ELEVENTH EDITION

Basic Technical Mathematics with Calculus

Allyn J. Washington
Dutchess Community College

Richard S. Evans
Corning Community College

Director, Courseware Portfolio Management: Deirdre Lynch
Executive Editor: Jeff Weidenaar
Editorial Assistant: Jennifer Snyder
Managing Producer: Karen Wernholm
Content Producer: Tamela Ambush
Producer: Jean Choe
Manager, Content Development: Kristina Evans
Math Content Developer: Megan M. Burns
Field Marketing Manager: Jennifer Crum
Product Marketing Manager: Alicia Frankel
Marketing Assistant: Hanna Lafferty
Senior Author Support/Technology Specialist: Joe Vetere
Manager, Rights Management: Gina M. Cheselka
Manufacturing Buyer: Carol Melville, LSC Communications
Composition: SPi Global
Associate Director of Design: Blair Brown
Cover Design: Barbara Atkinson
Cover Image: Daniel Schoenen/Image Broker/Alamy Stock Photo

Library of Congress Cataloging-in-Publication Data

Names: Washington, Allyn J. | Evans, Richard (Mathematics teacher)
Title: Basic technical mathematics with calculus / Allyn J. Washington, Dutchess Community College, Richard Evans, Corning Community College.
Description: 11th edition. | Boston : Pearson, [2018] | Includes indexes.
Identifiers: LCCN 2016020426| ISBN 9780134437736 (hardcover) | ISBN 013443773X (hardcover)
Subjects: LCSH: Mathematics–Textbooks. | Calculus–Textbooks.
Classification: LCC QA37.3 .W38 2018 | DDC 510–dc23
LC record available at **https://lccn.loc.gov/2016020426**

1 16

Student Edition:
ISBN 10: 0-13-443773-X
ISBN 13: 978-0-13-4437

Contents

Preface

Scope of the Book

Basic Technical Mathematics with Calculus, Eleventh Edition, is intended primarily for students in technical and pre-engineering technical programs or other programs for which coverage of mathematics is required. Chapters 1 through 20 provide the necessary background for further study with an integrated treatment of algebra and trigonometry. Chapter 21 covers the basic topics of analytic geometry, and Chapter 22 gives an introduction to statistics. Chapters 23 through 31 cover fundamental concepts of calculus including limits, derivatives, integrals, series representation of functions, and differential equations. In the examples and exercises, numerous applications from the various fields of technology are included, primarily to indicate where and how mathematical techniques are used. However, it is not necessary that the student have a specific knowledge of the technical area from which any given problem is taken. Most students using this text will have a background that includes some algebra and geometry. However, the material is presented in adequate detail for those who may need more study in these areas. The material presented here is sufficient for two to three semesters. One of the principal reasons for the arrangement of topics in this text is to present material in an order that allows a student to take courses concurrently in allied technical areas, such as physics and electricity. These allied courses normally require a student to know certain mathematics topics by certain definite times; yet the traditional order of topics in mathematics courses makes it difficult to attain this coverage without loss of continuity. However, the material in this book can be rearranged to fit any appropriate sequence of topics. The approach used in this text is not unduly rigorous mathematically, although all appropriate terms and concepts are introduced as needed and given an intuitive or algebraic foundation. The aim is to help the student develop an understanding of mathematical methods without simply providing a collection of formulas. The text material is developed recognizing that it is essential for the student to have a sound background in algebra and trigonometry in order to understand and succeed in any subsequent work in mathematics.

New to This Edition

You may have noticed something new on the cover of this book. Another author! Yes, after 50 years as a "solo act," Allyn Washington has a partner. New co-author Rich Evans is a veteran faculty member at Corning Community College (NY) and has brought a wealth of positive contributions to the book and accompanying MyMathLab course.

The new features of the eleventh edition include:

- **Refreshed design** – The book has been redesigned in full color to help students better use it and to help motivate students as they put in the hard work to learn the mathematics (because let's face it—a more modern looking book has more appeal).
- **Graphing calculator** – We have replaced the older TI-84 screens with those from the new TI-84 Plus-C (the color version). And Benjamin Rushing [Northwestern State University] has added graphing calculator help for students, accessible online via short URLs in the margins. If you'd like to see the complete listing of entries for the online graphing calculator manual, use the URL `goo.gl/eAUgW3`.
- **Applications** – The text features a wealth of new applications in the examples and exercises (over 200 in all!). Here is a sampling of the contexts for these new applications:
  - Power of a wind turbine (Section 3.4)
  - Height of One World Trade Center (Section 4.4)
  - GPS satellite velocity (Section 8.4)
  - Google's self-driving car laser distance (Section 9.6)
  - Phase angle for current/voltage lead and lag (Section 10.3)
  - Growth of computer processor transistor counts (Section 13.7)

CAUTION When you enter URLs for the Graphing Calculator Manual, take care to distinguish the following characters:

`l` = lowercase l

`I` = uppercase I

`1` = one

`O` = uppercase O

`0` = zero ■

   - Bezier curve roof design (Section 15.3)
   - Cardioid microphone polar pattern (Section 21.7)
   - Social networks usage (Section 22.1)
   - Video game system market share (Section 22.1)
   - Bluetooth headphone maximum revenue (Section 24.7)
   - Saddledome roof slopes (Section 29.3)
   - Weight loss differential equation (Section 31.6)
- **Exercises** – There are over 1000 new and updated exercises in the new edition. In creating new exercises, the authors analyzed aggregated student usage and performance data from MyMathLab for the previous edition of this text. The results of this analysis helped improve the quality and quantity of exercises that matter the most to instructors and students. There are a total of 14,000 exercises and 1400 examples in the eleventh edition.
- **Chapter Endmatter** – The exercises formerly called "Quick Chapter Review" are now labeled "Concept Check Exercises" (to better communicate their function within the chapter endmatter).
- **MyMathLab** – Features of the MyMathLab course for the new edition include:
   - Hundreds of new assignable algorithmic exercises help you address the homework needs of students. Additionally, all exercises are in the new HTML5 player, so they are accessible via mobile devices.
   - 223 new instructional videos (to augment the existing 203 videos) provide help for students as they do homework. These videos were created by Sue Glascoe (Mesa Community College) and Benjamin Rushing (Northwestern State University).
   - A new Graphing Calculator Manual, created specifically for this text, features instructions for the TI-84 and TI-89 family of calculators.
   - New PowerPoint® files feature animations that are designed to help you better teach key concepts.
   - Study skills modules help students with the life skills (e.g., time management) that can make the difference between passing and failing.

Content updates for the eleventh edition were informed by the extensive reviews of the text completed for this revision. These include:

- Unit analysis, including operations with units and unit conversions, has been moved from Appendix B to Section 1.4. Appendix B has been streamlined, but still contains the essential reference materials on units.
- In Section 1.3, more specific instructions have been provided for rounding combined operations with approximate numbers.
- Engineering notation has been added to Section 1.5.
- Finding the domain and range of a function *graphically* has been added to Section 3.4.
- The terms *input, output, piecewise defined functions,* and *practical domain and range* have been added to Chapter 3.
- In response to reviewer feedback, the beginning of Chapter 5 has been reorganized so that systems of equations has a strong introduction in Section 5.2. The prerequisite material needed for systems of equations (linear equations and graphs of linear functions) has been consolidated into Section 5.1. An example involving linear regression has also been added to Section 5.1.
- Solving systems using reduced row echelon form (*rref*) on a calculator has been added to Chapter 5.
- Several reviewers made the excellent suggestion to strengthen the focus on factoring in Chapter 6 by taking the contents of 6.1 (Special Products) and spreading it throughout the chapter. This change has been implemented. The terminology *greatest common factor* (GCF) has also been added to this chapter.

- In Chapter 7, the *square root property* is explicitly stated and illustrated.
- In Chapter 8, the unit circle definition of the trigonometric functions has been added.
- In Chapter 9, more emphasis had been given to solving equilibrium problems, including those that have more than one unknown.
- In Chapter 10, an example was added to show how the *phase angle* can be interpreted, and how it is different from the *phase shift*.
- In Chapter 16, the terminology *row echelon form* is used. Also, solving a system using *rref* is again illustrated. The material on using properties to evaluate determinants was deleted.
- The terminology *binomial coefficients* was added to Chapter 19.
- Chapter 22 (Introduction to Statistics) has undergone significant changes.
  - Section 22.1 now discusses common graphs used for both qualitative data (bar graphs and pie charts) and quantitative data (histograms, stem-and-leaf plots, and time series plots).
  - In Section 22.2, what was previously called the *arithmetic mean* is now referred to as simply the *mean*.
  - The *empirical rule* had been added to Section 22.4.
  - The sampling distribution of $\bar{x}$ has been formalized including the statement of the *central limit theorem*.
  - A discussion of interpolation and extrapolation has been added in the context of regression, as well as information on how to interpret the values of r and r^2.
  - The emphasis of Section 22.7 on nonlinear regression has been changed. Information on how to choose an appropriate type of model depending on the shape of the data has been added. However, a calculator is now used to obtain the actual regression equation.
- In Chapter 23, the terminology *direct substitution* has been introduced in the context of limits.
- Throughout the calculus chapters, many of the differentiation and integration rules have been given names so they can be easily referred to. These include, the *constant rule, power rule, constant multiple rule, product rule, quotient rule, general power rule, power rule for integration*, etc.
- In Chapter 30, the proof of the Fourier coefficients has been moved online.

Continuing Features

PAGE LAYOUT

Special attention has been given to the page layout. We specifically tried to avoid breaking examples or important discussions across pages. Also, all figures are shown immediately adjacent to the material in which they are discussed. Finally, we tried to avoid referring to equations or formulas by number when the referent is not on the same page spread.

CHAPTER INTRODUCTIONS

Each chapter introduction illustrates specific examples of how the development of technology has been related to the development of mathematics. In these introductions, it is shown that these past discoveries in technology led to some of the methods in mathematics, whereas in other cases mathematical topics already known were later very useful in bringing about advances in technology. Also, each chapter introduction contains a photo that refers to an example that is presented within that chapter.

WORKED-OUT EXAMPLES

EXAMPLE 3 Symbol in capital and in lowercase—forces on a beam

In the study of the forces on a certain beam, the equation $W = \frac{L(wL + 2P)}{8}$ is used. Solve for P.

$$8W = \frac{8L(wL + 2P)}{8} \quad \text{multiply both sides by 8}$$
$$8W = L(wL + 2P) \quad \text{simplify right side}$$
$$8W = wL^2 + 2LP \quad \text{remove parentheses}$$
$$8W - wL^2 = 2LP \quad \text{subtract } wL^2 \text{ from both sides}$$
$$P = \frac{8W - wL^2}{2L} \quad \text{divide both sides by } 2L \text{ and switch sides}$$
■

- **"HELP TEXT"** Throughout the book, special explanatory comments in blue type have been used in the examples to emphasize and clarify certain important points. Arrows are often used to indicate clearly the part of the example to which reference is made.
- **EXAMPLE DESCRIPTIONS** A brief descriptive title is given for each example. This gives an easy reference for the example, particularly when reviewing the contents of the section.
- **APPLICATION PROBLEMS** There are over 350 applied examples throughout the text that show complete solutions of application problems. Many relate to modern technology such as computer design, electronics, solar energy, lasers, fiber optics, the environment, and space technology. Others examples and exercises relate to technologies such as aeronautics, architecture, automotive, business, chemical, civil, construction, energy, environmental, fire science, machine, medical, meteorology, navigation, police, refrigeration, seismology, and wastewater. The Index of Applications at the end of the book shows the breadth of applications in the text.

KEY FORMULAS AND PROCEDURES

Throughout the book, important formulas are set off and displayed so that they can be easily referenced for use. Similarly, summaries of techniques and procedures consistently appear in color-shaded boxes.

"CAUTION" AND "NOTE" INDICATORS

CAUTION This heading is used to identify errors students commonly make or places where they frequently have difficulty. ■

The NOTE label in the side margin, along with accompanying blue brackets in the main body of the text, points out material that is of particular importance in developing or understanding the topic under discussion. [Both of these features have been clarified in the eleventh edition by adding a small design element to show where the CAUTION or NOTE feature *ends*.]

NOTE ▸

CHAPTER AND SECTION CONTENTS

A listing of learning outcomes for each chapter is given on the introductory page of the chapter. Also, a listing of the key topics of each section is given below the section number and title on the first page of the section. This gives the student and instructor a quick preview of the chapter and section contents.

PRACTICE EXERCISES

Most sections include some practice exercises in the margin. They are included so that a student is more actively involved in the learning process and can check his or her understanding of the material. They can also be used for classroom exercises. The answers to these exercises are given at the end of the exercises set for the section. There are over 450 of these exercises.

FEATURES OF EXERCISES

- **EXERCISES DIRECTLY REFERENCED TO TEXT EXAMPLES** The first few exercises in most of the text sections are referenced directly to a specific example of the section. These exercises are worded so that it is necessary for the student to refer to the example in order to complete the required solution. In this way, the student should be able to better review and understand the text material before attempting to solve the exercises that follow.

- **WRITING EXERCISES** There are over 270 writing exercises through the book (at least eight in each chapter) that require at least a sentence or two of explanation as part of the answer. These are noted by a pencil icon next to the exercise number.
- **APPLICATION PROBLEMS** There are about 3000 application exercises in the text that represent the breadth of applications that students will encounter in their chosen professions. The Index of Applications at the end of the book shows the breadth of applications in the text.

CHAPTER ENDMATTER

- **KEY FORMULAS AND EQUATIONS** Here all important formulas and equations are listed together with their corresponding equation numbers for easy reference.
- **CHAPTER REVIEW EXERCISES** These exercises consist of (a) Concept Check Exercises (a set of true/false exercises) and (b) Practice and Applications.
- **CHAPTER TEST** These are designed to mirror what students might see on the actual chapter test. Complete step-by-step solutions to all practice test problems are given in the back of the book.

MARGIN NOTES

Throughout the text, some margin notes point out relevant historical events in mathematics and technology. Other margin notes are used to make specific comments related to the text material. Also, where appropriate, equations from earlier material are shown for reference in the margin.

ANSWERS TO EXERCISES

The answers to odd-numbered exercises are given near the end of the book. The Student's Solution Manual contains solutions to every other odd-numbered exercise and the Instructor's Solution Manual contains solutions to all section exercises.

FLEXIBILITY OF COVERAGE

The order of coverage can be changed in many places and certain sections may be omitted without loss of continuity of coverage. Users of earlier editions have indicated successful use of numerous variations in coverage. Any changes will depend on the type of course and completeness required.

Technology and Supplements

MYMATHLAB® ONLINE COURSE (ACCESS CODE REQUIRED)

Built around Pearson's best-selling content, MyMathLab is an online homework, tutorial, and assessment program designed to work with this text to engage students and improve results. MyMathLab can be successfully implemented in any classroom environment—lab-based, hybrid, fully online, or traditional. **By addressing instructor and student needs, MyMathLab improves student learning.**

MOTIVATION

Students are motivated to succeed when they're engaged in the learning experience and understand the relevance and power of mathematics. MyMathLab's online homework offers students immediate feedback and tutorial assistance that motivates them to do more, which means they retain more knowledge and improve their test scores.

- **Exercises with immediate feedback**—over 7850 assignable exercises—are based on the textbook exercises, and regenerate algorithmically to give students unlimited opportunity for practice and mastery. MyMathLab provides helpful feedback when students enter incorrect answers and includes optional learning aids including Help Me Solve This, View an Example, videos, and the eText.

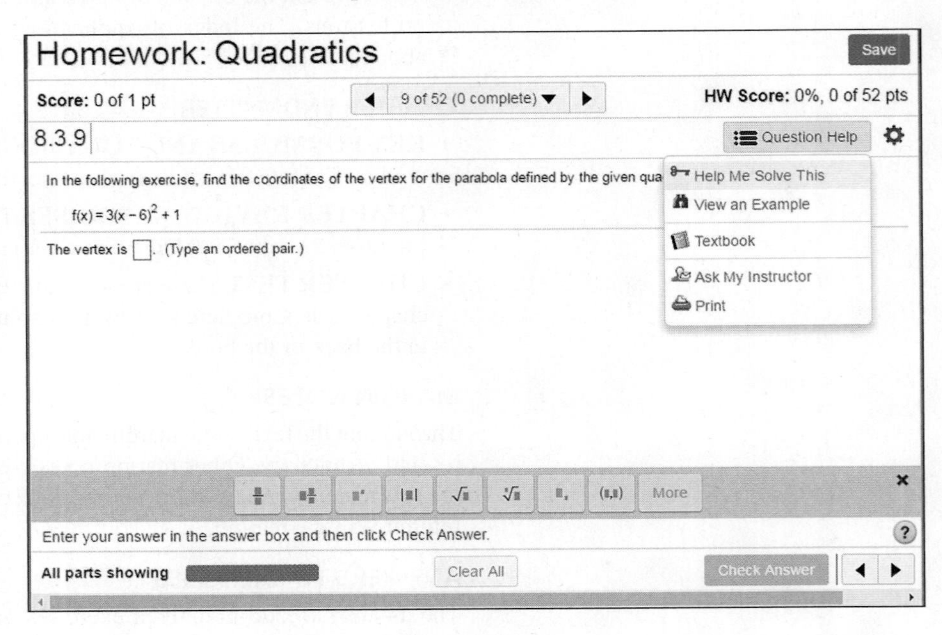

- **Learning Catalytics™** is a student response tool that uses students' smartphones, tablets, or laptops to engage them in more interactive tasks and thinking. Learning Catalytics fosters student engagement and peer-to-peer learning with real-time analytics.

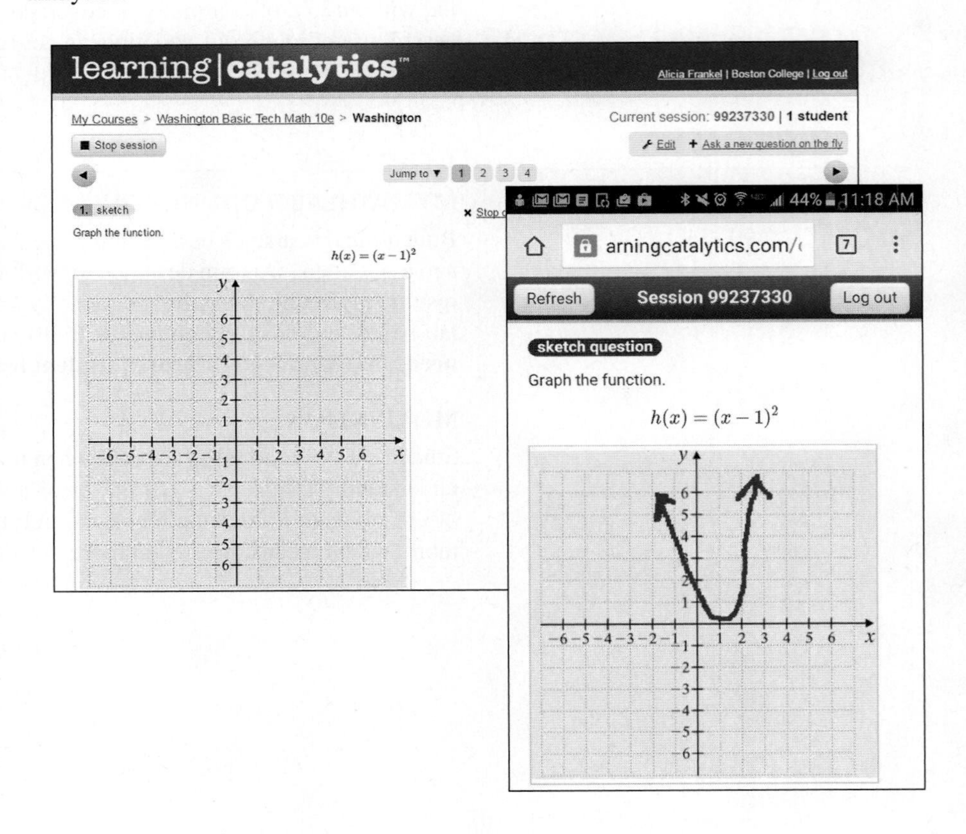

LEARNING TOOLS FOR STUDENTS

- **Instructional videos** - The nearly 440 videos in the 11th edition MyMathLab course provide help for students outside of the classroom. These videos are also available as learning aids within the homework exercises, for students to refer to at point-of-use.

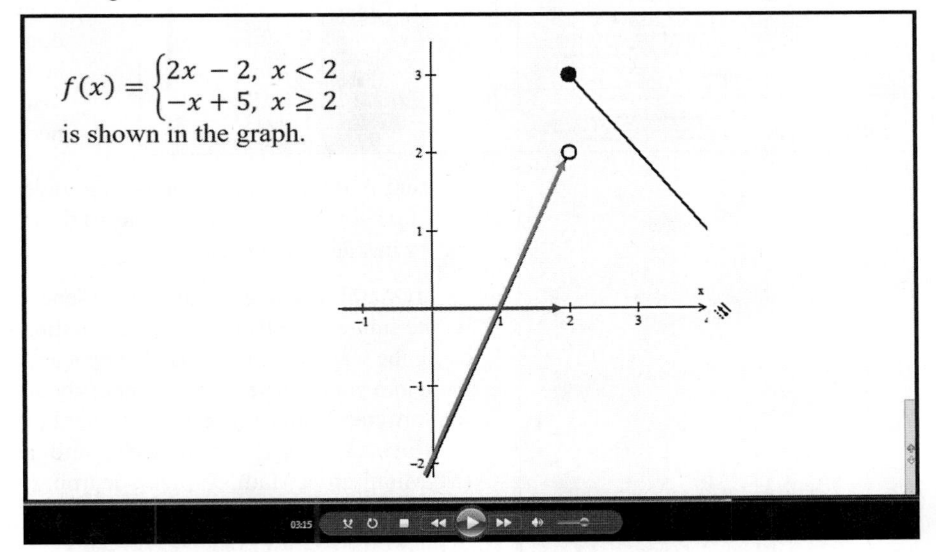

- **The complete eText** is available to students through their MyMathLab courses for the lifetime of the edition, giving students unlimited access to the eText within any course using that edition of the textbook. The eText includes links to videos.
- A new online **Graphing Calculator Manual**, created specifically for this text by Benjamin Rushing (Northwestern State University), features instructions for the TI-84 and TI-89 family of calculators.
- **Skills for Success Modules** help foster strong study skills in collegiate courses and prepare students for future professions. Topics include "Time Management" and "Stress Management".
- **Accessibility** and achievement go hand in hand. MyMathLab is compatible with the JAWS screen reader, and enables multiple-choice and free-response problem types to be read and interacted with via keyboard controls and math notation input. MyMathLab also works with screen enlargers, including ZoomText, MAGic, and SuperNova. And, all MyMathLab videos have closed-captioning. More information is available at **http://mymathlab.com/accessibility**.

SUPPORT FOR INSTRUCTORS

- New **PowerPoint®** files feature animations that are designed to help you better teach key concepts.

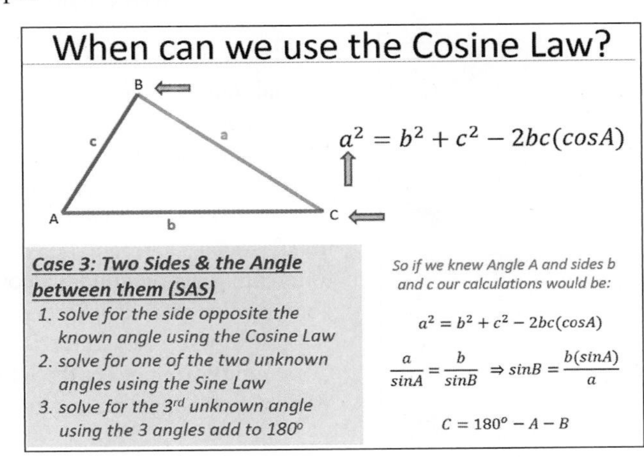

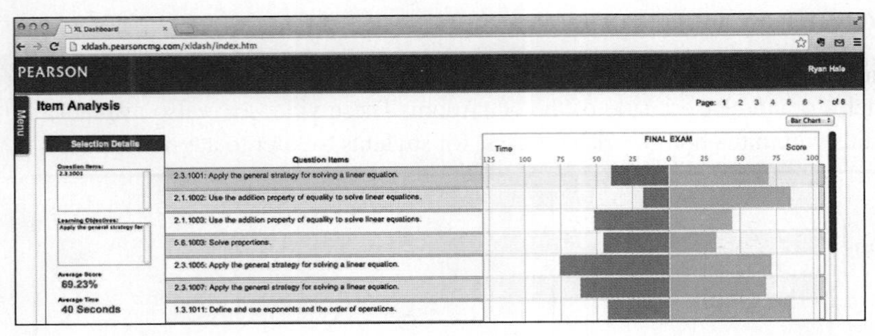

- **A comprehensive gradebook** with enhanced reporting functionality allows you to efficiently manage your course.
  - **The Reporting Dashboard** provides insight to view, analyze, and report learning outcomes. Student performance data is presented at the class, section, and program levels in an accessible, visual manner so you'll have the information you need to keep your students on track.
- **Item Analysis** tracks class-wide understanding of particular exercises so you can refine your class lectures or adjust the course/department syllabus. Just-in-time teaching has never been easier!

MyMathLab comes from an experienced partner with educational expertise and an eye on the future. Whether you are just getting started with MyMathLab, or have a question along the way, we're here to help you learn about our technologies and how to incorporate them into your course. To learn more about how MyMathLab helps students succeed, visit **www.mymathlab.com** or contact your Pearson rep.

MathXL® is the homework and assessment engine that runs MyMathLab. (MyMathLab is MathXL plus a learning management system.) MathXL access codes are also an option.

STUDENT'S SOLUTIONS MANUAL

ISBN-10: 0134434633 | ISBN-13: 9780134434636

The Student's Solutions Manual by Matthew Hudelson (Washington State University) includes detailed solutions for every other odd-numbered section exercise. The manual is available in print and is downloadable from within MyMathLab.

INSTRUCTOR'S SOLUTIONS MANUAL (DOWNLOADABLE)

ISBN-10: 0134435893 | ISBN-13: 9780134435893

The Instructor's Solution Manual by Matthew Hudelson (Washington State University) contains detailed solutions to every section exercise, including review exercises. The manual is available to qualified instructors for download in the Pearson Instructor Resource **www.pearsonhighered.com/irc** or within MyMathLab.

TESTGEN (DOWNLOADABLE)

ISBN-10: 0134435753 | ISBN-13: 9780134435756

TestGen enables instructors to build, edit, print, and administer tests using a bank of questions developed to cover all objectives in the text. TestGen is algorithmically based, allowing you to create multiple but equivalent versions of the same question or test. Instructors can also modify test bank questions or add new questions. The TestGen software and accompanying test bank are available to qualified instructors for download in the Pearson Web Catalog **www.pearsonhighered.com** or within MyMathLab.

Acknowledgments

Special thanks goes to Matthew Hudelson of Washington State University for preparing the Student's Solutions Manual and the Instructor's Solutions Manual. Thanks also to Bob Martin and John Garlow, both of Tarrant County College (TX) for their work on these manuals for previous editions. A special thanks to Ben Rushing of Northwestern State University of Louisiana for his work on the graphing calculator manual as well as instructional videos. Our gratitude is also extended to to Sue Glascoe (Mesa Community College) for creating instructional videos. We would also like to express appreciation for the work done by David Dubriske and Cindy Trimble in checking for accuracy in the text and exercises. Also, we again wish to thank Thomas Stark of Cincinnati State Technical and Community College for the RISERS approach to solving word problems in Appendix A. We also extend our thanks to Julie Hoffman, Personal Assistant to Allyn Washington.

We gratefully acknowledge the unwavering cooperation and support of our editor, Jeff Weidenaar. A warm thanks also goes to Tamela Ambush, Content Producer, for her help in coordinating many aspects of this project. A special thanks also to Julie Kidd, Project Manager at SPi Global, as well as the compositors Karthikeyan Lakshmikanthan and Vijay Sigamani, who set all the type for this edition.

The authors gratefully acknowledge the contributions of the following reviewers of the tenth edition in preparation for this revision. Their detailed comments and suggestions were of great assistance.

Bob Biega, *Kentucky Community and Technical College System*

Bill Burgin, *Gaston College*

Brian Carter, *St. Louis Community College*

Majid R. Chatsaz, *Penn State University Scranton*

Benjamin Falero, *Central Carolina Community College*

Kenny Fister, *Murray State University*

Joshua D. Hammond, *SUNY Jefferson Community College*

Harold Hayford, *Penn State Altoona*

Hengli Jiao, *Ferris State University*

Mohammad Kazemi, *University of North Carolina at Charlotte*

Mary Knappen, *Genesee Community College*

John F. Larson, *Southeastern Community College*

Michael Leonard, *Purdue University Calumet*

Jillian McMeans, *Asheville-Buncombe Technical Community College*

Cristal Miskovich, *Embry-Riddle Aeronautical University Worldwide*

Robert Mitchell, *Pennsylvania College of Technology*

M. Niebauer, *Penn State Erie, The Behrend College*

Kaan Ozmeral, *Central Carolina Community College*

Suzie Pickle, *College of Southern Nevada*

April Pritchett, *Murray State University*

Renee Quick, *Wallace State Community College*

Craig Rabatin, *West Virginia University – Parkersburg*

Ben Rushing Jr., *Northwestern State University of Louisiana*

Timothy Schoppert, *Embry-Riddle Aeronautical University*

Joshua Shelor, *Virginia Western Community College*

Natalie Sommer, *DeVry College of New York*

Tammy Sullivan, *Asheville-Buncombe Technical Community College*

Fereja Tahir, *Illinois Central College*

Tiffany Williams, *Hurry Georgetown Technical College*

Shirley Wilson, *Massachusetts Maritime Academy*

Tseng Y. Woo, *Durham Technical Community College*

Finally, we wish to sincerely thank again each of the over 375 reviewers of the eleven editions of this text. Their comments have helped further the education of more than two million students during since this text was first published in 1964.

Allyn Washington

Richard Evans

Basic Algebraic Operations

1

Interest in things such as the land on which they lived, the structures they built, and the motion of the planets led people in early civilizations to keep records and to create methods of counting and measuring.

In turn, some of the early ideas of arithmetic, geometry, and trigonometry were developed. From such beginnings, mathematics has played a key role in the great advances in science and technology.

Often, mathematical methods were developed from scientific studies made in particular areas, such as astronomy and physics. Many people were interested in the math itself and added to what was then known. Although this additional mathematical knowledge may not have been related to applications at the time it was developed, it often later became useful in applied areas.

In the chapter introductions that follow, examples of the interaction of technology and mathematics are given. From these examples and the text material, it is hoped you will better understand the important role that math has had and still has in technology. In this text, there are applications from technologies including (but not limited to) aeronautical, business, communications, electricity, electronics, engineering, environmental, heat and air conditioning, mechanical, medical, meteorology, petroleum, product design, solar, and space.

We begin by reviewing the concepts that deal with numbers and symbols. This will enable us to develop topics in algebra, an understanding of which is essential for progress in other areas such as geometry, trigonometry, and calculus.

LEARNING OUTCOMES

After completion of this chapter, the student should be able to:

- Identify real, imaginary, rational, and irrational numbers
- Perform mathematical operations on integers, decimals, fractions, and radicals
- Use the fundamental laws of algebra in numeric and algebraic expressions
- Employ mathematical order of operations
- Understand technical measurement, approximation, the use of significant digits, and rounding
- Use scientific and engineering notations
- Convert units of measurement
- Rearrange and solve basic algebraic equations
- Interpret word problems using algebraic symbols

◀ **From the Great Pyramid of Giza, built in Egypt 4500 years ago, to the modern technology of today, mathematics has played a key role in the advancement of civilization. Along the way, important discoveries have been made in areas such as architecture, navigation, transportation, electronics, communication, and astronomy. Mathematics will continue to pave the way for new discoveries.**

1.1 Numbers

Real Number System • Number Line • Absolute Value • Signs of Inequality • Reciprocal • Denominate Numbers • Literal Numbers

In technology and science, as well as in everyday life, we use the very familiar **counting numbers,** or **natural numbers** 1, 2, 3, and so on. The **whole numbers** include 0 as well as all the natural numbers. Because it is necessary and useful to use negative numbers as well as positive numbers in mathematics and its applications, the natural numbers are called the **positive integers,** and the numbers -1, -2, -3, and so on are the **negative integers.**

Therefore, *the* **integers** *include the positive integers, the negative integers,* and **zero,** *which is neither positive nor negative*. This means that the integers are the numbers . . . , $-3, -2, -1, 0, 1, 2, 3, \ldots$ and so on.

A **rational number** *is a number that can be expressed as the division of one integer* ***a*** *by another nonzero integer* ***b,*** *and can be represented by the fraction* ***a/b.*** Here ***a*** *is the* **numerator** *and* ***b*** *is the* **denominator.** Here we have used algebra by letting letters represent numbers.

■ Irrational numbers were discussed by the Greek mathematician Pythagoras in about 540 B.C.E.

Another type of number, an **irrational number,** *cannot be written in the form of a fraction that is the division of one integer by another integer*. The following example illustrates integers, rational numbers, and irrational numbers.

EXAMPLE 1 Identifying rational numbers and irrational numbers

The numbers 5 and -19 are integers. They are also rational numbers because they can be written as $\frac{5}{1}$ and $\frac{-19}{1}$, respectively. Normally, we do not write the 1's in the denominators.

■ For reference, $\pi = 3.14159265\ldots$

The numbers $\frac{5}{8}$ and $\frac{-11}{3}$ are rational numbers because the numerator and the denominator of each are integers.

The numbers $\sqrt{2}$ and π are irrational numbers. It is not possible to find two integers, one divided by the other, to represent either of these numbers. In decimal form, irrational numbers are nonterminating, nonrepeating decimals. It can be shown that square roots (and other roots) that cannot be expressed exactly in decimal form are irrational. Also, $\frac{22}{7}$ is sometimes used as an *approximation* for π, but it is not equal *exactly* to π. We must remember that $\frac{22}{7}$ is rational and π is irrational.

■ A notation that is often used for repeating decimals is to place a bar over the digits that repeat. Using this notation we can write $\frac{1121}{1665} = 0.6\overline{732}$ and $\frac{2}{3} = 0.\overline{6}$.

The decimal number 1.5 is rational since it can be written as $\frac{3}{2}$. Any such *terminating decimal* is rational. The number 0.6666 . . . , where the 6's continue on indefinitely, is rational because we may write it as $\frac{2}{3}$. In fact, any *repeating decimal* (in decimal form, a specific sequence of digits is repeated indefinitely) is rational. The decimal number 0.6732732732 . . . is a repeating decimal where the sequence of digits 732 is repeated indefinitely $(0.6732732732\ldots = \frac{1121}{1665})$. ■

The rational numbers together with the irrational numbers, including all such numbers that are positive, negative, or zero, make up the **real number system** (see Fig. 1.1). There are times we will encounter an **imaginary number,** *the name given to the square root of a*

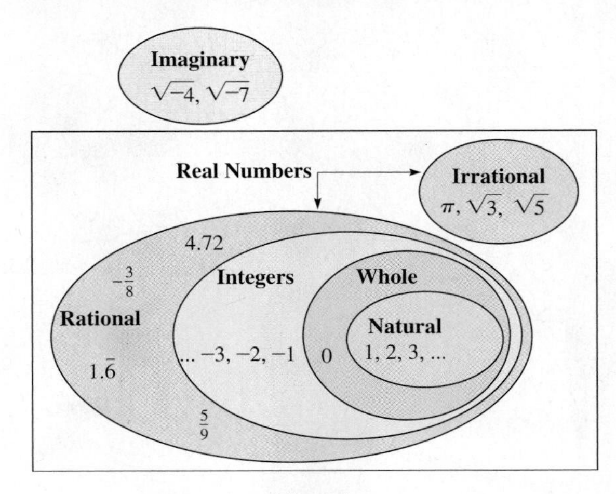

Fig. 1.1

negative number. Imaginary numbers are not real numbers and will be discussed in Chapter 12. However, unless specifically noted, we will use real numbers. Until Chapter 12, it will be necessary to only *recognize* imaginary numbers when they occur.

Also in Chapter 12, we will consider **complex numbers,** which include both the real numbers and imaginary numbers. See Exercise 39 of this section.

■ Real numbers and imaginary numbers are both included in the *complex number system*. See Exercise 39.

EXAMPLE 2 Identifying real numbers and imaginary numbers

(a) The number 7 is an integer. It is also rational because $7 = \frac{7}{1}$, and it is a real number since the real numbers include all the rational numbers.

(b) The number 3π is irrational, and it is real because the real numbers include all the irrational numbers.

(c) The numbers $\sqrt{-10}$ and $-\sqrt{-7}$ are imaginary numbers.

(d) The number $\frac{-3}{7}$ is rational and real. The number $-\sqrt{7}$ is irrational and real.

(e) The number $\frac{\pi}{6}$ is irrational and real. The number $\frac{\sqrt{-3}}{2}$ is imaginary. ■

■ Fractions were used by early Egyptians and Babylonians. They were used for calculations that involved parts of measurements, property, and possessions.

A **fraction** *may contain any number or symbol representing a number in its numerator or in its denominator*. The fraction indicates the division of the numerator by the denominator, as we previously indicated in writing rational numbers. Therefore, a fraction may be a number that is rational, irrational, or imaginary.

EXAMPLE 3 Fractions

(a) The numbers $\frac{2}{7}$ and $\frac{-3}{2}$ are fractions, and they are rational.

(b) The numbers $\frac{\sqrt{2}}{9}$ and $\frac{6}{\pi}$ are fractions, but they are not rational numbers. It is not possible to express either as one integer divided by another integer.

(c) The number $\frac{\sqrt{-5}}{6}$ is a fraction, and it is an imaginary number. ■

THE NUMBER LINE

Real numbers may be represented by points on a line. We draw a horizontal line and designate some point on it by O, which we call the **origin** (see Fig. 1.2). The integer *zero* is located at this point. Equal intervals are marked to the right of the origin, and the positive integers are placed at these positions. The other positive rational numbers are located between the integers. The points that cannot be defined as rational numbers represent irrational numbers. We cannot tell whether a given point represents a rational number or an irrational number unless it is specifically marked to indicate its value.

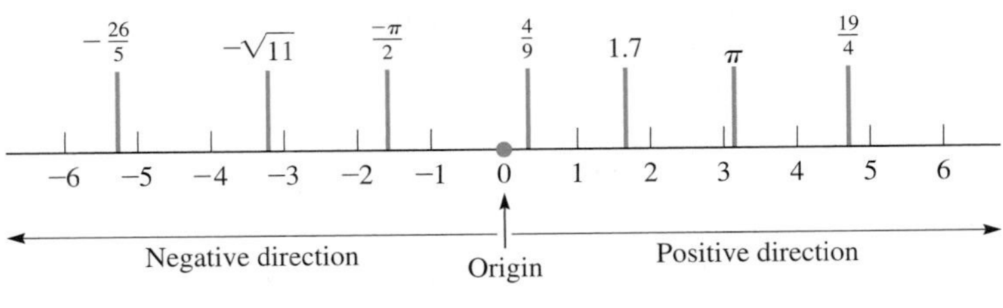

Fig. 1.2

The negative numbers are located on the number line by starting at the origin and marking off equal intervals *to the left, which is the* **negative direction.** As shown in Fig. 1.2, *the positive numbers are to the right of the origin and the negative numbers are to the left of the origin*. Representing numbers in this way is especially useful for graphical methods.

We next define another important concept of a number. *The* **absolute value** *of a positive number is the number itself, and the absolute value of a negative number is the corresponding positive number.* On the number line, we may interpret the absolute value of a number as the distance (which is always positive) between the origin and the number. Absolute value is denoted by writing the number between vertical lines, as shown in the following example.

EXAMPLE 4 Absolute value

The absolute value of 6 is 6, and the absolute value of -7 is 7. We write these as $|6| = 6$ and $|-7| = 7$. See Fig. 1.3.

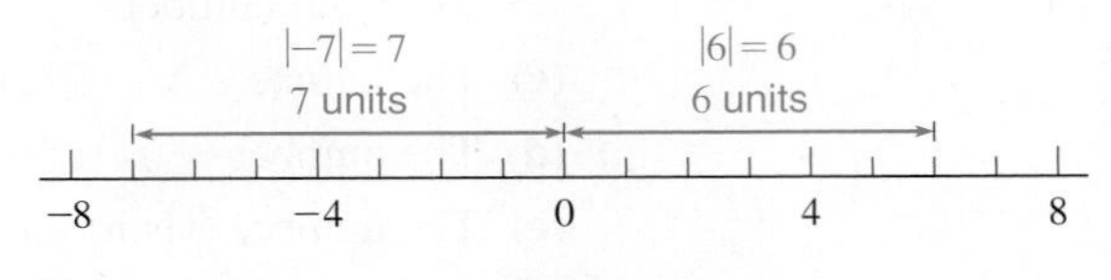

Fig. 1.3

Other examples are $\left|\frac{7}{5}\right| = \frac{7}{5}$, $|-\sqrt{2}| = \sqrt{2}$, $|0| = 0$, $-|\pi| = -\pi$, $|-5.29| = 5.29$, and $-|-9| = -9$ since $|-9| = 9$. ■

Practice Exercises

1. $|-4.2| = ?$ 2. $-\left|-\frac{3}{4}\right| = ?$

■ The symbols =, <, and > were introduced by English mathematicians in the late 1500s.

On the number line, *if a first number is to the right of a second number, then the first number is said to be* **greater than** *the second. If the first number is to the left of the second, it is* **less than** *the second number.* The symbol $>$ designates "is greater than," and the symbol $<$ designates "is less than." These are called **signs of inequality.** See Fig. 1.4.

EXAMPLE 5 Signs of inequality

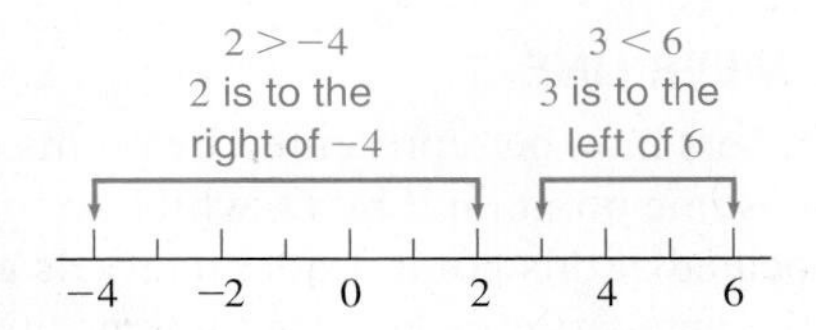

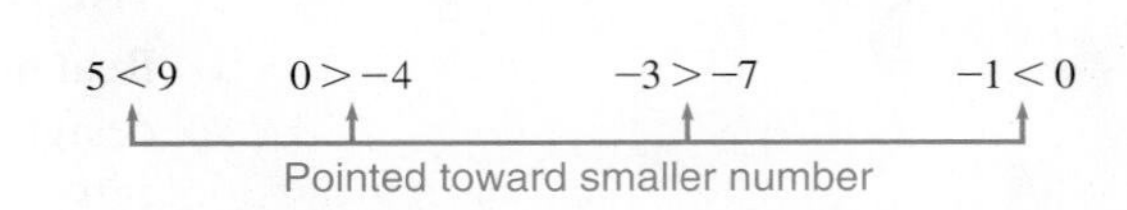

Fig. 1.4 ■

Practice Exercises

Place the correct sign of inequality (< or >) between the given numbers.

3. $-5 \quad 4$ 4. $0 \quad -3$

Every number, except zero, has a **reciprocal.** *The reciprocal of a number is 1 divided by the number.*

EXAMPLE 6 Reciprocal

The reciprocal of 7 is $\frac{1}{7}$. The reciprocal of $\frac{2}{3}$ is

$$\frac{1}{\frac{2}{3}} = 1 \times \frac{3}{2} = \frac{3}{2} \quad \text{invert denominator and multiply (from arithmetic)}$$

The reciprocal of 0.5 is $\frac{1}{0.5} = 2$. The reciprocal of $-\pi$ is $-\frac{1}{\pi}$. Note that the negative sign is retained in the reciprocal of a negative number.

We showed the multiplication of 1 and $\frac{3}{2}$ as $1 \times \frac{3}{2}$. We could also show it as $1 \cdot \frac{3}{2}$ or $1\left(\frac{3}{2}\right)$. We will often find the form with parentheses is preferable. ■

Practice Exercise

5. Find the reciprocals of

(a) -4 (b) $\frac{3}{8}$

In applications, *numbers that represent a measurement and are written with units of measurement are called* **denominate numbers.** The next example illustrates the use of units and the symbols that represent them.

■ For reference, see Appendix B for units of measurement and the symbols used for them.

EXAMPLE 7 Denominate numbers

(a) To show that a certain TV weighs 62 pounds, we write the weight as 62 lb.

(b) To show that a giant redwood tree is 330 feet high, we write the height as 300 ft.

(c) To show that the speed of a rocket is 1500 meters per second, we write the speed as 1500 m/s. (Note the use of s for second. We use s rather than sec.)

(d) To show that the area of a computer chip is 0.75 square inch, we write the area as 0.75 in.^{2}. (We will not use sq in.)

(e) To show that the volume of water in a glass tube is 25 cubic centimeters, we write the volume as 25 cm^3. (We will not use cu cm nor cc.) ■

It is usually more convenient to state definitions and operations on numbers in a general form. *To do this, we represent the numbers by letters, called* **literal numbers.** For example, if we want to say "If a first number is to the right of a second number on the number line, then the first number is greater than the second number," we can write "If a is to the right of b on the number line, then $a > b$." Another example of using a literal number is "The reciprocal of n is $1/n$."

Certain literal numbers may take on any allowable value, whereas other literal numbers represent the same value throughout the discussion. *Those literal numbers that may vary in a given problem are called* **variables,** and *those literal numbers that are held fixed are called* **constants.**

EXAMPLE 8 Variables and constants

(a) The resistance of an electric resistor is R. The current I in the resistor equals the voltage V divided by R, written as $I = V/R$. For this resistor, I and V may take on various values, and R is fixed. This means I and V are variables and R is a constant. For a *different* resistor, the value of R may differ.

(b) The fixed cost for a calculator manufacturer to operate a certain plant is b dollars per day, and it costs a dollars to produce each calculator. The total daily cost C to produce n calculators is

$$C = an + b$$

Here, C and n are variables, and a and b are constants, and the product of a and n is shown as an. For *another* plant, the values of a and b would probably differ.

If specific numerical values of a and b are known, say $a = \$7$ per calculator and $b = \$3000$, then $C = 7n + 3000$. Thus, constants may be numerical or literal. ■

EXERCISES 1.1

In Exercises 1–4, make the given changes in the indicated examples of this section, and then answer the given questions.

1. In the first line of Example 1, change the 5 to −7 and the −19 to 12. What other changes must then be made in the first paragraph?

2. In Example 4, change the 6 to −6. What other changes must then be made in the first paragraph?

3. In the left figure of Example 5, change the 2 to −6. What other changes must then be made?

4. In Example 6, change the $\frac{2}{3}$ to $\frac{3}{2}$. What other changes must then be made?

In Exercises 5–8, designate each of the given numbers as being an integer, rational, irrational, real, or imaginary. (More than one designation may be correct.)

5. $3, \sqrt{-4}$ **6.** $\frac{\sqrt{7}}{3}, -6$ **7.** $-\frac{\pi}{6}, \frac{1}{8}$ **8.** $-\sqrt{-6}, -2.33$

In Exercises 9 and 10, find the absolute value of each real number.

9. $3, -3, \frac{\pi}{4}, \sqrt{-1}$ **10.** $-0.857, \sqrt{2}, -\frac{19}{4}, \frac{\sqrt{-5}}{-2}$

In Exercises 11–18, insert the correct sign of inequality (> or <) between the given numbers.

11. 6 8 **12.** 7 5

13. π 3.1416 **14.** -4 0

15. -4 $-|-3|$ **16.** $-\sqrt{2}$ -1.42

17. $-\frac{2}{3}$ $-\frac{3}{4}$ **18.** -0.6 0.2

In Exercises 19 and 20, find the reciprocal of each number.

19. 3, $-\frac{4}{\sqrt{3}}$, $\frac{y}{b}$ **20.** $-\frac{1}{3}$, 0.25, $2x$

In Exercises 21 and 22, locate (approximately) each number on a number line as in Fig. 1.2.

21. 2.5, $-\frac{12}{5}$, $\sqrt{3}$, $-\frac{3}{4}$ **22.** $-\frac{\sqrt{2}}{2}$, 2π, $\frac{123}{19}$, $-\frac{7}{3}$

In Exercises 23–46, solve the given problems. Refer to Appendix B for units of measurement and their symbols.

23. Is an absolute value always positive? Explain.

24. Is -2.17 rational? Explain.

25. What is the reciprocal of the reciprocal of any positive or negative number?

26. Is the repeating decimal $2.\overline{72}$ rational or irrational?

27. True or False: A nonterminating, nonrepeating decimal is an irrational number.

28. If $b > a$ and $a > 0$, is $|b - a| < |b| - |a|$?

29. List the following numbers in numerical order, starting with the smallest: $-1, 9, \pi, \sqrt{5}, |-8|, -|-3|, -3.1$.

30. List the following numbers in numerical order, starting with the smallest: $\frac{1}{5}, -\sqrt{10}, -|-6|, -4, 0.25, |-\pi|$.

31. If a and b are positive integers and $b > a$, what type of number is represented by the following?

(a) $b - a$ **(b)** $a - b$ **(c)** $\frac{b - a}{b + a}$

32. If a and b represent positive integers, what kind of number is represented by (a) $a + b$, (b) a/b, and (c) $a \times b$?

33. For any positive or negative integer: (a) Is its absolute value always an integer? (b) Is its reciprocal always a rational number?

34. For any positive or negative rational number: (a) Is its absolute value always a rational number? (b) Is its reciprocal always a rational number?

35. Describe the location of a number x on the number line when (a) $x > 0$ and (b) $x < -4$.

36. Describe the location of a number x on the number line when (a) $|x| < 1$ and (b) $|x| > 2$.

37. For a number $x > 1$, describe the location on the number line of the reciprocal of x.

38. For a number $x < 0$, describe the location on the number line of the number with a value of $|x|$.

39. A *complex number* is defined as $a + bj$, where a and b are real numbers and $j = \sqrt{-1}$. For what values of a and b is the complex number $a + bj$ a real number? (All real numbers and all imaginary numbers are also complex numbers.)

40. A sensitive gauge measures the total weight w of a container and the water that forms in it as vapor condenses. It is found that $w = c\sqrt{0.1t + 1}$, where c is the weight of the container and t is the time of condensation. Identify the variables and constants.

41. In an electric circuit, the reciprocal of the total capacitance of two capacitors in series is the sum of the reciprocals of the capacitances $\left(\frac{1}{C_T} = \frac{1}{C_1} + \frac{1}{C_2}\right)$. Find the total capacitance of two capacitances of 0.0040 F and 0.0010 F connected in series.

42. Alternating-current (ac) voltages change rapidly between positive and negative values. If a voltage of 100 V changes to -200 V, which is greater in absolute value?

43. The memory of a certain computer has a bits in each byte. Express the number N of bits in n kilobytes in an equation. (A *bit* is a single digit, and bits are grouped in *bytes* in order to represent special characters. Generally, there are 8 bits per byte. If necessary,see Appendix B for the meaning of *kilo*.)

44. The computer design of the base of a truss is x ft long. Later it is redesigned and shortened by y in. Give an equation for the length L, in inches, of the base in the second design.

45. In a laboratory report, a student wrote "$-20°C > -30°C$." Is this statement correct? Explain.

46. After 5 s, the pressure on a valve is less than 60 lb/in.^{2} (pounds per square inch). Using t to represent time and p to represent pressure, this statement can be written "for $t > 5$ s, $p < 60$ lb/in.^{2}." In this way, write the statement "when the current I in a circuit is less than 4 A, the resistance R is greater than 12 Ω (ohms)."

Answers to Practice Exercises

1. 4.2 **2.** $-\frac{3}{4}$ **3.** < **4.** > **5.** **(a)** $-\frac{1}{4}$ **(b)** $\frac{8}{3}$

1.2 Fundamental Operations of Algebra

Fundamental Laws of Algebra • Operations on Positive and Negative Numbers • Order of Operations • Operations with Zero

If two numbers are added, it does not matter in which order they are added. (For example, $5 + 3 = 8$ and $3 + 5 = 8$, or $5 + 3 = 3 + 5$.) This statement, generalized and accepted as being correct for all possible combinations of numbers being added, is called the **commutative law** for addition. It states that *the sum of two numbers is the same,* *regardless of the order in which they are added.* We make no attempt to prove this law in general, but accept that it is true.

In the same way, we have the **associative law** for addition, which states that *the sum of three or more numbers is the same, regardless of the way in which they are grouped for addition.* For example, $3 + (5 + 6) = (3 + 5) + 6$.

The laws just stated for addition are also true for multiplication. Therefore, *the product of two numbers is the same, regardless of the order in which they are multiplied,* and *the product of three or more numbers is the same, regardless of the way in which they are grouped for multiplication.* For example, $2 \times 5 = 5 \times 2$, and $5 \times (4 \times 2) = (5 \times 4) \times 2$.

Another very important law is the **distributive law.** It states that *the product of one number and the sum of two or more other numbers is equal to the sum of the products of the first number and each of the other numbers of the sum.* For example,

■ Note carefully the difference:
associative law: $5 \times (4 \times 2)$
distributive law: $5 \times (4 + 2)$

$$5(4 + 2) = 5 \times 4 + 5 \times 2$$

In this case, it can be seen that the total is 30 on each side.

In practice, we use these **fundamental laws of algebra** naturally without thinking about them, except perhaps for the distributive law.

Not all operations are commutative and associative. For example, division is not commutative, because the order of division of two numbers does matter. For instance, $\frac{6}{5} \neq \frac{5}{6}$ ($\neq$ is read "does not equal"). (Also, see Exercise 54.)

Using literal numbers, the fundamental laws of algebra are as follows:

Commutative law of addition: $a + b = b + a$

Associative law of addition: $a + (b + c) = (a + b) + c$

Commutative law of multiplication: $ab = ba$

Associative law of multiplication: $a(bc) = (ab)c$

Distributive law: $a(b + c) = ab + ac$

■ Note the meaning of *identity.*

Each of these laws is an example of an *identity,* in that the expression to the left of the = sign equals the expression to the right for any value of each of a, b, and c.

OPERATIONS ON POSITIVE AND NEGATIVE NUMBERS

When using the basic operations (addition, subtraction, multiplication, division) on positive and negative numbers, we determine the result to be either positive or negative according to the following rules.

Addition of two numbers of the same sign *Add their absolute values and assign the sum their common sign.*

EXAMPLE 1 Adding numbers of the same sign

(a) $2 + 6 = 8$ — the sum of two positive numbers is positive

(b) $-2 + (-6) = -(2 + 6) = -8$ — the sum of two negative numbers is negative

■ From Section 1.1, we recall that a positive number is preceded by no sign. Therefore, in using these rules, we show the "sign" of a positive number by simply writing the number itself.

The negative number -6 is placed in parentheses because it is also preceded by a plus sign showing addition. It is not necessary to place the -2 in parentheses. ■

Addition of two numbers of different signs *Subtract the number of smaller absolute value from the number of larger absolute value and assign to the result the sign of the number of larger absolute value.*

EXAMPLE 2 Adding numbers of different signs

(a) $2 + (-6) = -(6 - 2) = -4$ ← the negative 6 has the larger absolute value

(b) $-6 + 2 = -(6 - 2) = -4$ ←

(c) $6 + (-2) = 6 - 2 = 4$ ← the positive 6 has the larger absolute value

(d) $-2 + 6 = 6 - 2 = 4$ ←

↑ the subtraction of absolute values ■

Subtraction of one number from another *Change the sign of the number being subtracted and change the subtraction to addition. Perform the addition.*

EXAMPLE 3 Subtracting positive and negative numbers

(a) $2 - 6 = 2 + (-6) = -(6 - 2) = -4$

Note that after changing the subtraction to addition, and changing the sign of 6 to make it -6, we have precisely the same illustration as Example 2(a).

(b) $-2 - 6 = -2 + (-6) = -(2 + 6) = -8$

Note that after changing the subtraction to addition, and changing the sign of 6 to make it -6, we have precisely the same illustration as Example 1(b).

(c) $-a - (-a) = -a + a = 0$

NOTE ▸ This shows that subtracting a number from itself results in zero, even if the number is negative. [Subtracting a negative number is equivalent to adding a positive number of the same absolute value.]

(d) $-2 - (-6) = -2 + 6 = 4$

(e) The change in temperature from $-12°C$ to $-26°C$ is
$-26°C - (-12°C) = -26°C + 12°C = -14°C$ ■

Multiplication and division of two numbers *The product (or quotient) of two numbers of the same sign is positive. The product (or quotient) of two numbers of different signs is negative.*

EXAMPLE 4 Multiplying and dividing positive and negative numbers

(a) $3(12) = 3 \times 12 = 36$ $\quad \frac{12}{3} = 4$ result is positive if both numbers are positive

(b) $-3(-12) = 3 \times 12 = 36$ $\quad \frac{-12}{-3} = 4$ result is positive if both numbers are negative

(c) $3(-12) = -(3 \times 12) = -36$ $\quad \frac{-12}{3} = -\frac{12}{3} = -4$ result is negative if one number is positive and the other is negative

(d) $-3(12) = -(3 \times 12) = -36$ $\quad \frac{12}{-3} = -\frac{12}{3} = -4$ ■

Practice Exercises

Evaluate: **1.** $-5 - (-8)$ **2.** $-5(-8)$

ORDER OF OPERATIONS

Often, how we are to combine numbers is clear by grouping the numbers using symbols such as **parentheses,** ( ); the **bar,** ____, between the numerator and denominator of a fraction; and **vertical lines** for absolute value. Otherwise, for an expression in which there are several operations, we use the following order of operations.

Order of Operations

1. Perform operations within grouping symbols (parentheses, brackets, or absolute value symbols).
2. Perform multiplications and divisions (from left to right).
3. Perform additions and subtractions (from left to right).

EXAMPLE 5 Order of operations

■ Note that $20 \div (2 + 3) = \frac{20}{2 + 3}$, whereas $20 \div 2 + 3 = \frac{20}{2} + 3$.

(a) $20 \div (2 + 3)$ is evaluated by first adding $2 + 3$ and then dividing. The grouping of $2 + 3$ is clearly shown by the parentheses. Therefore, $20 \div (2 + 3) = 20 \div 5 = 4$.

(b) $20 \div 2 + 3$ is evaluated by first dividing 20 by 2 and then adding. No specific grouping is shown, and therefore the division is done before the addition. This means $20 \div 2 + 3 = 10 + 3 = 13$.

NOTE ➧ **(c)** [$16 - 2 \times 3$ is evaluated by *first multiplying* 2 by 3 and then subtracting. We do not first subtract 2 from 16.] Therefore, $16 - 2 \times 3 = 16 - 6 = 10$.

(d) $16 \div 2 \times 4$ is evaluated by first dividing 16 by 2 and then multiplying. From left to right, the division occurs first. Therefore, $16 \div 2 \times 4 = 8 \times 4 = 32$.

(e) $|3 - 5| - |-3 - 6|$ is evaluated by first performing the subtractions within the absolute value vertical bars, then evaluating the absolute values, and then subtracting. This means that $|3 - 5| - |-3 - 6| = |-2| - |-9| = 2 - 9 = -7$. ■

Practice Exercises

Evaluate: **3.** $12 - 6 \div 2$ **4.** $16 \div (2 \times 4)$

When evaluating expressions, it is generally more convenient to change the operations and numbers so that the result is found by the addition and subtraction of positive numbers. When this is done, we must remember that

$$a + (-b) = a - b \qquad \textbf{(1.1)}$$

$$a - (-b) = a + b \qquad \textbf{(1.2)}$$

EXAMPLE 6 Evaluating numerical expressions

(a) $7 + (-3) - 6 = 7 - 3 - 6 = 4 - 6 = -2$ using Eq. (1.1)

(b) $\frac{18}{-6} + 5 - (-2)(3) = -3 + 5 - (-6) = 2 + 6 = 8$ using Eq. (1.2)

(c) $\frac{|3 - 15|}{-2} - \frac{8}{4 - 6} = \frac{12}{-2} - \frac{8}{-2} = -6 - (-4) = -6 + 4 = -2$

(d) $\frac{-12}{2 - 8} + \frac{5 - 1}{2(-1)} = \frac{-12}{-6} + \frac{4}{-2} = 2 + (-2) = 2 - 2 = 0$

In illustration (b), we see that the division and multiplication were done before the addition and subtraction. In (c) and (d), we see that the groupings were evaluated first. Then we did the divisions, and finally the subtraction and addition. ■

Practice Exercises

Evaluate: **5.** $2(-3) - \frac{4 - 8}{2}$ **6.** $\frac{|5 - 15|}{2} - \frac{-9}{3}$

EXAMPLE 7 Evaluating—velocity after collision

A 3000-lb van going at 40 mi/h ran head-on into a 2000-lb car going at 20 mi/h. An insurance investigator determined the velocity of the vehicles immediately after the collision from the following calculation. See Fig. 1.5.

$$\frac{3000(40) + (2000)(-20)}{3000 + 2000} = \frac{120{,}000 + (-40{,}000)}{3000 + 2000} = \frac{120{,}000 - 40{,}000}{5000}$$

$$= \frac{80{,}000}{5000} = 16 \text{ mi/h}$$

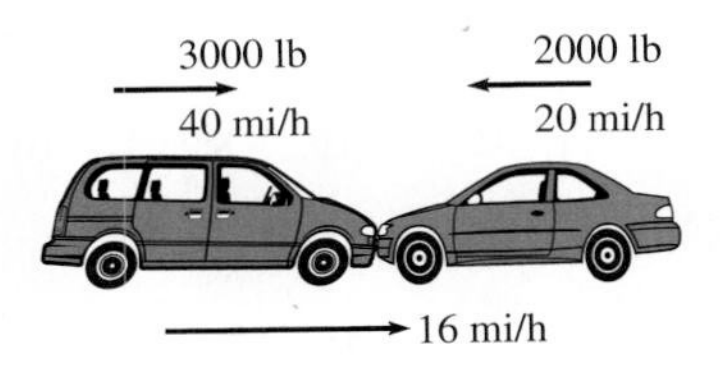

Fig. 1.5

The numerator and the denominator must be evaluated before the division is performed. The multiplications in the numerator are performed first, followed by the addition in the denominator and the subtraction in the numerator. ■

OPERATIONS WITH ZERO

Because operations with zero tend to cause some difficulty, we will show them here.

If a is a real number, the operations of addition, subtraction, multiplication, and division with zero are as follows:

$$a + 0 = a$$
$$a - 0 = a \qquad 0 - a = -a$$
$$a \times 0 = 0$$
$$0 \div a = \frac{0}{a} = 0 \qquad (\text{if } a \neq 0)$$

$\neq$ means "is not equal to"

EXAMPLE 8 Operations with zero

(a) $5 + 0 = 5$ **(b)** $-6 - 0 = -6$ **(c)** $0 - 4 = -4$

(d) $\frac{0}{6} = 0$ **(e)** $\frac{0}{-3} = 0$ **(f)** $\frac{5 \times 0}{7} = \frac{0}{7} = 0$ ■

Note that there is no result defined for division by zero. To understand the reason for this, consider the results for $\frac{6}{2}$ and $\frac{6}{0}$.

$$\frac{6}{2} = 3 \quad \text{since} \quad 2 \times 3 = 6$$

If $\frac{6}{0} = b$, then $0 \times b = 6$. This cannot be true because $0 \times b = 0$ for any value of b. Thus, ***division by zero is undefined***.

(The special case of $\frac{0}{0}$ is termed *indeterminate*. If $\frac{0}{0} = b$, then $0 = 0 \times b$, which is true for any value of b. Therefore, no specific value of b can be determined.)

EXAMPLE 9 Division by zero is undefined

$\frac{2}{5} \div 0$ is undefined $\quad \frac{8}{0}$ is undefined $\quad \frac{7 \times 0}{0 \times 6}$ is indeterminate ■

see above

CAUTION The operations with zero will not cause any difficulty if we remember to *never divide by zero*. ■

Division by zero is the only undefined basic operation. All the other operations with zero may be performed as for any other number.

EXERCISES 1.2

In Exercises 1–4, make the given changes in the indicated examples of this section, and then solve the resulting problems.

1. In Example 5(c), change 3 to (-2) and then evaluate.
2. In Example 6(b), change 18 to -18 and then evaluate.
3. In Example 6(d), interchange the 2 and 8 in the first denominator and then evaluate.
4. In the rightmost illustration in Example 9, interchange the 6 and the 0 above the 6. Is any other change needed?

In Exercises 5–38, evaluate each of the given expressions by performing the indicated operations.

5. $5 + (-8)$ **6.** $-4 + (-7)$ **7.** $-3 + 9$

8. $18 - 21$ **9.** $-19 - (-16)$ **10.** $-8 - (-10)$

11. $7(-4)$ **12.** $-9(3)$ **13.** $-7(-5)$

14. $\frac{-9}{3}$ **15.** $\frac{-6(20 - 10)}{-3}$ **16.** $\frac{-28}{-7(5 - 6)}$

17. $-2(4)(-5)$ 18. $-3(-4)(-6)$ 19. $2(2-7) \div 10$

20. $\dfrac{-64}{-2|4-8|}$ 21. $16 \div 2(-4)$ 22. $-20 \div 5(-4)$

23. $-9-|2-10|$ 24. $(7-7) \div (5-7)$ 25. $\dfrac{17-7}{7-7}$

26. $\dfrac{(7-7)(2)}{(7-7)(-1)}$ 27. $8-3(-4)$ 28. $-20+8 \div 4$

29. $-2(-6)+\left|\dfrac{8}{-2}\right|$ 30. $\dfrac{|-2|}{-2}-(-2)(-5)$

31. $10(-8)(-3) \div (10-50)$ 32. $\dfrac{7-|-5|}{-1(-2)}$

33. $\dfrac{24}{3+(-5)}-4(-9) \div (-3)$ 34. $\dfrac{-18}{3}-\dfrac{4-|-6|}{-1}$

35. $-7-\dfrac{|-14|}{2(2-3)}-3|6-8|$ 36. $-7(-3)+\dfrac{-6}{-3}-|-9|$

37. $\dfrac{3|-9-2(-3)|}{1+(-10)}$ 38. $\dfrac{20(-12)-40(-15)}{98-|-98|}$

In Exercises 39–46, determine which of the fundamental laws of algebra is demonstrated.

39. $6(7)=7(6)$ 40. $6+8=8+6$

41. $6(3+1)=6(3)+6(1)$ 42. $4(5 \times \pi)=(4 \times 5)(\pi)$

43. $3+(5+9)=(3+5)+9$ 44. $8(3-2)=8(3)-8(2)$

45. $(\sqrt{5} \times 3) \times 9=\sqrt{5} \times (3 \times 9)$

46. $(3 \times 6) \times 7=7 \times (3 \times 6)$

In Exercises 47–50, for numbers a and b, determine which of the following expressions equals the given expression.

(a) $a+b$ **(b)** $a-b$ **(c)** $b-a$ **(d)** $-a-b$

47. $-a+(-b)$ 48. $b-(-a)$

49. $-b-(-a)$ 50. $-a-(-b)$

In Exercises 51–66, solve the given problems. Refer to Appendix B for units of measurement and their symbols.

51. Insert the proper sign ($=, >, <$) to make the following true: $|5-(-2)| \quad |-5-|-2||$

52. Insert the proper sign ($=, >, <$) to make the following true: $|-3-|-7|| \quad ||-3|-7|$

53. (a) What is the sign of the product of an even number of negative numbers? (b) What is the sign of the product of an odd number of negative numbers?

54. Is subtraction commutative? Explain.

55. Explain why the following definition of the absolute value of a real number x is either correct or incorrect (the symbol $\geq$ means "is equal to or greater than"): If $x \geq 0$, then $|x|=x$; if $x<0$, then $|x|=-x$.

56. Explain what is the error if the expression $24-6 \div 2 \cdot 3$ is evaluated as 27. What is the correct value?

57. Describe the values of x and y for which (a) $-xy=1$ and (b) $\dfrac{x-y}{x-y}=1$.

58. Describe the values of x and y for which (a) $|x+y|=|x|+|y|$ and (b) $|x-y|=|x|+|y|$.

59. The changes in the price of a stock (in dollars) for a given week were -0.68, $+0.42$, $+0.06$, -0.11, and $+0.02$. What was the total change in the stock's price that week?

60. Using subtraction of signed numbers, find the difference in the altitude of the bottom of the Dead Sea, 1396 m below sea level, and the bottom of Death Valley, 86 m below sea level.

61. Some solar energy systems are used to supplement the utility company power supplied to a home such that the meter runs backward if the solar energy being generated is greater than the energy being used. With such a system, if the solar power averages 1.5 kW for a 3.0-h period and only 2.1 kW · h is used during this period, what will be the change in the meter reading for this period? *Hint:* Solar power generated makes the meter run in the negative direction while power used makes it run in the positive direction.

62. A baseball player's batting average (total number of hits divided by total number of at-bats) is expressed in decimal form from 0.000 (no hits for all at-bats) to 1.000 (one hit for each at-bat). A player's batting average is often shown as 0.000 before the first at-bat of the season. Is this a correct batting average? Explain.

63. The daily high temperatures (in °C) for Edmonton, Alberta, in the first week in March were recorded as $-7, -3, 2, 3, 1, -4$, and -6. What was the average daily temperature for the week? (Divide the algebraic sum of readings by the number of readings.)

64. A flare is shot up from the top of a tower. Distances above the flare gun are positive and those below it are negative. After 5 s the vertical distance (in ft) of the flare from the flare gun is found by evaluating $(70)(5)+(-16)(25)$. Find this distance.

65. Find the sum of the voltages of the batteries shown in Fig. 1.6. Note the directions in which they are connected.

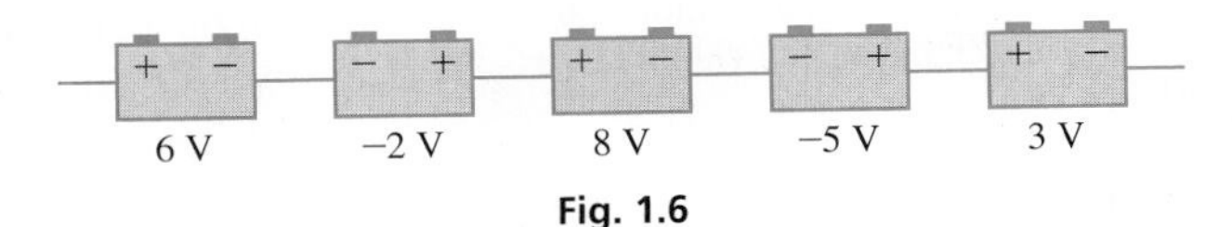

Fig. 1.6

66. A faulty gauge on a fire engine pump caused the apparent pressure in the hose to change every few seconds. The pressures (in lb/in.^{2} above and below the set pressure were recorded as: $+7, -2, -9, -6$. What was the change between (a) the first two readings, (b) between the middle two readings, and (c) the last two readings?

67. One oil-well drilling rig drills 100 m deep the first day and 200 m deeper the second day. A second rig drills 200 m deep the first day and 100 m deeper the second day. In showing that the total depth drilled by each rig was the same, state what fundamental law of algebra is illustrated.

68. A water tank leaks 12 gal each hour for 7 h, and a second tank leaks 7 gal each hour for 12 h. In showing that the total amount leaked is the same for the two tanks, what fundamental law of algebra is illustrated?

69. On each of the 7 days of the week, a person spends 25 min on Facebook and 15 min on Twitter. Set up the expression for the total time spent on these two sites that week. What fundamental law of algebra is illustrated?

70. A jet travels 600 mi/h relative to the air. The wind is blowing at 50 mi/h. If the jet travels with the wind for 3 h, set up the expression for the distance traveled. What fundamental law of algebra is illustrated?

Answers to Practice Exercises

1. 3 **2.** 40 **3.** 9 **4.** 2 **5.** −4 **6.** 8

1.3 Calculators and Approximate Numbers

Graphing Calculators • Approximate Numbers • Significant Digits • Accuracy and Precision • Rounding Off • Operations with Approximate Numbers • Estimating Results

You will be doing many of your calculations on a calculator, and a *graphing calculator* can be used for these calculations and many other operations. In this text, we will restrict our coverage of calculator use to graphing calculators because a *scientific calculator* cannot perform many of the required operations we will cover.

A brief discussion of the graphing calculator appears in Appendix C, and sample calculator screens appear throughout the book. Since there are many models of graphing calculators, *the notation and screen appearance for many operations will differ from one model to another*. You should *practice using your calculator and* ***review its manual*** *to be sure how it is used*. Following is an example of a basic calculation done on a graphing calculator.

■ The calculator screens shown with text material are for a TI-84 Plus. They are intended only as an illustration of a calculator screen for the particular operation. Screens for other models may differ.

EXAMPLE 1 Calculating on a graphing calculator

In order to calculate the value of $38.3 - 12.9(-3.58)$, the numbers are entered as follows. The calculator will perform the multiplication first, following the order of operations shown in Section 1.2. The sign of -3.58 is entered using the [(−)] key, before 3.58 is entered. The display on the calculator screen is shown in Fig. 1.7.

38.3 [−] 12.9 [×] [(−)] 3.58 [ENTER] keystrokes

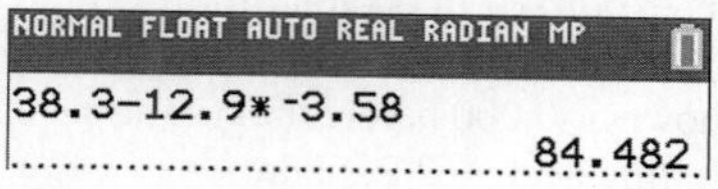

NORMAL FLOAT AUTO REAL RADIAN MP
38.3-12.9*-3.58
84.482

Fig. 1.7

This means that $38.3 - 12.9(-3.58) = 84.482$.

Note in the display that the negative sign of -3.58 is smaller and a little higher to distinguish it from the minus sign for subtraction. Also note the * shown for multiplication; the asterisk is the standard computer symbol for multiplication. ■

■ When less than half of a calculator screen is needed, a partial screen will be shown.

Looking back into Section 1.2, we see that *the minus sign is used in two different ways:* (1) to indicate subtraction and (2) to designate a negative number. This is clearly shown on a graphing calculator because there is a key for each purpose. The [−] key is used for subtraction, and the [(−)] key is used before a number to make it negative.

■ Some calculator keys on different models are labeled differently. For example, on some models, the EXE key is equivalent to the ENTER key.

We will first use a graphing calculator for the purpose of graphing in Section 3.5. Before then, we will show some calculational uses of a graphing calculator.

■ Calculator keystrokes for various operations can be found by using the URLs given in this text. A list of all the calculator instructions is at goo.gl/eAUgW3.

APPROXIMATE NUMBERS AND SIGNIFICANT DIGITS

Most numbers in technical and scientific work are **approximate numbers,** having been determined by some *measurement*. Certain other numbers are **exact numbers,** having been determined by a *definition* or *counting process*.

EXAMPLE 2 Approximate numbers and exact numbers

One person measures the distance between two cities on a map as 36 cm, and another person measures it as 35.7 cm. However, the distance cannot be measured *exactly*.

If a computer prints out the number of names on a list of 97, this 97 is exact. We know it is not 96 or 98. Since 97 was found from precise counting, it is exact.

By definition, 60 s = 1 min, and the 60 and the 1 are exact. ■

An approximate number may have to include some zeros to properly locate the decimal point. *Except for these zeros, all other digits are called* **significant digits.**

EXAMPLE 3 Significant digits

All numbers in this example are assumed to be approximate.

(a) 34.7 has three significant digits.

(b) 0.039 has two significant digits. The zeros properly locate the decimal point.

(c) 706.1 has four significant digits. The zero is not used for the location of the decimal point. It shows the number of tens in 706.1.

(d) 5.90 has three significant digits. ***The zero is not necessary as a placeholder*** and should not be written unless it is significant.

(e) 1400 has two significant digits, unless information is known about the number that makes either or both zeros significant. Without such information, we assume that the zeros are placeholders for proper location of the decimal point.

(f) Other approximate numbers with the number of significant digits are 0.0005 (one), 960,000 (two), 0.0709 (three), 1.070 (four), and 700.00 (five). ■

■ To show that zeros at the end of a whole number are significant, a notation that can be used is to place a bar over the last significant zero. Using this notation, $78{,}0\bar{0}0$ is shown to have four significant digits.

Practice Exercises

Determine the number of significant digits.

1. 1010 **2.** 0.1010

From Example 3, we see that *all nonzero digits are significant. Also, zeros not used as placeholders (for location of the decimal point) are significant.*

In calculations with approximate numbers, the number of significant digits and the position of the decimal point are important. *The* **accuracy** *of a number refers to the number of significant digits it has,* whereas *the* **precision** *of a number refers to the decimal position of the last significant digit.*

EXAMPLE 4 Accuracy and precision

One technician measured the thickness of a metal sheet as 3.1 cm and another technician measured it as 3.12 cm. Here, 3.12 is more precise since its last digit represents hundredths and 3.1 is expressed only to tenths. Also, 3.12 is more accurate since it has three significant digits and 3.1 has only two.

A concrete driveway is 230 ft long and 0.4 ft thick. Here, 230 is more accurate (two significant digits) and 0.4 is more precise (expressed to tenths). ■

The last significant digit of an approximate number is not exact. It has usually been determined by estimating or *rounding off.* However, it is not off by more than one-half of a unit in its place value.

EXAMPLE 5 Meaning of the last digit of an approximate number

When we write the measured distance on the map in Example 2 as 35.7 cm, we are saying that the distance is at least 35.65 cm and no more than 35.75 cm. Any value between these, rounded off to tenths, would be 35.7 cm.

In changing the fraction $\frac{2}{3}$ to the approximate decimal value 0.667, we are saying that the value is between 0.6665 and 0.6675. ■

■ On graphing calculators, it is possible to set the number of decimal places (to the right of the decimal point) to which results will be rounded off.

To **round off** *a number to a specified number of significant digits, discard all digits to the right of the last significant digit (replace them with zeros if needed to properly place the decimal point). If the first digit discarded is 5 or more, increase the last significant digit by 1 (round up). If the first digit discarded is less than 5, do not change the last significant digit (round down).*

EXAMPLE 6 Rounding off

(a) 70,360 rounded off to three significant digits is 70,400. Here, 3 is the third significant digit and the next digit is 6. Because $6 > 5$, the 3 is rounded up to 4 and the 6 is replaced with a zero to hold the place value.

(b) 70,430 rounded off to three significant digits, or to the nearest hundred, is 70,400. Here the 3 is replaced with a zero.

(c) 187.35 rounded off to four significant digits, or to tenths, is 187.4.

(d) 187.349 rounded off to four significant digits is 187.3. *We do not round up the* 4 *and then round up the* 3.

NOTE ➧ **(e)** 35.003 rounded off to four significant digits is 35.00. [We do not discard the zeros because they are significant and are not used only to properly place the decimal point.]

(f) 849,720 rounded off to three significant digits is $85\bar{0},000$. *The bar over the zero shows that digit is significant.* ■

Practice Exercises

Round off each number to three significant digits.

3. 2015 **4.** 0.3004

OPERATIONS WITH APPROXIMATE NUMBERS

NOTE ➧ [When performing operations on approximate numbers, we must express the result to an accuracy or precision that is valid.] Consider the following examples.

EXAMPLE 7 Precision—length of pipe

A pipe is made in two sections. One is measured as 16.3 ft long and the other as 0.927 ft long. What is the total length of the two sections together?

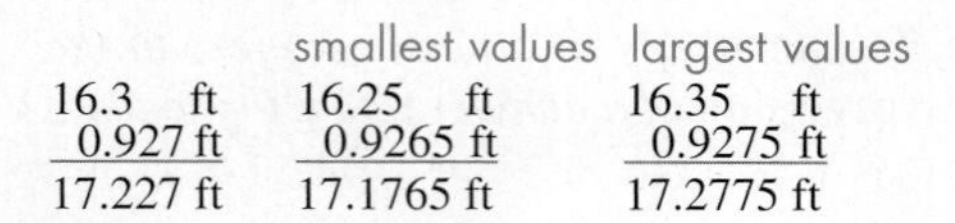

It may appear that we simply add the numbers as shown at the left. However, both numbers are approximate, and adding the smallest possible values and the largest possible values, the result differs by 0.1 (17.2 and 17.3) when rounded off to tenths. Rounded off to hundredths (17.18 and 17.28), they do not agree at all because the tenths digit is different. Thus, we get a good approximation for the total length if it is rounded off to *tenths,* the precision of the least precise length, and it is written as 17.2 ft. ■

EXAMPLE 8 Accuracy—area of land plot

We find the area of the rectangular piece of land in Fig. 1.8 by multiplying the length, 207.54 ft, by the width, 81.4 ft. Using a calculator, we find that $(207.54)(81.4) = 16{,}893.756$. This apparently means the area is 16,893.756 ft^2.

However, the area should not be expressed with this accuracy. Because the length and width are both approximate, we have

$$(207.535\text{ ft})(81.35\text{ ft}) = 16{,}882.97225\text{ ft}^2 \quad \text{least possible area}$$

$$(207.545\text{ ft})(81.45\text{ ft}) = 16{,}904.54025\text{ ft}^2 \quad \text{greatest possible area}$$

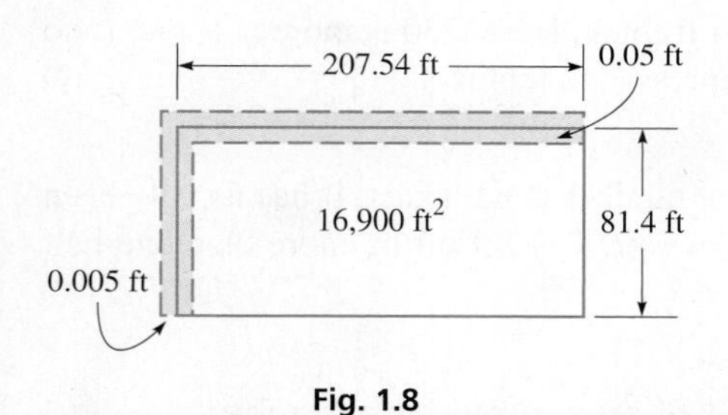

Fig. 1.8

These values agree when rounded off to three significant digits (16,900 ft^2) but do not agree when rounded off to a greater accuracy. Thus, we conclude that the result is accurate only to *three* significant digits, the accuracy of the least accurate measurement, and that the area is written as 16,900 ft^2. ■

■ The results of operations on approximate numbers shown at the right are based on reasoning that is similar to that shown in Examples 7 and 8.

The Result of Operations on Approximate Numbers

1. When approximate numbers are added or subtracted, the result is expressed with the precision of the least precise number.
2. When approximate numbers are multiplied or divided, the result is expressed with the accuracy of the least accurate number.
3. When the root of an approximate number is found, the result is expressed with the accuracy of the number.
4. When approximate numbers and exact numbers are involved, the accuracy of the result is limited only by the approximate numbers.

CAUTION Always express the result of a calculation with the proper accuracy or precision. When using a calculator, if additional digits are displayed, round off the *final* result (do not round off in any of the intermediate steps). ■

■ When rounding off a number, it may seem difficult to discard the extra digits. However, if you keep those digits, you show a number with too great an accuracy, and it is incorrect to do so.

EXAMPLE 9 Adding approximate numbers

Find the sum of the approximate numbers 73.2, 8.0627, and 93.57.

Showing the addition in the standard way and using a calculator, we have

$$\begin{array}{r} 73.2 \\ 8.0627 \\ 93.57 \\ \hline 174.8327 \end{array}$$

73.2 ← least precise number (expressed to tenths)

174.8327 ← final display must be rounded to tenths

Therefore, the sum of these approximate numbers is 174.8. ■

Practice Exercise

Evaluate using a calculator.

5. $40.5 + \frac{3275}{-60.041}$ (Numbers are approximate.)

EXAMPLE 10 Multiplying approximate numbers

In finding the product of the approximate numbers 2.4832 and 30.5 on a calculator, the final display shows 75.7376. However, since 30.5 has only three significant digits, the product is 75.7. ■

EXAMPLE 11 Combined operations

For problems with multiple operations, follow the correct order of operations as given in Section 1.2. Keep all the digits in the intermediate steps, but keep track of (perhaps by underlining) the significant digits that would be retained according to the appropriate rounding rule for each step. Then round off the final answer according to the last operation that is performed. For example,

$$(4.265 \times 2.60) \div (3.7 + 5.14) = \underline{11.0}89 \div \underline{8.8}4 = 1.3$$

Note that three significant digits are retained from the multiplication and one decimal place precision is retained from the addition. The final answer is rounded off to two significant digits, which is the accuracy of the least accurate number in the final division (based on the underlined significant digits). ■

EXAMPLE 12 Operations with exact numbers and approximate numbers

Using the exact number 600 and the approximate number 2.7, we express the result to tenths if the numbers are added or subtracted. If they are multiplied or divided, we express the result to two significant digits. Since 600 is exact, the accuracy of the result depends only on the approximate number 2.7.

$$600 + 2.7 = 602.7 \qquad 600 - 2.7 = 597.3$$
$$600 \times 2.7 = 1600 \qquad 600 \div 2.7 = 220$$

There are 16 pieces in a pile of lumber and the average length of a piece is 482 mm. Here 16 is exact, but 482 is approximate. To get the total length of the pieces in the pile, the product $16 \times 482 = 7712$ must be rounded off to three significant digits, the accuracy of 482. Therefore, we can state that the total length is about 7710 mm. ■

NOTE ➧ [A note regarding the equal sign (=) is in order. We will use it for its defined meaning of "equals exactly" and when the result is an approximate number that has been properly rounded off.] Although $\sqrt{27.8} \approx 5.27$, where ≈ means "equals approximately," we write $\sqrt{27.8} = 5.27$, since 5.27 has been properly rounded off.

You should *make a rough estimate* of the result when using a calculator. An estimation may prevent accepting an incorrect result after using an incorrect calculator sequence, particularly if the calculator result is far from the estimated value.

EXAMPLE 13 Estimating results

In Example 1, we found that

$$38.3 - 12.9(-3.58) = 84.482 \quad \text{using exact numbers}$$

When using the calculator, if we forgot to make 3.58 negative, the display would be −7.882, or if we incorrectly entered 38.3 as 83.3, the display would be 129.482.

However, if we estimate the result as

$$40 - 10(-4) = 80$$

we know that a result of −7.882 or 129.482 cannot be correct.

When estimating, we can often use one-significant-digit approximations. If the calculator result is far from the estimate, we should do the calculation again. ■

EXERCISES 1.3

In Exercises 1–4, make the given changes in the indicated examples of this section, and then solve the given problems.

1. In Example 3(b), change 0.039 to 0.390. Is there any change in the conclusion?

2. In Example 6(e), change 35.003 to 35.303 and then find the result.

3. In Example 10, change 2.4832 to 2.5 and then find the result.

4. In Example 13, change 12.9 to 21.9 and then find the estimated value.

In Exercises 5–10, determine whether the given numbers are approximate or exact.

5. A car with 8 cylinders travels at 55 mi/h.

6. A computer chip 0.002 mm thick is priced at $7.50.

7. In 24 h there are 1440 min.

8. A calculator has 50 keys, and its battery lasted for 50 h of use.

9. A cube of copper 1 cm on an edge has a mass of 9 g.

10. Of a building's 90 windows, 75 were replaced 15 years ago.

In Exercises 11–18, determine the number of significant digits in each of the given approximate numbers.

11. 107; 3004; 1040 **12.** 3600; 730; 2055

13. 6.80; 6.08; 0.068 **14.** 0.8730; 0.0075; 0.0305

15. 3000; 3000.1; 3000.10 **16.** 1.00; 0.01; 0.0100

17. 5000; 5000.0; $500\bar{0}$ **18.** 200; $20\bar{0}$; 200.00

In Exercises 19–24, determine which of the pair of approximate numbers is (a) more precise and (b) more accurate.

19. 30.8; 0.010 **20.** 0.041; 7.673

21. 0.1; 78.0 **22.** 7040; 0.004

23. 7000; 0.004 **24.** 50.060; $|-8.914|$

In Exercises 25–32, round off the given approximate numbers (a) to three significant digits and (b) to two significant digits.

25. 4.936 **26.** 80.53 **27.** −50.893 **28.** 7.004

29. 5968 **30.** 30.96 **31.** 0.9449 **32.** 0.9999

In Exercises 33–42, perform the indicated operations assuming all numbers are approximate. Round your answers using the procedure shown in Example 11.

33. $12.78 + 1.0495 - 1.633$ **34.** $3.64(17.06)$

35. $0.6572 \times 3.94 - 8.651$ **36.** $41.5 - 26.4 \div 3.7$

37. $8.75 + (1.2)(3.84)$ **38.** $28 - \dfrac{20.955}{2.2}$

39. $\dfrac{8.75(15.32)}{8.75 + 15.32}$ **40.** $\dfrac{8.97(4.003)}{2.0 + 4.78}$

41. $4.52 - \dfrac{2.056(309.6)}{395.2}$ **42.** $8.195 + \dfrac{14.9}{1.7 + 2.1}$

In Exercises 43–46, perform the indicated operations. The first number is approximate, and the second number is exact.

43. $0.9788 + 14.9$ **44.** $17.311 - 22.98$

45. $-3.142(65)$ **46.** $8.62 \div 1728$

In Exercises 47–50, answer the given questions. Refer to Appendix B for units of measurement and their symbols.

47. The manual for a heart monitor lists the frequency of the ultrasound wave as 2.75 MHz. What are the least possible and the greatest possible frequencies?

48. A car manufacturer states that the engine displacement for a certain model is 2400 cm^3. What should be the least possible and greatest possible displacements?

49. A flash of lightning struck a tower 3.25 mi from a person. The thunder was heard 15 s later. The person calculated the speed of sound and reported it as 1144 ft/s. What is wrong with this conclusion?

50. A technician records 4.4 s as the time for a robot arm to swing from the extreme left to the extreme right, 2.72 s as the time for the return swing, and 1.68 s as the difference in these times. What is wrong with this conclusion?

In Exercises 51–58, perform the calculations on a calculator without rounding.

51. Evaluate: (a) $2.2 + 3.8 \times 4.5$ (b) $(2.2 + 3.8) \times 4.5$

52. Evaluate: (a) $6.03 \div 2.25 + 1.77$ (b) $6.03 \div (2.25 + 1.77)$

53. Evaluate: (a) $2 + 0$ (b) $2 - 0$ (c) $0 - 2$ (d) 2×0 (e) $2 \div 0$ Compare with operations with zero on page 10.

54. Evaluate: (a) $2 \div 0.0001$ and $2 \div 0$ (b) $0.0001 \div 0.0001$ and $0 \div 0$ (c) Explain why the displays differ.

55. Show that π is not equal exactly to (a) 3.1416, or (b) $22/7$.

56. At some point in the decimal equivalent of a rational number, some sequence of digits will start repeating endlessly. An irrational number never has an endlessly repeating sequence of digits. Find the decimal equivalents of (a) $8/33$ and (b) π. Note the repetition for $8/33$ and that no such repetition occurs for π.

57. Following Exercise 56, show that the decimal equivalents of the following fractions indicate they are rational: (a) $1/3$ (b) $5/11$ (c) $2/5$. What is the repeating part of the decimal in (c)?

58. Following Exercise 56, show that the decimal equivalent of the fraction $124/990$ indicates that it is rational. Why is the last digit different?

In Exercises 59–64, assume that all numbers are approximate unless stated otherwise.

59. In 3 successive days, a home solar system produced 32.4 MJ, 26.704 MJ, and 36.23 MJ of energy. What was the total energy produced in these 3 days?

60. A shipment contains eight plasma televisions, each weighing 68.6 lb, and five video game consoles, each weighing 15.3 lb. What is the total weight of the shipment?

61. Certain types of iPhones and iPads weigh approximately 129 g and 298.8 g, respectively. What is the total weight of 12 iPhones and 16 iPads of these types? (*Source:* Apple.com.)

62. Find the voltage in a certain electric circuit by multiplying the sum of the resistances 15.2 Ω, 5.64 Ω, and 101.23 Ω by the current 3.55 A.

63. The percent of alcohol in a certain car engine coolant is found by performing the calculation $\dfrac{100(40.63 + 52.96)}{105.30 + 52.96}$. Find this percent of alcohol. The number 100 is exact.

64. The tension (in N) in a cable lifting a crate at a construction site was found by calculating the value of $\dfrac{50.45(9.80)}{1 + 100.9 \div 23}$, where the 1 is exact. Calculate the tension.

In Exercises 65 and 66, all numbers are approximate. (a) Estimate the result mentally using one-significant-digit approximations of all the numbers, and (b) compute the result using the appropriate rounding rules and compare with the estimate.

65. $7.84 \times 4.932 - 11.317$

66. $21.6 - 53.14 \div 9.64$

Answers to Practice Exercises

1. 3 **2.** 4 **3.** 2020 **4.** 0.300 **5.** -14.0

1.4 Exponents and Unit Conversions

Positive Integer Exponents • Zero and Negative Exponents • Order of Operations • Evaluating Algebraic Expressions • Converting Units

In mathematics and its applications, we often have a number multiplied by itself several times. To show this type of product, we use the notation a^n, where a is the number and n is the number of times it appears. *In the expression a^n, the number a is called the* **base,** *and n is called the* **exponent;** in words, *a^n is read as "the* ***n*th power of *a*.**"

EXAMPLE 1 Meaning of exponents

(a) $4 \times 4 \times 4 \times 4 \times 4 = 4^5$ — the fifth power of 4

(b) $(-2)(-2)(-2)(-2) = (-2)^4$ — the fourth power of -2

(c) $a \times a = a^2$ — the second power of a, called "a squared"

(d) $\left(\frac{1}{5}\right)\left(\frac{1}{5}\right)\left(\frac{1}{5}\right) = \left(\frac{1}{5}\right)^3$ — the third power of $\frac{1}{5}$, called "$\frac{1}{5}$ cubed" ■

We now state the basic operations with exponents using positive integers as exponents. Therefore, with m and n as positive integers, we have the following operations:

$$a^m \times a^n = a^{m+n} \tag{1.3}$$

$$\frac{a^m}{a^n} = a^{m-n} \, (m > n, a \neq 0) \qquad \frac{a^m}{a^n} = \frac{1}{a^{n-m}} \quad (m < n, a \neq 0) \tag{1.4}$$

$$(a^m)^n = a^{mn} \tag{1.5}$$

$$(ab)^n = a^n b^n \qquad \left(\frac{a}{b}\right)^n = \frac{a^n}{b^n} \quad (b \neq 0) \tag{1.6}$$

■ Two forms are shown for Eqs. (1.4) in order that the resulting exponent is a positive integer. We consider negative and zero exponents after the next three examples.

EXAMPLE 2 Illustrating Eqs. (1.3) and (1.4)

■ In a^3, which equals $a \times a \times a$, each a is called a factor. A more general definition of factor is given in Section 1.7.

Using Eq. (1.3): (add exponents)

$$a^3 \times a^5 = a^{3+5} = a^8$$

Using the meaning of exponents: (3 factors of a)(5 factors of a), 8 factors of a

$$a^3 \times a^5 = (a \times a \times a)(a \times a \times a \times a \times a) = a^8$$

■ Here we are using the fact that a (not zero) divided by itself equals 1, or $a/a = 1$.

Using first form Eq. (1.4): ($5 > 3$)

$$\frac{a^5}{a^3} = a^{5-3} = a^2$$

Using the meaning of exponents:

$$\frac{a^5}{a^3} = \frac{\overset{1}{\cancel{a}} \times \overset{1}{\cancel{a}} \times \overset{1}{\cancel{a}} \times a \times a}{\underset{1}{\cancel{a}} \times \underset{1}{\cancel{a}} \times \underset{1}{\cancel{a}}} = a^2$$

■ Note that Eq. (1.3) can be verified numerically, for example, by
$2^3 \times 2^5 = 8 \times 32 = 256$
$2^3 \times 2^5 = 2^{3+5} = 2^8 = 256$

Using second form Eq. (1.4): ($5 > 3$)

$$\frac{a^3}{a^5} = \frac{1}{a^{5-3}} = \frac{1}{a^2}$$

Using the meaning of exponents:

$$\frac{a^3}{a^5} = \frac{\overset{1}{\cancel{a}} \times \overset{1}{\cancel{a}} \times \overset{1}{\cancel{a}}}{\underset{1}{\cancel{a}} \times \underset{1}{\cancel{a}} \times \underset{1}{\cancel{a}} \times a \times a} = \frac{1}{a^2}$$ ■

EXAMPLE 3 Illustrating Eqs. (1.5) and (1.6)

Using Eq. (1.5): (multiply exponents)

$$(a^5)^3 = a^{5(3)} = a^{15}$$

Using the meaning of exponents:

$$(a^5)^3 = (a^5)(a^5)(a^5) = a^{5+5+5} = a^{15}$$

Using first form Eq. (1.6):

$$(ab)^3 = a^3b^3$$

Using the meaning of exponents:

$$(ab)^3 = (ab)(ab)(ab) = a^3b^3$$

Using second form Eq. (1.6):

$$\left(\frac{a}{b}\right)^3 = \frac{a^3}{b^3}$$

Using the meaning of exponents:

$$\left(\frac{a}{b}\right)^3 = \left(\frac{a}{b}\right)\left(\frac{a}{b}\right)\left(\frac{a}{b}\right) = \frac{a^3}{b^3}$$ ■

CAUTION When an expression involves a product or a quotient of different bases, only exponents of the same base may be combined. ■ Consider the following example.

EXAMPLE 4 Other illustrations of exponents

CAUTION In illustration (b), note that ax^2 means a times the square of x and does not mean a^2x^2, whereas $(ax)^3$ does mean a^3x^3. ■

(a) $(-x^2)^3 = [(-1)x^2]^3 = (-1)^3(x^2)^3 = -x^6$

(b) $ax^2(ax)^3 = ax^2(a^3x^3) = a^4x^5$ (exponent of 1; add exponents of a; add exponents of x)

(c) $\dfrac{(3 \times 2)^4}{(3 \times 5)^3} = \dfrac{3^4 2^4}{3^3 5^3} = \dfrac{3 \times 2^4}{5^3}$

(d) $\dfrac{(ry^3)^2}{r(y^2)^4} = \dfrac{r^2y^6}{ry^8} = \dfrac{r}{y^2}$ ■

Practice Exercises

Use Eqs. (1.3)–(1.6) to simplify the given expressions.

1. $ax^3(-ax)^2$ **2.** $\dfrac{(2c)^5}{(3cd)^2}$

EXAMPLE 5 Exponents—beam deflection

In analyzing the amount a beam bends, the following simplification may be used. (P is the force applied to a beam of length L; E and I are constants related to the beam.)

$$\frac{1}{2}\left(\frac{PL}{4EI}\right)\left(\frac{2}{3}\right)\left(\frac{L}{2}\right)^2 = \frac{1}{2}\left(\frac{PL}{4EI}\right)\left(\frac{2}{3}\right)\left(\frac{L^2}{2^2}\right)$$

$$= \frac{\overset{1}{\cancel{2}}PL(L^2)}{\underset{1}{\cancel{2}}(3)(4)(4)EI} = \frac{PL^3}{48EI}$$

In *simplifying* this expression, we combined exponents of L and divided out the 2 that was a factor common to the numerator and the denominator. ■

ZERO AND NEGATIVE EXPONENTS

If $n = m$ in Eqs. (1.4), we have $a^m/a^m = a^{m-m} = a^0$. Also, $a^m/a^m = 1$, since any nonzero quantity divided by itself equals 1. Therefore, for Eqs. (1.4) to hold for $m = n$,

$$a^0 = 1 \qquad (a \neq 0) \tag{1.7}$$

Equation (1.7) states that *any nonzero expression raised to the zero power is* 1. Zero exponents can be used with any of the operations for exponents.

EXAMPLE 6 Zero as an exponent

(a) $5^0 = 1$ **(b)** $(-3)^0 = 1$ **(c)** $-(-3)^0 = -1$ **(d)** $(2x)^0 = 1$

(e) $(ax + b)^0 = 1$ **(f)** $(a^2b^0c)^2 = a^4c^2$ ($b^0 = 1$) **(g)** $2t^0 = 2(1) = 2$

We note in illustration (g) that *only t is raised to the zero power*. If the quantity $2t$ were raised to the zero power, it would be written as $(2t)^0$, as in part (d). ■

Practice Exercise

3. Evaluate: $-(3x)^0$

Applying both forms of Eq. (1.4) to the case where $n > m$ leads to the definition of a negative exponent. For example, applying both forms to a^2/a^7, we have

$$\frac{a^2}{a^7} = a^{2-7} = a^{-5} \quad \text{and} \quad \frac{a^2}{a^7} = \frac{1}{a^{7-2}} = \frac{1}{a^5}$$

For these results to be equal, then $a^{-5} = 1/a^5$. Thus, if we define

$$a^{-n} = \frac{1}{a^n} \qquad (a \neq 0) \tag{1.8}$$

then all the laws of exponents will hold for negative integers.

■ Although positive exponents are generally preferred in a final result, there are some cases in which zero or negative exponents are to be used. Also, negative exponents are very useful in some operations that we will use later.

EXAMPLE 7 Negative exponents

■ Note carefully the difference in parts (d) and (e) of Example 7.

(a) $3^{-1} = \frac{1}{3}$ **(b)** $4^{-2} = \frac{1}{4^2} = \frac{1}{16}$ **(c)** $\frac{1}{a^{-3}} = a^3$ change signs of exponents

(d) $(3x)^{-1} = \frac{1}{3x}$ **(e)** $3x^{-1} = 3\left(\frac{1}{x}\right) = \frac{3}{x}$ **(f)** $\left(\frac{a^3}{x}\right)^{-2} = \frac{(a^3)^{-2}}{x^{-2}} = \frac{a^{-6}}{x^{-2}} = \frac{x^2}{a^6}$ ■

Practice Exercises

Simplify: **4.** $\frac{-7^0}{c^{-3}}$ **5.** $\frac{(3x)^{-1}}{2a^{-2}}$

ORDER OF OPERATIONS

We have seen that the basic operations on numbers must be performed in a particular order. Since raising a number to a power is actually multiplication, it is performed before additions and subtractions, and in fact, before multiplications and divisions.

■ The use of exponents is taken up in more detail in Chapter 11.

Order of Operations

1. Operations within grouping symbols
2. Exponents
3. Multiplications and divisions (from left to right)
4. Additions and subtractions (from left to right)

EXAMPLE 8 Using order of operations

(a) $8 - (-1)^2 - 2(-3)^2 = 8 - 1 - 2(9)$

$= 8 - 1 - 18 = -11$ apply exponents first, then multiply, and then subtract

(b) $806 \div (26.1 - 9.09)^2 = 806 \div (\underline{17.01})^2$

$= 806 \div \underline{289.3}401 = 2.79$ subtract inside parentheses first, then square the answer, and then divide

NOTE ▸ [In part (b), the significant digits retained from each intermediate step are underlined.] ■

EXAMPLE 9 Even and odd powers

Using the meaning of a power of a number, we have

$$(-2)^2 = (-2)(-2) = 4 \qquad (-2)^3 = (-2)(-2)(-2) = -8$$

$$(-2)^4 = 16 \qquad (-2)^5 = -32 \qquad (-2)^6 = 64 \qquad (-2)^7 = -128$$

NOTE ▸ [Note that a negative number raised to an even power gives a positive value, and a negative number raised to an odd power gives a negative value.] ■

EVALUATING ALGEBRAIC EXPRESSIONS

An algebraic expression is **evaluated** *by* **substituting** *given values of the literal numbers in the expression and calculating the result.* On a calculator, the x^2 key is used to square numbers, and the ∧ or x^y key is used for other powers.

To calculate the value of $20 \times 6 + 200/5 - 3^4$, we use the key sequence

20 × 6 + 200 ÷ 5 − 3 ∧ 4

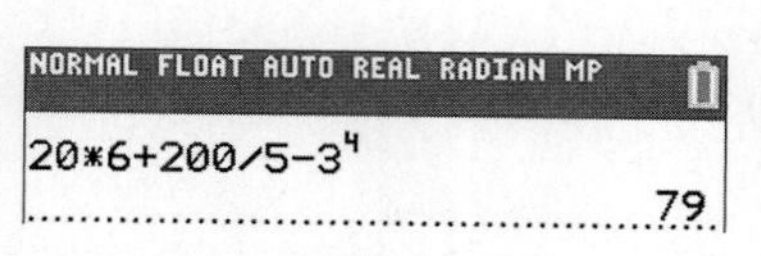

Fig. 1.9

with the result of 79 shown in the display of Fig. 1.9. **Note that calculators are programmed to follow the correct order of operations.**

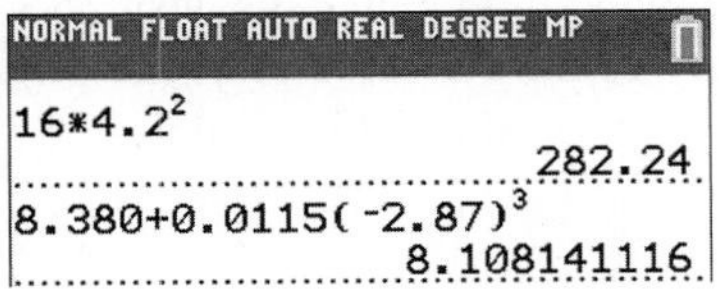

EXAMPLE 10 Evaluating an expression—free-fall distance

The distance (in ft) that an object falls in 4.2 s is found by substituting 4.2 for t in the expression $16.0t^2$ as shown below:

$$16.0(4.2)^2 = 280 \text{ ft}$$

```
NORMAL FLOAT AUTO REAL DEGREE MP
16*4.2²
                          282.24
8.380+0.0115(-2.87)³
                     8.108141116
```

Fig. 1.10

The calculator result from the first line of Fig. 1.10 has been rounded off to two significant digits, the accuracy of 4.2. ■

EXAMPLE 11 Evaluating an expression—length of a wire

A wire made of a special alloy has a length L (in m) given by $L = a + 0.0115T^3$, where T (in °C) is the temperature (between -4°C and 4°C). To find the wire length for L for $a = 8.380$ m and $T = -2.87$°C, we substitute these values to get

$$L = 8.380 + 0.0115(-2.87)^3 = 8.108 \text{ m}$$

The calculator result from the second line of Fig. 1.10 has been rounded to the nearest thousandth. ■

OPERATIONS WITH UNITS AND UNIT CONVERSIONS

Many problems in science and technology require us to perform operations on numbers with units. *For multiplication, division, powers, or roots, whatever operation is performed on the numbers also is performed on the units. For addition and subtraction, only numbers with the same units can be combined, and the answer will have the same units as the numbers in the problem.* Essentially, units are treated the same as any other algebraic symbol.

EXAMPLE 12 Algebraic operations with units

(a) $(2 \text{ ft})(4 \text{ lb}) = 8 \text{ ft} \cdot \text{lb}$ — the dot symbol represents multiplication

(b) $255 \text{ m} + 121 \text{ m} = 376 \text{ m}$ — note that the units are *not* added

(c) $(3.45 \text{ in.})^2 = 11.9 \text{ in.}^2$ — the unit is squared as well as the number

(d) $\left(65.0 \dfrac{\text{mi}}{\text{h}}\right)(3.52 \text{ h}) = 229 \text{ mi}$ — note that $\dfrac{\text{mi}}{\not{\text{h}}} \times \dfrac{\not{\text{h}}}{1} = \dfrac{\text{mi}}{1} = \text{mi}$

(e) $\dfrac{8.48\text{g}}{(1.69\text{m})^3} = \dfrac{8.48\text{g}}{(1.69)^3\text{m}^3} = 1.76 \text{ g/m}^3$ — the units are divided

(f) $\left(\dfrac{8.75 \text{ mi}}{1.32 \text{ min}^2}\right)\left(\dfrac{1 \text{ min}}{60 \text{ s}}\right)^2\left(\dfrac{5280 \text{ ft}}{1 \text{ mi}}\right) = \dfrac{(8.75 \text{ mi})(1 \text{ min}^2)(5280 \text{ ft})}{(1.32 \text{ min}^2)(3600 \text{ s}^2)(1 \text{ mi})}$ — the units min² and mi both cancel

$$= 9.72 \text{ ft/s}^2$$ ■

Often, it is necessary to **convert from one set of units to another.** This can be accomplished by using **conversion factors** (for example, 1 in. = 2.54 cm). Several useful conversion factors are shown in Table 1.1.

Metric prefixes are sometimes attached to units to indicate they are multiplied by a given power of ten. Table 1.2 (on the next page) shows some commonly used prefixes.

Table 1.1 Conversion Factors

Length	*Volume/Capacity*	*Weight/Mass*	*Energy/Power*
1 in. = 2.54 cm (exact)	1 ft³ = 28.32 L	1 lb = 453.6 g	1 Btu = 778.2 ft · lb
1 ft = 0.3048 m (exact)	1 L = 1.057 qt	1 kg = 2.205 lb	1 ft · lb = 1.356 J
1 km = 0.6214 mi	1 gal = 3.785 L	1 lb = 4.448 N	1 hp = 550 ft · lb/s (exact)
1 mi = 5280 ft (exact)			1 hp = 746.0 W

Table 1.2 Metric Prefixes

Prefix	*Factor*	*Symbol*
exa	10^{18}	E
peta	10^{15}	P
tera	10^{12}	T
giga	10^{9}	G
mega	10^{6}	M
kilo	10^{3}	k
hecto	10^{2}	h
deca	10^{1}	da
deci	10^{-1}	d
centi	10^{-2}	c
milli	10^{-3}	m
micro	10^{-6}	μ
nano	10^{-9}	n
pico	10^{-12}	p
femto	10^{-15}	f
atto	10^{-18}	a

■ See Appendix B for a description of the U.S. Customary and SI units, as well as a list of all the units used in this text and their symbols.

When a conversion factor is written in fractional form, *the fraction has a value of 1 since the numerator and denominator represent the same quantity*. For example, $\frac{1\text{ in.}}{2.54\text{ cm}} = 1$ or $\frac{1\text{ km}}{1000\text{ m}} = 1$. To convert units, we *multiply* the given number (including its units) by one or more of these fractions placed in such a way that the units we wish to eliminate will cancel and the units we wish to retain will remain in the answer. Since we are multiplying the given number by fractions that have a value of 1, the original quantity remains unchanged even though it will be expressed in different units.

EXAMPLE 13 Converting units

(a) The length of a certain smartphone is 13.8 cm. Convert this to inches.

$$13.8\text{ cm} = \underbrace{\frac{13.8\text{ cm}}{1}}_{\text{original number}} \times \underbrace{\frac{1\text{ in.}}{2.54\text{ cm}}}_{1} = \frac{(13.8)(1)\text{ in.}}{(1)(2.54)} = 5.43\text{ in.}$$

Notice that the unit cm appears in both the numerator and denominator and therefore cancels, leaving only the unit inches in the final answer.

(b) A car is traveling at 65.0 mi/h. Convert this speed to km/min (kilometers per minute).

From Table 1.1, we note that 1 km = 0.6214 mi, and we know that 1 h = 60 min. Using these values, we have

$$65.0\frac{\text{mi}}{\text{h}} = \frac{65.0\text{ mi}}{1\text{ h}} \times \frac{1\text{ km}}{0.6214\text{ mi}} \times \frac{1\text{ h}}{60\text{ min}} = \frac{(65.0)(1)(1)\text{ km}}{(1)(0.6214)(60)\text{ min}}$$
$$= 1.74\text{ km/min}$$

We note that the units mi and h appear in both numerator and denominator and therefore cancel out, leaving the units km and min. Also note that the 1's and 60 are exact.

(c) The density of iron is 7.86 g/cm^3 (grams per cubic centimeter). Express this density in kg/m^3 (kilograms per cubic meter).

From Table 1.2, 1 kg = 1000 g exactly, and 1 cm = 0.01 m exactly. Therefore,

$$7.86\frac{\text{g}}{\text{cm}^3} = \frac{7.86\text{ g}}{1\text{ cm}^3} \times \frac{1\text{ kg}}{1000\text{ g}} \times \left(\frac{1\text{ cm}}{0.01\text{ m}}\right)^3 = \frac{(7.86)(1)(1^3)\text{ kg}}{(1)(1000)(0.01^3)\text{ m}^3}$$
$$= \frac{(7.86)\text{ kg}}{0.001\text{ m}^3} = 7860\text{ kg/m}^3$$

Here, the units g and cm^3 are in both numerator and denominator and therefore cancel out, leaving units of kg and m^3. Also, all numbers are exact, except 7.86. ■

Practice Exercise

6. Convert 725 g/cm^2 to lb/in.^2

EXERCISES 1.4

In Exercises 1 and 2, make the given changes in the indicated examples of this section, and then simplify the resulting expression.

1. In Example 4(a), change $(-x^2)^3$ to $(-x^3)^2$.

2. In Example 6(d), change $(2x)^0$ to $2x^0$.

In Exercises 3–42, simplify the given expressions. Express results with positive exponents only.

3. x^3x^4 **4.** y^2y^7 **5.** $2b^4b^2$ **6.** $3k^5k$

7. $\frac{m^5}{m^3}$ **8.** $\frac{2x^6}{-x}$ **9.** $\frac{-n^5}{7n^9}$ **10.** $\frac{3s}{s^4}$

11. $(P^2)^4$ **12.** $(x^8)^3$ **13.** $(aT^2)^{30}$ **14.** $(3r^2)^3$

15. $\left(\frac{2}{b}\right)^3$ **16.** $\left(\frac{F}{t}\right)^{20}$ **17.** $\left(\frac{x^2}{-2}\right)^4$ **18.** $\left(\frac{3}{n^3}\right)^3$

19. $(8a)^0$ **20.** $-v^0$ **21.** $-3x^0$ **22.** $-(-2)^0$

23. 6^{-1} **24.** $-w^{-5}$ **25.** $\frac{1}{R^{-2}}$ **26.** $\frac{1}{-t^{-48}}$

27. $(-t^2)^7$ **28.** $(-y^3)^5$ **29.** $-\dfrac{L^{-3}}{L^{-5}}$ **30.** $2i^{40}i^{-70}$

31. $\dfrac{2v^4}{(2v)^4}$ **32.** $\dfrac{x^2x^3}{(x^2)^3}$ **33.** $\dfrac{(n^2)^4}{(n^4)^2}$ **34.** $\dfrac{(3t)^{-1}}{-3t^{-1}}$

35. $(\pi^0x^2a^{-1})^{-1}$ **36.** $(3m^{-2}n^4)^{-2}$ **37.** $(-8g^{-1}s^3)^2$

38. $ax^{-2}(-a^2x)^3$ **39.** $\left(\dfrac{4x^{-1}}{a^{-1}}\right)^{-3}$ **40.** $\left(\dfrac{2b^2}{-y^5}\right)^{-2}$

41. $\dfrac{15n^2T^5}{3n^{-1}T^6}$ **42.** $\dfrac{(nRT^{-2})^{32}}{R^{-2}T^{32}}$

In Exercises 43–50, evaluate the given expressions. In Exercises 45–50, all numbers are approximate.

43. $7(-4) - (-5)^2$ **44.** $6 - |-2|^5 - (-2)(8)$

45. $-(-26.5)^2 - (-9.85)^3$ **46.** $-0.711^2 - (-|-0.809|)^6$

47. $\dfrac{3.07(-|-1.86|)}{(-1.86)^4 + 1.596}$ **48.** $\dfrac{15.66^2 - (-4.017)^4}{1.044(-3.68)}$

49. $2.38(-10.7)^2 - \dfrac{254}{1.17^3}$ **50.** $4.2(2.6) + \dfrac{0.889}{1.89 - 1.09^2}$

In Exercises 51–62, perform the indicated operations.

51. Does $\left(\dfrac{1}{x^{-1}}\right)^{-1}$ represent the reciprocal of x?

52. Does $\left(\dfrac{0.2 - 5^{-1}}{10^{-2}}\right)^0$ equal 1? Explain.

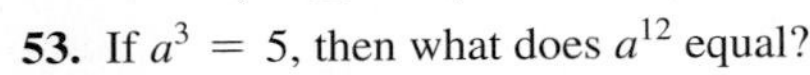

53. If $a^3 = 5$, then what does a^{12} equal?

54. Is $a^{-2} < a^{-1}$ for any negative value of a? Explain.

55. If a is a positive integer, simplify $(x^a \cdot x^{-a})^5$.

56. If a and b are positive integers, simplify $(-y^{a-b} \cdot y^{a+b})^2$.

57. In developing the "big bang" theory of the origin of the universe, the expression $(kT/(hc))^3(GkThc)^2c$ arises. Simplify this expression.

58. In studying planetary motion, the expression $(GmM)(mr)^{-1}(r^{-2})$ arises. Simplify this expression.

59. In designing a cam for a fire engine pump, the expression $\pi\left(\dfrac{r}{2}\right)^3\left(\dfrac{4}{3\pi r^2}\right)$ is used. Simplify this expression.

60. For a certain integrated electric circuit, it is necessary to simplify the expression $\dfrac{gM(2\pi fM)^{-2}}{2\pi fC}$. Perform this simplification.

61. If \$2500 is invested at 4.2% interest, compounded quarterly, the amount in the account after 6 years is $2500(1 + 0.042/4)^{24}$. Calculate this amount (the 1 is exact).

62. In designing a building, it was determined that the forces acting on an I beam would deflect the beam an amount (in cm), given by $\dfrac{x(1000 - 20x^2 + x^3)}{1850}$, where x is the distance (in m) from one end of the beam. Find the deflection for $x = 6.85$ m. (The 1000 and 20 are exact.)

63. Calculate the value of $\left(1 + \dfrac{1}{n}\right)^n$ for $n = 1, 10, 100, 1000$ on a calculator. Round to four decimal places. (For even larger values of n, the value will never exceed 2.7183. The limiting value is a number called e,which will be important in future chapters.)

64. For computer memory, the metric prefixes have an unusual meaning: 1 KB $= 2^{10}$ bytes, 1 MB $= 2^{10}$ KB, 1 GB $= 2^{10}$ MB, and 1 TB $= 2^{10}$ GB. How many bytes are there in 1 TB? (KB is kilobyte, MB is megabyte, GB is gigabyte, TB is terabyte)

In Exercises 65–68, perform the indicated operations and attach the correct units to your answers.

65. $\left(28.2\dfrac{\text{ft}}{\text{s}}\right)(9.81\text{ s})$

66. $\left(40.5\dfrac{\text{mi}}{\text{gal}}\right)(3.7\text{ gal})$

67. $\left(7.25\dfrac{\text{m}}{\text{s}^2}\right)\left(\dfrac{1\text{ ft}}{0.3048\text{ m}}\right)\left(\dfrac{60\text{ s}}{1\text{ min}}\right)^2$

68. $\left(238\dfrac{\text{kg}}{\text{m}^3}\right)\left(\dfrac{1000\text{ g}}{1\text{ kg}}\right)\left(\dfrac{1\text{ m}}{100\text{ cm}}\right)^3$

In Exercises 69–74, make the indicated conversions.

69. 15.7 qt to L **70.** 7.50 W to hp

71. 245 cm^2 to in.^{2} **72.** 85.7 mi^2 to km^2

73. $65.2\dfrac{\text{m}}{\text{s}}$ to $\dfrac{\text{ft}}{\text{min}}$ **74.** $25.0\dfrac{\text{mi}}{\text{gal}}$ to $\dfrac{\text{km}}{\text{L}}$

In Exercises 75–82, solve the given problems.

75. A laptop computer has a screen that measures 15.6 in. across its diagonal. Convert this to centimeters.

76. GPS satellites orbit the Earth at an altitude of about 12,500 mi. Convert this to kilometers.

77. A wastewater treatment plant processes 575,000 gal/day. Convert this to liters per hour.

78. Water flows from a fire hose at a rate of 85 gal/min. Convert this to liters per second.

79. The speed of sound is about 1130 ft/s. Change this to kilometers per hour.

80. A military jet flew at a rate of 7200 km/h. What is this speed in meters per second?

81. At sea level, atmospheric pressure is about 14.7 lb/in.^{2}. Express this in pascals (Pa). *Hint:* A pascal is a N/m^2 (see Appendix B).

82. The density of water is about 62.4 lb/ft^3. Convert this to kilograms per cubic meter.

Answers to Practice Exercises

1. a^3x^5 **2.** $\dfrac{2^5c^3}{3^2d^2} = \dfrac{32c^3}{9d^2}$ **3.** -1 **4.** $-c^3$ **5.** $\dfrac{a^2}{6x}$ **6.** 10.3 lb/in.^{2}

1.5 Scientific Notation

Meaning of Scientific Notation • Changing Numbers to and from Scientific Notation • Scientific Notation on a Calculator • Engineering Notation

In technical and scientific work, we encounter numbers that are inconvenient to use in calculations. Examples are: radio and television signals travel at 30,000,000,000 cm/s; the mass of Earth is 6,600,000,000,000,000,000,000 tons; a fiber in a fiber-optic cable has a diameter of 0.000005 m; some X-rays have a wavelength of 0.000000095 cm. Although calculators and computers can handle such numbers, a convenient and useful notation, called *scientific notation*, is used to represent these or any other numbers.

■ Television was invented in the 1920s and first used commercially in the 1940s.

The use of fiber optics was developed in the 1950s.

X-rays were discovered by Roentgen in 1895.

A number in **scientific notation** *is expressed as the product of a number greater than or equal to* 1 *and less than* 10, *and a power of* 10, *and is written as*

$$P \times 10^k$$

where $1 \leq P < 10$ and k is an integer. (The symbol $\leq$ means "is less than or equal to.")

EXAMPLE 1 Scientific notation

(a) $34{,}000 = 3.4 \times 10{,}000 = 3.4 \times 10^4$ (b) $6.82 = 6.82 \times 1 = 6.82 \times 10^0$

between 1 and 10

(c) $0.00503 = \dfrac{5.03}{1000} = \dfrac{5.03}{10^3} = 5.03 \times 10^{-3}$ ■

From Example 1, we see how a number is changed from ordinary notation to scientific notation. NOTE ▸ [The decimal point is moved so that only one nonzero digit is to its left. The number of places moved is the power of 10 (k), which is positive if the decimal point is moved to the left and negative if moved to the right.] To change a number from scientific notation to ordinary notation, this procedure is reversed. The next two examples illustrate these procedures.

EXAMPLE 2 Changing numbers to scientific notation

(a) $34{,}000 = 3.4 \times 10^4$ (4 places to left) (b) $6.82 = 6.82 \times 10^0$ (0 places) (c) $0.00503 = 5.03 \times 10^{-3}$ (3 places to right) ■

EXAMPLE 3 Changing numbers to ordinary notation

(a) To change 5.83×10^6 to ordinary notation, we move the decimal point six places to the right, including additional zeros to properly locate the decimal point.

$$5.83 \times 10^6 = 5{,}830{,}000$$

6 places to right

(b) To change 8.06×10^{-3} to ordinary notation, we must move the decimal point three places to the left, again including additional zeros to properly locate the decimal point.

$$8.06 \times 10^{-3} = 0.00806$$

3 places to left ■

Practice Exercises

1. Change 2.35×10^{-3} to ordinary notation.
2. Change 235 to scientific notation.

Scientific notation provides a practical way of handling very large or very small numbers. First, all numbers are expressed in scientific notation. Then the calculation can be done with numbers between 1 and 10, using the laws of exponents to find the power of ten of the result. Thus, scientific notation gives an important use of exponents.

■ See Exercise 43 of Exercises 1.1 for a brief note on computer data.

EXAMPLE 4 Scientific notation in calculations—processing rate

The processing rate of a computer processing 803,000 bytes of data in 0.00000525 s is

$$\frac{803{,}000}{0.00000525} = \frac{8.03 \times 10^5}{5.25 \times 10^{-6}} = \left(\frac{8.03}{5.25}\right) \times 10^{11} = 1.53 \times 10^{11} \text{ bytes/s}$$

$5 - (-6) = 11$

As shown, it is proper to leave the result (*rounded off*) in scientific notation. This method is useful when using a calculator and then estimating the result. In this case, the estimate is $(8 \times 10^5) \div (5 \times 10^{-6}) = 1.6 \times 10^{11}$. ■

Another advantage of scientific notation is that it clearly shows the precise number of significant digits when the final significant digit is 0, making it unnecessary to use the "bar" notation introduced in Section 1.3.

EXAMPLE 5 Scientific notation and significant digits—gravity

In determining the gravitational force between two stars 750,000,000,000 km apart, it is necessary to evaluate $750{,}000{,}000{,}000^2$. If 750,000,000,000 has *three* significant digits, we can show this by writing

$$75\bar{0}{,}000{,}000{,}000^2 = (7.50 \times 10^{11})^2 = 7.50^2 \times 10^{2\times 11} = 56.3 \times 10^{22}$$

Since 56.3 is not between 1 and 10, we can write this result in scientific notation as

$$56.3 \times 10^{22} = (5.63 \times 10)(10^{22}) = 5.63 \times 10^{23}$$ ■

We can enter numbers in scientific notation on a calculator, as well as have the calculator give results automatically in scientific notation. See the next example.

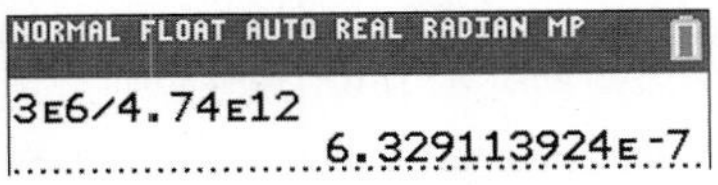

NORMAL FLOAT AUTO REAL RADIAN MP
3E6/4.74E12
6.329113924E-7

Fig. 1.11

EXAMPLE 6 Scientific notation on a calculator—wavelength

The wavelength λ (in m) of the light in a red laser beam can be found from the following calculation. Note the significant digits in the numerator.

$$\lambda = \frac{3{,}000{,}000}{4{,}740{,}000{,}000{,}000} = \frac{3.00 \times 10^6}{4.74 \times 10^{12}} = 6.33 \times 10^{-7} \text{ m}$$

The key sequence is 3 [EE] 6 [÷] 4.74 [EE] 12 [ENTER]. See Fig. 1.11. ■

Another commonly used notation, which is similar to scientific notation, is **engineering notation.** A number expressed in engineering notation is of the form

$$P \times 10^k$$

where $\mathbf{1 \le P < 1000}$ and **k is an integral multiple of 3.** Since the exponent k is a multiple of 3, the metric prefixes in Table 1.2 (Section 1.4) can be used to replace the power of ten. For example, an infrared wave that has a frequency of 850×10^9 Hz written in engineering notation can also be expressed as 850 GHz. The prefix *giga* (G) replaces the factor of 10^9.

EXAMPLE 7 Engineering notation and metric prefixes

Express each of the following quantities using engineering notation, and then replace the power of ten with the appropriate metric prefix.

less than 1000 — multiple of 3

(a) $48{,}000{,}000\ \Omega = 48 \times 10^6\ \Omega = 48\ \text{M}\Omega$

(b) $0.00000036\ \text{m} = 360 \times 10^{-9}\ \text{m} = 360\ \text{nm}$

(c) $1.3 \times 10^{-4}\ \text{A} = 0.00013\ \text{A} = 130 \times 10^{-6}\ \text{A} = 130\ \mu\text{A}$ ■

Practice Exercise

3. Write 0.0000728 s in engineering notation and using the appropriate metric prefix.

EXERCISES 1.5

In Exercises 1 and 2, make the given changes in the indicated examples of this section, and then rewrite the number as directed.

1. In Example 3(b), change the exponent -3 to 3 and then write the number in ordinary notation.

2. In Example 5, change the exponent 2 to -1 and then write the result in scientific notation.

In Exercises 3–10, change the numbers from scientific notation to ordinary notation.

3. 4.5×10^4 **4.** 6.8×10^7 **5.** 2.01×10^{-3} **6.** 9.61×10^{-5}

7. 3.23×10^0 **8.** 8×10^0 **9.** 1.86×10 **10.** 1×10^{-1}

In Exercises 11–18, change the numbers from ordinary notation to scientific notation.

11. 4000 **12.** 56,000 **13.** 0.0087 **14.** 0.00074

15. 609,000,000 **16.** 10 **17.** 0.0528 **18.** 0.0000908

In Exercises 19–22, perform the indicated calculations using a calculator and by first expressing all numbers in scientific notation. Assume that all numbers are exact.

19. 28,000(2,000,000,000) **20.** 50,000(0.006)

21. $\dfrac{88{,}000}{0.0004}$ **22.** $\dfrac{0.00003}{6{,}000{,}000}$

In Exercises 23–28, change the number from ordinary notation to engineering notation.

23. 35,600,000 **24.** 0.0000056 **25.** 0.0973

26. 925,000,000,000 **27.** 0.000000475 **28.** 370,000

In Exercises 29–32, perform the indicated calculations and then check the result using a calculator. Assume that all numbers are exact.

29. $2 \times 10^{-35} + 3 \times 10^{-34}$ **30.** $5.3 \times 10^{12} - 3.7 \times 10^{10}$

31. $(1.2 \times 10^{29})^3$ **32.** $(2 \times 10^{-16})^{-5}$

In Exercises 33–40, perform the indicated calculations using a calculator. All numbers are approximate.

33. 1320(649,000)(85.3) **34.** 0.0000569(3,190,000)

35. $\dfrac{0.0732(6710)}{0.00134(0.0231)}$ **36.** $\dfrac{0.00452}{2430(97{,}100)}$

37. $(3.642 \times 10^{-8})(2.736 \times 10^5)$ **38.** $\dfrac{(7.309 \times 10^{-1})^2}{5.9843(2.5036 \times 10^{-20})}$

39. $\dfrac{(3.69 \times 10^{-7})(4.61 \times 10^{21})}{0.0504}$ **40.** $\dfrac{(9.9 \times 10^7)(1.08 \times 10^{12})^2}{(3.603 \times 10^{-5})(2054)}$

In Exercises 41–50, change numbers in ordinary notation to scientific notation or change numbers in scientific notation to ordinary notation. See Appendix B for an explanation of symbols used.

41. The average number of tweets per day on Twitter in 2015 was 500,000,000.

42. A certain laptop computer has 17,200,000,000 bytes of memory.

43. A fiber-optic system requires 0.000003 W of power.

44. A red blood cell measures 0.0075 mm across.

45. The frequency of a certain cell phone signal is 1,200,000,000 Hz.

46. The PlayStation 4 game console has a graphic processing unit that can perform 1.84×10^{12} floating point operations per second (1.84 teraflops). (*Source:* playstation.com.)

47. The Gulf of Mexico oil spill in 2010 covered more than 12,000,000,000 m^2 of ocean surface.

48. A *parsec*, a unit used in astronomy, is about 3.086×10^{16} m.

49. The power of the signal of a laser beam probe is 1.6×10^{-12} W.

50. The electrical force between two electrons is about 2.4×10^{-43} times the gravitational force between them.

In Exercises 51–56, solve the given problems.

51. Write the following numbers in engineering notation and then replace the power of 10 with the appropriate metric prefix. (a) 2300 W (b) 0.23 W (c) 2,300,000 W (d) 0.00023 W

52. Write the following numbers in engineering notation and then replace the power of 10 with the appropriate metric prefix. (a) 8,090,000 Ω (b) 809,000 Ω (c) 0.0809 Ω

53. A *googol* is defined as 1 followed by 100 zeros. (a) Write this number in scientific notaion. (b) A *googolplex* is defined as 10 to the googol power. Write this number using powers of 10, and not the word *googol*. (Note the name of the Internet company.)

54. The number of electrons in the universe has been estimated at 10^{79}. How many times greater is a googol than the estimated number of electrons in the universe? (See Exercise 53.)

55. The diameter of the sun, 1.4×10^9 m, is about 109 times the diameter of Earth. Express the diameter of Earth in scientific notation.

56. GB means *gigabyte* where *giga* means *billion*, or 10^9. Actually, 1 GB $= 2^{30}$ bytes. Use a calculator to show that the use of *giga* is a reasonable choice of terminology.

In Exercises 57–60, perform the indicated calculations.

57. A computer can do an addition in 7.5×10^{-15} s. How long does it take to perform 5.6×10^6 additions?

58. Uranium is used in nuclear reactors to generate electricity. About 0.000000039% of the uranium disintegrates each day. How much of 0.085 mg of uranium disintegrates in a day?

59. If it takes 0.078 s for a GPS signal traveling at 2.998×10^8 m/s to reach the receiver in a car, find the distance from the receiver to the satellite.

60. (a) Determine the number of seconds in a day in scientific notation. (b) Using the result of part (a), determine the number of seconds in a century (assume 365.24 days/year).

In Exercises 61–64, perform the indicated calculations by first expressing all numbers in scientific notation.

61. One *atomic mass unit* (amu) is 1.66×10^{-27} kg. If one oxygen atom has 16 amu (an exact number), what is the mass of 125,000,000 oxygen atoms?

62. The rate of energy radiation (in W) from an object is found by evaluating the expression kT^4, where T is the thermodynamic temperature. Find this value for the human body, for which $k = 0.000000057\ \text{W/K}^4$ and $T = 303$ K.

63. In a microwave receiver circuit, the resistance R of a wire 1 m long is given by $R = k/d^2$, where d is the diameter of the wire. Find R if $k = 0.00000002196\ \Omega \cdot \text{m}^2$ and $d = 0.00007998$ m.

64. The average distance between the sun and Earth, 149,600,000 km, is called an *astronomical unit* (AU). If it takes light 499.0 s to travel 1 AU, what is the speed of light? Compare this with the speed of the GPS signal in Exercise 59.

Answers to Practice Exercises

1. 0.00235 **2.** 2.35×10^2 **3.** 72.8×10^{-6} s, 72.8 μs

1.6 Roots and Radicals

Principal *n*th Root • Simplifying Radicals • Using a Calculator • Imaginary Numbers

At times, we have to find the *square root* of a number, or maybe some other root of a number, such as a *cube root*. This means we must find a number that when squared, or cubed, and so on equals some given number. For example, to find the square root of 9, we must find a number that when squared equals 9. In this case, either 3 or -3 is an answer. Therefore, *either 3 or -3 is a square root of 9 since* $3^2 = 9$ and $(-3)^2 = 9$.

To have a general notation for the square root and have it represent *one* number, *we define the* **principal square root** *of a to be positive if a is positive and represent it by* $\sqrt{a}$. This means $\sqrt{9} = 3$ and not -3.

■ Unless we state otherwise, when we refer to the root of a number, it is the principal root.

The general notation for the **principal *n*th root** *of a is* $\sqrt[n]{a}$. (When $n = 2$, do not write the 2 for n.) *The* $\sqrt{\ }$ *sign is called a* **radical sign.**

EXAMPLE 1 Roots of numbers

(a) $\sqrt{2}$ (the square root of 2)

(b) $\sqrt[3]{2}$ (the cube root of 2)

(c) $\sqrt[4]{2}$ (the fourth root of 2)

(d) $\sqrt[7]{6}$ (the seventh root of 6) ■

NOTE ▸ [To have a single defined value for all roots (not just square roots) and to consider only real-number roots, we define the **principal *n*th root** of a to be positive if a is positive and to be negative if a is negative and n is odd.] (If a is negative and n is even, the roots are not real.)

EXAMPLE 2 Principal *n*th root

(a) $\sqrt{169} = 13 \quad (\sqrt{169} \neq -13)$

(b) $-\sqrt{64} = -8$

(c) $\sqrt[3]{27} = 3$ since $3^3 = 27$

(d) $\sqrt{0.04} = 0.2$ since $0.2^2 = 0.04$

(e) $-\sqrt[4]{256} = -4$

(f) $\sqrt[3]{-27} = -3$ (odd)

(g) $-\sqrt[3]{27} = -(+3) = -3$ ■

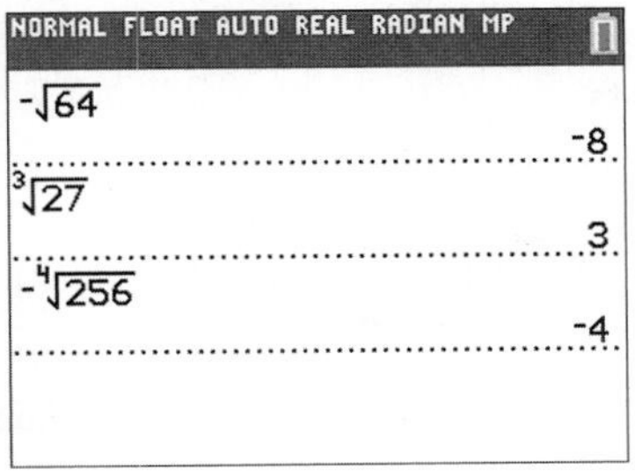

NORMAL FLOAT AUTO REAL RADIAN MP
-√64 -8
³√27 3
-⁴√256 -4

Fig. 1.12
Graphing calculator keystrokes: goo.gl/5HP7pF

The calculator evaluations of (b), (c), and (e) are shown in Fig. 1.12. The $\boxed{\sqrt{\ }}$ key is used for square roots and other roots are listed under the $\boxed{\text{MATH}}$ key.

Another property of square roots is developed by noting illustrations such as $\sqrt{36} = \sqrt{4 \times 9} = \sqrt{4} \times \sqrt{9} = 2 \times 3 = 6$. In general, this property states that *the square root of a product of positive numbers is the product of their square roots.*

$$\sqrt{ab} = \sqrt{a}\sqrt{b} \quad (a \text{ and } b \text{ positive real numbers}) \tag{1.9}$$

■ Try this one on your calculator: $\sqrt{12345678987654321}$

This property is used in simplifying radicals. It is most useful if either a or b is a **perfect square,** *which is the square of a rational number.*

EXAMPLE 3 Simplifying square roots

(a) $\sqrt{8} = \sqrt{(4)(2)} = \sqrt{4}\sqrt{2} = 2\sqrt{2}$

perfect squares — simplest form

(b) $\sqrt{75} = \sqrt{(25)(3)} = \sqrt{25}\sqrt{3} = 5\sqrt{3}$

(c) $\sqrt{4 \times 10^2} = \sqrt{4}\sqrt{10^2} = 2(10) = 20$

(Note that the square root of the square of a positive number is that number.) ■

In order to represent the square root of a number *exactly*, use Eq. (1.9) to write it in simplest form. However, in many applied problems, a decimal *approximation* obtained from a calculator is acceptable.

EXAMPLE 4 Approximating a square root—rocket descent

After reaching its greatest height, the time (in s) for a rocket to fall h ft is found by evaluating $0.25\sqrt{h}$. Find the time for the rocket to fall 1260 ft.

Using a calculator,

$$0.25\sqrt{1260} = 8.9 \text{ s}$$

The rocket takes 8.9 s to fall 1260 ft. The result from the calculator is rounded off to two significant digits, the accuracy of 0.25 (an approximate number). ■

In simplifying a radical, *all operations under a radical sign must be done before finding the root.*

EXAMPLE 5 More on simplifying square roots

(a) $\sqrt{16 + 9} = \sqrt{25}$ first perform the addition 16 + 9

$= 5$

NOTE ▸ [However, $\sqrt{16 + 9}$ is *not* $\sqrt{16} + \sqrt{9} = 4 + 3 = 7$.]

(b) $\sqrt{2^2 + 6^2} = \sqrt{4 + 36} = \sqrt{40} = \sqrt{4}\sqrt{10} = 2\sqrt{10}$,

NOTE ▸ [However, $\sqrt{2^2 + 6^2}$ is *not* $\sqrt{2^2} + \sqrt{6^2} = 2 + 6 = 8$.] ■

Practice Exercises

Simplify:

1. $\sqrt{12}$ 2. $\sqrt{36 + 144}$

In defining the principal square root, we did not define the square root of a negative number. However, in Section 1.1, we defined the square root of a negative number to be an **imaginary number.** More generally, *the even root of a negative number is an imaginary number, and the odd root of a negative number is a negative real number.*

EXAMPLE 6 Imaginary roots and real roots

even: $\sqrt{-64}$ is imaginary   even: $\sqrt[4]{-243}$ is imaginary   odd: $\sqrt[3]{-64} = -4$ (a real number) ■

Practice Exercise

3. Is $\sqrt[3]{-8}$ real or imaginary? If it is real, evaluate it.

A much more detailed coverage of roots, radicals, and imaginary numbers is taken up in Chapters 11 and 12.

EXERCISES 1.6

In Exercises 1–4, make the given changes in the indicated examples of this section and then solve the given problems.

1. In Example 2(b), change the square root to a cube root and then evaluate.
2. In Example 3(b), change $\sqrt{(25)(3)}$ to $\sqrt{(15)(5)}$ and explain whether or not this would be a better expression to use.
3. In Example 5(a), change the + to × and then evaluate.

4. In the first illustration of Example 6, place a − sign before the radical. Is there any other change in the statement?

In Exercises 5–38, simplify the given expressions. In each of 5–9 and 12–21, the result is an integer.

5. $\sqrt{49}$ **6.** $\sqrt{225}$ **7.** $-\sqrt{121}$ **8.** $-\sqrt{36}$

9. $-\sqrt{64}$ **10.** $\sqrt{0.25}$ **11.** $\sqrt{0.09}$ **12.** $-\sqrt{900}$

13. $\sqrt[3]{125}$ **14.** $\sqrt[4]{16}$ **15.** $\sqrt[4]{81}$ **16.** $-\sqrt[5]{-32}$

17. $(\sqrt{5})^2$ **18.** $(\sqrt[3]{31})^3$ **19.** $(-\sqrt[3]{-47})^3$ **20.** $(\sqrt[5]{-23})^5$

21. $(-\sqrt[4]{53})^4$ **22.** $\sqrt{75}$ **23.** $\sqrt{18}$ **24.** $-\sqrt{32}$

25. $\sqrt{1200}$ **26.** $\sqrt{50}$ **27.** $2\sqrt{84}$ **28.** $\dfrac{\sqrt{108}}{2}$

29. $\sqrt{\dfrac{80}{|3-7|}}$ **30.** $\sqrt{81 \times 10^2}$ **31.** $\sqrt[3]{-8^2}$ **32.** $\sqrt[4]{9^2}$

33. $\dfrac{7^2\sqrt{81}}{(-3)^2\sqrt{49}}$ **34.** $\dfrac{2^5\sqrt[5]{-243}}{-3\sqrt{144}}$ **35.** $\sqrt{36 + 64}$

36. $\sqrt{25 + 144}$ **37.** $\sqrt{3^2 + 9^2}$ **38.** $\sqrt{8^2 - 4^2}$

In Exercises 39–46, find the value of each square root by use of a calculator. Each number is approximate.

39. $\sqrt{85.4}$ **40.** $\sqrt{3762}$ **41.** $\sqrt{0.8152}$ **42.** $\sqrt{0.0627}$

43. (a) $\sqrt{1296 + 2304}$ (b) $\sqrt{1296} + \sqrt{2304}$

44. (a) $\sqrt{10.6276 + 2.1609}$ (b) $\sqrt{10.6276} + \sqrt{2.1609}$

45. (a) $\sqrt{0.0429^2 - 0.0183^2}$ (b) $\sqrt{0.0429^2} - \sqrt{0.0183^2}$

46. (a) $\sqrt{3.625^2 + 0.614^2}$ (b) $\sqrt{3.625^2} + \sqrt{0.614^2}$

In Exercises 47–58, solve the given problems.

47. The speed (in mi/h) of a car that skids to a stop on dry pavement is often estimated by $\sqrt{24s}$, where s is the length (in ft) of the skid marks. Estimate the speed if $s = 150$ ft.

48. The resistance in an amplifier circuit is found by evaluating $\sqrt{Z^2 - X^2}$. Find the resistance for $Z = 5.362\ \Omega$ and $X = 2.875\ \Omega$.

49. The speed (in m/s) of sound in seawater is found by evaluating $\sqrt{B/d}$ for $B = 2.18 \times 10^9$ Pa and $d = 1.03 \times 10^3$ kg/m^3. Find this speed, which is important in locating underwater objects using sonar.

50. The terminal speed (in m/s) of a skydiver can be approximated by $\sqrt{40m}$, where m is the mass (in kg) of the skydiver. Calculate the terminal speed (after reaching this speed, the skydiver's speed remains fairly constant before opening the parachute) of a 75-kg skydiver.

51. A TV screen is 52.3 in. wide and 29.3 in. high. The length of a diagonal (the dimension used to describe it—from one corner to the opposite corner) is found by evaluating $\sqrt{w^2 + h^2}$, where w is the width and h is the height. Find the diagonal.

52. A car costs \$38,000 new and is worth \$24,000 2 years later. The annual rate of depreciation is found by evaluating $100(1 - \sqrt{V/C})$, where C is the cost and V is the value after 2 years. At what rate did the car depreciate? (100 and 1 are exact.)

53. A *tsunami* is a very high ocean tidal wave (or series of waves) often caused by an earthquake. An Alaskan tsunami in 1958 measured over 500 m high; an Asian tsunami in 2004 killed over 230,000 people; a tsunami in Japan in 2011 killed over 10,000 people. An equation that approximates the speed v (in m/s) of a tsunami is $v = \sqrt{gd}$, where $g = 9.8$ m/s^2 and d is the average depth (in m) of the ocean floor. Find v (in km/h) for $d = 3500$ m (valid for many parts of the Indian Ocean and Pacific Ocean).

54. The greatest distance (in km) a person can see from a height h (in m) above the ground is $\sqrt{1.27 \times 10^4\, h + h^2}$. What is this distance for the pilot of a plane 9500 m above the ground?

55. Is it always true that $\sqrt{a^2} = a$? Explain.

56. For what values of x is **(a)** $x > \sqrt{x}$, **(b)** $x = \sqrt{x}$, and **(c)** $x < \sqrt{x}$?

57. A graphing calculator has a specific key sequence to find cube roots. Using a calculator, find $\sqrt[3]{2140}$ and $\sqrt[3]{-0.214}$.

58. A graphing calculator has a specific key sequence to find nth roots. Using a calculator, find $\sqrt[7]{0.382}$ and $\sqrt[7]{-382}$.

59. The resonance frequency f (in Hz) in an electronic circuit containing inductance L (in H) and capacitance C (in F) is given by $f = \dfrac{1}{2\pi\sqrt{LC}}$. Find the resonance frequency if $L = 0.250$ H and $C = 40.52 \times 10^{-6}$ F.

60. In statistics, the standard deviation is the square root of the variance. Find the standard deviation if the variance is 80.5 kg^2.

Answers to Practice Exercises

1. $2\sqrt{3}$ **2.** $6\sqrt{5}$ **3.** real, -2

1.7 Addition and Subtraction of Algebraic Expressions

Algebraic Expressions • Terms • Factors • Polynomials • Similar Terms • Simplifying • Symbols of Grouping

Because we use letters to represent numbers, we can see that all operations that can be used on numbers can also be used on literal numbers. In this section, we discuss the methods for adding and subtracting literal numbers.

Addition, subtraction, multiplication, division, and taking powers or roots are known as **algebraic operations.** *Any combination of numbers and literal symbols that results from algebraic operations is known as an* **algebraic expression.**

When an algebraic expression consists of several parts connected by plus signs and minus signs, each part (along with its sign) is known as a **term** *of the expression. If a given expression is made up of the product of a number of quantities, each of these quantities, or any product of them, is called a* **factor** *of the expression.*

CAUTION It is very important to distinguish clearly between *terms* and *factors*, because some operations that are valid for terms are not valid for factors, and conversely. Some of the common errors in handling algebraic expressions occur because these operations are not handled properly. ■

EXAMPLE 1 Terms and factors

In the study of the motion of a rocket, the following algebraic expression may be used.

terms

$$gt^2 - 2vt + 2s$$

factors

This expression has three terms: gt^2, $-2vt$, and $2s$. The first term, gt^2, has a factor of g and two factors of t. Any product of these factors is also a factor of gt^2. This means other factors are gt, t^2, and gt^2 itself. ■

EXAMPLE 2 Terms and factors

$7a(x^2 + 2y) - 6x(5 + x - 3y)$ is an expression with terms $7a(x^2 + 2y)$ and $-6x(5 + x - 3y)$.

The term $7a(x^2 + 2y)$ has individual factors of 7, a, and $(x^2 + 2y)$, as well as products of these factors. The factor $x^2 + 2y$ has two terms, x^2 and $2y$.

The term $-6x(5 + x - 3y)$ has factors 2, 3, x, and $(5 + x - 3y)$. The negative sign in front can be treated as a factor of -1. The factor $5 + x - 3y$ has three terms, 5, x, and $-3y$. ■

■ In Chapter 11, we will see that roots are equivalent to noninteger exponents.

A **polynomial** *is an algebraic expression with only nonnegative integer exponents on one or more variables, and has no variable in a denominator. The* **degree of a term** *is the sum of the exponents of the variables of that term, and the* **degree of the polynomial** *is the degree of the term of highest degree.*

A **multinomial** *is any algebraic expression of more than one term.* Terms like $1/x$ and $\sqrt{x}$ can be included in a multinomial, but not in a polynomial. (Since $1/x = x^{-1}$, the exponent is negative.)

EXAMPLE 3 Polynomials

Some examples of polynomials are as follows:

(a) $4x^2 - 5x + 3$ (degree 2) **(b)** $2x^6 - x$ (degree 6) **(c)** $3x$ (degree 1)

(d) $xy^3 + 7x - 3$ (degree 4) (add exponents of x and y)

(e) -6 (degree 0) $(-6 = -6x^0)$

From (c), note that a single term can be a polynomial, and from (e), note that a constant can be a polynomial. The expressions in (a), (b), and (d) are also multinomials.

The expression $x^2 + \sqrt{y + 2} - 8$ is a multinomial, but not a polynomial because of the square root term. ■

A polynomial with one term is called a **monomial.** *A polynomial with two terms is called a* **binomial,** *and one with three terms is called a* **trinomial.** *The numerical factor is called the* **numerical coefficient** (or simply **coefficient**) *of the term. All terms that differ at most in their numerical coefficients are known as* **similar** *or* **like** *terms.* That is, similar terms have the same variables with the same exponents.

EXAMPLE 4 Monomial, binomial, trinomial

(a) $7x^4$ is a monomial. The numerical coefficient is 7.

(b) $3ab - 6a$ is a binomial. The numerical coefficient of the first term is 3, and the numerical coefficient of the second term is -6. Note that the sign is attached to the coefficient.

(c) $8cx^3 - x + 2$ is a trinomial. The coefficients of the first two terms are 8 and -1.

(d) $x^2y^2 - 2x + 3y - 9$ is a polynomial with four terms (no special name). ■

EXAMPLE 5 Similar terms

(a) $8b - 6ab + 81b$ is a trinomial. The first and third terms are similar because they differ only in their numerical coefficients. The middle term is not similar to the others because it has a factor of a.

(b) $4x^2 - 3x$ is a binomial. The terms are not similar since the first term has two factors of x, and the second term has only one factor of x.

(c) $3x^2y^3 - 5y^3x^2 + x^2 - 2y^3$ is a polynomial. The commutative law tells us that $x^2y^3 = y^3x^2$, which means the first two terms are similar. ■

In adding and subtracting algebraic expressions, we combine similar terms into a single term. The ***simplified*** expression will contain only terms that are not similar.

EXAMPLE 6 Simplifying expressions

(a) $3x + 2x - 5y = 5x - 5y$ add similar terms—result has unlike terms

(b) $6a^2 - 7a + 8ax$ cannot be simplified since there are no like terms.

■ CAS (computer algebra system) calculators can display algebraic expressions and perform algebraic operations. The TI-89 graphing calculator is an example of such a calculator.

(c) $6a + 5c + 2a - c = 6a + 2a + 5c - c$ commutative law

$= 8a + 4c$ add like terms ■

To group terms in an algebraic expression, we use **symbols of grouping.** In this text, we use **parentheses,** ( ); **brackets,** [ ]; and **braces,** { }. All operations that occur in the numerator or denominator of a fraction are implied to be inside grouping symbols, as well as all operations under a radical symbol.

CAUTION In simplifying an expression using the distributive law, to remove the symbols of grouping if a MINUS sign precedes the grouping, change the sign of EVERY term in the grouping, or if a plus sign precedes the grouping retain the sign of every term. ■

EXAMPLE 7 Symbols of grouping

(a) $2(a + 2x) = 2a + 2(2x)$ use distributive law

$= 2a + 4x$

(b) $-(+a - 3c) = (-1)(+a - 3c)$ treat − sign as −1

$= (-1)(+a) + (-1)(-3c)$

$= -a + 3c$ note change of signs

Normally, $+a$ would be written simply as a. ■

Practice Exercises

Use the distributive law.

1. $3(2a + y)$ **2.** $-3(-2r + s)$

EXAMPLE 8 Simplifying: signs before parentheses

+ sign before parentheses

(a) $3c + (2b - c) = 3c + 2b - c = 2b + 2c$ use distributive law

$2b = +2b$ signs retained

− sign before parentheses

(b) $3c - (2b - c) = 3c - 2b + c = -2b + 4c$ use distributive law

$2b = +2b$ signs changed

(c) $3c - (-2b + c) = 3c + 2b - c = 2b + 2c$ use distributive law

signs changed

(d) $y(3 - y) - 2(y - 3) = 3y - y^2 - 2y + 6$ note the $-2(-3) = +6$

$= -y^2 + y + 6$ ■

■ Note in each case that the parentheses are removed and the sign before the parentheses is also removed.

Practice Exercise

3. Simplify $2x - 3(4y - x)$

EXAMPLE 9 Simplifying—machine part design

In designing a certain machine part, it is necessary to perform the following simplification.

$$16(8 - x) - 2(8x - x^2) - (64 - 16x + x^2) = 128 - 16x - 16x + 2x^2 - 64 + 16x - x^2$$
$$= 64 - 16x + x^2$$ ■

At times, we have expressions in which more than one symbol of grouping is to be removed in the simplification. [Normally, when several symbols of grouping are to be removed, it is more convenient to remove the innermost symbols first.]

NOTE ▸

CAUTION One of the most common errors made is changing the sign of only the first term when removing symbols of grouping preceded by a minus sign. Remember, if the symbols are preceded by a minus sign, we must change the sign of *every* term. ■

EXAMPLE 10 Several symbols of grouping

(a) $3ax - [ax - (5s - 2ax)] = 3ax - [ax - 5s + 2ax]$ ← remove parentheses

$= 3ax - ax + 5s - 2ax$ ← remove brackets

$= 5s$

(b) $3a^2b - \{[a - (2a^2b - a)] + 2b\} = 3a^2b - \{[a - 2a^2b + a] + 2b\}$ ← remove parentheses

$= 3a^2b - \{a - 2a^2b + a + 2b\}$ ← remove brackets

$= 3a^2b - a + 2a^2b - a - 2b$ ← remove braces

$= 5a^2b - 2a - 2b$ ■

Calculators use only parentheses for grouping symbols, and we often need to use one set of parentheses within another set. These are called **nested parentheses.** In the next example, note that the innermost parentheses are removed first.

EXAMPLE 11 Nested parentheses

$$2 - (3x - 2(5 - (7 - x))) = 2 - (3x - 2(5 - 7 + x))$$
$$= 2 - (3x - 10 + 14 - 2x)$$
$$= 2 - 3x + 10 - 14 + 2x = -x - 2$$ ■

TI-89 graphing calculator keystrokes for Example 11: goo.gl/sUCoav

EXERCISES 1.7

In Exercises 1–4, make the given changes in the indicated examples of this section, and then solve the resulting problems.

1. In Example 6(a), change $2x$ to $2y$.
2. In Example 8(a), change the sign before $(2b - c)$ from $+$ to $-$.
3. In Example 10(a), change $[ax - (5s - 2ax)]$ to $[(ax - 5s) - 2ax]$.
4. In Example 10(b), change $\{[a - (2a^2b - a)] + 2b\}$ to $\{a - [2a^2b - (a + 2b)]\}$.

In Exercises 5–51, simplify the given algebraic expressions.

5. $5x + 7x - 4x$
6. $6t - 3t - 4t$
7. $2y - y + 4x$
8. $-4C + L - 6C$
9. $3t - 4s - 3t - s$
10. $-8a - b + 12a + b$
11. $2F - 2T - 2 + 3F - T$
12. $x - 2y - 3x - y + z$
13. $a^2b - a^2b^2 - 2a^2b$
14. $-xy^2 - 3x^2y^2 + 2xy^2$
15. $2p + (p - 6 - 2p)$
16. $5 + (3 - 4n + p)$
17. $v - (7 - 9x + 2v)$
18. $-2a - \frac{1}{2}(b - a)$
19. $2 - 3 - (4 - 5a)$
20. $\sqrt{A} + (h - 2\sqrt{A}) - 3\sqrt{A}$
21. $(a - 3) + (5 - 6a)$
22. $(4x - y) - (-2x - 4y)$
23. $-(t - 2u) + (3u - t)$
24. $-2(6x - 3y) - (5y - 4x)$
25. $3(2r + s) - (-5s - r)$
26. $3(a - b) - 2(a - 2b)$
27. $-7(6 - 3j) - 2(j + 4)$
28. $-(5t + a^2) - 2(3a^2 - 2st)$
29. $-[(4 - 6n) - (n - 3)]$
30. $-[(A - B) - (B - A)]$
31. $2[4 - (t^2 - 5)]$
32. $-3\left[-3 - \frac{2}{3}(-a - 4)\right]$
33. $-2[-x - 2a - (a - x)]$
34. $-2[-3(x - 2y) + 4y]$
35. $aZ - [3 - (aZ + 4)]$
36. $9v - [6 - (-v - 4) + 4v]$
37. $5z - \{8 - [4 - (2z + 1)]\}$
38. $7y - \{y - [2y - (x - y)]\}$
39. $5p - (q - 2p) - [3q - (p - q)]$
40. $-(4 - \sqrt{LC}) - [(5\sqrt{LC} - 7) - (6\sqrt{LC} + 2)]$
41. $-2\{-(4 - x^2) - [3 + (4 - x^2)]\}$
42. $-\{-[-(x - 2a) - b] - (a - x)\}$
43. $5V^2 - (6 - (2V^2 + 3))$
44. $-2F + 2((2F - 1) - 5)$
45. $-(3t - (7 + 2t - (5t - 6)))$
46. $a^2 - 2(x - 5 - (7 - 2(a^2 - 2x) - 3x))$
47. $-4[4R - 2.5(Z - 2R) - 1.5(2R - Z)]$
48. $-3\{2.1e - 1.3[-f - 2(e - 5f)]\}$
49. In determining the size of a V belt to be used with an engine, the expression $3D - (D - d)$ is used. Simplify this expression.
50. When finding the current in a transistor circuit, the expression $i_1 - (2 - 3i_2) + i_2$ is used. Simplify this expression. (The numbers below the i's are *subscripts*. Different subscripts denote different variables.)
51. Research on a plastic building material leads to $[(B + \frac{4}{3}\alpha) + 2(B - \frac{2}{3}\alpha)] - [(B + \frac{4}{3}\alpha) - (B - \frac{2}{3}\alpha)]$. Simplify this expression.
52. One car goes 30 km/h for $t - 1$ hours, and a second car goes 40 km/h for $t + 2$ hours. Find the expression for the sum of the distances traveled by the two cars.
53. A shipment contains x hard drives with 4 terabytes of memory and $x + 25$ hard drives with 8 terabytes. Express the total number of terabytes of memory in the shipment as a variable expression and simplify.
54. Each of two suppliers has $2n + 1$ bundles of shingles costing \$30 each and $n - 2$ bundles costing \$20 each. How much more is the total value of the \$30 bundles than the \$20 bundles?
55. For the expressions $2x^2 - y + 2a$ and $3y - x^2 - b$ find (a) the sum, and (b) the difference if the second is subtracted from the first.
56. For the following expressions, subtract the third from the sum of the first two: $3a^2 + b - c^3$, $2c^3 - 2b - a^2$, $4c^3 - 4b + 3$.

In Exercises 57–60, answer the given questions.

57. Is the following simplification correct? Explain.

$$2x - 3y + 5 - (4x - y + 3) = 2x - 3y + 5 - 4x - y + 3$$
$$= -2x - 4y + 8$$

58. Is the following simplification correct? Explain.

$$2a - 3b - 4c - (-5a + 3b - 2c)$$
$$= 2a - 3b - 4c + 5a - 3b - 2c$$
$$= 7a - 6b - 6c$$

59. For any real numbers a and b, is it true that $|a - b| = |b - a|$? Explain.
60. Is subtraction associative? That is, in general, does $(a - b) - c$ equal $a - (b - c)$? Explain.

Answers to Practice Exercises

1. $6a + 3y$ 2. $6r - 3s$ 3. $5x - 12y$

1.8 Multiplication of Algebraic Expressions

Multiplying Monomials • Products of Monomials and Polynomials • Powers of Polynomials

To find the product of two or more monomials, we multiply the numerical coefficients to find the numerical coefficient of the product, and multiply the literal numbers, remembering that *the exponents may be combined only if the base is the same.*

EXAMPLE 1 Multiplying monomials

(a) $3c^5(-4c^2) = -12c^7$ multiply numerical coefficients and add exponents of c

(b) $(-2b^2y^3)(-9aby^5) = 18ab^3y^8$ add exponents of same base

(c) $2xy(-6cx^2)(3xcy^2) = -36c^2x^4y^3$ ■

NOTE ▸ [If a product contains a monomial that is raised to a power, we must first raise it to the indicated power before proceeding with the multiplication.]

EXAMPLE 2 Product containing power of a monomial

(a) $-3a(2a^2x)^3 = -3a(8a^6x^3) = -24a^7x^3$

(b) $2s^3(-st^4)^2(4s^2t) = 2s^3(s^2t^8)(4s^2t) = 8s^7t^9$ ■

We find the product of a monomial and a polynomial by using the distributive law, which states that we *multiply each term of the polynomial by the monomial.* In doing so, we must be careful to give the correct sign to each term of the product.

EXAMPLE 3 Product of monomial and polynomial

(a) $2ax(3ax^2 - 4yz) = 2ax(3ax^2) + (2ax)(-4yz) = 6a^2x^3 - 8axyz$

Practice Exercises

Perform the indicated multiplications.

1. $2a^3b(-6ab^2)$ **2.** $-5x^2y^3(2xy - y^4)$

(b) $5cy^2(-7cx - ac) = (5cy^2)(-7cx) + (5cy^2)(-ac) = -35c^2xy^2 - 5ac^2y^2$ ■

It is generally not necessary to write out the middle step as it appears in the preceding example. We write the answer directly. For instance, Example 3(a) would appear as $2ax(3ax^2 - 4yz) = 6a^2x^3 - 8axyz$.

We find the product of two polynomials by using the distributive law. The result is that *we multiply each term of one polynomial by each term of the other and add the results.* Again we must be careful to give each term of the product its correct sign.

EXAMPLE 4 Product of polynomials

■ Note that, using the distributive law, $(x - 2)(x + 3) = (x - 2)(x) + (x - 2)(3)$ leads to the same result.

$$(x - 2)(x + 3) = x(x) + x(3) + (-2)(x) + (-2)(3)$$
$$= x^2 + 3x - 2x - 6 = x^2 + x - 6$$ ■

Finding the power of a polynomial is equivalent to using the polynomial as a factor the number of times indicated by the exponent. It is sometimes convenient to write the power of a polynomial in this form before multiplying.

EXAMPLE 5 Power of a polynomial

two factors

(a) $(x + 5)^2 = (x + 5)(x + 5) = x^2 + 5x + 5x + 25$
$= x^2 + 10x + 25$

(b) $(2a - b)^3 = (2a - b)(2a - b)(2a - b)$ the exponent 3 indicates three factors

$$= (2a - b)(4a^2 - 2ab - 2ab + b^2)$$
$$= (2a - b)(4a^2 - 4ab + b^2)$$
$$= 8a^3 - 8a^2b + 2ab^2 - 4a^2b + 4ab^2 - b^3$$
$$= 8a^3 - 12a^2b + 6ab^2 - b^3$$ ■

TI-89 graphing calculator keystrokes for Example 5: goo.gl/GKgiFl

Practice Exercises

Perform the indicated multiplications.

3. $(2s - 5t)(s + 4t)$ **4.** $(3u + 2v)^2$

CAUTION We should note that in Example 5(a) $(x + 5)^2$ is *not* equal to $x^2 + 25$ because the term $10x$ is not included. We must follow the proper procedure and not simply square each of the terms within the parentheses. ■

EXAMPLE 6 Simplifying products—telescope lens

An expression used with a lens of a certain telescope is simplified as shown.

$$a(a + b)^2 + a^3 - (a + b)(2a^2 - s^2)$$
$$= a(a + b)(a + b) + a^3 - (2a^3 - as^2 + 2a^2b - bs^2)$$
$$= a(a^2 + ab + ab + b^2) + a^3 - 2a^3 + as^2 - 2a^2b + bs^2$$
$$= a^3 + a^2b + a^2b + ab^2 - a^3 + as^2 - 2a^2b + bs^2$$
$$= ab^2 + as^2 + bs^2$$ ■

EXERCISES 1.8

In Exercises 1–4, make the given changes in the indicated examples of this section, and then solve the resulting problems.

1. In Example 2(b), change the factor $(-st^4)^2$ to $(-st^4)^3$.
2. In Example 3(a), change the factor $2ax$ to $-2ax$.
3. In Example 4, change the factor $(x + 3)$ to $(x - 3)$.
4. In Example 5(b), change the exponent 3 to 2.

In Exercises 5–66, perform the indicated multiplications.

5. $(a^2)(ax)$ **6.** $(2xy)(x^2y^3)$ **7.** $-a^2c^2(a^2cx^3)$

8. $-2cs^2(-4cs)^2$ **9.** $(2ax^2)^2(-2ax)$ **10.** $6pq^3(3pq^2)^2$

11. $i^2(Ri + 2i)$ **12.** $2x(-p - q)$ **13.** $-3s(s^2 - 5t)$

14. $-3b(2b^2 - b)$ **15.** $5m(m^2n + 3mn)$ **16.** $a^2bc(2ac - 3b^2c)$

17. $3M(-M - N + 2)$ **18.** $-4c^2(-9gc - 2c + g^2)$

19. $xy(tx^2)(x + y^3)$ **20.** $-2(-3st^3)(3s - 4t)$

21. $(x - 3)(x + 5)$ **22.** $(a + 7)(a + 1)$

23. $(x + 5)(2x - 1)$ **24.** $(4t_1 + t_2)(2t_1 - 3t_2)$

25. $(y + 8)(y - 8)$ **26.** $(z - 4)(z + 4)$

27. $(2a - b)(-2b + 3a)$ **28.** $(-3 + 4w^2)(3w^2 - 1)$

29. $(2s + 7t)(3s - 5t)$ **30.** $(5p - 2q)(p + 8q)$

31. $(x^2 - 1)(2x + 5)$ **32.** $(3y^2 + 2)(2y - 9)$

33. $(x - 2y - 4)(x - 2y + 4)$

34. $(2a + 3b + 1)(2a + 3b - 1)$

35. $2(a + 1)(a - 9)$ **36.** $-5(y - 3)(y + 6)$

37. $-3(3 - 2T)(3T + 2)$ **38.** $2n(-n + 5)(6n + 5)$

39. $2L(L + 1)(4 - L)$ **40.** $ax(x + 4)(7 - x^2)$

41. $(3x - 7)^2$ **42.** $(x - 3y)^2$ **43.** $(x_1 + 3x_2)^2$

44. $(-7m - 1)^2$ **45.** $(xyz - 2)^2$ **46.** $(-6x^2 + b)^2$

47. $2(x + 8)^2$ **48.** $3(3R - 4)^2$

49. $(2 + x)(3 - x)(x - 1)$ **50.** $(-c^2 + 3x)^3$

51. $3T(T + 2)(2T - 1)$ **52.** $[(x - 2)^2(x + 2)]^2$

53. Let $x = 3$ and $y = 4$ to show that (a) $(x + y)^2 \neq x^2 + y^2$ and (b) $(x - y)^2 \neq x^2 - y^2$. ($\neq$ means "does not equal")

54. *Explain how* you would perform $(x + 3)^5$. Do not actually do the operations.

55. By multiplication, show that $(x + y)^3$ is not equal to $x^3 + y^3$.

56. By multiplication, show that $(x + y)(x^2 - xy + y^2) = x^3 + y^3$.

57. In finding the value of a certain savings account, the expression $P(1 + 0.01r)^2$ is used. Multiply out this expression.

58. A savings account of \$1000 that earns $r\%$ annual interest, compounded quarterly, has a value of $1000(1 + 0.0025r)^2$ after 6 months. Perform the indicated multiplication.

59. A contractor is designing a rectangular room that will have a pool table. The length of the pool table is twice its width. The contractor wishes to have 5 ft of open space between each wall and the pool table. See Fig. 1.13. Express the area of the room in terms of the width w of the pool table. Then perform the indicated operations.

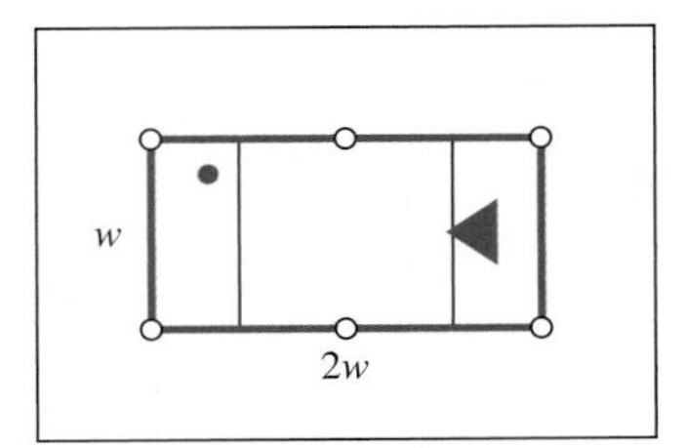

Fig. 1.13

60. The weekly revenue R (in dollars) of a flash drive manufacturer is given by $R = xp$, where x is the number of flash drives sold each week and p is the price (in dollars). If the price is given by the demand equation $p = 30 - 0.01x$, express the revenue in terms of x and simplify.

61. In using aircraft radar, the expression $(2R - X)^2 - (R^2 + X^2)$ arises. Simplify this expression.

62. In calculating the temperature variation of an industrial area, the expression $(2T^3 + 3)(T^2 - T - 3)$ arises. Perform the indicated multiplication.

63. In a particular computer design containing n circuit elements, n^2 switches are needed. Find the expression for the number of switches needed for $n + 100$ circuit elements.

64. Simplify the expression $(T^2 - 100)(T - 10)(T + 10)$, which arises when analyzing the energy radiation from an object.

65. In finding the maximum power in part of a microwave transmiter circuit, the expression $(R_1 + R_2)^2 - 2R_2(R_1 + R_2)$ is used. Multiply and simplify.

66. In determining the deflection of a certain steel beam, the expression $27x^2 - 24(x - 6)^2 - (x - 12)^3$ is used. Multiply and simplify.

Answers to Practice Exercises

1. $-12a^4b^3$ **2.** $-10x^3y^4 + 5x^2y^7$
3. $2s^2 + 3st - 20t^2$ **4.** $9u^2 + 12uv + 4v^2$

1.9 Division of Algebraic Expressions

Dividing Monomials • Dividing by a Monomial • Dividing One Polynomial by Another

To find the quotient of one monomial divided by another, we use the laws of exponents and the laws for dividing signed numbers. Again, the *exponents may be combined only if the base is the same.*

EXAMPLE 1 Dividing monomials

(a) $\dfrac{3c^7}{c^2} = 3c^{7-2} = 3c^5$

(b) $\dfrac{16x^3y^5}{4xy^2} = \dfrac{16}{4}(x^{3-1})(y^{5-2}) = 4x^2y^3$

(divide coefficients; subtract exponents)

(c) $\dfrac{-6a^2xy^2}{2axy^4} = -\left(\dfrac{6}{2}\right)\dfrac{a^{2-1}x^{1-1}}{y^{4-2}} = \dfrac{-3a}{y^2}$

As shown in illustration (c), we use only positive exponents in the final result unless there are specific instructions otherwise. ■

From arithmetic, we may show how a multinomial is to be divided by a monomial. When adding fractions (say $\frac{2}{7}$ and $\frac{3}{7}$), we have $\dfrac{2}{7} + \dfrac{3}{7} = \dfrac{2+3}{7}$.

Looking at this from *right to left,* we see that *the quotient of a multinomial divided by a monomial is found by dividing each term of the multinomial by the monomial and adding the results*. This can be shown as

■ This is an identity and is valid for all values of a and b, and all values of c except zero (which would make it undefined).

$$\frac{a+b}{c} = \frac{a}{c} + \frac{b}{c}$$

CAUTION Be careful: Although $\dfrac{a+b}{c} = \dfrac{a}{c} + \dfrac{b}{c}$, we must note that $\dfrac{c}{a+b}$ is *not* $\dfrac{c}{a} + \dfrac{c}{b}$. ■

EXAMPLE 2 Dividing by a monomial

(a) $\dfrac{4a^2 + 8a}{2a} = \dfrac{4a^2}{2a} + \dfrac{8a}{2a} = 2a + 4$

(b) $\dfrac{4x^3y - 8x^3y^2 + 2x^2y}{2x^2y} = \dfrac{4x^3y}{2x^2y} - \dfrac{8x^3y^2}{2x^2y} + \dfrac{2x^2y}{2x^2y}$

$= 2x - 4xy + 1$

(each term of numerator divided by denominator)

■ Until you are familiar with the method, it is recommended that you do write out the middle steps.

We usually do not write out the middle step as shown in these illustrations. The divisions of the terms of the numerator by the denominator are usually done by inspection (mentally), and the result is shown as it appears in the next example.

NOTE ▸ [Note carefully the last term 1 of the result. When all factors of the numerator are the same as those in the denominator, we are dividing a number by itself, which gives a result of 1.] ■

Practice Exercise

1. Divide: $\frac{4ax^2 - 6a^2x}{2ax}$

EXAMPLE 3 Dividing by a monomial—irrigation pump

The expression $\frac{2p + v^2d + 2ydg}{2dg}$ is used when analyzing the operation of an irrigation pump. Performing the indicated division, we have

$$\frac{2p + v^2d + 2ydg}{2dg} = \frac{p}{dg} + \frac{v^2}{2g} + y$$ ■

DIVISION OF ONE POLYNOMIAL BY ANOTHER

To divide one polynomial by another, use the following steps.

1. Arrange the dividend (the polynomial to be divided) and the divisor in descending powers of the variable.
2. Divide the first term of the dividend by the first term of the divisor. The result is the first term of the quotient.
3. Multiply the entire divisor by the first term of the quotient and *subtract* the product from the dividend.
4. Divide the first term of this difference by the first term of the divisor. This gives the second term of the quotient.
5. Multiply this term by the entire divisor and *subtract* the product from the first difference.
6. Repeat this process until the remainder is zero or of lower degree than the divisor.
7. Express the answer in the form quotient + $\frac{\text{remainder}}{\text{divisor}}$.

■ This is similar to long division of numbers.

EXAMPLE 4 Dividing one polynomial by another

Perform the division $(6x^2 + x - 2) \div (2x - 1)$.

(This division can also be indicated in the fractional form $\frac{6x^2 + x - 2}{2x - 1}$.)

We set up the division as we would for long division in arithmetic. Then, following the procedure outlined above, we have the following:

$$\begin{array}{r|l} & 3x + 2 \\ 2x - 1 & 6x^2 + x - 2 \\ & 6x^2 - 3x \\ & \quad\quad 4x - 2 \\ & \quad\quad 4x - 2 \\ & \quad\quad\quad 0 \end{array}$$

$\frac{6x^2}{2x}$ divide first term of dividend by first term of divisor

$3x(2x - 1)$

subtract

$6x^2 - 6x^2 = 0$

$x - (-3x) = 4x$

The remainder is zero and the quotient is $3x + 2$. Therefore, the answer is

$$3x + 2 + \frac{0}{2x - 1}, \text{ or simply } 3x + 2$$ ■

■ The answer to Example 4 can be checked by showing $(2x - 1)(3x + 2) = 6x^2 + x - 2$.

Practice Exercise

2. Divide: $(6x^2 + 7x - 3) \div (3x - 1)$

EXAMPLE 5 Quotient with a remainder

Perform the division $\frac{8x^3 + 4x^2 + 3}{4x^2 - 1}$. Because there is no x-term in the dividend, we should leave space for any x-terms that might arise (which we will show as $0x$).

$$\begin{array}{r l} & 2x + 1 \\ \text{divisor} \quad 4x^2 - 1 & \overline{)8x^3 + 4x^2 + 0x + 3} \quad \text{dividend} \quad \frac{8x^3}{4x^2} = 2x \\ & \underline{8x^3 \qquad\quad - 2x} \quad \text{subtract} \\ 0x - (-2x) = 2x & \quad 4x^2 + 2x + 3 \quad \frac{4x^2}{4x^2} = 1 \\ & \quad \underline{4x^2 \qquad\quad - 1} \quad \text{subtract} \\ & \qquad 2x + 4 \quad \text{remainder} \end{array}$$

TI-89 graphing calculator keystrokes for Example 5: goo.gl/WtS19R

■ The answer to Example 5 can be checked by showing $(4x^2 - 1)(2x + 1) + (2x + 4) = 8x^3 + 4x^2 + 3$

Because the degree of the remainder $2x + 4$ is less than that of the divisor, the long-division process is complete and the answer is $2x + 1 + \frac{2x + 4}{4x^2 - 1}$. ■

EXERCISES 1.9

In Exercises 1–4, make the given changes in the indicated examples of this section and then perform the indicated divisions.

1. In Example 1(c), change the denominator to $-2a^2xy^5$.
2. In Example 2(b), change the denominator to $2xy^2$.
3. In Example 4, change the dividend to $6x^2 - 7x + 2$.
4. In Example 5, change the sign of the middle term of the numerator from + to −.

In Exercises 5–24, perform the indicated divisions.

5. $\frac{8x^3y^2}{-2xy}$
6. $\frac{-18b^7c^3}{bc^2}$
7. $\frac{-16r^3t^5}{-4r^5t}$
8. $\frac{51mn^5}{17m^2n^2}$
9. $\frac{(15x^2y)(2xz)}{10xy}$
10. $\frac{(5sT)(8s^2T^3)}{10s^3T^2}$
11. $\frac{(4a^3)(2x)^2}{(4ax)^2}$
12. $\frac{12a^2b}{(3ab^2)^2}$
13. $\frac{3a^2x + 6xy}{3x}$
14. $\frac{2m^2n - 6mn}{-2m}$
15. $\frac{3rst - 6r^2st^2}{3rs}$
16. $\frac{-5a^2n - 10an^2}{5an}$
17. $\frac{4pq^3 + 8p^2q^2 - 16pq^5}{4pq^2}$
18. $\frac{a^2x_1x_2^2 + ax_1^3 - ax_1}{ax_1}$
19. $\frac{2\pi fL - \pi fR^2}{\pi fR}$
20. $\frac{9(aB)^4 - 6aB^4}{-3aB^3}$
21. $\frac{-7a^2b + 14ab^2 - 21a^3}{14a^2b^2}$
22. $\frac{2x^{n+2} + 4ax^n}{2x^n}$
23. $\frac{6y^{2n} - 4ay^{n+1}}{2y^n}$
24. $\frac{3a(F + T)b^2 - (F + T)}{a(F + T)}$

In Exercises 25–44, perform the indicated divisions. Express the answer as shown in Example 5 when applicable.

25. $(x^2 + 9x + 20) \div (x + 4)$
26. $(x^2 + 7x - 18) \div (x - 2)$
27. $(2x^2 + 7x + 3) \div (x + 3)$
28. $(3t^2 - 7t + 4) \div (t - 1)$
29. $(x^2 - 3x + 2) \div (x - 2)$
30. $(2x^2 - 5x - 7) \div (x + 1)$
31. $(x - 14x^2 + 8x^3) \div (2x - 3)$
32. $(6 + 7y + 6y^2) \div (2y + 1)$
33. $(4Z^2 - 5Z - 7) \div (4Z + 3)$
34. $(6x^2 - 5x - 9) \div (-4 + 3x)$
35. $\frac{x^3 + 3x^2 - 4x - 12}{x + 2}$
36. $\frac{3x^3 + 19x^2 + 13x - 20}{3x - 2}$
37. $\frac{2a^4 + 4a^2 - 16}{a^2 - 2}$
38. $\frac{6T^3 + T^2 + 2}{3T^2 - T + 2}$
39. $\frac{y^3 + 27}{y + 3}$
40. $\frac{D^3 - 1}{D - 1}$
41. $\frac{x^2 - 2xy + y^2}{x - y}$
42. $\frac{3r^2 - 5rR + 2R^2}{r - 3R}$
43. $\frac{t^3 - 8}{t^2 + 2t + 4}$
44. $\frac{a^4 + b^4}{a^2 - 2ab + 2b^2}$

In Exercises 45–56, solve the given problems.

45. When $2x^2 - 9x - 5$ is divided by $x + c$, the quotient is $2x + 1$. Find c.
46. When $6x^2 - x + k$ is divided by $3x + 4$, the remainder is zero. Find k.
47. By division show that $\frac{x^4 + 1}{x + 1}$ is not equal to x^3.
48. By division show that $\frac{x^3 + y^3}{x + y}$ is not equal to $x^2 + y^2$.
49. If a gas under constant pressure has volume V_1 at temperature T_1 (in kelvin), then the new volume V_2 when the temperature changes from T_1 to T_2 is given by $V_2 = V_1\left(1 + \frac{T_2 - T_1}{T_1}\right)$. Simplify the right-hand side of this equation.

50. The area of a certain rectangle can be represented by $6x^2 + 19x + 10$. If the length is $2x + 5$, what is the width? (Divide the area by the length.)

51. In the optical theory dealing with lasers, the following expression arises: $\dfrac{8A^5 + 4A^3\mu^2E^2 - A\mu^4E^4}{8A^4}$. ($\mu$ is the Greek letter mu.) Simplify this expression.

52. In finding the total resistance of the resistors shown in Fig. 1.14, the following expression is used.

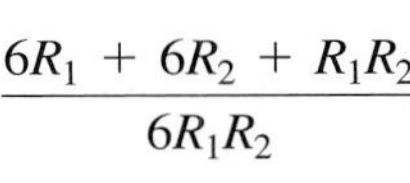

$$\frac{6R_1 + 6R_2 + R_1R_2}{6R_1R_2}$$

Simplify this expression.

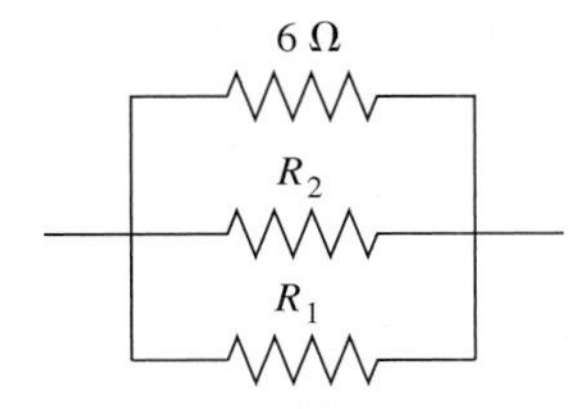

Fig. 1.14

53. When analyzing the potential energy associated with gravitational forces, the expression $\dfrac{GMm[(R + r) - (R - r)]}{2rR}$ arises. Simplify this expression.

54. A computer model shows that the temperature change T in a certain freezing unit is found by using the expression $\dfrac{3T^3 - 8T^2 + 8}{T - 2}$. Perform the indicated division.

55. In analyzing the displacement of a certain valve, the expression $\dfrac{s^2 - 2s - 2}{s^4 + 4}$ is used. Find the reciprocal of this expression and then perform the indicated division.

56. In analyzing a rectangular computer image, the area and width of the image vary with time such that the length is given by the expression $\dfrac{2t^3 + 94t^2 - 290t + 500}{2t + 100}$. By performing the indicated division, find the expression for the length.

Answers to Practice Exercises

1. $2x - 3a$ **2.** $2x + 3$

1.10 Solving Equations

Types of Equations • Solving Basic Types of Equations • Checking the Solution • First Steps • Ratio and Proportion

In this section, we show how algebraic operations are used in solving equations. In the following sections, we show some of the important applications of equations.

An **equation** *is an algebraic statement that two algebraic expressions are equal.* Any value of the **unknown** that produces equality when **substituted** in the equation is said to **satisfy** the equation and is called a **solution** of the equation.

EXAMPLE 1 Valid values for equations

The equation $3x - 5 = x + 1$ is true only if $x = 3$. Substituting 3 for x in the equation, we have $3(3) - 5 = 3 + 1$, or $4 = 4$; substituting $x = 2$, we have $1 = 3$, which is not correct.

This equation is valid for only one value of the unknown. *An equation valid only for certain values of the unknown is a* **conditional equation.** In this section, nearly all equations we solve will be conditional equations that are satisfied by only one value of the unknown. ■

EXAMPLE 2 Identity and contradiction

(a) The equation $x^2 - 4 = (x - 2)(x + 2)$ is true for all values of x. For example, substituting $x = 3$ in the equation, we have $3^2 - 4 = (3 - 2)(3 + 2)$, or $5 = 5$. Substituting $x = -1$, we have $(-1)^2 - 4 = (-1 - 2)(-1 + 2)$, or $-3 = -3$. *An equation valid for all values of the unknown is an* **identity.**

(b) The equation $x + 5 = x + 1$ is not true for any value of x. For any value of x we try, we find that the left side is 4 **greater** than the right side. *Such an equation is called a* **contradiction.** ■

■ Equations can be solved on most graphing calculators. An estimate (or guess) of the answer may be required to find the solution. See Exercises 47 and 48.

To **solve** *an equation, we find the values of the unknown that satisfy it.* There is one basic rule to follow when solving an equation:

Perform the same operation on both sides of the equation.

We do this to isolate the unknown and thus to find its value.

By performing the same operation on both sides of an equation, the two sides remain equal. Thus,

we may add the same number to both sides, subtract the same number from both sides, multiply both sides by the same number (not zero), or divide both sides by the same number (not zero).

EXAMPLE 3 Basic operations used in solving

In each of the following equations, we may isolate x, and thereby solve the equation, by performing the indicated operation.

■ The word *algebra* comes from Arabic and means "a restoration." It refers to the fact that when a number has been added to one side of an equation, the same number must be added to the other side to maintain equality.

$x - 3 = 12$	$x + 3 = 12$	$\frac{x}{3} = 12$	$3x = 12$
add 3 to both sides	subtract 3 from both sides	multiply both sides by 3	divide both sides by 3
$x - 3 + 3 = 12 + 3$	$x + 3 - 3 = 12 - 3$	$3\left(\frac{x}{3}\right) = 3(12)$	$\frac{3x}{3} = \frac{12}{3}$
$x = 15$	$x = 9$	$x = 36$	$x = 4$

NOTE ➧ [Each solution should be checked by substitution in the original equation.] ■

EXAMPLE 4 Operations used for solution; checking

Solve the equation $2t - 7 = 9$.

We are to perform basic operations to both sides of the equation to finally isolate t on one side. The steps to be followed are suggested by the form of the equation.

$2t - 7 = 9$	original equation
$2t - 7 + 7 = 9 + 7$	add 7 to both sides
$2t = 16$	combine like terms
$\frac{2t}{2} = \frac{16}{2}$	divide both sides by 2
$t = 8$	simplify

■ Note that the solution generally requires a combination of basic operations.

Therefore, we conclude that $t = 8$. Checking *in the original equation*, we have

$$2(8) - 7 \stackrel{?}{=} 9, \quad 16 - 7 \stackrel{?}{=} 9, \quad 9 = 9$$

The solution checks. ■

EXAMPLE 5 First remove parentheses

Solve the equation $x - 7 = 3x - (6x - 8)$.

■ With simpler numbers, many basic steps are done by inspection and not actually written down.

$x - 7 = 3x - 6x + 8$	parentheses removed
$x - 7 = -3x + 8$	x-terms combined on right
$4x - 7 = 8$	$3x$ added to both sides
$4x = 15$	7 added to both sides
$x = \frac{15}{4}$	both sides divided by 4

Checking in the original equation, we obtain (after simplifying) $-\frac{13}{4} = -\frac{13}{4}$. ■

CAUTION Note that we always check in the *original* equation. This is done since errors may have been made in finding the later equations. ■

Practice Exercises

Solve for x.

1. $3x + 4 = x - 6$

2. $2(5 - x) = x - 8$

■ Many other types of equations require more advanced methods for solving. These are considered in later chapters.

From these examples, we see that the following steps are used in solving the basic equations of this section.

Procedure for Solving Equations

1. Remove grouping symbols (distributive law).
2. Combine any like terms on each side (also after step 3).
3. Perform the same operations on both sides until $x =$ solution is obtained.
4. Check the solution in the original equation.

NOTE ➧ [If an equation contains numbers not easily combined by inspection, the best procedure is to first solve for the unknown and then perform the calculation.]

EXAMPLE 6 First solve for unknown—circuit current

When finding the current i (in A) in a certain radio circuit, the following equation and solution are used.

$$\begin{aligned} 0.0595 - 0.525i - 8.85(i + 0.00316) &= 0 \\ 0.0595 - 0.525i - 8.85i - 8.85(0.00316) &= 0 \quad \text{note how the above procedure is followed} \\ (-0.525 - 8.85)i &= 8.85(0.00316) - 0.0595 \\ i &= \frac{8.85(0.00316) - 0.0595}{-0.525 - 8.85} \quad \text{evaluate} \\ &= 0.00336 \text{ A} \end{aligned}$$

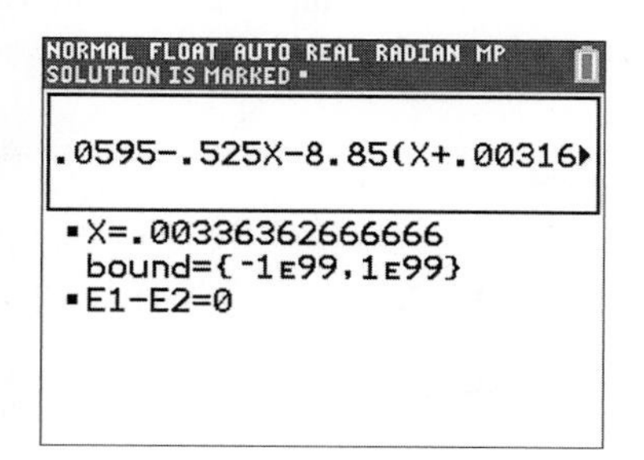

Fig. 1.15

Graphing calculator keystrokes: goo.gl/5wWTnp

The calculator solution of this equation, using the *Solver* feature, is shown in Fig. 1.15. When doing the calculation indicated above in the solution for i, be careful to group the numbers in the denominator for the division. Also, be sure to round off the result as shown above, but do not round off values before the final calculation. ■

RATIO AND PROPORTION

The quotient a/b is also called the **ratio** *of a to b. An equation stating that two ratios are equal is called a* **proportion.** Because a proportion is an equation, if one of the numbers is unknown, we can solve for its value as with any equation. Usually, this is done by noting the denominators and multiplying each side by a number that will clear the fractions.

■ In general, the proportion $\frac{a}{b} = \frac{c}{d}$ is equivalent to the equation $ad = bc$. Therefore, $\frac{x}{8} = \frac{3}{4}$ can be rewritten as $4x = 24$ in order to remove the fractions.

EXAMPLE 7 Ratio

If the ratio of x to 8 equals the ratio of 3 to 4, we have the proportion

$$\frac{x}{8} = \frac{3}{4}$$

We can solve this equation by multiplying both sides by 8. This gives

$$8\left(\frac{x}{8}\right) = 8\left(\frac{3}{4}\right), \quad \text{or} \quad x = 6$$

Substituting $x = 6$ into the original proportion gives the proportion $\frac{6}{8} = \frac{3}{4}$. Because these ratios are equal, the solution checks. ■

Practice Exercise

3. If the ratio of 2 to 5 equals the ratio of x to 30, find x.

■ Generally, units of measurement will not be shown in intermediate steps. The proper units will be shown with the data and final result.

■ If the result is required to be in feet, we have the following change of units (see Section 1.4):

$$6.08\text{ m}\left(\frac{1\text{ ft}}{0.3048\text{ m}}\right) = 19.9\text{ ft}$$

EXAMPLE 8 Proportion—roof truss

The supports for a roof are triangular trusses for which the longest side is 8/5 as long as the shortest side for all of the trusses. If the shortest side of one of the trusses is 3.80 m, what is the length of the longest side of that truss?

If we label the longest side L, since the ratio of sides is 8/5, we have

$$\frac{L}{3.80} = \frac{8}{5}$$

$$3.80\left(\frac{L}{3.80}\right) = 3.80\left(\frac{8}{5}\right)$$

$$L = 6.08\text{ m}$$

Checking, we note that $6.08/3.80 = 1.6$ and $8/5 = 1.6$. The solution checks. ■

The meanings of *ratio* and *proportion* (particularly ratio) will be of importance when studying trigonometry in Chapter 4. A detailed discussion of ratio and proportion is found in Chapter 18. A general method of solving equations involving fractions, such as we found in Examples 7 and 8, is given in Chapter 6.

EXERCISES 1.10

In Exercises 1–4, make the given changes in the indicated examples of this section and then solve the resulting problems.

1. In Example 3, change 12 to −12 in each of the four illustrations and then solve.
2. In Example 4, change $2t - 7$ to $7 - 2t$ and then solve.
3. In Example 5, change $(6x - 8)$ to $(8 - 6x)$ and then solve.
4. In Example 8, change 8/5 to 7/4 and then solve.

In Exercises 5–44, solve the given equations.

5. $x - 2 = 7$
6. $x - 4 = -1$
7. $x + 5 = 4$
8. $s + 6 = -3$
9. $-\frac{t}{2} = 5$
10. $\frac{x}{-4} = 2$
11. $\frac{y - 8}{3} = 4$
12. $\frac{7 - r}{6} = 3$
13. $4E = -20$
14. $2x = 12$
15. $5t + 9 = -1$
16. $5D - 2 = 13$
17. $5 - 2y = -3$
18. $-5t + 8 = 18$
19. $3x + 7 = x$
20. $6 + 4L = 5 - 3L$
21. $2(3q + 4) = 5q$
22. $3(4 - n) = -n$
23. $-(r - 4) = 6 + 2r$
24. $-(x + 2) + 5 = 5x$
25. $8(y - 5) = -2y$
26. $4(7 - F) = -7$
27. $0.1x - 0.5(x - 2) = 2$
28. $1.5x - 0.3(x - 4) = 6$
29. $-4 - 3(1 - 2p) = -7 + 2p$
30. $3 - 6(2 - 3t) = t - 5$
31. $\frac{4x - 2(x - 4)}{3} = 8$
32. $2x = \frac{-5(7 - 3x) + 2}{4}$
33. $|x| - 9 = 2$
34. $2 - |x| = 4$
35. $|2x - 3| = 5$
36. $|7 - x| = 1$

In Exercises 37–44, all numbers are approximate.

37. $5.8 - 0.3(x - 6.0) = 0.5x$
38. $1.9t = 0.5(4.0 - t) - 0.8$
39. $-0.24(C - 0.50) = 0.63$
40. $27.5(5.17 - 1.44x) = 73.4$
41. $\frac{x}{2.0} = \frac{17}{6.0}$
42. $\frac{3.0}{7.0} = \frac{R}{42}$
43. $\frac{165}{223} = \frac{13V}{15}$
44. $\frac{276x}{17.0} = \frac{1360}{46.4}$

In Exercises 45–56, solve the given problems.

45. Identify each of the following equations as a conditional equation, an identity, or a contradiction.
    (a) $2x + 3 = 3 + 2x$ (b) $2x - 3 = 3 - 2x$
46. Are there any values of a for which the equation $2x + a = 2x$ results in a conditional equation? Explain why or why not.
47. Solve the equation of Example 5 by using the Equation Solver of a graphing calculator.
48. Solve the equation of Example 6 by using the Equation Solver of a graphing calculator.
49. To find the amount of a certain investment of x dollars, it is necessary to solve the equation $0.03x + 0.06(2000 - x) = 96$. Solve for x.
50. In finding the rate v (in km/h) at which a polluted stream is flowing, the equation $15(5.5 + v) = 24(5.5 - v)$ is used. Find v.
51. In finding the maximum operating temperature T (in °C) for a computer integrated circuit, the equation $1.1 = (T - 76)/40$ is used. Find the temperature.
52. To find the voltage V in a circuit in a TV remote-control unit, the equation $1.12V - 0.67(10.5 - V) = 0$ is used. Find V.
53. In blending two gasolines of different octanes, in order to find the number n of gallons of one octane needed, the equation $0.14n + 0.06(2000 - n) = 0.09(2000)$ is used. Find n, given that 0.06 and 0.09 are exact and the first zero of 2000 is significant.

54. In order to find the distance x such that the weights are balanced on the lever shown in Fig. 1.16, the equation $210(3x) = 55.3x + 38.5(8.25 - 3x)$ must be solved. Find x. (3 is exact.)

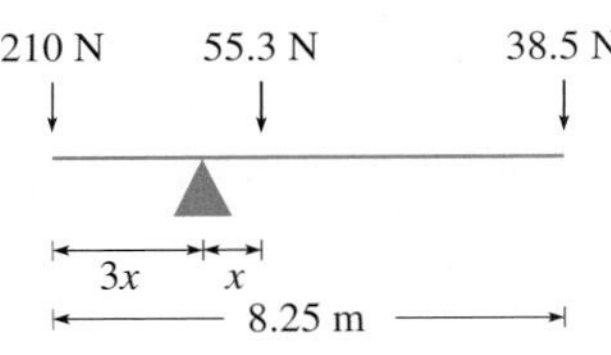

Fig. 1.16

55. The 2016 Nissan Leaf electric car can travel 107 mi on a fully charged 30-kW · h battery. How much electricity (in kW · h) is required to drive this car on a 350-mi trip? Assume all numbers are exact and round your answer to a whole number. (*Source*: www.nissanusa.com.)

56. An athlete who was jogging and wearing a Fitbit found that she burned 250 calories in 20 minutes. At that rate, how long will it take her to burn 400 calories? Assume all numbers are exact.

Answers to Practice Exercises

1. −5 **2.** 6 **3.** 12

1.11 Formulas and Literal Equations

Formulas • Literal Equations • Subscripts • Solve for Symbol before Substituting Numerical Values

■ Einstein published his first paper on relativity in 1905.

An important application of equations is in the use of *formulas* that are found in geometry and nearly all fields of science and technology. *A* **formula** (or **literal equation**) *is an equation that expresses the relationship between two or more related quantities.* For example, Einstein's famous formula $E = mc^2$ shows the equivalence of energy E to the mass m of an object and the speed of light c.

NOTE ➧ We can solve a formula for a particular symbol just as we solve any equation. [That is, we isolate the required symbol by using algebraic operations on literal numbers.]

EXAMPLE 1 Solving for symbol in formula—Einstein

In Einstein's formula $E = mc^2$, solve for m.

$$\frac{E}{c^2} = m \quad \text{divide both sides by } c^2$$

$$m = \frac{E}{c^2} \quad \text{switch sides to place } m \text{ at left}$$

The required symbol is usually placed on the left, as shown. ■

EXAMPLE 2 Symbol with subscript in formula—velocity

A formula relating acceleration a, velocity v, initial velocity v_0, and time is $v = v_0 + at$. Solve for t.

■ The subscript$_0$ makes v_0 a different literal symbol from v. (We have used subscripts in a few of the earlier exercises.)

$$v - v_0 = at \quad v_0 \text{ subtracted from both sides}$$

$$t = \frac{v - v_0}{a} \quad \text{both sides divided by } a \text{ and then sides switched}$$ ■

EXAMPLE 3 Symbol in capital and in lowercase—forces on a beam

TI-89 graphing calculator keystrokes for Example 3: goo.gl/1fYsOi

In the study of the forces on a certain beam, the equation $W = \dfrac{L(wL + 2P)}{8}$ is used. Solve for P.

$$8W = \frac{8L(wL + 2P)}{8} \quad \text{multiply both sides by 8}$$

$$8W = L(wL + 2P) \quad \text{simplify right side}$$

$$8W = wL^2 + 2LP \quad \text{remove parentheses}$$

$$8W - wL^2 = 2LP \quad \text{subtract } wL^2 \text{ from both sides}$$

$$P = \frac{8W - wL^2}{2L} \quad \text{divide both sides by } 2L \text{ and switch sides}$$ ■

CAUTION Be careful. Just as subscripts can denote different literal numbers, a capital letter and the same letter in lowercase are different literal numbers. In this example, W and w are different literal numbers. This is shown in several of the exercises in this section. ■

EXAMPLE 4 Formula with groupings—temperature and volume

The effect of temperature on measurements is important when measurements must be made with great accuracy. The volume V of a special precision container at temperature T in terms of the volume V_0 at temperature T_0 is given by $V = V_0[1 + b(T - T_0)]$, where b depends on the material of which the container is made. Solve for T.

Because we are to solve for T, we must isolate the term containing T. This can be done by first removing the grouping symbols.

$$V = V_0[1 + b(T - T_0)] \quad \text{original equation}$$
$$V = V_0[1 + bT - bT_0] \quad \text{remove parentheses}$$
$$V = V_0 + bTV_0 - bT_0V_0 \quad \text{remove brackets}$$
$$V - V_0 + bT_0V_0 = bTV_0 \quad \text{subtract } V_0 \text{ and add } bT_0V_0 \text{ to both sides}$$
$$T = \frac{V - V_0 + bT_0V_0}{bV_0} \quad \text{divide both sides by } bV_0 \text{ and switch sides}$$

■

Practice Exercises

Solve for the indicated letter. Each comes from the indicated area of study.

1. $\theta = kA + \lambda$, for λ (robotics)
2. $P = n(p - c)$, for p (economics)

NOTE ➧ [To determine the values of any literal number in an expression for which we know values of the other literal numbers, we should first solve for the required symbol and then evaluate.]

EXAMPLE 5 Solve for symbol before substituting—volume of sphere

The volume V (in mm^3) of a copper sphere changes with the temperature T (in °C) according to $V = V_0 + V_0\beta T$, where V_0 is the volume at 0°C. For a given sphere, $V_0 = 6715\ \text{mm}^3$ and $\beta = 5.10 \times 10^{-5}/°\text{C}$. Evaluate T for $V = 6908\ \text{mm}^3$.

We first solve for T and then substitute the given values.

$$V = V_0 + V_0\beta T$$
$$V - V_0 = V_0\beta T$$
$$T = \frac{V - V_0}{V_0\beta}$$

Now substituting, we have

$$T = \frac{6908 - 6715}{(6715)(5.10 \times 10^{-5})}$$
$$= 564°\text{C}$$

■

■ Note that copper melts at about 1100°C.

EXERCISES 1.11

In Exercises 1–4, solve for the given letter from the indicated example of this section.

1. For the formula in Example 2, solve for a.
2. For the formula in Example 3, solve for w.
3. For the formula in Example 4, solve for T_0.
4. For the formula in Example 5, solve for β. (Do not evaluate.)

In Exercises 5–42, each of the given formulas arises in the technical or scientific area of study shown. Solve for the indicated letter.

5. $E = IR$, for R (electricity)
6. $PV = nRT$, for T (chemistry)
7. $rL = g_2 - g_1$, for g_1 (surveying)
8. $W = S_dT - Q$, for Q (air conditioning)

9. $B = \frac{nTWL}{12}$, for n (construction management)

10. $P = 2\pi Tf$, for T (mechanics)

11. $p = p_a + dgh$, for g (hydrodynamics)

12. $2Q = 2I + A + S$, for I (nuclear physics)

13. $F_c = \frac{mv^2}{r}$, for r (centripetal force)

14. $P = \frac{4F}{\pi D^2}$, for F (automotive trades)

15. $S_T = \frac{A}{5T} + 0.05d$, for A (welding)

16. $u = -\frac{eL}{2m}$, for L (spectroscopy)

17. $ct^2 = 0.3t - ac$, for a (medical technology)

18. $2p + dv^2 = 2d(C - W)$, for W (fluid flow)

19. $T = \frac{c + d}{v}$, for d (traffic flow)

20. $B = \frac{\mu_0 I}{2\pi R}$, for R (magnetic field)

21. $\frac{K_1}{K_2} = \frac{m_1 + m_2}{m_1}$, for m_2 (kinetic energy)

22. $f = \frac{F}{d - F}$, for d (photography)

23. $a = \frac{2mg}{M + 2m}$, for M (pulleys)

24. $v = \frac{V(m + M)}{m}$, for M (ballistics)

25. $C_0^2 = C_1^2(1 + 2V)$, for V (electronics)

26. $A_1 = A(M + 1)$, for M (photography)

27. $N = r(A - s)$, for s (engineering)

28. $T = 3(T_2 - T_1)$, for T_1 (oil drilling)

29. $T_2 = T_1 - \frac{h}{100}$, for h (air temperature)

30. $p_2 = p_1 + rp_1(1 - p_1)$, for r (population growth)

31. $Q_1 = P(Q_2 - Q_1)$, for Q_2 (refrigeration)

32. $p - p_a = dg(y_2 - y_1)$, for y_1 (fire science)

33. $N = N_1T - N_2(1 - T)$, for N_1 (machine design)

34. $t_a = t_c + (1 - h)t_m$, for h (computer access time)

35. $L = \pi(r_1 + r_2) + 2x_1 + x_2$, for r_1 (pulleys)

36. $I = \frac{VR_2 + VR_1(1 + \mu)}{R_1R_2}$, for μ (electronics)

37. $P = \frac{V_1(V_2 - V_1)}{gJ}$, for V_2 (jet engine power)

38. $W = T(S_1 - S_2) - Q$, for S_2 (refrigeration)

39. $C = \frac{2eAk_1k_2}{d(k_1 + k_2)}$, for e (electronics)

40. $d = \frac{3LPx^2 - Px^3}{6EI}$, for L (beam deflection)

41. $V = C\left(1 - \frac{n}{N}\right)$, for n (property deprecation)

42. $\frac{p}{P} = \frac{AI}{B + AI}$, for B (atomic theory)

In Exercises 43–48, find the indicated values.

43. For a car's cooling system, the equation $p(C - n) + n = A$ is used. If $p = 0.25$, $C = 15.0$ L, and $A = 13.0$ L, solve for n (in L).

44. A formula used in determining the total transmitted power P_t in an AM radio signal is $P_t = P_c(1 + 0.500m^2)$. Find P_c if $P_t = 680$ W and $m = 0.925$.

45. A formula relating the Fahrenheit temperature F and the Celsius temperature C is $F = \frac{9}{5}C + 32$. Find the Celsius temperature that corresponds to 90.2°F.

46. In forestry, a formula used to determine the volume V of a log is $V = \frac{1}{2}L(B + b)$, where L is the length of the log and B and b are the areas of the ends. Find b (in ft²) if $V = 38.6$ ft³, $L = 16.1$ ft, and $B = 2.63$ ft². See Fig. 1.17.

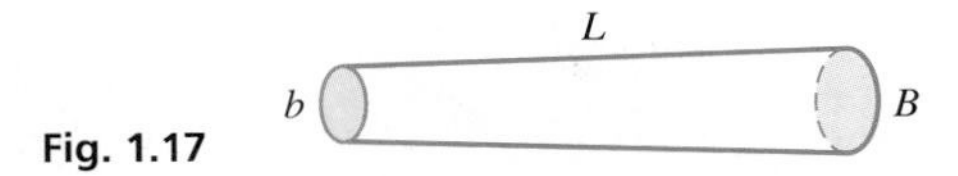

Fig. 1.17

47. The voltage V_1 across resistance R_1 is $V_1 = \frac{VR_1}{R_1 + R_2}$, where V is the voltage across resistances R_1 and R_2. See Fig. 1.18. Find R_2 (in Ω) if $R_1 = 3.56$ Ω, $V_1 = 6.30$ V, and $V = 12.0$ V.

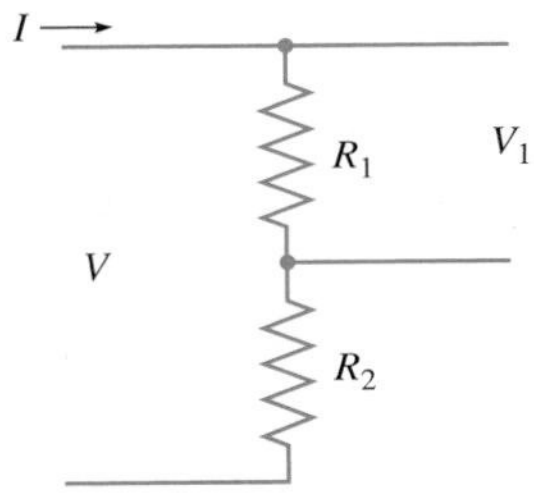

Fig. 1.18

48. The efficiency E of a computer multiprocessor compilation is given by $E = \frac{1}{q + p(1 - q)}$, where p is the number of processors and q is the fraction of the compilation that can be performed by the available parallel processors. Find p for $E = 0.66$ and $q = 0.83$.

In Exercises 49 and 50, set up the required formula and solve for the indicated letter.

49. One missile travels at a speed of v_2 mi/h for 4 h, and another missile travels at a speed of v_1 for $t + 2$ hours. If they travel a total of d mi, solve the resulting formula for t.

50. A microwave transmitter can handle x telephone communications, and 15 separate cables can handle y connections each. If the combined system can handle C connections, solve for y.

Answers to Practice Exercises

1. $\theta - kA$ **2.** $\frac{P + nc}{n}$

1.12 Applied Word Problems

Procedure for Solving Word Problems • Identifying the Unknown Quantities • Setting Up the Proper Equation • Examples of Solving Word Problems

Many applied problems are at first word problems, and we must put them into mathematical terms for solution. Usually, the most difficult part in solving a word problem is identifying the information needed for setting up the equation that leads to the solution. To do this, you must read the problem carefully to be sure that you understand all of the terms and expressions used. Following is an approach you should use.

■ See Appendix A, page A-1, for a variation to the method outlined in these steps. You might find it helpful.

Procedure for Solving Word Problems

1. Read the statement of the problem. First, read it quickly for a general overview. Then reread slowly and carefully, listing the information given.
2. Clearly identify the unknown quantities and then assign an appropriate letter to represent one of them, stating this choice clearly.
3. Specify the other unknown quantities in terms of the one in step 2.
4. If possible, make a sketch using the known and unknown quantities.
5. Analyze the statement of the problem and write the necessary equation. This is often the most difficult step because **some of the information may be implied and not explicitly stated.** Again, a very careful reading of the statement is necessary.
6. Solve the equation, clearly stating the solution.
7. Check the solution with the original statement of the problem.

Read the following examples very carefully and note just how the outlined procedure is followed.

EXAMPLE 1 Sum of forces on a beam

A 17-lb beam is supported at each end. The supporting force at one end is 3 lb more than at the other end. Find the forces.

Since the force at each end is required, we write

step 2 let $F =$ the smaller force (in lb)

as a way of establishing the unknown for the equation. Any appropriate letter could be used, and we could have let it represent the larger force.

■ Be sure to carefully identify your choice for the unknown. In most problems, there is really a choice. Using the word *let* clearly shows that a specific choice has been made.

Also, since the other force is 3 lb more, we write

step 3 $F + 3 =$ the larger force (in lb)

■ The statement after "let x (or some other appropriate letter) =" should be clear. It should completely define the chosen unknown.

step 4 We now draw the sketch in Fig. 1.19.

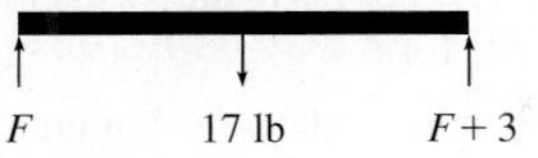

F 17 lb $F+3$

Fig. 1.19

Since the forces at each end of the beam support the weight of the beam, we have the equation

step 5 $F + (F + 3) = 17$

This equation can now be solved: $2F = 14$

step 6 $F = 7$ lb

step 7 Thus, the smaller force is 7 lb, and the larger force is 10 lb. This checks with the original statement of the problem. ■

CAUTION Always check a verbal problem with the original statement of the problem, not the first equation, because it was derived from the statement. ■

EXAMPLE 2 Office complex energy-efficient lighting

In designing an office complex, an architect planned to use 34 energy-efficient ceiling lights using a total of 1000 W. Two different types of lights, one using 25 W and the other using 40 W, were to be used. How many of each were planned?

Since we want to find the number of each type of light, we

$$\text{let } x = \text{number of 25 W lights}$$

■ "Let x = 25 W lights" is incomplete. We want to find out how many there are.

Also, since there are 34 lights in all,

$$34 - x = \text{number of 40 W lights}$$

We also know that the total wattage of all lights is 1000 W. This means

25 W each → number → 40 W each → number →

$$25x + 40(34 - x) = 1000 \quad \leftarrow \text{total wattage of all lights}$$

(total wattage of 25 W lights; total wattage of 40 W lights)

■ See Appendix A, page A-1, for a "sketch" that might be used with this example.

$$25x + 1360 - 40x = 1000$$
$$-15x = -360$$
$$x = 24$$

Therefore, there are 24 25-W lights and 10 40-W lights. The total wattage of these lights is $24(25) + 10(40) = 600 + 400 = 1000$. We see that this checks with the statement of the problem. ■

EXAMPLE 3 Number of medical slides

A medical researcher finds that a given sample of an experimental drug can be divided into 4 more slides with 5 mg each than with 6 mg each. How many slides with 5 mg each can be made up?

We are asked to find the number of slides with 5 mg, and therefore we

$$\text{let } x = \text{number of slides with 5 mg}$$

Because the sample may be divided into 4 more slides with 5 mg each than of 6 mg each, we know that

$$x - 4 = \text{number of slides with 6 mg}$$

Because *it is the same sample that is to be divided,* the total mass of the drug on each set of slides is the same. This means

5 mg each → number → 6 mg each → number →

$$5x = 6(x - 4)$$

(total mass 5-mg slides; total mass 6-mg slides)

$$5x = 6x - 24$$
$$-x = -24 \quad \text{or} \quad x = 24$$

Therefore, the sample can be divided into 24 slides with 5 mg each, or 20 slides with 6 mg each. Since the total mass, 120 mg, is the same for each set of slides, the solution checks with the statement of the problem. ■

Practice Exercise

1. Solve the problem in Example 3 by letting y = number of slides with 6 mg.

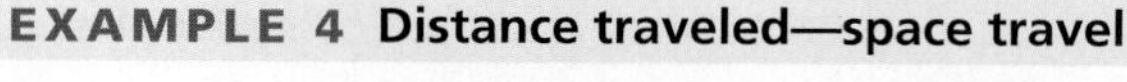

EXAMPLE 4 Distance traveled—space travel

■ A maneuver similar to the one in Example 4 was used on several servicing missions to the Hubble space telescope from 1999 to 2009.

A space shuttle maneuvers so that it may "capture" an already orbiting satellite that is 6000 km ahead. If the satellite is moving at 27,000 km/h and the shuttle is moving at 29,500 km/h, how long will it take the shuttle to reach the satellite? (All digits shown are significant.)

First, we let $t =$ the time for the shuttle to reach the satellite. Then, using the fact that the shuttle must go 6000 km farther in the same time, we draw the sketch in Fig. 1.20. Next, we use the formula *distance* = *rate* × *time* ($d = rt$). This leads to the following equation and solution.

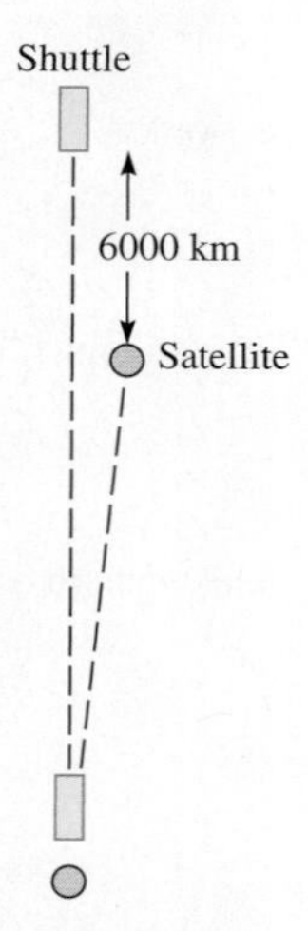

Fig. 1.20

speed of shuttle ↓ time ↓ — speed of satellite ↓ time ↓

$$29{,}500t = 6000 + 27{,}000t$$

distance traveled by shuttle — distance between at beginning — distance traveled by satellite

$$2500t = 6000$$
$$t = 2.400 \text{ h}$$

This means that it will take the shuttle 2.400 h to reach the satellite. In 2.400 h, the shuttle will travel 70,800 km, and the satellite will travel 64,800 km. We see that the solution checks with the statement of the problem. ■

EXAMPLE 5 Mixture—gasoline and methanol

■ "Let x = methanol" is incomplete. We want to find out the volume (in L) that is to be added.

A refinery has 7250 L of a gasoline-methanol blend that is 6.00% methanol. How much pure methanol must be added so that the resulting blend is 10.0% methanol?

First, let $x =$ the number of liters of methanol to be added. The total volume of methanol in the final blend is the volume in the original blend plus that which is added. This total volume is to be 10.0% of the final blend. See Fig. 1.21.

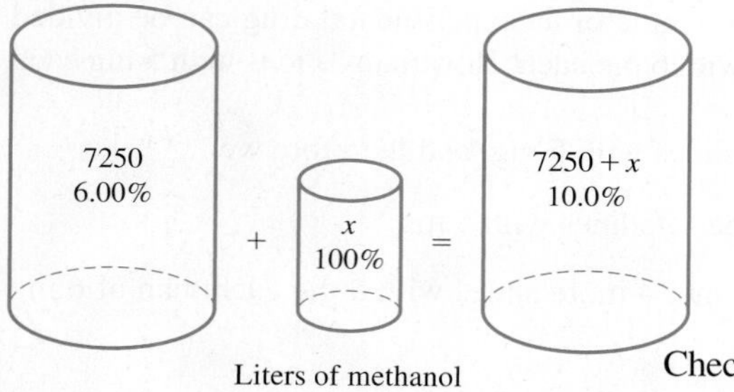

Fig. 1.21

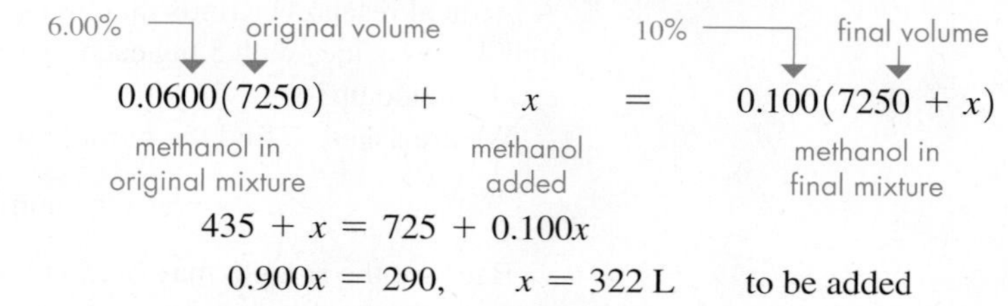

$$0.0600(7250) + x = 0.100(7250 + x)$$

$$435 + x = 725 + 0.100x$$
$$0.900x = 290, \quad x = 322 \text{ L} \quad \text{to be added}$$

Checking (to three significant digits), there would be 757 L of methanol of a total 7570 L. ■

EXERCISES 1.12

In Exercises 1–4, make the given changes in the indicated examples of this section and then solve the resulting problems.

1. In Example 2, in the first line, change 34 to 31.
2. In Example 3, in the second line, change "4 more slides" to "3 more slides."
3. In Example 4, in the second line, change 27,000 km/h to 27,100 km/h.
4. In Example 5, change "pure methanol" to "of a blend with 50.0% methanol."

In Exercises 5–34, set up an appropriate equation and solve. Data are accurate to two significant digits unless greater accuracy is given.

5. A certain new car costs \$5000 more than the same model new car cost 6 years ago. Together a new model today and 6 years ago cost \$64,000. What was the cost of each? (Assume all values are exact.)
6. The flow of one stream into a lake is 1700 ft^3/s more than the flow of a second stream. In 1 h, 1.98×10^7 ft^3 flow into the lake from the two streams. What is the flow rate of each?

7. Approximately 6.9 million wrecked cars are recycled in two consecutive years. There were 500,000 more recycled the second year than the first year. How many are recycled each year?

8. A business website had twice as many hits on the first day of a promotion as on the second day. If the total number of hits for both days was 495,000, find the number of hits on each day.

9. Petroleum rights to 140 acres of land are leased for $37,000. Part of the land leases for $200 per acre, and the reminder for $300 per acre. How much is leased at each price?

10. A vial contains 2000 mg, which is to be used for two dosages. One patient is to be administered 660 mg more than another. How much should be administered to each?

11. After installing a pollution control device, a car's exhaust contained the same amount of pollutant after 5.0 h as it had in 3.0 h. Before the installation the exhaust contained 150 ppm/h (parts per million per hour) of the pollutant. By how much did the device reduce the emission?

12. Three meshed spur gears have a total of 107 teeth. If the second gear has 13 more teeth than the first and the third has 15 more teeth than the second, how many teeth does each have?

13. A cell phone subscriber paid x dollars per month for the first 6 months. He then increased his data plan, and his bill increased by $10 per month for the next 4 months. If he paid a total of $890 for the 10-month period, find the amount of his bill before and after the increase.

14. A satellite television subscriber paid x dollars per month for the first year. Her monthly bill increased by $30 per month for the second and third years, and then another $20 for the fourth and fifth years. If the total amount paid for the 5-year period was $7320, find the three different monthly bill amounts.

15. The sum of three electric currents that come together at a point in a circuit is zero. If the second current is twice the first and the third current is 9.2 μA more than the first, what are the currents? (The sign indicates the direction of flow.)

16. A delivery firm uses one fleet of trucks on daily routes of 8 h. A second fleet, with five more trucks than the first, is used on daily routes of 6 h. Budget allotments allow for 198 h of daily delivery time. How many trucks are in each fleet?

17. A natural gas pipeline feeds into three smaller pipelines, each of which is 2.6 km longer than the main pipeline. The total length of the four pipelines is 35.4 km. How long is each section?

18. At 100% efficiency two generators would produce 750 MW of power. At efficiencies of 65% and 75%, they produce 530 MW. At 100% efficiency, what power would each produce?

19. A wholesaler sells three types of GPS systems. A dealer orders twice as many economy systems at $40 each, and 75 more econo-plus systems at $80 each, than deluxe systems at $140 each, for $42,000. How many of each were ordered?

20. A person won a state lottery prize of $20,000, from which 25% was deducted for taxes. The remainder was invested, partly for a 40% gain, and the rest for a 10% loss. How much was each investment if there was a $2000 net investment gain?

21. Train A is 520 ft long and traveling at 60.0 mi/h. Train B is 440 ft long and traveling at 40 mi/h in the opposite direction of train A on an adjacent track. How long does it take for the trains to completely pass each other? (*Footnote*: A law was once actually passed by the Wisconsin legislature that included "whenever two trains meet at an intersection . . . , neither shall proceed until the other has.")

22. A family has $3850 remaining of its monthly income after making the monthly mortgage payment, which is 23.0% of the monthly income. How much is the mortgage payment?

23. A ski lift takes a skier up a slope at 50 m/min. The skier then skis down the slope at 150 m/min. If a round trip takes 24 min, how long is the slope?

24. Before being put out of service, the supersonic jet Concorde made a trip averaging 120 mi/h less than the speed of sound for 1.0 h, and 410 mi/h more than the speed of sound for 3.0 h. If the trip covered 3990 mi, what is the speed of sound?

25. Trains at each end of the 50.0-km-long Eurotunnel under the English Channel start at the same time into the tunnel. Find their speeds if the train from France travels 8.0 km/h faster than the train from England and they pass in 17.0 min. See Fig. 1.22.

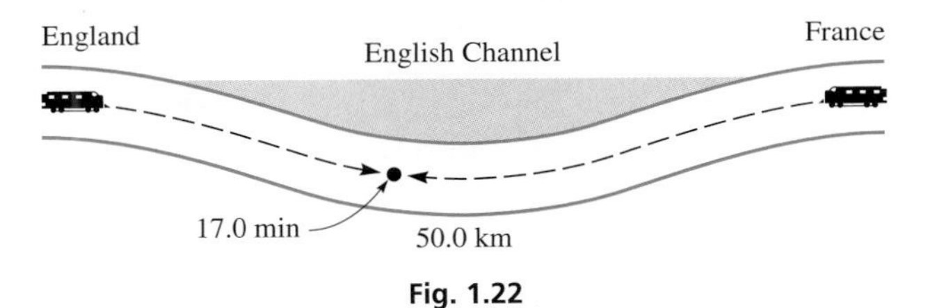

Fig. 1.22

26. An executive would arrive 10.0 min early for an appointment if traveling at 60.0 mi/h, or 5.0 min early if traveling at 45.0 mi/h. How much time is there until the appointment?

27. One lap at the Indianapolis Speedway is 2.50 mi. In a race, a car stalls and then starts 30.0 s after a second car. The first car travels at 260 ft/s, and the second car travels at 240 ft/s. How long does it take the first car to overtake the second, and which car will be ahead after eight laps?

28. A computer chip manufacturer produces two types of chips. In testing a total of 6100 chips of both types, 0.50% of one type and 0.80% of the other type were defective. If a total of 38 defective chips were found, how many of each type were tested?

29. Two gasoline distributors, A and B, are 228 mi apart on Interstate 80. A charges $2.90/gal and B charges $2.70/gal. Each charges 0.2¢/gal per mile for delivery. Where on Interstate 80 is the cost to the customer the same?

30. An outboard engine uses a gasoline-oil fuel mixture in the ratio of 15 to 1. How much gasoline must be mixed with a gasoline-oil mixture, which is 75% gasoline, to make 8.0 L of the mixture for the outboard engine?

31. A car's radiator contains 12 L of antifreeze at a 25% concentration. How many liters must be drained and then replaced by pure antifreeze to bring the concentration to 50% (the manufacturer's "safe" level)?

32. How much sand must be added to 250 lb of a cement mixture that is 22% sand to have a mixture that is 25% sand?

33. To pass a 20-m long semitrailer traveling at 70 km/h in 10 s, how fast must a 5.0-m long car go?

34. An earthquake emits primary waves moving at 8.0 km/s and secondary waves moving at 5.0 km/s. How far from the epicenter of the earthquake is the seismic station if the two waves arrive at the station 2.0 min apart?

Answer to Practice Exercise

1. 24 with 5 mg, 20 with 6 mg

CHAPTER 1 KEY FORMULAS AND EQUATIONS

Commutative law of addition: $a + b = b + a$

Commutative law of multiplication: $ab = ba$

Associative law of addition: $a + (b + c) = (a + b) + c$

Associative law of multiplication: $a(bc) = (ab)c$

Distributive law: $a(b + c) = ab + ac$

$$a + (-b) = a - b \quad (1.1)$$

$$a - (-b) = a + b \quad (1.2)$$

$$a^m \times a^n = a^{m+n} \quad (1.3)$$

$$\frac{a^m}{a^n} = a^{m-n} \quad (m > n, a \neq 0), \qquad \frac{a^m}{a^n} = \frac{1}{a^{n-m}} \quad (m < n, a \neq 0) \quad (1.4)$$

$$(a^m)^n = a^{mn} \quad (1.5)$$

$$(ab)^n = a^n b^n, \quad \left(\frac{a}{b}\right)^n = \frac{a^n}{b^n} \quad (b \neq 0) \quad (1.6)$$

$$a^0 = 1 \quad (a \neq 0) \quad (1.7)$$

$$a^{-n} = \frac{1}{a^n} \quad (a \neq 0) \quad (1.8)$$

$$\sqrt{ab} = \sqrt{a}\sqrt{b} \quad (a \text{ and } b \text{ positive real numbers}) \quad (1.9)$$

CHAPTER 1 REVIEW EXERCISES

CONCEPT CHECK EXERCISES

Determine each of the following as being either **true** *or* **false.** *If it is false, explain why.*

1. The absolute value of any real number is positive.
2. $16 - 4 \div 2 = 14$
3. For approximate numbers, $26.7 - 15 = 11.7$.
4. $2a^3 = 8a^3$
5. $0.237 = 2.37 \times 10^{-1}$
6. $-\sqrt{-4} = 2$
7. $4x - (2x + 3) = 2x + 3$
8. $(x - 7)^2 = 49 - 14x + x^2$
9. $\frac{6x + 2}{2} = 3x$
10. If $5x - 4 = 0$, $x = 5/4$
11. If $a - bc = d$, $c = (d - a)/b$
12. In setting up the solution to a word problem involving numbers of gears, it would be sufficient to "let $x =$ the first gear."

PRACTICE AND APPLICATIONS

In Exercises 13–24, evaluate the given expressions.

13. $-2 + (-5) - 3$
14. $6 - 8 - (-4)$
15. $\frac{(-5)(6)(-4)}{(-2)(3)}$
16. $\frac{(-9)(-12)(-4)}{24}$
17. $-5 - |2(-6)| + \frac{-15}{3}$
18. $3 - 5|-3 - 2| - \frac{|-4|}{-4}$
19. $\frac{18}{3 - 5} - (-4)^2$
20. $-(-3)^2 - \frac{-8}{(-2) - |-4|}$
21. $\sqrt{16} - \sqrt{64}$
22. $-\sqrt{81 + 144}$
23. $(\sqrt{7})^2 - \sqrt[3]{8}$
24. $-\sqrt[4]{16} + (\sqrt{6})^2$

In Exercises 25–32, simplify the given expressions. Where appropriate, express results with positive exponents only.

25. $(-2rt^2)^2$
26. $(3a^0b^{-2})^3$
27. $-3mn^{-5}t(8m^{-3}n^4)$
28. $\frac{15p^4q^2r}{5pq^5r}$
29. $\frac{-16N^{-2}(NT^2)}{-2N^0T^{-1}}$
30. $\frac{-35x^{-1}y(x^2y)}{5xy^{-1}}$
31. $\sqrt{45}$
32. $\sqrt{9 + 36}$

In Exercises 33–36, for each number, (a) determine the number of significant digits and (b) round off each to two significant digits.

33. 8000
34. 21,490
35. 9.050
36. 0.7000

In Exercises 37–40, evaluate the given expressions. All numbers are approximate.

37. $37.3 - 16.92(1.067)^2$
38. $\frac{8.896 \times 10^{-12}}{-3.5954 - 6.0449}$
39. $\frac{\sqrt{0.1958 + 2.844}}{3.142(65)^2}$
40. $\frac{1}{0.03568} + \frac{37,466}{29.63^2}$

In Exercises 41–46, make the indicated conversions.

41. 875 Btu to joules

42. 18.4 in. to meters

43. 65 km/h to ft/s

44. 12.25 g/L to lb/ft^3

45. 225 hp to joules per minute

46. 89.7 $lb/in.^2$ to N/cm^2

In Exercises 47–78, perform the indicated operations.

47. $a - 3ab - 2a + ab$

48. $xy - y - 5y - 4xy$

49. $6LC - (3 - LC)$

50. $-(2x - b) - 3(-x - 5b)$

51. $(2x - 1)(5 + x)$

52. $(C - 4D)(D - 2C)$

53. $(x + 8)^2$

54. $(2r - 9s)^2$

55. $\dfrac{2h^3k^2 - 6h^4k^5}{2h^2k}$

56. $\dfrac{4a^2x^3 - 8ax^4}{-2ax^2}$

57. $4R - [2r - (3R - 4r)]$

58. $-3b - [3a - (a - 3b)] + 4a$

59. $2xy - \{3z - [5xy - (7z - 6xy)]\}$

60. $x^2 + 3b + [(b - y) - 3(2b - y + z)]$

61. $(2x + 1)(x^2 - x - 3)$

62. $(x - 3)(2x^2 + 1 - 3x)$

63. $-3y(x - 4y)^2$

64. $-s(4s - 3t)^2$

65. $3p[(q - p) - 2p(1 - 3q)]$

66. $3x[2y - r - 4(x - 2r)]$

67. $\dfrac{12p^3q^2 - 4p^4q + 6pq^5}{2p^4q}$

68. $\dfrac{27s^3t^2 - 18s^4t + 9s^2t}{-9s^2t}$

69. $(2x^2 + 7x - 30) \div (x + 6)$

70. $(4x^2 - 41) \div (2x + 7)$

71. $\dfrac{3x^3 - 7x^2 + 11x - 3}{3x - 1}$

72. $\dfrac{w^3 + 7w - 4w^2 - 12}{w - 3}$

73. $\dfrac{4x^4 + 10x^3 + 18x - 1}{x + 3}$

74. $\dfrac{8x^3 - 14x + 3}{2x + 3}$

75. $-3\{(r + s - t) - 2[(3r - 2s) - (t - 2s)]\}$

76. $(1 - 2x)(x - 3) - (x + 4)(4 - 3x)$

77. $\dfrac{2y^3 - 7y + 9y^2 + 5}{2y - 1}$

78. $\dfrac{6x^2 + 5xy - 4y^2}{2x - y}$

In Exercises 79–90, solve the given equations.

79. $3x + 1 = x - 8$

80. $4y - 3 = 5y + 7$

81. $\dfrac{5x}{7} = \dfrac{3}{2}$

82. $\dfrac{2(4 - N)}{-3} = \dfrac{5}{4}$

83. $-6x + 5 = -3(x - 4)$

84. $-2(-4 - y) = 3y$

85. $2s + 4(3 - s) = 6$

86. $2|x| - 1 = 3$

87. $3t - 2(7 - t) = 5(2t + 1)$

88. $-(8 - x) = x - 2(2 - x)$

89. $2.7 + 2.0(2.1x - 3.4) = 0.1$

90. $0.250(6.721 - 2.44x) = 2.08$

In Exercises 91–100, change numbers in ordinary notation to scientific notation or change numbers in scientific notation to ordinary notation. (See Appendix B for an explanation of the symbols that are used.)

91. A certain computer has 60,000,000,000,000 bytes of memory.

92. The escape velocity (the velocity required to leave the Earth's gravitational field) is about 25,000 mi/h.

93. In 2015, the most distant known object in the solar system, a dwarf planet named V774104, was discovered. It was 15,400,000,000 km from the sun.

94. Police radar has a frequency of 1.02×10^9 Hz.

95. Among the stars nearest the Earth, Centaurus A is about 2.53×10^{13} mi away.

96. Before its destruction in 2001, the World Trade Center had nearly 10^7 ft^2 of office space.

97. The faintest sound that can be heard has an intensity of about 10^{-12} W/m^2.

98. An optical coating on glass to reduce reflections is about 0.00000015 m thick.

99. The maximum safe level of radiation in the air of a home due to radon gas is 1.5×10^{-1} Bq/L. (Bq is the symbol for bequerel, the metric unit of radioactivity, where 1 Bq = 1 decay/s.)

100. A certain virus was measured to have a diameter of about 0.00000018 m.

In Exercises 101–114, solve for the indicated letter. Where noted, the given formula arises in the technical or scientific area of study.

101. $V = \pi r^2 L$, for L (oil pipeline volume)

102. $R = \dfrac{2GM}{c^2}$, for G (astronomy: black holes)

103. $P = \dfrac{\pi^2 EI}{L^2}$, for E (mechanics)

104. $f = p(c - 1) - c(p - 1)$, for p (thermodynamics)

105. $Pp + Qq = Rr$, for q (moments of forces)

106. $V = IR + Ir$, for R (electricity)

107. $d = (n - 1)A$, for n (optics)

108. $mu = (m + M)v$, for M (physics: momentum)

109. $N_1 = T(N_2 - N_3) + N_3$, for N_2 (mechanics: gears)

110. $Q = \dfrac{kAt(T_2 - T_1)}{L}$, for T_1 (solar heating)

111. $R = \dfrac{A(T_2 - T_1)}{H}$, for T_2 (thermal resistance)

112. $Z^2\left(1 - \dfrac{\lambda}{2a}\right) = k$, for λ (radar design)

113. $d = kx^2[3(a + b) - x]$, for a (mechanics: beams)

114. $V = V_0[1 + 3a(T_2 - T_1)]$, for T_2 (thermal expansion)

In Exercises 115–120, perform the indicated calculations.

115. A computer's memory is 5.25×10^{13} bytes, and that of a model 30 years older is 6.4×10^4 bytes. What is the ratio of the newer computer's memory to the older computer's memory?

116. The time (in s) for an object to fall h feet is given by the expression $0.25\sqrt{h}$. How long does it take a person to fall 66 ft from a sixth-floor window into a net while escaping a fire?

117. The CN Tower in Toronto is 0.553 km high. The Willis Tower (formerly the Sears Tower) in Chicago is 442 m high. How much higher is the CN Tower than the Willis Tower?

118. The time (in s) it takes a computer to check n cells is found by evaluating $(n/2650)^2$. Find the time to check 4.8×10^3 cells.

119. The combined electric resistance of two parallel resistors is found by evaluating the expression $\frac{R_1R_2}{R_1 + R_2}$. Evaluate this for $R_1 = 0.0275\ \Omega$ and $R_2 = 0.0590\ \Omega$.

120. The distance (in m) from the Earth for which the gravitational force of the Earth on a spacecraft equals the gravitational force of the sun on it is found by evaluating $1.5 \times 10^{11}\sqrt{m/M}$, where m and M are the masses of the Earth and sun, respectively. Find this distance for $m = 5.98 \times 10^{24}$ kg and $M = 1.99 \times 10^{30}$ kg.

In Exercises 121–124, simplify the given expressions.

121. One transmitter antenna is $(x - 2a)$ ft long, and another is $(x + 2a)$ yd long. What is the sum, in feet, of their lengths?

122. In finding the value of an annuity, the expression $(Ai - R)(1 + i)^2$ is used. Multiply out this expression.

123. A computer analysis of the velocity of a link in an industrial robot leads to the expression $4(t + h) - 2(t + h)^2$. Simplify this expression.

124. When analyzing the motion of a communications satellite, the expression $\frac{k^2r - 2h^2k + h^2rv^2}{k^2r}$ is used. Perform the indicated division.

In Exercises 125–136, solve the given problems.

125. Does the value of $3 \times 18 \div (9 - 6)$ change if the parentheses are removed?

126. Does the value of $(3 \times 18) \div 9 - 6$ change if the parentheses are removed?

127. In solving the equation $x - (3 - x) = 2x - 3$, what conclusion can be made?

128. In solving the equation $7 - (2 - x) = x + 2$, what conclusion can be made?

129. For $x = 2$ and $y = -4$, evaluate (a) $2|x| - 2|y|$; (b) $2|x - y|$.

130. If $a < 0$, write the value of $|a|$ without the absolute value symbols.

131. If $3 - x < 0$, solve $|3 - x| + 7 = 2x$ for x.

132. Solve $|x - 4| + 6 = 3x$ for x. (Be careful!)

133. Show that $(x - y)^3 = -(y - x)^3$.

134. Is division associative? That is, is it true (if $b \neq 0, c \neq 0$) that $(a \div b) \div c = a \div (b \div c)$?

135. What is the ratio of 8×10^{-3} to 2×10^4?

136. What is the ratio of $\sqrt{4 + 36}$ to $\sqrt{4}$?

In Exercises 137–154, solve the given problems. All data are accurate to two significant digits unless greater accuracy is given.

137. A certain engine produces 250 hp. What is this power in kilowatts (kW)?

138. The pressure gauge for an automobile tire shows a pressure of 32 lb/in.^{2}. What is this pressure in N/m^2?

139. A certain automobile engine produces a maximum torque of 110 N · m. Convert this to foot pounds.

140. A typical electric current density in a wire is 1.2×10^6 A/m^2 (A is the symbol for ampere). Express this in mA/cm^2.

141. Two computer software programs cost \$190 together. If one costs \$72 more than the other, what is the cost of each?

142. A sponsor pays a total of \$9500 to run a commercial on two different TV stations. One station charges \$1100 more than the other. What does each charge to run the commercial?

143. Three chemical reactions each produce oxygen. If the first produces twice that of the second, the third produces twice that of the first, and the combined total is 560 cm^3, what volume is produced by each?

144. In testing the rate at which a polluted stream flows, a boat that travels at 5.5 mi/h in still water took 5.0 h to go downstream between two points, and it took 8.0 h to go upstream between the same two points. What is the rate of flow of the stream?

145. The voltage across a resistor equals the current times the resistance. In a microprocessor circuit, one resistor is 1200 Ω greater than another. The sum of the voltages across them is 12.0 mV. Find the resistances if the current is 2.4 μA in each.

146. An air sample contains 4.0 ppm (parts per million) of two pollutants. The concentration of one is four times the other. What are the concentrations?

147. One road crew constructs 450 m of road bed in 12 h. If another crew works at the same rate, how long will it take them to construct another 250 m of road bed?

148. The fuel for a two-cycle motorboat engine is a mixture of gasoline and oil in the ratio of 15 to 1. How many liters of each are in 6.6 L of mixture?

149. A ship enters Lake Superior from Sault Ste. Marie, moving toward Duluth at 17.4 km/h. Two hours later, a second ship leaves Duluth moving toward Sault Ste. Marie at 21.8 km/h. When will the ships pass, given that Sault Ste. Marie is 634 km from Duluth?

150. A helicopter used in fighting a forest fire travels at 105 mi/h from the fire to a pond and 70 mi/h with water from the pond to the fire. If a round-trip takes 30 min, how long does it take from the pond to the fire? See Fig. 1.23.

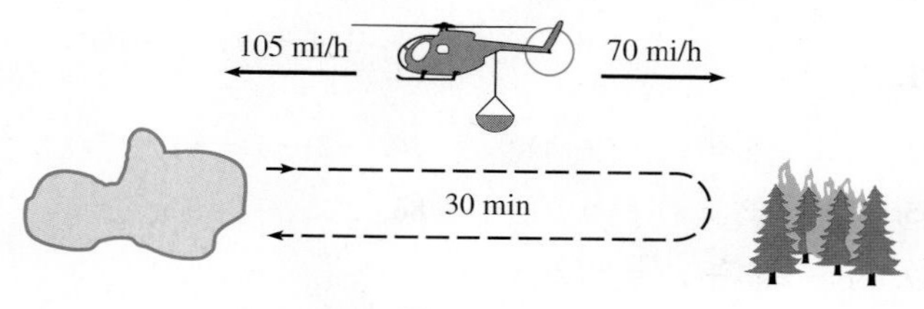

Fig. 1.23

151. One grade of oil has 0.50% of an additive, and a higher grade has 0.75% of the additive. How many liters of each must be used to have 1000 L of a mixture with 0.65% of the additive?

152. Each day a mining company crushes 18,000 Mg of shale-oil rock, some of it 72 L/Mg and the rest 150 L/Mg of oil. How much of each type of rock is needed to produce 120 L/Mg?

153. An architect plans to have 25% of the floor area of a house in ceramic tile. In all but the kitchen and entry, there are 2200 ft^2 of floor area, 15% of which is tile. What area can be planned for the kitchen and entry if each has an all-tile floor?

154. A karat equals 1/24 part of gold in an alloy (for example, 9-karat gold is 9/24 gold). How many grams of 9-karat gold must be mixed with 18-karat gold to get 200 g of 14-karat gold?

155. In calculating the simple interest earned by an investment, the equation $P = P_0 + P_0rt$ is used, where P is the value after an initial principal P_0 is invested for t years at interest rate r. Solve for r, and then evaluate r for $P = \$7625$, $P_0 = \$6250$, and $t = 4.000$ years. Write a paragraph or two explaining (a) your method for solving for r, and (b) the calculator steps used to evaluate r, noting the use of parentheses.

CHAPTER 1 PRACTICE TEST

As a study aid, we have included complete solutions for each Practice Test problem at the back of this book.

In Problems 1–5, evaluate the given expressions. In Problems 3 and 5, the numbers are approximate.

1. $\sqrt{9 + 16}$ **2.** $\dfrac{(7)(-3)(-2)}{(-6)(0)}$ **3.** $\dfrac{3.372 \times 10^{-3}}{7.526 \times 10^{12}}$

4. $\dfrac{(+6)(-2) - 3(-1)}{|2 - 5|}$ **5.** $\dfrac{346.4 - 23.5}{287.7} - \dfrac{0.944^3}{(3.46)(0.109)}$

In Problems 6–12, perform the indicated operations and simplify. When exponents are used, use only positive exponents in the result.

6. $(2a^0b^{-2}c^3)^{-3}$ **7.** $(2x + 3)^2$ **8.** $3m^2(am - 2m^3)$

9. $\dfrac{8a^3x^2 - 4a^2x^4}{-2ax^2}$ **10.** $\dfrac{6x^2 - 13x + 7}{2x - 1}$

11. $(2x - 3)(x + 7)$ **12.** $3x - [4x - (3 - 2x)]$

13. Solve for y: $5y - 2(y - 4) = 7$

14. Solve for x: $3(x - 3) = x - (2 - 3d)$

15. Convert 245 lb/ft^3 to kg/L.

16. Express 0.0000036 in scientific notation.

17. List the numbers -3, $|-4|$, $-\pi$, $\sqrt{2}$, and 0.3 in numerical order.

18. What fundamental law is illustrated by $3(5 + 8) = 3(5) + 3(8)$?

19. **(a)** How many significant digits are in the number 3.0450?
(b) Round it off to two significant digits.

20. If P dollars is deposited in a bank that compounds interest n times a year, the value of the account after t years is found by evaluating $P(1 + i/n)^{nt}$, where i is the annual interest rate. Find the value of an account for which $P = \$1000$, $i = 5\%$, $n = 2$, and $t = 3$ years (values are exact).

21. In finding the illuminance from a light source, the expression $8(100 - x)^2 + x^2$ is used. Simplify this expression.

22. The equation $L = L_0[1 + \alpha(t_2 - t_1)]$ is used when studying thermal expansion. Solve for t_2.

23. An alloy weighing 20 lb is 30% copper. How many pounds of another alloy, which is 80% copper, must be added for the final alloy to be 60% copper?

2 Geometry

LEARNING OUTCOMES

After completion of this chapter, the student should be able to:

- Identify perpendicular and parallel lines
- Identify supplementary, complementary, vertical, and corresponding angles
- Determine interior angles and sides of various triangles
- Use the Pythagorean theorem
- Identify and analyze different types of quadrilaterals and polygons
- Identify and analyze circles, arcs, and interior angles of a circle
- Calculate area and perimeter of a geometric shape
- Use an approximation method to estimate an irregular area
- Identify and analyze three-dimensional geometric figures, including volume, surface area, and dimensions

When building the pyramids nearly 5000 years ago, and today when using MRI (magnetic resonance imaging) to detect a tumor in a human being, the size and shape of an object are measured.

Because geometry deals with size and shape, the topics and methods of geometry are important in many of the applications of technology.

Many of the methods of measuring geometric objects were known in ancient times, and most of the geometry used in technology has been known for hundreds of years. In about 300 B.C.E., the Greek mathematician Euclid (who lived and taught in Alexandria, Egypt) organized what was known in geometry. He added many new ideas in a 13-volume set of writings known as the *Elements*. Centuries later it was translated into various languages, and today is second only to the Bible as the most published book in history.

The study of geometry includes the properties and measurements of angles, lines, and surfaces and the basic figures they form. In this chapter, we review the more important methods and formulas for calculating basic geometric measures, such as area and volume. Technical applications are included from areas such as architecture, construction, instrumentation, surveying and civil engineering, mechanical design, and product design of various types, as well as other areas of engineering.

Geometric figures and concepts are also basic to the development of many areas of mathematics, such as graphing and trigonometry. We will start our study of graphs in Chapter 3 and trigonometry in Chapter 4.

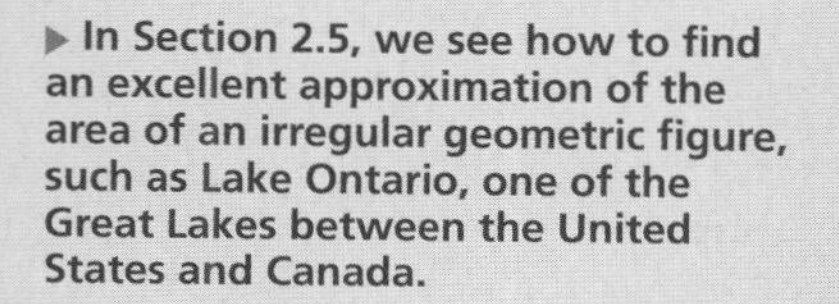

▶ In Section 2.5, we see how to find an excellent approximation of the area of an irregular geometric figure, such as Lake Ontario, one of the Great Lakes between the United States and Canada.

2.1 Lines and Angles

Basic Angles • Parallel Lines and Perpendicular Lines • Supplementary Angles and Complementary Angles • Angles Between Intersecting Lines • Segments of Transversals

■ The use of 360 comes from the ancient Babylonians and their number system based on 60 rather than 10, as we use today. However, the specific reason for the choice of 360 is not known. (One theory is that 360 is divisible by many smaller numbers and is close to the number of days in a year.)

It is not possible to define every term we use. In geometry, *the meanings of* **point, line,** *and* **plane** *are accepted without being defined.* These terms give us a starting point for the definitions of other useful geometric terms.

The *amount of rotation of a* **ray** *(or* **half-line***) about its endpoint is called an* **angle.** A ray is that part of a line (the word *line* means "straight line") to one side of a fixed point on the line. *The fixed point is the* **vertex** *of the angle. One complete rotation of a ray is an angle with a measure of* 360 **degrees,** *written as* 360°. Some special types of angles are as follows:

Name of angle	Measure of angle
Right angle	*90°*
Straight angle	*180°*
Acute angle	*Between 0° and 90°*
Obtuse angle	*Between 90° and 180°*

EXAMPLE 1 Basic angles

Figure 2.1(a) shows a right angle (marked as ⊾). The vertex of the angle is point B, and the ray is the half-line BA. Figure 2.1(b) shows a straight angle. Figure 2.1(c) shows an acute angle, denoted as $\angle E$ (or $\angle DEF$ or $\angle FED$). In Fig. 2.1(d), $\angle G$ is an obtuse angle.

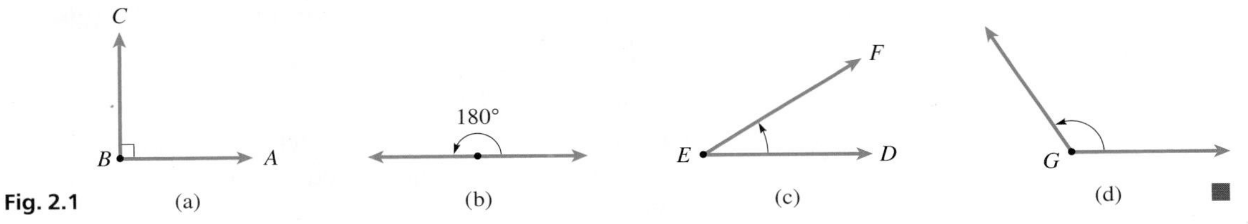

Fig. 2.1 ■

If two lines intersect such that the angle between them is a right angle, the lines are **perpendicular.** *Lines in the same plane that do not intersect are* **parallel.** These are illustrated in the following example.

EXAMPLE 2 Parallel lines and perpendicular lines

In Fig. 2.2(a), lines AC and DE are perpendicular (which is shown as $AC \perp DE$) because they meet in a right angle (again, shown as ⊾) at B.

In Fig. 2.2(b), lines AB and CD are drawn so they do not meet, even if extended. Therefore, these lines are parallel (which can be shown as $AB \parallel CD$).

In Fig. 2.2(c), $AB \perp BC$, $DC \perp BC$, and $AB \parallel DC$.

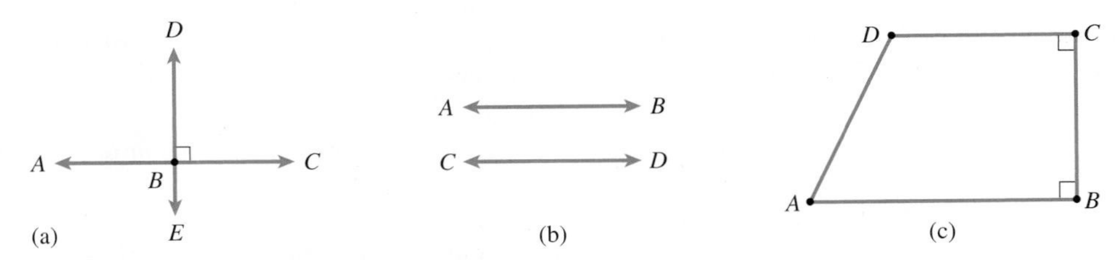

Fig. 2.2

It will be important to recognize perpendicular sides and parallel sides in many of the geometric figures in later sections. ■

■ Recognizing complementary angles is important in trigonometry.

If the sum of the measures of two angles is 180°, *then the angles are called* **supplementary angles.** Each angle is the **supplement** of the other. *If the sum of the measures of two angles is* 90°, *the angles are called* **complementary angles.** Each is the **complement** of the other.

EXAMPLE 3 Supplementary angles and complementary angles

(a) In Fig. 2.3, $\angle BAC = 55°$, and $\angle DAC = 125°$. Because $55° + 125° = 180°$, $\angle BAC$ and $\angle DAC$ are supplementary angles. $\angle BAD$ is a straight angle.

(b) In Fig. 2.4, we see that $\angle POQ$ is a right angle, or $\angle POQ = 90°$. Because $\angle POR + \angle ROQ = \angle POQ = 90°$, $\angle POR$ is the complement of $\angle ROQ$ (or $\angle ROQ$ is the complement of $\angle POR$).

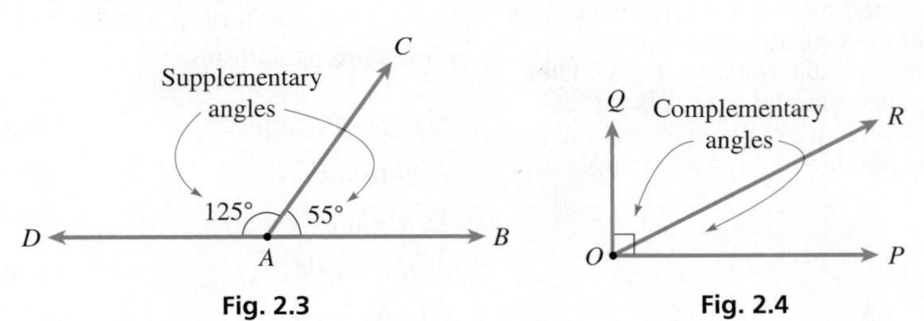

Fig. 2.3 Fig. 2.4 ■

Practice Exercise

1. What is the measure of the complement of $\angle BAC$ in Fig. 2.3?

It is often necessary to refer to certain special pairs of angles. *Two angles that have a common vertex and a side common between them are known as* **adjacent angles.** *If two lines cross to form equal angles on opposite sides of the point of intersection, which is the common vertex, these angles are called* **vertical angles.**

EXAMPLE 4 Adjacent angles and vertical angles

(a) In Fig. 2.5, $\angle BAC$ and $\angle CAD$ have a common vertex at A and the common side between them. This means that $\angle BAC$ and $\angle CAD$ are adjacent angles.

(b) In Fig. 2.6, lines AB and CD intersect at point O. Here, $\angle AOC$ and $\angle BOD$ are vertical angles, and they are equal. Also, $\angle BOC$ and $\angle AOD$ are vertical angles and are equal.

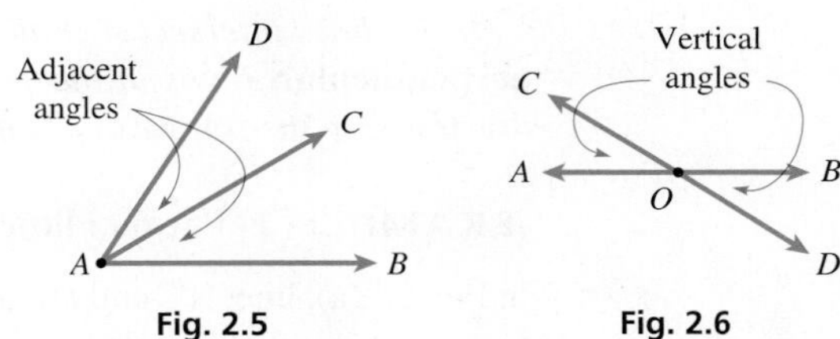

Fig. 2.5 Fig. 2.6 ■

Practice Exercise

2. In Fig. 2.7, if $\angle 2 = 42°$, then $\angle 5 = ?$

We should also be able to identify *the sides of an angle that are adjacent to the angle*. In Fig. 2.5, sides AB and AC are adjacent to $\angle BAC$, and in Fig. 2.6, sides OB and OD are adjacent to $\angle BOD$. Identifying sides adjacent and opposite an angle in a triangle is important in trigonometry.

In a plane, *if a line crosses two or more parallel or nonparallel lines, it is called a* **transversal.** In Fig. 2.7, $AB \parallel CD$, and the transversal of these two parallel lines is the line EF.

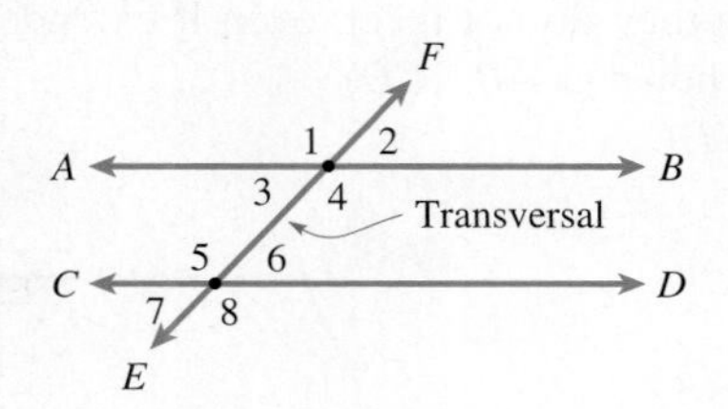

Fig. 2.7

When a transversal crosses a pair of parallel lines, certain pairs of equal angles result. In Fig. 2.7, the **corresponding angles** are equal (that is, $\angle 1 = \angle 5$, $\angle 2 = \angle 6$, $\angle 3 = \angle 7$, and $\angle 4 = \angle 8$). Also, the **alternate-interior angles** are equal ($\angle 3 = \angle 6$ and $\angle 4 = \angle 5$), and the **alternate-exterior angles** are equal ($\angle 1 = \angle 8$ and $\angle 2 = \angle 7$).

When more than two parallel lines are crossed by *two* transversals, such as is shown in Fig. 2.8, *the segments of the transversals between the same two parallel lines are called* **corresponding segments.** A useful theorem is that *the ratios of corresponding segments of the transversals are equal*. In Fig. 2.8, this means that

$$\frac{a}{b} = \frac{c}{d} \tag{2.1}$$

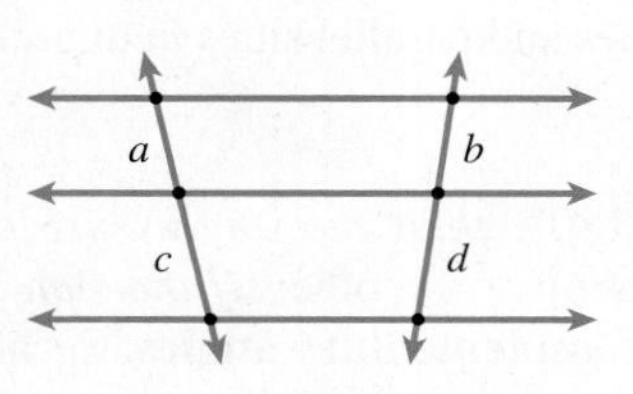

Fig. 2.8

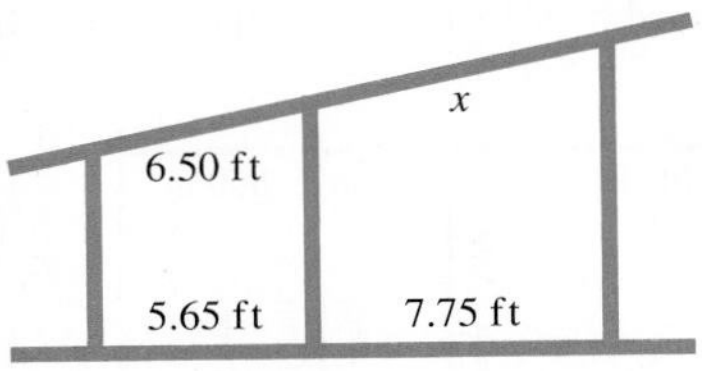

Fig. 2.9

EXAMPLE 5 Segments of transversals

In Fig. 2.9, part of the beam structure within a building is shown. The vertical beams are parallel. From the distances between beams that are shown, determine the distance x between the middle and right vertical beams.

Using Eq. (2.1), we have

$$\frac{6.50}{5.65} = \frac{x}{7.75}$$

$$x = \frac{6.50(7.75)}{5.65} = 8.92 \text{ ft} \quad \text{rounded off}$$

■

EXERCISES 2.1

In Exercises 1–4, answer the given questions about the indicated examples of this section.

1. In Example 2, what is the measure of $\angle ABE$ in Fig. 2.2(a)?
2. In Example 3(b), if $\angle POR = 35°$ in Fig. 2.4, what is the measure of $\angle QOR$?
3. In Example 4, how many different pairs of adjacent angles are there in Fig. 2.6?
4. In Example 5, if the segments of 6.50 ft and 7.75 ft are interchanged, (a) what is the answer, and (b) is the beam along which x is measured more nearly vertical or more nearly horizontal?

In Exercises 5–12, identify the indicated angles and sides in Fig. 2.10. In Exercises 9 and 10, also find the measures of the indicated angles.

5. Two acute angles
6. Two right angles
7. The straight angle
8. The obtuse angle
9. If $\angle CBD = 65°$, find its complement.
10. If $\angle CBD = 65°$, find its supplement.
11. The sides adjacent to $\angle DBC$
12. The acute angle adjacent to $\angle DBC$

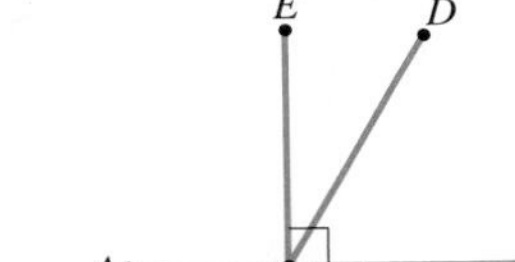

Fig. 2.10

In Exercises 13–15, use Fig. 2.11. In Exercises 16–18, use Fig. 2.12. Find the measures of the indicated angles.

13. $\angle AOB$
14. $\angle AOC$
15. $\angle BOD$
16. $\angle 3$
17. $\angle 4$
18. $\angle 5$

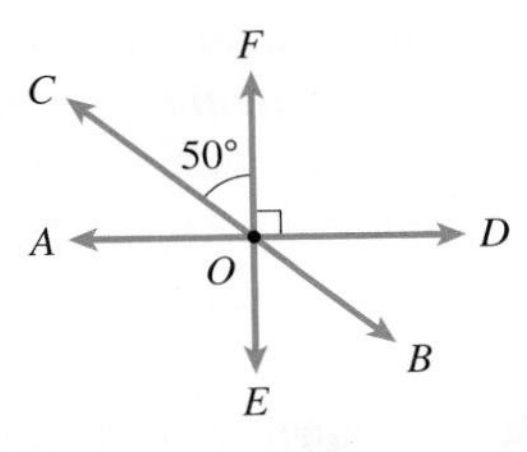

Fig. 2.11

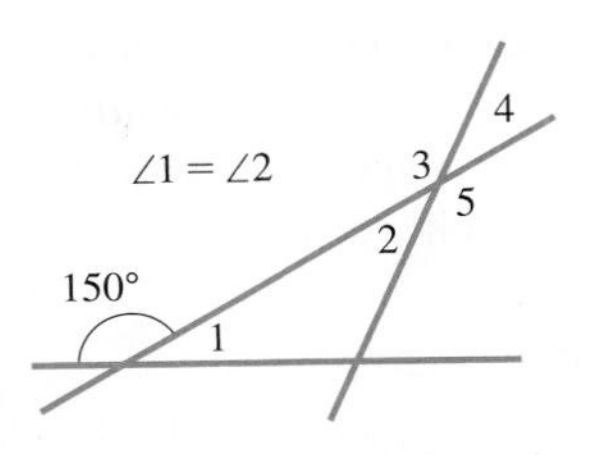

Fig. 2.12

In Exercises 19–24, find the measures of the angles in Fig. 2.13.

19. $\angle 1$
20. $\angle 2$
21. $\angle 3$
22. $\angle 4$
23. $\angle 5$
24. $\angle 6$

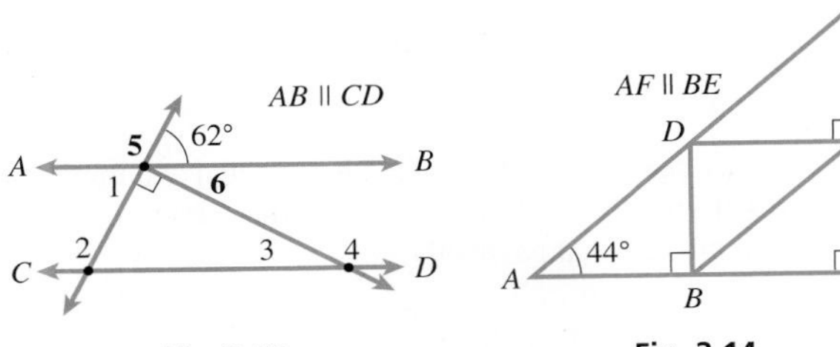

Fig. 2.13

Fig. 2.14

In Exercises 25–30, find the measures of the angles in the truss shown in Fig. 2.14. A truss is a rigid support structure that is used in the construction of buildings and bridges.

25. $\angle BDF$
26. $\angle ABE$
27. $\angle DEB$
28. $\angle DBE$
29. $\angle DFE$
30. $\angle ADE$

In Exercises 31–34, find the indicated distances between the straight irrigation ditches shown in Fig. 2.15. The vertical ditches are parallel.

31. a
32. b
33. c
34. d

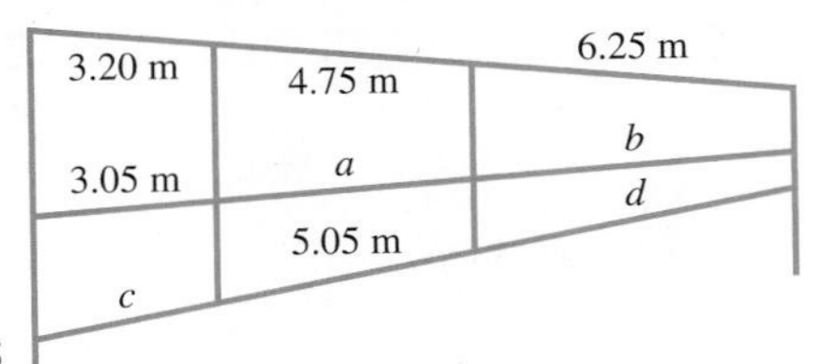

Fig. 2.15

In Exercises 35–40, find all angles of the given measures for the beam support structure shown in Fig. 2.16.

35. 25°
36. 45°
37. 65°
38. 70°
39. 110°
40. 115°

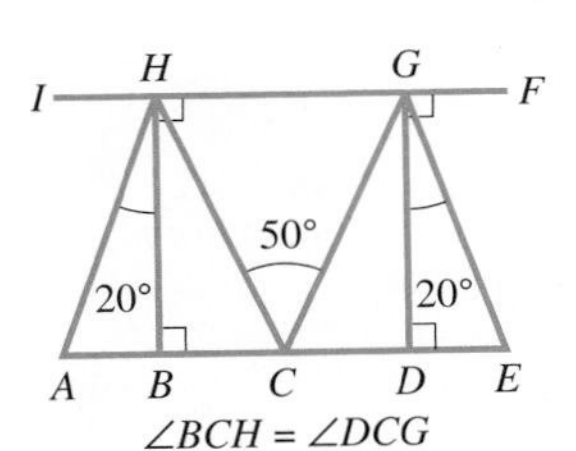

Fig. 2.16

In Exercises 41–46, solve the given problems

41. A plane was heading in a direction 58° east of directly north. It then turned and began to head in a direction 18° south of directly east. Find the measure of the obtuse angle formed between the two parts of the trip. See Fig. 2.17.

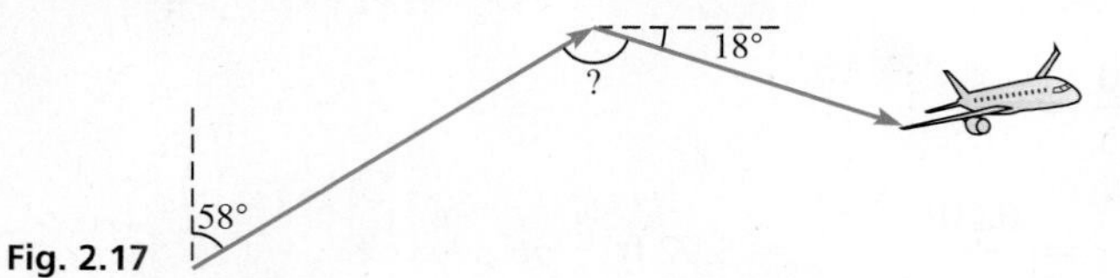

Fig. 2.17

42. $\angle A = (x + 20)°$ and $\angle B = (3x - 2)°$. Solve for x if (a) $\angle A$ and $\angle B$ are (a) complementary angles; (b) alternate-interior angles.

43. A steam pipe is connected in sections *AB, BC,* and *CD* as shown in Fig. 2.18. Find $\angle BCD$ if $AB \parallel CD$.

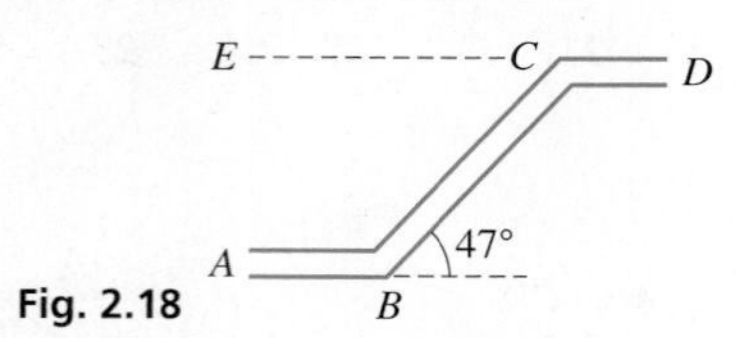

Fig. 2.18

44. Part of a laser beam striking a surface is reflected and the remainder passes through (see Fig. 2.19). Find the angle between the surface and the part that passes through.

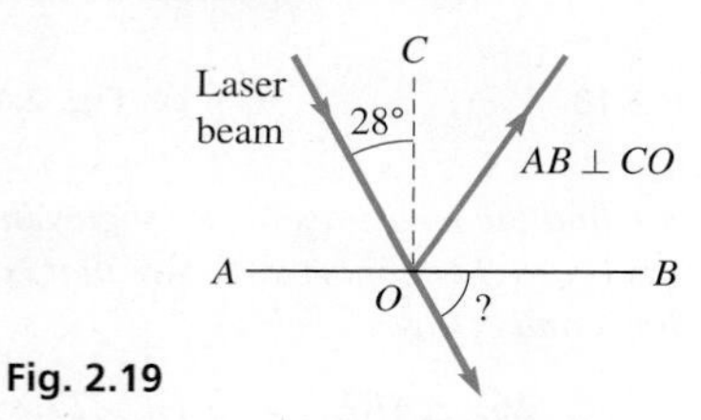

Fig. 2.19

45. An electric circuit board has equally spaced parallel wires with connections at points *A, B,* and *C*, as shown in Fig. 2.20. How far is *A* from *C*, if $BC = 2.15$ cm?

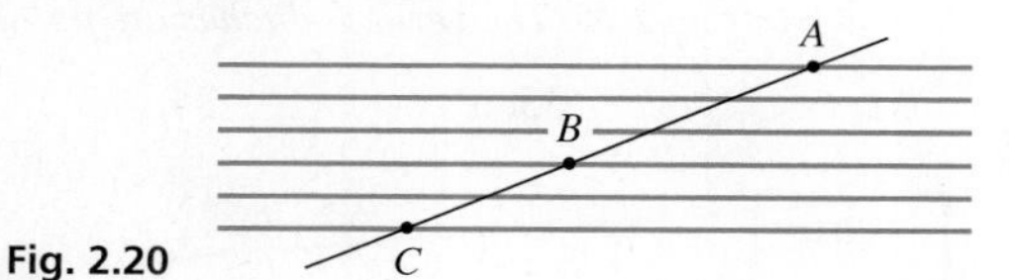

Fig. 2.20

46. Find the distance on Dundas St. W between Dufferin St. and Ossington Ave. in Toronto, as shown in Fig. 2.21. The north–south streets are parallel.

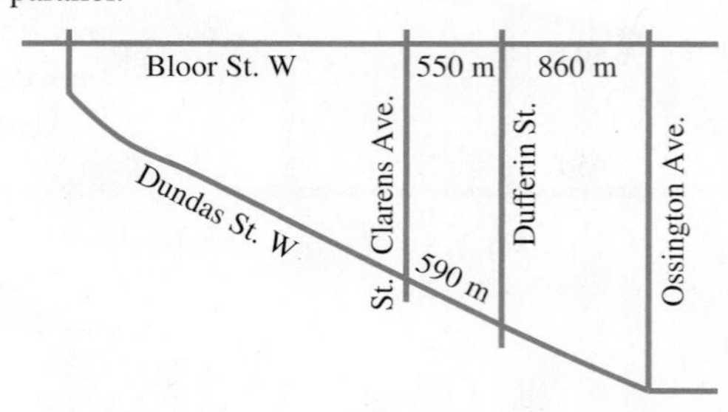

Fig. 2.21

In Exercises 47–50, solve the given problems related to Fig. 2.22.

47. $\angle 1 + \angle 2 + \angle 3 = ?$

48. $\angle 4 + \angle 2 + \angle 5 = ?$

49. Based on Exercise 48, what conclusion can be drawn about a closed geometric figure like the one with vertices at *A, B,* and *D*?

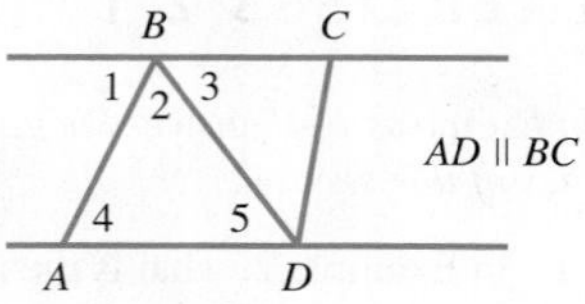

$AD \parallel BC$

Fig. 2.22

50. The *angle of elevation* is the angle above horizontal that an observer must look to see a higher object. The *angle of depression* is the angle below horizontal that an observer must look to see a lower object. See Fig. 2.23. Do the angle of elevation and the angle of depression always have the same measure? Explain why or why not.

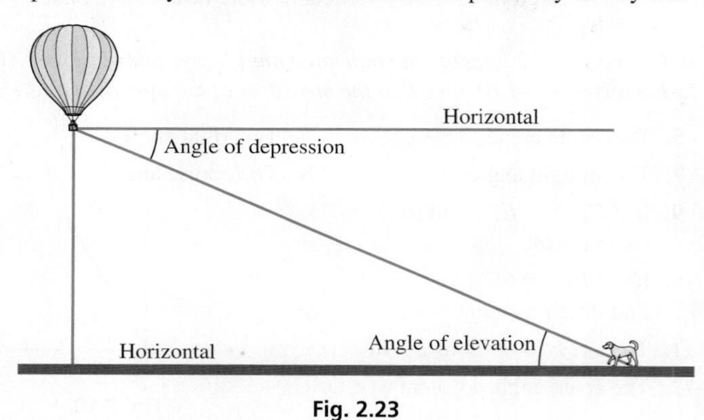

Fig. 2.23

Answers to Practice Exercises

1. 35° **2.** 138°

2.2 Triangles

Types and Properties of Triangles • Perimeter and Area • Hero's Formula • Pythagorean Theorem • Similar Triangles

When part of a plane is bounded and closed by straight-line segments, it is called a **polygon,** and it is named according to the number of sides it has. *A* **triangle** *has three sides, a* **quadrilateral** *has four sides, a* **pentagon** *has five sides, a* **hexagon** *has six sides, and so on.* The most important polygons are the triangle, which we consider in this section, and the quadrilateral, which we study in the next section.

TYPES AND PROPERTIES OF TRIANGLES

■ The properties of the triangle are important in the study of trigonometry, which we start in Chapter 4.

In a **scalene triangle,** no two sides are equal in length. In an **isosceles triangle,** two of the sides are equal in length, and the two *base angles* (the angles opposite the equal sides) are equal. In an **equilateral triangle,** the three sides are equal in length, and each of the three angles is 60°.

The most important triangle in technical applications is the **right triangle.** *In a right triangle, one of the angles is a right angle. The side opposite the right angle is the* **hypotenuse,** *and the other two sides are called* **legs.**

EXAMPLE 1 Types of triangles

Figure 2.24(a) shows a scalene triangle; each side is of a different length. Figure 2.24(b) shows an isosceles triangle with two equal sides of 2 in. and equal base angles of 40°. Figure 2.24(c) shows an equilateral triangle with equal sides of 5 cm and equal angles of 60°. Figure 2.24(d) shows a right triangle. The hypotenuse is side AB.

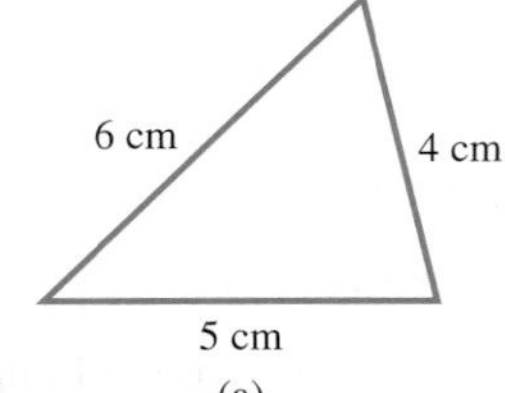

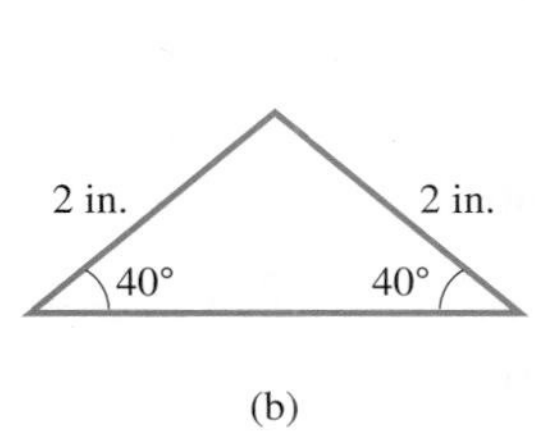

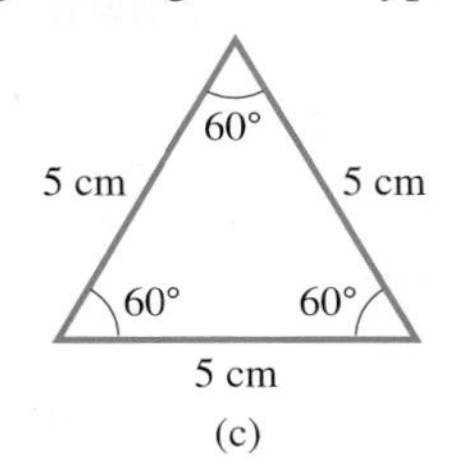

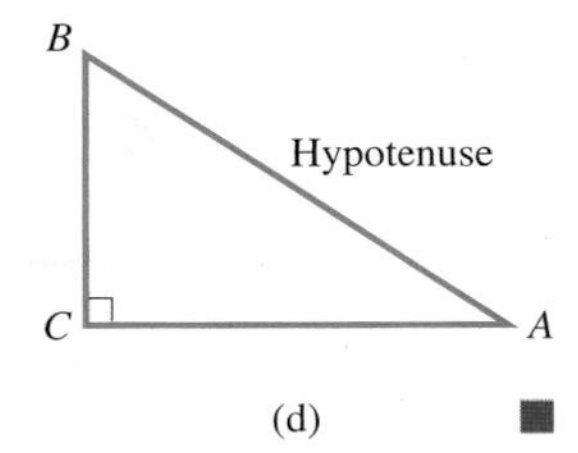

Fig. 2.24 ■

NOTE ➧ [One very important property of a triangle is that the sum of the measures of the three angles of a triangle is 180°.]

In the next example, we show this property by using material from Section 2.1.

EXAMPLE 2 Sum of angles of triangle

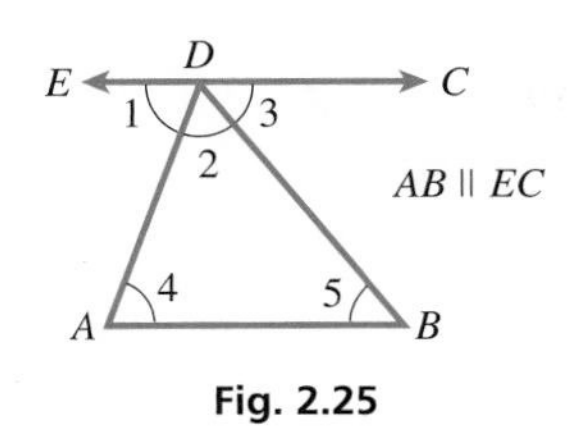

Fig. 2.25

In Fig. 2.25, because $\angle 1$, $\angle 2$, and $\angle 3$ constitute a straight angle,

$$\angle 1 + \angle 2 + \angle 3 = 180°$$

Also, by noting alternate interior angles, we see that $\angle 1 = \angle 4$ and $\angle 3 = \angle 5$. Therefore, by substitution, we have

$$\angle 4 + \angle 2 + \angle 5 = 180° \qquad \text{see Exercises 47–49 of Section 2.1}$$

Therefore, if two of the angles of a triangle are known, the third may be found by subtracting the sum of the first two from 180°. ■

EXAMPLE 3 Sum of angles—airplane flight

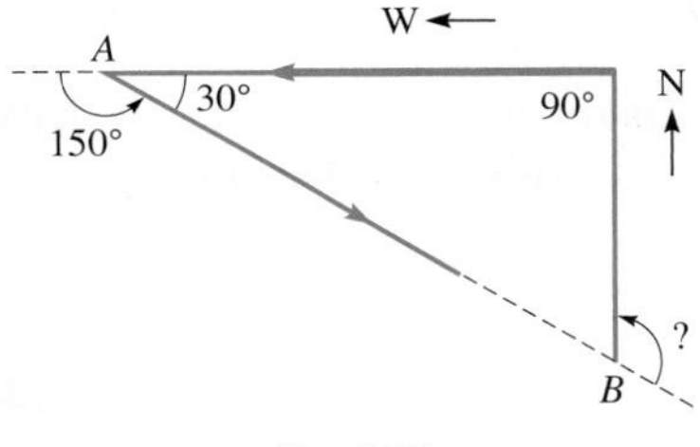

Fig. 2.26

An airplane is flying north and then makes a 90° turn to the west. Later, it makes another left turn of 150°. What is the angle of a third left turn that will cause the plane to again fly north? See Fig. 2.26.

From Fig. 2.26, we see that the interior angle of the triangle at A is the supplement of 150°, or 30°. Because the sum of the measures of the interior angles of the triangle is 180°, the interior angle at B is

$$\angle B = 180° - (90° + 30°) = 60°$$

The required angle is the supplement of 60°, which is 120°. ■

Practice Exercise

1. If the triangle in Fig. 2.27 is isosceles and the vertex angle (at the left) is 30°, what are the base angles?

A line segment drawn from a vertex of a triangle to the *midpoint* of the opposite side is called a **median** of the triangle. A basic property of a triangle is that *the three medians meet at a single point, called the* **centroid** *of the triangle*. See Fig. 2.27. Also, *the three* **angle bisectors** (lines from the vertices that divide the angles in half) *meet at a common point*. See Fig. 2.28.

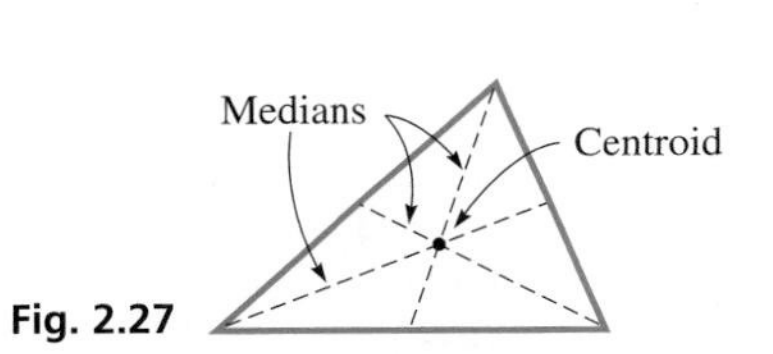

Fig. 2.27

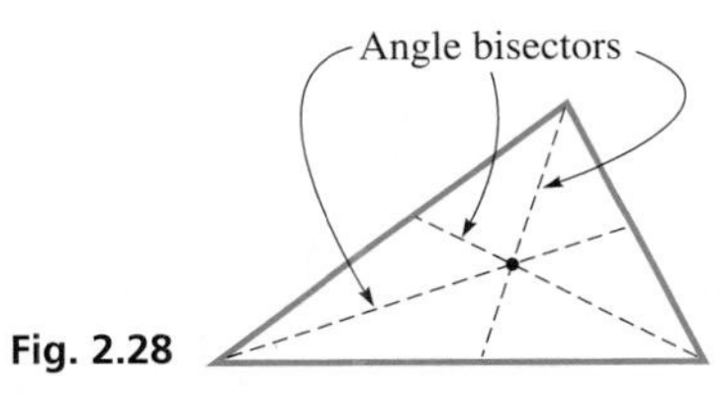

Fig. 2.28

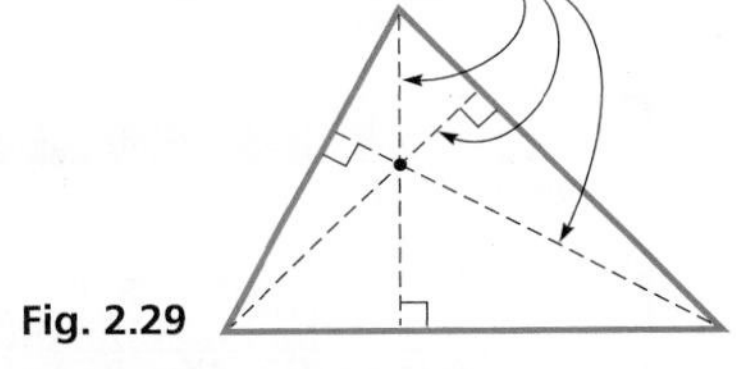

Fig. 2.29

An **altitude** (*or* **height**) *of a triangle is the line segment drawn from a vertex perpendicular to the opposite side (or its extension), which is called the* **base** *of the triangle. The three altitudes of a triangle meet at a common point*. See Fig. 2.29. The three common points of the medians, angle bisectors, and altitudes are generally not the same point for a given triangle.

PERIMETER AND AREA OF A TRIANGLE

We now consider two of the most basic measures of a plane geometric figure. The first of these is its **perimeter,** *which is the total distance around it*. In the following example, we find the perimeter of a triangle.

EXAMPLE 4 Perimeter of triangle

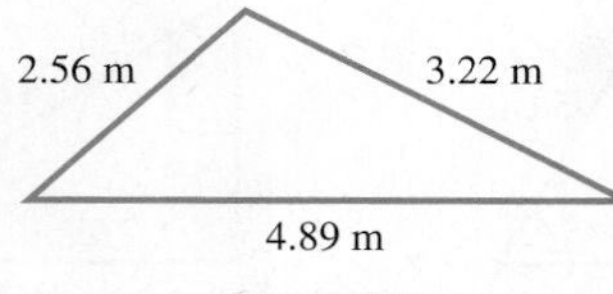

Fig. 2.30

A roof has triangular trusses with sides of 2.56 m, 3.22 m, and 4.89 m. Find the perimeter of one of these trusses. See Fig. 2.30.

Using the definition, the perimeter of one of these trusses is

$$p = 2.56 + 3.22 + 4.89 = 10.67 \text{ m}$$

The perimeter is 10.67 m (to hundredths, since each side is given to hundredths). ■

The second important measure of a geometric figure is its **area.** Although the concept of area is primarily intuitive, it is easily defined and calculated for the basic geometric figures. *Area gives a measure of the surface of the figure,* just as perimeter gives the measure of the distance around it.

The area A of a triangle of base b and altitude h is

$$A = \tfrac{1}{2}bh \tag{2.2}$$

The following example illustrates the use of Eq. (2.2).

EXAMPLE 5 Area of triangle

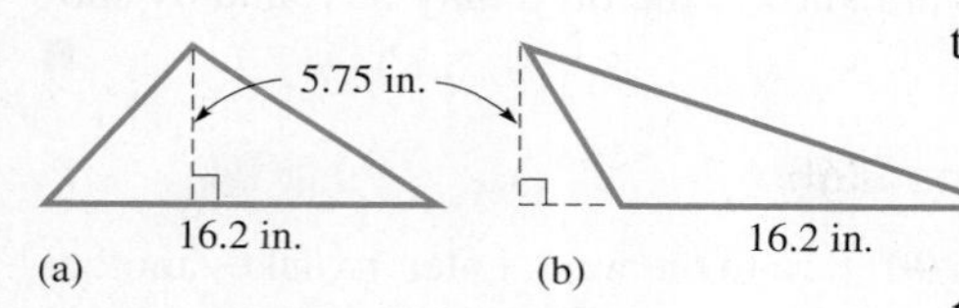

Fig. 2.31

Find the areas of the triangles in Fig. 2.31(a) and Fig. 2.31(b).

Even though the triangles are of different shapes, the base b of each is 16.2 in., and the altitude h of each is 5.75 in. Therefore, the area of each triangle is

$$A = \tfrac{1}{2}bh = \tfrac{1}{2}(16.2)(5.75) = 46.6 \text{ in.}^2$$ ■

■ Named for Hero (or Heron), a first-century Greek mathematician.

Another formula for the area of a triangle that is particularly useful when we have *a triangle with three known sides and no right angle is* **Hero's formula,** which is

$$A = \sqrt{s(s-a)(s-b)(s-c)}, \quad \text{where } s = \tfrac{1}{2}(a+b+c) \tag{2.3}$$

In Eq. (2.3), a, b, and c are the lengths of the sides, and s is one-half of the perimeter.

EXAMPLE 6 Hero's formula—area of land parcel

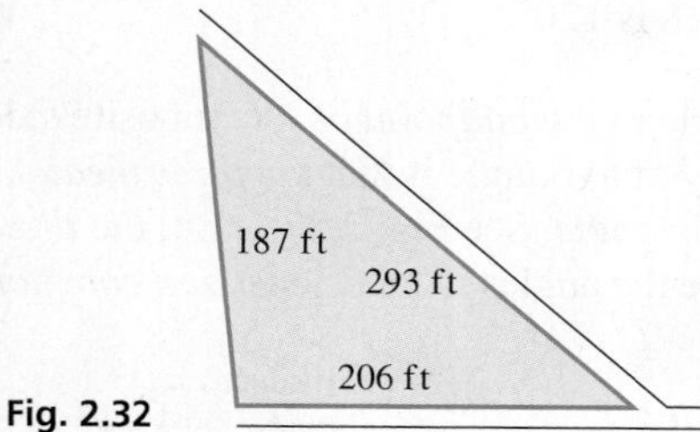

Fig. 2.32

A surveyor measures the sides of a triangular parcel of land between two intersecting straight roads to be 206 ft, 293 ft, and 187 ft. Find the area of this parcel (see Fig. 2.32).

To use Eq. (2.3), we first find s:

$$s = \tfrac{1}{2}(206 + 293 + 187)$$
$$= \tfrac{1}{2}(686) = 343 \text{ ft}$$

Now, substituting in Eq. (2.3), we have

$$A = \sqrt{343(343-206)(343-293)(343-187)}$$
$$= 19{,}100 \text{ ft}^2$$

```
NORMAL FLOAT AUTO REAL RADIAN MP
(206+293+187)/2
                         343
Ans→X
                         343
√X(X-206)(X-293)(X-187)
                 19144.96801
```
Fig. 2.33

The calculator solution is shown in Fig. 2.33. Note that the value of s was stored in memory (as x) using the STO▶ key and then used to find A. Using the calculator, it is not necessary to write down anything but the final result, properly *rounded off*. ■

THE PYTHAGOREAN THEOREM

As we have noted, one of the most important geometric figures in technical applications is the right triangle. A very important property of a right triangle is given by the **Pythagorean theorem,** which states that

■ Named for the Greek mathematician Pythagoras (sixth century B.C.E.)

> *in a **right triangle,** the square of the length of the hypotenuse equals the sum of the squares of the lengths of the other two sides.*

If c is the length of the hypotenuse and a and b are the lengths of the other two sides (see Fig. 2.34), the Pythagorean theorem is

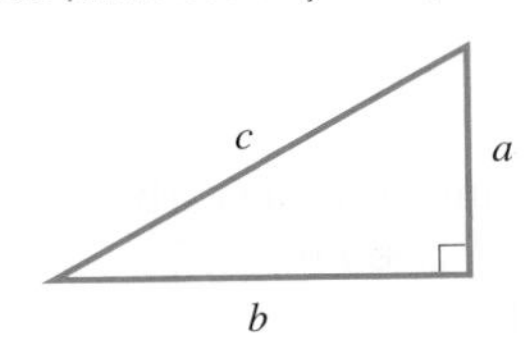

Fig. 2.34

$$c^2 = a^2 + b^2 \quad (2.4)$$

CAUTION The Pythagorean theorem applies only to right triangles. ■

EXAMPLE 7 Pythagorean theorem—length of wire

A pole is perpendicular to the level ground around it. A guy wire is attached 3.20 m up the pole and at a point on the ground, 2.65 m from the pole. How long is the guy wire?

We sketch the pole and guy wire as shown in Fig. 2.35. Using the Pythagorean theorem, and then substituting, we have

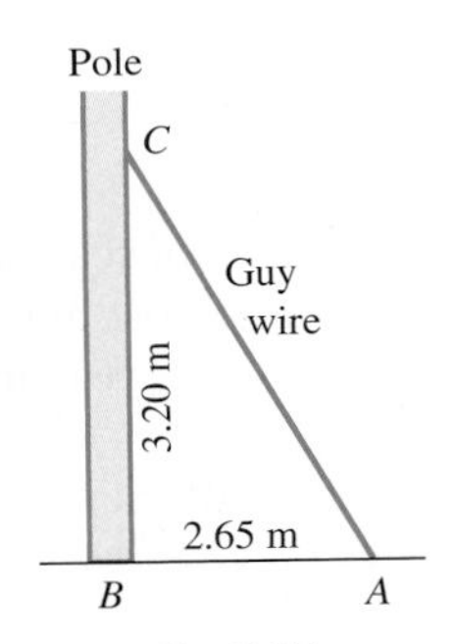

Fig. 2.35

$$AC^2 = AB^2 + BC^2$$
$$= 2.65^2 + 3.20^2$$
$$AC = \sqrt{2.65^2 + 3.20^2} = 4.15 \text{ m}$$

The guy wire is 4.15 m long. ■

SIMILAR TRIANGLES

The perimeter and area of a triangle are measures of its *size*. We now consider the shape of triangles.

Two triangles are **similar** *if they have the same shape (but not necessarily the same size).* There are two very important properties of similar triangles.

Properties of Similar Triangles

1. The corresponding angles of similar triangles are equal.
2. The corresponding sides of similar triangles are proportional.

If one property is true, then the other is also true, and therefore, the triangles are similar. In two similar triangles, the **corresponding sides** *are the sides, one in each triangle, that are between the same pair of equal corresponding angles.*

EXAMPLE 8 Similar triangles

In Fig. 2.36, a pair of similar triangles are shown. They are similar even though the corresponding parts are not in the same position relative to the page. Using standard symbols, we can write $\Delta ABC \sim \Delta A'B'C'$, where Δ means "triangle" and $\sim$ means "is similar to."

The pairs of corresponding angles are A and A', B and B', and C and C'. This means $A = A'$, $B = B'$, and $C = C'$.

The pairs of corresponding sides are AB and $A'B'$, BC and $B'C'$, and AC and $A'C'$. In order to show that these corresponding sides are proportional, we write

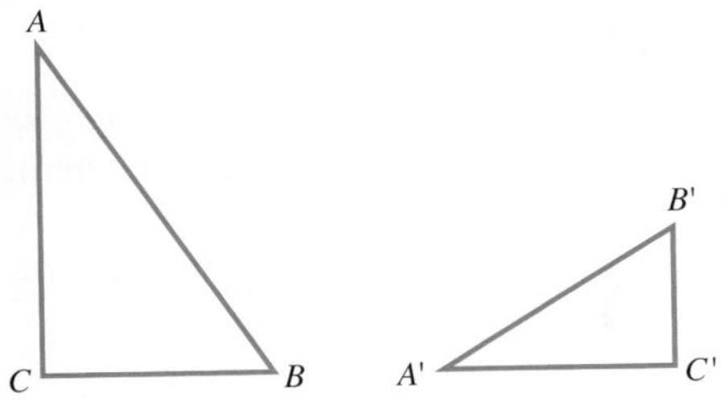

$$\frac{AB}{A'B'} = \frac{BC}{B'C'} = \frac{AC}{A'C'} \quad \begin{matrix} \leftarrow \text{sides of } \Delta ABC \\ \leftarrow \text{sides of } \Delta A'B'C' \end{matrix}$$ ■

Fig. 2.36

If we know that two triangles are similar, we can use the two basic properties of similar triangles to find the unknown parts of one triangle from the known parts of the other triangle. The next example illustrates this in a practical application.

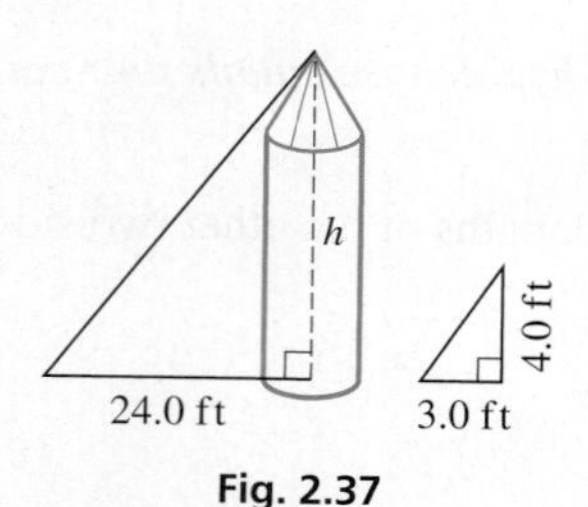

Fig. 2.37

Practice Exercise

2. In Fig. 2.37, knowing the value of h, what is the distance between the top of the silo and the end of its shadow?

EXAMPLE 9 Similar triangles—height of silo

On level ground, a silo casts a shadow 24 ft long. At the same time, a nearby vertical pole 4.0 ft high casts a shadow 3.0 ft long. How tall is the silo? See Fig. 2.37.

The rays of the sun are essentially parallel. The two triangles in Fig. 2.37 are similar since *each has a right angle and the angles at the tops are equal*. The other angles must be equal since the sum of the angles is 180°. The lengths of the hypotenuses are of no importance in this problem, so we use only the other sides in stating the ratios of corresponding sides. Denoting the height of the silo as h, we have

$$\frac{h}{4.0} = \frac{24}{3.0}, \qquad h = 32 \text{ ft}$$

We conclude that the silo is 32 ft high. ■

One of the most practical uses of similar geometric figures is that of **scale drawings.** Maps, charts, architectural blueprints, engineering sketches, and most drawings that appear in books (including many that have already appeared, and will appear, in this text) are familiar examples of scale drawings.

In any scale drawing, all distances are drawn at a certain ratio of the distances that they represent, and all angles are drawn equal to the angles they represent. Note the distances and angles shown in Fig. 2.38 in the following example.

EXAMPLE 10 Scale drawing

In drawing a map of the area shown in Fig. 2.38, a scale of 1 cm = 200 km is used. In measuring the distance between Chicago and Toronto on the map, we find it to be 3.5 cm. The actual distance x between Chicago and Toronto is found from the proportion

$$\begin{matrix}\text{actual distance} \rightarrow \\ \text{distance on map} \rightarrow\end{matrix} \frac{x}{3.5 \text{ cm}} = \overset{\text{scale} \downarrow}{\frac{200 \text{ km}}{1 \text{ cm}}} \quad \text{or} \quad x = 700 \text{ km}$$

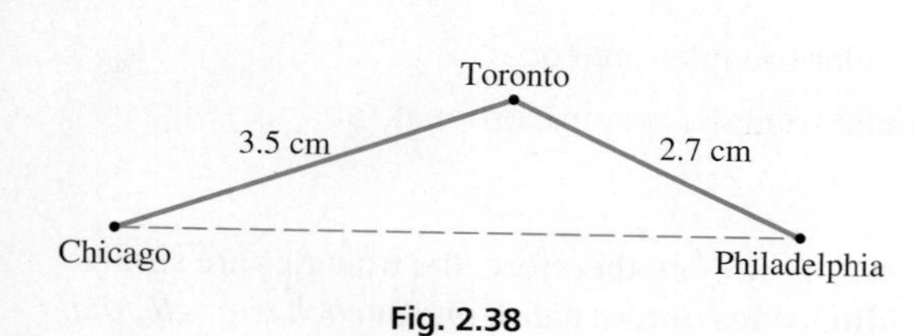

Fig. 2.38

If we did not have the scale but knew that the distance between Chicago and Toronto is 700 km, then by measuring distances on the map between Chicago and Toronto (3.5 cm) and between Toronto and Philadelphia (2.7 cm), we could find the distance between Toronto and Philadelphia. It is found from the following proportion, determined by use of similar triangles:

$$\frac{700 \text{ km}}{3.5 \text{ cm}} = \frac{y}{2.7 \text{ cm}}$$

$$y = \frac{2.7(700)}{3.5} = 540 \text{ km}$$ ■

Practice Exercise

3. On the same map, if the map distance from Toronto to Boston is 3.3 cm, what is the actual distance?

■ If the result is required to be in miles, we have the following change of units (see Section 1.4).

$$540 \text{ km}\left(\frac{0.6214 \text{ mi}}{1 \text{ km}}\right) = 340 \text{ mi}$$

Similarity requires *equal* angles and *proportional* sides. *If the corresponding angles and the corresponding sides of two triangles are equal, the two triangles are* **congruent.** As a result of this definition, the areas and perimeters of congruent triangles are also equal. Informally, we can say that similar triangles have the same shape, whereas congruent triangles have the same shape and same size.

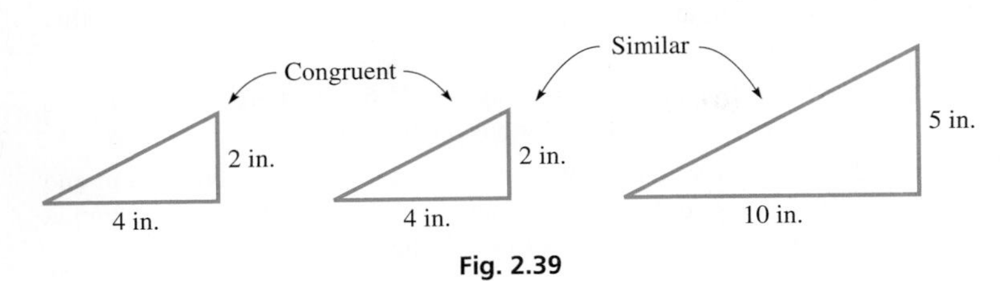

Fig. 2.39

EXAMPLE 11 Similar and congruent triangles

A right triangle with legs of 2 in. and 4 in. is congruent to any other right triangle with legs of 2 in. and 4 in. It is also similar to any right triangle with legs of 5 in. and 10 in., or any other right triangle that has legs in the ratio of 1 to 2, since the corresponding sides are proportional. See Fig. 2.39. ■

EXERCISES 2.2

In Exercises 1–4, answer the given questions about the indicated examples of this section.

1. In Example 2, if $\angle 1 = 70°$ and $\angle 5 = 45°$ in Fig. 2.25, what is the measure of $\angle 2$?

2. In Example 5, change 16.2 in. to 61.2 in. What is the answer?

3. In Example 7, change 2.65 m to 6.25 m. What is the answer?

4. In Example 9, interchange 4.0 ft and 3.0 ft. What is the answer?

In Exercises 5–8, determine $\angle A$ in the indicated figures.

5. Fig. 2.40 (a)

6. Fig. 2.40 (b)

7. Fig. 2.40 (c)

8. Fig. 2.40 (d)

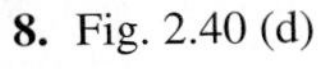

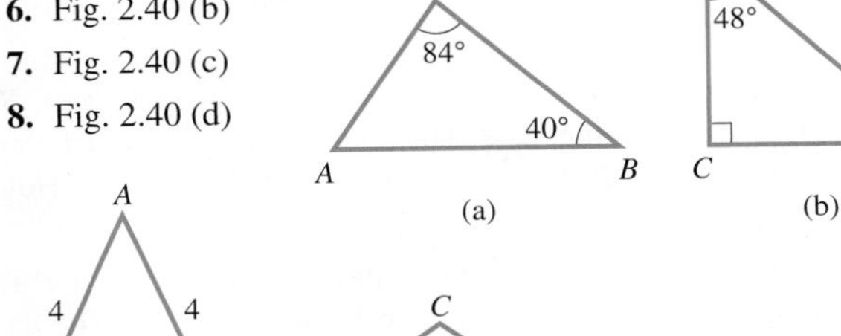

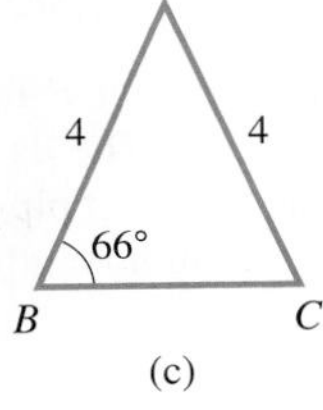

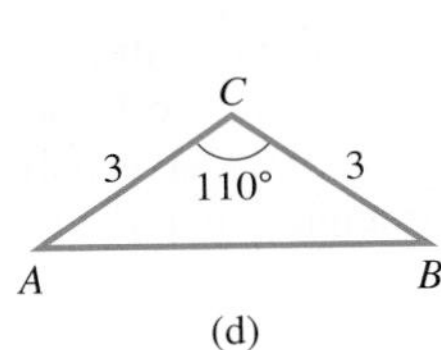

Fig. 2.40

In Exercises 9–16, find the area of each triangle.

9. Fig. 2.41(a)

10. Fig. 2.41(b)

11. Fig. 2.41(c)

12. Fig. 2.41(d)

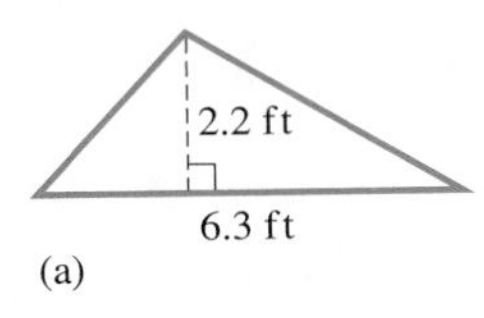

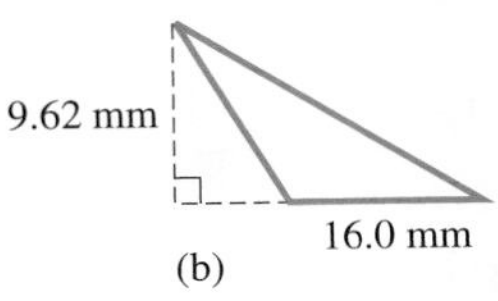

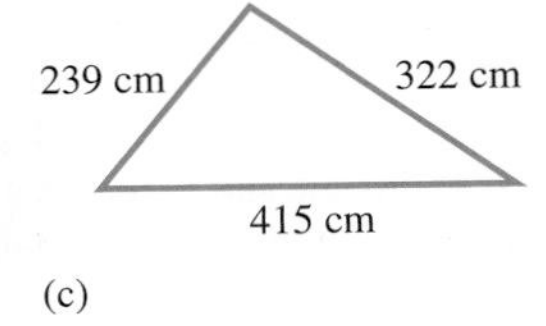

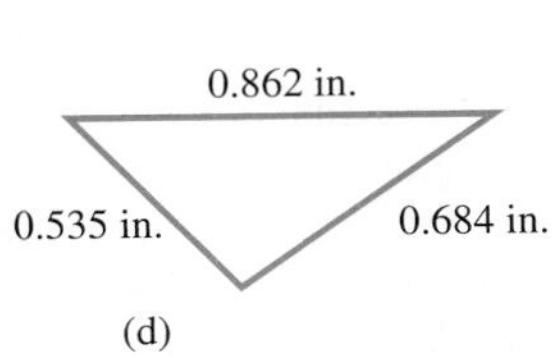

Fig. 2.41

13. Right triangle with legs 3.46 ft and 2.55 ft

14. Right triangle with legs 234 mm and 342 mm

15. Isosceles triangle, equal sides of 0.986 m, third side of 0.884 m

16. Equilateral triangle of sides 3200 yd

In Exercises 17–20, find the perimeter of each triangle.

17. Fig. 2.41(c)

18. Fig. 2.41(d)

19. An equilateral triangle of sides 21.5 cm

20. Isosceles triangle, equal sides of 2.45 in., third side of 3.22 in.

In Exercises 21–26, find the third side of the right triangle shown in Fig. 2.42 for the given values. The values in Exercises 21 and 22 are exact.

21. $a = 3$ in., $b = 4$ in.

22. $a = 5$ yd, $c = 13$ yd

23. $a = 13.8$ ft, $b = 22.7$ ft

24. $a = 2.48$ m, $b = 1.45$ m

25. $a = 175$ cm, $c = 551$ cm

26. $b = 0.474$ in., $c = 0.836$ in.

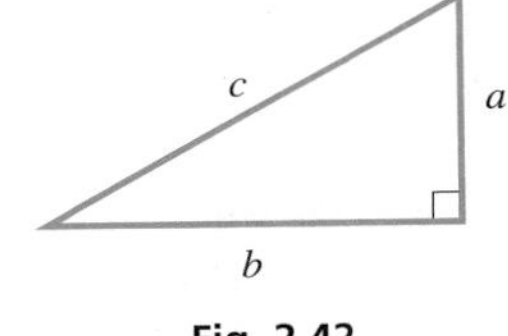

Fig. 2.42

In Exercises 27–30, use the right triangle in Fig. 2.43.

27. Find $\angle B$.

28. Find side c.

29. Find the perimeter.

30. Find the area.

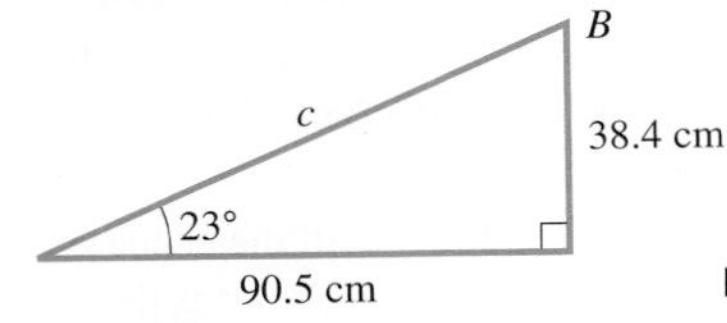

Fig. 2.43

In Exercises 31–58, solve the given problems.

31. What is the angle between the bisectors of the acute angles of a right triangle?

32. If the midpoints of the sides of an isosceles triangle are joined, another triangle is formed. What do you conclude about this inner triangle?

33. For what type of triangle is the centroid the same as the intersection of altitudes and the intersection of angle bisectors?

34. Is it possible that the altitudes of a triangle meet, when extended, outside the triangle? Explain.

35. The altitude to the hypotenuse of a right triangle divides the triangle into two smaller triangles. What do you conclude about the original triangle and the two new triangles? Explain.

36. If two triangles have the same three angles, can you conclude that the triangles are congruent? Explain why or why not.

37. In Fig. 2.44, show that $\Delta MKL \sim \Delta MNO$.

38. In Fig. 2.45, show that $\Delta ACB \sim \Delta ADC$.

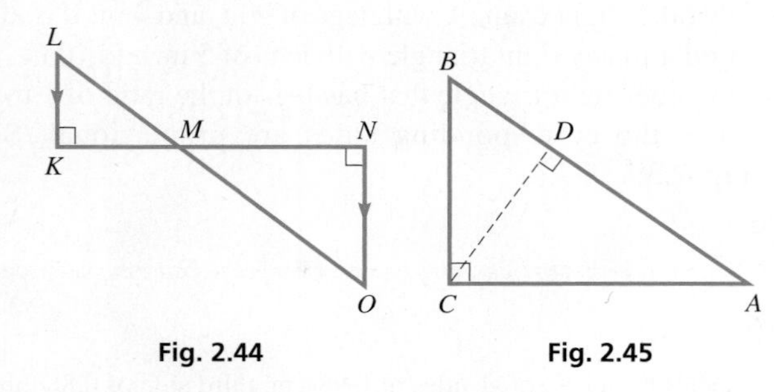

Fig. 2.44 Fig. 2.45

39. In Fig. 2.44, if $KN = 15$, $MN = 9$, and $MO = 12$, find LM.

40. In Fig. 2.45, if $AD = 9$ and $AC = 12$, find AB.

41. A *perfect triangle* is one that has sides that are integers and the perimeter and area are numerically equal integers. Is the triangle with sides 6, 25, and 29 a perfect triangle?

42. Government guidelines require that a sidewalk to street ramp be such that there is no more than 1.0 in. rise for each horizontal 20.0 in. of the ramp. How long should a ramp be for a curb that is 4.0 in. above the street?

43. The angle between the roof sections of an A-frame house is 50°. What is the angle between either roof section and a horizontal rafter?

44. A transmitting tower is supported by a wire that makes an angle of 52° with the level ground. What is the angle between the tower and the wire?

45. An 18.0-ft tall tree is broken in a wind storm such that the top falls and hits the ground 8.0 ft from the base. If the two sections of the tree are still connected at the break, how far up the tree (to the nearest tenth of a foot) was the break?

46. The Bermuda Triangle is sometimes defined as an equilateral triangle 1600 km on a side, with vertices in Bermuda, Puerto Rico, and the Florida coast. Assuming it is flat, what is its approximate area?

47. The sail of a sailboat is in the shape of a right triangle with sides of 8.0 ft, 15 ft, and 17 ft. What is the area of the sail?

48. An observer is 550 m horizontally from the launch pad of a rocket. After the rocket has ascended 750 m, how far is it from the observer?

49. In a practice fire mission, a ladder extended 10.0 ft just reaches the bottom of a 2.50-ft high window if the foot of the ladder is 6.00 ft from the wall. To what length must the ladder be extended to reach the top of the window if the foot of the ladder is 6.00 ft from the wall and cannot be moved?

50. The beach shade shown in Fig. 2.46 is made up of 30°-60°-90° triangular sections. Find x. (In a 30°-60°-90° triangle, the side opposite the 30° angle is one-half the hypotenuse.)

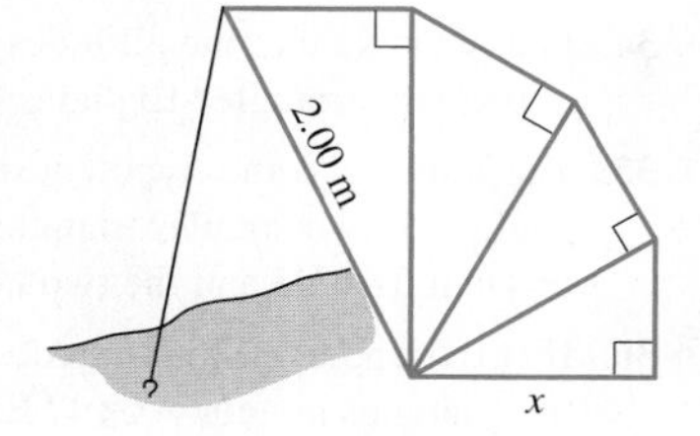

Fig. 2.46

51. A rectangular room is 18 ft long, 12 ft wide, and 8.0 ft high. What is the length of the longest diagonal from one corner to another corner of the room?

52. On a blueprint, a hallway is 45.6 cm long. The scale is 1.2 cm = 1.0 m. How long is the hallway?

53. Two parallel guy wires are attached to a vertical pole 4.5 m and 5.4 m above the ground. They are secured on the level ground at points 1.2 m apart. How long are the guy wires?

54. The two sections of a folding door, hinged in the middle, are at right angles. If each section is 2.5 ft wide, how far are the hinges from the far edge of the other section?

55. A 4.0-ft high wall stands 2.0 ft from a building. The ends of a straight pole touch the building and the ground 6.0 ft from the wall. A point on the pole touches the top of the wall. How long is the pole? See Fig. 2.47.

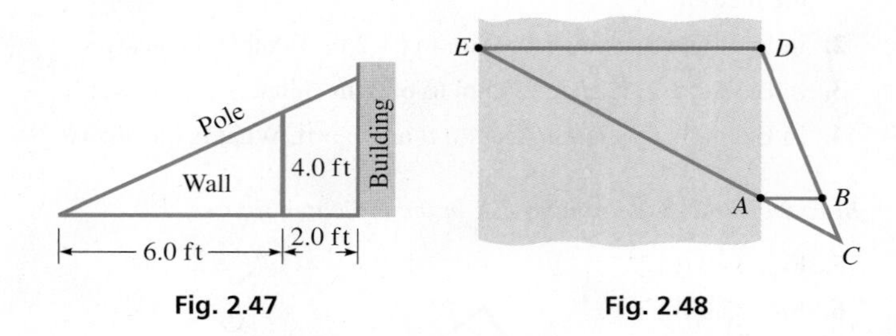

Fig. 2.47 Fig. 2.48

56. To find the width ED of a river, a surveyor places markers at A, B, C, and D, as shown in Fig. 2.48. The markers are placed such that $AB \parallel ED$, $BC = 50.0$ ft, $DC = 312$ ft, and $AB = 80.0$ ft. How wide is the river?

57. A water pumping station is to be built on a river at point P in order to deliver water to points A and B. See Fig. 2.49. The design requires that $\angle APD = \angle BPC$ so that the total length of piping that will be needed is a minimum. Find this minimum length of pipe.

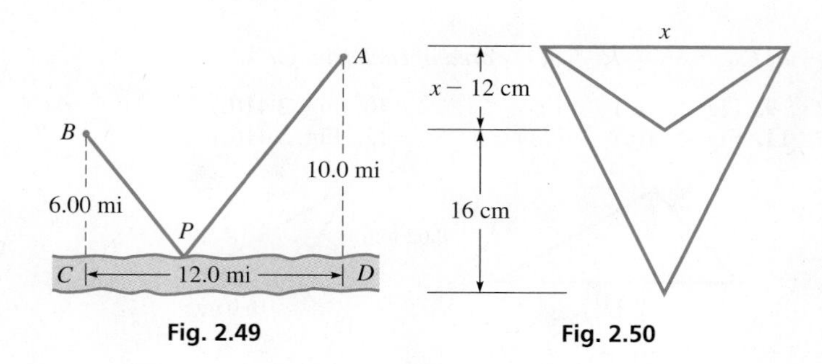

Fig. 2.49 Fig. 2.50

58. The cross section of a drainage trough has the shape of an isosceles triangle whose depth is 12 cm less than its width. If the depth is increased by 16 cm and the width remains the same, the area of the cross section is increased by 160 cm^2. Find the original depth and width. See Fig. 2.50.

Answers to Practice Exercises

1. 75° **2.** 40 ft **3.** 660 km

2.3 Quadrilaterals

Types and Properties of Quadrilaterals • Perimeter and Area

A **quadrilateral** *is a closed plane figure with four sides,* and these four sides form four interior angles. A general quadrilateral is shown in Fig. 2.51.

A **diagonal** *of a polygon is a straight line segment joining any two nonadjacent vertices.* The dashed line is one of two diagonals of the quadrilateral in Fig. 2.52.

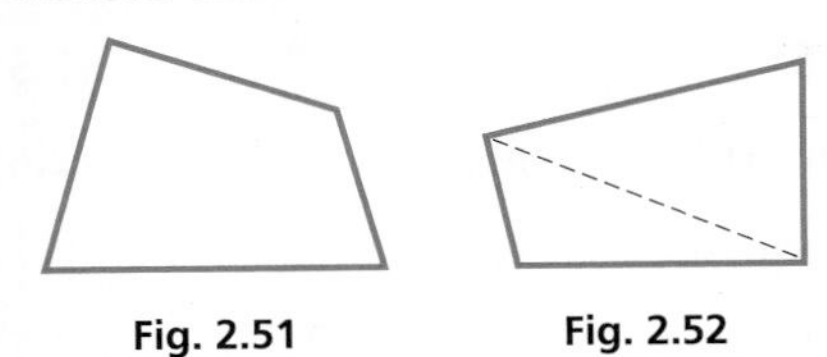

Fig. 2.51 **Fig. 2.52**

TYPES OF QUADRILATERALS

A **parallelogram** *is a quadrilateral in which opposite sides are parallel.* In a parallelogram, opposite sides are equal and opposite angles are equal. *A* **rhombus** *is a parallelogram with four equal sides.*

A **rectangle** *is a parallelogram in which intersecting sides are perpendicular,* which means that all four interior angles are right angles. In a rectangle, the longer side is usually called the **length,** and the shorter side is called the **width.** *A* **square** *is a rectangle with four equal sides.*

A **trapezoid** *is a quadrilateral in which two sides are parallel.* The parallel sides are called the **bases** of the trapezoid.

EXAMPLE 1 Types of quadrilaterals

A parallelogram is shown in Fig. 2.53(a). Opposite sides a are equal in length, as are opposite sides b. A rhombus with equal sides s is shown in Fig. 2.53(b). A rectangle is shown in Fig. 2.53(c). The length is labeled l, and the width is labeled w. A square with equal sides s is shown in Fig. 2.53(d). A trapezoid with bases b_1 and b_2 is shown in Fig. 2.53(e). ■

Practice Exercise

1. Develop a formula for the length d of a diagonal for the rectangle in Fig. 2.53(c).

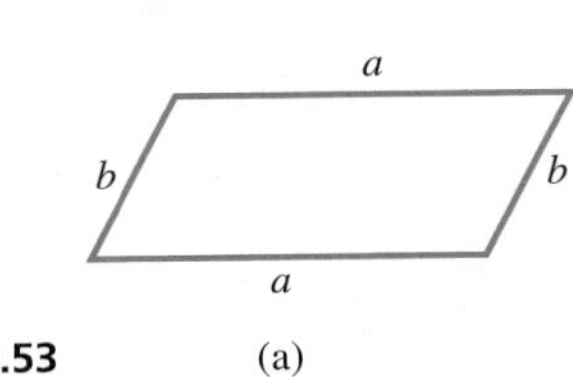

(a)

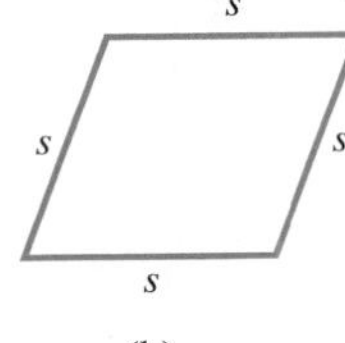

(b)

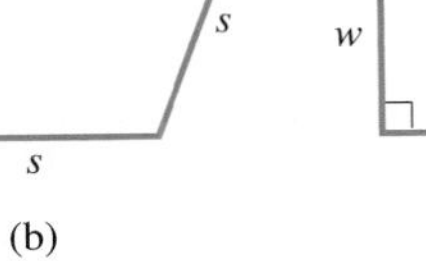

(c)

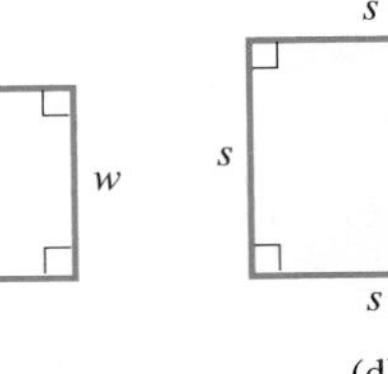

(d)

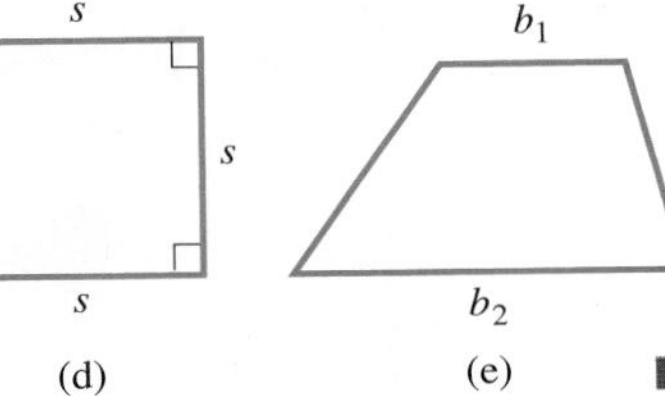

(e)

Fig. 2.53

PERIMETER AND AREA OF A QUADRILATERAL

The perimeter of a quadrilateral is the sum of the lengths of the four sides.

EXAMPLE 2 Perimeter—window molding

An architect designs a room with a rectangular window 36 in. high and 21 in. wide, with another window above in the shape of an equilateral triangle, 21 in. on a side. See Fig. 2.54. How much molding is needed for these windows?

The length of molding is the sum of the perimeters of the windows. For the rectangular window, the opposite sides are equal, which means the perimeter is twice the length l plus twice the width w. For the equilateral triangle, the perimeter is three times the side s. Therefore, the length L of molding is

$$\begin{aligned} L &= 2l + 2w + 3s \\ &= 2(36) + 2(21) + 3(21) \\ &= 177 \text{ in.} \end{aligned}$$

■

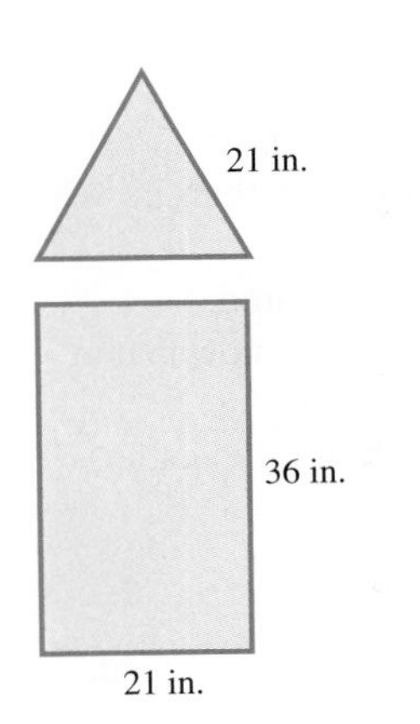

Fig. 2.54

We could write down formulas for the perimeters of the different kinds of triangles and quadrilaterals. However, if we *remember the meaning of perimeter as being the total distance around a geometric figure,* such formulas are not necessary.

Practice Exercise

2. If we did develop perimeter formulas, what is the formula for the perimeter p of a rhombus of side s?

For the areas of the square, rectangle, parallelogram, and trapezoid, we have the following formulas.

$A = s^2$	Square of side s (Fig. 2.55)	**(2.5)**
$A = lw$	Rectangle of length l and width w (Fig. 2.56)	**(2.6)**
$A = bh$	Parallelogram of base b and height h (Fig. 2.57)	**(2.7)**
$A = \frac{1}{2}h(b_1 + b_2)$	Trapezoid of bases b_1 and b_2 and height h (Fig. 2.58)	**(2.8)**

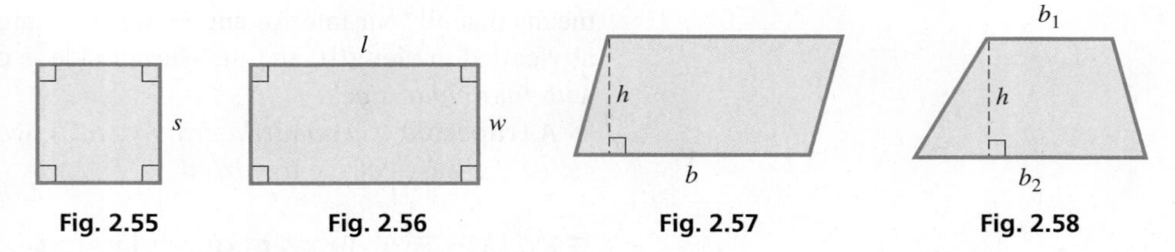

Fig. 2.55 Fig. 2.56 Fig. 2.57 Fig. 2.58

Because a rectangle, a square, and a rhombus are special types of parallelograms, the area of these figures can be found from Eq. (2.7). The area of a trapezoid is of importance when we find areas of irregular geometric figures in Section 2.5.

EXAMPLE 3 Area—park design

A city park is designed with lawn areas in the shape of a right triangle, a parallelogram, and a trapezoid, as shown in Fig. 2.59, with walkways between them. Find the area of each section of lawn and the total lawn area.

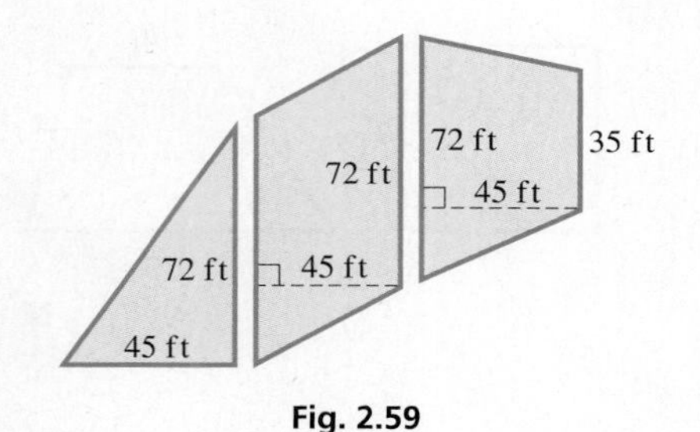

Fig. 2.59

$$A_1 = \tfrac{1}{2}bh = \tfrac{1}{2}(72)(45) = 1600 \text{ ft}^2$$
$$A_2 = bh = (72)(45) = 3200 \text{ ft}^2$$
$$A_3 = \tfrac{1}{2}h(b_1 + b_2) = \tfrac{1}{2}(45)(72 + 35) = 2400 \text{ ft}^2$$

The total lawn area is about 7200 ft^2. ■

EXAMPLE 4 Perimeter—computer chip dimensions

■ The computer microprocessor chip was first commercially available in 1971.

The length of a rectangular computer chip is 2.0 mm longer than its width. Find the dimensions of the chip if its perimeter is 26.4 mm.

Because the dimensions, the length and the width, are required, let w = the width of the chip. Because the length is 2.0 mm more than the width, we know that $w + 2.0$ = the length of the chip. See Fig. 2.60.

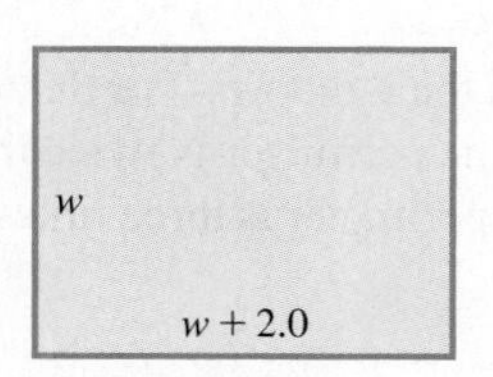

Fig. 2.60

Because the perimeter of a rectangle is twice the length plus twice the width, we have the equation

$$2(w + 2.0) + 2w = 26.4$$

because the perimeter is given as 26.4 mm. This is the equation we need.

Solving this equation, we have

$$2w + 4.0 + 2w = 26.4$$
$$4w = 22.4$$
$$w = 5.6 \text{ mm} \quad \text{and} \quad w + 2.0 = 7.6 \text{ mm}$$

Therefore, the length is 7.6 mm and the width is 5.6 mm. These values check with the statements of the original problem. ■

Practice Exercise

3. What is the area of the chip in Fig. 2.60?

EXERCISES 2.3

In Exercises 1–4, make the given changes in the indicated examples of this section and then solve the given problems.

1. In Example 1, interchange the lengths of b_1 and b_2 in Fig. 2.53(e). What type of quadrilateral is the resulting figure?
2. In Example 2, change the equilateral triangle of side 21 in. to a square of side 21 in. and then find the length of molding.
3. In Example 3, change the dimension of 45 ft to 55 ft in each figure and then find the area.
4. In Example 4, change 2.0 mm to 3.0 mm, and then solve.

In Exercises 5–12, find the perimeter of each figure.

5. Square: side of 85 m
6. Rhombus: side of 2.46 ft
7. Rectangle: $l = 9.200$ in., $w = 7.420$ in.
8. Rectangle: $l = 142$ cm, $w = 126$ cm
9. Parallelogram in Fig. 2.61
10. Parallelogram in Fig. 2.62
11. Trapezoid in Fig. 2.63
12. Trapezoid in Fig. 2.64

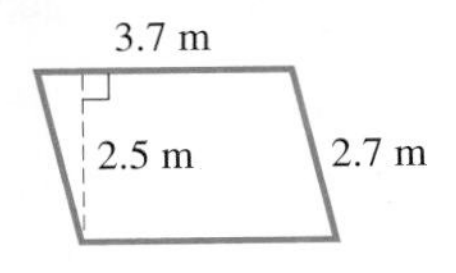

Fig. 2.61

27.3 in.
12.6 in.
14.2 in.

Fig. 2.62

0.730 ft
0.362 ft
0.298 ft
0.440 ft
0.612 ft

Fig. 2.63

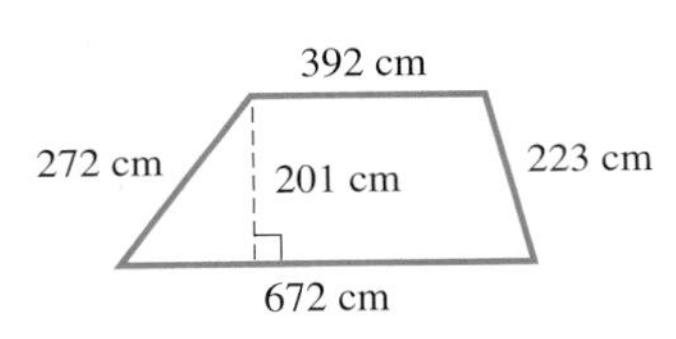

Fig. 2.64

In Exercises 13–20, find the area of each figure.

13. Square: $s = 6.4$ mm
14. Square: $s = 15.6$ ft
15. Rectangle: $l = 8.35$ in., $w = 2.81$ in.
16. Rectangle: $l = 142$ cm, $w = 126$ cm
17. Parallelogram in Fig. 2.61
18. Parallelogram in Fig. 2.62
19. Trapezoid in Fig. 2.63
20. Trapezoid in Fig. 2.64

In Exercises 21–24, set up a formula for the indicated perimeter or area. (Do not include dashed lines.)

21. The perimeter of the figure in Fig. 2.65 (a parallelogram and a square attached)
22. The perimeter of the figure in Fig. 2.66 (two trapezoids attached)
23. Area of figure in Fig. 2.65
24. Area of figure in Fig. 2.66

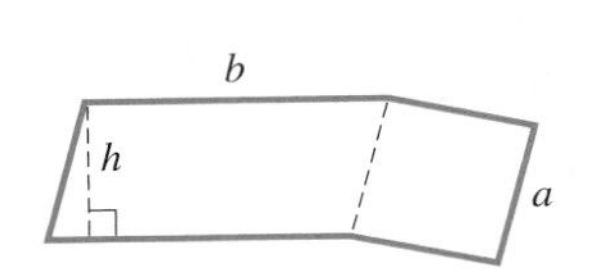

Fig. 2.65

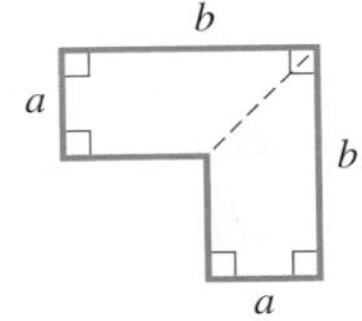

Fig. 2.66

In Exercises 25–46, solve the given problems.

25. If the angle between adjacent sides of a parallelogram is 90°, what conclusion can you make about the parallelogram?
26. What conclusion can you make about the two triangles formed by the sides and diagonal of a parallelogram? Explain.
27. Find the area of a square whose diagonal is 24.0 cm.
28. In a trapezoid, find the angle between the bisectors of the two angles formed by the bases and one nonparallel side.
29. Noting the quadrilateral in Fig. 2.67, determine the sum of the interior angles of a quadrilateral.

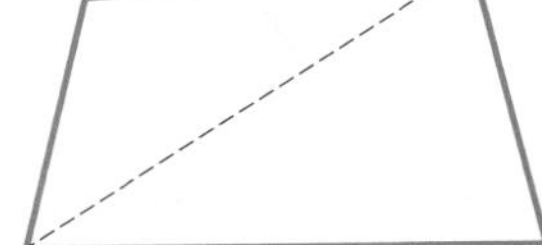
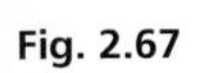
Fig. 2.67

30. The sum S of the measures of the interior angles of a polygon with n sides is $S = 180(n - 2)$. (a) Solve for n. (b) If $S = 3600°$, how many sides does the polygon have?
31. Express the area A of the large rectangle in Fig. 2.68 formed by the smaller rectangles in two ways. What property of numbers is illustrated by the results?

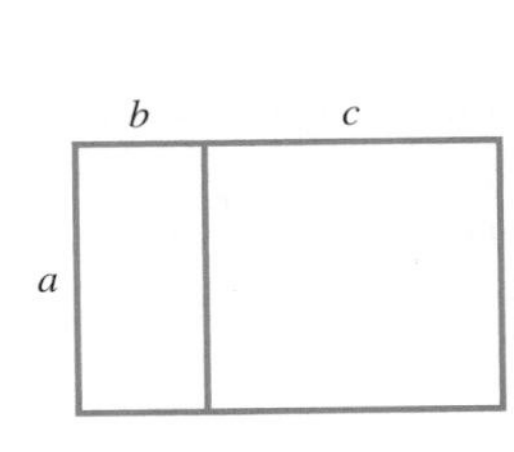

Fig. 2.68

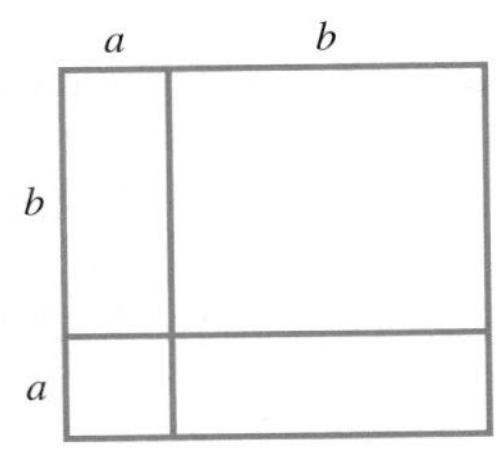

Fig. 2.69

32. Express the area of the square in Fig. 2.69 in terms of the smaller rectangles into which it is divided. What algebraic expression is illustrated by the results?
33. Noting how a diagonal of a rhombus divides an interior angle, explain why the automobile jack in Fig. 2.70 is in the shape of a rhombus.

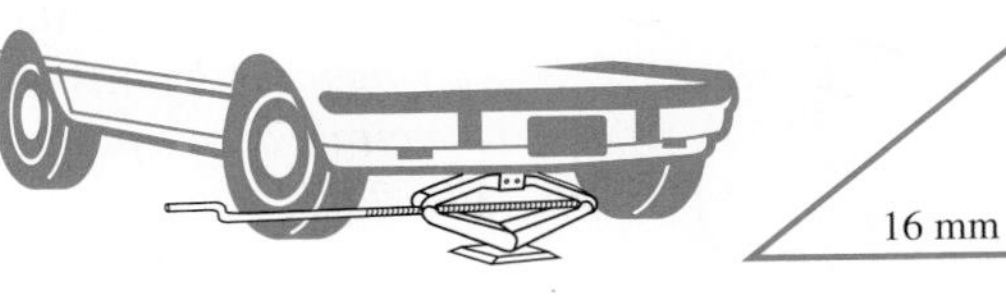
Fig. 2.70

12 mm
16 mm

Fig. 2.71

34. Part of an electric circuit is wired in the configuration of a rhombus and one of its altitudes as shown in Fig. 2.71. What is the length of wire in this part of the circuit?
35. A walkway 3.0 m wide is constructed along the outside edge of a square courtyard. If the perimeter of the courtyard is 320 m, (a) what is the perimeter of the square formed by the outer edge of the walkway? (b) What is the area of the walkway?
36. An architect designs a rectangular window such that the width of the window is 18 in. less than the height. If the perimeter of the window is 180 in., what are its dimensions?

37. Find the area of the cross section of concrete highway support shown in Fig. 2.72. All measurements are in feet and are exact.

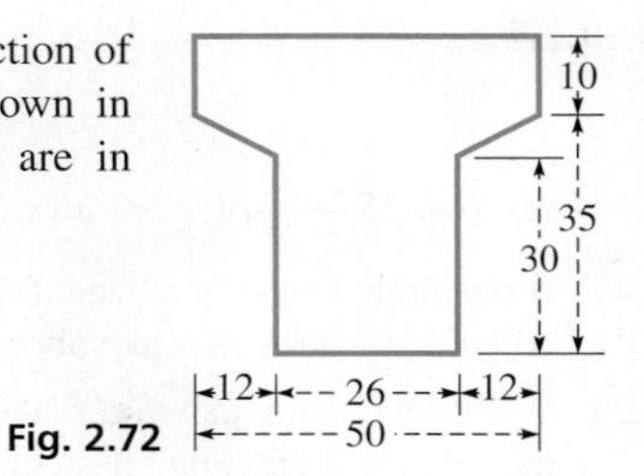

Fig. 2.72

38. A beam support in a building is in the shape of a parallelogram, as shown in Fig. 2.73. Find the area of the side of the beam shown.

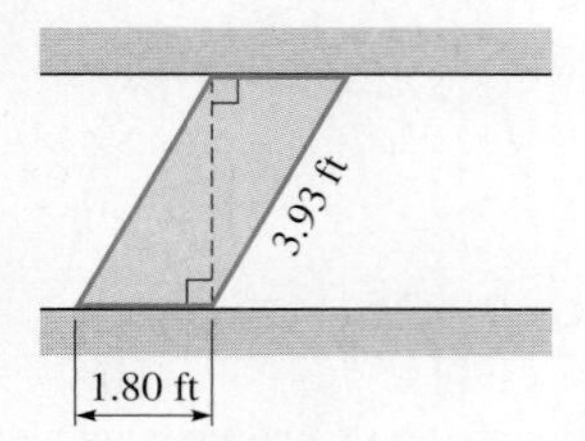

Fig. 2.73

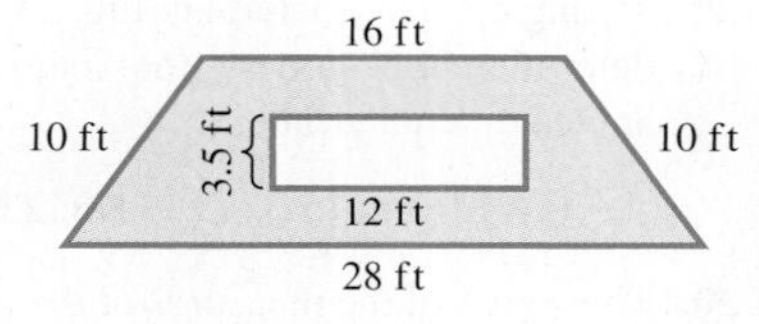

Fig. 2.74

39. Each of two walls (with rectangular windows) of an A-frame house has the shape of a trapezoid as shown in Fig. 2.74. If a gallon of paint covers 320 ft^2, how much paint is required to paint these walls? (All data are accurate to two significant digits.)

40. A 1080p high-definition widescreen television screen has 1080 pixels in the vertical direction and 1920 pixels in the horizontal direction. If the screen measures 15.8 in. high and 28.0 in. wide, find the number of pixels per square inch.

41. The ratio of the width to the height of a 43.3 cm (diagonal) laptop computer screen is 1.60. What is the width w and height h of the screen?

42. Six equal trapezoidal sections form a conference table in the shape of a hexagon, with a hexagonal opening in the middle. See Fig. 2.75. From the dimensions shown, find the area of the table top.

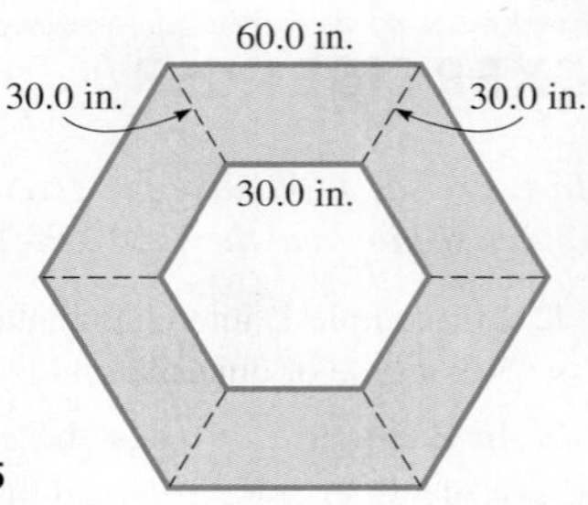

Fig. 2.75

43. A fenced section of a ranch is in the shape of a quadrilateral whose sides are 1.74 km, 1.46 km, 2.27 km, and 1.86 km, the last two sides being perpendicular to each other. Find the area of the section.

44. A rectangular security area is enclosed on one side by a wall, and the other sides are fenced. The length of the wall is twice the width of the area. The total cost of building the wall and fence is \$13,200. If the wall costs \$50.00/m and the fence costs \$5.00/m, find the dimensions of the area.

45. What is the sum of the measures of the interior angles of a quadrilateral? Explain.

46. Find a formula for the area of a rhombus in terms of its diagonals d_1 and d_2. (See Exercise 33.)

Answers to Practice Exercises

1. $d = \sqrt{l^2 + w^2}$ **2.** $p = 4s$ **3.** 43 mm^2

2.4 Circles

Properties of Circles • Tangent and Secant Lines • Circumference and Area • Circular Arcs and Angles • Radian Measure of an Angle

The next geometric figure we consider is the circle. *All points on a* **circle** *are at the same distance from a fixed point, the* **center** *of the circle. The distance from the center to a point on the circle is the* **radius** *of the circle. The distance between two points on the circle on a line through the center is the* **diameter.** Therefore, the diameter d is twice the radius r, or $d = 2r$. See Fig. 2.76.

There are also certain types of lines associated with a circle. *A* **chord** *is a line segment having its endpoints on the circle. A* **tangent** *is a line that touches (does not pass through) the circle at one point. A* **secant** *is a line that passes through two points of the circle.* See Fig. 2.77.

An important property of a tangent is that *a tangent is perpendicular to the radius drawn to the point of contact.* This is illustrated in the following example.

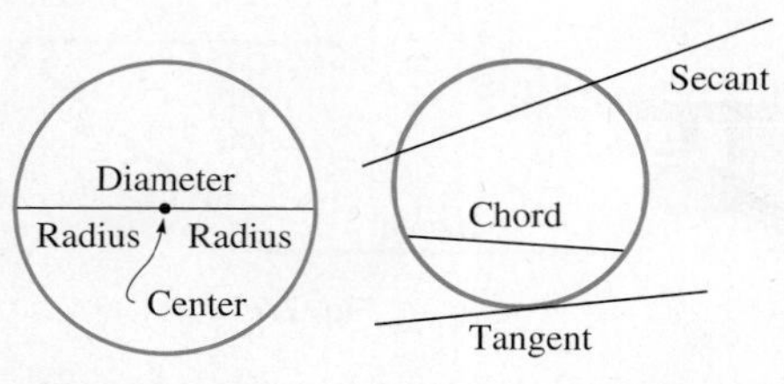

Fig. 2.76 Fig. 2.77

EXAMPLE 1 Tangent line perpendicular to radius

In Fig. 2.78, O is the center of the circle, and AB is tangent at B. If $\angle OAB = 25°$, find $\angle AOB$.

Because the center is O, OB is a radius of the circle. A tangent is perpendicular to a radius at the point of tangency, which means $\angle ABO = 90°$ so that

$$\angle OAB + \angle OBA = 25° + 90° = 115°$$

Because the sum of the angles of a triangle is 180°, we have

$$\angle AOB = 180° - 115° = 65°$$

■

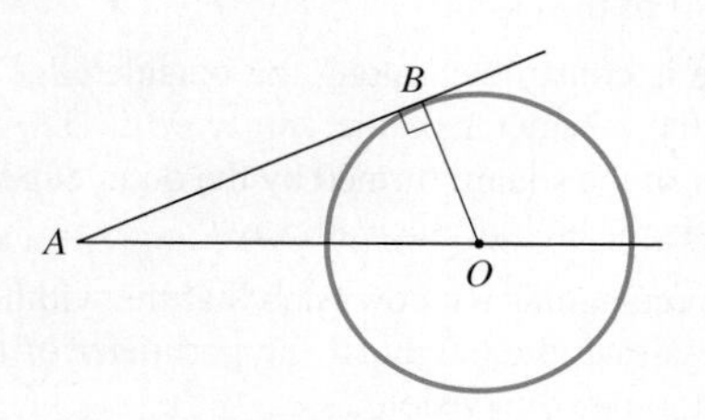

Fig. 2.78

CIRCUMFERENCE AND AREA OF A CIRCLE

The perimeter of a circle is called the **circumference.** The formulas for the circumference and area of a circle are as follows:

■ The symbol π (the Greek letter pi), which we use as a number, was first used in this way as a number in the 1700s.

$c = 2\pi r$	Circumference of a circle of radius r	**(2.9)**
$A = \pi r^2$	Area of a circle of radius r	**(2.10)**

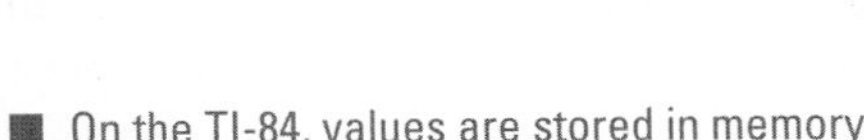

■ On the TI-84, values are stored in memory using up to 14 digits with a two-digit exponent.

Here, π equals approximately 3.1416. In using a calculator, π can be entered to a much greater accuracy by using the π key.

EXAMPLE 2 Area of circle—oil spill

A circular oil spill has a diameter of 2.4 km. It is to be enclosed within special flexible tubing. What is the area of the spill, and how long must the tubing be? See Fig. 2.79.

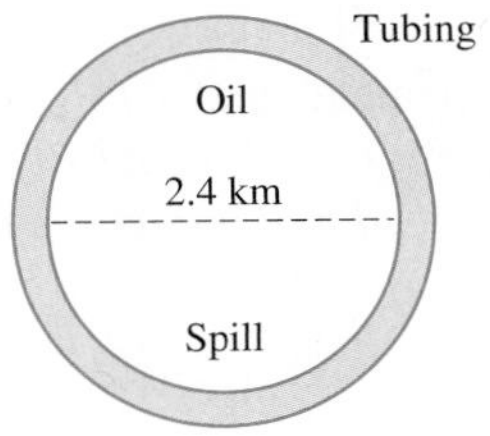

Since $d = 2r$, $r = d/2 = 1.2$ km. Using Eq. (2.10), the area is

$$\begin{aligned} A &= \pi r^2 = \pi(1.2)^2 \\ &= 4.5\text{ km}^2 \end{aligned}$$

Tubing
Oil
2.4 km
Spill

Fig. 2.79

The length of tubing needed is the circumference of the circle. Therefore,

$$\begin{aligned} c &= 2\pi r = 2\pi(1.2) && \text{note that } c = \pi d \\ &= 7.5\text{ km} && \text{rounded off} \end{aligned}$$

■

Many applied problems involve a combination of geometric figures. The following example illustrates one such combination.

EXAMPLE 3 Perimeter and area—machine part

A machine part is a square of side 3.25 in. with a quarter-circle removed (see Fig. 2.80). Find the perimeter and the area of the part.

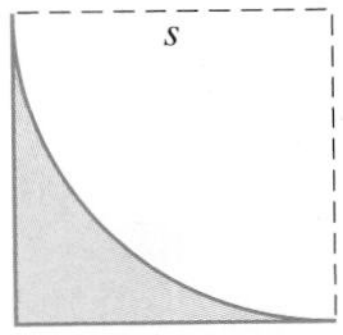

Setting up a formula for the perimeter, we add the two sides of length s to *one-fourth of the circumference of a circle with radius s*. For the area, we *subtract the area of one-fourth of a circle from the area of the square*. This gives

s
$s = 3.25$ in.

Fig. 2.80

$$p = \underset{\text{bottom and left}}{2s} + \underset{\text{circular section}}{\frac{2\pi s}{4}} = 2s + \frac{\pi s}{2} \qquad A = \underset{\text{square}}{s^2} - \underset{\text{quarter circle}}{\frac{\pi s^2}{4}}$$

where s is the side of the square and the radius of the circle. Evaluating, we have

$$p = 2(3.25) + \frac{\pi(3.25)}{2} = 11.6\text{ in.}$$

$$A = 3.25^2 - \frac{\pi(3.25)^2}{4} = 2.27\text{ in.}^2$$

■

Practice Exercises

1. Find the circumference of a circle with a radius of 20.0 cm.
2. Find the area of the circle in Practice Exercise 1.

CIRCULAR ARCS AND ANGLES

An **arc** *is part of a circle,* and *an angle formed at the center by two radii is a* **central angle.** The measure of an arc is the same as the central angle between the ends of the radii that define the arc. *A* **sector** *of a circle is the region bounded by two radii and the arc they intercept. A* **segment** *of a circle is the region bounded by a chord and its arc.* (There are two possible segments for a given chord. The smaller region is a *minor segment,* and the larger region is a *major segment.*) These are illustrated in the following example.

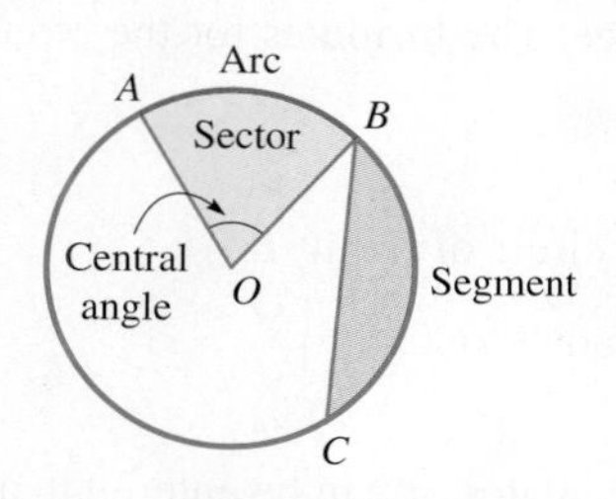

Fig. 2.81

EXAMPLE 4 Sector and segment

In Fig. 2.81, a sector of the circle is between radii OA and OB and arc AB (denoted as $\widehat{AB}$). If the measure of the central angle at O between the radii is 70°, the measure of $\widehat{AB}$ is 70°.

A segment of the circle is the region between chord BC and arc BC ($\widehat{BC}$). ■

An **inscribed angle** *of an arc is one for which the endpoints of the arc are points on the sides of the angle and for which the vertex is a point (not an endpoint) of the arc.* An important property of a circle is that *the measure of an inscribed angle is one-half of its intercepted arc.*

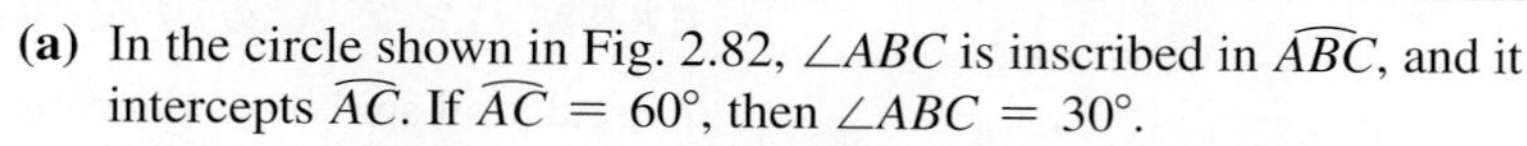

EXAMPLE 5 Inscribed angle

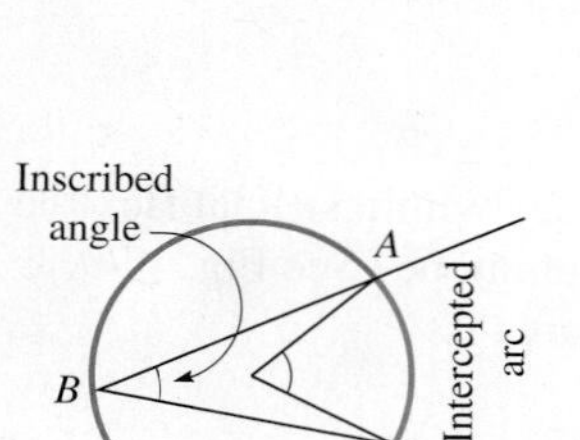

Fig. 2.82

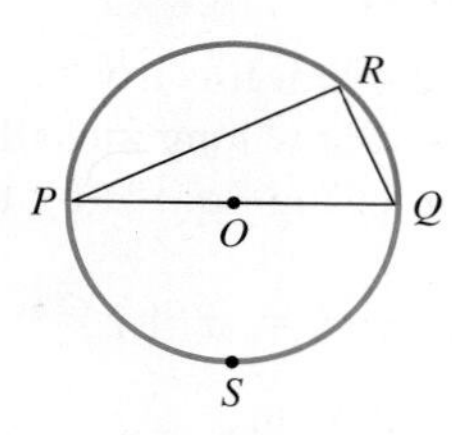

Fig. 2.83

(a) In the circle shown in Fig. 2.82, $\angle ABC$ is inscribed in $\widehat{ABC}$, and it intercepts $\widehat{AC}$. If $\widehat{AC} = 60°$, then $\angle ABC = 30°$.

(b) In the circle shown in Fig. 2.83, PQ is a diameter, and $\angle PRQ$ is inscribed in the semicircular $\widehat{PRQ}$. Since $\widehat{PSQ} = 180°$, $\angle PRQ = 90°$. From this we conclude that *an angle inscribed in a semicircle is a right angle.* ■

RADIAN MEASURE OF AN ANGLE

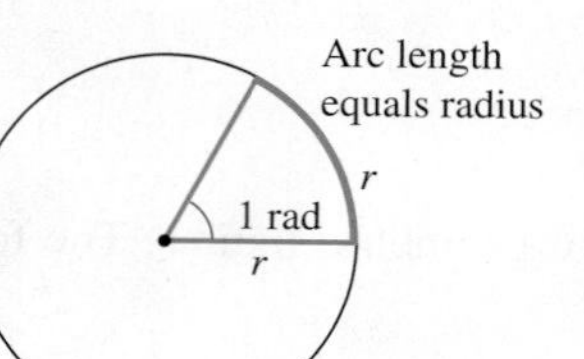

Fig. 2.84

To this point, we have measured all angles in degrees. There is another measure of an angle, the *radian,* that is defined in terms of an arc of a circle. We will find it of importance when we study trigonometry.

If a central angle of a circle intercepts an arc equal in length to the radius of the circle, the measure of the central angle is defined as **1 radian.** See Fig. 2.84. The radius can be marked off along the circumference 2π times (about 6.283 times). Thus, 2π rad $= 360°$ (where rad is the symbol for radian), and the basic relationship between radians and degrees is

$$\pi \text{ rad} = 180° \tag{2.11}$$

EXAMPLE 6 Radian measure of an angle

(a) If we divide each side of Eq. (2.11) by π, we get

$$1 \text{ rad} = 57.3°$$

where the result has been rounded off.

(b) To change an angle of 118.2° to radian measure, we have

$$118.2° = 118.2°\left(\frac{\pi \text{ rad}}{180°}\right) = 2.06 \text{ rad}$$

Multiplying 118.2° by π rad/180°, the unit that remains is rad, since degrees "cancel." We will review radian measure again when we study trigonometry. ■

Practice Exercise

3. Express the angle 85.0° in radian measure.

EXERCISES 2.4

In Exercises 1–4, answer the given questions about the indicated examples of this section.

1. In Example 1, if $\angle AOB = 72°$ in Fig. 2.78, then what is the measure of $\angle OAB$?

2. In the first line of Example 2, if "diameter" is changed to "radius," what are the results?

3. In Example 3, if the machine part is the unshaded part (rather than the shaded part) of Fig. 2.80, what are the results?

4. In Example 5(a), if $\angle ABC = 25°$ in Fig. 2.82, then what is the measure of $\widehat{AC}$?

In Exercises 5–8, refer to the circle with center at O in Fig. 2.85. Identify the following.

5. (a) A secant line
(b) A tangent line

6. (a) Two chords
(b) An inscribed angle

7. (a) Two perpendicular lines
(b) An isosceles triangle

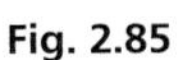

Fig. 2.85

8. (a) A segment
(b) A sector with an acute central angle

In Exercises 9–12, find the circumference of the circle with the given radius or diameter.

9. $r = 275$ ft **10.** $r = 0.563$ m
11. $d = 23.1$ mm **12.** $d = 8.2$ in.

In Exercises 13–16, find the area of the circle with the given radius or diameter.

13. $r = 0.0952$ yd **14.** $r = 45.8$ cm
15. $d = 2.33$ m **16.** $d = 1256$ ft

In Exercises 17 and 18, find the area of the circle with the given circumference.

17. $c = 40.1$ cm **18.** $c = 147$ m

In Exercises 19–22, refer to Fig. 2.86, where AB is a diameter, TB is a tangent line at B, and $\angle ABC = 65°$. Determine the indicated angles.

19. $\angle CBT$
20. $\angle BCT$
21. $\angle CAB$
22. $\angle BTC$

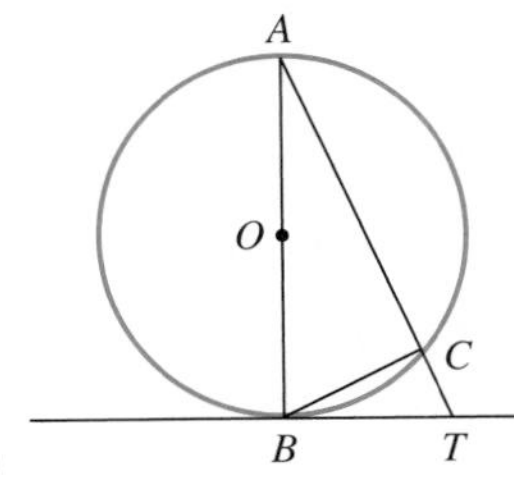

Fig. 2.86

In Exercises 23–26, refer to Fig. 2.87. Determine the indicated arcs and angles.

23. $\widehat{BC}$
24. $\widehat{AB}$
25. $\angle ABC$
26. $\angle ACB$

Fig. 2.87

In Exercises 27–30, change the given angles to radian measure.

27. 22.5°
28. 60.0°
29. 125.2°
30. 323.0°

In Exercises 31–34, find a formula for the indicated perimeter or area.

31. The perimeter of the quarter-circle in Fig. 2.88.

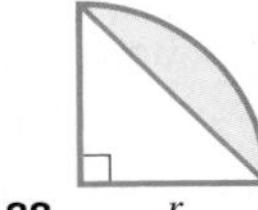

32. The perimeter of the figure in Fig. 2.89. A quarter-circle is attached to a triangle.

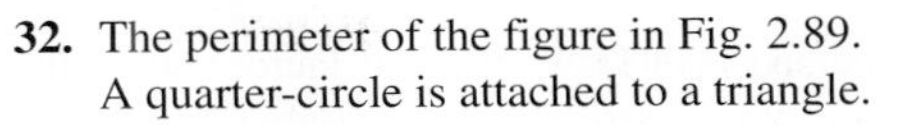

Fig. 2.88

33. The perimeter of the segment of the quarter-circle in Fig. 2.88.

34. The area of the figure in Fig. 2.89.

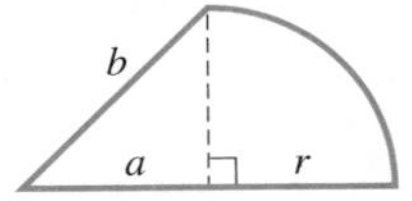

Fig. 2.89

In Exercises 35–58, solve the given problems.

35. Describe the location of the midpoints of a set of parallel chords of a circle.

36. The measure of $\widehat{AB}$ on a circle of radius r is 45°. What is the length of the arc in terms of r and π?

37. In a circle, a chord connects the ends of two perpendicular radii of 6.00 in. What is the area of the minor segment?

38. In Fig. 2.90, chords AB and DE are parallel. What is the relation between ΔABC and ΔCDE? Explain.

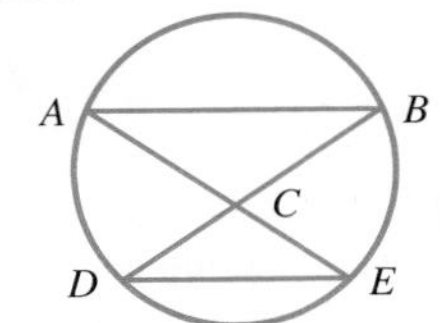

Fig. 2.90

39. Equation (2.9) is $c = 2\pi r$. Solve for π, and then use the equation $d = 2r$, where d is the diameter. State the meaning of the result.

40. In 1897, the Indiana House of Representatives passed unanimously a bill that included "the . . . important fact that the ratio of the diameter and circumference is as five-fourths to four." Under this definition, what would be the value of π? What is wrong with this House bill statement? (The bill also passed the Senate Committee and would have been enacted into law, except for the intervention of a Purdue professor.)

41. In Fig. 2.91, for the quarter-circle of radius r, find the formula for the segment area A in terms of r.

42. For the segment in Fig. 2.91, find the segment height h in terms of r.

Fig. 2.91

43. A person is in a plane 11.5 km above the shore of the Pacific Ocean. How far from the plane can the person see out on the Pacific? (The radius of Earth is 6378 km.)

44. The CN Tower in Toronto has an observation deck at 346 m above the ground. Assuming ground level and Lake Ontario level are equal, how far can a person see from the deck? (The radius of Earth is 6378 km.)

45. An FM radio station emits a signal that is clear within 85 km of the transmitting tower. Can a clear signal be received at a home 68 km west and 58 km south of the tower?

46. A circular pool 12.0 m in diameter has a sitting ledge 0.60 m wide around it. What is the area of the ledge?

47. The radius of the Earth's equator is 3960 mi. What is the circumference?

48. As a ball bearing rolls along a straight track, it makes 11.0 revolutions while traveling a distance of 109 mm. Find its radius.

49. The rim on a basketball hoop has an inside diameter of 18.0 in. The largest cross section of a basketball has a diameter of 12.0 in. What is the ratio of the cross-sectional area of the basketball to the area of the hoop?

50. With no change in the speed of flow, by what factor should the diameter of a fire hose be increased in order to double the amount of water that flows through the fire hose?

51. Using a tape measure, the circumference of a tree is found to be 112 in. What is the diameter of the tree (assuming a circular cross section)?

52. Suppose that a 5250-lb force is applied to a hollow steel cylindrical beam that has the cross section shown in Fig. 2.92. The *stress* on the beam is found by dividing the force by the cross-sectional area. Find the stress.

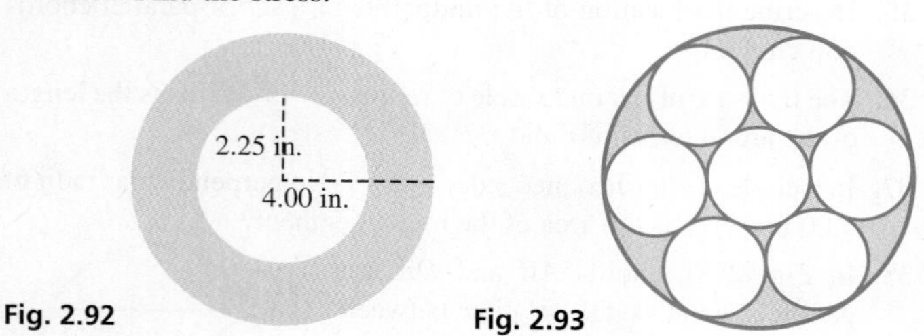

Fig. 2.92

Fig. 2.93

53. The cross section of a large circular conduit has seven smaller equal circular conduits within it. The conduits are tangent to each other as shown in Fig. 2.93. What fraction of the large conduit is occupied by the seven smaller conduits?

54. Find the area of the room in the plan shown in Fig. 2.94.

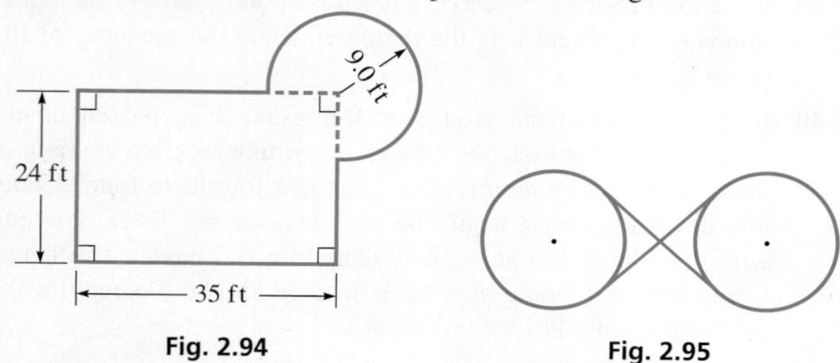

Fig. 2.94

Fig. 2.95

55. Find the length of the pulley belt shown in Fig. 2.95 if the belt crosses at right angles. The radius of each pulley wheel is 5.50 in.

56. Two pipes, each with a 25.0-mm-diameter hole, lead into a single larger pipe (see Fig. 2.96). In order to ensure proper flow, the cross-sectional area of the hole of the larger pipe is designed to be equal to the sum of the cross-sectional areas of the two smaller pipes. Find the inside diameter of the larger pipe.

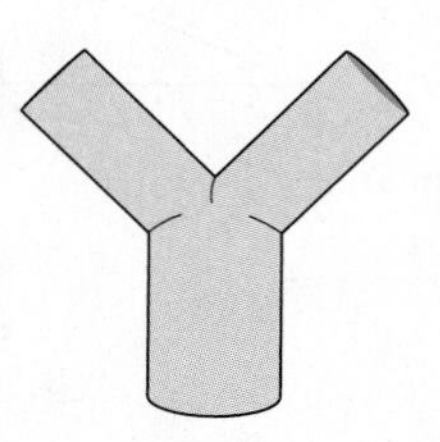

Fig. 2.96

57. The velocity of an object moving in a circular path is directed tangent to the circle in which it is moving. A stone on a string moves in a vertical circle, and the string breaks after 5.5 revolutions. If the string was initially in a vertical position, in what direction does the stone move after the string breaks? Explain.

58. Part of a circular gear with 24 teeth is shown in Fig. 2.97. Find the indicated angle.

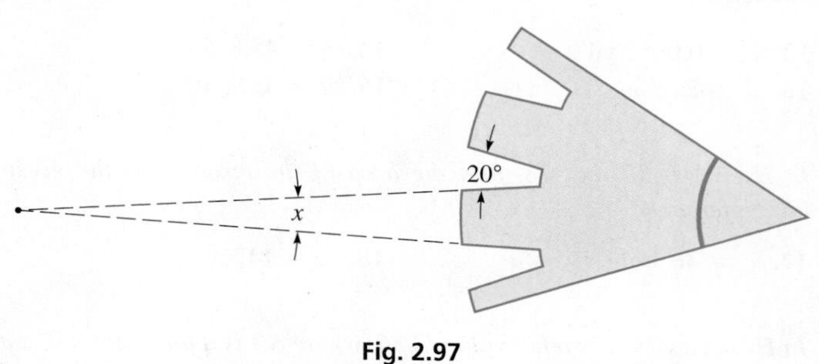

Fig. 2.97

Answers to Practice Exercises

1. 126 cm **2.** 1260 cm^2 **3.** 1.48 rad

2.5 Measurement of Irregular Areas

Trapezoidal Rule • Simpson's Rule

In practice it may be necessary to find the area of a figure with an irregular perimeter or one for which there is no specific formula. We now show two methods of finding a good *approximation* of such an area. These methods are particularly useful in technical areas such as surveying, architecture, and mechanical design.

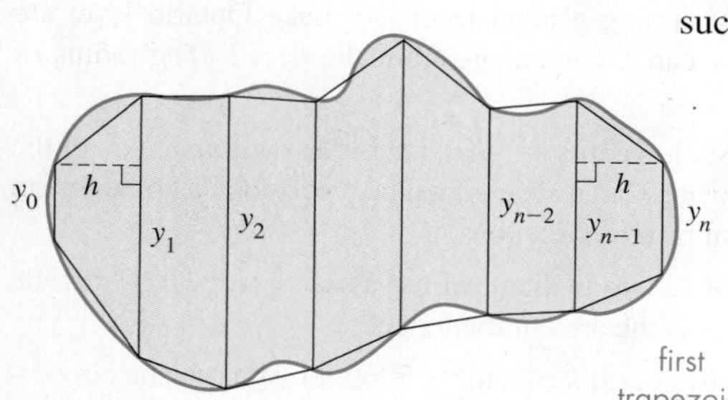

Fig. 2.98

THE TRAPEZOIDAL RULE

For the area in Fig. 2.98, we draw parallel lines at n equal intervals between the edges to form adjacent trapezoids. The sum of the areas of these trapezoids, all of equal height h, is a good approximation of the area. Now, labeling the lengths of the parallel lines $y_0, y_1, y_2, \ldots, y_n$, the total area of all trapezoids is

first trapezoid, second trapezoid, third trapezoid, next-to-last trapezoid, last trapezoid

$$A = \frac{h}{2}(y_0 + y_1) + \frac{h}{2}(y_1 + y_2) + \frac{h}{2}(y_2 + y_3) + \cdots + \frac{h}{2}(y_{n-2} + y_{n-1}) + \frac{h}{2}(y_{n-1} + y_n)$$

$$= \frac{h}{2}(y_0 + y_1 + y_1 + y_2 + y_2 + y_3 + \cdots + y_{n-2} + y_{n-1} + y_{n-1} + y_n)$$

Therefore, the approximate area is

■ Note carefully that the values of y_0 and y_n are *not* multiplied by 2.

$$A = \frac{h}{2}(y_0 + 2y_1 + 2y_2 + \cdots + 2y_{n-1} + y_n) \qquad \textbf{(2.12)}$$

which is known as the **trapezoidal rule.** The following examples illustrate its use.

EXAMPLE 1 Trapezoidal rule—area of cam

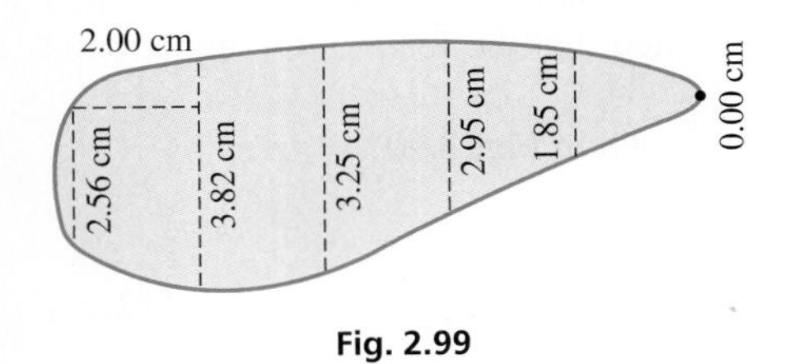

Fig. 2.99

A plate cam for opening and closing a valve is shown in Fig. 2.99. Widths of the face of the cam are shown at 2.00-cm intervals from one end of the cam. Find the area of the face of the cam.

From the figure, we see that

$$y_0 = 2.56 \text{ cm} \qquad y_1 = 3.82 \text{ cm} \qquad y_2 = 3.25 \text{ cm}$$
$$y_3 = 2.95 \text{ cm} \qquad y_4 = 1.85 \text{ cm} \qquad y_5 = 0.00 \text{ cm}$$

(In making such measurements, often a y-value at one end—or both ends—is zero. In such a case, the end "trapezoid" is actually a triangle.) From the given information in this example, $h = 2.00$ cm. Therefore, using the trapezoidal rule, Eq. (2.12), we have

$$\begin{aligned} A &= \frac{2.00}{2}[2.56 + 2(3.82) + 2(3.25) + 2(2.95) + 2(1.85) + 0.00] \\ &= 26.3 \text{ cm}^2 \end{aligned}$$

The area of the face of the cam is approximately 26.3 cm^2. ■

When approximating the area with trapezoids, we omit small parts of the area for some trapezoids and include small extra areas for other trapezoids. The omitted areas tend to compensate for the extra areas, which makes the approximation fairly accurate. Also, the use of smaller intervals improves the approximation because the total omitted area or total extra area is smaller.

EXAMPLE 2 Trapezoidal rule—Lake Ontario area

■ See the chapter introduction.

From a satellite photograph of Lake Ontario (as shown on page 54), one of the Great Lakes between the United States and Canada, measurements of the width of the lake were made along its length, starting at the west end, at 26.0-km intervals. The widths are shown in Fig. 2.100 and are given in the following table.

Distance from West End (km)	0.0	26.0	52.0	78.0	104	130	156
Width (km)	0.0	46.7	52.1	59.2	60.4	65.7	73.9

Distance from West End (km)	182	208	234	260	286	312
Width (km)	87.0	75.5	66.4	86.1	77.0	0.0

Fig. 2.100

Here, we see that $y_0 = 0.0$ km, $y_1 = 46.7$ km, $y_2 = 52.1$ km, . . . , and $y_n = 0.0$ km. Therefore, using the trapezoidal rule, the approximate area of Lake Ontario is found as follows:

$$\begin{aligned} A &= \frac{26.0}{2}[0.0 + 2(46.7) + 2(52.1) + 2(59.2) + 2(60.4) + 2(65.7) + 2(73.9) \\ &\quad + 2(87.0) + 2(75.5) + 2(66.4) + 2(86.1) + 2(77.0) + 0.0] \\ &= 19{,}500 \text{ km}^2 \end{aligned}$$

The area of Lake Ontario is actually 19,477 km^2. ■

Practice Exercise

1. In Example 2, use only the distances from west end of (in km) 0.0, 52.0, 104, 156, 208, 260, and 312. Calculate the area and compare with the answer in Example 2.

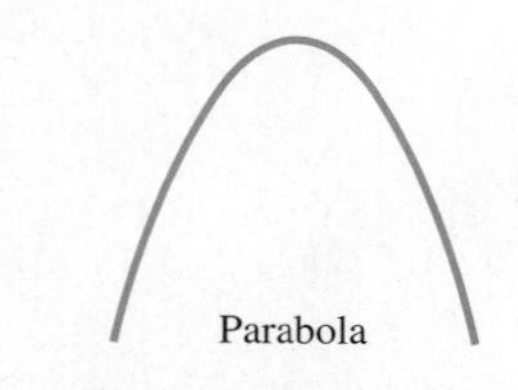

Fig. 2.101

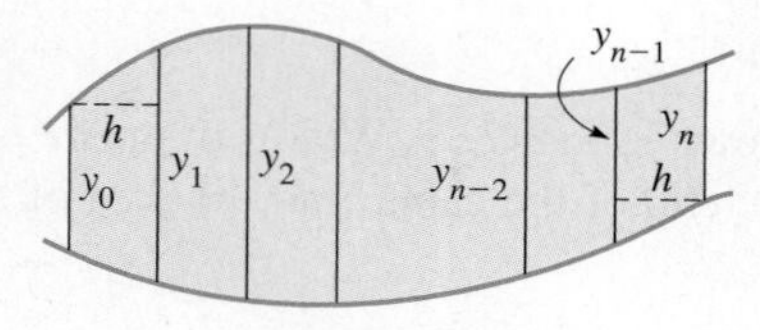

Fig. 2.102

SIMPSON'S RULE

For the second method of measuring an irregular area, we also draw parallel lines at equal intervals between the edges of the area. We then join the ends of these parallel lines with curved *arcs*. This takes into account the fact that the perimeters of most figures are curved. The arcs used in this method are not arcs of a circle, but arcs of a *parabola*. A parabola is shown in Fig. 2.101 and is discussed in detail in Chapter 21. [Examples of parabolas are (1) the path of a ball that has been thrown and (2) the cross section of a microwave "dish."]

The development of this method requires advanced mathematics. Therefore, we will simply state the formula to be used. It might be noted that the form of the equation is similar to that of the trapezoidal rule.

The approximate area of the geometric figure shown in Fig. 2.102 is given by

■ Named for the English mathematician Thomas Simpson (1710–1761).

$$A = \frac{h}{3}(y_0 + 4y_1 + 2y_2 + 4y_3 + \cdots + 2y_{n-2} + 4y_{n-1} + y_n) \tag{2.13}$$

Equation (2.13) *is known as* **Simpson's rule.** CAUTION In using Simpson's rule, the number n of intervals of width h must be *even*. ■

EXAMPLE 3 Simpson's rule—parking lot area

A parking lot is proposed for a riverfront area in a town. The town engineer measured the widths of the area at 100-ft (three sig. digits) intervals, as shown in Fig. 2.103. Find the area available for parking.

First, we see that there are six intervals, which means Eq. (2.13) may be used. With $y_0 = 407$ ft, $y_1 = 483$ ft, . . . , $y_6 = 495$ ft, and $h = 100$ ft, we have

$$A = \frac{100}{3}[407 + 4(483) + 2(382) + 4(378) + 2(285) + 4(384) + 495]$$

$$= 241{,}000 \text{ ft}^2$$ ■

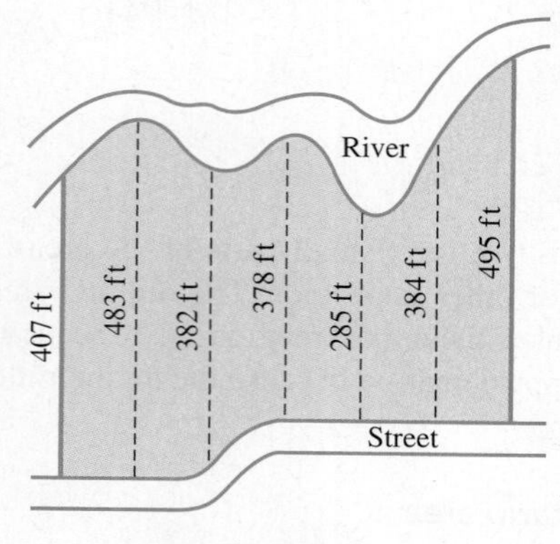

Fig. 2.103

For most areas, Simpson's rule gives a somewhat better approximation than the trapezoidal rule. The accuracy of Simpson's rule is also usually improved by using smaller intervals.

EXAMPLE 4 Simpson's rule—Easter Island area

From an aerial photograph, a cartographer determines the widths of Easter Island (in the Pacific Ocean) at 1.50-km intervals as shown in Fig. 2.104. The widths found are as follows:

Distance from South End (km)	0	1.50	3.00	4.50	6.00	7.50	9.00	10.5	12.0	13.5	15.0
Width (km)	0	4.8	5.7	10.5	15.2	18.5	18.8	17.9	11.3	8.8	3.1

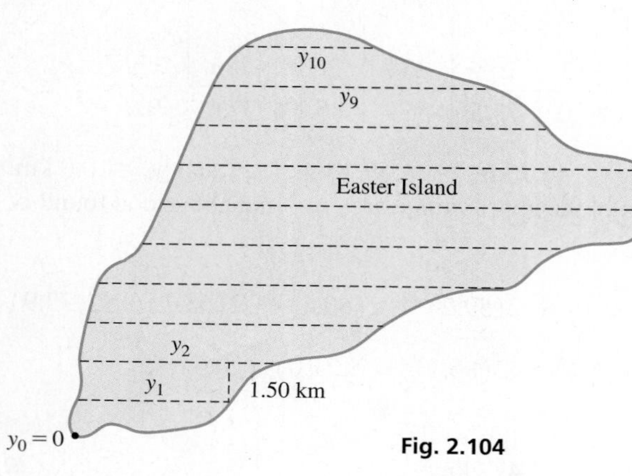

Fig. 2.104

Since there are ten intervals, Simpson's rule may be used. From the table, we have the following values: $y_0 = 0, y_1 = 4.8, y_2 = 5.7, \ldots,$ $y_9 = 8.8, y_{10} = 3.1$, and $h = 1.5$. Using Simpson's rule, the cartographer would approximate the area of Easter Island as follows:

$$A = \frac{1.50}{3}[0 + 4(4.8) + 2(5.7) + 4(10.5) + 2(15.2) + 4(18.5)$$
$$+ 2(18.8) + 4(17.9) + 2(11.3) + 4(8.8) + 3.1] = 174 \text{ km}^2$$ ■

EXERCISES 2.5

In Exercises 1 and 2, answer the given questions related to the indicated examples of this section.

1. In Example 1, if widths of the face of the same cam were given at 1.00-cm intervals (five more widths would be included), from the methods of this section, what is probably the most accurate way of finding the area? Explain.

2. In Example 4, if you use only the data from the south end of (in km) 0, 3.00, 6.00, 9.00, 12.0, and 15.0, would you choose the trapezoidal rule or Simpson's rule to calculate the area? Explain. Do not calculate the area for these data.

In Exercises 3 and 4, answer the given questions related to Fig. 2.105.

3. Which should be more accurate for finding the area, the trapezoidal rule or Simpson's rule? Explain.

4. If the trapezoidal rule is used to find the area, will the result probably be too high, about right, or too little? Explain.

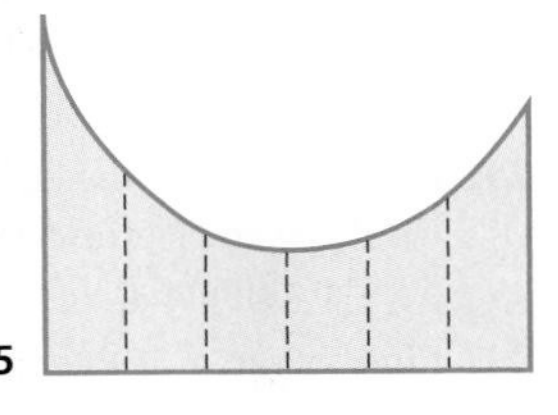
Fig. 2.105

In Exercises 5 and 6, answer the given questions related to Fig. 2.106.

5. If the trapezoidal rule was used to find the area of the region in Fig. 2.106, would the answer be approximate or exact? Explain.

6. Explain why Simpson's rule cannot be used to find the area of the region in Fig. 2.106.

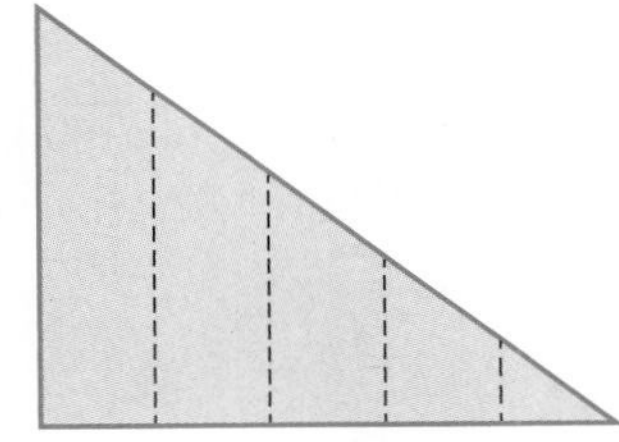
Fig. 2.106

In Exercises 7–18, calculate the indicated areas. All data are accurate to at least two significant digits.

7. The widths of a kidney-shaped swimming pool were measured at 2.0-m intervals, as shown in Fig. 2.107. Calculate the surface area of the pool, using the trapezoidal rule.

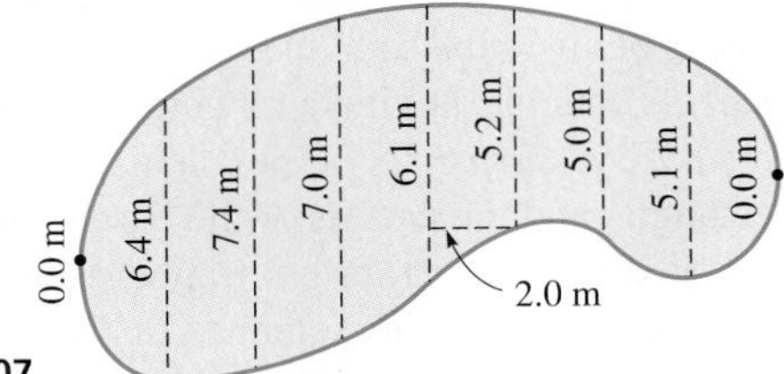

Fig. 2.107

8. Calculate the surface area of the swimming pool in Fig. 2.107, using Simpson's rule.

9. The widths of a cross section of an airplane wing are measured at 1.00-ft intervals, as shown in Fig. 2.108. Calculate the area of the cross section, using Simpson's rule.

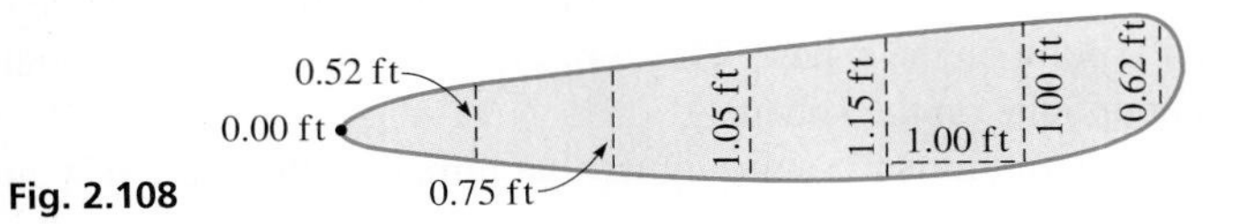

Fig. 2.108

10. Calculate the area of the cross section of the airplane wing in Fig. 2.108, using the trapezoidal rule.

11. Using aerial photography, the widths of an area burned by a forest fire were measured at 0.5-mi intervals, as shown in the following table:

Distance (mi)	0.0	0.5	1.0	1.5	2.0	2.5	3.0	3.5	4.0
Width (mi)	0.6	2.2	4.7	3.1	3.6	1.6	2.2	1.5	0.8

Determine the area burned by the fire by using the trapezoidal rule.

12. Find the area burned by the forest fire of Exercise 11, using Simpson's rule.

13. A cartographer measured the width of Bruce Peninsula in Ontario at 10-mm intervals on a map (scale 10 mm = 23 km), as shown in Fig. 2.109. The widths are shown in the following list. What is the area of Bruce Peninsula?

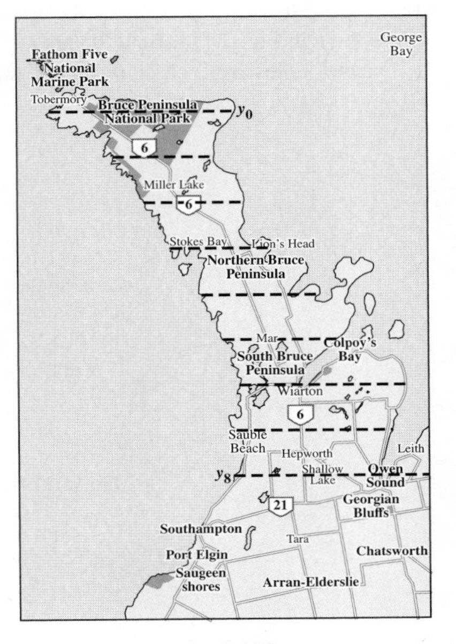

Fig. 2.109

$y_0 = 38$ mm $\quad y_1 = 24$ mm $\quad y_2 = 25$ mm
$y_3 = 17$ mm $\quad y_4 = 34$ mm $\quad y_5 = 29$ mm
$y_6 = 36$ mm $\quad y_7 = 34$ mm $\quad y_8 = 30$ mm

14. The widths (in m) of half the central arena in the Colosseum in Rome are shown in the following table, starting at one end and measuring from the middle to one side at 4.0-m intervals. Find the area of the arena by the trapezoidal rule. *Hint*: Remember to double the distances.

Dist. from middle (m)	0.0	4.0	8.0	12.0	16.0	20.0
Width (m)	55.0	54.8	54.0	53.6	51.2	49.0

Dist.	24.0	28.0	32.0	36.0	40.0	44.0
Width	45.8	42.0	37.2	31.1	21.7	0.0

15. The widths of the baseball playing area in Boston's Fenway Park at 45-ft intervals are shown in Fig. 2.110. Find the playing area using the trapezoidal rule.

16. Find the playing area of Fenway Park (see Exercise 15) by Simpson's rule.

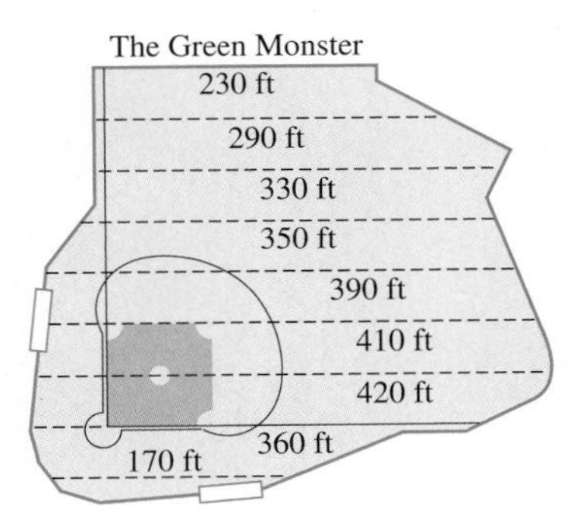

Fig. 2.110

17. Soundings taken across a river channel give the following depths with the corresponding distances from one shore.

Distance (ft)	0	50	100	150	200	250	300	350	400	450	500
Depth (ft)	5	12	17	21	22	25	26	16	10	8	0

Find the area of the cross section of the channel using Simpson's rule.

18. The widths of a bell crank are measured at 2.0-in. intervals, as shown in Fig. 2.111. Find the area of the bell crank if the two connector holes are each 2.50 in. in diameter.

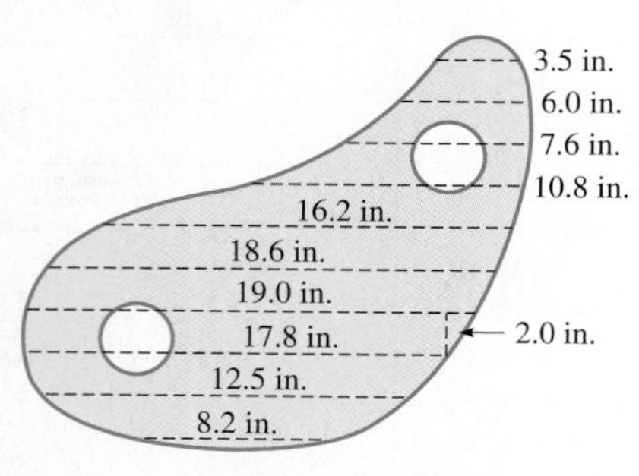

Fig. 2.111

In Exercises 19–22, calculate the area of the circle by the indicated method.

The lengths of parallel chords of a circle that are 0.250 in. apart are given in the following table. The diameter of the circle is 2.000 in. The distance shown is the distance from one end of a diameter.

Distance (in.)	0.000	0.250	0.500	0.750	1.000	1.250	1.500	1.750	2.000
Length (in.)	0.000	1.323	1.732	1.936	2.000	1.936	1.732	1.323	0.000

Using the formula $A = \pi r^2$, the area of the circle is 3.14 in.^{2}.

19. Find the area of the circle using the trapezoidal rule and only the values of distance of 0.000 in., 0.500 in., 1.000 in., 1.500 in., and 2.000 in. with the corresponding values of the chord lengths. Explain why the value found is less than 3.14 in.^{2}.

20. Find the area of the circle using the trapezoidal rule and all values in the table. Explain why the value found is closer to 3.14 in.^{2} than the value found in Exercise 19.

21. Find the area of the circle using Simpson's rule and the same table values as in Exercise 19. Explain why the value found is closer to 3.14 in.^{2} than the value found in Exercise 19.

22. Find the area of the circle using Simpson's rule and all values in the table. Explain why the value found is closer to 3.14 in.^{2} than the value found in Exercise 21.

Answer to Practice Exercise

1. 18,100 km^2

2.6 Solid Geometric Figures

Rectangular Solid • Cylinder • Prism • Cone • Pyramid • Sphere • Frustum

We now review the formulas for the *volume* and *surface area* of some basic solid geometric figures. Just as area is a measure of the surface of a plane geometric figure, **volume** is a measure of the space occupied by a solid geometric figure.

One of the most common solid figures is the **rectangular solid.** This figure has six sides (**faces**), and opposite sides are rectangles. All intersecting sides are perpendicular to each other. The **bases** of the rectangular solid are the top and bottom faces. A **cube** is a rectangular solid with all six faces being equal squares.

A **right circular cylinder** is *generated* by rotating a rectangle about one of its sides. Each **base** is a circle, and the *cylindrical surface* is perpendicular to each of the bases. The **height** is one side of the rectangle, and the **radius** of the base is the other side.

A **right circular cone** is generated by rotating a right triangle about one of its legs. The **base** is a circle, and the **slant height** is the hypotenuse of the right triangle. The **height** is one leg of the right triangle, and the **radius** of the base is the other leg.

The bases of a **right prism** are equal and parallel polygons, and the sides are rectangles. The **height** of a prism is the perpendicular distance between bases. The base of a **pyramid** is a polygon, and the other faces, the **lateral faces,** are triangles that meet at a common point, the **vertex.** A **regular pyramid** has congruent triangles for its lateral faces.

A **sphere** is generated by rotating a circle about a diameter. The **radius** is a line segment joining the center and a point on the sphere. The **diameter** is a line segment through the center and having its endpoints on the sphere.

The **frustum** of a cone or pyramid is the solid figure that remains after the top is cut off by a plane parallel to the base.

In the following formulas, V represents the *volume,* A represents the *total surface area,* S represents the *lateral surface area* (bases not included), B represents the *area of the base,* and p represents the *perimeter of the base.*

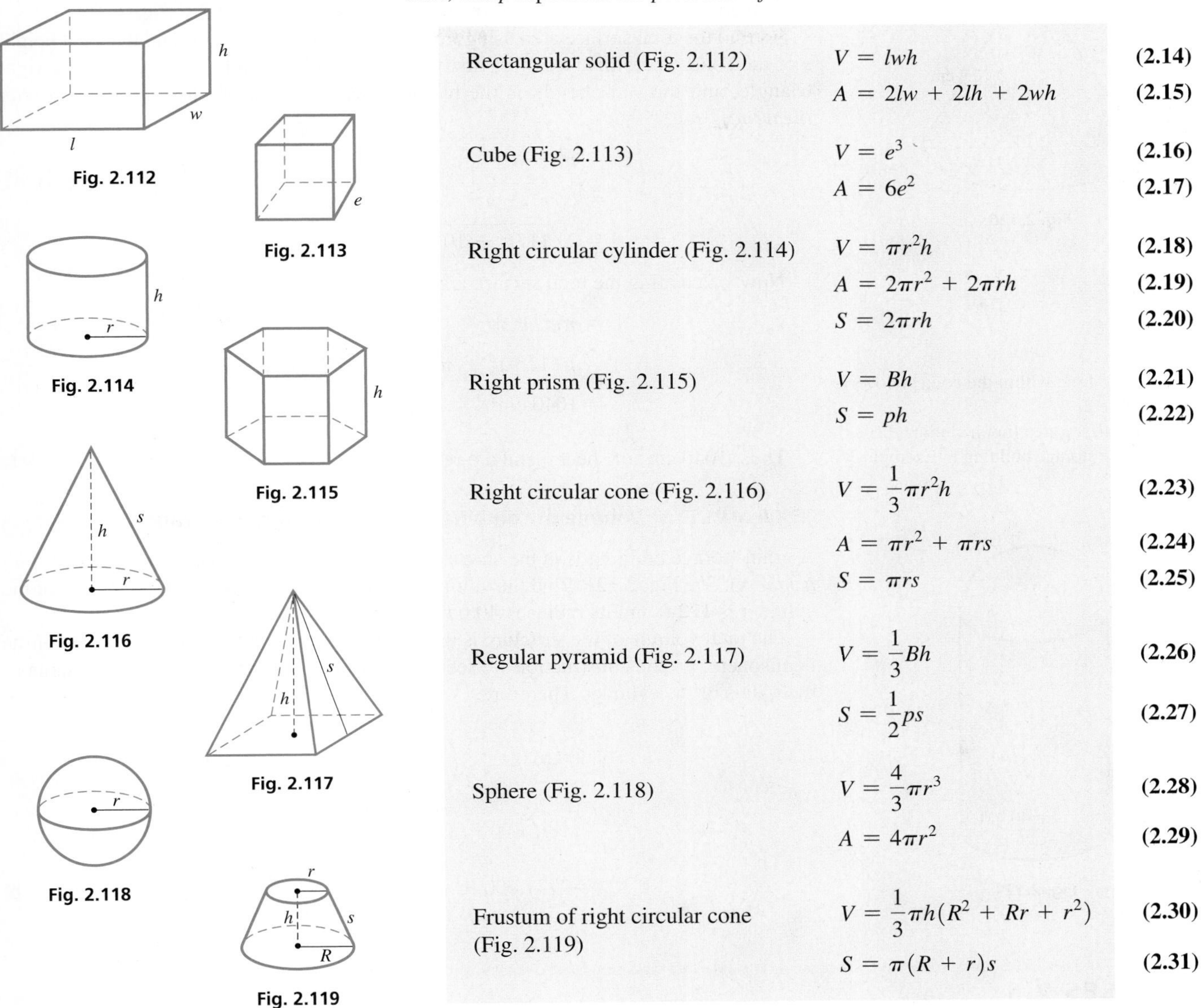

Fig. 2.112

Fig. 2.113

Fig. 2.114

Fig. 2.115

Fig. 2.116

Fig. 2.117

Fig. 2.118

Fig. 2.119

Rectangular solid (Fig. 2.112)
$$V = lwh \qquad (2.14)$$
$$A = 2lw + 2lh + 2wh \qquad (2.15)$$

Cube (Fig. 2.113)
$$V = e^3 \qquad (2.16)$$
$$A = 6e^2 \qquad (2.17)$$

Right circular cylinder (Fig. 2.114)
$$V = \pi r^2 h \qquad (2.18)$$
$$A = 2\pi r^2 + 2\pi rh \qquad (2.19)$$
$$S = 2\pi rh \qquad (2.20)$$

Right prism (Fig. 2.115)
$$V = Bh \qquad (2.21)$$
$$S = ph \qquad (2.22)$$

Right circular cone (Fig. 2.116)
$$V = \frac{1}{3}\pi r^2 h \qquad (2.23)$$
$$A = \pi r^2 + \pi rs \qquad (2.24)$$
$$S = \pi rs \qquad (2.25)$$

Regular pyramid (Fig. 2.117)
$$V = \frac{1}{3}Bh \qquad (2.26)$$
$$S = \frac{1}{2}ps \qquad (2.27)$$

Sphere (Fig. 2.118)
$$V = \frac{4}{3}\pi r^3 \qquad (2.28)$$
$$A = 4\pi r^2 \qquad (2.29)$$

Frustum of right circular cone (Fig. 2.119)
$$V = \frac{1}{3}\pi h(R^2 + Rr + r^2) \qquad (2.30)$$
$$S = \pi(R + r)s \qquad (2.31)$$

Equation (2.21) is valid for any prism, and Eq. (2.26) is valid for any pyramid. There are other types of cylinders and cones, but we restrict our attention to right circular cylinders and right circular cones, and we will often use "cylinder" or "cone" when referring to them.

EXAMPLE 1 Volume of rectangular solid—driveway construction

What volume of concrete is needed for a driveway 25.0 m long, 2.75 m wide, and 0.100 m thick?

The driveway is a rectangular solid for which $l = 25.0$ m, $w = 2.75$ m, and $h = 0.100$ m. Using Eq. (2.14), we have

$$V = (25.0)(2.75)(0.100)$$
$$= 6.88 \text{ m}^3$$

■

EXAMPLE 2 Total surface area of right circular cone—protective cover

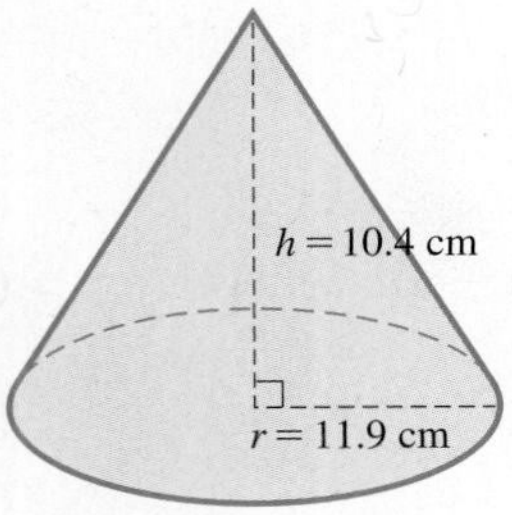

How many square centimeters of sheet metal are required to make a protective cone-shaped cover if the radius is 11.9 cm and the height is 10.4 cm? See Fig. 2.120.

$h = 10.4$ cm
$r = 11.9$ cm

Fig. 2.120

To find the total surface area using Eq. (2.24), we need the radius and the slant height s of the cone. Therefore, we must first find s. The radius and height are legs of a right triangle, and the slant height is the hypotenuse. To find s, we *use the Pythagorean theorem:*

$$s^2 = r^2 + h^2 \qquad \text{Pythagorean theorem}$$

$$s = \sqrt{r^2 + h^2} \qquad \text{solve for } s$$

$$= \sqrt{11.9^2 + 10.4^2} = 15.8 \text{ cm}$$

Now, calculating the total surface area, we have

$$A = \pi r^2 + \pi r s \qquad \text{Eq. (2.24)}$$

$$= \pi(11.9)^2 + \pi(11.9)(15.8) \qquad \text{substituting}$$

$$= 1040 \text{ cm}^2$$

Thus, 1040 cm^2 of sheet metal are required. ■

Practice Exercises

1. Find the volume within the conical cover in Example 2.
2. Find the surface area (not including the base) of the storage building in Example 3.

EXAMPLE 3 Volume of combination of solids—grain storage

A grain storage building is in the shape of a cylinder surmounted by a hemisphere (*half a sphere*). See Fig. 2.121. Find the volume of grain that can be stored if the height of the cylinder is 122 ft and its radius is 40.0 ft.

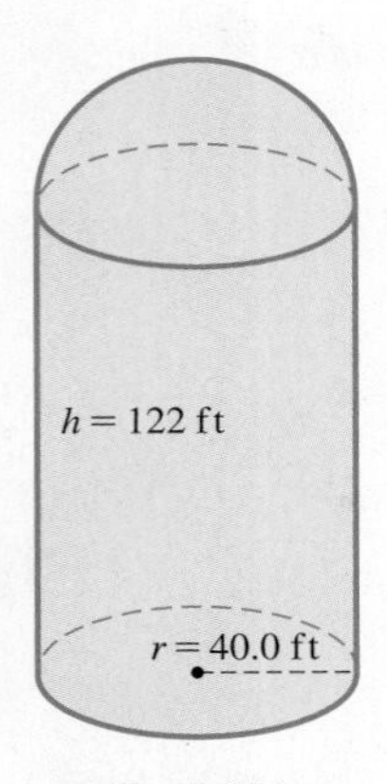

Fig. 2.121

The total volume of the structure is the volume of the cylinder plus the volume of the hemisphere. By the construction we see that the radius of the hemisphere is the same as the radius of the cylinder. Therefore,

$$V = \underbrace{\pi r^2 h}_{\text{cylinder}} + \underbrace{\frac{1}{2}\left(\frac{4}{3}\pi r^3\right)}_{\text{hemisphere}} = \pi r^2 h + \frac{2}{3}\pi r^3$$

$$= \pi(40.0)^2(122) + \frac{2}{3}\pi(40.0)^3$$

$$= 747{,}000 \text{ ft}^3$$

■

EXERCISES 2.6

In Exercises 1–4, answer the given questions about the indicated examples of this section.

1. In Example 1, if the length is doubled and the thickness is tripled, by what factor is the volume changed?
2. In Example 2, if the value of the slant height $s = 17.5$ cm is given instead of the height, what is the height?
3. In Example 2, if the radius is halved and the height is doubled, what is the volume?
4. In Example 3, if h is halved, what is the volume?

In Exercises 5–22, find the volume or area of each solid figure for the given values. See Figs. 2.112 to 2.119.

5. Volume of cube: $e = 6.95$ ft
6. Volume of right circular cylinder: $r = 23.5$ cm, $h = 48.4$ cm
7. Total surface area of right circular cylinder: $r = 689$ mm, $h = 233$ mm
8. Area of sphere: $r = 0.067$ in.
9. Volume of sphere: $r = 1.037$ yd
10. Volume of right circular cone: $r = 25.1$ m, $h = 5.66$ m
11. Lateral area of right circular cone: $r = 78.0$ cm, $s = 83.8$ cm
12. Lateral area of regular pyramid: $p = 345$ ft, $s = 272$ ft
13. Volume of regular pyramid: square base of side 0.76 in., $h = 1.30$ in.
14. Volume of right prism: square base of side 29.0 cm, $h = 11.2$ cm
15. Volume of frustum of right circular cone: $R = 37.3$ mm, $r = 28.2$ mm, $h = 45.1$ mm
16. Lateral area of frustum of right circular cone: $R = 3.42$ m, $r = 2.69$ m, $s = 3.25$ m

17. Lateral area of right prism: equilateral triangle base of side 1.092 m, $h = 1.025$ m

18. Lateral area of right circular cylinder: diameter = 250 ft, $h = 347$ ft

19. Volume of hemisphere: diameter = 0.65 yd

20. Volume of regular pyramid: square base of side 22.4 m, $s = 14.2$ m

21. Total surface area of right circular cone: $r = 3.39$ cm, $h = 0.274$ cm

22. Total surface area of regular pyramid: All faces and base are equilateral triangles of side 3.67 in. (This is often referred to as a *tetrahedron*.)

In Exercises 23–46, solve the given problems.

23. Equation (2.28) expresses the volume V of a sphere in terms of the radius r. Express V in terms of the diameter d.

24. Derive a formula for the total surface area A of a hemispherical volume of radius r (curved surface and flat surface).

25. The radius of a cylinder is twice as long as the radius of a cone, and the height of the cylinder is half as long as the height of the cone. What is the ratio of the volume of the cylinder to that of the cone?

26. The base area of a cone is one-fourth of the total area. Find the ratio of the radius to the slant height.

27. In designing a spherical weather balloon, it is decided to double the diameter of the balloon so that it can carry a heavier instrument load. What is the ratio of the final surface area to the original surface area?

28. During a rainfall of 1.00 in., what weight of water falls on an area of 1.00 mi^2? Each cubic foot of water weighs 62.4 lb.

29. A rectangular box is to be used to store radioactive materials. The inside of the box is 12.0 in. long, 9.50 in. wide, and 8.75 in. deep. What is the area of sheet lead that must be used to line the inside of the box?

30. A swimming pool is 50.0 ft wide, 78.0 ft long, 3.50 ft deep at one end, and 8.75 ft deep at the other end. How many cubic feet of water can it hold? (The slope on the bottom is constant.) See Fig. 2.122.

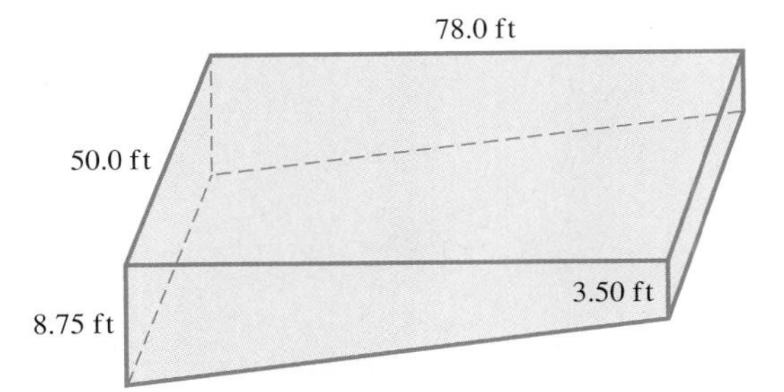

Fig. 2.122

31. The Alaskan oil pipeline is 750 mi long and has a diameter of 4.0 ft. What is the maximum volume of the pipeline?

32. The volume of a frustum of a pyramid is $V = \frac{1}{3}h(a^2 + ab + b^2)$ (see Fig. 2.123). (This equation was discovered by the ancient Egyptians.) If the base of a statue is the frustum of a pyramid, find its volume if $a = 2.50$ m, $b = 3.25$ m, and $h = 0.750$ m.

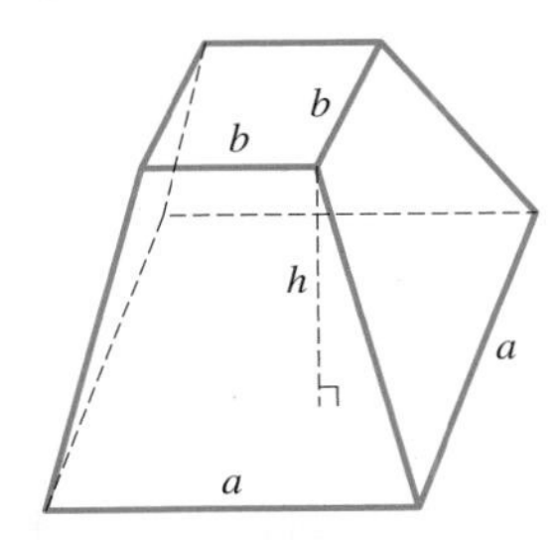

Fig. 2.123

33. A pole supporting a wind turbine is constructed of solid steel and is in the shape of a frustum of a cone. It measures 62.5 m high, and the diameter of the pole at the bottom and top are 3.88 m and 1.90 m, respectively. What is the volume of the pole?

34. A glass prism used in the study of optics has a right triangular base. The legs of the triangle are 3.00 cm and 4.00 cm. The prism is 8.50 cm high. What is the total surface area of the prism? See Fig. 2.124.

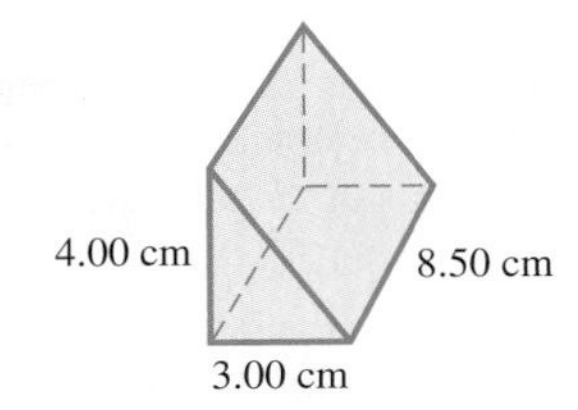

Fig. 2.124

35. The Great Pyramid of Egypt has a square base approximately 250 yd on a side. The height of the pyramid is about 160 yd. What is its volume? See Fig. 2.125.

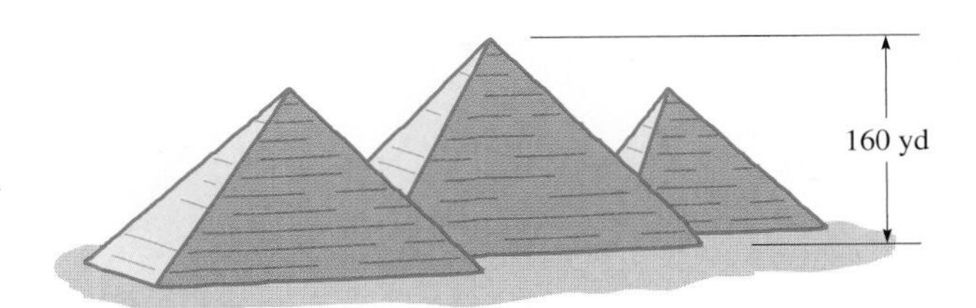

Fig. 2.125

36. A paper cup is in the shape of a cone as shown in Fig. 2.126. What is the surface area of the cup?

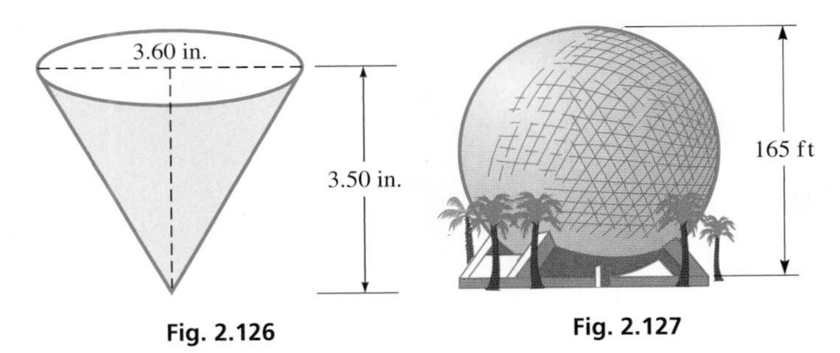

Fig. 2.126 Fig. 2.127

37. *Spaceship Earth* (shown in Fig. 2.127) at Epcot Center in Florida is a sphere of 165 ft in diameter. What is the volume of *Spaceship Earth*?

38. A propane tank is constructed in the shape of a cylinder with a hemisphere at each end, as shown in Fig. 2.128. Find the volume of the tank.

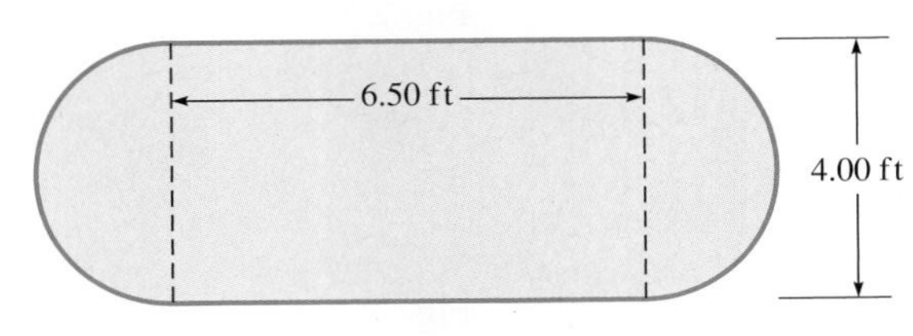

Fig. 2.128

39. A special wedge in the shape of a regular pyramid has a square base 16.0 mm on a side. The height of the wedge is 40.0 mm. What is the total surface area of the wedge (including the base)?

40. A lawn roller is a cylinder 0.96.0 m long and 0.60 m in diameter. How many revolutions of the roller are needed to roll 76 m^2 of lawn?

41. The circumference of a basketball is about 29.8 in. What is its volume?

42. What is the area of a paper label that is to cover the lateral surface of a cylindrical can 3.00 in. in diameter and 4.25 in. high? The ends of the label will overlap 0.25 in. when the label is placed on the can.

43. The side view of a rivet is shown in Fig. 2.129. It is a conical part on a cylindrical part. Find the volume of the rivet.

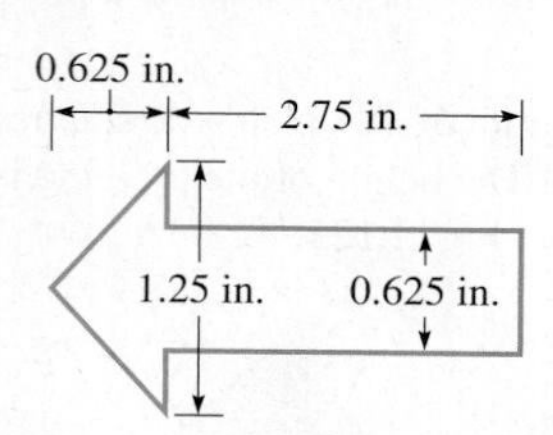

Fig. 2.129

44. A semicircular patio made of concrete 7.5 cm thick has a total perimeter of 18 m. What is the volume of concrete?

45. A ball bearing had worn down too much in a machine that was not operating properly. It remained spherical, but had lost 8.0% of its volume. By what percent had the radius decreased?

46. A dipstick is made to measure the volume remaining in the conical container shown in Fig. 2.130. How far below the full mark (at the top of the container) on the stick should the mark for half-full be placed?

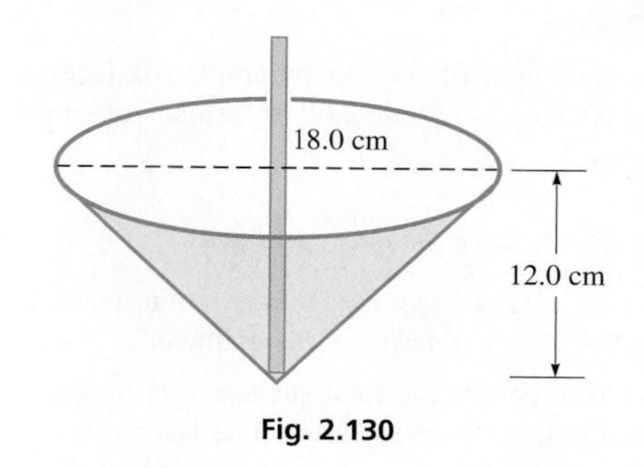

Fig. 2.130

Answers to Practice Exercises

1. 1540 cm^3 **2.** 40,700 ft^2

CHAPTER 2 KEY FORMULAS AND EQUATIONS

Line segments	Fig. 2.8	$\frac{a}{b} = \frac{c}{d}$	**(2.1)**
Triangle		$A = \frac{1}{2}bh$	**(2.2)**
Hero's formula		$A = \sqrt{s(s-a)(s-b)(s-c)}$, where $s = \frac{1}{2}(a+b+c)$	**(2.3)**
Pythagorean theorem	Fig. 2.34	$c^2 = a^2 + b^2$	**(2.4)**
Square	Fig. 2.55	$A = s^2$	**(2.5)**
Rectangle	Fig. 2.56	$A = lw$	**(2.6)**
Parallelogram	Fig. 2.57	$A = bh$	**(2.7)**
Trapezoid	Fig. 2.58	$A = \frac{1}{2}h(b_1 + b_2)$	**(2.8)**
Circle		$c = 2\pi r$	**(2.9)**
		$A = \pi r^2$	**(2.10)**
Radians	Fig. 2.84	π rad = 180°	**(2.11)**
Trapezoidal rule	Fig. 2.98	$A = \frac{h}{2}(y_0 + 2y_1 + 2y_2 + \cdots + 2y_{n-1} + y_n)$	**(2.12)**
Simpson's rule	Fig. 2.102	$A = \frac{h}{3}(y_0 + 4y_1 + 2y_2 + 4y_3 + \cdots + 2y_{n-2} + 4y_{n-1} + y_n)$	**(2.13)**

Rectangular solid	Fig. 2.112	$V = lwh$	**(2.14)**
		$A = 2lw + 2lh + 2wh$	**(2.15)**
Cube	Fig. 2.113	$V = e^3$	**(2.16)**
		$A = 6e^2$	**(2.17)**
Right circular cylinder	Fig. 2.114	$V = \pi r^2 h$	**(2.18)**
		$A = 2\pi r^2 + 2\pi rh$	**(2.19)**
		$S = 2\pi rh$	**(2.20)**
Right prism	Fig. 2.115	$V = Bh$	**(2.21)**
		$S = ph$	**(2.22)**
Right circular cone	Fig. 2.116	$V = \frac{1}{3}\pi r^2 h$	**(2.23)**
		$A = \pi r^2 + \pi rs$	**(2.24)**
		$S = \pi rs$	**(2.25)**
Regular pyramid	Fig. 2.117	$V = \frac{1}{3}Bh$	**(2.26)**
		$S = \frac{1}{2}ps$	**(2.27)**
Sphere	Fig. 2.118	$V = \frac{4}{3}\pi r^3$	**(2.28)**
		$A = 4\pi r^2$	**(2.29)**
Frustum of right circular cone	Fig. 2.119	$V = \frac{1}{3}\pi h(R^2 + Rr + r^2)$	**(2.30)**
		$S = \pi(R + r)s$	**(2.31)**

CHAPTER 2 REVIEW EXERCISES

CONCEPT CHECK EXERCISES

Determine each of the following as being either ***true*** *or* ***false****. If it is false, explain why.*

1. In Fig. 2.6, $\angle AOC$ is the complement of $\angle COB$.
2. A triangle of sides 9, 12, and 15 is a right triangle.
3. A quadrilateral with two sides of length a and two sides of length b is always a parallelogram.
4. The circumference of a circle of diameter d is πd.
5. Simpson's rule could be used to find the approximate area in Example 2 of Section 2.5.
6. The volume of a right circular cylinder is the base area times the height.

PRACTICE AND APPLICATIONS

In Exercises 7–10, use Fig. 2.131. Determine the indicated angles.

7. $\angle CGE$
8. $\angle EGF$
9. $\angle DGH$
10. $\angle EGI$

Fig. 2.131

In Exercises 11–18, find the indicated sides of the right triangle shown in Fig. 2.132.

11. $a = 12, b = 35, c = ?$
12. $a = 14, b = 48, c = ?$
13. $a = 400, b = 580, c = ?$
14. $b = 5600, c = 6500, a = ?$
15. $b = 0.380, c = 0.736, a = ?$
16. $b = 25.1, c = 128, a = ?$
17. $b = 38.3, c = 52.9, a = ?$
18. $a = 0.782, c = 0.885, b = ?$

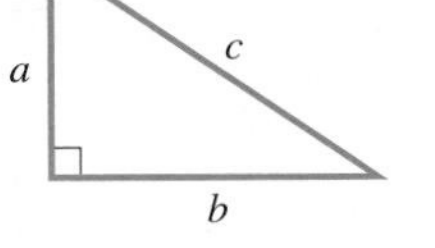

Fig. 2.132

In Exercises 19–26, find the perimeter or area of the indicated figure.

19. Perimeter: equilateral triangle of side 8.5 mm
20. Perimeter: rhombus of side 15.2 in.
21. Area: triangle: $b = 0.125$ ft, $h = 0.188$ ft
22. Area: triangle of sides 175 cm, 138 cm, 119 cm
23. Circumference of circle: $d = 74.8$ mm
24. Perimeter: rectangle, $l = 2980$ yd, $w = 1860$ yd
25. Area: trapezoid, $b_1 = 67.2$ in., $b_2 = 126.7$ in., $h = 34.2$ in.
26. Area: circle, $d = 0.328$ m

In Exercises 27–32, find the volume of the indicated solid geometric figure.

27. Prism: base is right triangle with legs 26.0 cm and 34.0 cm, height is 14.0 cm

28. Cylinder: base radius 36.0 in., height 2.40 in.

29. Pyramid: base area 3850 ft^2, height 125 ft

30. Sphere: diameter 2.21 mm

31. Cone: base radius 32.4 cm, height 50.7 cm

32. Frustum of a cone: base radius 2.336 ft, top radius 2.016 ft, height 4.890 ft

In Exercises 33–36, find the surface area of the indicated solid geometric figure.

33. Total area of cube of edge 0.520 m

34. Total area of cylinder: base diameter 12.0 ft, height 58.0 ft

35. Lateral area of cone: base radius 2.56 in., height 12.3 in.

36. Total area of sphere: $d = 12{,}760$ km

In Exercises 37–40; use Fig. 2.133. Line CT is tangent to the circle with center at O. Find the indicated angles.

37. $\angle BTA$ **38.** $\angle TAB$

39. $\angle BTC$ **40.** $\angle ABT$

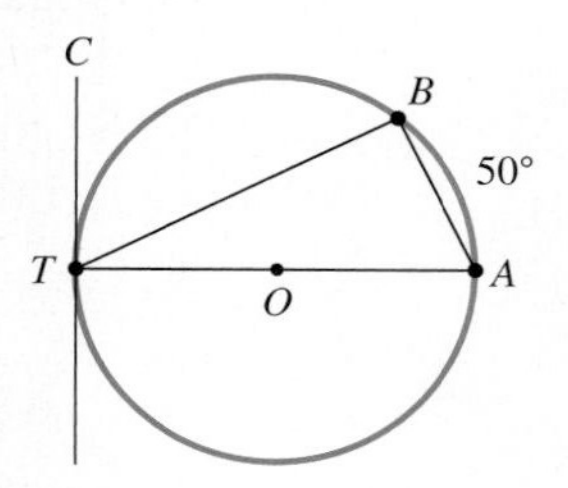

Fig. 2.133

In Exercises 41–44, use Fig. 2.134. Given that $AB = 4$, $BC = 4$, $CD = 6$, and $\angle ADC = 53°$, find the indicated angle and lengths.

41. $\angle ABE$

42. AD

43. BE

44. AE

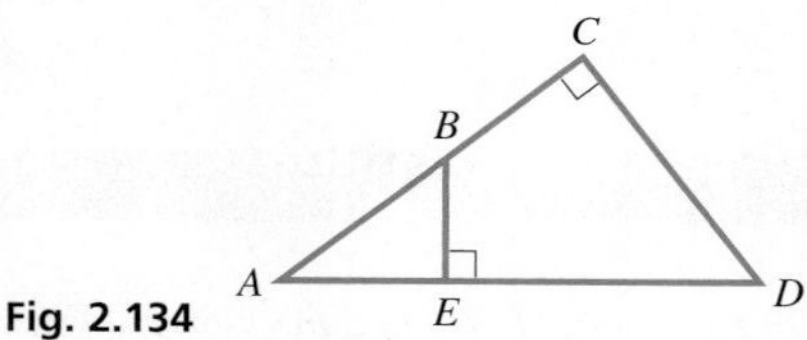

Fig. 2.134

In Exercises 45–48, find the formulas for the indicated perimeters and areas.

45. Perimeter of Fig. 2.135 (a right triangle and semicircle attached)

46. Perimeter of Fig. 2.136 (a square with a quarter circle at each end)

47. Area of Fig. 2.135 **48.** Area of Fig. 2.136

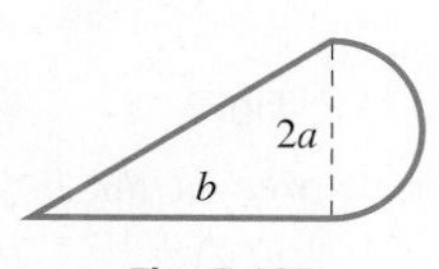

Fig. 2.135

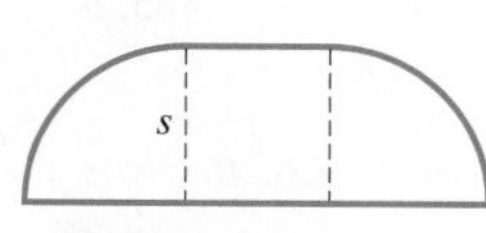

Fig. 2.136

In Exercises 49–54, answer the given questions.

49. Is a square also a rectangle, a parallelogram, and a rhombus?

50. If the measures of two angles of one triangle equal the measures of two angles of a second triangle, are the two triangles similar?

51. If the dimensions of a plane geometric figure are each multiplied by n, by how much is the area multiplied? Explain, using a circle to illustrate.

52. If the dimensions of a solid geometric figure are each multiplied by n, by how much is the volume multiplied? Explain, using a cube to illustrate.

53. What is an equation relating chord segments a, b, c, and d shown in Fig. 2.137. The dashed chords are an aid in the solution.

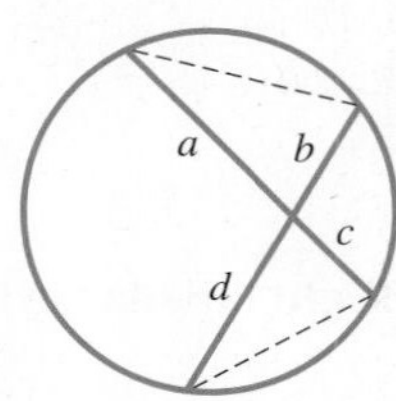

Fig. 2.137

54. From a common point, two line segments are tangent to the same circle. If the angle between the line segments is 36°, what is the angle between the two radii of the circle drawn from the points of tangency?

In Exercises 55–84, solve the given problems.

55. A tooth on a saw is in the shape of an isosceles triangle. If the angle at the point is 32°, find the two base angles.

56. A lead sphere 1.50 in. in diameter is flattened into a circular sheet 14.0 in. in diameter. How thick is the sheet?

57. A ramp for the disabled is designed so that it rises 0.48 m over a horizontal distance of 7.8 m. How long is the ramp?

58. An airplane is 2100 ft directly above one end of a 9500-ft runway. How far is the plane from the glide-slope indicator on the ground at the other end of the runway?

59. A machine part is in the shape of a square with equilateral triangles attached to two sides (see Fig. 2.138). Find the perimeter of the machine part.

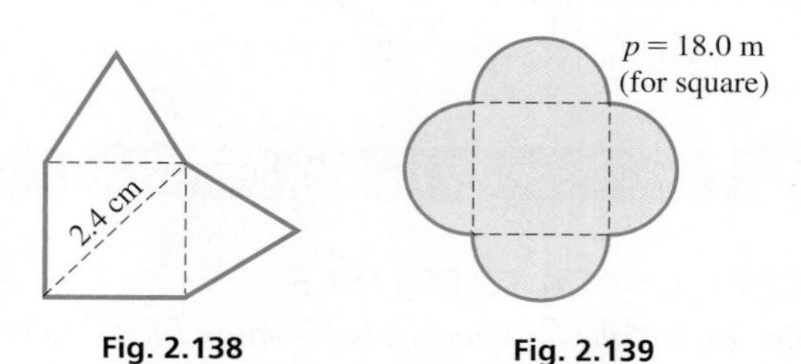

Fig. 2.138 Fig. 2.139

60. A patio is designed with semicircular areas attached to a square, as shown in Fig. 2.139. Find the area of the patio.

61. A cell phone transmitting tower is supported by guy wires. The tower and three parallel guy wires are shown in Fig. 2.140. Find the distance AB along the tower.

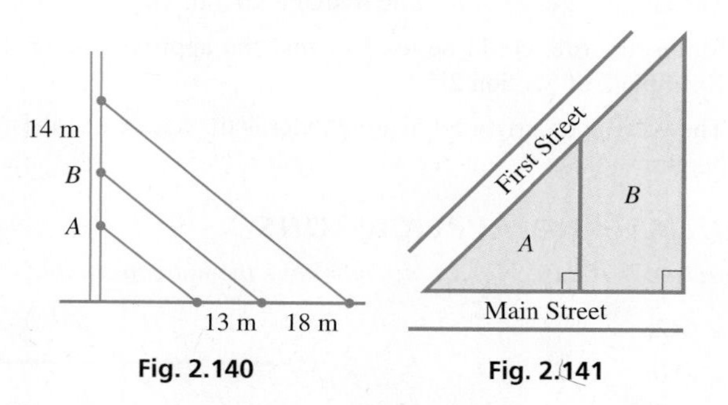

Fig. 2.140 Fig. 2.141

62. Find the areas of lots A and B in Fig. 2.141. A has a frontage on Main St. of 140 ft, and B has a frontage on Main St. of 84 ft. The boundary between lots is 120 ft.

63. To find the height of a flagpole, a person places a mirror at M, as shown in Fig. 2.142. The person's eyes at E are 160 cm above the ground at A. From physics, it is known that $\angle AME = \angle BMF$. If $AM = 120$ cm and $MB = 4.5$ m, find the height BF of the flagpole.

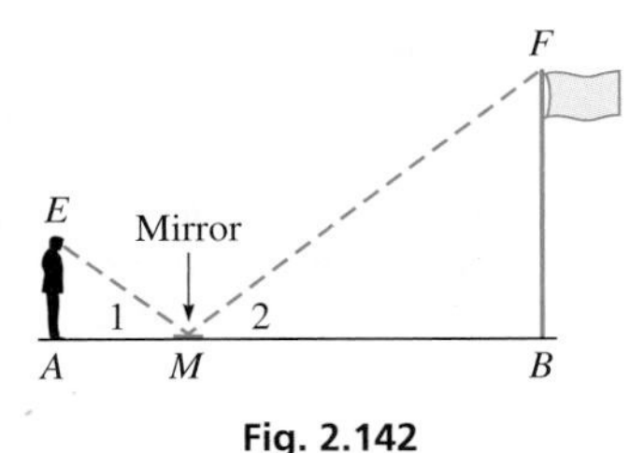

Fig. 2.142

64. A computer screen displays a circle inscribed in a square and a square inscribed in the circle. Find the ratio of (a) the area of the inner square to the area of the outer square, (b) the perimeter of the inner square to the perimeter of the outer square.

65. A typical scale for an aerial photograph is 1/18450. In an 8.00-by 10.0-in. photograph with this scale, what is the longest distance (in mi) between two locations in the photograph?

66. For a hydraulic press, the mechanical advantage is the ratio of the large piston area to the small piston area. Find the mechanical advantage if the pistons have diameters of 3.10 cm and 2.25 cm.

67. The diameter of the Earth is 7920 mi, and a satellite is in orbit at an altitude of 210 mi. How far does the satellite travel in one rotation about the Earth?

68. The roof of the Louisiana Superdome in New Orleans is supported by a circular steel tension ring 651 m in circumference. Find the area covered by the roof.

69. A rectangular piece of wallboard with two holes cut out for heating ducts is shown in Fig. 2.143. What is the area of the remaining piece?

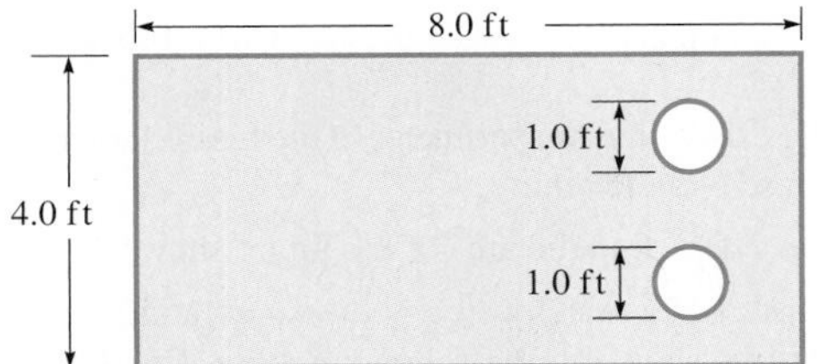

Fig. 2.143

70. The diameter of the sun is 1.38×10^6 km, the diameter of the Earth is 1.27×10^4 km, and the distance from the Earth to the sun (center to center) is 1.50×10^8 km. What is the distance from the center of the Earth to the end of the shadow due to the rays from the sun?

71. Using aerial photography, the width of an oil spill is measured at 250-m intervals, as shown in Fig. 2.144. Using Simpson's rule, find the area of the oil spill.

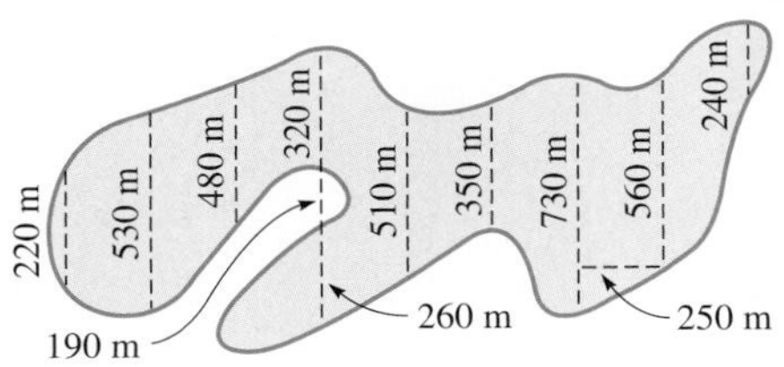

Fig. 2.144

72. To build a highway, it is necessary to cut through a hill. A surveyor measured the cross-sectional areas at 250-ft intervals through the cut as shown in the following table. Using the trapezoidal rule, determine the volume of soil to be removed.

Dist. (ft)	0	250	500	750	1000	1250	1500	1750
Area (ft^2)	560	1780	4650	6730	5600	6280	2260	230

73. The Hubble space telescope is within a cylinder 4.3 m in diameter and 13 m long. What is the volume within this cylinder?

74. A horizontal cross section of a concrete bridge pier is a regular hexagon (six sides, all equal in length, and all internal angles are equal), each side of which is 2.50 m long. If the height of the pier is 6.75 m, what is the volume of concrete in the pier?

75. A railroad track 1000.00 ft long expands 0.20 ft (2.4 in.) during the afternoon (due to an increase in temperature of about 30°F). Assuming that the track cannot move at either end and that the increase in length causes a bend straight up in the middle of the track, how high is the top of the bend?

76. On level ground a straight guy wire is attached to a vertical antenna. The guy wire is anchored in the ground 15.6 ft from the base of the antenna, and is 4.0 ft longer than the distance up the pole where it is attached. How long is the guy wire?

77. On a straight east-west road, a man walks 1500 m to the east and a woman walks 600 m to the west until they meet. They turn south and walk to a point that is 1700 m (on a straight line) from his starting point. How far is she (on a straight line) from her starting point?

78. A basketball court is 44 ft longer than it is wide. If the perimeter is 288 ft, what are the length and width?

79. A hot-water tank is in the shape of a right circular cylinder surmounted by a hemisphere as shown in Fig. 2.145. How many gallons does the tank hold? (1.00 ft^3 contains 7.48 gal.)

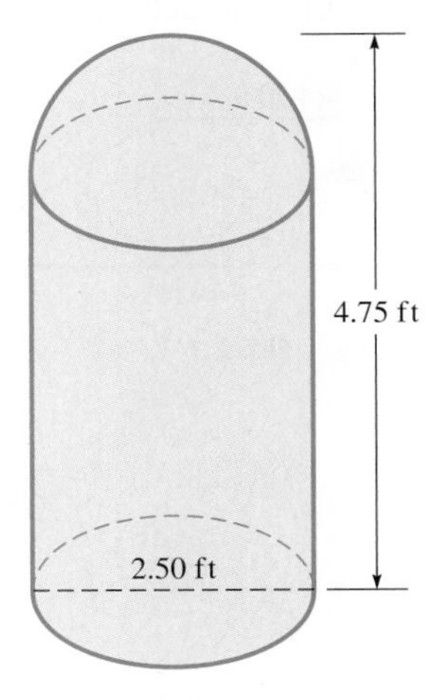

Fig. 2.145

80. A tent is in the shape of a regular pyramid surmounted on a cube. If the edge of the cube is 2.50 m and the total height of the tent is 3.25 m, find the area of the material used in making the tent (not including any floor area).

81. The diagonal of a TV screen is 152 cm. If the ratio of the width of the screen to the height of the screen is 16 to 9, what are the width and height?

82. An avid math student wrote to a friend, "My sailboat has a right triangular sail with edges (in ft) of $3k - 1$, $4k + 3$, and $5k + 2$. Can you tell me the area of the sail if $k > 1$?"

83. The friend in Exercise 82 replied, "Yes, and I have two cups that hold exactly the same amount of water, one of which is cylindrical and the other is hemispherical. If I tell you the height of the cylinder, can you tell me the radius of each, if the radii are equal?"

84. A satellite is 1590 km directly above the center of the eye of a circular (approximately) hurricane that has formed in the Atlantic Ocean. The distance from the satellite to the edge of the hurricane is 1620 km. What area does the hurricane cover? Neglect the curvature of the Earth and any possible depth of the hurricane.

85. The Pentagon, headquarters of the U.S. Department of Defense, is the world's largest office building. It is a regular pentagon (five sides, all equal in length, and all interior angles are equal) 921 ft on a side, with a diagonal of length 1490 ft. Using these data, draw a sketch and write one or two paragraphs to explain how to find the area covered within the outside perimeter of the Pentagon. (What is the area?)

CHAPTER 2 PRACTICE TEST

As a study aid, we have included complete solutions for each Practice Test problem at the back of this book.

1. In Fig. 2.146, determine $\angle 1$.
2. In Fig. 2.146, determine $\angle 2$.

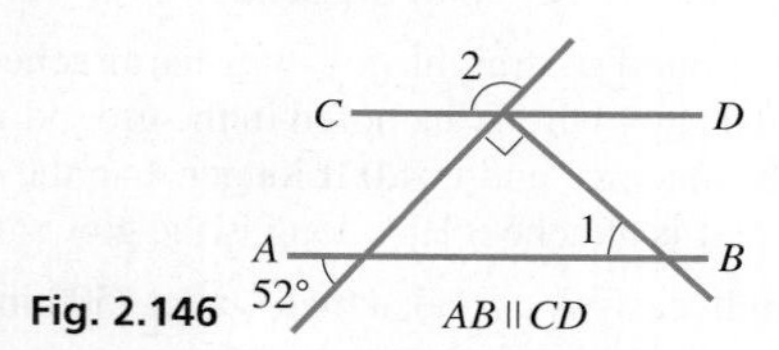

Fig. 2.146

3. A tree is 8.0 ft high and casts a shadow 10.0 ft long. At the same time, a telephone pole casts a shadow 25.0 ft long. How tall is the pole?
4. Find the area of a triangular wall pennant of sides 24.6 cm, 36.5 cm, and 40.7 cm.
5. What is the diagonal distance between corners of a rectangular field 125 ft wide and 170 ft long?
6. An office building hallway floor is designed in the trapezoidal shape shown in Fig. 2.147. What is the area of the hallway?

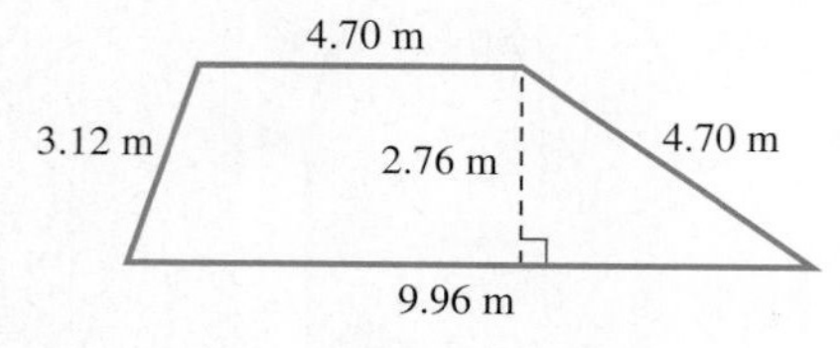

Fig. 2.147

7. What is (a) the mass (in kg) of a cubical block of ice, the edge of which is 0.40 m (the density of ice is 0.92×10^3 kg/m^3), and (b) the surface area of the block?
8. Find the surface area of a tennis ball whose circumference is 21.0 cm.
9. Find the volume of a right circular cone of radius 2.08 m and height 1.78 m.
10. In Fig. 2.148, find $\angle 1$.
11. In Fig. 2.148, find $\angle 2$.

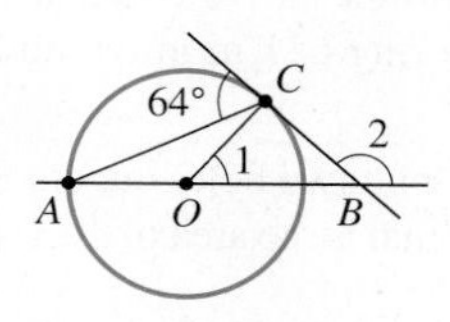

Fig. 2.148

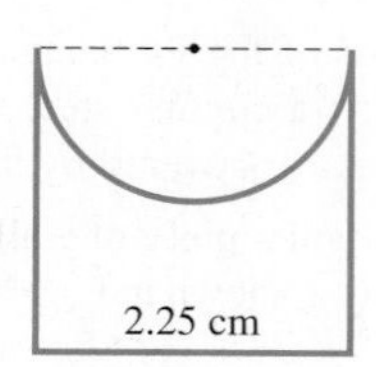

Fig. 2.149

12. In Fig. 2.149, find the perimeter of the figure shown. It is a square with a semicircle removed.
13. In Fig. 2.149, find the area of the figure shown.
14. The width of a marshy area is measured at 50-ft intervals, with the results shown in the following table. Using the trapezoidal rule, find the area of the marsh. (All data accurate to two or more significant digits.)

Distance (ft)	0	50	100	150	200	250	300
Width (ft)	0	90	145	260	205	110	20

Functions and Graphs

3

LEARNING OUTCOMES

After completion of this chapter, the student should be able to:

- Define a function and distinguish between dependent and independent variables
- Use mathematical functional notation
- Interpret a function as a process
- Determine domain and range for a function
- Graph a function using the rectangular coordinate system
- Use a graphing calculator to solve an equation graphically
- Graph and interpret sets of data values

By noting, for example, that stones fall faster than leaves, the Greek philosopher Aristotle (about 350 B.C.E.) reasoned that heavier objects fall faster than lighter ones.

For about 2000 years, this idea was generally accepted. Then in about 1600, the Italian scientist Galileo showed, by dropping objects from the Leaning Tower of Pisa, that the distance an object falls in a given time does not depend on its weight.

Galileo is generally credited with first using the *experimental method* by which controlled experiments are used to study natural phenomena. His aim was to find mathematical formulas that could be used to describe these phenomena. He realized that such formulas provided a way of showing a compact and precise relation between the variables.

In technology and science, determining how one quantity depends on others is a primary goal. A rule that shows such a relation is of great importance, and in mathematics such a rule is called a *function*. This chapter starts with a discussion of functions.

Examples of such relations in technology and science, as well as in everyday life, are numerous. Plant growth depends on sunlight and rainfall; traffic flow depends on roadway design; the sales tax on an item depends on the cost of an item; the time to access the Internet depends on how fast a computer processes data; distance traveled depends on time and speed of travel; electric voltage depends on the current and resistance. These are but a few of the innumerable possibilities.

A way of actually seeing how one quantity depends on another is by means of a *graph*. The basic method of graphing was devised by the French mathematicians Descartes and Fermat in the 1630s from their work to combine the methods of algebra and geometry. Their work was very influential in later developments in mathematics and technology. In this chapter, we discuss methods for graphing functions including the use of the graphing calculator.

◀ The power P that a wind turbine can extract from the wind depends on the velocity v of the wind according to the equation $P = \frac{1}{2}\rho C_p A v^3$. In Section 3.4, we will make a graph of this relationship for a specific wind turbine.

3.1 Introduction to Functions

Definition of Function • Independent and Dependent Variables • Functional Notation

In many applications, it is important to determine how two different variables are related, where the value of one variable depends on the other. For example, the experiments performed by Galileo on free-falling objects led to the discovery of the formula $s = 16t^2$, where s is the distance an object falls (in ft) and t is the time (in s). In this case, the distance depends on the time. Another example is that electrical measurements of voltage V and current I through a particular resistor would show that $V = kI$, where k is a constant. Here, the voltage depends on the current. In mathematics, whenever the value of one variable depends on the value of another, we call this relationship a ***function.***

Definition of a Function

A **function** is a relationship between two variables such that for every value of the first, there is only one corresponding value of the second.

The first variable is called the **independent variable,** and the second variable is called the **dependent variable.** We say that the dependent variable *is a function of* the independent variable.

CAUTION In a function, there may be only one value of the dependent variable for each value of the independent variable. ■

The first variable is termed *independent* because we can assign *any* reasonable value to it. The second variable is termed *dependent* because its value will depend on the choice of the independent variable. *Values of the independent variable and dependent variable are to be real numbers.* Therefore, there may be restrictions on their possible values. This is discussed in Section 3.2.

EXAMPLE 1 Examples of functions

(a) In the equation $y = 2x$, y is a function of x, because for each value of x there is only one value of y. For example, if $x = 3$, then $y = 6$ and no other value. By arbitrarily assigning values to x, and substituting, we see that the values of y we obtain *depend* on the values of x. Therefore, x is the independent variable, and y is the dependent variable.

(b) Figure 3.1 shows a cube of edge e. From Chapter 2, we know that the volume V in terms of the edge is $V = e^3$. Here, V is a function of e. The dependent variable is V and the independent variable is e. ■

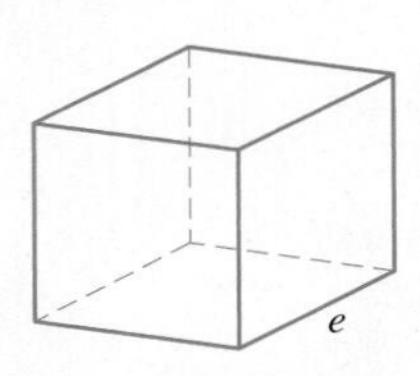

Fig. 3.1

EXAMPLE 2 Independent and dependent variables

The distance d (in km) a car travels in t hours at 75 km/h is $d = 75t$. Here t is the independent variable and d is the dependent variable, because we would choose values of t to find values of d.

If, however, the equation is written as $t = 75/d$, then d is the independent variable and t is the dependent variable, because we would choose values of d to find values of t. Note that by solving for t, the independent and dependent variables are interchanged. ■

■ A function is generally written in the form: dependent variable = an expression that involves the independent variable.

There are many ways to express functions, including formulas, tables, charts, and graphs. Functions will be important throughout this book, and we will use a number of different types of functions in later chapters.

FUNCTIONAL NOTATION

For convenience of notation, the phrase

"function of x" is written as $f(x)$.

This means that "y is a function of x" may be written as $y = f(x)$.

CAUTION Here, f denotes *dependence* and does not represent a quantity or a variable. Therefore, it also follows that $f(x)$ does *not* mean f times x. ■

EXAMPLE 3 Functional notation $f(x)$

If $y = 6x^3 - 5x$, we say that y is a function of x. It is also common to write such a function as $f(x) = 6x^3 - 5x$. Note that y and $f(x)$ both represent the same expression, $6x^3 - 5x$. However, the notation $f(x)$ is more descriptive because it shows the x-value that corresponds to y. For example, $f(2) = 38$ gives more information than $y = 38$ because it shows that when $x = 2$, $y = 38$. In general,

the value of the function $f(x)$ when $x = a$ is written as $f(a)$. ■

EXAMPLE 4 Meaning of $f(a)$

For a function $f(x)$, the value of $f(x)$ for $x = 2$ may be expressed as $f(2)$. Thus, substituting 2 for x in $f(x) = 3x - 7$, we have

$$f(2) = 3(2) - 7 = -1 \quad \text{substitute 2 for } x$$

The value of $f(x)$ for $x = -1.4$ is

$$f(-1.4) = 3(-1.4) - 7 = -11.2 \quad \text{substitute } -1.4 \text{ for } x$$ ■

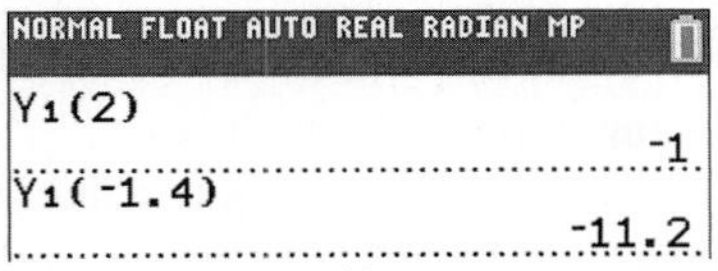

NORMAL FLOAT AUTO REAL RADIAN MP

Y1(2)
-1
Y1(-1.4)
-11.2

(a)

Graphing calculator keystrokes: goo.gl/CJQmwO

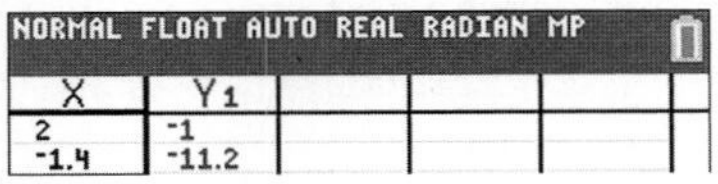

NORMAL FLOAT AUTO REAL RADIAN MP

X	Y1
2	-1
-1.4	-11.2

(b)

Graphing calculator keystrokes: goo.gl/McvaWD

Fig. 3.2

A calculator can be used to evaluate a function in several ways. One is to directly substitute the value into the function. A second is to enter the function as Y_1 and evaluate as shown in Fig. 3.2(a). A third way, which is very useful when many values are to be used, is to enter the function as Y_1 and use the *table* feature as shown in Fig. 3.2(b).

We must note that whatever number the variable a represents, *to find $f(a)$, we substitute a for x in $f(x)$*. This is true even if a is a literal number, as in the following examples.

EXAMPLE 5 Meaning of $f(a)$

■ TI-89 graphing calculator keystrokes for Example 5: goo.gl/2mz5nS

$$\text{If } g(t) = \frac{t^2}{2t + 1}, \text{ then } g(a^3) = \frac{(a^3)^2}{2(a^3) + 1} = \frac{a^6}{2a^3 + 1} \quad \text{substitute } a^3 \text{ for } t$$

$$\text{If } F(y) = 5 - 4y^2, \text{ then } F(a + 1) = 5 - 4(a + 1)^2 \quad \text{substitute } (a + 1) \text{ for } y$$

$$= 5 - 4(a^2 + 2a + 1) = -4a^2 - 8a + 1$$

$$\text{but } F(a) + 1 = (5 - 4a^2) + 1 = 6 - 4a^2$$

NOTE ▸ [Carefully note the difference between $F(a + 1)$ and $F(a) + 1$.] ■

Practice Exercise

1. For the function in Example 5, find $g(-1)$.

EXAMPLE 6 Functional notation—electric resistance

The electric resistance R of a certain resistor as a function of the temperature T (in °C is given by $R = 10.0 + 0.01T + 0.001T^2$. If a given temperature T is increased by 10°C, what is the value of R for the increased temperature as a function of T?

We are to determine R for a temperature of $T + 10$. Because

$$f(T) = 10.0 + 0.10T + 0.001T^2$$

then

$$f(T + 10) = 10.0 + 0.10(T + 10) + 0.001(T + 10)^2 \quad \text{substitute } T + 10 \text{ for } T$$

$$= 10.0 + 0.10T + 1.0 + 0.001T^2 + 0.02T + 0.1$$

$$= 11.1 + 0.12T + 0.001T^2$$ ■

At times, we need to define more than one function. We then use different symbols, such as $f(x)$ and $g(x)$, to denote these functions.

EXAMPLE 7 $f(x)$ and $g(x)$

For the functions $f(x) = 5x - 3$ and $g(x) = ax^2 + x$, where a is a constant, we have

$$f(-4) = 5(-4) - 3 = -23 \quad \text{substitute } -4 \text{ for } x \text{ in } f(x)$$

Practice Exercise

2. For the function in Example 7, find $f(-a^2x)$ and $g(-a^2x)$.

$$g(-4) = a(-4)^2 + (-4) = 16a - 4 \quad \text{substitute } -4 \text{ for } x \text{ in } g(x)$$ ■

A function may be viewed as a machine that takes an **input** (a value of the independent variable) and follows a set of instructions to transform it into an **output** (a value of the dependent variable). This is illustrated in the next example.

EXAMPLE 8 Function as a set of instructions

The function $f(x) = x^2 - 3x$ tells us to "square the input, multiply the input by 3, and subtract the second result from the first." An analogy would be a computer that was programmed so that when an input number was entered, it would square the number, then multiply the number by 3, and finally subtract the second result from the first. The result would be the output. This is diagramed in Fig. 3.3.

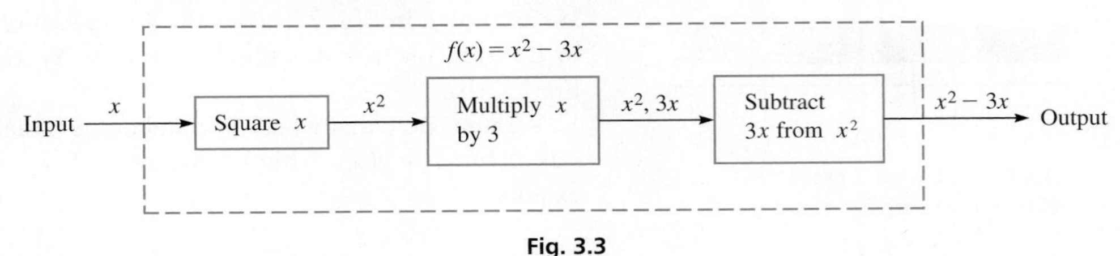

Fig. 3.3

The functions $f(t) = t^2 - 3t$ and $f(n) = n^2 - 3n$ are the same as the function $f(x) = x^2 - 3x$, because the operations performed on the independent variable are the same. [Although different literal symbols appear, this does not change the function.] ■

NOTE ▸

EXERCISES 3.1

In Exercises 1–4, solve the given problems related to the indicated examples of this section.

1. In Example 4, find the value of $f(-2)$.
2. In Example 5, evaluate $g(-a^2)$.
3. In Example 6, change "increased" to "decreased" in the second and third lines and then evaluate the function to find the proper expression.
4. In Example 8, change $x^2 - 3x$ to $x^3 + 4x$ and then determine the statements that should be placed in the three boxes in Fig. 3.3.

In Exercises 5–12, find the indicated functions.

5. Express the area A of a circle as a function of (a) its radius r and (b) its diameter d.
6. Express the circumference c of a circle as a function of (a) its radius r and (b) its diameter d.
7. Express the diameter d of a sphere as a function of its volume V.
8. Express the edge e of a cube as a function of its surface area A.
9. Express the area A of a square as a function of its diagonal d; express the diagonal d of a square as a function of its area A.
10. Express the perimeter p of a square as a function of its side s; express the side s of a square as a function of its perimeter p.
11. A circle is inscribed in a square (the circle is tangent to each side of the square). Express the total area A of the four corner regions bounded by the circle and the square as a function of the radius r of the circle.
12. Express the area A of an equilateral triangle as a function of its side s.

In Exercises 13–24, evaluate the given functions.

13. $f(x) = 2x + 1$; find $f(3)$ and $f(-5)$.
14. $f(x) = -x^2 - 9$; find $f(2)$ and $f(-2)$.
15. $f(x) = 6$; find $f(-2)$ and $f(0.4)$.
16. $f(T) = 7.2 - 2.5|T|$; find $f(2.6)$ and $f(-4)$.
17. $\phi(x) = \dfrac{6 - x^2}{2x}$; find $\phi(2\pi)$ and $\phi(-2)$.

18. $H(q) = \frac{8}{q} + 2\sqrt{q}$; find $H(4)$ and $H((-0.4)^2)$.

19. $g(t) = at^2 - a^2t$; find $g(\frac{1}{3})$ and $g(a)$.

20. $s(y) = 6\sqrt{y+11} - 3$; find $s(-2)$ and $s(a^2)$.

21. $K(s) = 3s^2 - s + 6$; find $K(-s+2)$ and $K(-s) + 2$.

22. $T(t) = 5t + 7$; find $T(2t + a - 1)$ and $T(2t) + a - 1$.

23. $f(x) = 8x + 3$; find $f(3x) - 3f(x)$.

24. $f(x) = 2x^2 + 1$; find $f(x+2) - [f(x) + 2]$.

In Exercises 25–28, evaluate the given functions. The values of the independent variable are approximate.

25. Given $f(x) = 5x^2 - 3x$, find $f(3.86)$ and $f(-6.92)$.

26. Given $g(t) = \sqrt{t + 1.0604} - 6t^3$, find $g(0.9261)$.

27. Given $F(H) = \frac{2H^2}{H + 36.85}$, find $F(-84.466)$.

28. Given $f(x) = \frac{x^4 - 2.0965}{6x}$, find $f(1.9654)$.

In Exercises 29–32, determine the function $y = f(x)$ that is represented by the given set of instructions.

29. Square the input, multiply this result by 3, and then subtract 4 from this result.

30. Cube the input, add 2 to this result, and then multiply this entire result by 5.

31. Cube the input, multiply the input by 7, and then subtract the second result from the first.

32. Square the input, divide the input by 4, and then add the second result to the first.

In Exercises 33–38, state the instructions of the function in words as in Example 8.

33. $f(x) = x^2 + 2$

34. $f(x) = 2x - 6$

35. $g(y) = 6y - y^3$

36. $\phi(s) = 8 - 5s + s^2$

37. $R(r) = 3(2r + 5)$

38. $f(z) = \frac{4z}{5 - z}$

In Exercises 39–42, write the equation as given by the statement. Then write the indicated function using functional notation.

39. The surface area A of a cubical open-top aquarium equals 5 times the square of an edge e of the aquarium.

40. A helicopter is at an altitude of 1000 m and is x m horizontally from a fire. Its distance d from the fire is the square root of the sum of 1000 squared and x squared.

41. The area A of the Bering Glacier in Alaska, given that its present area is 5200 km^2 and that it is melting at the rate of $120t$ km^2, where t is the time in centuries.

42. The electrical resistance R of a certain ammeter, in which the resistance of the coil is R_c, is the product of 10 and R_c divided by the sum of 10 and R_c.

In Exercises 43–52, solve the given problems.

43. A demolition ball is used to tear down a building. Its distance s (in m) above the ground as a function of time t (in s) after it is dropped is $s = 17.5 - 4.9t^2$. Because $s = f(t)$, find $f(1.2)$.

44. The length L (in ft) of a cable hanging between equal supports 100 ft apart is $L = 100(1 + 0.0003s^2)$, where s is the sag (in ft) in the middle of the cable. Because $L = f(s)$, find $f(15)$.

45. The stopping distance d (in ft) of a car going v mi/h is given by $d = v + 0.05v^2$. Because $d = f(v)$, find $f(30)$, $f(2v)$, and $f(60)$, using both $f(v)$ and $f(2v)$.

46. The electric power P (in W) dissipated in a resistor of resistance R (in Ω) is given by the function $P = \frac{200R}{(100 + R)^2}$. Because $P = f(R)$, find $f(R + 10)$ and $f(10R)$.

47. The pressure loss L (in lb/in.^{2}) of a fire hose as a function of its flow rate Q (in 100 gal/min) is $L = 1.2Q^2 + 1.5Q$. Find L for $Q = 450$ gal/min.

48. The sales tax in a city is 7.00%. The price tag on an item is D dollars. If the store has a 19% off price tag sale, express the total cost of the item as a function of D.

49. A motorist travels at 55 km/h for t hours and then at 65 km/h for $t + 1$ hours. Express the distance d traveled as a function of t.

50. A person considering the purchase of a car has two options: (1) purchase it for $35,000, or (2) lease it for three years for payments of $450/month plus $2000 down, with the option of buying the car at the end of the lease for $18,000. (a) For the lease option, express the amount P paid as a function of the numbers of months m. (b) What is the difference in the total cost if the person keeps the car for the three years, and then decides to buy it?

51. (a) Explain the meaning of $f[f(x)]$. (b) Find $f[f(x)]$ for $f(x) = 2x^2$.

52. If $f(x) = x$ and $g(x) = x^2$, find (a) $f[g(x)]$, and (b) $g[f(x)]$.

Answers to Practice Exercises

1. $f(-1) = -1$ 2. $f(-a^2x) = -5a^2x - 3$; $g(-a^2x) = a^5x^2 - a^2x$

3.2 More about Functions

Domain and Range • Functions from Verbal Statements • Relation

For a given function, *the complete set of possible values of the independent variable is called the* **domain** *of the function*, and the *complete set of all possible resulting values of the dependent variable is called the* **range** *of the function.*

NOTE ▶ As noted earlier, *we will use only* **real numbers** when working with functions. [This means there may be restrictions as to the values that may be used since values that lead to division by zero or to imaginary numbers may not be included in the domain or the range.]

EXAMPLE 1 Domain and range

The function $f(x) = x^2 + 2$ is defined for all real values of x. This means its domain is written as *all real numbers*. However, because x^2 is never negative, $x^2 + 2$ is never less than 2. We then write the range as *all real numbers* $f(x) \geq 2$, where the symbol $\geq$ means "is greater than or equal to."

The function $f(t) = \frac{1}{t+2}$ is not defined for $t = -2$, for this value would require division by zero. Also, no matter how large t becomes, $f(t)$ will never exactly equal zero. Therefore, the domain of this function is *all real numbers except* -2, and the range is *all real numbers except* 0. ■

■ Note that $f(2) = 6$ and $f(-2) = 6$. In a function, two different values of x may give the same value of y, but there may not be two different values of y for one value of x.

EXAMPLE 2 Domain and range

The function $g(s) = \sqrt{3 - s}$ is not defined for real numbers greater than 3, because such values make $3 - s$ negative and would result in imaginary values for $g(s)$. This means that the domain of this function is *all real numbers* $s \leq 3$, where the symbol $\leq$ means "is less than or equal to."

Also, because $\sqrt{3 - s}$ means the principal square root of $3 - s$ (see Section 1.6), we know that $g(s)$ cannot be negative. This tells us that the range of the function is *all real numbers* $g(s) \geq 0$. ■

Practice Exercise

1. Find the domain and range of the function $f(x) = \sqrt{x + 4}$.

In Examples 1 and 2, we found the domains by looking for values of the independent variable that cannot be used. The range was found through an inspection of the function. We generally use this procedure, although it may be necessary to use more advanced methods to find the range. This can be done by graphing the function as will be shown in future sections.

EXAMPLE 3 Find domain only

Find the domain of the function $f(x) = 16\sqrt{x} + \dfrac{1}{x}$.

From the term $16\sqrt{x}$, we see that x must be greater than or equal to zero in order to have real values. The term $\frac{1}{x}$ indicates that x cannot be zero, because of division by zero. Thus, putting these together, the domain is *all real numbers* $x > 0$.

As for the range, it is *all real numbers* $f(x) \geq 12$. More advanced methods are needed to determine this. ■

We have seen that the domain may be restricted because we do not use imaginary numbers or divide by zero. The domain may also be restricted by the definition of the function, or by practical considerations in an application.

EXAMPLE 4 Restricted domain

(a) A function defined as

$$f(x) = x^2 + 4 \qquad (\text{for } x > 2)$$

has a domain restricted to real numbers greater than 2 by definition. Thus, $f(5) = 29$, but $f(1)$ is not defined, because 1 is not in the domain. Also, the range is all real numbers greater than 8.

(b) The height h (in m) of a certain projectile as a function of the time t (in s) is

$$h = 20t - 4.9t^2$$

Negative values of time have no real meaning in this case. This is generally true in applications. Therefore, the domain is $t \geq 0$. Also, because we know the projectile will not continue in flight indefinitely, there is some upper limit on the value of t. These restrictions are not usually stated unless it affects the solution. There could be negative values of h if it is possible that the projectile is *below* the launching point at some time (such as a stone thrown from the top of a cliff). ■

Practice Exercise

2. For the function in Example 4(a), find $f(0)$ and $f(3)$ if defined.

The following example illustrates a function that is defined differently for different intervals of the domain. This kind of function is called a **piecewise-defined function.**

EXAMPLE 5 Piecewise-defined function

In a certain electric circuit, the current i (in mA) is a function of the time t (in s), which means $i = f(t)$. The function is

$$f(t) = \begin{cases} 8 - 2t & (\text{for } 0 \leq t \leq 4 \text{ s}) \\ 0 & (\text{for } t > 4 \text{ s}) \end{cases}$$

Because negative values of t are not usually meaningful, $f(t)$ is not defined for $t < 0$. Find the current for $t = 3$ s, $t = 6$ s, and $t = -1$ s.

We are to find $f(3)$, $f(6)$, and $f(-1)$, and we see that values of this function are determined differently depending on the value of t. Because 3 is between 0 and 4,

$$f(3) = 8 - 2(3) = 2 \quad \text{or} \quad i = 2 \text{ mA}$$

Because 6 is greater than 4, $f(6) = 0$, or $i = 0$ mA. We see that $i = 0$ mA for all values of t that are 4 or greater.

Because $f(t)$ is not defined for $t < 0$, $f(-1)$ is not defined. ■

FUNCTIONS FROM VERBAL STATEMENTS

To find a mathematical function from a verbal statement, we use methods like those for setting up equations in Chapter 1. In applied problems, the domain and range consist of the values of the independent and dependent variables that are *realistic* in the context of the problem. These are sometimes referred to as the **practical domain and range**. In this text, they will simply be called the domain and range.

EXAMPLE 6 Function from verbal statement

The fixed cost for a company to operate a certain plant is \$3000 per day. It also costs \$4 for each unit produced in the plant. Express the daily cost C of operating the plant as a function of the number n of units produced. Also find the domain and range of the function assuming that no more than 1000 units can be produced in one day.

The daily total cost C equals the fixed cost (\$3000) plus the cost of producing n units $(4n)$. Thus,

$$C = 3000 + 4n$$

Here, the domain is all values $0 \leq n \leq 1000$ since the number of units produced must be between 0 and 1000. The range is all values $3000 \leq C \leq 7000$ since the total cost is between \$3000 and \$7000. ■

EXAMPLE 7 Function from verbal statement

A metallurgist melts and mixes m grams (g) of solder that is 40% tin with n grams of another solder that is 20% tin to get a final solder mixture that contains 200 g of tin. Express n as a function of m. See Fig. 3.4.

The statement leads to the following equation:

$$\underset{\text{tin in first solder}}{0.40m} + \underset{\text{tin in second solder}}{0.20n} = \underset{\text{total amount of tin}}{200}$$

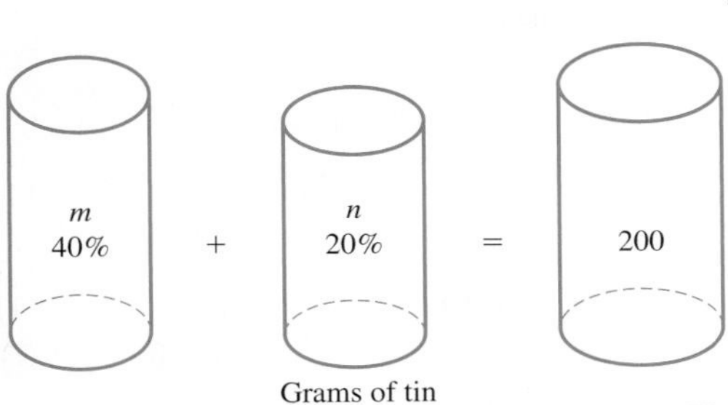

Fig. 3.4

Because we want $n = f(m)$, we now solve for n:

$$0.20n = 200 - 0.40m$$
$$n = 1000 - 2m$$

This is the required function. Note that neither m nor n can be negative. The domain is all values $0 \leq m \leq 500$ g, since m must be greater than or equal to 0 g and less than or equal to 500 g. The range is all values $0 \leq n \leq 1000$ g. ■

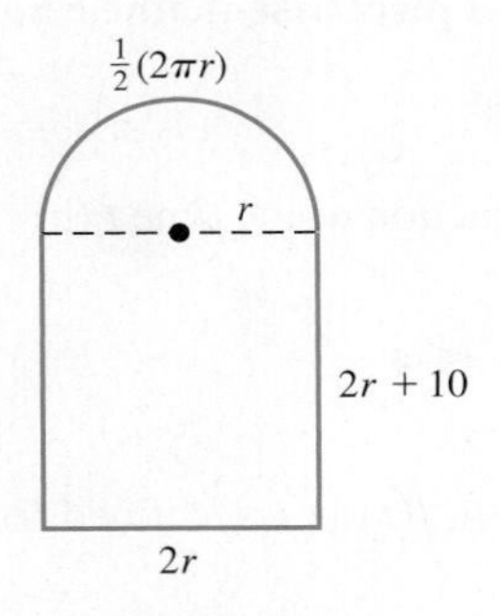

Fig. 3.5

EXAMPLE 8 Function from verbal statement

An architect designs a window in the shape of a rectangle with a semicircle on top, as shown in Fig. 3.5. The base of the window is 10 cm less than the height of the rectangular part. Express the perimeter p of the window as a function of the radius r of the circular part.

The perimeter is the distance around the window. Because the top part is a semicircle, the length of this top circular part is $\frac{1}{2}(2\pi r)$, and the base of the window is $2r$ because it is equal in length to the dashed line (the diameter of the circle). Finally, the base being 10 cm less than the height of the rectangular part tells us that each vertical side of the rectangle is $2r + 10$. Therefore, the perimeter p, where $p = f(r)$, is

$$\begin{aligned} p &= \tfrac{1}{2}(2\pi r) + 2r + 2(2r + 10) \\ &= \pi r + 2r + 4r + 20 \\ &= \pi r + 6r + 20 \end{aligned}$$

We see that the required function is $p = \pi r + 6r + 20$. Because the radius cannot be negative and there would be no window if $r = 0$, the domain of the function is all values $0 < r \leq R$, where R is a maximum possible value of r determined by design considerations. ■

Practice Exercise

3. In Example 8, find p as a function of r if there is to be a semicircular section of the window at the bottom, as well as at the top.

From the definition of a function, we know that any value of the independent variable must yield only one value of the dependent variable. *If a value of the independent variable yields one or more values of the dependent variable, the relationship is called a* **relation.** A function is a relation for which each value of the independent variable yields only one value of the dependent variable. Therefore, a function is a special type of relation. There are also relations that are not functions.

EXAMPLE 9 Relation

For $y^2 = 4x^2$, if $x = 2$, then y can be either 4 or -4. Because a value of x yields more than one value for y, we see that $y^2 = 4x^2$ is a relation, not a function. ■

EXERCISES 3.2

In Exercises 1–4, solve the given problems related to the indicated examples of this section.

1. In Example 1, in the first line, change x^2 to $-x^2$. What other changes must be made in the rest of the paragraph?
2. In Example 3, in the first line, change $\frac{1}{x}$ to $\frac{1}{x-1}$. What other changes must be made in the first paragraph?
3. In Example 5, find $f(2)$ and $f(5)$.
4. In Example 7, interchange 40% and 20% and then find the function.

In Exercises 5–14, find the domain and range of the given functions. In Exercises 11 and 12, explain your answers.

5. $f(x) = x + 5$
6. $g(u) = 3 - 4u^2$
7. $G(R) = \dfrac{3.2}{R}$
8. $F(r) = \sqrt{r + 4}$
9. $f(s) = \sqrt{s - 2}$
10. $T(t) = 2t^4 + t^2 - 1$
11. $H(h) = 2h + \sqrt{h} + 1$
12. $f(x) = \dfrac{-6}{\sqrt{2 - x}}$
13. $y = |x - 3|$
14. $y = x + |x|$

In Exercises 15–20, find the domain of the given functions.

15. $Y(y) = \dfrac{y + 1}{\sqrt{y - 2}}$
16. $f(n) = \dfrac{n^2}{6 - 2n}$
17. $f(D) = \sqrt{D} + \dfrac{1}{D - 2}$
18. $g(x) = \dfrac{\sqrt{x - 2}}{x - 3}$
19. For $x = \dfrac{3}{y - 1}$, find y as a function of x, and the domain of $f(x)$.
20. For $f(x) = \dfrac{1}{\sqrt{x}}$, what is the domain of $f(x + 4)$?

In Exercises 21–24, evaluate the indicated functions.

$F(t) = 3t - t^2 \text{ (for } t \leq 2)$ $\quad h(s) = \begin{cases} 2s & (\text{for } s < -1) \\ s + 1 & (\text{for } s \geq -1) \end{cases}$

$f(x) = \begin{cases} x + 1 & (\text{for } x < 1) \\ \sqrt{x + 3} & (\text{for } x \geq 1) \end{cases}$ $\quad g(x) = \begin{cases} \frac{1}{x} & (\text{for } x \neq 0) \\ 0 & (\text{for } x = 0) \end{cases}$

21. Find $F(1)$, $F(2)$, and $F(3)$.
22. Find $h(-8)$ and $h(-0.5)$.
23. Find $f(1)$ and $f(-0.25)$.
24. Find $g(0.2)$ and $g(0)$.

In Exercises 25–38, determine the appropriate functions.

25. A motorist travels at 40 mi/h for t h, and then continues at 55 mi/h for 2 h. Express the total distance d traveled as a function of t.

26. Express the cost C of insulating a cylindrical water tank of height 2 m as a function of its radius r, if the cost of insulation is \$3 per square meter.

27. A rocket burns up at the rate of 2 tons/min after falling out of orbit into the atmosphere. If the rocket weighed 5500 tons before reentry, express its weight w as a function of the time t, in minutes, of reentry.

28. A computer part costs \$3 to produce and distribute. Express the profit p made by selling 100 of these parts as a function of the price of c dollars each.

29. Upon ascending, a weather balloon ices up at the rate of 0.5 kg/m after reaching an altitude of 1000 m. If the mass of the balloon below 1000 m is 110 kg, express its mass m as a function of its altitude h if $h > 1000$ m.

30. A chemist adds x L of a solution that is 50% alcohol to 100 L of a solution that is 70% alcohol. Express the number n of liters of alcohol in the final solution as a function of x.

31. A company installs underground cable at a cost of \$500 for the first 50 ft (or up to 50 ft) and \$5 for each foot thereafter. Express the cost C as a function of the length l of underground cable if $l > 50$ ft.

32. The *mechanical advantage* of an inclined plane is the ratio of the length of the plane to its height. Express the mechanical advantage M of a plane of length 8 m as a function of its height h.

33. The capacities (in L) of two oil-storage tanks are x and y. The tanks are initially full; 1200 L is removed from them by taking 10% of the contents of the first tank and 40% of the contents of the second tank. (a) Express y as a function of x. (b) Find $f(400)$.

34. A city does not tax the first \$30,000 of a resident's income but taxes any amount over \$30,000 at 5%. (a) Find the tax T as a function of a resident's income I and (b) find $f(25{,}000)$ and $f(45{,}000)$.

35. A company finds that it earns a profit of \$15 on each cell phone and a profit of \$30 on each Blu-ray player that it sells. If x cell phones and y Blu-ray players are sold, the profit is \$2850. Express y as a function of x.

36. To subscribe to satellite radio, a customer is charged \$15 for the activation fee and then an additional \$18 per month. Express the total cost C as a function of the number of months of service x. If $C(x) = \$375$, find x.

37. For studying the electric current that is induced in wire rotating through a magnetic field, a piece of wire 60 cm long is cut into two pieces. One of these is bent into a circle and the other into a square. Express the total area A of the two figures as a function of the perimeter p of the square.

38. The cross section of an air-conditioning duct is in the shape of a square with semicircles on each side. See Fig. 3.6. Express the area A of this cross section as a function of the diameter d (in cm) of the circular part.

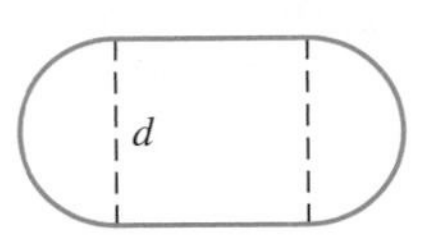

Fig. 3.6

In Exercises 39–52, solve the given problems.

39. A helicopter 120 m from a person takes off vertically. Express the distance d from the person to the helicopter as a function of the height h of the helicopter. What are the domain and the range of $d = f(h)$? See Fig. 3.7.

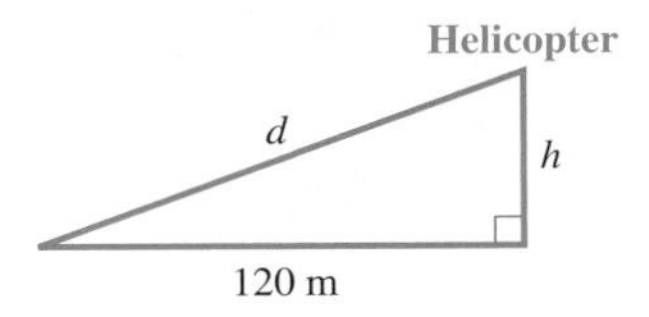

Fig. 3.7

40. A computer program displays a circular image of radius 6 in. If the radius is decreased by x in., express the area of the image as a function of x. What are the domain and range of $A = f(x)$?

41. A truck travels 300 km in $t - 3$ h. Express the average speed s of the truck as a function of t. What are the domain and range of $s = f(t)$?

42. A rectangular grazing range with an area of 8 mi² is to be fenced. Express the length l of the field as a function of its width w. What are the domain and range of $l = f(w)$?

43. The resonant frequency f of a certain electric circuit as a function of the capacitance is $f = \dfrac{1}{2\pi\sqrt{C}}$. Describe the domain.

44. A jet is traveling directly between Calgary, Alberta and Portland, Oregon, which are 550 mi apart. At any given time, x represents the jet's distance (in mi) from Calgary and y represents its distance from Portland. Find the domain and range of $y = f(x)$.

45. Using a piecewise-defined function, express the mass m of the weather balloon in Exercise 29 as a function of any height h.

46. Using a piecewise-defined function, express the cost C of installing any length l of the underground cable in Exercise 31.

47. A rectangular piece of cardboard twice as long as wide is to be made into an open box by cutting 2-in. squares from each corner and bending up the sides. (a) Express the volume V of the box as a function of the width w of the piece of cardboard. (b) Find the domain of the function.

48. A spherical buoy 36 in. in diameter is floating in a lake and is more than half above the water. (a) Express the circumference c of the circle of intersection of the buoy and water as a function of the depth d to which the buoy sinks. (b) Find the domain and range of the function.

49. For $f(x + 2) = |x|$, find $f(0)$.

50. For $f(x) = x^2$, find $\dfrac{f(x + h) - f(x)}{h}$.

51. What is the range of the function $f(x) = |x| + |x - 2|$?

52. For $f(x) = \sqrt{x - 1}$ and $g(x) = x^2$, find the domain of $g[f(x)]$. Explain.

Answers to Practice Exercises

1. Domain: all real numbers $x \geq -4$; Range: all real numbers $f(x) \geq 0$

2. $f(0)$ not defined, $f(3) = 13$ **3.** $p = 2\pi r + 4r + 20$

3.3 Rectangular Coordinates

Rectangular Coordinate System

■ Rectangular (Cartesian) coordinates were developed by the French mathematician René Descartes (1596–1650).

One of the most valuable ways of representing a function is by graphical representation. By using graphs, we can obtain a "picture" of the function; by using this picture, we can learn a great deal about the function.

To make a graphical representation of a function, recall from Chapter 1 that numbers can be represented by points on a line. For a function, we have values of the independent variable as well as the corresponding values of the dependent variable. Therefore, it is necessary to use two different lines to represent the values from each of these sets of numbers. We do this by placing the lines perpendicular to each other.

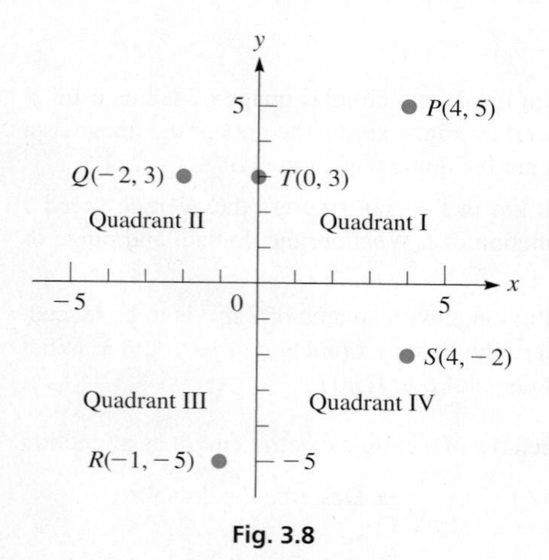

Fig. 3.8

Place one line horizontally and label it the **x-axis.** The values of the independent variable are normally placed on this axis. *The other line is placed vertically and labeled the* **y-axis.** Normally, the y-axis is used for values of the dependent variable. *The point of intersection is called the* **origin.** This is the **rectangular coordinate system.**

On the x-axis, positive values are to the right of the origin, and negative values are to the left of the origin. On the y-axis, positive values are above the origin, and negative values are below it. *The four parts into which the plane is divided are called* **quadrants,** which are numbered as in Fig. 3.8.

A point P in the plane is designated by the pair of numbers (x, y), where x is the value of the independent variable and y is the value of the dependent variable. *The x-value is the perpendicular distance of P from the y-axis, and the y-value is the perpendicular distance of P from the x-axis.* The values of x and y, written as (x, y), are the **coordinates** of the point P.

EXAMPLE 1 Locating points

(a) Locate the points $A(2, 1)$ and $B(-4, -3)$ on the rectangular coordinate system.

The coordinates $(2, 1)$ for A mean that the point is 2 units to the *right* of the y-axis and 1 unit *above* the x-axis, as shown in Fig. 3.9. The coordinates $(-4, -3)$ for B mean that the point is 4 units to the *left* of the y-axis and 3 units *below* the x-axis, as shown. The x-coordinate of A is 2, and the y-coordinate of A is 1. For point B, the x-coordinate is -4, and the y-coordinate is -3.

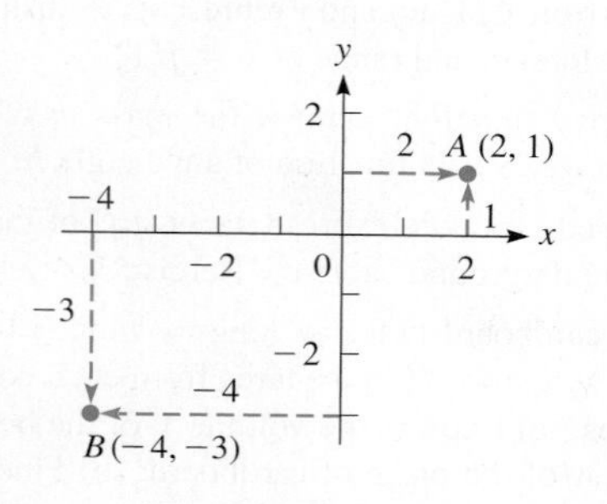

Fig. 3.9

(b) The positions of points $P(4, 5)$, $Q(-2, 3)$, $R(-1, -5)$, $S(4, -2)$, and $T(0, 3)$ are shown in Fig. 3.8. We see that this representation allows for *one point for any pair of values* (x, y). Also note that the point $T(0, 3)$ is on the y-axis. Any such point that is on either axis is not *in* any of the four quadrants. ■

NOTE ▸ [Note very carefully that the x-coordinate is always written first, and the y-coordinate is always written second. For this reason, (x, y) is called an **ordered pair.**] It is very important to keep the proper order when writing the coordinates of a point. (The x-coordinate is also known as the *abscissa,* and the y-coordinate is also known as the *ordinate.*)

EXAMPLE 2 Coordinates of vertices of rectangle

Three vertices of the rectangle in Fig. 3.10 are $A(-3, -2)$, $B(4, -2)$, and $C(4, 1)$. What is the fourth vertex?

We use the fact that opposite sides of a rectangle are equal and parallel to find the solution. Because both vertices of the base AB of the rectangle have a y-coordinate of -2, the base is parallel to the x-axis. Therefore, the top of the rectangle must also be parallel to the x-axis. Thus, the vertices of the top must both have a y-coordinate of 1, because one of them has a y-coordinate of 1. In the same way, the x-coordinates of the left side must both be -3. Therefore, the fourth vertex is $D(-3, 1)$. ■

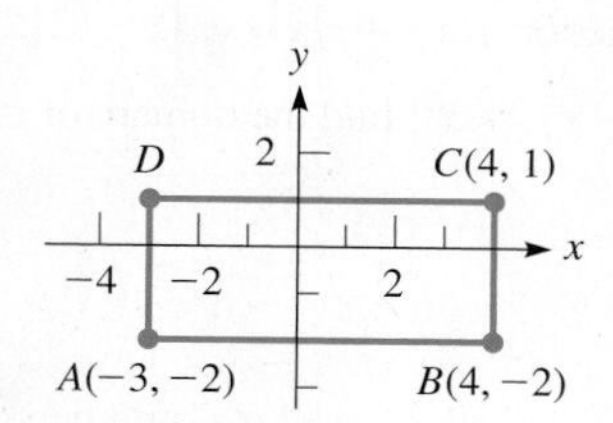

Fig. 3.10

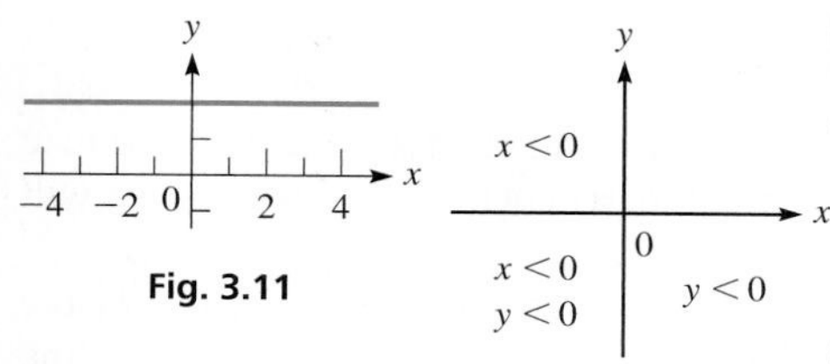

Fig. 3.11

Fig. 3.12

Practice Exercise

1. Where are all points for which $x = 0$ and $y > 0$?

EXAMPLE 3 Locating sets of points

(a) Where are all the points whose y-coordinates are 2?

We can see that the question could be stated as: "Where are all the points for which $y = 2$?" Because all such points are two units above the x-axis, the answer could be stated as "on a horizontal line 2 units above the x-axis." See Fig. 3.11.

(b) Where are all points (x, y) for which $x < 0$ and $y < 0$?

Noting that $x < 0$ means "x is less than zero," or "x is negative," and that $y < 0$ means the same for y, we want to determine where both x and y are negative. Our answer is "in the third quadrant," because both coordinates are negative for all points in the third quadrant, and this is the only quadrant for which this is true. See Fig. 3.12. ■

EXERCISES 3.3

In Exercises 1 and 2, make the given changes in the indicated examples of this section, and then solve the resulting problems.

1. In Example 2, change $A(-3, -2)$ to $A(0, -2)$ and then find the fourth vertex.

2. In Example 3(b), change $y < 0$ to $y > 0$ and then find the location of the points (x, y).

In Exercises 3 and 4, determine (at least approximately) the coordinates of the points shown in Fig. 3.13.

3. A, B, C **4.** D, E, F

In Exercises 5 and 6, plot the given points.

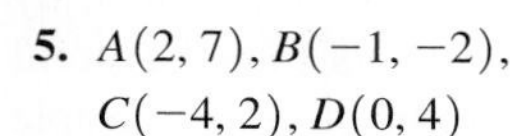

5. $A(2, 7), B(-1, -2), C(-4, 2), D(0, 4)$

6. $A(3, \frac{1}{2}), B(-6, 0), C(-\frac{5}{2}, -5), D(1, -3)$

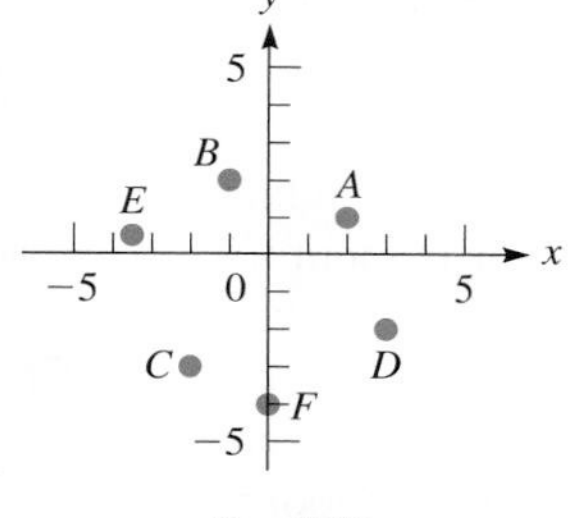

Fig. 3.13

In Exercises 7–10, plot the given points and then join these points, in the order given, by straight-line segments. Name the geometric figure formed.

7. $A(-1, 4), B(3, 4), C(2, -2), A(-1, 4)$

8. $A(0, 3), B(0, -1), C(4, -1), A(0, 3)$

9. $A(-2, -1), B(3, -1), C(3, 5), D(-2, 5), A(-2, -1)$

10. $A(-5, -2), B(4, -2), C(6, 3), D(-5, 3), A(-5, -2)$

In Exercises 11–14, find the indicated coordinates.

11. Three vertices of a rectangle are $(6, 3)$, $(-1, 3)$, and $(-1, -2)$. What are the coordinates of the fourth vertex?

12. Two vertices of an equilateral triangle are (7, 1) and (2, 1). What is the x-coordinate of the third vertex?

13. P is the point (3, 6). Locate point Q such that the x-axis is the perpendicular bisector of the line segment joining P and Q.

14. P is the point $(-4, 1)$. Locate point Q such that the line segment joining P and Q is bisected by the origin.

In Exercises 15–18, determine the quadrant in which the point (x, y) lies.

15. $x < 0$ and $y > 0$ **16.** $x > 0$ and $y < 0$

17. $x < 0$ and $y < 0$ **18.** $x > 0$ and $y > 0$

In Exercises 19–38, answer the given questions.

19. Where are all points whose x-coordinates are 1?

20. Where are all points whose y-coordinates are -3?

21. Where are all points such that $y = 3$?

22. Where are all points such that $|x| = 2$?

23. Where are all points whose x-coordinates equal their y-coordinates?

24. Where are all points whose x-coordinates equal the negative of their y-coordinates?

25. What is the x-coordinate of all points on the y-axis?

26. What is the y-coordinate of all points on the x-axis?

27. Where are all points for which $x > 0$?

28. Where are all points for which $y < 0$?

29. Where are all points for which $x < -1$?

30. Where are all points for which $y > 4$?

31. Where are all points for which $xy > 0$?

32. Where are all points for which $y/x < 0$?

33. Where are all points for which $xy = 0$?

34. Where are all points for which $x < y$?

35. If the point (a, b) is in the second quadrant, in which quadrant is $(a, -b)$?

36. The points $(3, -1)$, $(3, 0)$, and (x, y) are on the same straight line. Describe this line.

37. On a circular machine part, holes are to be drilled at the points $(2, 2)$, $(-2, 2)$, $(-2, -2)$, and $(2, -2)$, where $(0, 0)$ represents the center. Plot these points and find the distance between the points in quadrants I and III.

38. Join the points $(-1, -2)$, $(4, -2)$, $(7, 2)$, $(2, 2)$ and $(-1, -2)$ in order with straight-line segments. Find the distances between successive points and then identify the geometric figure formed.

Answer to Practice Exercise

1. On the positive y-axis

3.4 The Graph of a Function

Table of Values • Plotting Points • Be Careful about the Domain • Vertical-Line Test • Finding Domain and Range Graphically

Now that we have introduced the concepts of a function and the rectangular coordinate system, we are in a position to determine the graph of a function. In this way, we will obtain a visual representation of a function.

The graph of a function is the set of all points whose coordinates (x, y) satisfy the functional relationship $y = f(x)$. Because $y = f(x)$, we can write the coordinates of the points on the graph as $(x, f(x))$. Writing the coordinates in this manner tells us exactly how to find them. *We assume a certain value for x and then find the value of the function of x. These two numbers are the coordinates of the point.*

Because there is no limit to the possible number of points that can be chosen, we normally select a few values of x, obtain the corresponding values of the function, plot these points, and then join them. Therefore, we use the following basic procedure in plotting a graph.

■ As to just what values of x to choose and how many to choose, with a little experience you will usually be able to tell if you have enough points to plot an accurate graph.

Procedure for Plotting the Graph of a Function

1. Let x take on several values and calculate the corresponding values of y.
2. Tabulate these values, arranging the table so that values of x are increasing.
3. Draw the x- and y-axes, properly labeled, with an appropriate scale.
4. Plot the points and join them from left to right with a smooth curve.

EXAMPLE 1 Graphing a function by plotting points

Graph the function $f(x) = 3x - 5$.

For purposes of graphing, let $y = f(x)$, or $y = 3x - 5$. Then, let x take on various values and determine the corresponding values of y. Note that once we choose a given value of x, we have no choice about the corresponding y-value, as it is determined by evaluating the function. If $x = 0$, we find that $y = -5$. This means that the point $(0, -5)$ is on the graph of the function $3x - 5$. Choosing another value of x, for example $x = 1$, we find that $y = -2$. This means that the point $(1, -2)$ is on the graph of the function $3x - 5$. Continuing to choose a few other values of x, we tabulate the results, as shown in Fig. 3.14. It is best to arrange the table so that the values of x increase; then there is no doubt how they are to be connected, for they are then connected in the order shown. A graphing calculator can be used to quickly construct a table of ordered pairs by using its *table* feature (see Fig. 3.15). By plotting and connecting these points, we see that the graph of the function $3x - 5$ is a straight line. ■

■ We will show how a graph is displayed on a graphing calculator in the next section.

NORMAL FLOAT AUTO
PRESS + FOR △Tbl

X	Y1
-1	-8
0	-5
1	-2
2	1
3	4
4	7
5	10
6	13
7	16
8	19
9	22

X=-1

Fig. 3.15
Graphing calculator keystrokes: goo.gl/ii9YjS

Select values of x and calculate corresponding values of y

$f(x) = 3x - 5$
$f(-1) = 3(-1) - 5 = -8$
$f(0) = 3(0) - 5 = -5$
$f(1) = 3(1) - 5 = -2$
$f(2) = 3(2) - 5 = 1$
$f(3) = 3(3) - 5 = 4$

Tabulate with values of x increasing

x	y
−1	−8
0	−5
1	−2
2	1
3	4

Draw the axes with an appropriate scale, plot points, and join from left to right with smooth curve

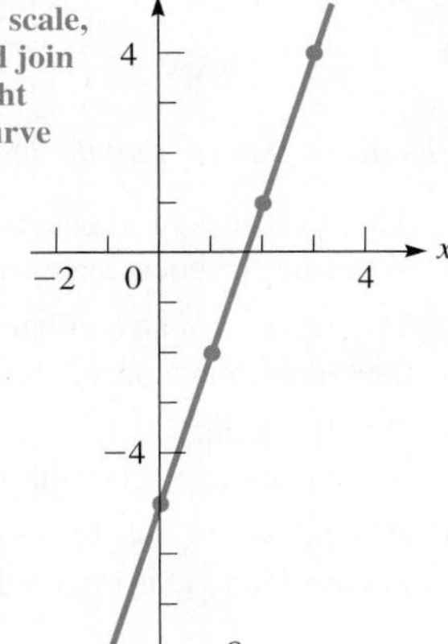

Fig. 3.14

EXAMPLE 2 Be careful: negative numbers

Graph the function $f(x) = 2x^2 - 4$.

First, let $y = 2x^2 - 4$ and tabulate the values as shown in Fig. 3.16. In determining the values in the table, take particular care to obtain the correct values of y for negative values of x. ***Mistakes are relatively common when dealing with negative numbers.*** We must carefully use the laws for signed numbers. For example, if $x = -2$, we have $y = 2(-2)^2 - 4 = 2(4) - 4 = 8 - 4 = 4$. Once the values are obtained, plot and connect the points with a smooth curve, as shown. ■

Select values of x and calculate corresponding values of y

$f(x) = 2x^2 - 4$
$f(-2) = 2(-2)^2 - 4 = 4$
$f(-1) = 2(-1)^2 - 4 = -2$
$f(0) = 2(0)^2 - 4 = -4$
$f(1) = 2(1)^2 - 4 = -2$
$f(2) = 2(2)^2 - 4 = 4$

Tabulate with values of x increasing

x	y
−2	4
−1	−2
0	−4
1	−2
2	4

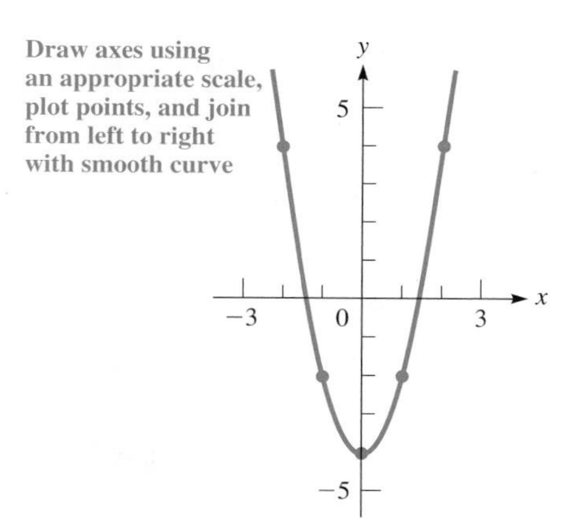

Fig. 3.16

When graphing a function, we must use extra care with certain parts of the graph. These include the following:

Special Notes on Graphing

1. Because the graphs of most common functions are smooth, *any place that the graph changes in a way that is not expected should be checked* with care. It usually helps to check values of x between which the question arises.
2. The domain of the function may not include all values of x. Remember, ***division by zero is not defined, and only real values of the variables may be used.***
3. In applications, we must *use only values of the variables that have meaning*. In particular, negative values of some variables, such as time, generally have no meaning.

The following examples illustrate these points.

EXAMPLE 3 Be careful: another point needed

Graph the function $y = x - x^2$.

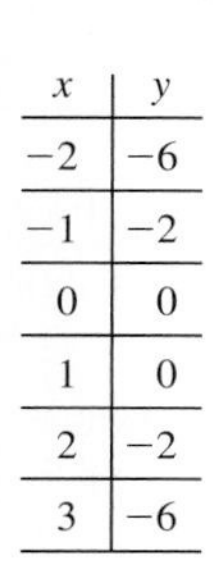

x	y
−2	−6
−1	−2
0	0
1	0
2	−2
3	−6

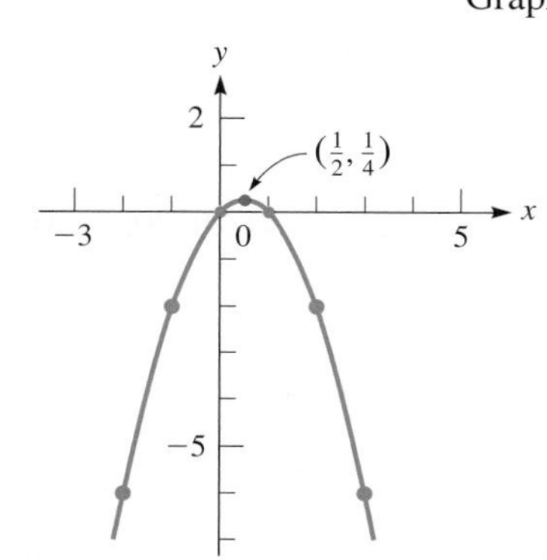

Fig. 3.17

First, we determine the values in the table, as shown with Fig. 3.17. Again, we must be careful when dealing with negative values of x. For the value $x = -1$, we have $y = (-1) - (-1)^2 = -1 - 1 = -2$. Once all values in the table have been found and plotted, note that $\boldsymbol{y = 0}$ ***for both*** $\boldsymbol{x = 0}$ ***and*** $\boldsymbol{x = 1}$**.** The question arises—***what happens between these values?*** Trying $x = \frac{1}{2}$, we find that $y = \frac{1}{4}$. Using this point completes the information needed to complete an accurate graph. ■

Note that in plotting these graphs, we do not stop the graph with the points we found, but indicate that the curve continues by drawing the graph past these points. However, this is not true for all graphs. As we will see in later examples, the graph should not include points for values of x not in the domain, and therefore may not continue past a particular point.

EXAMPLE 4 Division by zero is undefined

Graph the function $y = 1 + \frac{1}{x}$.

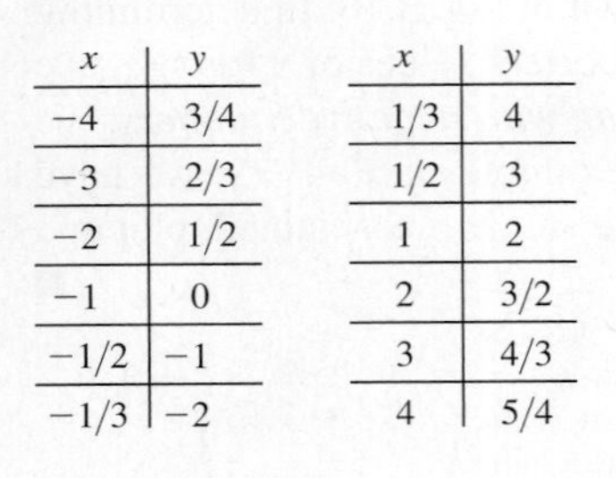

x	y
−4	3/4
−3	2/3
−2	1/2
−1	0
−1/2	−1
−1/3	−2

x	y
1/3	4
1/2	3
1	2
2	3/2
3	4/3
4	5/4

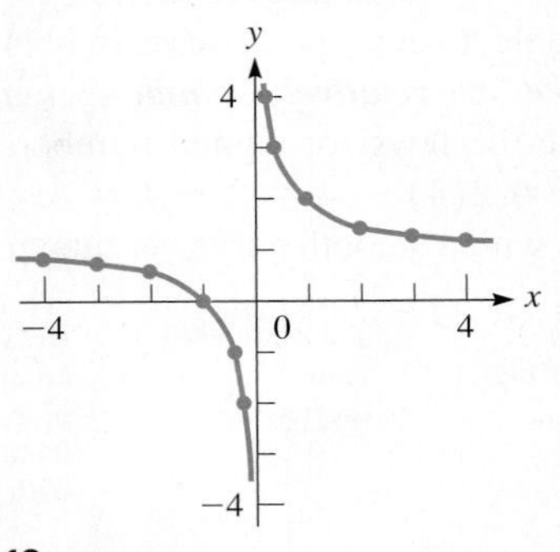

Fig. 3.18

In finding the points on this graph, as shown in Fig. 3.18, note that y is not defined for $x = 0$, due to division by zero. Thus, $x = 0$ is not in the domain, and no point on the graph will have an x-coordinate of zero. This means ***the curve will not cross the y-axis***. Although we cannot let $x = 0$, we can choose other values for x between -1 and 1 that are close to zero. In doing so, we find that as x gets closer to zero, the points get closer to the y-axis, although they do not reach or touch it. In this case, the y-axis is called an **asymptote** of the curve. ■

EXAMPLE 5 Be careful: Imaginary values

x	y
−1	0
0	1
1	1.4
2	1.7
3	2

x	y
4	2.2
5	2.4
6	2.6
7	2.8
8	3

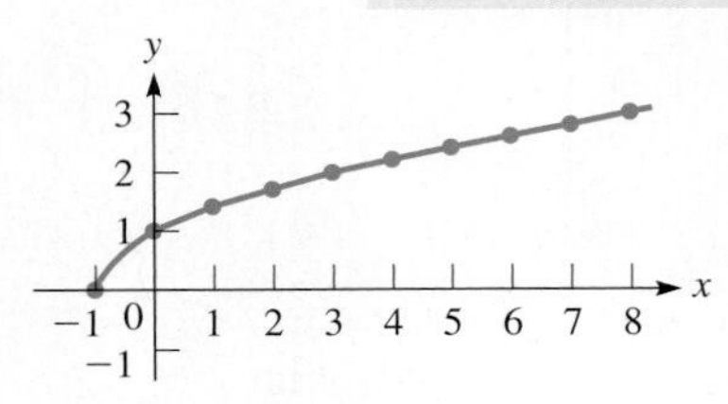

Fig. 3.19

Graph the function $y = \sqrt{x + 1}$.

When finding the points for the graph, we may not let x take on any value less than -1, for ***all such values would lead to imaginary values for y*** and are not in the domain. Also, because we have the positive square root indicated, the range consists of all values of y that are positive or zero $(y \geq 0)$. See Fig. 3.19. Note that the graph starts at $(-1, 0)$. ■

Practice Exercises

Determine for what value(s) of x there are no points on the graph.

1. $y = \frac{2x}{x + 5}$ **2.** $y = \sqrt{x - 3}$

Functions of a particular type have graphs with a certain basic shape, and many have been named. Three examples of this are the *straight line* (Example 1), the *parabola* (Examples 2 and 3), and the *hyperbola* (Example 4). We consider the straight line again in Chapter 5 and the parabola in Chapter 7. All of these graphs and others are studied in detail in Chapter 21. Other types of graphs are found in many of the later chapters.

EXAMPLE 6 Graph of wind turbine power

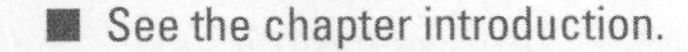

■ See the chapter introduction.

■ According to Betz's law, no turbine can capture more than 59.3% of the power in the wind. This is the maximum value of the power coefficient C_p.

The *extractable power* (in watts, W) in the wind hitting a wind turbine is given by $P = \frac{1}{2}\rho C_p A v^3$ where ρ is the air density (in kg/m^3), C_p is the power coefficient, A is the circular area swept by the blades (in m^2), and v is the wind velocity (in m/s). On a certain commercial wind turbine, $C_p = 0.400$ and the blades of the wind turbine are 51.5 m long. Assuming $\rho = 1.23$ kg/m^3, plot P as a function of v.

Substituting in the given values, the extractable power (in W) is given by

$$P = \tfrac{1}{2}\rho C_p A v^3 = \tfrac{1}{2}(1.23)(0.400)\pi(51.5)^2 v^3 = 2050v^3$$

Converting to kilowatts, we have $\boldsymbol{P = 2.050v^3}$ (in kW). Since the wind velocity cannot be negative, we will make a table of values beginning with a wind velocity of 0 m/s and going up to 10.0 m/s (a strong wind).

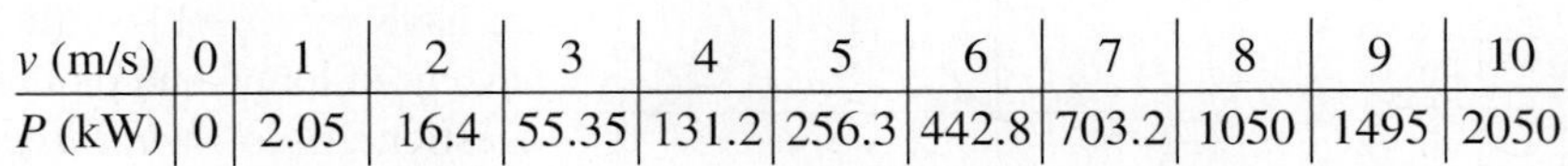

v (m/s)	0	1	2	3	4	5	6	7	8	9	10
P (kW)	0	2.05	16.4	55.35	131.2	256.3	442.8	703.2	1050	1495	2050

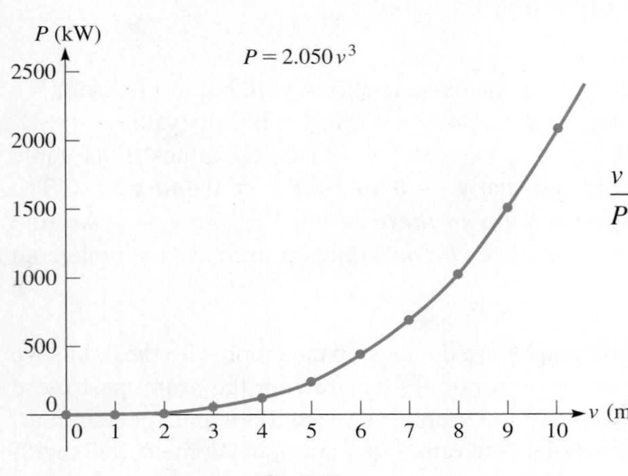

Fig. 3.20

The graph is shown in Fig. 3.20. Note that the scale on the P-axis is different from that on the v-axis. This is commonly done when the variables differ in magnitudes and ranges.

The graph and table both clearly show that for each successive increase in the wind velocity, the power increases by larger and larger amounts. For this reason, the wind velocity is an extremely important factor in determining the power generated by a wind turbine. ■

EXAMPLE 7 Piecewise-defined function

Graph the function $f(x) = \begin{cases} 2x + 1 & (\text{for } x \le 1) \\ 6 - x^2 & (\text{for } x > 1) \end{cases}$.

$f(-2) = 2(-2) + 1 = -3$
$f(-1) = 2(-1) + 1 = -1$
$f(0) = 2(0) + 1 = 1$
$f(1) = 2(1) + 1 = 3$
$f(2) = 6 - 2^2 = 2$
$f(3) = 6 - 3^2 = -3$

x	y
−2	−3
−1	−1
0	1
1	3
2	2
3	−3

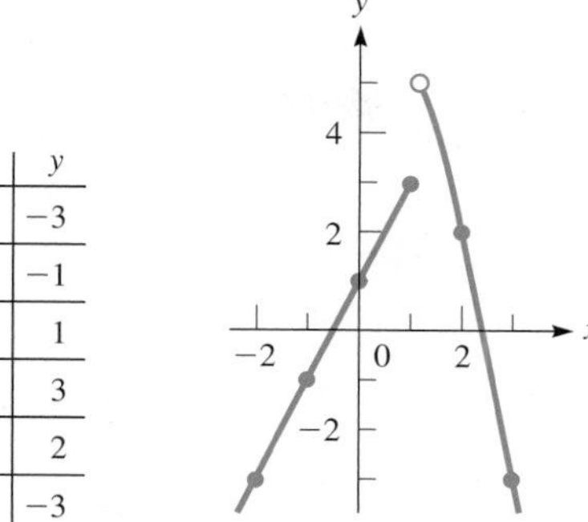

Fig. 3.21

First, let $y = f(x)$ and then tabulate the necessary values. In evaluating $f(x)$, we must be careful to use the proper part of the definition. To see where to start the curve for $x > 1$, we evaluate $6 - x^2$ for $x = 1$, but we must realize that ***the curve does not include this point*** (1, 5) and starts immediately to its right. To show that it is not part of the curve, draw it as an open circle. See Fig. 3.21.

A function such as this one, with a "break" in it, is called *discontinuous*. ■

A function is defined such that there is only one value of the dependent variable for each value of the independent variable, but a *relation* may have more than one such value of the dependent variable. The following test shows how to use a graph to determine whether or not a relation is also a function.

■ The graphs in Examples 1–7 in this section all represent functions since any vertical line crosses their graph at most once.

Vertical Line Test

The graph of a relation also represents a function if and only if **every vertical line crosses the graph at most once.**

EXAMPLE 8 Using the vertical line test

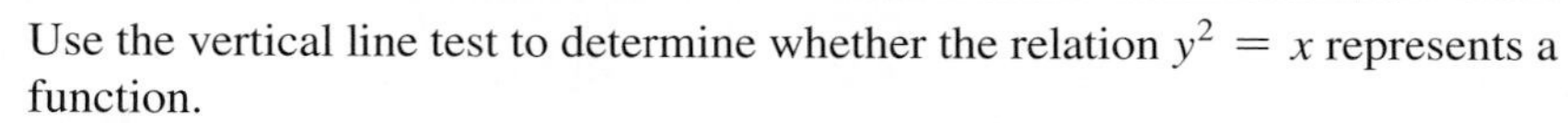

Use the vertical line test to determine whether the relation $y^2 = x$ represents a function.

By letting $x = 0$, 1, and 4, we get the table and graph shown in Fig. 3.22. Note that any positive value of x has *two* corresponding values of y. Since a vertical line crosses this graph twice, this relation is **not a function.** ■

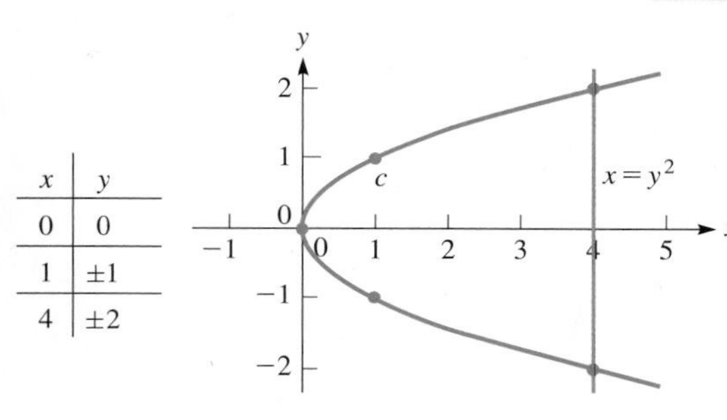

x	y
0	0
1	±1
4	±2

Fig. 3.22

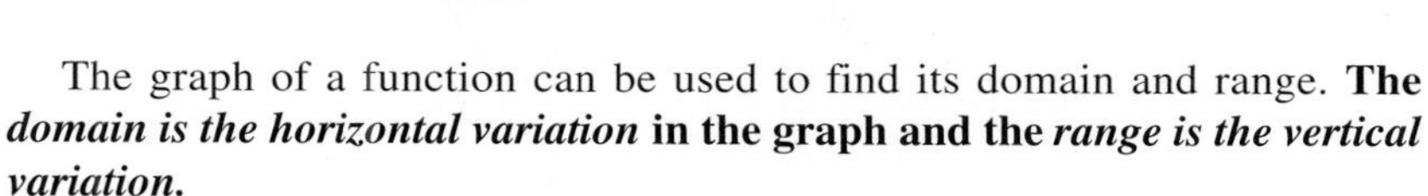

The graph of a function can be used to find its domain and range. **The *domain is the horizontal variation* in the graph and the *range is the vertical variation.***

EXAMPLE 9 Finding the domain and range graphically

For the functions graphed in blue in Fig. 3.23, the domain is shown in green and the range is shown in orange. A solid dot indicates the point is included, a hollow dot indicates the point is not included, and an arrow shows the graph continues indefinitely in that direction.

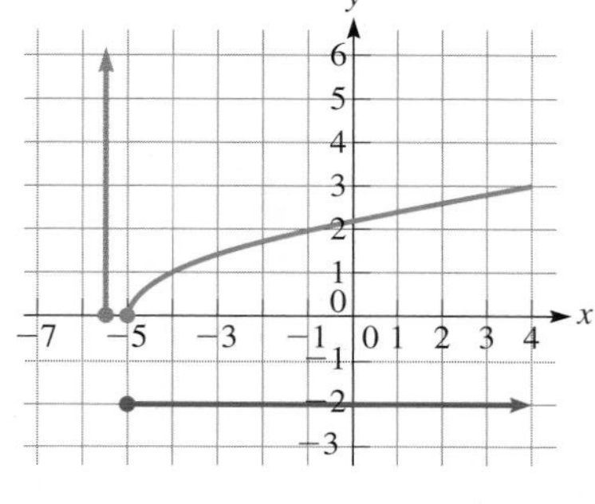

Domain: All values $x \ge -5$
Range: All values $y \ge 0$

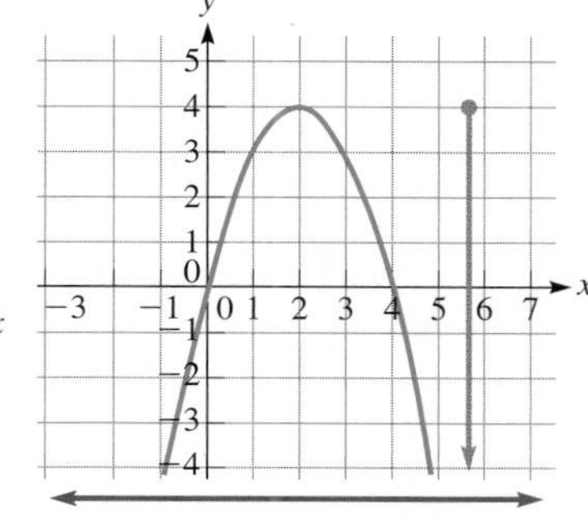

Domain: All real numbers
Range: All values $y \le 4$

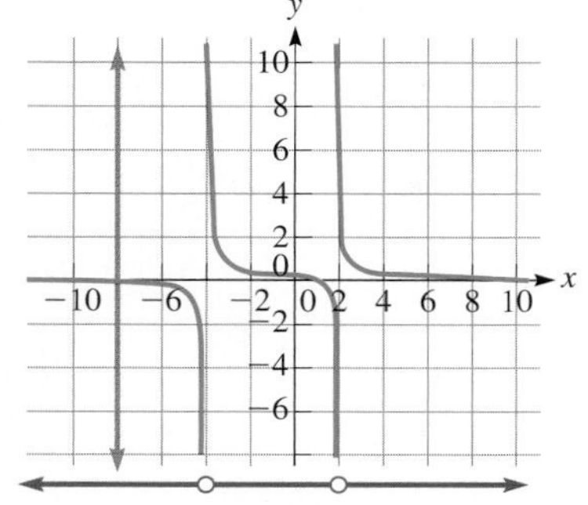

Domain: All values x except −4 and 2
Range: All real numbers

Fig. 3.23

NOTE ➧ [Note that a function's domain and range are the sets of x and y values respectively that its graph passes through.] ■

EXERCISES 3.4

In Exercises 1–4, make the given changes in the indicated examples of this section and then plot the graphs.

1. In Example 1, change the − sign to +.
2. In Example 2, change $2x^2 - 4$ to $4 - 2x^2$.
3. In Example 4, change the x in the denominator to $x - 1$.
4. In Example 5, change the + sign to −.

In Exercises 5–36, graph the given functions.

5. $y = 3x$
6. $y = -2x$
7. $y = 2x - 4$
8. $y = 4 - 3x$
9. $s = 7 - 2t$
10. $y = -3$
11. $y = \frac{1}{2}x - 3$
12. $A = 6 - \frac{1}{3}r$
13. $y = x^2$
14. $y = -2x^2$
15. $y = 6 - 2x^2$
16. $y = x^2 - 3$
17. $y = \frac{1}{2}x^2 + 2$
18. $y = 2x^2 + 1$
19. $y = x^2 + 2x$
20. $h = 20t - 5t^2$
21. $y = x^2 - 3x + 1$
22. $y = 2 + 3x + x^2$
23. $V = e^3$
24. $y = -2x^3$
25. $y = x^3 - x^2$
26. $L = 3e - e^3$
27. $D = v^4 - 4v^2$
28. $y = x^3 - x^4$
29. $P = \dfrac{8}{V} + 3$
30. $y = \dfrac{4}{x + 2}$
31. $y = \dfrac{4}{x^2}$
32. $p = \dfrac{1}{n^2 + 0.5}$
33. $y = \sqrt{9x}$
34. $y = \sqrt{4 - x}$
35. $v = \sqrt{16 - h^2}$
36. $y = \sqrt{x^2 - 16}$

In Exercises 37–40, use the graph to determine the domain and range of the given function.

37. Fig. 3.24(a)
38. Fig. 3.24(b)
39. Fig. 3.24(c)
40. Fig. 3.24(d)

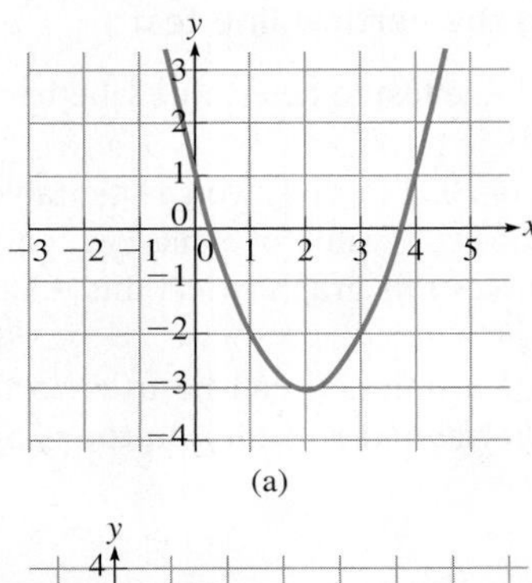

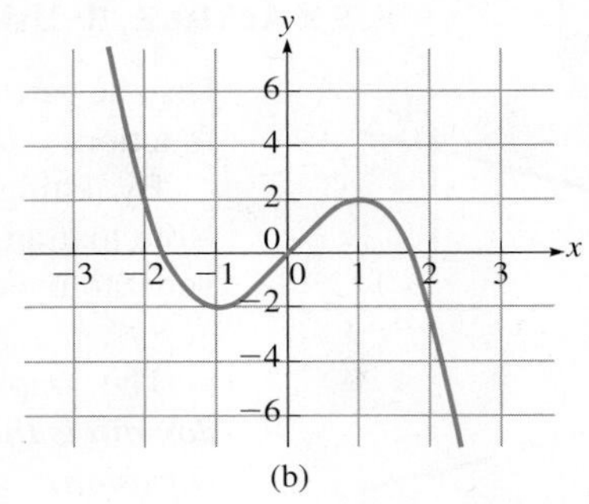

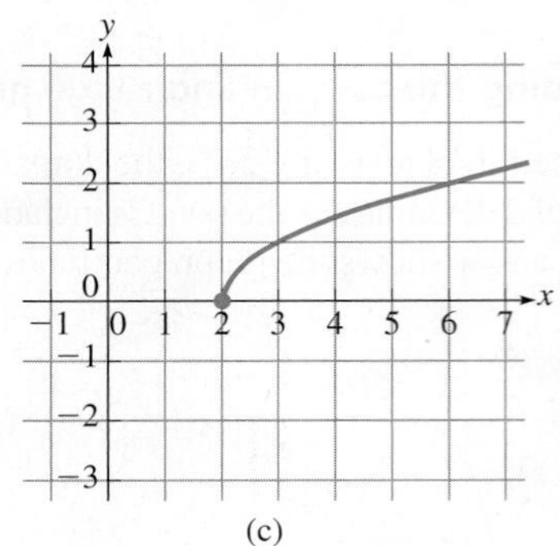

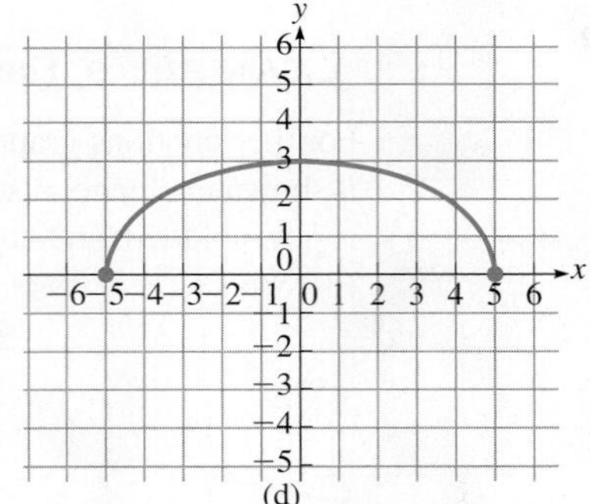

Fig. 3.24

In Exercises 41–70, graph the indicated functions.

41. In blending gasoline, the number of gallons n of 85-octane gas to be blended with m gal of 92-octane gas is given by the equation $n = 0.40m$. Plot n as a function of m.

42. The consumption of fuel c (in L/h) of a certain engine is determined as a function of the number r of r/min of the engine, to be $c = 0.011r + 40$. This formula is valid for 500 r/min to 3000 r/min. Plot c as a function of r. (r is the symbol for revolution.)

43. For a certain model of truck, its resale value V (in dollars) as a function of its mileage m is $V = 50{,}000 - 0.2m$. Plot V as a function of m for $m \leq 100{,}000$ mi.

44. The resistance R (in Ω) of a resistor as a function of the temperature T (in °C) is given by $R = 250(1 + 0.0032T)$. Plot R as a function of T.

45. The rate H (in W) at which heat is developed in the filament of an electric light bulb as a function of the electric current I (in A) is $H = 240I^2$. Plot H as a function of I.

46. The total annual fraction f of energy supplied by solar energy to a home as a function of the area A (in m^2) of the solar collector is $f = 0.065\sqrt{A}$. Plot f as a function of A.

47. The maximum speed v (in mi/h) at which a car can safely travel around a circular turn of radius r (in ft) is given by $r = 0.42v^2$. Plot r as a function of v.

48. The height h (in m) of a rocket as a function of the time t (in s) is given by the function $h = 1500t - 4.9t^2$. Plot h as a function of t, assuming level terrain.

49. The power P (in W/h) that a certain windmill generates is given by $P = 0.004v^3$, where v is the wind speed (in km/h). Plot the graph of P vs. v.

50. An astronaut weighs 750 N at sea level. The astronaut's weight at an altitude of x km above sea level is given by $w = 750\left(\frac{6400}{6400 + x}\right)^2$. Plot w as a function of x for $x = 0$ to $x = 8000$ km.

51. A formula used to determine the number N of board feet of lumber that can be cut from a 4-ft section of a log of diameter d (in in.) is $N = 0.22d^2 - 0.71d$. Plot N as a function of d for values of d from 10 in. to 40 in.

52. A guideline of the maximum affordable monthly mortgage M on a home is $M = 0.25(I - E)$, where I is the homeowner's monthly income and E is the homeowner's monthly expenses. If $E = \$600$, graph M as a function of I for $I = \$2000$ to $I = \$10{,}000$.

53. An airline requires that any carry-on bag has total dimensions (length + width + height) that do not exceed 45 in. For a carry-on that just meets this requirement and h as a length that is twice the width, express the volume V as a function of the width w. Draw the graph of $V = f(w)$.

54. A copper electrode with a mass of 25.0 g is placed in a solution of copper sulfate. An electric current is passed through the solution, and 1.6 g of copper is deposited on the electrode each hour. Express the total mass m on the electrode as a function of the time t and plot the graph.

55. A land developer is considering several options of dividing a large tract into rectangular building lots, many of which would have perimeters of 200 m. For these, the minimum width would be 30 m and the maximum width would be 70 m. Express the areas A of these lots as a function of their widths w and plot the graph.

56. The distance p (in m) from a camera with a 50-mm lens to the object being photographed is a function of the magnification m of the camera, given by $p = \frac{0.05(1 + m)}{m}$. Plot the graph for positive values of m up to 0.50.

57. A measure of the light beam that can be passed through an optic fiber is its numerical aperture N. For a particular optic fiber, N is a function of the index of refraction n of the glass in the fiber, given by $N = \sqrt{n^2 - 1.69}$. Plot the graph for $n \leq 2.00$.

58. $F = x^4 - 12x^3 + 46x^2 - 60x + 25$ is the force (in N) exerted by a cam on the arm of a robot, as shown in Fig. 3.25. Noting that x varies from 1 cm to 5 cm, plot the graph.

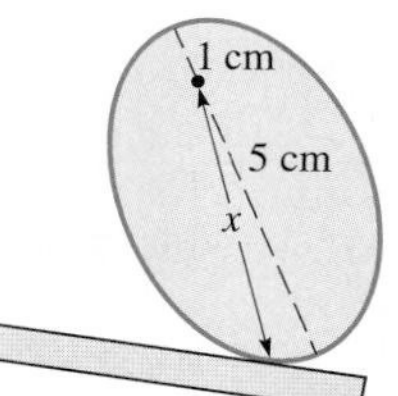

Fig. 3.25

59. The number of times S that a certain computer can perform a computation faster with a multiprocessor than with a uniprocessor is given by $S = \frac{5n}{4 + n}$, where n is the number of processors. Plot S as a function of n.

60. The voltage V across a capacitor in a certain electric circuit for a 2-s interval is $V = 2t$ during the first second and $V = 4 - 2t$ during the second second. Here, t is the time (in s). Plot V as a function of t.

61. Given that the point (1, 2) is on the graph of $y = f(x)$, must it be true that $f(2) = 1$? Explain.

62. In Fig. 3.26, part of the graph of $y = 2x^2 + 0.5$ is shown. What is the area of the rectangle, if its vertex is on the curve, as shown?

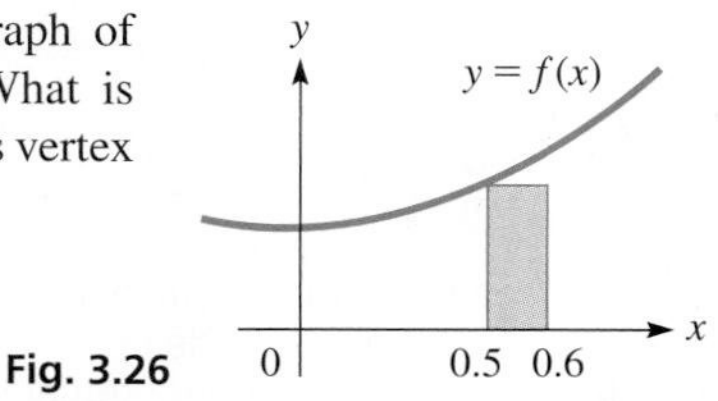

Fig. 3.26

63. On a taxable income of x dollars, a certain city's income tax T is defined as $T = 0.02x$ if $0 < x \leq 20{,}000$, $T = 400 + 0.03(x - 20{,}000)$ if $x > 20{,}000$. Graph $T = f(x)$ for $0 \leq x < 100{,}000$.

64. The temperature T (in °C) recorded on a day during which a cold front passed through a city was $T = 2 + h$ for $6 < h < 14$, $T = 16 - 0.5h$ for $14 \leq h < 20$, where h is the number of hours past midnight. Graph T as a function of h for $6 < h < 20$.

65. Plot the graphs of $y = x$ and $y = |x|$ on the same coordinate system. Explain why the graphs differ.

66. Plot the graphs of $y = 2 - x$ and $y = |2 - x|$ on the same coordinate system. Explain why the graphs differ.

67. Plot the graph of $f(x) = \begin{cases} 3 - x & (\text{for } x < 1) \\ x^2 + 1 & (\text{for } x \geq 1) \end{cases}$.

68. Plot the graph of $f(x) = \begin{cases} \frac{1}{x - 1} & (\text{for } x < 0) \\ \sqrt{x + 1} & (\text{for } x \geq 0) \end{cases}$.

69. Plot the graphs of (a) $y = x + 2$ and (b) $y = \dfrac{x^2 - 4}{x - 2}$. Explain the difference between the graphs.

70. Plot the graphs of (a) $y = x^2 - x + 1$ and (b) $y = \dfrac{x^3 + 1}{x + 1}$. Explain the difference between the graphs.

In Exercises 71–74, determine whether or not the indicated graph is that of a function.

71. Fig. 3.27(a)
72. Fig. 3.27(b)
73. Fig. 3.27(c)
74. Fig. 3.27(d)

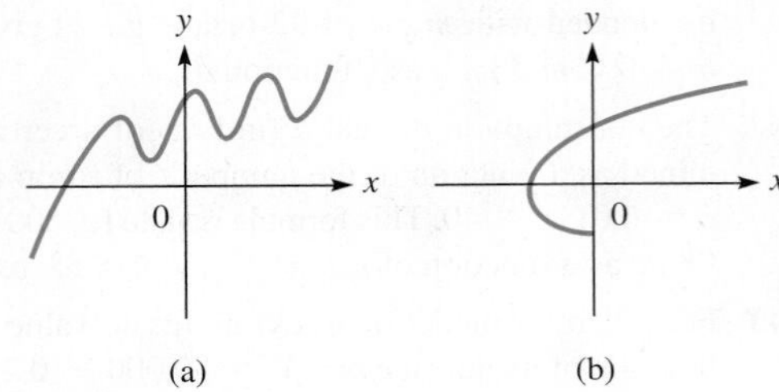

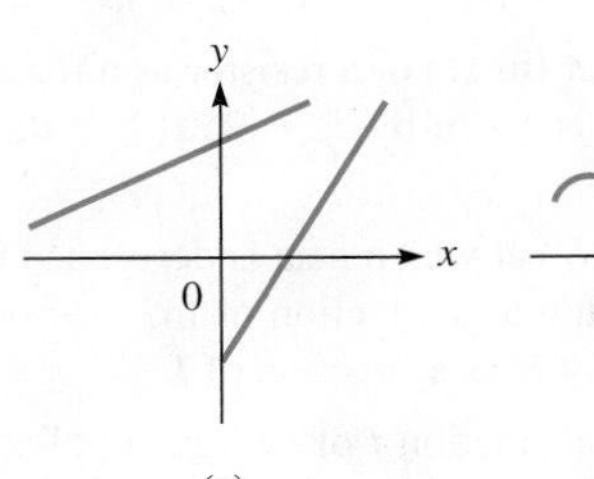

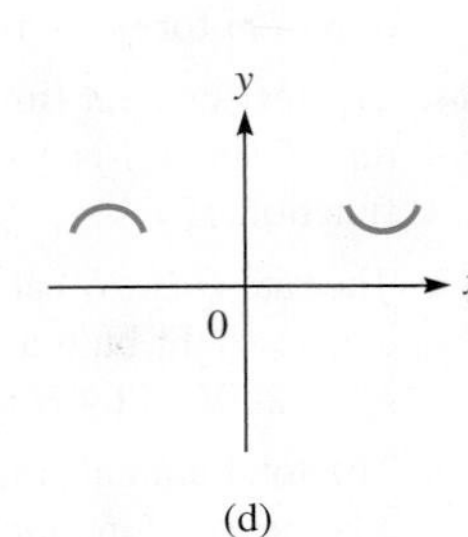

Fig. 3.27

Answers to Practice Exercises
1. $x = -5$ **2.** $x < 3$

3.5 Graphs on the Graphing Calculator

Calculator Setup for Graphing • Solving Equations Graphically • Finding the Maximum or Minimum from a Graph • Shifting a Graph

In this section, we will see that a graphing calculator can display a graph quickly and easily. As we have pointed out, to use your calculator effectively, *you must know the sequence of keys* for any operation you intend to use. The manual for any particular model should be used for a detailed coverage of its features.

EXAMPLE 1 Entering the function: window settings

To graph the function $y = 2x + 8$, first display $Y_1 =$ and then enter the $2x + 8$. The display for this is shown in Fig. 3.28(a).

Next use the *window* (or *range*) feature to set the part of the domain and the range that will be seen in the *viewing window*. For this function, set

$$\text{Xmin} = -6, \text{Xmax} = 2, \text{Xscl} = 1, \text{Ymin} = -2, \text{Ymax} = 10, \text{Yscl} = 1$$

in order to get a good view of the graph. The display for the *window* settings is shown in Fig. 3.28(b).

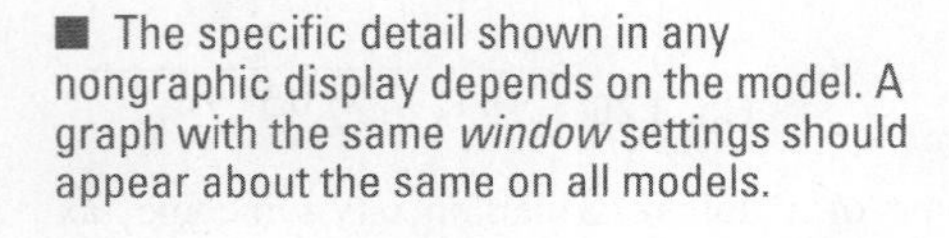
■ The specific detail shown in any nongraphic display depends on the model. A graph with the same *window* settings should appear about the same on all models.

Then display the graph, using the *graph* (or *exe*) key. The display showing the graph of $y = 2x + 8$ is shown in Fig. 3.28(c).

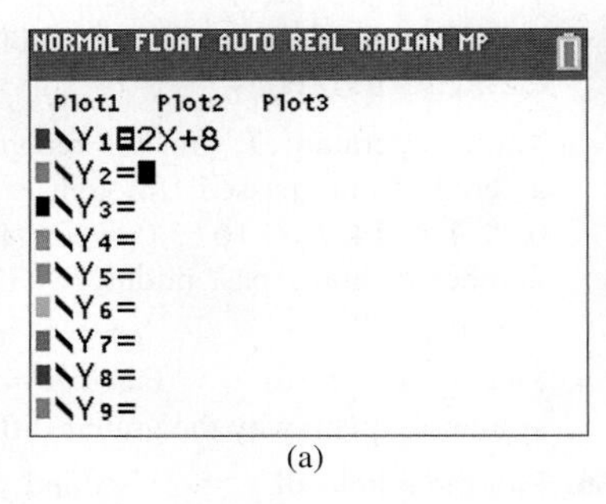

(a)

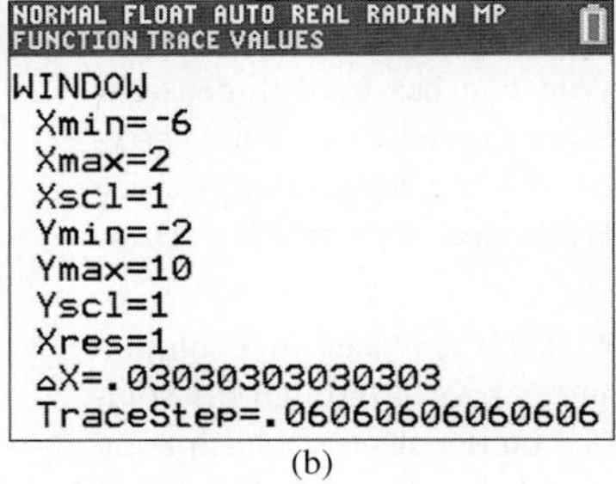

(b)

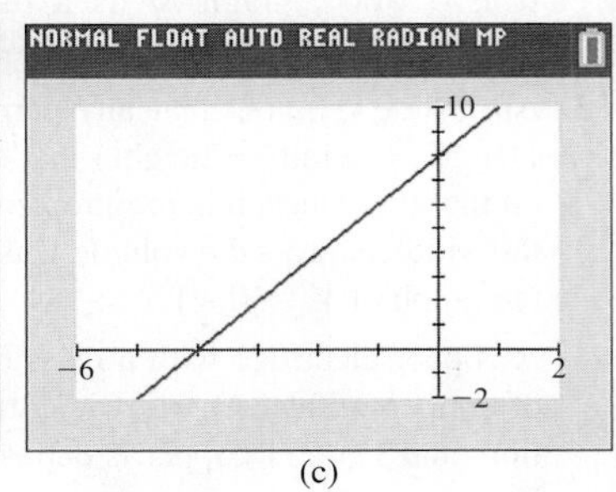

(c)

Fig. 3.28 ■

Practice Exercise
1. Use your calculator to determine good window settings for graphing $y = 6 - 3x$.

In Example 1, we noted that the *window* feature sets the intervals of the domain and range of the function that are seen in the viewing window. Unless the settings are appropriate, you may not get a good view of the graph.

EXAMPLE 2 Choose window settings carefully

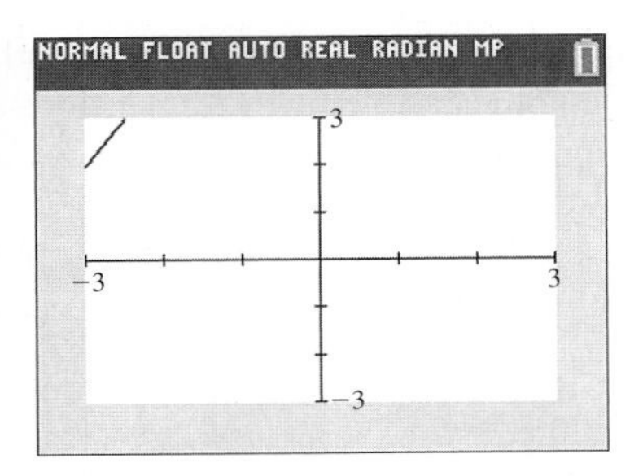

Fig. 3.29

■ In later examples, we generally will not show the *window* settings in the text. The Xmin, Xmax Ymin and Ymax will be labeled on the display, as in Fig. 3.29.

In Example 1, the *window* settings were chosen to give a good view of the graph of $y = 2x + 8$. However, if we had chosen settings of Xmin $= -3$, Xmax $= 3$, Ymin $= -3$, and Ymax $= 3$, we would get the view shown in Fig. 3.29. We see that very little of the graph can be seen. With some settings, it is possible that no part of the graph can be seen. Therefore, it is necessary to be careful in choosing the settings. This includes the scale settings (Xscl and Yscl), which should be chosen so that several (not too many or too few) axis markers are used.

Because the settings are easily changed, to get the general location of the graph, it is usually best to first choose intervals between the min and max values that are greater than is probably necessary. They can be reduced as needed.

Also, note that the calculator makes the graph just as we have been doing—by plotting points (square dots called *pixels*). It just does it a lot faster. ■

SOLVING EQUATIONS GRAPHICALLY

An equation can be solved by use of a graph. Most of the time, the solution will be approximate, but with a graphing calculator, it is possible to get good accuracy in the result. The procedure used is as follows:

Procedure for Solving an Equation Graphically

1. Collect all terms on one side of the equal sign. This gives us the equation $f(x) = 0$.
2. Set $y = f(x)$ and graph this function.
3. Find the points where the graph crosses the x-axis. These points are called the ***x*-intercepts** of the graph. At these points, $y = 0$.
4. The values of x for which $y = 0$ are the solutions of the equation. [These values are called the **zeros** of the function $f(x)$.]

EXAMPLE 3 Solving an equation graphically

Using a calculator, graphically solve the equation $x^2 - 2x = 1$.

Following the above procedure, we first rewrite the equation as $x^2 - 2x - 1 = 0$ and then set $y = x^2 - 2x - 1$. Next, we graph this function as shown in Fig. 3.30(a). Since the graph crosses the x-axis twice, the equation has two solutions which appear to be approximately $x \approx -0.5$ and $x \approx 2.5$.

A graphing calculator can be used to find the x-intercepts with a high level of accuracy by using the ***zero*** feature. When using the *zero* feature, one must move the crosshairs on the calculator screen to choose a left bound (to the left of the x-intercept), a right bound (to the right of the x-intercept), and then a guess (near the x-intercept), pressing Enter after each entry. The calculator will then display the x-intercept. This process must be repeated twice, once for each x-intercept. The solutions, which are shown in Fig. 3.30(b) and (c), are $x = -0.414$ and $x = 2.414$ (rounded to the nearest thousandth). Note that these solutions are close to our original estimates.

Graphing calculator keystrokes: goo.gl/9XiYoG

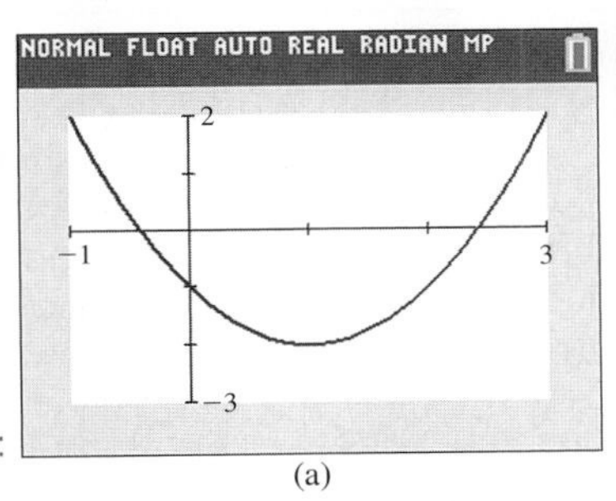

(a)

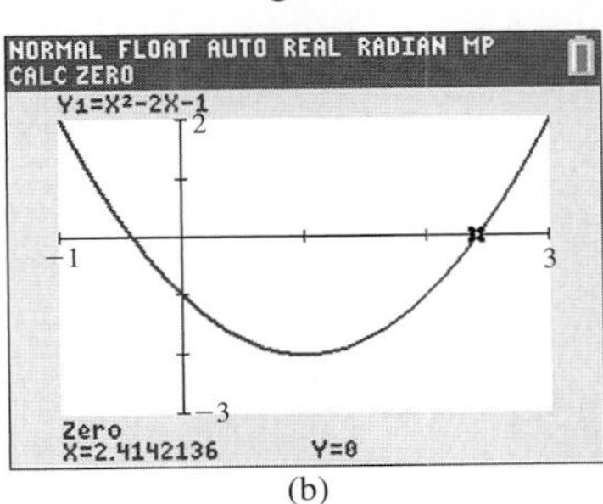

(b)

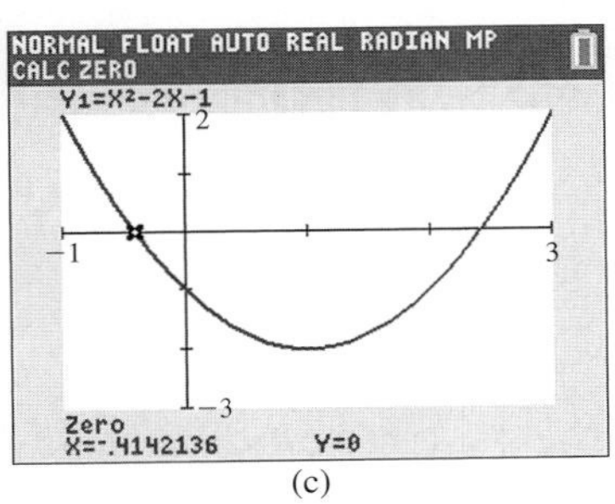

(c)

Fig. 3.30 ■

EXAMPLE 4 Solving graphically—container dimensions

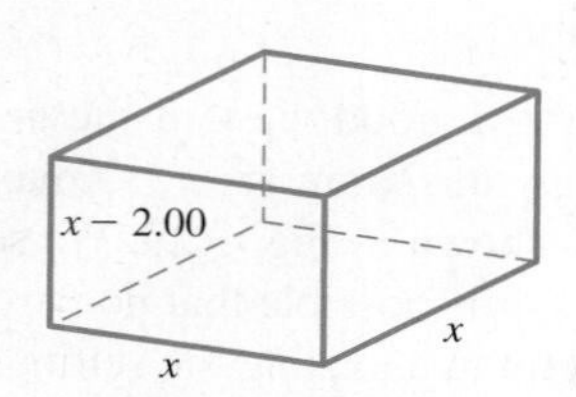

Fig. 3.31

A rectangular container whose volume is 30.0 cm^3 is made with a square base and a height that is 2.00 cm less than the length of the side of the base. Find the dimensions of the container by first setting up the necessary equation and then solving it graphically on a calculator.

Let $x =$ the length of the side of the square base (see Fig. 3.31); the height is then $x - 2.00$ cm. This means the volume is $(x)(x)(x - 2.00)$, or $x^3 - 2.00x^2$. Because the volume is 30.0 cm^3, we have the equation

$$x^3 - 2.00x^2 = 30.0$$

To solve it graphically on a calculator, first rewrite the equation as $x^3 - 2x^2 - 30 = 0$ and then set $y = x^3 - 2x^2 - 30$. The calculator view is shown in Fig. 3.32, where we use only positive values of x, because negative values have no meaning. Using the *zero* feature, we find that $x = 3.94$ cm is the approximate solution. Therefore, the dimensions are 3.94 cm, 3.94 cm, and 1.94 cm. Checking, these dimensions give a volume of 30.1 cm^3. The slight difference of 0.1 cm^3 is due to rounding off. A better check can be made by using the calculator value before rounding off. However, the final dimensions should be given only to three significant digits. ■

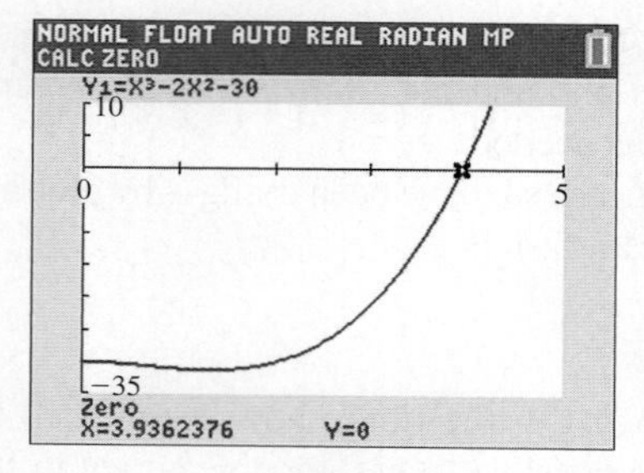

Fig. 3.32

Graphing calculator keystrokes: goo.gl/U0qeKP

EXAMPLE 5 Solving graphically—fuel cell power

The electric power P (in W) delivered by a certain fuel cell as a function of the resistance R (in Ω) in the circuit is given by $P = \frac{100R}{(0.50 + R)^2}$. Find (a) the maximum power, (b) the resistances that deliver a power of 40 W, and (c) the domain and range of the function.

■ The *minimum* and *maximum* features on a calculator are useful in determining a function's peaks and valleys. This is helpful in determining the range of a function.

(a) By using the ***maximum*** feature on a calculator (and selecting a left bound, a right bound, and a guess), the maximum power is found to be 50 W as shown in Fig. 3.33. This occurs when $R = 0.50$ Ω.

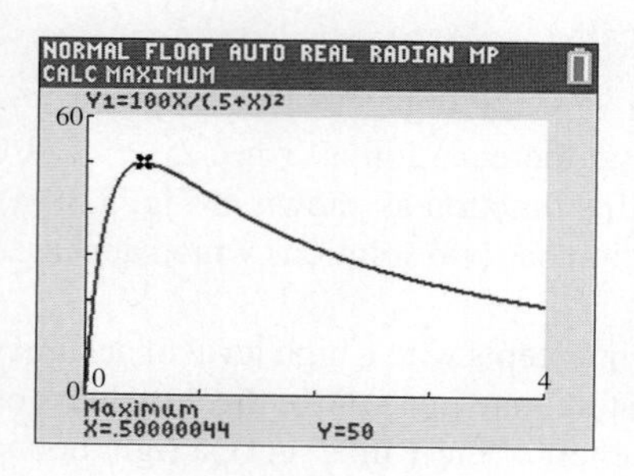

Fig. 3.33

Graphing calculator keystrokes: goo.gl/159pWS

(b) To find the resistances that deliver 40 W of power, we can find the points of intersection between the given function and the function $y = 40$. By using the ***intersect*** feature on a calculator (and choosing the first curve, the second curve, and a guess), the resistances are found to be 0.19 Ω and 1.31 Ω to the nearest hundredth. Note that the *intersect* feature must be applied twice, once for each intersecting point. See Fig. 3.34(a) and (b).

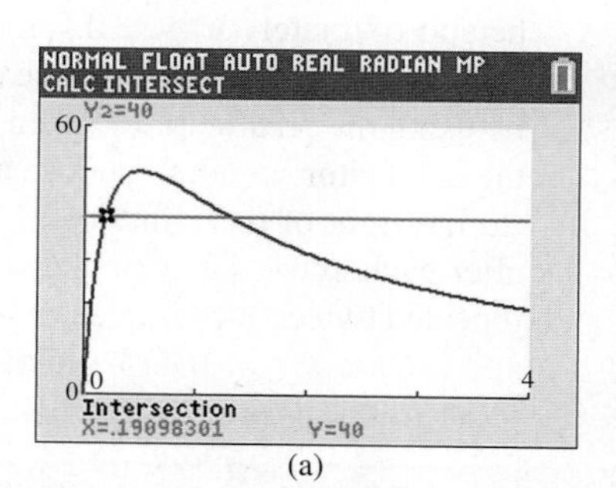

(a)

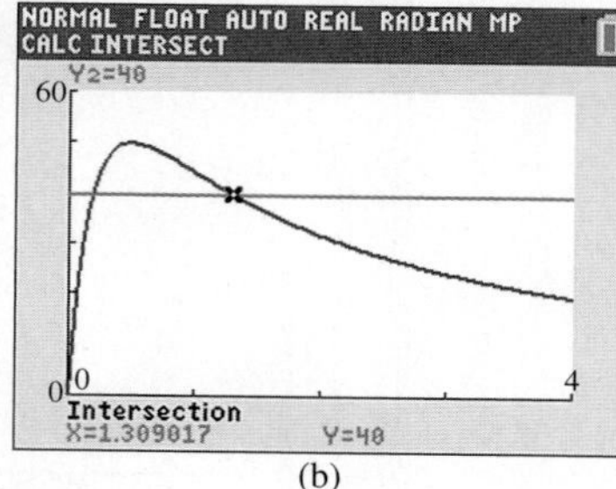

(b)

Fig. 3.34

Graphing calculator keystrokes: goo.gl/3aV11Z

(c) Since it is not reasonable for the resistance to be negative, the domain consists of all values $R \geq 0$. Since the power cannot be negative and the maximum power is 50 W, the range is all values $0 \leq P \leq 50$. ■

SHIFTING A GRAPH

By adding a positive constant to the right side of the function $y = f(x)$, the graph of the function is *shifted* straight up. If a negative constant is added to the right side of the function, the graph is shifted straight down.

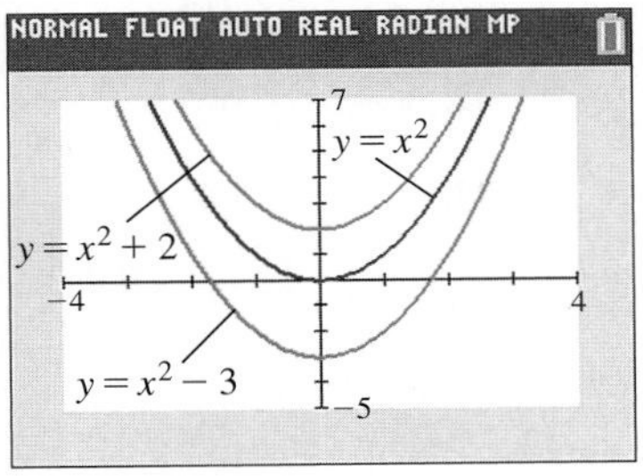

EXAMPLE 6 Shifting a graph vertically

If we add 2 to the right side of $y = x^2$, we get $y = x^2 + 2$. This will shift the graph of $y = x^2$ up 2 units.

If we add -3 to the right side of $y = x^2$, we get $y = x^2 - 3$. This will shift the graph of $y = x^2$ down 3 units. Figure 3.35 shows the original function $y = x^2$ (in blue) along with both vertical shifts. ■

Fig. 3.35

Practice Exercise

2. Describe the graph of $y = x^3 - 5$ as a shift of the graph of $y = x^3$.

By adding a constant to x in the function $y = f(x)$, the graph of the function is shifted to the right or to the left. **CAUTION** Adding a positive constant to x shifts the graph to the left, and adding a negative constant to x shifts the graph to the right. ■

EXAMPLE 7 Shifting a graph horizontally

For the function $y = x^2$, if we add 2 to x, we get $y = (x + 2)^2$, or if we add -3 to x, we get $y = (x - 3)^2$. The graphs of these three functions are displayed in Fig. 3.36.

We see that the graph of $y = (x + 2)^2$ is 2 units to the left of $y = x^2$ and that the graph of $y = (x - 3)^2$ is 3 units to the right of $y = x^2$. ■

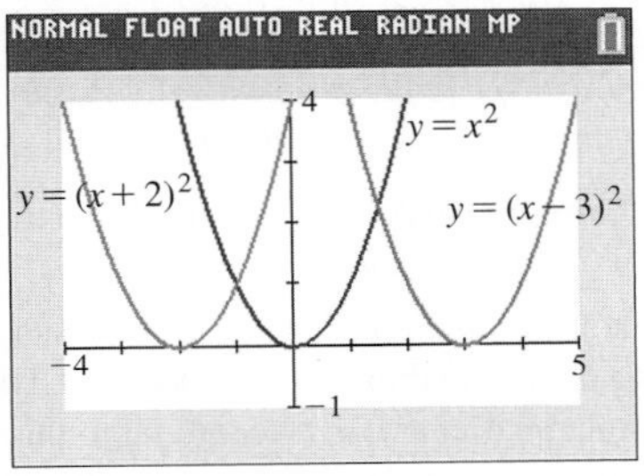

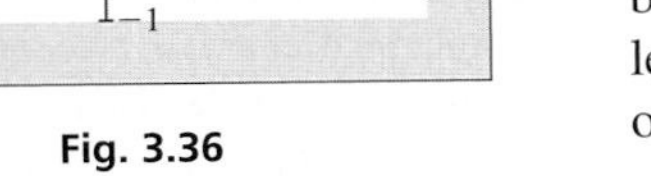

Fig. 3.36

Be very careful when shifting graphs horizontally. To check the direction and magnitude of a horizontal shift, find the value of x that makes the expression in parentheses equal to 0. For example, in Example 7, $y = (x - 3)^2$ is 3 units to the *right* of $y = x^2$ because $x = +3$ makes $x - 3$ equal to zero. The point (3, 0) on $y = (x - 3)^2$ is equivalent to the point (0, 0) on $y = x^2$. In the same way $x + 2 = 0$ for $x = -2$, and the graph of $y = (x + 2)^2$ is shifted 2 units to the *left* of $y = x^2$.

Summarizing how a graph is shifted, we have the following:

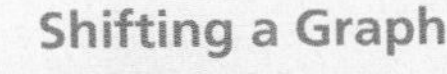

Shifting a Graph

Vertical shifts: $y = f(x) + k$ shifts the graph of $y = f(x)$ up k units if $k > 0$ and down k units if $k < 0$.

Horizontal shifts: $y = f(x + k)$ shifts the graph of $y = f(x)$ **left** k units if $k > 0$ and **right** k units if $k < 0$.

EXAMPLE 8 Shifting vertically and horizontally

The graph of a function can be shifted both vertically and horizontally. To shift the graph of $y = x^2$ to the right 3 units and down 2 units, add -3 to x and add -2 to the resulting function. In this way, we get $y = (x - 3)^2 - 2$. The graph of this function and the graph of $y = x^2$ are shown in Fig. 3.37.

Note that the point $(3, -2)$ on the graph of $y = (x - 3)^2 - 2$ is equivalent to the point (0, 0) on the graph of $y = x^2$. Checking, by setting $x - 3 = 0$, we get $x = 3$. This means the graph of $y = x^2$ has been shifted 3 units to the *right*. The vertical shift of -2 is clear. ■

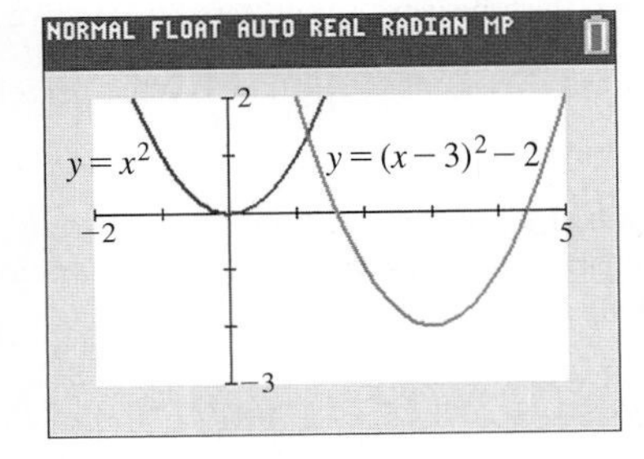

Fig. 3.37

Practice Exercise

3. Describe the graph of $y = (x + 2)^2 + 3$ relative to the graph of $y = x^2$.

EXERCISES 3.5

In Exercises 1 and 2, make the indicated changes in the given examples of this section and then solve.

1. In Example 3, change the sign on the left side of the equation from − to +.

2. In Example 8, in the second line, change "to the right 3 units and down 2 units" to "to the left 2 units and down 3 units."

In Exercises 3–18, display the graphs of the given functions on a graphing calculator. Use appropriate window settings.

3. $y = 3x - 1$

4. $y = 4 - 0.5x$

5. $y = x^2 - 4x$

6. $y = 8 - 2x^2$

7. $y = 6 - \frac{1}{2}x^3$

8. $y = x^4 - 6x^2$

9. $y = x^4 - 2x^3 - 5$

10. $y = x^2 - 3x^5 + 3$

11. $y = \frac{2x}{x - 2}$

12. $y = \frac{3}{x^2 - 4}$

13. $y = x + \sqrt{x + 3}$

14. $y = \sqrt[3]{2x + 1}$

15. $y = 3 + \frac{2}{x}$

16. $y = \frac{x^2}{\sqrt{3 - x}}$

17. $y = 2 - |x^2 - 4|$

18. $y = |4 - x^2| + 2$

In Exercises 19–28, use a graphing calculator to solve the given equations to the nearest 0.001.

19. $x^2 + x - 5 = 0$

20. $v^2 - 2v - 4 = 0$

21. $x^3 - 3 = 3x$

22. $x^4 - 2x = 0$

23. $s^3 - 4s^2 = 6$

24. $3x^2 - x^4 = 2 + x$

25. $\sqrt{5R + 2} = 3$

26. $\sqrt{x} + 3x = 7$

27. $\frac{1}{x^2 + 1} = 0$

28. $T - 2 = \frac{1}{T}$

In Exercises 29–40, use a graphing calculator to find the range of the given functions. Use the maximum or minimum feature when needed.

29. $y = -4x^2 + 8x + 3$

30. $y = 3x^2 + 12x - 4$

31. $y = \frac{10x}{1 + x^2}$

32. $y = 16\sqrt{x} + \frac{1}{x}$

33. $y = \frac{4}{x^2 - 4}$

34. $y = \frac{x + 1}{x^2}$

35. $y = \frac{x^2}{x + 1}$

36. $y = \frac{x}{x^2 - 4}$

37. $Y(y) = \frac{y + 1}{\sqrt{y - 2}}$

38. $f(n) = \frac{n^2}{6 - 2n}$

39. $f(D) = \frac{D^2 + 8D - 8}{D(D - 2) + 4(D - 2)}$

40. $g(x) = \frac{\sqrt{x - 2}}{x - 3}$

In Exercises 41–48, a function and how it is to be shifted is given. Find the shifted function, and then display the given function and the shifted function on the same screen of a graphing calculator.

41. $y = 3x$, up 1

42. $y = x^3$, down 2

43. $y = \sqrt{x}$, right 3

44. $y = \frac{2}{x}$, left 4

45. $y = -2x^2$, down 3, left 2

46. $y = -4x$, up 4, right 3

47. $y = \sqrt{2x + 1}$, up 1, left 1

48. $y = \sqrt{x^2 + 4}$, down 2, right 2

In Exercises 49–52, use the function for which the graph is shown in Fig. 3.38 to sketch graphs of the indicated functions.

49. $f(x + 2)$

50. $f(x - 2)$

51. $f(x - 2) + 2$

52. $f(x + 2) + 2$

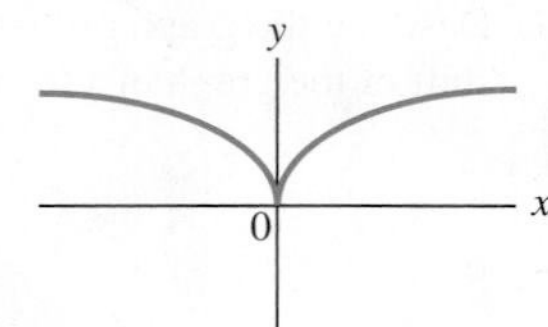

Fig. 3.38

In Exercises 53–60, solve the indicated equations graphically. Assume all data are accurate to two significant digits unless greater accuracy is given.

53. In an electric circuit, the current i (in A) as a function of voltage v is given by $i = 0.01v - 0.06$. Find v for $i = 0$.

54. For tax purposes, a corporation assumes that one of its computer systems depreciates according to the equation $V = 90{,}000 - 12{,}000t$, where V is the value (in dollars) after t years. According to this formula, when will the computer be fully depreciated (no value)?

55. Two cubical coolers together hold 40.0 L (40,000 cm^3). If the inside edge of one is 5.00 cm greater than the inside edge of the other, what is the inside edge of each?

56. The height h (in ft) of a rocket as a function of time t (in s) of flight is given by $h = 50 + 280t - 16t^2$. Determine when the rocket is at ground level. Also find the maximum height.

57. The length of a rectangular solar panel is 12 cm more than its width. If its area is 520 cm^2, find its dimensions.

58. A computer model shows that the cost (in dollars) to remove x percent of a pollutant from a lake is $C = \frac{8000x}{100 - x}$. What percent can be removed for $25,000?

59. In finding the illumination at a point x feet from one of two light sources that are 100 ft apart, it is necessary to solve the equation $9x^3 - 2400x^2 + 240{,}000x - 8{,}000{,}000 = 0$. Find x.

60. A rectangular storage bin is to be made from a rectangular piece of sheet metal 12 in. by 10 in., by cutting out equal corners of side x and bending up the sides. See Fig. 3.39. Find x if the storage bin is to hold 90 in.^{3}.

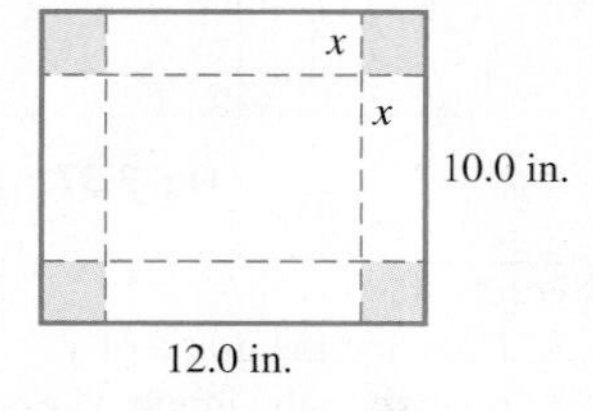

Fig. 3.39

In Exercises 61–68, solve the given problems.

61. The concentration C (in mg/L) of a drug in a patient's bloodstream t hours after taking a pill is given by $C = \frac{10.0\,t}{t^2 + 1.00}$. The patient should receive a second dose after the concentration has dropped below 1.50 mg/L. How long will this take?

62. Explain how to show the graph of the *relation* $y^2 = x$ on a graphing calculator, and then display it on the calculator. See Example 8 on page 99.

63. The cutting speed s (in ft/min) of a saw in cutting a particular type of metal piece is given by $s = \sqrt{t - 4t^2}$, where t is the time in seconds. What is the maximum cutting speed in this operation (to two significant digits)? (*Hint*: Find the range.)

64. Referring to Exercise 60, explain how to determine the maximum possible capacity for a storage bin constructed in this way. What is the maximum possible capacity (to three significant digits)?

65. A balloon is being blown up at a constant rate. (a) Sketch a reasonable graph of the radius of the balloon as a function of time. (b) Compare to a typical situation that can be described by $r = \sqrt[3]{3t}$, where r is the radius (in cm) and t is the time (in s).

66. A hot-water faucet is turned on. (a) Sketch a reasonable graph of the water temperature as a function of time. (b) Compare to a typical situation described by $T = \frac{t^3 + 80}{0.015t^3 + 4}$, where T is the water temperature (in °C) and t is the time (in s).

67. Display the graph of $y = cx^3$ with $c = -2$ and with $c = 2$. Describe the effect of the value of c.

68. Display the graph of $y = cx^4$, with $c = 4$ and with $c = \frac{1}{4}$. Describe the effect of the value of c.

Answers to Practice Exercises

1. x: -1 to 3: y: -2 to 8 is a good window setting.
2. 5 units down
3. 2 units left, 3 units up

3.6 Graphs of Functions Defined by Tables of Data

Straight-Line Segments Graph • Smooth Curve Graph • Reading a Graph • Linear Interpolation

As we noted in Section 3.1, there are ways other than formulas to show functions. One important way to show the relationship between variables is by means of a table of values found by observation or from an experiment.

Statistical data often give values that are taken for certain intervals or are averaged over various intervals, and there is no meaning to the intervals *between* the points. Such points should be connected by straight-line segments only to make them stand out better and make the graph easier to read. See Example 1.

EXAMPLE 1 Graph using straight-line segments

The electric energy usage (in kW • h) for a certain all-electric house for each month of a year is shown in the following table. Plot these data.

Month	Jan	Feb	Mar	Apr	May	Jun
Energy usage	2626	3090	2542	1875	1207	892

Month	July	Aug	Sep	Oct	Nov	Dec
Energy usage	637	722	825	1437	1825	2427

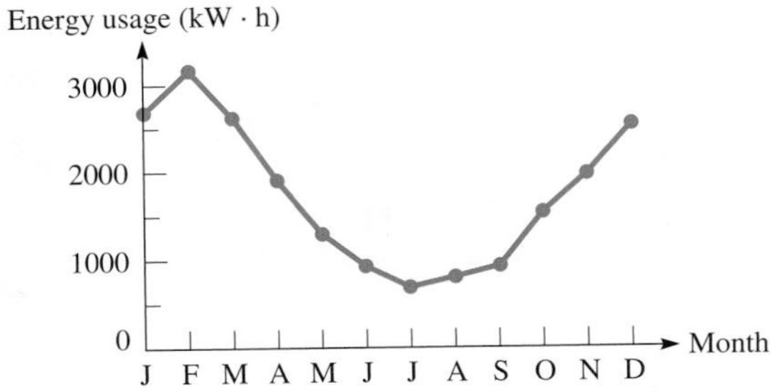

Fig. 3.40

We know that there is no meaning to the intervals between the months, because we have the total number of kilowatt-hours for each month. Therefore, we use straight-line segments, but only to make the points stand out better. See Fig. 3.40. ■

Data from experiments in science and technology often indicate that the variables could have a formula relating them, although the formula may not be known. In this case, when plotting the graph, the points should be connected by a smooth curve.

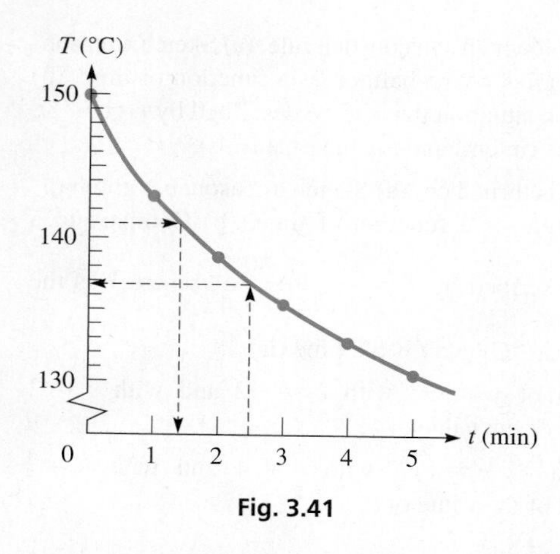

Fig. 3.41

■ The Celsius degree is named for the Swedish astronomer Andres Celsius (1701–1744). He designated 100° as the freezing point of water and 0° as the boiling point. These were later reversed.

■ In Chapters 5 and 22, we see how to find a *regression* equation that approximates the function relating the variables in a table of values.

EXAMPLE 2 Graph using smooth curve

Steam in a boiler was heated to 150°C and then allowed to cool. Its temperature T (in °C) was recorded each minute, as shown in the following table. Plot the graph.

Time (min)	0.0	1.0	2.0	3.0	4.0	5.0
Temperature (°C)	150.0	142.8	138.5	135.2	132.7	130.8

Because the temperature changes in a continuous way, there is meaning to the values in the intervals between points. Therefore, these points are joined by a smooth curve, as in Fig. 3.41. Note the indicated break in the vertical scale between 0 and 130. This is done so that the plotted points use up most of the space on the graph.

The dotted arrows in the graph show how we can estimate the value of one variable for a given value of the other. This is called **reading a graph.** For example, after 2.5 min, the temperature is approximately 136.8°C. Also, if the temperature is known to be 141°C, then the time is approximately 1.4 min. ■

LINEAR INTERPOLATION

In Example 2, we see that we can estimate values from a graph. However, unless a very accurate graph is drawn with expanded scales for both variables, only very approximate values can be found. There is a method, called *linear interpolation*, that uses the table itself to get more accurate results.

Linear interpolation *assumes that if a particular value of one variable lies between two of those listed in the table, then the corresponding value of the other variable is at the same proportional distance between the listed values.* On the graph, linear interpolation assumes that two points defined in the table are connected by a straight line. Although this is generally not correct, it is a good approximation if the values in the table are sufficiently close together.

EXAMPLE 3 Linear interpolation

■ Before the use of electronic calculators, interpolation was used extensively in mathematics textbooks for finding values from mathematics tables. It is still of use when using scientific and technical tables.

For the cooling steam in Example 2, we can use interpolation to find its temperature after 1.4 min. Because 1.4 min is $\frac{4}{10}$ of the way from 1.0 min to 2.0 min, we will assume that the value of T we want is $\frac{4}{10}$ of the way between 142.8 and 138.5, the values of T for 1.0 min and 2.0 min, respectively. The difference between these values is 4.3, and $\frac{4}{10}$ of 4.3 is 1.7 (rounded off to tenths). Subtracting (the values of T are decreasing) 1.7 from 142.8, we obtain 141.1. Thus, the required value of T is about 141.1°C. (Note that this agrees well with the result in Example 2.)

Another method of indicating the interpolation is shown in Fig. 3.42. From the figure, we have the proportion

$$\frac{0.4}{1.0} = \frac{x}{-4.3}$$

$$x = -1.7 \quad \text{rounded off}$$

		t	T		
1.0	0.4	1.0	142.8	x	−4.3
		1.4			
		2.0	138.5		

Fig. 3.42

Therefore,

$$142.8 + (-1.7) = 141.1°\text{C}$$

is the required value of T. If the values of T had been increasing, we would have added 1.7 to the value of T for 1.0 min. ■

Practice Exercise

1. In Example 3, interpolate to find T for $t = 4.2$ min.

EXERCISES 3.6

In Exercises 1–8, represent the data graphically.

1. The average monthly temperatures (in °C) for Washington, D.C., are as follows:

Month	J	F	M	A	M	J	J	A	S	O	N	D
Temp. (°C)	6	7	12	18	24	28	31	29	26	19	13	7

2. The average *exchange rate* for the number of Canadian dollars equal to one U.S. dollar for 2009–2016 is as follows:

Year	2009	2010	2011	2012	2013	2014	2015	2016
Can. Dol.	1.14	1.03	0.99	1.00	1.03	1.10	1.27	1.33

3. The amount of material necessary to make a cylindrical gallon container depends on the diameter, as shown in this table:

Diameter (in.)	3.0	4.0	5.0	6.0	7.0	8.0	9.0
Material (in.^{2})	322	256	224	211	209	216	230

4. An oil burner propels air that has been heated to 90°C. The temperature then drops as the distance from the burner increases, as shown in the following table:

Distance (m)	0.0	1.0	2.0	3.0	4.0	5.0	6.0
Temperature (°C)	90	84	76	66	54	46	41

5. A changing electric current in a coil of wire will induce a voltage in a nearby coil. Important in the design of transformers, the effect is called *mutual inductance*. For two coils, the mutual inductance (in H) as a function of the distance between them is given in the following table:

Distance (cm)	0.0	2.0	4.0	6.0	8.0	10.0	12.0
M. ind. (H)	0.77	0.75	0.61	0.49	0.38	0.25	0.17

6. The temperatures felt by the body as a result of the *wind-chill factor* for an outside temperature of 20°F (as determined by the National Weather Service) are given in the following table:

Wind speed (mi/h)	5	10	15	20	25	30	35	40
Temp. felt (°F)	13	9	6	4	3	1	0	−1

7. The time required for a sum of money to double in value, when compounded annually, is given as a function of the interest rate in the following table:

Rate (%)	4	5	6	7	8	9	10
Time (years)	17.7	14.2	11.9	10.2	9.0	8.0	7.3

8. The torque T of an engine, as a function of the frequency f of rotation, was measured as follows:

f (r/min)	500	1000	1500	2000	2500	3000	3500
T (ft · lb)	175	90	62	45	34	31	27

In Exercises 9–12, use the graph in Fig. 3.41, which relates the temperature of cooling steam and the time. Find the indicated values by reading the graph.

9. For $t = 4.3$ min, find T.

10. For $t = 1.8$ min, find T.

11. For $T = 145.0$°C, find t.

12. For $T = 133.5$°C, find t.

In Exercises 13 and 14, use the following table, which gives the valve lift L (in mm) of a certain cam as a function of the angle θ (in degrees) through which the cam is turned. Plot the values. Find the indicated values by reading the graph.

θ(°)	0	20	40	60	80	100	120	140
L (mm)	0	1.2	2.3	3.3	3.8	3.0	1.6	0

13. **(a)** For $\theta = 25°$, find L. **(b)** For $\theta = 96°$, find L.

14. For $L = 2.0$ mm, find θ.

In Exercises 15–18, find the indicated values by means of linear interpolation.

15. In Exercise 5, find the inductance for $d = 9.2$ cm.

16. In Exercise 6, find the temperature for $s = 12$ mi/h.

17. In Exercise 7, find the rate for $t = 10.0$ years.

18. In Exercise 8, find the torque for $f = 2300$ r/min.

In Exercises 19–22, use the following table that gives the rate R of discharge from a tank of water as a function of the height H of water in the tank. For Exercises 19 and 20, plot the graph and find the values from the graph. For Exercises 21 and 22, find the indicated values by linear interpolation.

Height (ft)	0	1.0	2.0	4.0	6.0	8.0	12
Rate (ft^3/s)	0	10	15	22	27	31	35

19. **(a)** for $R = 20$ ft^3/s, find H. **(b)** For $H = 2.5$ ft, find R.

20. **(a)** For $R = 34$ ft^3/s, find H. **(b)** For $H = 6.4$ ft, find R.

21. Find R for $H = 1.7$ ft.

22. Find H for $R = 25$ ft^3/s.

In Exercises 23–26, use the following table, which gives the fraction (as a decimal) of the total heating load of a certain system that will be supplied by a solar collector of area A (in m^2). Find the indicated values by linear interpolation.

f	0.22	0.30	0.37	0.44	0.50	0.56	0.61
A (m^2)	20	30	40	50	60	70	80

23. For $A = 36$ m^2, find f.

24. For $A = 52$ m^2, find f.

25. For $f = 0.59$, find A.

26. For $f = 0.27$, find A.

In Exercises 27–30, a method of finding values beyond those given is considered. By using a straight-line segment to extend a graph beyond the last known point, we can estimate values from the extension of the graph. The method is known as **linear extrapolation.** *Use this method to estimate the required values from the given graphs.*

27. Using Fig. 3.41, estimate T for $t = 5.3$ min.

28. Using the graph for Exercise 7, estimate T for $R = 10.4$%.

29. Using the graph for Exer. 19–22, estimate R for $H = 13$ ft.

30. Using the graph for Exer. 19–22, estimate R for $H = 16$ ft.

Answer to Practice Exercise

1. $T = 132.3$°C

CHAPTER 3 REVIEW EXERCISES

CONCEPT CHECK EXERCISES

*Determine each of the following as being either **true** or **false**. If it is false, explain why.*

1. $f(-x) = -f(x)$ for any function $f(x)$.
2. The domain of $f(x) = \sqrt{x - 1}$ is all real numbers $x > 1$.
3. All points for which $y > 1$ are in quadrants I and IV.
4. If $y^2 = x$, then y is a function of x.
5. If (2, 0) is an x-intercept of the graph of $y = f(x)$, then (2, 0) is also an intercept of the graph of $y = 2f(x)$.
6. The graph of $y = f(x + 4)$ is the graph of $y = f(x)$ shifted to the left 4 units.

PRACTICE AND APPLICATIONS

In Exercises 7–10, determine the appropriate function.

7. The radius of a circular water wave increases at the rate of 2 m/s. Express the area of the circle as a function of the time t (in s).
8. A conical sheet-metal hood is to cover an area 6 m in diameter. Find the total surface area A of the hood as a function of its height h.
9. A person invests x dollars at 5% APR (annual percentage rate) and y dollars at 4% APR. If the total annual income is $2000, solve for y as a function of x.
10. Fencing around a rectangular storage depot area costs twice as much along the front as along the other three sides. The back costs $10 per foot. Express the cost C of the fencing as a function of the width w if the length (along the front) is 20 ft longer than the width.

In Exercises 11–18, evaluate the given functions.

11. $f(x) = 7x - 5$; find $f(3)$ and $f(-6)$.
12. $g(I) = 8 - 3I$; find $g(\frac{1}{6})$ and $g(-4)$.
13. $H(h) = \sqrt{1 - 2h}$; find $H(-4)$ and $H(2h) + 2$.
14. $\phi(v) = \dfrac{6v - 9}{|v + 1|}$; find $\phi(-2)$ and $\phi(v + 1)$.
15. $F(x) = x^3 + 2x^2 - 3x$; find $F(3 + h) - F(3)$.
16. $f(x) = 3x^2 - 2x + 4$; find $\dfrac{f(x + h) - f(x)}{h}$.
17. $f(x) = 4 - 5x$; find $f(2x) - 2f(x)$.
18. $f(x) = 1 - 9x^2$; find $[f(x)]^2 - f(x^2)$.

In Exercises 19–22, evaluate the given functions. Values of the independent variable are approximate.

19. $f(x) = 8.07 - 2x$; find $f(5.87)$ and $f(-4.29)$.
20. $g(x) = 7x - x^2$; find $g(45.81)$ and $g(-21.85)$.
21. $G(S) = \dfrac{S - 0.087629}{3.0125S}$; find $G(0.17427)$ and $G(0.053206)$.
22. $h(t) = \dfrac{t^2 - 4t}{t^3 + 564}$; find $h(8.91)$ and $h(-4.91)$.

In Exercises 23–28, determine the domain and the range of the given functions.

23. $f(x) = x^4 + 1$
24. $G(z) = \dfrac{-4}{z^3}$
25. $g(t) = \dfrac{8}{\sqrt{t + 4}}$
26. $F(y) = 1 - 2\sqrt{y}$
27. $f(n) = 1 + \dfrac{2}{(n - 5)^2}$
28. $F(x) = 3 - |x|$

In Exercises 29–38, plot the graphs of the given functions. Check these graphs by using a calculator.

29. $y = 4x + 2$
30. $y = 5x - 10$
31. $s = 4t - t^2$
32. $y = x^2 - 8x - 5$
33. $y = 3 - x - 2x^2$
34. $V = 3 - 0.5s^3$
35. $A = 6 - s^4$
36. $y = x^4 - 4x$
37. $y = \dfrac{x}{x + 1}$
38. $Z = \sqrt{25 - 2R^2}$

In Exercises 39–46, use a calculator to solve the given equations to the nearest 0.1.

39. $7x - 3 = 0$
40. $3x + 11 = 0$
41. $x^2 + 1 = 6x$
42. $3t - 2 = t^2$
43. $x^3 - x^2 = 2 - x$
44. $5 - x^3 = 4x^2$
45. $\dfrac{5}{v^3} = v + 4$
46. $\sqrt{x} = 2x - 1$

In Exercises 47–50, use a calculator to find the range of the given function.

47. $y = x^4 - 5x^2$
48. $y = x\sqrt{4 - x^2}$
49. $A = w + \dfrac{2}{w}$
50. $y = 2x + \dfrac{3}{\sqrt{x}}$

In Exercises 51–70, solve the given problems.

51. Explain how $A(a, b)$ and $B(b, a)$ may be in different quadrants.
52. Determine the distance from the origin to the point (a, b).
53. Two vertices of an equilateral triangle are (0, 0) and (2, 0). What is the third vertex?
54. The points (1, 2) and (1, −3) are two adjacent vertices of a square. Find the other vertices.
55. Where are all points for which $|y/x| > 0$?
56. Describe the values of x and y for which $(1, -2)$, $(-1, -2)$, and (x, y) are on the same straight line.
57. Sketch the graph of a function for which the domain is $0 \le x \le 4$ and the range is $-5 \le y \le -2$.
58. Sketch the graph of a function for which the domain is all values of x and the range is $-2 < y < 2$.
59. If the function $y = \sqrt{x - 1}$ is shifted left 2 and up 1, what is the resulting function?

60. If the function $y = 3 - 2x$ is shifted right 1 and down 3, what is the resulting function?

61. For $f(x) = 2x^3 - 3$, display the graphs of $f(x)$ and $f(-x)$ on a graphing calculator. Describe the graphs in relation to the y-axis.

62. For $f(x) = 2x^3 - 3$, display the graphs of $f(x)$ and $-f(x)$ on a graphing calculator. Describe the graphs in relation to the x-axis.

63. Is it possible that the points (2, 3), (5, −1), and (−1, 3) are all on the graph of the same function? Explain.

64. For the functions $f(x) = |x|$ and $g(x) = \sqrt{x^2}$, use a graphing calculator to determine whether or not $f(x) = g(x)$ for all real x.

65. Express the area of a circle as a function of the side s of a square inscribed within the circle (all four corners of the square are on the circle.)

66. For $f(x) = x + 1$ and $g(x) = \dfrac{x^2 - 1}{x - 1}$ is $f(x) = g(x)$? Explain. Display the graph of each on a calculator (being very careful when choosing the window settings for x).

67. An equation used in electronics with a transformer antenna is $I = 12.5\sqrt{1 + 0.5m^2}$. For $I = f(m)$, find $f(0.55)$.

68. The percent p of wood lost in cutting it into boards 1.5 in. thick due to the thickness t (in in.) of the saw blade is $p = \dfrac{100t}{t + 1.5}$. Find p if $t = 0.4$ in. That is, since $p = f(t)$, find $f(0.4)$.

69. The angle A (in degrees) of a robot arm with the horizontal as a function of time t (for 0.0 s to 6.0 s) is given by $A = 8.0 + 12t^2 - 2.0t^3$. What is the greatest value of A to the nearest 0.1°? See Fig. 3.43. (*Hint*: Find the range.)

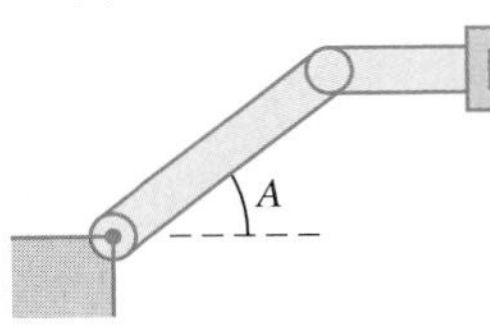

Fig. 3.43

70. The electric power it P (in W) produced by a certain battery is $P = \dfrac{24R}{R^2 + 1.40R + 0.49}$, where R is the resistance (in Ω) in the circuit. What is the maximum power produced? (*Hint*: Find the range.)

In Exercises 71–86, plot the graphs of the indicated functions.

71. When El Niño, a Pacific Ocean current, moves east and warms the water off South America, weather patterns in many parts of the world change significantly. Special buoys along the equator in the Pacific Ocean send data via satellite to monitoring stations. If the temperature T (in °C) at one of these buoys is $T = 28.0 + 0.15t$, where t is the time in weeks between Jan. 1 and Aug. 1 (30 weeks), plot the graph of $T = f(t)$.

72. The change C (in in.) in the length of a 100-ft steel bridge girder from its length at 40°F as a function of the temperature T is given by $C = 0.014(T - 40)$. Plot the graph for $T = 0$°F to $T = 120$°F.

73. The profit P (in dollars) a retailer makes in selling 50 tablets is given by $P = 50(p - 80)$, where p is the selling price. Plot P as a function of p for $p = \$70$ to $p = \$100$.

74. There are 500 L of oil in a tank that has the capacity of 100,000 L. It is filled at the rate of 7000 L/h. Determine the function relating the number of liters N and the time t while the tank is being filled. Plot N as a function of t.

75. The length L (in cm) of a pulley belt is 12 cm longer than the circumference of one of the pulley wheels. Express L as a function of the radius r of the wheel and plot L as a function of r.

76. The pressure loss P (in lb/in.^{2} per 100 ft) in a fire hose is given by $P = 0.00021Q^2 + 0.013Q$, where Q is the rate of flow (in gal/min). Plot the graph of P as a function of Q.

77. A *thermograph* measures the infrared radiation from each small area of a person's skin. Because the skin over a tumor radiates more than skin from nearby areas, a thermograph can help detect cancer cells. The emissivity e (in %) of radiation as a function of skin temperature T (in K) is $e = f(T) = 100(T^4 - 307^4)/307^4$, if nearby skin is at 34°C (307 K). Find e for $T = 309$ K.

78. If an amount A is due on a certain credit card, the minimum payment p is A, if A is \$20 or less, or \$20 plus 10% of any amount over \$20 that is owed. Find $p = f(A)$, and graph this function.

79. For a certain laser device, the laser output power P (in mW) is negligible if the drive current i is less than 80 mA. From 80 mA to 140 mA, $P = 1.5 \times 10^{-6}\, i^3 - 0.77$. Plot the graph of $P = f(i)$.

80. It is determined that a good approximation for the cost C (in cents/mi) of operating a certain car at a constant speed v (in mi/h) is given by $C = 0.025v^2 - 1.4v + 35$. Plot C as a function of v for $v = 10$ mi/h to $v = 60$ mi/h.

81. A medical researcher exposed a virus culture to an experimental vaccine. It was observed that the number of live cells N in the culture as a function of the time t (in h) after exposure was given by $N = \dfrac{1000}{\sqrt{t + 1}}$. Plot the graph of $N = f(t)$.

82. The electric field E (in V/m) from a certain electric charge is given by $E = 25/r^2$, where r is the distance (in m) from the charge. Plot the graph of $E = f(r)$ for values of r up to 10 cm.

83. To draw the approximate shape of an irregular shoreline, a surveyor measured the distances d from a straight wall to the shoreline at 20-ft intervals along the wall, as shown in the following table. Plot the graph of distance d as a function of the distance D along the wall.

D (ft)	0	20	40	60	80	100	120	140	160
d (ft)	15	32	56	33	29	47	68	31	52

84. The percent p of a computer network that is in use during a particular loading cycle as a function of the time t (in s) is given in the following table. Plot the graph of $p = f(t)$.

t (s)	0.0	0.2	0.4	0.6	0.8	1.0	1.2	1.4	1.6
P (%)	0	45	85	90	85	85	60	10	0

85. The vertical sag s (in ft) at the middle of an 800-ft power line as a function of the temperature T (in °F) is given in the following table. See Fig. 3.44. For the function $s = f(T)$, find $f(47)$ by linear interpolation.

T (°F)	0	20	40	60
s (ft)	10.2	10.6	11.1	11.7

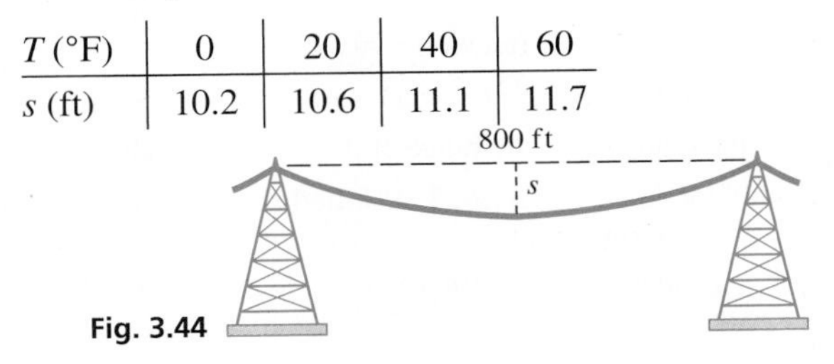

Fig. 3.44

86. In an experiment measuring the pressure p (in kPa) at a given depth d (in m) of seawater, the results in the following table were found. Plot the graph of $p = f(d)$ and from the graph determine $f(10)$.

d (m)	0.0	3.0	6.0	9.0	12	15
p (kPa)	101	131	161	193	225	256

In Exercises 87–96, solve the indicated equations graphically.

87. A person 250 mi from home starts toward home and travels at 60 mi/h for the first 2.0 h and then slows down to 40 mi/h for the rest of the trip. How long does it take the person to be 70 mi from home?

88. One industrial cleaner contains 30% of a certain solvent, and another contains 10% of the solvent. To get a mixture containing 50 gal of the solvent, 120 gal of the first cleaner is used. How much of the second must be used?

89. The solubility s (in kg/m^3 of water) of a certain type of fertilizer is given by $s = 135 + 4.9T + 0.19T^2$, where T is the temperature (in °C). Find T for $s = 500$ kg/m^3.

90. A 2.00-L (2000-cm^3) metal container is to be made in the shape of a right circular cylinder. Express the total area A of metal necessary as a function of the radius r of the base. Then find A for $r = 6.00$ cm, 7.00 cm, and 8.00 cm.

91. In an oil pipeline, the velocity v (in ft/s) of the oil as a function of the distance x (in ft) from the wall of the pipe is given by $v = 7.6x - 2.1x^2$. Find x for $v = 5.6$ ft/s. The diameter of the pipe is 3.50 ft.

92. For the health insurance of each employee, an insurance company charges a business with fewer than 2500 employees at an annual rate (in dollars) of $C(x) = 3000 - 20\sqrt{x - 1}$, where x is the number of employees. Find C for $x = 1500$.

93. A fountain is at the center of a circular pool of radius r ft, and the area is enclosed within a circular railing that is 45.0 ft from the edge of the pool. If the total area within the railing is 11,500 ft^2, what is the radius of the pool? See Fig. 3.45.

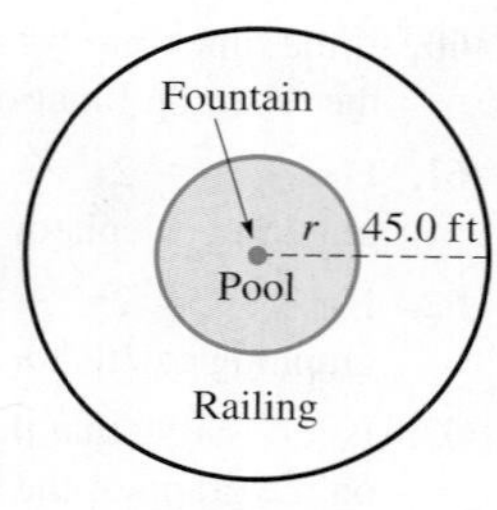

Fig. 3.45

94. One ball bearing is 1.00 mm more in radius and has twice the volume of another ball bearing. What is the radius of each?

95. A computer, using data from a refrigeration plant, estimates that in the event of a power failure, the temperature (in °C) in the freezers would be given by $T = \dfrac{4t^2}{t + 2} - 20$, where t is the number of hours after the power failure. How long would it take for the temperature to reach 0°C?

96. Two electrical resistors in parallel (see Fig. 3.46) have a combined resistance R_T given by $R_T = \dfrac{R_1R_2}{R_1 + R_2}$. If $R_2 = R_1 + 2.0$, express R_T as a function of R_1 and find R_1 if $R_T = 6.0\ \Omega$.

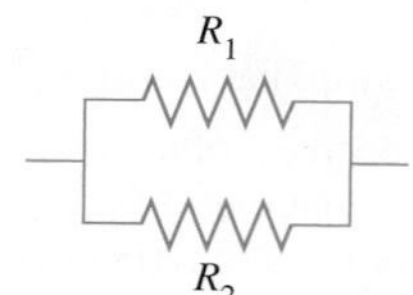

Fig. 3.46

97. Find the inner surface area A of a cylindrical 250.0-cm^3 cup as a function of the radius r of the base. Then if $A = 175.0$ cm^2, solve for r. Write one or two paragraphs explaining your method for setting up the function, and how you used a graphing calculator to solve for r with the given value of A.

CHAPTER 3 PRACTICE TEST

As a study aid, we have included complete solutions for each Practice Test problem at the back of this book.

1. Given $f(x) = \dfrac{8}{x} - 2x^2$, find (a) $f(-4)$ and (b) $f(x - 4)$.

2. A rocket has a mass of 2000 Mg at liftoff. If the first-stage engines burn fuel at the rate of 10 Mg/s, find the mass m of the rocket as a function of the time t (in s) while the first-stage engines operate. Sketch the graph for $t = 0$ s to $t = 60$ s.

3. Plot the graph of the function $f(x) = 4 - 2x$.

4. Use a graphing calculator to solve the equation $2x^2 - 3 = 3x$ to the nearest 0.1.

5. Plot the graph of the function $y = \sqrt{4 + 2x}$.

6. Locate all points (x, y) for which $x < 0$ and $y = 0$.

7. Find the domain and the range of the function $f(x) = \sqrt{6 - x}$.

8. If the function $y = 2x^2 - 3$ is shifted right 1 and up 3, what is the resulting function?

9. Use a graphing calculator to find the range of the function $y = \dfrac{x^2 + 2}{x + 2}$.

10. A window has the shape of a semicircle over a square, as shown in Fig. 3.47. Express the area of the window as a function of the radius of the circular part.

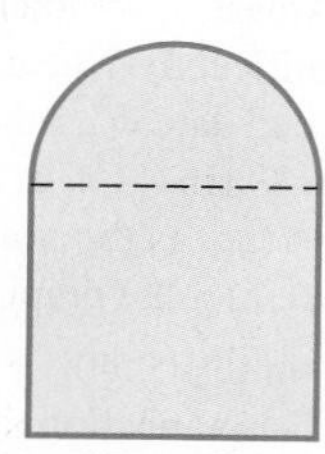

Fig. 3.47

11. The voltage V and current i (in mA) for a certain electrical experiment were measured as shown in the following table. Plot the graph of $i = f(V)$ and estimate $f(45.0)$ by reading the graph.

Voltage (V)	10.0	20.0	30.0	40.0	50.0	60.0
Current (mA)	145	188	220	255	285	315

12. From the table in Problem 11, find the voltage for $i = 200$ mA using linear interpolation.

The Trigonometric Functions

4

Triangles are often used in solving applied problems in technology. A short list of such problems includes those in air navigation, surveying, rocket motion, carpentry, structural design, electric circuits, and astronomy. In fact, it was because the Greek astronomer Hipparchus (about 150 B.C.E.) was interested in measuring distances such as that between the Earth and the moon that he started the study of **trigonometry.** Using records of earlier works, he organized and developed the real beginnings of this very important field of mathematics in order to make various astronomical measurements.

In trigonometry, we develop methods for calculating the sides and angles of triangles, and this in turn allows us to solve related applied problems. Because trigonometry has a great number of applications in many areas of study, it is considered one of the most practical branches of mathematics.

In this chapter, we introduce the basic trigonometric functions and show many applications of right triangles in science and technology. In later chapters, we will discuss other types of triangles and their applications.

As mathematics developed, it became clear, particularly in the 1800s and 1900s, that the trigonometric functions used for solving problems involving triangles were also very valuable in applications in which a triangle is not involved. This important use of the trigonometric functions is now essential in areas such as electronics, mechanical vibrations, acoustics, and optics, and these applications will be studied in later chapters.

LEARNING OUTCOMES

After completion of this chapter, the student should be able to:

- Define positive and negative angles
- Express an angle in degrees or radians and convert between the two measurements
- Determine a standard position angle
- Define, evaluate, and use the six trigonometric functions
- Use the Pythagorean theorem and trigonometric functions to solve a right triangle
- Employ the inverse trigonometric functions to solve for a missing angle
- Solve application problems involving right triangles

◀ **Using trigonometry, it is often possible to calculate distances that may not be directly measured. In Section 4.5, we show how we can measure the height of a missile at a given point in time.**

4.1 Angles

Positive and Negative Angles • Coterminal Angles • Angle Conversions • Standard Position of an Angle

In Chapter 2, we gave a basic definition of an *angle*. In this section, we extend this definition and also give some other important definitions related to angles.

An **angle** *is generated by rotating a ray about its fixed endpoint from an* ***initial position*** *to a* ***terminal position.*** *The initial position is called the* **initial side** *of the angle, the terminal position is called the* **terminal side,** *and the fixed endpoint is the* **vertex.** The angle itself is the amount of rotation from the initial side to the terminal side.

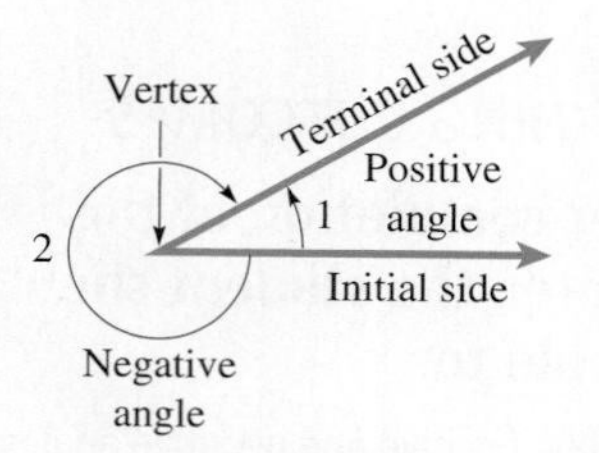

Fig. 4.1

If the rotation of the terminal side from the initial side is ***counterclockwise,*** *the angle is defined as* ***positive.*** *If the rotation is* ***clockwise,*** *the angle is* ***negative.*** In Fig. 4.1, $\angle 1$ is positive and $\angle 2$ is negative.

Many symbols are used to designate angles. Among the most widely used are certain Greek letters such as θ (theta), ϕ (phi), α (alpha), and β (beta). Capital letters representing the vertex (e.g., $\angle A$ or simply A) and other literal symbols, such as x and y, are also used commonly.

In Chapter 2, we introduced two measurements of an angle. These are the *degree* and the *radian*. Since degrees and radians are both important, we will briefly review the relationship between them in this section. However, we will not make use of radians until Chapter 8.

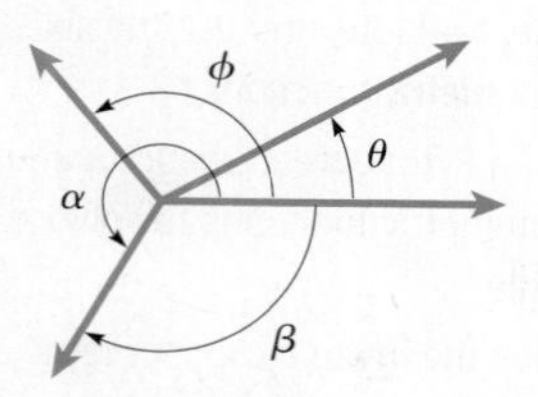

Fig. 4.2

From Section 2.1, we recall that *a* **degree** *is* $1/360$ *of one complete rotation.* In Fig. 4.2, $\angle\theta = 30°$, $\angle\phi = 140°$, $\angle\alpha = 240°$, and $\angle\beta = -120°$. Note that β is drawn in a clockwise direction to show that it is negative. The other angles are drawn in a counterclockwise direction to show that they are positive angles.

In Chapter 2, we used degrees and decimal parts of a degree. Most calculators use degrees in this decimal form. Another traditional way is to divide a degree into 60 equal parts called **minutes;** each minute is divided into 60 equal parts called **seconds.** The symbols ′ and ″ are used to designate minutes and seconds, respectively.

In Fig. 4.2, we note that angles α and β have the same initial and terminal sides. *Such angles are called* **coterminal angles.** An understanding of coterminal angles is important in certain concepts of trigonometry.

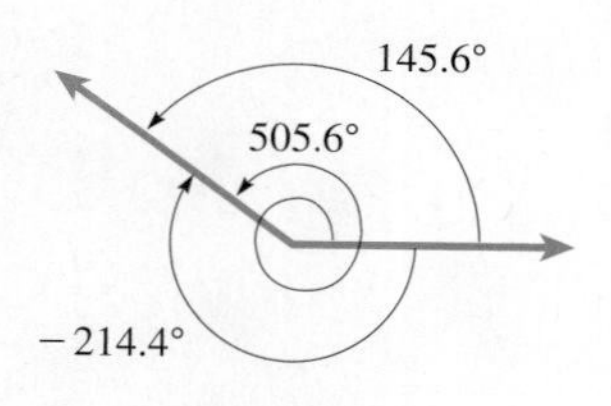

Fig. 4.3

EXAMPLE 1 Coterminal angles

Determine the measures of two angles that are coterminal with an angle of 145.6°.

Because there are 360° in one complete rotation, we can find a coterminal angle by adding 360° to the given angle of 145.6° to get 505.6°. Another coterminal angle can be found by subtracting 360° from 145.6° to get −214.4°. See the angles in Fig. 4.3. We could continue to add 360°, or subtract 360°, as many times as needed to get as many additional coterminal angles as may be required. ■

ANGLE CONVERSIONS

NOTE ▸ We will use only degrees as a measure of angles in most of this chapter. [Therefore, when using a calculator, be sure to use the ***mode*** feature to set the calculator for degrees.] In later chapters, we will use radians. We can convert between degrees and radians by using a calculator feature or by the definition (see Section 2.4) of π rad $= 180°$.

Before the extensive use of calculators, it was common to use degrees and minutes in tables, whereas calculators use degrees and decimal parts of a degree. Changing from one form to another can be done directly on a calculator by use of the *dms (degree-minute-second)* feature. The following examples illustrate angle conversions by using the definitions and by using the appropriate calculator features.

EXAMPLE 2 Convert radians to degrees

Express 1.36 rad in degrees.

We know that π rad = 180°, which means 1 rad = $180°/\pi$. Therefore,

$$1.36 \text{ rad} = 1.36\left(\frac{180°}{\pi}\right) = 77.9° \quad \text{to nearest } 0.1°$$

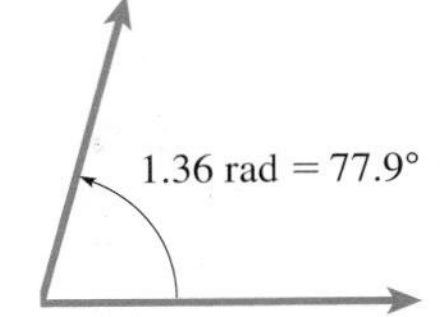

Fig. 4.4

This angle is shown in Fig. 4.4. We again note that degrees and radians are simply two different ways of measuring an angle.

In Fig. 4.5, a calculator display shows the conversions of 1.36 rad to degrees (calculator in *degree* mode) and 77.9° to radians (calculator in *radian* mode). ■

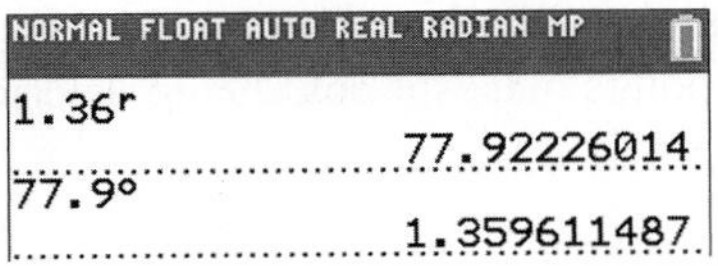

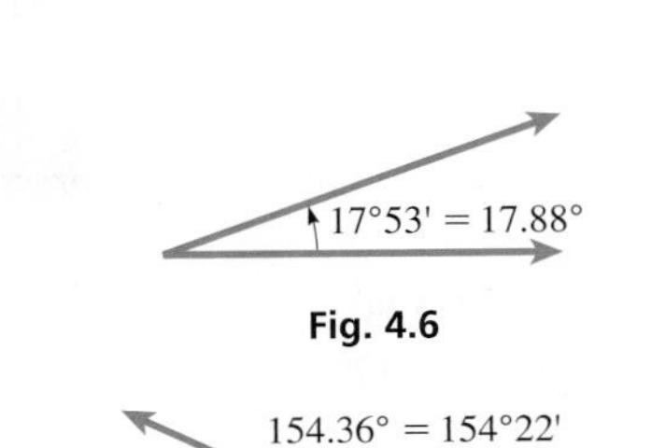

Fig. 4.5

Graphing calculator keystrokes: goo.gl/0lkBjy

EXAMPLE 3 Degrees, minutes—decimal form

(a) We change 17°53′ to decimal form by using the fact that 1° = 60′. This means that $53' = \left(\frac{53}{60}\right)° = 0.88°$ (to nearest 0.01°). Therefore, 17°53′ = 17.88°. See Fig. 4.6.

(b) The angle between two laser beams is 154.36°. To change this to an angle measured to the nearest minute, we have

$$0.36° = 0.36(60') = 22'$$

This means that 154.36° = 154°22′. See Fig. 4.7. ■

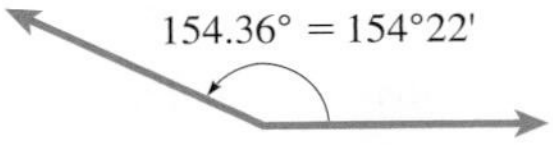

17°53′ = 17.88°

Fig. 4.6

154.36° = 154°22′

Fig. 4.7

Practice Exercises

1. Change 17°24′ to decimal form.
2. Change 38.25° to an angle measured in degrees and minutes.

STANDARD POSITION OF AN ANGLE

If the initial side of the angle is the positive x-axis and the vertex is the origin, the angle is said to be in **standard position.** The angle is then classified by the position of the terminal side. *If the terminal side is in the first quadrant, the angle is called a* **first-quadrant angle.** Similar terms are used when the terminal side is in the other quadrants. *If the terminal side coincides with one of the axes, the angle is a* **quadrantal angle.** For an angle in standard position, the terminal side can be determined if we know any point, except the origin, on the terminal side.

EXAMPLE 4 Angles in standard position

(a) A standard position angle of 60° is a first-quadrant angle with its terminal side 60° from the x-axis. See Fig. 4.8(a).

(b) A second-quadrant angle of 130° is shown in Fig. 4.8(b).

(c) A third-quadrant angle of 225° is shown in Fig. 4.8(c).

(d) A fourth-quadrant angle of 340° is shown in Fig. 4.8(d).

(e) A standard-position angle of −120° is shown in Fig. 4.8(e). Because the terminal side is in the third quadrant, it is a third-quadrant angle.

(f) A standard-position angle of 90° is a quadrantal angle since its terminal side is the positive y-axis. See Fig. 4.8(f).

Practice Exercise

3. In Fig. 4.8, which terminal side is that of a standard position angle of 240°?

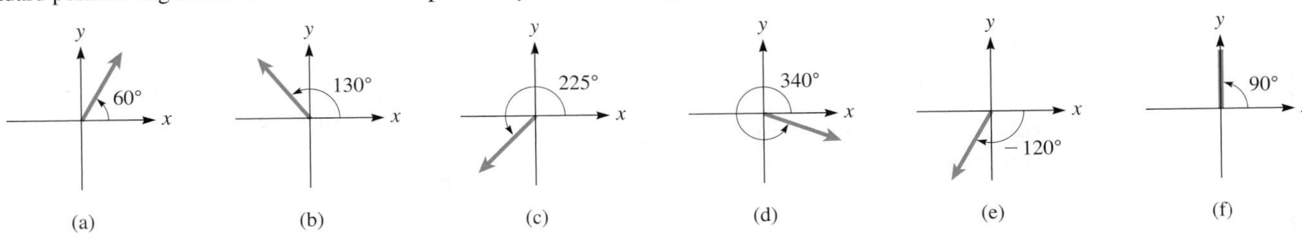

Fig. 4.8

■

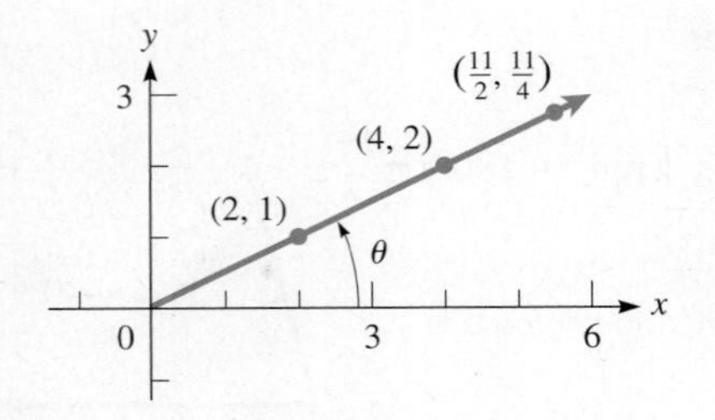

Fig. 4.9

EXAMPLE 5 Standard position—terminal side

In Fig. 4.9, θ is in standard position, and the terminal side is uniquely determined by knowing that it passes through the point (2, 1). The same terminal side passes through the points (4, 2) and $(\frac{11}{2}, \frac{11}{4})$, among an unlimited number of other points. Knowing that the terminal side passes through any one of these points makes it possible to determine the terminal side of the angle. ■

EXERCISES 4.1

In Exercises 1–4, find the indicated angles in the given examples of this section.

1. In Example 1, find another angle that is coterminal with the given angle.
2. In Example 3, change 53′ to 35′ and then find the decimal form.
3. In Example 4, find another standard-position angle that has the same terminal side as the angle in Fig. 4.8(c).
4. In Example 4, find another standard-position angle that has the same terminal side as the angle in Fig. 4.8(e).

In Exercises 5–8, draw the given angles in standard position.

5. 60°, 120°, −90° **6.** 330°, −150°, 450°
7. 50°, −120°, −30° **8.** 45°, 245°, −250°

In Exercises 9–14, determine one positive and one negative coterminal angle for each angle given.

9. 125° **10.** 173°
11. −150° **12.** 462°
13. 278.1° **14.** −197.6°

In Exercises 15–18, by means of the definition of a radian, change the given angles in radians to equal angles expressed in degrees to the nearest 0.01°.

15. 0.675 rad **16.** 0.838 rad
17. 4.447 rad **18.** −3.642 rad

In Exercises 19–22, use a calculator conversion sequence to change the given angles in radians to equal angles expressed in degrees to the nearest 0.01°.

19. 1.257 rad **20.** 2.089 rad
21. −4.110 rad **22.** 6.705 rad

In Exercises 23–26, use a calculator conversion sequence to change the given angles to equal angles expressed in radians to three significant digits.

23. 85.0° **24.** 237.4°
25. 384.8° **26.** −117.5°

In Exercises 27–30, change the given angles to equal angles expressed to the nearest minute.

27. 47.50° **28.** 715.80°
29. −5.62° **30.** 142.87°

In Exercises 31–34, change the given angles to equal angles expressed in decimal form to the nearest 0.01°.

31. 15°12′ **32.** 517°39′
33. 301°16′ **34.** −94°47′

In Exercises 35–42, draw angles in standard position such that the terminal side passes through the given point.

35. (4, 2) **36.** (−3, 8)
37. (−3, −5) **38.** (6, −1)
39. (−7, 5) **40.** (−4, −2)
41. (−2, 0) **42.** (0, 6)

In Exercises 43–50, the given angles are in standard position. Designate each angle by the quadrant in which the terminal side lies, or as a quadrantal angle.

43. 31°, 310° **44.** 180°, 92°
45. 435°, −270° **46.** −5°, 265°
47. 1 rad, 2 rad **48.** 3 rad, -3π rad
49. 4 rad, $\pi/3$ rad **50.** 12 rad, −2 rad

In Exercises 51 and 52, change the given angles to equal angles expressed in decimal form to the nearest 0.001°. *In Exercises 53 and 54, change the given angles to equal angles expressed to the nearest second.*

51. 21°42′36″ **52.** −107°16′23″
53. 86.274° **54.** 257.019°

55. A circular gear rotates clockwise by exactly 3.5 revolutions. By how many degrees does it rotate?
56. A windmill rotates 15.6 revolutions in a counterclockwise direction. By how many radians does it rotate?

Answers to Practice Exercises

1. 17.4° **2.** 38°15′ **3.** (e)

4.2 Defining the Trigonometric Functions

Definitions of the Trigonometric Functions • Evaluating the Trigonometric Functions • The Unit Circle

In this section, we define the six *trigonometric functions*. These functions are used in a wide variety of technological fields to solve problems involving triangles as well as quantities that oscillate repeatedly. The six trigonometric functions are defined below.

DEFINITION OF THE TRIGONOMETRIC FUNCTIONS

Let (x, y) be any point (other than the origin) on the terminal side of the angle θ in standard position, and let r be the distance between the origin and the given point (see Fig. 4.10).

Then the six trigonometric functions are defined as follows:

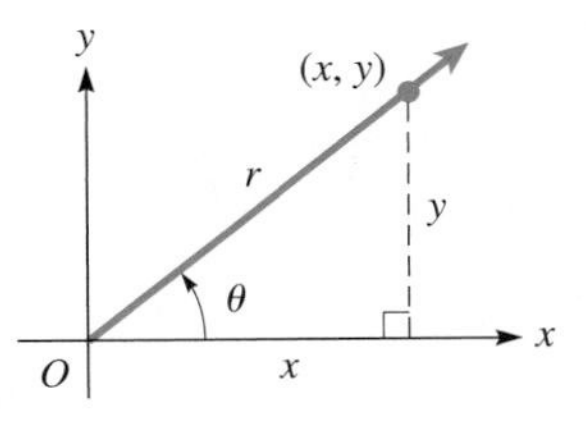

Fig. 4.10

sine of θ	$\sin\theta = \dfrac{y}{r}$	cosecant of θ	$\csc\theta = \dfrac{r}{y}$
cosine of θ	$\cos\theta = \dfrac{x}{r}$	secant of θ	$\sec\theta = \dfrac{r}{x}$
tangent of θ	$\tan\theta = \dfrac{y}{x}$	cotangent of θ	$\cot\theta = \dfrac{x}{y}$

(4.1)

NOTE ▸ [It is important to notice the reciprocal relationships that exist between sine and cosecant, cosine and secant, and tangent and cotangent.] If we know the value of one of the trigonometric functions, then we can find the value of the corresponding reciprocal function by interchanging the numerator and denominator.

The definition given above is completely general, which means it applies to *all angles*, no matter which of the four quadrants they terminate in. However, in this chapter, we will concentrate on angles between 0° and 90°. In Chapters 8 and 20, we will expand our scope and apply this definition to angles terminating in all four quadrants. In all cases, *the distance r from the origin to the point is always a positive number*, and it is called the ***radius vector.*** NOTE ▸ [It should also be noted that when either $x = 0$ or $y = 0$, some of the trigonometric ratios are undefined because their denominator is zero.] This will affect the domains of those functions, which will be discussed in Chapter 10 when we investigate their graphs.

EXAMPLE 1 Trigonometric ratios are independent of point

The definition of the trigonometric functions allows us to use *any point* on the terminal side of θ to determine the six trigonometric ratios. NOTE ▸ [It is important to realize that no matter which point we decide to use, we will get the same answer for these ratios].

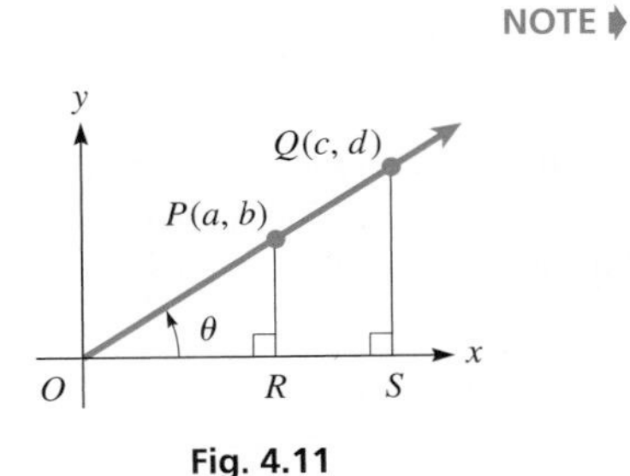

Fig. 4.11

For example, using point P in Fig. 4.11, we get $\tan\theta = \dfrac{b}{a}$. If we use point Q, we get $\tan\theta = \dfrac{d}{c}$. Since the triangles ORP and OSQ are ***similar triangles*** (their corresponding angles are equal), their corresponding sides must be proportional (see Section 2.2). This means $\dfrac{RP}{SQ} = \dfrac{OR}{OS}$, or equivalently, by multiplying both sides by $\dfrac{SQ}{OR}$, $\dfrac{RP}{OR} = \dfrac{SQ}{OS}$.

This proves that $\dfrac{b}{a} = \dfrac{d}{c}$, meaning that we will arrive at the same answer for $\tan\theta$ no matter which point we use on the terminal side. The same can be shown for the other trigonometric ratios. Therefore, each trigonometric ratio depends solely on the angle θ, and for each value of θ, there will be *only one* value of each trigonometric ratio. This means that the ratios are ***functions*** of the angle θ, and this is why we call them ***trigonometric functions.*** ■

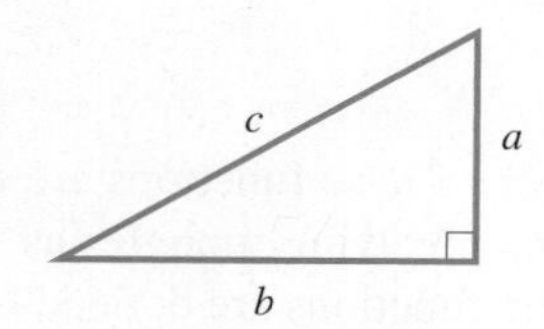

Fig. 4.12

EVALUATING THE TRIGONOMETRIC FUNCTIONS

The definitions in Eqs. (4.1) are used in evaluating the trigonometric functions. Also, we often use the *Pythagorean theorem*, which is discussed in Section 2.2, and which we restate here for reference. For the right triangle in Fig. 4.12, with hypotenuse c and legs a and b,

$$c^2 = a^2 + b^2 \tag{4.2}$$

Therefore, in Fig. 4.10 on the previous page, $r^2 = x^2 + y^2$, or $r = \sqrt{x^2 + y^2}$.

EXAMPLE 2 Values of trigonometric functions

Find the values of the trigonometric functions of the standard-position angle θ with its terminal side passing through the point (3, 4).

By placing the angle in standard position, as shown in Fig. 4.13, and drawing the terminal side through (3, 4), we find by use of the Pythagorean theorem that

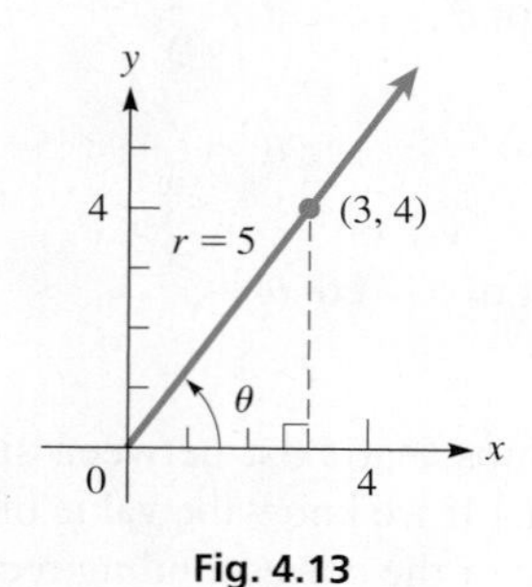

Fig. 4.13

$$r = \sqrt{3^2 + 4^2} = \sqrt{25} = 5$$

Using the values $x = 3$, $y = 4$, and $r = 5$, we find that

$$\sin\theta = \frac{4}{5} \qquad \csc\theta = \frac{5}{4}$$

$$\cos\theta = \frac{3}{5} \qquad \sec\theta = \frac{5}{3}$$

$$\tan\theta = \frac{4}{3} \qquad \cot\theta = \frac{3}{4}$$

We have left each of these results in the form of a fraction, which is considered to be an *exact form* in that there has been no approximation made. In writing decimal values, we find that $\tan\theta = 1.333$ and $\sec\theta = 1.667$, where these values have been rounded off and are therefore *approximate*. ■

Practice Exercise

1. In Example 2, change (3, 4) to (4, 3) and then find $\tan\theta$ and $\sec\theta$.

EXAMPLE 3 Values of trigonometric functions

Find the values of the trigonometric functions of the standard-position angle whose terminal side passes through (7.27, 4.19). The coordinates are approximate.

We show the angle and the given point in Fig. 4.14. From the Pythagorean theorem, we have

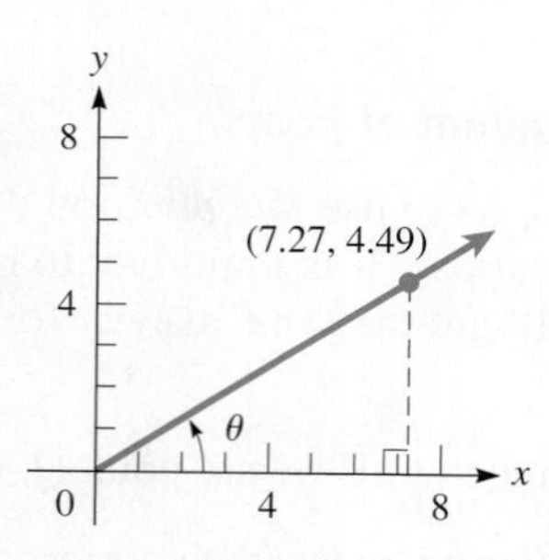

Fig. 4.14

$$r = \sqrt{7.27^2 + 4.49^2} = 8.545$$

(Here, we show a rounded-off value of r. It is not actually necessary to record the value of r because its value can be stored in the memory of a calculator. The reason for recording it here is to show the values used in the calculation of each of the trigonometric functions.) Therefore, we have the following values:

$$\sin\theta = \frac{4.49}{8.545} = 0.525 \qquad \csc\theta = \frac{8.545}{4.49} = 1.90$$

$$\cos\theta = \frac{7.27}{8.545} = 0.851 \qquad \sec\theta = \frac{8.545}{7.27} = 1.18$$

$$\tan\theta = \frac{4.49}{7.27} = 0.618 \qquad \cot\theta = \frac{7.27}{4.49} = 1.62$$

Because the coordinates are approximate, the results are rounded off.

The trigonometric functions can be evaluated directly on a calculator without rounding intermediate steps by storing the value of r and then using this stored value to compute the trigonometric ratios. The calculator evaluations of $\sin\theta$ and $\cos\theta$ are shown in Fig. 4.15. ■

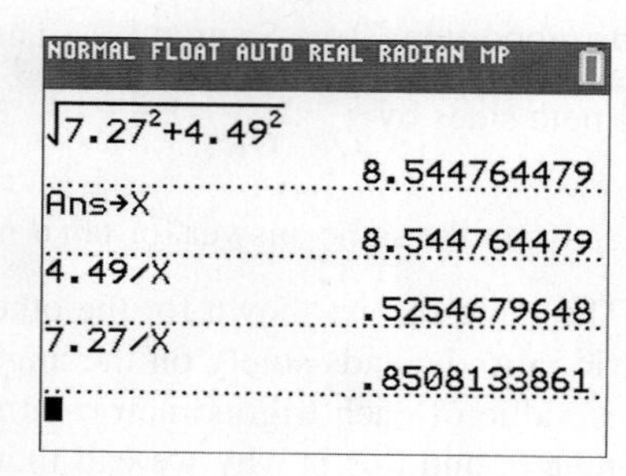

Fig. 4.15

Graphing calculator keystrokes: goo.gl/5qosro

CAUTION In Example 3, we expressed the result as $\sin\theta = 0.525$. A common error is to omit the angle and give the value as $\sin = 0.525$. This is a meaningless expression, for we must *show the angle* for which we have the value of a function. ■

If one of the trigonometric functions is known, it is possible to find the values of the other functions. The following example illustrates the method.

EXAMPLE 4 Given one function—find others

If we know that $\sin\theta = 3/7$, and that θ is a first-quadrant angle, we know the ratio of the ordinate to the radius vector (y to r) is 3 to 7. Therefore, the point on the terminal side for which $y = 3$ can be found by use of the Pythagorean theorem:

$$x^2 + 3^2 = 7^2$$

$$x = \sqrt{7^2 - 3^2} = \sqrt{49 - 9} = \sqrt{40} = 2\sqrt{10}$$

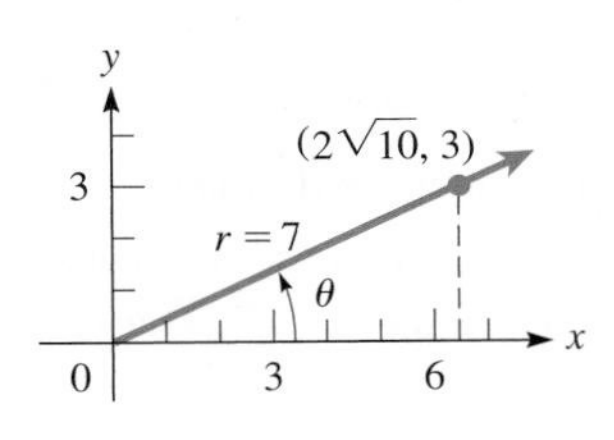

Fig. 4.16

Therefore, the point $(2\sqrt{10}, 3)$ is on the terminal side, as shown in Fig. 4.16.

Using the values $x = 2\sqrt{10}$, $y = 3$, and $r = 7$, we have the other trigonometric functions of θ. They are

$$\cos\theta = \frac{2\sqrt{10}}{7} \quad \tan\theta = \frac{3}{2\sqrt{10}} \quad \cot\theta = \frac{2\sqrt{10}}{3} \quad \sec\theta = \frac{7}{2\sqrt{10}} \quad \csc\theta = \frac{7}{3}$$

These values are *exact*. *Approximate* decimal values found on a calculator are

$$\cos\theta = 0.9035 \qquad \tan\theta = 0.4743 \qquad \cot\theta = 2.108$$

$$\sec\theta = 1.107 \qquad \csc\theta = 2.333$$

■

Practice Exercise

2. In Example 4, change $\sin\theta = 3/7$ to $\cos\theta = 3/7$, and then find approximate values of $\sin\theta$ and $\cot\theta$.

THE UNIT CIRCLE AND THE TRIGONOMETRIC FUNCTIONS

The circle that is centered at the origin and has a radius of 1 is called the ***unit circle***. Suppose the terminal side of θ in standard position intersects the unit circle at the point (x, y) as shown is Fig. 4.17.

Using the definition of the trigonometric functions and the fact that $r = 1$, $\sin\theta = \frac{y}{r} = \frac{y}{1} = y$ and $\cos\theta = \frac{x}{r} = \frac{x}{1} = x$. Thus, **the y-coordinate of the point on the unit circle equals $\sin\theta$ and the x-coordinate equals $\cos\theta$.** This fact can be very useful in many situations.

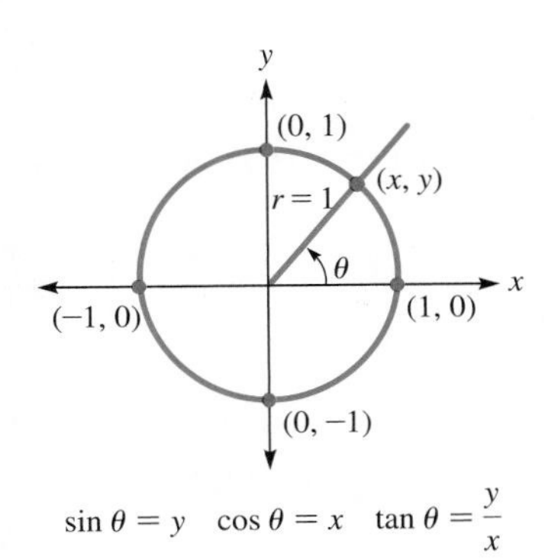

$\sin\theta = y \quad \cos\theta = x \quad \tan\theta = \frac{y}{x}$

Fig. 4.17

In terms of the unit circle, $\tan\theta = \frac{y}{x}$, the same as in the original definition. Also, the three reciprocal functions are still found by taking the reciprocal of the sine, cosine, or tangent.

EXAMPLE 5 The unit circle

(a) Use the unit circle to find cos 90° and sin 90°.

A 90° angle, when placed in standard position, intersects the unit circle at the point (0,1). Therefore, $\cos 90° = x = 0$ and $\sin 90° = y = 1$.

(b) Find $\cos\theta$ and $\csc\theta$ if the terminal side of θ (in standard position) intersects the unit circle at the point (0.6, 0.8).

We know that $\cos\theta = x = 0.6$. Since $\sin\theta = y = 0.8$, $\csc\theta$ is the reciprocal of 0.8.

Thus, $\csc\theta = \frac{1}{0.8} = \frac{10}{8} = \frac{5}{4}$. ■

EXERCISES 4.2

In Exercises 1 and 2, answer the given questions about the indicated examples of this section.

1. In Example 2, if (4, 3) replaces (3, 4), what are the values?
2. In Example 4, if 4/7 replaces 3/7, what are the values?

In Exercises 3–18, find values of the trigonometric functions of the angle (in standard position) whose terminal side passes through the given points. For Exercises 3–16, give answers in exact form. For Exercises 17 and 18, the coordinates are approximate.

3. (6, 8) 4. (5, 12) 5. (15, 8) 6. (240, 70)
7. (0.09, 0.40) 8. (1.1, 6.0) 9. (1.2, 3.5) 10. (1.2, 0.9)
11. $(1, \sqrt{15})$ 12. $(\sqrt{3}, 2)$ 13. (7, 7) 14. (840, 130)
15. (50, 20) 16. $(1, \frac{1}{2})$ 17. (0.687, 0.943) 18. (37.65, 21.87)

In Exercises 19–26, find the values of the indicated functions. In Exercises 19–22, give answers in exact form. In Exercises 23–26, the values are approximate.

19. Given $\cos\theta = 12/13$, find $\sin\theta$ and $\cot\theta$.
20. Given $\sin\theta = 1/2$, find $\cos\theta$ and $\csc\theta$.
21. Given $\tan\theta = 2$, find $\sin\theta$ and $\sec\theta$.
22. Given $\sec\theta = \sqrt{5}/2$, find $\tan\theta$ and $\cos\theta$.
23. Given $\sin\theta = 0.750$, find $\cot\theta$ and $\csc\theta$.
24. Given $\cos\theta = 0.0326$, find $\sin\theta$ and $\tan\theta$.
25. Given $\cot\theta = 0.254$, find $\cos\theta$ and $\tan\theta$.
26. Given $\csc\theta = 1.20$, find $\sec\theta$ and $\cos\theta$.

In Exercises 27–30, each given point is on the terminal side of an angle. Show that each of the given functions is the same for each point.

27. (3, 4), (6, 8), (4.5, 6), $\sin\theta$ and $\tan\theta$
28. (5, 12), (15, 36), (7.5, 18), $\cos\theta$ and $\cot\theta$
29. (0.3, 0.1), (9, 3), (33, 11), $\tan\theta$ and $\sec\theta$
30. (40, 30), (56, 42), (36, 27), $\csc\theta$ and $\cos\theta$

Use the unit circle to complete Exercises 31–34 (see Example 5).

31. Find $\cos 0°$.
32. Find $\sin 0°$.
33. If the angle θ intersects the unit circle at exactly (0.28, 0.96), find $\sin\theta$ and $\csc\theta$.
34. For the angle in Exercise 33, find $\cos\theta$ and $\sec\theta$.

In Exercises 35–42, answer the given questions.

35. For an acute angle A, which is greater, $\sin A$ or $\tan A$?
36. For an acute angle $A > 45°$, which is less, $\tan A$ or $\cot A$?
37. If $\tan\theta = 3/4$, what is the value of $\sin^2\theta + \cos^2\theta$? $[\sin^2\theta = (\sin\theta)^2]$
38. If $\sin\theta = 2/3$, what is the value of $\sec^2\theta - \tan^2\theta$?
39. If $y = \sin\theta$, what is $\cos\theta$ in terms of y?
40. If $x = \cos\theta$, what is $\tan\theta$ in terms of x?
41. From the definitions of the trigonometric functions, it can be seen that $\csc\theta$ is the reciprocal of $\sin\theta$. What function is the reciprocal of $\cos\theta$?
42. Refer to the definitions of the trigonometric functions in Eqs. (4.1). Is the quotient of one of the functions divided by $\cos\theta$ equal to $\tan\theta$? Explain.

Answers to Practice Exercises

1. $\tan\theta = 3/4$, $\sec\theta = 5/4$ 2. $\sin\theta = 0.9035$, $\cot\theta = 0.4743$

4.3 Values of the Trigonometric Functions

Function Values Using Geometry • Function Values from Calculator • Inverse Trigonometric Functions • Accuracy of Trigonometric Functions • Reciprocal Functions

We often need values of the trigonometric functions for angles measured in degrees. One way to find some of these values for particular angles is to use certain basic facts from geometry. This is illustrated in the next two examples.

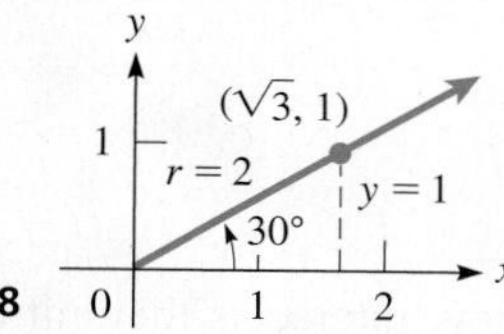

Fig. 4.18

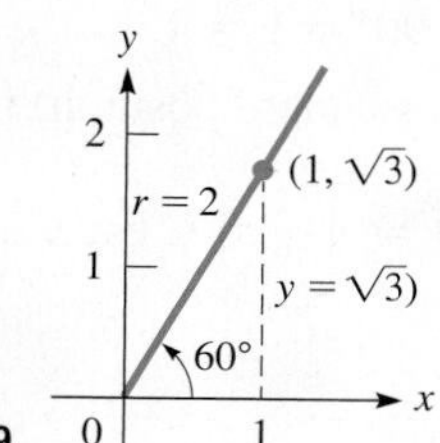

Fig. 4.19

EXAMPLE 1 Function values of 30° and 60°

From geometry, we find that in a right triangle, the side opposite a 30° angle is one-half of the hypotenuse. Therefore, in Fig. 4.18, letting $y = 1$ and $r = 2$, and using the Pythagorean theorem, $x = \sqrt{2^2 - 1^2} = \sqrt{3}$. Now with $x = \sqrt{3}$, $y = 1$, and $r = 2$,

$$\sin 30° = \frac{1}{2} \qquad \cos 30° = \frac{\sqrt{3}}{2} \qquad \tan 30° = \frac{1}{\sqrt{3}}$$

Using this same method, we find the functions of 60° to be (see Fig. 4.19)

$$\sin 60° = \frac{\sqrt{3}}{2} \qquad \cos 60° = \frac{1}{2} \qquad \tan 60° = \sqrt{3}$$

■

EXAMPLE 2 Function values of 45°

Find sin 45°, cos 45°, and tan 45°.

If we place an isosceles right triangle with one of its 45° angles in standard position and hypotenuse along the radius vector (see Fig. 4.20), the terminal side passes through (1, 1), since the legs of the triangle are equal. Using this point, $x = 1, y = 1$, and $r = \sqrt{2}$. Thus,

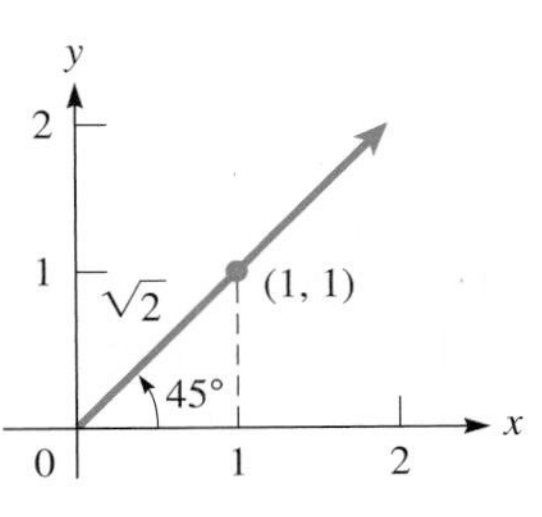

Fig. 4.20

$$\sin 45° = \frac{1}{\sqrt{2}} \qquad \cos 45° = \frac{1}{\sqrt{2}} \qquad \tan 45° = 1$$ ■

In Examples 1 and 2, we have given *exact* values. Decimal approximations are also given in the following table that summarizes the results for 30°, 45°, and 60°.

■ It is helpful to be familiar with these values, as they are used in later sections.

Trigonometric Functions of 30°, 45°, and 60°

	(exact values)			*(decimal approximations)*		
θ	30°	45°	60°	30°	45°	60°
$\sin\theta$	$\frac{1}{2}$	$\frac{1}{\sqrt{2}}$	$\frac{\sqrt{3}}{2}$	0.500	0.707	0.866
$\cos\theta$	$\frac{\sqrt{3}}{2}$	$\frac{1}{\sqrt{2}}$	$\frac{1}{2}$	0.866	0.707	0.500
$\tan\theta$	$\frac{1}{\sqrt{3}}$	1	$\sqrt{3}$	0.577	1.000	1.732

Another way to find values of the functions is to use a scale drawing. Measure the angle with a protractor, then measure directly the values of x, y, and r for some point on the terminal side, and finally use the proper ratios to evaluate the functions. However, this method is only approximate, and geometric methods work only for a limited number of angles. As it turns out, it is possible to find these values to any required accuracy through more advanced methods (using calculus and what are known as *power series*).

Values of $\sin\theta$, $\cos\theta$, and $\tan\theta$ are programmed into graphing calculators. For the remainder of this chapter, *be sure that your calculator is set for* ***degrees*** (not radians). The following examples illustrate using a calculator to find trigonometric values.

EXAMPLE 3 Values from calculator

Using a calculator to find the value of tan 67.36°, first enter the function and then the angle, just as it is written. The resulting display is shown in Fig. 4.21.

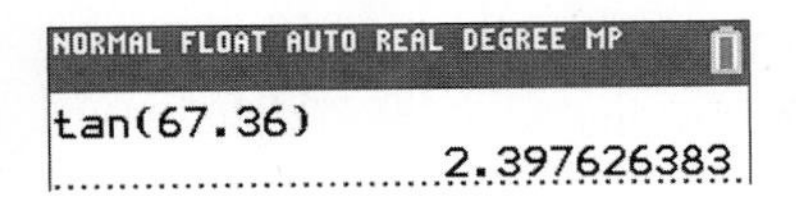

Fig. 4.21

Therefore, tan 67.36° = 2.397626383. ■

Not only are we able to find values of the trigonometric functions if we know the angle, but we can also find the angle if we know that value of a function. In doing this, we are actually using another important type of mathematical function, an **inverse trigonometric function.** They are discussed in detail in Chapter 20. For the purpose of using a calculator at this point, it is sufficient to recognize and understand the notation that is used.

■ Another notation that is used for $\sin^{-1} x$ is arcsin x.

The notation for "the angle whose sine is x" is $\sin^{-1} x$. This is called the *inverse sine function*. Equivalent meanings are given to $\cos^{-1} x$ (the angle whose cosine is x) and $\tan^{-1} x$ (the angle whose tangent is x). CAUTION Carefully note that the -1 used with a trigonometric function in this way shows an *inverse* trigonometric function and is *not* a negative exponent ($\sin^{-1} x$ represents an angle, not a function of an angle). ■ On a calculator, the $\sin^{-1}$ key (usually obtained by pressing [2ND] [SIN]) is used to find the angle when the sine of that angle is known. The following example illustrates the use of the similar $\cos^{-1}$ key.

EXAMPLE 4 Inverse function value from calculator

If $\cos \theta = 0.3527$, which means that $\theta = \cos^{-1} 0.3527$ (θ is the angle whose cosine is 0.3527), we can use a calculator to find θ. The display is shown in Fig. 4.22.

Therefore, $\theta = 69.35°$ (rounded off). ■

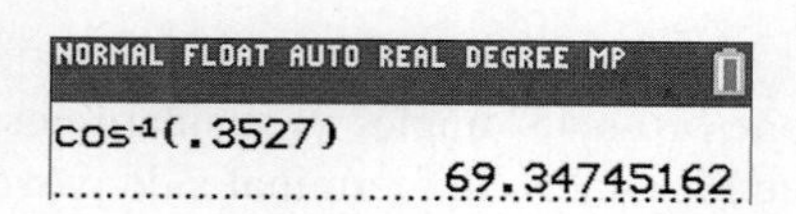

Fig. 4.22

When using the trigonometric functions, the angle is often *approximate*. Angles of 2.3°, 92.3°, and 182.3° are angles with equal accuracy, which shows that *the accuracy of an angle does not depend on the number of digits shown*. The measurement of an angle and the accuracy of its trigonometric functions are shown in Table 4.1:

Practice Exercises

1. Find the value of sin 12.5°.
2. Find θ if $\tan \theta = 1.039$.

Table 4.1 Angles and Accuracy of Trigonometric Functions

Measurements of Angle to Nearest	*Accuracy of Trigonometric Function*
1°	2 significant digits
0.1° or 10′	3 significant digits
0.01° or 1′	4 significant digits

We rounded off the result in Example 4 according to this table.

Although we can usually set up the solution of a problem in terms of the sine, cosine, or tangent, there are times when a value of the cotangent, secant, or cosecant is used. We now show how values of these functions are found on a calculator.

NOTE ▸ From the definitions, $\csc \theta = r/y$ and $\sin \theta = y/r$. This means *the value of* $\csc \theta$ *is the reciprocal of the value of* $\sin \theta$. Again, using the definitions, we find that *the value of* $\sec \theta$ *is the reciprocal of* $\cos \theta$ and *the value of* $\cot \theta$ *is the reciprocal of* $\tan \theta$. [Because the reciprocal of x equals x^{-1}, we can use the x^{-1} key along with the sin, cos, and tan keys, to find the values of $\csc \theta$, $\sec \theta$, and $\cot \theta$.]

EXAMPLE 5 Reciprocal functions value from calculator

To find the value of sec 27.82°, we use the fact that

$$\sec 27.82° = \frac{1}{\cos 27.82°} = 1.131 \quad \text{or} \quad \sec 27.82° = (\cos 27.82°)^{-1} = 1.131$$

Either of the above two options can be evaluated on a calculator to obtain the result 1.131, which has been rounded according to Table 4.1. ■

EXAMPLE 6 Given a function—find θ

To find the value of θ if $\cot \theta = 0.354$, we use the fact that

$$\tan \theta = \frac{1}{\cot \theta} = \frac{1}{0.354}$$

Therefore, $\theta = \tan^{-1}\left(\frac{1}{0.354}\right) = 70.5°$ (rounded off). ■

Practice Exercises

3. Find the values of cot 56.4°.
4. Find θ if $\csc \theta = 1.904$.

EXAMPLE 7 Given one function—find another

Find $\sin \theta$ if $\sec \theta = 2.504$.

Since the value of $\sec \theta$ is given, we know that $\cos \theta = 1/2.504$. This in turn tells us that $\theta = \cos^{-1}\left(\frac{1}{2.504}\right) \approx 66.46176°$ (extra significant digits are retained since this is an intermediate step).

Therefore, $\sin \theta = \sin(66.46176°) = 0.9168$ (see Fig. 4.23). ■

NORMAL FLOAT AUTO REAL DEGREE MP
cos⁻¹(1/2.504)
66.46176102
sin(66.46176102)
.9167937466

Fig. 4.23

The following example illustrates the use of the value of a trigonometric function in an applied problem. We consider various types of applications later in the chapter.

EXAMPLE 8 Evaluating cosine—rocket velocity

When a rocket is launched, its horizontal velocity v_x is related to the velocity v with which it is fired by the equation $v_x = v \cos \theta$. Here, θ is the angle between the horizontal and the direction in which it is fired (see Fig. 4.24). Find v_x if $v = 1250$ m/s and $\theta = 36.0°$.

Substituting the given values of v and θ in $v_x = v \cos \theta$, we have

$$v_x = 1250 \cos 36.0°$$
$$= 1010 \text{ m/s}$$

Therefore, the horizontal velocity is 1010 m/s (rounded off). ■

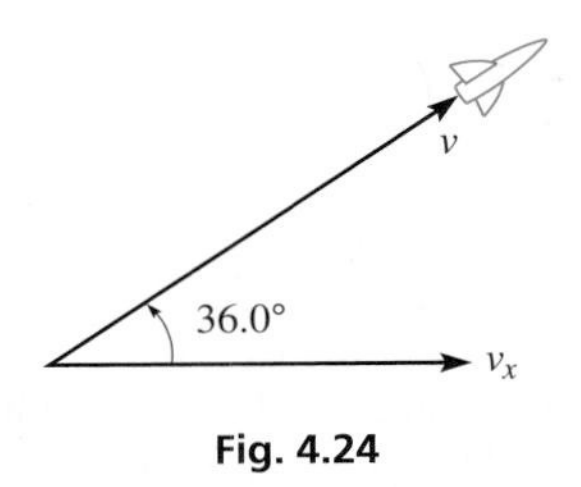

Fig. 4.24

EXERCISES 4.3

In Exercises 1–4, make the given changes in the indicated examples of this section and then find the indicated values.

1. In Example 4, change $\cos \theta$ to $\sin \theta$ and then find the angle.
2. In Example 5, change sec 27.82° to csc 27.82° and then find the value.
3. In Example 6, change 0.354 to 0.345 and then find the angle.
4. In Example 7, change $\sin \theta$ to $\tan \theta$ and then find the value.

In Exercises 5–20, find the values of the trigonometric functions. Round off results according to Table 4.1.

5. sin 34.9°
6. cos 72.5°
7. tan 57.6°
8. sin 36.0°
9. cos 15.71°
10. tan 8.653°
11. sin 88°
12. cos 0.7°
13. cot 57.86°
14. csc 22.81°
15. sec 80.4°
16. cot 41.8°
17. csc 0.49°
18. sec 7.8°
19. cot 85.96°
20. csc 76.30°

In Exercises 21–36, find θ for each of the given trigonometric functions. Round off results according to Table 4.1.

21. $\cos \theta = 0.3261$
22. $\tan \theta = 2.470$
23. $\sin \theta = 0.9114$
24. $\cos \theta = 0.0427$
25. $\tan \theta = 0.317$
26. $\sin \theta = 1.09$
27. $\cos \theta = 0.65007$
28. $\tan \theta = 5.7706$
29. $\csc \theta = 2.574$
30. $\sec \theta = 2.045$
31. $\cot \theta = 0.0606$
32. $\csc \theta = 1.002$
33. $\sec \theta = 0.305$
34. $\cot \theta = 14.4$
35. $\csc \theta = 8.26$
36. $\cot \theta = 0.1519$

In Exercises 37–40, use a protractor to draw the given angle. Measure off 10 units (centimeters are convenient) along the radius vector. Then measure the corresponding values of x and y. From these values, determine the trigonometric functions of the angle.

37. 40°
38. 75°
39. 15°
40. 53°

In Exercises 41–44, use a calculator to verify the given relationships or statements. $[\sin^2 \theta = (\sin \theta)^2]$

41. $\dfrac{\sin 43.7°}{\cos 43.7°} = \tan 43.7°$
42. $\sin^2 77.5° + \cos^2 77.5° = 1$
43. $\tan 70° = \dfrac{\tan 30° + \tan 40°}{1 - (\tan 30°)(\tan 40°)}$
44. $\sin 78.4° = 2(\sin 39.2°)(\cos 39.2°)$

In Exercises 45–50, explain why the given statements are true for an acute angle θ.

45. $\sin \theta$ is always between 0 and 1.
46. $\tan \theta$ can equal any positive real number.
47. $\cos \theta$ decreases in value as θ increases from 0° to 90°.
48. The value of $\sec \theta$ is never less than 1.
49. $\sin \theta + \cos \theta > 1$, if θ is acute.
50. If $\theta < 45°$, $\sin \theta < \cos \theta$.

In Exercises 51–54, find the values of the indicated trigonometric functions if θ is an acute angle.

51. Find $\sin \theta$, given $\tan \theta = 1.936$.
52. Find $\cos \theta$, given $\sin \theta = 0.6725$.
53. Find $\tan \theta$, given $\sec \theta = 1.3698$.
54. Find $\csc \theta$, given $\cos \theta = 0.1063$.

In Exercises 55–60, solve the given problems.

55. According to Snell's law, if a ray of light passes from air into water with an angle of incidence of 45.0°, then the angle of refraction θ_r is given by the equation $\sin 45.0° = 1.33 \sin \theta_r$. Find θ_r.
56. If the backup camera on a car is mounted at a height h above the road and is angled downward (from the horizontal) at an angle θ, then the distance x along the road between the car and the point at which the camera is directed is given by $x = \dfrac{h}{\tan \theta}$. Find x if $h = 24.0$ in. and $\theta = 15.0°$.

57. A brace is used in the structure shown in Fig. 4.25. Its length is $l = a(\sec\theta + \csc\theta)$. Find l if $a = 28.0$ cm and $\theta = 34.5°$.

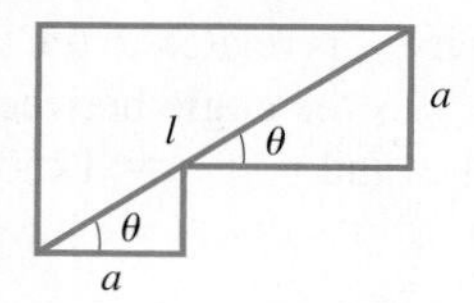

Fig. 4.25

58. The sound produced by a jet engine was measured at a distance of 100 m in all directions. The loudness d of the sound (in decibels) was found to be $d = 70.0 + 30.0\cos\theta$, where the 0° line was directed in front of the engine. Calculate d for $\theta = 54.5°$.

59. The signal from an AM radio station with two antennas d meters apart has a wavelength λ (in m). The intensity of the signal depends on the angle θ as shown in Fig. 4.26. An angle of minimum intensity is given by $\sin\theta = 1.50\,\lambda/d$. Find θ if $\lambda = 200$ m and $d = 400$ m.

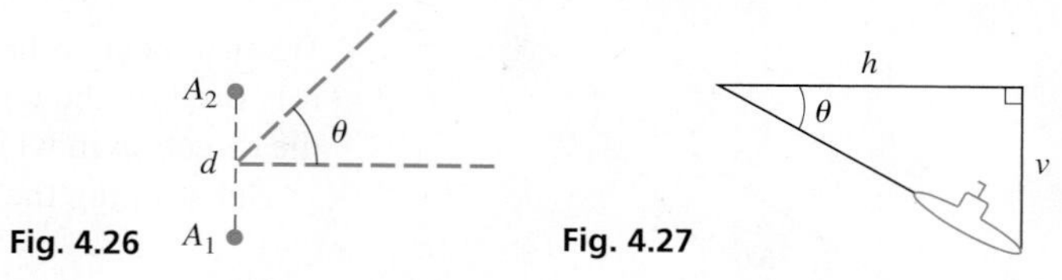

Fig. 4.26 Fig. 4.27

60. A submarine dives such that the horizontal distance h it moves and the vertical distance v it dives are related by $v = h\tan\theta$. Here, θ is the angle of the dive, as shown in Fig. 4.27. Find θ if $h = 2.35$ mi and $v = 1.52$ mi.

Answers to Practice Exercises

1. 0.216 **2.** 46.10° **3.** 0.664 **4.** 31.68°

4.4 The Right Triangle

Solving a Triangle • Cofunctions • Procedure for Solving a Right Triangle

We know that a triangle has three sides and three angles. If one side and any other two of these six parts are known, we can find the other three parts. One of the known parts must be a side, for if we know only the three angles, we know only that any triangle with these angles is similar to any other triangle with these angles.

EXAMPLE 1 Parts of a triangle

Assume that one side and two angles are known, such as the side of 5 and the angles of 35° and 90° in the triangle in Fig. 4.28. Then we may determine the third angle α by the fact that the sum of the angles of a triangle is always 180°. Of all possible similar triangles having the three angles of 35°, 90°, and 55° (which is α), we have the one with the particular side of 5 between angles of 35° and 90°. Only one triangle with these parts is possible (in the sense that all triangles with the given parts are *congruent* and have equal corresponding angles and sides). ■

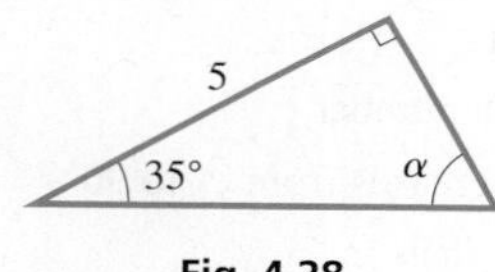

Fig. 4.28

To **solve a triangle** *means that, when we are given three parts of a triangle (at least one a side), we are to find the other three parts.* In this section, we are going to demonstrate the method of solving a right triangle. *Since one angle of the triangle is 90°, it is necessary to know one side and one other part.* Also, since the sum of the three angles is 180°, we know that *the sum of the other two angles is 90°, and they are acute angles.* This means they are **complementary angles.**

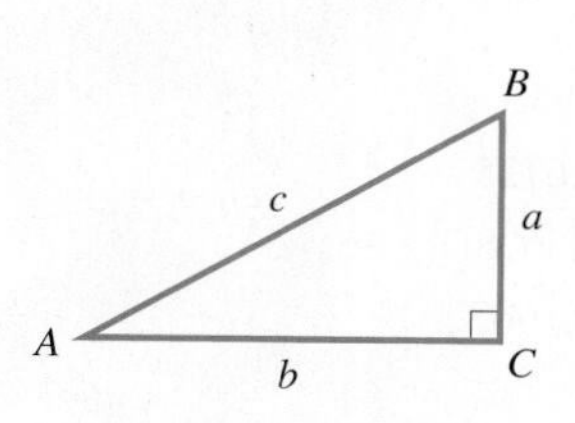

Fig. 4.29

For consistency, when we are labeling the parts of the right triangle, *we will use the letters A and B to denote the acute angles and C to denote the right angle. The letters a, b, and c will denote the sides opposite these angles, respectively. Thus, side c is the hypotenuse of the right triangle.* See Fig. 4.29.

In solving right triangles, we will find it convenient to express the trigonometric functions of the acute angles in terms of the sides. By placing the vertex of angle A at the origin and the vertex of right angle C on the positive x-axis, as shown in Fig. 4.30, we have the following ratios for angle A in terms of the sides of the triangle.

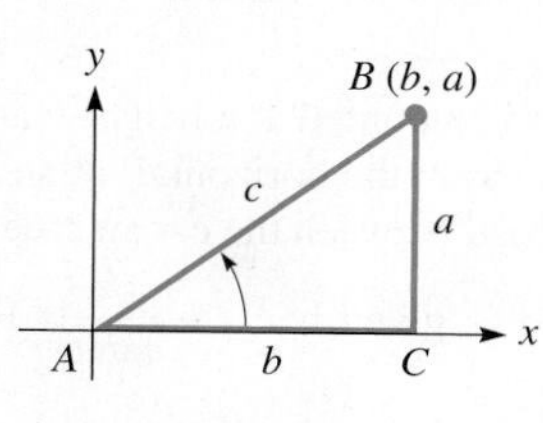

Fig. 4.30

$$\sin A = \frac{a}{c} \qquad \csc A = \frac{c}{a}$$
$$\cos A = \frac{b}{c} \qquad \sec A = \frac{c}{b} \qquad (4.3)$$
$$\tan A = \frac{a}{b} \qquad \cot A = \frac{b}{a}$$

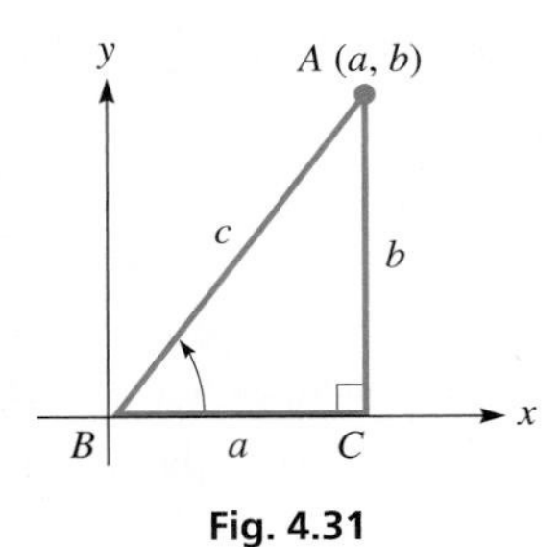

Fig. 4.31

If we should place the vertex of B at the origin, instead of the vertex of angle A, we would obtain the following ratios for the functions of angle B (see Fig. 4.31):

$$\sin B = \frac{b}{c} \qquad \csc B = \frac{c}{b}$$
$$\cos B = \frac{a}{c} \qquad \sec B = \frac{c}{a} \qquad \textbf{(4.4)}$$
$$\tan B = \frac{b}{a} \qquad \cot B = \frac{a}{b}$$

Equations (4.3) and (4.4) show that we may generalize our definitions of the trigonometric functions of an acute angle of a right triangle (we have chosen $\angle A$ in Fig. 4.32) to be as follows:

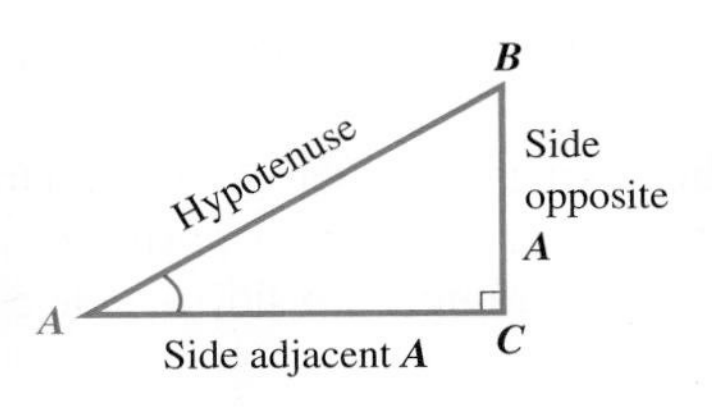

Fig. 4.32

$$\sin A = \frac{\text{side opposite } A}{\text{hypotenuse}} \qquad \csc A = \frac{\text{hypotenuse}}{\text{side opposite } A}$$
$$\cos A = \frac{\text{side adjacent A}}{\text{hypotenuse}} \qquad \sec A = \frac{\text{hypotenuse}}{\text{side adjacent } A} \qquad \textbf{(4.5)}$$
$$\tan A = \frac{\text{side opposite } A}{\text{side adjacent } A} \qquad \cot A = \frac{\text{side adjacent } A}{\text{side opposite } A}$$

■ Which side is adjacent or opposite depends on the angle being considered. In Fig. 4.32, the side opposite A is adjacent to B, and the side adjacent to A is opposite B.

Using the definitions in this form, we can solve right triangles without placing the angle in standard position. The angle need only be a part of any right triangle.

We note from the above discussion that $\sin A = \cos B$, $\tan A = \cot B$, and $\sec A = \csc B$. From this, we conclude that ***cofunctions of acute complementary angles are equal***. The sine function and cosine function are cofunctions, the tangent function and cotangent function are cofunctions, and the secant function and cosecant function are cofunctions.

EXAMPLE 2 Cofunctions of complementary angles

Given $a = 4$, $b = 7$, and $c = \sqrt{65}$ (see Fig. 4.33), find $\sin A$, $\cos A$, $\tan A$, $\sin B$, $\cos B$, and $\tan B$ in exact form and in approximate decimal form (to three significant digits).

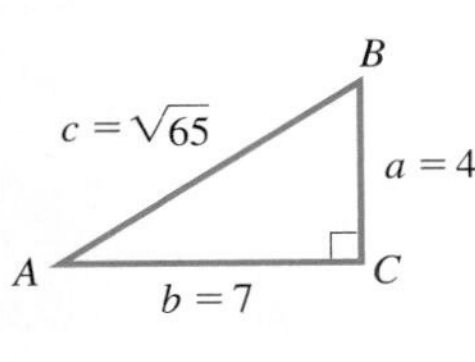

Fig. 4.33

$$\sin A = \frac{\text{side opposite angle } A}{\text{hypotenuse}} = \frac{4}{\sqrt{65}} = 0.496 \qquad \sin B = \frac{\text{side opposite angle } B}{\text{hypotenuse}} = \frac{7}{\sqrt{65}} = 0.868$$

$$\cos A = \frac{\text{side adjacent angle } A}{\text{hypotenuse}} = \frac{7}{\sqrt{65}} = 0.868 \qquad \cos B = \frac{\text{side adjacent angle } B}{\text{hypotenuse}} = \frac{4}{\sqrt{65}} = 0.496$$

$$\tan A = \frac{\text{side opposite angle } A}{\text{side adjacent angle } A} = \frac{4}{7} = 0.571 \qquad \tan B = \frac{\text{side opposite angle } B}{\text{side adjacent angle } B} = \frac{7}{4} = 1.75$$

We see that A and B are complementary angles. Comparing values of the functions of angles A and B, we see that $\sin A = \cos B$ and $\cos A = \sin \text{B}$. ■

We are now ready to solve right triangles, which can be done by using the following procedure.

Procedure for Solving a Right Triangle

1. Sketch a right triangle and label the known and unknown sides and angles.
2. Express each of the three unknown parts in terms of the known parts and solve for the unknown parts.
3. Check the results. The sum of the angles should be 180°. If only one side is given, check the computed side with the Pythagorean theorem. If two sides are given, check the angles and computed side by using appropriate trigonometric functions.

EXAMPLE 3 Given angle and side—find other parts

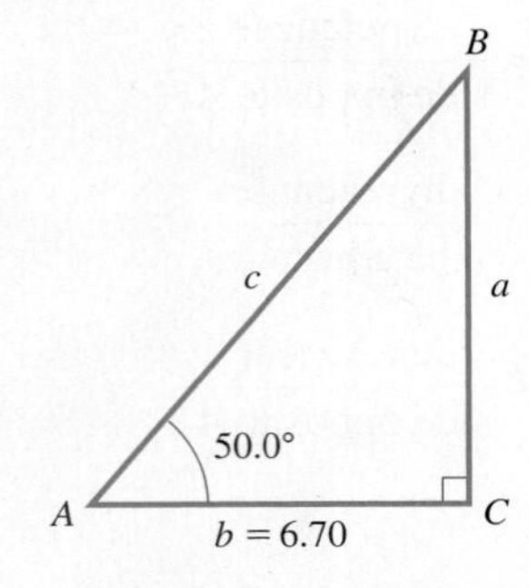

Fig. 4.34

Solve the right triangle with $A = 50.0°$ and $b = 6.70$.

We first sketch the triangle shown in Fig. 4.34. In making the sketch, we should make sure that angle C is the right angle, side c is opposite angle C, side a is opposite angle A, and side b is opposite angle B. The unknown parts of the triangle are side a, side c and angle B.

To find side a, we use the equation $\tan A = \dfrac{\text{side opposite } A}{\text{side adjacent } A}$.

$$\tan 50.0° = \frac{a}{6.70}.$$

$$a = 6.70 \tan 50.0° \qquad \text{multiply both sides by 6.70}$$

$$= 7.98$$

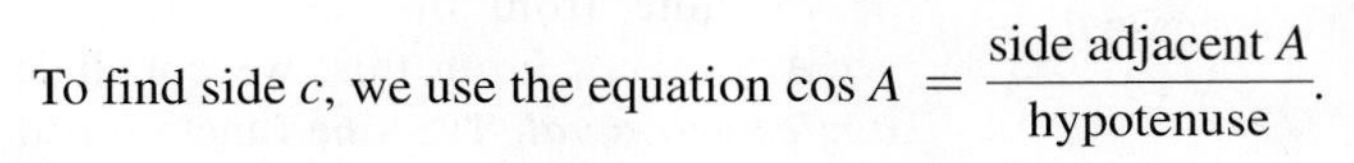

To find side c, we use the equation $\cos A = \dfrac{\text{side adjacent } A}{\text{hypotenuse}}$.

$$\cos 50.0° = \frac{6.70}{c}.$$

$$c(\cos 50.0°) = 6.70 \qquad \text{multiply both sides by } c$$

$$c = \frac{6.70}{\cos 50.0°} \qquad \text{divide both sides by cos 50.0°}$$

$$= 10.4$$

Now, solving for B, we know that $A + B = 90°$, or

$$B = 90° - A$$

$$= 90° - 50.0° = 40.0°$$

Therefore, $a = 7.98$, $c = 10.4$, and $B = 40.0°$.

$$\text{Checking the angles: } A + B + C = 50.0° + 40.0° + 90° = 180°$$

$$\text{Checking the sides: } c^2 = 108.16$$

$$a^2 + b^2 = 108.57$$

Because this is a right triangle, we should find that $a^2 + b^2 = c^2$. The sides check reasonably well, although not perfectly since the values of a and c have been rounded. The sides would check perfectly if nonrounded values were used. ■

Practice Exercises

1. In a right triangle, find a if $B = 20.0°$ and $c = 8.50$.
2. In a right triangle, find c if $B = 55.0°$ and $a = 228$.

CAUTION When solving for parts of a triangle, it is best to avoid using previously rounded answers to find other sides or angles because the round-off error can build up and deflate the accuracy of the final answer. In Example 3, we could have used the values of a and b, along with the Pythagorean theorem, to find the value of c. However, since the value of side a was rounded, this could decrease the accuracy of c. When it is absolutely necessary to use a rounded result to compute another answer, make sure to keep extra significant digits in the intermediate steps. ■

NOTE ➧ We should also point out that, by inspection, we can make a rough check on the sides and angles of any triangle. [The longest side is always opposite the largest angle, and the shortest side is always opposite the smallest angle.] In a right triangle, *the hypotenuse is always the longest side*. We see that this is true for the sides and angles for the triangle in Example 3, where c is the longest side (opposite the 90° angle) and b is the shortest side and is opposite the angle of 40°.

EXAMPLE 4 Given two sides—find other parts

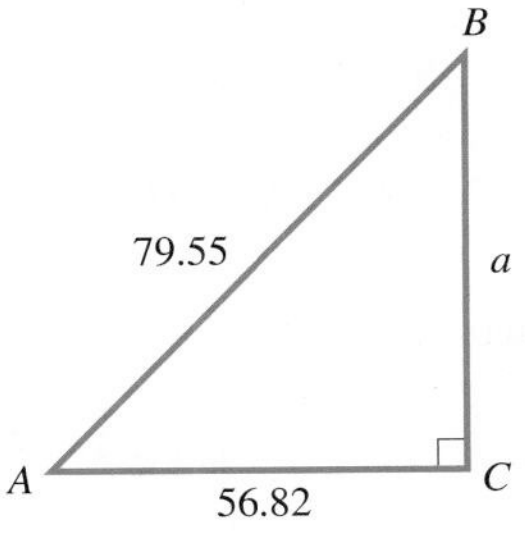

Solve the right triangle with $b = 56.82$ and $c = 79.55$.

We sketch the right triangle as shown in Fig. 4.35. Because two sides are given, we will use the Pythagorean theorem to find the third side a. Also, we will use the cosine to find $\angle A$.

B, A, C, 79.55, a, 56.82

Fig. 4.35

Because $c^2 = a^2 + b^2$, $a^2 = c^2 - b^2$. Therefore,

$$a = \sqrt{c^2 - b^2} = \sqrt{79.55^2 - 56.82^2}$$
$$= 55.67$$

Because $\cos A = \dfrac{\text{side adjacent } A}{\text{hypotenuse}}$, we have

$$\cos A = \frac{56.82}{79.55}$$
$$A = \cos^{-1}\left(\frac{56.82}{79.55}\right) = 44.42°$$

To find $\angle B$, we use the fact that $A + B = 90°$, which means that we have $B = 90° - A = 90° - 44.42° = 45.58°$.

We have now found that

$$a = 55.67 \qquad A = 44.42° \qquad B = 45.58°$$

Checking the sides and angles, we first note that side a is the shortest side and is opposite the smallest angle, $\angle A$. Also, the hypotenuse is the longest side. Next, using the sine function (we could use the cosine or tangent) to check the sides, we have

$$\sin 44.42° = \frac{55.67}{79.55}, \text{ or } 0.6999 \approx 0.6998 \qquad \sin 45.58° = \frac{56.82}{79.55}, \text{ or } 0.7142 \approx 0.7143$$

This shows that the values check. The small differences are due to round-off error. ■

Practice Exercise

3. In a right triangle, find B if $a = 20.0$ and $b = 28.0$.

■ As we noted earlier, the symbol $\approx$ means "equals approximately."

EXAMPLE 5 Unknown parts in terms of known parts

If A and a are known, express the unknown parts of a right triangle in terms of A and a.

We sketch a right triangle as in Fig. 4.36, and then set up the required expressions.

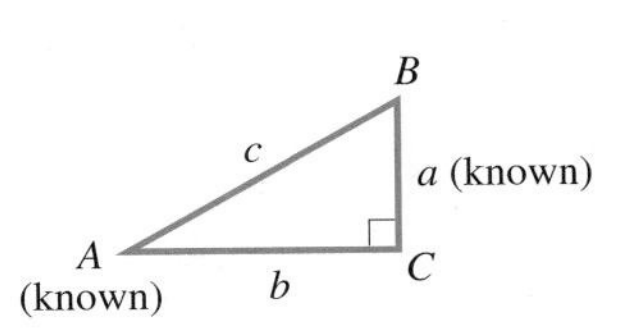

Fig. 4.36

Because $\dfrac{a}{b} = \tan A$, we have $a = b \tan A$, or $b = \dfrac{a}{\tan A}$.

Because $\dfrac{a}{c} = \sin A$, we have $a = c \sin A$, or $c = \dfrac{a}{\sin A}$.

Because A is known, $B = 90° - A$. ■

EXERCISES 4.4

In Exercises 1 and 2, make the given changes in the indicated examples of this section and then find the indicated values.

1. In Example 2, interchange a and b and then find the values.

2. In Example 4, change 56.82 to 65.82 and then solve the triangle.

In Exercises 3–6, draw appropriate figures and verify through observation that only one triangle may contain the given parts (that is, any others which may be drawn will be congruent).

3. A 60° angle included between sides of 3 in. and 6 in.

4. A side of 4 in. included between angles of 40° and 50°

5. A right triangle with a hypotenuse of 5 cm and a leg of 3 cm

6. A right triangle with a 70° angle between the hypotenuse and a leg of 5 cm

In Exercises 7–30, solve the right triangles with the given parts. Round off results according to Table 4.1. Refer to Fig. 4.37.

7. $A = 77.8°, a = 6700$

8. $A = 18.4°, c = 0.0897$

9. $a = 150, c = 345$

10. $a = 932, c = 1240$

11. $B = 32.1°, c = 238$

12. $B = 64.3°, b = 0.652$

13. $b = 82.0, c = 881$

14. $a = 5920, b = 4110$

15. $A = 32.10°, c = 56.85$

16. $B = 12.60°, c = 184.2$

17. $a = 56.73, b = 44.09$

18. $a = 9.908, c = 12.63$

19. $B = 37.5°, a = 0.862$

20. $A = 87.25°, b = 8.450$

21. $B = 74.18°, b = 1.849$

22. $A = 51.36°, a = 3692$

23. $a = 591.87, b = 264.93$

24. $b = 2.9507, c = 50.864$

25. $A = 2.975°, b = 14.592$

26. $B = 84.942°, a = 7413.5$

27. $B = 9.56°, c = 0.0973$

28. $a = 1.28, b = 16.3$

29. $a = 35.0, C = 90.0°$

30. $A = 25.7°, B = 64.3°$

Fig. 4.37

In Exercises 31–34, find the part of the triangle labeled either x or A in the indicated figure.

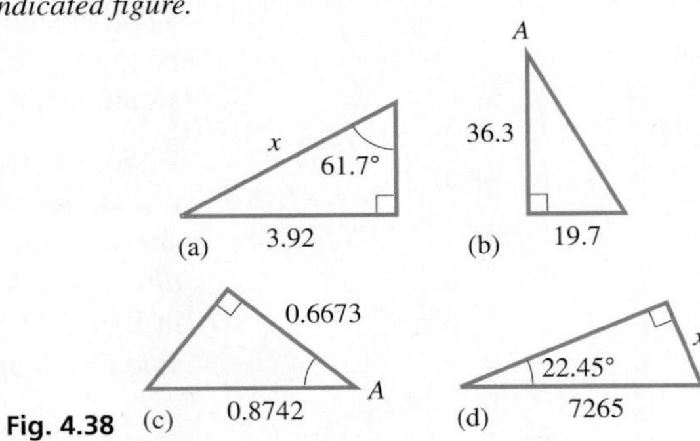

Fig. 4.38

31. Fig. 4.38(a)

32. Fig. 4.38(b)

33. Fig. 4.38(c)

34. Fig. 4.38(d)

In Exercises 35–38, find the indicated part of the right triangle that has the given parts.

35. One leg is 23.7, and the hypotenuse is 37.5. Find the smaller acute angle.

36. One leg is 8.50, and the angle opposite this leg is 52.3°. Find the other leg.

37. The hypotenuse is 964, and one angle is 17.6°. Find the longer leg.

38. The legs are 0.596 and 0.842. Find the larger acute angle.

In Exercises 39–42, solve the given problems.

39. Find the exact area of a circle inscribed in a regular hexagon (the circle is tangent to each of the six sides) of perimeter 72.

40. Find the exact perimeter of an equilateral triangle that is inscribed in a circle (each vertex is on the circle) of circumference 20π.

41. **One World Trade Center** refers to the main building of the rebuilt World Trade Center in New York City, which was opened in 2014. A person standing on level ground 350.0 ft from the base of the building must look upward (from the ground) at an angle of 78.85° to see the tip of the spire on top of the building. Use a right triangle to find the height of the building to 4 significant digits. (Does your answer remind you of a famous date in U.S. history?) *(This problem is included in memory of those who suffered and died as a result of the terrorist attack of September 11, 2001.)*

42. The screen on a certain Samsung Galaxy tablet has 2048 pixels along its length and 1536 pixels along its width. Using a right triangle, find the angle between the longer side and the diagonal of the screen. If the diagonal measures 9.7 in., find the length and width of the screen. (*Source*: www.samsung.com.)

Answers to Practice Exercises

1. $a = 7.99$ **2.** 398 **3.** 54.5°

4.5 Applications of Right Triangles

Angles of Elevation and Depression • Indirect Measurements of Distances and Angles

Many problems in science, technology, and everyday life can be solved by finding the missing parts of a right triangle. In this section, we illustrate a number of these in the examples and exercises.

EXAMPLE 1 Angle of elevation

Horseshoe Falls on the Canadian side of Niagara Falls can be seen from a small boat 2500 ft downstream. The **angle of elevation** *(the angle between the horizontal and the line of sight, when the object is above the horizontal)* from the observer to the top of Horseshoe Falls is 4°. How high are the Falls?

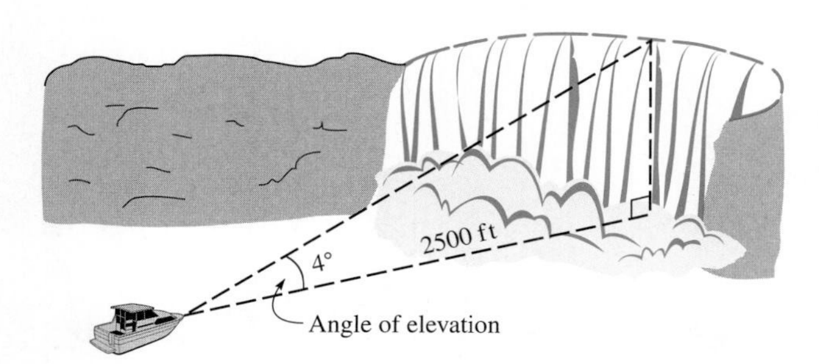

Fig. 4.39

By drawing an appropriate figure, as shown in Fig. 4.39, we show the given information and what we are to find. Here, we let h be the height of the Falls and label the horizontal distance from the boat to the base of the Falls as 2500 ft. From the figure, we see that

$$\tan 4° = \frac{h}{2500} \qquad \text{tangent of given angle} = \frac{\text{required opposite side}}{\text{given adjacent side}}$$

$$h = 2500 \tan 4° \qquad \text{multiply both sides by 2500}$$

$$= 170 \text{ ft}$$

We have rounded off the result since the data are good only to two significant digits. (Niagara Falls is on the border between the United States and Canada and is divided into the American Falls and the Horseshoe Falls. About 500,000 tons of water flow over the Falls each minute.) ■

EXAMPLE 2 Angle of depression

The Goodyear blimp is 1850 ft above the ground and south of the Rose Bowl in California during a Super Bowl game. The **angle of depression** *(the angle between the horizontal and the line of sight, when the object is below the horizontal)* of the north goal line from the blimp is 58.5°. How far is the observer in the blimp from the goal line?

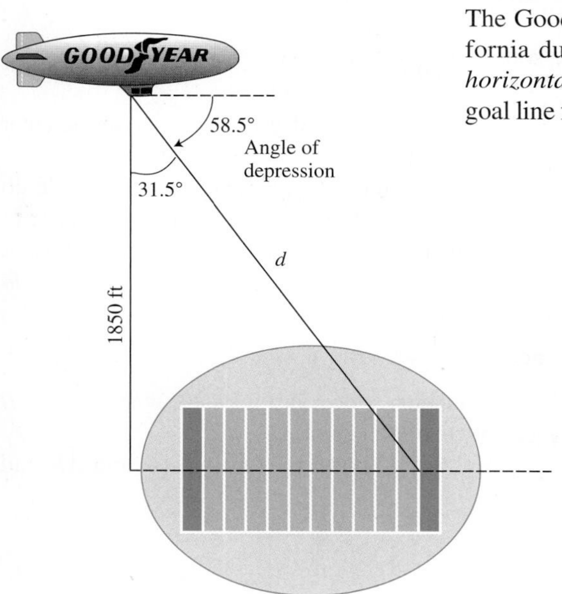

Fig. 4.40

Again, we sketch a figure as shown in Fig. 4.40. Here, we let d be the distance between the blimp and the north goal line. From the figure, we see that

$$\cos 31.5° = \frac{1850}{d} \qquad \text{cosine of known angle} = \frac{\text{given adjacent side}}{\text{required hypotenuse}}$$

$$d \cos 31.5° = 1850 \qquad \text{multiply both sides by } d$$

$$d = \frac{1850}{\cos 31.5°} \qquad \text{divide both sides by } \cos 31.5°$$

$$= 2170 \text{ ft}$$

In this case, we have rounded off the result to three significant digits, which is the accuracy of the given information. (The Super Bowl games in 1977, 1980, 1983, 1987, and 1993 were played in the Rose Bowl.) ■

Carefully note the difference between the angle of elevation and the angle of depression. NOTE ▸ [The angle of elevation is the angle through which an object is observed by *elevating* the line of sight above the horizontal. The angle of depression is the angle through which the object is observed by *depressing* (lowering) the line of sight below the horizontal.]

EXAMPLE 3 Height of missile

■ See the chapter introduction.

A missile is launched at an angle of 26.55° with respect to the horizontal. If it travels in a straight line over level terrain for 2.000 min and its average speed is 6355 km/h, what is its altitude at this time?

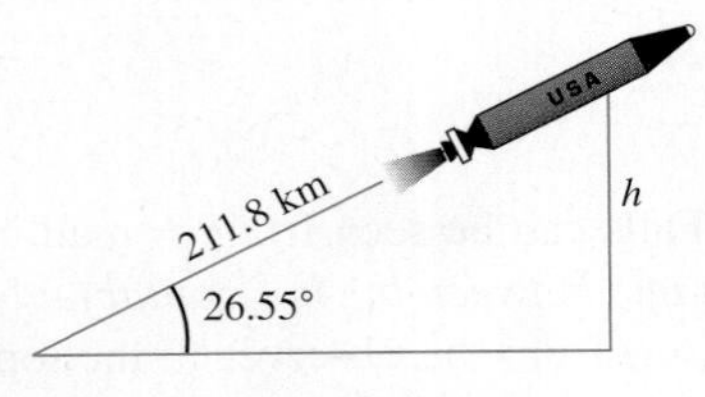

Fig. 4.41

In Fig. 4.41, we let h represent the altitude of the missile after 2.000 min (altitude is measured on a perpendicular line). Also, we determine that in this time the missile has flown 211.8 km in a direct line from the launching site. This is found from the fact that it travels at 6355 km/h for $\frac{1}{30.00}$ h (2.000 min), and distance = speed × time. We therefore have $\left(6355\frac{\text{km}}{\text{h}}\right)\left(\frac{1}{30.00}\text{h}\right) = 211.8$ km. This means

$$\sin 26.55° = \frac{h}{211.8} \qquad \text{sine of given angle} = \frac{\text{required opposite side}}{\text{known hypotenuse}}$$

$$h = 211.8(\sin 26.55°)$$

$$= 94.67 \text{ km}$$

■

EXAMPLE 4 Measurement of angle

One level at an airport is 22.5 ft above the level below. An escalator 48.0 ft long carries passengers between levels. What is the angle at which the escalator moves from one level to the other?

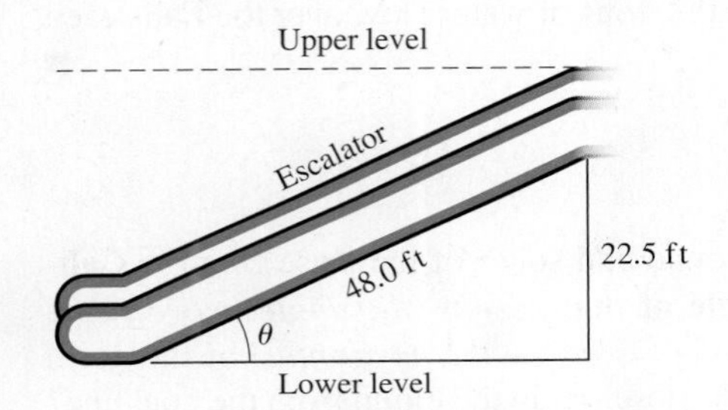

Fig. 4.42

In Fig. 4.42, we let θ be the angle at which the escalator rises (or descends) between levels. From the figure, we see that

$$\sin \theta = \frac{22.5}{48.0}$$

$$\theta = \sin^{-1}\left(\frac{22.5}{48.0}\right) = 28.0°$$

In finding this angle on the calculator, with the calculator in degree mode, we would enter $\sin^{-1}(22.5/48)$ and round off the displayed result of 27.95318688.

If we had been asked to find the horizontal distance the passengers move while on the escalator, we could use the angle we just calculated. However, since it is better to use given data, rather than derived results, that distance (42.4 ft) is better found using the Pythagorean theorem. ■

EXAMPLE 5 Surveyor—indirect measurement

Using lasers, a surveyor makes the measurements shown in Fig. 4.43, where points B and C are in a marsh. Find the distance between B and C.

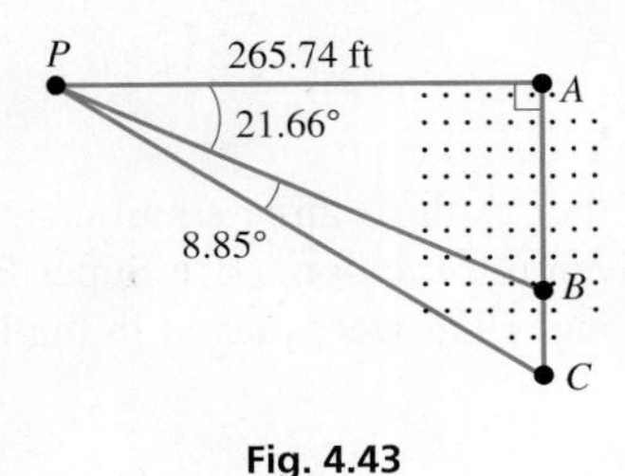

Fig. 4.43

Because the distance $BC = AC - AB$, BC is found by finding AC and AB and subtracting:

$$\frac{AB}{265.74} = \tan 21.66°$$

$$AB = 265.74 \tan 21.66°$$

$$\frac{AC}{265.74} = \tan(21.66° + 8.85°)$$

$$AC = 265.74 \tan 30.51°$$

$$BC = AC - AB = 265.74 \tan 30.51° - 265.74 \tan 21.66°$$

$$= 51.06 \text{ ft}$$

■

■ Lasers were first produced in the late 1950s.

EXAMPLE 6 Measurement of angle

A driver coming to an intersection sees the word STOP in the roadway. From the measurements shown in Fig. 4.44, find the angle θ that the letters make at the driver's eye.

From the figure, we know sides BS and BE in triangle BES and sides BT and BE in triangle BET. This means we can find $\angle TEB$ and $\angle SEB$ by use of the tangent. We then find θ from the fact that $\theta = \angle TEB - \angle SEB$.

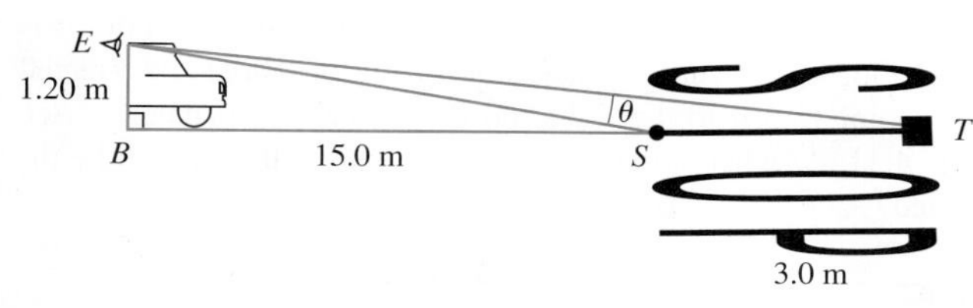

Fig. 4.44

$$\tan \angle TEB = \frac{18.0}{1.20} \qquad \angle TEB = 86.2°$$

$$\tan \angle SEB = \frac{15.0}{1.20} \qquad \angle SEB = 85.4°$$

$$\theta = 86.2° - 85.4° = 0.8°$$ ■

The indirect measurement of distances, as illustrated in the examples of this section, has been one of the most useful applications of trigonometry. As we mentioned in the chapter introduction, the Greek astronomer Hipparchus used some of the methods of trigonometry for measurements in astronomy. Since the time of Hipparchus, methods of indirect measurement have also been used in many other fields, such as surveying and navigation.

Practice Exercise

1. Find θ if the letters in the road are 2.0 m long, rather than 3.0 m long.

EXERCISES 4.5

In Exercises 1 and 2, make the given changes in the indicated examples of this section and then find the indicated values.

1. In Example 2, in line 4, change 58.5° to 62.1° and then find the distance.
2. In Example 3, in line 2, change 2.000 min to 3.000 min and then find the altitude.

In Exercises 3–44, solve the given problems. Sketch an appropriate figure, unless the figure is given.

3. A straight 12$\bar{0}$-ft culvert is built down a hillside that makes an angle of 25.0° with the horizontal. Find the height of the hill.
4. In 2000, about 70 metric tons of soil were removed from under the Leaning Tower of Pisa, and the angle the tower made with the ground was increased by about 0.5°. Before that, a point near the top of the tower was 50.5 m from a point at the base (measured along the tower), and this top point was directly above a point on the ground 4.25 m from the same base point. See Fig. 4.45. How much did the point on the ground move toward the base point?

Fig. 4.45

5. On level ground, a tree 12.0 m high has a shadow 85.0 m long. What is the angle of elevation of the sun?
6. The straight arm of a robot is 1.25 m long and makes an angle of 13.0° above a horizontal conveyor belt. How high above the belt is the end of the arm? See Fig. 4.46.

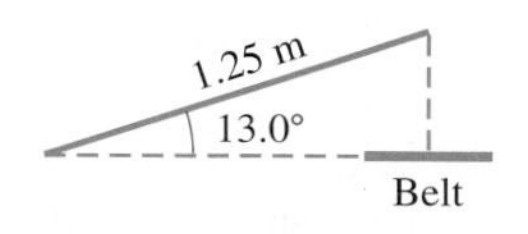

Fig. 4.46

7. The headlights of an automobile are set such that the beam drops 2.00 in. for each 25.0 ft in front of the car. What is the angle between the beam and the road?
8. A bullet was fired such that it just grazed the top of a table. It entered a wall, which is 12.60 ft from the graze point in the table, at a point 4.63 ft above the tabletop. At what angle was the bullet fired above the horizontal? See Fig. 4.47.

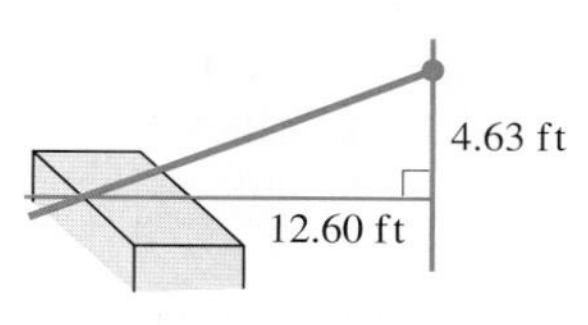

Fig. 4.47

9. In a series RL circuit, the *impedance* Z can be represented as the hypotenuse of a right triangle that has legs R (resistance) and X_L (inductive reactance). The angle ϕ between R and Z is called the *phase angle*. See Fig. 4.48. If $R = 12.0\ \Omega$ and $X_L = 15.0\ \Omega$, find the impedance and the phase angle.

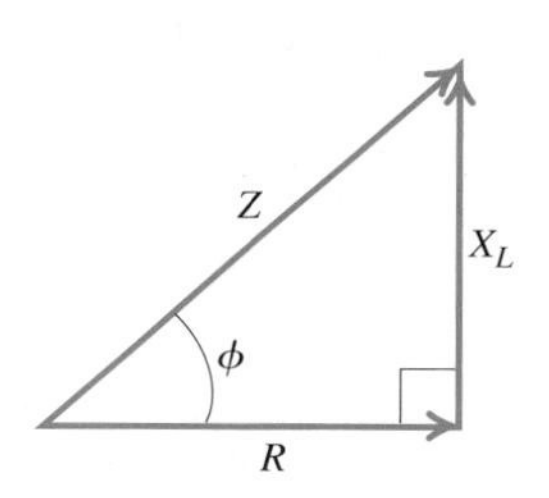

Fig. 4.48

10. In order to determine the length of a lake, a surveyor places poles in the ground at points A, B, and C and makes the measurements shown in Fig. 4.49. Determine the length of the lake, DE.

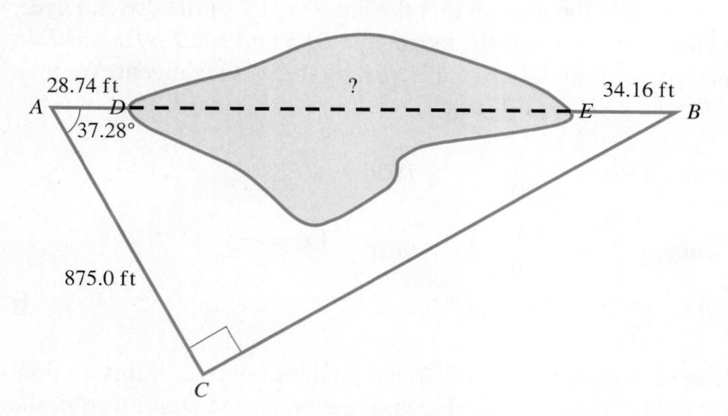

Fig. 4.49

11. The bottom of the doorway to a building is 2.65 ft above the ground, and a ramp to the door for the disabled is at an angle of 6.0° with the ground. How much longer must the ramp be in order to make the angle 3.0°?

12. On a test flight, during the landing of the space shuttle, the ship was 325 ft above the end of the landing strip. If it then came in at a constant angle of 6.5° with the landing strip, how far from the end of the landing strip did it first touch ground? (A successful reentry required that the angle of reentry be between 5.1° and 7.1°.)

13. From a point on the South Rim of the Grand Canyon, it is found that the angle of elevation of a point on the North Rim is 1.2°. If the horizontal distance between the points is 9.8 mi, how much higher is the point on the North Rim?

14. What is the steepest angle between the surface of a board 3.50 cm thick and a nail 5.00 cm long if the nail is hammered into the board such that it does not go through?

15. A robot is on the surface of Mars. The angle of depression from a camera in the robot to a rock on the surface of Mars is 13.33°. The camera is 196.0 cm above the surface. How far from the camera is the rock?

16. The Willis Tower in Chicago can be seen from a point on the ground known to be 5200 ft from the base of the tower. The angle of elevation from the observer to the top of the tower is 16°. How high is the Willis Tower? See Fig. 4.50.

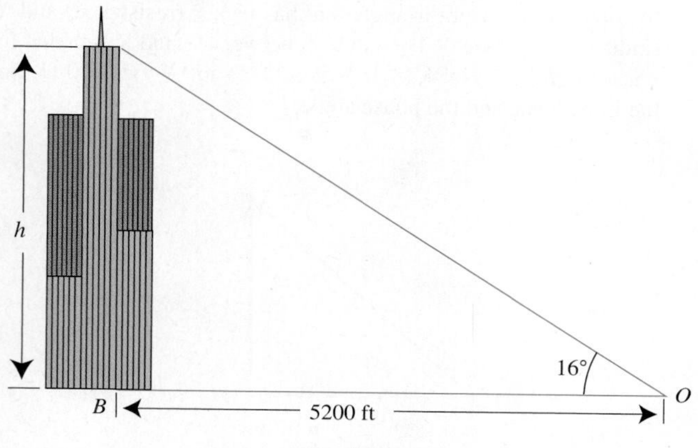

Fig. 4.50

17. A rectangular piece of plywood 4.00 ft by 8.00 ft is cut from one corner to an opposite corner. What are the angles between edges of the resulting pieces?

18. A guardrail is to be constructed around the top of a circular observation tower. The diameter of the observation area is 12.3 m. If the railing is constructed with 30 equal straight sections, what should be the length of each section?

19. To get a good view of a person in front of a teller's window, it is determined that a surveillance camera at a bank should be directed at a point 15.5 ft to the right and 6.75 ft below the camera. See Fig. 4.51. At what angle of depression should the camera be directed?

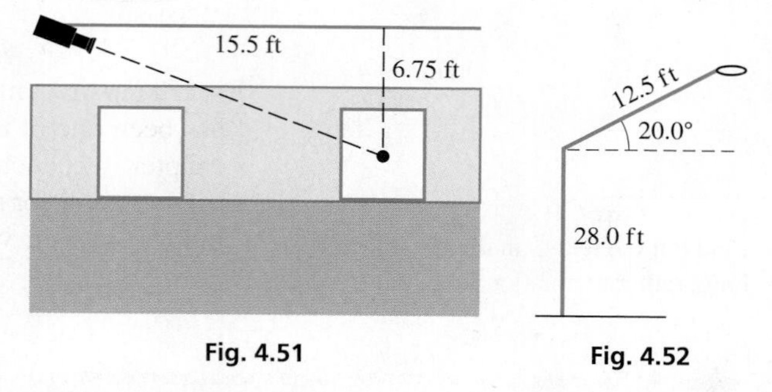

Fig. 4.51 Fig. 4.52

20. A street light is designed as shown in Fig. 4.52. How high above the street is the light?

21. A straight driveway is 85.0 ft long, and the top is 12.0 ft above the bottom. What angle does it make with the horizontal?

22. Part of the Tower Bridge in London is a drawbridge. This part of the bridge is 76.0 m long. When each half is raised, the distance between them is 8.0 m. What angle does each half make with the horizontal? See Fig. 4.53.

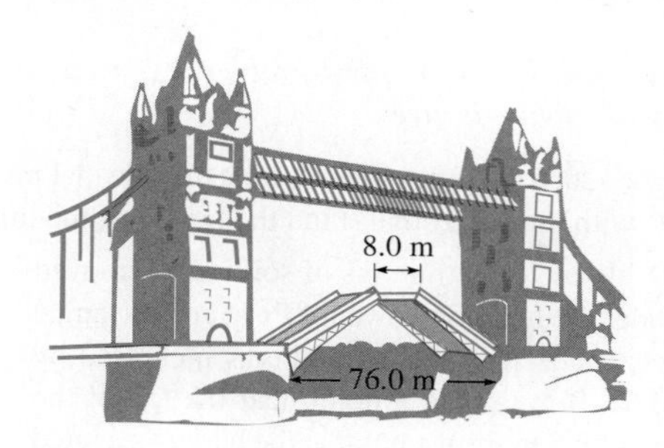

Fig. 4.53

23. The angle of inclination of a road is often expressed as *percent grade*, which is the vertical rise divided by the horizontal run (expressed as a percent). See Fig. 4.54. A 6.0% grade corresponds to a road that rises 6.0 ft for every 100 ft along the horizontal. Find the angle of inclination that corresponds to a 6.0% grade.

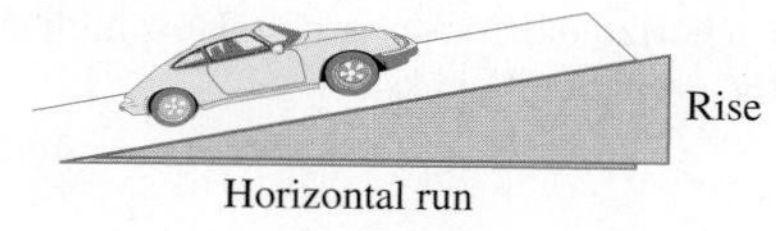

Fig. 4.54

24. A circular patio table of diameter 1.20 m has a regular octagon design inscribed within the outer edge (all eight vertices touch the circle). What is the perimeter of the octagon?

25. A square wire loop is rotating in the magnetic field between two poles of a magnet in order to induce an electric current. The axis of rotation passes through the center of the loop and is midway between the poles, as shown in the side view in Fig. 4.55. How far is the edge of the loop from either pole when the angle between the loop and the vertical is 78.0° if the side of the square is 7.30 cm and the poles are 7.66 cm apart?

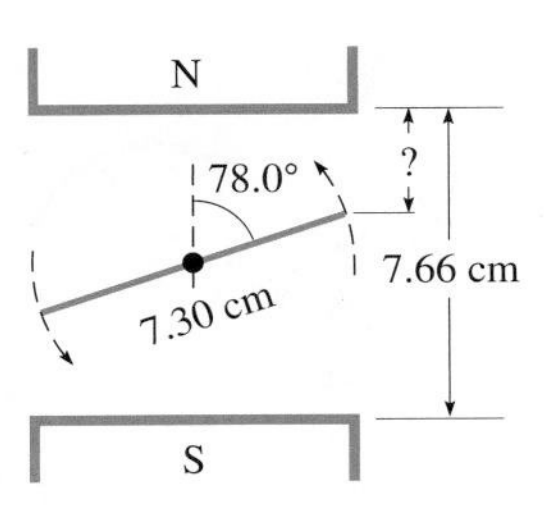

Fig. 4.55

26. From a space probe circling Io, one of Jupiter's moons, at an altitude of 552 km, it was observed that the angle of depression of the horizon was 39.7°. What is the radius of Io?

27. A manufacturing plant is designed to be in the shape of a regular pentagon with 92.5 ft on each side. A security fence surrounds the building to form a circle, and each corner of the building is to be 25.0 ft from the closest point on the fence. How much fencing is required?

28. A surveyor on the New York City side of the Hudson River wishes to find the height of a cliff (the Palisades) on the New Jersey side. Figure 4.56 shows the measurements that were made. How high is the cliff? (In the figure, the triangle containing the height h is vertical and perpendicular to the river.)

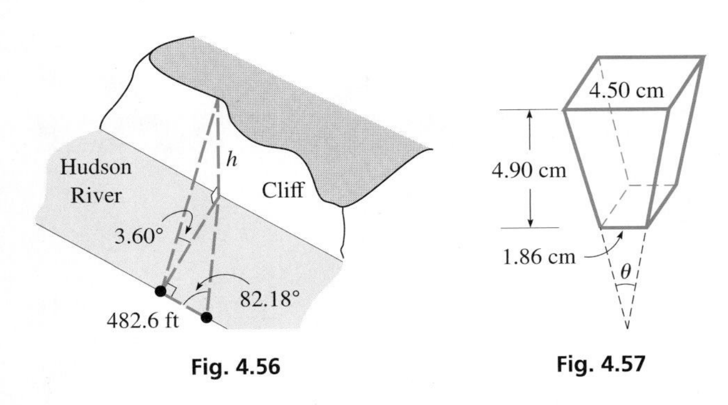

Fig. 4.56

Fig. 4.57

29. Find the angle θ in the taper shown in Fig. 4.57. (The front face is an isosceles trapezoid.)

30. The ratio of the width to the height of a TV screen is 16 to 9. What is the angle between the width and a diagonal of the screen?

31. When the distance from Earth to the sun is 94,500,000 mi, the angle that the sun subtends on Earth is 0.522°. Find the diameter of the sun.

32. Two ladders, each 6.50 m long are leaning against opposite walls of a level alley, with their feet touching. If they make angles of 38.0° and 68.0° with respect to the alley floor, how wide is the alley?

33. A TV screen is 116 cm wide and 65.3 cm high. A person is seated at an angle from the screen such that the near end of the screen is 224 cm away, and the far end is 302 cm away. If the person's eye level is at the bottom of the screen, what is the difference in the angles at the person's eye of the far end of the screen from the near end (angles α and β in Fig. 4.58)?

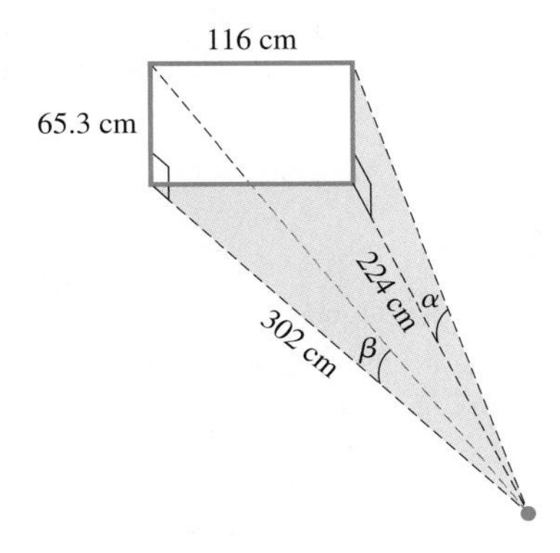

Fig. 4.58

34. A tract of land is in the shape of a trapezoid as shown in Fig. 4.59. Find the lengths of the nonparallel sides.

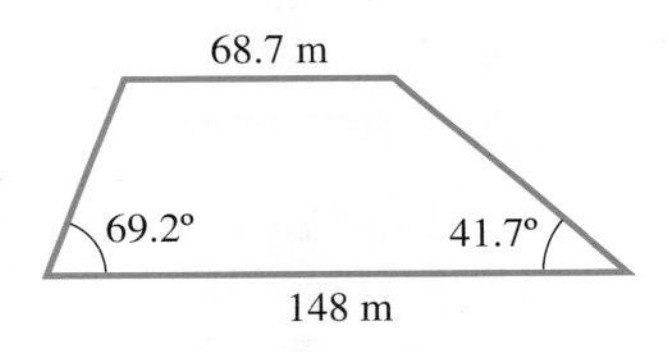

Fig. 4.59

35. A stairway 1.0 m wide goes from the bottom of a cylindrical storage tank to the top at a point halfway around the tank. The handrail on the outside of the stairway makes an angle of 31.8° with the horizontal, and the radius of the tank is 11.8 m. Find the length of the handrail. See Fig. 4.60.

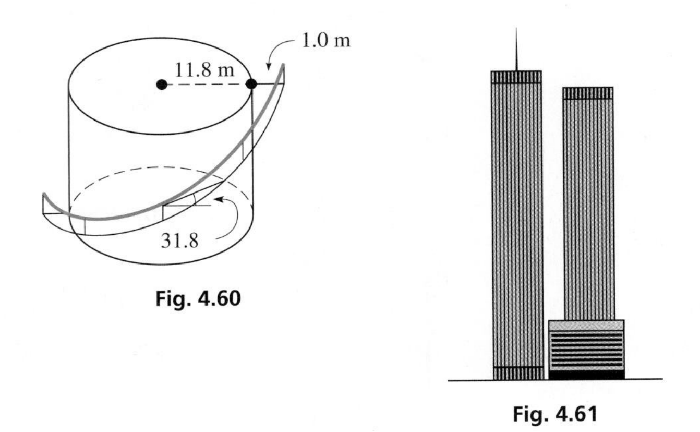

Fig. 4.60

Fig. 4.61

36. An antenna was on the top of the World Trade Center (before it was destroyed in 2001). From a point on the river 7800 ft from the Center, the angles of elevation of the top and bottom of the antenna were 12.1° and 9.9°, respectively. How tall was the antenna? (Disregard the small part of the antenna near the base that could not be seen.) The former World Trade Center is shown in Fig. 4.61.

37. Some of the streets of San Francisco are shown in Fig. 4.62. The distances between intersections A and B, A and D, and C and E are shown. How far is it between intersections C and D?

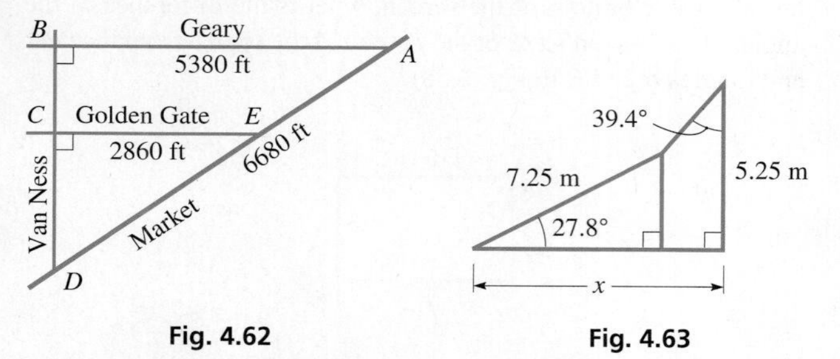

Fig. 4.62 **Fig. 4.63**

38. A supporting girder structure is shown in Fig. 4.63. Find the length x.

39. Find a formula for the area of the trapezoidal aqueduct cross section shown in Fig. 4.64.

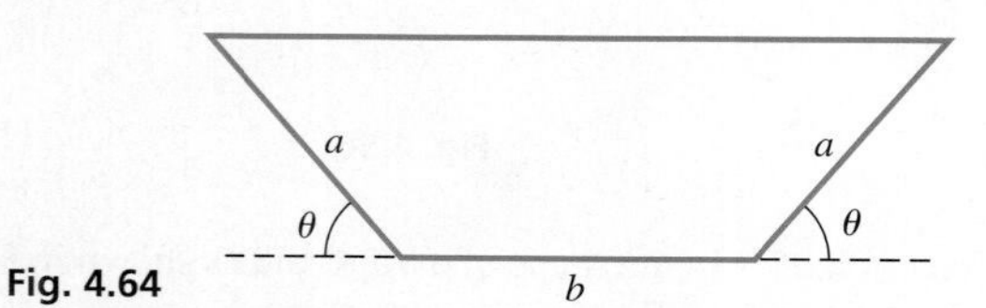

Fig. 4.64

40. The political banner shown in Fig. 4.65 is in the shape of a parallelogram. Find its area.

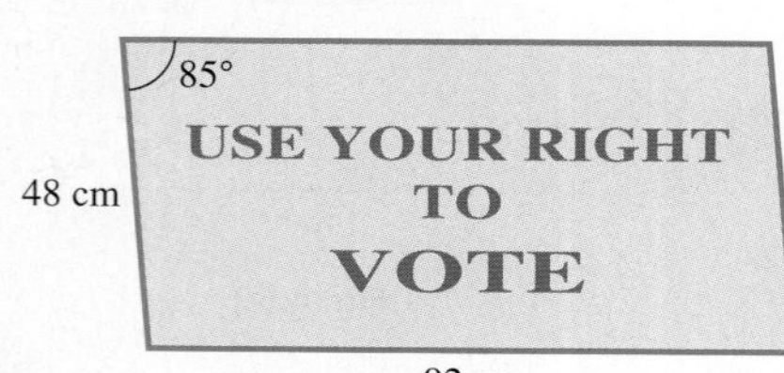

Fig. 4.65

41. The diameter d of a pipe can be determined by noting the distance x on the V-gauge shown in Fig. 4.66. Points A and B indicate where the pipe touches the gauge, and x equals either AV or VB. Find a formula for d in terms of x and θ.

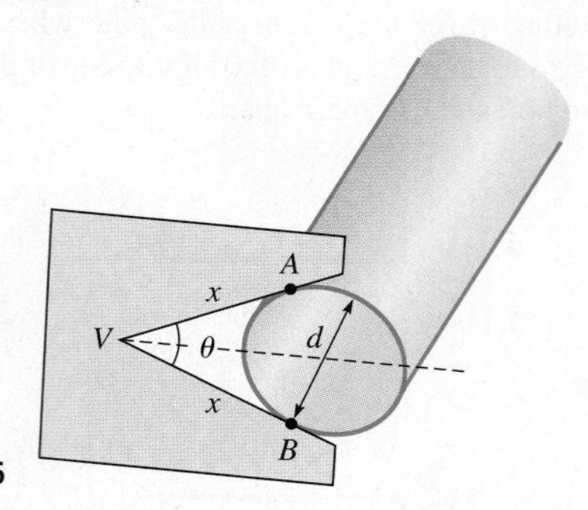

Fig. 4.66

42. A communications satellite is in orbit 35,300 km directly above the Earth's equator. What is the greatest latitude from which a signal can travel from the Earth's surface to the satellite in a straight line? The radius of the Earth is 6400 km.

43. What is the angle between the base of a cubical glass paperweight and a diagonal of the cube (from one corner to the opposite corner)?

44. Find the angle of view θ of the camera lens (see Fig. 4.67), given the measurements shown in the figure.

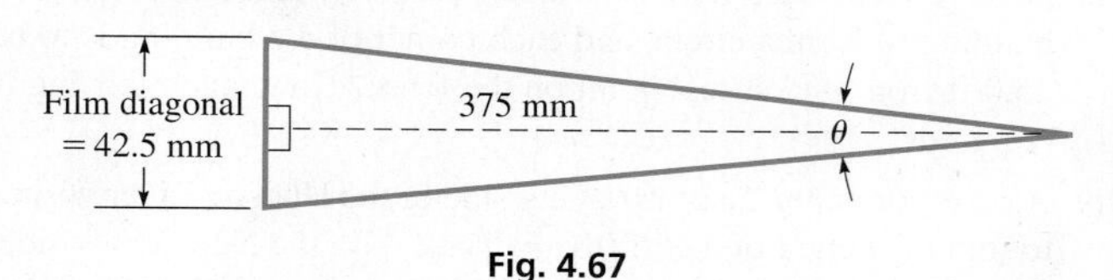

Fig. 4.67

Answers to Practice Exercise

1. $\theta = 0.6°$

CHAPTER 4 KEY FORMULAS AND EQUATIONS

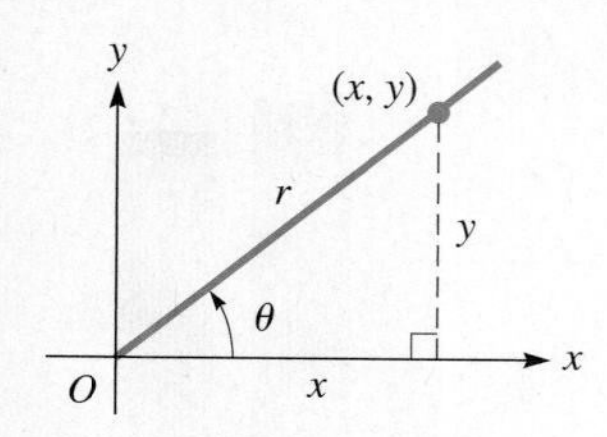

$$\sin\theta = \frac{y}{r} \qquad \csc\theta = \frac{r}{y}$$

$$\cos\theta = \frac{x}{r} \qquad \sec\theta = \frac{r}{x} \qquad (4.1)$$

$$\tan\theta = \frac{y}{x} \qquad \cot\theta = \frac{x}{y}$$

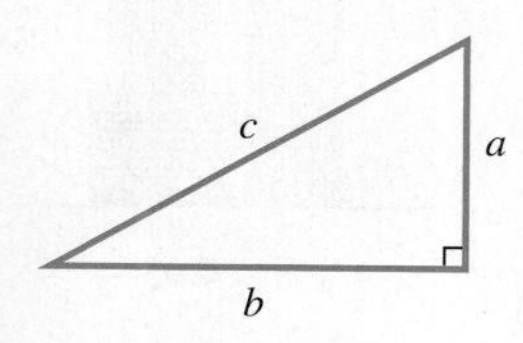

Pythagorean theorem

$$c^2 = a^2 + b^2 \qquad (4.2)$$

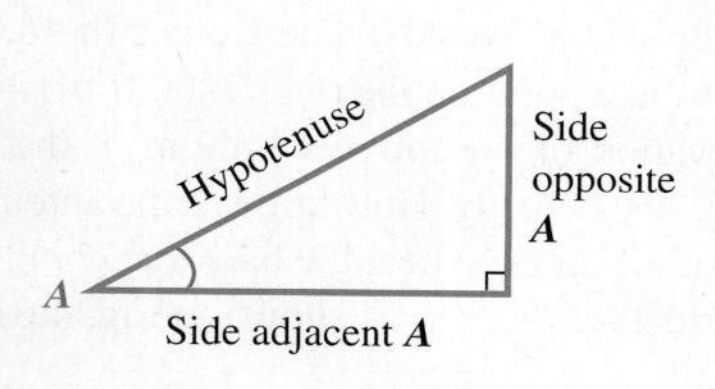

$$\sin A = \frac{\text{side opposite } A}{\text{hypotenuse}} \qquad \csc A = \frac{\text{hypotenuse}}{\text{side opposite } A}$$

$$\cos A = \frac{\text{side adjacent } A}{\text{hypotenuse}} \qquad \sec A = \frac{\text{hypotenuse}}{\text{side adjacent } A} \qquad (4.5)$$

$$\tan A = \frac{\text{side opposite } A}{\text{side adjacent } A} \qquad \cot A = \frac{\text{side adjacent } A}{\text{side opposite } A}$$

CHAPTER 4 REVIEW EXERCISES

CONCEPT CHECK EXERCISES

*Determine each of the following as being either **true** or **false**.*

1. A standard position angle of 205° is a second-quadrant angle.
2. In Example 2 of Section 4.2, if the terminal side passes through (6, 8) the values of the trigonometric functions are the same as those shown.
3. $\csc\theta$ and $\sin^{-1}\theta$ are the same.
4. In a right triangle with sides a, b, and c, and the standard angles opposite these sides, if $A = 45°$, then $a = b$.
5. In Example 2 of Section 4.5, if the angle of elevation from the north goal line to the blimp is 58.5°, the solution is the same.
6. If an angle has a cosine of 0.2, then the secant of the angle is 5.

PRACTICE AND APPLICATIONS

In Exercises 7–10, find the smallest positive angle and the smallest negative angle (numerically) coterminal with but not equal to the given angle.

7. 47.0° **8.** 338.8° **9.** −217.5° **10.** −0.72°

In Exercises 11–14, express the given angles in decimal form.

11. 31°54′ **12.** 574°45′ **13.** −83°21′ **14.** 321°27′

In Exercises 15–18, express the given angles to the nearest minute.

15. 17.5° **16.** −65.4° **17.** 749.75° **18.** 126.05°

In Exercises 19–22, determine the trigonometric functions of the angles (in standard position) whose terminal side passes through the given points. Give answers in exact form.

19. (24, 7) **20.** (5, 4) **21.** (48, 48) **22.** (0.36, 0.77)

In Exercises 23–28, find the indicated trigonometric functions. Give answers in decimal form, rounded off to three significant digits. Assume θ is an acute angle.

23. Given $\sin\theta = \frac{5}{13}$, find $\cos\theta$ and $\cot\theta$.

24. Given $\cos\theta = \frac{3}{8}$, find $\sin\theta$ and $\tan\theta$.

25. Given $\tan\theta = 2$, find $\cos\theta$ and $\csc\theta$.

26. Given $\cot\theta = 40$, find $\sin\theta$ and $\sec\theta$.

27. Given $\cos\theta = 0.327$, find $\sin\theta$ and $\csc\theta$.

28. Given $\csc\theta = 3.41$, find $\tan\theta$ and $\sec\theta$.

In Exercises 29–36, find the values of the trigonometric functions. Round off results.

29. sin 72.1° **30.** cos 40.3°

31. tan 85.68° **32.** sin 0.91°

33. sec 36.2° **34.** csc 82.4°

35. (cot 7.06°)(sin 7.06°) − cos 7.06°

36. (sec 79.36°)(sin 79.36°) − tan 79.36°

In Exercises 37–48, find θ for each of the given trigonometric functions. Assume θ is an acute angle. Round off results.

37. $\cos\theta = 0.850$ **38.** $\sin\theta = 0.63052$

39. $\tan\theta = 1.574$ **40.** $\cos\theta = 0.0135$

41. $\csc\theta = 4.713$ **42.** $\cot\theta = 0.7561$

43. $\sec\theta = 34.2$ **44.** $\csc\theta = 1.92$

45. $\cot\theta = 7.117$ **46.** $\sec\theta = 1.006$

47. $\sin\theta = 1.030$ **48.** $\tan\theta = 0.0052$

In Exercises 49 and 50, assume θ is an acute angle with the given trigonometric function value. Find the exact coordinates of the point where the terminal side of θ (in standard position) intersects the unit circle.

49. $\sin\theta = \dfrac{2}{5}$ **50.** $\tan\theta = 3$

In Exercises 51–60, solve the right triangles with the given parts. Refer to Fig. 4.68.

51. $A = 17.0°$, $b = 6.00$

52. $B = 68.1°$, $a = 1080$

53. $a = 81.0$, $b = 64.5$

54. $a = 106$, $c = 382$

55. $A = 37.5°$, $a = 12.0$

56. $B = 85.7°$, $b = 852.44$

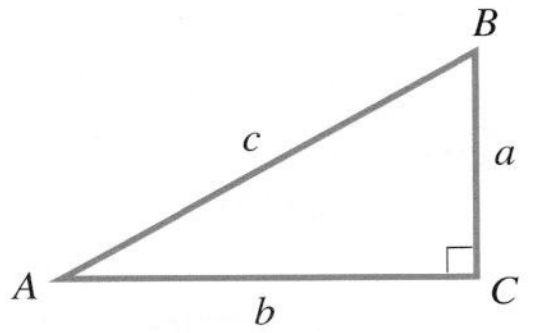

57. $b = 6.508$, $c = 7.642$ **58.** $a = 0.721$, $b = 0.144$

59. $A = 49.67°$, $c = 0.8253$ **60.** $B = 4.38°$, $b = 5682$

Fig. 4.68

In Exercises 61–105, solve the given problems.

61. Find the value of x for the triangle shown in Fig. 4.69.

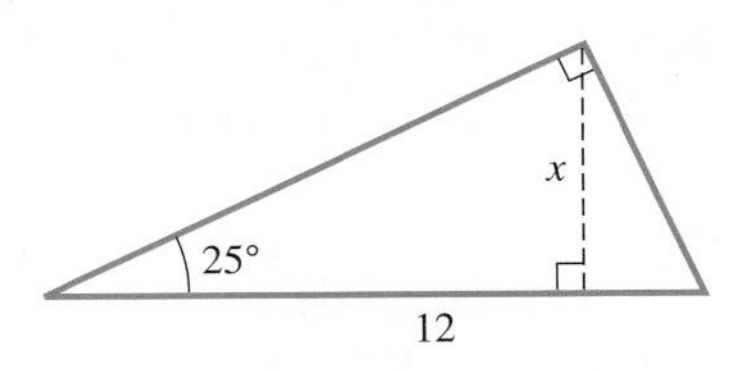

Fig. 4.69

62. Explain three ways in which the value of x can be found for the triangle shown in Fig. 4.70. Which of these methods is the easiest?

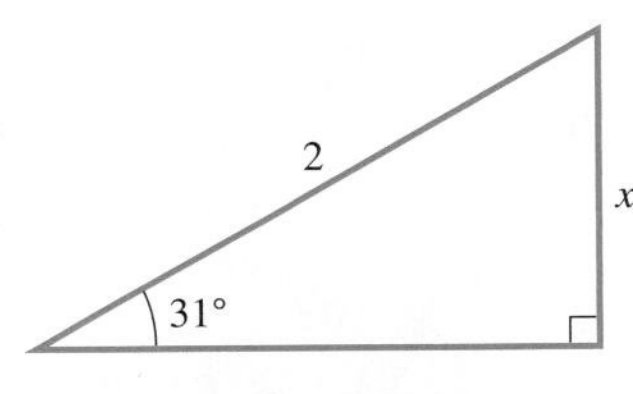

Fig. 4.70

63. Find the perimeter of a regular octagon (eight equal sides with equal interior angles) that is inscribed in a circle (all vertices of the octagon touch the circle) of radius 10.

64. Explain why values of $\sin\theta$ increase as θ increases from 0° to 90°.

65. What is x if (3, 2) and (x, 7) are on the same terminal side of an acute angle?

66. Two legs of a right triangle are 2.607 and 4.517. What is the smaller acute angle?

67. Show that the side c of any triangle ABC is related to the perpendicular h from C to side AB by the equation

$$c = h\cot A + h\cot B.$$

68. For the isosceles triangle shown in Fig. 4.71, show that $c = 2a\sin\frac{A}{2}$.

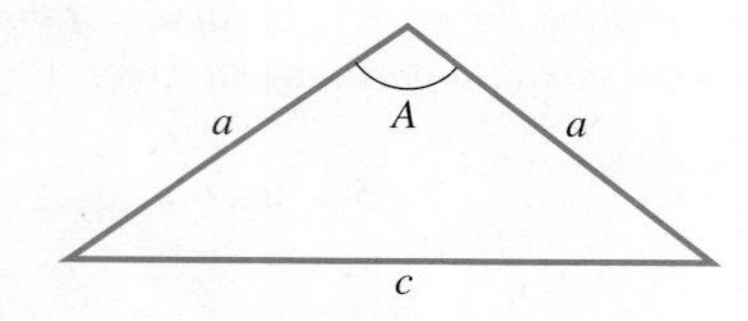

Fig. 4.71

69. In Fig. 4.72, find the length c of the chord in terms of r and the angle $\theta/2$.

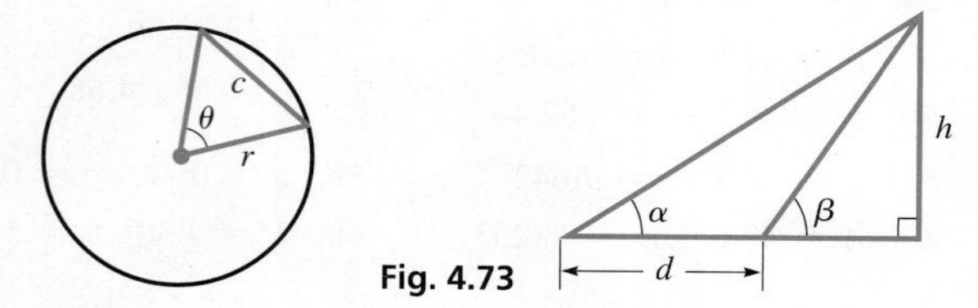

Fig. 4.72

Fig. 4.73

70. In Fig. 4.73, find a formula for h in terms of d, α, and β.

71. Find the angle between the line passing through the origin and (3, 2), and the line passing through the origin and (2, 3).

72. A sloped cathedral ceiling is between walls that are 7.50 ft high and 12.0 ft high. If the walls are 15.0 ft apart, at what angle does the ceiling rise?

73. The base of a 75-ft fire truck ladder is at the rear of the truck and is 4.8 ft above the ground. If the ladder is extended backward at an angle of 62° with the horizontal, how high up on the building does it reach, and how far from the building must the back of the truck be in order that the ladder just reach the building? See Fig. 4.74.

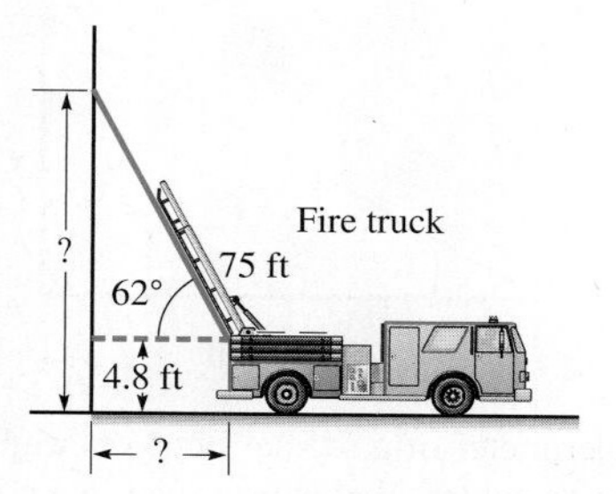

Fig. 4.74

74. A pendulum 1.25 m long swings through an angle of 5.6°. What is the distance between the extreme positions of the pendulum?

75. The voltage e at any instant in a coil of wire that is turning in a magnetic field is given by $e = E\cos\alpha$, where E is the maximum voltage and α is the angle the coil makes with the field. Find the acute angle α if $e = 56.9$ V and $E = 339$ V.

76. The area of a quadrilateral with diagonals d_1 and d_2 is $A = \frac{1}{2}d_1d_2\sin\theta$, where d_1 and d_2 are the diagonals and θ is the angle between them. Find the area of an approximately four-sided grass fire with diagonals of 320 ft and 440 ft and $\theta = 72°$.

77. For a car rounding a curve, the road should be banked at an angle θ according to the equation $\tan\theta = \frac{v^2}{gr}$. Here, v is the speed of the car and r is the radius of the curve in the road. See Fig. 4.75. Find θ for $v = 80.7$ ft/s (55.0 mi/h), $g = 32.2$ ft/s², and $r = 950$ ft.

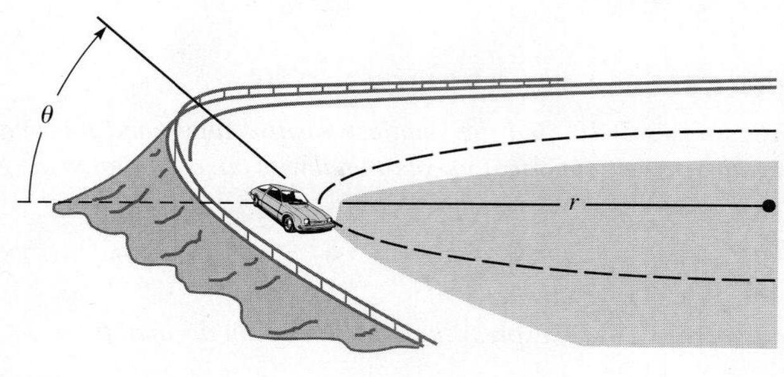

Fig. 4.75

78. The *apparent power* S in an electric circuit in which the power is P and the impedance phase angle is θ is given by $S = P\sec\theta$. Given $P = 12.0$ V · A and $\theta = 29.4°$, find S.

79. A surveyor measures two sides and the included angle of a triangular tract of land to be $a = 31.96$ m, $b = 47.25$ m, and $C = 64.09°$. (a) Show that a formula for the area A of the tract is $A = \frac{1}{2}ab\sin C$. (b) Find the area of the tract.

80. A water channel has the cross section of an isosceles trapezoid. See Fig. 4.76. (a) Show that a formula for the area of the cross section is $A = bh + h^2\cot\theta$. (b) Find A if $b = 12.6$ ft, $h = 4.75$ ft, and $\theta = 37.2°$.

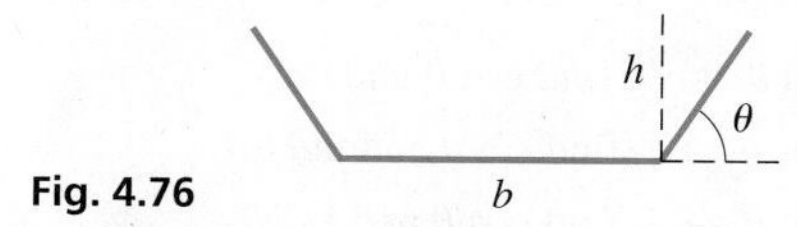

Fig. 4.76

81. In tracking an airplane on radar, it is found that the plane is 27.5 km on a direct line from the control tower, with an angle of elevation of 10.3°. What is the altitude of the plane?

82. A straight emergency chute for an airplane is 16.0 ft long. In being tested, the top of the chute is 8.5 ft above the ground. What angle does the chute make with the ground?

83. The windshield on an automobile is inclined 42.5° with respect to the horizontal. Assuming that the windshield is flat and rectangular, what is its area if it is 4.80 ft wide and the bottom is 1.50 ft in front of the top?

84. A water slide at an amusement park is 85 ft long and is inclined at an angle of 52° with the horizontal. How high is the top of the slide above the water level?

85. Find the area of the patio shown in Fig. 4.77.

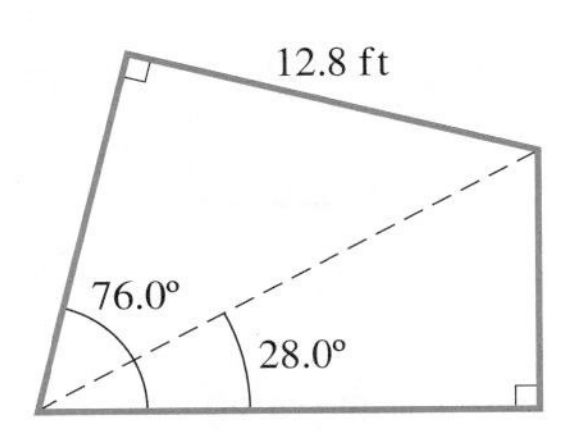

Fig. 4.77

86. The cross section (a regular trapezoid) of a levee to be built along a river is shown in Fig. 4.78. What is the volume of rock and soil that will be needed for a one-mile length of the levee?

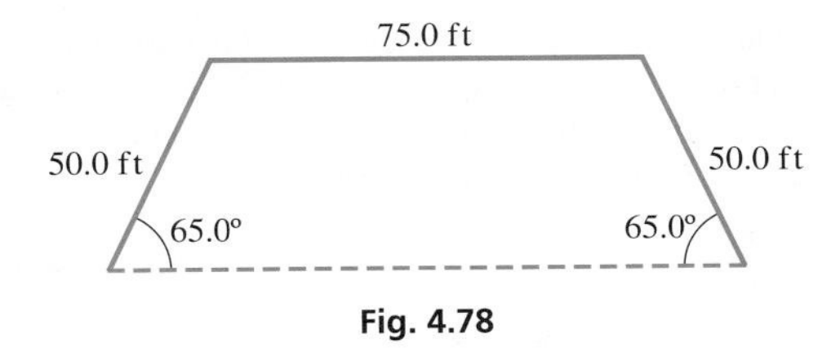

Fig. 4.78

87. The vertical cross section of an attic room in a house is shown in Fig. 4.79. Find the distance d across the floor.

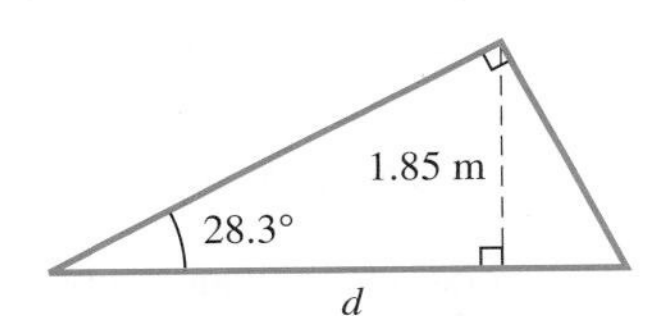

Fig. 4.79

88. The impedance Z and resistance R in an AC circuit may be represented by letting the impedance be the hypotenuse of a right triangle and the resistance be the side adjacent to the phase angle ϕ. If $R = 1.75 \times 10^3\ \Omega$ and $\phi = 17.38°$, find Z.

89. A typical aqueduct built by the Romans dropped on average at an angle of about 0.03° to allow gravity to move the water from the source to the city. For such an aqueduct of 65 km in length, how much higher was the source than the city?

90. The distance from the ground level to the underside of a cloud is called the *ceiling*. See Fig. 4.80. A ground observer 950 m from a searchlight aimed vertically notes that the angle of elevation of the spot of light on a cloud is 76°. What is the ceiling?

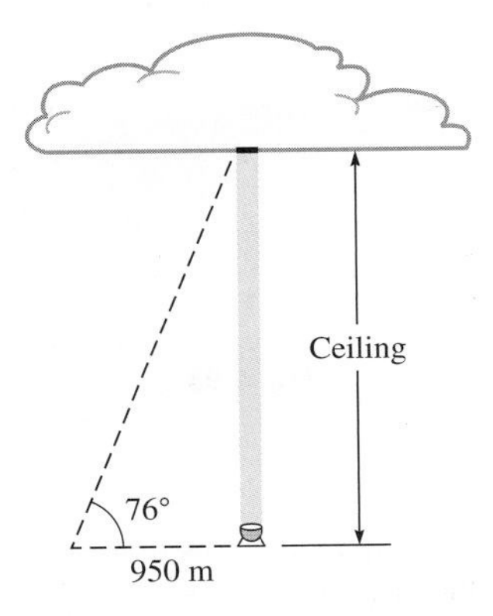

Fig. 4.80

91. The window of a house is shaded as shown in Fig. 4.81. What percent of the window is shaded when the angle of elevation θ of the sun is 65°?

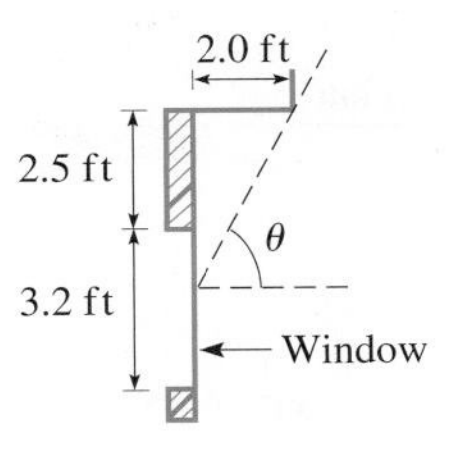

Fig. 4.81

92. A person standing on a level plain hears the sound of a plane, looks in the direction of the sound, but the plane is not there (familiar?). When the sound was heard, it was coming from a point at an angle of elevation of 25°, and the plane was traveling at 450 mi/h (660 ft/s) at a constant altitude of 2800 ft along a straight line. If the plane later passes directly over the person, at what angle of elevation should the person have looked directly to see the plane when the sound was heard? (The speed of sound is 1130 ft/s.) See Fig. 4.82.

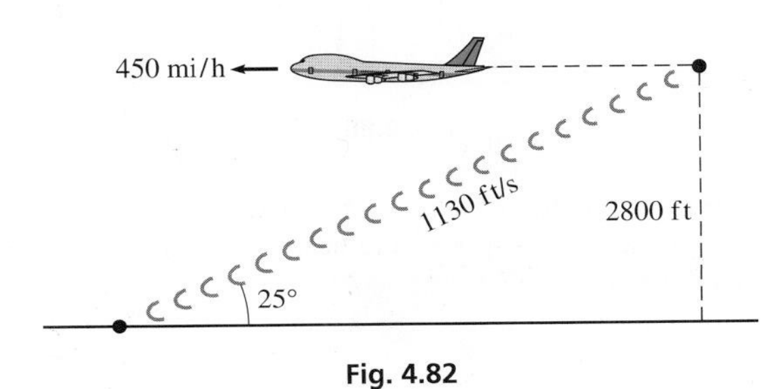

Fig. 4.82

93. In the structural support shown in Fig. 4.83, find x.

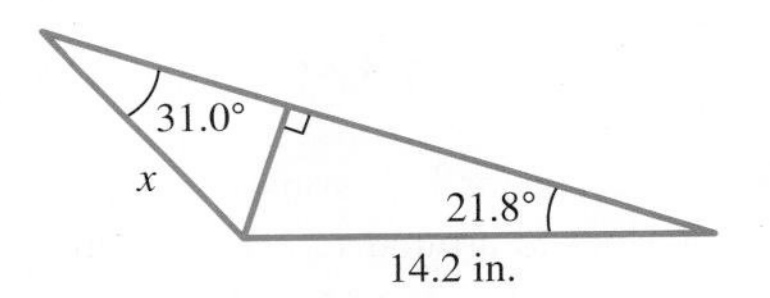

Fig. 4.83

94. The main span of the Mackinac Bridge (see Fig. 4.84) in northern Michigan is 1160 m long. The angle *subtended* by the span at the eye of an observer in a helicopter is 2.2°. Show that the distance calculated from the helicopter to the span is about the same if the line of sight is perpendicular to the end or to the middle of the span.

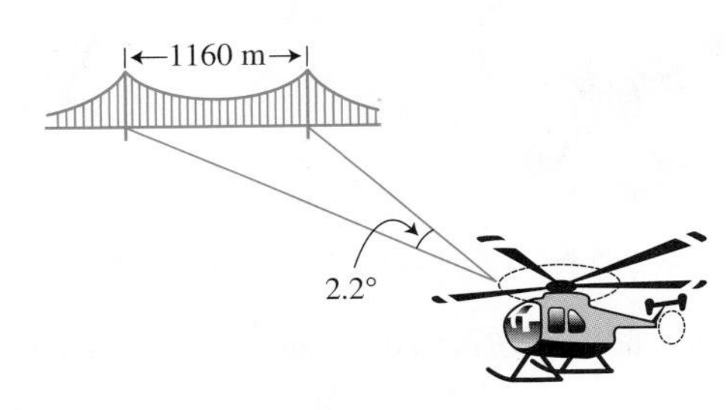

Fig. 4.84

95. A Coast Guard boat 2.75 km from a straight beach can travel at 37.5 km/h. By traveling along a line that is at 69.0° with the beach, how long will it take it to reach the beach? See Fig. 4.85.

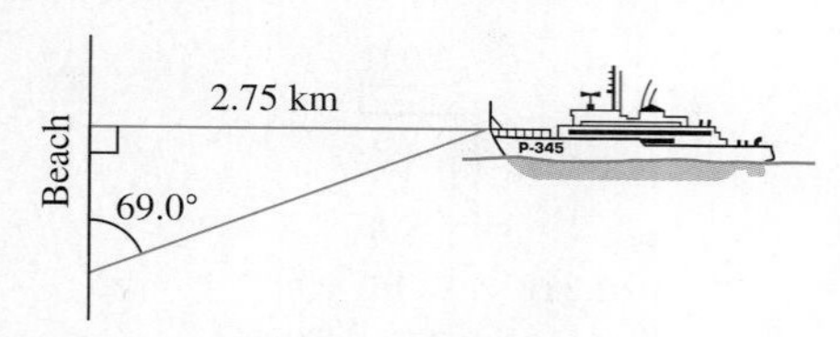

Fig. 4.85

96. Each side piece of the trellis shown in Fig. 4.86 makes an angle of 80.0° with the ground. Find the length of each side piece and the area covered by the trellis.

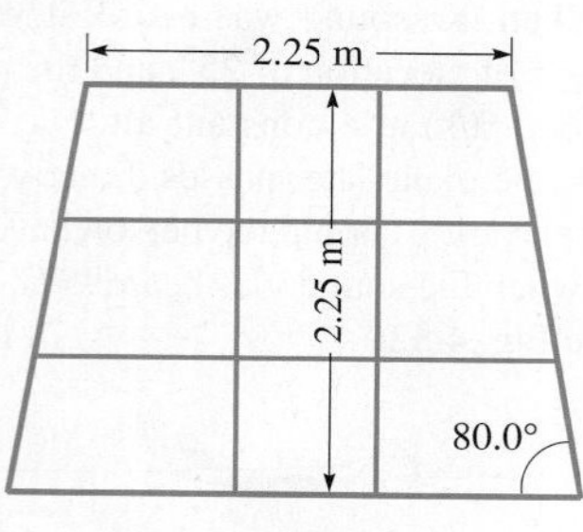

Fig. 4.86

97. A laser beam is transmitted with a "width" of 0.00200°. What is the diameter of a spot of the beam on an object 52,500 km distant? See Fig. 4.87.

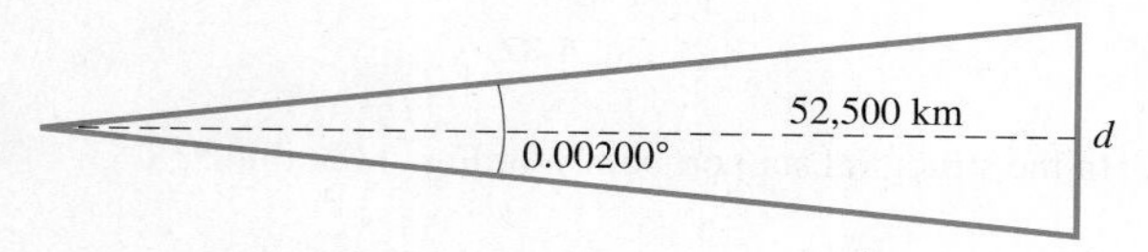

Fig. 4.87

98. The surface of a soccer ball consists of 20 regular hexagons (six sides) interlocked around 12 regular pentagons (five sides). See Fig. 4.88. (a) If the side of each hexagon and pentagon is 45.0 mm, what is the surface area of the soccer ball? (b) Find the surface area, given that the diameter of the ball is 222 mm, (c) Assuming that the given values are accurate, account for the difference in the values found in parts (a) and (b).

Fig. 4.88

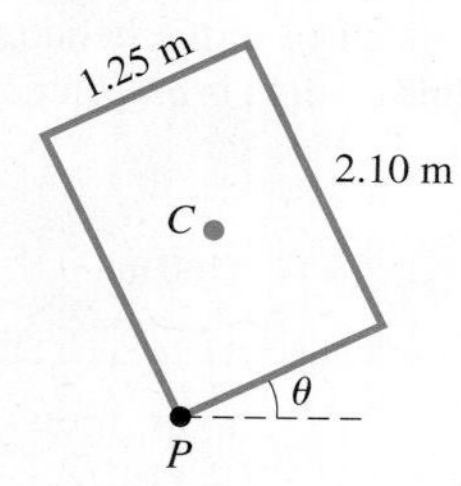

Fig. 4.89

99. Through what angle θ must the crate shown in Fig. 4.89 be tipped in order that its center of gravity C is directly above the pivot point P?

100. Find the gear angle θ in Fig. 4.90, if $t = 0.180$ in.

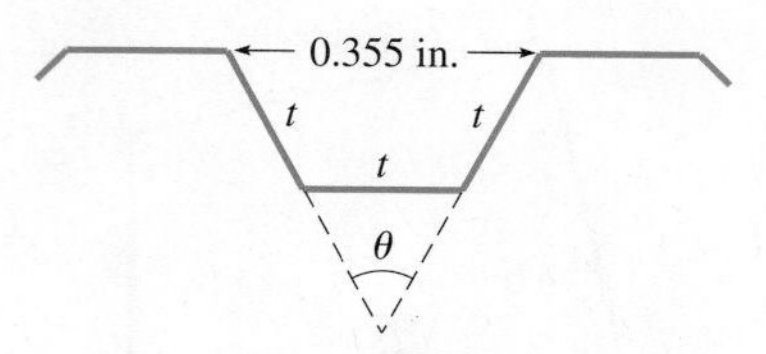

Fig. 4.90

101. A hang glider is directly above the shore of a lake. An observer on a hill is 375 m along a straight line from the shore. From the observer, the angle of elevation of the hang glider is 42.0°, and the angle of depression of the shore is 25.0°. How far above the shore is the hang glider?

102. A ground observer sights a weather balloon to the east at an angle of elevation of 15.0°. A second observer 2.35 mi to the east of the first also sights the balloon to the east at an angle of elevation of 24.0°. How high is the balloon? See Fig. 4.91.

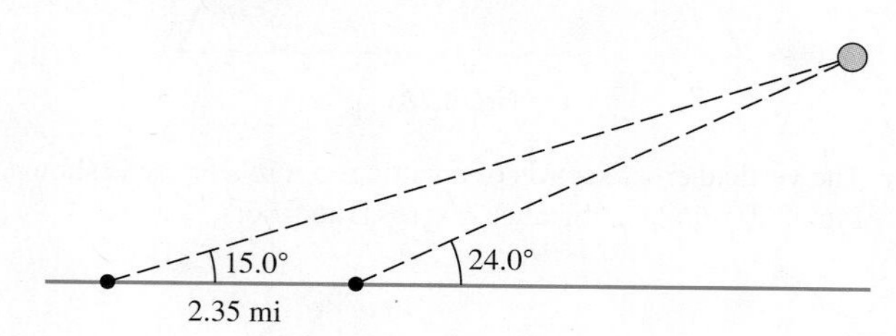

Fig. 4.91

103. A uniform strip of wood 5.0 cm wide frames a trapezoidal window, as shown in Fig. 4.92. Find the left dimension l of the outside of the frame.

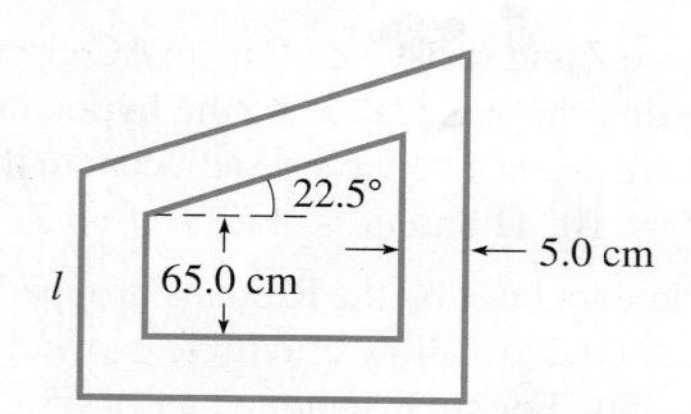

Fig. 4.92

104. A crop-dusting plane flies over a level field at a height of 25 ft. If the dust leaves the plane through a 30° angle and hits the ground after the plane travels 75 ft, how wide a strip is dusted? See Fig. 4.93.

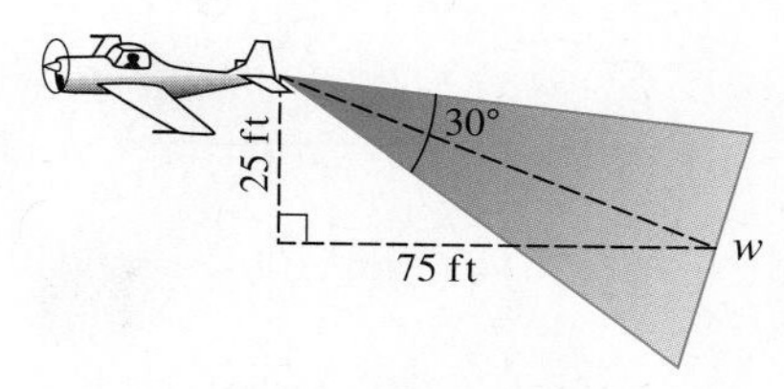

Fig. 4.93

105. A patio is designed in the shape of an isosceles trapezoid with bases 5.0 m and 7.0 m. The other sides are 6.0 m each. Write one or two paragraphs explaining how to use (a) the sine and (b) the cosine to find the internal angles of the patio, and (c) the tangent in finding the area of the patio.

CHAPTER 4 PRACTICE TEST

As a study aid, we have included complete solutions for each Practice Test problem at the end of this book.

1. Find the smallest positive angle and the smallest negative angle (numerically) coterminal with but not equal to $-165°$.
2. Express $37°39'$ in decimal form.
3. Find the value of tan 73.8°.
4. Find θ if $\cos\theta = 0.3726$ (assume θ is acute).
5. A ship's captain, desiring to travel due south, discovers due to an improperly functioning instrument, the ship has gone 22.62 km in a direction 4.05° east of south. How far from its course (to the east) is the ship?
6. Find $\tan\theta$ in fractional form if $\sin\theta = \frac{2}{3}$ (assume θ is acute).
7. Find $\csc\theta$ if $\tan\theta = 1.294$ (assume θ is acute).
8. Solve the right triangle in Fig. 4.94 if $A = 37.4°$ and $b = 52.8$.
9. Solve the right triangle in Fig. 4.94 if $a = 2.49$ and $c = 3.88$.

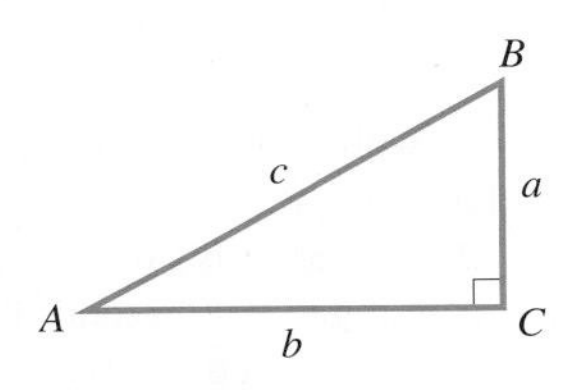

Fig. 4.94

10. The equal sides of an isosceles triangle are each 12.0, and each base angle is 42.0°. What is the length of the third side?
11. If $\tan\theta = 9/40$, find values of $\sin\theta$ and $\cos\theta$. Then evaluate $\sin\theta/\cos\theta$ (assume θ is acute).
12. In finding the wavelength λ (the Greek letter lambda) of light, the equation $\lambda = d\sin\theta$ is used. Find λ if $d = 30.05$ μm and $\theta = 1.167°$. (μ is the prefix for 10^{-6}.)
13. Determine the trigonometric functions of an angle in standard position if its terminal side passes through (5, 2). Give answers in exact and decimal forms.
14. The loading ramp at the back of a truck is 9.50 ft long, and the top of the ramp is 2.50 ft above the ground. How much longer should the ramp be to reduce the angle it makes with the ground to 4.50°?
15. A surveyor sights two points directly ahead. Both are at an elevation 18.525 m lower than the observation point. How far apart are the points if the angles of depression are 13.500° and 21.375°, respectively? See Fig. 4.95.

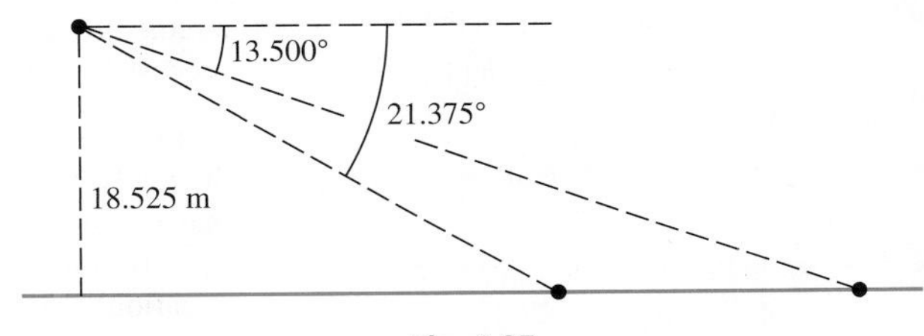

Fig. 4.95

5

Systems of Linear Equations; Determinants

LEARNING OUTCOMES

After completion of this chapter, the student should be able to:

- Identify linear equations
- Calculate the slope of a linear function
- Graph a linear function
- Use the slope-intercept form of a line
- Determine x-and y-intercepts
- Use a regression line to model data
- Test for a solution of a system of equations
- Solve a system of linear equations graphically
- Identify an inconsistent or dependent system of equations
- Solve a system of two or three linear equations algebraically
- Solve a system of two or three linear equations using determinants (Cramer's rule)
- Solve application problems involving systems of linear equations

As knowledge about electric circuits was first developing, in 1848 the German physicist Gustav Kirchhoff formulated what are now known as *Kirchhoff's current law* and *Kirchhoff's voltage law.*

These laws are still widely used today, and in using them, more than one equation is usually set up. To find the needed information about the circuit, it is necessary to find solutions that satisfy all equations at the same time. We will show the solution of circuits using Kirchhoff's laws in some of the exercises later in the chapter. Although best known for these laws of electric circuits, Kirchhoff is also credited in the study of optics as a founder of the modern chemical process known as *spectrum analysis.*

Methods of solving such *systems* of equations were well known to Kirchhoff, and this allowed the study of electricity to progress rapidly. In fact, 100 years earlier a book by the English mathematician Colin Maclaurin was published (2 years after his death) in which many well-organized topics in algebra were covered, including a general method of solving systems of equations. This method is now called *Cramer's rule* (named for the Swiss mathematician Gabriel Cramer, who popularized it in a book he wrote in 1750). We will explain some of the methods of solution, including Cramer's rule, later in the chapter.

Two or more equations that relate variables are found in many fields of science and technology. These include aeronautics, business, transportation, the analysis of forces on a structure, medical doses, and robotics, as well as electric circuits. These applications often require solutions that satisfy all equations simultaneously.

In this chapter, we restrict our attention to *linear* equations (variables occur only to the first power). We will consider systems of two equations with two unknowns and systems of three equations with three unknowns. Systems with other kinds of equations and systems with more unknowns are taken up later in the book.

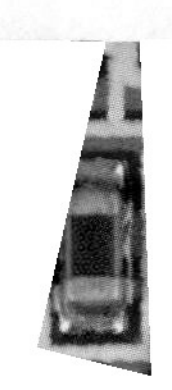

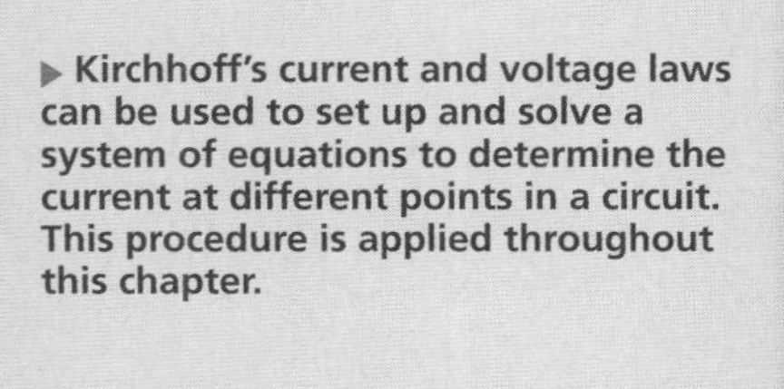

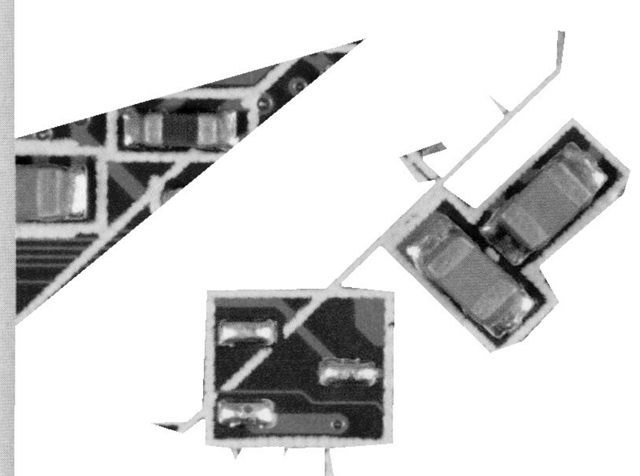

▶ **Kirchhoff's current and voltage laws can be used to set up and solve a system of equations to determine the current at different points in a circuit. This procedure is applied throughout this chapter.**

5.1 Linear Equations and Graphs of Linear Functions

Linear Equations • Solutions of Linear Equations • Linear Functions • Slope • Slope-Intercept Form • Sketching Lines by Intercepts • Linear Regression

In this chapter, we will discuss various techniques for solving *systems of equations*, which are groups of equations that contain more than one unknown. The methods we study assume that the equations are of a specific type, called *linear equations*. In this first section, we will define a linear equation and describe ways of graphing linear equations in two variables.

LINEAR EQUATIONS

An equation is called **linear** in a given set of variables if each term contains only one variable, *to the first power*, or is a constant. This is illustrated in the following example.

EXAMPLE 1 Identifying linear equations

(a) $4x - 2y = 12$ **is** a linear equation. — Both x and y are to the first power.

(b) $v = -32t + 24$ **is** a linear equation. — Both v and t are to the first power.

(c) $0.6F_1 + 0.8F_2 = F_3$ **is** a linear equation. — All three variables are to the first power.

(d) $xy = 150$ **is not** a linear equation. — The term on the left is a *product* of two different variables.

(e) $d = 16t^2$ **is not** a linear equation. — The variable t is squared.

(f) $u - \frac{6}{v} - 4w = 7$ **is not** a linear equation. — The variable v appears in the *denominator*. ■

Linear equations can have any number of unknowns (or variables). In Example 1, the equations in (a) and (b) are linear equations in two unknowns and the equation in (c) is a linear equation in three unknowns. The equations used throughout this chapter will be similar to the ones in (a)–(c).

EXAMPLE 2 Linear equations—Kirchhoff's current and voltage laws

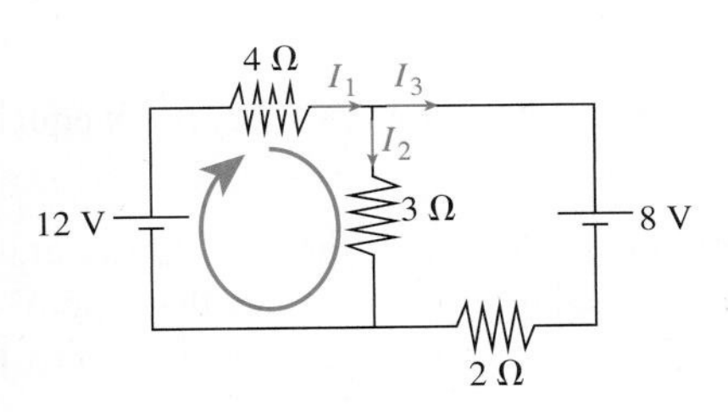

Fig. 5.1

■ See the chapter introduction.

Kirchhoff's current law states that the current going into any junction of an electric circuit equals the current going out. At the top junction in Fig. 5.1, this leads to

$$I_1 = I_2 + I_3$$

Kirchhoff's voltage law states that the sum of the voltages around any closed loop is zero. Going around the loop on the left in Fig. 5.1 (shown in blue) leads to

$$12 - 4I_1 - 3I_2 = 0$$

Both equations in this example are linear equations since the variables I_1, I_2, and I_3 occur only to the first power. Note that a third equation could be found by applying the voltage law to the closed loop on the right in Fig. 5.1. ■

The first method of solving a system of equations that we will discuss relies on graphing linear equations in two unknowns. Before taking this method up in Section 5.2, we will first develop additional ways of graphing these equations.

GRAPHING A LINEAR EQUATION IN TWO UNKNOWNS

A **linear equation in two unknowns** is an equation that can be written in the following form:

$$ax + by = c \tag{5.1}$$

A **solution** of such an equation is a pair of numbers, one for each variable, that when substituted into the equation, makes the equation true. Therefore, *each solution is an ordered pair* that can be plotted as a point in the rectangular coordinate system. There are an infinite number of these ordered-pair solutions, which together form the ***graph*** of the equation. This is illustrated in the following example.

NORMAL FLOAT AUTO REAL RADIAN MP
PRESS + FOR △Tbl

X	Y1
-2	-8
-1	-6
0	-4
1	-2
2	0
3	2
4	4
5	6
6	8
7	10
8	12

X=-2

Fig. 5.2

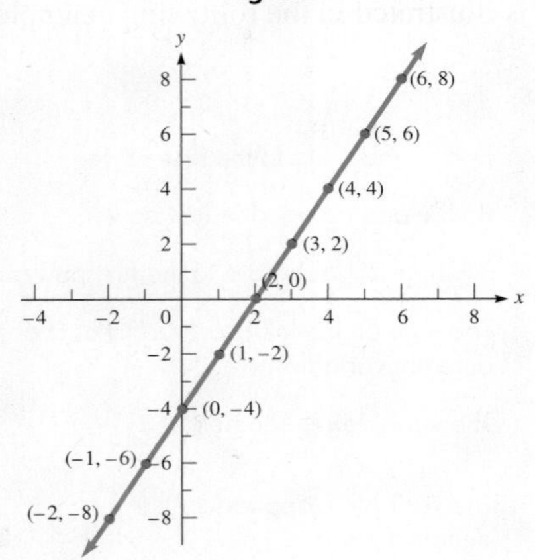

Fig. 5.3

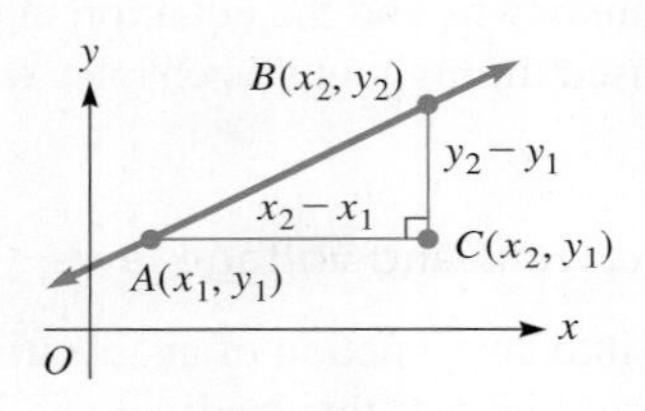

Fig. 5.4

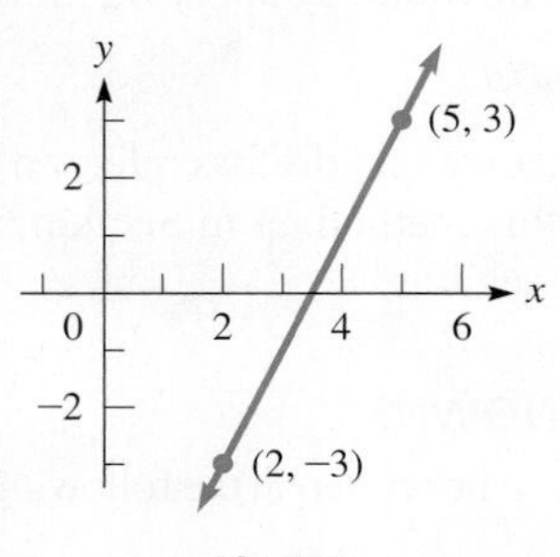

Fig. 5.5

EXAMPLE 3 Graphing a linear equation in two unknowns

Graph the equation $2x - y - 4 = 0$.

We first need to find several pairs of numbers (x, y) that are solutions to the equation, meaning they make the equation true when x and y are substituted in. If we first solve for y to get $y = 2x - 4$, the solutions are easier to find. We can substitute any number in for x and then find the corresponding value of y that makes the equation true. The resulting pair of numbers (x, y) is a *solution* to the equation, and when plotted, is one point on the graph. The *table* feature on a calculator can be used to find many solutions quickly, as shown in Fig. 5.2. All these pairs of values *satisfy* the original equation, meaning they make it true. When these points are plotted together, they form the graph of the equation as shown in Fig. 5.3. Note that the graph is a straight line. In general, all linear equations in two unknowns have graphs that are straight lines. ■

LINEAR FUNCTIONS AND SLOPE

For a linear equation in two unknowns, it is always possible to solve for one variable in terms of the other to put it in the form $\mathbf{y = mx + b}$, where m and b are constants (as we did in Example 3). Written this way, it is clear that y is a function of x, and thus, it is called a **linear function.** Another method for graphing a line is to interpret the values of m and b in this equation to obtain a graph of the line. To do this, one must first understand the concept of *slope*.

Consider the line that passes through points A, with coordinates (x_1, y_1), and B, with coordinates (x_2, y_2),in Fig. 5.4. Point C is horizontal from A and vertical from B. Thus, C has coordinates (x_2, y_1), and there is a right angle at C. One way of measuring the steepness of this line is to find the ratio of the vertical distance to the horizontal distance between two points. Therefore, *we define the* **slope** *of the line through two points as the difference in the y-coordinates divided by the difference in the x-coordinates.* For points A and B, the slope m is

$$\text{Slope} = m = \frac{y_2 - y_1}{x_2 - x_1} \tag{5.2}$$

CAUTION Be very careful not to reverse the order of subtraction. $x_1 - x_2$ is *not* equal to $x_2 - x_1$. ■

NOTE ➧ The slope is often referred to as the *rise* (vertical change) over the *run* (horizontal change). [The slope of a vertical line, for which $x_2 = x_1$, is undefined since the denominator of Eq. (5.2) is zero. Also, the slope of a horizontal line, for which $y_1 = y_2$, is zero.]

EXAMPLE 4 Slope of line through two points

Find the slope of the line through the points $(2, -3)$ and $(5, 3)$.

In Fig. 5.5, we draw the line through the two given points. By taking (5, 3) as (x_2, y_2), then (x_1, y_1) is (2, −3). We may choose either point as (x_2, y_2), but *once the choice is made the order must be maintained.* Using Eq. (5.2), the slope is

$$m = \frac{3 - (-3)}{5 - 2} = \frac{6}{3} = 2$$

The rise is 2 units for each unit (of run) in going from left to right. ■

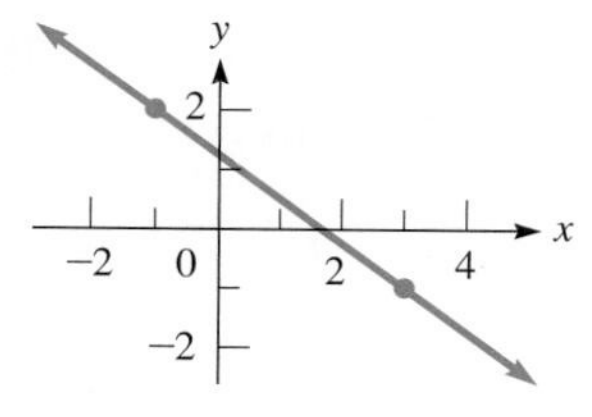

Fig. 5.6

Practice Exercise

1. Find the slope of the line through $(-2, 4)$ and $(-5, -6)$.

EXAMPLE 5 Slope of line through two points

Find the slope of the line through $(-1, 2)$ and $(3, -1)$.

In Fig. 5.6, we draw the line through these two points. By taking (x_2, y_2) as $(3, -1)$ and (x_1, y_1) as $(-1, 2)$, the slope is

$$m = \frac{-1 - 2}{3 - (-1)}$$

$$= \frac{-3}{3 + 1} = -\frac{3}{4}$$

The line *falls* 3 units for each 4 units in going from left to right. ■

We note in Example 1 that as *x increases, y increases and the slope is* ***positive.*** In Example 2, *as x increases, y decreases and the slope is* ***negative.*** Also, *the larger the absolute value of the slope, the steeper is the line.*

EXAMPLE 6 Slope and steepness of line

For each of the following lines shown in Fig. 5.7, we show the difference in the y-coordinates and in the x-coordinates between two points.

In Fig. 5.7(a), a line with a slope of 5 is shown. It rises sharply.
In Fig. 5.7(b), a line with a slope of $\frac{1}{2}$ is shown. It rises slowly.
In Fig. 5.7(c), a line with a slope of -5 is shown. It falls sharply.
In Fig. 5.7(d), a line with a slope of $-\frac{1}{2}$ is shown. It falls slowly.

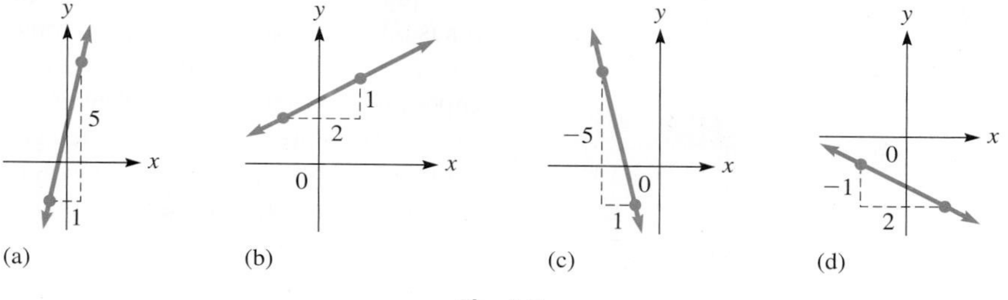

Fig. 5.7 ■

SLOPE-INTERCEPT FORM OF THE EQUATION OF A STRAIGHT LINE

We now show how the slope is related to the equation of a straight line. In Fig. 5.8, if we have two points, $(0, b)$ and a general point (x, y), the slope is

$$m = \frac{y - b}{x - 0}$$

Simplifying this, we have $mx = y - b$, or

$$y = mx + b \tag{5.3}$$

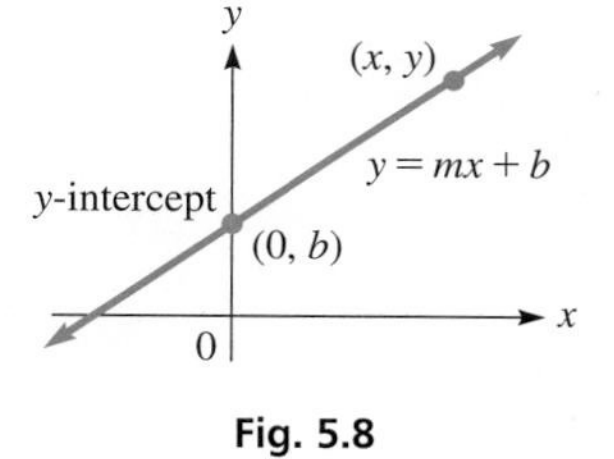

Fig. 5.8

In Eq. (5.3), ***m* is the slope,** and ***b* is the *y*-coordinate of the point where the line crosses the *y*-axis.** *This point is the* ***y-intercept*** *of the line,* and its coordinates are $(0, b)$.

NOTE ▸ [Equation (5.3) is the **slope-intercept form** of the equation of a straight line. The coefficient of x is the slope, and the constant is the ordinate of the y-intercept.] The point $(0, b)$ and simply b are both referred to as the y-intercept.

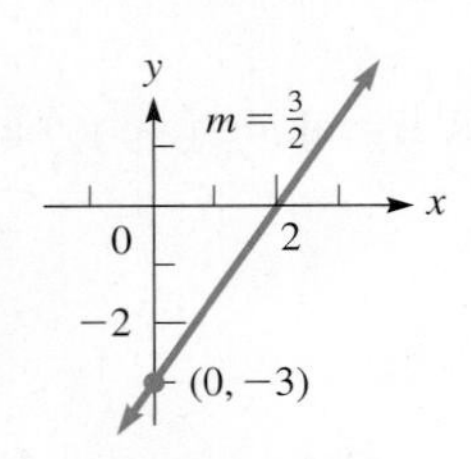

Fig. 5.9

EXAMPLE 7 Slope-intercept form of equation

Use the slope and the y-intercept of the line $y = \frac{3}{2}x - 3$ to sketch its graph.

Because we can write this as

slope — y-intercept ordinate

$$y = \frac{3}{2}x + (-3)$$

the slope is $\frac{3}{2}$ and the y-intercept is $(0, -3)$. To graph, we start at $(0, -3)$ and then move up 3 units and to the right by 2 units. See Fig. 5.9. ■

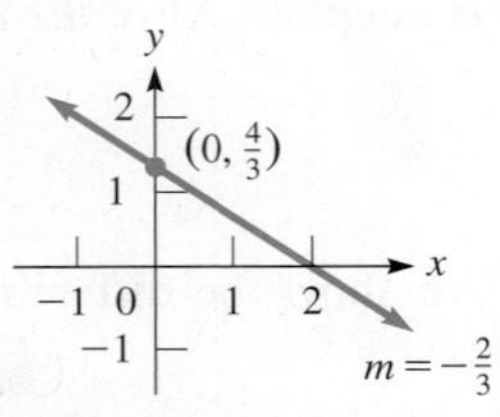

Fig. 5.10

EXAMPLE 8 Slope-intercept form of equation

Find the slope and the y-intercept of the line $2x + 3y = 4$.

We must first write the equation in slope-intercept form. Solving for y, we have

slope — y-intercept ordinate

$$y = -\frac{2}{3}x + \frac{4}{3}$$

Therefore, the slope is $-\frac{2}{3}$, and the y-intercept is the point $(0, \frac{4}{3})$. See Fig. 5.10. ■

Practice Exercise

2. Write the equation $3x - 5y - 15 = 0$ in slope-intercept form.

EXAMPLE 9 Slope-intercept form—Fahrenheit and Celsius temperature

The equation $F = \frac{9}{5}C + 32$ relates degrees Fahrenheit (F) and degrees Celsius (C). This is a linear function in slope-intercept form. The F-intercept is (0, 32), which represents the freezing point of water (0°C = 32°F). The slope is $\frac{9}{5}$, meaning the Fahrenheit temperature rises 9 degrees for every 5 degrees of Celsius temperature. To graph this by hand, we can start at (0, 32) and then go up 9 units and to the right 5 units (see Fig. 5.11). This can also be graphed on a calculator by entering $y_1 = \frac{9}{5}x + 32$ and then graphing with appropriate window settings (see Fig. 5.12). ■

■ Further details and discussion of slope and the graphs of linear equations are found in Chapter 21.

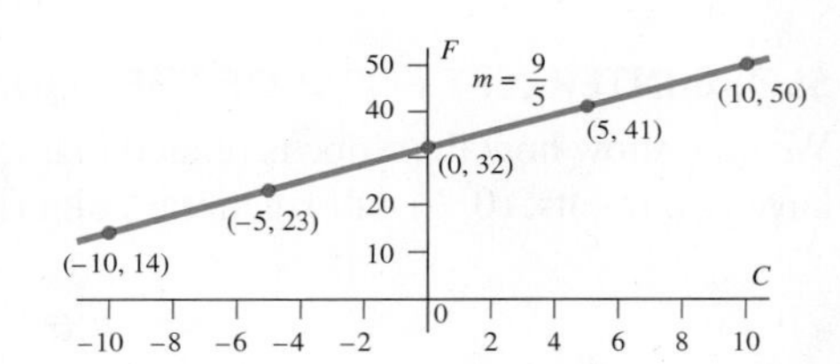

Fig. 5.11

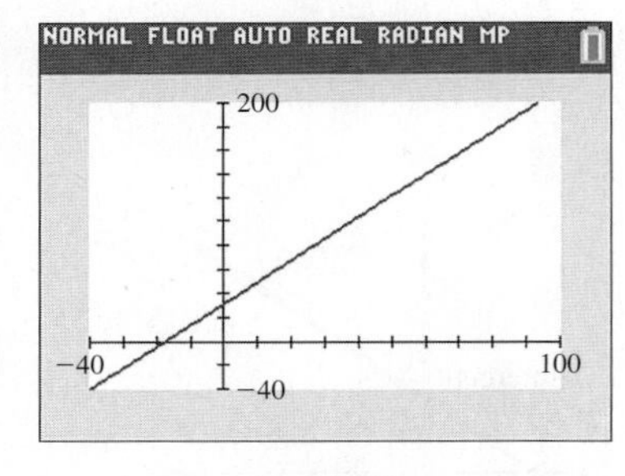

Fig. 5.12

SKETCHING LINES BY INTERCEPTS

Another way of sketching the graph of a straight line is to find two points on the line and then draw the line through these points. Two points that are easily determined are those where the line crosses the y-axis and the x-axis. We already know that the point where it crosses the y-axis is the y-intercept. [In the same way, the point where a line crosses the x-axis is called the ***x*-intercept,** and the coordinates of the x-intercept are $(a, 0)$.]

NOTE ▸

The intercepts are easily found because one of the coordinates is zero. By setting $x = 0$ and $y = 0$, in turn, and finding the value of the other variable, we get the coordinates of the intercepts. This method works except when both intercepts are at the origin, and we must find one other point, or use the slope-intercept method. In using the intercept method, a third point should be found as a check. The next example shows how a line is sketched by finding its intercepts.

■ a as well as $(a, 0)$ is often referred to as the x-intercept.

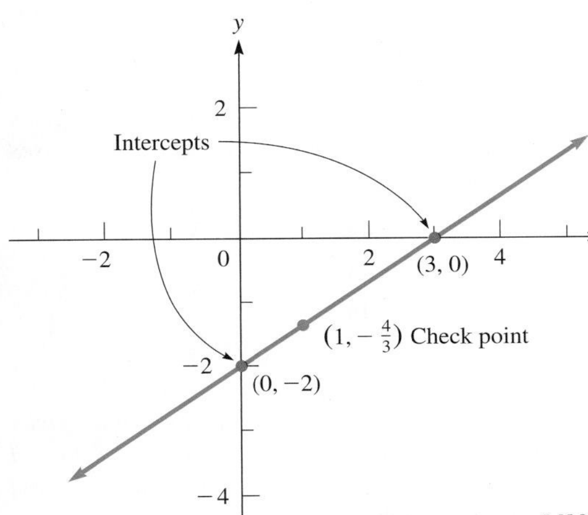

Fig. 5.13

EXAMPLE 10 Sketching line by intercepts

Sketch the graph of the line $2x - 3y = 6$ by finding its intercepts and one check point. See Fig. 5.13.

First, we let $x = 0$. This gives us $-3y = 6$, or $y = -2$. This gives us the y-intercept, which is the point $(0, -2)$. Next, we let $y = 0$, which gives us $2x = 6$, or $x = 3$. This means the x-intercept is the point (3, 0).

The intercepts are enough to sketch the line as shown in Fig. 5.13. To find a check point, we can use any value for x other than 3 or any value of y other than -2. Choosing $x = 1$, we find that $y = -\frac{4}{3}$. This means that the point $(1, -\frac{4}{3})$ should be on the line. In Fig. 5.13, we can see that it is on the line.

Note that in order to graph this on a calculator, we must first solve for y to get $y = \frac{2}{3}x - 2$. Then we can enter $y_1 = (2/3)x - 2$ and graph with appropriate window settings. ■

Practice Exercise

3. Find the intercepts of the line $3x - 5y - 15 = 0$.

LINEAR REGRESSION

■ Regression is discussed in detail in Chapter 22.

In Example 9, we saw that Fahrenheit and Celsius temperatures have an exact linear relationship. Sometimes when analyzing data, we find that two variables have an *approximate* linear relationship. This happens when the plotted data points are somewhat scattered, but generally follow a straight-line pattern. In these cases, a ***regression line*** can be used to *model* the data. A regression line is obtained from statistical methods that ensure it will be the line that best fits the data points. Many calculators can also be used to find regression lines, and the next example shows how this is done.

EXAMPLE 11 Linear regression—speed of sound versus temperature

In an experiment, the temperature (in °C) and the speed of sound in air (in m/s) were measured on seven days throughout the year as shown in the table:

Temperature, T (°C)	−7.2	−3.5	2.4	8.1	11.4	16.8	22.2
Measured speed of sound, v (m/s)	327	329	332	336	337	341	345

■ In addition to the temperature, other variables like humidity also influence the speed of sound in air.

The statistical graphing features on a calculator can be used to make a ***scatterplot*** of the data as shown in Fig. 5.14(a). Since the points generally follow a straight line, we can use the **LinReg($ax + b$)** feature to find the regression line, which is shown in Fig. 5.14(b). The result (after rounding to 3 decimal places) is $y = 0.604x + 330.957$. This equation can be entered for Y_1 and graphed through the scatterplot as shown in Fig. 5.14(c).

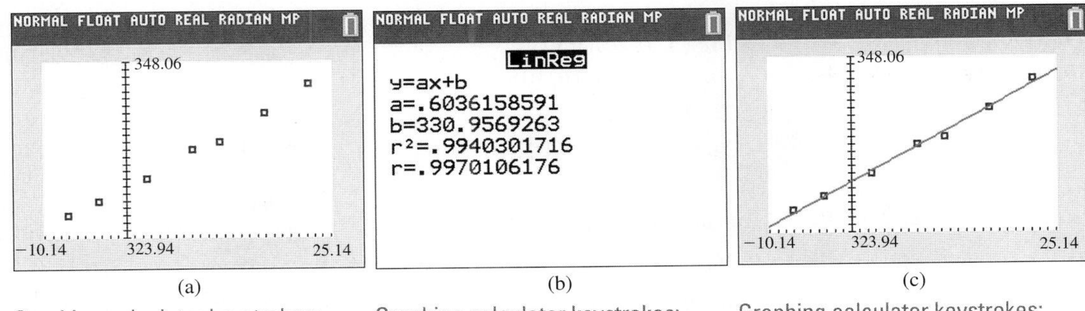

Fig. 5.14

(a) Graphing calculator keystrokes: goo.gl/awxJHG

(b) Graphing calculator keystrokes: goo.gl/ZvvFAo

(c) Graphing calculator keystrokes: goo.gl/dmQaN2

■ The values of r^2 and r shown in Fig. 5.14(b) measure how well the linear model fits the data. Values close to 1 or −1 indicate a very good fit.

The regression equation is in the form $y = mx + b$ with $m = 0.604$ and $b = 330.957$. Thus, it is a linear function in *slope-intercept form*. Using the variables in our example, we can rewrite the regression line as $\mathbf{v = 0.604T + 330.957}$.

The regression equation can be used to predict the value of either variable if the other variable is known. For example, to predict the speed of sound in air with temperature 20.0°C, we have $v = 0.604(20.0) + 330.957 = 343$ m/s. To predict the temperature that would result in a speed of sound of 334 m/s, we substitute 334 in for v and then solve for T, which yields $T = 5.0$ °C. ■

Practice Exercise

4. Find the linear regression model in Example 11 if the value 22.2 in the last pair of data is changed to 20.8.

EXERCISES 5.1

In Exercises 1–4, answer the given questions about the indicated examples of this section.

1. In Example 3, what is the solution if $x = -3$?

2. In Example 5, if the first y-coordinate is changed to -2, what changes result in the example?

3. In the first line of Example 8, if $+$ is changed to $-$, what changes result in the example?

4. In the first line of Example 10, if $-$ is changed to $+$, what changes occur in the graph?

In Exercises 5–8, determine whether or not the given equation is linear.

5. $8x - 3y = 12$ **6.** $2v + 3t^2 = 60$

7. $2l + 3w = 4lw$ **8.** $I_1 + I_3 = I_2$

In Exercises 9–12, determine whether or not the given pair of values is a solution to the given linear equation in two unknowns.

9. $2x + 3y = 9;\quad (3, 1), (5, \frac{1}{3})$

10. $5x + 2y = 1;\quad (0.2, -1); (1, -2)$

11. $-3x + 5y = 13;\quad (-1, 2), (4, 5)$

12. $4y - x = -10;\quad (2, -2), (2, 2)$

In Exercises 13–16, for each given value of x, determine the value of y that gives a solution to the given linear equation in two unknowns.

13. $3x - 2y = 12;\quad x = 2, x = -3$

14. $6y - 5x = 60;\quad x = -10, x = 8$

15. $x - 4y = 2;\quad x = 3, x = -0.4$

16. $3x - 2y = 9;\quad x = \frac{2}{3}, x = -3$

In Exercises 17–24, find the slope of the line that passes through the given points.

17. $(1, 0), (3, 8)$ **18.** $(3, 1), (2, -7)$

19. $(-1, 2), (-4, 17)$ **20.** $(-1, -2), (6, 10)$

21. $(5, -3), (-2, -5)$ **22.** $(-3, 4), (-7, -4)$

23. $(0.4, 0.5), (-0.2, 0.2)$ **24.** $(-2.8, 3.4), (1.2, 4.2)$

In Exercises 25–32, sketch the line with the given slope and y-intercept.

25. $m = 2, (0, -1)$ **26.** $m = 3, (0, 1)$

27. $m = 0, (0, 5)$ **28.** $m = -4, (0, -2)$

29. $m = \frac{1}{2}, (0, 0)$ **30.** $m = \frac{2}{3}, (0, -1)$

31. $m = -9, (0, 20)$ **32.** $m = -0.3, (0, -1.4)$

In Exercises 33–40, find the slope and y-intercept of the line with the given equation and sketch the graph using the slope and y-intercept. A calculator can be used to check your graph.

33. $y = -2x + 1$ **34.** $y = -4x$

35. $y = x - 4$ **36.** $y = \frac{4}{5}x + 2$

37. $5x - 2y = 40$ **38.** $-2y = 7$

39. $24x + 40y = 15$ **40.** $1.5x - 2.4y = 3.0$

In Exercises 41–48, find the x- and y-intercepts of the line with the given equation. Sketch the line using the intercepts. A calculator can be used to check the graph.

41. $x + 2y = 4$ **42.** $3x + y = 3$

43. $4x - 3y = 12$ **44.** $5y - x = 5$

45. $y = 3x + 6$ **46.** $y = -2x - 4$

47. $12x + y = 30$ **48.** $y = 0.25x + 4.5$

In Exercises 49–52, find the equation of the regression line for the given data. Then use this equation to make the indicated estimate. Round decimals in the regression equation to three decimal places. Round estimates to the same accuracy as the given data.

49. The following table gives the weld diameters d (in mm) and the shear strengths s (in kN/mm) for a sample of spot welds. Find the equation of the regression line, and then estimate the shear strength of a spot weld that has a diameter of 4.5 mm.

Diameter, d (mm)	4.2	4.4	4.6	4.8	5.0	5.2	5.4
Shear strength, s (kN/mm)	51	54	69	76	75	85	89

50. The following table gives the percentage p of adults living in households with only wireless telephone services, where t is the number of years after the year 2000. Find the equation of the regression line, and then estimate the percentage of adults living in households with only wireless telephone services in the year 2016.

Number of years after 2000, t	6	8	10	12	14
Percentage with only wireless phones, p (%)	10	16	25	34	43

51. The following data gives the diameter in inches (measured 4.5 ft above ground) and the volume of wood (in ft^3) for a sample of black cherry trees. Find the equation of the regression line, and then estimate the volume of wood in a black cherry tree that has a diameter of 15.0 in.

Diameter, d (in.)	8.8	11.0	11.2	11.7	13.7	16.0	17.9	20.6
Volume, v (ft^3)	10.2	15.6	19.9	21.3	25.7	38.3	58.3	77.0

52. The following table gives the fraction f (as a decimal) of the total heating load of a certain system that will be supplied by a solar collector of area A (in m^2). Find the equation of the regression line, and then estimate the fraction of the heating load that will be supplied by a solar collector with area 38 m^2.

Area, A (m^2)	20	30	40	50	60	70	80
Fraction, f	0.22	0.30	0.37	0.44	0.50	0.56	0.61

In Exercises 53–56, solve the given problems.

53. Find the intercepts of the line $\frac{x}{a} + \frac{y}{b} = 1$.

54. Find the intercepts of the line $y = mx + b$.

55. Do the points $(1, -2)$, $(3, -3)$, $(5, -4)$, $(7, -6)$ and $(11, -7)$ lie on the same straight line?

56. The points $(-2, 5)$ and $(3, 7)$ are on the same straight line.
(a) Find another point on the line for which $x = -12$.
(b) Find another point on the line for which $y = -3$.

In Exercises 57–62, sketch the indicated lines.

57. The diameter of the large end, d (in in.), of a certain type of machine tool can be found from the equation $d = 0.2l + 1.2$, where l is the length of the tool. Sketch d as a function of l, for values of l to 10 in. See Fig. 5.15

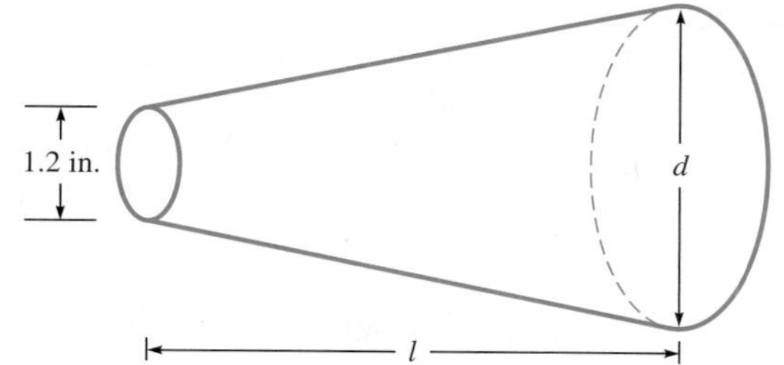

Fig. 5.15

58. In mixing 85-octane gasoline and 93-octane gasoline to produce 91-octane gasoline, the equation $0.85x + 0.93y = 910$ is used. Sketch the graph.

59. Two electric currents, I_1 and I_2 (in mA), in part of a circuit in a computer are related by the equation $4I_1 - 5I_2 = 2$. Sketch I_2 as a function of I_1. These currents can be negative.

60. In 2009, it was estimated that about 4.1 billion text messages were sent each day in the United States, and in 2012 it was estimated that this number was 6.2 billion. Assuming the growth in texting is linear, set up a function for the number N of daily text messages and the time t in years, with $t = 0$ being 2009. Sketch the graph for $t = 0$ to $t = 8$.

61. A plane cruising at 220 m/min starts its descent from 2.5 km at 150 m/min. Find the equation for its altitude h as a function of time t and sketch the graph for $t = 0$ to $t = 10$ min.

62. A straight ski run declines 80 m over a horizontal distance of 480 m. Find the equation for the height h of a skier above the bottom as a function of the horizontal distance d moved. Sketch the graph.

Answers to Practice Exercises

1. 10/3 **2.** $y = \frac{3}{5}x - 3$ **3.** $(5, 0), (0, -3)$
4. $v = 0.620T + 331$

5.2 Systems of Equations and Graphical Solutions

System of Linear Equations • Solution of a System • Solution by Graphing • Using a Graphing Calculator • Inconsistent and Dependent Systems

In many applications, there is *more than one unknown* that we wish to solve for. To accomplish this, we must also have more than one equation that involves these unknowns. In fact, we need as many equations as there are unknowns. A group of two or more equations involving two or more unknowns is called a **system of equations**. In the remaining sections of this chapter, we will discuss various techniques for solving systems of linear equations. We begin by studying systems of two equations with two unknowns.

Two linear equations, each containing the same two unknowns,

$$\begin{aligned} a_1x + b_1y &= c_1 \\ a_2x + b_2y &= c_2 \end{aligned} \tag{5.4}$$

NOTE ▶ *are said to form a* **system of linear equations in two unknowns**. [*A* **solution of the system** *is a pair of numbers (x, y) that make* ***both equations true***.] Except in special circumstances, there will only be one pair of numbers that make *both* equations true. The following two examples illustrate the meaning of a solution of a system.

EXAMPLE 1 Testing for solution of a system

Is (2, 8) a solution of the system $\begin{aligned} x + 2y &= 18 \\ -5x + 2y &= 6 \end{aligned}$?

Substituting $x = 2$ and $y = 8$ into each equation, we get

$$2 + 2(8) \stackrel{?}{=} 18 \qquad -5(2) + 2(8) \stackrel{?}{=} 6$$
$$18 = 18 \qquad 6 = 6$$

Since both equations are true, $x = 2$, $y = 8$ is a solution of the system. ■

EXAMPLE 2 Testing for solution of a system—finding forces

Are the forces in the following system given by $F_1 = 18$ lb and $F_2 = 41$ lb?

$$\begin{aligned} 2F_1 + 4F_2 &= 200 \\ F_2 &= 2F_1 \end{aligned}$$

Substituting into the equations, we have

$$2(18) + 4(41) \stackrel{?}{=} 200 \qquad 41 \stackrel{?}{=} 2(18)$$
$$200 = 200 \qquad 41 \neq 36$$

Since one of these equations is not true, $F_1 = 18$ lb, $F_2 = 41$ lb is *not* a solution of the system. ■

SOLUTION BY GRAPHING

In Section 5.1, we saw that the graph of a linear equation in two unknowns is a straight line. Because a solution of a system of linear equations in two unknowns is a pair of values (x, y) that satisfies *both* equations, graphically *the solution is given by the coordinates of the point of intersection of the two lines.* This must be the case, for the coordinates of this point constitute the only pair of values to satisfy *both* equations. (In some special cases, there may be no solution; in others, there may be many solutions.See Examples 7 and 8.)

Therefore, when we solve a system of equations in two unknowns graphically, *we must graph each line and determine the point of intersection.* When doing this by hand, we often need to estimate the point of intersection. However, we will show in Example 5 that a calculator can be used to find this point very accurately.

EXAMPLE 3 Determine the point of intersection

Solve the system of equations

$$y = x - 3$$
$$y = -2x + 1$$

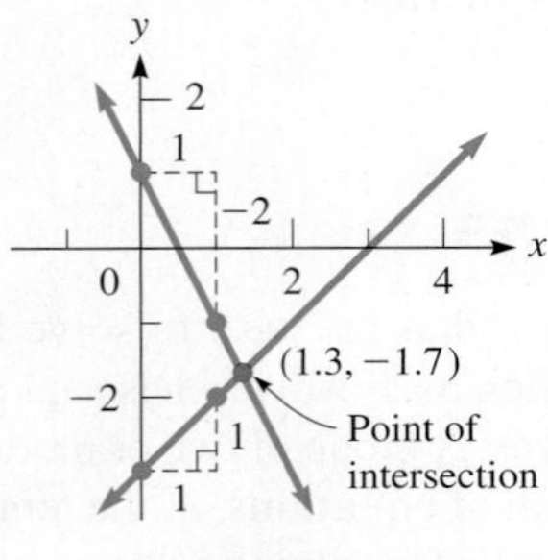

Fig. 5.16

Because each of the equations is in slope-intercept form, note that $m = 1$ and $b = -3$ for the first line and that $m = -2$ and $b = 1$ for the second line. Using these values, we sketch the lines, as shown in Fig. 5.16.

From the figure, it can be seen that *the lines cross at about the point* $(1.3, -1.7)$. This means that the solution is approximately

$$x = 1.3 \qquad y = -1.7$$

(The exact solution is $x = \frac{4}{3}, y = -\frac{5}{3}$.)

NOTE ▸ [In checking the solution, be careful to substitute the values in both equations.] Making these substitutions gives us

$$-1.7 \stackrel{?}{=} 1.3 - 3 \quad \text{and} \quad -1.7 \stackrel{?}{=} -2(1.3) + 1$$
$$= -1.7 \qquad\qquad \approx -1.6$$

These values show that the solution checks. [The point $(1.3, -1.7)$ is *on* the first line and *almost on* the second line. The difference in values when checking the values for the second line is due to the fact that the solution is *approximate*.] ■

EXAMPLE 4 Determine the point of intersection

Solve the system of equations

$$2x + 5y = 10$$
$$3x - y = 6$$

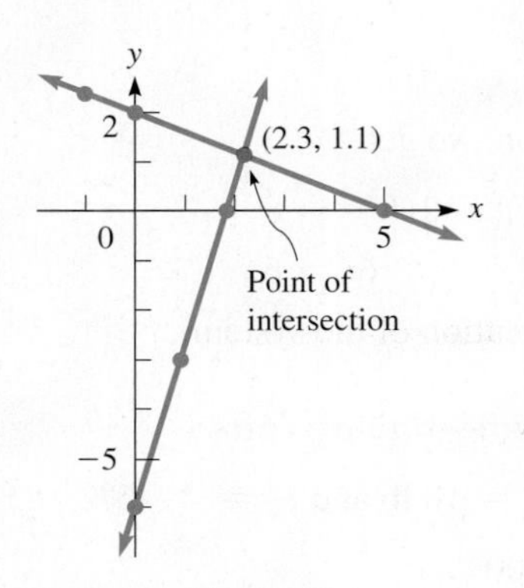

Fig. 5.17

We could write each equation in slope-intercept form in order to sketch the lines. Also, we could use the form in which they are written to find the intercepts. Choosing to find the intercepts and draw lines through them, let $y = 0$; then $x = 0$. Therefore, the intercepts of the first line are the points (5, 0) and (0, 2). A third point is $(-1, \frac{12}{5})$. The intercepts of the second line are (2, 0) and $(0, -6)$. A third point is $(-1, -3)$. Plotting these points and drawing the proper straight lines, we see that the lines cross at about (2.3, 1.1). [The exact values are $(\frac{40}{17}, \frac{18}{17})$.] The solution of the system of equations is approximately $x = 2.3$, $y = 1.1$ (see Fig. 5.17).

Checking, we have

$$2(2.3) + 5(1.1) \stackrel{?}{=} 10 \quad \text{and} \quad 3(2.3) - 1.1 \stackrel{?}{=} 6$$
$$10.1 \approx 10 \qquad\qquad 5.8 \approx 6$$

This shows the solution is correct to the accuracy we can get from the graph. ■

Practice Exercise

1. In Example 4, change the 6 to 3, and then solve.

SOLVING SYSTEMS OF EQUATIONS USING A GRAPHING CALCULATOR

A calculator can be used to find the point of intersection with much greater accuracy than is possible by hand-sketching the lines. Once we locate the point of intersection, the features of the calculator allow us to get the accuracy we need.

EXAMPLE 5 Using a calculator to find points of intersection

Using a calculator, we now solve the systems of equations in Examples 3 and 4.

Solving the system for Example 3, let $y_1 = x - 3$ and $y_2 = -2x + 1$. Then display the lines as shown in Fig. 5.18(a). Then using the *intersect* feature of the calculator, we find (rounded to the nearest 0.001) that the solution is $x = 1.333$, $y = -1.667$.

Solving the system for Example 4, let $y_1 = -2x/5 + 2$ and $y_2 = 3x - 6$. Then display the lines as shown in Fig. 5.18(b). Then using the *intersect* feature, we find (rounded to the nearest 0.001) that the solution is $x = 2.353$, $y = 1.059$.

■ We could use the *trace* and *zoom* features, but the *intersect* feature is meant for this type of problem, and is more accurate.

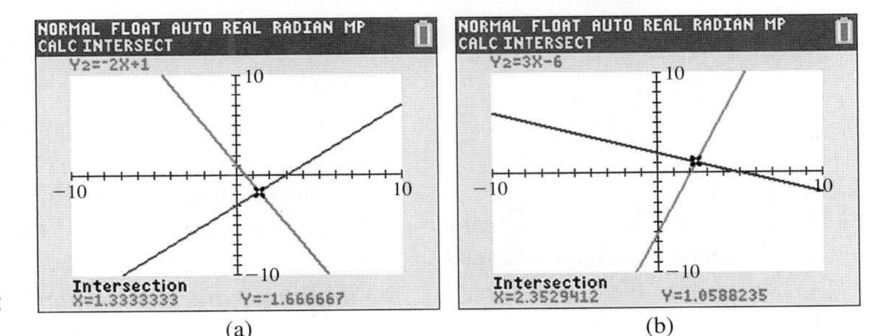

Fig. 5.18
Graphing calculator keystrokes: goo.gl/EqBrUk ■

APPLICATIONS INVOLVING TWO LINEAR EQUATIONS

Linear equations in two unknowns are often useful in solving applied problems. Just as in Section 1.12, *we must read the statement carefully in order to identify the unknowns and the information for setting up the equations.*

EXAMPLE 6 Solving a system—speed of a car

A driver traveled for 1.5 h at a constant speed along a highway. Then, through a construction zone, the driver reduced the car's speed by 20 mi/h for 30 min. If 100 mi were covered in the 2.0 h, what were the two speeds?

First, let v_h = the highway speed and v_c = the speed in the construction zone. Two equations are found by using

1. distance = rate × time [for units, mi $= (\frac{\text{mi}}{\text{h}})\text{h}$], and
2. the fact that "the driver reduced the car's speed by 20 mi/h for 30 min."

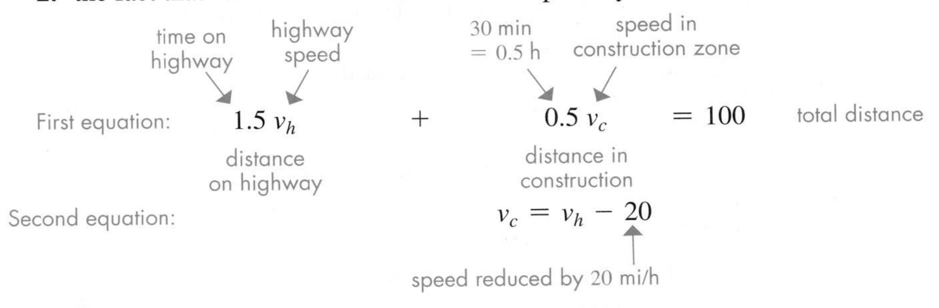

Since the second equation is solved for v_c, we will treat v_c as the dependent variable (y). Solving the first equation for v_c (by subtracting $1.5v_h$ from both sides and multiplying both sides by 2), we get $v_c = -3v_h + 200$. The sketch of the two equations is shown in Fig. 5.19(a). To graph these equations on a calculator, we must use x for v_h and graph $y_1 = -3x + 200$ and $y_2 = x - 20$. Then the *intersect* feature can be used to find the point of intersection as shown in Fig. 5.19(b).

We see that the point of intersection is (55, 35), which means that the solution is $y_1 = 55$ mi/h and $v_c = 35$ mi/h. Checking *in the statement of the problem,* we have $(1.5\text{ h})(55\text{ mi/h}) + (0.5\text{ h})(35\text{ mi/h}) = 100\text{ mi}$. ■

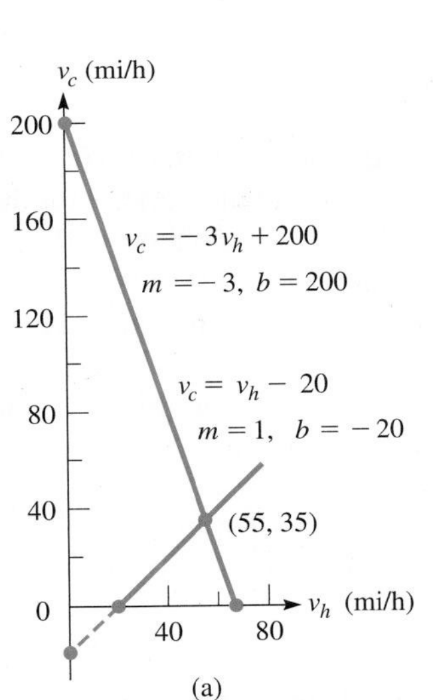

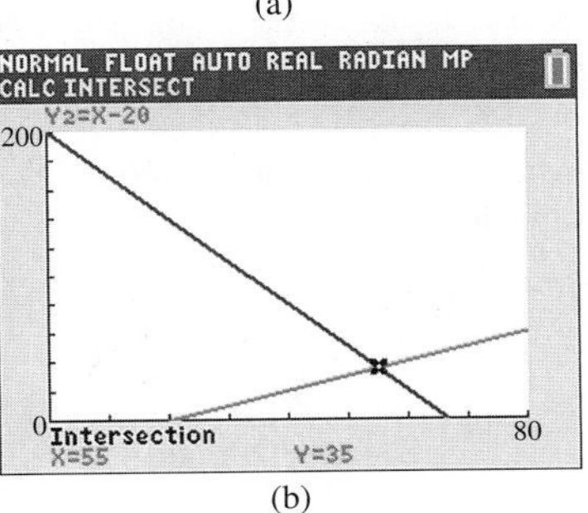

Fig. 5.19
Graphing calculator keystrokes: goo.gl/bMavGh

INCONSISTENT AND DEPENDENT SYSTEMS

The lines of each system in the previous examples intersect in a single point, and each system has *one* solution. Such systems are called *consistent* and *independent.* Most systems that we will encounter have just one solution. However, as we now show, *not all systems have just one such solution.* The following examples illustrate a system that has *no solution* and a system that has an *unlimited number of solutions.*

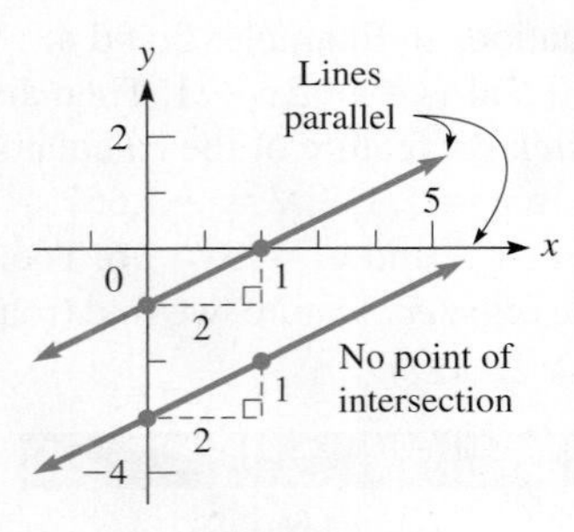

Fig. 5.20

EXAMPLE 7 Inconsistent system

Solve the system of equations

$$x = 2y + 6$$
$$6y = 3x - 6$$

Writing each of these equations in slope-intercept form, we have for the first equation

$$y = \tfrac{1}{2}x - 3$$

For the second equation, we have

$$y = \tfrac{1}{2}x - 1$$

From these, we see that each line has a slope of $\frac{1}{2}$ and that the y-intercepts are $(0, -3)$ and $(0, -1)$. Therefore, we know that the y-intercepts are different, but the *slopes are the same.* Because the slope indicates that each line rises $\frac{1}{2}$ unit for y for each unit x increases, *the lines are parallel and do not intersect,* as shown in Fig. 5.20. [This means that there are *no solutions* for this system of equations. Such a system is called **inconsistent**.] ■

NOTE ▸

Practice Exercise

2. Is the system in Example 7 inconsistent if the −6 in the second equation is changed to −12?

EXAMPLE 8 Dependent system

Solve the system of equations

$$x - 3y = 9$$
$$-2x + 6y = -18$$

We find that the intercepts and a third point for the first line are (9, 0), $(0, -3)$, and $(3, -2)$. For the second line, we then find that the intercepts are the same as for the first line. We also find that the check point $(3, -2)$ also satisfies the equation of the second line. This means that the two lines are really the same line.

Another check is to write each equation in slope-intercept form. This gives us the equation $y = \frac{1}{3}x - 3$ for each line. See Fig. 5.21.

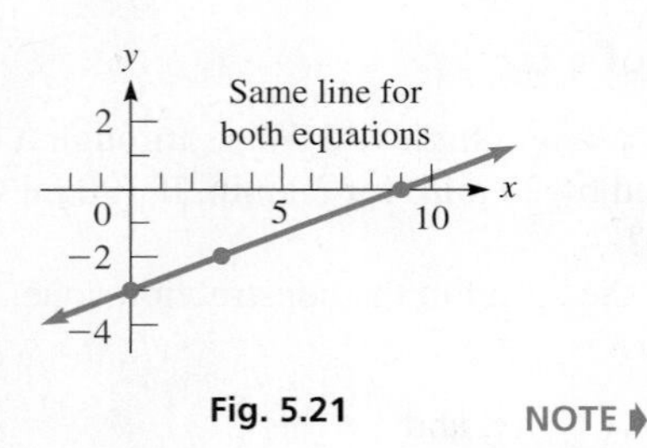

Fig. 5.21

NOTE ▸ [Because the lines are the same, the coordinates of any point on this common line constitute a solution of the system. Since no unique solution can be determined, the system is called **dependent**.] ■

Later in the chapter, we will show algebraic ways of finding out whether a given system is consistent, inconsistent, or dependent.

EXERCISES 5.2

In Exercises 1–4, answer the given questions about the indicated examples of this section.

1. In Example 2, is $F_1 = 20$ lb, $F_2 = 40$ lb a solution?
2. In the second equation of Example 4, if − is replaced with +, what is the solution?
3. In Example 7, what changes occur if the 6 in the first equation is changed to a 2?
4. In Example 8, by changing what one number in the first equation does the system become (a) inconsistent? (b) consistent?

In Exercises 5–12, determine whether or not the given pair of values is a solution of the given system of linear equations.

5. $x - y = 5$ $\quad x = 4, y = -1$
   $2x + y = 7$
6. $2x + y = 8$ $\quad x = -1, y = 10$
   $3x - y = -13$
7. $A + 5B = -7$ $\quad A = -2, B = 1$
   $3A - 4B = -4$

8. $3y - 6x = 4$ $x = \frac{1}{3}, y = 2$
$6x - 3y = -4$

9. $2x - 5y = 0$ $x = \frac{1}{2}, y = -\frac{1}{5}$
$4x + 10y = 4$

10. $6i_1 + i_2 = 5$ $i_1 = 1, i_2 = -1$
$3i_1 - 4i_2 = -1$

11. $3x - 2y = 2.2$ $x = 0.6, y = -0.2$
$5x + y = 2.8$

12. $7t - s = 3.2$ $s = -1.1, t = 0.3$
$2s + t = 2.5$

In Exercises 13–28, solve each system of equations by sketching the graphs. Use the slope and the y-intercept or both intercepts. Estimate the result to the nearest 0.1 if necessary.

13. $y = -x + 4$
$y = x - 2$

14. $y = \frac{1}{2}x - 1$
$y = -x + 8$

15. $y = 2x - 6$
$y = -\frac{1}{3}x + 1$

16. $2y = x - 8$
$y = 2x + 2$

17. $3x + 2y = 6$
$x - 3y = 3$

18. $4R - 3V = -8$
$6R + V = 6$

19. $2x - 5y = 10$
$3x + 4y = -12$

20. $-5x + 3y = 15$
$2x + 7y = 14$

21. $s - 4t = 8$
$2s = t + 4$

22. $y = 4x - 6$
$2x = y - 4$

23. $y = -x + 3$
$2y = 6 - 4x$

24. $p - 6 = 6v$
$v = 3 - 3p$

25. $x - 4y = 6$
$2y = x + 4$

26. $x + y = 3$
$3x - 2y = 14$

27. $-2r_1 + 2r_2 = 7$
$4r_1 - 2r_2 = 1$

28. $2x - 3y = -5$
$3x + 2y = 12$

In Exercises 29–40, solve each system of equations to the nearest 0.001 for each variable by using a calculator.

29. $x = 4y + 2$
$3y = 2x + 3$

30. $1.2x - 2.4y = 4.8$
$3.0x = -2.0y + 7.2$

31. $4.0x - 3.5y = 1.5$
$1.4y + 0.2x = 1.4$

32. $5F - 2T = 7$
$4T + 3F = 8$

33. $x - 5y = 10$
$2x - 10y = 20$

34. $18x - 3y = 7$
$2y = 1 + 12x$

35. $1.9v = 3.2t$
$1.2t - 2.6v = 6$

36. $3y = 14x - 9$
$12x + 23y = 0$

37. $5x = y + 3$
$4x = 2y - 3$

38. $0.75u + 0.67v = 5.9$
$2.1u - 3.9v = 4.8$

39. $7R = 18V + 13$
$-1.4R + 3.6V = 2.6$

40. $y = 6x + 2$
$12x - 2y = -4$

In Exercises 41–44, solve the given problems.

41. In planning a search pattern from an aircraft carrier, a pilot plans to fly at p mi/h relative to a wind that is blowing at w mi/h. Traveling with the wind, the ground speed would be 300 mi/h, and against the wind the ground speed would be 220 mi/h. This leads to two equations:

$p + w = 300$
$p - w = 220$

Are the speeds 260 mi/h and 40 mi/h?

42. The electric resistance R of a certain resistor is a function of the temperature T given by the equation $R = aT + b$, where a and b are constants. If $R = 1200\ \Omega$ when $T = 10.0°C$ and $R = 1280\ \Omega$ when $T = 50.0°C$, we can find the constants a and b by substituting and obtaining the equations

$1200 = 10.0a + b$
$1280 = 50.0a + b$

Are the constants $a = 4.00\ \Omega/°C$ and $b = 1160\ \Omega$?

43. The forces acting on part of a structure are shown in Fig. 5.22. An analysis of the forces leads to the equations

$0.80F_1 + 0.50F_2 = 50$
$0.60F_1 - 0.87F_2 = 12$

Are the forces $F_1 = 45$ N and $F_2 = 28$ N?

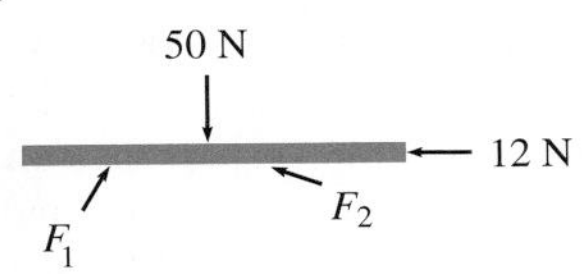

Fig. 5.22

44. A student earned $4000 during the summer and decided to put half into an IRA (Individual Retirement Account). If the IRA was invested in two accounts earning 4.0% and 5.0%, the total income for the first year is $92. The equations to determine the amounts of x and y are

$x + y = 2000$
$0.040x + 0.050y = 92$

Are the amounts $x = \$1200$ and $y = \$800$?

In Exercises 45–50, graphically solve the given problems. A calculator may be used.

45. Chains support a crate, as shown in Fig. 5.23. The equations relating tensions T_1 and T_2 are given below. Determine the tensions to the nearest 1 N from the graph.

$0.8T_1 - 0.6T_2 = 12$
$0.6T_1 + 0.8T_2 = 68$

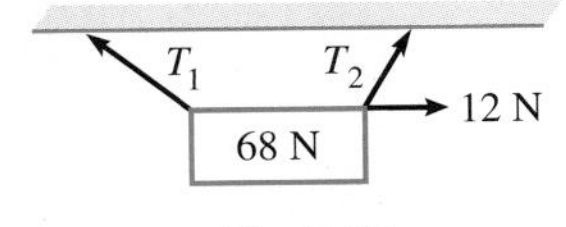

Fig. 5.23

46. An architect designing a parking lot has a row 202 ft wide to divide into spaces for compact cars and full-size cars. The architect determines that 16 compact car spaces and 6 full-size car spaces use the width, or that 12 compact car spaces and 9 full-size car spaces use all but 1 ft of the width. What are the widths (to the nearest 0.1 ft) of the spaces being planned? See Fig. 5.24.

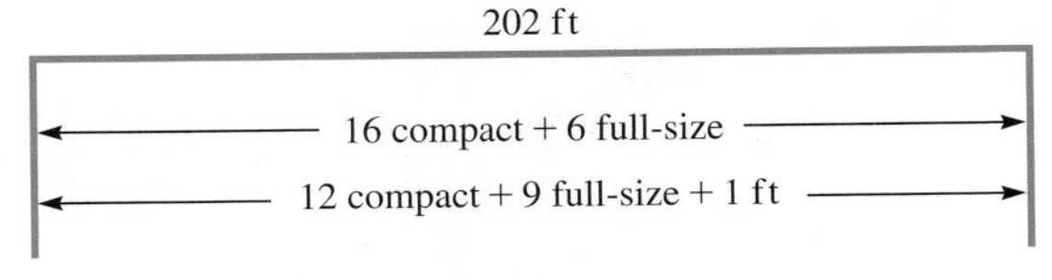

Fig. 5.24

47. A small plane can fly 780 km with a tailwind in 3 h. If the tailwind is 50% stronger, it can fly 700 km in 2 h 30 min. Find the speed p of the plane in still air, and the speed w of the wind.

48. The equations relating the currents i_1 and i_2 shown in Fig. 5.25 are given below. Find the currents to the nearest 0.1 A.

$$2i_1 + 6(i_1 + i_2) = 12$$
$$4i_2 + 6(i_1 + i_2) = 12$$

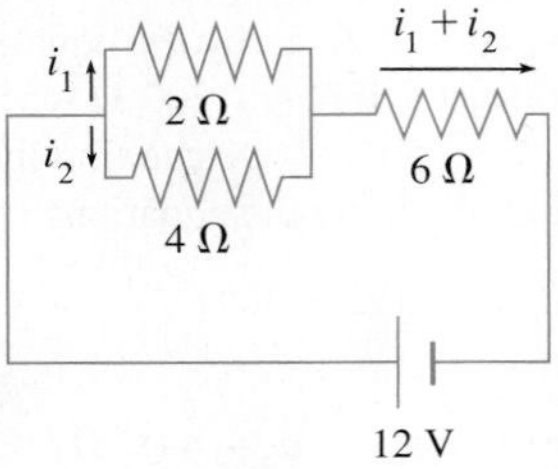

Fig. 5.25

49. A construction company placed two orders, at the same prices, with a lumber retailer. The first order was for 8 sheets of plywood and 40 framing studs for a total cost of $304. The second order was for 25 sheets of plywood and 12 framing studs for a total cost of $498. Find the price of one sheet of plywood and one framing stud.

50. A certain car gets 21 mi/gal in city driving and 28 mi/gal in highway driving. If 18 gal of gas are used in traveling 448 mi, how many miles were driven in the city, and how many were driven on the highway (assuming that only the given rates of usage were actually used)?

In Exercises 51–52, find the required values.

51. Determine the values of m and b that make the following system have no solution:

$$y = 3x - 7$$
$$y = mx + b$$

52. What, if any, value(s) of m and b result in the following system having (a) one solution, (b) an unlimited number of solutions?

$$y = -3x + b$$
$$y = mx + 8$$

Answer to Practice Exercises

1. $x = 1.5, y = 1.4$ **2.** Yes

5.3 Solving Systems of Two Linear Equations in Two Unknowns Algebraically

Solution by Substitution • Solution by Addition or Subtraction

The graphical method of solving two linear equations in two unknowns is good for getting a "picture" of the solution. One problem is that graphical methods usually give *approximate* results, although the accuracy obtained when using a calculator is excellent. If *exact* solutions are required, we turn to other methods. In this section, we present two algebraic methods of solution.

SOLUTION BY SUBSTITUTION

The first method involves the *elimination of one variable by* **substitution.** The basic idea of this method is to *solve one of the equations for one of the unknowns and substitute this into the other equation.* The result is an equation with only one unknown, and this equation can then be solved for the unknown. Following is the basic procedure to be used.

Solution of Two Linear Equations by Substitution

1. Solve one equation for one of the unknowns.
2. Substitute this solution into the other equation. At this point, we have a linear equation in one unknown.
3. Solve the resulting equation for the value of the unknown it contains.
4. Substitute this value into the equation of step 1 and solve for the other unknown.
5. Check the values in both original equations.

In following this procedure, you must first choose an unknown for which to solve. Often it makes little difference, but if it is easier to solve a particular equation for one of the unknowns, that is the one to use.

EXAMPLE 1 Solution by substitution

Solve the following system of equations by substitution.

$$x - 3y = 6$$
$$2x + 3y = 3$$

Here, it is easiest to solve the first equation for x:

step 1 $$x = 3y + 6 \quad \textbf{(A1)}$$

in second equation, x replaced by $3y + 6$

step 2 $$2(3y + 6) + 3y = 3 \quad \text{substituting}$$

step 3 $$6y + 12 + 3y = 3 \quad \text{solving for } y$$
$$9y = -9$$
$$y = -1$$

Now, put the value $y = -1$ into Eq. (A1) since this is already solved for x in terms of y. Solving for x, we have

step 4 $$x = 3(-1) + 6 = 3 \quad \text{solving for } x$$

step 5 Therefore, the solution of the system is $x = 3, y = -1$. As a check, substitute these values into each of the original equations. This gives us $3 - 3(-1) = 6$ and $2(3) + 3(-1) = 3$, which verifies the solution. ■

EXAMPLE 2 Solution by substitution

Solve the following system of equations by substitution.

$$-5x + 2y = -4$$
$$10x + 6y = 3$$

It makes little difference which equation or which unknown is chosen. Therefore,

$$2y = 5x - 4 \quad \text{solving first equation for } y$$
$$y = \frac{5x - 4}{2} \quad \textbf{(B1)}$$

in second equation, y replaced by $\frac{5x-4}{2}$

TI-89 graphing calculator keystrokes for Example 2: goo.gl/cJChPo

$$10x + 6\left(\frac{5x - 4}{2}\right) = 3 \quad \text{substituting}$$
$$10x + 3(5x - 4) = 3 \quad \text{solving for } x$$
$$10x + 15x - 12 = 3$$
$$25x = 15$$
$$x = \frac{3}{5}$$

Substituting this value into the expression for y, Eq. (B1), we obtain

$$y = \frac{5(3/5) - 4}{2} = \frac{3 - 4}{2} = -\frac{1}{2} \quad \text{solving for } y$$

Therefore, the solution of this system is $x = \frac{3}{5}, y = -\frac{1}{2}$. Substituting these values in both original equations shows that the solution checks. ■

Practice Exercise

1. Solve the system in Example 2 by first solving for x and then substituting.

SOLUTION BY ADDITION OR SUBTRACTION

The method of substitution is useful if one of the equations can easily be solved for one of the unknowns. However, often a fraction results (such as in Example 2), and this leads to additional algebraic steps to find the solution. Even worse, if the coefficients are themselves decimals or fractions, the algebraic steps are even more involved.

Therefore, we now present another algebraic method of solving a system of equations, *elimination of a variable by* **addition** *or* **subtraction.** Following is the basic procedure to be used.

■ For reference, Eqs. (5.4) are

$$a_1x + b_1y = c_1$$
$$a_2x + b_2y = c_2$$

Solution of Two Linear Equations by Addition or Subtraction

1. Write the equations in the form of Eqs. (5.4), if they are not already in this form.
2. If necessary, multiply all terms of each equation by a constant chosen so that the coefficients of one unknown will be numerically the same in both equations. (They can have the same or different signs.)
3. **(a)** If the numerically equal coefficients have *different* signs, *add* the terms on each side of the resulting equations.
   (b) If the numerically equal coefficients have the *same* sign, *subtract* the terms on each side of one equation from the terms of the other equation.
4. Solve the resulting linear equation in the other unknown.
5. Substitute this value into one of the original equations to find the value of the other unknown.
6. Check by substituting both values into both original equations.

■ Some prefer always to use addition of the terms of the resulting equations. This avoids possible errors that may be caused when subtracting.

EXAMPLE 3 Solution by addition

Use the method of elimination by addition or subtraction to solve the system of equations.

$$x - 3y = 6 \quad \text{already in the form of Eqs. (5.4)}$$
$$2x + 3y = 3$$

We look at the coefficients to determine the best way to eliminate one of the unknowns. Since the coefficients of the y-terms are *numerically the same and opposite in sign,* we may immediately *add* terms of the two equations together to eliminate y. Adding the terms of the left sides and adding terms of the right sides, we obtain

$$x + 2x - 3y + 3y = 6 + 3$$
$$3x = 9$$
$$x = 3$$

Substituting this value into the first equation, we obtain

$$3 - 3y = 6$$
$$-3y = 3$$
$$y = -1$$

The solution $x = 3$, $y = -1$ agrees with the results obtained for the same problem illustrated in Example 1. ■

Practice Exercise

2. Solve the system in Example 3 by first multiplying the terms of the first equation by 2, and then subtracting.

The following example illustrates the solution of a system of equations by subtracting the terms of one equation from the terms of the other equation. CAUTION Be very careful when subtracting, particularly when dealing with negative numbers. Remember that subtracting a negative number is the same as adding a positive number. ■

EXAMPLE 4 Solution by subtraction

Use the method of addition or subtraction to solve the following system of equations.

$$3x - 2y = 4$$
$$x + 3y = 2$$

Looking at the coefficients of x and y, we see that we must multiply the second equation by 3 to make the coefficients of x the same. To make the coefficients of y numerically the same, we must multiply the first equation by 3 and the second equation by 2. Thus, the best method is to multiply the second equation by 3 and eliminate x. **CAUTION** Be careful to multiply the terms on *both* sides: a common error is to forget to multiply the value on the right. ■ After multiplying, the coefficients of x have the *same sign*. Therefore, we *subtract* terms of the second equation from those of the first equation:

$$\begin{aligned} 3x - 2y &= 4 \\ 3x + 9y &= 6 \quad \text{each term of second equation multiplied by 3} \\ \hline -11y &= -2 \quad \text{subtract} \\ y &= \frac{2}{11} \end{aligned}$$

$3x - 3x = 0$; $-2y - (+9y) = -11y$; $4 - 6 = -2$

In order to find the value of x, substitute $y = \frac{2}{11}$ into one of the original equations. Choosing the second equation (its form is somewhat simpler), we have

$$\begin{aligned} x + 3\left(\frac{2}{11}\right) &= 2 \\ 11x + 6 &= 22 \quad \text{multiply each term by 11} \\ x &= \frac{16}{11} \end{aligned}$$

Therefore, the solution is $x = \frac{16}{11}$, $y = \frac{2}{11}$. Substituting these values into both of the original equations shows that the solution checks. ■

EXAMPLE 5 System of Example 4 solved by addition

As noted in Example 4, we can solve that system of equations by first multiplying the terms of the first equation by 3 and those of the second equation by 2, and thereby eliminating y. Doing this, we have

$$\begin{aligned} 9x - 6y &= 12 \quad \text{each term of first equation multiplied by 3} \\ 2x + 6y &= 4 \quad \text{each term of second equation multiplied by 2} \\ \hline 11x \quad &= 16 \quad \text{add} \\ x &= \frac{16}{11} \\ 3\left(\tfrac{16}{11}\right) - 2y &= 4 \quad \text{substituting } x = 16/11 \text{ in the first original equation} \\ 48 - 22y &= 44 \quad \text{multiply each term by 11} \\ -22y &= -4 \\ y &= \frac{2}{11} \end{aligned}$$

$9x + 2x = 11x$; $-6y + 6y = 0$; $12 + 4 = 16$

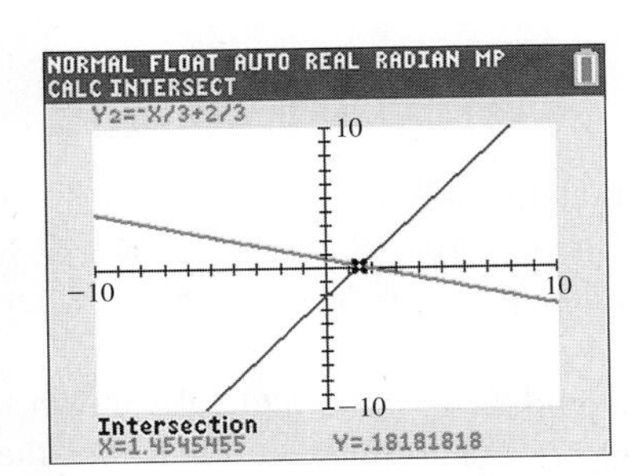

Fig. 5.26

Therefore, the solution is $x = \frac{16}{11}$, $y = \frac{2}{11}$, as found in Example 4.

A calculator solution of this system is shown in Fig. 5.26, where $y_1 = 3x/2 - 2$ and $y_2 = -x/3 + 2/3$. The point of intersection is (1.455, 0.182). This solution is the same as the algebraic solutions because $\frac{16}{11} = 1.455$ and $\frac{2}{11} = 0.182$. ■

EXAMPLE 6 Solution by subtraction—mixing metal alloys

■ The second equation is based on the amount of copper. We could have used the amount of zinc, which would have led to the equation $0.30A + 0.60B = 0.40(300)$. Because we need only two equations, we may use any two of these three equations to find the solution.

By weight, one alloy is 70% copper and 30% zinc. Another alloy is 40% copper and 60% zinc. How many grams of each are required to make 300 g of an alloy that is 60% copper and 40% zinc?

Let A = the required number of grams of the first alloy and B = the required number of grams of the second alloy. Our equations are determined from:

1. The total weight of the final alloy is 300 g: $A + B = 300$.
2. The final alloy will have 180 g of copper (60% of 300 g), and this comes from 70% of A ($0.70A$) and 40% of B ($0.40B$): $0.70A + 0.40B = 180$.

These two equations can now be solved simultaneously:

$$A + B = 300 \quad \text{sum of weights is 300 g}$$

$$\text{copper} \rightarrow 0.70A + 0.40B = 180 \leftarrow 60\% \text{ of } 300 \text{ g}$$

(70% — weight of first alloy; 40% — weight of second alloy)

$$\begin{aligned} 4A + 4B &= 1200 && \text{multiply each term of first equation by 4} \\ 7A + 4B &= 1800 && \text{multiply each term of second equation by 10} \\ \hline 3A &= 600 && \text{subtract first equation from second equation} \\ A &= 200 \text{ g} \\ B &= 100 \text{ g} && \text{by substituting into first equation} \end{aligned}$$

Checking with the statement of the problem, using the percentages of zinc, we have $0.30(200) + 0.60(100) = 0.40(300)$, or $60 \text{ g} + 60 \text{ g} = 120 \text{ g}$. We use zinc here since we used the percentages of copper for the equation. ■

Practice Exercise

3. Solve the problem in Example 6 by using the equation shown in the note above.

EXAMPLE 7 Inconsistent system

In solving the system of equations

$$\begin{aligned} 4x &= 2y + 3 \\ -y + 2x - 2 &= 0 \end{aligned}$$

first note that the equations are not in the correct form. Therefore, writing them in the form of Eqs. (5.4), we have

$$\begin{aligned} 4x - 2y &= 3 \\ 2x - y &= 2 \end{aligned}$$

Now, multiply the second equation by 2 and subtract to get

$$\begin{aligned} 4x - 2y &= 3 \\ 4x - 2y &= 4 \\ \hline 0 &= -1 \end{aligned}$$

($4x - 4x = 0$, $-2y - (-2y) = 0$, $3 - 4 = -1$)

NORMAL FLOAT AUTO REAL RADIAN MP

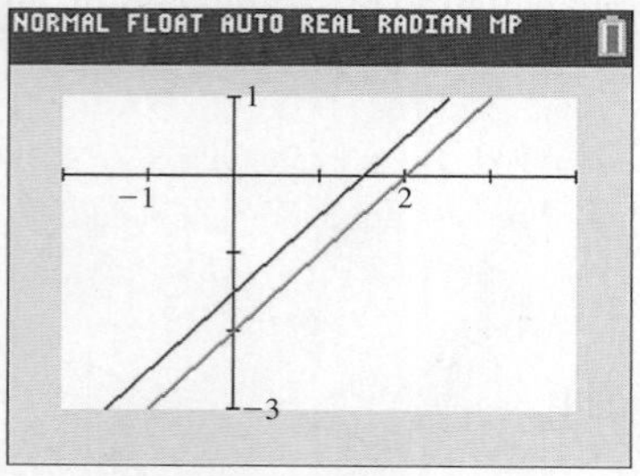

Fig. 5.27

NOTE ▶ [Because 0 does not equal -1, we conclude that there is no solution. When the result is $0 = a$ $(a \neq 0)$, the system is *inconsistent*.] From Section 5.2, we know this means the lines representing the equations are parallel. See Fig. 5.27, where $y_1 = 2x - 3/2$, $y_2 = 2x - 2$.

NOTE ▶ [If the result of solving a system is $0 = 0$, the system is *dependent*.] As shown in Section 5.2, this means there is an unlimited number of solutions, and the lines that represent the equations are really the same line. ■

EXERCISES 5.3

In Exercises 1–4, make the given changes in the indicated examples of this section and then solve the resulting problems.

1. In Example 1, change the + to − in the second equation and then solve the system of equations.

2. In Example 3, change the + to − in the second equation and then solve the system of equations.

3. In Example 4, change x to $2x$ in the second equation and then solve the system of equations.

4. In Example 7, change 3 to 4 in the first equation and then find if there is any change in the conclusion that is drawn.

In Exercises 5–14, solve the given systems of equations by the method of elimination by substitution.

5. $x = y + 3$
$x - 2y = 5$

6. $x = 2y + 1$
$2x - 3y = 4$

7. $p = V - 4$
$V + p = 10$

8. $y = 2x + 10$
$2x + y = -2$

9. $x + y = -5$
$2x - y = 2$

10. $3x + y = 1$
$2y = 3x - 16$

11. $2x + 3y = 7$
$6x - y = 1$

12. $6s + 6t = 3$
$4s - 2t = 17$

13. $33x + 2y = 34$
$40y = 9x + 11$

14. $3A + 3B = -1$
$5A = -6B - 1$

In Exercises 15–24, solve the given systems of equations by the method of elimination by addition or subtraction.

15. $x + 2y = 5$
$x - 2y = 1$

16. $x + 3y = 7$
$2x + 3y = 5$

17. $2x - 3y = 4$
$2x + y = -4$

18. $R - 4r = 17$
$4r + 3R = 3$

19. $12t + 9y = 14$
$6t = 7y - 16$

20. $3x - y = 3$
$4x = 3y + 14$

21. $v + 2t = 7$
$2v + 4t = 9$

22. $3x - y = 5$
$3y - 9x = -15$

23. $2x - 3y - 4 = 0$
$3x + 2 = 2y$

24. $3i_1 + 5 = -4i_2$
$3i_2 = 5i_1 - 2$

In Exercises 25–36, solve the given systems of equations by either method of this section.

25. $2x - y = 5$
$6x + 2y = -5$

26. $3x + 2y = 4$
$6x - 6y = 13$

27. $6x + 3y + 4 = 0$
$5y + 9x + 6 = 0$

28. $1 + 6q = 5p$
$3p - 4q = 7$

29. $15x + 10y = 11$
$20x - 25y = 7$

30. $2x + 6y = -3$
$-6x - 18y = 5$

31. $12V + 108 = -84C$
$36C + 48V + 132 = 0$

32. $66x + 66y = -77$
$33x - 132y = 143$

33. $44A = 1 - 15B$
$5B = 22 + 7A$

34. $60x - 40y = 80$
$2.9x - 2.0y = 8.0$

35. $2b = 6a - 16$
$33a = 4b + 39$

36. $30P = 55 - Q$
$19P + 14Q + 32 = 0$

In Exercises 37–42, in order to make the coefficients easier to work with, first multiply each term of the equation or divide each term of the equation by a number selected by inspection. Then proceed with the solution of the system by an appropriate algebraic method.

37. $0.3x - 0.7y = 0.4$
$0.2x + 0.5y = 0.7$

38. $250R + 225Z = 400$
$375R - 675Z = 325$

39. $40s - 30t = 60$
$20s - 40t = -50$

40. $0.060x + 0.048y = -0.084$
$0.065y - 0.13x = 0.078$

41. $\frac{x}{3} + \frac{2y}{3} = 2$
$\frac{x}{2} - 2y = \frac{5}{2}$

42. $\frac{2x}{5} - \frac{y}{5} = 1$
$\frac{3x}{4} - y = \frac{5}{4}$

In Exercises 43–50, solve the indicated or given systems of equations by an appropriate algebraic method.

43. Find the function $f(x) = ax + b$, if $f(2) = 1$ and $f(-1) = -5$.

44. Find the function $f(x) = ax + b$, if $f(6) = -1$ and $f(-6) = 11$.

45. Solve the following system of equations by (a) solving the first equation for x and substituting, and (b) solving the first equation for y and substituting.

$2x + y = 4$
$3x - 4y = -5$

46. Solve for x and y: $\frac{5}{x + y} + \frac{2}{x - y} = 3$

$\frac{20}{x + y} - \frac{2}{x - y} = 2$

47. Find the voltages V_1 and V_2 of the batteries shown in Fig. 5.28. The terminals are aligned in the same direction in Fig. 5.28(a) and in the opposite directions in Fig. 5.28(b).

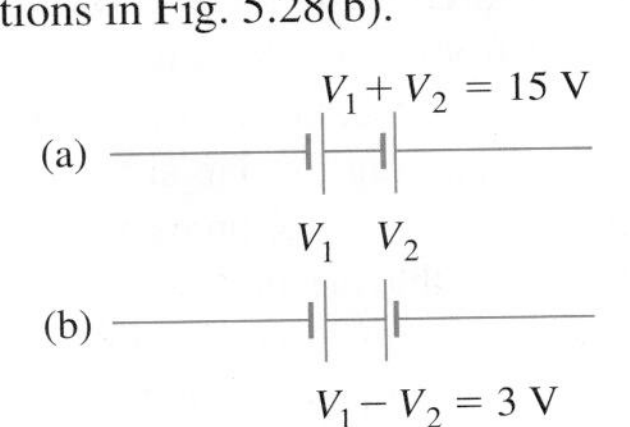

Fig. 5.28

48. A spring of length L is stretched x cm for each newton of weight hung from it. Weights of 3 N and then 5 N are hung from the spring, leading to the equations

$L + 3x = 18$
$L + 5x = 22$

Solve for L and x.

49. Two grades of gasoline are mixed to make a blend with 1.50% of a special additive. Combining x liters of a grade with 1.80% of the additive to y liters of a grade with 1.00% of the additive gives 10,000 L of the blend. The equations relating x and y are

$$x + y = 10{,}000$$
$$0.0180x + 0.0100y = 0.0150(10{,}000)$$

Find x and y (to three significant digits).

50. A 6.0% solution and a 15.0% solution of a drug are added to 200 mL of a 20.0% solution to make 1200 mL of a 12.0% solution for a proper dosage. The equations relating the number of milliliters of the added solutions are

$$x + y + 200 = 1200$$
$$0.060x + 0.150y + 0.200(200) = 0.120(1200)$$

Find x and y (to three significant digits).

In Exercises 51–64, set up appropriate systems of two linear equations and solve the systems algebraically. All data are accurate to at least two significant digits.

51. A person's email for a day contained a total of 78 messages. The number of spam messages was two less than four times the other messages. How many were spam?

52. A 150-m cable is cut into two pieces such that one piece is four times as long as the other. How long is each piece?

53. The weight W_f supported by the front wheels of a certain car and the weight W_r supported by the rear wheels together equal the weight of the car, 17,700 N. See Fig. 5.29. Also, the ratio of W_r to W_f is 0.847. What are the weights supported by each set of wheels?

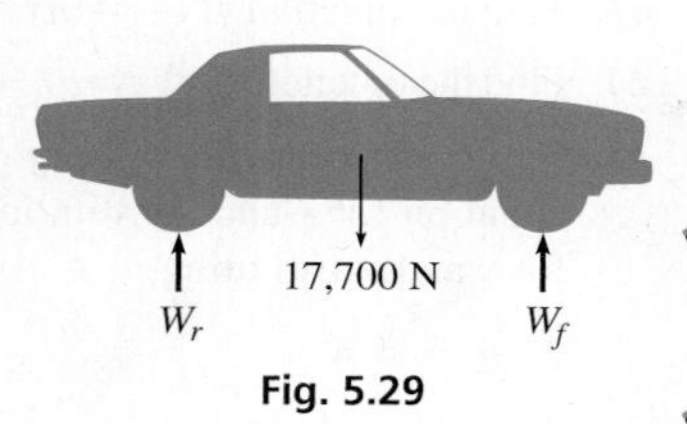

Fig. 5.29

54. A sprinkler system is used to water two areas. If the total water flow is 980 L/h and the flow through one sprinkler is 65% as much as the other, what is the flow in each?

55. In a test of a heat-seeking rocket, a first rocket is launched at 2000 ft/s, and the heat-seeking rocket is launched along the same flight path 12 s later at a speed of 3200 ft/s. Find the times t_1 and t_2 of flight of the rockets until the heat-seeking rocket destroys the first rocket.

56. The *torque* of a force is the product of the force and the perpendicular distance from a specified point. If a lever is supported at only one point and is in balance, the sum of the torques (about the support) of forces acting on one side of the support must equal the sum of the torques of the forces acting on the other side. Find the forces F_1 and F_2 that are in the positions shown in Fig. 5.30(a) and then move to the positions in Fig. 5.30(b). The lever weighs 20 N and is in balance in each case.

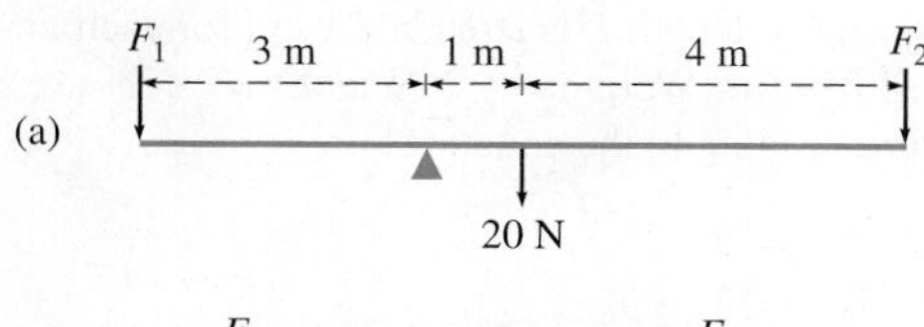

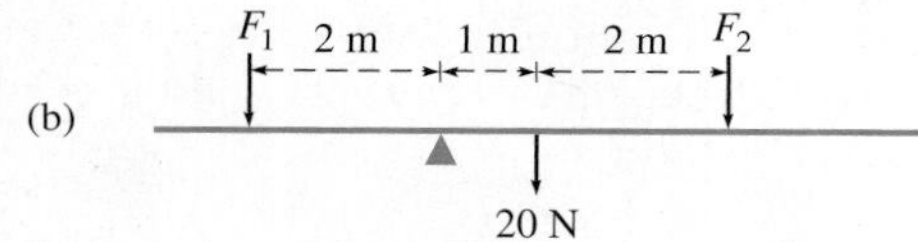

Fig. 5.30

57. There are two types of offices in an office building, and a total of 54 offices. One type rents for \$900/month and the other type rents for \$1250/month. If all offices are occupied and the total rental income is \$55,600/month, how many of each type are there?

58. An airplane flies into a headwind with an effective ground speed of 140 mi/h. On the return trip it flies with the tailwind and has an effective ground speed of 240 mi/h. Find the speed p of the plane in still air, and the speed w of the wind.

59. In an election, candidate A defeated candidate B by 2000 votes. If 1.0% of those who voted for A had voted for B, B would have won by 1000 votes. How many votes did each receive?

60. An underwater (but near the surface) explosion is detected by sonar on a ship 30 s before it is heard on the deck. If sound travels at 5000 ft/s in water and 1100 ft/s in air, how far is the ship from the explosion? See Fig. 5.31.

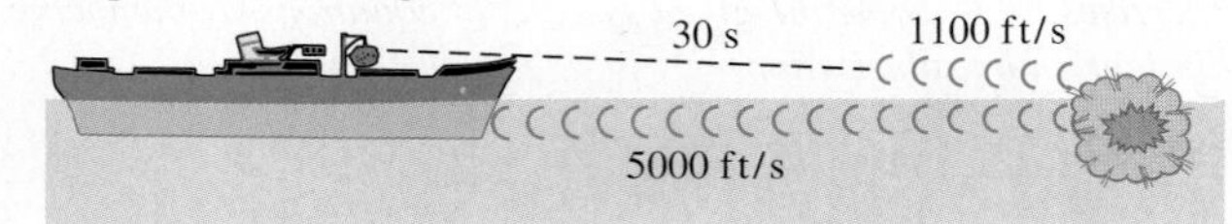

Fig. 5.31

61. A small isolated farm uses a windmill and a gas generator for power. During a 10-day period, they produced 3010 kW · h of power, with the windmill operating at 45.0% of capacity and the generator at capacity. During the following 10-day period, they produced 2900 kW · h with the windmill at 72.0% of capacity and the generator down 60 h for repairs (at capacity otherwise). What is the capacity (in kW) of each?

62. In mixing a weed-killing chemical, a 40% solution of the chemical is mixed with an 85% solution to get 20 L of a 60% solution. How much of each solution is needed?

63. What conclusion can you draw from a sales report that states that "sales this month were \$8000 more than last month, which means that total sales for both months are \$4000 more than twice the sales last month"?

64. Regarding the forces on a truss, a report stated that force F_1 is twice force F_2 and that twice the sum of the two forces less 6 times F_2 is 6 N. Explain your conclusion about the magnitudes of the forces found from this support.

In Exercises 65–68, answer the given questions.

65. What condition(s) must be placed on the constants of the system of equations

$$ax + y = c$$
$$bx + y = d$$

such that there is a unique solution for x and y?

66. What conditions must be placed on the constants of the system of equations in Exercise 65 such that the system is (a) inconsistent? (b) Dependent?

67. For the dependent system of Example 8 on page 150, both equations can be written as $y = \frac{1}{3}x - 3$. The solution for the system can then be shown in terms of a general point on the graph as $(x, \frac{1}{3}x - 3)$. This is referred to as the solution with arbitrary x. Find the solutions in this form for this system for $x = -3$, and $x = 9$.

68. For the dependent system of Example 8 on page 150, write the form for the solution with arbitrary y. See Exercise 67.

Answers to Practice Exercises

1. $x = 3/5, y = -1/2$ **2.** $x = 3, y = -1$
3. $A = 200$ g, $B = 100$ g

5.4 Solving Systems of Two Linear Equations in Two Unknowns by Determinants

Determinant of the Second Order • Cramer's Rule • Solving Systems of Equations by Determinants

Consider two linear equations in two unknowns, as given in Eqs. (5.4):

$$\begin{aligned} a_1x + b_1y &= c_1 \\ a_2x + b_2y &= c_2 \end{aligned} \quad (5.4)$$

If we multiply the first of these equations by b_2 and the second by b_1, we obtain

$$\begin{aligned} a_1b_2x + b_1b_2y &= c_1b_2 \\ a_2b_1x + b_2b_1y &= c_2b_1 \end{aligned} \quad (5.5)$$

We see that the coefficients of y are the same. Thus, subtracting the second equation from the first, we can solve for x. The solution can be shown to be

$$x = \frac{c_1b_2 - c_2b_1}{a_1b_2 - a_2b_1} \quad (5.6)$$

In the same manner, we may show that

$$y = \frac{a_1c_2 - a_2c_1}{a_1b_2 - a_2b_1} \quad (5.7)$$

■ Determinants were invented by the German mathematician Gottfried Wilhelm Leibniz (1646–1716).

The expression $a_1b_2 - a_2b_1$, which appears in each of the denominators of Eqs. (5.6) and (5.7), is an example of a special kind of expression called a *determinant of the second order*. The determinant $a_1b_2 - a_2b_1$ is denoted by

$$\begin{vmatrix} a_1 & b_1 \\ a_2 & b_2 \end{vmatrix}$$

Therefore, by definition, *a* **determinant of the second order** *is*

$$\begin{vmatrix} a_1 & b_1 \\ a_2 & b_2 \end{vmatrix} = a_1b_2 - a_2b_1 \quad (5.8)$$

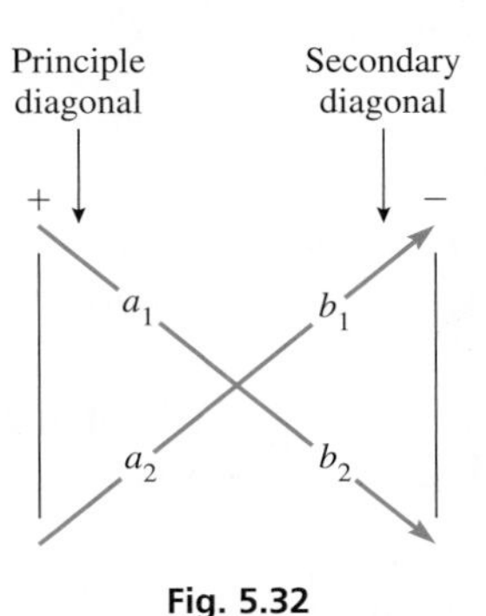

Fig. 5.32

The numbers a_1 and b_1 are called the **elements** *of the first* **row** *of the determinant. The numbers a_1 and a_2 are the elements of the first* **column** *of the determinant.* In the same manner, the numbers a_2 and b_2 are the elements of the second row, and the numbers b_1 and b_2 are the elements of the second column. *The numbers a_1 and b_2 are the elements of the* **principal diagonal,** *and the numbers a_2 and b_1 are the elements of the* **secondary diagonal.** Thus, one way of stating the definition indicated in Eq. (5.8) is that *the value of a determinant of the second order is found by taking the product of the elements of the principal diagonal and subtracting the product of the elements of the secondary diagonal.*

A diagram that is often helpful for remembering the expansion of a second-order determinant is shown in Fig. 5.32. CAUTION It is very important in following Eq. (5.8), or the diagram, that you remember to subtract the product a_2b_1 of the secondary diagonal from the product a_1b_2 of the principal diagonal. ■ The following examples illustrate the evaluation of determinants of the second order.

EXAMPLE 1 Evaluating a second-order determinant

$$\begin{vmatrix} -5 & 8 \\ 3 & 7 \end{vmatrix} = (-5)(7) - 3(8) = -35 - 24 = -59$$

■

EXAMPLE 2 Evaluating second-order determinants

(a) $\begin{vmatrix} 4 & 6 \\ 3 & 17 \end{vmatrix} = 4(17) - (3)(6) = 68 - 18 = 50$

CAUTION Be careful. If a diagonal contains a zero, the product is zero. ■

(b) $\begin{vmatrix} 4 & 0 \\ -3 & 17 \end{vmatrix} = 4(17) - (-3)(0) = 68 - 0 = 68$

(c) $\begin{vmatrix} 3.6 & 6.1 \\ -3.2 & -17.2 \end{vmatrix} = 3.6(-17.2) - (-3.2)(6.1) = -42.4$

Note the signs of the terms being combined. ■

Practice Exercise

1. Evaluate the determinant $\begin{vmatrix} 2 & -4 \\ 3 & 5 \end{vmatrix}$.

We note that the numerators and denominators of Eqs. (5.6) and (5.7) may be written as determinants. The numerators of the equations are

$$\begin{vmatrix} c_1 & b_1 \\ c_2 & b_2 \end{vmatrix} \quad \text{and} \quad \begin{vmatrix} a_1 & c_1 \\ a_2 & c_2 \end{vmatrix}$$

Therefore, the solutions for x and y can be expressed directly in terms of determinants, which is called **Cramer's rule.**

Cramer's Rule

For a system of equations of the form of Eqs. (5.4), the solutions for x and y are given by

$$x = \frac{\begin{vmatrix} c_1 & b_1 \\ c_2 & b_2 \end{vmatrix}}{\begin{vmatrix} a_1 & b_1 \\ a_2 & b_2 \end{vmatrix}} \quad \text{and} \quad y = \frac{\begin{vmatrix} a_1 & c_1 \\ a_2 & c_2 \end{vmatrix}}{\begin{vmatrix} a_1 & b_1 \\ a_2 & b_2 \end{vmatrix}} \qquad \textbf{(5.9)}$$

provided the determinant in the denominator is not equal to zero.

■ Note carefully the location of c_1 and c_2.

It is important to notice three key features of the determinants in Cramer's rule:

1. The determinants in the denominators are the same and are formed by using the coefficients of x and y.
2. In the **solution for x,** the determinant in the numerator is obtained from the determinant in the denominator by **replacing the x-coefficients with the constant terms.**
3. In the **solution for y,** the determinant in the numerator is obtained from the determinant in the denominator by **replacing the y-coefficients with the constant terms.**

■ Named for the Swiss mathematician Gabriel Cramer (1704–1752).

CAUTION In Using Cramer's rule, we must be very sure that the equations are written in the form of Eqs. (5.4) before setting up the determinants. That is, be sure the unknowns are in the same order in each equation. ■

The following examples illustrate the method of solving systems of equations by determinants.

EXAMPLE 3 Solving a system using Cramer's rule

Solve the following system of equations by determinants:

$$2x + y = 1$$
$$5x - 2y = -11$$

NOTE ▸ [First, note that the equations are in the proper form of Eqs. (5.4) for solution by determinants.] Next, set up the determinant for the denominator, which consists of the four coefficients in the system, written as shown. It is

$$\begin{vmatrix} 2 & 1 \\ 5 & -2 \end{vmatrix}$$

x-coefficients — *y*-coefficients

For finding x, the determinant in the numerator is obtained from this determinant by replacing the first column by the constants that appear on the right sides of the equations. Thus, *the numerator for the solution for x* is

$$\begin{vmatrix} 1 & 1 \\ -11 & -2 \end{vmatrix}$$

replace *x*-coefficients with the constants

For finding y, the determinant in the numerator is obtained from the determinant of the denominator by replacing the second column by the constants that appear on the right sides of the equations. Thus, *the numerator for the solution for y* is

$$\begin{vmatrix} 2 & 1 \\ 5 & -11 \end{vmatrix}$$

replace *y*-coefficients with the constants

Now, set up the solutions for x and y using the determinants above:

■ Determinants can be evaluated on a graphing calculator. This is shown in Section 5.6.

$$x = \frac{\begin{vmatrix} 1 & 1 \\ -11 & -2 \end{vmatrix}}{\begin{vmatrix} 2 & 1 \\ 5 & -2 \end{vmatrix}} = \frac{1(-2) - (-11)(1)}{2(-2) - (5)(1)} = \frac{-2 + 11}{-4 - 5} = \frac{9}{-9} = -1$$

$$y = \frac{\begin{vmatrix} 2 & 1 \\ 5 & -11 \end{vmatrix}}{\begin{vmatrix} 2 & 1 \\ 5 & -2 \end{vmatrix}} = \frac{2(-11) - (5)(1)}{-9} = \frac{-22 - 5}{-9} = 3$$

Therefore, the solution to the system of equations is $x = -1, y = 3$.

Substituting these values into the equations, we have

$$2(-1) + 3 \stackrel{?}{=} 1 \quad \text{and} \quad 5(-1) - 2(3) \stackrel{?}{=} -11$$
$$1 = 1 \qquad\qquad -11 = -11$$

which shows that they check.

NOTE ▸ [Because the same determinant appears in each denominator, it needs to be evaluated only once.] This means that three determinants are to be evaluated in order to solve the system. ■

Practice Exercise

2. Solve the following system by determinants.

$$x + 2y = 4$$
$$3x - y = -9$$

EXAMPLE 4 Solving a system using Cramer's rule

Solve the following system of equations by determinants. Numbers are approximate.

$$5.3x + 7.2y = 4.5$$
$$3.2x - 6.9y = 5.7$$

constants (→ first column of numerator); coefficients (→ denominator)

$$x = \frac{\begin{vmatrix} 4.5 & 7.2 \\ 5.7 & -6.9 \end{vmatrix}}{\begin{vmatrix} 5.3 & 7.2 \\ 3.2 & -6.9 \end{vmatrix}} = \frac{4.5(-6.9) - 5.7(7.2)}{5.3(-6.9) - 3.2(7.2)} = \frac{-72.09}{-59.61} = 1.2$$

constants (→ second column of numerator); coefficients (→ denominator)

$$y = \frac{\begin{vmatrix} 5.3 & 4.5 \\ 3.2 & 5.7 \end{vmatrix}}{\begin{vmatrix} 5.3 & 7.2 \\ 3.2 & -6.9 \end{vmatrix}} = \frac{5.3(5.7) - 3.2(4.5)}{-59.61} = \frac{15.81}{-59.61} = -0.27$$

It is important to round only the final answers when performing these calculations. When substituted into the original equations, the solutions check reasonably well. The small differences are due to rounding the final answers. ■

EXAMPLE 5 Solving a system—investment income

Two investments totaling \$18,000 yield an annual income of \$700. If the first investment has an interest rate of 5.5% and the second a rate of 3.0%, what is the value of each?

Let x = the value of the first investment and y = the value of the second investment. We know that the total of the two investments is \$18,000. This leads to the equation $x + y = \$18{,}000$. The first investment yields $0.055x$ dollars annually, and the second yields $0.030y$ dollars annually. This leads to the equation $0.055x + 0.030y = 700$. These two equations are then solved simultaneously:

$$x + y = 18{,}000 \quad \text{sum of investments}$$
$$0.055x + 0.030y = 700 \quad \leftarrow \text{income}$$

(↑ 5.5% value under $0.055x$; ↑ 3.0% value under $0.030y$)

$$x = \frac{\begin{vmatrix} 18{,}000 & 1 \\ 700 & 0.030 \end{vmatrix}}{\begin{vmatrix} 1 & 1 \\ 0.055 & 0.030 \end{vmatrix}} = \frac{540 - 700}{0.030 - 0.055} = \frac{-160}{-0.025} = 6400$$

The value of y can be found most easily by substituting this value of x into the first equation, $y = 18{,}000 - x = 18{,}000 - 6400 = 11{,}600$.

Therefore, the values invested are \$6400 and \$11,600. Checking, the total income is $\$6400(0.055) + \$11{,}600(0.030) = \$700$, which agrees with the statement of the problem. ■

Practice Exercise

3. In Example 5, change 5.5% to 5.0% and solve for the investments.

CAUTION The equations must be in the form of Eqs. (5.4) before the determinants are set up. ■ The specific positions of the values in the determinants are based on that form of writing the system. If either unknown is missing from an equation, a zero must be placed in the proper position. Also, from Example 4, we see that determinants are easier to use than other algebraic methods when the coefficients are decimals.

CAUTION If the determinant of the denominator is zero, we do not have a unique solution because this would require division by zero. If the determinant of the denominator is zero and that of the numerator is not zero, the system is inconsistent. If the determinants of both numerator and denominator are zero, the system is dependent. ■

EXERCISES 5.4

In Exercises 1–4, make the given changes in the indicated examples of this section and then solve the resulting problems.

1. In Example 2(a), change the 6 to −6 and then evaluate.
2. In Example 2(a), change the 4 to −4 and the 6 to −6 and then evaluate.
3. In Example 3, change the + to − in the first equation and then solve the system of equations.
4. In Example 5, change \$700 to \$830 and then solve for the values of the investments.

In Exercises 5–18, evaluate the given determinants.

5. $\begin{vmatrix} 8 & 3 \\ 4 & 1 \end{vmatrix}$ 6. $\begin{vmatrix} -1 & 3 \\ 2 & 6 \end{vmatrix}$ 7. $\begin{vmatrix} 3 & -5 \\ 7 & -2 \end{vmatrix}$

8. $\begin{vmatrix} -4 & 7 \\ 1 & -3 \end{vmatrix}$ 9. $\begin{vmatrix} 15 & -9 \\ 12 & 0 \end{vmatrix}$ 10. $\begin{vmatrix} -20 & -15 \\ -8 & -6 \end{vmatrix}$

11. $\begin{vmatrix} -20 & 110 \\ -70 & -80 \end{vmatrix}$ 12. $\begin{vmatrix} -6.5 & 12.2 \\ -15.5 & 34.6 \end{vmatrix}$ 13. $\begin{vmatrix} 0.75 & -1.32 \\ 0.15 & 1.18 \end{vmatrix}$

14. $\begin{vmatrix} 0.20 & -0.05 \\ 0.28 & 0.09 \end{vmatrix}$ 15. $\begin{vmatrix} -8 & -4 \\ -32 & 16 \end{vmatrix}$ 16. $\begin{vmatrix} 43 & -7 \\ -81 & 16 \end{vmatrix}$

17. $\begin{vmatrix} 2 & a-1 \\ a+2 & a \end{vmatrix}$ 18. $\begin{vmatrix} x+y & y-x \\ 2x & 2y \end{vmatrix}$

In Exercises 19–28, solve the given systems of equations by determinants. (These are the same as those for Exercises 15–24 of Section 5.3.)

19. $x + 2y = 5$
$x - 2y = 1$

20. $x + 3y = 7$
$2x + 3y = 5$

21. $2x - 3y = 4$
$2x + y = -4$

22. $R - 4r = 17$
$4r + 3R = 3$

23. $12t + 9y = 14$
$6t = 7y - 16$

24. $3x - y = 3$
$4x = 3y + 14$

25. $v + 2t = 7$
$2v + 4t = 9$

26. $3x - y = 5$
$3y - 9x = -15$

27. $2x - 3y - 4 = 0$
$3x + 2 = 2y$

28. $3i_1 + 5 = -4i_2$
$3i_2 = 5i_1 - 2$

In Exercises 29–36, solve the given systems of equations by determinants. All numbers are approximate. (Exercises 29–32 are the same as Exercises 37–40 of Section 5.3.)

29. $0.3x - 0.7y = 0.4$
$0.2x + 0.5y = 0.7$

30. $250R + 225Z = 400$
$375R - 675Z = 325$

31. $40s - 30t = 60$
$20s - 40t = -50$

32. $0.060x + 0.048y = -0.084$
$0.065y - 0.13x = 0.078$

33. $301x - 529y = 1520$
$385x - 741y = 2540$

34. $0.25d + 0.63n = -0.37$
$-0.61d - 1.80n = 0.55$

35. $1.2y + 10.8 = -8.4x$
$3.5x + 4.8y + 12.9 = 0$

36. $6541x + 4397y = -7732$
$3309x - 8755y = 7622$

In Exercises 37–40, answer the given questions about the determinant to the right. $\begin{vmatrix} a & b \\ c & d \end{vmatrix}$

37. What is the value of the determinant if $c = d = 0$?
38. What change in value occurs if both the rows and the columns are interchanged?
39. What is the value of the determinant if $a = kb$ and $c = kd$?
40. How does the value change if a and c are doubled?

In Exercises 41–44, solve the given systems of equations by determinants. All numbers are accurate to at least two significant digits.

41. The forces acting on a link of an industrial robot are shown in Fig. 5.33. The equations for finding forces F_1 and F_2 are

$F_1 + F_2 = 21$
$2F_1 = 5F_2$

Find F_1 and F_2.

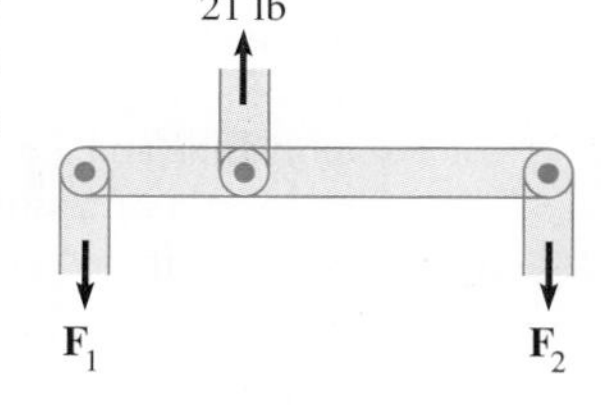

Fig. 5.33

42. The area of a quadrilateral is

$$A = \frac{1}{2}\left(\begin{vmatrix} x_0 & x_1 \\ y_0 & y_1 \end{vmatrix} + \begin{vmatrix} x_1 & x_2 \\ y_1 & y_2 \end{vmatrix} + \begin{vmatrix} x_2 & x_3 \\ y_2 & y_3 \end{vmatrix} + \begin{vmatrix} x_3 & x_0 \\ y_3 & y_0 \end{vmatrix}\right)$$

where (x_0, y_0), (x_1, y_1), (x_2, y_2), and (x_3, y_3) are the rectangular coordinates of the vertices of the quadrilateral, listed counterclockwise. (This *surveyor's formula* can be generalized to find the area of any polygon.)

A surveyor records the locations of the vertices of a quadrilateral building lot on a rectangular coordinate system as (12.79, 0.00), (67.21, 12.30), (53.05, 47.12), and (10.09, 53.11), where distances are in meters. Find the area of the lot.

43. An airplane begins a flight with a total of 36.0 gal of fuel stored in two separate wing tanks. During the flight, 25.0% of the fuel in one tank is used, and in the other tank 37.5% of the fuel is used. If the total fuel used is 11.2 gal, the amounts x and y used from each tank can be found by solving the system of equations

$x + y = 36.0$
$0.250x + 0.375y = 11.2$

Find x and y.

44. In applying Kirchhoff's laws (see the chapter introduction; the equations can be found in most physics textbooks) to the electric circuit shown in Fig. 5.34, the following equations are found. Find the indicated currents I_1 and I_2 (in A).

$$52I_1 - 27I_2 = -420$$
$$-27I_1 + 76I_2 = 210$$

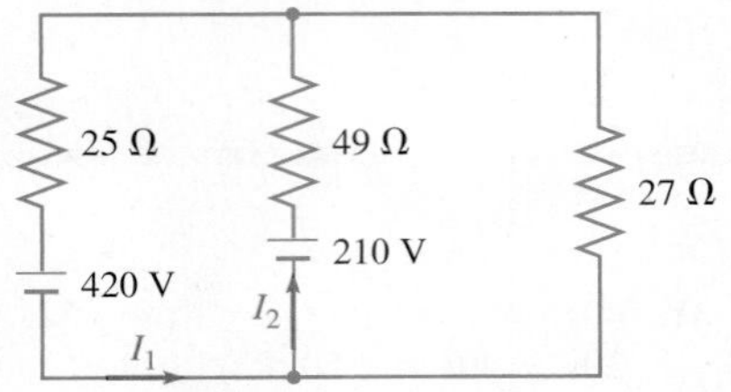

25 Ω 49 Ω 27 Ω 210 V 420 V I_2 I_1

Fig. 5.34

In Exercises 45–56, set up appropriate systems of two linear equations in two unknowns and then solve the systems by determinants. All numbers are accurate to at least two significant digits.

45. A new development has 3-bedroom homes and 4-bedroom homes. The developer's profit was \$25,000 from each 3-br home, and \$35,000 from each 4-br home, totaling \$6,800,000. Total annual property taxes are \$560,000, with \$2000 from each 3-br home and \$3000 from each 4-br home. How many of each were built?

46. Two joggers are 2.0 mi apart. If they jog toward each other, they will meet in 12 min. If they jog in the same direction, the faster one will overtake the slower one in 2.0 h. At what rate does each jog?

47. A shipment of 320 cell phones and radar detectors was destroyed due to a truck accident. On the insurance claim, the shipper stated that each phone was worth \$110, each detector was worth \$160, and their total value was \$40,700. How many of each were in the shipment?

48. Two types of electromechanical carburetors are being assembled and tested. Each of the first type requires 15 min of assembly time and 2 min of testing time. Each of the second type requires 12 min of assembly time and 3 min of testing time. If 222 min of assembly time and 45 min of testing time are available, how many of each type can be assembled and tested, if all the time is used?

49. Since ancient times, a rectangle for which the length L is approximately 1.62 times the width w has been considered the most pleasing to view, and is called a *golden rectangle*. If a painting in the shape of a golden rectangle has a perimeter of 4.20 m, find the dimensions.

50. The ratio of men to women on a bus was 5/7. Then two women and one man boarded, and the ratio was 7/10. How many men and women were on the bus before the last three passengers boarded?

51. A machinery sales representative receives a fixed salary plus a sales commission each month. If \$6200 is earned on sales of \$70,000 in one month and \$4700 is earned on sales of \$45,000 in the following month, what are the fixed salary and the commission percent?

52. A moving walkway at an airport is 65.0 m long. A child running at a constant speed takes 20.0 s to run along the walkway in the direction it is moving, and then 52.0 s to run all the way back. What are the speed of the walkway and the speed of the child?

53. A boat carrying illegal drugs leaves a port and travels at 42 mi/h. A Coast Guard cutter leaves the port 24 min later and travels at 50 mi/h in pursuit of the boat. Find the times each has traveled when the cutter overtakes the boat with drugs. See Fig. 5.35.

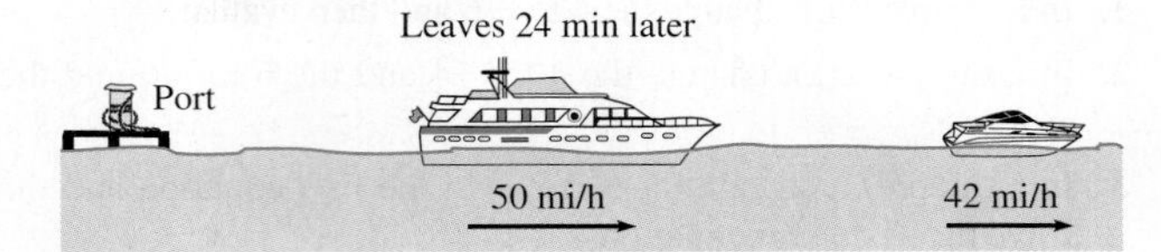

Fig. 5.35

54. Sterling silver is 92.5% silver and 7.5% copper. One silver-copper alloy is 94.0% silver, and a second silver-copper alloy is 85.0% silver. How much of each should be used in order to make 100 g of sterling silver?

55. A surveyor measures the angle of elevation to the top of a hill to be 15.5°. He then moves 345 ft closer, on level ground, and remeasures the angle of elevation to be 21.4° (see Fig. 5.36). Find the height h of the hill and the distance d. *Hint:* Use the tangent function to set up a system of equations.

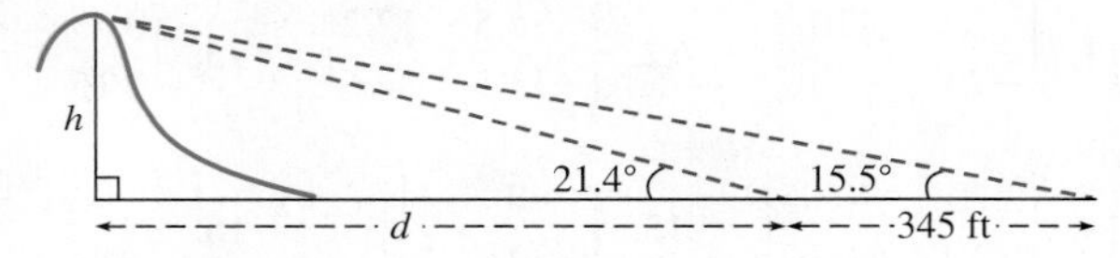

Fig. 5.36

56. The velocity of sound in steel is 15,900 ft/s faster than the velocity of sound in air. One end of a long steel bar is struck, and an instrument at the other end measures the time it takes for the sound to reach it. The sound in the bar takes 0.0120 s, and the sound in the air takes 0.180 s. What are the velocities of sound in air and in steel?

Answers to Practice Exercises

1. 22 **2.** $x = -2, y = 3$ **3.** $x = \$8000, y = \$10,000$

5.5 Solving Systems of Three Linear Equations in Three Unknowns Algebraically

Algebraic Method Using Addition or Subtraction • Solving a System Using Reduced Row Echelon Form

Many technical problems involve systems of linear equations with more than two unknowns. In this section, we solve systems with three unknowns, and in Chapter 16 we will show how systems with even more unknowns are solved.

Solving such systems is very similar to solving systems in two unknowns. In this section, we will show the algebraic method, and in the next section we will show how determinants are used. Graphical solutions are not used since a linear equation in three unknowns represents a plane in space. We will, however, briefly show graphical interpretations of systems of three linear equations at the end of this section.

A system of three linear equations in three unknowns written in the form

$$\begin{aligned} a_1x + b_1y + c_1z &= d_1 \\ a_2x + b_2y + c_2z &= d_2 \\ a_3x + b_3y + c_3z &= d_3 \end{aligned} \qquad \textbf{(5.10)}$$

has as its solution the set of values x, y, and z that satisfy all three equations simultaneously. The method of solution involves multiplying ***two*** of the equations by the proper numbers to eliminate **one** of the unknowns between these equations. We then repeat this process, using a ***different pair*** of the original equations, being sure that we eliminate the same unknown as we did between the first pair of equations. At this point we have two linear equations in two unknowns that can be solved by any of the methods previously discussed.

EXAMPLE 1 Algebraically solving a system

Solve the following system of equations:

■ We can choose to first eliminate any one of the three unknowns. We have chosen y.

■ At this point we could have used Eqs. (1) and (3) or Eqs. (2) and (3), but we must set them up to eliminate y.

(1)	$4x + y + 3z = 1$	
(2)	$2x - 2y + 6z = 11$	
(3)	$-6x + 3y + 12z = -4$	
(4)	$8x + 2y + 6z = 2$	(1) multiplied by 2
	$2x - 2y + 6z = 11$	(2)
(5)	$10x + 12z = 13$	adding
(6)	$12x + 3y + 9z = 3$	(1) multiplied by 3
	$-6x + 3y + 12z = -4$	(3)
(7)	$18x - 3z = 7$	subtracting
	$10x + 12z = 13$	(5)
(8)	$72x - 12z = 28$	(7) multiplied by 4
(9)	$82x = 41$	adding
(10)	$x = \frac{1}{2}$	
(11)	$18(\frac{1}{2}) - 3z = 7$	substituting (10) in (7)
(12)	$-3z = -2$	
(13)	$z = \frac{2}{3}$	
(14)	$4(\frac{1}{2}) + y + 3(\frac{2}{3}) = 1$	substituting (13) and (10) in (1)
(15)	$2 + y + 2 = 1$	
(16)	$y = -3$	

Thus, the solution is $x = \frac{1}{2}$, $y = -3$, $z = \frac{2}{3}$. Substituting in the equations, we have

$$4(\tfrac{1}{2}) + (-3) + 3(\tfrac{2}{3}) \stackrel{?}{=} 1 \qquad 2(\tfrac{1}{2}) - 2(-3) + 6(\tfrac{2}{3}) \stackrel{?}{=} 11 \qquad -6(\tfrac{1}{2}) + 3(-3) + 12(\tfrac{2}{3}) \stackrel{?}{=} -4$$
$$1 = 1 \qquad\qquad 11 = 11 \qquad\qquad -4 = -4$$

TI-89 graphing calculator keystrokes for Example 1: goo.gl/cDclHw

We see that the solution checks.

If we had first eliminated x, we would then have had to solve two equations in y and z. If we had first eliminated z, we would then have had to solve two equations in x and y. In each case, the basic procedure is the same. ■

Practice Exercise

1. Solve the system in Example 1 by first eliminating x.

EXAMPLE 2 Setting up and solving a system—voltage

Three voltages, e_1, e_2, and e_3, where e_3 is three times e_1, are in series with the same polarity [see Fig. 5.37(a)] and have a total voltage of 85 mV. If e_2 is reversed in polarity [see Fig. 5.37(b)], the voltage is 35 mV. Find the voltages.

Because the voltages are in series with the same polarity, $e_1 + e_2 + e_3 = 85$. Then, since e_3 is three times e_1, we have $e_3 = 3e_1$. Then, with the reversed polarity of e_2, we have $e_1 - e_2 + e_3 = 35$. Writing these equations in standard form, we have the following solution:

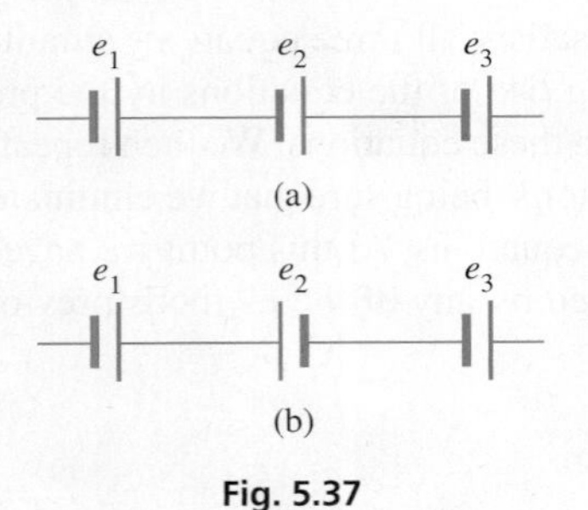

Fig. 5.37

$$
\begin{array}{lrl}
(1) & e_1 + e_2 + e_3 = 85 & \\
(2) & 3e_1 \quad - e_3 = 0 & \text{rewriting second equation} \\
(3) & e_1 - e_2 + e_3 = 35 & \\
\hline
(4) & 2e_1 \quad + 2e_3 = 120 & \text{adding (1) and (3)} \\
(5) & 6e_1 \quad - 2e_3 = 0 & \text{(2) multiplied by 2} \\
\hline
(6) & 8e_1 \quad = 120 & \text{adding} \\
(7) & e_1 = 15\text{ mV} & \\
(8) & 3(15) - e_3 = 0 & \text{substituting (7) in (2)} \\
(9) & e_3 = 45\text{ mV} & \\
(10) & 15 + e_2 + 45 = 85 & \text{substituting (7) and (9) in (1)} \\
(11) & e_2 = 25\text{ mV} &
\end{array}
$$

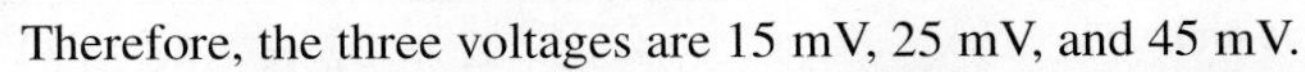

Therefore, the three voltages are 15 mV, 25 mV, and 45 mV.

Checking the solution, the sum of the three voltages is 85 mV, e_3 is three times e_1, and the sum of e_1 and e_3 less e_2 is 35 mV. ■

A calculator can be used to solve a system of equations using the ***reduced row echelon form*** **(rref)** feature. It is programmed to perform the same row operations that we use, and it continues until *all* the unknowns are isolated, making the solutions clear. The difference is that the calculator performs the operations on the numerical values only (without the variables), which are entered into an array with rows and columns called a *matrix*. The following example illustrates this process.

EXAMPLE 3 Solving a system using rref—robotic forces

The forces acting on the main link of an industrial robotic arm are shown in Fig. 5.38. An analysis of the forces leads to the following system of equations. Determine the forces.

$$
\begin{aligned}
A + 60 &= 0.8T \\
B &= 0.6T \\
8A + 6B + 80 &= 5T
\end{aligned}
$$

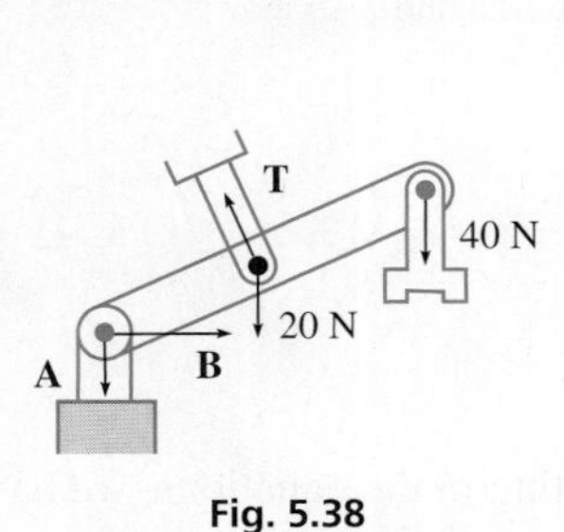

Fig. 5.38

$$
\begin{aligned}
A \qquad - 0.8T &= -60 \\
B - 0.6T &= 0 \\
8A + 6B \quad - 5T &= -80
\end{aligned}
$$

1. Write the system in the form of Eqs. (5.10), lining up the variables, equal signs, and constants.

$$
\left[\begin{array}{ccc:c}
1 & 0 & -0.8 & -60 \\
0 & 1 & -0.6 & 0 \\
8 & 6 & -5 & -80
\end{array}\right]
$$

2. Construct a matrix with 3 rows and 4 columns that includes the coefficients and constants, but not the variables. This is called an **augmented matrix.** Insert zeros for missing terms. The dotted line represents the equal sign. Enter this into a calculator using the *Matrix > Edit* feature. See Fig. 5.39(a) on the next page.

$$
\left[\begin{array}{ccc:c}
1 & 0 & 0 & 4 \\
0 & 1 & 0 & 48 \\
0 & 0 & 1 & 80
\end{array}\right]
$$

3. Use the *Matrix > Math > rref* feature to convert the matrix to **reduced row echelon form.** In this form, the solutions for A, B, and T are evident. See Fig. 5.39(b) on the next page.

Therefore, the forces are $A = 4$ N, $B = 48$ N, and $T = 80$ N.

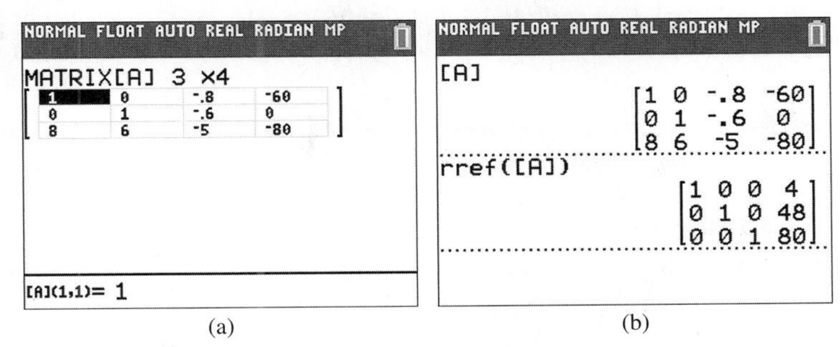

Fig. 5.39 (a) (b)

Graphing calculator keystrokes: goo.gl/6gLiuy ■

■ Matrices are discussed in detail in Chapter 16.

Systems with four or more unknowns are solved in a manner similar to that used for three unknowns. With four unknowns, one is eliminated between three different pairs of equations, and the resulting three equations are then solved.

Linear systems with more than two unknowns may have an unlimited number of solutions or be inconsistent. After eliminating unknowns, if we have $0 = 0$, there is an unlimited number of solutions. If we have $0 = a (a \neq 0)$, the system is inconsistent, and there is no solution. (See Exercises 33–36.)

As noted earlier, a linear equation in three unknowns represents a plane in space. For three linear equations in three unknowns, if the planes intersect at a point, there is a unique solution [Fig. 5.40(a)]; if they intersect in a line, there is an unlimited number of solutions [Fig. 5.40(b)]. If the planes do not have a common intersection, the system is inconsistent. In inconsistent systems, the planes can be parallel [Fig. 5.41(a)], two can be parallel [Fig. 5.41(b)], or they can intersect in three parallel lines [Fig. 5.41(c)]. If one plane is coincident with another plane, the system has an unlimited number of solutions if they intersect with the third plane, or is inconsistent otherwise.

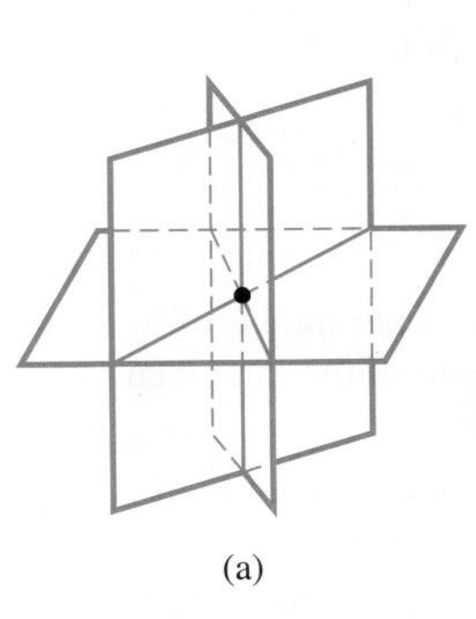

(a)

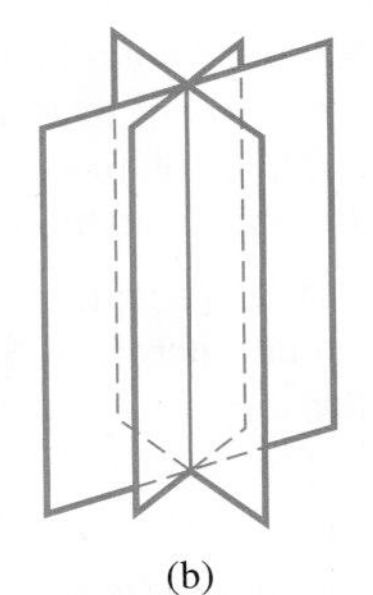

(b)

Fig. 5.40

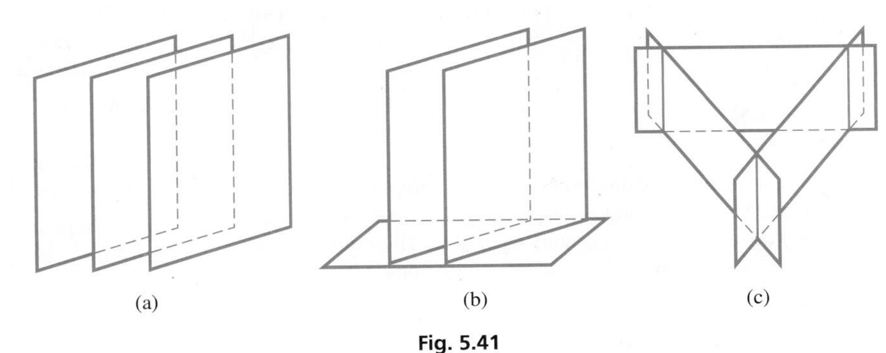

(a) (b) (c)

Fig. 5.41

EXERCISES 5.5

In Exercises 1 and 2, make the given changes in Example 1 of this section and then solve the resulting system of equations.

1. In the second equation, change the constant to the right of the = sign from 11 to 12, and in the third equation, change the constant to the right of the = sign from −4 to −14.
2. Change the second equation to $8x + 9z = 10$ (no y-term).

In Exercises 3–14, solve the given systems of equations.

3. $x + y + z = 2$
   $x - z = 1$
   $x + y = 1$

4. $x + y - z = -3$
   $x + z = 2$
   $2x - y + 2z = 3$

5. $2x + 3y + z = 2$
   $-x + 2y + 3z = -1$
   $-3x - 3y + z = 0$

6. $2x + y - z = 4$
   $4x - 3y - 2z = -2$
   $8x - 2y - 3z = 3$

7. $5l + 6w - 3h = 6$
   $4l - 7w - 2h = -3$
   $3l + w - 7h = 1$

8. $3r + s - t = 2$
   $r + t - 2s = 0$
   $4r - s + t = 3$

9. $2x - 2y + 3z = 5$
   $2x + y - 2z = -1$
   $3z + y - 4x = 0$

10. $2u + 2v + 3w = 0$
    $3u + v + 4w = 21$
    $-u - 3v + 7w = 15$

11. $3x - 7y + 3z = 6$
    $3x + 3y + 6z = 1$
    $5x - 5y + 2z = 5$

12. $18x + 24y + 4z = 46$
    $63x + 6y - 15z = -75$
    $-90x + 30y - 20z = -55$

13. $10x + 15y - 25z = 35$
$40x - 30y - 20z = 10$
$16x - 2y + 8z = 6$

14. $2i_1 - 4i_2 - 4i_3 = 3$
$3i_1 + 8i_2 + 2i_3 = -11$
$4i_1 + 6i_2 - i_3 = -8$

In Exercises 15–18, solve the given systems of equations using the reduced row echelon form (rref) feature on a calculator. The decimals in Exercises 17–18 are approximate.

15. $-x + 3y - 2z = 7$
$3x + 4y - 7z = -8$
$x + 2y + z = 2$

16. $3a + 2b + c = 20$
$4a - 10c = -10$
$-a - 2b + 2c = -1$

17. $I_2 = I_1 + I_3$
$8.00I_1 + 10.0I_2 = 80.0$
$6.00I_3 + 10.0I_2 = 60.0$

18. $1.21x + 1.32y + 1.20z = 6.81$
$4.93x - 1.25y + 3.65z = 22.0$
$2.85x + 3.25y - 2.70z = 2.76$

In Exercises 19 and 20, find the indicated functions.

19. Find the function $f(x) = ax^2 + bx + c$, if $f(1) = 3$, $f(-2) = 15$, and $f(3) = 5$.

20. Find the function $f(x) = ax^2 + bx + c$, if $f(1) = -3$, $f(-3) = -35$, and $f(3) = -11$.

In Exercises 21–32, solve the systems of equations. In Exercises 25–32, it is necessary to set up the appropriate equations. All numbers are accurate to at least three significant digits.

21. A medical supply company has 1150 worker-hours for production, maintenance, and inspection. Using this and other factors, the number of hours used for each operation, P, M, and I, respectively, is found by solving the following system of equations:

$P + M + I = 1150$
$P = 4I - 100$
$P = 6M + 50$

22. Three oil pumps fill three different tanks. The pumping rates of the pumps (in L/h) are r_1, r_2, and r_3, respectively. Because of malfunctions, they do not operate at capacity each time. Their rates can be found by solving the following system of equations:

$r_1 + r_2 + r_3 = 14{,}000$
$r_1 + 2r_2 = 13{,}000$
$3r_1 + 3r_2 + 2r_3 = 36{,}000$

23. The forces acting on a certain girder, as shown in Fig. 5.42, can be found by solving the following system of equations:

$0.707F_1 - 0.800F_2 = 0$
$0.707F_1 + 0.600F_2 - F_3 = 10.0$
$3.00F_2 - 3.00F_3 = 20.0$

Find the forces, in newtons.

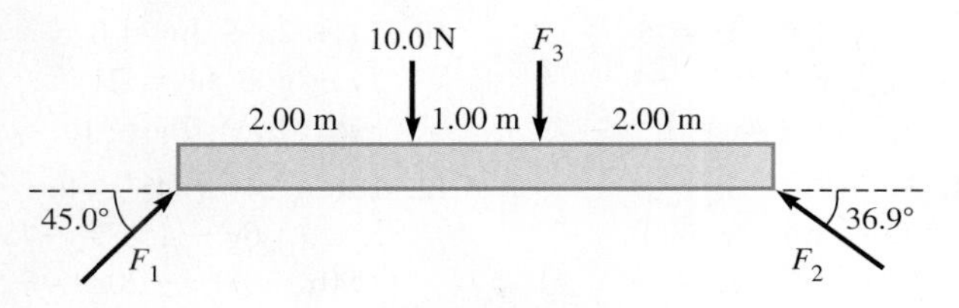

Fig. 5.42

24. Using Kirchhoff's laws (see the chapter introduction; the equations can be found in most physics textbooks) with the electric circuit shown in Fig. 5.43, the following equations are found. Find the indicated currents (in A) I_1, I_2, and I_3.

$1.0I_1 + 3.0(I_1 - I_3) = 12$
$2.0I_2 + 4.0(I_2 + I_3) = 12$
$1.0I_1 - 2.0I_2 + 3.0I_3 = 0$

Fig. 5.43

25. Find angles A, B, and C in the roof truss shown in Fig. 5.44.

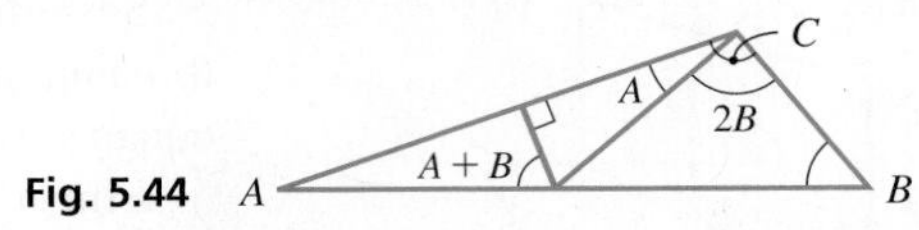

Fig. 5.44

26. Under certain conditions, the cost per mile C of operating a car is a function of the speed v (in mi/h) of the car, given by $C = av^2 + bv + c$. If $C = 28$ ¢/mi for $v = 10$ mi/h, $C = 22$ ¢/mi for $v = 30$ mi/h, and $C = 24$ ¢/mi for $v = 50$ mi/h, find C as a function of v.

27. In an election with three candidates for mayor, the initial vote count gave A 200 more votes than B, and 500 more votes than C. An error was found, and 1.0% of A's initial votes went to B, and 2.0% of A's initial votes went to C, such that B had 100 more votes than C. How many votes did each have in the final tabulation?

28. A person invests \$22,500, partly at 5.00%, partly at 6.00%, the remainder at 6.50%, with a total annual interest of \$1308. If the interest received at 5.00% equals the interest received at 6.00%, how much is invested at each rate?

29. A university graduate school conferred 420 advanced academic degrees at graduation. There were 100 more MA degrees than MS and PhD degrees combined, and 3 times as many MS degrees as PhD degrees. How many of each were awarded?

30. The computer systems at three weather bureaus have a combined hard-disk memory capacity of 8.0 TB (terabytes). The memory capacity of systems A and C have 0.2 TB more memory than twice that of system B, and twice the sum of the memory capacities of systems A and B is three times that of system C. What are the memory capacities of each of these computer systems?

31. By weight, one fertilizer is 20% potassium, 30% nitrogen, and 50% phosphorus. A second fertilizer has percents of 10, 20, and 70, respectively, and a third fertilizer has percents of 0, 30, and 70, respectively. How much of each must be mixed to get 200 lb of fertilizer with percents of 12, 25, and 63, respectively?

32. The average traffic flow (number of vehicles) from noon until 1 P.M. in a certain section of one-way streets in a city is shown in Fig. 5.45. Explain why an analysis of the flow through these intersections is not sufficient to obtain unique values for x, y, and z. *Hint*: The total traffic going into each intersection equals the traffic going out.

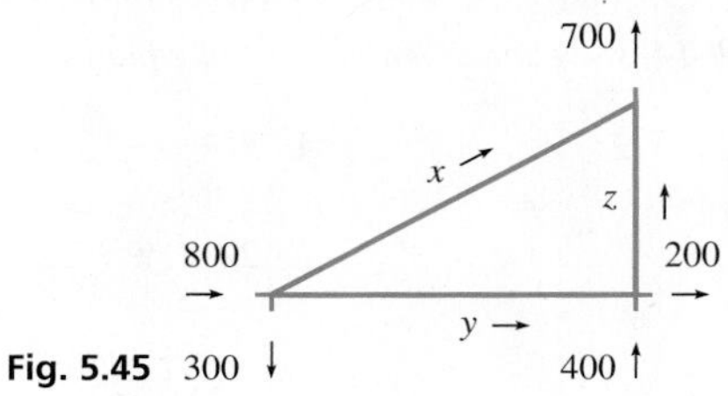

Fig. 5.45

In Exercises 33–36, show that the given systems of equations have either an unlimited number of solutions or no solution. If there is an unlimited number of solutions, find one of them.

33. $x - 2y - 3z = 2$
$x - 4y - 13z = 14$
$-3x + 5y + 4z = 0$

34. $x - 2y - 3z = 2$
$x - 4y - 13z = 14$
$-3x + 5y + 4z = 2$

35. $3x + 3y - 2z = 2$
$2x - y + z = 1$
$x - 5y + 4z = -3$

36. $3x + y - z = -3$
$x + y - 3z = -5$
$-5x - 2y + 3z = -7$

Answer to Practice Exercise

1. $x = 1/2, y = -3, z = 2/3$

5.6 Solving Systems of Three Linear Equations in Three Unknowns by Determinants

Determinant of the Third Order • Cramer's Rule • Solving Systems of Equations by Determinants • Determinants on the Calculator

Just as systems of two linear equations in two unknowns can be solved by determinants, so can systems of three linear equations in three unknowns. The system

$$\begin{aligned} a_1x + b_1y + c_1z &= d_1 \\ a_2x + b_2y + c_2z &= d_2 \\ a_3x + b_3y + c_3z &= d_3 \end{aligned} \tag{5.10}$$

can be solved in general terms by the method of elimination by addition or subtraction. This leads to the following solutions for x, y, and z.

$$\begin{aligned} x &= \frac{d_1b_2c_3 + d_3b_1c_2 + d_2b_3c_1 - d_3b_2c_1 - d_1b_3c_2 - d_2b_1c_3}{a_1b_2c_3 + a_3b_1c_2 + a_2b_3c_1 - a_3b_2c_1 - a_1b_3c_2 - a_2b_1c_3} \\ y &= \frac{a_1d_2c_3 + a_3d_1c_2 + a_2d_3c_1 - a_3d_2c_1 - a_1d_3c_2 - a_2d_1c_3}{a_1b_2c_3 + a_3b_1c_2 + a_2b_3c_1 - a_3b_2c_1 - a_1b_3c_2 - a_2b_1c_3} \\ z &= \frac{a_1b_2d_3 + a_3b_1d_2 + a_2b_3d_1 - a_3b_2d_1 - a_1b_3d_2 - a_2b_1d_3}{a_1b_2c_3 + a_3b_1c_2 + a_2b_3c_1 - a_3b_2c_1 - a_1b_3c_2 - a_2b_1c_3} \end{aligned} \tag{5.11}$$

The expressions that appear in the numerators and denominators of Eqs. (5.11) are examples of a **determinant of the third order.** *This determinant is defined by*

$$\begin{vmatrix} a_1 & b_1 & c_1 \\ a_2 & b_2 & c_2 \\ a_3 & b_3 & c_3 \end{vmatrix} = a_1b_2c_3 + a_3b_1c_2 + a_2b_3c_1 - a_3b_2c_1 - a_1b_3c_2 - a_2b_1c_3 \tag{5.12}$$

The elements, rows, columns, and diagonals of a third-order determinant are defined just as are those of a second-order determinant. For example, the principal diagonal is made up of the elements a_1, b_2, and c_3.

Probably the easiest way of remembering the method of finding the value of a third-order determinant is as follows: *Rewrite the first two columns to the right of the determinant. The products of the elements of the principal diagonal and the two parallel diagonals to the right of it are then added. The products of the elements of the secondary diagonal and the two parallel diagonals to the right of it are subtracted from the first sum. The algebraic sum of these six products gives the value of the determinant.* These products are indicated in Fig. 5.46.

CAUTION This method is used only for third-order determinants. It does not work for determinants of order higher than three. ■

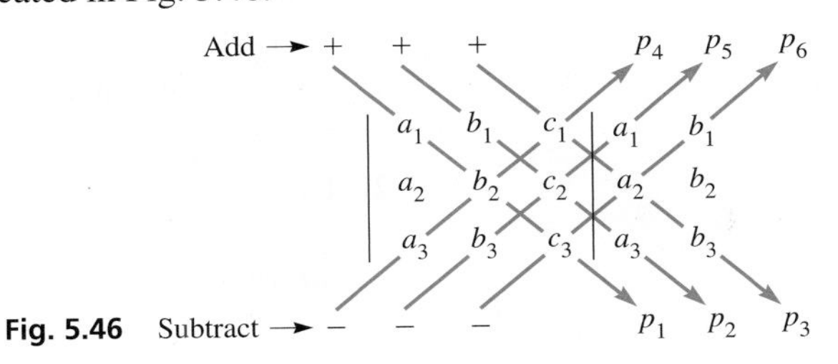

Fig. 5.46

Examples 1 and 2 illustrate this method of evaluating third-order determinants.

EXAMPLE 1 Evaluating a third-order determinant

$$\begin{vmatrix} 1 & 5 & 4 \\ -2 & 3 & -1 \\ 2 & -1 & 5 \end{vmatrix}\begin{matrix} 1 & 5 \\ -2 & 3 \\ 2 & -1 \end{matrix} = \overset{p_1}{15} + \overset{p_2}{(-10)} + \overset{p_3}{(+8)} - \underset{p_4}{(24)} - \underset{p_5}{(1)} - \underset{p_6}{(-50)} = 38$$

■

Practice Exercise

1. Interchange the rows and columns of the determinant in Example 1 (make the first row 1 −2 2 and the first column 1 5 4, etc.) and then evaluate.

EXAMPLE 2 Evaluating a determinant with and without a calculator

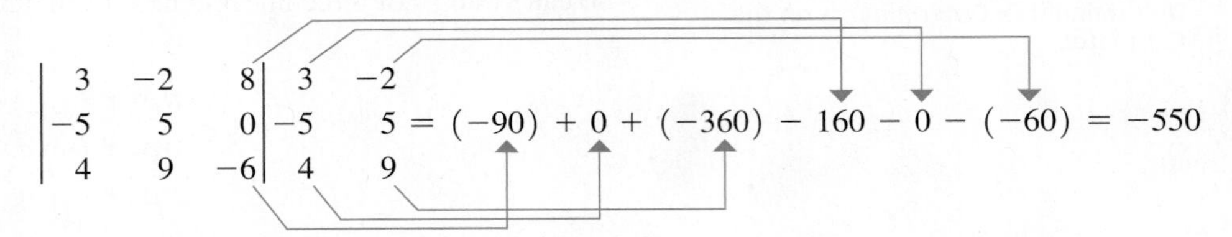

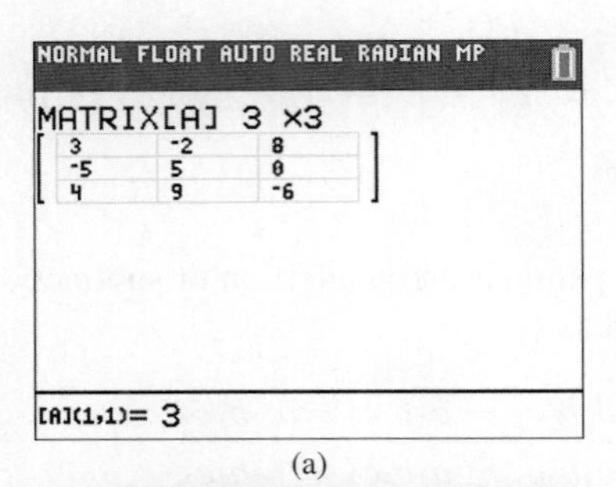

(a)

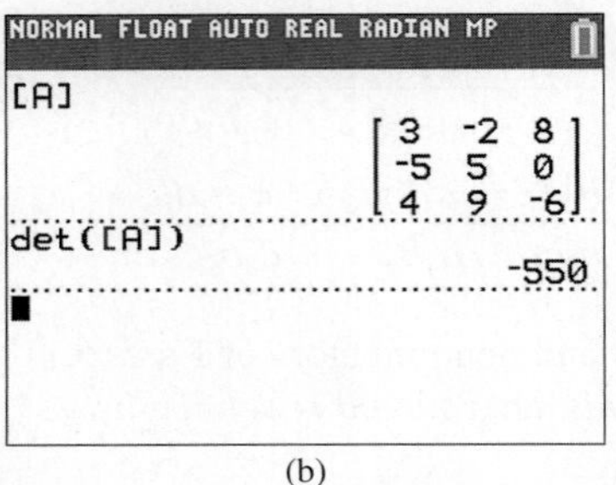

(b)

Fig. 5.47

Graphing calculator keystrokes: goo.gl/rIVzBM

This determinant can also be evaluated on a calculator using the *matrix* feature. Figure 5.47 shows two windows for the determinant. The first shows the window for entering the numbers in the *matrix*. The second shows the matrix displayed and the evaluation of the determinant. ■

Practice Exercise

2. Using a calculator, evaluate the determinant in Example 1.

Inspection of Eqs. (5.11) reveals that the numerators of these solutions may also be written in terms of determinants. Thus, we may write the general solution to a system of three equations in three unknowns using determinants, which again is Cramer's rule.

Cramer's Rule: Three Unknowns

For a system of equations in three unknowns of the form of Eqs. (5.10), the solutions are given by

$$x = \frac{\begin{vmatrix} d_1 & b_1 & c_1 \\ d_2 & b_2 & c_2 \\ d_3 & b_3 & c_3 \end{vmatrix}}{\begin{vmatrix} a_1 & b_1 & c_1 \\ a_2 & b_2 & c_2 \\ a_3 & b_3 & c_3 \end{vmatrix}} \quad y = \frac{\begin{vmatrix} a_1 & d_1 & c_1 \\ a_2 & d_2 & c_2 \\ a_3 & d_3 & c_3 \end{vmatrix}}{\begin{vmatrix} a_1 & b_1 & c_1 \\ a_2 & b_2 & c_2 \\ a_3 & b_3 & c_3 \end{vmatrix}} \quad z = \frac{\begin{vmatrix} a_1 & b_1 & d_1 \\ a_2 & b_2 & d_2 \\ a_3 & b_3 & d_3 \end{vmatrix}}{\begin{vmatrix} a_1 & b_1 & c_1 \\ a_2 & b_2 & c_2 \\ a_3 & b_3 & c_3 \end{vmatrix}} \qquad (5.13)$$

provided the determinant in the denominator is not equal to zero.

If the determinant of the denominator is not zero, there is a unique solution to the system of equations. If all determinants are zero, there is an *unlimited number of solutions.* If the determinant of the denominator is zero and any of the determinants of the numerators is not zero, the system is *inconsistent,* and there is *no solution.*

An analysis of Eqs. (5.13) shows that the situation is precisely the same as it was when we were using determinants to solve systems of two linear equations. That is,

1. The determinants in the denominators are the same and are formed by using the coefficients of x, y, and z.
2. The determinants in the numerators are the same as the one in the denominators except that **the constant terms replace the column of coefficients of the unknown for which we are solving.**

EXAMPLE 3 Solving a system using Cramer's rule

Solve the following system by determinants.

$$\begin{aligned} 3x + 2y - 5z &= -1 \\ 2x - 3y - z &= 11 \\ 5x - 2y + 7z &= 9 \end{aligned}$$

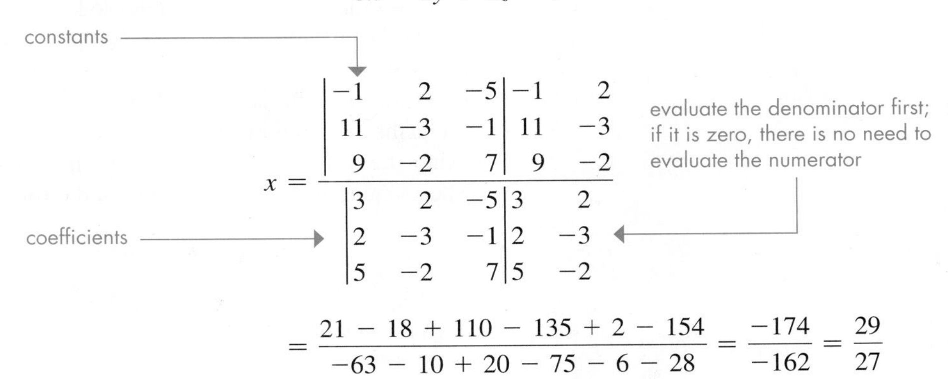

$$x = \frac{\begin{vmatrix} -1 & 2 & -5 \\ 11 & -3 & -1 \\ 9 & -2 & 7 \end{vmatrix} \begin{matrix} -1 & 2 \\ 11 & -3 \\ 9 & -2 \end{matrix}}{\begin{vmatrix} 3 & 2 & -5 \\ 2 & -3 & -1 \\ 5 & -2 & 7 \end{vmatrix} \begin{matrix} 3 & 2 \\ 2 & -3 \\ 5 & -2 \end{matrix}}$$

$$= \frac{21 - 18 + 110 - 135 + 2 - 154}{-63 - 10 + 20 - 75 - 6 - 28} = \frac{-174}{-162} = \frac{29}{27}$$

constants

$$y = \frac{\begin{vmatrix} 3 & -1 & -5 \\ 2 & 11 & -1 \\ 5 & 9 & 7 \end{vmatrix} \begin{matrix} 3 & -1 \\ 2 & 11 \\ 5 & 9 \end{matrix}}{-162} = \frac{231 + 5 - 90 + 275 + 27 + 14}{-162}$$

denominator $= -162$ from solution for x

$$= \frac{462}{-162} = -\frac{77}{27}$$

■ Figure 5.48 shows the calculator solution using determinants. The decimal solutions can be converted to fractions as shown in the bottom line for z.

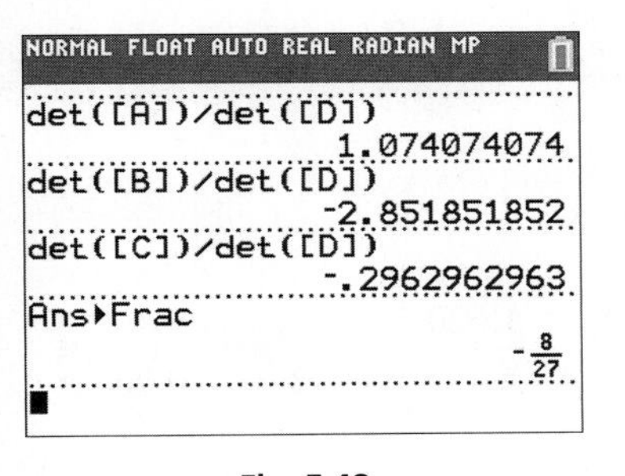

Fig. 5.48
Graphing calculator keystrokes: goo.gl/H4aQYg

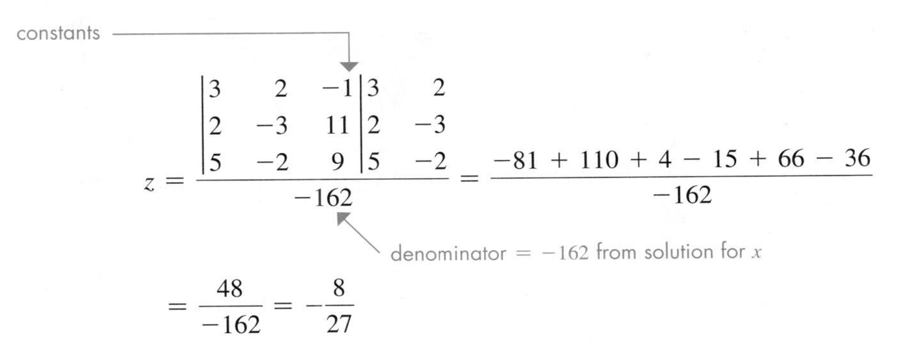

$$z = \frac{\begin{vmatrix} 3 & 2 & -1 \\ 2 & -3 & 11 \\ 5 & -2 & 9 \end{vmatrix} \begin{matrix} 3 & 2 \\ 2 & -3 \\ 5 & -2 \end{matrix}}{-162} = \frac{-81 + 110 + 4 - 15 + 66 - 36}{-162}$$

$$= \frac{48}{-162} = -\frac{8}{27}$$

Substituting in each of the original equations shows that the solution checks:

$$3\left(\frac{29}{27}\right) + 2\left(-\frac{77}{27}\right) - 5\left(-\frac{8}{27}\right) = \frac{87 - 154 + 40}{27} = \frac{-27}{27} = -1$$

$$2\left(\frac{29}{27}\right) - 3\left(-\frac{77}{27}\right) - \left(-\frac{8}{27}\right) = \frac{58 + 231 + 8}{27} = \frac{297}{27} = 11$$

$$5\left(\frac{29}{27}\right) - 2\left(-\frac{77}{27}\right) + 7\left(-\frac{8}{27}\right) = \frac{145 + 154 - 56}{27} = \frac{243}{27} = 9$$

After the values of x and y were determined, we could have evaluated z by substituting the values of x and y into one of the original equations. ■

EXAMPLE 4 Setting up and solving a system—mixing acid solutions

An 8.0% solution, an 11% solution, and an 18% solution of nitric acid are to be mixed to get 150 mL of a 12% solution. If the volume of acid from the 8.0% solution equals half the volume of acid from the other two solutions, how much of each is needed?

Let x = volume of 8.0% solution needed, y = volume of 11% solution needed, and z = volume of 18% solution needed.

The fact that the sum of the volumes of the three solutions is 150 mL leads to the equation $x + y + z = 150$. Because there are $0.080x$ mL of pure acid from the first solution, $0.11y$ mL from the second solution, and $0.18z$ mL from the third solution, and $0.12(150)$ mL in the final solution, we are led to the equation $0.080x + 0.11y + 0.18z = 18$. Finally, using the last stated condition, we have the equation $0.080x = 0.5(0.11y + 0.18z)$. These equations are then written in the form of Eqs. (5.10) and solved.

$$\begin{aligned} x + y + z &= 150 \\ 0.080x + 0.11y + 0.18z &= 18 \\ 0.080x &= 0.055y + 0.090z \end{aligned}$$

sum of volumes; volumes of pure acid; one-half of acid in others (acid in 8.0% solution: $0.080x$)

$$\begin{aligned} x + y + z &= 150 \\ 0.080x + 0.11y + 0.18z &= 18 \\ 0.080x - 0.055y - 0.090z &= 0 \end{aligned}$$

standard form of Eqs. (5.10) by rewriting third equation

$$x = \frac{\left|\begin{array}{ccc|cc} 150 & 1 & 1 & 150 & 1 \\ 18 & 0.11 & 0.18 & 18 & 0.11 \\ 0 & -0.055 & -0.090 & 0 & -0.055 \end{array}\right.}{\left|\begin{array}{ccc|cc} 1 & 1 & 1 & 1 & 1 \\ 0.080 & 0.11 & 0.18 & 0.080 & 0.11 \\ 0.080 & -0.055 & -0.090 & 0.080 & -0.055 \end{array}\right.}$$

$$= \frac{-1.485 + 0 - 0.990 - 0 + 1.485 + 1.620}{-0.0099 + 0.0144 - 0.0044 - 0.0088 + 0.0099 + 0.0072} = \frac{0.630}{0.0084} = 75$$

$$y = \frac{\left|\begin{array}{ccc|cc} 1 & 150 & 1 & 1 & 150 \\ 0.080 & 18 & 0.18 & 0.080 & 18 \\ 0.080 & 0 & -0.090 & 0.080 & 0 \end{array}\right.}{0.0084}$$

$$= \frac{-1.620 + 2.160 + 0 - 1.440 - 0 + 1.080}{0.0084} = \frac{0.180}{0.0084} = 21$$

$$z = \frac{\left|\begin{array}{ccc|cc} 1 & 1 & 150 & 1 & 1 \\ 0.080 & 0.11 & 18 & 0.080 & 0.11 \\ 0.080 & -0.055 & 0 & 0.080 & -0.055 \end{array}\right.}{0.0084}$$

the value of z can also be found by substituting $x = 75$ and $y = 21$ into the first equation

$$= \frac{0 + 1.440 - 0.660 - 1.320 + 0.990 - 0}{0.0084} = \frac{0.450}{0.0084} = 54$$

Therefore, 75 mL of the 8.0% solution, 21 mL of the 11% solution, and 54 mL of the 18% solution are required to make the 12% solution. Results have been rounded off to two significant digits, the accuracy of the data. Checking with the statement of the problem, we see that these volumes total 150 mL. ■

EXERCISES 5.6

In Exercises 1 and 2, make the given change in the indicated examples of this section and then solve the resulting problems.

1. In Example 1, interchange the first and second rows of the determinant and then evaluate it.
2. In Example 3, change the constant to the right of the = sign in the first equation from −1 to −3, change the constant to the right of the = sign in the third equation from 9 to 11, and then solve the resulting system of equations.

In Exercises 3–14, evaluate the given third-order determinants.

3. $\begin{vmatrix} 5 & 4 & -1 \\ -2 & -6 & 3 \\ 7 & 1 & 1 \end{vmatrix}$

4. $\begin{vmatrix} -7 & 0 & 0 \\ 2 & 4 & 5 \\ 1 & 4 & 2 \end{vmatrix}$

5. $\begin{vmatrix} 8 & 9 & -6 \\ -3 & 7 & 2 \\ 4 & -2 & 5 \end{vmatrix}$

6. $\begin{vmatrix} -2 & 4 & -1 \\ 5 & -1 & 4 \\ 4 & -8 & 2 \end{vmatrix}$

7. $\begin{vmatrix} -8 & -4 & -6 \\ 5 & -1 & 0 \\ 2 & 10 & -1 \end{vmatrix}$

8. $\begin{vmatrix} 10 & 2 & -7 \\ -2 & -3 & 6 \\ 6 & 5 & -2 \end{vmatrix}$

9. $\begin{vmatrix} 4 & -3 & -11 \\ -9 & 2 & -2 \\ 0 & 1 & -5 \end{vmatrix}$

10. $\begin{vmatrix} 9 & -2 & 0 \\ -1 & 3 & -6 \\ -4 & -6 & -2 \end{vmatrix}$

11. $\begin{vmatrix} 20 & 18 & -50 \\ -15 & 24 & -12 \\ -20 & 55 & -22 \end{vmatrix}$

12. $\begin{vmatrix} 20 & 0 & -15 \\ -4 & 30 & 1 \\ 6 & -1 & 40 \end{vmatrix}$

13. $\begin{vmatrix} 0.1 & -0.2 & 0 \\ -0.5 & 1 & 0.4 \\ -2 & 0.8 & 2 \end{vmatrix}$

14. $\begin{vmatrix} 0.25 & -0.54 & -0.42 \\ 1.20 & 0.35 & 0.28 \\ -0.50 & 0.12 & -0.44 \end{vmatrix}$

In Exercises 15–28, solve the given systems of equations by use of determinants. (Exercises 17–26 are the same as Exercises 3–12 of Section 5.5.)

15. $2x + 3y + z = 4$
 $3x - z = -3$
 $x - 2y + 2z = -5$

16. $4x + y + z = 2$
 $2x - y - z = 4$
 $3y + z = 2$

17. $x + y + z = 2$
 $x - z = 1$
 $x + y = 1$

18. $x + y - z = -3$
 $x + z = 2$
 $2x - y + 2z = 3$

19. $2x + 3y + z = 2$
 $-x + 2y + 3z = -1$
 $-3x - 3y + z = 0$

20. $2x + y - z = 4$
 $4x - 3y - 2z = -2$
 $8x - 2y - 3z = 3$

21. $5l + 6w - 3h = 6$
 $4l - 7w - 2h = -3$
 $3l + w - 7h = 1$

22. $3r + s - t = 2$
 $r + t - 2s = 0$
 $4r - s + t = 3$

23. $2x - 2y + 3z = 5$
 $2x + y - 2z = -1$
 $3z + y - 4x = 0$

24. $2u + 2v + 3w = 0$
 $3u + v + 4w = 21$
 $-u - 3v + 7w = 15$

25. $3x - 7y + 3z = 6$
 $3x + 3y + 6z = 1$
 $5x - 5y + 2z = 5$

26. $18x + 24y + 4z = 46$
 $63x + 6y - 15z = -75$
 $-90x + 30y - 20z = -55$

27. $p + 2q + 2r = 0$
 $2p + 6q - 3r = -1$
 $4p - 3q + 6r = -8$

28. $9x + 12y + 2z = 23$
 $21x + 2y - 5z = -25$
 $6y - 18x - 4z = -11$

The three points (x_1, y_1), (x_2, y_2), and (x_3, y_3) are collinear (they lie on a single straight line) if

$\begin{vmatrix} x_1 & y_1 & 1 \\ x_2 & y_2 & 1 \\ x_3 & y_3 & 1 \end{vmatrix} = 0$. *In Exercises 29 and 30, use this fact to decide if the given points are colliner.*

29. $(-2, 1)(2, 3)$, $(6, 5)$
30. $(-8, 20)$, $(-3, 6)$, $(1, -8)$

In Exercises 31–34, use the determinant at the right. Answer the questions about the determinant for the changes given in each exercise. $\begin{vmatrix} 2 & 4 & 1 \\ 3 & 6 & 5 \\ 7 & 9 & 8 \end{vmatrix} = 35$

31. How does the value change if the first two rows are interchanged?
32. What is the value if the second row is replaced with the first row (the first row remains unchanged—the first and second rows are the same)?
33. How does the value change if the elements of the first row are added to the corresponding elements of the second row (the first row remains unchanged)?
34. How does value change if each element of the first row is multiplied by 2?

In Exercises 35–46, solve the given problems by determinants. In Exercises 40–46, set up appropriate systems of equations. All numbers are accurate to at least two significant digits.

35. In analyzing the forces on the bell-crank mechanism shown in Fig. 5.49, the following equations are found. Find the forces.

 $A - 0.60F = 80$
 $B - 0.80F = 0$
 $6.0A - 10F = 0$

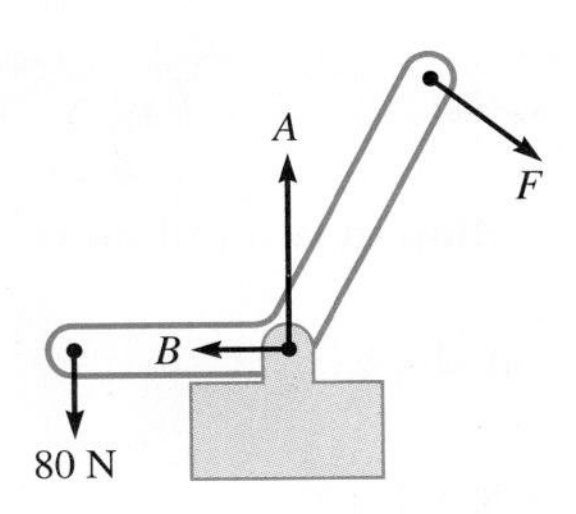

Fig. 5.49

36. Using Kirchhoff's laws (see the chapter introduction; the equations can be found in most physics textbooks) with the circuit shown in Fig. 5.50, the following equations are found. Find the indicated currents (in A) i_1, i_2, and i_3.

$$\begin{aligned} 19I_1 - 12I_2 \quad\quad &= 60 \\ 12I_1 - 18I_2 + 6.0I_3 &= 0 \\ 6.0I_2 - 18I_3 &= 0 \end{aligned}$$

Fig. 5.50

37. In a laboratory experiment to measure the acceleration of an object, the distances traveled by the object were recorded for three different time intervals. These data led to the following equations:

$$\begin{aligned} s_0 + 2v_0 + 2a &= 20 \\ s_0 + 4v_0 + 8a &= 54 \\ s_0 + 6v_0 + 18a &= 104 \end{aligned}$$

Here, s_0 is the initial displacement (in ft), v_0 is the initial velocity (in ft/s), and a is the acceleration (in ft/s^2). Find s_0, v_0, and a.

38. The angle θ between two links of a robot arm is given by $\theta = at^3 + bt^2 + ct$, where t is the time during an 11.8-s cycle. If $\theta = 19.0°$ for $t = 1.00$ s, $\theta = 30.9°$ for $t = 3.00$ s, and $\theta = 19.8°$ for $t = 5.00$ s, find the equation $\theta = f(t)$. See Fig. 5.51.

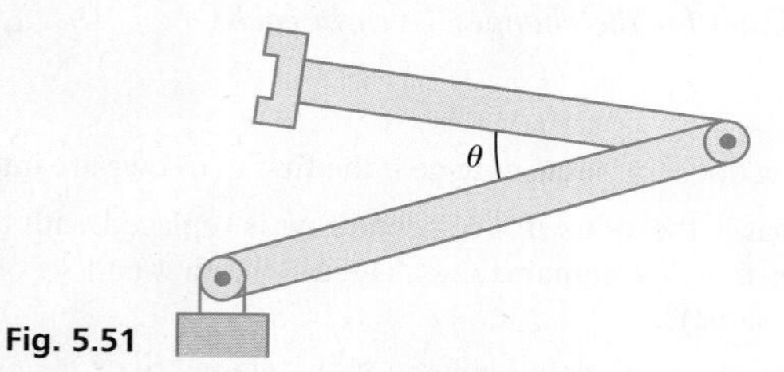

Fig. 5.51

39. The angles of a quadrilateral shaped parcel of land with two equal angles (see Fig. 5.52) have measures such that $\angle A = \angle C + \angle B$ and $\angle A + 2\angle B + \angle C = 280°$. Find the measures of these angles.

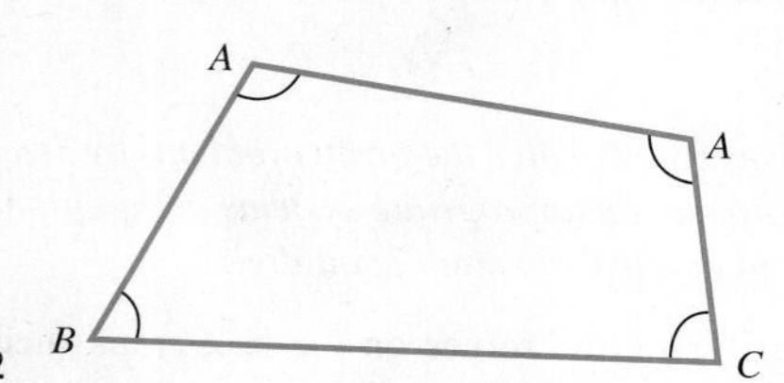

Fig. 5.52

40. A certain 18-hole golf course has par-3, par-4, and par-5 holes, and there are twice as many par-4 holes as par-5 holes. How many holes of each type are there if a golfer has par on every hole for a score of 70?

41. The increase L in length of a long metal rod is a function of the temperature T (in °C) given by $L = a + bT + cT^2$, where a, b, and c are constants. By evaluating a, b, and c, find $L = f(T)$, if $L = 6.4$ mm for $T = 2.0$°C, $L = 8.6$ mm for $T = 4.0$°C, and $V = 11.6$ mm for $T = 6.0$°C.

42. An online retailer requires three different size containers to package its products for shipment. The costs of these containers, A, B, and C, and their capacities are shown as follows:

Container	A	B	C
Cost ($ each)	4	6	7
Capacity (in.^{3})	200	400	600

If the retailer orders 2500 containers with a total capacity of 1.1×10^6 in.^{3} at a cost of $15,000, how many of each are in the order?

43. An alloy used in electrical transformers contains nickel (Ni), iron (Fe), and molybdenum (Mo). The percent of Ni is 1% less than five times the percent of Fe. The percent of Fe is 1% more than three times the percent of Mo. Find the percent of each in the alloy.

44. A company budgets $750,000 in salaries, hardware, and computer time for the design of a new product. The salaries are as much as the others combined, and the hardware budget is twice the computer budget. How much is budgeted for each?

45. A person spent 1.10 h in a car going to an airport, 1.95 h flying in a jet, and 0.520 h in a taxi to reach the final destination. The jet's speed averaged 12.0 times that of the car, which averaged 15.0 mi/h more than the taxi. What was the average speed of each if the trip covered 1140 mi?

46. An intravenous aqueous solution is made from three mixtures to get 500 mL with 6.0% of one medication, 8.0% of a second medication, and 86% water. The percents in the mixtures are, respectively, 5.0, 20, 75 (first), 0, 5.0, 95 (second), and 10, 5.0, 85 (third). How much of each is used?

Answers to Practice Exercises

1. 38 **2.** 38

CHAPTER 5 KEY FORMULAS AND EQUATIONS

Linear equation in two unknowns $\quad ax + by = c \quad$ **(5.1)**

Definition of slope $\quad m = \dfrac{y_2 - y_1}{x_2 - x_1} \quad$ **(5.2)**

Slope-intercept form $\quad y = mx + b \quad$ **(5.3)**

System of two linear equations

$$\begin{aligned} a_1x + b_1y &= c_1 \\ a_2x + b_2y &= c_2 \end{aligned} \quad \textbf{(5.4)}$$

Second-order determinant

$$\begin{vmatrix} a_1 & b_1 \\ a_2 & b_2 \end{vmatrix} = a_1b_2 - a_2b_1 \quad (5.8)$$

Cramer's rule

$$x = \frac{\begin{vmatrix} c_1 & b_1 \\ c_2 & b_2 \end{vmatrix}}{\begin{vmatrix} a_1 & b_1 \\ a_2 & b_2 \end{vmatrix}} \quad \text{and} \quad y = \frac{\begin{vmatrix} a_1 & c_1 \\ a_2 & c_2 \end{vmatrix}}{\begin{vmatrix} a_1 & b_1 \\ a_2 & b_2 \end{vmatrix}} \quad (5.9)$$

System of three linear equations

$$\begin{aligned} a_1x + b_1y + c_1z &= d_1 \\ a_2x + b_2y + c_2z &= d_2 \\ a_3x + b_3y + c_3z &= d_3 \end{aligned} \quad (5.10)$$

Third-order determinant

$$\begin{vmatrix} a_1 & b_1 & c_1 \\ a_2 & b_2 & c_2 \\ a_3 & b_3 & c_3 \end{vmatrix} = a_1b_2c_3 + a_3b_1c_2 + a_2b_3c_1 - a_3b_2c_1 - a_1b_3c_2 - a_2b_1c_3 \quad (5.12)$$

Cramer's rule

$$x = \frac{\begin{vmatrix} d_1 & b_1 & c_1 \\ d_2 & b_2 & c_2 \\ d_3 & b_3 & c_3 \end{vmatrix}}{\begin{vmatrix} a_1 & b_1 & c_1 \\ a_2 & b_2 & c_2 \\ a_3 & b_3 & c_3 \end{vmatrix}} \quad y = \frac{\begin{vmatrix} a_1 & d_1 & c_1 \\ a_2 & d_2 & c_2 \\ a_3 & d_3 & c_3 \end{vmatrix}}{\begin{vmatrix} a_1 & b_1 & c_1 \\ a_2 & b_2 & c_2 \\ a_3 & b_3 & c_3 \end{vmatrix}} \quad z = \frac{\begin{vmatrix} a_1 & b_1 & d_1 \\ a_2 & b_2 & d_2 \\ a_3 & b_3 & d_3 \end{vmatrix}}{\begin{vmatrix} a_1 & b_1 & c_1 \\ a_2 & b_2 & c_2 \\ a_3 & b_3 & c_3 \end{vmatrix}} \quad (5.13)$$

CHAPTER 5 REVIEW EXERCISES

CONCEPT CHECK EXERCISES

Determine each of the following as being either **true** *or* **false**. *If it is false, explain why.*

1. $x = 2, y = -3$ is a solution of the linear equation $4x - 3y = -1$.
2. The slope of the line having intercepts $(0, -3)$ and $(2, 0)$ is $3/2$.
3. The system of equations $3x + y = 5, 2y + 6x = 10$ is inconsistent.
4. One method of finding the solution of the system of equations $2x + 3y = 8, x - y = 3$, would be by multiplying each term of the second equation by 2, and then adding the left sides and adding the right sides.
5. $\begin{vmatrix} 2 & 4 \\ -1 & 3 \end{vmatrix} = 2$
6. One method of finding the solution of the following system of equations would be to subtract the left side and the right side of the first equation from the respective sides of the second and third equations to create a system of two equations in x and z.

$$\begin{aligned} 3x - 2y - z &= 1 \\ 4x - 2y + 3z &= -6 \\ x - 2y + z &= -5 \end{aligned}$$

7. $\begin{vmatrix} 1 & 0 & 3 \\ 0 & 4 & -1 \\ -1 & 7 & 0 \end{vmatrix} = 0$
8. If the graphs of the equations in a linear system are parallel lines with different y-intercepts, the system is called dependent.

PRACTICE AND APPLICATIONS

In Exercises 9–12, evaluate the given determinants.

9. $\begin{vmatrix} -2 & 5 \\ 3 & 1 \end{vmatrix}$
10. $\begin{vmatrix} 40 & 10 \\ -20 & -60 \end{vmatrix}$
11. $\begin{vmatrix} -18 & -33 \\ -21 & 44 \end{vmatrix}$
12. $\begin{vmatrix} 0.91 & -1.2 \\ 0.73 & -5.0 \end{vmatrix}$

In Exercises 13–16, find the slopes of the lines that pass through the given points.

13. $(2,0), (4, -8)$
14. $(-1, -5), (-4, 4)$
15. $(40, -20), (-30, -40)$
16. $(-6, \frac{1}{2}), (1, -\frac{7}{2})$

In Exercises 17–20, find the slope and the y-intercepts of the lines with the given equations. Sketch the graphs.

17. $y = -2x + 4$
18. $2y = \frac{2}{3}x - 3$
19. $8x - 2y = 5$
20. $6x = 16 + 6y$

In Exercises 21–28, solve the given systems of equations graphically.

21. $y = 2x - 4$
    $y = -\frac{3}{2}x + 3$
22. $y + 3x = 3$
    $y = 2x - 6$
23. $4A - B = 6$
    $3A + 2B = 12$
24. $2x - 5y = 10$
    $3x + y = 6$
25. $7x = 2y + 14$
    $y = -4x + 4$
26. $5v = 15 - 3t$
    $t = 6v - 12$
27. $3M + 4N = 6$
    $2M - 3N = 2$
28. $5x + 2y = 5$
    $4x - 8y = 6$

In Exercises 29–38, solve the given systems of equations by using an appropriate algebraic method.

29. $x + 2y = 5$
$x + 3y = 7$

30. $2x - y = 7$
$x + y = 2$

31. $4x + 3y = -4$
$y = 2x - 3$

32. $r = -3s - 2$
$-2r - 9s = 2$

33. $10i - 27v = 29$
$40i + 33v = 69$

34. $3x - 6y = 5$
$2y + 7x = 4$

35. $7x = 2y - 6$
$7y = 12 - 4x$

36. $3R = 8 - 5I$
$6I = 8R + 11$

37. $90x - 110y = 40$
$30x - 15y = 25$

38. $0.42x - 0.56y = 1.26$
$0.98x - 1.40y = -0.28$

In Exercises 39–48, solve the given systems of equations by determinants. (These are the same as for Exercises 29–38.)

39. $x + 2y = 5$
$x + 3y = 7$

40. $2x - y = 7$
$x + y = 2$

41. $4x + 3y = -4$
$y = 2x - 3$

42. $r = -3s - 2$
$-2r - 9s = 2$

43. $10i - 27v = 29$
$40i + 33v = 69$

44. $3x - 6y = 5$
$2y + 7x = 4$

45. $7x = 2y - 6$
$7y = 12 - 4x$

46. $3R = 8 - 5I$
$6I = 8R + 11$

47. $90x - 110y = 40$
$30x - 15y = 25$

48. $0.42x - 0.56y = 1.26$
$0.98x - 1.40y = -0.28$

In Exercises 49–52, explain your answers. In Exercises 49–51, choose an exercise from among Exercises 39–48 that you think is most easily solved by the indicated method.

49. Substitution **50.** Addition or subtraction **51.** Determinants

52. Considering the slope m and the y-intercept b, explain how to determine if there is (a) a unique solution, (b) an inconsistent solution, or (c) a dependent solution.

In Exercises 53–56, evaluate the given determinants.

53. $\begin{vmatrix} 4 & -1 & 8 \\ -1 & 6 & -2 \\ 2 & 1 & -1 \end{vmatrix}$

54. $\begin{vmatrix} -500 & 0 & -500 \\ 250 & 300 & -100 \\ -300 & 200 & 200 \end{vmatrix}$

55. $\begin{vmatrix} -2.2 & -4.1 & 7.0 \\ 1.2 & 6.4 & -3.5 \\ -7.2 & 2.4 & -1.0 \end{vmatrix}$

56. $\begin{vmatrix} 30 & 22 & -12 \\ 0 & -34 & 44 \\ 35 & -41 & -27 \end{vmatrix}$

In Exercises 57–62, solve the given systems of equations algebraically. In Exercises 61 and 62, the numbers are approximate.

57. $2x + y + z = 4$
$x - 2y - z = 3$
$3x + 3y - 2z = 1$

58. $x + 2y + z = 2$
$3x - 6y + 2z = 2$
$2x - z = 8$

59. $2r + s + 2t = 8$
$3r - 2s - 4t = 5$
$-2r + 3s + 4t = -3$

60. $4u + 4v - 2w = -4$
$20u - 15v + 10w = -10$
$24u - 12v - 9w = 39$

61. $3.6x + 5.2y - z = -2.2$
$3.2x - 4.8y + 3.9z = 8.1$
$6.4x + 4.1y + 2.3z = 5.1$

62. $32t + 24u + 63v = 32$
$19v - 31u + 42t = 132$
$48t + 12u + 11v = 0$

In Exercises 63 and 64, solve the given systems of equations using the rref feature on a calculator.

63. $5x + y - 4z = -5$
$3x - 5y - 6z = -20$
$x - 3y + 8z = -27$

64. $x + y + z = 80$
$2x - 3y = -20$
$2x + 3z = 115$

In Exercises 65–70, solve the given systems of equations by determinants. In Exercises 69 and 70, the numbers are approximate. (These systems are the same as for Exercises 57–62.)

65. $2x + y + z = 4$
$x - 2y - z = 3$
$3x + 3y - 2z = 1$

66. $x + 2y + z = 2$
$3x - 6y + 2z = 2$
$2x - z = 8$

67. $2r + s + 2t = 8$
$3r - 2s - 4t = 5$
$-2r + 3s + 4t = -3$

68. $4u + 4v - 2w = -4$
$20u - 15v + 10w = -10$
$24u - 12v - 9w = 39$

69. $3.6x + 5.2y - z = -2.2$
$3.2x - 4.8y + 3.9z = 8.1$
$6.4x + 4.1y + 2.3z = 5.1$

70. $32t + 24u + 63v = 32$
$19v - 31u + 42t = 132$
$48t + 12u + 11v = 0$

In Exercises 71–74, solve for x. Here, we see that we can solve an equation in which the unknown is an element of a determinant.

71. $\begin{vmatrix} 2 & 5 \\ 1 & x \end{vmatrix} = 3$

72. $\begin{vmatrix} -1 & x \\ 3 & 4 \end{vmatrix} = 7$

73. $\begin{vmatrix} x & 1 & 2 \\ 0 & -1 & 3 \\ -2 & 2 & 1 \end{vmatrix} = 5$

74. $\begin{vmatrix} 1 & 2 & -1 \\ -2 & 3 & x \\ -1 & 2 & -2 \end{vmatrix} = -3$

In Exercises 75 and 76, find the equation of the regression line for the given data. Then use this equation to make the indicated estimate. Round decimals in the regression equation to three decimal places. Round estimates to the same accuracy as the given data.

75. The weight (in lb) and the highway fuel efficiency [in miles per gallon (mpg)] are shown for eight different cars in the table below. Find the equation of the regression line, and then estimate the highway mpg of a car that weighs 3500 lb.

Weight, lb	2595	3135	3315	3630	3860	4020	4205	4415
Highway mpg	38	32	31	27	28	27	25	25

76. The data given in the table below are the dosage (in mg) of a certain drug and the pulse rate for eight different patients. Find the equation of the regression line, and then estimate the pulse rate of a patient who received a dosage of 0.25 mg of the drug.

Dosage, mg	0.12	0.22	0.29	0.34	0.38	0.40	0.45	0.48
Pulse rate, beats/s	96	89	91	90	88	82	80	79

In Exercises 77 and 78, determine the value of k that makes the system dependent. In Exercises 79 and 80, determine the value of k that makes the system inconsistent.

77. $3x - ky = 6$
$x + 2y = 2$

78. $5x + 20y = 15$
$2x + ky = 6$

79. $kx - 2y = 5$
$4x + 6y = 1$

80. $2x - 5y = 7$
$kx + 10y = 2$

In Exercises 81 and 84, solve the given systems of equations by any appropriate method. All numbers in 81 are accurate to two significant digits, and in 82 they are accurate to three significant digits.

81. A 20-ft crane arm with a supporting cable and with a 100-lb box suspended from its end has forces acting on it, as shown in Fig. 5.53. Find the forces (in lb) from the following equations.

$$F_1 + 2.0F_2 = 280$$
$$0.87F_1 - F_3 = 0$$
$$3.0F_1 - 4.0F_2 = 600$$

Fig. 5.53

82. In applying Kirchhoff's laws to the electric circuit shown in Fig. 5.54, the following equations result. Find the indicated currents (in A).

$$i_1 + i_2 + i_3 = 0$$
$$5.20i_1 - 3.25i_2 = 8.33 - 6.45$$
$$3.25i_2 - 2.62i_3 = 6.45 - 9.80$$

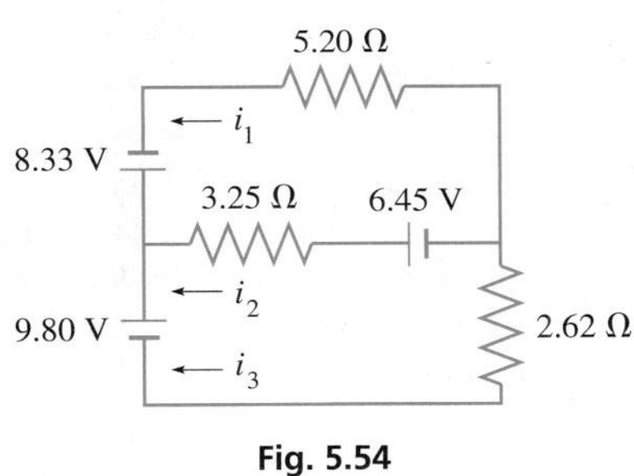

Fig. 5.54

In Exercises 83–106, set up systems of equations and solve by any appropriate method. All numbers are accurate to at least two significant digits.

83. A study found that the fuel consumption for transportation contributes a percent p_1 that is 16%, less than the percent p_2 of all other sources combined. Find p_1 and p_2.

84. A certain amount of a fuel contains 150,000 Btu of potential heat. Part is burned at 80% efficiency, and the rest is burned at 70%, efficiency, such that the total amount of heat actually delivered is 114,000 Btu. Find the amounts burned at each efficiency.

85. The sales representatives of a company have a choice of being paid 10%, of their sales in a month, or \$2400 plus 4%, of their sales in the month. For what monthly sales is the income the same, and what is that income?

86. A person invested a total of \$20,900 into two bonds, one with an annual interest rate of 6.00%, and the other with an annual interest rate of 5.00% per year. If the total annual interest from the bonds is \$1170, how much is invested in each bond?

87. A total of 42 tons of two types of ore is to be loaded into a smelter. The first type contains 6.0% copper, and the second contains 2.4% copper. Find the necessary amounts of each ore (to the nearest 1 ton) to produce 2 tons of copper.

88. As a 40-ft pulley belt makes one revolution, one of the two pulley wheels makes one more revolution than the other. Another wheel of half the radius replaces the smaller wheel and makes six more revolutions than the larger wheel for one revolution of the belt. Find the circumferences of the wheels. See Fig. 5.55.

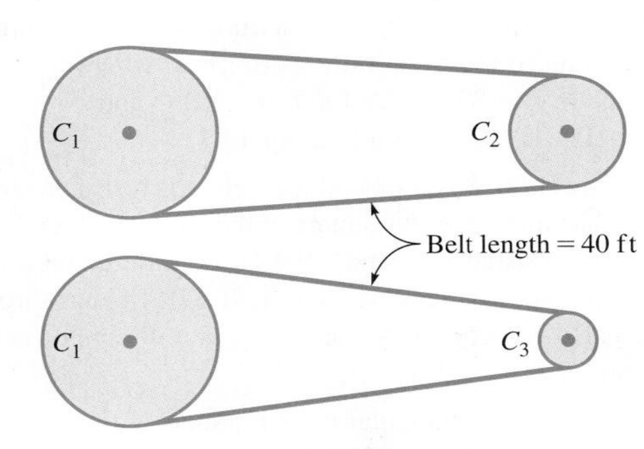

Fig. 5.55

89. A computer analysis showed that the temperature T of the ocean water within 1000 m of a nuclear-plant discharge pipe was given by $T = \frac{a}{x + 100} + b$, where x is the distance from the pipe and a and b are constants. If $T = 14°C$ for $x = 0$ and $T = 10°C$ for $x = 900$ m, find a and b.

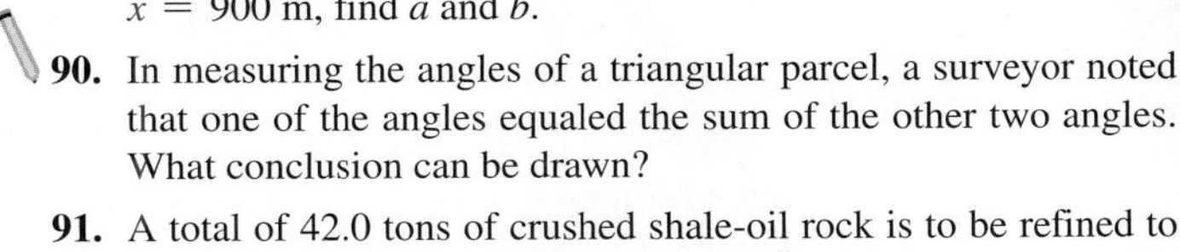

90. In measuring the angles of a triangular parcel, a surveyor noted that one of the angles equaled the sum of the other two angles. What conclusion can be drawn?

91. A total of 42.0 tons of crushed shale-oil rock is to be refined to extract the oil. The first contains 18.0 gal/ton and the second contains 30.0 gal/ton. How much of each must be refined to produce 1050 gal of oil?

92. An online merchandise company charges \$4 for shipping orders of less than \$50, \$6 for orders from \$50 to \$200, and \$8 for orders over \$200. One day the total shipping charges were \$2160 for 384 orders. Find the number of orders shipped at each rate if the number of orders under \$50 was 12 more than twice the number of orders over \$200.

93. In an office building one type of office has 800 ft^2 and rents for \$900/month. A second type of office has 1100 ft^2 and rents for \$1250/month. How many of each are there if they have a total of 49,200 ft^2 of office space and rent for a total of \$55,600/month?

94. A study showed that the percent of persons 25 to 29 years old who completed at least four years of college was 31.7% in 2010 and 34.0% in 2014. Assuming the increase to be linear, find an equation relating the percent p and number of years n after 2010. Based on this equation, what would be the percent in 2020? (*Source:* nces.ed.gov.)

95. A satellite is to be launched from a space shuttle. It is calculated that the satellite's speed will be 24,200 km/h if launched directly ahead of the shuttle or 21,400 km/h if launched directly to the rear of the shuttle. What is the speed of the shuttle and the launching speed of the satellite relative to the shuttle? See Fig. 5.56.

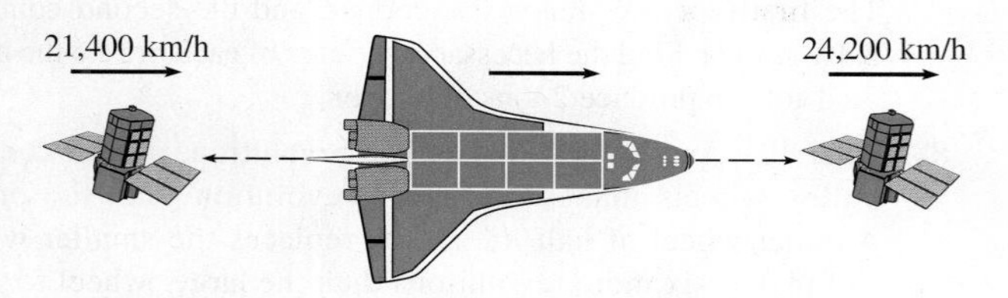

Fig. 5.56

96. The velocity v of sound is a function of the temperature T according to the function $v = aT + b$, where a and b are constants. If $v = 337.5$ m/s for $T = 10.0°C$ and $v = 346.6$ m/s for $T = 25.0°C$, find v as a function of T.

97. The power (in W) dissipated in an electric resistance (in Ω) equals the resistance times the square of the current (in A). If 1.0 A flows through resistance R_1 and 3.0 A flows through resistance R_2, the total power dissipated is 14.0 W. If 3.0 A flows through R_1 and 1.0 A flows through R_2, the total power dissipated is 6.0 W. Find R_1 and R_2.

98. Twelve equal rectangular ceiling panels are placed as shown in Fig. 5.57. If each panel is 6.0 in. longer than it is wide and a total of 132.0 ft of edge and middle strips is used, what are the dimensions of the room?

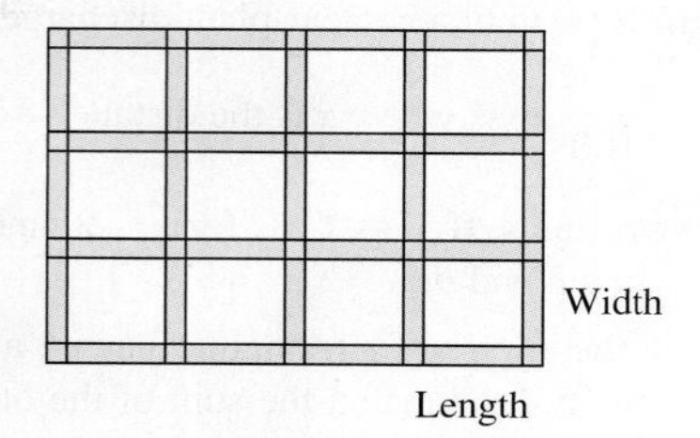

Fig. 5.57

99. The weight of a lever may be considered to be at its center. A 20-ft lever of weight w is balanced on a fulcrum 8 ft from one end by a load L at that end. If a load of $4L$ is placed at that end, it requires a 20-lb weight at the other end to balance the lever. What are the initial load L and the weight w of the lever? (See Exercise 56 of Section 5.3.) See Fig. 5.58.

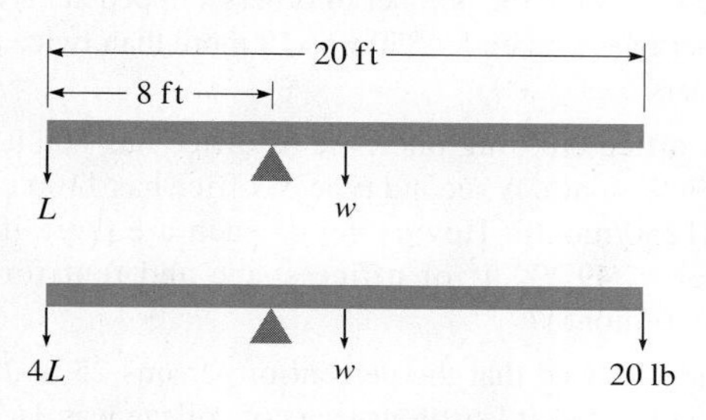

Fig. 5.58

100. Two fuel mixtures, one of 2.0% oil and 98.0% gasoline and another of 8.0% oil and 92.0% gasoline, are to be used to make 10.0 L of a fuel that is 4.0% oil and 96.0% gasoline for use in a chain saw. How much of each mixture is needed?

101. For the triangular truss shown in Fig. 5.59, $\angle A$ is 55° less than twice $\angle B$, and $\angle C$ is 25° less than $\angle B$. Find the measures of the three angles.

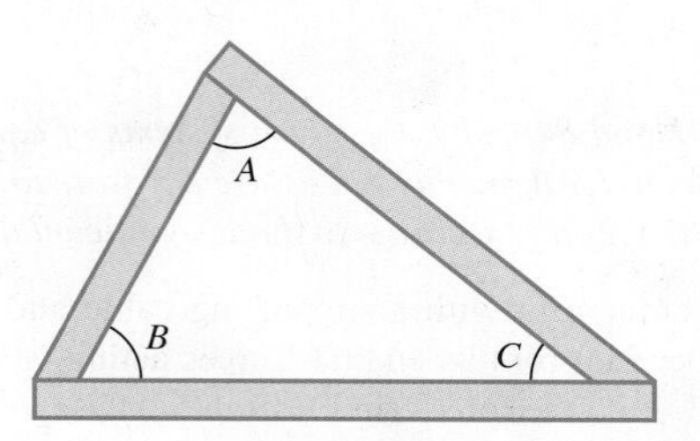

Fig. 5.59

102. A manufacturer produces three models of DVD players in a year. Four times as many of model A are produced as model C, and 7000 more of model B than model C. If the total production for the year is 97,000 units, how many of each are produced?

103. Three computer programs A, B, and C require a total of 140 MB (megabytes) of hard-disk memory. If three other programs, two requiring the same memory as B and one the same as C, are added to a disk with A, B, and C, a total of 236 MB are required. If three other programs, one requiring the same memory as A and two the same memory as C, are added to a disk with A, B, and C, a total of 304 MB are required. How much memory is required for each of A, B, and C?

104. In a laboratory, electrolysis was used on a solution of sulfuric acid, silver nitrate, and cupric sulfate, releasing hydrogen gas, silver, and copper. A total mass of 1.750 g is released. The mass of silver deposited is 3.40 times the mass of copper deposited, and the mass of copper and 70.0 times the mass of hydrogen combined equals the mass of silver deposited less 0.037 g. How much of each is released?

105. Gold loses about 5.3%, and silver about 10%, of its weight when immersed in water. If a gold-silver alloy weighs 6.0 N in air and 5.6 N in water, find the weight in air of the gold and the silver in the alloy.

106. A person engaged a workman for 48 days. For each day that he labored, he received 24 cents, and for each day he was idle, he paid 12 cents for his board. At the end of the 48 days, the account was settled, when the laborer received 504 cents. Determine the number of working days and the number of days he was idle. (Source: *Elementary Algebra: Embracing the First Principles of the Science* by Charles Davies, A.S. Barnes & Company, 1852)

107. Write one or two paragraphs giving reasons for choosing a particular method of solving the following problem. If a first pump is used for 2.2 h and a second pump is used for 2.7 h, 1100 ft^3 can be removed from a wastewater-holding tank. If the first pump is used for 1.4 h and the second for 2.5 h, 840 ft^3 can be removed. How much can each pump remove in 1.0 h? (What is the result to two significant digits?)

CHAPTER 5 PRACTICE TEST

As a study aid, we have included complete solutions for each Practice Test problem at the back of this book.

1. Find the slope of the line through $(2, -5)$ and $(-1, 4)$.

2. Is $x = -2, y = 3$ a solution to the system of equations

   $2x + 5y = 11$

   $y - 5x = 12$?

   Explain.

3. Evaluate: $\begin{vmatrix} -1 & 3 & -2 \\ 4 & -3 & 0 \\ 5 & -4 & 2 \end{vmatrix}$

4. Solve by substitution:

   $x + 2y = 5$

   $4y = 3 - 2x$

5. Solve by determinants:

   $3x - 2y = 4$

   $2x + 5y = -1$

6. By finding the slope and y-intercept, sketch the graph of $2x + y = 4$.

7. The perimeter of a rectangular ranch is 24 km, and the length is 6.0 km more than the width. Set up equations relating the length l and the width w, and then solve for l and w.

8. Solve by addition or subtraction:

   $6N - 2P = 13$

   $4N + 3P = -13$

9. In testing an anticholesterol drug, it was found that each milligram of drug administered reduced a person's blood cholesterol level by 2 units. Set up the function relating the cholesterol level C as a function of the dosage d for a person whose cholesterol level is 310 before taking the drug. Sketch the graph.

10. Solve the following system of equations graphically. Determine the values of x and y to the nearest 0.1.

    $2x - 3y = 6$

    $4x + y = 4$

11. Solve algebraically:

    $x + y + z = 4$

    $x + y - z = 6$

    $2x + y + 2z = 5$

12. By volume, one alloy is 60% copper, 30% zinc, and 10% nickel. A second alloy has percents of 50, 30, and 20, respectively. A third alloy is 30% copper and 70% nickel. How much of each alloy is needed to make 100 cm^3 of a resulting alloy with percents of 40, 15, and 45, respectively?

13. Solve for y by determinants:

    $3x + 2y - z = 4$

    $2x - y + 3z = -2$

    $x + 4z = 5$

14. Use the *rref* feature on a calculator to find the three currents given in the following system of equations. Round answers to three significant digits.

    $I_1 + I_2 = I_3$

    $3.00I_1 - 4.00I_2 = 10.0$

    $4.00I_2 + 5.00I_3 = 5.00$

6

Factoring and Fractions

LEARNING OUTCOMES

After completion of this chapter, the student should be able to:

- Factor algebraic expressions using common factors, difference of squares, sum and difference of cubes, grouping, and techniques for trinomials
- Simplify algebraic fractions
- Add, subtract, multiply, and divide algebraic fractions
- Solve equations involving algebraic fractions
- Simplify complex fractions

In this chapter, we further develop operations with products, quotients, and fractions and show how they can be utilized to solve certain types of equations. These algebraic methods are extremely useful in many applied problems as well as in higher-level mathematics.

The development of the symbols now used in algebra in itself led to advances in mathematics and science. Until about 1500, most problems and their solutions were stated in words, which made them very lengthy and often difficult to follow.

As time went on, writers abbreviated some of the words in a problem, but they still essentially wrote out the solution in words. Eventually, some symbols did start to come into use. For example, the $+$ and $-$ signs first appeared in a published book in the late 1400s, the square root symbol $\sqrt{\ }$ was first used in 1525, and the $=$ sign was introduced in 1557. Then, in the late 1500s, the French lawyer François Viète wrote several articles in which he used symbols, including letters, to represent numbers. He so improved the symbolism of algebra that he is often called "the father of algebra." By the mid-1600s, the notation being used was reasonably similar to that we use today.

The use of symbols in algebra made it more useful in all fields of mathematics. It was important in the development of calculus in the 1600s and 1700s. In turn, this gave scientists a powerful tool that allowed for much greater advancement in all areas of science and technology.

Although the primary purpose of this chapter is to develop additional algebraic methods, many technical applications of these operations will be shown. The important applications in optics are noted by the picture of the Hubble telescope shown below. Other areas of application include electronics, mechanical design, thermodynamics, and physics.

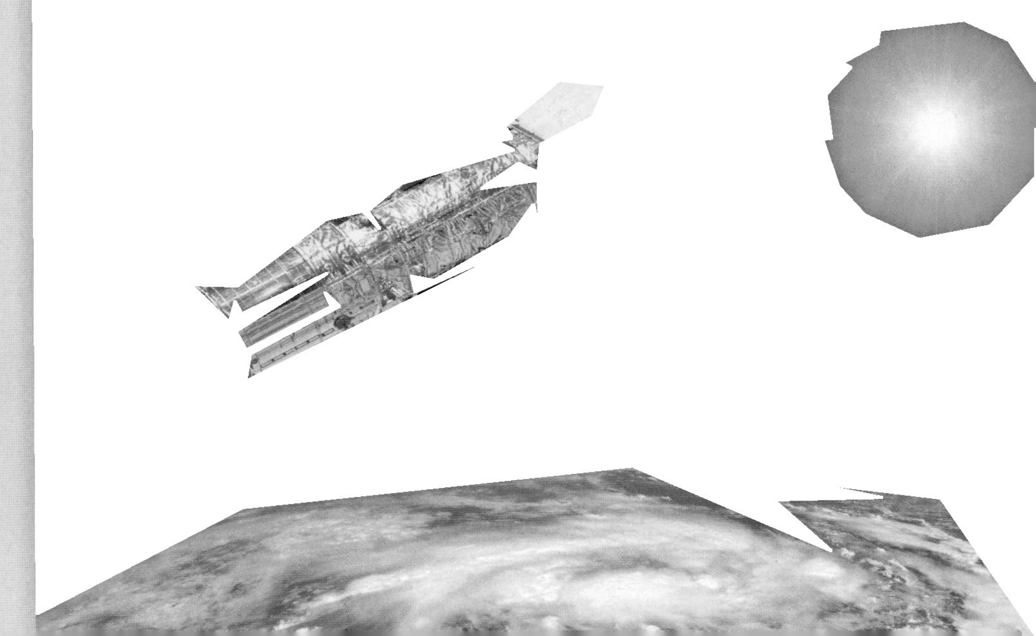

▶ **Great advances in our knowledge of the universe have been made through the use of the Hubble space telescope since the mid-1990s. In Section 6.7, we illustrate the use of algebraic operations in the design of lenses and reflectors in telescopes.**

6.1 Factoring: Greatest Common Factor and Difference of Squares

Factoring • Factoring Out the GCF • Factoring Difference of Two Squares • Prime Factors • Factoring Completely • Factoring by Grouping

When an algebraic expression is written as the **product** of two or more quantities, each of these quantities is called a **factor** of the expression. In practice, we often have an expression and we need to find its factors. Finding these factors, which is essentially reversing the process of multiplication, is called **factoring.** Factoring is needed in many technical applications that require us to solve higher-degree equations, perform operations on algebraic fractions, or solve certain literal equations.

Since factoring is the reversal of multiplying, it is important that we understand certain types of special products. Each special product, when used in reverse, gives us a method for factoring. The first three sections of this chapter are devoted to discussing different techniques that can be used to factor expressions. In this section, we discuss the first two of these techniques: factoring out the *greatest common factor* and factoring the *difference of squares.*

FACTORING OUT THE GREATEST COMMON FACTOR (GCF)

From the distributive law, we know that $a(x + y) = ax + ay$. When written in reverse, this becomes

$$ax + ay = a(x + y) \qquad \textbf{(6.1)}$$

Equation (6.1) provides us with a technique of factoring called **factoring out the greatest common factor** (or GCF, for short). In this case, a is the GCF and it has been *factored out* of the expression.

To factor out the GCF, we first find the greatest monomial that is common to each term of the expression. This is the GCF. We then use Eq. (6.1) to write the expression as the product of the GCF and the remaining expression. Although we will be discussing several other factoring techniques in this chapter, **factoring out the GCF is always the first step in factoring.** The following examples illustrate this method.

EXAMPLE 1 Factoring out the GCF

In factoring $6x - 2y$, we note each term contains a factor of 2:

$$6x - 2y = 2(3x) - 2y = 2(3x - y)$$

Here, 2 is the greatest common factor, and $2(3x - y)$ is the required factored form of $6x - 2y$.

NOTE ▸ [We check the result by multiplication.] In this case,

$$2(3x - y) = 6x - 2y$$

Since the result of the multiplication gives the original expression, the factored form is correct. ■

In Example 1, we determined the greatest common factor of 2 by inspection. This is normally the way in which it is found. Once the GCF has been found, the other factor can be determined by *dividing* the original expression by the GCF.

The next example illustrates the case where the GCF is the same as one of the terms, and special care must be taken to complete the factoring correctly.

EXAMPLE 2 Greatest common factor same as term

Factor: $4ax^2 + 2ax$.

The numerical factor 2 and the literal factors a and x are common to each term. Therefore, the greatest common factor of $4ax^2 + 2ax$ is $2ax$. This means that

$$4ax^2 + 2ax = 2ax(2x) + 2ax(1) = 2ax(2x + 1)$$

Note the presence of the 1 in the factored form. When we divide $4ax^2 + 2ax$ by $2ax$, we get

$$\frac{4ax^2 + 2ax}{2ax} = \frac{4ax^2}{2ax} + \frac{2ax}{2ax}$$
$$= 2x + 1$$

■

Practice Exercises

Factor:

1. $3cx^3 - 9cx$ **2.** $9cx^3 - 3cx$

CAUTION Note that in Example 2, $2ax$ divided by $2ax$ is 1 and does not simply cancel out leaving nothing. In cases like this, where the GCF is the same as one of the terms, it is a common error to omit the 1. However, we must include the 1. Without it, when the factored form is multiplied out, we would not obtain the original expression. ■

Usually, the division shown in Example 2 is done mentally. However, we show it here to emphasize the actual operation that is being performed when we factor out a GCF.

EXAMPLE 3 Greatest common factor by inspection

Factor: $6a^5x^2 - 9a^3x^3 + 3a^3x^2$.

After inspecting each term, we determine that each contains a factor of 3, a^3, and x^2. Thus, the greatest common factor is $3a^3x^2$. This means that

$$6a^5x^2 - 9a^3x^3 + 3a^3x^2 = 3a^3x^2(2a^2 - 3x + 1)$$

■

When factoring, it is important to factor out the *greatest* common factor, not just any common factor. Factoring out something other than the GCF will lead to an expression that is not factored completely. To be **factored completely,** the factors must be **prime,** meaning they contain no factor other than ± 1 and $\pm$ itself. The following example illustrates this point.

EXAMPLE 4 Factoring completely

If we factor the common factor of 2 out of the expression $12x + 6x^2$, we get $12x + 6x^2 = 2(6x + 3x^2)$.

However, this is not factored completely. The factor $6x + 3x^2$ is not prime because it may be further factored as $6x + 3x^2 = 3x(2 + x)$.

If we instead factor out the *greatest* common factor $6x$, we get $12x + 6x^2 = 6x(2 + x)$.

This is now factored completely since the factor $2 + x$ is prime. ■

In these examples, note that factoring an expression does not actually change the expression, although it does change the *form* of the expression. In equating the expression to its factored form, we write an *identity*.

It is often necessary to use factoring when solving an equation. This is illustrated in the following example.

EXAMPLE 5 Using factoring to solve a literal equation—FM reception

An equation used in the analysis of FM reception is $R_F = \alpha(2R_A + R_F)$. Solve for R_F.

The steps in the solution are as follows:

■ FM radio was developed in the early 1930s.

$$R_F = \alpha(2R_A + R_F) \quad \text{original equation}$$
$$R_F = 2\alpha R_A + \alpha R_F \quad \text{use distributive law}$$
$$R_F - \alpha R_F = 2\alpha R_A \quad \text{subtract } \alpha R_F \text{ from both sides}$$
$$R_F(1 - \alpha) = 2\alpha R_A \quad \text{factor out } R_F \text{ on left}$$
$$R_F = \frac{2\alpha R_A}{1 - \alpha} \quad \text{divide both sides by } 1 - \alpha$$

We see that we collected both terms containing R_F on the left so that we could factor and thereby solve for R_F. ■

FACTORING THE DIFFERENCE OF TWO SQUARES

From Section 1.8, we know that in order to multiply two binomials, we multiply each term of the first binomial by each term of the second. When we apply this procedure to the special product $(a + b)(a - b)$, we get $(a + b)(a - b) = a^2 - ab + ab - b^2$. Since the two terms in the middle add up to zero, this simplifies to the following:

$$(a + b)(a - b) = a^2 - b^2 \quad (6.2)$$

We can use Eq. (6.2) to find this type of special product directly, as shown in the following example:

EXAMPLE 6 Finding the product $(a + b)(a - b)$

(a) $(2x + 5)(2x - 5) = (2x)^2 - 5^2 = 4x^2 - 25$ Eq. (6.2)

(b) To find the product $Fa(L - a)(L + a)$, which arises from analyzing forces on a beam, we have

$$Fa(L - a)(L + a) = Fa(L^2 - a^2) \quad \text{Eq. (6.2)}$$
$$= FaL^2 - Fa^3 \quad \text{distributive property}$$ ■

An important consequence of Eq. (6.2) is that, when written in reverse, it gives us a method of factoring called **factoring the difference of two squares**.

$$a^2 - b^2 = (a + b)(a - b) \quad (6.3)$$

Equation (6.3) can be used to factor the difference between two perfect squares as shown in the following examples.

EXAMPLE 7 Factoring difference of two squares

In factoring $x^2 - 16$, note that x^2 is the square of x and 16 is the square of 4. Therefore,

■ Usually, in factoring an expression of this type, where it is very clear what numbers are squared, we do not actually write out the middle step as shown. However, if in doubt, write it out.

squares

$$x^2 - 16 = x^2 - 4^2 = (x + 4)(x - 4)$$

difference sum difference ■

EXAMPLE 8 Factoring difference of two squares

(a) Because $4x^2$ is the square of $2x$ and 9 is the square of 3, we may factor $4x^2 - 9$ as

$$4x^2 - 9 = (2x)^2 - 3^2 = (2x + 3)(2x - 3)$$

(b) In the same way,

$$(y - 3)^2 - 16x^4 = [(y - 3) + 4x^2][(y - 3) - 4x^2]$$
$$= (y - 3 + 4x^2)(y - 3 - 4x^2)$$

where we note that $16x^4 = (4x^2)^2$. ■

FACTORING COMPLETELY

NOTE ▸ As noted before, the **greatest common factor should be factored out first.** [However, we must be careful to see if the other factor can itself be factored.] It is possible, for example, that the other factor is the difference of squares. CAUTION This means that **complete factoring** often requires more than one step. Be sure to include only prime factors in the final result. ■

EXAMPLE 9 Factoring completely

(a) In factoring $20x^2 - 45$, note a common factor of 5 in each term. Therefore, $20x^2 - 45 = 5(4x^2 - 9)$. However, the factor $4x^2 - 9$ itself is the difference of squares. Therefore, $20x^2 - 45$ is completely factored as

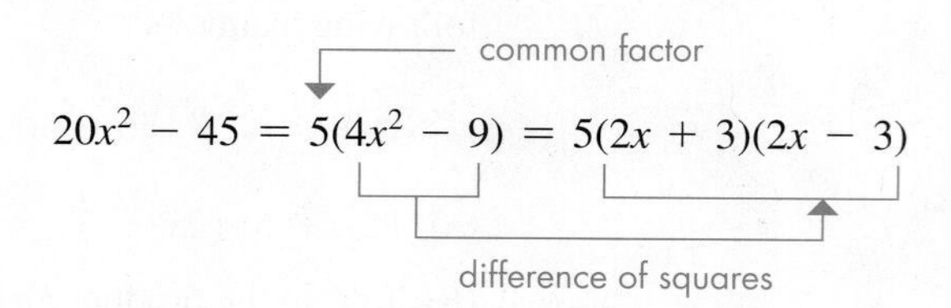

(b) In factoring $x^4 - y^4$, note that we have the difference of two squares. Therefore, $x^4 - y^4 = (x^2 + y^2)(x^2 - y^2)$. However, the factor $x^2 - y^2$ is also the difference of squares. This means that

$$x^4 - y^4 = (x^2 + y^2)(x^2 - y^2) = (x^2 + y^2)(x + y)(x - y)$$

(c) In analyzing the energy collected by different circular solar cells, the expression $45R^2 - 20r^2$ arises. In factoring this expression, we note the factor 5 in each term. Therefore, $45R^2 - 20r^2 = 5(9R^2 - 4r^2)$. However, $9R^2 - 4r^2$ is the difference of squares. Therefore, factoring the original expression, we have

$$45R^2 - 20r^2 = 5(9R^2 - 4r^2) = 5(3R - 2r)(3R + 2r)$$ ■

Practice Exercises

Factor: **3.** $9c^2 - 64$ **4.** $18c^2 - 128$

CAUTION In Example 9(b), the factor $x^2 + y^2$ is *prime*. It is *not* equal to $(x + y)^2$ since $(x + y)^2 = (x + y)(x + y) = x^2 + 2xy + y^2$. ■

FACTORING BY GROUPING

Terms in an expression can sometimes be grouped and then factored by methods of this section. The following example illustrates this method of *factoring by grouping*. In the next section, we discuss another type of expression that can be factored by grouping.

EXAMPLE 10 Factoring by grouping

Factor: $2x - 2y + ax - ay$.

We see that there is no common factor to all four terms, but that each of the first two terms contains a factor of 2, and each of the third and fourth terms contains a factor of a. Grouping terms this way and then factoring each group, we have

$$\begin{aligned} 2x - 2y + ax - ay &= (2x - 2y) + (ax - ay) \\ &= 2(x - y) + a(x - y) \quad \text{now note the common factor of } (x - y) \\ &= (x - y)(2 + a) \end{aligned}$$

■

EXERCISES 6.1

In Exercises 1–4, make the given changes in the indicated examples of this section and then solve the indicated problems.

1. In Example 2, change the + sign to − and then factor.
2. In Example 2, set the given expression equal to B and then solve for a.
3. In Example 9(a), change the coefficient of the first term from 20 to 5 and then factor.
4. In Example 10, change both − signs to + and then factor.

In Exercises 5–8, find the indicated special products directly by inspection.

5. $(T + 6)(T - 6)$
6. $(s + 5t)(s - 5t)$
7. $(4x - 5y)(4x + 5y)$
8. $(3v - 7y)(3v + 7y)$

In Exercises 9–48, factor the given expressions completely.

9. $7x + 7y$
10. $3a - 3b$
11. $5a - 5$
12. $2x^2 + 2$
13. $3x^2 - 9x$
14. $20s + 4s^2$
15. $7b^2h - 28b$
16. $5a^2 - 20ax$
17. $72n^3 + 24n$
18. $90p^3 - 15p^2$
19. $2x + 4y - 8z$
20. $23a - 46b + 69c$
21. $3ab^2 - 6ab + 12ab^3$
22. $4pq - 14q^2 - 16pq^2$
23. $12pq^2 - 8pq - 28pq^3$
24. $27a^2b - 24ab - 9a$
25. $2a^2 - 2b^2 + 4c^2 - 6d^2$
26. $5a + 10ax - 5ay - 20az$
27. $x^2 - 9$
28. $r^2 - 25$
29. $100 - 4A^2$
30. $49 - Z^4$
31. $36a^4 + 1$
32. $324z^2 + 4$
33. $162s^2 - 50t^2$
34. $36s^2 - 121t^4$
35. $144n^2 - 169p^4$
36. $36a^2b^2 + 169c^2$
37. $(x + y)^2 - 9$
38. $(a - b)^2 - 1$
39. $8 - 2x^2$
40. $5a^4 - 125a^2$
41. $300x^2 - 2700z^2$
42. $28x^2 - 700y^2$
43. $2(I - 3)^2 - 8$
44. $a(x + 2)^2 - ay^2$
45. $x^4 - 16$
46. $81 - y^4$
47. $x^{10} - x^2$
48. $2x^4 - 8y^4$

In Exercises 49–54, solve for the indicated variable.

49. $2a - b = ab + 3$, for a
50. $n(x + 1) = 5 - x$, for x
51. $3 - 2s = 2(3 - st)$, for s
52. $k(2 - y) = y(2k - 1)$, for y
53. $(x + 2k)(x - 2) = x^2 + 3x - 4k$, for k
54. $(2x - 3k)(x + 1) = 2x^2 - x - 3$, for k

In Exercises 55–62, factor the given expressions by grouping as illustrated in Example 10.

55. $3x - 3y + bx - by$
56. $am + an + cn + cm$
57. $a^2 + ax - ab - bx$
58. $2y - y^2 - 6y^4 + 12y^3$
59. $x^3 + 3x^2 - 4x - 12$
60. $S^3 - 5S^2 - S + 5$
61. $x^2 - y^2 + x - y$
62. $4p^2 - q^2 + 2p + q$

In Exercises 63 and 64, evaluate the given expressions by using factoring. The results may be checked with a calculator.

63. $\dfrac{8^9 - 8^8}{7}$
64. $\dfrac{5^9 - 5^7}{7^2 - 5^2}$

In Exercises 65 and 66, give the required explanations.

65. Factor $n^2 + n$, and then explain why it represents a positive even integer if n is a positive integer.
66. Factor $n^3 - n$, and then explain why it represents a multiple of 6 if n is an integer greater than 1.

In Exercises 67–76, factor the expressions completely. In Exercises 73 and 74, it is necessary to set up the proper expression. Each expression comes from the technical area indicated.

67. $2Q^2 + 2$ (fire science)
68. $4d^2D^2 - 4d^3D - d^4$ (machine design)

69. $81s - s^3$ (rocket path)

70. $12(4 - x^2) - 2x(4 - x^2) - (4 - x^2)^2$ (container design)

71. $rR^2 - r^3$ (pipeline flow)

72. $p_1R^2 - p_1r^2 - p_2R^2 + p_2r^2$ (fluid flow)

73. A square as large as possible is cut from a circular metal plate of radius r. Express in factored form the area of the metal pieces that are left.

74. A pipe of outside diameter d is inserted into a pipe of inside radius r. Express in factored form the cross-sectional area within the larger pipe that is outside the smaller pipe.

75. A spherical float has a volume of air within it of radius r_1, and the outer radius of the float is r_2. Express in factored form the difference in areas of the outer surface and the inner surface.

76. The kinetic energy of an object of mass m traveling at velocity v is given by $\frac{1}{2}mv^2$. Suppose a car of mass m_0 equipped with a crash-avoidance system automatically applies the brakes to avoid a collision and slows from a velocity of v_1 to a velocity of v_2. Find an expression, in factored form, for the difference between the original and final kinetic energy.

In Exercises 77–82, solve for the indicated variable. Each equation comes from the technical area indicated.

77. $i_1R_1 = (i_2 - i_1)R_2$, for i_1 (electricity: ammeter)

78. $nV + n_1v = n_1V$, for n_1 (acoustics)

79. $3BY + 5Y = 9BS$, for B (physics: elasticity)

80. $Sq + Sp = Spq + p$, for q (computer design)

81. $ER = AtT_0 - AtT_1$, for t (energy conservation)

82. $R = kT_2^4 - kT_1^4$, for k (factor resulting denominator) (radiation)

Answers to Practice Exercises

1. $3cx(x^2 - 3)$ **2.** $3cx(3x^2 - 1)$
3. $(3c - 8)(3c + 8)$ **4.** $2(3c - 8)(3c + 8)$

6.2 Factoring Trinomials

Factoring Trinomials with Leading Coefficient of 1 • Factoring General Trinomials by Trial and Error • Factoring General Trinomials by Grouping • Factoring Perfect Square Trinomials • Factoring Completely

In Section 6.1, we saw that factoring expressions can be viewed as the reverse of multiplication. In this section, we apply this same idea to expressions with three terms, called **trinomials.** In order to factor trinomials, we must first understand the process by which two binomials are multiplied, and then reverse that process. In other words, we have to find two binomials that, when multiplied together, will equal the given trinomial. We begin with the simplest case, which is factoring trinomials in which the coefficient of x^2, called the **leading coefficient,** is 1.

FACTORING TRINOMIALS WITH A LEADING COEFFICIENT OF 1

Recall that when two binomials are multiplied, each term in the first binomial is multiplied by each term in the second. Therefore,

$$(x + a)(x + b) = x^2 + ax + bx + ab = x^2 + (a + b)x + ab$$

This result gives us the following special product:

$$(x + a)(x + b) = x^2 + (a + b)x + ab \tag{6.4}$$

Notice that the **coefficient of x is the *sum* of a and b** and **the last term is the *product* of a and b.** Reversing this process will allow us to factor a trinomial when the coefficient of x^2 is 1 as shown in the diagram below:

sum

coefficient $= 1 \longrightarrow x^2 + (a + b)x + ab = (x + a)(x + b)$

product

Essentially, *we need to find two integers that have a product equal to the last term of a given trinomial and a sum equal to the coefficient of x.* Once we find these two integers, we insert them into the following product to get the factored form: $(x\square)(x\square)$.

EXAMPLE 1 Factoring trinomial $x^2 + (a + b)x + ab$

In factoring $x^2 + 3x + 2$, we set it up as

$$x^2 + 3x + 2 = (x\square)(x\square)$$

The constant 2 tells us that the product of the required integers is 2. Thus, the only possibilities are 2 and 1 (or 1 and 2). The + sign before the 2 indicates that the sign before the 1 and 2 in the factors must be the same. The + sign before the 3, the sum of the integers, tells us that both signs are positive. Therefore,

$$x^2 + 3x + 2 = (x + 2)(x + 1)$$

In factoring $x^2 - 3x + 2$, the analysis is the same until we note that the middle term is negative. This tells us that both signs are negative in this case. Therefore,

$$x^2 - 3x + 2 = (x - 2)(x - 1)$$

For a trinomial with first term x^2 and constant $+2$ to be factorable, the middle term must be $+3x$ or $-3x$. No other middle terms are possible. This means, for example, the expressions $x^2 + 4x + 2$ and $x^2 - x + 2$ cannot be factored. ■

EXAMPLE 2 Factoring trinomials $x^2 + (a + b)x + ab$

(a) In order to factor $x^2 + 7x - 8$, *we must find two integers whose product is* -8 *and whose sum is* $+7$. The possible factors of -8 are

-8 and $+1$ $\quad$ $+8$ and -1 $\quad$ -4 and $+2$ $\quad$ $+4$ and -2

Inspecting these, we see that only $+8$ and -1 have the sum of $+7$. Therefore,

$$x^2 + 7x - 8 = (x + 8)(x - 1)$$

In choosing the correct values for the integers, it is usually fairly easy to find a pair for which the product is the final term. However, choosing the pair of integers that correctly fits the middle term is the step that often is not done properly. *Special attention must be given to choosing the integers so that the expansion of the resulting factors has the correct* ***middle term*** *of the original expression.*

■ Always multiply the factors together to check to see that you get the **correct middle term.**

(b) In the same way, we have

$$x^2 - x - 12 = (x - 4)(x + 3)$$

because -4 and $+3$ is the only pair of integers whose product is -12 and whose sum is -1.

(c) Also,

$$x^2 - 5xy + 6y^2 = (x - 3y)(x - 2y)$$

because -3 and -2 is the only pair of integers whose product is $+6$ and whose sum is -5. Here, we find second terms of each factor with a product of $6y^2$ and sum of $-5xy$, which means that each second term must have a factor of y, as we have shown above. ■

Practice Exercises

Factor: **1.** $x^2 - x - 2$ **2.** $x^2 - 4x - 12$

FACTORING GENERAL TRINOMIALS BY TRIAL AND ERROR

We will now turn our attention to factoring *general trinomials*, where *the coefficient of the squared term can be any integer* (not just 1 as in the previous examples). Let us first review the process by which two general binomials are multiplied:

$$(ax + b)(cx + d) = acx^2 + adx + bcx + bd$$

or

$$(ax + b)(cx + d) = acx^2 + (ad + bc)x + bd \tag{6.5}$$

Rewriting Eq. (6.5) with the sides reversed gives us a strategy for factoring general trinomials by trial and error as shown in the diagram below:

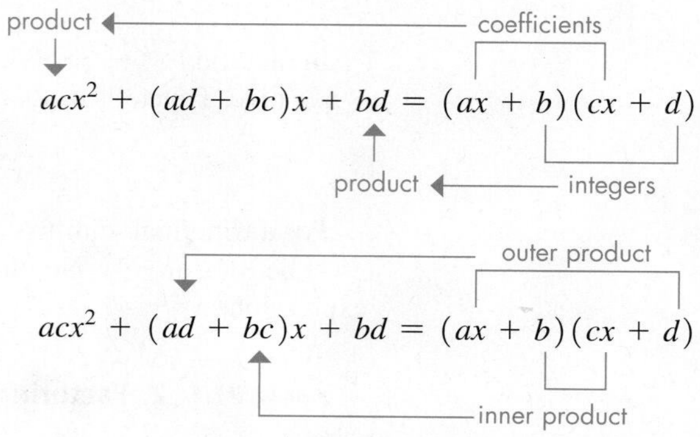

product ← coefficients

$$acx^2 + (ad + bc)x + bd = (ax + b)(cx + d)$$

product ← integers

outer product

$$acx^2 + (ad + bc)x + bd = (ax + b)(cx + d)$$

inner product

This diagram shows us that

1. *the coefficient of x^2 is the product of the coefficients a and c in the factors,*
2. *the final constant is the product of the constants b and d in the factors, and*
3. ***the coefficient of x is the sum of the inner and outer products.***

CAUTION In finding the factors, we must try possible combinations of *a, b, c,* and *d* that give the proper inner and outer products for the middle term of the given expression. This requires some trial and error as shown in the following examples. ■

EXAMPLE 3 Factoring a general trinomial by trial and error

To factor $2x^2 + 11x + 5$, we take the factors of 2 to be +2 and +1 (we use only positive coefficients a and c when the coefficient of x^2 is positive). We set up the factoring as

$$2x^2 + 11x + 5 = (2x \square)(x \square)$$

NOTE ▸ Because the product of the integers to be found is +5, only integers of the same sign need to be considered. [Also because the sum of the outer and inner products is +11, the integers are positive.] The factors of +5 are +1 and +5, and −1 and −5, which means that +1 and +5 is the only possible pair. Now, trying the factors

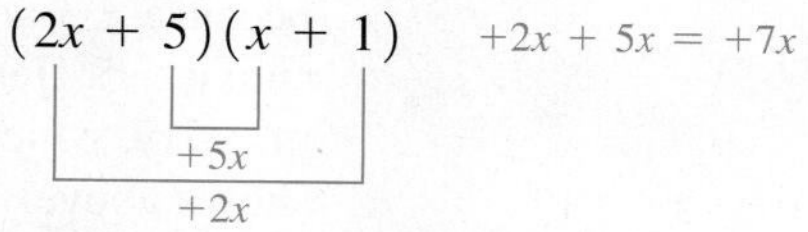

$$(2x + 5)(x + 1) \qquad +2x + 5x = +7x$$

(inner: $+5x$; outer: $+2x$)

we see that $7x$ is not the correct middle term.

■ In Example 3, if we were given $5 + 11x + 2x^2$, we can factor it as $(5 + x)(1 + 2x)$, or rewrite it as $2x^2 + 11x + 5$ before factoring.

Therefore, we now try

$$(2x + 1)(x + 5) \quad x + 10x = +11x$$

(inner product $+x$, outer product $+10x$)

and we have the correct sum of $+11x$. Therefore,

$$2x^2 + 11x + 5 = (2x + 1)(x + 5)$$

For a trinomial with a first term $2x^2$ and a constant $+5$ to be factorable, we can now see that the middle term must be either $\pm 11x$ or $\pm 7x$. This means that $2x^2 + 7x + 5 = (2x + 5)(x + 1)$, but a trinomial such as $2x^2 + 8x + 5$ is not factorable. ■

EXAMPLE 4 Factoring a general trinomial by trial and error

In factoring $4x^2 + 4x - 3$, the coefficient 4 in $4x^2$ shows that the possible coefficients of x in the factors are 4 and 1, or 2 and 2. [The 3 shows that the only possible constants in the factors are 1 and 3, and the minus sign with the 3 tells us that these integers have different signs.] This gives us the following possible combinations of factors, along with the resulting sum of the outer and inner products:

NOTE ➧

$$(4x + 3)(x - 1): -4x + 3x = -x$$
$$(4x + 1)(x - 3): -12x + x = -11x$$
$$(4x - 3)(x + 1): 4x - 3x = +x$$
$$(4x - 1)(x + 3): 12x - x = +11x$$
$$(2x + 3)(2x - 1): -2x + 6x = +4x$$
$$(2x - 3)(2x + 1): 2x - 6x = -4x$$

TI-89 graphing calculator keystrokes for Example 4: goo.gl/h084Fd

We see that the factors that have the correct middle term of $+4x$ are $(2x + 3)(2x - 1)$. This means that

$$4x^2 + 4x - 3 = (2x + 3)(2x - 1)$$

(inner product $+6x$, outer product $-2x$; $-2x + 6x = +4x$)

Expressing the result with the factors reversed is an equally correct answer.

NOTE ➧ [Another hint given by the coefficients of the original expression is that the plus sign with the $4x$ tells us that the larger of the outer and inner products must be positive.] ■

EXAMPLE 5 Factoring a general trinomial—beam deflection

An expression that arises when analyzing the deflection of beams is $9x^2 - 32Lx + 28L^2$. When factoring this expression, we get

$$9x^2 - 32Lx + 28L^2 = (x - 2L)(9x - 14L)$$

NOTE ➧ [There are numerous possible combinations for 9 and 28, but we must carefully check that the middle term is the **sum of the inner and outer products** of the factors we have chosen.]

$$-14Lx - 18Lx = -32Lx \quad \text{checking middle term}$$ ■

Practice Exercise

3. Factor $6s^2 + 19st - 20t^2$ by trial and error.

FACTORING GENERAL TRINOMIALS BY GROUPING

Sometimes when using trial and error, there are many possible combinations of values to try and it can be difficult to find the ones that result in the correct middle term. We now discuss another alternative for factoring general trinomials, which utilizes factoring by grouping. In Section 6.1, we saw that we could sometimes factor an expression containing four terms by grouping the first two terms and the last two terms. This same technique can be applied to trinomials of the form $ax^2 + bx + c$ if we first split up the x-term into two terms using the method described below:

1. Find the product ac.
2. Find two numbers whose product is ac and whose sum is b.
3. Replace the middle term bx with two x-terms having these numbers as coefficients.
4. Complete the factorization by grouping.

This method is sometimes referred to as "***ac* splits *b*.**" The following example illustrates how this is done.

EXAMPLE 6 Factoring a general trinomial by grouping

Factor the trinomial $3x^2 - 10xy + 8y^2$.

We start by finding the product $ac = 3(8) = 24$. Next, we find two numbers that have a product of 24 and a sum of -10. The two numbers that work are -6 and -4. We use these numbers as coefficients to split up the middle term and then factor by grouping.

$$3x^2 - 10xy + 8y^2 = 3x^2 - 6xy - 4xy + 8y^2$$ $-10xy = -6xy - 4xy$

$$= 3x(x - 2y) \underline{\ ?\ } (x - 2y)$$ Factor out the GCF $3x$ from the first two terms. Since the second set of parentheses must match the first, determine the expression to factor out of the second two terms.

$$= 3x(x - 2y) - 4y(x - 2y)$$ The required expression is $-4y$.

$$= (x - 2y)(3x - 4y)$$ Factor out the common binomial factor $(x - 2y)$.

NOTE ▸ [It is important to note that when splitting the middle term, the two terms can be inserted in either order.] For example, replacing $-10xy$ with $-4xy - 6xy$ would lead to an equivalent answer, although the factors may be in a different order. ■

Practice Exercise

4. Factor $3x^2 - 2x - 16$ by grouping.

To summarize, a general trinomial can be factored either by trial and error or by grouping. Either method can be used, depending on one's preference.

FACTORING PERFECT SQUARE TRINOMIALS

Some trinomials factor into two identical factors. For example, $x^2 + 6x + 9 = (x + 3)(x + 3) = (x + 3)^2$. Trinomials that factor in this way are called ***perfect square trinomials.*** If we recognize a perfect square trinomial, then it is much easier to factor since we know the two factors must be the same. The two special products below show the form of a perfect square trinomial.

$$(a + b)^2 = a^2 + 2ab + b^2 \quad \textbf{(6.6)}$$
$$(a - b)^2 = a^2 - 2ab + b^2 \quad \textbf{(6.7)}$$

Notice that the first and last terms of Eqs. (6.6) and (6.7) are both perfect squares. When we encounter a trinomial in which the first and last terms are perfect squares, there is a *possibility* that it is a perfect square trinomial, but we must also check the middle term as demonstrated in the following example.

EXAMPLE 7 Checking for perfect square trinomials

(a) In factoring $x^2 + 12x + 36$, we should notice that the first term is the square of x and the last term is the square of 6. Therefore, this *might* be a perfect square trinomial and factor into two identical factors. It is wise to try this possibility first and see if the inner and outer products combine to equal the middle term:

$$x^2 + 12x + 36 \stackrel{?}{=} (x + 6)(x + 6) \qquad 6x + 6x = 12x$$

Since the middle term checks, we have the correct factorization, which can be written as $(x + 6)^2$.

(b) In factoring $36x^2 - 84xy + 49y^2$, notice that the first term is the square of $6x$ and the last term is the square of $7y$. We will check the middle term to see if this is a perfect square trinomial:

$$36x^2 - 84xy + 49y^2 \stackrel{?}{=} (6x - 7y)(6x - 7y) \qquad -42xy - 42xy = -84xy$$

Since the middle term checks, we have the correct factorization, which is $(6x - 7y)^2$. This expression would have been much more difficult to factor if we had used either the trial-and-error method (without recognizing it as a perfect square trinomial) or the grouping method.

(c) In factoring $4x^2 - 25x + 25$, notice the first term is the square of $2x$ and the last term is the square of 5. To see if this is a perfect square trinomial, we again check the middle term:

$$4x^2 - 25x + 25 \stackrel{?}{=} (2x - 5)(2x - 5) \qquad -10x - 10x = -20x \neq -25x$$

The middle term does *not* check, *so we do not have the correct factorization.* When factored correctly using either trial and error or grouping, we find that $4x^2 - 25x + 25 = (4x - 5)(x - 5)$. ■

Practice Exercise

5. Factor: $9x^2 + 24x + 16$

CAUTION Example 7(c) demonstrates an important point. Just because the first and last terms of a trinomial are perfect squares, we can't *assume* the expression is a perfect square trinomial. We must check the middle term. ■

FACTORING COMPLETELY

As mentioned in the previous section, it is important that we factor expressions *completely*. To do this, *first factor out the greatest common factor* (if possible), and then attempt to further factor the remaining expression.

EXAMPLE 8 Factoring completely

When factoring $2x^2 + 6x - 8$, first note the common factor of 2. This leads to

$$2x^2 + 6x - 8 = 2(x^2 + 3x - 4)$$

Now, notice that $x^2 + 3x - 4$ is also factorable. Therefore,

$$2x^2 + 6x - 8 = 2(x + 4)(x - 1)$$

■

CAUTION If you attempt to factor the previous example without first factoring out the common factor of 2, it becomes much more difficult, and also the result will not be factored completely. There would still be a common factor inside the parentheses, so additional steps would be needed. **It is always easier to factor out the greatest common factor first.** ■

EXAMPLE 9 Factoring completely—rocket flight

A study of the path of a certain rocket leads to the expression $16t^2 + 240t - 1600$, where t is the time of flight. Factor this expression.

An inspection shows that there is a common factor of 16. Factoring out 16 leads to

■ Liquid-fuel rockets were designed in the United States in the 1920s but were developed by German engineers. They were first used in the 1940s during World War II.

$$\begin{aligned} 16t^2 + 240t - 1600 &= 16(t^2 + 15t - 100) \\ &= 16(t + 20)(t - 5) \end{aligned}$$

Note that by first factoring out the common factor of 16, the resulting trinomial $t^2 + 15t - 100$ is *much easier* to factor than the original trinomial $16t^2 + 240t - 1600$. ■

Practice Exercise

Factor: **6.** $6x^2 + 9x - 6$

EXERCISES 6.2

In Exercises 1–4, make the given changes in the indicated examples of this section and then factor.

1. In Example 1, change the 3 to 4 and the 2 to 3.

2. In Example 2(a), change the + before $7x$ to −.

3. In Example 3, change the + before $11x$ to −.

4. In Example 8, change the 8 to 36.

In Exercises 5–8, find each product.

5. $(x - 7)^2$ **6.** $(y + 6)^2$

7. $(a + 3b)^2$ **8.** $(2n + 5m)^2$

In Exercises 9–54, factor the given expressions completely.

9. $x^2 + 4x + 3$ **10.** $x^2 - 5x - 6$

11. $s^2 - s - 42$ **12.** $a^2 + 14a - 32$

13. $t^2 + 5t - 24$ **14.** $r^3 - 11r^2 + 18r$

15. $x^2 + 8x + 16$ **16.** $D^2 + 8D + 16$

17. $a^2 - 6ab + 9b^2$ **18.** $b^2 - 12bc + 36c^2$

19. $3x^2 - 5x - 2$ **20.** $6n^2 - 39n - 21$

21. $12y^2 - 32y - 12$ **22.** $25x^2 + 45x - 10$

23. $2s^2 + 13s + 11$ **24.** $5 - 12y + 7y^2$

25. $3z^2 - 19z + 6$ **26.** $10R^4 - 6R^2 - 4$

27. $2t^2 + 7t - 15$ **28.** $20 - 20n + 3n^2$

29. $3t^2 - 7tu + 4u^2$ **30.** $3x^2 + xy - 14y^2$

31. $6x^2 + x - 5$ **32.** $2z^2 + 13z - 5$

33. $9x^2 + 7xy - 2y^2$ **34.** $4r^2 + 11rs - 3s^2$

35. $12m^2 + 60m + 75$ **36.** $48q^2 + 72q + 27$

37. $8x^2 - 24x + 18$ **38.** $3a^2c^2 - 6ac + 3$

39. $9t^2 - 15t + 4$ **40.** $6t^4 + t^2 - 12$

41. $8b^6 + 31b^3 - 4$ **42.** $12n^4 + 8n^2 - 15$

43. $4p^2 - 25pq + 6q^2$ **44.** $12x^2 + 4xy - 5y^2$

45. $12x^2 + 47xy - 4y^2$ **46.** $8r^2 - 14rs - 9s^2$

47. $12 - 14x + 2x^2$ **48.** $6y^2 - 33y - 18$

49. $4x^5 + 14x^3 - 8x$ **50.** $12B^2 + 22BH - 4H^2$

51. $ax^3 + 4a^2x^2 - 12a^3x$ **52.** $15x^2 - 39x^3 + 18x^4$

53. $4x^{2n} + 13x^n - 12$ **54.** $12B^{2n} + 19B^nH - 10H^2$

In Exercises 55–66, factor the given expressions completely. Each is from the technical area indicated.

55. $16t^2 - 80t + 64$ (projectile motion)

56. $9x^2 - 33Lx + 30L^2$ (civil engineering)

57. $d^4 - 10d^2 + 16$ (magnetic field)

58. $3e^2 + 18e - 1560$ (fuel efficiency)

59. $3Q^2 + Q - 30$ (fire science)

60. $bT^2 - 40bT + 400b$ (thermodynamics)

61. $V^2 - 2nBV + n^2B^2$ (chemistry)

62. $a^4 + 8a^2\pi^2f^2 + 16\pi^4f^4$ (periodic motion: energy)

63. $wx^4 - 5wLx^3 + 6wL^2x^2$ (beam design)

64. $1 - 2r^2 + r^4$ (lasers)

65. $3Adu^2 - 4Aduv + Adv^2$ (water power)

66. $k^2A^2 + 2k\lambda A + \lambda^2 - \alpha^2$ (robotics)

In Exercises 67–70, solve the given problems.

67. Find the two integer values of k that make $4x^2 + kx + 9$ a perfect square trinomial.

68. Find the two integer values of k that make $16y^2 + ky + 25$ a perfect square trinomial.

69. Find six values of k such that $x^2 + kx + 18$ can be factored.

70. Explain why most students would find $24x^2 - 23x - 12$ more difficult to factor than $23x^2 - 18x - 5$.

When an object is thrown upward with an initial velocity of v_0 (in ft/s) from an initial height of s_0 (in ft), its height after t seconds is given by $-16t^2 + v_0t + s_0$. In Exercises 71 and 72, find an expression for the height and write it in factored form.

71. $v_0 = 32$ ft/s, $s_0 = 128$ ft **72.** $v_0 = 24$ ft/s, $s_0 = 40$ ft

Answers to Practice Exercises

1. $(x - 2)(x + 1)$ **2.** $(x - 6)(x + 2)$
3. $(6s - 5t)(s + 4t)$ **4.** $(3x - 8)(x + 2)$
5. $(3x + 4)^2$ **6.** $3(2x - 1)(x + 2)$

6.3 The Sum and Difference of Cubes

Factoring the Sum of Cubes • Factoring the Difference of Cubes • Summary of Methods of Factoring

We have seen that the difference of squares can be factored, but that the sum of squares often cannot be factored. We now turn our attention to the sum and difference of cubes, both of which can be factored. The following two equations show us how either a sum or difference of cubes can be factored into the product of a binomial and a trinomial. (The proof is left as an exercise.)

■ Note carefully the positions of the + and − signs.

$$a^3 + b^3 = (a + b)(a^2 - ab + b^2) \quad (6.8)$$

$$a^3 - b^3 = (a - b)(a^2 + ab + b^2) \quad (6.9)$$

In these equations, the second factors are usually prime.

EXAMPLE 1 Factoring sum and difference of cubes

Table of Cubes
$1^3 = 1$
$2^3 = 8$
$3^3 = 27$
$4^3 = 64$
$5^3 = 125$
$6^3 = 216$

(a) $x^3 + 8 = x^3 + 2^3$

$$= (x + 2)[(x)^2 - 2x + 2^2]$$

$$= (x + 2)(x^2 - 2x + 4)$$

(b) $x^3 - 1 = x^3 - 1^3$

$$= (x - 1)[(x)^2 + (1)(x) + 1^2]$$

$$= (x - 1)(x^2 + x + 1)$$

(c) $8 - 27x^3 = 2^3 - (3x)^3$ $\quad$ $8 = 2^3$ and $27x^3 = (3x)^3$

$$= (2 - 3x)[2^2 + 2(3x) + (3x)^2]$$

$$= (2 - 3x)(4 + 6x + 9x^2)$$ ■

EXAMPLE 2 First, factor out the common factor

In factoring $ax^5 - ax^2$, we first note that each term has a common factor of ax^2. This is factored out to get $ax^2(x^3 - 1)$. However, the expression is not completely factored because $1 = 1^3$, which means that $x^3 - 1$ is the difference of cubes [see Example 1(b)]. We complete the factoring by the use of Eq. (6.9). Therefore,

$$ax^5 - ax^2 = ax^2(x^3 - 1)$$

$$= ax^2(x - 1)(x^2 + x + 1)$$ ■

Practice Exercises

Factor:

1. $x^3 + 216$ **2.** $x^3 - 216$

EXAMPLE 3 Factoring difference of cubes—volume of steel bearing

The volume of material used to make a steel bearing with a hollow core is given by $\frac{4}{3}\pi R^3 - \frac{4}{3}\pi r^3$. Factor this expression.

$$\frac{4}{3}\pi R^3 - \frac{4}{3}\pi r^3 = \frac{4}{3}\pi(R^3 - r^3) \quad \text{common factor of } \frac{4}{3}\pi$$

$$= \frac{4}{3}\pi(R - r)(R^2 + Rr + r^2) \quad \text{using Eq. (6.9)}$$ ■

SUMMARY OF METHODS OF FACTORING

In the first three sections of this chapter, we discussed several methods of factoring. It is important to know which methods to use and the order in which to use them. The diagram below shows a general strategy that can be used to factor most expressions.

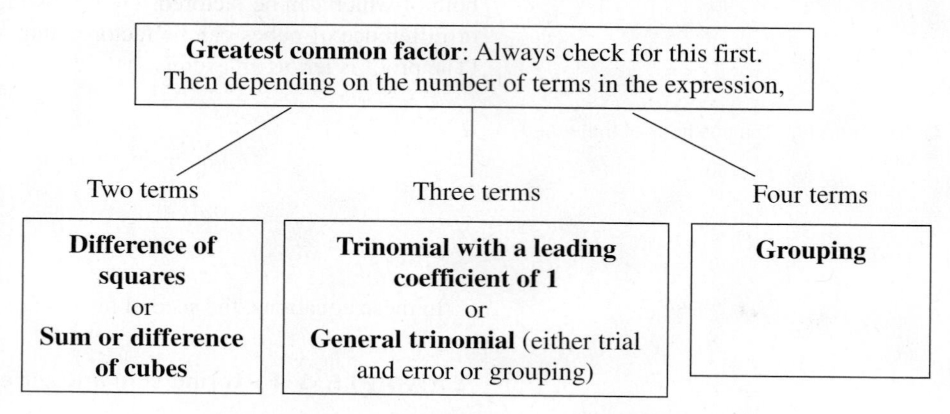

CAUTION Always be sure the expression is factored **completely.** ■

EXERCISES 6.3

In Exercises 1 and 2, make the given changes in the indicated examples of this section and then factor.

1. In Example 1(a), change the + before the 8 to −.
2. In Example 2, change $-ax^2$ to $+a^4x^2$.

In Exercises 3–30, factor the given expressions completely.

3. $x^3 + 1$
4. $R^3 + 27$
5. $y^3 - 125$
6. $z^3 - 8$
7. $27 - t^3$
8. $8r^3 - 1$
9. $8a^3 - 27b^3$
10. $64x^4 + 125x$
11. $4x^3 + 32$
12. $3y^3 - 81$
13. $7n^5 - 7n^2$
14. $64 - 8s^9$
15. $54x^3y - 6x^3y^4$
16. $12a^3 + 96a^3b^3$
17. $x^6y^3 + x^3y^6$
18. $16r^3 - 432$
19. $3a^6 - 3a^2$
20. $81y^2 - x^6$
21. $\frac{4}{3}\pi R^3 - \frac{4}{3}\pi r^3$
22. $0.027x^3 + 0.125$
23. $27L^6 + 216L^3$
24. $a^3s^5 - 8000a^3s^2$
25. $(a + b)^3 + 64$
26. $125 + (2x + y)^3$
27. $64 - x^6$
28. $2a^6 - 54b^6$
29. $x^6 - 2x^3 + 1$
30. $n^6 + 4n^3 + 4$

In Exercises 31–36, factor the given expressions completely. Each is from the technical area indicated.

31. $32x - 4x^4$ (rocket trajectory)
32. $kT^3 - kT_0^3$ (thermodynamics)
33. $D^4 - d^3D$ (machine design)
34. $(h + 2t)^3 - h^3$ (container design)
35. $QH^4 + Q^4H$ (thermodynamics)
36. $\left(\frac{s}{r}\right)^{12} - \left(\frac{s}{r}\right)^6$ (molecular interaction)

In Exercises 37 and 38, perform the indicated operations.

37. Using multiplication, verify the identity $a^3 + b^3 = (a + b)(a^2 - ab + b^2)$.
38. Using multiplication, verify the identity $a^3 - b^3 = (a - b)(a^2 + ab + b^2)$.

In Exercises 39–42, solve the given problems.

39. Factor $x^6 - y^6$ as the difference of cubes.
40. Factor $x^6 - y^6$ as the difference of squares. Then explain how the result of Exercise 39 can be shown to be the same as the result in this exercise.
41. By factoring $(n^3 + 1)$, explain why this expression represents a number that is not prime if n is an integer greater than one.
42. A certain wind turbine produces a power of $1200v^3$ (in watts) where v is the wind speed. Find an expression, in factored form, for the difference in power for the wind speeds v_1 and v_2.

Answers to Practice Exercises

1. $(x + 6)(x^2 - 6x + 36)$ 2. $(x - 6)(x^2 + 6x + 36)$

6.4 Equivalent Fractions

Fundamental Principle of Fractions • Equivalent Fractions • Simplest Form, or Lowest Terms, of a Fraction • Factors That Differ Only in Sign

When we deal with algebraic expressions, we must be able to work effectively with fractions. Because algebraic expressions are representations of numbers, the basic operations on fractions from arithmetic form the basis of our algebraic operations. In this section, we demonstrate a very important property of fractions, and in the following two sections, we establish the basic algebraic operations with fractions.

The following property of fractions, often referred to as the **fundamental principle of fractions,** is very important when working with fractions.

Fundamental Principle of Fractions

The value of a fraction is unchanged if both the numerator and denominator are multiplied or divided by the same nonzero number.

Two fractions are said to be **equivalent** if one can be obtained from the other by use of the fundamental principle.

EXAMPLE 1 Equivalent arithmetic fractions

If we multiply the numerator and the denominator of the fraction $\frac{6}{8}$ by 2, we obtain the equivalent fraction $\frac{12}{16}$. If we divide the numerator and the denominator of $\frac{6}{8}$ by 2, we obtain the equivalent fraction $\frac{3}{4}$. This means that the fractions $\frac{6}{8}$, $\frac{3}{4}$, and $\frac{12}{16}$ are equivalent. Obviously, there is an unlimited number of other fractions that are equivalent to these fractions. ■

EXAMPLE 2 Equivalent algebraic fractions

We may write

$$\frac{ax}{2} = \frac{3a^2x}{6a}$$

because we get the fraction on the right by multiplying the numerator and the denominator of the fraction on the left by $3a$. This means that the fractions are equivalent. ■

SIMPLEST FORM, OR LOWEST TERMS, OF A FRACTION

One of the most important operations with a fraction is that of ***reducing*** it to its **simplest form,** or **lowest terms.**

Simplest Form of a Fraction

A fraction is said to be in its simplest form if the numerator and the denominator have no common factors other than $+1$ or -1.

In reducing a fraction to its simplest form, we use the fundamental principle of fractions and *divide* both the numerator and the denominator by all ***factors*** that are common to each.

NOTE ▸ [It will be assumed throughout this text that if any of the literal symbols were to be evaluated, numerical values would be such that none of the denominators would be equal to zero. Therefore, all operations are done properly since the undefined operation of division by zero is avoided.]

EXAMPLE 3 Reducing fraction to lowest terms

In order to reduce the fraction

$$\frac{16ab^3c^2}{24ab^2c^5}$$

to its lowest terms, note that both the numerator and the denominator contain the factor $8ab^2c^2$. Therefore,

$$\frac{16ab^3c^2}{24ab^2c^5} = \frac{2b(8ab^2c^2)}{3c^3(8ab^2c^2)} = \frac{2b}{3c^3}$$

common factor

Here, we divided out the common factor of $8ab^2c^2$. The resulting fraction is in its lowest terms, because there are no common factors in the numerator and the denominator other than $+1$ or -1. ■

Practice Exercise

1. Reduce to lowest terms: $\frac{9xy^5}{15x^3y^2}$

As shown in the previous example, we reduce a fraction to its simplest form by **dividing both its numerator and denominator by the common factor.** This process, which produces an equivalent fraction, is often referred to as **cancellation.**

CAUTION It is very important that you cancel only **factors** that appear in the numerator and denominator. This means that the numerator and denominator must both be in **factored form** before making any cancellations. It is a common error to cancel expressions without first factoring the numerator and denominator. In particular, **terms** (expressions separated by + or − signs) **cannot be cancelled.** ■

EXAMPLE 4 Cancel factors only

When simplifying the expression

$$\frac{x^2(x-2)}{x^2-4}$$

a term, but not a factor, of the denominator

NOTE ▸ many students would "cancel" the x^2 from the numerator and the denominator. [This is incorrect, because x^2 ***is a term***, not a factor, of the denominator.] In order to simplify the above fraction properly, we should factor the denominator. We then get

$$\frac{x^2(x-2)}{(x-2)(x+2)} = \frac{x^2}{x+2}$$

Here, the common ***factor*** $x - 2$ has been divided out. ■

Practice Exercise

2. Reduce to lowest terms: $\frac{4a}{4a-2x}$

EXAMPLE 5 Distinguishing between term and factor

(a) $\frac{2a}{2ax} = \frac{1}{x}$ — $2a$ is a factor of the numerator and the denominator

We divide out the common factor of $2a$. Note that the resulting numerator is 1.

(b) $\frac{2a}{2a+x}$ — $2a$ is a term, but not a factor, of the denominator

NOTE ▸ [This cannot be reduced, because there are no common *factors* in the numerator and the denominator.] ■

EXAMPLE 6 Remaining factor of 1 in denominator

$$\frac{2x^2 + 8x}{x + 4} = \frac{2x\overset{1}{\cancel{(x + 4)}}}{\underset{1}{\cancel{(x + 4)}}} = \frac{2x}{1}$$

$$= 2x$$

The numerator and the denominator were each divided by $x + 4$ after factoring the numerator. The only remaining factor in the denominator is 1, and it is generally not written in the final result. Another way of writing the denominator is $1(x + 4)$, which shows the factor of 1 more clearly. ■

EXAMPLE 7 Cancel factors only

$$\frac{x^2 - 4x + 4}{x^2 - 4} = \frac{(x - 2)\overset{1}{\cancel{(x - 2)}}}{(x + 2)\underset{1}{\cancel{(x - 2)}}}$$

$$= \frac{x - 2}{x + 2}$$ — x is a term but not a factor

Practice Exercise

3. Reduce to lowest terms: $\frac{x^2 - x - 2}{x^2 + 3x + 2}$

Here, the numerator and the denominator have each been *factored first and then the common factor $x - 2$ has been divided out.* [In the final form, neither the x's nor the 2's may be canceled, because they are terms, not *factors.*] ■

NOTE ➧

EXAMPLE 8 Reducing fraction—mechanical vibration

In the mathematical analysis of the vibrations in a certain mechanical system, the following expression and simplification are used:

$$\frac{8s + 12}{4s^2 + 26s + 30} = \frac{4(2s + 3)}{2(2s^2 + 13s + 15)} = \frac{\overset{2}{\cancel{4}}\overset{1}{\cancel{(2s + 3)}}}{\underset{1}{\cancel{2}}\underset{1}{\cancel{(2s + 3)}}(s + 5)}$$

$$= \frac{2}{s + 5}$$

TI-89 graphing calculator keystrokes for Example 8: goo.gl/PcU1xM

In the third fraction, note that the factors common to both the numerator and the denominator are 2 and $(2s + 3)$. ■

FACTORS THAT DIFFER ONLY IN SIGN

In simplifying fractions, we must be able to identify factors that differ only in ***sign.*** Because $-(y - x) = -y + x = x - y$, we have

$$x - y = -(y - x) \tag{6.10}$$

NOTE ➧ [The factors $x - y$ and $y - x$ differ only in sign.] The following examples illustrate the simplification of fractions where a change of signs is necessary.

EXAMPLE 9 Factors differ only in sign

$$\frac{x^2-1}{1-x} = \frac{(x-1)(x+1)}{-(x-1)} = \frac{x+1}{-1} = -(x+1)$$

NOTE ▶ [In the second fraction, we replaced $1 - x$ with the equivalent expression $-(x - 1)$.] In the third fraction, the common factor $x - 1$ was divided out. Finally, we expressed the result in the more convenient form by dividing $x + 1$ by -1. Replacing $1 - x$ with $-(x - 1)$ is the same as factoring -1 from the terms of $1 - x$.

It should be clear that the factors $1 - x$ and $x - 1$ differ only in sign. However, in simplifying fractions, the fact that they can provide a cancellation often goes unnoticed, and a simplification may be incomplete. ■

Practice Exercise

4. Reduce to lowest terms: $\frac{y-2}{4-y^2}$

EXAMPLE 10 Factors differ only in sign

$$\frac{2x^4-128x}{20+7x-3x^2} = \frac{2x(x^3-64)}{(4-x)(5+3x)} = \frac{2x(x-4)(x^2+4x+16)}{-(x-4)(3x+5)}$$

$$= -\frac{2x(x^2+4x+16)}{3x+5}$$

Again, the factor $4 - x$ has been replaced by the equivalent expression $-(x - 4)$. This allows us to recognize the common factor of $x - 4$.

Also, note that the order of the terms of the factor $5 + 3x$ was changed in writing the third fraction. This was done only to write the terms in the more standard form with the x-term first. However, because both terms are *positive,* it is simply an application of the commutative law of addition, and the factor itself is not actually changed. ■

EXERCISES 6.4

In Exercises 1 and 2, make the given changes in the indicated examples of this section and then solve the resulting problems.

1. In Example 7, change the numerator to $x^2 - 4x - 12$ and then simplify.

2. In Example 10, change the numerator to $2x^4 - 32x^2$ and then simplify.

In Exercises 3–10, multiply the numerator and the denominator of each fraction by the given factor and obtain an equivalent fraction.

3. $\frac{2}{3}$ (by 7)

4. $\frac{7}{5}$ (by 9)

5. $\frac{ax}{y}$ (by $3a$)

6. $\frac{2x^2y}{3n}$ (by $2xn^2$)

7. $\frac{2}{x+3}$ (by $x - 2$)

8. $\frac{7}{a-1}$ (by $a + 2$)

9. $\frac{a}{x-y}$ (by $x + y$)

10. $\frac{B-1}{B+1}$ (by $1 - B$)

In Exercises 11–18, divide the numerator and the denominator of each fraction by the given factor and obtain an equivalent fraction.

11. $\frac{28}{44}$ (by 4)

12. $\frac{25}{65}$ (by 5)

13. $\frac{4x^2y}{8xy^2}$ (by $2x$)

14. $\frac{6a^3b^2}{9a^5b^4}$ (by $3a^2b^2$)

15. $\frac{4(R-2)}{(R-2)(R+2)}$ (by $R - 2$)

16. $\frac{(x+5)(x-3)}{3(x+5)}$ (by $x + 5$)

17. $\frac{s^2-3s-10}{2s^2+3s-2}$ (by $s + 2$)

18. $\frac{6x^2+13x-5}{6x^3-2x^2}$ (by $1 - 3x$)

In Exercises 19–26, replace the A with the proper expression such that the fractions are equivalent.

19. $\frac{3x}{2y} = \frac{A}{6y^2}$

20. $\frac{2R}{R+T} = \frac{2R^2T}{A}$

21. $\frac{6}{a+4} = \frac{6a-24}{A}$

22. $\frac{a+1}{5a^2c} = \frac{A}{5a^3c - 5a^2c}$

23. $\frac{2x^3+2x}{x^4-1} = \frac{A}{x^2-1}$

24. $\frac{n^2-1}{n^3+1} = \frac{A}{n^2-n+1}$

25. $\frac{x^2+3bx-4b^2}{x-b} = \frac{x+4b}{A}$

26. $\frac{4y^2-1}{4y^2+6y-4} = \frac{A}{2y+4}$

In Exercises 27–62, reduce each fraction to simplest form.

27. $\frac{5a}{9a}$

28. $\frac{6x}{15x}$

29. $\frac{18x^2y}{24xy}$

30. $\frac{2a^2xy}{6axyz^2}$

31. $\frac{b+8}{5ab+40a}$

32. $\frac{t-a}{t^2-a^2}$

33. $\frac{4a-4b}{4a-2b}$

34. $\frac{20s-5r}{10r-5s}$

35. $\frac{4x^2+1}{4x^2-1}$

36. $\frac{x^2-y^2}{x^2+y^2}$

37. $\frac{3x^2-6x}{x-2}$

38. $\frac{10T^2+15T}{3+2T}$

39. $\frac{3+2y}{4y^3+6y^2}$

40. $\frac{6-3t}{4t^3-8t^2}$

41. $\frac{x^2-10x+25}{x^2-25}$

42. $\frac{4a^2+12ab+9b^2}{4a^2+6ab}$

43. $\frac{2w^4+5w^2-3}{w^4+11w^2+24}$

44. $\frac{3y^3+7y^2+4y}{4+5y+y^2}$

45. $\frac{5x^2-6x-8}{x^3+x^2-6x}$

46. $\frac{5s^2+8rs-4s^2}{6r^2-17rs+5s^2}$

47. $\frac{N^4-16}{8N-16}$

48. $\frac{3+x(4+x)}{3+x}$

49. $\frac{t+4}{(2t+9)t+4}$

50. $\frac{2A^3+8A^4+8A^5}{4A+2}$

51. $\frac{(x-1)(3+x)}{(3-x)(1-x)}$

52. $\frac{(2x-1)(x+6)}{(x-3)(1-2x)}$

53. $\frac{y^2-x^2}{2x-2y}$

54. $\frac{x^2-y^2-4x+4y}{x^2-y^2+4x-4y}$

55. $\frac{n^3+n^2-n-1}{n^3-n^2-n+1}$

56. $\frac{3a^2-13a-10}{5+4a-a^2}$

57. $\frac{(x+5)(x-2)(x+2)(3-x)}{(2-x)(5-x)(3+x)(2+x)}$

58. $\frac{(2x-3)(3-x)(x-7)(3x+1)}{(3x+2)(3-2x)(x-3)(7+x)}$

59. $\frac{x^3+y^3}{2x+2y}$

60. $\frac{w^3-8}{w^2+2w+4}$

61. $\frac{6x^2+2x}{1+27x^3}$

62. $\frac{24-3a^3}{a^2-4a+4}$

In Exercises 63–66, after finding the simplest form of each fraction, explain why it cannot be simplified more.

63. (a) $\frac{x^2(x+2)}{x^2+4}$ (b) $\frac{x^4+4x^2}{x^4-16}$

64. (a) $\frac{2x+3}{2x+6}$ (b) $\frac{2(x+6)}{2x+6}$

65. (a) $\frac{x^2-x-2}{x^2-x}$ (b) $\frac{x^2-x-2}{x^2+x}$

66. (a) $\frac{x^3-x}{1-x}$ (b) $\frac{2x^2+4x}{2x^2+4}$

In Exercises 67 and 68, answer the given questions.

67. If $3 - x < 0$, is $\frac{x^2-9}{3-x} > 0$?

68. If $x - 4 < 0$, is $\frac{x^2-16}{x^3+64} < 0$?

In Exercises 69–74, reduce each fraction to simplest form. Each is from the indicated area of application.

69. $\frac{v^2-v_0^2}{vt-v_0t}$ (rectilinear motion)

70. $\frac{a^3-b^3}{a^3-ab^2}$ (beam design)

71. $\frac{mu^2-mv^2}{mu-mv}$ (nuclear energy)

72. $\frac{16(t^2-2tt_0+t_0^2)(t-t_0-3)}{3t-3t_0}$ (rocket motion)

73. $\frac{E^2R^2-E^2r^2}{(R^2+2Rr+r^2)^2}$ (electric power)

74. $\frac{r_0^3-r_i^3}{r_0^2-r_i^2}$ (machine design)

Answers to Practice Exercises

1. $\frac{3y^3}{5x^2}$ **2.** $\frac{2a}{2a-x}$ **3.** $\frac{x-2}{x+2}$ **4.** $-\frac{1}{y+2}$

6.5 Multiplication and Division of Fractions

Multiplication of Fractions • Division of Fractions • Using Calculator for Checking Answers

From arithmetic, recall that *the product of two fractions is a fraction whose numerator is the product of the numerators and whose denominator is the product of the denominators of the given fractions.* Also, *the quotient of two fractions is found by inverting the second fraction and then proceeding as in multiplication.* Symbolically, these operations are shown below:

Multiplication and Division of Fractions

$$\frac{a}{b} \times \frac{c}{d} = \frac{ac}{bd}$$

$$\frac{a}{b} \div \frac{c}{d} = \frac{\frac{a}{b}}{\frac{c}{d}} = \frac{a}{b} \times \frac{d}{c} = \frac{ad}{bc}$$

The first three examples illustrate the multiplication of fractions.

EXAMPLE 1 Multiplying basic fractions

(a) $\frac{3}{5} \times \frac{2}{7} = \frac{(3)(2)}{(5)(7)} = \frac{6}{35}$ ← multiply numerators / ← multiply denominators

(b) $\frac{3a}{5b} \times \frac{15b^2}{a} = \frac{(3a)(15b^2)}{(5b)(a)}$

$= \frac{45ab^2}{5ab} = \frac{9b}{1}$

$= 9b$

(c) $(6x)\left(\frac{2y}{3x^2}\right) = \left(\frac{6x}{1}\right)\left(\frac{2y}{3x^2}\right)$

$= \frac{12xy}{3x^2} = \frac{4y}{x}$

In (b), we divided out the common factor $5ab$ to reduce the resulting fraction to its lowest terms. In (c), note that we treated $6x$ as $\frac{6x}{1}$. ■

When multiplying fractions, we usually want to express the final result in simplest form. This means we will have to express the numerator and denominator of the final answer in factored form and cancel any common factors. NOTE ▸ [For this reason, when multiplying fractions, it is best to only *indicate* the multiplications between the numerators and denominators rather than actually perform them.] By doing this, the numerator and denominator of the answer will already be partially factored. We can then factor them completely and cancel any common factors to simplify the answers. If we actually perform the multiplications, the answer will be very difficult to simplify. The following example illustrates this point.

EXAMPLE 2 First indicate the multiplications

In performing the multiplication

$$\frac{3(x-y)}{(x-y)^2} \times \frac{(x^2-y^2)}{6x+9y}$$

if we multiplied out the numerators and the denominators before performing any factoring, we would have to simplify the fraction

$$\frac{3x^3 - 3x^2y - 3xy^2 + 3y^3}{6x^3 - 3x^2y - 12xy^2 + 9y^3}$$

It is possible to factor the resulting numerator and denominator, but it would be very difficult. However, as stated previously, we should ***first indicate the multiplications*** and not actually perform them, and ***the solution is much easier.*** Then we factor the resulting numerator and denominator, divide out (cancel) any common factors, and the solution is then complete. Doing this, we have

$$\frac{3(x-y)}{(x-y)^2} \times \frac{(x^2-y^2)}{6x+9y} = \frac{3(x-y)(x^2-y^2)}{(x-y)^2(6x+9y)} \quad \text{indicate multiplication}$$

$$= \frac{3(x-y)(x+y)(x-y)}{(x-y)^2(3)(2x+3y)} \quad \text{factor completely}$$

$$= \frac{\cancel{3}\cancel{(x-y)^2}(x+y)}{\cancel{3}\cancel{(x-y)^2}(2x+3y)} \quad \text{cancel common factors}$$

$$= \frac{x+y}{2x+3y}$$

The common factor $3(x-y)^2$ is readily recognized using this procedure. ■

Practice Exercise

1. Multiply: $\frac{2x}{4x+8} \times \frac{x+2}{xy-x}$

■ It is possible to factor and indicate the product of the factors, showing only a single step, as we have done here.

EXAMPLE 3 Multiplying algebraic fractions

$$\left(\frac{2x-4}{4x+12}\right)\left(\frac{2x^2+x-15}{3x-1}\right) = \frac{\overset{}{\cancel{2}}(x-2)(2x-5)\cancel{(x+3)}}{\underset{2}{\cancel{4}}\cancel{(x+3)}(3x-1)} \quad \text{multiplications indicated}$$

$$= \frac{(x-2)(2x-5)}{2(3x-1)}$$

TI-89 graphing calculator keystrokes for Example 3: goo.gl/hEjz0z

Here, the common factor is $2(x+3)$. It is permissible to multiply out the final form of the numerator and the denominator, but it is often preferable to leave the numerator and the denominator in factored form, as indicated. ■

The following examples illustrate the division of fractions.

EXAMPLE 4 Dividing basic fractions

(a) $\frac{6x}{7} \div \frac{5}{3} = \frac{6x}{7} \times \frac{3}{5} = \frac{18x}{35}$ (multiply; invert)

(b) $\dfrac{\frac{3a^2}{5c}}{\frac{2c^2}{a}} = \frac{3a^2}{5c} \times \frac{a}{2c^2} = \frac{3a^3}{10c^3}$ (multiply; invert) ■

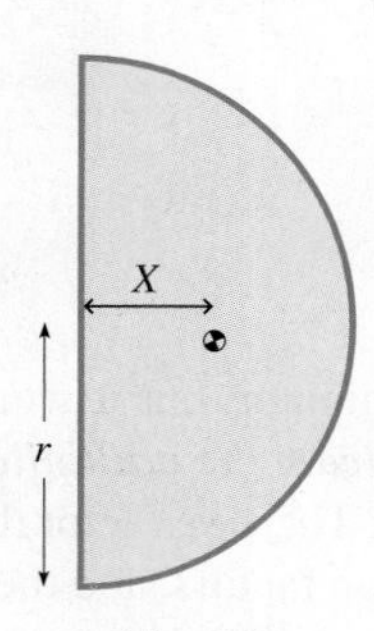

Fig. 6.1

EXAMPLE 5 Division by a fraction—center of gravity

When finding the center of gravity (shown as ⊕ in Fig. 6.1) of a uniform flat semicircular metal plate, the equation $X = \frac{4\pi r^3}{3} \div \left(\frac{\pi r^2}{2} \times 2\pi\right)$ is derived. Simplify the right side of this equation to find X as a function of r in simplest form.

The parentheses indicate that we should perform the multiplication first:

$$X = \frac{4\pi r^3}{3} \div \left(\frac{\pi r^2}{2} \times 2\pi\right) = \frac{4\pi r^3}{3} \div \left(\frac{2\pi^2 r^2}{2}\right) \qquad 2\pi = \frac{2\pi}{1}$$

$$= \frac{4\pi r^3}{3} \div \frac{\pi^2 r^2}{1} = \frac{4\pi r^3}{3} \times \frac{1}{\pi^2 r^2}$$

$$= \frac{4\pi r^3}{3\pi^2 r^2} = \frac{4r}{3\pi} \qquad \text{divide out the common factor of } \pi r^2$$

This is the exact solution. Approximately, $X = 0.424r$. ■

EXAMPLE 6 Dividing algebraic fractions

$$(x + y) \div \frac{2x + 2y}{6x + 15y} = \frac{x + y}{1} \times \frac{6x + 15y}{2x + 2y} = \frac{\cancel{(x + y)}(3)(2x + 5y)}{2\cancel{(x + y)}} \qquad \text{indicate multiplication}$$

invert

$$= \frac{3(2x + 5y)}{2} \qquad \text{simplify}$$ ■

Practice Exercise

2. Divide: $\frac{3x}{a + 1} \div \frac{x^2 + 2x}{a^2 + a}$

EXAMPLE 7 Dividing algebraic fractions

invert

$$\frac{\frac{4 - x^2}{x^2 - 3x + 2}}{\frac{x + 2}{x^2 - 9}} = \frac{4 - x^2}{x^2 - 3x + 2} \times \frac{x^2 - 9}{x + 2} = \frac{(2 - x)(2 + x)(x - 3)(x + 3)}{(x - 2)(x - 1)(x + 2)} \qquad \text{factor and indicate multiplications}$$

$$= \frac{-\cancel{(x - 2)}\cancel{(x + 2)}(x - 3)(x + 3)}{\cancel{(x - 2)}(x - 1)\cancel{(x + 2)}} \qquad \text{replace } (2 - x) \text{ with } -(x - 2) \text{ and } (2 + x) \text{ with } (x + 2)$$

$$= -\frac{(x - 3)(x + 3)}{x - 1} \quad \text{or} \quad \frac{(x - 3)(x + 3)}{1 - x} \qquad \text{simplify}$$

Note the use of Eq. (6.10) when the factor $(2 - x)$ was replaced by $-(x - 2)$ to get the first form of the answer. As shown, it can also be used to get the alternate form of the answer, although it is not necessary to give this form. ■

USE OF A CALCULATOR FOR CHECKING ANSWERS GRAPHICALLY

A graphing calculator can be used to check that two algebraic expressions are equivalent by showing that their graphs are the same. To do this, we enter the two expressions as y_1 and y_2 in a calculator and then graph them using appropriate window settings. If the graphs overlap each other, then the expressions are equivalent (at least throughout their common domains). The table feature on a calculator can also confirm that the two functions pass through the same points. This is shown in the next example.

EXAMPLE 8 Using a calculator to check answer

To graphically check the result of Example 7, we first let y_1 equal the original expression and y_2 equal the final result. In this case,

$$y_1 = [(4 - x^2)/(x^2 - 3x + 2)]/[(x + 2)/(x^2 - 9)]$$
$$y_2 = -(x - 3)(x + 3)/(x - 1)$$

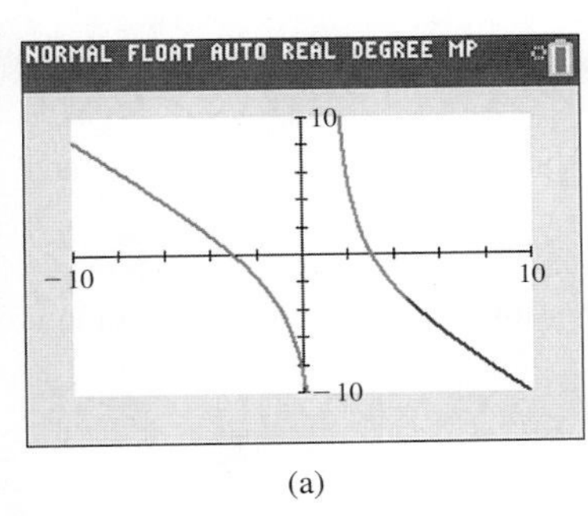

(a)

NORMAL FLOAT AUTO REAL DEGREE MP
PRESS + FOR ΔTbl

X	Y1	Y2
0	-9	-9
1	ERROR	ERROR
2	ERROR	5
3	ERROR	0
4	-2.333	-2.333
5	-4	-4
6	-5.4	-5.4
7	-6.667	-6.667
8	-7.857	-7.857
9	-9	-9
10	-10.11	-10.11

X=0

(b)

Fig. 6.2
Graphing calculator keystrokes: goo.gl/xnU441

Figure 6.2(a) shows y_2 (in red) being graphed right over y_1 (in blue). Since the graphs are the same, we can consider this as a *check* (but not as a *proof*) that the expressions are equivalent.

The table in Fig. 6.2(b) also shows the expressions are equivalent throughout their common domains since the y-values, when they exist, are the same for y_1 and y_2. The x-values that cause division by zero in either function have *undefined* y-values and are shown as "ERROR" in the table. It is important to note that although the two functions agree with each other, their domains are different. ■

EXERCISES 6.5

In Exercises 1 and 2, make the given changes in the indicated examples of this section and then solve the resulting problems.

1. In Example 3, change the first denominator to $4x - 10$ and do the multiplication.
2. In Example 7, change the numerator $x + 2$ of the divisor to $x + 3$ and then simplify.

In Exercises 3–38, simplify the given expressions involving the indicated multiplications and divisions.

3. $\frac{3}{10} \times \frac{2}{9}$
4. $11 \times \frac{13}{33}$
5. $\frac{4x}{3y} \times \frac{9y^2}{2}$
6. $\frac{18sy^3}{ax^2} \times \frac{(ax)^2}{3s}$
7. $\frac{2}{9} \div \frac{4}{7}$
8. $\frac{5}{16} \div \frac{25}{-13}$
9. $\frac{yz}{ab} \div \frac{bz}{ay}$
10. $\frac{sr^2}{2t} \div \frac{st}{4}$
11. $\frac{4x + 16}{5y} \times \frac{y^2}{2x + 8}$
12. $\frac{2y^2 + 6y}{6z} \times \frac{z^3}{y^2 - 9}$
13. $\frac{u^2 - v^2}{u + 2v}(3u + 6v)$
14. $(x - y)\frac{x + 2y}{x^2 - y^2}$
15. $\frac{2a + 8}{15} \div \frac{16 + 8a + a^2}{125}$
16. $\frac{a^2 - a}{3a + 9} \div \frac{a^2 - 2a + 1}{18 - 2a^2}$
17. $\frac{x^4 - 9}{x^3} \div \frac{x^2 + 3}{x}$
18. $\frac{9B^2 - 16}{B + 1} \div (4 - 3B)$
19. $\frac{3ax^2 - 9ax}{10x^2 + 5x} \times \frac{2x^2 + x}{a^2x - 3a^2}$
20. $\frac{4R^2 - 36}{R^3 - 25R} \times \frac{7R - 35}{3R^2 + 9R}$
21. $\left(\frac{x^4 - 1}{8x + 16}\right)\left(\frac{2x^2 - 8x}{x^3 + x}\right)$
22. $\left(\frac{2x^2 - 4x - 6}{3x - x^2}\right)\left(\frac{4x^2 - x^3}{4x^2 - 4x - 8}\right)$
23. $\dfrac{\frac{x^2 + ax}{2b - cx}}{\frac{a^2 + 2ax + x^2}{2bx - cx^2}}$
24. $\dfrac{\frac{x^4 - 11x^2 + 28}{2x^2 + 6}}{\frac{4 - x^2}{2x^2 + 3}}$
25. $\frac{35a + 25}{12a + 33} \div \frac{28a + 20}{36a + 99}$
26. $\frac{2a^3 + a^2}{2b^3 + b^2} \div \frac{2ab + a}{2ab + b}$
27. $\frac{x^2 - 6x + 5}{4x^2 - 5x - 6} \times \frac{12x + 9}{x^2 - 1}$
28. $\frac{n^2 + 5n}{3n^2 + 8n + 4} \times \frac{2n^2 - 8}{n^3 + 3n^2 - 10n}$
29. $\dfrac{\frac{6T^2 - NT - N^2}{2V^2 - 9V - 35}}{\frac{8T^2 - 2NT - N^2}{20V^2 + 26V - 60}}$
30. $\dfrac{\frac{4L^3 - 9L}{8L^2 + 10L - 3}}{3L^2 - 2L^3}$
31. $\frac{7x^2}{3a} \div \left(\frac{a}{x} \times \frac{a^2x}{x^2}\right)$
32. $\left(\frac{3u}{8v^2} \div \frac{9u^2}{2w^2}\right) \times \frac{2u^4}{15vw}$
33. $\left(\frac{4t^2 - 1}{t - 5} \div \frac{2t + 1}{2t}\right) \times \frac{2t^2 - 50}{1 + 4t + 4t^2}$
34. $\frac{2x^2 - 5x - 3}{x - 4} \div \left(\frac{x - 3}{x^2 - 16} \times \frac{1}{3 - x}\right)$

35. $\dfrac{x^3 - y^3}{2x^2 - 2y^2} \times \dfrac{y^2 + 2xy + x^2}{x^2 + xy + y^2}$

36. $\dfrac{2M^2 + 4M + 2}{6M - 6} \div \dfrac{5M + 5}{M^2 - 1}$

37. $\left(\dfrac{ax + bx + ay + by}{p - q}\right)\left(\dfrac{3p^2 + 4pq - 7q^2}{a + b}\right)$

38. $\dfrac{x^4 + x^5 - 1 - x}{x - 1} \div \dfrac{x - 1}{x}$

In Exercises 39–42, simplify the given expressions and then check your answers with a calculator as in Example 8.

39. $\dfrac{x}{2x + 4} \times \dfrac{x^2 - 4}{3x^2}$ **40.** $\dfrac{4t^2 - 25}{4t^2} \div \dfrac{10 + 4t}{8}$

41. $\dfrac{2x^2 + 3x - 2}{2 + 3x - 2x^2} \div \dfrac{5x + 10}{4x + 2}$

42. $\dfrac{16x^2 - 8x + 1}{9x} \times \dfrac{12x + 3}{1 - 16x^2}$

In Exercises 43–46, simplify the given expressions. The technical application of each is indicated.

43. $\dfrac{2\pi}{\lambda}\left(\dfrac{a + b}{2ab}\right)\left(\dfrac{ab\lambda}{2a + 2b}\right)$ (optics)

44. $\left(\dfrac{8\pi n^2 eu}{mv^2 - mvu^2}\right)\left(\dfrac{mv^2}{2mv^2 - 2\pi ne^2}\right)$ (electromagnetism)

45. $\dfrac{c\lambda^2 - c\lambda_0^2}{\lambda_0^2} \div \dfrac{\lambda^2 + \lambda_0^2}{\lambda_0^2}$ (cosmology)

46. $(p_1 - p_2) \div \left(\dfrac{\pi a^4 p_1 - \pi a^4 p_2}{81u}\right)$ (hydrodynamics)

47. The speed v of a satellite can be found from the equation $v^2 = \dfrac{GmM}{r^2} \div \dfrac{m}{r}$. Simplify the right side of the equation and then solve for v.

48. Simplify the following expression involving units (the number 2 is exact):

$$\frac{\left(28\frac{m}{s}\right)^2}{2(0.90)\left(9.8\frac{m}{s^2}\right)}$$

Answers to Practice Exercises

1. $\dfrac{1}{2(y - 1)}$ **2.** $\dfrac{3a}{x + 2}$

6.6 Addition and Subtraction of Fractions

Lowest Common Denominator • Addition and Subtraction of Fractions • Complex Fractions

From arithmetic, recall that *the sum (or difference) of a set of fractions that all have the same denominator is the sum (or difference) of the numerators divided by the common denominator.*

Addition and Subtraction of Fractions

$$\frac{a}{c} + \frac{b}{c} = \frac{a + b}{c}$$

$$\frac{a}{c} - \frac{b}{c} = \frac{a - b}{c}$$

Because algebraic expressions represent numbers, this fact is also true in algebra. Addition and subtraction of such fractions are illustrated in the following example.

EXAMPLE 1 Combining basic fractions

(a) $\dfrac{5}{9} + \dfrac{2}{9} - \dfrac{4}{9} = \dfrac{5 + 2 - 4}{9}$ ← sum of numerators; ← same denominators

$= \dfrac{3}{9} = \dfrac{1}{3}$ ← final result in lowest terms

use parentheses to show subtraction of both terms

(b) $\dfrac{b}{ax} + \dfrac{1}{ax} - \dfrac{2b - 1}{ax} = \dfrac{b + 1 - (2b - 1)}{ax} = \dfrac{b + 1 - 2b + 1}{ax}$

$= \dfrac{2 - b}{ax}$ ■

CAUTION When subtracting a fraction with more than one term in its numerator, be sure to distribute the minus sign as was done in Example 1(b). ■

LOWEST COMMON DENOMINATOR

If the fractions to be combined do not all have the same denominator, we must first change each to an equivalent fraction so that the resulting fractions do have the same denominator. NOTE ▸ [Normally, the denominator that is most convenient and useful is the **lowest common denominator** (**LCD**). This is the product of all the prime factors that appear in the denominators, with each factor raised to the *highest power* to which it appears *in any one* of the denominators.] This means that the lowest common denominator is the *simplest* algebraic expression into which all given denominators will divide exactly. Following is the procedure for finding the lowest common denominator of a set of fractions.

Procedure for Finding the Lowest Common Denominator

1. Factor each denominator into its prime factors.
2. For each different prime factor that appears, note the highest power to which it is raised in any one of the denominators.
3. Form the product of all the different prime factors, each raised to the power found in step 2. This product is the lowest common denominator.

The two examples that follow illustrate the method of finding the LCD.

EXAMPLE 2 Lowest common denominator (LCD)

Find the LCD of the fractions

$$\frac{3}{4a^2b} \qquad \frac{5}{6ab^3} \qquad \frac{1}{4ab^2}$$

We first express each denominator in terms of powers of its prime factors:

highest powers — already seen to be highest power of 2

$$4a^2b = 2^2a^2b \qquad 6ab^3 = 2 \times 3 \times ab^3 \qquad 4ab^2 = 2^2ab^2$$

The prime factors to be considered are 2, 3, a, and b. The largest exponent of 2 that appears is 2. Therefore, 2^2 is a factor of the LCD. NOTE ▸ [What matters is that the highest power of 2 that appears is 2, not the fact that 2 appears in all three denominators with a total of five factors.] The largest exponent of 3 that appears is 1 (understood in the second denominator). Therefore, 3 is a factor of the LCD. The largest exponent of a that appears is 2, and the largest exponent of b that appears is 3. Thus, a^2 and b^3 are factors of the LCD. Therefore, the LCD of the fractions is

$$2^2 \times 3 \times a^2b^3 = 12a^2b^3$$

This is the simplest expression into which ***each*** of the denominators above will divide exactly. ■

NOTE ▸ Finding the LCD of a set of fractions can be a source of difficulty. [Remember, it is necessary to find all of the prime factors that are present in any denominator and then find the highest power to which each is raised in any denominator. The product of these prime factors, each raised to its highest power, is the LCD.]

The following example illustrates this procedure.

EXAMPLE 3 Lowest common denominator

Find the LCD of the following fractions:

$$\frac{x-4}{x^2-2x+1} \qquad \frac{1}{x^2-1} \qquad \frac{x+3}{x^2-x}$$

Factoring each of the denominators, we find that the fractions are

$$\frac{x-4}{(x-1)^2} \qquad \frac{1}{(x-1)(x+1)} \qquad \frac{x+3}{x(x-1)}$$

The factor $(x - 1)$ appears in all the denominators. It is squared in the first fraction and appears only to the first power in the other two fractions. Thus, we must have $(x - 1)^2$ as a factor in the LCD. We do not need a higher power of $(x - 1)$ because, as far as this factor is concerned, each denominator will divide into it evenly. Next, the second denominator has a factor of $(x + 1)$. Therefore, the LCD must also have a factor of $(x + 1)$; otherwise, the second denominator would not divide into it exactly. Finally, the third denominator shows that a factor of x is also needed. The LCD is therefore $x(x + 1)(x - 1)^2$. All three denominators will divide exactly into this expression, and there is no simpler expression for which this is true. ■

Practice Exercise

1. Find the LCD of the following fractions:

$$\frac{x}{2x^2+2x}, \frac{1}{x^2+2x+1}, \frac{2x+4}{4x^2}$$

ADDITION AND SUBTRACTION OF FRACTIONS

Once we have found the LCD for the fractions, we multiply the numerator and the denominator of each fraction by the proper quantity to make the resulting denominator in each case the lowest common denominator. After this step, it is necessary only to add the numerators, place this result over the common denominator, and simplify.

EXAMPLE 4 Finding LCD—combining fractions

Combine: $\dfrac{2}{3r^2} + \dfrac{4}{rs^3} - \dfrac{5}{3s}$.

By looking at the denominators, notice that the factors necessary in the LCD are 3, r, and s. The 3 appears only to the first power, the largest exponent of r is 2, and the largest exponent of s is 3. Therefore, the LCD is $3r^2s^3$. Now, write each fraction with this quantity as the denominator. NOTE ▸ [Since the denominator of the first fraction already contains factors of 3 and r^2, it is necessary to introduce the factor of s^3. In other words, we must multiply the numerator and the denominator of this fraction by s^3.] For similar reasons, we must multiply the numerators and the denominators of the second and third fractions by $3r$ and r^2s^2, respectively. This leads to

$$\frac{2}{3r^2} + \frac{4}{rs^3} - \frac{5}{3s} = \frac{2(s^3)}{(3r^2)(s^3)} + \frac{4(3r)}{(rs^3)(3r)} - \frac{5(r^2s^2)}{(3s)(r^2s^2)}$$ change to equivalent fractions with LCD

factors needed in each

$$= \frac{2s^3}{3r^2s^3} + \frac{12r}{3r^2s^3} - \frac{5r^2s^2}{3r^2s^3}$$

$$= \frac{2s^3 + 12r - 5r^2s^2}{3r^2s^3}$$ combine numerators over LCD ■

Practice Exercise

2. Combine: $\dfrac{5}{2ab} - \dfrac{3}{4a^2}$

EXAMPLE 5 Find LCD—adding fractions

$$\frac{a}{x-1}+\frac{a}{x+1}=\frac{a(x+1)}{(x-1)(x+1)}+\frac{a(x-1)}{(x+1)(x-1)}$$ change to equivalent fraction with LCD

factors needed

$$=\frac{ax+a+ax-a}{(x+1)(x-1)}$$ combine numerators over LCD

$$=\frac{2ax}{(x+1)(x-1)}$$ simplify

When we multiply each fraction by the quantity required to obtain the proper denominator, we do not actually have to write the common denominator under each numerator. Placing all the products that appear in the numerators over the common denominator is sufficient. Hence, the illustration in this example would appear as

$$\frac{a}{x-1}+\frac{a}{x+1}=\frac{a(x+1)+a(x-1)}{(x-1)(x+1)}=\frac{ax+a+ax-a}{(x-1)(x+1)}$$

$$=\frac{2ax}{(x-1)(x+1)}$$ ■

Practice Exercise

3. Combine: $\frac{5}{x-3}-\frac{2}{x+2}$

EXAMPLE 6 Combining fractions

$$\frac{3x}{2x^2-2x-24}-\frac{x-1}{x^2-8x+16}=\frac{3x}{2(x-4)(x+3)}-\frac{x-1}{(x-4)^2}$$ factor denominators

$$=\frac{3x(x-4)-(x-1)(2)(x+3)}{2(x-4)^2(x+3)}$$ change to equivalent fractions with LCD

$$=\frac{(3x^2-12x)-(2x^2+4x-6)}{2(x-4)^2(x+3)}$$ expand in numerator

$$=\frac{3x^2-12x-2x^2-4x+6}{2(x-4)^2(x+3)}=\frac{x^2-16x+6}{2(x-4)^2(x+3)}$$ simplify ■

CAUTION In doing a problem like the one in Example 6, many errors may arise in the use of the minus sign. Remember, if a minus sign precedes an expression inside parentheses, the signs of *all* terms must be changed before they can be combined with other terms. ■

EXAMPLE 7 Combining fractions—missile firing

The following expression is found in the analysis of the dynamics of missile firing. The indicated addition is performed as shown.

$$\frac{1}{s}-\frac{1}{s+4}+\frac{8}{s^2+8s+16}=\frac{1}{s}-\frac{1}{s+4}+\frac{8}{(s+4)^2}$$ factor third denominator

$$=\frac{1(s+4)^2-1(s)(s+4)+8s}{s(s+4)^2}$$ LCD has one factor of s and two factors of $(s+4)$

$$=\frac{s^2+8s+16-s^2-4s+8s}{s(s+4)^2}$$ expand terms of the numerator

$$=\frac{12s+16}{s(s+4)^2}=\frac{4(3s+4)}{s(s+4)^2}$$ simplify/factor

We factored the numerator in the final result to see whether or not there were any factors common to the numerator and the denominator. Because there are none, either form of the result is acceptable. ■

Practice Exercise

4. Combine:

$\frac{3}{2x+2}-\frac{2}{x^2+2x+1}$

COMPLEX FRACTIONS

A **complex fraction** *is one in which the numerator, the denominator, or both the numerator and the denominator contain fractions.* The following examples illustrate the simplification of complex fractions.

EXAMPLE 8 Complex fraction

■ The original complex fraction can be written as a division as follows:

$$\frac{2}{x} \div \left(1 - \frac{4}{x}\right)$$

$$\frac{\frac{2}{x}}{1 - \frac{4}{x}} = \frac{\frac{2}{x}}{\frac{x-4}{x}} \quad \text{first, perform subtraction in denominator}$$

$$= \frac{2}{x} \times \frac{x}{x-4} = \frac{2\cancel{x}}{\cancel{x}(x-4)} \quad \text{invert divisor and multiply}$$

$$= \frac{2}{x-4} \quad \text{simplify}$$

■

EXAMPLE 9 Complex fraction—calculator check

TI-89 graphing calculator keystrokes for Example 9: goo.gl/vyDyAB

$$\frac{1 - \frac{2}{x}}{\frac{1}{x} + \frac{2}{x^2 + 4x}} = \frac{1 - \frac{2}{x}}{\frac{1}{x} + \frac{2}{x(x+4)}} = \frac{\frac{x-2}{x}}{\frac{x+4+2}{x(x+4)}} \quad \text{perform subtraction and addition}$$

$$= \frac{x-2}{x} \times \frac{x(x+4)}{x+6} \quad \text{invert divisor and multiply}$$

$$= \frac{(x-2)(\cancel{x})(x+4)}{\cancel{x}(x+6)} \quad \text{indicate multiplication}$$

$$= \frac{(x-2)(x+4)}{x+6} \quad \text{simplify}$$

With $y_1 = (1 - 2/x)/[1/x + 2/(x^2 + 4x)]$ and $y_2 = (x-2)(x+4)/(x+6)$, a calculator check shows that both graphs are the same. In Fig. 6.3, y_2 (the red curve) is being plotted right over y_1 (the blue curve). ■

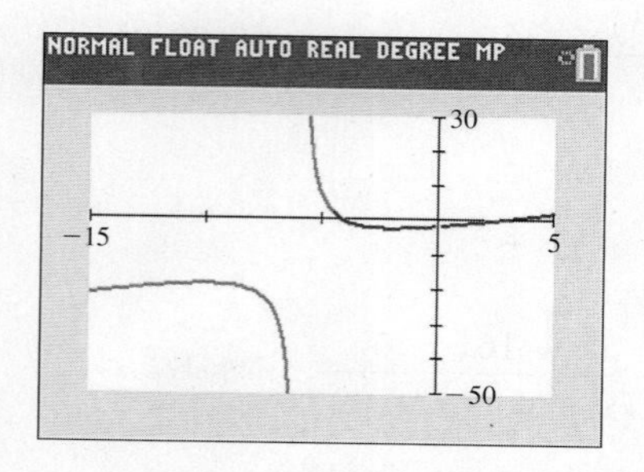

Fig. 6.3

EXERCISES 6.6

In Exercises 1–4, make the given changes in the indicated examples of this section and then solve the resulting problems.

1. In Example 2, change the denominator of the third fraction to $4a^2b^2$ and then find the LCD.
2. In Example 5, add the fraction $\frac{2}{x^2 - 1}$ to those being added and then find the result.
3. In Example 7, change the denominator of the third fraction to $s^2 + 4s$ and then find the result.
4. In Example 8, change the fraction in the numerator to $2/x^2$ and then simplify.

In Exercises 5–46, perform the indicated operations and simplify. For Exercises 35, 36, 41, and 42, check the solution with a graphing calculator.

5. $\frac{5}{6} + \frac{7}{6}$
6. $\frac{2}{13} + \frac{6}{13}$
7. $\frac{1}{x} + \frac{7}{x}$
8. $\frac{2}{a} + \frac{3}{a}$
9. $\frac{1}{2} + \frac{3}{4}$
10. $\frac{5}{9} - \frac{1}{3}$
11. $\frac{4}{3x} + \frac{x}{3} + \frac{1}{x}$
12. $\frac{a+3}{a} - \frac{a}{2}$
13. $\frac{a}{x} - \frac{b}{x^2}$
14. $\frac{3}{2s^2} + \frac{5}{4s}$
15. $\frac{6}{5x^3} + \frac{a}{25x}$
16. $\frac{a}{6y} - \frac{2b}{3y^4}$

17. $\frac{2}{5a} + \frac{1}{a} - \frac{a}{10}$

18. $\frac{1}{2A} - \frac{6}{B} - \frac{9}{4C}$

19. $\frac{x + 1}{2x} - \frac{y - 3}{4y}$

20. $\frac{1 - x}{6y} - \frac{3 + x}{4y}$

21. $\frac{y^2}{y + 3} - \frac{2y + 15}{y + 3}$

22. $\frac{t^2 + 4}{t - 4} - \frac{5t}{t - 4}$

23. $\frac{x}{2x - 2} + \frac{4}{3x - 3}$

24. $\frac{5}{6y + 3} - \frac{a}{4 + 8y}$

25. $\frac{4}{x(x + 1)} - \frac{3}{2x}$

26. $\frac{3}{ax + ay} - \frac{1}{a^2}$

27. $\frac{s}{2s - 6} + \frac{1}{4} - \frac{3s}{4s - 12}$

28. $\frac{2}{x + 2} - \frac{3 - x}{x^2 + 2x} + \frac{1}{x}$

29. $\frac{3R}{R^2 - 9} - \frac{2}{3R + 9}$

30. $\frac{2}{n^2 + 4n + 4} - \frac{3}{4 + 2n}$

31. $\frac{3}{x^2 - 8x + 16} - \frac{2}{4 - x}$

32. $\frac{2a - b}{c - 3d} - \frac{b - 2a}{3d - c}$

33. $\frac{v + 4}{v^2 + 5v + 4} - \frac{v - 2}{v^2 - 5v + 6}$

34. $\frac{N - 1}{2N^3 - 4N^2} - \frac{5}{2 - N}$

35. $\frac{x - 1}{3x^2 - 13x + 4} - \frac{3x + 1}{4 - x}$

36. $\frac{x}{4x^2 - 12x + 5} + \frac{2x - 1}{4x^2 - 4x - 15}$

37. $\frac{t}{t^2 - t - 6} - \frac{2t}{t^2 + 6t + 9} + \frac{t}{9 - t^2}$

38. $\frac{5}{2x^3 - 3x^2 + x} - \frac{x}{x^4 - x^2} + \frac{2 - x}{2x^2 + x - 1}$

39. $\frac{1}{w^3 + 1} + \frac{1}{w + 1} - 2$

40. $\frac{2}{8 - x^3} + \frac{1}{x^2 - x - 2}$

41. $\dfrac{\frac{3}{x}}{\frac{1}{x} - 1}$

42. $\dfrac{a - \frac{1}{a}}{1 - \frac{1}{a}}$

43. $\dfrac{\frac{x}{y} - \frac{y}{x}}{1 + \frac{y}{x}}$

44. $\dfrac{\frac{V^2 - 9}{V}}{\frac{1}{V} - \frac{1}{3}}$

45. $\dfrac{\frac{3}{x} + \frac{1}{x^2 + x}}{\frac{1}{x + 1} - \frac{1}{x - 1}}$

46. $1 + \dfrac{1}{1 + \dfrac{1}{1 + \frac{1}{x}}}$

The expression $f(x + h) - f(x)$ is frequently used in the study of calculus. (If necessary, refer to Section 3.1 for a review of functional notation.) In Exercises 47–50, determine and then simplify this expression for the given functions.

47. $f(x) = \frac{x}{x + 1}$

48. $f(x) = \frac{3}{1 - 2x}$

49. $f(x) = \frac{1}{x^2}$

50. $f(x) = \frac{2}{x^2 + 4}$

In Exercises 51–59, simplify the given expressions. In Exercise 60, answer the given question.

51. Using the definitions of the trigonometric functions given in Section 4.2, find an expression that is equivalent to $(\tan\theta)(\cot\theta) + (\sin\theta)^2 - \cos\theta$, in terms of x, y, and r.

52. Using the definitions of the trigonometric functions given in Section 4.2, find an expression that is equivalent to $\sec\theta - (\cot\theta)^2 + \csc\theta$, in terms of x, y, and r.

53. If $f(x) = 2x - x^2$, find $f\left(\frac{1}{a}\right)$.

54. If $f(x) = x^2 + x$, find $f\left(a + \frac{1}{a}\right)$.

55. If $f(x) = x - \frac{1}{x}$, find $f(a + 1)$.

56. If $f(x) = 2x - 3$, find $f[1/f(x)]$.

57. The sum of two numbers a and b is divided by the sum of their reciprocals. Simplify the expression for this quotient.

58. Find A and B if $\frac{2x - 9}{x^2 - x - 6} = \frac{A}{x - 3} + \frac{B}{x + 2}$.

59. If $x = \frac{mn}{m + n}$ and $y = \frac{mn}{m - n}$, show that $\frac{y^2 - x^2}{y^2 + x^2} = \frac{2mn}{m^2 + n^2}$.

60. When adding fractions, explain why it is better to find the lowest common denominator rather than any denominator that is common to the fractions.

In Exercises 61–70, perform the indicated operations. Each expression occurs in the indicated area of application.

61. $\frac{3}{4\pi} - \frac{3H_0}{4\pi H}$ (transistor theory)

62. $1 + \frac{9}{128T} - \frac{27P}{64T^3}$ (thermodynamics)

63. $\frac{2n^2 - n - 4}{2n^2 + 2n - 4} + \frac{1}{n - 1}$ (optics)

64. $\frac{b}{x^2 + y^2} - \frac{2bx^2}{x^4 + 2x^2y^2 + y^4}$ (magnetic field)

65. $\left(\frac{3Px}{2L^2}\right)^2 + \left(\frac{P}{2L}\right)^2$ (force of a weld)

66. $\dfrac{a}{b^2h} + \dfrac{c}{bh^2} - \dfrac{1}{6bh}$ (strength of materials)

67. $\dfrac{\dfrac{L}{C} + \dfrac{R}{sC}}{sL + R + \dfrac{1}{sC}}$ (electronics)

68. $\dfrac{\dfrac{m}{c}}{1 - \dfrac{p^2}{c^2}}$ (airfoil design)

69. $\dfrac{1}{R^2} + \left(\omega C - \dfrac{1}{\omega L}\right)^2$ (electricity)

70. $\dfrac{\dfrac{x}{h_1} + \dfrac{x - L}{h_2}}{1 + \dfrac{x(L - x)}{h_1h_2}}$ (optics)

In Exercises 71 and 72, find the required expressions.

71. A boat that travels at v km/h in still water travels 5 km upstream and then 5 km downstream in a stream that flows at w km/h. Write an expression for the total time the boat traveled as a single fraction.

72. A person travels a distance d at an average speed v_1, and then returns over the same route at an average speed v_2. Write an expression in simplest form for the average speed of the round trip.

Answers to Practice Exercises

1. $4x^2(x + 1)^2$ **2.** $\dfrac{10a - 5b}{4a^2b}$

3. $\dfrac{3x + 16}{(x - 3)(x + 2)}$ **4.** $\dfrac{3x - 1}{2(x + 1)^2}$

6.7 Equations Involving Fractions

Multiply Each Term by the LCD • Solving Equations Involving Fractions • Applications • Extraneous Solutions

Many important equations in science and technology contain fractions. In order to solve these types of equations, it is beneficial to first rewrite them in a form that contains no fractions. This can be done by using the following procedure:

Solving Equations Involving Fractions

1. Multiply each term of the equation by the LCD of all the denominators. This should eliminate all fractions from the equation.
2. Solve the resulting equation using previous methods.
3. Check the solution in the original equation.

The following examples illustrate this procedure.

EXAMPLE 1 Multiply each term by LCD

Solve for x: $\dfrac{x}{12} - \dfrac{1}{8} = \dfrac{x + 2}{6}$.

First, note that the LCD of the terms of the equation is 24. Therefore, multiply each term by 24. This gives

■ Note carefully that we are multiplying both sides of the equation by the LCD and not combining terms over a common denominator as when adding fractions.

$$\frac{24(x)}{12} - \frac{24(1)}{8} = \frac{24(x + 2)}{6} \quad \text{each term multiplied by LCD}$$

Reduce each term to its lowest terms and solve the resulting equation:

$$2x - 3 = 4(x + 2) \quad \text{each term reduced}$$
$$2x - 3 = 4x + 8$$
$$-2x = 11$$
$$x = -\frac{11}{2}$$

When we check this solution in the original equation, we obtain $-7/12$ on each side of the equal sign. Therefore, the solution is correct. ■

EXAMPLE 2 Solving a literal equation

Solve for x: $\frac{x}{2} - \frac{1}{b^2} = \frac{x}{2b}$.

First, determine that the LCD of the terms of the equation is $2b^2$. Then multiply each term by $2b^2$ and continue with the solution:

$$\frac{2b^2(x)}{2} - \frac{2b^2(1)}{b^2} = \frac{2b^2(x)}{2b} \quad \text{each term multiplied by LCD}$$

$$b^2x - 2 = bx \quad \text{each term reduced}$$

$$b^2x - bx = 2$$

$$x(b^2 - b) = 2 \quad \text{factor}$$

$$x = \frac{2}{b^2 - b}$$

NOTE ➧ [Note the use of factoring in arriving at the final result.] Checking shows that each side of the original equation is equal to $\frac{1}{b^2(b-1)}$. ■

EXAMPLE 3 Solving an equation—focal length of a lens

An equation relating the focal length f of a lens with the object distance p and the image distance q is given below. See Fig. 6.4. Solve for q.

■ See the chapter introduction.

$$f = \frac{pq}{p + q} \quad \text{given equation}$$

■ The first telescope was invented by Lippershay, a Dutch lens maker, in about 1608.

Because the only denominator is $p + q$, the LCD is also $p + q$. By first multiplying each term by $p + q$, the solution is completed as follows:

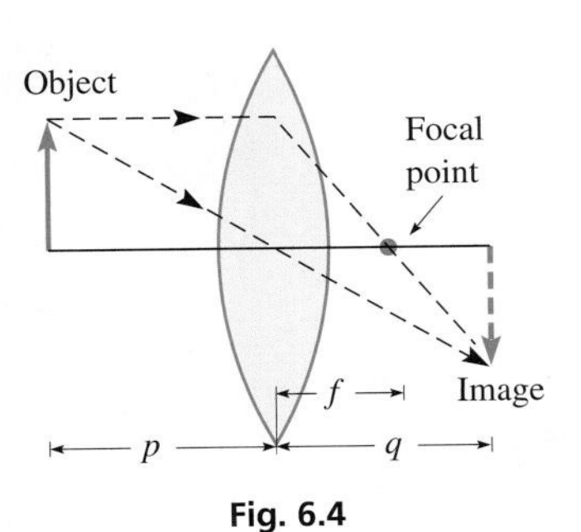

Fig. 6.4

$$f(p + q) = \frac{pq(p + q)}{p + q} \quad \text{each term multiplied by LCD}$$

$$fp + fq = pq \quad \text{reduce term on right}$$

$$fq - pq = -fp \quad \text{rearrange terms}$$

$$q(f - p) = -fp \quad \text{factor}$$

$$q = \frac{-fp}{f - p} \quad \text{divide by } f - p$$

$$q = -\frac{fp}{-(p - f)} \quad \text{use } x - y = -(y - x)$$

$$q = \frac{fp}{p - f}$$

The last form is preferred because there is no minus sign before the fraction. However, either form of the result is correct. ■

EXAMPLE 4 Solving an equation—planetary motion

In astronomy, when developing the equations that describe the motion of the planets, the equation

$$\frac{1}{2}v^2 - \frac{GM}{r} = -\frac{GM}{2a}$$

is found. Solve for M.

■ In 1609, the Italian scientist Galileo (1564–1642) learned of the invention of the telescope and developed it for astronomical observations. Among his first discoveries were the four largest moons of the planet Jupiter.

First, determine that the LCD of the terms of the equation is $2ar$. Multiplying each term by $2ar$ and proceeding, we have

$$\frac{2ar(v^2)}{2} - \frac{2ar(GM)}{r} = -\frac{2ar(GM)}{2a} \quad \text{mulitply each term by the LCD}$$

$$arv^2 - 2aGM = -rGM \quad \text{reduce each term}$$

$$rGM - 2aGM = -arv^2 \quad \text{rearrange terms}$$

$$M(rG - 2aG) = -arv^2 \quad \text{factor}$$

$$M = -\frac{arv^2}{rG - 2aG} \quad \text{divide by } rG - 2aG$$

$$= \frac{arv^2}{2aG - rG} \quad \text{use } x - y = -(y - x)$$

Practice Exercise

1. Solve for x: $\frac{x-1}{3a} = \frac{5}{6} + \frac{x}{a^2}$

Again, note the use of factoring to arrive at the final result. ■

CAUTION Note very carefully that the method of multiplying each term of the equation by the LCD to eliminate the denominators may be used only when there is an *equation*. If there is no equation, do not multiply by the LCD to eliminate denominators. ■

EXAMPLE 5 Extraneous solution

Solve for x: $\frac{2}{x+1} - \frac{1}{x} = -\frac{2}{x^2+x}$.

The solution is as follows:

$$\frac{2(x)(x+1)}{x+1} - \frac{x(x+1)}{x} = -\frac{2x(x+1)}{x(x+1)} \quad \text{multiply each side by the LCD of } x(x+1)$$

$$2x - (x+1) = -2 \quad \text{simplify each fraction}$$

$$2x - x - 1 = -2 \quad \text{complete the solution}$$

$$x = -1$$

Checking this solution *in the original equation,* notice the zero in the denominators of the first and third terms of the equation. **Because division by zero is undefined, $x = -1$ cannot be a solution.** *Thus, there is* **no solution** *to this equation.* ■

CAUTION The previous example points out clearly why it is necessary to check solutions in the original equation. It also shows that whenever we multiply each term by a common denominator that contains the unknown, it is possible to obtain a value that is not a solution of the original equation. Such a value is termed an **extraneous solution.** Only certain equations will lead to extraneous solutions, but we must be careful to identify them when they occur. ■

A number of stated problems give rise to equations involving fractions. The following examples illustrate the solution of such problems.

■ The first large-scale electronic computer was the ENIAC. It was constructed at the Univ. of Pennsylvania in the mid-1940s and used until 1955. It had 18,000 vacuum tubes and occupied 15,000 ft^2 of floor space.

■ A military programmable computer called *Colossus* was used to break the German codes in World War II. It was developed, in secret, before ENIAC.

EXAMPLE 6 Setting up an equation—computer processing

An industrial firm uses a computer system that processes and prints out its data for an average day in 20 min. To process the data more rapidly and to handle increased future computer needs, the firm plans to add new components to the system. One set of new components can process the data in 12 min, without the present system. How long would it take the new system, a combination of the present system and the new components, to process the data?

First, let $x =$ the number of minutes for the new system to process the data. Next, we know that it takes the present system 20 min to do it. This means that it processes $\frac{1}{20}$ of the data in 1 min, or $\frac{1}{20}x$ of the data in x min. In the same way, the new components can process $\frac{1}{12}x$ of the data in x min. When x min have passed, the new system (including the present system and the new components) will have processed all of the data. Therefore,

$$\underbrace{\frac{x}{20}}_{\text{part of data processed by present system}} + \underbrace{\frac{x}{12}}_{\text{part of data processed by new components}} = \underbrace{1}_{\text{one complete processing (all of data)}}$$

$$\frac{60x}{20} + \frac{60x}{12} = 60(1) \quad \text{each term multiplied by LCD of 60}$$

$$3x + 5x = 60 \quad \text{each term reduced}$$

$$8x = 60$$

$$x = \frac{60}{8} = 7.5 \text{ min}$$

Therefore, the new system should take about 7.5 min to process the data. ■

Practice Exercise

2. What is the result in Example 6, if the new system can process the data in 6.0 min?

EXAMPLE 7 Setting up an equation—rate of travel

A bus averaging 80.0 km/h takes 1.50 h longer to travel from Pittsburgh to Charlotte, NC, than a train that averages 96.0 km/h. How far apart are Pittsburgh and Charlotte?

Let $d =$ the distance from Pittsburgh and Charlotte. Since distance = rate × time, then distance ÷ rate = time. Therefore, the time the train takes is $d/96.0$ h, and the time the bus takes is $d/80.0$ h. This means

$$\frac{d}{80.0} - \frac{d}{96.0} = 1.50 \qquad \begin{aligned} 80 &= 16(5) = (2^4)(5) \\ 96 &= 32(3) = (2^5)(3) \\ \text{LCD} &= (2^5)(3)(5) = 480 \end{aligned}$$

$$\frac{480d}{80.0} - \frac{480d}{96.0} = 480(1.50)$$

$$6.00d - 5.00d = 720, \qquad d = 720 \text{ km}$$ ■

EXERCISES 6.7

In Exercises 1–4, make the given changes in the indicated examples of this section and then solve for the indicated variable.

1. In Example 2, change the second denominator from b^2 to b and then solve for x.

2. In Example 3, solve for p.

3. In Example 4, solve for G.

4. In Example 5, change the numerator on the right to 1 and then solve for x.

In Exercises 5–32, solve the given equations and check the results.

5. $\frac{x}{2} + 6 = 2x$

6. $\frac{x}{5} + 2 = \frac{15 + x}{10}$

7. $\frac{x}{6} - \frac{1}{2} = \frac{x}{3}$

8. $\frac{3N}{8} - \frac{3}{4} = \frac{N - 4}{2}$

9. $\frac{1}{4} - \frac{t - 3}{8} = \frac{1}{6}$

10. $\frac{2x - 7}{3} + 5 = \frac{1}{5}$

11. $\frac{3x}{7} - \frac{5}{21} = \frac{2 - x}{14}$

12. $\frac{F - 3}{12} - \frac{2}{3} = \frac{1 - 3F}{2}$

13. $\frac{3}{T} + 2 = \frac{5}{3}$

14. $\frac{1}{2y} - \frac{1}{2} = 4$

15. $\frac{4}{a} + \frac{1}{5} = \frac{3}{a}$

16. $\frac{2}{3} + \frac{3}{y} = \frac{5}{2y}$

17. $3 - \frac{x - 2}{5x} = \frac{1}{5}$

18. $\frac{1}{2R} = \frac{2}{3R} + \frac{1}{3}$

19. $\frac{3a}{a - 3} = 5$

20. $\frac{x}{2x - 3} = 4$

21. $\frac{2}{s} = \frac{3}{s - 1}$

22. $\frac{5}{2n + 4} = \frac{3}{4n}$

23. $\frac{5}{2x + 4} + \frac{3}{6x + 12} = 2$

24. $\frac{6}{4x - 6} + \frac{2}{4} = \frac{6}{3 - 2x}$

25. $\frac{2}{z - 5} - \frac{7}{10 - 2z} = 4$

26. $\frac{4}{4 - x} + 2 = \frac{2}{12 - 3x} + \frac{1}{3}$

27. $\frac{1}{4x} + \frac{3}{2x} = \frac{2}{x + 1}$

28. $\frac{3}{t + 3} - \frac{1}{t} = \frac{5}{6 + 2t}$

29. $\frac{5}{y} = \frac{2}{y - 3} + \frac{7}{2y^2 - 6y}$

30. $\frac{1}{2x + 3} = \frac{5}{2x} - \frac{4}{2x^2 + 3x}$

31. $\frac{1}{x^2 - x} - \frac{1}{x} + \frac{1}{1 - x} = 0$

32. $\frac{2}{x^2 - 1} - \frac{2}{x + 1} = \frac{1}{x - 1}$

33. $\frac{2}{B^2 - 4} - \frac{1}{B - 2} = \frac{1}{2B + 4}$

34. $\frac{2}{2x^2 + 5x - 3} - \frac{1}{4x - 2} + \frac{3}{2x + 6} = 0$

In Exercises 35–50, solve for the indicated letter. In Exercises 39–50, each of the given formulas arises in the technical or scientific area of study listed.

35. $2 - \frac{1}{b} + \frac{3}{c} = 0$, for c

36. $\frac{2}{3} = \frac{h}{x} + \frac{1}{6x}$, for x

37. $\frac{t - 3}{b} - \frac{t}{2b - 1} = \frac{1}{2}$, for t

38. $\frac{1}{a^2 + 2a} - \frac{y}{2a} = \frac{2y}{a + 2}$, for y

39. $\frac{s - s_0}{t} = \frac{v + v_0}{2}$, for v (velocity of object)

40. $S = \frac{P}{A} + \frac{Mc}{I}$, for A (machine design)

41. $V = 1.2\left(5.0 + \frac{8.0R}{8.0 + R}\right)$, for R

(electric resistance in Fig. 6.5)

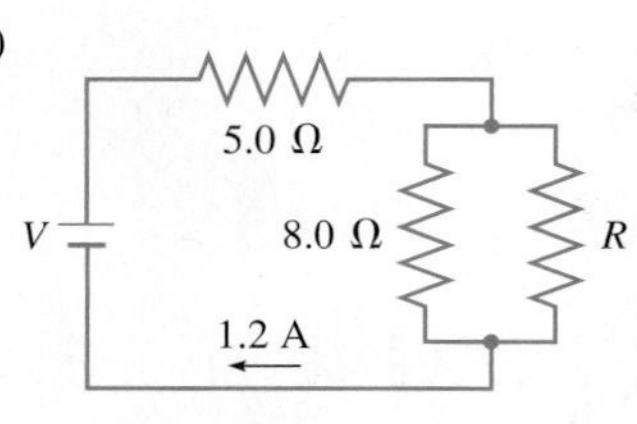

Fig. 6.5

42. $C = \frac{7p}{100 - p}$, for p (environmental pollution)

43. $z = \frac{1}{g_m} - \frac{jX}{g_m R}$, for R (FM transmission)

44. $A = \frac{1}{2}wp - \frac{1}{2}w^2 - \frac{\pi}{8}w^2$, for p (architecture)

45. $P = \frac{RT}{V - b} - \frac{a}{V^2}$, for T (thermodynamics)

46. $\frac{1}{x} + \frac{1}{nx} = \frac{1}{f}$, for f (photography)

47. $\frac{1}{C} = \frac{1}{C_2} + \frac{1}{C_1 + C_3}$, for C_1 (electricity: capacitors)

48. $D = \frac{wx^4}{24EI} - \frac{wLx^3}{6EI} + \frac{wL^2x^2}{4EI}$, for w (beam design)

49. $f(n - 1) = \frac{1}{\frac{1}{R_1} + \frac{1}{R_2}}$, for R_1 (optics)

50. $P = \frac{\frac{1}{1 + i}}{1 - \frac{1}{1 + i}}$, for i (business)

In Exercises 51–64, set up appropriate equations and solve the given stated problems. All numbers are accurate to at least two significant digits.

51. One pump can empty an oil tanker in 5.0 h, and a second pump can empty the tanker in 8.0 h. How long would it take the two pumps working together to empty the tanker?

52. One company determines that it will take its crew 450 h to clean up a chemical dump site, and a second company determines that it will take its crew 600 h to clean up the site. How long will it take the two crews working together?

53. One automatic packaging machine can package 100 boxes of machine parts in 12 min, and a second machine can do it in 10 min. A newer model machine can do it in 8.0 min. How long will it take the three machines working together?

54. A painting crew can paint a structure in 12 h, or the crew can paint it in 7.2 h when working with a second crew. How long would it take the second crew to do the job if working alone?

55. An elevator traveled from the first floor to the top floor of a building at an average speed of 2.0 m/s and returned to the first floor at 2.2 m/s. If it was on the top floor for 90 s and the total elapsed time was 5.0 min, how far above the first floor is the top floor?

56. A beam of light travels from the source to a water tank and then through the tank in 1.67×10^{-8} s. The total distance from the light source to the back of the tank is 4.50 m. If the speed of light in air is 3.00×10^8 m/s, and in water is 2.25×10^8 m/s, how far does the beam travel through the water tank (neglect the glass walls)? see Fig. 6.6.

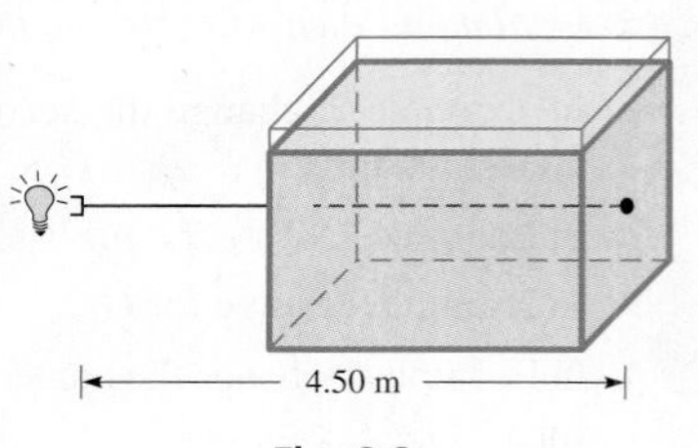

Fig. 6.6

57. A total of 276 m^3 of water was pumped from a basement for 4 h 50 min. The pumping rate for part of the time was 60.0 m^3/h, and was then reduced to 18 m^3/h. What volume was pumped at the lower rate?

58. A commuter traveled 36.0 km to work at an average speed of v_1 km/h and later returned over the same route at an average speed of 8.00 km/h less. If the total time for the two trips was 2.00 h, find v_1.

59. A jet travels 75% of the way to a destination at a speed of Mach 2 (about 2400 km/h), and then the rest of the way at Mach 1 (about 1200 km/h). What was the jet's average Mach speed for the trip?

60. A commuter rapid transit train travels 24 km farther between stops A and B than between stops B and C. If it averages 60 km/h from A to B and 30 km/h between B and C, and an express averages 50 km/h between A and C (not stopping at B), how far apart are stops A and C?

61. A jet takes the same time to travel 2580 km with the wind as it does to travel 1800 km against the wind. If its speed relative to the air is 450 km/h, what is the speed of the wind?

62. An engineer travels from Aberdeen, Scotland, to the Montrose oil field in the North Sea on a ship that averages 28 km/h. After spending 6.0 h at the field, the engineer returns to Aberdeen in a helicopter that averages 140 km/h. If the total trip takes 15.0 h, how far is the Montrose oil field from Aberdeen?

63. The current through each of the resistances R_1 and R_2 in Fig. 6.7 equals the voltage V divided by the resistance. The sum of the currents equals the current i in the rest of the circuit. Find the voltage if $i = 1.2$ A, $R_1 = 2.7\ \Omega$, and $R_2 = 6.0\ \Omega$.

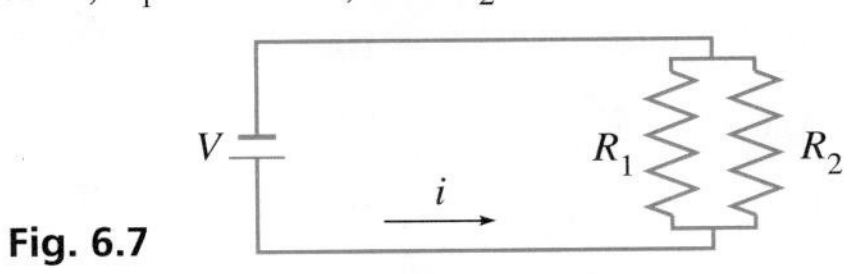

Fig. 6.7

64. A fox, pursued by a greyhound, has a start of 60 leaps. He makes 9 leaps while the greyhound makes but 6; but, 3 leaps of the greyhound are equivalent to 7 of the fox. How many leaps must the greyhound make to overcome the fox? (Source: *Elementary Algebra: Embracing the First Principles of the Science* by Charles Davies, A.S. Barnes & Company, 1852) (*Hint:* Let the unit of distance be one fox leap.)

In Exercises 65 and 66, find constants A and B such that the equation is true.

65. $\frac{x - 12}{x^2 + x - 6} = \frac{A}{x + 3} + \frac{B}{x - 2}$

66. $\frac{23 - x}{2x^2 + 7x - 4} = \frac{A}{2x - 1} - \frac{B}{x + 4}$

Answers to Practice Exercises

1. $x = \frac{5a^2 + 2a}{2a - 6}$ **2.** 4.6 min

CHAPTER 6 KEY FORMULAS AND EQUATIONS

$$ax + ay = a(x + y) \quad (6.1)$$

$$(a + b)(a - b) = a^2 - b^2 \quad (6.2)$$

$$a^2 - b^2 = (a + b)(a - b) \quad (6.3)$$

$$(x + a)(x + b) = x^2 + (a + b)x + ab \quad (6.4)$$

$$(ax + b)(cx + d) = acx^2 + (ad + bc)x + bd \quad (6.5)$$

$$(a + b)^2 = a^2 + 2ab + b^2 \quad (6.6)$$

$$(a - b)^2 = a^2 - 2ab + b^2 \quad (6.7)$$

$$a^3 + b^3 = (a + b)(a^2 - ab + b^2) \quad (6.8)$$

$$a^3 - b^3 = (a - b)(a^2 + ab + b^2) \quad (6.9)$$

$$x - y = -(y - x) \quad (6.10)$$

CHAPTER 6 REVIEW EXERCISES

CONCEPT CHECK EXERCISES

*Determine each of the following as being either **true** or **false**. If it is false, explain why.*

1. $(2a - 3)^2 = 4a^2 - 9$

2. $2x^2 - 8 = 2(x - 2)(x + 2)$

3. $6x^2 - 7x + 1 = (3x - 1)(2x - 1)$

4. $(x + 2)^3 = x^3 + 8$

5. $\frac{x^2 + x - 2}{x^2 + 3x - 4} = \frac{x - 2}{3x - 4}$

6. $\frac{x^2 - y^2}{2x + 1} \div (x + y) = \frac{x - y}{2x + 1}$

7. $\frac{x}{y} - \frac{x + 1}{y + 1} = \frac{x + y}{y^2 + y}$

8. $x = -1$ is a solution to the equation $\frac{1}{x + 1} - \frac{2}{x} = -\frac{1}{x^2 + x}$

PRACTICE AND APPLICATIONS

In Exercises 9–14, find the products by inspection. No intermediate steps should be necessary.

9. $3a(4x + 5a)$

10. $-7xy(4x^2 - 7y)$

11. $(3a + 4b)(3a - 4b)$

12. $(x - 4z)(x + 4z)$

13. $(b - 2)(b + 8)$

14. $(5 - y)(7 - y)$

In Exercises 15–46, factor the given expressions completely.

15. $5s + 20t$

16. $7x - 28y$

17. $a^2x^2 + a^2$

18. $3ax - 6ax^4 - 9a$

19. $W^2b^{x+2} - 144b^x$

20. $900n^a - n^{a+4}$

21. $16(x + 2)^2 - t^4$

22. $25s^4 - 36t^2$

23. $36t^2 - 24t + 4$

24. $9 - 12x + 4x^2$

25. $25t^2 + 10t + 1$

26. $4c^2 + 36cd + 81d^2$

27. $x^2 + x - 42$

28. $x^2 - 4x - 45$

29. $t^4 - 5t^2 - 36$

30. $3N^4 - 33N^2 + 30$

31. $2k^2 - k - 36$

32. $3 - 2x - 5x^2$

33. $8x^2 - 8x - 70$

34. $27F^3 + 21F^2 - 48F$

35. $10b^2 + 23b - 5$

36. $12x^2 - 7xy - 12y^2$

37. $4x^2 - 16y^2$

38. $4a^2x^2 + 26a^2x + 36a^2$

39. $250 - 16y^6$

40. $8a^6 + 64a^3$

41. $8x^3 + 27$

42. $2a^6 + 4a^3 + 2$

43. $ab^2 - 3b^2 + a - 3$

44. $axy - ay + ax - a$

45. $nx + 5n - x^2 + 25$

46. $ty - 4t + y^2 - 16$

In Exercises 47–70, perform the indicated operations and express results in simplest form.

47. $\dfrac{48ax^3y^6}{9a^3xy^6}$

48. $\dfrac{-39r^2s^4t^8}{52rs^5t}$

49. $\dfrac{6x^2 - 7x - 3}{2x^2 - 5x + 3}$

50. $\dfrac{p^4 - 4p^2 - 4}{12 + p^2 - p^4}$

51. $\dfrac{4x + 4y}{35x^2} \times \dfrac{28x}{x^2 - y^2}$

52. $\left(\dfrac{6x - 3}{x^2}\right)\left(\dfrac{4x^2 - 12x}{6 - 12x}\right)$

53. $\dfrac{18 - 6L}{L^2 - 6L + 9} \div \dfrac{L^2 - 2L - 15}{L^2 - 9}$

54. $\dfrac{6x^2 - xy - y^2}{2x^2 + xy - y^2} \div \dfrac{16y^2 - 4x^2}{2x^2 + 6xy + 4y^2}$

55. $\dfrac{\dfrac{3x}{7x^2 + 13x - 2}}{\dfrac{6x^2}{x^2 + 4x + 4}}$

56. $\dfrac{\dfrac{3y - 3x}{2x^2 + 3xy - 2y^2}}{\dfrac{3x^2 - 3y^2}{x^2 + 4xy + 4y^2}}$

57. $\dfrac{x + \dfrac{1}{x} + 1}{x^2 - \dfrac{1}{x}}$

58. $\dfrac{\dfrac{4}{y} - 4y}{2 - \dfrac{2}{y}}$

59. $\dfrac{4}{9x} - \dfrac{5}{12x^2}$

60. $\dfrac{3}{10a^2} + \dfrac{1}{4a^3}$

61. $\dfrac{6}{x} - \dfrac{7}{2x} + \dfrac{3}{xy}$

62. $\dfrac{T}{T^2 + 2} - \dfrac{1}{2T + T^3}$

63. $\dfrac{a + 1}{a} - \dfrac{a - 3}{a + 2}$

64. $\dfrac{y}{y + 2} - \dfrac{1}{y^2 + 2y}$

65. $\dfrac{2x}{x^2 + 2x - 3} - \dfrac{1}{6x + 2x^2}$

66. $\dfrac{x}{4x^2 + 4x - 3} + \dfrac{3}{9 - 4x^2}$

67. $\dfrac{3x}{2x^2 - 2} - \dfrac{2}{4x^2 - 5x + 1}$

68. $\dfrac{2n - 1}{4 - n} + \dfrac{n + 2}{20 - 5n}$

69. $\dfrac{3x}{x^2 + 2x - 3} - \dfrac{2}{x^2 + 3x} + \dfrac{x}{x - 1}$

70. $\dfrac{3}{y^4 - 2y^3 - 8y^2} + \dfrac{y - 1}{y^2 + 2y} - \dfrac{y - 3}{y^2 - 4y}$

In Exercises 71–74, graphically check the results for the indicated exercises of this set on a calculator.

71. Exercise 49

72. Exercise 52

73. Exercise 57

74. Exercise 66

In Exercises 75–82, solve the given equations.

75. $\dfrac{x}{2} - 3 = \dfrac{x - 10}{4}$

76. $\dfrac{4x}{c} - \dfrac{2}{2c} = \dfrac{6}{c} - x$, for x

77. $\dfrac{2}{t} - \dfrac{1}{at} = 2 + \dfrac{a}{t}$, for t

78. $\dfrac{3}{a^2y} - \dfrac{1}{ay} = \dfrac{9}{a}$, for y

79. $\dfrac{2x}{2x^2 - 5x} - \dfrac{3}{x} = \dfrac{1}{4x - 10}$

80. $\dfrac{3}{x^2 + 3x} - \dfrac{1}{x} = \dfrac{1}{3 + x}$

81. Given $f(x) = \dfrac{1}{x + 2}$, solve for x if $f(x + 2) = 2f(x)$.

82. Given $f(x) = \dfrac{x}{x + 1}$, solve for x if $3f(x) + f\left(\dfrac{1}{x}\right) = 2$.

In Exercises 83–100, solve the given problems. Where indicated, the expression is found in the stated technical area.

83. In an algebraic fraction, what effect on the result is made if the sign is changed of: (a) an odd number of factors? (b) an even number of factors?

84. Algebraically show that the reciprocal of the reciprocal of the number x is x.

85. Show that $xy = \dfrac{1}{4}[(x + y)^2 - (x - y)^2]$.

86. Show that $x^2 + y^2 = \dfrac{1}{2}[(x + y)^2 + (x - y)^2]$.

87. Multiply: $Pb(L + b)(L - b)$ (architecture)

88. Multiply: $kr(R - r)$ (blood flow)

89. Expand: $[2b + (n - 1)\lambda]^2$ (optics)

90. Factor: $\pi r_1^2 l - \pi r_2^2 l$ (jet plane fuel supply)

91. Factor: $cT_2 - cT_1 + RT_2 - RT_1$ (pipeline flow)

92. Factor: $9600t + 8400t^2 - 1200t^3$ (solar energy)

93. Simplify and express in factored form: $(2R - r)^2 - (r^2 + R^2)$ (aircraft radar)

94. Express in factored form: $2R(R + r) - (R + r)^2$ (electricity: power)

95. Expand and simplify: $(n + 1)^3(2n + 1)^3$ (fluid flow in pipes)

96. Expand and simplify: $2(e_1 - e_2)^2 + 2(e_2 - e_3)^2$ (mechanical design)

97. Expand and simplify: $10a(T - t) + a(T - t)^2$ (instrumentation)

98. Expand the third term and then factor by grouping: $pa^2 + (1 - p)b^2 - [pa + (1 - p)b]^2$ (nuclear physics)

99. A metal cube of edge x is heated and each edge increases by 4 mm. Express the increase in volume in factored form.

100. Express the difference in volumes of two ball bearings of radii r mm and 3 mm in factored form. $(r > 3\text{ mm})$

In Exercises 101–112, perform the indicated operations and simplify the given expressions. Each expression is from the indicated technical area of application.

101. $\left(\dfrac{2wtv^2}{Dg}\right)\left(\dfrac{b\pi^2D^2}{n^2}\right)\left(\dfrac{6}{bt^2}\right)$ (machine design)

102. $\dfrac{m}{c} \div \left[1 - \left(\dfrac{p}{c}\right)^2\right]$ (airfoil design)

103. $\dfrac{\dfrac{\pi ka}{2}(R^4 - r^4)}{\pi ka(R^2 - r^2)}$ (flywheel rotation)

104. $\dfrac{V}{kp} - \dfrac{RT}{k^2p^2}$ (electric motors)

105. $1 - \dfrac{d^2}{2} + \dfrac{d^4}{24} - \dfrac{d^6}{120}$ (aircraft emergency locator transmitter)

106. $\dfrac{wx^2}{2T_0} + \dfrac{kx^4}{12T_0}$ (bridge design)

107. $\dfrac{4k - 1}{4k - 4} + \dfrac{1}{2k}$ (spring stress)

108. $\dfrac{Am}{k} - \dfrac{g}{2}\left(\dfrac{m}{k}\right)^2 + \dfrac{AML}{k}$ (rocket fuel)

109. $1 - \dfrac{3a}{4r} - \dfrac{a^3}{4r^3}$ (hydrodynamics)

110. $\dfrac{1}{F} + \dfrac{1}{f} - \dfrac{d}{fF}$ (optics)

111. $\dfrac{\dfrac{u^2}{2g} - x}{\dfrac{1}{2gc^2} - \dfrac{u^2}{2g} + x}$ (mechanism design)

112. $\dfrac{V}{\dfrac{1}{2R} + \dfrac{1}{2R + 2}}$ (electricity)

In Exercises 113–122, solve for the indicated letter. Each equation is from the indicated technical area of application.

113. $W = mgh_2 - mgh_1$, for m (work done on object)

114. $h(T_1 + T_2) = T_2$, for T_1 (engine efficiency)

115. $R = \dfrac{wL}{H(w + L)}$, for L (architecture)

116. $\dfrac{J}{T} = \dfrac{t}{\omega_1 - \omega_2}$, for ω_2 (rotational torque)

117. $E = V_0 + \dfrac{(m + M)V^2}{2} + \dfrac{p^2}{2I}$, for I (nuclear physics)

118. $\dfrac{q_2 - q_1}{d} = \dfrac{f + q_1}{D}$, for q_1 (photography)

119. $s^2 + \dfrac{cs}{m} + \dfrac{kL^2}{mb^2} = 0$, for c (mechanical vibrations)

120. $I = \dfrac{A}{x^2} + \dfrac{B}{(10 - x)^2}$, for A (optics)

121. $F = \dfrac{m}{sC} + \dfrac{F_0}{s}$, for C (mechanics)

122. $C = \dfrac{k}{n(1 - k)}$, for k (reinforced concrete design)

In Exercises 123–130, set up appropriate equations and solve the given stated problems. All numbers are accurate to at least two significant digits.

123. If a certain car's lights are left on, the battery will be dead in 4.0 h. If only the radio is left on, the battery will be dead in 24 h. How long will the battery last if both the lights and the radio are left on?

124. Two pumps are being used to fight a fire. One pumps 5000 gal in 20 min, and the other pumps 5000 gal in 25 min. How long will it take the two pumps together to pump 5000 gal?

125. A number m is the *harmonic mean* of number x and y if $1/m$ equals the average of $1/x$ and $1/y$ (divide $1/x + 1/y$ by 2). Find the harmonic mean of the musical notes that have frequencies of 400 Hz and 1200 Hz.

126. An auto mechanic can do a certain motor job in 3.0 h, and with an assistant he can do it in 2.1 h. How long would it take the assistant to do the job alone?

127. The *relative density* of an object may be defined as its weight in air w_a, divided by the difference of its weight in air and its weight when submerged in water, w_w. For a lead weight, $w_a = 1.097\ w_w$. Find the relative density of lead.

128. A car travels halfway to its destination at 80.0 km/h and the remainder of the distance at 60.0 km/h. What is the average speed of the car for the trip?

129. For electric resistors in parallel, the reciprocal of the combined resistance equals the sum of the reciprocals of the individual resistances. For three resistors of 12 Ω, R ohms, and $2R$ ohms, in parallel, the combined resistance is 6.0 Ω. Find R.

130. A fire engine averaged 48 mi/h going to a fire, and 36 mi/h on its return to the station. If the total traveling time was 35 min, how far was the station from the fire?

131. In analyzing a certain electric circuit, the expression

$$\frac{\left(1+\frac{1}{s}\right)\left(1+\frac{1}{s/2}\right)}{3+\frac{1}{s}+\frac{1}{s/2}}$$

is used. Simplify this expression, and write a paragraph describing your procedure. When you "cancel," explain what basic operation is being performed.

CHAPTER 6 PRACTICE TEST

As a study aid, we have included complete solutions for each Practice Test problem at the end of this book.

1. Find the product: $2x(2x-3)^2$.

2. The following equation is used in electricity.

Solve for R_1: $\frac{1}{R}=\frac{1}{R_1+r}+\frac{1}{R_2}$

3. Reduce to simplest form: $\frac{2x^2+5x-3}{2x^2+12x+18}$

4. Factor: $4x^2-16y^2$

5. Factor: pb^3+8a^3p (business application)

6. Factor: $2a-4T-ba+2bT$

7. Factor: $36x^2+14x-16$

In problems 8–10, perform the indicated operations and simplify.

8. $\frac{3}{4x^2}-\frac{2}{x^2-x}-\frac{x}{2x-2}$

9. $\frac{x^2+x}{2-x}\div\frac{x^2}{x^2-4x+4}$

10. $\frac{1-\frac{3}{2x+2}}{\frac{x}{5}-\frac{1}{2}}$

11. If one riveter can do a job in 12 days, and a second riveter can do it in 16 days, how long would it take for them to do it together?

12. Solve for x: $\frac{3}{2x^2-3x}+\frac{1}{x}=\frac{3}{2x-3}$

Quadratic Equations

7

LEARNING OUTCOMES

After completion of this chapter, the student should be able to:

- Identify quadratic functions
- Solve quadratic functions by factoring, by completing the square, by using the quadratic formula, and by graphing
- Solve application problems involving quadratic equations
- Graph a quadratic function and identify important points on the curve

In this chapter, we develop algebraic and graphical methods for solving an important type of equation called a *quadratic equation*. We will also see that the graph of a *quadratic function* is a *parabola*, which has many modern applications in science and technology.

One important application of a parabola is a television satellite dish. Because of its parabolic design, incoming signals are reflected off the surface of the dish and directed to a common point called the *focus*, where the receiver is placed.

The parabolic shape also occurs in the supporting cables of suspension bridges. It can be shown that when a hanging cable is subjected to a uniformly distributed load, such as the deck of a suspension bridge, the cable takes the form of a parabola.

Another important application of a parabola (and thereby of quadratic functions) is that of a projectile, examples of which are a baseball, an artillery shell, and a rocket. When Galileo showed that the distance an object falls does not depend on its weight, he also discovered that this distance depends on the square of the time of fall. This, in turn, was shown to mean that the path of a projectile is parabolic (not considering air resistance). To find the location of a projectile for any given time of flight requires the solution of a quadratic equation.

The Babylonians developed methods of solving quadratic equations nearly 4000 years ago. However, their use in applied situations was limited, as they had little real-world use for them. Today, in addition to those mentioned above, quadratic equations have applications in architecture, electric circuits, mechanical systems, forces on structures, and product design.

◀ **The Golden Gate Bridge is a parabolic suspension bridge. In Section 7.4, we will use a quadratic function to represent the height of its curved supporting cables.**

7.1 Quadratic Equations; Solution by Factoring

General Quadratic Equation • Solutions of a Quadratic Equation • Solving a Quadratic Equation by Factoring • Equations with Fractions

Given that a, b, and c are constants ($a \neq 0$), *the equation*

$$ax^2 + bx + c = 0 \quad \textbf{(7.1)}$$

is called the **general quadratic equation in** $\boldsymbol{x}$. The left side of Eq. (7.1) is a polynomial function of degree 2. *This function,* $f(x) = ax^2 + bx + c$, *is known as the* **quadratic function.** Any equation that can be simplified and then written in the form of Eq. (7.1) is a quadratic equation in one unknown.

Among the applications of quadratic equations and functions are the following examples: In finding the flight time t of a projectile, the equation $s_0 + v_0t - 16t^2 = 0$ occurs; in analyzing the electric current i in a circuit, the function $f(i) = Ei - Ri^2$ is found; and in determining the forces at a distance x along a beam, the function $f(x) = ax^2 + bLx + cL^2$ is used.

NOTE ▸ [Because it is the x^2-term in Eq. (7.1) that distinguishes the quadratic equation from other types of equations, the equation is *not quadratic* if $a = 0$. However, either b or c (or both) may be zero, and the equation is still quadratic.] No power of x higher than the second may be present in a quadratic equation. Also, we should be able to properly identify a quadratic equation even when it does not initially appear in the form of Eq. (7.1). The following two examples illustrate how we may recognize quadratic equations.

■ It is usually easier to have integer coefficients and $a > 0$ in a quadratic equation. This can be done by multiplying each term by the LCD if there are fractions, and to change the sign of all terms if $a < 0$.

EXAMPLE 1 Examples of quadratic equations

The following are quadratic equations.

$x^2 - 4x - 5 = 0$ ($a = 1$, $b = -4$, $c = -5$) — To show this equation in the form of Eq. (7.1), it can be written as $1x^2 + (-4)x + (-5) = 0$.

$3x^2 - 6 = 0$ ($a = 3$, $c = -6$) — Because there is no x-term, $b = 0$.

$2x^2 + 7x = 0$ ($a = 2$, $b = 7$) — Because no constant appears, $c = 0$.

$(m - 3)x^2 - mx + 7 = 0$ — The constants in Eq. (7.1) may include literal expressions. In this case, $m - 3$ takes the place of a, $-m$ takes the place of b, and $c = 7$.

$4x^2 - 2x = x^2$ — After all nonzero terms have been collected on the left side, the equation becomes $3x^2 - 2x = 0$.

$(x + 1)^2 = 4$ — Expanding the left side and collecting all nonzero terms on the left, we have $x^2 + 2x - 3 = 0$. ■

EXAMPLE 2 Examples of equations not quadratic

The following are not quadratic equations.

$bx - 6 = 0$ — There is no x^2-term.

$x^3 - x^2 - 5 = 0$ — There should be no term of degree higher than 2. Thus, there can be no x^3-term in a quadratic equation.

$x^2 + x - 7 = x^2$ — When terms are collected, there will be no x^2-term. ■

SOLUTIONS OF A QUADRATIC EQUATION

Recall that *the* **solution** *of an equation consists of all numbers* (**roots**), *which, when substituted in the equation, give equality.* There are *two* roots for a quadratic equation. At times, these roots are equal (see Example 3), so only one number is actually a solution. Also, the roots can contain *imaginary* numbers. If this happens, all we wish to do at this point is to recognize that there are no real roots.

EXAMPLE 3 Solutions (roots) of a quadratic equation

(a) The quadratic equation $3x^2 - 7x + 2 = 0$ has roots $x = 1/3$ and $x = 2$. This is seen by substituting these numbers in the equation.

$$3(\tfrac{1}{3})^2 - 7(\tfrac{1}{3}) + 2 = 3(\tfrac{1}{9}) - \tfrac{7}{3} + 2 = \tfrac{1}{3} - \tfrac{7}{3} + 2 = \tfrac{0}{3} = 0$$

$$3(2)^2 - 7(2) + 2 = 3(4) - 14 + 2 = 14 - 14 = 0$$

■ Until the 1600s, most mathematicians did not accept negative, irrational, or imaginary roots of an equation. It was also generally accepted that an equation had only one root.

(b) The quadratic equation $4x^2 - 4x + 1 = 0$ has a **double root** (*both roots are the same*) of $x = 1/2$. Showing that this number is a solution, we have

$$4(\tfrac{1}{2})^2 - 4(\tfrac{1}{2}) + 1 = 4(\tfrac{1}{4}) - 2 + 1 = 1 - 2 + 1 = 0$$

(c) The quadratic equation $x^2 + 9 = 0$ has roots $x = \sqrt{-9}$ and $x = -\sqrt{-9}$. These are **imaginary roots** because they involve taking the square root of a negative number. There are **no real roots.** ■

SOLVING A QUADRATIC EQUATION BY FACTORING

We now describe a method for solving quadratic equations that relies on factoring.

NOTE ▸ [Using the **zero product rule,** which states that a product is zero if and only if any of its factors is zero, we have the following steps in solving a quadratic equation.]

Procedure for Solving a Quadratic Equation by Factoring

1. Collect all terms on the left and simplify to the form $ax^2 + bx + c = 0$.
2. Factor the quadratic expression.
3. Set each factor equal to zero.
4. Solve the resulting linear equations. These solutions are the roots of the quadratic equation.
5. Check the solutions in the original equation.

EXAMPLE 4 Solving quadratic equation by factoring

$$x^2 - x - 12 = 0$$

$$(x - 4)(x + 3) = 0 \quad \text{factor}$$

$$x - 4 = 0 \qquad x + 3 = 0 \quad \text{set each factor equal to zero}$$

$$x = 4 \qquad x = -3 \quad \text{solve}$$

The roots are $x = 4$ and $x = -3$. We can check them in the original equation by substitution. Therefore,

$$(4)^2 - (4) - 12 \stackrel{?}{=} 0 \qquad (-3)^2 - (-3) - 12 \stackrel{?}{=} 0$$

$$0 = 0 \qquad 0 = 0$$

Both roots satisfy the original equation. ■

Practice Exercise

1. Solve for x: $x^2 + 4x - 21 = 0$

EXAMPLE 5 First put equation in proper form

$$x^2 + 4 = 4x \quad \text{equation not in form of Eq. (7.1)}$$
$$x^2 - 4x + 4 = 0 \quad \text{subtract } 4x \text{ from both sides}$$
$$(x - 2)^2 = 0 \quad \text{factor}$$
$$x - 2 = 0, \qquad x = 2 \quad \text{solve}$$

Because $(x - 2)^2 = (x - 2)(x - 2)$, both factors are the same. This means there is a double root of $x = 2$. Substitution shows that $x = 2$ satisfies the original equation. ■

Practice Exercise

2. Solve for x: $9x^2 + 1 = 6x$

EXAMPLE 6 Quadratic equations with $b = 0$ or $c = 0$

(a) In solving the equation $3x^2 - 12 = 0$, we note that $b = 0$ (there is no x-term). However, we can solve it by factoring. First we note the common factor of 3. Because it is a constant, we can first divide all terms by 3, and proceed with the solution.

$$3x^2 - 12 = 0$$
$$x^2 - 4 = 0 \quad \text{divide each term by 3}$$
$$(x - 2)(x + 2) = 0 \quad \text{factor}$$
$$x - 2 = 0 \qquad x + 2 = 0 \quad \text{set each factor equal to zero}$$
$$x = 2 \qquad x = -2 \quad \text{solve}$$

The roots 2 and -2 check. We could also have first factored the 3 from each term, but the results would be the same since 3 is a constant, and the only two factors that can be set equal to zero are $x - 2$ and $x + 2$.

NOTE ▶ **(b)** In solving the equation $3x^2 - 12x = 0$, we note that $c = 0$ (there is no constant term). However, we can solve it by factoring. We note the factor of 3, and because 3 is a constant, we can divide each term by 3. [We also note the common factor x, but because we are solving for x, we *cannot* divide each term by x. If we divide by x, we lose one of the two roots.] Therefore,

$$3x^2 - 12x = 0$$
$$x^2 - 4x = 0 \quad \text{divide each term by 3, but not by } x$$
$$x(x - 4) = 0 \quad \text{factor}$$
$$x = 0, 4$$

These roots check. Again, if we had divided out the x, we would not have found the root $x = 0$, and therefore our solution would be incomplete. ■

EXAMPLE 7 Quadratic equation—fire hose flow rate

For a certain fire hose, the pressure loss P (in lb/in.^{2} per 100 ft of hose) is $P = 2Q^2 + Q$, where Q is the flow rate (in 100 gal/min). Find Q for $P = 15$ lb/in.^{2} per 100 ft.

Substituting, we have the following equation and solution.

CAUTION It is essential for the quadratic expression on the left to be equal to zero (on the right). The first step must be to write the equation in the form $ax^2 + bx + c = 0$. ■

$$2Q^2 + Q = 15$$
$$2Q^2 + Q - 15 = 0$$
$$(2Q - 5)(Q + 3) = 0$$
$$2Q - 5 = 0, \qquad Q = 2.5 \text{ 100 gal/min}$$
$$Q + 3 = 0, \qquad Q = -3 \text{ 100 gal/min} \quad \text{not realistically possible}$$

These roots check, but the negative root is not realistically possible, which means the only solution is $Q = 2.5$ 100 gal/min, (or 250 gal/min). ■

A number of equations involving fractions lead to quadratic equations after the fractions are eliminated. The following two examples illustrate the process of solving such equations with fractions. As discussed in Section 6.7, the fractions can be eliminated by multiplying each term of the equation by the LCD of all the denominators.

EXAMPLE 8 Fractional equation solved as quadratic

Solve for x: $\frac{1}{x} + 3 = \frac{2}{x+2}$.

$$\frac{x(x+2)}{x} + 3x(x+2) = \frac{2x(x+2)}{x+2} \quad \text{multiply each term by the LCD, } x(x+2)$$

$$x + 2 + 3x^2 + 6x = 2x \quad \text{reduce each term}$$

$$3x^2 + 5x + 2 = 0 \quad \text{collect terms on left}$$

$$(3x+2)(x+1) = 0 \quad \text{factor}$$

$$3x + 2 = 0, \quad x = -\frac{2}{3} \quad \text{set each factor equal to zero and solve}$$

$$x + 1 = 0, \quad x = -1$$

■

CAUTION Remember, if either value gives division by zero, the root is extraneous, and must be excluded from the solution. ■

Both of these solutions check when substituted into the original equation.

EXAMPLE 9 Fractional equation—speed of truck

A lumber truck travels 60 mi from a sawmill to a lumber camp and then back in 7 h travel time. If the truck averages 5 mi/h less on the return trip than on the trip to the camp, find its average speed to the camp. See Fig. 7.1.

Let $v =$ the average speed (in mi/h) of the truck going to the camp. This means that the average speed of the return trip was $(v - 5)$ mi/h.

We also know that $d = vt$ (distance equals speed times time), which tells us that $t = d/v$. Thus, the time for each part of the trip is the distance divided by the speed.

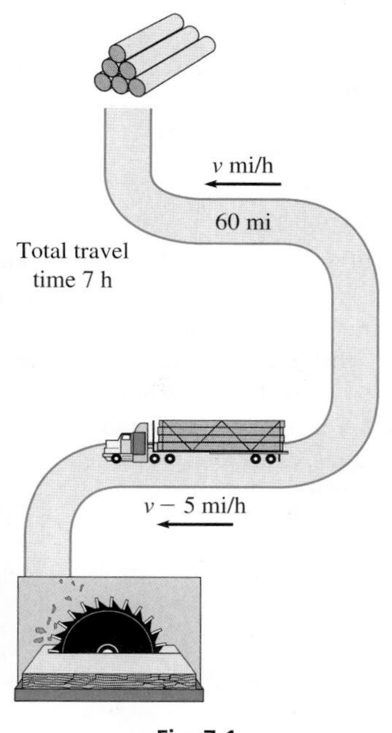

Fig. 7.1

$$\underset{\text{time to camp}}{\frac{60}{v}} + \underset{\text{time from camp}}{\frac{60}{v-5}} = \underset{\text{total time}}{7}$$

$$60(v-5) + 60v = 7v(v-5) \quad \text{multiply each term by } v(v-5)$$

$$7v^2 - 155v + 300 = 0 \quad \text{collect terms on the left}$$

$$(7v - 15)(v - 20) = 0 \quad \text{factor}$$

$$7v - 15 = 0, \quad v = \frac{15}{7} \quad \text{set each factor equal to zero and solve}$$

$$v - 20 = 0, \quad v = 20$$

The value $v = 15/7$ mi/h cannot be a solution because the return speed of 5 mi/h less would be negative. Thus, the solution is $v = 20$ mi/h, which means the return speed was 15 mi/h. The trip to the camp took 3 h, and the return took 4 h, which shows the solution checks. ■

The procedure used to solve quadratic equations in this section only works *if we are able to factor* the quadratic expression. In these cases, the solutions are rational numbers. Since not all quadratic expressions are factorable, other methods for solving quadratic equations will be discussed in the remaining sections of this chapter.

EXERCISES 7.1

In Exercises 1 and 2, make the given changes in the indicated examples of this section and then solve the resulting quadratic equations.

1. In Example 3(a), change the − sign before $7x$ to + and then solve.

2. In Example 8, change the numerator of the first term to 2 and the numerator of the term on the right to 1 and then solve.

In Exercises 3–10, determine whether or not the given equations are quadratic. If the resulting form is quadratic, identify a, b, and c, with $a > 0$. Otherwise, explain why the resulting form is not quadratic.

3. $x(x - 2) = 4$ **4.** $(3x - 2)^2 = 2$

5. $x^2 = (x + 2)^2$ **6.** $x(2x^2 + 5) = 7 + 2x^2$

7. $n(n^2 + n - 1) = n^3$ **8.** $(T - 7)^2 = (2T + 3)^2$

9. $y^2(y - 2) = 3(y - 2)$ **10.** $z(z + 4) = (z + 1)(z + 5)$

In Exercises 11–48, solve the given quadratic equations by factoring.

11. $x^2 - 25 = 0$ **12.** $B^2 - 400 = 0$

13. $4y^2 = 9$ **14.** $2x^2 = 0.32$

15. $x^2 - 5x - 14 = 0$ **16.** $x^2 + x - 6 = 0$

17. $R^2 + 12 = 7R$ **18.** $x^2 + 30 = 11x$

19. $40x - 16x^2 = 0$ **20.** $15L = 20L^2$

21. $12m^2 = 3$ **22.** $9 = a^2x^2$

23. $3x^2 - 13x + 4 = 0$ **24.** $A^2 + 8A + 16 = 0$

25. $4x = 3 - 7x^2$ **26.** $4x^2 + 25 = 20x$

27. $6x^2 = 13x - 6$ **28.** $6z^2 = 6 + 5z$

29. $4x(x + 1) = 3$ **30.** $t(43 + t) = 9 - 9t^2$

31. $6y^2 + by = 2b^2$ **32.** $2x^2 - 7ax + 4a^2 = a^2$

33. $8s^2 + 16s = 90$ **34.** $18t^2 = 48t - 32$

35. $(x + 2)^3 = x^3 + 8$ **36.** $V(V^2 - 4) = V^2(V - 1)$

37. $(x + a)^2 - b^2 = 0$ **38.** $bx^2 - b = x - b^2x$

39. $x^2 + 2ax = b^2 - a^2$ **40.** $x^2(a^2 + 2ab + b^2) = x(a + b)$

41. In Eq. (7.1), for $a = 2$, $b = -7$, and $c = 3$, show that the sum of the roots is $-b/a$.

42. For the equation of Exercise 41 show that the product of the roots is c/a.

43. In finding the dimensions of a crate, the equation $12x^2 - 64x + 64 = 0$ is used. Solve for x, if $x > 2$.

44. If a rocket is launched with an initial velocity of 320 ft/s, its height above ground after t seconds is given by $-16t^2 + 320t$ (in ft). Find the times when the height is 0.

45. The voltage V across a semiconductor in a computer is given by $V = \alpha I + \beta I^2$, where I is the current (in A). If a 6-V battery is conducted across the semiconductor, find the current if $\alpha = 2\ \Omega$ and $\beta = 0.5\ \Omega/\text{A}$.

46. The mass m (in Mg) of the fuel supply in the first-stage booster of a rocket is $m = 135 - 6t - t^2$, where t is the time (in s) after launch. When does the booster run out of fuel?

47. The power P (in MW) produced between midnight and noon by a nuclear power plant is $P = 4h^2 - 48h + 744$, where h is the hour of the day. At what time is the power 664 MW?

48. In determining the speed s (in mi/h) of a car while studying its fuel economy, the equation $s^2 - 16s = 3072$ is used. Find s.

In Exercises 49 and 50, find the indicated quadratic equations.

49. Find a quadratic equation for which the solutions are 0.5 and 2.

50. Find a quadratic equation for which the solutions are a and b.

In Exercises 50 and 51, although the equations are not quadratic, factoring will lead to one quadratic factor and the solution can be completed by factoring as with a quadratic equation. Find the three roots of each equation.

51. $x^3 - x = 0$ **52.** $x^3 - 4x^2 - x + 4 = 0$

In Exercises 53–56, solve the given equations involving fractions.

53. $\dfrac{1}{x - 3} + \dfrac{4}{x} = 2$ **54.** $2 - \dfrac{1}{x} = \dfrac{3}{x + 2}$

55. $\dfrac{1}{2x} - \dfrac{3}{4} = \dfrac{1}{2x + 3}$ **56.** $\dfrac{x}{2} + \dfrac{1}{x - 3} = 3$

In Exercises 57–60, set up the appropriate quadratic equations and solve.

57. The spring constant k is the force F divided by the amount x the spring stretches ($k = F/x$). See Fig. 7.2(a). For two springs in series [see Fig. 7.2(b)], the reciprocal of the spring constant k_c for the combination equals the sum of the reciprocals of the individual spring constants. Find the spring constants for each of two springs in series if $k_c = 2$ N/cm and one spring constant is 3 N/cm more than the other.

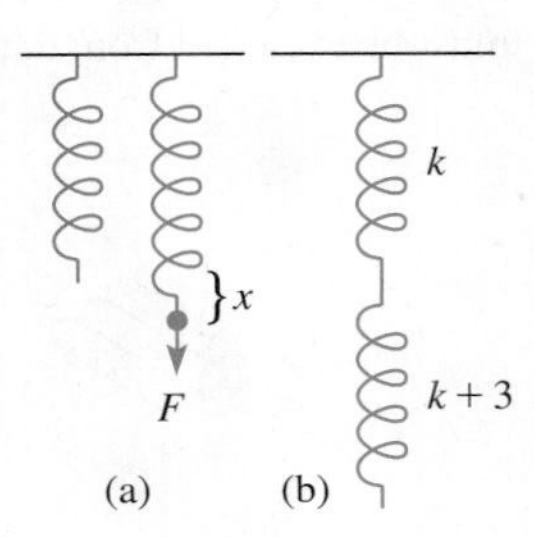

Fig. 7.2

58. The combined resistance R of two resistances R_1 and R_2 connected in parallel [see Fig. 7.3(a)] is equal to the product of the individual resistances divided by their sum. If the two resistances are connected in series [see Fig. 7.3(b)], their combined resistance is the sum of their individual resistances. If two resistances connected in parallel have a combined resistance of 3.0 Ω and the same two resistances have a combined resistance of 16 Ω when connected in series, what are the resistances?

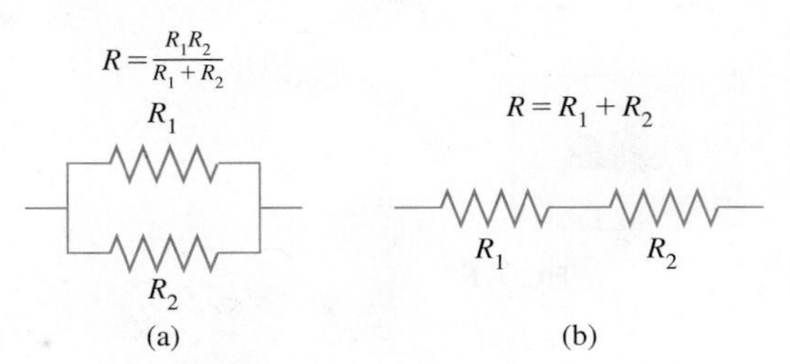

Fig. 7.3

59. A hydrofoil made the round-trip of 120 km between two islands in 3.5 h of travel time. If the average speed going was 10 km/h less than the average speed returning, find these speeds.

60. A rectangular solar panel is 20 cm by 30 cm. By adding the same amount to each dimension, the area is doubled. How much is added?

Answers to Practice Exercises

1. $x = -7, x = 3$ **2.** $x = 1/3, x = 1/3$

7.2 Completing the Square

Using the Square Root Property • Solving a Quadratic Equation by Completing the Square

Most quadratic equations that arise in applications cannot be solved by factoring. Therefore, we now develop a method called **completing the square,** which can be used to solve *any* quadratic equation. In the next section, this method is used to develop a general formula that can be used to solve any quadratic equation. Solving an equation by completing the square relies on the following property:

Square Root Property

If $x^2 = k$, then $x = \pm\sqrt{k}$.

This property allows us to take the square root on both sides of an equation, ***but we must remember to include the positive and negative square root***. There are actually two solutions, $x = \sqrt{k}$ and $x = -\sqrt{k}$. The following example illustrates how this property is used.

EXAMPLE 1 Using the square root property

(a) If $x^2 = 16$, then $x = \pm\sqrt{16} = \pm 4$. The solutions are $x = 4$ and $x = -4$.

(b) To solve $3y^2 - 4 = 6$, we must *first isolate the squared expression* to get $y^2 = \frac{10}{3}$. We then have $y = \pm\sqrt{\frac{10}{3}}$. The two solutions are $y = \sqrt{\frac{10}{3}}$ and $y = -\sqrt{\frac{10}{3}}$.

(c) If $(x - 2)^2 = 5$, then $x - 2 = \pm\sqrt{5}$. Adding 2 to both sides yields $x = 2 \pm \sqrt{5}$. Separately, the solutions are $x = 2 + \sqrt{5}$ and $x = 2 - \sqrt{5}$. ■

We now describe how to solve a quadratic equation by completing the square. Our goal is to rewrite the given equation in the form of Example 1(c) so we can finish solving it using the square root property. The following examples illustrate this method.

EXAMPLE 2 Method of completing the square

To find the roots of the quadratic equation

$$x^2 - 6x - 8 = 0$$

NOTE ▸ first note that the left side is not factorable. [However, $x^2 - 6x$ is part of the special product $(x - 3)^2 = x^2 - 6x + 9$ and this special product is a *perfect square.*] By adding 9 to $x^2 - 6x$, we have $(x - 3)^2$. Therefore, we rewrite the original equation as

$$x^2 - 6x = 8$$

$$x^2 - 6x + 9 = 17 \quad \text{add 9 to both sides}$$

$$(x - 3)^2 = 17 \quad \text{factor left side as a square}$$

$$x - 3 = \pm\sqrt{17} \quad \text{use square root property}$$

The $\pm$ sign means that $x - 3 = \sqrt{17}$ or $x - 3 = -\sqrt{17}$.

By adding 3 to each side, we obtain

$$x = 3 \pm \sqrt{17}$$

which means $x = 3 + \sqrt{17}$ and $x = 3 - \sqrt{17}$ are the two roots of the equation.

Therefore, by adding 9 to $x^2 - 6x$ on the left, we have $x^2 - 6x + 9$, which is the perfect square of $x - 3$. This means, by adding 9 to each side, we have *completed the square of* $x - 3$ on the left side. Then, by equating $x - 3$ to the positive and negative square root of the number on the right side, we were able to find the two roots of the quadratic equation by solving two linear equations. ■

■ The method of completing the square is used later in Section 7.4 and also in Chapter 21.

Practice Exercise

1. Solve by completing the square: $x^2 + 4x - 12 = 0$

How we determine the number to be added to complete the square is based on the special products in Eqs. (6.6) and (6.7). We rewrite these as

$$(x + a)^2 = x^2 + 2ax + a^2 \tag{7.2}$$

$$(x - a)^2 = x^2 - 2ax + a^2 \tag{7.3}$$

We must be certain that the coefficient of the x^2-term is 1 before we start to complete the square. The coefficient of x in each case is numerically $2a$, and the number added to complete the square is a^2. [Thus, if we take half the coefficient of the x-term and square this result, we have the number that completes the square.] In our example, the numerical coefficient of the x-term was 6, and 9 was added to complete the square. Therefore, the procedure for solving a quadratic equation by completing the square is as follows:

NOTE ▸

Solving a Quadratic Equation by Completing the Square

1. Divide each side by a (the coefficient of x^2).
2. Rewrite the equation with the constant on the right side.
3. Complete the square: Add the square of one-half of the coefficient of x to both sides.
4. Write the left side as a square and simplify the right side.
5. Equate the square root of the left side to the positive and negative square root of the right side.
6. Solve the two resulting linear equations.

(a)

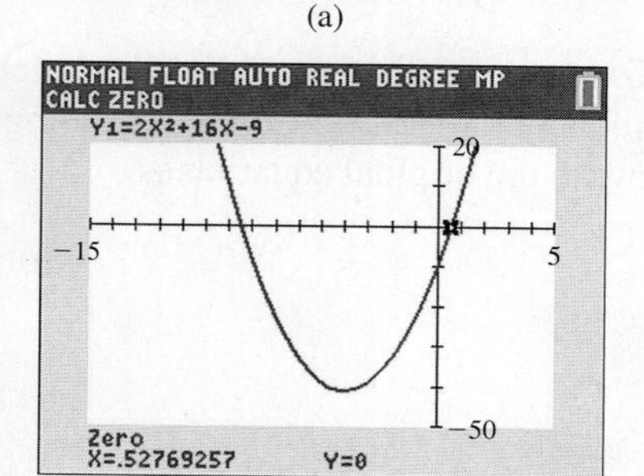

(b)

Fig. 7.4
Graphing calculator keystrokes:
goo.gl/IzJpFL

EXAMPLE 3 Method of completing the square

Solve $2x^2 + 16x - 9 = 0$ by completing the square.

$$x^2 + 8x - \frac{9}{2} = 0$$
1. Divide each term by 2 to make coefficient of x^2 equal to 1.

$$x^2 + 8x = \frac{9}{2}$$
2. Put constant on right by adding 9/2 to each side.

$$\frac{1}{2}(8) = 4; \quad 4^2 = 16$$

$$x^2 + 8x + 16 = \frac{9}{2} + 16 = \frac{41}{2}$$
3. Divide coefficient 8 of x by 2, square the 4, and add to both sides.

$$(x + 4)^2 = \frac{41}{2}$$
4. Write left side as $(x + 4)^2$.

$$x + 4 = \pm\sqrt{\frac{41}{2}}$$
5. Take the square root of each side and equate the $x + 4$ to the positive and negative square root of 41/2.

$$x = -4 \pm \sqrt{\frac{41}{2}}$$
6. Solve for x.

Therefore, the roots are $-4 + \sqrt{\frac{41}{2}}$ and $-4 - \sqrt{\frac{41}{2}}$. The calculator solution, using the *zero* feature, is shown in Fig. 7.4(a) and (b). We use the *zero* feature twice, since we can find only one root at a time. We see from the calculator display that the decimal approximations of the roots are 0.5277 and −8.5277. ■

EXERCISES 7.2

In Exercises 1 and 2, make the given changes in the indicated examples of this section and then solve the resulting quadratic equations by completing the square.

1. In Example 2, change the $-$ sign before $6x$ to $+$.
2. In Example 3, change the coefficient of the second term from 16 to 12.

In Exercises 3–12, solve the given quadratic equations by using the square root property.

3. $x^2 = 25$
4. $x^2 = 100$
5. $x^2 = 7$
6. $s^2 = 15$
7. $2y^2 - 5 = 1$
8. $4x^2 - 7 = 2$
9. $(x - 2)^2 = 25$
10. $(x + 2)^2 = 10$
11. $(y + 3)^2 = 7$
12. $(x - \frac{5}{2})^2 = 100$

In Exercises 13–32, solve the given quadratic equations by completing the square.

13. $x^2 + 2x - 15 = 0$
14. $x^2 - 8x - 20 = 0$
15. $D^2 + 3D + 2 = 0$
16. $t^2 + 5t - 6 = 0$
17. $n^2 = 6n - 4$
18. $(R + 9)(R + 1) = 13$
19. $v(v + 4) = 6$
20. $12 = 8Z - Z^2$
21. $2s^2 + 5s = 3$
22. $8x^2 + 2x = 6$
23. $3y^2 = 3y + 2$
24. $3x^2 = 3 - 4x$
25. $2y^2 - y - 2 = 0$
26. $2 + 6v = 9v^2$
27. $10T - 5T^2 = 4$
28. $\pi^2 y^2 + 2\pi y = 3$
29. $9x^2 + 6x + 1 = 0$
30. $2x^2 = 3x - 2a$
31. $x^2 + 2bx + c = 0$
32. $px^2 + qx + r = 0$

In Exercises 33–36, use completing the square to solve the given problems.

33. The voltage V across a certain electronic device is related to the temperature T (in °C) by $V = 4.0T - 0.2T^2$. For what temperature(s) is $V = 15$ V?
34. A flare is shot vertically into the air such that its distance s (in ft) above the ground is given by $s = 64t - 16t^2$, where t is the time (in s) after it was fired. Find t for $s = 48$ ft.
35. A woman is holding a selfie stick so her cell phone camera is exactly 30 in. from her face. The horizontal distance between the woman's face and the cell phone is exactly 6 in. more than the vertical distance. How far above her face is the cell phone?
36. A rectangular storage area is 8.0 m longer than it is wide. If the area is 28 m^2, what are its dimensions?

Answer to Practice Exercise

1. $x = -6, \quad x = 2$

7.3 The Quadratic Formula

The Quadratic Formula • Character of the Roots of a Quadratic Equation

We now use the method of completing the square to derive a general formula that may be used for the solution of any quadratic equation.

Consider the general quadratic equation:

$$ax^2 + bx + c = 0 \quad (a \neq 0)$$

When we divide through by a, we obtain

$$x^2 + \frac{b}{a}x + \frac{c}{a} = 0$$

Subtracting c/a from each side, we have

$$x^2 + \frac{b}{a}x = -\frac{c}{a}$$

Half of b/a is $b/2a$, which squared is $b^2/4a^2$. Adding $b^2/4a^2$ to each side gives us

$$x^2 + \frac{b}{a}x + \frac{b^2}{4a^2} = -\frac{c}{a} + \frac{b^2}{4a^2}$$

Writing the left side as a perfect square and combining fractions on the right side,

$$\left(x + \frac{b}{2a}\right)^2 = \frac{b^2 - 4ac}{4a^2}$$

Equating $x + \frac{b}{2a}$ to the positive and negative square root of the right side,

$$x + \frac{b}{2a} = \frac{\pm\sqrt{b^2 - 4ac}}{2a}$$

■ The quadratic formula, with the $\pm$ sign, means that the solutions to the quadratic equation

$$ax^2 + bx + c = 0$$

are

$$x = \frac{-b + \sqrt{b^2 - 4ac}}{2a}$$

and

$$x = \frac{-b - \sqrt{b^2 - 4ac}}{2a}$$

When we subtract $b/2a$ from each side and simplify the resulting expression, we obtain the **quadratic formula:**

$$x = \frac{-b \pm \sqrt{b^2 - 4ac}}{2a} \qquad (7.4)$$

The quadratic formula gives us a quick general way of solving any quadratic equation. We need only write the equation in the standard form $ax^2 + bx + c = 0$; substitute the values of a, b, and c into the formula; and simplify.

EXAMPLE 1 Quadratic formula—rational roots

Solve: $x^2 - 5x + 6 = 0$.

($a = 1$, $b = -5$, $c = 6$)

Here, using the indicated values of a, b, and c in the quadratic formula, we have

■ The fact the square root in the quadratic formula simplified to a whole number tells us that it is also possible to solve this equation by factoring.

$$x = \frac{-(-5) \pm \sqrt{(-5)^2 - 4(1)(6)}}{2(1)} = \frac{5 \pm \sqrt{25 - 24}}{2} = \frac{5 \pm 1}{2}$$

$$x = \frac{5 + 1}{2} = 3 \quad \text{or} \quad x = \frac{5 - 1}{2} = 2$$

The roots $x = 3$ and $x = 2$ check when substituted in the original equation. ■

CAUTION It must be emphasized that, in using the quadratic formula, the entire expression $-b \pm \sqrt{b^2 - 4ac}$ is divided by $2a$. It is a relatively common error to divide only the radical $\sqrt{b^2 - 4ac}$. ■

EXAMPLE 2 Quadratic formula—irrational roots

Solve: $2x^2 - 7x - 5 = 0$.

($a = 2$, $b = -7$, $c = -5$)

TI-89 graphing calculator keystrokes for Example 2: goo.gl/zxbcnN

Substituting the values for a, b, and c in the quadratic formula, we have

$$x = \frac{-(-7) \pm \sqrt{(-7)^2 - 4(2)(-5)}}{2(2)} = \frac{7 \pm \sqrt{49 + 40}}{4} = \frac{7 \pm \sqrt{89}}{4}$$

$$x = \frac{7 + \sqrt{89}}{4} = 4.108 \quad \text{or} \quad x = \frac{7 - \sqrt{89}}{4} = -0.6085$$

The exact roots are $x = \frac{7 \pm \sqrt{89}}{4}$ (this form is often used when the roots are irrational). Approximate decimal values are $x = 4.108$ and $x = -0.6085$. ■

Practice Exercise

1. Solve using the quadratic formula: $3x^2 + x - 5 = 0$

EXAMPLE 3 Quadratic formula—double root

Graphing calculator program for solving a quadratic equation: `goo.gl/gTefFl`

Solve: $9x^2 + 24x + 16 = 0$.

In this example, $a = 9$, $b = 24$, and $c = 16$. Thus,

$$x = \frac{-24 \pm \sqrt{24^2 - 4(9)(16)}}{2(9)} = \frac{-24 \pm \sqrt{576 - 576}}{18} = \frac{-24 \pm 0}{18} = -\frac{4}{3}$$

NOTE ➧ [Here, both roots are $-\frac{4}{3}$, so $x = -\frac{4}{3}$ is called a *double root*. We will always get a double root when $b^2 - 4ac = 0$, as in this case.] ■

EXAMPLE 4 Quadratic formula—imaginary roots

Solve: $3x^2 - 5x + 4 = 0$.

In this example, $a = 3$, $b = -5$, and $c = 4$. Therefore,

■ Recall that the square root of a negative number is called an *imaginary number*.

$$x = \frac{-(-5) \pm \sqrt{(-5)^2 - 4(3)(4)}}{2(3)} = \frac{5 \pm \sqrt{25 - 48}}{6} = \frac{5 \pm \sqrt{-23}}{6}$$

NOTE ➧ [These roots contain imaginary numbers. This happens if $b^2 - 4ac < 0$.] ■

The previous examples illustrate the *character of the roots* of a quadratic equation. If a, b, and c are rational numbers, by noting the value of $b^2 - 4ac$ (called the **discriminant**), we have the following:

■ If $b^2 - 4ac$ is positive and a perfect square, $ax^2 + bx + c$ is factorable.

■ Examples 1–4 illustrate each of the four cases listed to the right.

Character of the Roots of a Quadratic Equation

1. If $b^2 - 4ac$ is positive and a perfect square, the roots are real, rational, and unequal.
2. If $b^2 - 4ac$ is positive but not a perfect square, the roots are real, irrational, and unequal.
3. If $b^2 - 4ac = 0$, the roots are real, rational, and equal.
4. If $b^2 - 4ac < 0$, the roots contain imaginary numbers and are unequal. Therefore, there are no real roots.

We can use the value of $b^2 - 4ac$ to help in checking the roots or in finding the character of the roots without having to solve the equation completely.

EXAMPLE 5 Quadratic formula—literal numbers

The equation $s = s_0 + v_0t - \frac{1}{2}gt^2$ is used in the analysis of projectile motion (see Fig. 7.5). Solve for t.

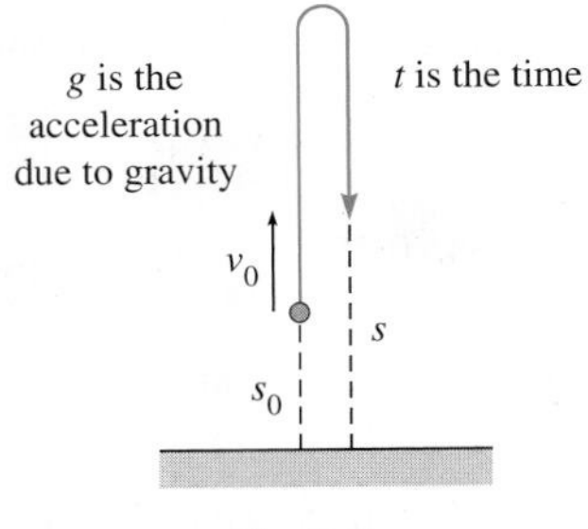

Fig. 7.5

$$gt^2 - 2v_0t - 2(s_0 - s) = 0 \quad \text{multiply by } -2\text{, put in form } ax^2 + bx + c = 0$$

In this form, we see that $a = g$, $b = -2v_0$, and $c = -2(s_0 - s)$:

$$t = \frac{-(-2v_0) \pm \sqrt{(-2v_0)^2 - 4g(-2)(s_0 - s)}}{2g}$$

$$= \frac{2v_0 \pm \sqrt{4(v_0^2 + 2gs_0 - 2gs)}}{2g}$$

$$= \frac{2v_0 \pm 2\sqrt{v_0^2 + 2gs_0 - 2gs}}{2g}$$

$$= \frac{v_0 \pm \sqrt{v_0^2 + 2gs_0 - 2gs}}{g}$$ ■

EXAMPLE 6 Quadratic formula—patio dimensions

A rectangular area 17.0 m long and 12.0 m wide is to be used for a patio with a rectangular pool. One end and one side of the patio area around the pool (the chairs, sunning, etc.) are to be the same width. The other end with the diving board is to be twice as wide, and the other side is to be three times as wide as the narrow side. The pool area is to be 96.5 m^2. What are the widths of the patio ends and sides, and the dimensions of the pool? See Fig. 7.6.

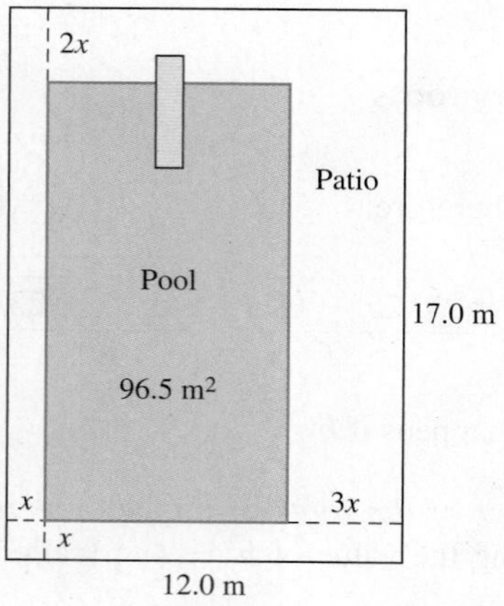

First, let $x =$ the width of the narrow end and side of the patio. The other end is then $2x$ in width, and the other side is $3x$ in width. Because the pool area is 96.5 m^2, we have

$$\underbrace{(17.0 - 3x)}_{\text{pool length}}\underbrace{(12.0 - 4x)}_{\text{pool width}} = \underbrace{96.5}_{\text{pool area}}$$

$$204 - 68.0x - 36.0x + 12x^2 = 96.5$$

$$12x^2 - 104.0x + 107.5 = 0$$

$$x = \frac{-(-104.0) \pm \sqrt{(-104.0)^2 - 4(12)(107.5)}}{2(12)} = \frac{104.0 \pm \sqrt{5656}}{24}$$

Fig. 7.6

Evaluating, we get $x = 7.5$ m and $x = 1.2$ m. The value 7.5 m cannot be the required result because the width of the patio would be greater than the width of the entire area. For $x = 1.2$ m, the pool would have a length 13.4 m and width 7.2 m. These give an area of 96.5 m^2, which checks. The widths of the patio area are then 1.2 m, 1.2 m, 2.4 m, and 3.6 m. ■

EXERCISES 7.3

In Exercises 1 and 2, make the given changes in the indicated examples of this section and then solve the resulting equations by the quadratic formula.

1. In Example 1, change the $-$ sign before $5x$ to $+$.

2. In Example 2, change the coefficient of x^2 from 2 to 3.

In Exercises 3–34, solve the given quadratic equations using the quadratic formula. If there are no real roots, state this as the answer. Exercises 3–6 are the same as Exercises 13–16 of Section 7.2.

3. $x^2 + 2x - 15 = 0$

4. $x^2 - 8x - 20 = 0$

5. $D^2 + 3D + 2 = 0$

6. $t^2 + 5t - 6 = 0$

7. $x^2 - 5x + 3 = 0$

8. $x^2 + 10x - 4 = 0$

9. $v^2 = 15 - 2v$

10. $16V - 24 = 2V^2$

11. $8s^2 + 20s = 12$

12. $4x^2 + x = 3$

13. $3y^2 = 3y + 2$

14. $3x^2 = 3 - 4x$

15. $z + 2 = 2z^2$

16. $2 + 6v = 9v^2$

17. $30y^2 + 23y - 40 = 0$

18. $62x + 63 = 40x^2$

19. $5t^2 + 3 = 7t$

20. $2d(d - 2) = -7$

21. $s^2 = 9 + s(1 - 2s)$

22. $20r^2 = 20r + 1$

23. $25y^2 = 81$

24. $37T = T^2$

25. $15 + 4z = 32z^2$

26. $4x^2 - 12x = 7$

27. $x^2 - 0.20x - 0.40 = 0$

28. $3.2x^2 = 2.5x + 7.6$

29. $0.29Z^2 - 0.18 = 0.63Z$

30. $13.2x = 15.5 - 12.5x^2$

31. $x^2 + 2cx - 1 = 0$

32. $x^2 - 7x + (6 + a) = 0$

33. $b^2x^2 + 1 - a = (b + 1)x$

34. $c^2x^2 - x - 1 = x^2$

In Exercises 35–38, without solving the given equations, determine the character of the roots.

35. $2x^2 - 7x = -8$

36. $3x^2 = 14 - 19x$

37. $3.6t^2 + 2.1 = 7.7t$

38. $0.45s^2 + 0.33 = 0.12s$

In Exercises 39–62, solve the given problems. All numbers are accurate to at least two significant digits.

39. Find k if the equation $x^2 + 4x + k = 0$ has a real double root.

40. Find the smallest positive integer value of k if the equation $x^2 + 3x + k = 0$ has roots with imaginary numbers.

41. Solve the equation $x^4 - 5x^2 + 4 = 0$ for x. [*Hint:* The equation can be written as $(x^2)^2 - 5(x^2) + 4 = 0$. First solve for x^2.]

42. Without drawing the graph or completely solving the equation, explain how to find the number of x-intercepts of a quadratic function.

43. Use the discriminant $b^2 - 4ac$ to determine if the equation $90x^2 - 123x + 40 = 0$ can be solved by factoring. Explain why or why not. Do not solve.

44. Solve $6x^2 - x = 15$ for x by (a) factoring, (b) completing the square, and (c) the quadratic formula. Which is (a) longest? (b) shortest?

45. In machine design, in finding the outside diameter D_0 of a hollow shaft, the equation $D_0^2 - DD_0 - 0.25D^2 = 0$ is used. Solve for D_0 if $D = 3.625$ cm.

46. A missile is fired vertically into the air. The distance s (in ft) above the ground as a function of time t (in s) is given by $s = 300 + 500t - 16t^2$. (a) When will the missile hit the ground? (b) When will the missile be 1000 ft above the ground?

47. In analyzing the deflection of a certain beam, the equation $8x^2 - 15Lx + 6L^2 = 0$ is used. Solve for x, if $x < L$.

48. A homeowner wants to build a rectangular patio with an area of 20.0 m^2, such that the length is 2.0 m more than the width. What should the dimensions be?

49. Two cars leave an intersection at the same time, one going due east and the other due south. After one has gone 2.0 km farther than the other, they are 6.0 km apart on a direct line. How far did each go?

50. A student cycled 3.0 km/h faster to college than when returning, which took 15 min longer. If the college is 4.0 km from home, what were the speeds to and from college?

51. For a rectangle, if the ratio of the length to the width equals the ratio of the length plus the width to the length, the ratio is called the *golden ratio.* Find the value of the golden ratio, which the ancient Greeks thought had the most pleasing properties to look at.

52. When focusing a camera, the distance r the lens must move from the infinity setting is given by $r = f^2/(p - f)$, where p is the distance from the object to the lens, and f is the focal length of the lens. Solve for f.

53. In calculating the current in an electric circuit with an inductance L, a resistance R, and a capacitance C, it is necessary to solve the equation $Lm^2 + Rm + 1/C = 0$. Solve for m in the terms of L, R, and C. See Fig. 7.7.

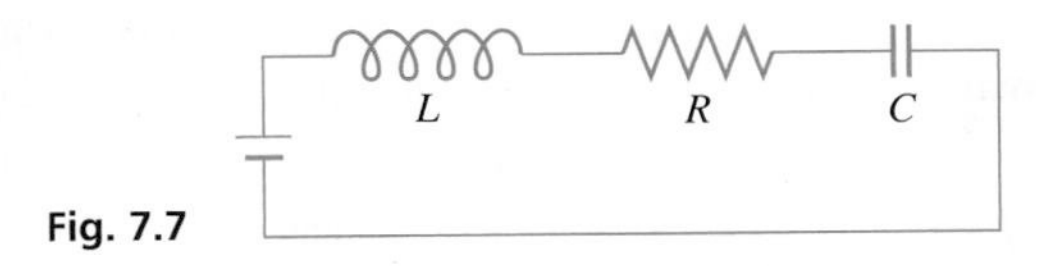

Fig. 7.7

54. In finding the radius r of a circular arch of height h and span b, we use the formula shown below. Solve for h.

$$r = \frac{b^2 + 4h^2}{8h}$$

55. A computer monitor has a viewing screen that is 33.8 cm wide and 27.3 cm high, with a uniform edge around it. If the edge covers 20.0% of the monitor front, what is the width of the edge?

56. An investment of \$2000 is deposited at a certain annual interest rate. One year later, \$3000 is deposited in another account at the same rate. At the end of the second year, the accounts have a total value of \$5319.05. The interest rate r can be found by solving $2000(1 + r)^2 + 3000(1 + r) = \$5319.05$. What is the interest rate?

57. In remodeling a house, an architect finds that by adding the same amount to each dimension of a 12-ft by 16-ft rectangular room, the area would be increased by 80 ft^2. How much must be added to each dimension?

58. Two pipes together drain a wastewater-holding tank in 6.00 h. If used alone to empty the tank, one takes 2.00 h longer than the other. How long does each take to empty the tank if used alone?

59. In order to have the proper strength, the angle iron shown in Fig. 7.8 must have a cross-sectional area of 53.5 cm^2. Find the required thickness x.

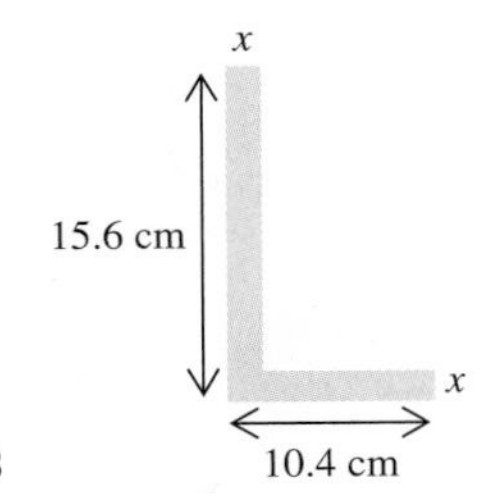

Fig. 7.8

60. For an optical lens, the sum of the reciprocals of p, the distance of the object from the lens, and q, the distance of the image from the lens, equals the reciprocal of f, the focal length of the lens. If p is 5.0 cm greater than q $(q > 0)$ and $f = 4.0$ cm, find p and q.

61. The length of a tennis court is 12.8 m more than its width. If the area of the tennis court is 262 m^2, what are its dimensions? See Fig. 7.9.

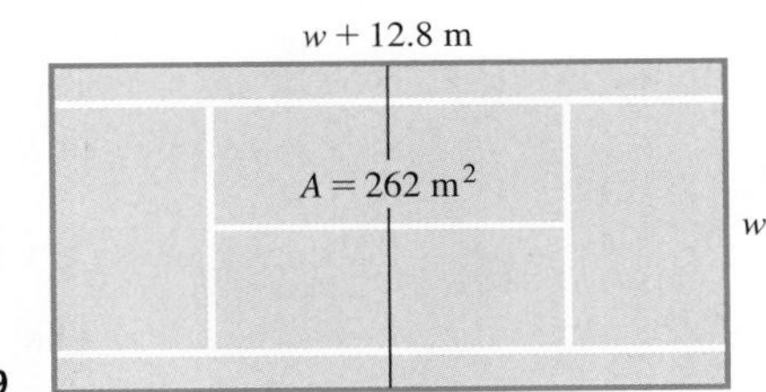

Fig. 7.9

62. Two circular oil spills are tangent to each other. If the distance between centers is 800 m and they cover a combined area of 1.02×10^6 m^2, what is the radius of each? See Fig. 7.10.

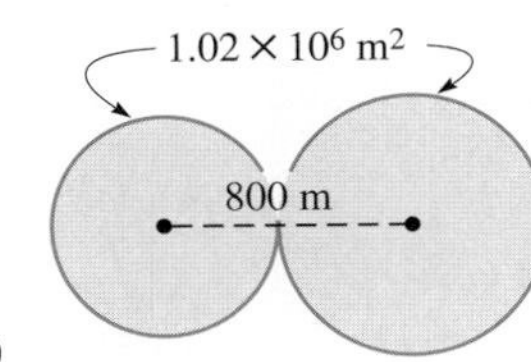

Fig. 7.10

Answer to Practice Exercise

1. $x = \dfrac{-1 \pm \sqrt{61}}{6}$

7.4 The Graph of the Quadratic Function

The Parabola • Vertex and y-intercept • Solving Quadratic Equations Graphically

In this section, we discuss the graph of the quadratic function $ax^2 + bx + c$ and show the graphical solution of a quadratic equation. By letting $y = ax^2 + bx + c$, we can graph this function, as in Chapter 3.

EXAMPLE 1 Graphing a quadratic function

Graph the function $f(x) = x^2 + 2x - 3$.

First, let $y = x^2 + 2x - 3$. Then set up a table of values and graph the function as shown in Fig. 7.11. We can also display it on a calculator as shown in Fig. 7.12.

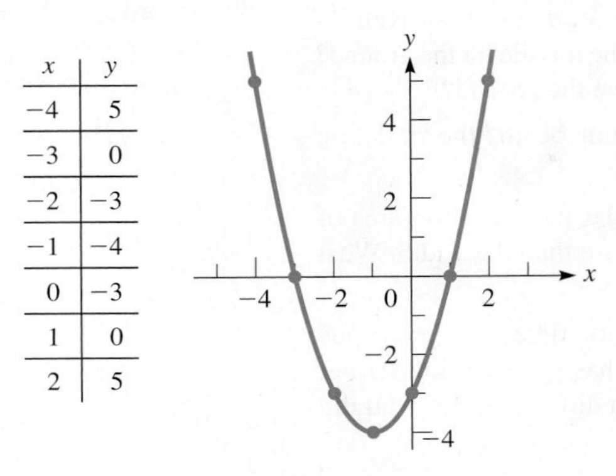

x	y
−4	5
−3	0
−2	−3
−1	−4
0	−3
1	0
2	5

Fig. 7.11

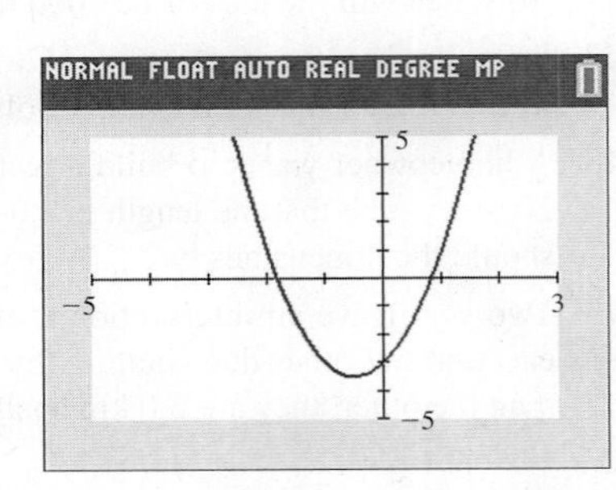

Fig. 7.12 ■

The shape of the graph in Figs. 7.11 and 7.12 is called a **parabola,** and *the graph of any quadratic function* $y = ax^2 + bx + c$ *will have the same basic shape.* A parabola can open upward (as in Example 1) or downward. The location of the parabola and how it opens depends on the values of a, b, and c.

In Example 1, the parabola has a minimum point at $(-1, -4)$, and the curve opens upward. *All parabolas have an* **extreme point** *of this type.* [For $y = ax^2 + bx + c$, if $a > 0$, the parabola has a minimum point and opens upward. If $a < 0$, the parabola has a maximum point and opens downward.] *The extreme point of the parabola is known as its* **vertex.**

NOTE ▶

EXAMPLE 2 Parabola—extreme points

The graph of $y = 2x^2 - 8x + 6$ is shown in Fig. 7.13(a). For this parabola, $a = 2$ $(a > 0)$ and it opens upward. The vertex (a minimum point) is $(2, -2)$.

The graph of $y = -2x^2 + 8x - 6$ is shown in Fig. 7.13(b). For this parabola, $a = -2$ $(a < 0)$ and it opens downward. The vertex (a maximum point) is $(2, 2)$.

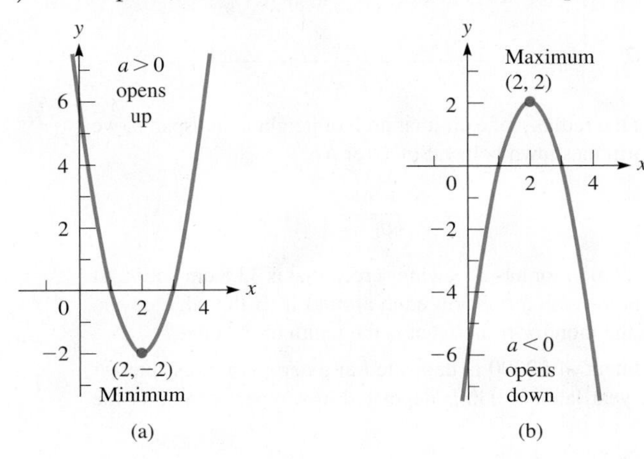

Fig. 7.13 ■

We can sketch the graphs of a parabola by using its basic shape and knowing two or three points, including the vertex. Even when using a graphing calculator, we can get a check on the graph by knowing the vertex and how the parabola opens.

In order to find the coordinates of the extreme point, start with the quadratic function

$$y = ax^2 + bx + c$$

Then factor a from the two terms containing x, obtaining

$$y = a\left(x^2 + \frac{b}{a}x\right) + c$$

Now, completing the square of the terms within parentheses, we have

$$y = a\left(x^2 + \frac{b}{a}x + \frac{b^2}{4a^2}\right) + c - \frac{b^2}{4a}$$

$$= a\left(x + \frac{b}{2a}\right)^2 + c - \frac{b^2}{4a}$$

NOTE ▶ [This form of the function shows a *horizontal shift of* $-b/2a$, which means **the vertex is at $x = -b/2a$.**] *The y-coordinate of the vertex can be found by substituting this x-value into the original function.*

■ For a review of horizontal and vertical shifts in a function, see Examples 7 and 8 in Section 3.5.

Another easily found point is the y-intercept. By substituting $x = 0$ into $y = ax^2 + bx + c$, we get $y = c$. *This means that the point* $(0, c)$ *is the y-intercept.* This information is summarized below.

Vertex and y-Intercept of the Quadratic Function $y = ax^2 + bx + c$

Vertex: $\left(-\frac{b}{2a}, f\left(-\frac{b}{2a}\right)\right)$ **(7.5)**

y-Intercept: $(0, c)$

EXAMPLE 3 Graphing a parabola—vertex—y-intercept

For the graph of the function $y = 2x^2 - 8x + 6$, find the vertex and y-intercept and sketch the graph. (This function was also used in Example 2.)

First, $a = 2$ and $b = -8$. This means that the x-coordinate of the vertex is

$$\frac{-b}{2a} = \frac{-(-8)}{2(2)} = \frac{8}{4} = 2$$

and the y-coordinate is

$$y = 2(2^2) - 8(2) + 6 = -2$$

Thus, the vertex is $(2, -2)$. Because $a > 0$, it is a minimum point.

Because $c = 6$, the y-intercept is (0,6).

We can use the minimum point $(2, -2)$ and the y-intercept (0, 6), along with the fact that the graph is a parabola, to get an approximate sketch of the graph. Noting that a parabola increases (or decreases) away from the vertex in the same way on each side of it (it is *symmetric* with respect to a vertical line through the vertex), we sketch the graph in Fig. 7.14. It is the same graph as that shown in Fig. 7.13(a). ■

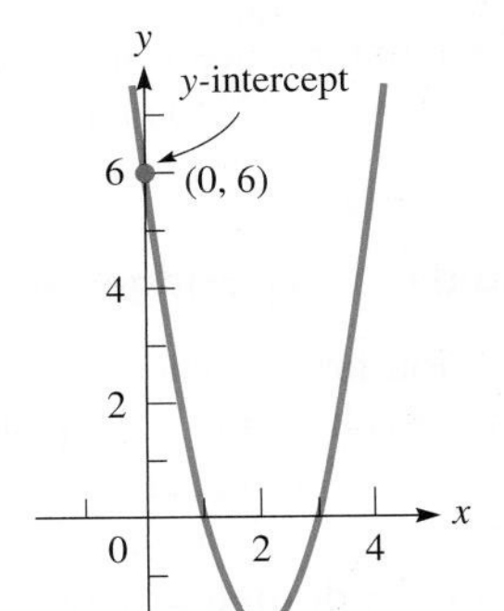

Fig. 7.14

Practice Exercise

1. (a) Find the vertex and y-intercept for the graph of the function $y = 3x^2 + 12x - 4$.
   (b) Is the vertex a minimum or a maximum point?

If we are not using a graphing calculator, we may need one or two additional points to get a good sketch of a parabola. This would be especially true if the y-intercept is close to the vertex. Two points we can find are the x-intercepts, if the parabola crosses the x-axis (one point if the vertex is on the x-axis). They are found by setting $y = 0$ and solving the quadratic equation $ax^2 + bx + c = 0$. Also, we may simply find one or two points other than the vertex and the y-intercept. Sketching a parabola in this way is shown in the following two examples.

EXAMPLE 4 Graphing a parabola using the vertex and intercepts

Sketch the graph of $y = -x^2 + x + 6$.

We first note that $a = -1$ and $b = 1$. Therefore, the x-coordinate of the maximum point $(a < 0)$ is $-\frac{1}{2(-1)} = \frac{1}{2}$. The y-coordinate is $-(\frac{1}{2})^2 + \frac{1}{2} + 6 = \frac{25}{4}$. This means that the maximum point is $(\frac{1}{2}, \frac{25}{4})$.

The y-intercept is (0,6).

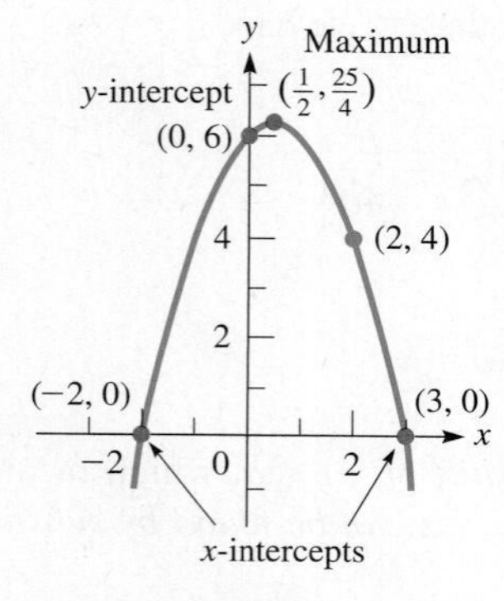

Fig. 7.15

We can see from Fig. 7.15 that the vertex and y-intercept are too close together for us to get a good idea of the shape of the parabola, so additional points are needed. We will find the x-intercepts.

$$
\begin{aligned}
-x^2 + x + 6 &= 0 && \text{set } y = 0 \\
x^2 - x - 6 &= 0 && \text{multiply both sides by } -1 \\
(x - 3)(x + 2) &= 0 && \text{factor} \\
x - 3 = 0 \quad & x + 2 = 0 && \text{set each factor equal to 0} \\
x = 3 \quad & x = -2 && \text{solve}
\end{aligned}
$$

This means that the x-intercepts are (3, 0) and $(-2, 0)$, as shown in Fig. 7.15.

Also, rather than finding the x-intercepts, we can let $x = 2$ (or some value to the right of the vertex) and then use the point (2, 4).

Note that the y-coordinate of the vertex can be used to find the range of a quadratic function. In this case, the range is all values $y \leq \frac{25}{4}$. The domain is all real numbers. ■

EXAMPLE 5 Graphing a parabola—no x-term

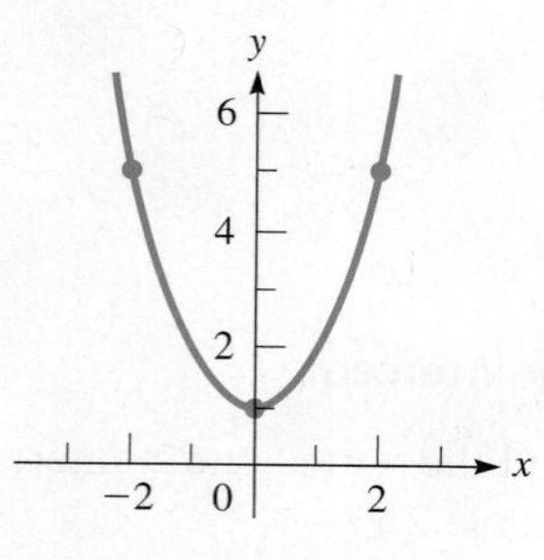

Fig. 7.16

Sketch the graph of $y = x^2 + 1$.

Because there is no x-term, $b = 0$. This means that the x-coordinate of the minimum point $(a > 0)$ is 0 and that the minimum point and the y-intercept are both (0, 1). We know that the graph opens upward, because $a > 0$, which in turn means that it does not cross the x-axis. Now, letting $x = 2$ and $x = -2$, find the points (2, 5) and $(-2, 5)$ on the graph, which is shown in Fig. 7.16. ■

EXAMPLE 6 Finding a quadratic function—Golden Gate Bridge

■ See the chapter introduction.

The Golden Gate Bridge in San Francisco is a suspension bridge, and its supporting cables are parabolic. With the origin at the low point of the cable, find the equation that represents the height of the parabolic cables if the towers are 4200 ft apart and the maximum sag is 300 ft? See Fig. 7.17.

Since both the y-intercept and vertex are at the origin, the desired equation must be of the form $y = ax^2$. We also know that the point (2100, 300) is on the parabola. By substituting these values into the equation, we can solve for a:

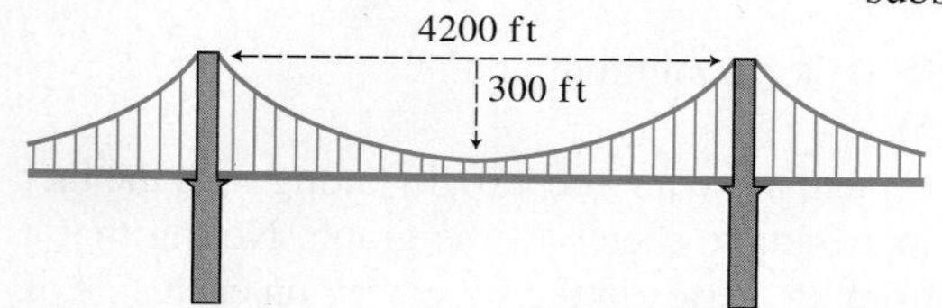

Fig. 7.17

$$300 = a(2100^2)$$

$$a = \frac{300}{2100^2} = 0.000068$$

Thus, the equation for the cable is $y = 0.000068x^2$. This can be used to find the height of the vertical supports at any desired location. ■

SOLVING QUADRATIC EQUATIONS GRAPHICALLY

In Section 3.5, we showed that to solve an equation graphically, we (1) collect all terms on the left side of the equation, with zero on the right side, (2) graph the function on the left side, and (3) use the ***zero*** feature to find the x-coordinate(s) of the x-intercepts. This procedure works for *all* equations, including quadratic equations. We used this to check the solutions obtained in Example 3 of Section 7.2. Another method for solving an equation graphically is to use the ***intersect*** feature to find the point(s) of intersection between the left and right sides of the equation. This method does not require having a zero on the right side. The following example illustrates these methods.

EXAMPLE 7 Solving equation graphically—height of a projectile

A projectile is fired vertically upward from the ground with a velocity of 38 m/s. Its distance above the ground is given by $s = -4.9t^2 + 38t$, where s is the distance (in m) and t is the time (in s). Graph the function, and from the graph determine (1) when the projectile will hit the ground, (2) how long it takes to reach 45 m above the ground, and (3) the maximum height of the projectile.

1. To find when the projectile hits the ground, we set the height $s = 0$ and then solve for t. This results in the equation $-4.9t^2 + 38t = 0$. Graphically, this is the x-intercept of the function $y = -4.9t^2 + 38t$, which can be found using the *zero* feature on a calculator. The solution is $t = 7.76$ s as shown in Fig. 7.18(a).
2. To find when the projectile is 45 m above the ground, we can use the *intersect* feature to find the points of intersection between the given function and the horizontal line $y = 45$. There are two solutions: $t = 1.46$ s and $t = 6.30$ s. The first of these solutions is shown in Fig. 7.18(b).
3. The maximum height occurs at the vertex, which can be found using the *maximum* feature on a calculator. From Fig. 7.18(c), we see that the maximum height is 73.7 m, which occurs after 3.88 s.

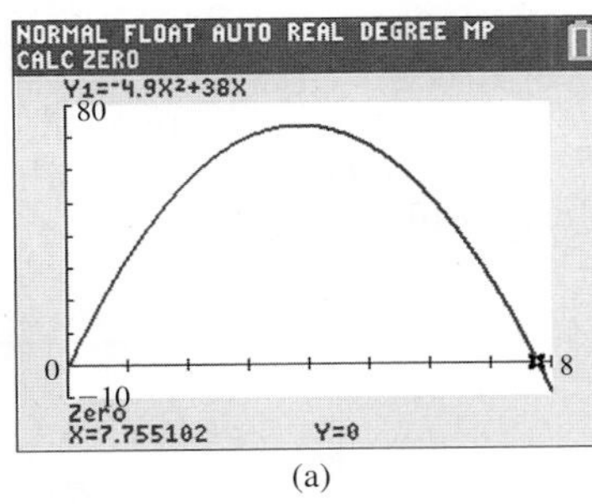

(a)

Graphing calculator keystrokes: goo.gl/GLlbOC

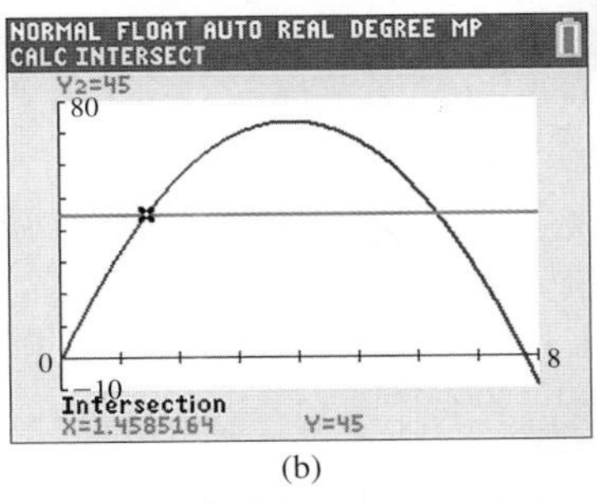

(b)

Graphing calculator keystrokes: goo.gl/pM6cv9

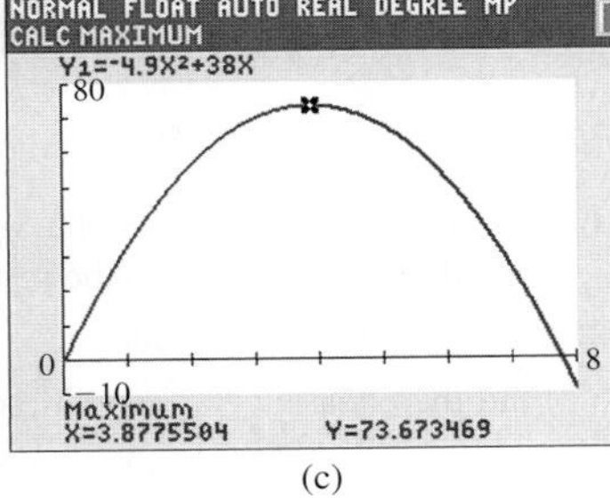

(c)

Graphing calculator keystrokes: goo.gl/Qlfejw ■

Fig. 7.18

EXERCISES 7.4

In Exercises 1 and 2, make the given changes in the indicated examples of this section and then solve the resulting problems.

1. In Example 3, change the $-$ sign before $8x$ to $+$ and then sketch the graph.
2. In Example 7, change 38 to 42 and then answer the questions.

In Exercises 3–8, sketch the graph of each parabola by using only the vertex and the y-intercept. Check the graph using a calculator.

3. $y = x^2 - 6x + 5$
4. $y = -x^2 - 4x - 3$
5. $y = -3x^2 + 10x - 4$
6. $s = 2t^2 + 8t - 5$
7. $R = v^2 - 4v$
8. $y = -2x^2 - 5x$

In Exercises 9–12, sketch the graph of each parabola by using the vertex, the y-intercept, and the x-intercepts. Check the graph using a calculator.

9. $y = x^2 - 4$ **10.** $y = x^2 + 3x$

11. $y = -2x^2 - 6x + 8$ **12.** $u = -3v^2 + 12v - 9$

In Exercises 13–16, sketch the graph of each parabola by using the vertex, the y-intercept, and two other points, not including the x-intercepts. Check the graph using a calculator.

13. $y = 2x^2 + 3$ **14.** $s = t^2 + 2t + 2$

15. $y = -2x^2 - 2x - 6$ **16.** $y = -3x^2 - x$

In Exercises 17–24, use a calculator to solve the given equations. Round solutions to the nearest hundredth. If there are no real roots, state this.

17. $2x^2 - 7 = 0$ **18.** $5 - x^2 = 0$

19. $-3x^2 + 9x - 5 = 0$ **20.** $2t^2 = 7t + 4$

21. $x(2x - 1) = -3$ **22.** $6w - 15 = 3w^2$

23. $4R^2 = 12 - 7R$ **24.** $3x^2 - 25 = 20x$

In Exercises 25–30, use a calculator to graph all three parabolas on the same coordinate system. In Exercises 25–28, describe (a) the shifts (see page 105) of $y = x^2$ that occur and (b) how each parabola opens. In Exercises 29 and 30, describe (a) the shifts and (b) the stretching and shrinking.

25. (a) $y = x^2$ (b) $y = x^2 + 3$ (c) $y = x^2 - 3$

26. (a) $y = x^2$ (b) $y = (x - 3)^2$ (c) $y = (x + 3)^2$

27. (a) $y = x^2$ (b) $y = (x - 2)^2 + 3$ (c) $y = (x + 2)^2 - 3$

28. (a) $y = x^2$ (b) $y = -x^2$ (c) $y = -(x - 2)^2$

29. (a) $y = x^2$ (b) $y = 3x^2$ (c) $y = \frac{1}{3}x^2$

30. (a) $y = x^2$ (b) $y = -3(x - 2)^2$ (c) $y = \frac{1}{3}(x + 2)^2$

In Exercises 31–50, solve the given applied problem.

31. Use a calculator to find the vertex of $s = -9.8t^2 + 25t + 4$. Round the coordinates to the nearest hundredth.

32. Find the range of the function $s = -16t^2 + 64t + 6$.

33. Find the equation of the quadratic function that has vertex (0, 0) and passes through the point (25, 125).

34. A parabolic satellite dish is 8.40 in. deep and 36.0 in. across its opening. If the dish is positioned so it opens directly upward with its vertex at the origin, find the equation of its parabolic cross section.

35. Find the smallest integer value of c such that $y = 2x^2 - 4x - c$ has at least one real root.

36. Find the smallest integer value of c such that $y = 3x^2 - 12x + c$ has no real roots.

37. Find the equation of the parabola that contains the points $(-2, -3)$, $(0, -3)$, and $(2, 5)$.

38. Find the equation of the parabola that contains the points $(-1, 14)(1, 9)$, and $(2, 8)$.

39. The vertical distance d (in cm) of the end of a robot arm above a conveyor belt in its 8-s cycle is given by $d = 2t^2 - 16t + 47$. Sketch the graph of $d = f(t)$.

40. When mineral deposits form a uniform coating 1 mm thick on the inside of a pipe of radius r (in mm), the cross-sectional area A through which water can flow is $A = \pi(r^2 - 2r + 1)$. Sketch $A = f(r)$.

41. The shape of the Gateway Arch in St. Louis can be approximated by the parabola $y = 192 - 0.0208\,x^2$ (in meters) if the origin is at ground level, under the center of the Arch. Display the equation representing the Arch on a calculator. How high and wide is the Arch?

42. Under specified conditions, the pressure loss L (in lb/in.^{2} per 100 ft), in the flow of water through a fire hose in which the flow is q gal/min, is given by $L = 0.0002q^2 + 0.005q$. Sketch the graph of L as a function of q, for $q < 100$ gal/min.

43. A computer analysis of the power P (in W) used by a pressing machine shows that $P = 50i - 3i^2$, where i is the current (in A). Sketch the graph of $P = f(i)$.

44. Tests show that the power P (in hp) of an automobile engine as a function of r (in r/min) is given by $P = -5.0 \times 10^{-6}r^2 + 0.050r - 45$ $(1500 < r < 6000 \text{ r/min})$. Sketch the graph of P vs. r and find the maximum power that is produced.

45. The height h (in m) of a fireworks shell shot vertically upward as a function of time t (in s) is $h = -4.9t^2 + 68t + 2$. How long should the fuse last so that the shell explodes at the top of its trajectory?

46. In a certain electric circuit, the resistance R (in Ω) that gives resonance is found by solving the equation $25R = 3(R^2 + 4)$. Solve this equation graphically (to 0.1 Ω).

47. If the radius of a circular solar cell is increased by 1.00 cm, its area is 96.0 cm^2. What was the original radius?

48. The diagonal of a rectangular floor is 3.00 ft less than twice the length of one of the sides. If the other side is 15.0 ft long, what is the area of the floor?

49. A security fence is to be built around a rectangular parking area of 20,000 ft^2. If the front side of the fence costs \$20/ft and the other three sides cost \$10/ft, solve graphically for the dimensions (to 1 ft) of the parking area if the fence is to cost \$7500. See Fig. 7.19.

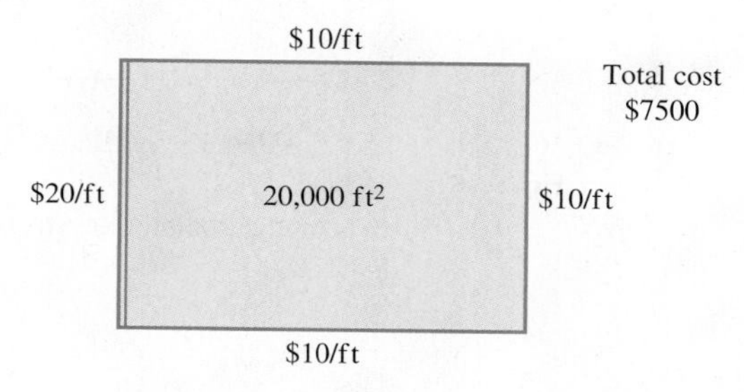

Fig. 7.19

50. An airplane pilot could decrease the time t (in h) needed to travel the 630 mi from Ottawa to Milwaukee by 20 min if the plane's speed v is increased by 40 mi/h. Set up the appropriate equation and solve graphically for v (to two significant digits).

Answers to Practice Exercise

1. (a) $(-2, -16)$, $(0, -4)$ (b) minimum

CHAPTER 7 KEY FORMULAS AND EQUATIONS

Quadratic equation	$ax^2 + bx + c = 0$	**(7.1)**
Quadratic formula	$x = \dfrac{-b \pm \sqrt{b^2 - 4ac}}{2a}$	**(7.4)**
Vertex	$\left(-\dfrac{b}{2a}, f\left(-\dfrac{b}{2a}\right)\right)$	**(7.5)**

CHAPTER 7 REVIEW EXERCISES

CONCEPT CHECK EXERCISES

Determine each of the following as being either **true** *or* **false**. *If it is false, explain why.*

1. The solution of the equation $x^2 - 2x = 0$ is $x = 2$.
2. The first steps in solving the equation $2x^2 + 4x - 7 = 0$ by completing the square is to divide 4 by 2, square the result, which is then added to each side of the equation.
3. The quadratic formula is $x = \dfrac{-b \pm \sqrt{b^2 - 4ac}}{2a}$.
4. The graph of $y = ax^2 + bx + c$ has a minimum point if $a > 0$.

PRACTICE AND APPLICATIONS

In Exercises 5–16, solve the given quadratic equations by factoring.

5. $x^2 + 3x - 4 = 0$
6. $x^2 + 3x - 10 = 0$
7. $x^2 - 10x + 21 = 0$
8. $P^2 - 27 = 6P$
9. $3x^2 + 11x = 4$
10. $11y = 6y^2 + 3$
11. $6t^2 = 13t - 5$
12. $3x^2 + 5x + 2 = 0$
13. $4s^2 = 18s$
14. $23n + 35 = 6n^2$
15. $4B^2 = 8B + 21$
16. $6\pi^2x^2 = 8 - 47\pi x$

In Exercises 17–28, solve the given quadratic equations by using the quadratic formula.

17. $x^2 - 4x - 96 = 0$
18. $x^2 + 3x - 18 = 0$
19. $m^2 + 2m = 6$
20. $1 + 7D = D^2$
21. $2x^2 - x = 36$
22. $6x^2 = 28 - 2x$
23. $18s + 12 = 24s^2$
24. $2 - 7x = 5x^2$
25. $2.1x^2 + 2.3x + 5.5 = 0$
26. $0.30R^2 - 0.42R = 0.15$
27. $4x = 9 - 6x^2$
28. $25t = 24t^2 - 20$

In Exercises 29–42, solve the given quadratic equations by any appropriate algebraic method. If there are no real roots, state this as the answer.

29. $4x^2 - 5 = 15$
30. $12y^2 = 20y$
31. $x^2 + 4x - 4 = 0$
32. $x^2 + 3x + 1 = 0$
33. $3x^2 + 8x + 2 = 0$
34. $3p^2 = 28 - 5p$
35. $4v^2 + v = 3$
36. $3n - 6 = 18n^2$
37. $7 + 3C = -2C^2$
38. $5y = 4y^2 - 8$
39. $a^2x^2 + 2ax + 2 = 0$
40. $16r^2 = 8r - 1$
41. $ay^2 = a - 3y$
42. $2bx = x^2 - 3b$

In Exercises 43–46, solve the given quadratic equations by completing the square.

43. $x^2 - x - 30 = 0$
44. $x^2 = 2x + 5$
45. $2t^2 = t + 4$
46. $4x^2 - 8x = 3$

In Exercises 47–50, solve the given equations.

47. $\dfrac{x - 4}{x - 1} = \dfrac{2}{x}$
48. $\dfrac{V - 1}{3} = \dfrac{5}{V} + 1$
49. $\dfrac{x^2 - 3x}{x - 3} = \dfrac{x^2}{x + 2}$
50. $\dfrac{x - 2}{x - 5} = \dfrac{15}{x^2 - 5x}$

In Exercises 51–54, sketch the graphs of the given functions by using the vertex, the y-intercept, and one or two other points.

51. $y = 2x^2 - x - 1$
52. $y = -4x^2 - 1$
53. $y = x - 3x^2$
54. $y = 2x^2 + 8x - 10$

In Exercises 55–58, solve the given equations by using a calculator. Round solutions to the nearest hundredth. If there are no real roots, state this as the answer.

55. $2x^2 + x - 18 = 0$
56. $-4s^2 - 5s = 1$
57. $3x^2 = -x - 2$
58. $x(15x - 12) = 8$

In Exercises 59 and 60, solve the given problems.

59. A quadratic equation $f(x) = 0$ has a solution $x = 2$. Its graph has its vertex at $(-1, 6)$. What is the other solution?
60. Find c such that $y = 2x^2 + 16x + c$ has exactly one real root.

In Exercises 61–76, solve the given quadratic equations by any appropriate method. All numbers are accurate to at least two significant digits.

61. The bending moment M of a simply supported beam of length L with a uniform load of w kg/m at a distance x from one end is $M = 0.5wLx - 0.5wx^2$. For what values of x is $M = 0$?
62. A nuclear power plant supplies a fixed power level at a constant voltage. The current I (in A) is found by solving the equation $I^2 - 17I - 12 = 0$. Solve for $I > 0$.
63. A computer design of a certain rectangular container uses the equation $12x^2 - 80x + 96 = 0$. Solve for x, if $x < 4$.
64. In fighting a fire, it is necessary to spray over nearby trees such that the height h (in m) of the spray is $h = 1.5 + 7.2x - 1.2x^2$, where x (in m) is the horizontal distance from the nozzle. What is the maximum height of the spray?

65. A computer analysis shows that the cost C (in dollars) for a company to make x units of a certain product is given by $C = 0.1x^2 + 0.8x + 7$. How many units can be made for \$50?

66. For laminar flow of fluids, the coefficient K used to calculate energy loss due to sudden enlargements is given by $K = 1.00 - 2.67R + R^2$, where R is the ratio of cross-sectional areas. If $K = 0.500$, what is the value of R?

67. At an altitude h (in ft) above sea level, the boiling point of water is lower by T°F than the boiling point at sea level, which is 212°F. The difference can be approximated by solving the equation $T^2 + 520T - h = 0$. What is the boiling point in Boulder, Colorado (altitude 5300 ft)?

68. In a natural gas pipeline, the velocity v (in m/s) of the gas as a function of the distance x (in cm) from the wall of the pipe is given by $v = 5.2x - x^2$. Determine x for $v = 4.8$ m/s.

69. The height h of an object ejected at an angle θ from a vehicle moving with velocity v is given by $h = vt \sin\theta - 16t^2$, where t is the time of flight. Find t (to 0.1 s) if $v = 44$ ft/s, $\theta = 65°$, and $h = 18$ ft.

70. In studying the emission of light, in order to determine the angle at which the intensity is a given value, the equation $\sin^2 A - 4\sin A + 1 = 0$ must be solved. Find angle A (to 0.1°). $[\sin^2 A = (\sin A)^2.]$

71. A computer analysis shows that the number n of electronic components a company should produce for supply to equal demand is found by solving.

$$\frac{n^2}{500{,}000} = 144 - \frac{n}{500}.\text{ Find } n.$$

72. To determine the resistances of two resistors that are to be in parallel in an electric circuit, it is necessary to solve the equation $\frac{20}{R} + \frac{20}{R + 10} = \frac{1}{5}$. Find R (to nearest 1 Ω).

73. In designing a cylindrical container, the formula $A = 2\pi r^2 + 2\pi rh$ is used. Solve for r.

74. In determining the number of bytes b that can be stored on a hard disk, the equation $b = kr(R - r)$ is used. Solve for r.

75. In the study of population growth, the equation $p_2 = p_1 + rp_1(1 - p_1)$ occurs. Solve for p_1.

76. In the study of the velocities of deep-water waves, the equation $v^2 = k^2\left(\frac{L}{C} + \frac{C}{L}\right)$ occurs. Solve for L.

In Exercises 77–90, set up the necessary equation where appropriate and solve the given problems. All numbers are accurate to at least two significant digits.

77. In testing the effects of a drug, the percent of the drug in the blood was given by $p = 0.090t - 0.015t^2$, where t is the time (in h) after the drug was administered. Sketch the graph of $p = f(t)$.

78. A machinery pedestal is made of two concrete cubes, one on top of the other. The pedestal is 8.00 ft high and contains 152 ft³ of concrete. Find the edge of each cube.

79. Concrete contracts as it dries. If the volume of a cubical concrete block is 29 cm³ less and each edge is 0.10 cm less after drying, what was the original length of an edge of the block?

80. The change p (in lb/in.²) in pressure in a certain fire hose, as a function of time (in min), is found to be $p = 2.3t^2 - 9.2t$. Sketch the graph of $p = f(t)$ for $t < 5$ min.

81. By adding the same amount to its length and its width, a developer increased the area of a rectangular lot by 3000 m² to make it 80 m by 100 m. What were the original dimensions of the lot? See Fig. 7.20.

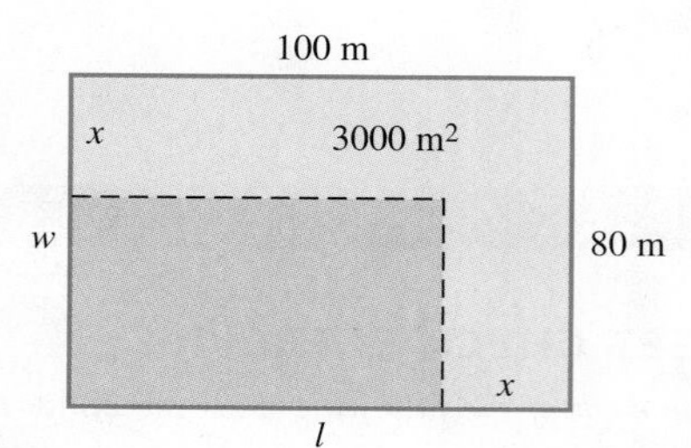

Fig. 7.20

82. A jet flew 1200 mi with a tailwind of 50 mi/h. The tailwind then changed to 20 mi/h for the remaining 570 mi of the flight. If the total time of the flight was 3.0 h, find the speed of the jet relative to the air.

83. A food company increased the profit on a package of frozen vegetables by decreasing the volume, and keeping the price the same. The thickness of the rectangular package remained at 4.00 cm, but the length and width were reduced by equal amounts from 16.0 cm and 12.0 cm such that the volume was reduced by 10.0%. What are the dimensions of the new container?

84. A military jet flies directly over and at right angles to the straight course of a commercial jet. The military jet is flying at 200 mi/h faster than four times the speed of the commercial jet. How fast is each going if they are 2050 mi apart (on a direct line) after 1 h?

85. The width of a rectangular LCD television screen is 22.9 in. longer than the height. If the diagonal is 60.0 in., find the dimensions of the screen. See Fig. 7.21.

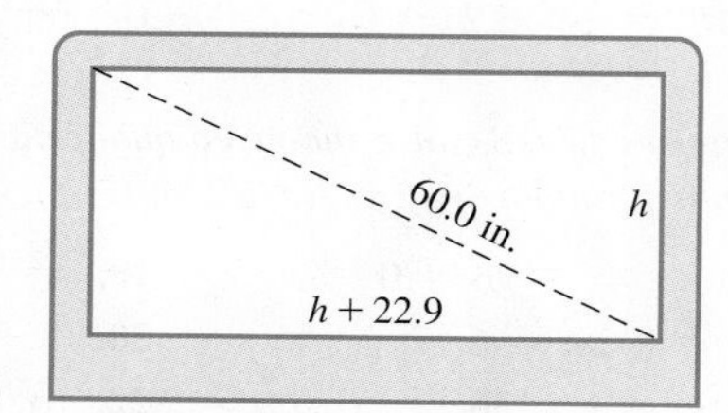

Fig. 7.21

86. Find the exact value of x that is defined in terms of the *continuing fraction* at the right (the pattern continues endlessly). Explain your method of solution.

$$x = 2 + \cfrac{1}{2 + \cfrac{1}{2 + \cfrac{1}{2 + \cdots}}}$$

87. An electric utility company is placing utility poles along a road. It is determined that five fewer poles per kilometer would be necessary if the distance between poles were increased by 10 m. How many poles are being placed each kilometer?

88. An architect is designing a Norman window (a semicircular part over a rectangular part) as shown in Fig. 7.22. If the area of the window is to be 16.0 ft^2 and the height of the rectangular part is 4.00 ft, find the radius of the circular part.

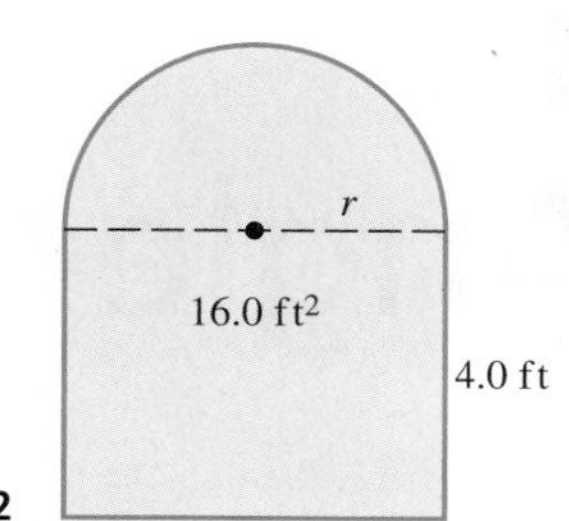

Fig. 7.22

89. A testing station found p parts per million (ppm) of sulfur dioxide in the air as a function of the hour h of the day to be $p = 0.00174(10 + 24h - h^2)$. Sketch the graph of $p = f(h)$ and, from the graph, find the time when $p = 0.205$ ppm.

90. A compact disc (CD) is made such that it is 53.0 mm from the edge of the center hole to the edge of the disc. Find the radius of the hole if 1.36% of the disc is removed in making the hole. See Fig. 7.23.

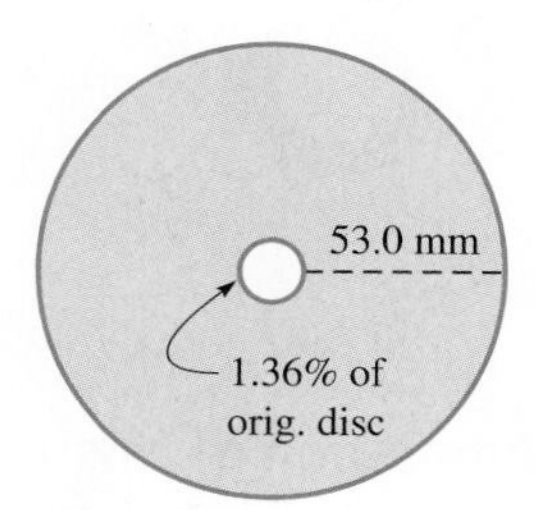

Fig. 7.23

91. An electronics student is asked to solve the equation $\frac{1}{2} = \frac{1}{R} + \frac{1}{R+1}$ for R. Write one or two paragraphs explaining your procedure for the solution, including a discussion of what methods of the chapter may be used in completing the solution.

CHAPTER 7 PRACTICE TEST

As a study aid, we have included complete solutions for each Practice Test problem at the end of this book.

1. Solve by factoring: $2x^2 + 5x = 12$
2. Solve by using the quadratic formula: $x^2 = 3x + 5$
3. Solve graphically using a calculator: $4x^2 - 5x - 3 = 0$
4. Solve algebraically: $2x^2 - x = 6 - 2x(3 - x)$
5. Solve algebraically: $\frac{3}{x} - \frac{2}{x+2} = 1$
6. Sketch the graph of $y = 2x^2 + 8x + 5$ using the extreme point and the y-intercept.
7. In electricity, the formula $P = EI - RI^2$ is used. Solve for I in terms of E, P, and R.
8. Solve by completing the square: $x^2 = 6x + 9$
9. The perimeter of a rectangular window is 8.4 m, and its area is 3.8 m^2. Find its dimensions.
10. If $y = x^2 - 8x + 8$, sketch the graph using the vertex and any other useful points and find graphically the values of x for which $y = 0$.

8

Trigonometric Functions of Any Angle

LEARNING OUTCOMES

After completion of this chapter, the student should be able to:

- Determine the magnitude and sign of any trigonometric function of any angle
- Identify reference angles and use them to evaluate trigonometric functions of angles in any quadrant
- Express an angle in degrees or radians and convert between the two measurements
- Given the value of a trigonometric function of an angle, find the angle(s)
- Use radian measure of an angle to find circular arc length
- Solve problems involving area of a sector of a circle
- Solve problems involving angular velocity
- Solve application problems involving trigonometric functions of any angle

When we introduced the trigonometric functions in Chapter 4, we defined them in general but used them only with acute angles.

In this chapter, we show how these functions are used with angles of any size and with angles measured in radians.

By the mid-1700s, the trigonometric functions had been used for many years as ratios, as in Chapter 4. It was also known that they are useful in describing periodic functions (functions for which values repeat at specific intervals) without reference to triangles. In about 1750, this led the Swiss mathematician Leonhard Euler to include, for the first time in a textbook, the trigonometric functions of numbers (not angles). As we will see in this chapter, this is equivalent to using these functions on angles measured in radians.

Euler wrote over 70 volumes in mathematics and applied subjects such as astronomy, mechanics, and music. (Many of these volumes were dictated, as he was blind for the last 17 years of his life.) He is noted as one of the great mathematicians of all time. His interest in applied subjects often led him to study and develop topics in mathematics that were used in those applications.

Today, trigonometric functions of numbers are of importance in many areas of application such as electric circuits, mechanical vibrations, and rotational motion. Although electronics were unknown in the 1700s, the trigonometric functions of numbers developed at that time played an important role in leading us into the electronic age of today.

▶ Satellites orbiting the Earth travel at very high speeds. In Section 8.4, we will see how this speed can be calculated by knowing the angular velocity of the satellite and the radius of its orbit.

8.1 Signs of the Trigonometric Functions

Signs of the Trigonometric Functions in Each of the Four Quadrants • Evaluating Functions Knowing Point on Terminal Side

Recall the definitions of the trigonometric functions that were given in Section 4.2. *Here, the point (x, y) is a point on the terminal side of angle θ, and r is the radius vector.* See Fig. 8.1.

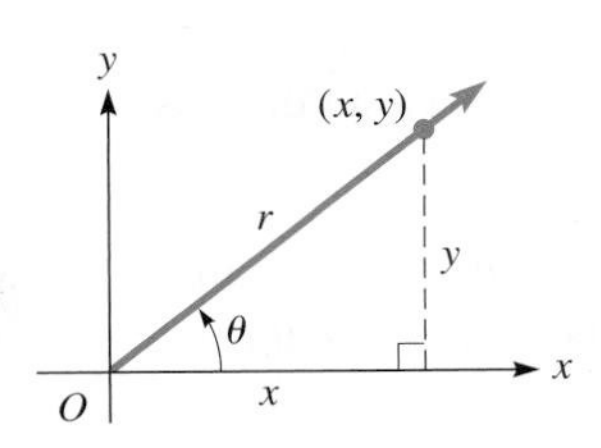

Fig. 8.1

$$\sin\theta = \frac{y}{r} \qquad \csc\theta = \frac{r}{y}$$
$$\cos\theta = \frac{x}{r} \qquad \sec\theta = \frac{r}{x} \qquad \textbf{(8.1)}$$
$$\tan\theta = \frac{y}{x} \qquad \cot\theta = \frac{x}{y}$$

As noted before, these definitions are valid for a standard-position angle of any size. In this section, we determine the *sign* of the trigonometric functions in each of the four quadrants.

We can find the values of the trigonometric functions if we know the coordinates (x, y) of a point on the terminal side and the radius vector r. *Because r is always taken to be positive, the functions will vary in sign, depending on the values of x and y.* If either x or y is zero in the denominator, the function is undefined. We will consider this in the next section.

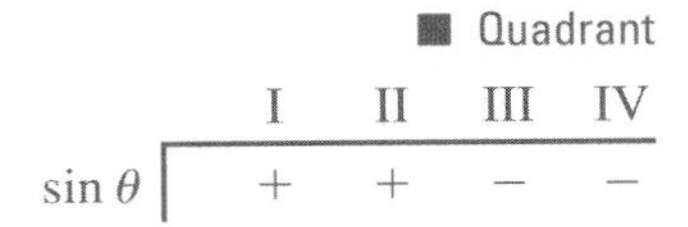

■ Quadrant	I	II	III	IV
$\sin\theta$	+	+	−	−

Because $\sin\theta = y/r$, *the sign of* $\sin\theta$ *depends on the sign of* y. Because $y > 0$ in the first and second quadrants and $y < 0$ in the third and fourth quadrants, $\sin\theta$ *is positive if the terminal side is in the first or second quadrant,* and $\sin\theta$ *is negative if the terminal side is in the third or fourth quadrant.* See Fig. 8.2.

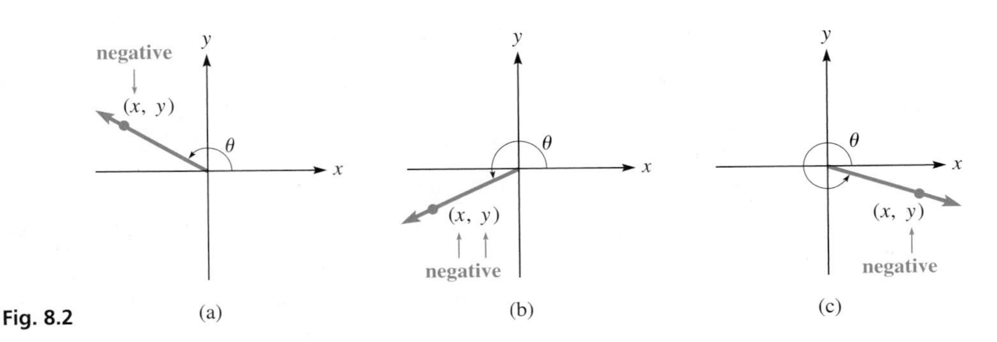

Fig. 8.2

EXAMPLE 1 Sign of $\sin\theta$ in each quadrant

The value of sin 20° is positive, because the terminal side of 20° is in the first quadrant. The value of sin 160° is positive, because the terminal side of 160° is in the second quadrant. The values of sin 200° and sin 340° are negative, because the terminal sides are in the third and fourth quadrants, respectively. ■

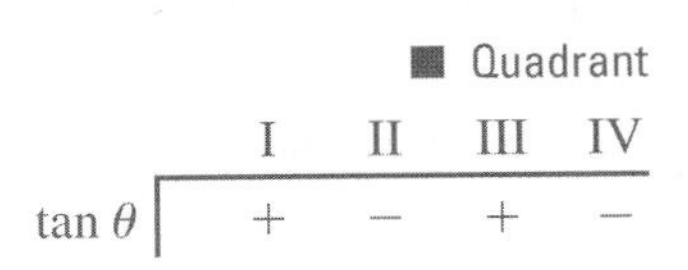

■ Quadrant	I	II	III	IV
$\tan\theta$	+	−	+	−

Because $\tan\theta = y/x$, *the sign of* $\tan\theta$ *depends on the ratio of y to x.* Because x and y are both positive in the first quadrant, both negative in the third quadrant, and have different signs in the second and fourth quadrants, $\tan\theta$ *is positive if the terminal side is in the first or third quadrant,* and $\tan\theta$ *is negative if the terminal side is in the second or fourth quadrant.* See Fig. 8.2.

EXAMPLE 2 Sign of $\tan\theta$ in each quadrant

The values of tan 20° and tan 200° are positive, because the terminal sides of these angles are in the first and third quadrants, respectively. The values of tan 160° and tan 340° are negative, because the terminal sides of these angles are in the second and fourth quadrants, respectively. ■

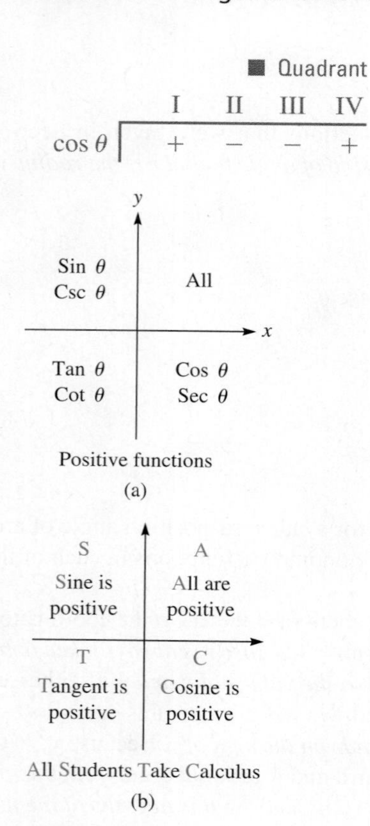

Fig. 8.3

Because $\cos\theta = x/r$, *the sign of* $\cos\theta$ *depends on the sign of* x. Because $x > 0$ in the first and fourth quadrants and $x < 0$ in the second and third quadrants, $\cos\theta$ *is positive if the terminal side is in the first or fourth quadrant,* and $\cos\theta$ *is negative if the terminal side is in the second or third quadrant.* See Fig. 8.2 on the previous page.

EXAMPLE 3 Sign of $\cos\theta$ in each quadrant

The values of cos 20° and cos 340° are positive, because the terminal sides of these angles are in the first and fourth quadrants, respectively. The values of cos 160° and cos 200° are negative, because the terminal sides of these angles are in the second and third quadrants, respectively. ■

Because $\csc\theta$ is defined in terms of y and r, as in $\sin\theta$, $\csc\theta$ has the same sign as $\sin\theta$. For similar reasons, $\cot\theta$ has the same sign as $\tan\theta$, and $\sec\theta$ has the same sign as $\cos\theta$. Therefore, as shown in Fig. 8.3(a),

All functions of first-quadrant angles are positive. Sin θ and csc θ are positive for second-quadrant angles. Tan θ and cot θ are positive for third-quadrant angles. Cos θ and sec θ are positive for fourth-quadrant angles. All others are negative.

Figure 8.3(b) shows a helpful memory aid for the signs of sine, cosine, and tangent. This discussion does not include the *quadrantal angles,* which are angles with terminal sides on one of the axes. They will be discussed in the next section.

EXAMPLE 4 Positive and negative functions

(a) The following are ***positive:***

sin 150° cos 290° tan 190° cot 260° sec 350° csc 100°

(b) The following are ***negative:***

sin 300° cos 150° tan 100° cot 300° sec 200° csc 250° ■

EXAMPLE 5 Determining the quadrant of θ

(a) If $\sin\theta > 0$, then the terminal side of θ is in either the first or second quadrant.

(b) If $\sec\theta < 0$, then the terminal side of θ is in either the second or third quadrant.

(c) If $\cos\theta > 0$ and $\tan\theta < 0$, then the terminal side of θ is in the fourth quadrant. Only in the fourth quadrant are both signs correct. ■

EXAMPLE 6 Evaluating functions

Determine the trigonometric functions of θ if the terminal side of θ passes through $(-1, \sqrt{3})$. See Fig. 8.4.

We know that $x = -1$, $y = \sqrt{3}$, and from the Pythagorean theorem, we find that $r = 2$. Therefore, the trigonometric functions of θ are

$$\sin\theta = \frac{\sqrt{3}}{2} = 0.8660 \qquad \cos\theta = -\frac{1}{2} = -0.5000 \qquad \tan\theta = -\sqrt{3} = -1.732$$

$$\cot\theta = -\frac{1}{\sqrt{3}} = -0.5774 \qquad \sec\theta = -2 = -2.000 \qquad \csc\theta = \frac{2}{\sqrt{3}} = 1.155$$

The point $(-1, \sqrt{3})$ is in the second quadrant, and the signs of the functions of θ are those of a second-quadrant angle. ■

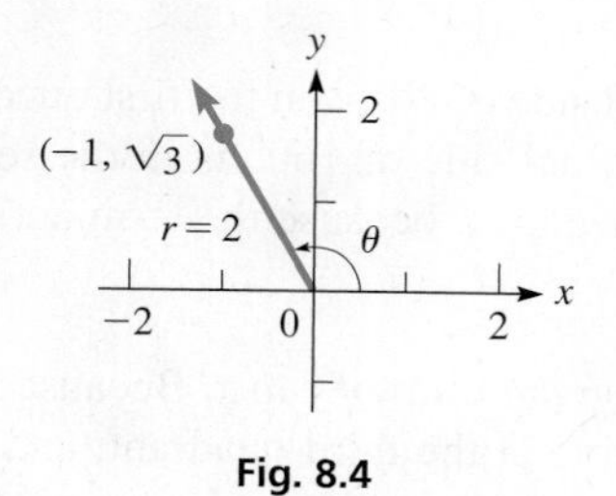

Fig. 8.4

When dealing with negative angles, or angles greater than 360°, the sign of the function is still determined by the location of the terminal side of the angle. For example, $\sin(-120°) < 0$ because the terminal side of $-120°$ is in the third quadrant. Also, for the same reason, $\sin 600° < 0$ (same terminal side as $-120°$ or $240°$).

Practice Exercise

1. Determine the sign of the given functions:
   (a) sin 140° (b) tan 255°
   (c) sec 175°

Practice Exercise

2. Determine the value of $\cos\theta$ if the terminal side of θ passes through $(-1, 4)$.

EXERCISES 8.1

In Exercises 1 and 2, answer the given questions about the indicated examples of this section.

1. In Example 4, if 90° is added to each angle, what is the sign of each resulting function?
2. In Example 6, if the point $(-1, \sqrt{3})$ is replaced with the point $(1, -\sqrt{3})$, what are the resulting values?

In Exercises 3–16, determine the sign of the given functions.

3. tan 135°, sec 50° **4.** sin 240°, cos 300°
5. sin 290°, cos 200° **6.** tan 320°, sec 185°
7. csc 98°, cot 82° **8.** cos 260°, csc 290°
9. sec 150°, tan 220° **10.** sin 335°, cot 265°
11. cos 348°, csc 238° **12.** cot 110°, sec 309°
13. tan 460°, sin (−185°) **14.** csc(−200°), cos 550°
15. cot(−95°), cos 710° **16.** sin 539°, tan(−480°)

In Exercises 17–24, find the trigonometric functions of θ if the terminal side of θ passes through the given point. All coordinates are exact.

17. (2, 1) **18.** (−5, 5) **19.** (−2, −3) **20.** (16, −12)
21. (−0.5, 1.2) **22.** (−39, −80) **23.** (20, −8) **24.** (0.9, 4)

In Exercises 25–30, for the given values, determine the quadrant(s) in which the terminal side of the angle lies.

25. $\sin\theta = 0.5000$ **26.** $\cos\theta = 0.8666$
27. $\tan\theta = -2.500$ **28.** $\sin\theta = -0.8666$
29. $\cos\theta = -0.5000$ **30.** $\tan\theta = 0.4270$

In Exercises 31–40, determine the quadrant in which the terminal side of θ lies, subject to both given conditions.

31. $\sin\theta > 0, \cos\theta < 0$ **32.** $\tan\theta > 0, \cos\theta < 0$
33. $\sec\theta < 0, \cot\theta < 0$ **34.** $\cos\theta > 0, \csc\theta < 0$
35. $\csc\theta < 0, \tan\theta < 0$ **36.** $\tan\theta < 0, \cos\theta > 0$
37. $\sin\theta > 0, \tan\theta > 0$ **38.** $\sec\theta > 0, \csc\theta < 0$
39. $\sin\theta > 0, \cot\theta < 0$ **40.** $\tan\theta > 0, \csc\theta < 0$

In Exercises 41–44, with (x, y) in the given quadrant, determine whether the given ratio is positive or negative.

41. III, $\frac{x}{r}$ **42.** II, $\frac{y}{r}$ **43.** IV, $\frac{y}{x}$ **44.** III, $\frac{x}{y}$

Answers to Practice Exercises

1. (a) + (b) + (c) − **2.** $-1/\sqrt{17}$

8.2 Trigonometric Functions of Any Angle

Reference Angle • Evaluating Trigonometric Functions • Evaluations on a Calculator • Quadrantal Angles • Negative Angles

```
NORMAL FLOAT AUTO REAL DEGREE MP
sin(20)
                .3420201433
sin(160)
                .3420201433
sin(200)
               -.3420201433
sin(340)
               -.3420201433
```

Fig. 8.5

In the last section, we saw that the signs of the trigonometric functions depend on the quadrant of θ. In this section, we will investigate an important connection between certain angles in different quadrants. To illustrate this, let's start by taking the sine of the angles 20°, 160°, 200°, and 340°. These values are shown in Fig. 8.5.

The signs of the function values agree with what we would expect; $\sin\theta$ is positive in quadrants I and II and negative in quadrants III and IV. The important thing to notice, however, is that **the numerical values (disregarding the signs) are the same for all four angles.** The reason for this is that these angles have something in common, called a **reference angle,** which is defined below:

The **reference angle,** labeled θ_{ref}, of a given angle θ is the **positive, acute angle formed between the terminal side of θ and the x-axis.**

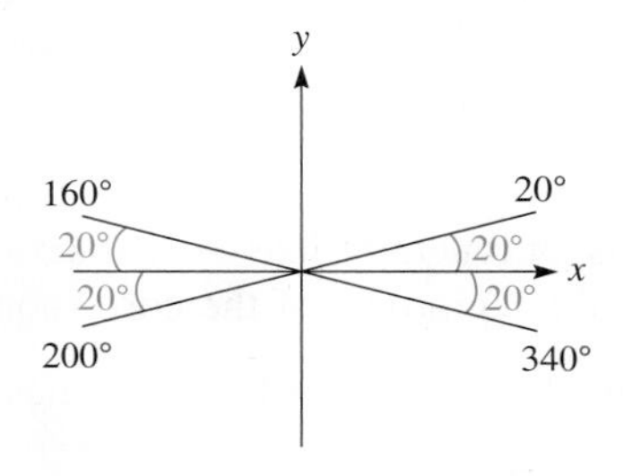

Fig. 8.6

■ In diagrams, we may choose to draw the arcs for reference angles without arrows since we know reference angles are always positive.

In our example, all four angles have a reference angle of 20°. Figure 8.6 shows the terminal sides of these angles drawn in standard position along with their reference angles. **Since the angles share a common reference angle, they will also have the same trigonometric function values, neglecting the sign.** Our example demonstrated this for sine, and the same is true for the other five functions. This is because the values of x, y, and r in the definition of the trigonometric functions will be the same, except for the signs of x and y, which will depend on the quadrant.

For this reason, reference angles are very important in trigonometry. To find the reference angle of a given angle, we sketch the angle in standard position and then determine the acute angle between its terminal side and the x-axis. This is illustrated in the following example.

EXAMPLE 1 Finding reference angles

Figure 8.7 shows the reference angles of 150°, 230°, 285°, and 420°. Pay careful attention to how the reference angle is calculated in each of the four quadrants.

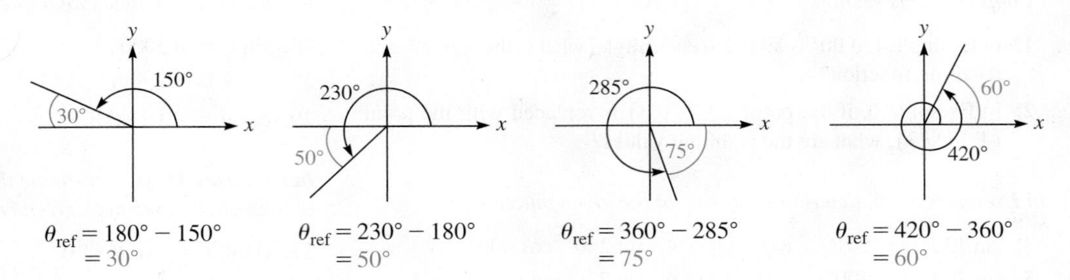

Fig. 8.7

We have observed that the trigonometric functions of different angles with the same reference angle will be the same, except for possibly the sign. Since the reference angle is an acute angle, it will always have a positive function value. The signs in the other quadrants will follow the rules stated in Section 8.1. This is summarized below:

Let θ be any angle in standard position, and let θ_{ref} be its reference angle. If we let "trig" stand for any of the six trigonometric functions, then

$$\text{trig } \theta = \pm \text{ trig } \theta_{ref} \quad \textbf{(8.2)}$$

where the sign is determined by the quadrant of θ.

According to the statement above, $\sin\theta = \pm\sin\theta_{ref}$, $\cos\theta = \pm\cos\theta_{ref}$, $\tan\theta = \pm\tan\theta_{ref}$, etc. This means that a trigonometric function of any angle can be found by first finding the trigonometric function of its reference angle and then attaching the correct sign, depending on the quadrant. The following example illustrates this process.

EXAMPLE 2 Evaluating using reference angles

same function — reference angle

$$\sin 160° = +\sin(180° - 160°) = \sin 20° = 0.3420$$
$$\tan 110° = -\tan(180° - 110°) = -\tan 70° = -2.747$$
$$\cos 225° = -\cos(225° - 180°) = -\cos 45° = -0.7071$$
$$\cot 260° = +\cot(260° - 180°) = \cot 80° = 0.1763$$
$$\sec 304° = +\sec(360° - 304°) = \sec 56° = 1.788$$
$$\sin 357° = -\sin(360° - 357°) = -\sin 3° = -0.0523$$

determines quadrant — proper sign for function in quadrant

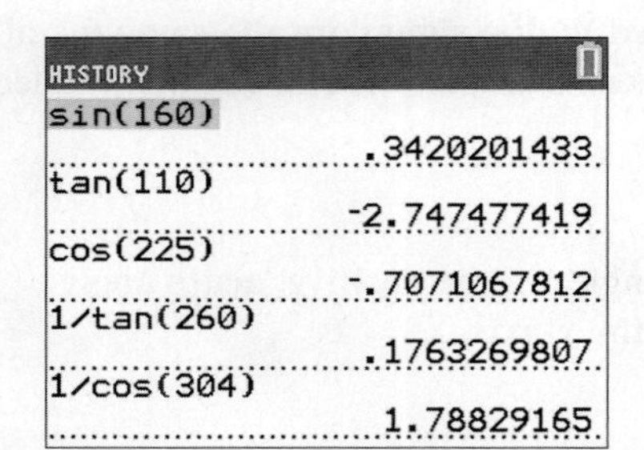
```
HISTORY
sin(160)
                .3420201433
tan(110)
               -2.747477419
cos(225)
               -.7071067812
1/tan(260)
                .1763269807
1/cos(304)
                 1.78829165
```

Fig. 8.8

A calculator will give values, with the proper signs, of functions like those in Example 2. To find csc θ, sec θ, and cot θ, we must take the reciprocal of the corresponding function. For example, $\sec 304° = \frac{1}{\cos 304°}$. The display for the first five lines of Example 2 is shown in Fig. 8.8. NOTE ▸ [In most examples, we will round off values to four significant digits (as in Example 2). However, if the angle is approximate, use the guidelines in Table 4.1 of Section 4.3 for rounding off values.]

Practice Exercises

Express each function in terms of the reference angle.

1. sin 165°; csc 345°
2. sec 195°; cos 345°
3. tan 165°; cot 195°

EXAMPLE 3 Quadrant II angle—area of triangle

A formula for finding the area of a triangle, knowing sides a and b and the included $\angle C$, is $A = \frac{1}{2}ab \sin C$. A surveyor uses this formula to find the area of a triangular tract of land for which $a = 173.2$ m, $b = 156.3$ m, and $C = 112.51°$. See Fig. 8.9.

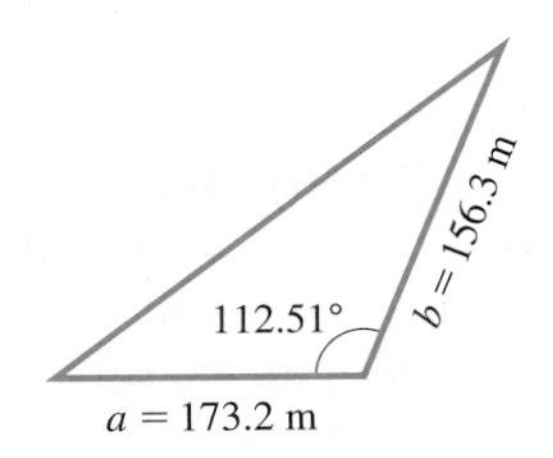

Fig. 8.9

$$A = \tfrac{1}{2}(173.2)(156.3)\sin 112.51°$$
$$= 12{,}500 \text{ m}^2 \qquad \text{rounded to four significant digits}$$

The calculator automatically uses a positive value for sin 112.51°. ■

Reference angles are especially important when we wish to find an angle that has a given trigonometric function value. This is because the inverse trigonometric functions on a calculator are programmed to give angles within the specific intervals shown below:

■ The reason that the calculator displays these angles is shown in Chapter 20, when the inverse trigonometric functions are discussed in detail.

$\sin^{-1}x$ always returns an angle between $-90°$ and $90°$

$\cos^{-1}x$ always returns an angle between $0°$ and $180°$

$\tan^{-1}x$ always returns and angle between $-90°$ and $90°$

CAUTION Sometimes we wish to find an angle that is *not* within the intervals given above. This is when reference angles must be used. ■ For example, if we need to find a second quadrant angle for which $\sin\theta = 0.853$, then taking $\sin^{-1}(0.853)$ **will not return the correct angle.** In this type of situation, we must use the reference angle to find the correct angle. This is illustrated in the following examples.

EXAMPLE 4 Finding angles given sin θ

If $\sin\theta = 0.2250$, we see from the first line of the calculator display shown in Fig. 8.10 that $\theta = 13.00°$ (rounded off).

This result is correct, but remember that there is also a second quadrant angle (where sine is also positive) that has a sine of 0.2550. This angle has a *reference angle* of 13.00° and is therefore given by $180° - 13.00° = 167.00°$.

If we need only an acute angle, $\theta = 13.00°$ is correct. However, if a second-quadrant angle is required, we see that $\theta = 167.00°$ is the angle (see Fig. 8.11). These can be checked by finding the values of sin 13.00° and sin 167.00°.

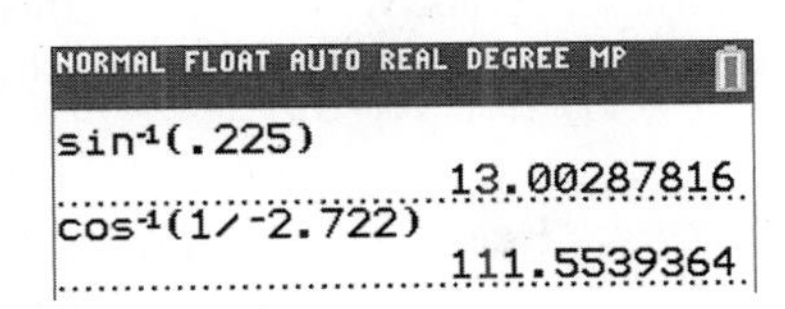

Fig. 8.10

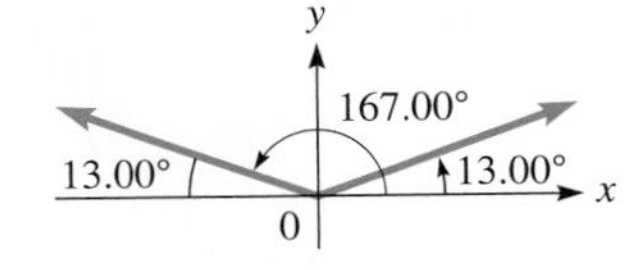

Fig. 8.11 ■

EXAMPLE 5 Finding angles given sec θ

For $\sec\theta = -2.722$ and $0° \le \theta < 360°$ (this means θ may equal 0° or be between 0° and 360°), we see from the second line of the calculator display shown in Fig. 8.10 that $\theta = 111.55°$ (rounded off).

The angle 111.55° is the second-quadrant angle, but $\sec\theta < 0$ in the third quadrant as well. The reference angle is $\theta_{ref} = 180° - 111.55° = 68.45°$, and the third-quadrant angle is $180° + 68.45° = 248.45°$. Therefore, the two angles between 0° and 360° for which $\sec\theta = -2.722$ are 111.55° and 248.45° (see Fig. 8.12). These angles can be checked by finding sec 111.55° and sec 248.45°. ■

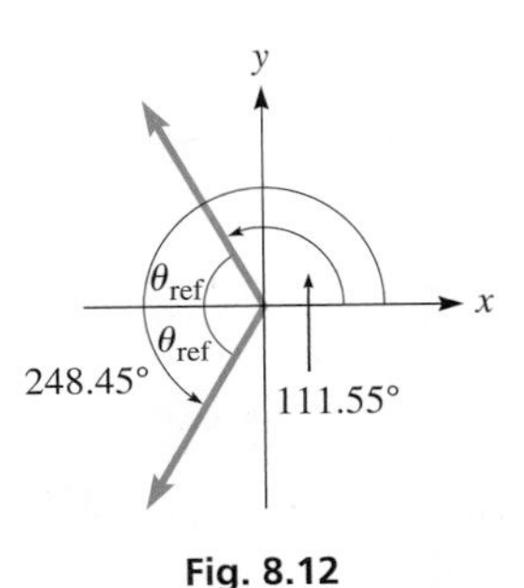

Fig. 8.12

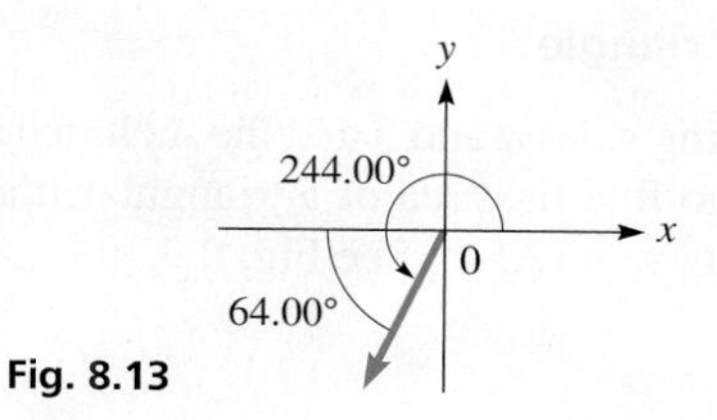

Fig. 8.13

Practice Exercise

4. If $\cos\theta = 0.5736$, find θ for $0° \le \theta < 360°$.

EXAMPLE 6 Finding angles given $\tan\theta$

Given that $\tan\theta = 2.050$ and $\cos\theta < 0$, find θ for $0° \le \theta < 360°$.

Because $\tan\theta$ is positive and $\cos\theta$ is negative, θ must be a third-quadrant angle. A calculator will display an angle of 64.00° (rounded off) for $\tan^{-1}(2.050)$. However, because we need a third-quadrant angle, *we must add* 64.00° (*the reference angle*) *to* 180°. Thus, the required angle is 244.00° (see Fig. 8.13). Check by finding tan 244.00°.

If $\tan\theta = -2.050$ and $\cos\theta < 0$, the calculator will display an angle of $-64.00°$ for $\tan^{-1}(-2.050)$. We would then have to *recognize that the reference angle is* 64.00° *and subtract it from* 180° *to get* 116.00°, the required second-quadrant angle. This can be checked by finding tan 116.00°. ■

NOTE ▸ The calculator gives the reference angle (disregarding any minus signs) in all cases except when $\cos\theta$ is negative. [To avoid confusion from the angle displayed by the calculator, **a good procedure is to find the reference angle first.**] Then it can be used to determine the angle required by the problem.

We can *find the reference angle by entering the absolute value of the function. The displayed angle will be the reference angle.* The required angle θ is then found by using the reference angle along with the quadrant of θ as shown below:

$$\begin{aligned} \theta &= \theta_{\text{ref}} && \text{(first quadrant)} \\ \theta &= 180° - \theta_{\text{ref}} && \text{(second quadrant)} \\ \theta &= 180° + \theta_{\text{ref}} && \text{(third quadrant)} \\ \theta &= 360° - \theta_{\text{ref}} && \text{(fourth quadrant)} \end{aligned} \tag{8.3}$$

EXAMPLE 7 Using reference angle

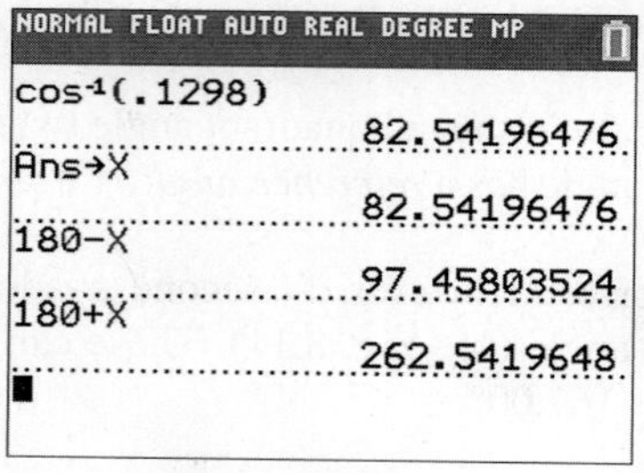

```
NORMAL FLOAT AUTO REAL DEGREE MP
cos⁻¹(.1298)
                    82.54196476
Ans→X
                    82.54196476
180-X
                    97.45803524
180+X
                    262.5419648
```

Fig. 8.14

Given that $\cos\theta = -0.1298$, find θ for $0° \le \theta < 360°$.

Because $\cos\theta$ is negative, θ is either a second-quadrant angle or a third-quadrant angle. Using 0.1298, the calculator tells us that the reference angle is 82.54°.

For the second-quadrant angle, we subtract 82.54° from 180° to get 97.46°. For the third-quadrant angle, we add 82.54° to 180° to get 262.54°. See Fig. 8.14. Therefore, the two solutions are 97.46° and 262.54°. ■

In Section 4.2, we showed how the unit circle could be used to evaluate the trigonometric functions. We now restate this as a definition, which is a special case of the general definition in Eq. (8.1) with the restriction that $r = 1$.

Unit Circle Definition of the Trigonometric Functions

Suppose the terminal side of θ in standard position intersects the unit circle at the point (x, y) as shown is Fig. 8.15. Then the six trigonometric functions are defined as follows:

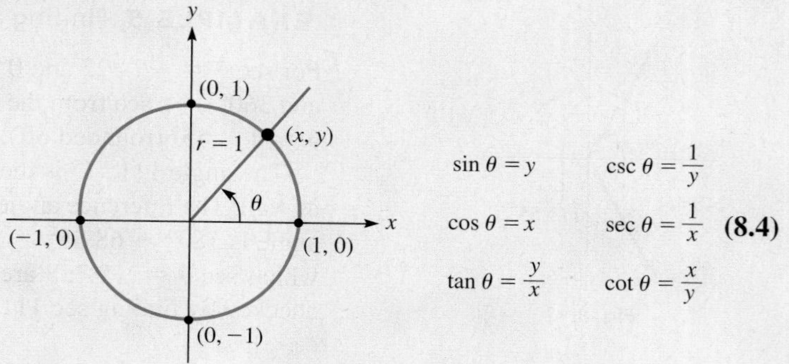

Fig. 8.15

There are many situations when the unit circle definition is useful, including evaluating trigonometric functions of **quadrantal angles,** which have their terminal side along one of the axes (see Fig. 8.16). The sine of θ is simply the y-coordinate on the unit circle, and the cosine of θ is the x-coordinate. The other functions are found by taking ratios, and **if the denominator is zero, the function is *undefined*.**

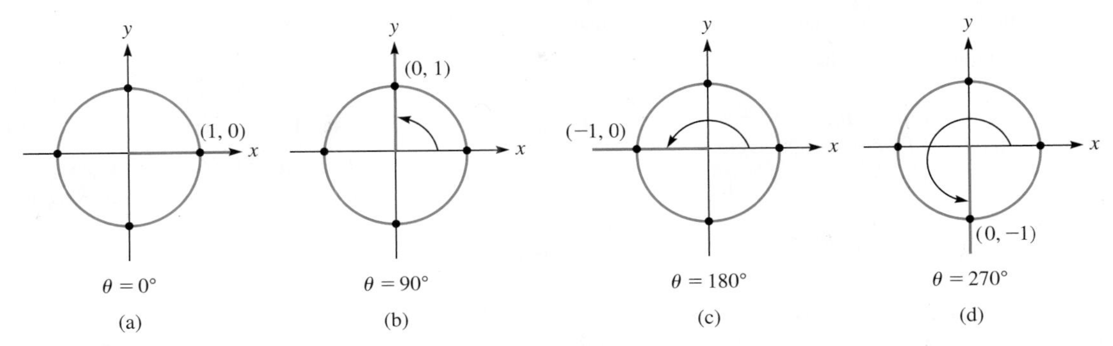

Fig. 8.16

EXAMPLE 8 Quadrantal angles and the unit circle

(a) Find the six trigonometric functions of 0°.

As shown in Fig. 8.16(a), $\theta = 0°$ intersects the unit circle at (1, 0). Therefore,

$$\sin 0° = 0 \qquad \csc 0° = \frac{1}{0} \text{ (undefined)}$$

$$\cos 0° = 1 \qquad \sec 0° = \frac{1}{1} = 1$$

$$\tan 0° = \frac{0}{1} = 0 \qquad \cot 0° = \frac{1}{0} \text{ (undefined)}$$

(b) Find the six trigonometric functions of 270°.

As shown in Fig. 8.16(d), $\theta = 270°$ intersects the unit circle at $(0, -1)$. Thus,

$$\sin 270° = -1 \qquad \csc 270° = \frac{1}{-1} = -1$$

$$\cos 270° = 0 \qquad \sec 270° = \frac{1}{0} \text{ (undefined)}$$

$$\tan 270° = \frac{-1}{0} \text{ (undefined)} \qquad \cot 270° = \frac{0}{-1} = 0$$

■

Practice Exercise

5. Find cos 180° and tan 90°, or state that they are undefined.

To evaluate functions of **negative angles,** we can use functions of corresponding positive angles, if we use the correct ***sign.*** In Fig. 8.17, note that $\sin\theta = y/r$, and $\sin(-\theta) = -y/r$, which means $\sin(-\theta) = -\sin\theta$. In the same way, we can get all of the relations between functions of $-\theta$ and the functions of θ. Therefore,

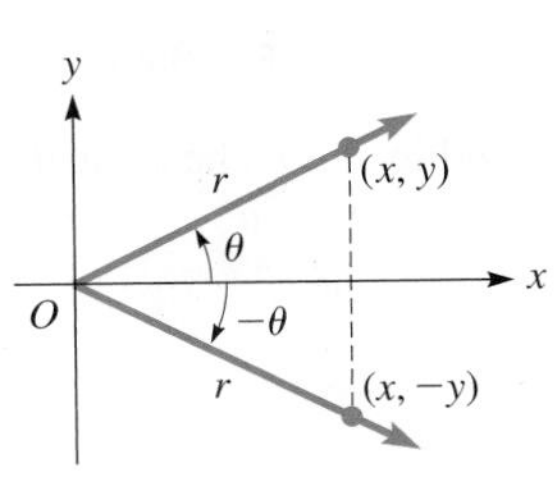

Fig. 8.17

$$\sin(-\theta) = -\sin\theta \qquad \cos(-\theta) = \cos\theta \qquad \tan(-\theta) = -\tan\theta$$
$$\csc(-\theta) = -\csc\theta \qquad \sec(-\theta) = \sec\theta \qquad \cot(-\theta) = -\cot\theta \tag{8.5}$$

EXAMPLE 9 Negative angles

$$\sin(-60°) = -\sin 60° = -0.8660 \qquad \cos(-60°) = \cos 60° = 0.5000$$

$$\tan(-60°) = -\tan 60° = -1.732 \qquad \cot(-60°) = -\cot 60° = -0.5774$$

$$\sec(-60°) = \sec 60° = 2.000 \qquad \csc(-60°) = -\csc 60° = -1.155$$

■

EXERCISES 8.2

In Exercises 1 and 2, make the given changes in the indicated examples of this section and then solve the resulting problems.

1. In Example 2, add 40° to each angle, express in terms of the same function of a positive acute angle, and then evaluate.
2. In Example 7, change the − to + and then find θ.

In Exercises 3–8, express the given trigonometric function in terms of the same function of the reference angle.

3. sin 155°, cos 220°
4. tan 91°, sec 345°
5. tan 105°, csc 328°
6. cos 190°, tan 290°
7. sec 425°, sin(−520°)
8. tan 920°, csc(−550°)

In Exercises 9–42, the given angles are approximate. In Exercises 9–16, find the values of the given trigonometric functions by finding the reference angle and attaching the proper sign.

9. sin 195°
10. tan 311°
11. cos 106.3°
12. sin 93.4°
13. sec 328.33°
14. cot 516.53°
15. tan(−109.1°)
16. csc(−108.4°)

In Exercises 17–24, find the values of the given trigonometric functions directly from a calculator.

17. cos(−62.7°)
18. cos 141.4°
19. sin 310.36°
20. tan 242.68°
21. csc 194.82°
22. sec 441.08°
23. tan 148.25°
24. sin(−215.5°)

In Exercises 25–38, find θ for $0° \leq \theta < 360°$.

25. $\sin\theta = -0.8480$
26. $\tan\theta = -1.830$
27. $\cos\theta = 0.4003$
28. $\sec\theta = -1.637$
29. $\cot\theta = -0.012$
30. $\csc\theta = -8.09$
31. $\sin\theta = 0.870$, $\cos\theta < 0$
32. $\tan\theta = 0.932$, $\sin\theta < 0$
33. $\cos\theta = -0.12$, $\tan\theta > 0$
34. $\sin\theta = -0.192$, $\tan\theta < 0$
35. $\csc\theta = -1.366$, $\cos\theta > 0$
36. $\cos\theta = 0.0726$, $\sin\theta < 0$
37. $\sec\theta = 2.047$, $\cot\theta < 0$
38. $\cot\theta = -0.3256$, $\csc\theta > 0$

In Exercises 39–42, find the exact value of each expression without the use of a calculator. (Hint: Start by expressing each quantity in terms of its reference angle.)

39. cos 60° + cos 70° + cos 110°
40. sin 200° − sin 150° + sin 160°
41. tan 40° + tan 135° − tan 220°
42. sec 130° − sec 230° + sec 300°

In Exercises 43–46, determine the function that satisfies the given conditions.

43. Find $\tan\theta$ when $\sin\theta = -0.5736$ and $\cos\theta > 0$.
44. Find $\sin\theta$ when $\cos\theta = 0.422$ and $\tan\theta < 0$.
45. Find $\cos\theta$ when $\tan\theta = -0.809$ and $\csc\theta > 0$.
46. Find $\cot\theta$ when $\sec\theta = 6.122$ and $\sin\theta < 0$.

In Exercises 47–50, insert the proper sign, $>$ or $<$ or $=$, between the given expressions. Explain your answers.

47. sin 90° 2 sin 45°
48. cos 360° 2 cos 180°
49. tan 180° tan 0°
50. sin 270° 3 sin 90°

In Exercises 51–56, solve the given problems. In Exercises 53 and 54, assume $0° < \theta < 90°$. (Hint: Review cofunctions on page 125.)

51. Using the fact that sin 75° = 0.9659, evaluate cos 195°.
52. Using the fact that cot 20° = 2.747, evaluate tan 290°.
53. Express $\tan(270° - \theta)$ in terms of the $\cot\theta$.
54. Express $\cos(90° + \theta)$ in terms of $\sin\theta$.
55. For a triangle with angles A, B, and C, evaluate $\tan A + \tan(B + C)$.
56. (a) Is sin 180° = 2 sin 90°? (b) Is sin 360° = 2 sin 180°?

In Exercises 57–60, evaluate the given expressions.

57. The current i in an alternating-current circuit is given by $i = i_m \sin\theta$, where i_m is the maximum current in the circuit. Find i if $i_m = 0.0259$ A and $\theta = 495.2°$.
58. The force F that a rope exerts on a crate is related to force F_x directed along the x-axis by $F = F_x \sec\theta$, where θ is the standard-position angle for F. See Fig. 8.18. Find F if $F_x = -365$ N and $\theta = 127.0°$.

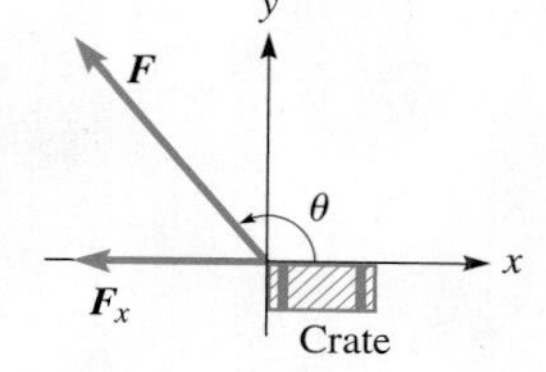

Fig. 8.18

59. For the slider mechanism shown in Fig. 8.19, $y \sin\alpha = x \sin\beta$. Find y if $x = 6.78$ in., $\alpha = 31.3°$, and $\beta = 104.7°$.

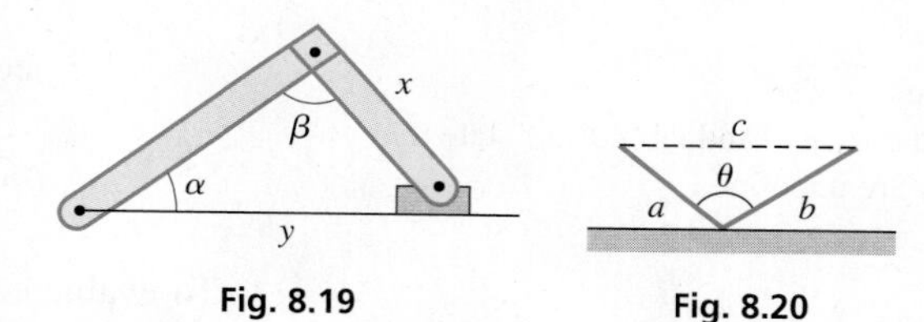

Fig. 8.19 Fig. 8.20

60. A laser follows the path shown in Fig. 8.20. The angle θ is related to the distances a, b, and c by $2ab\cos\theta = a^2 + b^2 - c^2$. Find θ if $a = 12.9$ cm, $b = 15.3$ cm, and $c = 24.5$ cm.

Answers to Practice Exercises

1. sin 15°; −csc 15° 2. −sec 15°; cos 15° 3. −tan 15°; cot 15° 4. 55.0°, 305.0° 5. −1, undefined

8.3 Radians

Definition of a Radian • Converting Angle Measurements • Evaluations on a Calculator

For many problems in which trigonometric functions are used, particularly those involving the solution of triangles, degree measurements of angles are convenient and quite sufficient. However, division of a circle into 360 equal parts is by definition, and it is arbitrary and artificial (see the margin comment on page 55).

In numerous other types of applications and in more theoretical discussions, the *radian* is a more meaningful measure of an angle. We defined the radian in Chapter 2 and reviewed it briefly in Chapter 4. In this section, we discuss the radian in detail and start by reviewing its definition.

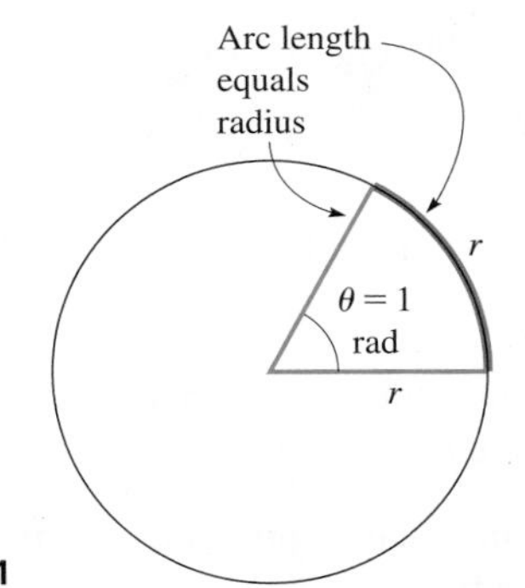

Fig. 8.21

A **radian** *is the measure of an angle with its vertex at the center of a circle and with an intercepted arc on the circle equal in length to the radius of the circle.* See Fig. 8.21.

Because the circumference of any circle in terms of its radius is given by $c = 2\pi r$, the ratio of the circumference to the radius is 2π. This means that the radius may be laid off 2π (about 6.28) times along the circumference, regardless of the length of the radius. Therefore, note that radian measure is independent of the radius of the circle. The definition of a radian is based on an important property of a circle and is therefore a more natural measure of an angle. In Fig. 8.22, the numbers on each of the radii indicate the number of radians in the angle measured in standard position. The circular arrow shows an angle of 6 radians.

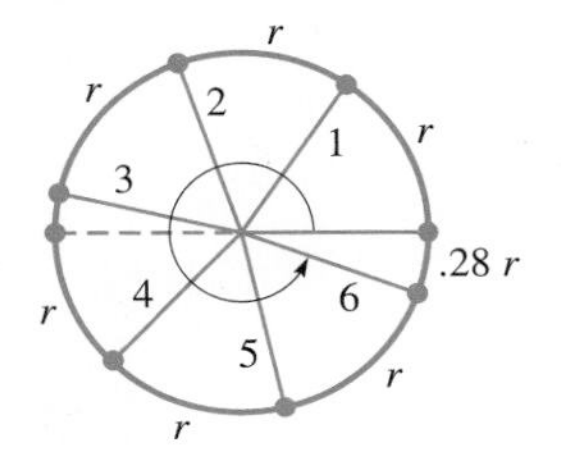

Fig. 8.22

Because the radius may be laid off 2π times along the circumference, it follows that there are 2π radians in one complete rotation. Also, there are 360° in one complete rotation. Therefore, 360° is *equivalent* to 2π radians. It then follows that the relation between degrees and radians is $2\pi \text{ rad} = 360°$. The following equations can be used to convert between degrees and radians:

Converting Angles

$$\pi \text{ rad} = 180° \quad \textbf{(8.6)}$$

Degrees to Radians

$$1° = \frac{\pi}{180} \text{ rad} = 0.01745 \text{ rad} \quad \textbf{(8.7)}$$

Radians to Degrees

$$1 \text{ rad} = \frac{180°}{\pi} = 57.30° \quad \textbf{(8.8)}$$

From Eqs. (8.6), (8.7), and (8.8), note that we convert angle measurements from degrees to radians or radians to degrees using the following procedure:

Procedure for Converting Angle Measurements

1. To convert an angle measured in degrees to the same angle measured in radians, multiply the number of degrees by $\pi/180°$.
2. To convert an angle measured in radians to the same angle measured in degrees, multiply the number of radians by $180°/\pi$.

EXAMPLE 1 Converting to and from radians

(a) $18.0° = \left(\frac{\pi}{180°}\right)(18.0°) = \frac{\pi}{10.0} = 0.314$ rad (See Fig. 8.23.)

(converting degrees to radians; degrees cancel)

(b) $2.00 \text{ rad} = \left(\frac{180°}{\pi}\right)(2.00) = \frac{360°}{\pi} = 114.6°$ (See Fig. 8.23.)

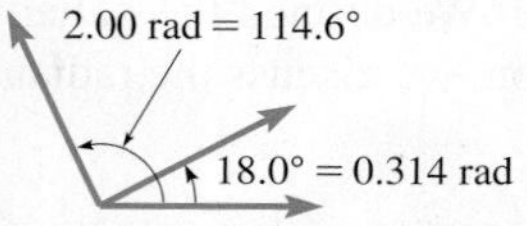

(converting radians to degrees)

2.00 rad = 114.6°

18.0° = 0.314 rad

Fig. 8.23

Multiplying by $\pi/180°$ or $180°/\pi$ is actually multiplying by 1, because π rad $= 180°$. The unit of measurement is different, but *the angle is the same.* ■

■ See Section 1.4 on unit conversions.

Because of the definition of the radian, it is common to express radians in terms of π, particularly for angles whose degree measure is a fraction of 180°.

EXAMPLE 2 Radians in terms of π

(a) Converting 30° to radian measure, we have

$$30° = \left(\frac{\pi}{180°}\right)(30°) = \frac{\pi}{6} \text{ rad} \qquad \text{(See Fig. 8.24.)}$$

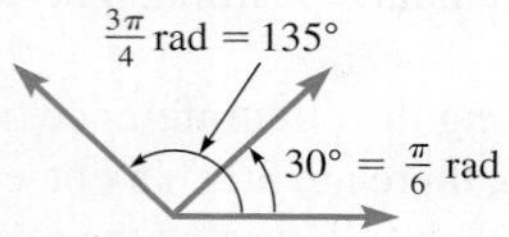

(b) Converting $3\pi/4$ rad to degrees, we have

$$\frac{3\pi}{4} \text{ rad} = \left(\frac{180°}{\pi}\right)\left(\frac{3\pi}{4}\right) = 135° \qquad \text{(See Fig. 8.24.)}$$ ■

$\frac{3\pi}{4}$ rad = 135°

30° = $\frac{\pi}{6}$ rad

Fig. 8.24

Practice Exercises

1. Convert 36° to radians in terms of π.
2. Convert $7\pi/9$ to degrees.

We wish now to make a very important point. Because π is a number (a little greater than 3) that is the ratio of the circumference of a circle to its diameter, it is the ratio of one length to another. This means *radians have no units,* and *radian measure amounts to measuring angles in terms of real numbers.* It is this property that makes radians useful in many applications.

CAUTION When the angle is measured in radians, it is customary that no units are shown. The radian is understood to be the unit of measurement. If the symbol rad is used, it is only to emphasize that the angle is in radians. ■

EXAMPLE 3 No angle units indicates radians

(a) $60° = \left(\frac{\pi}{180°}\right)(60.0°) = \frac{\pi}{3.00} = 1.05$ 1.05 = 1.05 rad

(no units indicates radian measure)

(b) $3.80 = \left(\frac{180°}{\pi}\right)(3.80) = 218°$

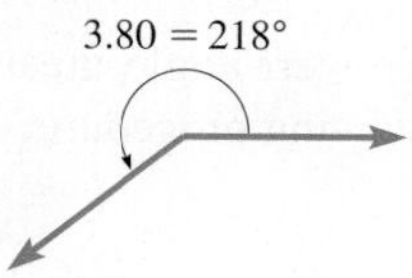

3.80 = 218°

Fig. 8.25

Because no units are shown for 1.05 and 3.80, they are known to be measured in radians. See Fig. 8.25. ■

As shown in Section 4.1, the *angle* feature on a calculator can be used to directly convert from degrees to radians or from radians to degrees. In Fig. 8.26, we show the calculator display for changing $\pi/2$ rad to degrees (calculator in degree mode) and for changing 60° to radians (calculator in radian mode). It is important to set the calculator to the correct mode when using these features.

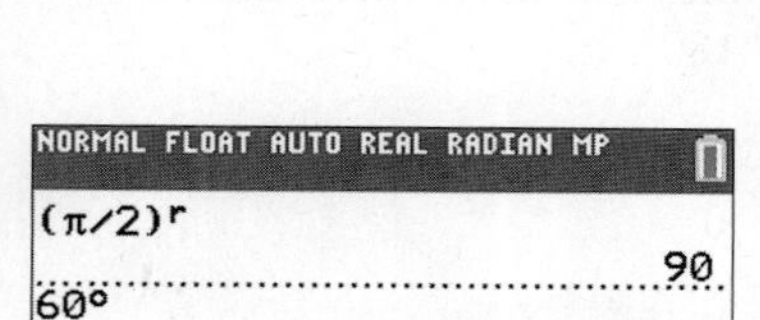

Fig. 8.26

A calculator can also be used to find a trigonometric function of an angle that is given in radians. If the calculator is in radian mode, it then uses values in radians directly and will *consider any angle entered to be in radians.*

CAUTION Always be careful to have your calculator in the proper mode. Check the setting in the *mode* feature. If you are working in degrees, use the degree mode, but if you are working in radians, use the radian mode. ■

EXAMPLE 4 Calculator evaluations with radians

In each of the following, we assume the angles are in radians since no unit is indicated. Using a calculator in the radian mode, we get

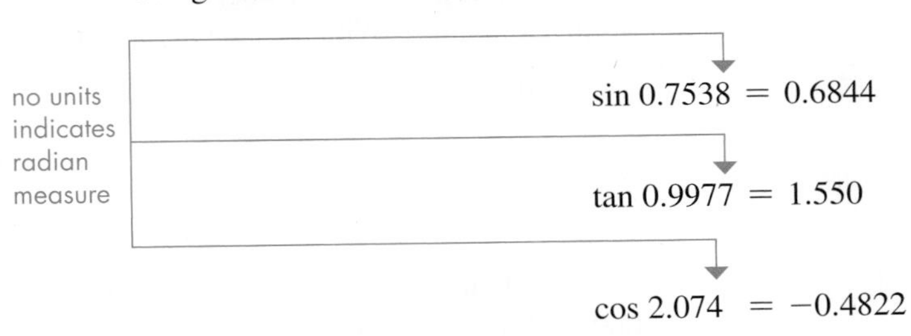

$$\sin 0.7538 = 0.6844$$

$$\tan 0.9977 = 1.550$$

$$\cos 2.074 = -0.4822$$ ■

Practice Exercise

3. Find the value of sin 3.56.

In the following application, the resulting angle is a unitless number, and it is therefore in radian measure.

EXAMPLE 5 Application of radian measure

The velocity v of an object undergoing simple harmonic motion at the end of a spring is given by

$$v = A\sqrt{\frac{k}{m}}\cos\sqrt{\frac{k}{m}}t$$

the angle is $\left(\sqrt{\frac{k}{m}}\right)(t)$

Here, m is the mass of the object (in g), k is a constant depending on the spring, A is the maximum distance the object moves, and t is the time (in s). Find the velocity (in cm/s) after 0.100 s of a 36.0-g object at the end of a spring for which $k = 400$ g/s^2, if $A = 5.00$ cm.

Substituting, we have

$$v = 5.00\sqrt{\frac{400}{36.0}}\cos\sqrt{\frac{400}{36.0}}(0.100)$$

Using calculator memory for $\sqrt{\frac{400}{36.0}}$, and with the calculator in radian mode, we have

$$v = 15.7 \text{ cm/s}$$

See Fig. 8.27. ■

NORMAL FLOAT AUTO REAL RADIAN MP
√400/36→X
3.333333333
5Xcos(.1X)
15.74928244

Fig. 8.27

If we need a reference angle in radians, recall that $\pi/2 = 90°$, $\pi = 180°$, $3\pi/2 = 270°$, and $2\pi = 360°$ (see Fig. 8.28). These and their decimal approximations are shown in Table 8.1.

Table 8.1 Quadrantal Angles

Degrees	*Radians*	*Radians (decimal)*
90°	$\frac{\pi}{2}$	1.571
180°	π	3.142
270°	$\frac{3\pi}{2}$	4.712
360°	2π	6.283

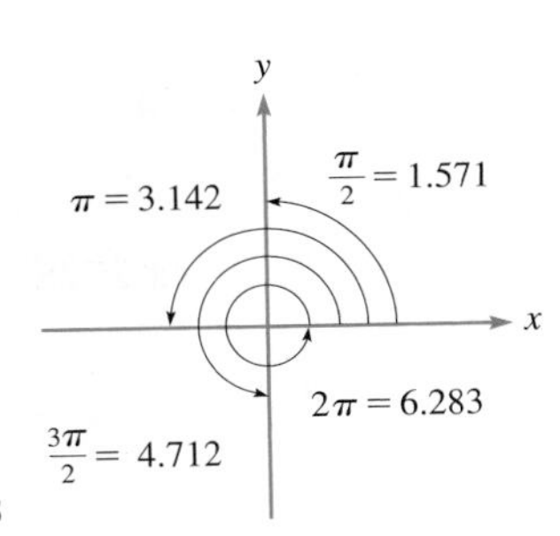

Fig. 8.28

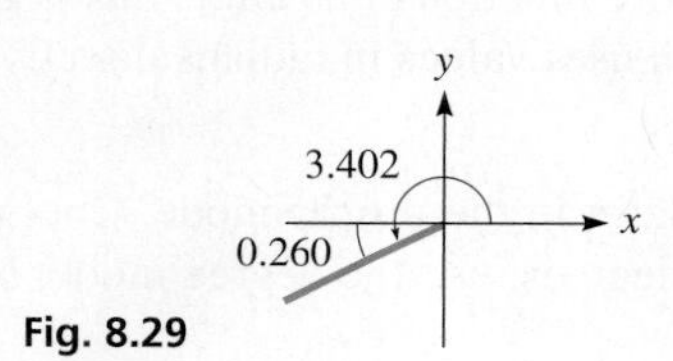

Fig. 8.29

EXAMPLE 6 Reference angles in radians

An angle of 3.402 is greater than 3.142 but less than 4.712. Thus, it is a third-quadrant angle, and the reference angle is $3.402 - \pi = 0.260$. The calculator π key can be used. See Fig. 8.29.

An angle of 5.210 is between 4.712 and 6.283. Therefore, it is in the fourth quadrant and the reference angle is $2\pi - 5.210 = 1.073$. ■

EXAMPLE 7 Find angle in radians for given function

Find θ in radians, such that $\cos\theta = 0.8829$ and $0 \leq \theta < 2\pi$.

Because $\cos\theta$ is positive and θ between 0 and 2π, we want a first-quadrant angle and a fourth-quadrant angle. With the calculator in radian mode,

$$\cos^{-1} 0.8829 = 0.4888 \quad \text{first-quadrant angle}$$

$$2\pi - 0.4888 = 5.794 \quad \text{fourth-quadrant angle}$$

$$\theta = 0.4888 \quad \text{or} \quad \theta = 5.794 \quad \text{see Fig. 8.30}$$ ■

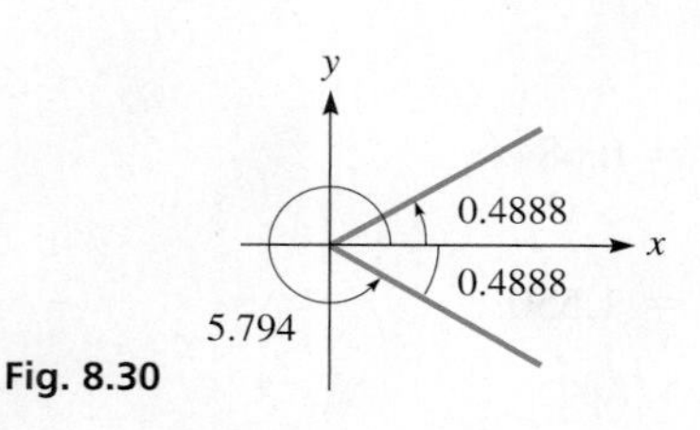

Fig. 8.30

CAUTION When one first encounters radian measure, expressions such as sin 1 and $\sin\theta = 1$ are often confused. The first is equivalent to $\sin 57.30°$ (because 1 radian = 57.30°). The second means θ is the angle for which the sine is 1. Because $\sin 90° = 1$, we can say that $\theta = 90°$ or $\theta = \pi/2$. ■ The following example gives additional illustrations.

Practice Exercise

4. If $\sin\theta = 0.4235$, find the smallest positive θ (in radians).

EXAMPLE 8 Angle in radians—value of function

(a) $\sin\pi/3 = \sqrt{3}/2$ **(b)** $\cos\theta = 0.5$, $\theta = 60° = \pi/3$ (smallest positive θ)

(c) $\tan 2 = -2.185$ **(d)** $\tan\theta = 2$, $\theta = 1.107$ (smallest positive θ) ■

EXERCISES 8.3

In Exercises 1 and 2, make the given changes in the indicated examples of this section and then solve the resulting problems.

1. In Example 3(b), change 3.80 to 2.80.

2. In Example 7, change cos to sin.

In Exercises 3–10, express the given angle measurements in radian measure in terms of π.

3. 15°, 120° **4.** 12°, 225° **5.** 75°, 330° **6.** 36°, 315°

7. 210°, 99° **8.** 5°, 300° **9.** 720°, −9° **10.** −66°, 540°

In Exercises 11–18, the given numbers express angle measure. Express the measure of each angle in terms of degrees.

11. $\frac{3\pi}{5}, \frac{3\pi}{2}$ **12.** $\frac{3\pi}{10}, \frac{11\pi}{6}$ **13.** $\frac{5\pi}{9}, \frac{7\pi}{4}$ **14.** $\frac{8\pi}{15}, \frac{4\pi}{3}$

15. $\frac{7\pi}{18}, \frac{5\pi}{6}$ **16.** $\frac{\pi}{40}, \frac{5\pi}{4}$ **17.** $-\frac{\pi}{15}, \frac{3\pi}{20}$ **18.** $\frac{9\pi}{2}, -\frac{4\pi}{15}$

In Exercises 19–26, express the given angles in radian measure. Round off results to the number of significant digits in the given angle.

19. 84.0° **20.** 54.3° **21.** 252° **22.** 104°

23. −333.5° **24.** 268.7° **25.** 478.5° **26.** −86.1°

In Exercises 27–34, the given numbers express the angle measure. Express the measure of each angle in terms of degrees, with the same accuracy as the given value.

27. 0.750 **28.** 0.240 **29.** 3.407 **30.** 1.703

31. 12.4 **32.** −34.4 **33.** −16.42 **34.** 100.0

In Exercises 35–42, evaluate the given trigonometric functions by first changing the radian measure to degree measure. Round off results to four significant digits.

35. $\sin\frac{\pi}{4}$ **36.** $\cos\frac{\pi}{6}$ **37.** $\tan\frac{5\pi}{12}$

38. $\sin\frac{71\pi}{36}$ **39.** $\cos\frac{5\pi}{6}$ **40.** $\tan\left(-\frac{7\pi}{3}\right)$

41. sec 4.5920 **42.** cot 3.2732

In Exercises 43–50, evaluate the given trigonometric functions directly, without first changing to degree measure. Round answers to three significant digits.

43. tan 0.7359 **44.** cos 1.4308 **45.** sin 4.24

46. tan 3.47 **47.** sec 2.07 **48.** sin(−1.34)

49. cot(−4.86) **50.** csc 6.19

In Exercises 51–58, find θ to four significant digits for $0 \leq \theta < 2\pi$.

51. $\sin\theta = 0.3090$ **52.** $\cos\theta = -0.9135$

53. $\tan\theta = -0.2126$ **54.** $\sin\theta = -0.0436$

55. $\cos\theta = 0.6742$ **56.** $\cot\theta = 1.860$

57. $\sec\theta = -1.307$ **58.** $\csc\theta = 3.940$

In Exercises 59–64, solve the given problems. (Hint: For problems 61–64, review cofunctions on page 125.)

59. Find the radian measure of an angle at the center of a circle of radius 12 cm that intercepts an arc of 15 cm on the circle.

60. Find the length of arc of a circle of radius 10 in. that is intercepted from the center of the circle by an angle of 3 radians.

61. Using the fact that $\sin\frac{\pi}{8} = 0.3827$, find the value of $\cos\frac{5\pi}{8}$. (A calculator should be used only to check the result.)

62. Using the fact that $\tan\frac{\pi}{6} = 0.5774$, find the value of $\cot\frac{5\pi}{3}$. (A calculator should be used only to check the result.)

63. Express $\tan\left(\frac{\pi}{2} + \theta\right)$ in terms of $\cot\theta$. $\left(0 < \theta < \frac{\pi}{2}\right)$

64. Express $\cos\left(\frac{3\pi}{2} + \theta\right)$ in terms of $\sin\theta$. $\left(0 < \theta < \frac{\pi}{2}\right)$

In Exercises 65–72, evaluate the given problems.

65. A unit of angle measurement used in artillery is the *mil,* which is defined as a central angle of a circle that intercepts an arc equal in length to 1/6400 of the circumference. How many mils are in a central angle of 34.4°?

66. Through how many radians does the minute hand of a clock move in 25 min?

67. After the brake was applied, a bicycle wheel went through 1.60 rotations. Through how many radians did a spoke rotate?

68. The *London Eye* is a Ferris wheel erected in London in 1999. It is 135 m high and has 32 air-conditioned passenger capsules, each able to hold 25 people. Through how many radians does it move after loading capsule 1, until capsule 25 is reached?

69. A flat plate of weight W oscillates as shown in Fig. 8.31. Its potential energy V is given by $V = \frac{1}{2}Wb\theta^2$, where θ is measured in radians. Find V if $W = 8.75$ lb, $b = 0.75$ ft, and $\theta = 5.5°$.

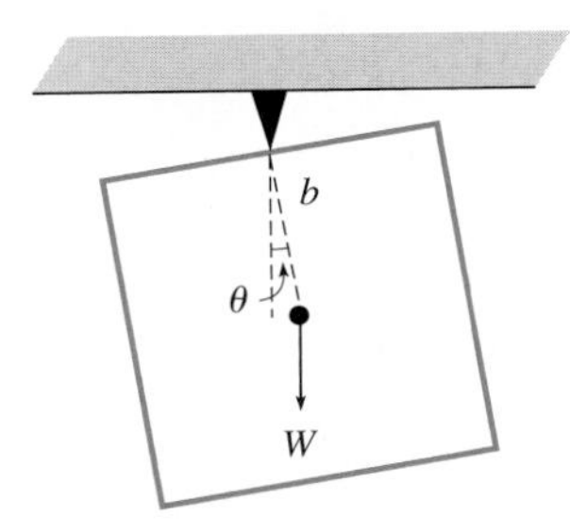

Fig. 8.31

70. The charge q (in C) on a capacitor as a function of time is $q = A\sin\omega t$. If t is measured in seconds, in what units is ω measured? Explain.

71. The height h of a rocket launched 1200 m from an observer is found to be $h = 1200\tan\dfrac{5t}{3t + 10}$ for $t < 10$ s, where t is the time after launch. Find h for $t = 8.0$ s.

72. The electric intensity I (in W/m^2) from the two radio antennas in Fig. 8.32 is a function of θ given by $I = 0.023\cos^2(\pi\sin\theta)$. Find I for $\theta = 40.0°$. $[\cos^2\alpha = (\cos\alpha)^2.]$ (*Hint*: You will need to use both the degree mode and the radian mode on your calculator.)

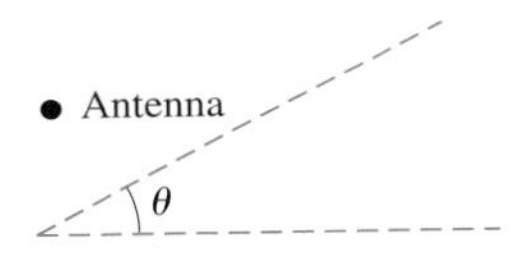

Fig. 8.32

Answers to Practice Exercises

1. $\pi/5$ **2.** 140° **3.** -0.4063 **4.** 0.4373

8.4 Applications of Radian Measure

Arc Length on a Circle • Area of a Sector of a Circle • Linear and Angular Velocity

Radian measure has numerous applications in mathematics and technology. In this section, we discuss applications involving circular arc length, the area of a sector, and the relationship between linear and angular velocity.

ARC LENGTH ON A CIRCLE

From geometry, we know that *the length of an arc on a circle is proportional to the central angle* formed by the radii that intercept the arc. The length of arc of a complete circle is the circumference. Letting s represent the length of arc, we may state that $s = 2\pi r$ for a complete circle. Because 2π is the central angle (in radians) of the complete circle, the length s of any circular arc with central angle θ (in radians) is given by

$$s = \theta r \quad (\theta \text{ in radians}) \tag{8.9}$$

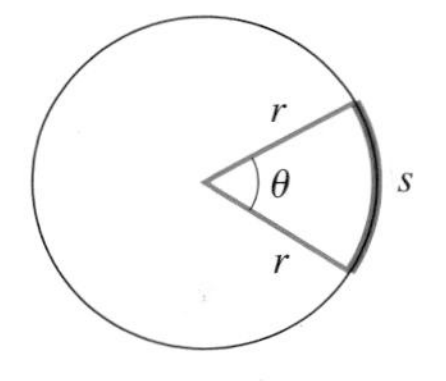

Fig. 8.33

Therefore, if we know the central angle θ ***in radians*** and the radius of a circle, we can find the length of a circular arc directly by using Eq. (8.9). See Fig. 8.33.

EXAMPLE 1 Arc length

Find the length of arc on a circle of radius $r = 3.00$ in., for which the central angle $\theta = \pi/6$. See Fig. 8.34.

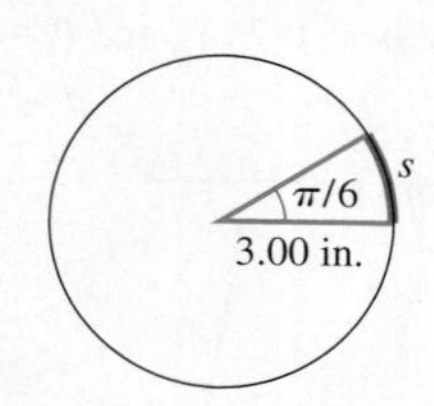

Fig. 8.34

θ in radians

$$s = \left(\frac{\pi}{6}\right)(3.00) = \frac{\pi}{2.00}$$
$$= 1.57 \text{ in.}$$

Therefore, the length of arc s is 1.57 in. ■

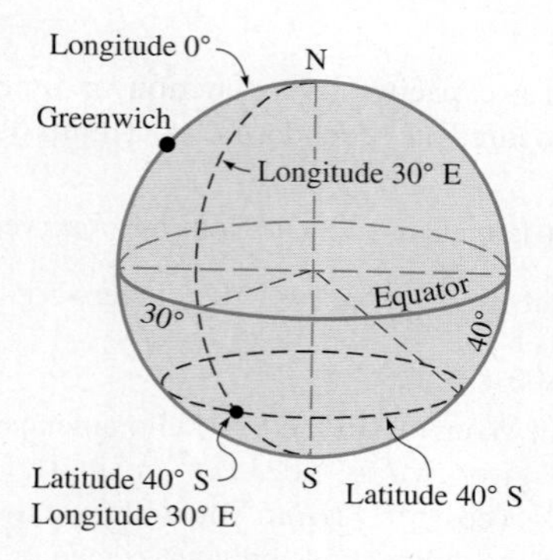

Fig. 8.35

Among the important applications of arc length are distances on the Earth's surface. For most purposes, the Earth may be regarded as a sphere (the diameter at the equator is slightly greater than the distance between the poles). A *great circle* of the Earth (or any sphere) is the circle of intersection of the surface of the sphere and a plane that passes through the center.

The equator is a great circle and is designated as 0° *latitude*. Other *parallels of latitude* are parallel to the equator with diameters decreasing to zero at the poles, which are 90° N and 90° S. See Fig. 8.35.

Meridians of longitude are half great circles between the poles. The *prime meridian* through Greenwich, England, is designated as 0°, with meridians to 180° measured east and west from Greenwich. Positions on the surface of the Earth are designated by longitude and latitude.

EXAMPLE 2 Arc length—nautical mile

The traditional definition of a *nautical mile* is the length of arc along a great circle of the Earth for a central angle of 1′. The modern international definition is a distance of 1852 m. What measurement of the Earth's radius does this definition use?

Here, $\theta = 1' = (1/60)°$, and $s = 1852$ m. Solving for r, we have

$$r = \frac{s}{\theta} = \frac{1852}{\left(\frac{1}{60}\right)^{\circ}\left(\frac{\pi}{180^{\circ}}\right)} = 6.367 \times 10^6 \text{ m}$$
$$= 6367 \text{ km}$$

θ in radians

Historically, the fact that the Earth is not a perfect sphere has led to many variations in the distance used for a nautical mile. ■

Practice Exercise

1. Find θ in degrees if $s = 2.50$ m and $r = 1.75$ m.

AREA OF A SECTOR OF A CIRCLE

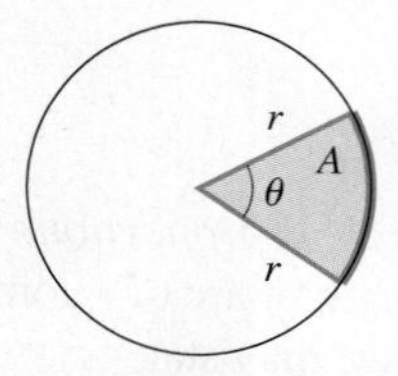

Fig. 8.36

Another application of radians is finding the area of a sector of a circle (see Fig. 8.36). Recall from geometry that areas of sectors of circles are proportional to their central angles. The area of a circle is $A = \pi r^2$, which can be written as $A = \frac{1}{2}(2\pi)r^2$. Because the angle for a complete circle is 2π, *the area of any sector of a circle in terms of the radius and central angle (in radians) is*

$$A = \frac{1}{2}\theta r^2 \quad (\theta \text{ in radians}) \tag{8.10}$$

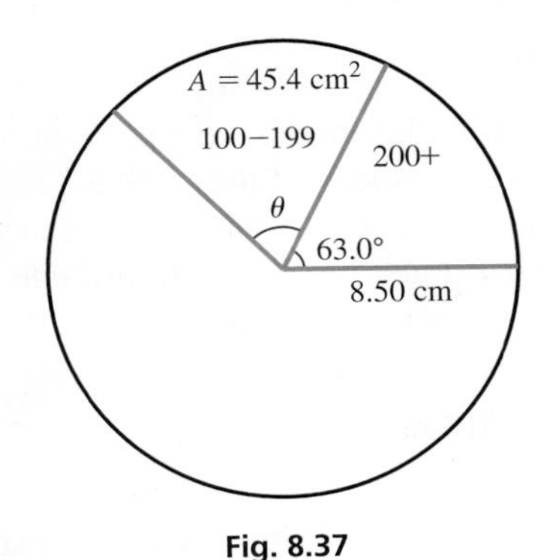

Fig. 8.37

EXAMPLE 3 Area of sector of circle—text messages

(a) Showing that 17.5% of high school students send at least 200 text messages per day on a pie chart of radius 8.50 cm means that the central angle of the sector is 63.0° (17.5% of 360°). The area of this sector (see Fig. 8.37) is

$$A = \frac{1}{2}(63.0)\left(\frac{\pi}{180}\right)(8.50^2) = 39.7 \text{ cm}^2$$

θ in radians

(b) Given that the area of the pie chart in Fig. 8.37 that shows the percent of students who send 100–199 text messages per day is 45.4 cm², we can find the central angle of this sector by first solving for θ. This gives

$$\theta = \frac{2A}{r^2} = \frac{2(45.4)}{8.50^2} = 1.26$$

no units indicates radian measure

This means the central angle is 1.26 rad, or 72.2°. ■

Practice Exercise

2. Find A if $r = 17.5$ in. and $\theta = 125°$.

CAUTION We should note again that the equations in this section require that the angle θ is expressed in radians. A common error is to use θ in degrees. ■

LINEAR AND ANGULAR VELOCITY

The average velocity of a moving object is defined by $v = s/t$, where v is the average velocity, s is the distance traveled, and t is the elapsed time. For an object moving in a circular path with constant speed, the distance traveled is the length of arc through which it moves. Therefore, if we divide both sides of Eq. (8.9) by t, we obtain

$$\frac{s}{t} = \frac{\theta r}{t} = \frac{\theta}{t}r$$

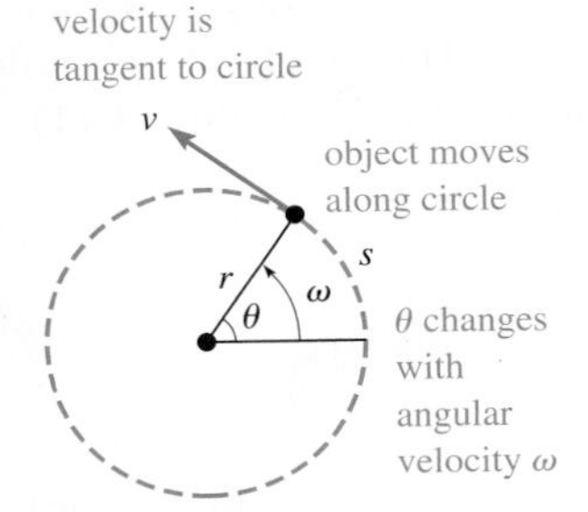

Fig. 8.38

where θ/t is called the *angular velocity* and is designated by ω. Therefore,

$$v = \omega r \qquad \textbf{(8.11)}$$

Equation (8.11) expresses the relationship between the **linear velocity** v *and the* **angular velocity** ω *of an object moving around a circle of radius r.* See Fig. 8.38. In the figure, v is shown directed tangent to the circle, for that is its direction for the position shown. The direction of v changes constantly.

■ Some of the typical units used for angular velocity are

rad/s	rad/min	rad/h
°/s	°/min	
r/s	r/min	

(r represents revolutions.)
(r/min is the same as rpm. This text does not use rpm.)

CAUTION In order to use Eq. (8.11), *the units for* ω *must be radians per unit time.* However, in practice, ω is often given in revolutions per minute or in some similar unit. In these cases, it is necessary to convert the units of ω to radians per unit of time before substituting in Eq. (8.11). ■

EXAMPLE 4 Angular velocity—hang glider motion

A person on a hang glider is moving in a horizontal circular arc of radius 90.0 m with an angular velocity of 0.125 rad/s. The person's linear velocity is

$$v = (0.125 \text{ rad/s})(90.0 \text{ m}) = 11.3 \text{ m/s}$$

(Remember that radians are numbers and are not included in the final set of units.) This means that the person is moving along the circumference of the arc at 11.3 m/s (40.7 km/h). ■

EXAMPLE 5 Angular velocity—satellite orbit

A communications satellite remains at an altitude of 22,320 mi above a point on the equator. If the radius of the Earth is 3960 mi, what is the velocity of the satellite?

■ See chapter introduction.

In order for the satellite to remain over a point on the equator, it must rotate exactly once each day around the center of the Earth (and it must remain at an altitude of 22,320 mi). Because there are 2π radians in each revolution, the angular velocity is

$$\omega = \frac{1 \text{ r}}{1 \text{ day}} = \frac{2\pi \text{ rad}}{24 \text{ h}} = 0.2618 \text{ rad/h}$$

The radius of the circle through which the satellite moves is its altitude plus the radius of the Earth, or $22{,}320 + 3960 = 26{,}280$ mi. Thus, the velocity is

$$v = 0.2618(26{,}280) = 6880 \text{ mi/h}$$ ■

Practice Exercise

3. Find r if $v = 25.0$ ft/s and the angular velocity is 6.0°/min.

EXAMPLE 6 Angular velocity—pulley rotation

A pulley belt 10.0 ft long takes 2.00 s to make one complete revolution. The radius of the pulley is 6.00 in. What is the angular velocity (in revolutions per minute) of a point on the rim of the pulley? See Fig. 8.39.

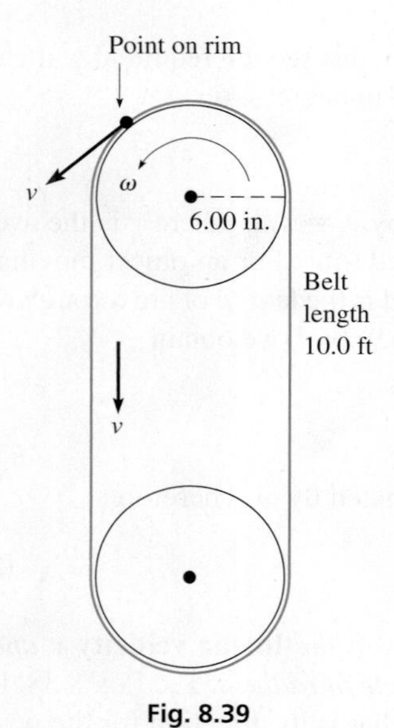

Fig. 8.39

Because the linear velocity of a point on the rim of the pulley is the same as the velocity of the belt, $v = 10.0$ ft/2.00 s $= 5.00$ ft/s. The radius of the pulley is $r = 6.00$ in. $= 0.500$ ft, and we can find ω by substituting into Eq. (8.11). This gives us

$$v = \omega r$$
$$5.00 = \omega(0.500)$$
$$\omega = 10.0 \text{ rad/s}$$

To convert this answer to revolutions per minute, we use the procedure described in Section 1.4:

$$10.0\,\frac{\text{rad}}{\text{s}} = \underbrace{\frac{10.0 \text{ rad}}{1 \text{ s}}}_{\text{original number}} \times \underbrace{\frac{60 \text{ s}}{1 \text{ min}}}_{1} \times \underbrace{\frac{1 \text{ r}}{2\pi \text{ rad}}}_{1} = \frac{10.0(60)\text{r}}{2\pi \text{ min}} = 95.5\,\frac{\text{r}}{\text{min}}$$ ■

Many types of problems use radian measure. Following is one involving electric current.

EXAMPLE 7 Application to electric current

The current at any time in a certain alternating-current electric circuit is given by $i = I \sin 120\pi t$, where I is the maximum current and t is the time in seconds. Given that $I = 0.0685$ A, find i for $t = 0.00500$ s.

Substituting, with the calculator in radian mode, we get

$$i = 0.0685 \sin[(120\pi)(0.00500)]$$
$$= 0.0651 \text{ A}$$

NOTE ➧ [$(120\pi)(0.00500)$ is a pure number, and therefore is an angle in radians.] ■

EXERCISES 8.4

In Exercises 1–4, make the given changes in the indicated examples of this section, and then solve the resulting problems.

1. In Example 1, change $\pi/6$ to $\pi/4$.
2. In Example 3(a), change 17.5% to 18.5%.
3. In Example 4, change 90.0 m to 115 m.
4. In Example 6, change 2.00 s to 2.50 s.

In Exercises 5–16, for an arc length s, area of sector A, and central angle θ of a circle of radius r, find the indicated quantity for the given values.

5. $r = 5.70$ in., $\theta = \pi/4$, $s = ?$
6. $r = 21.2$ cm, $\theta = 2.65$, $s = ?$
7. $s = 915$ mm, $\theta = 136.0°$, $r = ?$
8. $s = 0.3456$ ft, $\theta = 73.61°$, $A = ?$
9. $s = 0.3913$ mi, $r = 0.9449$ mi, $A = ?$
10. $s = 319$ m, $r = 229$ m, $\theta = ?$
11. $r = 3.8$ cm, $\theta = 6.7$, $A = ?$
12. $r = 46.3$ in., $\theta = 2\pi/5$, $A = ?$
13. $A = 0.0119$ ft^2, $\theta = 326.0°$, $r = ?$
14. $A = 1200$ mm^2, $\theta = 17°$, $s = ?$
15. $A = 165$ m^2, $r = 40.2$ m, $s = ?$
16. $A = 67.8$ mi^2, $r = 67.8$ mi, $\theta = ?$

In Exercises 17–58, solve the given problems.

17. In traveling three-fourths of the way around a traffic circle a car travels 0.203 mi. What is the radius of the traffic circle? See Fig. 8.40.

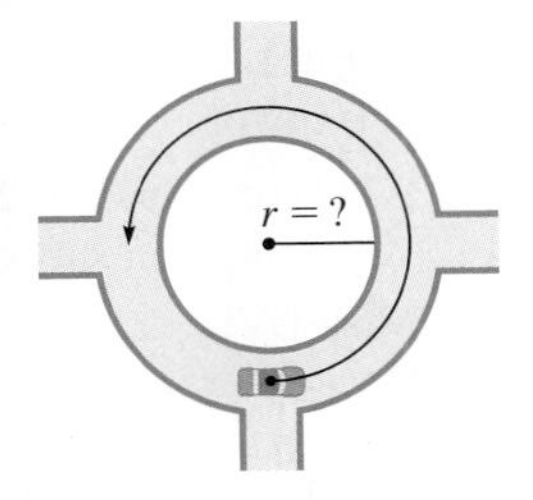

Fig. 8.40

18. The latitude of Miami is 26° N, and the latitude of the north end of the Panama Canal is 9° N. Both are at a longitude of 80° W. What is the distance between Miami and the Canal? Explain how the angle used in the solution is found. The radius of the Earth is 3960 mi.

19. The central angle corresponding to the part of a belt making contact with the larger pulley in Fig. 8.41 is 213°. Find the length of the belt in contact with the pulley if its radius is 4.50 in.

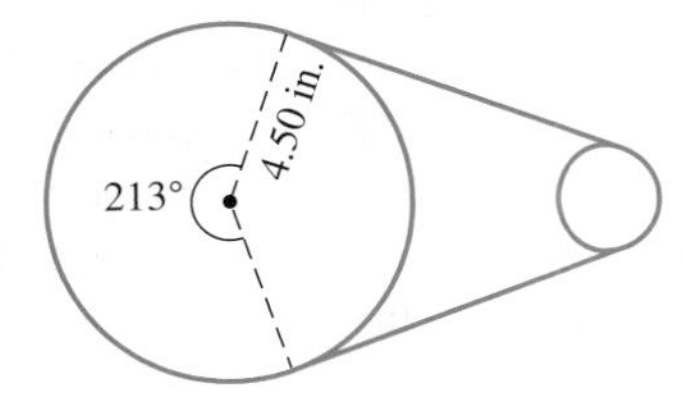

Fig. 8.41

20. A pizza is cut into eight equal pieces, the area of each being 88 cm^2. What was the diameter of the original pizza?

21. When between 12:00 noon and 1:00 P.M. are the minute and hour hands of a clock 180° apart?

22. A cam is in the shape of a circular sector, as shown in Fig. 8.42. What is the perimeter of the cam?

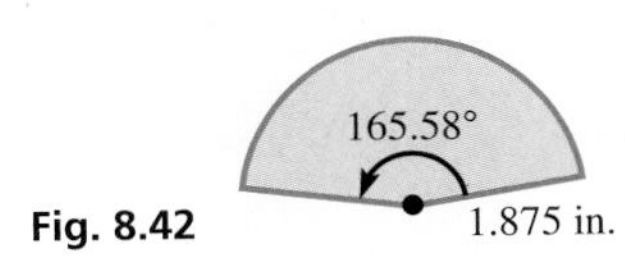

Fig. 8.42

23. A lawn sprinkler can water up to a distance of 65.0 ft. It turns through an angle of 115.0°. What area can it water?

24. A spotlight beam sweeps through a horizontal angle of 75.0°. If the range of the spotlight is 250 ft, what area can it cover?

25. If a car makes a U-turn in 6.0 s, what is its average angular velocity in the turn, expressed in rad/s?

26. A ceiling fan has blades 61.0 cm long. What is the linear velocity of the tip of a blade when the fan is rotating at 8.50 r/s?

27. What is the floor area of the hallway shown in Fig. 8.43? The outside and inside of the hallway are circular arcs.

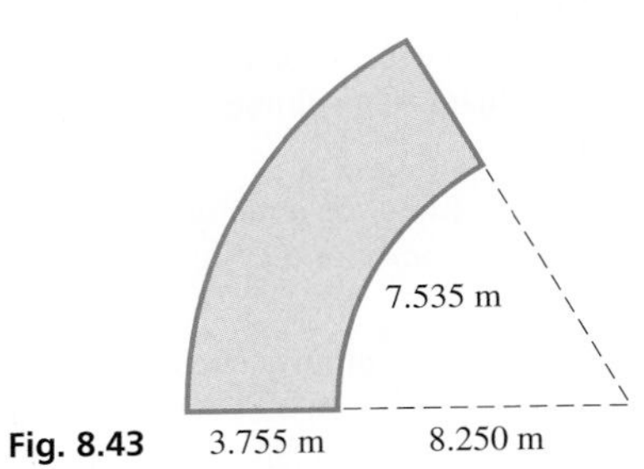

Fig. 8.43

28. The arm of a car windshield wiper is 12.75 in. long and is attached at the middle of a 15.00-in. blade. (Assume that the arm and blade are in line.) What area of the windshield is cleaned by the wiper if it swings through 110.0° arcs?

29. Part of a railroad track follows a circular arc with a central angle of 28.0°. If the radius of the arc of the inner rail is 93.67 ft and the rails are 4.71 ft apart, how much longer is the outer rail than the inner rail?

30. A wrecking ball is dropped as shown in Fig. 8.44. Its velocity at the bottom of its swing is $v = \sqrt{2gh}$, where g is the acceleration due to gravity. What is its angular velocity at the bottom if $g = 9.80$ m/s^2 and $h = 4.80$ m?

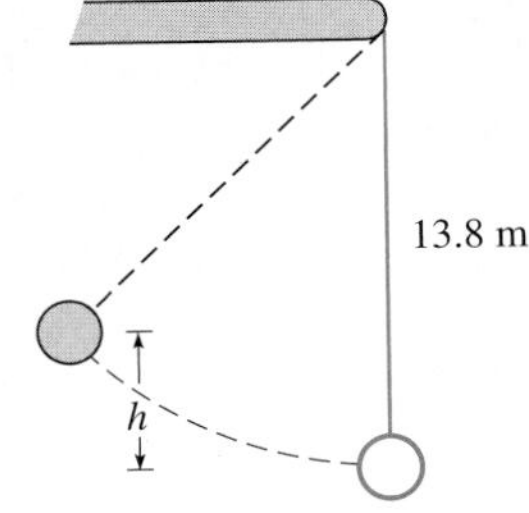

Fig. 8.44

31. Part of a security fence is built 2.50 m from a cylindrical storage tank 11.2 m in diameter. What is the area between the tank and this part of the fence if the central angle of the fence is 75.5°? See Fig. 8.45.

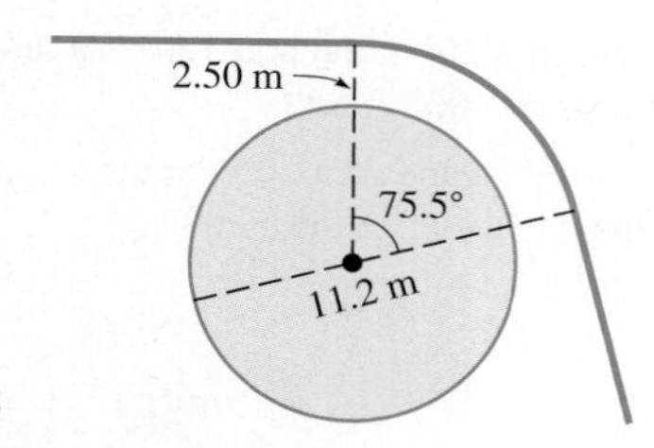

Fig. 8.45

32. Through what angle does the drum in Fig. 8.46 turn in order to lower the crate 18.5 ft?

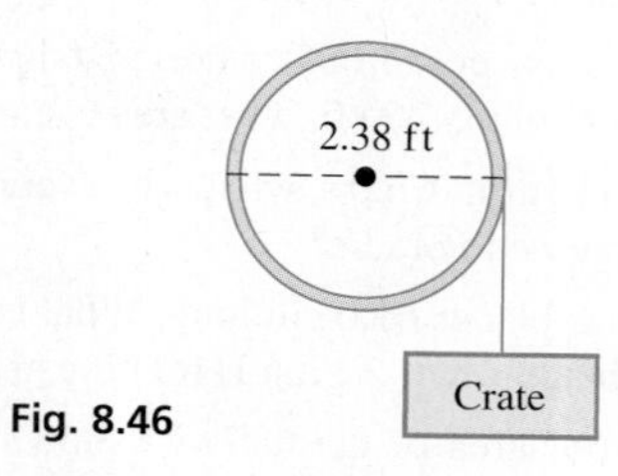

Fig. 8.46

33. A section of road follows a circular arc with a central angle of 15.6°. The radius of the inside of the curve is 285.0 m, and the road is 15.2 m wide. What is the volume of the concrete in the road if it is 0.305 m thick?

34. The propeller of the motor on a motorboat is rotating at 130 rad/s. What is the linear velocity of a point on the tip of a blade if it is 22.5 cm long?

35. A storm causes a pilot to follow a circular-arc route, with a central angle of 12.8°, from city A to city B rather than the straight-line route of 185.0 km. How much farther does the plane fly due to the storm? (*Hint*: First find the radius of the circle.)

36. An interstate route exit is a circular arc 330 m long with a central angle of 79.4°. What is the radius of curvature of the exit?

37. The paddles of a riverboat have a radius of 8.50 ft and revolve at 20.0 r/min. What is the speed of a tip of one of the paddles?

38. The sweep second hand of a watch is 15.0 mm long. What is the linear velocity of the tip?

39. A DVD has a diameter of 4.75 in. and rotates at 360.0 r/min. What is the linear velocity of a point on the outer edge?

40. The *Singapore Flyer* is a Ferris wheel that has 28 air-conditioned capsules, each able to hold 28 people. It is 165 m high, with a 150-m-diameter wheel, and makes one revolution in 37 min. Find the speed (in cm/s) of a capsule.

41. GPS satellites orbit the Earth twice per day with an orbital radius of 26,600 km. What is the linear velocity (in km/s) of these satellites?

42. Assume that the Earth rotates around the sun in a circular orbit of radius 93,000,000 mi. (which is approximately correct). What is the Earth's linear velocity (in mi/h)?

43. The sprocket assembly for a 28.0-in. bike is shown in Fig. 8.47. How fast (in r/min) does the rider have to pedal in order to go 15.0 mi/h on level ground?

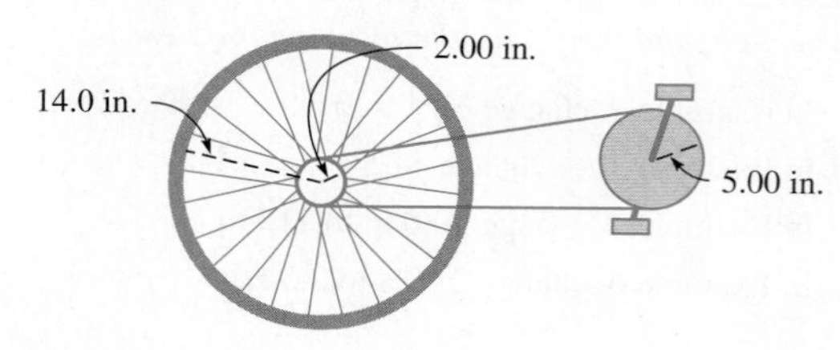

Fig. 8.47

44. The flywheel of a car engine is 0.36 m in diameter. If it is revolving at 750 r/min, through what distance does a point on the rim move in 2.00 s?

45. Two streets meet at an angle of 82.0°. What is the length of the piece of curved curbing at the intersection if it is constructed along the arc of a circle 15.0 ft in radius? See Fig. 8.48.

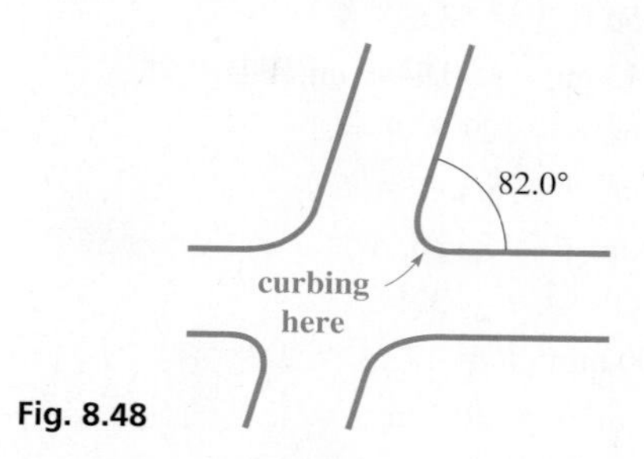

Fig. 8.48

46. An ammeter needle is deflected 52.00° by a current of 0.2500 A. The needle is 3.750 in. long, and a circular scale is used. How long is the scale for a maximum current of 1.500 A?

47. A drill bit $\frac{3}{8}$ in. in diameter rotates at 1200 r/min. What is the linear velocity of a point on its circumference?

48. A helicopter blade is 2.75 m long and is rotating at 420 r/min. What is the linear velocity of the tip of the blade?

49. Most DVD players use *constant linear velocity*. This means that the angular velocity of the disk continually changes, but the linear velocity of the point being read by the laser remains the same. If a DVD rotates at 1590 r/min for a point 2.25 cm from the center, find the number of revolutions per minute for a point 5.51 cm from the center.

50. A jet is traveling westward with the sun directly overhead (the jet is on a line between the sun and the center of the Earth). How fast must the jet fly in order to keep the sun directly overhead? (Assume that the Earth's radius is 3960 mi, the altitude of the jet is low, and the Earth rotates about its axis once in 24.0 h.)

51. A 1500-kW wind turbine (windmill) rotates at 40.0 r/min. What is the linear velocity of a point on the end of a blade, if the blade is 35 ft long (from the center of rotation)?

52. What is the linear velocity of a point in Charleston, South Carolina, which is at a latitude of 32°46′ N? The radius of the Earth is 3960 mi.

53. Through what total angle does the drive shaft of a car rotate in 1 s when the tachometer reads 2400 r/min?

54. A baseball field is designed such that the outfield fence is along the arc of a circle with its center at second base. If the radius of the circle is 280 ft, what is the playing area of the field? See Fig. 8.49.

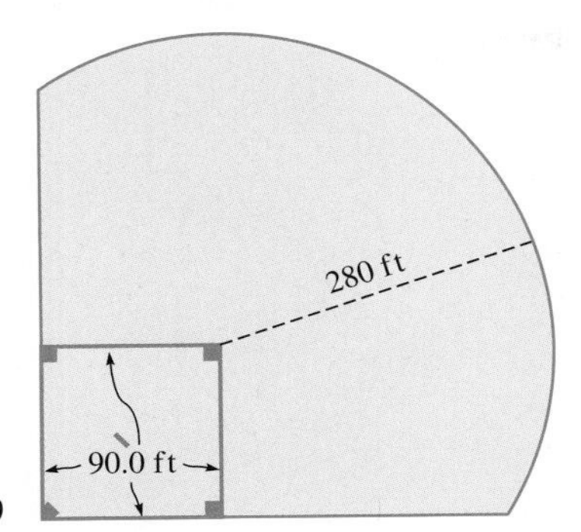

Fig. 8.49

55. The turbine fan blade of a turbojet engine is 1.2 m in diameter and rotates at 250 r/s. How fast is the tip of a blade moving?

56. A patio is in the shape of a circular sector with a central angle of 160.0°. It is enclosed by a railing of which the circular part is 11.6 m long. What is the area of the patio?

57. An oil storage tank 4.25 m long has a flat bottom as shown in Fig. 8.50. The radius of the circular part is 1.10 m. What volume of oil does the tank hold?

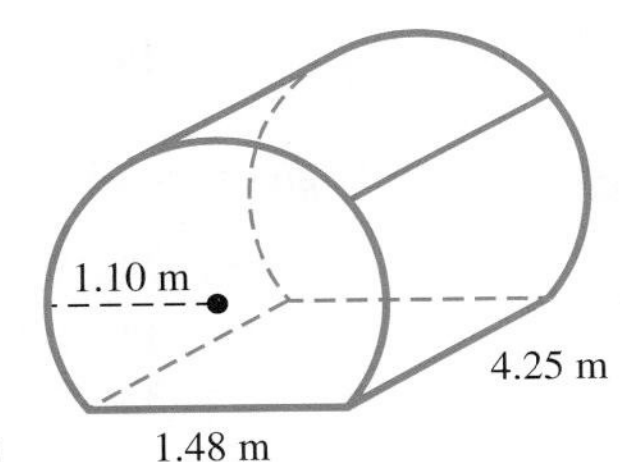

Fig. 8.50

58. Two equal beams of light illuminate the area shown in Fig. 8.51. What area is lit by both beams?

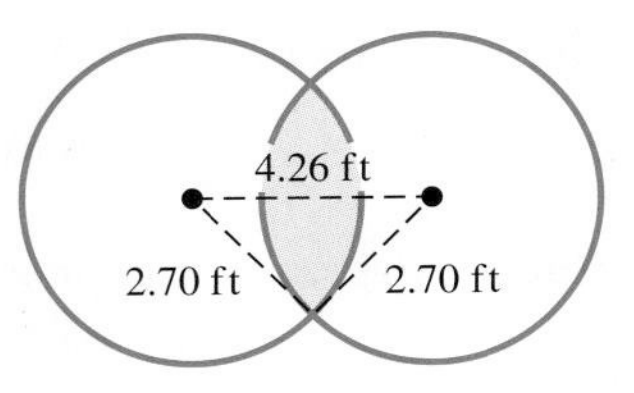

Fig. 8.51

In Exercises 59–62, another use of radians is illustrated.

59. Use a calculator (in radian mode) to evaluate the ratios $(\sin\theta)/\theta$ and $(\tan\theta)/\theta$ for $\theta = 0.1, 0.01, 0.001$, and 0.0001. From these values, explain why it is possible to say that

$$\sin\theta = \tan\theta = \theta \qquad \textbf{(8.12)}$$

approximately for very small angles measured in radians.

60. Using Eq. (8.12), evaluate tan 0.001°. Compare with a calculator value.

61. An astronomer observes that a star 12.5 light-years away moves through an angle of 0.2″ in 1 year. Assuming it moved in a straight line perpendicular to the initial line of observation, how many miles did the star move? ($1 \text{ light-year} = 5.88 \times 10^{12}$ mi.) Use Eq. (8.12).

62. In calculating a back line of a lot, a surveyor discovers an error of 0.05° in an angle measurement. If the lot is 136.0 m deep, by how much is the back-line calculation in error? See Fig. 8.52. Use Eq. (8.12).

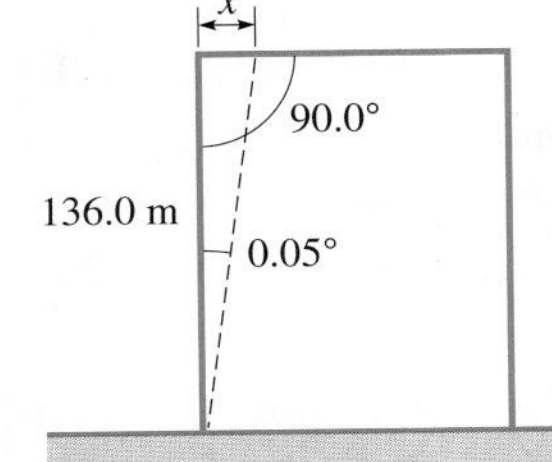

Fig. 8.52

Answers to Practice Exercises

1. 81.9° **2.** 334 in.^{2} **3.** 14,300 ft

CHAPTER 8 KEY FORMULAS AND EQUATIONS

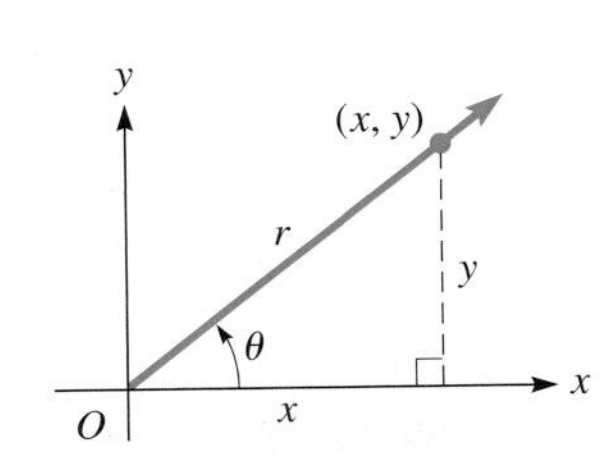

Fig. 1

$$\sin\theta = \frac{y}{r} \quad \csc\theta = \frac{r}{y}$$
$$\cos\theta = \frac{x}{r} \quad \sec\theta = \frac{r}{x} \qquad \textbf{(8.1)}$$
$$\tan\theta = \frac{y}{x} \quad \cot\theta = \frac{x}{y}$$

$$\text{trig }\theta = \pm\text{trig }\theta_{\text{ref}} \text{ ("trig" is any trigonometric function)} \qquad \textbf{(8.2)}$$

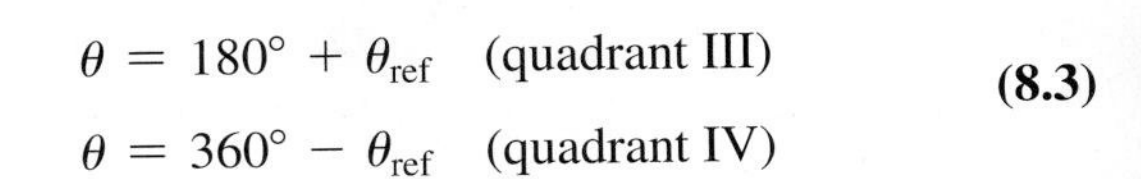

$$\theta = \theta_{\text{ref}} \text{ (quadrant I)} \qquad \theta = 180° + \theta_{\text{ref}} \text{ (quadrant III)}$$
$$\theta = 180° - \theta_{\text{ref}} \text{ (quadrant II)} \qquad \theta = 360° - \theta_{\text{ref}} \text{ (quadrant IV)} \qquad \textbf{(8.3)}$$

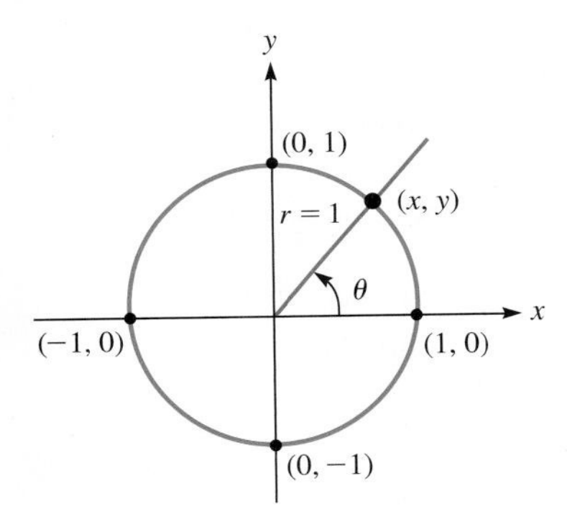

$$\sin\theta = y \quad \csc\theta = \frac{1}{y}$$
$$\cos\theta = x \quad \sec\theta = \frac{1}{x} \qquad \textbf{(8.4)}$$
$$\tan\theta = \frac{y}{x} \quad \cot\theta = \frac{x}{y}$$

Negative angles

$$\sin(-\theta) = -\sin\theta \quad \cos(-\theta) = \cos\theta \quad \tan(-\theta) = -\tan\theta$$
$$\csc(-\theta) = -\csc\theta \quad \sec(-\theta) = \sec\theta \quad \cot(-\theta) = -\cot\theta \tag{8.5}$$

Radian-degree conversions

$$\pi \text{ rad} = 180° \tag{8.6}$$

$$1° = \frac{\pi}{180} \text{ rad} = 0.01745 \text{ rad} \tag{8.7}$$

$$1 \text{ rad} = \frac{180°}{\pi} = 57.30° \tag{8.8}$$

Circular arc length

$$s = \theta r \qquad (\theta \text{ in radians}) \tag{8.9}$$

Circular sector area

$$A = \frac{1}{2}\theta r^2 \qquad (\theta \text{ in radians}) \tag{8.10}$$

Linear and angular velocity

$$v = \omega r \tag{8.11}$$

CHAPTER 8 REVIEW EXERCISES

CONCEPT CHECK EXERCISES

Determine each of the following as being either **true** *or* **false**. *If it is false, explain why.*

1. If θ is a fourth-quadrant angle, then $\cos\theta > 0$.
2. $\sin 40° = -\sin 220°$
3. For $0° < \theta < 360°$, $\sin\theta > 0$, and $\tan\theta < 0$, then $\sec\theta > 0$.
4. To convert an angle measured in radians to an angle measured in degrees, multiply the number of radians by $180°/\pi$.
5. The length of arc s of a circular arc of radius r and central angle θ (in radians) is $s = \frac{1}{2}\theta r$.
6. If the sine and cosine of an angle are both negative, then the angle must terminate in the third quadrant.

PRACTICE AND APPLICATIONS

In Exercises 7–10, find the trigonometric functions of θ. The terminal side of θ passes through the given point.

7. (6, 8)
8. (−12, 5)
9. (42, −12)
10. (−0.2, −0.3)

In Exercises 11–14, express the given trigonometric functions in terms of the same function of the reference angle.

11. cos 132°, tan 194°
12. sin 243°, cot 318°
13. sin 289°, sec(−15°)
14. cos 463°, csc(−100°)

In Exercises 15–18, express the given angle measurements in terms of π.

15. 40°, 153°
16. 22.5°, 324°
17. 408°, −202.5°
18. 12°, −162°

In Exercises 19–26, the given numbers represent angle measure. Express the measure of each angle in degrees.

19. $\frac{7\pi}{5}, \frac{13\pi}{18}$
20. $\frac{3\pi}{8}, \frac{7\pi}{20}$
21. $\frac{\pi}{15}, \frac{11\pi}{6}$
22. $\frac{27\pi}{10}, \frac{5\pi}{4}$
23. 0.560
24. −1.354
25. −36.07
26. 14.5

In Exercises 27–32, express the given angles in radians (not in terms of π).

27. 102°
28. 305°
29. 20.25°
30. 148.38°
31. −636.2°
32. 385.4°

In Exercises 33–36, express the given angles in radian measure in terms of π.

33. 270°
34. 210°
35. −300°
36. 75°

In Exercises 37–56, determine the values of the given trigonometric functions directly on a calculator. The angles are approximate. Express answers to Exercises 49–52, to four significant digits.

37. cos 237.4°
38. sin 141.3°
39. cot 295°
40. tan 184°
41. csc 247.82°
42. sec 96.17°
43. sin 542.8°
44. cos 326.72°
45. tan 301.4°
46. sin 703.9°
47. tan 436.42°
48. cos(−162.32°)
49. $\sin\frac{9\pi}{5}$
50. $\sec\frac{5\pi}{8}$
51. $\cos\left(-\frac{7\pi}{6}\right)$
52. $\tan\frac{23\pi}{12}$
53. sin 0.5906
54. tan 0.8035
55. csc 2.153
56. cos(−7.190)

In Exercises 57–60, find θ in degrees for $0° \le \theta < 360°$.

57. $\tan\theta = 0.1817$
58. $\sin\theta = -0.9323$
59. $\cos\theta = -0.4730$
60. $\cot\theta = 1.196$

In Exercises 61–64, find θ in radians for $0 \le \theta < 2\pi$.

61. $\cos\theta = 0.8387$

62. $\csc\theta = 9.569$

63. $\sin\theta = -0.8650$

64. $\tan\theta = 8.480$

In Exercises 65–68, find θ in degrees for $0° \le \theta < 360°$.

65. $\cos\theta = -0.672$, $\sin\theta < 0$

66. $\tan\theta = -1.683$, $\cos\theta < 0$

67. $\cot\theta = 0.4291$, $\cos\theta < 0$

68. $\sin\theta = 0.2626$, $\tan\theta < 0$

In Exercises 69–76, for an arc of length s, area of sector A, and central angle θ of circle of radius r, find the indicated quantity for the given values.

69. $s = 20.3$ in., $\theta = 107.5°$, $r = ?$

70. $s = 5840$ ft, $r = 1060$ ft, $\theta = ?$

71. $A = 265\text{ mm}^2$, $r = 12.8$ mm, $\theta = ?$

72. $A = 0.908\text{ km}^2$, $\theta = 234.5°$, $r = ?$

73. $r = 4.62$ m, A $= 32.8\text{ m}^2$, $s = ?$

74. $\theta = 98.5°$, $A = 0.493\text{ ft}^2$, $s = ?$

75. $\theta = 0.85°$, $s = 7.94$ in., $A = ?$

76. $r = 254$ cm, $s = 7.61$ cm, $A = ?$

In Exercises 77–103, solve the given problems.

77. Without a calculator, evaluate $\tan 200° + 2\cot 110° + \tan(-160°)$.

78. Without a calculator, evaluate $2\cos 40° + \cos 140° + \sin 230°$.

79. In Fig. 8.53, show that the area of a design label (a segment of a circle) intercepted by angle θ is $A = \frac{1}{2}r^2(\theta - \sin\theta)$. Find the area if $r = 4.00$ cm and $\theta = 1.45$.

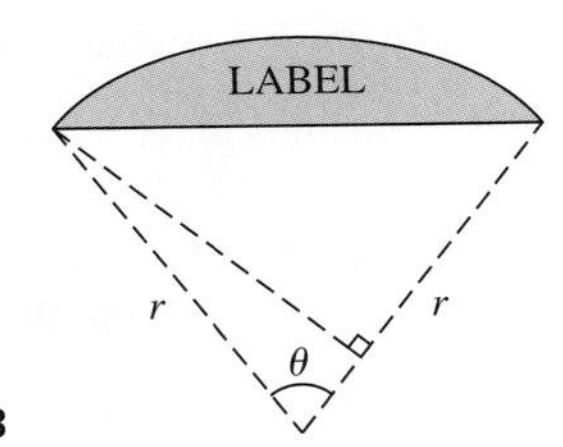

Fig. 8.53

80. The cross section of a tunnel is the major segment of a circle of radius 12.0 ft wide. The base of the tunnel is 20.0 ft wide. What is the area of the cross section? See Exercise 79.

81. Find (a) the area and (b) the perimeter of the parcel of land shown in Fig. 8.54. Its shape is a right triangle attached to a circular sector.

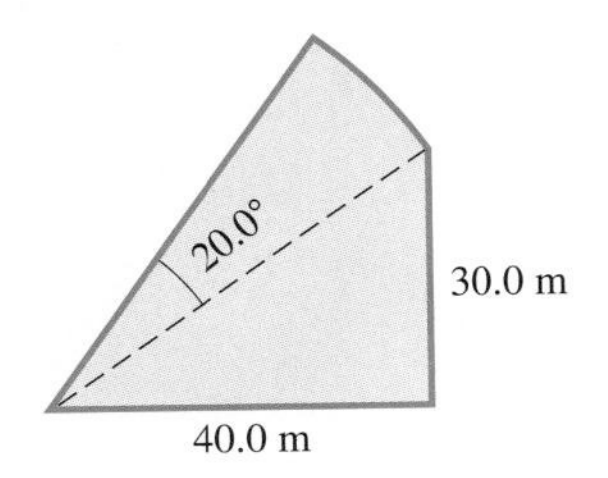

Fig. 8.54

82. The speedometer of a car is designed to be accurate with tires that are 14.0 in. in radius. If the tires are changed to 15.0 in. in radius, and the speedometer shows 55 mi/h, how fast is the car actually going?

83. The instantaneous power p (in W) input to a resistor in an alternating-current circuit is $p = p_m \sin^2 377t$, where p_m is the maximum power input and t is the time (in s). Find p for $p_m = 0.120$ W and $t = 2.00$ ms. $[\sin^2\theta = (\sin\theta)^2.]$

84. The horizontal distance x through which a pendulum moves is given by $x = a(\theta + \sin\theta)$, where a is a constant and θ is the angle between the vertical and the pendulum. Find x for $a = 45.0$ cm and $\theta = 0.175$.

85. A sector gear with a pitch radius of 8.25 in. and a 6.60-in. arc of contact is shown in Fig. 8.55. What is the sector angle θ?

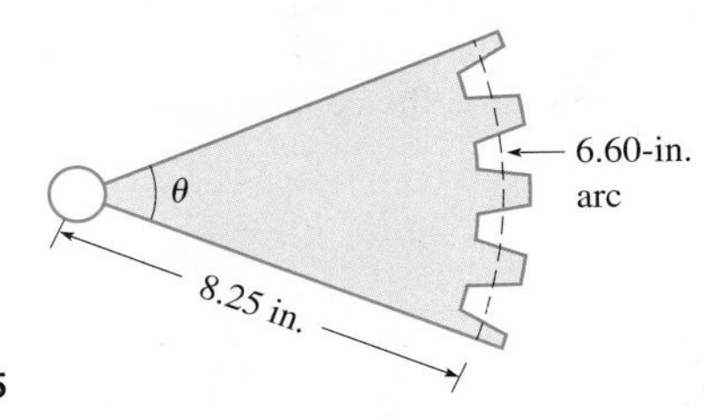

Fig. 8.55

86. Two pulleys have radii of 10.0 in. and 6.00 in., and their centers are 40.0 in. apart. If the pulley belt is uncrossed, what must be the length of the belt?

87. A special vehicle for traveling on glacial ice in *Banff National Park* in the Canadian Rockies has tires 4.8 ft in diameter. If the vehicle travels at 3.5 mi/h, what is the angular velocity (in r/min) of a tire?

88. A rotating circular restaurant at the top of a hotel has a diameter of 32.5 m. If it completes one revolution in 24.0 min, what is the velocity of the outer surface?

89. Find the velocity (in mi/h) of the moon as it revolves about the Earth. Assume it takes 28 days for one revolution at a distance of 240,000 mi from the Earth.

90. The stopboard of a shot-put circle is a circular arc 1.22 m in length. The radius of the circle is 1.06 m. What is the central angle?

91. The longitude of Anchorage, Alaska, is 150° W, and the longitude of St. Petersburg, Russia, is 30° E. Both cities are at a latitude of 60° N. (a) Find the great circle distance (see page 254) from Anchorage to St. Petersburg over the north pole. (b) Find the distance between them along the 60° N latitude arc. The radius of the Earth is 3960 mi. What do the results show?

92. A piece of circular filter paper 15.0 cm in diameter is folded such that its effective filtering area is the same as that of a sector with central angle of 220°. What is the filtering area?

93. To produce an electric current, a circular loop of wire of diameter 25.0 cm is rotating about its diameter at 60.0 r/s in a magnetic field. What is the greatest linear velocity of any point on the loop?

94. Find the area of the decorative glass panel shown in Fig. 8.56. The panel is made up of two equal circular sectors and an isosceles triangle.

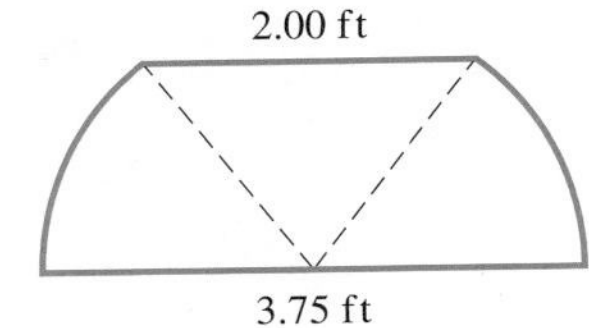

Fig. 8.56

95. A circular hood is to be used over a piece of machinery. It is to be made from a circular piece of sheet metal 3.25 ft in radius. A hole 0.75 ft in radius and a sector of central angle 80.0° are to be removed to make the hood. What is the area of the top of the hood?

96. The chain on a chain saw is driven by a sprocket 7.50 cm in diameter. If the chain is 108 cm long and makes one revolution in 0.250 s, what is the angular velocity (in r/s) of the sprocket?

97. An *ultracentrifuge*, used to observe the sedimentation of particles such as proteins, may rotate as fast as 80,000 r/min. If it rotates at this rate and is 7.20 cm in diameter, what is the linear velocity of a particle at the outer edge?

98. A computer is programmed to shade in a sector of a pie chart 2.44 cm in radius. If the perimeter of the shaded sector is 7.32 cm, what is the central angle (in degrees) of the sector? See Fig. 8.57.

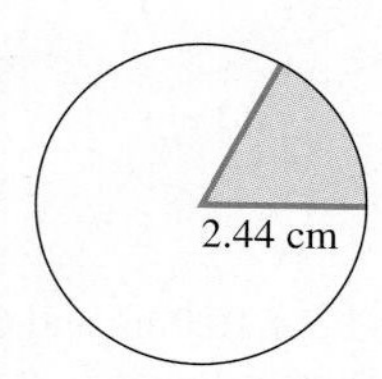

Fig. 8.57

99. A Gothic arch, commonly used in medieval European structures, is formed by two circular arcs. In one type, each arc is one-sixth of a circle, with the center of each at the base on the end of the other arc. See Fig. 8.58. Therefore, the width of the arch equals the radius of each arc. For such an arch, find the area of the opening if the width is 15.0 m.

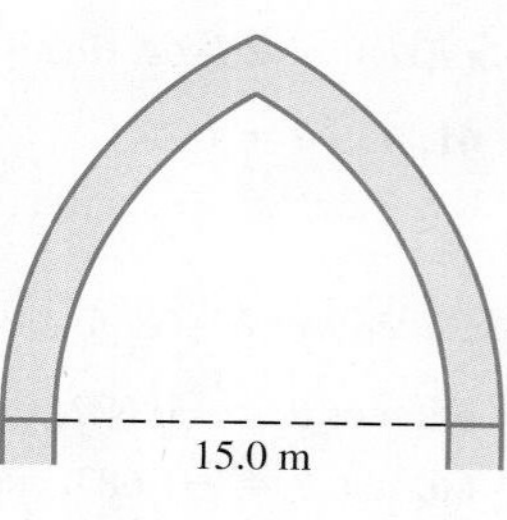

Fig. 8.58

100. The Trans-Alaska Pipeline was assembled in sections 40.0 ft long and 4.00 ft in diameter. If the depth of the oil in one horizontal section is 1.00 ft, what is the volume of oil in this section?

101. A laser beam is transmitted with a "width" of 0.0008° and makes a circular spot of radius 2.50 km on a distant object. How far is the object from the source of the laser beam? Use Eq. (8.12) on page 259.

102. The planet Venus subtends an angle of 15″ to an observer on Earth. If the distance between Venus and Earth is 1.04×10^8 mi, what is the diameter of Venus? Use Eq. (8.12) on page 259.

103. Write a paragraph explaining how you determine the units for the result of the following problem: An astronaut in a spacecraft circles the moon once each 1.95 h. If the altitude of the spacecraft is constant at 70.0 mi, what is its velocity? The radius of the moon is 1080 mi. (What is the answer?)

CHAPTER 8 PRACTICE TEST

As a study aid, we have included complete solutions for each Practice Test problem at the end of this book.

1. Change 150° to radians in terms of π.

2. Determine the sign of (a) sec 285°, (b) sin 245°, (c) cot 265°.

3. Express sin 205° in terms of the sine of the reference angle. Do not evaluate.

4. Find $\sin\theta$ and $\sec\theta$ if θ is in standard position and the terminal side passes through $(-9, 12)$.

5. An airplane propeller blade is 2.80 ft long and rotates at 2200 r/min. What is the linear velocity of a point on the tip of the blade?

6. Given that 3.572 is the measure of an angle, express the angle in degrees.

7. If $\tan\theta = 0.2396$, find θ, in degrees, for $0° \le \theta < 360°$.

8. If $\cos\theta = -0.8244$ and $\csc\theta < 0$, find θ in radians for $0 \le \theta < 2\pi$.

9. If $\sin\theta = -\dfrac{6}{7}$ and $\cos\theta > 0$, find $\tan\theta$.

10. The floor of a sunroom is in the shape of a circular sector of arc length 32.0 ft and radius 8.50 ft. What is the area of the floor?

11. The face of a flat wedge is a circular sector with an area of 38.5 cm^2 and diameter of 12.2 cm. What is the arc length of the wedge?

Vectors and Oblique Triangles

LEARNING OUTCOMES

After completion of this chapter, the student should be able to:

- Distinguish between a scalar and a vector
- Add vectors graphically by the polygon method and the parallelogram method
- Resolve a vector into its x- and y-components
- Add vectors by components
- Solve application problems involving vectors
- Solve oblique triangles using the law of sines and/or the law of cosines
- Solve application problems involving oblique triangles

In many applications, we often deal with such things as forces and velocities. To study them, both their *magnitudes* and *directions* must be known. In general, a quantity for which we must specify *both magnitude and direction* is called a *vector.*

In basic applications, a vector is usually represented by an arrow showing its magnitude and direction, although this was not common before the 1800s. In 1743, the French mathematician d'Alembert published a paper on dynamics in which he used some diagrams, but most of the text was algebraic. In 1788, the French mathematician Lagrange wrote a classic work on *Analytical Mechanics,* but the text was all algebraic and included no diagrams.

In the 1800s, calculus was used to greatly advance the use of vectors, and in turn, these advancements became very important in further developments in scientific fields such as electromagnetic theory. Also in the 1800s, mathematicians defined a vector more generally and opened up study in new areas of advanced mathematics. A vector is an excellent example of a math concept that came from a basic physics concept.

In addition to studying vectors in this chapter, we also develop methods of solving *oblique* triangles (triangles that are not right triangles). Although obviously not known in their modern forms, one of these methods, the *law of cosines,* was known to the Greek astronomer Ptolemy (90–168), and the other method, the *law of sines,* was known to Islamic mathematicians of the 1100s. In solving oblique triangles, we often use the trig functions of obtuse angles.

Vectors are of great importance in many fields of science and technology, including physics, engineering, structural design, and navigation. The law of sines and law of cosines are also applied in those fields when working with triangles that don't contain a right angle.

◀ Vectors can be used to find certain forces when a system is in equilibrium. In Section 9.4, we use vectors to find the tension in a rope that is holding up a mountain climber.

9.1 Introduction to Vectors

Scalars and Vectors • Addition of Vectors • Scalar Multiple of a Vector • Subtraction of Vectors

We deal with many quantities that may be described by a number that shows only the magnitude. These include lengths, areas, time intervals, monetary amounts, and temperatures. *Quantities such as these, described only by the* ***magnitude,*** *are known as* **scalars.**

Many other quantities, called **vectors,** *are fully described only when both the* ***magnitude*** *and* ***direction*** *are specified.* The following example shows the difference between scalars and vectors.

EXAMPLE 1 Scalars and vectors—speed versus velocity

A jet is traveling at 600 mi/h. From this statement alone, we know only the *speed* of the jet. *Speed is a scalar quantity,* and it tells us only the *magnitude* of the rate. Knowing only the speed of the jet, we know the rate at which it is moving, but we do not know where it is headed.

If the phrase "in a direction 10° south of west" is added to the sentence about the jet, we specify the direction of travel as well as the speed. We then know the *velocity* of the jet; that is the *direction* of travel as well as the rate at which it is moving. ***Velocity*** *is a* ***vector*** *quantity.* ■

For an example of the action of two vectors, consider a boat moving on a river. We will assume that the boat is driven by a motor that can move it at 8 mi/h in still water and that the river's current is going 6 mi/h downstream, as shown in Fig. 9.1. We quickly see that the movement of the boat depends on the direction in which it is headed. If it heads downstream, it moves at 14 mi/h, for the water is moving at 6 mi/h and the boat moves at 8 mi/h with respect to the water. If it heads upstream, however, it moves only at 2 mi/h, because the river is acting directly against the motor. If the boat heads directly across the river, the point it reaches on the other side is not directly opposite the point from which it started. This is so because the river is moving the boat downstream *at the same time* the boat moves across the river.

Checking this last case further, assume that the river is 0.4 mi wide where the boat is crossing. It takes 0.05 h $(0.4 \text{ mi} \div 8 \text{ mi/h} = 0.05 \text{ h})$ to cross. In 0.05 h, the river will carry the boat 0.3 mi $(0.05 \text{ h} \times 6 \text{ mi/h} = 0.3 \text{ mi})$ downstream. This means the boat went 0.3 mi downstream as it went 0.4 mi across the river. From the Pythagorean theorem, we see that it went 0.5 mi from its starting point to its finishing point:

$$d^2 = 0.4^2 + 0.3^2 = 0.25$$
$$d = 0.5 \text{ mi}$$

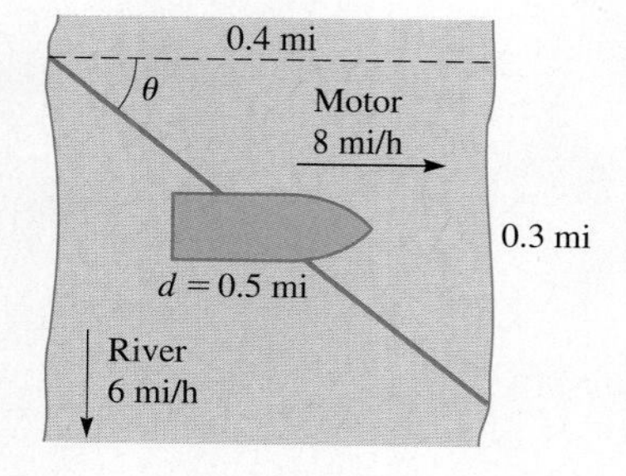

Fig. 9.1

Because the 0.5 mi was traveled in 0.05 h, the magnitude of the velocity (the *speed*) of the boat was actually

$$v = \frac{d}{t} = \frac{0.5 \text{ mi}}{0.05 \text{ h}} = 10 \text{ mi/h}$$

Also, note that the direction of this velocity can be represented along a line that makes an angle θ with the line directed directly across the river, as shown in Fig. 9.1. We can find this angle by noting that

$$\tan\theta = \frac{0.3 \text{ mi}}{0.4 \text{ mi}} = 0.75$$
$$\theta = \tan^{-1} 0.75 = 37°$$

Therefore, when headed directly across the river, the boat's velocity is 10 mi/h directed at an angle of 37° downstream from a line directly across the river.

ADDITION OF VECTORS

We have just seen two velocity vectors being *added.* Note that these vectors are not added the way numbers are added. We must take into account their directions as well as their magnitudes. Reasoning along these lines, let us now define the sum of two vectors.

We will represent a vector quantity by a letter printed in **boldface** type. The same letter in *italic* (lightface) type represents the magnitude only. Thus, **A** is a vector of magnitude A. In handwriting, one usually places an arrow over the letter to represent a vector, such as $\vec{A}$.

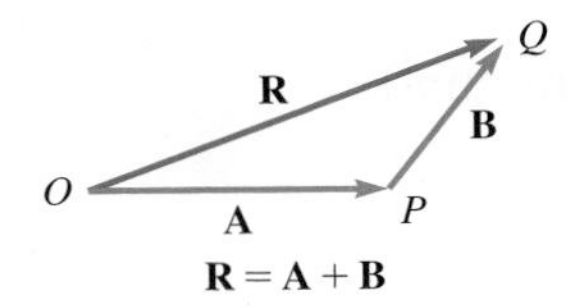

Fig. 9.2

Let **A** and **B** represent vectors directed from O to P and P to Q, respectively (see Fig. 9.2). *The vector sum* **A** + **B** *is the vector* **R,** *from the* **initial point** *O to the* **terminal point** *Q. Here, vector* **R** *is called the* **resultant.** *In general, a resultant is a single vector that is the vector sum of any number of other vectors.*

There are two common methods of adding vectors by means of a diagram. The first method, called the **polygon method,** is illustrated in Fig. 9.3. To add **B** to **A,** shift **B** parallel to itself until its tail touches the head of **A.** *The vector sum* **A** + **B** *is the resultant vector* **R,** *which is drawn from the tail of* **A** *to the head of* **B.** NOTE ▸ [In using this method, we can move a vector for addition as long as its magnitude and direction remain unchanged. (Because the magnitude and direction specify a vector, two vectors in different locations are considered the same if they have the same magnitude and direction.)] When using a diagram to add vectors, it must be drawn with reasonable accuracy.

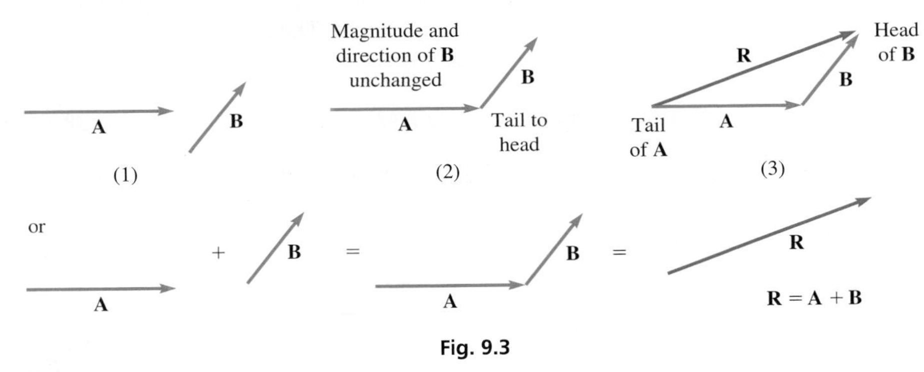

Fig. 9.3

Three or more vectors are added in the same general manner. We place the initial point of the second vector at the terminal point of the first vector, the initial point of the third vector at the terminal point of the second vector, and so on. The resultant is the vector from the initial point of the first vector to the terminal point of the last vector. NOTE ▸ [The order in which they are added does not matter.]

EXAMPLE 2 Adding vectors—polygon method

The addition of vectors **A, B,** and **C** is shown in Fig. 9.4.

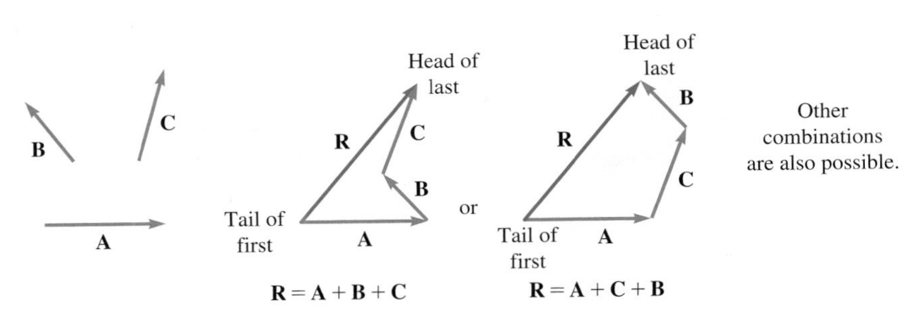

Fig. 9.4

■

Practice Exercise

1. For the vectors in Example 2, show that **R** = **B** + **C** + **A.**

Another method that is convenient when only two vectors are being added is the **parallelogram method.** To use this method, we place the two vectors being added *tail to tail* and then form a parallelogram that includes those two vectors as sides. The resultant is then the *diagonal of the parallelogram*, with its initial point being the common initial point of the two vectors being added. This method is illustrated in the following example.

EXAMPLE 3 Adding vectors—parallelogram method

The addition of vectors **A** and **B** is shown in Fig. 9.5.

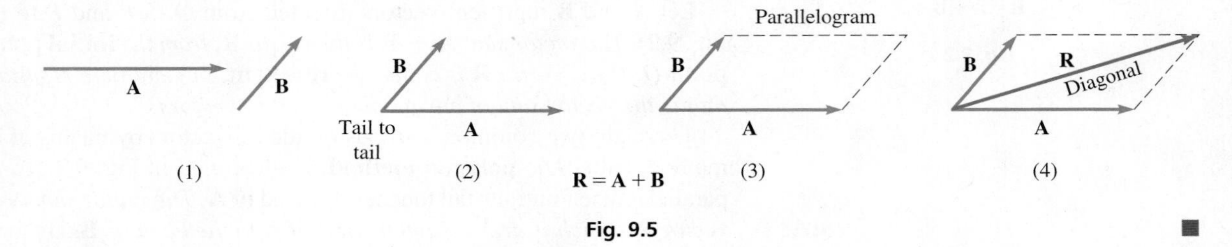

Fig. 9.5

■

If vector **C** is in the same direction as vector **A** and **C** has a magnitude n times that of **A**, then **C** = n**A**, where *the vector* n**A** *is called the* **scalar multiple** *of vector* **A**. [This means that 2**A** is a vector that is twice as long as **A** but is in the same direction.] Note carefully that only the magnitudes of **A** and 2**A** are different. Their directions are the same. The addition of scalar multiples of vectors is illustrated in the following example.

NOTE ▸

EXAMPLE 4 Scalar multiple of a vector

For vectors **A** and **B** in Fig. 9.6, find vector 3**A** + 2**B**.

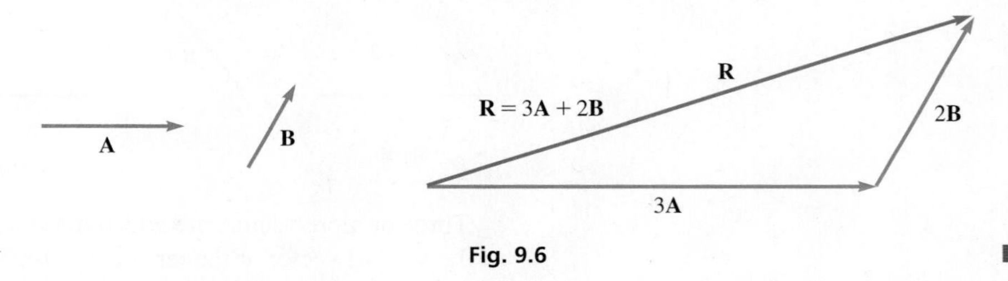

Fig. 9.6

■

Vector **B** is **subtracted** from vector **A** by reversing the direction of **B** and proceeding as in vector addition. [Thus, **A** − **B** = **A** + (−**B**), where the minus sign indicates that vector −**B** has the opposite direction of vector **B**.] Vector subtraction is illustrated in the following example.

NOTE ▸

EXAMPLE 5 Subtracting vectors

For vectors **A** and **B** in Fig. 9.7, find vector 2**A** − **B**.

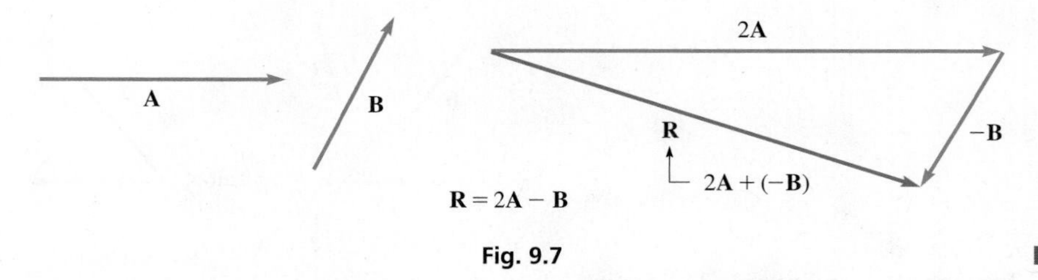

Fig. 9.7

■

Among the most important applications of vectors is that of the forces acting on a structure or object. The next example shows the addition of forces by using the parallelogram method.

Practice Exercise

2. For the vectors in Example 5, find vector **B** − 3**A**.

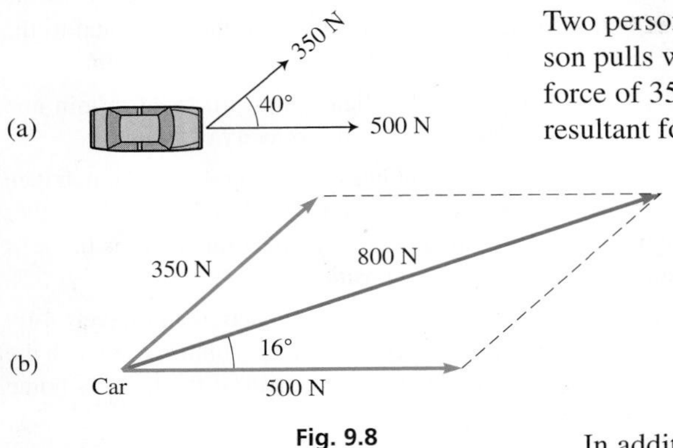

Fig. 9.8

EXAMPLE 6 Adding vectors—pulling a car

Two persons pull horizontally on ropes attached to a car that is stuck in mud. One person pulls with a force of 500 N directly to the right, while the other person pulls with a force of 350 N at an angle of 40° from the first force, as shown in Fig. 9.8(a). Find the resultant force on the car.

We make a scale drawing of the forces as shown in Fig. 9.8(b), measuring the magnitudes of the forces with a ruler and the angles with a protractor. [The scale drawing of the forces is made larger and with a different scale than that in Fig. 9.8(a) in order to get better accuracy.] We then complete the parallelogram and draw in the diagonal that represents the resultant force. Finally, we find that the resultant force is about 800 N and that it acts at an angle of about 16° from the first force. ■

In addition to force, two other important vector quantities are *velocity* and *displacement*. Velocity as a vector is illustrated in Example 1. NOTE ➧ [The **displacement** of an object is the ***change*** *in its position*. Displacement is given by the distance from a reference point and the angle from a reference direction.] The following example illustrates the difference between *distance* and *displacement*.

EXAMPLE 7 Adding vectors—jet displacement

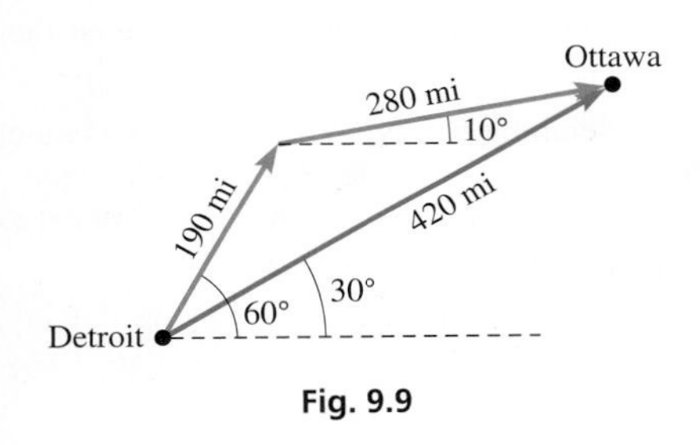

Fig. 9.9

To avoid a storm, a jet travels at 60° north of east from Detroit for 190 mi and then turns to a direction of 10° north of east for 280 mi to Ottawa. Find the displacement of Ottawa from Detroit.

We make a scale drawing in Fig. 9.9 to show the route taken by the jet. Measuring distances with a ruler and angles with a protractor, we find that Ottawa is about 420 mi from Detroit, at an angle of about 30° north of east. By giving both the magnitude and the ***direction,*** we have given the displacement.

If the jet returned directly from Ottawa to Detroit, its *displacement* from Detroit would be *zero,* although it traveled a *distance* of about 890 mi. ■

EXERCISES 9.1

In Exercises 1–4, find the resultant vectors if the given changes are made in the indicated examples of this section.

1. In Example 2, what is the resultant of the three vectors if the direction of vector **A** is reversed?

2. In Example 4, for vectors **A** and **B**, what is vector 2**A** + 3**B**?

3. In Example 5, for vectors **A** and **B**, what is vector 2**B** − **A**?

4. In Example 6, if 20° replaces 40°, what is the resultant force?

In Exercises 5–8, determine whether a scalar or a vector is described in (a) and (b). Explain your answers.

5. (a) A soccer player runs 15 m from the center of the field.
(b) A soccer player runs 15 m from the center of the field toward the opponents' goal.

6. (a) A small-craft warning reports 25 mi/h winds.
(b) A small-craft warning reports 25 mi/h winds from the north.

7. (a) An arm of an industrial robot pushes with a 10-lb force downward on a part.
(b) A part is being pushed with a 10-lb force by an arm of an industrial robot.

8. (a) A ballistics test shows that a bullet hit a wall at a speed of 400 ft/s.
(b) A ballistics test shows that a bullet hit a wall at a speed of 400 ft/s perpendicular to the wall.

In Exercises 9–14, add the given vectors by drawing the appropriate resultant.

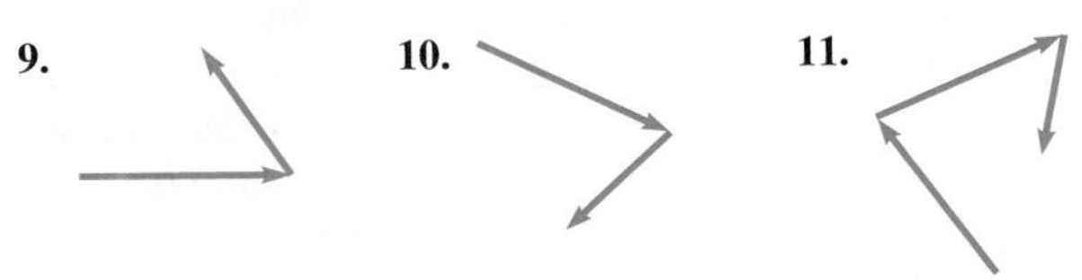

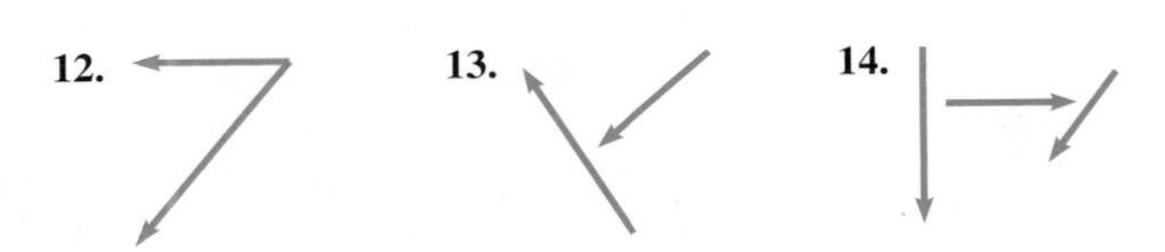

In Exercises 15–18, draw the given vectors and find their sum graphically. The magnitude is shown first, followed by the direction as an angle in standard position.

15. 3.6 cm, 0°; 4.3 cm, 90°
16. 2.3 cm, 45°; 5.2 cm, 120°
17. 6.0 cm, 150°; 1.8 cm, 315°
18. 7.5 cm, 240°; 2.3 cm, 30°

In Exercises 19–40, find the indicated vector sums and differences with the given vectors by means of diagrams. (You might find graph paper to be helpful.)

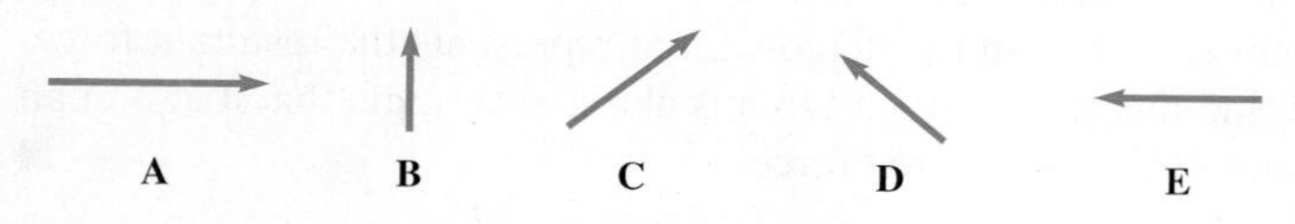

19. A + B
20. B + C
21. C + D
22. D + E
23. A + C + E
24. B + D + A
25. 2A + D + E
26. B + E + A
27. B + 3E
28. A + 2C
29. 3C + E
30. 2D + A
31. A − B
32. C − D
33. E − B
34. C − 2A
35. 3B + $\frac{1}{2}$A
36. 2B − $\frac{3}{2}$E
37. B + 2C − E
38. A + 2D − 3B
39. C − B − $\frac{3}{4}$A
40. D − 2C − $\frac{1}{2}$E

In Exercises 41–48, solve the given problems. Use a ruler and protractor as in Examples 6 and 7.

41. Two forces that act on an airplane wing are called the *lift* and the *drag*. Find the resultant of these forces acting on the airplane wing in Fig. 9.10.

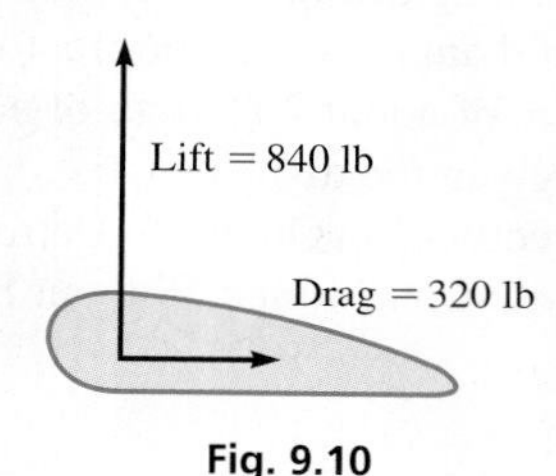

Fig. 9.10

42. Two electric charges create an electric field intensity, a vector quantity, at a given point. The field intensity is 30 kN/C to the right and 60 kN/C at an angle of 45° above the horizontal to the right. Find the resultant electric field intensity at this point.

43. A ski tow is moving skiers vertically upward at 24 m/min and horizontally at 44 m/min. What is the velocity of the tow?

44. A jet travels 17 km in a straight line as it also descends 8 km. It then turns upward and travels 10 km in a straight line until it reaches its original altitude, all in an easterly direction. What is the jet's displacement from its original position?

45. A driver takes the wrong road at an intersection and travels 4 mi north, then 6 mi east, and finally 10 mi to the southeast to reach the home of a friend. What is the displacement of the friend's home from the intersection?

46. A ship travels 20 km in a direction of 30° south of east, then turns and goes due south for another 40 km, and finally turns again 30° south of east and goes another 20 km. What is the ship's displacement from its original position?

47. A rope holds a helium-filled balloon in place with a tension of 510 N at an angle of 80° with the ground due to a wind. The weight (a vertical force) of the balloon and contents is 400 N, and the upward buoyant force is 900 N. The wind creates a horizontal force of 90 N on the balloon. What is the resultant force on the balloon?

48. While unloading a crate weighing 610 N, the chain from a crane supports it with a force of 650 N at an angle of 20° from the vertical. What force must a horizontal rope exert on the crate so that the total force (including its weight) on the crate is zero?

Answers to Practice Exercises

1. Same as **R** in Fig. 9.4. **2.** (Half scale)

9.2 Components of Vectors

Resolving a Vector into Components • Finding x- and y-Components • Meaning of Sign of Component

Using diagrams is useful in developing an understanding of vectors. However, unless the diagrams are drawn with great care, the results we get are not too accurate. Therefore, other methods are needed in order to get more accurate results.

In this section, we show how a given vector can be made to be the sum of two other vectors, with any required degree of accuracy. In the next section, we show how this lets us add vectors to get their sum with the required accuracy in the result.

Two vectors that, when added together, have a resultant equal to the original vector, are called **components** *of the original vector.* In the illustration of the boat in Section 9.1, the velocities of 8 mi/h across the river and 6 mi/h downstream are components of the 10 mi/h vector directed at the angle θ.

Certain components of a vector are of particular importance. If the initial point of a vector is placed at the origin of a rectangular coordinate system and its direction is given by an angle in standard position, we may find its **x- and y- components.** NOTE ▸ [These components are vectors directed along the axes that, when added together, equal the given vector.] The initial points of these components are at the origin, and the terminal points are at the points where perpendicular lines from the terminal point of the given vector cross the axes. *Finding these component vectors is called* **resolving the vector into its components.**

EXAMPLE 1 Components of first-quadrant vector

Find the x- and y-components of the vector **A** shown in Fig. 9.11. The magnitude of **A** is 7.25.

From the figure, we see that A_x, the magnitude of the x-component of $\mathbf{A}_x$, is related to **A** by

$$\frac{A_x}{A} = \cos 62.0°$$
$$A_x = A \cos 62.0°$$

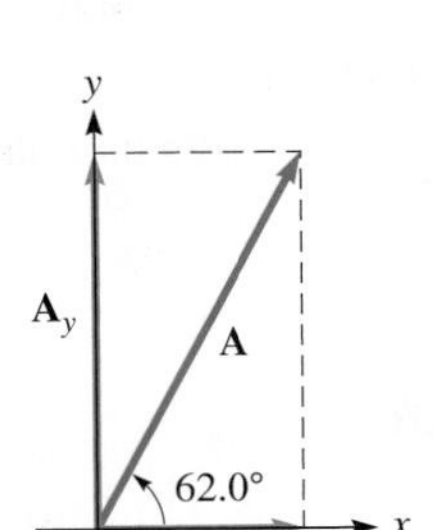

Fig. 9.11

In the same way, A_y, the magnitude of the y-component of $\mathbf{A}_y$, is related to **A** ($\mathbf{A}_y$ could be placed along the vertical dashed line) by

$$\frac{A_y}{A} = \sin 62.0°$$
$$A_y = A \sin 62.0°$$

From these relations, knowing that $A = 7.25$, we have

$$A_x = 7.25 \cos 62.0° = 3.40$$
$$A_y = 7.25 \sin 62.0° = 6.40$$

This means that the x-component is directed along the x-axis to the right and has a magnitude of 3.40. Also, the y-component is directed along the y-axis upward and its magnitude is 6.40. These two component vectors can replace vector **A,** because the effect they have is the same as **A.** ■

Practice Exercise

1. For the vector in Example 1, change the angle to 25.0° and find the x- and y-components.

EXAMPLE 2 Components of second-quadrant vector—land survey

A surveyor's marker is located 14.4 m from the set pin at a 126.0° standard position angle as shown in Fig. 9.12. Find the x- and y-components of the displacement of the marker from the pin.

Let the pin be at the origin, the initial point of the displacement vector, with the angle in standard position. The displacement vector directed along the x-axis, $\mathbf{d}_x$, is related to the displacement vector **d**, of magnitude d by

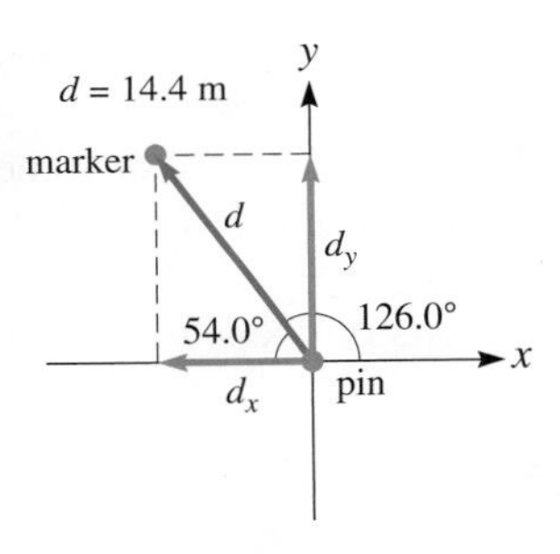

Fig. 9.12

$$d_x = d \cos 126.0°$$

(magnitude; standard-position angle)

Because the displacement vector directed along the y-axis, $\mathbf{d}_y$, could also be placed along the vertical dashed line, it is related to the displacement vector **d** by

$$d_y = d \sin 126.0°$$

Thus, the vectors $\mathbf{d}_x$ and $\mathbf{d}_y$ have the magnitudes

$$d_x = 14.4 \cos 126.0° = -8.46 \text{ m} \qquad d_y = 14.4 \sin 126.0° = 11.6 \text{ m}$$

Therefore, we have resolved the given displacement vector into two components: one directed along the negative x-axis, of magnitude 8.46 m, and the other, directed along the positive y-axis, of magnitude 11.6 m.

It is also possible to use the reference angle, as long as *the proper sign is attached to each component*. In this case, the reference angle is 54.0°, and therefore

$$d_x = -14.4 \cos 54.0° = -8.46 \text{ m} \qquad d_y = +14.4 \sin 54.0° = 11.6 \text{ m}$$

(directed along negative x-axis; directed along positive y-axis)

The minus sign shows that the x-component is directed to the left. ■

Practice Exercise

2. For the vector in Example 2, change the angle to 306.0° and find the x- and y-components.

From Examples 1 and 2, we can see that the steps used in finding the x- and y-components of a vector are as follows:

Steps Used in Finding the x- and y-Components of a Vector

1. Place vector **A** such that θ is in standard position.
2. Calculate A_x and A_y from $A_x = A\cos\theta$ and $A_y = A\sin\theta$. We may use the reference angle if we note the direction of the component.
3. Check the components to see if each is in the correct direction and has a magnitude that is proper for the reference angle.

EXAMPLE 3 Components of third-quadrant vector

Resolve vector **A**, of magnitude 375.4 and direction $\theta = 205.32°$, into its x- and y-components. See Fig. 9.13.

By placing **A** such that θ is in standard position, we see that

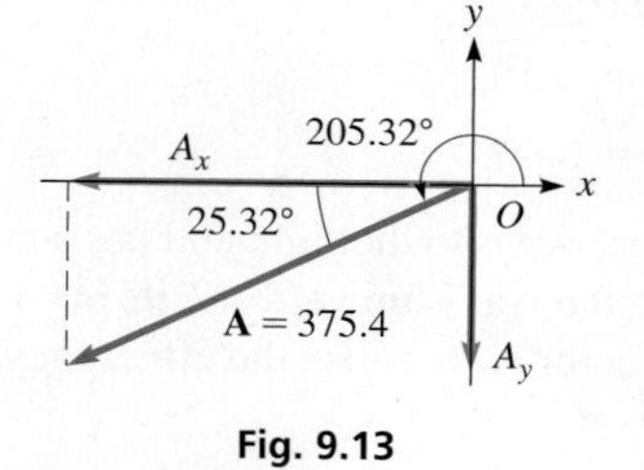

Fig. 9.13

$$A_x = A\cos 205.32° = 375.4\cos 205.32° = -339.3$$

and

directed along negative axis

$$A_y = A\sin 205.32° = 375.4\sin 205.32° = -160.5$$

The angle 205.32° places the vector in the third quadrant, and each of the components is directed along the negative axis. This must be the case for a third-quadrant angle. Also, the reference angle is 25.32°, and we see that the magnitude of A_x is greater than the magnitude of A_y, which must be true for a reference angle that is less than 45°. ■

EXAMPLE 4 Vector components—tension in cable

The tension **T** in a cable supporting the sign shown in Fig. 9.14(a) is 85.0 lb. If the cable makes an angle of 53.5° with the horizontal, find the horizontal and vertical components of the tension.

Practice Exercise

3. Find the components of the tension in Example 4, if the cable makes an angle of 143.5° with the horizontal.

The tension is the force the cable exerts on the sign. Showing the tension in Fig. 9.14(b), we see that

$$T_y = T\sin 53.5° = 85.0\sin 53.5°$$
$$= 68.3 \text{ lb}$$
$$T_x = T\cos 53.5° = 85.0\cos 53.5°$$
$$= 50.6 \text{ lb}$$

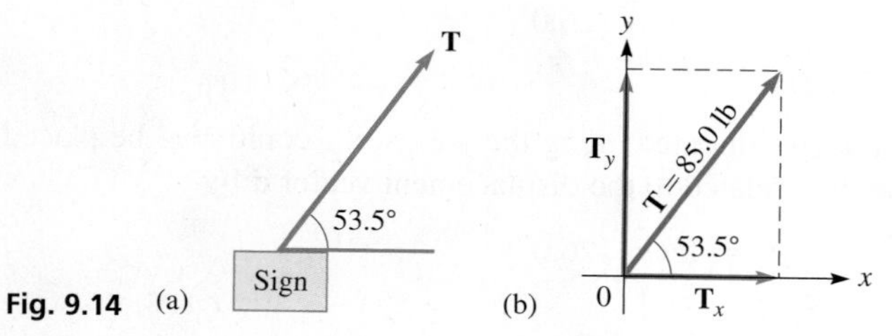

Fig. 9.14 (a) (b)

Note that $T_y > T_x$, which should be the case for an acute angle greater than 45°. ■

EXERCISES 9.2

In Exercises 1–4, find the component vectors if the given changes are made in the indicated examples of this section.

1. In Example 2, change 126.0° to 216.0°.
2. In Example 3, change 205.32° to 295.32°.
3. In Example 3, change 205.32° to 270.00°.
4. In Example 4, change 53.5° to 45.0°.

In Exercises 5–10, find the horizontal and vertical components of the vectors shown in the given figures. In each, the magnitude of the vector is 750.

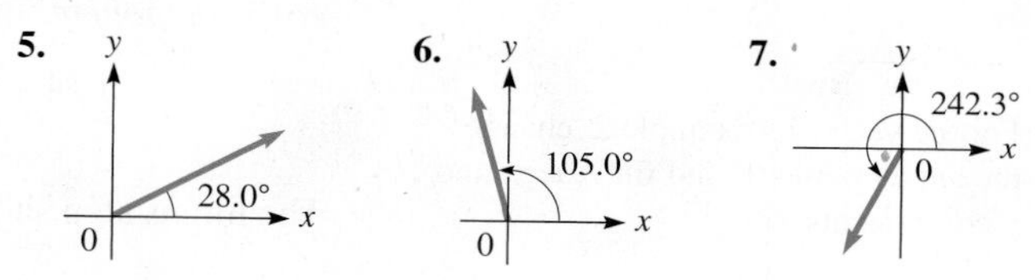

8.

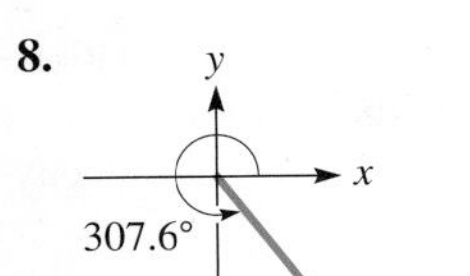

9.

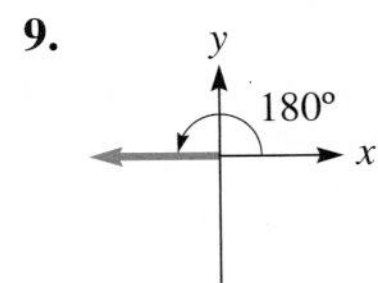

10.

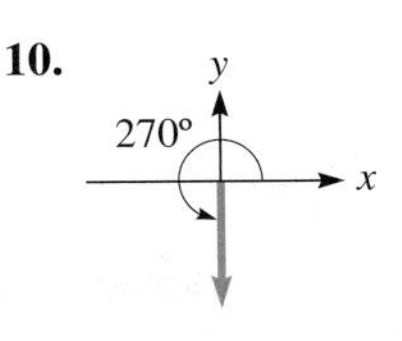

In Exercises 11–20, find the x- and y-components of the given vectors by use of the trigonometric functions. The magnitude is shown first, followed by the direction as an angle in standard position.

11. 15.8 lb, $\theta = 76.0°$

12. 9750 N, $\theta = 243.0°$

13. 76.8 m/s, $\theta = 145.0°$

14. 0.0998 ft/s, $\theta = 296.0°$

15. 8.17 ft/s^2, $\theta = 270.0°$

16. 16.4 cm/s^2, $\theta = 156.5°$

17. 2.65 mN, $\theta = 197.3°$

18. 6.78 lb, $\theta = 22.5°$

19. 38.47 ft, $\theta = 145.82°$

20. 509.4 m, $\theta = 360.0°$

In Exercises 21–34, find the required horizontal and vertical components of the given vectors.

21. A nuclear submarine approaches the surface of the ocean at 25.0 km/h and at an angle of 17.3° with the surface. What are the components of its velocity? See Fig. 9.15.

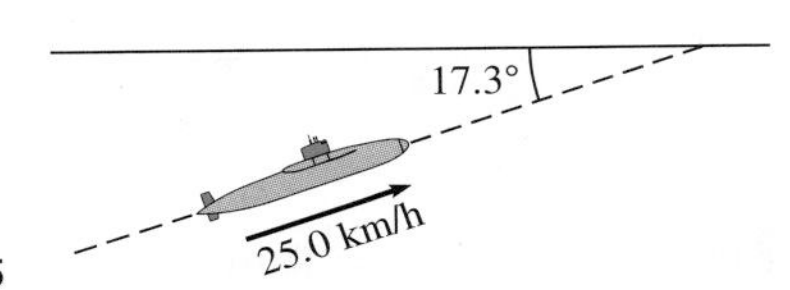

Fig. 9.15

22. A wind-blown fire with a speed of 18 ft/s is moving toward a highway at an angle of 66° with the highway. What is the component of the velocity toward the highway?

23. A car is being unloaded from a ship. It is supported by a cable from a crane and guided into position by a horizontal rope. If the tension in the cable is 2790 lb and the cable makes an angle of 3.5° with the vertical, what are the weight W of the car and the tension T in the rope? (*Hint:* Find the components of the cable vector.) See Fig. 9.16.

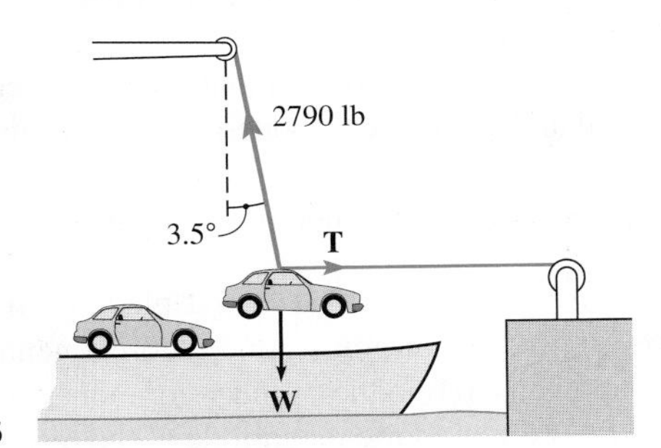

Fig. 9.16

24. A car traveling at 25 mi/h hits a stone wall at an angle of 65° with the wall. What is the component of the car's velocity that is perpendicular to the wall?

25. With the sun directly overhead, a plane is taking off at 125 km/h at an angle of 22.0° above the horizontal. How fast is the plane's shadow moving along the runway?

26. The end of a robot arm is 3.50 ft on a line 78.6° above the horizontal from the point where it does a weld. What are the components of the displacement from the end of the robot arm to the welding point?

27. Vehicles on either side of a canal are towing a barge in the canal. If one pulls with a force of 760 N at 18° with respect to a line in the center of the canal, at what angle should the other pull with a force of 820 N so that the barge moves down the center of the canal? See Fig. 9.17.

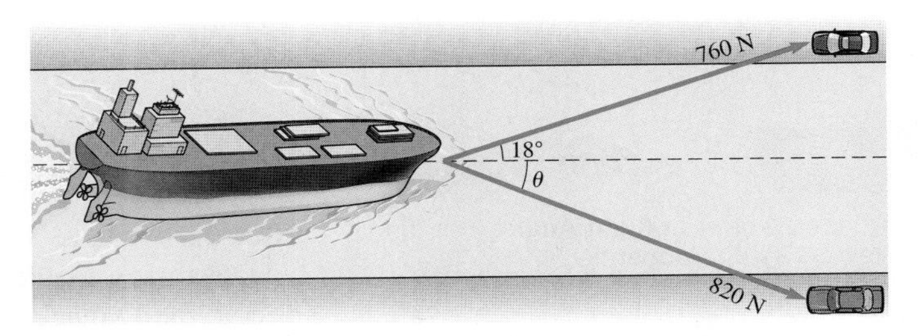

Fig. 9.17

28. The tension in a rope attached to a boat is 55.0 lb. The rope is attached to the boat 12.0 ft below the level at which it is being drawn in. At the point where there are 36.0 ft of rope out, what force tends to bring the boat toward the wharf, and what force tends to raise the boat? See Fig. 9.18.

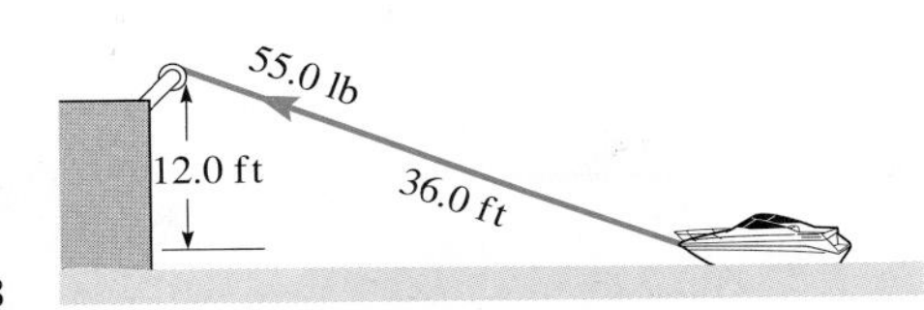

Fig. 9.18

29. A person applies a force of 210 N perpendicular to a jack handle that is at an angle of 25° above the horizontal. What are the horizontal and vertical components of the force?

30. A water skier is pulled up the ramp shown in Fig. 9.19 at 28 ft/s. How fast is the skier rising when leaving the ramp?

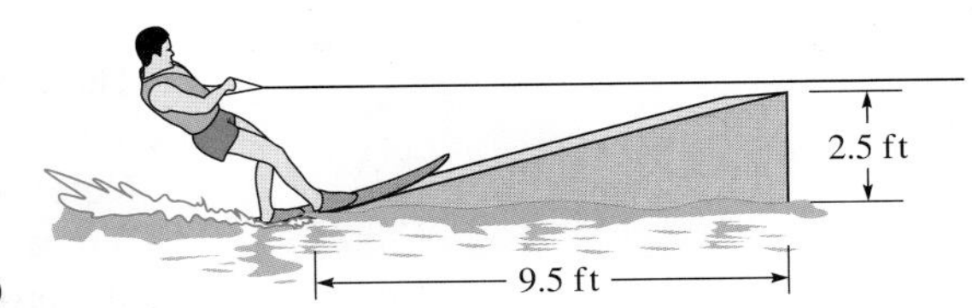

Fig. 9.19

31. A wheelbarrow is being pushed with a force of 80 lb directed along the handles, which are at an angle of 20° above the horizontal. What is the effective force for moving the wheelbarrow forward?

32. Two upward forces are acting on a bolt. One force of 60.5 lb acts at an angle of 82.4° above the horizontal, and the other force of 37.2 lb acts at an angle of 50.5° below the first force. What is the total upward force on the bolt? See Fig. 9.20.

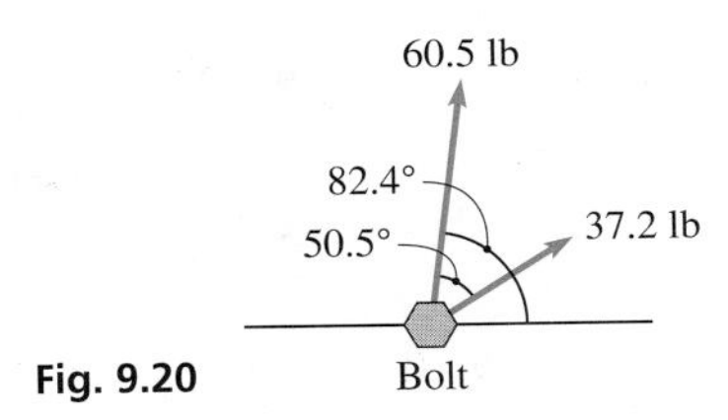

Fig. 9.20

33. Vertical wind sheer in the lowest 100 m above the ground is of great importance to aircraft when taking off or landing. It is defined as the rate at which the wind velocity changes per meter above ground. If the vertical wind sheer at 50 m above the ground is 0.75(km/h)/m directed at angle of 40° above the ground, what are its vertical and horizontal components?

34. At one point, the *Pioneer* space probe was entering the gravitational field of Jupiter at an angle of 2.55° below the horizontal with a velocity of 18,550 mi/h. What were the components of its velocity?

Answers to Practice Exercises

1. $A_x = 6.57, A_y = 3.06$ **2.** $d_x = 8.46$ m, $d_y = -11.6$ m
3. $T_x = -68.3$ lb, $T_y = 50.6$ lb

9.3 Vector Addition by Components

Adding Vectors at Right Angles • Adding Vectors by Components

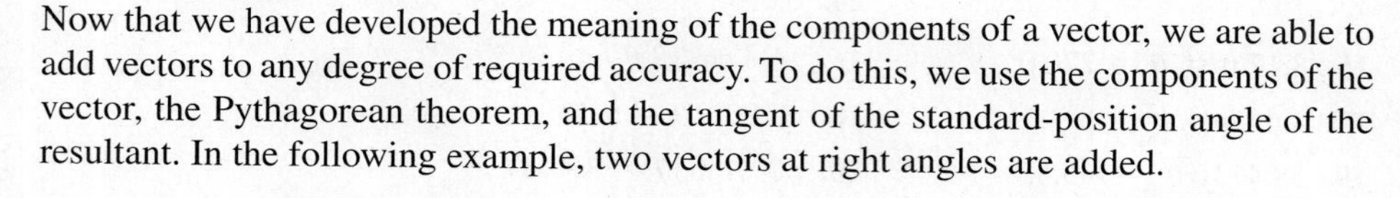

Now that we have developed the meaning of the components of a vector, we are able to add vectors to any degree of required accuracy. To do this, we use the components of the vector, the Pythagorean theorem, and the tangent of the standard-position angle of the resultant. In the following example, two vectors at right angles are added.

EXAMPLE 1 Adding vectors at right angles

Add vectors **A** and **B,** with $A = 14.5$ and $B = 9.10$. The vectors are at right angles, as shown in Fig. 9.21.

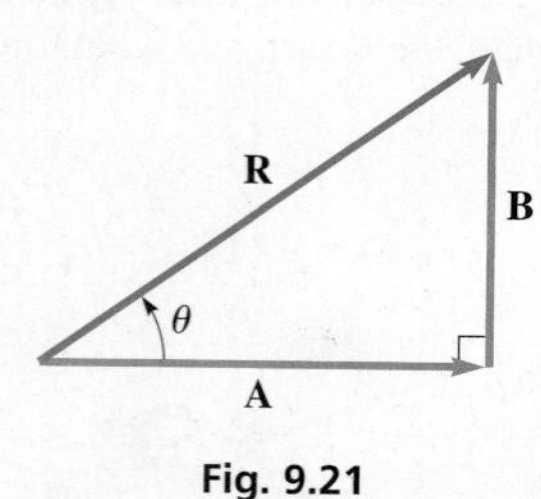

Fig. 9.21

We can find the magnitude R of the resultant vector **R** by use of the Pythagorean theorem. This leads to

$$R = \sqrt{A^2 + B^2} = \sqrt{(14.5)^2 + (9.10)^2}$$
$$= 17.1$$

We now determine the direction of the resultant vector **R** by specifying its direction as the angle θ in Fig. 9.21, that is, the angle that **R** makes with vector **A.** Therefore,

$$\tan\theta = \frac{B}{A} = \frac{9.10}{14.5}$$
$$\theta = \tan^{-1}\left(\frac{9.10}{14.5}\right)$$
$$\theta = 32.1°$$

Therefore, we see that **R** is a vector of magnitude $R = 17.1$ and in a direction 32.1° from vector **A.**

Note that Fig. 9.21 shows vectors **A** and **B** as horizontal and vertical, respectively. This means they are the horizontal and vertical components of the resultant vector **R.** A similar procedure can be used when the vectors are not at right angles by first determining the horizontal and vertical components of **R** and then proceeding as above. ■

NOTE ▸ [If the vectors being added are not at right angles, first place each vector with its tail at the origin. Next, resolve each vector into its x- and y-components. Then add the x-components and add the y-components to find the x- and y-components of the resultant. Then, by using the Pythagorean theorem, find the magnitude of the resultant, and by use of the tangent, find the angle that gives the direction of the resultant.]

CAUTION Remember, a vector is not completely specified unless both its magnitude and its direction are specified. A common error is to determine the magnitude, but not to find the angle θ that is used to define its direction. ■

The following examples illustrate the addition of vectors by first finding the components of the given vectors.

EXAMPLE 2 Adding by first combining components

Find the resultant of two vectors **A** and **B** such that $A = 12\bar{0}0$, $\theta_A = 270.0°$, $B = 1750$, and $\theta_B = 115.0°$.

First, place the vectors on a coordinate system with the tail of each at the origin as shown in Fig. 9.22(a). Then resolve each vector into its x- and y-components, as shown in Fig. 9.22(b) and as calculated below. (Note that **A** is vertical and has no horizontal component.) Next, the components are combined, as in Fig. 9.22(c) and as calculated. Finally, the magnitude of the resultant and the angle θ (to determine the direction), as shown in Fig. 9.22(d), are calculated. The reference angle is found first and then used to determine the standard position angle θ.

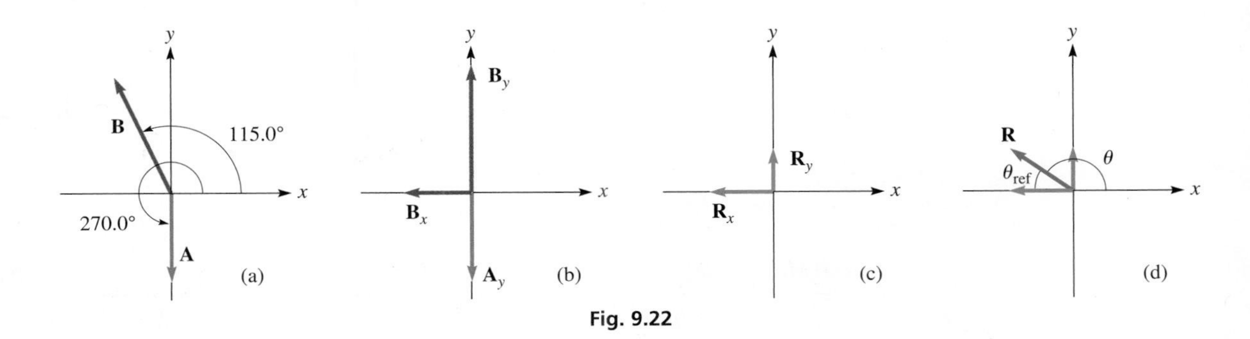

Fig. 9.22

$$\begin{aligned}
A_x &= A \cos 270.0° = 1200 \cos 270.0° = 0 \\
B_x &= B \cos 115.0° = 1750 \cos 115.0° = -739.6 \\
A_y &= A \sin 270.0° = 1200 \sin 270.0° = -1200 \\
B_y &= B \sin 115.0° = 1750 \sin 115.0° = 1586
\end{aligned}$$

— Fig. 9.23(b)

$$\begin{aligned}
R_x &= A_x + B_x = 0 - 739.6 = -739.6 \\
R_y &= A_y + B_y = -1200 + 1586 = 386
\end{aligned}$$

— Fig. 9.23(c)

$$R = \sqrt{R_x^2 + R_y^2} = \sqrt{(-739.6)^2 + 386^2} = 834$$

$$\tan\theta_{ref} = \left|\frac{R_y}{R_x}\right| = \frac{386}{739.6}, \quad \theta_{ref} = 27.6°, \quad \theta = 152.4° \leftarrow 180° - 27.6°$$

— Fig. 9.23(d)

Note that in the last line above, we subtracted θ_{ref} from 180° since the resultant lies in quadrant II (R_x is negative and R_y is positive). Thus, the resultant has a magnitude of 834 and is directed at a standard-position angle of 152.4°.

CAUTION If we obtained $\tan\theta$ directly from R_y/R_x, the calculator would give us an angle of $-27.6°$, and we would have to recognize 27.6° as the reference angle. For this reason, when either R_x or R_y is negative, ***it is best to find the reference angle first*** by disregarding the signs of the components. Then the standard position angle can be found by using the reference angle and the quadrant in which the resultant lies. ■

The values shown in this example have been rounded off. In using the calculator, R_x and R_y can each be calculated in one step and stored for the calculation of R and θ in order to reduce round-off error. We will show these steps in the next example. ■

Practice Exercise

1. In Example 2, change 270° to 90° and then add the vectors.

EXAMPLE 3 Adding vectors by components—swimmer velocity

A person who swims 2.5 ft/s in still water is swimming at an angle of 57° north of east in a stream flowing 38° south of east at 1.7 ft/s. Find the person's resultant velocity.

Practice Exercise

2. In Example 3, change 322° to 142° and find the resultant velocity.

Letting north be along the positive y-axis, and east along the positive x-axis, we have standard position angles of 57° and 322°. Figure 9.23(a) shows the person's vector **P** and the stream vector **S** with the standard position angles. Figure 9.23(b) shows the components of vectors **P** and **S**, and Fig. 9.23(c) shows the resultant **R** and its components.

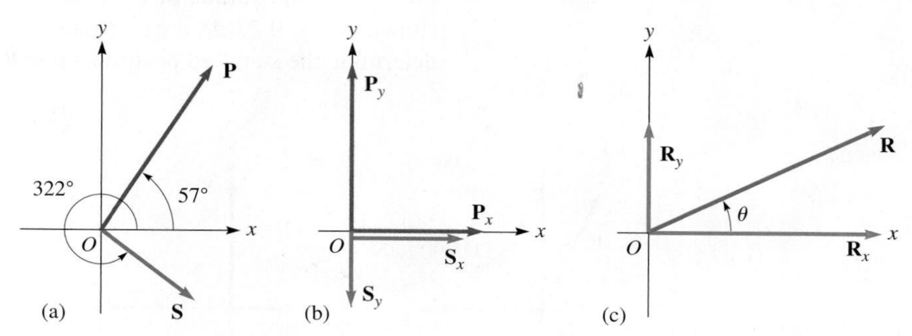

Fig. 9.23

Vector	*Magnitude*	*Angle*	*x-Component*		*y-Component*	
P	2.5 ft/s	57°	$P_x = 2.5 \cos 57°$	$= 1.36$ ft/s	$P_y = 2.5 \sin 57°$	$= 2.10$ ft/s
S	1.7 ft/s	322°	$S_x = 1.7 \cos 322°$	$= 1.34$ ft/s	$S_y = 1.7 \sin 322°$	$= -1.05$ ft/s
R			$R_x = P_x + S_x$	$= 2.70$ ft/s	$R_y = P_y + S_y$	$= 1.05$ ft/s

$$R = \sqrt{R_x^2 + R_y^2} = \sqrt{2.70^2 + 1.05^2} = 2.9 \text{ ft/s}$$

$$\theta = \tan^{-1}\frac{R_y}{R_x} = \tan^{-1}\left(\frac{1.05}{2.70}\right) = 21°$$ ← don't forget the direction

```
NORMAL FLOAT AUTO REAL DEGREE MP
2.5cos(57)+1.7cos(322)→X
                  2.701215869
2.5sin(57)+1.7sin(322)→Y
                  1.050051912
√(X²+Y²)
                  2.898133224
tan⁻¹(Y/X)
                  21.24274968
```

Fig. 9.24

The person's resultant velocity is 2.9 ft/s at a standard position angle of 21°, as shown in Fig. 9.23(c). We know the resultant is in the first quadrant because both R_x and R_y are positive.

In the table, we have carried some extra digits and rounded off all final values. However, when using a calculator, it is necessary only to calculate, and store, R_x and R_y in one step each. We then use them to calculate the values of R and θ, which is shown in Fig. 9.24. In the calculator solution, $X = R_x$ and $Y = R_y$. In finding this solution, we see that both R_x and R_y are positive, which means that θ is a first-quadrant angle. ■

Some general formulas can be derived from the previous examples. For a given vector **A,** directed at an angle θ, of magnitude A,

$$A_x = A \cos \theta \quad A_y = A \sin \theta \tag{9.1}$$

If R_x and R_y are the components of the resultant vector, then

$$R = \sqrt{R_x^2 + R_y^2} \tag{9.2}$$

$$\theta_{\text{ref}} = \tan^{-1}\frac{|R_y|}{|R_x|} \tag{9.3}$$

The standard position angle θ is found by using the reference angle from Eq. (9.3) and the quadrant in which the resultant lies.

From the previous examples, we see that the following procedure is used for adding vectors.

Graphing calculator program for adding vectors: goo.gl/tPdGhn

Procedure for Adding Vectors by Components

1. Add the x-components of the given vectors to obtain R_x.
2. Add the y-components of the given vectors to obtain R_y.
3. Find the magnitude of the resultant **R.** Use Eq. (9.2)

$$R = \sqrt{R_x^2 + R_y^2}$$

4. Find the standard-position angle θ for the resultant **R.** First, find the reference angle θ_{ref} for the resultant **R** by using Eq. (9.3)

$$\theta_{ref} = \tan^{-1}\frac{|R_y|}{|R_x|}.$$

Then use the reference angle and the quadrant of **R** (as determined by the signs of R_x and R_y) to determine the standard position angle θ.

CAUTION The formulas we use to find the x- and y-components assume that the angles are in *standard position* (reference angles can also be used if we manually attach the correct sign to each component). ***Make sure to convert all angles to standard-position angles (or reference angles) before calculating the components.*** ■

EXAMPLE 4 Standard-position angle of resultant

Find the resultant of the three given vectors in Fig. 9.25. The magnitudes of these vectors are $T = 422$, $U = 405$, and $V = 210$.

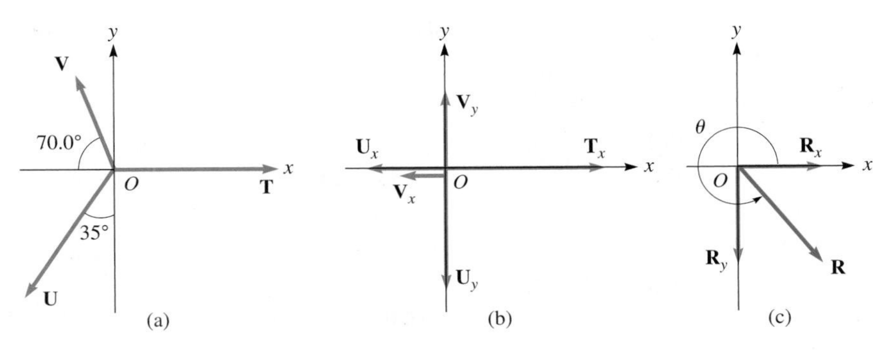

Fig. 9.25

We will start by changing each given angle to a standard-position angle (abbreviated SP Angle). In the following table, we show the x- and y-components of the given vectors, and the sums of these components give us the components of **R.**

Vector	*Magnitude*	*SP Angle*	*x-Component*	*y-Component*
T	422	0°	$422 \cos 0° = 422.0$	$422 \sin 0° = 0.0$
U	405	235.0°	$405 \cos 235.0° = -232.3$	$405 \sin 235.0° = -331.8$
V	210	110.0°	$210 \cos 110° = -71.8$	$210 \sin 110.0° = 197.3$
R			$R_x = 117.9$	$R_y = -134.5$

We now use the components of **R** to find its magnitude and reference angle:

$$R = \sqrt{117.9^2 + (-134.5)^2} = 179$$ make sure to include parentheses around negative numbers being squared

$$\theta_{ref} = \tan^{-1}\left(\frac{134.5}{117.9}\right) = 48.8°$$ we used positive 134.5 since we took the absolute value

Because R_x is positive and R_y is negative, we know that θ is **a fourth-quadrant angle.** Therefore, to find θ, we subtract θ_{ref} from 360°. This means

$$\theta = 360° - 48.8° = 311.2°$$ ■

Practice Exercise

3. In Example 4, find the resultant of vectors **U** and **V.** (Do not include vector **T.**)

EXERCISES 9.3

In Exercises 1 and 2, find the resultant vectors if the given changes are made in the indicated examples of this section.

1. In Example 2, find the resultant if θ_A is changed to 0°.

2. In Example 4, find the resultant if 70° is changed to 20°.

*In Exercises 3–6, vectors **A** and **B** are at right angles. Find the magnitude and direction (the angle from vector **A**) of the resultant.*

3. $A = 14.7$, $B = 19.2$

4. $A = 592$, $B = 195$

5. $A = 3.086$, $B = 7.143$

6. $A = 1734$, $B = 3297$

In Exercises 7–14, with the given sets of components, find R and θ.

7. $R_x = 5.18, R_y = 8.56$

8. $R_x = 89.6, R_y = -52.0$

9. $R_x = -0.982, R_y = 2.56$

10. $R_x = -729, R_y = -209$

11. $R_x = -646, R_y = 2030$

12. $R_x = -31.2, R_y = -41.2$

13. $R_x = 6941, R_y = -1246$

14. $R_x = 7.627, R_y = -6.353$

In Exercises 15–32, add the given vectors by components.

15. $A = 18.0, \theta_A = 0.0°$; $B = 12.0, \theta_B = 27.0°$

16. $F = 154, \theta_F = 90.0°$; $T = 128, \theta_T = 43.0°$

17. $A = 368, \theta_A = 235.3°$; $B = 227, \theta_B = 295.0°$

18. $A = 30.7, \theta_A = 18.2°$; $B = 45.2, \theta_B = 251.0°$

19. $C = 5650, \theta_C = 76.0°$; $D = 1280, \theta_D = 160.0°$

20. $A = 6.89, \theta_A = 123.0°$; $B = 29.0, \theta_B = 260.0°$

21. $A = 9.821, \theta_A = 34.27°$; $B = 17.45, \theta_B = 752.50°$

22. $E = 1653, \theta_E = 36.37°$; $F = 9807, \theta_F = 253.06°$

23. $A = 21.9, \theta_A = 236.2°$; $B = 96.7, \theta_B = 11.5°$; $C = 62.9, \theta_C = 143.4°$

24. $R = 630, \theta_R = 189.6°$; $F = 176, \theta_F = 320.1°$; $T = 324, \theta_T = 75.4°$

25. $U = 0.364, \theta_U = 175.7°$; $V = 0.596, \theta_V = 319.5°$; $W = 0.129, \theta_W = 100.6°$

26. $A = 64, \theta_A = 126°$; $B = 59, \theta_B = 238°$; $C = 32, \theta_C = 72°$

27. The displacement vectors in Fig. 9.26.

28. The velocity vectors in Fig. 9.27.

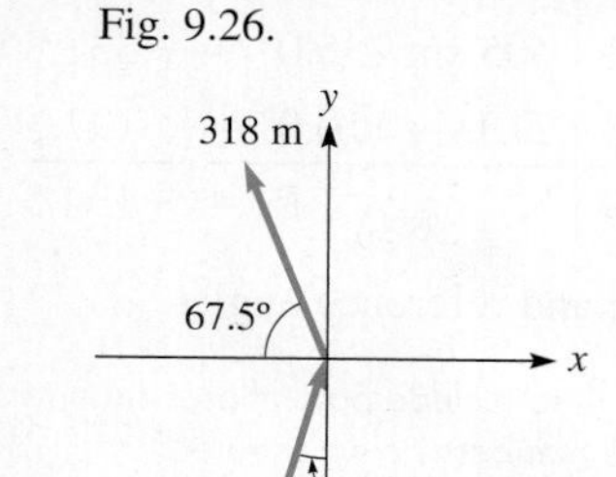

Fig. 9.26

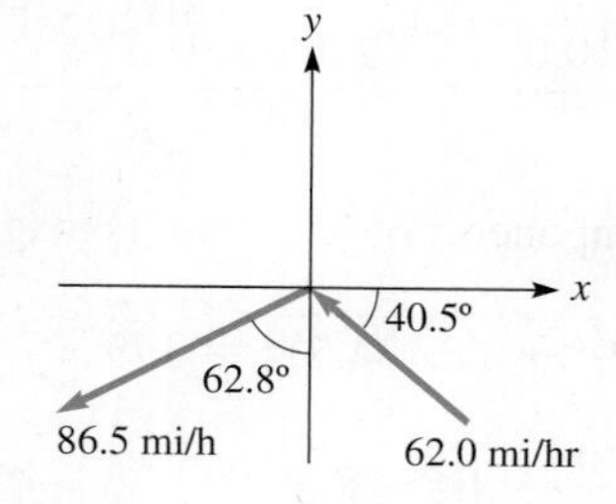

Fig. 9.27

Answers to Practice Exercises

1. $R = 2880, \theta = 104.9°$ **2.** $R = 3.1$ ft/s, $\theta = 90°$
3. $R = 333, \theta = 203.8°$

29. The force vectors in Fig. 9.28.

30. The displacement vectors in Fig. 9.29.

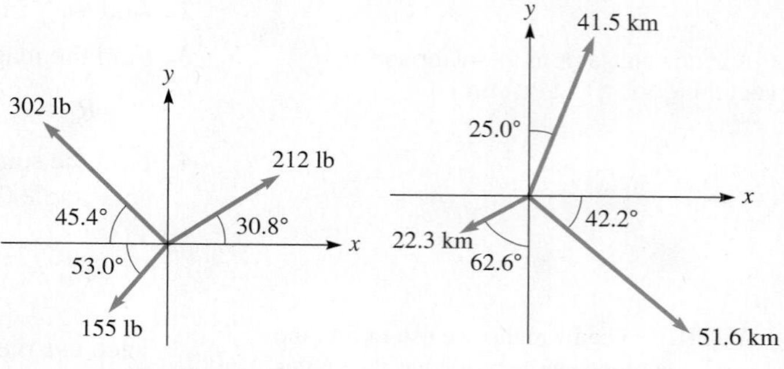

Fig. 9.28

Fig. 9.29

31. In order to move an ocean liner into the channel, three tugboats exert the forces shown in Fig. 9.30. What is the resultant of these forces?

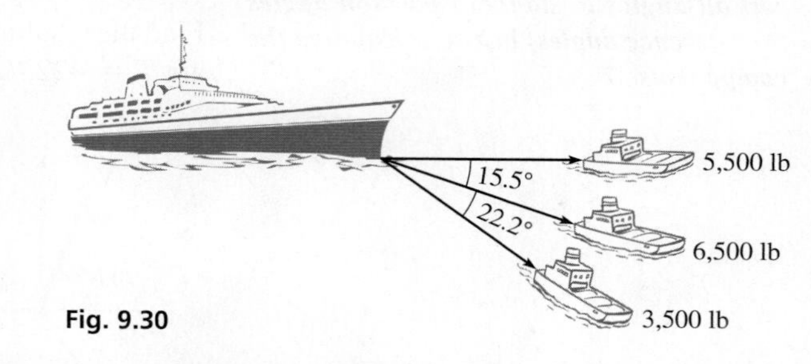

Fig. 9.30

32. A naval cruiser on maneuvers travels 54.0 km at 18.7° west of north, then turns and travels 64.5 km at 15.6° south of east, and finally turns to travel 72.4 km at 38.1° east of south. Find its displacement from its original position. See Fig. 9.31.

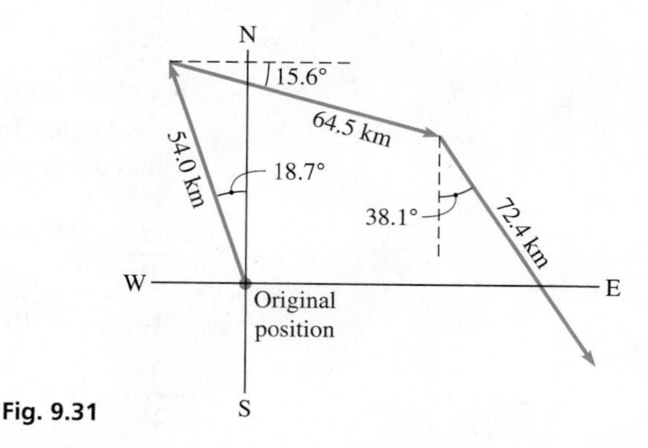

Fig. 9.31

In Exercises 33–36, solve the given problems.

33. The angle between two forces of 1700 N and 2500 N is 37° when placed tail to tail. What is the magnitude of the resultant force?

34. What is the magnitude of a velocity vector that makes an angle of 18.0° with its horizontal component of 420 mi/h?

35. The angle between two equal-momentum vectors of 15.0 kg · m/s in magnitude is 72.0° when placed tail to tail. What is the magnitude of the resultant?

36. The resultant **R** of displacements **A** and **B** is perpendicular to **A.** If $A = 25$ m and $R = 25$ m, find the angle between **A** and **B.**

9.4 Applications of Vectors

Forces • Displacement • Velocity • Other Applications

In Section 9.1, we introduced the important vector quantities of force, velocity, and displacement, and we found vector sums by use of diagrams. Now, we can use the method of Section 9.3 to find sums of these kinds of vectors and others and to use them in various types of applications.

EXAMPLE 1 Forces at right angles

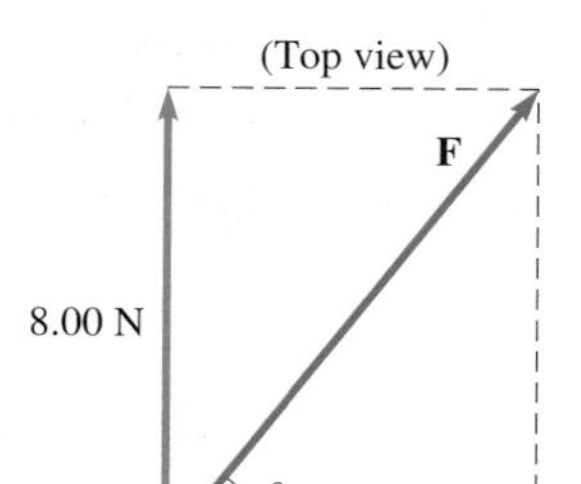

Fig. 9.32

■ The metric unit of force, the newton (N), is named for the great English mathematician and physicist Sir Isaac Newton (1642–1727). His name will appear on other pages of this text, as some of his many accomplishments are noted.

In centering a figurine on a table, two persons apply forces on it. These forces are at right angles and have magnitudes of 6.00 N and 8.00 N. The angle between their lines of action is 90.0°. What is the resultant of these forces on the figurine?

By means of an appropriate diagram (Fig. 9.32), we may better visualize the actual situation. Note that a good choice of axes (unless specified, it is often convenient to choose the x- and y-axes to fit the problem) is to have the x-axis in the direction of the 6.00-N force and the y-axis in the direction of the 8.00-N force. (This is possible because the angle between them is 90°.) With this choice, note that the two given forces will be the x- and y-components of the resultant. Therefore, we arrive at the following results:

$$F_x = 6.00\text{ N},\ F_y = 8.00\text{ N}$$

$$F = \sqrt{(6.00)^2 + (8.00)^2} = 10.0\text{ N}$$

$$\theta = \tan^{-1}\frac{F_y}{F_x} = \tan^{-1}\frac{8.00}{6.00}$$

$$= 53.1°$$

The resultant has a magnitude of 10.0 N and acts at an angle of 53.1° from the 6.00-N force. ■

EXAMPLE 2 Ship displacement

A ship sails 32.50 mi due east and then turns 41.25° north of east. After sailing another 16.18 mi, where is it with reference to the starting point?

In this problem, we are to find the resultant displacement of the ship from the two given displacements. The problem is diagrammed in Fig. 9.33, where the first displacement is labeled vector **A** and the second as vector **B.**

Because east corresponds to the positive x-direction, note that the x-component of the resultant is $\mathbf{A} + \mathbf{B}_x$ and the y-component of the resultant is $\mathbf{B}_y$. Therefore, we have the following results:

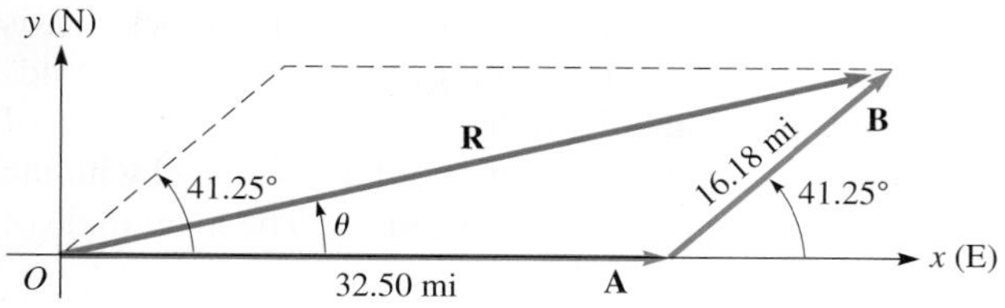

Fig. 9.33

$$R_x = A + B_x = 32.50 + 16.18\cos 41.25°$$

$$= 32.50 + 12.16$$

$$= 44.66\text{ mi}$$

$$R_y = 16.18\sin 41.25° = 10.67\text{ mi}$$

$$R = \sqrt{(44.66)^2 + (10.67)^2} = 45.92\text{ mi}$$

$$\theta = \tan^{-1}\frac{10.67}{44.66}$$

$$= 13.44°$$

Therefore, the ship is 45.92 mi from the starting point, in a direction 13.44° north of east. ■

Practice Exercise

1. In Example 2, change "41.25° north of east" to "in the direction of 41.25° north of west," and then find the resultant displacement of the ship.

EXAMPLE 3 Airplane velocity

■ An aircraft's *heading* is the direction in which it is pointed. Its *air speed* is the speed at which it travels with respect to the air surrounding it. Due to the wind, the heading and air speed will be different than the direction and speed relative to the ground.

■ Recall that a bar over a digit indicates it is significant.

An airplane headed due east is in a wind blowing from the southeast. What is the resultant velocity of the plane with respect to the ground if the velocity of the plane with respect to the air is $6\bar{0}0$ km/h and that of the wind is $10\bar{0}$ km/h? See Fig. 9.34.

If **A** is the plane's air vector (air speed and directional heading) and **W** is the wind vector (wind speed and wind direction), the calculations below can be used to find the magnitude and direction of the resultant ground vector **R** (ground speed and ground course).

Vector	*Magnitude*	*SP Angle*	*x-Component*	*y-Component*
A	600 km/h	0°	$600 \cos 0° = 600$ km/h	$600 \sin 0° = 0.0$ km/h
W	100 km/h	135.0°	$100 \cos 135.0° = -70.7$ km/h	$100 \sin 135.0° = 70.7$ km/h
R			529 km/h	70.7 km/h

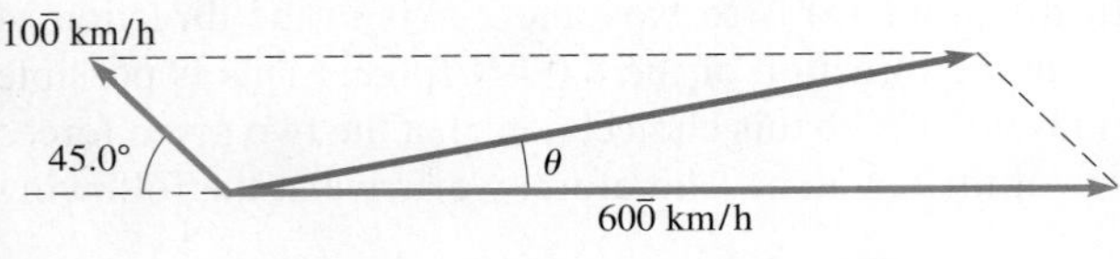

Fig. 9.34

$$R = \sqrt{529^2 + 70.7^2} = 534 \frac{\text{km}}{\text{h}}$$

$$\theta = \tan^{-1}\left(\frac{70.7}{529}\right) = 7.6°$$

Practice Exercise

2. In Example 3, find the plane's resultant velocity if the wind is from the southwest.

Therefore, the plane is traveling 534 km/h with respect to the ground and is flying on a ground course of 7.6° north of east. From this, we observe that a plane does not necessarily head in the direction of its destination. ■

EQUILIBRIUM OF FORCES

As we have seen, an important vector quantity is the force acting on an object. One of the most important applications of vectors involves forces that are in **equilibrium.** NOTE ▸ [For an object to be in equilibrium, the net force acting on it in any direction must be zero. This condition is satisfied if the sum of the x-components of the forces is zero and the sum of the y-components of the forces is also zero.] The following two examples illustrate forces in equilibrium.

EXAMPLE 4 Forces on block on inclined plane

A cement block is resting on a straight inclined plank that makes an angle of 30.0° with the horizontal. If the block weighs 80.0 lb, what is the force of friction between the block and the plank?

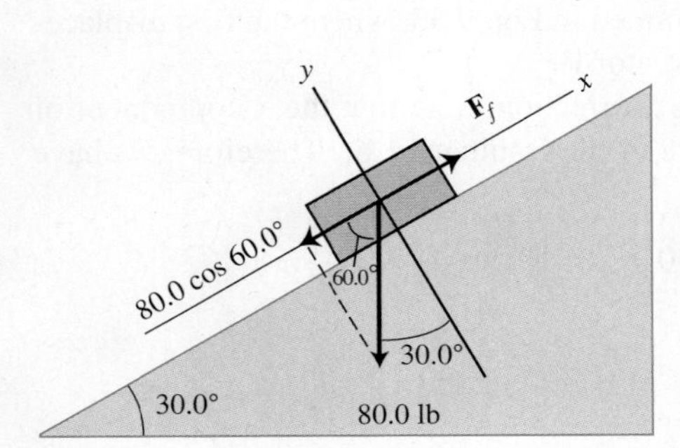

Fig. 9.35

The weight of the cement block is the force exerted on the block due to gravity. Therefore, the weight is directed vertically downward. The frictional force tends to oppose the motion of the block and is directed upward along the plank. The frictional force must be sufficient to counterbalance that component of the weight of the block that is directed down the plank for the block to be at rest (not moving). The plank itself "holds up" that component of the weight that is perpendicular to the plank. A convenient set of coordinates (see Fig. 9.35) is one with the origin at the center of the block and with the x-axis directed up the plank and the y-axis perpendicular to the plank. The magnitude of the frictional force $\mathbf{F}_f$ is given by

$$F_f = 80.0 \cos 60.0° \quad \text{component of weight down plank equals frictional force}$$

$$= 40.0 \text{ lb}$$

■ Newton's *third law of motion* states that when an object exerts a force on another object, the second object exerts on the first object a force of the same magnitude but in the opposite direction. The force exerted by the plank on the block is an illustration of this law.

We have used the 60.0° angle since it is the reference angle. We could have expressed the frictional force as $F_f = 80.0 \sin 30.0°$.

Here, it is assumed that the block is small enough that we may calculate all forces as though they act at the center of the block (although we know that the frictional force acts at the surface between the block and the plank). ■

■ See chapter introduction.

EXAMPLE 5 Equilibrium—forces on a climber

A 165-lb mountain climber is suspended by a rope that is angled 60.0° from the horizontal. See Fig. 9.36. If **T** is the tension in the rope and **F** is the horizontal force with which the climber pushes on the side of a cliff, find the magnitudes of the two forces, T and F.

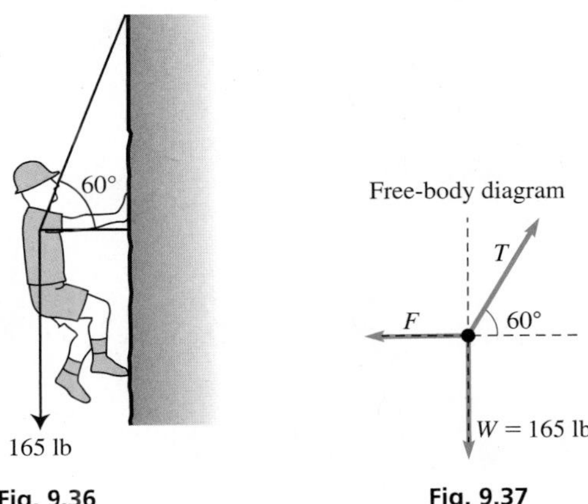

Fig. 9.36 **Fig. 9.37**

We will use a ***free-body diagram*** (Fig. 9.37) to analyze all the forces acting on the climber's body. *A free-body diagram shows a body (an object) removed from its environment along with all the external forces acting on the body.* In this case, the large dot represents the climber's body and the vectors represent the three forces acting on it (the climber's weight, the tension in the rope, and the reactive force of the cliff pushing against the climber's body). Note that the force **F** of the cliff tends to push the climber's body *toward the left*.

Using standard position angles, the x- and y-components of the three vectors are shown below:

Vector	*Magnitude (lb)*	*SP Angle*	*x-Component (lb)*	*y-Component (lb)*
T	T	60.0°	$T\cos 60.0° = 0.5T$	$T\sin 60.0° = 0.866T$
F	F	180°	$F\cos 180° = -F$	$F\sin 180° = 0.0$
W	165	270°	$165\cos 270° = 0.0$	$165\sin 270° = -165$
R			0	0

Note that since the forces are in equilibrium, both the x- and y-components must have a sum of zero:

$$0.5T - F = 0$$
$$0.866T - 165 = 0$$

The second equation can be solved for T, and this solution can be substituted into the first equation to find F:

$$T = \frac{165}{0.866} = 190.5 \text{ lb}$$
$$0.5(190.5) - F = 0$$
$$F = 95.3 \text{ lb}$$

Therefore, the tension in the rope is $T = 191$ lb and the force with which the climber pushes against the cliff is $F = 95.3$ lb. ■

Practice Exercise

3. In Example 5, find T and F if 60.0° is changed to 75.0°.

EXAMPLE 6 Equilibrium involving a system—cable tension

An 865-lb crate of building supplies is hanging from two cables as shown in Fig. 9.38. Find the tensions T_1 and T_2 assuming that the forces are in equilibrium.

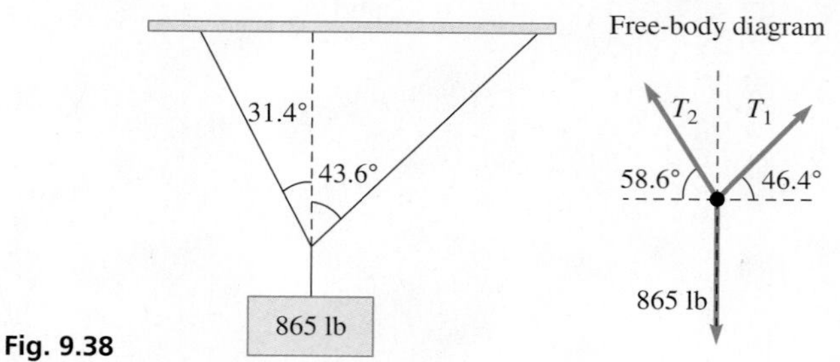

Fig. 9.38

The table below shows the x- and y-components of the three vectors:

Vector	*Magnitude (lb)*	*SP Angle*	*x-Component (lb)*	*y-Component (lb)*
A	T_1	46.4°	$T_1 \cos 46.4° = 0.6896T_1$	$T_1 \sin 46.4° = 0.7242T_1$
B	T_2	121.4°	$T_2 \cos 121.4° = -0.5210T_2$	$T_2 \sin 121.4° = 0.8536T_2$
C	865	270°	$865 \cos 270° = 0$	$865 \sin 270° = -865$
R			0	0

Setting the sums of the x- and y-components equal to zero gives us the following system of equations:

$$0.6896T_1 - 0.5210T_2 = 0$$

$$0.7242T_1 + 0.8536T_2 - 865 = 0$$

This system can be solved by the *substitution method* as described in Section 5.3. We solve the first equation for T_1 and then substitute the result into the second equation to solve for T_2.

$$T_1 = \frac{0.5210T_2}{0.6896} = 0.7555T_2$$

$$0.7242(0.7555T_2) + 0.8536T_2 - 865 = 0$$

$$0.5471T_2 + 0.8536T_2 - 865 = 0$$

$$1.401T_2 - 865 = 0$$

$$T_2 = \frac{865}{1.401} = 617.4 \text{ lb}$$

Lastly, the solution for T_2 can be used to find T_1:

$$T_1 = 0.7555(617.4) = 466 \text{ lb}$$

Therefore, the tensions in the two cables are $T_1 = 466$ lb and $T_2 = 617$ lb. ■

CAUTION Be careful about rounding intermediate steps because it may diminish the accuracy of the final answers. In the last example, one extra significant digit was included in the intermediate steps in order to minimize the round-off error. ■

EXERCISES 9.4

In Exercises 1 and 2, find the necessary quantities if the given changes are made in the indicated examples of this section.

1. In Example 2, find the resultant location if the 41.25° angle is changed to 31.25°.

2. In Example 3, find the resultant velocity if the wind is from the northwest, rather than southeast.

In Exercises 3–36, solve the given problems.

3. Two hockey players strike the puck at the same time, hitting it with horizontal forces of 6.85 lb and 4.50 lb that are perpendicular to each other. Find the resultant of these forces.

4. A jet is 115 mi east and 88.3 mi north of Niagara Falls. What is its displacement from Niagara Falls?

5. In lifting a heavy piece of equipment from the mud, a cable from a crane exerts a vertical force of 6520 N, and a cable from a truck exerts a force of 8280 N at 15.0° above the horizontal. Find the resultant of these forces.

6. At a point in the plane, two electric charges create an electric field (a vector quantity) of 25.9 kN/C at 10.8° above the horizontal to the right and 12.6 kN/C at 83.4 below the horizontal to the right. Find the resultant electric field.

7. A motorboat leaves a dock and travels 1580 ft due west; then it turns 35.0° to the south and travels another 1640 ft to a second dock. What is the displacement of the second dock from the first dock?

8. Toronto is 650 km at 19.0° north of east from Chicago. Cincinnati is 390 km at 48.0° south of east from Chicago. What is the displacement of Cincinnati from Toronto?

9. From a fixed point, a surveyor locates a pole at 215.6 ft due east and a building corner at 358.2 ft at 37.72° north of east. What is the displacement of the building from the pole?

10. A rocket is launched with a vertical component of velocity of 2840 km/h and a horizontal component of velocity of 1520 km/h. What is its resultant velocity?

11. In testing the behavior of a tire on ice, a force of 520 lb is exerted to the side, and a force of 780 lb is exerted to the front. What is the resultant force on the tire?

12. To raise a crate, two ropes are attached to its top. If the force in one rope is 240 lb at 25° from the vertical, what must be the force in the second rope at 35° from the vertical in order to lift the crate straight upward?

13. A storm front is moving east at 18.0 km/h and south at 12.5 km/h. Find the resultant velocity of the front.

14. To move forward, a helicopter pilot tilts the helicopter forward. If the rotor generates a force of 3200 lb, with a horizontal component (thrust) of 420 lb, what is the vertical component (lift)?

15. The acceleration (a vector quantity) of gravity on a sky diver is 9.8 m/s^2. If the force of the wind also causes an acceleration of 1.2 m/s^2 at an angle of 15° above the horizontal, what is the resultant acceleration of the sky diver?

16. In an accident, a truck with momentum (a vector quantity) of 22,100 kg · m/s strikes a car with momentum of 17,800 kg · m/s from the rear. The angle between their directions of motion is 25.0°. What is the resultant momentum?

17. In an automobile safety test, a shoulder and seat belt exerts a force of 95.0 lb directly backward and a force of 83.0 lb backward at an angle of 20.0° below the horizontal on a dummy. If the belt holds the dummy from moving farther forward, what force did the dummy exert on the belt? See Fig. 9.39.

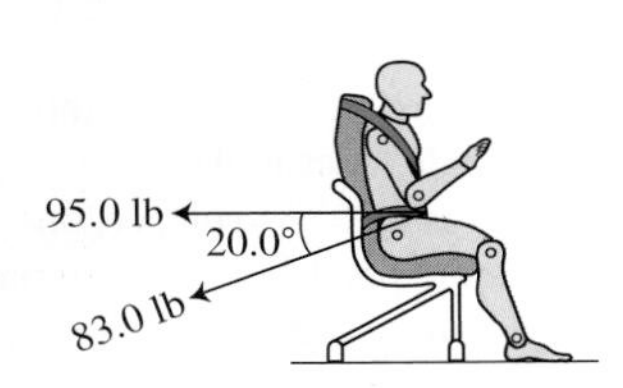

Fig. 9.39

18. Two perpendicular forces act on a ring at the end of a chain that passes over a pulley and holds an automobile engine. If the forces have the values shown in Fig. 9.40, what is the weight of the engine?

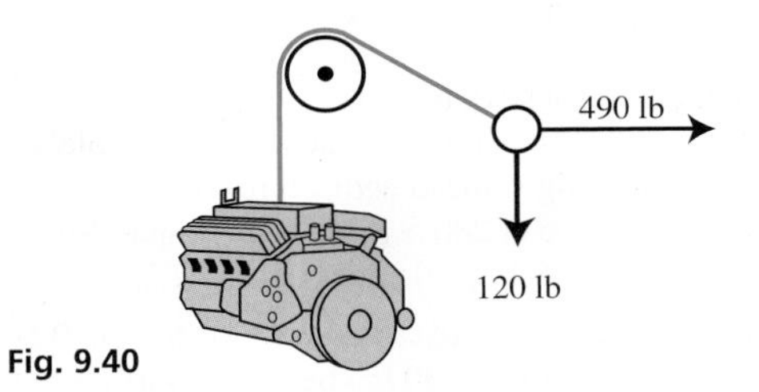

Fig. 9.40

19. A plane flies at 550 km/h into a head wind of 60 km/h at 78° with the direction of the plane. Find the resultant velocity of the plane with respect to the ground. See Fig. 9.41.

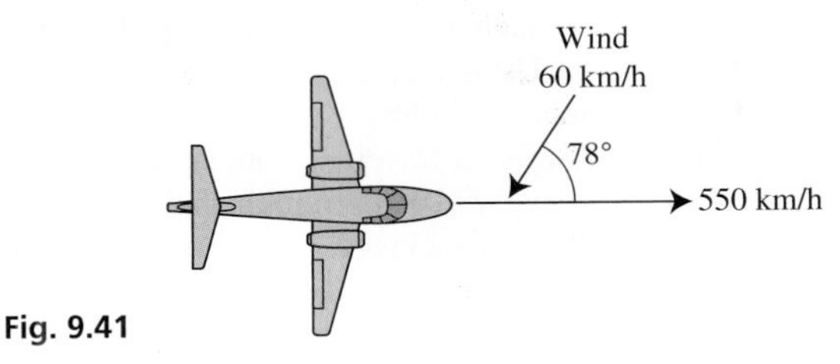

Fig. 9.41

20. A ship's navigator determines that the ship is moving through the water at 17.5 mi/h with a heading of 26.3° north of east, but that the ship is actually moving at 19.3 mi/h in a direction of 33.7° north of east. What is the velocity of the current?

21. Assuming the vectors in Fig. 9.42 are in equilibrium, find T_1 and T_2.

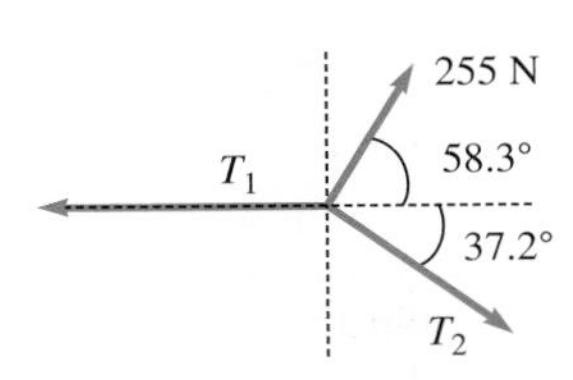

Fig. 9.42

22. A block of ice slides down a (frictionless) ramp with an acceleration of 5.3 m/s^2. If the ramp makes an angle of 32.7° with the horizontal, find g, the acceleration due to gravity. See Fig. 9.43.

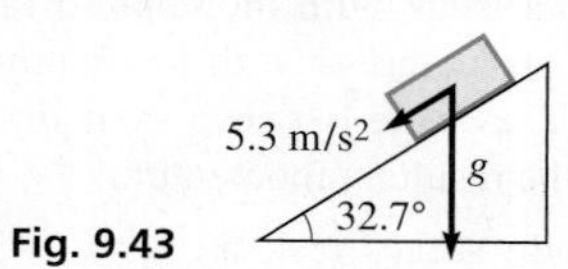

Fig. 9.43

23. In navigation, one method of expressing *bearing* is to give a single angle measured clockwise from due north. A radar station notes the bearing of two planes, one 15.5 km away on a bearing of 22.0°, and the other 12.0 km away on a bearing of 180.0°. Find the displacement of the second plane from the first. See Fig. 9.44.

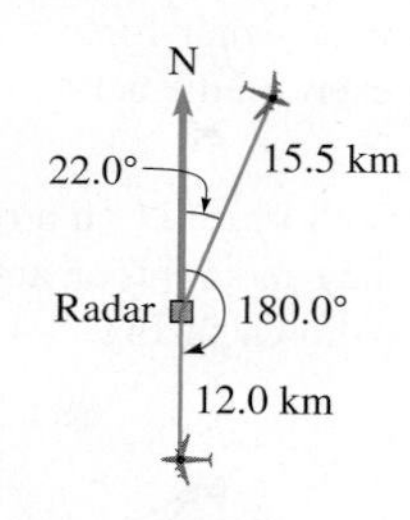

Fig. 9.44

24. On a mountain trek, a pack mule becomes obstinate and refuses to move. One man pulls on a rope attached to the mule's harness at 15° to the left of straight ahead with a force of 320 N, and a second man pulls with a force of 280 N at 12° to the right. With what force does the mule resist if the mule does not move?

25. As a nuclear submarine moves eastward, it travels 0.50 km in a straight line as it descends 0.30 km below the surface. It then turns southward and travels 0.90 km in a straight line while resurfacing. What is its displacement from its original position?

26. As a satellite travels around Earth at an altitude of 22,320 mi, the magnitude of its linear velocity (see Section 8.4) remains unchanged, but its *direction* changes continually, which means there is a change in velocity. Using a diagram based on Fig. 9.45, draw the difference in velocities (subtract the first velocity from the second) for the satellite as it travels (a) one-sixth, and (b) one-quarter of its orbit. Draw the vectors from the point at which their lines of action cross. What conclusion can be drawn about the direction of the change in velocity?

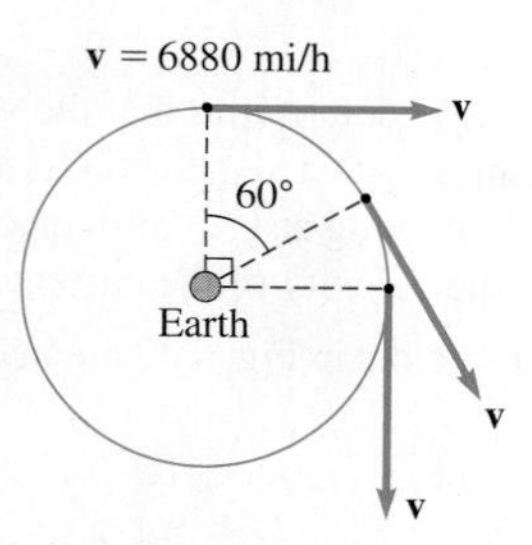

Fig. 9.45

27. A passenger on a cruise ship traveling due east at a speed of 32 km/h notes that the smoke from the ship's funnels makes an angle of 15° with the ship's wake. If the wind is from the southwest, find the speed of the wind.

28. A mine shaft goes due west 75 m from the opening at an angle of 25° below the horizontal surface. It then becomes horizontal and turns 30° north of west and continues for another 45 m. What is the displacement of the end of the tunnel from the opening?

29. A crowbar 1.5 m long is supported underneath 1.2 m from the end by a rock, with the other end under a boulder. If the crowbar is at an angle of 18° with the horizontal and a person pushes down, perpendicular to the crowbar, with a force of 240 N, what vertical force is applied to the boulder? See Fig. 9.46. (The forces are related by $1.2\mathbf{F}_1 = 0.3\mathbf{F}_2$.)

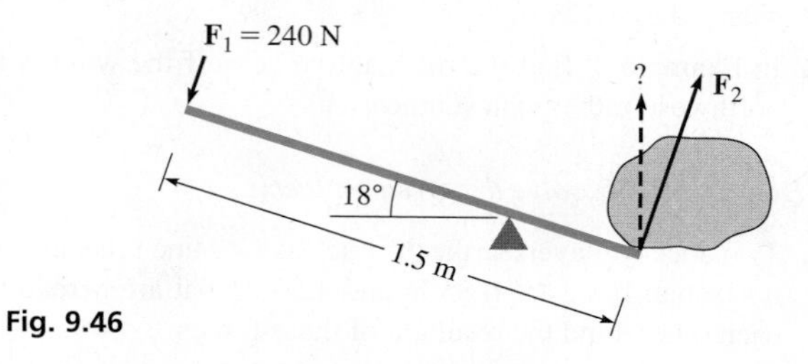

Fig. 9.46

30. A scuba diver's body is directed downstream at 75° to the bank of a river. If the diver swims at 25 m/min, and the water is moving at 5.0 m/min, what is the diver's velocity?

31. A sign that weighs 275 lb hangs from the side of a building. The sign is supported by a cable angled at 45° from a rigid horizontal bar as shown in Fig. 9.47. Make a free-body diagram, and then find the tension T in the upper cable and the force F exerted on the horizontal bar. See Example 5.

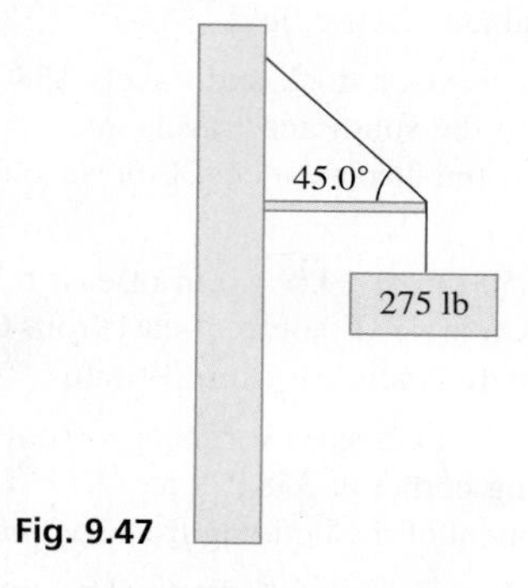

Fig. 9.47

32. A fire boat that travels 24.0 km/h in still water crosses a river to reach the opposite bank at a point directly opposite that from which it left. If the river flows at 5.0 km/h, what is the velocity of the boat while crossing?

33. In searching for a boat lost at sea, a Coast Guard cutter leaves a port and travels 75.0 mi due east. It then turns 65° north of east and travels another 75.0 mi, and finally turns another 65.0° toward the west and travels another 75.0 mi. What is its displacement from the port?

34. A car is held stationary on a ramp by two forces. One is the force of 480 lb by the brakes, which hold it from rolling down the ramp. The other is a reaction force by the ramp of 2250 lb, perpendicular to the ramp. This force keeps the car from going through the ramp. See Fig. 9.48. What is the weight of the car, and at what angle with the horizontal is the ramp inclined?

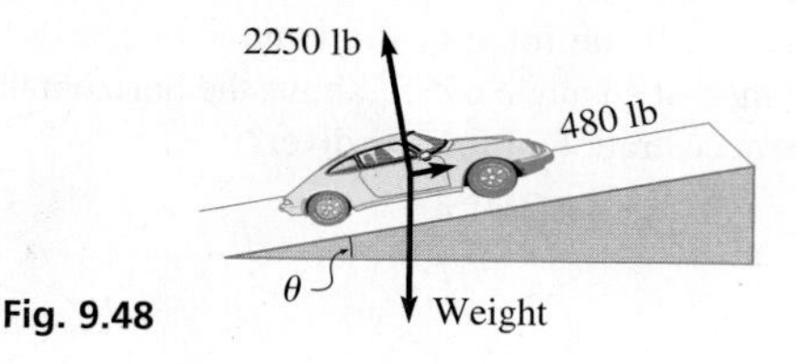

Fig. 9.48

35. A plane is moving at 75.0 m/s, and a package with weather instruments is ejected horizontally from the plane at 15.0 m/s, perpendicular to the direction of the plane. If the vertical velocity v_v (in m/s), as a function of time t (in s) of fall, is given by $v_v = 9.80t$, what is the velocity of the package after 2.00 s (before its parachute opens)?

36. A flat rectangular barge, 48.0 m long and 20.0 m wide, is headed directly across a stream at 4.5 km/h. The stream flows at 3.8 km/h. What is the velocity, relative to the riverbed, of a person walking diagonally across the barge at 5.0 km/h while facing the opposite upstream bank?

37. Assuming the vectors in Fig. 9.49 are in equilibrium, find T_1 and T_2. See Example 6.

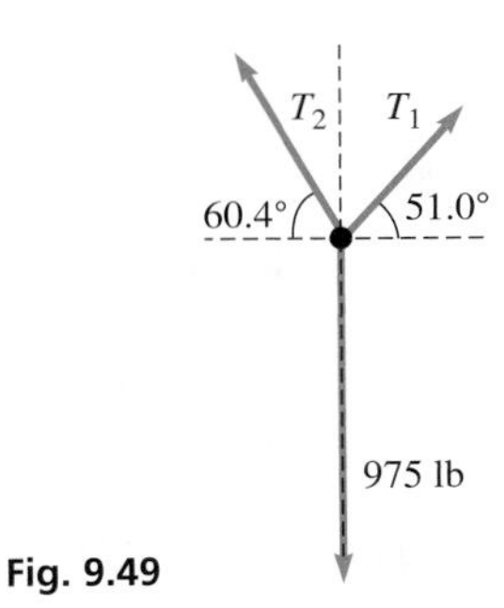

Fig. 9.49

38. A traffic light that weighs 215 N is suspended from two cables which are angled 30.0° and 53.0° from the vertical as shown in Fig. 9.50. Make a free-body diagram, and then find the tensions T_L and T_R in the left and right cables, respectively. See Example 6.

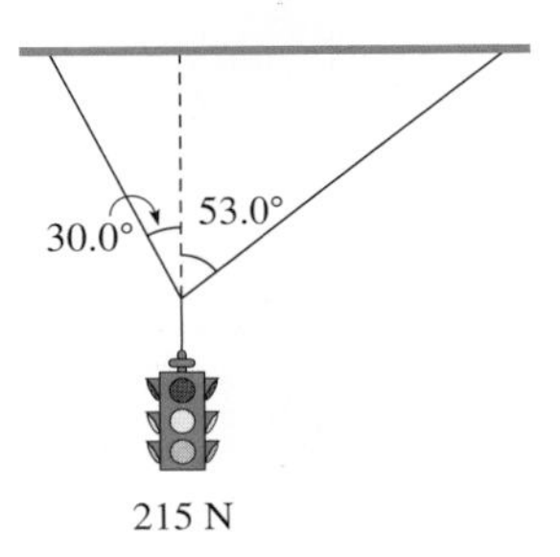

Fig. 9.50

Answers to Practice Exercises

1. $R = 22.97$ mi, $\theta = 27.68°$ **2.** 674 km/h, 6.0° north of east
3. $T = 171$ lb, $F = 44.2$ lb

9.5 Oblique Triangles, the Law of Sines

Oblique Triangles • Law of Sines • Case 1: Two Angles and One Side • Case 2: Two Sides and the Angle Opposite One of Them • Ambiguous Case

To this point, we have limited our study of triangle solution to right triangles. However, *many triangles that require solution do not contain a right angle. Such triangles are called* **oblique triangles.** We now discuss solutions of oblique triangles.

In Section 4.4, we stated that we need to know three parts, at least one of them a side, to solve a triangle. There are four possible such combinations of parts, and these combinations are as follows:

Case 1. *Two angles and one side*

Case 2. *Two sides and the angle opposite one of them*

Case 3. *Two sides and the included angle*

Case 4. *Three sides*

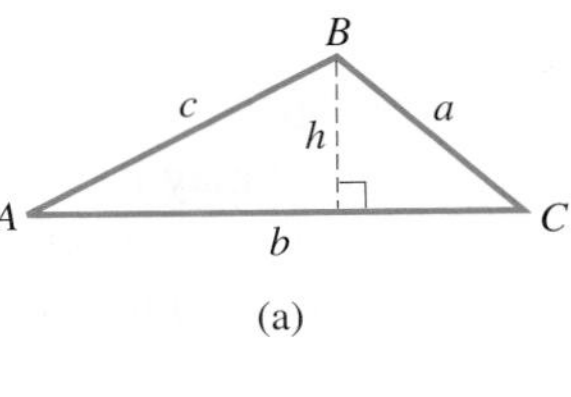

(a)

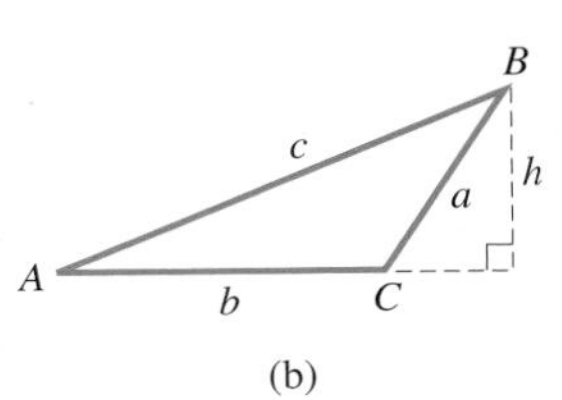

(b)

Fig. 9.51

There are several ways in which oblique triangles may be solved, but we restrict our attention to the two most useful methods, the **law of sines** and the **law of cosines.** In this section, we discuss the law of sines and show that it may be used to solve Case 1 and Case 2.

Let ABC be an oblique triangle with sides a, b, and c opposite angles A, B, and C, respectively. By drawing a perpendicular h from B to side b, as shown in Fig. 9.51(a), note that $h/c = \sin A$ and $h/a = \sin C$, and therefore

$$h = c \sin A \quad \text{or} \quad h = a \sin C \tag{9.4}$$

This relationship also holds if one of the angles is obtuse. By drawing a perpendicular h to the extension of side b, as shown in Fig. 9.51(b), we see that $h/c = \sin A$.

Also, in Fig. 9.51(b), $h/a = \sin(180° - C)$, where $180° - C$ is also the reference angle for the obtuse angle C. For a second-quadrant reference angle $180° - C$, we have $\sin[180° - (180° - C)] = \sin C$. Putting this all together, $\sin(180° - C) = \sin C$. This means that

$$h = c \sin A \quad \text{or} \quad h = a \sin(180° - C) = a \sin C \tag{9.5}$$

The results are precisely the same in Eqs. (9.4) and (9.5). Setting the results for h equal to each other, we have

$$c \sin A = a \sin C$$

Dividing each side by $\sin A \sin C$ and reversing sides gives us

$$\frac{a}{\sin A} = \frac{c}{\sin C} \tag{9.6}$$

By dropping a perpendicular from A to a, we also derive the result

$$c \sin B = b \sin C$$

which, when each side is divided by $\sin B \sin C$, and reversing sides, gives us

$$\frac{b}{\sin B} = \frac{c}{\sin C} \tag{9.7}$$

If we had dropped a perpendicular to either side a or side c, equations similar to Eqs. (9.6) and (9.7) would have resulted.

Next we combine Eqs. (9.6) and (9.7) into an important basic equation that allows us to solve oblique triangles for which the given information fits either Case 1 or Case 2.

Combining Eqs. (9.6) and (9.7), *for any triangle with sides a, b, and c, opposite angles A, B, and C, respectively,* such as the one shown in Fig. 9.52, *we have the following*:

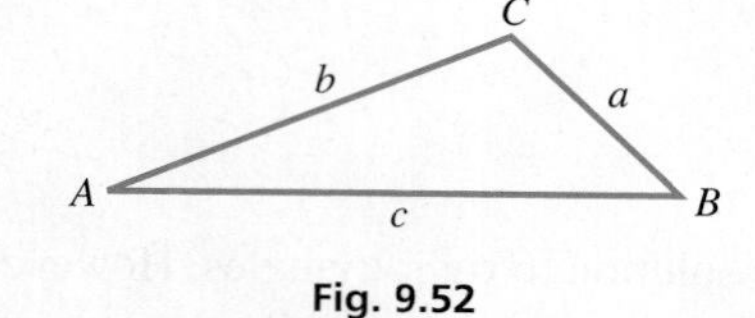

Fig. 9.52

Law of Sines

$$\frac{a}{\sin A} = \frac{b}{\sin B} = \frac{c}{\sin C} \tag{9.8}$$

Another form of the law of sines can be obtained by equating the reciprocals of each of the fractions in Eq. (9.8). The law of sines is a statement of proportionality between the sides of a triangle and the sines of the angles opposite them. NOTE ▸ [Note that there are actually three equations combined in Eq. (9.8). Of these, we use the one with three known parts of the triangle and we solve for the fourth part.] In finding the complete solution of a triangle, it may be necessary to use two of the three equations.

CASE 1: TWO ANGLES AND ONE SIDE

Now, we see how the law of sines is used in the solution of a triangle in which two angles and one side are known. If two angles are known, the third may be found from the fact that the sum of the angles in a triangle is 180°. Then, since the given side and its opposite angle are both known, we can use the law of sines (twice) to find each of the other two sides.

EXAMPLE 1 Case 1: Two angles and one side

Given $c = 6.00$, $A = 60.0°$, and $B = 40.0°$, find a, b, and C.

First, we can see that

$$C = 180.0° - (60.0° + 40.0°) = 80.0°$$

We now know side c and angle C, which allows us to use Eq. (9.8). Therefore, using the equation relating a, A, c, and C, we have

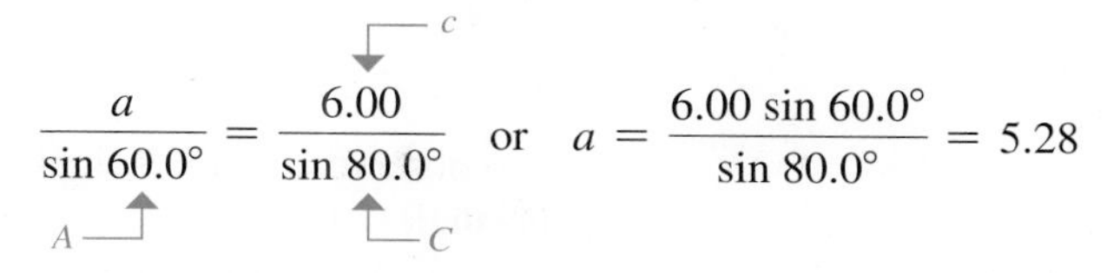

$$\frac{a}{\sin 60.0^\circ} = \frac{6.00}{\sin 80.0^\circ} \quad \text{or} \quad a = \frac{6.00 \sin 60.0^\circ}{\sin 80.0^\circ} = 5.28$$

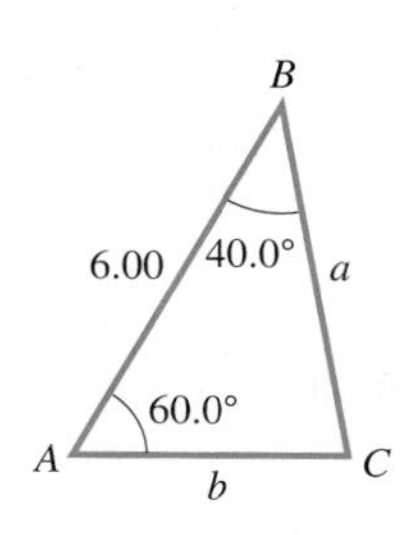

Fig. 9.53

Now, using the equation relating b, B, c, and C, we have

$$\frac{b}{\sin 40.0^\circ} = \frac{6.00}{\sin 80.0^\circ} \quad \text{or} \quad b = \frac{6.00 \sin 40.0^\circ}{\sin 80.0^\circ} = 3.92$$

Thus, $a = 5.28$, $b = 3.92$, and $C = 80.0^\circ$. See Fig. 9.53. We could also have used the form of Eq. (9.8) relating a, A, b, and B in order to find b, but any error in calculating a would make b in error as well. ■

Practice Exercise

1. In Example 1, change the value of B to 65.0°, and then find a.

NOTE ▸ [The solution of a triangle can be checked approximately by noting that the smallest angle is opposite the shortest side, and the largest angle is opposite the longest side.] Note that this is the case in Example 1, where b (shortest side) is opposite B (smallest angle), and c (longest side) is opposite C (largest angle).

EXAMPLE 2 Case 1: Two angles and one side

Solve the triangle with the following given parts: $a = 63.71$, $A = 56.29^\circ$, and $B = 97.06^\circ$. See Fig. 9.54.

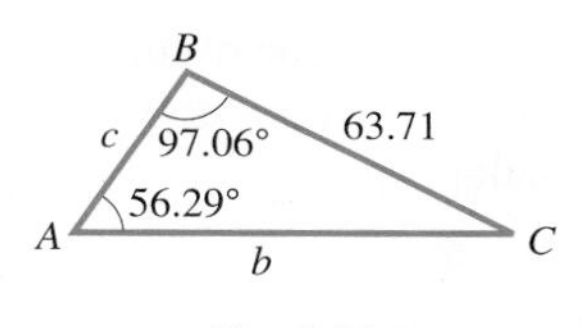

Fig. 9.54

From the figure, we see that we are to find angle C and sides b and c. We first determine angle C:

$$C = 180^\circ - (A + B) = 180^\circ - (56.29^\circ + 97.06^\circ)$$
$$= 26.65^\circ$$

Noting the three angles, we know that c is the shortest side (C is the smallest angle) and b is the longest side (B is the largest angle). This means that the length of a is between c and b, or $c < 63.71$ and $b > 63.71$. Now using the ratio $a/\sin A$ of Eq. (9.8) (the law of sines) to find sides b and c, we have

$$\frac{b}{\sin 97.06^\circ} = \frac{63.71}{\sin 56.29^\circ} \quad \text{or} \quad b = \frac{63.71 \sin 97.06^\circ}{\sin 56.29^\circ} = 76.01$$

$$\frac{c}{\sin 26.65^\circ} = \frac{63.71}{\sin 56.29^\circ} \quad \text{or} \quad c = \frac{63.71 \sin 26.65^\circ}{\sin 56.29^\circ} = 34.35$$

Thus, $b = 76.01$, $c = 34.35$, and $C = 26.65^\circ$. Note that $c < a$ and $b > a$, as expected. ■

If the given information is appropriate, the law of sines may be used to solve applied problems. The following example illustrates the use of the law of sines in such a problem.

EXAMPLE 3 Case 1—distance to helicopter

Two observers A and B sight a helicopter due east. The observers are 3540 ft apart, and the angles of elevation they each measure to the helicopter are 32.0° and 44.0°, respectively. How far is observer A from the helicopter? See Fig. 9.55.

■ The first successful helicopter was made in the United States by Igor Sikorsky in 1939.

Letting H represent the position of the helicopter, we see that angle B within the triangle ABH is $180° - 44.0° = 136.0°$. This means that the angle at H within the triangle is

$$H = 180° - (32.0° + 136.0°) = 12.0°$$

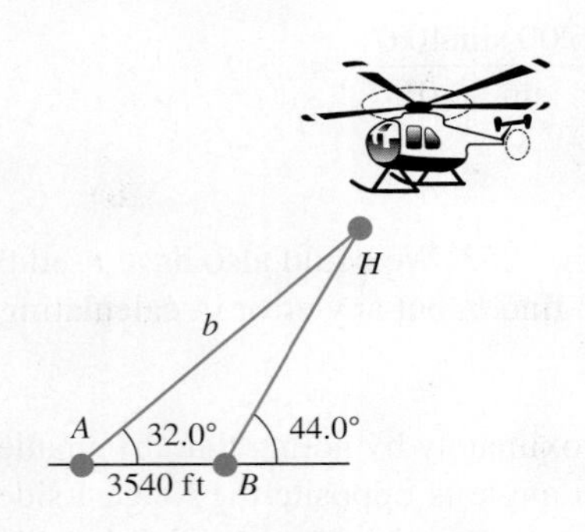

Fig. 9.55

Now, using the law of sines to find required side b, we have

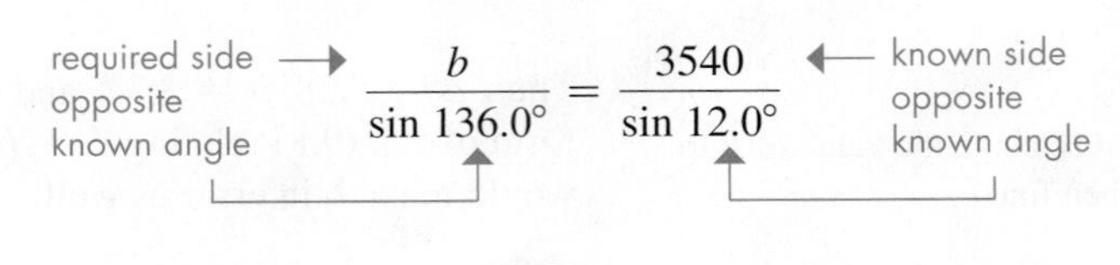

$$\frac{b}{\sin 136.0°} = \frac{3540}{\sin 12.0°}$$

or

$$b = \frac{3540 \sin 136.0°}{\sin 12.0°} = 11{,}800 \text{ ft}$$

Thus, observer A is about 11,800 ft from the helicopter. ■

CASE 2: TWO SIDES AND THE ANGLE OPPOSITE ONE OF THEM

For a triangle in which we know two sides and the angle opposite one of the given sides, the solution will be either *one triangle*, or *two triangles*, or even possibly *no triangle*. The following examples illustrate how each of these results is possible.

EXAMPLE 4 Case 2: Two sides and angle opposite

Solve the triangle with the following given parts: $a = 40.0$, $b = 60.0$, and $A = 30.0°$.

First, make a good scale drawing [Fig. 9.56(a)] by drawing angle A and measuring off 60 for b. This will more clearly show that side $a = 40.0$ will intersect side c at either position B or B'. This means there are two triangles that satisfy the given values. Using the law of sines, we solve the case for which B is an acute angle:

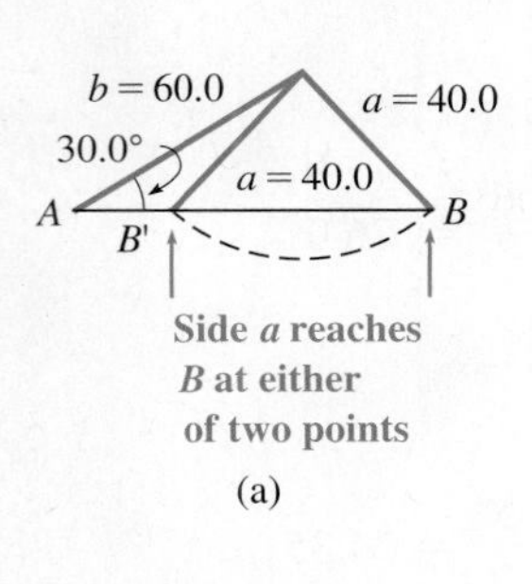

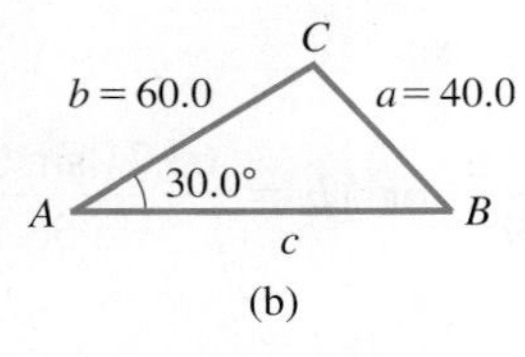

Fig. 9.56

$$\frac{60.0}{\sin B} = \frac{40.0}{\sin 30.0°} \quad \text{or} \quad \sin B = \frac{60.0 \sin 30.0°}{40.0}$$

$$B = \sin^{-1}\left(\frac{60.0 \sin 30.0°}{40.0}\right) = 48.6°$$

$$C = 180° - (30.0° + 48.6°) = 101.4°$$

Therefore, $B = 48.6°$ and $C = 101.4°$. Using the law of sines again to find c, we have

$$\frac{c}{\sin 101.4°} = \frac{40.0}{\sin 30.0°}$$

$$c = \frac{40.0 \sin 101.4°}{\sin 30.0°}$$

$$= 78.4$$

Thus, $B = 48.6°$, $C = 101.4°$, and $c = 78.4$. See Fig. 9.56(b).

The other solution is the case in which B', opposite side b, is an obtuse angle. Therefore,

$$B' = 180° - B = 180° - 48.6°$$
$$= 131.4°$$
$$C' = 180° - (30.0° + 131.4°)$$
$$= 18.6°$$

Using the law of sines to find c', we have

$$\frac{c'}{\sin 18.6°} = \frac{40.0}{\sin 30.0°}$$
$$c' = \frac{40.0 \sin 18.6°}{\sin 30.0°}$$
$$= 25.5$$

This means that the second solution is $B' = 131.4°$, $C' = 18.6°$, and $c' = 25.5$. See Fig. 9.56(c). ■

EXAMPLE 5 Case 2: Possible solutions

In Example 4, if $a > 60.0$, only one solution would result. In this case, side a would intercept side c at B. It also intercepts the extension of side c, but this would require that angle A not be included in the triangle (see Fig. 9.57). Thus, only one solution may result if $a > b$.

In Example 4, there would be *no solution* if side a were not at least 30.0. If this were the case, side a would not be long enough to even touch side c. It can be seen that a must at least equal $b \sin A$. If it is just equal to $b \sin A$, there is *one solution,* a right triangle. See Fig. 9.58.

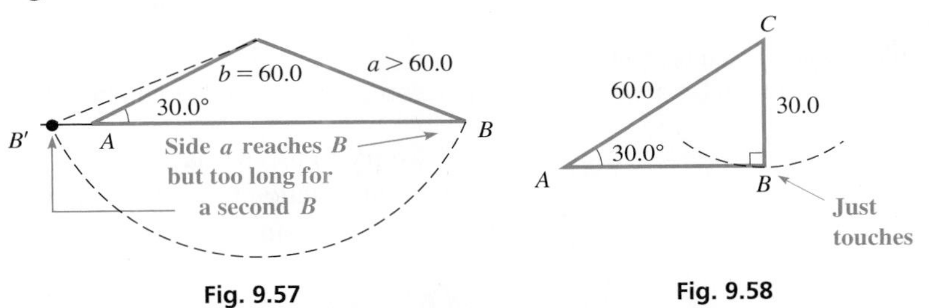

Fig. 9.57

Fig. 9.58 ■

Practice Exercise

2. Determine which of the four possible solution types occurs if $c = 28$, $b = 48$, and $C = 30°$.

Summarizing the results for Case 2 as illustrated in Examples 4 and 5, we make the following conclusions. Given sides a and b and angle A [assuming here that a and A $(A < 90°)$ are corresponding parts], we have the following summary of solutions for Case 2.

Summary of Solutions: Two Sides and the Angle Opposite One of Them

1. No solution if $a < b \sin A$. See Fig. 9.59(a).
2. A right triangle solution if $a = b \sin A$. See Fig. 9.59(b).
3. Two solutions if $b \sin A < a < b$. See Fig. 9.59(c).
4. One solution if $a > b$. See Fig. 9.59(d).

CAUTION When two sides and an opposite angle are given, there may be **two solutions** of the triangle. ■

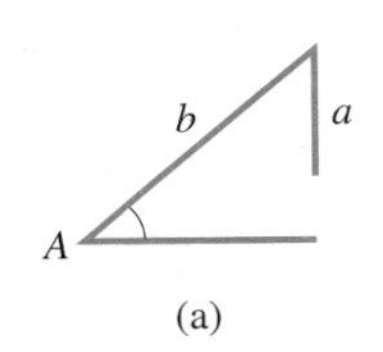

(a)

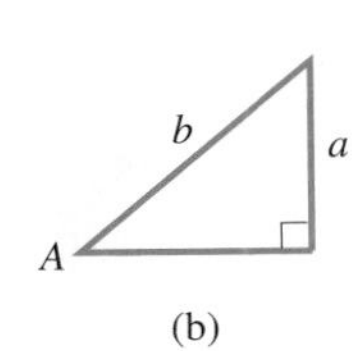

(b)

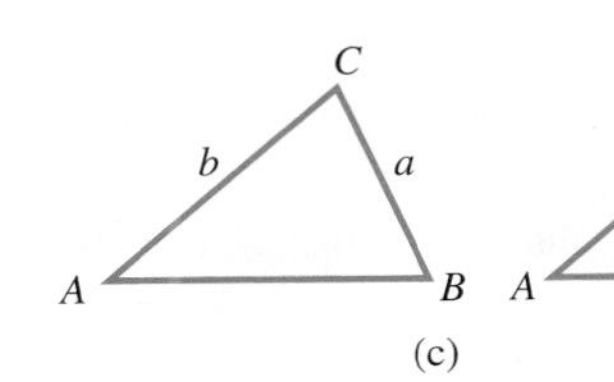

(c)

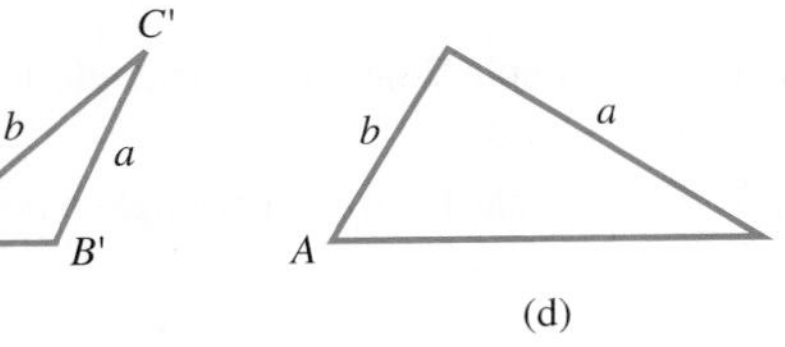

(d)

Fig. 9.59

NOTE ▸ [Note that in order to have two solutions, we must know two sides and the angle opposite one of the sides, and the shorter side must be opposite the known angle.]

If there is *no solution,* the calculator will indicate an *error*. If the solution is a *right triangle,* the calculator will show an angle of *exactly* 90° (no extra decimal digits will be displayed).

NOTE ▸ [For the reason that two solutions may result for Case 2, it is called the **ambiguous case.**] However, it must also be kept in mind that *there may only be one solution*. A *careful check of the given parts* must be made in order to determine whether there is one solution or two solutions. The following example illustrates Case 2 in an applied problem.

EXAMPLE 6 Case 2—heading of plane

Kingston, Jamaica, is 43.2° south of east of Havana, Cuba. What should be the heading of a plane from Havana to Kingston if the wind is from the west at 40.0 km/h and the plane's speed with respect to the air is 300 km/h?

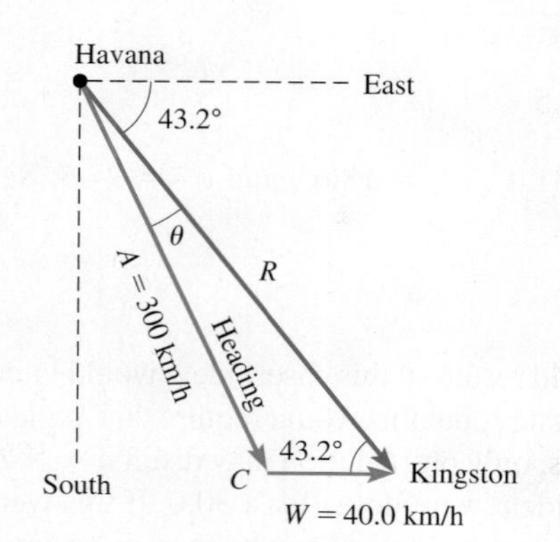

Fig. 9.60

The heading should be set so that the resultant of the plane's velocity with respect to the air **A** and the velocity of the wind **W** will be in the direction from Havana to Kingston. This means that the resultant velocity **R** of the plane with respect to the ground must be at an angle of 43.2° south of east from Havana.

Using the given information, we draw the vector triangle shown in Fig. 9.60. In the triangle, we know that the angle at Kingston is 43.2° by noting the alternate-interior angles (see page 56). By finding θ, the required heading can be found. There can be only one solution, because $A > W$. Using the law of sines, we have

$$\frac{40.0}{\sin\theta} = \frac{300}{\sin 43.2°}$$

(known side opposite required angle → 40.0, sin θ; known side opposite known angle → 300, sin 43.2°)

$$\sin\theta = \frac{40.0 \sin 43.2°}{300}, \qquad \theta = 5.2°$$

Therefore, the heading should be 43.2° + 5.2° = 48.4° south of east. ■

Practice Exercise

3. In Example 6, what should be the heading if 300 km/h is changed to 500 km/h?

If we try to use the law of sines for Case 3 or Case 4, we find that we do not have enough information to complete any of the ratios. These cases can, however, be solved by the law of cosines as shown in the next section.

EXAMPLE 7 Cases 3 and 4 not solvable by law of sines

Given (Case 3) two sides and the included angle of a triangle $a = 2$, $b = 3$, $C = 45°$, and (Case 4) the three sides (Case 4) $a = 5$, $b = 6$, $c = 7$, we set up the ratios

$$\text{(Case 3)}\frac{2}{\sin A} = \frac{3}{\sin B} = \frac{c}{\sin 45°} \quad \text{and} \quad \text{(Case 4)}\frac{5}{\sin A} = \frac{6}{\sin B} = \frac{7}{\sin C}$$

The solution cannot be found because each of the three possible equations in either Case 3 or Case 4 contains two unknowns. ■

EXERCISES 9.5

In Exercises 1 and 2, solve the resulting triangles if the given changes are made in the indicated examples of this section.

1. In Example 2, solve the triangle if the value of B is changed to 82.94°.
2. In Example 4, solve the triangle if the value of b is changed to 70.0.

In Exercises 3–20, solve the triangles with the given parts.

3. $a = 45.7$, $A = 65.0°$, $B = 49.0°$
4. $b = 3.07$, $A = 26.0°$, $C = 120.0°$
5. $c = 4380$, $A = 37.4°$, $B = 34.6°$
6. $a = 932$, $B = 0.9°$, $C = 82.6°$

7. $a = 4.601, b = 3.107, A = 18.23°$
8. $b = 362.2, c = 294.6, B = 110.63°$
9. $b = 7751, c = 3642, B = 20.73°$
10. $a = 150.4, c = 250.9, C = 76.43°$
11. $a = 63.8, B = 58.4°, C = 22.2°$
12. $a = 0.130, A = 55.2°, B = 117.5°$
13. $b = 4384, B = 47.43°, C = 64.56°$
14. $b = 283.2, B = 13.79°, C = 103.62°$
15. $a = 5.240, b = 4.446, B = 48.13°$
16. $a = 89.45, c = 37.36, C = 15.62°$
17. $b = 2880, c = 3650, B = 31.4°$
18. $a = 0.841, b = 0.965, A = 57.1°$
19. $a = 450, b = 1260, A = 64.8°$
20. $a = 20, c = 10, C = 30°$

In Exercises 21–38, use the law of sines to solve the given problems.

21. A small island is approximately a triangle in shape. If the longest side of the island is 520 m, and two of the angles are 45° and 55°, what is the length of the shortest side?

22. A boat followed a triangular route going from dock A, to dock B, to dock C, and back to dock A. The angles turned were 135° at B and 125° at C. If B is 875 m from A, how far is it from B to C?

23. The loading ramp at a delivery service is 12.5 ft long and makes a 18.0° angle with the horizontal. If it is replaced with a ramp 22.5 ft long, what angle does the new ramp make with the horizontal?

24. In an aerial photo of a triangular field, the longest side is 86.0 cm, the shortest side is 52.5 cm, and the largest angle is 82.0°. The scale is 1 cm = 2 m. Find the actual length of the third side of the field.

25. The Pentagon (headquarters of the U.S. Department of Defense) is the largest office building in the world. It is a regular pentagon (five sides), 921 ft on a side. Find the greatest straight-line distance from one point on the outside of the building to another outside point (the length of a diagonal).

26. Two ropes hold a 175-lb crate as shown in Fig. 9.61. Find the tensions $\mathbf{T}_1$ and $\mathbf{T}_2$ in the ropes. (*Hint:* Move vectors so that they are tail to head to form a triangle. The vector sum $\mathbf{T}_1 + \mathbf{T}_2$ must equal 175 lb for equilibrium. See page 278.)

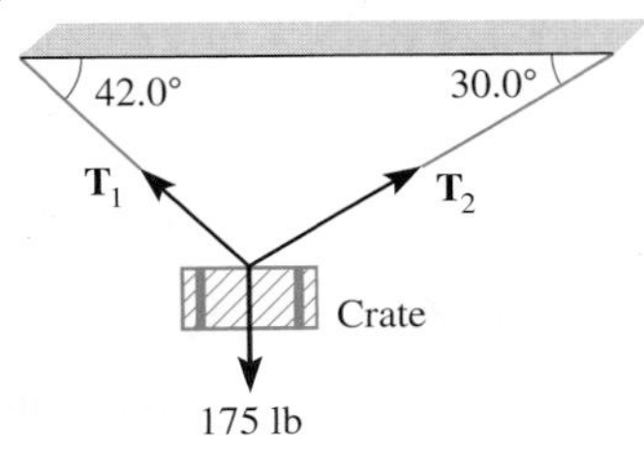

Fig. 9.61

27. Find the tension **T** in the left guy wire attached to the top of the tower shown in Fig. 9.62. (*Hint:* The horizontal components of the tensions must be equal and opposite for equilibrium. See page 278. Thus, move the tension vectors tail to head to form a triangle with a vertical resultant. This resultant equals the upward force at the top of the tower for equilibrium. This last force is not shown and does not have to be calculated.)

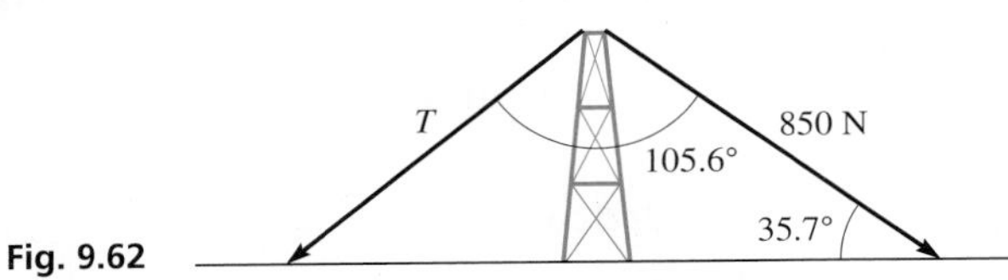

Fig. 9.62

28. Find the distance from Atlanta to Raleigh, North Carolina, from Fig. 9.63.

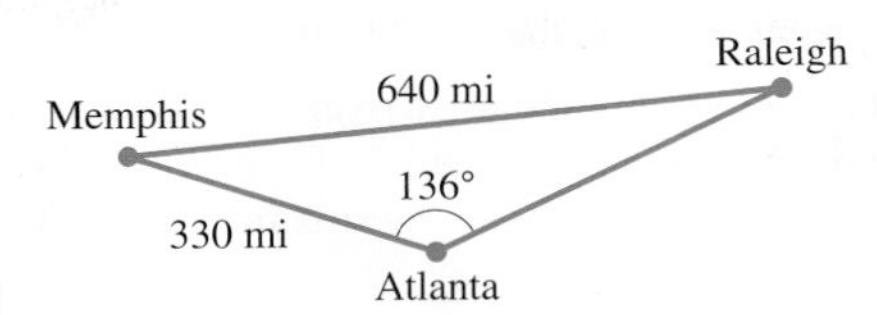

Fig. 9.63

29. An Amazon robot begins at point O and travels in a straight line across a warehouse floor to point A where it picks up some merchandise. It then turns a 35° corner and travels 32.5 m to point B where it drops off the merchandise. See Fig. 9.64. If the robot must now turn a 29.0° corner to return to its original position, how far must it travel to get there?

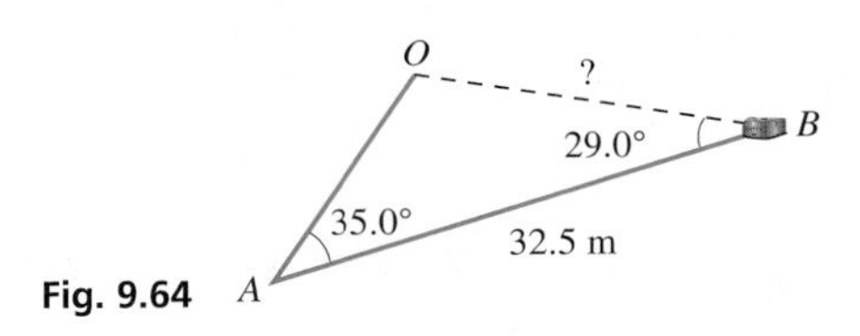

Fig. 9.64

30. When an airplane is landing at an 8250-ft runway, the angles of depression to the ends of the runway are 10.0° and 13.5°. How far is the plane from the near end of the runway?

31. Find the total length of the path of the laser beam that is shown in Fig. 9.65.

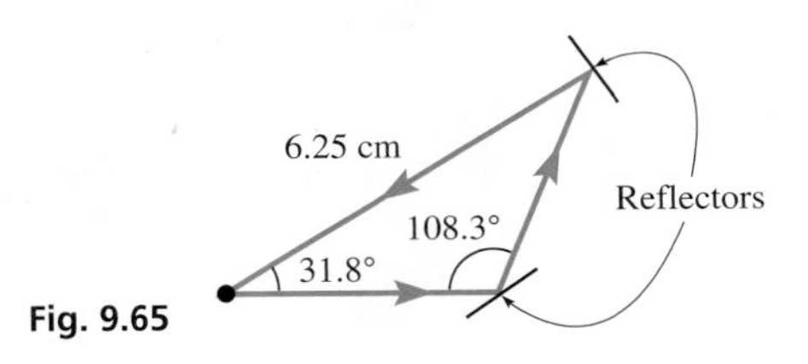

Fig. 9.65

32. In widening a highway, it is necessary for a construction crew to cut into the bank along the highway. The present angle of elevation of the straight slope of the bank is 23.0°, and the new angle is to be 38.5°, leaving the top of the slope at its present position. If the slope of the present bank is 220 ft long, how far horizontally into the bank at its base must they dig?

33. A communications satellite is directly above the extension of a line between receiving towers A and B. It is determined from radio signals that the angle of elevation of the satellite from tower A is 89.2°, and the angle of elevation from tower B is 86.5°. See Fig. 9.66. If A and B are 1290 km apart, how far is the satellite from A? (Neglect the curvature of Earth.)

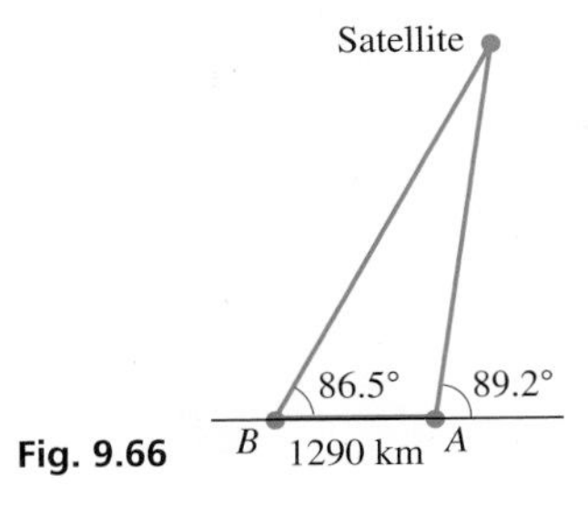

Fig. 9.66

34. A Mars rover travels 569 m in a straight line and then stops to take a soil sample. It then changes direction and travels another 642 m where it takes another sample. If it must turn on a 47.2° corner to return to its starting point, what are the possible distances from its starting point?

35. A boat owner wishes to cross a river 2.60 km wide and go directly to a point on the opposite side 1.75 km downstream. The boat goes 8.00 km/h in still water, and the stream flows at 3.50 km/h. What should the boat's heading be?

36. A motorist traveling along a level highway at 75 km/h directly toward a mountain notes that the angle of elevation of the mountain top changes from about 20° to about 30° in a 20-min period. How much closer on a direct line did the mountain top become?

37. A person standing on level ground from the base of the Skylon Tower in Niagara Falls observes an elevator going up the side of the tower. At one instant, the angle of elevation to the elevator is 42.5° and then 15.0 s later, the angle of elevation is 60.7°. If the elevator travels at a rate of 10.0 ft/s, how high above ground is the elevator at the second instant?

38. Point P on the mechanism shown in Fig. 9.67 is driven back and forth horizontally. If the minimum value of angle θ is 32.0°, what is the distance between extreme positions of P? What is the maximum possible value of angle θ?

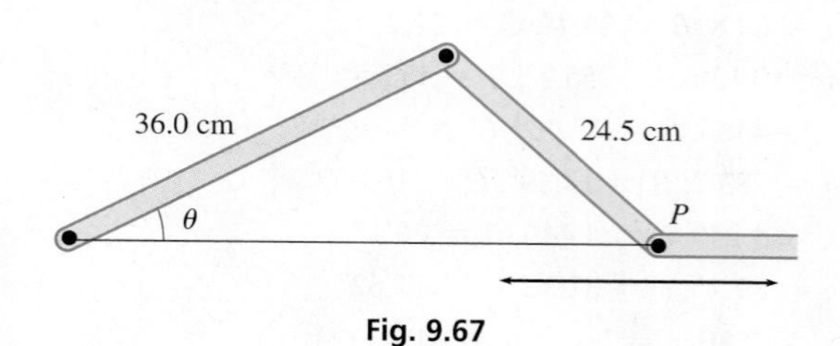

Fig. 9.67

Answers to Practice Exercises
1. $a = 6.34$ **2.** Two solutions **3.** 46.3°

9.6 The Law of Cosines

Law of Cosines • Case 3: Two Sides & Included Angle • Case 4: Three Sides • Summary of Solving Oblique Triangles

As noted in the last section, the law of sines cannot be used for Case 3 (two sides and the included angle) and Case 4 (three sides). In this section, we develop the *law of cosines,* which can be used for Cases 3 and 4. After finding another part of the triangle using the law of cosines, we will often find it easier to complete the solution using the law of sines.

Consider any oblique triangle—for example, either triangle shown in Fig. 9.68. For each triangle, $h/b = \sin A$, or $h = b \sin A$. Also, using the Pythagorean theorem, we obtain $a^2 = h^2 + x^2$ for each triangle. Therefore [with $(\sin A)^2 = \sin^2 A$],

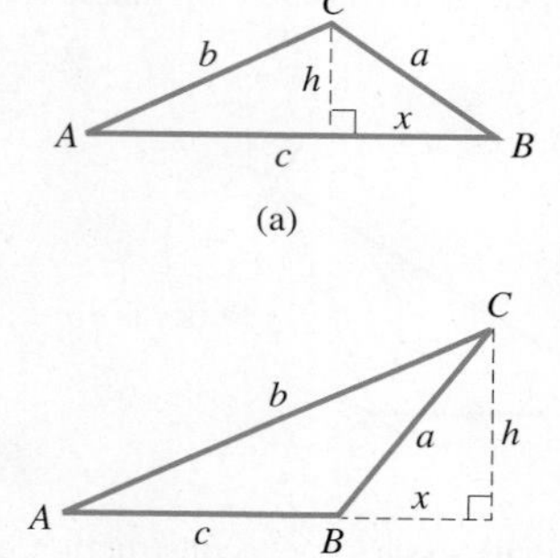

Fig. 9.68

$$a^2 = b^2 \sin^2 A + x^2 \tag{9.9}$$

In Fig. 9.68(a), note that $(c - x)/b = \cos A$, or $c - x = b \cos A$. Solving for x, we have $x = c - b \cos A$. In Fig. 9.68(b), $c + x = b \cos A$, and solving for x, we have $x = b \cos A - c$. Substituting these relations into Eq. (9.9), we obtain

$$a^2 = b^2 \sin^2 A + (c - b \cos A)^2$$

and

$$a^2 = b^2 \sin^2 A + (b \cos A - c)^2 \tag{9.10}$$

respectively. When expanded, these both give

$$\begin{aligned} a^2 &= b^2 \sin^2 A + b^2 \cos^2 A + c^2 - 2bc \cos A \\ &= b^2(\sin^2 A + \cos^2 A) + c^2 - 2bc \cos A \end{aligned} \tag{9.11}$$

Recalling the definitions of the trigonometric functions, we know that $\sin \theta = y/r$ and $\cos \theta = x/r$. Thus, $\sin^2\theta + \cos^2\theta = (y^2 + x^2)/r^2$. However, $x^2 + y^2 = r^2$, which means

$$\sin^2 \theta + \cos^2 \theta = 1 \tag{9.12}$$

This equation is valid for any angle θ, since we have made no assumptions as to the properties of θ. Thus, by substituting Eq. (9.12) into Eq. (9.11), we arrive at the following:

Law of Cosines

$$a^2 = b^2 + c^2 - 2bc \cos A \tag{9.13}$$

The law of cosines can also be written in the following forms:

$$b^2 = a^2 + c^2 - 2ac \cos B$$
$$c^2 = a^2 + b^2 - 2ab \cos C$$

CASE 3: TWO SIDES AND THE INCLUDED ANGLE

If two sides and the included angle of a triangle are known, the forms of the law of cosines show that we may directly solve for the side opposite the given angle. Then, as noted earlier, the solution may be completed using the law of sines.

EXAMPLE 1 Case 3: Two sides and included angle

Solve the triangle with $a = 45.0$, $b = 67.0$, and $C = 35.0°$. See Fig. 9.69.

Because angle C is known, first solve for side c, using the law of cosines in the form $c^2 = a^2 + b^2 - 2ab\cos C$. Substituting, we have

unknown side opposite known angle

$$c^2 = 45.0^2 + 67.0^2 - 2(45.0)(67.0)\cos 35.0°$$

known sides

$$c = \sqrt{45.0^2 + 67.0^2 - 2(45.0)(67.0)\cos 35.0°} = 39.7$$

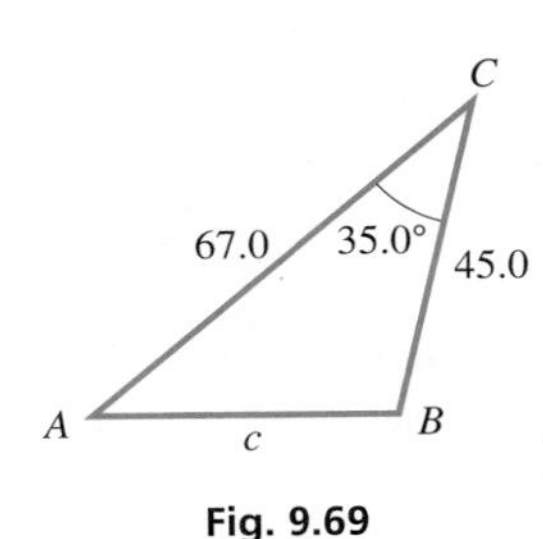

Fig. 9.69

Now we will use the law of sines to find angle A (the *smaller* remaining angle).

$$\frac{45.0}{\sin A} = \frac{39.7}{\sin 35.0°}$$ ← sides opposite angles

$$\sin A = \frac{45.0 \sin 35.0°}{39.7}, \qquad A = 40.6°$$

Angle B can be found by using the fact that the sum of the angles is 180°:

$$B = 180° - 35.0° - 40.6° = 104.4°$$

Therefore, $c = 39.7$, $A = 40.6°$, and $B = 104.4°$.

After finding side c, solving for angle B rather than angle A, the calculator would show $B = 75.5°$. Then, when subtracting the sum of angles B and C from 180°, we would get $A = 69.5°$. Although this appears to be correct, it is not. CAUTION Because the largest angle of a triangle might be greater than 90°, first solve for the smaller unknown angle. ■ This smaller angle (opposite the shorter known side) cannot be greater than 90°, and the larger unknown angle can be found by subtraction as is done in this example. ■

Practice Exercise

1. In Example 1, change the value of a to 95.0, and find A and B.

EXAMPLE 2 Case 3—finding a resultant force

Two forces are acting on a bolt. One is a 78.0-N force acting horizontally to the right, and the other is a force of 45.0 N acting upward to the right, 15.0° from the vertical. Find the resultant force **F**. See Fig. 9.70.

Moving the 45.0-N vector to the right and using the lower triangle with the 105.0° angle, the magnitude of **F** is

$$F = \sqrt{78.0^2 + 45.0^2 - 2(78.0)(45.0)\cos 105.0°}$$
$$= 99.6 \text{ N}$$

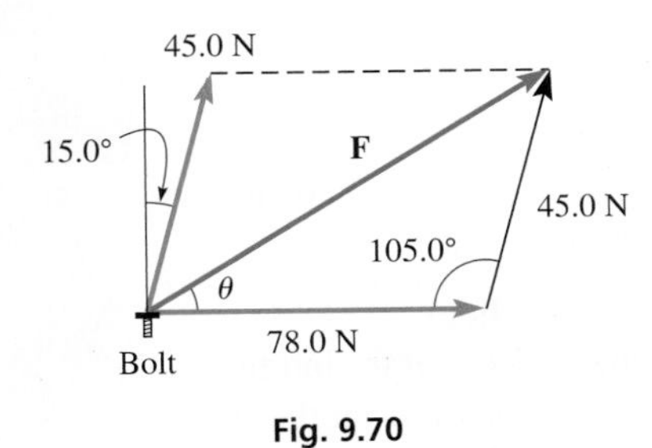

Fig. 9.70

To find θ, use the law of sines:

$$\frac{45.0}{\sin\theta} = \frac{99.6}{\sin 105.0°}, \qquad \sin\theta = \frac{45.0\sin 105.0°}{99.6}$$

This gives us $\theta = 25.9°$.

We could also solve this problem using vector components. For the x-component of the resultant we would add the 78.0-N vector to the x-component of the 45.0-N vector, and the y-component of the resultant is the y-component of the 45.0-N vector. ■

CASE 4: THREE SIDES

Given the three sides of a triangle, we may solve for the angle opposite any side using the law of cosines. **CAUTION** The best procedure is to find the largest angle first. This avoids the ambiguous case if we switch to the law of sines and there is an obtuse angle. ■ *The largest angle is opposite the longest side.* Another procedure is to use the law of cosines to find two angles.

EXAMPLE 3 Case 4: Three sides—roof angle

For a triangular roof truss with sides 49.3 ft, 21.6 ft, and 42.6 ft, find the angles between the sides. See Fig. 9.71.

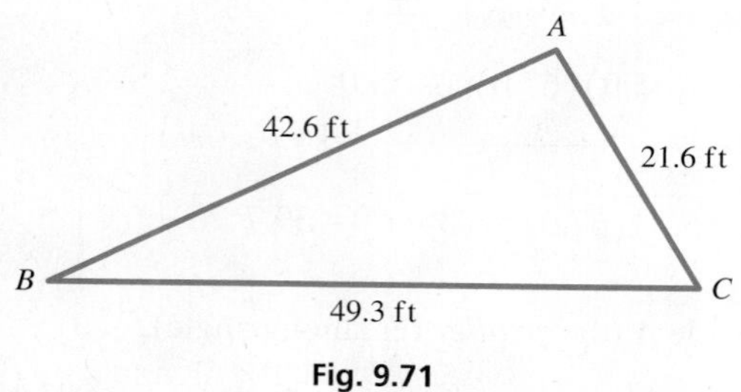

Fig. 9.71

First, we let the sides be a, b, and c, with opposite angles of A, B, and C. Then, because the longest side is 49.3 ft, we first solve for angle A. Because $a^2 = b^2 + c^2 - 2bc \cos A$,

$$\cos A = \frac{b^2 + c^2 - a^2}{2bc} = \frac{21.6^2 + 42.6^2 - 49.3^2}{2(21.6)(42.6)}$$

$$A = 94.6°$$

Now we can use the law of sines to find angle B:

$$\frac{21.6}{\sin B} = \frac{49.3}{\sin 94.65°}$$ ← extra significant digit is included for intermediate step

$$\sin B = \frac{21.6 \sin 94.65°}{49.3}$$

$$B = 25.9°$$

Angle C can be found by subtracting from 180°: $C = 180° - 94.6° - 25.9° = 59.5°$. ■

EXAMPLE 4 Case 4—antenna angle

A vertical radio antenna is to be built on a hillside with a constant slope. A guy wire is to be attached at a point 145 ft up the antenna, and at a point 110 ft from the base of the antenna up the hillside. If the guy wire is 170 ft long, what angle does the antenna make with the hillside?

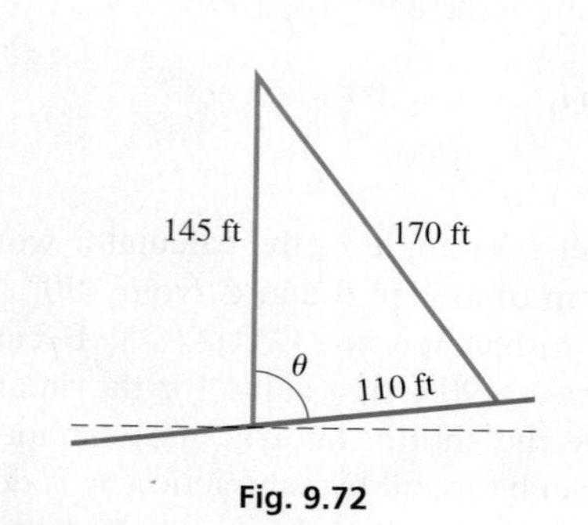

Fig. 9.72

From Fig. 9.72, we can set up the equation necessary for the solution.

$$170^2 = 110^2 + 145^2 - 2(110)(145)\cos\theta$$

$$\theta = \cos^{-1}\frac{110^2 + 145^2 - 170^2}{2(110)(145)} = 82.4°$$ ■

Practice Exercise

2. Using Fig. 9.72, and only the data given in Example 4, find the angle between the guy wire and the hillside.

Summary of Solving Oblique Triangles

Case 1: Two Angles and One Side
Find the unknown angle by subtracting the sum of the known angles from 180°. Use the **law of sines** to find the unknown sides.

Case 2: Two Sides and the Angle Opposite One of Them
Use the known side and the known angle opposite it to find the angle opposite the other known side. Find the third angle from the fact that the sum of the angles is 180°. Use the **law of sines** to find the third side. **CAUTION:** There may be two solutions. See page 287 for a summary of Case 2, the *ambiguous case*. ■

Case 3: Two Sides and the Included Angle
Find the third side by using the **law of cosines.** Find the *smaller* unknown angle (opposite the shorter side) by using the **law of sines.** Complete the solution using the fact that the sum of the angles is 180°.

Case 4: Three Sides
Find the *largest angle* (opposite the longest side) by using the **law of cosines.** Find a second angle by using the **law of sines.** Complete the solution by using the fact that the sum of the angles is 180°.

Other variations in finding the solutions can be used. For example, after finding the third side in Case 3, it is possible to find an angle using the law of cosines. Also, in Case 4, all angles can be found by using the law of cosines. The methods shown above are those normally used.

EXERCISES 9.6

In Exercises 1 and 2, solve the resulting triangles if the given changes are made in the indicated examples of this section.

1. In Example 1, solve the triangle if the value of C is changed to 145°.
2. In Example 3, solve the triangle if 49.3 is changed to 54.3.

In Exercises 3–20, solve the triangles with the given parts.

3. $a = 6.00, b = 7.56, C = 54.0°$
4. $b = 87.3, c = 34.0, A = 130.0°$
5. $a = 4530, b = 924, C = 98.0°$
6. $a = 0.0845, c = 0.116, B = 85.0°$
7. $a = 395.3, b = 452.2, c = 671.5$
8. $a = 2.331, b = 2.726, c = 2.917$
9. $a = 385.4, b = 467.7, c = 800.9$
10. $a = 0.2433, b = 0.2635, c = 0.1538$
11. $a = 320, b = 847, C = 158.0°$
12. $b = 18.3, c = 27.1, A = 8.7°$
13. $a = 2140, c = 428, B = 86.3°$
14. $a = 1.13, b = 0.510, C = 77.6°$
15. $b = 103.7, c = 159.1, C = 104.67°$
16. $a = 49.32, b = 54.55, B = 114.36°$
17. $a = 723, b = 598, c = 158$
18. $a = 1.78, b = 6.04, c = 4.80$
19. $a = 1500, A = 15°, B = 140°$
20. $a = 17, b = 24, c = 42$. Explain your answer.

In Exercises 21–40, use the law of cosines to solve the given problems.

21. An iRobot Roomba vacuum cleaner travels 12.5 ft across the floor where it then hits a wall and turns on a 47.0° corner. If it travels 15.0 ft in this new direction, how far is it from its starting point?
22. Set up equations (do not solve) to solve the triangle in Fig. 9.73 by the law of cosines. Why is the law of sines easier to use?

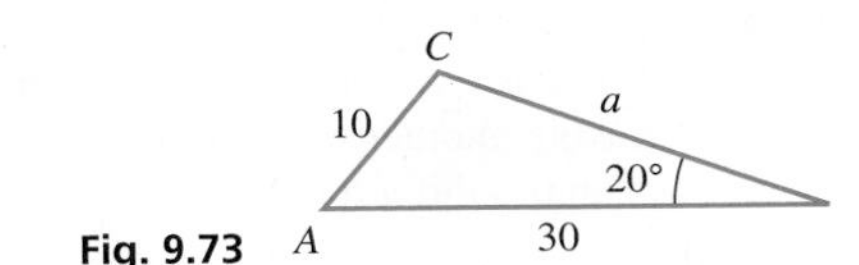

Fig. 9.73

23. Write the form of the law of cosines, given sides a and b, and angle $C = 90°$.
24. Find the length of the chord intercepted by a central angle of 54.2° in a circle or radius 18.0 cm.
25. An airplane leaves an airport traveling 385 mi/h on a course 27.3° east of due north. Fifteen minutes later, a second plane leaves the same airport traveling 455 mi/h on a course 19.4° west of due north. What is the distance between the two planes one hour after the first plane departed?
26. Three circles of radii 24 in., 32 in., and 42 in. are externally tangent to each other (each is tangent to the other two). Find the largest angle of the triangle formed by joining their centers.
27. In viewing an island from a ship, at a point 3.15 km from one end of the island, and 7.25 km from the other end, the island subtends an angle of 33.9°. What is the length of the island?
28. The robot arm shown in Fig. 9.74 places packages on a conveyor belt. What is the distance x?

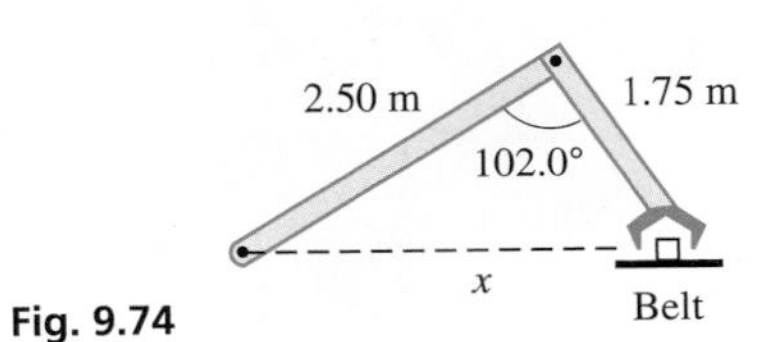

Fig. 9.74

29. Find the angle between the front legs and the back legs of the folding chair shown in Fig. 9.75.

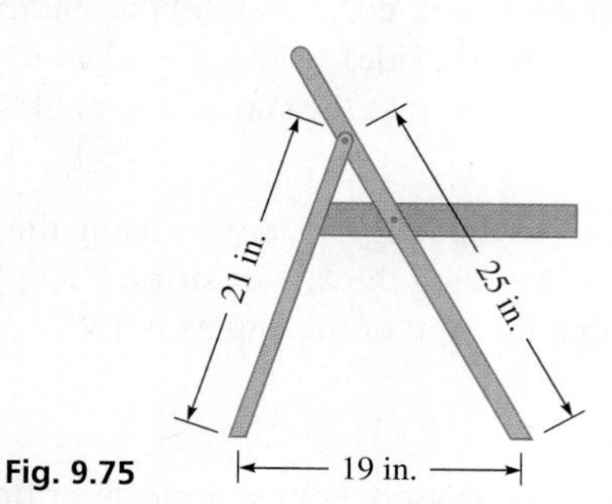

Fig. 9.75

30. In a baseball field, the four bases are at the vertices of a square 90.0 ft on a side. The pitching rubber is 60.5 ft from home plate. See Fig. 9.76. How far is it from the pitching rubber to first base?

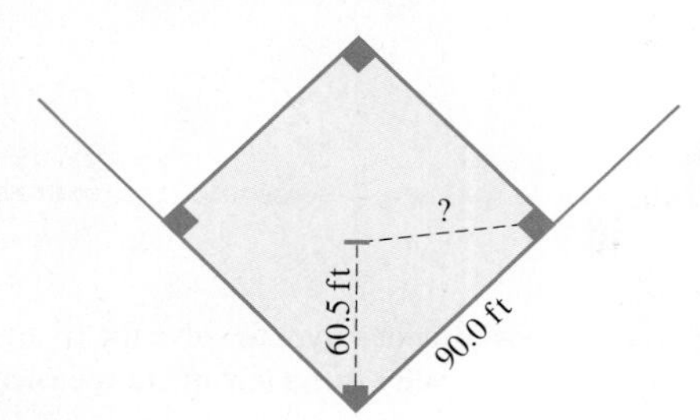

Fig. 9.76

31. A plane leaves an airport and travels 624 km due east. It then turns toward the north and travels another 326 km. It then turns again less than 180° and travels another 846 km directly back to the airport. Through what angles did it turn?

32. The apparent depth of an object submerged in water is less than its actual depth. A coin is actually 5.00 in. from an observer's eye just above the surface, but it appears to be only 4.25 in. The real light ray from the coin makes an angle with the surface that is 8.1° greater than the angle the apparent ray makes. How much deeper is the coin than it appears to be? See Fig. 9.77.

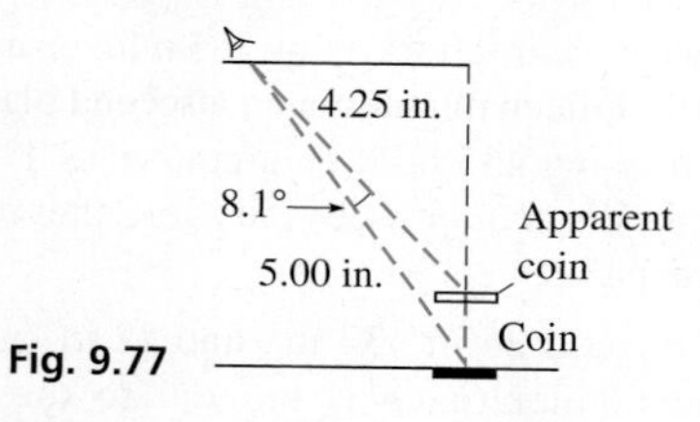

Fig. 9.77

33. A nut is in the shape of a regular hexagon (six sides). If each side is 9.53 mm, what opening on a wrench is necessary to tighten the nut? See Fig. 9.78.

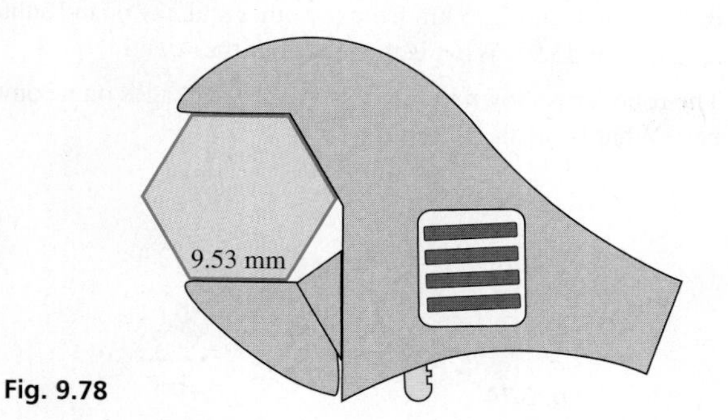

Fig. 9.78

34. The GPS of a plane over Lake Erie indicates that the plane's position is 190 km from Detroit and 110 km from London, Ontario, which are known to be 160 km apart. What is the angle between the plane's directions to Detroit and London?

35. A ferryboat travels at 11.5 km/h with respect to the water. Because of the river current, it is traveling at 12.7 km/h with respect to the land in the direction of its destination. If the ferryboat's heading is 23.6° from the direction of its destination, what is the velocity of the current?

36. Two persons are talking to each other on cell phones. The angle between their signals at the tower is 132°, as shown in Fig. 9.79. How far apart are they?

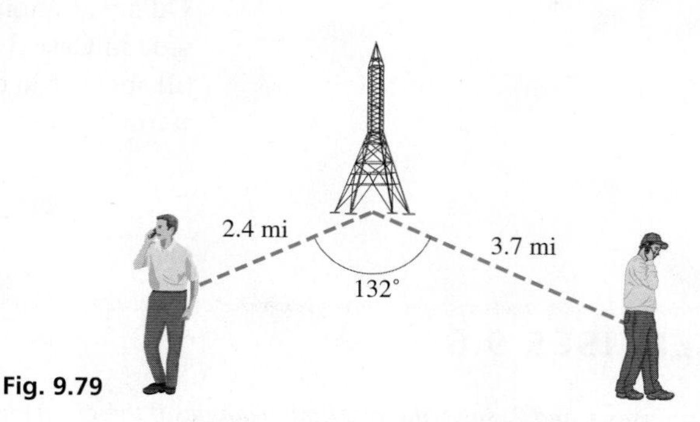

Fig. 9.79

37. An air traffic controller sights two planes that are due east from the control tower and headed toward each other. One is 15.8 mi from the tower at an angle of elevation of 26.4°, and the other is 32.7 mi from the tower at an angle of elevation of 12.4°. How far apart are the planes?

38. A ship's captain notes that a second ship is 14.5 km away at a bearing measured clockwise from true north of 46.3°, and that a third ship was at a distance of 21.7 km at a bearing of 201.0°. How far apart are the second and third ships?

39. Google's self-driving car uses a laser and detects a pedestrian a distance of 15.8 m away. A split second later (assume the car has not moved), the laser rotates by 56.3° and detects a street light pole 12.1 m away. How far is the pedestrian from the street light pole?

40. A triangular machine part has sides of 5 cm and 8 cm. Explain why the law of sines, or the law of cosines, is used to start the solution of the triangle if the third known part is (a) the third side, (b) the angle opposite the 5-cm side, or (c) the angle between the 5-cm and 8-cm sides.

Answers to Practice Exercises

1. $B = 43.8°, A = 101.2°$ **2.** 57.7°

CHAPTER 9 KEY FORMULAS AND EQUATIONS

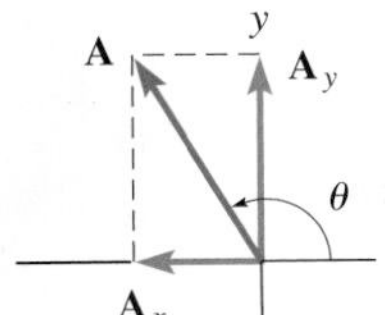

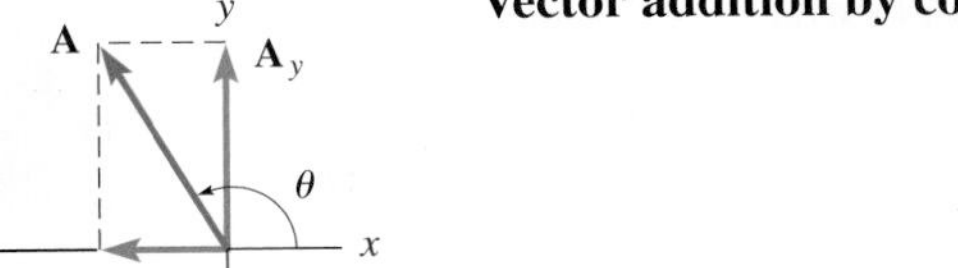

Vector addition by components $A_x = A\cos\theta \qquad A_y = A\sin\theta$ (9.1)

$$R = \sqrt{R_x^2 + R_y^2} \quad (9.2)$$

$$\theta_{ref} = \tan^{-1}\frac{|R_y|}{|R_x|} \quad (9.3)$$

Law of sines

$$\frac{a}{\sin A} = \frac{b}{\sin B} = \frac{c}{\sin C} \quad (9.8)$$

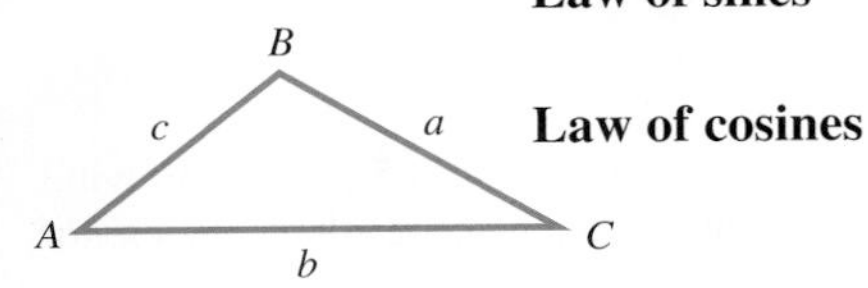

Law of cosines

$$a^2 = b^2 + c^2 - 2bc\cos A$$
$$b^2 = a^2 + c^2 - 2ac\cos B \quad (9.13)$$
$$c^2 = a^2 + b^2 - 2ab\cos C$$

CHAPTER 9 REVIEW EXERCISES

CONCEPT CHECK EXERCISES

Determine each of the following as being either ***true*** or ***false***.

1. To add two vectors by use of a diagram, place the tail of the second at the head of the first, keeping magnitudes and directions. The sum is the vector from the tail of the first to the head of the second.
2. For vector **A**, in standard position at angle θ, of magnitude A, the magnitude of the x-component is $A_x = A\sin\theta$.
3. In adding vectors, the answer is the magnitude of the resultant.
4. The Law of Sines is used in solving a triangle for which two sides and the included angle are known.
5. The Law of Cosines is used in solving a triangle for which two sides and the angle opposite one of them are known.
6. The ambiguous case occurs when two sides of a triangle and one of their opposite angles are given.

PRACTICE AND APPLICATIONS

In Exercises 7–10, find the x- and y-components of the given vectors by use of the trigonometric functions.

7. $A = 72.0,\ \theta_A = 24.0°$
8. $A = 8050,\ \theta_A = 149.0°$
9. $A = 5.716,\ \theta_A = 215.59°$
10. $A = 657.1,\ \theta_A = 343.74°$

In Exercises 11–14, vectors **A** *and* **B** *are at right angles. Find the magnitude and direction (from* **A**) *of the resultant.*

11. $A = 327$, $B = 505$
12. $A = 6.8$, $B = 2.9$
13. $A = 5296$, $B = 3298$
14. $A = 26.52$, $B = 89.86$

In Exercises 15–22, add the given vectors by using components.

15. $A = 780,\ \theta_A = 28.0°$; $B = 346,\ \theta_B = 320.0°$
16. $J = 0.0120,\ \theta_J = 370.5°$; $K = 0.00781,\ \theta_K = 260.0°$
17. $A = 22.51,\ \theta_A = 130.16°$; $B = 7.604,\ \theta_B = 200.09°$
18. $A = 18{,}760,\ \theta_A = 110.43°$; $B = 4835,\ \theta_B = 350.20°$
19. $Y = 51.33,\ \theta_Y = 12.25°$; $Z = 42.61,\ \theta_Z = 291.77°$
20. $A = 7.031,\ \theta_A = 122.54°$; $B = 3.029,\ \theta_B = 214.82°$
21. $A = 0.750,\ \theta_A = 15.0°$; $B = 0.265,\ \theta_B = 192.4°$; $C = 0.548,\ \theta_C = 344.7°$
22. $S = 8120,\ \theta_S = 141.9°$; $T = 1540,\ \theta_T = 165.2°$; $U = 3470,\ \theta_U = 296.0°$

In Exercises 23–40, solve the triangles with the given parts.

23. $A = 48.0°,\ B = 68.0°,\ a = 145$
24. $A = 132.0°,\ b = 0.750,\ C = 32.0°$
25. $a = 22.8,\ B = 33.5°,\ C = 125.3°$
26. $A = 71.0°,\ B = 48.5°,\ c = 8.42$
27. $b = 7607,\ c = 4053,\ B = 110.09°$
28. $A = 77.06°,\ a = 12.07,\ c = 5.104$
29. $b = 14.5,\ c = 13.0,\ C = 56.6°$
30. $B = 40.6°,\ b = 7.00,\ c = 18.0$
31. $a = 186,\ B = 130.0°,\ c = 106$
32. $b = 750,\ c = 1100,\ A = 56°$
33. $a = 7.86,\ b = 2.45,\ C = 2.5°$
34. $a = 0.208,\ c = 0.697,\ B = 165.4°$
35. $A = 67.16°,\ B = 96.84°,\ c = 532.9$
36. $A = 43.12°,\ a = 7.893,\ b = 4.113$
37. $a = 17,\ b = 12,\ c = 25$
38. $a = 4114,\ b = 9110,\ c = 5016$
39. $a = 0.530,\ b = 0.875,\ c = 1.25$
40. $a = 47.4,\ b = 40.0,\ c = 45.5$

In Exercises 41–74, solve the given problems.

41. Three straight streets intersect each other such that two of the angles of intersection are 22° and 112°, and the shortest distances between any two intersections is 540 m. What is the longest distance between any two intersections?

42. A triangular brace is designed such that sides meet at angles of 42.0° and 59.5°, with the longest side being 5.00 cm longer than the shortest side. What is the perimeter of the brace?

43. An architect determines the two acute angles and one of the legs of a right triangular wall panel. Show that the area A_t is

$$A_t = \frac{a^2 \sin B}{2 \sin A}$$

44. If $A = 30°$ and $b = 18$, what values of a will result in the ambiguous case with two triangle solutions? Explain.

45. Find the horizontal and vertical components of the force shown in Fig. 9.80.

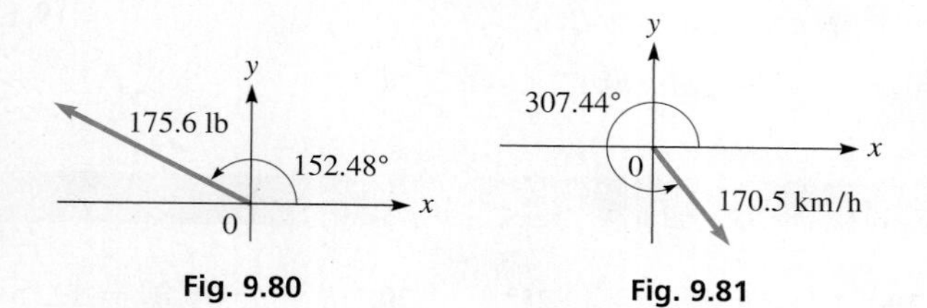

Fig. 9.80 Fig. 9.81

46. Find the horizontal and vertical components of the velocity shown in Fig. 9.81.

47. In a ballistics test, a bullet was fired into a block of wood with a velocity of 2200 ft/s and at an angle of 71.3° with the surface of the block. What was the component of the velocity perpendicular to the surface?

48. A storm cloud is moving at 15 mi/h from the northwest. A television tower is 60° south of east of the cloud. What is the component of the cloud's velocity toward the tower?

49. Assuming the vectors in Fig. 9.82 are in equilibrium, find T_1 and T_2.

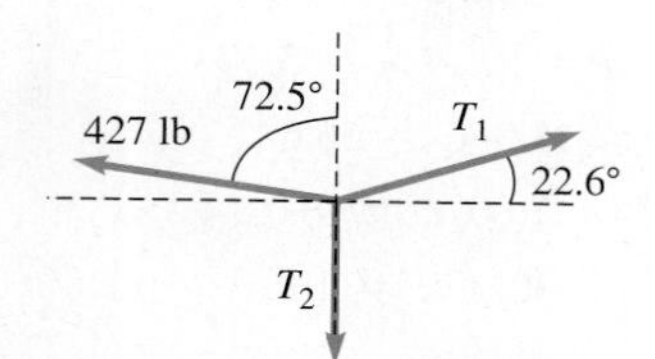

Fig. 9.82

50. A rocket is launched at an angle of 42.0° with the horizontal and with a speed of 2500 ft/s. What are its horizontal and vertical components of velocity?

51. Three forces of 3200 lb, 1300 lb, and 2100 lb act on a bolt as shown in Fig. 9.83. Find the resultant force.

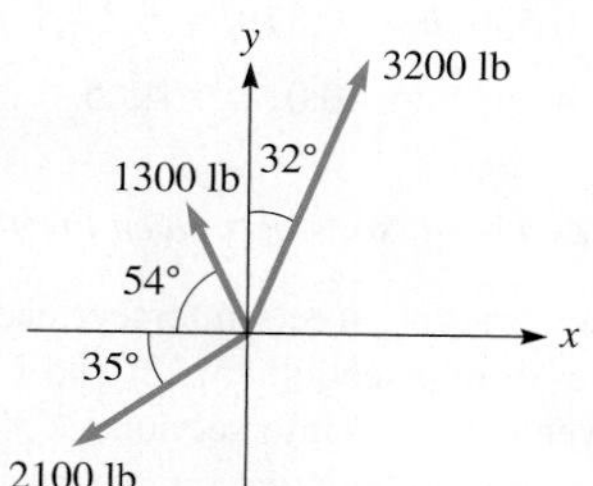

Fig. 9.83

52. In Fig. 9.84, **F** represents the total surface tension force around the circumference on the liquid in the capillary tube. The vertical component of **F** holds up the liquid in the tube above the liquid surface outside the tube. What is the vertical component of **F**?

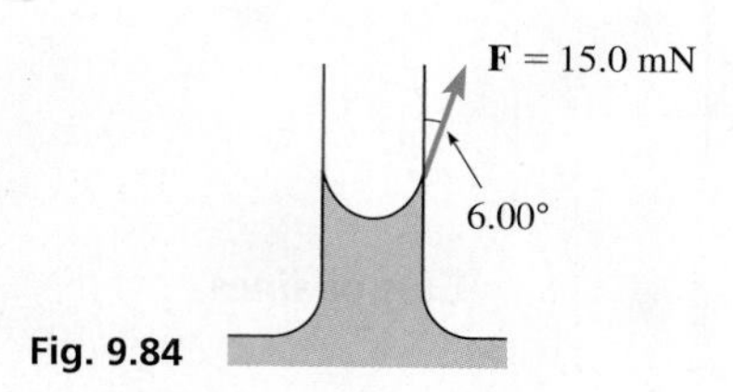

Fig. 9.84

53. A helium-filled balloon rises vertically at 3.5 m/s as the wind carries it horizontally at 5.0 m/s. What is the resultant velocity of the balloon?

54. A shearing pin is designed to break and disengage gears before damage is done to a machine. In a test, a vertically upward force of 8250 lb and a 7520 lb at 35.0° below the horizontal are applied to a shearing pin. What is the resultant force?

55. A magnetic force of 0.15 N is applied at an angle of 22.5° above the horizontal on an iron bar. A second magnetic force of 0.20 N is applied from the opposite side at an angle of 15.0° above the horizontal. What is the upward force on the bar?

56. A boat is tied to a dock at two points using ropes as shown in Fig. 9.85. The boat exerts a force of 325 N at an angle of 25° from the line perpendicular to the dock. Make a free-body diagram, and then find the tensions T_p and T_s in the ropes perpendicular to and slanted from the dock, respectively.

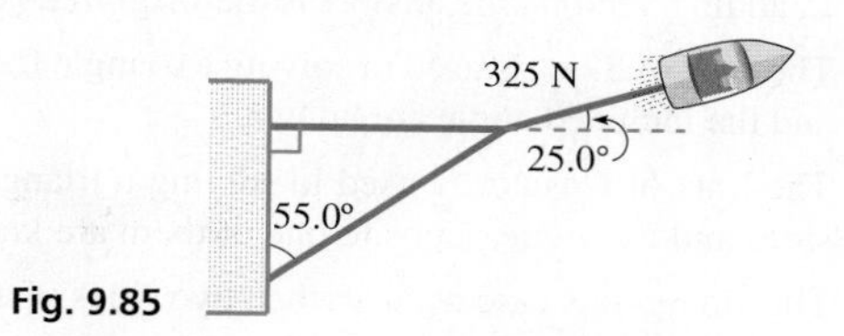

Fig. 9.85

57. A water molecule (H_2O) consists of two hydrogen atoms and one oxygen atom. The distance from the nucleus of each hydrogen atom to the nucleus of the oxygen atom is 0.96 pm (see Section 1.4) and the bond angle (see Fig. 9.86) is 105°. How far is one hydrogen nucleus from the other?

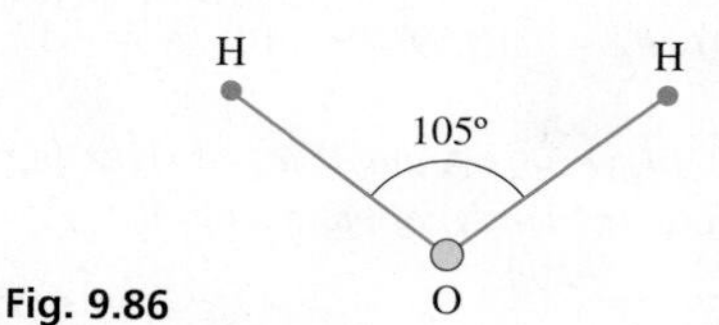

Fig. 9.86

58. A crate weighing 562 lb hangs from two ropes which are angled 41.5° and 37.2° from the vertical as shown in Fig. 9.87. Make a free-body diagram, and then find the tensions T_L and T_R in the left and right ropes, respectively.

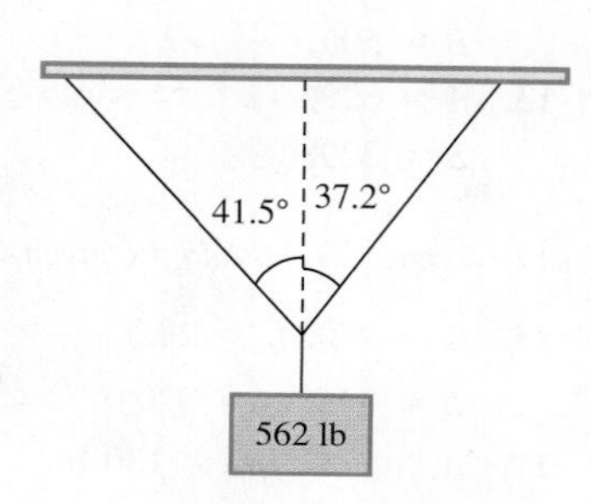

Fig. 9.87

59. In Fig. 9.88, a damper mechanism in an air-conditioning system is shown. If $\theta = 27.5°$ when the spring is at its shortest and longest lengths, what are these lengths?

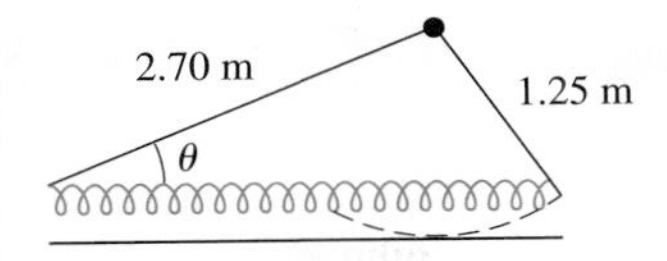

Fig. 9.88

60. A bullet is fired from the ground of a level field at an angle of 39.0° above the horizontal. It travels in a straight line at 2200 ft/s for 0.20 s when it strikes a target. The sound of the strike is recorded 0.32 s later on the ground. If sound travels at 1130 ft/s, where is the recording device located?

61. In order to get around an obstruction, an oil pipeline is constructed in two straight sections, one 3.756 km long and the other 4.675 km long, with an angle of 168.85° between the sections where they are joined. How much more pipeline was necessary due to the obstruction?

62. Three pipes of radii 2.50 cm, 3.25 cm, and 4.25 cm are welded together lengthwise. See Fig. 9.89. Find the angles between the center-to-center lines.

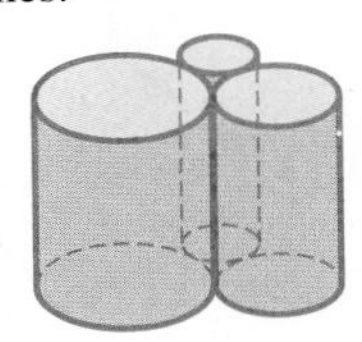
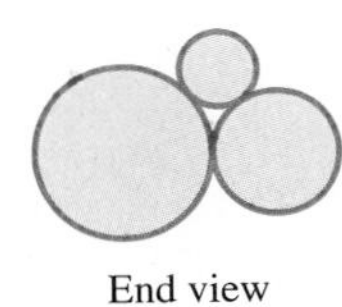

Fig. 9.89

63. Two satellites are being observed at the same observing station. One is 22,500 mi from the station, and the other is 18,700 mi away. The angle between their lines of observation is 105.4°. How far apart are the satellites?

64. Find the side x in the truss in Fig. 9.90.

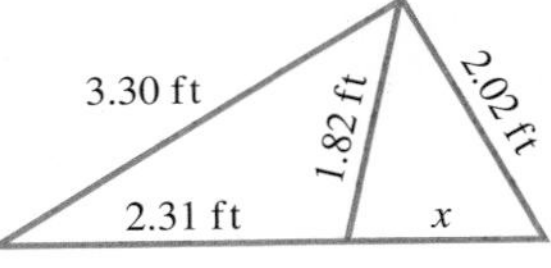

Fig. 9.90

65. A water skier is towed 65 m to the south of the starting point, and then is turned 40° east of south for another 35 m. What is the skier's displacement from the starting point?

66. A 116-cm wide TV screen is viewed at an angle such that the near end of the screen is 224 cm from the viewer, and the far end is 302 cm distant. What angle does the width of the screen subtend at the viewer's eye? See Fig. 9.91.

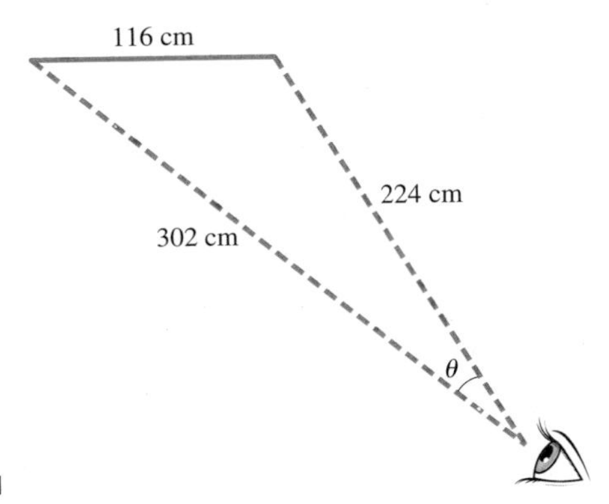

Fig. 9.91

67. The angle of depression of a fire noticed west of a fire tower is 6.2°. The angle of depression of a pond, also west of the tower, is 13.5°. If the fire and pond are at the same altitude, and the tower is 2.25 km from the pond on a direct line, how far is the fire from the pond?

68. A surveyor wishes to find the distance between two points between which there is a security-restricted area. The surveyor measures the distance from each of these points to a third point and finds them to be 226.73 m and 185.12 m. If the angle between the lines of sight from the third point to the other points is 126.724°, how far apart are the two points?

69. Atlanta is 290 mi and 51.0° south of east from Nashville. The pilot of an airplane due north of Atlanta radios Nashville and finds the plane is on a line 10.5° south of east from Nashville. How far is the plane from Nashville?

70. In going around a storm, a plane flies 125 mi south, then 140 mi at 30.0° south of west, and finally 225 mi at 15.0° north of west. What is the displacement of the plane from its original position?

71. One end of a 1450-ft bridge is sighted from a distance of 3250 ft. The angle between the lines of sight of the ends of the bridge is 25.2°. From these data, how far is the observer from the other end of the bridge?

72. A plane is traveling horizontally at 1250 ft/s. A missile is fired horizontally from it 30.0° from the direction in which the plane is traveling. If the missile leaves the plane at 2040 ft/s, what is its velocity 10.0 s later if the vertical component is given by $v_V = -32.0t$ (in ft/s)?

73. A sailboat is headed due north, and its sail is set perpendicular to the wind, which is from the south of west. The component of the force of the wind in the direction of the heading is 480 N, and the component perpendicular to the heading (the *drift* component) is 650 N. What is the force exerted by the wind, and what is the direction of the wind? See Fig. 9.92.

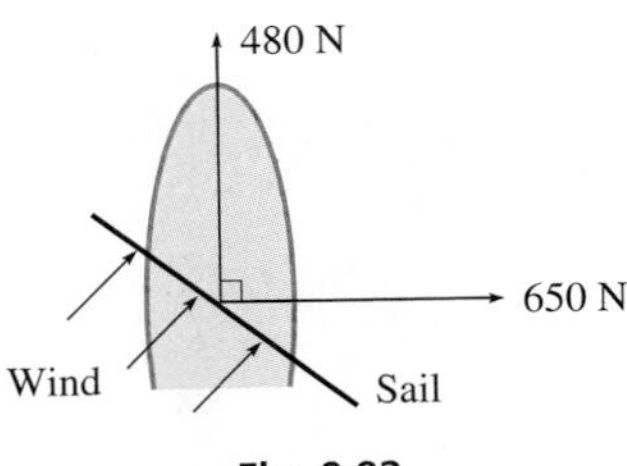

Fig. 9.92

74. Boston is 650 km and 21.0° south of west from Halifax, Nova Scotia. Radio signals locate a ship 10.5° east of south from Halifax and 5.6° north of east from Boston. How far is the ship from each city?

75. The resultant of three horizontal forces, 45 N, 35 N, and 25 N, that act on a bolt is zero. Write a paragraph or two explaining how to find the angles between the forces.

CHAPTER 9 PRACTICE TEST

As a study aid, we have included complete solutions for each Practice Test problem at the end of this book.

In all triangle solutions, sides a, b, c, are opposite angles A, B, C, respectively.

1. By use of a diagram, find the vector sum $2\mathbf{A} + \mathbf{B}$ for the given vectors.

2. For the triangle in which $a = 22.5$, $B = 78.6°$, and $c = 30.9$, find b.
3. A surveyor locates a tree 36.50 m to the northeast of a set position. The tree is 21.38 m north of a utility pole. What is the displacement of the utility pole from the set position?
4. For the triangle in which $A = 18.9°$, $B = 104.2°$, and $a = 426$, find c.
5. Solve the triangle in which $a = 9.84$, $b = 3.29$, and $c = 8.44$.
6. For vector **R**, find R and standard position θ if $R_x = -235$ and $R_y = 152$.
7. Find the horizontal and vertical components of a force of 871 kN that is directed at a standard-position angle of 284.3°.
8. A ship leaves a port and travels due west. At a certain point it turns 31.5° north of west and travels an additional 42.0 mi to a point 63.0 mi on a direct line from the port. How far from the port is the point where the ship turned?
9. Find the sum of the vectors for which $A = 449$, $\theta_A = 74.2°$, $B = 285$, and $\theta_B = 208.9°$ by using components.
10. Solve the triangle for which $a = 22.3$, $b = 29.6$, and $A = 36.5°$.
11. Assuming the vectors in Fig. 9.93 are in equilibrium, find T_1 and T_2.

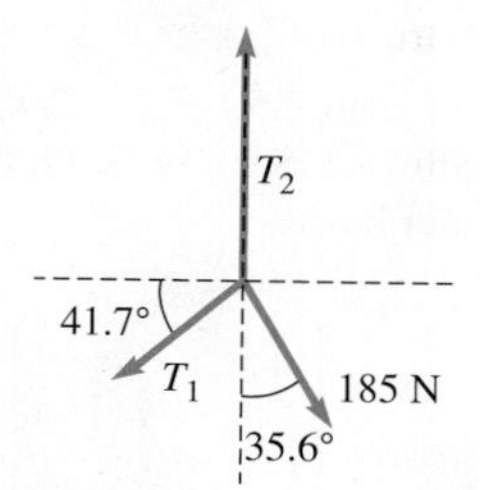

Fig. 9.93

Graphs of the Trigonometric Functions

LEARNING OUTCOMES

After completion of this chapter, the student should be able to:

- Graph the functions $y = \sin x$ and $y = \cos x$
- Determine amplitude, period, and displacement
- Graph the functions $y = a \sin x$ and $y = a \cos x$
- Graph the functions $y = a \sin bx$ and $y = a \cos bx$
- Graph the functions $y = a \sin(bx + c)$ and $y = a \cos(bx + c)$
- Graph the functions $y = \tan x$, $y = \cot x$, $y = \sec x$, and $y = \csc x$
- Solve application problems involving graphs of trigonometric functions
- Graph composite trigonometric curves by addition of ordinates
- Graph Lissajous figures

The electronics era is thought by many to have started in the 1880s with the discovery of the vacuum tube by the American inventor Thomas Edison and the discovery of radio waves by the German physicist Heinrich Hertz.

Then in the 1890s, the cathode-ray tube was developed and, as an oscilloscope, has been used since that time to analyze various types of wave forms, such as sound waves and radio waves. Since the mid-1900s, devices similar to a cathode-ray tube have been used in TV picture tubes and computer displays.

What is seen on the screen of an oscilloscope are electric signals that are represented by graphs of trigonometric functions. As noted earlier, the basic method of graphing was developed in the mid-1600s, and using trigonometric functions of numbers has been common since the mid-1700s. Therefore, the graphs of the trigonometric functions were well known in the late 1800s and became very useful in the development of electronics.

The graphs of the trigonometric functions are useful in many areas of application, particularly those that involve wave motion and periodic values. Filtering electronic signals in communications, mixing musical sounds in a recording studio, studying the seasonal temperatures of an area, and analyzing ocean waves and tides illustrate some of the many applications of periodic motion.

As well as their use in applications, the graphs of the trigonometric functions give us one of the clearest ways of showing the properties of the various functions. Therefore, in this chapter, we show graphs of these functions, with emphasis on the sine and cosine functions.

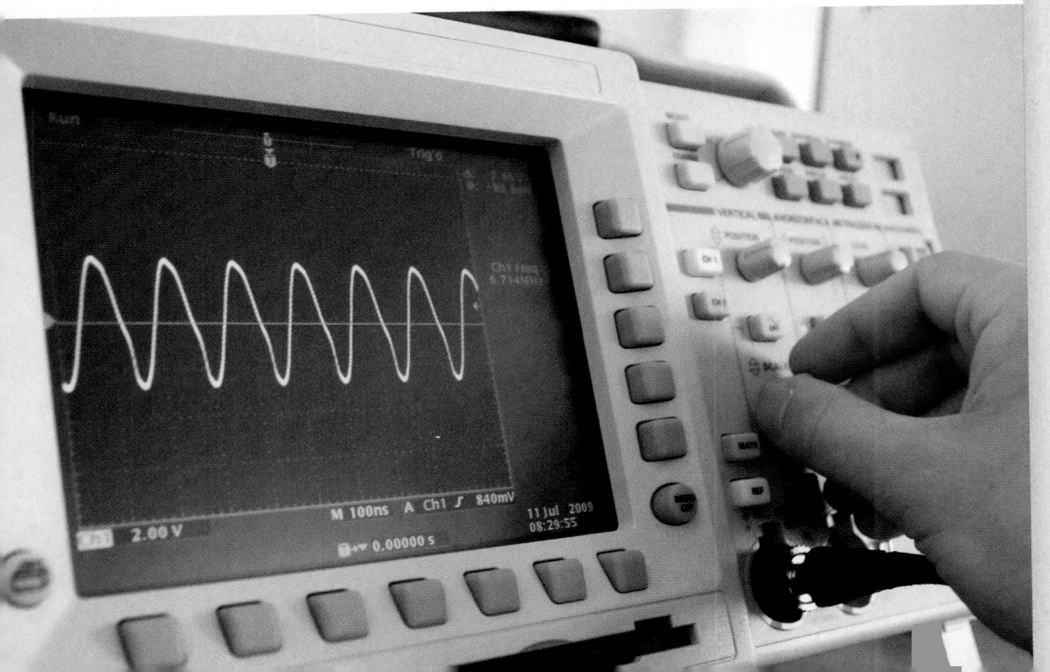

◀ **In Section 10.6, we show the resulting curve when an oscilloscope is used to combine and display electric signals.**

10.1 Graphs of $y = a \sin x$ and $y = a \cos x$

Graphs of $y = \sin x$ and $y = a \cos x$ • Amplitude • Graphs of $y = a \sin x$ and $y = a \cos x$

When plotting and sketching the graphs of the trigonometric functions, it is normal to *express the angle in radians.* By using radians, both the independent variable and the dependent variable are real numbers. NOTE ▶ [Therefore, it is necessary to be able to readily use angles expressed in radians.] If necessary, review Section 8.3 on radian measure of angles.

In this section, the graphs of the sine and cosine functions are shown. We begin by making a table of values of x and y for the function $y = \sin x$, where x and y are used in the standard way as the *independent variable* and *dependent variable.* We plot the points to obtain the graph in Fig. 10.1.

x	0	$\frac{\pi}{6}$	$\frac{\pi}{3}$	$\frac{\pi}{2}$	$\frac{2\pi}{3}$	$\frac{5\pi}{6}$	π
y	0	0.5	0.87	1	0.87	0.5	0

x	$\frac{7\pi}{6}$	$\frac{4\pi}{3}$	$\frac{3\pi}{2}$	$\frac{5\pi}{3}$	$\frac{11\pi}{6}$	2π
y	−0.5	−0.87	−1	−0.87	−0.5	0

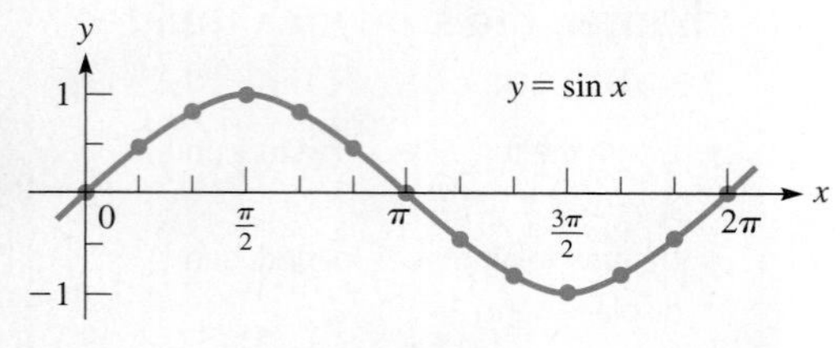

Fig. 10.1

The graph of $y = \cos x$ may be drawn in the same way. The next table gives values for plotting the graph of $y = \cos x$, and the graph is shown in Fig. 10.2.

x	0	$\frac{\pi}{6}$	$\frac{\pi}{3}$	$\frac{\pi}{2}$	$\frac{2\pi}{3}$	$\frac{5\pi}{6}$	π
y	1	0.87	0.5	0	−0.5	−0.87	−1

x	$\frac{7\pi}{6}$	$\frac{4\pi}{3}$	$\frac{3\pi}{2}$	$\frac{5\pi}{3}$	$\frac{11\pi}{6}$	2π
y	−0.87	−0.5	0	0.5	0.87	1

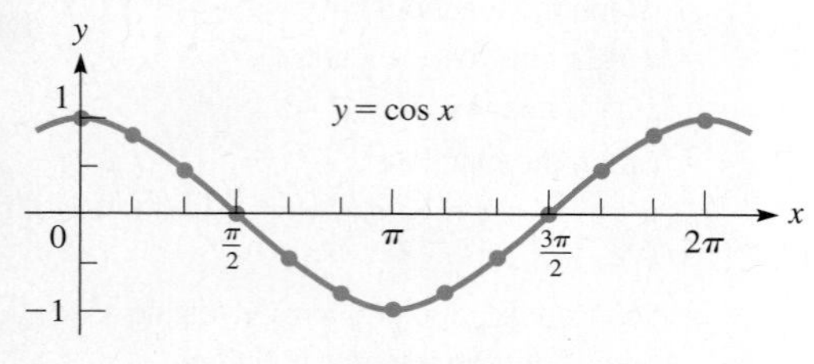

Fig. 10.2

The graphs are continued beyond the values shown in the tables to indicate that *they continue indefinitely in each direction.* To show this more clearly, in Figs. 10.3 and 10.4, note the graphs of $y = \sin x$ and $y = \cos x$ from $x = -10$ to $x = 10$.

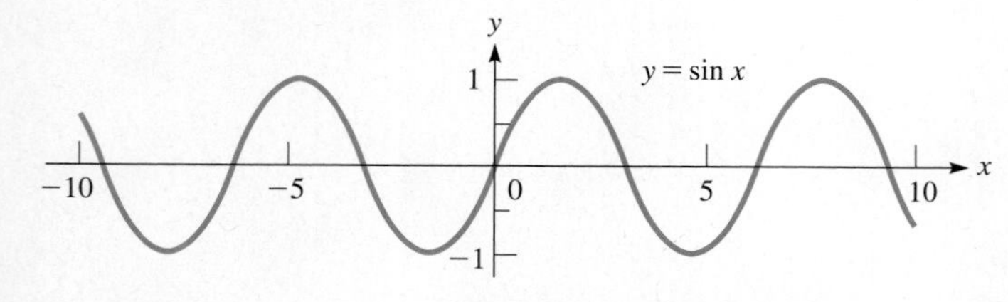

Fig. 10.3

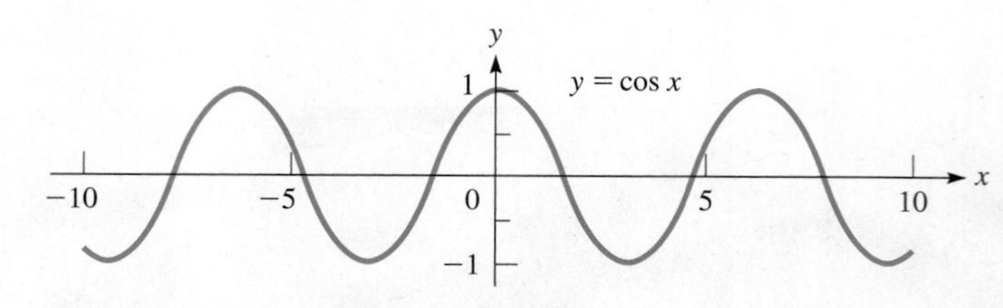

Fig. 10.4

From these tables and graphs, it can be seen that *the graphs of $y = \sin x$ and $y = \cos x$ are of exactly the same shape (called* **sinusoidal**), *with the cosine curve displacement $\pi/2$ units to the left of the sine curve.* The shape of these curves should be recognized readily, with special note as to the points at which they cross the axes. This information will be especially valuable in *sketching* similar curves, because the basic sinusoidal shape remains the same.

NOTE ▶ [It will not be necessary to plot numerous points every time we wish to sketch such a curve. A few key points will be enough.]

AMPLITUDE

To obtain the graph of $y = a \sin x$, note that all the y-values obtained for the graph of $y = \sin x$ are to be multiplied by the number a. In this case, the greatest value of the sine function is $|a|$. *The number* $|a|$ *is called the* **amplitude** *of the curve and represents the greatest y-value of the curve.* Also, the curve will have no value less than $-|a|$. This is true for $y = a \cos x$ as well as $y = a \sin x$.

EXAMPLE 1 Plot graph of $y = a \sin x$

Plot the graph of $y = 2 \sin x$.

Because $a = 2$, the amplitude of this curve is $|2| = 2$. This means that the maximum value of y is 2 and the minimum value is $y = -2$. The table of values follows, and the curve is shown in Fig. 10.5.

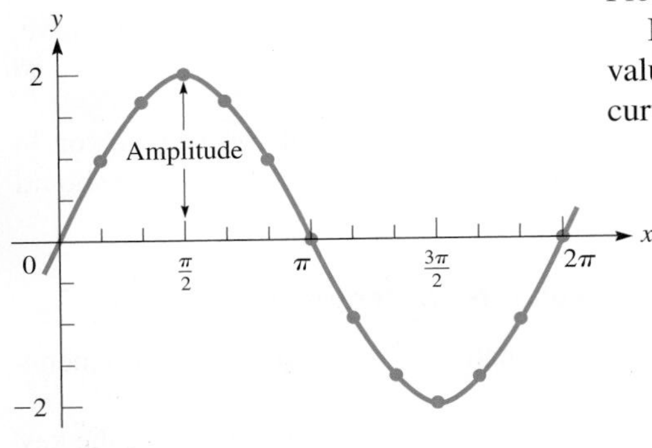

Fig. 10.5

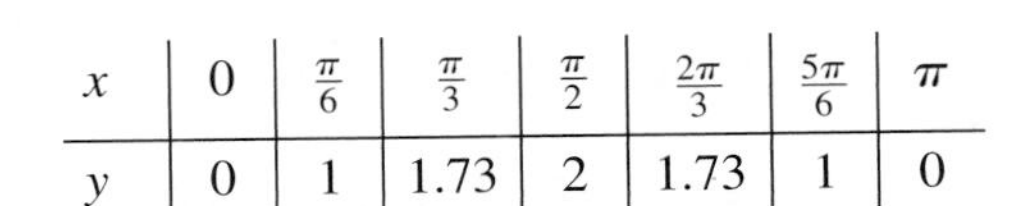

x	0	$\frac{\pi}{6}$	$\frac{\pi}{3}$	$\frac{\pi}{2}$	$\frac{2\pi}{3}$	$\frac{5\pi}{6}$	π
y	0	1	1.73	2	1.73	1	0

x	$\frac{7\pi}{6}$	$\frac{4\pi}{3}$	$\frac{3\pi}{2}$	$\frac{5\pi}{3}$	$\frac{11\pi}{6}$	2π
y	−1	−1.73	−2	−1.73	−1	0

■

EXAMPLE 2 Plot graph of $y = a \cos x$

Plot the graph of $y = -3 \cos x$.

In this case, $a = -3$, and this means that the amplitude is $|-3| = 3$. Therefore, the maximum value of y is 3, and the minimum value of y is -3. The table of values follows, and the curve is shown in Fig. 10.6.

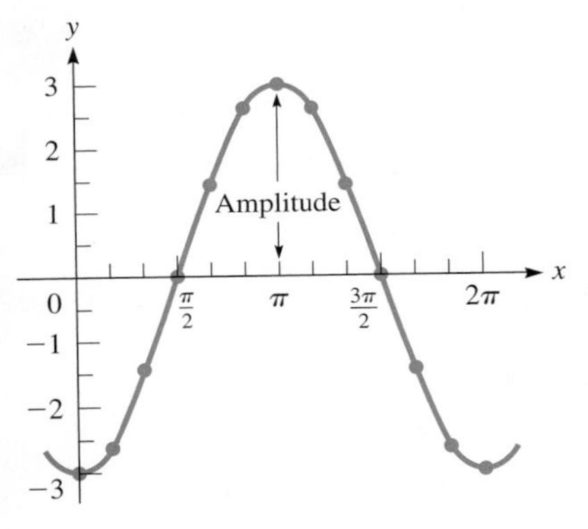

Fig. 10.6

x	0	$\frac{\pi}{6}$	$\frac{\pi}{3}$	$\frac{\pi}{2}$	$\frac{2\pi}{3}$	$\frac{5\pi}{6}$	π
y	−3	−2.6	−1.5	0	1.5	2.6	3

x	$\frac{7\pi}{6}$	$\frac{4\pi}{3}$	$\frac{3\pi}{2}$	$\frac{5\pi}{3}$	$\frac{11\pi}{6}$	2π
y	2.6	1.5	0	−1.5	−2.6	−3

NOTE ▶ [Note that the effect of the negative sign with the number a is to *invert* the curve about the x-axis.] ■

From the previous examples, note that the function $y = a \sin x$ has zeros for $x = 0, \pi, 2\pi$ and that it has its maximum or minimum values for $x = \pi/2, 3\pi/2$. The function $y = a \cos x$ has its zeros for $x = \pi/2, 3\pi/2$ and its maximum or minimum values for $x = 0, \pi, 2\pi$. This is summarized in Table 10.1. Therefore, by knowing the general shape of the sine curve, where it has its zeros, and what its amplitude is, *we can rapidly* ***sketch*** *curves of the form* $y = a \sin x$ *and* $y = a \cos x$.

Table 10.1

	$x = 0, \pi, 2\pi$	$\frac{\pi}{2}, \frac{3\pi}{2}$
$y = a \sin x$	zeros	max. or min.
$y = a \cos x$	max. or min.	zeros

Because the graphs of $y = a \sin x$ and $y = a \cos x$ can extend indefinitely to the right and to the left, we see that the domain of each is all real numbers. We should note that the key values of $x = 0, \pi/2, \pi, 3\pi/2$, and 2π are those only for x from 0 to 2π. Corresponding values ($x = 5\pi/2, 3\pi$, and their negatives) could also be used. Also, from the graphs, we can readily see that the range of these functions is $-|a| \le f(x) \le |a|$.

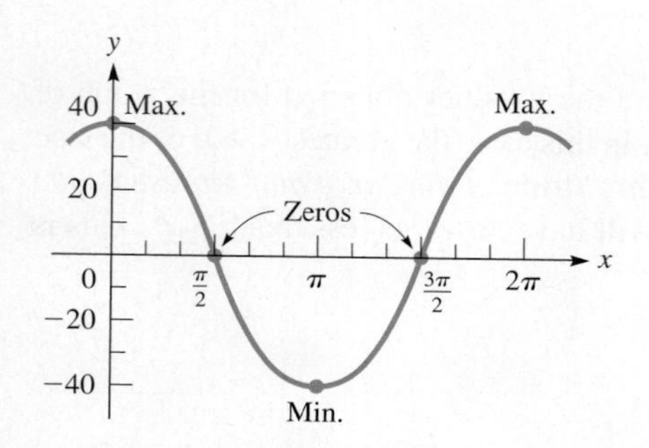

Fig. 10.7

Practice Exercise

1. For the graph of $y = -6 \sin x$, set up a table of key values for $0 \leq x \leq 2\pi$.

EXAMPLE 3 Using key values to sketch graph

Sketch the graph of $y = 40 \cos x$.

First, we set up a table of values for the points where the curve has its zeros, maximum points, and minimum points:

x	0	$\frac{\pi}{2}$	π	$\frac{3\pi}{2}$	2π
y	40 max.	0	−40 min.	0	40 max.

Now, we plot these points and join them, knowing the basic sinusoidal shape of the curve. See Fig. 10.7. ■

The graphs of $y = a \sin x$ and $y = a \cos x$ can be displayed easily on a calculator. In the next example, we see that the calculator displays the features of the curve we should expect from our previous discussion.

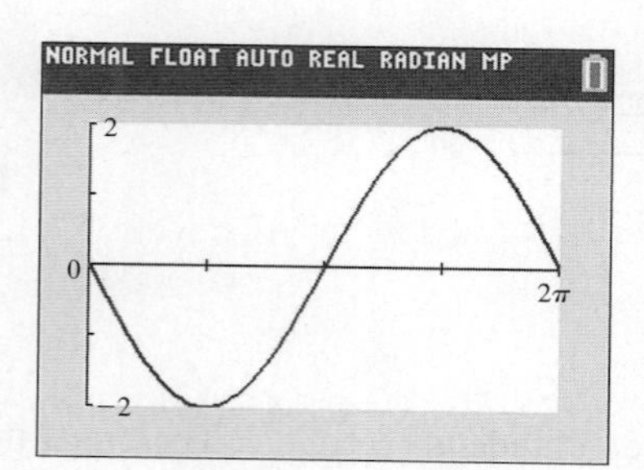

Fig. 10.8

EXAMPLE 4 Displaying graphs on calculator—water wave

A certain water wave can be represented by the equation $y = -2 \sin x$ (measurements in ft). Display the graph on a calculator.

Using radian mode on the calculator, Fig. 10.8 shows the calculator view for the key values in the following table:

x	0	$\frac{\pi}{2}$	π	$\frac{3\pi}{2}$	2π
y	0	−2 min.	0	2 max.	0

We see the amplitude of 2 and the effect of the negative sign in inverting the curve. ■

EXERCISES 10.1

In Exercises 1 and 2, graph the function if the given changes are made in the indicated examples of this section.

1. In Example 2, if the sign of the coefficient of $\cos x$ is changed, plot the graph of the resulting function.
2. In Example 4, if the sign of the coefficient of $\sin x$ is changed, display the graph of the resulting function.

In Exercises 3–6, complete the following table for the given functions and then plot the resulting graphs.

x	$-\pi$	$-\frac{3\pi}{4}$	$-\frac{\pi}{2}$	$-\frac{\pi}{4}$	0	$\frac{\pi}{4}$	$\frac{\pi}{2}$	$\frac{3\pi}{4}$	π
y									

x	$\frac{5\pi}{4}$	$\frac{3\pi}{2}$	$\frac{7\pi}{4}$	2π	$\frac{9\pi}{4}$	$\frac{5\pi}{2}$	$\frac{11\pi}{4}$	3π
y								

3. $y = \sin x$
4. $y = \cos x$
5. $y = -3 \cos x$
6. $y = -4 \sin x$

In Exercises 7–22, give the amplitude and sketch the graphs of the given functions. Check each using a calculator.

7. $y = 3 \sin x$
8. $y = 15 \sin x$
9. $y = \frac{5}{2} \sin x$
10. $y = 35 \sin x$
11. $y = 200 \cos x$
12. $y = 0.25 \cos x$
13. $y = 0.8 \cos x$
14. $y = \frac{3}{2} \cos x$
15. $y = -\sin x$
16. $y = -30 \sin x$
17. $y = -1500 \sin x$
18. $y = -0.2 \sin x$
19. $y = -4 \cos x$
20. $y = -8 \cos x$
21. $y = -50 \cos x$
22. $y = -0.4 \cos x$

Although units of π are convenient, we must remember that π is only a number. Numbers that are not multiples of π may be used. In Exercises 23–26, plot the indicated graphs by finding the values of y that correspond to values of x of 0, 1, 2, 3, 4, 5, 6, and 7 on a calculator. (Remember, the numbers 0, 1, 2, and so on represent radian measure.)

23. $y = \sin x$
24. $y = -30 \sin x$
25. $y = -12 \cos x$
26. $y = 2 \cos x$

In Exercises 27–32, solve the given problems.

27. Find the function and graph it for a function of the form $y = a \sin x$ that passes through $(\pi/2, -2)$.
28. Find the function and graph it for a function of the form $y = a \sin x$ that passes through $(3\pi/2, -2)$.

29. The height (in ft) of a person on a ferris wheel above its center is given by $h = -32 \cos t$ where t is the time (in s). Graph one complete cycle of this function.

30. The horizontal displacement d (in m) of the bob on a large pendulum is $d = 5 \sin t$, where t is the time (in s). Graph two cycles of this function.

31. The graph displayed on an oscilloscope can be represented by $y = -0.05 \sin x$. Display this curve on a calculator.

32. The displacement y (in cm) of the end of a robot arm for welding is $y = 4.75 \cos t$, where t is the time (in s). Display this curve on a calculator.

In Exercises 33–36, the graph of a function of the form $y = a \sin x$ or $y = a \cos x$ is shown. Determine the specific function of each.

33.

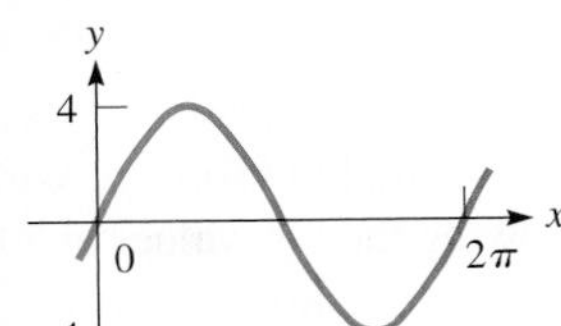

34.

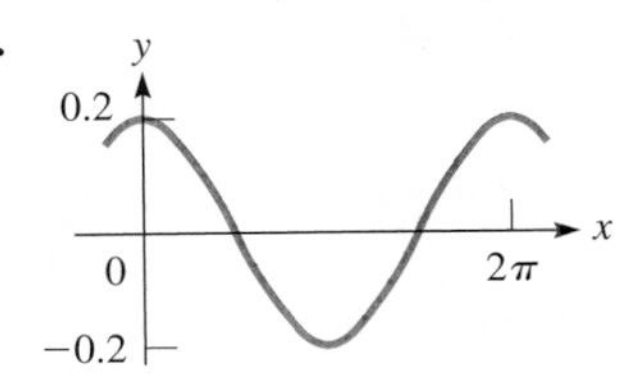

35.

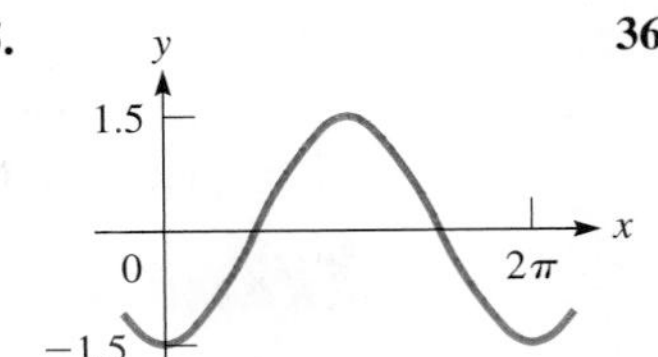

36.

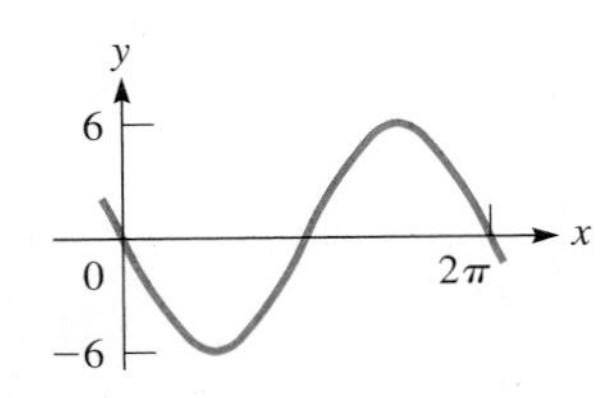

Each ordered pair given in Exercises 37–40 is located on the graph of either $y = a \sin x$ or $y = a \cos x$ where the amplitude is $|a| = 2.50$. Use the trace feature on a calculator to determine the equation of the function that contains the given point. (All points are located such that the x value is between $-\pi$ and π.)

37. $(0.67, -1.55)$

38. $(-1.20, 0.90)$

39. $(2.07, 1.20)$

40. $(-2.47, -1.55)$

Answers to Practice Exercise

1.

x	0	$\frac{\pi}{2}$	π	$\frac{3\pi}{2}$	2π
y	0	-6 min.	0	6 max.	0

10.2 Graphs of $y = a \sin bx$ and $y = a \cos bx$

Period of a Function • Graphs of $y = a \sin bx$ and $y = a \cos bx$ • Important Values for Sketching

In graphing the function $y = \sin x$, we see that the values of y repeat every 2π units of x. This is because $\sin x = \sin(x + 2\pi) = \sin(x + 4\pi)$, and so forth. For any function F, we say that it has a *period P* if $F(x) = F(x + P)$. For functions that are periodic, such as the sine and cosine, *the* **period** *is the x-distance between a point and the next corresponding point for which the value of y repeats.*

Let us now plot the curve $y = \sin 2x$. This means that we choose a value of x, multiply this value by 2, and find the sine of the result. This leads to the following table of values for this function:

x	0	$\frac{\pi}{8}$	$\frac{\pi}{4}$	$\frac{3\pi}{8}$	$\frac{\pi}{2}$	$\frac{5\pi}{8}$	$\frac{3\pi}{4}$	$\frac{7\pi}{8}$	π	$\frac{9\pi}{8}$	$\frac{5\pi}{4}$
$2x$	0	$\frac{\pi}{4}$	$\frac{\pi}{2}$	$\frac{3\pi}{4}$	π	$\frac{5\pi}{4}$	$\frac{3\pi}{2}$	$\frac{7\pi}{4}$	2π	$\frac{9\pi}{4}$	$\frac{5\pi}{2}$
y	0	0.7	1	0.7	0	−0.7	−1	−0.7	0	0.7	1

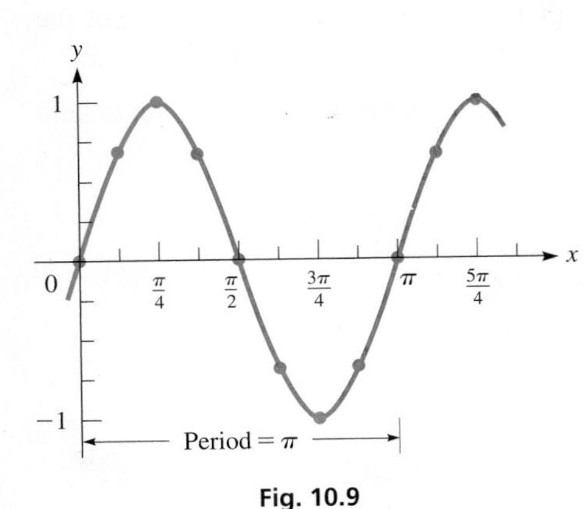

Fig. 10.9

Plotting these points, we have the curve shown in Fig. 10.9.

From the table and Fig. 10.9, note that $y = \sin 2x$ repeats after π units of x. The effect of the 2 is that the period of $y = \sin 2x$ is half the period of the curve of $y = \sin x$. We then conclude that if the period of a function $F(x)$ is P, then the period of $F(bx)$ is P/b.

NOTE ▸ [Because both $\sin x$ and $\cos x$ have a period of 2π, each of the functions $\sin bx$ and $\cos bx$ has a period of $2\pi/b$.]

EXAMPLE 1 Finding period of a function

(a) The period of $\cos 4x$ is $\frac{2\pi}{4} = \frac{\pi}{2}$.

(b) The period of $\sin 3\pi x$ is $\frac{2\pi}{3\pi} = \frac{2}{3}$.

(c) The period of $\sin \frac{1}{2}x$ is $\frac{2\pi}{\frac{1}{2}} = 4\pi$.

(d) The period of $\cos \frac{\pi}{4}x$ is $\frac{2\pi}{\frac{\pi}{4}} = 8$.

In (a), the period tells us that the curve of $y = \cos 4x$ will repeat every $\pi/2$ (approximately 1.57) units of x. In (b), we see that the curve of $y = \sin 3\pi x$ will repeat every 2/3 of a unit. In (c) and (d), the periods are longer than those of $y = \sin x$ and $y = \cos x$. ■

Practice Exercises

Find the period of each function.

1. $y = \sin \pi x$

2. $y = \cos \frac{1}{3}x$

NOTE ➧ [Combining the value of the period with the value of the amplitude from Section 10.1, we conclude that the functions $y = a \sin bx$ and $y = a \cos bx$ each has an amplitude of $|a|$ and a period of $2\pi/b$.] These properties are very useful in sketching these functions.

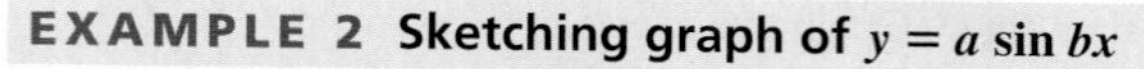

EXAMPLE 2 Sketching graph of $y = a \sin bx$

Sketch the graph of $y = 3 \sin 4x$ for $0 \le x \le \pi$.

Because $a = 3$, the amplitude is 3. The $4x$ tells us that the period is $2\pi/4 = \pi/2$. This means that $y = 0$ for $x = 0$ and for $y = \pi/2$. Because this sine function is zero halfway between $x = 0$ and $x = \pi/2$, we find that $y = 0$ for $x = \pi/4$. Also, the fact that the graph of the sine function reaches its maximum and minimum values halfway between zeros means that $y = 3$ for $x = \pi/8$, and $y = -3$ for $x = 3\pi/8$. Note that the values of x in the following table are those for which $4x = 0, \pi/2, \pi, 3\pi/2, 2\pi$, and so on.

x	0	$\frac{\pi}{8}$	$\frac{\pi}{4}$	$\frac{3\pi}{8}$	$\frac{\pi}{2}$	$\frac{5\pi}{8}$	$\frac{3\pi}{4}$	$\frac{7\pi}{8}$	π
y	0	3	0	−3	0	3	0	−3	0

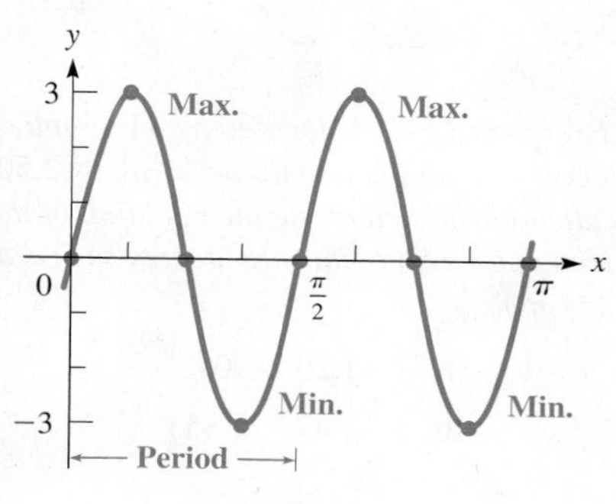

Fig. 10.10

Practice Exercise

3. Find the amplitude and period of the function $y = -8 \sin \frac{\pi x}{4}$

Using the values from the table and the fact that the curve is sinusoidal in form, we sketch the graph of this function in Fig. 10.10. We see again that knowing the key values and the basic shape of the curve allows us to sketch the graph of the curve quickly and easily. ■

Note from Example 2 that *an important distance in sketching a sine curve or a cosine curve is one-fourth of the period.* For $y = a \sin bx$, it is one-fourth of the period from the origin to the first value of x where y is at its maximum (or minimum) value. Then we proceed another one-fourth period to a zero, another one-fourth period to the next minimum (or maximum) value, another to the next zero (this is where the period is completed), and so on.

Similarly, one-fourth of the period is useful in sketching the graph of $y = \cos bx$. For this function, the maximum (or minimum) value occurs for $x = 0$. At the following one-fourth-period values, there is a zero, a minimum (or maximum), a zero, and a maximum (or minimum) at the start of the next period. NOTE ➧ [Therefore, by finding one-fourth of the period, we can easily find the important values for sketching the curve.]

We now summarize the important values for sketching the graphs of $y = a \sin bx$ and $y = a \cos bx$.

Important Values for Sketching $y = a \sin bx$ and $y = a \cos bx$

1. The amplitude: $|a|$
2. The period: $2\pi/b$
3. Values of the function for each one-fourth period

EXAMPLE 3 Using important values to sketch graph

Sketch the graph of $y = -2 \cos 3x$ for $0 \le x \le 2\pi$.

Note that the amplitude is 2 and the period is $\frac{2\pi}{3}$. This means that one-fourth of the period is $\frac{1}{4} \times \frac{2\pi}{3} = \frac{\pi}{6}$. Because the cosine curve is at a maximum or minimum for $x = 0$, we find that $y = -2$ for $x = 0$ (the negative value is due to the minus sign before the function), which means it is a minimum point. The curve then has a zero at $x = \frac{\pi}{6}$, a maximum value of 2 at $x = 2(\frac{\pi}{6}) = \frac{\pi}{3}$, a zero at $x = 3(\frac{\pi}{6}) = \frac{\pi}{2}$, and its next value of -2 at $x = 4(\frac{\pi}{6}) = \frac{2\pi}{3}$, and so on. Therefore, we have the following table:

x	0	$\frac{\pi}{6}$	$\frac{\pi}{3}$	$\frac{\pi}{2}$	$\frac{2\pi}{3}$	$\frac{5\pi}{6}$	π	$\frac{7\pi}{6}$	$\frac{4\pi}{3}$	$\frac{3\pi}{2}$	$\frac{5\pi}{3}$	$\frac{11\pi}{6}$	2π
y	−2	0	2	0	−2	0	2	0	−2	0	2	0	−2

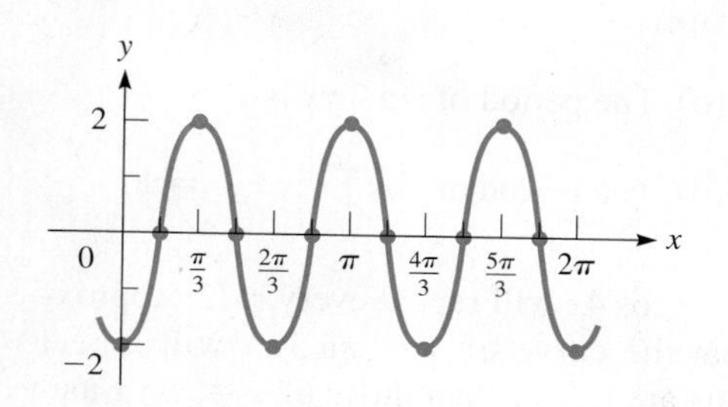

Fig. 10.11

Using this table and the sinusoidal shape of the cosine curve, we sketch the function in Fig. 10.11. ■

EXAMPLE 4 Graph on calculator—generator voltage

A generator produces a voltage $V = 200 \cos 50\pi t$, where t is the time in seconds (50π has units of rad/s; thus, $50\pi t$ is an angle in radians). Use a calculator to display the graph of V as a function of t for $0 \leq t \leq 0.06$ s.

The amplitude is 200 V and the period is $2\pi/(50\pi) = 0.04$ s. Because the period is not in terms of π, it is more convenient to use decimal units for t rather than to use units in terms of π as in the previous graphs. Thus, we have the following table of values:

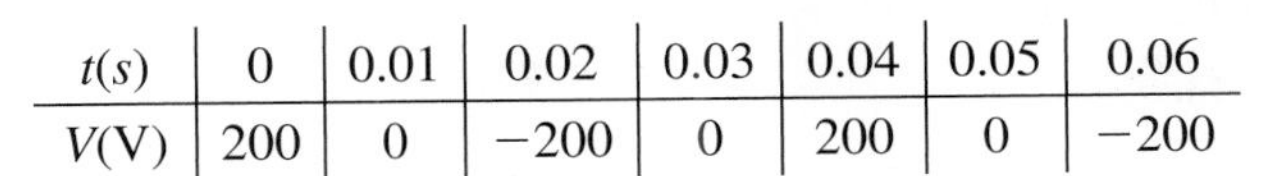

t(s)	0	0.01	0.02	0.03	0.04	0.05	0.06
V(V)	200	0	−200	0	200	0	−200

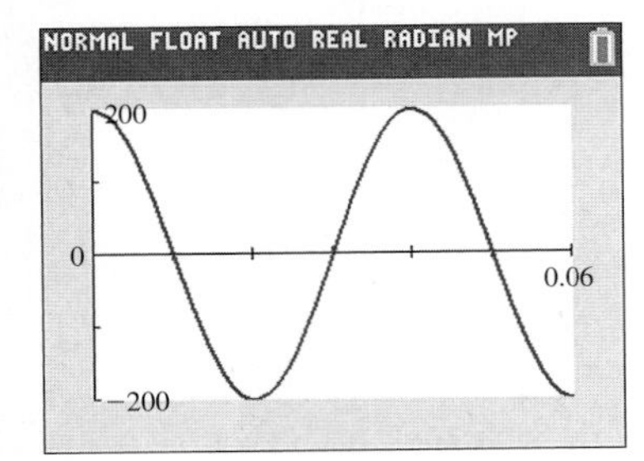

Fig. 10.12

For the calculator, use x for t and y for V. This means we graph the function $y_1 = 200 \cos 50\pi x$ (in the radian mode), as shown in Fig. 10.12. From the amplitude of 200 V and the above table, we choose the *window* values shown in Fig. 10.12. We do not consider negative values of t, for they have no real meaning in this problem. ■

EXERCISES 10.2

In Exercises 1 and 2, graph the function if the given changes are made in the indicated examples of this section.

1. In Example 2, if the coefficient of x is changed from 4 to 6, graph one complete cycle of the resulting function.
2. In Example 3, if the coefficient of x is changed from 3 to 4, sketch the graph of the resulting function.

In Exercises 3–22, find the amplitude and period of each function and then sketch its graph.

3. $y = 2 \sin 6x$
4. $y = 4 \sin 2x$
5. $y = 3 \cos 8x$
6. $y = 28 \cos 10x$
7. $y = -2 \sin 12x$
8. $y = -\frac{1}{5} \sin 5x$
9. $y = -\cos 16x$
10. $y = -4 \cos 3x$
11. $y = 520 \sin 2\pi x$
12. $y = 2 \sin 3\pi x$
13. $y = 3 \cos 4\pi x$
14. $y = 4 \cos 10\pi x$
15. $y = 15 \sin \frac{1}{3}x$
16. $y = -25 \sin 0.4x$
17. $y = -\frac{1}{2} \cos \frac{2}{3}x$
18. $y = \frac{1}{3} \cos 0.75x$
19. $y = 0.4 \sin \frac{2\pi x}{9}$
20. $y = 15 \cos \frac{\pi x}{10}$
21. $y = 3.3 \cos \pi^2 x$
22. $y = -12.5 \sin \frac{2x}{\pi}$

In Exercises 23–26, the period is given for a function of the form $y = \sin bx$. Write the function corresponding to the given period.

23. $\frac{\pi}{3}$
24. $\frac{5\pi}{2}$
25. $\frac{1}{3}$
26. 6

In Exercises 27–30, graph the given functions. In Exercises 27 and 28, use the equations for negative angles in Section 8.2 to first rewrite the function with a positive angle, and then graph the resulting function.

27. $y = 3 \sin(-2x)$
28. $y = -5 \cos(-4\pi x)$
29. $y = 8|\cos(\frac{\pi}{2}x)|$
30. $y = 0.4|\sin 6x|$

In Exercises 31–36, solve the given problems.

31. Display the graphs of $y = 2 \sin 3x$ and $y = 2 \sin(-3x)$ on a calculator. What conclusion do you draw from the graphs?
32. Display the graphs of $y = 2 \cos 3x$ and $y = \cos(-3x)$ on a calculator. What conclusion do you draw from the graphs?
33. By noting the periods of $\sin 2x$ and $\sin 3x$, find the period of the function $y = \sin 2x + \sin 3x$ by finding the least common multiple of the individual periods.
34. By noting the period of $\cos \frac{1}{2}x$ and $\cos \frac{1}{3}x$, find the period of the function $y = \cos \frac{1}{2}x + \cos \frac{1}{3}x$ by finding the least common multiple of the individual periods.
35. Find the function and graph it for a function of the form $y = -2 \sin bx$ that passes through $(\pi/4, -2)$ and for which b has the smallest possible positive value.
36. Find the function and graph it for a function of the form $y = 2 \sin bx$ that passes through $(\pi/6, 2)$ and for which b has the smallest possible positive value.

In Exercises 37 and 38, use the fact that the frequency, in cycles/s (or Hz), is the reciprocal of the period (in s). (Frequency is discussed in more detail in Section 10.5.)

37. The carrier signal transmitted by a certain FM radio station is given by $y = \sin(6.60 \times 10^8 t)$. Find the frequency and express it in MHz (1 MHz $= 10^6$ Hz).
38. The current in a certain alternating-current circuit is given by $i = 2.5 \sin(120\pi t)$. Find the period and the frequency.

In Exercises 39–42, graph the indicated functions.

39. The standard electric voltage in a 60-Hz alternating-current circuit is given by $V = 170 \sin 120\pi t$, where t is the time in seconds. Sketch the graph of V as a function of t for $0 \leq t \leq 0.05$ s.
40. To tune the instruments of an orchestra before a concert, an A note is struck on a piano. The piano wire vibrates with a displacement y (in mm) given by $y = 3.20 \cos 880\pi t$, where t is in seconds. Sketch the graph of y vs. t for $0 \leq t \leq 0.01$ s.

41. The velocity v (in in./s) of a piston is $v = 450 \cos 3600t$, where t is in seconds. Sketch the graph of v vs. t for $0 \leq t \leq 0.006$ s.

42. On a Florida beach, the tides have water levels about 4 m between low and high tides. The period is about 12.5 h. Find a cosine function that describes these tides if high tide is at midnight of a given day. Sketch the graph.

In Exercises 43–46, the graph of a function of the form $y = a \sin bx$ or $y = a \cos bx$ is shown. Determine the specific function of each.

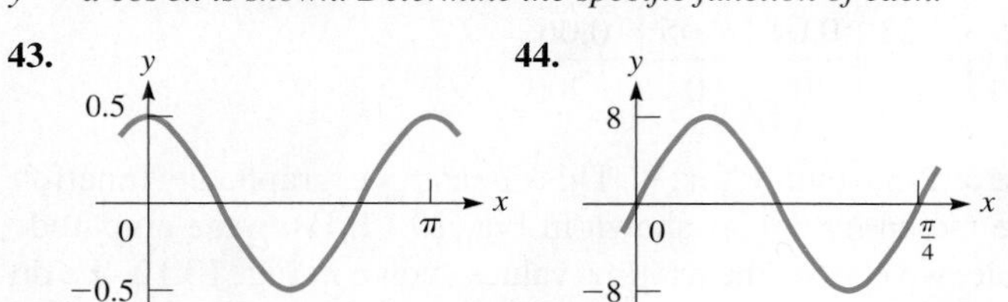

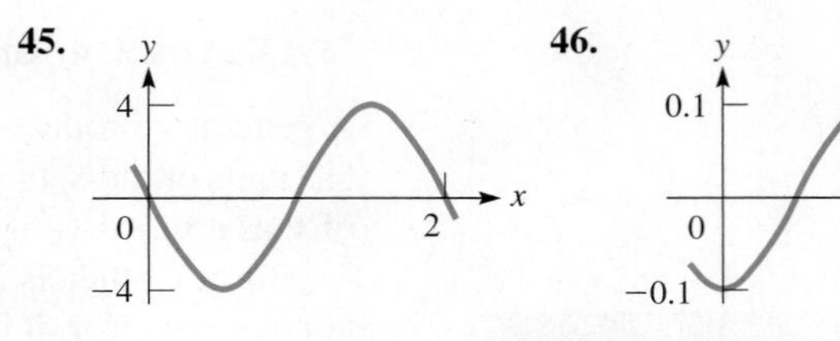

Answers to Practice Exercises

1. 2 **2.** 6π **3.** amp.: 8; per.: 8

10.3 Graphs of $y = a \sin (bx + c)$ and $y = a \cos (bx + c)$

Phase Angle • Displacement • Graphs of $y = a \sin (bx + c)$ and $y = a \cos (bx + c)$ • Cycle

In the function $y = a \sin (bx + c)$, c represents the **phase angle**. It is another very important quantity in graphing the sine and cosine functions. Its meaning is illustrated in the following example.

EXAMPLE 1 Sketch function with phase angle

Sketch the graph of $y = \sin(2x + \frac{\pi}{4})$.

Here, $c = \pi/4$. Therefore, in order to obtain values for the table, we assume a value for x, multiply it by 2, add $\pi/4$ to this value, and then find the sine of the result. The values shown are those for which $2x + \pi/4 = 0, \pi/4, \pi, 2, 3\pi/4, \pi$, and so on, which are the important values for $y = \sin 2x$.

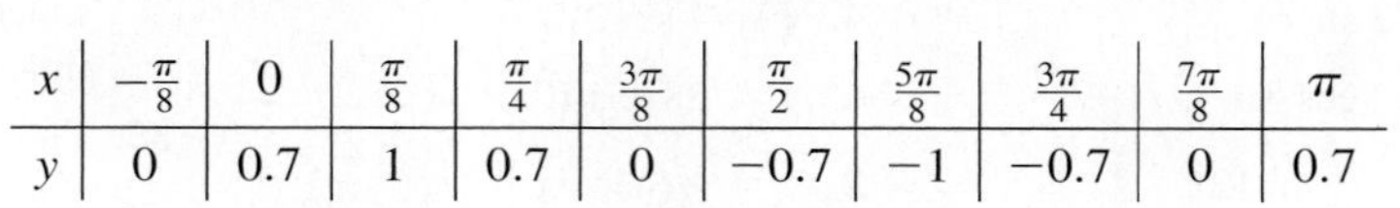

x	$-\frac{\pi}{8}$	0	$\frac{\pi}{8}$	$\frac{\pi}{4}$	$\frac{3\pi}{8}$	$\frac{\pi}{2}$	$\frac{5\pi}{8}$	$\frac{3\pi}{4}$	$\frac{7\pi}{8}$	π
y	0	0.7	1	0.7	0	−0.7	−1	−0.7	0	0.7

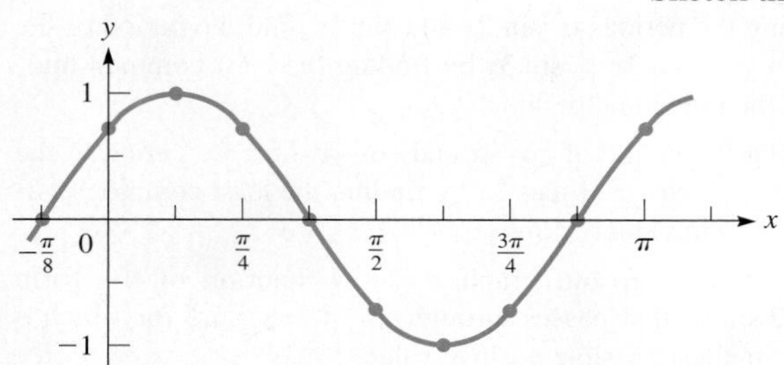

Fig. 10.13

Solving $2x + \pi/4 = 0$, we get $x = -\pi/8$, which gives $y = \sin 0 = 0$. The other values for y are found in the same way. See Fig. 10.13. ■

Note from Example 1 that *the graph of $y = \sin(2x + \frac{\pi}{4})$ is precisely the same as the graph of $y = \sin 2x$, except that it is **shifted** $\pi/8$ units to the left.* In Fig. 10.14, a calculator view shows the graphs of $y = \sin 2x$ and $y = \sin(2x + \frac{\pi}{4})$. We see that the shapes are the same except that the graph of $y = \sin(2x + \frac{\pi}{4})$ is $\pi/8$ units to the left of the graph of $y = \sin 2x$.

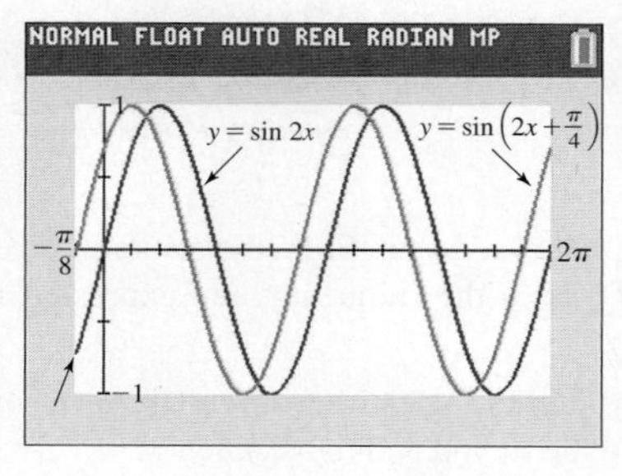

Fig. 10.14

In general, the effect of c in the equation $y = a \sin(bx + c)$ is to shift the curve of $y = a \sin bx$ to the left if $c > 0$, or shift the curve to the right if $c < 0$. The amount of this shift is given by $-c/b$. Due to its importance in sketching curves, *the quantity $-c/b$ is called the* **displacement** (*or* **phase shift**).

We can see the reason that the displacement is $-c/b$ by noting corresponding points on the graphs of $y = \sin bx$ and $y = \sin(bx + c)$. For $y = \sin bx$, when $x = 0$, then $y = 0$. For $y = \sin(bx + c)$, when $x = -c/b$, then $y = 0$. The point $(-c/b, 0)$ on the graph of $y = \sin(bx + c)$ is $-c/b$ units horizontally from the point (0, 0) on the graph of $y = \sin x$. In Fig. 10.14, $-c/b = -\pi/8$.

Therefore, we use the displacement combined with the amplitude and the period along with the other information from the previous sections to sketch curves of the functions $y = a \sin(bx + c)$ and $y = a \cos(bx + c)$, where $b > 0$.

■ Note that Eqs. (10.1) can be used in two different ways. If we are given an equation, we can find the amplitude, period, and displacement. Also, **if we are given the amplitude, period, and displacement, we can solve for a, b, and c to find the equation.**

Important Quantities to Determine for Sketching Graphs of $y = a\sin(bx + c)$ and $y = a\cos(bx + c)$

$$\text{Amplitude} = |a|$$

$$\text{Period} = \frac{2\pi}{b} \qquad \textbf{(10.1)}$$

$$\text{Displacement} = -\frac{c}{b}$$

By use of these quantities and the one-fourth period distance, the graphs of the sine and cosine functions can be readily sketched.

A general illustration of the graph of $y = a\sin(bx + c)$ is shown in Fig. 10.15.

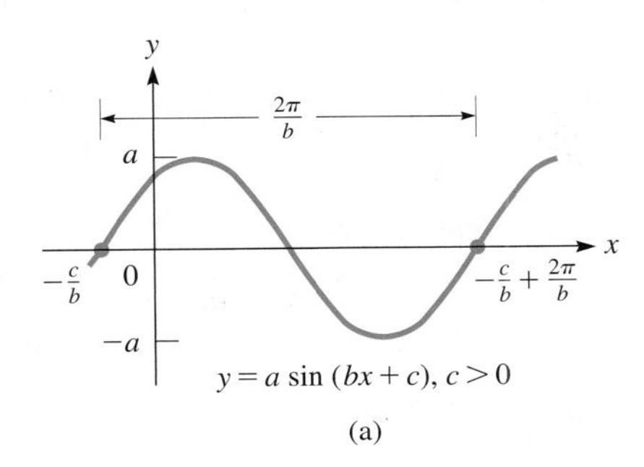

Since $c < 0$, $-c/b$ is positive

For each $a > 0, b > 0$

$y = a\sin(bx + c), c < 0$

(b)

Fig. 10.15

CAUTION Note that the displacement is negative (to the left) for $c > 0$ and positive (to the right) for $c < 0$ as shown in Figs. 10.15(a) and (b), respectively. ■

NOTE ➧ [Note also that we can find the displacement for the graphs of $y = a\sin(bx + c)$ by solving $bx + c = 0$ for x. We see that $x = -c/b$.] This is the same as the horizontal shift for graphs of functions that is shown on page 105.

EXAMPLE 2 Sketching graph of $y = a\sin(bx + c)$

Sketch the graph of $y = 2\sin(3x - \pi)$.

First, note that $a = 2$, $b = 3$, and $c = -\pi$. Therefore, the amplitude is 2, the period is $2\pi/3$, and the displacement is $-(-\pi/3) = \pi/3$. (We can also get the displacement from $3x - \pi = 0$, $x = \pi/3$.)

Note that the curve "starts" at $x = \pi/3$ and starts repeating $2\pi/3$ units to the right of this point. Be sure to grasp this point well. *The period tells us the number of units along the x-axis between such corresponding points.* One-fourth of the period is $\frac{1}{4}\left(\frac{2\pi}{3}\right) = \frac{\pi}{6}$.

Important values are at $\frac{\pi}{3}, \frac{\pi}{3} + \frac{\pi}{6} = \frac{\pi}{2}, \frac{\pi}{3} + 2\left(\frac{\pi}{6}\right) = \frac{2\pi}{3}$, and so on. We now make the table of important values and sketch the graph shown in Fig. 10.16.

x	0	$\frac{\pi}{6}$	$\frac{\pi}{3}$	$\frac{\pi}{2}$	$\frac{2\pi}{3}$	$\frac{5\pi}{6}$	π
y	0	−2	0	2	0	−2	0

← $\frac{1}{4}$(period) $= \frac{\pi}{6}$ ■

Fig. 10.16

EXAMPLE 3 Sketching graph of $y = a \cos(bx + c)$

Sketch the graph of the function $y = -\cos(2x + \frac{\pi}{6})$.

First, we determine that

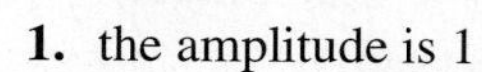

1. the amplitude is 1
2. the period is $\frac{2\pi}{2} = \pi$
3. the displacement is $-\frac{\pi}{6} \div 2 = -\frac{\pi}{12}$

We now make a table of important values, noting that the curve starts repeating π units to the right of $-\frac{\pi}{12}$.

x	$-\frac{\pi}{12}$	$\frac{\pi}{6}$	$\frac{5\pi}{12}$	$\frac{2\pi}{3}$	$\frac{11\pi}{12}$
y	-1	0	1	0	-1

← $\frac{1}{4}$(period) $= \frac{\pi}{4}$

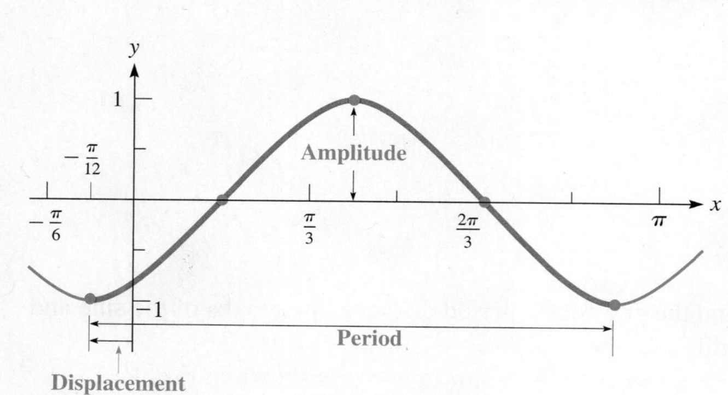

Fig. 10.17

From this table, we sketch the graph in Fig. 10.17. ■

Practice Exercise

1. For the graph of $y = 8 \sin(2x - \pi/3)$, determine amplitude, period, and displacement.

Each of the heavy portions of the graphs in Figs. 10.16 and 10.17 is called a *cycle* of the curve. A **cycle** *is any section of the graph that includes exactly one period.*

EXAMPLE 4 Graph on calculator

View the graph of $y = 2 \cos(\frac{1}{2}x - \frac{\pi}{6})$ on a calculator.

From the values $a = 2$, $b = 1/2$, and $c = -\pi/6$, we determine that

1. the amplitude is 2
2. the period is $2\pi \div \frac{1}{2} = 4\pi$
3. the displacement is $-(-\frac{\pi}{6}) \div \frac{1}{2} = \frac{\pi}{3}$

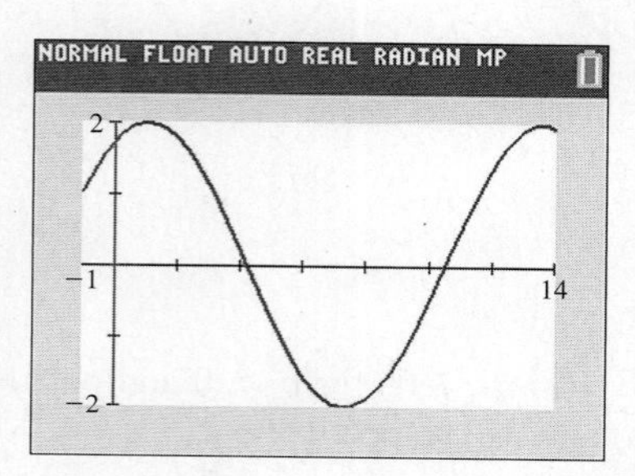

Fig. 10.18

This curve's cycle starts at $x = \frac{\pi}{3}$ (≈ 1.05) and ends at $\frac{\pi}{3} + 4\pi = \frac{13\pi}{3}$ (≈ 13.6). When choosing the window settings, we should make sure these two values are between the Xmin and Xmax. The amplitude determines the Ymin and Ymax. The graph is shown in Fig. 10.18. ■

EXAMPLE 5 Phase angle—ac voltage/current lead and lag

In alternating-current (ac) circuits, voltage (v) and current (i) can both be represented by sine waves with the same period. Depending on the existence of resistors, capacitors, or inductors in the circuit, these waves can be *in phase* (maximums and minimums occur at the same time) or one can *lead or lag* the other. See Fig. 10.19.

■ In Chapter 12, we show how the phase angle depends on the resistance, capacitance, and inductance in a circuit.

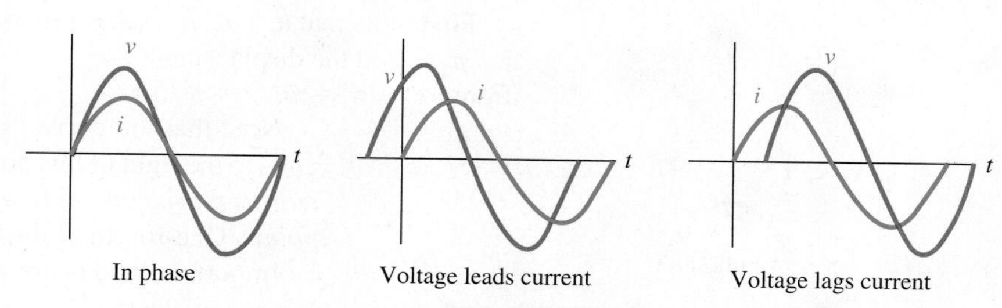

Fig. 10.19

If we assume that the sine wave for current passes through the origin, it can be represented as $i = I_m \sin(2\pi ft)$, where f is the frequency (in cycles/s) and t is time (in s). The voltage can then be represented by $v = V_m \sin(2\pi ft + \phi)$, where ϕ is the phase angle. [The **phase angle,** often written in degrees, tells us how many degrees (assuming 360° is one complete cycle) the voltage leads (if ϕ is positive) or lags (if ϕ is negative) the current by.]

NOTE ▶

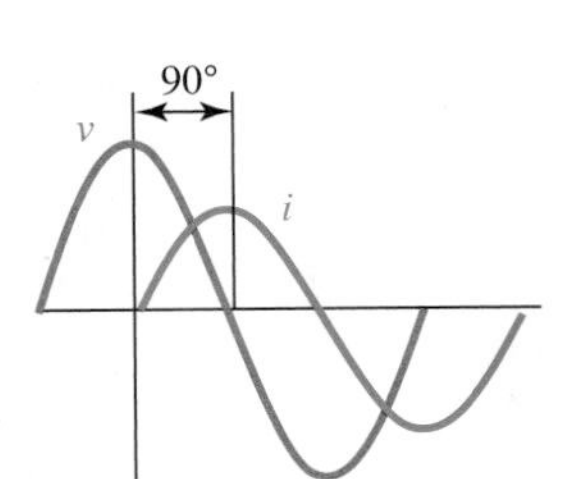

Fig. 10.20 Voltage leads current by 90°

For example, if $i = 15 \sin(120\pi t)$ and $v = 170 \sin(120\pi t + 90°)$, then voltage leads current by 90°, or one-quarter of the period. See Fig. 10.20. It is somewhat unusual that the voltage function combines radians ($2\pi ft$) and degrees (90°), but this is commonly done in this particular application. If we convert the phase angle to $\frac{\pi}{2}$ radians, we can then find the displacement, which gives the time shift in seconds between the two waves:

$$\text{Displacement} = -\frac{c}{b} = -\frac{\pi/2}{120\pi} = -\frac{1}{240}\ \text{s}$$

Therefore, voltage leads current by $\frac{1}{240}$ s. This is one-quarter of the period, with the period equal to $\frac{2\pi}{120\pi} = \frac{1}{60}$ s. NOTE ▶ [Note that in order to graph the voltage function on a calculator, the phase angle must be expressed in radians.] ■

Practice Exercise

2. In Example 5, find the function for voltage if voltage lags current by 45° and all other information remains the same.

EXERCISES 10.3

In Exercises 1 and 2, graph the function if the given changes are made in the indicated examples of this section.

1. In Example 3, if the sign before $\pi/6$ is changed, sketch the graph of the resulting function.
2. In Example 4, if the sign before $\pi/6$ is changed, sketch the graph of the resulting function.

In Exercises 3–26, determine the amplitude, period, and displacement for each function. Then sketch the graphs of the functions. Check each using a calculator.

3. $y = \sin\left(x - \frac{\pi}{6}\right)$
4. $y = 3\sin\left(x + \frac{\pi}{4}\right)$
5. $y = \cos\left(x + \frac{\pi}{6}\right)$
6. $y = 2\cos\left(x - \frac{\pi}{8}\right)$
7. $y = 0.2\sin\left(2x + \frac{\pi}{2}\right)$
8. $y = -\sin\left(3x - \frac{\pi}{2}\right)$
9. $y = -\cos(2x - \pi)$
10. $y = 0.4\sin\left(3x + \frac{\pi}{3}\right)$
11. $y = \frac{1}{2}\sin\left(\frac{1}{2}x - \frac{\pi}{4}\right)$
12. $y = 2\sin\left(\frac{1}{4}x + \frac{\pi}{2}\right)$
13. $y = 30\cos\left(\frac{1}{3}x + \frac{\pi}{3}\right)$
14. $y = -\frac{1}{3}\cos\left(\frac{1}{2}x - \frac{\pi}{8}\right)$
15. $y = -\sin\left(\pi x + \frac{\pi}{8}\right)$
16. $y = -2\sin(2\pi x - \pi)$
17. $y = 0.08\cos\left(4\pi x - \frac{\pi}{5}\right)$
18. $y = 25\cos\left(3\pi x + \frac{\pi}{4}\right)$
19. $y = -0.6\sin(2\pi x - 1)$
20. $y = 1.8\sin\left(\pi x + \frac{1}{3}\right)$
21. $y = 40\cos(3\pi x + 1)$
22. $y = 360\cos(6\pi x - 3)$
23. $y = \sin(\pi^2 x - \pi)$
24. $y = -\frac{1}{2}\sin\left(2x - \frac{1}{\pi}\right)$
25. $y = -\frac{3}{2}\cos\left(\pi x + \frac{\pi^2}{6}\right)$
26. $y = \pi\cos\left(\frac{1}{\pi}x + \frac{1}{3}\right)$

In Exercises 27–30, write the equation for the given function with the given amplitude, period, and displacement, respectively. [Hint: Use Eqs. (10.1) to determine a, b, and c.]

27. sine, 4, 3π, $-\pi/4$
28. cosine, 8, $2\pi/3$, $\pi/3$
29. cosine, 12, 1/2, 1/8
30. sine, 18, 4, -1

In Exercises 31 and 32, show that the given equations are identities. The method using a calculator is indicated in Exercise 31.

31. By viewing the graphs of $y_1 = \sin x$ and $y_2 = \cos(x - \pi/2)$, show that $\cos(x - \pi/2) = \sin x$.
32. Show that $\cos(2x - 3\pi/8) = \cos(3\pi/8 - 2x)$.

In Exercises 33–40, solve the given problems.

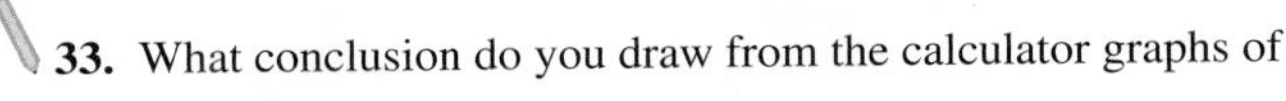

33. What conclusion do you draw from the calculator graphs of
$$y_1 = 2\sin\left(3x + \frac{\pi}{6}\right) \text{ and } y_2 = -2\sin\left[-\left(3x + \frac{\pi}{6}\right)\right]?$$
34. What conclusion do you draw from the calculator graphs of
$$y_1 = 2\cos\left(3x + \frac{\pi}{6}\right) \text{ and } y_2 = -2\cos\left[-\left(3x + \frac{\pi}{6}\right)\right]?$$
35. The current in an alternating-current circuit is given by $i = 12\sin(120\pi t)$. Find a function for the voltage v if the amplitude is 18V and voltage lags current by 60°. Then find the displacement. See Example 5.
36. The current in an alternating-current circuit is given by $i = 8.50\sin(120\pi t)$. Find a function for the voltage v if the amplitude is 28V and voltage leads current by 45°. Then find the displacement. See Example 5.
37. A wave traveling in a string may be represented by the equation $y = A\sin 2\pi\left(\frac{t}{T} - \frac{x}{\lambda}\right)$. Here, A is the amplitude, t is the time the wave has traveled, x is the distance from the origin, T is the time required for the wave to travel one *wavelength* λ (the Greek letter lambda). Sketch three cycles of the wave for which $A = 2.00$ cm, $T = 0.100$ s, $\lambda = 20.0$ cm, and $x = 5.00$ cm.
38. The electric current i (in μA) in a certain circuit is given by $i = 3.8\cos 2\pi(t + 0.20)$, where t is the time in seconds. Sketch three cycles of this function.

39. A certain satellite circles Earth such that its distance y, in miles north or south (altitude is not considered) from the equator, is $y = 4500\cos(0.025t - 0.25)$, where t is the time (in min) after launch. Use a calculator to graph two cycles of the curve.

40. In performing a test on a patient, a medical technician used an ultrasonic signal given by the equation $I = A\sin(\omega t + \phi)$. Use a calculator to view two cycles of the graph of I vs. t if $A = 5\ \text{nW/m}^2$, $\omega = 2 \times 10^5$ rad/s, and $\phi = 0.4$. Explain how you chose your calculator's window settings.

In Exercises 41–44, give the specific form of the equation by evaluating a, b, and c through an inspection of the given curve. Explain how a, b, and c are found.

41. $y = a\sin(bx + c)$ Fig. 10.21

42. $y = a\cos(bx + c)$ Fig. 10.21

43. $y = a\cos(bx + c)$ Fig. 10.22

44. $y = a\sin(bx + c)$ Fig. 10.22

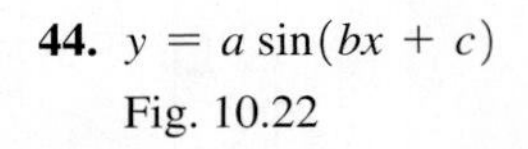

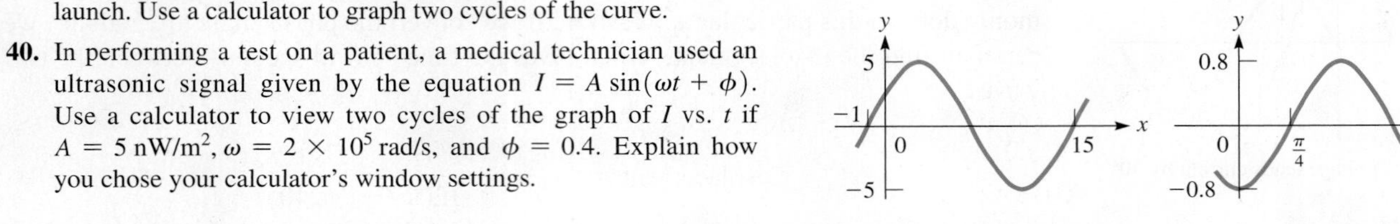

Fig. 10.21

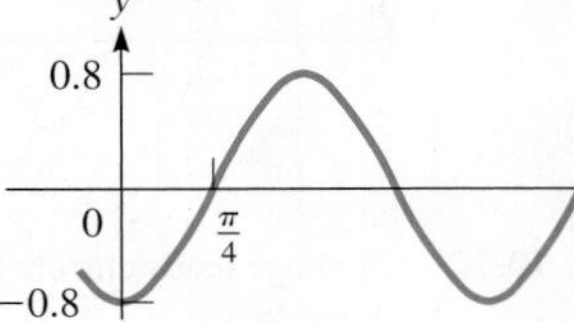

Fig. 10.22

Answers to Practice Exercises

1. amp. = 8, per. = π, disp. = $\pi/6$ **2.** $v = 170\sin(120\pi t - 45°)$

10.4 Graphs of $y = \tan x$, $y = \cot x$, $y = \sec x$, $y = \csc x$

Graph of $y = \tan x$ • Reciprocal Functions • Asymptotes • Graphs of $y = \cot x$, $y = \sec x$, $y = \csc x$

The graph of $y = \tan x$ is displayed in Fig. 10.23. Because from Section 4.3 we know that $\csc x = 1/\sin x$, $\sec x = 1/\cos x$, and $\cot x = 1/\tan x$, we are able to find values of $y = \csc x$, $y = \sec x$, and $y = \cot x$ and graph these functions, as shown in Figs. 10.24–10.26.

■ The following reciprocal relationships are the basis for graphing $y = \cot x$, $y = \sec x$, and $y = \csc x$.

$$\csc x = \frac{1}{\sin x} \qquad \sec x = \frac{1}{\cos x}$$

$$\cot x = \frac{1}{\tan x} \qquad \textbf{(10.2)}$$

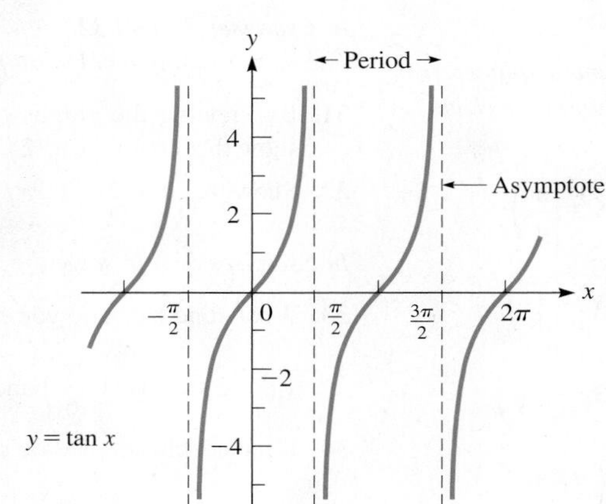

Fig. 10.23

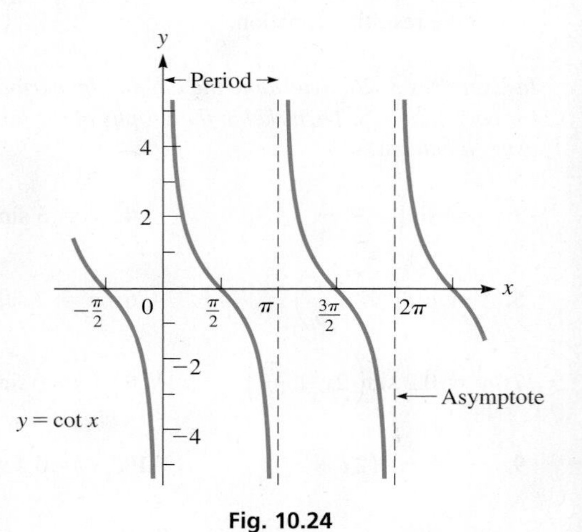

Fig. 10.24

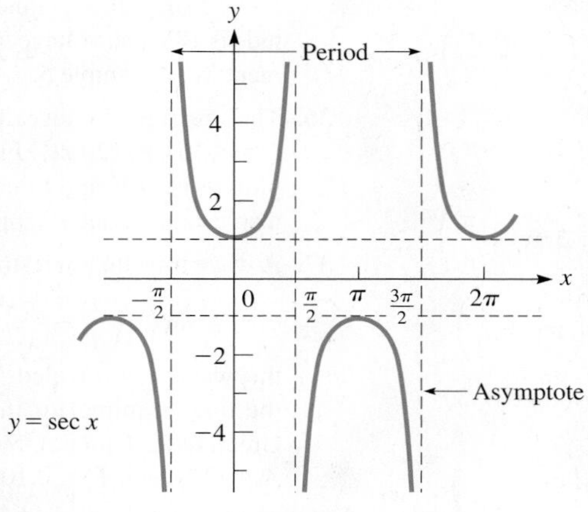

Fig. 10.25

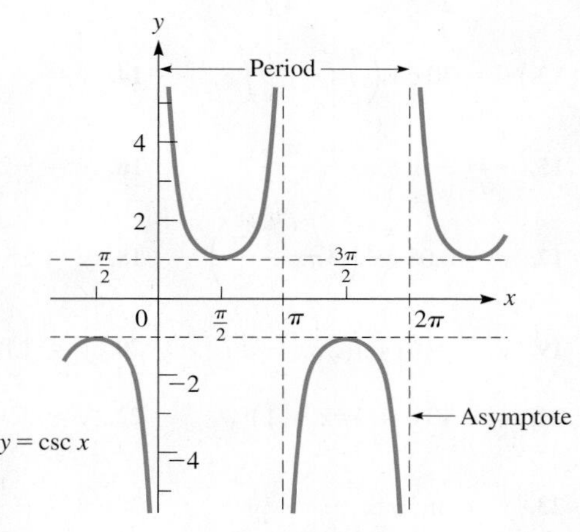

Fig. 10.26

From these graphs, note that the period for $y = \tan x$ and $y = \cot x$ is π, and that the period of $y = \sec x$ and $y = \csc x$ is 2π. *The vertical dashed lines are* **asymptotes** (see Sections 3.4 and 21.6). The curves *approach* these lines, but never actually reach them.

The functions are not defined for the values of x for which the curve has asymptotes. This means that the domains do not include these values of x. Thus, we see that the domains of $y = \tan x$ and $y = \sec x$ include all real numbers, except the values $x = -\pi/2, \pi/2, 3\pi/2$, and so on. The domain of $y = \cot x$ and $y = \csc x$ include all real numbers except $x = -\pi, 0, \pi, 2\pi$, and so on.

From the graphs, we see that the ranges of $y = \tan x$ and $y = \cot x$ are all real numbers, but that the ranges of $y = \sec x$ and $y = \csc x$ do not include the real numbers between -1 and 1.

To sketch functions such as $y = a \sec x$, first sketch $y = \sec x$ and then multiply the y-values by a. [Here, a is not an amplitude, because the ranges of these functions are not limited in the same way they are for the sine and cosine functions.] NOTE ▸

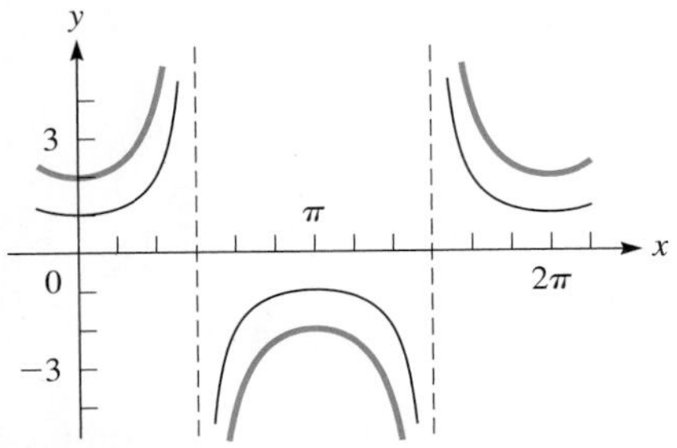

Fig. 10.27

EXAMPLE 1 Sketching graph of $y = a \sec x$

Sketch the graph of $y = 2 \sec x$.

First, we sketch in $y = \sec x$, the curve shown in black in Fig. 10.27. Then we multiply the y-values of this secant function by 2. Although we can only estimate these values and do this approximately, a reasonable graph can be sketched this way. The desired curve is shown in blue in Fig. 10.27. ■

Using a calculator, we can display the graphs of these functions more easily and more accurately than by sketching them. By knowing the general shape and period of the function, the values for the *window* settings can be determined.

EXAMPLE 2 Calculator graph of $y = a \cot bx$

View at least two cycles of the graph of $y = 0.5 \cot 2x$ on a calculator.

Because the period of $y = \cot x$ is π, the period of $y = \cot 2x$ is $\pi/2$. Therefore, we choose the *window* settings as follows:

Xmin $= 0$ ($x = 0$ is one asymptote of the curve)
Xmax $= \pi$ (the period is $\pi/2$; two periods is π)
Ymin $= -5$, Ymax $= 5$ (the range is all x; this shows enough of the curve)

We must remember to enter the function as $y_1 = 0.5(\tan 2x)^{-1}$, because $\cot x = (\tan x)^{-1}$. The graphing calculator view is shown in Fig. 10.28. We can view many more cycles of the curve with appropriate *window* settings. ■

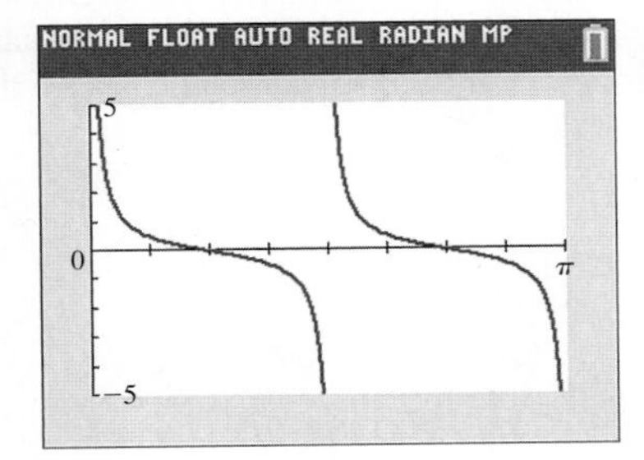

Fig. 10.28

EXAMPLE 3 Calculator graph of $y = a \csc(bx + c)$

View at least two periods of the graph of $y = 2 \csc(2x + \pi/4)$ on a calculator.

Because the period of $\csc x$ is 2π, the period of $\csc(2x + \pi/4)$ is $2\pi/2 = \pi$. Recalling that $\csc x = (\sin x)^{-1}$, the curve will have the same displacement as $y = \sin(2x + \pi/4)$. Therefore, displacement is $-\frac{\pi/4}{2} = -\frac{\pi}{8}$. There is some flexibility in choosing the *window* settings, and as an example, we choose the following settings.

Xmin $= -0.5$ (the displacement is $-\pi/8 = -0.4$)
Xmax $= 6$ (displacement $= -\pi/8$; period $= \pi$, $-\pi/8 + 2\pi = 15\pi/8 = 5.9$)
Ymin $= -6$, Ymax $= 6$ (there is no curve between $y = -2$ and $y = 2$)

With $y_1 = 2[\sin(2x + \pi/4)]^{-1}$, the function is graphed in blue in Fig. 10.29. The graph of $y = 2 \sin(2x + \pi/4)$ is shown in red for reference. ■

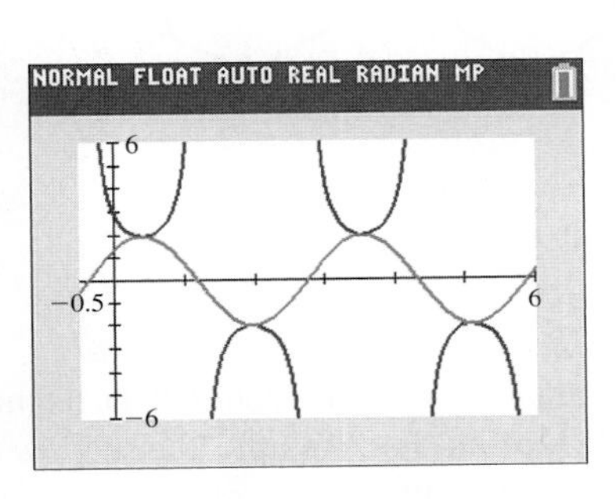

Fig. 10.29

EXERCISES 10.4

In Exercises 1 and 2, view the graphs on a calculator if the given changes are made in the indicated examples of this section.

1. In Example 2, change 0.5 to 5.

2. In Example 3, change the sign before $\pi/4$.

In Exercises 3–6, fill in the following table for each function and plot the graph from these points.

x	$-\frac{\pi}{2}$	$-\frac{\pi}{3}$	$-\frac{\pi}{4}$	$-\frac{\pi}{6}$	0	$\frac{\pi}{6}$	$\frac{\pi}{4}$	$\frac{\pi}{3}$	$\frac{\pi}{2}$	$\frac{2\pi}{3}$	$\frac{3\pi}{4}$	$\frac{5\pi}{6}$	π	y

3. $y = \tan x$

4. $y = \cot x$

5. $y = \sec x$

6. $y = \csc x$

In Exercises 7–14, sketch the graphs of the given functions by use of the basic curve forms (Figs. 10.23, 10.24, 10.25, and 10.26). See Example 1.

7. $y = 2 \tan x$

8. $y = 3 \cot x$

9. $y = \frac{1}{2} \sec x$

10. $y = \frac{3}{2} \csc x$

11. $y = -8 \cot x$

12. $y = -0.1 \tan x$

13. $y = -3 \csc x$

14. $y = -60 \sec x$

In Exercises 15–24, view at least two cycles of the graphs of the given functions on a calculator.

15. $y = \tan 2x$

16. $y = 2 \cot 3x$

17. $y = \frac{1}{2} \sec 3x$

18. $y = -0.4 \csc 2x$

19. $y = -2 \cot\left(2x + \frac{\pi}{6}\right)$

20. $y = \tan\left(3x - \frac{\pi}{2}\right)$

21. $y = 18 \csc\left(3x - \frac{\pi}{3}\right)$

22. $y = 12 \sec\left(2x + \frac{\pi}{4}\right)$

23. $y = 75 \tan\left(0.5x - \frac{\pi}{16}\right)$

24. $y = 0.5 \sec\left(0.2x + \frac{\pi}{25}\right)$

In Exercises 25–28, solve the given problems. In Exercises 29–32, sketch the appropriate graphs, and check each on a calculator.

25. Using the graph of $y = \tan x$, explain what happens to $\tan x$ as x gets closer to $\pi/2$ (a) from the left and (b) from the right.

26. Using a calculator, graph $y = \sin x$ and $y = \csc x$ in the same window. When $\sin x$ reaches a maximum or minimum, explain what happens to $\csc x$.

27. Write the equation of a secant function with zero displacement, a period of 4π, and that passes through $(0, -3)$.

28. Use a graphing calculator to show that $\sin x < \tan x$ for $0 < x < \pi/2$, although $\sin x$ and $\tan x$ are nearly equal for the values near zero.

29. Near Antarctica, an iceberg with a vertical face 200 m high is seen from a small boat. At a distance x from the iceberg, the angle of elevation θ of the top of the iceberg can be found from the equation $x = 200 \cot \theta$. Sketch x as a function of θ.

30. In a laser experiment, two mirrors move horizontally in equal and opposite distances from point A. The laser path from and to point B is shown in Fig. 10.30. From the figure, we see that $x = a \tan \theta$. Sketch the graph of $x = f(\theta)$ for $a = 5.00$ cm.

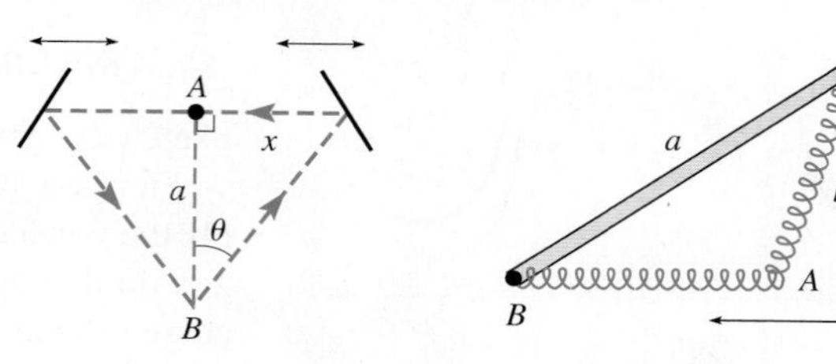

Fig. 10.30

Fig. 10.31

31. A mechanism with two springs is shown in Fig. 10.31, where point A is restricted to move horizontally. From the law of sines, we see that $b = (a \sin B) \csc A$. Sketch the graph of b as a function of A for $a = 4.00$ cm and $B = \pi/4$.

32. A cantilever column of length L will buckle if too large a downward force P is applied d units off center. The horizontal deflection x (see Fig. 10.32) is $x = d[\sec(kL) - 1]$, where k is a constant depending on P, and $0 < kL < \pi/2$. For a constant d, sketch the graph of x as a function of kL.

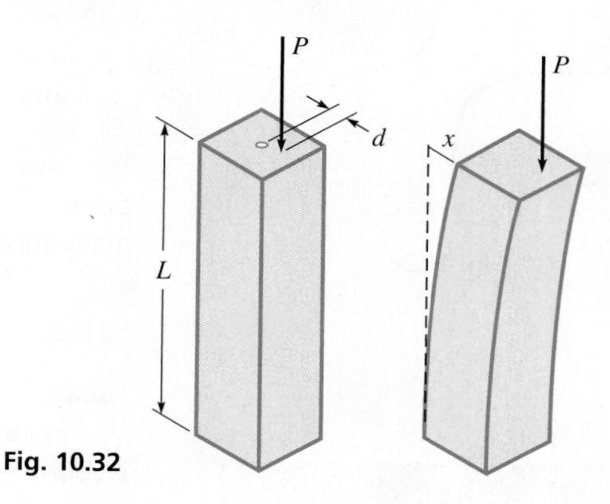

Fig. 10.32

10.5 Applications of the Trigonometric Graphs

Simple Harmonic Motion • Alternating Electric Current • Frequency

When an object moves in a circular path with constant velocity (see Section 8.4), its *projection* on a diameter moves with **simple harmonic motion.** For example, the shadow of a ball at the end of a string and moving at a constant rate, moves with simple harmonic motion. We now consider this physical concept and some of its applications.

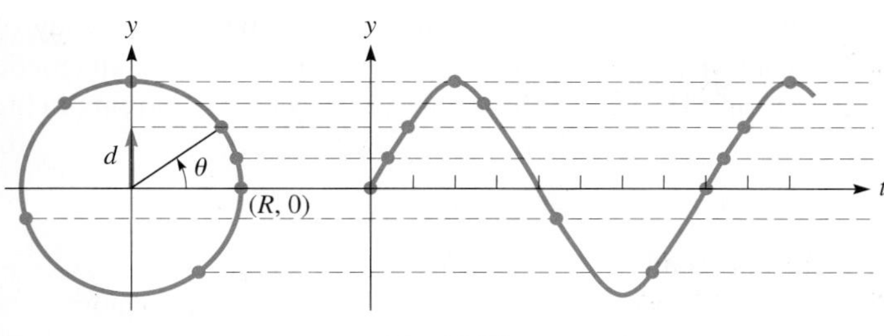

Fig. 10.33

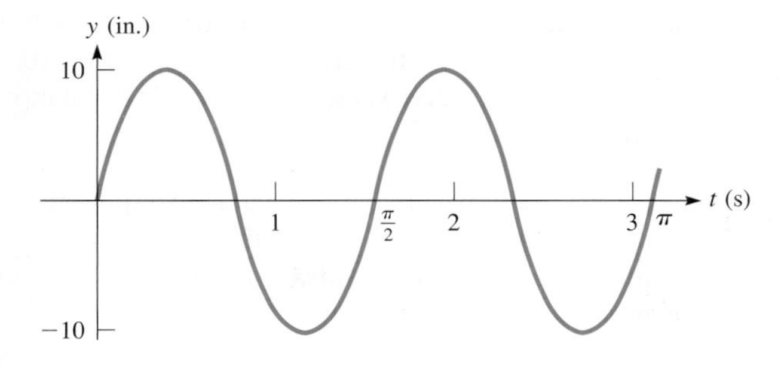

Fig. 10.34

EXAMPLE 1 Simple harmonic motion

In Fig. 10.33, assume that a particle starts at the end of the radius at $(R, 0)$ and moves counterclockwise around the circle with constant angular velocity ω. *The displacement of the projection on the y-axis is d* and is given by $d = R\sin\theta$. The displacement is shown for a few different positions of the end of the radius.

Because $\theta/t = \omega$, or $\theta = \omega t$, we have

$$d = R\sin\omega t \qquad \textbf{(10.3)}$$

as the equation for the displacement of this projection, with time t as the independent variable.

For the case where $R = 10.0$ in. and $\omega = 4.00$ rad/s, we have

$$d = 10.0\sin 4.00t$$

By sketching or viewing the graph of this function, we can find the displacement d of the projection for a given time t. The graph is shown in Fig. 10.34. ■

In Example 1, note that *time is the independent variable.* This is motion for which the object (the end of the projection) remains at the same horizontal position $(x = 0)$ and moves only vertically according to a sinusoidal function. In the previous sections, we dealt with functions in which y is a sinusoidal function of the horizontal displacement x. Think of a water wave. At *one point* of the wave, the motion is only vertical and sinusoidal with time. At *one given time,* a picture would indicate a sinusoidal movement from one horizontal position to the next.

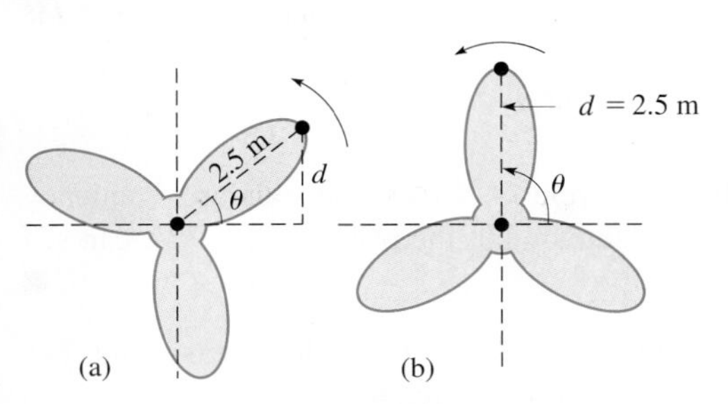

Fig. 10.35

EXAMPLE 2 Simple harmonic motion—windmill displacement

A windmill is used to pump water. The radius of the blade is 2.5 m, and it is moving with constant angular velocity. If the vertical displacement of the end of the blade is timed from the point it is at an angle of 45° $(\pi/4 \text{ rad})$ from the horizontal [see Fig. 10.35(a)], the displacement d is given by

$$d = 2.5\sin\left(\omega t + \frac{\pi}{4}\right)$$

If the blade makes an angle of 90° $(\pi/2 \text{ rad})$ when $t = 0$ [see Fig. 10.35(b)], the displacement d is given by

$$d = 2.5\sin\left(\omega t + \frac{\pi}{2}\right) \quad \text{or} \quad d = 2.5\cos\omega t$$

If timing started at the first maximum for the displacement, the resulting curve for the displacement would be that of the cosine function. ■

Practice Exercise

1. If the windmill blade in Example 2 starts at an angle of 135°, what equation gives its displacement in terms of the cosine function?

Other examples of simple harmonic motion are (1) the movement of a pendulum bob through its arc (a very close approximation to simple harmonic motion), (2) the motion of an object "bobbing" in water, (3) the movement of the end of a vibrating rod (which we hear as sound), and (4) the displacement of a weight moving up and down on a spring. Other phenomena that give rise to equations like those for simple harmonic motion are found in the fields of optics, sound, and electricity. The equations for such phenomena have the same mathematical form because they result from vibratory movement or motion in a circle.

EXAMPLE 3 Alternating current

As shown in Section 10.3, the sine and cosine functions are important in the study of alternating current, which is caused by the motion of a wire passing through a magnetic field. If the wire is moving in a circular path, with angular velocity ω, the current i in the wire at time t is given by an equation of the form

$$i = I_m \sin(\omega t + \phi)$$

where I_m is the maximum current attainable and ϕ is the phase angle.

Given that $I_m = 6.00$ A, $\omega = 120\pi$ rad/s, and $\phi = \pi/6$, we have the equation

$$i = 6.00 \sin\left(120\pi t + \frac{\pi}{6}\right)$$

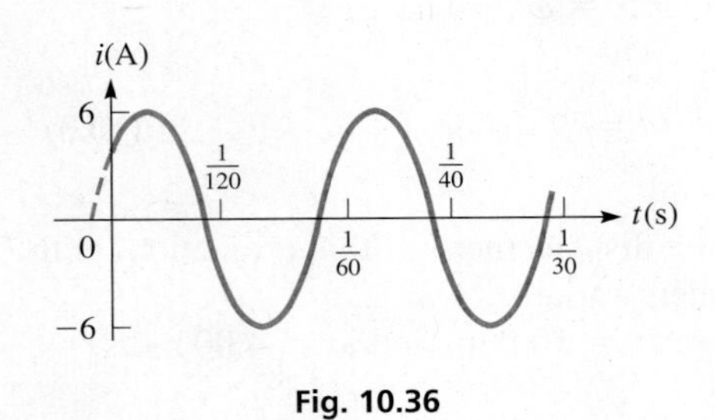

Fig. 10.36

From this equation, note that the amplitude is 6.00 A, the period is $\frac{1}{60}$ s, and the displacement is $-\frac{1}{720}$ s. From these values, we draw the graph as shown in Fig. 10.36. Because the current takes on both positive and negative values, we conclude that it moves alternately in one direction and then the other. ■

It is a common practice to express the rate of rotation in terms of *the* **frequency** f*, the number of cycles per unit of time*, rather than directly in terms of the angular velocity ω, the number of radians per unit of time. *The SI unit for frequency is the* **hertz** (Hz), *and* 1 Hz = 1 cycle/s. Because there are 2π rad in one cycle, we have

■ Named for the German physicist Heinrich Hertz (1857–1894).

$$\omega = 2\pi f \tag{10.4}$$

It is the frequency f that is referred to in electric current, on radio stations, for musical tones, and so on.

It is important to note that **the frequency and the period are reciprocals of each other.** The following example illustrates this point.

EXAMPLE 4 Frequency—hertz

For the electric current in Example 3, $\omega = 120\pi$ rad/s. The corresponding frequency f is

$$f = \frac{120\pi}{2\pi} = 60 \text{ Hz}$$

This means that 120π rad/s corresponds to 60 cycles/s. This is the standard frequency used for alternating current. We have already shown that the period is $\frac{1}{60}$ s. We can see here that the frequency and the period are reciprocals. ■

Practice Exercise

2. In Example 1, what is the frequency?

Sometimes the frequency is given in terms of a unit of time other than the second. For example, suppose a wheel rotates at 1500 r/min. Here the frequency is 1500 cycles/min and the period is $\frac{1}{1500}$ min.

EXERCISES 10.5

In Exercises 1 and 2, answer the given questions about the indicated examples of this section.

1. In Example 1, what is the equation relating d and t if the end of the radius starts at $(0, R)$?
2. In Example 2, if the blade starts at an angle of $-45°$, what is the equation relating d and t as (a) a sine function? (b) A cosine function?

In Exercises 3–6, solve the given problems.

3. If $f = 40$ cycles/min, find the period.
4. If the period is $\frac{1}{10}$ s, find f.
5. If $\omega = 40\pi$ rad/s, find the period.
6. If the period is $\frac{1}{5}$ s, find ω.

A graphing calculator may be used in the following exercises.

In Exercises 7 and 8, sketch two cycles of the curve given by $d = R \sin \omega t$ for the given values.

7. $R = 2.40$ cm, $\omega = 2.00$ rad/s
8. $R = 18.5$ ft, $f = 0.250$ Hz

In Exercises 9 and 10, a point on a cam is 8.30 cm from the center of rotation. The cam is rotating with a constant angular velocity, and the vertical displacement $d = 8.30$ cm for $t = 0$ s. See Fig. 10.37. Sketch two cycles of d as a function of t for the given values.

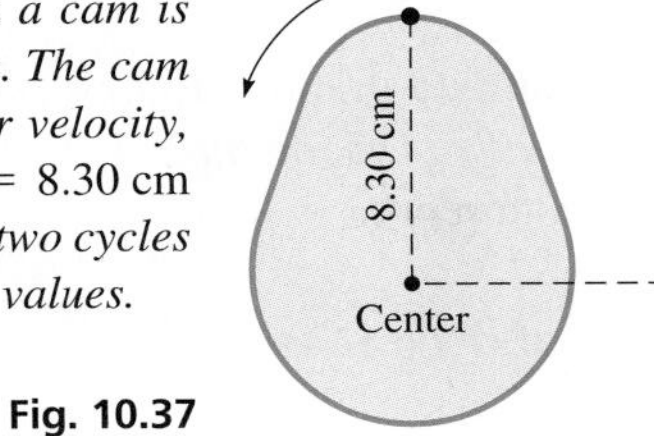

Fig. 10.37

9. $f = 3.20$ Hz

10. $\omega = 3.20$ rad/s

In Exercises 11 and 12, a satellite is orbiting Earth such that its displacement D north of the equator (or south if $D < 0$) is given by $D = A\sin(\omega t + \phi)$. Sketch two cycles of D as a function of t for the given values.

11. $A = 500$ mi, $\omega = 3.60$ rad/h, $\phi = 0$

12. $A = 850$ km, $f = 1.6 \times 10^{-4}$ Hz, $\phi = \pi/3$

In Exercises 13 and 14, for an alternating-current circuit in which the voltage e is given by $e = E\cos(\omega t + \phi)$, sketch two cycles of the voltage as a function of time for the given values.

13. $E = 170$ V, $f = 60.0$ Hz, $\phi = -\pi/3$

14. $E = 80$ mV, $\omega = 377$ rad/s, $\phi = \pi/2$

In Exercises 15 and 16, refer to the wave in the string described in Exercise 37 of Section 10.3. For a point on the string, the displacement y is given by $y = A\sin 2\pi\left(\frac{t}{T} - \frac{x}{\lambda}\right)$. We see that each point on the string moves with simple harmonic motion. Sketch two cycles of y as a function of t for the given values.

15. $A = 3.20$ mm, $T = 0.050$ s, $\lambda = 40.0$ mm, $x = 5.00$ mm

16. $A = 0.350$ in., $T = 0.250$ s, $\lambda = 24.0$ in., $x = 20.0$ in.

In Exercises 17 and 18, the air pressure within a plastic container changes above and below the external atmospheric pressure by $p = p_0 \sin 2\pi ft$. Sketch two cycles of $p = f(t)$ for the given values.

17. $p_0 = 2.80$ lb/in.^{2}, $f = 2.30$ Hz

18. $p_0 = 45.0$ kPa, $f = 0.450$ Hz

In Exercises 19–26, sketch the required curves.

19. The vertical position y (in m) of the tip of a high speed fan blade is given by $y = 0.10\cos 360t$, where t is in seconds. Use a calculator to graph two complete cycles of this function.

20. A study found that, when breathing normally, the increase in volume V (in L) of air in a person's lungs as a function of the time t (in s) is $V = 0.30\sin 0.50\pi t$. Sketch two cycles.

21. Sketch two cycles of the radio signal $e = 0.014\cos(2\pi ft + \pi/4)$ (e in volts, f in hertz, and t in seconds) for a station broadcasting with $f = 950$ kHz ("95" on the AM radio dial).

22. Sketch two cycles of the acoustical intensity I of the sound wave for which $I = A\cos(2\pi ft - \phi)$, given that t is in seconds, $A = 0.027$ W/cm^2, $f = 240$ Hz, and $\phi = 0.80$.

23. The rotating beacon of a parked police car is 12 m from a straight wall. (a) Sketch the graph of the length L of the light beam, where $L = 12\sec \pi t$, for $0 \leq t \leq 2.0$ s. (b) Which part(s) of the graph show meaningful values? Explain.

24. The motion of a piston of a car engine approximates simple harmonic motion. Given that the stroke (twice the amplitude) is 0.100 m, the engine runs at 2800 r/min, and the piston starts by moving upward from the middle of its stroke, find the equation for the displacement d as a function of t (in min). Sketch two cycles.

25. A riverboat's paddle has a 12-ft radius and rotates at 18 r/min. Find the equation of motion of the vertical displacement (from the center of the wheel) y of the end of a paddle as a function of the time t (in min) if the paddle is initially horizontal and moves upward when the wheel begins to turn. Sketch two cycles.

26. The sinusoidal electromagnetic wave emitted by an antenna in a cellular phone system has a frequency of 7.5×10^9 Hz and an amplitude of 0.045 V/m. Find the equation representing the wave if it starts at the origin. Sketch two cycles.

Answer to Practice Exercises

1. $d = 2.5\cos(\omega t + \pi/4)$ **2.** $f = 2/\pi$ Hz

10.6 Composite Trigonometric Curves

Additional of Ordinates • Parametric Equations • Lissajous Figures

Many applications involve functions that in themselves are a combination of two or more simpler functions. In this section, we discuss methods by which the curve of such a function can be found by combining values from the simpler functions.

EXAMPLE 1 Function that is sum of simpler functions

Sketch the graph of $y = 2 + \sin 2x$.

This function is the sum of the simpler functions $y_1 = 2$ and $y_2 = \sin 2x$. We may find values for y by adding 2 to each important value of $y_2 = \sin 2x$.

For $y_2 = \sin 2x$, the amplitude is 1, and the period is $2\pi/2 = \pi$. Therefore, we obtain the values in the following table and sketch the graph in Fig. 10.38.

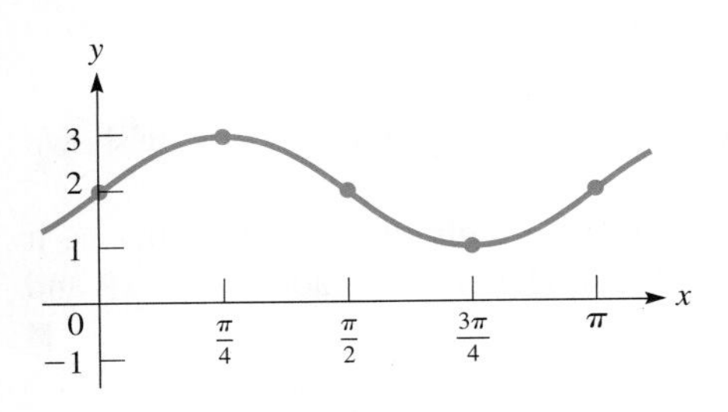

Fig. 10.38

x	0	$\frac{\pi}{4}$	$\frac{\pi}{2}$	$\frac{3\pi}{4}$	π
$\sin 2x$	0	1	0	−1	0
$2 + \sin 2x$	2	3	2	1	2

Note that this is a vertical shift of 2 units of the graph of $y = \sin 2x$, in the same way as discussed on page 105. ■

ADDITION OF ORDINATES

Another way to sketch the resulting graph is to *first sketch the two simpler curves and then add the y-values graphically. This method is called* **addition of ordinates** and is illustrated in the following example.

EXAMPLE 2 Addition of ordinates

Sketch the graph of $y = 2\cos x + \sin 2x$.

On the same set of coordinate axes, we sketch the curves $y = 2\cos x$ and $y = \sin 2x$. These are the dashed and solid curves shown in black in Fig. 10.39. For various values of x, we determine the distance above or below the x-axis of each curve and add these distances, noting that those above the axis are positive and those below the axis are negative. We thereby *graphically* ***add*** *the y-values* of these curves to get points on the resulting curve, shown in blue in Fig. 10.39.

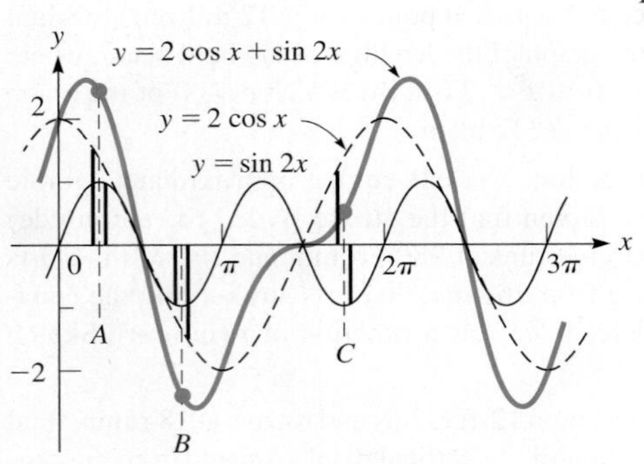

Fig. 10.39

At A, add the two lengths (shown side-by-side for clarity) to get the length for y. At B, both lengths are negative, and the value for y is the sum of these negative values. At C, one is positive and the other negative, and we must subtract the lower length from the upper one to get the length for y.

We combine these lengths for enough x-values to get a good curve. Some points are easily found. Where one curve crosses the x-axis, its value is zero, and the resulting curve has its point on the other curve. Here, where $\sin 2x$ is zero, the points for the resulting curve lie on the curve of $2\cos x$.

We should also add values where each curve is at its maximum or its minimum. *Extra care should be taken for those values of x for which one curve is* ***positive*** *and the other is* ***negative.*** ■

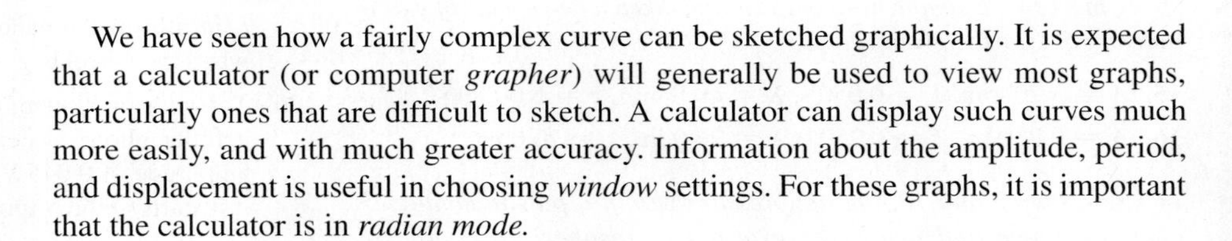

We have seen how a fairly complex curve can be sketched graphically. It is expected that a calculator (or computer *grapher*) will generally be used to view most graphs, particularly ones that are difficult to sketch. A calculator can display such curves much more easily, and with much greater accuracy. Information about the amplitude, period, and displacement is useful in choosing *window* settings. For these graphs, it is important that the calculator is in *radian mode.*

EXAMPLE 3 Calculator addition of ordinates

Use a calculator to display the graph of $y = \frac{x}{2} - \cos x$.

Here, we note that the curve is a combination of the straight line $y = x/2$ and the trigonometric curve $y = \cos x$. There are several good choices for the *window* settings, depending on how much of the curve is to be viewed. To see a little more than one period of $\cos x$, we can make the following choices:

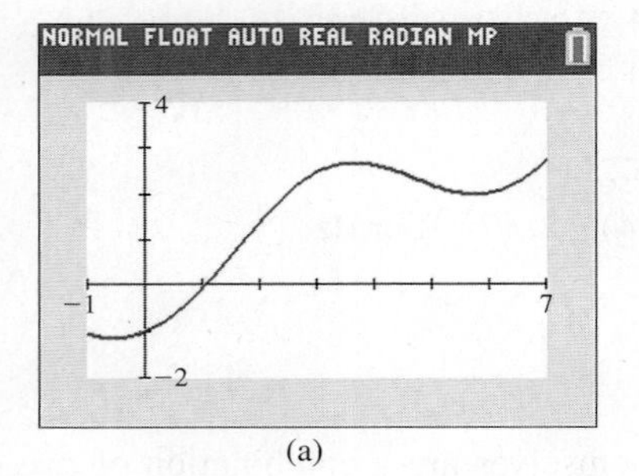

(a)

Xmin = −1 (to start to left of y-axis)

Xmax = 7 (period of $\cos x$ is $2\pi = 6.3$)

Ymin = −2 [line passes through (0, 0); amplitude of $y = \cos x$ is 1]

Ymax = 4 (slope of line is 1/2)

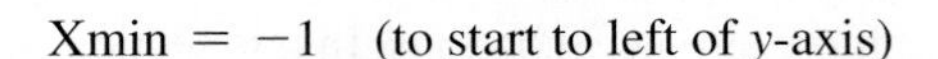

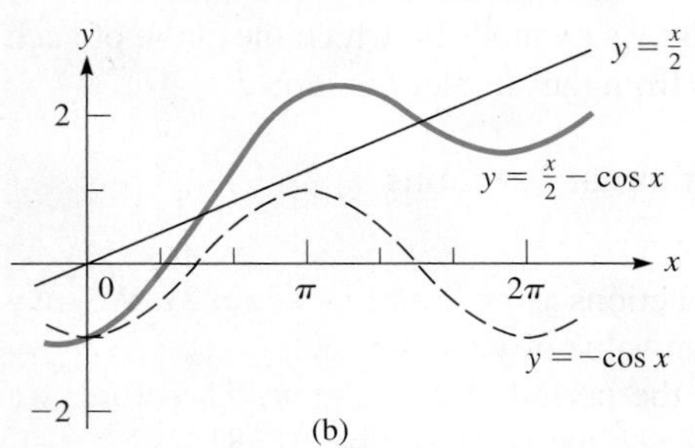

(b)

Fig. 10.40

See Fig. 10.40(a). The graphs of $y = x/2 - \cos x$, $y = x/2$, and $y = -\cos x$ are shown in Fig. 10.40(b).

We show $y = -\cos x$ rather than $y = \cos x$ because with addition of ordinates, it is easier to add graphic values than to subtract them. Here we can add $y_1 = x/2$ and $y_2 = -\cos x$ in order to get $y = x/2 - \cos x$. ■

EXAMPLE 4 Calculator addition of ordinates—string vibration

In a laboratory test, an instantaneous photo of a vibrating string showed that the vertical displacement y (in mm) of the string could be represented by $y = \cos \pi x - 2 \sin 2x$, where x is the distance (in mm) along the string. Display the graph of this function on a calculator.

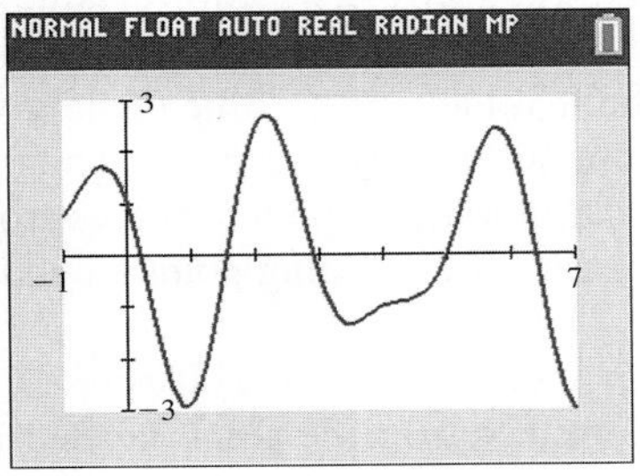

The combination of $y = \cos \pi x$ and $y = 2 \sin 2x$ leads to the following choices for the *window* settings:

Xmin $= -1$ (to start to the left of the y-axis)

Xmax $= 7$ (the periods are 2 and π; this shows at least two periods of each)

Ymin $= -3$, Ymax $= 3$ (the sum of the amplitudes is 3)

Fig. 10.41

There are many possible choices for Xmin and Xmax to get a good view of the graph on a calculator. However, because the sum of the amplitudes is 3, note that the curve cannot be below $y = -3$ or above $y = 3$. The calculator view is shown in Fig. 10.41.

This graph can be constructed by using addition of ordinates, although it is difficult to do very accurately. ■

LISSAJOUS FIGURES

An important application of trigonometric curves is made when they are added at *right angles.* The methods for doing this are shown in the following examples.

EXAMPLE 5 Graphing parametric equations

Plot the graph for which the values of x and y are given by the equations $y = \sin 2\pi t$ and $x = 2 \cos \pi t$. *Equations given in this form, x and y in terms of a third variable, are called* **parametric equations.**

Because both x and y are in terms of t, by assuming values of t, we find corresponding values of x and y and use these values to plot the graph. Because the periods of $\sin 2\pi t$ and $2 \cos \pi t$ are $t = 1$ and $t = 2$, respectively, we will use values of $t = 0, 1/4, 1/2, 3/4, 1$, and so on. These give us convenient values of $0, \pi/4, \pi/2, 3\pi/4, \pi$, and so on to use in the table. We plot the points in Fig. 10.42.

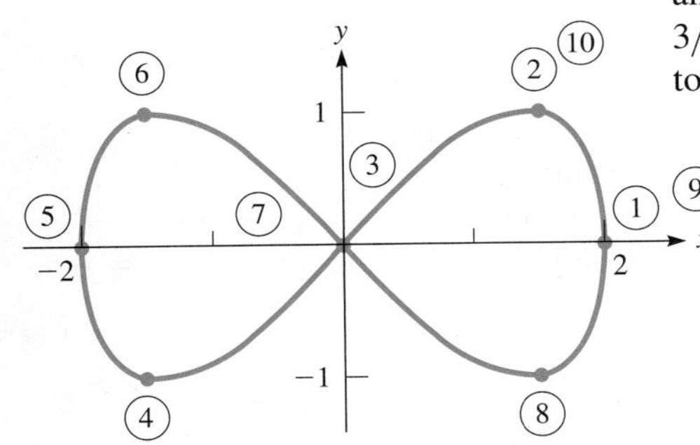

Fig. 10.42

t	0	$\frac{1}{4}$	$\frac{1}{2}$	$\frac{3}{4}$	1	$\frac{5}{4}$	$\frac{3}{2}$	$\frac{7}{4}$	2	$\frac{9}{4}$
x	2	1.4	0	−1.4	−2	−1.4	0	1.4	2	1.4
y	0	1	0	−1	0	1	0	−1	0	1
Point number	1	2	3	4	5	6	7	8	9	10

■

Because x and y are trigonometric functions of a third variable t, and because the x-axis is at right angles to the y-axis, values of x and y obtained in this way result in a combination of two trigonometric curves at right angles. *Figures obtained in this way are called* **Lissajous figures.** Note that the Lissajous figure in Fig. 10.42 *is not a function* since there are *two* values of y for each value of x (except $x = -2, 0, 2$) in the domain.

■ Named for the French physicist Jules Lissajous (1822–1880).

In practice, Lissajous figures can be displayed on an *oscilloscope* by applying an electric signal between a pair of horizontal plates and another signal between a pair of vertical plates. These signals are then seen on the screen of the oscilloscope. This type of screen (a cathode-ray tube) is similar to that used on most older TV sets.

There is additional coverage of parametric equations in Chapter 21. They are also used in some topics in the study of calculus.

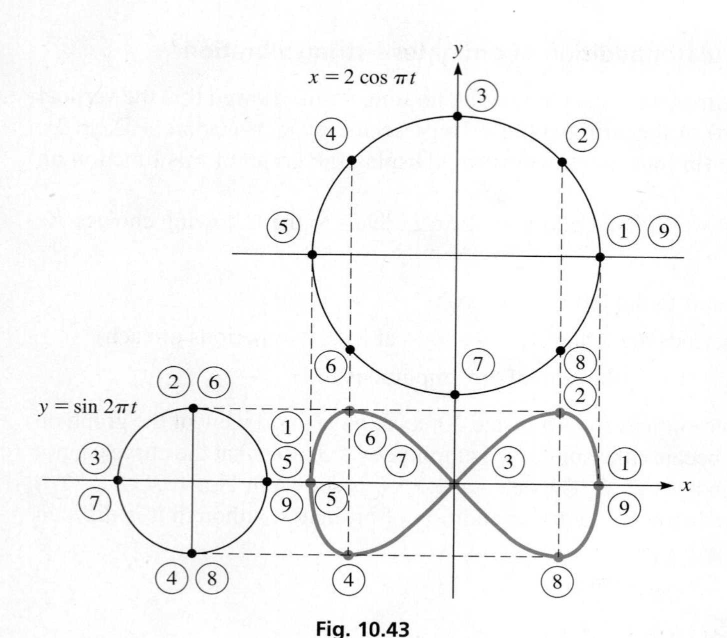

Fig. 10.43

■ See the chapter introduction.

EXAMPLE 6 Graphing Lissajous figures

If a circle is placed on the x-axis and another on the y-axis, we may represent the coordinates (x, y) for the curve of Example 5 by the lengths of the projections (see Example 1 of Section 10.5) of a point moving around each circle. A careful study of Fig. 10.43 will clarify this. We note that the radius of the circle giving the x-values is 2 and that the radius of the circle giving the y-values is 1. This is due to the way in which x and y are defined. Also, due to these definitions, the point revolves around the y-circle twice as fast as the corresponding point around the x-circle.

When $t = 0$, we get the point labeled 1 on the blue curve, with its x- and y-coordinates determined from the points labeled 1 on the top and left circle respectively. Points labeled 2, 3, 4, . . . are found similarly for $t = \frac{1}{4}, \frac{1}{2}, \frac{3}{4}$, etc.

On an oscilloscope, this curve would result when two electric signals are used, with the first having twice the amplitude and one-half the frequency of the second. ■

Most graphing calculators can be used to display a curve defined by parametric equations. It is necessary to use the *mode* feature and select *parametric equations.* Use the manual for the calculator, as there are some differences as to how this is done on various calculators.

EXAMPLE 7 Calculator Lissajous figures

Use a calculator to display the graph defined by the parametric equations $x = 2 \cos \pi t$ and $y = \sin 2\pi t$. These are the same equations as those used in Examples 5 and 6.

First, select the parametric equation option from the *mode* feature and enter the parametric equations $x_{1T} = 2 \cos \pi t$ and $y_{1T} = \sin 2\pi t$. Then make the following *window* settings:

Tmin = 0 (standard default settings, and the usual choice)
Tmax = 2 (the periods are 2 and 1; the longer period is 2)
Tstep = 0.01 (small enough to give a nice smooth curve)
Xmin = −2, Xmax = 2 (smallest and largest possible values of x), Xscl = 1
Ymin = −1, Ymax = 1 (smallest and largest possible values of y), Yscl = 0.5

The calculator graph is shown in Fig. 10.44. ■

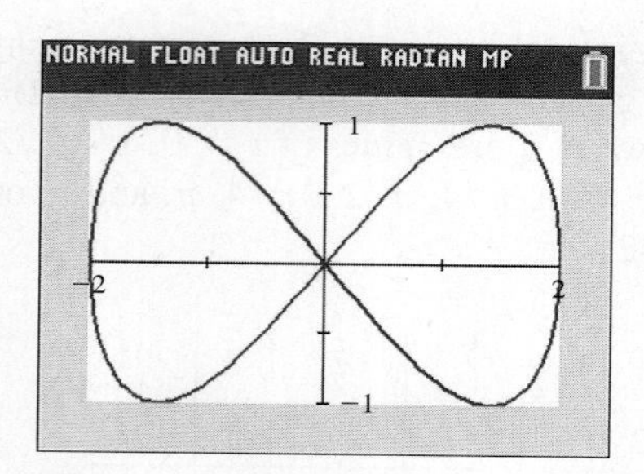

Fig. 10.44

EXERCISES 10.6

In Exercises 1–8, sketch the curves of the given functions by addition of ordinates.

1. $y = 1 + \sin x$
2. $y = 3 - 2 \cos x$
3. $y = \frac{1}{3}x + \sin 2x$
4. $y = x - \sin x$
5. $y = \frac{1}{10}x^2 - \sin \pi x$
6. $y = \frac{1}{4}x^2 + \cos 3x$
7. $y = \sin x + \cos x$
8. $y = \sin x + \sin 2x$

In Exercises 9–20, display the graphs of the given functions on a calculator.

9. $y = x^3 + 10 \sin 2x$
10. $y = \dfrac{1}{x^2 + 1} - \cos \pi x$
11. $y = \sin x - 1.5 \sin 2x$
12. $y = \cos 3x - 3 \sin x$
13. $y = 20 \cos 2x + 30 \sin x$
14. $y = \frac{1}{2} \sin 4x + \cos 2x$
15. $y = 2 \sin x - \cos 1.5x$
16. $y = 8 \sin 0.5x - 12 \sin x$
17. $y = \sin \pi x - \cos 2x$
18. $y = 2 \cos 4x - \cos\left(x - \dfrac{\pi}{4}\right)$
19. $y = 2 \sin\left(2x - \dfrac{\pi}{6}\right) + \cos\left(2x + \dfrac{\pi}{3}\right)$
20. $y = 3 \cos 2\pi x + \sin \frac{\pi}{2}x$

In Exercises 21–26, plot the Lissajous figures.

21. $x = 8 \cos t,\ y = 5 \sin t$

22. $x = 5 \cos t + 2,\ y = 5 \sin t - 3$

23. $x = 2 \sin t,\ y = 3 \sin t$

24. $x = 2 \cos t,\ y = \cos(t + 4)$

25. $x = \sin \pi t,\ y = 2 \cos 2\pi t$

26. $x = \cos\left(t + \frac{\pi}{4}\right),\ y = \sin 2t$

In Exercises 27–34, use a calculator to display the Lissijous figures defined by the given parametric equations.

27. $x = \cos \pi\left(t + \frac{1}{6}\right),\ y = 2 \sin \pi t$

28. $x = \sin^2 \pi t,\ y = \cos \pi t$

29. $x = 4 \cos 3t,\ y = \cos 2t$

30. $x = 2 \sin \pi t,\ y = 3 \sin 3\pi t$

31. $x = t \cos t,\ y = t \sin t\ (t \geq 0)$

32. $x = \frac{3t}{1 + t^3},\ y = \frac{3t^2}{1 + t^3}$

33. $x = 2 \cot t,\ y = 1 - \cos 2t$

34. $x = 2 \sin 2t,\ y = \frac{2 \sin^3 2t}{\cos t}$

Parametric equations can be defined for functions other than trigonometric functions. In Exercises 35 and 36, draw the graphs of the indicated parametric equations. In Exercises 37 and 38, display the graphs on a calculator.

35. $x = \sqrt{t} + 4,\ y = 3\sqrt{t} - 1$

36. $x = t^2 + 1,\ y = t - 1$

37. $x = 4 - t,\ y = t^3$

38. $x = \sqrt[5]{2t - 1},\ y = 2t^2 - 3$

In Exercises 39–50, sketch the appropriate curves. A calculator may be used.

39. An object oscillating on a spring has a displacement (in ft) given by $y = 0.4 \sin 4t + 0.3 \cos 4t$, where t is the time (in s). Sketch the graph.

40. The voltage e in a certain electric circuit is given by $e = 50 \sin 50\pi t + 80 \sin 60\pi t$, where t is the time (in s). Sketch the graph.

41. An analysis of the temperature records of Louisville, Kentucky, indicates that the average daily temperature T (in °F) during the year is approximately $T = 56 - 22 \cos\left[\frac{\pi}{6}(x - 0.5)\right]$, where x is measured in months ($x = 0.5$ is Jan. 15, etc.). Sketch the graph of T vs. x for one year.

42. An analysis of data shows that the mean density d (in mg/cm^3) of a calcium compound in the bones of women is given by $d = 139.3 + 48.6 \sin(0.0674x - 0.210)$, where x represents the ages of women ($20 \leq x \leq 80$ years). (A woman is considered to be osteoporotic if $d < 115$ mg/cm^3.) Sketch the graph.

43. A normal person with a pulse rate of 60 beats/min has a blood pressure of "120 over 80." This means the pressure is oscillating between a high (systolic) of 120 mm of mercury (shown as mmHg) and a low (diastolic) of 80 mmHg. Assuming a sinusoidal type of function, find the pressure p as a function of the time t if the initial pressure is 120 mmHg. Sketch the graph for the first 5 s.

44. The strain e (dimensionless) on a cable caused by vibration is $e = 0.0080 - 0.0020 \sin 30t + 0.0040 \cos 10t$, where t is measured in seconds. Sketch two cycles of e as a function of t.

45. The intensity I of an alarm (in dB—decibel) signal is given by $I = 40 + 50 \sin t - 20 \cos 2t$, where t is measured in seconds. Display two cycles of I as a function of t on a calculator.

46. The available solar energy depends on the amount of sunlight, and the available time in a day for sunlight depends on the time of the year. An approximate correction factor (in min) to standard time is $C = 10 \sin \frac{1}{29}(n - 80) - 7.5 \cos \frac{1}{58}(n - 80)$, where n is the number of the day of the year. Sketch C as a function of n.

47. Two signals are seen on an oscilloscope as being at right angles. The equations for the displacements of these signals are $x = 4 \cos \pi t$ and $y = 2 \sin 3\pi t$. Sketch the figure that appears on the oscilloscope.

48. In the study of optics, light is said to be *elliptically polarized* if certain optic vibrations are out of phase. These may be represented by Lissajous figures. Determine the Lissajous figure for two light waves given by $w_1 = \sin \omega t$ and $w_2 = \sin\left(\omega t + \frac{\pi}{4}\right)$.

49. In checking electric circuit elements, a *square wave* such as that shown in Fig. 10.45 may be displayed on an oscilloscope. Display the graph of $y = 1 + \frac{4}{\pi} \sin\left(\frac{\pi x}{4}\right) + \frac{4}{3\pi} \sin\left(\frac{3\pi x}{4}\right)$ on a calculator, and compare it with Fig. 10.45. This equation gives the first three terms of a *Fourier series*. As more terms of the series are added, the approximation to a square wave is better.

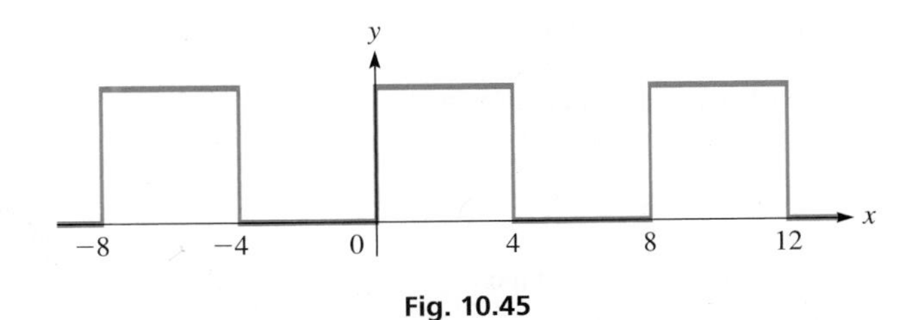

Fig. 10.45

50. Another type of display on an oscilloscope may be a *sawtooth wave* such as that shown in Fig. 10.46. Display the graph of $y = 1 - \frac{8}{\pi^2}\left(\cos \frac{\pi x}{2} + \frac{1}{9} \cos \frac{3\pi x}{2}\right)$ on a calculator and compare it with Fig. 10.46. See Exercise 49.

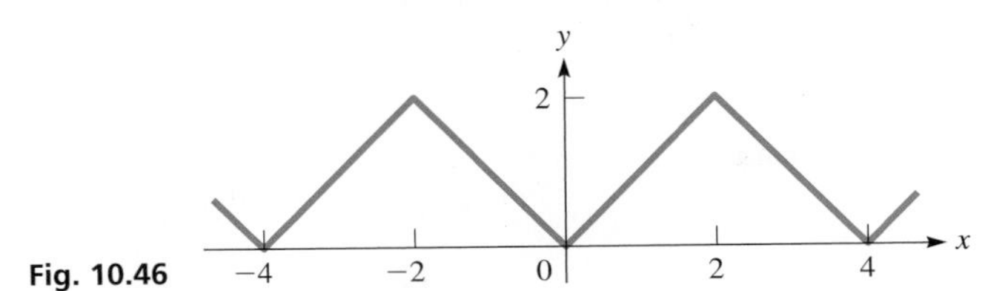

Fig. 10.46

CHAPTER 10 KEY FORMULAS AND EQUATIONS

For the graphs of $y = a \sin(bx + c)$ and $y = a \cos(bx + c)$

$$\text{Amplitude} = |a| \qquad \text{Period} = \frac{2\pi}{b} \qquad \text{Displacement} = -\frac{c}{b} \qquad (10.1)$$

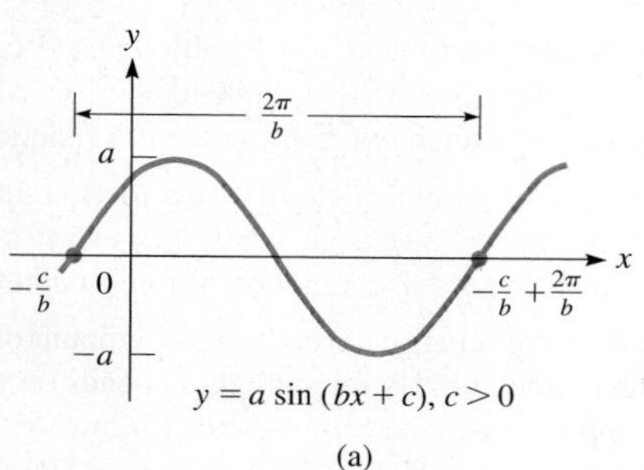

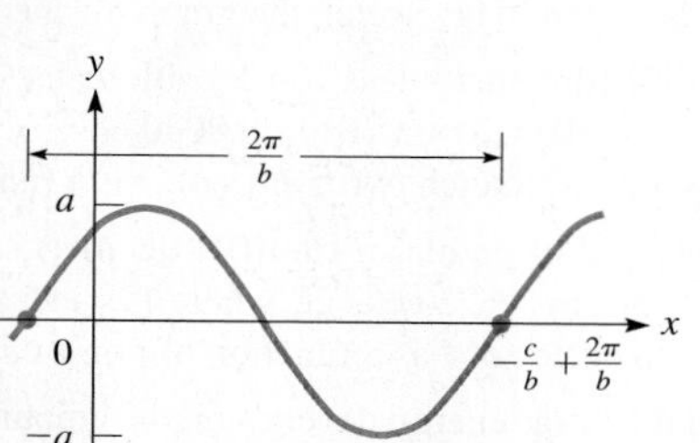

$y = a \sin(bx + c),\ c > 0$

(a)

For each $a > 0,\ b > 0$

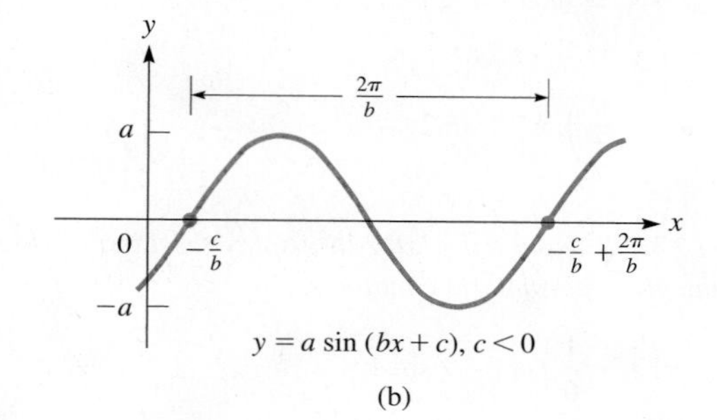

$y = a \sin(bx + c),\ c < 0$

(b)

Reciprocal relationships

$$\csc x = \frac{1}{\sin x} \qquad \sec x = \frac{1}{\cos x} \qquad \cot x = \frac{1}{\tan x} \qquad (10.2)$$

Simple harmonic motion

$$d = R \sin \omega t \qquad (10.3)$$

Angular velocity and frequency

$$\omega = 2\pi f \qquad (10.4)$$

CHAPTER 10 REVIEW EXERCISES

CONCEPT CHECK EXERCISES

*Determine each of the following as being either **true** or **false**. If it is false, explain why.*

1. A point on the graph of $y = 2 \cos x$ is $(\pi, -2)$.
2. The period of the graph of $y = 4 \sin \frac{1}{2}\pi x$ is 4.
3. The displacement of the graph of $y = -2 \cos\left(2x - \frac{\pi}{4}\right)$ is $-\frac{\pi}{8}$.
4. The amplitude of the graph of $y = 3 \tan 2x$ is 3.
5. The frequency of an alternating current described by the equation $i = 8 \sin 60\pi t$ (t in s) is 120 Hz.
6. The graph of $y = \sin x - 2 \cos \frac{1}{2}x$ passes through $(\pi, 0)$.

PRACTICE AND APPLICATIONS

In Exercises 7–34, sketch the curves of the given trigonometric functions. Check each using a calculator.

7. $y = \frac{2}{3} \sin x$
8. $y = -4 \sin x$
9. $y = -2 \cos x$
10. $y = 2.3 \cos(-x)$
11. $y = 2 \sin 3x$
12. $y = 4.5 \sin 12x$
13. $y = 0.4 \cos 4x$
14. $y = 24 \cos 6x$
15. $y = -3 \cos \frac{1}{3}x$
16. $y = 3 \sin(-0.5x)$
17. $y = \sin \pi x$
18. $y = 36 \sin 4\pi x$
19. $y = 5 \cos\left(\frac{\pi x}{2}\right)$
20. $y = -\cos 6\pi x$
21. $y = -0.5 \sin\left(-\frac{\pi x}{6}\right)$
22. $y = 8 \sin \frac{\pi}{4}x$
23. $y = 12 \sin\left(3x - \frac{\pi}{2}\right)$
24. $y = 3 \sin\left(\frac{x}{2} + \frac{\pi}{2}\right)$
25. $y = -2 \cos(4x + \pi)$
26. $y = 0.8 \cos\left(\frac{x}{6} - \frac{\pi}{3}\right)$
27. $y = -\sin\left(\pi x + \frac{\pi}{6}\right)$
28. $y = 250 \sin(3\pi x - \pi)$
29. $y = 8 \cos\left(4\pi x - \frac{\pi}{2}\right)$
30. $y = 3 \cos(2\pi x + \pi)$
31. $y = 0.3 \tan 0.5x$
32. $y = \frac{1}{4} \sec x$
33. $y = -3 \csc x$
34. $y = -5 \cot \pi x$

In Exercises 35–38, sketch the curves of the given functions by addition of ordinates.

35. $y = 2 + \frac{1}{2} \sin 2x$
36. $y = \frac{1}{2}x - \cos \frac{1}{3}x$
37. $y = \sin 2x + 3 \cos x$
38. $y = \sin 3x + 2 \cos 2x$

In Exercises 39–46, display the curves of the given functions on a calculator.

39. $y = 2 \sin x - \cos 2x$
40. $y = 10 \sin 3x - 20 \cos x$
41. $y = \cos\left(x + \frac{\pi}{4}\right) - 0.4 \sin 2x$
42. $y = 2 \cos \pi x + \cos(2\pi x - \pi)$
43. $y = \frac{\sin x}{x}$
44. $y = \sqrt{x} \sin 0.5x$

45. $y = \sin^2 x + \cos^2 x \qquad (\sin^2 x = (\sin x)^2)$
What conclusion can be drawn from the graph?

46. $y = \sin\left(x + \frac{\pi}{4}\right) - \cos\left(x - \frac{\pi}{4}\right) + 1$

What conclusion can be drawn from the graph?

In Exercises 47–50, give the specific form of the indicated equation by determining a, b, and c through an inspection of the given curve.

47. $y = a \sin(bx + c)$
(Figure 10.47)

48. $y = a \cos(bx + c)$
(Figure 10.47)

49. $y = a \cos(bx + c)$
(Figure 10.48)

50. $y = a \sin(bx + c)$
(Figure 10.48)

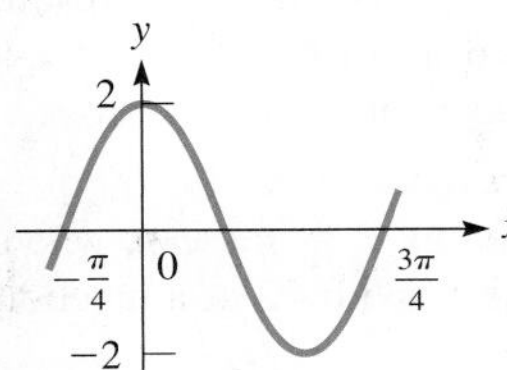

Fig. 10.47

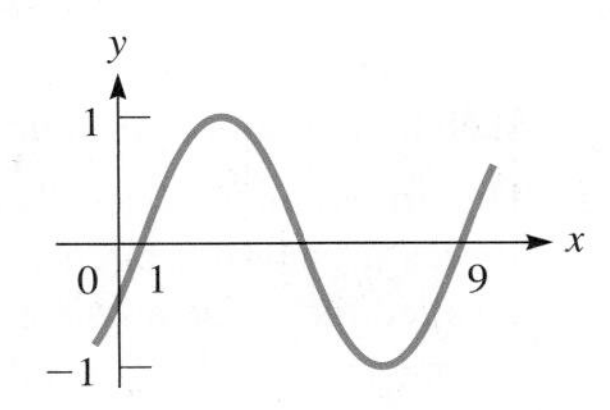

Fig. 10.48

In Exercises 51–56, display the Lissajous figures on a calculator.

51. $x = 5 \cos t + 3, y = 9 \sin t - 1$

52. $x = 10 \cos^3 t, y = 10 \sin^3 t$

53. $x = -\cos 2\pi t, y = 2 \sin \pi t$

54. $x = \sin\left(t + \frac{\pi}{6}\right), y = \sin t$

55. $x = 2 \cos\left(2\pi t + \frac{\pi}{4}\right), y = \cos \pi t$

56. $x = \cos\left(t - \frac{\pi}{6}\right), y = \cos\left(2t + \frac{\pi}{3}\right)$

In Exercises 57–68, solve the given problems.

57. Display the function $y = 2|\sin 0.2\pi x| - |\cos 0.4\pi x|$ on a graphing calculator.

58. Display the function $y = 0.2|\tan 2x|$ on a calculator.

59. Show that $\cos(x + \frac{\pi}{4}) = \sin(\frac{\pi}{4} - x)$ on a calculator.

60. Show that $\tan(x - \frac{\pi}{3}) = -\tan(\frac{\pi}{3} - x)$ on a calculator.

61. What is the period of the function $y = 2 \cos 0.5x + \sin 3x$?

62. What is the period of the function $y = \sin \pi x + 3 \sin 0.25\pi x$?

63. Find the function and graph it if it is of the form $y = a \sin x$ and passes through $(5\pi/2, 3)$.

64. Find the function and graph it if it is of the form $y = a \cos x$ and passes through $(4\pi, -3)$.

65. Find the function and graph it if it is of the form $y = 3 \cos bx$ and passes through $(\pi/3, -3)$ and b has the smallest possible positive value.

66. Find the function and graph it if it is of the form $y = 3 \sin bx$ and passes through $(\pi/3, 0)$ and b has the smallest possible positive value.

67. Find the function and graph it for a function of the form $y = 3 \sin(\pi x + c)$ that passes through $(-0.25, 0)$ and for which c has the smallest possible positive value.

68. Write the equation of the cosecant function with zero displacement, a period of 2, and that passes through (0.5, 4).

In Exercises 69–94, sketch the appropriate curves. A calculator may be used.

69. The range R of a rocket is given by $R = \frac{v_0^2 \sin 2\theta}{g}$. Sketch R as a function of θ for $v_0 = 1000$ m/s and $g = 9.8$ m/s^2. See Fig. 10.49.

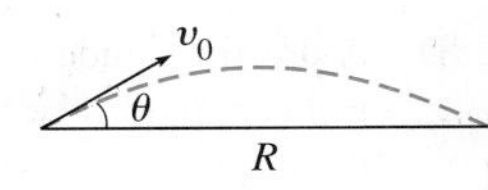

Fig. 10.49

70. The blade of a saber saw moves vertically up and down at 18 strokes per second. The vertical displacement y (in cm) is given by $y = 1.2 \sin 36\pi t$, where t is in seconds. Sketch at least two cycles of the graph of y vs. t.

71. The velocity v (in cm/s) of a piston in a certain engine is given by $v = \omega D \cos \omega t$, where ω is the angular velocity of the crankshaft in radians per second and t is the time in seconds. Sketch the graph of v vs. t if the engine is at 3000 r/min and $D = 3.6$ cm.

72. A light wave for the color yellow can be represented by the equation $y = A \sin 3.4 \times 10^{15}\, t$. With A as a constant, sketch two cycles of y as a function of t (in s).

73. The electric current i (in A) in a circuit in which there is a *full-wave rectifier* is $i = 10|\sin 120\pi t|$. Sketch the graph of $i = f(t)$ for $0 \le t \le 0.05$s. What is the period of the current?

74. A circular disk suspended by a thin wire attached to the center of one of its flat faces is twisted through an angle θ. Torsion in the wire tends to turn the disk back in the opposite direction (thus, the name *torsion pendulum* is given to this device). The angular displacement θ (in rad) as a function of time t (in s) is $\theta = \theta_0 \cos(\omega t + \phi)$, where θ_0 is the maximum angular displacement, ω is a constant that depends on the properties of the disk and wire, and ϕ is the phase angle. Sketch the graph of θ vs. t if $\theta_0 = 0.100$ rad, $\omega = 2.50$ rad/s, and $\phi = \pi/4$. See Fig. 10.50.

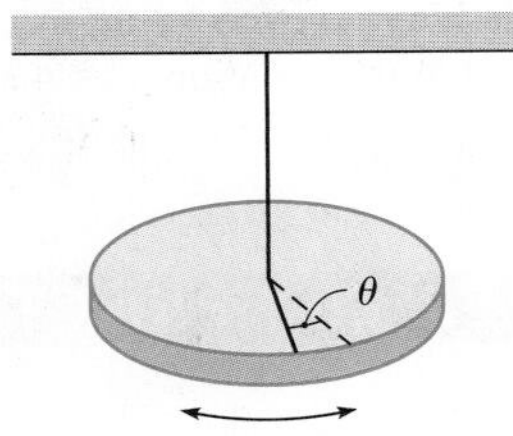

Fig. 10.50

75. The vertical displacement y of a point at the end of a propeller blade of a small boat is $y = 14.0 \sin 40.0\pi t$. Sketch two cycles of y (in cm) as a function of t (in s).

76. In optics, two waves are said to interfere destructively if, when they both pass through a medium, the amplitude of the resulting wave is zero. Sketch the graph of $y = \sin x + \cos(x + \pi/2)$ and find whether or not it would represent destructive interference of two waves.

77. The vertical displacement y (in ft) of a buoy floating in water is given by $y = 3.0 \cos 0.2t + 1.0 \sin 0.4t$, where t is in seconds. Sketch the graph of y as a function of t for the first 40 s.

78. Find the simplest function that represents the vertical projection y of a 12-mm sweep second hand on a watch as a function of time t (in s).

79. The *London Eye* Ferris wheel has a circumference of 424 m, and it takes 30 min for one complete revolution. Find the equation for the height y (in m) above the bottom as a function of the time t (in min). Sketch the graph.

80. The vertical motion of a rubber raft on a lake approximates simple harmonic motion due to the waves. If the amplitude of the motion is 0.250 m and the period is 3.00 s, find an equation for the vertical displacement y as a function of the time t. Sketch two cycles.

81. A drafting student draws a circle through the three vertices of a right triangle. The hypotenuse of the triangle is the diameter d of the circle, and from Fig. 10.51, we see that $d = a \sec \theta$. Sketch the graph of d as a function of θ for $a = 3.00$ in.

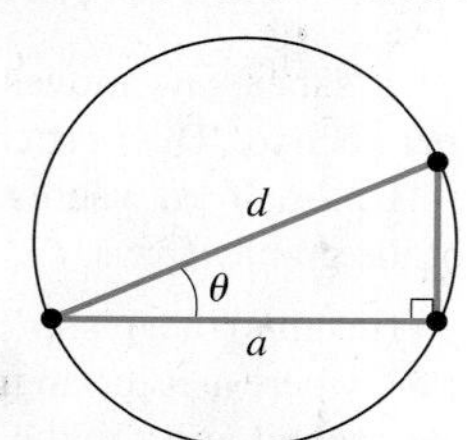

Fig. 10.51

82. The height h (in m) of a certain rocket ascending vertically is given by $h = 800 \tan \theta$, where θ is the angle of elevation from an observer 800 m from the launch pad. Sketch h as a function of θ.

83. At 40°N latitude the number of hours h of daylight each day during the year is approximately $h = 12.2 + 2.8 \sin\left[\frac{\pi}{6}(x - 2.7)\right]$, where x is measured in months ($x = 0.5$ is Jan. 15, etc.). Sketch the graph of h vs. x for one year. (Some of the cities near 40°N are Philadelphia, Madrid, Naples, Ankara, and Beijing.)

84. The equation in Exercise 83 can be used for the number of hours of daylight at 40°S latitude with the appropriate change. Explain what change is necessary and determine the proper equation. Sketch the graph. (This would be appropriate for southern Argentina and Wellington, New Zealand.)

85. If the upper end of a spring is not fixed and is being moved with a sinusoidal motion, the motion of the bob at the end of the spring is affected. Sketch the curve if the motion of the upper end of a spring is being moved by an external force and the bob moves according to the equation $y = 4 \sin 2t - 2 \cos 2t$.

86. One wave in a medium will affect another wave in the same medium depending on the difference in displacements. For water waves $y_1 = 2 \cos x$, $y_2 = 4 \sin(x + \pi/2)$, and $y_3 = 4 \sin(x + 3\pi/2)$, show this effect by graphing $y_1 + y_2$, $y_1 + y_3$, and $y_2 + y_3$.

87. The height y (in cm) of an irregular wave in a string as a function of time t (in s) is $y = 0.15 - \cos 0.25t + 0.50 \sin 0.5t$. Sketch the graph.

88. The loudness L (in decibels) of a fire siren as a function of the time t (in s) is approximately $L = 40 - 35 \cos 2t + 60 \sin t$. Sketch this function for $0 \le t \le 10$ s.

89. The path of a roller mechanism used in an assembly-line process is given by $x = \theta - \sin \theta$ and $y = 1 - \cos \theta$. Sketch the path for $0 \le \theta \le 2\pi$.

90. The equations for two microwave signals that give a resulting curve on an oscilloscope are $x = 6 \sin \pi t$ and $y = 4 \cos 4\pi t$. Sketch the graph of the curve displayed on the oscilloscope.

91. The impedance Z (in Ω) and resistance R (in Ω) for an alternating-current circuit are related by $Z = R \sec \phi$, where ϕ is called the phase angle. Sketch the graph for Z as a function of ϕ for $-\pi/2 < \phi < \pi/2$.

92. The current in a certain alternating-current circuit is given by $i = 2.0 \sin(120\pi t)$. Find the function for voltage if the amplitude is 5.0 V and voltage lags current by 30°. Graph both functions in the same window for $0 \le t \le 0.02$ s.

93. The charge q (in C) on a certain capacitor as a function of the time t (in s) is given by $q = 0.0003(3 - 2 \sin 100t \cos 100t)$. Sketch one complete cycle of q vs. t.

94. The instantaneous power p (in W) in an electric circuit is defined as the product of the instantaneous voltage e and the instantaneous current i (in A). If we have $e = 100 \cos 200t$ and $i = 2 \cos(200t + \frac{\pi}{4})$, plot the graph e vs. t and the graph of i vs. t on the same coordinate system. Then sketch the graph of p vs. t by multiplying appropriate values of e and i.

95. A wave passing through a string can be described at any instant by the equation $y = a \sin(bx + c)$. Write one or two paragraphs explaining the change in the wave (a) if a is doubled, (b) if b is doubled, and (c) if c is doubled.

CHAPTER 10 PRACTICE TEST

As a study aid, we have included complete solutions for each Chapter Test problem at the back of this book.

1. Determine the amplitude, period, and displacement of the function $y = -3 \sin(4\pi x - \pi/3)$.

In Problems 2–5, sketch the graphs of the given functions.

2. $y = 0.5 \cos \frac{\pi}{2}x$

3. $y = 2 + 3 \sin x$

4. $y = 3 \sec x$

5. $y = 2 \sin(2x - \frac{\pi}{3})$

6. A wave is traveling in a string. The displacement y (in in.) as a function of the time t (in s) from its equilibrium position is given by $y = A \cos(2\pi/T)t$. T is the period (in s) of the motion. If $A = 0.200$ in. and $T = 0.100$ s, sketch two cycles of y vs t.

7. Sketch the graph of $y = 2 \sin x + \cos 2x$ by addition of ordinates.

8. Use a calculator to display the Lissajous figure for which $x = \sin \pi t$ and $y = 2 \cos 2\pi t$.

9. Sketch two cycles of the curve of a projection of the end of a radius on the y-axis. The radius is of length R and it is rotating counterclockwise about the origin at 2.00 rad/s. It starts at an angle of $\pi/6$ with the positive x-axis.

10. Find the function of the form $y = 2 \sin bx$ if its graph passes through $(\pi/3, 2)$ and b is the smallest possible positive value. Then graph the function.

Exponents and Radicals

11

For use in later chapters, we now further develop the use of exponents and radicals.

In previous chapters, we have used only exponents that are integers, and by introducing exponents that are fractions we will show the relationship between exponents and radicals (for example, we will show that $\sqrt{x} = x^{1/2}$). In more advanced math and in applications, it is more convenient to use fractional exponents rather than radicals.

As we noted in Chapter 6, the use of symbols led to advances in mathematics and science. As variables became commonly used in the 1600s, it was common to write, for example, x^3 as xxx. For larger powers, this is obviously inconvenient, and the modern use of exponents came into use. The first to use exponents consistently was the French mathematician René Descartes in the 1630s.

The meaning of negative and fractional exponents was first found by the English mathematician Wallis in the 1650s, although he did not write them as we do today. In the 1670s, it was the great English mathematician and physicist Isaac Newton who first used all exponents (positive, negative, and fractional) with modern notation. This improvement in notation made the development of many areas of mathematics, particularly calculus, easier. In this way, it led to many advances in the applications of mathematics.

As we develop the various operations with exponents and radicals, we will show their uses in some technical areas of application. They are used in a number of formulas in areas such as electronics, hydrodynamics, optics, solar energy, and machine design.

LEARNING OUTCOMES

After completion of this chapter, the student should be able to:

- Use the laws of exponents to simplify expressions
- Convert radicals to fractional exponents and vice versa
- Perform mathematical operations with exponents and radicals
- Simplify expressions containing radicals
- Rationalize a denominator containing radicals

◀ In finding the rate at which solar radiation changes at a solar-energy collector, the following expression is found:

$$\frac{(t^4 + 100)^{1/2} - 2t^3(t + 6)(t^4 + 100)^{-1/2}}{[(t^4 + 100)^{1/2}]^2}$$

In Section 11.2, we show that this can be written in a much simpler form.

11.1 Simplifying Expressions with Integer Exponents

Laws of Exponents • Simplifying Expressions • Zero and Negative Exponents

The laws of exponents were given in Section 1.4. For reference, they are

$$a^m \times a^n = a^{m+n} \quad \textbf{(11.1)}$$

$$\frac{a^m}{a^n} = a^{m-n} \quad \text{or} \quad \frac{a^m}{a^n} = \frac{1}{a^{n-m}} \quad (a \neq 0) \quad \textbf{(11.2)}$$

$$(a^m)^n = a^{mn} \quad \textbf{(11.3)}$$

$$(ab)^n = a^n b^n, \quad \left(\frac{a}{b}\right)^n = \frac{a^n}{b^n} \quad (b \neq 0) \quad \textbf{(11.4)}$$

$$a^0 = 1 \quad (a \neq 0) \quad \textbf{(11.5)}$$

$$a^{-n} = \frac{1}{a^n} \quad (a \neq 0) \quad \textbf{(11.6)}$$

Although Eqs. (11.1) to (11.4) were originally defined for positive integers as exponents, we showed in Section 1.4 that with the definitions in Eqs. (11.5) and (11.6), they are valid for all integer exponents. Later in the chapter, we will show how fractions are used as exponents. These equations are very important to the developments of this chapter and should be reviewed and learned thoroughly.

EXAMPLE 1 Simplifying basic expressions

(a) Applying Eq. (11.1) and then Eq. (11.6), we have

$$a^3 \times a^{-5} = a^{3-5} = a^{-2} = \frac{1}{a^2}$$

Negative exponents are generally not used in the expression of the final result, unless specified otherwise. However, they are often used in intermediate steps.

(b) Applying Eq. (11.1), then (11.6), and then (11.4), we have

$$(10^3 \times 10^{-4})^2 = (10^{3-4})^2 = (10^{-1})^2 = \left(\frac{1}{10}\right)^2 = \frac{1}{10^2} = \frac{1}{100}$$

NOTE ➧ The result is in proper form as either $1/10^2$ or $1/100$. If the exponent is large, it is common to leave the exponent in the answer. [If the power of ten is being used as part of scientific notation, then the form with the negative exponent 10^{-2} is used.] ■

EXAMPLE 2 Simplifying basic expressions

(a) Applying Eqs. (11.2) and (11.5), we have

$$\frac{a^2b^3c^0}{ab^7} = \frac{a^{2-1}(1)}{b^{7-3}} = \frac{a}{b^4}$$

(b) Applying Eqs. (11.4) and (11.3), and then (11.6), we have

$$(x^{-2}y)^3 = (x^{-2})^3(y^3) = x^{-6}y^3 = \frac{y^3}{x^6}$$ ■

Practice Exercise

1. Simplify: $a(x^{-1}y^3)^2$

Often, *several different combinations of Eqs. (11.1) to (11.6) can be used to simplify an expression.* This is illustrated in the next example.

EXAMPLE 3 Simplification done in different ways

$$\text{(a)}\quad (x^2y)^2\left(\frac{2}{x}\right)^{-2} = \frac{(x^4y^2)}{\left(\frac{2}{x}\right)^2} = \frac{x^4y^2}{\frac{4}{x^2}} = \frac{x^4y^2}{1} \times \frac{x^2}{4} = \frac{x^6y^2}{4}$$

$$\text{(b)}\quad (x^2y)^2\left(\frac{2}{x}\right)^{-2} = (x^4y^2)\left(\frac{2^{-2}}{x^{-2}}\right) = (x^4y^2)\left(\frac{x^2}{2^2}\right) = \frac{x^6y^2}{4}$$

In (a), we first used Eq. (11.6) and then (11.4). The simplification was completed by changing the division of a fraction to multiplication and using Eq. (11.1). In (b), we first used Eq. (11.3), then (11.6), and finally (11.1). ■

Practice Exercise

2. Simplify: $x\left(\frac{3}{x}\right)^{-3}$

EXAMPLE 4 Exponents and units of measurement

When writing a denominate number, if units of measurement appear in the denominator, they can be written using negative exponents. For example, the metric unit for pressure is the *pascal,* where $1\text{ Pa} = 1\text{ N/m}^2$. This can be written as

■ Named for the French mathematician and scientist Blaise Pascal (1623–1662).

$$1\text{ Pa} = 1\text{ N/m}^2 = 1\text{ N}\cdot\text{m}^{-2}$$

where $1/\text{m}^2 = \text{m}^{-2}$.

The metric unit for energy is the *joule,* where $1\text{ J} = 1\text{ kg}\cdot(\text{m}\cdot\text{s}^{-1})^2$, or

■ Named for the English physicist James Prescott Joule (1818–1899).

$$1\text{ J} = 1\text{ kg}\cdot\text{m}^2\cdot\text{s}^{-2} = 1\text{ kg}\cdot\text{m}^2/\text{s}^2$$ ■

Care must be taken to apply the laws of exponents properly. Certain common problems are pointed out in the following examples.

EXAMPLE 5 Be careful with zero exponents

The expression $(-5x)^0$ equals 1, whereas the expression $-5x^0$ equals -5. For $(-5x)^0$, the parentheses show that the expression $-5x$ is raised to the zero power, whereas for $-5x^0$, only x is raised to the zero power and we have

$$-5x^0 = -5(1) = -5$$

Also, $(-5)^0 = 1$ but $-5^0 = -1$. Again, for $(-5)^0$, parentheses show -5 raised to the zero power, whereas for -5^0, only 5 is raised to the zero power.

Similarly, $(-2)^2 = 4$ and $-2^2 = -4$.

For the same reasons, $2x^{-1} = \frac{2}{x}$, whereas $(2x)^{-1} = \frac{1}{2x}$. ■

Practice Exercise

3. Simplify: $3/x^0$

EXAMPLE 6 Simplification done in different ways

$$\text{(a)}\quad (2a + b^{-1})^{-2} = \frac{1}{(2a + b^{-1})^2} = \frac{1}{\left(2a + \frac{1}{b}\right)^2} = \frac{1}{\left(\frac{2ab + 1}{b}\right)^2}$$

$$= \frac{1}{\frac{(2ab + 1)^2}{b^2}} = \frac{b^2}{(2ab + 1)^2}$$

not necessary to expand the denominator

$$\text{(b)}\quad (2a + b^{-1})^{-2} = \left(2a + \frac{1}{b}\right)^{-2} = \left(\frac{2ab + 1}{b}\right)^{-2}$$

$$= \frac{(2ab + 1)^{-2}}{b^{-2}} = \frac{b^2}{(2ab + 1)^2}$$

positive exponents used in the final result ■

EXAMPLE 7 Be careful not to make a common error

There is an error that is commonly made in simplifying the type of expression in Example 6. We must be careful to see that

$$(2a + b^{-1})^{-2} \quad \text{is } \textit{not} \text{ equal to} \quad (2a)^{-2} + (b^{-1})^{-2}, \quad \text{or} \quad \frac{1}{4a^2} + b^2$$

CAUTION When raising a binomial (or any multinomial) to a power, we cannot simply raise each term to the power to obtain the result. ■

However, when raising a product of factors to a power, we use the equation $(ab)^n = a^n b^n$. Thus,

$$(2ab^{-1})^{-2} = (2a)^{-2}(b^{-1})^{-2} = \frac{b^2}{(2a)^2} = \frac{b^2}{4a^2}$$

We see that we must be careful to distinguish between the power of a sum of terms and the power of a product of factors. ■

CAUTION From the preceding examples, we see that when a *factor* is moved from the denominator to the numerator of a fraction, or conversely, the sign of the exponent is changed. We should carefully note the word *factor;* this rule does not apply to moving *terms* in the numerator or the denominator. ■

EXAMPLE 8 Factor moved from numerator to denominator

(a) $$3L^{-1} - (2L)^{-2} = \frac{3}{L} - \frac{1}{(2L)^2} = \frac{3}{L} - \frac{1}{4L^2}$$
$$= \frac{12L - 1}{4L^2}$$

(b) $$3^{-1}\left(\frac{4^{-2}}{3 - 3^{-1}}\right) = \frac{1}{3}\left(\frac{1}{4^2}\right)\left(\frac{1}{3 - \frac{1}{3}}\right) = \frac{1}{3 \times 4^2}\left(\frac{1}{\frac{9 - 1}{3}}\right)$$
$$= \frac{1}{3 \times 4^2}\left(\frac{3}{8}\right) = \frac{1}{128}$$ ■

Practice Exercise

4. Simplify: $(3a)^{-1} - 3a^{-2}$

EXAMPLE 9 Simplifying a complex expression

terms

$$\frac{1}{x^{-1}}\left(\frac{x^{-1} - y^{-1}}{x^2 - y^2}\right) = \frac{x}{1}\left(\frac{\frac{1}{x} - \frac{1}{y}}{x^2 - y^2}\right) = x\left(\frac{\frac{y - x}{xy}}{x^2 - y^2}\right)$$
$$= \frac{\frac{x(y - x)}{xy}}{(x - y)(x + y)}$$
$$= \frac{x(y - x)}{xy} \times \frac{1}{(x - y)(x + y)}$$
$$= \frac{x(y - x)}{xy(x - y)(x + y)} = \frac{-(x - y)}{y(x - y)(x + y)}$$
$$= -\frac{1}{y(x + y)}$$

CAUTION Note that in this example, the x^{-1} and y^{-1} in the numerator could *not* be moved directly to the denominator with positive exponents because they are only *terms* of the original numerator, not *factors*. ■ ■

EXAMPLE 10 Simplifying an expression from calculus

In finding out the rate at which a quantity is changing, it may be necessary to simplify an expression found by using the advanced mathematics of calculus. Simplify the following expression, which is derived using calculus.

$$\begin{aligned}&3(x+4)^2(x-3)^{-2} - 2(x-3)^{-3}(x+4)^3\\&= \frac{3(x+4)^2}{(x-3)^2} - \frac{2(x+4)^3}{(x-3)^3} = \frac{3(x-3)(x+4)^2 - 2(x+4)^3}{(x-3)^3}\\&= \frac{(x+4)^2[3(x-3) - 2(x+4)]}{(x-3)^3}\\&= \frac{(x+4)^2(x-17)}{(x-3)^3}\end{aligned}$$

■

EXERCISES 11.1

In Exercises 1–4, solve the resulting problems if the given changes are made in the indicated examples of this section.

1. In Example 3, change the factor x^2 to x^{-2} and then find the result.
2. In Example 6, change the term $2a$ to $2a^{-1}$ and then find the result.
3. In Example 8(b), change the 3^{-1} in the denominator to 3^{-2} and then find the result.
4. In Example 9, change the sign in the numerator from $-$ to $+$ and then find the result.

In Exercises 5–56, express each of the given expressions in simplest form with only positive exponents.

5. x^8x^{-3}
6. y^9y^{-2}
7. $2a^2a^{-6}$
8. $5ss^{-5}$
9. $\dfrac{c^7}{c^{-2}}$
10. $\dfrac{t^{-8}}{t^{-3}}$
11. $\dfrac{x^5}{y^{-2}}$
12. $\dfrac{n^{-6}}{m^{-4}}$
13. $5^0 \times 5^{-3}$
14. $(3^2 \times 4^{-3})^3$
15. $(2\pi x^{-1})^2$
16. $(3xy^{-2})^3$
17. $2(5an^{-2})^{-1}$
18. $4(6s^2t^{-1})^{-2}$
19. $(-4)^0$
20. -4^0
21. $-9x^0$
22. $(-7x)^0$
23. $3x^{-2}$
24. $(3x)^{-2}$
25. $(7a^{-1}x)^{-3}$
26. $7a^{-1}x^{-3}$
27. $\left(\dfrac{2}{n^3}\right)^{-4}$
28. $\left(\dfrac{x^3}{-3}\right)^{-2}$
29. $3\left(\dfrac{a}{b^{-2}}\right)^{-3}$
30. $5\left(\dfrac{2n^{-2}}{D^{-1}}\right)^{-2}$
31. $(a+b)^{-1}$
32. $a^{-1} + b^{-1}$
33. $2x^{-3} + 4y^{-2}$
34. $(3x - 2y)^{-2}$
35. $(2a^{-n})^2\left(\dfrac{3}{2a^n}\right)^{-1}$
36. $(7 \times 3^{-a})\left(\dfrac{3^a}{7}\right)^2$
37. $\left(\dfrac{3a^2}{4b}\right)^{-3}\left(\dfrac{4}{a}\right)^{-5}$
38. $(2np^{-2})^{-2}(4^{-1}p^2)^{-1}$
39. $\left(\dfrac{V^{-1}}{2t}\right)^{-2}\left(\dfrac{t^2}{V^{-2}}\right)^{-3}$
40. $ab\left(\dfrac{a^{-2}}{b^2}\right)^{-3}\left(\dfrac{a^{-3}}{b^5}\right)^2$
41. $3a^{-2} + (3a^{-2})^4$
42. $3(a^{-1}z^2)^{-3} + c^{-2}z^{-1}$
43. $2 \times 3^{-1} + 4 \times 3^{-2}$
44. $5 \times 8^{-2} - 3^{-1} \times 2^3$
45. $(R_1^{-1} + R_2^{-1})^{-1}$
46. $6^{-2}(2a - b^{-2})^{-1}$
47. $(n^{-2} - 2n^{-1})^2$
48. $2(3^{-3} - 9^{-1})^{-2}$
49. $\dfrac{6^{-1}}{4^{-2} + 2}$
50. $\dfrac{x - y^{-1}}{x^{-1} - y}$
51. $\dfrac{x^{-2} - y^{-2}}{x^{-1} - y^{-1}}$
52. $\dfrac{ax^{-2} + a^{-2}x}{a^{-1} + x^{-1}}$
53. $2t^{-2} + t^{-1}(t+1)$
54. $3x^{-1} - x^{-3}(y+2)$
55. $(D-1)^{-1} + (D+1)^{-1}$
56. $4(2x-1)(x+2)^{-1} - (2x-1)^2(x+2)^{-2}$

In Exercises 57–78, solve the given problems.

57. If $x < 0$, is it ever true that $x^{-2} < x^{-1}$?
58. Is it true that $(a+b)^0 = 1$ for all values of a and b?
59. Express $4^2 \times 64$ (a) as a power of 4 and (b) as a power of 2.
60. Express $1/81$ (a) as a power of 9 and (b) as a power of 3.
61. (a) By use of Eqs. (11.4) and (11.6), show that
$$\left(\frac{a}{b}\right)^{-n} = \left(\frac{b}{a}\right)^n$$
(b) Verify the equation in part (a) by evaluating each side with $a = 3.576$, $b = 8.091$, and $n = 7$.
62. For what integer values of n is $(-3)^{-n} = -3^{-n}$? Explain.
63. For what integer value(s) of n is $n^\pi > \pi^n$?
64. Evaluate $(8^{19})^{12}/(8^{16})^{14}$. What happens when you try to evaluate this on a calculator?

65. Is it true that $[-2^0 - (-1)^0]^0 = 1$?

66. Is it true that, if $x \neq 0$, $[(-x^{-2})^{-2}]^{-2} = 1/x^2$?

67. If $f(x) = 4^x$, find $f(a + 2)$.

68. If $f(x) = 2(9^x)$, find $f(2 - a)$.

69. Solve for x: $2^{5x} = 2^7(2^{2x})^2$.

70. In analyzing the tuning of an electronic circuit, the expression $[\omega\omega_0^{-1} - \omega_0\omega^{-1}]^2$ is used. Expand and simplify this expression.

71. The metric unit of energy, the *joule* (J), can be expressed as $\text{kg}\cdot\text{s}^{-2}\cdot\text{m}^2$. Simplify these units and include *newtons* (see Appendix B) and only positive exponents in the final result.

72. The units for the electric quantity called *permittivity* are $\text{C}^2\cdot\text{N}^{-1}\cdot\text{m}^{-2}$. Given that $1\ \text{F} = 1\ \text{C}^2\cdot\text{J}^{-1}$, show that the units of permittivity are F/m. See Appendix B.

73. When studying a solar energy system, the units encountered are $\text{kg}\cdot\text{s}^{-1}(\text{m}\cdot\text{s}^{-2})^2$. Simplify these units and include *joules* (see Exercise 71) and only positive exponents in the final result.

74. The metric units for the velocity v of an object are $\text{m}\cdot\text{s}^{-1}$, and the units for the acceleration a of the object are $\text{m}\cdot\text{s}^{-2}$. What are the units for v/a?

75. Given that $v = a^p t^r$, where v is the velocity of an object, a is its acceleration, and t is the time, use the metric units given in Exercise 74 to show that $p = r = 1$.

76. In optics, the combined focal length F of two lenses is given by $F = [f_1^{-1} + f_2^{-1} + d(f_1f_2)^{-1}]^{-1}$, where f_1 and f_2 are the focal lengths of the lenses and d is the distance between them. Simplify the right side of this equation.

77. The monthly loan payment P for loan amount A with an annual interest rate r (as a decimal) for t years is

$$P = \frac{A\left(\frac{r}{12}\right)}{1 - \left(1 + \frac{r}{12}\right)^{-12t}}$$

Find the monthly payment for a \$20,000 car loan if it is a 5-year loan with an annual interest rate of 4%.

78. An idealized model of the thermodynamic process in a gasoline engine is the *Otto cycle*. The efficiency e of the process is

$$e = \frac{\frac{T_1r^\gamma}{r} - \frac{T_2r^\gamma}{r} - T_1 + T_2}{\frac{T_1r^\gamma}{r} - \frac{T_2r^\gamma}{r}}.\quad \text{Show that } e = 1 - \frac{1}{r^{\gamma-1}}.$$

Answers to Practice Exercises

1. $\frac{ay^6}{x^2}$ **2.** $\frac{x^4}{27}$ **3.** 3 **4.** $\frac{a-9}{3a^2}$

11.2 Fractional Exponents

Meaning of Fractional Exponents • Evaluations • Graphs • Simplifying Expressions

In Section 11.1, we reviewed the use of integer exponents, including exponents that are negative integers and zero. We now show how rational numbers may be used as exponents. With the appropriate definitions, all the laws of exponents are valid for all rational numbers as exponents.

Equation (11.3) states that $(a^m)^n = a^{mn}$. If we were to let $m = \frac{1}{2}$ and $n = 2$, we would have $(a^{1/2})^2 = a^1$. However, we already have a way of writing a quantity that, when squared, equals a. This is written as $\sqrt{a}$. To be consistent with previous definitions and to allow the laws of exponents to hold, we define

■ The radical in Eq. (11.7) is the symbol for the nth root of a number. Do not confuse it with $\sqrt{a}$, the square root of a.

$$a^{1/n} = \sqrt[n]{a} \tag{11.7}$$

CAUTION a^{-n} is not equal to $a^{1/n}$. ■

In order that Eqs. (11.3) and (11.7) may hold at the same time, we define

$$a^{m/n} = \sqrt[n]{a^m} = (\sqrt[n]{a})^m \tag{11.8}$$

These definitions are valid for all the laws of exponents. We must note that Eqs. (11.7) and (11.8) are valid as long as $\sqrt[n]{a}$ does not involve the even root of a negative number. Such numbers are imaginary and are considered in Chapter 12.

EXAMPLE 1 Meaning of fractional exponent

We now verify that Eq. (11.1) ($a^m \times a^n = a^{m+n}$) holds for the above definitions:

$$a^{1/4}a^{1/4}a^{1/4}a^{1/4} = a^{(1/4)+(1/4)+(1/4)+(1/4)} = a^1$$

Now, $a^{1/4} = \sqrt[4]{a}$ by definition. Also, by definition, $\sqrt[4]{a}\,\sqrt[4]{a}\,\sqrt[4]{a}\,\sqrt[4]{a} = a$. Equation (11.1) is thereby verified for $n = 4$ in Eq. (11.7).

Equation (11.3) is verified by the following:

$$(a^{1/4})(a^{1/4})(a^{1/4})(a^{1/4}) = (a^{1/4})^4 = a^1 = (\sqrt[4]{a})^4$$ ■

We may interpret $a^{m/n}$ in Eq. (11.8) as the mth power of the nth root of a, as well as the nth root of the mth power of a. This is illustrated in the following example.

EXAMPLE 2 Interpretation of fractional exponent

$$8^{2/3} = (\sqrt[3]{8})^2 = (2)^2 = 4 \quad \text{or} \quad 8^{2/3} = \sqrt[3]{8^2} = \sqrt[3]{64} = 4$$ ■

NOTE ▸ [Although the power and the root may be applied in either order when evaluating a fractional exponent, it is usually easier to find the root first, as indicated by the denominator of the exponent.] This allows us to find the root of the smaller number, which is normally easier to find.

EXAMPLE 3 Evaluating

To evaluate $(64)^{5/2}$, we should proceed as follows:

$$(64)^{5/2} = [(64)^{1/2}]^5 = 8^5 = 32{,}768$$

If we raised 64 to the fifth power first, we would have

$$(64)^{5/2} = (64^5)^{1/2} = (1{,}073{,}741{,}824)^{1/2}$$

We would now have to evaluate the indicated square root. This demonstrates why it is preferable to find the indicated root first. ■

EXAMPLE 4 Evaluating

(a) $(16)^{3/4} = (16^{1/4})^3 = 2^3 = 8$

(b) $4^{-1/2} = \dfrac{1}{4^{1/2}} = \dfrac{1}{2}$ **(c)** $9^{3/2} = (9^{1/2})^3 = 3^3 = 27$

We note in (b) that Eq. (11.6) must also hold for negative rational exponents. In writing $4^{-1/2}$ as $1/4^{1/2}$, the *sign* of the exponent is changed. ■

Practice Exercises

Evaluate: **1.** $81^{5/4}$ **2.** $27^{-4/3}$

Fractional exponents allow us to find roots of numbers on a calculator. By use of the appropriate key (on most calculators x^y or ∧), we may raise any positive number to any power. For roots, we use the equivalent fractional exponent. Powers that are fractions or decimal in form are entered directly.

EXAMPLE 5 Fractional exponents and calculator

The thermodynamic temperature T [in kelvins, (K)] is related to the pressure P (in kPa) of a gas by the equation $T = 80.5P^{2/7}$. Find the value of T for $P = 750$ kPa.

■ Named for the Scottish physicist Lord Kelvin (1824–1907).

Substituting, we have

$$T = 80.5(750)^{2/7}$$

The calculator display for this calculation is shown in Fig. 11.1. Therefore, $T = 534$ K. ■

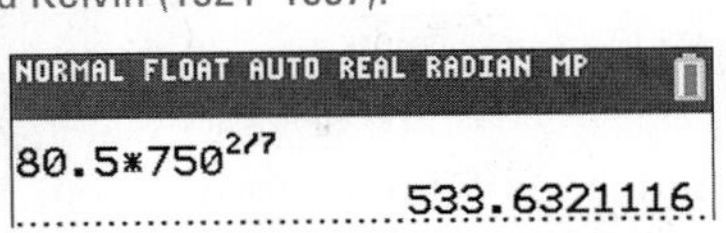

Fig. 11.1

When finding powers of negative numbers, some calculators will show an error. If this is the case, enter the positive value of the number and then enter a negative sign for the result when appropriate. NOTE ▸ [From Section 1.6, we recall that an even root of a negative number is imaginary, and an odd root of a negative number is negative.]

EXAMPLE 6 Fractional exponents and graphs

Plot the graph of the function $y = 2x^{1/3}$.

In obtaining points for the graph, we use $(1/3)$ as the power when using the calculator. If your calculator does not evaluate powers of negative numbers, enter positive values for the negative values of x and then make the results negative. Because $x^{1/3} = \sqrt[3]{x}$, we know that $x^{1/3}$ is negative for negative values of x because we have an odd root of a negative number. We get the following table of values:

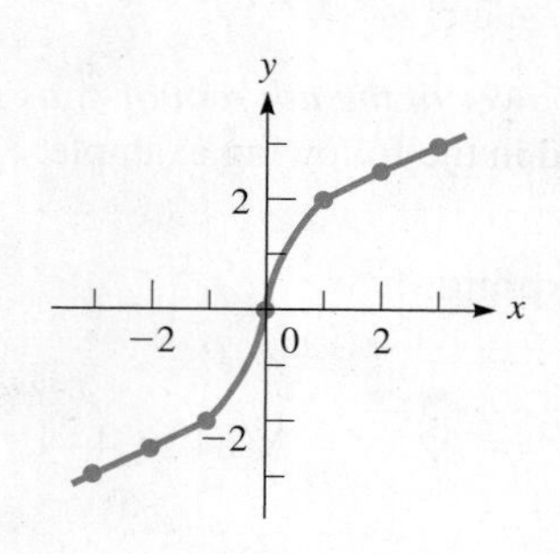

Fig. 11.2

x	-3	-2	-1	0	1	2	3
y	-2.9	-2.5	-2.0	0	2.0	2.5	2.9

The graph, shown in Fig. 11.2, can also be displayed on a calculator. ■

Fractional exponents are often easier to use in more complex expressions involving roots. This is true in algebra and in topics from more advanced mathematics. Any expression with radicals can also be expressed with fractional exponents and then simplified.

EXAMPLE 7 Simplifying expressions

(a) $(8a^2b^4)^{1/3} = [(8^{1/3})(a^2)^{1/3}(b^4)^{1/3}]$ using $(ab)^n = a^nb^n$

$= 2a^{2/3}b^{4/3}$ using $a^{1/n} = \sqrt[n]{a}$ and $(a^m)^n = a^{mn}$

(b) $a^{3/4}a^{4/5} = a^{3/4+4/5} = a^{31/20}$ using $a^m \times a^n = a^{m+n}$

(c) $\left(\dfrac{4^{-9/2}x^{2/3}}{2^{3/2}x^{-1/3}}\right)^{2/3} = \left(\dfrac{x^{2/3}x^{1/3}}{2^{3/2}4^{9/2}}\right)^{2/3}$ using $a^{-n} = \dfrac{1}{a^n}$

$= \left(\dfrac{x^{2/3+1/3}}{2^{3/2}4^{9/2}}\right)^{2/3}$ using $a^m \times a^n = a^{m+n}$

$= \dfrac{x^{(1)(2/3)}}{2^{(3/2)(2/3)}4^{(9/2)(2/3)}}$ using $(ab)^n = a^nb^n$ and $\left(\dfrac{a}{b}\right)^n = \dfrac{a^n}{b^n}$

$= \dfrac{x^{2/3}}{(2)(4^3)} = \dfrac{x^{2/3}}{128}$

(d) $(4x^4)^{-1/2} - 3x^{-3} = \dfrac{1}{(4x^4)^{1/2}} - \dfrac{3}{x^3}$ using $a^{-n} = \dfrac{1}{a^n}$

$= \dfrac{1}{2x^2} - \dfrac{3}{x^3} = \dfrac{x-6}{2x^3}$ using $a^{1/n} = \sqrt[n]{a}$, common denominator ■

Practice Exercises

Simplify:

3. $(64a^2c^3)^{2/3}$ **4.** $2x^{-4} - (8x^3)^{-1/3}$

■ See the chapter introduction.

EXAMPLE 8 Simplifying expression—solar radiation

During a day, the rate R at which the radiation changes at a solar-radiation collector is

$$R = \frac{(t^4 + 100)^{1/2} - 2t^3(t+6)(t^4+100)^{-1/2}}{[(t^4+100)^{1/2}]^2}$$

■ Solar collectors supply the power for most space satellites.

Here, R is measured in $\text{kW}/(\text{m}^2 \cdot \text{h})$, t is the number of hours from noon, and $-6\text{ h} \le t \le 8\text{ h}$. Express the right side of this equation in simpler form and find R for $t = 0$ (noon) and for $t = 4\text{ h}$ (4 P.M.)

Performing the simplification, we have the following steps:

$$R = \frac{(t^4+100)^{1/2} - \dfrac{2t^3(t+6)}{(t^4+100)^{1/2}}}{t^4+100} \quad \leftarrow \text{using } a^{-n} = \frac{1}{a^n} \quad \leftarrow \text{using } (a^m)^n = a^{mn}$$

$$= \frac{\dfrac{(t^4+100)^{1/2}(t^4+100)^{1/2} - 2t^3(t+6)}{(t^4+100)^{1/2}}}{t^4+100} \quad \leftarrow \text{common denominator}$$

$$= \frac{(t^4+100) - 2t^3(t+6)}{(t^4+100)^{1/2}} \times \frac{1}{t^4+100} \quad \leftarrow \text{invert divisor and multiply}$$

$$= \frac{100 - 12t^3 - t^4}{(t^4+100)^{1/2}(t^4+100)} = \frac{100 - 12t^3 - t^4}{(t^4+100)^{3/2}} \quad \leftarrow \text{using } a^m \times a^n = a^{m+n}$$

For $t = 0$: $\quad R = \dfrac{100 - 12(0^3) - 0^4}{(0^4+100)^{3/2}} = 0.10 \text{ kW/(m}^2 \cdot \text{h)}$

For $t = 4$ h: $\quad R = \dfrac{100 - 12(4^3) - 4^4}{(4^4+100)^{3/2}} = -0.14 \text{ kW/(m}^2 \cdot \text{h)}$

This shows that the radiation is increasing at noon and decreasing at 4 P.M. ■

EXERCISES 11.2

In Exercises 1–4, solve the resulting problems if the given changes are made in the indicated examples of this section.

1. In Example 2, change the exponent to $4/3$.

2. In Example 4(b), change the exponent to $-3/2$.

3. In Example 6, change the exponent to $-1/3$.

4. In Example 7(d), change the exponent $-1/2$ to $-3/2$.

In Exercises 5–26, evaluate the given expressions.

5. $36^{1/2}$ **6.** $27^{1/3}$ **7.** $81^{1/4}$ **8.** $125^{2/3}$

9. $100^{3/2}$ **10.** $-16^{5/4}$ **11.** $8^{-2/3}$ **12.** $16^{-7/4}$

13. $64^{-2/3}$ **14.** $-32^{-4/5}$ **15.** $5^{1/2}5^{3/2}$ **16.** $(4^4)^{3/2}$

17. $(3^6)^{2/3}$ **18.** $\dfrac{121^{-1/2}}{100^{1/2}}$ **19.** $\dfrac{1000^{1/3}}{-400^{-1/2}}$ **20.** $\dfrac{-7^{-1/2}}{6^{-1}7^{1/2}}$

21. $\dfrac{15^{2/3}}{5^2 15^{-1/3}}$ **22.** $\dfrac{(-27)^{4/3}}{6}$ **23.** $\dfrac{(-8)^{2/3}}{-2}$ **24.** $\dfrac{-4^{-1/2}}{(-64)^{-4/3}}$

25. $125^{-2/3} - 100^{-3/2}$ **26.** $32^{0.4} + 25^{-0.5}$

In Exercises 27–30, use a calculator to evaluate each expression.

27. $17.98^{1/4}$ **28.** $(-750.81)^{2/3}$

29. $4.0187^{-4/9}$ **30.** $0.1863^{-7/6}$

In Exercises 31–54, simplify the given expressions. Express all answers with positive exponents.

31. $m^{1/3}m^{1/2}$ **32.** $x^{5/6}x^{-1/3}$ **33.** $\dfrac{4y^{-1/2}}{-y^{2/5}}$

34. $\dfrac{s^{1/4}s^{2/3}}{5s^{-1}}$ **35.** $\dfrac{x^{3/10}}{x^{-1/5}x^2}$ **36.** $\dfrac{R^{-2/5}R^2}{R^{-3/10}}$

37. $(8a^3b^6)^{1/3}$ **38.** $(8b^{-4}c^2)^{4/3}$ **39.** $(16a^4b^3)^{-3/4}$

40. $(32C^5D^4)^{-2/5}$ **41.** $\left(\dfrac{a^{5/7}}{a^{2/3}}\right)^{7/4}$ **42.** $\left(\dfrac{4a^{5/6}b^{-1/5}}{a^{2/3}b^2}\right)^{-1/2}$

43. $\dfrac{1}{2}(4x^2+1)^{-1/2}(8x)$ **44.** $\dfrac{(x+x^{1/2})(x-x^{1/2})}{x}$

45. $\dfrac{y^{3/8}(y^{5/8} - y^{13/8})}{y^{1/2}(y^{1/2} - y^{-1/2})}$ **46.** $\dfrac{3^{-1}a^{1/2}}{4^{-1/2}b} \div \dfrac{9^{1/2}a^{-1/3}}{2b^{-1/4}}$

47. $(T^{-1} + 2T^{-2})^{-1/2}$ **48.** $(a^{-2} - a^{-4})^{-1/4}$

49. $(a^3)^{-4/3} + a^{-2}$ **50.** $(4N^6)^{-1/2} - 2N^{-1}$

51. $[(a^{1/2} - a^{-1/2})^2 + 4]^{1/2}$ **52.** $4x^{1/2} + \dfrac{1}{2}x^{-1/2}(4x+1)$

53. $x^2(2x-1)^{-1/2} + 2x(2x-1)^{1/2}$

54. $(3n-1)^{-2/3}(1-n) - (3n-1)^{1/3}$

In Exercises 55–58, graph the given functions.

55. $f(x) = 3x^{1/2}$ **56.** $f(x) = 2x^{2/3}$

57. $f(t) = t^{-4/5}$ **58.** $f(V) = 4V^{3/2}$

In Exercises 59–72, perform the indicated operations.

59. Is $\left[(4^{1/2})^{-3/4} + \left(\frac{1}{4^3}\right)^{1/8}\right]^4 = 2$?

60. Is $\left[\left(-\frac{1}{8}\right)^{-1/3} + \frac{1}{8}\left(\frac{1}{32}\right)^{-4/5}\right]^0 = 1$?

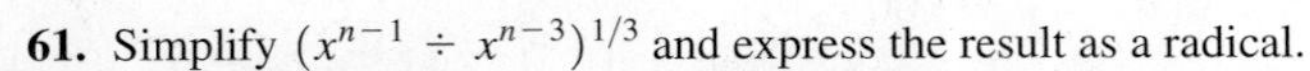

61. Simplify $(x^{n-1} \div x^{n-3})^{1/3}$ and express the result as a radical.

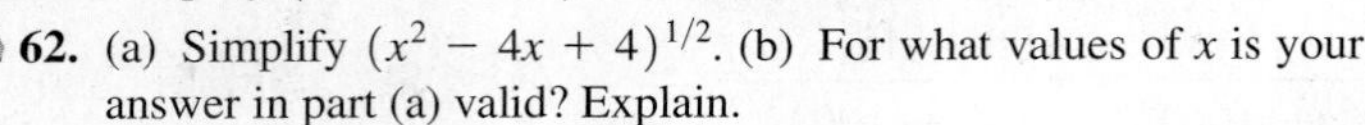

62. (a) Simplify $(x^2 - 4x + 4)^{1/2}$. (b) For what values of x is your answer in part (a) valid? Explain.

63. A factor used in determining the performance of a solar-energy storage system is $(A/S)^{-1/4}$, where A is the actual storage capacity and S is a standard storage capacity. If this factor is 0.5, explain how to find the ratio A/S.

64. A factor used in measuring the loudness sensed by the human ear is $(I/I_0)^{0.3}$, where I is the intensity of the sound and I_0 is a reference intensity. Evaluate this factor for $I = 3.2 \times 10^{-6}$ W/m^2 (ordinary conversation) and $I_0 = 10^{-12}$ W/m^2.

65. The period T of a satellite circling Earth is given by $T^2 = kR^3\left(1 + \frac{d}{R}\right)^3$, where R is the radius of Earth, d is the distance of the satellite above Earth, and k is a constant. Solve for R, using fractional exponents in the result.

66. (a) Express the radius of a sphere as a function of its volume v using fractional exponents. (b) If the volume of the moon is 2.19×10^{19} m^3, find its radius.

67. A plastic cup is in the shape of a right circular cone for which the base radius equals the height. (a) Express the base radius r as a function of the volume V using fractional exponents. (b) Find the radius if the cup holds 125 cm^3 of liquid.

68. The withdrawal resistance R of a nail of diameter d indicates its holding power. One formula for R is $R = ks^{5/2}dh$, where k is a constant, s is the specific gravity of the wood, and h is the depth of the nail in the wood. Solve for s using fractional exponents in the result.

69. Carbon-14 has a half-life of approximately 5730 years. If A_0 is the original amount present, then the amount present after t years is given by $A(t) = A_0 2^{-t/5730}$. If 55.0 mg is present initially, how much is present after 3250 years?

70. Living organisms contain a certain amount of carbon-14. After death, this amount decays according to the equation in Exercise 69 (this is used in carbon-14 dating). What *percentage* of an organism's carbon-14 is still present 2580 years after death?

71. For a heat-seeking rocket in pursuit of an aircraft, the distance d (in km) from the rocket to the aircraft is $d = \frac{500(\sin\theta)^{1/2}}{(1 - \cos\theta)^{3/2}}$, where θ is shown in Fig. 11.3. Find d for $\theta = 125.0°$.

72. The electric current i (in A) in a circuit with a battery of voltage E, a resistance R, and an inductance L, is $i = \frac{E}{R}(1 - e^{-Rt/L})$, where t is the time after the circuit is closed. See Fig. 11.4. Find i for $E = 6.20$ V, $R = 1.20\ \Omega$, $L = 3.24$ H, and $t = 0.00100$ s. (The number e is irrational and can be found from the calculator.)

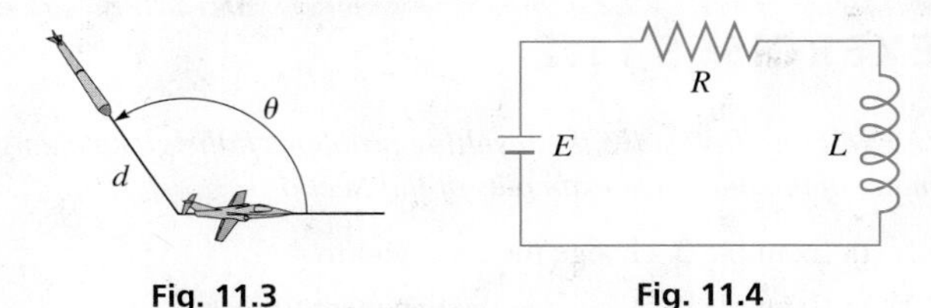

Fig. 11.3 Fig. 11.4

Answers to Practice Exercises

1. 243 **2.** $\frac{1}{81}$ **3.** $16a^{4/3}c^2$ **4.** $\frac{4 - x^3}{2x^4}$

11.3 Simplest Radical Form

Operations with Radicals • Removing Perfect *n*th-power Factors • Reducing the Order of a Radical • Rationalizing the Denominator.

As we have said, any expression with radicals can also be expressed with fractional exponents. For adding or subtracting radicals, there is little advantage to changing form, but with multiplication or division of radicals, fractional exponents have some advantages. Therefore, we now define operations with radicals so that they are consistent with the laws of exponents. This will let us use the form that is more convenient for the operation being performed.

$$\sqrt[n]{a^n} = (\sqrt[n]{a})^n = a \quad \textbf{(11.9)}$$

$$\sqrt[n]{a}\sqrt[n]{b} = \sqrt[n]{ab} \quad \textbf{(11.10)}$$

$$\sqrt[m]{\sqrt[n]{a}} = \sqrt[mn]{a} \quad \textbf{(11.11)}$$

$$\frac{\sqrt[n]{a}}{\sqrt[n]{b}} = \sqrt[n]{\frac{a}{b}} \quad (b \neq 0) \quad \textbf{(11.12)}$$

NOTE ▶ *The number under the radical is called the* **radicand,** *and the number indicating the root being taken is called the* **order** (*or* **index**) *of the radical.* [To avoid difficulties with imaginary numbers (which are considered in the next chapter), we will assume that all letters represent positive numbers.]

EXAMPLE 1 Operation with radicals

Following are illustrations using Eqs. (11.9) to (11.12).

(a) $\sqrt[5]{4^5} = (\sqrt[5]{4})^5 = 4$ using Eq. (11.9)

(b) $\sqrt[3]{2}\sqrt[3]{3} = \sqrt[3]{2 \times 3} = \sqrt[3]{6}$ using Eq. (11.10)

(c) $\sqrt[3]{\sqrt{5}} = \sqrt[3 \times 2]{5} = \sqrt[6]{5}$ using Eq. (11.11)

(d) $\dfrac{\sqrt{7}}{\sqrt{3}} = \sqrt{\dfrac{7}{3}}$ using Eq. (11.12) ■

EXAMPLE 2 Root of sum and root of product

In Example 5 of Section 1.6, we saw that

$$\sqrt{16 + 9} \quad \text{is } not \text{ equal to} \quad \sqrt{16} + \sqrt{9}$$

However, using Eq. (11.10),

$$\sqrt{16 \times 9} = \sqrt{16} \times \sqrt{9}$$
$$= 4 \times 3 = 12$$

CAUTION We must be careful to distinguish between the root of a sum of terms and the root of a product of factors. ■ This is the same as with powers of sums and powers of products, as shown in Example 7 of Section 11.1. It should be the same, as a root can be interpreted as a fractional exponent. ■

EXAMPLE 3 Removing perfect-square factors

To simplify $\sqrt{75}$, we know that $75 = (25)(3)$ and that $\sqrt{25} = 5$. As in Section 1.6 and now using Eq. (11.10), we write

$$\sqrt{75} = \sqrt{(25)(3)} = \sqrt{25}\sqrt{3} = 5\sqrt{3}$$

(25): perfect square

NOTE ▶ This illustrates one step that should always be carried out in simplifying radicals: [Always remove all perfect nth-power factors from the radicand of a radical of order n]. ■

EXAMPLE 4 Removing perfect nth-power factors

(a) $\sqrt{72} = \sqrt{(36)(2)} = \sqrt{36}\sqrt{2} = 6\sqrt{2}$

(36): perfect square

(b) $\sqrt{a^3b^2} = \sqrt{(a^2)(a)(b^2)} = \sqrt{a^2}\sqrt{a}\sqrt{b^2} = ab\sqrt{a}$

(a^2), (b^2): perfect squares

(c) cube root → $\sqrt[3]{40} = \sqrt[3]{(8)(5)} = \sqrt[3]{8}\sqrt[3]{5} = 2\sqrt[3]{5}$

(8): perfect cube

(d) fifth root → $\sqrt[5]{64x^8y^{12}} = \sqrt[5]{(32)(2)(x^5)(x^3)(y^{10})(y^2)}$

(32), (x^5), (y^{10}): perfect fifth powers

$$= \sqrt[5]{(32)(x^5)(y^{10})}\sqrt[5]{2x^3y^2}$$
$$= 2xy^2\sqrt[5]{2x^3y^2}$$ ■

Practice Exercises

Simplify: **1.** $\sqrt{8x^5}$ **2.** $\sqrt[3]{16a^7b^2}$

The next example illustrates another procedure used to simplify radicals. It is to *reduce the order of the radical,* when it is possible to do so.

EXAMPLE 5 Reducing order of radical

(a) $\sqrt[6]{8} = \sqrt[6]{2^3} = 2^{3/6} = 2^{1/2} = \sqrt{2}$

Here, we started with a sixth root and ended with a square root, thereby reducing the order of the radical. Fractional exponents are often helpful for this.

(b) $\sqrt[8]{16} = \sqrt[8]{2^4} = 2^{4/8} = 2^{1/2} = \sqrt{2}$

(c) $\dfrac{\sqrt[4]{9}}{\sqrt{3}} = \dfrac{\sqrt[4]{3^2}}{\sqrt{3}} = \dfrac{3^{2/4}}{3^{1/2}} = 1$ **(d)** $\dfrac{\sqrt[6]{8}}{\sqrt{7}} = \dfrac{\sqrt[6]{2^3}}{\sqrt{7}} = \dfrac{2^{1/2}}{7^{1/2}} = \sqrt{\dfrac{2}{7}}$

(e) $\sqrt[9]{27x^6y^{12}} = \sqrt[9]{3^3x^6y^9y^3} = 3^{3/9}x^{6/9}y^{9/9}y^{3/9} = 3^{1/3}x^{2/3}yy^{1/3} = y\sqrt[3]{3x^2y}$ ■

Practice Exercise

3. Reduce the order of the radical: $\sqrt[6]{27}$

If a radical is to be written in its *simplest form,* the two operations illustrated in the last four examples must be performed. Therefore, we have the following:

Steps to Reduce a Radical to Simplest Form

1. Remove all perfect nth-power factors from a radical of order n.
2. If possible, reduce the order of the radical.

When working with fractions, it has traditionally been the practice to write a fraction with radicals in a form in which the denominator contains no radicals. This step of simplification was performed primarily for ease of calculation, but with a calculator, it does not matter to any extent that there is a radical in the denominator. However, the procedure of writing a radical in this form, called **rationalizing the denominator,** is at times useful for other purposes. Therefore, the following examples show how the process of rationalizing the denominator is carried out.

EXAMPLE 6 Rationalizing the denominator

To write $\sqrt{\frac{2}{5}}$ in an equivalent form in which the denominator is not included under the radical sign, we *create a perfect square in the denominator* by multiplying the numerator and the denominator under the radical by 5. The steps are written as follows:

$$\sqrt{\frac{2}{5}} = \sqrt{\frac{2 \times 5}{5 \times 5}} = \sqrt{\frac{10}{25}} = \frac{\sqrt{10}}{\sqrt{25}} = \frac{\sqrt{10}}{5}$$

perfect square ■

EXAMPLE 7 Rationalizing the denominator

(a) $\dfrac{5}{\sqrt{18}} = \dfrac{5}{3\sqrt{2}} = \dfrac{5(\sqrt{2})}{3\sqrt{2}(\sqrt{2})} = \dfrac{5\sqrt{2}}{3\sqrt{4}} = \dfrac{5\sqrt{2}}{6}$

perfect square

(b) cube root → $\sqrt[3]{\dfrac{2}{3}} = \sqrt[3]{\dfrac{2 \times 9}{3 \times 9}} = \sqrt[3]{\dfrac{18}{27}} = \dfrac{\sqrt[3]{18}}{\sqrt[3]{27}} = \dfrac{\sqrt[3]{18}}{3}$

perfect cube

In (a), a perfect square was made by multiplying the numerator and denominator by $\sqrt{2}$. In (b), we made a perfect cube, because a cube root was being found. ■

Practice Exercise

4. Rationalize the denominator: $\sqrt{\dfrac{a}{3b}}$

EXAMPLE 8 Rationalizing expression—pendulum period

The period T (in s) for one cycle of a simple pendulum is given by $T = 2\pi\sqrt{L/g}$, where L is the length of the pendulum and g is the acceleration due to gravity. Rationalize the denominator on the right side of this equation if $L = 3.0$ ft and $g = 32$ ft/s^2.

Substituting, and then rationalizing, we have

$$T = 2\pi\sqrt{\frac{3.0}{32}} = 2\pi\sqrt{\frac{3.0 \times 2}{32 \times 2}}$$

$$= 2\pi\sqrt{\frac{6.0}{64}} = \frac{2\pi}{8}\sqrt{6.0}$$

$$= \frac{\pi}{4}\sqrt{6.0} = 1.9 \text{ s}$$

■

EXAMPLE 9 Simplifying expression

Simplify $\sqrt{\frac{1}{2a^2} + 2b^{-2}}$ and rationalize the denominator.

We will write the expression using only positive exponents, and then perform the required operations.

$$\sqrt{\frac{1}{2a^2} + \frac{2}{b^2}} = \sqrt{\frac{b^2 + 4a^2}{2a^2b^2}}$$ ← first combine fractions over lowest common denominator

$$= \frac{\sqrt{b^2 + 4a^2}}{ab\sqrt{2}}$$ ← sum of squares—radical cannot be simplified

$$= \frac{\sqrt{b^2 + 4a^2}\sqrt{2}}{ab\sqrt{2 \times 2}}$$

$$= \frac{\sqrt{2(b^2 + 4a^2)}}{2ab}$$

■

EXERCISES 11.3

In Exercises 1–4, simplify the resulting expressions if the given changes are made in the indicated examples of this section.

1. In Example 4(b), change the exponent of b to 4 and then find the resulting expression.
2. In Example 5(d), change the 8 to 27 and then find the resulting expression.
3. In Example 7(b), change the root to a fourth root and then find the resulting expression.
4. In Example 9, replace b^{-2} with b^{-4} and then find the resulting expression.

In Exercises 5–66, write each expression in simplest radical form. If a radical appears in the denominator, rationalize the denominator.

5. $\sqrt{75}$
6. $\sqrt{40}$
7. $\sqrt{108}$
8. $\sqrt{98}$
9. $\sqrt{x^2y^5}$
10. $\sqrt{pq^2r^7}$
11. $\sqrt{x^4y^3z^6}$
12. $\sqrt{75ab^2}$
13. $\sqrt{80R^5TV^4}$
14. $\sqrt{132M^2N^3}$
15. $\sqrt[3]{16}$
16. $\sqrt[4]{48}$
17. $\sqrt[5]{96}$
18. $\sqrt[3]{-512}$
19. $\sqrt[3]{8a^2}$
20. $\sqrt[3]{25a^4b^2}$
21. $\sqrt[4]{64r^3s^4t^5}$
22. $\sqrt[5]{16x^5y^3z^{11}}$
23. $\sqrt[5]{8}\sqrt[5]{-4}$
24. $\sqrt[3]{4}\sqrt[7]{64}$
25. $\sqrt[3]{P}\sqrt[3]{P^2V}$
26. $\sqrt[6]{3m^5n^8}\sqrt[6]{9mn}$
27. $\frac{2}{\sqrt{3}}$
28. $\frac{1}{\sqrt{2}}$
29. $\sqrt{\frac{3}{2}}$
30. $\sqrt{\frac{11}{12}}$
31. $\sqrt[3]{\frac{9}{12}}$
32. $\sqrt[4]{\frac{2}{125}}$
33. $\sqrt[5]{\frac{1}{9}}$
34. $\sqrt[6]{\frac{5}{256}}$
35. $\sqrt[4]{400}$
36. $\sqrt[8]{81}$
37. $\sqrt[6]{64}$
38. $\sqrt[9]{-27}$
39. $\sqrt{4 \times 10^4}$
40. $\sqrt{4 \times 10^5}$
41. $\sqrt{4 \times 10^6}$
42. $\sqrt[3]{16 \times 10^5}$
43. $\sqrt[4]{9y^2}$

44. $\sqrt[6]{b^3c^4}$ **45.** $\sqrt[4]{\frac{1}{4}}$ **46.** $\frac{\sqrt[4]{640}}{\sqrt[4]{5}}$

47. $\sqrt[4]{\sqrt[3]{16}}$ **48.** $\sqrt[5]{\sqrt[4]{9}}$ **49.** $\sqrt{\sqrt{\sqrt{n}}}$

50. $\sqrt{b^4\sqrt{a}}$ **51.** $\sqrt{28u^3v^{-5}}$ **52.** $\sqrt{98x^6y^{-7}}$

53. $\sqrt{64 + 144}$ **54.** $\sqrt{9 + 81}$ **55.** $\sqrt{\frac{2x}{3c^4}}$

56. $\sqrt{\frac{n}{m^3}}$ **57.** $\sqrt{\frac{5}{4} - \frac{1}{8}}$ **58.** $\sqrt{\frac{1}{a^2} + \frac{1}{b}}$

59. $\sqrt{xy^{-1} + x^{-1}y}$ **60.** $\sqrt{x^2 + 8^{-1}}$ **61.** $\sqrt{\frac{C - 2}{C + 2}}$

62. $\sqrt{a^2 + 2ab + b^2}$ **63.** $\sqrt{a^2 + b^2}$ **64.** $\sqrt{4x^2 - 1}$

65. $\sqrt{9x^2 - 6x + 1}$ **66.** $\sqrt{\frac{1}{2} + 2r + 2r^2}$

In Exercises 67–76, perform the required operation.

67. Change to radicals of the same order: $\sqrt{a}$; $\sqrt[3]{b}$; $\sqrt[6]{c}$.

68. Change to radicals of the same order: $3x^{2/3}$; $2y^{1/2}$; $(5z)^{1/4}$.

69. Display the graphs of $y_1 = \sqrt{x + 2}$ and $y_2 = \sqrt{x} + \sqrt{2}$ on a calculator to show that $\sqrt{x + 2}$ is not equal to $\sqrt{x} + \sqrt{2}$.

70. The *escape velocity* required to overcome the gravitational pull of a planet or star is $v_{esc} = \sqrt{\frac{2GM}{R}}$. Write this equation with a rationalized denominator.

71. In designing musical instruments, the equation $f = \sqrt{\frac{T}{4\mu L^2}}$ arises for the frequency of vibration of strings. Write this equation with a rationalized denominator.

72. An approximate equation for the efficiency E (in percent) of an engine is $E = 100(1 - 1/\sqrt[5]{R^2})$, where R is the compression ratio. Explain how this equation can be written with fractional exponents and then find E for $R = 7.35$.

73. When analyzing the velocity of an object falling through a great distance, the expression $a\sqrt{2g/a}$ arises. Show by rationalizing the denominator that this expression takes on a simpler form.

74. According to the Doppler effect, the frequency f_0 observed of a sound of frequency f, when the observer is moving away from the source with velocity u, is given by $f_0 = \frac{f_v}{v - u}\sqrt{1 - \left(\frac{u}{v}\right)^2}$, where v is the velocity of sound. Simplify the right side.

75. In analyzing an electronic filter circuit, the expression $\frac{8A}{\pi^2\sqrt{1 + (f_0/f)^2}}$ is used. Rationalize the denominator, expressing the answer without the fraction f_0/f.

76. The expression $\frac{1}{2L}\sqrt{R^2 - \frac{4L}{C}}$ occurs in the study of electric circuits. Simplify this expression by combining terms under the radical and rationalizing the denominator.

Answers to Practice Exercises

1. $2x^2\sqrt{2x}$ **2.** $2a^2\sqrt[3]{2ab^2}$ **3.** $\sqrt{3}$ **4.** $\frac{\sqrt{3ab}}{3b}$

11.4 Addition and Subtraction of Radicals

Similar Radicals • Adding and Subtracting Radicals

When adding or subtracting algebraic expressions, including those with radicals, we combine similar terms. Thus, *radicals must be* **similar,** *differing only in numerical coefficients, to be added.* This means they must have the same order and same radicand.

In order to add radicals, we first express each radical in its simplest form, rationalize any denominators, and then combine those that are similar. For those that are not similar, we can only indicate the addition.

EXAMPLE 1 Adding similar radicals

(a) $2\sqrt{7} - 5\sqrt{7} + \sqrt{7} = -2\sqrt{7}$ all similar radicals

This result follows the distributive law, as it should. We can write

$$2\sqrt{7} - 5\sqrt{7} + \sqrt{7} = (2 - 5 + 1)\sqrt{7} = -2\sqrt{7}$$

We can also see that the terms combine just as

$$2x - 5x + x = -2x$$

(b) $\sqrt[5]{6} + 4\sqrt[5]{6} - 2\sqrt[5]{6} = 3\sqrt[5]{6}$ all similar radicals

(c) $\sqrt{5} + 2\sqrt{3} - 5\sqrt{5} = 2\sqrt{3} - 4\sqrt{5}$ answer contains two terms

similar radicals

We note in (c) that we are able only to indicate the final subtraction because the radicals are not similar. ■

Practice Exercise

1. Add: $\sqrt{10} + 2\sqrt{7} - 2\sqrt{10}$

EXAMPLE 2 Adding radicals

(a) $\sqrt{2} + \sqrt{8} = \sqrt{2} + \sqrt{4 \times 2} = \sqrt{2} + \sqrt{4}\sqrt{2}$
$= \sqrt{2} + 2\sqrt{2} = 3\sqrt{2}$

(b) $\sqrt[3]{24} + \sqrt[3]{81} = \sqrt[3]{8 \times 3} + \sqrt[3]{27 \times 3} = \sqrt[3]{8}\sqrt[3]{3} + \sqrt[3]{27}\sqrt[3]{3}$
$= 2\sqrt[3]{3} + 3\sqrt[3]{3} = 5\sqrt[3]{3}$

CAUTION Notice that $\sqrt{8}$, $\sqrt[3]{24}$, and $\sqrt[3]{81}$ were simplified before performing the additions. We also note that $\sqrt{2} + \sqrt{8}$ is *not* equal to $\sqrt{2 + 8}$. ■ ■

Practice Exercise

2. Combine: $5\sqrt{44} - \sqrt{99}$

We note in the illustrations of Example 2 that the radicals do not initially appear to be similar. However, after each is simplified, we are able to recognize the similar radicals.

EXAMPLE 3 Simplify each—combine similar radicals

(a) $3\sqrt{125} - \sqrt{20} + \sqrt{27} = 3\sqrt{25 \times 5} - \sqrt{4 \times 5} + \sqrt{9 \times 3}$
$= 3(5\sqrt{5}) - 2\sqrt{5} + 3\sqrt{3}$ ← not similar to others
$= 13\sqrt{5} + 3\sqrt{3}$

(b) $\sqrt{24} + \sqrt{\frac{3}{2}} = \sqrt{4 \times 6} + \sqrt{\frac{3 \times 2}{2 \times 2}} = \sqrt{4}\sqrt{6} + \sqrt{\frac{6}{4}}$ rationalize denominator
$= 2\sqrt{6} + \frac{\sqrt{6}}{2} = \frac{4\sqrt{6} + \sqrt{6}}{2} = \frac{5}{2}\sqrt{6}$ combine similar radicals ■

EXAMPLE 4 Radicals with literal numbers

$$\sqrt{\frac{2}{3a}} - 2\sqrt{\frac{3}{2a}} = \sqrt{\frac{2(3a)}{3a(3a)}} - 2\sqrt{\frac{3(2a)}{2a(2a)}} = \sqrt{\frac{6a}{9a^2}} - 2\sqrt{\frac{6a}{4a^2}}$$
$$= \frac{1}{3a}\sqrt{6a} - \frac{2}{2a}\sqrt{6a} = \frac{1}{3a}\sqrt{6a} - \frac{1}{a}\sqrt{6a}$$
$$= \frac{\sqrt{6a} - 3\sqrt{6a}}{3a} = \frac{-2\sqrt{6a}}{3a} = -\frac{2}{3a}\sqrt{6a}$$ ■

Practice Exercise

3. Combine: $\sqrt{\frac{8x}{5y}} - \sqrt{10xy}$

EXAMPLE 5 Adding radicals—roof truss

A roof truss for a house has been designed as shown in Fig. 11.5. Find an expression for the exact number of linear feet of wood needed to construct the truss.

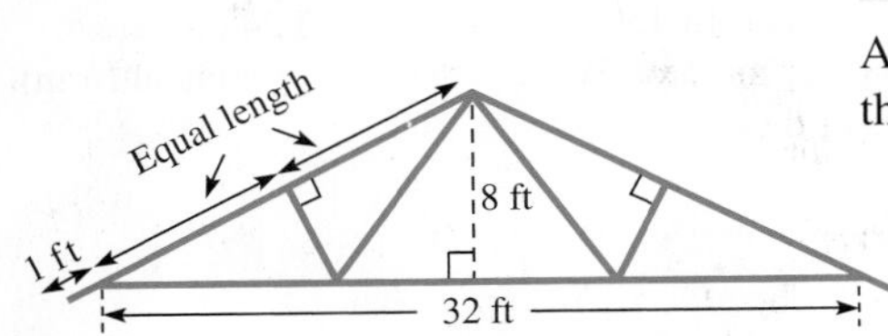

Fig. 11.5

The left side of the truss is shown in Fig. 11.6. By the Pythagorean theorem,

$$x^2 = 16^2 + 8^2$$
$$x = \sqrt{320} = \sqrt{64(5)} = 8\sqrt{5}$$

Therefore, $a = \frac{8\sqrt{5}}{2} = 4\sqrt{5}$.

Since triangles ADE and ACB are similar triangles, $\frac{b}{8} = \frac{4\sqrt{5}}{16}$, or $b = 2\sqrt{5}$. Side c can then be found using the Pythagorean theorem:

$$c^2 = a^2 + b^2 = (4\sqrt{5})^2 + (2\sqrt{5})^2 = 4^2(\sqrt{5})^2 + 2^2(\sqrt{5})^2 = 16(5) + 4(5) = 100$$

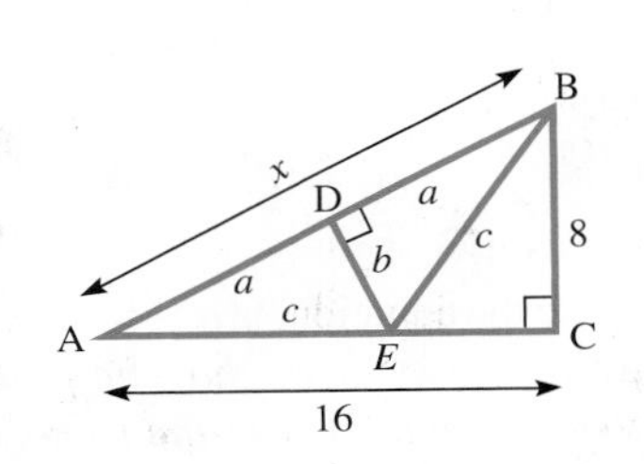

Fig. 11.6

Thus, $c = \sqrt{100} = 10$. The total amount of linear feet needed for the truss is

$$32 + 1 + 1 + 10 + 10 + 8\sqrt{5} + 8\sqrt{5} + 2\sqrt{5} + 2\sqrt{5}$$
$$= 54 + 20\sqrt{5} \text{ ft (approximately 99 ft).}$$ ■

EXERCISES 11.4

In Exercises 1 and 2, simplify the resulting expressions if the given changes are made in the indicated examples of this section.

1. In Example 3(a), change $\sqrt{27}$ to $\sqrt{45}$ and then find the resulting simplified expression.

2. In Example 4, change $\sqrt{\dfrac{2}{3a}}$ to $\sqrt{\dfrac{1}{6a}}$ and then find the resulting simplified expression.

In Exercises 3–42, express each radical in simplest form, rationalize denominators, and perform the indicated operations.

3. $3\sqrt{7} + 5\sqrt{7}$
4. $8\sqrt{11} - 3\sqrt{11}$
5. $\sqrt{28} + \sqrt{5} - 3\sqrt{7}$
6. $8\sqrt{6} - \sqrt{12} - 5\sqrt{6}$
7. $2\sqrt{18} - \sqrt{27} + \sqrt{50}$
8. $\sqrt{32} + 5\sqrt{24} - \sqrt{54}$
9. $\sqrt{5} + \sqrt{16 + 4}$
10. $\sqrt{7} + \sqrt{36 + 27}$
11. $2\sqrt{3t^2} - 3\sqrt{12t^2}$
12. $4\sqrt{2n^2} - \sqrt{50n^2}$
13. $\sqrt{18y} - 3\sqrt{8y}$
14. $\sqrt{27x} + 3\sqrt{18x}$
15. $2\sqrt{28} + 3\sqrt{175}$
16. $\sqrt{100 + 25} - 7\sqrt{80}$
17. $3\sqrt{200} - \sqrt{162} - \sqrt{288}$
18. $2\sqrt{44} - \sqrt{99} + \sqrt{2}\sqrt{88}$
19. $3\sqrt{75R} + 2\sqrt{48R} - 2\sqrt{18R}$
20. $2\sqrt{28} - \sqrt{108} - 6\sqrt{175}$
21. $\sqrt{40} + \sqrt{\frac{5}{2}}$
22. $3\sqrt{84} - \sqrt{\frac{3}{7}}$
23. $\sqrt{\frac{1}{2}} + \sqrt{\frac{25}{2}} - 4\sqrt{18}$
24. $\sqrt{6} - \sqrt{\frac{2}{3}} - \sqrt{\frac{1}{24}}$
25. $\sqrt[3]{81} + \sqrt[3]{3000}$
26. $\sqrt[3]{-16} + \sqrt[3]{54}$
27. $\sqrt[4]{32} - \sqrt[8]{4}$
28. $\sqrt[6]{\sqrt{2}} - \sqrt[12]{2^{13}}$
29. $5\sqrt{a^3b} - \sqrt{4ab^5}$
30. $\sqrt{2R^2I} + \sqrt{8}\sqrt{I^3}$
31. $\sqrt{6}\sqrt{5}\sqrt{3} - \sqrt{40a^2}$
32. $3\sqrt{60b^2n} - b\sqrt{135n}$
33. $\sqrt[3]{24a^2b^4} - \sqrt[3]{3a^5b}$
34. $\sqrt[5]{32a^6b^4} + 3a\sqrt[5]{243ab^9}$
35. $\sqrt{\dfrac{a}{c^5}} - \sqrt{\dfrac{c}{a^3}}$
36. $\sqrt{\dfrac{14x}{21y}} + \sqrt{\dfrac{27y}{8x}}$
37. $\sqrt[3]{ab^{-1}} - \sqrt[3]{8a^{-2}b^2}$
38. $\sqrt{\dfrac{2xy^{-1}}{3}} + \sqrt{\dfrac{27x^{-1}y}{8}}$
39. $\sqrt{\dfrac{T-V}{T+V}} - \sqrt{\dfrac{T+V}{T-V}}$
40. $\sqrt{\dfrac{16}{x} + 8 + x} - \sqrt{1 - \dfrac{1}{x}}$
41. $\sqrt{4x + 8} + 2\sqrt{9x + 18}$
42. $3\sqrt{50y - 75} - \sqrt{8y - 12}$

In Exercises 43–48, express each radical in simplest form, rationalize denominators, and perform the indicated operations. Then use a calculator to verify the result.

43. $3\sqrt{45} + 3\sqrt{75} - 2\sqrt{500}$
44. $2\sqrt{40} + 3\sqrt{90} - 5\sqrt{250}$
45. $2\sqrt{\frac{2}{3}} + \sqrt{24} - 5\sqrt{\frac{3}{2}}$
46. $\sqrt{\frac{2}{7}} - 2\sqrt{\frac{7}{2}} + 5\sqrt{56}$
47. $2\sqrt[3]{16} - \sqrt[3]{\frac{1}{4}}$
48. $5\sqrt[4]{810} - \sqrt[4]{\frac{5}{8}}$

In Exercises 49–56, solve the given problems.

49. Find the exact sum of the positive roots of $x^2 - 2x - 2 = 0$ and $x^2 + 2x - 11 = 0$.

50. For the quadratic equation $ax^2 + bx + c = 0$, if a, b, and c are integers, the sum of the roots is a rational number. Explain.

51. Without calculating the actual value, determine whether $10\sqrt{11} - \sqrt{1000}$ is positive or negative. Explain.

52. The adjacent sides of a parallelogram are $\sqrt{12}$ and $\sqrt{27}$ units long. What is the perimeter of the parallelogram?

53. The two legs of a right triangle are $2\sqrt{2}$ and $2\sqrt{6}$ units long. What is the perimeter of the triangle?

54. The current I (in A) passing through a resistor R (in Ω) in which P watts of power are dissipated is $I = \sqrt{P/R}$. If the power dissipated in the resistors shown in Fig. 11.7 is W watts, what is the sum of the currents in radical form?

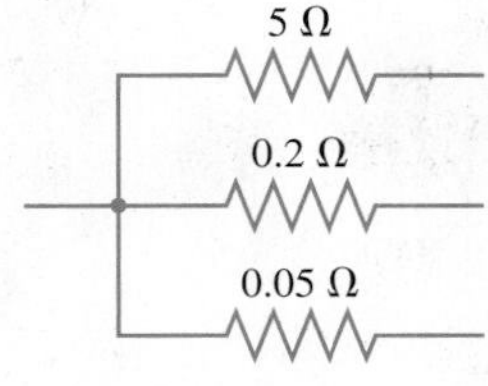

Fig. 11.7

55. A rectangular piece of plywood 4 ft by 8 ft has corners cut from it, as shown in Fig. 11.8. Find the perimeter of the remaining piece in exact form and in decimal form.

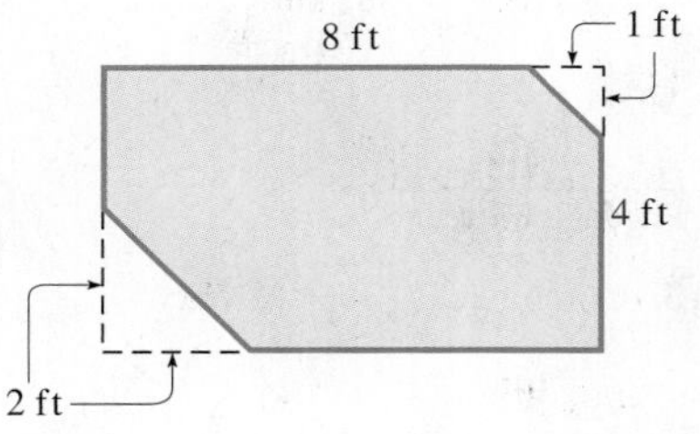

Fig. 11.8

56. Three squares with areas of 150 cm², 54 cm², and 24 cm² are displayed on a computer monitor. What is the sum (in radical form) of the perimeters of these squares?

Answers to Practice Exercises

1. $2\sqrt{7} - \sqrt{10}$ 2. $7\sqrt{11}$ 3. $\dfrac{(2 - 5y)\sqrt{10xy}}{5y}$

11.5 Multiplication and Division of Radicals

Multiplication of Radicals • Power of a Radical • Division of Radicals • Rationalizing the Denominator

When multiplying expressions containing radicals, we use the equation $\sqrt[n]{a}\sqrt[n]{b} = \sqrt[n]{ab}$, along with the normal procedures of algebraic multiplication. *Note that the product of two radicals can be combined only if the orders of the radicals are the same.* The following examples illustrate the method.

EXAMPLE 1 Multiplying monomial radicals

(a) $\sqrt{5}\sqrt{2} = \sqrt{5 \times 2} = \sqrt{10}$

(b) $\sqrt{33}\sqrt{3} = \sqrt{33 \times 3} = \sqrt{99} = \sqrt{9 \times 11} = \sqrt{9}\sqrt{11}$ (perfect square)

$= 3\sqrt{11}$

or $\sqrt{33}\sqrt{3} = \sqrt{33 \times 3} = \sqrt{11 \times 3 \times 3}$

$= 3\sqrt{11}$

Note that we express the resulting radical in simplest form.

(c) $\sqrt[3]{6}\sqrt[3]{4} = \sqrt[3]{6(4)} = \sqrt[3]{24} = \sqrt[3]{8}\sqrt[3]{3}$ (perfect cube)

$= 2\sqrt[3]{3}$

(d) $\sqrt[5]{8a^3b^4}\sqrt[5]{8a^2b^3} = \sqrt[5]{(8a^3b^4)(8a^2b^3)} = \sqrt[5]{64a^5b^7} = \sqrt[5]{32a^5b^5}\sqrt[5]{2b^2}$ (perfect fifth power)

$= 2ab\sqrt[5]{2b^2}$ ■

Practice Exercise

1. Multiply: $\sqrt{7}\sqrt{14}$

EXAMPLE 2 Multiplying binomial radicals

(a) $\sqrt{2}(3\sqrt{5} - 4\sqrt{2}) = 3\sqrt{2}\sqrt{5} - 4\sqrt{2}\sqrt{2} = 3\sqrt{10} - 4\sqrt{4}$

$= 3\sqrt{10} - 4(2) = 3\sqrt{10} - 8$

(b) $(5\sqrt{7} - 2\sqrt{3})(4\sqrt{7} + 3\sqrt{3}) = (5\sqrt{7})(4\sqrt{7}) + (5\sqrt{7})(3\sqrt{3})$

$- (2\sqrt{3})(4\sqrt{7}) - (2\sqrt{3})(3\sqrt{3})$

$= (5)(4)\sqrt{7}\sqrt{7} + (5)(3)\sqrt{7}\sqrt{3}$

$- (2)(4)\sqrt{3}\sqrt{7} - (2)(3)\sqrt{3}\sqrt{3}$

$= 20(7) + 15\sqrt{21} - 8\sqrt{21} - 6(3)$

$= 140 + 7\sqrt{21} - 18$

$= 122 + 7\sqrt{21}$ ■

NOTE ▸ [When raising a single-term radical expression to a power, the power can be applied to each factor of the term separately. However, when raising a binomial to a power, the entire binomial must be multiplied by itself the indicated number of times.] This is illustrated in the following example.

EXAMPLE 3 Power of radical

(a) $(2\sqrt{7})^2 = 2^2(\sqrt{7})^2 = 4(7) = 28$

(b) $(2\sqrt{7})^3 = 2^3(\sqrt{7})^3 = 8(\sqrt{7})^2(\sqrt{7}) = 8(7)\sqrt{7}$

$= 56\sqrt{7}$

(c) $(3 + \sqrt{5})^2 = (3 + \sqrt{5})(3 + \sqrt{5}) = 9 + 3\sqrt{5} + 3\sqrt{5} + (\sqrt{5})^2$

$= 9 + 6\sqrt{5} + 5$

$= 14 + 6\sqrt{5}$

(d) $(\sqrt{a} - \sqrt{b})^2 = (\sqrt{a} - \sqrt{b})(\sqrt{a} - \sqrt{b}) = (\sqrt{a})^2 - \sqrt{ab} - \sqrt{ab} + (\sqrt{b})^2$

$= a + b - 2\sqrt{ab}$ ■

CAUTION Again, we note that to multiply radicals and combine them under one radical sign, it is necessary that the order of the radicals be the same. ■ If necessary, we can make the order of each radical the same by appropriate operations on each radical separately. Fractional exponents are frequently useful for this purpose.

Practice Exercises

Multiply: 2. $2\sqrt{5}(\sqrt{15} - 3\sqrt{3})$

3. $(2\sqrt{6} - \sqrt{3})(3\sqrt{6} + 2\sqrt{3})$

EXAMPLE 4 Use of fractional exponents

(a) $\sqrt[3]{2}\sqrt{5} = 2^{1/3}5^{1/2} = 2^{2/6}5^{3/6} = (2^2 5^3)^{1/6} = \sqrt[6]{500}$

(b) $$\begin{aligned}\sqrt[3]{4a^2b}\sqrt[4]{8a^3b^2} &= (2^2a^2b)^{1/3}(2^3a^3b^2)^{1/4} = (2^2a^2b)^{4/12}(2^3a^3b^2)^{3/12}\\ &= (2^8a^8b^4)^{1/12}(2^9a^9b^6)^{1/12} = (2^{17}a^{17}b^{10})^{1/12}\\ &= (2^{12}a^{12})^{1/12}(2^5a^5b^{10})^{1/12}\\ &= 2a(2^5a^5b^{10})^{1/12}\\ &= 2a\sqrt[12]{32a^5b^{10}}\end{aligned}$$

■

DIVISION OF RADICALS

If a fraction involving a radical is to be changed in form, *rationalizing the denominator* or *rationalizing the numerator* is the principal step. Although calculators have made the rationalization of denominators unnecessary for calculation, this process often makes the form of the fraction simpler. Also, rationalizing numerators is useful at times in more advanced mathematics. We now consider rationalizing when the denominator (or numerator) to be rationalized has more than one term.

NOTE ▸ [If the denominator (or numerator) is the sum (or difference) of two terms, at least one of which is a radical, the fraction is rationalized by multiplying both the numerator and the denominator by the difference (or sum) of the same two terms, if the radicals are square roots.]

EXAMPLE 5 Rationalizing the denominator

The fraction $\frac{1}{\sqrt{3}-\sqrt{2}}$ can be rationalized by multiplying the numerator and the denominator by $\sqrt{3}+\sqrt{2}$. In this way, the radicals will be removed from the denominator.

$$\frac{1}{\sqrt{3}-\sqrt{2}} \times \frac{\sqrt{3}+\sqrt{2}}{\sqrt{3}+\sqrt{2}} = \frac{\sqrt{3}+\sqrt{2}}{(\sqrt{3})^2+\sqrt{6}-\sqrt{6}-(\sqrt{2})^2} = \frac{\sqrt{3}+\sqrt{2}}{3-2} = \sqrt{3}+\sqrt{2}$$

change sign

■

The reason this technique works is that an expression of the form $a^2 - b^2$ is created in the denominator, where a or b (or both) is a radical. We see that the result is a denominator free of radicals.

EXAMPLE 6 Rationalizing the denominator

Rationalize the denominator of $\frac{4\sqrt{6}-\sqrt{2}}{\sqrt{6}+\sqrt{2}}$ and simplify the result.

$$\frac{4\sqrt{6}-\sqrt{2}}{\sqrt{6}+\sqrt{2}}\left(\frac{\sqrt{6}-\sqrt{2}}{\sqrt{6}-\sqrt{2}}\right) = \frac{24-4\sqrt{12}-\sqrt{12}+2}{6-\sqrt{12}+\sqrt{12}-2} = \frac{24-5\sqrt{12}+2}{4}$$

change sign

$$= \frac{26-5(2\sqrt{3})}{4} = \frac{13-5\sqrt{3}}{2}$$

TI-89 graphing calculator keystrokes for Example 6: goo.gl/bsu8HF

We note that, after rationalizing the denominator, the result has a much simpler form than the original expression.

■

Practice Exercise

4. Rationalize the denominator and simplify: $\frac{3-2\sqrt{3}}{2+7\sqrt{3}}$

As we noted earlier, in certain types of algebraic operations, it may be necessary to rationalize the numerator of an expression. This procedure is illustrated in the following example.

EXAMPLE 7 Rationalizing the numerator—semiconductor expression

In studying the properties of a semiconductor, the expression

$$\frac{C_1 + C_2\sqrt{1 + 2V}}{\sqrt{1 + 2V}}$$

is used. Here, C_1 and C_2 are constants and V is the voltage across a junction of the semiconductor. Rationalize the numerator of this expression.

Multiplying numerator and denominator by $C_1 - C_2\sqrt{1 + 2V}$, we have

$$\begin{aligned}\frac{C_1 + C_2\sqrt{1 + 2V}}{\sqrt{1 + 2V}} &= \frac{(C_1 + C_2\sqrt{1 + 2V})(C_1 - C_2\sqrt{1 + 2V})}{\sqrt{1 + 2V}(C_1 - C_2\sqrt{1 + 2V})} \\ &= \frac{C_1^2 - C_2^2(\sqrt{1 + 2V})^2}{C_1\sqrt{1 + 2V} - C_2\sqrt{1 + 2V}\sqrt{1 + 2V}} \\ &= \frac{C_1^2 - C_2^2(1 + 2V)}{C_1\sqrt{1 + 2V} - C_2(1 + 2V)}\end{aligned}$$

■

EXERCISES 11.5

In Exercises 1–4, perform the indicated operations on the resulting expressions if the given changes are made in the indicated examples of this section.

1. In Example 2(a), change $4\sqrt{2}$ to $4\sqrt{8}$ and then perform the multiplication.
2. In Example 4(a), change $\sqrt{5}$ to $\sqrt[6]{5}$ and then perform the multiplication.
3. In Example 5, change the sign in the denominator from − to + and then rationalize the denominator.
4. In Example 6, rationalize the numerator of the given expression.

In Exercises 5–48, perform the indicated operations, expressing answers in simplest form with rationalized denominators.

5. $\sqrt{5}\sqrt{7}$
6. $\sqrt{2}\sqrt{51}$
7. $\sqrt{2}\sqrt{6}$
8. $\sqrt{7}\sqrt{35}$
9. $\sqrt[3]{4}\sqrt[3]{2}$
10. $\sqrt[5]{4}\sqrt[5]{16}$
11. $(4\sqrt{3})^2$
12. $2(3\sqrt{5})^3$
13. $\sqrt{72}\sqrt{\frac{5}{2}}$
14. $\sqrt{\frac{6}{7}}\sqrt{\frac{2}{3}}$
15. $\sqrt{3}(\sqrt{6} - \sqrt{5})$
16. $3\sqrt{5}(\sqrt{15} - 2\sqrt{5})$
17. $(2 - \sqrt{5})(2 + \sqrt{5})$
18. $4(1 - \sqrt{7})^2$
19. $(\sqrt{30} - 2\sqrt{3})(\sqrt{30} + 7\sqrt{3})$
20. $(3\sqrt{7a} - \sqrt{50})(\sqrt{7a} + \sqrt{2})$
21. $(3\sqrt{11} - \sqrt{x})(2\sqrt{11} + 5\sqrt{x})$
22. $(2\sqrt{10} + 3\sqrt{15})(\sqrt{10} - 7\sqrt{15})$
23. $\sqrt{a}(\sqrt{ab} + \sqrt{c^3})$
24. $\sqrt{3x}(\sqrt[3]{3x} - \sqrt{xy})$
25. $\dfrac{\sqrt{6} - 9}{\sqrt{6}}$
26. $\dfrac{5 - \sqrt{10}}{\sqrt[4]{10}}$
27. $(\sqrt{x} - 4\sqrt{y})^2$
28. $(\sqrt{c} + \sqrt{5d})(\sqrt{c} - \sqrt{5d})$
29. $(\sqrt{2a} - \sqrt{b})(\sqrt{2a} + 3\sqrt{b})$
30. $(2\sqrt{mn} + 3\sqrt{n})^2$
31. $\sqrt{2}\sqrt[3]{3}$
32. $\sqrt[5]{64}\sqrt[3]{16}$
33. $\dfrac{1}{\sqrt{7} + \sqrt{3}}$
34. $\dfrac{6\sqrt[4]{25}}{5 - 6\sqrt{5}}$
35. $\dfrac{\sqrt{2} - 1}{\sqrt{7} - 3\sqrt{2}}$
36. $\dfrac{2\sqrt{15} - 3}{\sqrt{15} + 4}$
37. $\dfrac{2\sqrt{3} - 5\sqrt{5}}{\sqrt{3} + 2\sqrt{5}}$
38. $\dfrac{\sqrt{15} - 3\sqrt{5}}{2\sqrt{15} - \sqrt{5}}$
39. $\dfrac{6\sqrt{x}}{\sqrt{x} - \sqrt{5}}$
40. $\dfrac{\sqrt{5c} + 3d}{\sqrt{5c} - d}$
41. $(\sqrt[5]{\sqrt{6} - \sqrt{5}})(\sqrt[5]{\sqrt{6} + \sqrt{5}})$
42. $\sqrt[3]{5 - \sqrt{17}}\sqrt[3]{5 + \sqrt{17}}$
43. $\left(\sqrt{\dfrac{2}{R}} + \sqrt{\dfrac{R}{2}}\right)\left(\sqrt{\dfrac{2}{R}} - 2\sqrt{\dfrac{R}{2}}\right)$
44. $(3 + \sqrt{6 - 7a})(2 - \sqrt{6 - 7a})$
45. $\dfrac{\sqrt{x + y}}{\sqrt{x - y} - \sqrt{x}}$
46. $\dfrac{\sqrt{1 + a}}{a - \sqrt{1 - a}}$
47. $\dfrac{\sqrt{a} + \sqrt{a - 2}}{\sqrt{a} - \sqrt{a - 2}}$
48. $\dfrac{\sqrt{T^4 - V^4}}{\sqrt{V^{-2} - T^{-2}}}$

In Exercises 49–52, perform the indicated operations, expressing answers in simplest form with rationalized denominators. Then verify the result with a calculator.

49. $(\sqrt{11} + \sqrt{6})(\sqrt{11} - 2\sqrt{6})$

50. $(2\sqrt{5} - \sqrt{7})(3\sqrt{5} + \sqrt{7})$

51. $\dfrac{2\sqrt{6} - \sqrt{5}}{3\sqrt{6} - 4\sqrt{5}}$ **52.** $\dfrac{3\sqrt{7} - 12\sqrt{2}}{15\sqrt{7} - 12\sqrt{2}}$

In Exercises 53–56, combine the terms into a single fraction, but do not rationalize the denominators.

53. $2\sqrt{x} + \dfrac{1}{\sqrt{x}}$ **54.** $\dfrac{3}{2\sqrt{3x - 4}} - \sqrt{3x - 4}$

55. $\dfrac{x^2}{\sqrt{2x + 1}} + 2x\sqrt{2x + 1}$ **56.** $4\sqrt{x^2 + 1} - \dfrac{4x}{\sqrt{x^2 + 1}}$

In Exercises 57–60, rationalize the numerator of each fraction.

57. $\dfrac{\sqrt{5} + 2\sqrt{2}}{3\sqrt{10}}$ **58.** $\dfrac{\sqrt{19} - 3}{5}$

59. $\dfrac{\sqrt{x + h} - \sqrt{x}}{h}$ **60.** $\dfrac{\sqrt{3x + 4} + \sqrt{3x}}{8}$

In Exercises 61–76, solve the given problems.

61. Are there any values of x or y such that $\sqrt{x} + \sqrt{y} = \sqrt{x + y}$?

62. For what real values of x and y is $\sqrt[3]{x} + \sqrt[3]{y} = \sqrt[3]{x + y}$?

63. By substitution, show that $x = 1 - \sqrt{2}$ is a solution of the equation $x^2 - 2x - 1 = 0$.

64. For the quadratic equation $ax^2 + bx + c = 0$, if a, b, and c are integers, the product of the roots is a rational number. Explain.

65. Determine the relationship between a and c in $ax^2 + bx + c = 0$ if the roots of the equation are reciprocals.

66. Evaluate $r^2 - s^2$ if $r = -b + \sqrt{b^2 - 4ac}$ and $s = -b - \sqrt{b^2 - 4ac}$.

67. Rationalize the denominator of $\dfrac{1}{\sqrt[3]{x^2} + \sqrt[3]{x} + 1}$ by using the equation $a^3 - b^3 = (a - b)(a^2 + ab + b^2)$. (*Hint:* The denominator is of the form $a^2 + ab + b^2$.)

68. One leg of a right triangle is $2\sqrt{7}$ and the hypotenuse is 6. What is the area of the triangle?

69. For an object oscillating at the end of a spring and on which there is a force that retards the motion, the equation $m^2 + bm + k^2 = 0$ must be solved. Here, b is a constant related to the retarding force, and k is the spring constant. By substitution, show that $m = \frac{1}{2}(\sqrt{b^2 - 4k^2} - b)$ is a solution.

70. Among the products of a specialty furniture company are tables with tops in the shape of a regular octagon (eight sides). Express the area A of a table top as a function of the side s of the octagon.

71. An expression used in determining the characteristics of a spur gear is $\dfrac{50}{50 + \sqrt{V}}$. Rationalize the denominator.

72. When studying the orbits of Earth satellites, the expression $\left(\dfrac{GM}{4\pi^2}\right)^{1/3} T^{2/3}$ arises. Express it in simplest rationalized radical form.

73. In analyzing a tuned amplifier circuit, the expression $\dfrac{2Q}{\sqrt{\sqrt{2} - 1}}$ is used. Rationalize the denominator.

74. When analyzing the ratio of resultant forces when forces **F** and **T** act on a structure, the expression $\sqrt{F^2 + T^2}/\sqrt{F^{-2} + T^{-2}}$ arises. Simplify this expression.

75. The resonant frequency ω of a capacitance C in parallel with a resistance R and inductance L (see Fig. 11.9) is

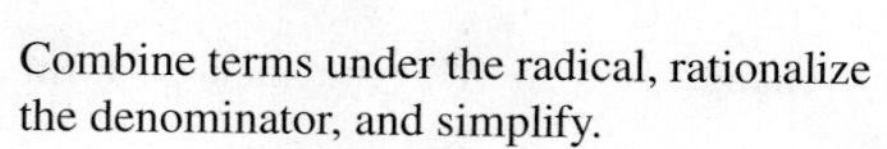

$$\omega = \frac{1}{\sqrt{LC}}\sqrt{1 - \frac{R^2C}{L}}.$$

Combine terms under the radical, rationalize the denominator, and simplify.

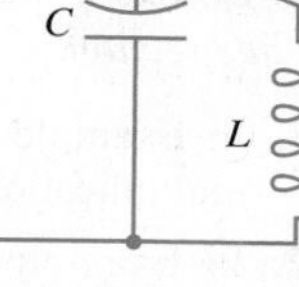

Fig. 11.9

76. In fluid dynamics, the expression $\dfrac{a}{1 + m/\sqrt{r}}$ arises. Combine terms in the denominator, rationalize the denominator, and simplify.

Answers to Practice Exercises

1. $7\sqrt{2}$ **2.** $10\sqrt{3} - 6\sqrt{15}$ **3.** $30 + 3\sqrt{2}$ **4.** $\dfrac{-48 + 25\sqrt{3}}{143}$

CHAPTER 11 KEY FORMULAS AND EQUATIONS

Exponents

$$a^m \times a^n = a^{m+n} \quad (11.1)$$

$$\frac{a^m}{a^n} = a^{m-n} \quad \text{or} \quad \frac{a^m}{a^n} = \frac{1}{a^{n-m}} \quad (a \neq 0) \quad (11.2)$$

$$(a^m)^n = a^{mn} \quad (11.3)$$

$$(ab)^n = a^nb^n, \quad \left(\frac{a}{b}\right)^n = \frac{a^n}{b^n} \quad (b \neq 0) \quad (11.4)$$

$$a^0 = 1 \quad (a \neq 0) \quad (11.5)$$

$$a^{-n} = \frac{1}{a^n} \quad (a \neq 0) \quad (11.6)$$

Fractional exponents

$$a^{1/n} = \sqrt[n]{a} \quad (11.7)$$

$$a^{m/n} = \sqrt[n]{a^m} = (\sqrt[n]{a})^m \quad (11.8)$$

Radicals

$$\sqrt[n]{a^n} = (\sqrt[n]{a})^n = a \quad (11.9)$$

$$\sqrt[n]{a}\sqrt[n]{b} = \sqrt[n]{ab} \quad (11.10)$$

$$\sqrt[m]{\sqrt[n]{a}} = \sqrt[mn]{a} \quad (11.11)$$

$$\frac{\sqrt[n]{a}}{\sqrt[n]{b}} = \sqrt[n]{\frac{a}{b}} \quad (b \neq 0) \quad (11.12)$$

CHAPTER 11 REVIEW EXERCISES

CONCEPT CHECK EXERCISES

Determine each of the following as being either **true** *or* **false.** *If it is false, explain why.*

1. $2x^0 = 1$

2. $\sqrt{16+4} = 6$

3. $(16^{-1/2})\left(\frac{1}{4^{-2}}\right) = 2$

4. $\sqrt{48} - \sqrt{27} = \sqrt{3}$

5. $\frac{1}{\sqrt{3}-\sqrt{2}} = \sqrt{3}+\sqrt{2}$

6. $\sqrt{5}\sqrt{7} = \sqrt{35}$

PRACTICE AND APPLICATIONS

In Exercises 7–34, express each expression in simplest form with only positive exponents.

7. $8x^{-3}y^0$ **8.** $(2c)^{-1}z^{-2}$ **9.** $\frac{9c^{-1}}{d^{-3}}$ **10.** $\frac{-5x^0}{3y^{-1}}$

11. $3(25)^{3/2}$ **12.** $32^{2/5}$ **13.** $8(400^{-3/2})$ **14.** $7(1000)^{-2/3}$

15. $\left(\frac{3}{t^2}\right)^{-2}$ **16.** $\left(\frac{2x^3}{3}\right)^{-3}$ **17.** $\frac{-8^{2/3}}{49^{-1/2}}$ **18.** $\frac{4^{-45}}{2^{-90}}$

19. $(2a^{2/3}b^{1/6})^6$ **20.** $(ax^{-1/2}y^{1/4})^8$ **21.** $(-32m^{15}n^{10})^{3/5}$

22. $(27x^{-6}y^9)^{2/3}$ **23.** $6L^{-2} - 4C^{-1}$ **24.** $C^4C^{-1} + (C^4)^{-1}$

25. $\frac{3x^{-1}}{3x^{-1}+y^{-1}}$ **26.** $\frac{9a}{(2a)^{-1}-a}$ **27.** $(a - 3b^{-1})^{-1}$

28. $(2s^{-2}+t)^{-2}$ **29.** $(x^3 - y^{-3})^{1/3}$

30. $(27a^3)^{2/3}(a^{-2}+1)^{1/2}$ **31.** $(W^2 - 2WH + H^2)^{-1/2}$

32. $\left[\frac{(9a)^0(4x^2)^{1/3}(3b^{1/2})}{(2b^0)^2}\right]^{-6}$

33. $2x(x-1)^{-2} - 2(x^2+1)(x-1)^{-3}$

34. $4(1-T^2)^{1/2} - (1-T^2)^{-1/2}$

In Exercises 35–84, perform the indicated operations and express the result in simplest radical form with rationalized denominators.

35. $\sqrt{60}$ **36.** $4\sqrt{96}$ **37.** $\sqrt{ab^5c^2}$

38. $\sqrt{x^3y^4z^6}$ **39.** $\sqrt{9a^3b^4}$ **40.** $\sqrt{8x^5y^2}$

41. $\sqrt{84st^3u^{-2}}$ **42.** $\sqrt{52L^2C^{-5}}$ **43.** $\frac{5}{\sqrt{2s}}$

44. $\frac{3a}{\sqrt{5x}}$ **45.** $\sqrt{\frac{8}{27}}$ **46.** $\sqrt{\frac{63}{8V}}$

47. $\sqrt[4]{8m^6n^9}$ **48.** $\sqrt[3]{81a^7b^{-3}}$ **49.** $\sqrt[4]{\sqrt[3]{64}}$

50. $\sqrt{a^{-3}\sqrt[5]{b^{12}}}$ **51.** $\sqrt{36+4} - 2\sqrt{10}$

52. $2\sqrt{68x} - \sqrt{153x}$ **53.** $\sqrt{63} - 2\sqrt{112} - \sqrt{28}$

54. $7\sqrt{20} - \sqrt{80} - 2\sqrt{125}$ **55.** $a\sqrt{2x^3} + \sqrt{8a^2x^3}$

56. $2\sqrt{m^2n^3} - \sqrt{n^5}$ **57.** $\sqrt[3]{8a^4} + b\sqrt[3]{a}$

58. $\sqrt[4]{2xy^5} - \sqrt[4]{32xy}$ **59.** $5\sqrt{5}(6\sqrt{5} - \sqrt{35})$

60. $7\sqrt{8}(5\sqrt{2} - \sqrt{6})$ **61.** $2\sqrt{2}(\sqrt{6} - \sqrt{10})$

62. $3\sqrt{5}(\sqrt{15m^2} + 2\sqrt{35m^2})$

63. $(2 - 3\sqrt{17B})(3 + \sqrt{17B})$

64. $(5\sqrt{6} - 4)(3\sqrt{6} + 5)$

65. $(2\sqrt{7} - 3\sqrt{a})(3\sqrt{7} + \sqrt{a})$

66. $(3\sqrt{2} - \sqrt{13})(5\sqrt{2} + 3\sqrt{13})$

67. $(6\sqrt{x} - 4\sqrt{y})^2$

68. $(\sqrt{8a} - 2\sqrt{b})(\sqrt{8a} + 2\sqrt{b})$

69. $\frac{\sqrt{3x}}{2\sqrt{3x}-\sqrt{y}}$ **70.** $\frac{10\sqrt{a}}{4\sqrt{a}-2c}$ **71.** $\frac{\sqrt{2}}{\sqrt{3}-4\sqrt{2}}$

72. $\frac{4}{6-2\sqrt{7}}$ **73.** $\frac{\sqrt{7}-\sqrt{5}}{\sqrt{5}+3\sqrt{7}}$ **74.** $\frac{7-4\sqrt{6}}{3+4\sqrt{6}}$

75. $\frac{2\sqrt{x}-a}{3\sqrt{x}+5a}$ **76.** $\frac{2\sqrt{3y}-\sqrt{2}}{9\sqrt{3y}-5\sqrt{2}}$

77. $\sqrt{4b^2+1}$ **78.** $x^{3/8}(x^{5/8} - 8x^{13/8})$

79. $(3x^{1/2} - 2x)(3x^{1/2} + 2x)$ **80.** $(x^{3n-1}/x^{n-1})^{1/n}$

81. $(1 + 6^{1/2})(3^{1/2} + 2^{1/2})(3^{1/2} - 2^{1/2})$

82. $\sqrt{a^{-2} + \frac{1}{b^2}}$

83. $\left(\frac{2-\sqrt{15}}{2}\right)^2 - \left(\frac{2-\sqrt{15}}{2}\right)$

84. $\sqrt{3+n}(\sqrt{3+n} - \sqrt{n})^{-1}$

In Exercises 85 and 86, perform the indicated operations and express the result in simplified radical form with rationalized denominators. In Exercises 87 and 88, without a calculator, show that the given equations are true. Then verify each result with a calculator.

85. $(\sqrt{7} - 2\sqrt{15})(3\sqrt{7} - \sqrt{15})$

86. $\frac{2\sqrt{3}-7\sqrt{14}}{3\sqrt{3}+2\sqrt{14}}$

87. $\sqrt{\sqrt{2}-1}(\sqrt{2}+1) = \sqrt{\sqrt{2}+1}$

88. $\sqrt{\sqrt{3}+1}(\sqrt{3}-1) = \sqrt{2(\sqrt{3}-1)}$

In Exercises 89–106, perform the indicated operations.

89. The legs of a right triangle are $\sqrt{18}$ and $\sqrt{32}$. Find the perimeter of the triangle.

90. One of the legs of a right triangle is $3\sqrt{5}$ and the hypotenuse is $5\sqrt{3}$. Find the area of the triangle.

91. Evaluate $3x^2 - 2x + 5$ for $x = \frac{1}{2}(2 - \sqrt{3})$.

92. Evaluate x if $x^{-1/3} = 0.2$.

93. The average annual increase i (in %) of the cost of living over n years is given by $i = 100[(C_2/C_1)^{1/n} - 1]$, where C_1 is the cost of living index for the first year and C_2 is the cost of living index for the last year of the period. Evaluate i if $C_1 = 218.1$ and $C_2 = 237.0$ are the values for 2010 and 2015, respectively.

94. Kepler's third law of planetary motion may be given as $T = kr^{3/2}$, where T is the time for one revolution of a planet around the sun, r is its mean radius from the sun, and $k = 1.115 \times 10^{-12}$ year/mi$^{3/2}$. Find the time for one revolution of Venus about the sun if $r = 6.73 \times 10^7$ mi.

95. The speed v of a ship of weight W whose engines produce power P is $v = k\sqrt[3]{P/W}$. Express this equation (a) with a fractional exponent and (b) as a radical with the denominator rationalized.

96. In analyzing the orbit of an Earth satellite, the expression $\sqrt{1 + \frac{2E}{m}\left(\frac{h}{GM}\right)^2}$ is used. Combine terms under the radical and express the result with the denominator rationalized.

97. A square plastic sheet of side x is stretched by an amount equal to $\sqrt{x}$ horizontally and vertically. Find the expression for the percent increase in the area of the sheet.

98. The value P of an object originally worth P_0 and that appreciates at an annual rate of r for t years is given by $P = P_0(1 + r)^t$. Solve for r.

99. The expression $(1 - v^2/c^2)^{-1/2}$ arises in the theory of relativity. Simplify, using only positive exponents in the result.

100. The compression ratio r of a certain gasoline engine is related to the efficiency e (in %) of the engine by $r = \left(\frac{100 - e}{100}\right)^{-2.5}$. What compression ratio is necessary to have an efficiency of 55%?

101. A square is decreasing in size on a computer screen. To find how fast the side of the square changes, we must rationalize the numerator of $\frac{\sqrt{A + h} - \sqrt{A}}{h}$. Perform this operation.

102. A plane takes off 200 mi east of its destination, and a 50 mi/h wind is from the south. If the plane's velocity is 250 mi/h and its heading is always toward its destination, its path can be described as $y = 100[(0.005x)^{0.8} - (0.005x)^{1.2}]$, $0 \leq x \leq 200$ mi. Sketch the path and check using a graphing calculator.

103. In an experiment, a laser beam follows the path shown in Fig. 11.10. Express the length of the path in simplest radical form.

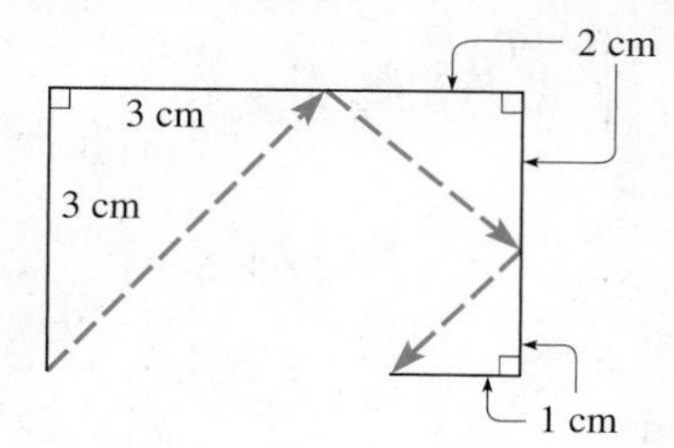

Fig. 11.10

104. The flow rate V (in ft/s) through a storm drain pipe is found from $V = 110(0.55)^{2/3}(0.0018)^{1/2}$. Find the value of V.

105. The frequency of a certain electric circuit is given by $\frac{1}{2\pi\sqrt{\frac{LC_1C_2}{C_1 + C_2}}}$. Express this in simplest rationalized radical form.

106. A computer analysis of an experiment showed that the fraction f of viruses surviving X-ray dosages was given by $f = \frac{20}{d + \sqrt{3d + 400}}$, where d is the dosage. Express this with the denominator rationalized.

107. In calculating the forces on a tower by the wind, it is necessary to evaluate $0.018^{0.13}$. Write a paragraph explaining why this form is preferable to the equivalent radical form.

CHAPTER 11 PRACTICE TEST

As a study aid, we have included complete solutions for each Practice Test problem at the end of this book.

In Problems 1–14, simplify the given expressions. For those with exponents, express each result with only positive exponents. For the radicals, rationalize the denominator where applicable.

1. $-5y^0$ **2.** $(3\pi x^{-4})^{-2}$ **3.** $\frac{100^{3/2}}{8^{-2/3}}$

4. $(as^{-1/3}t^{3/4})^{12}$ **5.** $2\sqrt{20} - \sqrt{125}$ **6.** $(2x^{-1} + y^{-2})^{-1}$

7. $(\sqrt{2x} - 3\sqrt{y})^2$ **8.** $\sqrt[3]{\sqrt{4}}$ **9.** $\frac{3 - 2\sqrt{2}}{2\sqrt{x}}$

10. $\sqrt{27a^4b^3}$ **11.** $2\sqrt{2}(3\sqrt{10} - \sqrt{6})$

12. $(2x + 3)^{1/2} + (x + 1)(2x + 3)^{-1/2}$

13. $\left(\frac{4a^{-1/2}b^{3/4}}{b^{-2}}\right)\left(\frac{b^{-1}}{2a}\right)$ **14.** $\frac{2\sqrt{15} + \sqrt{3}}{\sqrt{15} - 2\sqrt{3}}$

15. Express $\frac{3^{-1/2}}{2}$ in simplest radical form with a rationalized denominator.

16. In the study of fluid flow in pipes, the expression $0.220N^{-1/6}$ is found. Evaluate this expression for $N = 64 \times 10^6$.

Complex Numbers

12

Among the important areas of research in the mid-nineteenth century was the study of light, which was determined to be a form of electromagnetic radiation.

Today, extensive use of electromagnetic waves is made in a great variety of situations. These include the transmission of signals for radio, television, and cell phones. Also, radar, microwave ovens, and X-ray machines are among the many other applications of electromagnetic waves.

In the 1860s, James Clerk Maxwell, a Scottish physicist, used the research of others—and his own—to develop a set of very important equations for electromagnetic radiation. In doing so, he actually predicted *mathematically* the existence of electromagnetic waves, such as radio waves.

It was not until 1887 that Heinrich Hertz, a German physicist, produced and observed radio waves in the laboratory. We see that mathematics was used in predicting the existence of one of the most important devices in use today, over 20 years before it was actually observed.

Important in the study of electromagnetic waves and many areas of electricity and electronics are *complex numbers,* which we study in this chapter. Complex numbers include imaginary numbers as well as real numbers. Other than briefly noting imaginary numbers in Chapters 1 and 7, we have purposely avoided any extended discussion of them until now.

Despite their names, complex numbers and imaginary numbers have very real and useful applications, as we have noted. Besides electricity and electronics, they are used in the studies of mechanical vibrations, optics, acoustics, and signal processing.

LEARNING OUTCOMES

After completion of this chapter, the student should be able to:

- Express complex numbers in rectangular form, polar form, exponential form, and graphically
- Convert a complex number from one form to another
- Perform mathematical operations on complex numbers in rectangular, polar, and exponential forms
- Add complex numbers graphically
- Solve algebraic equations involving complex numbers
- Find nth roots of a complex number using DeMoivre's theorem
- Solve application problems involving complex numbers, including the analysis of alternating-current circuits

◀ In Section 12.7, we see that tuning in a radio station involves a basic application of complex numbers to electricity.

12.1 Basic Definitions

Imaginary Unit • Rectangular Form of a Complex Number • Equality of Complex Numbers • Conjugate

Because the square of a positive number or a negative number is positive, it is not possible to square any real number and have a negative result. Therefore, we must define a number system to include square roots of negative numbers. We will find that such numbers can be used to great advantage in certain applications.

If the radicand in a square root is negative, we can express the indicated root as the product of $\sqrt{-1}$ and the square root of a positive number. *The symbol* $\sqrt{-1}$ *is defined as the* **imaginary unit** *and is denoted by the symbol j.* Therefore, we have

$$j = \sqrt{-1} \quad \text{and} \quad j^2 = -1 \qquad \textbf{(12.1)}$$

Generally, mathematicians use the symbol i for $\sqrt{-1}$, and therefore most nontechnical textbooks use i. However, one of the major technical applications of complex numbers is in electronics, where i represents electric current. Therefore, we will use j for $\sqrt{-1}$, which is also the standard symbol in electronics textbooks.

EXAMPLE 1 Square roots in terms of j

Express the following square roots in terms of j.

(a) $\sqrt{-9} = \sqrt{(9)(-1)} = \sqrt{9}\sqrt{-1} = 3j$

(b) $\sqrt{-0.25} = \sqrt{0.25}\sqrt{-1} = 0.5j$

(c) $\sqrt{-5} = \sqrt{5}\sqrt{-1} = \sqrt{5}j = j\sqrt{5}$ — the form $j\sqrt{5}$ is better than $\sqrt{5}j$

NOTE ▸ [When a radical appears, we write the result in a form with the j before the radical as in the last illustration. This clearly shows that the j is not under the radical.] ■

Practice Exercise

1. Write $\sqrt{-25}$ in terms of j.

EXAMPLE 2 Be careful in simplifying

$(\sqrt{-4})^2 = (\sqrt{4}\sqrt{-1})^2 = (j\sqrt{4})^2 = 4j^2 = -4$

NOTE ▸ [The simplification of this expression does not follow Eq. (11.10), which states that $\sqrt{ab} = \sqrt{a}\sqrt{b}$ for square roots. This is the reason it was noted as being valid only if a and b are not negative. Therefore, we see that Eq. (11.10) does not always hold for negative values of a and b.]

In simplifying $(\sqrt{-4})^2$, we can write it as $(\sqrt{-4})(\sqrt{-4})$. However, we *cannot now write this product as* $\sqrt{(-4)(-4)}$, for this leads to an incorrect result of $+4$. In fact, *we cannot use Eq. (11.10) if both a and b are negative.*

CAUTION It cannot be overemphasized that $(\sqrt{-4})^2$ *is not equal to* $\sqrt{(-4)(-4)}$. As shown at the beginning of this example, for $(\sqrt{-4})^2$, we must first write $\sqrt{-4}$ as $j\sqrt{4}$ before continuing with the simplification. ■ ■

EXAMPLE 3 First, express number in terms of j

To further illustrate the method of handling square roots of negative numbers, consider the difference between $\sqrt{-3}\sqrt{-12}$ and $\sqrt{(-3)(-12)}$. For these expressions, we have

$$\sqrt{-3}\sqrt{-12} = (j\sqrt{3})(j\sqrt{12}) = (\sqrt{3}\sqrt{12})j^2 = (\sqrt{36})j^2$$
$$= 6(-1) = -6$$
$$\sqrt{(-3)(-12)} = \sqrt{36} = 6$$

For $\sqrt{-3}\sqrt{-12}$, we have the product of square roots of negative numbers, whereas for $\sqrt{(-3)(-12)}$, we have the product of negative numbers under the radical. We must be careful to note the difference. ■

NOTE ▶ [From Examples 2 and 3, we see that when we are dealing with square roots of negative numbers, each should be expressed in terms of j before proceeding.] To do this, for any positive real number a we write

$$\sqrt{-a} = j\sqrt{a} \qquad (a > 0) \tag{12.2}$$

EXAMPLE 4 Expressing numbers in terms of j

(a) $\sqrt{-6} = \sqrt{(6)(-1)} = \sqrt{6}\sqrt{-1} = j\sqrt{6}$

this step is correct if only one is negative, as in this case

(b) $-\sqrt{-75} = -\sqrt{(25)(3)(-1)} = -\sqrt{(25)(3)}\sqrt{-1} = -5j\sqrt{3}$

Practice Exercises

Simplify: **2.** $\sqrt{-5}\sqrt{-20}$ **3.** $-\sqrt{-48}$

CAUTION We note that $-\sqrt{-75}$ is *not* equal to $\sqrt{75}$. ■ ■

At times, we need to raise imaginary numbers to some power. Using the definitions of exponents and of j, we have the following results:

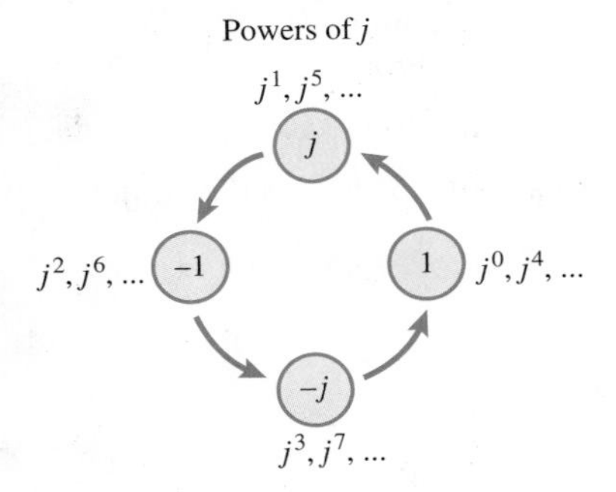

Fig. 12.1

$$j = j \qquad j^5 = j^4 j = j$$
$$j^2 = -1 \qquad j^6 = j^4 j^2 = (1)(-1) = -1$$
$$j^3 = j^2 j = -j \qquad j^7 = j^4 j^3 = (1)(-j) = -j$$
$$j^4 = j^2 j^2 = (-1)(-1) = 1 \qquad j^8 = j^4 j^4 = (1)(1) = 1$$

The powers of j go through the cycle, $j, -1, -j, 1, j, -1, -j, 1$, and so forth. See Fig. 12.1. Noting this—and the fact that j raised to a power that is a multiple of 4 equals 1—allows us to raise j to any integral power almost on sight.

EXAMPLE 5 Powers of j

(a) $j^{10} = j^8 j^2 = (1)(-1) = -1$

(b) $j^{45} = j^{44} j = (1)(j) = j$

(c) $j^{531} = j^{528} j^3 = (1)(-j) = -j$

exponents 8, 44, 528 are multiples of 4 ■

Practice Exercise

4. Simplify: $-j^{25}$

RECTANGULAR FORM OF A COMPLEX NUMBER

Using real numbers and the imaginary unit j, we define a new kind of number.

NOTE ▶ [A **complex number** is any number that can be written in the form $a + bj$ where a and b are real numbers.] If $a = 0$ and $b \neq 0$, we have the number bj, *which is a* **pure imaginary number.** *If $b = 0$, then $a + bj$ is a real number. The form $a + bj$ is known as the* **rectangular form** *of a complex number, where a is known as the* **real part** *and b is known as the* **imaginary part.** We see that complex numbers include all real numbers and all pure imaginary numbers.

A comment here about the words *imaginary* and *complex* is in order. The choice of the names of these numbers is historical in nature, and unfortunately it leads to some misconceptions about the numbers. The use of *imaginary* does not imply that the numbers do not exist. Imaginary numbers do in fact exist, as they are defined above. In the same way, the use of *complex* does not imply that the numbers are complicated and therefore difficult to understand. With the proper definitions and operations, we can work with complex numbers, just as with any other type of number.

■ Even negative numbers were not widely accepted by mathematicians until late in the sixteenth century.

■ Imaginary numbers were so named because the French mathematician René Descartes (1596–1650) referred to them as "imaginaries." Most of the mathematical development of them occurred in the eighteenth century.

■ Complex numbers were named by the German mathematician Karl Friedrich Gauss (1777–1855).

The real part of a complex number is positive or negative (or zero), and the same is true of the imaginary part. However, the complex number itself is not positive or negative in the usual sense. [Also, two complex numbers are equal if and only if the real parts are equal and the imaginary parts are equal, or $a + bj = x + yj$ if $a = x$ and $b = y$.]

NOTE ▸

EXAMPLE 6 Equality of complex numbers

(a) $a + bj = 3 + 4j$ if $a = 3$ and $b = 4$

real part — imaginary part

(b) $x + 3j = yj - 5$ if $x = -5$ and $y = 3$

(c) What values of x and y satisfy the equation $x + 3(xj + y) = 5 - j - jy$?

Rewriting each side in the form $a + bj$, and equating real parts and equating imaginary parts, we have

$$(x + 3y) + 3xj = 5 + (-1 - y)j$$

For equality, $x + 3y = 5$ and $3x = -1 - y$. The solution of this system of equations is $x = -1$ and $y = 2$. ■

Practice Exercise

5. Evaluate x and y: $4 - 6j - x = j + jy$

The **conjugate** *of the complex number* $a + bj$ *is the complex number* $a - bj$. We see that the sign of the imaginary part is changed to obtain the conjugate.

EXAMPLE 7 Conjugates

(a) $3 - 2j$ is the conjugate of $3 + 2j$. We may also say that $3 + 2j$ is the conjugate of $3 - 2j$. Thus, each is the conjugate of the other.

(b) $-2 - 5j$ and $-2 + 5j$ are conjugates.

(c) $6j$ and $-6j$ are conjugates.

(d) 3 is the conjugate of 3 (imaginary part is zero). ■

EXERCISES 12.1

In Exercises 1–4, perform the indicated operations on the resulting expressions if the given changes are made in the indicated examples of this section.

1. In Example 3, put a − sign before the first radical of the first illustration and then simplify.
2. In Example 4(a), put a j in front of the radical and then simplify.
3. In Example 5(a), add 40 to the exponent and then evaluate.
4. In Example 6(c), change the 5 on the right side of the equation to −5 and then solve.

In Exercises 5–16, express each number in terms of j.

5. $\sqrt{-81}$ **6.** $\sqrt{-121}$ **7.** $-\sqrt{-4}$ **8.** $-\sqrt{-49}$

9. $\sqrt{-0.36}$ **10.** $-\sqrt{-0.01}$ **11.** $4\sqrt{-8}$ **12.** $3\sqrt{-48}$

13. $\sqrt{-\frac{7}{4}}$ **14.** $-\sqrt{-\frac{5}{9}}$ **15.** $-\sqrt{-4\pi^3}$ **16.** $\sqrt{-\pi^4}$

In Exercises 17–32, simplify each of the given expressions.

17. (a) $(\sqrt{-7})^2$ (b) $\sqrt{(-7)^2}$

18. (a) $\sqrt{(-15)^2}$ (b) $(\sqrt{-15})^2$

19. (a) $\sqrt{(-2)(-8)}$ (b) $\sqrt{-2}\sqrt{-8}$

20. (a) $\sqrt{-9}\sqrt{-16}$ (b) $\sqrt{(-9)(-16)}$

21. $\sqrt{-\frac{1}{15}}\sqrt{-\frac{27}{5}}$ **22.** $-\sqrt{(-\frac{4}{7})(-\frac{49}{16})}$

23. $-\sqrt{-5}\sqrt{-2}\sqrt{-10}$ **24.** $-\sqrt{(-3)(-7)}\sqrt{-21}$

25. (a) $-j^6$ (b) $(-j)^6$ **26.** (a) $-j^{21}$ (b) $(-j)^{21}$

27. $j^2 - j^6$ **28.** $2j^5 - 1/j^{-2}$

29. $j^{15} - j^{13}$ **30.** $3j^{48} + j^{200}$ **31.** $-\sqrt{(-j)^2}$ **32.** $-\sqrt{-j^2}$

In Exercises 33–50, perform the indicated operations and simplify each complex number to its rectangular form.

33. $2 + \sqrt{-9}$ **34.** $-26 + \sqrt{-64}$ **35.** $3j - \sqrt{-100}$

36. $-\sqrt{1} - \sqrt{-400}$ **37.** $\sqrt{-4j^2} + \sqrt{-4}$ **38.** $5 - 2\sqrt{25j^2}$

39. $2j^2 + 3j$ **40.** $j^3 - 6$ **41.** $\sqrt{18} - \sqrt{-8}$

42. $\sqrt{-27} + \sqrt{12}$ **43.** $(\sqrt{-2})^2 + j^4$ **44.** $(2\sqrt{2})^2 - j^6$

45. $5j(-3j)(j^2)$ 46. $-7(2j)(-j^3)$ 47. $\frac{-3 + \sqrt{18}}{6}$

48. $\frac{10 - \sqrt{75}}{10}$ 49. $\frac{\sqrt{-9} - 6}{3}$ 50. $\frac{12 - 9\sqrt{-4}}{4}$

In Exercises 51–54, find the conjugate of each complex number.

51. (a) $6 - 7j$ (b) $8 + j$ 52. (a) $2j - 3$ (b) $-9 - j$

53. (a) $2j$ (b) -4 54. (a) 6 (b) $-5j$

In Exercises 55–60, find the values of x and y that satisfy the given equations.

55. $7x - 2yj = 14 + 4j$ 56. $2x + 3jy = -6 + 12j$

57. $6j - 7 = 3 - x - yj$ 58. $9 - j = xj + 1 - y$

59. $x - 2j^2 + 7j = yj = 2xj^3$ 60. $2x - 6xj^3 - 3j^2 = yj - y + 7j^5$

In Exercises 61–64, solve the given equations for x. Express the answer in simplified form in terms of j.

61. $x^2 + 32 = 0$ 62. $x^2 - 2x + 2 = 0$

63. $x^2 + 2x + 7 = 0$ 64. $3x^2 - 6x + 4 = 0$

In Exercises 65–74, answer the given questions.

65. How is a number changed if it is multiplied by (a) j^4; (b) j^2?

66. Evaluate: (a) j^{-8}; (b) j^{-6}

67. Are $8j$ and $-8j$ the solutions to the equation $x^2 + 64 = 0$?

68. Are $2j$ and $-2j$ solutions to the equation $x^4 + 16 = 0$?

69. Evaluate $j + j^2 + j^3 + j^4 + j^5 + j^6 + j^7 + j^8$.

70. What is the smallest positive value of n for which $j^{-1} = j^n$?

71. Is it possible that a given complex number and its conjugate are equal? Explain.

72. Is it possible that a given complex number and the negative of its conjugate are equal? Explain.

73. Show that the real part of $x + yj$ equals the imaginary part of $j(x + yj)$.

74. Explain why a real number is a complex number, but a complex number may not be a real number.

Answers to Practice Exercises

1. $5j$ 2. -10 3. $-4j\sqrt{3}$ 4. $-j$ 5. $x = 4, y = -7$

12.2 Basic Operations with Complex Numbers

Addition, Subtraction, Multiplication, and Division of Complex Numbers • Complex Numbers on Calculator

The basic operations of addition, subtraction, multiplication, and division of complex numbers are based on the operations for binomials with real coefficients. In performing these operations, we treat j as we would any other literal number, although we must properly handle any powers of j that might occur. [However, we must be careful to express all complex numbers in terms of j before performing these operations.] Therefore, we now have the definitions for these operations.

NOTE ▸

Basic Operations on Complex Numbers

Addition:

$$(a + bj) + (c + dj) = (a + c) + (b + d)j \quad (12.3)$$

Subtraction:

$$(a + bj) - (c + dj) = (a - c) + (b - d)j \quad (12.4)$$

Multiplication:

$$(a + bj)(c + dj) = ac + adj + bcj + bdj^2 = (ac - bd) + (ad + bc)j \quad (12.5)$$

Division:

$$\frac{a + bj}{c + dj} = \frac{(a + bj)(c - dj)}{(c + dj)(c - dj)} = \frac{(ac + bd) + (bc - ad)j}{c^2 + d^2} \quad (12.6)$$

Equations (12.3) and (12.4) show that we *add and subtract complex numbers by combining the real parts and combining the imaginary parts.*

EXAMPLE 1 Adding, subtracting complex numbers

(a) $(3 - 2j) + (-5 + 7j) = (3 - 5) + (-2 + 7)j$
$= -2 + 5j$

(b) $(7 + 9j) - (6 - 4j) = (7 - 6) + (9 - (-4))j$
$= 1 + 13j$ ■

Practice Exercise

1. Simplify: $(-6 + j) - (-3 + 2j)$

When complex numbers are multiplied, Eq. (12.5) indicates that *we express numbers in terms of j, proceed as with any algebraic multiplication, and note that* $j^2 = -1$.

EXAMPLE 2 Multiplying complex numbers

(a) $(6 - \sqrt{-4})(\sqrt{-9}) = (6 - 2j)(3j)$ write in terms of j

$$= 18j - 6j^2 = 18j - 6(-1)$$
$$= 6 + 18j$$

(b) $(-9.4 - 6.2j)(2.5 + 1.5j) = (-9.4)(2.5) + (-9.4)(1.5j) + (-6.2j)(2.5) + (-6.2j)(1.5j)$

$$= -23.5 - 14.1j - 15.5j - 9.3j^2$$
$$= -23.5 - 29.6j - 9.3(-1)$$
$$= -14.2 - 29.6j$$ ■

Practice Exercise

2. Multiply: $(3 - 7j)(9 + 2j)$

The procedure shown in Eq. (12.6) for dividing by a complex number is the same as that used for rationalizing the denominator of a fraction. The result is in the proper form of a complex number. Therefore, *to divide by a complex number, multiply the numerator and the denominator by the conjugate of the denominator.*

EXAMPLE 3 Dividing complex numbers

(a) $\dfrac{7 - 2j}{3 + 4j} = \dfrac{(7 - 2j)(3 - 4j)}{(3 + 4j)(3 - 4j)}$ multiply by conjugate of denominator

$$= \frac{21 - 28j - 6j + 8j^2}{9 - 16j^2} = \frac{21 - 34j + 8(-1)}{9 - 16(-1)}$$
$$= \frac{13 - 34j}{25}$$

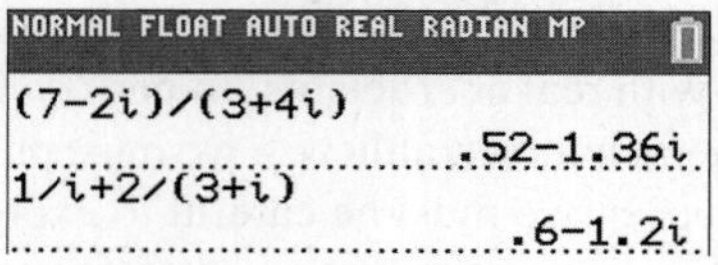

This could be written in the form $a + bj$ as $\frac{13}{25} - \frac{34}{25}j$, but this type of result is generally left as a single fraction. In decimal form, the result would be expressed as $0.52 - 1.36j$.

(b) $\dfrac{1}{j} + \dfrac{2}{3 + j} = \dfrac{(3 + j) + 2j}{j(3 + j)} = \dfrac{3 + 3j}{3j - 1} = \dfrac{3 + 3j}{-1 + 3j} \cdot \dfrac{-1 - 3j}{-1 - 3j}$

$$= \frac{-3 - 12j - 9j^2}{1 - 9j^2} = \frac{6 - 12j}{10}$$
$$= \frac{3 - 6j}{5}$$

```
NORMAL FLOAT AUTO REAL RADIAN MP
(7-2i)/(3+4i)
                         .52-1.36i
1/i+2/(3+i)
                           .6-1.2i
```

Fig. 12.2

Note the use of i instead of j, located above the decimal point key on a TI-84.

Practice Exercise

3. Divide: $\dfrac{5 - 3j}{2 + 7j}$

Most calculators are programmed for complex numbers. The solutions for (a) and (b) are shown in Fig. 12.2 with the results in decimal form. ■

EXAMPLE 4 Dividing complex numbers—alternating current

In an alternating-current circuit, the voltage E is given by $E = IZ$, where I is the current (in A) and Z is the impedance (in Ω). Each of these can be represented by complex numbers. Find the complex number representation for I if $E = 4.20 - 3.00j$ volts and $Z = 5.30 + 2.65j$ ohms. (This type of circuit is discussed in Section 12.7.)

Because $I = E/Z$, we have

$$I = \frac{4.20 - 3.00j}{5.30 + 2.65j} = \frac{(4.20 - 3.00j)(5.30 - 2.65j)}{(5.30 + 2.65j)(5.30 - 2.65j)}$$ multiply by conjugate of denominator

$$= \frac{22.26 - 11.13j - 15.90j + 7.95j^2}{5.30^2 - 2.65^2j^2} = \frac{22.26 - 7.95 - 27.03j}{5.30^2 + 2.65^2}$$
$$= \frac{14.31 - 27.03j}{35.11} = 0.408 - 0.770j \text{ amperes}$$ ■

EXERCISES 12.2

In Exercises 1–4, perform the indicated operations on the resulting expressions if the given changes are made in the indicated examples of this section.

1. In Example 1(b), change the sign in the first parentheses from + to − and then perform the addition.
2. In Example 2(b), change the sign before $6.2j$ from − to +, and then perform the multiplication.
3. In Example 3(a), change the sign in the denominator from + to − and then simplify.
4. In Example 3(b), change the sign in the second denominator from + to − and then simplify.

In Exercises 5–38, perform the indicated operations, expressing all answers in the form $a + bj$.

5. $(3 - 7j) + (2 - j)$
6. $(-4 - j) + (-7 - 4j)$
7. $(7j - 6) - (19 - 3j)$
8. $(5.4 - 3.4j) - (2.9j + 5.5)$
9. $0.23 - (0.46 - 0.19j) + 0.67j$
10. $(7 - j) - (4 - 4j) - (j - 6)$
11. $(12j - 21) - (15 - 18j) - 9j$
12. $(0.062j - 0.073) - 0.030j - (0.121 - 0.051j)$
13. $(7 - j)(7j)$
14. $(-2.2j)(1.5j - 4.0)$
15. $(4 - j)(5 + 2j)$
16. $(8j - 5)(7 + 4j)$
17. $(\sqrt{-18}\sqrt{-4})(3j)$
18. $\sqrt{-6}\sqrt{-12}\sqrt{30}$
19. $(2 - 3j)(5 + 4j)$
20. $(9 - 2j)(6 + j)$
21. $j\sqrt{-7} - j^6\sqrt{112} + 3j$
22. $j^2\sqrt{-7} - \sqrt{-28} + 8j^3$
23. $(3 - 7j)^2$
24. $(8j + 20)^2$
25. $(8 + 3j)(8 - 3j)$
26. $(6 + 8j)(6 - 8j)$
27. $\dfrac{6j}{2 - 5j}$
28. $\dfrac{0.25}{3 - \sqrt{-1}}$
29. $\dfrac{1 - j}{3j}$
30. $\dfrac{12 + 10j}{6 - 8j}$
31. $\dfrac{j\sqrt{2} - 5}{j\sqrt{2} + 3}$
32. $\dfrac{j^5 - j^3}{3 + j}$
33. $\dfrac{j^2 - j}{2j - j^8}$
34. $\dfrac{3}{2j} - \dfrac{5}{j - 6}$
35. $\dfrac{4j}{1 - j} - \dfrac{j + 8}{2 + 3j}$
36. $\dfrac{(6j + 5)(2 - 4j)}{(5 - j)(4j + 1)}$
37. $(4j^5 - 5j^4 + 2j^3 - 3j^2)^2$
38. $(2j^2 - 3j^3 + 2j^4 - 2j^5)^6$

In Exercises 39–42, evaluate each expression on a calculator. Express answers in the form $a + bj$.

39. $(3j^9 - 5j^3)(4j^6 - 6j^8)$
40. $(5j - 4j^2 + 3j^7)(2j^{12} - j^{13})$
41. $\dfrac{(2 - j^3)^4}{(j^8 - j^6)^3} + j$
42. $(1 + j)^{-3}(2 - j)^{-2}$

In Exercises 43–56, solve the given problems.

43. Show that $-1 - j$ is a solution to the equation $x^2 + 2x + 2 = 0$.
44. Show that $1 - j\sqrt{3}$ is a solution to the equation $x^2 + 4 = 2x$.
45. What is the sum of the solutions for the equation $x^2 - 4x + 13 = 0$?
46. What is the product of the solutions to the equation in Exercise 45?
47. Multiply $-3 + j$ by its conjugate.
48. Divide $2 - 3j$ by its conjugate.
49. Write the reciprocal of $3 - j$ in rectangular form.
50. Write the reciprocal of $2 + 5j$ in rectangular form.
51. Write $j^{-2} + j^{-3}$ in rectangular form.
52. Solve for x: $(x + 2j)^2 = 5 + 12j$
53. Solve for x: $(x + 3j)^2 = 7 - 24j$
54. For $\frac{3}{5} + \frac{4}{5}j$, find: (a) the conjugate; (b) the reciprocal.
55. If $f(x) = x + \frac{1}{x}$, find $f(1 + 3j)$.
56. When finding the current in a certain electric circuit, the expression $(s + 1 + 4j)(s + 1 - 4j)$ occurs. Simplify this expression.

In Exercises 57–60, solve the given problems. Refer to Example 4.

57. If $I = 0.835 - 0.427j$ amperes and $Z = 250 + 170j$ ohms, find the complex-number representation for E.
58. If $E = 5.70 - 3.65j$ mV and $I = 0.360 - 0.525j$ μA, find the complex-number representation for Z.
59. If $E = 85 + 74j$ volts and $Z = 2500 - 1200j$ ohms, find the complex-number representation for I.
60. In an alternating-current circuit, two impedances Z_1 and Z_2 have a total impedance Z_T of $Z_T = \dfrac{Z_1Z_2}{Z_1 + Z_2}$. Find Z_T for $Z_1 = 3.2 + 4.8j$ mΩ and $Z_2 = 4.8 - 6.4j$ mΩ.

In Exercises 61–64, answer or explain as indicated.

61. What type of number is the result of (a) adding a complex number to its conjugate and (b) subtracting a complex number from its conjugate?
62. If the reciprocal of $a + bj$ equals $a - bj$, what condition must a and b satisfy?
63. Explain why the product of a complex number and its conjugate is real and nonnegative.
64. Explain how to show that the reciprocal of the imaginary unit is the negative of the imaginary unit.

Answers to Practice Exercises

1. $-3 - j$ 2. $41 - 57j$ 3. $\dfrac{-11 - 41j}{53}$

12.3 Graphical Representation of Complex Numbers

Complex Plane • Complex Number as a Point • Adding and Subtracting Complex Numbers Graphically

Graphically, we represent a complex number as a ***point***, designated as $a + bj$, in the rectangular coordinate system. *The **real part** is the **x-value** of the point, and the **imaginary part** is the **y-value*** of the point. Used in this way, *the coordinate system is called the* **complex plane,** *the horizontal axis is the* **real axis,** and *the vertical axis is the* **imaginary axis.**

EXAMPLE 1 Complex numbers in the complex plane

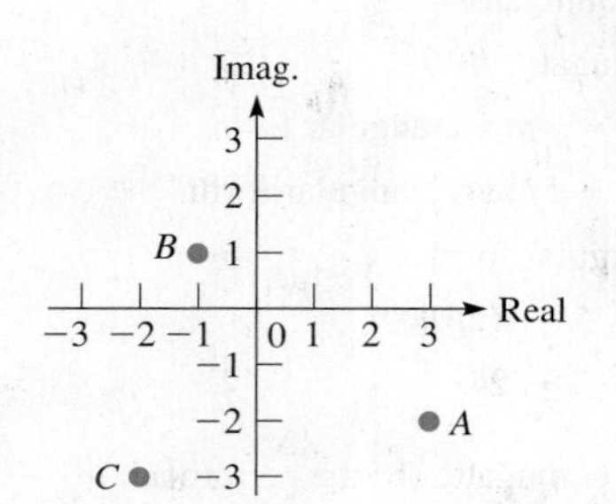

Fig. 12.3

In Fig. 12.3, A represents the complex number $3 - 2j$, B represents $-1 + j$, and C represents $-2 - 3j$. These complex numbers are represented by the points $(3, -2)$, $(-1, 1)$, and $(-2, -3)$, respectively, of the standard rectangular coordinate system.

Keep in mind that the meaning given to the points representing complex numbers in the complex plane is different from the meaning given to the points in the standard rectangular coordinate system. *A point in the complex plane represents a **single complex number,*** whereas *a point in the rectangular coordinate system represents a **pair of real numbers.*** ■

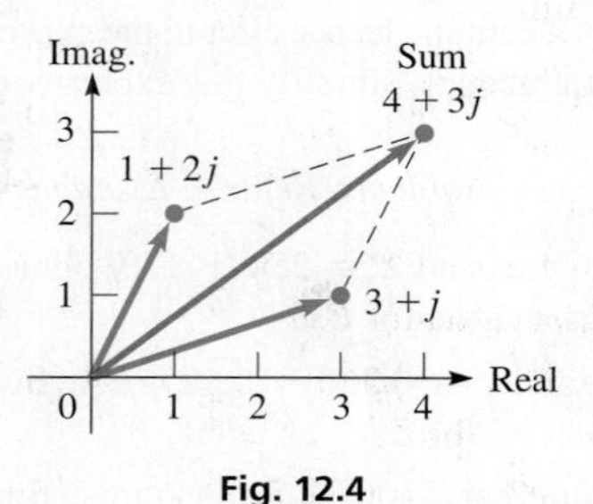

Fig. 12.4

In the complex plane, consider two complex numbers—for example, $1 + 2j$ and $3 + j$—and their sum $4 + 3j$. Drawing lines from the origin to these points (see Fig. 12.4), note that if we think of the complex numbers as being vectors, their sum is the vector sum. Because complex numbers can be used to represent vectors, the numbers are particularly important. *Any complex number can be thought of as representing a vector from the origin to its point in the complex plane.* This leads to the method used to add complex numbers graphically.

Steps to Add Complex Numbers Graphically

1. Find the point corresponding to one of the numbers and draw a line from the origin to this point.
2. Repeat step 1 for the second number.
3. Complete a parallelogram with the lines drawn as adjacent sides. The resulting fourth vertex is the point representing the sum.

NOTE ▸ [Note that this is equivalent to adding vectors by graphical means.]

EXAMPLE 2 Adding complex numbers graphically

(a) Add the complex numbers $5 - 2j$ and $-2 - j$ graphically.

The solution is shown in Fig. 12.5. We see that the fourth vertex of the parallelogram is at $3 - 3j$. This is, of course, the algebraic sum of these two complex numbers.

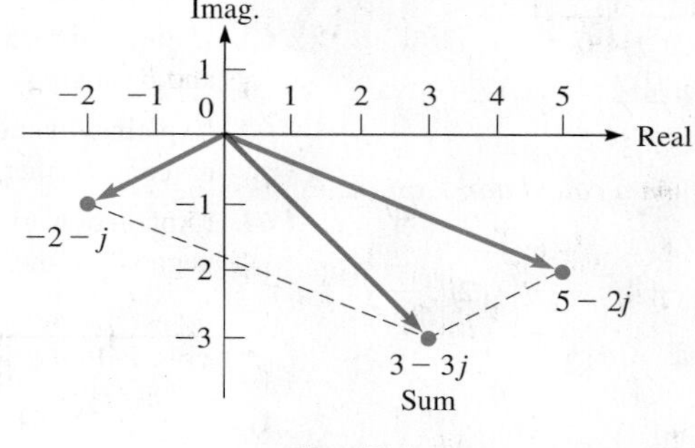

Fig. 12.5

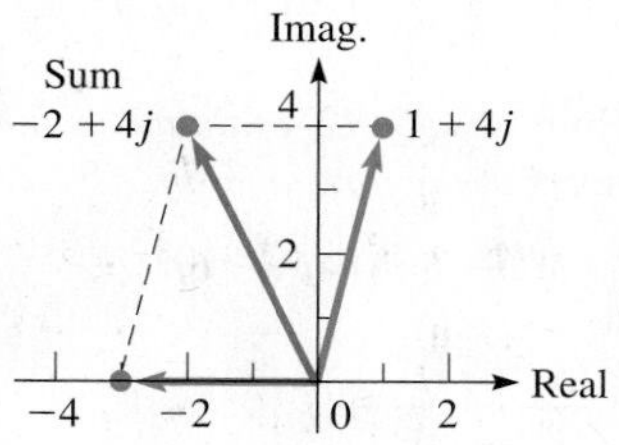

Fig. 12.6

(b) Add the complex numbers -3 and $1 + 4j$.

First, note that $-3 = -3 + 0j$, which means that the point representing -3 is on the negative real axis. In Fig. 12.6, we show the numbers -3 and $1 + 4j$ on the graph and complete the parallelogram. From the graph, we see that the sum is $-2 + 4j$. ■

EXAMPLE 3 Subtracting complex numbers graphically

Subtract $4 - 2j$ from $2 - 3j$ graphically.

Subtracting $4 - 2j$ is equivalent to adding $-4 + 2j$. Therefore, we complete the solution by adding $-4 + 2j$ and $2 - 3j$, as shown in Fig. 12.7. The result is $-2 - j$. ■

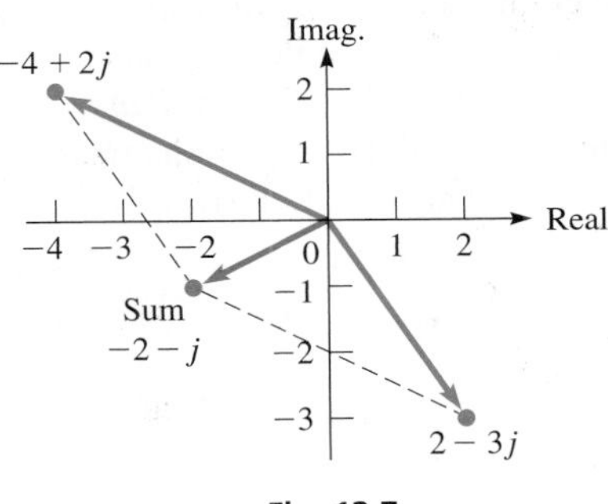

Fig. 12.7

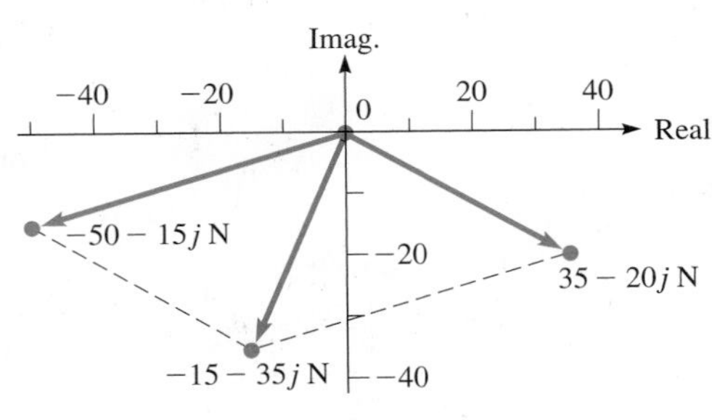

Fig. 12.8

EXAMPLE 4 Adding complex numbers—forces on a bolt

Two forces acting on an overhead bolt can be represented by $35 - 20j$ N and $-50 - 15j$ N. Find the resultant force graphically.

The forces are shown in Fig. 12.8. From the graph, we can see that the sum of the forces, which is the resultant force, is $-15 - 35j$ N. ■

EXERCISES 12.3

In Exercises 1 and 2, perform the indicated operations for the resulting complex numbers if the given changes are made in the indicated examples of this section.

1. In Example 2(a), change the sign of the imaginary part of the second complex number and then add the numbers graphically.

2. In Example 3, change the sign of the imaginary part of the second complex number and do the subtraction graphically.

In Exercises 3–8, locate the given numbers in the complex plane.

3. $2 + 6j$ **4.** $-5 + j$ **5.** $-4 - 3j$

6. 10 **7.** $-3j$ **8.** $3 - 4j$

In Exercises 9–26, perform the indicated operations graphically. Check them algebraically.

9. $2 + (3 + 4j)$ **10.** $2j + (-2 + 3j)$

11. $(5 - j) + (3 + 2j)$ **12.** $(3 - 2j) + (-1 - j)$

13. $5j - (1 - 4j)$ **14.** $(0.2 - 0.1j) - 0.1$

15. $(2 - 4j) + (j - 2)$ **16.** $(-1 - 6j) + (3j + 6)$

17. $(3 - 2j) - (4 - 6j)$ **18.** $(-25 - 40j) - (20 - 55j)$

19. $(80 + 300j) - (260 + 150j)$

20. $(-j - 2) - (-1 - 3j)$

21. $(3 - j) + (6 + 5j)$ **22.** $(7 + 4j) - (3j - 8)$

23. $(3 - 6j) - (-1 - 8j)$ **24.** $(-6 - 3j) + (2 - 7j)$

25. $(2j + 1) - 3j - (j + 1)$ **26.** $(6 - j) - 9 - (2j - 3)$

In Exercises 27–30, show the given number, its negative, and its conjugate on the same coordinate system.

27. $3 + 2j$ **28.** $4j - 2$ **29.** $-3 - 5j$ **30.** $5 - j$

In Exercises 31 and 32, show the numbers $a + bj$, $3(a + bj)$, and $-3(a + bj)$ on the same coordinate system. The multiplication of a complex number by a real number is called **scalar multiplication** *of the complex number.*

31. $3 - j$ **32.** $-10 - 30j$

In Exercises 33–36, perform the indicated vector operations graphically on the complex number $2 + 4j$.

33. Graph the complex number and its conjugate. Describe the relative positions.

34. Add the number and its conjugate. Describe the result.

35. Subtract the conjugate from the number. Describe the result.

36. Graph the number, the number multiplied by j, the number multiplied by j^2, and the number multiplied by j^3 on the same graph. Describe the result of multiplying a complex number by j.

In Exercises 37 and 38, perform the indicated vector additions graphically. Check them algebraically.

37. Two ropes hold a boat at a dock. The tensions in the ropes can be represented by $40 + 10j$ lb and $50 - 25j$ lb. Find the resultant force.

38. Relative to the air, a plane heads north of west with a velocity that can be represented by $-480 + 210j$ km/h. The wind is blowing from south of west with a velocity that can be represented by $60 + 210j$ km/h. Find the resultant velocity of the plane.

12.4 Polar Form of a Complex Number

Polar Form $r(\cos\theta + j\sin\theta)$ • Expressing Numbers in Polar Form

In this section, we use the fact that a complex number can be represented by a vector to write complex numbers in another form. This form has advantages when multiplying and dividing complex numbers, and we will discuss these operations later in this chapter.

In the complex plane, by drawing a vector from the origin to the point that represents the number $x + yj$, an angle θ in standard position is formed. The point $x + yj$ is r units from the origin. In fact, *we can find any point in the complex plane by knowing the angle θ and the value of r.* The equations relating x, y, r, and θ are similar to those developed for vectors in Chapter 9. Referring to Fig. 12.9, we see that

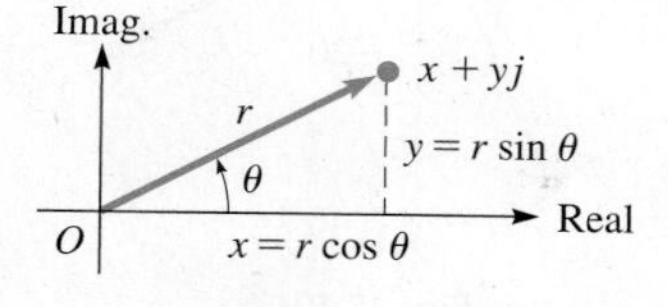

Fig. 12.9

$$x = r\cos\theta \qquad y = r\sin\theta \tag{12.7}$$

$$r^2 = x^2 + y^2 \qquad \tan\theta = \frac{y}{x} \tag{12.8}$$

Substituting Eq. (12.7) into the rectangular form $x + yj$ of a complex number, we have $x + yj = r\cos\theta + j(r\sin\theta)$, or

■ r cis θ is sometimes used as a shorthand version of $r(\cos\theta + i\sin\theta)$, where i is being used instead of j.

$$x + yj = r(\cos\theta + j\sin\theta) \tag{12.9}$$

The right side of Eq. (12.9) is called the **polar form** (sometimes the **trigonometric form**). *The length r is the* **absolute value,** or *the* **modulus,** and *θ is the* **argument** *of the complex number.* Equations (12.7)–(12.9) define the polar form of a complex number.

EXAMPLE 1 Representing a number in polar form

Represent the complex number $3 + 4j$ graphically and give its polar form.

From the rectangular form $3 + 4j$, we see that $x = 3$ and $y = 4$. Using Eqs. (12.8), we have

$$r = \sqrt{3^2 + 4^2} = 5 \qquad \theta = \tan^{-1}\frac{4}{3} = 53.1^\circ$$

Thus, the polar form is $5(\cos 53.1^\circ + j\sin 53.1^\circ)$. See Fig. 12.10. ■

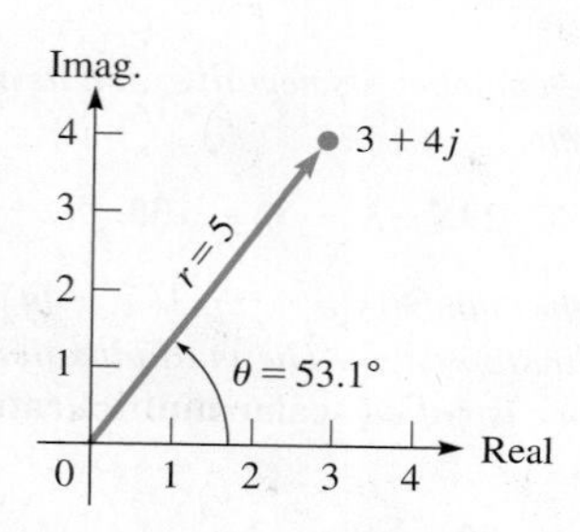

Fig. 12.10

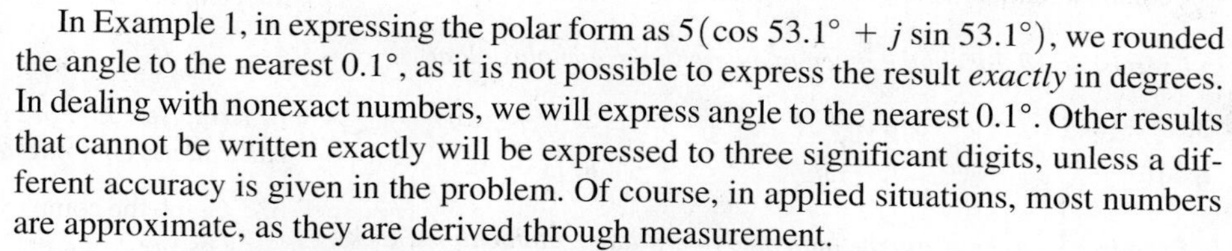
In Example 1, in expressing the polar form as $5(\cos 53.1^\circ + j\sin 53.1^\circ)$, we rounded the angle to the nearest 0.1°, as it is not possible to express the result *exactly* in degrees. In dealing with nonexact numbers, we will express angle to the nearest 0.1°. Other results that cannot be written exactly will be expressed to three significant digits, unless a different accuracy is given in the problem. Of course, in applied situations, most numbers are approximate, as they are derived through measurement.

Another convenient and widely used notation for the polar form is $r\underline{/\theta}$. We must remember in using this form that it represents a complex number and is simply a shorthand way of writing $r(\cos\theta + j\sin\theta)$. Therefore,

Practice Exercise

1. Write in polar form: $15 - 8j$

$$r\underline{/\theta} = r(\cos\theta + j\sin\theta) \tag{12.10}$$

EXAMPLE 2 Polar form $r\underline{/\theta}$

(a) $3(\cos 40^\circ + j\sin 40^\circ) = 3\underline{/40^\circ}$

(b) $6.26(\cos 217.3^\circ + j\sin 217.3^\circ) = 6.26\underline{/217.3^\circ}$

(c) $5\underline{/120^\circ} = 5(\cos 120^\circ + j\sin 120^\circ)$

(d) $14.5\underline{/306.2^\circ} = 14.5(\cos 306.2^\circ + j\sin 306.2^\circ)$ ■

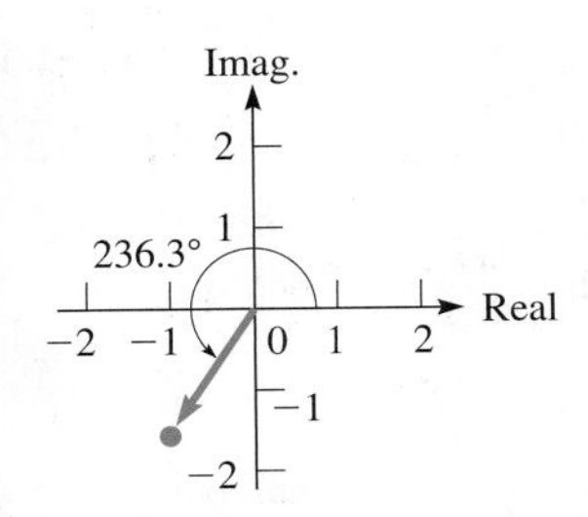

Fig. 12.11

EXAMPLE 3 Rectangular form to polar form

Represent the complex number $-1.04 - 1.56j$ graphically and give its polar form.

The graphical representation is shown in Fig. 12.11. From Eqs. (12.8), we have

$$r = \sqrt{(-1.04)^2 + (-1.56)^2} = 1.87$$

$$\theta_{\text{ref}} = \tan^{-1}\frac{1.56}{1.04} = 56.3° \qquad \theta = 180° + 56.3° = 236.3°$$

Because both the real and imaginary parts are negative, θ is a third-quadrant angle. Therefore, we found the reference angle before finding θ. This means the polar forms are

$$1.87(\cos 236.3° + j\sin 236.3°) = 1.87\underline{/236.3°}$$ ■

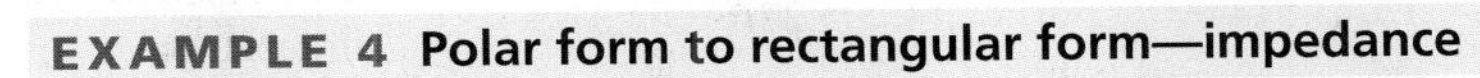

EXAMPLE 4 Polar form to rectangular form—impedance

The impedance Z (in Ω) in an alternating-current circuit is given by $Z = 3560\underline{/-32.4°}$. Express this in rectangular form.

From the polar form, we have $r = 3560\ \Omega$ and $\theta = -32.4°$ (it is common to use negative angles in this type of application). This means that we can also write

$$Z = 3560[(\cos(-32.4°) + j\sin(-32.4°)]$$

See Fig. 12.12. This means that

$$x = 3560\cos(-32.4°) = 3010$$

$$y = 3560\sin(-32.4°) = -1910$$

Therefore, the rectangular form is $Z = 3010 - 1910j\ \Omega$. ■

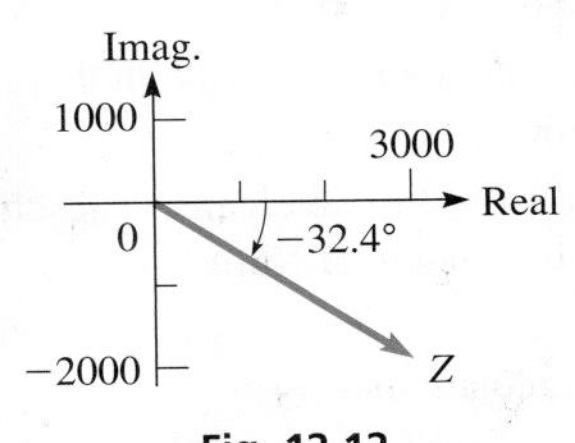

Fig. 12.12

Practice Exercise

2. Write in rectangular form: $2.50\underline{/120°}$

Complex numbers can be converted between rectangular and polar forms on most calculators. Figure 12.13 shows the calculator conversions for Examples 3 and 4. Note that for Example 3, the calculator returns an angle of $-123.7°$. This is coterminal to 236.3°, the angle we found.

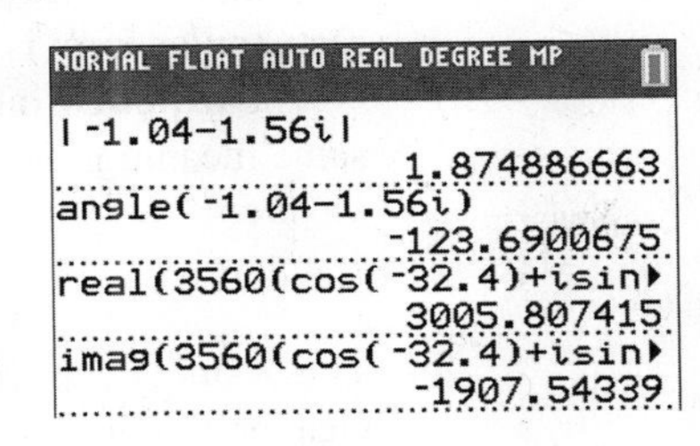

Fig. 12.13
Graphing calculator keystrokes for Examples 3 & 4: goo.gl/5iR10K

EXAMPLE 5 Polar form—real and imaginary numbers

Represent the numbers 5, −5, $7j$, and $-7j$ in polar form. See Fig. 12.14.

Real numbers lie on the real axis in the complex plane, while imaginary numbers lie on the imaginary axis. Therefore, the polar form angle of any real or imaginary number is a multiple of 90° as shown below:

$$5 = 5(\cos 0° + j\sin 0°) = 5\underline{/0°}$$ 5 units from the origin at an angle of 0°

$$-5 = 5(\cos 180° + j\sin 180°) = 5\underline{/180°}$$ 5 units from the origin at an angle of 180°

$$7j = 7(\cos 90° + j\sin 90°) = 7\underline{/90°}$$ 7 units from the origin at an angle of 90°

$$-7j = 7(\cos 270° + j\sin 270°) = 7\underline{/270°}$$ 7 units from the origin at an angle of 270° ■

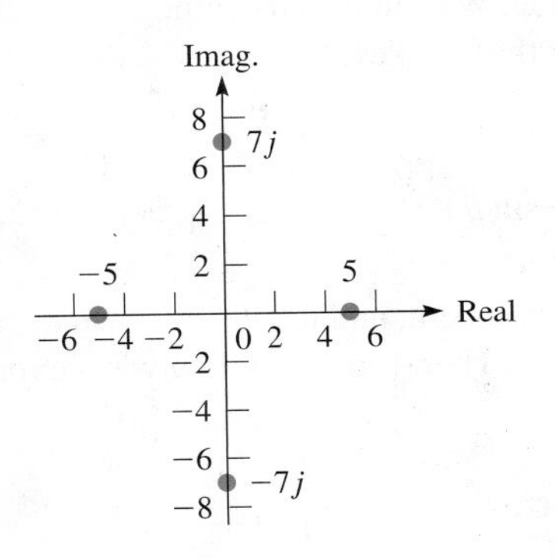

Fig. 12.14

Practice Exercise

3. Represent $-10j$ in polar form.

EXERCISES 12.4

In Exercises 1 and 2, change the sign of the real part of the complex number in the indicated example of this section and then perform the indicated operations for the resulting complex number.

1. Example 1 **2.** Example 3

In Exercises 3–18, represent each complex number graphically and give the polar form of each.

3. $8 + 6j$ **4.** $-8 - 15j$ **5.** $30 - 40j$

6. $12j - 5$ **7.** $3.00j - 2.00$ **8.** $7.00 - 5.00j$

9. $-0.55 - 0.24j$ **10.** $460 - 460j$ **11.** $1 + j\sqrt{3}$

12. $\sqrt{2} - j\sqrt{2}$ **13.** $3.514 - 7.256j$ **14.** $95.27j + 62.31$

15. -3 **16.** 60 **17.** $9j$ **18.** $-2j$

In Exercises 19–36, represent each complex number graphically and give the rectangular form of each.

19. $5.00(\cos 54.0° + j \sin 54.0°)$ **20.** $6(\cos 180° + j \sin 180°)$

21. $160(\cos 150.0° + j \sin 150.0°)$ **22.** $2.50(\cos 315.0° + j \sin 315.0°)$

23. $3.00(\cos 232.0° + j \sin 232.0°)$

24. $220.8(\cos 155.13° + j \sin 155.13°)$

25. $0.08(\cos 360° + j \sin 360°)$ **26.** $15(\cos 0° + j \sin 0°)$

27. $120(\cos 270° + j \sin 270°)$ **28.** $\cos 600.0° + j \sin 600.0°$

29. $4.75\angle 172.8°$ **30.** $1.50\angle 897.7°$ **31.** $0.9326\angle 229.54°$

32. $277.8\angle -342.63°$ **33.** $7.32\angle -270°$ **34.** $18.3\angle 540.0°$

35. $86.42\angle 94.62°$ **36.** $4629\angle 182.44°$

In Exercises 37–44, solve the given problems.

37. What is the argument for any negative real number?

38. For $x + yj$, what is the argument if $x = y < 0$?

39. Show that the conjugate of $r\angle\theta$ is $r\angle -\theta$.

40. The impedance in a certain circuit is $Z = 8.5\angle -36°\ \Omega$. Write this in rectangular form.

41. The voltage of a certain generator is represented by $2.84 - 1.06j$ kV. Write this voltage in polar form.

42. Find the magnitude and direction of a force on a bolt that is represented by $40.5 + 24.5j$ newtons.

43. The electric field intensity of a light wave can be described by $12.4\angle 78.3°$ V/m. Write this in rectangular form.

44. The current in a certain microprocessor circuit is represented by $3.75\angle 15.0°\ \mu$A. Write this in rectangular form.

Answers to Practice Exercises

1. $17(\cos 331.9° + j \sin 331.9°)$ **2.** $-1.25 + 2.17j$ **3.** $10\angle 270°$

12.5 Exponential Form of a Complex Number

Exponential Form $re^{j\theta}$ • Expressing Numbers in Exponential Form • Summary of Important Forms

Another important form of a complex number is the *exponential form.* It is commonly used in electronics, engineering, and physics applications. As we will see in the next section, it is also convenient for multiplication and division of complex numbers, as the rectangular form is for addition and subtraction.

The **exponential form** *of a complex number is written as* $re^{j\theta}$, where r and θ have the same meanings as given in the previous section, although θ is expressed in radians. *The number e is a special irrational number and has an approximate value*

■ The number e is named for the Swiss mathematician Leonhard Euler (1707–1783). His works in mathematics and other fields filled about 70 volumes.

$$e = 2.7182818284590452$$

This number e is very important in mathematics, and we will see it again in the next chapter. For now, it is necessary to accept the value for e, although in calculus its meaning is shown along with the reason it has the above value. We can find its value on a calculator by using the e^x key, with $x = 1$, or by using the e key.

In advanced mathematics, it is shown that

$$re^{j\theta} = r(\cos\theta + j\sin\theta) \tag{12.11}$$

NOTE ▸ By expressing θ in radians, the expression $j\theta$ is an exponent, and it can be shown to obey all the laws of exponents as discussed in Chapter 11. [Therefore, we will always express θ in radians when using exponential form.]

EXAMPLE 1 Polar form to exponential form

Express the number $8.50\angle 136.3°$ in exponential form.

Because this complex number is in polar form, we note that $r = 8.50$ and that we must express 136.3° in radians. Changing 136.3° to radians, we have

$$\frac{136.3\pi}{180} = 2.38 \text{ rad}$$

Therefore, the required exponential form is $8.50e^{2.38j}$. This means that

$$8.50\angle 136.3^\circ = 8.50e^{2.38j}$$

degrees to radians

value of r

We see that the principal step in changing from polar form to exponential form is to change θ from degrees to radians. ■

Practice Exercise

1. Express $25.0(\cos 127.0^\circ + j \sin 127.0^\circ)$ in exponential form.

Complex numbers are often used to represent vectors, where r is the magnitude and θ is the direction. The following example illustrates this.

EXAMPLE 2 Rectangular to exponential form—robotic displacement

The displacement d of a welding point from the end of a robot arm can be expressed as $2.00 - 4.00j$ ft. Express the displacement in exponential form and find its magnitude.

From the rectangular form, we have $x = 2.00$ and $y = -4.00$. Therefore,

$$d = \sqrt{(2.00)^2 + (-4.00)^2} = 4.47 \text{ ft}$$

$$\theta = \tan^{-1}\frac{-4.00}{2.00} = -63.4^\circ$$

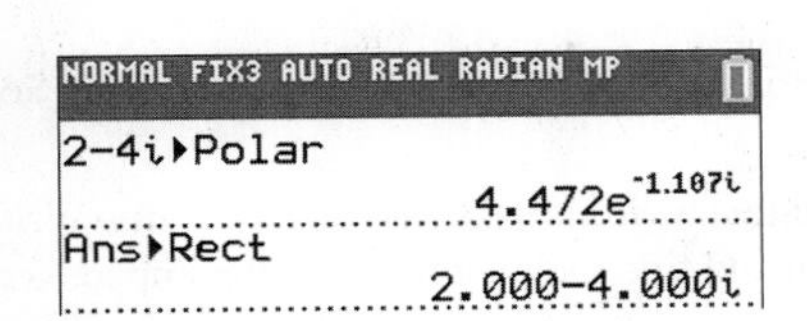

Fig. 12.15
Graphing calculator keystrokes: goo.gl/NOHI3j

■ Recall that π rad $= 180^\circ$.

It is common to express the direction in terms of negative angles when the imaginary part is negative. Since $63.4^\circ = 1.11$ rad, the exponential form is $4.47e^{-1.11j}$. The modulus of 4.47 means the magnitude of the displacement is 4.47 ft. Figure 12.15 shows the calculator features for directly converting between rectangular and polar forms. The calculator treats exponential and polar forms as being the same since each uses r and θ. ■

Practice Exercise

2. Express $-20.5 - 16.8j$ in exponential form.

EXAMPLE 3 Exponential form to other forms

Express the complex number $2.00e^{4.80j}$ in polar and rectangular forms.

We first express 4.80 rad as 275.0°. From the exponential form, we know that $r = 2.00$. Thus, the polar form is

$$2.00(\cos 275.0^\circ + j \sin 275.0^\circ)$$

Using the distributive law, we rewrite the polar form and then evaluate. Thus,

$$\begin{aligned} 2.00e^{4.80j} &= 2.00(\cos 275.0^\circ + j \sin 275.0^\circ) \\ &= 2.00 \cos 275.0^\circ + (2.00 \sin 275.0^\circ)j \\ &= 0.174 - 1.99j \end{aligned}$$ ■

Practice Exercise

3. Represent $3.00e^{2.66j}$ in rectangular form.

We now summarize the three important forms of a complex number. See Fig. 12.16.

Rectangular:	$x + yj$
Polar:	$r(\cos\theta + j\sin\theta) = r\angle\theta$
Exponential:	$re^{j\theta}$

It follows that

$$x + yj = r(\cos\theta + j\sin\theta) = r\angle\theta = re^{j\theta} \tag{12.12}$$

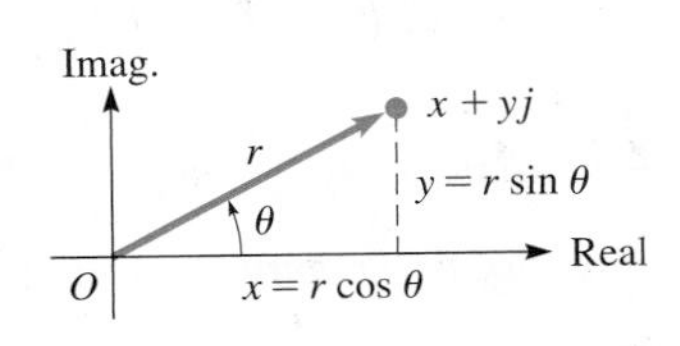

Fig. 12.16

where

$$r^2 = x^2 + y^2 \qquad \tan\theta = \frac{y}{x} \tag{12.8}$$

In Eq. (12.12), the argument θ is the same for exponential and polar forms. It is usually expressed in radians in exponential form and in degrees in polar form.

EXERCISES 12.5

In Exercises 1 and 2, perform the indicated operations for the resulting complex numbers if the given changes are made in the indicated examples of this section.

1. In Example 1, change 136.3° to 226.3° and then find the exponential form.

2. In Example 3, change the exponent to $3.80j$ and then find the polar and rectangular forms.

In Exercises 3–22, express the given numbers in exponential form.

3. $3.00(\cos 60.0° + j \sin 60.0°)$ **4.** $575(\cos 135.0° + j \sin 135°)$

5. $0.450(\cos 282.3° + j \sin 282.3°)$

6. $2.10(\cos 588.7° + j \sin 588.7°)$

7. $375.5[\cos(-95.46°) + j \sin(-95.46°)]$

8. $1672[\cos(-7.14°) + j \sin(-7.14°)]$

9. $0.515\underline{/198.3°}$ **10.** $4650\underline{/326.5°}$ **11.** $4.06\underline{/-61.4°}$

12. $0.0192\underline{/76.7°}$ **13.** $9245\underline{/296.32°}$ **14.** $82.76\underline{/470.09°}$

15. $3 - 4j$ **16.** $-1 - 5j$ **17.** $-30 + 20j$

18. $100j + 600$ **19.** $5.90 + 2.40j$ **20.** $47.3 - 10.9j$

21. $-634.6 - 528.2j$ **22.** $5477j - 8573$

In Exercises 23–30, express the given complex numbers in polar and rectangular forms.

23. $3.00e^{0.500j}$ **24.** $20.0e^{1.00j}$ **25.** $464e^{1.85j}$

26. $2.50e^{3.84j}$ **27.** $3.20e^{-5.41j}$ **28.** $0.800e^{3.00j}$

29. $1724e^{2.391j}$ **30.** $820.7e^{-3.492j}$

In Exercises 31–34, perform the indicated operations by using properties of exponents and express results in rectangular and polar forms.

31. $(4.55e^{1.32j})^2$ **32.** $(0.926e^{0.253j})^3$

33. $(625e^{3.46j})(4.40e^{1.22j})$ **34.** $(18.0e^{5.13j})(25.5e^{0.77j})$

In Exercises 35–40, perform the indicated operations.

35. Using a calculator, express $3.73 + 5.24j$ in exponential form. See Fig. 12.15.

36. Using a calculator, express $75.6e^{1.25j}$ in rectangular form. See Fig. 12.15.

37. The impedance in an antenna circuit is $3.75 + 1.10j$ ohms. Write this in exponential form and find the magnitude of the impedance.

38. The intensity of the signal from a radar microwave signal is $37.0[\cos(-65.3°) + j \sin(-65.3°)]$ V/m. Write this in exponential form.

39. In an electric circuit, the *admittance* is the reciprocal of the impedance. If the impedance is $2800 - 1450j$ ohms in a certain circuit, find the exponential form of the admittance.

40. If $E = 115e^{0.315j}$ V and $I = 28.6e^{-0.723j}$ A, find the exponential form of Z given that $E = IZ$.

Answers to Practice Exercises

1. $25.0e^{2.22j}$ **2.** $26.5e^{3.83j}$ **3.** $-2.66 + 1.39j$

12.6 Products, Quotients, Powers, and Roots of Complex Numbers

Multiplication, Division, and Powers in Polar and Exponential Forms • DeMoivre's Theorem • Roots

The operations of multiplication and division can be performed with complex numbers in polar and exponential forms, as well as rectangular form. It is convenient to use polar form for multiplication and division as well as for finding powers or roots of complex numbers.

Using exponential form and laws of exponents, we multiply two complex numbers as

$$[r_1e^{j\theta_1}] \times [r_2e^{j\theta_2}] = r_1r_2e^{j\theta_1 + j\theta_2} = r_1r_2e^{j(\theta_1+\theta_2)}$$

Rewriting this result in polar form, we have

$$[r_1e^{j\theta_1}] \times [r_2e^{j\theta_2}] = [r_1(\cos\theta_1 + j\sin\theta_1)] \times [r_2(\cos\theta_2 + j\sin\theta_2)]$$

and

$$r_1r_2e^{j(\theta_1+\theta_2)} = r_1r_2[\cos(\theta_1+\theta_2) + j\sin(\theta_1+\theta_2)]$$

The polar expressions are equal, which means *the product of two complex numbers is*

$$(r_1\underline{/\theta_1})(r_2\underline{/\theta_2}) = r_1r_2\underline{/\theta_1 + \theta_2}, \text{ or}$$
$$r_1(\cos\theta_1 + j\sin\theta_1)r_2(\cos\theta_2 + j\sin\theta_2)$$
$$= r_1r_2[\cos(\theta_1+\theta_2) + j\sin(\theta_1+\theta_2)] \quad (12.13)$$

NOTE ▸ [The magnitudes are multiplied, and the angles are added.]

EXAMPLE 1 Multiplication in polar form

To multiply the complex numbers $3.61\underline{/56.3°}$ and $1.41\underline{/315.0°}$, we have

multiply

$$(3.61\underline{/56.3°})(1.41\underline{/315.0°}) = (3.61)(1.41)\underline{/56.3° + 315.0°}$$

add

$$= 5.09\underline{/371.3°}$$
$$= 5.09\underline{/11.3°}$$

Practice Exercise

1. Find the polar form product: $(3\underline{/50°})(5\underline{/65°})$

Note that the angle in the final result is between 0° and 360°. This is usually the case, although in some applications, it is expressed as a negative angle. ■

If we wish to *divide* one complex number in exponential form by another, we arrive at the following result:

$$\frac{r_1 e^{j\theta_1}}{r_2 e^{j\theta_2}} = \frac{r_1}{r_2} e^{j(\theta_1 - \theta_2)} \tag{12.14}$$

Therefore, *the result of dividing one complex number in polar form by another is given by*

$$\frac{r_1\underline{/\theta_1}}{r_2\underline{/\theta_2}} = \frac{r_1}{r_2}\underline{/\theta_1 - \theta_2}\text{, or}$$

$$\frac{r_1(\cos\theta_1 + j\sin\theta_1)}{r_2(\cos\theta_2 + j\sin\theta_2)} = \frac{r_1}{r_2}[\cos(\theta_1 - \theta_2) + j\sin(\theta_1 - \theta_2)] \tag{12.15}$$

NOTE ➧ [The magnitudes are divided, and the angles are subtracted.]

EXAMPLE 2 Division in polar form

Divide the first complex number of Example 1 by the second.

Using the $r(\cos\theta + j\sin\theta)$ notation, we have

divide

$$\frac{3.61(\cos 56.3° + j\sin 56.3°)}{1.41(\cos 315.0° + j\sin 315.0°)} = \frac{3.61}{1.41}[\cos(56.3° - 315.0°) + j\sin(56.3° - 315.0°)]$$

subtract

$$= 2.56[\cos(-258.7°) + j\sin(-258.7°)]$$
$$= 2.56(\cos 101.3° + j\sin 101.3°)$$ ■

Practice Exercise

2. Find the polar form quotient:

$$\frac{3(\cos 50° + j\sin 50°)}{5(\cos 65° + j\sin 65°)}$$

We have just seen that multiplying and dividing numbers in polar form can be easily performed. **CAUTION** However, if we are to add or subtract numbers given in polar form, we must convert them to rectangular form before we do the addition or subtraction. ■

EXAMPLE 3 Addition in polar form

Perform the addition $1.563\angle 37.56° + 3.827\angle 146.23°$.

In order to do this addition, we must change each number to rectangular form:

$$\begin{aligned}1.563\angle 37.56° + 3.827\angle 146.23° &= 1.563(\cos 37.56° + j\sin 37.56°) + 3.827(\cos 146.23° + j\sin 146.23°)\\ &= 1.2390 + 0.9528j - 3.1813 + 2.1273j\\ &= -1.9423 + 3.0801j\end{aligned}$$

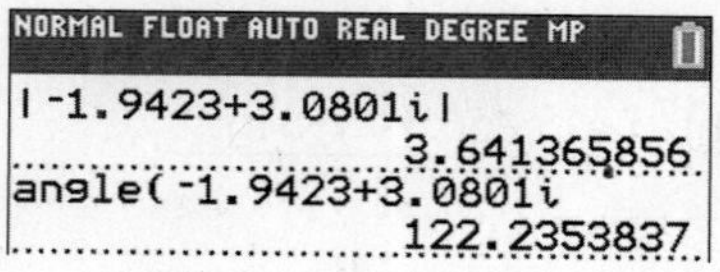

```
NORMAL FLOAT AUTO REAL DEGREE MP
|-1.9423+3.0801i|
                    3.641365856
angle(-1.9423+3.0801i
                    122.2353837
```

Fig. 12.17
Graphing calculator keystrokes: goo.gl/NiY4h2

Converting back to polar form, we get $3.641\angle 122.24°$ (see Fig. 12.17). ■

DEMOIVRE'S THEOREM

To raise a complex number to a power, we use the exponential form of the number along with the properties of exponents $(ab)^n = a^n b^n$ and $(a^m)^n = a^{mn}$. This leads to

$$(re^{j\theta})^n = r^n e^{jn\theta} \quad \textbf{(12.16)}$$

Extending this to polar form, we have

$$(r\angle\theta)^n = r^n\angle n\theta, \text{ or}$$
$$[r(\cos\theta + j\sin\theta)]^n = r^n(\cos n\theta + j\sin n\theta) \quad \textbf{(12.17)}$$

■ Named for the mathematician Abraham DeMoivre (1667–1754).

Equation (12.17) is known as **DeMoivre's theorem** *and is valid for all real values of n. It is also used for finding roots of complex numbers if n is a fractional exponent.* We note that *the magnitude is raised to the power,* and *the angle is multiplied by the power.*

EXAMPLE 4 Power by DeMoivre's theorem

Using DeMoivre's theorem, find $(2 + 3j)^3$.

Converting to polar form, $2 + 3j = 3.61\angle 56.3°$. Therefore,

$$\begin{aligned}[3.61(\cos 56.3° + j\sin 56.3°)]^3 &= (3.61)^3[\cos(3\times 56.3°) + j\sin(3\times 56.3°)]\\ &= 47.0(\cos 168.9° + j\sin 168.9°)\\ &= 47.0\angle 168.9°\end{aligned}$$

Expressing θ in radians, we have $\theta = 56.3° = 0.983$ rad. Therefore,

$$(3.61e^{0.983j})^3 = (3.61)^3 e^{3\times 0.983j} = 47.0e^{2.95j}$$

$$\begin{aligned}(2 + 3j)^3 &= 47.0(\cos 168.9° + j\sin 168.9°)\\ &= 47.0\angle 168.9°\\ &= 47.0e^{2.95j} = -46 + 9j\end{aligned}$$ ■

Practice Exercise

3. Find the polar form power: $(3\ \text{cis}\ 50°)^8$

EXAMPLE 5 Cube roots by DeMoivre's theorem

Find the cube root of -1.

Because -1 is a real number, we can find its cube root by means of the definition. Since $(-1)^3 = -1$, $\sqrt[3]{-1} = -1$. We check this by DeMoivre's theorem. Writing -1 in polar form, we have

$$-1 = 1(\cos 180° + j\sin 180°)$$

Applying DeMoivre's theorem, with $n = \frac{1}{3}$, we obtain

$$(-1)^{1/3} = 1^{1/3}(\cos \tfrac{1}{3}180° + j \sin \tfrac{1}{3}180°) = \cos 60° + j \sin 60°$$

$$= \frac{1}{2} + j\frac{\sqrt{3}}{2} \quad \text{exact answer}$$

$$= 0.5000 + 0.8660j \quad \text{decimal approximation}$$

Observe that we did not obtain -1 as the answer. If we check the answer, in the form $\frac{1}{2} + j\frac{\sqrt{3}}{2}$, by actually cubing it, we obtain -1! Therefore, it is a correct answer.

We should note that it is possible to take $\frac{1}{3}$ of any angle up to 1080° and still have an angle less than 360°. Because 180° and 540° have the same terminal side, let us try writing -1 as $1(\cos 540° + j \sin 540°)$. Using DeMoivre's theorem, we have

$$(-1)^{1/3} = 1^{1/3}(\cos \tfrac{1}{3}540° + j \sin \tfrac{1}{3}540°) = \cos 180° + j \sin 180° = -1$$

We have found the answer we originally anticipated.

Angles of 180° and 900° also have the same terminal side, so we try

$$(-1)^{1/3} = 1^{1/3}(\cos \tfrac{1}{3}900° + j \sin \tfrac{1}{3}900°) = \cos 300° + j \sin 300°$$

$$= \frac{1}{2} - j\frac{\sqrt{3}}{2} \quad \text{exact answer}$$

$$= 0.5000 - 0.8660j \quad \text{decimal approximation}$$

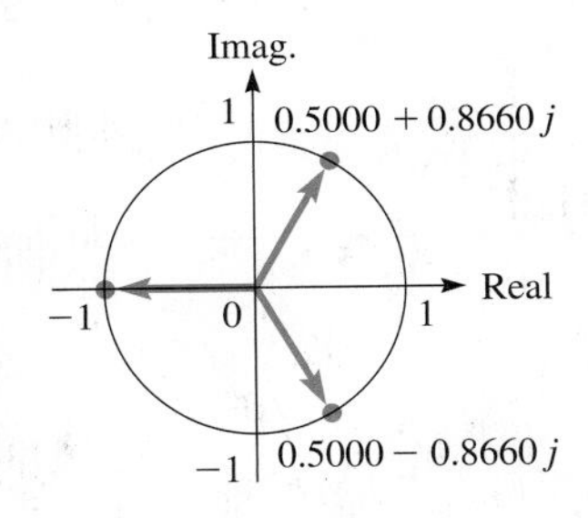

Fig. 12.18

Checking this, we find that it is also a correct root. We may try 1260°, but $\frac{1}{3}(1260°) = 420°$, which has the same functional values as 60°, and would give us the answer $0.5000 + 0.8660j$ again.

We have found, therefore, *three cube roots* of -1. They are

$$-1, \quad \frac{1}{2} + j\frac{\sqrt{3}}{2}, \quad \frac{1}{2} - j\frac{\sqrt{3}}{2}$$

These roots are graphed in Fig. 12.18. Note that they are equally spaced on the circumference of a circle of radius 1. ■

NOTE ➧ [When the results of Example 5 are generalized, it can be proven that there are n nth roots of a complex number.] When graphed, these roots are on a circle of radius $r^{1/n}$ and are equally spaced $360°/n$ apart. Following is the method for finding these n roots.

Using DeMoivre's Theorem to Find the n nth Roots of a Complex Number

1. Express the number in polar form.
2. Express the root as a fractional exponent.
3. Use Eq. (12.17) with θ to find one root.
4. Use Eq. (12.17) and add 360° to θ, $n - 1$ times, to find the other roots.

EXAMPLE 6 Square roots by DeMoivre's theorem

Find the two square roots of $2j$.

First, we write $2j$ in polar form as $2j = 2(\cos 90° + j \sin 90°)$. To find square roots, we use the exponent $1/2$. The first square root is

$$(2j)^{1/2} = 2^{1/2}\left(\cos\frac{90°}{2} + j\sin\frac{90°}{2}\right) = \sqrt{2}(\cos 45° + j\sin 45°) = 1 + j$$

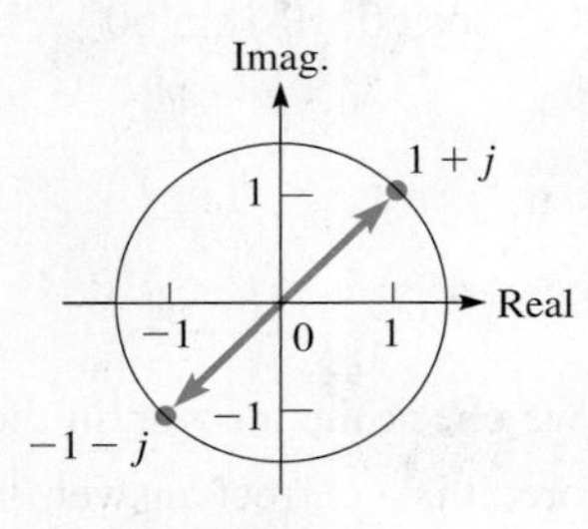

Fig. 12.19

To find the other square root, we add 360° to 90°. This gives us

$$(2j)^{1/2} = 2^{1/2}\left(\cos\frac{450°}{2} + j\sin\frac{450°}{2}\right) = \sqrt{2}(\cos 225° + j\sin 225°) = -1 - j$$

Therefore, the two square roots of $2j$ are $1 + j$ and $-1 - j$. We see in Fig. 12.19 that they are on a circle of radius $\sqrt{2}$ and 180° apart. ■

Practice Exercise

4. Find the two square roots of $8j$.

EXAMPLE 7 Sixth roots by DeMoivre's theorem

Find all the roots of the equation $x^6 - 64 = 0$.

Solving for x, we have $x^6 = 64$, or $x = \sqrt[6]{64}$. Therefore, we have to find the six sixth roots of 64. Writing 64 in polar form, we have $64 = 64(\cos 0° + j\sin 0°)$. Using the exponent $1/6$ for the sixth root, we have the following solutions:

Graphing calculator program for finding complex roots: goo.gl/1C05Pr

First root: $$64^{1/6} = 64^{1/6}\left(\cos\frac{0°}{6} + j\sin\frac{0°}{6}\right) = 2(\cos 0° + j\sin 0°) = 2$$

add 360°

Second root: $$64^{1/6} = 64^{1/6}\left(\cos\frac{0° + 360°}{6} + j\sin\frac{0° + 360°}{6}\right)$$
$$= 2(\cos 60° + j\sin 60°) = 1 + j\sqrt{3}$$

add 2 × 360°

TI-89 graphing calculator keystrokes for Example 7: goo.gl/g0r3Rx

Third root: $$64^{1/6} = 64^{1/6}\left(\cos\frac{0° + 720°}{6} + j\sin\frac{0° + 720°}{6}\right)$$
$$= 2(\cos 120° + j\sin 120°) = -1 + j\sqrt{3}$$

add 3 × 360°

Fourth root: $$64^{1/6} = 64^{1/6}\left(\cos\frac{0° + 1080°}{6} + j\sin\frac{0° + 1080°}{6}\right)$$
$$= 2(\cos 180° + j\sin 180°) = -2$$

add 4 × 360°

Fifth root: $$64^{1/6} = 64^{1/6}\left(\cos\frac{0° + 1440°}{6} + j\sin\frac{0° + 1440°}{6}\right)$$
$$= 2(\cos 240° + j\sin 240°) = -1 - j\sqrt{3}$$

add 5 × 360°

Sixth root: $$64^{1/6} = 64^{1/6}\left(\cos\frac{0° + 1800°}{6} + j\sin\frac{0° + 1800°}{6}\right)$$
$$= 2(\cos 300° + j\sin 300°) = 1 - j\sqrt{3}$$

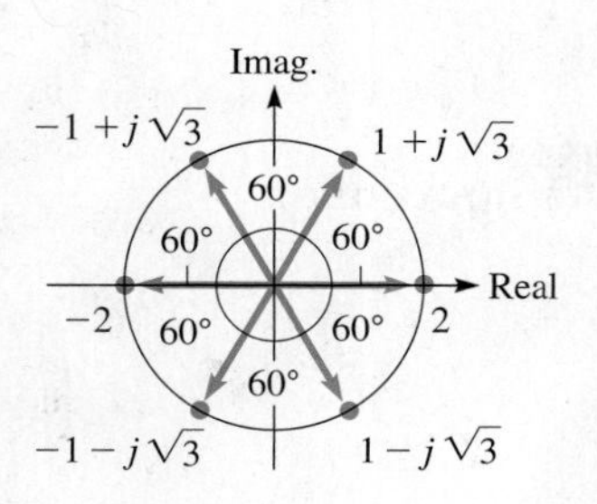

Fig. 12.20

These roots are graphed in Fig. 12.20. Note that they are equally spaced 60° apart on the circumference of a circle of radius 2. ■

From the text and examples of this and previous sections, we are able see the uses and advantages of the different forms of complex numbers. These can be summarized as follows:

Rectangular form: Used for all operations; best for addition and subtraction.

Polar form: Used for multiplication, division, powers, roots.

Exponential form: Used for multiplication, division, powers, and theoretical purposes (e.g., deriving DeMoivre's theorem)

EXERCISES 12.6

In Exercises 1–4, perform the indicated operations for the resulting complex numbers if the given changes are made in the indicated examples of this section.

1. In Example 1, change the sign of the angle in the first complex number and then perform the multiplication.

2. In Example 2, change the sign of the angle in the second complex number and then divide.

3. In Example 4, change the exponent to 5 and then find the result.

4. In Example 6, replace $2j$ with $-2j$ and then find the roots.

In Exercises 5–20, perform the indicated operations. Leave the result in polar form.

5. $[4(\cos 60° + j\sin 60°)][2(\cos 20° + j\sin 20°)]$

6. $[3(\cos 120° + j\sin 120°)][5(\cos 45° + j\sin 45°)]$

7. $(0.5\angle 140°)(6\angle 110°)$

8. $(0.4\angle 320°)(5.5\angle -150°)$

9. $\dfrac{8(\cos 100° + j\sin 100°)}{4(\cos 65° + j\sin 65°)}$

10. $\dfrac{[3(\cos 115° + j\sin 115°)]^2}{45(\cos 80° + j\sin 80°)}$

11. $\dfrac{12\angle 320°}{5\angle -210°}$

12. $\dfrac{2\angle 90°}{4\angle 75°}$

13. $[0.2(\cos 35° + j\sin 35°)]^3$

14. $[3(\cos 120° + j\sin 120°)]^4$

15. $(2/135°)^8$

16. $(1/142°)^{10}$

17. $\dfrac{(50\angle 236°)(2\angle 84°)}{125\angle 47°}$

18. $\dfrac{(6\angle 137°)^2}{(2\angle 141°)(6\angle 195°)}$

19. $\dfrac{(4\angle 24°)(10\angle 326°)}{(1\angle 62°)^3(8\angle 77°)}$

20. $\dfrac{(25\angle 194°)(6\angle 239°)}{(30\angle 17°)(10\angle 29°)}$

In Exercises 21–24, perform the indicated operations. Express results in polar form. See Example 3.

21. $2.78\angle 56.8° + 1.37\angle 207.3°$

22. $15.9\angle 142.6° - 18.5\angle 71.4°$

23. $7085\angle 115.62° - 4667\angle 296.34°$

24. $307.5\angle 326.54° + 726.3\angle 96.41°$

In Exercises 25–34, change each number to polar form and then perform the indicated operations. Express the result in rectangular and polar forms. Check by performing the same operation in rectangular form.

25. $(3 + 4j)(5 - 12j)$

26. $(5j - 2)(-1 - j)$

27. $(7 - 3j)(8 + j)$

28. $(1 + 5j)(4 + 2j)$

29. $\dfrac{21}{3 - 9j}$

30. $\dfrac{40j}{2j + 7}$

31. $\dfrac{30 + 40j}{5 - 12j}$

32. $\dfrac{5j - 2}{-1 - j}$

33. $(3 + 4j)^4$

34. $(-1 - j)^3$

In Exercises 35–40, use DeMoivre's theorem to find all the indicated roots. Be sure to find all roots.

35. The two square roots of $4(\cos 60° + j\sin 60°)$

36. The three cube roots of $27(\cos 120° + j\sin 120°)$

37. The three cube roots of $3 - 4j$

38. The two square roots of $-5 + 12j$

39. The square roots of $1 + j$

40. The cube roots of $\sqrt{3} + j$

In Exercises 41–46, find all of the roots of the given equations.

41. $x^4 - 1 = 0$

42. $x^3 - 8 = 0$

43. $x^3 + 27j = 0$

44. $x^4 - j = 0$

45. $x^5 + 32 = 0$

46. $x^6 + 8 = 0$

In Exercises 47–56, solve the given problems.

47. Using the results of Example 5, find the cube roots of -125.

48. Using the results of Example 6, find the square roots of $32j$.

49. In Example 5, we showed that one cube root of -1 is $\frac{1}{2} - \frac{1}{2}j\sqrt{3}$. Cube this number in rectangular form and show that the result is -1.

50. Explain why the two square roots of a complex number are negatives of each other.

51. The cube roots of -1 can be found by solving the equation $x^3 + 1 = 0$. Find these roots by factoring $x^3 + 1$ as the sum of cubes and compare with Example 5.

52. The cube roots of 8 can be found by solving the equation $x^3 - 8 = 0$. Find these roots by factoring $x^3 - 8$ as the difference of cubes and compare with Exercise 42.

53. The electric power p (in W) supplied to an element in a circuit is the product of the voltage e and the current i (in A). Find the expression for the power supplied if $e = 6.80\angle 56.3^\circ$ volts and $i = 0.0705\angle -15.8^\circ$ amperes.

54. The displacement d (in in.) of a weight suspended on a system of two springs is $d = 6.03\angle 22.5^\circ + 3.26\angle 76.0^\circ$ in. Perform the addition and express the answer in polar form.

55. The voltage across a certain inductor is $V = (8.66\angle 90.0^\circ)(50.0\angle 135.0^\circ)/(10.0\angle 60.0^\circ)$ volts. Simplify this expression and find the magnitude of the voltage.

56. In a microprocessor circuit, the current is $I = 3.75\angle 15.0^\circ$ μA and the impedance is $Z = 2500\angle -35.0^\circ$ ohms. Find the voltage E in rectangular form. Use $E = IZ$.

Answers to Practice Exercises

1. $15\angle 115^\circ$ **2.** $0.6\angle 345^\circ$ **3.** $6561\angle 40^\circ$ **4.** $2 + 2j, -2 - 2j$

12.7 An Application to Alternating-current (ac) Circuits

Basic Circuit with Resistance, Inductance, and Capacitance • Impedance • Phase Angle • Phasor • Resonance

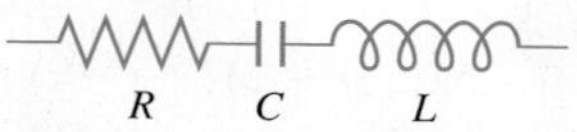

R C L

Fig. 12.21

We now show an application of complex numbers in a basic type of alternating-current circuit. We show how the voltage is measured between any two points in a circuit containing a resistance, a capacitance, and an inductance. This circuit is similar to one noted in earlier examples and exercises in this chapter.

A *resistance* is any part of a circuit that tends to obstruct the flow of electric current through the circuit. It is denoted by R (units in ohms, Ω) and in diagrams by —WWW—, as shown in Fig. 12.21. A *capacitance* is two nonconnected plates in a circuit; no current actually flows across the gap between them. In an ac circuit, an electric charge is continually going to and from each plate and, therefore, the current in the circuit is not effectively stopped. It is denoted by C (units in farads, F) and in diagrams by —||— (see Fig. 12.21). An *inductance* is basically a coil of wire in which current is induced because the current is continually changing in the circuit. It is denoted by L (units in henrys, H) and in diagrams by a coil symbol (see Fig. 12.21). All these elements affect the voltage in an ac circuit. We state here the relation each has to the voltage and current in the circuit.

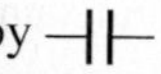

In Chapter 10, we noted that the current and voltage in an ac circuit could be represented by a sine or a cosine curve. Therefore, each reaches peak values periodically. *If they reach their respective peak values at the same time, they are* ***in phase.*** *If the voltage reaches its peak before the current, the voltage* ***leads*** *the current. If the voltage reaches its peak after the current, the voltage* ***lags*** *the current* (see Example 5 in Section 10.3).

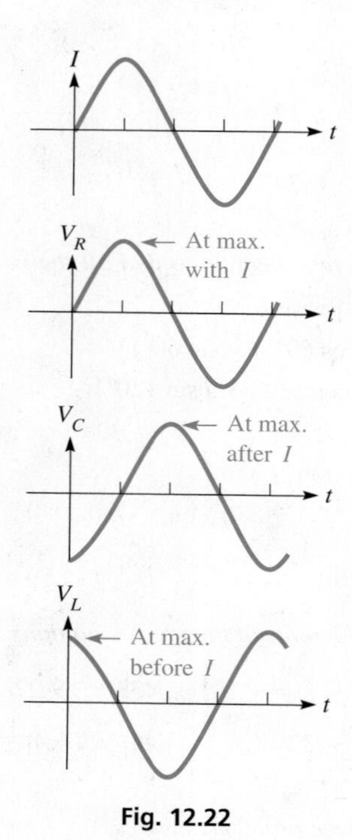

Fig. 12.22

In the study of electricity, it is shown that the voltage across a resistance is in phase with the current. The voltage across a capacitor lags the current by 90°, and the voltage across an inductance leads the current by 90°. This is shown in Fig. 12.22, where, in a given circuit, I represents the current, V_R is the voltage across a resistor, V_C is the voltage across a capacitor, V_L is the voltage across an inductor, and t represents time.

Each element in an ac circuit tends to offer a type of resistance to the flow of current. *The effective resistance of any part of the circuit is called the* **reactance,** and it is denoted by X. The voltage across any part of the circuit whose reactance is X is given by $V = IX$, where I is the current (in amperes) and V is the voltage (in volts). Therefore,

the voltage V_R across a resistor with resistance R,

the voltage V_C across a capacitor with reactance X_C, and

the voltage V_L across an inductor with reactance X_L

are, respectively,

■ Equations (12.18) are based on Ohm's law, which states that the current is proportional to the voltage for a constant resistance. It is named for the German physicist Georg Ohm (1787–1854). The ohm (Ω) is named for him.

$$V_R = IR \qquad V_C = IX_C \qquad V_L = IX_L \tag{12.18}$$

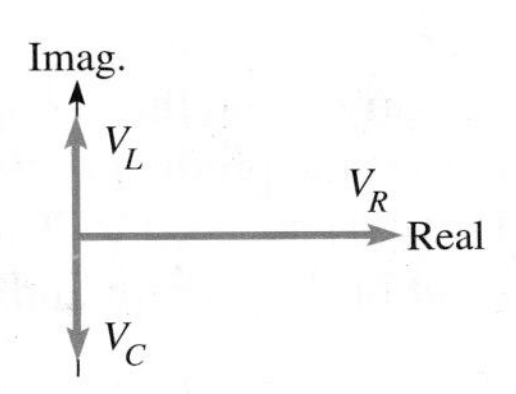

Fig. 12.23

To determine the voltage across a combination of these elements of a circuit, we must account for the reactance, as well as the phase of the voltage across the individual elements. Because the voltage across a resistor is in phase with the current, we represent V_R along the positive real axis as a real number. Because the voltage across an inductance leads the current by 90°, we represent this voltage as a positive, pure imaginary number. In the same way, by representing the voltage across a capacitor as a negative, pure imaginary number, we show that the voltage *lags* the current by 90°. These representations are meaningful since the positive imaginary axis is +90° from the positive real axis, and the negative imaginary axis is −90° from the positive real axis. See Fig. 12.23.

The circuit elements shown in Fig. 12.23 are in *series,* and all circuits we consider (except Exercises 22 and 23) are series circuits. The total voltage across a series of all three elements is given by $V_R + V_L + V_C$, which we represent by V_{RLC}. Therefore,

■ In the 1880s, it was decided that alternating current (favored by George Westinghouse) would be used to distribute electric power. Thomas Edison had argued for the use of direct current.

$$V_{RLC} = IR + IX_Lj - IX_Cj = I[R + j(X_L - X_C)]$$

This expression is also written as

$$V_{RLC} = IZ \tag{12.19}$$

■ The Greek letter phi, denoted ϕ, represents the phase angle. It is pronounced fi, like "fly."

where the symbol Z is called the **impedance** *of the circuit. It is the total effective resistance to the flow of current by a combination of the elements in the circuit,* taking into account the phase of the voltage in each element. From its definition, we see that Z is a complex number.

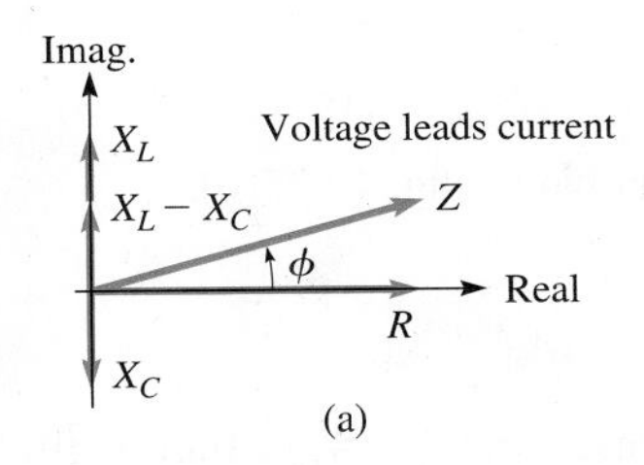

$$Z = R + j(X_L - X_C) \tag{12.20}$$

with a magnitude

$$|Z| = \sqrt{R^2 + (X_L - X_C)^2} \tag{12.21}$$

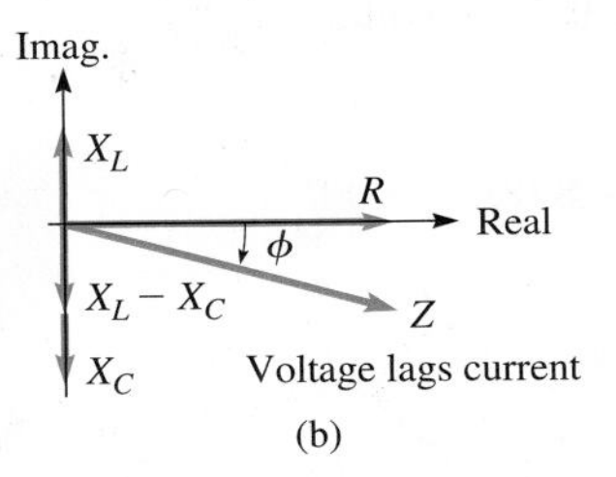

Fig. 12.24

Also, as a complex number, it makes an angle ϕ with the x axis, given by

$$\phi = \tan^{-1}\frac{X_L - X_C}{R} \tag{12.22}$$

All these equations are based on phase relations of voltages with respect to the current. Therefore, *the angle ϕ represents the phase angle between the current and the voltage.* [The standard way of expressing ϕ is to use a positive angle if the voltage leads the current and use a negative angle if the voltage lags the current.] Using Eq. (12.22), a calculator will give the correct angle even when $\tan\phi < 0$ since Z would be in the fourth quadrant and have a negative phase angle.

NOTE ▸

If the voltage leads the current, then $X_L > X_C$ as shown in Fig. 12.24(a). If the voltage lags the current, then $X_L < X_C$ as shown in Fig. 12.24(b).

In the examples and exercises of this section, the commonly used units and symbols for electrical quantities are used. For a summary of these units and symbols, see Appendix B. For common metric prefixes, see Section 1.4.

EXAMPLE 1 Finding impedance and voltage

In the series circuit shown in Fig. 12.25(a), $R = 12.0\ \Omega$ and $X_L = 5.00\ \Omega$. A current of 2.00 A is in the circuit. Find the voltage across each element, the impedance, the voltage across the combination, and the phase angle between the current and the voltage.

The voltage across the resistor (between points *a* and *b*) is the product of the current and the resistance ($V = IR$). This means $V_R = (2.00)(12.0) = 24.0$ V. The voltage across the inductor (between points *b* and *c*) is the product of the current and the reactance, or $V_L = (2.00)(5.00) = 10.0$ V.

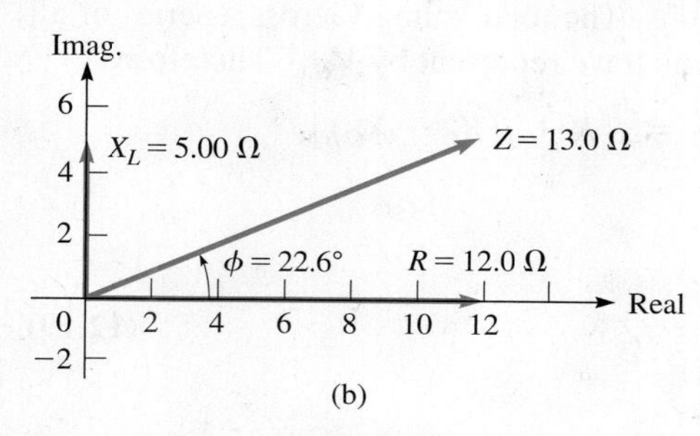

Fig. 12.25

To find the voltage across the combination, between points *a* and *c*, we must first find the magnitude of the impedance. Note that ***the voltage is not the arithmetic sum of V_R and V_L***, as we must account for the phase.

By Eq. (12.20), the impedance is (there is no capacitor)

$$Z = 12.0 + 5.00j$$

with magnitude

$$|Z| = \sqrt{R^2 + X_L^2} = \sqrt{(12.0)^2 + (5.00)^2} = 13.0\ \Omega$$

Thus, the magnitude of the voltage across the combination of the resistor and the inductance is

$$|V_{RL}| = (2.00)(13.0) = 26.0\ \text{V}$$

The phase angle between the voltage and the current is found by Eq. (12.22). This gives

$$\phi = \tan^{-1}\frac{5.00}{12.0} = 22.6°$$

The voltage *leads* the current by 22.6°, and this is shown in Fig. 12.25(b). ■

Practice Exercise

1. In Example 1, replace $X_L = 5.00\ \Omega$ with $X_C = 5.00\ \Omega$, and find the magnitude of the voltage and the phase angle ϕ.

EXAMPLE 2 Finding impedance and phase angle

For a circuit in which $R = 8.00\ \text{m}\Omega$, $X_L = 7.00\ \text{m}\Omega$, and $X_C = 13.0\ \text{m}\Omega$, find the impedance and the phase angle between the current and the voltage.

By the definition of impedance, Eq. (12.20), we have

$$Z = 8.00 + (7.00 - 13.0)j = 8.00 - 6.00j$$

where the magnitude of the impedance is

$$|Z| = \sqrt{(8.00)^2 + (-6.00)^2} = 10.0\ \text{m}\Omega$$

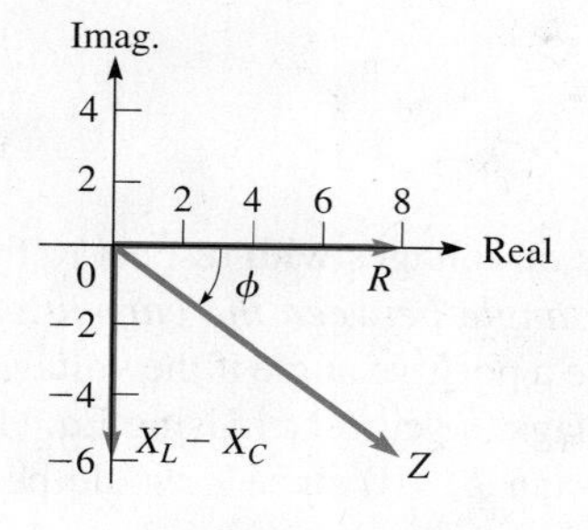

Fig. 12.26

The phase angle is found by

$$\phi = \tan^{-1}\frac{-6.00}{8.00} = -36.9°$$

The angle $\phi = -36.9°$ is given directly by the calculator, and it is the angle we want. As we noted after Eq. (12.22), we express ϕ as a negative angle if the voltage lags the current, as it does in this example. See Fig. 12.26.

From the values above, we write the impedance in polar form as $Z = 10.0\underline{/-36.9°}\ \text{m}\Omega$. ■

Note that the resistance is represented in the same way as a vector along the positive x-axis. Actually, resistance is not a vector quantity but is represented in this manner in order to assign an angle as the phase of the current. The important concept in this analysis is that *the phase* ***difference*** *between the current and voltage is constant,* and therefore any direction may be chosen arbitrarily for one of them. Once this choice is made, other phase angles are measured with respect to this direction. A common choice, as above, is to make the phase angle of the current zero. If an arbitrary angle is chosen, it is necessary to treat the current, voltage, and impedance as complex numbers.

EXAMPLE 3 Finding voltage

In a particular circuit, the current is $2.00 - 3.00j$ A and the impedance is $6.00 + 2.00j$ ohms. The voltage across this part of the circuit is

$$\begin{aligned} V &= (2.00 - 3.00j)(6.00 + 2.00j) = 12.0 - 14.0j - 6.00j^2 \\ &= 12.0 - 14.0j + 6.00 \\ &= 18.0 - 14.0j \text{ volts} \end{aligned}$$

The magnitude of the voltage is

$$|V| = \sqrt{(18.0)^2 + (-14.0)^2} = 22.8 \text{ V}$$

■

■ The ampere (A) is named for the French physicist André Ampere (1775–1836).

■ The volt (V) is named for the Italian physicist Alessandro Volta (1745–1827).

■ The farad (F) is named for the British physicist Michael Faraday (1791–1867).

■ The henry (H) is named for the U.S. physicist Joseph Henry (1797–1878).

■ The coulomb (C) is named for the French physicist Charles Coulomb (1736–1806).

Because the voltage across a resistor is in phase with the current, this voltage can be represented as having a phase difference of zero with respect to the current. Therefore, the resistance is indicated as an arrow in the positive real direction, denoting the fact that the current and the voltage are in phase. *Such a representation is called a* **phasor.** The arrow denoted by R, as in Fig. 12.25(b) and Fig. 12.26, is actually the phasor representing the voltage across the resistor. Remember, the positive real axis is arbitrarily chosen as the direction of the phase of the current.

To show properly that the voltage across an inductance leads the current by 90°, its reactance (effective resistance) is multiplied by j. We know that there is a positive 90° angle between a positive real number and a positive imaginary number. In the same way, by multiplying the capacitive reactance by $-j$, we show the 90° difference in phase between the voltage and the current in a capacitor, with the current leading. Therefore, jX_L represents the phasor for the voltage across an inductor and $-jX_C$ is the phasor for the voltage across the capacitor. The phasor for the voltage across the combination of the resistance, inductance, and capacitance is Z, where the phase difference between the voltage and the current for the combination is the angle ϕ.

NOTE ➧ [From this, we see that multiplying a phasor by j means to perform the operation of rotating it through 90°. For this reason, j is also called the ***j*-operator.**]

EXAMPLE 4 Multiplication by j

Multiplying a positive real number A by j, we have $A \times j = Aj$, which is a positive imaginary number. In the complex plane, Aj is 90° from A, which means that by multiplying A by j we rotated A by j 90°. Similarly, we see that $Aj \times j = Aj^2 = -A$, which is a negative real number, rotated 90° from Aj. Therefore, successive multiplications of A by j give us

$A \times j = Aj$	positive imaginary number
$Aj \times j = Aj^2 = -A$	negative real number
$-A \times j = -Aj$	negative imaginary number
$-Aj \times j = -Aj^2 = A$	positive real number

See Fig. 12.27. (See Exercise 36 in Section 12.3.) ■

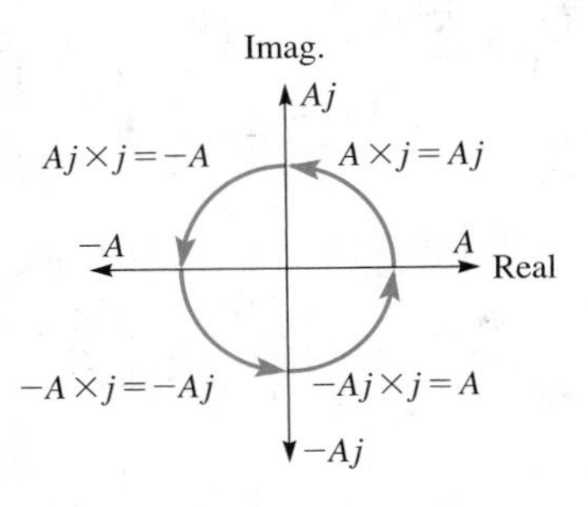

Fig. 12.27

An alternating current is produced by a coil of wire rotating through a magnetic field. If the angular velocity of the wire is ω, the capacitive and inductive reactances are given by

$$X_C = \frac{1}{\omega C} \quad \text{and} \quad X_L = \omega L \tag{12.23}$$

Therefore, if ω, C, and L are known, the reactance of the circuit can be found.

EXAMPLE 5 Finding voltage-current phase difference

If $R = 12.0\ \Omega$, $L = 0.300$ H, $C = 250\ \mu$F, and $\omega = 80.0$ rad/s, find the impedance and the phase difference between the current and the voltage.

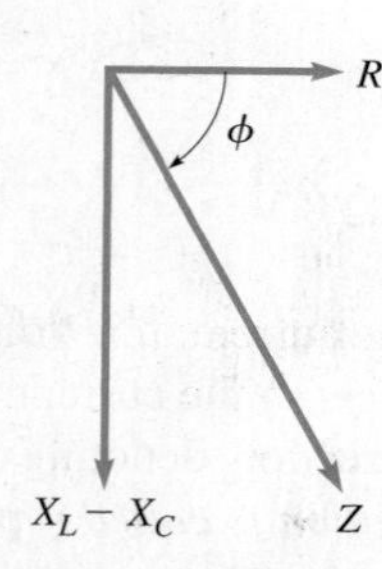

Fig. 12.28

$$X_C = \frac{1}{(80.0)(250 \times 10^{-6})} = 50.0\ \Omega$$

$$X_L = (0.300)(80.0) = 24.0\ \Omega$$

$$Z = 12.0 + (24.0 - 50.0)j = 12.0 - 26.0j$$

$$|Z| = \sqrt{(12.0)^2 + (-26.0)^2} = 28.6\ \Omega$$

$$\phi = \tan^{-1}\frac{-26.0}{12.0} = -65.2°$$

$$Z = 28.6\underline{/-65.2°}\ \Omega$$

The voltage *lags* the current (see Fig. 12.28). ■

Recall from Section 10.5 that the angular velocity ω is related to the frequency f by the relation $\omega = 2\pi f$, where ω is in rad/s and f is in Hz. It is very common to use frequency when discussing alternating current.

An important concept in the application of this theory is that of **resonance.** *For resonance, the impedance of any circuit is a minimum, or the total impedance is R.* Thus, $X_L - X_C = 0$. Also, it can be seen that the current and the voltage are in phase under these conditions. Resonance is required for the tuning of radio and television receivers.

EXAMPLE 6 Resonance

■ See the chapter introduction.

In the antenna circuit of a radio, the inductance is 4.20 mH, and the capacitance is variable. What range of values of capacitance is necessary for the radio to receive the AM band of radio stations, with frequencies from 530 kHz to 1600 kHz?

For proper tuning, the circuit should be in resonance, or $X_L = X_C$. This means that

$$2\pi fL = \frac{1}{2\pi fC} \quad \text{or} \quad C = \frac{1}{(2\pi f)^2 L}$$

For $f_1 = 530$ kHz $= 5.30 \times 10^5$ Hz and $L = 4.20$ mH $= 4.20 \times 10^{-3}$ H,

$$C_1 = \frac{1}{(2\pi)^2(5.30 \times 10^5)^2(4.20 \times 10^{-3})} = 2.15 \times 10^{-11}\ \text{F} = 21.5\ \text{pF}$$

and for $f_2 = 1600$ kHz $= 1.60 \times 10^6$ Hz and $L = 4.20 \times 10^{-3}$ H, we have

$$C_2 = \frac{1}{(2\pi)^2(1.60 \times 10^6)^2(4.20 \times 10^{-3})} = 2.36 \times 10^{-12}\ \text{F} = 2.36\ \text{pF}$$

The capacitance should be capable of varying from 2.36 pF to 21.6 pF. ■

■ From Section 1.4, the following prefixes are defined as follows:

Prefix	*Factor*	*Symbol*
pico	10^{-12}	p
milli	10^{-3}	m
kilo	10^{3}	k

EXERCISES 12.7

In Exercises 1 and 2, perform the indicated operations if the given changes are made in the indicated examples of this section.

1. In Example 1, change the value of X_L to 16.0 Ω and then solve the given problem.

2. In Example 5, double the values of L and C and then solve the given problem.

In Exercises 3–6, use the circuit shown in Fig. 12.29. The current in the circuit is 5.75 mA. Determine the indicated quantities.

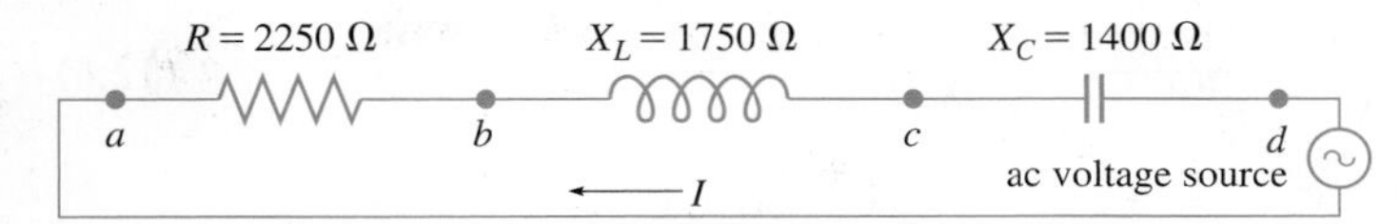

Fig. 12.29

3. The voltage across the resistor (between points a and b).

4. The voltage across the inductor (between points b and c).

5. **(a)** The magnitude of the impedance across the resistor and the inductor (between points a and c).
   (b) The phase angle between the current and the voltage for the combination in (a).
   (c) The magnitude of the voltage across the combination in (a).

6. **(a)** The magnitude of the impedance across the resistor, inductor, and capacitor (between points a and d).
   (b) The phase angle between the current and the voltage for the combination in (a).
   (c) The magnitude of the voltage across the combination in (a).

In Exercises 7–10, an ac circuit contains the given combination of circuit elements from among a resistor ($R = 45.0$ Ω), a capacitor ($C = 86.2\ \mu$F), and an inductor ($L = 42.9$ mH). If the frequency in the circuit is $f = 60.0$ Hz, find (a) the magnitude of the impedance and (b) the phase angle between the current and the voltage.

7. The circuit has the inductor and the capacitor (an LC circuit).

8. The circuit has the resistor and the capacitor (an RC circuit).

9. The circuit has the resistor and the inductor (an RL circuit).

10. The circuit has the resistor, the inductor, and the capacitor (an RLC circuit).

In Exercises 11–24, solve the given problems.

11. Given that the current in a given circuit is $3.90 - 6.04j$ mA and the impedance is $5.16 + 1.14j$ kΩ, find the magnitude of the voltage.

12. Given that the voltage in a given circuit is $8.375 - 3.140j$ V and the impedance is $2.146 - 1.114j$ Ω, find the magnitude of the current.

13. A resistance ($R = 25.3$ Ω) and a capacitance ($C = 2.75$ nF) are in an AM radio circuit. If $f = 1200$ kHz, find the impedance across the resistor and the capacitor.

14. A resistance ($R = 64.5$ Ω) and an inductance ($L = 1.08$ mH) are in a telephone circuit. If $f = 8.53$ kHz, find the impedance across the resistor and inductor.

15. The reactance of an inductor is 1200 Ω for $f = 280$ Hz. What is the inductance?

16. A resistor, an inductor, and a capacitor are connected in series across an ac voltage source. A voltmeter measures 12.0 mV, 15.5 mV, and 10.5 mV, respectively, when placed across each element separately. What is the voltage of the source?

17. An inductance of 12.5 μH and a capacitance of 47.0 nF are in series in an amplifier circuit. Find the frequency for resonance.

18. A capacitance ($C = 95.2$ nF) and an inductance are in series in the circuit of a receiver for navigation signals. Find the inductance if the frequency for resonance is 50.0 kHz.

19. In Example 6, what should be the capacitance in order to receive a 680-kHz radio signal?

20. A 220-V source with $f = 60.0$ Hz is connected in series to an inductance ($L = 2.05$ H) and a resistance R in an electric-motor circuit. Find R if the current is 0.250 A.

21. The power P (in W) supplied to a series combination of elements in an ac circuit is $P = VI\cos\phi$, where V is the effective voltage, I is the effective current, and ϕ is the phase angle between the current and voltage. If $V = 225$ mV across the resistor, capacitor, and inductor combination in Exercise 10, determine the power supplied to these elements.

22. For two impedances Z_1 and Z_2 in parallel, the combined impedance Z_C is given by $Z_C = \dfrac{Z_1 Z_2}{Z_1 + Z_2}$. Find the combined impedance (in rectangular form) for the parallel circuit elements in Fig. 12.30 if the current in the circuit has a frequency of 60.0 Hz.

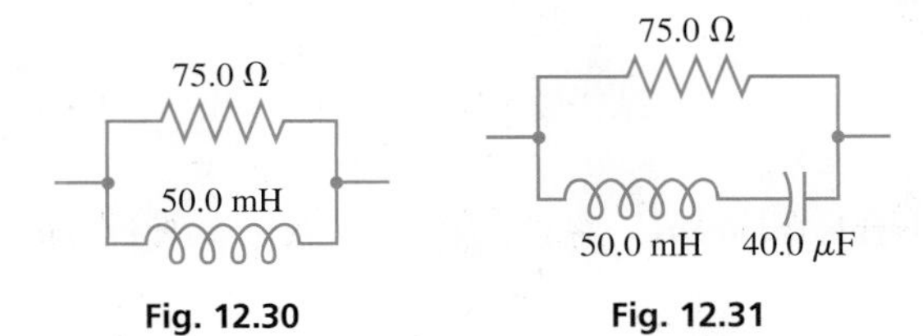

Fig. 12.30 **Fig. 12.31**

23. Find the combined impedance (in rectangular form) of the circuit elements in Fig. 12.31. The frequency of the current in the circuit is 60.0 Hz. See Exercise 22.

24. (a) If the complex number j, in polar form, is multiplied by itself, what is the resulting number in polar and rectangular forms? (b) In the complex plane, where is the resulting complex number in relation to j?

Answer to Practice Exercise

1. $V_{RC} = 26.0$ V, $\theta = -22.6°$

CHAPTER 12 KEY FORMULAS AND EQUATIONS

Chapter Equations for Complex Numbers

Imaginary unit

$$j = \sqrt{-1} \quad \text{and} \quad j^2 = -1 \tag{12.1}$$

$$\sqrt{-a} = j\sqrt{a} \quad (a > 0) \tag{12.2}$$

Basic operations

$$(a + bj) + (c + dj) = (a + c) + (b + d)j \tag{12.3}$$

$$(a + bj) - (c + dj) = (a - c) + (b - d)j \tag{12.4}$$

$$(a + bj)(c + dj) = (ac - bd) + (ad + bc)j \tag{12.5}$$

$$\frac{a + bj}{c + dj} = \frac{(a + bj)(c - dj)}{(c + dj)(c - dj)} = \frac{(ac + bd) + (bc - ad)j}{c^2 + d^2} \tag{12.6}$$

Complex number forms

Rectangular: $x + yj$

Polar: $r(\cos\theta + j\sin\theta) = r\underline{/\theta}$

Exponential: $re^{j\theta}$

$$x = r\cos\theta \qquad y = r\sin\theta \tag{12.7}$$

$$r^2 = x^2 + y^2 \qquad \tan\theta = \frac{y}{x} \tag{12.8}$$

$$x + yj = r(\cos\theta + j\sin\theta) = r\underline{/\theta} = re^{j\theta} \tag{12.12}$$

Product in polar form

$$\begin{aligned} r_1(\cos\theta_1 + j\sin\theta_1)r_2(\cos\theta_2 + j\sin\theta_2) &= r_1r_2[\cos(\theta_1 + \theta_2) + j\sin(\theta_1 + \theta_2)] \\ (r_1\underline{/\theta_1})(r_2\underline{/\theta_2}) &= r_1r_2\underline{/\theta_1 + \theta_2} \end{aligned} \tag{12.13}$$

Quotient in polar form

$$\frac{r_1(\cos\theta_1 + j\sin\theta_1)}{r_2(\cos\theta_2 + j\sin\theta_2)} = \frac{r_1}{r_2}[\cos(\theta_1 - \theta_2) + j\sin(\theta_1 - \theta_2)] \tag{12.15}$$

$$\frac{r_1\underline{/\theta_1}}{r_2\underline{/\theta_2}} = \frac{r_1}{r_2}\underline{/\theta_1 - \theta_2}$$

DeMoivre's theorem

$$[r(\cos\theta + j\sin\theta)]^n = r^n(\cos n\theta + j\sin n\theta) \tag{12.17}$$

$$(r\underline{/\theta})^n = r^n\underline{/n\theta}$$

Chapter Equations for Alternating-current Circuits

Voltage, current, reactance

$$V_R = IR \quad V_C = IX_C \quad V_L = IX_L \tag{12.18}$$

Impedance

$$V_{RLC} = IZ \tag{12.19}$$

$$Z = R + j(X_L - X_C) \tag{12.20}$$

$$|Z| = \sqrt{R^2 + (X_L - X_C)^2} \tag{12.21}$$

Phase angle

$$\phi = \tan^{-1}\frac{X_L - X_C}{R} \tag{12.22}$$

Capacitive reactance and inductive reactance

$$X_C = \frac{1}{\omega C} \quad \text{and} \quad X_L = \omega L \tag{12.23}$$

CHAPTER 12 REVIEW EXERCISES

CONCEPT CHECK EXERCISES

*Determine each of the following as being either **true** or **false**. If it is false, explain why.*

1. $(\sqrt{-9})^2 = 9$

2. $\dfrac{1+j}{1-j} = j$

3. $4\angle 90^\circ = 4j$

4. $2e^{\pi j} = -2$

5. $(2\angle 120^\circ)^3 = 8$

6. The phase angle ϕ for an impedance $Z = 4 - 4j$ is 45°.

PRACTICE AND APPLICATIONS

In Exercises 7–20, perform the indicated operations, expressing all answers in simplest rectangular form.

7. $(6 - 2j) + (4 + j)$

8. $(12 + 7j) + (-8 + 6j)$

9. $(18 - 3j) - (12 - 5j)$

10. $(-4 - 2j) - \sqrt{-49}$

11. $5j(6 - 5j)$

12. $-3j(4 - 7j)$

13. $(6 - 3j)(4 + 3j)$

14. $(4 - 9j)(1 + 2j)$

15. $\dfrac{3}{7 - 6j}$

16. $\dfrac{48j}{4 + 18j}$

17. $\dfrac{6 - \sqrt{-16}}{\sqrt{-4}}$

18. $\dfrac{3 + \sqrt{-4}}{4 - j}$

19. $\dfrac{5j - (3 - j)}{4 - 2j}$

20. $\dfrac{2 - (6 - j)}{1 - 2j}$

In Exercises 21–24, find the values of x and y for which the equations are valid.

21. $3x - 2j = yj - 9$

22. $2xj - 2y = (y + 3)j - 3$

23. $(3 + 2j^3)(x + jy) = 4 + j^9$

24. $(x + jy)(7j - 4) = j(x - 5)$

In Exercises 25–28, perform the indicated operations graphically. Check them algebraically.

25. $(-1 + 5j) + (4 + 6j)$

26. $(7 - 2j) + (-5 + 4j)$

27. $(9 + 2j) - (5 - 6j)$

28. $(4j + 8) - (11 - 3j)$

In Exercises 29–36, give the polar and exponential forms of each of the complex numbers.

29. $1 - j$

30. $4 - 3j^3$

31. $-22 - 77j$

32. $60 - 20j$

33. $1.07 + 4.55j$

34. $158j - 327$

35. 5000

36. $-4j^5$

In Exercises 37–48, give the rectangular form of each number:

37. $2(\cos 225^\circ + j \sin 225^\circ)$

38. $48(\cos 60^\circ + j \sin 60^\circ)$

39. $5.011(\cos 123.82^\circ + j \sin 123.82^\circ)$

40. $2.417(\cos 656.26^\circ + j \sin 656.26^\circ)$

41. $0.62\angle -72^\circ$

42. $20\angle 160^\circ$

43. $27.08\angle 346.27^\circ$

44. $1.689\angle 194.36^\circ$

45. $2.00e^{0.25j}$

46. $e^{-3.62j}$

47. $(35.37e^{1.096j})^2$

48. $(13.6e^{2.158j})(3.27e^{3.888j})$

In Exercises 49–64, perform the indicated operations. Leave the result in polar form.

49. $[3(\cos 32^\circ + j \sin 32^\circ)][5(\cos 52^\circ + j \sin 52^\circ)]$

50. $[2.5(\cos 162^\circ + j \sin 162^\circ)][8(\cos 115^\circ + j \sin 115^\circ)]$

51. $(40\angle 18^\circ)(0.5\angle 245^\circ)$

52. $(0.1254\angle 172.38^\circ)(27.17\angle 204.34^\circ)$

53. $\dfrac{24(\cos 165^\circ + j \sin 165^\circ)}{[3(\cos 55^\circ + j \sin 55^\circ)]^3}$

54. $\dfrac{18(\cos 403^\circ + j \sin 403^\circ)}{[2(\cos 96^\circ + j \sin 96^\circ)]^2}$

55. $\dfrac{245.6\angle 326.44^\circ}{17.19\angle 192.83^\circ}$

56. $\dfrac{4\angle 206^\circ}{100\angle -320^\circ}$

57. $0.983\angle 47.2^\circ + 0.366\angle 95.1^\circ$

58. $17.8\angle 110.4^\circ - 14.9\angle 226.3^\circ$

59. $7644\angle 294.36^\circ - 6871\angle 17.86^\circ$

60. $4.944\angle 327.49^\circ + 8.009\angle 7.37^\circ$

61. $[2(\cos 16^\circ + j \sin 16^\circ)]^{10}$

62. $[3(\cos 36^\circ + j \sin 36^\circ)]^6$

63. $(7\angle 110.5^\circ)^3$

64. $(536\angle 220.3^\circ)^4$

In Exercises 65–68, change each number to polar form and then perform the indicated operations. Express the final result in rectangular and polar forms. Check by performing the same operation in rectangular form using a calcultor.

65. $(1 - j)^{10}$

66. $(\sqrt{3} + j)^8(1 + j)^5$

67. $\dfrac{(5 + 5j)^4}{(-1 - j)^6}$

68. $(\sqrt{3} - j)^{-8}$

In Exercises 69–72, find all the roots of the given equations.

69. $x^3 + 8 = 0$

70. $x^3 - 1 = 0$

71. $x^4 + j = 0$

72. $x^5 - 32j = 0$

In Exercises 73–76, determine the rectangular form and the polar form of the complex number for which the graphical representation is shown in the given figure.

73.

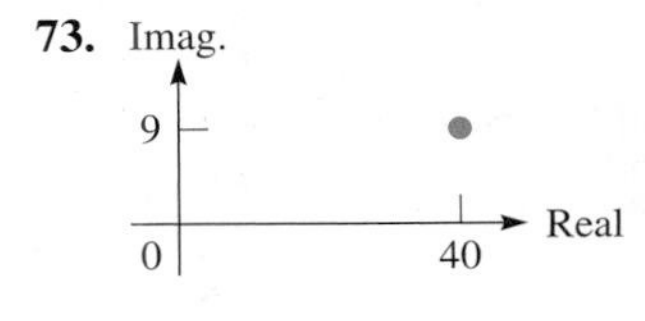

74.

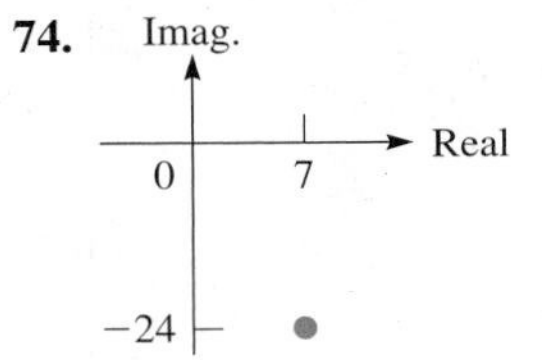

75. **76.**

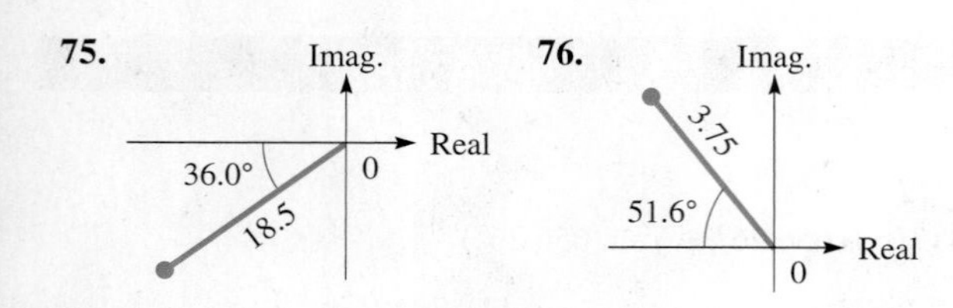

In Exercises 77–88, solve the given problems.

77. Evaluate $x^2 - 2x + 4$ for $x = 5 - 2j$.

78. Evaluate $2x^2 + 5x - 7$ for $x = -8 + 7j$.

79. Solve for x: $x^2 = 8x - 41$ (Express the solutions in simplified form in terms of j.)

80. Solve for x: $2x^2 = 6x - 9$ (Express the solutions in simplified form in terms of j.)

81. Are $1 - j$ and $-1 - j$ solutions to the equation $x^2 - 2x + 2 = 0$?

82. Show that $\frac{1}{2}(1 + j\sqrt{3})$ is the reciprocal of its conjugate.

83. Solve for x: $(1 + jx)^2 = 1 + j - x^2$

84. What is the argument for any negative imaginary number?

85. If $f(x) = 2x - (x - 1)^{-1}$, find $f(1 + 2j)$.

86. If $f(x) = x^{-2} + 3x^{-1}$, find $f(4 + j)$.

87. Using a calculator, express $5 - 3j$ in polar form.

88. Using a calculator, express $25.0e^{2.25j}$ in rectangular form.

In Exercises 89–100, find the required quantities.

89. A 60-V ac voltage source is connected in series across a resistor, an inductor, and a capacitor. The voltage across the inductor is 60 V, and the voltage across the capacitor is 60 V. What is the voltage across the resistor?

90. In a series ac circuit with a resistor, an inductor, and a capacitor, $R = 6.50\ \Omega$, $X_C = 3.74\ \Omega$, and $Z = 7.50\ \Omega$. Find X_L.

91. In a series ac circuit with a resistor, an inductor, and a capacitor, $R = 6250\ \Omega$, $Z = 6720\ \Omega$, and $X_L = 1320\ \Omega$. Find the phase angle ϕ.

92. A coil of wire rotates at 120.0 r/s. If the coil generates a current in a circuit containing a resistance of 12.07 Ω, an inductance of 0.1405 H, and an impedance of 22.35 Ω, what must be the value of a capacitor (in F) in the circuit?

93. What is the frequency f for resonance in a circuit for which $L = 2.65$ H and $C = 18.3\ \mu$F?

94. The displacement of an electromagnetic wave is given by $d = A(\cos \omega t + j \sin \omega t) + B(\cos \omega t - j \sin \omega t)$. Find the expressions for the magnitude and phase angle of d.

95. Two cables lift a crate. The tensions in the cables can be represented by $2100 - 1200j$ N and $1200 + 5600j$ N. Express the resultant tension in polar form.

96. A boat is headed across a river with a velocity (relative to the water) that can be represented as $6.5 + 1.7j$ mi/h. The velocity of the river current can be represented as $-1.1 - 4.3j$ mi/h. Express the resultant velocity of the boat in polar form.

97. In the study of shearing effects in the spinal column, the expression $\dfrac{1}{\mu + j\omega n}$ is found. Express this in rectangular form.

98. In the theory of light reflection on metals, the expression $\dfrac{\mu(1 - kj) - 1}{\mu(1 - kj) + 1}$ is encountered. Simplify this expression.

99. Show that $e^{j\pi} = -1$.

100. Show that $(e^{j\pi})^{1/2} = j$.

101. A computer programmer is writing a program to determine the n nth roots of a real number. Part of the program is to find the number of real roots and the number of pure imaginary roots. Write one or two paragraphs explaining how these numbers of roots can be determined without actually finding the roots.

CHAPTER 12 PRACTICE TEST

As a study aid, we have included complete solutions for each Practice Test problem at the end of this book.

1. Add, expressing the result in rectangular form:
$(3 - \sqrt{-4}) + (5\sqrt{-9} - 1)$.

2. Multiply, expressing the result in polar form:
$(2/130^\circ)(3/45^\circ)$.

3. Express $2 - 7j$ in polar form.

4. Express in terms of j: (a) $-\sqrt{-64}$ (b) $-j^{15}$.

5. Add graphically: $(4 - 3j) + (-1 + 4j)$.

6. Simplify, expressing the result in rectangular form: $\dfrac{2 - 4j}{5 + 3j}$.

7. Express $2.56(\cos 125.2^\circ + j \sin 125.2^\circ)$ in exponential form.

8. For an ac circuit in which $R = 3.50\ \Omega$, $X_L = 6.20\ \Omega$, and $X_C = 7.35\ \Omega$, find the impedance and the phase angle between the current and the voltage.

9. Express $3.47 - 2.81j$ in exponential form.

10. Find the values of x and y: $x + 2j - y = yj - 3xj$.

11. What is the capacitance of the circuit in a radio that has an inductance of 8.75 mH if it is to receive a station with frequency 600 kHz?

12. Find the cube roots of j.

Exponential and Logarithmic Functions

13

By the early 1600s, astronomy had progressed to the point of finding accurate information about the motion of the heavenly bodies.

Also, navigation had led to a more systematic exploration of Earth. In making the accurate measurements needed in astronomy and navigation, many lengthy calculations had to be performed, and all such calculations had to be done by hand.

Noting that astronomers' calculations usually involved sines of angles, John Napier (1550–1617), a Scottish mathematician, constructed a table of values that allowed multiplication of these sines by addition of values from the table. These tables of *logarithms* first appeared in 1614. Therefore, logarithms were essentially invented to make multiplications by means of addition, thereby making them easier.

Napier's logarithms were not in base 10, and the English mathematician Henry Briggs (1561–1631) realized that logarithms in base 10 would make the calculations even easier. He spent many years laboriously developing a table of base 10 logarithms, which was not completed until after his death.

Logarithms were enthusiastically received by mathematicians and scientists as a long-needed tool for lengthy calculations. The great French mathematician Pierre Laplace (1749–1827) stated "by shortening the labors doubled the life of the astronomer." Logarithms were commonly used for calculations until the 1970s when the scientific calculator came into use.

In this chapter, we study the *logarithmic function* and the *exponential function*. Although logarithms are no longer used directly for calculations, they are of great importance in many scientific and technical applications and in advanced mathematics. For example, they are used to measure the intensity of sound, the intensity of earthquakes, the power gains and losses in electrical transmission lines, and to distinguish between a base and an acid. Exponential functions are used in electronics, mechanical systems, thermodynamics, nuclear physics, biology in studying population growth, and in business to calculate compound interest.

LEARNING OUTCOMES

After completion of this chapter, the student should be able to:

- Evaluate and graph an exponential function
- Apply properties of logarithms
- Change equations from exponential form to logarithmic form and vice versa
- Evaluate and graph a logarithmic function
- Identify the exponential and logarithmic functions as inverse functions
- Solve logarithmic and exponential equations
- Change a logarithm in one base to a logarithm in another base
- Solve application problems involving logarithmic and exponential functions
- Graph functions on logarithmic or semilogarithmic paper

◀ The growth of population can often be measured using an exponential function. This is illustrated in Section 13.6.

13.1 Exponential Functions

The Exponential Function • Graphing Exponential Functions • Features of Exponential Functions

In Chapter 11, we showed that any rational number can be used as an exponent. Now letting *the exponent be a variable,* we define the **exponential function** *as*

$$y = b^x \tag{13.1}$$

where $b > 0$, $b \neq 1$, and x is any real number. *The number b is called the* **base.**

For an exponential function, we use only real numbers. Therefore, $b > 0$, because if b were negative and x were a fractional exponent with an even-number denominator, y would be imaginary. Also, $b \neq 1$, since 1 to any real power is 1 (y would be constant).

EXAMPLE 1 Exponential functions

From the definition, $y = 3^x$ is an exponential function, but $y = (-3)^x$ is not because the base is negative. However, $y = -3^x$ is an exponential function, because it is -1 times 3^x, and any real-number multiple of an exponential function is also an exponential function.

Also, $y = (\sqrt{3})^x$ is an exponential function since it can be written as $y = 3^{x/2}$. As long as x is a real number, so is $x/2$. Therefore, the exponent of 3 is real.

The function $y = 3^{-x}$ is an exponential function. If x is real, so is $-x$.

Other exponential functions are: $y = -2(8^{-0.55x})$ and $y = 35(1.0001)^x$. ■

EXAMPLE 2 Evaluating an exponential function

Evaluate the function $y = -2(4^x)$ for the given values of x.

(a) If $x = 2$, $y = -2(4^2) = -2(16) = -32$.

(b) If $x = -2$, $y = -2(4^{-2}) = -2/16 = -1/8$.

(c) If $x = 3/2$, $y = -2(4^{3/2}) = -2(8) = -16$.

(d) If $x = \sqrt{2}$, $y = -2(4^{\sqrt{2}}) = -14.206$ (calculator evaluation).

(e) If $x = \pi$, $y = -2(4^{\pi}) = -155.76$ (calculator evaluation). ■

GRAPHING EXPONENTIAL FUNCTIONS

We now show some representative graphs of the exponential function.

EXAMPLE 3 Graphing an exponential function

Plot the graph of $y = 2^x$.

For this function, we have the values in the following table:

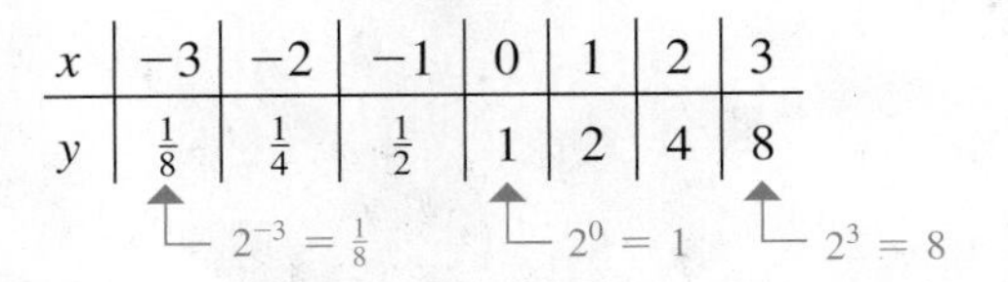

x	-3	-2	-1	0	1	2	3
y	$\frac{1}{8}$	$\frac{1}{4}$	$\frac{1}{2}$	1	2	4	8

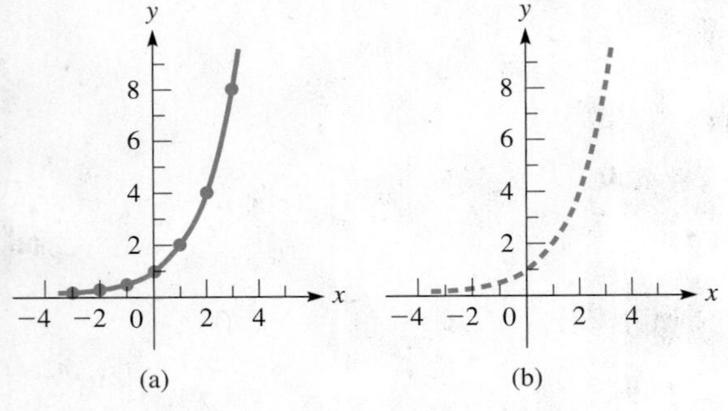

Fig. 13.1

The curve is shown in Fig. 13.1(a). In Chapter 1, we used only integer exponents, and the enlarged points are for these values. In Chapter 11, we introduced rational exponents, and using them would fill in many points between those for integers, but all the points for irrational numbers would be missing and the curve would be dotted [see Fig. 13.1(b)]. Using all real numbers for exponents, including the irrational numbers, we have all points on the curve shown in Fig. 13.1(a).

We see that the x-axis is an *asymptote* of the curve. The points on the left side of the graph get closer and closer to the x-axis, but they never touch it. ■

Practice Exercises

Evaluate $y = 16^x$ for:

1. $x = 3/2$ **2.** $x = -0.5$

The bases b of the exponential function of greatest importance in applications are greater than 1. However, in order to understand how the exponential function differs somewhat if $b < 1$, we now use a calculator to display such a graph.

EXAMPLE 4 Exponential function on a calculator

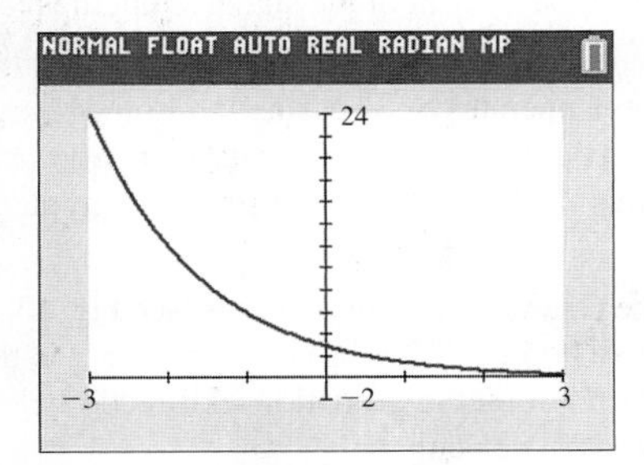

Fig. 13.2

Display the graph of $y = 3(\frac{1}{2})^x$ on a calculator.

For this function, $(1/2)^x = 1/2^x = 2^{-x}$. Because x may be any real number, *as x becomes more negative, y increases more rapidly*. Therefore, on a calculator, we let $y_1 = 3(1/2)^x$ [or $y_1 = 3(.5^x)$] and have the display in Fig. 13.2 with the *window* settings shown. ■

Any exponential curve where $b > 1$ will be similar in shape to that shown in Fig. 13.1, and if $b < 1$, it will be similar to the curve in Fig. 13.2. From these examples, we can see that exponential functions have the following basic features.

Basic Features of the Exponential Function $y = b^x$

1. The domain is all values of x; the range is $y > 0$.
2. The x-axis is an asymptote of the graph.
3. As x increases, b^x increases if $b > 1$, and b^x decreases if $b < 1$.

As we have noted, exponential functions are important in many applications. We now illustrate an application in the next example, and others are shown in the exercises.

EXAMPLE 5 Exponential function—rocket trajectory

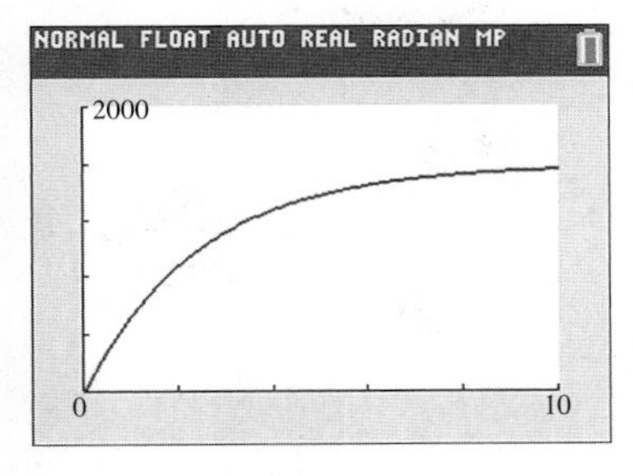

Fig. 13.3

A computer analysis of a rocket trajectory showed its height h (in m) as a function of time t (in s) was $h = 1600(1 - e^{-0.40t})$. Display the graph of this function on a calculator. Here e is the number introduced on page 356 and is equal to approximately 2.718.

Here, $e^{-0.40t} = 1$ for $t = 0$, which means $h = 0$ for $t = 0$. Also, $e^{-0.40t}$ becomes smaller as t increases, and this means h cannot be greater than 1600 m. Therefore, $h = 1600$ m is a horizontal asymptote. This leads to the settings and curve shown in Fig. 13.3. Here we have used y for h, x for t, and $y_1 = 1600(1 - e^{-.4x})$. ■

EXERCISES 13.1

In Exercises 1 and 2, perform the indicated operations if the given changes are made in the indicated examples of this section.

1. In Example 2(c), change the sign of x and then evaluate.
2. In Example 3, change the sign of the exponent and then plot the graph.

In Exercises 3–6, use a calculator to evaluate (to three significant digits) the given numbers.

3. $3^{\sqrt{5}}$ **4.** 1.5^{π} **5.** $(2\pi)^{-e}$ **6.** $(2e)^{-\sqrt{2}}$

In Exercises 7–10, determine if the given functions are exponential functions.

7. (a) $y = 5^x$ (b) $y = 5^{-x}$ **8.** (a) $y = -7^x$ (b) $y = (-7)^{-x}$

9. (a) $y = -7(-5)^{-x}$ (b) $y = -7(5^{-x})$

10. (a) $y = (\sqrt{5})^{-x}$ (b) $y = -(\sqrt{-5})^x$

In Exercises 11–16, evaluate the exponential function $y = 4^x$ for the given values of x.

11. $x = 0.5$ **12.** $x = 4$ **13.** $x = -2$

14. $x = -0.5$ **15.** $x = -3/2$ **16.** $x = 5/2$

In Exercises 17–22, plot the graphs of the given functions.

17. $y = 4^x$ **18.** $y = 0.25^x$ **19.** $y = 0.2(10^{-x})$

20. $y = -5(1.6^{-x})$ **21.** $y = 0.5\pi^x$ **22.** $y = 2e^x$

In Exercises 23–30, display the graphs of the given functions on a calculator.

23. $y = 0.3(2.55)^x$ **24.** $y = -1.5(4.15)^x$

25. $y = 0.1(0.25^{2x})$ **26.** $y = 0.4(0.95)^{3x}$

27. $i = 1.2(2 + 6^{-t})$ **28.** $y = 0.5e^{-x}$

29. (a) $y = 10^x$ (b) $y = 2^x$ (c) $y = 1.1^x$

30. (a) $y = 0.1^x$ (b) $y = 0.5^x$ (c) $y = 0.9^x$

In Exercises 31–46, solve the given problems.

31. Find the base b of the function $y = b^x$ if its graph passes through the point (3, 64).

32. Find the base b of the function $y = b^x$ if its graph passes through the point $(-2, 64)$.

33. For the function $f(x) = b^x$, show that $f(c + d) = f(c) \cdot f(d)$.

34. For the function $f(x) = b^x$, show that $f(c - d) = f(c)/f(d)$.

35. Use a calculator to graph the function $y = 2^{|x|}$.

36. To show the *damping effect* of an exponential function, use a calculator to display the graph of $y = (x^3)(2^{-x})$. Be sure to use appropriate window settings.

37. Use a calculator to find the value(s) for which $x^2 = 2^x$.

38. Use a calculator to find the value(s) for which $x^3 = 3^x$.

39. For $f(x) = Ae^{kx}$, find the expression for $\dfrac{f(x + 1) - f(x)}{f(x)}$.

40. A medical research lab is growing a virus for a vaccine that grows at a rate of 2.3% per hour. If there are 500.0 units of the virus originally, the amount present after t hours is given by $N = 500.0(1.023)^t$. How many units of the virus are present after two days?

41. The value V of a bank account in which \$250 is invested at 5.00% interest, compounded annually, is $V = 250(1.0500)^t$, where t is the time in years. Find the value of the account after 4 years.

42. The intensity I of an earthquake is given by $I = I_0(10)^R$, where I_0 is a minimum intensity for comparison and R is the Richter scale magnitude of the earthquake. Evaluate I in terms of I_0 if $R = 5.5$.

43. The electric current i (in mA) in the circuit shown in Fig. 13.4 is $i = 2.5(1 - e^{-0.10t})$, where t is the time (in s). Evaluate i for $t = 5.0$ ms (1 ms $= 10^{-3}$ s).

44. The strength I of a certain cable signal is given by $I = I_0 e^{-0.0015x}$, where I_0 is the signal strength at the source and x is the distance (in km) from the source. What percent of the signal strength is lost 15 km from the source?

45. The flash unit on a camera operates by releasing the stored charge on a capacitor. For a particular unit, the charge q (in μC) as a function of the time t (in s) is $q = 100e^{-10t}$. Display the graph on a calculator.

46. The height y (in m) of the Gateway Arch in St. Louis (see Fig. 13.5) is given by $y = 230.9 - 19.5(e^{x/38.9} + e^{-x/38.9})$, where x is the distance (in m) from the point on the ground level directly below the top. Display the graph on a calculator.

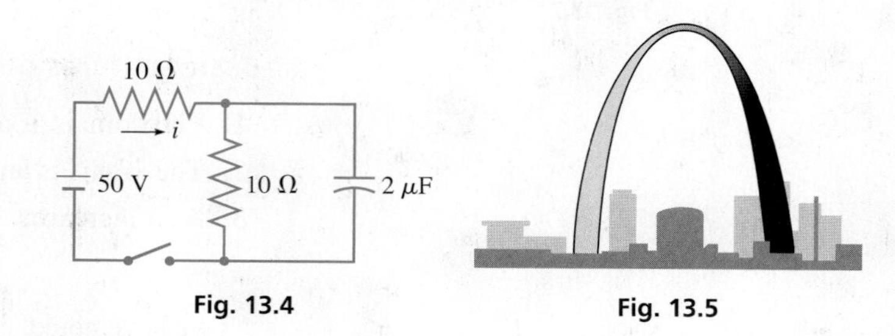

Fig. 13.4

Fig. 13.5

Answers to Practice Exercises

1. 64 **2.** 1/4

13.2 Logarithmic Functions

Exponential Form • Logarithmic Form • Logarithmic Function • Graphing Logarithmic Functions • Features of Logarithmic Functions • Inverse Functions

For many uses in mathematics and for many applications, it is necessary to express the exponent x in the exponential function $y = b^x$ in terms of y and the base b. This is done by defining a *logarithm*. Therefore, *if* $y = b^x$, *the exponent* x *is the* **logarithm** *of the number* y *to the base* b. We write this as

$$\text{If } y = b^x, \text{ then } x = \log_b y. \tag{13.2}$$

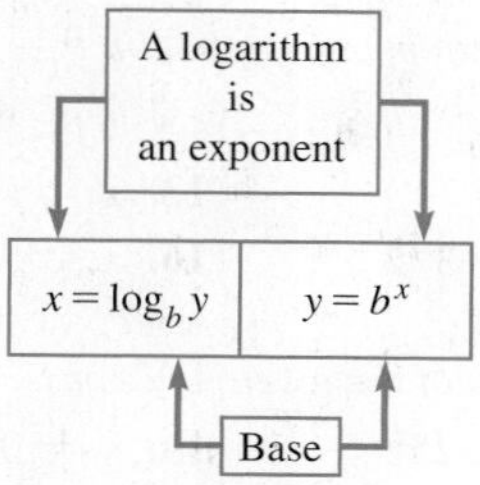

A logarithm is an exponent

$x = \log_b y$ | $y = b^x$

Base

Fig. 13.6

This means that x is the power to which the base b must be raised in order to equal the number y. That is, x is a logarithm, and ***a logarithm is an exponent.*** As with the exponential function, for the equation $x = \log_b y$, x may be any real number, b is a positive number other than 1, and y is a positive real number. In Eq. (13.2),

$y = b^x$ *is the* **exponential form,** and $x = \log_b y$ *is the* **logarithmic form**

See Fig. 13.6.

■ When you see the word *logarithm*, think *exponent.*

EXAMPLE 1 Exponential form and logarithmic form

The equation $y = 2^x$ is written as $x = \log_2 y$ when written in logarithmic form. When we choose values of y to find corresponding values of x from this equation, we ask ourselves "2 raised to what power x gives y?"

This means that if $y = 8$, we ask "what power of 2 gives us 8?" Then knowing that $2^3 = 8$, we know that $x = 3$. Therefore, $3 = \log_2 8$. ■

EXAMPLE 2 Changing to logarithmic form

(a) $3^2 = 9$ in logarithmic form is $2 = \log_3 9$.

(b) $4^{-1} = 1/4$ in logarithmic form is $-1 = \log_4(1/4)$.

CAUTION Remember, the exponent may be negative. The *base* must be positive. ■ ■

EXAMPLE 3 Changing between forms

(a) $(64)^{1/3} = 4$ in logarithmic form is $\frac{1}{3} = \log_{64} 4$.

(b) $(32)^{3/5} = 8$ in logarithmic form is $\frac{3}{5} = \log_{32} 8$.

(c) $\log_2 32 = 5$ in exponential form is $32 = 2^5$.

(d) $\log_6\left(\frac{1}{36}\right) = -2$ in exponential form is $\frac{1}{36} = 6^{-2}$.

(e) To change $4 \log_{16} 8 = 3$ to exponential form, first write it as $\log_{16} 8 = 3/4$. Then we can write the exponential form $8 = 16^{3/4}$. ■

EXAMPLE 4 Evaluating by changing form

(a) Find b, given that $-4 = \log_b\left(\frac{1}{81}\right)$.

Writing this in exponential form, we have $\frac{1}{81} = b^{-4}$. Thus, $\frac{1}{81} = \frac{1}{b^4}$ or $\frac{1}{3^4} = \frac{1}{b^4}$. Therefore, $b = 3$.

(b) Given $\log_4 y = 1/2$, in exponential form it becomes $y = 4^{1/2}$, or $y = 2$. ■

Practice Exercises

1. Change $125^{2/3} = 25$ to logarithmic form.

2. Change $\log_3(1/3) = -1$ to exponential form.

We see that exponential form is very useful for determining values written in logarithmic form. For this reason, it is important that you learn to transform readily from one form to the other.

CAUTION In order to change a function of the form $y = ab^x$ into logarithmic form, we must first write it as $y/a = b^x$. The coefficient of b^x must be equal to 1. In the same way, the coefficient of $\log_b y$ must be 1 in order to change it into exponential form. ■

EXAMPLE 5 Evaluating—satellite power

The power supply P (in W) of a certain satellite is given by $P = 75e^{-0.005t}$, where t is the time (in days) after launch. By writing this equation in logarithmic form, solve for t.

We start by dividing both sides of the equation by 75 to isolate the expression $e^{-0.005t}$:

$$\frac{P}{75} = e^{-0.005t}$$

Writing this in logarithmic form, we have $\log_e(P/75) = -0.005t$, or

$$t = \frac{\log_e\left(\frac{P}{75}\right)}{-0.005} = -200 \log_e\left(\frac{P}{75}\right) \qquad \frac{1}{-0.005} = -200$$

We recall from Section 12.5 that e is the special irrational number equal to about 2.718. It is important as a base of logarithms, and this is discussed in Section 13.5. ■

THE LOGARITHMIC FUNCTION

When we are working with functions, we must keep in mind that a function is defined by the operation being performed on the independent variable, and not by the letter chosen to represent it. However, for consistency, it is standard practice to let y represent the dependent variable and x represent the independent variable. Therefore, *the* **logarithmic function** *is*

$$y = \log_b x \tag{13.3}$$

As with the exponential function, $b > 0$ and $b \neq 1$.

NOTE ▸ [Equations (13.2) and (13.3) do not represent different functions, due to the difference in location of the variables, because they represent the same operation on the independent variable that appears in each.] However, Eq. (13.3) expresses the function with the standard dependent and independent variables.

EXAMPLE 6 Logarithmic function

For the logarithmic function $y = \log_2 x$, we have the standard independent variable x and the standard dependent variable y.

If $x = 16$, $y = \log_2 16$, which means that $y = 4$, because $2^4 = 16$.
If $x = \frac{1}{16}$, $y = \log_2(\frac{1}{16})$, which means that $y = -4$, because $2^{-4} = \frac{1}{16}$. ■

Practice Exercise

3. For the function in Example 6, evaluate y for $x = 8$.

GRAPHING LOGARITHMIC FUNCTIONS

We now show some representative graphs for the logarithmic function. From these graphs, we can see the basic properties of the logarithmic function.

EXAMPLE 7 Graphing a logarithmic function

Plot the graph of $y = \log_2 x$.

We can find the points for this graph more easily if we first put the equation in exponential form: $x = 2^y$. By assuming values for y, we can find the corresponding values for x.

$2^{-2} = \frac{1}{4}$ $\quad$ $2^2 = 4$

x	$\frac{1}{4}$	$\frac{1}{2}$	1	2	4	8
y	−2	−1	0	1	2	3

Using these values, we construct the graph seen in Fig. 13.7. ■

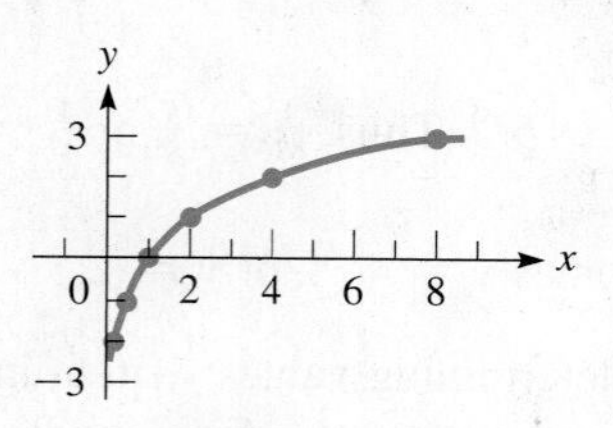

Fig. 13.7

Practice Exercise

4. In Example 7, find y if $x = 64$.

The log keys standard to all scientific and graphing calculators are [log] (for $\log_{10} x$) and [ln] (for $\log_e x$). Therefore, for now, we restrict graphing logarithmic functions to bases 10 and e on the graphing calculator. In Section 13.5, we will see how to use the calculator to graph a logarithmic function with any positive real-number base.

EXAMPLE 8 Logarithmic graph on calculator—sound intensity

The loudness b (in dB, or decibels) of a sound is defined as $b = 10 \log_{10}(I/I_0)$, where I/I_0 is the ratio of the intensity of the sound to the intensity of sound of minimum intensity that is detectable. On a calculator, display the graph of $y = 10 \log_{10} x$, where y represents b, the loudness of a sound, and x represents I/I_0, the ratio of intensities.

On a calculator, we set $y_1 = 10 \log x$, and the curve is displayed as shown in Fig. 13.8. The settings for X_{min} and X_{max} were selected since the domain is $x > 0$. The realistic values of x are $x \geq 1$, since sound for $x < 1$ is not detectable. ■

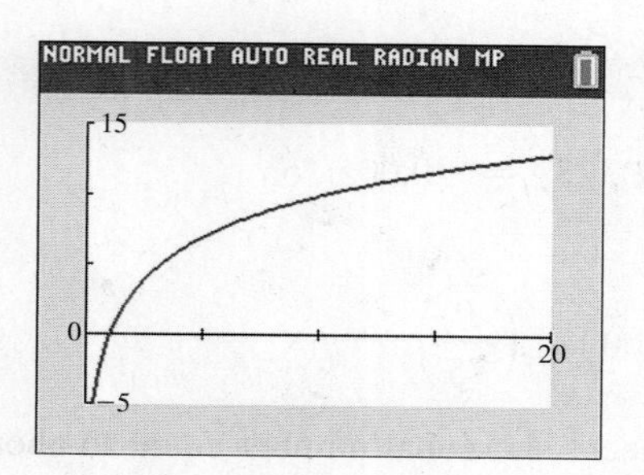

Fig. 13.8

From these graphs, we can see that logarithmic functions have the following features. We consider only the features for which $b > 1$, for these are the bases of greatest importance.

Basic Features of Logarithmic Functions ($b > 1$)

1. The domain is $x > 0$; the range is all values of y.
2. The negative y-axis is an asymptote of the graph of $y = \log_b x$.
3. If $0 < x < 1$, $\log_b x < 0$; if $x = 1$, $\log_b x = 0$; if $x > 1$, $\log_b x > 0$.
4. If $x > 1$, x increases more rapidly than $\log_b x$.

Table 1

x	1	4	16	64
$\log_2 x$	0	2	4	6
2^x	2	16	65,536	1.8×10^{19}

We just noted that if $b > 1$ and $x > 1$, x increases more rapidly than $\log_b x$. It is also true that b^x increases more rapidly than x. Actually, as x becomes larger, $\log_b x$ increases very slowly, and b^x increases very rapidly. Using $\log_2 x$ and 2^x and a calculator, we have the table of values shown to the left. This shows that we must choose values of x carefully when graphing these functions.

INVERSE FUNCTIONS

For the exponential function $y = b^x$ and the logarithmic function $y = \log_b x$, if we solve for the independent variable in one of the functions by changing the form, then interchange the variables, we obtain the other function. *Such functions are called* **inverse functions.**

This means that the x- and y-coordinates of inverse functions are interchanged. As a result, the graphs of inverse functions are mirror images of each other across the line $y = x$. This is illustrated in the following example.

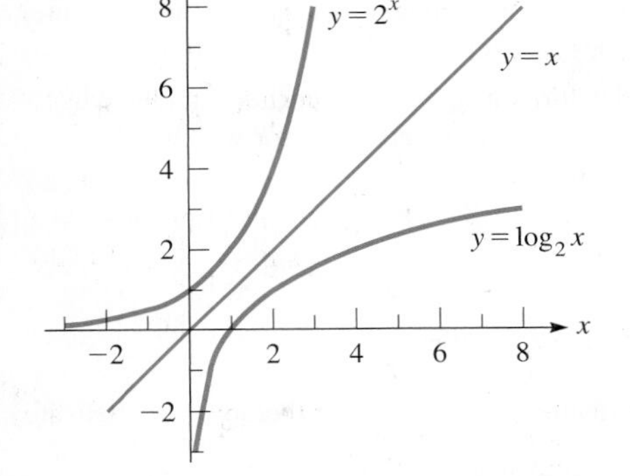

Fig. 13.9

EXAMPLE 9 Inverse functions

The functions $y = 2^x$ and $y = \log_2 x$ are inverse functions. We show this by solving $y = 2^x$ for x and then interchanging x and y.

Writing $y = 2^x$ in logarithmic form gives us $x = \log_2 y$. Then interchanging x and y, we have $y = \log_2 x$, which is the inverse function.

Making a table of values for each function, we have

$y = 2^x$:

x	−3	−2	−1	0	1	2	3
y	$\frac{1}{8}$	$\frac{1}{4}$	$\frac{1}{2}$	1	2	4	8

$y = \log_2 x$:

x	$\frac{1}{8}$	$\frac{1}{4}$	$\frac{1}{2}$	1	2	4	8
y	−3	−2	−1	0	1	2	3

We see that the coordinates are interchanged. In Fig. 13.9, note that the graphs of these two functions reflect each other across the line $y = x$. ■

Practice Exercise

5. What is the inverse function of $y = 10^x$?

For a function, there is exactly one value of y in the range for each value of x in the domain. This must also hold for the inverse function. Thus, for a function to have an inverse function, there must be only one x for each y. This is true for $y = b^x$ and $y = \log_b x$, as we have seen earlier in this section.

EXERCISES 13.2

In Exercises 1–4, perform the indicated operations if the given changes are made in the indicated examples of this section.

1. In Example 3(b), change the exponent to 4/5 and then make any other necessary changes.

2. In Example 4(b), change the 1/2 to 5/2 and then make any other necessary changes.

3. In Example 6, change the logarithm base to 4 and then make any other necessary changes.

4. In Example 7, change the logarithm base to 4 and then plot the graph.

In Exercises 5–16, express the given equations in logarithmic form.

5. $3^4 = 81$ **6.** $5^2 = 25$ **7.** $7^3 = 343$

8. $2^7 = 128$ **9.** $4^{-2} = \frac{1}{16}$ **10.** $3^{-2} = \frac{1}{9}$

11. $2^{-6} = \frac{1}{64}$ **12.** $(12)^0 = 1$ **13.** $8^{1/3} = 2$

14. $(81)^{3/4} = 27$ **15.** $(\frac{1}{4})^2 = \frac{1}{16}$ **16.** $(\frac{1}{2})^{-3} = 8$

In Exercises 17–28, express the given equations in exponential form.

17. $\log_2 32 = 5$ **18.** $\log_{11} 121 = 2$

19. $\log_9 9 = 1$ **20.** $\log_{15} 1 = 0$

21. $\log_{25} 5 = \frac{1}{2}$ **22.** $3 \log_8 16 = 4$

23. $5 \log_{243} 3 = 1$ **24.** $\log_{32}(\frac{1}{8}) = -0.6$

25. $\log_{10} 0.01 = -2$ **26.** $\log_7(\frac{1}{49}) = -2$

27. $\log_{0.5} 16 = -4$ **28.** $\log_{1/3} 3 = -1$

In Exercises 29–44, determine the value of the unknown.

29. $\log_4 16 = x$

30. $\log_5 125 = x$

31. $\log_{10} 0.01 = x$

32. $\log_{16}(\frac{1}{4}) = x$

33. $\log_7 y = 3$

34. $\log_8(N + 1) = 3$

35. $3 \log_8(A - 2) = -2$

36. $\log_7 y = -2$

37. $\log_b 3 = 2$

38. $\log_b 625 = 4$

39. $\log_b 4 = -\frac{1}{3}$

40. $3 \log_b 4 = 2$

41. $\log_{10} 10^{0.2} = x$

42. $\log_5 5^{2.3} = R + 1$

43. $\log_3 27^{-1} = x + 1$

44. $\log_b(\frac{1}{4}) = -0.5$

In Exercises 45–50, plot the graphs of the given functions.

45. $y = \log_3 x$

46. $y = -\log_4(-x)$

47. $y = \log_{0.5} x$

48. $y = 3 \log_2 x$

49. $N = 0.2 \log_4 v$

50. $A = 2.4 \log_{10}(2r)$

In Exercises 51–54, display the graphs of the given functions on a graphing calculator.

51. $y = 3 \log_e x$

52. $y = 5 \log_{10}|x|$

53. $y = -\log_{10}(-x)$

54. $y = -\log_e(-2x)$

In Exercises 55–76, perform the indicated operations.

55. Evaluate $\log_4 x$ for (a) $x = 1/64$ and (b) $x = -1/2$.

56. Evaluate: (a) $\log_b b$ (b) $\log_b 1$

57. If $f(x) = \log_5 x$, find: (a) $f(\sqrt{5})$ (b) $f(0)$

58. If $f(x) = \log_b x$ and $f(3) = 2$, find $f(9)$.

59. Find b such that the point (2, 2) is on the graph of $y = \log_b x$.

60. Find b such that the point $(8, -3)$ is on the graph of $y = \log_b x$.

61. On a calculator, display the graphs of: (a) $y_1 = \log_{10} x$ (b) $y_2 = \log_{10}(x + 2)$ (c) $y_3 = \log_{10}(x - 2)$ (See p. 105.)

62. On a calculator, display the graphs of $y_1 = 2 \log_{10} x$ and $y_2 = \log_{10} x^2$. Describe any similarities or differences.

63. Using a calculator, solve the equation $\log_{10} x = x - 2$.

64. Use a calculator to display the graphs (on the same screen) of $y = \log_e(x^2 + c)$, with $c = -4, 0, 4$. Describe the results.

65. Find the domain of the function $f(x) = \log_e(2 - x)$.

66. The time t (in years) for an investment to double is $t = \dfrac{\ln 2}{\ln(1 + r)}$, where r is the annual interest rate. How long does it take an investment to double if $r =$ (a) 3.00%, (b) 6.00%, (c) 9.00%?

67. The magnitudes (visual brightness), m_1 and m_2, of two stars are related to their (actual) brightnesses, b_1 and b_2, by the equation $m_1 - m_2 = 2.5 \log_{10}(b_2/b_1)$. Solve for b_2.

68. The velocity v of a rocket when its fuel is completely burned is given by $v = u \log_e(w_0/w)$, where u is the exhaust velocity, w_0 is the liftoff weight, and w is the burnout weight. Solve for w.

69. An equation relating the number N of atoms of radium at any time t in terms of the number N_0 of atoms at $t = 0$ is $\log_e(N/N_0) = -kt$, where k is a constant. Solve for N.

70. The capacitance C of a cylindrical capacitor is given by $C = k/\log_e(R_2/R_1)$, where R_1 and R_2 are the inner and outer radii. Solve for R_1.

71. The work W done by a sample of nitrogen gas during an isothermal (constant temperature) change from volume V_1 to volume V_2 is given by $W = k \log_e(V_2/V_1)$. Solve for V_1.

72. An equation used in measuring the flow of water in a channel is $C = -a \log_{10}(b/R)$. Solve for R.

73. The time t (in ps) required for N calculations by a certain computer design is $t = N + \log_2 N$. Sketch the graph of this function.

74. When a tractor-trailer turns a right angle corner, its rear wheels follow a curve called a *tractrix*, the equation for which is

$$y = \log_e\left(\frac{1 + \sqrt{1 - x^2}}{x}\right) - \sqrt{1 - x^2}.$$

Display the curve on a graphing calculator.

75. For a radioactive isotope, the constant of proportionality k is given by $k = \dfrac{\log_e(^1/_2)}{\text{half-life}}$. Cobalt-60, used in cancer therapy, has a half-life of 5.27 years. What is the value of k?

76. In Exercise 47, the graph of $y = \log_{0.5} x$ is plotted. By inspecting the graph and noting the properties of $\log_{0.5} x$, describe some of the differences of logarithms to a base less than 1 from those to a base greater than 1.

In Exercises 77–80, show that the given functions are inverse functions of each other. Then display the graphs of each function and the line $y = x$ on a graphing calculator and note that each is the mirror image of the other across $y = x$.

77. $y = 10^{x/2}$ and $y = 2 \log_{10} x$

78. $y = e^x$ and $y = \log_e x$

79. $y = 3x$ and $y = x/3$

80. $y = 2x + 4$ and $y = 0.5x - 2$

Answers to Practice Exercise

1. $2/3 = \log_{125} 25$ **2.** $1/3 = 3^{-1}$ **3.** 3 **4.** 6 **5.** $y = \log_{10} x$

13.3 Properties of Logarithms

Sum of Logarithms for Product • Difference of Logarithms for Quotient • Multiple of Logarithm for Power • Logarithms of 1 and b

Because a logarithm is an exponent, it must follow the laws of exponents. The laws used in this section to derive the very useful properties of logarithms are listed here for reference.

$$b^u b^v = b^{u+v} \tag{13.4}$$

$$\frac{b^u}{b^v} = b^{u-v} \tag{13.5}$$

$$(b^u)^n = b^{nu} \tag{13.6}$$

■ A logarithm is an exponent. However, a historical curiosity is that logarithms were developed before exponents were used.

The next example shows the reasoning used in deriving the properties of logarithms.

EXAMPLE 1 Sum of logarithms for product

We know that $8 \times 16 = 128$. Writing these numbers as powers of 2, we have

$$8 = 2^3 \qquad 16 = 2^4 \qquad 128 = 2^7 = 2^{3+4}$$

The logarithmic forms can be written as

$$3 = \log_2 8 \qquad 4 = \log_2 16 \qquad 3 + 4 = \log_2 128$$

This means that

$$\log_2 8 + \log_2 16 = \log_2 128$$

where

$$8 \times 16 = 128$$

The *sum of the logarithms* of 8 and 16 equals the logarithm of 128, where the *product* of 8 and 16 equals 128. ■

■ For reference, Eqs. (13.4) to (13.6) are

$$b^u b^v = b^{u+v}$$
$$\frac{b^u}{b^v} = b^{u-v}$$
$$(b^u)^n = b^{nu}$$

Following Example 1, if we let $u = \log_b x$ and $v = \log_b y$ and write these equations in exponential form, we have $x = b^u$ and $y = b^v$. Therefore, forming the product of x and y, we obtain

$$xy = b^u b^v = b^{u+v} \quad \text{or} \quad xy = b^{u+v}$$

Writing this last equation in logarithmic form yields

$$u + v = \log_b xy$$

This means that the **logarithm of a product** can be written as

$$\log_b xy = \log_b x + \log_b y \tag{13.7}$$

Equation (13.7) states the property that *the logarithm of the product of two numbers is equal to the sum of the logarithms of the numbers.*

Using the same definitions of u and v to form the quotient of x and y, we then have

$$\frac{x}{y} = \frac{b^u}{b^v} = b^{u-v} \quad \text{or} \quad \frac{x}{y} = b^{u-v}$$

Writing this last equation in logarithmic form, we have

$$u - v = \log_b\left(\frac{x}{y}\right).$$

Therefore, the **logarithm of a quotient** is given by

$$\log_b\left(\frac{x}{y}\right) = \log_b x - \log_b y \tag{13.8}$$

Equation (13.8) states the property that *the logarithm of the quotient of two numbers is equal to the logarithm of the numerator minus the logarithm of the denominator.*

CAUTION Noting Equations (13.7) and (13.8), it is very clear that

$\log x + \log y$ is *not* equal to $\log(x + y)$

$\log x - \log y$ is *not* equal to $\log(x - y)$ ■

If we again let $u = \log_b x$ and write this in exponential form, we have $x = b^u$. To find the nth power of x, we write

$$x^n = (b^u)^n = b^{nu}$$

Expressing this equation in logarithmic form yields

$$nu = \log_b(x^n).$$

Thus, the **logarithm of a power** is given by

$$\log_b(x^n) = n \log_b x \qquad \textbf{(13.9)}$$

Equation (13.9) states that *the logarithm of the nth power of a number is equal to n times the logarithm of the number.* The exponent n may be any real number, which, of course, includes all rational and irrational numbers.

In Section 13.2, we showed that the base b of logarithms must be a positive number. Because $x = b^u$ and $y = b^v$, this means that x and y are also positive numbers. Therefore, the *properties of logarithms that have just been derived are valid only for positive values of x and y.*

■ In advanced mathematics, the logarithms of negative and imaginary numbers are defined.

EXAMPLE 2 Logarithms of product, quotient, power

(a) Using Eq. (13.7), we may express $\log_4 15$ as a sum of logarithms:

$$\log_4 15 = \log_4(3 \times 5) = \log_4 3 + \log_4 5$$

logarithm of product
sum of logarithms

(b) Using Eq. (13.8), we may express $\log_4(\frac{5}{3})$ as the difference of logarithms:

$$\log_4\left(\frac{5}{3}\right) = \log_4 5 - \log_4 3$$

logarithm of quotient
difference of logarithms

(c) Using Eq. (13.9), we may express $\log_4(t^2)$ as twice $\log_4 t$:

$$\log_4(t^2) = 2 \log_4 t$$

logarithm of power
multiple of logarithm

(d) Using Eq. (13.8) and then Eq. (13.7), we have

$$\log_4\left(\frac{xy}{z}\right) = \log_4(xy) - \log_4 z$$

$$= \log_4 x + \log_4 y - \log_4 z$$ ■

Practice Exercises

Express as a sum or difference of logarithms.

1. $\log_3 10$

2. $\log_3(2a/5)$

EXAMPLE 3 Sum or difference of logarithms as a single quantity

We may also express a sum or difference of logarithms as the logarithm of a single quantity.

(a) $\log_4 3 + \log_4 x = \log_4(3 \times x) = \log_4 3x$ using Eq. (13.7)

(b) $\log_4 3 - \log_4 x = \log_4\left(\frac{3}{x}\right)$ using Eq. (13.8)

(c) $\log_4 3 + 2\log_4 x = \log_4 3 + \log_4(x^2) = \log_4 3x^2$ using Eqs. (13.7) and (13.9)

(d) $\log_4 3 + 2\log_4 x - \log_4 y = \log_4\left(\frac{3x^2}{y}\right)$ using Eqs. (13.7), (13.8), and (13.9) ■

Practice Exercise

3. Express as a single logarithm: $\log_3 5 - 3\log_3 x$

In Section 13.2, we noted that $\log_b 1 = 0$. Also, because $b = b^1$ in logarithmic form is $\log_b b = 1$, we have $\log_b(b^x) = x \log_b b = x(1) = x$. In addition, the logarithmic form of $\log_b x = \log_b x$ is $b^{\log_b x} = x$.

Summarizing these properties, we have

$$\log_b 1 = 0 \qquad \log_b b = 1 \tag{13.10}$$

$$\log_b(b^x) = x \tag{13.11}$$

$$b^{\log_b x} = x \tag{13.12}$$

These equations can be used to simplify certain expressions.

EXAMPLE 4 Exact values for certain logarithms

(a) We may evaluate $\log_3 9$ using Eq. (13.11):

$$\log_3 9 = \log_3(3^2) = 2$$

We can establish the exact value since the base of logarithms and the number being raised to the power are the same. Of course, this could have been evaluated directly from the definition of a logarithm.

(b) Using Eq. (13.11), we can write $\log_3(3^{0.4}) = 0.4$. Although we did not evaluate $3^{0.4}$, we can evaluate $\log_3(3^{0.4})$. ■

Practice Exercise

4. Find the exact value of $2 \log_2 8$.

EXAMPLE 5 Using the properties of logarithms

(a) $\log_2 6 = \log_2(2 \times 3) = \log_2 2 + \log_2 3 = 1 + \log_2 3$

(b) $\log_5 \frac{1}{5} = \log_5 1 - \log_5 5 = 0 - 1 = -1$

(c) $\log_7 \sqrt{7} = \log_7(7^{1/2}) = \frac{1}{2} \log_7 7 = \frac{1}{2}$

(d) $3^{\log_3 8} + 4^{\log_4 7} = 8 + 7 = 15$ ■

EXAMPLE 6 Evaluation in two ways

The following illustration shows the evaluation of a logarithm in two different ways. Either method is appropriate.

(a) $\log_5(\frac{1}{25}) = \log_5 1 - \log_5 25 = 0 - \log_5(5^2) = -2$

(b) $\log_5(\frac{1}{25}) = \log_5(5^{-2}) = -2$ ■

EXAMPLE 7 Solving equation with logarithms

Use the basic properties of logarithms to solve the following equation for y in terms of x: $\log_b y = 2 \log_b x + \log_b a$.

Using Eq. (13.9) and then Eq. (13.7), we have

$$\log_b y = \log_b(x^2) + \log_b a = \log_b(ax^2)$$

Because we have the logarithm to the base b of different expressions on each side of the resulting equation, the expressions must be equal. Therefore,

$$y = ax^2$$ ■

EXAMPLE 8 Solving equation—radioactive decay

An equation encountered in the study of radioactive elements is $\log_e N - \log_e N_0 = kt$. Here, N is the amount of the element present at any time t, and N_0 is the original amount. Solve for N as a function of t.

Using Eq. (13.8), we rewrite the left side of this equation, obtaining

$$\log_e\left(\frac{N}{N_0}\right) = kt$$

Rewriting this in exponential form, we have

$$\frac{N}{N_0} = e^{kt}, \quad \text{or} \quad N = N_0 e^{kt}$$

■

EXERCISES 13.3

In Exercises 1–8, perform the indicated operations on the resulting expressions if the given changes are made in the original expressions of the indicated examples of this section.

1. In Example 2(a), change the 15 to 21.
2. In Example 2(d), change the x to 2 and the z to 3.
3. In Example 3(b), change the 3 to 5.
4. In Example 3(c), change the 2 to 3.
5. In Example 4(a), change the 9 to 27.
6. In Example 5(a), change the 6 to 10.
7. In Example 5(c), change the 7's to 5's.
8. In Example 7, change the 2 to 3.

In Exercises 9–20, express each as a sum, difference, or multiple of logarithms. See Example 2.

9. $\log_5 33$
10. $\log_3 14$
11. $\log_7\left(\frac{9}{2}\right)$
12. $\log_3\left(\frac{2}{11}\right)$
13. $\log_2(a^3)$
14. $2\log_8(n^5)$
15. $\log_6(abc^2)$
16. $\log_2\left(\frac{xy}{z^2}\right)$
17. $10\log_5\sqrt{t}$
18. $\log_4\sqrt[7]{x}$
19. $\log_2\left(\frac{\sqrt{x}}{a^2}\right)$
20. $\log_3\left(\frac{\sqrt[3]{y}}{7x}\right)$

In Exercises 21–28, express each as the logarithm of a single quantity. See Example 3.

21. $\log_b a + \log_b c$
22. $\log_2 3 + \log_2 x$
23. $\log_5 9 - \log_5 3$
24. $-\log_8 R + \log_8 V$
25. $\log_b\sqrt{x} + \log_b x^2$
26. $\log_4 3^3 + \log_4 9$
27. $2\log_e 2 + 3\log_e \pi - \log_e 3$
28. $\frac{1}{2}\log_b a - 2\log_b 5 - 3\log_b x$

In Exercises 29–36, determine the exact value of each of the given expressions.

29. $\log_2\left(\frac{1}{32}\right)$
30. $\log_3\left(\frac{2}{162}\right)$
31. $\log_2(2^{2.5})$
32. $\log_5(5^{0.1})$
33. $6\log_7\sqrt{7}$
34. $\pi\log_6\sqrt[3]{6}$
35. $4^{\log_4 8}$
36. $10^{2\log_{10}3}$

In Exercises 37–44, express each as a sum, difference, or multiple of logarithms. In each case, part of the logarithm may be determined exactly.

37. $\log_3 18$
38. $\log_5 375$
39. $\log_2\left(\frac{4}{7}\right)$
40. $\log_{10}(0.05)$
41. $\log_3\sqrt{6}$
42. $\log_2\sqrt[3]{24}$
43. $\log_{10}3000$
44. $3\log_{10}(40^2)$

In Exercises 45–56, solve for y in terms of x.

45. $\log_b y = \log_b 2 + \log_b x$
46. $\log_b y = \log_b 6 - \log_b x$
47. $\log_4 y = \log_4 x - \log_4 10 + \log_4 6$
48. $\log_3 y = -2\log_3(x + 1) + \log_3 7$
49. $\log_{10} y = 2\log_{10}7 - 3\log_{10}x$
50. $\log_b y = 3\log_b\sqrt{x} + 2\log_b 10$
51. $5\log_2 y - \log_2 x = 3\log_2 4 + \log_2 a$
52. $4\log_2 x - 3\log_2 y = \log_2 27$
53. $\log_2 x + \log_2 y = 1$
54. $\pi\log_4 x + \log_4 y = 1$
55. $2\log_5 x - \log_5 y = 2$
56. $\log_8 x = 2\log_8 y + 4$

In Exercises 57–70, solve the given problems.

57. Explain why $\log_{10}(x + 3)$ is not equal to $\log_{10}x + \log_{10}3$.
58. Express as the logarithm of a single quantity: $2\log_2(2x) - \log_2 x^2$. For what values of x is the value of this expression valid? Explain.

59. Display the graphs of $y = \log_e(e^2x)$ and $y = 2 + \log_e x$ on a calculator and explain why they are the same.

60. If $x = \log_b 2$ and $y = \log_b 3$, express $\log_b 12$ in terms of x and y.

61. If $\log_b x = 2$ and $\log_b y = 3$, find $\log_b \sqrt{x^2y^4}$.

62. Is it true that $\log_b(ab)^x = x \log_b a + x$?

63. Simplify: $\log_b(1 + b^{2x}) - \log_b(1 + b^{-2x})$

64. If $f(x) = \log_b x$, express $\frac{f(x + h) - f(x)}{h}$ as a single logarithm.

65. On the same screen of a calculator, display the graphs of $y_1 = \log_{10} x - \log_{10}(x^2 + 1)$ and $y_2 = \log_{10} \frac{x}{x^2 + 1}$. What conclusion can be drawn from the display?

66. The use of the insecticide DDT was banned in the United States in 1972. A computer analysis shows that an expression relating the amount A still present in an area, the original amount A_0, and the time t (in years) since 1972 is $\log_{10} A = \log_{10} A_0 + 0.1t \log_{10} 0.8$. Solve for A as a function of t.

67. A study of urban density shows that the population density D (in persons/mi^2) is related to the distance r (in mi) from the city center by $\log_e D = \log_e a - br + cr^2$, where a, b, and c are positive constants. Solve for D as a function of r.

68. When a person ingests a medication capsule, it is found that the rate R (in mg/min) that it enters the bloodstream in time t (in min) is given by $\log_{10} R - \log_{10} 5 = t \log_{10} 0.95$. Solve for R as a function of t.

69. A container of water is heated to 90°C and then placed in a room at 0°C. The temperature T of the water is related to the time t (in min) by $\log_e T = \log_e 90.0 - 0.23\,t$. Find T as a function of t.

70. In analyzing the power gain in a microprocessor circuit, the equation $N = 10(2 \log_{10} I_1 - 2 \log_{10} I_2 + \log_{10} R_1 - \log_{10} R_2)$ is used. Express this with a single logarithm on the right side.

Answers to Practice Exercises

1. $\log_3 2 + \log_3 5$ **2.** $\log_3 2 + \log_3 a - \log_3 5$
3. $\log_4(5/x^3)$ **4.** 6

13.4 Logarithms to the Base 10

Common Logarithms • Antilogarithm • Calculations Using Base 10 Logarithms

In Section 13.2, we stated that a base of logarithms must be a positive number, not equal to 1. In the examples and exercises of the previous sections, we used a number of different bases. However, there are two particular bases that are especially important in many applications. They are 10 and e, where e is the irrational number approximately equal to 2.718 that we introduced in Section 12.5 and have used in the previous sections of this chapter.

Base 10 logarithms were developed for calculational purposes and were used a great deal for making calculations until the 1970s, when the modern scientific calculator became widely available. Base 10 logarithms are still used in several scientific measurements, and therefore a need still exists for them. Base e logarithms are used extensively in technical and scientific work; we consider them in detail in the next section.

Logarithms to the base 10 are called **common logarithms.** They may be found directly on a calculator using the [log] key. This calculator key indicates the common notation. [When no base is shown, it is assumed to be the base 10.] NOTE ▸

EXAMPLE 1 Base 10 logarithm on a calculator

Using a calculator to find log 426, as shown in the display in Fig. 13.10, we find that

$$\log 426 = 2.629$$

no base shown means base is 10

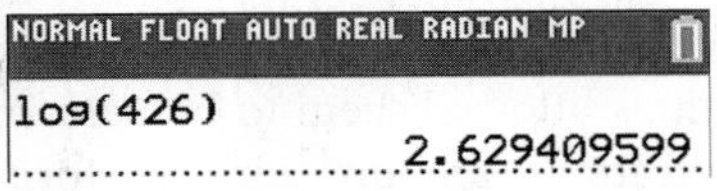

NORMAL FLOAT AUTO REAL RADIAN MP
log(426)
2.629409599

Fig. 13.10

NOTE ▸ when the result is rounded off. [The decimal part of a logarithm is normally expressed to the same accuracy as that of the number of which it is the logarithm.] In this case, since 426 has three significant digits, the decimal part of the answer (629) is rounded to three significant digits.

Because $10^2 = 100$ and $10^3 = 1000$, and in this case

$$10^{2.629} = 426$$

we see that the 2.629 power of 10 gives a number between 100 and 1000. ■

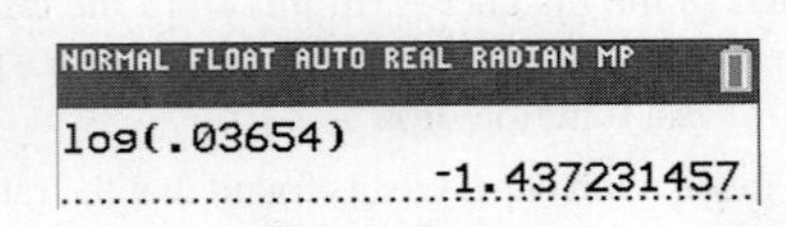

Fig. 13.11

EXAMPLE 2 Negative base 10 logarithm

Finding log 0.03654, as shown in the calculator display in Fig. 13.11, we see that

$$\log 0.03654 = -1.4372$$

We note that the logarithm here is negative. This should be the case when we recall the meaning of a logarithm. Raising 10 to a negative power gives us a number between 0 and 1, and here we have

$$10^{-1.4372} = 0.03654$$ ■

We may also use a calculator to find a number N if we know $\log N$. *In this case, we refer to N as the* **antilogarithm** *of $\log N$.* On the calculator, we use the 10^x key. We note that it shows the basic definition of a logarithm. (On many scientific calculators, the key sequence inv log is used. Note that this sequence shows that the exponential and logarithmic functions are inverse functions.)

EXAMPLE 3 Antilogarithm—inverse logarithm

Given $\log N = 1.1854$, we find N as shown in the first line of the calculator display in Fig. 13.12. Therefore,

$$N = 15.32$$

where the result has been rounded off. Because the given logarithm has four significant digits in its decimal part (1854), the antilogarithm is rounded to four significant digits. Since $10^1 = 10$ and $10^2 = 100$, we see that $N = 10^{1.1854}$ and is a number between 10 and 100. ■

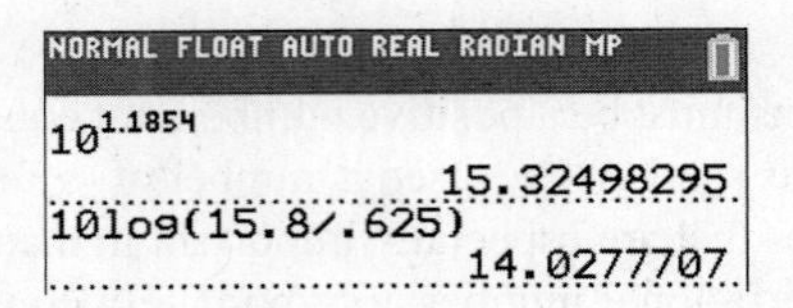

Fig. 13.12

The following example illustrates an application in which a measurement requires the direct use of the value of a logarithm.

EXAMPLE 4 Base 10 logarithm—power gain

The power gain G [in decibels (dB)] of an electronic device is given by $G = 10\log(P_0/P_i)$, where P_0 is the output power (in W) and P_i is the input power. Determine the power gain for an amplifier for which $P_0 = 15.8$ W and $P_i = 0.625$ W.

Substituting the given values, we have

$$G = 10\log\frac{15.8}{0.625} = 14.0 \text{ dB}$$

The second line of the display in Fig. 13.12 shows this evaluation on a calculator. ■

■ The unit of sound intensity level (used for power gain), the bel (B), is named for the U.S. inventor Alexander Graham Bell (1847–1922). The decibel is the commonly used unit.

Practice Exercises

Evaluate x using a calculator.

1. $x = \log 0.5392$ **2.** $\log x = 2.2901$

As noted in the chapter introduction, logarithms were developed for calculational purposes. They were first used in the seventeenth century for making tedious and complicated calculations that arose in astronomy and navigation. These complicated calculations were greatly simplified, because logarithms allowed them to be performed by means of basic additions, subtractions, multiplications, and divisions. Performing calculations in this way provides an opportunity to better understand the meaning and properties of logarithms. Also, certain calculations cannot be done directly on a calculator but can be done by logarithms.

EXAMPLE 5 Calculation using logarithms

A certain computer design has 64 different sequences of ten binary digits (either 0 or 1) so that the total number of possible states is $(2^{10})^{64} = 1024^{64}$. Evaluate 1024^{64} using logarithms. (It is very possible that your calculator cannot do this calculation directly.)

Because $\log x^n = n \log x$, we know that $\log 1024^{64} = 64 \log 1024$. Although most calculators will not directly evaluate 1024^{64}, we can use one to find the value of $64 \log 1024$. Because 1024^{64} is *exact,* we will show ten calculator digits until we round off the result. We therefore evaluate 1024^{64} as follows:

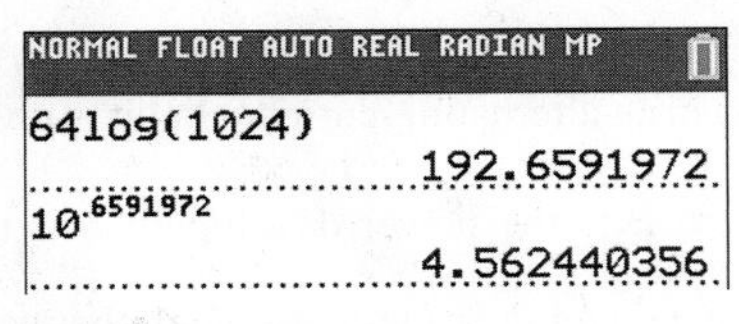

Fig. 13.13

$$\text{Let } N = 1024^{64}$$

$$\log N = \log 1024^{64} = 64 \log 1024 \quad \text{using } \log_b x^n = n \log_b x$$

$$= 192.6591972 \quad \text{calculator evaluation—see Fig. 13.13}$$

$$N = 10^{192.6591972} \quad \text{meaning of logarithm}$$

$$= 10^{192} \times 10^{0.6591972} \quad \text{using } b^u b^v = b^{u+v}$$

$$= (10^{192}) \times (4.5624) \quad \text{antilogarithm of 0.6591972 is 4.5624 (rounded off)—see Fig. 13.13}$$

$$= 4.5624 \times 10^{192}$$

Note that by using logarithms, we were able to obtain the answer and immediately express it in scientific notation. Calculating this directly (without logarithms) will cause an overflow error on many calculators due to the extreme size of the answer. ■

Example 5 shows that calculations using logarithms are based on Eqs. (13.7), (13.8), and (13.9). Multiplication is performed by addition of logarithms, division is performed by the subtraction of logarithms, and a power is found by a multiple of a logarithm. A root of a number is found by using the fractional exponent form of the power.

EXERCISES 13.4

In Exercises 1 and 2, find the indicated values if the given changes are made in the indicated examples of this section.

1. In Example 2, change 0.03654 to 0.3654 and then find the required value.

2. In Example 3, change 1.1854 to 2.1854 and then find the required value.

In Exercises 3–12, find the common logarithm of each of the given numbers by using a calculator.

3. 278 **4.** 0.0640

5. 9.24×10^6 **6.** 3.19^3

7. 1.174^{-4} **8.** 8.043×10^{-8}

9. $\sin 45.2°$ **10.** $\tan 12.6$

11. $\sqrt{274}$ **12.** $\log_2 16$

In Exercises 13–20, find the antilogarithm of each of the given logarithms by using a calculator.

13. 1.257 **14.** 0.929 **15.** −1.2154 **16.** −6.9788

17. 3.30112 **18.** 0.02436 **19.** −2.23746 **20.** −10.336

In Exercises 21–24, use logarithms to evaluate the given expressions.

21. 185^{100} **22.** $\dfrac{895}{73.4^{86}}$

23. $\dfrac{1}{247^{50}}$ **24.** $\dfrac{126{,}000^{20}}{2.63^{2.5}}$

In Exercises 25–28, use a calculator to verify the given values.

25. $\log 14 + \log 0.5 = \log 7$ **26.** $\log 500 - \log 20 = \log 25$

27. $\log 81 = 4 \log 3$ **28.** $\log 6 = 0.5 \log 36$

In Exercises 29–32, find the logarithms of the given numbers.

29. The signal used by some cell phones has a frequency of 9.00×10^8 Hz.

30. The diameter of the planet Jupiter is 1.43×10^8 m.

31. About $1.3 \times 10^{-14}\%$ of carbon atoms are carbon-14, the radioactive isotope used in determining the age of samples from ancient sites.

32. In an air sample taken in an urban area, 0.000005 of the air was carbon dioxide.

In Exercises 33–36, find the indicated values.

33. Find T (in K) if $\log T = 8$, where T is the temperature sufficient for nuclear fission.

34. Find v (in m/s) if $\log v = 8.4768$, where v is the speed of light.

35. Find e if $\log e = -0.35$, where e is the efficiency of a certain gasoline engine.

36. Find E (in J) if $\log E = -18.49$, where E is the energy of a photon of visible light.

In Exercises 37–42, solve the given problems by evaluating the appropriate logarithms.

37. Simplify: $\dfrac{\log_b x^2}{\log 100}$.

38. Evaluate: $\log(\log 10^{100})$ (10^{100} is called a googol.)

39. Evaluate: $2(10^{\log 0.1}) + 3(10^{\log 0.01})$.

40. A stereo amplifier has an input power of 0.750 W and an output power of 25.0 W. What is the power gain? (See Example 4.)

41. Measured on the Richter scale, the magnitude of an earthquake of intensity I is defined as $R = \log(I/I_0)$, where I_0 is a minimum level for comparison. What is the Richter scale reading for the 1964 Alaska earthquake for which $I = 1{,}600{,}000{,}000\ I_0$?

42. How many more times was the intensity of the 2011 earthquake in Japan ($R = 9.0$) than the 2015 earthquake in Nepal ($R = 7.8$)? (See Exercise 41.)

In Exercises 43 and 44, use logarithms to perform the indicated calculations.

43. A certain type of optical switch in a fiber-optic system allows a light signal to continue in either of two fibers. How many possible paths could a light signal follow if it passes through 400 such switches?

44. The peak current I_m (in A) in an alternating-current circuit is given by $I_m = \sqrt{\dfrac{2P}{Z\cos\phi}}$, where P is the power developed, Z is the magnitude of the impedance, and ϕ is the phase angle between the current and voltage. Evaluate I_m for $P = 5.25$ W, $Z = 320\ \Omega$, and $\phi = 35.4°$.

Answers to Practice Exercises

1. −0.2683 **2.** 195.0

13.5 Natural Logarithms

Natural Logarithm (ln x) • Change of Base • Calculator Values and Graphs

As we have noted, another number important as a base of logarithms is the number e. *Logarithms to the base e are called* **natural logarithms.** Since e is an irrational number equal to about 2.718, it may appear to be a very unnatural choice as a base of logarithms. However, in calculus, the reason for its choice and the fact that it is a very natural number for a base of logarithms are shown.

Just as $\log x$ refers to logarithms to the base 10, the notation $\ln x$ is used to denote logarithms to the base e. We briefly noted this in Section 13.2 in discussing the calculator. Due to the extensive use of natural logarithms, the notation **ln *x*** is more convenient than **log$_e$ *x***, although they mean the same thing.

Because more than one base is important, at times it is useful to change a logarithm from one base to another. If $u = \log_b x$, then $b^u = x$. Taking logarithms of both sides of this last expression to the base a, we have

$$\log_a b^u = \log_a x$$
$$u \log_a b = \log_a x$$
$$u = \frac{\log_a x}{\log_a b}$$

However, $u = \log_b x$. This leads us to the **change-of-base formula** shown below:

$$\log_b x = \frac{\log_a x}{\log_a b} \tag{13.13}$$

■ Equation (13.13) enables us to express the following relationships between common and natural logarithms:

$$\ln x = \frac{\log x}{\log e}$$

$$\log x = \frac{\ln x}{\ln 10}$$

NOTE ▶ [Equation (13.13) allows us to change a logarithm in one base to a logarithm in another base.] The following examples illustrate the method of performing this operation.

EXAMPLE 1 Change of base to find natural log

Change log 20 to a logarithm with base e; that is, find ln 20.

Using Eq. (13.13) with $a = 10$, $b = e$, and $x = 20$, we have

$$\log_e 20 = \frac{\log_{10} 20}{\log_{10} e}$$

or $$\ln 20 = \frac{\log 20}{\log e} = 2.996$$ see the calculator display in Fig. 13.14

```
NORMAL FLOAT AUTO REAL RADIAN MP
log(20)/log(e)
                    2.995732274
```

Fig. 13.14

■ On a TI-84, e is located above the division key.

This means that $e^{2.996} = 20$. ■

EXAMPLE 2 Change of base to find log to base 5

Find $\log_5 560$.

In Eq. (13.13), if we let $a = 10$, $b = 5$, and $x = 560$, we have

$$\log_5 560 = \frac{\log 560}{\log 5} = 3.932$$ see first line of calculator display in Fig. 13.15

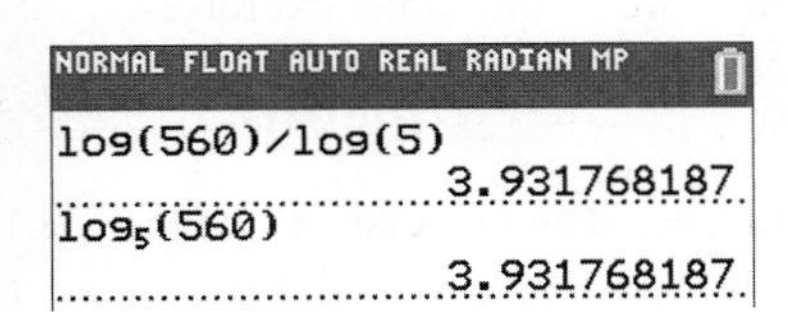

Fig. 13.15
Graphing calculator keystrokes for Examples 1 and 2: goo.gl/2fX7Y6

From the definition of a logarithm, this means that

$$5^{3.932} = 560$$

Note that some calculators allow for direct calculation of logarithms to any base as shown in the second line of Fig. 13.15. ■

Values of natural logarithms can be found directly on a calculator. The ln key is used for this purpose. In order to find the antilogarithm of a natural logarithm, we use the e^x key. The following example illustrates finding a natural logarithm and an antilogarithm on a graphing calculator.

Practice Exercise

1. Find $\log_3 23$.

EXAMPLE 3 Natural log on calculator

(a) From the first line of the calculator display in Fig. 13.16, we find that

$$\ln 236.5 = 5.4659$$

which means that $e^{5.4659} = 236.5$.

(b) Given that $\ln N = -0.8729$, we determine N by finding $e^{-0.8729}$ on the calculator. This gives us (see the second line of the display in Fig. 13.16)

$$N = 0.4177$$ ■

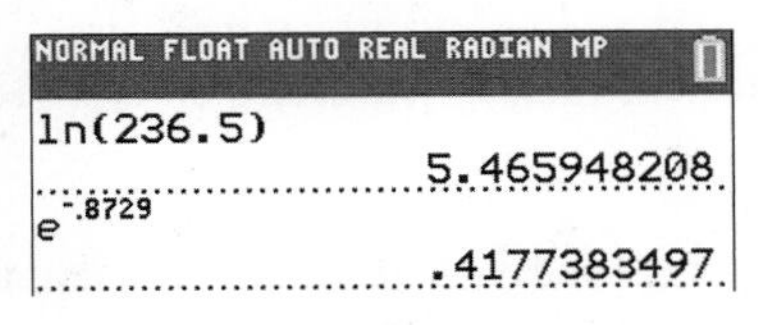

Fig. 13.16

Practice Exercise

2. If $\ln N = 1.8081$, find N.

For graphing calculators that don't allow direct entry of logarithms to any base, the change-of-base formula can be used to enter the function and display its graph. This is demonstrated in the next example.

EXAMPLE 4 Calculator graph of log to base 2

Display the graph of $y = 3 \log_2 x$ on a calculator.

To display this graph, we can use the fact that

$$\log_2 x = \frac{\log x}{\log 2} \quad \text{or} \quad \log_2 x = \frac{\ln x}{\ln 2}$$

Therefore, we enter the function

$$y = \frac{3 \log x}{\log 2} \quad \left(\text{or } y = \frac{3 \ln x}{\ln 2}\right)$$

in the calculator. Some calculators allow the function to be entered directly as $y = \log_2 x$. Figure 13.17(a) shows three equivalent forms of the function. They all produce the same graph, which is shown in Fig. 13.17(b). ■

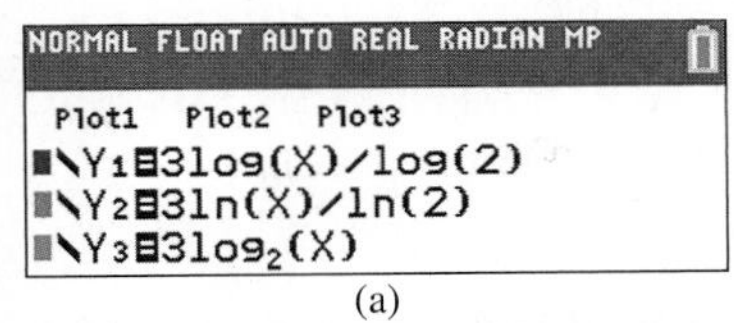

(a)

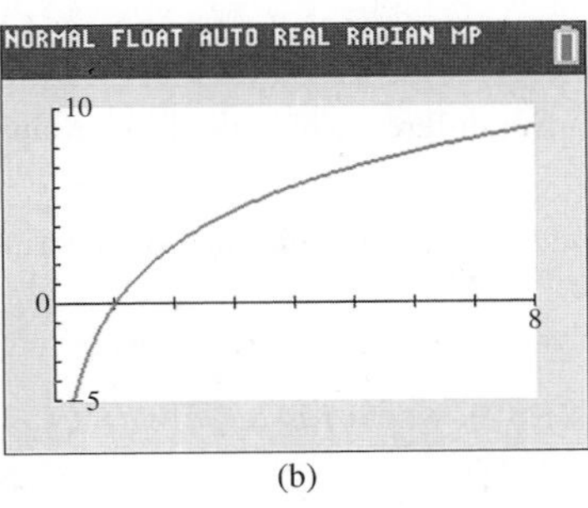

(b)

Fig. 13.17

In Section 13.3, we introduced the properties $\log_b(b^x) = x$ and $b^{\log_b x} = x$. If we let $b = e$, we get the following special cases of these properties, which can be used to simplify certain expressions:

$$\ln(e^x) = x \tag{13.14}$$

$$e^{\ln x} = x \tag{13.15}$$

Practice Exercise

3. Use a calculator to evaluate $\ln 57.2$.

EXAMPLE 5 Simplifying with base e

Using Eq. (13.14), $\ln(e^2) = 2$ and $\ln(e^{0.03t}) = 0.03t$

Using Eq. (13.15), $e^{\ln 7} = 7$ and $e^{\ln 2y} = 2y$ ■

Applications of natural logarithms are found in many fields of technology. One such application is shown in the next example, and others are found in the exercises.

EXAMPLE 6 Natural log—electric current

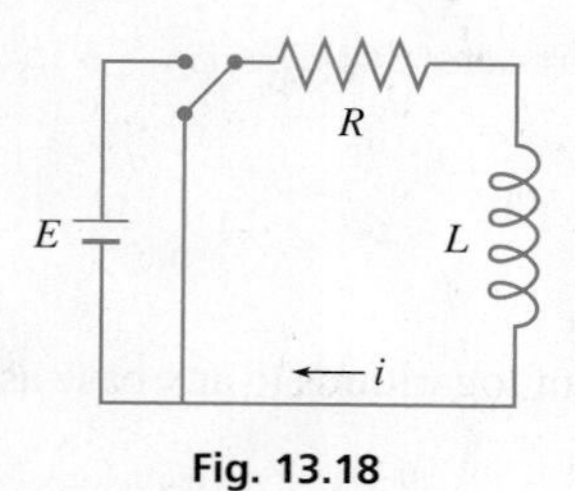

Fig. 13.18

The electric current i in a circuit containing a resistance and an inductance (see Fig. 13.18) is given by $\ln(i/I) = -Rt/L$, where I is the current at $t = 0$, R is the resistance, t is the time, and L is the inductance. Calculate how long (in s) it takes i to reach 0.430 A, if $I = 0.750$ A, $R = 7.50\ \Omega$, and $L = 1.25$ H.

Solving for t and then evaluating, we have

$$t = -\frac{L\ln(i/I)}{R} = -\frac{L(\ln i - \ln I)}{R} \quad \text{either form can be used}$$

$$= -\frac{1.25(\ln 0.430 - \ln 0.750)}{7.50} = 0.0927 \text{ s} \quad \text{evaluating}$$

Therefore, the current changes from 0.750 A to 0.430 A in 0.0927 s. ■

EXERCISES 13.5

In Exercises 1 and 2, find the indicated values if the given changes are made in the indicated examples of this section.

1. In Example 1, change 20 to 200 and then evaluate.

2. In Example 2, change the base to 4 and then evaluate.

In Exercises 3–8, use logarithms to the base 10 to find the natural logarithms of the given numbers.

3. 26.0 **4.** 6310 **5.** 1.562

6. 0.5017 **7.** 0.0073267 **8.** 4.438×10^{-4}

In Exercises 9–14, use logarithms to the base 10 to find the indicated logarithms.

9. $\log_7 52$ **10.** $\log_2 86$ **11.** $\log_\pi 245$

12. $\log_{12} 0.122$ **13.** $\log_{40} 7.50 \times 10^2$ **14.** $\log_{100} 3720$

In Exercises 15–22, find the natural logarithms of the given numbers.

15. 76.1 **16.** 293 **17.** 1.394

18. 6552 **19.** 0.8926 **20.** 2.086×10^{-3}

21. $(0.012937)^4$ **22.** $\sqrt{0.000060808}$

In Exercises 23–30, find the natural antilogarithms of the given logarithms.

23. 2.190 **24.** 5.420 **25.** 0.0084210

26. 0.632 **27.** −0.7429 **28.** −2.94218

29. −23.504 **30.** -8.04×10^{-3}

In Exercises 31–34, use a calculator to display the indicated graphs.

31. The graph of $y = \log_5 x$ **32.** The graph of $y = 2\log_8 x$

33. Graphically show that $y = 2^x$ and $y = \log_2 x$ are inverse functions. (See Example 9 of Section 13.2.)

34. Explain what is meant by the expression $\ln \ln x$. Display the graph of $y = \ln \ln x$ on a calculator.

In Exercises 35–38, use a calculator to verify the given values.

35. $\ln 5 + \ln 8 = \ln 40$ **36.** $2\ln 6 - \ln 3 = \ln 12$

37. $4\ln 3 = \ln 81$ **38.** $\ln 5 - 0.5\ln 25 = \ln 1$

In Exercises 39–54, solve the given problems.

39. Evaluate: $\sqrt{\ln e^9}$.

40. Solve for y in terms of x: $\ln y + 2\ln x = 1 + \ln 5$.

41. Solve for x: $\ln(\log x) = 0$.

42. If $\ln x = 3$ and $\ln y = 4$, find $\sqrt{x^2 y}$.

43. If $x = \ln 4$ and $y = \ln 5$, express $\ln 80$ in terms of x and y.

44. Express $\ln(xe^{-x})$ as the sum or difference of logarithms, evaluating where possible.

45. Express $\ln(e^2\sqrt{1 - x})$ as the sum or difference of logarithms, evaluating where possible.

46. Simplify: $2\ln(e^{3x+1}) - e^{\ln(2x-3)}$.

47. Find f (in Hz) if $\ln f = 21.619$, where f is the frequency of the microwaves in a microwave oven.

48. Find k (in 1/Pa) if $\ln k = -21.504$, where k is the compressibility of water.

49. If interest is compounded continuously (daily compounded interest closely approximates this), with an interest rate i, a bank account will double in t years according to $i = (\ln 2)/t$. Find i if the account is to double in 8.5 years.

50. World population is currently growing by 1.1% annually. If it continues at this rate, the time (in years) for the population to double in size is given by $t = \frac{\ln 2}{\ln 1.011}$. How many years is this (to the nearest year)?

51. For the electric circuit of Example 6, find how long it takes the current to reach 0.1 of the initial value of 0.750 A.

52. The heat loss rate Q (in W/m) through a certain cylindrical pipe insulation is given by $Q = \frac{2\pi(0.040)(T_1 - T_2)}{\ln(r_2/r_1)}$, where T_1 is the temperature inside the insulation, T_2 is the temperature outside the insulation, and r_1 and r_2 are the inside and outside radii of the insulation, respectively. Find the heat loss rate if $T_1 = 145°\text{C}$, $T_2 = 15.0°\text{C}$, $r_1 = 1.50$ in., and $r_2 = 2.50$ in.

53. The distance x traveled by a motorboat in t seconds after the engine is cut off is given by $x = k^{-1}\ln(kv_0t + 1)$, where v_0 is the velocity of the boat at the time the engine is cut and k is a constant. Find how long it takes a boat to go 150 m if $v_0 = 12.0$ m/s and $k = 6.80 \times 10^{-3}$/m.

54. The electric current i (in A) in a circuit that has a 1-H inductor, a 10-Ω resistor, and a 6-V battery, and the time t (in s) are related by the equation $10t = -\ln(1 - i/0.6)$. Solve for i.

Answers to Practice Exercises

1. 2.854 **2.** 6.099 **3.** 4.05

13.6 Exponential and Logarithmic Equations

Exponential Equations • Logarithmic Equations

EXPONENTIAL EQUATIONS

An equation in which the variable occurs in an exponent is called an **exponential equation.** Although some may be solved by changing to logarithmic form, they are generally solved by *taking the logarithm of each side* and then using the properties of logarithms.

EXAMPLE 1 Exponential equation—solved two ways

(a) We can solve the exponential equation $2^x = 8$ by writing it in logarithmic form. This gives

$$x = \log_2 8 = 3 \qquad 2^3 = 8$$

This method is good if we can directly evaluate the resulting logarithm.

(b) Because 2^x and 8 are equal, the logarithms of 2^x and 8 are also equal. Therefore, we can also solve $2^x = 8$ in a more general way by taking logarithms (to any proper base) of both sides and equating these logarithms. This gives us

$$\log 2^x = \log 8 \quad \text{or} \quad \ln 2^x = \ln 8$$

$$x\log 2 = \log 8 \qquad x\ln 2 = \ln 8 \qquad \text{using } \log_b(x^n) = n\log_b x$$

$$x = \frac{\log 8}{\log 2} = 3 \qquad x = \frac{\ln 8}{\ln 2} = 3 \qquad \text{using a calculator}$$

■

EXAMPLE 2 Solving an exponential equation

Solve the equation $3^{x-2} = 5$.

Taking logarithms of each side and equating them, we have

$$\log 3^{x-2} = \log 5$$

$$(x - 2)\log 3 = \log 5 \qquad \text{using } \log_b(x^n) = n\log_b x$$

$$x = 2 + \frac{\log 5}{\log 3} = 3.465$$

This solution means that

$$3^{3.465-2} = 3^{1.465} = 5$$

which can be checked by a calculator. ■

Practice Exercise

1. Solve for x: $2^{x+1} = 7$

EXAMPLE 3 Solving an exponential equation

Solve the equation $2(4^{x-1}) = 17^x$.

TI-89 graphing calculator keystrokes for Example 3: goo.gl/4Vv5Gn

By taking logarithms of each side, we have the following:

$$\log 2 + \log(4^{x-1}) = \log 17^x \qquad \text{using } \log_b xy = \log_b x + \log_b y$$

$$\log 2 + (x-1)\log 4 = x \log 17 \qquad \text{using } \log_b(x^n) = n \log_b x$$

$$x \log 4 - x \log 17 = \log 4 - \log 2$$

$$x(\log 4 - \log 17) = \log 4 - \log 2$$

$$x = \frac{\log 4 - \log 2}{\log 4 - \log 17} = -0.479$$

■

EXAMPLE 4 Exponential equation—atmospheric pressure

At constant temperature, the atmospheric pressure p (in Pa) at an altitude h (in m) is given by $p = p_0 e^{kh}$, where p_0 is the pressure where $h = 0$ (usually taken as sea level). Given that $p_0 = 101.3$ kPa (atmospheric pressure at sea level) and $p = 68.9$ kPa for $h = 3050$ m, find the value of k.

Because the equation is defined in terms of e, we can solve it most easily by taking natural logarithms of *each side*. By doing this, we have the following solution:

$$\ln p = \ln(p_0 e^{kh}) = \ln p_0 + \ln e^{kh} \qquad \text{using } \log_b xy = \log_b x + \log_b y$$

$$= \ln p_0 + kh \qquad \text{using } \ln e^x = x$$

$$\ln p - \ln p_0 = kh$$

$$k = \frac{\ln p - \ln p_0}{h}$$

Substituting the given values, we have

$$k = \frac{\ln(68.9 \times 10^3) - \ln(101.3 \times 10^3)}{3050} = -0.000126/\text{m}$$

■

LOGARITHMIC EQUATIONS

Some of the important measurements in scientific and technical work are defined in terms of logarithms. Using these formulas can lead to solving a **logarithmic equation,** *which is an equation with the logarithm of an expression involving the variable.* To solve logarithmic equations, we usually wish to isolate the logarithm and then convert to exponential form. If more than one logarithm appears in the equation, then the properties of logarithms are used to combine them into a single logarithm before proceeding. The following examples illustrate this process.

EXAMPLE 5 Logarithmic equation—loudness of sound

The human ear responds to sound on a scale that is approximately proportional to the logarithm of the intensity of the sound. Therefore, the loudness of sound (measured in dB) is defined by the equation $b = 10 \log(I/I_0)$, where I is the intensity of the sound and I_0 is the minimum intensity detectable.

■ Decibel levels of some everyday sounds:

Sound	Decibels (dB)
Whisper	20
Normal speech	60
Lawn mower	100
Aircraft at takeoff	180

A busy street has a loudness of 70 dB, and riveting has a loudness of 100 dB. To find how many times greater the intensity I_r of the sound of riveting is than the intensity I_c of the sound of the city street, we substitute in the above equation. This gives

$$70 = 10\log\left(\frac{I_c}{I_0}\right) \quad \text{and} \quad 100 = 10\log\left(\frac{I_r}{I_0}\right)$$

To solve for I_c and I_r, we divide each side by 10 and then use exponential form:

$$7.0 = \log\left(\frac{I_c}{I_0}\right) \quad \text{and} \quad 10 = \log\left(\frac{I_r}{I_0}\right)$$

$$\frac{I_c}{I_0} = 10^{7.0} \qquad \frac{I_r}{I_0} = 10^{10}$$

$$I_c = I_0(10^{7.0}) \qquad I_r = I_0(10^{10})$$

■ This demonstrates that sound intensity levels change much more than loudness levels. (See Exercise 56.)

Because we want the number of times I_r is greater than I_c, we divide I_r by I_c:

$$\frac{I_r}{I_c} = \frac{I_0(10^{10})}{I_0(10^{7.0})} = \frac{10^{10}}{10^{7.0}} = 10^{3.0} \quad \text{or} \quad I_r = 10^{3.0}I_c = 1000I_c$$

Thus, the sound of riveting is 1000 times as intense as the sound of the city street. ■

Practice Exercise

2. How many times greater is the sound of a loud car radio (80 dB) than ordinary conversation (60 dB)?

EXAMPLE 6 Logarithmic equation—population growth

■ See the chapter introduction.

An analysis of the population of Canada led to the equation $\log_2 P = \log_2 35.9 + 0.0115t$, where P is the projected population (in millions) and t is the number of years after 2015. Determine the projected population in 2025.

The solution is as follows:

$$\log_2 P - \log_2 35.9 = 0.0115(10)$$

$$\log_2(P/35.9) = 0.115 \qquad \text{using } \log_b\left(\frac{x}{y}\right) = \log_b x - \log_b y$$

$$P/35.9 = 2^{0.115} \qquad \text{exponential form}$$

$$P = 35.9(2^{0.115}) = 38.9 \text{ million} \qquad \text{projected 2025 population}$$

■

EXAMPLE 7 Solving a logarithmic equation

TI-89 graphing calculator keystrokes for Example 7: goo.gl/dnceMY

Solve the logarithmic equation $2\ln 2 + \ln x = \ln 3$.

Using the properties of logarithms, we have the following solution:

$$2\ln 2 + \ln x = \ln 3$$

$$\ln 2^2 + \ln x - \ln 3 = 0 \qquad \text{using } \log_b(x^n) = n\log_b x$$

$$\ln\frac{4x}{3} = 0 \qquad \text{using } \log_b xy = \log_b x + \log_b y \text{ and } \log_b\left(\frac{x}{y}\right) = \log_b x - \log_b y$$

$$\frac{4x}{3} = e^0 = 1 \qquad \text{exponential form}$$

$$4x = 3, \qquad x = 3/4$$

Because $\ln(3/4) = \ln 3 - \ln 4$, this solution checks in the *original equation*. ■

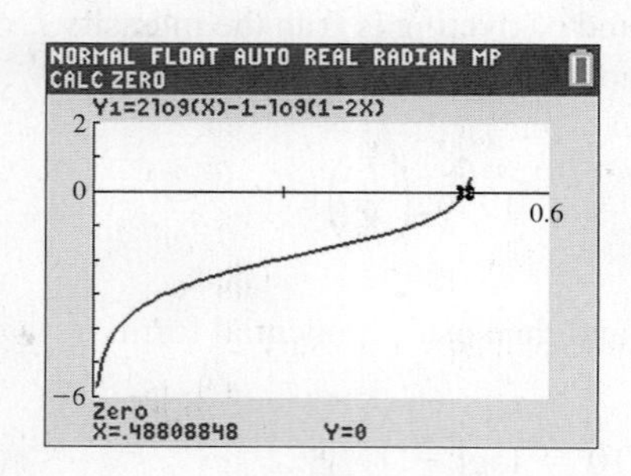

Fig. 13.19
Graphing calculator keystrokes: goo.gl/m06FdH

EXAMPLE 8 Solving a logarithmic equation

Solve the logarithmic equation $2 \log x - 1 = \log(1 - 2x)$.

$$\log x^2 - \log(1 - 2x) = 1$$

$$\log \frac{x^2}{1 - 2x} = 1 \qquad \text{using } \log_b\left(\frac{x}{y}\right) = \log_b x - \log_b y$$

$$\frac{x^2}{1 - 2x} = 10^1 \qquad \text{exponential form}$$

$$x^2 = 10 - 20x$$

$$x^2 + 20x - 10 = 0$$

$$x = \frac{-20 \pm \sqrt{400 + 40}}{2} = -10 \pm \sqrt{110}$$

NOTE ➧ [Note that logarithms of negative numbers are not defined and $-10 - \sqrt{110}$ is negative and cannot be used in the first term of the original equation.] Therefore, the only solution is

$$x = -10 + \sqrt{110} = 0.488.$$

The calculator graph of $y = 2 \log x - 1 - \log(1 - 2x)$ is shown in Fig. 13.19. The zero displayed on the screen shows that our answer is correct. ■

Practice Exercise

3. Solve for x: $3 \log 2 - \log(x + 1) = 1$

EXERCISES 13.6

In Exercises 1 and 2, find the indicated values if the given changes are made in the indicated examples of this section.

1. In Example 2, change the sign in the exponent from − to + and then solve the equation.

2. In Example 7, change ln 3 to ln 6 and then solve the equation.

In Exercises 3–32, solve the given equations.

3. $2^x = 16$ **4.** $3^x = 1/81$ **5.** $3.50^x = 82.9$

6. $\pi^x = 15$ **7.** $3^{-x} = 0.525$ **8.** $e^{-x} = 17.54$

9. $6^{x+2} = 85$ **10.** $5^{x-1} = 0.07$ **11.** $3(14^x) = 400$

12. $5(0.8^x) = 2$ **13.** $0.6^x = 2^{x^2}$ **14.** $15.6^{x+2} = 23^x$

15. $3^{x^2+8} = 27^{2x}$ **16.** $8^x = 4^{x^2-1}$ **17.** $3 \log_8 x = -2$

18. $5 \log_{32} x = -3$ **19.** $3 \log_2(x - 1) = 12$

20. $5 \log_6(7x + 1) = 10$ **21.** $\log_2 x + \log_2 7 = \log_2 21$

22. $2 \log_2 3 - \log_2 x = \log_2 45$ **23.** $2 \log(3 - x) = 1$

24. $9 \log(2x - 1) = 3$ **25.** $\log 12x^2 - \log 3x = 3$

26. $\ln x - \ln(1/3) = 1$ **27.** $3 \ln 2 + \ln(x - 1) = \ln 24$

28. $\log_2 x + \log_2(x + 2) = 3$ **29.** $\frac{1}{2} \log(x + 2) + \log 5 = 1$

30. $2 \log_x 2 + \log_2 x = 3$

31. $\log(2x - 1) + \log(x + 4) = 1$

32. $\ln(2x - 1) - 2 \ln 4 = 3 \ln 2$

In Exercises 33–42, use a calculator to solve the given equations.

33. $15^{-x} = 1.326$ **34.** $e^{2x} = 3.625$

35. $4(3^x) = 5$ **36.** $5^{x+2} = 3e^{2x}$

37. $3 \ln 2x = 2$ **38.** $\log 4x + \log x = 2$

39. $2 \ln 2 - \ln x = -1$ **40.** $\log(x - 3) + \log x = \log 4$

41. $2^{2x} - 2^x - 6 = 0$ **42.** $9^x - 3^x - 12 = 0$

In Exercises 43–62, solve the given problems.

43. If $4^x = 5$, find y if $y = 4^{-3x}$.

44. If $2^{-x} = 7$, find y if $y = 4^{2x}$.

45. Solve for x: $e^x + e^{-x} = 3$. (*Hint:* Multiply each term by e^x and then it can be treated as a quadratic equation in e^x.)

46. Solve for x: $3^x + 3^{-x} = 4$. See Exercise 45.

47. If $y = 1.5e^{-0.90x}$, find y when $x = 7.1$.

48. What values of x cannot be solutions of the equation $y = \log(2x - 5) + \log(x^2 + 1)$?

49. Use logarithms to find the x-intercept of the graph of $y = 3 - 4^{x+2}$.

50. Use a calculator to find the point of intersection of the curves of $3x + 5y = 6$ and $y = 1.5 \ln(x + 3)$.

51. According to one model, the number N of Americans (in millions) age 65 and older that will have Alzheimer's disease t years after 2015 is given by $N = 5.1(1.03)^t$. In what year will this number reach 8.0 million?

52. Referring to Exercise 66 in Section 13.3, in what year will the amount of DDT be 25% of the original amount?

53. Forensic scientists determine the temperature T (in °C) of a body t hours after death from the equation $T = T_0 + (37 - T_0)0.97^t$, where T_0 is the air temperature. If a body is discovered at midnight with a body temperature of 27°C in a room at 22°C, at what time did death occur?

54. When a camera flash goes off, the batteries recharge the flash's capacitor to a charge Q according to $Q = Q_0(1 - e^{-kt})$, where Q_0 is the maximum charge. How long does it take to recharge the capacitor to 90% of capacity if $k = 0.5$?

55. In chemistry, the pH value of a solution is a measure of its acidity. The pH value is defined by $\text{pH} = -\log(\text{H}^+)$, where H^+ is the hydrogen-ion concentration. If the pH of a sample of rainwater is 4.764, find the hydrogen-ion concentration. (If $\text{pH} < 7$, the solution is acid. If $\text{pH} > 7$, the solution is basic.) Acid rain has a pH between 4 and 5, and normal rain is slightly acidic with a pH of about 5.6.

56. Referring to Example 5, show that if the difference in loudness of two sounds is d decibels, the louder sound is $10^{d/10}$ more intense than the quieter sound.

57. Measured on the Richter scale, the magnitude of an earthquake of intensity I is defined as $R = \log(I/I_0)$, where I_0 is a minimum level for comparison. How many times I_0 was the 1906 San Francisco earthquake whose magnitude was 8.3 on the Richter scale?

58. How many more times intense was the 1906 San Francisco earthquake ($R = 8.3$) than the 1935 Timiskaming earthquake (felt in Ontario, Quebec, and northeast United States) ($R = 6.1$)?

59. Studies have shown that the concentration c (in mg/cm^3 of blood) of aspirin in a typical person is related to the time t (in h) after the aspirin reaches maximum concentration by the equation $\ln c = \ln 15 - 0.20t$. Solve for c as a function of t.

60. In an electric circuit containing a resistor and a capacitor with an initial charge q_0, the charge q on the capacitor at any time t after closing the switch can be found by solving the equation $\ln q = -\dfrac{t}{RC} + \ln q_0$. Here, R is the resistance, and C is the capacitance. Solve for q as a function of t.

61. An Earth satellite loses 0.1% of its remaining power each week. An equation relating the power P, the initial power P_0, and the time t (in weeks) is $\ln P = t \ln 0.999 + \ln P_0$. Solve for P as a function of t.

62. The atmospheric pressure P (in bars) at an altitude h (in km) above Earth's surface can be estimated using the equation $P = e^{-h/7}$. Solve for h as a function of P. At what elevation is the pressure 0.35 bars?

Many exponential and logarithmic equations cannot be solved algebraically as we did in this section. However, they can be solved graphically as was shown in the check of Example 8. In Exercises 63–66, solve the given equations graphically by use of a calculator.

63. $2^x + 3^x = 50$

64. $4^x + x^2 = 25$

65. The curve in which a uniform wire or rope hangs under its own weight is called a *catenary*. An example of a catenary that we see every day is a wire strung between utility poles, as shown in Fig. 13.20. For a particular wire, the equation of the catenary it forms is $y = 2(e^{x/4} + e^{-x/4})$, where (x, y) is a point on the curve. Find x for $y = 5.8$ m.

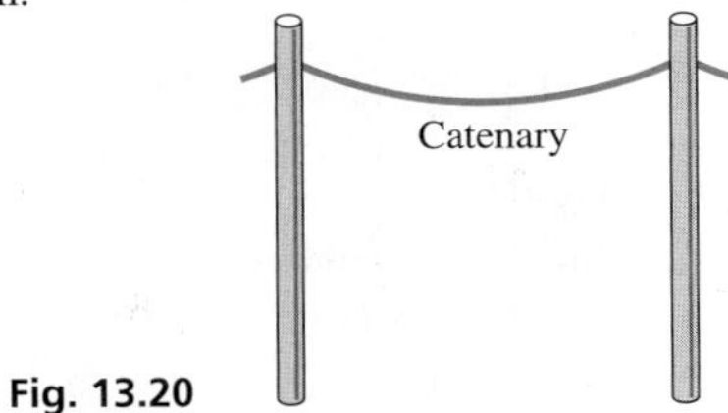

Fig. 13.20

66. A computer analysis of sales of a product showed that the net profit p (in \$1000) during a given year was $p = 2\ln(t + 2) - 0.5t$, where t is in months. Graphically determine the date when $p = 0$.

Answers to Practice Exercises

1. 1.807 **2.** 100 **3.** $-1/5$

13.7 Graphs on Logarithmic and Semilogarithmic Paper

Logarithmic Scale • Semilogarithmic (Semilog) Paper • Logarithmic (Log-Log) Paper

When constructing the graphs of some functions, one of the variables changes much more rapidly than the other. We saw this in graphing the exponential and logarithmic functions in Sections 13.1 and 13.2. The following example illustrates this point.

EXAMPLE 1 Graph of exponential function

Plot the graph of $y = 4(3^x)$.

Constructing the following table of values,

x	−1	0	1	2	3	4	5
y	1.3	4	12	36	108	324	972

we then plot these values as shown in Fig. 13.21(a).

We see that as x changes from −1 to 5, y changes much more rapidly, from about 1 to nearly 1000. Also, because of the scale that must be used, we see that it is not possible to show accurately the differences in the y-values on the graph.

Even a calculator cannot show the graph accurately for the values near $x = 0$, if we wish to view all of this part of the curve. In fact, the calculator view shows the curve as being on the axis for these values, as we see in Fig. 13.21(b). Note in this figure that we have shown the same values of the range of the function as we did in Fig. 13.21(a). ■

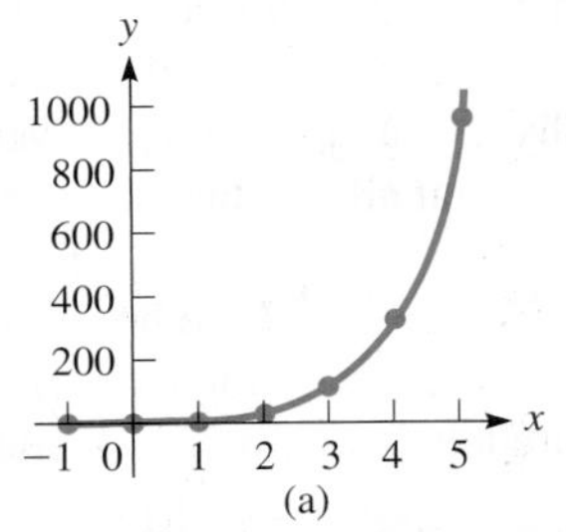

(a)

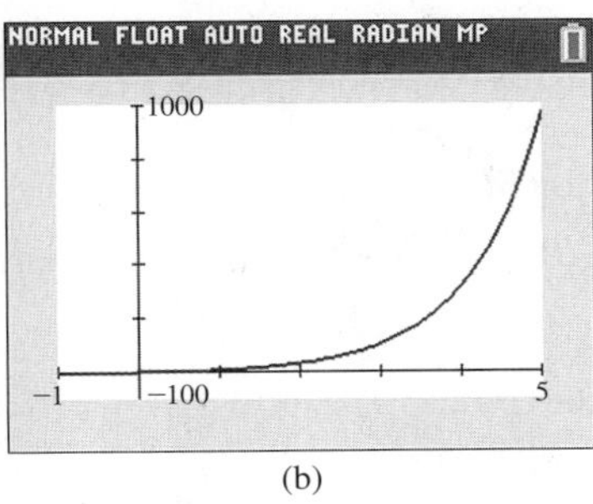

(b)

Fig. 13.21

It is possible to graph a function with a large change in values, for one or both variables, more accurately than can be done on the standard rectangular coordinate system. This is done by using a scale marked off in distances proportional to the logarithms of the values being represented. *Such a scale is called a* **logarithmic scale.** For example, $\log 1 = 0$, $\log 2 = 0.301$, and $\log 10 = 1$. Thus, on a logarithmic scale, the 2 is placed 0.301 unit of distance from the 1 to the 10. Figure 13.22 shows a logarithmic scale with the numbers represented and the distance used for each.

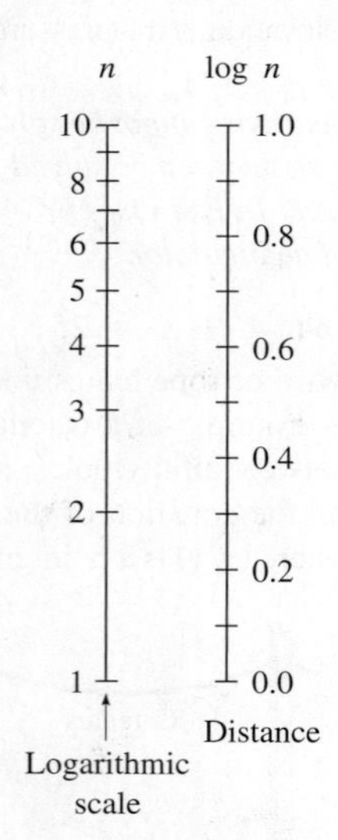

Fig. 13.22

On a logarithmic scale, the distances between the integers are not equal, but this scale does allow for a much greater range of values and much greater accuracy for many of the values. There is another advantage to using logarithmic scales. Many equations that would have more complex curves when graphed on the standard rectangular coordinate system will have simpler curves, often straight lines, when graphed using logarithmic scales. In many cases, this makes the analysis of the curve much easier.

Zero and negative numbers do not appear on the logarithmic scale. In fact, all numbers used on the logarithmic scale must be positive, because the domain of the logarithmic function includes only positive real numbers. Thus, the logarithmic scale must start at some number greater than zero. This number is a power of 10 and can be very small, say, $10^{-6} = 0.000001$, but it is positive.

If we wish to use a large range of values for only one of the variables, we use what is known as **semilogarithmic,** or **semilog,** graph paper. On this graph paper, only one axis (usually the y-axis) uses a logarithmic scale. If we wish to use a large range of values for both variables, we use **logarithmic,** or **log-log,** graph paper. Both axes are marked with logarithmic scales.

The following examples illustrate semilog and log-log graphs, which are used in many technical and scientific areas to display coordinates of plotted points more clearly.

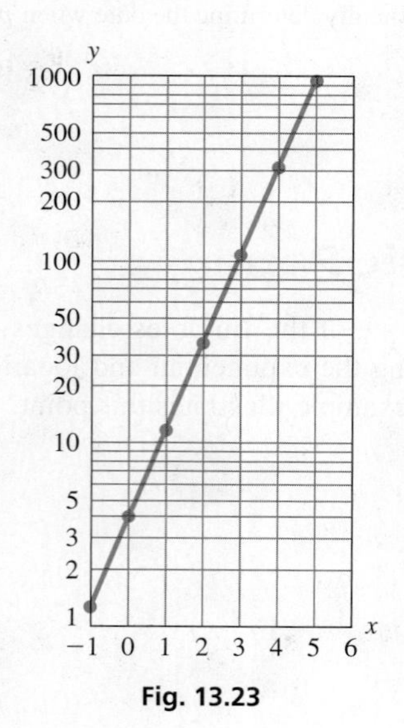

Fig. 13.23

EXAMPLE 2 Graph on semilogarithmic paper

Construct the graph of $y = 4(3^x)$ on semilogarithmic graph paper.

This is the same function as in Example 1, and we repeat the table of values:

x	−1	0	1	2	3	4	5
y	1.3	4	12	36	108	324	972

Again, we see that the range of y-values is large. When we plotted this curve on the rectangular coordinate system in Example 1, we had to use large units along the y-axis. This made the values of 1.3, 4, 12, and 36 appear at practically the same level. However, when we use semilog graph paper, we can label each axis such that all y-values are accurately plotted as well as the x-values.

The logarithmic scale is shown in **cycles,** and we must label the base line of the first cycle as 1 times a power of 10 (0.01, 0.1, 1, 10, 100, and so on) with the following cycle labeled with the next power of 10. The lines between are labeled with 2, 3, 4, and so on, times the proper power of 10. See the vertical scale in Fig. 13.23. We now plot the points in the table on the graph. The resulting graph is a straight line, as we see in Fig. 13.23. Taking logarithms of each side of the equation, we have

$$\log y = \log[4(3^x)] = \log 4 + \log 3^x \quad \text{using } \log_b xy = \log_b x + \log_b y$$
$$= \log 4 + x \log 3 \quad \text{using } \log_b(x^n) = n \log_b x$$

However, because $\log y$ was plotted automatically (because we used semilogarithmic paper), the graph really represents

$$u = \log 4 + x \log 3$$

where $u = \log y$; $\log 3$ and $\log 4$ are constants, and therefore this equation is of the form $u = mx + b$, which is a straight line (see Section 5.1).

The logarithmic scale in Fig. 13.23 has *three cycles,* because all values of three powers of 10 are represented. ■

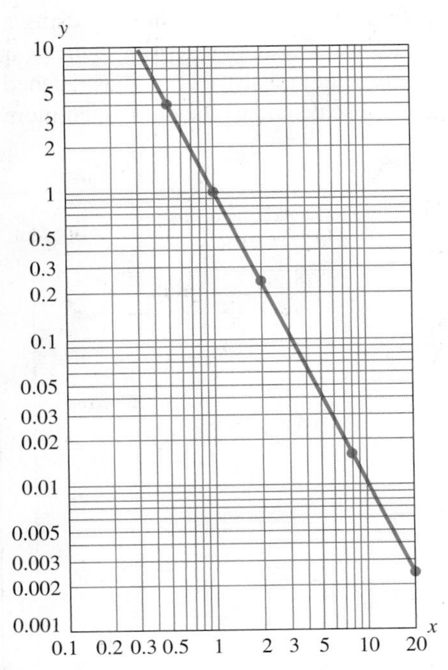

Fig. 13.24

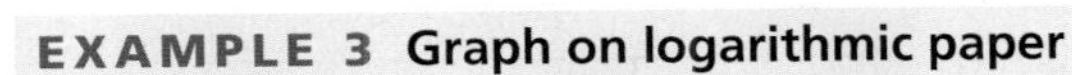

EXAMPLE 3 Graph on logarithmic paper

Construct the graph of $x^4y^2 = 1$ on logarithmic paper.

First, we solve for y and make a table of values. Considering positive values of x and y, we have

$$y = \sqrt{\frac{1}{x^4}} = \frac{1}{x^2}$$

x	0.5	1	2	8	20
y	4	1	0.25	0.0156	0.0025

We plot these values on log-log paper on which both scales are logarithmic, as shown in Fig. 13.24. We again see that we have a straight line. Taking logarithms of both sides of the equation, we have

$$\log(x^4y^2) = \log 1$$

$$\log x^4 + \log y^2 = 0 \qquad \text{using } \log_b xy = \log_b x + \log_b y$$

$$4 \log x + 2 \log y = 0 \qquad \text{using } \log_b(x^n) = n \log_b x$$

If we let $u = \log y$ and $v = \log x$, we then have

$$4v + 2u = 0 \quad \text{or} \quad u = -2v$$

which is the equation of a straight line, as shown in Fig. 13.24. Note, however, that not all graphs on logarithmic paper are straight lines. ■

EXAMPLE 4 Graph on log-log paper—beam deflection

The deflection (in ft) of a certain cantilever beam as a function of the distance x (in ft) from one end is

$$d = 0.0001(30x^2 - x^3)$$

If the beam is 20.0 ft long, plot a graph of d as a function of x on log-log paper.

Constructing a table of values, we have

x (ft)	1.00	1.50	2.00	3.00	4.00
d (ft)	0.00290	0.00641	0.0112	0.0243	0.0416

x (ft)	5.00	10.0	15.0	20.0
d (ft)	0.0625	0.200	0.338	0.400

Because the beam is 20.0 ft long, there is no meaning to values of x greater than 20.0 ft. The graph is shown in Fig. 13.25. ■

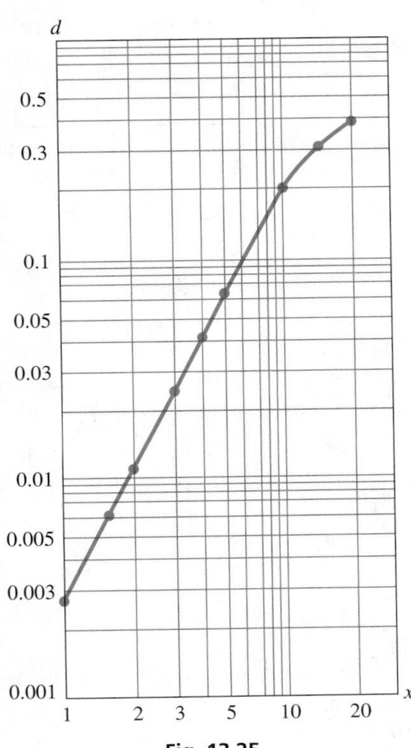

Fig. 13.25

Logarithmic and semilogarithmic paper may be useful for plotting data. Often, the data cover too large a range of values to be plotted on ordinary graph paper. The next example illustrates the use of semilogarithmic paper to plot data.

EXAMPLE 5 Excel scatterplot on semilog scale—transistors

The number of transistors in computer processors has increased dramatically over the years. The Excel spreadsheet in Fig. 13.26 shows some selected processors, the date of release, and the number of transistors. Also shown is a scatterplot of the data using a semilog scale. If plotted on a standard scale (see Fig. 13.27), the points prior to the year 2000 all appear to lie on the x-axis due to the fact that the transistor count has reached extremely high numbers in recent years. The semilog scale shows the data much more clearly.

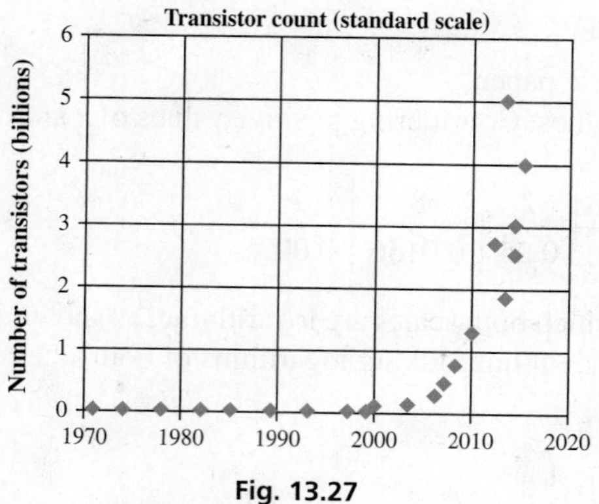

Fig. 13.27

Processor	Date	Transistor count
Intel 4004	1971	2,300
Intel 8080	1974	4,500
Intel 8086	1978	29,000
Intel 80286	1982	134,000
Intel 80386	1985	275,000
Intel 80486	1989	1,200,000
Intel Pentium	1993	3,100,000
AMD K6	1997	8,800,000
Intel Pentium III	1999	9,500,000
Intel Pentium 4	2000	42,000,000
AMD K8	2003	105,900,000
Intel Core 2 Duo	2006	291,000,000
AMD K10	2007	463,000,000
Intel Core i7	2008	731,000,000
IBM Power 7	2010	1,200,000,000
IBM zEC12	2012	2,750,000,000
Intel Ivy Bridge E	2013	1,860,000,000
Xbox One	2013	5,000,000,000
Intel Haswell E	2014	2,600,000,000
Apple A8X	2014	3,000,000,000
IBM z13	2015	3,990,000,000

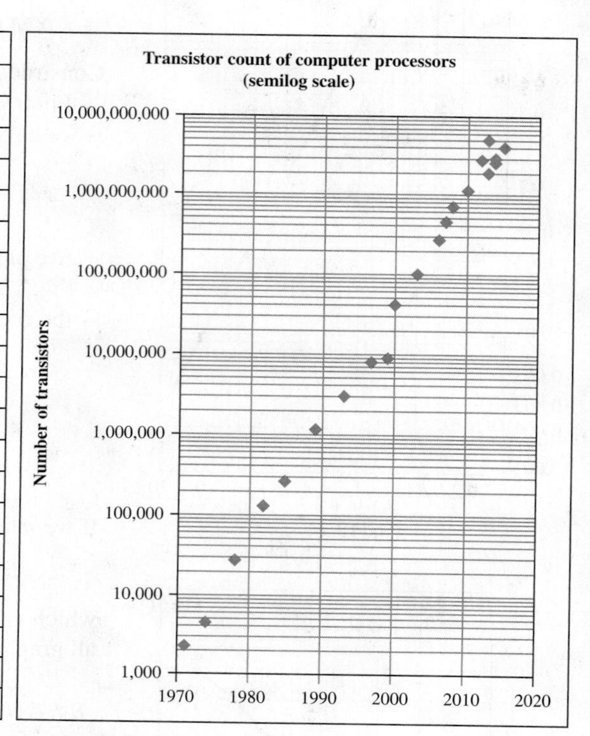

Fig. 13.26

EXERCISES 13.7

In Exercises 1 and 2, make the given changes in the indicated examples, and then draw the graphs.

1. In Example 2, change the 4 to 2 and then make the graph.
2. In Example 3, change the 1 to 4 and then make the graph.

In Exercises 3–10, plot the graphs of the given functions on semilogarithmic paper.

3. $y = 2^x$
4. $y = 3(5^x)$
5. $y = 5(4^{-x})$
6. $y = 6^{-x}$
7. $y = x^3$
8. $y = 2x^4$
9. $y = 2x^3 + 6x$
10. $y = 4x^3 + 2x^2$

In Exercises 11–18, plot the graphs of the given functions on log-log paper.

11. $y = 0.01x^4$
12. $y = \sqrt{x}$
13. $y = x^{2/3}$
14. $y = 8x^{0.25}$
15. $xy = 40$
16. $x^2y^3 = 16$
17. $x^2y^2 = 25$
18. $x^3y = 8$

In Exercises 19–26, determine the type of graph paper on which the graph of the given function is a straight line. Using the appropriate paper, sketch the graph.

19. $y = 3^{-x}$
20. $y = 0.2x^3$
21. $y = 3x^6$
22. $y = 5(10^{-x})$
23. $y = 4^{x/2}$
24. $xy^3 = 10$
25. $x\sqrt{y} = 4$
26. $y(2^x) = 3$

In Exercises 27–38, plot the indicated graphs.

27. On the moon, the distance s (in ft) a rock will fall due to gravity is $s = 2.66t^2$, Where t is the time (in s) of fall. Plot the graph of s as a function of t for $0 \le t \le 10$ s on (a) a regular rectangular coordinate system and (b) a semilogarithmic coordinate system.
28. By pumping, the air pressure in a tank is reduced by 18% each second. Thus, the pressure p (in kPa) in the tank is given by $p = 101(0.82)^t$, where t is the time (in s). Plot the graph of p as a function of t for $0 \le t \le 30$ s on (a) a regular rectangular coordinate system and (b) a semilogarithmic coordinate system.

29. Strontium-90 decays according to the equation $N = N_0e^{-0.028t}$, where N is the amount present after t years and N_0 is the original amount. Plot N as a function of t on semilog paper if $N_0 = 1000$ g.

30. The electric power P (in W) in a certain battery as a function of the resistance R (in Ω) in the circuit is given by $P = \dfrac{100R}{(0.50 + R)^2}$.

Plot P as a function of R on semilog paper, using the logarithmic scale for R and values of R from 0.01 Ω to 10 Ω.

31. The acceleration g (in m/s^2) produced by the gravitational force of Earth on a spacecraft is given by $g = 3.99 \times 10^{14}/r^2$, where r is the distance from the center of Earth to the spacecraft. On log-log paper, graph g as a function of r from $r = 6.37 \times 10^6$ m (Earth's surface) to $r = 3.91 \times 10^8$ m (the distance to the moon).

32. In undergoing an adiabatic (no *heat* gained or lost) expansion of a gas, the relation between the pressure p (in kPa) and the volume v (in m^3) is $p^2v^3 = 850$. On log-log paper, graph p as a function of v from $v = 0.10$ m^3 to $v = 10$ m^3.

33. The number of cell phone subscribers in the United States from 1994 to 2015 is shown in the following table. Plot N as a function of the year on semilog paper.

Year	1994	1997	2000	2003	2006	2009	2012	2015
$N(\times 10^6)$	24.1	55.3	109	159	233	275	300	359

34. The period T (in years) and mean distance d (given as a ratio of that of Earth) from the sun to the planets (Mercury, Venus, Earth, Mars, Jupiter, Saturn, Uranus, Neptune, Pluto) are given below. Plot T as a function of d on log-log paper. (Note that Pluto is currently considered to be a *dwarf planet*.)

Planet	M	V	E	M	J	S	U	N	P
d	0.39	0.72	1.00	1.52	5.20	9.54	19.2	30.1	39.5
T	0.24	0.62	1.00	1.88	11.9	29.5	84.0	165	249

35. The intensity level B (in dB) and the frequency f (in Hz) for a sound of constant loudness were measured as shown in the table that follows. Plot the data for B as a function of f on semilog paper, using the log scale for f.

f (Hz)	100	200	500	1000	2000	5000	10,000
B (dB)	40	30	22	20	18	24	30

36. The atmospheric pressure p (in kPa) at a given altitude h (in km) is given in the following table. On semilog paper, plot p as a function of h.

h (km)	0	10	20	30	40
p (kPa)	101	25	6.3	2.0	0.53

37. One end of a very hot steel bar is sprayed with a stream of cool water. The rate of cooling R (in °F/s) as a function of the distance d (in in.) from one end of the bar is then measured, with the results shown in the following table. On log-log paper, plot R as a function of d. Such experiments are made to determine the hardness of steel.

d (in.)	0.063	0.13	0.19	0.25
R (°F/s)	600	190	100	72

d (in.)	0.38	0.50	0.75	1.0	1.5
R (°F/s)	46	29	17	10	6.0

38. The magnetic intensity H (in A/m) and flux density B (in teslas) of annealed iron are given in the following table. Plot H as a function of B on log-log paper. (The unit tesla is named in honor of Nikola Tesla. See Chapter 20 introduction.)

B (T)	0.0042	0.043	0.67	1.01
H (A/m)	10	50	100	150

B (T)	1.18	1.44	1.58	1.72
H (A/m)	200	500	1000	10,000

In Exercises 39 and 40, plot the indicated semilogarithmic graphs for the following application.

In a particular electric circuit, called a low-pass filter, the input voltage V_i is across a resistor and a capacitor, and the output voltage V_0 is across the capacitor (see Fig. 13.28). The voltage gain G (in dB) is given by

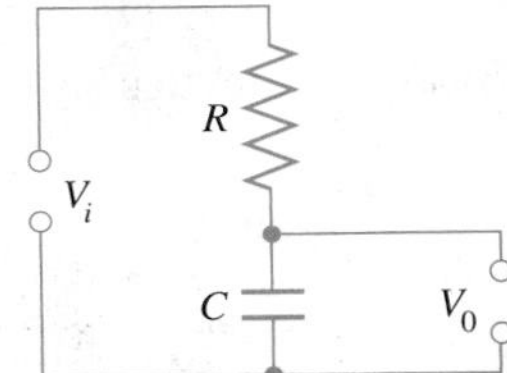

Fig. 13.28

$$G = 20 \log \frac{1}{\sqrt{1 + (\omega T)^2}}$$

where $\tan \phi = -\omega T$.

Here, ϕ is the phase angle of V_0/V_i. For values of ωT of 0.01, 0.1, 0.3, 1.0, 3.0, 10.0, 30.0, and 100, plot the indicated graphs. These graphs are called a Bode diagram for the circuit.

39. Calculate values of G for the given values of ωT and plot a semilogarithmic graph of G vs. ωT.

40. Calculate values of ϕ (as negative angles) for the given values of ωT and plot a semilogarithmic graph of ϕ vs. ωT.

CHAPTER 13 KEY FORMULAS AND EQUATIONS

Exponential function $y = b^x$ (13.1)

Logarithmic form $x = \log_b y$ (13.2)

Logarithmic function $y = \log_b x$ (13.3)

Laws of exponents $b^u b^v = b^{u+v}$ (13.4)

$\dfrac{b^u}{b^v} = b^{u-v}$ (13.5)

$(b^u)^n = b^{nu}$ (13.6)

Properties of logarithms $\log_b xy = \log_b x + \log_b y$ (13.7)

$\log_b\left(\dfrac{x}{y}\right) = \log_b x - \log_b y$ (13.8)

$\log_b(x^n) = n \log_b x$ (13.9)

$\log_b 1 = 0 \quad \log_b b = 1$ (13.10)

$\log_b(b^x) = x$ (13.11)

$b^{\log_b x} = x$ (13.12)

Change-of-base-formula $\log_b x = \dfrac{\log_a x}{\log_a b}$ (13.13)

Base *e* properties $\ln(e^x) = x$ (13.14)

$e^{\ln x} = x$ (13.15)

CHAPTER 13 REVIEW EXERCISES

CONCEPT CHECK EXERCISES

Determine each of the following as being either **true** *or* **false**. *If it is false, explain why.*

1. $y = -6^{-x}$ is an exponential function.
2. The logarithmic form of $y = -2^{-x}$ is $x = -\log_{-2} y$.
3. $\log\dfrac{4}{5} = \dfrac{\log 4}{\log 5}$
4. $\log\left(\dfrac{1}{100}\right) = -2$
5. $2 \ln x = \dfrac{\ln x^2}{-\ln e}$
6. If $2 \ln x - 1 = \ln x$, then $x = e$.

PRACTICE AND APPLICATIONS

In Exercises 7–18, determine the value of x.

7. $\log_{10} x = 4$
8. $\log_9 x = 3$
9. $\log_5 x = -1$
10. $\log_4(\sin x) = -0.5$
11. $2 \log_{1/2} 8 = x$
12. $\log_{12} 144 = x - 3$
13. $\log_8 32 = x + 1$
14. $\log_9 27 = x$
15. $\log_x 36 = 2$
16. $\log_x(1/243) = 5$
17. $\log_x 10 = \frac{1}{3}$
18. $\log_x 8 = 0$

In Exercises 19–30, express each as a sum, difference, or multiple of logarithms. Wherever possible, evaluate logarithms of the result.

19. $\log_3 6x$
20. $\log_5\left(\dfrac{7}{a^3}\right)$
21. $\log_3(t^2)$
22. $\log_6 \sqrt{5}$
23. $\log_2 56$
24. $\log_7 196$
25. $\log_4 \sqrt{48}$
26. $\log \sqrt[4]{32y}$
27. $\log_3\left(\dfrac{9}{x}\right)$
28. $\log_6\left(\dfrac{5}{36}\right)$
29. $\log_{10}(1000x^4)$
30. $\log_3(9^2 \times 6^3)$

In Exercises 31–42, solve for y in terms of x.

31. $\log_6 y = \log_6 4 - \log_6 x$
32. $6 \ln y = 3 \ln e^2 - 9 \ln x$
33. $3 \ln y = 2 + 3 \ln x$
34. $2(\log_9 y + 2 \log_9 x) = 1$
35. $\log_3 y = \frac{1}{2} \log_3 7 + \frac{1}{2} \log_3 x$
36. $(\log_2 3)(\log_2 y) - \log_2 x = 3$
37. $\log_5 x + \log_5 y = \log_5 3 + 1$
38. $\log_7 y = 2 \log_7 5 + \log_7 x + 3$

39. $2(\log_4 y - 3\log_4 x) = 3$

40. $\dfrac{\log_7 x}{\log_7 4} - \log_7 y = 1$

41. $\log_x y = \ln e^3$

42. $10^y = 3^{x+1}$

In Exercises 43–50, display the graphs of the given functions on a calculator.

43. $y = 0.5(5^x)$

44. $y = 3(2^{-x})$

45. $R = 0.2\log_4 r$

46. $y = 10\log_{16} x$

47. $y = \log_{3.15} x$

48. $s = 0.1\log_{4.05} t$

49. $y = 1 - e^{-|x|}$

50. $s = 2(1 - e^{-0.2t})$

In Exercises 51–54, use a calculator to find each logarithm.

51. $\log(0.0824)$

52. $\log(3.16 \times 10^5)$

53. $\ln(87.9)$

54. $\ln(0.0052)$

In Exercises 55–58, use the change-of-base formula to find each logarithm.

55. $\log_3(594)$

56. $\log_7(0.019)$

57. $\log_{4.25}(0.0067)$

58. $\log_{5.02}(852.1)$

In Exercises 59–66, solve the given equations.

59. $e^{2x} = 5$

60. $2(5^x) = 15$

61. $3^{x+2} = 5^x$

62. $2^x/3^{1-x} = 12^{x+1}$

63. $\log_4 z + \log_4 6 = \log_4 12$

64. $\log_8(x + 2) = 2 - \log_8 2$

65. $2\log_3 2 - \log_3(x + 1) = \log_3 5$

66. $\log(n + 2) + \log n = 0.4771$

In Exercises 67 and 68, plot the graphs of the given functions on semilogarithmic paper. In Exercises 69 and 70, plot the graphs of the given functions on log-log paper.

67. $y = 8^x$

68. $y = 5x^3$

69. $y = \sqrt[3]{x}$

70. $xy^4 = 16$

In Exercises 71–74, evaluate the given expressions using the property $b^{\log_b x} = x$.

71. $10^{\log 4}$

72. $2e^{\ln 7.5}$

73. $3e^{2\ln 2}$

74. $5(10^{2\log 3})$

In Exercises 76–112, solve the given problems.

75. If $\log N = 1.513$ and $\ln P = 3.796$, find $N + P$.

76. Use a calculator to verify that $3\ln 2 + 0.5\ln 64 = 3\ln 4$.

77. Evaluate $\sqrt[3]{\ln e^8} - \sqrt{\log 10^4}$.

78. Solve for x: $2^x + 32(2^{-x}) = 12$.

79. If $f(x) = 2\log_b x$ and $f(8) = 3$, find $f(4)$.

80. Evaluate: $\log(2\log 1000)$.

81. Evaluate: $7(10^{\log 0.1}) + 6000(100^{\log 0.001})$.

82. If $x = \log_b 7$ and $y = \log_b 2$, express $\log_b 42$ in terms of x and y.

83. Use a calculator to solve the equation $\log_5 x = 2x - 7$.

84. Use logarithms to find the x-intercept of the graph of $y = 2 - 5^{2x-3}$.

85. The bending moment M (in N · m) of a particular concrete column is given by $\ln M = 15.34$. What is the value of M?

86. The temperature T (in °C) of the coffee in a cup t min after being heated in a microwave was $T = 22.5 + 76.0e^{-0.15t}$. Display the graph of this function on a calculator.

87. If A_0 dollars are invested at an annual interest rate of r (in %), the value of A after n years is $A = A_0(1 + r)^n$. (a) Solve for n. (b) How many years does it take A to double if $r = 4.00\%$?

88. The current i (in μA) in a microchip circuit is $i = 25(2^{-2500t})$, where t is the time (in ns) after the current is turned off. How long does it take the current to be 0.25 μA?

89. The formula $\ln(I/I_0) = -\beta h$ is used in estimating the thickness of the ozone layer. Here, I_0 is the intensity of a wavelength of sunlight before reaching Earth's atmosphere, I is the intensity of the light after passing through h cm of the ozone layer, and β is a constant. Solve for I.

90. The amount A of cesium-137 (a dangerous radioactive element) remaining after t years is given by $A = A_0(0.5^{t/30.3})$, where A_0 is the initial amount. In what year will the cesium-137 be 10% of that released at the Chernobyl disaster in 1986?

91. The amount A of alcohol in a person's bloodstream is given by $A = A_0e^{kt}$, where A_0 is an initial amount (mg alcohol/mL blood), t is the time (in h) after drinking the alcohol, and k is a constant depending on the person. If $A_0 = 0.24$ mg/mL, and $A = 0.16$ mg/mL in 2.0 h, how long should the person wait to drive if the legal limit is 0.08 mg/mL?

92. A state lottery pays \$500 for a \$1 ticket if a person picks the correct three-digit number determined by the random draw of three numbered balls. To find the number of drawings x required to have a 50% chance of winning, we must solve the equation $1 - 0.999^x = 0.5$. For how many drawings does a person have to buy a ticket to have a 50% chance of winning?

93. An approximate formula for the population (in millions) of the United States since 2000 is $P = 283e^{0.0085t}$, where t is the number of years since 2000. According to this model, in what year will the U.S. population reach 400 million?

94. A computer analysis of the luminous efficiency E (in lumens/W) of a tungsten lamp as a function of its input power P (in W) is given by $E = 22.0(1 - 0.65e^{-0.008P})$. Sketch the graph of E as a function of P for $0 \leq P \leq 1000$ W.

95. An original amount of 100 mg of radium radioactively decomposes such that N mg remain after t years. The function relating t and N is $t = 2350(\ln 100 - \ln N)$. Sketch the graph.

96. The time t (in s) to chemically change 5 kg of a certain substance into another is given by $t = -5\log\left(\dfrac{5 - x}{5}\right)$, where x is the number of kilograms that have been changed at any time. Sketch the graph.

97. An equation that may be used for the angular velocity ω of the slider mechanism in Fig. 13.29 is $2\ln\omega = \ln 3g + \ln\sin\theta - \ln l$, where g is the acceleration due to gravity. Solve for $\sin\theta$.

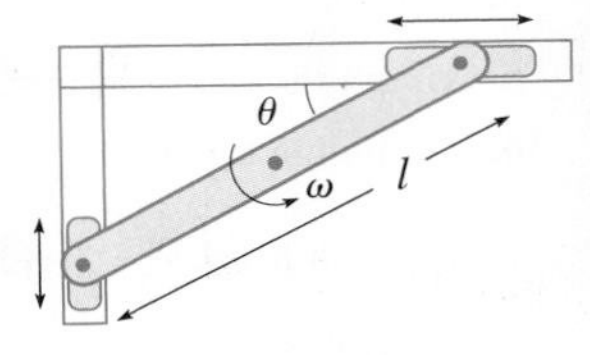

Fig. 13.29

98. Taking into account the weight loss of fuel, the maximum velocity v_m of a rocket is $v_m = u(\ln m_0 - \ln m_s) - gt_f$, where m_0 is the initial mass of the rocket and fuel, m_s is the mass of the rocket shell, t_f is the time during which fuel is expended, u is the velocity of the expelled fuel, and g is the acceleration due to gravity. Solve for m_0.

99. An equation used to calculate the capacity C (in bits/s) of a telephone channel is $C = B\log_2(1 + R)$. Solve for R.

100. To find the number n of years that an initial value A of equipment takes to have salvage value S, the equation $n = \dfrac{\log S - \log A}{\log(1 - d)}$ is used. Here, d is the annual rate of depreciation. Solve for d.

101. The magnitudes (visual brightnesses), m_1 and m_2, of two stars are related to their (actual) brightnesses, b_1 and b_2, by the equation $m_1 - m_2 = 2.5\log(b_2/b_1)$. As a result of this definition, magnitudes may be negative, and *magnitudes decrease as brightnesses increase.* The magnitude of the brightest star, Sirius, is -1.4, and the magnitudes of the faintest stars observable with the naked eye are about 6.0. How much brighter is Sirius than these faintest stars?

102. The power gain of an electronic device such as an amplifier is defined as $n = 10\log(P_0/P_i)$, where n is measured in decibels, P_0 (in W) is the power output, and P_i (in W) is the power input. If $P_0 = 10.0$ W and $P_i = 0.125$ W, calculate the power gain. (See Example 4 in Section 13.4.)

103. In studying the frictional effects on a flywheel, the revolutions per minute R that it makes as a function of the time t (in min) is given by $R = 4520(0.750)^{2.50t}$. Find t for $R = 1950$ r/min.

104. The efficiency e of a gasoline engine as a function of its compression ratio r is given by $e = 1 - r^{1-\gamma}$, where γ is a constant. Find γ for $e = 0.55$ and $r = 7.5$.

105. The intensity I of light decreases from its value I_0 as it passes a distance x through a medium. Given that $x = k(\ln I_0 - \ln I)$, where k is a constant depending on the medium, find x for $I = 0.850I_0$ and $k = 5.00$ cm.

106. Under certain conditions, the temperature T and pressure p are related by the following equation, where T_0 is the temperature at pressure p_0: $\dfrac{T}{T_0} = \left(\dfrac{p}{p_0}\right)^{\frac{k-1}{k}}$. Solve for k.

107. The temperature T of an object with an initial temperature T_1 in water at temperature T_0 as a function of the time t is given by $T = T_1 + (T_0 - T_1)e^{-kt}$. Solve for t.

108. According to one projection, the number of users (in millions) of the Internet in North America is given by $y = 200\log(3.47 + 1.80t)$, where t is the number of years after 2000. On a calculator, display the graph of this function from 2000 to 2020.

109. Pure water is running into a brine solution, and the same amount of solution is running out. The number n of kilograms of salt in the solution after t min is found by solving the equation $\ln n = -0.04t + \ln 20$. Solve for n as a function of t.

110. For the circuit in Fig. 13.30, the current i (in mA) is given by $i = 1.6e^{-100t}$. Plot the graph of i as a function of t for the first 0.05 s on semilog paper.

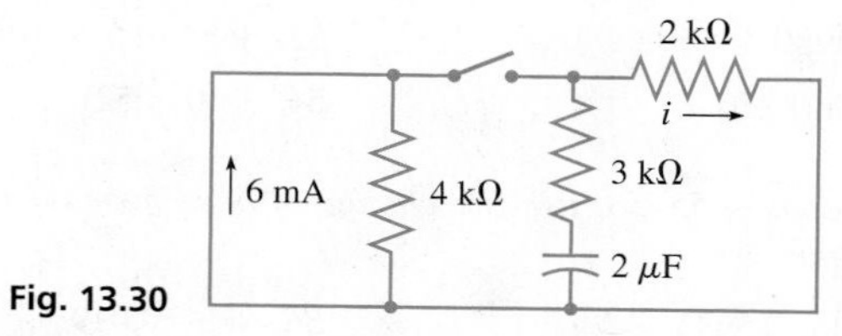

Fig. 13.30

111. For a particular solar-energy system, the collector area A required to supply a fraction F of the total energy is given by $A = 480\,F^{2.2}$. Plot A (in m^2) as a function of F, from $F = 0.1$ to $F = 0.9$, on semilog paper.

112. The current I (in μA) and resistance R (in Ω) were measured as follows in a certain microcomputer circuit:

$R(\Omega)$	100	200	500	1000	2000	5000	10,000
$I(\mu A)$	81	41	16	8.2	4.0	1.6	0.8

Plot I as a function of R on log-log paper.

113. While checking logarithmic curves on a calculator, a machine-design student noted that a certain robotic arm was shaped like part of the graph of $y = \ln(2x^{2/3})$. As a check, the student re-wrote the equation as $y = (2\ln x)/3 + \ln 4 - \ln(\ln e^2)$. Write a paragraph explaining (a) whether the second equation is equivalent to the first, and (b) if the graphs of the two equations are identical.

CHAPTER 13 PRACTICE TEST

As a study aid, we have included complete solutions for each Practice Test problem at the end of this book.

In Problems 1–4, determine the value of x.

1. $\log_9 x = -\frac{1}{2}$

2. $\log_3 x - \log_3 2 = 2$

3. $\log_x 64 = 3$

4. $3^{3x+1} = 8$

5. Graph the function $y = 2\log_4 x$.

6. Graph the function $y = 2(3^x)$ on semilog paper.

7. Express $\log_5\left(\dfrac{4a^3}{7}\right)$ as a combination of a sum, difference, and multiple of logarithms, including $\log_5 2$.

8. Solve for y in terms of x: $3\log_7 x - \log_7 y = 2$.

9. An equation used for a certain electric circuit is $\ln i - \ln I = -t/RC$. Solve for i.

10. Evaluate: $\dfrac{2\ln 0.9523}{\log 6066}$.

11. Evaluate: $\log_5 732$.

12. If A_0 dollars are invested at 8%, compounded continuously for t years, the value A of the investment is given by $A = A_0e^{0.08t}$. Determine how long it takes for the investment to double in value.

Additional Types of Equations and Systems of Equations

14

LEARNING OUTCOMES

After completion of this chapter, the student should be able to:

- Graph conic sections using a calculator
- Solve systems involving conics and other nonlinear equations graphically
- Solve systems involving nonlinear equations algebraically by the method of substitution or by the method of addition and subtraction
- Solve application problems involving systems of nonlinear equations
- Solve equations in quadratic form
- Solve equations with radicals
- Solve application problems involving equations in quadratic form or equations with radicals

In this chapter, we discuss graphical and algebraic solutions of systems of equations of types different from those of earlier chapters. We also consider solutions of two special types of equations.

One of the methods includes solving systems of equations by finding points of intersection between two graphs. The intersection of curves was very much part of the basic method of solution of equations used in the mid-1600s by Rene Descartes, who developed the coordinate system. Since that time, graphical solutions of equations have been very common and very useful in science and technology.

In the study of optics in the 1800s, it was found that light traveled much faster in free space than in other mediums, such as glass and water. Scientists then defined n, the *index of refraction*, to be the ratio of the speed of light in free space to the speed of light in a particular substance. In 1836, the French mathematician Cauchy developed the equation $n = A + B\lambda^{-2} + C\lambda^{-4}$ that related the index of refraction with the wavelength of light λ in a medium. Once the constants A, B, and C were found for a particular medium (by solving three simultaneous equations as in Chapter 5), this equation can be solved for λ for a particular value of n using a method known at the time that is used in this chapter (see Exercise 38 on page 414). Again, we see that an earlier mathematical method was useful in dealing with a new scientific discovery.

Applications of the types of equations and systems of equations of this chapter are found in many fields of science and technology. These include physics, electricity, business, and structural design.

◀ **In most cell phone specifications, the screen size is given by its diagonal length. In Section 14.3, we will use this information along with the area to find the dimensions of the screen.**

14.1 Graphical Solution of Systems of Equations

Conic Sections • Parabola, Circle, Ellipse, Hyperbola • Solving Systems of Equations Graphically

In this section, we first discuss the graphs of the *circle, parabola, ellipse*, and *hyperbola*, which are known as the **conic sections.** We saw the parabola earlier when discussing quadratic functions in Chapter 7. Then we will find graphical solutions of systems of equations involving these and other nonlinear equations.

EXAMPLE 1 Calculator graph—parabola

Graph the equation $y = 3x^2 - 6x$.

We graphed equations of this form in Section 7.4. Because the general quadratic function is $y = ax^2 + bx + c$, for $y = 3x^2 - 6x$, we have $a = 3$, $b = -6$, and $c = 0$. Therefore, $-b/(2a) = 1$, which means the x-coordinate of the vertex is $-(-6)/6 = 1$. Because $y = -3$ for $x = 1$, the vertex is $(1, -3)$. It is a minimum point since $a > 0$.

Knowing the vertex and the fact that the graph goes through the origin ($c = 0$), we choose appropriate *window* settings and have the display shown in Fig. 14.1.

As we showed in Section 7.4, the curve is a **parabola,** and a parabola always results if the equation is of the form of the quadratic function $y = ax^2 + bx + c$. ■

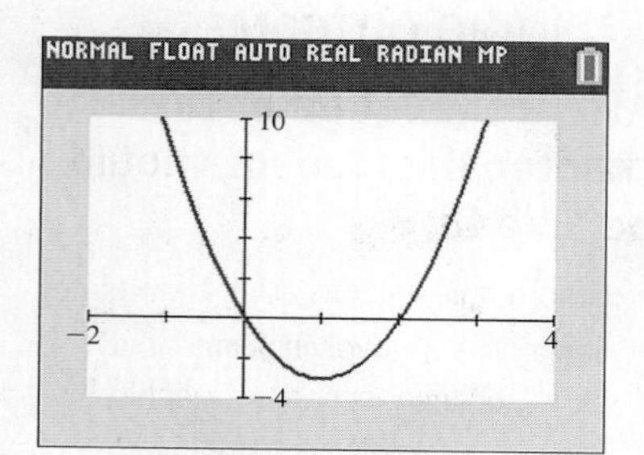

Fig. 14.1

EXAMPLE 2 Plotted graph—circle

Plot the graph of the equation $x^2 + y^2 = 25$.

We first solve this equation for y, and we obtain $y = \sqrt{25 - x^2}$, or $y = -\sqrt{25 - x^2}$, which we write as $y = \pm\sqrt{25 - x^2}$. We now assume values for x and find the corresponding values for y.

x	0	±1	±2	±3	±4	±5
y	±5	±4.9	±4.6	±4	±3	0

If $x > 5$, values of y are imaginary. We cannot plot these because x and y must both be real. When we show $y = \pm 3$ for $x = \pm 4$, this is a short way of representing four points. These points are $(4, 3)$, $(4, -3)$, $(-4, 3)$, and $(-4, -3)$.

In Fig. 14.2, *the resulting curve is a* **circle.** A circle with its center at the origin results from an equation of the form $x^2 + y^2 = r^2$, where r is the radius. ■

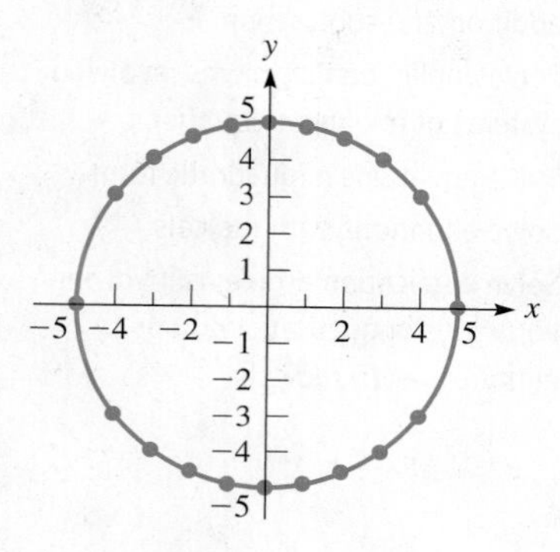

Fig. 14.2

From the graph of the circle in Fig. 14.2, we see that *the equation of a circle does not represent a function.* There are *two* values of y for most of the values of x in the domain. We must take this into account when displaying the graph of such an equation on a graphing calculator. This is illustrated in the next example.

■ See the *vertical-line test* on page 99.

EXAMPLE 3 Calculator graph—ellipse

Display the graph of the equation $2x^2 + 5y^2 = 10$ on a calculator.

First, solving for y, we get $y = \pm\sqrt{\dfrac{10 - 2x^2}{5}}$. To display the graph of this equation on a calculator, *we must enter both functions*, one as $y_1 = \sqrt{(10 - 2x^2)/5}$, and the other as $y_2 = -\sqrt{(10 - 2x^2)/5}$.

Trying some *window* settings (or noting that the domain is from $-\sqrt{5}$ to $\sqrt{5}$ and the range is from $-\sqrt{2}$ to $\sqrt{2}$), we get the graphing calculator display that is shown in Fig. 14.3.

The curve is an **ellipse.** An ellipse will be the resulting curve if the equation is of the form $ax^2 + by^2 = c$, where the constants a, b, and c have the same sign, and for which $a \neq b$. ■

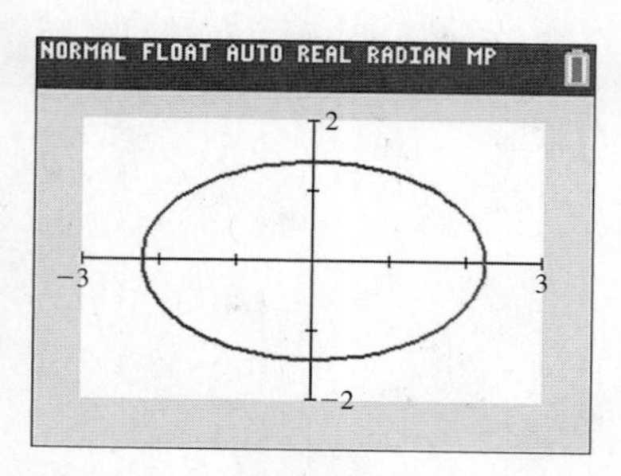

Fig. 14.3

■ A more complete discussion of the conic sections is found in Chapter 21.

EXAMPLE 4 Calculator graph—hyperbola

Display the graph of $2x^2 - y^2 = 4$ on a calculator.

Solving for y, we get

$$y = \pm\sqrt{2x^2 - 4}$$

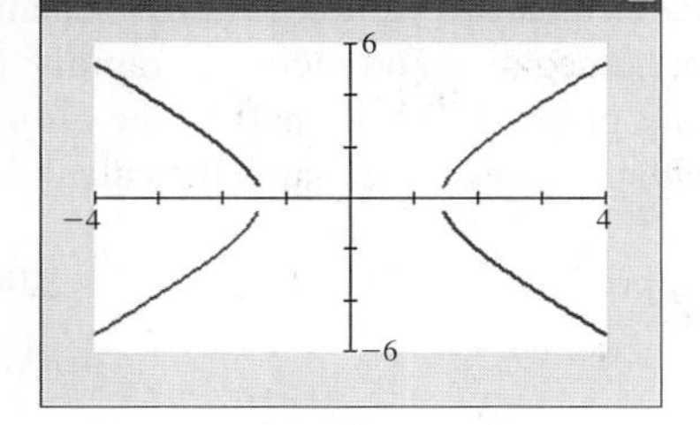

Fig. 14.4

As in Example 3, we enter two functions in the calculator, one with the plus sign and the other with the minus sign, and we have the display shown in Fig. 14.4. The *window* settings are chosen by noting that the values $-\sqrt{2} < x < \sqrt{2}$ are not in the domain of either $y_1 = \sqrt{2x^2 - 4}$ or $y_2 = -\sqrt{2x^2 - 4}$. These values of x would lead to imaginary values of y.

The curve is a **hyperbola,** which results when we have an equation of the form $ax^2 + by^2 = c$, if a and b have *different* signs.

Note that the calculator graph in Fig. 14.4 shows small gaps in the graph where it crosses the x-axis. This sometimes happens when conic sections are graphed on a calculator because of pixel limitations. In reality, these gaps do not exist. ■

SOLVING SYSTEMS OF EQUATIONS

As in solving systems of linear equations, we solve any system by finding the values of x and y that satisfy both equations at the same time. To solve a system graphically, we graph the equations and find the coordinates of all points of intersection. If the curves do not intersect, the system has no real solutions.

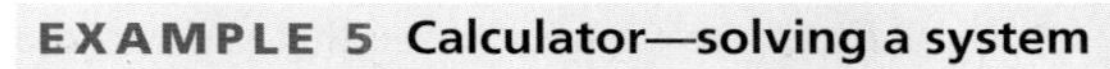

EXAMPLE 5 Calculator—solving a system

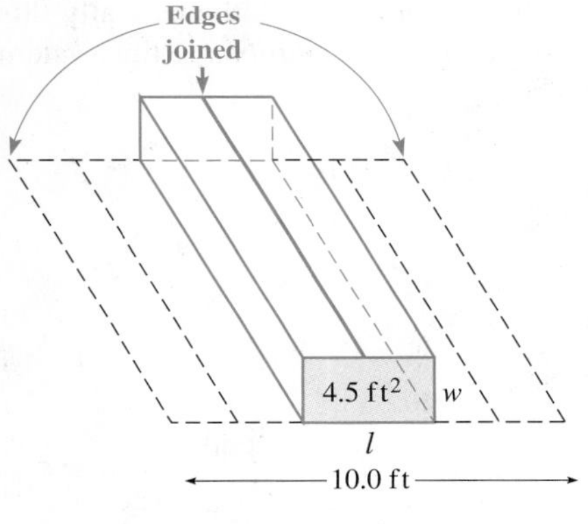

Fig. 14.5

For proper ventilation, the vent for a hot-air heating system is to have a rectangular cross-sectional area of 4.5 ft² and is to be made from sheet metal 10.0 ft wide. Find the dimensions of this cross-sectional area of the vent.

In Fig. 14.5, we have let l = the length and w = the width of the area. Because the area is 4.5 ft², we have $lw = 4.5$. Also, since the sheet metal is 10.0 ft wide, this is the perimeter of the area. This gives us $2l + 2w = 10.0$, or $l + w = 5.0$. This means that the system of equations to be solved is

$$lw = 4.5$$
$$l + w = 5.0$$

Solving each equation for l, we have

$$l = 4.5/w \quad \text{and} \quad l = 5.0 - w$$

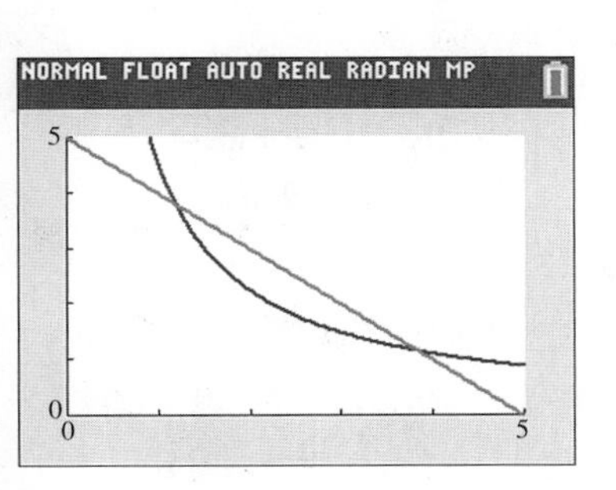

Fig. 14.6

We now display the graphs of these two equations on a calculator, using x for w and y for l. Because negative values of l and w have no meaning to the solution and since the straight line $l = 5.0 - w$ has intercepts of (5.0, 0) and (0, 5.0), we choose the *window* settings as shown in Fig. 14.6.

Using the *intersect* feature with the graph in Fig. 14.6, we find that the solutions are approximately (1.2, 3.8) and (3.8, 1.2). Using the length as the longer dimension, we have the solution of

$$l = 3.8 \text{ ft} \quad \text{and} \quad w = 1.2 \text{ ft}$$

We see that this checks with the statement of the problem. ■

In Example 5, we graphed the equation $xy = 4.5$ (having used x for w and y for l). The graph of this equation is also a *hyperbola*, another form of which is $xy = c$.

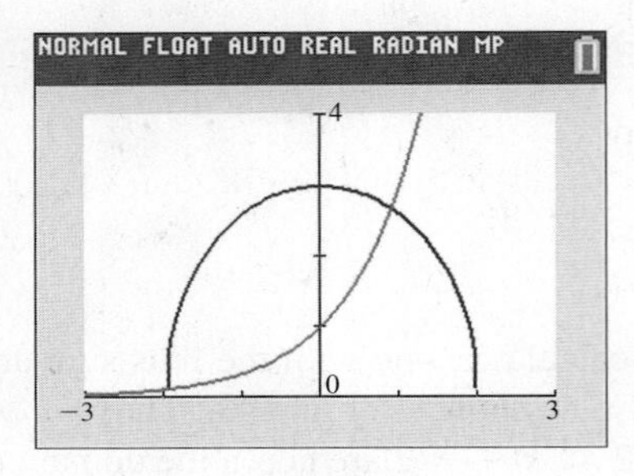

Fig. 14.7

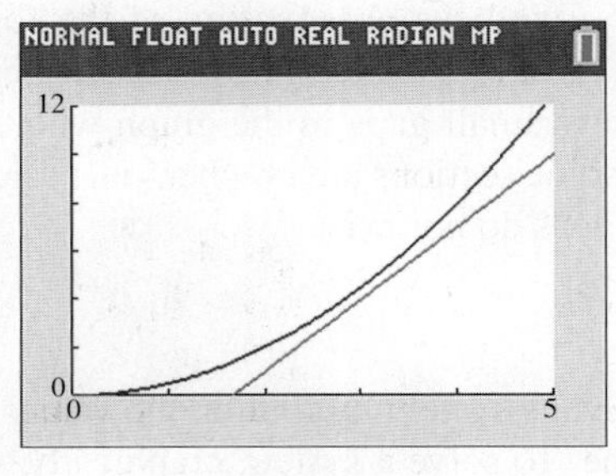

Fig. 14.8

Practice Exercise

1. In Example 7, determine how many times the rocket paths cross if the 5 in the second equation is changed to 4.

EXAMPLE 6 Calculator—solving a system

Graphically solve the system of equations:

$$9x^2 + 4y^2 = 36$$
$$y = 3^x$$

Solving the first equation for y, we have $y = \pm\frac{1}{2}\sqrt{36 - 9x^2}$. Because this is an ellipse, as shown in Example 3, its domain extends from $x = -2$ to $x = 2$ ($9x^2$ cannot be greater than 36). The exponential curve cannot be negative and increases rapidly, as discussed in Chapter 13. This means we need only graph the upper part of the ellipse (using the + sign). Therefore, we choose the *window* settings as shown in the calculator display in Fig. 14.7.

Using the *intersection* feature, we find the approximate solutions of $x = -2.00$, $y = 0.11$ and $x = 0.90$, $y = 2.68$. ■

EXAMPLE 7 System of equations with no solution—rocket paths

In a computer game, two rockets follow paths described by the equations $x^2 = 2y$ and $3x - y = 5$. Determine if the rocket paths ever cross.

Solving the equation of the path of the first rocket, we get $y = 0.5x^2$. This parabola has its vertex at the origin and opens upward since $a > 0$. This means there cannot be a point of intersection for $y < 0$.

The straight line rocket path has intercepts of $(0.00, -5.00)$ and $(1.67, 0.00)$. This means any possible point of intersection must be in the first quadrant. Therefore, we have the *window* settings as shown in Fig. 14.8.

From the figure, we see that the two rocket paths do not cross. Mathematically this means that the curves do not intersect and that there are *no real solutions* to this system of equations. ■

EXERCISES 14.1

In Exercises 1–4, make the given changes in the indicated examples of this section, and then perform the indicated operations.

1. In Example 1, change the − sign before $6x$ to +.
2. In Example 2, change the + sign before y^2 to −.
3. In Example 6, change the coefficient of x^2 to 25.
4. In Example 7, change the coefficient of y in the first equation to 3.

In Exercises 5–30, solve the given systems of equations graphically by using a calculator. Find all values to at least the nearest 0.1.

5. $y = \frac{x}{2}$, $x^2 + y^2 = 16$
6. $5x - y = 5$, $y = x^2 - 13$
7. $x^2 + 2y^2 = 8$, $x - 2y = 4$
8. $y = 3x - 6$, $xy = 6$
9. $y = x^2 - 2$, $4y = 12x - 17$
10. $4x^2 + 25y^2 = 21$, $10y = 31 - 9x$
11. $8y = 15x^2$, $xy = 20$
12. $y = -2x^2$, $y = x^2 - 6$
13. $y = x^2 - 3$, $x^2 + y^2 = 25$
14. $y = x^3$, $2x^2 + 4y^2 = 32$
15. $x^2 - 4y^2 = 16$, $x^2 + y^2 = 1$
16. $y = 2x^2 - 4x$, $x^2y = -4$
17. $2x^2 + 3y^2 = 19$, $x^2 + y^2 = 9$
18. $x^2 - 3y^2 = 2$, $2x^2 + y^2 = 16$
19. $x^2 + y^2 = 7$, $y(x + 2) = 3$
20. $x^2 + y^2 = 4$, $y^2 = x + 4$
21. $y = \frac{x^2}{4}$, $y = \sin x$
22. $y = 3 + 2x - x^2$, $y = 2\cos 2x$
23. $y = e^{-x}$, $y = x^{2/3}$
24. $y = 2^{x+1}$, $x^2 + y^2 = 4$
25. $x^2 - y^2 = 7$, $y = 5\log x$
26. $2x^2 + 8y^2 = 32$, $y = 2\ln x$
27. $y = \ln(x - 1)$, $y = \sin\frac{1}{2}x$
28. $y = \cos x$, $y = \log_3 x$
29. $10^{x+y} = 150$, $y = x^2$
30. $e^{x^2+y^2} = 20$, $xy = 4$

In Exercise 31, draw the appropriate figures. In Exercises 32–38, set up systems of equations and solve them graphically.

31. By drawing rough sketches, show that a parabola and an ellipse can have 0, 1, 2, 3, or 4 possible points of intersection.

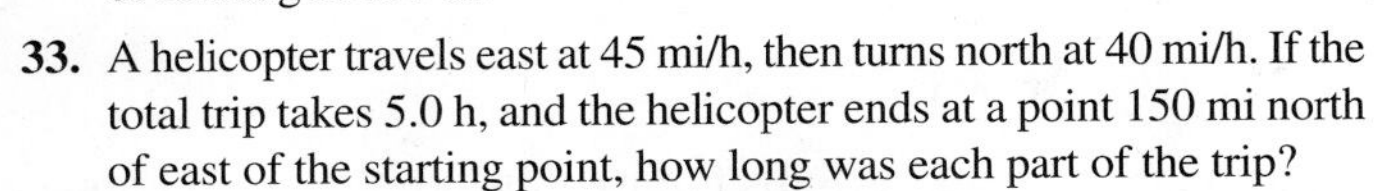

32. A rectangular security area is to be enclosed by fencing and divided in two equal parts of 1600 m^2 each by a fence parallel to the shorter sides. Find the dimensions of the security area if the total amount of fencing is 280 m.

33. A helicopter travels east at 45 mi/h, then turns north at 40 mi/h. If the total trip takes 5.0 h, and the helicopter ends at a point 150 mi north of east of the starting point, how long was each part of the trip?

34. A 4.60-m insulating strip is placed completely around a rectangular solar panel with an area of 1.20 m^2. What are the dimensions of the panel?

35. The power developed in an electric resistor is i^2R, where i is the current. If a first current passes through a 2.0-Ω resistor and a second current passes through a 3.0-Ω resistor, the total power produced is 12 W. If the resistors are reversed, the total power produced is 16 W. Find the currents (in A) if $i > 0$.

36. A circular hot tub is located on the square deck of a home. The side of the deck is 24 ft more than the radius of the hot tub, and there are 780 ft^2 of deck around the tub. Find the radius of the hot tub and the length of the side of the deck. Explain your answer.

37. Assume Earth is a sphere, with $x^2 + y^2 = 41$ as the equation of a circumference (distance in thousands of km). If a meteorite approaching Earth has a path described as $y^2 = 20x + 140$, will the meteorite strike Earth? If so, where?

38. Two people, one walking 1.0 km/h faster than the other, are on straight roads that are perpendicular. After meeting at an intersection, each continues on straight. How fast is each walking if they are 7.0 km apart (on a straight line) 1.0 h after meeting?

Answer to Practice Exercise

1. Two

14.2 Algebraic Solution of Systems of Equations

Solution by Substitution • Solution by Addition or Subtraction

Often, the graphical method is the easiest way to solve a system of equations. With a graphing calculator, it is possible to find the result with good accuracy. However, the graphical method does not usually give the *exact* answer. Using algebraic methods to find exact solutions for some systems of equations is either not possible or quite involved. There are systems, however, for which there are relatively simple algebraic solutions. In this section, we consider two useful methods, both of which we discussed before when we were studying systems of linear equations.

SOLUTION BY SUBSTITUTION

The first method is *substitution*. If we can solve one of the equations for one of its variables, we can substitute this solution into the other equation. We then have only one unknown in the resulting equation, and we can then solve this equation by methods discussed in earlier chapters.

EXAMPLE 1 Solution by substitution

By substitution, solve the system of equations

$$\begin{aligned} 2x - y &= 4 \\ x^2 - y^2 &= 4 \end{aligned}$$

We solve the first equation for y, obtaining $y = 2x - 4$. We now substitute $2x - 4$ for y in the second equation, getting

$$x^2 - (2x - 4)^2 = 4 \quad \text{in second equation, } y \text{ replaced by } 2x - 4$$

When simplified, this gives a quadratic equation.

$$x^2 - (4x^2 - 16x + 16) = 4$$

$$-3x^2 + 16x - 20 = 0$$

$$x = \frac{-16 \pm \sqrt{256 - 4(-3)(-20)}}{-6} = \frac{-16 \pm \sqrt{16}}{-6} = \frac{-16 \pm 4}{-6} = \frac{10}{3}, 2$$

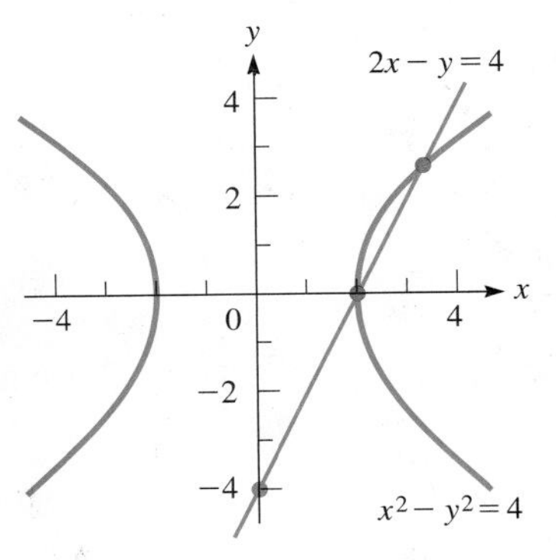

Fig. 14.9

We now find the corresponding values of y by substituting into $y = 2x - 4$. Thus, we have the solutions $x = \frac{10}{3}$, $y = \frac{8}{3}$, and $x = 2$, $y = 0$. As a check, we find that these values also satisfy the equation $x^2 - y^2 = 4$. Compare these solutions with those that would be obtained from Fig. 14.9. ■

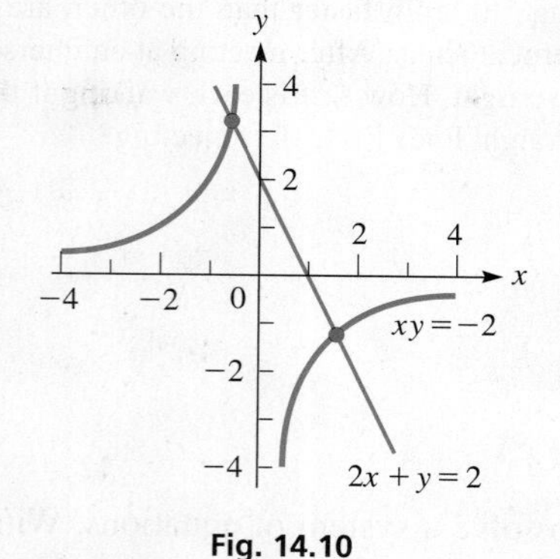

Fig. 14.10

EXAMPLE 2 Solution by substitution

By substitution, solve the system of equations

$$xy = -2$$
$$2x + y = 2$$

From the first equation, we have $y = -2/x$. Substituting this into the second equation, we have

in second equation, y replaced by $-\frac{2}{x}$

$$2x + \left(-\frac{2}{x}\right) = 2$$
$$2x^2 - 2 = 2x$$
$$x^2 - x - 1 = 0$$
$$x = \frac{1 \pm \sqrt{1 + 4}}{2} = \frac{1 \pm \sqrt{5}}{2}$$

By substituting these values for x into either of the original equations, we find the corresponding values of y, and we have the solutions

$$x = \frac{1 + \sqrt{5}}{2}, y = 1 - \sqrt{5} \quad \text{and} \quad x = \frac{1 - \sqrt{5}}{2}, y = 1 + \sqrt{5}$$

These can be checked by substituting in the original equations. In decimal form, they are

$$x \approx 1.618, y \approx -1.236 \quad \text{and} \quad x \approx -0.618, y \approx 3.236$$

The graphical solutions are shown in Fig. 14.10.

NOTE ▸ [The solutions can also be found by first solving the second equation for y, or either equation for x, and then substituting in the other equation.] ■

Practice Exercises

1. Solve by substitution:
$$2x^2 + y^2 = 3$$
$$x - y = 2$$

SOLUTION BY ADDITION OR SUBTRACTION

The other algebraic method is that of elimination by *addition or subtraction*. This method is most useful if both equations have only squared terms and constants.

EXAMPLE 3 Solution by addition

By addition or subtraction, solve the system of equations

$$2x^2 + y^2 = 9$$
$$x^2 - y^2 = 3$$

We note that if we add the corresponding sides of each equation, y^2 is eliminated. This leads to the solution.

$$2x^2 + y^2 = 9$$
$$\underline{x^2 - y^2 = 3}$$
$$3x^2 \quad = 12 \quad \text{add}$$
$$x^2 = 4$$
$$x = \pm 2$$

For $x = 2$, we have two corresponding y-values, $y = \pm 1$. Also, for $x = -2$, we have two corresponding y-values, $y = \pm 1$. Thus, we have four solutions:

$$x = 2, y = 1 \qquad x = 2, y = -1 \qquad x = -2, y = 1 \qquad x = -2, y = -1$$

Each solution checks in the original equations. The graphical solutions are shown in Fig. 14.11. ■

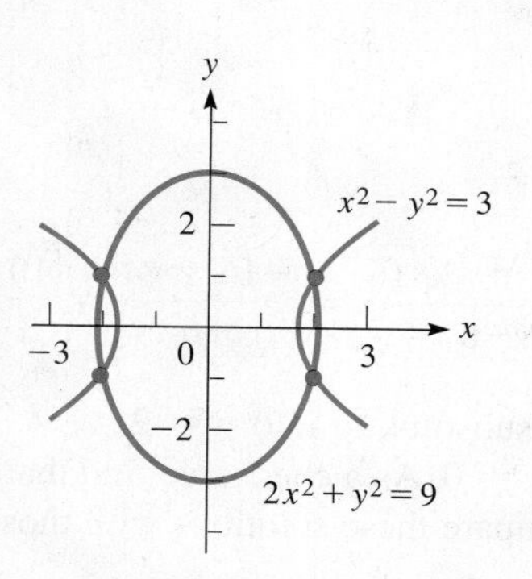

Fig. 14.11

EXAMPLE 4 Solution by subtraction

By addition or subtraction, solve the system of equations

$$5x^2 + 2y^2 = 17$$
$$x^2 + y^2 = 4$$

Multiplying the second equation by 2, and subtracting the resulting equations, we have

$$5x^2 + 2y^2 = 17$$
$$2x^2 + 2y^2 = 8 \quad \text{each term of second equation multiplied by 2}$$
$$3x^2 \qquad = 9 \quad \text{subtract}$$
$$x^2 = 3, \quad x = \pm\sqrt{3}$$

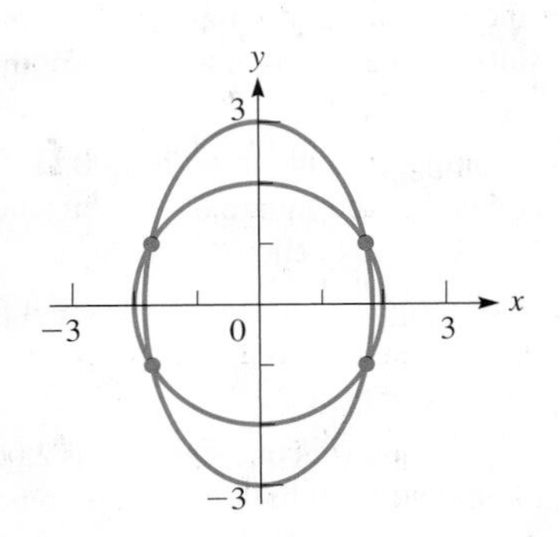

Fig. 14.12

The corresponding values of y for each value of x are ± 1. Again, we have four solutions:

$$x = \sqrt{3}, y = 1 \qquad x = \sqrt{3}, y = -1$$
$$x = -\sqrt{3}, y = 1 \qquad x = -\sqrt{3}, y = -1$$

Each solution checks when substituted in the original equations. The graphical solutions are shown in Fig. 14.12. ■

Practice Exercise

2. Solve by addition or subtraction:

$$x^2 + y^2 = 6$$
$$2x^2 - y^2 = 6$$

EXAMPLE 5 Solution by substitution— cost of machine parts

A certain number of machine parts cost \$1000. If they cost \$5 less per part, ten additional parts could be purchased for the same amount of money. What is the cost of each part?

Let c = the cost per part, and n = the number of parts. From the first statement, we see that $cn = 1000$. From the second statement, $(c - 5)(n + 10) = 1000$. Rewriting these equations, we have

$$n = 1000/c$$
$$cn + 10c - 5n - 50 = 1000$$

$$c\left(\frac{1000}{c}\right) + 10c - 5\left(\frac{1000}{c}\right) - 50 = 1000 \quad \text{substituting first equation in second equation}$$

$$1000 + 10c - \frac{5000}{c} - 50 = 1000$$

$$10c - \frac{5000}{c} - 50 = 0$$

$$c^2 - 5c - 500 = 0$$
$$(c + 20)(c - 25) = 0$$
$$c = -20, 25$$

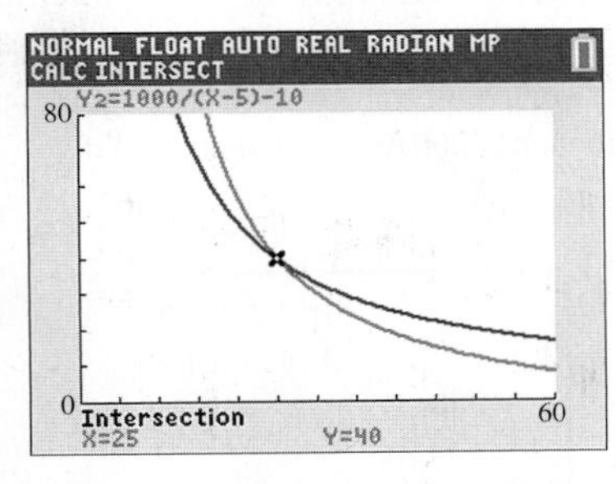

Fig. 14.13

Graphing calculator keystrokes: goo.gl/pkrirm

Because a negative answer has no significance in this particular situation, we see that the solution is c = \$25 per part. Checking with the original statement of the problem, we see that this is correct. The calculator solution is shown in Fig. 14.13. Here, we have let $y_1 = 1000/x$ and $y_2 = 1000/(x - 5) - 10$. ■

EXERCISES 14.2

In Exercises 1–4, make the given changes in the indicated examples of this section and then solve the resulting systems of equations

1. In Example 1, change the sign before y in the first equation from − to + and then solve the system.
2. In Example 2, change the right side of the second equation from 2 to 3 and then solve the system.
3. In Example 3, change the coefficient of x^2 in the first equation from 2 to 1 and then solve the system.
4. In Example 4, change the left side of the first equation from $5x^2 + 2y^2$ to $2x^2 + 5y^2$ and then solve the system.

In Exercises 5–28, solve the given systems of equations algebraically

5. $y = 2x + 9$
$y = x^2 + 1$

6. $y = 2x - 1$
$y = 2x^2 + 2x - 3$

7. $x + 2y = 3$
$x^2 + y^2 = 26$

8. $p^2 + 4h^2 = 4$
$h = p + 1$

9. $x + y = 1$
$x^2 - y^2 = 1$

10. $3x + 3y = 6$
$2x^2 - y^2 = 1$

11. $2x - y = 2$
$2x^2 + 3y^2 = 4$

12. $6y - x = 6$
$x^2 + 3y^2 = 36$

13. $wh = 9$
$w + h = 6$

14. $xy = 3$
$y - x = 2$

15. $xy = 4$
$12x - 9y = 6$

16. $xy = -4$
$2x + y = -2$

17. $y = x^2$
$y = 3x^2 - 50$

18. $M = L^2 - 1$
$2L^2 - M^2 = 2$

19. $x^2 - 3y = -8$
$x^2 + y^2 = 10$

20. $s^2 + t^2 = 8$
$st = 4$

21. $D^2 - 1 = R$
$D^2 - 2R^2 = 1$

22. $2x^2 + y^2 = 2$
$2y^2 = x^2 + 8$

23. $x^2 + y^2 = 25$
$x^2 - 2y^2 = 7$

24. $3x^2 - y^2 = 4$
$x^2 + 4y^2 = 10$

25. $x^2 + 3y^2 = 37$
$2x^2 - 9y^2 = 14$

26. $5x^2 - 4y^2 = 15$
$3y^2 + 4x^2 = 12$

27. $x^2 + y^2 + 4x = 1$
$x^2 + y^2 - 2y = 9$

28. $x^2 + y^2 - 4x - 2y + 4 = 0$
$x^2 + y^2 - 2x - 4y + 4 = 0$

(Hint for Exercises 27 and 28: First, subtract one equation from the other to get an equation relating x and y. Then substitute this equation in either given equation.)

In Exercises 29–46, solve the indicated systems of equations algebraically. In Exercises 33–46, it is necessary to set up the systems of equations properly

29. Solve for x and y: $x^2 - y^2 = a^2 - b^2; x - y = a - b$.

30. For what value of b are the two solutions of the system $x^2 - 2y = 5; y = 3x + b$ equal to each other? For this value of b, what is true of the graphs of the two functions?

31. A rocket is fired from behind a ship and follows the path given by $h = 3x - 0.05x^2$, where h is its altitude (in mi) and x is the horizontal distance traveled (in mi). A missile fired from the ship follows the path given by $h = 0.8x - 15$. For $h > 0$ and $x > 0$, find where the paths of the rocket and missile cross.

32. A 2-kg block collides with an 8-kg block. Using the physical laws of conservation of energy and conservation of momentum, along with given conditions, the following equations involving the velocities are established:

$$v_1^2 + 4v_2^2 = 41$$

$$2v_1 + 8v_2 = 12$$

Find these velocities (in m/s) if $v_2 > 0$.

33. One face of a washer has an area of 37.7 cm^2. The inner radius is 2.00 cm less than the outer radius. What are the radii?

34. A triangular sail has sides of 10.5 m, 8.5 m, and 5.0 m. Find the height of the sail to the 10.5-m side.

35. The edges of a rectangular piece of plastic sheet are joined together to make a plastic tube. If the area of the plastic sheet is 216 cm^2, and the volume of the resulting tube is 224 cm^3, what are the dimensions of the plastic sheet?

36. The impedance Z in an alternating-current circuit is 2.00 Ω. If the resistance R is numerically equal to the square of the reactance X, find R and X. Use $Z^2 = R^2 + X^2$ (See Section 12.7).

37. A roof truss is in the shape of a right triangle. If there are 4.60 m of lumber in the truss and the longest side is 2.20 m long, what are the lengths of the other two sides of the truss?

38. In a certain roller mechanism, the radius of one steel ball is 2.00 cm greater than the radius of a second steel ball. If the difference in their masses is 7100 g, find the radii of the balls. The density of steel is 7.70 g/cm^3.

39. A set of equal electrical resistors in series has a total resistance (the sum of the resistances) of 78.0 Ω. Another set of two fewer equal resistors in series also has a total resistance of 78.0 Ω. If each resistor in the second set is 1.3 Ω greater than each of the first set, how many are in each set?

40. Security fencing encloses a rectangular storage area of 1600 m^2 that is divided into two sections by additional fencing parallel to the shorter sides. Find the dimensions of the storage area if 220 m of fencing are used.

41. An open liner for a carton is to be made from a rectangular sheet of cardboard of area 216 in.^{2} by cutting equal 2.00-in. squares from each corner and bending up the sides. If the volume within the liner is 224 in.^{3}, what are dimensions of the cardboard sheet?

42. Two guy wires, one 140 ft long and the other 120 ft long, are attached at the same point of a TV tower with the longer one secured in the (level) ground 30 ft farther from the base of the tower than the shorter one. How high up on the tower are they attached?

43. A workbench is in the shape of a trapezoid, as shown in Fig. 14.14. If the perimeter of the workbench is 260 cm, what is its area if $y > x$?

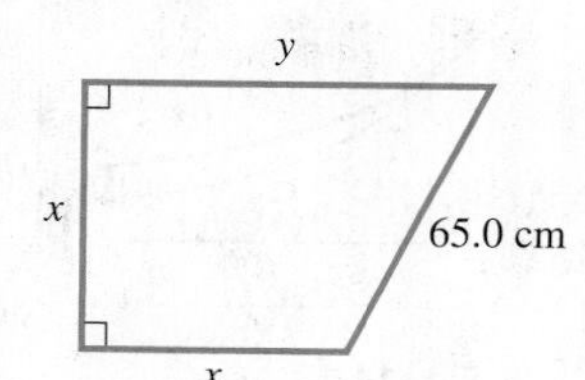

Fig. 14.14

44. A rectangular play area is twice as long as it is wide. If the area is 648 m^2, what are its dimensions?

45. A jet travels at 610 mi/h relative to the air. It takes the jet 1.6 h longer to travel the 3660 mi from London to Washington, D.C., against the wind than it takes from Washington to London with the wind. Find the velocity of the wind.

46. In a marketing survey, a company found that the total gross income for selling t tables at a price of p dollars each was \$35,000. It then increased the price of each table by \$100 and found that the total income was only \$27,000 because 40 fewer tables were sold. Find p and t.

Answers to Practice Exercises

1. $x = 1/3, y = -5/3; x = 1, y = -1$

2. $x = 2, y = \sqrt{2}; x = 2, y = -\sqrt{2};$
$x = -2, y = \sqrt{2}; x = -2, y = -\sqrt{2}$

14.3 Equations in Quadratic Form

Substituting to Fit Quadratic Form • Solving Equations in Quadratic Form • Extraneous Roots

Often, we encounter equations that can be solved by methods applicable to quadratic equations, even though these equations are not actually quadratic. [They do have the property, however, that with a proper substitution they may be written in the form of a quadratic equation.] All that is necessary is that the equation have terms including some variable quantity, its square, and perhaps a constant term. The following example illustrates these types of equations.

NOTE ▶

EXAMPLE 1 Identifying quadratic form

(a) The equation $x - 2\sqrt{x} - 5 = 0$ is an equation in quadratic form, because if we let $y = \sqrt{x}$, we have $x = (\sqrt{x})^2 = y^2$, and the resulting equation is $y^2 - 2y - 5 = 0$.

(b) $t^{-4} - 5t^{-2} + 3 = 0$ ($(t^{-2})^2$)

By letting $y = t^{-2}$, we have $y^2 - 5y + 3 = 0$.

(c) $t^3 - 3t^{3/2} - 7 = 0$ ($(t^{3/2})^2$)

By letting $y = t^{3/2}$, we have $y^2 - 3y - 7 = 0$.

(d) $(x + 1)^4 - (x + 1)^2 - 1 = 0$ ($[(x+1)^2]^2$)

By letting $y = (x + 1)^2$, we have $y^2 - y - 1 = 0$.

(e) $x^{10} - 2x^5 + 1 = 0$ ($(x^5)^2$)

By letting $y = x^5$, we have $y^2 - 2y + 1 = 0$. ■

The following examples illustrate the method of solving equations in quadratic form.

EXAMPLE 2 Solving an equation in quadratic form

Solve the equation $2x^4 + 7x^2 = 4$.

We first let $y = x^2$ to write the equation in quadratic form. We will then solve the resulting quadratic equation for y. However, solutions for x are required, so we again let $y = x^2$ to solve for x.

$$2y^2 + 7y - 4 = 0 \quad \text{let } y = x^2$$

$$(2y - 1)(y + 4) = 0 \quad \text{factor and solve for } y$$

$$y = \tfrac{1}{2} \quad \text{or} \quad y = -4$$

$$x^2 = \tfrac{1}{2} \quad \text{or} \quad x^2 = -4 \quad y = x^2 \text{ (to solve for } x)$$

$$x = \pm\frac{1}{\sqrt{2}} \quad \text{or} \quad x = \pm 2j$$

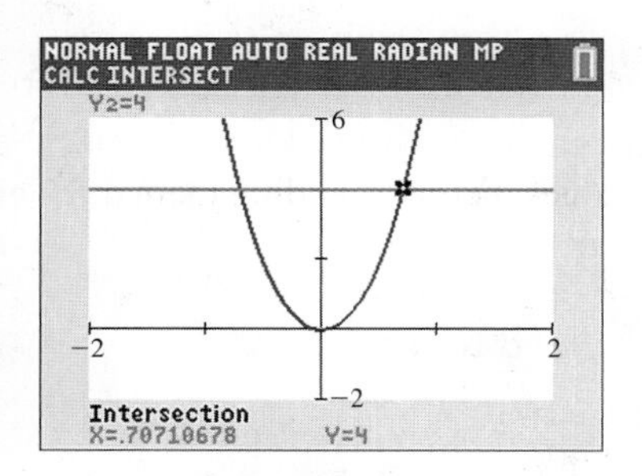

Fig. 14.15
Graphing calculator keystrokes: goo.gl/Akg6eq

We can let $y_1 = 2x^4 + 7x^2$ and $y_2 = 4$ to solve the system on a calculator. The display is shown in Fig. 14.15, and we see that there are two points of intersection, one for $x = 0.7071$ (shown on screen) and the other for $x = -0.7071$ (not shown). Because $1/\sqrt{2} = 0.7071$, this verifies the real solutions. The imaginary solutions cannot be found graphically. Substitution of each value *in the original equation* shows each value to be a solution. ■

Two of the solutions in Example 2 are complex numbers. We were able to find these solutions directly from the definition of the square root of a negative number. In some cases (see Exercise 22 of this section), it is necessary to use the method of Section 12.6 to find such complex-number solutions.

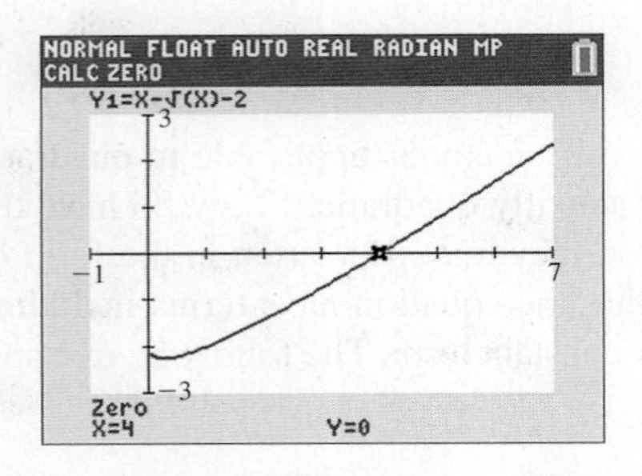

Fig. 14.16

Graphing calculator keystrokes: goo.gl/8fqPDi

Practice Exercise

1. Solve for x: $x + \sqrt{x} - 2 = 0$

EXAMPLE 3 Solving an equation with a square root

Solve the equation $x - \sqrt{x} - 2 = 0$.

By letting $y = \sqrt{x}$, we have

$$y^2 - y - 2 = 0$$
$$(y - 2)(y + 1) = 0$$
$$y = 2 \quad \text{or} \quad y = -1$$
$$\sqrt{x} = 2 \quad \text{or} \quad \sqrt{x} = -1 \qquad y = \sqrt{x}$$
$$x = 4 \qquad \text{no soultion}$$

Since $\sqrt{x}$ cannot be negative, the only solution is $x = 4$. This solution checks when substituted in the original equation.

The graph of $y_1 = x - \sqrt{x} - 2$ is shown in the calculator display in Fig. 14.16. Note that $x = 4$ is the only solution shown. ■

CAUTION Example 3 illustrates a very important point. Whenever an operation involving the unknown is performed on an equation, this operation may introduce roots into a subsequent equation that are not roots of the original equation. Therefore, we must check all answers in the original equation. ■ Only operations involving constants—that is, adding, subtracting, multiplying by, or dividing by constants—are certain not to introduce the **extraneous roots.** We first encountered the concept of an extraneous root in Section 6.7, when we discussed equations involving fractions.

EXAMPLE 4 Solving an equation with negative exponents

Solve the equation $x^{-2} + 3x^{-1} + 1 = 0$.

By substituting $y = x^{-1}$, we have $y^2 + 3y + 1 = 0$. To solve this equation, we may use the quadratic formula:

$$y = \frac{-3 \pm \sqrt{9 - 4}}{2} = \frac{-3 \pm \sqrt{5}}{2}$$

Since $x = 1/y$, we have

$$x = \frac{2}{-3 + \sqrt{5}} \quad \text{or} \quad x = \frac{2}{-3 - \sqrt{5}}$$

These answers in decimal form are

$$x \approx -2.618 \quad \text{or} \quad x \approx -0.382$$

These results check when substituted in the original equation. In checking these decimal answers, it is more accurate to use the calculator values, before rounding them off. This can be done by storing the calculator values in memory. ■

Practice Exercise

2. Solve for x: $x^{-4} - 8x^{-2} + 16 = 0$

EXAMPLE 5 Solving an equation with grouped terms

Solve the equation $(x^2 - x)^2 - 8(x^2 - x) + 12 = 0$.

By substituting $y = x^2 - x$, we have

$$y^2 - 8y + 12 = 0$$
$$(y - 2)(y - 6) = 0$$
$$y = 2 \quad \text{or} \quad y = 6$$
$$x^2 - x = 2 \quad \text{or} \quad x^2 - x = 6 \qquad y = x^2 - x$$

Solving each of these equations, we have

$$x^2 - x - 2 = 0 \qquad\qquad x^2 - x - 6 = 0$$
$$(x - 2)(x + 1) = 0 \qquad (x - 3)(x + 2) = 0$$
$$x = 2 \quad \text{or} \quad x = -1 \qquad x = 3 \quad \text{or} \quad x = -2$$

Each value checks when substituted in the original equation. ■

■ See chapter introduction.

EXAMPLE 6 Quadratic form—cell phone screen

An advertisement for a cell phone states that it has a 106-mm screen (diagonal) with an area of 5040 mm^2. Find the length l and width w of the screen. See Fig. 14.17.

Fig. 14.17

Because the required quantities are the length and width, let l = the length of the screen and let w = its width. Because the area is 5040 mm^2, $lw = 5040$. Also, using the Pythagorean theorem and the fact that the diagonal is 106 mm, we have the equation $l^2 + w^2 = 106^2$. Therefore, we are to solve the system of equations

$$lw = 5040 \qquad l^2 + w^2 = 11{,}236$$

Solving the first equation for l, we have $l = 5040/w$. Substituting this into the second equation, we have

$$\left(\frac{5040}{w}\right)^2 + w^2 = 11{,}236$$

$$\frac{5040^2}{w^2} + w^2 = 11{,}236$$

$$25{,}401{,}600 + w^4 = 11{,}236w^2$$

Let $x = w^2$,

$$x^2 - 11{,}236x + 25{,}401{,}600 = 0$$

$$x = \frac{-(-11{,}236) \pm \sqrt{(-11{,}236)^2 - 4(1)(25{,}401{,}600)}}{2(1)}$$

$$x = 8100 \quad \text{or} \quad x = 3136$$

Therefore, $w^2 = 8100$, or $w^2 = 3136$.

Solving for w, we get $w = \pm 90$, or $w = \pm 56$. Only positive values are meaningful in this problem, which means if $w = 56$ mm, then $l = 90$ mm (or $l = 56$ mm, $w = 90$ mm). Checking with the statement of the problem, we see that these dimensions for the screen give an area of 5040 mm^2 and a diagonal of 106 mm. ■

■ The first permanent photograph was taken in 1816 by the French inventor Joseph Niepce (1765 – 1833).

EXERCISES 14.3

In Exercises 1 and 2, make the given changes in the indicated examples of this section and then solve the resulting equations

1. In Example 2, change the + before the $7x^2$ to − and then solve the equation.
2. In Example 3, change the 2 to 6 and then solve the equation.

In Exercises 3–28, solve the given equations algebraically. In Exercise 10, explain your method

3. $x^4 - 10x^2 + 9 = 0$
4. $4R^4 + 15R^2 = 4$
5. $3x^{-2} - 7x^{-1} - 6 = 0$
6. $10x^{-2} + 3x^{-1} - 1 = 0$
7. $x^{-4} + 2x^{-2} = 24$
8. $x^{-1} - x^{-1/2} = 2$
9. $2x - 5\sqrt{x} + 3 = 0$
10. $4x + 3\sqrt{x} = 1$
11. $3\sqrt[3]{x} - 5\sqrt[6]{x} + 2 = 0$
12. $\sqrt{x} + 3\sqrt[4]{x} = 28$
13. $x^{2/3} - 2x^{1/3} - 15 = 0$
14. $y^3 + 2y^{3/2} = 80$
15. $8n^{1/2} - 20n^{1/4} = 12$
16. $3x^{4/3} + 5x^{2/3} = 2$
17. $(x - 1) - \sqrt{x - 1} = 20$
18. $(C + 1)^{-2/3} + 5(C + 1)^{-1/3} - 6 = 0$
19. $(x^2 - 2x)^2 - 11(x^2 - 2x) + 24 = 0$
20. $(x^2 - 1)^2 + (x^2 - 1)^{-2} = 2$
21. $x - 3\sqrt{x - 2} = 6$ (Let $y = \sqrt{x - 2}$.)
22. $x^6 + 7x^3 = 8$
23. $\dfrac{1}{(x - 1)^2} + \dfrac{1}{x - 1} = 12$
24. $x^2 - 3x = \sqrt{x^2 - 3x + 2}$
25. $\dfrac{1}{s^2 + 1} + \dfrac{2}{s^2 + 3} = 1$
26. $(x + \frac{2}{x})^2 - 6x - \frac{12}{x} = -9$
27. $e^{2x} - 3e^x + 2 = 0$
28. $10^{2x} - 2(10^x) = 0$

In Exercises 29–34, solve the given equations algebraically and check the solutions with a calculator

29. $x^4 - 20x^2 + 64 = 0$
30. $x^{-2} - x^{-1} - 42 = 0$
31. $x + 2 = 3\sqrt{x}$
32. $3x^{2/3} = 12x^{1/3} + 36$
33. $(\log x)^2 - 3\log x + 2 = 0$
34. $2^x + 32(2^{-x}) = 12$

In Exercises 35–42, solve the given problems algebraically.

35. Solve for x: $\log(x^4 + 4) - \log(5x^2) = 0$.

36. A paper drinking cup in the shape of a cone is constructed from 6π in.^{2} of paper. If the height of the cone is 4 in., find the radius. (*Hint*: Lateral surface area $S = \pi r\sqrt{r^2 + h^2}$.)

37. The equivalent resistance R_T of two resistors R_1 and R_2 in parallel is given by $R_T^{-1} = R_1^{-1} + R_2^{-1}$. If $R_T = 1.00\ \Omega$ and $R_2 = \sqrt{R_1}$, find R_1 and R_2.

38. An equation used in the study of the dispersion of light is $\mu = A + B\lambda^{-2} + C\lambda^{-4}$. Solve for λ. (See the chapter introduction.)

39. In the theory dealing with optical interferometers, the equation $\sqrt{F} = 2\sqrt{p}/(1 - p)$ is used. Solve for p if $F = 16$.

40. A special washer is made from a circular disc 3.50 cm in radius by removing a rectangular area of 12.0 cm^2 from the center. If each corner of the rectangular area is 0.50 cm from the outer edge of the washer, what are the dimensions of the area that is removed?

41. A rectangular TV screen has an area of 1540 in.^{2} and a diagonal of 60.0 in. Find the dimensions of the screen.

42. A roof truss in the shape of a right triangle has a perimeter of 90 ft. If the hypotenuse is 1 ft longer than one of the other sides, what are the sides of the truss?

Answers to Practice Exercises

1. 1 **2.** 1/2, 1/2, −1/2, −1/2

14.4 Equations with Radicals

Solving by Squaring Both Sides • Isolating a Radical • Solving a Nested Radical Equation

Equations containing radicals are normally solved by squaring both sides of the equation if the radical represents a square root or by a similar operation for the other roots. However, when we do this, we often introduce *extraneous roots*. **CAUTION** Thus, it is very important that all solutions be checked in the original equation. ■

EXAMPLE 1 Solve by squaring both sides

Solve the equation $\sqrt{x - 4} = 2$.

By squaring both sides of the equation, we have

$$(\sqrt{x-4})^2 = 2^2$$
$$x - 4 = 4$$
$$x = 8$$

This solution checks when put into the original equation. ■

EXAMPLE 2 Solve by squaring both sides

Solve the equation $2\sqrt{3x - 1} = 3x$.

Squaring both sides of the equation gives us

$$(2\sqrt{3x-1})^2 = (3x)^2 \quad \text{don't forget to square the 2}$$
$$4(3x - 1) = 9x^2$$
$$12x - 4 = 9x^2$$
$$9x^2 - 12x + 4 = 0$$
$$(3x - 2)^2 = 0$$
$$x = \frac{2}{3} \quad \text{(double root)}$$

Checking this solution in the original equation, we have

$$2\sqrt{3(\tfrac{2}{3}) - 1} \stackrel{?}{=} 3(\tfrac{2}{3}), \quad 2\sqrt{2 - 1} \stackrel{?}{=} 2, \quad 2 = 2$$

Therefore, the solution $x = \frac{2}{3}$ checks.

We can check this solution graphically by letting $y_1 = 2\sqrt{3x - 1}$ and $y_2 = 3x$. The calculator display is shown in Fig. 14.18. The *intersection* feature shows that the only x-value that the curves have in common is $x = 0.6667$, which agrees with the solution of $x = 2/3$. This also means the line y_1 is tangent to the curve of y_2. ■

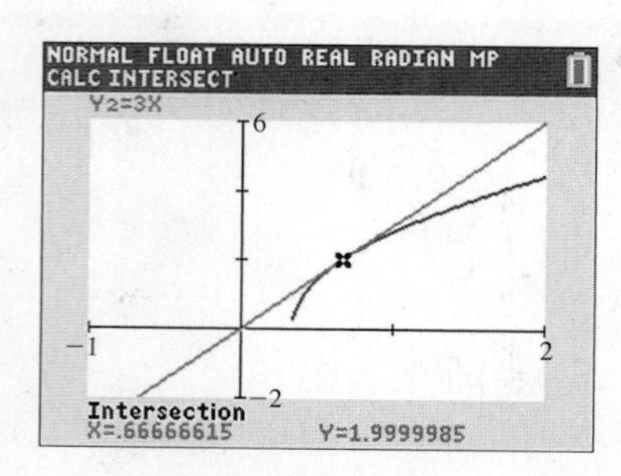

Fig. 14.18

Graphing calculator keystrokes: goo.gl/qJsVbI

Practice Exercise

1. Solve for x: $\sqrt{2x + 3} = x$

EXAMPLE 3 Solve by cubing both sides

Solve the equation $\sqrt[3]{x-8}=2$.

Cubing both sides of the equation, we have

$$x-8=8$$
$$x=16$$

Checking this solution in the original equation, we get

$$\sqrt[3]{16-8}\stackrel{?}{=}2, \qquad 2=2$$

Therefore, the solution checks. ■

If a radical and other terms are on one side of the equation, we *first isolate the radical.* That is, we rewrite the equation with the radical on one side and all other terms on the other side.

EXAMPLE 4 Solve by isolating the radical

Solve the equation $\sqrt{x-1}+3=x$.

We first isolate the radical by subtracting 3 from each side. This gives us

$$\sqrt{x-1}=x-3$$

We now square both sides and proceed with the solution:

$$\begin{aligned}(\sqrt{x-1})^2&=(x-3)^2\\ x-1&=x^2-6x+9\\ x^2-7x+10&=0\\ (x-5)(x-2)&=0\\ x=5 \quad &\text{or} \quad x=2\end{aligned}$$

square the expression on each side, not just the terms separately

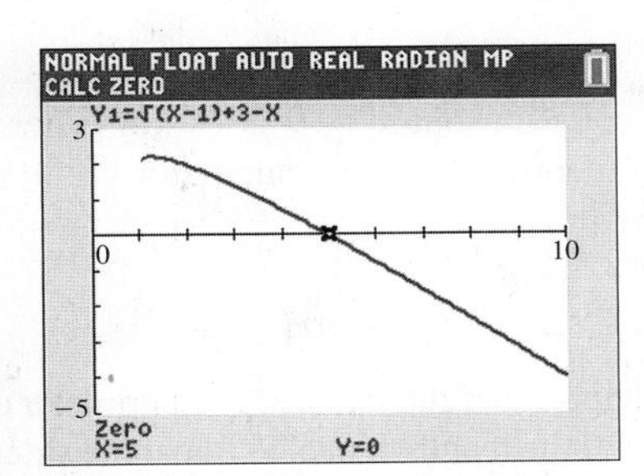

Fig. 14.19

Graphing calculator keystrokes: goo.gl/MUOaE1

The solution $x=5$ checks, but the solution $x=2$ gives $4=2$. Thus, the only solution is $x=5$. The value $x=2$ is an extraneous root. The graph of $y=\sqrt{x-1}+3-x$, shown in Fig. 14.19, verifies that $x=5$ is a solution but $x=2$ is not. ■

Practice Exercise

2. Solve for x: $\sqrt{x+4}+2=x$

EXAMPLE 5 Solve by isolating the radical

Solve the equation $\sqrt{x+1}+\sqrt{x-4}=5$.

This is most easily solved by first isolating one of the radicals by placing the other radical on the right side of the equation. We then square both sides of the resulting equation.

$$\begin{aligned}\sqrt{x+1}&=5-\sqrt{x-4} \quad \text{two terms}\\ (\sqrt{x+1})^2&=(5-\sqrt{x-4})^2\\ x+1&=25-10\sqrt{x-4}+(\sqrt{x-4})^2 \quad \text{be carefull}\\ x+1&=25-10\sqrt{x-4}+x-4\end{aligned}$$

Now, isolating the radical on one side of the equation and squaring again, we have

$$\begin{aligned}10\sqrt{x-4}&=20\\ \sqrt{x-4}&=2 \quad \text{divide by 10}\\ x-4&=4 \quad \text{square both sides}\\ x&=8\end{aligned}$$

This solution checks.

We note again that in squaring $5-\sqrt{x-4}$, we do not simply square 5 and $\sqrt{x-4}$. We must square the entire expression. ■

Practice Exercise

3. Solve for x: $\sqrt{x+3}+\sqrt{x}=3$

EXAMPLE 6 Nested radical equation

Solve the equation $\sqrt{7 + \sqrt{x}} - 1 = \sqrt{x}$.

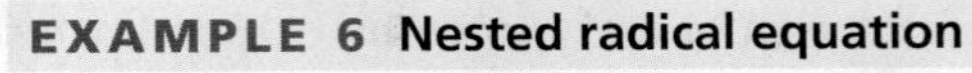

We first **isolate the outer radical** and then square both sides. We then isolate the remaining radical and square both sides again. The steps are shown below.

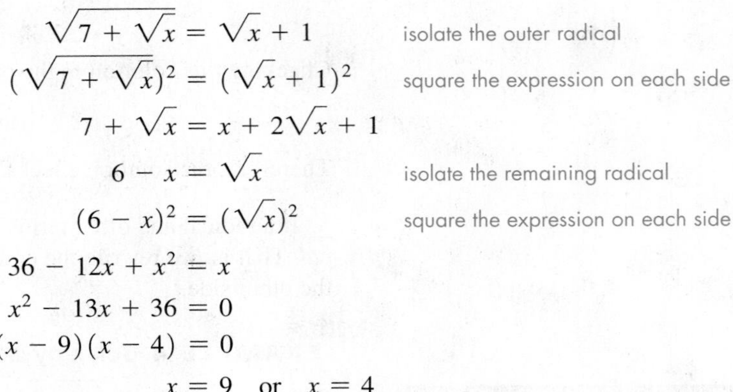

$$\sqrt{7 + \sqrt{x}} = \sqrt{x} + 1 \qquad \text{isolate the outer radical}$$
$$(\sqrt{7 + \sqrt{x}})^2 = (\sqrt{x} + 1)^2 \qquad \text{square the expression on each side}$$
$$7 + \sqrt{x} = x + 2\sqrt{x} + 1$$
$$6 - x = \sqrt{x} \qquad \text{isolate the remaining radical}$$
$$(6 - x)^2 = (\sqrt{x})^2 \qquad \text{square the expression on each side}$$
$$36 - 12x + x^2 = x$$
$$x^2 - 13x + 36 = 0$$
$$(x - 9)(x - 4) = 0$$
$$x = 9 \quad \text{or} \quad x = 4$$

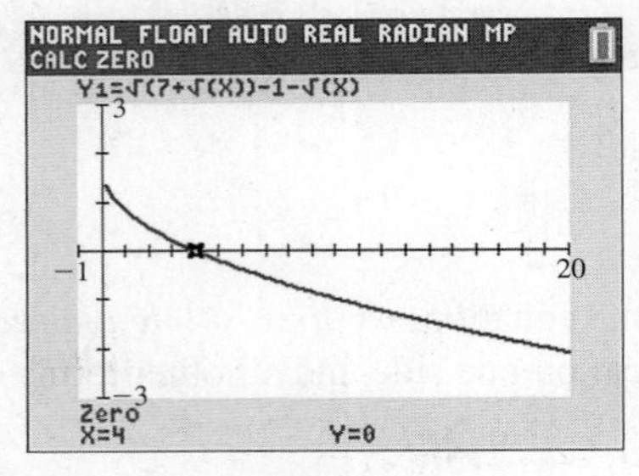

Fig. 14.20

When checking, $x = 4$ satisfies the original equation but $x = 9$ does not. Therefore, the only solution is $x = 4$ ($x = 9$ is an extraneous root). The calculator graph in Fig. 14.20 verifies that $x = 4$ is the only solution. ■

EXAMPLE 7 System with a radical—holograph dimensions

Each cross section of a holographic image is in the shape of a right triangle. The perimeter of the cross section is 60 cm, and its area is 120 cm^2. Find the length of each of the three sides.

If we let the two legs of the triangle be x and y, as shown in Fig. 14.21, from the formulas for the perimeter p and the area A of a triangle, we have

$$p = x + y + \sqrt{x^2 + y^2} \quad \text{and} \quad A = \tfrac{1}{2}xy$$

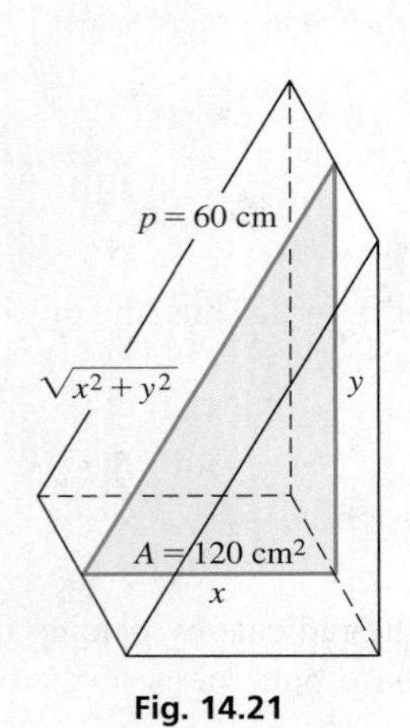

Fig. 14.21

where the hypotenuse was found by use of the Pythagorean theorem. Using the information given in the statement of the problem, we arrive at the equations

$$x + y + \sqrt{x^2 + y^2} = 60 \quad \text{and} \quad xy = 240$$

Isolating the radical in the first equation and then squaring both sides, we have

$$\sqrt{x^2 + y^2} = 60 - x - y$$
$$x^2 + y^2 = 3600 - 120x - 120y + x^2 + 2xy + y^2$$
$$0 = 3600 - 120x - 120y + 2xy$$

Solving the second of the original equations for y, we have $y = 240/x$. Substituting, we have

$$0 = 3600 - 120x - 120\left(\frac{240}{x}\right) + 2x\left(\frac{240}{x}\right)$$
$$0 = 3600x - 120x^2 - 120(240) + 480x \qquad \text{multiply by } x$$
$$0 = 30x - x^2 - 240 + 4x \qquad \text{divide by 120}$$
$$x^2 - 34x + 240 = 0 \qquad \text{collect terms on left}$$
$$(x - 10)(x - 24) = 0$$
$$x = 10 \text{ cm} \quad \text{or} \quad x = 24 \text{ cm}$$

■ Holography is a method of producing a three-dimensional image without the use of a lens. The theory of holography was developed in the late 1940s by the British engineer Dennis Gabor (1900–1979). After the invention of lasers, the first holographs were produced in the early 1960s.

If $x = 10$ cm, then $y = 24$ cm, or if $x = 24$ cm, then $y = 10$ cm. Therefore, the legs of the holographic cross section are 10 cm and 24 cm, and the hypotenuse is 26 cm. For these sides, $p = 60$ cm and $A = 120$ cm^2. We see that these values check with the statement of the problem. ■

EXERCISES 14.4

In Exercises 1–4, make the given changes in the indicated examples of this section, and then solve the resulting equations.

1. In Example 2, change the $3x$ on the right to 3.
2. In Example 3, change the 8 under the radical to 19.
3. In Example 4, change the 3 on the left to 7.
4. In Example 5, change the 4 under the second radical to 14.

In Exercises 5–34, solve the given equations. In Exercises 19 and 22, explain how the extraneous root is introduced.

5. $\sqrt{x - 8} = 2$
6. $\sqrt{x + 4} = 3$
7. $\sqrt{15 - 2x} = x$
8. $2\sqrt{2P + 5} = P$
9. $2\sqrt{3x + 2} = 6x$
10. $\sqrt{5x - 1} + 3 = x$
11. $2\sqrt{3 - x} - x = 5$
12. $x - 3\sqrt{2x + 1} = -5$
13. $\sqrt[3]{y - 7} = 2$
14. $\sqrt[4]{5 - x} = 2$
15. $5\sqrt{s} - 6 = s$
16. $6\sqrt{x} + 4 = 4x$
17. $\sqrt{x^2 - 11} = 5$
18. $t^2 = 3 - \sqrt{2t^2 - 3}$
19. $\sqrt{x + 4} + 8 = x$
20. $\sqrt{x^2 - x - 4} = x + 2$
21. $\sqrt{5 + \sqrt{x}} = \sqrt{x} - 1$
22. $\sqrt{13 + \sqrt{x}} = \sqrt{x} + 1$
23. $x - 3x\sqrt{x} = 0$
24. $\sqrt{3 - 2\sqrt{x}} = \sqrt{x}$
25. $2\sqrt{x + 2} - \sqrt{3x + 4} = 1$
26. $\sqrt{x - 1} + \sqrt{x + 2} = 3$
27. $\sqrt{5x + 1} - 1 = 3\sqrt{x}$
28. $\sqrt{x - 7} = \sqrt{x} - 7$
29. $\sqrt{6x - 5} - \sqrt{x + 4} = 2$
30. $\sqrt{5x - 4} - \sqrt{x} = 2$
31. $\sqrt{x - 9} = \dfrac{36}{\sqrt{x - 9}} - \sqrt{x}$
32. $\sqrt[4]{x + 10} = \sqrt{x - 2}$
33. $\sqrt{x - \sqrt{2x}} = 2$
34. $\sqrt{3x + \sqrt{3x + 4}} = 4$

In Exercises 35–38, solve the given equations algebraically and check the solutions with a graphing calculator.

35. $\sqrt{3x + 4} = x$
36. $\sqrt{x - 2} + 3 = x$
37. $\sqrt{2x + 1} + 3\sqrt{x} = 9$
38. $\sqrt{2x + 1} - \sqrt{x + 4} = 1$

In Exercises 39–52, solve the given problems.

39. If $f(x) = \sqrt{x + 3}$, find x if $f(x + 6) = 5$.
40. Solve $\sqrt{x - 2} = \sqrt[4]{x - 2} + 12$ by first writing it in quadratic form as shown in Section 14.3.
41. Solve $\sqrt{x - 1} + x = 3$ algebraically. Then compare the solution with that of Example 4. Noting that the algebraic steps after isolating the radical are identical, why is the solution different?
42. The resonant frequency f in an electric circuit with an inductance L and a capacitance C is given by $f = \dfrac{1}{2\pi\sqrt{LC}}$. Solve for L.
43. A formula used in calculating the range R for radio communication is $R = \sqrt{2rh + h^2}$. Solve for h.
44. The distance d (in km) to the horizon from a height h (in km) above the surface of Earth is $d = \sqrt{1.28 \times 10^4 h + h^2}$. Find h for $d = 980$ km.
45. In the study of spur gears in contact, the equation $kC = \sqrt{R_1^2 - R_2^2} + \sqrt{r_1^2 - r_2^2} - A$ is used. Solve for r_1^2.
46. The speed s (in m/s) at which a tsunami wave moves is related to the depth d (in m) of the ocean according to $s = \sqrt{gd}$, where g is the acceleration of gravity (9.8 m/s^2). If a wave from the 2004 Indian Ocean tsunami was traveling at 195 m/s, estimate the depth of the ocean at that point.
47. If the value of a home increases from v_1 to v_2 over n years, the average annual rate of growth (as a decimal) is given by $r = \sqrt[n]{\dfrac{v_2}{v_1}} - 1$. Suppose the value of a home increases on average by 3.6% per year over 10 years. If its value at the end of the 10-year period is \$325,000, find its value at the beginning of the period.
48. A smaller of two cubical boxes is centered on the larger box, and they are taped together with a wide adhesive that just goes around both boxes (see Fig. 14.22). If the edge of the larger box is 1.00 in. greater than that of the smaller box, what are the lengths of the edges of the boxes if 100.0 in. of tape is used?

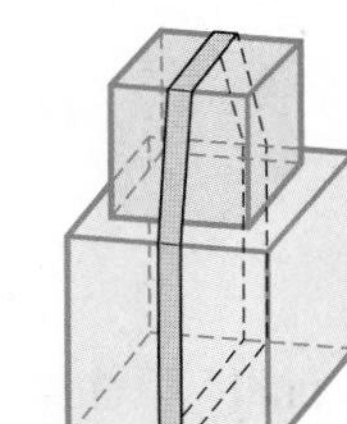

Fig. 14.22

49. A freighter is 5.2 km farther from a Coast Guard station on a straight coast than from the closest point A on the coast. If the station is 8.3 km from A, how far is it from the freighter?
50. The velocity v of an object that falls through a distance h is given by $v = \sqrt{2gh}$, where g is the acceleration due to gravity. Two objects are dropped from heights that differ by 10.0 m such that the sum of their velocities when they strike the ground is 20.0 m/s. Find the heights from which they are dropped if $g = 9.80$ m/s^2.
51. A point D on Denmark's largest island is 2.4 mi from the nearest point S on the coast of Sweden (assume the coast is straight, which is nearly the case). A person in a motorboat travels straight from D to a point on the beach x mi from S and then travels x mi farther along the beach away from S. Find x if the person traveled a total of 4.5 mi. See Fig. 14.23.

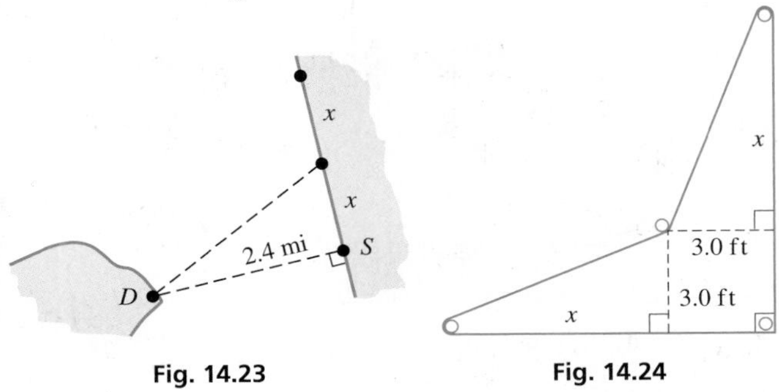

Fig. 14.23

Fig. 14.24

52. The length of the roller belt in Fig. 14.24 is 28.0 ft. Find x.

Answers to Practice Exercises

1. 3 **2.** 5 **3.** 1

CHAPTER 14 REVIEW EXERCISES

CONCEPT CHECK EXERCISES

*Determine each of the following as being either **true** or **false**. If it is false, explain why.*

1. To get the calculator display of the equation $2x^2 + y^2 = 4$, let $y_1 = \sqrt{4 - 2x^2}$.
2. A solution of the system $x^2 + 2y^2 = 9, 2x^2 - y = 3$, is $(-1, 2)$.
3. A solution of the equation $8x^{-4} - 2x^{-2} - 1 = 0$ is $(\sqrt{2}, 1)$.
4. A solution of the equation $\sqrt{3 - x} + 1 = 0$ is $x = 2$.

PRACTICE AND APPLICATIONS

In Exercises 5–14, solve the given systems of equations by use of a calculator.

5. $x + 2y = 9$, $y = 3x^2$
6. $x + y = 3$, $x^2 + y^2 = 25$
7. $\frac{1}{2}x + \frac{1}{3}y = 1$, $x^2 + 4y^2 = 4$
8. $x^2 - 2y = 0$, $y = 3x - 5$
9. $y = x^2 + 1$, $4x^2 + 16y^2 = 35$
10. $\frac{x^2}{4} + y^2 = 1$, $x^2 - y^2 = 1$
11. $y = 11 - x^2$, $y = 2x^2 - 1$
12. $x^2y = 63$, $y = 25 - 2x^2$
13. $y = x^2 - 2x$, $y = 1 - e^{-x} + x$
14. $y = 2 \ln x$, $y = \sin x$

In Exercises 15–24, solve each of the given systems of equations algebraically.

15. $y = 4x^2$, $y = 8x$
16. $x + y = 12$, $xy = 20$
17. $3R = L^2$, $R^2 + L^2 = 4$
18. $y = x^2$, $2x^2 - y^2 = 1$
19. $4u^2 + v = 3$, $6u + 9v = 3$
20. $x^2 + 7y^2 = 56$, $2x^2 - 8y^2 = 90$
21. $4x^2 - 7y^2 = 21$, $x^2 + 2y^2 = 99$
22. $s - t = 6$, $\sqrt{s} - \sqrt{t} = 1$
23. $4x^2 + 3xy = 4$, $x + 3y = 4$
24. $\frac{6}{x} + \frac{3}{y} = 4$, $\frac{36}{x^2} + \frac{36}{y^2} = 13$

In Exercises 25–46, solve the given equations.

25. $x^4 - 20x^2 + 64 = 0$
26. $t^6 - 26t^3 - 27 = 0$
27. $x^{3/2} - 9x^{3/4} + 8 = 0$
28. $x^{1/2} + 3x^{1/4} - 28 = 0$
29. $D^{-2} + 4D^{-1} - 21 = 0$
30. $6x - 9\sqrt{x} - 15 = 0$
31. $4(\ln x)^2 - \ln x^2 = 0$
32. $e^x + e^{-x} = 2$
33. $2(x + 1)^2 - 5(x + 1) = 3$
34. $\left(\frac{x}{x+1}\right)^2 + \frac{2x}{x+1} = 8$
35. $2t + 7 = 15\sqrt{t}$
36. $\sqrt{10 + 3\sqrt{x}} = \sqrt{x}$
37. $\frac{4}{r^2 + 1} + \frac{7}{2r^2 + 1} = 2$
38. $(x^2 + 5x)^2 - 5(x^2 + 5x) = 6$
39. $3\sqrt{2Z + 4} = 2Z$
40. $3\sqrt[3]{x - 2} = 9$
41. $\sqrt{5x + 9} + 1 = x$
42. $2\sqrt{5x - 3} - 1 = 2x$
43. $\sqrt{x + 1} + \sqrt{x} = 2$
44. $\sqrt{3x^2 - 2} - \sqrt{x^2 + 7} = 1$
45. $\sqrt{n + 4} + 2\sqrt{n + 2} = 3$
46. $\sqrt{3x - 2} - \sqrt{x + 7} = 1$

*In Exercises 47 and 48, find the value of x. (In each, the expression on the right is called a **continued radical.** Also, . . . means that the pattern continues indefinitely.) (Hint: Square both sides.) Noting the result, complete the solution and explain your method.*

47. $x = \sqrt{2 + \sqrt{2 + \sqrt{2 + \cdots}}}$
48. $x = \sqrt{6 - \sqrt{6 - \sqrt{6 - \cdots}}}$

In Exercises 49–54, solve the given equations algebraically and check the solutions with a calculator.

49. $x^3 - 2x^{3/2} - 48 = 0$
50. $(x + 1)^4 - 54 = 3(x + 1)^2$
51. $2\sqrt{3x + 1} - \sqrt{x - 1} = 6$
52. $3\sqrt{x} + \sqrt{x - 9} = 11$
53. $\sqrt[3]{x^3 - 7} = x - 1$
54. $\sqrt{x^2 + 7} + \sqrt[4]{x^2 + 7} = 6$

In Exercises 55–66, solve the given problems.

55. Solve for x and y: $x^2 - y^2 = 2a + 1$; $x - y = 1$
56. Solve for x: $\log(\sqrt{x} + 38) - \log x = 1$
57. Solve $\sqrt{\sqrt{x} - 1} = 2$ for x. Check using the graph on a calculator.
58. If $f(x) = \sqrt{8 - 2x}$ and $f(x + 1) = 2$, find x.
59. Use a graphing calculator to solve the following system of three equations: $x^2 + y^2 = 13, y = x - 1, xy = 6$.
60. Algebraically solve the following system of three equations: $y = -x^2, y = x - 1, xy = 1$. Explain the results.
61. Doctors can estimate the surface area A of an adult body (in m^2) using the equation $A = \sqrt{\frac{HW}{3600}}$, where H is height (in cm) and W is weight (in kg). Estimate the height of a person who has weight of 81.2 kg and a surface area of 1.25 m^2.
62. The frequency ω of a certain RLC circuit is given by $\omega = \frac{\sqrt{R^2 + 4(L/C)} + R}{2L}$. Solve for C.
63. In the theory dealing with a suspended cable, the equation $y = \sqrt{s^2 - m^2} - m$ is used. Solve for m.
64. The equation $V = e^2cr^{-2} - e^2Zr^{-1}$ is used in spectroscopy. Solve for r.

65. In an experiment, an object is allowed to fall, stop, and then fall for twice the initial time. The total distance the object falls is 45 ft. The equations relating the times t_1 and t_2 (in s) of fall are $16t_1^2 + 16t_2^2 = 45$ and $t_2 = 2t_1$. Find the times of fall.

66. If two objects collide and the kinetic energy remains constant, the collision is termed perfectly elastic. Under these conditions, if an object of mass m_1 and initial velocity u_1 strikes a second object (initially at rest) of mass m_2, such that the velocities after collision are v_1 and v_2, the following equations are found:

$$m_1u_1 = m_1v_1 + m_2v_2$$

$$\tfrac{1}{2}m_1u_1^2 = \tfrac{1}{2}m_1v_1^2 + \tfrac{1}{2}m_2v_2^2$$

Solve these equations for m_2 in terms of u_1, v_1, and m_1.

In Exercises 67–78, set up the appropriate equations and solve them.

67. A wrench is dropped by a worker at a construction site. Four seconds later the worker hears it hit the ground below. How high is the worker above the ground? (The velocity of sound is 1100 ft/s, and the distance the wrench falls as a function of time is $s = 16t^2$.)

68. The rectangular screen for a laptop computer has an area of 840 cm² and a perimeter of 119 cm. What are the dimensions of the screen?

69. The perimeter of a banner in the shape of an isosceles triangle is 72 in., and its area is 240 in.². Using a calculator, graphically find the lengths of the sides of the banner.

70. A rectangular field is enclosed by fencing and a wall along one side and half of an adjacent side. See Fig. 14.25. If the area of the field is 9000 ft² and 240 ft of fencing are used, what are the dimensions of the field?

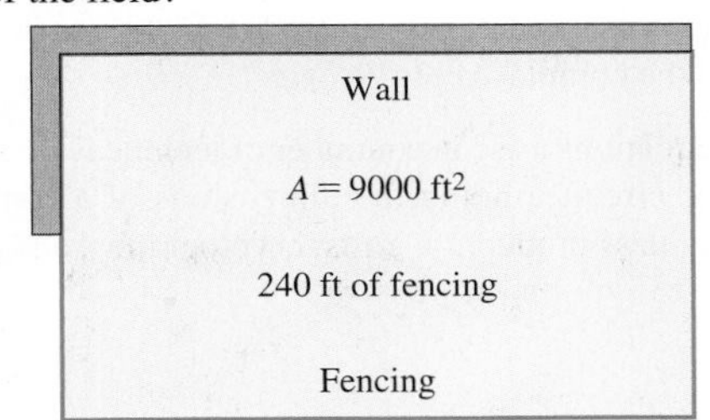

Fig. 14.25

71. A circuit on a computer chip is designed to be within the area shown in Fig. 14.26. If this part of the chip has an area of 9.0 mm² and a perimeter of 16 mm, find x and y.

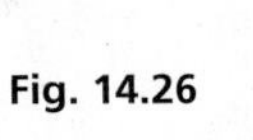

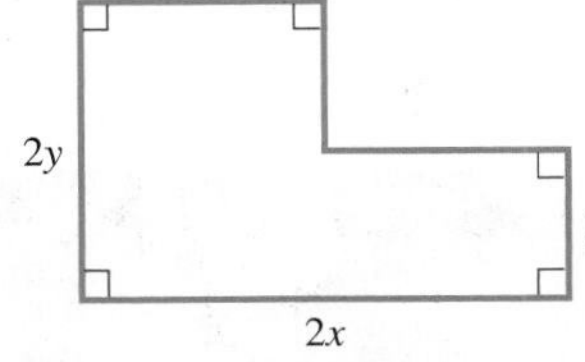

Fig. 14.26

72. For the plywood piece shown in Fig. 14.27, find x and y.

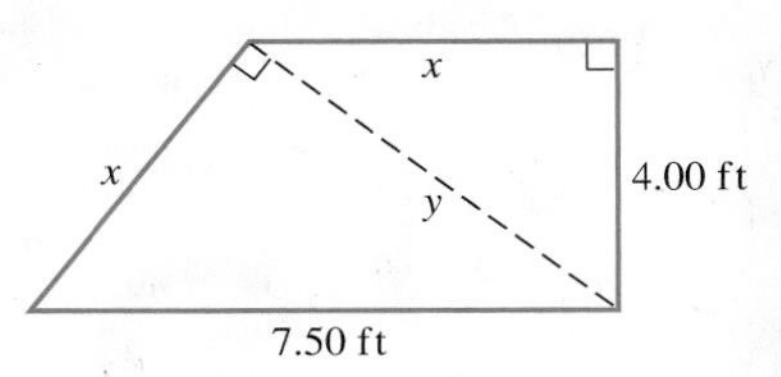

Fig. 14.27

73. The viewing window on a graphing calculator has an area of 1770 mm² and a diagonal of 62 mm. What are the length and width of the rectangle?

74. A trough is made from a piece of sheet metal 12.0 in. wide. The cross section of the trough is shown in Fig. 14.28. Find x.

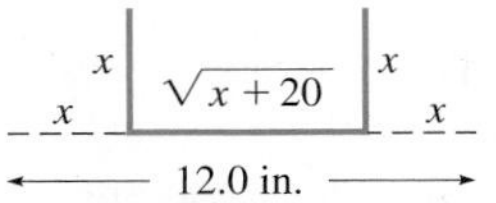

x $\quad \sqrt{x + 20} \quad$ x

x $\qquad\qquad$ x

← 12.0 in. →

Fig. 14.28

75. The circular solar cell and square solar cell shown in Fig. 14.29 have a combined surface area of 40.0 cm². Find the radius of the circular cell and the side of the square cell.

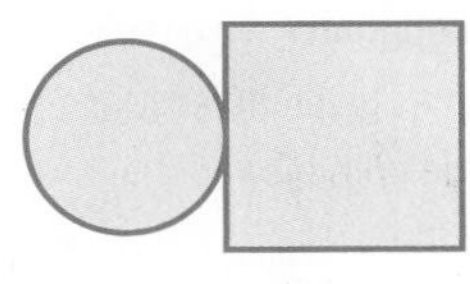

Fig. 14.29 ← 7.00 cm →

76. A plastic band 19.0 cm long is bent into the shape of a triangle with sides $\sqrt{x - 1}$, $\sqrt{5x - 1}$, and 9. Find x.

77. A Coast Guard ship travels from Houston to Mobile, and later it returns to Houston at a speed that is 6.0 mi/h faster. If Houston is 510 mi from Mobile and the total travel time is 35 h, find the speed of the ship in each direction.

78. Two trains are approaching the same crossing on tracks that are at right angles to each other. Each is traveling at 60.0 km/h. If one is 6.00 km from the crossing when the other is 3.00 km from it, how much later will they be 4.00 km apart (on a direct line)?

79. Using a computer, an engineer designs a triangular support structure with sides (in m) of x, $\sqrt{x - 1}$, and 4.00 m. If the perimeter is to be 9.00 m, the equation to be solved is $x + \sqrt{x - 1} + 4.00 = 9.00$. This equation can be solved by either of two methods used in this chapter. Write one or two paragraphs identifying the methods and explaining how they are used to solve the equation.

CHAPTER 14 PRACTICE TEST

As a study aid, we have included complete solutions for each Practice Test problem at the end of this book.

1. Solve for x: $x^{1/2} - 2x^{1/4} = 3$.
2. Solve for x: $3\sqrt{x - 2} - \sqrt{x + 1} = 1$.
3. Solve for x: $x^4 - 17x^2 + 16 = 0$.
4. Solve for x and y algebraically: $x^2 - 2y = 5$
   $2x + 6y = 1$
5. Solve for x: $\sqrt[3]{2x + 5} = 5$.
6. The velocity v of an object falling under the influence of gravity in terms of its initial velocity v_0, the acceleration due to gravity g, and the height h fallen is given by $v = \sqrt{v_0^2 + 2gh}$. Solve for h.
7. Solve for x and y graphically: $x^2 - y^2 = 4$
   $xy = 2$
8. A rectangular desktop has a perimeter of 14.0 ft and an area of 10.0 ft². Find the length and the width of the desktop.

15 Equations of Higher Degree

LEARNING OUTCOMES

After completion of this chapter, the student should be able to:

- Use the remainder theorem to evaluate polynomials and to find remainders
- Use the factor theorem to identify factors and zeros of polynomials
- Perform synthetic division
- Use the fundamental theorem of algebra to determine the number of roots of an equation
- Solve equations given at least one root
- Determine the possible rational roots of an equation
- Determine the maximum possible number of positive and negative roots by using Descartes' rule of signs
- Solve polynomial equations of degree three and higher
- Solve application problems involving polynomial equations

The errors in the hand-compiled mathematical tables used by the British mathematician Charles Babbage (1792–1871) led him to design a machine for solving polynomial equations.

A model of his *difference engine* was well received, but even after many years with government financing, a full-scale model was never successfully produced. Despite this setback, he then designed an *analytical engine,* which he hoped would perform many kinds of calculations. Although never built, it did have the important features of a modern computer: input, storage, control unit, and output. Because of this design, Babbage is often considered the father of the computer. (In 1992, his design was used successfully to make the device.)

We see that polynomials played an important role in the development of computers. Today, among many things, computers are used to solve equations, including polynomial equations, which include linear equations (first degree) and quadratic equations (second degree).

Various methods of solving higher-degree polynomial equations were developed in the 1500s and 1600s. Although known prior to 1650, the *fundamental theorem of algebra* (which we state later in the chapter) was proven by Karl Friedrich Gauss in 1799, and this was a big advancement in the solution of these equations. Today, Gauss is considered by many as the greatest mathematician of all time.

In this chapter, we study some of the methods that have been developed for solving higher-degree polynomial equations. The solutions we find include all possible roots, including complex-number roots. Some calculators are programmed to find all such roots, but graphical methods cannot be used to find the complex roots.

Applications of higher-degree equations arise in a number of technical areas. Included in these are the resistance in an electric circuit, finding the dimensions of a container or structure, robotics, calculating various business production costs, cryptography, and graphics design.

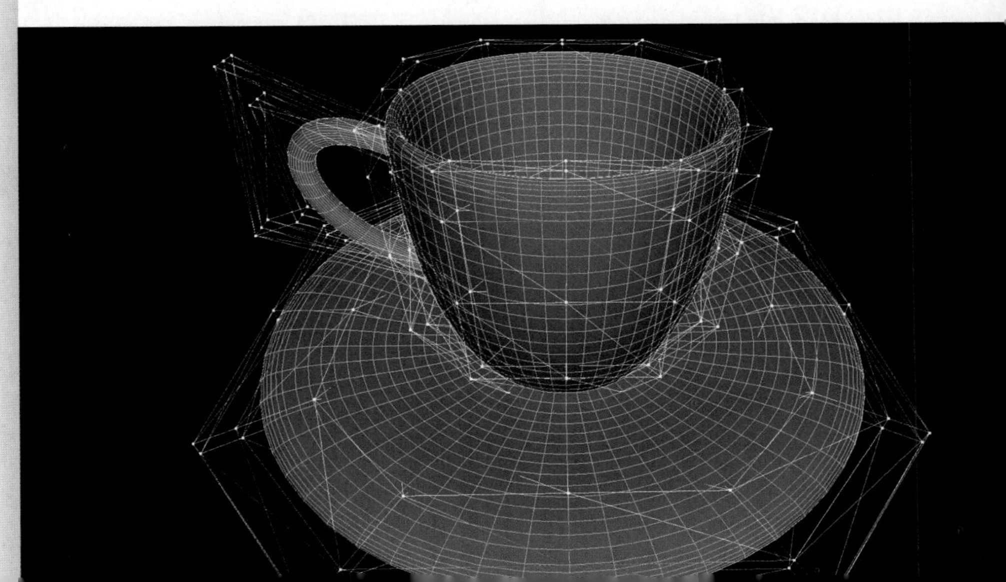

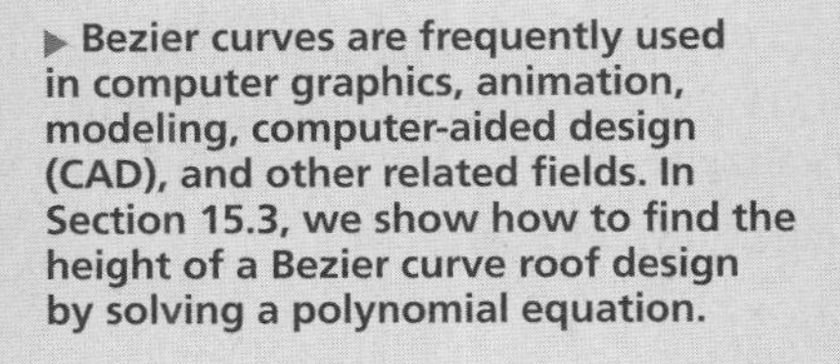

▶ Bezier curves are frequently used in computer graphics, animation, modeling, computer-aided design (CAD), and other related fields. In Section 15.3, we show how to find the height of a Bezier curve roof design by solving a polynomial equation.

15.1 The Remainder and Factor Theorems; Synthetic Division

Polynomial Function • Remainder Theorem • Factor Theorem • Synthetic Division

In solving higher-degree polynomial equations, the quadratic formula can be used for second-degree equations, and methods have been found for certain third- and fourth-degree equations. It can also be proven that polynomial equations of degree higher than 4 cannot in general be solved algebraically.

In this section, we present two theorems and a simplified method for algebraic division. These will help us in solving polynomial equations later in the chapter.

Any function of the form

$$f(x) = a_n x^n + a_{n-1}x^{n-1} + \cdots + a_0 \qquad (15.1)$$

where $a_n \neq 0$ and n is a positive integer or zero is called a **polynomial function.** We will be considering only polynomials in which the coefficients $a_0, a_1, \ldots, a_n$ are real numbers.

If we divide a polynomial by $x - r$, we find a result of the form

$$f(x) = (x - r)q(x) + R \qquad (15.2)$$

where $q(x)$ is the quotient and R is the remainder.

EXAMPLE 1 Division with remainder

$$\begin{array}{r} 3x + 11 \\ x - 2\overline{)3x^2 + 5x - 8} \\ -(3x^2 - 6x) \\ \hline 11x - 8 \\ -(11x - 22) \\ \hline 14 \end{array}$$

Divide $f(x) = 3x^2 + 5x - 8$ by $x - 2$.

The division is shown at the left, and it shows that

$$3x^2 + 5x - 8 = (x - 2)(3x + 11) + 14$$

where, for this function $f(x)$ with $r = 2$, we identify $q(x)$ and R as

$$q(x) = 3x + 11 \qquad R = 14$$

THE REMAINDER THEOREM

If we now set $x = r$ in Eq. (15.2), we have $f(r) = q(r)(r - r) + R = q(r)(0) + R$, or

$$f(r) = R \qquad (15.3)$$

This leads us to the **remainder theorem,** which is stated below.

Remainder Theorem

If a polynomial $f(x)$ is divided by $x - r$ until a constant remainder R is obtained, then $f(r) = R$.

This means the remainder equals the value of the function at $x = r$.

EXAMPLE 2 Verifying the remainder theorem

In Example 1, $f(x) = 3x^2 + 5x - 8$, $R = 14$, and $r = 2$.

We find that

$$\begin{aligned} f(2) &= 3(2^2) + 5(2) - 8 \\ &= 12 + 10 - 8 \\ &= 14 \end{aligned}$$

Therefore, $f(2) = 14$ verifies that $f(r) = R$ for this example.

EXAMPLE 3 Using the remainder theorem

By using the remainder theorem, determine the remainder when $3x^3 - x^2 - 20x + 5$ is divided by $x + 4$.

In using the remainder theorem, we determine the remainder when the function is divided by $x - r$ by evaluating the function for $x = r$. To have $x + 4$ in the proper form to identify r, we write it as $x - (-4)$. *This means that* $r = -4$, and we therefore evaluate the function $f(x) = 3x^3 - x^2 - 20x + 5$ for $x = -4$, or find $f(-4)$:

$$\begin{aligned} f(-4) &= 3(-4)^3 - (-4)^2 - 20(-4) + 5 = -192 - 16 + 80 + 5 \\ &= -123 \end{aligned}$$

The remainder is -123 when $3x^3 - x^2 - 20x + 5$ is divided by $x + 4$. ■

THE FACTOR THEOREM

The remainder theorem leads to another important theorem known as the **factor theorem.** It states that *if* $f(r) = R = 0$, *then* $x - r$ *is a factor of* $f(x)$. We see in Eq. (15.2) that if the remainder $R = 0$, then $f(x) = (x - r)q(x)$, and this shows that $x - r$ is a factor of $f(x)$. Therefore, we have the following meanings for $f(r) = 0$.

Zero, Factor, and Root of the Function $f(x)$

If $f(r) = 0$, then $x = r$ is a zero of $f(x)$,

$x - r$ is a factor of $f(x)$, and

$x = r$ is a root of the equation $f(x) = 0$

EXAMPLE 4 Using the factor theorem

(a) We determine that $t + 1$ is a factor of $f(t) = t^3 + 2t^2 - 5t - 6$ because $f(-1) = 0$, as we now show:

$$f(-1) = (-1)^3 + 2(-1)^2 - 5(-1) - 6 = -1 + 2 + 5 - 6 = 0$$

(b) However, $t + 2$ is not a factor of $f(t)$ because $f(-2)$ is not zero, as we now show:

$$f(-2) = (-2)^3 + 2(-2)^2 - 5(-2) - 6 = -8 + 8 + 10 - 6 = 4$$ ■

Practice Exercises

1. Use the factor theorem to determine whether $t - 2$ is a factor of $t^3 + 2t^2 - 5t - 4$.
2. Use the remainder theorem to determine whether $1/2$ is a zero of $f(x) = 2x^3 + 9x^2 - 11x + 3$.

SYNTHETIC DIVISION

In the sections that follow, we will find that division of a polynomial by the factor $x - r$ is also useful in solving polynomial equations. Therefore, we now develop a simplified form of long division, known as **synthetic division.** It allows us to easily find the coefficients of the quotient and the remainder. If the degree of the equation is high, it is easier to use synthetic division than to calculate $f(r)$. The method for synthetic division is developed in the following example.

EXAMPLE 5 Developing synthetic division

Divide $x^4 + 4x^3 - x^2 - 16x - 14$ by $x - 2$.

We first perform this division in the usual manner:

$$\begin{array}{r|l} & x^3 + 6x^2 + 11x + 6 \\ \hline x - 2 & x^4 + 4x^3 - x^2 - 16x - 14 \\ & x^4 - 2x^3 \\ & \quad 6x^3 - x^2 \\ & \quad 6x^3 - 12x^2 \\ & \qquad 11x^2 - 16x \\ & \qquad 11x^2 - 22x \\ & \qquad\qquad 6x - 14 \\ & \qquad\qquad 6x - 12 \\ & \qquad\qquad\quad -2 \end{array}$$

In doing the division, notice that we repeat many terms and that the only important numbers are the coefficients. This means there is no need to write in the powers of x. To the left of the division, we write it without x's and without identical terms.

$$\begin{array}{r|rrrrr} & 1 & 6 & 11 & 6 & \\ \hline -2 & 1 & 4 & -1 & -16 & -14 \\ & & -2 & & & \\ & & 6 & & & \\ & & & -12 & & \\ & & & 11 & & \\ & & & & -22 & \\ & & & & 6 & \\ & & & & & -12 \\ & & & & & -2 \end{array}$$

All numbers below the dividend may be written in two lines. Then all coefficients of the quotient, except the first, appear in the bottom line. Therefore, the line above the dividend is omitted, and we have the form at the left.

$$\begin{array}{r|rrrrr} -2 & 1 & 4 & -1 & -16 & -14 \\ & & -2 & -12 & -22 & -12 \\ \hline & & 6 & 11 & 6 & -2 \end{array}$$

Now, write the first coefficient (in this case, 1) in the bottom line. Also, change the -2 to 2, which is the actual value of r. Then in the form at the left, write the 2 on the right. *In this form, the* 1, 6, 11, *and* 6 *are the coefficients of the* x^3, x^2, x, *and constant term of the quotient. The* -2 *is the remainder.*

$$\begin{array}{rrrrr|l} 1 & 4 & -1 & -16 & -14 & 2 \\ & -2 & -12 & -22 & -12 & \\ \hline 1 & 6 & 11 & 6 & -2 & \end{array}$$

Finally, it is easier to use addition rather than subtraction in the process, so we change the signs of the numbers in the middle row. Remember that originally the bottom line was found by subtraction. Therefore, we have the last form on the left.

$$\begin{array}{rrrrr|l} 1 & 4 & -1 & -16 & -14 & 2 \\ & 2 & 12 & 22 & 12 & \\ \hline 1 & 6 & 11 & 6 & -2 & \end{array}$$

In the last form at the left, we have 1 (of the bottom row) $\times\ 2(= r) = 2$, the first number of the middle row. In the second column, $4 + 2 = 6$, the second number in the bottom row. Then, $6 \times 2\ (= r) = 12$, the second number of the second row; $-1 + 12 = 11$; $11 \times 2 = 22$; $22 + (-16) = 6$; $6 \times 2 = 12$; and $12 + (-14) = -2$.

We read the bottom line of the last form, the one we use in *synthetic division,* as

$$1x^3 + 6x^2 + 11x + 6 \text{ with a remainder of } -2$$

The method of synthetic division shown in the last form is outlined below. ■

Procedure for Synthetic Division

1. Write the coefficients of $f(x)$. Be certain that the powers are in descending order and that zeros are inserted for missing powers.
2. Carry down the left coefficient, then multiply it by r, and place this product under the second coefficient of the top line.
3. Add the two numbers in the second column and place the result below. Multiply this sum by r and place the product under the third coefficient of the top line.
4. Continue this process until the bottom row has as many numbers as the top row.

Using synthetic division, the last number in the bottom row is the remainder, and the other numbers are the respective coefficients of the quotient. The first term of the quotient is of degree one less than the first term of the dividend.

EXAMPLE 6 Using synthetic division

Divide $x^5 + 2x^4 - 4x^2 + 3x - 4$ by $x + 3$ using synthetic division.

Because the powers of x are in descending order, write down the coefficients of $f(x)$. In doing so, we must be certain to include a zero for the missing x^3 term. Next, note that the divisor is $x + 3$, which means that $r = -3$. The -3 is placed to the right. This gives us a top line of

coefficients → 1 2 0 −4 3 −4 |−3 ← r

Next, we carry the left coefficient, 1, to the bottom line and multiply it by r, -3, placing the product, -3, in the middle line under the second coefficient, 2. We then add the 2 and the -3 and place the result, -1, below. This gives

add

1	2	0	−4	3	−4	|−3
	−3	1 × −3 = −3				
1	−1					

Now, multiply the -1 by -3 $(= r)$ and place the result, 3, in the middle line under the zero. Now, add and continue the process, obtaining the following result:

1	2	0	−4	3	−4	|−3
	−3	3	−9	39	−126	
1	−1	3	−13	42	−130	

coefficients and constant of quotient; remainder

Because the degree of the dividend is 5, the degree of the quotient is 4. This means that the quotient is $x^4 - x^3 + 3x^2 - 13x + 42$ and the remainder is -130. In turn, this means that for $f(x) = x^5 + 2x^4 - 4x^2 + 3x - 4$, we have $f(-3) = -130$. ■

EXAMPLE 7 Using synthetic division

By synthetic division, divide $3x^4 - 5x + 6$ by $x - 4$.

3	0	0	−5	6	|4
	12	48	192	748	
3	12	48	187	754	

The quotient is $3x^3 + 12x^2 + 48x + 187$, and the remainder is 754. ■

Practice Exercise

3. Use synthetic division to divide $3x^3 - 5x + 6$ by $x + 2$.

EXAMPLE 8 Checking factor with synthetic division

By synthetic division, determine whether or not $t + 4$ is a factor of $t^4 + 2t^3 - 15t^2 - 32t - 16$.

1	2	−15	−32	−16	|−4
	−4	8	28	16	
1	−2	−7	−4	0	

Because the remainder is zero, $t + 4$ is a factor. We may also conclude that

$$f(t) = (t + 4)(t^3 - 2t^2 - 7t - 4)$$ ■

EXAMPLE 9 Checking rational factor

By using synthetic division, determine whether $2x - 3$ is a factor of $2x^3 - 3x^2 + 8x - 12$.

CAUTION We first note that the coefficient of x in the possible factor is not 1. Thus, we cannot use $r = 3$, because the factor is not of the form $x - r$. ■ However, $2x - 3 = 2(x - \frac{3}{2})$, which means that if $2(x - \frac{3}{2})$ is a factor of the function, $2x - 3$ is a factor. If we use $r = \frac{3}{2}$ and find that the remainder is zero, then $x - \frac{3}{2}$ is a factor.

$$\begin{array}{rrrr|l} 2 & -3 & 8 & -12 & \frac{3}{2} \\ & 3 & 0 & 12 & \\ \hline 2 & 0 & 8 & 0 & \end{array}$$

Because the remainder is zero, $x - \frac{3}{2}$ is a factor. Also, the quotient is $2x^2 + 8$, which may be factored into $2(x^2 + 4)$. Thus, 2 is also a factor of the function. This means that $2(x - \frac{3}{2})$ is a factor of the function, and this in turn means that $2x - 3$ is a factor. This tells us that

$$2x^3 - 3x^2 + 8x - 12 = (2x - 3)(x^2 + 4)$$ ■

EXAMPLE 10 Checking a zero

Determine whether or not -12.5 is a zero of the function $f(x) = 6x^3 + 61x^2 - 171x + 100$.

If -12.5 is a zero of $f(x)$, then $x - (-12.5)$, or $x + 12.5$, is a factor of $f(x)$, and $f(-12.5) = 0$. We can find the remainder by direct use of the remainder theorem or by synthetic division.

Using synthetic division and a calculator to make the calculations, we have the following setup and calculator sequence:

$$\begin{array}{rrrr|l} 6 & 61 & -171 & 100 & -12.5 \\ & -75 & 175 & -50 & \\ \hline 6 & -14 & 4 & 50 & \end{array}$$

Because the remainder is 50, and not zero, -12.5 is not a zero of $f(x)$. ■

Practice Exercise

4. Use synthetic division to determine whether $x + 2$ is a factor of $2x^3 + x^2 - 12x - 8$.

EXERCISES 15.1

In Exercises 1–4, make the given changes in the indicated examples of this section, and then perform the indicated operations.

1. In Example 3, change the $x + 4$ to $x + 3$ and then find the remainder.
2. In Example 4(a), change the $t + 1$ to $t - 1$ and then determine if $t - 1$ is a factor.
3. In Example 6, change the $x + 3$ to $x + 2$ and then perform the synthetic division.
4. In Example 9, change the $2x - 3$ to $2x + 3$ and then determine whether $2x + 3$ is a factor.

In Exercises 5–10, find the remainder by long division.

5. $(x^3 + 2x - 8) \div (x - 2)$
6. $(x^4 - 4x^3 - x^2 + x - 100) \div (x + 3)$
7. $(2x^5 - x^2 + 8x + 44) \div (x + 1)$
8. $(4s^3 - 9s^2 - 24s - 17) \div (s - 5)$
9. $(2x^4 - 3x^3 - 2x^2 - 15x - 16) \div (2x - 3)$
10. $(2x^4 - 11x^2 - 15x - 17) \div (2x + 1)$

In Exercises 11–16, find the remainder using the remainder theorem. Do not use synthetic division.

11. $(R^4 + R^3 - 9R^2 + 3) \div (R - 3)$
12. $(4x^4 - x^3 + 5x - 7) \div (x - 5)$
13. $(2x^4 - 7x^3 - x^2 + 8) \div (x + 1)$
14. $(3n^4 - 13n^2 + 10n - 10) \div (n + 4)$
15. $(x^5 - 3x^3 + 5x^2 - 10x + 6) \div (x + 2)$
16. $(3x^4 - 12x^3 - 60x + 4) \div (x - 0.5)$

In Exercises 17–22, use the factor theorem to determine whether or not the second expression is a factor of the first expression. Do not use synthetic division.

17. $8x^3 + 2x^2 - 32x - 8,\ x - 2$
18. $3x^3 + 14x^2 + 7x - 4,\ x + 4$
19. $3V^4 - 7V^3 + V + 8,\ V - 2$
20. $x^5 - 2x^4 + 3x^3 - 6x^2 - 4x + 8,\ x - 1$
21. $x^{51} - 2x - 1,\ x + 1$
22. $x^7 - 128^{-1},\ x + 2^{-1}$

In Exercises 23–32, perform the indicated divisions by synthetic division.

23. $(x^3 + 2x^2 - x - 2) \div (x - 1)$

24. $(x^3 - 3x^2 - x + 2) \div (x - 3)$

25. $(x^3 + 2x^2 - 3x + 4) \div (x + 4)$

26. $(2x^3 - 4x^2 + x - 1) \div (x + 2)$

27. $(p^6 - 6p^3 - 2p^2 - 6) \div (p - 2)$

28. $(x^5 + 4x^4 - 8) \div (x + 1)$

29. $(x^7 - 128) \div (x - 2)$

30. $(20x^4 + 11x^3 - 89x^2 + 60x - 77) \div (x + 2.75)$

31. $(2x^4 + x^3 + 3x^2 - 1) \div (2x - 1)$

32. $(6t^4 + 5t^3 - 10t + 4) \div (3t - 2)$

In Exercises 33–40, use the factor theorem and synthetic division to determine whether or not the second expression is a factor of the first.

33. $2x^5 - x^3 + 3x^2 - 4;\quad x + 1$

34. $t^5 - 3t^4 - t^2 - 6;\quad t - 3$

35. $4x^3 - 9x^2 + 2x - 2;\quad x - \frac{1}{4}$

36. $3x^3 - 5x^2 + x + 1;\quad x + \frac{1}{3}$

37. $2Z^4 - Z^3 - 4Z^2 + 1;\quad 2Z - 1$

38. $6x^4 + 5x^3 - x^2 + 6x - 2;\quad 3x - 1$

39. $4x^4 + 2x^3 - 8x^2 + 3x + 12;\quad 2x + 3$

40. $3x^4 - 2x^3 + x^2 + 15x + 4;\quad 3x + 4$

In Exercises 41–44, use synthetic division to determine whether or not the given numbers are zeros of the given functions.

41. $x^4 - 5x^3 - 15x^2 + 5x + 14;\quad 7$

42. $r^4 + 5r^3 - 18r - 8;\quad -4$

43. $85x^3 + 348x^2 - 263x + 120;\quad -4.8$

44. $2x^3 + 13x^2 + 10x - 4;\quad \frac{1}{2}$

In Exercises 45–60, solve the given problems.

45. If $f(x) = 2x^3 + 3x^2 - 19x - 4$, and $f(x) = (x + 4)g(x)$, find $g(x)$.

46. Using synthetic division, divide $ax^2 + bx + c$ by $x + 1$.

47. By division, show that $2x - 1$ is a factor of $f(x) = 4x^3 + 8x^2 - x - 2$. May we therefore conclude that $f(1) = 0$? Explain.

48. By division, show that $x^2 + 2$ is a factor of $f(x) = 3x^3 - x^2 + 6x - 2$. May we therefore conclude that $f(-2) = 0$? Explain.

49. For what value of k is $x - 2$ a factor of $f(x) = 2x^3 + kx^2 - x + 14$?

50. For what value of k is $x + 1$ a factor of $f(x) = 3x^4 + 3x^3 + 2x^2 + kx - 4$?

51. Use synthetic division: $(x^3 - 3x^2 + x - 3) \div (x + j)$.

52. Use synthetic division: $(2x^3 - 7x^2 + 10x - 6) \div [x - (1 + j)]$.

53. If $f(x) = -g(x)$, do the functions have the same zeros? Explain.

54. Do the functions $f(x)$ and $f(-x)$ have the same zeros? Explain.

55. If $f(x) = 3x^3 - 5ax^2 - 3a^2x + 5a^3$, find $f(x) \div (x + a)$.

56. The length of a rectangular box is 3 cm longer than its width. If the volume as a function of the width is $f(w) = 2w^3 + 5w^2 - 3w$, find the height if the box.

57. In finding the electric current in a certain circuit, it is necessary to factor the denominator of $\frac{2s}{s^3 + 5s^2 + 4s + 20}$. Is (a) $(s - 2)$ or (b) $(s + 5)$ a factor?

58. In the theory of the motion of a sphere moving through a fluid, the function $f(r) = 4r^3 - 3ar^2 - a^3$ is used. Is (a) $r = a$ or (b) $r = 2a$ a zero of $f(r)$?

59. In finding the volume V (in cm^3) of a certain gas in equilibrium with a liquid, it is necessary to solve the equation $V^3 - 6V^2 + 12V = 8$. Use synthetic division to determine if $V = 2$ cm^3.

60. An architect is designing a window in the shape of a segment of a circle. An approximate formula for the area is $A = \frac{h^3}{2w} + \frac{2wh}{3}$, where A is the area, w is the width, and h is the height of the segment. If the width is 1.500 m and the area is 0.5417 m^2, use synthetic division to show that $h = 0.500$ m.

Answers to Practice Exercises

1. No $(R = 2)$ **2.** Yes $(R = 0)$
3. Quotient: $3x^2 - 6x + 7$, $R = -8$ **4.** No $(R = 4)$

15.2 The Roots of an Equation

The Fundamental Theorem of Algebra • Linear Factors • Complex Roots • Remaining Quadratic Factor

In this section, we present certain theorems that are useful in determining the number of roots in the equation $f(x) = 0$ and the nature of some of these roots. In dealing with polynomial equations of higher degree, it is helpful to have as much of this kind of information as we can find before actually solving for all of the roots, including any possible complex roots. A graph does not show the complex roots an equation may have, but we can verify the real roots, as we show by use of a graphing calculator in some of the examples that follow.

The first of these theorems is so important that it is called the **fundamental theorem of algebra.** It states:

Every polynomial equation has at least one (real or complex) root.

■ This theorem was first proven in 1799 by the German mathematician Karl Gauss (1777–1855) for his doctoral thesis. See the chapter introduction on page 420.

The proof of this theorem is of an advanced nature, and therefore we accept its validity at this time. However, using the fundamental theorem, we can show the validity of other theorems that are useful in solving equations.

Let us now assume that we have a polynomial equation $f(x) = 0$ and that we are looking for its roots. By the fundamental theorem, we know that it has at least one root. Assuming that we can find this root by some means (the factor theorem, for example), we call this root r_1. Thus,

$$f(x) = (x - r_1)f_1(x)$$

where $f_1(x)$ is the polynomial quotient found by dividing $f(x)$ by $(x - r_1)$. However, because the fundamental theorem states that any polynomial equation has at least one root, this must apply to $f_1(x) = 0$ as well. Let us assume that $f_1(x) = 0$ has the root r_2. Therefore, this means that $f(x) = (x - r_1)(x - r_2)f_2(x)$. Continuing this process until one of the quotients is a constant a, we have

$$f(x) = a(x - r_1)(x - r_2)\cdots(x - r_n)$$

Note that one linear factor appears each time a root is found and that the degree of the quotient is one less each time. Thus, there are n linear factors, if the degree of $f(x)$ is n.

Therefore, based on the fundamental theorem of algebra, we have the following two related theorems.

Consequences of the Fundamental Theorem of Algebra

1. A polynomial of the *nth* degree can be factored into n linear factors.
2. A polynomial equation of degree n has exactly n roots.

These theorems are illustrated in the following example.

EXAMPLE 1 Illustrating fundamental theorem

For the function $f(x) = 2x^4 - 3x^3 - 12x^2 + 7x + 6 = 0$, we are given the factors that we show. In the next section, we will see how to find these factors.

For the function $f(x)$, we have

$$2x^4 - 3x^3 - 12x^2 + 7x + 6 = (x - 3)(2x^3 + 3x^2 - 3x - 2)$$
$$2x^3 + 3x^2 - 3x - 2 = (x + 2)(2x^2 - x - 1)$$
$$2x^2 - x - 1 = (x - 1)(2x + 1)$$
$$2x + 1 = 2(x + \tfrac{1}{2})$$

Therefore,

$$2x^4 - 3x^3 - 12x^2 + 7x + 6 = 2(x - 3)(x + 2)(x - 1)(x + \tfrac{1}{2}) = 0$$

The degree of $f(x)$ is 4. There are four linear factors: $(x - 3)$, $(x + 2)$, $(x - 1)$, and $(x + \frac{1}{2})$. There are four roots of the equation: 3, -2, 1, and $-\frac{1}{2}$. Thus, we have verified each of the theorems above for this example. ■

It is not necessary for each root of an equation to be different from the other roots. Actually, there is no limit as to the number of roots an equation can have that are the same. For example, the equation $(x - 1)^{16} = 0$ has sixteen roots, all of which are 1. Such roots are referred to as *multiple* (or *repeated*) roots. In this case, the root of 1 has a *multiplicity* of 16.

When we solve the equation $x^2 + 1 = 0$, the roots are j and $-j$. In fact, for any equation (with real coefficients) that has a root of the form $a + bj$ $(b \neq 0)$, there is also a root of the form $a - bj$. This is summarized below.

Complex Conjugate Root Theorem

If the coefficients of the equation $f(x) = 0$ are real and $a + bj$ $(b \neq 0)$ is a complex root, then its conjugate, $a - bj$, is also a root.

EXAMPLE 2 Illustrating complex roots

Consider the equation $f(x) = (x - 1)^3(x^2 + x + 1) = 0$.

The factor $(x - 1)^3$ shows that there is a triple root of 1, and there is a total of five roots, because the highest-power term would be x^5 if we were to multiply out the function. To find the other two roots, we use the quadratic formula on the *factor* $(x^2 + x + 1)$. This is permissible, because we are finding the values of x for

$$x^2 + x + 1 = 0$$

For this, we have

$$x = \frac{-1 \pm \sqrt{1 - 4}}{2}$$

Thus,

$$x = \frac{-1 + j\sqrt{3}}{2} \quad \text{and} \quad x = \frac{-1 - j\sqrt{3}}{2}$$

Fig. 15.1

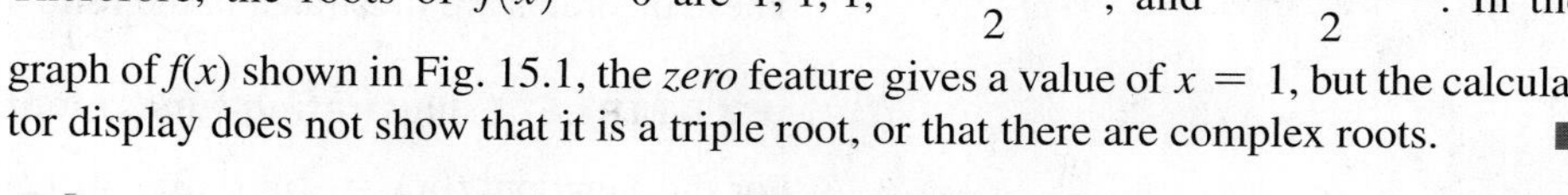

Therefore, the roots of $f(x) = 0$ are $1, 1, 1, \frac{-1 + j\sqrt{3}}{2}$, and $\frac{-1 - j\sqrt{3}}{2}$. In the graph of $f(x)$ shown in Fig. 15.1, the *zero* feature gives a value of $x = 1$, but the calculator display does not show that it is a triple root, or that there are complex roots. ■

NOTE ▸ [From Example 2, we can see that whenever enough roots are known so that the remaining factor is quadratic, it is possible to find the remaining roots from the quadratic formula.] This is true for finding real or complex roots.

EXAMPLE 3 Solving equation—given one root

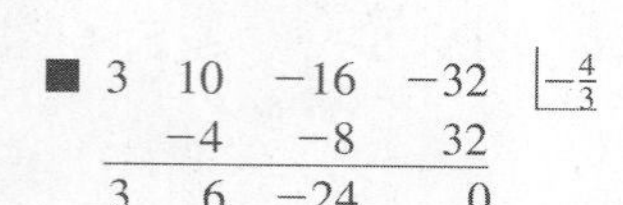

■ 3	10	−16	−32	$\lfloor -\frac{4}{3}$
	−4	−8	32	
3	6	−24	0	

Solve the equation $3x^3 + 10x^2 - 16x - 32 = 0$; $-\frac{4}{3}$ is a root.

Using synthetic division and the given root, we have the setup shown at the left. From this, we see that

$$3x^3 + 10x^2 - 16x - 32 = \left(x + \tfrac{4}{3}\right)(3x^2 + 6x - 24)$$

We know that $x + \frac{4}{3}$ is a factor from the given root and that $3x^2 + 6x - 24$ is a factor found from synthetic division. This second factor can be factored as

$$3x^2 + 6x - 24 = 3(x^2 + 2x - 8) = 3(x + 4)(x - 2)$$

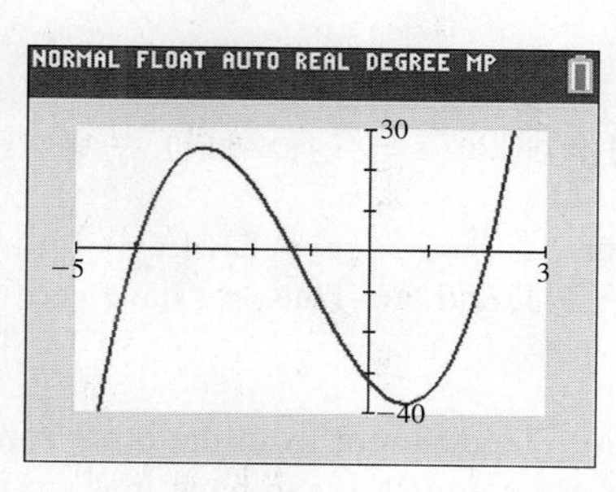

Fig. 15.2

Therefore, we have

$$3x^3 + 10x^2 - 16x - 32 = 3\left(x + \tfrac{4}{3}\right)(x + 4)(x - 2)$$

This means the roots are $-\frac{4}{3}$, -4, and 2. Because the three roots are real, they can be found from a calculator display, as shown in Fig. 15.2. ■

Practice Exercise

1. Solve the equation in Example 3, given that -4 is a root.

EXAMPLE 4 Solving equation—given two roots

1	3	−4	−10	−4	−1
	−1	−2	6	4	
1	2	−6	−4	0	

Solve $x^4 + 3x^3 - 4x^2 - 10x - 4 = 0$; -1 and 2 are roots.

Using synthetic division and the root -1, the first setup at the left shows that

$$x^4 + 3x^3 - 4x^2 - 10x - 4 = (x + 1)(x^3 + 2x^2 - 6x - 4)$$

We now know that $x - 2$ must be a factor of $x^3 + 2x^2 - 6x - 4$, because it is a factor of the original function. Again, using synthetic division and this time the root 2, we have the second setup at the left. Thus,

1	2	−6	−4	2
	2	8	4	
1	4	2	0	

$$x^4 + 3x^3 - 4x^2 - 10x - 4 = (x + 1)(x - 2)(x^2 + 4x + 2)$$

Because the original equation can now be written as

$$(x + 1)(x - 2)(x^2 + 4x + 2) = 0$$

the remaining two roots are found by solving

$$x^2 + 4x + 2 = 0$$

by the quadratic formula. This gives us

$$x = \frac{-4 \pm \sqrt{16 - 8}}{2} = \frac{-4 \pm 2\sqrt{2}}{2} = -2 \pm \sqrt{2}$$

Therefore, the roots are $-1, 2, -2 + \sqrt{2}$, and $-2 - \sqrt{2}$. ■

Practice Exercise

2. Solve $x^4 - x^3 - 2x^2 - 4x - 24 = 0$, given that -2 and 3 are roots.

EXAMPLE 5 Solving equation—given double root

Solve the equation $3x^4 - 26x^3 + 63x^2 - 36x - 20 = 0$, given that 2 is a double root.

Using synthetic division, we have the first setup at the left. It tells us that

$$3x^4 - 26x^3 + 63x^2 - 36x - 20 = (x - 2)(3x^3 - 20x^2 + 23x + 10)$$

3	−26	63	−36	−20	2
	6	−40	46	20	
3	−20	23	10	0	

Also, because 2 is a double root, it must be a root of $3x^3 - 20x^2 + 23x + 10 = 0$. Using synthetic division again, we have the second setup at the left. This second quotient $3x^2 - 14x - 5$ factors into $(3x + 1)(x - 5)$. The roots are 2, 2, $-\frac{1}{3}$, and 5.

3	−20	23	10	2
	6	−28	−10	
3	−14	−5	0	

NOTE ➧ [Because the quotient of the first division is the dividend for the second division, both divisions can be done without rewriting the first quotient as follows:]

	3	−26	63	−36	−20	2
first division →		6	−40	46	20	
	3	−20	23	10	0	2
second division →		6	−28	−10		
	3	−14	−5	0		

■

EXAMPLE 6 Solving equation—given complex root

Solve the equation $2x^4 - 5x^3 + 11x^2 - 3x - 5 = 0$, given that $1 + 2j$ is a root.

Because $1 + 2j$ is a root, we know that $1 - 2j$ is also a root. Using synthetic division twice, we can then reduce the remaining factor to a quadratic function.

2	−5	11	−3	−5	$1 + 2j$
	$2 + 4j$	$-11 - 2j$	$4 - 2j$	5	
2	$-3 + 4j$	$-2j$	$1 - 2j$	0	$1 - 2j$
	$2 - 4j$	$-1 + 2j$	$-1 + 2j$		
2	−1	−1	0		

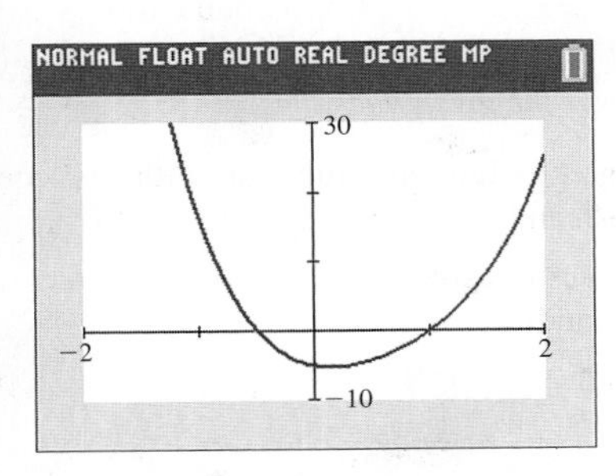

Fig. 15.3

The quadratic factor $2x^2 - x - 1$ factors into $(2x + 1)(x - 1)$. Therefore, the roots of the equation are $1 + 2j$, $1 - 2j$, 1, and $-\frac{1}{2}$. As we can see, the calculator display in Fig. 15.3 shows only the real roots. ■

GRAPHICAL FEATURES INDICATING NUMBER AND TYPES OF ROOTS

If $f(x)$ is of degree n and the roots of $f(x) = 0$ are real and different, the graph of $f(x)$ crosses the x-axis n times and $f(x)$ changes signs as it crosses. See Fig. 15.4, where $n = 4$, the equation has four changes of sign, and the graph crosses four times.

For each pair of complex roots, the number of times the graph crosses the x-axis is reduced by two (Fig. 15.5). For multiple roots, the graph crosses the x-axis once if the multiple is odd, or is tangent to the x-axis if the multiple is even (Fig. 15.6).

The graph must cross the x-axis at least once if the degree of $f(x)$ is odd, because the range includes all real numbers (Figs. 15.5 and 15.6). The curve may not cross the x-axis if the degree of $f(x)$ is even, because the range is bounded at a minimum point or a maximum point (Fig. 15.7), as we have seen for a quadratic function.

$y = f(x)$
$= x^4 - 5x^3 + 5x^2 + 5x - 6$
$= (x+1)(x-1)(x-2)(x-3)$

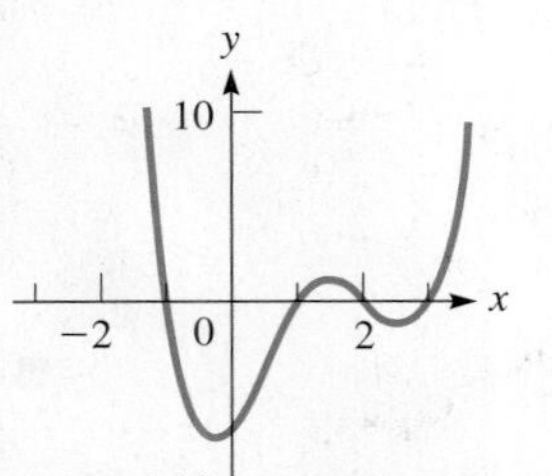

Fig. 15.4

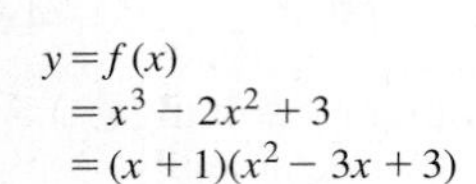

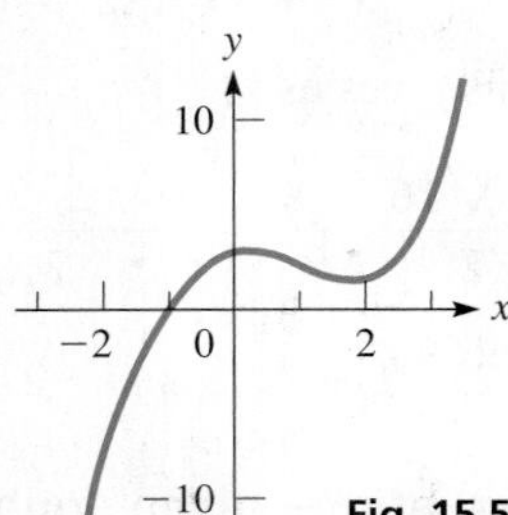

Fig. 15.5

$y = f(x)$
$= x^5 + x^4 - 2x^3 - 2x^2 + x + 1$
$= (x+1)^3(x-1)^2$

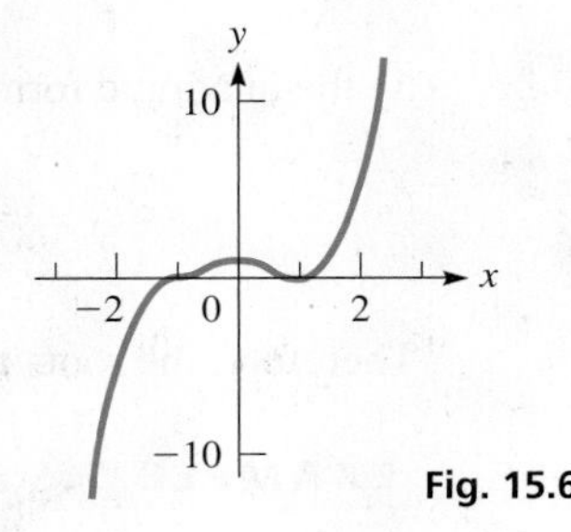

Fig. 15.6

$y = f(x)$
$= x^4 + 4$
$= (x^2 - 2x + 2)(x^2 + 2x + 2)$

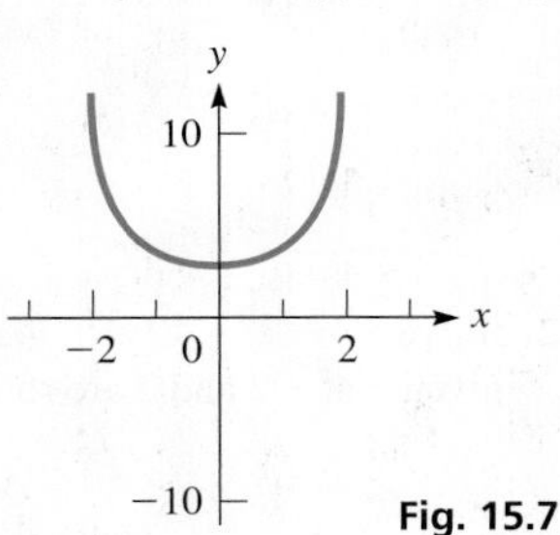

Fig. 15.7

EXERCISES 15.2

In Exercises 1 and 2, make the given changes in the indicated examples of this section and then solve the resulting equation.

1. In Example 2, change the middle term of the second factor to $2x$.
2. In Example 6, change the middle three terms to the left of the $=$ sign to $-3x^3 + 7x^2 + 7x$, given the same root.

In Exercises 3–6, find the roots of the given equations by inspection.

3. $(x + 3)(x^2 - 4) = 0$
4. $x(2x + 5)^2(x^2 - 64) = 0$
5. $(x - 5)(x^2 + 9)$
6. $(4y^2 + 9)(25y^2 - 10y + 1) = 0$

In Exercises 7–26, find the remaining roots of the given equations using synthetic division, given the roots indicated.

7. $x^3 - 5x^2 + 2x + 8 = 0 \quad (r_1 = 2)$
8. $R^3 + 1 = 0 \quad (r_1 = -1)$
9. $2x^3 + 11x^2 + 20x + 12 = 0 \quad (r_1 = -\frac{3}{2})$
10. $4x^3 + 6x^2 - 2x - 1 = 0 \quad (r_1 = \frac{1}{2})$
11. $5x^3 - 2x^2 + 5x - 2 = 0 \quad (r_1 = j)$
12. $3x^3 + 15x^2 + 27x + 15 = 0 \quad (r_1 = -2 + j)$
13. $t^3 - 7t^2 + 17t - 15 = 0 \quad (r_1 = 2 + j)$
14. $x^4 - 2x^3 - 20x^2 - 8x - 96 = 0 \quad (r_1 = 6, r_2 = -4)$
15. $2x^4 - 19x^3 + 39x^2 + 35x - 25 = 0$ (5 is a double root)
16. $4n^4 + 28n^3 + 61n^2 + 42n + 9 = 0$ (-3 is a double root)
17. $6x^4 + 5x^3 - 15x^2 + 4 = 0 \quad (r_1 = -\frac{1}{2}, r_2 = \frac{2}{3})$
18. $6x^4 - 5x^3 - 14x^2 + 14x - 3 = 0 \quad (r_1 = \frac{1}{3}, r_2 = \frac{3}{2})$
19. $2x^4 - x^3 - 4x^2 + 10x - 4 = 0 \quad (r_1 = 1 + j)$
20. $s^4 - 8s^3 - 72s - 81 = 0 \quad (r_1 = 3j)$
21. $x^5 - 3x^4 + 4x^3 - 4x^2 + 3x - 1 = 0$ (1 is a triple root)
22. $12x^5 - 7x^4 + 41x^3 - 26x^2 - 28x + 8 = 0$ $(r_1 = 1, r_2 = \frac{1}{4}, r_3 = -\frac{2}{3})$
23. $P^5 - 3P^4 - P + 3 = 0 \quad (r_1 = 3, r_2 = j)$
24. $4x^5 + x^3 - 4x^2 - 1 = 0 \quad (r_1 = 1, r_2 = \frac{1}{2}j)$
25. $x^6 + 2x^5 - 4x^4 - 10x^3 - 41x^2 - 72x - 36 = 0$ (-1 is a double root; $2j$ is a root)
26. $x^6 - x^5 - 2x^3 - 3x^2 - x - 2 = 0$ (j is a double root)

In Exercises 27 and 28, answer the given questions.

27. Why cannot a third-degree polynomial function with real coefficients have zeros of 1, 2, and j?
28. How can the graph of a fourth-degree polynomial equation have its only x-intercepts as 0, 1, and 2?

In Exercises 29–32, solve the given problems.

29. Find k such that $x - 2$ is a factor of $2x^3 + kx^2 - kx - 2$.

30. Find k such that $x - 1$ is a factor of $x^3 - 4x^2 - kx + 2$.

31. Equations of the form $y^2 = x^3 + ax + b$ are called *elliptic curves* and are used in cryptography. If $y = 3$ for the curve $y^2 = x^3 - 4x + 6$, use synthetic division to show that one possible value of x is $x = -1$. Then find any other possible values of x.

32. Form a polynomial equation of the smallest possible degree and with integer coefficients, having a double root of 3, and a root of j.

Answers to Practice Exercises

1. $-4, -4/3, 2$ **2.** $-2, 3, 2j, -2j$

15.3 Rational and Irrational Roots

Rational Root Theorem • Descartes' Rule of Signs • Roots of a Polynomial Equation

The product $(x + 2)(x - 4)(x + 3)$ equals $x^3 + x^2 - 14x - 24$. Here, we find that the constant 24 is determined only by the 2, 4, and 3. These numbers represent the roots of the equation if the given function is set equal to zero. In fact, if we find all the integer roots of an equation with integer coefficients and represent the equation in the form

$$f(x) = (x - r_1)(x - r_2) \cdots (x - r_k)f_{k+1}(x) = 0$$

where all the roots indicated are integers, the constant term of $f(x)$ must have factors of $r_1, r_2, \ldots, r_k$. This leads us to the theorem that states that

in a polynomial equation $f(x) = 0$, if the coefficient of the highest power is 1, then any integer roots are factors of the constant term of $f(x)$.

EXAMPLE 1 Possible integer roots if $a_0 = 1$

The equation $x^5 - 4x^4 - 7x^3 + 14x^2 - 44x + 120 = 0$ can be written as

$$(x - 5)(x + 3)(x - 2)(x^2 + 4) = 0$$

We now note that $5(3)(2)(4) = 120$. Thus, the roots 5, -3, and 2 are numerical factors of $|120|$. The theorem states nothing about the signs involved. ■

If the coefficient a_n of the highest-power term of $f(x)$ is an integer not equal to 1, the polynomial equation $f(x) = 0$ may have rational roots that are not integers. We can factor a_n from every term of $f(x)$. Thus, any polynomial equation $f(x) = a_nx^n + a_{n-1}x^{n-1} + \cdots + a_0 = 0$ with integer coefficients can be written in the form

$$f(x) = a_n\left(x^n + \frac{a_{n-1}}{a_n}x^{n-1} + \cdots + \frac{a_0}{a_n}\right) = 0$$

Because a_0 and a_n are integers, a_0/a_n is a rational number. Using the same reasoning as with integer roots applied to the polynomial within the parentheses, we see that any rational roots are factors of a_0/a_n This leads to the following theorem:

Rational Root Theorem

Any rational root r_r of a polynomial equation (with integer coefficients)

$$f(x) = a_nx^n + a_{n-1}x^{n-1} + \cdots + a_0 = 0$$

is an integer factor of a_0 divided by an integer factor of a_n. Therefore,

$$r_r = \frac{\text{integer factor of } a_0}{\text{integer factor of } a_n} \qquad \textbf{(15.4)}$$

EXAMPLE 2 Possible rational roots

If $f(x) = 4x^3 - 3x^2 - 25x - 6$, any possible rational roots must be integer factors of 6 divided by integer factors of 4. These factors of 6 are 1, 2, 3, and 6, and these factors of 4 are 1, 2, and 4. Forming all possible positive and negative quotients, any possible rational roots that exist will be found in the following list: $\pm 1, \pm\frac{1}{2}, \pm\frac{1}{4}, \pm 2, \pm 3, \pm\frac{3}{2}, \pm\frac{3}{4}, \pm 6$.

The roots of this equation are -2, 3, and $-\frac{1}{4}$. ■

There are 16 different possible rational roots in Example 2, but we cannot tell which of these are the actual roots. Therefore we now present a rule, known as *Descartes' rule of signs,* which will help us to find these roots.

■ Named for the French mathematician René Descartes (1596–1650). (See page 94).

Descartes' Rule of Signs

1. The number of positive roots of a polynomial equation $f(x) = 0$ cannot exceed the number of changes in sign in $f(x)$ in going from one term to the next in $f(x)$.
2. The number of negative roots cannot exceed the number of sign changes in $f(-x)$.

We can reason this way: If $f(x)$ has all positive terms, then any positive number substituted in $f(x)$ must give a positive value for $f(x)$. This indicates that the number substituted in the function is not a root. Thus, there must be at least one negative and one positive term in the function for any positive number to be a root. This is not a proof, but it does indicate the type of reasoning used in developing the theorem.

EXAMPLE 3 Using Descartes' rule of signs

By Descartes' rule of signs, determine the maximum possible number of positive and negative roots of $3x^3 - x^2 - x + 4 = 0$.

Here, $f(x) = 3x^3 - x^2 - x + 4$. Reading the terms from left to right, there are two changes in sign, which we can show as follows:

$$f(x) = 3x^3 - x^2 - x + 4$$

1 2 ← two sign changes

Because there are *two* changes of sign in $f(x)$, there are *no more than two* positive roots of $f(x) = 0$.

To find the maximum possible number of negative roots, we must find the number of sign changes in $f(-x)$. Thus,

$$\begin{aligned} f(-x) &= 3(-x)^3 - (-x)^2 - (-x) + 4 \\ &= -3x^3 - x^2 + x + 4 \end{aligned}$$

← one sign change

Practice Exercise

1. Determine the maximum possible number of positive and negative roots of $9x^4 - 12x^3 - 5x^2 + 12x - 4 = 0$.

NOTE ▸ There is only one change of sign in $f(-x)$; therefore, there is one negative root. [When there is just one change of sign in $f(x)$, there is a positive root, and when there is just one change of sign in $f(-x)$, there is a negative root.] ■

EXAMPLE 4 Using Descartes' rule of signs

For the equation $4x^5 - x^4 - 4x^3 + x^2 - 5x - 6 = 0$, we write

$$f(x) = 4x^5 - x^4 - 4x^3 + x^2 - 5x - 6$$

← three sign changes

$$f(-x) = -4x^5 - x^4 + 4x^3 + x^2 + 5x - 6$$

← two sign changes

Thus, there are no more than three positive and two negative roots. ■

At this point, let us summarize the information we can determine about the roots of a polynomial equation $f(x) = 0$ of degree n and with real coefficients.

Roots of a Polynomial Equation of Degree *n*

1. There are n roots.
2. Complex roots appear in conjugate pairs.
3. Any rational roots must be factors of the constant term divided by factors of the coefficient of the highest-power term.
4. The maximum number of positive roots is the number of sign changes in $f(x)$, and the maximum number of negative roots is the number of sign changes in $f(-x)$.
5. Once we find $n - 2$ of the roots, we can find the remaining roots by the quadratic formula.

Because synthetic division is easy to perform, it is usually used to try possible roots. When a root is found, the quotient is of one degree less than the degree of the dividend. Each root found makes the ensuing work easier. The following examples show the complete method, as well as two other helpful rules.

EXAMPLE 5 Finding roots of equation

Find the roots of the equation $2x^3 + x^2 + 5x - 3 = 0$.

Because $n = 3$, there are three roots. If we can find one of these roots, we can use the quadratic formula to find the other two. We have

$$f(x) = 2x^3 + x^2 + 5x - 3 \quad \text{and} \quad f(-x) = -2x^3 + x^2 - 5x - 3$$

which shows there is one positive root and no more than two negative roots, which may or may not be rational. The *possible* rational roots are $\pm 1, \pm\frac{1}{2}, \pm\frac{3}{2}, \pm 3$.

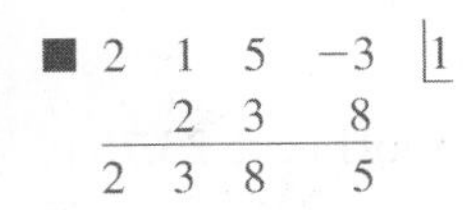

2	1	5	−3	1
	2	3	8	
2	3	8	5	

2	1	5	−3	$\frac{1}{2}$
	1	1	3	
2	2	6	0	

First, trying the root 1 (always a possibility if there are positive roots), we have the synthetic division shown at the left. The remainder of 5 tells us that 1 is not a root, but we have gained some additional information, if we observe closely. If we try any positive number larger than 1, the results in the last row will be larger positive numbers than we now have. The products will be larger, and therefore the sums will also be larger. Thus, there is no positive root larger than 1. This leads to the following rule: [When we are trying a positive root, if the bottom row contains all positive numbers, then there are no roots larger than the value tried.] This rule tells us that there is no reason to try $+\frac{3}{2}$ and $+3$ as roots. (It is also true that when trying a negative root, if the signs alternate in the bottom row, then there are no roots less than the value tried.)

NOTE ▶

Now, let us try $+\frac{1}{2}$, as shown at the left. The zero remainder tells us that $+\frac{1}{2}$ is a root, and the remaining factor is $2x^2 + 2x + 6$, which itself factors to $2(x^2 + x + 3)$. By the quadratic formula, we find the remaining roots by solving the equation $x^2 + x + 3 = 0$. This gives us

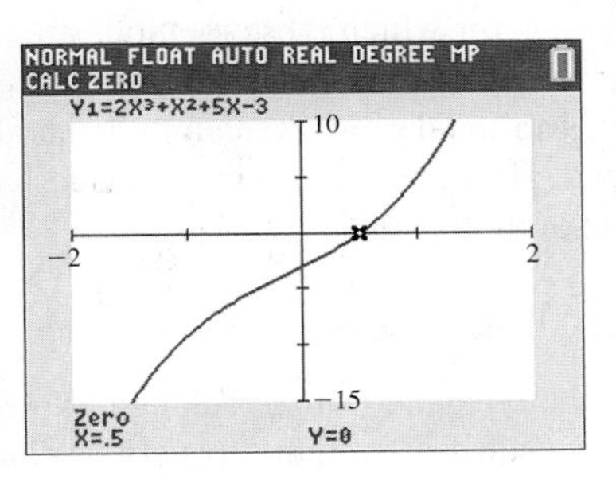

Fig. 15.8

$$x = \frac{-1 \pm \sqrt{1 - 12}}{2} = \frac{-1 \pm j\sqrt{11}}{2}$$

The three roots are $+\frac{1}{2}$, $\frac{-1 + j\sqrt{11}}{2}$, and $\frac{-1 - j\sqrt{11}}{2}$. There are no negative roots because the nonpositive roots are complex. Proceeding this way, we did not have to try any negative roots. The calculator display in Fig. 15.8 verifies the one real root at $x = 1/2$. ■

Practice Exercise

2. Find the roots of the equation $9x^4 - 12x^3 - 5x^2 + 12x - 4 = 0$.

TI-89 graphing calculator keystrokes for Example 5: goo.gl/t8ejzJ

■ See chapter introduction. Bezier curves were made popular by French engineer Pierre Bézier, who used them to design automobile bodies at Renault in the 1960s.

EXAMPLE 6 Finding roots—Bezier curve design

Bezier curves are used frequently in computer graphics to generate smooth curves and surfaces. In particular, cubic Bezier curves are formed by four points, two that define the curve's start and end (at $t = 0$ and $t = 1$) and two that "stretch" the middle part of the graph. The red curve in Fig. 15.9 shows a Bezier curve design for the roofline of a new art museum. It is defined by the parametric equations shown below, where the units for x and y are in tens of feet.

$$x = 2t^3 - t^2 + 6t$$
$$y = t^3 - 9t^2 + 9t$$
$$0 \leq t \leq 1$$

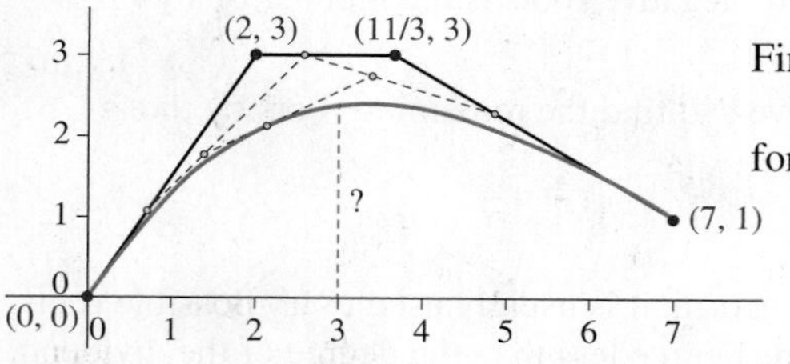

Fig. 15.9

Find the height of the roofline at the point that is 30 ft to the right of the left edge.

To find this height, we will substitute 3 for x in the top parametric equation, solve for t, and then substitute the result into the bottom equation to find y.

$$3 = 2t^3 - t^2 + 6t$$
$$2t^3 - t^2 + 6t - 3 = 0$$

If the above equation has any rational roots, they must be of the form $\dfrac{\pm 1, \pm 3}{\pm 1, \pm 2}$. Since we know $0 \leq t \leq 1$, the only possibilities are $t = \frac{1}{2}$ or $t = 1$. We also know that $t = 1$ represents an endpoint, so it is not the solution we desire. Thus, we will try $t = \frac{1}{2}$ (or 0.5) using synthetic division.

$$\begin{array}{rrrr|l} 2 & -1 & 6 & -3 & \underline{0.5} \\ & 1 & 0 & 3 & \\ \hline 2 & 0 & 6 & 0 & \end{array}$$

Since the remainder is zero, $t = 0.5$ is a root (the quotient $2x^2 + 6$ indicates that the other roots are not real numbers). To find the value of y, we substitute $t = 0.5$ into the second parametric equation:

■ The parametric equations of a cubic Bezier curve are of the form $(1 - t)^3P_0 + 3(1 - t)^2tP_1 + 3(1 - t)t^2P_2 + t^3P_3$. The polynomials multiplied by each of the four points are called *Bernstein polynomials.*

$$y = (0.5)^3 - 9(0.5)^2 + 9(0.5) = 2.375$$

Therefore, $y = 2.375$ (in tens of ft.), which means the height of the roofline is 23.75 ft at the desired location. ■

By the methods we have presented, we can look for *all roots* of a polynomial equation. These include any possible *complex roots* and *exact values of the rational and irrational roots,* if they exist. These methods allow us to solve a great many polynomial equations for these roots, but there are numerous other equations for which these methods are not sufficient.

When a polynomial equation has more than two irrational roots, we cannot generally find these roots by the methods we have developed. However, approximate values can be found on a calculator in situations where exact solutions are not needed.

EXAMPLE 7 Finding roots with calculator—box design

The bottom part of a box to hold a jigsaw puzzle is to be made from a rectangular piece of cardboard 37.0 cm by 31.0 cm by cutting out equal squares from the corners, bending up the sides, and taping the corners. See Fig. 15.10. If the volume of the box is to be 2770 cm^3, find the side of the square that is to be cut out.

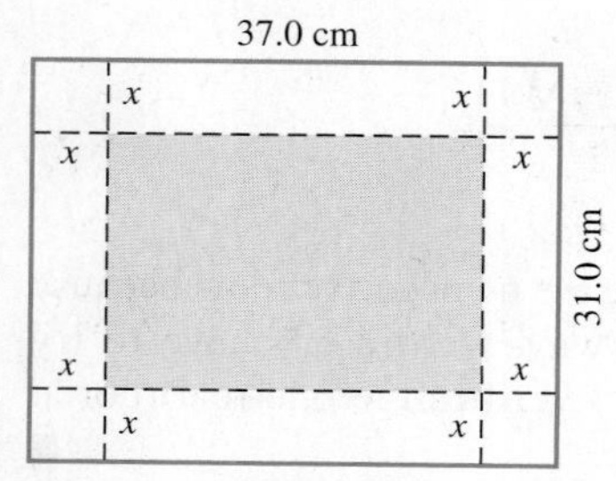

Fig. 15.10

Let $x =$ the side of the square to be cut out. This means

$$2770 = x(37.0 - 2x)(31.0 - 2x) \quad \text{volume} = 2770 \text{ cm}^3$$
$$= 4x^3 - 136x^2 + 1147x$$
$$4x^3 - 136x^2 + 1147x - 2770 = 0 \quad \text{simplify with terms on the left}$$

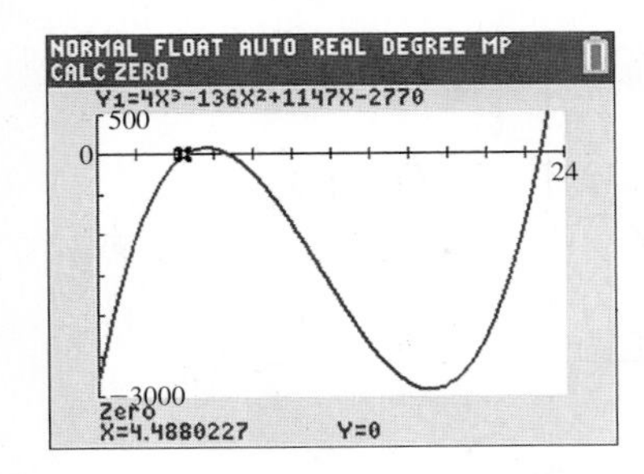

Fig. 15.11
Graphing calculator keystrokes: goo.gl/hM0AjP

The graph of $y = 4x^3 - 136x^2 + 1147x - 2770$ is shown in Fig. 15.11. Note that there are three positive real roots. This is consistent with there being three changes in sign in the function. However, the domain of the function is $0 < x < 15.5$ since cutting out squares larger than 15.5 cm would leave nothing for the width of the box. Therefore, there are two possible sizes for the squares to be cut from each corner: $x = 4.49$ cm (this zero is displayed in Fig. 15.11) or $x = 6.79$ cm. The third solution, $x = 22.7$ cm, is not in the domain. ■

EXERCISES 15.3

In Exercises 1 and 2, make the given changes in the indicated examples of this section and then perform the indicated operation.

1. In Example 4, change all signs between terms. (Leave the first term positive.)

2. In Example 5, change the + sign before the $5x$ to −.

In Exercises 3–20, solve the given equations without using a calculator.

3. $x^3 + 5x^2 + 2x - 8 = 0$

4. $2x^3 + 5x^2 - x + 6 = 0$

5. $x^3 + 2x^2 - 5x - 6 = 0$

6. $t^3 - 12t - 16 = 0$

7. $3x^4 - x^2 - 2x = 0$

8. $21t^3 + 56t^2 - 7 = 0$

9. $2x^3 - 3x^2 - 3x + 2 = 0$

10. $4x^3 - 16x^2 + 21x - 9 = 0$

11. $x^4 - 11x^2 - 12x + 4 = 0$

12. $8x^4 - 32x^3 - x + 4 = 0$

13. $5n^4 - 2n^3 + 40n - 16 = 0$

14. $8n^4 - 34n^2 + 28n - 6 = 0$

15. $12x^4 + 44x^3 + 21x^2 - 11x = 6$

16. $9x^4 - 3x^3 + 34x^2 - 12x = 8$

17. $D^5 + D^4 - 9D^3 - 5D^2 + 16D + 12 = 0$

18. $x^6 - x^4 - 14x^2 + 24 = 0$

19. $4x^5 - 24x^4 + 49x^3 - 38x^2 + 12x - 8 = 0$

20. $2x^5 + 5x^4 - 4x^3 - 19x^2 - 16x = 4$

In Exercises 21–24, use a calculator to solve the given equations to the nearest 0.01.

21. $2x^3 - 8x + 3 = 0$

22. $2x^4 - 15x^2 - 7x + 3 = 0$

23. $8x^4 + 36x^3 + 35x^2 - 4x - 4 = 0$

24. $2x^5 - 3x^4 + 8x^3 - 4x^2 - 4x + 2 = 0$

In Exercises 25–44, solve the given problems. Use a calculator to solve if necessary.

25. Solve the following system algebraically:
$y = x^4 - 11x^2;\quad y = 12x - 4$

26. Find rational values of a such that $(x - a)$ will divide into $x^3 + x^2 - 4x - 4$ with a remainder of zero.

27. Where does the graph of the function $f(x) = 4x^3 + 3x^2 - 20x - 15$ cross the x-axis?

28. Where does the graph of the function $f(s) = 2s^4 - s^3 - 5s^2 + 7s - 6$ cross the s-axis?

29. By checking only the equation and the coefficients, determine the smallest and largest possible rational roots of the equation $2x^4 + x^2 - 22x + 8 = 0$.

30. By checking only the equation and the coefficients, determine the smallest and largest possible rational roots of the equation $2x^4 + x^2 + 22x + 26 = 0$.

31. The angular acceleration α (in rad/s^2) of the wheel of a car is given by $\alpha = -0.2t^3 + t^2$, where t is the time (in s). For what values of t is $\alpha = 2.0$ rad/s^2?

32. In finding one of the dimensions d (in in.) of the support columns of a building, the equation $3d^3 + 5d^2 - 400d - 18{,}000 = 0$ is found. What is this dimension?

33. The deflection y of a beam at a horizontal distance x from one end is given by $y = k(x^4 - 2Lx^3 - L^3x)$, where L is the length of the beam and k is a constant. For what values of x is the deflection zero?

34. The specific gravity s of a sphere of radius r that sinks to a depth h in water is given by $s = \dfrac{3rh^2 - h^3}{4r^3}$. Find the depth to which a spherical buoy of radius 4.0 cm sinks if $s = 0.50$.

35. Cubic Bezier curves are commonly used to control the timing of animations. A certain "ease in" curve is given by $x = 0.1t^3 - 1.2t^2 + 2.1t$, $y = -2t^3 + 3t^2$ for $0 \le t \le 1$, where x represents the percentage of elapsed time for the animation and y represents the percentage of the progression of the animation (both as decimals). What percentage of the animation will be completed after 50% of the time has elapsed? (See chapter introduction.)

36. The pressure difference p (in kPa) at a distance x (in km) from one end of an oil pipeline is given by $p = x^5 - 3x^4 - x^2 + 7x$. If the pipeline is 4 km long, where is $p = 0$?

37. A rectangular tray is made from a square piece of sheet metal 10.0 cm on a side by cutting equal squares from each corner, bending up the sides, and then welding them together. How long is the side of the square that must be cut out if the volume of the tray is 70.0 cm^3?

38. The angle θ (in degrees) of a robot arm with the horizontal as a function of time t (in s) is given by $\theta = 15 + 20t^2 - 4t^3$ for $0 \le t \le 5$ s. Find t for $\theta = 40°$.

39. The radii of four different-sized ball bearings differ by 1.00 mm in radius from one size to the next. If the volume of the largest equals the volumes of the other three combined, find the radii.

40. A rectangular safe is to be made of steel of uniform thickness, including the door. The inside dimensions are 1.20 m, 1.20 m, and 2.00 m. If the volume of steel is 1.25 m^3, find its thickness.

41. Elliptic curve cryptography uses equations of the form $y^2 = x^3 + ax + b$ and a type of point addition where the "sum" R of points P and Q is found by extending a line through P and Q, determining the third point where the line intersects the curve, and then reflecting that point across the x-axis. For the curve $y^2 = x^3 - 6x + 9$, find the coordinates of point R for $P(-3, 0)$ and $Q(0, 3)$. See Fig. 15.12.

42. Each of three revolving doors has a perimeter of 6.60 m and revolves through a volume of 9.50 m^3 in one revolution about their common vertical side. What are the door's dimensions?

43. If a, b, and c are positive integers, find the combinations of the possible positive, negative, and nonreal complex roots if $f(x) = ax^3 - bx^2 + c = 0$.

44. An equation $f(x) = 0$ involves only odd powers of x with positive coefficients. Explain why this equation has no real root except $x = 0$.

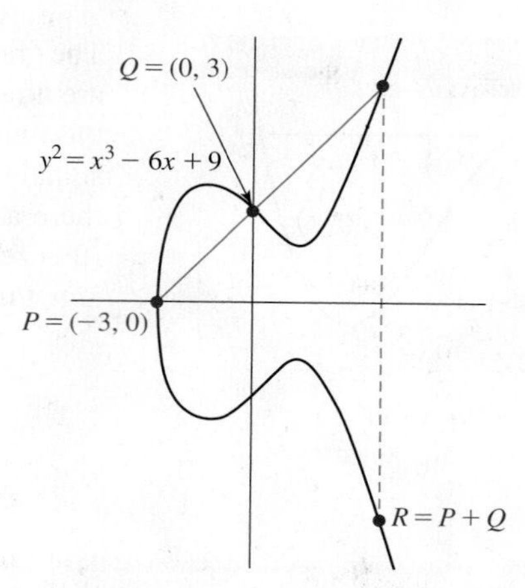

Fig. 15.12

Answers to Practice Exercises

1. 3 positive, 1 negative **2.** −1, 2/3, 2/3, 1

CHAPTER 15 KEY FORMULAS AND EQUATIONS

Polynomial function $$f(x) = a_nx^n + a_{n-1}x^{n-1} + \cdots + a_0 \quad (15.1)$$

Remainder theorem $$f(x) = (x - r)q(x) + R \quad (15.2)$$

$$f(r) = R \quad (15.3)$$

Rational root theorem $$r_r = \frac{\text{integer factor of } a_0}{\text{integer factor of } a_n} \quad (15.4)$$

CHAPTER 15 REVIEW EXERCISES

CONCEPT CHECK EXERCISES

Determine each of the following as being either ***true*** *or* ***false.*** *If it is false, explain why.*

1. If $3x^2 + 5x - 8$ is divided by $x - 2$, the remainder is 12.

2. Using synthetic division to divide $2x^3 - 3x^2 - 23x + 12$ by $x + 3$, the bottom row of numbers is 2 −9 −4 0.

3. Without solving, it can be determined that 1/8 is a possible rational root of the equation $4x^4 - 3x^3 + 5x^2 - x + 8 = 0$.

4. Without solving, it can be determined that there is no more than one possible negative root of the equation $x^4 - 3x^2 + x + 1 = 0$.

PRACTICE AND APPLICATIONS

In Exercises 5–8, find the remainder of the indicated division by the remainder theorem.

5. $(2x^3 - 4x^2 - x + 7) \div (x - 1)$

6. $(x^3 - 2x^2 + 9) \div (x + 2)$

7. $(3n^3 + n + 4) \div (n + 3)$

8. $(x^4 - 5x^3 + 8x^2 + 15x - 2) \div (x - 3)$

In Exercises 9–12, use the factor theorem to determine whether or not the second expression is a factor of the first.

9. $x^4 + x^3 + x^2 - 2x - 3;\ x + 1$

10. $2s^3 - 6s - 4;\ s - 2$

11. $2t^4 - 10t^3 - t^2 - 3t + 10;\ t + 5$

12. $9v^3 + 6v^2 + 4v + 2;\ 3v + 1$

In Exercises 13–20, use synthetic division to perform the indicated divisions.

13. $(x^3 + 4x^2 + 5x + 1) \div (x - 1)$

14. $(3x^3 - 2x^2 + 7) \div (x - 3)$

15. $(2x^3 - 3x^2 - 4x + 3) \div (x + 2)$

16. $(3D^3 + 8D^2 - 16) \div (D + 4)$

17. $(x^4 + 3x^3 - 20x^2 - 2x + 56) \div (x + 6)$

18. $(x^4 - 6x^3 + x - 8) \div (x - 3)$

19. $(2m^5 - 48m^3 + m^2 - 9) \div (m - 5)$

20. $(x^6 + 63x^3 + 5x^2 - 9x - 8) \div (x + 4)$

In Exercises 21–24, use synthetic division to determine whether or not the given numbers are zeros of the given functions.

21. $y^3 + 4y^2 - 9;\ \ -3$

22. $8y^4 - 32y^3 - y + 4;\ \ 4$

23. $2x^4 - x^3 + 2x^2 + x - 1;\ \ -1, \frac{1}{2}$

24. $6W^4 + 9W^3 - 2W^2 + 6W - 4;\ \ -\frac{3}{2}, -\frac{1}{2}$

In Exercises 25–36, find all the roots of the given equations, using synthetic division and the given roots.

25. $x^3 - 4x^2 - 7x + 10 = 0 \quad (r_1 = 5)$

26. $3B^3 - 10B^2 + B + 14 = 0 \quad (r_1 = 2)$

27. $x^4 - 10x^3 + 35x^2 - 50x + 24 = 0 \quad (r_1 = 1, r_2 = 2)$

28. $2x^4 - 2x^3 - 10x^2 - 2x - 12 = 0 \quad (r_1 = 3, r_2 = -2)$

29. $4p^4 - p^2 - 18p + 9 = 0 \quad (r_1 = \frac{1}{2}, r_2 = \frac{3}{2})$

30. $15x^4 + 4x^3 + 56x^2 + 16x - 16 = 0 \quad (r_1 = 2/5, r_2 = -2/3)$

31. $4x^4 + 4x^3 + x^2 + 4x - 3 = 0 \quad (r_1 = j)$

32. $x^4 + 2x^3 - 4x - 4 = 0 \quad (r_1 = -1 + j)$

33. $s^5 + 3s^4 - s^3 - 11s^2 - 12s - 4 = 0 \quad (-1 \text{ is a triple root})$

34. $24x^5 + 10x^4 + 7x^2 - 6x + 1 = 0$
$(r_1 = -1, r_2 = \frac{1}{4}, r_3 = \frac{1}{3})$

35. $V^5 + 4V^4 + 5V^3 - V^2 - 4V - 5 = 0$
$(r_1 = 1, r_2 = -2 + j)$

36. $2x^6 - x^5 + 8x^2 - 4x = 0 \quad (r_1 = \frac{1}{2}, r_2 = 1 + j)$

In Exercises 37–44, solve the given equations.

37. $x^3 + x^2 - 10x + 8 = 0$

38. $x^3 - 8x^2 + 20x = 16$

39. $6r^3 - 9r^2 + 3 = 0$

40. $2x^3 - 3x^2 - 11x + 6 = 0$

41. $6x^3 - x^2 - 12x = 5$

42. $6y^3 + 19y^2 + 2y = 3$

43. $4t^4 - 17t^2 + 14t - 3 = 0$

44. $2x^4 + 5x^3 - 14x^2 - 23x + 30 = 0$

In Exercises 45–70, solve the given problems. Where appropriate, set up the required equations.

45. Graph the function $f(x) = 3x^4 - 7x^3 - 26x^2 + 16x + 32$, and use the graph as an assist in factoring the function.

46. Graph the function $f(x) = 8x^4 - 22x^3 - 11x^2 + 52x - 12$, and use the graph as an assist in factoring the function.

47. What are the possible number of real zeros (double roots count as two, etc.) for a polynomial with real coefficients and of degree 5?

48. What are the possible combinations of real and nonreal complex zeros (double roots count as two, etc.) of a fourth-degree polynomial?

49. If a calculator shows a real root, how many nonreal complex roots are possible for a sixth-degree polynomial equation $f(x) = 0$?

50. Find rational values of a such that $(x - a)$ will divide into $x^3 - 13x + 3$ with a remainder of -9.

51. Explain how to find k if $x + 2$ is a factor of $f(x) = 3x^3 + kx^2 - 8x - 8$. What is k?

52. Explain how to find k if $x - 3$ is a factor of $f(x) = kx^4 - 15x^2 - 5x - 12$. What is k?

53. Form a polynomial equation of degree 3 with integer coefficients and having roots of j and 5.

54. If $f(x) = 3x^4 - 18x^3 - 2x^2 + 13x - 6$, and $f(x) = g(x)(x - 6)$, find $g(x)$.

55. Solve the following system algebraically: $x^2 = y + 3; xy = 2$

56. Where does the graph of the function $f(x) = 2x^4 - 7x^3 + 11x^2 - 28x + 12$ cross the x-axis?

57. A silo is to be constructed in the shape of a cylinder with a hemisphere as its top. Because of design constraints, the total height is to be 40.0 ft. Find the radius that would be required in order for the silo to hold 15,500 ft^3 of wood chips. Solve graphically.

58. The edge of a cube is 10 cm greater than the radius of a sphere. If the volumes of the figures are equal, what is this volume?

59. A computer analysis of the number of crimes committed each month in a certain city for the first 10 months of a year showed that $n = x^3 - 9x^2 + 15x + 600$. Here, n is the number of monthly crimes and x is the number of the month (as of the last day). In what month were 580 crimes committed?

60. A company determined that the number s (in thousands) of computer chips that it could supply at a price p of less than \$5 is given by $s = 4p^2 - 25$, whereas the demand d (in thousands) for the chips is given by $d = p^3 - 22p + 50$. For what price is the supply equal to the demand?

61. In order to find the diameter d (in cm) of a helical spring subject to given forces, it is necessary to solve the equation $64d^3 - 144d^2 + 108d - 27 = 0$. Solve for d.

62. A cubical tablet for purifying water is wrapped in a sheet of foil 0.500 mm thick. The total volume of the tablet and foil is 33.1% greater than the volume of the tablet alone. Find the length of the edge of the tablet.

63. For the mirror shown in Fig. 15.13, the reciprocal of the focal distance f equals the sum of the reciprocals of the object distance p and image distance q (in in.). Find p, if $q = p + 4$ and $f = (p + 1)/p$.

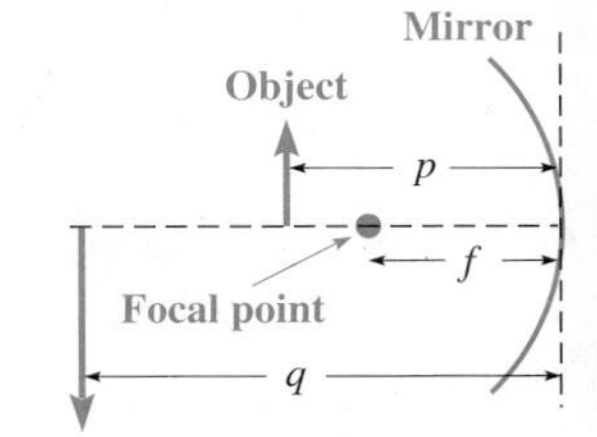

Fig. 15.13 Image

64. Three electric capacitors are connected in series. The capacitance of the second is 1 μF more than the first, and the third is 2 μF more than the second. The capacitance of the combination is 1.33 μF. The equation used to determine C, the capacitance of the first capacitor, is

$$\frac{1}{C} + \frac{1}{C + 1} + \frac{1}{C + 3} = \frac{3}{4}$$

Find the values of the capacitances.

65. The height of a cylindrical oil tank is 3.2 m more than the radius. If the volume of the tank is 680 m^3, what are the radius and the height of the tank?

66. A grain storage bin has a square base, each side of which is 5.5 m longer than the height of the bin. If the bin holds 160 m^3 of grain, find its dimensions.

67. A rectangular door has a diagonal brace that is 0.900 ft longer than the height of the door. If the area of the door is 24.3 ft^2, find its dimensions.

68. The radius of one ball bearing is 1.0 mm greater than the radius of a second ball bearing. If the sum of their volumes is 100 mm^3, find the radius of each.

69. The entrance to a garden area is a parabolic portal that can be described by $y = 4 - x^2$ (in m). Find the largest area of a rectangular gate that can be installed by graphing the function for area and finding its maximum. This is similar to finding the maximum point of a parabola as in Section 7.4.

70. An open container (no top) is to be made from a square piece of sheet metal, 20.0 cm on a side, by cutting equal squares from the corners and bending up the sides. Find the side of each cut-out square such that the volume is a maximum (see Exercise 69.)

71. A computer science student is to write a computer program that will print out the values of n for which $x + r$ is a factor of $x^n + r^n$. Write a paragraph that states which are the values of n and explains how they are found.

CHAPTER 15 PRACTICE TEST

As a study aid, we have included complete solutions for each Practice Test problem at the end of this book.

1. Is -3 a zero for the function $2x^3 + 3x^2 + 7x - 6$? Explain.
2. Find the remaining roots of the equation $x^4 - 2x^3 - 7x^2 + 20x - 12 = 0$; 2 is a double root.
3. Use synthetic division to perform the division $(x^3 - 5x^2 + 4x - 9) \div (x - 3)$.
4. Use the factor theorem and synthetic division to determine whether or not $2x + 1$ is a factor of $2x^4 + 15x^3 + 23x^2 - 16$.
5. Use the remainder theorem to find the remainder of the division $(x^3 + 4x^2 + 7x - 9) \div (x + 4)$.
6. Solve for x: $2x^4 - x^3 + 5x^2 - 4x - 12 = 0$.
7. The ends of a 10-ft beam are supported at different levels. The deflection y of the beam is given by $y = kx^2(x^3 + 436x - 4000)$, where x is the horizontal distance from one end and k is a constant. Find the values of x for which the deflection is zero.
8. A cubical metal block is heated such that its edge increases by 1.0 mm and its volume is doubled. Find the edge of the cube to the nearest tenth. Solve graphically using a calculator.

Matrices; Systems of Linear Equations

16

While working with systems of linear equations in the 1850s and 1860s, the English mathematician Arthur Cayley developed the use of a *matrix* (as we will show, a matrix is simply a rectangular array of numbers).

His interest was in the mathematical methods involved, and for many years matrices were used by mathematicians with little or no reference to possible applications other than solving systems of equations.

It was not until the 1920s that one of the first major applications of matrices was made when physicists used them in developing theories about the elementary particles within the atom. Their work is still very important in atomic and nuclear physics.

Since about 1950, matrices have become a very important and useful tool in many other areas of application, such as social science and economics. They are also used extensively in business and industry in making appropriate decisions in research, development, and production.

Here, in the case of matrices, we have a mathematical method that was developed only for its usefulness in mathematics, but has been found later to be very useful in many areas of application.

Prior to the development of matrices, other methods of solving systems of linear equations were used, but they often involved a great deal of calculation. This led the German mathematician Karl Friedrich Gauss in the early 1800s to develop a systematic method of solving systems of equations, now known as *Gaussian elimination*. This led to a matrix method that is now used in programming computers to solve systems of equations.

In the final two sections of this chapter, we develop the method of Gaussian elimination, and show the methods that were used to evaluate determinants before the extensive use of calculators and computers.

LEARNING OUTCOMES

After completion of this chapter, the student should be able to:

- Add and subtract matrices
- Multiply a matrix by a scalar
- Perform matrix multiplication
- Find the inverse of a matrix
- Solve a system of linear equations by using the inverse of the matrix of coefficients
- Solve a system of linear equations by using Gaussian elimination
- Identify a system of linear equations as having a unique solution, no solution, or an unlimited number of solutions
- Evaluate determinants of any order
- Solve a system of linear equations by determinants

◀ **In Section 16.2, we see how to determine the amount of material and worker time required for the production of automobile transmissions.**

16.1 Matrices: Definitions and Basic Operations

Elements of a Matrix • Square Matrix • Equality of Matrices • Matrix Addition and Subtraction • Scalar Multiplication

In this section, we introduce the definitions and some basic operations with matrices. In the three sections that follow, we develop additional operations and show how they are used in solving systems of linear equations.

A **matrix** *is an ordered rectangular array of numbers.* To distinguish such an array from a determinant, we enclose it within brackets. As with a determinant, *the individual members are called* **elements** *of the matrix.*

■ We first introduced the term *matrix* in Section 5.5, where we used them to solve systems of equations.

EXAMPLE 1 Illustrations of matrices

Some examples of matrices are shown here.

$$\begin{bmatrix} 2 & 8 \\ 1 & 0 \end{bmatrix} \quad \begin{bmatrix} 2 & -4 & 6 \\ -1 & 0 & 5 \end{bmatrix} \quad \begin{bmatrix} 4 & 6 \\ 0 & -1 \\ -2 & 5 \\ 3 & 0 \end{bmatrix}$$

$$\begin{bmatrix} -1 & 8 & 6 & 7 & 9 \\ 2 & 6 & 0 & 4 & 3 \\ 5 & -1 & 8 & 10 & 2 \end{bmatrix} \quad \begin{bmatrix} -1 & 2 & 0 & 9 \end{bmatrix}$$

■

As we can see, it is not necessary for the number of columns and number of rows to be the same, although such is the case for a determinant. However, *if the number of rows does equal the number of columns, the matrix is called a* **square matrix.** We will find that square matrices are of special importance. *If all the elements of a matrix are zero, the matrix is called a* **zero matrix.** It is convenient to designate a given matrix by a capital letter. The **size** of a matrix is given by the number of rows (first) and columns (second). For example, a 3×4 matrix has three rows and four columns.

CAUTION We must be careful to distinguish between a *matrix* and a *determinant.* A matrix is simply any rectangular array of numbers, whereas a determinant is a specific value associated with a square matrix. ■

EXAMPLE 2 Illustrating rows and columns

Consider the following matrices:

$$A = \begin{bmatrix} 5 & 0 & -1 \\ 1 & 2 & 6 \\ 0 & -4 & -5 \end{bmatrix} \quad B = \begin{bmatrix} 9 \\ 8 \\ 1 \\ 5 \end{bmatrix} \quad C = \begin{bmatrix} -1 & 6 & 8 & 9 \end{bmatrix} \quad O = \begin{bmatrix} 0 & 0 \\ 0 & 0 \end{bmatrix}$$

Matrix A is a 3×3 square matrix, matrix B is a 4×1 matrix (four rows and one column), matrix C is a 1×4 matrix (one row and four columns), and matrix O is a 2×2 zero matrix. ■

To be able to refer to specific elements of a matrix and to give a general representation, a double-subscript notation is usually employed. That is,

$$A = \begin{bmatrix} a_{11} & a_{12} & a_{13} \\ a_{21} & a_{22} & a_{23} \\ a_{31} & a_{32} & a_{33} \end{bmatrix}$$

row ↗ ↖ column

We see that the first subscript refers to the row in which the element lies and the second subscript refers to the column in which the element lies.

NOTE ➧ [Two matrices are said to be **equal** if and only if they are *identical*. That is, they must have the same number of columns, the same number of rows, and the corresponding elements must be equal.]

EXAMPLE 3 Equality of matrices

(a) $$\begin{bmatrix} a_{11} & a_{12} & a_{13} \\ a_{21} & a_{22} & a_{23} \end{bmatrix} = \begin{bmatrix} 1 & -5 & 0 \\ 4 & 6 & -3 \end{bmatrix}$$

if and only if $a_{11} = 1, a_{12} = -5, a_{13} = 0, a_{21} = 4, a_{22} = 6$, and $a_{23} = -3$.

(b) The matrices

$$\begin{bmatrix} 1 & 2 & 3 \\ -1 & -2 & -5 \end{bmatrix} \text{ and } \begin{bmatrix} 1 & 2 & -5 \\ -1 & -2 & 3 \end{bmatrix}$$

are not equal, because the elements in the third column are reversed.

(c) The matrices

$$\begin{bmatrix} 2 & 3 \\ -1 & 5 \end{bmatrix} \text{ and } \begin{bmatrix} 2 & 3 & 0 \\ -1 & 5 & 0 \end{bmatrix}$$

are not equal, because the number of columns is different. This is true despite the fact that both elements of the third column are zeros. ■

EXAMPLE 4 Matrix equation—equilibrium forces

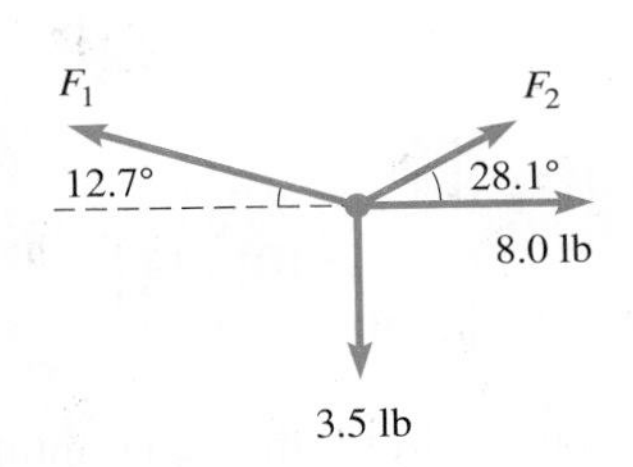

Fig. 16.1

The forces acting on a bolt are in equilibrium, as shown in Fig. 16.1. Analyzing the horizontal and vertical components as in Section 9.4, we find the following matrix equation. Find forces F_1 and F_2.

$$\begin{bmatrix} 0.98F_1 - 0.88F_2 \\ 0.22F_1 + 0.47F_2 \end{bmatrix} = \begin{bmatrix} 8.0 \\ 3.5 \end{bmatrix}$$

From the equality of matrices, we know that $0.98F_1 - 0.88F_2 = 8.0$ and $0.22F_1 + 0.47F_2 = 3.5$. Therefore, to find the forces F_1 and F_2, we must solve the system of equations

$$\begin{aligned} 0.98F_1 - 0.88F_2 &= 8.0 \\ 0.22F_1 + 0.47F_2 &= 3.5 \end{aligned}$$

Using determinants, as shown in Section 5.4, we have

$$F_1 = \frac{\begin{vmatrix} 8.0 & -0.88 \\ 3.5 & 0.47 \end{vmatrix}}{\begin{vmatrix} 0.98 & -0.88 \\ 0.22 & 0.47 \end{vmatrix}} = \frac{8.0(0.47) - 3.5(-0.88)}{0.98(0.47) - 0.22(-0.88)} = 10.5 \text{ lb}$$

Using determinants again, or by substituting this value into either equation, we find that $F_2 = 2.6$ lb. These values check when substituted into the original matrix equation. ■

MATRIX ADDITION AND SUBTRACTION

If two matrices have the same number of rows and the same number of columns, their **sum** is defined as the matrix consisting of the sums of the corresponding elements.

NOTE ➧ [If the number of rows or the number of columns of the two matrices are not equal, they cannot be added.]

EXAMPLE 5 Adding matrices

(a)

$$\begin{bmatrix} 8 & 1 & -5 & 9 \\ 0 & -2 & 3 & 7 \end{bmatrix} + \begin{bmatrix} -3 & 4 & 6 & 0 \\ 6 & -2 & 6 & 5 \end{bmatrix} = \begin{bmatrix} 8+(-3) & 1+4 & -5+6 & 9+0 \\ 0+6 & -2+(-2) & 3+6 & 7+5 \end{bmatrix}$$

$$= \begin{bmatrix} 5 & 5 & 1 & 9 \\ 6 & -4 & 9 & 12 \end{bmatrix}$$

(b) The matrices

$$\begin{bmatrix} 3 & -5 & 8 \\ 2 & 9 & 0 \\ 4 & -2 & 3 \end{bmatrix} \quad \text{and} \quad \begin{bmatrix} 3 & -5 & 8 & 0 \\ 2 & 9 & 0 & 0 \\ 4 & -2 & 3 & 0 \end{bmatrix}$$

cannot be added since the second matrix has one more column than the first matrix. This is true even though the extra column contains only zeros. ■

Practice Exercise

1. For matrices A and B, find $A + B$.

$$A = \begin{bmatrix} 2 & -4 \\ -7 & 5 \end{bmatrix} \quad B = \begin{bmatrix} -9 & -6 \\ 7 & 3 \end{bmatrix}$$

The product of a number and a matrix (known as **scalar multiplication** *of a matrix) is defined as the matrix whose elements are obtained by multiplying each element of the given matrix by the given number.* Thus, we obtain matrix kA by multiplying the elements of matrix A by k. In this way, $A + A$ and $2A$ are the same matrix.

EXAMPLE 6 Scalar multiplication

For the matrix A, where

$$A = \begin{bmatrix} -5 & 7 \\ 3 & 0 \end{bmatrix} \quad \text{we have} \quad 2A = \begin{bmatrix} 2(-5) & 2(7) \\ 2(3) & 2(0) \end{bmatrix} = \begin{bmatrix} -10 & 14 \\ 6 & 0 \end{bmatrix}$$ ■

Practice Exercise

2. In Example 6, find matrix $-A$.

By combining the definitions for the addition of matrices and for the scalar multiplication of a matrix, we can define the subtraction of matrices. That is, the **difference** of matrices A and B is given by $A - B = A + (-B)$. Therefore, we change the sign of each element of B, and proceed as in addition.

EXAMPLE 7 Subtracting matrices

$$\begin{bmatrix} 7 & -4 \\ -9 & 3 \end{bmatrix} - \begin{bmatrix} -2 & 6 \\ -8 & 5 \end{bmatrix} = \begin{bmatrix} 7 & -4 \\ -9 & 3 \end{bmatrix} + \begin{bmatrix} 2 & -6 \\ 8 & -5 \end{bmatrix} = \begin{bmatrix} 9 & -10 \\ -1 & -2 \end{bmatrix}$$ ■

Practice Exercise

3. For matrices A and B in Practice Exercise 1, find $A - 2B$.

The operations of addition, subtraction, and multiplication of a matrix by a number are like those for real numbers. For these operations, the algebra of matrices is like the algebra of real numbers. We see that the following laws hold for matrices.

$$A + B = B + A$$ (commutative law) **(16.1)**

$$A + (B + C) = (A + B) + C$$ (associative law) **(16.2)**

$$k(A + B) = kA + kB$$ **(16.3)**

$$A + O = A$$ **(16.4)**

Here, we have let O represent the zero matrix. We will find in the next section that not all laws for matrix operations are like those for real numbers.

EXERCISES 16.1

In Exercises 1 and 2, make the given changes in the indicated examples of this section and then perform the indicated operations.

1. In Example 5(a), interchange the second and third columns of the second matrix and then add the matrices.
2. In Example 6, find the matrix $-2A$.

In Exercises 3–10, determine the value of the literal numbers in each of the given matrix equalities. If the matrices cannot be equal, explain why.

3. $\begin{bmatrix} a & b \\ c & d \end{bmatrix} = \begin{bmatrix} 1 & -3 \\ 4 & 7 \end{bmatrix}$

4. $\begin{bmatrix} x \\ x+y \end{bmatrix} = \begin{bmatrix} 2 \\ 5 \end{bmatrix}$

5. $\begin{bmatrix} x & 2y & z \\ r/4 & -s & -5t \end{bmatrix} = \begin{bmatrix} -2 & 10 & -9 \\ 12 & -4 & 5 \end{bmatrix}$

6. $[a + bj \quad 2c - dj \quad 3e + fj] = [5j \quad a + 6 \quad 3b + c]$ $(j = \sqrt{-1})$

7. $\begin{bmatrix} C + D \\ D - 2E \\ 3E \end{bmatrix} = \begin{bmatrix} 3 \\ 2 \\ 6 \end{bmatrix}$

8. $\begin{bmatrix} 2x - 3y \\ x + 4y \end{bmatrix} = \begin{bmatrix} 13 \\ 1 \end{bmatrix}$

9. $\begin{bmatrix} x - 3 & x + y \\ x - z & y + z \\ x + t & y - t \end{bmatrix} = \begin{bmatrix} 5 & 3 \\ 4 & -1 \end{bmatrix}$

10. $\begin{bmatrix} x \\ x + y \end{bmatrix} = \begin{bmatrix} 2 & 0 \\ 4 & -3 \end{bmatrix}$

In Exercises 11–14, find the indicated sums of matrices.

11. $\begin{bmatrix} 6 & 3 \\ -5 & -4 \end{bmatrix} + \begin{bmatrix} -1 & 7 \\ 5 & -2 \end{bmatrix}$

12. $\begin{bmatrix} 1 & 0 & 9 \\ 3 & -5 & -2 \end{bmatrix} + \begin{bmatrix} 4 & -1 & 7 \\ 2 & 0 & -3 \end{bmatrix}$

13. $\begin{bmatrix} 50 & -82 \\ -34 & 57 \\ -15 & 62 \end{bmatrix} + \begin{bmatrix} -55 & 82 \\ 30 & 14 \\ 26 & -70 \end{bmatrix}$

14. $\begin{bmatrix} 4.7 & 2.1 & -9.6 \\ -6.8 & 4.8 & 7.4 \\ -1.9 & 0.7 & 5.9 \end{bmatrix} + \begin{bmatrix} -4.9 & -9.6 & -2.1 \\ 3.4 & 0.7 & 0.0 \\ 5.6 & 10.1 & -1.6 \end{bmatrix}$

In Exercises 15–26, use matrices A, B, C, and D to find the indicated matrices. If the operations cannot be performed, explain why.

$$A = \begin{bmatrix} 6 & -3 \\ 4 & -5 \end{bmatrix} \quad B = \begin{bmatrix} 3 & 12 \\ -9 & -6 \end{bmatrix}$$

$$C = \begin{bmatrix} -1 & 4 & -7 \\ 2 & -6 & 11 \end{bmatrix} \quad D = \begin{bmatrix} 7 & 9 & -6 \\ -4 & 0 & 8 \end{bmatrix}$$

15. $A + B$ 16. $A - B$ 17. $A + C$ 18. $2A + B$
19. $C - D$ 20. $2C + D$ 21. $-2C + D$ 22. $2B - D$
23. $D - 4C$ 24. $-C - 2D$ 25. $B - 3A$ 26. $2A - \frac{1}{3}B$

In Exercises 27–34, use the matrices for Exercises 15–26 and find the indicated matrices using a calculator.

27. $3A$ 28. $-2C$ 29. $C + 3D$ 30. $\frac{1}{3}B - \frac{1}{2}A$
31. $5A - 4B$ 32. $-4C - 3D$ 33. $-6B - 4A$ 34. $3C - 2D$

In Exercises 35–38, use matrices A and B to show that the indicated laws hold for these matrices. In Exercise 35, explain the meaning of the result.

$$A = \begin{bmatrix} -1 & 2 & 3 & 7 \\ 0 & -3 & -1 & 4 \\ 9 & -1 & 0 & -2 \end{bmatrix} \quad B = \begin{bmatrix} 4 & -1 & -3 & 0 \\ 5 & 0 & -1 & 1 \\ 1 & 11 & 8 & 2 \end{bmatrix}$$

35. $A + B = B + A$ 36. $A + O = A$
37. $-(A - B) = B - A$ 38. $3(A + B) = 3A + 3B$

In Exercises 39 and 40, find the unknown quantities in the given matrix equations.

39. An airplane is flying in a direction 21.0° north of east at 235 km/h but is headed 14.5° north of east. The wind is from the southeast. Find the speed of the wind v_w and the speed of the plane v_p relative to the wind from the given matrix equation. See Fig. 16.2.

$$\begin{bmatrix} v_p \cos 14.5° - v_w \cos 45.0° \\ v_p \sin 14.5° + v_w \sin 45.0° \end{bmatrix} = \begin{bmatrix} 235 \cos 21.0° \\ 235 \sin 21.0° \end{bmatrix}$$

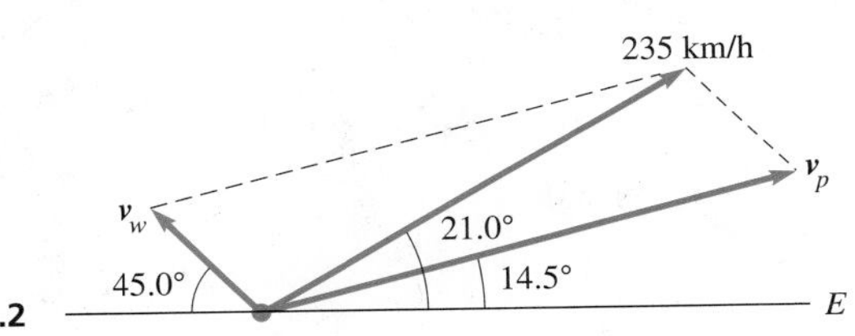

Fig. 16.2

40. Find the electric currents shown in Fig. 16.3 by solving the following matrix equation:

$$\begin{bmatrix} I_1 + I_2 + I_3 \\ -2I_1 + 3I_2 \\ -3I_2 + 6I_3 \end{bmatrix} = \begin{bmatrix} 0 \\ 24 \\ 0 \end{bmatrix}$$

Fig. 16.3

In Exercises 41–44, perform the indicated matrix operations.

41. The contractor of a housing development constructs four different types of houses, with either a carport, a one-car garage, or a two-car garage. The following matrix shows the number of houses of each type and the type of garage.

	Type A	Type B	Type C	Type D
Carport	96	75	0	0
1-car garage	62	44	24	0
2-car garage	0	35	68	78

If the contractor builds two additional identical developments, find the matrix showing the total number of each house-garage type built in the three developments.

42. The inventory of a drug supply company shows that the following numbers of cases of bottles of vitamins C and B_3 (niacin) are in stock: Vitamin C — 25 cases of 100-mg bottles, 10 cases of 250-mg bottles, and 32 cases of 500-mg bottles; vitamin B_3 — 30 cases of 100-mg bottles, 18 cases of 250-mg bottles, and 40 cases of 500-mg bottles. This is represented by matrix A below. After two shipments are sent out, each of which can be represented by matrix B below, find the matrix that represents the remaining inventory.

$$A = \begin{bmatrix} 25 & 10 & 32 \\ 30 & 18 & 40 \end{bmatrix} \qquad B = \begin{bmatrix} 10 & 5 & 6 \\ 12 & 4 & 8 \end{bmatrix}$$

43. One serving of brand K breakfast cereal provides the given percentages of the given vitamins and minerals: vitamin A, 15%; vitamin C, 25%; calcium, 10%; iron, 25%. One serving of brand G provides: vitamin A, 10%; vitamin C, 10%; calcium, 10%; iron, 45%. One serving of tomato juice provides: vitamin A, 15%; vitamin C, 30%; calcium, 3%; iron, 3%. One serving of orange-pineapple juice provides vitamin A, 0%; vitamin C, 100%; calcium, 2%; iron, 2%. Set up a two-row, four-column matrix B to represent the data for the cereals and a similar matrix J for the juices.

44. Referring to Exercise 43, find the matrix $B + J$ and explain the meaning of its elements.

Answers to Practice Exercises

1. $\begin{bmatrix} -7 & -10 \\ 0 & 8 \end{bmatrix}$ **2.** $\begin{bmatrix} 5 & -7 \\ -3 & 0 \end{bmatrix}$ **3.** $\begin{bmatrix} 20 & 8 \\ -21 & -1 \end{bmatrix}$

16.2 Multiplication of Matrices

Multiplication of Matrices • Identity Matrix • Inverse of a Matrix

The definition for the multiplication of matrices does not have an intuitive basis. However, through the solution of a system of linear equations we can, at least in part, show why multiplication is defined as it is. Consider Example 1.

EXAMPLE 1 Reasoning for definition of multiplication

If we solve the system of equations

$$\begin{aligned} 2x + y &= 1 \\ 7x + 3y &= 5 \end{aligned}$$

we get $x = 2, y = -3$. Checking this solution in each of the equations, we get

$$\begin{aligned} 2(2) + 1(-3) &= 1 \\ 7(2) + 3(-3) &= 5 \end{aligned}$$

Let us represent the coefficients of the equations by the matrix $\begin{bmatrix} 2 & 1 \\ 7 & 3 \end{bmatrix}$ and the solutions by the matrix $\begin{bmatrix} 2 \\ -3 \end{bmatrix}$. If we now indicate the multiplications of these matrices and perform it as shown

$$\begin{bmatrix} 2 & 1 \\ 7 & 3 \end{bmatrix}\begin{bmatrix} 2 \\ -3 \end{bmatrix} = \begin{bmatrix} 2(2) + 1(-3) \\ 7(2) + 3(-3) \end{bmatrix} = \begin{bmatrix} 1 \\ 5 \end{bmatrix}$$

we note that we obtain a matrix that properly represents the right-side values of the equations. (Note the products and sums in the resulting matrix.) ■

Following reasons along the lines indicated in Example 1, we now define the **multiplication of matrices.** If the number of columns in a first matrix equals the number of rows in a second matrix, the product of these matrices is formed as follows: *The element in a specified row and a specified column of the product matrix is the sum of the products formed by multiplying each element in the specified row of the first matrix by the corresponding element in the specific column of the second matrix.* The product matrix will have the same number of rows as the first matrix and the same number of columns as the second matrix. Consider the following examples.

■ We can multiply two matrices if the inside numbers of their sizes are the same:

$$(m \times \underbrace{n)(n}_{\text{same}} \times p) = m \times p$$

The size of the product is given by the outside numbers.

EXAMPLE 2 Multiplying matrices

Find the product AB, where

$$A = \begin{bmatrix} 2 & 1 \\ -3 & 0 \\ 1 & 2 \end{bmatrix} \qquad B = \begin{bmatrix} -1 & 6 & 5 & -2 \\ 3 & 0 & 1 & -4 \end{bmatrix}$$

With two columns in matrix A and two rows in matrix B, the product can be formed. The element in the first row and first column of the product is the sum of the products of the corresponding elements of the first row of A and first column of B. The elements in the first row and second column of the product is the sum of the products of corresponding elements of the first row of A and second column of B. We continue until we have three rows (the number in A) and four columns (the number in B).

Practice Exercise

1. Find the product AB.

$$A = \begin{bmatrix} -1 & 4 \\ 8 & -2 \\ 0 & 12 \end{bmatrix} \qquad B = \begin{bmatrix} 5 & -3 \\ -7 & 10 \end{bmatrix}$$

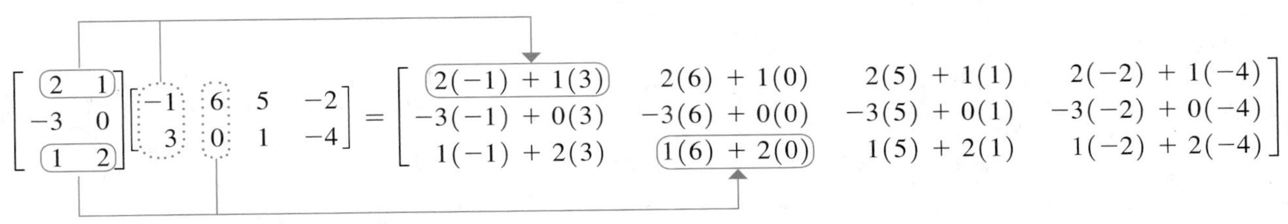

$$\begin{bmatrix} 2 & 1 \\ -3 & 0 \\ 1 & 2 \end{bmatrix}\begin{bmatrix} -1 & 6 & 5 & -2 \\ 3 & 0 & 1 & -4 \end{bmatrix} = \begin{bmatrix} 2(-1)+1(3) & 2(6)+1(0) & 2(5)+1(1) & 2(-2)+1(-4) \\ -3(-1)+0(3) & -3(6)+0(0) & -3(5)+0(1) & -3(-2)+0(-4) \\ 1(-1)+2(3) & 1(6)+2(0) & 1(5)+2(1) & 1(-2)+2(-4) \end{bmatrix}$$

$$= \begin{bmatrix} 1 & 12 & 11 & -8 \\ 3 & -18 & -15 & 6 \\ 5 & 6 & 7 & -10 \end{bmatrix}$$

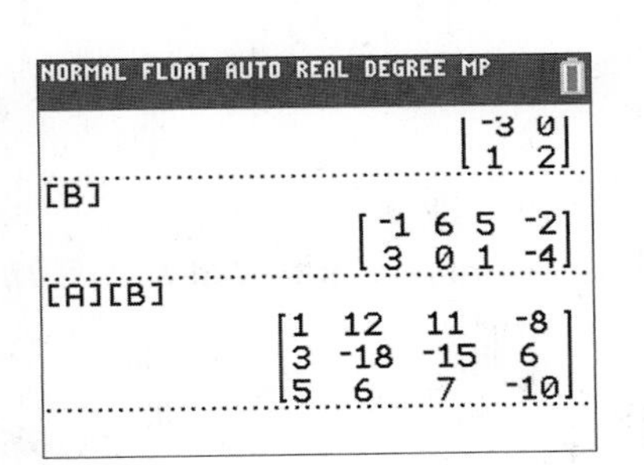

Fig. 16.4
Graphing calculator keystrokes: goo.gl/iS5GKz

The elements used to form the element in the first row and first column and the element in the third row and second column of the product are outlined in color. Figure 16.4 shows the calculator display for the matrix product AB.

In trying to form the product BA, we see that B has four columns and A has three rows. Because these numbers are not the same, the product cannot be formed. **CAUTION** Therefore, $AB \neq BA$, which means matrix multiplication is not commutative (except in special cases), and therefore differs from multiplication of real numbers. ■

EXAMPLE 3

The product of two matrices below may be formed because the first matrix has four columns and the second matrix has four rows. The matrix is formed as shown.

$$\begin{bmatrix} -1 & 9 & 3 & -2 \\ 2 & 0 & -7 & 1 \end{bmatrix}\begin{bmatrix} 6 & -2 \\ 1 & 0 \\ 3 & -5 \\ 3 & 9 \end{bmatrix} = \begin{bmatrix} -1(6)+9(1)+3(3)+(-2)(3) & -1(-2)+9(0)+3(-5)+(-2)(9) \\ 2(6)+0(1)+(-7)(3)+1(3) & 2(-2)+0(0)+(-7)(-5)+1(9) \end{bmatrix}$$

$$= \begin{bmatrix} -6+9+9-6 & 2+0-15-18 \\ 12+0-21+3 & -4+0+35+9 \end{bmatrix}$$

$$= \begin{bmatrix} 6 & -31 \\ -6 & 40 \end{bmatrix}$$

■

Practice Exercise

2. In Example 3, how many rows and columns does the product matrix have if the matrices switch positions?

IDENTITY MATRIX

There are two special matrices of particular importance in the multiplication of matrices. The first of these is the **identity matrix I,** *which is a square matrix with 1's for elements of the principal diagonal with all other elements zero.* (The principal diagonal starts with the element a_{11}.) It has the property that if it is multiplied by another square matrix with the same number of rows and columns, then the second matrix equals the product matrix.

EXAMPLE 4 Identity matrix

Show that $AI = IA = A$ for the matrix

$$A = \begin{bmatrix} 2 & -3 \\ 4 & 1 \end{bmatrix}$$

Because A has two rows and two columns, we choose I with two rows and two columns. Therefore, for this case

$$I = \begin{bmatrix} 1 & 0 \\ 0 & 1 \end{bmatrix}$$ ← elements of principal diagonal are 1's

Forming the indicated products, we have results as follows:

$$AI = \begin{bmatrix} 2 & -3 \\ 4 & 1 \end{bmatrix}\begin{bmatrix} 1 & 0 \\ 0 & 1 \end{bmatrix}$$

$$= \begin{bmatrix} 2(1) + (-3)(0) & 2(0) + (-3)(1) \\ 4(1) + 1(0) & 4(0) + 1(1) \end{bmatrix} = \begin{bmatrix} 2 & -3 \\ 4 & 1 \end{bmatrix}$$

$$IA = \begin{bmatrix} 1 & 0 \\ 0 & 1 \end{bmatrix}\begin{bmatrix} 2 & -3 \\ 4 & 1 \end{bmatrix}$$

$$= \begin{bmatrix} 1(2) + 0(4) & 1(-3) + 0(1) \\ 0(2) + 1(4) & 0(-3) + 1(1) \end{bmatrix} = \begin{bmatrix} 2 & -3 \\ 4 & 1 \end{bmatrix}$$

Therefore, we see that $AI = IA = A$. ■

INVERSE OF A MATRIX

For a given square matrix A, its **inverse** A^{-1} is the other important special matrix. *The matrix A and its inverse A^{-1} have the property that*

$$AA^{-1} = A^{-1}A = I \qquad \textbf{(16.5)}$$

If the product of two square matrices equals the identity matrix, the matrices are called inverses of each other. In the next section, we develop the procedure for finding the inverse of a square matrix, and the section that follows shows how the inverse is used in the solution of systems of equations. At this point, we simply show that the product of certain matrices equals the identity matrix and that therefore these matrices are inverses of each other. CAUTION Only square matrices can have inverses, and if the determinant of a matrix is zero, then the inverse *does not exist.* ■

EXAMPLE 5 Inverse matrix

For the given matrices A and B, show that $AB = BA = I$, and therefore that $B = A^{-1}$.

$$A = \begin{bmatrix} 1 & -3 \\ -2 & 7 \end{bmatrix} \qquad B = \begin{bmatrix} 7 & 3 \\ 2 & 1 \end{bmatrix}$$

Forming the products AB and BA, we have the following:

$$AB = \begin{bmatrix} 1 & -3 \\ -2 & 7 \end{bmatrix}\begin{bmatrix} 7 & 3 \\ 2 & 1 \end{bmatrix} = \begin{bmatrix} 7 - 6 & 3 - 3 \\ -14 + 14 & -6 + 7 \end{bmatrix} = \begin{bmatrix} 1 & 0 \\ 0 & 1 \end{bmatrix}$$

$$BA = \begin{bmatrix} 7 & 3 \\ 2 & 1 \end{bmatrix}\begin{bmatrix} 1 & -3 \\ -2 & 7 \end{bmatrix} = \begin{bmatrix} 7 - 6 & -21 + 21 \\ 2 - 2 & -6 + 7 \end{bmatrix} = \begin{bmatrix} 1 & 0 \\ 0 & 1 \end{bmatrix}$$

Since $AB = I$ and $BA = I$, $B = A^{-1}$ and $A = B^{-1}$. ■

Practice Exercise

3. Show that $AB = BA = I$.

$$A = \begin{bmatrix} 5 & -7 \\ -2 & 3 \end{bmatrix} \qquad B = \begin{bmatrix} 3 & 7 \\ 2 & 5 \end{bmatrix}$$

EXAMPLE 6 Matrix multiplication—transmission production

A company makes three types of automobile transmissions: 4-gear manual (type X), 4-gear automatic (type Y), and 5-gear automatic (type Z). In one day, it produces 40 of type X, 50 of type Y, and 80 of type Z. Required for production are 4 units of parts (some preassembled) and 1 worker-hour for type X, 5 units of parts and 2 worker-hours for type Y, and 3 units of parts and 2 worker-hours for type Z. Letting matrix A represent the number of each type produced, and matrix B represent the parts and time requirements, we have

■ See the chapter introduction.

$$A = \begin{bmatrix} 40 & 50 & 80 \end{bmatrix} \qquad B = \begin{bmatrix} 4 & 1 \\ 5 & 2 \\ 3 & 2 \end{bmatrix}$$

(In A: columns type X, type Y, type Z; number of each type produced. In B: columns units of parts, worker-hours; rows type X, type Y, type Z; parts and time required for each.)

The product AB gives the total number of units of parts and the total number of worker-hours needed for the day's production in a one-row, two-column matrix:

$$AB = \begin{bmatrix} 40 & 50 & 80 \end{bmatrix} \begin{bmatrix} 4 & 1 \\ 5 & 2 \\ 3 & 2 \end{bmatrix}$$

$$= \begin{bmatrix} 160 + 250 + 240 & 40 + 100 + 160 \end{bmatrix} = \begin{bmatrix} 650 & 300 \end{bmatrix}$$

(650: total units of parts; 300: total worker-hours)

Therefore, 650 units of parts and 300 worker-hours are required. ■

We have seen that *matrix multiplication is not commutative* (see Example 2); that is, $AB \neq BA$ in general. This differs from the multiplication of real numbers. Another difference is that it is possible that $AB = O$, even though neither A nor B is O. There are also some similarities. For example, $AI = A$, where I and the number 1 are analogous. Also, *the distributive property* $A(B + C) = AB + AC$ *holds for matrix multiplication.*

EXERCISES 16.2

In Exercises 1 and 2, make the given changes in the indicated examples of this section and then perform the indicated multiplications.

1. In Example 2, interchange columns 1 and 2 in matrix A and then do the multiplication.
2. In Example 5, in A change -2 to 2 and -3 to 3, in B change 2 to -2 and 3 to -3, and then do the multiplications.

In Exercises 3–10, perform the indicated multiplications.

3. $\begin{bmatrix} 4 & -2 \end{bmatrix} \begin{bmatrix} -1 & 0 \\ 2 & 6 \end{bmatrix}$

4. $\begin{bmatrix} -\frac{1}{2} & 6 & -\frac{2}{3} \end{bmatrix} \begin{bmatrix} 4 & 8 \\ \frac{1}{3} & -\frac{1}{2} \\ 0 & 9 \end{bmatrix}$

5. $\begin{bmatrix} 9 & -1 & 3 \\ 7 & 0 & 2 \end{bmatrix} \begin{bmatrix} 40 \\ -15 \\ 20 \end{bmatrix}$

6. $\begin{bmatrix} 0 & -1 & 2 \\ 4 & 11 & 2 \end{bmatrix} \begin{bmatrix} 3 & -1 \\ 1 & 2 \\ 6 & 1 \end{bmatrix}$

7. $\begin{bmatrix} -8 & \frac{3}{2} \\ \frac{1}{3} & -6 \\ 2 & 8 \end{bmatrix} \begin{bmatrix} 2 & -3 \\ 4 & 5 \end{bmatrix}$

8. $\begin{bmatrix} 12 & -47 \\ 43 & -18 \\ 36 & -22 \end{bmatrix} \begin{bmatrix} 1 & 2 \\ -1 & 1 \end{bmatrix}$

9. $\begin{bmatrix} -1 & 7 \\ 3 & 5 \\ 10 & -1 \\ -5 & 12 \end{bmatrix} \begin{bmatrix} 2 & 1 \\ 5 & -3 \end{bmatrix}$

10. $\begin{bmatrix} 5 & 4 \end{bmatrix} \begin{bmatrix} 4 & -4 \\ -5 & 5 \end{bmatrix}$

In Exercises 11–14, use a graphing calculator to perform the indicated multiplications.

11. $\begin{bmatrix} 2 & -3 \\ 8 & -1 \end{bmatrix} \begin{bmatrix} 3 & 0 & -1 \\ 7 & -5 & 6 \end{bmatrix}$

12. $\begin{bmatrix} -7 & 8 \\ 5 & 0 \end{bmatrix} \begin{bmatrix} -90 & 100 \\ 10 & 40 \end{bmatrix}$

13. $\begin{bmatrix} -7.1 & 2.3 & 0.5 \\ -3.8 & -2.4 & 4.9 \end{bmatrix} \begin{bmatrix} 6.5 & -5.2 \\ 4.9 & 1.7 \\ -1.8 & 6.9 \end{bmatrix}$

14. $\begin{bmatrix} 1 & 2 & -6 & -6 & 1 \\ -2 & 4 & 0 & 1 & 2 \end{bmatrix} \begin{bmatrix} 1 \\ -1 \\ 0 \\ -5 \\ 2 \end{bmatrix}$

In Exercises 15–18, find, if possible, AB and BA. If it is not possible, explain why.

15. $A = \begin{bmatrix} 1 & -3 & 8 \end{bmatrix}$ $\quad B = \begin{bmatrix} -1 \\ 5 \\ 7 \end{bmatrix}$

16. $A = \begin{bmatrix} -3 & 2 & 0 \\ 1 & -4 & 5 \end{bmatrix}$ $\quad B = \begin{bmatrix} -2 & 0 \\ 4 & -6 \\ 5 & -1 \end{bmatrix}$

17. $A = \begin{bmatrix} -10 & 25 & 40 \\ 42 & -5 & 0 \end{bmatrix}$ $\quad B = \begin{bmatrix} 6 \\ -15 \\ 12 \end{bmatrix}$

18. $A = \begin{bmatrix} -2 & 1 & 7 \\ 3 & -1 & 0 \\ 0 & 2 & -1 \end{bmatrix}$ $\quad B = \begin{bmatrix} 4 & -1 & 5 \end{bmatrix}$

In Exercises 19–22, show that $AI = IA = A$.

19. $A = \begin{bmatrix} 1 & 8 \\ -2 & 2 \end{bmatrix}$ $\quad$ **20.** $A = \begin{bmatrix} -15 & 28 \\ -5 & 64 \end{bmatrix}$

21. $A = \begin{bmatrix} 3 & 9 & -15 \\ 8 & 0 & 4 \\ 6 & -12 & 24 \end{bmatrix}$ $\quad$ **22.** $A = \begin{bmatrix} -1 & 2 & 0 \\ 4 & -3 & 1 \\ 2 & 1 & 3 \end{bmatrix}$

In Exercises 23–26, determine whether or not $B = A^{-1}$.

23. $A = \begin{bmatrix} 5 & -2 \\ -2 & 1 \end{bmatrix}$ $\quad B = \begin{bmatrix} 1 & 2 \\ 2 & 5 \end{bmatrix}$

24. $A = \begin{bmatrix} 3 & -4 \\ 5 & -7 \end{bmatrix}$ $\quad B = \begin{bmatrix} 7 & -4 \\ 5 & -2 \end{bmatrix}$

25. $A = \begin{bmatrix} 1 & -2 & 3 \\ 2 & -5 & 7 \\ -1 & 3 & -5 \end{bmatrix}$ $\quad B = \begin{bmatrix} 4 & -1 & 1 \\ 3 & -2 & -1 \\ 1 & -1 & -1 \end{bmatrix}$

26. $A = \begin{bmatrix} 1 & -1 & 3 \\ 3 & -4 & 8 \\ -2 & 3 & -4 \end{bmatrix}$ $\quad B = \begin{bmatrix} 8 & -5 & -4 \\ 4 & -2 & -1 \\ -1 & 1 & 1 \end{bmatrix}$

In Exercises 27–30, determine by matrix multiplication whether or not A is the proper matrix of solution values.

27. $3x - 2y = -1$
$4x + y = 6$ $\quad A = \begin{bmatrix} 1 \\ 2 \end{bmatrix}$

28. $4x + y = -5$
$3x + 4y = 8$ $\quad A = \begin{bmatrix} -2 \\ 3 \end{bmatrix}$

29. $3x + y + 2z = 1$
$x - 3y + 4z = -3$
$2x + 2y + z = 1$ $\quad A = \begin{bmatrix} -1 \\ 2 \\ 1 \end{bmatrix}$

30. $2x - y + z = 7$
$x - 3y + 2z = 7$
$3x + y = 7$ $\quad A = \begin{bmatrix} 3 \\ -2 \\ -1 \end{bmatrix}$

In Exercises 31–34, perform the indicated matrix multiplications on a calculator, using the following matrices. For matrix A, $A^2 = A \times A$.

$$A = \begin{bmatrix} 2 & -3 & -5 \\ -1 & 4 & 5 \\ 1 & -3 & -4 \end{bmatrix} \quad B = \begin{bmatrix} 1 & -2 & -6 \\ -3 & 2 & 9 \\ 2 & 0 & -3 \end{bmatrix} \quad C = \begin{bmatrix} 1 & -3 & -4 \\ -1 & 3 & 4 \\ 1 & -3 & -4 \end{bmatrix}$$

31. Show that $A^2 = A$.

32. Show that $C^2 = O$.

33. Show that $B^3 = B$.

34. Show that $A^3B^3 = AB$.

In Exercises 35–48, solve the given problems.

35. For matrices $A = \begin{bmatrix} a & b \\ b & a \end{bmatrix}$ and $B = \begin{bmatrix} c & d \\ d & c \end{bmatrix}$, show that $AB = BA$.

36. For matrix $A = \begin{bmatrix} 1 & 2 & 2 \\ 2 & 1 & 2 \\ 2 & 2 & 1 \end{bmatrix}$, show that $A^2 - 4A - 5I = O$.

37. Using two rows and columns, show that $(-I)^2 = I$.

38. For $J = \begin{bmatrix} j & 0 \\ 0 & j \end{bmatrix}$, where $j = \sqrt{-1}$, show that $J^2 = -I$, $J^3 = -J$, and $J^4 = I$. Explain the similarity with j^2, j^3, and j^4.

39. Show that $A^2 - I = (A + I)(A - I)$ for $A = \begin{bmatrix} 2 & 4 \\ 3 & 5 \end{bmatrix}$.

40. By using $A = \begin{bmatrix} 0 & 2 \\ 0 & 0 \end{bmatrix}$, $B = \begin{bmatrix} 3 & 1 \\ 2 & 0 \end{bmatrix}$, $C = \begin{bmatrix} 6 & 3 \\ 2 & 0 \end{bmatrix}$, show that $AB = AC$ does not necessarily mean that $B = C$.

41. From past records, satellite television Company A finds that among its current customers, 92% will still be a customer one year from now and the remaining 8% will be lost to other competing companies. Also, among customers who are currently with other companies, 14% will switch to Company A within one year and the remaining 86% will remain with other companies. This is represented by the following *transition matrix T*:

		One year from now	
		A	Other
Now	A	0.92	0.08
	Other	0.14	0.86

$= T$

Company A currently has 40% of the market share, given by
A Other
$[0.40 \quad 0.60]$. Find the market share of Company A after one year by multiplying $[0.40 \quad 0.60]\,T$.

42. In Exercise 41, the market share of Company A in future years can be found by multiplying by the transition matrix T repeatedly. What would the market share of Company A be after 3 years?

43. In studying the motion of electrons, one of the Pauli spin matrices used is $s_y = \begin{bmatrix} 0 & -j \\ j & 0 \end{bmatrix}$, where $j = \sqrt{-1}$. Show that $s_y^2 = I$.

44. To rotate a set of points $(x_1, y_1), (x_2, y_2), (x_3, y_3), \ldots$ counterclockwise about the origin by angle θ, we multiply $\begin{bmatrix} \cos\theta & -\sin\theta \\ \sin\theta & \cos\theta \end{bmatrix}\begin{bmatrix} x_1 & x_2 & x_3 & \cdots \\ y_1 & y_2 & y_3 & \end{bmatrix}$. If a photo on a computer screen has corners at $(4, 2)$, $(-4, 2)$, $(-4, -2)$, and $(4, -2)$, find the coordinates of the corners after the photo has been rotated counterclockwise about the origin by 30°.

45. In an *ammeter,* nearly all the electric current flows through a *shunt,* and the remaining known fraction of current is measured by the meter. See Fig. 16.5. From the given matrix equation, find voltage v_2 and current i_2 in terms of v_1, i_1, and resistance R, whichever may be applicable.

$$\begin{bmatrix} v_2 \\ i_2 \end{bmatrix} = \begin{bmatrix} 1 & 0 \\ -\frac{1}{R} & 1 \end{bmatrix}\begin{bmatrix} v_1 \\ i_1 \end{bmatrix}$$

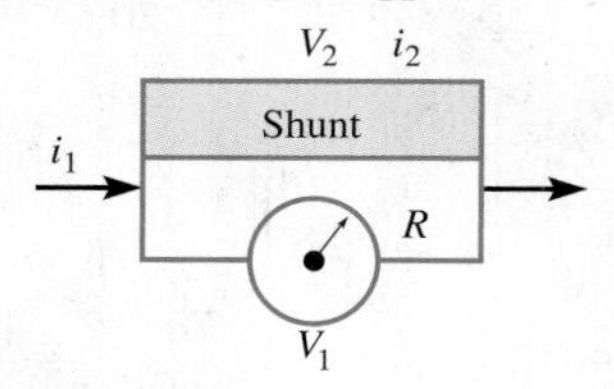

Fig. 16.5

46. In the theory related to the reproduction of color photography, the equation

$$\begin{bmatrix} X \\ Y \\ Z \end{bmatrix} = \begin{bmatrix} 1.0 & 0.1 & 0 \\ 0.5 & 1.0 & 0.1 \\ 0.3 & 0.4 & 1.0 \end{bmatrix}\begin{bmatrix} x \\ y \\ z \end{bmatrix}$$

is found. The X, Y, and Z represent the red, green, and blue densities of the reproductions, respectively, and the x, y, and z represent the red, green, and blue densities, respectively, of the subject. Give the equations relating X, Y, and Z and x, y, and z.

47. The path of an Earth satellite can be written as

$$[x \quad y]\begin{bmatrix} 2.76 & -1 \\ 1 & 2.81 \end{bmatrix}\begin{bmatrix} x \\ y \end{bmatrix} = [7.76 \times 10^7]$$

where distances are in miles. What type of curve is represented? (See Section 14.1.)

48. Using Kirchhoff's laws on the circuit shown in Fig. 16.6, the following matrix equation is found. By matrix multiplication, find the resulting system of equations.

$$\begin{bmatrix} R_1 + R_2 & -R_2 & 0 \\ -R_2 & R_2 + R_3 + R_4 & -R_4 \\ 0 & -R_4 & R_4 + R_5 \end{bmatrix}\begin{bmatrix} I_1 \\ I_2 \\ I_3 \end{bmatrix} = \begin{bmatrix} V_1 \\ 0 \\ -V_2 \end{bmatrix}$$

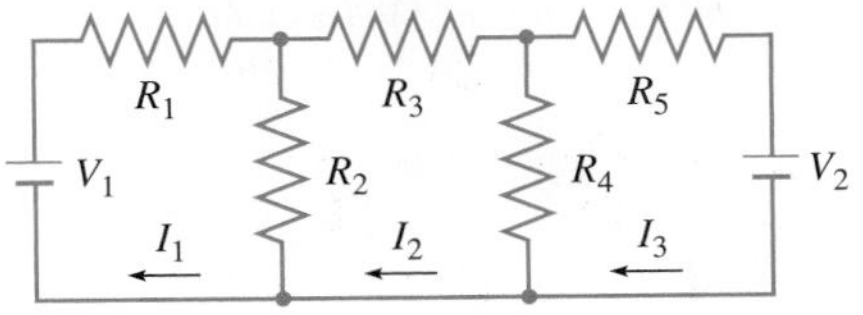

Fig. 16.6

Answers to Practice Exercises

1. $\begin{bmatrix} -33 & 43 \\ 54 & -44 \\ -84 & 120 \end{bmatrix}$ **2.** 4 rows, 4 columns **3.** $AB = BA = \begin{bmatrix} 1 & 0 \\ 0 & 1 \end{bmatrix}$

16.3 Finding the Inverse of a Matrix

Inverse of a 2 × 2 Matrix • Gauss–Jordan Method • Inverse on a Calculator

We now show how to find the inverse of a matrix, and in the next section, we show how it is used in solving a system of linear equations. First, we show two methods of finding the **inverse of a 2 × 2 matrix.** The first method is as follows:

1. *Interchange the elements on the principal diagonal.*
2. *Change the signs of the off-diagonal elements.*
3. *Divide each resulting element by the determinant of the given matrix.*

CAUTION This method can be used with second-order square matrices *but not with higher-order matrices.* ■ The following example illustrates this method.

CAUTION If the value of the determinant is zero, the inverse matrix *does not exist.* It would cause division by zero in the method used in Example 1. ■

■ In Exercise 37, you are asked to verify the method used in Example 1.

EXAMPLE 1 Inverse of 2 × 2 matrix—method 1

Find the inverse of the matrix $A = \begin{bmatrix} 2 & -3 \\ 4 & -7 \end{bmatrix}$.

First, we interchange the elements on the principal diagonal and change the signs of the off-diagonal elements. This gives us the matrix

$$\begin{bmatrix} -7 & 3 \\ -4 & 2 \end{bmatrix}$$

← signs changed
← elements interchanged

Now, we find the determinant of the original matrix, which means we evaluate

$$\begin{vmatrix} 2 & -3 \\ 4 & -7 \end{vmatrix} = -14 - (-12) = -2$$

We now divide each element of the second matrix by -2. This gives

$$A^{-1} = \frac{1}{-2}\begin{bmatrix} -7 & 3 \\ -4 & 2 \end{bmatrix} = \begin{bmatrix} \frac{-7}{-2} & \frac{3}{-2} \\ \frac{-4}{-2} & \frac{2}{-2} \end{bmatrix} = \begin{bmatrix} \frac{7}{2} & -\frac{3}{2} \\ 2 & -1 \end{bmatrix}$$ ← inverse

Checking by multiplication gives

$$AA^{-1} = \begin{bmatrix} 2 & -3 \\ 4 & -7 \end{bmatrix}\begin{bmatrix} \frac{7}{2} & -\frac{3}{2} \\ 2 & -1 \end{bmatrix} = \begin{bmatrix} 7 - 6 & -3 + 3 \\ 14 - 14 & -6 + 7 \end{bmatrix} = \begin{bmatrix} 1 & 0 \\ 0 & 1 \end{bmatrix} = I$$

Because $AA^{-1} = I$, the matrix A^{-1} is the proper inverse matrix. ■

Practice Exercise

1. Find the inverse:

$$A = \begin{bmatrix} 3 & -8 \\ 1 & -2 \end{bmatrix}$$

GAUSS–JORDAN METHOD

The second method, called the *Gauss–Jordan method,* involves *transforming the given matrix into the identity matrix while* ***transforming the identity matrix into the inverse.*** There are three types of steps allowable in making these transformations:

■ Named for the German mathematician Karl Gauss (1777–1855) and the German geodesist Wilhelm Jordan (1842–1899).

1. *Any two rows may be interchanged.*
2. *Every element in any row may be multiplied by any number other than zero.*
3. *Any row may be replaced by a row whose elements are the sum of a nonzero multiple of itself and a nonzero multiple of another row.*

NOTE ▸ [Note that these are **row operations,** not column operations, and are the same operations used in solving a system of equations by addition and subtraction.]

EXAMPLE 2 2 × 2 Inverse—Gauss-Jordan method

Find the inverse of the matrix

$$A = \begin{bmatrix} 2 & -3 \\ 4 & -7 \end{bmatrix}$$ this is the same matrix as in Example 1

First, we set up the given matrix with the identity matrix as follows:

$$\left[\begin{array}{cc|cc} 2 & -3 & 1 & 0 \\ 4 & -7 & 0 & 1 \end{array}\right]$$

The vertical line simply shows the separation of the two matrices.

We wish to transform the left matrix into the identity matrix. Therefore, the first requirement is a 1 for element a_{11}. Therefore, we divide all elements of the first row by 2. This gives the following setup:

$$\left[\begin{array}{cc|cc} 1 & -\frac{3}{2} & \frac{1}{2} & 0 \\ 4 & -7 & 0 & 1 \end{array}\right]$$

Next, we want to have a zero for element a_{21}. Therefore, we subtract 4 times each element of row 1 from the corresponding element in row 2, replacing the elements of row 2. This gives us the following setup:

■ As in this example, (1) always work one column at a time, from left to right and (2) never undo the work in a previously completed column.

$$\left[\begin{array}{cc|cc} 1 & -\frac{3}{2} & \frac{1}{2} & 0 \\ 4-4(1) & -7-4(-\frac{3}{2}) & 0-4(\frac{1}{2}) & 1-4(0) \end{array}\right] \text{ or } \left[\begin{array}{cc|cc} 1 & -\frac{3}{2} & \frac{1}{2} & 0 \\ 0 & -1 & -2 & 1 \end{array}\right]$$

Next, we want to have 1, not −1, for element a_{22}. Therefore, we multiply each element of row 2 by −1. This gives

$$\left[\begin{array}{cc|cc} 1 & -\frac{3}{2} & \frac{1}{2} & 0 \\ 0 & 1 & 2 & -1 \end{array}\right]$$

Finally, we want zero for element a_{12}. Therefore, we add $\frac{3}{2}$ times each element of row 2 to the corresponding elements of row 1, replacing row 1. This gives

$$\left[\begin{array}{cc|cc} 1+\frac{3}{2}(0) & -\frac{3}{2}+\frac{3}{2}(1) & \frac{1}{2}+\frac{3}{2}(2) & 0+\frac{3}{2}(-1) \\ 0 & 1 & 2 & -1 \end{array}\right] \text{ or } \left[\begin{array}{cc|cc} 1 & 0 & \frac{7}{2} & -\frac{3}{2} \\ 0 & 1 & 2 & -1 \end{array}\right]$$

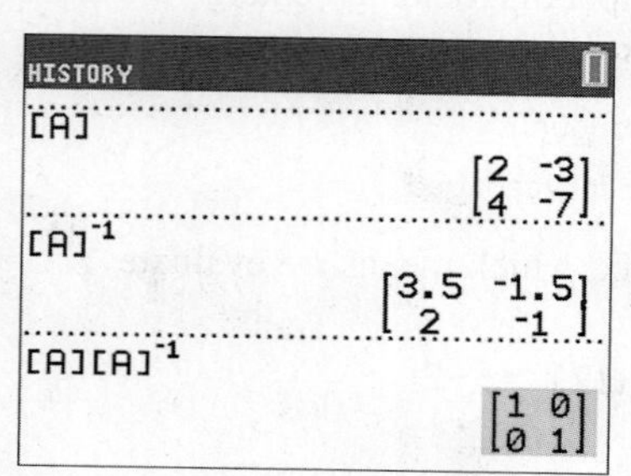

Fig. 16.7
Graphing calculator keystrokes: goo.gl/6a8wvW

At this point, we have transformed the given matrix into the identity matrix, and the identity matrix into the inverse. Therefore, the matrix to the right of the vertical bar in the last setup is the required inverse. Thus,

$$A^{-1} = \begin{bmatrix} \frac{7}{2} & -\frac{3}{2} \\ 2 & -1 \end{bmatrix}$$

To find the inverse on a calculator, we first enter the matrix A and then use the (x^{-1}) key. Figure 16.7 shows matrix A, its inverse A^{-1}, and the product $A^{-1}A$, which equals the identity matrix. ■

Practice Exercise

2. Find the inverse using the Gauss–Jordan method: $A = \begin{bmatrix} 3 & -8 \\ 1 & -2 \end{bmatrix}$

In transforming a matrix into the identity matrix, we work on one column at a time, transforming the columns in order from left to right. It is generally best to make the element on the principal diagonal for the column 1 first and then make all other elements in the column 0. The method is applicable for any square matrix. The following two examples illustrate this method, with the row operations shown in red. For example, the notation $-4R_1 + R_2$ means that a row is being replaced with -4 times row 1 plus row 2.

■ Matrices and inverse matrices can be used to send and decode encrypted messages. See Exercises 43 and 44.

EXAMPLE 3 2 × 2 Inverse—Gauss-Jordan method

Find the inverse of the matrix $\begin{bmatrix} -3 & 6 \\ 4 & 5 \end{bmatrix}$.

original setup; make top left entry 1; make other entry in first column 0

$$\left[\begin{array}{cc|cc} -3 & 6 & 1 & 0 \\ 4 & 5 & 0 & 1 \end{array}\right] \xrightarrow{-\frac{1}{3}R_1} \left[\begin{array}{cc|cc} 1 & -2 & -\frac{1}{3} & 0 \\ 4 & 5 & 0 & 1 \end{array}\right] \xrightarrow{-4R_1 + R_2} \left[\begin{array}{cc|cc} 1 & -2 & -\frac{1}{3} & 0 \\ 0 & 13 & \frac{4}{3} & 1 \end{array}\right]$$

make second endtry in bottom row 1; make other entry in second column 0

$$\xrightarrow{\frac{1}{13}R_2} \left[\begin{array}{cc|cc} 1 & -2 & -\frac{1}{3} & 0 \\ 0 & 1 & \frac{4}{39} & \frac{1}{13} \end{array}\right] \xrightarrow{2R_2 + R_1} \left[\begin{array}{cc|cc} 1 & 0 & -\frac{5}{39} & \frac{2}{13} \\ 0 & 1 & \frac{4}{39} & \frac{1}{13} \end{array}\right]$$

Therefore, $A^{-1} = \begin{bmatrix} -\frac{5}{39} & \frac{2}{13} \\ \frac{4}{39} & \frac{1}{13} \end{bmatrix}$, which can be checked by multiplication. ■

EXAMPLE 4 3 × 3 Inverse—Gauss-Jordan method

Find the inverse of the matrix $\begin{bmatrix} 1 & 2 & -1 \\ 3 & 5 & -1 \\ -2 & -1 & -2 \end{bmatrix}$.

original setup: top left entry is already 1; make other entries in first column 0; make second entry in middle row 1

$$\left[\begin{array}{ccc|ccc} 1 & 2 & -1 & 1 & 0 & 0 \\ 3 & 5 & -1 & 0 & 1 & 0 \\ -2 & -1 & -2 & 0 & 0 & 1 \end{array}\right] \xrightarrow[2R_1 + R_3]{-3R_1 + R_2} \left[\begin{array}{ccc|ccc} 1 & 2 & -1 & 1 & 0 & 0 \\ 0 & -1 & 2 & -3 & 1 & 0 \\ 0 & 3 & -4 & 2 & 0 & 1 \end{array}\right] \xrightarrow{-1R_2} \left[\begin{array}{ccc|ccc} 1 & 2 & -1 & 1 & 0 & 0 \\ 0 & 1 & -2 & 3 & -1 & 0 \\ 0 & 3 & -4 & 2 & 0 & 1 \end{array}\right]$$

make other entries in second column 0; make third entry in bottom row 1; make other entries in third column 0

$$\xrightarrow[-3R_2 + R_3]{-2R_2 + R_1} \left[\begin{array}{ccc|ccc} 1 & 0 & 3 & -5 & 2 & 0 \\ 0 & 1 & -2 & 3 & -1 & 0 \\ 0 & 0 & 2 & -7 & 3 & 0 \end{array}\right] \xrightarrow{\frac{1}{2}R_3} \left[\begin{array}{ccc|ccc} 1 & 0 & 3 & -5 & 2 & 0 \\ 0 & 1 & -2 & 3 & -1 & 0 \\ 0 & 0 & 1 & -\frac{7}{2} & \frac{3}{2} & \frac{1}{2} \end{array}\right] \xrightarrow[2R_3 + R_2]{-3R_3 + R_1} \left[\begin{array}{ccc|ccc} 1 & 0 & 0 & \frac{11}{2} & -\frac{5}{2} & -\frac{3}{2} \\ 0 & 1 & 0 & -4 & 2 & 1 \\ 0 & 0 & 1 & -\frac{7}{2} & \frac{3}{2} & \frac{1}{2} \end{array}\right]$$

Therefore, the required inverse matrix is

$$\begin{bmatrix} \frac{11}{2} & -\frac{5}{2} & -\frac{3}{2} \\ -4 & 2 & 1 \\ -\frac{7}{2} & \frac{3}{2} & \frac{1}{2} \end{bmatrix}$$

which may be checked by multiplication. See Fig. 16.8 for a calculator window showing A^{-1} in decimal and fractional form. ■

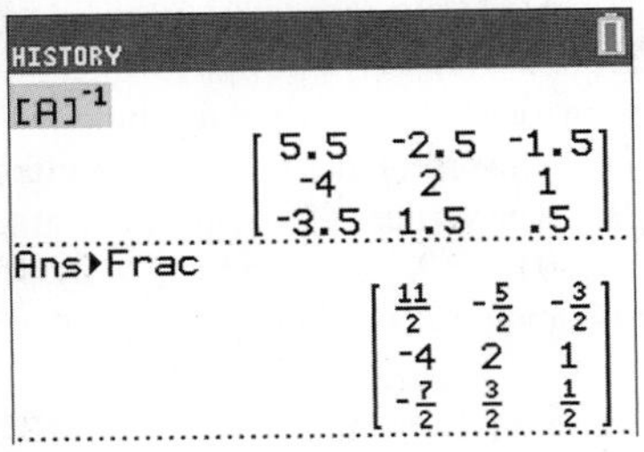

```
HISTORY
[A]^-1
        [ 5.5  -2.5  -1.5 ]
        [ -4    2     1   ]
        [ -3.5  1.5   .5  ]
Ans▸Frac
        [ 11/2  -5/2  -3/2 ]
        [ -4     2     1   ]
        [ -7/2   3/2   1/2 ]
```

Fig. 16.8
Graphing calculator keystrokes: goo.gl/4gGEkb

EXERCISES 16.3

In Exercises 1 and 2, make the given changes in the indicated examples of this section and then find the matrix inverses.

1. In Example 1, change the element -7 to -5 and then find the inverse using the same method.
2. In Example 3, change the element -3 to -2 and then find the inverse using the same method.

In Exercises 3–10, find the inverse of each of the given matrices by the method of Example 1 of this section.

3. $\begin{bmatrix} 2 & -3 \\ -2 & 4 \end{bmatrix}$ 4. $\begin{bmatrix} -6 & 3 \\ 3 & -2 \end{bmatrix}$ 5. $\begin{bmatrix} -1 & 5 \\ 4 & 10 \end{bmatrix}$

6. $\begin{bmatrix} 0.8 & -0.1 \\ -0.4 & -0.5 \end{bmatrix}$ 7. $\begin{bmatrix} 0 & -4 \\ 2 & 6 \end{bmatrix}$ 8. $\begin{bmatrix} 70 & -30 \\ -60 & 40 \end{bmatrix}$

9. $\begin{bmatrix} -30 & -45 \\ 26 & 50 \end{bmatrix}$ 10. $\begin{bmatrix} 7.2 & -3.6 \\ -1.3 & -5.7 \end{bmatrix}$

In Exercises 11–20, find the inverse of each of the given matrices by transforming the identity matrix, as in Examples 2–4.

11. $\begin{bmatrix} 1 & 2 \\ 2 & 5 \end{bmatrix}$ 12. $\begin{bmatrix} 1 & 5 \\ -1 & -4 \end{bmatrix}$

13. $\begin{bmatrix} 2 & 4 \\ -1 & -1 \end{bmatrix}$ 14. $\begin{bmatrix} -2 & 6 \\ 3 & -4 \end{bmatrix}$

15. $\begin{bmatrix} -2 & 5 \\ -1 & 2 \end{bmatrix}$ 16. $\begin{bmatrix} -2 & 3 \\ -3 & 5 \end{bmatrix}$

17. $\begin{bmatrix} 1 & -3 & -2 \\ -2 & 7 & 3 \\ 1 & -1 & -3 \end{bmatrix}$ 18. $\begin{bmatrix} 1 & 2 & -1 \\ 3 & 7 & -5 \\ -1 & -2 & 0 \end{bmatrix}$

19. $\begin{bmatrix} 1 & 3 & 2 \\ -2 & -5 & -1 \\ 2 & 4 & 0 \end{bmatrix}$ 20. $\begin{bmatrix} 1 & 3 & 4 \\ -1 & -4 & -2 \\ 4 & 9 & 20 \end{bmatrix}$

In Exercises 21–28, find the inverse of each of the given matrices by using a calculator. The matrices in Exercises 23 and 24 are the same as those in Exercises 19 and 20.

21. $\begin{bmatrix} -2 & 8 \\ -1 & 6 \end{bmatrix}$ 22. $\begin{bmatrix} 20 & -45 \\ -12 & 24 \end{bmatrix}$

23. $\begin{bmatrix} 1 & 3 & 2 \\ -2 & -5 & -1 \\ 2 & 4 & 0 \end{bmatrix}$ 24. $\begin{bmatrix} 1 & 3 & 4 \\ -1 & -4 & -2 \\ 4 & 9 & 20 \end{bmatrix}$

25. $\begin{bmatrix} 2 & 4 & 0 \\ 3 & 4 & -2 \\ -1 & 1 & 2 \end{bmatrix}$ 26. $\begin{bmatrix} 10 & -5 & 30 \\ -2 & 4 & -5 \\ 20 & 5 & 5 \end{bmatrix}$

27. $\begin{bmatrix} 1 & -2 & 1 & 0 \\ 1 & -2 & 2 & -3 \\ 0 & 1 & -1 & 1 \\ -2 & 3 & -2 & 3 \end{bmatrix}$ 28. $\begin{bmatrix} 12.5 & -2.6 & 1.2 & 7.6 \\ -4.6 & 10.0 & -4.7 & -6.8 \\ 5.7 & -3.7 & 7.3 & 11.0 \\ 8.8 & 6.8 & 14.0 & 4.7 \end{bmatrix}$

In Exercises 29–31, find BA^{-1}. In Exercises 32–34, find CA^{-1}.

$$B = \begin{bmatrix} 8 & -2 \\ 3 & 4 \end{bmatrix} \qquad C = \begin{bmatrix} 5 & -1 & 0 \\ 2 & -2 & 1 \\ -3 & 0 & 4 \end{bmatrix}$$

29. $A = \begin{bmatrix} 2 & -4 \\ -1 & 3 \end{bmatrix}$ 30. $A = \begin{bmatrix} -4 & 1 \\ 6 & -2 \end{bmatrix}$

31. $A = \begin{bmatrix} 5 & -3 \\ 2 & -1 \end{bmatrix}$ 32. $A = \begin{bmatrix} -3 & 1 & -1 \\ 1 & -4 & -7 \\ 1 & 2 & 5 \end{bmatrix}$

33. $A = \begin{bmatrix} 1 & -1 & 1 \\ 0 & -2 & 1 \\ -2 & -3 & 0 \end{bmatrix}$

34. $A = \begin{bmatrix} 1 & 0 & 2 \\ -1 & 2 & 3 \\ 1 & -1 & 0 \end{bmatrix}$

In Exercises 35–44, solve the given problems.

35. Show that the matrix $\begin{bmatrix} 1 & 1 \\ 1 & 1 \end{bmatrix}$ has no inverse.

36. Find the determinant of the matrix $\begin{bmatrix} 1 & -2 & 0 \\ -2 & 4 & 8 \\ 3 & -6 & 6 \end{bmatrix}$. Explain what this tells us about its inverse.

37. For the matrix $A = \begin{bmatrix} a & b \\ c & d \end{bmatrix}$, show that

$$\frac{1}{ad - bc}\begin{bmatrix} a & b \\ c & d \end{bmatrix}\begin{bmatrix} d & -b \\ -c & a \end{bmatrix} = \begin{bmatrix} 1 & 0 \\ 0 & 1 \end{bmatrix}.$$

This verifies the method of Example 1.

38. Describe the relationship between the elements of the matrix $\begin{bmatrix} a & 0 & 0 \\ 0 & b & 0 \\ 0 & 0 & c \end{bmatrix}$ and the elements of its inverse.

39. In Exercise 44 of Section 16.2, we saw that the matrix multiplication $\begin{bmatrix} \cos\theta & -\sin\theta \\ \sin\theta & \cos\theta \end{bmatrix}\begin{bmatrix} x_1 & x_2 & x_3 & \cdots \\ y_1 & y_2 & y_3 & \end{bmatrix}$ rotates the points in the second matrix counterclockwise about the origin by angle θ. If we replace the first matrix with its inverse, it has the opposite effect of rotating *clockwise* by angle θ. The vertices of a triangle in a graphic design program are at $(\sqrt{2}, 0)$, $(-\sqrt{2}, 0)$, and $(0, \sqrt{2})$. If the triangle is rotated clockwise about the origin by 45°, find the new coordinates of the vertices.

40. The matrix $A = \begin{bmatrix} 2 & -1 & 1 \\ -1 & 4 & -3 \\ 1 & -3 & 2 \end{bmatrix}$ is *symmetric* (note the elements on opposite sides of the main diagonal are equal). Show that A^{-1} is also symmetric.

41. For the *four-terminal network* shown in Fig. 16.9, it can be shown that the voltage matrix V is related to the coefficient matrix A and the current matrix I by $V = A^{-1}I$, where

$$V = \begin{bmatrix} v_1 \\ v_2 \end{bmatrix} \quad A = \begin{bmatrix} a_{11} & a_{12} \\ a_{21} & a_{22} \end{bmatrix} \quad I = \begin{bmatrix} i_1 \\ i_2 \end{bmatrix}$$

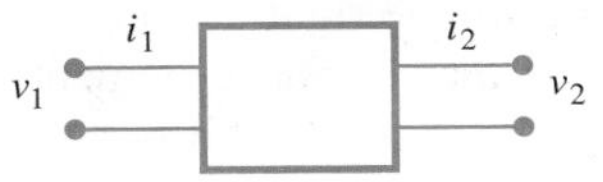

Fig. 16.9

Find the equations for v_1 and v_2 that give each in terms of i_1 and i_2.

42. The rotations of a robot arm such as that shown in Fig. 16.10 are often represented by matrices. The values represent trigonometric functions of the angles of rotation. For the following rotation matrix R, find R^{-1}.

$$R = \begin{bmatrix} 0.8 & 0.0 & -0.6 \\ 0.0 & 1.0 & 0.0 \\ 0.6 & 0.0 & 0.8 \end{bmatrix}$$

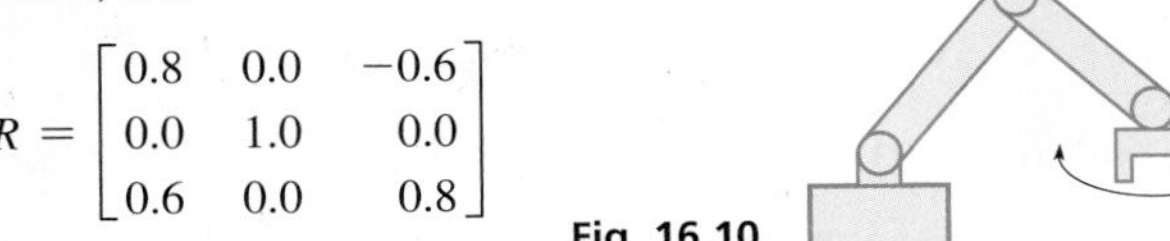

Fig. 16.10

43. An avid math student sends an encrypted matrix message E to a friend who has the coding matrix C. The friend can decode E to get message M by multiplying E by C^{-1}. If

$$E = \begin{bmatrix} 42 & 61 & 47 & 62 \\ 18 & 16 & 4 & 19 \\ 7 & 10 & 7 & 12 \\ 41 & 34 & 47 & 55 \end{bmatrix} \quad \text{and} \quad C = \begin{bmatrix} 1 & 1 & 2 & 1 \\ 1 & 0 & 1 & 1 \\ 1 & 2 & 1 & 2 \\ 1 & 1 & 0 & 1 \end{bmatrix},$$

what is the message? ($0 =$ space, $1 = A$, $2 = B$, $3 = C$, etc.)

44. (See Exercise 43.) Using the same encryption code C, the student's friend answered with

$$R = \begin{bmatrix} 19 & 9 & 20 & 23 \\ 70 & 83 & 67 & 88 \\ 45 & 43 & 40 & 64 \\ 39 & 30 & 45 & 51 \end{bmatrix}.$$ What was the return message R?

Answers to Practice Exercises

1. $\begin{bmatrix} -1 & 4 \\ -\frac{1}{2} & \frac{3}{2} \end{bmatrix}$ **2.** $\begin{bmatrix} -1 & 4 \\ -\frac{1}{2} & \frac{3}{2} \end{bmatrix}$

16.4 Matrices and Linear Equations

Multiply Matrix of Constants by Inverse of Matrix of Coefficients • Solving Systems of Equations • Matrix Solution on a Calculator

As we stated at the beginning of Section 16.1, matrices can be used to solve systems of equations. In this section, we show one method by which this is done. As we develop this method, it will be apparent that there is a great deal of numerical work involved. However, methods such as this one are easily programmed for use on a computer, which can do the arithmetic work very rapidly. Also, most calculators can perform these operations, and we will show an example at the end of the section in which a calculator is used to solve a system of four equations more readily than with earlier methods. It is the *method* of solving the system of equations that is of primary importance here.

Let us consider the system of equations

$$a_1x + b_1y = c_1$$
$$a_2x + b_2y = c_2$$

Recalling the definition of equality of matrices, we can write this system as

$$\begin{bmatrix} a_1x + b_1y \\ a_2x + b_2y \end{bmatrix} = \begin{bmatrix} c_1 \\ c_2 \end{bmatrix} \tag{16.6}$$

If we let

$$A = \begin{bmatrix} a_1 & b_1 \\ a_2 & b_2 \end{bmatrix} \quad X = \begin{bmatrix} x \\ y \end{bmatrix} \quad C = \begin{bmatrix} c_1 \\ c_2 \end{bmatrix} \tag{16.7}$$

the left side of Eq. (16.6) can be written as the product of matrices A and X.

$$AX = C \tag{16.8}$$

If we now multiply (on the left) each side of this matrix equation by A^{-1}, we have

$$A^{-1}AX = A^{-1}C$$

Because $A^{-1}A = I$, we have

$$IX = A^{-1}C$$

However, $IX = X$. Therefore,

$$X = A^{-1}C \tag{16.9}$$

NOTE ▸ [Equation (16.9) states that we can solve a system of linear equations by multiplying the one-column matrix of the constants on the right by the inverse of the matrix of the coefficients.] The result is a one-column matrix whose elements are the required values for the solution. **CAUTION** Note that $X = A^{-1}C$, and **not** CA^{-1}. The order of matrix multiplication must be carefully followed. ■

EXAMPLE 1 Matrix solution—two equations

Use matrices to solve the system of equations

$$2x - y = 7$$
$$5x - 3y = 18$$

The matrices for Eq. (16.7), and the matrix equation are

$$A = \begin{bmatrix} 2 & -1 \\ 5 & -3 \end{bmatrix} \quad X = \begin{bmatrix} x \\ y \end{bmatrix} \quad C = \begin{bmatrix} 7 \\ 18 \end{bmatrix} \quad \begin{bmatrix} 2 & -1 \\ 5 & -3 \end{bmatrix}\begin{bmatrix} x \\ y \end{bmatrix} = \begin{bmatrix} 7 \\ 18 \end{bmatrix}$$

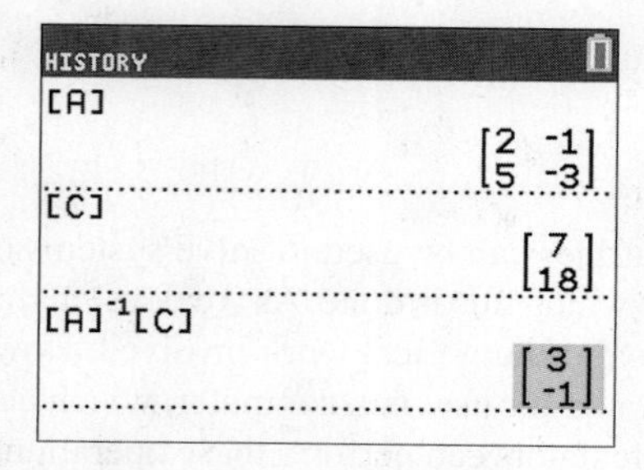

Fig. 16.11
Graphing calculator keystrokes: goo.gl/7CT1W2

By either of the methods of the previous section, we can determine the inverse of matrix A to be

$$A^{-1} = \begin{bmatrix} 3 & -1 \\ 5 & -2 \end{bmatrix}$$

We now form the matrix product $A^{-1}C$.

$$A^{-1}C = \begin{bmatrix} 3 & -1 \\ 5 & -2 \end{bmatrix}\begin{bmatrix} 7 \\ 18 \end{bmatrix} = \begin{bmatrix} 21 - 18 \\ 35 - 36 \end{bmatrix} = \begin{bmatrix} 3 \\ -1 \end{bmatrix}$$

Because $X = A^{-1}C$, this means that

$$\begin{bmatrix} x \\ y \end{bmatrix} = \begin{bmatrix} 3 \\ -1 \end{bmatrix}$$

Therefore, the required solution is $x = 3$ and $y = -1$, which checks when these values are substituted into the original equations. See the calculator matrix solution shown in Fig. 16.11. ■

Practice Exercise

1. Use matrices to solve the system of equations
$x - 3y = 6$
$2x + y = 5$

EXAMPLE 2 Matrix solution two equations—electric current

For the electric circuit shown in Fig. 16.12, the equations used to find the currents (in amperes) i_1 and i_2 are

$$\begin{aligned} 2.30i_1 + 6.45(i_1 + i_2) &= 15.0 \\ 1.25i_2 + 6.45(i_1 + i_2) &= 12.5 \end{aligned} \quad \text{or} \quad \begin{aligned} 8.75i_1 + 6.45i_2 &= 15.0 \\ 6.45i_1 + 7.70i_2 &= 12.5 \end{aligned}$$

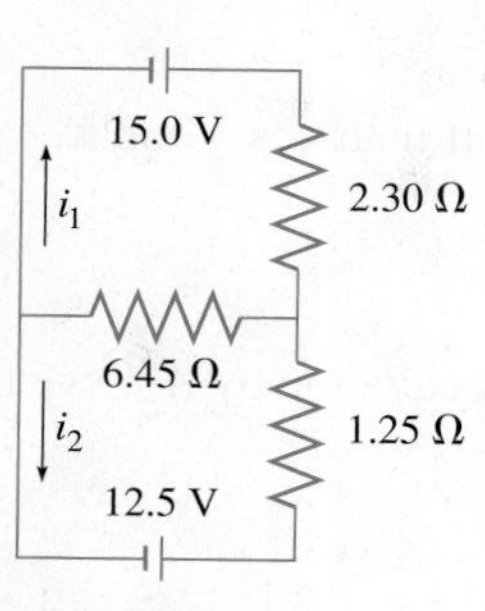

Fig. 16.12

Using matrices to solve this system of equations, we set up the matrix A of coefficients, the matrix C of constants, and the matrix X of currents as

$$A = \begin{bmatrix} 8.75 & 6.45 \\ 6.45 & 7.70 \end{bmatrix} \quad C = \begin{bmatrix} 15.0 \\ 12.5 \end{bmatrix} \quad X = \begin{bmatrix} i_1 \\ i_2 \end{bmatrix}$$

We now find the inverse of A as

$$A^{-1} = \frac{1}{8.75(7.70) - 6.45(6.45)}\begin{bmatrix} 7.70 & -6.45 \\ -6.45 & 8.75 \end{bmatrix} = \begin{bmatrix} 0.2988 & -0.2503 \\ -0.2503 & 0.3395 \end{bmatrix}$$

Therefore,

$$X = A^{-1}C = \begin{bmatrix} 0.2988 & -0.2503 \\ -0.2503 & 0.3395 \end{bmatrix}\begin{bmatrix} 15.0 \\ 12.5 \end{bmatrix}$$

$$= \begin{bmatrix} 0.2988(15.0) - 0.2503(12.5) \\ -0.2503(15.0) + 0.3395(12.5) \end{bmatrix} = \begin{bmatrix} 1.35 \\ 0.49 \end{bmatrix}$$

Therefore, the required currents are $i_1 = 1.35$ A and $i_2 = 0.49$ A. These values check when substituted into the original equations. ■

EXAMPLE 3 Matrix solution—three equations

Use matrices to solve the system of equations

$$\begin{aligned} x + 2y - z &= -4 \\ 3x + 5y - z &= -5 \\ -2x - y - 2z &= -5 \end{aligned}$$

Setting up matrices A, C, and X, we have

$$A = \begin{bmatrix} 1 & 2 & -1 \\ 3 & 5 & -1 \\ -2 & -1 & -2 \end{bmatrix} \quad C = \begin{bmatrix} -4 \\ -5 \\ -5 \end{bmatrix} \quad X = \begin{bmatrix} x \\ y \\ z \end{bmatrix}$$

Finding A^{-1} (see Example 4 of Section 16.3) and solving for X, we have

$$A^{-1} = \begin{bmatrix} \frac{11}{2} & -\frac{5}{2} & -\frac{3}{2} \\ -4 & 2 & 1 \\ -\frac{7}{2} & \frac{3}{2} & \frac{1}{2} \end{bmatrix}$$

$$X = A^{-1}C = \begin{bmatrix} \frac{11}{2} & -\frac{5}{2} & -\frac{3}{2} \\ -4 & 2 & 1 \\ -\frac{7}{2} & \frac{3}{2} & \frac{1}{2} \end{bmatrix}\begin{bmatrix} -4 \\ -5 \\ -5 \end{bmatrix} = \begin{bmatrix} -2 \\ 1 \\ 4 \end{bmatrix}$$

Thix means that $x = -2$, $y = 1$, and $z = 4$. ■

EXAMPLE 4 Matrix solution on a calculator—4 equations

Use a calculator to perform the necessary matrix operations in solving the following

$$\begin{aligned} 2r + 4s - t + u &= 5 \\ r - 2s + 3t - u &= -4 \\ 3r + s + 2t - 4u &= 8 \\ 4r + 5s - t + 3u &= -1 \end{aligned}$$

First, we set up matrices A, X, and C:

$$A = \begin{bmatrix} 2 & 4 & -1 & 1 \\ 1 & -2 & 3 & -1 \\ 3 & 1 & 2 & -4 \\ 4 & 5 & -1 & 3 \end{bmatrix} \quad X = \begin{bmatrix} r \\ s \\ t \\ u \end{bmatrix} \quad C = \begin{bmatrix} 5 \\ -4 \\ 8 \\ -1 \end{bmatrix}$$

It is now necessary only to enter matrices A and C in the calculator and find the matrix product $A^{-1}C$, as shown in Fig. 16.13(a). (There is no need to record or display A^{-1}.) This shows that the solution is

$$r = -2 \qquad s = 3 \qquad t = 0.5 \qquad u = -2.5$$

This solution can be checked on the calculator by storing the resulting matrix X as matrix B and finding the matrix product AB, which should equal matrix C as shown in Fig. 16.13(b). The product AB is equivalent to substituting each value into the original equations, as shown in Eq. (16.8). ■

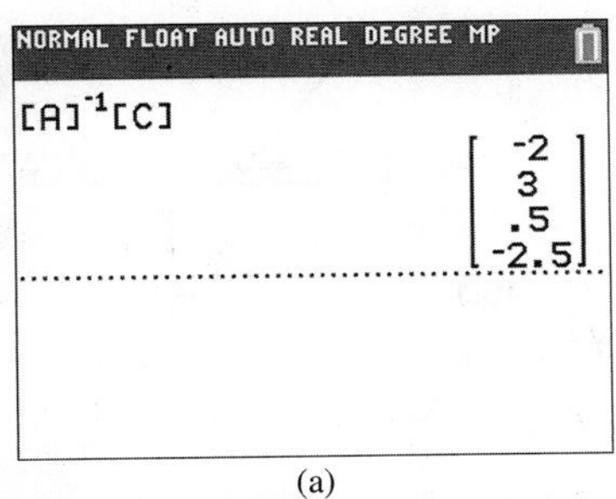

(a)

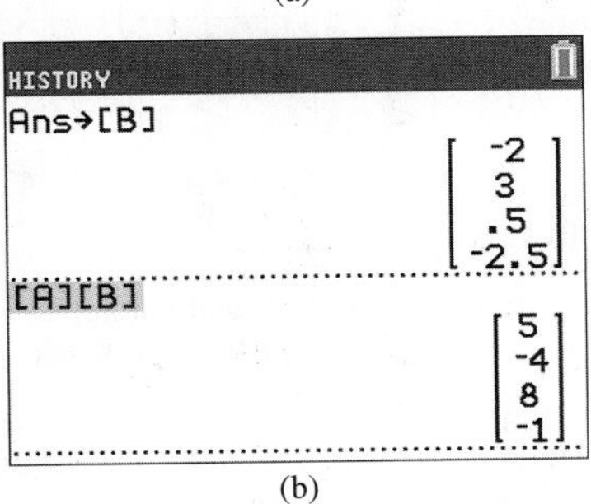

(b)

Fig. 16.13
Graphing calculator keystrokes: goo.gl/yJxmrF

EXERCISES 16.4

In Exercises 1 and 2, make the given changes in the indicated examples of this section and then solve the systems of equations.

1. In Example 1, change the 18 to 19 and then solve the system of equations.

2. In Example 3, change the -4 in the top equation to -2 and then solve the system of equations.

In Exercises 3–8, solve the given systems of equations by using the inverse of the coefficient matrix. The numbers in parentheses refer to exercises from Section 16.3, where the inverses may be checked.

3. $2x - 3y = -6$ (3)
$-2x + 4y = 11$

4. $-x + 5y = 4$ (5)
$4x + 10y = -4$

5. $x + 2y = 7$ (11)
$2x + 5y = 11$

6. $2x + 4y = -9$ (13)
$-x - y = 2$

7. $x - 3y - 2z = -8$ (17)
$-2x + 7y + 3z = 19$
$x - y - 3z = -3$

8. $x + 3y + 2z = 5$ (19)
$-2x - 5y - z = -1$
$2x + 4y = -2$

In Exercises 9–16. solve the given systems of equations by using the inverse of the coefficient matrix

9. $2x - 3y = 3$
$4x - 5y = 4$

10. $x + 2y = 3$
$3x + 4y = 11$

11. $2.5x + 2.8y = -3.0$
$3.5x - 1.6y = 9.6$

12. $24x - 10y = -800$
$31x + 25y = 180$

13. $x + 2y + 2z = -4$
$4x + 9y + 10z = -18$
$-x + 3y + 7z = -7$

14. $x - 4y - 2z = -7$
$-x + 5y + 5z = 18$
$3x - 7y + 10z = 38$

15. $2x + 4y + z = 5$
$-2x - 2y - z = -6$
$-x + 2y + z = 0$

16. $4x + y = 2$
$-2x - y + 3z = -18$
$2x + y - z = 8$

In Exercises 17–28, solve the given systems of equations by using the inverse of the coefficient matrix. Use a calculator to perform the necessary matrix operations and display the results and the check. See Example 4.

17. $3x - y = 4$
$7x + 2y = 18$

18. $5s + 2t = 1$
$s - 3t = 7$

19. $3x - 2y = 5$
$9x + 4y = -5$

20. $4u + 3v = 7$
$6u - 2v = -9$

21. $2x - y - z = 7$
$4x - 3y + 2z = 4$
$3x + 5y + z = -10$

22. $6x + 2y + 9z = 13$
$7x + 6y - 6z = 6$
$5x - 4y + 3z = 15$

23. $u - 3v - 2w = 9$
$3u + 2v + 6w = 20$
$4u - v + 3w = 25$

24. $4x + 2y - 2z = 2$
$3x - 2y - 8z = -3$
$x + 3y + z = 10$

25. $x - 5y + 2z - t = -18$
$3x + y - 3z + 2t = 17$
$4x - 2y + z - t = -1$
$-2x + 3y - z + 4t = 11$

26. $2p + q + 5r + s = 5$
$p + q - 3r - 4s = -1$
$3p + 6q - 2r + s = 8$
$2p + 2q + 2r - 3s = 2$

27. $2v + 3w + x - y - 2z = 6$
$6v - 2w - x + 3y - z = 21$
$v + 3w - 4x + 2y + 3z = -9$
$3v - w - x + 7y + 4z = 5$
$v + 6w + 6x - 4y - z = -4$

28. $4x - y + 2z - 2t + u = -15$
$8x + y - z + 4t - 2u = 26$
$2x - 6y - 2z + t - u = 10$
$2x + 5y + z - 3t + 8u = -22$
$4x - 3y + 2z + 4t + 2u = -4$

In Exercises 29–40, solve the indicated systems of equations using the inverse of the coefficient matrix. In Exercises 35–40, it is necessary to set up the appropriate equations.

29. For the following system of equations, solve for x^2 and y using the matrix methods of this section, and then solve for x and y.
$x^2 + y = 2$
$2x^2 - y = 10$

30. For the following system of equations, solve for x^2 and y^2 using the matrix methods of this scction, and then solve for x and y.
$x^2 - y^2 = 8$
$x^2 + y^2 = 10$

31. Solve any pair of the system of equations $2x - y = 4$, $3x + y = 1$, and $x - 2y = 5$. Show that the solution is valid for any pair chosen. What conclusions can be drawn about the graphs of the three equations?

32. In solving the system of equations $3x - 4y = 5$, $8y - 6x = 7$, when conclusion can be drawn?

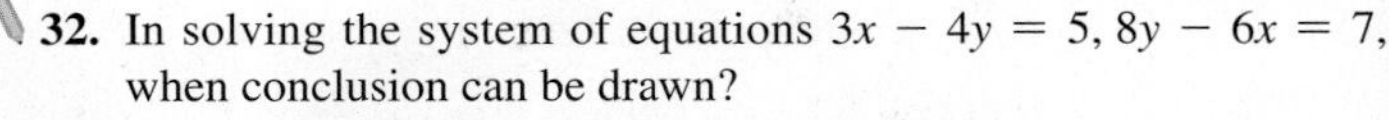

33. Forces $\mathbf{F_1}$ and $\mathbf{F_2}$ hold up a beam that weights 2540 N, as shown in Fig. 16.14. The equations used to find the forces are
$A \sin 47.2° + B \sin 64.4° = 2540$
$A \cos 47.2° - B \cos 64.4° = 0$
Find the magnitude of each force.

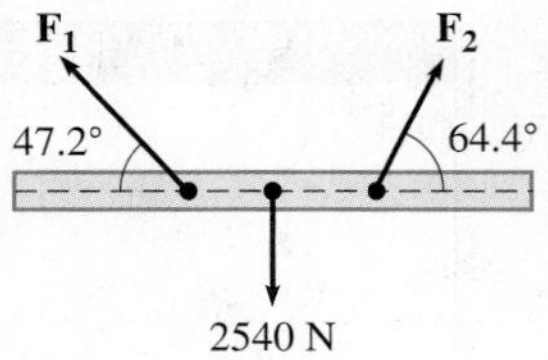

$\mathbf{F_1}$ $\mathbf{F_2}$ 47.2° 64.4° 2540 N

Fig. 16.14

34. In applying Kirchhoff's laws to the circuit shown in Fig. 16.15, following equations are found. Determine the indicated currents (in A).
$I_A + I_B + I_C = 0$
$2I_A - 5I_B = 6$
$5I_B - I_C = -3$

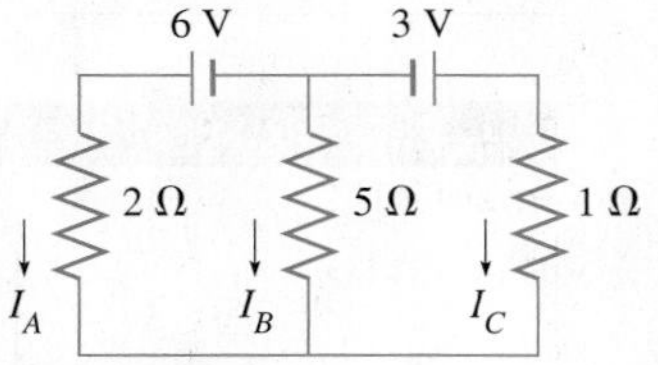

6 V 3 V 2 Ω 5 Ω 1 Ω I_A I_B I_C

Fig. 16.15

35. Two batteries in an electric circuit have a combined voltage of 18 V, and one battery produces 6 V less than twice the other. What is the voltage of each?

36. For college expenses, a student took out a loan at 4.00%, and a semester later took out a second loan at 3.00%. The total annual interest for the two loans was \$245. If the second loan had been for twice as much, the annual interest would have been \$290. How much was each loan?

37. What volume of each of a 20% acid solution and a 50% acid solution should be combined to form 48 mL of a 25% solution?

38. Two computer programs together require 236 MB (megabytes) of memory. If one program requires 20.0 MB of memory more than twice the memory of the other, what are the memory requirements of each program?

39. A research chemist wants to make 10.0 L of gasoline containing 2.0% of a new experimental additive. Gasoline without additive and two mixtures of gasoline with additive, one with 5.0% and the other with 6.0%, are to be used. If four times as much gasoline without additive as the 5.0% mixture is to be used, how much of each is needed?

40. A river tour boat takes 5.0 h to cruise downstream and 7.0 h for the return upstream. If the river flows at 4.0 mi/h, how fast does the boat travel in still water, and how far downstream does the boat go before starting the return trip?

Answer to Practice Exercise

1. $x = 3, y = -1$

16.5 Gaussian Elimination

Row Echelon Form • Gaussian Elimination • Number of Possible Solutions • Consistent and Inconsistent Systems

We now show a general method that can be used to solve a system of linear equations. The procedure used in this method is similar to that used in finding the inverse of a matrix in Section 16.3. It is known as **Gaussian elimination,** and as noted in the chapter introduction, it was developed in the early 1800s by Karl Gauss. Today, it is commonly used in computer programs for the solutions of systems of linear equations.

When using this method, we use certain row operations to convert the system of equations to *row echelon form*. Then the solution can be found by substituting back into the equations from the bottom up. The following examples illustrate this method.

EXAMPLE 1 Row echelon form

$$\begin{aligned} x - 3y - z &= 1 \\ y + 2z &= 5 \\ z &= 2 \end{aligned}$$

The system of equations to the left is said to be in **row echelon form.** We see that the third equation directly gives us the value $z = 2$. Since the second equation contains only y and z, we can now substitute $z = 2$ into the second equation to get $y = 1$. Then we can find x by substituting $y = 1$ and $z = 2$ into the first equation, and get $x = 6$. Therefore, after writing the system of equations in the form to the left, the solution is completed by substituting back, starting with the last equation. ■

The method of Gaussian elimination is based on first rewriting the given system of equations in row echelon form like in Example 1, so that the last equation shows the value of one of the unknowns, and the others are found by substituting back. The method to obtain the proper form is based on the use of any of the following three operations on the equations of the original system of equations.

Operations Used to Obtain Row Echelon Form

1. Any two equations may be interchanged.
2. Both sides of an equation may be multiplied by a nonzero constant.
3. A nonzero multiple of one equation may be added to another equation.

Using three linear equations in three unknowns as an example, by using the above operations, we can change the system

$$\begin{aligned} a_1x + b_1y + c_1z &= d_1 \\ a_2x + b_2y + c_2z &= d_2 \\ a_3x + b_3y + c_3z &= d_3 \end{aligned} \tag{16.10}$$

into the equivalent system in row echelon form

$$\begin{aligned} x + b_4y + c_4z &= d_4 \\ y + c_5z &= d_5 \\ z &= d_6 \end{aligned} \tag{16.11}$$

The solution is completed by substituting the value of z into the second equation, and then substituting the values of y and z into the first equation, as in Example 1.

EXAMPLE 2 Gaussian elimination with two equations

Solve the following system of equations using Gaussian elimination:

$$\begin{aligned} 2x \;+ y &= 4 \\ 3x - 2y &= 3 \end{aligned}$$

The steps are given below, with the necessary row operations shown in red:

this is in the form of Eq. (16.10)		make coefficient of x in first equation 1		eliminate x from second equation		make coefficient of y in second equation 1
$\begin{aligned} 2x + y &= 4 \\ 3x - 2y &= 3 \end{aligned}$	$\xrightarrow{\frac{1}{2}R_1}$	$\begin{aligned} x + \tfrac{1}{2}y &= 2 \\ 3x - 2y &= 3 \end{aligned}$	$\xrightarrow{-3R_1 + R_2}$	$\begin{aligned} x + \tfrac{1}{2}y &= 2 \\ -\tfrac{7}{2}y &= -3 \end{aligned}$	$\xrightarrow{-\frac{2}{7}R_2}$	$\begin{aligned} x + \tfrac{1}{2}y &= 2 \\ y &= \tfrac{6}{7} \end{aligned}$

To find the value of x, we substitute $y = \frac{6}{7}$ into the first equation: $x + \frac{1}{2}\left(\frac{6}{7}\right) = 2$, or $x = \frac{11}{7}$.

Thus, the solution is $x = \frac{11}{7}$, $y = \frac{6}{7}$, which checks when substituted into the original equations. ■

EXAMPLE 3 Gaussian elimination with three equations

Solve the following system of equations using Gaussian elimination:

$$\begin{aligned} x + 3y - 2z &= -5 \\ 2x - \;\; y + 4z &= \;\;\, 7 \\ -3x + 2y - 3z &= -1 \end{aligned}$$

In this example, we will perform the row operations on the ***augmented matrix,*** which includes the coefficients and constant terms, but not the variables. The equal signs are replaced with dotted lines. The steps are shown below.

top left entry is already 1 — make other entries in first column 0 — make second entry in middle row 1

$$\left[\begin{array}{ccc:c} 1 & 3 & -2 & -5 \\ 2 & -1 & 4 & 7 \\ -3 & 2 & -3 & -1 \end{array}\right] \xrightarrow[3R_1 + R_3]{-2R_1 + R_2} \left[\begin{array}{ccc:c} 1 & 3 & -2 & -5 \\ 0 & -7 & 8 & 17 \\ 0 & 11 & -9 & -16 \end{array}\right] \xrightarrow{-\frac{1}{7}R_2} \left[\begin{array}{ccc:c} 1 & 3 & -2 & -5 \\ 0 & 1 & -\frac{8}{7} & -\frac{17}{7} \\ 0 & 11 & -9 & -16 \end{array}\right]$$

make second entry in bottom row 0 — make third entry in bottom row 1

$$\xrightarrow{-11R_2 + R_3} \left[\begin{array}{ccc:c} 1 & 3 & -2 & -5 \\ 0 & 1 & -\frac{8}{7} & -\frac{17}{7} \\ 0 & 0 & \frac{25}{7} & \frac{75}{7} \end{array}\right] \xrightarrow{\frac{7}{25}R_3} \left[\begin{array}{ccc:c} 1 & 3 & -2 & -5 \\ 0 & 1 & -\frac{8}{7} & -\frac{17}{7} \\ 0 & 0 & 1 & 3 \end{array}\right] \longrightarrow z = 3$$

The last line above shows that $z = 3$. This can be substituted into the second equation to find y, and then the values of x and y can be substituted into the first equation to find x.

$$\begin{aligned} y - \tfrac{8}{7}(3) &= -\tfrac{17}{7} \\ y &= 1 \end{aligned} \qquad\qquad \begin{aligned} x + 3(1) - 2(3) &= -5 \\ x &= -2 \end{aligned}$$

Thus, the solution is $x = -2$, $y = 1$, $z = 3$. This solution checks in the original equations.

Note that in this example, it is possible to continue performing row operations until the left side of the augmented matrix is the identity matrix. This is called **reduced row echelon form** (rref). In this form, the system is completely solved and there is no need to back-substitute from the bottom up. Many calculators have a *rref* feature that can be used to solve systems of equations (see Example 3 in Section 5.5). Figure 16.16 shows the calculator solution for this example. ■

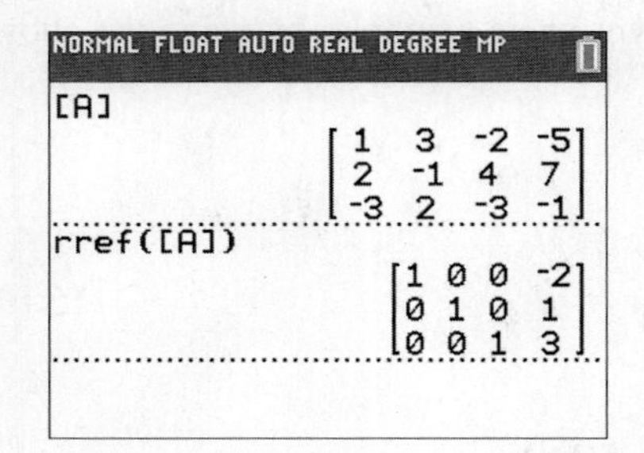

Fig. 16.16
Graphing calculator keystrokes: goo.gl/d40wzr

EXAMPLE 4 Gaussian elimination leading to infinite solutions

Solve the following system of equations using Gaussian elimination:

$$\begin{aligned} 4y + z &= 2 \\ 2x + 6y - 2z &= 3 \\ 4x + 8y - 5z &= 4 \end{aligned}$$

We will again work with the augmented matrix. Since the first equation doesn't contain x, we start by interchanging the first two equations. The steps are shown below.

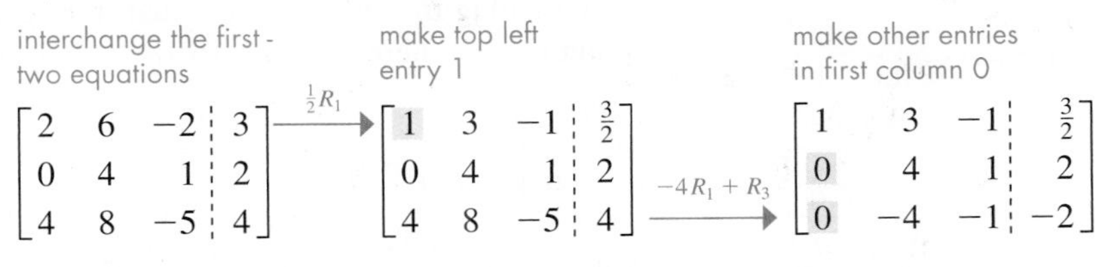

$$\left[\begin{array}{ccc|c} 2 & 6 & -2 & 3 \\ 0 & 4 & 1 & 2 \\ 4 & 8 & -5 & 4 \end{array}\right] \xrightarrow{\frac{1}{2}R_1} \left[\begin{array}{ccc|c} 1 & 3 & -1 & \frac{3}{2} \\ 0 & 4 & 1 & 2 \\ 4 & 8 & -5 & 4 \end{array}\right] \xrightarrow{-4R_1 + R_3} \left[\begin{array}{ccc|c} 1 & 3 & -1 & \frac{3}{2} \\ 0 & 4 & 1 & 2 \\ 0 & -4 & -1 & -2 \end{array}\right]$$

make second entry in middle row 1 — make second entry in bottom row 0

$$\xrightarrow{\frac{1}{4}R_2} \left[\begin{array}{ccc|c} 1 & 3 & -1 & \frac{3}{2} \\ 0 & 1 & \frac{1}{4} & \frac{1}{2} \\ 0 & -4 & -1 & -2 \end{array}\right] \xrightarrow{4R_2 + R_3} \left[\begin{array}{ccc|c} 1 & 3 & -1 & \frac{3}{2} \\ 0 & 1 & \frac{1}{4} & \frac{1}{2} \\ 0 & 0 & 0 & 0 \end{array}\right] \longrightarrow$$

$0 = 0$ is true, so there are an infinite number of solutions

The last row of zeros indicates that z can be any number. But it is still possible to express both x and y in terms of z. We first solve the equation in the second row for y in terms of z, which results in $y = \frac{1}{2} - \frac{1}{4}z$. Then this expression can be substituted into the equation in the top row to solve for x in terms of z.

$$\begin{aligned} x + 3\left(\tfrac{1}{2} - \tfrac{1}{4}z\right) - z &= \tfrac{3}{2} \\ x &= \tfrac{7}{4}z \end{aligned}$$

Therefore, the solution is given by $x = \frac{7}{4}z$ and $y = \frac{1}{2} - \frac{1}{4}z$, where z can be *any number*. NOTE ▶ [This means there are an infinite number of solutions.] For example, if $z = 4$, then $x = 7$ and $y = -\frac{1}{2}$.

NOTE ▶ [Note that the original matrix of coefficients on the left of the dotted line has a determinant of zero, and therefore its inverse does not exist. Thus, the method described in Section 16.4 cannot be used to solve this system. This underscores the importance of Gaussian elimination.] ■

POSSIBLE NUMBER OF SOLUTIONS

When we solved systems of linear equations in Chapter 5, we found that not all systems have unique solutions, as they do in Examples 2 and 3. In Example 4, we illustrated the use of Gaussian elimination on a system of equations for which the solution is not unique.

In Example 4, one of the equations became $0 = 0$, and there was an unlimited number of solutions. If any of the equations of a system becomes $0 = a$, $a \neq 0$, then the system is inconsistent and there is no solution. Example 5 illustrates such a system.

If a system has more unknowns than equations, or it can be written this way, as in Example 4, it usually has an unlimited number of solutions. It is possible, however, that such a system is inconsistent.

If a system has more equations than unknowns, it is inconsistent unless enough equations become $0 = 0$ such that at least one solution is found. The following example illustrates two systems in which there are more equations than unknowns.

Practice Exercise

1. Use Gaussian elimination to solve the system of equations

$$\begin{aligned} 2x - y &= 7 \\ 4x + 3y &= -1 \end{aligned}$$

EXAMPLE 5 Consistent and inconsistent systems

Solve the following systems of equations by Gaussian elimination.

First system	Second system
$x + 2y = 5$	$x + 2y = 5$
$3x - y = 1$	$3x - y = 1$
$4x + y = 6$	$4x + y = 2$
$x + 2y = 5$	$x + 2y = 5$
$-7y = -14$	$-7y = -14$
$-7y = -14$	$-7y = -18$
$x + 2y = 5$	$x + 2y = 5$
$y = 2$	$y = 2$
$-7y = -14$	$-7y = -18$
$x + 2y = 5$	$x + 2y = 5$
$y = 2$	$y = 2$
$0 = 0$	$0 = -4$
$x = 1$	

The solutions are shown at the left. We note that each system has three equations and two unknowns.

In the solution of the first system, the third equation becomes $0 = 0$, and only two equations are needed to find the solution $x = 1, y = 2$.

In the solution of the second system, the third equation becomes $0 = -4$, which means the system is inconsistent and there is no solution.

Graphing the systems, note that in Fig. 16.17, each of the lines passes through the point (1, 2), whereas in Fig. 16.18, there is no point common to the three lines. ■

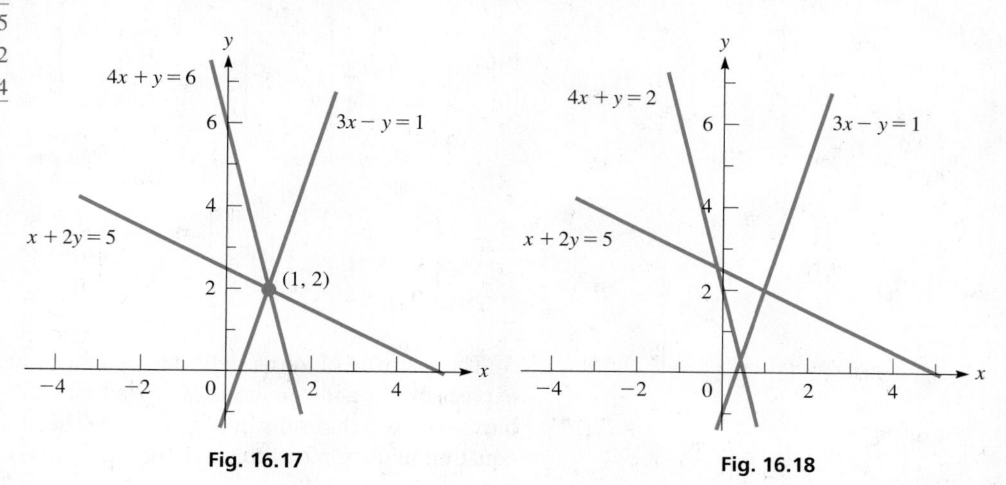

Fig. 16.17

Fig. 16.18

EXERCISES 16.5

In Exercises 1 and 2, make the given changes in the indicated examples of this section, and then solve the resulting systems using Gaussian elimination.

1. In Example 2, interchange the two equations and then solve the system with the equations in this order.

2. In Example 5, change the third equation of the second system to $5x + 3y = 11$.

In Exercises 3–24, solve the given systems of equations by Gaussian elimination. If there is an unlimited number of solutions, find two of them.

3. $x + 2y = 4$
$3x - y = 5$

4. $2x + y = 1$
$5x + 2y = 1$

5. $4x - 3y = 2$
$-2x + 4y = 3$

6. $-3x + 2y = 4$
$4x + y = -5$

7. $2x - 3y + z = 4$
$6y - 4x - 2z = 9$

8. $3s + 4t - u = -5$
$2u - 6s - 8t = 10$

9. $x + 3y + 3z = -3$
$2x + 2y + z = -5$
$-2x - y + 4z = 6$

10. $9x - 3y + 6z = 9$
$4x - 2y + z = 3$
$6x + 6y + 3z = 4$

11. $w + 2x - y + 3z = 12$
$2w - 2y - z = 3$
$3x - y - z = -1$
$-w + 2x + y + 2z = 3$

12. $3x - 2y = -11$
$5x + y = -1$
$2x + 3y = 10$
$x - 5y = -21$

13. $x - 4y + z = 5$
$2x - y + 3z = 4$

14. $4x + z = 6$
$2x - y - 2z = -2$

15. $2x - y + z = 5$
$3x + 2y - 2z = 4$
$5x + 8y - 8z = 5$

16. $3u + 6v + 2w = -2$
$u + 3v - 4w = 2$
$2u - 3v - 2w = -2$

17. $x + 3y + z = 4$
$2x - 6y - 3z = 10$
$4x - 9y + 3z = 4$

18. $30x + 20y - 10z = 30$
$4x - 2y - 6z = 4$
$-5x + 20y - 25z = -5$

19. $2x - 4y = 7$
$3x + 5y = -6$
$9x - 7y = 15$

20. $4x - y = 5$
$8x + 8y = 12$
$6x - 4y = 7$
$2x + y = 4$

21. $6x + 10y = -4$
$24x - 18y = 13$
$15x - 33y = 19$
$6x + 68y = -33$

22. $x + 3y - z = 1$
$3x - y + 4z = 4$
$-2x + 2y + 3z = 17$
$3x + 7y + 5z = 23$

23. $4x - 8y - 8z = 12$
$10x + 5y + 15z = 20$
$-6x - 3y - 3z = 15$
$3x + 3y - 2z = 2$

24. $2x - y - 2z - t = 4$
$4x + 2y + 3z + 2t = 3$
$-2x - y + 4z = -2$

In Exercises 25–28, solve the given system of equations by using the rref feature on a calculator.

25. $7x + 5y - 3z = 16$
$3x - 5y + 2z = -8$
$5x + 3y - 7z = 0$

26. $x + y + z = 6$
$2y + 5z = -4$
$2x + 5y - z = 27$

27. $-2r + 3t + 5u = 1$
$-4r + s - 2t + u = 0$
$r - 4s + 2t + u = 4$
$-5r - 3s + 8u = 0$

28. $2x + 5y - 9z + 3w = 151$
$5x + 6y - 4z + 2w = 103$
$3x - 4y + 2z + 7w = 16$
$11x + 7y + 4z - 8w = -32$

In Exercises 29 and 30, solve the given problems using Gaussian elimination.

29. Solve the system $a_1x + b_1y = c_1$, $a_2x + b_2y = c_2$ and show that the result is the same as that obtained using Cramer's rule as in Section 5.4. See page 160.

30. Solve the system $x + 2y = 6$, $2x + ay = 4$ and show that the solution depends on the value of a. What value of a does the solution show may not be used?

In Exercises 31–34, set up systems of equations and solve by Gaussian elimination.

31. Two jets are 2370 km apart and traveling toward each other, one at 720 km/h and the other at 860 km/h. How far does each travel before they pass?

32. The voltage across an electric resistor equals the current (in A) times the resistance (in Ω). If a current of 3.00 A passes through each of two resistors, the sum of the voltages is 10.5 V. If 2.00 A passes through the first resistor and 4.00 A passes through the second resistor, the sum of the voltages is 13.0 V. Find the resistances.

33. Three machines together produce 650 parts each hour. Twice the production of the second machine is 10 parts/h more than the sum of the production of the other two machines. If the first operates for 3.00 h and the others operate for 2.00 h, 1550 parts are produced. Find the production rate of each machine.

34. A business bought three types of computer programs to be used in offices at different locations. The first costs $35 each and uses 190 MB of memory; the second costs $50 each and uses 225 MB of memory; and the third costs $60 each and uses 130 MB of memory. If as many of the third type were purchased as the other two combined, with a total cost of $2600 and total memory requirement of 8525 MB, how many of each were purchased?

Answer to Practice Exercise

1. $x = 2, y = -3$

16.6 Higher-order Determinants

Minors • Expansion by Minors • Properties of Determinants • Solving Systems of Equations

In Chapter 5, we limited our discussion of determinants to those of the second and third orders. We now show some methods of evaluating higher-order determinants, and use these methods to solve systems of equations.

From Section 5.7, we recall the definition of a third-order determinant. By rearranging the terms and factoring out a_1, $-a_2$, and a_3, we have

$$\begin{vmatrix} a_1 & b_1 & c_1 \\ a_2 & b_2 & c_2 \\ a_3 & b_3 & c_3 \end{vmatrix} = a_1b_2c_3 + a_3b_1c_2 + a_2b_3c_1 - a_3b_2c_1 - a_1b_3c_2 - a_2b_1c_3$$

$$= a_1(b_2c_3 - b_3c_2) - a_2(b_1c_3 - b_3c_1) + a_3(b_1c_2 - b_2c_1)$$

$$= a_1\begin{vmatrix} b_2 & c_2 \\ b_3 & c_3 \end{vmatrix} - a_2\begin{vmatrix} b_1 & c_1 \\ b_3 & c_3 \end{vmatrix} + a_3\begin{vmatrix} b_1 & c_1 \\ b_2 & c_2 \end{vmatrix} \quad \textbf{(16.12)}$$

In Eq. (16.12), the third-order determinant is expanded as products of the elements of the first column and second-order determinants, known as *minors*. In general, *the* **minor** *of an element of a determinant is the determinant that results by deleting the row and column in which the element lies.*

EXAMPLE 1 Minors

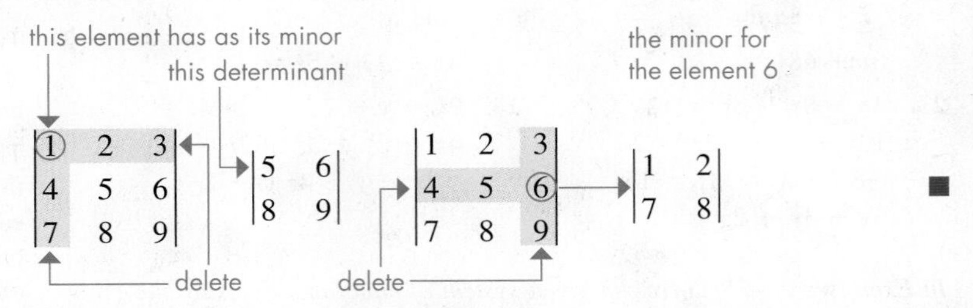

Equation (16.12) is only one way of expressing the expansion of the third-order determinant, but it does lead to a general theorem for the expansion of a determinant of any order. It is not a proof of the following theorem, but does provide a basis for it.

Sign chart for the expansion row or column

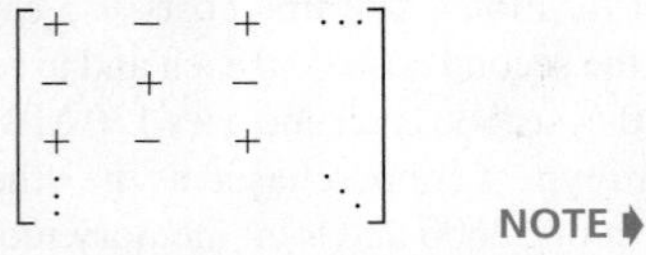

Fig. 16.19

EXPANSION OF A DETERMINANT BY MINORS

The value of a determinant of order n may be found by forming the n products of the elements of any column (or row) and their minors. A product is given a plus sign if the sum of the number of the column and the number of the row in which the element lies is even, and a minus sign if this sum is odd. (See the sign chart in Fig. 16.19.) The algebraic sum of these terms is the value of the determinant. NOTE ▶ [Since we can expand by *any* row or column, we usually choose one that has one or more zeros in order to make the calculations easier.]

EXAMPLE 2 Expansion by minors

In evaluating the following determinant, note that the third column has two zeros. This means that expanding by the third column will require less numerical work. Therefore,

$$\begin{vmatrix} 3 & -2 & 0 & 2 \\ 1 & 0 & -1 & 4 \\ -3 & 1 & 2 & -2 \\ 2 & -1 & 0 & -1 \end{vmatrix} = +(0)\begin{vmatrix} 1 & 0 & 4 \\ -3 & 1 & -2 \\ 2 & -1 & -1 \end{vmatrix} - (-1)\begin{vmatrix} 3 & -2 & 2 \\ -3 & 1 & -2 \\ 2 & -1 & -1 \end{vmatrix} + (2)\begin{vmatrix} 3 & -2 & 2 \\ 1 & 0 & 4 \\ 2 & -1 & -1 \end{vmatrix} - (0)\begin{vmatrix} 3 & -2 & 2 \\ 1 & 0 & 4 \\ -3 & 1 & -2 \end{vmatrix}$$

$$= \begin{vmatrix} 3 & -2 & 2 \\ -3 & 1 & -2 \\ 2 & -1 & -1 \end{vmatrix} + 2\begin{vmatrix} 3 & -2 & 2 \\ 1 & 0 & 4 \\ 2 & -1 & -1 \end{vmatrix}$$

$$= \left[3\begin{vmatrix} 1 & -2 \\ -1 & -1 \end{vmatrix} - (-3)\begin{vmatrix} -2 & 2 \\ -1 & -1 \end{vmatrix} + 2\begin{vmatrix} -2 & 2 \\ 1 & -2 \end{vmatrix}\right] + 2\left[-(-2)\begin{vmatrix} 1 & 4 \\ 2 & -1 \end{vmatrix} + 0\begin{vmatrix} 3 & 2 \\ 2 & -1 \end{vmatrix} - (-1)\begin{vmatrix} 3 & 2 \\ 1 & 4 \end{vmatrix}\right]$$

expanding first determinant by first column — expanding second determinant by second column

$$= [3(-1 - 2) + 3(2 + 2) + 2(4 - 2)] + 2[2(-1 - 8) + (12 - 2)]$$

$$= [-9 + 12 + 4] + 2[-18 + 10] = 7 + 2(-8) = -9$$

As we noted in Chapter 5, most calculators can be used to quickly evaluate determinants. Figure 16.20 shows the calculator evaluation of this determinant.

```
NORMAL FLOAT AUTO REAL DEGREE MP
[A]
      [ 3 -2  0  2 ]
      [ 1  0 -1  4 ]
      [-3  1  2 -2 ]
      [ 2 -1  0 -1 ]
det([A])
                 -9
```

Fig. 16.20

Graphing calculator keystrokes: goo.gl/xH0CwQ ■

Practice Exercise

1. Use expansion by minors to evaluate the determinant.

$$\begin{vmatrix} -4 & -1 & 0 & 2 \\ 2 & -3 & -2 & 1 \\ 1 & 0 & 0 & -2 \\ 3 & 0 & 5 & 4 \end{vmatrix}$$

PROPERTIES OF DETERMINANTS

We now state some properties which allow us to find certain determinants quickly and to see how certain changes to a matrix affect the value of the determinant.

1. *If each element above or each element below the principal diagonal of a determinant is zero, then the product of the elements of the principal diagonal is the value of the determinant.*
2. *If all corresponding rows and columns of a determinant are interchanged, the value of the determinant is unchanged.*
3. *If two columns (or rows) of a determinant are identical, the value of the determinant is zero.*
4. *If two columns (or rows) of a determinant are interchanged, the value of the determinant is changed in sign.*
5. *If all elements of a column (or row) are multiplied by the same number k, the value of the determinant is multiplied by k.*
6. *If all the elements of any column (or row) are multiplied by the same number k, and the resulting numbers are added to the corresponding elements of another column (or row), the value of the determinant is unchanged.*

EXAMPLE 3 Using properties of determinants

(a) $$\begin{vmatrix} 2 & 1 & 8 \\ 0 & -5 & 9 \\ 0 & 0 & -6 \end{vmatrix} = 2(-5)(-6) = 60$$ property 1

(b) identical $$\begin{bmatrix} 3 & 5 & 2 \\ -4 & 6 & 9 \\ -4 & 6 & 9 \end{bmatrix} = 0$$ property 3

(c) $$\begin{vmatrix} -1 & 0 & 6 \\ 6 & 3 & -6 \\ 0 & 5 & 3 \end{vmatrix} = 3\begin{vmatrix} -1 & 0 & 6 \\ 2 & 1 & -2 \\ 0 & 5 & 3 \end{vmatrix}$$ property 5
the value of the first is 141,
and the value of the second is 47
$141 = 3(47)$ ■

The properties given here can be used to simplify the evaluation of determinants using expansion by minors. For example, property 6 can be used to get all zeros under the principal diagonal. Then, by property 1, the determinant is the product of the elements of the principal diagonal. However, due to the ability of calculators and computers to evaluate determinants, this is not as necessary as it was in the past.

SOLVING SYSTEMS OF LINEAR EQUATIONS BY DETERMINANTS

We can use the expansion of determinants by minors to solve systems of linear equations. **Cramer's rule** *for solving systems of linear equations, as stated in Section 5.6, is valid for any system of n equations in n unknowns.*

EXAMPLE 4 Solving a system of four equations

Solve the following system of equations:

$$\begin{aligned} x + 2y + z \qquad\quad &= 5 \\ 2x \qquad + z + 2t &= 1 \\ x - y + 3z + 4t &= -6 \\ 4x - y \qquad - 2t &= 0 \end{aligned}$$

constants

expanding by fourth row

$$x = \frac{\begin{vmatrix} 5 & 2 & 1 & 0 \\ 1 & 0 & 1 & 2 \\ -6 & -1 & 3 & 4 \\ 0 & -1 & 0 & -2 \end{vmatrix}}{\begin{vmatrix} 1 & 2 & 1 & 0 \\ 2 & 0 & 1 & 2 \\ 1 & -1 & 3 & 4 \\ 4 & -1 & 0 & -2 \end{vmatrix}} = \frac{-(0)\begin{vmatrix} 2 & 1 & 0 \\ 0 & 1 & 2 \\ -1 & 3 & 4 \end{vmatrix} + (-1)\begin{vmatrix} 5 & 1 & 0 \\ 1 & 1 & 2 \\ -6 & 3 & 4 \end{vmatrix} - (0)\begin{vmatrix} 5 & 2 & 0 \\ 1 & 0 & 2 \\ -6 & -1 & 4 \end{vmatrix} + (-2)\begin{vmatrix} 5 & 2 & 1 \\ 1 & 0 & 1 \\ -6 & -1 & 3 \end{vmatrix}}{(1)\begin{vmatrix} 0 & 1 & 2 \\ -1 & 3 & 4 \\ -1 & 0 & -2 \end{vmatrix} - 2\begin{vmatrix} 2 & 1 & 2 \\ 1 & 3 & 4 \\ 4 & 0 & -2 \end{vmatrix} + (1)\begin{vmatrix} 2 & 0 & 2 \\ 1 & -1 & 4 \\ 4 & -1 & -2 \end{vmatrix} - (0)\begin{vmatrix} 2 & 0 & 1 \\ 1 & -1 & 3 \\ 4 & -1 & 0 \end{vmatrix}}$$

expanding by first row

$$= \frac{-(-26) - 2(-14)}{1(0) - 2(-18) + 1(18)} = \frac{26 + 28}{36 + 18} = \frac{54}{54} = 1$$

Note that we chose to expand the determinant in the numerator by the fourth row since it contains two zeros. In the denominator, we expanded by the first row since it contains a zero, and no other row or column contains more than one zero.

In solving for y, we again note the two zeros in the fourth row:

expanding by fourth row

$$y = \frac{\begin{vmatrix} 1 & 5 & 1 & 0 \\ 2 & 1 & 1 & 2 \\ 1 & -6 & 3 & 4 \\ 4 & 0 & 0 & -2 \end{vmatrix}}{54} = \frac{-4\begin{vmatrix} 5 & 1 & 0 \\ 1 & 1 & 2 \\ -6 & 3 & 4 \end{vmatrix} + (-2)\begin{vmatrix} 1 & 5 & 1 \\ 2 & 1 & 1 \\ 1 & -6 & 3 \end{vmatrix}}{54}$$

$$= \frac{-4(-26) - 2(-29)}{54} = \frac{104 + 58}{54} = \frac{162}{54} = 3$$

Substituting $x = 1$ and $y = 3$ in the first equation gives us $z = -2$. Then substituting $x = 1$ and $y = 3$ in the fourth equation gives us $t = 1/2$. Therefore, the required solution is $x = 1, y = 3, z = -2, t = 1/2$. We can check the solution by substituting in the second or third equation (we used the first and fourth to *find* values of z and t.) ■

Practice Exercise

2. Solve the system in Example 4 if the 5 in the first equation is changed to a 2.

EXERCISES 16.6

In Exercises 1 and 2, make the given changes in the indicated examples of this section. Then evaluate the resulting determinants.

1. In Example 2, change the -2 in the first row to 0 and then find the determinant.

2. In Example 3(a), change the 2 in the first row to a 3 and then find the determinant.

In Exercises 3–6, evaluate each determinant by inspection. Observation will allow evaluation by using the properties of this section.

3. $\begin{vmatrix} 10 & -5 & 8 \\ 0 & 3 & -8 \\ 0 & 0 & -3 \end{vmatrix}$

4. $\begin{vmatrix} -3 & 0 & 0 \\ 0 & 10 & 0 \\ -9 & -1 & -5 \end{vmatrix}$

5. $\begin{vmatrix} 3 & -2 & 4 & 2 \\ 5 & -1 & 2 & -1 \\ 3 & -2 & 4 & 2 \\ 0 & 3 & -6 & 0 \end{vmatrix}$

6. $\begin{vmatrix} -12 & -24 & -24 & 15 \\ 12 & 32 & 32 & -35 \\ -22 & 18 & 18 & 18 \\ 44 & 0 & 0 & -26 \end{vmatrix}$

In Exercises 7–10, use the given value of the determinant at the right and the properties of this section to evaluate the following determinants.

$$\begin{vmatrix} 2 & -3 & 1 \\ -4 & 1 & 3 \\ 1 & -3 & -2 \end{vmatrix} = 40$$

7. $\begin{vmatrix} 2 & 1 & -3 \\ -4 & 3 & 1 \\ 1 & -2 & -3 \end{vmatrix}$

8. $\begin{vmatrix} 2 & -3 & 1 \\ -4 & 1 & 3 \\ 2 & -6 & -4 \end{vmatrix}$

9. $\begin{vmatrix} 2 & -3 & 3 \\ -4 & 1 & 9 \\ 1 & -3 & -6 \end{vmatrix}$

10. $\begin{vmatrix} 2 & -4 & 1 \\ -3 & 1 & -3 \\ 1 & 3 & -2 \end{vmatrix}$

In Exercises 11–20, evaluate the given determinants by expansion by minors.

11. $\begin{vmatrix} 2 & 0 & 0 \\ -2 & 1 & 4 \\ 4 & -2 & 3 \end{vmatrix}$

12. $\begin{vmatrix} 10 & 0 & -3 \\ -2 & -4 & 1 \\ 3 & 0 & 2 \end{vmatrix}$

13. $\begin{vmatrix} 3 & 1 & 0 \\ -2 & 3 & -1 \\ 4 & 2 & 5 \end{vmatrix}$

14. $\begin{vmatrix} -40 & 30 & -20 \\ -8 & 8 & 16 \\ -15 & 75 & -45 \end{vmatrix}$

15. $\begin{vmatrix} 4 & 3 & 6 & 0 \\ 3 & 0 & 0 & 4 \\ 5 & 0 & 1 & 2 \\ 2 & 1 & 1 & 7 \end{vmatrix}$

16. $\begin{vmatrix} 6 & -3 & -6 & 3 \\ -2 & 1 & 2 & -1 \\ 18 & 7 & -1 & 5 \\ 0 & -1 & 10 & 10 \end{vmatrix}$

17. $\begin{vmatrix} 5 & 3 & 0 & 5 \\ 4 & 2 & 1 & 2 \\ 3 & 2 & -2 & 2 \\ 0 & 1 & 2 & -1 \end{vmatrix}$

18. $\begin{vmatrix} -2 & 2 & 1 & 3 \\ 1 & 4 & 3 & 1 \\ 4 & 3 & -2 & -2 \\ 3 & -2 & 1 & 5 \end{vmatrix}$

19. $\begin{vmatrix} 1 & 2 & 0 & 1 & 0 \\ 0 & 2 & 1 & 0 & 1 \\ 1 & 0 & -1 & 1 & -1 \\ -2 & 0 & -1 & 2 & 1 \\ 1 & 0 & 2 & -1 & -2 \end{vmatrix}$

20. $\begin{vmatrix} -1 & 3 & 5 & 0 & -5 \\ 0 & 1 & 7 & 3 & -2 \\ 5 & -2 & -1 & 0 & 3 \\ -3 & 0 & 2 & -1 & 3 \\ 6 & 2 & 1 & -4 & 2 \end{vmatrix}$

In Exercises 21–24, solve the given systems of equations by determinants. Evaluate by expansion by minors.

21. $2x + 2t = 0$
$3x + y + z = -1$
$2y - z + 3t = 1$
$2z - 3t = 1$

22. $2x + y + z = 4$
$2y - 2z - t = 3$
$3y - 3z + 2t = 1$
$6x - y + t = 0$

23. $x + 2y - z = 6$
$y - 2z - 3t = -5$
$3x - 2y + t = 2$
$2x + y + z - t = 0$

24. $2p + 3r + s = 4$
$p - 2r - 3s + 4t = -1$
$3p + r + s - 5t = 3$
$-p + 2r + s + 3t = 2$

In Exercises 25–28, solve the given systems of equations by determinants. Evaluate by using a calculator.

25. $2x + y + z = 2$
$3y - z + 2t = 4$
$y + 2z + t = 0$
$3x + 2z = 4$

26. $6x + 3y + 3z = 0$
$x - y + 2t = 2$
$2y + z + 4t = 2$
$5x + 2z + 2t = 4$

27. $D + E + 2F = 1$
$2D - E + G = -2$
$D - E - F - 2G = 4$
$2D - E + 2F - G = 0$

28. $3x + y + t = 0$
$3z + 2t = 8$
$6x + 2y + 2z + t = 3$
$3x - y - z - t = 0$

In Exercises 29–32, make the indicated changes in the determinant at the right, and then solve the indicated problem. Assume the elements are nonzero, unless otherwise specified.

$$\begin{vmatrix} a & b & c \\ d & e & f \\ g & h & i \end{vmatrix}$$

29. Evaluate the determinant if $a = c$, $d = f$, and $g = i$.

30. Evaluate the determinant if $b = c = f = 0$.

31. By what factor is the value of the determinant changed if all elements are doubled?

32. How is the value changed if a is added to g, b added to h, and c added to i?

In Exercises 33–38, solve the given problems by using determinants.

33. In applying Kirchhoff's laws (see Example 2 on page 141) to the circuit shown in Fig. 16.21, the following equations are found. Determine the indicated currents (in A).

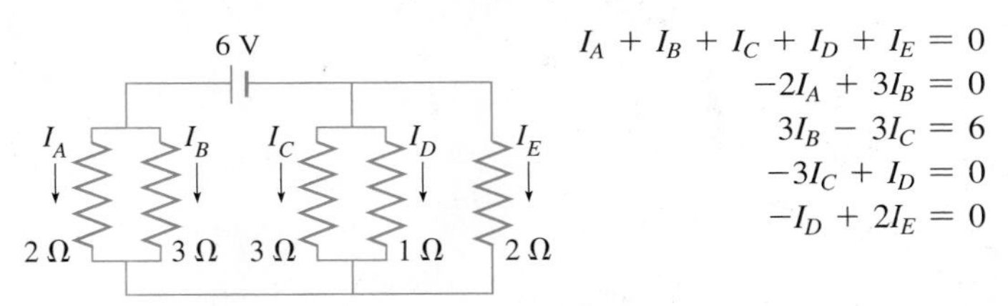

$$\begin{aligned} I_A + I_B + I_C + I_D + I_E &= 0 \\ -2I_A + 3I_B &= 0 \\ 3I_B - 3I_C &= 6 \\ -3I_C + I_D &= 0 \\ -I_D + 2I_E &= 0 \end{aligned}$$

Fig. 16.21

34. In analyzing the forces A, B, C, and D shown on the beam in Fig. 16.22, the following equations are used. Find these forces.

$$\begin{aligned} A + B &= 850 \\ A + B + 400 &= 0.8C + 0.6D \\ 0.6C &= 0.8D \\ 5A - 5B + 4C - 3D &= 0 \end{aligned}$$

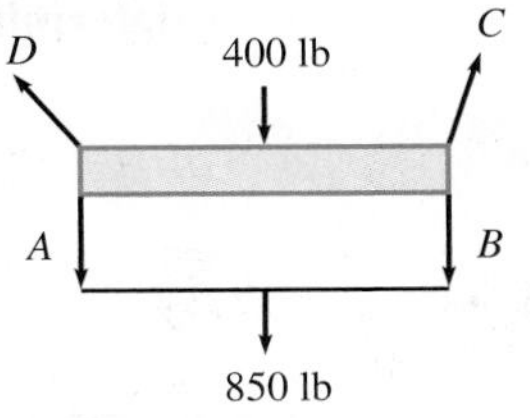

D 400 lb C
A B
850 lb

Fig. 16.22

35. The area of a triangle with vertices (x_1, y_1), (x_2, y_2), and (x_3, y_3) is given by $A = \pm\frac{1}{2}\begin{vmatrix} x_1 & y_1 & 1 \\ x_2 & y_2 & 1 \\ x_3 & y_3 & 1 \end{vmatrix}$. (If the answer comes out negative, take the absolute value).
A natural gas company locates the three corners of a triangular-shaped shale natural gas reserve at (0,0), (8.45, 3.64), and (1.82, 5.70), where all measurements are in kilometers. Find the area of the reserve.

36. An alloy is to be made from four other alloys containing copper (Cu), nickel (Ni), zinc (Zn), and iron (Fe). The first is 80% Cu and 20% Ni. The second is 60% Cu, 20% Ni, and 20% Zn. The third is 30% Cu, 60% Ni, and 10% Fe. The fourth is 20% Ni, 40% Zn, and 40% Fe. How much of each is needed so that the final alloy has 56 g Cu, 28 g Ni, 10 g Zn, and 6 g Fe?

37. In testing for air pollution, a given air sample contained a total of 6.0 parts per million (ppm) of four pollutants, sulfur dioxide (SO_2), nitric oxide (NO), nitrogen dioxide (NO_2), and carbon monoxide (CO). The ppm of CO was 10 times that of SO_2, which in turn equaled those of NO and NO_2. There was a total of 0.8 ppm of SO_2 and NO. How many ppm of each were present in the air sample?

38. A tablet with a 32-GB hard drive starts out with three different apps A, B, and C, which use up 4% of the tablet's memory. Two more apps are added, each using the same amount of memory as app A, and the apps then use a total of 6% of the tablet's memory. Then, in addition to those, three more apps requiring the same memory as app B are added. All eight apps combined use up 13.5% of the tablet's memory. Find the number of megabytes (MB) of memory required for each app A, B, and C. (1 GB = 1000 MB.)

Answers to Practice Exercises

1. 135 **2.** $x = 1, y = 2, z = -3, t = 1$

CHAPTER 16 KEY FORMULAS AND EQUATIONS

Basic laws for matrices

$$A + B = B + A \quad \text{(commutative law)} \tag{16.1}$$

$$A + (B + C) = (A + B) + C \quad \text{(associative law)} \tag{16.2}$$

$$k(A + B) = kA + kB \tag{16.3}$$

$$A + O = A \tag{16.4}$$

Inverse matrix

$$AA^{-1} = A^{-1}A = I \tag{16.5}$$

Solving systems of equations by matrices

$$\begin{bmatrix} a_1x + b_1y \\ a_2x + b_2y \end{bmatrix} = \begin{bmatrix} c_1 \\ c_2 \end{bmatrix} \tag{16.6}$$

$$A = \begin{bmatrix} a_1 & b_1 \\ a_2 & b_2 \end{bmatrix} \quad X = \begin{bmatrix} x \\ y \end{bmatrix} \quad C = \begin{bmatrix} c_1 \\ c_2 \end{bmatrix} \tag{16.7}$$

$$AX = C \tag{16.8}$$

$$X = A^{-1}C \tag{16.9}$$

Gaussian elimination

$$\begin{aligned} a_1x + b_1y + c_1z &= d_1 \\ a_2x + b_2y + c_2z &= d_2 \\ a_3x + b_3y + c_3z &= d_3 \end{aligned} \qquad \begin{aligned} x + b_4y + c_4z &= d_4 \\ y + c_5z &= d_5 \\ z &= d_6 \end{aligned} \qquad \text{(16.10) (16.11)}$$

Higher-order determinants

$$\begin{vmatrix} a_1 & b_1 & c_1 \\ a_2 & b_2 & c_2 \\ a_3 & b_3 & c_3 \end{vmatrix} = a_1\begin{vmatrix} b_2 & c_2 \\ b_3 & c_3 \end{vmatrix} - a_2\begin{vmatrix} b_1 & c_1 \\ b_3 & c_3 \end{vmatrix} + a_3\begin{vmatrix} b_1 & c_1 \\ b_2 & c_2 \end{vmatrix} \tag{16.12}$$

CHAPTER 16 REVIEW EXERCISES

CONCEPT CHECK EXERCISES

Determine each of the following as being either **true** *or* **false.** *If it is false, explain why.*

1. $2\begin{bmatrix} 3 & -1 \\ 0 & 2 \end{bmatrix} = \begin{bmatrix} 6 & -2 \\ 0 & 2 \end{bmatrix}$

2. $\begin{bmatrix} 3 & -1 \\ 0 & 2 \end{bmatrix}\begin{bmatrix} 1 & 3 \\ -2 & -1 \end{bmatrix} = \begin{bmatrix} 5 & 10 \\ -4 & -2 \end{bmatrix}$

3. If $A = \begin{bmatrix} -1 & 1 \\ 4 & -3 \end{bmatrix}$, then $A^{-1} = \begin{bmatrix} -3 & -1 \\ -4 & -1 \end{bmatrix}$

4. $\begin{vmatrix} 3 & 1 & -1 \\ 2 & -2 & 3 \\ 0 & 1 & 2 \end{vmatrix} = 3\begin{vmatrix} -2 & 3 \\ 1 & 2 \end{vmatrix} - 2\begin{vmatrix} 1 & -1 \\ 1 & 2 \end{vmatrix}$

5. $\begin{bmatrix} 2 & -1 \\ 5 & -3 \end{bmatrix}\begin{bmatrix} 3 \\ 7 \end{bmatrix} = \begin{bmatrix} -1 \\ -6 \end{bmatrix}$ is the matrix solution of $\begin{aligned} 2x - y &= 3 \\ 5x - 3y &= 7 \end{aligned}$.

6. After using two steps of the Gaussian elimination method for solving the system $\begin{aligned} 2x + 3y &= 4 \\ 3x + 2y &= 1 \end{aligned}$, we would have $\begin{aligned} x + \frac{3}{2}y &= 1 \\ -\frac{5}{2}y &= -5 \end{aligned}$.

PRACTICE AND APPLICATIONS

In Exercises 7–12, determine the values of the literal numbers.

7. $\begin{bmatrix} 4a \\ a - b \end{bmatrix} = \begin{bmatrix} 8 \\ 5 \end{bmatrix}$

8. $\begin{bmatrix} x - y \\ 2x + 2z \\ 4y + z \end{bmatrix} = \begin{bmatrix} 1 \\ 3 \\ -1 \end{bmatrix}$

9. $\begin{bmatrix} 2x & 3y & 2z \\ x + a & y + b & c - z \end{bmatrix} = \begin{bmatrix} 4 & -9 & 8 \\ 7 & -4 & 2 \end{bmatrix}$

10. $\begin{bmatrix} a + bj & b \\ aj & b - aj \end{bmatrix} = \begin{bmatrix} 6j & 2d \\ 2cj & ej^2 \end{bmatrix} \quad (j = \sqrt{-1})$

11. $\begin{bmatrix} \cos \pi & \sin \frac{\pi}{6} \\ x + y & x - y \end{bmatrix} = \begin{bmatrix} x & y \\ a & b \end{bmatrix}$

12. $\begin{bmatrix} \ln e & \log 100 \\ a^2 & b^2 \end{bmatrix} = \begin{bmatrix} a + b & a - b \\ x & y \end{bmatrix}$

In Exercises 13–18, use the given matrices and perform the indicated operations.

$$A = \begin{bmatrix} 2 & -3 \\ 4 & 1 \\ -5 & 0 \\ 2 & -3 \end{bmatrix} \quad B = \begin{bmatrix} -1 & 0 \\ 4 & -6 \\ -3 & -2 \\ 1 & -7 \end{bmatrix} \quad C = \begin{bmatrix} 5 & -6 \\ 2 & 8 \\ 0 & -2 \end{bmatrix}$$

13. $A + B$ **14.** $2C$

15. $B - A$ **16.** $2A - B$

17. $2A - 3B$ **18.** $2(A - B)$

In Exercises 19–22, perform the indicated matrix multiplications.

19. $\begin{bmatrix} 2 & -1 \\ -2 & 1 \end{bmatrix}\begin{bmatrix} 1 & -1 \\ 2 & -2 \end{bmatrix}$

20. $\begin{bmatrix} 6 & -4 & 1 & 0 \\ 2 & 0 & -4 & 3 \end{bmatrix}\begin{bmatrix} 7 & -1 & 6 \\ 4 & 0 & 1 \\ 3 & -2 & 5 \\ 9 & 1 & 0 \end{bmatrix}$

21. $\begin{bmatrix} 0 & 0.6 \\ 0.2 & 0.0 \\ 0.4 & -0.1 \end{bmatrix}\begin{bmatrix} 0.1 & -0.4 & 0.5 \\ 0.5 & 0.1 & 0.0 \end{bmatrix}$

22. $\begin{bmatrix} 0 & -1 & 6 \\ 8 & 1 & 4 \\ 7 & -2 & -1 \end{bmatrix}\begin{bmatrix} 5 & -1 & 7 & 1 & 5 \\ 0 & 1 & 0 & 4 & 1 \\ 1 & -2 & 3 & 0 & 1 \end{bmatrix}$

In Exercises 23–30, find the inverses of the given matrices. Check each by using a calculator.

23. $\begin{bmatrix} 2 & -5 \\ 2 & -4 \end{bmatrix}$ **24.** $\begin{bmatrix} -5 & -30 \\ 10 & 50 \end{bmatrix}$

25. $\begin{bmatrix} -0.8 & -0.1 \\ 0.4 & -0.7 \end{bmatrix}$ **26.** $\begin{bmatrix} 50 & -12 \\ 42 & -80 \end{bmatrix}$

27. $\begin{bmatrix} 1 & 1 & -2 \\ -1 & -2 & 1 \\ 0 & 3 & 4 \end{bmatrix}$ **28.** $\begin{bmatrix} -1 & -1 & 2 \\ 2 & 3 & 0 \\ 1 & 4 & 1 \end{bmatrix}$

29. $\begin{bmatrix} 2 & -4 & 3 \\ 4 & -6 & 5 \\ -2 & 1 & -1 \end{bmatrix}$ **30.** $\begin{bmatrix} 3 & 1 & -4 \\ -3 & 1 & -2 \\ -6 & 0 & 3 \end{bmatrix}$

In Exercises 31–38, solve the given systems of equations using the inverse of the coefficient matrix.

31. $2x - 3y = -9$
$4x - y = -13$

32. $5A - 7B = 62$
$6A + 5B = -6$

33. $33x + 52y = -450$
$45x - 62y = 1380$

34. $0.24x - 0.26y = -3.1$
$0.40x + 0.34y = -1.3$

35. $2u - 3v + 2w = 7$
$3u + v - 3w = -6$
$u + 4v + w = -13$

36. $6x + 6y - 3z = 24$
$x + 4y + 2z = 5$
$3x - 2y + z = 17$

37. $x + 2y + 3z = 1$
$3x - 4y - 3z = 2$
$7x - 6y + 6z = 2$

38. $3x + 2y + z = 2$
$2x + 3y - 6z = 3$
$x + 3y + 3z = 1$

In Exercises 39–46, solve the given systems of equations by Gaussian elimination. For Exercises 39–44, use those that are indicated from Exercises 31–38.

39. Exercise 31 **40.** Exercise 32 **41.** Exercise 35

42. Exercise 36 **43.** Exercise 37 **44.** Exercise 38

45. $2x + 3y - z = 10$
$x - 2y + 6z = -6$
$5x + 4y + 4z = 14$

46. $x - 3y + 4z - 2t = 6$
$2x + y - 2z + 3t = 7$
$3x - 9y + 12z - 6t = 12$

In Exercises 47–50, solve the systems of equations by determinants. As indicated, use the systems from Exercises 35–38.

47. Exercise 35 **48.** Exercise 36

49. Exercise 37 **50.** Exercise 38

In Exercises 51–58, solve the given systems of equations by using the coefficient matrix. Use a calculator to perform the necessary matrix operations and display the results and the check.

51. $3x - 2y + z = 6$
$2x + 3z = 3$
$4x - y + 5z = 6$

52. $7n + p + 2r = 3$
$4n - 2p + 4r = -2$
$2n + 3p - 6r = 3$

53. $2x - 3y + z - t = -8$
$4x + 3z + 2t = -3$
$2y - 3z - t = 12$
$x - y - z + t = 3$

54. $6x + 4y - 4z - 4t = 0$
$5y + 3z + 4t = 3$
$6y - 3z + 4t = 9$
$6x - y + 2z - 2t = -3$

55. $3x - y + 6z - 2t = 8$
$2x + 5y + z + 2t = 7$
$4x - 3y + 8z + 3t = -17$
$3x + 5y - 3z + t = 8$

56. $A + B + 2C - 3D = 15$
$3A + 3B - 8C - 2D = 9$
$6A - 4B + 6C + D = -6$
$2A + 2B - 4C - 2D = 8$

57. $4r - s + 8t - 2u + 4v = -1$
$3r + 2s - 4t + 3u - v = 4$
$3r + 3s + 2t + 5u + 6v = 13$
$6r - s + 2t - 2u + v = 0$
$r - 2s + 4t - 3u + 3v = 1$

58. $2v + 3w + 2x - 2y + 5z = -1$
$7v + 8w + 3x + y - 4z = 3$
$v - 2w - 4x - 4y - 8z = -9$
$3v - w + 7x + 5y - 3z = -18$
$4v + 5w + x + 3y - 6z = 7$

In Exercises 59–62, use matrices A and B.

$$A = \begin{bmatrix} 1 & 0 \\ 3 & 4 \end{bmatrix} \qquad B = \begin{bmatrix} 0 & 1 & 0 \\ 0 & 0 & 1 \\ 1 & 0 & 0 \end{bmatrix}$$

59. Find A^2, A^3, and A^4.

60. Show that $(A^2)^2 = A^4$.

61. Show that $B^3 = I$.

62. Show that $B^4 = B$.

In Exercises 63–66, evaluate the given determinants by expansion by minors.

63. $\begin{vmatrix} 7 & 2 & 3 \\ 1 & -5 & 0 \\ -6 & 4 & -3 \end{vmatrix}$

64. $\begin{vmatrix} 4 & -5 & -1 \\ 6 & 1 & 6 \\ -2 & 4 & -3 \end{vmatrix}$

65. $\begin{vmatrix} 1 & 4 & 0 & -3 \\ 3 & 1 & 2 & 5 \\ -2 & -2 & -4 & 1 \\ -1 & 6 & 3 & -4 \end{vmatrix}$

66. $\begin{vmatrix} -2 & 6 & 6 & -1 \\ 1 & -2 & -5 & 2 \\ 5 & -4 & 4 & 3 \\ -3 & 1 & -2 & -3 \end{vmatrix}$

In Exercises 67–70, use the determinants for Exercises 63–66, and evaluate each using a calculator.

In Exercises 71 and 72, use the matrix N.

$$N = \begin{bmatrix} 0 & -1 \\ 1 & 0 \end{bmatrix}$$

71. Show that $N^{-1} = -N$.

72. Show that $N^2 = -I$.

In Exercises 73–76, solve the given problems.

73. For any real number n, show that $\begin{bmatrix} n & 1+n \\ 1-n & -n \end{bmatrix}^2 = I$.

74. For the matrix $N = \begin{bmatrix} 1 & 1 \\ 1 & 1 \end{bmatrix}$, find (a) N^2, (b) N^3, (c) N^4. What is N^{20}? Explain.

75. If $B^{-1} = \begin{bmatrix} 1 & 2 \\ 3 & 4 \end{bmatrix}$ and $AB = \begin{bmatrix} -1 & 3 \\ 0 & 2 \end{bmatrix}$, find A.

76. For $A = \begin{bmatrix} 1 & 2 \\ 3 & 4 \end{bmatrix}$, $B = \begin{bmatrix} 2 & -3 \\ 3 & 5 \end{bmatrix}$, $C = \begin{bmatrix} 0 & 1 \\ 2 & 4 \end{bmatrix}$, verify the associative law of multiplication, $A(BC) = (AB)C$.

In Exercises 77–80, use matrices A and B.

$$A = \begin{bmatrix} 1 & -2 \\ 0 & 3 \end{bmatrix} \qquad B = \begin{bmatrix} -3 & 1 \\ 2 & -1 \end{bmatrix}$$

77. Show that $(A + B)(A - B) \neq A^2 - B^2$.

78. Show that $(A + B)^2 \neq A^2 + 2AB + B^2$.

79. Show that the inverse of $2A$ is $A^{-1}/2$.

80. Show that the inverse of $B/2$ is $2B^{-1}$.

In Exercises 81–84, solve the given systems of equations by use of matrices as in Section 16.4.

81. Two electric resistors, R_1 and R_2, are tested with currents and voltages such that the following equations are found:

$$2R_1 + 3R_2 = 26$$
$$3R_1 + 2R_2 = 24$$

Find the resistances R_1 and R_2 (in Ω).

82. A company produces two products, each of which is processed in two departments. Considering the worker time available, the numbers x and y of each product produced each week can be found by solving the system of equations

$$4.0x + 2.5y = 1200$$
$$3.2x + 4.0y = 1200$$

Find x and y.

83. A beam is supported as shown in Fig. 16.23. Find the force F and the tension T by solving the following system of equations:

$$0.500F = 0.866T$$
$$0.866F + 0.500T = 350$$

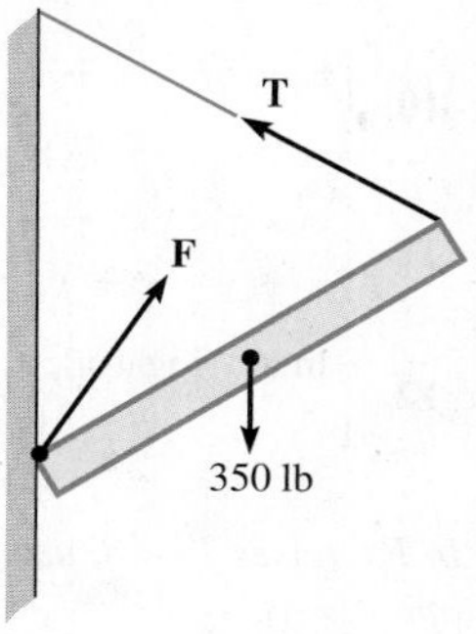

Fig. 16.23

84. To find the electric currents (in A) indicated in Fig. 16.24, it is necessary to solve the following equations.

$$I_A + I_B + I_C = 0$$
$$5I_A - 2I_B = -4$$
$$2I_B - I_C = 0$$

Find I_A, I_B, and I_C.

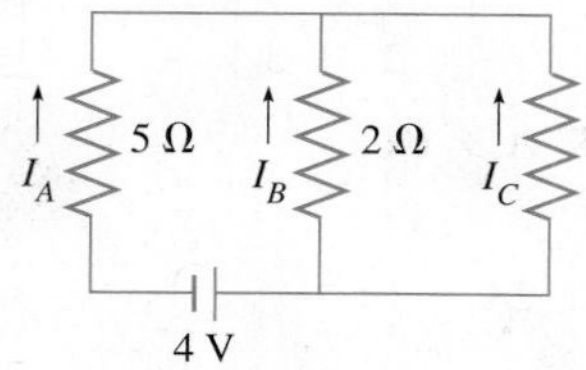

Fig. 16.24

In Exercises 85–88, solve the systems of equations in Exercises 81–84, by Gaussian elimination.

In Exercises 89–92, solve the given problems by setting up the necessary equations and solving them by any appropriate method of this chapter.

89. A crime suspect passes an intersection in a car traveling at 110 mi/h. The police pass the intersection 3.0 min later in a car traveling at 135 mi/h. How long is it before the police overtake the suspect?

90. A contractor needs a backhoe and a generator for two different jobs. Renting the backhoe for 5.0 h and the generator for 6.0 h costs $425 for one job. On the other job, renting the backhoe for 2.0 h and the generator for 8.0 h costs $310. What are the hourly charges for the backhoe and the generator?

91. By mass, three alloys have the following percentages of lead, zinc, and copper.

	Lead	*Zinc*	*Copper*
Alloy A	60%	30%	10%
Alloy B	40%	30%	30%
Alloy C	30%	70%	

How many grams of each of alloys A, B, and C must be mixed to get 100 g of an alloy that is 44% lead, 38% zinc, and 18% copper?

92. On a 750-mi trip from Salt Lake City to San Francisco that took a total of 5.5 h, a person took a limousine to the airport, then a plane, and finally a car to reach the final destination. The limousine took as long as the final car trip and the time for connections. The limousine averaged 55 mi/h, the plane averaged 400 mi/h, and the car averaged 40 mi/h. The plane traveled four times as far as the limousine and car combined. How long did each part of the trip and the connections take?

In Exercises 93–96, perform the indicated matrix operations.

93. An automobile maker has two assembly plants at which cars with either 4, 6, or 8 cylinders and with either standard or automatic transmission are assembled. The annual production at the first plant of cars with the number of cylinders–transmission type (standard, automatic) is as follows:

4: 12,000, 15,000; 6: 24,000, 8000; 8: 4000, 30,000

At the second plant the annual production is

4: 15,000, 20,000; 6: 12,000, 3000; 8: 2000, 22,000

Set up matrices for this production, and by matrix addition find the matrix for the total production by the number of cylinders and type of transmission.

94. Set up a matrix representing the information given in Exercise 91. A given shipment contains 500 g of alloy A, 800 g of alloy B, and 700 g of alloy C. Set up a matrix for this information. By multiplying these matrices, obtain a matrix that gives the total weight of lead, zinc, and copper in the shipment.

95. The matrix equation

$$\left(\begin{bmatrix} R_1 & -R_2 \\ -R_2 & R_1 \end{bmatrix} + R_2\begin{bmatrix} 1 & 0 \\ 0 & 1 \end{bmatrix}\right)\begin{bmatrix} i_1 \\ i_2 \end{bmatrix} = \begin{bmatrix} 6 \\ 0 \end{bmatrix}$$

may be used to represent the system of equations relating the currents and resistances of the circuit in Fig. 16.25. Find this system of equations by performing the indicated matrix operations.

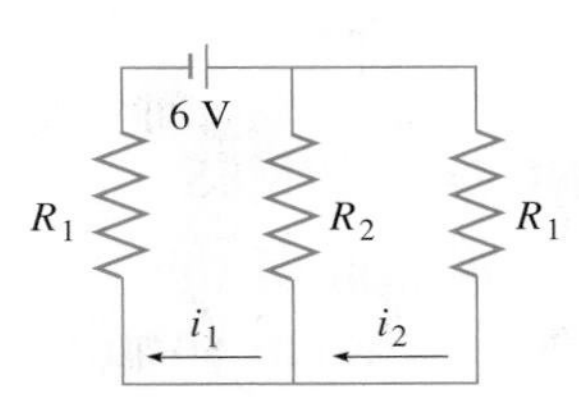

Fig. 16.25

96. A person prepared a meal of the following items, each having the given number of grams of protein, carbohydrates, and fat, respectively. Beef stew: 25, 21, 22; coleslaw: 3, 10, 10; (light) ice cream: 7, 25, 6. If the calorie count of each gram of protein, carbohydrate, and fat is 4.1 Cal/g, 3.9 Cal/g, and 8.9 Cal/g, respectively, find the total calorie count of each item by matrix multiplication. (1 Cal = 1 kcal.)

97. A hardware company has 60 different retail stores in which 1500 different products are sold. Write a paragraph or two explaining why matrices provide an efficient method of inventory control, and what matrix operations in this chapter would be of use.

CHAPTER 16 PRACTICE TEST

As a study aid, we have included complete solutions for each Practice Test problem at the end of this book.

1. For matrices A and B, find $A - 2B$.

$$A = \begin{bmatrix} 3 & -1 & 4 \\ 2 & 0 & -2 \end{bmatrix} \quad B = \begin{bmatrix} 1 & 4 & 5 \\ -1 & -2 & 3 \end{bmatrix}$$

2. Evaluate the literal symbols.

$$\begin{bmatrix} 2x & x - y & z \\ x + z & 2y & y + z \end{bmatrix} = \begin{bmatrix} 6 & -2 & 4 \\ a & b & c \end{bmatrix}$$

3. For matrices C and D, find CD and DC, if possible.

$$C = \begin{bmatrix} 1 & 0 & 4 \\ 2 & -2 & 1 \\ -1 & 3 & 2 \end{bmatrix} \quad D = \begin{bmatrix} 2 & -2 \\ 4 & -5 \\ 6 & 1 \end{bmatrix}$$

4. Determine whether or not $B = A^{-1}$.

$$A = \begin{bmatrix} 2 & -5 \\ 1 & -2 \end{bmatrix} \quad B = \begin{bmatrix} -2 & 5 \\ -1 & 2 \end{bmatrix}$$

5. For matrix C of Problem 3, find C^{-1}.

6. Solve by using the inverse of the coefficient matrix.

$$\begin{aligned} 2x - 3y &= 11 \\ x + 2y &= 2 \end{aligned}$$

7. Solve the system of equations in Problem 6 by Gaussian elimination.

8. Evaluate the following determinant by expansion by minors.

$$\begin{vmatrix} 2 & -4 & -3 \\ -3 & 6 & 2 \\ 5 & -1 & 5 \end{vmatrix}$$

9. Solve the following system of equations by using the inverse of the coefficient matrix. Use a calculator to perform the necessary matrix operations and display the result and check. Write down the display shown in the window with the solutions.

$$\begin{aligned} 7x - 2y + z &= 6 \\ 2x + 3y - 4z &= 6 \\ 4x - 5y + 2z &= 10 \end{aligned}$$

10. Fifty shares of stock A and 30 shares of stock B cost $2600. Thirty shares of stock A and 40 shares of stock B cost $2000. What is the price per share of each stock? Solve by setting up the appropriate equations and then using the inverse of the coefficient matrix.

17 Inequalities

LEARNING OUTCOMES

After completion of this chapter, the student should be able to:

- Graph solutions of inequalities on the number line
- Apply the properties of inequalities to solve linear inequalities
- Solve polynomial and rational inequalities using critical values
- Solve inequalities involving absolute value
- Solve inequalities and systems of inequalities with two variables graphically
- Solve linear programming problems involving constraint and objective functions
- Solve application problems involving inequalities

Having devoted a great deal of time to the solution of equations and systems of equations, we now turn our attention to solving inequalities and systems of inequalities.

In doing so, we will find it necessary to find *all values* of the variable or variables that satisfy the inequality or system of inequalities.

There are numerous technical applications of inequalities. For example, in electricity it might be necessary to find the values of a current that are *greater than* a specified value. In designing a link in a robotic mechanism, it might be necessary to find the forces that are *less than* a specified value. Computers can be programmed to switch from one part of a program to another, based upon a result that is greater than (or less than) some given value.

Systems of linear *equations* have been studied for more than 2000 years, but almost no attention was given to systems of linear *inequalities* until World War II in the 1940s. Problems of deploying personnel and aircraft effectively and allocating supplies efficiently led the U.S. Air Force to have a number of scientists, economists, and mathematicians look for solutions. From this, a procedure for analyzing such problems was devised in 1947 by George Danzig and his colleagues. Their method involved using systems of linear inequalities and is today called *linear programming*. We introduce the basic method of linear programming in the final section of this chapter.

Here, we see that a mathematical method was developed as a result of a military need. Today, linear programming is widely used in business and industry in order to set production levels for maximizing profits and minimizing costs.

▶ **In Section 17.6, we use inequalities to show how an airline can minimize operating costs when determining how many planes it needs.**

17.1 Properties of Inequalities

Solution of an Inequality • Conditional Inequality • Absolute Inequality • Properties of Inequalities

In Chapter 1, we first introduced the signs of inequality. To this point, only a basic understanding of their meanings has been necessary to show certain intervals associated with a variable. In this section, we review the meanings and develop certain basic properties of inequalities. We also show the meaning of the solution of an inequality and how it is shown on the number line.

The expression $a < b$ is read as "a is less than b," and the expression $a > b$ is read as "a is greater than b." *These signs define what is known as the* **sense** *(indicated by the direction of the sign) of the inequality.* Two inequalities are said to have the same sense if the signs of inequality point in the same direction. They are said to have the opposite sense if the signs of inequality point in opposite directions. *The two sides of the inequality are called* **members** *of the inequality.*

EXAMPLE 1 Sense of inequality

The inequalities $x + 3 > 2$ and $x + 1 > 0$ have the same sense, as do the inequalities $3x - 1 < 4$ and $x^2 - 1 < 3$.

The inequalities $x - 4 < 0$ and $x > -4$ have the opposite sense, as do the inequalities $2x + 4 > 1$ and $3x^2 - 7 < 1$. ■

The **solution** *of an inequality consists of all values of the variable that make the inequality a true statement.* Most inequalities with which we deal are **conditional inequalities,** *which are true for some, but not all, values of the variable.* Also, *some inequalities are true for all values of the variable, and they are called* **absolute inequalities.** A solution of an inequality consists of only real numbers, as the terms *greater than* or *less than* have not been defined for complex numbers.

EXAMPLE 2 Conditional and absolute inequalities

The inequality $x + 1 > 0$ is true for all values of x greater than -1. Therefore, the values of x that satisfy this inequality are written as $x > -1$. This illustrates the difference between the solution of an equation and the solution of an inequality. NOTE ➧ [The solution of an equation normally consists of a few specific numbers, whereas the solution to an inequality normally consists of an *interval of values* of the variable.] Any and all values within this interval are termed solutions of the inequality. Because the inequality $x + 1 > 0$ is satisfied only by the values of x in the interval $x > -1$, it is a *conditional inequality*.

The inequality $x^2 + 1 > 0$ is true for all real values of x, since x^2 is never negative. Therefore, it is an *absolute inequality*. ■

There are occasions when it is convenient to combine an inequality with an equality. For such purposes, the symbols $\leq$, meaning *less than or equal to,* and $\geq$, meaning *greater than or equal to,* are used.

EXAMPLE 3 Greater (or less) than or equal to

If we wish to state that x is positive, we can write $x > 0$. However, the value zero is not included in the solution. If we wish to state that x is not negative, we write $x \geq 0$. Here, zero is part of the solution.

In order to state that x is less than or equal to -5, we write $x \leq -5$. ■

In the sections that follow, we will solve inequalities. It is often useful to show the solution on the number line. The next example shows how this is done.

EXAMPLE 4 Graphing solutions of inequalities

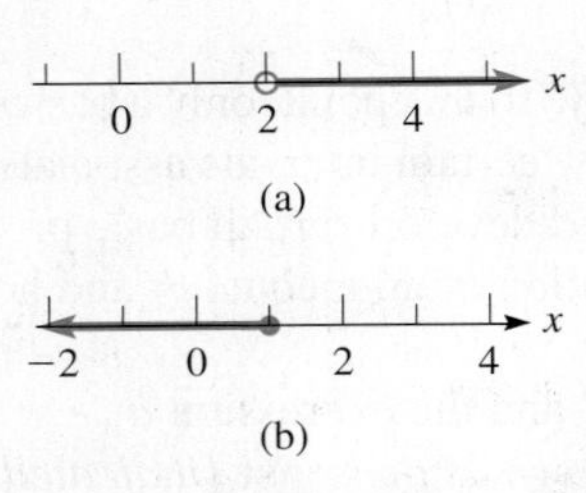

Fig. 17.1

(a) To graph $x > 2$, we draw a small open circle at 2 on the number line (which is equivalent to the x-axis). Then we draw a solid line to the right of the point and with an arrowhead pointing to the right, indicating all values greater than 2. See Fig. 17.1(a). The *open circle* shows that the point is not part of the indicated solution.

(b) To graph $x \leq 1$, we follow the same basic procedure as in part (a), except that we use a solid circle and the arrowhead points to the left. See Fig. 17.1(b). The *solid circle* shows that the point is part of the indicated solution. ■

PROPERTIES OF INEQUALITIES

We now show the basic operations performed on inequalities. These are the same operations as those performed on equations, but in certain cases the results take on a different form. *The following are the* **properties of inequalities.**

1. *The sense of an inequality is not changed when the same number is added to—or subtracted from—both members of the inequality.* Symbolically, this may be stated as "if $a > b$, then $a + c > b + c$ and $a - c > b - c$."

EXAMPLE 5 Illustrations of property 1

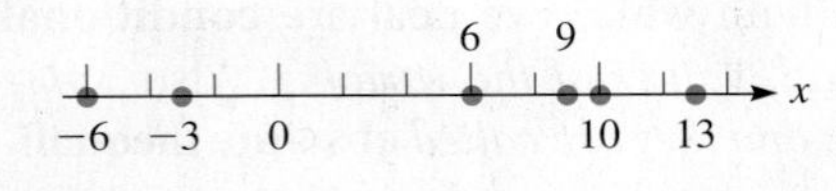

Fig. 17.2

Using property 1 on the inequality $9 > 6$, we have the following results:

$9 > 6$	$9 > 6$
add 4 to each member	subtract 12 from each member
$9 + 4 > 6 + 4$	$9 - 12 > 6 - 12$
$13 > 10$	$-3 > -6$

In Fig. 17.2, we see that 9 is to the right of 6, 13 is to the right of 10, and -3 is to the right of -6. ■

Practice Exercises

For $-6 < 3$, determine the inequality if
1. 8 is added to each member.
2. 8 is subtracted from each member.

2. *The sense of an inequality is not changed if both members are multiplied or divided by the same positive number.* Symbolically, this is stated as "if $a > b$, then $ac > bc$, and $a/c > b/c$, provided that $c > 0$."

EXAMPLE 6 Illustrations of property 2

Using property 2 on the inequality $8 < 15$, we have the following results:

$8 < 15$	$8 < 15$
multiply both members by 2	divide both members by 2
$2(8) < 2(15)$	$\frac{8}{2} < \frac{15}{2}$
$16 < 30$	$4 < \frac{15}{2}$

■

3. ***The sense of an inequality is reversed if both members are multiplied or divided by the same negative number.*** Symbolically, this is stated as "if $a > b$, then $ac < bc$, and $a/c < b/c$, provided that $c < 0$."

CAUTION In using this property of inequalities, be very careful to note that the inequality sign remains the same if both members are multiplied or divided by a positive number, but that **the inequality sign is reversed if both members are multiplied or divided by a negative number.** Most of the errors made in dealing with inequalities occur when using this property. ■

EXAMPLE 7 Be careful in using property 3

Using property 3 on the inequality $4 > -2$, we have the following results:

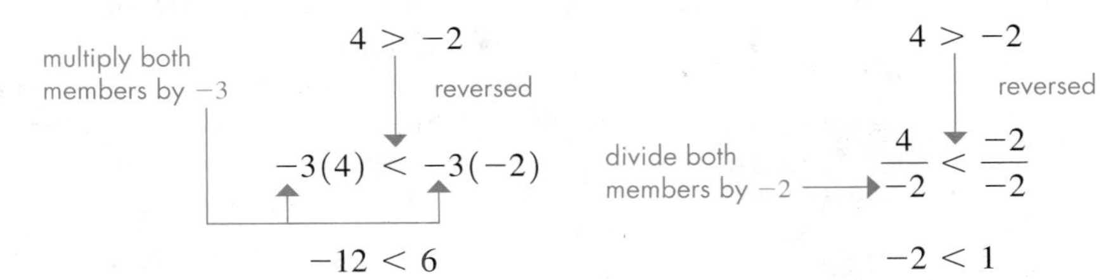

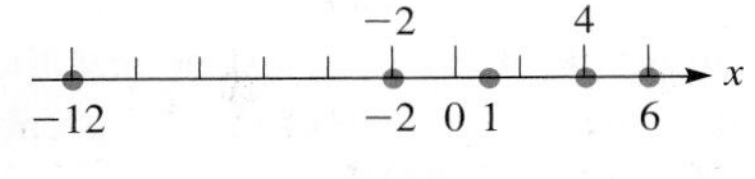

Fig. 17.3

In Fig. 17.3, we see that 4 is to the right of -2, but that -12 *is to the left of* 6 and that -2 *is to the left of* 1. This is consistent with reversing the sense of the inequality when it is multiplied by -3 and when it is divided by -2. ■

Practice Exercises

For the inequality $-6 < 3$, state the inequality that results.

3. Multiply both members by 4.

4. Divide both members by -3.

4. *If both members of an inequality are positive numbers and n is a positive integer, then the inequality formed by taking the nth power of each member, or the nth root of each member, is in the same sense as the given inequality.* Symbolically, this is stated as "if $a > b$, then $a^n > b^n$, and $\sqrt[n]{a} > \sqrt[n]{b}$, provided that $n > 0$, $a > 0$, $b > 0$."

EXAMPLE 8 Illustrations of property 4

Using property 4 on the inequality $16 > 9$, we have

$16 > 9$	$16 > 9$
square both members	take square root of both members
$16^2 > 9^2$	$\sqrt{16} > \sqrt{9}$
$256 > 81$	$4 > 3$

■

Many inequalities have more than two members. In fact, inequalities with three members are common, and care must be used in stating these inequalities.

EXAMPLE 9 Inequalities with three members

(a) To state that 5 is less than 6, and also greater than 2, we may write $2 < 5 < 6$, or $6 > 5 > 2$. (Generally, the *less than* form is preferred.)

(b) To state that a number x may be greater than -1 *and* also less than or equal to 3, we write $-1 < x \leq 3$. (It can also be written as $x > -1$ *and* $x \leq 3$.) This is shown in Fig. 17.4(a). Note the use of the open circle and the solid circle.

(c) By writing $x \leq -4$ *or* $x > 2$, we state that x is less than or equal to -4, *or* greater than 2. ***It may not be shown as*** $2 < x \leq -4$, for this shows x as being less than -4, and also greater than 2, and *no such numbers exist.* See Fig. 17.4(b). ■

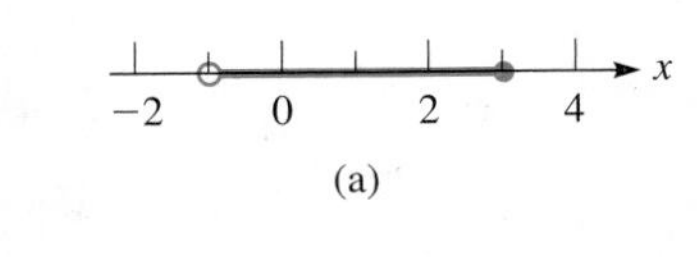

(a)

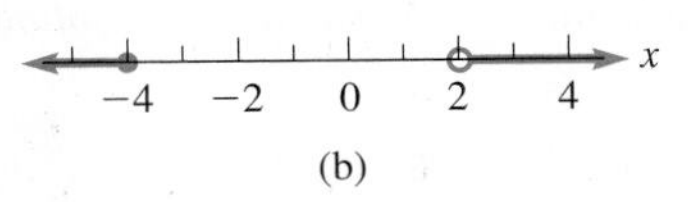

(b)

Fig. 17.4

NOTE ▸ [Note carefully that **and** is used when the solution consists of values that make **both** statements true. The word **or** is used when the solution consists of values that make **either** statement true.] (In everyday speech, *or* can sometimes mean that one statement is true or another statement is true, but *not* that both are true.)

EXAMPLE 10 Meaning of *and*—meaning of *or*

The inequality $x^2 - 3x + 2 > 0$ is satisfied if x is either greater than 2 *or* less than 1. This is written as $x > 2$ *or* $x < 1$, but it is incorrect to state it as $1 > x > 2$. (If we wrote it this way, we would be saying that the same value of x is less than 1 *and* at the same time greater than 2. Of course, as noted for this type of situation in Example 9, no such number exists.) However, we could say that the inequality is not satisfied for $1 \leq x \leq 2$, which means those values of x greater than *or* equal to 1 *and* less than *or* equal to 2 (between or equal to 1 and 2). ■

Practice Exercise

5. Graph the inequality $-2 < x \leq 2$ on the number line.

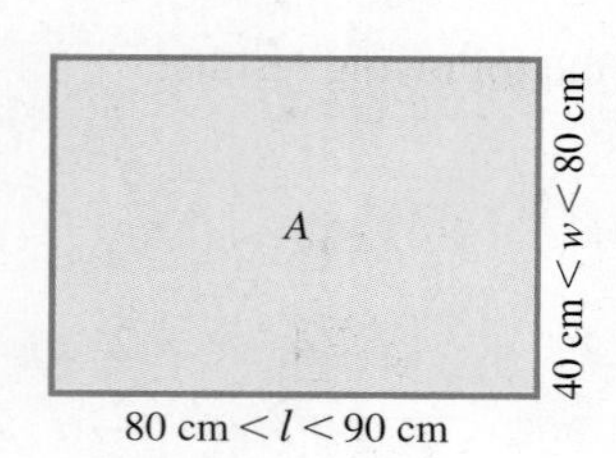

Fig. 17.5

EXAMPLE 11 Setting up an inequality— solar panel design

The design of a rectangular solar panel shows that the length l is between 80 cm and 90 cm and the width w is between 40 cm and 80 cm. See Fig. 17.5. Find the values of area the panel may have.

Because l is to be less than 90 cm and w less than 80 cm, the area must be less than $(90 \text{ cm})(80 \text{ cm}) = 7200 \text{ cm}^2$. Also, because l is to be greater than 80 cm and w greater than 40 cm, the area must be greater than $(80 \text{ cm})(40 \text{ cm}) = 3200 \text{ cm}^2$. Therefore, the area A may be represented as

$$3200 \text{ cm}^2 < A < 7200 \text{ cm}^2$$

This means the area is greater than 3200 cm^2 *and* less than 7200 cm^2. ■

EXAMPLE 12 Setting up an inequality—current through diode

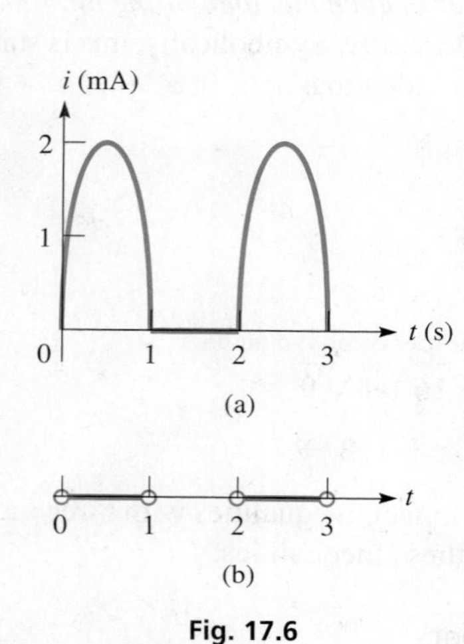

Fig. 17.6

A semiconductor *diode* has the property that an electric current flows through it in only one direction. If it is an alternating-current circuit, the current in the circuit flows only during the half-cycle when the diode allows it to flow. If a source of current given by $i = 2 \sin \pi t$ (i in mA, t in seconds) is connected in series with a diode, write the inequalities for the current and the time. Assume the source is on for 3.0 s and a positive current passes through the diode.

We are to find the values of t that correspond to $i > 0$. From the properties of the sine function, we know that $2 \sin \pi t$ has a period of $2\pi/\pi = 2.0$ s. Therefore, the current is zero for $t = 0$, 1.0 s, 2.0 s, and 3.0 s.

The source current is positive for $0 < t < 1.0$ s and for $2.0\ s < t < 3.0$ s.
The source current is negative for $1.0 \text{ s} < t < 2.0$ s.
Therefore, in the circuit

$$i > 0 \quad \text{for} \quad 0 < t < 1.0 \text{ s} \quad \text{and} \quad 2.0 \text{ s} < t < 3.0 \text{ s}$$
$$i = 0 \quad \text{for} \quad t = 0,\ 1.0 \text{ s} \leq t \leq 2.0 \text{ s}$$

A graph of the current in the circuit as a function of time is shown in Fig. 17.6(a). In Fig. 17.6(b), the values of t for which $i > 0$ are shown. ■

EXERCISES 17.1

In Exercises 1–4, make the given changes in the indicated examples of this section and then perform the indicated operations.

1. In Example 2, in the first paragraph, change the $>$ to $<$ and then complete the meaning of the resulting inequality as in the first sentence. Rewrite the meaning as in the second line.
2. In Example 4(b), change $\leq$ to $>$ and then graph the resulting inequality.
3. In Example 7, change the inequality to $-2 > -4$ and then perform the two operations shown in color.
4. In Example 9(b), change the -1 to -3 and the 3 to 1 and then write the two forms in which an inequality represents the statement.

In Exercises 5–12, for the inequality $4 < 9$, state the inequality that results when the given operations are performed on both members.

5. Add 5.
6. Subtract 16.
7. Multiply by 4.
8. Multiply by -2.
9. Divide by -1.
10. Divide by 0.5.
11. Square both.
12. Take square roots.

In Exercises 13–24, give the inequalities equivalent to the following statements about the number x.

13. Greater than -2
14. Less than 0.7
15. Less than or equal to 38
16. Greater than or equal to -6
17. Greater than 1 and less than 7
18. Greater than or equal to -200 and less than 650
19. Less than -5, or greater than or equal to 3
20. Less than or equal to 8, or greater than or equal to 12
21. Less than 1, or greater than 3 and less than or equal to 5
22. Greater than or equal to 0 and less than or equal to 2, or greater than 5
23. Greater than -2 and less than 2, or greater than or equal to 3 and less than 4
24. Less than -4, or greater than or equal to 0 and less than or equal to 1, or greater than or equal to 5

In Exercises 25–28, give verbal statements equivalent to the given inequalities involving the number x.

25. $0 < x \leq 9$

26. $x < 5$ or $x > 7$

27. $x < -10$ or $10 \leq x < 20$

28. $-1 \leq x < 3$ or $5 < x < 7$

In Exercises 29–44, graph the given inequalities on the number line.

29. $x < 3$

30. $x \geq -1$

31. $x \leq -1$ or $x > 0.5$

32. $x < -300$ or $x \geq 0$

33. $0 \leq x < 5$

34. $-4 < y < -2$

35. $x \geq -3$ and $x < 5$

36. $x > 4$ and $x < 3$

37. $x < -1$ or $1 \leq x < 4$

38. $-3 < x < 0$ or $x > 3$

39. $-3 < x < -1$ or $0.5 < x \leq 3$

40. $x \leq 4$ or $x > -4$

41. $t \leq -5$ and $t \geq -5$

42. $x < 1$ or $1 < x \leq 4$

43. $(x \leq 5$ or $x \geq 8)$ and $(3 < x < 10)$

44. $(x < 7$ and $x > 2)$ or $(x \geq 10$ or $x < 1)$

In Exercises 45–48, answer the given questions about the inequality $0 < a < b$.

45. Is $a^2 < b^2$ a conditional inequality or an absolute inequality?

46. Is $|a - b| < b - a$?

47. If each member of the inequality $2 > 1$ is multiplied by $a - b$, is the result $2(a - b) > (a - b)$?

48. What is wrong with the following sequence of steps?

$$a < b, ab < b^2, ab - b^2 < 0, b(a - b) < 0, b < 0$$

In Exercises 49–52, solve the given problems.

49. Write the relationship between $(|x| + |y|)$ and $|x + y|$ if $x > 0$ and $y < 0$.

50. Write the relationship between $|xy|$ and $|x||y|$ if $x > 0$ and $y < 0$.

51. Explain the error in the following "proof" that $3 < 2$:

(1) $1/8 < 1/4$ (2) $0.5^3 < 0.5^2$ (3) $\log 0.5^3 < \log 0.5^2$
(4) $3 \log 0.5 < 2 \log 0.5$ (5) $3 < 2$

52. If $x \neq y$, show that $x^2 + y^2 > 2xy$.

In Exercises 53–62, some applications of inequalities are shown.

53. The length L and width w (in yd) of a rectangular soccer field should satisfy the inequalities $110 \leq L \leq 120$ and $70 \leq w \leq 80$. Express the possible diagonal lengths d as an inequality.

54. A breakfast cereal company guarantees the calorie count shown for each serving is accurate within 5%. If the package shows a serving has 200 cal, write an inequality for the possible calorie counts.

55. An electron microscope can magnify an object from 2000 times to 1,000,000 times. Assuming these values are exact, express these magnifications M as an inequality and graph them.

56. A busy person glances at a digital clock that shows 9:36. Another glance a short time later shows the clock at 9:44. Express the amount of time t (in min) that could have elapsed between glances by use of inequalities. Graph these values of t.

57. An Earth satellite put into orbit near Earth's surface will have an elliptic orbit if its velocity v is between 18,000 mi/h and 25,000 mi/h. Write this as an inequality and graph these values of v.

58. Fossils found in Jurassic rocks indicate that dinosaurs flourished during the Jurassic geological period, 140 MY (million years ago) to 200 MY. Write this as an inequality, with t representing past time. Graph the values of t.

59. A DVD player spins at 1530 r/min at the innermost edge and gradually slows to a rate of 630 r/min at the outer edge. Use an inequality to express the angular velocity ω of the DVD player.

60. The velocity v of an ultrasound wave in soft human tissue may be represented as 1550 ± 60 m/s, where the ± 60 m/s gives the possible variation in the velocity. Express the possible velocities by an inequality.

61. A driver using the Google Maps app finds that it is 300 mi to her destination. If her speed always stays between 50 mi/h and 60 mi/h, use an inequality to express the required time t for the trip.

62. If the current from the source in Example 12 is $i = 5 \cos 4\pi t$ and the diode allows only negative current to flow, write the inequalities and draw the graph for the current in the circuit as a function of time for $0 \leq t \leq 1$ s.

Answers to Practice Exercises

1. $2 < 11$ **2.** $-14 < -5$ **3.** $-24 < 12$ **4.** $2 > -1$

5. −2 −1 0 1 2

17.2 Solving Linear Inequalities

Linear Inequalities • Inequalities with Three Members • Solving an Inequality with a Calculator

Using the properties and definitions discussed in Section 17.1, we can now proceed to solve inequalities. In this section, we solve linear inequalities in one variable. Similar to linear functions as defined in Chapter 5, *a* **linear inequality** *is one in which each term contains only one variable and the exponent of each variable is* 1. We will consider linear inequalities in two variables in Section 17.5.

The procedure for solving a linear inequality in one variable is similar to what we used in solving basic equations in Chapter 1. We solve the inequality by isolating the variable, and to do this we perform the same operations on each member of the inequality. The operations are based on the properties given in Section 17.1.

EXAMPLE 1 Solutions using basic operations

In each of the following inequalities, by performing the indicated operation, we isolate x and thereby solve the inequality.

$x + 2 < 4$	$\frac{x}{2} > 4$	$2x \leq 4$
Subtract 2 from each member.	Multiply each member by 2.	Divide each member by 2.
$x < 2$	$x > 8$	$x \leq 2$

Each solution can be checked by substituting any number in the indicated interval into the original inequality. For example, any value less than 2 will satisfy the first inequality, whereas any number greater than 8 will satisfy the second inequality. ■

EXAMPLE 2 Solving a linear inequality

Solve the following inequality: $3 - 2x \geq 15$.

We have the following solution:

$$3 - 2x \geq 15 \quad \text{original inequality}$$
$$-2x \geq 12 \quad \text{subtract 3 from each member}$$
inequality reversed →
$$x \leq -6 \quad \text{divide each member by } -2$$

CAUTION Again, carefully note that the sign of inequality was reversed when each number was divided by -2. ■ We check the solution by substituting -7 in the original inequality, obtaining $17 \geq 15$. ■

Practice Exercise

1. Solve the inequality: $4 + 3x < 10$

EXAMPLE 3 Solving a linear inequality

Solve the inequality $2x \leq 3 - x$.

The solution proceeds as follows:

$$2x \leq 3 - x \quad \text{original inequality}$$
$$3x \leq 3 \quad \text{add } x \text{ to each member}$$
$$x \leq 1 \quad \text{divide each member by 3}$$

This solution checks and is represented in Fig. 17.7, as we showed in Section 17.1.

This inequality could have been solved by combining x-terms on the right. In doing so, we would obtain $1 \geq x$. Because this might be misread, it is best to combine the variable terms on the left, as we did above. ■

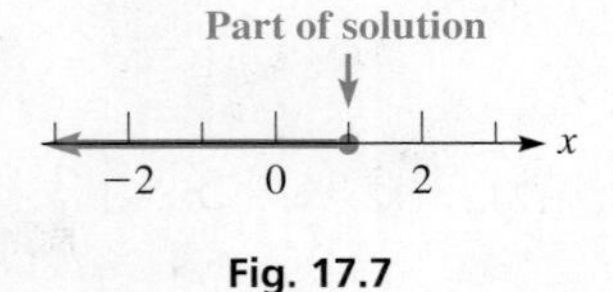

Fig. 17.7

EXAMPLE 4 Solving a linear inequality

Solve the inequality $\frac{3}{2}(1 - x) > \frac{1}{4} - x$.

$$\frac{3}{2}(1 - x) > \frac{1}{4} - x \quad \text{original inequality}$$
$$6(1 - x) > 1 - 4x \quad \text{multiply each member by 4}$$
$$6 - 6x > 1 - 4x \quad \text{remove parentheses}$$
$$-6x > -5 - 4x \quad \text{subtract 6 from each member}$$
$$-2x > -5 \quad \text{add } 4x \text{ to each member}$$
$$x < \frac{5}{2} \quad \text{divide each member by } -2$$

Note that the sense of the inequality was reversed when we divided by -2. This solution is shown in Fig. 17.8. Any value of $x < 5/2$ checks when substituted into the original inequality. ■

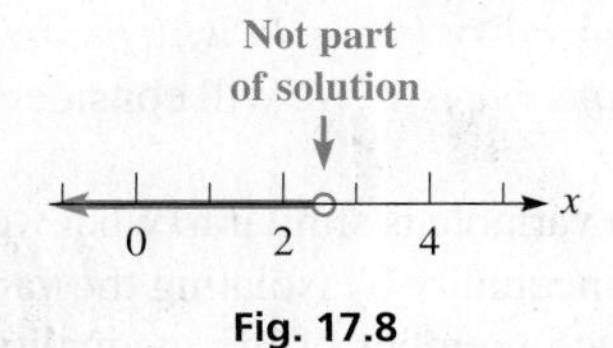

Fig. 17.8

Practice Exercise

2. Solve the inequality: $2(3 - x) > 5 + 4x$

The following example illustrates an application that involves the solution of an inequality.

EXAMPLE 5 Linear inequality—missile velocity

The velocity v (in ft/s) of a missile in terms of the time t (in s) is given by $v = 960 - 32t$. For how long is the velocity positive? (Because velocity is a vector, this can also be interpreted as asking "how long is the missile moving upward?")

In terms of inequalities, we are asked to find the values of t for which $v > 0$. This means that we must solve the inequality $960 - 32t > 0$. The solution is as follows:

$$960 - 32t > 0 \quad \text{original inequality}$$
$$-32t > -960 \quad \text{subtract 960 from each member}$$
$$t < 30 \text{ s} \quad \text{divide each member by } -32$$

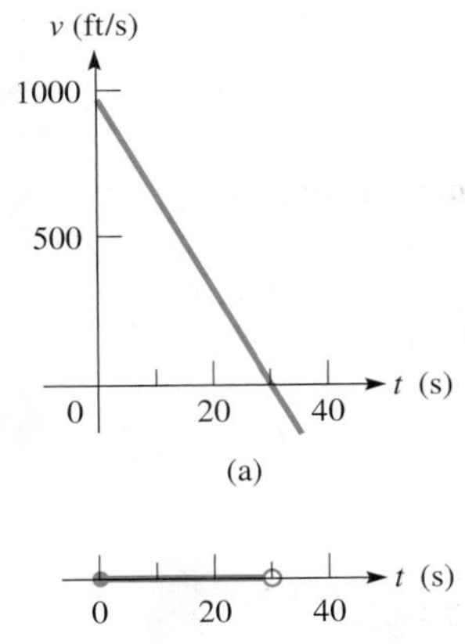

(a)

(b)

Fig. 17.9

Negative values of t have no meaning in this problem. Checking $t = 0$, we find that $v = 960$ ft/s. Therefore, the complete solution is $0 \leq t < 30$ s.

In Fig. 17.9(a), we show the graph of $v = 960 - 32t$, and in Fig. 17.9(b), we show the solution $0 \leq t < 30$ s on the number line (which is really the t-axis in this case). Note that the values of v are above the t-axis for those values of t that are part of the solution. This shows the relationship of the graph of v as a function of t, and the solution as graphed on the number line (the t-axis). ■

INEQUALITIES WITH THREE MEMBERS

EXAMPLE 6 Solving inequality with three members

Solve: $-1 < 2x + 3 < 6$.

We have the following solution.

$$-1 < 2x + 3 < 6 \quad \text{original inequality}$$
$$-4 < 2x < 3 \quad \text{subtract 3 from each member}$$
$$-2 < x < \frac{3}{2} \quad \text{divide each member by 2}$$

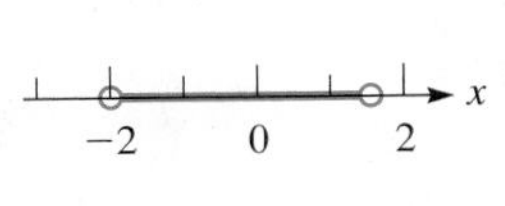

Fig. 17.10

The solution is shown in Fig. 17.10. ■

EXAMPLE 7 Solving inequality with three members

TI-89 graphing calculator keystrokes for Example 7: goo.gl/AeXOiN

Solve the inequality $2x < x - 4 \leq 3x + 8$.

Since we cannot isolate x in the middle member (or in any member), we rewrite the inequality as

$$2x < x - 4 \quad \text{and} \quad x - 4 \leq 3x + 8$$

We then solve each of the inequalities, keeping in mind that the solution must satisfy both of them. Therefore, we have

$$2x < x - 4 \quad \text{and} \quad x - 4 \leq 3x + 8$$
$$-2x \leq 12$$
$$x < -4 \qquad\qquad x \geq -6$$

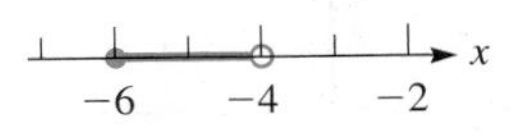

Fig. 17.11

We see that the solution is $x < -4$ *and* $x \geq -6$, which can also be written as $-6 \leq x < -4$. This second form is generally preferred since it is more concise and more easily interpreted. The solution checks and is shown in Fig. 17.11. ■

Practice Exercise

3. Solve the inequality: $-2 \leq 4x - 3 < 5$

EXAMPLE 8 Inequality with three members—pump rates

In emptying a wastewater tank, one pump can remove no more than 40 L/min. If it operates for 8.0 min and a second pump operates for 5.0 min, what must be the pumping rate of the second pump if 480 L are to be removed?

Let x = the pumping rate of the first pump and y = the pumping rate of the second pump. Because the first operates for 8.0 min and the second for 5.0 min to remove 480 L, we have

$$\underset{\text{first pump}}{8.0x} + \underset{\text{second pump}}{5.0y} = \underset{\text{total}}{480} \quad \longleftarrow \text{amounts pumped}$$

Because we know that the first pump can remove no more than 40 L/min, which means that $0 \leq x \leq 40$ L/min, we solve for x and then substitute in this inequality:

$$x = 60 - 0.625y \qquad \text{solve for } x$$
$$0 \leq 60 - 0.625y \leq 40 \qquad \text{substitute in inequality}$$
$$-60 \leq -0.625y \leq -20 \qquad \text{subtract 60 from each member}$$
$$96 \geq y \geq 32 \qquad \text{divide each member by } -0.625$$
$$32 \leq y \leq 96 \text{ L/min} \qquad \text{use} \leq \text{symbol (optional step)}$$

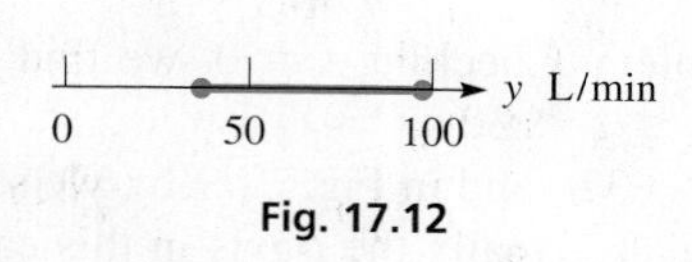

Fig. 17.12

This means that the second pump must be able to pump at least 32 L/min and no more than 96 L/min. See Fig. 17.12.

Although this was a three-member inequality combined with equalities, the solution was done in the same way as with a two-member inequality. ■

Practice Exercise

4. Solve the problem in Example 8 if the second pump operates for 4.0 min.

SOLVING AN INEQUALITY WITH A CALCULATOR

A calculator can be used to show the solution of an inequality. On most calculators, a value of 1 is shown if the inequality is satisfied, and a value of 0 is shown if the inequality is not satisfied. This means that if, for example, we enter $8 > 3$ (see the manual to see how $>$ is entered), the calculator displays 1, and if we enter $3 > 8$, it shows 0, as in the display in Fig. 17.13. We can also use an inequality feature to show graphically the solution of an inequality.

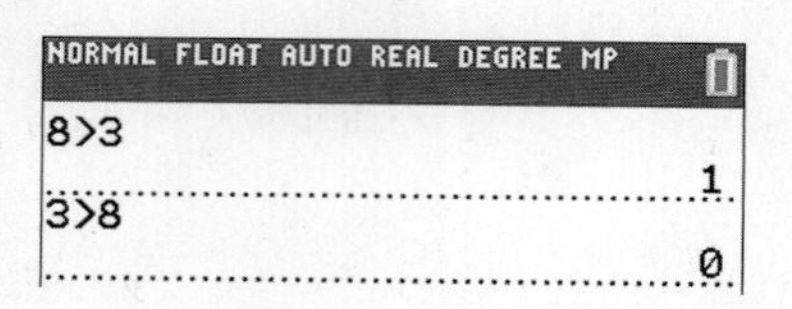

Fig. 17.13
Graphing calculator keystrokes: goo.gl/65nifo

EXAMPLE 9 Inequality solution on calculator

Display the solution of the inequality $2x \leq 3 - x$ (see Example 3) on a calculator.

We set y_1 *equal to the inequality,* as shown in Fig. 17.14(a) and then graph y_1. From Fig. 17.14(b), we see that $y_1 = 1$ up to $x = 1$. Using the *value* feature, we can also see that $y_1 = 1$ right at $x = 1$. This means the inequality is satisfied for all values of x less than or equal to 1. Thus, the solution set is $x \leq 1$. Compare Fig. 17.14(b) with Fig. 17.7.

■ Depending on the calculator model and the settings, the solution may show a vertical connection where the solution changes from 0 to 1, or 1 to 0.

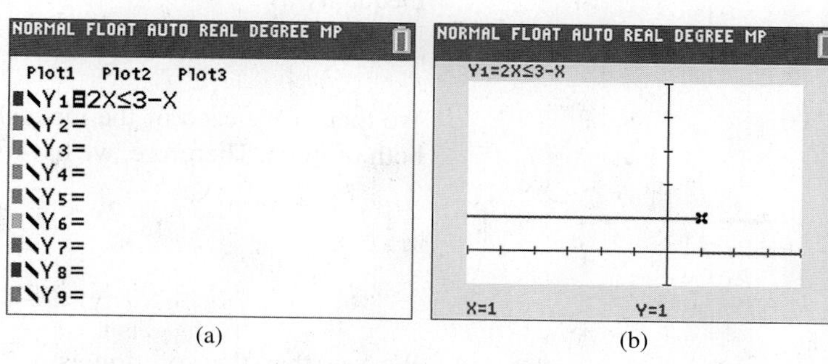

Fig. 17.14
Graphing calculator keystrokes: goo.gl/ssgQrs

■

```
NORMAL FLOAT AUTO REAL DEGREE MP
Plot1  Plot2  Plot3
\Y1= -1<2X+3 and 2X+3<6
\Y2=
\Y3=
\Y4=
\Y5=
\Y6=
\Y7=
\Y8=
\Y9=
```

(a)

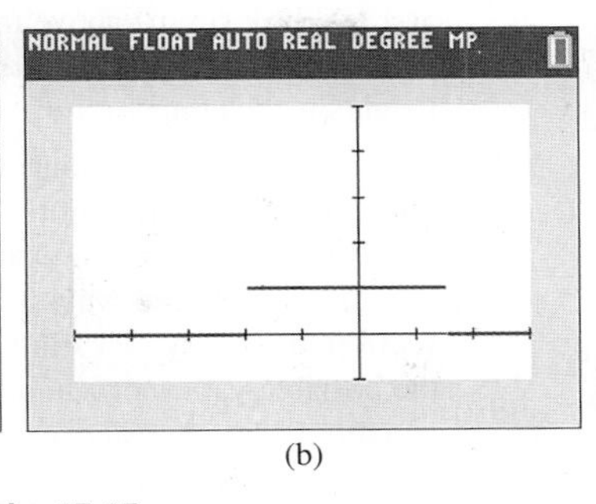

(b)

Fig. 17.15
Graphing calculator keystrokes:
goo.gl/pJq8uJ

EXAMPLE 10 Inequality solution on calculator

Display the solution of the inequality $-1 < 2x + 3 < 6$ (see Example 6) on a calculator.

In order to display the solution, we must write the inequality as

$$-1 < 2x + 3 \quad \text{and} \quad 2x + 3 < 6 \qquad \text{see Example 7}$$

Then enter $y_1 = -1 < 2x + 3$ and $2x + 3 < 6$ (consult the manual to determine how "and" is entered) in the calculator, as shown in Fig. 17.15(a). From Fig. 17.15(b), we see that the solution is $-2 < x < 1.5$. The *trace* and *zoom* features can be used to get accurate values. Compare Fig. 17.15(b) with Fig. 17.10. ■

EXERCISES 17.2

In Exercises 1–4, make the given changes in the indicated examples of this section and then perform the indicated operations.

1. In Example 2, change 3 to 25 and solve the resulting inequality.
2. In Example 4, change the 1/4 to 7/4 and then solve and display the resulting inequality.
3. In Example 6, change the + in the middle member to − and then solve the resulting inequality. Graph the solution.
4. In Example 10, change the + in the middle member to − and then display the solution on a graphing calculator.

In Exercises 5–28, solve the given inequalities. Graph each solution.

5. $x - 3 > -4$
6. $3x + 2 \le 11$
7. $\frac{1}{2}x < 32$
8. $-4t > 12$
9. $3x - 5 \le -11$
10. $\frac{1}{3}x + 2 \ge 1$
11. $12 - 2y > 16$
12. $32 - 5x < -8$
13. $\frac{4x - 5}{2} \le x$
14. $1.5 - 5.2x \ge 3.7 + 2.3x$
15. $180 - 6(T + 12) > 14T + 285$
16. $-2[x - (3 - 2x)] > \frac{1 - 5x}{3} + 2$
17. $2.50(1.50 - 3.40x) < 3.84 - 8.45x$
18. $(2x - 7)(x + 1) \le 4 - x(1 - 2x)$
19. $\frac{1}{3} - \frac{L}{2} < L + \frac{3}{2}$
20. $\frac{x}{5} - 2 > \frac{2}{3}(x + 3)$
21. $-1 \le 2x + 1 \le 3$
22. $4 < 6R + 2 \le 16$
23. $-4 \le 1 - x < -1$
24. $0 \le 3 - 2x \le 6$
25. $2x < x - 1 \le 3x + 5$
26. $x + 19 \le 25 - x < 2x$
27. $2s - 3 < s - 5 < 3s - 3$
28. $0 < 1 - x \le 3$ or $-1 < 2x - 3 < 5$

In Exercises 29–38, solve the inequalities by displaying the solutions on a calculator. See Examples 9 and 10.

29. $3x - 2 < 8 - x$
30. $40(x - 2) > x + 60$
31. $\frac{1}{2}(x + 15) \ge 5 - 2x$
32. $\frac{1}{3}x - 2 \le \frac{1}{2}x + 1$
33. $0.1 < 0.5 - 0.2t < 0.9$
34. $-3 < 2 - \frac{s}{3} \le -1$
35. $x - 3 < 2x + 5 < 6x + 7$
36. $n - 3 < 2n + 4 \le 1 - n$
37. $-2(2.5^x) + 5 \ge 3$
38. $\ln(x - 3) \ge 1$

In Exercises 39–60, solve the given problems by setting up and solving appropriate inequalities. Graph each solution.

39. Determine the values of x that are in the domain of the function $f(x) = \sqrt{2x - 10}$.
40. Determine the values of x that are in the domain of the function $f(x) = 1/\sqrt{3 - 0.5x}$.
41. For what values of k are the roots of the equation $x^2 - kx + 9 = 0$ imaginary?
42. For what values of k are the roots of the equation $2x^2 + 3x + k = 0$ real and unequal?
43. For $-6 < x < 2$, find a and b if $a < 5 - x < b$.
44. For $8 > -x > -4$, find a and b such that $a < x + 1 < b$.
45. Insert the proper sign $(=, >, <)$ for the ? such that $|5 - (-2)| \ ? \ |-5 - |-2||$ is true.
46. Insert the proper sign $(=, >, <)$ for the ? such that $|-3 - |-7|| \ ? \ ||-3|-7|$ is true.
47. What range of annual interest I will give between \$240 and \$360 annual income from an investment of \$7500?
48. Parking at an airport costs \$3.00 for the first hour, or any part thereof, and \$2.50 for each additional hour, or any part thereof. What range of hours costs at least \$28 and no more than \$78?
49. A contractor is considering two similar jobs, each of which is estimated to take n hours to complete. One pays \$350 plus \$15 per hour, and the other pays \$25 per hour. For what values of n will the contractor make more at the second position?

50. In designing plastic pipe, if the inner radius r is increased by 5.00 cm, and the inner cross-sectional area is increased by between 125 cm^2 and 175 cm^2, what are the possible inner radii of the pipe?

51. The relation between the temperature in degrees Fahrenheit F and degrees Celsius C is $9C = 5(F - 32)$. What temperatures F correspond to temperatures between 10°C and 20°C?

52. The voltage drop V across a resistor is the product of the current i (in A) and the resistance R (in Ω). Find the possible voltage drops across a variable resistor R, if the minimum and maximum resistances are 1.6 kΩ and 3.6 kΩ, respectively, and the current is constant at 2.5 mA.

53. A rectangular PV (photovoltaic) solar panel is designed to be 1.42 m long and supply 130 W/m^2 of power. What must the width of the panel be in order to supply between 100 W and 150 W?

54. A beam is supported at each end, as shown in Fig. 17.16. Analyzing the forces leads to the equation $F_1 = 13 - 3d$. For what values of d is F_1 more than 6 N?

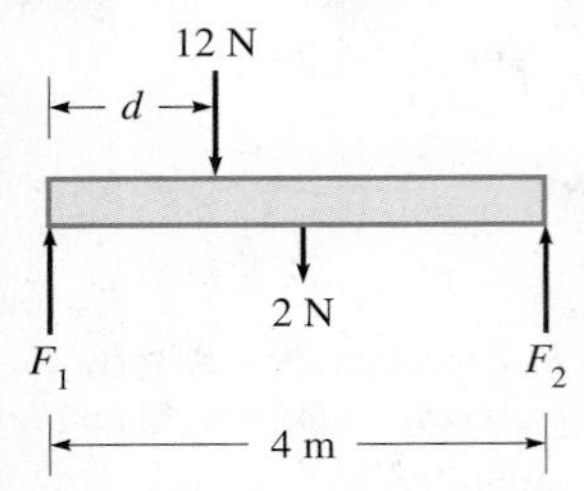

Fig. 17.16

55. The mass m (in g) of silver plate on a dish is increased by electroplating. The mass of silver on the plate is given by $m = 125 + 15.0t$, where t is the time (in h) of electroplating. For what values of t is m between 131 g and 164 g?

56. For a ground temperature of T_0 (in °C), the temperature T (in° C) at a height h (in m) above the ground is given approximately by $T = T_0 - 0.010h$. If the ground temperature is 25°C, for what heights is the temperature above 10°C?

57. During a given rush hour, the numbers of vehicles shown in Fig. 17.17 go in the indicated directions in a one-way-street section of a city. By finding the possible values of x and the equation relating x and y, find the possible values of y.

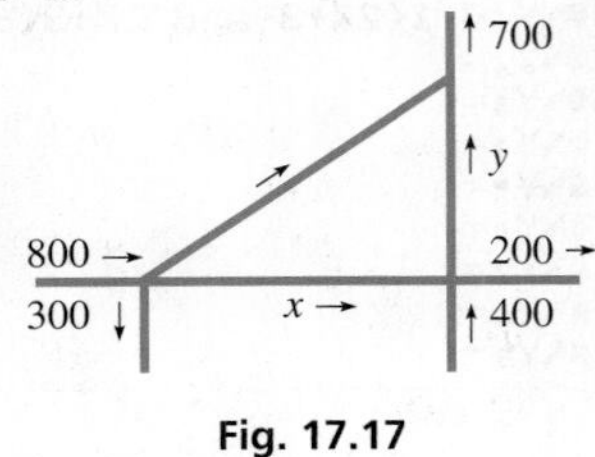

Fig. 17.17

58. The minimum legal speed on a certain interstate highway is 45 mi/h, and the maximum legal speed is 65 mi/h. What legal distances can a motorist travel in 4 h on this highway without stopping?

59. The route of a rapid transit train is 40 km long, and the train makes five stops of equal length. If the train is actually moving for 1 h and each stop must be at least 2 min, what are the lengths of the stops if the train maintains an average speed of at least 30 km/h, including stop times?

60. An oil company plans to install eight storage tanks, each with a capacity of x liters, and five additional tanks, each with a capacity of y liters, such that the total capacity of all tanks is 440,000 L. If capacity y will be at least 40,000 L, what are the possible values of capacity x?

Answers to Practice Exercises

1. $x < 2$ **2.** $x < 1/6$ **3.** $1/4 \le x < 2$ **4.** $40 \le y \le 120$ L/min

17.3 Solving Nonlinear Inequalities

Critical Values • Solving Polynomial and Rational Inequalities • Solving Inequalities Graphically

In this section, we develop methods of solving inequalities with polynomials, rational expressions (expressions involving fractions), and nonalgebraic expressions. To develop the basic method for solving these types of inequalities, we now take another look at a linear inequality.

In Fig. 17.18, we see that all values of the linear function $f(x) = ax + b$ $(a \neq 0)$ are positive on one side of the point at which $f(x) = 0$, and all values of $f(x)$ are negative on the opposite side of the same point. The means that *we can solve a linear inequality by expressing it with* ***zero*** *on the right and then finding the* ***sign*** *of the resulting function on either side of zero.*

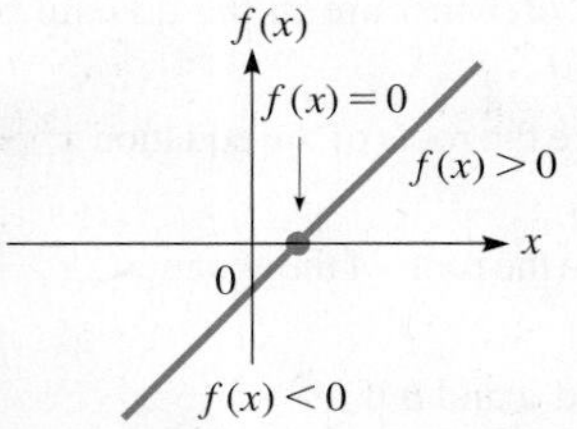

Fig. 17.18

EXAMPLE 1 Sign of function with zero on right

Solve the inequality $2x - 5 > 1$.

Finding the equivalent inequality with zero on the right, we have $2x - 6 > 0$. Setting the left member equal to zero, we have

$$2x - 6 = 0 \quad \text{for} \quad x = 3$$

which means $f(x) = 2x - 6$ has one sign for $x < 3$, and the other sign for $x > 3$. Testing values in these intervals, we find, for example, that

$$f(x) = -2 \quad \text{for} \quad x = 2 \qquad \text{and} \qquad f(x) = +2 \quad \text{for} \quad x = 4$$

Therefore, the solution to the original inequality is $x > 3$. The solution in Fig. 17.19(b) corresponds to the positive values of $f(x)$ in Fig. 17.19(a).

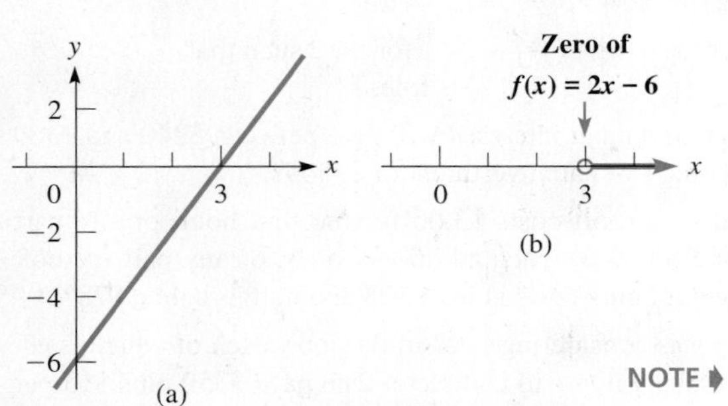

Fig. 17.19

NOTE ▸ [We could have solved this inequality by methods of the previous section, but the important idea here is to use the **sign** of the function, with zero on the right.] ■

We can extend this method to solving inequalities with polynomials of higher degree. We first find the equivalent inequality with zero on the right and then find the zeros of the function on the left (often by factoring and setting each factor equal to zero). The function can only change sign at these zeros. Thus, we can solve the inequality by determining the sign of the function on either side of each of the zeros.

This method is especially useful in solving rational inequalities (inequalities involving fractions). A fraction is zero if its numerator equals zero and is undefined if its denominator equals zero. *The values of x for which a function is zero or undefined are called the* **critical values** *of the function.* As all negative and positive values of x are considered (starting with negative numbers of large absolute value, proceeding through zero, and ending with large positive numbers), a function can change sign only at a critical value.

We now outline the method of solving an inequality by using the critical values of the function. Several examples of the method follow.

Using Critical Values to Solve a Polynomial or Rational Inequality

1. Determine the equivalent inequality with zero on the right.
2. Find the **critical values** of the function on the left side of the inequality. These are the values of x for which the function is zero or undefined.
3. Using **test values,** determine the sign of the function to the left of the leftmost critical value, between critical values, and to the right of the rightmost critical value.
4. Those intervals in which the function has the proper sign satisfy the inequality.

EXAMPLE 2 Solving a quadratic inequality

Solve the inequality $x^2 - 3 > 2x$.

We first find the equivalent inequality with zero on the right. Therefore, we have $x^2 - 2x - 3 > 0$. We then factor the left member and have

$$(x + 1)(x - 3) > 0$$

Setting each factor equal to zero, we find the critical values are -1 and 3. These critical values are the only possible places where the function can change in sign. Therefore, ***we must determine the sign of $f(x)$ for each of the intervals $x < -1$, $-1 < x < 3$, and $x > 3$.***

To do this, we will choose a ***test value*** from each interval and substitute it into the function to determine its sign. The following table shows each interval, the test value, the sign of each factor on each interval, and the resulting sign of the function.

Interval	*Test value*	$(x + 1)$	$(x - 3)$	*Sign of* $(x + 1)(x - 3)$
$x < -1$	-2	$-$	$-$	$+$
$-1 < x < 3$	0	$+$	$-$	$-$
$x > 3$	4	$+$	$+$	$+$

Because we want values for which the product is *greater than zero* (*or positive*), the solution is $x < -1$ or $x > 3$.

The solution that is shown in Fig. 17.20(b) corresponds to the positive values of the function $f(x) = x^2 - 2x - 3$, shown in Fig. 17.20(a). ■

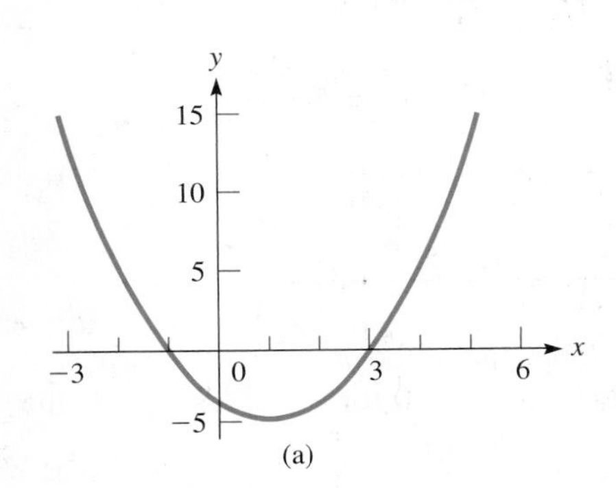

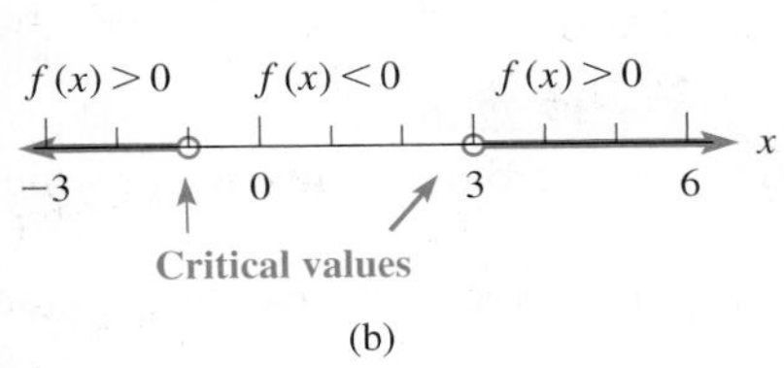

Fig. 17.20

EXAMPLE 3 Solving a cubic inequality

Solve the inequality $x^3 - 4x^2 + x + 6 < 0$.

By methods developed in Chapter 15, we factor the function on the left and obtain $(x + 1)(x - 2)(x - 3) < 0$. The critical values are $-1, 2, 3$. We wish to determine the sign of the left member for the intervals $x < -1$, $-1 < x < 2$, $2 < x < 3$, and $x > 3$. The following table shows this information.

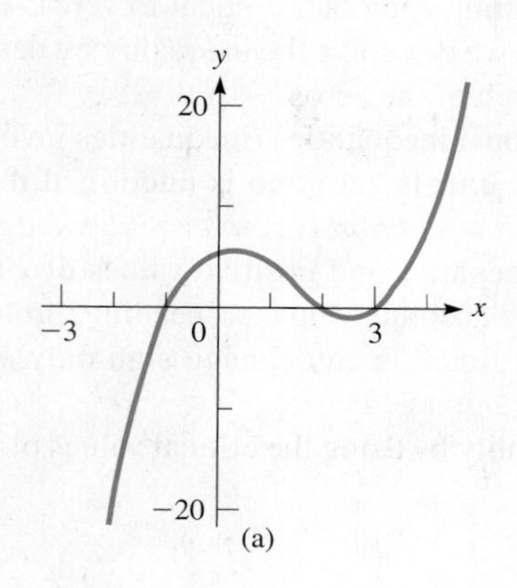

$$f(x) = (x + 1)(x - 2)(x - 3)$$

Interval	*Test value*	$(x + 1)$	$(x - 2)$	$(x - 3)$	*Sign of f(x)*
$x < -1$	-2	$-$	$-$	$-$	$-$
$-1 < x < 2$	0	$+$	$-$	$-$	$+$
$2 < x < 3$	2.5	$+$	$+$	$-$	$-$
$x > 3$	4	$+$	$+$	$+$	$+$

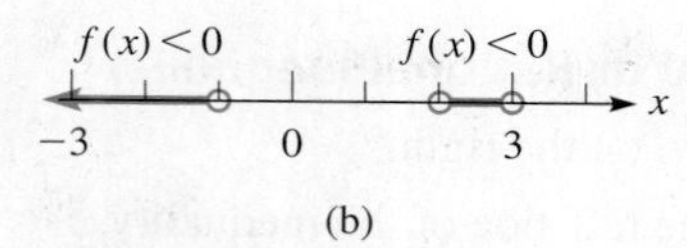

Fig. 17.21

Because we want values for $f(x) < 0$, the solution is $x < -1$ or $2 < x < 3$. The solution shown in Fig. 17.21(b) corresponds to the values of $f(x) < 0$ in Fig. 17.21(a). ■

The following example illustrates an applied situation that involves the solution of an inequality.

EXAMPLE 4 Solving a quadratic inequality—force on a cam

The force F (in N) acting on a cam varies according to the time t (in s), and it is given by the function $F = 2t^2 - 12t + 20$. For what values of t, $0 \leq t \leq 6$ s, is the force at least 4 N?

For a force of at least 4 N, we know that $F \geq 4$ N, or $2t^2 - 12t + 20 \geq 4$. This means we are to solve the inequality $2t^2 - 12t + 16 \geq 0$, and the solution is as follows:

$$2t^2 - 12t + 16 \geq 0$$
$$t^2 - 6t + 8 \geq 0$$
$$(t - 2)(t - 4) \geq 0$$

The critical values are $t = 2$ and $t = 4$, which leads to the following table:

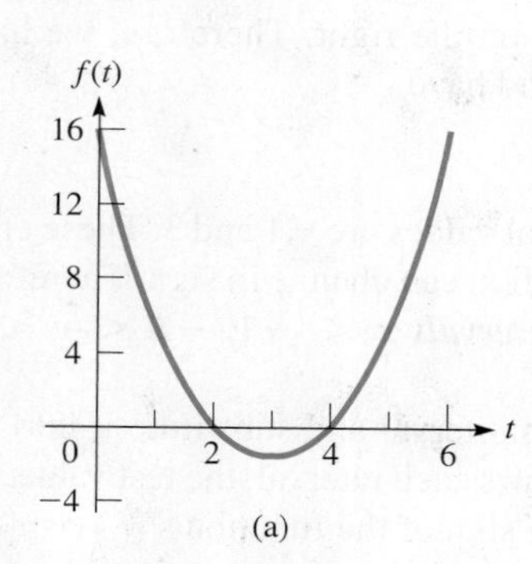

Interval	*Test value*	$(t - 2)$	$(t - 4)$	*Sign of* $(t - 2)(t - 4)$
$0 \leq t < 2$	1	$-$	$-$	$+$
$2 < t < 4$	3	$+$	$-$	$-$
$4 < t \leq 6$	5	$+$	$+$	$+$

We see that the values of t that satisfy the *greater than* part of the problem are $0 \leq t < 2$ and $4 < t \leq 6$. Because we know that $(t - 2)(t - 4) = 0$ for $t = 2$ and $t = 4$, the solution is

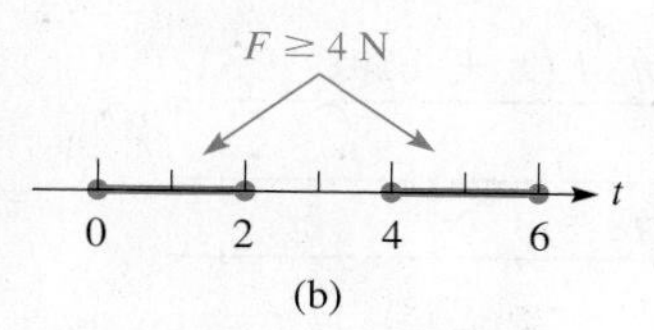

Fig. 17.22

$$0 \leq t \leq 2 \text{ s} \quad \text{or} \quad 4 \text{ s} \leq t \leq 6 \text{ s}$$

The graph of $f(t) = 2t^2 - 12t + 16$ is shown in Fig. 17.22(a). The solution, shown in Fig. 17.22(b), corresponds to the values of $f(t)$ that are zero or positive or for which $F \geq 4$ N.

We note here that if the cam rotates in 6-s intervals, the force on the cam is periodic, varying from 2 N to 20 N. ■

Practice Exercise

1. Solve the inequality: $x^2 - x - 42 < 0$

EXAMPLE 5 Solving a rational inequality

Solve the inequality $\dfrac{(x-2)^2(x+3)}{4-x} < 0$.

Since this function is a fraction, the critical values are the values of x that make either the numerator or denominator equal to zero. Thus, the critical values are -3, 2, and 4, and we have the following table:

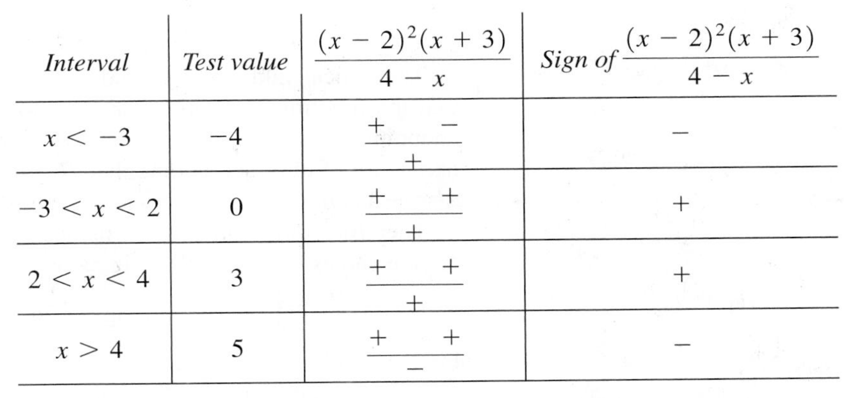

Interval	*Test value*	$\dfrac{(x-2)^2(x+3)}{4-x}$	*Sign of* $\dfrac{(x-2)^2(x+3)}{4-x}$
$x < -3$	-4	$\dfrac{+\quad -}{+}$	$-$
$-3 < x < 2$	0	$\dfrac{+\quad +}{+}$	$+$
$2 < x < 4$	3	$\dfrac{+\quad +}{+}$	$+$
$x > 4$	5	$\dfrac{+\quad +}{-}$	$-$

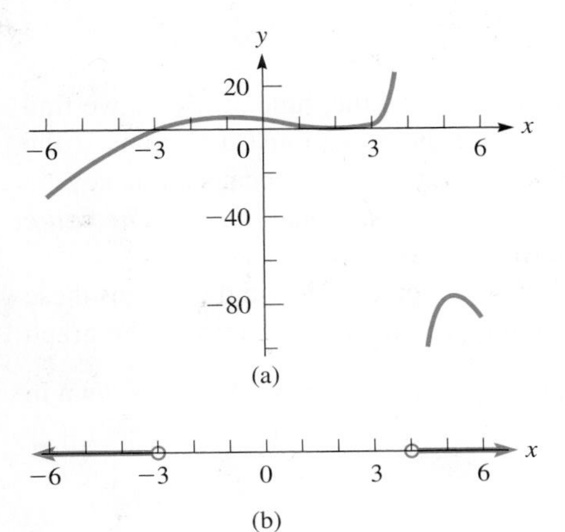

Fig. 17.23

Thus, the solution is $x < -3$ or $x > 4$. The graph of the function is shown in Fig. 17.23(a), and the graph of the solution is shown in Fig. 17.23(b). ■

EXAMPLE 6 Solving a rational inequality

Solve the inequality $\dfrac{x+3}{x-1} > 2$.

CAUTION It seems that all we have to do is multiply both members by $x - 1$ and solve the resulting linear inequality. However, this will lead to an incorrect result. The reason is that we assume $x - 1$ is positive when we multiply, but $x - 1$ is negative for $x < 1$. ■

To avoid this problem, *first subtract* 2 *from each member* and combine terms on the left. [This procedure should be used with any inequality involving fractions with the variable in the denominator.] Therefore, the solution is

NOTE ▸

$$\frac{x+3}{x-1} - 2 > 0 \qquad \text{subtract 2 from both members}$$

$$\frac{x+3-2(x-1)}{x-1} > 0 \qquad \text{combine over the common denominator}$$

$$\frac{5-x}{x-1} > 0 \qquad \text{this is the form to use}$$

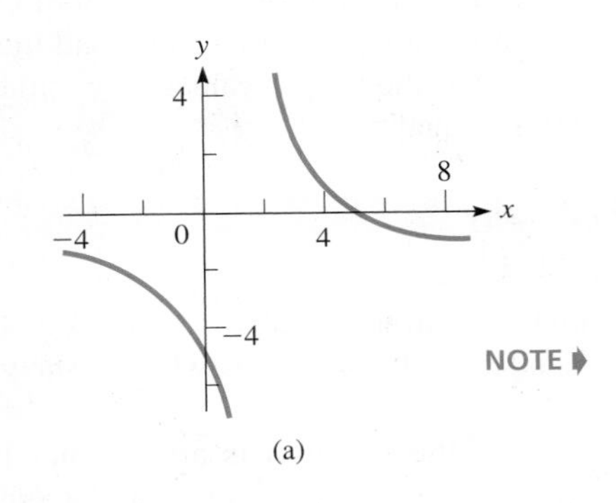

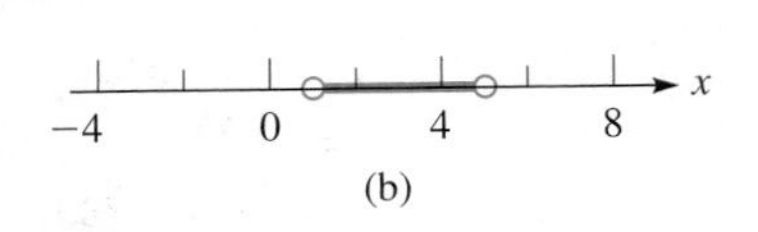

Fig. 17.24

The critical values are 1 and 5, and we have the following table of signs:

Interval	*Test value*	$(5-x)/(x-1)$	*Sign of* $(5-x)/(x-1)$
$x < 1$	0	$+\ /\ -$	$-$
$1 < x < 5$	3	$+\ /\ +$	$+$
$x > 5$	6	$-\ /\ +$	$-$

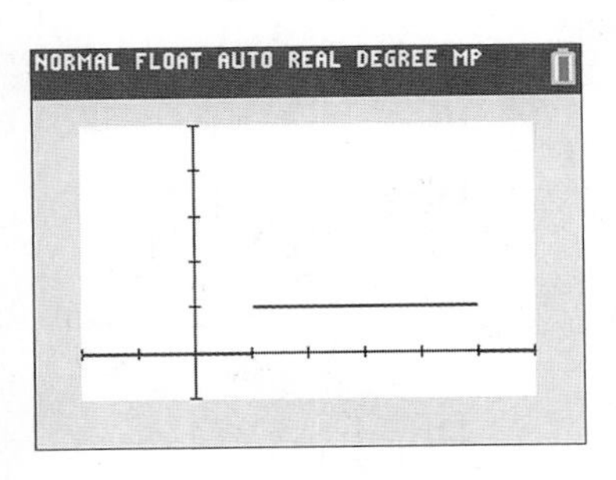

Fig. 17.25

Graphing calculator keystrokes: goo.gl/EICmmh

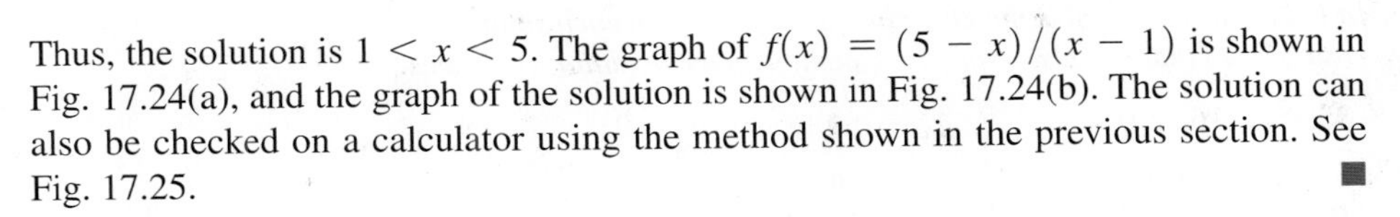

Thus, the solution is $1 < x < 5$. The graph of $f(x) = (5-x)/(x-1)$ is shown in Fig. 17.24(a), and the graph of the solution is shown in Fig. 17.24(b). The solution can also be checked on a calculator using the method shown in the previous section. See Fig. 17.25. ■

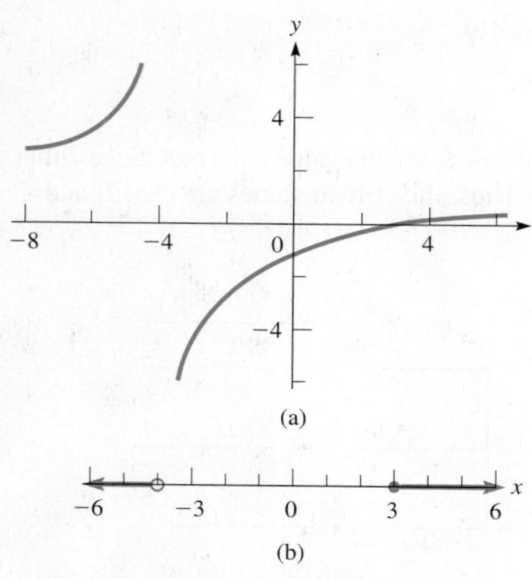

Fig. 17.26

Practice Exercise

2. Solve the inequality: $\frac{x+3}{2x-5} \geq 0$

EXAMPLE 7 Values representing real numbers

Find the values of x for which $\sqrt{\frac{x-3}{x+4}}$ represents a real number.

For the expression to represent a real number, the fraction under the radical must be greater than or equal to zero. This means we must solve the inequality.

$$\frac{x-3}{x+4} \geq 0$$

The critical values are -4 and 3. After testing each of the three intervals, we find that the function is positive for $x < -4$ or $x > 3$. Now we must determine if the endpoints -4 and 3 are solutions. The value $x = 3$ is a solution because it makes the numerator of the fraction (and therefore the whole fraction) equal to zero. ***The value $x = -4$ is not a solution since it causes division by zero.***

Therefore, the solution of the inequality is $x < -4$ or $x \geq 3$, and this means these are the values for which the original expression represents a real number. The graph of $f(x) = \frac{x-3}{x+4}$ is shown in Fig. 17.26(a), and the graph of the solution is shown in Fig. 17.26(b). ■

SOLVING INEQUALITIES GRAPHICALLY

We can get an approximate solution of an inequality from the graph of a function, including functions that are not factorable or not algebraic. The method follows several of the earlier examples. First, write the equivalent inequality with zero on the right and then graph this function. Those values of x corresponding to the proper values of y (either above or below the x-axis) are those that satisfy the inequality.

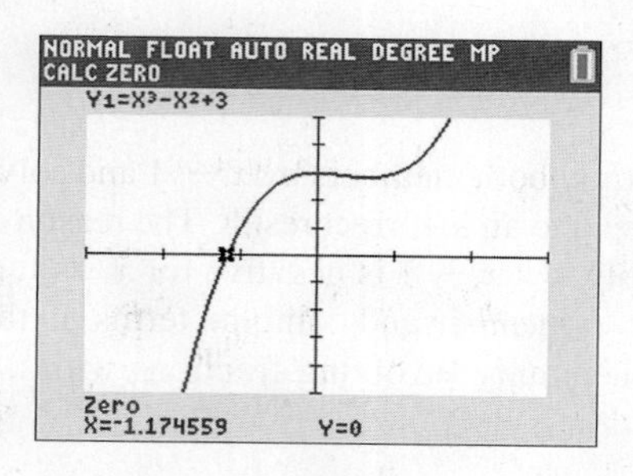

Fig. 17.27

Graphing calculator keystrokes: goo.gl/91Iu6l

EXAMPLE 8 Solving an inequality graphically

Use a calculator to solve the inequality $x^3 > x^2 - 3$.

Finding the equivalent inequality with zero on the right, we have $x^3 - x^2 + 3 > 0$. On the calculator, we then let $y_1 = x^3 - x^2 + 3$ and graph this function, which is shown in Fig. 17.27.

Using the *zero* feature, we find that the zero of the function is approximately $x = -1.17$. Because we want values of x that correspond to positive values of y, we see that the solution is $x > -1.17$. ■

EXERCISES 17.3

In Exercises 1 and 2, make the given changes in the indicated examples of this section and then solve the resulting inequalities.

1. In Example 2, change the − sign before the 3 to + and change $2x$ to $4x$, then solve the resulting inequality, and graph the solution.
2. In Example 5, change the exponent on $(x - 2)$ from 2 to 3, then solve the resulting inequality, and graph the solution.

In Exercises 3–22, solve the given inequalities. Graph each solution. It is suggested that you also graph the function on a calculator as a check.

3. $x^2 - 16 < 0$
4. $x^2 + 3x \geq 0$
5. $2x^2 \geq 4x$
6. $x^2 - 4x < 21$
7. $2x^2 - 12 \leq -5x$
8. $9t^2 + 6t > -1$
9. $x^2 + 4x \leq -4$
10. $6x^2 + 1 < 5x$
11. $R^2 + 4 > 0$
12. $2x^4 + 4 < 2$
13. $x^3 + x^2 - 2x < 0$
14. $3x^3 - 6x^2 + 3x \geq 0$
15. $s^3 + 2s^2 - s \geq 2$
16. $n^4 - 2n^3 + 8n + 12 \leq 7n^2$
17. $\frac{2x-3}{x+6} \leq 0$
18. $\frac{x+15}{x-9} > 0$
19. $\frac{x^2-6x-7}{x+5} > 0$
20. $\frac{(x-2)^2(5-x)}{(4-x)^3} \leq 0$
21. $\frac{x}{x+3} > 1$
22. $\frac{2p}{p-1} > 3$

In Exercises 23–30, solve the inequalities by displaying the solutions on a calculator. See Examples 9 and 10 in Section 17.2.

23. $3x^2 + 5x \geq 2$
24. $12x^2 + x < 1$
25. $\frac{T-8}{3-T} \leq 0$
26. $\frac{6x+2}{x+3} \geq 0$

27. $\dfrac{6-x}{3-x-4x^2} \geq 0$

28. $\dfrac{4-x}{3+2x-x^2} > 0$

29. $\dfrac{x^4(9-x)(x-5)(2-x)}{(4-x)^5} < 0$

30. $\dfrac{2}{x-3} < 4$

In Exercises 31–34, determine the values for x for which the radicals represent real numbers.

31. $\sqrt{(x-4)(x+1)}$

32. $\sqrt{x^2-3x}$

33. $\sqrt{-2x-x^2}$

34. $\sqrt{\dfrac{x^3+6x^2+8x}{3-x}}$

In Exercises 35–42, solve the given inequalities graphically by using a calculator. See Example 8. Round all decimals to the nearest hundredth.

35. $x^3 - x > 2$

36. $0.5x^3 < 3 - 2x^2$

37. $x^4 < x^2 - 2x - 1$

38. $3x^4 + x + 1 > 5x^2$

39. $2^x > x + 2$

40. $\log x < 1 - 2x^2$

41. $\sin x < 0.1x^2 - 1$

42. $4 \cos 2x > 2x - 3$

In Exercises 43–50, use inequalities to solve the given problems.

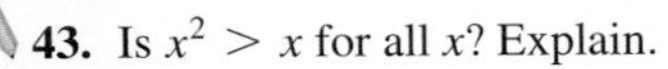

43. Is $x^2 > x$ for all x? Explain.

44. Is $x > 1/x$ for all x? Explain.

45. Find an inequality of the form $ax^2 + bx + c < 0$ with $a > 0$ for which the solution is $-1 < x < 4$.

46. Find an inequality of the form $ax^3 + bx < 0$ with $a > 0$ for which the solution is $x < -1$ or $0 < x < 1$.

47. Algebraically find the values of x for which $2^{x+2} > 3^{2x-3}$.

48. Graphically find the values of x for which $2 \log_2 x < \log_3(x+1)$.

49. For what values of real numbers a and b does the inequality $(x-a)(x-b) < 0$ have real solutions?

50. Algebraically find the intervals for which $f(x) = 2x^4 - 5x^3 + 3x^2$ is positive and those for which it is negative. Using only this information, draw a rough sketch of the graph of the function.

In Exercises 51–62, answer the given questions by solving the appropriate inequalities.

51. The power p (in W) used by a motor is given by $p = 9 + 5t - t^2$, where t is the time (in min). For what values of t is the power greater than 15 W?

52. The weight w (in tons) of fuel in a rocket after launch is $w = 2000 - t^2 - 140t$, where t is the time (in min). During what period of time is the weight of fuel greater than 500 tons?

53. The weekly sales S (in thousands of units) t weeks after a new smartwatch is released on the market is given by $S = \dfrac{100t}{t^2 + 100}$. When will the sales be 4000 units or more?

54. The object distance p (in cm) and image distance q (in cm) for a camera of focal length 3.00 cm is given by $p = 3.00q/(q - 3.00)$. For what values of q is $p > 12.0$ cm?

55. The total capacitance C of capacitors C_1 and C_2 in series is $C^{-1} = C_1^{-1} + C_2^{-1}$. If $C_2 = 4.00\ \mu\text{F}$, find C_1 if $C > 1.00\ \mu\text{F}$.

56. A rectangular field is to be enclosed by a fence and divided down the middle by another fence. The middle fence costs \$4/ft and the other fence cost \$8/ft. If the area of the field is to be 8000 ft^2, and the cost of the fence cannot exceed \$4000, what are the possible dimensions of the field?

57. The weight w (in N) of an object h meters above the surface of Earth is $w = r^2 w_0/(r+h)^2$, where r is the radius of Earth and w_0 is the weight of the object at sea level. Given that $r = 6380$ km, if an object weighs 200 N at sea level, for what altitudes is its weight less than 100 N?

58. The value V after two years of an amount A invested at an annual interest rate r is $V = A(1+r)^2$. If \$10,000 is invested in order that the value V is between \$11,000 and \$11,500, what rates of interest (to 0.1%) will provide this?

59. A triangular postage stamp is being designed such that the height h is 1.0 cm more than the base b. Find the possible height h such that the area of the stamp is at least 3.0 cm^2.

60. A laser source is 2.0 in. from the nearest point P on a flat mirror, and the laser beam is directed at a point Q that is on the mirror and is x in. from P. The beam is then reflected to the receiver, which is x in. from Q. What is x if the total length of the beam is greater than 6.5 in.? See Fig. 17.28.

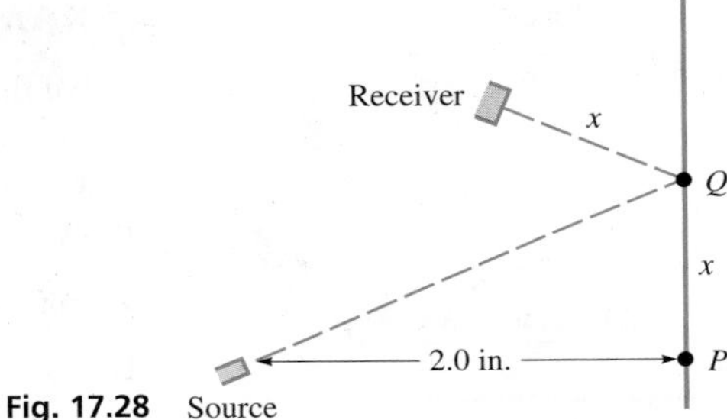

Fig. 17.28

61. A plane takes off from Winnipeg and flies due east at 620 km/h. At the same time, a second plane takes off from the surface of Lake Winnipeg 310 km due north of Winnipeg and flies due north at 560 km/h. For how many hours are the planes less than 1000 km apart?

62. An open box (no top) is formed from a piece of cardboard 8.00 in. square by cutting equal squares from the corners, turning up the resulting sides, and taping the edges together. Find the edges of the squares that are cut out in order that the volume of the box is greater than 32.0 in^3. See Fig. 17.29.

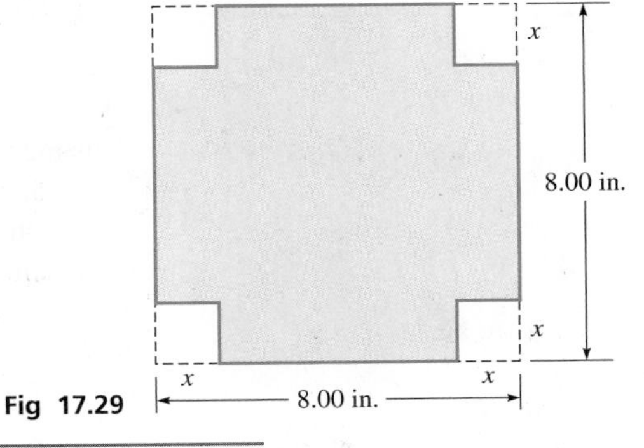

Fig 17.29

Answers to Practice Exercises

1. $-6 < x < 7$ **2.** $x \leq -3$ or $x > 5/2$

17.4 Inequalities Involving Absolute Values

Absolute Value Greater Than Given Value • Absolute Value Less Than Given Value

Inequalities involving absolute values are often useful in later topics in mathematics such as calculus and in applications such as the accuracy of measurements. In this section, we show the meaning of such inequalities and how they are solved.

If we wish to write the inequality $|x| > 1$ without absolute-value signs, we must note that we are considering values of x that are *numerically* larger than 1. Thus, we may write this inequality in the equivalent form $x < -1$ or $x > 1$. We now note that *the original inequality, with an absolute-value sign, can be written in terms of two equivalent inequalities, neither involving absolute values.* If we are asked to write the inequality $|x| < 1$ without absolute-value signs, we write $-1 < x < 1$, since we are considering values of x numerically less than 1.

Following reasoning similar to this, whenever absolute values are involved in inequalities, the following two relations allow us to write equivalent inequalities without absolute values. For $n > 0$,

■ It is sometimes necessary to isolate the absolute value before using Eqs. (17.1) and (17.2).

If $|f(x)| > n$, then $f(x) < -n$ or $f(x) > n$. **(17.1)**

If $|f(x)| < n$, then $-n < f(x) < n$. **(17.2)**

EXAMPLE 1 Absolute value less than

Solve the inequality $|x - 3| < 2$.

Here, we want values of x such that $x - 3$ is numerically smaller than 2, or the values of x within 2 units of $x = 3$. These are given by the inequality $1 < x < 5$. Now, using Eq. (17.2), we have

$$-2 < x - 3 < 2$$

By adding 3 to all three members of this inequality, we have

$$1 < x < 5$$

which is the proper interval. See Fig. 17.30. ■

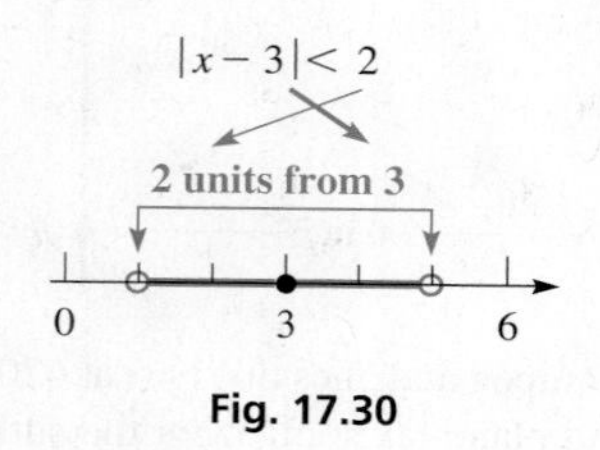

Fig. 17.30

EXAMPLE 2 Absolute value greater than

Solve the inequality $|2x - 1| > 5$.

By using Eq. (17.1), we have

$$2x - 1 < -5 \quad \text{or} \quad 2x - 1 > 5$$

Completing the solution, we have

$$2x < -4 \quad \text{or} \quad 2x > 6 \quad \text{add 1 to each member}$$

$$x < -2 \qquad\quad x > 3 \quad \text{divide each member by 2}$$

TI-89 graphing calculator keystrokes for Example 2: goo.gl/52M0nF

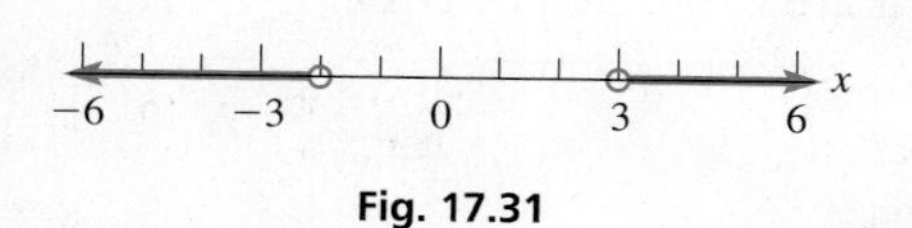

Fig. 17.31

NOTE ▶ This means that the given inequality is satisfied for $x < -2$ or for $x > 3$. [We must be very careful to remember that we cannot write this as $3 < x < -2$.] The solution is shown in Fig. 17.31.

The meaning of this inequality is that the numerical value of $2x - 1$ is greater than 5. By considering values in these intervals, we can see that this is true for values of x less than -2 or greater than 3. ■

EXAMPLE 3 Absolute value greater than or equal to

Solve the inequality $2\left|\frac{2x}{3}+1\right| \geq 4$.

The solution is as follows:

$$2\left|\frac{2x}{3}+1\right| \geq 4 \quad \text{original inequality}$$

$$\left|\frac{2x}{3}+1\right| \geq 2 \quad \text{divide each member by 2 to isolate the absolute value}$$

$$\frac{2x}{3}+1 \leq -2 \quad \text{or} \quad \frac{2x}{3}+1 \geq 2 \quad \text{using Eq. (17.1)}$$

$$2x+3 \leq -6 \qquad 2x+3 \geq 6$$

$$2x \leq -9 \qquad 2x \geq 3$$

$$x \leq -\frac{9}{2} \qquad x \geq \frac{3}{2} \quad \text{solution}$$

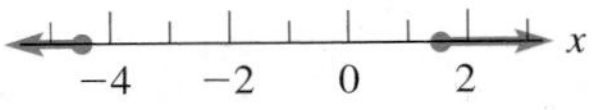

Fig. 17.32

This solution is shown in Fig. 17.32. Note that the sign of equality does not change the method of solution. It simply indicates that $-\frac{9}{2}$ and $\frac{3}{2}$ are included in the solution. ■

Practice Exercise

1. Solve the inequality: $|2x - 9| > 3$

EXAMPLE 4 Absolute value less than

Solve the inequality $|3 - 2x| < 3$.

We have the following solution.

$$|3-2x| < 3 \quad \text{original inequality}$$

$$-3 < 3-2x < 3 \quad \text{using Eq. (17.2)}$$

$$-6 < -2x < 0$$

$$3 > x > 0 \quad \text{divide by } -2 \text{ and reverse signs of inequality}$$

$$0 < x < 3 \quad \text{solution}$$

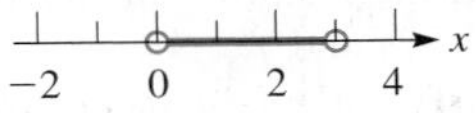

Fig. 17.33

The meaning of the inequality is that the numerical value of $3 - 2x$ is less than 3. This is true for values of x between 0 and 3. The solution is shown in Fig. 17.33. ■

Practice Exercise

2. Solve the inequality: $|4 - x| \leq 2$

EXAMPLE 5 Absolute value—fire hose pressure

The pressure from a fire truck pump is 120 lb/in.^{2} with a possible variation of $\pm$ 10 lb/in.^{2} Express this pressure p in terms of an inequality with absolute values.

This statement tells us that the pressure is no less than 110 lb/in.^{2} and no more than 130 lb/in.^{2}. Another way of stating this is that the numerical difference between the actual value of p (unknown exactly) and the measured value, 120 lb/in.^{2}, is less than or equal to 10 lb/in.^{2}. Using an absolute value inequality, this is written as

$$|p - 120| \leq 10$$

where values are in lb/in.^{2}.

We can see that this is the correct way of writing the inequality by using Eq. (17.2), which gives us

$$-10 \leq p - 120 \leq 10$$

$$110 \leq p \leq 130$$

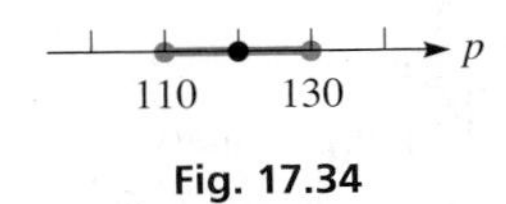

Fig. 17.34

This verifies that p should not be less than 110 lb/in.^{2} or more than 130 lb/in.^{2}. This solution is shown in Fig. 17.34. ■

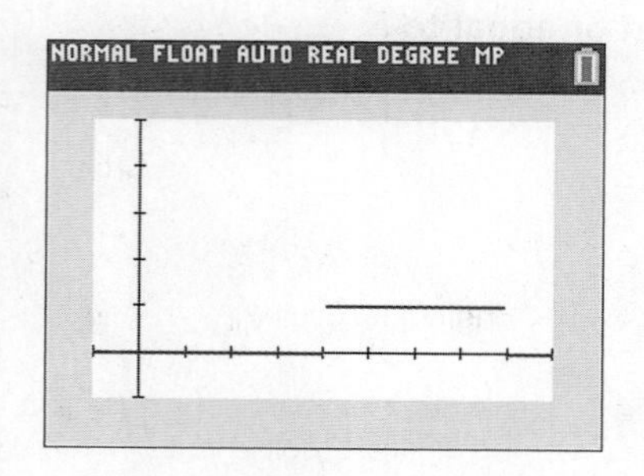

Fig. 17.35
Graphing calculator keystrokes: goo.gl/ES0dlj

Most calculators can be used to display the solution of an inequality that involves absolute values just as they can display the solutions to other inequalities. Because calculators differ in their operation, check the manual of your calculator to see how an absolute value is entered and displayed. The following example shows the display of the solution of an inequality on a graphing calculator.

EXAMPLE 6 Calculator solution of absolute-value inequality

Display the solution to the inequality $\left|\frac{x}{2} - 3\right| < 1$ on a calculator.

On the calculator, set $y_1 = \text{abs}(x/2 - 3) < 1$ and obtain the display shown in Fig. 17.35. From this display, we see that the solution is $4 < x < 8$. ■

EXERCISES 17.4

In Exercises 1 and 2, make the given changes in the indicated examples of this section and then solve the resulting inequalities.

1. In Example 2, change the $>$ to $<$, solve the resulting inequality, and graph the solution.
2. In Example 4, change the $<$ to $>$, solve the resulting inequality, and graph the solution.

In Exercises 3–24, solve the given inequalities. Graph each solution.

3. $|x - 4| < 1$
4. $|x + 4| < 6$
5. $|5x + 4| > 6$
6. $\left|\frac{1}{2}N - 1\right| > 3$
7. $1 + |6x - 5| \leq 5$
8. $|30 - 42x| \leq 0$
9. $|3 - 4x| > 3$
10. $3 + |3x + 1| \geq 5$
11. $\left|\frac{t + 1}{5}\right| < 5$
12. $\left|\frac{2x - 9}{4}\right| < 3$
13. $|20x + 85| \leq 46$
14. $|2.6x - 9.1| > 10.4$
15. $2|x - 24| > 84$
16. $3|4 - 3x| \leq 10$
17. $8 + 3|3 - 2x| < 11$
18. $5 - 4|1 - 7x| > 13$
19. $4|2 - 5x| \geq 6$
20. $2.5|7.1 - 2.0x| \leq 6.5$
21. $\left|\frac{3R}{5} + 1\right| < 8$
22. $\left|\frac{4x}{3} - 5\right| \geq 7$
23. $\left|6.5 - \frac{x}{2}\right| \geq 2.3$
24. $\left|\frac{2w + 5}{w + 1}\right| \geq 1$

In Exercises 25–28, solve the given inequalities by displaying the solutions on a calculator. See Example 6.

25. $|2x - 5| < 3$
26. $2|6 - T| > 5$
27. $\left|4 - \frac{x}{2}\right| \geq 1$
28. $\left|\frac{y}{3} + 12\right| - 3 \leq 22$

In Exercises 29–32, solve the given quadratic inequalities. Check each by displaying the solution on a calculator.

29. $|x^2 + x - 4| > 2$
    [After using Eq. (17.1), you will have two inequalities. The solution includes the values of x that satisfy *either* of the inequalities.]
30. $|x^2 + 3x - 1| > 3$ (See Exercise 29.)
31. $|x^2 + x - 4| < 2$
    [Use Eq. (17.2), and then treat the resulting inequality as two inequalities of the form $f(x) > -n$ and $f(x) < n$. The solution includes the values of x that satisfy *both* of the inequalities.]
32. $|x^2 + 3x - 1| < 3$ (See Exercise 31.)

In Exercises 33–40, solve the given problems.

33. Solve for x if $|x| < a$ and $a \leq 0$. Explain.
34. Solve for x if $|x - 1| < 4$ and $x \geq 0$.
35. Solve for x: $|x - 5| < 3$ and $|x - 7| < 2$.
36. Solve for x: $1 < |x - 2| < 3$.
37. If $|x - 1| < 4$, find a and b if $a < x + 4 < b$.
38. Solve for x if $|x - 1| > 4$ and $|x - 3| < 5$.
39. The thickness t (in km) of Earth's crust varies and can be described as $|t - 27| \leq 23$. What are the minimum and maximum values of the thickness of Earth's crust?
40. A motorist notes the gasoline gauge and estimates there are about 9 gal in the tank, but knows the estimate may be off by as much as 1 gal. This means we can write $|n - 9| \leq 1$, where n is the number of gallons in the tank. Using this inequality, what distance can the car go on this gas, if it gets 25 mi/gal?

In Exercises 41–48, use inequalities involving absolute values to solve the given problems.

41. The production p (in barrels) of oil at a refinery is estimated at $2{,}000{,}000 \pm 200{,}000$. Express p using an inequality with absolute values and describe the production in a verbal statement.

42. According to the Waze navigation app, the time required for a driver to reach his destination is 52 min. If this time is accurate to ± 3 min, express the travel time t using an inequality with absolute values.

43. The temperature T (in °C) at which a certain machine can operate properly is 70 ± 20. Express the temperature T for proper operation using an inequality with absolute values.

44. The *Mach number* M of a moving object is the ratio of its velocity v to the velocity of sound v_s, and v_s varies with temperature. A jet traveling at 1650 km/h changes its altitude from 500 m to 5500 m. At 500 m (with the temperature at 27° C), $v_s = 1250$ km/h, and at 5500 m (-3°C), $v_s = 1180$ km/h. Express the range of M, using an inequality with absolute values.

45. The diameter d of a certain type of tubing is 3.675 cm with a tolerance of 0.002 cm. Express this as an inequality with absolute values.

46. The velocity v (in ft/s) of a projectile launched upward from the ground is given by $v = -32t + 56$, where t is given in seconds. Given that speed $= |\text{velocity}|$, find the times at which the speed is greater than 8 ft/s.

47. The voltage v in a certain circuit is given by $v = 6.0 - 200i$, where i is the current (in A). For what values of the current is the absolute value of the voltage less than 2.0 V?

48. A rocket is fired from a plane flying horizontally at 9000 ft. The height h (in ft) of the rocket above the plane is given by $h = 560t - 16t^2$, where t is the time (in s) of flight of the rocket. When is the rocket more than 4000 ft above or below the plane? See Fig. 17.36.

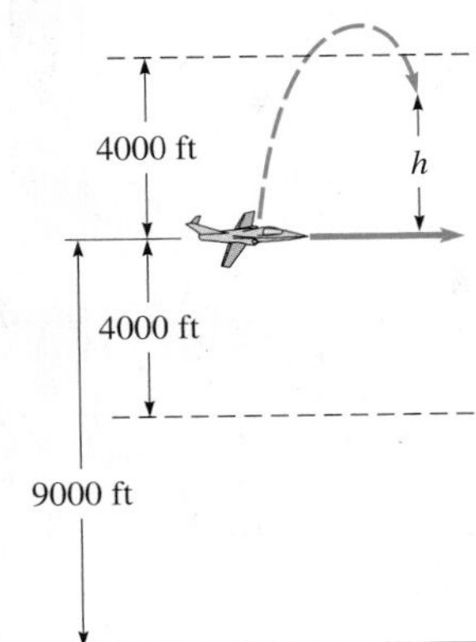

Fig. 17.36

Answers to Practice Exercises

1. $x < 3$ or $x > 6$ **2.** $2 \le x \le 6$

17.5 Graphical Solution of Inequalities with Two Variables

Points Above and Below Curve • Use of Dashed Curve or Solid Curve • Solution Using Calculator

To this point, we have considered inequalities with one variable and certain methods of solving them. We may also graphically solve inequalities involving two variables, such as x and y. In this section, we consider the solution of such inequalities.

Let us consider the function $y = f(x)$. We know that the coordinates of points on the graph satisfy the equation $y = f(x)$. However, for points above the graph of the function, we have $y > f(x)$, and for points below the graph of the function, we have $y < f(x)$. Consider the following example.

EXAMPLE 1 Checking points above and below line

Consider the linear function $y = 2x - 1$, the graph of which is shown in Fig. 17.37. This equation is satisfied for points on the line. For example, the point (2, 3) is on the line, and we have $3 = 2(2) - 1 = 3$. Therefore, for points on the line, we have $y = 2x - 1$, or $y - 2x + 1 = 0$.

The point (2, 4) is above the line, because we have $4 > 2(2) - 1$, or $4 > 3$. Therefore, for points above the line, we have $y > 2x - 1$, or $y - 2x + 1 > 0$. In the same way, for points below the line, $y < 2x - 1$ or $y - 2x + 1 < 0$. We note this is true for the point (2, 1), because $1 < 2(2) - 1$, or $1 < 3$.

The line for which $y = 2x - 1$ and the regions for which $y > 2x - 1$ and for which $y < 2x - 1$ are shown in Fig. 17.37.

Summarizing,

$y > 2x - 1$ for points *above* the line

$y = 2x - 1$ for points *on* the line

$y < 2x - 1$ for points *below* the line ■

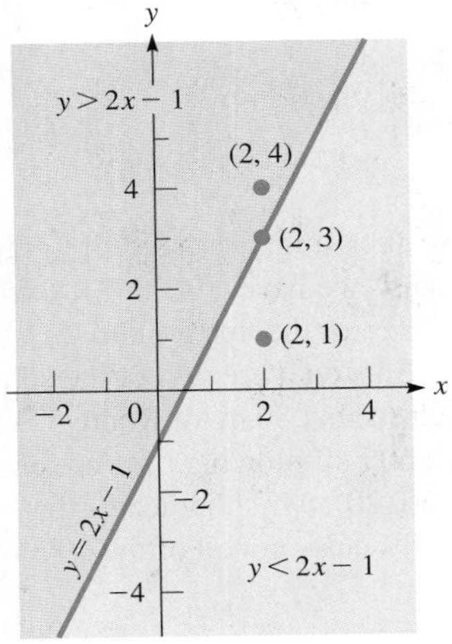

Fig. 17.37

The illustration in Example 1 leads us to the graphical method of indicating the points that satisfy an inequality with two variables. First, we solve the inequality for y and then determine the graph of the function $y = f(x)$. NOTE ▸ [If we wish to solve the inequality $y > f(x)$, we indicate the appropriate points by shading in the region *above* the curve. For the inequality $y < f(x)$, we indicate the appropriate points by shading in the region *below* the curve.] We note that the complete solution to the inequality consists of all points in an entire region of the plane.

EXAMPLE 2 Sketching $y < f(x)$ inequality

Draw a sketch of the graph of the inequality $y < x + 3$.

First, we draw the function $y = x + 3$, as shown by the dashed line in Fig. 17.38. Because we wish to find all the points that satisfy the inequality $y < x + 3$, we show these points by shading in the region below the line. The line is shown as a *dashed line* to indicate that points on it do not satisfy the inequality.

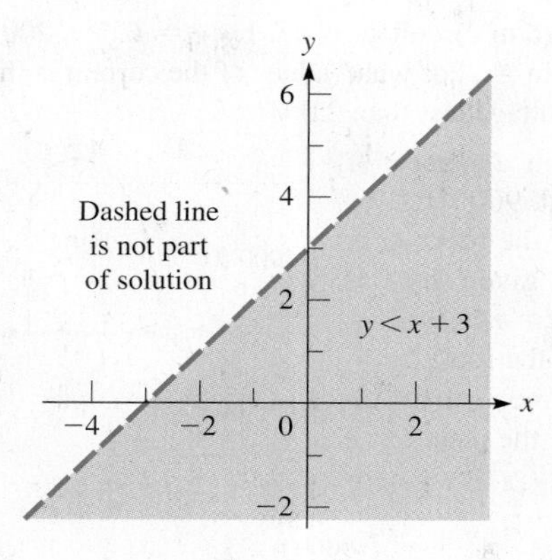

Fig. 17.38

Most calculators can be used to display the solution of an inequality involving two variables by shading a region above a curve, below a curve, or between curves. The manner in which this is done varies according to the model of the calculator. Therefore, the manual should be used to determine how this is done on any particular model of calculator. In Fig. 17.39, such a calculator display is shown for this inequality.

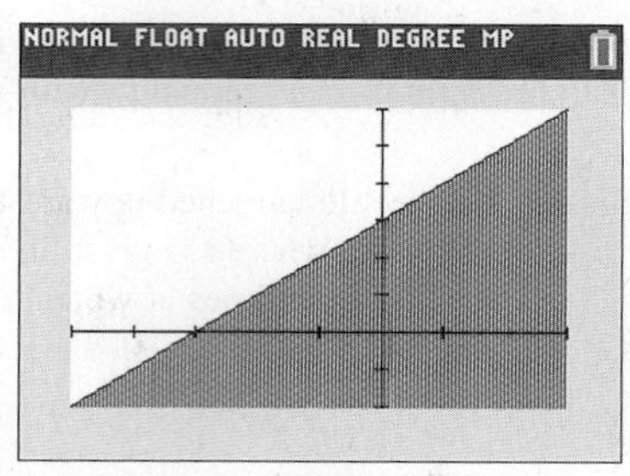

Fig. 17.39

Graphing calculator keystrokes: goo.gl/koaQzU ■

Practice Exercise

1. Does the point $(-2, 1)$ satisfy the inequality in Example 2?

EXAMPLE 3 Sketching $y \leq f(x)$ inequality—plowing distances

After a snowstorm in the Rochester, MN area, it is estimated it will take 30 min to plow each mile of Route 14, and 45 min to plow each mile of Route 52. If no more than 60 plowing-hours are available, what combinations of Routes 14 and 52 can be plowed?

Let $x =$ miles of Route 14 that can be plowed and $y =$ miles of Route 52 that can be plowed. The time to plow along each route is the product of the time for each mile and the number of miles to be plowed. This gives us

$$\underbrace{(0.50\text{ h/mi})(x\text{ mi})}_{\text{time to plow Rt. 14}} + \underbrace{(0.75\text{ h/mi})(y\text{ mi})}_{\text{time to plow Rt. 52}} \leq \underbrace{60\text{ h}}_{\text{max. available time}}$$

(30 min → 0.50 h/mi; 45 min → 0.75 h/mi)

$$0.50x + 0.75y \leq 60$$

$$y \leq 80 - 0.67x$$

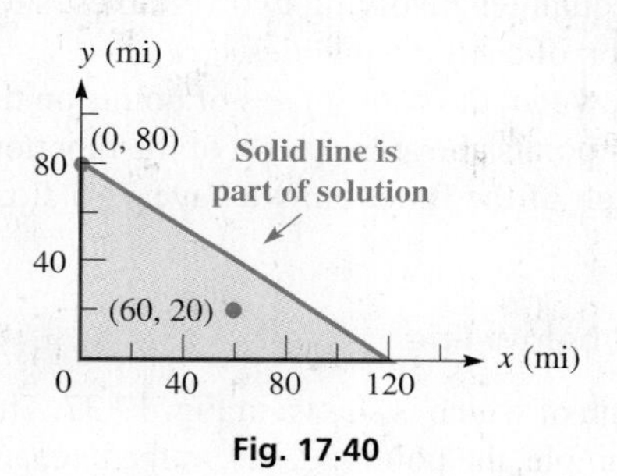

Fig. 17.40

Noting that negative values of x and y do not have meaning, we have the graph in Fig. 17.40, shading in the region below the line because we have $y < 80 - 0.67x$ for that region. Any point in the shaded region, or on the axes or the line around the shaded region, gives a solution. The *solid line* indicates that points on it are part of the solution.

The point (0, 80), for example, is a solution and tells us that 80 mi of Route 52 can be plowed if none of Route 14 is plowed. In this case, all 60 h of plowing time are used for Route 52. Another possibility is shown by the point (60, 20), which indicates that 60 mi of Route 14 and 20 mi of Route 52 can be plowed. In this case, not all of the 60 plowing hours are used. ■

Practice Exercise

2. In Example 3, is it possible to plow 80 mi of Route 14 and 30 mi of Route 52?

EXAMPLE 4 Draw a sketch of the graph of the inequality $y > x^2 - 4$

Although the graph of $y = x^2 - 4$ is not a straight line, the method of solution is the same. We graph the function $y = x^2 - 4$ as a dashed curve, since it is not part of the solution, as shown in Fig. 17.41. We then shade in the region above the curve to indicate the points that satisfy the inequality. ■

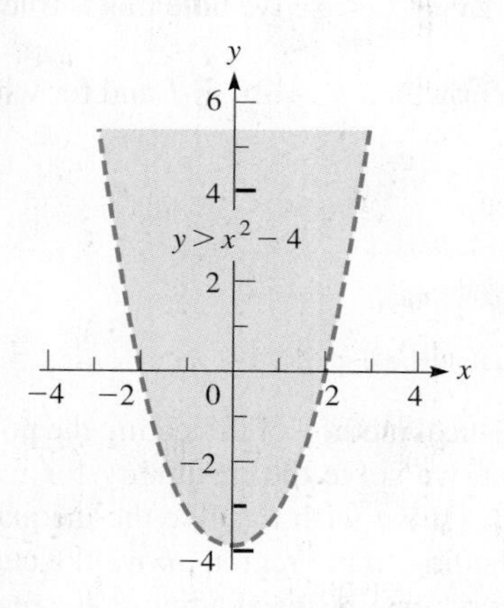

Fig. 17.41

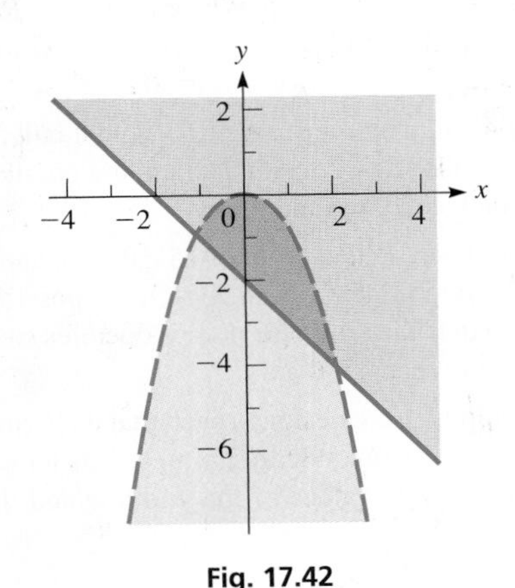

Fig. 17.42

■ Some calculators can show solid lines and dashed lines when graphing inequalities.

EXAMPLE 5 Solution of system of inequalities

Draw a sketch of the region that is defined by the system of inequalities $y \geq -x - 2$ and $y + x^2 < 0$.

Similar to the solution of a system of equations, *the solution of a system of inequalities is any pair of values (x, y) that satisfies both inequalities.* This means we want the region common to both inequalities. In Fig. 17.42, we first shade in the region above the line $y = -x - 2$ and then shade in the region below the parabola $y = -x^2$. The region defined by this system is the darkly shaded region below the parabola that is also above and on the line. The calculator display for this region is shown in Fig. 17.43.

If we are asked to find the region defined by $y \geq -x - 2$ *or* $y + x^2 < 0$, it consists of both shaded regions and all points on the line.

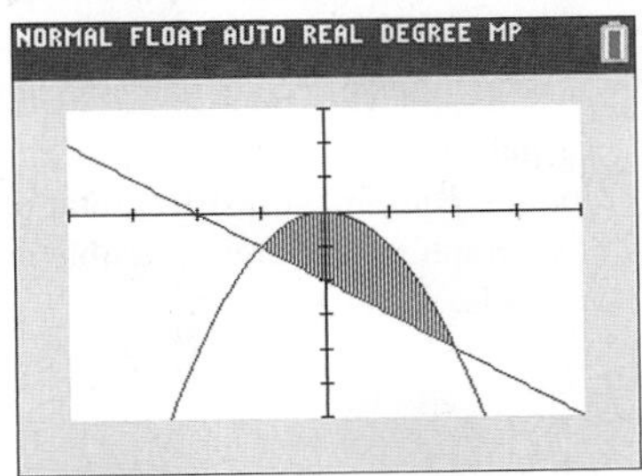

Fig. 17.43

Graphing calculator keystrokes: goo.gl/Tgv1gH ■

EXERCISES 17.5

In Exercises 1 and 2, make the given changes in the indicated examples of this section and then draw the graph of the resulting inequality.

1. In Example 2, change $x + 3$ to $3 - x$ and then draw the graph of the resulting inequality.
2. In Example 4, change $x^2 - 4$ to $4 - x^2$ and then draw the graph of the resulting inequality.

In Exercises 3–22, draw a sketch of the graph of the given inequality.

3. $y > x - 1$
4. $2y < 3x - 2$
5. $y \geq 2x + 5$
6. $y \leq 15 - 3x$
7. $3x + 2y + 6 > 0$
8. $x + 4y - 8 < 0$
9. $4y < x^2$
10. $y \leq 2x^2 - 3$
11. $2x^2 - 4x - y > 0$
12. $y \leq x^3 - 8$
13. $y < 32x - x^4$
14. $y \leq \sqrt{2x + 5}$
15. $y > \dfrac{10}{x^2 + 1}$
16. $y < \ln x$
17. $y > 1 + \sin 2x$
18. $y > |x| - 3$
19. $-20 < 5y \leq -10$
20. $y > |x + 3|$
21. $|x| < |y|$
22. $2|x - 3y| > 4$

In Exercises 23–34, draw a sketch of the graph of the region in which the points satisfy the given system of inequalities.

23. $y > x$, $y > 1 - x$
24. $y \leq 2x$, $y \geq x - 1$
25. $y \leq 2x^2$, $y > x - 2$
26. $y > x^2$, $y < x + 4$
27. $y > \frac{1}{2}x^2$, $y \leq 4x - x^2$
28. $y < 4 - x$, $y < \sqrt{16 - x^2}$
29. $y \geq 0$, $y \leq \sin x$, $0 \leq x \leq 3\pi$
30. $y > 0$, $y > 1 - x$, $y < e^x$
31. $|y + 2| < 5$, $|x - 3| \leq 2$
32. $16x + 3y - 12 > 0$, $y > x^2 - 2x - 3$, $|2x - 3| < 3$
33. $y > 0, x < 0, y \leq x$
34. $y \leq 0, x \geq 0, y \geq x$

In Exercises 35–44, use a calculator to display the solution of the given inequality or system of inequalities.

35. $2x + y < 5$
36. $4x - 2y > 1$
37. $y \geq 1 - x^2$
38. $y < |4 - 2x|$
39. $y > 2x - 1$, $y < x^4 - 8$
40. $y < 3 - x$, $y > 3x - x^3$
41. $y > x^2 + 2x - 8$, $y < \dfrac{1}{x} - 2$
42. $2y > -4x^2$, $y < 1 - e^{-x}$
43. $y \leq |2x - 3|$, $y > 1 - 2x^2$
44. $y \geq |4 - x^2|$, $y < 2 \ln|x|$

In Exercises 45–50, solve the given problems.

45. By an inequality, define the region below the line $9x - 3y + 12 = 0$.
46. By an inequality, define the region that is bounded by or includes the parabola $x^2 - 2y = 0$, and that contains the point (1, 0.4).
47. For $Ax + By < C$, if $B < 0$, would you shade above or below the line?

48. Find a system of inequalities that would describe the region within the triangle with vertices (0, 0) (0, 4), and (2, 0).

49. Draw a graph of the solution of the system $y \geq 2x^2 - 6$ and $y = x - 3$.

50. Draw a graph of the solution of the system $y < |x + 2|$ and $y = x^2$.

In Exercises 51–56, set up the necessary inequalities and sketch the graph of the region in which the points satisfy the indicated inequality or system of inequalities.

51. A telephone company is installing two types of fiber-optic cable in an area. It is estimated that no more than 300 m of type A cable, and at least 200 m but no more than 400 m of type B cable, are needed. Graph the possible lengths of cable that are needed.

52. A refinery can produce gasoline and diesel fuel, in amounts of any combination, except that equipment restricts total production to 2.5×10^5 barrels/day. Graph the different possible production combinations of the two fuels.

53. The elements of an electric circuit dissipate p watts of power. The power p_R dissipated by a resistor in the circuit is given by $p_R = Ri^2$, where R is the resistance (in Ω) and i is the current (in A). Graph the possible values of p and i for $p > p_R$ and $R = 0.5\ \Omega$.

54. The cross-sectional area A (in m^2) of a certain trapezoidal culvert in terms of its depth d (in m) is $A = 2d + d^2$. Graph the possible values of d and A if A is between 1 m^2 and 2 m^2.

55. One pump can remove wastewater at the rate of 75 gal/min, and a second pump works at the rate of 45 gal/min. Graph the possible values of the time (in min) that each of these pumps operates such that together they pump more than 4500 gal.

56. A rectangular computer chip is being designed such that its perimeter is no more than 15 mm, its width at least 2 mm, and its length at least 3 mm. Graph the possible values of the width w and the length l.

Answers to Practice Exercises

1. No **2.** No

17.6 Linear Programming

Constraints • Objective Function • Feasible Point • Vertices of Region of Feasible Points • Maximizing or Minimizing Objective Function

An important area in which graphs of inequalities with two or more variables are used is in the branch of mathematics known as **linear programming** (in this context, "programming" does not mean computer programming). This subject, which we mentioned in the chapter introduction, is widely applied in industry, business, economics, and technology. The analysis of many social problems can also be made by use of linear programming.

■ This section is an introduction to linear programming. Other methods are developed in a more complete coverage.

Linear programming is used to analyze problems such as those related to maximizing profits, minimizing costs, or the use of materials with certain constraints of production. First, we look at a similar type of mathematical problem.

EXAMPLE 1 Maximum value of F

Find the maximum value of F, where $F = 2x + 3y$ and x and y are subject to the conditions that

$$x \geq 0, y \geq 0$$
$$x + y \leq 6$$
$$x + 2y \leq 8$$

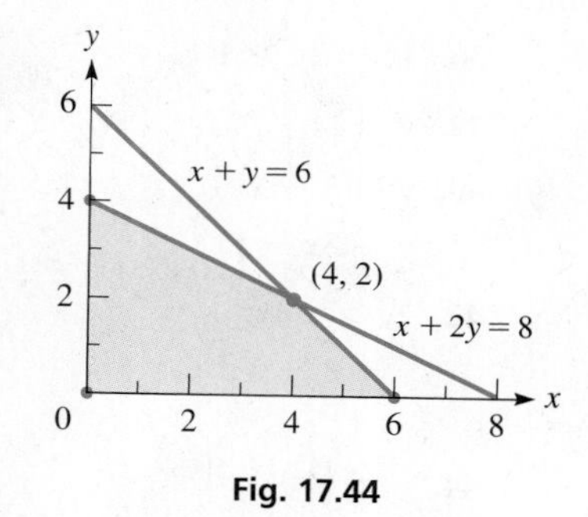

Fig. 17.44

These four inequalities that define the conditions on x and y are known as the **constraints** of the problem, and F is known as the **objective function.**

We now graph this set of inequalities, as shown in Fig. 17.44. Each point in the shaded region (including the line segments on the edges) satisfies all the constraints and is known as a **feasible point.**

The maximum value of F must be found at one of the feasible points. Testing for values at the vertices of the region, we have the values in the following table:

Point	(0, 0)	(6, 0)	(4, 2)	(0, 4)
Value of F	0	12	14	12

If we evaluate F at any other feasible point, we will find that $F < 14$. Therefore, the maximum value of F under the given constraints is 14. ■

Practice Exercise

1. In Example 1, what is the maximum value of F if the third constraint is changed to $x + y \leq 5$?

In Example 1, we found the maximum value of a *linear* objective function, subject to *linear* constraints (thus the name *linear programming*). We found the maximum value of F, subject to the given constraints, to be at one of the vertices of the region of feasible points. [In fact, in the theory of linear programming it is established that the maximum and minimum values of the objective function will always occur at a *vertex* of the region of feasible points] (or at all points along a line segment connecting two vertices).

NOTE ➧

■ The vertices of the region of feasible points are found by determining the points of intersection between the individual equations.

EXAMPLE 2 Maximum and minimum values of F

Find the maximum and minimum values of the objective function $F = 3x + y$, subject to the constraints

$$x \geq 2, y \geq 0$$
$$x + y \leq 8$$
$$2y - x \leq 1$$

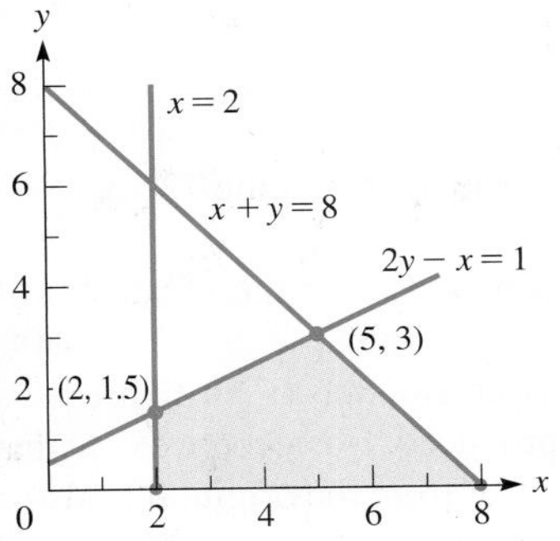

The constraints are graphed as shown in Fig. 17.45. We then locate the vertices, and evaluate F at these vertices as follows.

Vertex	(2, 0)	(8, 0)	(5, 3)	(2, 1.5)
Value of F	6	24	18	7.5

Fig. 17.45

Therefore, we see that the maximum value of F is 24, and the minimum value of F is 6. If we check the value of F at any other feasible point, we will find a value between 6 and 24. ■

The following two examples show the use of linear programming in finding a maximum value and a minimum value in applied situations.

EXAMPLE 3 Linear programming—speaker assembly

A company makes two types of stereo-speaker systems, their good-quality system and their highest-quality system. The production of these systems requires assembly of the speaker system itself and the production of the cabinets in which they are installed. The good-quality system requires 3 worker-hours for speaker assembly and 2 worker-hours for cabinet production for each complete system. The highest-quality system requires 4 worker-hours for speaker assembly and 6 worker-hours for cabinet production for each complete system. Available skilled labor allows for a maximum of 480 worker-hours per week for speaker assembly and a maximum of 540 worker-hours per week for cabinet production. It is anticipated that all systems will be sold and that the profit will be $30 for each good-quality system and $75 for each highest-quality system. How many of each system should be produced to provide the greatest profit?

■ The loudspeaker was developed in the early 1920s by U.S. inventors Edward Kellogg and Chester Rice.

First, let x = the number of good-quality systems and y = the number of highest-quality systems made in one week. Thus, the profit P is given by

$$P = 30x + 75y$$

We know that negative numbers are not valid for either x or y, and therefore we have $x \geq 0$ and $y \geq 0$.

The number of available worker-hours per week for each part of the production also restricts the number of systems that can be made. In the speaker-assembly shop, 3 worker-hours are needed for each good-quality system and 4 worker-hours are needed for each highest-quality system. The constraint that only 480 hours are available in the speaker-assembly shop means that $3x + 4y \leq 480$.

In the cabinet shop, it takes 2 worker-hours for each good-quality system and 6 worker-hours for each highest-quality system. The constraint that only 540 hours are available in the cabinet shop means that $2x + 6y \leq 540$.

Therefore, we want to maximize the profit $P = 30x + 75y$, which is the objective function, under the constraints

$x \geq 0, y \geq 0$	number of systems produced cannot be negative
$3x + 4y \leq 480$	worker-hours for speaker assembly
$2x + 6y \leq 540$	worker-hours for cabinet production

The constraints are graphed as shown in Fig. 17.46. We locate the vertices and evaluate the profit P at these points as follows:

Vertex	(0, 0)	(160, 0)	(72, 66)	(0, 90)
Profit ($)	0	4800	7110	6750

Therefore, we see that the greatest profit of $7110 is made by producing 72 good-quality systems and 66 highest-quality systems.

A way of showing the number of each system to be made for the greatest profit is to assume values of the profit P and graph these lines. For example, for $P = \$3000$ or $P = \$6000$, we have the lines shown in Fig. 17.46. Both amounts of profit are possible with various combinations of speaker systems being produced. However, we note that the line for $P = \$6000$ passes through feasible points farther from the origin. It is also clear that the lines for the profits are parallel and that *the greatest profit attainable is given by the line passing through A,* where $3x + 4y = 480$ and $2x + 6y = 540$ intersect. This also illustrates why the greatest profit is found at one of the vertices of the region of feasible points.

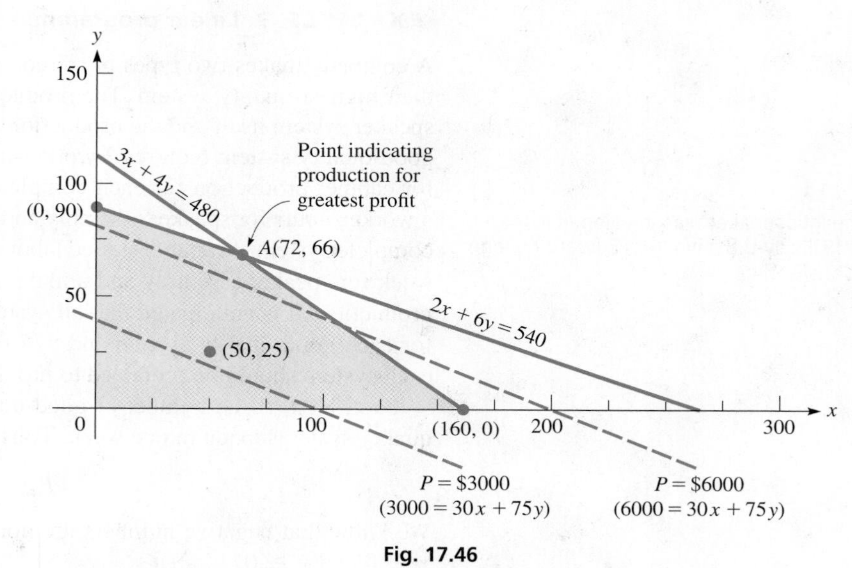

Fig. 17.46

Practice Exercise

2. Is it possible to make 100 good- and 50 highest-quality systems?

The problems in linear programming that arise in business and industry involve many more variables than in the simplified examples in this text. They are solved by computers using matrices, but the basic idea is the same as that shown in this section.

EXAMPLE 4 Linear programming—minimizing airline costs

■ See the chapter introduction.

An airline plans to open new routes and use two types of planes, A and B, on these routes. It is expected there would be at least 400 first-class passengers and 2000 economy-class passengers on these routes each day. Plane A costs $18,000/day to operate and has seats for 40 first-class and 80 economy-class passengers. Plane B costs $16,000/day to operate and has seats for 20 first-class and 160 economy-class passengers. How many of each type of plane should be used for these routes to minimize operating costs? (Assume that the planes will be used only on these routes.)

We first let $x =$ the number of A planes and $y =$ the number of B planes to be used. Then the daily operating cost $C = 18{,}000x + 16{,}000y$ is the objective function. The constraints are

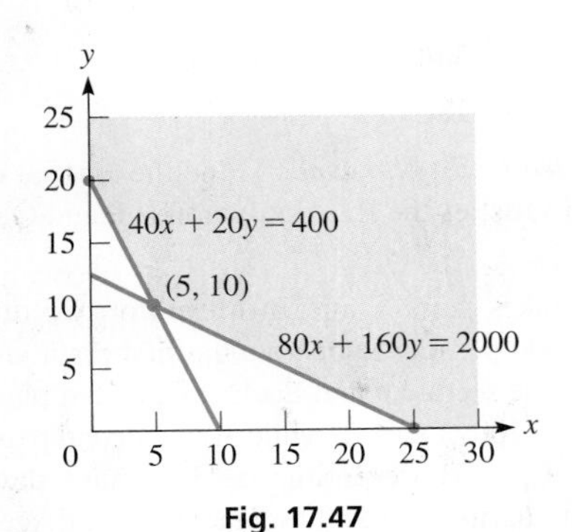

Fig. 17.47

$x \geq 0, y \geq 0$	number of planes cannot be negative
$40x + 20y \geq 400$	at least 400 first-class passengers
$80x + 160y \geq 2000$	at least 2000 economy-class passengers

The constraints are graphed as shown in Fig. 17.47. We see that the region of feasible points is unlimited (there could be more than 2400 passengers), but we still want to evaluate the operating cost C at the vertices. Evaluating C at these points, we have

Vertex	(25, 0)	(5, 10)	(0, 20)
Cost ($)	450,000	250,000	320,000

Therefore, five type-A planes and ten type-B planes should be used on these routes to keep the operating costs at a minimum of $250,000. Obviously, if the number of passengers does not meet the numbers expected, these numbers of planes should be changed accordingly. ■

EXERCISES 17.6

In Exercises 1 and 2, make the given changes in the indicated examples of this section and then find the indicated values.

1. In Example 1, change the last constraint to $2x + y \leq 8$. Then graph the feasible points and find the maximum value of F.
2. In Example 2, change the last constraint to $3y - x \leq 4$. Then graph the feasible points and find the maximum and minimum values of F.

In Exercises 3–16, find the indicated maximum and minimum values by the linear programming method of this section. For Exercises 5–16, the constraints are shown below the objective function.

3. Graphing the constraints of a linear programming problem shows the consecutive vertices of the region of feasible points to be (0, 0), (12, 0), (10, 7), (0, 5), and (0, 0). What are the maximum and minimum values of the objective function $F = 2x + 5y$ in this region?
4. Graphing the constraints of a linear programming problem shows the consecutive vertices of the region of feasible points to be (1, 3), (8, 0), (9, 7), (5, 8), (0, 6), and (1, 3). What are the maximum and minimum values of the objective function $F = 2x + 5y$ in this region?

5. Maximum P:
$P = 5x + 2y$
$x \geq 0, y \geq 0$
$2x + y \leq 6$

6. Maximum P:
$P = 2x + 7y$
$x \geq 1, y \geq 0$
$x + 4y \leq 8$

7. Maximum P:
$P = 5x + 9y$
$x \geq 0, y \geq 0$
$x + 2y \leq 6$

8. Minimum C:
$C = 10x + 20y$
$x \geq 0, y \geq 0$
$3x + 5y \geq 30$

9. Minimum C:
$C = 5x + 6y$
$x \geq 0, y \geq 0$
$x + y \geq 5$
$x + 2y \geq 7$

10. Minimum C:
$C = 6x + 4y$
$x \geq 0, y \geq 0$
$2x + y \geq 6$
$x + y \geq 5$

11. Maximum and minimum F:
$F = x + 3y$
$x \geq 1, y \geq 2$
$y - x \leq 3$
$y + 2x \leq 8$

12. Maximum and minimum F:
$F = 3x - y$
$x \geq 1, y \geq 0$
$x + 4y \leq 8$
$4x + y \leq 8$

13. Maximum P:
$P = 8x + 6y$
$x \geq 0, y \geq 0$
$2x + 5y \leq 10$
$4x + 2y \leq 12$

14. Minimum C:
$C = 3x + 8y$
$x \geq 0, y \geq 0$
$6x + y \geq 6$
$x + 4y \geq 4$

15. Minimum C:
$C = 6x + 4y$
$y \geq 2$
$x + y \leq 12$
$x + 2y \geq 12$
$2x + y \geq 12$

16. Maximum P:
$P = 3x + 4y$
$2x + y \geq 2$
$x + 2y \geq 2$
$x + y \leq 2$

In Exercises 17–22, solve the given linear programming problems.

17. A political candidate plans to spend no more than \$9000 on newspaper and radio advertising, with no more than twice being spent on newspaper ads at \$50 each than radio ads at \$150 each. It is assumed each newspaper ad is read by 8000 people, and each radio ad is heard by 6000 people. How many of each should be used to maximize the number of people who hear or see the message?

18. An oil refinery refines types A and B of crude oil and can refine as much as 4000 barrels each week. Type A crude has 2 kg of impurities per barrel, type B has 3 kg of impurities per barrel, and the refinery can handle no more than 9000 kg of these impurities each week. How much of each type should be refined in order to maximize profits, if the profit is \$4/barrel for type A and \$5/barrel for type B?

19. A manufacturer produces a business calculator and a graphing calculator. Each calculator is assembled in two sets of operations, where each operation is in production 8 h during each day. The average time required for a business calculator in the first operation is 3 min, and 6 min is required in the second operation. The graphing calculator averages 6 min in the first operation and 4 min in the second operation. All calculators can be sold; the profit for a business calculator is \$8, and the profit for a graphing calculator is \$10. How many of each type of calculator should be made each day in order to maximize profit?

20. Using the information given in Example 3, with the one change that the profit on each good-quality speaker system is \$60 (instead of \$30), how many of each system should be made? Explain why this one change in data makes such a change in the solution.

21. Brands A and B of breakfast cereal are both enriched with vitamins P and Q. The necessary information about these cereals is as follows:

	Cereal A	*Cereal B*	*RDA*
Vitamin P	1 unit/oz	2 units/oz	10 units
Vitamin Q	5 units/oz	3 units/oz	30 units
Cost	12¢/oz	18¢/oz	

(RDA is the *Recommended Daily Allowance*.) Find the amount of each cereal that together satisfies the RDA of vitamins P and Q at the lowest cost.

22. A computer company makes parts A and B in each of two different plants. It costs \$4000 per day to operate the first plant and \$5000 per day to operate the second plant. Each day the first plant produces 100 of part A and 200 of part B, while at the second plant 250 of part A and 100 of part B are produced. How many days should each plant operate to produce 2000 of each part and keep operating costs at a minimum?

Answers to Practice Exercise

1. 13 [at (2, 3)] **2.** No

CHAPTER 17 KEY FORMULAS AND EQUATIONS

If $|f(x)| > n$, then $f(x) < -n$ or $f(x) > n$. **(17.1)**

If $|f(x)| < n$, then $-n < f(x) < n$. **(17.2)**

CHAPTER 17 REVIEW EXERCISES

CONCEPT CHECK EXERCISES

Determine each of the following as being either **true** *or* **false**. *If it is false, explain why.*

1. $x < -3$ or $x > 1$ may also be written as $1 < x < -3$.

2. The solution of the inequality $-3x > 6$ is $x > -2$.

3. The solution of the inequality $x^2 - 2x - 8 > 0$ is $x < -2$ or $x > 4$.

4. The solution of the inequality $|x - 2| < 5$ is $x < -3$ or $x > 7$.

5. The graphical solution of the inequality $y \geq x + 1$ is shown as all points above the line $y = x + 1$.

6. The maximum value of the objective function $F = x + 2y$, subject to the conditions $x \geq 0$, $y \geq 0$, $2x + 3y \leq 6$, is 4.

PRACTICE AND APPLICATIONS

In Exercises 7–22, solve each of the given inequalities algebraically. Graph each solution.

7. $2x - 12 > 0$

8. $2.4(T - 4.0) \geq 5.5 - 2.4T$

9. $4 < 2x - 1 < 11$

10. $2x < x + 1 < 4x + 7$

11. $5x^2 + 9x < 2$

12. $x^2 + 2x > 63$

13. $6n^2 - n > 35$

14. $2x^3 + 4 \leq x^2 + 8x$

15. $\dfrac{(2x - 1)(3 - x)}{x + 4} > 0$

16. $x^4 + x^2 \leq 0$

17. $\dfrac{8}{x} < 2$

18. $\dfrac{1}{x - 2} > \dfrac{1}{4}$

19. $|3x + 2| \leq 4$

20. $|4 - 3x| \geq 7$

21. $|3 - 5x| > 7$

22. $3\left|2 - \dfrac{y}{2}\right| \leq 0$

In Exercises 23–30, solve the given inequalities on a calculator such that the display is the graph of the solution.

23. $5 - 3x > 0$

24. $6 \leq 2 - 4x < 9$

25. $2 \leq \dfrac{4n - 2}{3} < 3$

26. $3x < 2x + 1 < x - 5$

27. $\dfrac{8 - R}{2R + 1} \leq 0$

28. $\dfrac{(3 - x)^2}{2x + 7} \leq 0$

29. $|x - 30| > 48$

30. $2|2x - 9| < 8$

In Exercises 31–34, use a calculator to solve the given inequalities. Graph the appropriate function and from the graph determine the solution.

31. $x^3 + x + 1 < 0$

32. $\frac{4}{R+2} > 6$

33. $e^{-t} > 0.5$

34. $\sin 2x < 0.8 \quad (0 < x < 4)$

In Exercises 35–46, draw a sketch of the region in which the points satisfy the given inequality or system of inequalities.

35. $y > 12 - 3x$

36. $y < \frac{1}{2}x + 2$

37. $4y - 6x - 8 \leq 0$

38. $3y - x + 6 \geq 0$

39. $2x > 6 - y$

40. $y \leq \frac{6}{x^2 - 49}$

41. $y - |x + 1| < 0$

42. $2y + 2x^3 + 6x > 3$

43. $y > x + 1$
$y < 4 - x^2$

44. $y > 2x - x^2$
$y \geq -2$

45. $y \leq \frac{4}{x^2 + 1}$
$y < x - 3$

46. $y < \cos\frac{1}{2}x$
$y > \frac{1}{2}e^x$
$-\pi < x < \pi$

In Exercises 47–54, use a calculator to display the region in which the points satisfy the given inequality or system of inequalities.

47. $y < 3x + 5$

48. $x > 8 - 4y$

49. $y > 8 + 7x - x^2$

50. $y < x^3 + 4x^2 - x - 4$

51. $y < 32x - x^4$

52. $y > 2x - 1$
$y < 6 - 3x^2$

53. $y > 1 - x \sin 2x$
$y < 5 - x^2$

54. $y > |x - 1|$
$y < 4 + \ln x$

In Exercises 55–58, determine the values of x for which the given radicals represent real numbers.

55. $\sqrt{4 - x}$

56. $\sqrt{x + 5}$

57. $\sqrt{x^2 + 3x}$

58. $\sqrt{\frac{x - 1}{2x + 5}}$

In Exercises 59–62, find the indicated maximum and minimum values by the method of linear programming. The constraints are shown below the objective function.

59. Maximum P:
$P = 2x + 9y$
$x \geq 0, y \geq 0$
$x + 4y \leq 13$
$3y - x \leq 8$

60. Maximum P:
$P = x + 2y$
$x \geq 1, y \geq 0$
$3x + y \leq 6$
$2x + 3y \leq 8$

61. Minimum C:
$C = 3x + 4y$
$x \geq 0, y \geq 1$
$2x + 3y \geq 6$
$4x + 2y \geq 5$

62. Minimum C:
$C = 2x + 4y$
$x \geq 0, y \geq 0$
$x + 3y \geq 6$
$4x + 7y \geq 18$

In Exercises 63–90, solve the given problems using inequalities. (All data are accurate to at least two significant digits.)

63. Under what conditions is $|a + b| < |a| + |b|$?

64. Is $|a - b| < |a| + |b|$ always true? Explain.

65. For what values of x is the graph of $y = x^2 + 3x$ above the graph of $y = 2x + 6$?

66. Solve for x: $a + |bx| < c$ given that $a - c < 0$.

67. Form an inequality of the form $ax^2 + bx + c < 0$ with $a > 0$ for which the solution is $-3 < x < 5$.

68. By means of an inequality, define the region above the line $x - 3y - 6 = 0$.

69. Draw a graph of the system $y < 1 - x^2$ and $y = x^2$.

70. Find a system of inequalities that would describe the region within the quadrilateral with vertices (0, 0), (4, 4), (0, −3), and (4, −3).

71. Find the values for which $f(x) = (x - 2)(x - 3)$ is positive, zero, and negative. Use this information along with $f(0)$ and $f(5)$ to make a rough sketch of the graph of $f(x)$.

72. Follow the same instructions as in Exercise 71 for the function $f(x) = (x - 2)/(x - 3)$.

73. Describe the region that satisfies the system $y \geq 2x - 3, y < 2x - 3$.

74. Describe the region defined by $y \geq 2x - 3$ or $y < 2x - 3$.

75. If two adjacent sides of a square design on a TV screen expand 6.0 cm and 10.0 cm, respectively, how long is each side of the original square if the perimeter of the resulting rectangle is at least twice that of the original square? See Fig. 17.48.

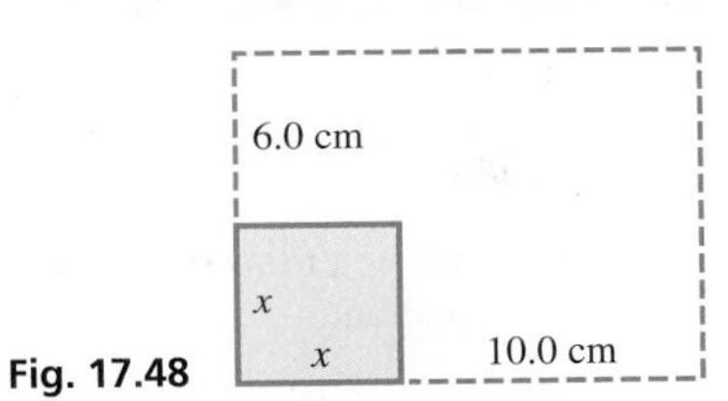

Fig. 17.48

76. The value V (in \$) of each building lot in a development is estimated as $V = 65{,}000 + 5000t$, where t is the time in years from now. For how long is the value of each lot no more than \$90,000?

77. The cost C of producing two of one type of calculator and five of a second type is \$50. If the cost of producing each of the second type is between \$5 and \$8, what are the possible costs of producing each of the first type?

78. City A is 600 km from city B. One car starts from A for B 1 h before a second car. The first car averages 60 km/h, and the second car averages 80 km/h for the trip. For what times after the first car starts is the second car ahead of the first car?

79. The pressure p (in kPa) at a depth d (in m) in the ocean is given by $p = 101 + 10.1d$. For what values of d is $p > 500$ kPa?

80. After conducting tests, it was determined that the stopping distance x (in ft) of a car traveling 60 mi/h was $|x - 290| \leq 35$. Express this inequality without absolute values and find the interval of stopping distances that were found in the tests.

81. A heating unit with 80% efficiency and a second unit with 90% efficiency deliver 360,000 Btu of heat to an office complex. If the first unit consumes an amount of fuel that contains no more than 261,000 Btu, what is the Btu content of the fuel consumed by the second unit?

82. A rectangular parking lot is to have a perimeter of 180 m and an area of at least 2000 m^2. What are the possible dimensions of the lot?

83. The electric power p (in W) dissipated in a resistor is given by $p = Ri^2$, where R is the resistance (in Ω) and i is the current (in A). For a given resistor, $R = 12.0\ \Omega$, and the power varies between 2.50 W and 8.00 W. Find the values of the current.

84. The reciprocal of the total resistance of two electric resistances in parallel equals the sum of the reciprocals of the resistances. If a 2.0-Ω resistance is in parallel with a resistance R, with a total resistance greater than 0.5 Ω, find R.

85. The efficiency e (in %) of a certain gasoline engine is given by $e = 100(1 - r^{-0.4})$, where r is the *compression ratio* for the engine. For what values of r is $e > 50\%$?

86. A rocket is fired such that its height h (in mi) is given by $h = 41t - t^2$. For what values of t (in min) is the height greater than 400 mi?

87. In developing a new product, a company estimates that it will take no more than 1200 min of computer time for research and no more than 1000 min of computer time for development. Graph the possible combinations of the computer times that are needed.

88. A natural gas supplier has a maximum of 120 worker-hours per week for delivery and for customer service. Graph the possible combinations of times available for these two services.

89. A company produces two types of cell phones, the regular model and the deluxe model. For each regular model produced, there is a profit of $8, and for each deluxe model the profit is $15. The same amount of materials is used to make each model, but the supply is sufficient only for 450 cell phones per day. The deluxe model requires twice the time to produce as the regular model. If only regular models were made, there would be time enough to produce 600 per day. Assuming all cell phones will be sold, how many of each model should be produced if the profit is to be a maximum?

90. A company that manufactures DVD/CD players gets two different parts, A and B, from two different suppliers. Each package of parts from the first supplier costs $2.00 and contains 6 of each type of part. Each package of parts from the second supplier costs $1.50 and contains 4 of A and 8 of B. How many packages should be bought from each supplier to keep the total cost to a minimum, if production requirements are 600 of A and 900 of B?

91. In planning a new city development, an engineer uses a rectangular coordinate system to locate points within the development. A park in the shape of a quadrilateral has corners at (0, 0), (0, 20), (40, 20), and (20, 40) (measurements in meters). Write two or three paragraphs explaining how to describe the park region with inequalities and find these inequalities.

CHAPTER 17 PRACTICE TEST

As a study aid, we have included complete solutions for each Practice Test problem at the end of this book.

1. State conditions on x and y in terms of inequalities if the point (x, y) is in the second quadrant.

In Problems 2–7, solve the given inequalities algebraically and graph each solution.

2. $\dfrac{-x}{2} \geq 3$

3. $3x + 1 < -5$

4. $-1 < 1 - 2x < 5$

5. $\dfrac{x^2 + x}{x - 2} \leq 0$

6. $|2x + 1| \geq 3$

7. $|2 - 3x| < 8$

8. Sketch the region in which the points satisfy the following system of inequalities:

$$y < x^2$$
$$y \geq x + 1$$

9. Determine the values of x for which $\sqrt{x^2 - x - 6}$ represents a real number.

10. The length of a rectangular lot is 20 m more than its width. If the area is to be at least 4800 m^2, what values may the width be?

11. Type A wire costs $0.10 per foot, and type B wire costs $0.20 per foot. Use a graph to show the possible combinations of lengths of wire that can be purchased for less than $5.00.

12. The range of the visible spectrum in terms of the wavelength λ of light ranges from about $\lambda = 400$ nm (violet) to about $\lambda = 700$ nm (red). Express these values using an inequality with absolute values.

13. Solve the inequality $x^2 > 12 - x$ on a graphing calculator such that the display is the graph of the solution.

14. By using linear programming, find the maximum value of the objective function $P = 5x + 3y$ subject to the following constraints: $x \geq 0$, $y \geq 0$, $2x + 3y \leq 12$, $4x + y \leq 8$.

Variation

18

As millions watched on television, men walked on the moon for the first time in 1969. Numerous discoveries in science, engineering, and mathematics helped make it possible for a spacecraft to travel to the moon. However, it can be argued that the beginning came with the formulation of the *universal law of gravitation* by Newton in the 1680s, which was based on earlier discoveries regarding the motion of the planets and the moon.

The universal law of gravitation is stated using the terminology of *variation,* the principal topic of this chapter. Using the language of variation, we state how one variable changes as other related variables change. There are many examples in various fields of study where the value of one variable is affected by changes in other variables.

We begin this chapter by reviewing the meanings of *ratio* and *proportion,* which were first introduced in Chapter 1. Then we see how ratio and proportion lead to variation and setting up relationships between variables. Depending on the situation, an increase in one variable can lead to either an increase or decrease in another variable. This leads to two types of variation: direct variation and inverse variation.

Applications of variation are found in all areas of technology. It is used in acoustics, biology, chemistry, computer technology, economics, electronics, environmental technology, hydrodynamics, mechanics, navigation, optics, physics, space technology, thermodynamics, and other fields.

LEARNING OUTCOMES

After completion of this chapter, the student should be able to:

- Solve application problems involving ratios and proportions
- Set up relationships between variables in terms of direct variation, inverse variation, or joint variation
- Solve application problems involving the different types of variation

◀ In Section 18.2, we show how a person's body mass index is related to their weight and height.

18.1 Ratio and Proportion

Ratio • Proportion

In order to develop the meaning of *variation,* we now review and expand our discussion of *ratio* and *proportion.* First, from Chapter 1, recall that *the quotient a/b is the* **ratio** *of a to b*. Therefore, a fraction is a ratio.

A measurement is the ratio of the measured magnitude to an accepted unit of measurement. For example, measuring the length of an object as 5 cm means it is five times as long as the accepted unit of length, the centimeter. Other examples of ratios are density (weight/volume), relative density (density of object/density of water), and pressure (force/area). Thus, ratios compare quantities of the same kind (for example, the trigonometric ratios) or express the division of magnitudes of different quantities (such a ratio is also called a **rate**).

EXAMPLE 1 Ratio—units of measurement

The approximate airline distance from Ottawa to Milwaukee is 600 mi, and the approximate airline distance from Ottawa to Toledo is 400 mi. The ratio of these distances is

$$\frac{600\text{ mi}}{400\text{ mi}} = \frac{3}{2}$$

■ The first jet-propelled airplane was flown in Germany in 1928.

Because both units are in miles, the resulting ratio is a dimensionless number.

If a jet travels from Ottawa to Milwaukee in 2 h, its average speed is

$$\frac{600\text{ mi}}{2\text{ h}} = 300\text{ mi/h}$$

In this case, we must attach the proper units to the ratio. ■

As we noted in Example 1, we must be careful to attach the proper units to the resulting ratio. Generally, the ratio of measurements of the same kind should be expressed as a dimensionless number. Consider the following example.

EXAMPLE 2 Ratio of measurements of same kind

The length of a certain room is 24 ft, and the width of the room is 18 ft. Therefore, the ratio of the length to the width is $\frac{24}{18}$, or $\frac{4}{3}$.

If the width of the room is expressed as 6 yd, we have the ratio $24\text{ ft}/6\text{ yd} = 4\text{ ft}/1\text{ yd}$. However, this does not clearly show the ratio. It is better and more meaningful first to change the units of one of the measurements to the units of the other measurement. Changing the length from 6 yd to 18 ft, we express the ratio as $\frac{4}{3}$, as we saw above. From this ratio, we can easily see that the length is $\frac{4}{3}$ as long as the width. ■

Dimensionless ratios are often used in definitions in mathematics and in technology. For example, the irrational number π is the dimensionless ratio of the circumference of a circle to its diameter. The specific gravity (or relative density) of a substance is the ratio of its density to the density of water. Other illustrations are found in the exercises for this section.

From Chapter 1, also recall that *an equation stating that two ratios are equal is called a* **proportion.** By this definition, a proportion is

$$\frac{a}{b} = \frac{c}{d} \tag{18.1}$$

Consider the following example.

EXAMPLE 3 Proportion—scale of map

On a certain map, 1 in. represents 31 mi, which means that we have a ratio of distances of 1 in./31 mi. Also on this map, Cooperstown, NY (home of the baseball Hall of Fame), is 11.5 in. from Canton, OH (home of the football Hall of Fame). Therefore, to find the distance from Cooperstown to Canton, we can set up the proportion

map distances

$$\frac{11.5 \text{ in.}}{x} = \frac{1 \text{ in.}}{31 \text{ mi.}}$$

land distances

$$(31x)\left(\frac{11.5}{x}\right) = (31x)\left(\frac{1}{31}\right) \quad \text{multiply each side by LCD} = 31x$$

$$356.5 = x \quad \text{or} \quad x = 360 \text{ mi} \quad \text{rounded}$$

NOTE ➧ [The ratio 1 in./31 mi is the *scale* of the map and has a special meaning, relating map distances in inches to land distances in miles. In a case like this, we should not change either unit to the other, even though they are both units of length.] ■

EXAMPLE 4 Proportion—change units of measurement

Given that 1 in. = 2.54 cm, what is the length in centimeters of the diagonal of a flat computer screen that is 17.0 in. long? See Fig. 18.1.

If we equate the ratio of known lengths to the ratio of the given length to the required length, we can find the required length by solving the resulting proportion (which is an equation). This gives us

Fig. 18.1

$$\frac{1 \text{ in.}}{2.54 \text{ cm}} = \frac{17.0 \text{ in.}}{x \text{ cm}}$$

$$x = (17.0)(2.54)$$

$$= 43.2 \text{ cm}$$

Therefore, the diagonal of the flat computer screen is 17.0 in., or 43.2 cm. ■

EXAMPLE 5 Proportion—magnitude of electric field

The magnitude of an electric field E is the ratio of the force F on a charge q to the magnitude of q. We can write this as $E = F/q$. If we know the force exerted on a particular charge at some point in the field, we can determine the force that would be exerted on another charge placed at the same point. For example, if we know that a force of 10 nN is exerted on a charge of 4.0 nC, we can then determine the force that would be exerted on a charge of 6.0 nC by the proportion

forces at point

$$\frac{10 \times 10^{-9}}{4.0 \times 10^{-9}} = \frac{F}{6.0 \times 10^{-9}}$$

charges at point

$$F = \frac{(6.0 \times 10^{-9})(10 \times 10^{-9})}{4.0 \times 10^{-9}}$$

$$= 15 \times 10^{-9} = 15 \text{ nN}$$ ■

Practice Exercise

1. In a certain electric field a force of 21 nN is exerted on a charge of 6.0 nC. At the same point, what is the force on a charge of 16 nC?

EXAMPLE 6 Proportion—metal alloy

An alloy is 5 parts tin and 3 parts lead. How many grams of each are in 40 g of the alloy?

First, let $x =$ the number of grams of tin in the 40 g of the alloy. Next, we note that there are 8 total parts of alloy, of which 5 are tin. Thus, 5 is to 8 as x is to 40. Therefore,

$$\frac{5}{8} = \frac{x}{40}$$

(parts tin → 5; total parts → 8; x ← grams of tin; 40 ← total grams)

$$x = 40\left(\frac{5}{8}\right) = 25 \text{ g}$$

There are 25 g of tin and 15 g of lead. The ratio 25/15 is the same as 5/3. ■

Practice Exercise

2. An alloy is 7 parts zinc and 5 parts lead. How much lead is there in 48 g of the alloy?

EXERCISES 18.1

In Exercises 1 and 2, make the given changes in the indicated examples of this section and then solve the indicated problem.

1. In Example 2, change 18 ft to 16 ft.
2. In Example 6, change 5 parts tin to 7 parts tin.

In Exercises 3–10, express the ratios in the simplest form.

3. 12 V to 3 V
4. 63 ft^2 to 18 ft^2
5. 96 h to 3 days
6. 120 s to 4 min
7. 20 qt to 2.5 gal
8. 25 cm^2 to 75 mm^2
9. 0.175 kg to 3500 mg
10. 2000 μm to 6 mm

In Exercises 11–26, find the required ratios.

11. The *P/E ratio (price-to-earnings ratio)* of a company's stock is the ratio of the price of the stock to the company's earnings (or profits) per share of stock, both measured in dollars. Find the P/E ratio of a stock that sells for \$17.59 and has an earnings of \$0.76 per share.
12. For a particular acute angle in a right triangle, the ratio of the opposite side to the hypotenuse is 12/13. Find the ratio of the adjacent side to the hypotenuse.
13. The *efficiency* of a power amplifier is defined as the ratio of the power output to the power input. Find the efficiency of an amplifier for which the power output is 2.6 W and the power input is 9.6 W.
14. A virus 3.0×10^{-5} cm long appears to be 1.2 cm long through a microscope. What is the *magnification* (ratio of image length to object length) of the microscope?
15. The *coefficient of friction* for two contacting surfaces is the ratio of the frictional force between them to the perpendicular force that presses them together. If it takes 450 N to overcome friction to move a 1.10-kN crate along the floor, what is the coefficient of friction between the crate and the floor? See Fig. 18.2.

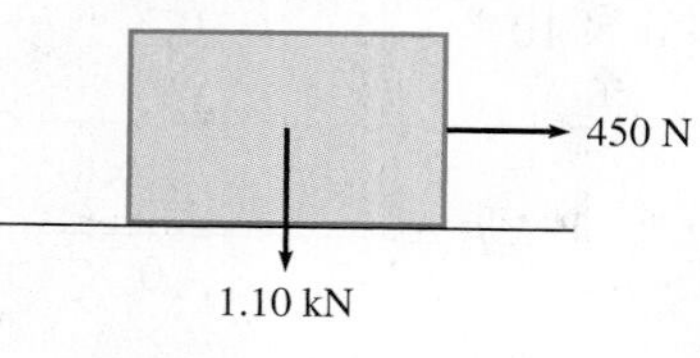

Fig. 18.2

16. The *Mach number* of a moving object is the ratio of its speed to the speed of sound (1200 km/h). Find the Mach number of a military jet that flew at 7200 km/h.
17. The *capacitance C* of a capacitor is defined as the ratio of its charge q (in C) to the voltage V. Find C (in F) for which $q = 5.00\ \mu$C and $V = 200$ V. (1 F = 1 C/1 V.)
18. The *atomic mass* of an atom of carbon is defined to be 12 u. The ratio of the atomic mass of an atom of oxygen to that of an atom of carbon is $\frac{4}{3}$. What is the atomic mass of an atom of oxygen? (The symbol u represents the *unified atomic mass unit,* where 1 u = 1.66×10^{-27} kg.)
19. An important design feature of an aircraft wing is its *aspect ratio.* It is defined as the ratio of the square of the span of the wing (wingtip to wingtip) to the total area of the wing. If the span of the wing for a certain aircraft is 32.0 ft and the area is 195 ft^2, find the aspect ratio.
20. For an automobile engine, the ratio of the cylinder volume to compressed volume is the *compression ratio.* If the cylinder volume of 820 cm^3 is compressed to 110 cm^3, find the compression ratio.
21. The *specific gravity* of a substance is the ratio of its density to the density of water. If the density of steel is 487 lb/ft^3 and that of water is 62.4 lb/ft^3, what is the specific gravity of steel?
22. The *percent grade* of a road is the ratio of vertical rise to the horizontal change in distance (expressed in percent). If a highway rises 75 m for each 1.2 km along the horizontal, what is the percent grade?
23. The *percent error* in a measurement is the ratio of the error in the measurement to the actual correct value, expressed as a percent. When writing a computer program, the memory remaining is determined as 2450 MB and then it is correctly found to be 2540 MB. What is the percent error in the first reading?
24. The electric *current* in a given circuit is the ratio of the voltage to the resistance. What is the current (1 V/1 Ω = 1 A) for a circuit where the voltage is 24.0 mV and the resistance is 10.0 Ω?
25. The *mass* of an object is the ratio of its weight to the acceleration g due to gravity. If a space probe weighs 8.46 kN on Earth, where $g = 9.80\ \text{m/s}^2$, find its mass in kg. (See Appendix B for the definition of a newton.)
26. *Power* is defined as the ratio of work done to the time required to do the work. If an engine performs 3.65 kJ of work in 15.0 s, find the power (in W) developed by the engine. (See Appendix B for the definition of a watt.)

In Exercises 27–30, find the required quantities from the given proportions.

27. According to Boyle's law, the relation $p_1/p_2 = V_2/V_1$ holds for pressures p_1 and p_2 and volumes V_1 and V_2 of a gas at constant temperature. Find V_1 if $p_1 = 36.6$ kPa, $p_2 = 84.4$ kPa, and $V_2 = 0.0447$ m^3.

28. For two connected gears, the relation

$$\frac{d_1}{d_2} = \frac{N_1}{N_2}$$

holds, where d is the diameter of the gear and N is the number of teeth. Find N_1 if $d_1 = 2.60$ in., $d_2 = 11.7$ in., and $N_2 = 45$. The ratio N_2/N_1 is called the *gear ratio*. See Fig. 18.3.

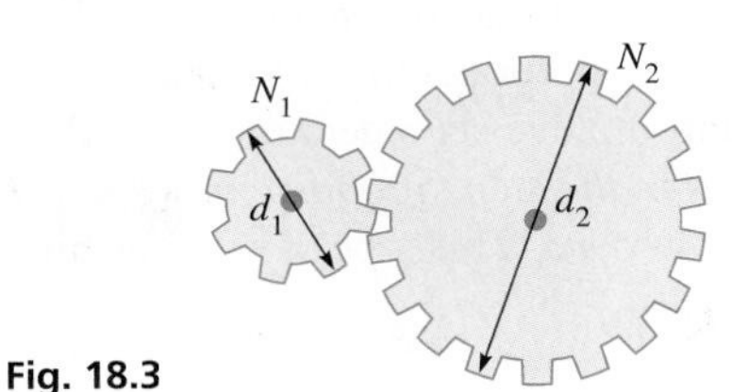

Fig. 18.3

29. In an electric instrument called a "Wheatstone bridge," electric resistances are related by

$$\frac{R_1}{R_2} = \frac{R_3}{R_4}$$

Find R_2 if
$R_1 = 6.00\ \Omega$,
$R_3 = 62.5\ \Omega$, and
$R_4 = 15.0\ \Omega$.
See Fig. 18.4.

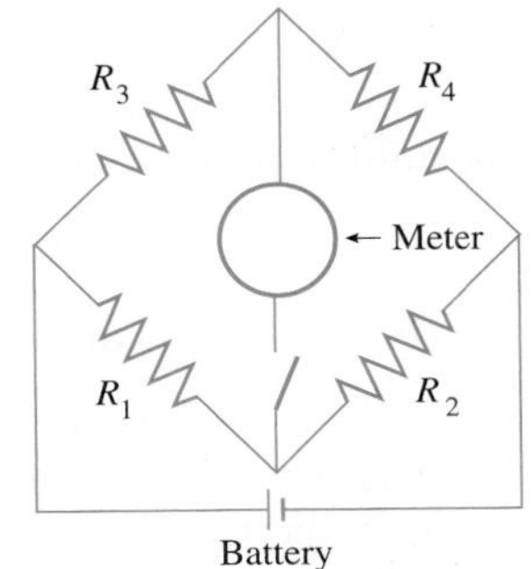

Fig. 18.4

30. In a transformer, an electric current in one coil of wire induces a current in a second coil. For a transformer,

$$\frac{i_1}{i_2} = \frac{t_2}{t_1}$$

where i is the current and t is the number of windings in each coil. In a neon sign amplifier, $i_1 = 1.2$ A and the *turns ratio* $t_2/t_1 = 160$. Find i_2. See Fig. 18.5.

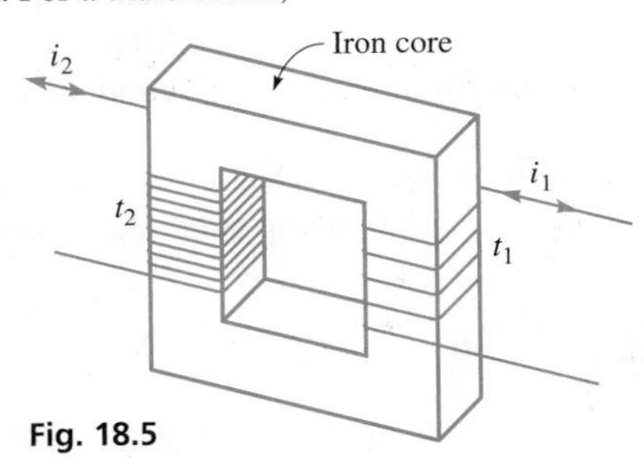

Fig. 18.5

In Exercises 31–52, answer the given questions by setting up and solving the appropriate proportions.

31. Given that 1.00 in.^{2} = 6.45 cm^2, what area in square inches is 36.3 cm^2?

32. Given that 1.000 kg = 2.205 lb, what mass in kilograms is equivalent to 175.5 lb?

33. Given that 1.00 hp = 746 W, what power in horsepower is 250 W?

34. Given that 1.50 L = 1.59 qt, what capacity in quarts is 2.75 L?

35. Given that 2.00 km = 1.24 mi, what distance in kilometers is 750 mi?

36. Given that 10^4 cm^2 = 10^6 mm^2, what area in square centimeters is 2.50×10^5 mm^2?

37. How many meters per second are equivalent to 45.0 km/h?

38. How many gallons per hour are equivalent to 540 mL/min?

39. A particular type of automobile engine produces 62,500 cm^3 of carbon monoxide in 2.00 min. How much carbon monoxide is produced in 45.0 s?

40. An airplane consumes 36.0 gal of gasoline in flying 425 mi. Under similar conditions, how far can it fly on 52.5 gal?

41. After a race, a runner's GPS watch shows the length of the race course to be 5058 m. If the actual length of the course is 4998 m, find the percent error in the watch's reading. See Exercise 23.

42. The weight of a person on Earth and the weight of the same person on Mars are proportional. If an astronaut weighs 920 N on Earth and 350 N on Mars, what is the weight of another astronaut on Mars if the astronaut weighs 640 N on Earth?

43. Two separate sections of a roof have the same slope. If the rise and run on one section are, respectively, 3.0 m and 6.3 m, what is the run on the other section if its rise is 4.2 m?

44. When a bullet is fired from a loosely held rifle, the ratio of the mass of the bullet to that of the rifle equals the negative of the reciprocal of the ratio of the velocity of the bullet to that of the rifle. If a 3.0 kg rifle fires a 5.0-g bullet and the velocity of the bullet is 300 m/s, what is the recoil velocity of the rifle?

45. A TV screen has an area of 1160 in.^{2}. A PIP (picture in picture) has an area that is one-third as large as the remaining area seen. What are the areas of the PIP and the remaining area on the screen?

46. If c/d is in *inverse ratio* to a/b, then $a/b = d/c$ (see Exercises 29 and 30). The current i (in A) in an electric circuit is in inverse ratio to the resistance R (in Ω). If $i = 0.25$ mA when $R = 2.8\ \Omega$, what is i when $R = 7.2\ \Omega$?

47. By weight, the ratio of chlorine to sodium in table salt is 35.46 to 23.00. How much sodium is contained in 50.00 kg of salt?

48. How much 100% fire-fighting foam concentrate must be added to 1000 L of a 2.0% solution to make a 4.0% solution?

49. In testing for quality control, it was found that 17 of every 500 computer chips produced by a company in a day were defective. If a total of 595 defective parts were found, what was the total number of chips produced during that day?

50. The TV revenue of $2.50 billion is divided in the ratio of 1.85 to 1.65 between the players and owners of a football league as a result of the contract between them. What does each receive?

51. The ratio of the width to height of an HDTV screen is 16 to 9, and for an older traditional screen the ratio is 4 to 3. For a 60.0 in. (diagonal) screen of each type, which has the greater area, and by what percent?

52. Of Earth's water area, the Pacific Ocean covers 46.0%, and the Atlantic Ocean covers 23.9%. Together they cover a total of 2.53×10^8 km^2. What is the area of each?

Answers to Practice Exercises

1. 56 nN **2.** 20 g

18.2 Variation

Direct Variation • Inverse Variation • Joint Variation • Calculating the Constant of Proportionality

Scientific laws are often stated in terms of ratios and proportions. For example, Charles' law can be stated as "for a perfect gas under constant pressure, the ratio of any two volumes this gas may occupy equals the ratio of the absolute temperatures." Symbolically, this could be stated as $V_1/V_2 = T_1/T_2$. Thus, if the ratio of the volumes and one of the values of the temperature are known, we can easily find the other temperature.

■ Named for the French physicist, Jacques Charles (1746–1823).

By multiplying both sides of the proportion of Charles' law by V_2/T_1, we can change the form of the proportion to $V_1/T_1 = V_2/T_2$. This statement says that the ratio of the volume to the temperature (for constant pressure) is constant. Thus, if any pair of values of volume and temperature is known, this ratio of V_1/T_1 can be calculated. This ratio of V_1/T_1 can be called a constant k, which means that Charles' law can be written as $V/T = k$. We now have the statement that the ratio of the volume to temperature is always constant; or, as it is normally stated, "The volume is proportional to the temperature." Therefore, we write $V = kT$, the clearest and most informative statement of Charles' law.

Thus, *for any two quantities always in the same proportion, we say that one is* **proportional to** (*or* **varies directly as**) *the second.* This type of relationship is called **direct variation.**

Direct Variation

We say that ***y* is proportional to *x*** (or ***y* varies directly as *x***) if

$$y = kx \tag{18.2}$$

where k is called the **constant of proportionality.**

EXAMPLE 1 Direct variation—applications

(a) The circumference c of a circle is proportional to (varies directly as) the radius r. We write this as $c = kr$. Because we know that $c = 2\pi r$ for a circle, we know that in this case the constant of proportionality is $k = 2\pi$.

(b) The fact that the volume V of paint varies directly as (is proportional to) the area A being painted is written as $V = kA$. If the area increases (or decreases), this equation tells us that the volume of paint increases (or decreases) proportionally. [In this case, the constant of proportionality is different for different types of paints or surfaces, although it remains constant for any given paint and surface being painted.] ■

NOTE ➧

It is very common that, when two quantities are related, the product of the two quantities remains constant. This type of relationship is called **inverse variation.**

Inverse Variation

We say that ***y* is inversely proportional to *x*** (or ***y* varies inversely as *x***) if

$$y = \frac{k}{x} \tag{18.3}$$

where k is the constant of proportionality.

EXAMPLE 2 Inverse variation—Boyle's law

Boyle's law states that "at a given temperature, the pressure p of an ideal gas varies inversely as the volume V." We write this as $p = k/V$. In this case, as the volume of the gas increases, the pressure decreases, or as the volume decreases, the pressure increases. ■

■ Named for the English physicist, Robert Boyle (1627–1691).

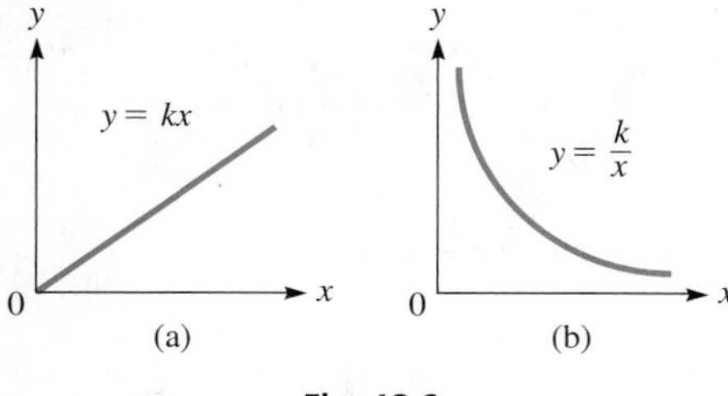

Fig. 18.6

In Fig. 18.6(a), the graph of the equation for direct variation $y = kx$, where $(x \geq 0)$, is shown. It is a straight line of slope k $(k > 0)$ and y-intercept of 0. We see that y increases as x increases, or y decreases as x decreases. In Fig. 18.6(b), the graph of the equation for inverse variation $y = k/x$ $(k > 0, x > 0)$ is shown. (It is a *hyperbola*.) As x increases, y decreases, or as x decreases, y increases.

For many relationships, one quantity varies as a specific power of another quantity. The terms *varies directly* and *varies inversely as* are used in the following example with a specific power of the independent variable.

EXAMPLE 3 Vary as specific power—applications

(a) The statement that the volume V of a sphere varies directly as the cube of its radius is written as $V = kr^3$. In this case, we know that $k = 4\pi/3$. We see that as the radius increases, the volume increases much more rapidly. For example, if $r = 2.00$ cm, $V = 33.5\text{ cm}^3$, and if $r = 3.00$ cm, $V = 113\text{ cm}^3$.

(b) A company finds that the number n of units of a product that are sold is inversely proportional to the square of the price p of the product. This is written as $n = k/p^2$. As the price is raised, the number of units that are sold decreases much more rapidly. ■

One quantity may vary as the product of two or more other quantities. Such variation is called **joint variation.**

Joint Variation

We say that **y varies jointly as x and** z if

$$y = kxz \tag{18.4}$$

where k is the constant of proportionality.

EXAMPLE 4 Joint variation—cost of sheet metal

The cost C of a piece of sheet metal varies jointly as the area A of the piece and the cost c per unit area. This is written as $C = kAc$. Here, C increases as the product Ac increases, and decreases as the product Ac decreases. ■

Direct, inverse, and joint variations may be combined. A given relationship may be a combination of two or all three of these types of variation.

EXAMPLE 5 Combined variation—law of gravitation

■ Formulated by the great English mathematician and physicist, Isaac Newton (1642–1727).

Newton's *universal law of gravitation* can be stated as follows: "The force F of gravitation between two objects varies jointly as the masses m_1 and m_2 of the objects, and inversely as the square of the distance r between their centers." We write this as

$$F = \frac{Gm_1m_2}{r^2}$$

← force varies jointly as masses and
← inversely as the square of the distance

where G is the constant of proportionality. ■

Practice Exercise

1. Express the relationship that y varies directly as the square of x and inversely as z.

CALCULATING THE CONSTANT OF PROPORTIONALITY

Once we have used the given statement to set up a general equation in terms of the variables and the constant of proportionality, we may calculate the value of the constant of proportionality if *one complete set of values* of the variables is known. *This value can then be substituted into the general equation to find the specific equation* relating the variables. We can then find the value of any one of the variables for any set of the others.

EXAMPLE 6 Find one value—given others

If y varies inversely as x, and $x = 15$ when $y = 4$, find the value of y when $x = 12$.

The solution is as follows:

$$y = \frac{k}{x} \quad \text{general equation showing inverse variation}$$

$$4 = \frac{k}{15}, \quad k = 60 \quad \text{evaluate constant of proportionality}$$

$$y = \frac{60}{x} \quad \text{specific equation relating } y \text{ and } x$$

$$y = \frac{60}{12} = 5 \quad \text{evaluating } y \text{ for } x = 12$$

■

Practice Exercise

2. If y varies directly as x, and $y = 5$ when $x = 20$, find the value of y when $x = 12$.

EXAMPLE 7 Vary as square root—frequency of wire vibration

The frequency f of vibration of a wire varies directly as the square root of the tension T of the wire. If $f = 420$ Hz when $T = 1.14$ N, find f when $T = 3.40$ N.

The steps in making this evaluation are outlined below:

$$f = k\sqrt{T} \quad \text{set up general equation: } f \text{ varies directly as } \sqrt{T}$$

$$420 \text{ Hz} = k\sqrt{1.14 \text{ N}} \quad \text{substitute given set of values and evaluate } k$$

$$k = 393 \text{ Hz/N}^{1/2}$$

$$f = 393\sqrt{T} \quad \text{substitute value of } k \text{ to get specific equation}$$

$$f = 393\sqrt{3.40} \quad \text{evaluate } f \text{ for } T = 3.40 \text{ N}$$

$$= 725 \text{ Hz}$$

We note that k has a set of units associated with it, and this usually will be the case in applied situations. As long as we do not change the units that are used for any of the variables, the units for the final variable that is evaluated will remain the same. ■

EXAMPLE 8 Joint variation—body mass index

■ See chapter introduction.

A person's body mass index (BMI) is directly proportional to their weight and inversely proportional to the square of their height. If weight and height both increase by 5%, by what factor does the BMI change?

The original BMI is given by

■ The constant of proportionality for BMI depends on the units chosen for weight and height. For kilograms and meters, $k = 1$. For pounds and inches, $k = 703$.

$$\text{BMI}_{\text{original}} = \frac{kw_0}{h_0^{\,2}}$$

where w_0 and h_0 represent the original weight and height, respectively. After the weight and height increase by 5%, their new values will be $1.05w_0$ and $1.05h_0$. Thus, the new BMI is

$$\text{BMI}_{\text{new}} = \frac{k(1.05w_0)}{(1.05h_0)^2} = \left(\frac{1.05}{1.05^2}\right)\frac{kw_0}{h_0^{\,2}} = 0.952(\text{BMI}_{\text{original}})$$

Therefore, the BMI changes by a factor of 0.952. This means the new BMI is 95.2% of the original BMI. ■

Practice Exercise

3. In Example 8, by what factor does the BMI change if weight and height both increase by 10%?

EXAMPLE 9 Joint variation—gravitational forces on a spacecraft

In Example 5, we stated Newton's universal law of gravitation. This law was formulated in the late seventeenth century, but it has numerous modern space-age applications. Use this law to solve the following problem.

■ See the chapter introduction.

A spacecraft is traveling from Earth to the moon, which are 240,000 mi apart. The mass of the moon is 0.0123 that of Earth. How far from Earth is the gravitational force of Earth on the spacecraft equal to the gravitational force of the moon on the spacecraft?

■ The first landing on the moon was by the crew of the U.S. spacecraft *Apollo 11* in July 1969.

From Example 5, we have the gravitational force between two objects as

$$F = \frac{Gm_1m_2}{r^2}$$

where the constant of proportionality G is the same for any two objects. Because we want the force between Earth and the spacecraft to equal the force between the moon and the spacecraft, we have

$$\frac{Gm_sm_e}{r^2} = \frac{Gm_sm_m}{(240{,}000 - r)^2}$$

where m_s, m_e, and m_m are the masses of the spacecraft, Earth, and the moon, respectively; r is the distance from Earth to the spacecraft; and $240{,}000 - r$ is the distance from the moon to the spacecraft. Because $m_m = 0.0123m_e$, we have

$$\frac{Gm_sm_e}{r^2} = \frac{Gm_s(0.0123m_e)}{(240{,}000 - r)^2}$$

$$\frac{1}{r^2} = \frac{0.0123}{(240{,}000 - r)^2} \quad \text{divide each side by } Gm_sm_e$$

$$(240{,}000 - r)^2 = 0.0123r^2 \quad \text{multiply each side by LCD}$$

$$240{,}000 - r = 0.111r \quad \text{take square roots}$$

$$1.111r = 240{,}000$$

$$r = 216{,}000 \text{ mi}$$

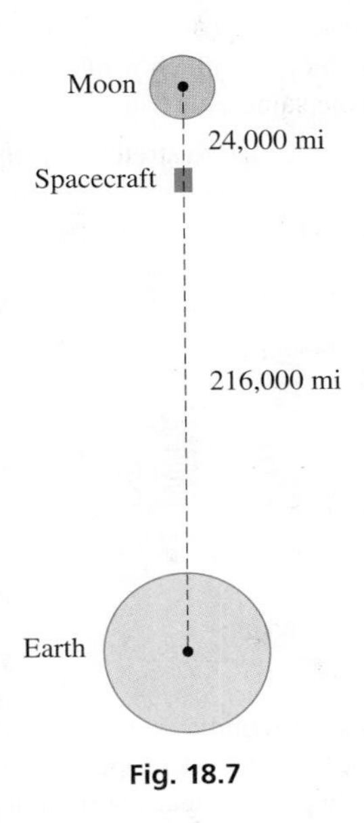

Fig. 18.7

Therefore, the spacecraft is 216,000 mi from Earth and 24,000 mi from the moon when the gravitational forces are equal. See Fig. 18.7. ■

Practice Exercise

4. What is the effect on the gravitational force F between two masses if the distance d is increased to $10d$?

EXERCISES 18.2

In Exercises 1–4, make the given changes in the indicated examples of this section and solve the indicated problems.

1. In Example 1(a), change radius r to diameter d, write the appropriate equation, and find the value of the constant of proportionality.
2. In Example 6, change "inversely as x" to "inversely as the square of x" and then solve the resulting problem.
3. In Example 7, change 1.14 N to 1.35 N and then solve the resulting problem.
4. In Example 8, change "weight and height both increase by 5%" to "weight increases by 10% and height increases by 5%" and then solve the resulting problem.

In Exercises 5–12, set up the general equations from the given statements.

5. The speed v at which a galaxy is moving away from Earth varies directly as its distance r from Earth.
6. The demand D for a product varies inversely as its price P.
7. The electric resistance R of a wire varies inversely as the square of its diameter d.
8. The volume V of silt carried by a river is proportional to the sixth power of the velocity v of the river.
9. In a tornado, the pressure P that a roof will withstand is inversely proportional to the square root of the area A of the roof.
10. During an adiabatic (no heat loss or gain) expansion of a gas, the pressure p is inversely proportional to the 3/2 power of the volume V.
11. The stiffness S of a beam varies jointly as its width w and the cube of its depth d.
12. The average electric power P entering a load varies jointly as the resistance R of the load and the square of the effective voltage V, and inversely as the square of the impedance Z.

In Exercises 13–16, express the meaning of the given equation in a verbal statement, using the language of variation. (k and π are constants.)

13. $A = \pi r^2$

14. $s = \frac{k}{\sqrt[3]{t}}$

15. $f = \frac{kL}{\sqrt{m}}$

16. $V = \pi r^2 h$

In Exercises 17–20, give the specific equation relating the variables after evaluating the constant of proportionality for the given set of values.

17. V varies directly as the square of H, and $V = 48$ when $H = 4$.

18. n is inversely proportional to the cube root of p, and $n = 4$ when $p = 27$.

19. p is proportional to q and inversely proportional to the cube of r, and $p = 6$ when $q = 3$ and $r = 2$.

20. v is proportional to t and the square root of s, and $v = 80$ when $s = 4$ and $t = 5$.

In Exercises 21–28, find the required value by setting up the general equation and then evaluating.

21. Find y when $x = 10$ if y varies directly as x, and $y = 200$ when $x = 16$.

22. Find y when $x = 5$ if y varies directly as the square of x, and $y = 6$ when $x = 8$.

23. Find s when $t = 8.50$ if s is inversely proportional to t, and $s = 820$ when $t = 0.200$.

24. Find p for $q = 0.8$ if p is inversely proportional to the square of q, and $p = 18$ when $q = 0.2$.

25. Find y for $x = 6$ and $z = 0.5$ if y varies directly as x and inversely as z, and $y = 60$ when $x = 4$ and $z = 10$.

26. Find r when $n = 160$ if r varies directly as the square root of n and $r = 4$ when $n = 250$.

27. Find f when $p = 2$ and $c = 4$ if f varies jointly as p and the cube of c, and $f = 8$ when $p = 4$ and $c = 0.1$.

28. Find v when $r = 2$, $s = 3$, and $t = 4$ if v varies jointly as r and s and inversely as the square of t, and $v = 8$ when $r = 8$, $s = 6$, and $t = 2$.

In Exercises 29 and 30, A varies directly as x, and B varies directly as x, although not in the same proportion as A. All numbers are positive.

29. Show that $A + B$ varies directly as x.

30. Show that $\sqrt{AB}$ varies directly as x.

In Exercises 31–64, solve the given applied problems involving variation.

31. The amount of hydroelectric power produced by a dam is proportional to the flow rate of the water. If a particular dam produces 35,800 kW of power when the flow rate is 2250 ft^3/s, how much power will be produced if the flow rate is 2150 ft^3/s?

32. The *blade tip speed* of a wind turbine is directly proportional to the rotation rate ω. For a certain wind turbine, the blade tip speed is 125 mi/h when $\omega = 12.0$ r/min. Find the blade tip speed when $\omega = 17.0$ r/min.

33. The amount of heat H required to melt ice is proportional to the mass m of ice that is melted. If it takes 2.93×10^5 J to melt 875 g of ice, how much heat is required to melt 625 g?

34. In electroplating, the mass m of the material deposited varies directly as the time t during which the electric current is on. Set up the equation for this relationship if 2.50 g are deposited in 5.25 h.

35. The velocity v of an Earth satellite varies directly as the square root of its mass m, and inversely as the square root of its distance r from the center of Earth. If the mass is halved and the distance is doubled, how is the speed affected?

36. The flow of water Q through a fire hose is proportional to the cross-sectional area A of the hose. If 250 gal flows through a hose of diameter 2.00 in. in a given time, how much would flow through a hose 3.00 in. in diameter in the same time?

37. Hooke's law states that the force needed to stretch a spring is proportional to the amount the spring is stretched. If 10.0 lb stretches a certain spring 4.00 in, how much will the spring be stretched by a force of 6.00 lb? See Fig. 18.8.

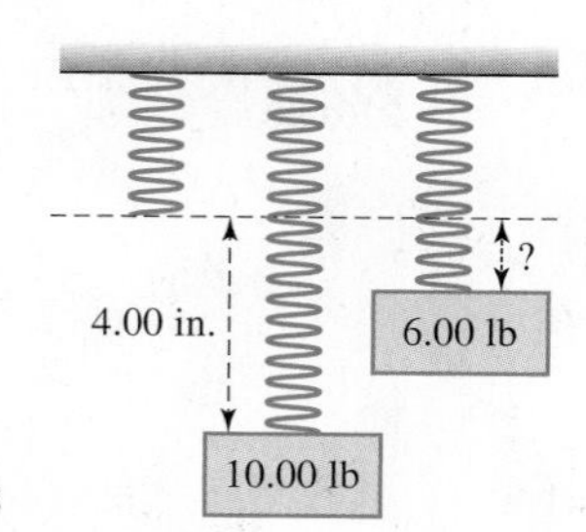

Fig. 18.8

38. The rate H of heat removal by an air conditioner is proportional to the electric power input P. The constant of proportionality is the *performance coefficient*. Find the performance coefficient of an air conditioner for which $H = 1.8$ kW and $P = 720$ W.

39. The energy E available daily from a solar collector varies directly as the percent p that the sun shines during the day. If a collector provides 1200 kJ for 75% sunshine, how much does it provide for a day during which there is 35% sunshine?

40. The distance d that can be seen from horizon to horizon from an airplane varies directly as the square root of the altitude h of the airplane. If $d = 133$ mi for $h = 12{,}000$ ft, find d for $h = 16{,}000$ ft.

41. The time t required to empty a wastewater-holding tank is inversely proportional to the cross-sectional area A of the drainage pipe. If it takes 2.0 h to empty a tank with a drainage pipe for which $A = 48$ in.^{2}, how long will it take to empty the tank if $A = 68$ in.^{2}?

42. The time t required to make a particular trip is inversely proportional to the average speed v. If a jet takes 2.75 h at an average speed of 520 km/h, how long will it take at an average speed of 620 km/h? Explain the meaning of the constant of proportionality.

43. In a physics experiment, a given force was applied to three objects. The mass m and the resulting acceleration a were recorded as follows:

m (g)	2.0	3.0	4.0
a (cm/s^2)	30	20	15

(a) Is the relationship $a = f(m)$ one of direct or inverse variation? Explain. (b) Find $a = f(m)$.

44. The lift L of each of three model airplane wings of width w was measured and recorded as follows:

w (cm)	20	40	60
L (N)	10	40	90

If L varies directly as the square of w, find $L = f(w)$. Does it matter which pair of values is used to find the constant of proportionality? Explain.

45. The power P required to propel a ship varies directly as the cube of the speed s of the ship. If 5200 hp will propel a ship at 12.0 mi/h, what power is required to propel it at 15.0 mi/h?

46. The f-number lens setting of a camera varies directly as the square root of the time t that the film is exposed. If the f-number is 8 (written as $f/8$) for $t = 0.0200$ s, find the f-number for $t = 0.0098$ s.

47. The force F on the blade of a wind generator varies jointly as the blade area A and the square of the wind velocity v. Find the equation relating F, A, and v if $F = 19.2$ lb when $A = 3.72\ \text{ft}^2$ and $v = 31.4$ ft/s.

48. The escape velocity v a spacecraft needs to leave the gravitational field of a planet varies directly as the square root of the product of the planet's radius R and its acceleration due to gravity g. For Mars and Earth, $R_M = 0.533R_e$ and $g_M = 0.400g_e$. Find v_M for Mars if $v_e = 11.2$ km/s.

49. The force F between two parallel wires carrying electric currents is inversely proportional to the distance d between the wires. If a force of 0.750 N exists between wires that are 1.25 cm apart, what is the force between them if they are separated by 1.75 cm?

50. The velocity v of a pulse traveling in a string varies directly as the square root of the tension T in the string. If the velocity of a pulse in a string is 450 ft/s when the tension is 20.0 lb, find the velocity when the tension is 30.0 lb.

51. The average speed s of oxygen molecules in the air is directly proportional to the square root of the absolute temperature T. If the speed of the molecules is 460 m/s at 273 K, what is the speed at 300 K?

52. The time t required to test a computer memory unit varies directly as the square of the number n of memory cells in the unit. If a unit with 4800 memory cells can be tested in 15.0 s, how long does it take to test a unit with 8400 memory cells?

53. The electric resistance R of a wire varies directly as its length l and inversely as its cross-sectional area A. Find the relation between resistance, length, and area for a wire that has a resistance of 0.200 Ω for a length of 225 ft and cross-sectional area of 0.0500 in.^{2}.

54. The general gas law states that the pressure P of an ideal gas varies directly as the thermodynamic temperature T and inversely as the volume V. If $P = 610$ kPa for $V = 10.0\ \text{cm}^3$ and $T = 290$ K, find V for $P = 400$ kPa and $T = 400$ K.

55. Noting the general gas law in Exercise 54, if V remains constant, (a) express P as a function of T. (b) A car tire is filled with air at 10.0°C with a pressure of 225 kPa, and after a while the air temperature in the tire is 60.0°C. Assuming no change in volume, what is the pressure? [*Hint*: You must convert the temperatures to thermodynamic temperatures (in K) using $T_K = T_C + 273.15$.]

56. The weight of an object is the force F on the object due to gravity, where $F = mg$. State how g (the acceleration due to gravity) varies with respect to the mass m_b of the spherical body on which the object lies, and the radius of that body (assume all of the mass is at its center.). (See Example 5.)

57. The power P in an electric circuit varies jointly as the resistance R and the square of the current I. If the power is 10.0 W when the current is 0.500 A and the resistance is 40.0 Ω, find the power if the current is 2.00 A and the resistance is 20.0 Ω.

58. The difference $m_1 - m_2$ in magnitudes (visual brightnesses) of two stars varies directly as the base 10 logarithm of the ratio b_2/b_1 of their actual brightnesses. For two particular stars, if $b_2 = 100b_1$ for $m_1 = 7$ and $m_2 = 2$, find the equation relating m_1, m_2, b_1, and b_2.

59. The power gain G by a parabolic microwave dish varies directly as the square of the diameter d of the opening and inversely as the square of the wavelength λ of the wave carrier. Find the equation relating G, d, and λ if $G = 5.5 \times 10^4$ for $d = 2.9$ m and $\lambda = 3.0$ cm.

60. The intensity I of sound varies directly as the power P of the source and inversely as the square of the distance r from the source. Two sound sources are separated by a distance d, and one has twice the power output of the other. Where should an observer be located on a line between them such that the intensity of each sound is the same?

61. The x-component of the acceleration of an object moving around a circle with constant angular velocity ω varies jointly as $\cos \omega t$ and the square of ω. If the x-component of the acceleration is $-11.4\ \text{ft/s}^2$ when $t = 1.00$ s for $\omega = 0.524$ rad/s, find the x-component of the acceleration when $t = 2.00$ s.

62. The tangent of the proper banking angle θ of the road for a car making a turn is directly proportional to the square of the car's velocity v and inversely proportional to the radius r of the turn. If 7.75° is the proper banking angle for a car traveling at 20.0 m/s around a turn of radius 300 m, what is the proper banking angle for a car traveling at 30.0 m/s around a turn of radius 250 m? See Fig. 18.9.

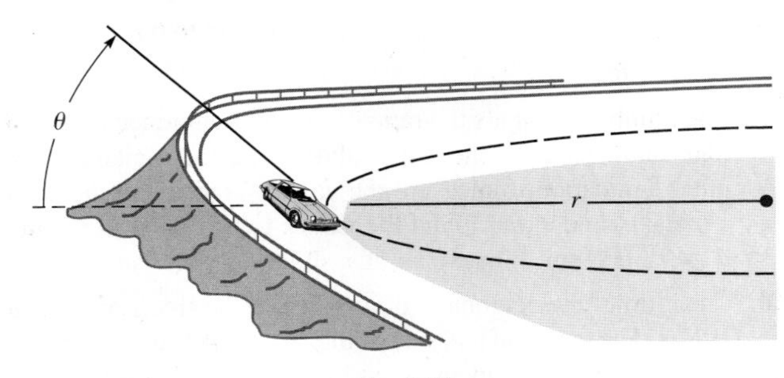

Fig. 18.9

63. The acoustical intensity I of a sound wave is proportional to the square of the pressure amplitude P and inversely proportional to the velocity v of the wave. If $I = 0.474\ \text{W/m}^2$ for $P = 20.0$ Pa and $v = 346$ m/s, find I if $P = 15.0$ Pa and $v = 320$ m/s.

64. To cook a certain vegetable mix in a microwave oven, the instructions are to cook 4.0 oz for 2.5 min or 8.0 oz for 3.5 min. Assuming the cooking time t is proportional to some power (not necessarily an integer) of the weight w, use logarithms to find t as a function of w.

Answers to Practice Exercises

1. $y = kx^2/z$ **2.** $y = 3$ **3.** $\text{BMI}_{\text{new}} = 0.909(\text{BMI}_{\text{original}})$
4. 0.01 F

CHAPTER 18 KEY FORMULAS AND EQUATIONS

Proportion	$\frac{a}{b} = \frac{c}{d}$	**(18.1)**
Direct variation	$y = kx$	**(18.2)**
Inverse variation	$y = \frac{k}{x}$	**(18.3)**
Joint variation	$y = kxz$	**(18.4)**

CHAPTER 18 REVIEW EXERCISES

CONCEPT CHECK EXERCISES

Determine each of the following as being either **true** *or* **false.** *If it is false, explain why.*

1. The ratio of 25 cm to 50 mm is 5.

2. If 20 m is divided into two parts in the ratio of 3/2, the parts are 14 m and 6 m.

3. If y varies inversely as the square of x, and $y = 4$ when $x = 1/4$, then $y = 4/x^2$.

4. If R varies jointly with s and the square of t, and inversely as the square of r, then $k = \left(\frac{R}{s}\right)\left(\frac{r}{t}\right)^2$.

PRACTICE AND APPLICATIONS

In Exercises 5–18, find the indicated ratios.

5. 840 mg to 3g

6. 300 nm to 6 μm

7. 375 mL to 25 cL

8. 12 ks to 2 h

9. The number π equals the ratio of the circumference c of a circle to its diameter d. To check the value of π, a technician used computer simulation to measure the circumference and diameter of a metal cylinder and found the values to be $c = 4.2736$ cm and $d = 1.3603$ cm. What value of π did the technician get?

10. The ratio of the diagonal d of a square to the side s of the square is $\sqrt{2}$. The diagonal and side of the face of a glass cube are found to be $d = 35.375$ mm and $s = 25.014$ mm. What is the value of $\sqrt{2}$ found from these measurements?

11. The mechanical advantage of a lever is the ratio of the output force F_0 to the input force F_i. Find the mechanical advantage if $F_0 = 24$ kN and $F_i = 5000$ N. See Fig. 18.10.

Fig. 18.10

12. For an automobile, the ratio of the number n_1 of teeth on the ring gear to the number n_2 of teeth on the pinion gear is the *rear axle ratio* of the car. Find this ratio if $n_1 = 64$ and $n_2 = 20$.

13. The pressure p exerted on a surface is the ratio of the force F on the surface to its area A. Find the pressure on a square patch, 2.25 in. on a side, on a tank if the force on the patch is 42.3 lb.

14. The electric resistance R of a resistor is the ratio of the voltage V across the resistor to the current i in the resistor. Find R if $V = 0.632$ V and $i = 2.03$ mA.

15. The *heat of vaporization* of a substance is the amount of heat needed to change a unit amount from liquid to vapor. Experimentation shows 7910 J are needed to change 3.50 g of water to steam. What is the heat of vaporization of water?

16. What is the ratio of the volume of Earth $(R = 6380 \text{ km})$ to the volume of a bacterium $(r = 0.0040 \text{ mm})$?

17. The ratio of the commission for selling a home to the selling price of the home is the commission rate. What is this rate if the commission of $17,100 is charged for selling a home priced at $380,000?

18. A total cholesterol level of 200 mg/dL is considered too high, but many doctors consider the ratio of the total cholesterol to HDL (high density lipids—the "good" cholesterol) a more important measure of risk to coronary heart disease. What is this ratio if the total cholesterol is 179 mg/dL and the HDL is 39 mg/dL? (Average for women is about 4.5, and for men is about 5.0.)

In Exercises 19 and 20, given that $a/b = c/d$ (b and d not zero), show that indicated proportions are correct.

19. $\frac{a+b}{b} = \frac{c+d}{d}$

20. $\frac{a-b}{b} = \frac{c-d}{d}$

In Exercises 21–36, answer the given questions by setting up and solving the appropriate proportions.

21. On a map of Hawaii, 1.3 in. represents 20.0 mi. If the distance on the map between Honolulu and Kaanapali on Maui is 6.0 in., how far is Kaanapali from Honolulu?

22. Given that 1.000 lb $=$ 453.6 g, what is the weight in pounds of a 14.0-g computer disk?

23. Given that 1.00 Btu $=$ 1060 J, how much heat in joules is produced by a heating element that produces 2660 Btu?

24. Given that 1.00 L $=$ 61.0 in.^{3}, what is the capacity in liters of a cubical box that is 3.23 in. along an edge?

25. A computer printer can print 60 pages in 45 s. How many pages can it print in 3 min?

26. A solar heater with a collector area of 58.0 m^2 is required to heat 2560 kg of water. Under the same conditions, how much water can be heated by a rectangular solar collector 9.50 m by 8.75 m?

27. The dosage of a certain medicine is 25 mL for each 10 lb of the patient's weight. What is the dosage for a person weighing 56 kg?

28. A woman invests \$50,000 and a man invests \$20,000 in a partnership. If profits are to be shared in the ratio that each invested in the partnership, how much does each receive from \$10,500 in profits?

29. On a certain blueprint, a measurement of 25.0 ft is represented by 2.00 in. What is the actual distance between two points if they are 5.75 in. apart on the blueprint?

30. The chlorine concentration in a water supply is 0.12 part per million. How much chlorine is there in a cylindrical holding tank 4.22 m in radius and 5.82 m high filled from the water supply?

31. One fiber-optic cable carries 60.0% as many messages as another fiber-optic cable. Together they carry 12,000 messages. How many does each carry?

32. Two types of roadbed material, one 50% rock and the other 100% rock, are used in the ratio of 4 to 1 to form a roadbed. If a total of 150 tons are used, how much rock is in the roadbed?

33. To neutralize 80.0 kg of sodium hydroxide, 98.0 kg of sulfuric acid are needed. How much sodium hydroxide can be neutralized with 37.0 kg of sulfuric acid?

34. For analysis, 105 mg of DNA sample is divided into two vials in the ratio of 2 to 5. How many milligrams are in each vial?

35. A total of 322 bolts are in two containers. The ratio of the number of bolts in the first container to the number in the second container is $5/9$. How many are in each container?

36. A gasoline company sells octane-87 gas and octane-91 gas in the ratio of 9 to 2. How many of each octane are sold of a total of 16.5 million gallons?

In Exercises 37–40, give the specific equation relating the variables after evaluating the constant of proportionality for the given set of values.

37. y varies directly as the square of x, and $y = 27$ when $x = 3$.

38. f varies inversely as l, and $f = 500$ when $l = 800$.

39. v is directly proportional to x and inversely proportional to the cube of y, and $v = 10$ when $x = 5$ and $y = 4$.

40. r varies jointly as u, v, and the square of w, and $r = 1/8$ when $u = 1/2$, $v = 1/4$, and $w = 1/3$.

In Exercises 41–82, solve the given applied problems.

41. For the lever balanced at the fulcrum, the relation

$$\frac{F_1}{F_2} = \frac{L_2}{L_1}$$

holds, where F_1 and F_2 are forces on opposite sides of the fulcrum at distances L_1 and L_2 (see Fig. 18.11). Find L_2 if $F_1 = 4.50$ lb, $F_2 = 6.75$ lb, and $L_1 = 17.5$ in.

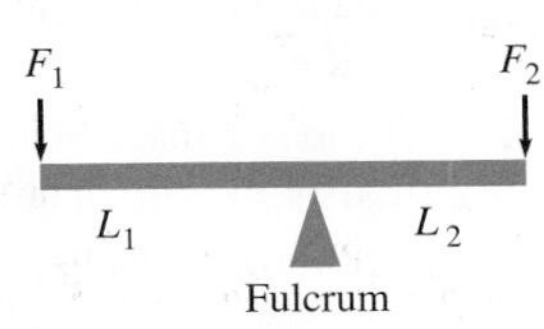

Fig. 18.11

42. A company finds that the volume V of sales of a certain item and the price P of the item are related by

$$\frac{P_1}{P_2} = \frac{V_2}{V_1}$$

Find V_2 if $P_1 = \$8.00$, $P_2 = \$6.00$, and $V_1 = 3000$ per week.

43. The image height h and object height H for the lens shown in Fig. 18.12 are related to the image distance q and object distance p by

$$\frac{h}{H} = \frac{q}{p}$$

Find q if

$h = 24$ cm,

$H = 84$ cm, and

$p = 36$ cm.

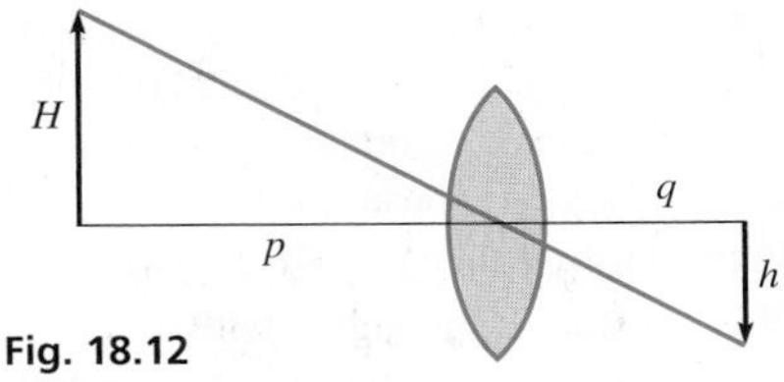

Fig. 18.12

44. For two pulleys connected by a belt, the relation

$$\frac{d_1}{d_2} = \frac{n_2}{n_1}$$

holds, where d is the diameter of the pulley and n is the number of revolutions per unit time it makes. Find n_2 if $d_1 = 4.60$ in., $d_2 = 8.30$ in., and $n_1 = 18.0$ r/min.

45. An apartment owner charges rent R proportional to the floor area A of the apartment. Find the equation relating R and A if an apartment of 900 ft^2 rents for \$850/month.

46. The number r of aluminum cans that can be made by recycling n used cans is proportional to n. How many cans can be made from 50,000 used cans if $r = 115$ cans for $n = 125$ cans?

47. Under certain conditions, the rate of increase v of bacteria is proportional to the number N of bacteria present. Find v for $N = 7500$ bacteria, if $v = 800$ bacteria/h for $N = 4000$ bacteria.

48. The index of refraction n of a medium varies inversely as the velocity of light v within it. For quartz, $n = 1.46$ and $v = 2.05 \times 10^8$ m/s. What is n for a diamond, in which $v = 1.24 \times 10^8$ m/s?

49. The force F needed to tighten a bolt varies as the length L of the wrench handle. See Fig. 18.13. If $F = 250$ N for $L = 22$ cm, and $F = 550$ N for $L = 10$ cm, determine the equation relating F and L.

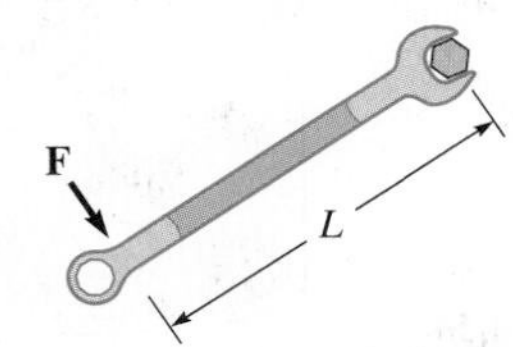

Fig. 18.13

50. The value V of a precious stone varies directly as the square of its weight w. What is the loss in value if a stone for which $V = \$27,000$ is cut into two pieces with weights in the ratio of 2 to 1?

51. The period T of a pendulum varies directly as the square root of its length L. If $T = \pi/2$ s for $L = 2.00$ ft, find T for $L = 4.00$ ft.

52. The monthly payment p on a mortgage is directly proportional to amount A that is borrowed. If $p = \$635$ for $A = \$100,000.00$, find p if $A = \$345,000$.

53. The velocity v (in ft/s) of water from a fire hose is directly proportional to the square root of the pressure p (in lb/in.^{2}). If $v = 120$ ft/s for $p = 100$ lb/in.^{2}, find v for $p = 64$ lb/in.^{2}.

54. The component of velocity v_x of an object moving in a circle with constant angular velocity ω varies jointly with ω and $\sin \omega t$. If $\omega = \pi/6$ rad/s, and $v_x = -4\pi$ ft/s when $t = 1.00$ s, find v_x when $t = 9.00$ s.

55. The charge C on a capacitor varies directly as the voltage V across it. If the charge is 6.3 μC with a voltage of 220 V across a capacitor, what is the charge on it with a voltage of 150 V across it?

56. The amount of natural gas burned is proportional to the amount of oxygen consumed. If 24.0 lb of oxygen is consumed in burning 15.0 lb of natural gas, how much air, which is 23.2% oxygen by weight, is consumed to burn 50.0 lb of natural gas?

57. The power P of a gas engine is proportional to the area A of the piston. If an engine with a piston area of 8.00 in.^{2} can develop 30.0 hp, what power is developed by an engine with a piston area of 6.0 in.^{2}?

58. The decrease in temperature at a point above a region is directly proportional to the altitude of the point above the region. If the temperature T on the ground in Toronto is 22°C and a plane 3.50 km above notes that the temperature is 1.0°C, what is the temperature at the top of the CN Tower, which is 522 m high? (Assume there are no other temperature effects.)

59. The distance d an object falls under the influence of gravity varies directly as the square of the time t of fall. If an object falls 64.0 ft in 2.00 s, how far will it fall in 3.00 s?

60. The kinetic energy E of a moving object varies jointly as the mass m of the object and the square of its velocity v. If a 5.00-kg object, traveling at 10.0 m/s, has a kinetic energy of 250 J, find the kinetic energy of an 8.00-kg object moving at 50.0 m/s.

61. In a particular computer design, N numbers can be sorted in a time proportional to the square of log N. How many times longer does it take to sort 8000 numbers than to sort 2000 numbers?

62. The velocity v of a jet of fluid flowing from an opening in the side of a container is proportional to the square root of the depth d of the opening. If the velocity of the jet from an opening at a depth of 1.22 m is 4.88 m/s, what is the velocity of a jet from an opening at a depth of 7.62 m? See Fig. 18.14.

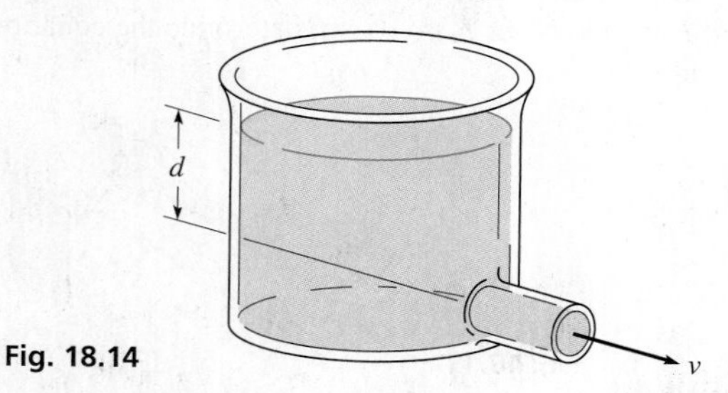

Fig. 18.14

63. In an electric circuit containing an inductance L and a capacitance C, the resonant frequency f is inversely proportional to the square root of the capacitance. If the resonant frequency in a circuit is 25.0 Hz and the capacitance is 95.0 μF, what is the resonant frequency of this circuit if the capacitance is 25.0 μF?

64. The rate of emission R of radiant energy from the surface of a body is proportional to the fourth power of the theromodynamic temperature T. Given that a 25.0-W (the rate of emission) lamp has an operating temperature of 2500 K, what is the operating temperature of a similar 40.0-W lamp?

65. The frequency f of a radio wave is inversely proportional to its wavelength λ. The constant of proportionality is the velocity of the wave, which equals the speed of light. Find this velocity if an FM radio wave has a frequency of 90.9 MHz and a wavelength of 3.29 m.

66. The acceleration of gravity g on a satellite in orbit around Earth varies inversely as the square of its distance r from the center of Earth. If $g = 8.7$ m/s^2 for a satellite at an altitude of 400 km above the surface of Earth, find g if it is 1000 km above the surface. The radius of the Earth is 6.4×10^6 m.

67. If the weight w of an airplane varies directly as the cube of its length L, and $w = 3520$ lb for $L = 42.0$ ft, what would be the weight of a plane that is 50.0 ft long?

68. The thrust T of a propeller varies jointly as the square of the number n of revolutions per second and the fourth power of its diameter d. What is the effect on T if n is doubled and d is halved?

69. Using *holography* (a method of producing an image without using a lens), an image of concentric circles is formed. The radius r of each circle varies directly as the square root of the wavelength λ of the light used. If $r = 3.56$ cm for $\lambda = 575$ nm, find r if $\lambda = 483$ nm.

70. A metal circular ring has a circular cross section of radius r. If R is the radius of the ring (measured to the middle of the cross section), the volume V of metal in the ring varies directly as R and the square of r. If $V = 2550$ mm^3 for $r = 2.32$ mm and $R = 24.0$ mm, find V for $r = 3.50$ mm and $R = 32.0$ mm. See Fig. 18.15.

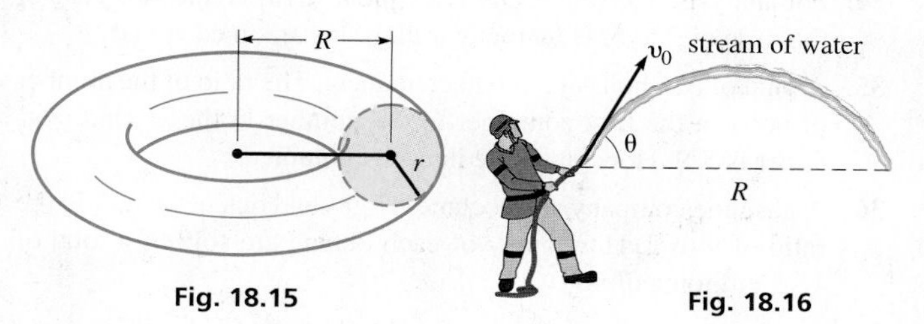

Fig. 18.15

Fig. 18.16

71. The range R of a stream of water from a fire hose varies jointly as the square of the initial velocity v_0 and the sine of twice the angle θ from the horizontal from which it is released. See Fig. 18.16. A stream for which $v_0 = 35$ m/s and $\theta = 22°$ has a range of 85 m. Find the range if $v_0 = 28$ m/s and $\theta = 43°$.

72. Kepler's third law of planetary motion states that the square of the period of any planet is proportional to the cube of the mean radius (about the sun) of that planet, with the constant of proportionality being the same for all planets. Using the fact that the period of Earth is 1 year and its mean radius is 93.0 million miles, calculate the mean radius for Venus, given that its period is 7.38 months.

73. The stopping distance d of a car varies directly as the square of the velocity v of the car when the brakes are applied. A car moving at 32 mi/h can stop in 52 ft. What is the stopping distance for the car if it is moving at 55 mi/h?

74. The load L that a helical spring can support varies directly as the cube of its wire diameter d and inversely as its coil diameter D. A spring for which $d = 0.120$ in. and $D = 0.953$ in. can support 45.0 lb. What is the coil diameter of a similar spring that supports 78.5 lb and for which $d = 0.156$ in.?

75. The volume rate of flow R of blood through an artery varies directly as the fourth power of the radius r of the artery and inversely as the distance d along the artery. If an operation is successful in effectively increasing the radius of an artery by 25% and decreasing its length by 2%, by how much is the volume rate of flow increased?

76. The safe, uniformly distributed load L on a horizontal beam, supported at both ends, varies jointly as the width w and the square of the depth d and inversely as the distance D between supports. Given that one beam has double the dimensions of another, how many times heavier is the safe load it can support than the first can support?

77. A bank statement exactly 30 years old is discovered. It states, "This 10-year-old account is now worth \$185.03 and pays 4% interest compounded annually." An investment with annual compound interest varies directly as $1 + r$ to the power n, where r is the interest rate expressed as a decimal and n is the number of years of compounding. What was the value of the original investment, and what is it worth now?

78. The distance s that an object falls due to gravity varies jointly as the acceleration g due to gravity and the square of the time t of fall. The acceleration due to gravity on the moon is 0.172 of that on Earth. If a rock falls for t_0 seconds on Earth, how many times farther would the rock fall on the moon in $3t_0$ seconds?

79. The heat loss L through fiberglass insulation varies directly as the time t and inversely as the thickness d of the fiberglass. If the loss through 8.0 in. of fiberglass is 1200 Btu in 30 min, what is the loss through 6.0 in. in 1 h 30 min?

80. A quantity important in analyzing the rotation of an object is its *moment of inertia I*. For a ball bearing, the moment of inertia varies directly as its mass m and the square of its radius r. Find the general expression for I if $I = 39.9\ \text{g} \cdot \text{cm}^2$ for $m = 63.8$ g and $r = 1.25$ cm.

81. In the study of polarized light, the intensity I is proportional to the square of the cosine of the angle θ of transmission. If $I = 0.025\ \text{W/m}^2$ for $\theta = 12.0°$, find I for $\theta = 20.0°$.

82. The force F that acts on a pendulum bob is proportional to the mass m of the bob and the sine of the angle θ the pendulum makes with the vertical. If $F = 0.120$ N for $m = 0.350$ kg and $\theta = 2.00°$, find F for $m = 0.750$ kg and $\theta = 3.50°$.

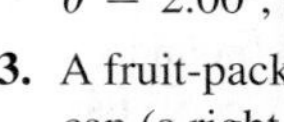

83. A fruit-packing company plans to reduce the size of its fruit juice can (a right circular cylinder) by 10% and keep the price of each can the same (effectively raising the price). The radius and the height of the new can are to be equally proportional to those of the old can. Write one or two paragraphs explaining how to determine the percent decrease in the radius and the height of the old can that is required to make the new can.

CHAPTER 18 PRACTICE TEST

As a study aid, we have included complete solutions for each Practice Test problem at the end of this book.

1. Express the ratio of 180 s to 4 min in simplest form.

2. A person 1.8 m tall is photographed, and the film image is 20.0 mm high. Under the same conditions, how tall is a person whose film image is 14.5 mm high?

3. The change L in length of a copper rod varies directly as the change T in temperature of the rod. Set up an equation for this relationship if $L = 2.7$ cm for $T = 150°\text{C}$.

4. Given that 1.00 in. = 2.54 cm, what is the length (in in.) of a computer screen image that is 7.24 cm long?

5. The perimeter of a rectangular solar panel is 210.0 in. The ratio of the length to the width is 7 to 3. What are the dimensions of the panel?

6. The difference p in pressure in a fluid between that at the surface and that at a point below varies jointly as the density d of the fluid and the depth h of the point. The density of water is $1000\ \text{kg/m}^3$, and the density of alcohol is $800\ \text{kg/m}^3$. This difference in pressure at a point 0.200 m below the surface of water is 1.96 kPa. What is the difference in pressure at a point 0.300 m below the surface of alcohol? (All data are accurate to three significant digits.)

7. The crushing load L of a pillar varies directly as the fourth power of its radius r and inversely as the square of its length l. If one pillar has twice the radius and three times the length of a second pillar, what is the ratio of the crushing load of the first pillar to that of the second pillar?

19

Sequences and the Binomial Theorem

LEARNING OUTCOMES

After completion of this chapter, the student should be able to:

- Find the terms, common difference, number of terms, or sum of an arithmetic sequence
- Find the terms, common ratio, number of terms, or sum of a finite geometric sequence
- Find the sum of an infinite geometric series
- Evaluate expressions involving factorials
- Expand binomials to any power using Pascal's triangle and the binomial theorem
- Obtain terms of a binomial series
- Solve application problems involving sequences and series

If a person is saving for the future and invests \$1000 at 5%, compounded annually, the value of the investment 40 years later would be about \$7040. Even better, if the interest is compounded daily, which is a common method today, it would be worth about \$7390 in 40 years. A person saving for retirement would do much better by putting aside \$1000 each year and letting the interest accumulate. If \$1000 is invested each year at 5%, compounded annually, the total investment would be worth about \$126,840 after 40 years. If the interest is compounded daily, it would be worth about \$131,000 in 40 years. Obviously, if more is invested, or the interest rate is higher, the value of these investments would be higher.

Each of these values can be found quickly since the values of the annual investments form what is called a *geometric sequence,* and formulas can be formed for such sums. Such formulas involving compound interest are widely used in calculating values such as monthly car payments, home mortgages, and annuities.

A *sequence* is a set of numbers arranged in some specific way and usually follows a pattern. Sequences have been of interest to people for centuries. There are records that date back to at least 1700 B.C.E. showing calculations involving sequences. Euclid, in his *Elements*, dealt with sequences in about 300 B.C.E. More advanced forms of sequences were used extensively in the study of advanced mathematics in the 1700s and 1800s. These advances in mathematics have been very important in many areas of science and technology.

Of the many types of sequences, we study certain basic ones in this chapter. Included are those used in the expansion of a binomial to a power. We show applications in areas such as physics and chemistry in studying radioactivity, biology in studying population growth, and, of course, in business when calculating interest.

▶ **Sequences are basic to many calculations in business, including compound interest. In Section 19.2, we show such a calculation.**

19.1 Arithmetic Sequences

Arithmetic Sequence • Common Difference• Finite and Infinite Sequences • Sum of n Terms

A *sequence* of numbers may consist of numbers chosen in any way we may wish to select. This could include a random selection of numbers, although the sequences that are useful follow a pattern. We consider only those sequences that include real numbers or literal numbers that represent real numbers.

An **arithmetic sequence** (*or* **arithmetic progression**) *is a set of numbers in which each number after the first can be obtained from the preceding one by adding to it a fixed number called the* **common difference.** This definition can be expressed in terms of the *recursion formula*

$$a_n = a_{n-1} + d \tag{19.1}$$

where a_n is any term, a_{n-1} is the preceding term, and d is the common difference.

EXAMPLE 1 Illustrations of arithmetic sequences

(a) The sequence 2, 5, 8, 11, 14, . . . , is an arithmetic sequence with a common difference $d = 3$. We can obtain any term by adding 3 to the previous term. We see that the fifth term is $a_5 = a_4 + d$, or $14 = 11 + 3$.

(b) The sequence 7, 2, −3, −8, . . . , is an arithmetic sequence with $d = -5$. We can get any term after the first by adding −5 to the previous term.

The three dots after the 14 in part (a) and after −8 in part (b) mean that the sequences continue. ■

If we know the first term of an arithmetic sequence, we can find any other term by adding the common difference enough times to get the desired term. This, however, is very inefficient, and there is a general way of finding a particular term.

If a_1 is the first term and d is the common difference, the second term is $a_1 + d$, the third term is $a_1 + 2d$, and so on. For the nth term, we need to add d to the first term $n - 1$ times. Therefore, the **nth term, a_n, of the arithmetic sequence** is given by

■ Another form of Eq. (19.2) is $d = \frac{a_n - a_1}{n - 1}$.

$$a_n = a_1 + (n - 1)d \tag{19.2}$$

Equation (19.2) can be used to find any given term in any arithmetic sequence. We can refer to a_n as the *last term* of an arithmetic sequence if no terms beyond it are included in the sequence. Such a sequence is called a ***finite*** sequence. If the terms in a sequence continue without end, the sequence is called an ***infinite*** sequence.

EXAMPLE 2 Finding a specified term

Find the tenth term of the arithmetic sequence 2, 5, 8, ...

By subtracting any term from the following term, we find the common difference $d = 3$, and see that the first term $a_1 = 2$. Therefore, the tenth term, a_{10}, is

$$\begin{aligned} a_{10} &= \underset{a_1}{2} + (\underset{n}{10} - 1)\underset{d}{3} = 2 + (9)(3) \\ &= 29 \end{aligned}$$

NOTE ▸ [The three dots after the 8 show that the sequence continues. With no additional information given, this indicates that it is an infinite arithmetic sequence.] ■

Practice Exercise

1. Find the 20th term of the arithmetic sequence 2, 8, 14, . . .

EXAMPLE 3 Finding the common difference

Find the common difference between successive terms of the arithmetic sequence for which the third term is 5 and the 34th term is -119.

In order to use Eq. (19.2), we could calculate a_1, but we can also treat the third term as a_1 and the 34th term as a_{32}. This gives us the same sequence from 5 to -119. Therefore, using the values $a_1 = 5$, $a_{32} = -119$, and $n = 32$, we can find the value of d. Substituting these values in Eq. (19.2) gives

$$-119 = 5 + (32 - 1)d$$
$$31d = -124$$
$$d = -4$$

There is no information as to whether this is a finite or an infinite sequence. The solution is the same in either case. ■

Practice Exercise

2. Find d if $a_1 = 4$ and $a_{13} = 34$.

EXAMPLE 4 Finding the number of terms—velocity of a package

A package delivery company uses a metal (very low friction) ramp to slide packages from the sorting area to the loading area below. If a given package is pushed such that it starts down the ramp at 25 cm/s, and the package accelerates as it slides down the ramp such that it gains 35 cm/s during each second, after how many seconds is its velocity 305 cm/s? See Fig. 19.1.

Here, we see that the velocity (in cm/s) of the package after each second is

$$60, 95, 130, \ldots, 305, \ldots$$

Therefore, $a_1 = 60$ (the 25 cm/s was at the beginning, that is, after 0 s), $d = 35$, $a_n = 305$, and we are to find n, which in this case, represents the number of seconds during which the velocity increases. Therefore,

$$305 = 60 + (n - 1)(35)$$
$$245 = 35n - 35$$
$$35n = 280$$
$$n = 8.0$$

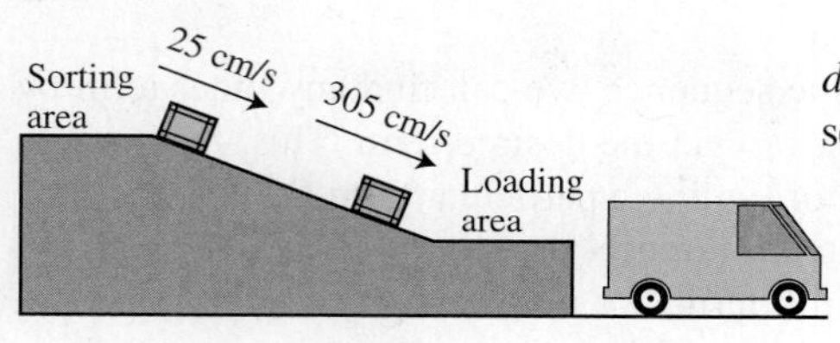

Fig. 19.1

This means that the velocity of a package sliding down the ramp is 305 cm/s after 8.0 s. ■

SUM OF *n* TERMS

Another important quantity related to an arithmetic sequence is the sum of the first n terms. We can indicate this sum by starting the sum either with the first term or with the last term, as shown by these two equations:

$$S_n = a_1 + (a_1 + d) + (a_1 + 2d) + \cdots + (a_n - d) + a_n$$

or

$$S_n = a_n + (a_n - d) + (a_n - 2d) + \cdots + (a_1 + d) + a_1$$

If we now add the corresponding members of these two equations, we obtain the result

$$2S_n = (a_1 + a_n) + (a_1 + a_n) + (a_1 + a_n) + \cdots + (a_1 + a_n) + (a_1 + a_n)$$

Each term on the right in parentheses has the same expression $(a_1 + a_n)$, and there are n such terms. This tells us that *the sum of the first n terms is given by*

$$S_n = \frac{n}{2}(a_1 + a_n) \tag{19.3}$$

The use of Eq. (19.3) is illustrated in the following examples.

Karl Friedrich Gauss (1777–1856) is considered one of the greatest mathematicians of all time. A popular story about him is that at the age of 10 his schoolmaster asked his class to add all the numbers from 1 to 100, assuming it would take some time to do. He had barely finished stating the problem when Gauss handed in the correct answer of 5050. He had done it essentially as we did the solution in Example 5.

EXAMPLE 5 Finding the sum of terms

Find the sum of the first 1000 positive integers.

The first 1000 positive integers form a finite arithmetic sequence for which $a_1 = 1, a_{1000} = 1000, n = 1000$, and $d = 1$. Substituting these values in Eq. (19.3) (in which we do not use the value of d), we have

$$S_{1000} = \frac{1000}{2}(1 + 1000) = 500(1001)$$
$$= 500{,}500$$

Practice Exercise

3. Find S_9 if $d = -2$ and $a_9 = 1$.

If any three of the five values a_1, a_n, n, d, and S_n are given for a particular arithmetic sequence, the other two may be found from Eqs. (19.2) and (19.3). Consider the following example.

EXAMPLE 6 Finding the sum of terms—increasing voltage

The voltage across a resistor increases such that during each second the increase is 0.002 mV less than during the previous second. Given that the increase during the first second is 0.350 mV, what is the total voltage increase during the first 10.0 s?

We are asked to find the sum of the voltage increases 0.350 mV, 0.348 mV, 0.346 mV, . . . , so as to include ten increases. This means we want the sum of an arithmetic sequence for which $a_1 = 0.350$, $d = -0.002$, and $n = 10$. Because we need a_n to use Eq. (19.3), we first calculate it, using Eq. (19.2):

$$a_{10} = 0.350 + (10 - 1)(-0.002) = 0.332 \text{ mV}$$

Now, we use Eq. (19.3) to find the sum with $a_1 = 0.350$, $a_{10} = 0.332$, and $n = 10$:

$$S_{10} = \frac{10}{2}(0.350 + 0.332) = 3.410 \text{ mV}$$

Thus, the total voltage increase is 3.410 mV.

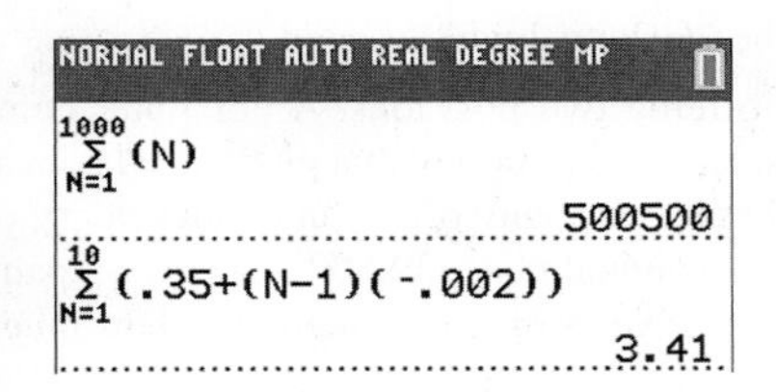

Fig. 19.2
Graphing calculator keystrokes: goo.gl/AMe122

The *summation* feature on a calculator can also be used to find the sum of the terms of a sequence. The sums found in Examples 5 and 6 are shown in Fig. 19.2. The symbol $\sum$ (the Greek capital letter sigma) is used to indicate the *sum* of the terms as the *index* N ranges between the values shown below and above the $\sum$ symbol.

EXERCISES 19.1

In Exercises 1 and 2, make the given changes in the indicated examples of this section and then solve the resulting problems.

1. In Example 3, change −119 to −88 and then find the common difference.
2. In Example 5, change 1000 to 500 and then find the sum.

In Exercises 3–6, write the first five terms of the arithmetic sequence with the given values.

3. $a_1 = 4, d = 2$
4. $a_1 = 6, d = -\frac{1}{2}$
5. $a_1 = 2.5, a_5 = -1.5$
6. $a_2 = -2, a_5 = 43$

In Exercises 7–14, find the nth term of the arithmetic sequence with the given values.

7. $1, 4, 7, \ldots; n = 8$
8. $-6, -4, -2, \ldots; n = 10$
9. $\frac{8}{3}, \frac{5}{3}, \frac{2}{3}, \ldots; n = 9$
10. $2, 0.5, -1, \ldots; n = 25$
11. $a_1 = -0.7, d = 0.4, n = 80$
12. $a_1 = \frac{3}{2}, d = \frac{1}{6}, n = 601$
13. $a_1 = b, a_3 = 5b, n = 25$
14. $a_1 = -c, a_4 = 8c, n = 30$

In Exercises 15–18, find the sum of the n terms of the indicated arithmetic sequence.

15. $4 + 8 + 12 + \cdots + 64$
16. $27 + 24 + 21 + \cdots + (-9)$
17. $-2, -\frac{5}{2}, -3, \ldots; n = 10$
18. $3k, \frac{10}{3}k, \frac{11}{3}k, \ldots; n = 40$

In Exercises 19–26, find any of the values of a_1, d, a_n, n, or S_n that are missing for an arithmetic sequence.

19. $5, 13, 21, \ldots, 45$
20. $-2, -1.5, -1, \ldots, 28$
21. $a_1 = \frac{5}{3}, n = 20, S_{20} = \frac{40}{3}$
22. $d = -3, n = 3, a_3 = -5.9$
23. $d = 3, n = 30, S_{30} = 1875$
24. $d = 9, n = 18, S_{18} = 1521$
25. $a_1 = 7.4, d = -0.5, a_n = -23.1$
26. $a_1 = -\frac{9}{7}, n = 19, a_{19} = -\frac{36}{7}$

In Exercises 27–56, find the indicated quantities for the appropriate arithmetic sequence.

27. $a_6 = 560, a_{10} = 720$ (find a_1, d, S_n for $n = 10$)

28. $a_{17} = -91, a_2 = -73$ (find a_1, d, S_n for $n = 40$)

29. Is ln 3, ln 6, ln 12, . . . an arithmetic sequence? Explain. If it is, what is the fifth term?

30. Is sin 2°, sin 4°, sin 6°, . . . an arithmetic sequence? Explain. If it is, what is the fifth term?

31. Find a formula with variable n for the nth term of the arithmetic sequence with $a_1 = 3$ and $a_{n+1} = a_n + 2$ for $n = 1, 2, 3, \ldots$.

32. In the equation $a_n = a_1 + (n - 1)d$, solve for n.

33. Write the first five terms of the arithmetic sequence for which the second term is b and the third term is c.

34. If the sum of the first two terms of an arithmetic sequence equals the sum of the first three terms, find the sum of the first five terms.

35. Find the sum of the first 100 positive integers. (See the margin note on page 517.)

36. If x, 5, and $x + 8$ are the first three terms of an arithmetic sequence, find the sum of the first 20 terms.

37. Find x if $3 - x$, $-x$, and $\sqrt{9 - 2x}$ are the first three terms of an arithmetic sequence.

38. If $x, x + 2y, 2x + 3y, \ldots$ is an arithmetic sequence, express S_{100} in terms of x.

39. The sum of the angles inside a triangle, quadrilateral, and pentagon are 180°, 360°, and 540°, respectively. Assuming this pattern continues, what is the sum of the angles inside a dodecagon (12 sides)?

40. A person begins an exercise program of jogging 10 min each day for the first week. Each week thereafter, the person must increase their daily jogging time by 3 min. During which week will the person be jogging 55 min per day?

41. A beach now has an area of 9500 m^2 but is eroding such that it loses 100 m^2 more of its area each year than during the previous year. If it lost 400 m^2 during the last year, what will be its area 8 years from now?

42. During a period of heavy rains, on a given day 110,000 ft^3/s of water was being released from a dam. In order to minimize downstream flooding, engineers then reduced the releases by 10,000 ft^3/s each day thereafter. How much water was released during the first week of these releases?

43. At a logging camp, 15 layers of logs are so piled that there are 20 logs in the bottom layer, and each layer has 1 less log than the layer below it. How many logs are in the pile?

44. In order to prevent an electric current surge in a circuit, the resistance R in the circuit is stepped down by 4.0 Ω after each 0.1 s. If the voltage V is constant at 120 V, do the resulting currents I (in A) form an arithmetic sequence if $V = IR$?

45. There are 12 seats in the first row around a semicircular stage. Each row behind the first has 4 more seats than the row in front of it. How many rows of seats are there if there is a total of 300 seats?

46. A bank loan of \$8000 is repaid in annual payments of \$1000 plus 10% interest on the unpaid balance. What is the total amount of interest paid?

47. A car depreciates \$1800 during the first year after it is bought. Each year thereafter it depreciates \$150 less than the year before. How many years after it was bought will it be considered to have no value, and what was the original cost?

48. The sequence of ships' bells is as follows: 12:30 A.M. one bell is rung, and each half hour later one more bell is rung than the previous time until eight bells are rung. The sequence is then repeated starting at 4:30 A.M., again until eight bells are rung. This pattern is followed throughout the day. How many bells are rung in one day?

49. When a stone is dropped from the edge of a cliff at the Grand Canyon, it falls 16 ft during the first second, 48 ft during the second second, 80 ft during the third second, 112 ft during the fourth second, and so on. Find (a) the distance the stone falls during the tenth second and (b) the total distance the stone falls during the first 10 seconds.

50. In preparing a bid for constructing a new building, a contractor determines that the foundation and basement will cost \$605,000 and the first floor will cost \$360,000. Each floor above the first will cost \$15,000 more than the one below it. How much will the building cost if it is to be 18 floors high?

51. A college graduate is offered two positions. A computer company offers an annual salary of \$42,000 with a guaranteed annual raise of \$1200. A marketing company offers an annual salary of \$44,000 with a guaranteed annual raise of \$600. Which company will pay more for the first six years of employment, and how much more?

52. A person has a \$5000 balance due on a credit card account that charges 1% interest per month on the unpaid balance B_n. Assuming no extra charges, if \$250 is paid each month, find (a) a recursion formula [similar to Eq. (19.1)] for B_n and (b) B_n after two months.

53. Derive a formula for S_n in terms of n, a_1, and d.

54. A *harmonic sequence* is a sequence of numbers whose reciprocals form an arithmetic sequence. Is a harmonic sequence also an arithmetic sequence? Explain.

55. Show that the sum of the first n positive integers is $\frac{1}{2}n(n + 1)$.

56. Show that the sum of the first n positive odd integers is n^2.

Answers to Practice Exercises

1. 116 **2.** $d = 2.5$ **3.** $S_9 = 81$

19.2 Geometric Sequences

Geometric Sequence • Common Ratio • Sum of n Terms

A second type of important sequence of numbers is the **geometric sequence** (or **geometric progression**). *In a geometric sequence, each number after the first can be obtained from the preceding one by multiplying it by a fixed number, called the* **common ratio.** We can express this definition in terms of the *recursion formula*

$$a_n = ra_{n-1} \tag{19.4}$$

where a_n is any term, a_{n-1} is the preceding term, and r is the common ratio. One important application of geometric sequences is in computing compound interest on savings accounts. Other applications are found in areas such as biology and physics.

EXAMPLE 1 Illustrations of geometric sequences

(a) The sequence 2, 4, 8, 16, . . . , is a geometric sequence with a common ratio of 2. Any term after the first can be obtained by multiplying the previous term by 2. We see that the fourth term $a_4 = ra_3$, or $16 = 2(8)$.

(b) The sequence 9, -3, 1, $-1/3$, . . . , is a geometric sequence with a common ratio of $-1/3$. We can obtain any term after the first by multiplying the previous term by $-1/3$. ■

If we know the first term, we can find any other term by multiplying by the common ratio a sufficient number of times. When we do this for a general geometric sequence, we can find the nth term in terms of the first term a_1, the common ratio r, and n. Thus, the second term is a_1r, the third term is a_1r^2, and so forth. In general, the expression for the **nth term of a geometric sequence is**

$$a_n = a_1r^{n-1} \tag{19.5}$$

EXAMPLE 2 Finding a specified term

Find the eighth term of the geometric sequence 8, 4, 2, ...

By dividing any term by the previous term, we find the common ratio to be $\frac{1}{2}$. From the terms given, we see that $a_1 = 8$. From the statement of the problem, we know that $n = 8$. Thus, we substitute into Eq. (19.5) to find a_8:

$$a_8 = \underset{a_1}{8}\left(\underset{r}{\frac{1}{2}}\right)^{\overset{n}{8-1}} = \frac{8}{2^7} = \frac{1}{16}$$ ■

Practice Exercise

1. Find the sixth term of the geometric sequence 8, 2, 1/2, . . .

EXAMPLE 3 Finding the common ratio

Find the common ratio r if $a_1 = \frac{8}{625}$ and $a_{10} = -\frac{3125}{64}$.

Using Eq. (19.5), we have

$$-\frac{3125}{64} = \frac{8}{625}r^{10-1} = \frac{8}{625}r^9 \qquad \text{substituting in Eq. (19.5)}$$

$$r^9 = -\frac{(3125)(625)}{(64)(8)} \qquad \text{solving for } r$$

$$r = \left[-\frac{(3125)(625)}{(64)(8)}\right]^{1/9} = -2.5$$ ■

EXAMPLE 4 Finding a specified term

Find the seventh term of the geometric sequence for which the second term is 3, the fourth term is 9, and $r > 0$.

To get from the second term to the fourth term, we must multiply by the common ratio twice. Thus,

$$9 = 3r^2, \qquad r = \sqrt{3} \qquad \text{(because } r > 0\text{)}$$

Practice Exercise

2. Find the sixth term of the geometric sequence for which $a_1 = 81$ and $r = -1/3$.

Now, to get from the fourth term to the seventh term, we must multiply by the common ratio 3 times. Therefore,

$$a_7 = 9r^3 = 9(\sqrt{3})^3 = 27\sqrt{3}$$

■

EXAMPLE 5 Geometric sequence—chemical changes

In an experiment, 22.0% of a substance changes chemically each 10.0 min. If there is originally 120 g of the substance, how much will remain after 45.0 min?

Let $P =$ the portion of the substance remaining after each minute. From the statement of the problem, we know that $r = 0.780$, since 78.0% remains after each 10.0-min period. We also know that $a_1 = 120$ g, and we let n represent the number of minutes of elapsed time. This means that $P = 120(0.780)^{n/10.0}$. It is necessary to divide by 10.0 because the ratio is given for a 10.0-min period. In order to find P when $n = 45.0$ min, we write

$$P = 120(0.780)^{45.0/10.0} = 120(0.780)^{4.50}$$
$$= 39.2 \text{ g}$$

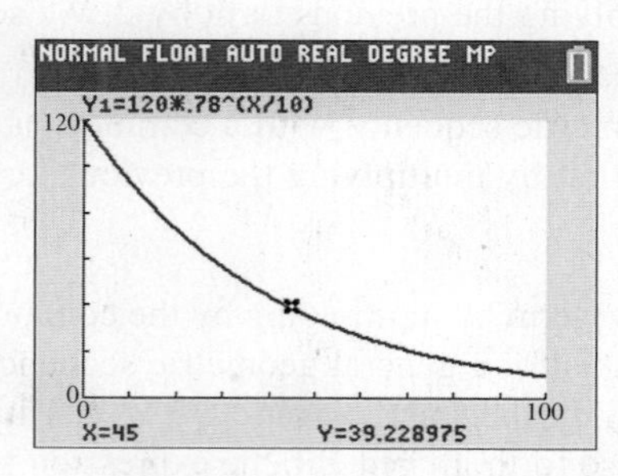

Fig. 19.3

This means that 39.2 g remain after 45.0 min. Note that the power 4.50 represents 4.50 ten-minute periods. See Fig. 19.3. ■

SUM OF n TERMS

A general expression for the sum S_n of the first n terms of a geometric sequence may be found by directly forming the sum and multiplying this equation by r:

$$S_n = a_1 + a_1r + a_1r^2 + \cdots + a_1r^{n-1}$$
$$rS_n = a_1r + a_1r^2 + a_1r^3 + \cdots + a_1r^n$$

If we now subtract the second of these equations from the first, we get $S_n - rS_n = a_1 - a_1r^n$. All other terms cancel by subtraction. Now, factoring S_n from the terms on the left and a_1 from the terms on the right, we solve for S_n. Thus, the **sum S_n of the first n terms of a geometric sequence** is

$$S_n = \frac{a_1(1 - r^n)}{1 - r} \qquad (r \neq 1) \tag{19.6}$$

EXAMPLE 6 Finding the sum of terms

Find the sum of the first seven terms of the geometric sequence for which the first term is 2 and the common ratio is 1/2.

We are to find S_n, given that $a_1 = 2$, $r = \frac{1}{2}$, and $n = 7$. Using Eq. (19.6), we have

Practice Exercise

3. Find the sum of the six terms of the geometric sequence in Practice Exercise 1.

$$S_7 = \frac{2(1 - (\frac{1}{2})^7)}{1 - \frac{1}{2}} = \frac{2(1 - \frac{1}{128})}{\frac{1}{2}} = 4\left(\frac{127}{128}\right) = \frac{127}{32}$$

■

EXAMPLE 7 Finding the sum of terms—compound interest

If \$100 is invested each year at 5% interest compounded annually, what would be the total amount of the investment after 10 years (before the 11th deposit is made)?

■ See the chapter introduction.

After 1 year, the amount invested will have added to it the interest for the year. Therefore, for the last (10th) \$100 invested, its value will become

$$\$100(1 + 0.05) = \$100(1.05) = \$105$$

■ It is reported that Albert Einstein was once asked what was the greatest discovery ever made, and his reply was "compound interest."

The next to last \$100 will have interest added twice. After 1 year, its value becomes \$100(1.05), and after 2 years it is $\$100(1.05)(1.05) = \$100(1.05)^2$. In the same way, the value of the first \$100 becomes $\$100(1.05)^{10}$, since it will have interest added 10 times. This means that we are to find the sum of the sequence

$$\underset{\text{1 year in account}}{100(1.05)} + \underset{\text{2 years in account}}{100(1.05)^2} + \underset{\text{3 years in account}}{100(1.05)^3} + \cdots + \underset{\text{10 years in account}}{100(1.05)^{10}}$$

or

$$100[1.05 + (1.05)^2 + (1.05)^3 + \cdots + (1.05)^{10}]$$

For the sequence in the brackets, we have $a_1 = 1.05$, $r = 1.05$, and $n = 10$. Thus,

$$S_{10} = \frac{1.05[1 - (1.05)^{10}]}{1 - 1.05} = 13.2068$$

The total value of the \$100 investments is $100(13.2068) = \$1320.68$. We see that \$320.68 in interest has been earned. ■

EXERCISES 19.2

In Exercises 1 and 2, make the given changes in the indicated examples of this section and then solve the given problems.

1. In Example 4, change "seventh" to "tenth."

2. In Example 6, change $\frac{1}{2}$ to $\frac{1}{3}$ and then find the sum.

In Exercises 3–6, write down the first five terms of the geometric sequence with the given values.

3. $a_1 = 6400$, $r = 0.25$

4. $a_1 = 0.09$, $r = -\frac{2}{3}$

5. $a_1 = \frac{1}{6}$, $r = 3$

6. $a_3 = -12$, $r = 2$

In Exercises 7–14, find the nth term of the geometric sequence with the given values.

7. $\frac{1}{2}, 1, 2, \ldots; n = 9$

8. $10, 1, 0.1, \ldots; n = 8$

9. $125, -25, 5, \ldots; n = 7$

10. $0.1, 0.3, 0.9, \ldots; n = 5$

11. $a_1 = -2700$, $r = -\frac{1}{3}$, $n = 10$

12. $a_2 = 24$, $r = \frac{1}{2}$, $n = 12$

13. $10^{100}, -10^{98}, 10^{96}, \ldots; n = 51$

14. $-2, 4k, -8k^2, \ldots; n = 6$

In Exercises 15–20, find the sum of the first n terms of the indicated geometric sequence with the given values.

15. $\frac{1}{8} + \frac{1}{2} + 2 + \cdots + 32$

16. $162 - 54 + 18 - \cdots - \frac{2}{3}$

17. $384, 192, 96, \ldots, n = 7$

18. $a_1 = 9$, $a_n = -243$, $n = 4$

19. $a_1 = 96$, $r = -\frac{k}{2}$, $n = 10$

20. $\log 2, \log 4, \log 16, \ldots; n = 6$

In Exercises 21–28, find any of the values of a_1, r, a_n, n, or S_n that are missing.

21. $\frac{1}{16}, \frac{1}{4}, 1, \ldots, 64$

22. $5, 1, 0.2, \ldots, 0.00032$

23. $r = \frac{3}{2}$, $n = 5$, $S_5 = 211$

24. $r = -\frac{1}{2}$, $a_n = \frac{1}{8}$, $n = 7$

25. $a_n = 27$, $n = 4$, $S_4 = 40$

26. $a_1 = 3$, $n = 7$, $a_7 = 192$

27. $a_2 = 15$, $r = \frac{1}{5}$, $a_n = \frac{3}{625}$

28. $r = -2$, $n = 6$, $S_6 = 42$

In Exercises 29–56, find the indicated quantities.

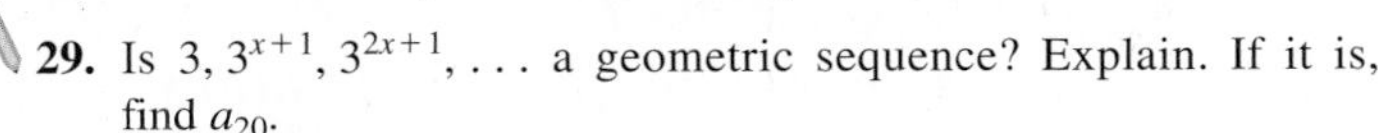

29. Is $3, 3^{x+1}, 3^{2x+1}, \ldots$ a geometric sequence? Explain. If it is, find a_{20}.

30. Show that a, $\sqrt{ab}$, b are three successive terms of a geometric sequence $(a > 0, b > 0)$.

31. Find x if 2, 6, $2x + 8$ are successive terms of a geometric sequence.

32. Write the first five terms of a geometric sequence of which the first two terms are a and b.

33. Find the sum of the first n terms of $x^2 - x^3 + x^4 - x^5 + \cdots$

34. Do the term-by-term products of two geometric sequences form a geometric sequence?

35. Each stroke of a pump removes 8.2% of the remaining air from a container. What percent of the air remains after 50 strokes?

36. From 2010 to 2015, Canada had a 1.08% average growth rate of population, compounded annually. If the population in 2010 was 34.1 million, what was the 2015 population?

37. An electric current decreases by 12.5% each 1.00 μs. If the initial current is 3.27 mA, what is the current after 8.20 μs?

38. A copying machine is set to reduce the dimensions of material copied by 10%. A drawing 12.0 cm wide is reduced, and then the copies are in turn reduced. What is the width of the drawing on the sixth reduction?

39. How much is an investment of $250 worth after 8 years if it earns annual interest of 4.8% compounded monthly? [4.8% annual interest compounded monthly means that 0.4% (4.8%/12) interest is added each month.]

40. A chemical spill pollutes a stream. A monitoring device finds 620 ppm (parts per million) of the chemical 1.0 mi below the spill, and the readings decrease by 12.5% for each mile farther downstream. How far downstream is the reading 100 ppm?

41. Measurements show that the temperature of a distant star is presently 9800°C and is decreasing by 10% every 800 years. What will its temperature be in 4000 years?

42. A swimming pool has an automatic cleaning system. When debris has entered the pool, the system can remove 15% of it in an hour. How long does it take for this system to remove 80% of the debris from the pool?

43. The strength of a signal in a fiber-optic cable decreases 12% for every 15 km along the cable. What percent of the signal remains after 100 km?

44. A series of deposits, each of value A and made at equal time intervals, earns an interest rate of i for the time interval. The deposits have a total value of

$$A(1 + i) + A(1 + i)^2 + A(1 + i)^3 + \cdots + A(1 + i)^n$$

after n time intervals (just before the next deposit). Find a formula for this sum.

45. A new car that was purchased for $40,000 depreciates by 10% each year. Find the value of the car (to the nearest dollar) when it is 10 years old.

46. The power on a space satellite is supplied by a radioactive isotope. On a given satellite, the power decreases by 0.2% each day. What percent of the initial power remains after 1 year?

47. If you decided to save money by putting away 1¢ on a given day, 2¢ one week later, 4¢ a week later, and so on, how much would you have to put away 6 months (26 weeks) after putting away the 1¢?

48. How many direct ancestors (parents, grandparents, and so on) does a person have in the ten generations that preceded him or her (assuming that no ancestor appears in more than one line of descent)?

49. A professional baseball player is offered a contract for an annual salary of $5,000,000 for six years. Also offered is a bonus (based on performance) of either $400,000 each year, or a 5.00% increase in salary each year. Which bonus option pays more over the term of the contract, and how much more?

50. If 85% of an aspirin remains in the bloodstream after 3.0 h, how long after an 80-mg aspirin is taken does it take for there to be 35 mg in the bloodstream?

51. A certain tennis ball was dropped from a height of 100 ft. On each bounce, the ball reached a height equal to 55% of the previous height. Find the height of the ball (to the nearest hundredth) after the tenth bounce.

52. Write down several terms of a general geometric sequence. Then take the logarithm of each term. Explain why the resulting sequence is an arithmetic sequence.

53. Do the squares of the terms of a geometric sequence also form a geometric sequence? Explain.

54. If $a_1, a_2, a_3, \ldots$ is an arithmetic sequence, explain why $2^{a_1}, 2^{a_2}, 2^{a_3}, \ldots$ is a geometric sequence.

55. The numbers 8, x, y form an arithmetic sequence, and the numbers x, y, 36 form a geometric sequence. Find all of the possible sequences.

56. For $f(x) = ab^x$, is $f(1), f(3), f(5)$ an arithmetic sequence or a geometric sequence? What is the common difference (or common ratio)?

Answers to Practice Exercises

1. 1/128 **2.** $a_6 = -1/3$ **3.** $S_6 = 1365/128$

19.3 Infinite Geometric Series

Series • Infinity • Limit • Sum of an Infinite Geometric Series • Repeating Decimal

In the previous sections, we developed formulas for the sum of the first n terms of an arithmetic sequence and of a geometric sequence. *The indicated sum of the terms of a sequence is called a* **series.**

EXAMPLE 1 Illustrations of series

(a) The indicated sum of the terms of the arithmetic sequence 2, 5, 8, 11, 14, . . . is the series $2 + 5 + 8 + 11 + 14 + \cdots$.

(b) The indicated sum of terms of the geometric sequence $1, \frac{1}{2}, \frac{1}{4}, \frac{1}{8}, \ldots$ is the series $1 + \frac{1}{2} + \frac{1}{4} + \frac{1}{8} + \cdots$. ■

The series associated with a finite sequence will sum up to a real number. The series associated with an infinite arithmetic sequence will not sum up to a real number, as the terms being added become larger and larger numerically. The sum is unbounded, as we can see in Example 1(a). The series associated with an infinite geometric sequence may or may not sum up to a real number, as we now show.

Let us now consider the sum of the first n terms of the infinite geometric sequence $1, \frac{1}{2}, \frac{1}{4}, \ldots$. This is the sum of the n terms of the associated geometric series

$$1 + \frac{1}{2} + \frac{1}{4} + \cdots + \frac{1}{2^{n-1}}$$

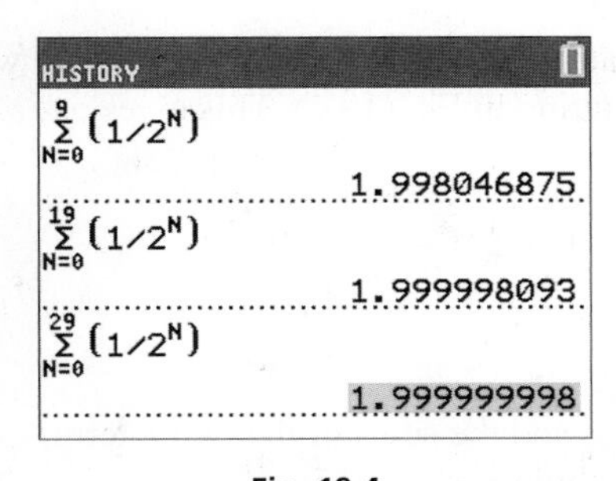

Fig. 19.4
Graphing calculator keystrokes: goo.gl/zVSbSH

Here, $a_1 = 1$ and $r = \frac{1}{2}$, and we find that we get the values of S_n for the given values of n in the following table:

n	2	3	4	5	6	7	8	9	10
S_n	$\frac{3}{2}$	$\frac{7}{4}$	$\frac{15}{8}$	$\frac{31}{16}$	$\frac{63}{32}$	$\frac{127}{64}$	$\frac{255}{128}$	$\frac{511}{256}$	$\frac{1023}{512}$

$1 + \frac{1}{2} + \frac{1}{4} + \frac{1}{8}$ (for $n = 4$)

$1 + \frac{1}{2} + \frac{1}{4} + \frac{1}{8} + \frac{1}{16} + \frac{1}{32} + \frac{1}{64}$ (for $n = 7$)

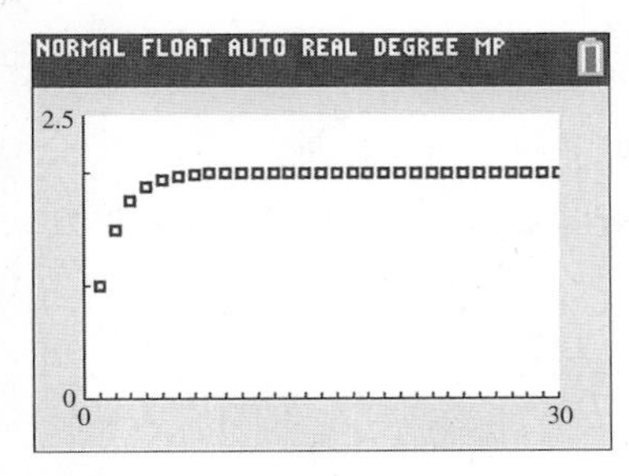

Fig. 19.5
Graphing calculator keystrokes: goo.gl/Dx9WEn

The series for $n = 4$ and $n = 7$ are shown. We see that as n gets larger, the value of S_n gets closer and closer to 2. Figure 19.4 shows the values of S_{10}, S_{20}, and S_{30}. Figure 19.5 shows the graph of S_1 through S_{30}. We can see numerically and graphically that S_n *approaches* 2 as n gets larger. For the sum of the first n terms of a geometric sequence

$$S_n = a_1\frac{1 - r^n}{1 - r}$$

the term r^n becomes exceedingly small if $|r| < 1$, and if n is sufficiently large, this term is effectively zero. *If this term were exactly zero,* the sum would be

$$S_n = 1\frac{1 - 0}{1 - \frac{1}{2}} = 2$$

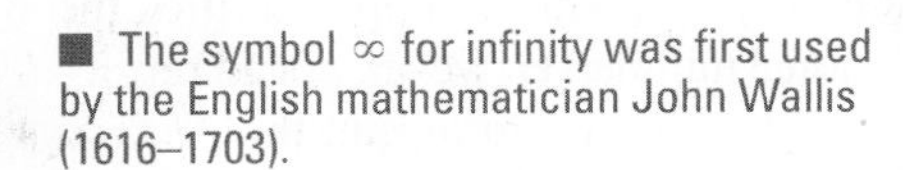
■ The symbol ∞ for infinity was first used by the English mathematician John Wallis (1616–1703).

The only problem is that we cannot find any number large enough for n to make $(\frac{1}{2})^n$ zero. There is, however, an accepted notation for this. This notation is

$$\lim_{n \to \infty} r^n = 0 \qquad (\text{if } |r| < 1)$$

and it is read as "the limit, as n *approaches* infinity, of r to the nth power is zero."

CAUTION The symbol ∞ is read as **infinity,** but it must *not* be thought of as a number. It is simply a symbol that stands for a process of considering numbers that become large without bound. ■ The number called the **limit** of the sums is simply the number the sums get closer and closer to, as n is considered to approach infinity. This notation and terminology are of particular importance in calculus.

If we consider values of r such that $|r| < 1$ and let the values of n become unbounded, we find that $\lim_{n \to \infty} r^n = 0$. The formula for *the sum of the terms of an infinite geometric series then becomes*

$$S = \frac{a_1}{1 - r} \quad (|r| < 1) \tag{19.7}$$

where a_1 is the first term and r is the common ratio. If $|r| \geq 1$, S is unbounded in value.

NORMAL FLOAT AUTO REAL DEGREE MP

L1	L2	L3	L4	L5
1	4	4	------	------
2	-.5	3.5		
3	.0625	3.5625		
4	-.0078	3.5547		
5	9.8E-4	3.5557		
6	-1E-4	3.5555		
7	1.5E-5	3.5556		
8	-2E-6	3.5556		
9	2.4E-7	3.5556		
10	-3E-8	3.5556		
11	3.7E-9	3.5556		

L3(1)=4

Fig. 19.6
Graphing calculator keystrokes:
goo.gl/sGrU7q

EXAMPLE 2 Sum of an infinite geometric series

Find the sum of the infinite geometric series

$$4 - \frac{1}{2} + \frac{1}{16} - \frac{1}{128} + \cdots$$

Here, we see that $a_1 = 4$. We find r by dividing any term by the previous term, and we find that $r = -\frac{1}{8}$. We then find the sum by substituting in Eq. (19.7). This gives us

$$S = \frac{4}{1 - \left(-\frac{1}{8}\right)} = \frac{4}{1 + \frac{1}{8}}$$
$$= \frac{4}{1} \times \frac{8}{9} = \frac{32}{9}$$

In Fig. 19.6, the terms of the series are shown in L_2 and the sums of the first n terms are shown in L_3. We can see that the sums are approaching $\frac{32}{9} = 3.555\ldots$. ■

Practice Exercise

1. Find the sum of the infinite geometric series $9 + 3 + 1 + 1/3 + \cdots$.

EXAMPLE 3 Decimal form to fraction

Find the fraction that has its decimal form 0.121212. . . .

This decimal form can be considered as being

$$0.12 + 0.0012 + 0.000012 + \cdots$$

which means that we have an infinite geometric series in which $a_1 = 0.12$ and $r = 0.01$. Thus,

$$S = \frac{0.12}{1 - 0.01} = \frac{0.12}{0.99}$$
$$= \frac{4}{33}$$

Therefore, the decimal 0.121212 . . . and the fraction $\frac{4}{33}$ represent the same number. ■

■ Note that the series $1 + \frac{3}{2} + \frac{9}{4} + \frac{27}{8} + \cdots$ does *not* have a finite sum since $|r| = \frac{3}{2} \geq 1$.

Practice Exercise

2. Find the fraction that has as its decimal form 0.272727. . . .

NOTE ▸ [The decimal in Example 3 is called a **repeating decimal,** because a particular sequence of digits in the decimal form repeats endlessly.] This example verifies the theorem that any repeating decimal represents a rational number. However, not all repeating decimals start repeating immediately. If the numbers never do repeat, the decimal represents an irrational number. For example, there are no repeating decimals that represent π, $\sqrt{2}$, or e. The decimal form of the number e does repeat at one point, but the repetition stops. As we noted in Section 12.5, the decimal form of e to 16 decimal places is 2.7 1828 1828 4590 452. We see that the sequence of digits 1828 repeats only once.

EXAMPLE 4 Repeating decimal

Find the fraction that has its decimal form the repeating decimal 0.50345345345. . . .

We first separate the decimal into the beginning, nonrepeating part, and the infinite repeating decimal, which follows. Thus, we have

$$0.50345345345\ldots = 0.50 + 0.00345345345\ldots$$

This means that we are to add $\frac{50}{100}$ to the fraction that represents the sum of the terms of the infinite geometric series $0.00345 + 0.00000345 + \cdots$. For this series, $a_1 = 0.00345$ and $r = 0.001$. We find this sum to be

$$S = \frac{0.00345}{1 - 0.001} = \frac{0.00345}{0.999} = \frac{115}{33{,}300} = \frac{23}{6660}$$

Therefore,

$$0.50345345\ldots = \frac{5}{10} + \frac{23}{6660} = \frac{5(666) + 23}{6660} = \frac{3353}{6660}$$ ■

■ A calculator can be used to change a repeating decimal to a fraction. Figure 19.7 shows this for the decimals in Examples 3 and 4.

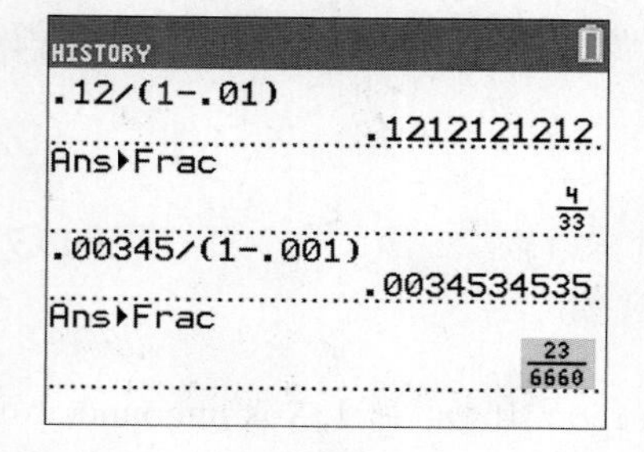

Fig. 19.7
Graphing calculator keystrokes:
goo.gl/OdbYXF

EXAMPLE 5 Infinite geometric series—pendulum distance

Each swing of a certain pendulum bob is 95% as long as the previous swing. How far does the bob travel if the first swing is 40.0 in. long?

We are to find the sum of the terms of an infinite geometric series for which $a_1 = 40.0$ and $r = 95\% = \frac{19}{20}$. Substituting these values into Eq. (19.7), we obtain

$$S = \frac{40.0}{1 - \frac{19}{20}} = \frac{40.0}{\frac{1}{20}} = (40.0)(20) = 800 \text{ in.}$$

The pendulum bob travels 800 in. (about 67 ft). ■

Practice Exercise

3. In Example 5, how far does the bob travel if each swing is 99% as long as the previous swing?

EXERCISES 19.3

In Exercises 1 and 2, make the given changes in the indicated examples of this section, and then solve the resulting problems.

1. In Example 2, change all − signs to + in the series.
2. In Example 3, change the decimal form to 0.012012012 . . . .

In Exercises 3–6, find the indicated quantity for an infinite geometric series.

3. $a_1 = 4, r = \frac{1}{2}, S = ?$
4. $a_3 = 68, r = -\frac{1}{3}, S = ?$
5. $a_1 = 0.5, S = 0.625, r = ?$
6. $S = 4 + 2\sqrt{2}, r = \dfrac{1}{\sqrt{2}}, a_1 = ?$

In Exercises 7–14, find the sums of the given infinite geometric series.

7. $20 - 1 + 0.05 - \cdots$
8. $9 + 8.1 + 7.29 + \cdots$
9. $4 + \frac{7}{2} + \frac{49}{16} + \cdots$
10. $6 - 4 + \frac{8}{3} - \cdots$
11. $1 + 10^{-4} + 10^{-8} + \cdots$
12. $1000 - 300 + 90 - \cdots$
13. $(2 + \sqrt{3}) + 1 + (2 - \sqrt{3}) + \cdots$
14. $(1 + \sqrt{2}) - 1 + (\sqrt{2} - 1) - \cdots$

In Exercises 15–24, find the fractions equal to the given decimals.

15. 0.33333 . . .
16. 0.272727 . . .
17. 0.499999 . . .
18. 0.999999 . . .
19. 0.404040 . . .
20. 0.070707 . . .
21. 0.0273273273 . . .
22. 0.82222 . . .
23. 0.1181818 . . .
24. 0.3336336336 . . .

In Exercises 25–36, solve the given problems by use of the sum of an infinite geometric series.

25. When two dice are rolled repeatedly, the probability of rolling a sum of 6 before a sum of 7 is $\frac{5}{36} + \left(\frac{25}{36}\right)\frac{5}{36} + \left(\frac{25}{36}\right)^2\frac{5}{36} + \left(\frac{25}{36}\right)^3\frac{5}{36} + \cdots$. Find this probability.
26. Explain why there is no infinite geometric series with $a_1 = 5$ and $S = 2$.
27. Liquid is continuously collected in a wastewater-holding tank such that during a given hour only 92.0% as much liquid is collected as in the previous hour. If 28.0 gal are collected in the first hour, what must be the minimum capacity of the tank?
28. If 75% of all aluminum cans are recycled, what is the total number of recycled cans that can be made from 400,000 cans that are recycled over and over until all the aluminum from these cans is used up? (Assume no aluminum is lost in the recycling process.)
29. The amounts of plutonium-237 that decay each day because of radioactivity form a geometric sequence. Given that the amounts that decay during each of the first 4 days are 5.882 g, 5.782 g, 5.684 g, and 5.587 g, respectively, what total amount will decay?
30. A helium-filled balloon rose 120 ft in 1.0 min. Each minute after that, it rose 75% as much as in the previous minute. What was its maximum height?
31. A bicyclist traveling at 10 m/s then begins to coast. The bicycle travels 0.90 as far each second as in the previous second. How far does the bicycle travel while coasting?
32. If a major league baseball team can make it to the World Series, it can be a great financial boost to the economy of their city. Let us assume, if their team plays in the World Series, that tourists will spend $20,000,000 for hotels, restaurants, local transporation, tickets for the Series and city attractions, and so forth. We now assume that 75% of this money will be spent in a second round of spending in the city by those who received it. A third round, fourth round, fifth round, and so on of 75% spending will follow. Now assuming this continues indefinitely, what is the total amount of this spending, which in effect is added to the economy of the city because of the World Series? *This problem illustrates one of the major reasons a city wants to host major events that attract many tourists.*
33. A double-pane window has a thin air space between two parallel panes of glass. If 90.0% of the incoming light is transmitted through a pane, and 10.0% is reflected, what percent of the light is transmitted through the second pane? See Fig. 19.8, which illustrates one incident ray of light and its unlimited reflections and transmissions through the two panes of glass.

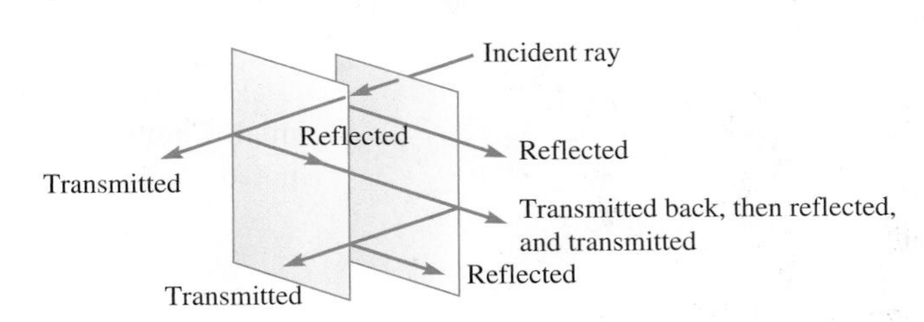

Fig. 19.8

34. A square has sides of 20 cm. Another square is inscribed in the first square by joining the midpoints of the sides. Assuming that such inscribed squares can be formed endlessly, find the sum of the areas of all the squares and explain how the sum is found. See Fig. 19.9.

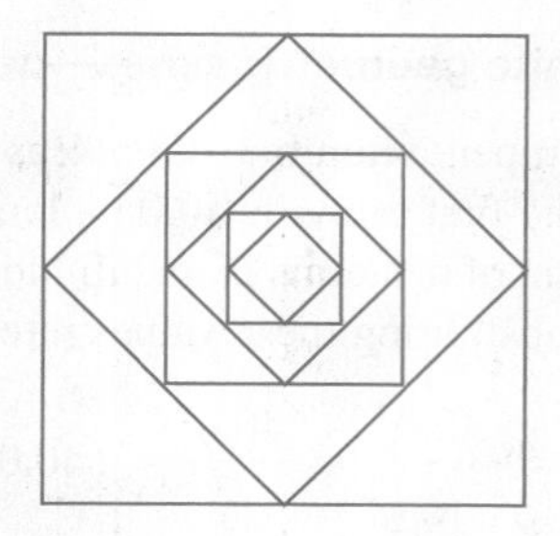

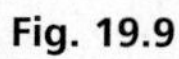

Fig. 19.9

35. Find x if the sum of the terms of the infinite geometric series $1 + 2x + 4x^2 + \cdots$ is 2/3.

36. Find the sum of the terms of the infinite series $1 + 2x + 3x^2 + 4x^3 + \cdots$ for $|x| < 1$. (*Hint:* Use $S - xS$.)

Answers to Practice Exercises

1. $S = 27/2$ **2.** $3/11$ **3.** 4000 in.

19.4 The Binomial Theorem

Properties of $(a + b)^n$ • Factorial Notation • Binomial Theorem • Pascal's Triangle • Binomial Series

To expand $(x + 2)^5$ would require a number of multiplications that would be a tedious operation. We now develop *the* **binomial theorem,** *by which it is possible to expand binomials to any given power without direct multiplication.* Using this type of expansion, it is also possible to expand some expressions for which direct multiplication is not possible. The binomial theorem is used to develop expressions needed in certain mathematics topics and in technical applications.

By direct multiplication, we may obtain the following expansions of the binomial $a + b$:

$$(a + b)^0 = 1$$
$$(a + b)^1 = a + b$$
$$(a + b)^2 = a^2 + 2ab + b^2$$
$$(a + b)^3 = a^3 + 3a^2b + 3ab^2 + b^3$$
$$(a + b)^4 = a^4 + 4a^3b + 6a^2b^2 + 4ab^3 + b^4$$
$$(a + b)^5 = a^5 + 5a^4b + 10a^3b^2 + 10a^2b^3 + 5ab^4 + b^5$$

Inspection shows these expansions have certain properties, and we assume that these properties are valid for the expansion of $(a + b)^n$, where n is any positive integer.

Properties of the Binomial $(a + b)^n$

1. There are $n + 1$ terms.
2. The first term is a^n, and the final term is b^n.
3. Progressing from the first term to the last, the exponent of a decreases by 1 from term to term, the exponent of b increases by 1 from term to term, and the sum of the exponents of a and b in each term is n.
4. If the coefficient of any term is multiplied by the exponent of a in that term and this product is divided by the number of that term, we obtain the coefficient of the next term.
5. The coefficients of terms equidistant from the ends are equal.

The following example illustrates the use of the basic properties in expanding a binomial. Carefully note the diagram in which each of these five properties is specifically noted.

EXAMPLE 1 Using basic binomial properties

Using the basic properties, develop the expansion for $(a + b)^5$.

Because the exponent of the binomial is 5, we have $n = 5$.

From property 1, we know that there are six terms.

From property 2, we know that the first term is a^5 and the final term is b^5.

From property 3, we know that the factors of a and b in terms 2, 3, 4, and 5 are a^4b, a^3b^2, a^2b^3, and ab^4, respectively.

From property 4, we obtain the coefficients of terms 2, 3, 4, and 5. In the first term, a^5, the coefficient is 1. Multiplying by 5, the power of a, and dividing by 1, the number of the term, we obtain 5, which is the coefficient of the second term. Thus, the second term is $5a^4b$. Again using property 4, we obtain the coefficient of the third term. The coefficient of the second term is 5. Multiplying by 4, and dividing by 2, we obtain 10. This means that the third term is $10a^3b^2$.

From property 5, we know that the coefficient of the fifth term is the same as the second and that the coefficient of the fourth term is the same as the third. These properties are illustrated in the following diagram:

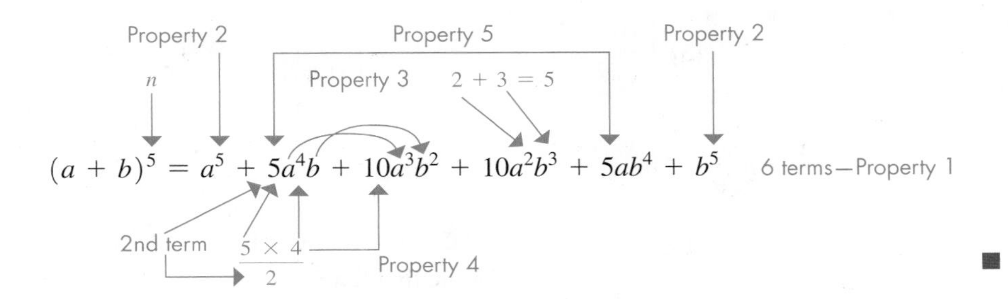

■

It is not necessary to use the above properties directly to expand a given binomial. If they are applied to $(a + b)^n$, a general formula for the expansion of a binomial may be obtained. In developing and stating the general formula, it is convenient to use the **factorial notation $n!$**, where

$$n! = n(n - 1)(n - 2) \cdots (2)(1) \tag{19.8}$$

We see that $n!$, read "n factorial," represents the product of the first n positive integers.

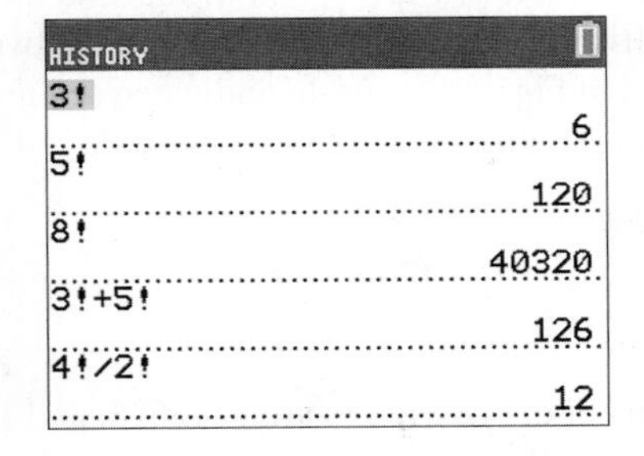

Fig. 19.10

Graphing calculator keystrokes: goo.gl/W9BKSd

EXAMPLE 2 Evaluating factorials

(a) $3! = (3)(2)(1) = 6$

(b) $5! = (5)(4)(3)(2)(1) = 120$

(c) $8! = (8)(7)(6)(5)(4)(3)(2)(1) = 40{,}320$

(d) $3! + 5! = 6 + 120 = 126$

(e) $\dfrac{4!}{2!} = \dfrac{(4)(3)(2)(1)}{(2)(1)} = 12$

In evaluating factorials, we must remember that they represent products of numbers. **CAUTION** In parts (d) and (e), we see that $3! + 5!$ *is not* $(3 + 5)!$ and $4!/2!$ *is not* $2!$. ■ The evaluation of the factorials in Example 2 are shown in the calculator display in Fig. 19.10. ■

Practice Exercise

1. Evaluate $9!/7!$.

THE BINOMIAL FORMULA

Based on the binomial properties, *the* **binomial theorem** *states that the following* **binomial formula** *is valid for all positive integer values of n* (the binomial theorem is proven through advanced methods).

$$(a+b)^n = a^n + na^{n-1}b + \frac{n(n-1)}{2!}a^{n-2}b^2 + \frac{n(n-1)(n-2)}{3!}a^{n-3}b^3 + \cdots + b^n \qquad \textbf{(19.9)}$$

EXAMPLE 3 Using the binomial formula

Using the binomial formula, expand $(2x+3)^6$.

In using the binomial formula for $(2x+3)^6$, we use $2x$ for a, 3 for b, and 6 for n. Thus,

$$(2x+3)^6 = (2x)^6 + 6(2x)^5(3) + \frac{(6)(5)}{2}(2x)^4(3)^2 + \frac{(6)(5)(4)}{(2)(3)}(2x)^3(3)^3 + \frac{(6)(5)(4)(3)}{(2)(3)(4)}(2x)^2(3)^4 + \frac{(6)(5)(4)(3)(2)}{(2)(3)(4)(5)}(2x)(3)^5 + (3)^6$$

($n = 6$; $n-1$; n; $n-1$; $n-2$; $2!$; $3!$; $4!$; $5!$)

$$= 64x^6 + 576x^5 + 2160x^4 + 4320x^3 + 4860x^2 + 2916x + 729$$

■

PASCAL'S TRIANGLE

For the first few integer powers of a binomial $a + b$, the coefficients can be obtained by setting them up in the following pattern, known as **Pascal's triangle.**

■ Named for the French scientist and mathematician Blaise Pascal (1623–1662).

$n = 0$							1						
$n = 1$						1		1					
$n = 2$					1		2		1				
$n = 3$				1		3		3		1			
$n = 4$			1		4		6		4		1		
$n = 5$		1		5		10		10		5		1	
$n = 6$	1		6		15		20		15		6		1

see expansions on page 526

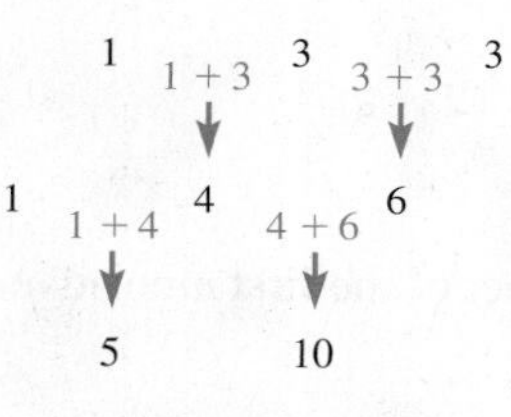

Fig. 19.11

We note that the first and last coefficients shown in each row are 1, and the second and next-to-last coefficients are equal to n. Other coefficients are obtained by adding the two nearest coefficients in the row above, as illustrated in Fig. 19.11 for the indicated section of Pascal's triangle. This pattern may be continued indefinitely, although use of Pascal's triangle is cumbersome for high values of n.

EXAMPLE 4 Using Pascal's triangle

Using Pascal's triangle, expand $(5s - 2t)^4$.

Here, we note that $n = 4$. Thus, the coefficients of the five terms are 1, 4, 6, 4, and 1, respectively. Also, here we use $5s$ for a and $-2t$ for b. We are expanding this expression as $[(5s) + (-2t)]^4$. Therefore,

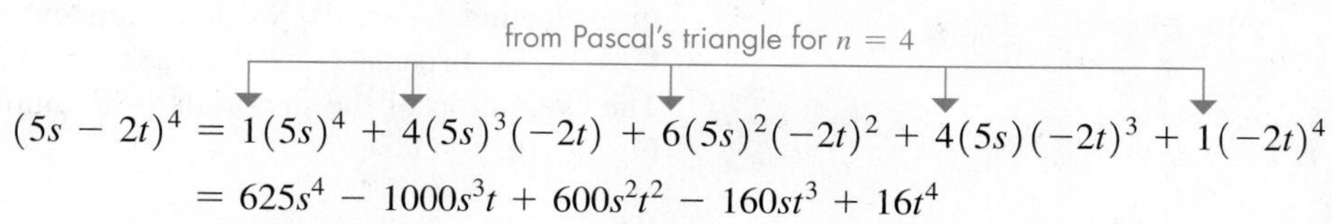

$$(5s - 2t)^4 = 1(5s)^4 + 4(5s)^3(-2t) + 6(5s)^2(-2t)^2 + 4(5s)(-2t)^3 + 1(-2t)^4$$
$$= 625s^4 - 1000s^3t + 600s^2t^2 - 160st^3 + 16t^4$$

The coefficients 1, 4, 6, 4, and 1 are called **binomial coefficients.** In addition to using Pascal's triangle, these coefficients can be found on a calculator as shown in Fig. 19.12.

■

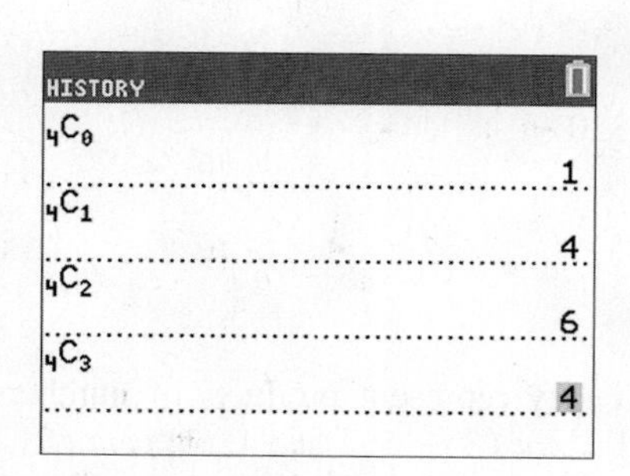

Fig. 19.12
Graphing calculator keystrokes: goo.gl/L8TMfu

Practice Exercise

2. Using Pascal's triangle, expand $(5s + t)^4$.

In certain uses of a binomial expansion, it is not necessary to obtain all terms. Only the first few terms are required. The following example illustrates finding the first four terms of an expansion.

EXAMPLE 5 Using the binomial formula

Find the first four terms of the expansion of $(x + 7)^{12}$.

Here, we use x for a, 7 for b, and 12 for n. Thus, from the binomial formula, we have

$$(x + 7)^{12} = x^{12} + 12x^{11}(7) + \frac{(12)(11)}{2}x^{10}(7^2) + \frac{(12)(11)(10)}{(2)(3)}x^9(7^3) + \cdots$$

$$= x^{12} + 84x^{11} + 3234x^{10} + 75{,}460x^9 + \cdots$$ ■

Practice Exercise

3. Find the first three terms of the expansion of $(x - 4)^9$.

If we let $a = 1$ and $b = x$ in the binomial formula, we obtain the **binomial series.**

Binomial Series

$$(1 + x)^n = 1 + nx + \frac{n(n-1)}{2!}x^2 + \frac{n(n-1)(n-2)}{3!}x^3 + \cdots \quad \textbf{(19.10)}$$

Through advanced methods, the binomial series can be shown to be valid for any real number n if $|x| < 1$. When n is either negative or a fraction, we obtain an infinite series. In such a case, we calculate as many terms as may be needed although such a series is not obtainable through direct multiplication. The binomial series may be used to develop important expressions that are used in applications and more advanced mathematics topics.

EXAMPLE 6 Binomial series—forces on a beam

In the analysis of forces on beams, the expression $1/(1 + m^2)^{3/2}$ is used. Use the binomial series to find the first four terms of the expansion.

Using negative exponents, we have

$$1/(1 + m^2)^{3/2} = (1 + m^2)^{-3/2}$$

Now, in using Eq. (19.10), we have $n = -3/2$ and $x = m^2$:

$$(1 + m^2)^{-3/2} = 1 + \left(-\frac{3}{2}\right)(m^2) + \frac{(-\frac{3}{2})(-\frac{3}{2} - 1)}{2!}(m^2)^2 + \frac{(-\frac{3}{2})(-\frac{3}{2} - 1)(-\frac{3}{2} - 2)}{3!}(m^2)^3 + \cdots$$

Therefore,

$$\frac{1}{(1 + m^2)^{3/2}} = 1 - \frac{3}{2}m^2 + \frac{15}{8}m^4 - \frac{35}{16}m^6 + \cdots$$ ■

EXAMPLE 7 Evaluation using binomial series

Approximate the value of 0.97^7 by use of the binomial series.

We note that $0.97 = 1 - 0.03$, which means $0.97^7 = [1 + (-0.03)]^7$. Using four terms of the binomial series, we have

$$0.97^7 = [1 + (-0.03)]^7$$

$$= 1 + 7(-0.03) + \frac{7(6)}{2!}(-0.03)^2 + \frac{7(6)(5)}{3!}(-0.03)^3$$

$$= 1 - 0.21 + 0.0189 - 0.000945 = 0.807955$$

From a calculator, we find that $0.97^7 = 0.807983$ (to six decimal places), which means these values agree to four significant digits with a value 0.8080. Greater accuracy can be found using the binomial series if more terms are used. ■

Practice Exercise

4. Using the binomial series, approximate the value of 1.03^7.

EXERCISES 19.4

In Exercises 1 and 2, make the given changes in the indicated examples of this section and then solve the given problems.

1. In Example 3, change the exponent from 6 to 5.

2. In Example 7, change 0.97 to 0.98.

In Exercises 3–12, expand and simplify the given expressions by use of the binomial formula.

3. $(t + 4)^3$ **4.** $(x - 2)^3$ **5.** $(2x - 3)^4$

6. $(4x^2 + 5)^4$ **7.** $(6 + 0.1)^5$ **8.** $(xy^2 - z)^5$

9. $(n + 2\pi^3)^5$ **10.** $(1 - j)^6$ $(j = \sqrt{-1})$

11. $(2a - b^2)^6$ **12.** $\left(\frac{a}{x} + x\right)^6$

In Exercises 13–16, expand and simplify the given expressions by use of Pascal's triangle.

13. $(5x - 3)^4$ **14.** $(x - 4)^5$ **15.** $(2a + 1)^6$ **16.** $(x - 3)^7$

In Exercises 17–24, find the first four terms of the indicated expansions.

17. $(x + 2)^{10}$ **18.** $(x - 4)^8$ **19.** $(2a - 1)^7$

20. $(3b + 2)^9$ **21.** $(x^{1/2} - 4y)^{12}$ **22.** $(2a - x^{-1})^{11}$

23. $\left(b^2 + \frac{1}{2b}\right)^{20}$ **24.** $\left(2x^2 + \frac{y}{3}\right)^{15}$

In Exercises 25–28, approximate the value of the given expression to three decimal places by using three terms of the appropriate binomial series. Check using a calculator.

25. 1.05^6 **26.** $\sqrt{0.927}$ **27.** $\sqrt[3]{1.045}$ **28.** 0.98^{-7}

In Exercises 29–36, find the first four terms of the indicated expansions by use of the binomial series.

29. $(1 + x)^8$ **30.** $(1 + x)^{-1/3}$ **31.** $(1 - 3x)^{-2}$

32. $(1 - 2\sqrt{x})^9$ **33.** $\sqrt{1 + x}$ **34.** $\frac{1}{\sqrt{1 - x}}$

35. $\frac{1}{\sqrt{9 - 9x}}$ **36.** $\sqrt{4 + x^2}$

In Exercises 37–40, solve the given problems involving factorials.

37. Using a calculator, evaluate (a) $17! + 4!$, (b) $21!$, (c) $17! \times 4!$, and (d) $68!$.

38. Using a calculator, evaluate (a) $8! - 7!$, (b) $8!/7!$, (c) $8! \times 7!$, and (d) $56!$.

39. Show that $n! = n \times (n - 1)!$ for $n \geq 2$. To use this equation for $n = 1$, explain why it is necessary to define $0! = 1$ (this is a standard definition of $0!$).

40. Show that $\frac{(n + 1)!}{(n - 2)!} = n^3 - n$ for $n \geq 2$. See Exercise 39.

In Exercises 41–44, find the indicated terms by use of the following information. The $r + 1$ term of the expansion of $(a + b)^n$ is given by

$$\frac{\overbrace{n(n - 1)(n - 2) \cdots (n - r + 1)}^{r \text{ factors}}}{r!} a^{n-r} b^r$$

41. The term involving b^5 in $(a + b)^8$

42. The term involving y^6 in $(x + y)^{10}$

43. The fifth term of $(2x - 3b)^{12}$

44. The sixth term of $(\sqrt{a} - \sqrt{b})^{14}$

In Exercises 45–58, solve the given problems.

45. Without expanding, evaluate

$$(2x - 1)^3 + 3(2x - 1)^2(3 - 2x) + 3(2x - 1)(3 - 2x)^2 + (3 - 2x)^3$$

46. Add up the coefficients of the first few powers of $(a + b)^n$ by using Pascal's triangle. Then by noting the pattern, determine the sum of the coefficients of $(a + b)^{79}$.

47. Explain why $n!$ ends in a zero if $n > 4$.

48. Expand $[(a + b) + c]^3$ using the binomial theorem. Group terms as indicated.

49. A mechanical system has 4 redundant engines, each with a 95% probability of functioning properly. When the binomial theorem is used to expand $(0.95 + 0.05)^4$, the resulting terms give the probabilities of exactly 4, 3, 2, 1, and 0 engines functioning properly. Find the probability of exactly 3 engines functioning properly (the term containing 0.95^3). Round to the third decimal place.

50. In Exercise 49, find the probability that *at least one* engine functions properly (the sum of the terms containing 0.95^4, 0.95^3, 0.95^2, and 0.95^1). Round to the third decimal place.

51. Approximate $\sqrt{6}$ to hundredths by noting that $\sqrt{6} = \sqrt{4(1.5)} = 2\sqrt{1 + 0.5}$ and using four terms of the appropriate binomial series.

52. Approximate $\sqrt[3]{10}$ by using the method of Exercise 51.

53. A company purchases a piece of equipment for A dollars, and the equipment depreciates at a rate of r each year. Its value V after n years is $V = A(1 - r)^n$. Expand this expression for $n = 5$.

54. In finding the rate of change of emission of energy from the surface of a body at temperature T, the expression $(T + h)^4$ is used. Expand this expression.

55. In the theory associated with the magnetic field due to an electric current, the expression $1 - \frac{x}{\sqrt{a^2 + x^2}}$ is found. By expanding $(a^2 + x^2)^{-1/2}$, find the first three nonzero terms that could be used to approximate the given expression.

56. In the theory related to the dispersion of light, the expression $1 + \frac{A}{1 - \lambda_0^2/\lambda^2}$ arises. (a) Let $x = \lambda_0^2/\lambda^2$ and find the first four terms of the expansion of $(1 - x)^{-1}$. (b) Find the same expansion by using long division. (c) Write the original expression in expanded form, using the results of (a) and (b).

57. A cubic Bezier curve is defined by four points (P_0 through P_3) from the parametric equations given by $(1-t)^3P_0 + 3(1-t)^2tP_1 + 3(1-t)t^2P_2 + t^3P_3$, where $0 \le t \le 1$. Note that the points P_0 through P_3 are multiplied by terms of the binomial expansion $[(1-t)+t]^3$. Using a similar process, find the equation of the fourth-degree Bezier curve defined by the five points P_0 through P_4 (see Example 6 in Section 15.3).

58. Find the first four terms of the expansion of $(1+x)^{-1}$ and then divide $1+x$ into 1. Compare the results.

Answers to Practice Exercises

1. 72 **2.** $625s^4 + 500s^3t + 150s^2t^2 + 20st^3 + t^4$
3. $x^9 - 36x^8 + 576x^7 - \cdots$ **4.** 1.230 (rounded off)

CHAPTER 19 KEY FORMULAS AND EQUATIONS

Arithmetic sequences

Recursion formula $\quad a_n = a_{n-1} + d \qquad$ **(19.1)**

nth term $\quad a_n = a_1 + (n-1)d \qquad$ **(19.2)**

Sum of n terms $\quad S_n = \frac{n}{2}(a_1 + a_n) \qquad$ **(19.3)**

Geometric sequences

Recursion formula $\quad a_n = ra_{n-1} \qquad$ **(19.4)**

nth term $\quad a_n = a_1r^{n-1} \qquad$ **(19.5)**

Sum of n terms $\quad S_n = \frac{a_1(1-r^n)}{1-r} \quad (r \ne 1) \qquad$ **(19.6)**

Sum of an infinite geometric series $\quad S = \frac{a_1}{1-r} \quad (|r| < 1) \qquad$ **(19.7)**

Factorial notation $\quad n! = n(n-1)(n-2)\cdots(2)(1) \qquad$ **(19.8)**

Binomial formula

$$(a+b)^n = a^n + na^{n-1}b + \frac{n(n-1)}{2!}a^{n-2}b^2 + \frac{n(n-1)(n-2)}{3!}a^{n-3}b^3 + \cdots + b^n \qquad \textbf{(19.9)}$$

Binomial series

$$(1+x)^n = 1 + nx + \frac{n(n-1)}{2!}x^2 + \frac{n(n-1)(n-2)}{3!}x^3 + \cdots \qquad \textbf{(19.10)}$$

CHAPTER 19 REVIEW EXERCISES

CONCEPT CHECK EXERCISES

Determine each of the following as being either **true** *or* **false**. *If it is false, explain why.*

1. For an arithmetic sequence, if $a_1 = 5$, $d = 3$, and $n = 10$, then $a_{10} = 35$.

2. For a geometric sequence, if $a_3 = 8$, $a_5 = 2$, then $r = \frac{1}{2}$.

3. The sum of the geometric series $6 - 2 + \frac{4}{3} - \cdots$ is $\frac{9}{2}$.

4. Evaluating the first three terms, $(x+y)^{10} = x^{10} + 10x^9y + 45x^8y^2 + \cdots$

PRACTICE AND APPLICATIONS

In Exercises 5–12, find the indicated term of each sequence.

5. $1, 6, 11, \ldots$ (17th) **6.** $1, -3, -7, \ldots$ (21st)

7. $500, 100, 20, \ldots$ (9th) **8.** $0.025, 0.01, 0.004, \ldots$ (7th)

9. $-1, 3.5, 8, \ldots$ (16th) **10.** $-1, -\frac{5}{3}, -\frac{7}{3}, \ldots$ (25th)

11. $\frac{3}{4}, \frac{1}{2}, \frac{1}{3}, \ldots$ (7th) **12.** $5^{-2}, 5^0, 5^2, \ldots$ (18th)

In Exercises 13–16, find the sum of each sequence with the indicated values.

13. $a_1 = -4$, $n = 15$, $a_{15} = 17$ (arith.)

14. $a_1 = 300$, $d = -\frac{20}{3}$, $n = 10$

15. $a_1 = 16$, $r = -\frac{1}{2}$, $n = 14$

16. $a_1 = 64$, $a_n = 729$, $n = 7$ (geom., $r > 0$)

In Exercises 17–26, find the indicated quantities for the appropriate sequences.

17. $a_1 = 17$, $d = -2$, $n = 9$, $S_9 = ?$

18. $d = \frac{4}{3}$, $a_1 = -3$, $a_n = 17$, $n = ?$

19. $a_1 = 4, r = \sqrt{2}, n = 7, a_7 = ?$

20. $a_n = \frac{49}{8}, r = -\frac{2}{7}, S_n = \frac{1911}{32}, a_1 = ?$

21. $a_1 = 80, a_n = -25, S_n = 220, d = ?$

22. $a_6 = 3, d = 0.2, n = 11, S_{11} = ?$

23. $n = 6, r = -0.25, S_6 = 204.75, a_6 = ?$

24. $a_1 = 10, r = 0.1, S_n = 11.111, n = ?$

25. $a_1 = -1, a_n = 32, n = 12, S_{12} = ?$ (arith.)

26. $a_2 = 800, a_n = 6400, S_n = 32{,}500, d = 700, n = ?$ (arith.)

In Exercises 27–30, find the sums of the given infinite geometric series.

27. $0.9 + 0.6 + 0.4 + \cdots$

28. $1280 - 320 + 80 - \cdots$

29. $1 + 1.02^{-1} + 1.02^{-2} + \cdots$

30. $3 - \sqrt{3} + 1 - \cdots$

In Exercises 31–34, find the fractions equal to the given decimals.

31. 0.030303 . . .

32. 0.363363 . . .

33. 0.0727272 . . .

34. 0.25399399399 . . .

In Exercises 35–38, expand and simplify the given expression. In Exercises 39–42, find the first four terms of the appropriate expansion.

35. $(x - 5)^4$

36. $(3 + 0.1)^4$

37. $(x^2 + 4)^5$

38. $(3n^{1/2} - 2a)^6$

39. $(a + 3e)^{10}$

40. $\left(\frac{x}{4} - y\right)^{12}$

41. $\left(p^2 - \frac{q}{6}\right)^9$

42. $(2s^2 - \frac{3}{2}t^{-1})^{14}$

In Exercises 43–50, find the first four terms of the indicated expansions by use of the binomial series.

43. $(1 + x)^{12}$

44. $(1 - 2x)^{10}$

45. $\sqrt{1 + x^2}$

46. $(4 - 4\sqrt{x})^{-1}$

47. $\sqrt{1 - a^2}$

48. $\sqrt{1 + 2b^4}$

49. $(2 - 4x)^{-3}$

50. $(1 + 4x)^{-1/4}$

In Exercises 51–98, solve the given problems by use of an appropriate sequence or series. All numbers are accurate to at least two significant digits.

51. Find the sum of the first 1000 positive even integers.

52. Find the sum of the first 300 positive odd integers.

53. What is the fifth term of the arithmetic sequence in which the first term is a and the second term is b?

54. Find three consecutive numbers in an arithmetic sequence such that their sum is 15 and the sum of their squares is 77.

55. For a geometric sequence, is it possible that $a_3 = 6$, $a_5 = 9$, and $a_7 = 12$?

56. Find three consecutive numbers in a geometric sequence such that their product is 64 and the sum of their squares is 84.

57. Find a formula with variable n of the arithmetic sequence with $a_1 = -5, a_{n+1} = a_n - 3$, for $n = 1, 2, 3, \ldots$.

58. Find the ninth term of the sequence $(a + 2b), b, -a, \ldots$.

59. Approximate the value of $(1.06)^{-6}$ by using three terms of the appropriate binomial series. Check using a calculator.

60. Approximate the value of $\sqrt[4]{0.94}$ by using three terms of the appropriate binomial series. Check using a calculator.

61. Approximate the value of $\sqrt{30}$ by noting that $\sqrt{30} = \sqrt{25(1.2)} = 5\sqrt{1 + 0.2}$ and using three terms of the appropriate binomial series.

62. Approximate $\sqrt[3]{29.7}$ by using the *method* of Exercise 61.

63. Without expanding, evaluate

$$(3x - 2)^4 + 4(3x - 2)^3(5 - 3x) + 6(3x - 2)^2(5 - 3x)^2 + 4(3x - 2)(5 - 3x)^3 + (5 - 3x)^4$$

64. If $x, x + 2, x + 3$ forms a geometric sequence, find x.

65. If $x + 1, x^2 - 2, 4x - 2$ forms an arithmetic sequence, find x.

66. If $x + x^2 + x^3 + \cdots = 2$, find x.

67. Each stroke of a pile driver moves a post 2 in. less than the previous stroke. If the first stroke moves the post 24 in., which stroke moves the post 4 in.?

68. During each hour, an exhaust fan removes 15.0% of the carbon dioxide present in the air in a room at the beginning of the hour. What percent of the carbon dioxide remains after 10.0 h?

69. Each 1.0 mm of a filter through which light passes reduces the intensity of the light by 12%. How thick should the filter be to reduce the intensity of the light to 20%?

70. A pile of dirt and ten holes are in a straight line. It is 20 ft from the dirt pile to the nearest hole, and the holes are 8 ft apart. If a backhoe takes two trips to fill each hole, how far must it travel in filling all the holes if it starts and ends at the dirt pile?

71. A roof support with equally spaced vertical pieces is shown in Figure 19.13. Find the total length of the vertical pieces if the shortest one is 10.0 in. long.

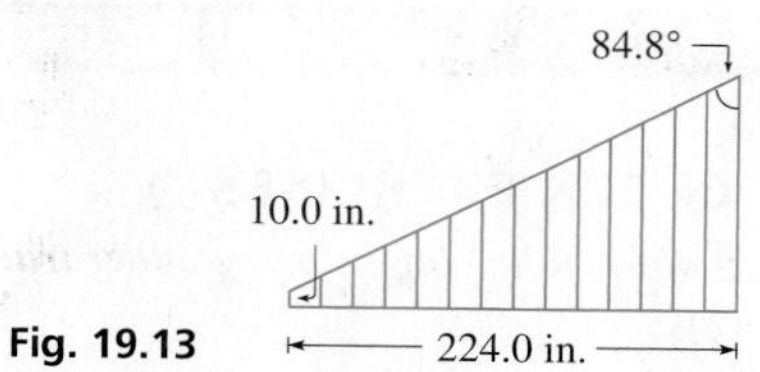

Fig. 19.13

72. During each microsecond, the current in an electric circuit decreases by 9.3%. If the initial current is 2.45 mA, how long does it take to reach 0.50 mA?

73. A machine that costs \$8600 depreciates 1.0% in value each month. What is its value 5 years after it was purchased?

74. The level of chemical pollution in a lake is 4.50 ppb (parts per billion). If the level increases by 0.20 ppb in the following month and by 5.0% less each month thereafter, what will be the maximum level?

75. A piece of paper 0.0040 in. thick is cut in half. These two pieces are then placed one on the other and cut in half. If this is repeated such that the paper is cut in half 40 times, how high will the pile be?

76. After the power is turned off, an object on a nearly frictionless surface slows down such that it travels 99.9% as far during 1 s as during the previous second. If it travels 100 cm during the first second after the power is turned off, how far does it travel while sliding?

77. Under gravity, an object falls 16 ft during the first second, 48 ft during the second second, 80 ft during the third second, and so on. How far will it fall during the 20th second?

78. For the object in Exercise 77, what is the total distance fallen during the first 20 s?

79. An object suspended on a spring is oscillating up and down. If the first oscillation is 10.0 cm and each oscillation thereafter is 9/10 of the preceding one, find the total distance the object travels if it bounces indefinitely.

80. During each oscillation, a pendulum swings through 85% of the distance of the previous oscillation. If the pendulum swings through 80.8 cm in the first oscillation, through what total distance does it move in 12 oscillations?

81. A person invests \$1000 each year at the beginning of the year. What is the total value of these investments after 20 years if they earn 7.5% annual interest, compounded semiannually?

82. A well driller charges \$10.00 for drilling the first meter of a well and for each meter thereafter charges 0.20% more than for the preceding meter. How much is charged for drilling a 150-m well?

83. When new, an article cost \$250. It was then sold at successive yard sales at 40% of the previous price. What was the price at the fourth yard sale?

84. Each side of an equilateral triangle is 2 cm in length. The midpoints are joined to form a second equilateral triangle. The midpoints of the second triangle are joined to form a third equilateral triangle. Find the sum of all of the perimeters of the triangles if this process is continued indefinitely. See Fig. 19.14.

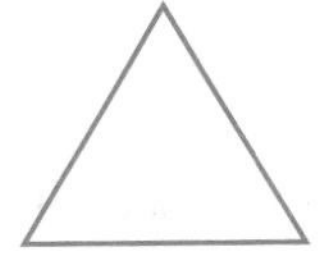
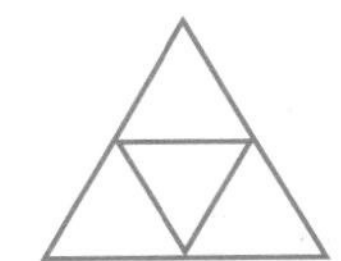
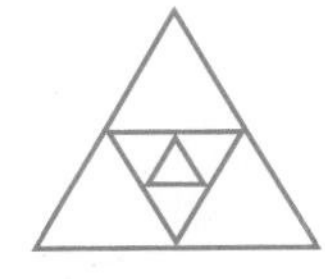

Fig. 19.14

85. In testing a type of insulation, the temperature in a room was made to fall to 2/3 of the initial temperature after 1.0 h, to 2/5 of the initial temperature after 2.0 h, to 2/7 of the initial temperature after 3.0 h, and so on. If the initial temperature was 50.0°C, what was the temperature after 12.0 h? (This is an illustration of a *harmonic sequence.*)

86. Two competing businesses make the same item and sell it initially for \$100. One increases the price by \$8 each year for 5 years, and the other increases the price by 8% each year for 5 years. What is the difference in price after 5 years?

87. In hydrodynamics, while studying compressible fluid flow, the expression $\left(1 + \frac{a-1}{2}m^2\right)^{a/(a-1)}$ arises. Find the first three terms of the expansion of this expression.

88. In finding the partial pressure P_F of the fluorine gas under certain conditions, the equation

$$P_F = \frac{(1 + 2 \times 10^{-10}) - \sqrt{1 + 4 \times 10^{-10}}}{2} \text{ atm}$$

is found. By using three terms of the expansion for $\sqrt{1 + x}$, approximate the value of this expression.

89. During 1 year a beach eroded 1.2 m to a line 48.3 m from the wall of a building. If the erosion is 0.1 m more each year than the previous year, when will the waterline reach the wall?

90. On a highway with a steep incline, the runaway truck ramp is constructed so that a vehicle that has lost its brakes can stop. The ramp is designed to slow a truck in succeeding 20-m distances by 10 km/h, 12 km/h, 14 km/h, ... If the ramp is 160 m long, will it stop a truck moving at 120 km/h when it reaches the ramp?

91. Each application of an insecticide destroys 75% of a certain insect. How many applications are needed to destroy at least 99.9% of the insects?

92. At the end of 2005 a house was valued at \$375,000, and then increased in value by 5.00%, on average, each year 2006 through 2011. Its value did not change in 2012, but then decreased by 8.00%, on average, each year 2013 through 2016. What was its value at the end of 2016?

93. Oil pumped from a certain oil field decreases 10% each year. How long will it take for production to be 10% of the first year's production?

94. A wire hung between two poles is parabolic in shape. To find the length of wire between two points on the wire, the expression $\sqrt{1 + 0.08x^2}$ is used. Find the first three terms of the binomial expansion of this expression.

95. Show that the middle term of a finite arithmetic sequence equals S/n, if n is odd.

96. The sum of three terms of a geometric sequence is −3, and the second is 6 more than the first. Find these terms.

97. Do the reciprocals of the terms of a geometric sequence form a geometric sequence? Explain.

98. The terms $a, a + 12, a + 24$ form an arithmetic sequence, and the terms $a, a + 24, a + 12$ form a geometric sequence. Find these sequences.

99. Derive a formula for the value V after 1 year of an amount A invested at r% (as a decimal) annual interest, compounded n times during the year. If A = \$1000 and $r = 0.10$ (10%), write two or three paragraphs explaining why the amount of interest increases as n increases and stating your approach to finding the maximum possible amount of interest.

CHAPTER 19 PRACTICE TEST

As a study aid, we have included complete solutions for each Practice Test problem at the end of this book.

1. Write the first five terms of the sequence for which (a) $a_1 = 8$ and $d = -1/2$; (b) $a_1 = 8$ and $r = -1/2$.
2. Find the sum of the first seven terms of the sequence $6, -2, \frac{2}{3}, \ldots$.
3. If 4, $x + 6$, and $3x + 2$ are the first three terms of an arithmetic sequence, find the sum of the first 10 terms.
4. Find the fraction equal to the decimal 0.454545. . . .
5. Find the first three terms of the expansion of $\sqrt{1 - 4x}$.
6. Expand and simply the expression $(2x - y)^5$.
7. What is the value after 20 years of an investment of \$2500 if it draws 5% annual interest compounded annually?
8. Find the sum of the first 100 even integers.
9. A ball is dropped from a height of 8.00 ft, and on each rebound, it rises to 1/2 of the height it last fell. If it bounces indefinitely, through what total distance will it move?

Additional Topics in Trigonometry

20

As the use of electricity became widespread in the 1880s, there was a serious debate over the best way to distribute electric power. The American inventor Thomas Edison favored the use of direct current because it was safer and did not vary with time. Another American inventor and engineer, George Westinghouse, favored alternating current because the voltage could be stepped up and down with transformers during transmission.

Also favoring the use of alternating current was Nikola Tesla, an American (born in Croatia) electrical engineer, and he had a strong influence on the fact that alternating current came to be used for transmission. Tesla developed many electrical devices, among them the *polyphase generator* that allowed alternating current to be transmitted with constant instantaneous power. Using this type of generator, power losses are greatly reduced in transmission lines, which allows for smaller conductors, and the power can be generated far from where it is used. Three-phase systems are used in most commercial electric generators as well as some electric cars, including those produced by Tesla Motors, a company named in honor of Nikola Tesla.

The trigonometric relationships that we develop in this chapter are important for a number of reasons. In fact, we already made use of some of them in Section 10.4 when we graphed certain trigonometric functions, and in Chapter 9 in deriving the law of cosines. In calculus, certain problems use trigonometric relationships, even including some in which these functions do not appear in the initial problem or final answer. Also, they are useful in a number of technical applications in areas such as electronics, optics, solar energy, and robotics.

Later in the chapter, we see that various trigonometric relationships are used in solving equations with trigonometric functions. Also, we develop the concept of the inverse trigonometric functions that were introduced in Chapter 4.

LEARNING OUTCOMES

After completion of this chapter, the student should be able to:

- Recognize the basic trigonometric identities and use them to prove other trigonometric identities and simplify trigonometric expressions
- Recognize and apply the formulas for trignonometric functions of sums and differences of angles as well as half and double angles
- Solve trigonometric equations
- Evaluate inverse trignonometric functions
- Find algebraic expressions for expressions involving inverse trigonometric functions
- Solve application problems involving trigonometric identities and inverse trigonometric functions

◀ **Tesla electric cars are powered by a three-phase alternating-current engine. In Section 20.2, we see how certain trigonometric identities can be used to show an important property of a three-phase generator.**

20.1 Fundamental Trigonometric Identities

Trigonometric Identities • Basic Identities • Proving Trigonometric Identities

From the definitions in Chapters 4 and 8, recall that $\sin\theta = y/r$ and $\csc\theta = r/y$ (see Fig. 20.1). Because $y/r = 1/(r/y)$, we see that $\sin\theta = 1/\csc\theta$. These definitions hold for *any* angle, which means that $\sin\theta = 1/\csc\theta$ is true for *any* angle. *This type of relation, which is true for any value of the variable, is called an* **identity.** Of course, values where division by zero would be indicated are excluded.

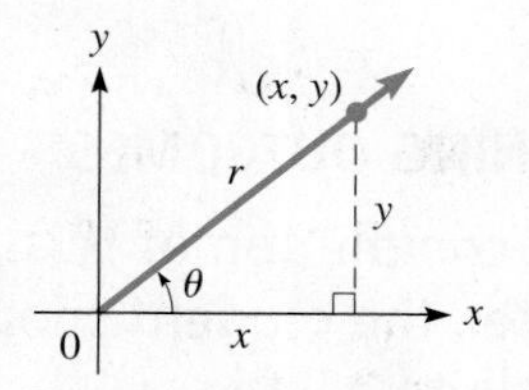

Fig. 20.1

In this section, we develop several important identities involving the trigonometric functions. We also show how the basic identities are used to verify other identities.

From the definitions, we have

$$\sin\theta\csc\theta = \frac{y}{r}\times\frac{r}{y} = 1 \quad \text{or} \quad \sin\theta = \frac{1}{\csc\theta} \quad \text{or} \quad \csc\theta = \frac{1}{\sin\theta}$$

$$\cos\theta\sec\theta = \frac{x}{r}\times\frac{r}{x} = 1 \quad \text{or} \quad \cos\theta = \frac{1}{\sec\theta} \quad \text{or} \quad \sec\theta = \frac{1}{\cos\theta}$$

$$\tan\theta\cot\theta = \frac{y}{x}\times\frac{x}{y} = 1 \quad \text{or} \quad \tan\theta = \frac{1}{\cot\theta} \quad \text{or} \quad \cot\theta = \frac{1}{\tan\theta}$$

$$\frac{\sin\theta}{\cos\theta} = \frac{y/r}{x/r} = \frac{y}{x} = \tan\theta \qquad \frac{\cos\theta}{\sin\theta} = \frac{x/r}{y/r} = \frac{x}{y} = \cot\theta$$

Also, from the definitions and the Pythagorean theorem in the form of $x^2 + y^2 = r^2$, we arrive at the following identities.

By dividing the Pythagorean relation through by r^2, we have

$$\left(\frac{x}{r}\right)^2 + \left(\frac{y}{r}\right)^2 = 1 \quad \text{which leads us to} \quad \cos^2\theta + \sin^2\theta = 1$$

By dividing the Pythagorean relation by x^2, we have

$$1 + \left(\frac{y}{x}\right)^2 = \left(\frac{r}{x}\right)^2 \quad \text{which leads us to} \quad 1 + \tan^2\theta = \sec^2\theta$$

By dividing the Pythagorean relation by y^2, we have

$$\left(\frac{x}{y}\right)^2 + 1 = \left(\frac{r}{y}\right)^2 \quad \text{which leads us to} \quad \cot^2\theta + 1 = \csc^2\theta$$

BASIC IDENTITIES

NOTE ➧ [The term $\cos^2\theta$ is the common way of writing $(\cos\theta)^2$, and it means to square the value of the cosine of the angle.] Obviously, the same holds true for the other functions.

Summarizing these results, we have the following important identities.

$$\sin\theta = \frac{1}{\csc\theta} \quad (20.1)$$

$$\cos\theta = \frac{1}{\sec\theta} \quad (20.2)$$

$$\tan\theta = \frac{1}{\cot\theta} \quad (20.3)$$

$$\tan\theta = \frac{\sin\theta}{\cos\theta} \quad (20.4)$$

$$\cot\theta = \frac{\cos\theta}{\sin\theta} \quad (20.5)$$

$$\sin^2\theta + \cos^2\theta = 1 \quad (20.6)$$

$$1 + \tan^2\theta = \sec^2\theta \quad (20.7)$$

$$1 + \cot^2\theta = \csc^2\theta \quad (20.8)$$

In using the basic identities, θ may stand for any angle or number or expression representing an angle or a number.

EXAMPLE 1 Using basic identities

(a) $\sin ax = \dfrac{1}{\csc ax}$ using Eq. (20.1) **(b)** $\tan 157° = \dfrac{\sin 157°}{\cos 157°}$ using Eq. (20.4)

(c) $\sin^2 \frac{\pi}{4} + \cos^2 \frac{\pi}{4} = 1$ using Eq. (20.6) ■

EXAMPLE 2 Checking identities with numerical values

Let us check the last two illustrations of Example 1 for the given values of θ.

(a) Using a calculator, we find that

$$\frac{\sin 157°}{\cos 157°} = \frac{0.3907311285}{-0.9205048535} = -0.4244748162 \text{ and } \tan 157° = -0.4244748162$$

We see that $\sin 157°/\cos 157° = \tan 157°$.

(b) Checking Example 1(c), refer to Fig. 20.2.

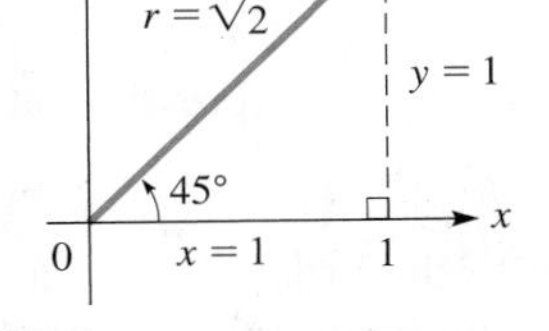

Fig. 20.2

$$\sin \frac{\pi}{4} = \sin 45° = \frac{1}{\sqrt{2}} = \frac{\sqrt{2}}{2} \quad \text{and} \quad \cos \frac{\pi}{4} = \cos 45° = \frac{\sqrt{2}}{2}$$

$$\sin^2 \frac{\pi}{4} + \cos^2 \frac{\pi}{4} = \left(\frac{\sqrt{2}}{2}\right)^2 + \left(\frac{\sqrt{2}}{2}\right)^2 = \frac{1}{2} + \frac{1}{2} = 1$$

We see that this checks with Eq. (20.6) for these values. ■

EXAMPLE 3 Simplifying basic expressions

(a) Multiply and simplify the expression $\sin\theta \tan\theta(\csc\theta + \cot\theta)$.

$$\begin{aligned} \sin\theta \tan\theta(\csc\theta + \cot\theta) &= \sin\theta \tan\theta \csc\theta + \sin\theta \tan\theta \cot\theta \\ &= \left(\frac{1}{\csc\theta}\right) \tan\theta \csc\theta + \sin\theta\left(\frac{1}{\cot\theta}\right) \cot\theta \\ &= \tan\theta + \sin\theta \end{aligned}$$

using Eqs. (20.1) and (20.4)

(b) Factor and simplify the expression $\tan^3\alpha + \tan\alpha$.

$$\begin{aligned} \tan^3\alpha + \tan\alpha &= \tan\alpha(\tan^2\alpha + 1) \\ &= \tan\alpha \sec^2\alpha \end{aligned}$$

using Eq. (20.7)

We see that when algebraic operations are performed on trigonometric terms, they are handled in just the same way as with algebraic terms. This is true for the basic operations of addition, subtraction, multiplication, and division, as well as other operations used in simplifying expressions, such as factoring or taking roots. ■

Practice Exercise

1. Multiply and simplify: $\cos x(\sec x + \tan x)$

PROVING TRIGONOMETRIC IDENTITIES

A great many identities exist among the trigonometric functions. We are going to use the basic identities that have been developed in Eqs. (20.1) through (20.8), along with a few additional ones developed in later sections to prove the validity of still other identities.

CAUTION The ability to prove trigonometric identities depends to a large extent on being very familiar with the basic identities so that you can recognize them in somewhat different forms. ■

If you do not learn these basic identities and learn them well, you will have difficulty in following the examples and doing the exercises. The more readily you recognize these forms, the more easily you will be able to prove such identities.

In proving identities, look for combinations that appear in, or are very similar to, those in the basic identities. This is illustrated in the following examples.

EXAMPLE 4 Proving a trig identity

In proving the identity

$$\sin x = \frac{\cos x}{\cot x}$$

we know that $\cot x = \dfrac{\cos x}{\sin x}$. Because sin x appears on the left, substituting for cot x on the right will eliminate cot x and introduce sin x. This should help us proceed in proving the identity. Thus, starting with the right-hand side, we have

$$\frac{\cos x}{\cot x} = \frac{\cos x}{\dfrac{\cos x}{\sin x}} = \frac{\cos x}{1} \times \frac{\sin x}{\cos x} = \sin x \quad \text{cancel } \cos x \text{ factors}$$

Eq. (20.5) invert

By showing that the right side may be changed exactly to sin x, the expression on the left side, we have proved the identity. ■

Practice Exercise

2. Prove the identity $\dfrac{\cos x \csc x}{\cot^2 x} = \tan x$.

CAUTION Performing the algebraic operations carefully and correctly is very important when working with trigonometric expressions. Operations such as substituting, factoring, and simplifying fractions are frequently used. ■

Some important points should be made in relation to the proof of the identity of Example 4. We must recognize which basic identities may be useful. The proof of an identity requires the use of the basic algebraic operations, and these must be done carefully and correctly. Although in Example 4, we changed the right side to the form on the left, we could have changed the left to the form on the right. From this and the fact that various substitutions are possible, we see that a variety of procedures can be used to prove any given identity.

EXAMPLE 5 Trig identity—two different solutions

Prove that $\tan\theta \csc\theta = \sec\theta$.

In proving this identity, we know that $\tan\theta = \dfrac{\sin\theta}{\cos\theta}$ and also that $\dfrac{1}{\cos\theta} = \sec\theta$. Thus, by substituting for $\tan\theta$, we introduce $\cos\theta$ in the denominator, which is equivalent to introducing $\sec\theta$ in the numerator. Therefore, changing only the left side, we have

$$\tan\theta \csc\theta = \frac{\sin\theta}{\cos\theta}\csc\theta = \frac{\sin\theta}{\cos\theta}\,\frac{1}{\sin\theta}$$

Eq. (20.4) Eq. (20.1)

$$= \frac{1}{\cos\theta} \quad \text{cancel } \sin\theta \text{ factors}$$

$$= \sec\theta \quad \text{using Eq. (20.2)}$$

Having changed the left side into the form on the right side, we have proven the identity.

Many variations of the preceding steps are possible. We could have changed the right side to obtain the form on the left. For example,

$$\sec\theta = \frac{1}{\cos\theta} \quad \text{using Eq. (20.2)}$$

$$= \frac{\sin\theta}{\cos\theta \sin\theta} = \frac{\sin\theta}{\cos\theta}\,\frac{1}{\sin\theta} \quad \text{multiply numerator and denominator by } \sin\theta \text{ and rewrite}$$

$$= \tan\theta \csc\theta \quad \text{using Eqs. (20.4) and (20.1)}$$

■

In proving the identities in Examples 4 and 5, we showed that the expression on one side of the equal sign can be changed into the expression on the other side. Because we want to *prove* that the expressions are equal, we cannot transpose terms or multiply each side by the same quantity as we do when solving equations. Each side must be treated separately. NOTE ▸ [Although we could make changes to each side independently to hopefully arrive at the same expression, we will restrict the method of proof to changing only one side into the same form as the other side.] In this way, we know the form we are attempting to attain, and by looking ahead, we are better able to make the necessary changes.

There is no set procedure for working with identities. The most important factors are to (1) ***recognize the proper forms,*** (2) ***try to identify what effect a change may have*** before performing it, and (3) ***perform all steps correctly.*** NOTE ▸ [Normally, it is easier to change the form of the more complicated side to the same form as the less complicated side.] If the forms are about the same, a first close look often suggests possible steps to use.

EXAMPLE 6 Proving a trig identity

Prove the identity $\dfrac{\sec^2 y}{\cot y} - \tan^3 y = \tan y$.

Because the more complicated side is on the left, we will change the left side to tan y, the form on the right. Because we want tan y as the final form, we will use the basic identities to introduce tan y where possible. Thus, noting the cot y in the denominator, we substitute $1/\tan y$ for cot y. This gives

$$\frac{\sec^2 y}{\cot y} - \tan^3 y = \frac{\sec^2 y}{\frac{1}{\tan y}} - \tan^3 y = \sec^2 y \tan y - \tan^3 y$$

In the form on the right we now note the factor of tan y in each term. Factoring tan y from each of these terms, and then using the basic identity $1 + \tan^2 y = \sec^2 y$ in the form $\sec^2 y - \tan^2 y = 1$, we can complete the proof. Therefore,

$$\sec^2 y \tan y - \tan^3 y = \tan y(\sec^2 y - \tan^2 y) = (\tan y)(1)$$
$$= \tan y$$

NOTE ▸ [Note carefully that we had to use the identity $1 + \tan^2 y = \sec^2 y$ in a somewhat different form from that shown in Eq. (20.7).] The use of such variations in form to prove identities is a very common procedure. ■

Practice Exercise

3. Prove the identity:

$$\frac{\cos^2 x}{\tan x} + \sin^2 x \cot x = \cot x$$

EXAMPLE 7 Proving a trig identity

Prove the identity $\dfrac{1 - \sin x}{\sin x \cot x} = \dfrac{\cos x}{1 + \sin x}$.

The combination $1 - \sin x$ also suggests $1 - \sin^2 x$, since multiplying $(1 - \sin x)$ by $(1 + \sin x)$ gives $1 - \sin^2 x$, which can then be replaced by $\cos^2 x$. Thus, changing only the left side, we have

$$\frac{1 - \sin x}{\sin x \cot x} = \frac{(1 - \sin x)(1 + \sin x)}{\sin x \cot x(1 + \sin x)}$$ multiply numerator and denominator by $1 + \sin x$

$$= \frac{1 - \sin^2 x}{\sin x\left(\frac{\cos x}{\sin x}\right)(1 + \sin x)} = \frac{\cos^2 x}{\cos x(1 + \sin x)}$$ Eq. (20.6); cancel $\sin x$

$$= \frac{\cos x}{1 + \sin x}$$ cancel $\cos x$ ■

EXAMPLE 8 Trig identity—radiation rate of electric charge

In finding the radiation rate of an accelerated electric charge, it is necessary to show that $\sin^3\theta = \sin\theta - \sin\theta\cos^2\theta$. Show this by changing the left side.

Because each term on the right has a factor of $\sin\theta$, we see that we can proceed by writing $\sin^3\theta$ as $\sin\theta(\sin^2\theta)$. Then the factor $\sin^2\theta$ and the $\cos^2\theta$ on the right suggest the use of Eq. (20.6). Thus, we have

$$\sin^3\theta = \sin\theta(\sin^2\theta) = \sin\theta(1 - \cos^2\theta)$$

$$= \sin\theta - \sin\theta\cos^2\theta \qquad \text{multiplying}$$

Because we substituted for $\sin^2\theta$, we used Eq. (20.6) in the form $\sin^2\theta = 1 - \cos^2\theta$. ■

EXAMPLE 9 Simplifying a trig expression

Simplify the expression $\dfrac{\csc x}{\tan x + \cot x}$.

We proceed with a simplification in a manner similar to proving an identity, although we do not know what the result should be. Following is one procedure for this simplification:

TI-89 graphing calculator keystrokes for Example 9: goo.gl/8AiqR8

$$\frac{\csc x}{\tan x + \cot x} = \frac{\csc x}{\tan x + \dfrac{1}{\tan x}} = \frac{\csc x}{\dfrac{\tan^2 x + 1}{\tan x}}$$

$$= \frac{\csc x \tan x}{\tan^2 x + 1} = \frac{\csc x \tan x}{\sec^2 x} = \frac{\dfrac{1}{\sin x}\dfrac{\sin x}{\cos x}}{\dfrac{1}{\cos^2 x}}$$

$$= \frac{1}{\sin x}\frac{\sin x}{\cos x}\frac{\cos^2 x}{1} = \cos x$$

■

A calculator can be used to check an identity or a simplification. This is done by graphing the function on each side of an identity, or the initial expression and the final expression for simplification. If the two graphs are the same, the identity or simplification is probably shown to be correct, although this is not strictly a proof.

EXAMPLE 10 Trig identity—verifying by calculator

Use a calculator to verify the identity of Example 7.

Noting this identity as $\dfrac{1 - \sin x}{\sin x \cot x} = \dfrac{\cos x}{1 + \sin x}$, on a calculator, we let

$$y_1 = (1 - \sin x)/(\sin x/\tan x) \qquad \text{noting that } \cot x = 1/\tan x$$

$$y_2 = \cos x/(1 + \sin x)$$

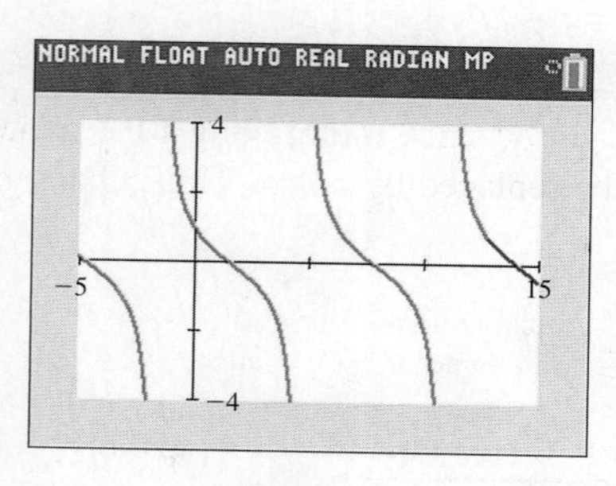

Fig. 20.3

We then graph these two functions as shown in Fig. 20.3. The screenshot shows y_2 (in red) in the process of being plotted right on top of y_1 (in blue). Because these curves are the same, the identity appears to be verified (although it has not been *proven*). ■

Working with trigonometric identities has often been considered a difficult topic. However, as we emphasized earlier, knowing the basic identities very well and performing the necessary algebraic steps properly should make proving these identities a much easier task.

EXERCISES 20.1

In Exercises 1 and 2, make the given changes in the indicated examples of this section and then prove the resulting identities.

1. In Example 4, change the right side to $\tan x/\sec x$.
2. In Example 6, change the first term on the left to $\sin y/\cos^3 y$.

In Exercises 3–6, use a calculator to check the indicated basic identities for the given angles.

3. Eq. (20.4) for $\theta = 38°$
4. Eq. (20.5) for $\theta = 280°$
5. Eq. (20.6) for $\theta = 4\pi/3$
6. Eq. (20.7) for $\theta = 5\pi/6$

In Exercises 7–12, multiply and simplify. In Exercises 13–18, factor and simplify.

7. $\cos x(\tan x - \sec x)$
8. $\csc y(\sin y + 3\cos y)$
9. $\cos\theta\cot\theta(\sec\theta - 2\tan\theta)$
10. $(\csc x - 1)(\csc x + 1)$
11. $\tan u(\cot u + \tan u)$
12. $\cos^2 t(1 + \tan^2 t)$
13. $\sin x + \sin x\tan^2 x$
14. $\sin\theta - \sin\theta\cos^2\theta$
15. $\sin^3 t\cos t + \sin t\cos^3 t$
16. $\tan^2 u\sec^2 u - \tan^4 u$
17. $\csc^4 y - 1$
18. $\sin x + \sin x\cot^2 x$

In Exercises 19–38, prove the given identities.

19. $\dfrac{\sin x}{\tan x} = \cos x$
20. $\dfrac{\csc\theta}{\sec\theta} = \cot\theta$
21. $\sin x\sec x = \tan x$
22. $\cot\theta\sec\theta = \csc\theta$
23. $\csc^2 x(1 - \cos^2 x) = 1$
24. $\sec\theta(1 - \sin^2\theta) = \cos\theta$
25. $\sin x(1 + \cot^2 x) = \csc x$
26. $\csc x(\csc x - \sin x) = \cot^2 x$
27. $\cos\theta\cot\theta + \sin\theta = \csc\theta$
28. $\dfrac{\csc x}{\cos x} - \tan x = \cot x$
29. $\cot\theta\sec^2\theta - \dfrac{1}{\tan\theta} = \tan\theta$
30. $\sin y + \sin y\cot^2 y = \csc y$
31. $\tan x + \cot x = \sec x\csc x$
32. $\tan x + \cot x = \tan x\csc^2 x$
33. $\cos^2 x - \sin^2 x = 1 - 2\sin^2 x$
34. $\dfrac{1 + \cos x}{\sin x} = \dfrac{\sin x}{1 - \cos x}$
35. $\dfrac{\sin\theta}{\csc\theta} + \dfrac{\cos\theta}{\sec\theta} = 1$
36. $\dfrac{\sec\theta}{\cos\theta} - \dfrac{\tan\theta}{\cot\theta} = 1$
37. $2\sin^4 x - 3\sin^2 x + 1 = \cos^2 x(1 - 2\sin^2 x)$
38. $\dfrac{\sin^2\theta + 2\cos\theta - 1}{\sin^2\theta + 3\cos\theta - 3} = \dfrac{1}{1 - \sec\theta}$

In Exercises 39–46, simplify the given expressions. The result will be one of sin x, cos x, tan x, cot x, sec x, or csc x.

39. $\dfrac{\tan x\csc^2 x}{1 + \tan^2 x}$
40. $\dfrac{\cos x - \cos^3 x}{\sin x - \sin^3 x}$
41. $\cot x(\sec x - \cos x)$
42. $\sin x(\tan x + \cot x)$
43. $\dfrac{\tan x + \cot x}{\csc x}$
44. $\dfrac{1 + \tan x}{\sin x} - \sec x$
45. $\dfrac{\cos x + \sin x}{1 + \tan x}$
46. $\dfrac{\sec x - \cos x}{\tan x}$

In Exercises 47–50, for a first-quadrant angle, express the first function listed in terms of the second function listed.

47. $\sin x$, $\sec x$
48. $\cos x$, $\csc x$
49. $\tan x$, $\csc x$
50. $\cot x$, $\sec x$

In Exercises 51–54, use a calculator to verify the given identities by comparing the graphs of each side.

51. $\sin x(\csc x - \sin x) = \cos^2 x$
52. $\cos y(\sec y - \cos y) = \sin^2 y$
53. $\dfrac{\sec x + \csc x}{1 + \tan x} = \csc x$
54. $\dfrac{\cot x + 1}{\cot x} = 1 + \tan x$

In Exercises 55–58, use a calculator to determine whether the given equations are identities.

55. $\sec\theta\tan\theta\csc\theta = \tan^2\theta + 1$
56. $\sin x\cos x\tan x = \cos^2 x - 1$
57. $\dfrac{2\cos^2 x - 1}{\sin x\cos x} = \tan x - \cot x$
58. $\cos^3 x\csc^3 x\tan^3 x = \csc^2 x - \cot^2 x$

In Exercises 59–62, solve the given problems involving trigonometric identities.

59. When designing a solar-energy collector, it is necessary to account for the latitude and longitude of the location, the angle of the sun, and the angle of the collector. In doing this, the equation $\cos\theta = \cos A\cos B\cos C + \sin A\sin B$ is used. If $\theta = 90°$, show that $\cos C = -\tan A\tan B$.

60. In determining the path of least time between two points under certain conditions, it is necessary to show that
$$\sqrt{\frac{1 + \cos\theta}{1 - \cos\theta}}\sin\theta = 1 + \cos\theta$$
Show this by transforming the left-hand side.

61. Show that the length l of the straight brace shown in Fig. 20.4 can be found from the equation
$$l = \frac{a(1 + \tan\theta)}{\sin\theta}$$

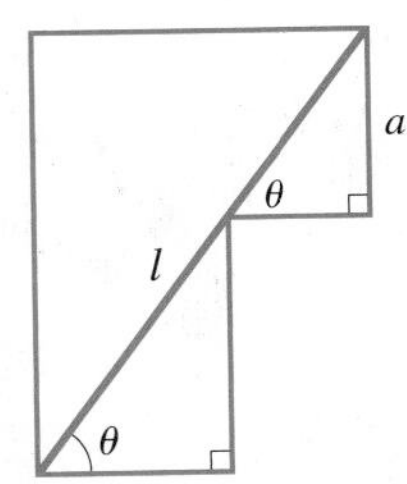

Fig. 20.4

62. The path of a point on the circumference of a circle, such as a point on the rim of a bicycle wheel as it rolls along, tracks out a curve called a *cycloid*. See Fig. 20.5. To find the distance through which a point moves, it is necessary to simplify the expression $(1 - \cos\theta)^2 + \sin^2\theta$. Perform this simplification.

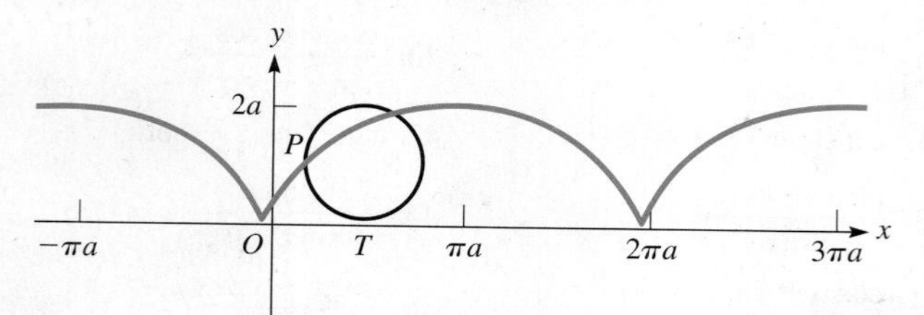

Fig. 20.5

In Exercises 63–70, solve the given problems.

63. Explain how to transform $\sin\theta\tan\theta + \cos\theta$ into $\sec\theta$.

64. Explain how to transform $\tan^2\theta\cos^2\theta + \cot^2\theta\sin^2\theta$ into 1.

65. Show that $\sin^2 x(1 - \sec^2 x) + \cos^2 x(1 + \sec^4 x)$ has a constant value.

66. Show that $\cot y \csc y \sec y - \csc y \cos y \cot y$ has a constant value.

67. If $\tan x + \cot x = 2$, evaluate $\tan^2 x + \cot^2 x$.

68. If $\sec x + \cos x = 2$, evaluate $\sec^2 x + \cos^2 x$.

69. Prove that $\sec^2\theta + \csc^2\theta = \sec^2\theta\csc^2\theta$ by expressing each function in terms of its x, y, and r definition.

70. Prove that $\dfrac{\csc\theta}{\tan\theta + \cot\theta} = \cos\theta$ by expressing each function in terms of its x, y, and r definition.

In Exercises 71–74, use the given substitutions to show that the given equations are valid. In each, $0 < \theta < \pi/2$.

71. If $x = \cos\theta$, show that $\sqrt{1 - x^2} = \sin\theta$.

72. If $x = 3\sin\theta$, show that $\sqrt{9 - x^2} = 3\cos\theta$.

73. If $x = 2\tan\theta$, show that $\sqrt{4 + x^2} = 2\sec\theta$.

74. If $x = 4\sec\theta$, show that $\sqrt{x^2 - 16} = 4\tan\theta$.

Answers to Practice Exercises

1. $1 + \sin x$ 2. $\tan x$ 3. $\cot x(\cos^2 x + \sin^2 x) = \cot x$

20.2 The Sum and Difference Formulas

Formulas for $\sin(\alpha + \beta)$ and $\cos(\alpha + \beta)$ • Formulas for $\sin(\alpha - \beta)$ and $\cos(\alpha - \beta)$ • Formulas for $\tan(\alpha + \beta)$ and $\tan(\alpha - \beta)$

There are other important relations among the trigonometric functions. The most important and useful relations are those that involve twice an angle and half an angle. To obtain these relations, in this section, we derive the expressions for the sine and cosine of the sum and difference of two angles. These expressions will lead directly to the desired relations of double and half angles that we will derive in the following sections.

■ For reference, Eq. (12.13) is
$r_1(\cos\theta_1 + j\sin\theta_1)r_2(\cos\theta_2 + j\sin\theta_2) = r_1r_2[\cos(\theta_1 + \theta_2) + j\sin(\theta_1 + \theta_2)]$

Equation (12.13), shown in the margin, gives the polar (or trigonometric) form of the product of two complex numbers. We can use this formula to derive the expressions for the sine and cosine of the sum and difference of two angles.

Using Eq. (12.13) to find the product of the complex numbers $\cos\alpha + j\sin\alpha$ and $\cos\beta + j\sin\beta$, which are represented in Fig. 20.6, we have

$$(\cos\alpha + j\sin\alpha)(\cos\beta + j\sin\beta) = \cos(\alpha + \beta) + j\sin(\alpha + \beta)$$

Expanding the left side, and then switching sides, we have

$$\cos(\alpha + \beta) + j\sin(\alpha + \beta) = (\cos\alpha\cos\beta - \sin\alpha\sin\beta) + j(\sin\alpha\cos\beta + \cos\alpha\sin\beta)$$

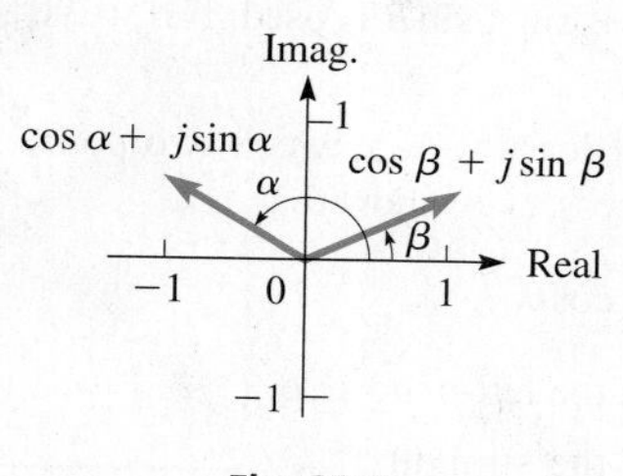

Fig. 20.6

Because two complex numbers are equal if their real parts are equal and their imaginary parts are equal, we have the following formulas:

$$\sin(\alpha + \beta) = \sin\alpha\cos\beta + \cos\alpha\sin\beta \quad \textbf{(20.9)}$$

$$\cos(\alpha + \beta) = \cos\alpha\cos\beta - \sin\alpha\sin\beta \quad \textbf{(20.10)}$$

EXAMPLE 1 Verifying the sin ($\alpha + \beta$) formula

Verify that $\sin 90° = 1$, by finding $\sin(60° + 30°)$.

$$\sin 90° = \sin(60° + 30°) = \sin 60° \cos 30° + \cos 60° \sin 30° \quad \text{using Eq. (20.9)}$$

$$= \frac{\sqrt{3}}{2} \times \frac{\sqrt{3}}{2} + \frac{1}{2} \times \frac{1}{2} \quad \text{for values, see Section 4.3}$$

$$= \frac{3}{4} + \frac{1}{4} = 1$$

CAUTION It should be obvious from this example that $\sin(\alpha + \beta)$ is *not* equal to $\sin\alpha + \sin\beta$. If we used such a formula, we would get $\sin 90° = \frac{1}{2}\sqrt{3} + \frac{1}{2} = 1.366$ for the combination $(60° + 30°)$. This is not possible, because the values of the sine function never exceed 1 in value. ■ ■

EXAMPLE 2 Using cos ($\alpha + \beta$) with numerical values

Given that $\sin\alpha = \frac{5}{13}$ (α in the first quadrant) and $\sin\beta = -\frac{3}{5}$ (for β in the third quadrant), find $\cos(\alpha + \beta)$.

Because $\sin\alpha = \frac{5}{13}$ for α in the first quadrant, from Fig. 20.7, we have $\cos\alpha = \frac{12}{13}$. Also, because $\sin\beta = -\frac{3}{5}$ for β in the third quadrant, from Fig. 20.7, we also have $\cos\beta = -\frac{4}{5}$.

Then, by using Eq. (20.10), we have

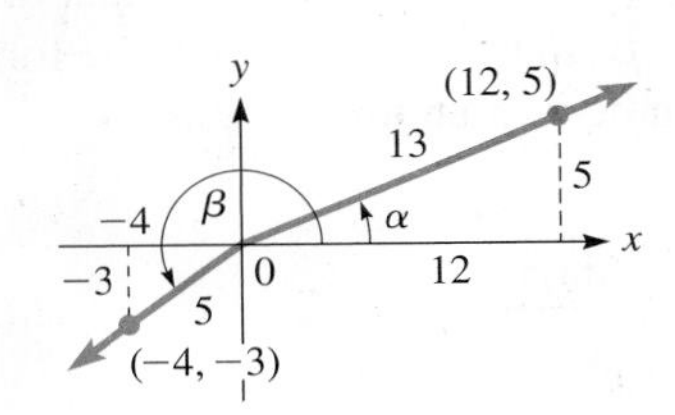

Fig. 20.7

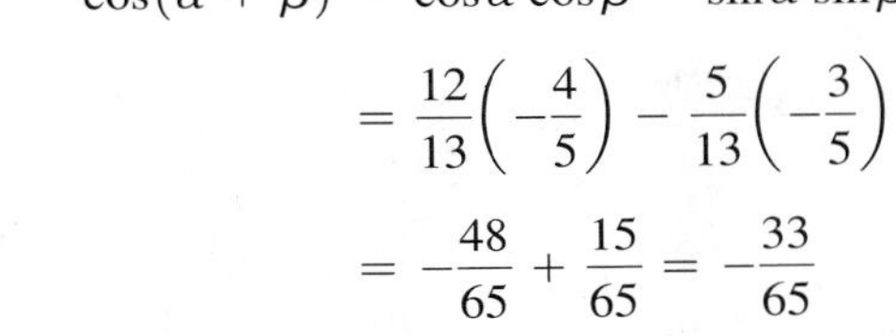

$$\cos(\alpha + \beta) = \cos\alpha \cos\beta - \sin\alpha \sin\beta$$

$$= \frac{12}{13}\left(-\frac{4}{5}\right) - \frac{5}{13}\left(-\frac{3}{5}\right)$$

$$= -\frac{48}{65} + \frac{15}{65} = -\frac{33}{65}$$

■

From Eqs. (20.9) and (20.10), we can easily find expressions for $\sin(\alpha - \beta)$ and $\cos(\alpha - \beta)$. This is done by finding $\sin[\alpha + (-\beta)]$ and $\cos[\alpha + (-\beta)]$. Thus, we have

$$\sin(\alpha - \beta) = \sin[\alpha + (-\beta)] = \sin\alpha \cos(-\beta) + \cos\alpha \sin(-\beta)$$

Because $\cos(-\beta) = \cos\beta$ and $\sin(-\beta) = -\sin\beta$ [see Eq. (8.5) on page 247], we have

$$\sin(\alpha - \beta) = \sin\alpha \cos\beta - \cos\alpha \sin\beta \qquad \textbf{(20.11)}$$

In the same manner, we find that

$$\cos(\alpha - \beta) = \cos\alpha \cos\beta + \sin\alpha \sin\beta \qquad \textbf{(20.12)}$$

EXAMPLE 3 Using cos ($\alpha - \beta$) formula

Find $\cos 15°$ from $\cos(45° - 30°)$.

$$\cos 15° = \cos(45° - 30°) = \cos 45° \cos 30° + \sin 45° \sin 30° \quad \text{using Eq. (20.12)}$$

$$= \frac{\sqrt{2}}{2} \times \frac{\sqrt{3}}{2} + \frac{\sqrt{2}}{2} \times \frac{1}{2} = \frac{\sqrt{6} + \sqrt{2}}{4} \quad \text{(exact)}$$

$$= 0.9659$$

■

EXAMPLE 4 Sin $(\alpha - \beta)$ formula—oscillating spring

In analyzing the motion of an object oscillating up and down at the end of a spring, the expression $\sin(\omega t + \alpha)\cos\alpha - \cos(\omega t + \alpha)\sin\alpha$ occurs. Simplify this expression.

If we let $x = \omega t + \alpha$, the expression becomes $\sin x \cos\alpha - \cos x \sin\alpha$, which is the form for $\sin(x - \alpha)$. Therefore,

$$\sin(\omega t + \alpha)\cos\alpha - \cos(\omega t + \alpha)\sin\alpha = \sin x\cos\alpha - \cos x\sin\alpha = \sin(x - \alpha)$$
$$= \sin(\omega t + \alpha - \alpha) = \sin\omega t$$ ■

EXAMPLE 5 Using cos $(\alpha + \beta)$ formula

Evaluate $\cos 23^\circ \cos 67^\circ - \sin 23^\circ \sin 67^\circ$.

We note that this expression fits the form of the right side of Eq. (20.10), so

$$\cos 23^\circ\cos 67^\circ - \sin 23^\circ\sin 67^\circ = \cos(23^\circ + 67^\circ)$$
$$= \cos 90^\circ$$
$$= 0$$

Practice Exercise

1. Evaluate: $\sin 115^\circ \cos 25^\circ - \cos 115^\circ \sin 25^\circ$

Again, we are able to evaluate this expression by ***recognizing the form*** of the given expression. Evaluation by a calculator will verify the result. ■

By dividing the right side of Eq. (20.9) by that of Eq. (20.10), we can determine expressions for $\tan(\alpha + \beta)$, and by dividing the right side of Eq. (20.11) by that of Eq. (20.12), we can determine an expression for $\tan(\alpha - \beta)$. The derivation of these formulas is Exercise 33 of this section. These formulas can be written together, as

$$\tan(\alpha \pm \beta) = \frac{\tan\alpha \pm \tan\beta}{1 \mp \tan\alpha\tan\beta} \qquad \textbf{(20.13)}$$

The formula for $\tan(\alpha + \beta)$ uses the upper signs, and the formula for $\tan(\alpha - \beta)$ uses the lower signs.

Certain trigonometric identities can be proven by the formulas derived in this section. The following examples illustrate this use of these formulas.

EXAMPLE 6 Sin $(\alpha + \beta)$—calculator verification

Prove that $\sin(180^\circ + x) = -\sin x$.

$$\sin(180^\circ + x) = \sin 180^\circ \cos x + \cos 180^\circ \sin x \qquad \text{using Eq. (20.9)}$$
$$= (0)\cos x + (-1)\sin x \qquad \sin 180^\circ = 0,\ \cos 180^\circ = -1$$
$$= -\sin x$$

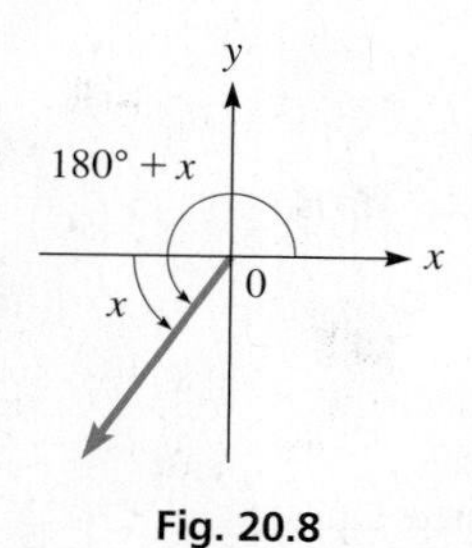

Fig. 20.8

Although x may or may not be an acute angle, this agrees with the results for the sine of a third-quadrant angle as discussed in Section 8.2. See Fig. 20.8. The calculator check of this identity, with $y_1 = \sin(\pi + x)$ and $y_2 = -\sin x$ is shown in Fig. 20.9. This screen was captured while y_2 (in red) was being graphed right over y_1 (in blue).

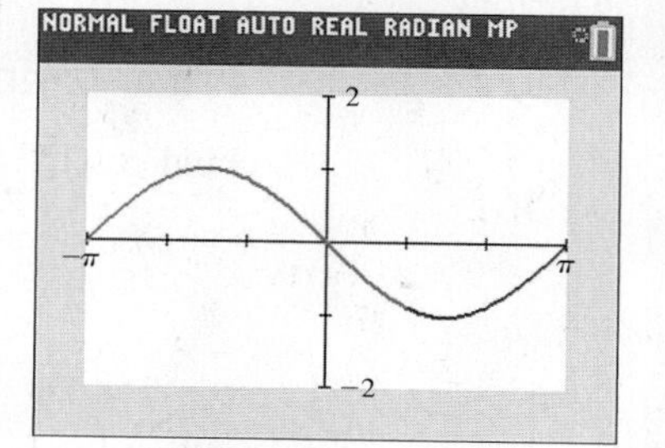

Fig. 20.9

■

EXAMPLE 7 Proving a trig identity with tan $(\alpha \pm \beta)$

Show that $\tan(\alpha + \beta)\tan(\alpha - \beta) = \dfrac{\tan^2\alpha - \tan^2\beta}{1 - \tan^2\alpha\tan^2\beta}$.

Using both of Eqs. (20.13), we have

$$\tan(\alpha + \beta)\tan(\alpha - \beta) = \left(\frac{\tan\alpha + \tan\beta}{1 - \tan\alpha\tan\beta}\right)\left(\frac{\tan\alpha - \tan\beta}{1 + \tan\alpha\tan\beta}\right)$$

$$= \frac{\tan^2\alpha - \tan^2\beta}{1 - \tan^2\alpha\tan^2\beta}$$

■

EXAMPLE 8 Trig identity using sin $(\alpha - \beta)$

Simplify the expression $\dfrac{\sin(\alpha - \beta)}{\sin\alpha\sin\beta}$.

$$\frac{\sin(\alpha - \beta)}{\sin\alpha\sin\beta} = \frac{\sin\alpha\cos\beta - \cos\alpha\sin\beta}{\sin\alpha\sin\beta} \quad \text{using Eq. (20.11)}$$

$$= \frac{\sin\alpha\cos\beta}{\sin\alpha\sin\beta} - \frac{\cos\alpha\sin\beta}{\sin\alpha\sin\beta} = \frac{\cos\beta}{\sin\beta} - \frac{\cos\alpha}{\sin\alpha}$$

$$= \cot\beta - \cot\alpha \quad \text{using Eq. (20.5)}$$

■

Practice Exercise

2. Simplify: $\tan(180° + x)$

EXAMPLE 9 Using sin $(\alpha - \beta)$—three-phase generator

■ See chapter introduction.

Alternating electric current is produced essentially by a coil of wire rotating in a magnetic field, and this is the basis for designing generators of alternating current. A three-phase generator uses three coils of wire and thereby produces three electric currents at the same time. This is the most widely used type of *polyphase generator* as mentioned in the chapter introduction on page 535.

The voltages induced in a three-phase generator can be represented as

$$E_1 = E_0\sin\omega t \qquad E_2 = E_0\sin\left(\omega t - \tfrac{2\pi}{3}\right) \qquad E_3 = E_0\sin\left(\omega t - \tfrac{4\pi}{3}\right)$$

where E_0 is the maximum voltage and ω is the angular velocity of rotation. Show that the sum of these voltages at any time t is zero.

Setting up the sum $E_1 + E_2 + E_3$ and using Eq. (20.11), we have

$$E_1 + E_2 + E_3$$
$$= E_0\left[\sin\omega t + \sin\left(\omega t - \tfrac{2\pi}{3}\right) + \sin\left(\omega t - \tfrac{4\pi}{3}\right)\right]$$
$$= E_0\left(\sin\omega t + \sin\omega t\cos\tfrac{2\pi}{3} - \cos\omega t\sin\tfrac{2\pi}{3} + \sin\omega t\cos\tfrac{4\pi}{3} - \cos\omega t\sin\tfrac{4\pi}{3}\right)$$
$$= E_0\left[\sin\omega t + (\sin\omega t)\left(-\tfrac{1}{2}\right) - (\cos\omega t)\left(\tfrac{1}{2}\sqrt{3}\right) + (\sin\omega t)\left(-\tfrac{1}{2}\right) - (\cos\omega t)\left(-\tfrac{1}{2}\sqrt{3}\right)\right]$$
$$= E_0\left[(\sin\omega t)\left(1 - \tfrac{1}{2} - \tfrac{1}{2}\right) + (\cos\omega t)\left(\tfrac{1}{2}\sqrt{3} - \tfrac{1}{2}\sqrt{3}\right)\right] = 0$$

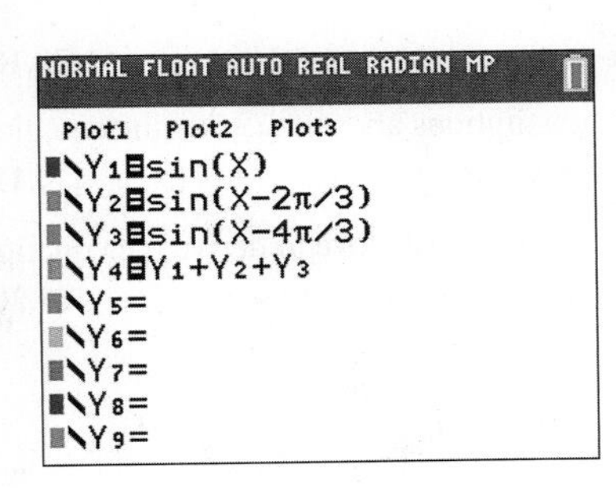

Fig. 20.10
Graphing calculator keystrokes: goo.gl/VJ1NCJ

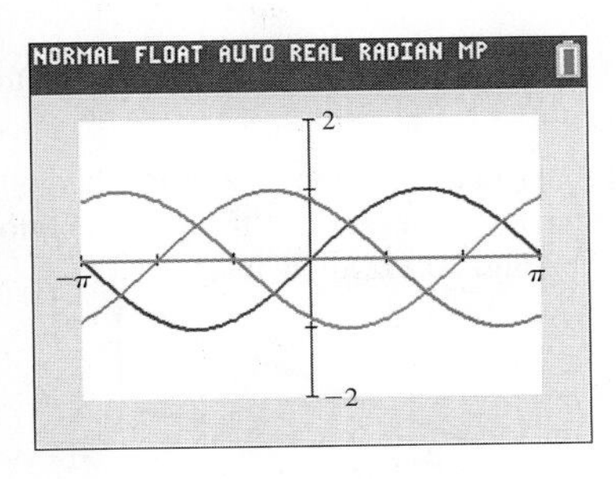

Fig. 20.11

To show that $E_1 + E_2 + E_3 = 0$ on a calculator, in Fig. 20.10 we let $y_1 = \sin x$, $y_2 = \sin(x - 2\pi/3)$, $y_3 = \sin(x - 4\pi/3)$, and $y_4 = y_1 + y_2 + y_3$. In Fig. 20.11, y_1, y_2, and y_3 are the three sine curves, and y_4 is the pink line along the x-axis that shows $y_1 + y_2 + y_3 = 0$. This shows us that the sum of the three voltages is zero at any point in time. ■

EXERCISES 20.2

In Exercises 1 and 2, make the given changes in the indicated examples of this section and then solve the given problems.

1. In Example 2, change $\frac{5}{13}$ to $\frac{12}{13}$ and then find the value of $\cos(\alpha + \beta)$.

2. In Example 6, change $180° + x$ to $180° - x$ and then determine what other changes result.

In Exercises 3–6, determine the values of the given functions as indicated.

3. Find sin 105° by using $105° = 60° + 45°$.

4. Find tan 75° by using $75° = 30° + 45°$.

5. Find cos 15° by using $15° = 60° - 45°$.

6. Find sin 15° by using $15° = 225° - 210°$.

In Exercises 7–10, evaluate the given functions with the following information: $\sin\alpha = 4/5$ (α in first quadrant) and $\cos\beta = -12/13$ (β in second quadrant).

7. $\sin(\alpha + \beta)$ **8.** $\tan(\beta - \alpha)$

9. $\cos(\alpha + \beta)$ **10.** $\sin(\alpha - \beta)$

In Exercises 11–20, simplify the given expressions.

11. $\sin x \cos 2x + \sin 2x \cos x$ **12.** $\sin 3x \cos x - \sin x \cos 3x$

13. $\cos 5x \cos x + \sin 5x \sin x$ **14.** $\dfrac{\tan(x - y) + \tan x}{1 - \tan(x - y)\tan y}$

15. $\sin(90° - x)$ **16.** $\cos(\frac{3}{2}\pi - x)$

17. $\tan(x - \pi)$ **18.** $\sin(x + \pi/2)$

19. $\sin 3x \cos(3x - \pi) - \cos 3x \sin(3x - \pi)$

20. $\cos(x + \pi)\cos(x - \pi) + \sin(x + \pi)\sin(x - \pi)$

In Exercises 21–24, evaluate each expression by first changing the form. Verify each by use of a calculator.

21. $\sin 122° \cos 32° - \cos 122° \sin 32°$

22. $\cos 250° \cos 70° + \sin 250° \sin 70°$

23. $\cos\frac{\pi}{5}\cos\frac{3\pi}{10} - \sin\frac{\pi}{5}\sin\frac{3\pi}{10}$ **24.** $\dfrac{\tan\frac{\pi}{10} + \tan\frac{3\pi}{20}}{1 - \tan\frac{\pi}{10}\tan\frac{3\pi}{20}}$

In Exercises 25–28, prove the given identities.

25. $\sin(x + y)\sin(x - y) = \sin^2 x - \sin^2 y$

26. $\cos(x + y)\cos(x - y) = \cos^2 x - \sin^2 y$

27. $\cos(\alpha + \beta) + \cos(\alpha - \beta) = 2\cos\alpha\cos\beta$

28. $\tan(90° + x) = -\cot x$ [Explain why Eq. (20.13) cannot be used for this, but Eqs. (20.9) and (20.10) can be used.]

In Exercises 29–32, verify each identity by comparing the graph of the left side with the graph of the right side on a calculator.

29. $\cos(\frac{\pi}{6} + x) = \dfrac{\sqrt{3}\cos x - \sin x}{2}$

30. $\sin(120° - x) = \dfrac{\sqrt{3}\cos x + \sin x}{2}$

31. $\tan(\frac{3\pi}{4} + x) = \dfrac{\tan x - 1}{\tan x + 1}$ **32.** $\cos(\frac{\pi}{2} - x) = \sin x$

In Exercises 33–36, derive the given equations as indicated. Equations (20.14)–(20.16) are known as the product formulas.

33. By dividing the right side of Eq. (20.9) by that of Eq. (20.10), and dividing the right side of Eq. (20.11) by that of Eq. (20.12), derive Eq. (20.13).

$$\tan(\alpha \pm \beta) = \frac{\tan\alpha \pm \tan\beta}{1 \mp \tan\alpha\tan\beta} \quad \textbf{(20.13)}$$

(*Hint:* Divide numerator and denominator by $\cos\alpha\cos\beta$.)

34. By adding Eqs. (20.9) and (20.11), derive the equation
$$\sin\alpha\cos\beta = \tfrac{1}{2}[\sin(\alpha + \beta) + \sin(\alpha - \beta)] \quad \textbf{(20.14)}$$

35. By adding Eqs. (20.10) and (20.12), derive the equation
$$\cos\alpha\cos\beta = \tfrac{1}{2}[\cos(\alpha + \beta) + \cos(\alpha - \beta)] \quad \textbf{(20.15)}$$

36. By subtracting Eq. (20.10) from Eq. (20.12), derive
$$\sin\alpha\sin\beta = \tfrac{1}{2}[\cos(\alpha - \beta) - \cos(\alpha + \beta)] \quad \textbf{(20.16)}$$

In Exercises 37–40, derive the given equations by letting $\alpha + \beta = x$ and $\alpha - \beta = y$, which leads to $\alpha = \frac{1}{2}(x + y)$ and $\beta = \frac{1}{2}(x - y)$. The resulting equations are known as the factor formulas.

37. Use Eq. (20.14) and the substitutions above to derive the equation
$$\sin x + \sin y = 2\sin\tfrac{1}{2}(x + y)\cos\tfrac{1}{2}(x - y) \quad \textbf{(20.17)}$$

38. Use Eqs. (20.9) and (20.11) and the substitutions above to derive the equation
$$\sin x - \sin y = 2\sin\tfrac{1}{2}(x - y)\cos\tfrac{1}{2}(x + y) \quad \textbf{(20.18)}$$

39. Use Eq. (20.15) and the substitutions above to derive the equation
$$\cos x + \cos y = 2\cos\tfrac{1}{2}(x + y)\cos\tfrac{1}{2}(x - y) \quad \textbf{(20.19)}$$

40. Use Eq. (20.16) and the substitutions above to derive the equation
$$\cos x - \cos y = -2\sin\tfrac{1}{2}(x + y)\sin\tfrac{1}{2}(x - y) \quad \textbf{(20.20)}$$

In Exercises 41–54, solve the given problems.

41. Evaluate exactly: $\sin(x + 30°)\cos x - \cos(x + 30°)\sin x$

42. Simplify: $\sin(\frac{\pi}{6} - \theta) + \cos(\frac{\pi}{3} - \theta)$

43. Show that $\frac{\sin 2x}{\sin x} = 2\cos x$. [*Hint:* $\sin 2x = \sin(x + x)$.]

44. Using graphs displayed on a calculator, verify the identity in Exercise 43.

45. Explain how the exact value of sin 75° can be found using either Eq. (20.9) or Eq. (20.11).

46. A vertical pole of length L is placed on top of a hill of height h. From the plain below the angles of elevation of the top and bottom of the pole are α and β. See Fig. 20.12. Show that

$$h = \frac{L\cos\alpha\sin\beta}{\sin(\alpha - \beta)}$$

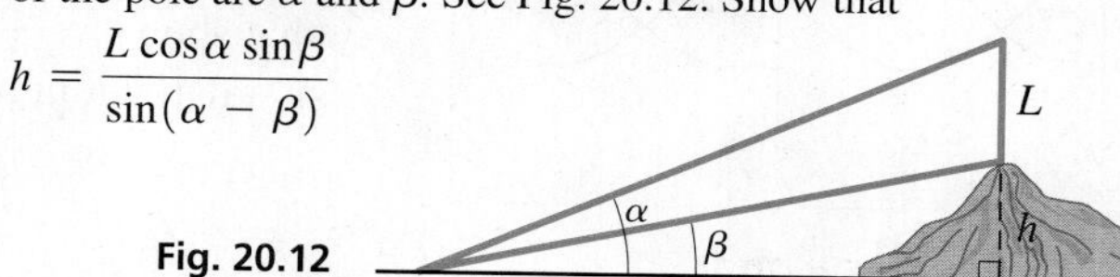

Fig. 20.12

47. The design of a certain three-phase alternating-current generator uses the fact that the sum of the currents $I\cos(\theta + 30°)$, $I\cos(\theta + 150°)$, and $I\cos(\theta + 270°)$ is zero. Verify this.

48. The current (in A) in a certain AC circuit is given by $i = 4\sin(120\pi t + \frac{\pi}{2})$. Use the sum formula for sine to write this in a different form and then simplify.

49. For voltages $V_1 = 20\sin 120\pi t$ and $V_2 = 20\cos 120\pi t$, show that $V = V_1 + V_2 = 20\sqrt{2}\sin(120\pi t + \pi/4)$. Use a calculator to verify this result.

50. The displacements y_1 and y_2 of two waves traveling through the same medium are given by $y_1 = A\sin 2\pi(t/T - x/\lambda)$ and $y_2 = A\sin 2\pi(t/T + x/\lambda)$. Find an expression for the displacement $y_1 + y_2$ of the combination of the waves.

51. The displacement d of a water wave is given by the equation $d = d_0\sin(\omega t + \alpha)$. Show that this can be written as $d = d_1\sin\omega t + d_2\cos\omega t$, where $d_1 = d_0\cos\alpha$ and $d_2 = d_0\sin\alpha$.

52. A weight **w** is held in equilibrium by forces **F** and **T** as shown in Fig. 20.13. Equations relating w, F, and T are

$$F\cos\theta = T\sin\alpha$$
$$w + F\sin\theta = T\cos\alpha$$

Show that $w = \dfrac{T\cos(\theta + a)}{\cos\theta}$.

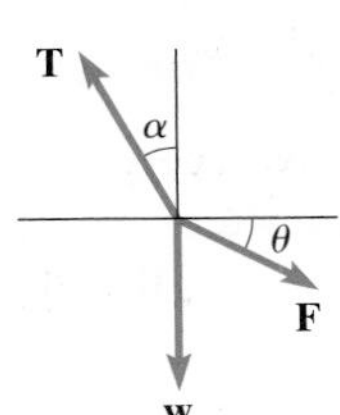

Fig. 20.13

53. For the two bevel gears shown in Fig. 20.14, the equation

$\tan\alpha = \dfrac{\sin\beta}{R + \cos\beta}$ is used.

Here, R is the ratio of gear 1 to gear 2. Show that

$$R = \frac{\sin(\beta - \alpha)}{\sin\alpha}.$$

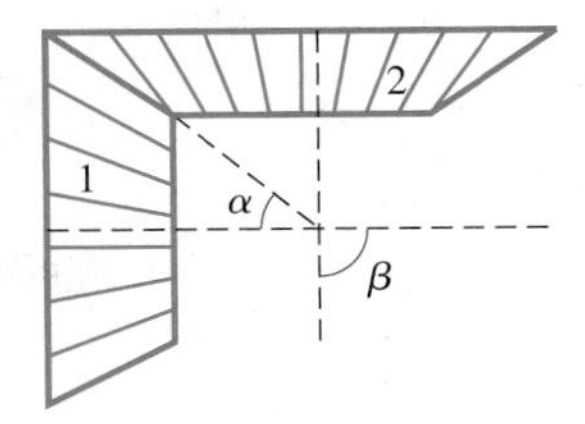

Fig. 20.14

54. In the analysis of the angles of incidence i and reflection r of a light ray subject to certain conditions, the following expression is found:

$$E_2\left(\frac{\tan r}{\tan i} + 1\right) = E_1\left(\frac{\tan r}{\tan i} - 1\right)$$

Show that $E_2 = E_1\dfrac{\sin(r - i)}{\sin(r + i)}$.

Answers to Practice Exercises

1. 1 **2.** $\tan x$

20.3 Double-Angle Formulas

Formula for $\sin 2\alpha$ • Formulas for $\cos 2\alpha$ • Formula for $\tan 2\alpha$

If we let $\beta = \alpha$ in the sum formulas for sine, cosine, and tangent (given in Section 20.2), we can derive the important double-angle formulas:

$$\sin(\alpha + \alpha) = \sin(2\alpha) = \sin\alpha\cos\alpha + \cos\alpha\sin\alpha = 2\sin\alpha\cos\alpha$$

$$\cos(\alpha + \alpha) = \cos\alpha\cos\alpha - \sin\alpha\sin\alpha = \cos^2\alpha - \sin^2\alpha$$

$$\tan(\alpha + \alpha) = \frac{\tan\alpha + \tan\alpha}{1 - \tan\alpha\tan\alpha} = \frac{2\tan\alpha}{1 - \tan^2\alpha}$$

Then using the basic identity $\sin^2 x + \cos^2 x = 1$, other forms of the equation for $\cos 2\alpha$ may be derived. Summarizing these forms, we have

$$\sin 2\alpha = 2\sin\alpha\cos\alpha \quad (20.21)$$
$$\cos 2\alpha = \cos^2\alpha - \sin^2\alpha \quad (20.22)$$
$$= 2\cos^2\alpha - 1 \quad (20.23)$$
$$= 1 - 2\sin^2\alpha \quad (20.24)$$
$$\tan 2\alpha = \frac{2\tan\alpha}{1 - \tan^2\alpha} \quad (20.25)$$

These double-angle formulas are widely used in applications of trigonometry, especially in calculus. They should be recognized quickly in any of the above forms.

CAUTION Note carefully that $\sin 2\alpha$ **IS NOT** $2\sin\alpha$. ■

EXAMPLE 1 Using double-angle formulas

(a) If $\alpha = 30°$, we have

$$\cos 60° = \cos 2(30°) = \cos^2 30° - \sin^2 30° = \left(\frac{\sqrt{3}}{2}\right)^2 - \left(\frac{1}{2}\right)^2 = \frac{1}{2} \quad \text{using Eq. (20.22)}$$

(b) If $\alpha = 3x$, we have

$$\sin 6x = \sin 2(3x) = 2 \sin 3x \cos 3x \quad \text{using Eq. (20.21)}$$

(c) If $2\alpha = x$, we may write $\alpha = x/2$, which means that

$$\sin x = \sin 2\left(\frac{x}{2}\right) = 2 \sin\frac{x}{2} \cos\frac{x}{2} \quad \text{using Eq. (20.21)}$$

(d) If $\alpha = \frac{\pi}{6}$, we have

$$\tan\frac{\pi}{3} = \tan 2\left(\frac{\pi}{6}\right) = \frac{2 \tan\frac{\pi}{6}}{1 - \tan^2(\frac{\pi}{6})} = \frac{2(\sqrt{3}/3)}{1 - (\sqrt{3}/3)^2} = \sqrt{3} \quad \text{using Eq. (20.25)}$$ ■

EXAMPLE 2 Simplification using $\cos 2\alpha$ formula

Simplify the expression $\cos^2 2x - \sin^2 2x$.

Since this is the difference of the square of the cosine of an angle and the square of the sine of the same angle, it fits the right side of Eq. (20.22). Therefore, letting $\alpha = 2x$, we have

$$\cos^2 2x - \sin^2 2x = \cos 2(2x) = \cos 4x$$ ■

EXAMPLE 3 Using $\sin 2\alpha$ formula—area of land

To find the area A of a right triangular tract of land, a surveyor may use the formula $A = \frac{1}{4}c^2 \sin 2\theta$, where c is the hypotenuse and θ is *either* of the acute angles. Derive this formula.

In Fig. 20.15, we see that $\sin\theta = a/c$ and $\cos\theta = b/c$, which gives us

$$a = c \sin\theta \quad \text{and} \quad b = c \cos\theta$$

The area is given by $A = \frac{1}{2}ab$, which leads to the solution

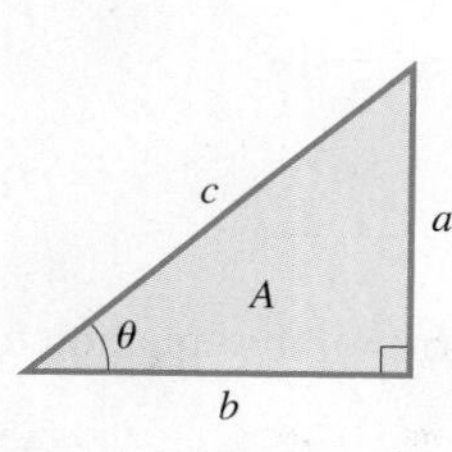

Fig. 20.15

$$A = \frac{1}{2}ab = \frac{1}{2}(c \sin\theta)(c \cos\theta)$$

$$= \frac{1}{2}c^2 \sin\theta \cos\theta = \frac{1}{2}c^2\left(\frac{1}{2}\sin 2\theta\right) \quad \text{using Eq. (20.21)}$$

$$= \frac{1}{4}c^2 \sin 2\theta$$

In using Eq. (20.21), we divided both sides by 2 to get $\sin\theta \cos\theta = \frac{1}{2}\sin 2\theta$.

If we had labeled the upper acute angle in Fig. 20.15 as θ, we would have $a = c \cos\theta$ and $b = c \sin\theta$. Using these values in the formula for the area gives the same solution. ■

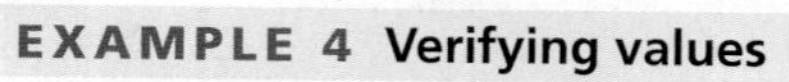

EXAMPLE 4 Verifying values

(a) Verifying the values of sin 90°, using the functions of 45°, we have

$$\sin 90° = \sin 2(45°) = 2 \sin 45° \cos 45° = 2\left(\frac{\sqrt{2}}{2}\right)\left(\frac{\sqrt{2}}{2}\right) = 1 \quad \text{using Eq. (20.21)}$$

(b) Using Eq. (20.25), $\tan 142° = \dfrac{2 \tan 71°}{1 - \tan^2 71°}$. Using a calculator, we have

$$\tan 142° = -0.7812856265 \quad \text{and} \quad \frac{2 \tan 71°}{1 - \tan^2 71°} = -0.7812856265$$ ■

Practice Exercise

1. Evaluate cos 90° using values for 45°.

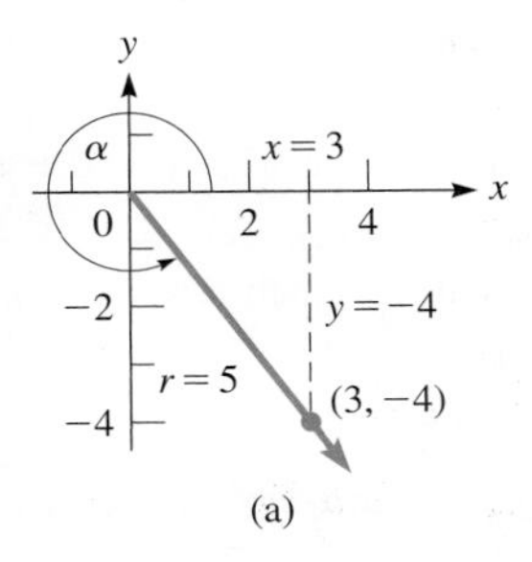

(a)

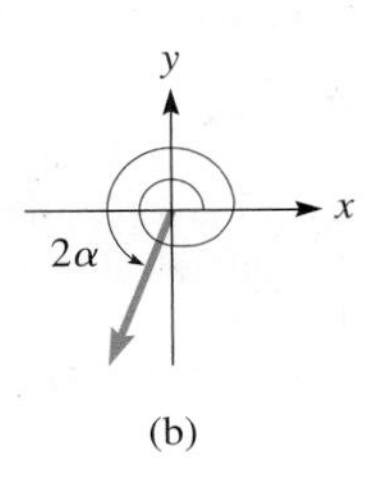

(b)

Fig. 20.16

EXAMPLE 5 Evaluation using $\sin 2\alpha$ formula

If $\cos\alpha = \frac{3}{5}$ for a fourth-quadrant angle, from Fig. 20.16(a) we see that $\sin\alpha = -\frac{4}{5}$. Thus,

$$\sin 2\alpha = 2\sin\alpha\cos\alpha \qquad \text{Eq. (20.21)}$$
$$= 2\left(-\frac{4}{5}\right)\left(\frac{3}{5}\right) = -\frac{24}{25}$$

In Fig. 20.16(b), angle 2α is shown to be in the third-quadrant, verifying the sign of the result. ($\cos\alpha = 3/5$, $\alpha = 307°$, $2\alpha = 614°$, which is a third-quadrant angle.) ■

EXAMPLE 6 Simplification using $\cos 2\alpha$

Simplify the expression $\dfrac{2}{1 + \cos 2x}$.

$$\frac{2}{1+\cos 2x} = \frac{2}{1 + (2\cos^2 x - 1)} \qquad \text{using Eq. (20.2)}$$
$$= \frac{2}{2\cos^2 x} = \sec^2 x \qquad \text{using Eq. (20.23)}$$ ■

Practice Exercise

2. Simplify: $\dfrac{\sin 2x}{\cos^2 x - \sin^2 x}$

EXAMPLE 7 Trig identity—calculator verification

Prove the identity $\dfrac{\sin 3x}{\sin x} + \dfrac{\cos 3x}{\cos x} = 4\cos 2x$.

Because the left side is the more complex side, we change it to the form on the right:

$$\frac{\sin 3x}{\sin x} + \frac{\cos 3x}{\cos x} = \frac{\sin 3x\cos x + \cos 3x\sin x}{\sin x\cos x} \qquad \text{combining fractions}$$
$$= \frac{\sin(3x + x)}{\frac{1}{2}\sin 2x} \qquad \begin{array}{l}\leftarrow \text{using Eq. (20.9)}\\ \leftarrow \text{using Eq. (20.21)}\end{array}$$
$$= \frac{2\sin 4x}{\sin 2x} = \frac{2(2\sin 2x\cos 2x)}{\sin 2x} \qquad \text{using Eq. (20.21)}$$
$$= 4\cos 2x$$

We can check this identity by comparing the graphs of

$$y_1 = (\sin 3x)/(\sin x) + (\cos 3x)/(\cos x) \quad \text{and} \quad y_2 = 4\cos 2x.$$

Figure 20.17 shows the graph of y_2 (in red) being plotted over the graph of y_1 (in blue). ■

TI-89 graphing calculator keystrokes for Example 7: goo.gl/Ou20NX

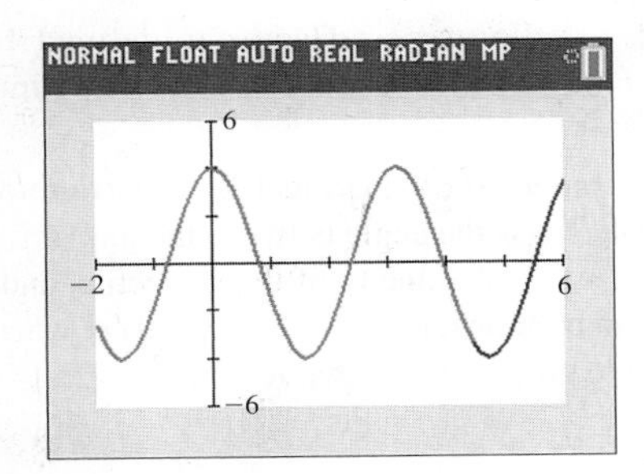

Fig. 20.17

EXERCISES 20.3

In Exercises 1–4, make the given changes in the indicated examples of this section and then solve the resulting problems.

1. In Example 1(d), change $\frac{\pi}{6}$ to $\frac{\pi}{3}$ and then evaluate $\tan\frac{2\pi}{3}$.
2. In Example 2, change $2x$ to $3x$ and then simplify.
3. In Example 5, change 3/5 to 4/5 and then evaluate $\sin 2\alpha$.
4. In Example 6, change the + in the denominator to − and then simplify the expression on the left.

In Exercises 5–8, determine the values of the indicated functions in the given manner.

5. Find sin 60° by using the functions of 30°.
6. Find sin 120° by using the functions of 60°.
7. Find tan 120° by using the functions of 60°.
8. Find cos 60° by using the functions of 30°.

In Exercises 9–14, use a calculator to verify the values found by using the double-angle formulas.

9. Find sin 100° directly and by using functions of 50°.
10. Find tan 184° directly and by using functions of 92°.
11. Find cos 96° directly and by using functions of 48°.
12. Find cos 276° directly and by using functions of 138°.
13. Find $\tan\frac{2\pi}{7}$ directly and by using functions of $\frac{\pi}{7}$.
14. Find $\sin 1.2\pi$ directly and by using functions of 0.6π.

In Exercises 15–18, evaluate the indicated functions with the given information.

15. Find $\sin 2x$ if $\cos x = \frac{4}{5}$ (in first quadrant).

16. Find $\cos 2x$ if $\sin x = -\frac{12}{13}$ (in third quadrant).

17. Find $\tan 2x$ if $\sin x = 0.5$ (in second quadrant).

18. Find $\sin 4x$ if $\tan x = -0.6$ (in fourth quadrant).

In Exercises 19–30, simplify the given expressions.

19. $6 \sin 5x \cos 5x$

20. $4 \sin^2 x \cos^2 x$

21. $1 - 2 \sin^2 4x$

22. $\dfrac{4 \tan 4\theta}{1 - \tan^2 4\theta}$

23. $2 \cos^2 \frac{1}{2}x - 1$

24. $4 \sin \frac{1}{2}x \cos \frac{1}{2}x$

25. $8 \sin^2 2x - 4$

26. $6 \cos 3x \sin 3x$

27. $\dfrac{\sin 6\theta}{\cos 3\theta}$

28. $\cos^4 u - \sin^4 u$

29. $\dfrac{\sin 3x}{\sin x} - \dfrac{\cos 3x}{\cos x}$

30. $\dfrac{\cos 3x}{\sin x} + \dfrac{\sin 3x}{\cos x}$

In Exercises 31–40, prove the given identities.

31. $\cos^2 \alpha - \sin^2 \alpha = 2 \cos^2 \alpha - 1$

32. $\cos^2 \alpha - \sin^2 \alpha = 1 - 2 \sin^2 \alpha$

33. $\dfrac{\cos x - \tan x \sin x}{\sec x} = \cos 2x$

34. $2 + \dfrac{\cos 2\theta}{\sin^2 \theta} = \csc^2 \theta$

35. $\dfrac{\sin 2\theta}{1 + \cos 2\theta} = \tan \theta$

36. $\dfrac{2 \tan \alpha}{1 + \tan^2 \alpha} = \sin 2\alpha$

37. $1 - \cos 2\theta = \dfrac{2}{1 + \cot^2 \theta}$

38. $\dfrac{\cos^3 \theta + \sin^3 \theta}{\cos \theta + \sin \theta} = 1 - \dfrac{1}{2} \sin 2\theta$

39. $\ln(1 - \cos 2x) - \ln(1 + \cos 2x) = 2 \ln \tan x$

40. $\log(20 \sin^2 \theta + 10 \cos 2\theta) = 1$

In Exercises 41–44, verify each identity by comparing the graph of the left side with the graph of the right side on a calculator.

41. $\tan 2\theta = \dfrac{2}{\cot \theta - \tan \theta}$

42. $\dfrac{1 - \tan^2 x}{\sec^2 x} = \cos 2x$

43. $(\sin x + \cos x)^2 = 1 + \sin 2x$

44. $2 \csc 2x \tan x = \sec^2 x$

In Exercises 45–62, solve the given problems.

45. Express $\sin 3x$ in terms of $\sin x$ only.

46. Express $\cos 3x$ in terms of $\cos x$ only.

47. Express $\cos 4x$ in terms of $\cos x$ only.

48. Express $\sin 4x$ in terms of $\sin x$ and $\cos x$.

49. Find the exact value of $\cos 2x + \sin 2x \tan x$.

50. Find the exact value of $\cos^4 x - \sin^4 x - \cos 2x$.

51. Simplify: $\log(\cos x - \sin x) + \log(\cos x + \sin x)$.

52. For an acute angle θ, show that $2 \sin \theta > \sin 2\theta$.

53. Without graphing, determine the amplitude and period of the function $y = 4 \sin x \cos x$. Explain.

54. Without graphing, determine the amplitude and period of the function $y = \cos^2 x - \sin^2 x$.

55. The path of a bouncing ball is given by $y = \sqrt{(\sin x + \cos x)^2}$. Show that this path can also be shown as $y = \sqrt{1 + \sin 2x}$. Use a calculator to show that this can also be shown as $y = |\sin x + \cos x|$.

56. The equation for the trajectory of a missile fired into the air at an angle α with velocity v_0 is $y = x \tan \alpha - \dfrac{g}{2v_0^2 \cos^2 \alpha} x^2$. Here, g is the acceleration due to gravity. On the right of the equal sign, combine terms and simplify.

57. The CN Tower in Toronto is 553 m high, and has an observation deck at the 335-m level. How far from the top of the tower must a 553-m high helicopter be so that the angle subtended at the helicopter by the part of the tower above the deck equals the angle subtended at the helicopter below the deck? In Fig. 20.18 these are the angles α and β.

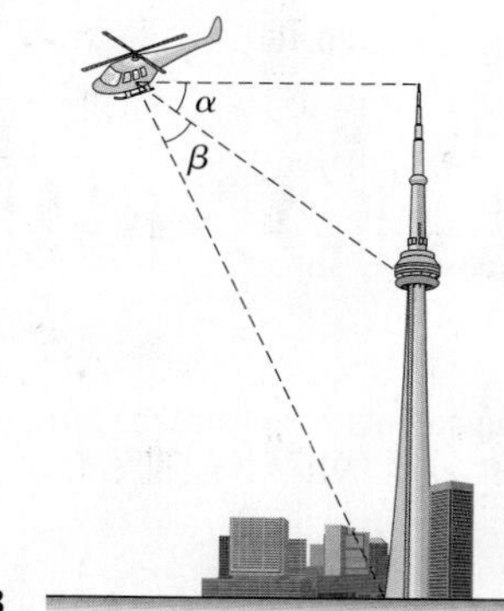

Fig. 20.18

58. The cross section of a radio-wave reflector is defined by $x = \cos 2\theta$, $y = \sin \theta$. Find the relation between x and y by eliminating θ.

59. To find the horizontal range R of a projectile, the equation $R = vt \cos \alpha$ is used, where α is the angle between the line of fire and the horizontal, v is the initial velocity of the projectile, and t is the time of flight. It can be shown that $t = (2v \sin \alpha)/g$, where g is the acceleration due to gravity. Show that $R = (v^2 \sin 2\alpha)/g$. See Fig. 20.19.

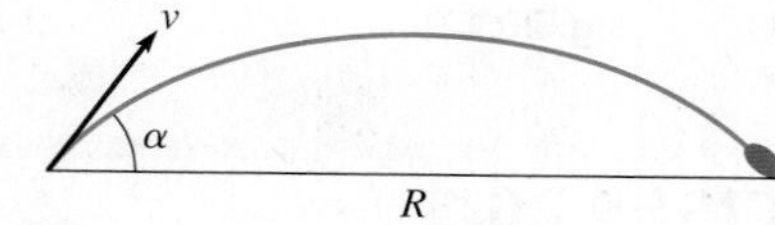

Fig. 20.19

60. In analyzing light reflection from a cylinder onto a flat surface, the expression $3 \cos \theta - \cos 3\theta$ arises. Show that this equals $2 \cos \theta \cos 2\theta + 4 \sin \theta \sin 2\theta$.

61. The instantaneous electric power p in an inductor is given by the equation $p = vi \sin \omega t \sin(\omega t - \pi/2)$. Show that this equation can be written as $p = -\frac{1}{2} vi \sin 2\omega t$.

62. In the study of the stress at a point in a bar, the equation $s = a \cos^2 \theta + b \sin^2 \theta - 2t \sin \theta \cos \theta$ arises. Show that this equation can be written as
$s = \frac{1}{2}(a + b) + \frac{1}{2}(a - b) \cos 2\theta - t \sin 2\theta$.

Answers to Practice Exercises

1. $\cos 90° = \cos^2 45° - \sin^2 45° = (\frac{1}{2}\sqrt{2})^2 - (\frac{1}{2}\sqrt{2})^2 = 0$

2. $\tan 2x$

20.4 Half-Angle Formulas

Formula for $\sin(\alpha/2)$ • Formula for $\cos(\alpha/2)$

If we let $\theta = \alpha/2$ in the identity $\cos 2\theta = 1 - 2\sin^2\theta$ and then solve for $\sin(\alpha/2)$,

$$\sin\frac{\alpha}{2} = \pm\sqrt{\frac{1 - \cos\alpha}{2}} \tag{20.26}$$

Also, with the same substitution in the identity $\cos 2\theta = 2\cos^2\theta - 1$, which is then solved for $\cos(\alpha/2)$, we have

$$\cos\frac{\alpha}{2} = \pm\sqrt{\frac{1 + \cos\alpha}{2}} \tag{20.27}$$

CAUTION In each of Eqs. (20.26) and (20.27), the sign chosen depends on the quadrant in which $\frac{\alpha}{2}$ lies. ■

EXAMPLE 1 Evaluation using $\cos(\alpha/2)$ formula

We can find cos 165° by using the relation

$$\cos 165° = -\sqrt{\frac{1 + \cos 330°}{2}} \quad \text{using Eq. (20.27)}$$

$$= -\sqrt{\frac{1 + 0.8660}{2}} = -0.9659$$

Here, the minus sign is used, since 165° is in the second quadrant, and the cosine of a second-quadrant angle is negative. ■

EXAMPLE 2 Evaluation using $\sin(\alpha/2)$ formula

Simplify $\sqrt{\frac{1 - \cos 114°}{2}}$ by expressing the result in terms of one-half the given angle. Then, using a calculator, show that the values are equal.

We note that the given expression fits the form of the right side of Eq. (20.26), which means that

$$\sqrt{\frac{1 - \cos 114°}{2}} = \sin\tfrac{1}{2}(114°) = \sin 57°$$

Using a calculator shows that

$$\sqrt{\frac{1 - \cos 114°}{2}} = 0.8386705679 \quad \text{and} \quad \sin 57° = 0.8386705679$$

which verifies the equation for these values. ■

EXAMPLE 3 Simplification using $\cos(\alpha/2)$ formula

Simplify the expression $\sqrt{\frac{9 + 9\cos 6x}{2}}$.

$$\sqrt{\frac{9 + 9\cos 6x}{2}} = \sqrt{\frac{9(1 + \cos 6x)}{2}} = 3\sqrt{\frac{1 + \cos 6x}{2}}$$

$$= 3\cos\tfrac{1}{2}(6x) \quad \text{using Eq. (20.27) with } \alpha = 6x$$

$$= 3\cos 3x$$

Noting the original expression, we see that $\cos 3x$ cannot be negative. ■

Practice Exercise

1. Simplify: $\sqrt{\frac{25 - 25\cos 4x}{2}}$

EXAMPLE 4 Trig identity with $\sin(\alpha/2)$—kinetic theory of gases

In the kinetic theory of gases, the expression $\sqrt{(1 - \cos\alpha)^2 + \sin^2\alpha}$ is found. Show that this expression equals $2\sin\frac{1}{2}\alpha$:

$$\begin{aligned}\sqrt{(1 - \cos\alpha)^2 + \sin^2\alpha} &= \sqrt{1 - 2\cos\alpha + \cos^2\alpha + \sin^2\alpha} && \text{expanding}\\ &= \sqrt{1 - 2\cos\alpha + 1} && \text{using Eq. (20.26)}\\ &= \sqrt{2 - 2\cos\alpha}\\ &= \sqrt{2(1 - \cos\alpha)} && \text{factoring}\end{aligned}$$

This last expression is very similar to that for $\sin\frac{1}{2}\alpha$, except that no 2 appears in the denominator. Therefore, multiplying the numerator and the denominator under the radical by 2 leads to the solution:

$$\begin{aligned}\sqrt{2(1 - \cos\alpha)} &= \sqrt{\frac{4(1 - \cos\alpha)}{2}} = 2\sqrt{\frac{1 - \cos\alpha}{2}}\\ &= 2\sin\tfrac{1}{2}\alpha && \text{using Eq. (20.26)}\end{aligned}$$

Noting the original expression, we see that $\sin\frac{1}{2}\alpha$ cannot be negative. ■

EXAMPLE 5 Evaluation using $\cos(\alpha/2)$ formula

Given that $\tan\alpha = \frac{8}{15}$ $(180° < \alpha < 270°)$, find $\cos\left(\frac{\alpha}{2}\right)$.

Knowing that $\tan\alpha = \frac{8}{15}$ for a third-quadrant angle, we determine from Fig. 20.20 that $\cos\alpha = -\frac{15}{17}$. This means

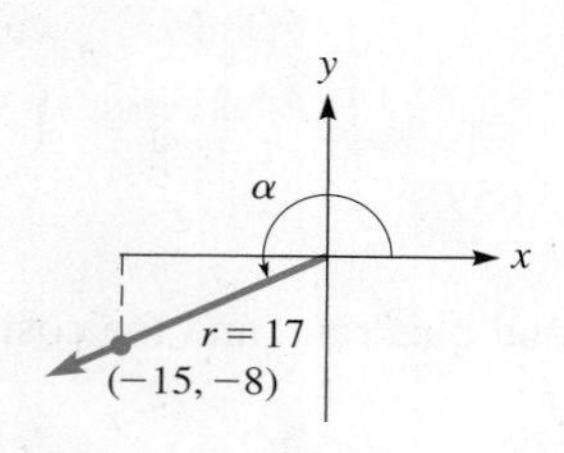

Fig. 20.20

$$\begin{aligned}\cos\frac{\alpha}{2} &= -\sqrt{\frac{1 + (-15/17)}{2}} = -\sqrt{\frac{2}{34}} && \text{using Eq. (20.27)}\\ &= -\frac{1}{17}\sqrt{17} = -0.2425\end{aligned}$$

Because $180° < \alpha < 270°$, we know that $90° < \frac{\alpha}{2} < 135°$ and therefore $\frac{\alpha}{2}$ is in the second quadrant. Because the cosine is negative for second-quadrant angles, we use the negative value of the radical. ■

EXAMPLE 6 Simplification—calculator verification

Show that $2\cos^2\frac{x}{2} - \cos x = 1$.

The first step is to substitute for $\cos\frac{x}{2}$, which will result in each term on the left being in terms of x and no $\frac{x}{2}$ terms will exist. This might allow us to combine terms. So we perform this operation, and we have for the left side

$$\begin{aligned}2\cos^2\frac{x}{2} - \cos x &= 2\left(\frac{1 + \cos x}{2}\right) - \cos x && \text{using Eq. (20.27) with both sides squared}\\ &= 1 + \cos x - \cos x = 1\end{aligned}$$

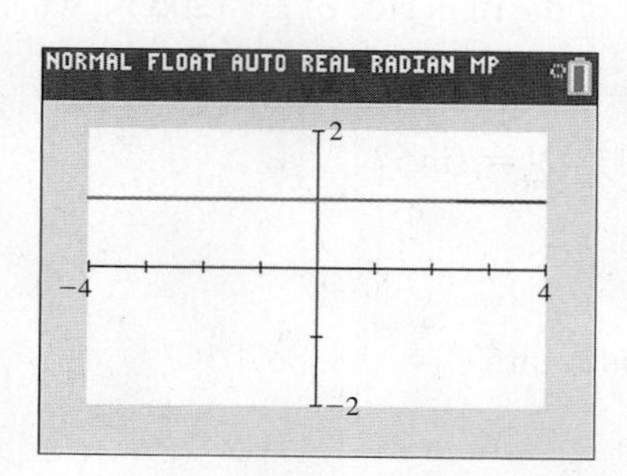

Fig. 20.21

From Fig. 20.21, we verify that the graph of $y_1 = 2[\cos(x/2)]^2 - \cos x$ is the same as the graph of $y_2 = 1$. ■

EXAMPLE 7 Formulas for other functions of $\alpha/2$

We can find relations for other functions of $\frac{\alpha}{2}$ by expressing these functions in terms of $\sin\left(\frac{\alpha}{2}\right)$ and $\cos\left(\frac{\alpha}{2}\right)$. For example,

$$\begin{aligned}\sec\frac{\alpha}{2} &= \frac{1}{\cos\frac{\alpha}{2}} = \pm\frac{1}{\sqrt{\frac{1 + \cos\alpha}{2}}} && \text{using Eq. (20.27)}\\ &= \pm\sqrt{\frac{2}{1 + \cos\alpha}}\end{aligned}$$ ■

Practice Exercise

2. Find the formula for $\csc(x/2)$.

EXERCISES 20.4

In Exercises 1 and 2, make the given changes in the indicated examples of this section and then solve the given problem.

1. In Example 2, change the − sign in the numerator to +.

2. In Example 5, change $\frac{8}{15}$ $(180^\circ < \alpha < 270^\circ)$ to $-\frac{8}{15}$ $(270^\circ < \alpha < 360^\circ)$.

In Exercises 3–8, use the half-angle formulas to evaluate the given functions.

3. $\cos 15^\circ$ **4.** $\sin 22.5^\circ$ **5.** $\sin 105^\circ$

6. $\cos 112.5^\circ$ **7.** $\cos\frac{3\pi}{8}$ **8.** $\sin\frac{11\pi}{12}$

In Exercises 9–12, simplify the given expressions by giving the results in terms of one-half the given angle. Then use a calculator to verify the result.

9. $\sqrt{\dfrac{1-\cos 236^\circ}{2}}$ **10.** $\sqrt{\dfrac{1+\cos 98^\circ}{2}}$

11. $\sqrt{\dfrac{1+\cos 96^\circ}{2}}$ **12.** $\sqrt{\dfrac{1-\cos 328^\circ}{2}}$

In Exercises 13–20, simplify the given expressions.

13. $\sqrt{\dfrac{1-\cos 8x}{2}}$ **14.** $\sqrt{\dfrac{4+4\cos 8\beta}{8}}$

15. $\sqrt{\dfrac{1+\cos 6x}{2}}$ **16.** $\sqrt{2-2\cos 64x}$

17. $\sqrt{4-4\cos 10\theta}$ **18.** $\sqrt{18+18\cos 1.4x}$

19. $2\sin^2\dfrac{x}{2}+\cos x$ **20.** $2\cos^2\dfrac{\theta}{2}\sec\theta$

In Exercises 21–24, evaluate the indicated functions.

21. Find the value of $\sin\left(\frac{\alpha}{2}\right)$ if $\cos\alpha = \frac{12}{13}(0^\circ < \alpha < 90^\circ)$.

22. Find the value of $\cos\left(\frac{\alpha}{2}\right)$ if $\sin\alpha = -\frac{4}{5}(180^\circ < \alpha < 270^\circ)$.

23. Find the value of $\cos\left(\frac{\alpha}{2}\right)$ if $\tan\alpha = -0.2917(90^\circ < \alpha < 180^\circ)$.

24. Find the value of $\sin\left(\frac{\alpha}{2}\right)$ if $\cos\alpha = 0.4706(270^\circ < \alpha < 360^\circ)$.

In Exercises 25–28, derive the required expressions.

25. Derive an expression for $\csc\left(\frac{\alpha}{2}\right)$ in terms of $\sec\alpha$

26. Derive an expression for $\sec\left(\frac{\alpha}{2}\right)$ in terms of $\sec\alpha$.

27. Derive an expression for $\tan\left(\frac{\alpha}{2}\right)$ in terms of $\sin\alpha$ and $\cos\alpha$.

28. Derive an expression for $\cot\left(\frac{\alpha}{2}\right)$ in terms of $\sin\alpha$ and $\cos\alpha$.

In Exercises 29–32, prove the given identities.

29. $\sin\dfrac{\alpha}{2} = \dfrac{1-\cos\alpha}{2\sin\frac{\alpha}{2}}$ **30.** $\cos\dfrac{\theta}{2} = \dfrac{\sin\theta}{2\sin\frac{\theta}{2}}$

31. $2\cos\dfrac{x}{2} = (1+\cos x)\sec\dfrac{x}{2}$

32. $\cos^2\dfrac{x}{2}\left[1+\left(\dfrac{\sin x}{1+\cos x}\right)^2\right] = 1$

In Exercises 33–36, verify each identity by comparing the graph of the left side with the graph of the right side on a calculator.

33. $2\sin^2\dfrac{\alpha}{2} - \cos^2\dfrac{\alpha}{2} = \dfrac{1-3\cos\alpha}{2}$ **34.** $\cos^2\dfrac{A}{2} - \sin^2\dfrac{A}{2} = \dfrac{\sin 2A}{2\sin A}$

35. $2\sin^2\dfrac{\theta}{2} = \dfrac{\sin^2\theta}{1+\cos\theta}$ **36.** $\tan\dfrac{\alpha}{2} = \dfrac{\sin\alpha}{1+\cos\alpha}$

In Exercises 37–48, use the half-angle formulas to solve the given problems.

37. Find $\tan\theta$ if $\sin(\theta/2) = 3/5$.

38. In a right triangle with sides and angles as shown in Fig. 20.22, show that $\sin^2\frac{A}{2} = \frac{c-b}{2c}$.

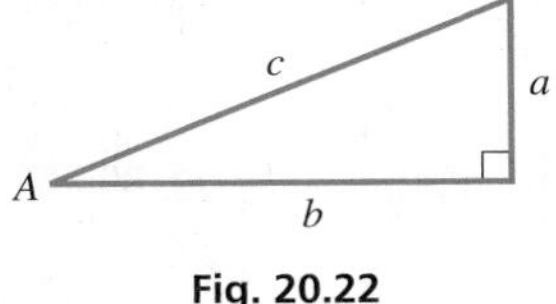

Fig. 20.22

39. Find the exact value of tan 22.5° using half-angle formulas.

40. If $180^\circ < \theta < 270^\circ$ and $\tan(\theta/2) = -\pi/3$, find $\sin\theta$.

41. If $90^\circ < \theta < 180^\circ$ and $\sin\theta = 4/5$, find $\cos(\theta/2)$.

42. Find the area of the segment of the circle in Fig. 20.23, expressing the result in terms of $\theta/2$.

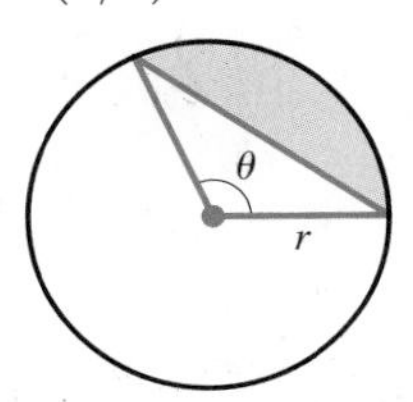

Fig. 20.23

43. In finding the path of a sliding particle, the expression $\sqrt{8-8\cos\theta}$ is used. Simplify this expression.

44. In designing track for a railway system, the equation $d = 4r\sin^2\frac{A}{2}$ is used. Solve for d in terms of $\cos A$.

45. In electronics, in order to find the *root-mean-square current* in a circuit, it is necessary to express $\sin^2\omega t$ in terms of $\cos 2\omega t$. Show how this is done.

46. In studying interference patterns of radio signals, the expression $2E^2 - 2E^2\cos(\pi-\theta)$ arises. Show that this can be written as $4E^2\cos^2(\theta/2)$.

47. The index of refraction n, the angle A of a prism, and the minimum angle of deflection ϕ are related

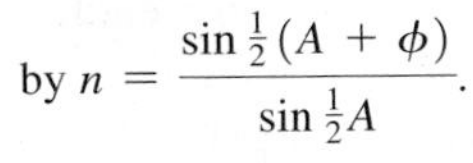

by $n = \dfrac{\sin\frac{1}{2}(A+\phi)}{\sin\frac{1}{2}A}$.

See Fig. 20.24. Show that an equivalent expression is

$$n = \sqrt{\frac{1-\cos A\cos\phi + \sin A\sin\phi}{1-\cos A}}$$

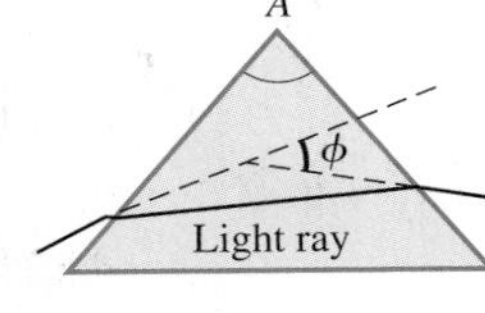

Fig. 20.24

48. For the structure shown in Fig. 20.25, show that $x = 2l\sin^2\frac{1}{2}\theta$.

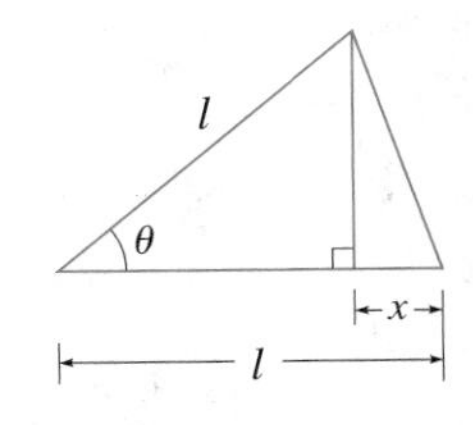

Fig. 20.25

Answers to Practice Exercises

1. $5\sin 2x$ **2.** $\pm\sqrt{\frac{2}{1-\cos x}}$

20.5 Solving Trigonometric Equations

Solve for Trig Function - Then Angle • Use Algebraic Methods and Trigonometric Identities • Use of Calculator

One of the most important uses of the trigonometric identities is in the solution of equations involving trigonometric functions. The solution of this type of equation consists of the angles that satisfy the equation. When solving for the angle, we generally first solve for a value of a function of the angle and then find the angle from this value of the function.

When equations are written in terms of more than one function, the identities provide a way of changing many of them to equations or factors involving only one function of the same angle. Thus, *the solution is found by using algebraic methods and trigonometric identities and values.* From Chapter 8, recall that we must be careful regarding the sign of the value of a trigonometric function in finding the angle. Figure 20.26 shows again the quadrants in which the functions are positive. Functions not listed are negative.

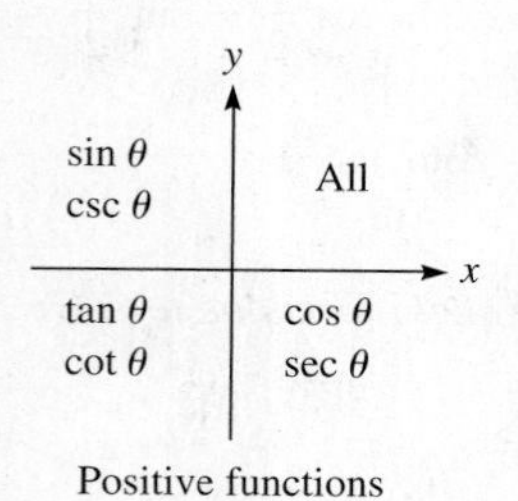

Fig. 20.26

EXAMPLE 1 First solve for $\cos\theta$

Solve the equation $2\cos\theta - 1 = 0$ for all values of θ such that $0 \leq \theta \leq 2\pi$.

Solving the equation for $\cos\theta$, we obtain $\cos\theta = \frac{1}{2}$. The problem asks for all values of θ from 0 to 2π that satisfy the equation. We know that the cosines of angles in the first and fourth quadrants are positive. Also, we know that $\cos\frac{\pi}{3} = \frac{1}{2}$, which means that $\frac{\pi}{3}$ is the reference angle. Therefore, the solution proceeds as follows:

$$2\cos\theta - 1 = 0$$

$$2\cos\theta = 1 \quad \text{solve for } \cos\theta$$

$$\cos\theta = \frac{1}{2}$$

$$\theta = \frac{\pi}{3}, \frac{5\pi}{3} \quad \theta \text{ in quadrants I and IV}$$

■

Practice Exercise

1. Solve for x $(0 \leq x < 2\pi)$: $2\sin x + 1 = 0$

EXAMPLE 2 Use identity—factor—solve for $\sin x$

Solve the equation $2\cos^2 x - \sin x - 1 = 0 \quad (0 \leq x < 2\pi)$.

By use of the identity $\sin^2 x + \cos^2 x = 1$, this equation may be put in terms of $\sin x$ only. Thus, we have

$$2(1 - \sin^2 x) - \sin x - 1 = 0 \quad \text{use identity}$$

$$-2\sin^2 x - \sin x + 1 = 0 \quad \text{solve for } \sin x$$

$$2\sin^2 x + \sin x - 1 = 0$$

$$(2\sin x - 1)(\sin x + 1) = 0 \quad \text{factor}$$

Setting each factor equal to zero, we find $\sin x = 1/2$, or $\sin x = -1$. For the domain 0 to 2π, $\sin x = 1/2$ gives $x = \pi/6, 5\pi/6$, and $\sin x = -1$ gives $x = 3\pi/2$. Therefore,

$$x = \frac{\pi}{6}, \frac{5\pi}{6}, \frac{3\pi}{2}$$

These values check when substituted in the original equation. ■

GRAPHICAL SOLUTIONS

As with algebraic equations, graphical solutions of trigonometric equations are approximate, whereas algebraic solutions often give exact solutions. As before, *we collect all terms on the left of the equal sign, with zero on the right.* We then *graph the function on the left to find its zeros by finding the values of x where the graph crosses (or is tangent to) the x-axis.*

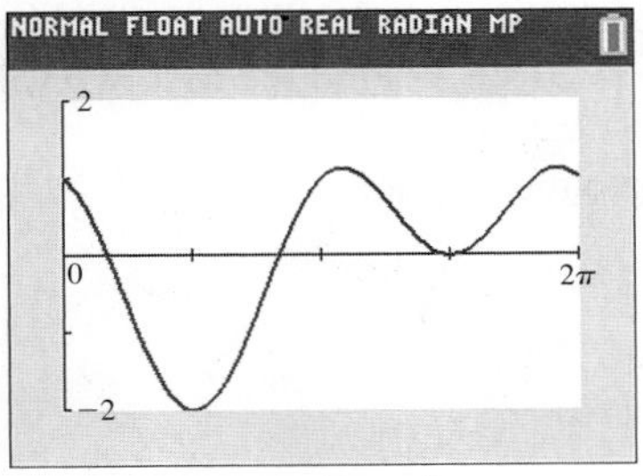

Fig. 20.27

EXAMPLE 3 Solution using calculator

Graphically solve the equation $2\cos^2 x - \sin x - 1 = 0$ $(0 \le x < 2\pi)$ by using a calculator. (This is the same equation as in Example 2.)

Because all the terms of the equation are on the left, with zero on the right, we now set $y = 2\cos^2 x - \sin x - 1$. We then enter this function in the calculator as Y_1, and the graph is displayed in Fig. 20.27. Using the *zero* feature of a calculator, we find that $y = 0$ for

$$x = 0.52, 2.62, 4.71$$

These values are the same as in Example 2. We note that for $x = 4.71$, the curve *touches* the x-axis but does not cross it. This means it is *tangent* to the x-axis. ■

EXAMPLE 4 Solve—check for extraneous solutions

Solve the equation $\cos(x/2) = 1 + \cos x$ $(0 \le x < 2\pi)$.

By using the half-angle formula for $\cos(x/2)$ and then squaring both sides of the resulting equation, this equation can be solved:

$$\pm\sqrt{\frac{1 + \cos x}{2}} = 1 + \cos x \qquad \text{using identity}$$

$$\frac{1 + \cos x}{2} = 1 + 2\cos x + \cos^2 x \qquad \text{squaring both sides}$$

$$2\cos^2 x + 3\cos x + 1 = 0 \qquad \text{simplifying}$$

$$(2\cos x + 1)(\cos x + 1) = 0 \qquad \text{factoring}$$

$$\cos x = -\frac{1}{2}, -1$$

$$x = \frac{2\pi}{3}, \frac{4\pi}{3}, \pi$$

CAUTION In finding this solution, we squared both sides of the original equation. In doing this, we may have introduced **extraneous solutions** (see Section 14.3). Thus, we must check each solution in the *original equation* to see if it is valid. ■ Hence,

$$x = \frac{2\pi}{3}: \quad \cos\frac{\pi}{3} \stackrel{?}{=} 1 + \cos\frac{2\pi}{3} \quad \text{or} \quad \frac{1}{2} \stackrel{?}{=} 1 + \left(-\frac{1}{2}\right) \quad \text{or} \quad \frac{1}{2} = \frac{1}{2}$$

$$x = \frac{4\pi}{3}: \quad \cos\frac{2\pi}{3} \stackrel{?}{=} 1 + \cos\frac{4\pi}{3} \quad \text{or} \quad -\frac{1}{2} \stackrel{?}{=} 1 + \left(-\frac{1}{2}\right) \quad \text{or} \quad -\frac{1}{2} \ne \frac{1}{2}$$

$$x = \pi: \quad \cos\frac{\pi}{2} \stackrel{?}{=} 1 + \cos\pi \quad \text{or} \quad 0 \stackrel{?}{=} 1 - 1 \quad \text{or} \quad 0 = 0$$

Thus, the apparent solution $x = \frac{4\pi}{3}$ is not a solution of the original equation. The correct solutions are $x = \frac{2\pi}{3}$ and $x = \pi$. We can see that these values agree with the values of x for which the graph of $y_1 = \cos(x/2) - 1 - \cos x$ crosses the x-axis in Fig. 20.28. ■

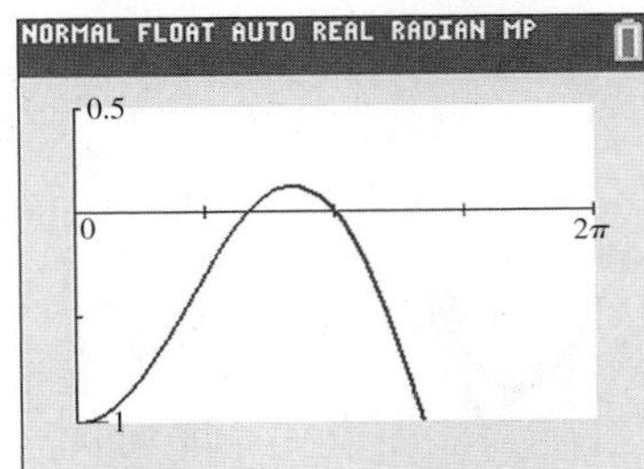

Fig. 20.28

EXAMPLE 5 Trig equation—spring displacement

The vertical displacement y of an object at the end of a spring, which itself is being moved up and down, is given by $y = 3.50 \sin t + 1.20 \sin 2t$. Find the first two values of t (in seconds) for which $y = 0$.

Using the double-angle formula for sin $2t$ leads to the solution.

$$
\begin{aligned}
3.50 \sin t + 1.20 \sin 2t &= 0 && \text{setting } y = 0 \\
3.50 \sin t + 2.40 \sin t \cos t &= 0 && \text{using identities} \\
\sin t(3.50 + 2.40 \cos t) &= 0 && \text{factoring} \\
\sin t &= 0 \quad \text{or} \quad \cos t = -1.46 \\
t &= 0.00, 3.14, \ldots
\end{aligned}
$$

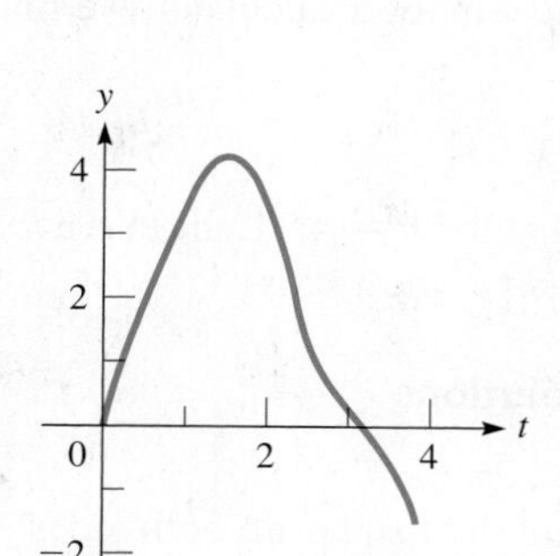

Fig. 20.29

Because $\cos t$ cannot be numerically larger than 1, there are no values of t for which $\cos t = -1.46$. Thus, the required times are $t = 0.00$ s, 3.14 s.

We can see that these values agree with the values of t for which the graph of $y = 3.50 \sin t + 1.20 \sin 2t$ crosses the t-axis in Fig. 20.29. (In using a graphing calculator, use x for t.) ■

EXAMPLE 6 Solve for 2θ—then θ

Solve the equation $\tan 2\theta - \cot 2\theta = 0 \; (0 \leq \theta < 2\pi)$.

$$
\begin{aligned}
\tan 2\theta - \frac{1}{\tan 2\theta} &= 0 && \text{using } \cot 2\theta = \frac{1}{\tan 2\theta} \\
\tan^2 2\theta &= 1 && \text{multiplying by } \tan 2\theta \text{ and adding 1 to each side} \\
\tan 2\theta &= \pm 1 && \text{taking square roots}
\end{aligned}
$$

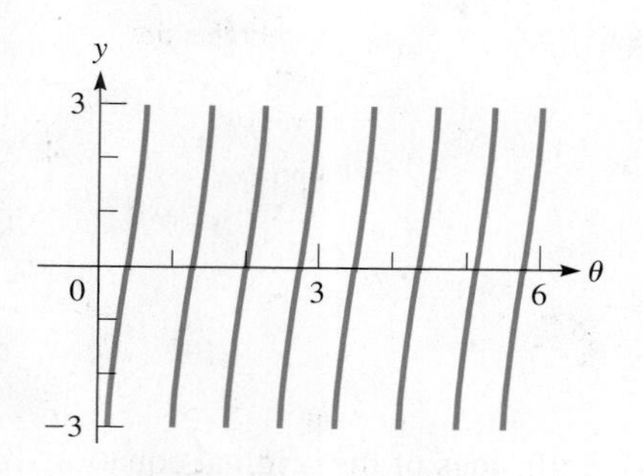

Fig. 20.30

For $0 \leq \theta < 2\pi$, we must have values of 2θ such that $0 \leq 2\theta < 4\pi$. Therefore,

$$2\theta = \frac{\pi}{4}, \frac{3\pi}{4}, \frac{5\pi}{4}, \frac{7\pi}{4}, \frac{9\pi}{4}, \frac{11\pi}{4}, \frac{13\pi}{4}, \frac{15\pi}{4}$$

This means that the solutions are

$$\theta = \frac{\pi}{8}, \frac{3\pi}{8}, \frac{5\pi}{8}, \frac{7\pi}{8}, \frac{9\pi}{8}, \frac{11\pi}{8}, \frac{13\pi}{8}, \frac{15\pi}{8}$$

These values satisfy the original equation. Because we multiplied through by $\tan 2\theta$ in the solution, any value of θ that leads to $\tan 2\theta = 0$ would not be valid, because this would indicate division by zero in the original equation.

We see that these solutions agree with the values of θ for which the graph of $y = \cos 2\theta - \cot 2\theta$ crosses the θ-axis in Fig. 20.30. (In using a graphing calculator, use x for θ. ■

Practice Exercise

2. Solve for x $(0 \leq x < 2\pi)$: $\sec^2 x + 2 \tan x = 0$

EXAMPLE 7 Recognize trigonometric form—solve

Solve the equation $\cos 3x \cos x + \sin 3x \sin x = 1 (0 \leq x < 2\pi)$.

The left side of this equation is of the general form $\cos(A - x)$, where $A = 3x$. Therefore,

$$\cos 3x \cos x + \sin 3x \sin x = \cos(3x - x) = \cos 2x$$

The original equation becomes

$$\cos 2x = 1$$

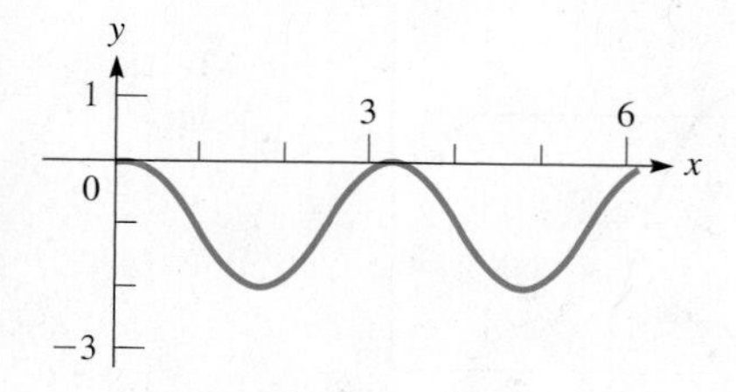

Fig. 20.31

This equation is satisfied if $2x = 0$ or $2x = 2\pi$. The solutions are $x = 0$ and $x = \pi$. Only through recognition of the proper trigonometric form can we readily solve this equation.

We see that these solutions agree with the two values of x for which the graph of $y = \cos 3x \cos x + \sin 3x \sin x - 1$ touches the x-axis in Fig. 20.31. ■

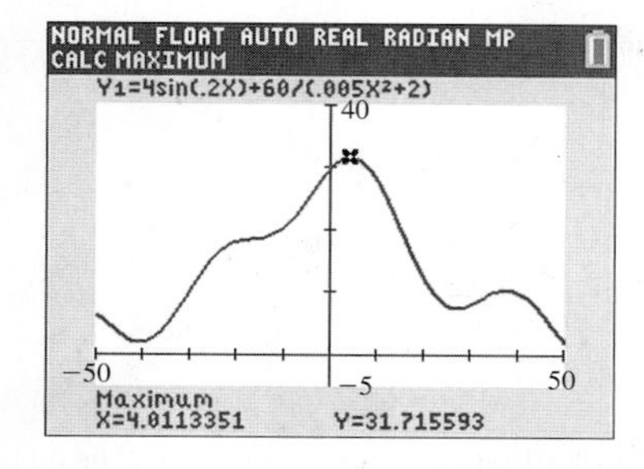

Fig. 20.32

Graphing calculator keystrokes: goo.gl/mYkgmM

EXAMPLE 8 Solve using calculator—height of tsunami wave

Aerial photographs and computer analysis show that a certain tsunami wave could be represented as $y = 4\sin 0.2x + \dfrac{60}{0.005x^2 + 2}$, where y (in m) is the height of the wave for a horizontal displacement x (in m). Find the height of the peak of the wave and the displacement x for the peak.

Entering the function in the calculator as y_1, the graph is shown in Fig. 20.32. We can find the coordinates of the peak of the wave using the *maximum* feature, as shown in the display. Thus, the peak of the wave is 31.7 m when $x = 4.0$ m. ■

EXERCISES 20.5

In Exercises 1–4, make the given changes in the indicated examples of this section and then solve the resulting problems.

1. In Example 1, change $2\cos\theta$ to $\tan\theta$.
2. In Example 2, change $2\cos^2 x$ to $2\sin^2 x$.
3. In Example 4, change $\cos\frac{1}{2}x$ to $\sin\frac{1}{2}x$ and on the right of the equal sign change the $+$ to $-$.
4. In Example 7, change the $+$ to $-$.

In Exercises 5–20, solve the given trigonometric equations analytically (using identities when necessary for exact values when possible) for values of x for $0 \le x < 2\pi$.

5. $\sin x - 1 = 0$
6. $2\cos x + 1 = 0$
7. $1 - 2\cos x = 0$
8. $4\tan x + 2 = 3(1 + \tan x)$
9. $4 - \sec^2 x = 0$
10. $3\tan x - \cot x = 0$
11. $2\sin^2 x - \sin x = 0$
12. $\sin 4x - \sin 2x = 0$
13. $\sin 2x \sin x + \cos x = 0$
14. $\sin x - \sin\dfrac{x}{2} = 0$
15. $2\cos^2 x - 2\cos 2x - 1 = 0$
16. $\csc^2 x + 2 = 3\csc x$
17. $4\tan x - \sec^2 x = 0$
18. $\sin\left(x - \frac{\pi}{4}\right) = \cos\left(x - \frac{\pi}{4}\right)$
19. $\sin 2x \cos x - \cos 2x \sin x = 0$
20. $\cos 3x \cos x - \sin 3x \sin x = 0$

In Exercises 21–38, solve the given trigonometric equations analytically and by use of a calculator. Compare results. Use values of x for $0 \le x < 2\pi$.

21. $\tan x + 1 = 0$
22. $2\sin x + 1 = 0$
23. $3 - 4\cos x = 7 - (2 - \cos x)$
24. $7\sin x - 2 = 3(2 - \sin x)$
25. $4 - 3\csc^2 x = 0$
26. $|\sin x| = \frac{1}{2}$
27. $\sin 4x - \cos 2x = 0$
28. $3\cos x - 4\cos^2 x = 0$
29. $2\sin x = \tan x$
30. $\cos 2x + \sin^2 x = 0$
31. $\sin^2 x - 2\sin x = 1$
32. $2\cos^2 2x + 1 = 3\cos 2x$
33. $\tan x + 3\cot x = 4$
34. $\tan^2 x + 4 = 2\sec^2 x$
35. $\sin 2x + \cos 2x = 0$
36. $2\sin 4x + \csc 4x = 3$
37. $2\sin 2x - \cos x \sin^3 x = 0$
38. $\sec^4 x = \csc^4 x$

In Exercises 39–54, solve the indicated equations analytically.

39. $\sin 3x + \sin x = 0$ [*Hint*: See Eq. (20.17).]
40. $\cos 3x - \cos x = 0$ [*Hint*: See Eq. (20.20).]
41. Is there any positive acute angle θ for which $\sin\theta + \cos\theta + \tan\theta + \cot\theta + \sec\theta + \csc\theta = 1$? Explain.
42. Use a calculator to determine the minimum value of the function to the left of the equal sign in Exercise 41 (for a positive acute angle).
43. Solve the system of equations $r = \sin\theta$, $r = \sin 2\theta$, for $0 \le \theta < 2\pi$.
44. Solve the system of equations $r = \sin\theta$, $r = \cos 2\theta$, for $0 \le \theta < 2\pi$.
45. Find the angles of a triangle if one side is twice another side and the angles opposite these sides differ by 60°.
46. If two musical tones of frequencies 220 Hz and 223 Hz are played together, beats will be heard. This can be represented by $y = \sin 440\pi t + \sin 446\pi t$. Graph this function and estimate t (in s) when $y = 0$ between beats for $0.15 < t < 1.15$ s.
47. The acceleration due to gravity g (in m/s^2) varies with latitude, approximately given by $g = 9.7805(1 + 0.0053\sin^2\theta)$, where θ is the latitude in degrees. Find θ for $g = 9.8000$.
48. Under certain conditions, the electric current i (in A) in the circuit shown in Fig. 20.33 is given below. For what value of t (in s) is the current first equal to zero?

$$i = -e^{-100t}(32.0\sin 624.5t + 0.200\cos 624.5t)$$

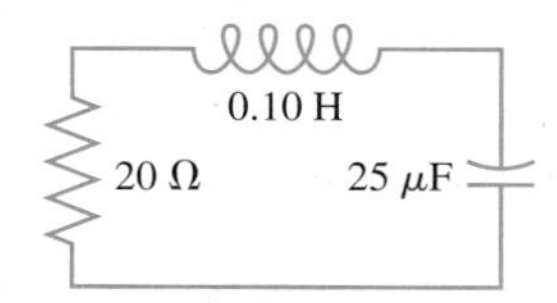

Fig. 20.33

49. The vertical displacement y (in m) of the end of a robot arm is given by $y = 2.30\cos 0.1t - 1.35\sin 0.2t$. Find the first four values of t (in s) for which $y = 0$.
50. In finding the maximum illuminance from a point source of light, it is necessary to solve the equation $\cos\theta \sin 2\theta - \sin^3\theta = 0$. Find θ if $0 < \theta < 90°$.

51. To find the angle θ subtended by a certain object on a camera film, it is necessary to solve the equation $\frac{p^2 \tan\theta}{0.0063 + p\tan\theta} = 1.6$, where p is the distance from the camera to the object. Find θ if $p = 4.8$ m.

52. The velocity of a certain piston is maximum when the acute crank angle θ satisfies the equation $8\cos\theta + \cos 2\theta = 0$. Find this angle.

53. Resolve a force of 500.0 N into two components, perpendicular to each other, for which the sum of their magnitudes is 700.0 N, by using the angle between a component and the resultant.

54. Looking for a lost ship in the North Atlantic Ocean, a plane flew from Reykjavik, Iceland, 160 km west. It then turned and flew due north and then made a final turn to fly directly back to Reykjavik. If the total distance flown was 480 km, how long were the final two legs of the flight? Solve by setting up and solving an appropriate trigonometric equation. (The Pythagorean theorem may be used only as a check.)

In Exercises 55–62, solve the given equations graphically.

55. $3\sin x - x = 0$

56. $4\cos x + 3x = 0$

57. $2\sin 2x = x^2 + 1$

58. $\sqrt{x} - \sin 3x = 1$

59. $2\ln x = 1 - \cos 2x$

60. $e^x = 1 + \sin x$

61. In finding the frequencies of vibration of a vibrating wire, the equation $x\tan x = 2.00$ occurs. Find x if $0 < x < \pi/2$.

62. An equation used in astronomy is $\theta - e\sin\theta = M$. Solve for θ for $e = 0.25$ and $M = 0.75$.

Answers to Practice Exercises

1. $x = 7\pi/6, 11\pi/6$ **2.** $x = 3\pi/4, 7\pi/4$

20.6 The Inverse Trigonometric Functions

Inverse Functions • Inverse Trigonometric Functions • Ranges of the Inverse Trig Functions

When we studied the exponential and logarithmic functions, we often changed an expression from one form to the other. The exponential function $y = b^x$, written in logarithmic form with x as a function of y, is $x = \log_b y$. Then we wrote the logarithmic function as $y = \log_b x$, because it is standard practice to use y as the dependent variable and x as the independent variable.

These two functions, the exponential function $y = b^x$ and the logarithmic function $y = \log_b x$, are called **inverse functions** (see page 379). *This means that if we solve for the independent variable in terms of the dependent variable in one function, we will arrive at the functional relationship expressed by the other.* It also means that, for every value of x, there is only one corresponding value of y.

Just as we are able to solve $y = b^x$ for the exponent by writing it in logarithmic form, at times it is necessary to solve for the independent variable (the angle) in trigonometric functions. Therefore, we define the **inverse sine function**

$$y = \sin^{-1} x \qquad \left(-\frac{\pi}{2} \le y \le \frac{\pi}{2}\right) \qquad \textbf{(20.28)}$$

where y is the angle whose sine is x. This means that x is the value of the sine of the angle y, or $x = \sin y$. (It is necessary to show the range as $-\pi/2 \le y \le \pi/2$, as we will see shortly.)

CAUTION In Eq. (20.28), the -1 is *not* an exponent. The -1 in $\sin^{-1} x$ is the notation showing the *inverse function*. ■ We introduced this notation in Chapter 4 when finding the angle with a known value of one of the trigonometric functions.

The notations Arcsin x, arcsin x, $\text{Sin}^{-1} x$ are also used to designate the inverse sine. Some calculators use the (inv) key and then the (sin) key to find values of the inverse sine. However, because most calculators use the notation $\sin^{-1}$ as a second function on the (sin) key, we will continue to use $\sin^{-1} x$ for the inverse sine.

Similar definitions are used for the other inverse trigonometric functions. They also have meanings similar to that of Eq. (20.28).

EXAMPLE 1 Meaning of inverse trig functions

(a) $y = \cos^{-1}x$ is read as "y is the angle whose cosine is x." In this case, $x = \cos y$.

(b) $y = \tan^{-1}2x$ is read as "y is the angle whose tangent is $2x$." In this case, $2x = \tan y$.

(c) $y = \csc^{-1}(1 - x)$ is read as "y is the angle whose cosecant is $1 - x$." In this case, $1 - x = \csc y$, or $x = 1 - \csc y$. ■

We have seen that $y = \sin^{-1}x$ means that $x = \sin y$. From our previous work with the trigonometric functions, we know that there is an unlimited number of possible values of y for a given value of x in $x = \sin y$. Consider the following example.

EXAMPLE 2 Values of trig functions

(a) For $x = \sin y$, we know that

$$\sin\frac{\pi}{6} = \frac{1}{2} \quad \text{and} \quad \sin\frac{5\pi}{6} = \frac{1}{2}$$

In fact, $x = \frac{1}{2}$ also for values of y of $-\frac{7\pi}{6}, \frac{13\pi}{6}, \frac{17\pi}{6}$, and so on.

(b) For $x = \cos y$, we know that

$$\cos 0 = 1 \quad \text{and} \quad \cos 2\pi = 1$$

In fact, $\cos y = 1$ for y equal to any even multiple of π. ■

From Chapter 3, we know that *to have a properly defined function, there must be only one value of the dependent variable for a given value of the independent variable.* Therefore, as in Eq. (20.28), in order to have only one value of y for each value of x in the domain of the inverse trigonometric functions, it is not possible to include all values of y in the range. For this reason, *the range of each of the* **inverse trigonometric functions** *is defined as follows:*

$$\begin{array}{llll} -\frac{\pi}{2} \le \sin^{-1}x \le \frac{\pi}{2} & 0 \le \cos^{-1}x \le \pi & -\frac{\pi}{2} < \tan^{-1}x < \frac{\pi}{2} & 0 < \cot^{-1}x < \pi \\ 0 \le \sec^{-1}x \le \pi & \left(\sec^{-1}x \ne \frac{\pi}{2}\right) & -\frac{\pi}{2} \le \csc^{-1}x \le \frac{\pi}{2} & (\csc^{-1}x \ne 0) \end{array} \tag{20.29}$$

We must choose a value of y in the range as defined in Eqs. (20.29) that corresponds to a given value of x in the domain. We will discuss the domains and the reasons for these definitions following the next two examples.

EXAMPLE 3 Ranges of inverse trig functions

(a) $\sin^{-1}\left(\frac{1}{2}\right) = \frac{\pi}{6}$ first-quadrant angle

This is the only value of the function that lies within the defined range. The value $\frac{5\pi}{6}$ is not correct, even though $\sin\left(\frac{5\pi}{6}\right) = \frac{1}{2}$, since $\frac{5\pi}{6}$ lies outside the defined range.

(b) $\cos^{-1}\left(-\frac{1}{2}\right) = \frac{2\pi}{3}$ second-quadrant angle

Other values such as $\frac{4\pi}{3}$ and $-\frac{2\pi}{3}$ are not correct, since they are not within the defined range for the function $\cos^{-1}x$.

(c) $\tan^{-1}(1) = \frac{\pi}{4}$ first-quadrant angle

Other values such as $\frac{5\pi}{4}$ and $-\frac{3\pi}{4}$ are not correct since they are not within the defined range of $\tan^{-1}x$. ■

EXAMPLE 4 Range of inverse tangent

$$\tan^{-1}(-1) = -\frac{\pi}{4} \quad \text{fourth-quadrant angle}$$

This is the only value within the defined range for the function $\tan^{-1}x$. **CAUTION** We must remember that when x is negative for $\sin^{-1}x$ and $\tan^{-1}x$, the value of y is a fourth-quadrant angle, expressed as a negative angle. ■ This is a direct result of the definition. [The single exception is $\sin^{-1}(-1) = -\pi/2$, which is a quadrantal angle and is not in the fourth quadrant.] ■

In choosing these values to be the ranges of the inverse trigonometric functions, we first note that the *domain* of $y = \sin^{-1}x$ and $y = \cos^{-1}x$ are each $-1 \leq x \leq 1$, because the sine and cosine functions take on only these values. Therefore, for each value in this domain, we use only one value of y in the range of the function. Although the domain of $y = \tan^{-1}x$ is all real numbers, we still use only one value of y in the range.

The ranges of the inverse trigonometric functions are chosen so that if x is positive, the resulting value is an angle in the first quadrant. However, care must be taken in choosing the range for negative values of x.

Because the sine of a second-quadrant angle is positive, we cannot choose these angles for $\sin^{-1}x$ for negative values of x. Therefore, we chose fourth-quadrant angles in the form of negative angles in order to have a continuous range of values for $\sin^{-1}x$. The range for $\tan^{-1}x$ is chosen in the same way for similar reasons. However, because the cosine of a fourth-quadrant angle is positive, the range for $\cos^{-1}x$ cannot be the same. To keep a continuous range of values for $\cos^{-1}x$, second-quadrant angles are used.

Values for the other functions are chosen such that the result is also an angle in the first quadrant if x is positive. As for negative values of x, it rarely makes any difference, since either positive values of x arise, or we can use one of the other functions. Our definitions, however, are those that are generally used.

The graphs of the inverse trigonometric functions can be used to show the domains and ranges. We can obtain the graph of the inverse sine function by first sketching the sine curve $x = \sin y$ *along the y-axis*. We then mark the specific part of this curve for which $-\frac{\pi}{2} \leq y \leq \frac{\pi}{2}$ as the graph of the inverse sine function. The graphs of the other inverse trigonometric functions are found in the same manner. In Figs. 20.34, 20.35, and 20.36, the graphs $x = \sin y$, $x = \cos y$, and $x = \tan y$, respectively, are shown. The heavier, colored portions indicate the graphs of the respective inverse trigonometric functions.

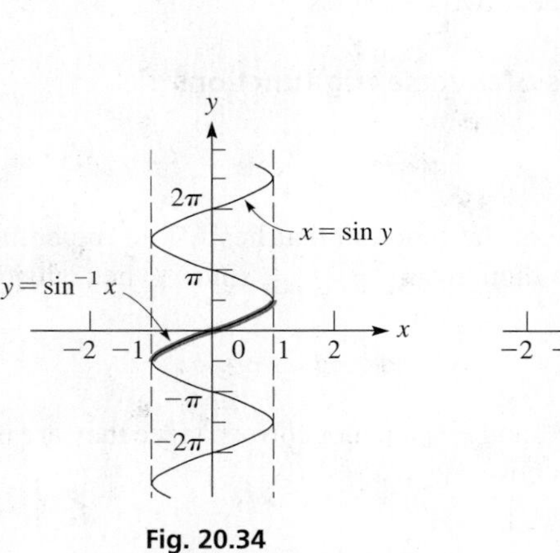

Fig. 20.34

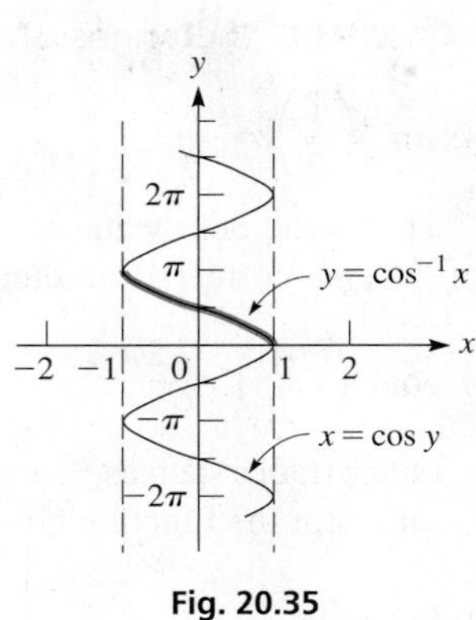

Fig. 20.35

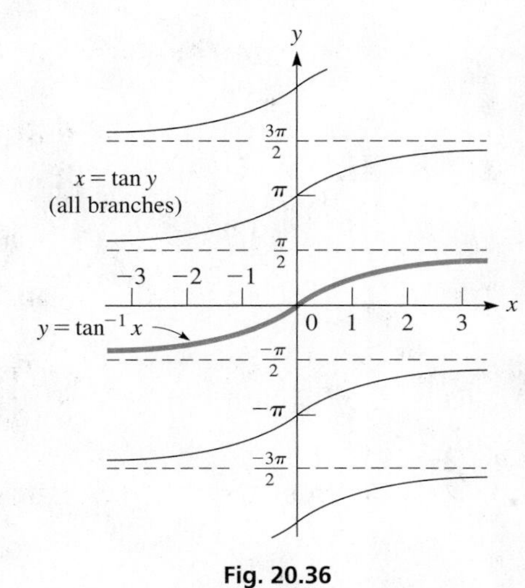

Fig. 20.36

The following examples further illustrate the values and meanings of the inverse trigonometric functions.

EXAMPLE 5 Values of inverse trig functions

(a) $\sin^{-1}(-\sqrt{3}/2) = -\pi/3$ **(b)** $\cos^{-1}(-1) = \pi$

(c) $\tan^{-1}0 = 0$ **(d)** $\tan^{-1}(\sqrt{3}) = \pi/3$

Using a calculator in radian mode, we find the following values:

(e) $\sin^{-1}0.6294 = 0.6808$ **(f)** $\sin^{-1}(-0.1568) = -0.1574$

(g) $\cos^{-1}(-0.8026) = 2.5024$ **(h)** $\tan^{-1}(-1.9268) = -1.0921$

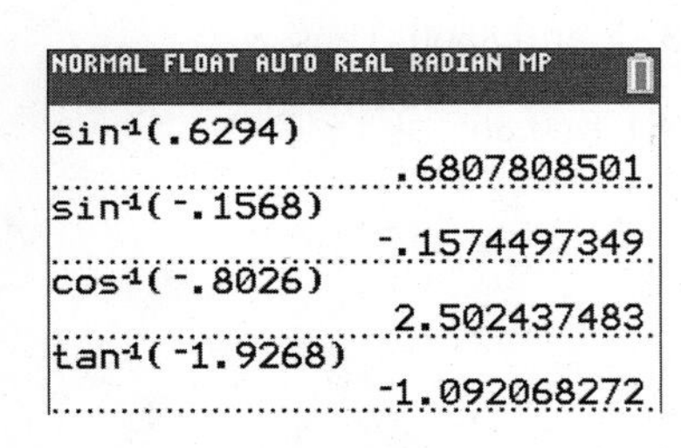

Fig. 20.37

In Fig. 20.37, we show the evaluations for parts (e)–(h) on a calculator. We note that in each case the calculator gives the value of the inverse function in the defined range for the given function. ■

EXAMPLE 6 Given an inverse function—solve for *x*

Given that $y = \pi - \sec^{-1}2x$, solve for x.

We first find the expression for $\sec^{-1}2x$ and then use the meaning of the inverse secant. The solution follows:

$$y = \pi - \sec^{-1}2x$$
$$\sec^{-1}2x = \pi - y \quad \text{solve for } \sec^{-1}2x$$
$$2x = \sec(\pi - y) \quad \text{use meaning of inverse secant}$$
$$x = -\frac{1}{2}\sec y \quad \sec(\pi - y) = -\sec y$$

As sec $2x$ and 2 sec x are different functions, $\sec^{-1}2x$ and $2\sec^{-1}x$ are also different functions. Because the values of $\sec^{-1}2x$ are restricted, so are the resulting values of y. ■

Practice Exercise

1. Evaluate: $\tan^{-1}(-\sqrt{3}/3)$

EXAMPLE 7 Solve for an angle—power in an electric inductor

The instantaneous power p in an electric inductor is given by the equation $p = vi\sin\omega t\cos\omega t$. Solve for t.

Noting the product $\sin\omega t\cos\omega t$ suggests using $\sin 2\alpha = 2\sin\alpha\cos\alpha$. Then, using the meaning of the inverse sine, we can complete the solution:

$$p = vi\sin\omega t\cos\omega t$$
$$= \frac{1}{2}vi\sin 2\omega t \quad \text{using double-angle formula}$$
$$\sin 2\omega t = \frac{2p}{vi}$$
$$2\omega t = \sin^{-1}\left(\frac{2p}{vi}\right) \quad \text{using meaning of inverse sine}$$
$$t = \frac{1}{2\omega}\sin^{-1}\left(\frac{2p}{vi}\right)$$
■

If we know the value of one of the inverse functions, we can find the trigonometric functions of the angle. If general relations are desired, a representative triangle is very useful. The following examples illustrate these methods.

EXAMPLE 8 Angle in terms of inverse functions

(a) Find $\cos(\sin^{-1}0.5)$.

Knowing that the values of inverse trigonometric functions are *angles,* we see that $\sin^{-1}0.5$ is a first-quadrant angle. Thus, we find $\sin^{-1}0.5 = \pi/6$. The problem is now to find $\cos(\pi/6)$. This is, of course, $\sqrt{3}/2$, or 0.8660. Thus,

$$\cos(\sin^{-1}0.5) = \cos(\pi/6) = 0.8660$$

(b) $$\sin(\cot^{-1}1) = \sin(\pi/4) \quad \text{first-quadrant angle}$$
$$= \frac{\sqrt{2}}{2} = 0.7071$$

(c) $$\tan[\cos^{-1}(-1)] = \tan\pi \quad \text{quadrantal angle}$$
$$= 0$$

(d) $$\cos[\csc^{-1}(-4.1754)] = \cos[\sin^{-1}(-4.1754^{-1})] = 0.9709 \quad \text{using a calculator}$$

Note the different uses of -1 in (d). For $\sin^{-1}(-4.1754^{-1})$ the $\sin^{-1}$ denotes the inverse sine function, whereas the -4.1759^{-1} denotes the reciprocal of -4.1759. ■

Practice Exercise

2. Evaluate: $\sin[\cos^{-1}(-0.5)]$

EXAMPLE 9 Trig function of inverse trig function

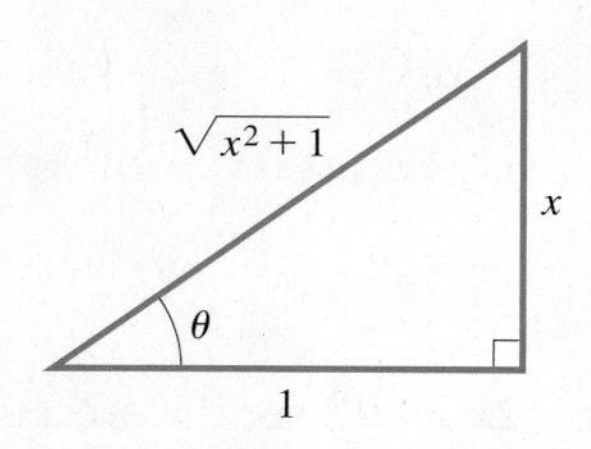

Fig. 20.38

TI-89 graphing calculator keystrokes for Example 9: goo.gl/zoZao5

Find $\sin(\tan^{-1}x)$.

We know that $\tan^{-1}x$ is another say of stating "the angle whose tangent is x." Thus, let us draw a right triangle (as in Fig. 20.38) and label one of the acute angles as θ, the side opposite θ as x, and the side adjacent to θ as 1. In this way, we see that, by definition, $\tan\theta = \frac{x}{1}$, or $\theta = \tan^{-1}x$, which means θ is the desired angle. By the Pythagorean theorem, the hypotenuse of this triangle is $\sqrt{x^2 + 1}$. Now, we find that $\sin\theta$, which is the same as $\sin(\tan^{-1}x)$, is $x/\sqrt{x^2 + 1}$. Thus,

$$\sin(\tan^{-1}x) = \frac{x}{\sqrt{x^2 + 1}}$$ ■

EXAMPLE 10 Trig function of inverse trig function

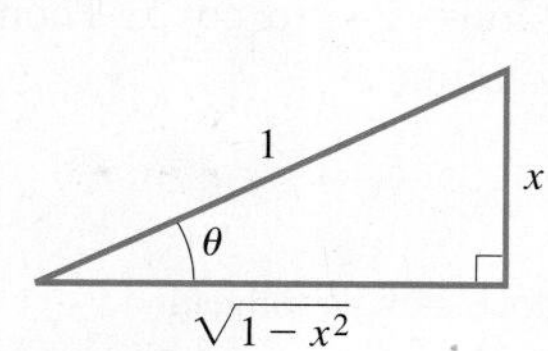

Fig. 20.39

Find $\cos(2\sin^{-1}x)$.

From Fig. 20.39, we see that $\theta = \sin^{-1}x$. From the double-angle formulas, we have

$$\cos 2\theta = 1 - 2\sin^2\theta$$

Thus, because $\sin\theta = x$, we have

$$\cos(2\sin^{-1}x) = 1 - 2x^2$$ ■

EXAMPLE 11 Inverse trig function—shelf support angles

A triangular brace of sides a, b, and c supports a shelf, as shown in Fig. 20.40. Find the expression for the angle between sides b and c.

The law of cosines leads to the solution:

$$a^2 = b^2 + c^2 - 2bc\cos A \quad \text{law of cosines}$$
$$2bc\cos A = b^2 + c^2 - a^2 \quad \text{solving for } \cos A$$
$$\cos A = \frac{b^2 + c^2 - a^2}{2bc}$$
$$A = \cos^{-1}\left(\frac{b^2 + c^2 - a^2}{2bc}\right) \quad \text{using meaning of inverse cosine}$$ ■

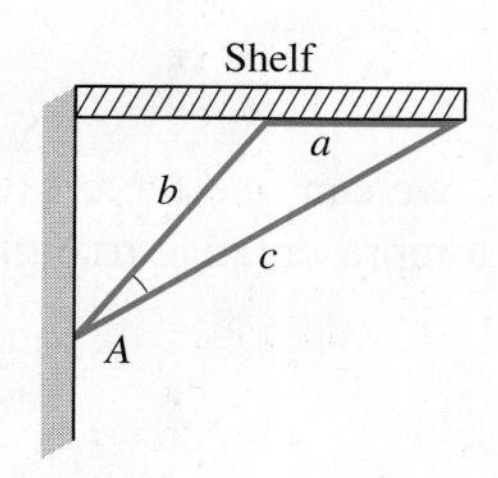

Fig. 20.40

EXERCISES 20.6

In Exercises 1–4, make the given changes in the indicated examples of this section and then solve the resulting problems.

1. In Example 1(b), change $2x$ to $3A$.
2. In Example 3(a), change $1/2$ to -1.
3. In Example 8(a), change 0.5 to 1.
4. In Example 9, change sin to cos.

In Exercises 5–10, write down the meaning of each of the given equations. See Example 1.

5. $y = \tan^{-1}5x$
6. $y = \csc^{-1}4x$
7. $y = 2\sin^{-1}x$
8. $y = 3\tan^{-1}2x$
9. $y = 5\cos^{-1}(2x - 1)$
10. $y = 4\sec^{-1}(3x + 2)$

In Exercises 11–28, evaluate exactly the given expressions if possible.

11. $\cos^{-1}0.5$
12. $\sin^{-1}1$
13. $\tan^{-1}1$
14. $\cos^{-1}2$
15. $\tan^{-1}(-\sqrt{3})$
16. $\sec^{-1}(-2)$
17. $\sec^{-1}0.5$
18. $\cot^{-1}\sqrt{3}$
19. $\sin^{-1}(-\sqrt{2}/2)$
20. $\cos^{-1}(-\sqrt{3}/2)$
21. $\sin(\tan^{-1}\sqrt{3})$
22. $\tan[\sin^{-1}(2/3)]$
23. $\cos^{-1}[\cos(-\pi/4)]$
24. $\tan^{-1}[\tan(2\pi/3)]$
25. $\cos[\tan^{-1}(-5)]$
26. $\sec[\cos^{-1}(-0.5)]$
27. $\cos(2\csc^{-1}1)$
28. $\sin(2\tan^{-1}2)$

In Exercises 29–36, use a calculator to evaluate the given expressions.

29. $\tan^{-1}(-2.8229)$
30. $\cos^{-1}(-0.6561)$
31. $\sin^{-1}0.0219$
32. $\cot^{-1}0.2846$
33. $\tan[\cos^{-1}(-0.6281)]$
34. $\cos[\tan(-7.2256)]$
35. $\sin[\tan^{-1}(-0.2297)]$
36. $\tan[\sin^{-1}(-0.3019)]$

In Exercises 37–42, solve the given equations for x.

37. $y = \sin 3x$
38. $y = \cos(2x - \pi)$
39. $y = \tan^{-1}(x/4)$
40. $5y = 2\sin^{-1}(x/6)$
41. $1 - y = \cos^{-1}(1 - x)$
42. $2y = \cot^{-1}3x - 5$

In Exercises 43–50, find an algebraic expression for each of the given expressions.

43. $\tan(\sin^{-1}x)$
44. $\sin(\cos^{-1}x)$
45. $\sin(\sec^{-1}\frac{x}{4})$
46. $\cos(\tan^{-1}\frac{x}{3})$
47. $\sec(\csc^{-1}3x)$
48. $\tan(\sin^{-1}2x)$
49. $\sin(2\sin^{-1}x)$
50. $\cos(2\tan^{-1}x)$

In Exercises 51–56, solve the given problems with the use of the inverse trigonometric functions.

51. Is $\sin^{-1}(\sin x) = x$ for all x? Explain.
52. In the analysis of ocean tides, the equation $y = A\cos 2(\omega t + \phi)$ is used. Solve for t.
53. Show that the area A of a segment of a circle of radius r, bounded by a chord at a distance d from the center, is given by $A = r^2\cos^{-1}(d/r) - d\sqrt{r^2 - d^2}$. See Fig. 20.41.

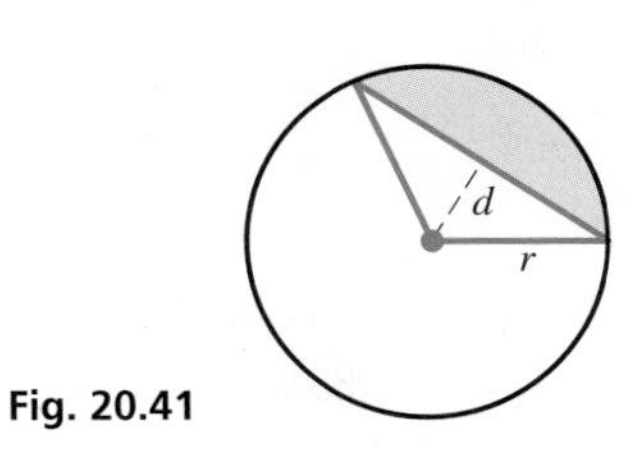

Fig. 20.41

54. For an object of weight w on an inclined plane that is at an angle θ to the horizontal, the equation relating w and θ is $\mu w\cos\theta = w\sin\theta$, where μ is the coefficient of friction between the surfaces in contact. Solve for θ.
55. The electric current in a certain circuit is given by $i = I_m[\sin(\omega t + \alpha)\cos\phi + \cos(\omega t + \alpha)\sin\phi]$. Solve for t.
56. The time t as a function of the displacement d of a piston is given by $t = \frac{1}{2\pi f}\cos^{-1}\frac{d}{A}$. Solve for d.

In Exercises 57 and 58, prove that the given expressions are equal. In Exercise 57, use the relation for $\sin(\alpha + \beta)$ and show that the sine of the sum of the angles on the left equals the sine of the angle on the right. In Exercise 58, use the relation for $\tan(\alpha + \beta)$.

57. $\sin^{-1}\frac{3}{5} + \sin^{-1}\frac{5}{13} = \sin^{-1}\frac{56}{65}$
58. $\tan^{-1}\frac{1}{3} + \tan^{-1}\frac{1}{2} = \frac{\pi}{4}$

In Exercises 59 and 60, evaluate the given expressions. In Exercises 61 and 62, find an equivalent algebraic expression.

59. $\cos\left(\sin^{-1}\frac{\sqrt{2}}{2} + \cos^{-1}\frac{3}{5}\right)$
60. $\sin\left(\sin^{-1}\frac{1}{2} + \cos^{-1}\frac{4}{5}\right)$
61. $\sin(\sin^{-1}x + \cos^{-1}y)$
62. $\cos(\sin^{-1}x - \cos^{-1}y)$

In Exercises 63–66, evaluate the given expressions.

63. $\sin^{-1}0.5 + \cos^{-1}0.5$
64. $\tan^{-1}\sqrt{3} + \cot^{-1}\sqrt{3}$
65. $\sin^{-1}x + \sin^{-1}(-x)$
66. $\sin^{-1}x + \cos^{-1}x$

In Exercises 67–70, solve for the angle A for the given triangles in the given figures in terms of the given sides and angles. For Exercises 67 and 68, explain your method.

67.

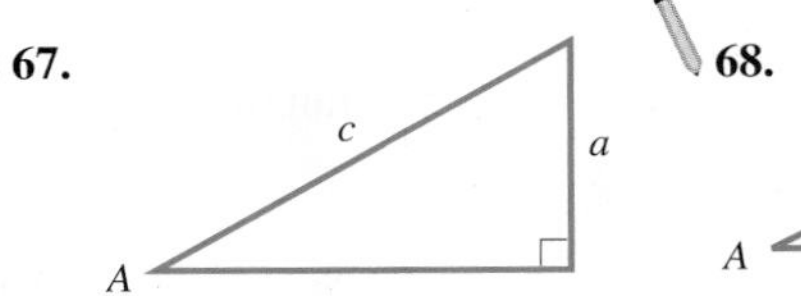

Fig. 20.42

68.

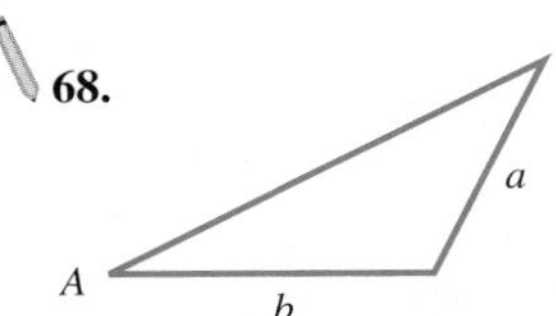

Fig. 20.43

69.

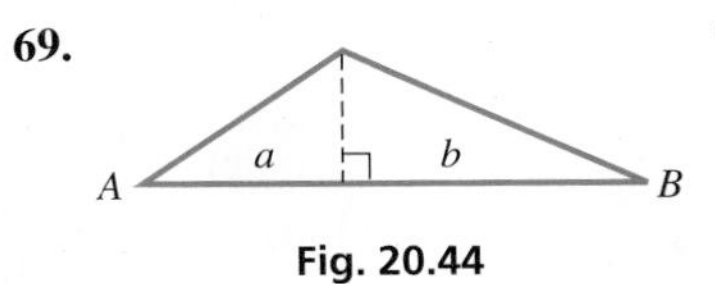

Fig. 20.44

70.

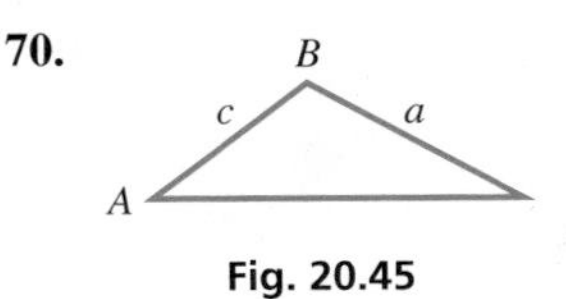

Fig. 20.45

In Exercises 71–76, solve the given problems.

71. The height of the Statue of Liberty is 151 ft. See Fig. 20.46. From the deck of a boat at a horizontal distance d from the statue, the angles of elevation of the top of the statue and the top of its pedestal are α and β, respectively. Show that

$$\alpha = \tan^{-1}\left(\frac{151}{d} + \tan\beta\right)$$

Fig. 20.46

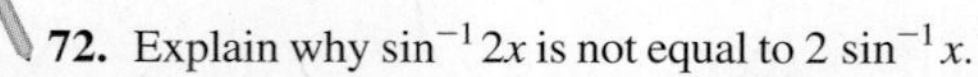

72. Explain why $\sin^{-1}2x$ is not equal to $2\sin^{-1}x$.

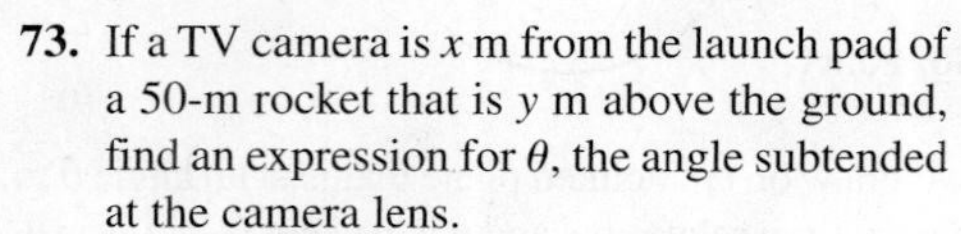

73. If a TV camera is x m from the launch pad of a 50-m rocket that is y m above the ground, find an expression for θ, the angle subtended at the camera lens.

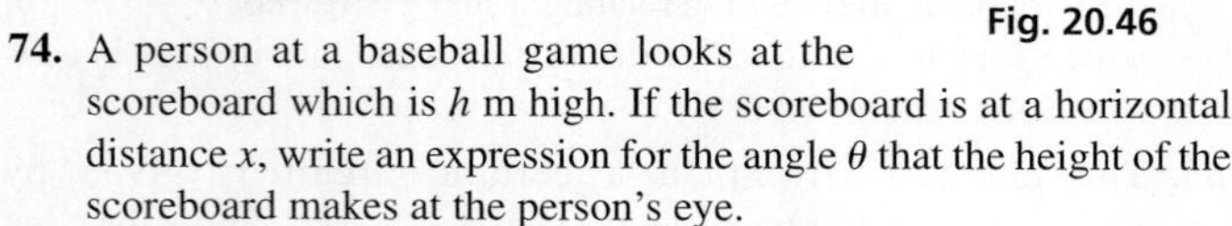

74. A person at a baseball game looks at the scoreboard which is h m high. If the scoreboard is at a horizontal distance x, write an expression for the angle θ that the height of the scoreboard makes at the person's eye.

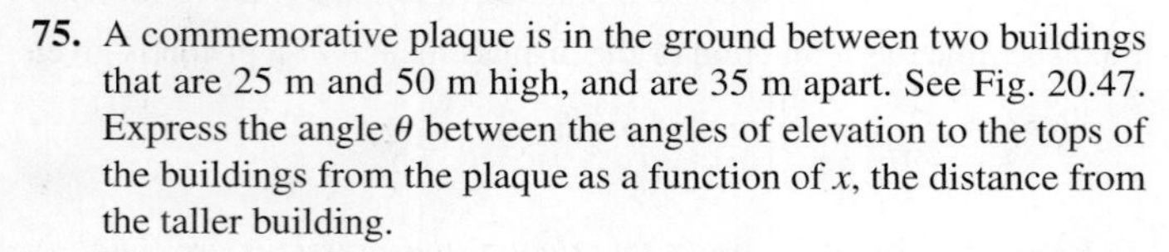

75. A commemorative plaque is in the ground between two buildings that are 25 m and 50 m high, and are 35 m apart. See Fig. 20.47. Express the angle θ between the angles of elevation to the tops of the buildings from the plaque as a function of x, the distance from the taller building.

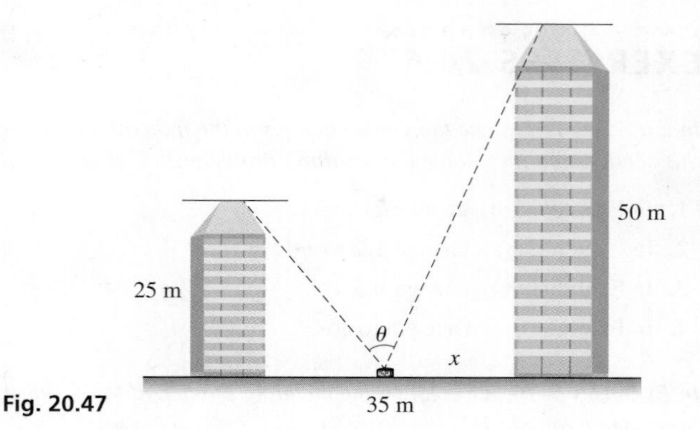

Fig. 20.47

76. Show that the length L of the pulley belt shown in Fig. 20.48 is $L = 24 + 11\pi + 10\sin^{-1}\frac{5}{13}$.

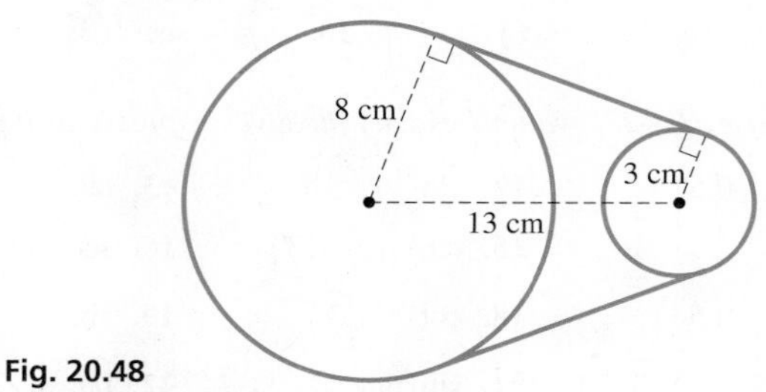

Fig. 20.48

Answers to Practice Exercises

1. $-\pi/6$ **2.** $\sqrt{3}/2$

CHAPTER 20 KEY FORMULAS AND EQUATIONS

Basic trigonometric identities

$$\sin\theta = \frac{1}{\csc\theta} \quad \textbf{(20.1)} \qquad \tan\theta = \frac{\sin\theta}{\cos\theta} \quad \textbf{(20.4)} \qquad \sin^2\theta + \cos^2\theta = 1 \quad \textbf{(20.6)}$$

$$\cos\theta = \frac{1}{\sec\theta} \quad \textbf{(20.2)} \qquad \cot\theta = \frac{\cos\theta}{\sin\theta} \quad \textbf{(20.5)} \qquad 1 + \tan^2\theta = \sec^2\theta \quad \textbf{(20.7)}$$

$$\tan\theta = \frac{1}{\cot\theta} \quad \textbf{(20.3)} \qquad 1 + \cot^2\theta = \csc^2\theta \quad \textbf{(20.8)}$$

Sum and difference identities

$$\sin(\alpha + \beta) = \sin\alpha\cos\beta + \cos\alpha\sin\beta \quad \textbf{(20.9)}$$

$$\cos(\alpha + \beta) = \cos\alpha\cos\beta - \sin\alpha\sin\beta \quad \textbf{(20.10)}$$

$$\sin(\alpha - \beta) = \sin\alpha\cos\beta - \cos\alpha\sin\beta \quad \textbf{(20.11)}$$

$$\cos(\alpha - \beta) = \cos\alpha\cos\beta + \sin\alpha\sin\beta \quad \textbf{(20.12)}$$

$$\tan(\alpha \pm \beta) = \frac{\tan\alpha \pm \tan\beta}{1 \mp \tan\alpha\tan\beta} \quad \textbf{(20.13)}$$

Double-angle formulas

$$\sin 2\alpha = 2\sin\alpha\cos\alpha \quad \textbf{(20.21)}$$

$$\cos 2\alpha = \cos^2\alpha - \sin^2\alpha \quad \textbf{(20.22)}$$

$$= 2\cos^2\alpha - 1 \quad \textbf{(20.23)}$$

$$= 1 - 2\sin^2\alpha \quad \textbf{(20.24)}$$

$$\tan 2\alpha = \frac{2\tan\alpha}{1 - \tan^2\alpha} \quad \textbf{(20.25)}$$

Half-angle formulas

$$\sin\frac{\alpha}{2} = \pm\sqrt{\frac{1-\cos\alpha}{2}} \quad (20.26)$$

$$\cos\frac{\alpha}{2} = \pm\sqrt{\frac{1+\cos\alpha}{2}} \quad (20.27)$$

Inverse trigonometric functions

$$y = \sin^{-1}x \quad \left(-\frac{\pi}{2} \le y \le \frac{\pi}{2}\right) \quad (20.28)$$

$$-\frac{\pi}{2} \le \sin^{-1}x \le \frac{\pi}{2} \quad 0 \le \cos^{-1}x \le \pi \quad -\frac{\pi}{2} < \tan^{-1}x < \frac{\pi}{2} \quad 0 < \cot^{-1}x < \pi$$
$$0 \le \sec^{-1}x \le \pi \quad \left(\sec^{-1}x \ne \frac{\pi}{2}\right) \quad -\frac{\pi}{2} \le \csc^{-1}x \le \frac{\pi}{2} \quad (\csc^{-1}x \ne 0) \quad (20.29)$$

CHAPTER 20 REVIEW EXERCISES

CONCEPT CHECK EXERCISES

*Determine each of the following as being either **true** or **false**. If it is false, explain why.*

1. $\tan 2\theta = \dfrac{\cos 2\theta}{\sin 2\theta}$
2. $\cos^2 2\alpha = 1 - \sin^2 2\alpha$
3. $\cos(\alpha - \beta) = \cos\alpha\cos\beta - \sin\alpha\sin\beta$
4. $\sin\dfrac{\alpha}{2} = \sqrt{\dfrac{1-\cos\alpha}{2}}$
5. $\sin^{-1}(-1) = \dfrac{3\pi}{2}$
6. For $0 \le \theta < 2\pi$, the solution of the equation $2\cos\theta + 1 = 0$ is $\theta = 2\pi/3, 4\pi/3$.

PRACTICE AND APPLICATIONS

In Exercises 7–14, determine the values of the indicated functions in the given manner.

7. Find sin 120° by using 120° = 90° + 30°.
8. Find cos 30° by using 30° = 90° − 60°.
9. Find sin 315° by using 315° = 360° − 45°.
10. Find $\tan\frac{5\pi}{4}$ by using $\frac{5\pi}{4} = \pi + \frac{\pi}{4}$.
11. Find $\cos\pi$ by using $\pi = 2(\frac{\pi}{2})$.
12. Find $\sin\frac{3\pi}{2}$ by using $\frac{3\pi}{2} = 2(\frac{3\pi}{4})$.
13. Find tan 60° by using 60° = 2(30°).
14. Find cos 45° by using $45° = \frac{1}{2}(90°)$.

In Exercises 15–22, simplify the given expressions by using one of the basic formulas of the chapter. Then use a calculator to verify the result by finding the value of the original expression and the value of the simplified expression.

15. $\sin 12°\cos 38° + \cos 12°\sin 38°$
16. $\cos^2 148° - \sin^2 148°$
17. $2\sin\frac{\pi}{14}\cos\frac{\pi}{14}$
18. $1 - 2\sin^2\frac{\pi}{8}$
19. $\cos 73°\cos(-142°) + \sin 73°\sin(-142°)$
20. $\cos 3°\cos 215° - \sin 3°\sin 215°$
21. $\dfrac{2\tan 18°}{1 - \tan^2 18°}$
22. $\sqrt{\dfrac{1 - \cos 166°}{8}}$

In Exercises 23–30, simplify each of the given expressions. Expansion of any term is not necessary; recognition of the proper form leads to the proper result.

23. $\sin 2x\cos 3x + \cos 2x\sin 3x$
24. $\cos 7x\cos 3x + \sin 7x\sin 3x$
25. $4\sin 7x\cos 7x$
26. $\dfrac{\tan x + \tan 2x}{1 - \tan x\tan 2x}$
27. $2 - 4\sin^2 6x$
28. $\cos^2 2x - \sin^2 2x$
29. $\sqrt{2 + 2\cos 2x}$
30. $\sqrt{32 - 32\cos 4x}$

In Exercises 31–38, evaluate the given expressions.

31. $\sin^{-1}(-1)$
32. $\sec^{-1}\sqrt{2}$
33. $\cos^{-1}0.8629$
34. $\tan^{-1}(-6.249)$
35. $\tan[\sin^{-1}(-0.5)]$
36. $\cos[\tan^{-1}(-\sqrt{3})]$
37. $\sin^{-1}[\cos(7\pi/6)]$
38. $\cos^{-1}[\cot(-\pi/4)]$

In Exercises 39–42, simplify the given expressions.

39. $\dfrac{\sec y}{\cos y} - \dfrac{\tan y}{\cot y}$
40. $\dfrac{\sin 2\theta}{2\csc\theta} - \cos^3\theta$
41. $\sin x(\csc x - \sin x)$
42. $\cos y(\sec y - \cos y)$

In Exercises 43–50, prove the given identities.

43. $\dfrac{\sec^4 x - 1}{\tan^2 x} = 2 + \tan^2 x$
44. $\cos^2 y - \sin^2 y = \dfrac{1 - \tan^2 y}{1 + \tan^2 y}$
45. $2\csc 2x\cot x = 1 + \cot^2 x$
46. $\sin x\cot^2 x = \csc x - \sin x$
47. $\dfrac{1 - \sin^2\theta}{1 - \cos^2\theta} = \cot^2\theta$
48. $\dfrac{\cos 2\theta}{\cos^2\theta} = 1 - \tan^2\theta$
49. $\sin\dfrac{\theta}{2}\cos\dfrac{\theta}{2} = \dfrac{\sin\theta}{2}$
50. $\cos(x - y)\cos y - \sin(x - y)\sin y = \cos x$

In Exercises 51–58, simplify the given expressions. The result will be one of sin *x*, cos *x*, tan *x*, cot *x*, sec *x*, *or* csc *x*.

51. $\frac{\sec x}{\sin x} - \sec x \sin x$

52. $\cos x \cot x + \sin x$

53. $\sin x \tan x + \cos x$

54. $\frac{\tan x \csc x}{\sin x} - \cot x$

55. $\frac{\sin x \cot x + \cos x}{2 \cot x}$

56. $\frac{1 + \cos 2x}{2 \cos x}$

57. $\frac{\sin 2x \sec x}{2}$

58. $(\sec x + \tan x)(1 - \sin x)$

In Exercises 59–66, verify each identity by comparing the graph of the left side with the graph of the right side on a calculator.

59. $\frac{\cos\theta - \sin\theta}{\cos\theta + \sin\theta} = \frac{\cot\theta - 1}{\cot\theta + 1}$

60. $\sin 3y \cos 2y - \cos 3y \sin 2y = \sin y$

61. $\sin 4x(\cos^2 2x - \sin^2 2x) = \frac{\sin 8x}{2}$

62. $\csc 2x + \cot 2x = \cot x$

63. $\frac{\sin x}{\csc x - \cot x} = 1 + \cos x$

64. $\cos x - \sin\frac{x}{2} = \left(1 - 2\sin\frac{x}{2}\right)\left(1 + \sin\frac{x}{2}\right)$

65. $\tan\frac{\alpha}{2} = \csc\alpha - \cot\alpha$

66. $\sec\frac{x}{2} + \csc\frac{x}{2} = \frac{2(\sin\frac{x}{2} + \cos\frac{x}{2})}{\sin x}$

In Exercises 67–70, solve for x.

67. $y = 2\cos 2x$

68. $y - 2 = 2\tan\left(x - \frac{\pi}{2}\right)$

69. $y = \frac{\pi}{4} - 3\sin^{-1}5x$

70. $2y = \sec^{-1}4x - 2$

In Exercises 71–82, solve the given equations for x $(0 \le x < 2\pi)$.

71. $3(\tan x - 2) = 1 + \tan x$

72. $5\sin x = 3 - (\sin x + 2)$

73. $2(1 - 2\sin^2 x) = 1$

74. $\sec x = 2\tan^2 x$

75. $2\sin^2\theta + 3\cos\theta - 3 = 0$

76. $2\sin 2x + 1 = 0$

77. $\sin x = \sin\frac{x}{2}$

78. $\cos 2x = \sin(-x)$

79. $\sin 2x = \cos 3x$

80. $\cos 3x \cos x + \sin 3x \sin x = 0$

81. $\sin^2\left(\frac{x}{2}\right) - \cos x + 1 = 0$

82. $\sin x + \cos x = 1$

In Exercises 83–86, determine whether the equality is an identity or a conditional equation. If it is an identity, prove it. If it is a conditional equation, solve it for $0 \le x < 2\pi$.

83. $\tan x + \cot x = \csc x \sec x$

84. $\tan x - \sin^2 x = \cos^2 x - \sec^2 x$

85. $\sin x \cos x - 1 = \cos x - \sin x$

86. $2\tan x = \sin 2x \sec^2 x$

In Exercises 87–90, solve the given equations graphically.

87. $x + \ln x - 3\cos^2 x = 2$

88. $e^{\sin x} - 2 = x\cos^2 x$

89. $2\tan^{-1}x + x^2 = 3$

90. $3\sin^{-1}x = 6\sin x + 1$

In Exercises 91–96, find an algebraic expression for each of the given expressions.

91. $\tan(\cot^{-1}x)$

92. $\sec(\sin^{-1}x)$

93. $\cos(\sin^{-1}\frac{x}{5})$

94. $\tan(\sec^{-1}\frac{x}{2})$

95. $\cos(\sin^{-1}x + \tan^{-1}y)$

96. $\sin(2\cos^{-1}x)$

In Exercises 97–100, use the given substitutions to show that the equations are valid for $0 \le \theta < \pi/2$.

97. If $x = 2\cos\theta$, show that $\sqrt{4 - x^2} = 2\sin\theta$.

98. If $x = 2\sec\theta$, show that $\sqrt{x^2 - 4} = 2\tan\theta$.

99. If $x = \tan\theta$, show that $\frac{x}{\sqrt{1 + x^2}} = \sin\theta$.

100. If $x = \cos\theta$, show that $\frac{\sqrt{1 - x^2}}{x} = \tan\theta$.

In Exercises 101–104, find the exact value of the given expression for the triangle in Fig. 20.49.

101. $\sin 2\theta$

102. $\sec 2\theta$

103. $\cos(\theta/2)$

104. $\tan(\theta/2)$

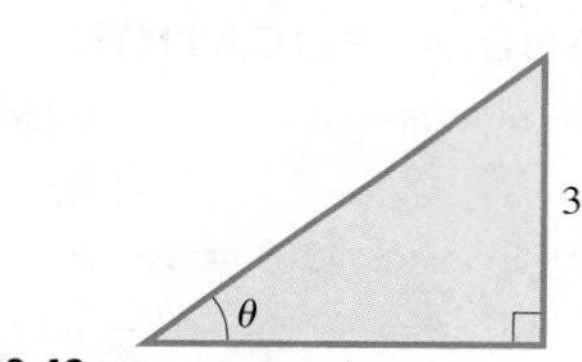

Fig. 20.49

In Exercises 105–130, use the methods and formulas of this chapter to solve the given problems.

105. Prove: $(\cos\theta + j\sin\theta)^2 = \cos 2\theta + j\sin 2\theta \quad (j = \sqrt{-1})$

106. Prove: $(\cos\theta + j\sin\theta)^3 = \cos 3\theta + j\sin 3\theta \quad (j = \sqrt{-1})$

107. Evaluate exactly: $2(\tan^2 x + \sin^2 x - \sec^2 x) + \cos 2x$

108. Evaluate exactly: $\sec^4\theta - \sec^2\theta - \tan^4\theta - \tan^2\theta$

109. Solve the inequality $\sin 2x > 2\cos x$ for $0 \le x < 2\pi$.

110. Show that $(\cos 2\alpha + \sin^2\alpha)\sec^2\alpha$ has a constant value.

111. Show that $y = A\sin 2t + B\cos 2t$ may be written as $y = C\sin(2t + \alpha)$, where $C = \sqrt{A^2 + B^2}$ and $\tan\alpha = B/A$. (*Hint:* Let $A/C = \cos\alpha$ and $B/C = \sin\alpha$.)

112. For the triangle in Fig. 20.50, find the expression for sin $(A/2)$.

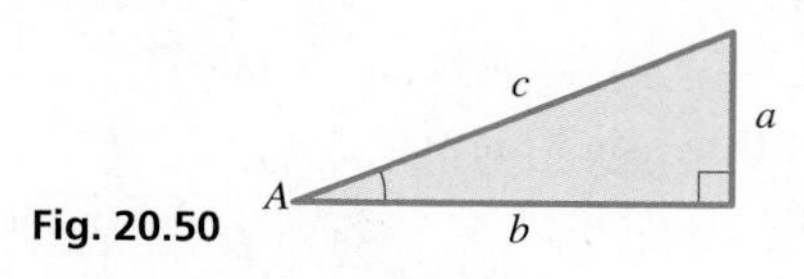

Fig. 20.50

113. For the triangle in Fig. 20.50 find the expression for sin 2*A*.

114. Express $\cos(A + B + C)$ in terms of $\sin A$, $\sin B$, $\sin C$, $\cos A$, $\cos B$, and $\cos C$.

115. Forces **A** and **B** act on a bolt such that **A** makes an angle θ with the x-axis and **B** makes an angle θ with the y-axis as shown in Fig. 20.51. The resultant **R** has components $R_x = A\cos\theta - B\sin\theta$ and $R_y = A\sin\theta + B\cos\theta$. Using these components, show that $R = \sqrt{A^2 + B^2}$.

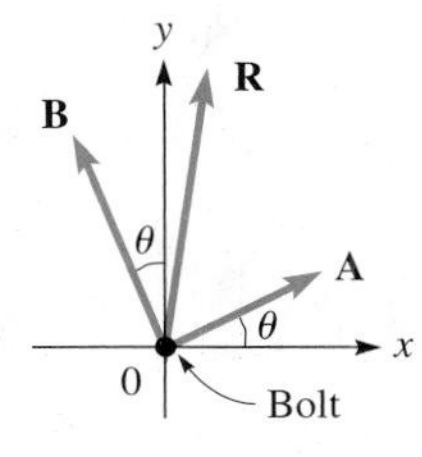

Fig. 20.51

116. For a certain alternating-current generator, the expression $I\cos\theta + I\cos(\theta + 2\pi/3) + I\cos(\theta + 4\pi/3)$ arises. Simplify this expression.

117. Some comets follow a parabolic path that can be described by the equation $r = (k/2)\csc^2(\theta/2)$, where r is the distance to the sun and k is a constant. Show that this equation can be written as $r = k/(1 - \cos\theta)$.

118. In studying the interference of light waves, the identity

$\dfrac{\sin\frac{3}{2}x}{\sin\frac{1}{2}x}\sin x = \sin x + \sin 2x$ is used. Prove this identity.

[*Hint:* $\sin\frac{3}{2}x = \sin(x + \frac{1}{2}x)$.]

119. In the study of chemical spectroscopy, the equation

$\omega t = \sin^{-1}\dfrac{\theta - \alpha}{R}$ arises. Solve for θ.

120. The power p in a certain electric circuit is given by $p = 2.5[\cos\alpha\sin(\omega t + \phi) - \sin\alpha\cos(\omega t + \phi)]$. Solve for t.

121. In surveying, when determining an azimuth (a measure used for reference purposes), it might be necessary to simplify the expression $(2\cos\alpha\cos\beta)^{-1} - \tan\alpha\tan\beta$. Perform this operation by expressing it in the simplest possible form when $\alpha = \beta$.

122. In analyzing the motion of an automobile universal joint, the equation $\sec^2 A - \sin^2 B\tan^2 A = \sec^2 C$ is used. Show that this equation is true if $\tan A\cos B = \tan C$.

123. The instantaneous power p used by a certain motor is given by $p = VI\cos\phi\cos^2\omega t - VI\sin\phi\cos\omega t\sin\omega t$. Simplify this expression.

124. In studying waveforms, a sawtooth wave may often be approximated by $y = \sin\pi x + \frac{1}{2}\sin 2\pi x + \frac{1}{3}\sin 3\pi x$. For what values of x, $0 \le x < 2$, is the sum of the first two terms equal to zero?

125. A person in the stands of a football field is "sitting" on the 50-yd line (which is 53.3 yd long), y yd above it and x yd horizontally from it. (a) Find the expression for θ, the angle subtended at the person's eye by the 50-yd line, and (b) find θ if $x = 28.0$ yd and $y = 12.0$ yd.

126. A person is 100 m above the ground and x m on a direct line from an object on the ground. Find an expression for 2θ, where θ is the angle of depression from the person to the object.

127. A roof truss is in the shape of an isosceles triangle of height 6.4 ft. If the total length of the three members is 51.2 ft, what is the length of a rafter? Solve by setting up and solving an appropriate trigonometric equation. (The Pythagorean theorem may be used only as a check.)

128. The angle of elevation of the top of the Washington Monument from a point on level ground 250 ft from the center of its base is twice the angle of elevation of the top from a point 610 ft farther away. Find the height of the monument. See Fig. 20.52.

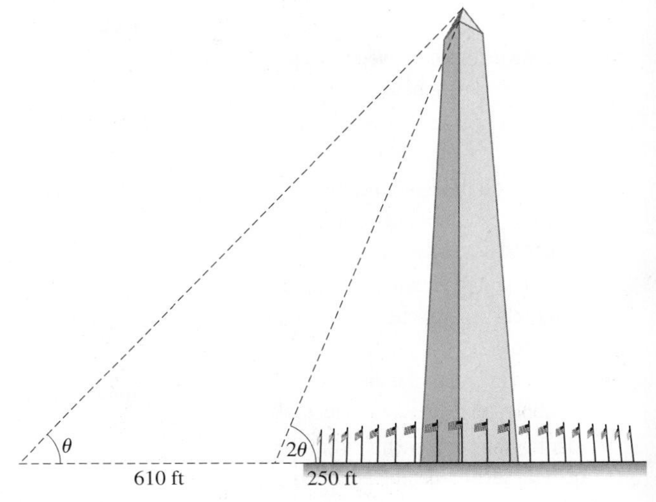

Fig. 20.52

129. If a plane surface inclined at angle θ moves horizontally, the angle for which the lifting force of the air is a maximum is found by solving the equation $2\sin\theta\cos^2\theta - \sin^3\theta = 0$, where $0 < \theta < 90°$. Solve for θ.

130. To determine the angle between two sections of a certain robot arm, the equation $1.20\cos\theta + 0.135\cos 2\theta = 0$ is to be solved. Find the required angle θ if $0° < \theta < 180°$.

131. In checking the angles of a section of a bridge support, an engineer finds the expression $\cos(2\sin^{-1}0.40)$. Write a paragraph explaining how the value of this expression can be found without the use of a calculator.

CHAPTER 20 PRACTICE TEST

As a study aid, we have included complete solutions for each Practice Test problem at the end of this book.

1. Prove that $\sec\theta - \dfrac{\tan\theta}{\csc\theta} = \cos\theta$.

2. Solve for x $(0 \le x < 2\pi)$ analytically, using trigonometric relations where necessary: $\sin 2x + \sin x = 0$.

3. Find an algebraic expression for $\cos(\sin^{-1}x)$.

4. The electric current as a function of the time for a particular circuit is given by $i = 8.00e^{-20t}(1.73\cos 10.0t - \sin 10.0t)$. Find the time (in s) when the current is first zero.

5. Prove that $\dfrac{\tan\alpha + \tan\beta}{\tan\alpha - \tan\beta} = \dfrac{\sin(\alpha + \beta)}{\sin(\alpha - \beta)}$.

6. Prove that $\cot^2 x - \cos^2 x = \cot^2 x\cos^2 x$.

7. Find $\cos\frac{1}{2}x$ if $\sin x = -\frac{3}{5}$ and $270° < x < 360°$.

8. The intensity of a certain type of polarized light is given by $I = I_0\sin 2\theta\cos 2\theta$. Solve for θ.

9. Solve graphically: $x - 2\cos x = 5$.

10. Find the exact value of x: $\cos^{-1}x = -\tan^{-1}(-1)$.

21

Plane Analytic Geometry

LEARNING OUTCOMES

After completion of this chapter, the student should be able to:

- Find the distance between two points in the coordinate plane
- Find the slope and the inclination of a line
- Undersatnd the relationship between the slopes of parallel and perpendicular lines
- Determine the equation and properties of a circle, a parabola, an ellipse, and a hyperbola
- Sketch the graph of a line, a circle, a parabola, an ellipse, or a hyperbola
- Identify the conic section from the general second-degree equation
- Obtain a new coordinate system by translation and/or rotation of axes
- Convert between rectangular and polar coordinates
- Graph functions in polar coordinates
- Solve application problems involving lines, conics, and polar coordinates

The development of geometry made little progress from the time of the ancient Greeks until the 1600s. Then in the 1630s, two French mathematicians, Rene Descartes and Pierre de Fermat, independently introduced algebra into the study of geometry. Each used algebraic notation and a coordinate system, which allowed them to analyze the properties of curves. Fermat briefly indicated his methods in a letter he wrote in 1636. However, in 1637 Descartes included a much more complete work as an appendix to his *Discourse on Method,* and he is therefore now considered as the founder of *analytic geometry.*

At the time, it was known, for example, that a projectile follows a parabolic path and that the orbits of the planets are ellipses, and there was a renewed interest in curves of various kinds because of their applications. However, Descartes was primarily interested in studying the relationship of algebra to geometry. In doing so, he started the development of analytic geometry, which has many applications and also was very important in the invention of calculus that came soon thereafter. In turn, through these advances in mathematics, many more areas of science and technology were able to be developed.

The underlying principle of analytic geometry is the relationship of an algebraic equation and the geometric properties of the curve that represents the equation. In this chapter, we develop equations for a number of important curves and find their properties through an analysis of their equations. Most important among these curves are the *conic sections,* which we briefly introduced in Chapter 14.

As we have noted, analytic geometry has applications in the study of projectile motion and planetary orbits. Other important applications range from the design of gears, airplane wings, and automobile headlights to the construction of bridges and nuclear towers.

▶ With improving technology, satellite dishes used today have become smaller and even portable. In Section 21.4, we will show the basic reflective property of a parabola that makes a satellite dish work.

21.1 Basic Definitions

The Distance Formula • The Slope of a Line • Inclination • Parallel Lines • Perpendicular Lines

As we have noted, analytic geometry deals with the relationship between an algebraic equation and the geometric curve it represents. In this section, we develop certain basic concepts that will be needed for future use in establishing the proper relationships between an equation and a curve.

THE DISTANCE FORMULA

The first of these concepts involves the distance between any two points in the coordinate plane. If these points lie on a line parallel to the x-axis, the distance from the first point (x_1, y) to the second point (x_2, y) is $|x_2 - x_1|$. The absolute value is used because we are interested only in the magnitude of the distance. Therefore, we could also denote the distance as $|x_1 - x_2|$. Similarly, the distance between two points (x, y_1) and (x, y_2) that lie on a line parallel to the y-axis is $|y_2 - y_1|$ or $|y_1 - y_2|$.

EXAMPLE 1 Distance between points

The line segment joining $A(-1, 5)$ and $B(-4, 5)$ in Fig. 21.1 is parallel to the x-axis. Therefore, the distance between these points is

$$d = |-4 - (-1)| = 3 \quad \text{or} \quad d = |-1 - (-4)| = 3$$

Also, in Fig. 21.1, the line segment joining $C(2, -3)$ and $D(2, 6)$ is **parallel** to the y-axis. The distance d between these points is

$$d = |6 - (-3)| = 9 \quad \text{or} \quad d = |-3 - 6| = 9$$

■

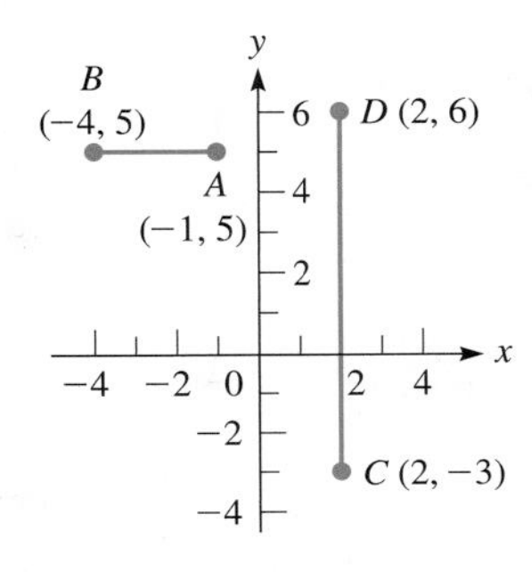

Fig. 21.1

We now wish to find the length of a line segment joining any two points in the plane. If these points are on a line that is not parallel to either axis (see Fig. 21.2), we use the Pythagorean theorem to find the distance between them. By making a right triangle with the line segment joining the points as the hypotenuse and line segments parallel to the axes as legs, we have $d^2 = (x_2 - x_1)^2 + (y_2 - y_1)^2$. Solving for d, we get *the* **distance formula,** *which gives the distance between any two points in the plane.* This formula is

$$d = \sqrt{(x_2 - x_1)^2 + (y_2 - y_1)^2} \tag{21.1}$$

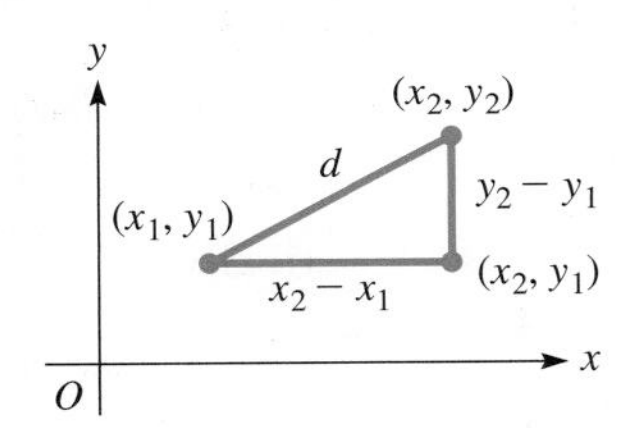

Fig. 21.2

Here, we choose the positive square root because we are concerned only with the magnitude of the length of the line segment.

EXAMPLE 2 Distance formula—either point as (x_1, y_1)

The distance between $(3, -1)$ and $(-2, -5)$ is given by

$$\begin{aligned} d &= \sqrt{[(-2) - 3]^2 + [(-5) - (-1)]^2} \\ &= \sqrt{(-5)^2 + (-4)^2} = \sqrt{25 + 16} \\ &= \sqrt{41} = 6.403 \end{aligned}$$

See Fig. 21.3.

NOTE ▸ [It makes no difference which point is chosen as (x_1, y_1) and which is chosen as (x_2, y_2), because the differences in the x-coordinates and the y-coordinates are squared.] We obtain the same value for the distance when we calculate it as

$$\begin{aligned} d &= \sqrt{[3 - (-2)]^2 + [(-1) - (-5)]^2} \\ &= \sqrt{5^2 + 4^2} = \sqrt{41} = 6.403 \end{aligned}$$

■

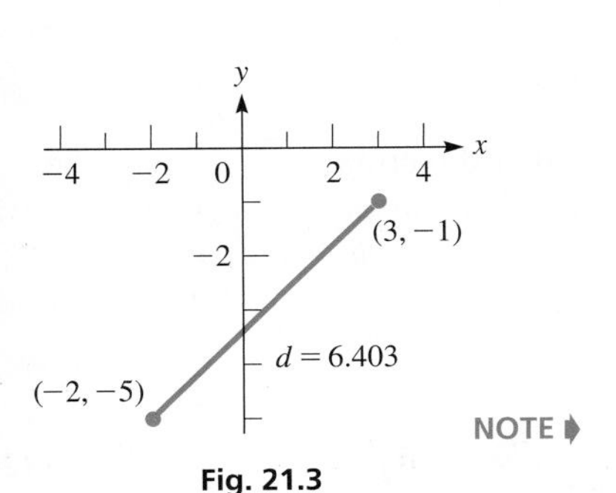

Fig. 21.3

Practice Exercise

1. Find the distance between $(-2, 6)$ and $(5, -3)$.

THE SLOPE OF A LINE

Another important quantity for a line is its *slope,* which we defined in Chapter 5. Here, we review its definition and develop its meaning in more detail.

The **slope** *of a line through two points is defined as the difference in the y-coordinates (rise) divided by the difference in the x-coordinates (run).* Therefore, the slope, m, which gives a measure of the direction of a line, is defined as

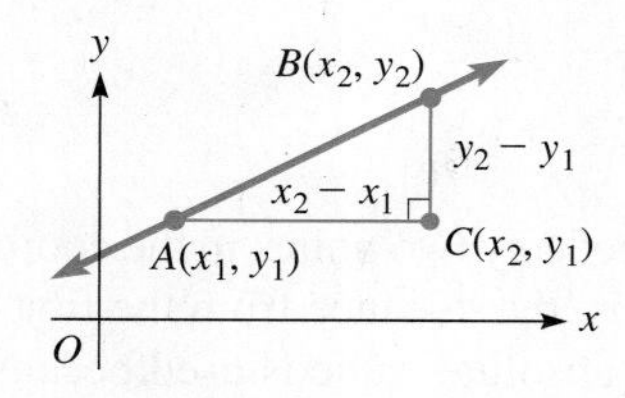

Fig. 21.4

$$m = \frac{y_2 - y_1}{x_2 - x_1} \quad (21.2)$$

See Fig. 21.4. **CAUTION** We may interpret either of the points as (x_1, y_1) and the other as (x_2, y_2), although we must be careful to place the $y_2 - y_1$ in the numerator and $x_2 - x_1$ in the denominator. ■ When the line is horizontal, $y_2 - y_1 = 0$ and $m = 0$. When the line is vertical, $x_2 - x_1 = 0$ and the slope is undefined.

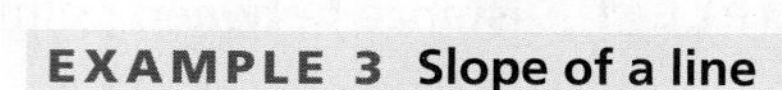

EXAMPLE 3 Slope of a line

The slope of a line through $(3, -5)$ and $(-2, -6)$ is

$$m = \frac{-6 - (-5)}{-2 - 3} = \frac{-6 + 5}{-5} = \frac{1}{5}$$

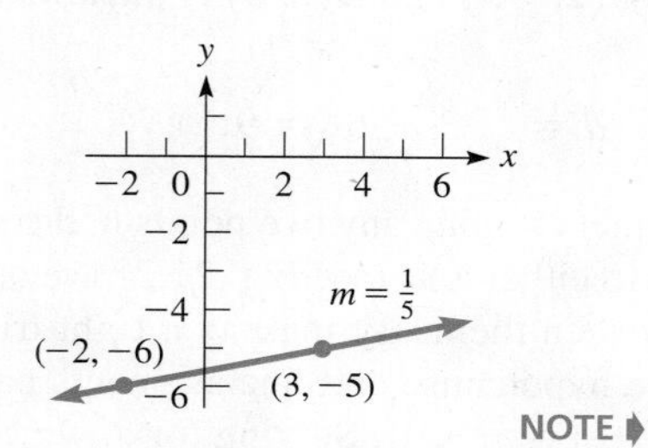

Fig. 21.5

See Fig. 21.5. Again, we may interpret either of the points as (x_1, y_1) and the other as (x_2, y_2). We can also obtain the slope of this same line from

$$m = \frac{-5 - (-6)}{3 - (-2)} = \frac{1}{5}$$ ■

The larger the numerical value of the slope of a line, the more nearly vertical is the line. NOTE ▸ [Also, a line rising to the right has a positive slope, and a line falling to the right has a negative slope.]

EXAMPLE 4 Magnitude and sign of slope

(a) The line in Example 3 has a positive slope, which is numerically small. From Fig. 21.5, it can be seen that the line rises slightly to the right.

(b) The line joining $(3, 4)$ and $(4, -6)$ has a slope of

$$m = \frac{4 - (-6)}{3 - 4} = -10 \quad \text{or} \quad m = \frac{-6 - 4}{4 - 3} = -10$$

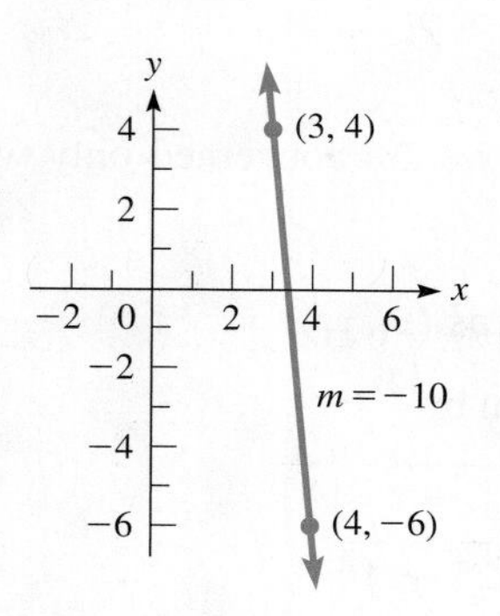

Fig. 21.6

This line falls sharply to the right, as shown in Fig. 21.6. ■

If a given line is extended indefinitely in either direction, it must cross the x-axis at some point unless it is parallel to the x-axis. *The angle measured from the x-axis in a positive direction to the line is called the* **inclination** *of the line* (see Fig. 21.7). The inclination of a line parallel to the x-axis is defined to be zero. *An alternate definition of slope, in terms of the inclination* α, *is*

$$m = \tan\alpha \quad (0^\circ \le \alpha < 180^\circ) \quad (21.3)$$

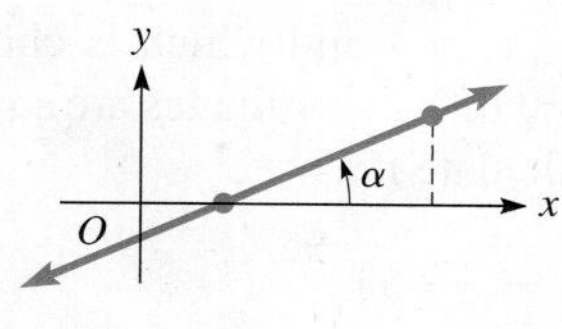

Fig. 21.7

Because the slope can be defined in terms of any two points on the line, we can choose the x-intercept and any other point. Therefore, from the definition of the tangent of an angle, we see that Eq. (21.3) is in agreement with Eq. (21.2).

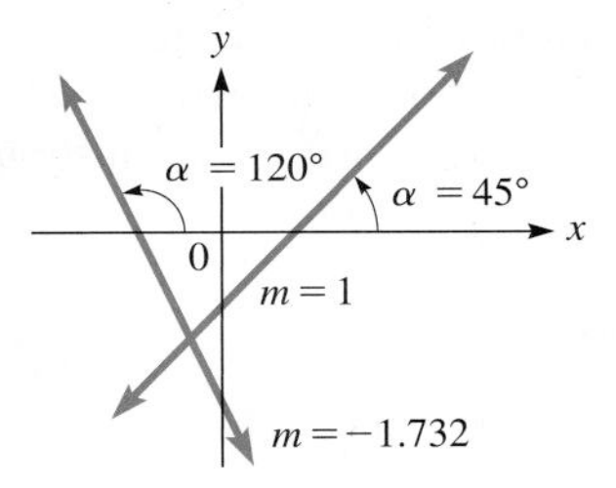

Fig. 21.8

EXAMPLE 5 Inclination

(a) The slope of a line with an inclination of 45° is

$$m = \tan 45° = 1.000$$

(b) If a line has a slope of -1.732, we know that $\tan\alpha = -1.732$. Because $\tan\alpha$ is negative, α must be a second-quadrant angle. Therefore, using a calculator to show that $\tan 60° = 1.732$, we find that $\alpha = 180° - 60° = 120°$. See Fig. 21.8.

We see that if the inclination is an acute angle, the slope is positive and the line rises to the right. If the inclination is an obtuse angle, the slope is negative and the line falls to the right. ■

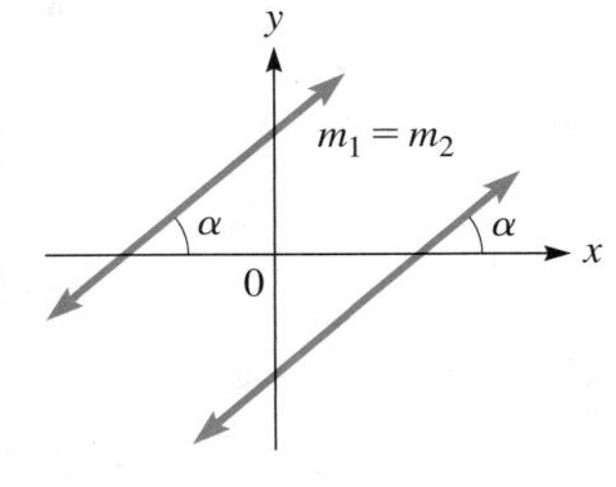

Fig. 21.9

Any two parallel lines crossing the x-axis have the same inclination. Therefore, as shown in Fig. 21.9, the *slopes of parallel lines are equal.* This can be stated as

$$m_1 = m_2 \quad \text{(for } \| \text{ lines)} \qquad (21.4)$$

If two lines are perpendicular, this means that there must be 90° between their inclinations (Fig. 21.10). The relation between their inclinations is

$$\alpha_2 = \alpha_1 + 90°$$
$$90° - \alpha_2 = -\alpha_1$$

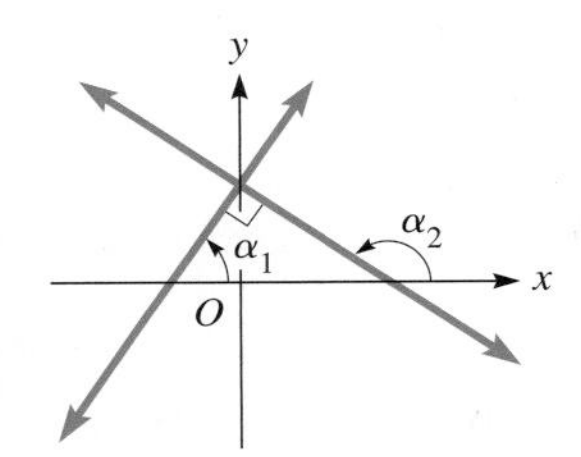

Fig. 21.10

If neither line is vertical (the slope of a vertical line is undefined) and we take the tangent in this last relation, we have

$$\tan(90° - \alpha_2) = \tan(-\alpha_1)$$

or

$$\cot\alpha_2 = -\tan\alpha_1$$

because a function of the complement of an angle equals the cofunction of that angle (see page 125) and because $\tan(-\alpha) = -\tan\alpha$ (see page 247). But $\cot\alpha = 1/\tan\alpha$, which means $1/\tan\alpha_2 = -\tan\alpha_1$. Using the inclination definition of slope, we have *as the relation between slopes of perpendicular lines,*

$$m_2 = -\frac{1}{m_1} \quad \text{or} \quad m_1 m_2 = -1 \quad \text{for } \perp \text{ lines} \qquad (21.5)$$

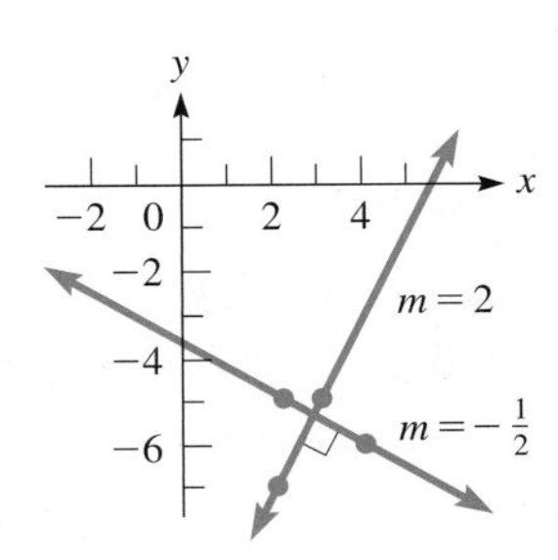

Fig. 21.11

EXAMPLE 6 Perpendicular lines

The line through $(3, -5)$ and $(2, -7)$ has a slope of

$$m_1 = \frac{-5 + 7}{3 - 2} = 2$$

The line through $(4, -6)$ and $(2, -5)$ has a slope of

$$m_2 = \frac{-6 - (-5)}{4 - 2} = -\frac{1}{2}$$

Because the slopes of the two lines are negative reciprocals, we know that the lines are perpendicular. See Fig. 21.11. ■

Practice Exercise

2. What is the slope of a line perpendicular to the line through $(-1, 6)$ and $(-3, -3)$?

Using the formulas for distance and slope, we can show certain basic geometric relationships. The following examples illustrate the use of the formulas and thereby show the use of algebra in solving problems that are basically geometric. This illustrates the methods of analytic geometry.

EXAMPLE 7 Use of algebra in geometric problems

(a) Show that the line segments joining $A(-5, 3)$, $B(6, 0)$, and $C(5, 5)$ form a right triangle. See Fig. 21.12.

If these points are vertices of a right triangle, the slopes of two of the sides must be negative reciprocals. This would show perpendicularity. These slopes are

$$m_{AB} = \frac{3-0}{-5-6} = -\frac{3}{11} \qquad m_{AC} = \frac{3-5}{-5-5} = \frac{1}{5} \qquad m_{BC} = \frac{0-5}{6-5} = -5$$

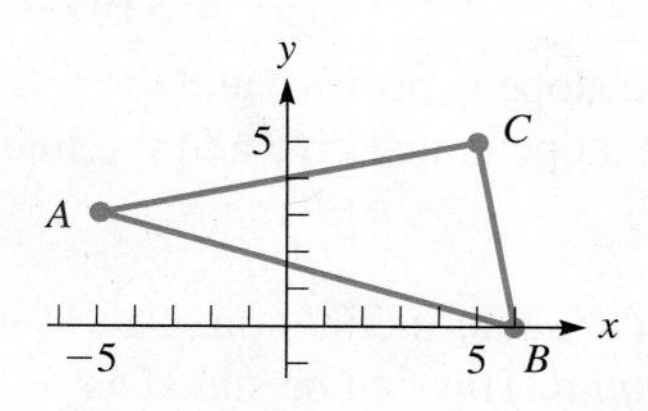

Fig. 21.12

We see that the slopes of AC and BC are negative reciprocals, which means $AC \perp BC$. From this we conclude that the triangle is a right triangle.

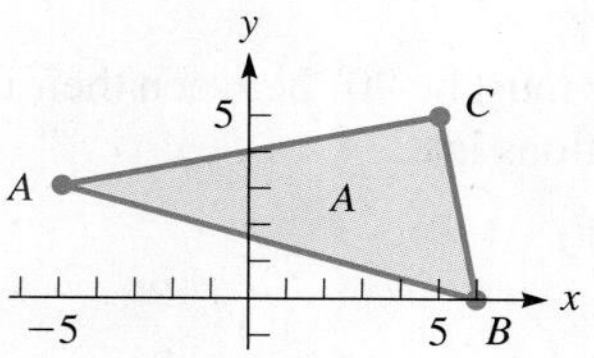

(b) Find the area of the triangle in part (a). See Fig. 21.13.

Because the right angle is at C, the legs of the triangle are AC and BC. The area is one-half the product of the lengths of the legs of a right triangle. The lengths of the legs are

$$d_{AC} = \sqrt{(-5-5)^2 + (3-5)^2} = \sqrt{104} = 2\sqrt{26}$$
$$d_{BC} = \sqrt{(6-5)^2 + (0-5)^2} = \sqrt{26}$$

Fig. 21.13

Therefore, the area is $A = \frac{1}{2}(2\sqrt{26})(\sqrt{26}) = 26$. ■

EXERCISES 21.1

In Exercises 1–4, make the given changes in the indicated examples of this section and then solve the resulting problems.

1. In Example 2, change $(-2, -5)$ to $(-2, 5)$.
2. In Example 3, change $(3, -5)$ to $(-3, -5)$.
3. In Example 5(b), change -1.732 to -0.5774.
4. In Example 7(a), change $B(6, 0)$ to $B(-4, -2)$.

In Exercises 5–14, find the distance between the given pairs of points.

5. $(3, 8)$ and $(-1, -2)$
6. $(-1, 3)$ and $(-8, -4)$
7. $(4, -5)$ and $(4, -8)$
8. $(-3, -7)$ and $(2, 10)$
9. $(-12, 20)$ and $(32, -13)$
10. $(23, -9)$ and $(-25, 11)$
11. $(\sqrt{32}, \sqrt{18})$ and $(-\sqrt{50}, \sqrt{8})$
12. $(e, -\pi)$ and $(-2e, -\pi)$
13. $(1.22, -3.45)$ and $(-1.07, -5.16)$
14. (a, h^2) and $[a + h, (a + h)^2]$

In Exercises 15–24, find the slopes of the lines through the points in Exercises 5–14.

In Exercises 25–28, find the slopes of the lines with the given inclinations.

25. 30°
26. 62.5°
27. 176.2°
28. 93.5°

In Exercises 29–32, find the inclinations of the lines with the given slopes.

29. 0.364
30. 1.903
31. -6.691
32. -0.0721

In Exercises 33–36, determine whether the lines through the two pairs of points are parallel or perpendicular.

33. $(6, -1)$ and $(4, 3)$; $(-5, 2)$ and $(-7, 6)$
34. $(-3, 9)$ and $(4, 4)$; $(9, -1)$ and $(4, -8)$
35. $(-1, -4)$ and $(2, 3)$; $(-5, 2)$ and $(-19, 8)$
36. $(-a, -2b)$ and $(3a, 6b)$; $(2a, -6b)$ and $(5a, 0)$

In Exercises 37–40, determine the value of k.

37. The distance between $(-1, 3)$ and $(11, k)$ is 13.
38. The distance between $(k, 0)$ and $(0, 2k)$ is 10.
39. Points $(6, -1)$, $(3, k)$, and $(-3, -7)$ are on the same line.
40. The points in Exercise 39 are the vertices of a right triangle, with the right angle at $(3, k)$.

In Exercises 41–44, show that the given points are vertices of the given geometric figures.

41. $(2, 3)$, $(4, 9)$, and $(-2, 7)$ are vertices of an isosceles triangle.
42. $(-1, 3)$, $(3, 5)$, and $(5, 1)$ are the vertices of a right triangle.
43. $(-5, -4)$, $(7, 1)$, $(10, 5)$, and $(-2, 0)$ are the vertices of a parallelogram.
44. $(-5, 6)$, $(0, 8)$, $(-3, 1)$, and $(2, 3)$ are the vertices of a square.

In Exercises 45–48, find the indicated areas and perimeters.

45. Find the area of the triangle in Exercise 42.

46. Find the area of the square in Exercise 44.

47. Find the perimeter of the triangle in Exercise 41.

48. Find the perimeter of the parallelogram in Exercise 43.

In Exercises 49–52, use the following definition to find the midpoints between the given points on a straight line.

The *midpoint* between points (x_1, y_1) and (x_2, y_2) on a straight line is the point.

$$\left(\frac{x_1 + x_2}{2}, \frac{y_1 + y_2}{2}\right)$$

49. $(-4, 9)$ and $(6, 1)$

50. $(-1, 6)$ and $(-13, -8)$

51. $(-12.4, 25.7)$ and $(6.8, -17.3)$

52. $(2.6, 5.3)$ and $(-4.2, -2.7)$

In Exercises 53–66, solve the given problems.

53. Find the relation between x and y such that (x, y) is always 3 units from the origin.

54. Find the relation between x and y such that (x, y) is always equidistant from the y-axis and $(2, 0)$.

55. Show that the diagonals of a square are perpendicular to each other. [*Hint:* Use $(0, 0)$, $(a, 0)$, and $(0, a)$ as three of the vertices.]

56. The center of a circle is $(2, -3)$, and one end of a diameter is $(-1, 2)$. What are the coordinates of the other end of the diameter?

57. A line segment has a slope of 3 and one endpoint at $(-2, 5)$. If the other endpoint is on the x-axis, what are its coordinates?

58. The points $(-1, 3)$, $(5, y)$, and $(2, 4)$ are collinear (on the same line). Find y.

59. Find the coordinates of the point on the y-axis that is equidistant from $(-3, -5)$ and $(2, 4)$.

60. Are the points $(3, 5)$, $(1, -1)$, and $(6, 14)$ collinear?

61. Are the points $(-1, 6)$ and $(5, 5)$ equidistant from $(2, 1)$?

62. The *grade* of a highway is its slope expressed as a percent (a 5% grade means the slope is $\frac{5}{100}$). If the grade of a certain highway is 6%, find (a) its angle of inclination and (b) the change in elevation (in ft) of a car driving for 2.00 mi uphill along this highway (1 mi = 5280 ft).

63. On a computer drawing showing the specifications for a mounting bracket, holes are to be drilled at the points (32.5, 25.5) and (88.0, 62.5), where all measurements are in mm. Find the distance between the centers of the two holes.

64. A triangular machine part has vertices at $A(-3, 6)$, $B(-1, 1)$, and $C(5, 2)$. What is the length of the *median* from $(-1, 1)$? See Fig. 21.14. (A median is a line segment that connects a vertex with the midpoint of the opposite side. See Exercises 49–52.)

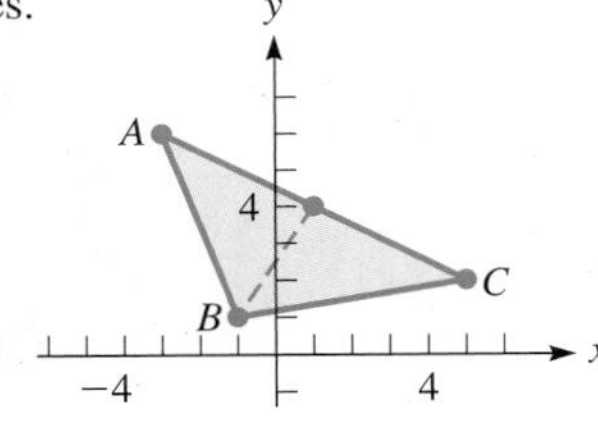

Fig. 21.14

65. Denver is 1350 km east and 900 km south of Seattle. Edmonton is 620 km east and 640 km north of Seattle. How far is Denver from Edmonton?

66. A person is working out on a treadmill inclined at 12% (the slope of the treadmill expressed in percent). What is the angle between the treadmill and the horizontal?

Answers to Practice Exercises

1. $d = \sqrt{130} = 11.4$ **2.** $m = -2/9$

21.2 The Straight Line

Point-slope Form of Equation • Vertical Line • Horizontal Line • Slope-intercept Form of Equation • General Form of Equation

In Chapter 5, we derived the *slope-intercept form* of the equation of the straight line. Here, we extend the development to include other forms of the equation of a straight line. Also, other methods of finding and applying these equations are shown. For completeness, we review some of the material in Chapter 5.

Using the definition of slope, we can derive the general type of equation that represents a straight line. This is another basic method of analytic geometry. That is, equations of a particular form can be shown to represent a particular type of curve. When we recognize the form of the equation, we know the kind of curve it represents. As we have seen, this is of great assistance in sketching the graph.

A straight line can be defined as a *curve* with a constant slope. This means that the value for the slope is the same for any two different points on the line that might be chosen. Thus, considering point (x_1, y_1) on a line to be fixed (Fig. 21.15) and another point $P(x, y)$ that *represents* any other point on the line, we have

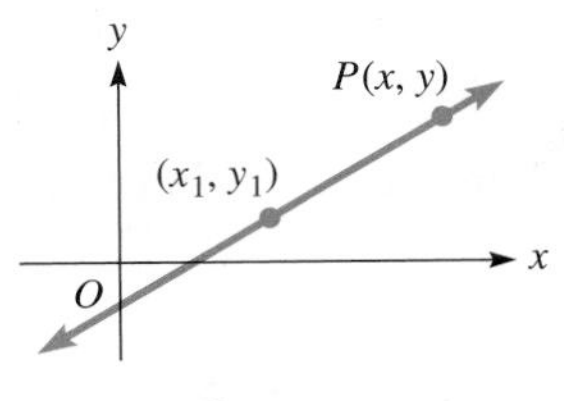

Fig. 21.15

$$m = \frac{y - y_1}{x - x_1}$$

which can be written as

$$y - y_1 = m(x - x_1). \quad (21.6)$$

Equation (21.6) is the **point-slope form** *of the equation of a straight line.* It is useful when we know the slope of a line and some point through which the line passes.

EXAMPLE 1 Point-slope form

Find the equation of the line that passes through $(-4, 1)$ with a slope of $-1/2$. See Fig. 21.16.

Substituting in Eq. (21.6), we find that

slope

$$y - 1 = (-\tfrac{1}{2})[x - (-4)]$$

coordinates

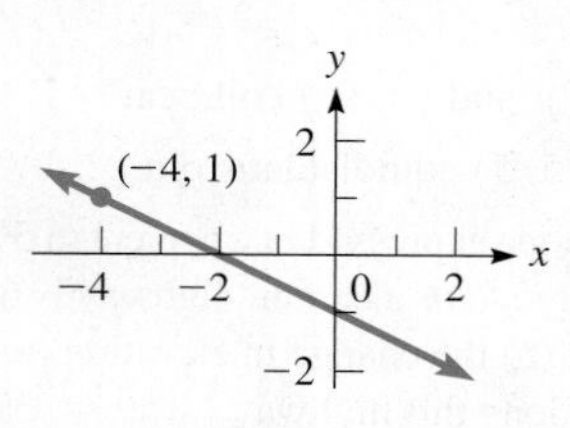

Fig. 21.16

Simplifying, we have

$$2y - 2 = -x - 4$$
$$x + 2y + 2 = 0$$

■

EXAMPLE 2 Equation of a line through two points

Find the equation of the line through $(2, -1)$ and $(6, 2)$.

We first find the slope of the line through these points:

$$m = \frac{2 - (-1)}{6 - 2} = \frac{3}{4}$$

Then by using either of the two known points and Eq. (21.6), we can find the equation of the line (see Fig. 21.17):

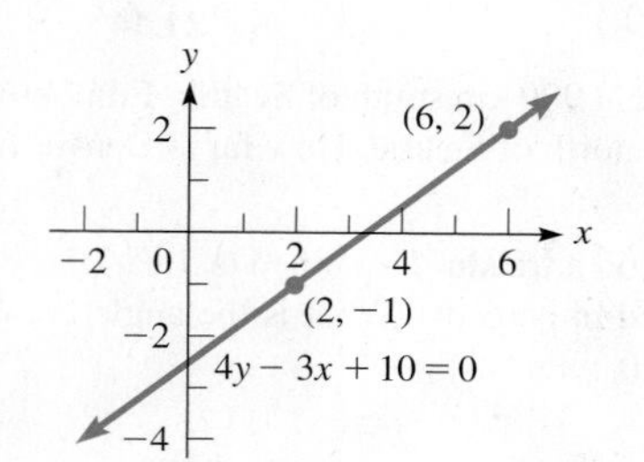

Fig. 21.17

$$y - (-1) = \frac{3}{4}(x - 2)$$
$$4y + 4 = 3x - 6$$
$$-3x + 4y + 10 = 0$$

■

Equation (21.6) can be used for any line except for one parallel to the y-axis. Such a line has an undefined slope. However, it does have the property that all points on it have the same x-coordinate, regardless of the y-coordinate. *We represent a line parallel to the y-axis as*

$$x = a \qquad \text{see Fig. 21.18} \qquad \textbf{(21.7)}$$

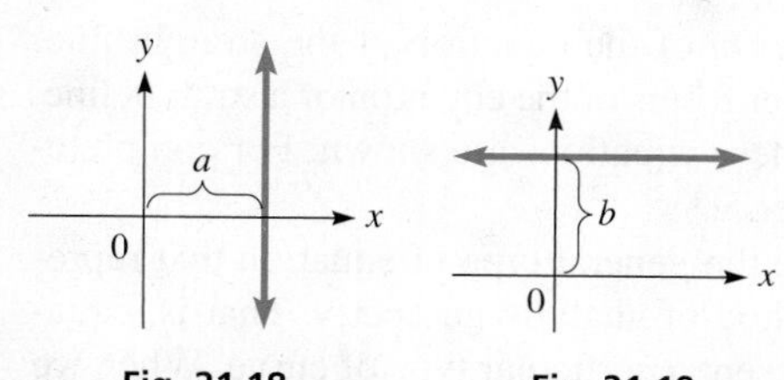

Fig. 21.18 Fig. 21.19

A line parallel to the x-axis has a slope of zero. From Eq. (21.6), we find its equation is $y = y_1$. To keep the same form as Eq. (21.7), we write this as

$$y = b \qquad \text{see Fig. 21.19} \qquad \textbf{(21.8)}$$

EXAMPLE 3 Vertical line—horizontal line

(a) The line $x = 2$ is a line parallel to the y-axis and 2 units to the right of it. This line is shown in Fig. 21.20.

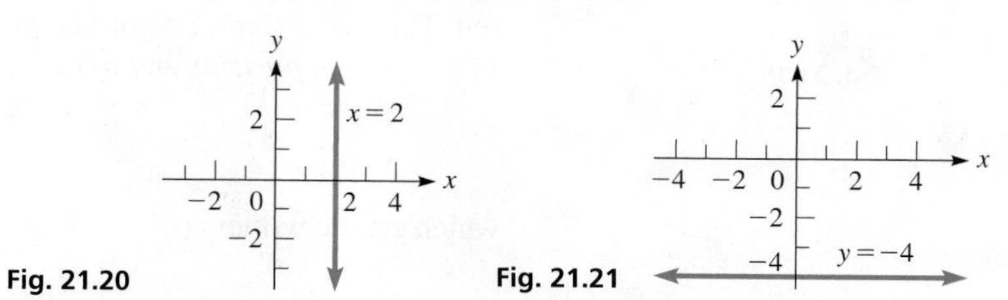

Fig. 21.20 Fig. 21.21

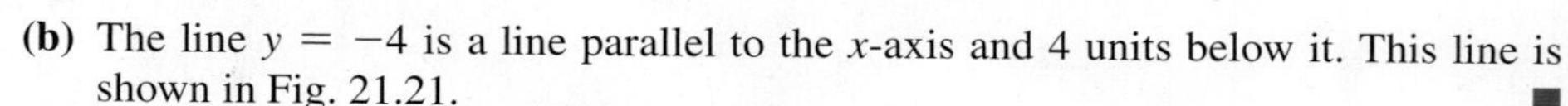

(b) The line $y = -4$ is a line parallel to the x-axis and 4 units below it. This line is shown in Fig. 21.21.

■

Practice Exercise

1. Find the equation of the straight line that passes through $(-5, 2)$ and $(-1, -6)$.

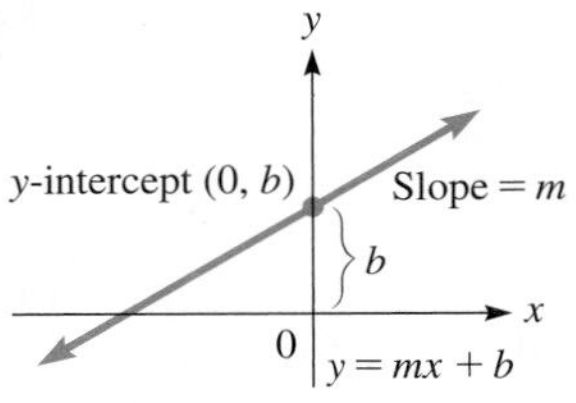

If we choose the special point $(0, b)$, which is the y-intercept of the line, as the point to use in Eq. (21.6), we have $y - b = m(x - 0)$, or

Fig. 21.22

$$y = mx + b \tag{21.9}$$

Equation (21.9) is the **slope-intercept form** *of the equation of a straight line,* and we first derived it in Chapter 5. Its primary usefulness lies in the fact that once we find the equation of a line and then write it in slope-intercept form, we know that the slope of the line is the coefficient of the x-term and that it crosses the y-axis at the coordinate indicated by the constant term. See Fig. 21.22.

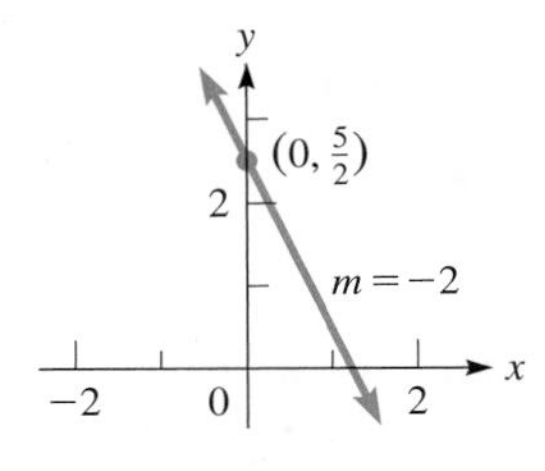

Fig. 21.23

Practice Exercise

2. What are the slope and y-intercept of the line $4x - 3y = 6$?

EXAMPLE 4 Slope-intercept form

Find the slope and the y-intercept of the straight line for which the equation is $2y + 4x - 5 = 0$.

We write this equation in slope-intercept form:

$$2y = -4x + 5$$

$$y = -2x + \frac{5}{2}$$

(slope: -2; y-coordinate of intercept: $\frac{5}{2}$)

Because the coefficient of x in this form is -2, the slope is -2. The constant on the right is $5/2$, which means that the y-intercept is $(0, 5/2)$. See Fig. 21.23. ■

An equation of a line can also be written in the form

■ In Eq. (21.10), A and B cannot both be zero.

$$Ax + By + C = 0 \tag{21.10}$$

which is known as the **general form** *of the equation of the straight line.*

EXAMPLE 5 General form

Find the general form of the equation of the line parallel to the line $3x + 2y - 6 = 0$ and that passes through the point $(-1, 2)$.

Because the line whose equation we want is parallel to the line $3x + 2y - 6 = 0$, it has the same slope. Thus, writing $3x + 2y - 6 = 0$ in slope-intercept form,

$$2y = -3x + 6 \qquad \text{solving for } y$$

$$y = -\frac{3}{2}x + 3$$

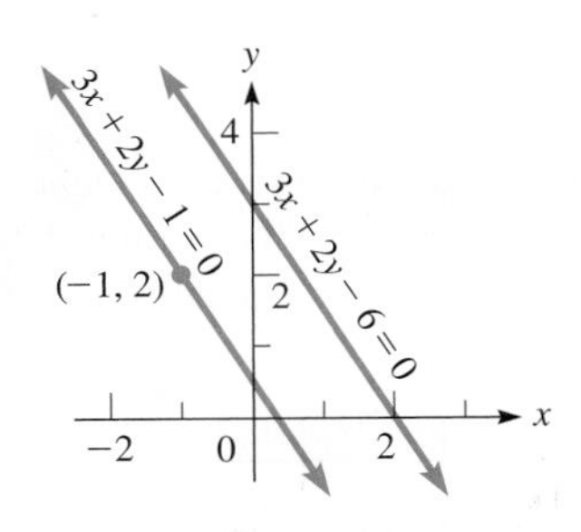

Fig. 21.24

Because the slope of $3x + 2y - 6 = 0$ is $-3/2$, the slope of the required line is also $-3/2$. Using $m = -3/2$, the point $(-1, 2)$, and the point-slope form, we have

$$y - 2 = -\frac{3}{2}(x + 1)$$

$$2y - 4 = -3(x + 1)$$

$$3x + 2y - 1 = 0$$

This is the general form of the equation. Both lines are shown in Fig. 21.24. ■

Practice Exercise

3. Write the general form of the equation of the straight line that passes through $(0, -1)$ with a slope of 3.

In many physical situations, a linear relationship exists between variables as shown in the following two examples.

EXAMPLE 6 Straight line—slope as acceleration

For a period of 6.0 s, the velocity v of a car varies linearly with the elapsed time t. If $v = 40$ ft/s when $t = 1.0$ s and $v = 55$ ft/s when $t = 4.0$ s, find the equation relating v and t and graph the function. From the graph, find the initial velocity and the velocity after 6.0 s. What is the meaning of the slope of the line?

With v as the dependent variable and t as the independent variable, the slope is

$$m = \frac{v_2 - v_1}{t_2 - t_1}$$

Using the information given in the statement of the problem, we have

$$m = \frac{55 - 40}{4.0 - 1.0} = 5.0$$

Then using the point-slope form of the equation of a straight line, we have

$$v - 40 = 5.0(t - 1.0)$$
$$v = 5.0t + 35$$

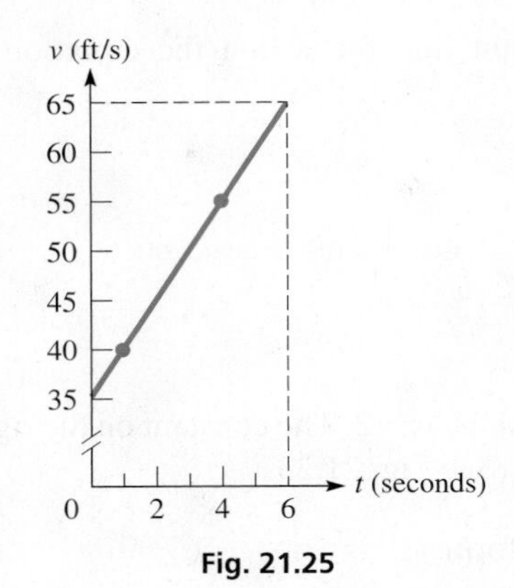

Fig. 21.25

The given values are sufficient to graph the line in Fig. 21.25. There is no need to include negative values of t, since they have no physical meaning. We see that the line crosses the v-axis at 35. This means that the initial velocity (for $t = 0$) is 35 ft/s. Also, when $t = 6.0$ s, we see that $v = 65$ ft/s.

The slope is the ratio of the change in velocity to the change in time. This is the car's *acceleration*. Here, the speed of the car increases 5.0 ft/s each second. We can express this acceleration as 5.0 (ft/s)/s, or 5.0 ft/s^2. ■

In Chapter 5, we showed how a calculator can be used to find a **linear regression line** to fit a set of data points that are approximately linear (see Example 11 of Section 5.1). The following example illustrates this process once again. The mathematics behind linear regression will be further developed in Chapter 22.

EXAMPLE 7 Linear regression—weight of a bear

The following table shows the chest girth (the distance around the chest) g (in inches) and the weight w (in lb) for a group of wild bears. Find the linear regression line for this data and use it to estimate the weight of a bear that has a chest girth of (a) 45.0 in. and (b) 60.0 in.

Chest girth, g (in.)	23.0	28.5	34.5	36.0	41.0	49.0	54.5
Weight, w (lb)	40.0	128	148	190	220	398	476

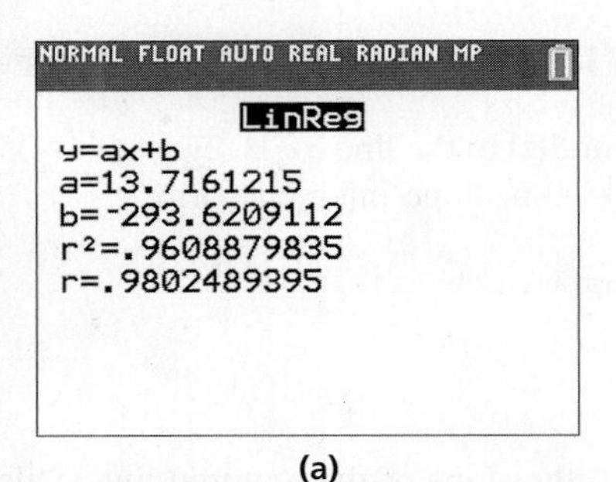

(a)

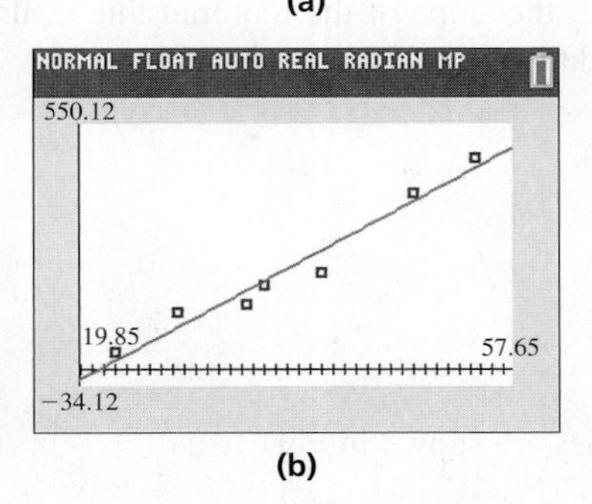

(b)

Fig. 21.26
Graphing calculator keystrokes: goo.gl/9lhmWj

We first enter the chest girth and weight values into the calculator lists L_1 and L_2, respectively, and make a *scatterplot* of the data. We then use the **LinReg($ax+b$)** feature to get the equation of the regression line [see Fig. 21.26(a)]. Thus, the regression line is given by

$$w = 13.716g - 293.621$$

The slope is 13.716 and the w-intercept is $(0, -293.621)$. Figure 21.26(b) shows the regression line plotted through the scatterplot.

To estimate the weight of a bear with chest girth 45.0 in., we evaluate $13.716(45.0) - 293.621 \approx 324$ lb. This is called **interpolation** because the chest girth of 45.0 in. is *inside* the range of the given data. To estimate the weight of a bear with a chest girth of 60.0 in., we get $13.716(60.0) - 293.621 \approx 529$ lb. This is called **extrapolation** because the chest girth 60.0 in. is *outside* the range of the given data. ■

EXERCISES 21.2

In Exercises 1–4, make the given changes in the indicated examples of this section and then solve the resulting problems.

1. In Example 1, change $(-4, 1)$ to $(4, -1)$.
2. In Example 2, change $(2, -1)$ to $(-2, 1)$.
3. In Example 4, change the $+$ before $4x$ to $-$.
4. In Example 5, change the $+$ before $2y$ to $-$.

In Exercises 5–20, find the equation of each of the lines with the given properties. Sketch the graph of each line.

5. Passes through $(-3, 8)$ with a slope of 4.
6. Passes through $(-2, -1)$ with a slope of -2.
7. Passes through $(2, -5)$ and $(4, 2)$.
8. Has an x-intercept $(4, 0)$ and a y-intercept of $(0, -6)$.
9. Passes through $(-7, 12)$ with an inclination of 45°.
10. Has a y-intercept $(0, -2)$ and an inclination of 120°.
11. Passes through $(5.3, -2.7)$ and is parallel to the x-axis.
12. Passes through $(-15, 9)$ and is perpendicular to the x-axis.
13. Is parallel to the y-axis and is 3 units to the left of it.
14. Is parallel to the x-axis and is 4.1 units below it.
15. Perpendicular to line with slope of -3; passes through $(1, -2)$.
16. Parallel to line through $(-1, 7)$ and $(3, 1)$; passes through $(-3, -4)$.
17. Has equal intercepts and passes through $(5, 2)$.
18. Is perpendicular to the line $6.0x - 2.4y - 3.9 = 0$ and passes through $(7.5, -4.7)$.
19. Has a slope of -3 and passes through the intersection of the lines $5x - y = 6$ and $x + y = 12$.
20. Passes through the point of intersection of $2x + y - 3 = 0$ and $x - y - 3 = 0$ and through the point $(4, -3)$.

In Exercises 21–28, reduce the equations to slope-intercept form and find the slope and the y-intercept. Sketch each line.

21. $4x - y = 8$
22. $2x - 3y - 6 = 0$
23. $3x + 5y - 10 = 0$
24. $4y = 6x - 9$
25. $3x - 2y - 1 = 0$
26. $2y + 4x - 5 = 0$
27. $11.2x + 1.6 = 3.2y$
28. $11.5x + 4.60y = 5.98$

In Exercises 29–36, determine whether the given lines are parallel, perpendicular, or neither.

29. $3x - 2y + 5 = 0$ and $4y = 6x - 1$
30. $8x - 4y + 1 = 0$ and $4x + 2y - 3 = 0$
31. $6x - 3y - 2 = 0$ and $2y + x - 4 = 0$
32. $3y - 2x = 4$ and $6x - 9y = 5$
33. $5x + 2y - 3 = 0$ and $10y = 7 - 4x$
34. $48y - 36x = 71$ and $52x = 17 - 39y$
35. $4.5x - 1.8y = 1.7$ and $2.4x + 6.0y = 0.3$
36. $3.5y = 4.3 - 1.5x$ and $3.6x + 8.4y = 1.7$

In Exercises 37–62, solve the given problems.

37. Find k if the lines $4x - ky = 6$ and $6x + 3y + 2 = 0$ are parallel.
38. Find k if the lines given in Exercise 37 are perpendicular.
39. Find k if the lines $3x - y = 9$ and $kx + 3y = 5$ are perpendicular. Explain how this value is found.
40. Find k such that the line through $(k, 2)$ and $(3, 1 - k)$ is perpendicular to the line $x - 2y = 5$. Explain your method.
41. Find the shortest distance from $(4, 1)$ to the line $4x - 3y + 12 = 0$.
42. Find the acute angle between the lines $x + y = 3$ and $2x - 5y = 4$.
43. Show that the following lines intersect to form a parallelogram. $8x + 10y = 3$; $2x - 3y = 5$; $4x - 6y = -3$; $5y + 4x = 1$.
44. For nonzero values of a, b, and c, find the intercepts of the line $ax + by + c = 0$.
45. For nonzero values of a, b, c, and d, show that (a) lines $ax + by + c = 0$ and $ax + by + d = 0$ are parallel, and (b) lines $ax + by + c = 0$ and $bx - ay + d = 0$ are perpendicular.
46. Find the equation of the line with positive intercepts that passes through $(3, 2)$ and forms with the axes a triangle of area 12.
47. The equation $4x - 2y = k$ defines a *family of lines*, one for each value of k. On a calculator display the lines for $k = -4$, $k = 0$, and $k = 4$. What conclusion do you draw about this family of lines?
48. The equation $kx - 2y = 4$ defines a family of lines, one for each value of k. On a calculator display the lines for $k = -4$, $k = 0$, and $k = 4$. What conclusion do you draw about this family of lines?
49. Show that the determinant equation at the right defines a straight line. $$\begin{vmatrix} y & x & b \\ m & 1 & 0 \\ 1 & 0 & 1 \end{vmatrix} = 0$$
50. Find the equation of the line that has one negative intercept, passes through $(3, 2)$, and forms with an axis and the line $y = 2$ a triangle with an area of 12.
51. In the 1700s, the French physicist Reaumur established a temperature scale on which the freezing point of water was 0° and the boiling point was 80°. Set up an equation for the Fahrenheit temperature F (freezing point 32°, boiling point 212°) as a function of the Reaumur temperature R.
52. The voltage V across part of an electric circuit is given by $V = E - iR$, where E is a battery voltage, i is the current, and R is the resistance. If $E = 6.00$ V and $V = 4.35$ V for $i = 9.17$ mA, find V as a function of i. Sketch the graph (i and V may be negative).
53. The velocity of sound v increases 0.607 m/s for each increase in temperature T of 1.00°C. If $v = 343$ m/s for $T = 20.0$°C, express v as a function of T.
54. An acid solution is made from x liters of a 20% solution and y liters of a 30% solution. If the final solution contains 20 L of acid, find the equation relating x and y.
55. A wall is 15 cm thick. At the outside, the temperature is 3°C, and at the inside, it is 23°C. If the temperature changes at a constant rate through the wall, write an equation of the temperature T in the wall as a function of the distance x from the outside to a point inside the wall. What is the meaning of the slope of the line?

56. An oil-storage tank is emptied at a constant rate. At 10 A.M., 1800 barrels remain, and at 2 P.M., 600 barrels remain. If pumping started at 8 A.M., find the equation relating the number of barrels n at time t (in h) from 8 A.M. When will the tank be empty?

57. In a research project on cancer, a tumor was determined to weigh 30 mg when first discovered. While being treated, it grew smaller by 2 mg each month. Find the equation relating the weight w of the tumor as a function of the time t in months. Graph the equation.

58. The length of a rectangular solar cell is 10 cm more than the width w. Express the perimeter p of the cell as a function of w. What is the meaning of the slope of the line?

59. A light beam is reflected off the edge of an optic fiber at an angle of 0.0032°. The diameter of the fiber is 48 μm. Find the equation of the reflected beam with the x-axis (at the center of the fiber) and the y-axis as shown in Fig. 21.27.

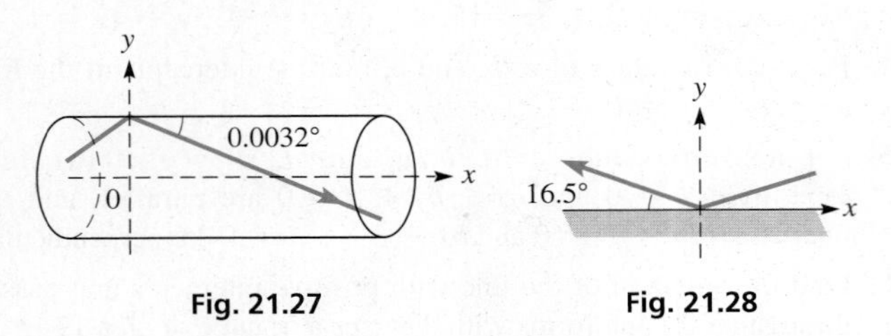

Fig. 21.27

Fig. 21.28

60. A police report stated that a bullet caromed upward off a floor at an angle of 16.5° with the floor, as shown in Fig. 21.28. What is the equation of the bullet's path after impact?

61. A survey of the traffic on a particular highway showed that the number of cars passing a particular point each minute varied linearly from 6:30 A.M. to 8:30 A.M. on workday mornings. The study showed that an average of 45 cars passed the point in 1 min at 7 A.M. and that 115 cars passed in 1 min at 8 A.M. If n is the number of cars passing the point in 1 min, and t is the number of minutes after 6:30 A.M., find the equation relating n and t, and graph the equation. From the graph, determine n at 6:30 A.M. and at 8:30 A.M. What is the meaning of the slope of the line?

62. After taking off, a plane gains altitude at 600 m/min for 5.0 min and then continues to gain altitude at 300 m/min for 15 min. It then continues at a constant altitude. Find the altitude h as a function of time t for the first 20 min, and sketch the graph of $h = f(t)$.

In Exercises 63 and 64, find the equation of the regression line for the given data. Then use this equation to make the indicated estimate. Round decimals in the regression equation to three decimal places. Round estimates to the same accuracy as the given data.

63. Measurements were taken to record the underwater pressure (in kPa) at various depths (in m). The resulting data is shown in the table. Find the equation of the regression line and then estimate the pressure at a depth of 35 ft. Is this interpolation or extrapolation?

Depth, d (m)	0	10	20	30	40	50
Pressure, p (kPa)	101	201	292	396	500	586

64. The shear strength of the bond between two propellants is important in rocket engines. The following table shows the age of the propellant t (in days) and the shear strength s (in psi). Find the equation of the regression line and then estimate the shear strength for propellant that is 7 days old. Is this interpolation or extrapolation?

Age of propellant, t (days)	14	56	88	133	150	167
Shear strength, s (psi)	2654	2316	2200	1708	1754	1678

In Exercises 65–68, treat the given nonlinear functions as linear functions in order to sketch their graphs. At times, this can be useful in showing certain values of a function. For example, $y = 2 + 3x^2$ can be shown as a straight line by graphing y as a function of x^2. A table of values for this graph is shown along with the corresponding graph in Fig. 21.29.

x	0	1	2	3	4	5
x^2	0	1	4	9	16	25
y	2	5	14	29	50	77

Fig. 21.29

65. The number n of memory cells of a certain computer that can be tested in t seconds is given by $n = 1200\sqrt{t}$. Sketch n as a function of $\sqrt{t}$.

66. The force F (in lb) applied to a lever to balance a certain weight on the opposite side of the fulcrum is given by $F = 40/d$, where d is the distance (in ft) of the force from the fulcrum. Sketch F as a function of $1/d$.

67. A spacecraft is launched such that its altitude h (in km) is given by $h = 300 + 2t^{3/2}$ for $0 \leq t < 100$ s. Sketch this as a linear function.

68. The current i (in A) in a certain electric circuit is given by $i = 6(1 - e^{-t})$. Sketch this as a linear function.

Answers to Practice Exercises

1. $2x + y + 8 = 0$ **2.** $m = 4/3, b = -2$ **3.** $3x - y - 1 = 0$

21.3 The Circle

Standard Equation • Symmetry • General Equation • Calculator Graph

We have found that we can obtain a general equation that represents a straight line by considering a fixed point on the line and then a general point $P(x, y)$ that can represent any other point on the same line. Mathematically, we can state this as "the line is the **locus** of a point $P(x, y)$ that *moves* from a fixed point with constant slope along the line." That is, the point $P(x, y)$ can be considered as a variable point that moves along the line.

In this way, we can define a number of important curves. *A* **circle** *is defined as the locus of a point $P(x, y)$ that moves so that it is always equidistant from a fixed point. We call this fixed distance the* **radius,** *and we call the fixed point the* **center** *of the circle.* Thus, using this definition, calling the fixed point (h, k) and the radius r, we have

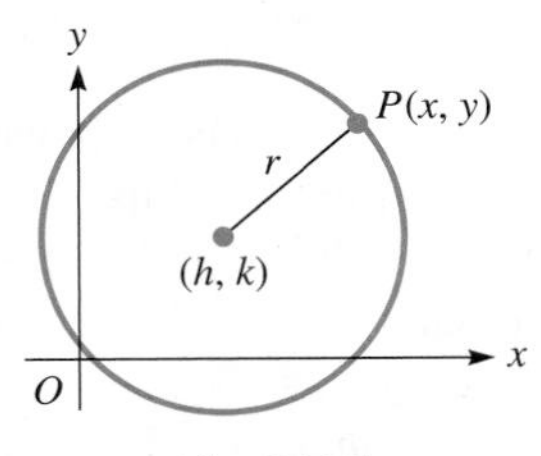

Fig. 21.30

$$\sqrt{(x-h)^2 + (y-k)^2} = r$$

or, by squaring both sides, we have

$$(x-h)^2 + (y-k)^2 = r^2 \tag{21.11}$$

Equation (21.11) is called the **standard equation of a circle** *with center at (h, k) and radius r.* See Fig. 21.30.

EXAMPLE 1 Standard equation

The equation $(x-1)^2 + (y+2)^2 = 16$ represents a circle with center at $(1, -2)$ and a radius of 4. We determine these values by considering the equation of the circle to be in the form of Eq. (21.11) as

$$(x - 1)^2 + [y - (-2)]^2 = 4^2$$

form requires − signs; coordinates of center; radius

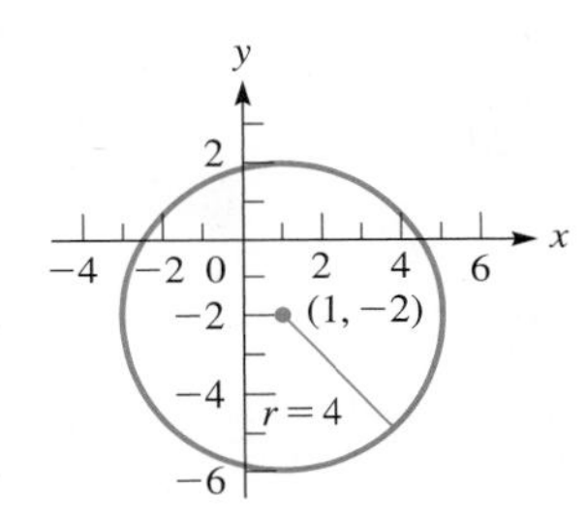

Fig. 21.31

Note carefully the way in which we found the y-coordinate of the center. *We must have a minus sign before each of the coordinates.* Here, to get the y-coordinate, we had to write $+2$ as $-(-2)$. This circle is shown in Fig. 21.31. ■

EXAMPLE 2 Find equation of circle

Find the equation of the circle with center at $(2, 1)$ and that passes through $(4, -6)$.

In Eq. (21.11), we can determine the equation of this circle if we can find h, k, and r. From the given information, the center is $(2, 1)$, which means $h = 2$ and $k = 1$. To find r, we use the fact that *all points on the circle must satisfy the equation of the circle.* The point $(4, -6)$ must satisfy Eq. (21.11), with $h = 2$ and $k = 1$. This means

$$(4-2)^2 + (-6-1)^2 = r^2 \quad \text{or} \quad r^2 = 53$$

Therefore, the equation of the circle is

$$(x-2)^2 + (y-1)^2 = 53$$

This circle is shown in Fig. 21.32. ■

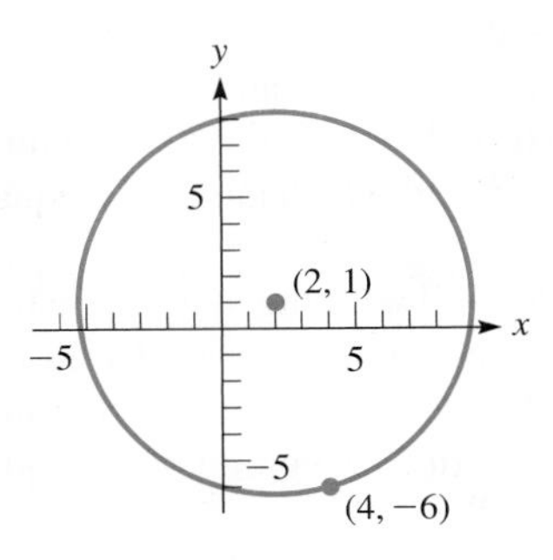

Fig. 21.32

Practice Exercise

1. Find the center and radius of the circle $(x+7)^2 + (y-2)^2 = 1$

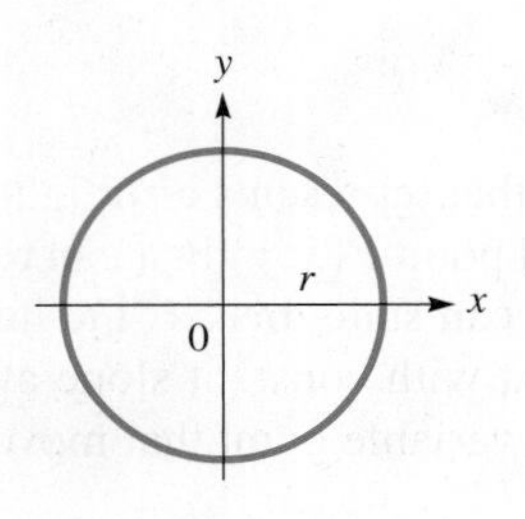

Fig. 21.33

If the center of the circle is at the origin, which means that the coordinates of the center are $(0, 0)$, the equation of the circle (see Fig. 21.33) becomes

$$x^2 + y^2 = r^2 \tag{21.12}$$

The following example illustrates an application using this type of circle and one with its center not at the origin.

EXAMPLE 3 Circle—friction drive

A student is drawing a friction drive in which two circular disks are in contact with each other. They are represented by circles in the drawing. The first has a radius of 10.0 cm, and the second has a radius of 12.0 cm. What is the equation of each circle if the origin is at the center of the first circle and the positive x-axis passes through the center of the second circle? See Fig. 21.34.

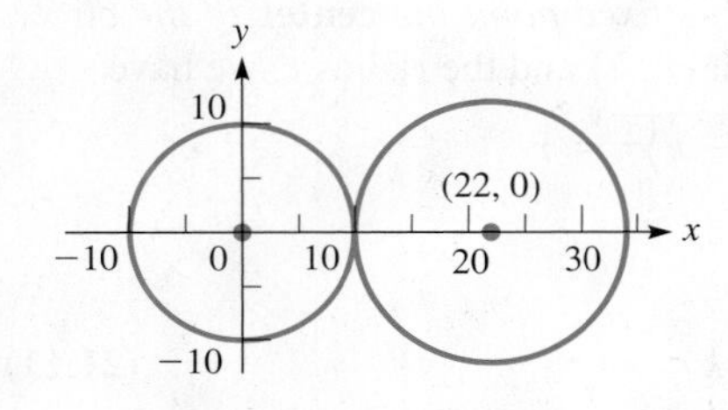

Fig. 21.34

Because the center of the smaller circle is at the origin, we can use Eq. (21.12). Given that the radius is 10.0 cm, we have as its equation

$$x^2 + y^2 = 100$$

The fact that the two disks are in contact tells us that they meet at the point $(10.0, 0)$. Knowing that the radius of the larger circle is 12.0 cm tells us that its center is at $(22.0, 0)$. Thus, using Eq. (21.11) with $h = 22.0$, $k = 0$, and $r = 12.0$,

$$(x - 22.0)^2 + (y - 0)^2 = 12.0^2$$

or

$$(x - 22.0)^2 + y^2 = 144$$

as the equation of the larger circle. ■

SYMMETRY

A circle with its center at the origin exhibits an important property of the graphs of many equations. *It is* **symmetric** *to the x-axis and also to the y-axis.* Symmetry to the x-axis can be thought of as meaning that the lower half of the curve is a reflection of the upper half, and conversely. NOTE ▶ [It can be shown that if $-y$ can replace y in an equation without changing the equation, the graph of the equation is **symmetric to the x-axis.** Symmetry to the y-axis is similar. If $-x$ can replace x in the equation without changing the equation, the graph is **symmetric to the y-axis.**]

This type of circle is also symmetric to the origin as well as being symmetric to both axes. The meaning of symmetry to the origin is that the origin is the midpoint of any two points (x, y) and $(-x, -y)$ that are on the curve. NOTE ▶ [Thus, if $-x$ can replace x and $-y$ can replace y at the same time, without changing the equation, the graph of the equation is **symmetric to the origin.**]

EXAMPLE 4 Symmetry

The equation of the circle with its center at the origin and with a radius of 6 is $x^2 + y^2 = 36$.

The symmetry of this circle can be shown analytically by the substitutions mentioned above. Replacing x by $-x$, we obtain $(-x)^2 + y^2 = 36$. Because $(-x)^2 = x^2$, this equation can be rewritten as $x^2 + y^2 = 36$. Because this substitution did not change the equation, the graph is symmetric to the y-axis.

Replacing y by $-y$, we obtain $x^2 + (-y)^2 = 36$, which is the same as $x^2 + y^2 = 36$. This means that the curve is symmetric to the x-axis.

Replacing x by $-x$ and simultaneously replacing y by $-y$, we obtain $(-x)^2 + (-y)^2 = 36$, which is the same as $x^2 + y^2 = 36$. This means that the curve is symmetric to the origin. This circle is shown in Fig. 21.35. ■

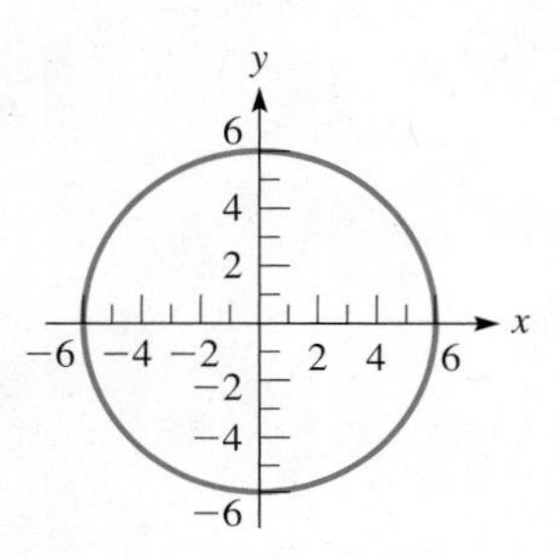

Fig. 21.35

If we multiply out each of the terms in Eq. (21.11), we may combine the resulting terms to obtain

$$x^2 - 2hx + h^2 + y^2 - 2ky + k^2 = r^2$$
$$x^2 + y^2 - 2hx - 2ky + (h^2 + k^2 - r^2) = 0 \quad \textbf{(21.13)}$$

Because each of h, k, and r is constant for any given circle, the coefficients of x and y and the term within parentheses in Eq. (21.13) are constants. Equation (21.13) can then be written as

$$x^2 + y^2 + Dx + Ey + F = 0 \quad \textbf{(21.14)}$$

Equation (21.14) is called the **general equation** *of the circle.* It tells us that any equation that can be written in that form will represent a circle (except in special cases where it may represent a single point or the empty set).

EXAMPLE 5 General equation

Find the center and radius of the circle.

$$x^2 + y^2 - 6x + 8y - 24 = 0$$

NOTE ▶ We can find this information if we write the given equation in standard form. [To do so, we must complete the square in the x-terms and also in the y-terms.] This is done by first writing the equation in the form

$$(x^2 - 6x \quad) + (y^2 + 8y \quad) = 24$$

To complete the square of the x-terms, we take half of -6, which is -3, square it, and add the result, 9, to each side of the equation. In the same way, we complete the square of the y-terms by adding 16 to each side of the equation, which gives

$$\left(\frac{-6}{2}\right)^2 \qquad \left(\frac{8}{2}\right)^2 \qquad \text{add to both sides}$$
$$(x^2 - 6x + 9) + (y^2 + 8y + 16) = 24 + 9 + 16$$
$$(x - 3)^2 + (y + 4)^2 = 49$$
$$(x - 3)^2 + [y - (-4)]^2 = 7^2$$

coordinates of center; radius

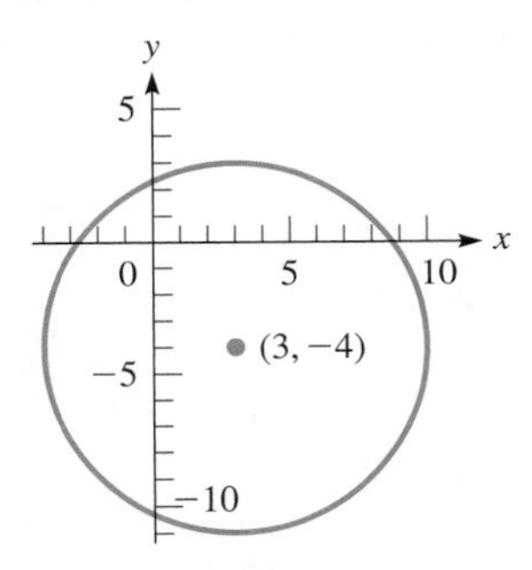

Fig. 21.36

Practice Exercise

2. Find the center and radius of the circle $x^2 + y^2 - 8x + 6y + 21 = 0$

Thus, the center is $(3, -4)$, and the radius is 7 (see Fig. 21.36). ■

EXAMPLE 6 Circle—pendulum motion

A certain pendulum is found to swing through an arc of the circle $3x^2 + 3y^2 - 9.60y - 2.80 = 0$. What is the length (in m) of the pendulum, and from what point is it swinging?

We see that this equation represents a circle by dividing through by 3. This gives us $x^2 + y^2 - 3.20y - 2.80/3 = 0$. The length of the pendulum is the radius of the circle, and the point from which it swings is the center. These are found as follows:

$$x^2 + (y^2 - 3.20y + 1.60^2) = 1.60^2 + 2.80/3 \quad \text{complete squares in both } x\text{- and } y\text{-terms}$$
$$x^2 + (y - 1.60)^2 = 3.493 \quad \text{standard form}$$

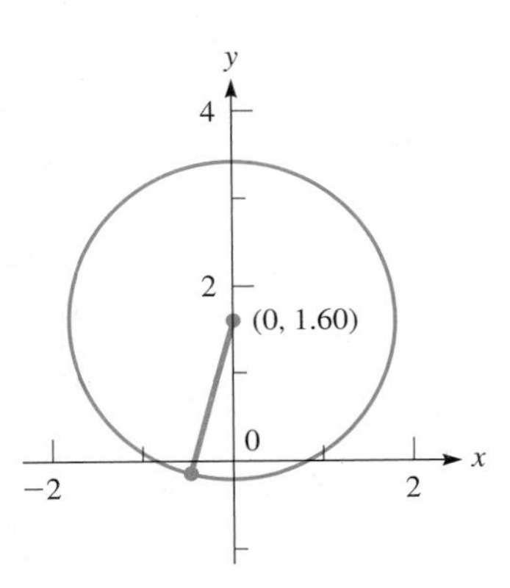

Fig. 21.37

Because $\sqrt{3.493} = 1.87$, the length of the pendulum is 1.87 m. The point from which it is swinging is (0, 1.60). See Fig. 21.37.

Replacing x by $-x$, the equation does not change. Replacing y by $-y$, the equation does change (the 3.20 y term changes sign). Thus, the circle is symmetric only to the y-axis. ■

In Section 14.1, we noted that the equation of a circle does not represent a *function* because there are two values of y for most values of x in the domain. [In fact, it might be necessary to use the quadratic formula to find the two functions to enter into a calculator in order to view the curve.] This is illustrated in the following example.

NOTE ▸

EXAMPLE 7 Calculator graph of circle

Display the graph of the circle $3x^2 + 3y^2 + 6y - 20 = 0$ on a calculator.

To fit the form of a quadratic equation in y, we write

$$3y^2 + 6y + (3x^2 - 20) = 0$$

Now, using the quadratic formula to solve for y, we let

$$a = 3 \quad b = 6 \quad c = 3x^2 - 20$$

$$y = \frac{-6 \pm \sqrt{6^2 - 4(3)(3x^2 - 20)}}{2(3)}$$

which means we get the *two functions*

$$y_1 = \frac{-6 + \sqrt{276 - 36x^2}}{6} \quad \text{and} \quad y_2 = \frac{-6 - \sqrt{276 - 36x^2}}{6}$$

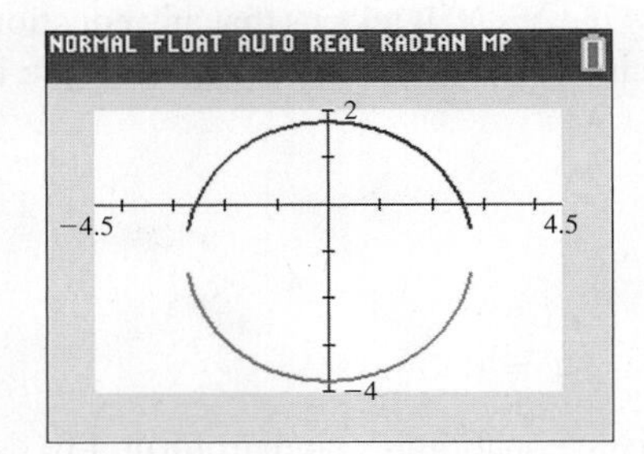

Fig. 21.38

which are entered into the calculator to get the view shown in Fig. 21.38.

The *window* values were chosen so that the length along the x-axis is about 1.5 times that along the y-axis so as to have less distortion in the circle. The gaps shown on the left and right sides of the circle should not be there, but graphing utilities sometimes have difficulty graphing parts of curves that are nearly vertical. ■

Practice Exercise

3. Is the circle in Example 7 symmetric to the x-axis? the y-axis? the origin?

EXERCISES 21.3

In Exercises 1–4, make the given changes in the indicated examples of this section and then solve the resulting problems.

1. In Example 1, change $(y + 2)^2$ to $(y + 1)^2$.

2. In Example 2, change $(2, 1)$ to $(-2, 1)$.

3. In Example 5, change the + before $8y$ to −.

4. In Example 7, change the + before $6y$ to −.

In Exercises 5–8, determine the center and the radius of each circle.

5. $(x - 2)^2 + (y - 1)^2 = 25$ **6.** $(x - 3)^2 + (y + 4)^2 = 49$

7. $4(x + 1)^2 + 4y^2 = 121$ **8.** $9x^2 + 9(y - 6)^2 = 64$

In Exercises 9–24, find the equation of each of the circles from the given information.

9. Center at $(0, 0)$, radius 3 **10.** Center at $(0, 0)$, radius 12

11. Center at $(2, 3)$, radius 4 **12.** Center at $(\frac{3}{2}, -2)$, radius $\frac{5}{2}$

13. Center at $(12, -15)$, radius 18

14. Center at $(-3, -5)$, radius $2\sqrt{3}$

15. The origin and $(-6, 8)$ are the ends of a diameter

16. The points $(3, 8)$ and $(-3, 0)$ are the ends of a diameter.

17. Concentric with the circle $(x - 2)^2 + (y - 1)^2 = 4$ and passes through $(4, -1)$

18. Concentric with the circle $x^2 + y^2 + 2x - 8y + 8 = 0$ and passes through $(-2, 3)$

19. Center at $(-3, 5)$ and tangent to line $y = 10$

20. Center at $(-7, 1)$, tangent to y-axis

21. Tangent to both axes and the lines $y = 4$ and $x = -4$

22. Tangent to lines $y = 2$ and $y = 8$, center on line $y = x$

23. Center at the origin, tangent to the line $x + y = 2$

24. Center at $(5, 12)$, tangent to the line $y = 2x - 3$

In Exercises 25–36, determine the center and radius of each circle. Sketch each circle.

25. $x^2 + (y - 3)^2 = 4$ **26.** $(x - 2)^2 + (y + 3)^2 = 49$

27. $4(x + 1)^2 + 4(y - 5)^2 = 81$

28. $4(x + 7)^2 + 4(y + 11)^2 = 169$

29. $2x^2 + 2y^2 - 16 = 4x$

30. $y^2 + x^2 - 4x = 6y + 12$

31. $x^2 + y^2 + 4.20x - 2.60y = 3.51$

32. $2x^2 + 2y^2 + 44x + 28y = 52$

33. $4x^2 + 4y^2 - 9 = 16y$ **34.** $9x^2 + 9y^2 + 18y = 7$

35. $2x^2 + 2y^2 = 4x + 8y + 1$ **36.** $9x^2 + 9y^2 = 36x - 12$

In Exercises 37–40, determine whether the circles with the given equations are symmetric to either axis or the origin.

37. $x^2 + y^2 = 100$ **38.** $x^2 + y^2 - 4x - 5 = 0$

39. $3x^2 + 3y^2 + 24y = 8$ **40.** $5x^2 + 5y^2 - 10x + 20y = 3$

In Exercises 41–68, solve the given problems.

41. A square is inscribed in the circle $x^2 + y^2 = 32$ (all four vertices are on the circle). Find the area of the square.

42. For the point $(-2, 5)$, find the point that is symmetric to it with respect to (a) the x-axis, (b) the y-axis, (c) the origin.

43. Find the intercepts of the circle $x^2 + y^2 - 6x + 5 = 0$.

44. Find the intercepts of the circle $x^2 + y^2 + 2x - 2y = 0$.

45. Find the area bounded between the two circles given by $x^2 + y^2 = 9$ and $x^2 + y^2 = 25$.

46. If the equation of the small right circle in Fig. 21.39 is $x^2 + y^2 = a^2$, what is the equation of the large circle? (Circles are tangent as shown with centers on the x-axis; the small circles are congruent.)

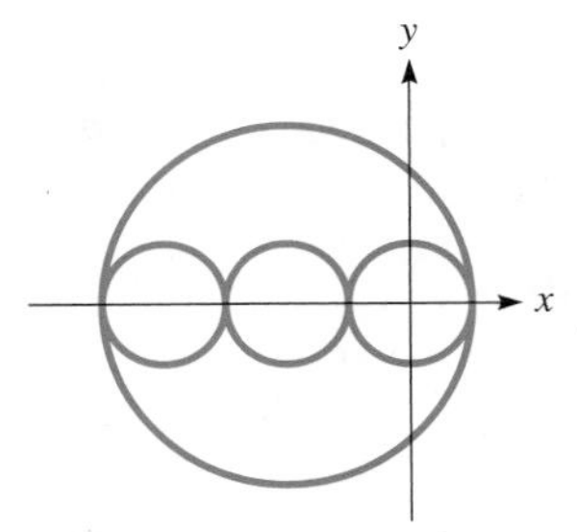

Fig. 21.39

47. Determine whether the circle $x^2 - 6x + y^2 - 7 = 0$ crosses the x-axis.

48. Find the points of intersection of the circle $x^2 + y^2 - x - 3y = 0$ and the line $y = x - 1$.

49. When graphed on a calculator, a circle sometimes looks like it is longer in one direction than it is in the other. Explain why this can happen.

50. When a circle is graphed on a calculator, there are sometimes gaps in the graph on the left and right sides of the circle. Why does this happen? Does the graph really have gaps?

51. Use a calculator to view the circle $x^2 + y^2 + 5y - 4 = 0$.

52. Use a calculator to view the circle $2x^2 + 2y^2 + 2y - x - 1 = 0$.

53. What type of graph is represented by the equations
(a) $y = \sqrt{9 - (x - 2)^2}$? (b) $y = -\sqrt{9 - (x - 2)^2}$?
(c) Are the equations in parts (a) and (b) functions? Explain.

54. What type of graph is represented by the equations
(a) $x^2 + (y - 1)^2 = 0$? (b) $x^2 + (y - 1)^2 = -1$?

55. Is the point $(0.1, 3.1)$ inside, outside, or on the circle $x^2 + y^2 - 2x - 4y + 3 = 0$?

56. Find the equation of the line along which the diameter of the circle $x^2 + y^2 - 2x - 4y - 4 = 0$ lies, if the diameter is parallel to the line $3x + 5y = 4$.

57. For the equation $(x - h)^2 + (y - k)^2 = p$, how does the value of p indicate whether the graph is a circle, a point, or does not exist? Explain.

58. Determine whether the graph of $x^2 + y^2 - 2x + 10y + 29 = 0$ is a circle, a point, or does not exist.

59. The inner and outer circles of the cross section of a pipe are represented by the equations $2.00x^2 + 2.00y^2 = 5.73$ and $2.80x^2 + 2.80y^2 = 8.91$. How thick (in in.) is the pipe wall?

60. A 12-ft pole leaning against a wall slips to the ground. Find the equation that represents the path of the midpoint of the pole. See Fig. 21.40.

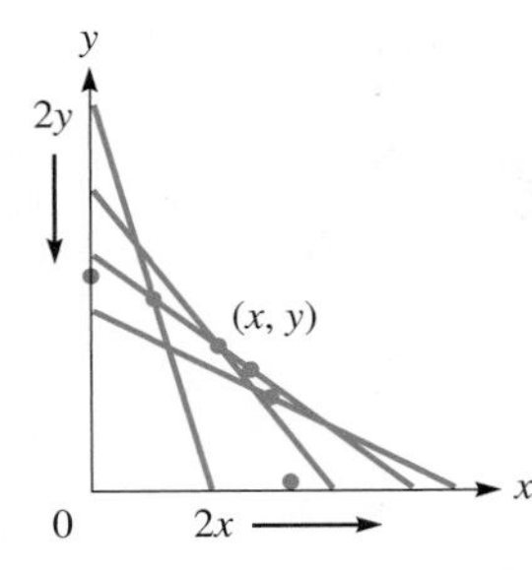

Fig. 21.40

61. In a hoisting device, two of the pulley wheels may be represented by $x^2 + y^2 = 14.5$ and $x^2 + y^2 - 19.6y + 86.0 = 0$. How far apart (in in.) are the wheels?

62. A rose garden has two intersecting circular paths with a fountain in the center of the longer path. The paths can be represented by the equations $x^2 + y^2 - 20x + 75 = 0$ and $x^2 + y^2 = 100$ (all distances in meters) with the fountain at the origin. What are (a) the closest and (b) the farthest the shorter path is to the fountain? (c) How far apart are the intersections of the paths? See Fig. 21.41.

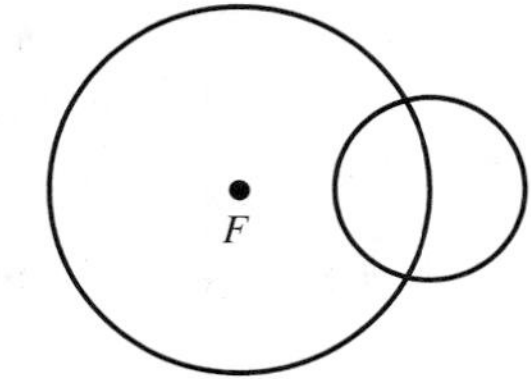

Fig. 21.41

63. A wire is rotating in a circular path through a magnetic field to induce an electric current in the wire. The wire is rotating at 60.0 Hz with a constant velocity of 37.7 m/s. Taking the origin at the center of the circle of rotation, find the equation of the path of the wire.

64. A communications satellite remains stationary at an altitude of 22,500 mi over a point on Earth's equator. It therefore rotates once each day about Earth's center. Its velocity is constant, but the horizontal and vertical components, v_H and v_V, of the velocity constantly change. Show that the equation relating v_H and v_V (in mi/h) is that of a circle. The radius of Earth is 3960 mi.

65. Find the equation describing the rim of a circular porthole 2 ft in diameter if the top is 6 ft below the surface of the water. Take the origin at the water surface directly above the center of the porthole.

66. An earthquake occurred 37° north of east of a seismic recording station. If the tremors travel at 4.8 km/s and were recorded 25 s later at the station, find the equation of the circle that represents the tremor recorded at the station. Take the station to be at the center of the coordinate system.

67. In analyzing the strain on a beam, *Mohr's circle* is often used. To form it, normal strain is plotted as the x-coordinate and shear strain is plotted as the y-coordinate. The center of the circle is midway between the minimum and maximum values of normal strain on the x-axis. Find the equation of Mohr's circle if the minimum normal strain is 100×10^{-6} and the maximum normal strain is 900×10^{-6} (strain is unitless). Sketch the graph.

68. An architect designs a Norman window, which has the form of a semicircle surmounted on a rectangle, as in Fig. 21.42. Find the area (in m²) of the window if the circular part is on the circle $x^2 + y^2 - 3.00y + 1.25 = 0$.

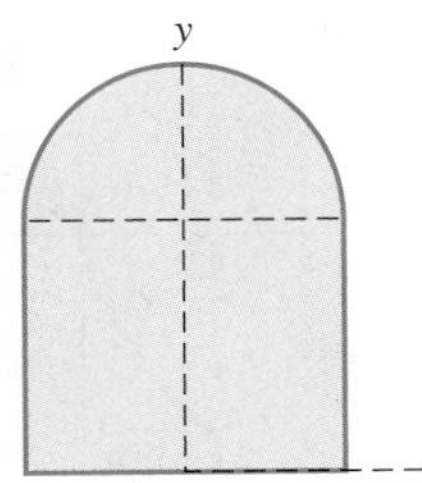

Fig. 21.42

Answers to Practice Exercises

1. $C(-7, 2), r = 1$ **2.** $C(4, -3), r = 2$ **3.** No, yes, no

21.4 The Parabola

Directrix • Focus • Axis • Vertex • Standard Form of Equation • Axis along x-axis • Axis along y-axis • Calculator Display

In Chapter 7, we showed that the graph of a quadratic function is a *parabola.* We now define the parabola more generally and find the general form of its equation.

A **parabola** *is defined as the locus of a point $P(x, y)$ that moves so that it is always equidistant from a given line (the* **directrix***) and a given point (the* **focus***). The line through the focus that is perpendicular to the directrix is the* **axis** *of the parabola. The point midway between the focus and directrix is the* **vertex.**

Using the definition, we now find the equation of the parabola with the focus at $(p, 0)$ and the directrix $x = -p$. With these choices, we find a general equation of a parabola with its vertex at the origin.

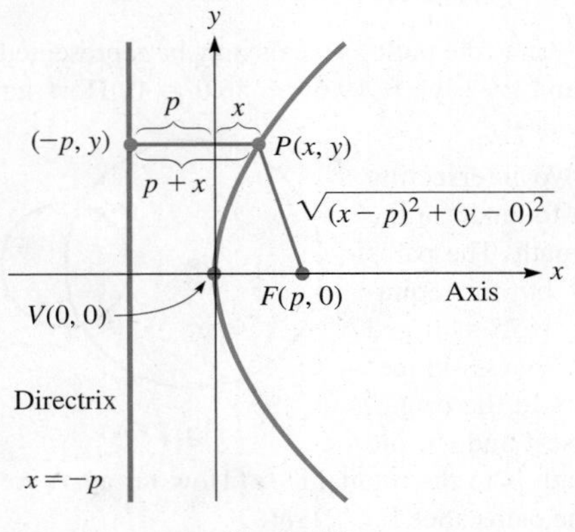

Fig. 21.43

From the definition, the distance from $P(x, y)$ on the parabola to the focus $(p, 0)$ must equal the distance from $P(x, y)$ to the directrix $x = -p$. The distance from P to the focus is found from the distance formula. The distance from P to the directrix is the perpendicular distance and is along a line parallel to the x-axis. These distances are shown in Fig. 21.43. Therefore, we have

$$\sqrt{(x-p)^2 + (y-0)^2} = x + p$$

$$(x-p)^2 + y^2 = (x+p)^2 \qquad \text{squaring both sides}$$

$$x^2 - 2px + p^2 + y^2 = x^2 + 2px + p^2$$

Simplifying, we obtain

$$y^2 = 4px \tag{21.15}$$

Equation (21.15) is called the **standard form of the equation of a parabola** *with its axis along the x-axis and the vertex at the origin.* The value of p is the *directed distance* from the vertex to the focus. The parabola is symmetric to the x-axis because $(-y)^2 = 4px$ is the same as $y^2 = 4px$.

EXAMPLE 1 Equation of parabola opening right

Find the coordinates of the focus and the equation of the directrix and sketch the graph of the parabola $y^2 = 12x$.

Because the equation of this parabola fits the form of Eq. (21.15), we know that the vertex is at the origin. The coefficient of 12 tells us that

$$4p = 12, \quad p = 3$$

The focus is the point $(3, 0)$, and the directrix is the line $x = -3$, as shown in Fig. 21.44. ■

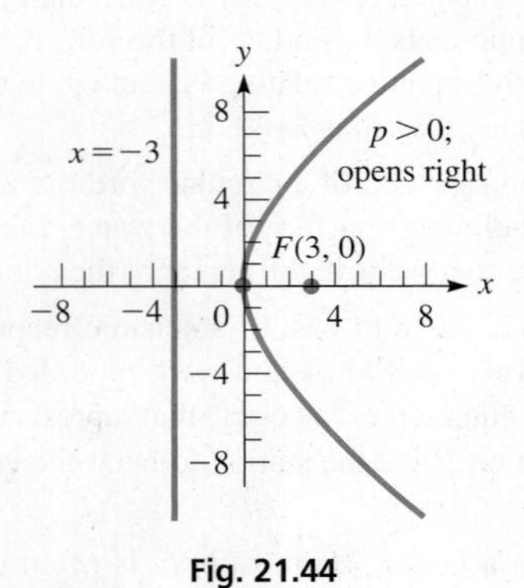

Fig. 21.44

EXAMPLE 2 Equation of parabola opening left

If the focus is to the left of the origin, with the directrix an equal distance to the right, the coefficient of the x-term is negative. This tells us that the parabola opens to the left, rather than to the right, as in the case when the focus is to the right of the origin. For example, the parabola $y^2 = -8x$ has its vertex at the origin, its focus at $(-2, 0)$, and the line $x = 2$ as its directrix. We determine this from the equation as follows:

$$y^2 = -8x \qquad 4p = -8, \qquad p = -2$$

Because $p = -2$, we find

the focus is $(-2, 0)$
the directrix is the line $x = -(-2)$, or $x = 2$

The parabola opens to the left, as shown in Fig. 21.45. ■

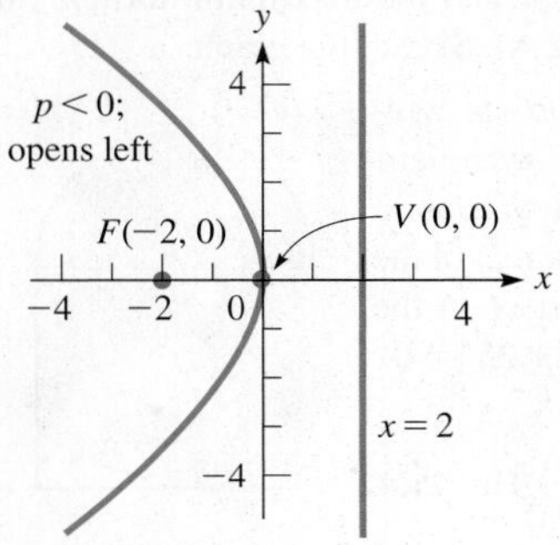

Fig. 21.45

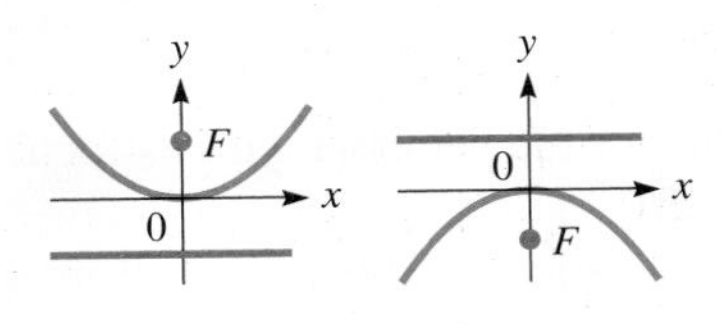

Fig. 21.46

If we chose the focus as the point $(0, p)$ and the directrix as the line $y = -p$ (see Fig. 21.46), we would find that the resulting equation is

$$x^2 = 4py \tag{21.16}$$

This is the standard form of the equation of a parabola with the y-axis as its axis and the vertex at the origin. Its symmetry to the y-axis can be proved, because $(-x)^2 = 4py$ is the same as $x^2 = 4py$. [We note that the difference between this equation and Eq. (21.15) is that x is squared and y appears to the first power in Eq. (21.16), rather than the reverse, as in Eq. (21.15).]

NOTE ▶

EXAMPLE 3 Standard form—axis along y-axis

(a) The parabola $x^2 = 4y$ fits the form of Eq. (21.16). Therefore, its axis is along the y-axis and its vertex is at the origin. From the equation, we find the value of p, which in turn tells us the location of the vertex and the directrix.

$$x^2 = 4y \qquad 4p = 4, \qquad p = 1$$

Focus $(0, p)$ is $(0, 1)$; directrix $y = -p$ is $y = -1$. The parabola is shown in Fig. 21.47, and we see in this case that it opens upward.

(b) The parabola $2x^2 = -9y$ fits the form of Eq. (21.16) if we write it in the form

$$x^2 = -\tfrac{9}{2}y$$

Here, we see that $4p = -9/2$. Therefore, its axis is along the y-axis, and its vertex is at the origin. Because $4p = -9/2$, we have

$$p = -\frac{9}{8} \qquad \text{focus}\left(0, -\frac{9}{8}\right) \qquad \text{directrix } y = \frac{9}{8}$$

The parabola opens downward, as shown in Fig. 21.48. ■

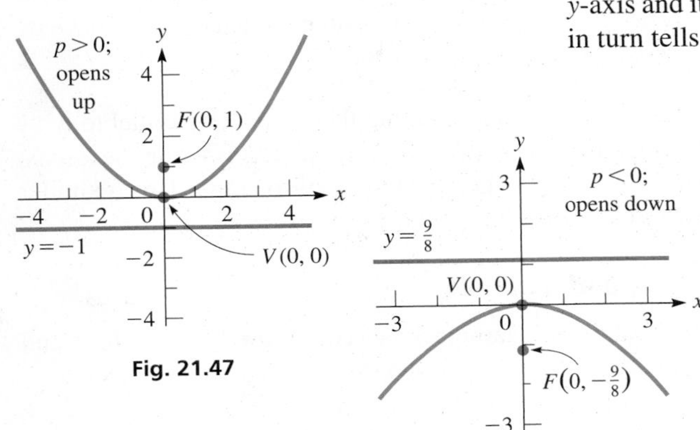

Fig. 21.47

Fig. 21.48

Practice Exercise

Find the coordinates of the focus and the equation of the directrix for each parabola.

1. $y^2 = -24x$ **2.** $x^2 = 40y$

■ See chapter introduction.

EXAMPLE 4 Parabola—satellite dish

In calculus, it can be shown that an electromagnetic wave (light wave, television signal, etc.) parallel to the axis of a parabolic reflector will pass through the focus of the parabola. Applications of this property of a parabola are many, including a satellite television dish and a spotlight (the light from the focus is reflected as a beam).

A parabolic satellite television dish is 46.0 cm across and 4.50 cm deep, as shown in Fig. 21.49. Find where the receiver should be located to receive all of the waves reflected off the parabolic surface.

With the vertex at the origin and the axis along the x-axis, we will use the general form $y^2 = 4px$ of the parabola. Because the parabolic opening is 46.0 cm and is 4.50 cm deep, the point $(4.50, 23.0)$ will be on any of the parabolic cross sections. Therefore, we find p by substituting $(4.50, 23.0)$ in the equation. This means

$$23.0^2 = 4p(4.50), \qquad p = 29.4 \text{ cm}$$

and the receiver should be placed 29.4 cm from the vertex as shown in Fig. 21.50. The equation of the parabolic cross section is $y^2 = 118x$. ■

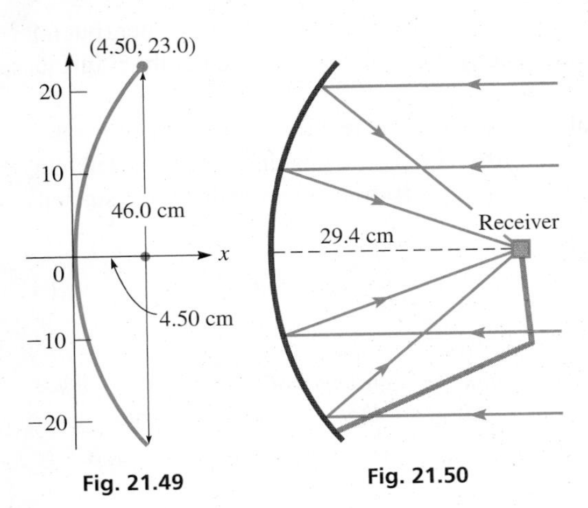

Fig. 21.49

Fig. 21.50

Equations (21.15) and (21.16) give the general forms of the equation of a parabola with vertex at the origin and focus on one of the coordinate axes. We now use the definition to find the equation of a parabola that has its vertex at a point other than the origin.

EXAMPLE 5 Find equation from definition

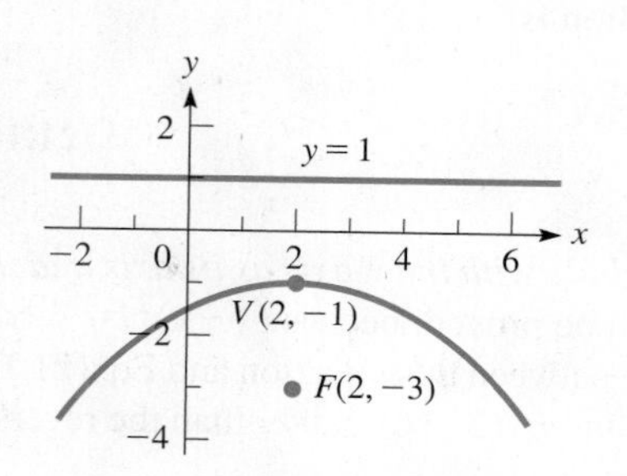

Fig. 21.51

Using the definition of the parabola, find the equation of the parabola with its focus at $(2, -3)$ and its directrix the line $y = 1$. See Fig. 21.51.

Choosing a general point $P(x, y)$ on the parabola and equating the distance from this point to $(2, -3)$ and to the line $y = 1$, we have

$$\sqrt{(x-2)^2 + (y+3)^2} = 1 - y$$

distance P to F = distance P to $y = 1$

$$(x-2)^2 + (y+3)^2 = (1-y)^2 \qquad \text{squaring both sides}$$

$$x^2 - 4x + 4 + y^2 + 6y + 9 = 1 - 2y + y^2$$

$$8y = -x^2 + 4x - 12$$

We note that this type of equation has appeared frequently in earlier chapters. The x-term and the constant (-12 in this case) are characteristic of a parabola that does not have its vertex at the origin if the directrix is parallel to the x-axis. ■

To view a parabola on a calculator, if the axis is along the y-axis, or parallel to it, as in Examples 3 and 5, we solve for y and use this function. If the axis is along the x-axis, or parallel to it, as in Examples 1, 2, and 4, we get *two* functions to graph. This is similar to Example 7 on page 582. These cases are shown in the next example.

EXAMPLE 6 Calculator graph of parabola

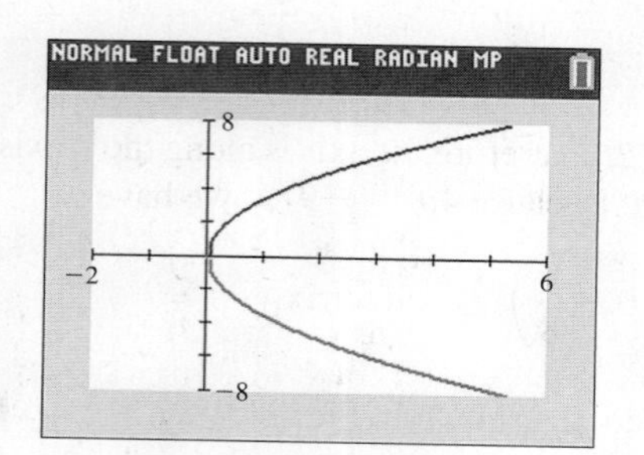

Fig. 21.52

(a) To display the graph of the parabola in Example 5 on a calculator, we solve for y and enter this function in the calculator. Therefore, we graph the function

$$y_1 = (-12 + 4x - x^2)/8$$

and this graph should match the one shown in Fig. 21.51.

(b) To display the graph of the parabola in Example 1, where $y^2 = 12x$, we must enter the two separate functions

$$y_1 = \sqrt{12x} \quad \text{and} \quad y_2 = -\sqrt{12x}$$

in the calculator, and we get the display in Fig. 21.52. ■

NOTE ▸ [We see that the equation of a parabola is characterized by the square of either (but not both) x and y and a first power term in the other.] We will consider the parabola further in Sections 21.7, 21.8, and 21.9.

The parabola has numerous technical applications. The reflection property illustrated in Example 4 has other important applications, such as the design of a radar antenna. Other examples of parabolas include the path of a projectile and the cables of a suspension bridge.

EXERCISES 21.4

In Exercises 1–4, make the given changes in the indicated examples of this section and then solve the resulting problems. In each, find the focus and directrix, and sketch the parabola.

1. In Example 1, change $12x$ to $20x$.
2. In Example 2, change $-8x$ to $-20x$.
3. In Example 3(a), change $4y$ to $-6y$.
4. In Example 3(b), change $-9y$ to $7y$.

In Exercises 5–16, determine the coordinates of the focus and the equation of the directrix of the given parabolas. Sketch each curve.

5. $y^2 = 4x$
6. $y^2 = 16x$
7. $y^2 = -64x$
8. $y^2 = -36x$
9. $x^2 = 72y$
10. $x^2 = y$
11. $x^2 = -4y$
12. $x^2 + 12y = 0$
13. $2y^2 - 5x = 0$
14. $3x^2 + 8y = 0$
15. $y = 0.48x^2$
16. $x = -7.6y^2$

In Exercises 17–30, find the equations of the parabolas satisfying the given conditions. The vertex of each is at the origin.

17. Focus $(3, 0)$
18. Focus $(0, 0.4)$
19. Focus $(0, -0.5)$
20. Focus $(-2.5, 0)$
21. Directrix $y = -0.16$
22. Directrix $x = 20$
23. Directrix $x = -84$
24. Directrix $y = 2.3$
25. Axis $x = 0$, passes through $(-1, 8)$
26. Symmetric to x-axis, passes through $(2, -1)$
27. Passes through $(3, 5)$ and $(3, -5)$
28. Passes through $(6, -1)$ and $(-6, -1)$
29. Passes through $(3, 3)$ and $(12, 6)$
30. Passes through $(-5, -5)$ and $(-10, -20)$

In Exercises 31–58, solve the given problems.

31. At what point(s) do the parabolas $y^2 = 2x$ and $x^2 = -16y$ intersect?

32. Using trigonometric identities, show that the parametric equations $x = \sin t$, $y = 2(1 - \cos^2 t)$ are the equations of a parabola.

33. Find the equation of the parabola with focus $(6, 1)$ and directrix $x = 0$, by use of the definition. Sketch the curve.

34. Find the equation of the parabola with focus $(1, 1)$ and directrix $y = 5$, by use of the definition. Sketch the curve.

35. Use a calculator to view the parabola $y^2 + 2x + 8y + 13 = 0$.

36. Use a calculator to view the parabola $y^2 - 2x - 6y + 19 = 0$.

37. The equation of a parabola with vertex (h, k) and axis parallel to the x-axis is $(y - k)^2 = 4p(x - h)$. (This is shown in Section 21.7.) Sketch the parabola for which (h, k) is $(2, -3)$ and $p = 2$.

38. The equation of a parabola with vertex (h, k) and axis parallel to the y-axis is $(x - h)^2 = 4p(y - k)$. (This is shown in Section 21.7.) Sketch the parabola for which (h, k) is $(-1, 2)$ and $p = -3$.

39. The chord of a parabola that passes through the focus and is parallel to the directrix is called the *latus rectum* of the parabola. Find the length of the latus rectum of the parabola $y^2 = 4px$.

40. Find the equation of the circle that has the focus and the vertex of the parabola $x^2 = 8y$ as the ends of a diameter.

41. Find the standard equation of the parabola with vertex $(0, 0)$ and that passes through $(2, 2)$ and $(8, 4)$.

42. For either standard form of the equation of a parabola, describe what happens to the shape of the parabola as $|p|$ increases.

43. A television satellite dish measures 80.0 cm across its opening and is 12.5 cm deep. Find the distance between the vertex and the focus (where the receiver is placed).

44. The rate of development of heat H (in W) in a resistor of resistance R (in Ω) of an electric circuit is given by $H = Ri^2$, where i is the current (in A) in the resistor. Sketch the graph of H vs. i, if $R = 6.0\ \Omega$.

45. A solar furnace uses a parabolic reflector to direct the sun's rays to a common focal point. The largest solar furnace in the world, located in Odeillo, France, has a parabolic reflector with a vertical cross section that measures 130 ft high and 17.9 ft deep. Find the focal length (the distance between the vertex and the focus).

46. The entrance to a building is a parabolic arch 5.6 m high at the center and 7.4 m wide at the base. What equation represents the arch if the vertex is at the top of the arch?

47. The sun is at the focus of a comet's parabolic orbit. When the comet is 4×10^7 km from the sun, the angle between the axis of the parabola and the line between the sun and comet is 60°. What is the closest distance the comet comes to the sun if $p > 0$?

48. What is the length of the horizontal bar across the parabolically shaped window shown in Fig. 21.53?

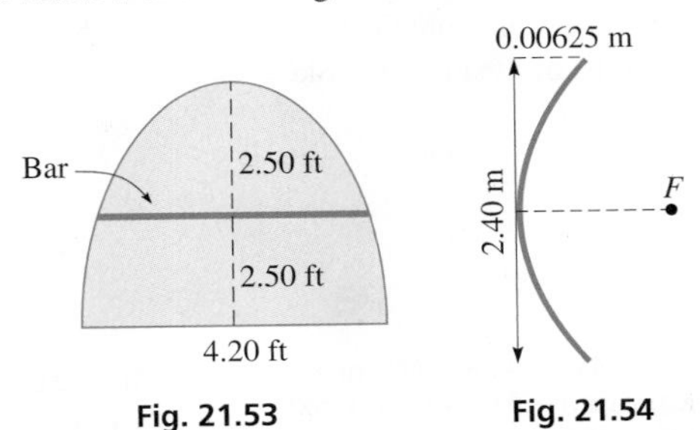

Fig. 21.53 Fig. 21.54

49. The primary mirror in the Hubble space telescope has a parabolic cross section, which is shown in Fig. 21.54. What is the focal length (vertex to focus) of the mirror?

50. A rocket is fired horizontally from a plane. Its horizontal distance x and vertical distance y from the point at which it was fired are given by $x = v_0 t$ and $y = \frac{1}{2}gt^2$, where v_0 is the initial velocity of the rocket, t is the time, and g is the acceleration due to gravity. Express y as a function of x and show that it is the equation of a parabola.

51. To launch a spacecraft to the moon, it is first put into orbit around Earth and then into a parabolic path toward the moon. Assume the parabolic path is represented by $x^2 = 4py$ and the spacecraft is later observed at $(10, 3)$ (units in thousands of km) after launch. Will this path lead directly to the moon at $(110, 340)$?

52. Under certain load conditions, a beam fixed at both ends is approximately parabolic in shape. If a beam is 4.0 m long and the deflection in the middle is 2.0 cm, find an equation to represent the shape of the beam. See Fig. 21.55.

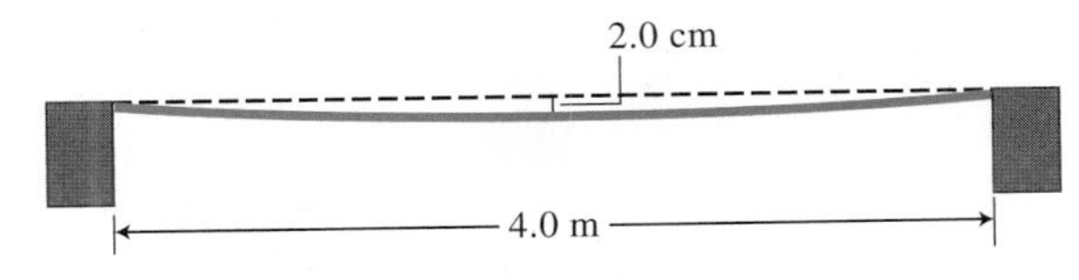

Fig. 21.55

53. A spotlight with a parabolic reflector is 15.0 cm wide and is 6.50 cm deep. See Fig. 21.56 and Example 4. Where should the filament of the bulb be located so as to produce a beam of light?

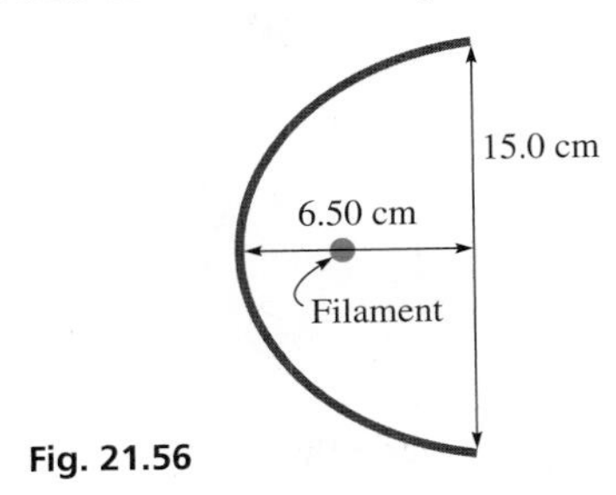

Fig. 21.56

54. A wire is fastened 36.0 ft up on each of two telephone poles that are 200 ft apart. Halfway between the poles the wire is 30.0 ft above the ground. Assuming the wire is parabolic, find the height of the wire 50.0 ft from either pole.

55. The total annual fraction f of energy supplied by solar energy to a home is given by $f = 0.065\sqrt{A}$, where A is the area of the solar collector. Sketch the graph of f as a function of A $(0 < A \leq 200 \text{ m}^2)$.

56. The velocity v (in ft/s) of a jet of water flowing from an opening in the side of a certain container is given by $v = 8\sqrt{h}$, where h is the depth (in ft) of the opening. Sketch a graph of v vs. h.

57. A small island is 4 km from a straight shoreline. A ship channel is equidistant between the island and the shoreline. Write an equation for the channel.

58. Under certain circumstances, the maximum power P (in W) in an electric circuit varies as the square of the voltage of the source E_0 and inversely as the internal resistance R_i (in Ω) of the source. If 10 W is the maximum power for a source of 2.0 V and internal resistance of 0.10 Ω, sketch the graph of P vs. E_0 if R_i remains constant.

Answers to Practice Exercises

1. $F(-6, 0)$; dir. $x = 6$ **2.** $F(0, 10)$; dir. $y = -10$

21.5 The Ellipse

Foci • Vertices • Major Axis • Minor Axis • Standard Equation – Axis along x-axis • Standard Equation – Axis along y-axis • Calculator Display

The next important curve is the ellipse. *An* **ellipse** *is defined as the locus of a point $P(x, y)$ that moves so that the sum of its distances from two fixed points is constant. These fixed points are the* **foci** *of the ellipse.* Letting this sum of distances be $2a$ and the foci be the points $(-c, 0)$ and $(c, 0)$, we have

$$\sqrt{(x-c)^2 + y^2} + \sqrt{[x-(-c)]^2 + y^2} = 2a$$

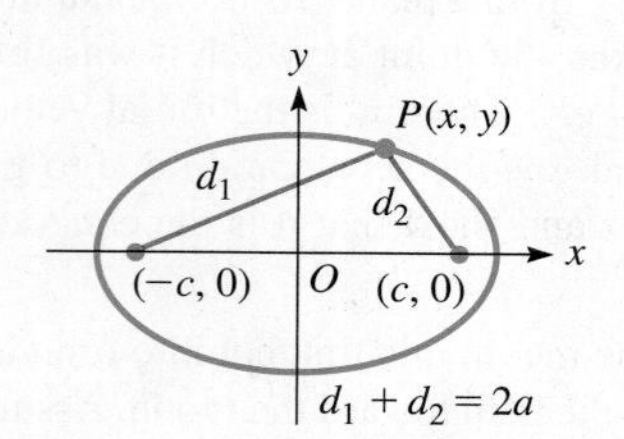

Fig. 21.57

See Fig. 21.57. The ellipse has its center at the origin such that c is the length of the line segment from the center to a focus. We will also see that a has a special meaning. Now, from Section 14.4, we see that we should move one radical to the right and then square each side. This leads to the following steps:

$$\sqrt{(x+c)^2 + y^2} = 2a - \sqrt{(x-c)^2 + y^2}$$

$$(x+c)^2 + y^2 = 4a^2 - 4a\sqrt{(x-c)^2 + y^2} + \left(\sqrt{(x-c)^2 + y^2}\right)^2$$

$$x^2 + 2cx + c^2 + y^2 = 4a^2 - 4a\sqrt{(x-c)^2 + y^2} + x^2 - 2cx + c^2 + y^2$$

$$4a\sqrt{(x-c)^2 + y^2} = 4a^2 - 4cx$$

$$a\sqrt{(x-c)^2 + y^2} = a^2 - cx$$

$$a^2(x^2 - 2cx + c^2 + y^2) = a^4 - 2a^2cx + c^2x^2$$

$$(a^2 - c^2)x^2 + a^2y^2 = a^2(a^2 - c^2)$$

We now define $a^2 - c^2 = b^2$ (this reason will be shown presently). Therefore,

$$b^2x^2 + a^2y^2 = a^2b^2$$

Dividing through by a^2b^2, we have

$$\frac{x^2}{a^2} + \frac{y^2}{b^2} = 1 \qquad \textbf{(21.17)}$$

A graphical analysis of this equation is found on the next page.

The x-intercepts are $(-a, 0)$ and $(a, 0)$. This means that $2a$ (the sum of distances used in the derivation) is also the distance between the x-intercepts. *The points* $(a, 0)$ *and* $(-a, 0)$ *are the* **vertices** *of the ellipse, and the line between them is the* **major axis** [see Fig. 21.58(a)]. Thus, *a is the length of the* **semimajor axis.**

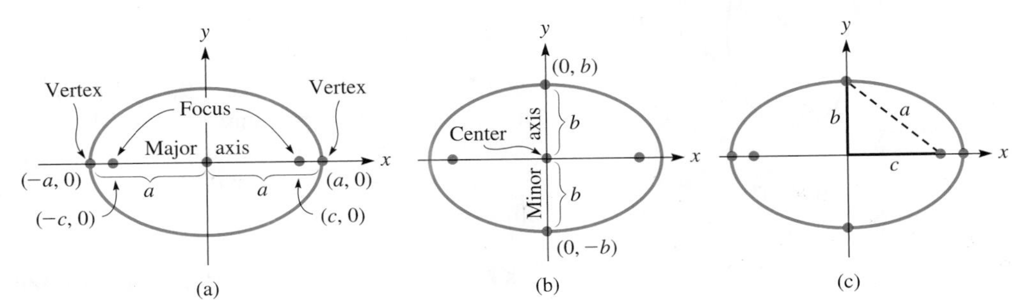

Fig. 21.58

We can now state that *Eq. (21.17) is called the* **standard equation of the ellipse** *with its major axis along the x-axis and its center at the origin.*

The y-intercepts of this ellipse are $(0, -b)$ and $(0, b)$. *The line joining these intercepts is called the* **minor axis** *of the ellipse* [Fig. 21.58(b)], *which means b is the length of the* **semiminor** axis. The intercept $(0, b)$ is equidistant from $(-c, 0)$ and $(c, 0)$. Since the sum of the distances from these points to $(0, b)$ is $2a$, the distance $(c, 0)$ to $(0, b)$ must be a. Thus, we have a right triangle with line segments of lengths a, b, and c, with a as hypotenuse [Fig. 21.58(c)]. Therefore,

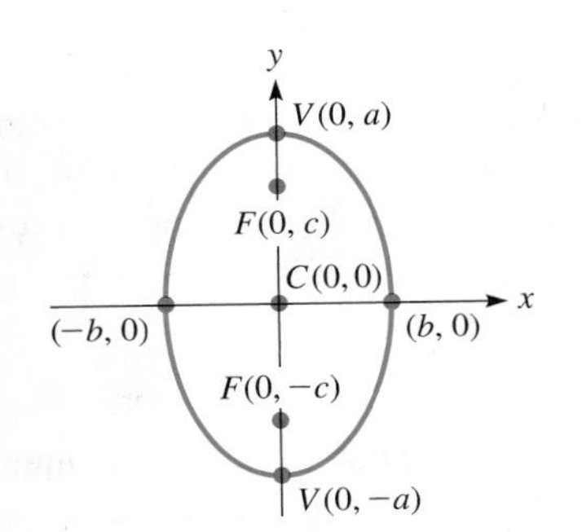

Fig. 21.59

$$a^2 = b^2 + c^2 \qquad (21.18)$$

is the relation between distances a, b, and c. This also shows why b was defined as it was in the derivation of Eq. (21.17).

If we choose points on the y-axis as the foci, *the standard equation of the ellipse, with its center at the origin and its major axis along the y-axis, is*

$$\frac{y^2}{a^2} + \frac{x^2}{b^2} = 1 \qquad (21.19)$$

In this case, the vertices are $(0, a)$ and $(0, -a)$, the foci are $(0, c)$ and $(0, -c)$, and the ends of the minor axis are $(b, 0)$ and $(-b, 0)$. See Fig. 21.59.

NOTE ➧ [For any ellipse, $a > b$.]

The ellipses represented by Eqs. (21.17) and (21.19) are both symmetric to both axes and to the origin.

EXAMPLE 1 Standard equation—major axis along *x*-axis

The ellipse $\frac{x^2}{25} + \frac{y^2}{9} = 1$ seems to fit the form of either Eq. (21.17) or Eq. (21.19).

CAUTION Because $a^2 = b^2 + c^2$, we know that a is always larger than b. ■ Because the square of the larger number appears under x^2, we know the equation is in the form of Eq. (21.17). Therefore, $a^2 = 25$ and $b^2 = 9$, or $a = 5$ and $b = 3$. This means that the vertices are $(5, 0)$ and $(-5, 0)$, and the minor axis extends from $(0, -3)$ to $(0, 3)$. See Fig. 21.60.

We find c from the relation $c^2 = a^2 - b^2$. This means that $c^2 = 16$ and the foci are $(4, 0)$ and $(-4, 0)$. ■

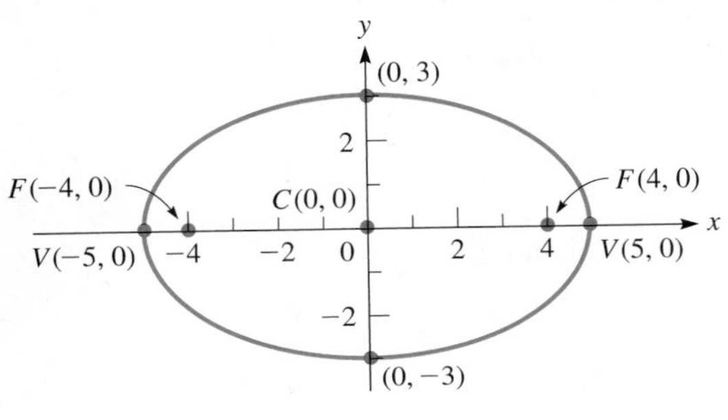

Fig. 21.60

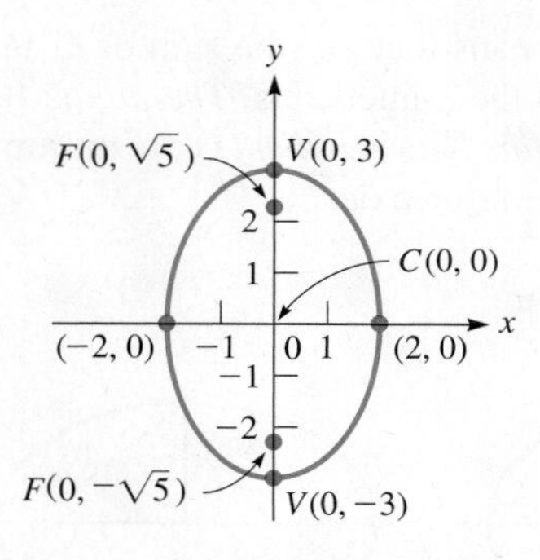

Fig. 21.61

EXAMPLE 2 Standard equation—major axis along y-axis

The equation of the ellipse

$$\frac{x^2}{4} + \frac{y^2}{9} = 1$$

(b^2 under x^2, a^2 under y^2)

has the larger denominator, 9, under the y^2. Therefore, it fits the form of Eq. (21.19) with $a^2 = 9$ and $b^2 = 4$. This means that the vertices are $(0, 3)$ and $(0, -3)$, and the minor axis extends from $(-2, 0)$ to $(2, 0)$. In turn, we know that $c^2 = a^2 - b^2 = 9 - 4 = 5$ and that the foci are $(0, \sqrt{5})$ and $(0, -\sqrt{5})$. This ellipse is shown in Fig. 21.61. ■

EXAMPLE 3 Find vertices, foci, minor axis

Find the coordinates of the vertices, the ends of the minor axis, and the foci of the ellipse $4x^2 + 16y^2 = 64$.

This equation must be put in standard form first, which we do by dividing through by 64. When this is done, we obtain

$$\frac{x^2}{16} + \frac{y^2}{4} = 1$$

(form requires + and 1)

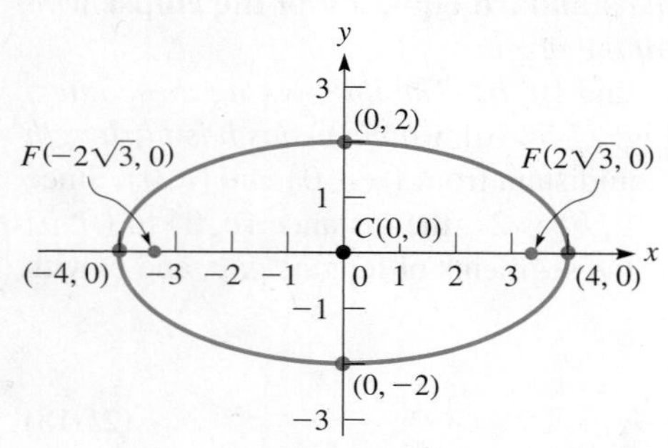

Fig. 21.62

We see that $a^2 = 16$ and $b^2 = 4$, which tells us that $a = 4$ and $b = 2$. Then $c = \sqrt{16 - 4} = \sqrt{12} = 2\sqrt{3}$. Because a^2 appears under x^2, the vertices are $(4, 0)$ and $(-4, 0)$. The ends of the minor axis are $(0, 2)$ and $(0, -2)$, and the foci are $(2\sqrt{3}, 0)$ and $(-2\sqrt{3}, 0)$. See Fig. 21.62. ■

EXAMPLE 4 Ellipse—satellite orbit

A satellite to study Earth's atmosphere has a minimum altitude of 600 mi and a maximum altitude of 2000 mi. If the path of the satellite about Earth is an ellipse with the center of Earth at one focus, what is the equation of its path? Assume that the radius of Earth is 4000 mi.

We set up the coordinate system such that the center of the ellipse is at the origin and the center of Earth is at the right focus, as shown in Fig. 21.63. We know that the distance between vertices is

$$2a = 2000 + 4000 + 4000 + 600 = 10{,}600 \text{ mi}$$
$$a = 5300 \text{ mi}$$

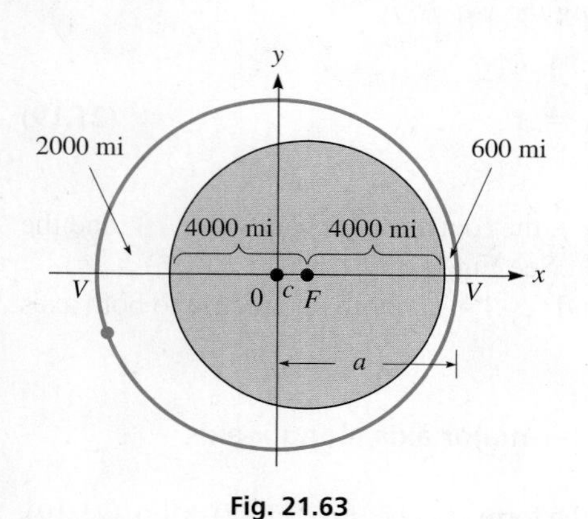

Fig. 21.63

From the right focus to the right vertex is 4600 mi. This tells us

$$c = a - 4600 = 5300 - 4600 = 700 \text{ mi}$$

We can now calculate b^2 as

$$b^2 = a^2 - c^2 = 5300^2 - 700^2 = 2.76 \times 10^7 \text{ mi}^2$$

Because $a^2 = 5300^2 = 2.81 \times 10^7 \text{ mi}^2$, the equation is

$$\frac{x^2}{2.81 \times 10^7} + \frac{y^2}{2.76 \times 10^7} = 1$$

or

$$2.76x^2 + 2.81y^2 = 7.76 \times 10^7$$

■

Practice Exercises

Find the vertices and foci of each ellipse.

1. $\frac{x^2}{9} + y^2 = 1$ **2.** $25x^2 + 4y^2 = 25$

3. In Example 4, find the equation if the altitudes are changed to 800 mi and 2200 mi.

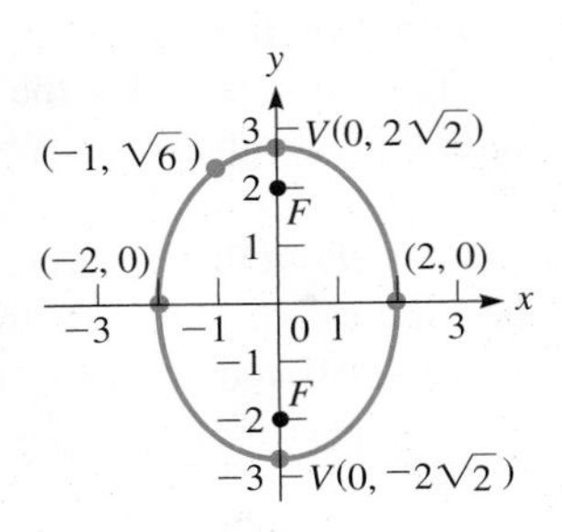

Fig. 21.64

EXAMPLE 5 Find equation—given points

Find the equation of the ellipse with its center at the origin and an end of its minor axis at $(2, 0)$ and which passes through $(-1, \sqrt{6})$.

Because the center is at the origin and an end of the minor axis is at $(2, 0)$, we know that the ellipse is of the form of Eq. (21.19) and that $b = 2$. Thus, we have

$$\frac{y^2}{a^2} + \frac{x^2}{2^2} = 1$$

In order to find a^2, we use the fact that the ellipse passes through $(-1, \sqrt{6})$. This means that these coordinates satisfy the equation of the ellipse. This gives

$$\frac{(\sqrt{6})^2}{a^2} + \frac{(-1)^2}{4} = 1, \qquad \frac{6}{a^2} = \frac{3}{4}, \qquad a^2 = 8$$

Therefore, the equation of the ellipse, shown in Fig. 21.64, is

$$\frac{y^2}{8} + \frac{x^2}{4} = 1$$

■

The following example illustrates the use of the definition of the ellipse to find the equation of an ellipse with its center at a point other than the origin.

EXAMPLE 6 Find equation from definition—calculator

Using the definition, find the equation of the ellipse with foci at $(1, 3)$ and $(9, 3)$, with a major axis of 10.

Recalling that the sum of distances in the definition equals the length of the major axis, we now use the same method as in the derivation of Eq. (21.17).

$$\sqrt{(x-1)^2 + (y-3)^2} + \sqrt{(x-9)^2 + (y-3)^2} = 10 \quad \text{use definition of ellipse}$$

$$\sqrt{(x-1)^2 + (y-3)^2} = 10 - \sqrt{(x-9)^2 + (y-3)^2} \quad \text{isolate radical}$$

$$x^2 - 2x + 1 + y^2 - 6y + 9 = 100 - 20\sqrt{(x-9)^2 + (y-3)^2} + x^2 - 18x + 81 + y^2 - 6y + 9 \quad \text{square both sides and simplify}$$

$$20\sqrt{(x-9)^2 + (y-3)^2} = 180 - 16x \quad \text{isolate radical}$$

$$5\sqrt{(x-9)^2 + (y-3)^2} = 45 - 4x \quad \text{divide by 4}$$

$$25(x^2 - 18x + 81 + y^2 - 6y + 9) = 2025 - 360x + 16x^2 \quad \text{square both sides}$$

$$9x^2 - 90x + 25y^2 - 150y + 225 = 0 \quad \text{simplify}$$

The additional x- and y-terms are characteristic of the equation of an ellipse whose center is not at the origin (see Fig. 21.65).

To view this ellipse on a calculator as shown in Fig. 21.66, we solve for y to get the two functions needed. The solutions are $y = \dfrac{15 \pm 3\sqrt{10x - x^2}}{5}$.

■

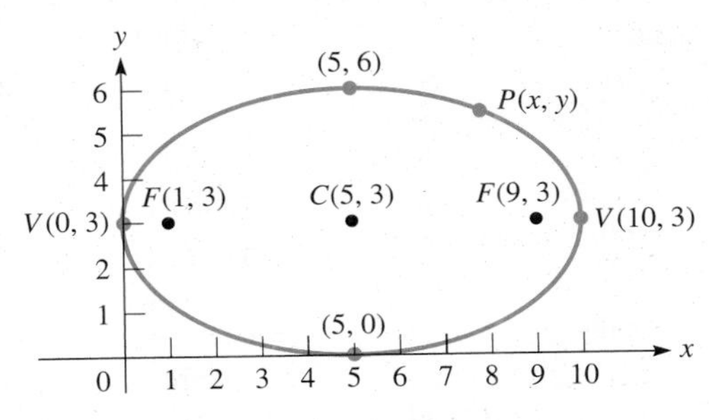

Fig. 21.65

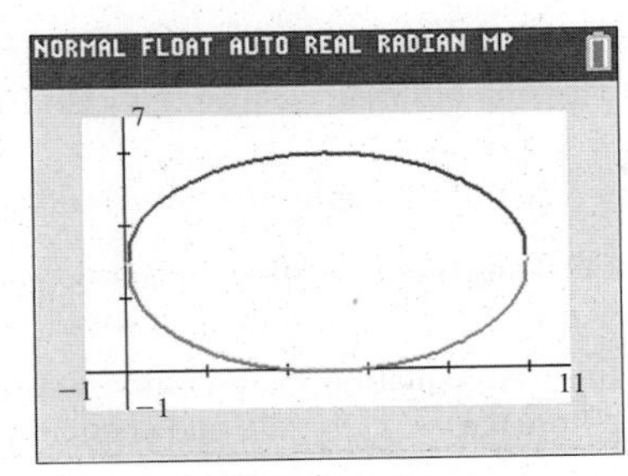

Fig. 21.66

NOTE ▸ [We see that the equation of an ellipse is characterized by both an x^2-term and a y^2-term, having different coefficients (in value but not in sign).] We note that the coefficients of the squared terms differ, whereas for the circle, they are the same. We will consider the ellipse further in Sections 21.7, 21.8, and 21.9.

■ The Polish astronomer Nicolaus Copernicus (1473–1543) is credited as being the first to suggest that Earth revolved about the sun, rather than the previously held belief that Earth was the center of the universe.

The ellipse has many applications. The orbits of the planets about the sun are elliptical. Gears, cams, and springs are often elliptical in shape. Arches are often constructed in the form of a semiellipse. These and other applications are illustrated in the exercises.

EXERCISES 21.5

In Exercises 1 and 2, make the given changes in the indicated examples of this section and then solve the given problems.

1. In Example 1, change the 9 to 36; find the vertices, ends of the minor axis, and foci; and sketch the ellipse.
2. In Example 2, interchange the 4 and 9; find the vertices, ends of the minor axis, and foci; and sketch the ellipse.

In Exercises 3–16, find the coordinates of the vertices and foci of the given ellipses. Sketch each curve.

3. $\frac{x^2}{4} + y^2 = 1$
4. $\frac{x^2}{100} + \frac{y^2}{64} = 1$
5. $\frac{x^2}{25} + \frac{y^2}{144} = 1$
6. $\frac{x^2}{49} + \frac{y^2}{81} = 1$
7. $\frac{4x^2}{25} + \frac{9y^2}{4} = 1$
8. $16x^2 + \frac{9y^2}{25} = 1$
9. $4x^2 + 9y^2 = 324$
10. $9x^2 + 36y^2 = 144$
11. $49x^2 + 8y^2 = 196$
12. $y^2 = 25(1 - x^2)$
13. $y^2 = 8(2 - x^2)$
14. $2x^2 + 3y^2 = 600$
15. $4x^2 + 25y^2 = 0.25$
16. $9x^2 + 4y^2 = 0.09$

In Exercises 17–28, find the equations of the ellipses satisfying the given conditions. The center of each is at the origin.

17. Vertex $(15, 0)$, focus $(9, 0)$
18. Minor axis 8, vertex $(0, -5)$
19. Focus $(0, 8)$, major axis 34
20. Vertex $(0, 13)$, focus $(0, -12)$
21. End of minor axis $(0, 12)$, focus $(8, 0)$
22. Sum of lengths of major and minor axes is 18, focus $(3, 0)$
23. Vertex $(8, 0)$, passes through $(2, 3)$
24. Focus $(0, 2)$, passes through $(-1, \sqrt{3})$
25. Passes through $(2, 2)$ and $(1, 4)$
26. Passes through $(-2, 2)$ and $(1, \sqrt{6})$
27. The sum of distances from (x, y) to $(6, 0)$ and $(-6, 0)$ is 20.
28. The sum of distances from (x, y) to $(0, 2)$ and $(0, -2)$ is 5.

In Exercises 29–56, solve the given problems.

29. Find any point(s) of intersection of the graphs of the ellipse $4x^2 + 9y^2 = 40$ and the parabola $y^2 = 4x$.
30. Find the equation of the circle that has the same center as the ellipse $4x^2 + 9y^2 = 36$ and is internally tangent to the ellipse.
31. Find the equation of the ellipse with foci $(-2, 1)$ and $(4, 1)$ and a major axis of 10, by use of the definition. Sketch the curve.
32. Find the equation of the ellipse with foci $(1, 4)$ and $(1, 0)$ that passes through $(4, 4)$, by use of the definition. Sketch the curve.
33. Use a calculator to view the ellipse $4x^2 + 3y^2 + 16x - 18y + 31 = 0$.
34. Use a calculator to view the ellipse $4x^2 + 8y^2 + 4x - 24y + 1 = 0$.
35. The equation of an ellipse with center (h, k) and major axis parallel to the x-axis is $\frac{(x-h)^2}{a^2} + \frac{(y-k)^2}{b^2} = 1$. (This is shown in Section 21.7.) Sketch the ellipse that has a major axis of 6, a minor axis of 4, and for which (h, k) is $(2, -1)$.
36. The equation of an ellipse with center (h, k) and major axis parallel to the y-axis is $\frac{(y-k)^2}{a^2} + \frac{(x-h)^2}{b^2} = 1$. (This is shown in Section 21.7.) Sketch the ellipse that has a major axis of 8, a minor axis of 6, and for which (h, k) is $(1, 3)$.
37. For what values of k does the ellipse $x^2 + ky^2 = 1$ have its vertices on the y-axis? Explain how these values are found.
38. For what value of k does the ellipse $x^2 + k^2y^2 = 25$ have a focus at $(3, 0)$? Explain how this value is found.
39. Show that the ellipse $2x^2 + 3y^2 - 8x - 4 = 0$ is symmetric to the x-axis.
40. Show that the ellipse $5x^2 + y^2 - 3y - 7 = 0$ is symmetric to the y-axis.
41. Show that the parametric equations $x = 2\sin t$ and $y = 3\cos t$ define an ellipse.
42. For the ellipse given by $\frac{x^2}{25} + \frac{y^2}{9} = 1$, find the length of the line segment perpendicular to the major axis that passes through a focus and spans the width of the ellipse.
43. A person exercising on an elliptical trainer is moving her feet along an elliptical path with a horizontal major axis of 32 in. and a vertical minor axis of 10 in. Find the equation of the ellipse if the center is at the origin.
44. An Australian football field is elliptical. If a field can be represented by the equation $\frac{x^2}{1500} + \frac{y^2}{1215} = 15$, what are the dimensions (in m) of the field?
45. The electric power P (in W) dissipated in a resistance R (in Ω) is given by $P = Ri^2$, where i is the current (in A) in the resistor. Find the equation for the total power of 64 W dissipated in two resistors, with resistances 2.0 Ω and 8.0 Ω, respectively, and with currents i_1 and i_2, respectively. Sketch the graph, assuming that negative values of current are meaningful.

46. The *eccentricity* e of an ellipse is defined as $e = c/a$. A cam in the shape of an ellipse can be described by the equation $x^2 + 9y^2 = 81$. Find the eccentricity of this elliptical cam.

47. A space object (dubbed 2003 UB313), larger and more distant than Pluto, was discovered in 2003. In its elliptical orbit with the sun at one focus, it is 3.5 billion miles from the sun at the closest, and 9.0 billion miles at the farthest. What is the eccentricity of its orbit? See Exercise 46.

48. Halley's Comet has an elliptical orbit with $a = 17.94$ AU (AU is astronomical unit, 1 AU $= 9.3 \times 10^7$ mi) and $b = 4.552$ AU, with the sun at one focus. What is the closest that the comet comes to the sun? Explain your method.

49. A draftsman draws a series of triangles with a base from $(-3, 0)$ to $(3, 0)$ and a perimeter of 14 cm (all measurements in centimeters). Find the equation of the curve on which all of the third vertices of the triangles are located.

50. A lithotripter is used to break up a kidney stone by placing the kidney stone at one focus of an ellipsoid end-section and a source of shock waves at the focus of the other end-section. If the vertices of the end-sections are 30.0 cm apart and a minor axis of the ellipsoid is 6.0 cm, how far apart are the foci? In a lithotripter, the shock waves are reflected as the sound waves noted in Exercise 51.

51. An ellipse has a focal property such that a light ray or sound wave emanating from one focus will be reflected through the other focus. Many buildings, such as Statuary Hall in the U.S. Capitol and the Taj Mahal, are built with elliptical ceilings with the property that a sound from one focus is easily heard at the other focus. If a building has a ceiling whose cross sections are part of an ellipse that can be described by the equation $36x^2 + 225y^2 = 8100$ (measurements in meters), how far apart must two persons stand in order to whisper to each other using this focal property?

52. An airplane wing is designed such that a certain cross section is an ellipse 8.40 ft wide and 1.20 ft thick. Find the equation that can be used to describe the perimeter of this cross section.

53. A road passes through a tunnel with a semielliptical cross section 64 ft wide and 18 ft high at the center. What is the height of the tallest vehicle that can pass through the tunnel at a point 22 ft from the center? See Fig. 21.67.

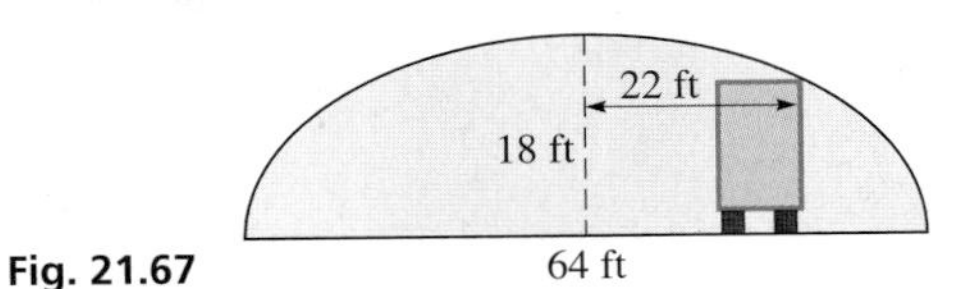

Fig. 21.67

54. An architect designs a window in the shape of an ellipse 4.50 ft wide and 3.20 ft high. Find the perimeter of the window from the formula $p = \pi(a + b)$. This formula gives a good *approximation* for the perimeter when a and b are nearly equal.

55. The ends of a horizontal tank 20.0 ft long are ellipses, which can be described by the equation $9x^2 + 20y^2 = 180$, where x and y are measured in feet. The area of an ellipse is $A = \pi ab$. Find the volume of the tank.

56. A laser beam 6.80 mm in diameter is incident on a plane surface at an angle of 62.0°, as shown in Fig. 21.68. What is the elliptical area that the laser covers on the surface? (See Exercise 55.)

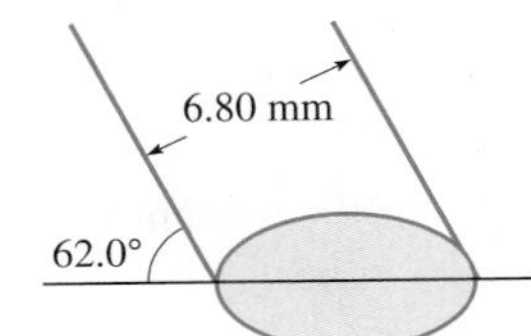

Fig. 21.68

Answers to Practice Exercises

1. $V(3, 0), V(-3, 0); F(2\sqrt{2}, 0), F(-2\sqrt{2}, 0)$
2. $V(0, 5/2), V(0, -5/2); F(0, \sqrt{21}/2), F(0, -\sqrt{21}/2)$
3. $2.98x^2 + 3.03y^2 = 9.00 \times 10^7$

21.6 The Hyperbola

Foci • Vertices • Asymptotes • Transverse Axis • Conjugate Axis • Standard Equation–Axis along x-axis • Standard Equation–Axis along y-axis • Calculator Display • $xy = c$ Hyperbola

A **hyperbola** *is defined as the locus of a point $P(x, y)$ that moves so that the difference of the distances from two fixed points (the* **foci***) is constant.* We choose the foci to be $(-c, 0)$ and $(c, 0)$ (see Fig. 21.69), and the constant difference to be $2a$. As with the ellipse, c is the length of the line segment from the center to a focus, and a (as we will see) the length of the line segment from the center to a vertex. Therefore,

$$\sqrt{(x + c)^2 + y^2} - \sqrt{(x - c)^2 + y^2} = 2a$$

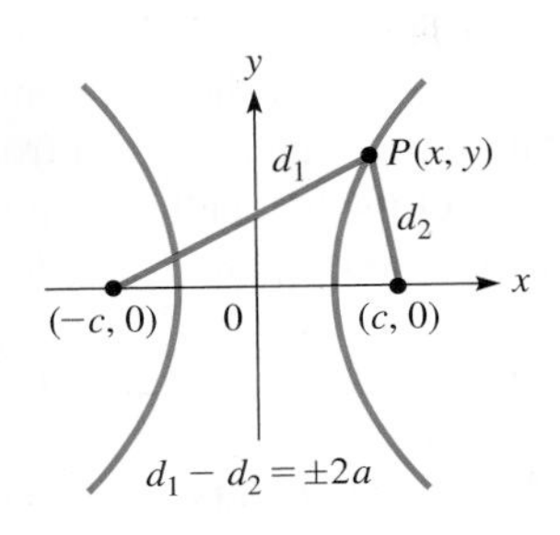

Fig. 21.69

Following the same procedure as with the ellipse, the **standard equation of the hyperbola** with its center at the origin is

$$\frac{x^2}{a^2} - \frac{y^2}{b^2} = 1 \qquad (21.20)$$

CAUTION When we derive this equation, we have a definition of the relation between a, b, and c that is different from that for the ellipse. ■ This relation is

$$c^2 = a^2 + b^2 \qquad (21.21)$$

In Eq. (21.20), if we let $y = 0$, we find that the x-intercepts are $(-a, 0)$ and $(a, 0)$, just as they are for the ellipse. *These are the* **vertices** *of the hyperbola.* For $x = 0$, we find that we have imaginary solutions for y, which means there are no points on the curve that correspond to a value of $x = 0$.

To find the meaning of b, we solve Eq. (21.20) for y in the special form:

$$\frac{y^2}{b^2} = \frac{x^2}{a^2} - 1$$

$$= \frac{x^2}{a^2} - \frac{a^2x^2}{a^2x^2} = \frac{x^2}{a^2}\left(1 - \frac{a^2}{x^2}\right)$$

$$y^2 = \frac{b^2x^2}{a^2}\left(1 - \frac{a^2}{x^2}\right)$$

multiply through by b^2 and take square root of each side

$$y = \pm\frac{bx}{a}\sqrt{1 - \frac{a^2}{x^2}} \tag{21.22}$$

We note that, if large values of x are assumed in Eq. (21.22), the quantity under the radical becomes approximately 1. In fact, the larger x becomes, the nearer 1 this expression becomes, because the x^2 in the denominator of a^2/x^2 makes this term nearly zero. Thus, for large values of x, Eq. (21.22) is approximately

$$y = \pm\frac{bx}{a} \tag{21.23}$$

NOTE ▸ [The slopes of the asymptotes are *negatives* of each other.]

Equation (21.23) is seen to represent two straight lines, each of which passes through the origin. One has a slope of b/a, and the other has a slope of $-b/a$. NOTE ▸ [These lines are called the **asymptotes** of the hyperbola. An **asymptote** is a line that the curve approaches as one of the variables approaches some particular value.] The graph of the tangent function also has asymptotes, as we saw in Fig. 10.23. We can designate this limiting procedure with notation introduced in Chapter 19 by

$$y \to \frac{bx}{a} \quad \text{as} \quad x \to \pm\infty$$

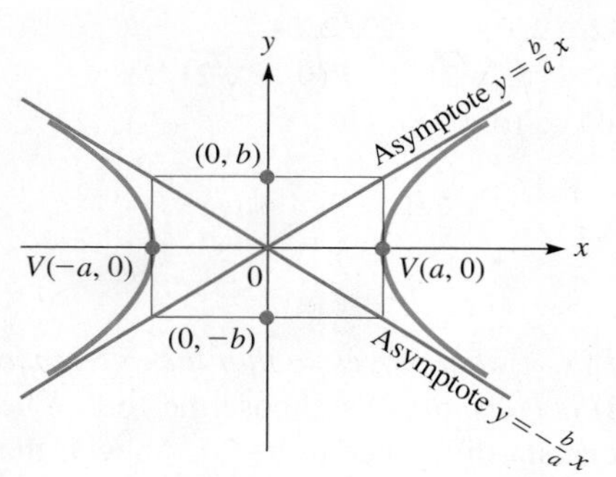

Fig. 21.70

The easiest way to sketch a hyperbola is to draw its asymptotes and then draw the hyperbola out from each vertex so that it comes closer and closer to each asymptote as x becomes numerically larger. To draw the asymptotes, first draw a small rectangle $2a$ by $2b$ with the origin at the center, as shown in Fig. 21.70. Then straight lines, the asymptotes, are drawn through opposite vertices of the rectangle. This shows us that the significance of the value of b is in the slopes of the asymptotes.

A hyperbola in the form of Eq. (21.20) *has a* **transverse axis** *of length* $2a$ *along the* x*-axis and a* **conjugate axis** *of length* $2b$ *along the* y*-axis.* This means that a represents the length of the semitransverse axis and b represents the length of the semiconjugate axis. See Fig. 21.71. From the definition of c, it is the length of the line segment from the center to a focus. Also, c is the length of the semidiagonal of the rectangle, as shown in Fig. 21.71. This shows us the geometric meaning of the relationship among a, b, and c given in Eq. (21.21).

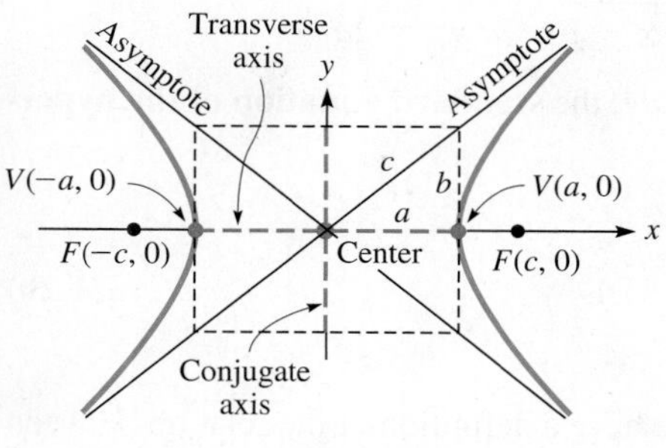

Fig. 21.71

When sketching the graph of a hyperbola, it is clear that the important quantities are a, the semitransverse axis, b, the semiconjugate axis, and c, the distance from the origin to a focus. In doing so, it is important to keep in mind that $c > a$ and $c > b$, but that either a or b can be greater than the other, or they can be equal.

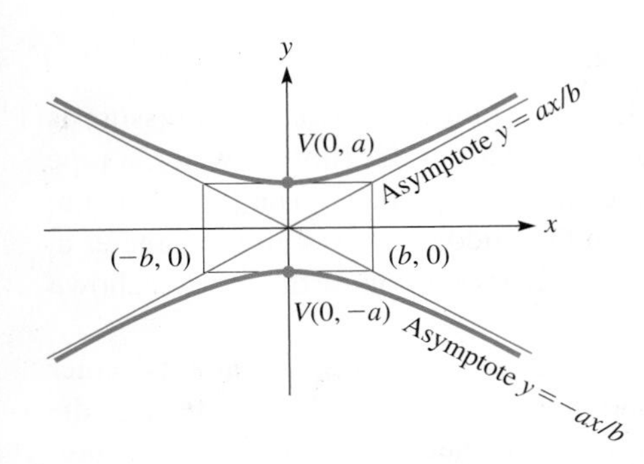

Fig. 21.72

If the transverse axis is along the y-axis and the conjugate axis is along the x-axis, the equation of the hyperbola with its center at the origin (see Fig. 21.72) is

$$\frac{y^2}{a^2} - \frac{x^2}{b^2} = 1 \qquad \textbf{(21.24)}$$

The hyperbolas represented by Eqs. (21.20) and (21.24) are both symmetric to both axes and to the origin.

EXAMPLE 1 Standard equation—transverse axis on x-axis

The hyperbola $\frac{x^2}{16} - \frac{y^2}{9} = 1$ (a^2 = 16, b^2 = 9)

fits the form of Eq. (21.20). We know that it fits Eq. (21.20) and not Eq. (21.24) because the x^2-term is the positive term with 1 on the right. From the equation, we see that $a^2 = 16$ and $b^2 = 9$, or $a = 4$ and $b = 3$. In turn, this means the vertices are $(4, 0)$ and $(-4, 0)$ and the conjugate axis extends from $(0, -3)$ to $(0, 3)$.

Because $c^2 = a^2 + b^2$, we find that $c^2 = 25$, or $c = 5$. The foci are $(-5, 0)$ and $(5, 0)$.

Drawing the rectangle and the asymptotes in Fig. 21.73, we then sketch in the hyperbola from each vertex toward each asymptote. ■

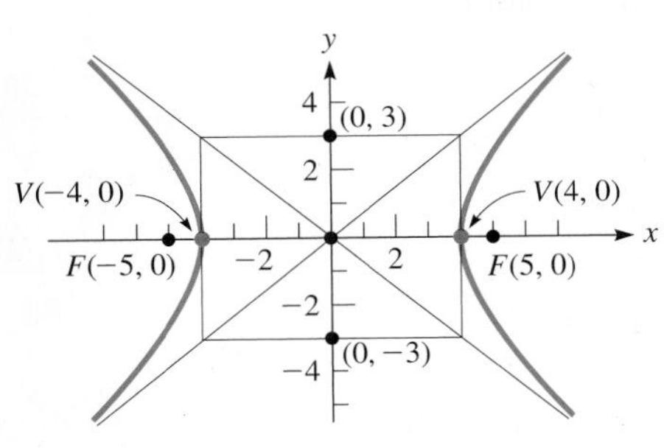

Fig. 21.73

EXAMPLE 2 Standard equation—transverse axis on y-axis

The hyperbola $\frac{y^2}{4} - \frac{x^2}{16} = 1$ (a^2 = 4, b^2 = 16)

has vertices at $(0, -2)$ and $(0, 2)$. Its conjugate axis extends from $(-4, 0)$ to $(4, 0)$. The foci are $(0, -2\sqrt{5})$ and $(0, 2\sqrt{5})$. We find this directly from the equation because the y^2-term is the positive term with 1 on the right. This means the equation fits the form of Eq. (21.24) with $a^2 = 4$ and $b^2 = 16$. Also, $c^2 = 20$, which means that $c = \sqrt{20} = 2\sqrt{5}$.

Because $2a$ extends along the y-axis, we see that the equations of the asymptotes are $y = \pm(a/b)x$. This is not a contradiction of Eq. (21.23) but the extension of it for a hyperbola with its transverse axis along the y-axis. The ratio a/b gives the slope of the asymptote. The hyperbola is shown in Fig. 21.74. ■

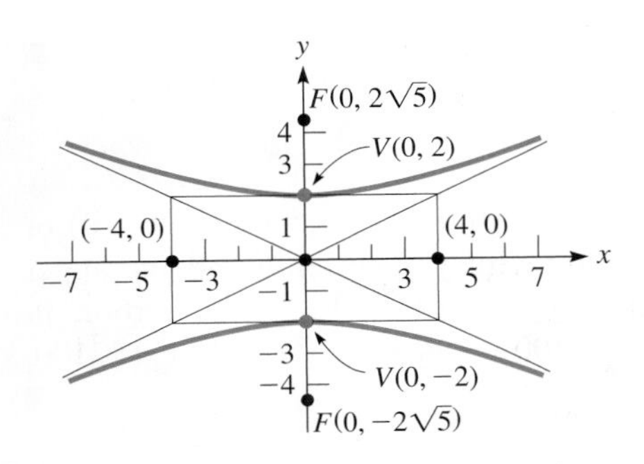

Fig. 21.74

EXAMPLE 3 Find vertices, foci

Determine the coordinates of the vertices and foci of the hyperbola

$$4x^2 - 9y^2 = 36$$

First, by dividing through by 36, we have

$$\frac{x^2}{9} - \frac{y^2}{4} = 1$$

(form requires − and 1)

From this form, we see that $a^2 = 9$ and $b^2 = 4$. In turn, this tells us that $a = 3$, $b = 2$, and $c = \sqrt{9 + 4} = \sqrt{13}$. Because a^2 appears under x^2, the equation fits the form of Eq. (21.20). Therefore, the vertices are $(-3, 0)$ and $(3, 0)$ and the foci are $(-\sqrt{13}, 0)$ and $(\sqrt{13}, 0)$. The hyperbola is shown in Fig. 21.75. ■

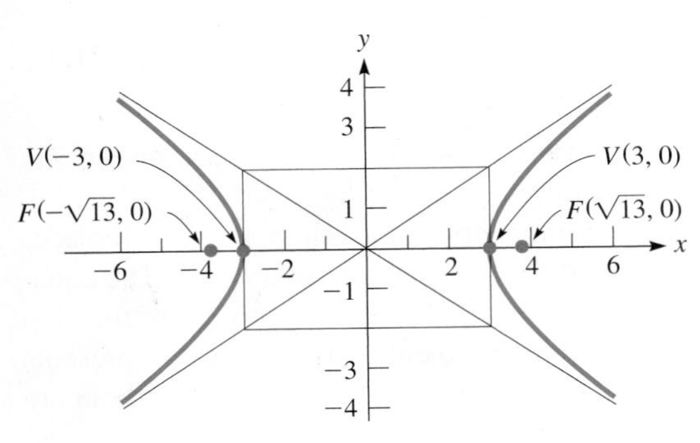

Fig. 21.75

Practice Exercises

Find the vertices and foci of each hyperbola.

1. $\frac{x^2}{36} + \frac{y^2}{13} = 1$ **2.** $4y^2 - x^2 = 4$

EXAMPLE 4 Hyperbola—water pipe

In physics, it is shown that where the velocity of a fluid is greatest, the pressure is the least. In designing an experiment to study this effect in the flow of water, a pipe is constructed such that its lengthwise cross section is hyperbolic. The pipe is 1.0 m long, 0.2 m in diameter at the narrowest point in the middle, and 0.4 m in diameter at each end. What is the equation that represents the cross section of the pipe as shown in Fig. 21.76?

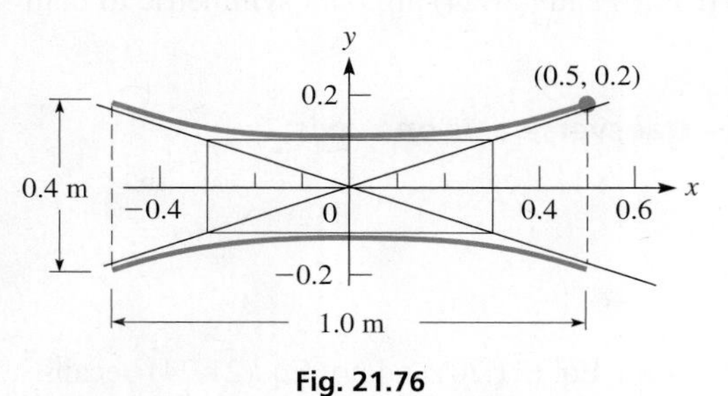

Fig. 21.76

As shown, the hyperbola has its transverse axis along the y-axis and its center at the origin. This means the general equation is given by Eq. (21.24). Because the radius at the middle of the pipe is 0.1 m, we know that $a = 0.1$ m. Also, because it is 1.0 m long and the radius at the end is 0.2 m, we know the point $(0.5, 0.2)$ is on the hyperbola. This point must satisfy the equation.

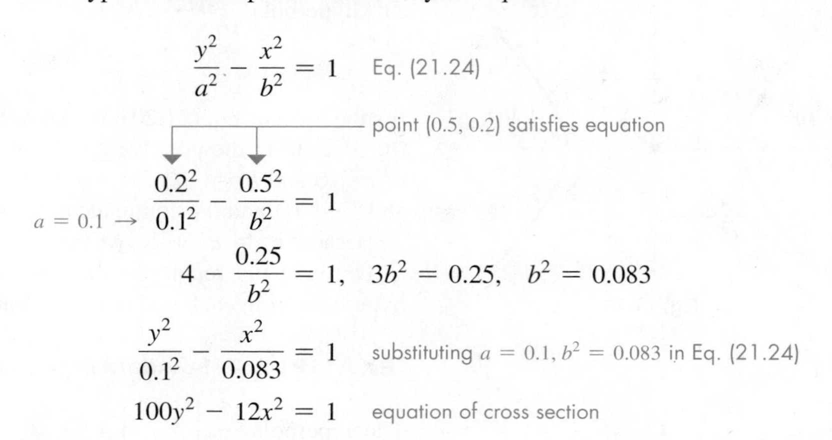

$$\frac{y^2}{a^2} - \frac{x^2}{b^2} = 1 \quad \text{Eq. (21.24)}$$

point (0.5, 0.2) satisfies equation

$$a = 0.1 \rightarrow \frac{0.2^2}{0.1^2} - \frac{0.5^2}{b^2} = 1$$

$$4 - \frac{0.25}{b^2} = 1, \quad 3b^2 = 0.25, \quad b^2 = 0.083$$

$$\frac{y^2}{0.1^2} - \frac{x^2}{0.083} = 1 \quad \text{substituting } a = 0.1, b^2 = 0.083 \text{ in Eq. (21.24)}$$

$$100y^2 - 12x^2 = 1 \quad \text{equation of cross section}$$

■

Practice Exercise

3. In Example 4, find the equation if 0.4 m is changed to 0.6 m.

EXAMPLE 5 Hyperbola—calculator display

If we use a calculator to display a hyperbola represented by either Eq. (21.20) or Eq. (21.24), we have two functions when we solve for y. One represents the upper half of the hyperbola, and the other represents the lower half. For the hyperbola in Example 4, the functions are $y_1 = \sqrt{(12x^2 + 1)/100}$ and $y_2 = -\sqrt{(12x^2 + 1)/100}$. See Fig. 21.77. ■

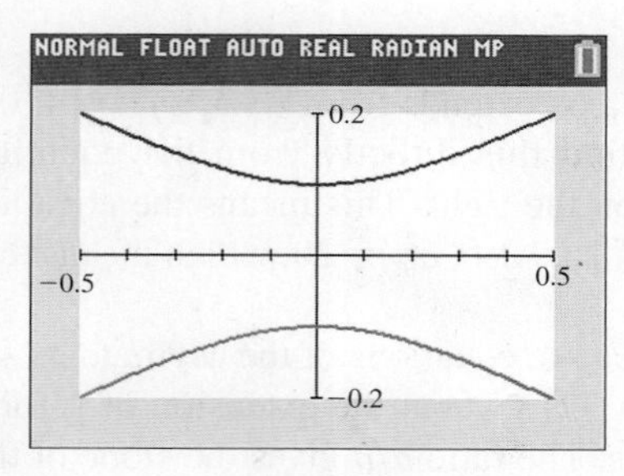

Fig. 21.77

Equations (21.20) and (21.24) give us the standard forms of the equation of the hyperbola with its center at the origin and its foci on one of the coordinate axes. There is another important equation form that represents a hyperbola, and it is

$$xy = c \qquad \textbf{(21.25)}$$

The asymptotes of this hyperbola are the coordinate axes, and the foci are on the line $y = x$ *if* c *is positive or on the line* $y = -x$ *if* c *is negative.*

The hyperbola represented by Eq. (21.25) is symmetric to the origin, for if $-x$ replaces x and $-y$ replaces y at the same time, we obtain $(-x)(-y) = c$, or $xy = c$. The equation is unchanged. However, if $-x$ replaces x or $-y$ replaces y, but not both, the sign on the left is changed. This means it is not symmetric to either axis. Here, c represents a constant and is not related to the focus. Two examples of this type of hyperbola are shown on the next page.

EXAMPLE 6 $xy = c$ **Hyperbola**

Plot the graph of the hyperbola $xy = 4$.

We find the values in the table below and then plot the appropriate points. Here, it is permissible to use a limited number of points, because we know the equation represents a hyperbola. Therefore, using $y = 4/x$, we obtain the values

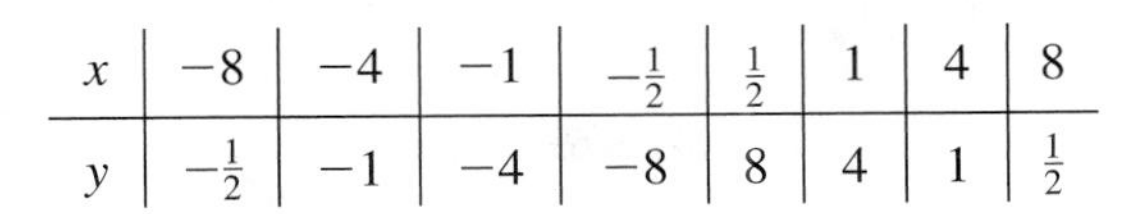

x	-8	-4	-1	$-\frac{1}{2}$	$\frac{1}{2}$	1	4	8
y	$-\frac{1}{2}$	-1	-4	-8	8	4	1	$\frac{1}{2}$

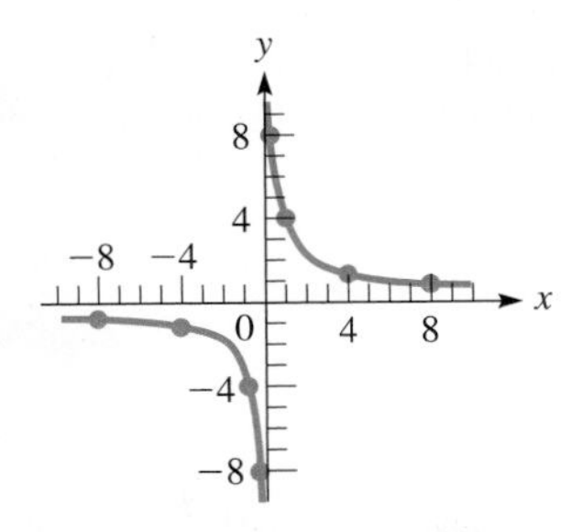

Fig. 21.78

Note that neither x nor y may equal zero. The hyperbola is shown in Fig. 21.78.

If the constant on the right is negative (for example, if $xy = -4$), then the two branches of the hyperbola are in the second and fourth quadrants. ■

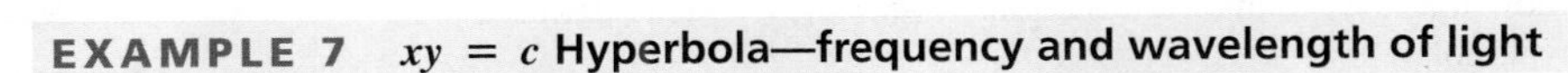

EXAMPLE 7 $xy = c$ **Hyperbola—frequency and wavelength of light**

For a light wave, the product of its frequency f of vibration and its wavelength λ is a constant, and this constant is the speed of light c. For green light, for which $f = 600$ THz, $\lambda = 500$ nm. Graph λ as a function of f for any light wave.

From the statement above, we know that $f\lambda = c$, and from the given values, we have

$$(600 \text{ THz})(500 \text{ nm}) = (6.0 \times 10^{14} \text{ Hz})(5.0 \times 10^{-7} \text{ m}) = 3.0 \times 10^8 \text{ m/s}$$

which means $c = 3.0 \times 10^8$ m/s. We are to sketch $f\lambda = 3.0 \times 10^8$. Solving for λ as $\lambda = (3.0 \times 10^8)/f$ **we have the table at the left** (only positive values have meaning). See Fig. 21.79. (Violet light has wavelengths of about 400 nm, orange light has wavelengths of about 600 nm, and red light has wavelengths of about 700 nm.) ■

f (THz)	750	600	500	430
λ (nm)	400	500	600	700

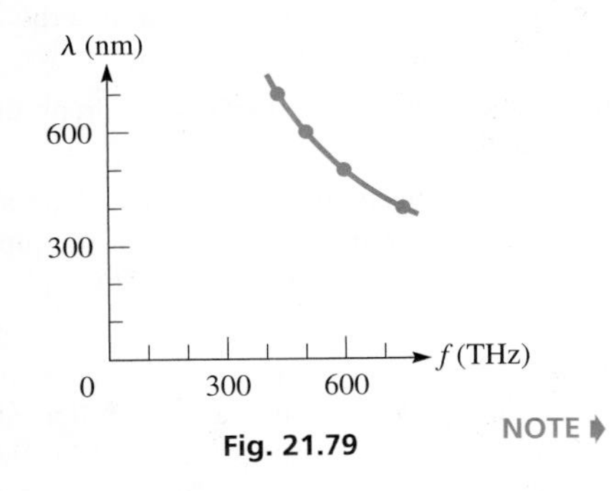

Fig. 21.79

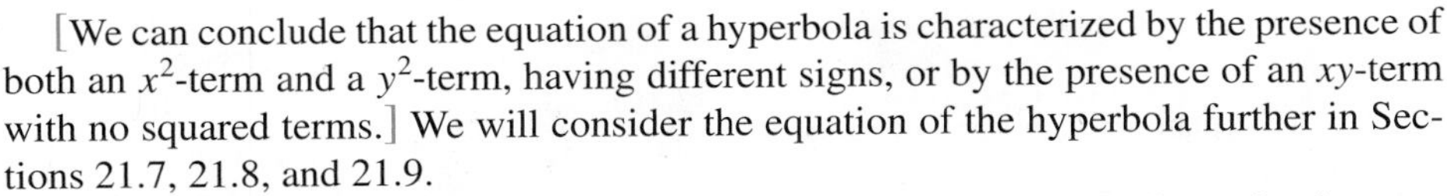

NOTE ▸ [We can conclude that the equation of a hyperbola is characterized by the presence of both an x^2-term and a y^2-term, having different signs, or by the presence of an xy-term with no squared terms.] We will consider the equation of the hyperbola further in Sections 21.7, 21.8, and 21.9.

The hyperbola has some very useful applications. The LORAN radio navigation system is based on the use of hyperbolic paths. Some reflecting telescopes use hyperbolic mirrors. The paths of comets that never return to pass by the sun are hyperbolic. Some additional applications are illustrated in the exercises.

■ The first reasonable measurement of the speed of light was made by the Danish astronomer Olaf Roemer (1644–1710). He measured the time required for light to come from the moons of Jupiter across Earth's orbit.

EXERCISES 21.6

In Exercises 1 and 2, make the given changes in the indicated examples of this section and then solve the resulting problems.

1. In Example 2, interchange the denominators of 4 and 16; find the vertices, ends of the conjugate axis, and foci. Sketch the curve.
2. In Example 3, change $-9y^2$ to $-y^2$ and then follow the same instructions as in Exercise 1.

In Exercises 3–16, find the coordinates of the vertices and the foci of the given hyperbolas. Sketch each curve.

3. $\dfrac{x^2}{25} - \dfrac{y^2}{144} = 1$
4. $\dfrac{x^2}{16} - \dfrac{y^2}{4} = 1$
5. $\dfrac{y^2}{9} - \dfrac{x^2}{1} = 1$
6. $\dfrac{y^2}{4} - \dfrac{x^2}{21} = 1$
7. $\dfrac{4x^2}{25} - \dfrac{y^2}{4} = 1$
8. $\dfrac{9y^2}{25} - x^2 = 1$
9. $9x^2 - y^2 = 4$
10. $4x^2 - 9y^2 = 6$
11. $2y^2 - 5x^2 = 20$
12. $9y^2 - 16x^2 = 9$
13. $y^2 = 4(x^2 + 1)$
14. $y^2 = 9(x^2 - 1)$
15. $4x^2 - y^2 = 0.64$
16. $9y^2 - x^2 = 0.72$

In Exercises 17–28, find the equations of the hyperbolas satisfying the given conditions. The center of each is at the origin.

17. Vertex $(3, 0)$, focus $(5, 0)$
18. Vertex $(0, -1)$, focus $(0, -\sqrt{3})$
19. Conjugate axis $= 48$, vertex $(0, 10)$
20. Sum of lengths of transverse and conjugate axes 28, focus $(10, 0)$
21. Passes through $(2, 3)$, focus $(2, 0)$

22. Passes through $(8, \sqrt{3})$, vertex $(4, 0)$
23. Passes through $(5, 4)$ and $(3, \frac{4}{5}\sqrt{5})$
24. Passes through $(1, 2)$ and $(2, 2\sqrt{2})$
25. Asymptote $y = 2x$, vertex $(1, 0)$
26. Asymptote $y = -4x$, vertex $(0, 4)$
27. The difference of distances to (x, y) from $(10, 0)$ and $(-10, 0)$ is 12.
28. The difference of distances to (x, y) from $(0, 4)$ and $(0, -4)$ is 6.

In Exercises 29–54, solve the given problems.

29. Sketch the graph of the hyperbola $xy = 2$.
30. Sketch the graph of the hyperbola $xy = -4$.
31. Show that the parametric equations $x = \sec t$, $y = \tan t$ define a hyperbola.
32. Show that all hyperbolas $\dfrac{x^2}{\cos^2\theta} - \dfrac{y^2}{\sin^2\theta} = 1$ have foci at $(\pm 1, 0)$ for all values of θ.
33. Find any points of intersection of the ellipse $2x^2 + y^2 = 17$ and the hyperbola $y^2 - x^2 = 5$.
34. Find any points of intersection of the hyperbolas $x^2 - 3y^2 = 22$ and $xy = 5$.
35. Find the equation of the hyperbola with foci $(1, 2)$ and $(11, 2)$, and a transverse axis of 8, by use of the definition. Sketch the curve.
36. Find the equation of the hyperbola with vertices $(-2, 4)$ and $(-2, -2)$, and a conjugate axis of 4, by use of the definition. Sketch the curve.
37. Use a calculator to view the hyperbola $x^2 - 4y^2 + 4x + 32y - 64 = 0$.
38. Use a calculator to view the hyperbola $5y^2 - 4x^2 + 8x + 40y + 56 = 0$.
39. The equation of a hyperbola with center (h, k) and transverse axis parallel to the x-axis is $\dfrac{(x - h)^2}{a^2} - \dfrac{(y - k)^2}{b^2} = 1$. (This is shown in Section 21.7.) Sketch the hyperbola that has a transverse axis of 4, a conjugate axis of 6, and for which (h, k) is $(-3, 2)$.
40. The equation of a hyperbola with center (h, k) and transverse axis parallel to the y-axis is $\dfrac{(y - k)^2}{a^2} - \dfrac{(x - h)^2}{b^2} = 1$. (This is shown in Section 21.7.) Sketch the hyperbola that has a transverse axis of 2, a conjugate axis of 8, and for which (h, k) is $(5, 0)$.
41. Two concentric (same center) hyperbolas are called conjugate hyperbolas if the transverse and conjugate axes of one are, respectively, the conjugate and transverse axes of the other. What is the equation of the hyperbola conjugate to the hyperbola given by $\dfrac{x^2}{9} - \dfrac{y^2}{16} = 1$?
42. As with an ellipse, the *eccentricity* e of a hyperbola is defined as $e = c/a$. Find the eccentricity of the hyperbola $2x^2 - 3y^2 = 24$.
43. Find the equation of the hyperbola that has the same foci as the ellipse $\frac{x^2}{169} + \frac{y^2}{144} = 1$ and passes through $(4\sqrt{2}, 3)$.
44. Explain how a branch of a hyperbola differs from a parabola.
45. Determine the equation of the hyperbola for which the difference in distances from $(-6, 0)$ and $(6,0)$ is (a) 4, (b) 8.
46. Two persons, 1000 m apart, heard an explosion, one hearing it 4.0 s before the other. Explain why the location of the explosion can be on one of the points of a hyperbola.
47. The cross section of the roof of a storage building shown in Fig. 21.80 is hyperbolic with the horizontal beam passing through the focus. Find the equation of the hyperbola such that its center is at the origin.

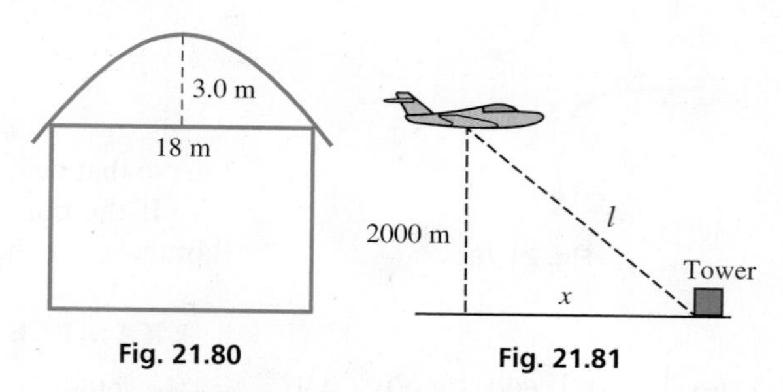

Fig. 21.80 Fig. 21.81

48. A plane is flying at a constant altitude of 2000 m. Show that the equation relating the horizontal distance x and the direct-line distance l from a control tower to the plane is that of a hyperbola. Sketch the graph of l as a function of x. See Fig. 21.81.
49. A jet travels 600 km at a speed of v km/h for t hours. Graph the equation relating v as a function of t.
50. A drain pipe 100 m long has an inside diameter d (in m) and an outside diameter D (in m). If the volume of material of the pipe itself is 0.50 m^3, what is the equation relating d and D? Graph D as a function of d.
51. Ohm's law in electricity states that the product of the current i and the resistance R equals the voltage V across the resistance. If a battery of 6.00 V is placed across a variable resistor R, find the equation relating i and R and sketch the graph of i as a function of R.
52. A ray of light directed at one focus of a hyperbolic mirror is reflected toward the other focus. Find the equation that represents the hyperbolic mirror shown in Fig. 21.82.

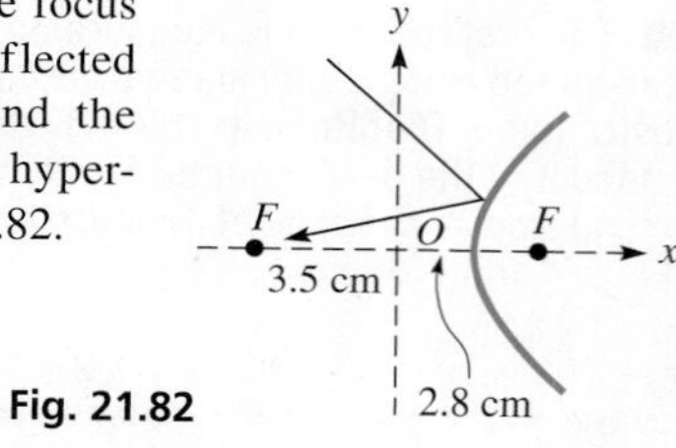

Fig. 21.82

53. A radio signal is sent simultaneously from stations A and B 600 km apart on the Carolina coast. A ship receives the signal from A 1.20 ms before it receives the signal from B. Given that radio signals travel at 300 km/ms, draw a graph showing the possible locations of the ship. This problem illustrates the basis of LORAN.
54. Maximum intensity for monochromatic (single-color) light from two sources occurs where the difference in distances from the sources is an integer number of wavelengths. Find the equation of the curves of maximum intensity in a thin film between the sources where the difference in paths is two wavelengths and the sources are four wavelengths apart. Let the sources be on the x-axis and the origin midway between them. Use units of one wavelength for both x and y.

Answers to Practice Exercises

1. $V(6, 0)$, $V(-6, 0)$; $F(7, 0)$, $F(-7, 0)$
2. $V(0, 1)$, $V(0, -1)$; $F(0, \sqrt{5})$, $F(0, -\sqrt{5})$
3. $100y^2 - 32x^2 = 1$

21.7 Translation of Axes

Centers of Curves Not at Origin • Axes of Curves Parallel to Coordinate Axes • Translation of Axes

The equations we have considered for the parabola, the ellipse, and the hyperbola are those for which the center of the ellipse or hyperbola, or vertex of the parabola, is at the origin. In this section, we consider, without specific use of the definition, the equations of these curves for the cases in which the axis of the curve is parallel to one of the coordinate axes. This is done by **translation of axes**.

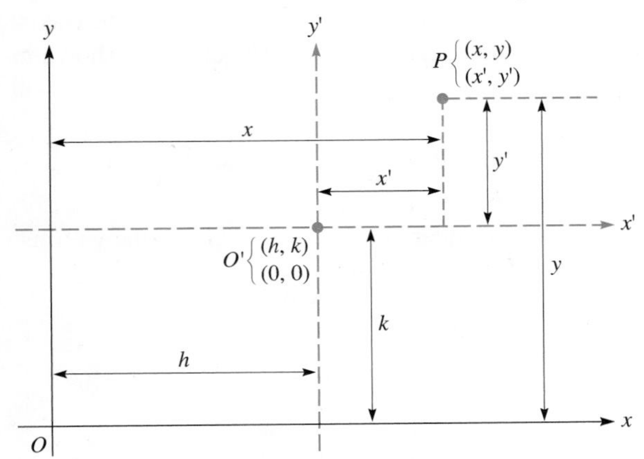

Fig. 21.83

In Fig. 21.83, we choose a point (h, k) in the xy-coordinate plane as the origin of another coordinate system, the $x'y'$-coordinate system. The x'-axis is parallel to the x-axis and the y'-axis is parallel to the y-axis. Every point now has two sets of coordinates (x, y) and (x', y'). We see that

$$x = x' + h \quad \text{and} \quad y = y' + k \tag{21.26}$$

Equation (21.26) can also be written in the form

$$x' = x - h \quad \text{and} \quad y' = y - k \tag{21.27}$$

EXAMPLE 1 Find an equation—given vertex and focus

Find the equation of the parabola with vertex $(2, 4)$ and focus $(4, 4)$.

If we let the origin of the $x'y'$-coordinate system be the point $(2, 4)$, then the point $(4, 4)$ is the point $(2, 0)$ in the $x'y'$-system. This means $p = 2$ and $4p = 8$. See Fig. 21.84. In the $x'y'$-system, the equation is

$$(y')^2 = 8(x')$$

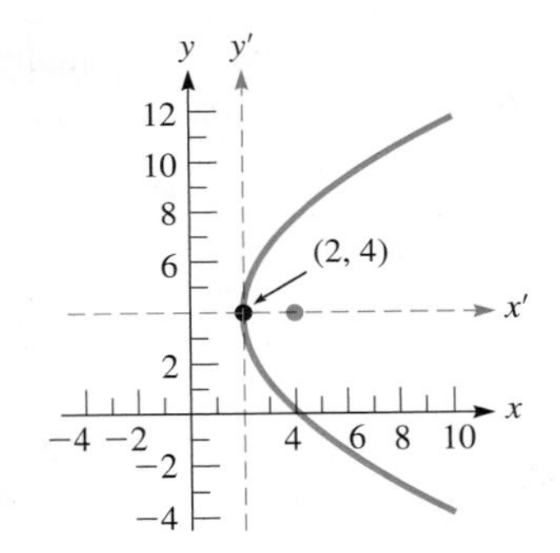

Fig. 21.84

Using Eqs. (21.27), we have

$$(y - 4)^2 = 8(x - 2)$$

coordinates of vertex (2, 4)

as the equation of the parabola in the xy-coordinate system. ■

Following the method of Example 1, by writing the equation of the curve in the $x'y'$-system and then using Eqs. (21.27), we have the following more general forms of the equations of the parabola, ellipse, and hyperbola.

Parabola, vertex (h, k): $(y - k)^2 = 4p(x - h)$ (axis parallel to x-axis) **(21.28)**

$(x - h)^2 = 4p(y - k)$ (axis parallel to y-axis) **(21.29)**

Ellipse, center (h, k): $\dfrac{(x - h)^2}{a^2} + \dfrac{(y - k)^2}{b^2} = 1$ (major axis parallel to x-axis) **(21.30)**

$\dfrac{(y - k)^2}{a^2} + \dfrac{(x - h)^2}{b^2} = 1$ (major axis parallel to y-axis) **(21.31)**

Hyperbola, center (h, k): $\dfrac{(x - h)^2}{a^2} - \dfrac{(y - k)^2}{b^2} = 1$ (transverse axis parallel to x-axis) **(21.32)**

$\dfrac{(y - k)^2}{a^2} - \dfrac{(x - h)^2}{b^2} = 1$ (transverse axis parallel to y-axis) **(21.33)**

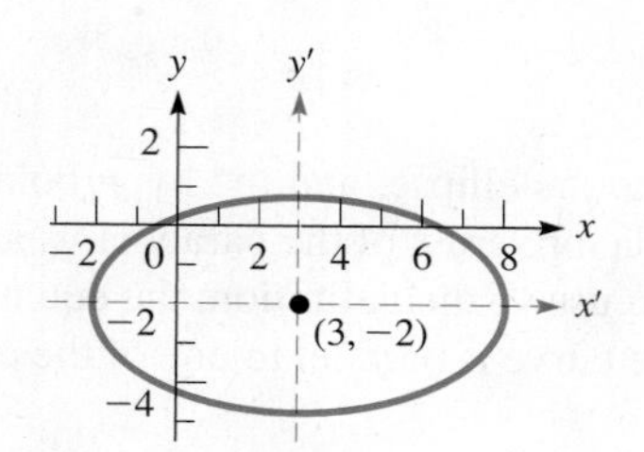

Fig. 21.85

EXAMPLE 2 Describing a curve—given equation

Describe the curve of the equation

$$\frac{(x-3)^2}{25}+\frac{(y+2)^2}{9}=1$$

We see that this equation fits the form of Eq. (21.30) with $h = 3$ and $k = -2$. It is the equation of an ellipse with its center at $(3, -2)$ and its major axis parallel to the x-axis. The semimajor axis is $a = 5$, and the semiminor axis is $b = 3$. The ellipse is shown in Fig. 21.85. ■

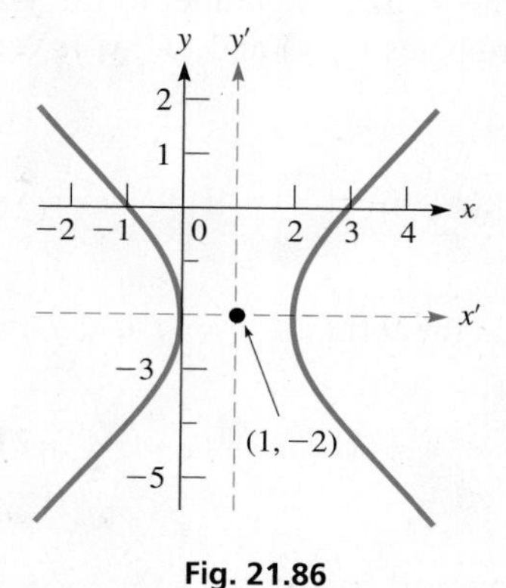

Fig. 21.86

EXAMPLE 3 Finding the center of a hyperbola

Find the center of the hyperbola $2x^2 - y^2 - 4x - 4y - 4 = 0$.

To analyze this curve, we first complete the square in the x-terms and in the y-terms. This will allow us to recognize the values of h and k.

$$2x^2 - 4x - y^2 - 4y = 4$$
$$2(x^2 - 2x \quad) - (y^2 + 4y \quad) = 4$$
$$2(x^2 - 2x + 1) - (y^2 + 4y + 4) = 4 + 2 - 4$$

NOTE ▸ [We note here that when we added 1 to complete the square of the x-terms within the parentheses, we were actually adding 2 to the left side. Thus, we added 2 to the right side. Similarly, when we added 4 to the y-terms within the parentheses, we were actually subtracting 4 from the left side. Thus, we subtract 4 from the right side.] Continuing, we have

$$2(x-1)^2 - (y+2)^2 = 2$$

coordinates of center $(1, -2)$

$$\frac{(x-1)^2}{1}-\frac{(y+2)^2}{2}=1$$

Therefore, the center of the hyperbola is $(1, -2)$. See Fig. 21.86. ■

Practice Exercise

1. Find the center and foci of ellipse $2x^2 + 3y^2 - 8x + 18y + 29 = 0$.

EXAMPLE 4 Translation of axes—surface area of a beaker

Cylindrical glass beakers are to be made with a height of 3 in. Express the surface area in terms of the radius of the base and sketch the curve.

The total surface area S of a beaker is the sum of the area of the base and the lateral surface area of the side. In general, S in terms of the radius r of the base and height h of the side is $S = \pi r^2 + 2\pi rh$. Because $h = 3$ in., we have

$$S = \pi r^2 + 6\pi r$$

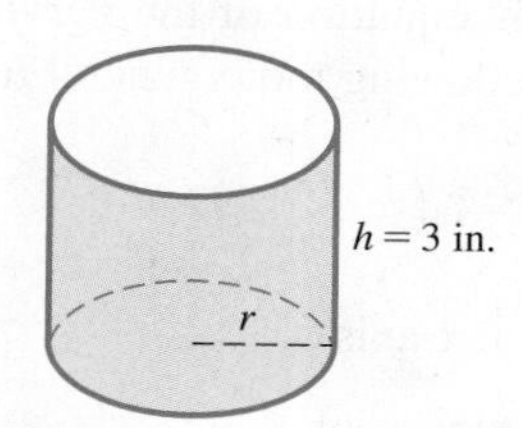

Fig. 21.87

which is the desired relationship. See Fig. 21.87.

To analyze the equation relating S and r, we complete the square of the r terms:

$$S = \pi(r^2 + 6r)$$
$$S + 9\pi = \pi(r^2 + 6r + 9) \quad \text{complete the square}$$
$$S + 9\pi = \pi(r+3)^2$$

vertex $(-3, -9\pi)$

$$(r+3)^2 = \frac{1}{\pi}(S + 9\pi)$$

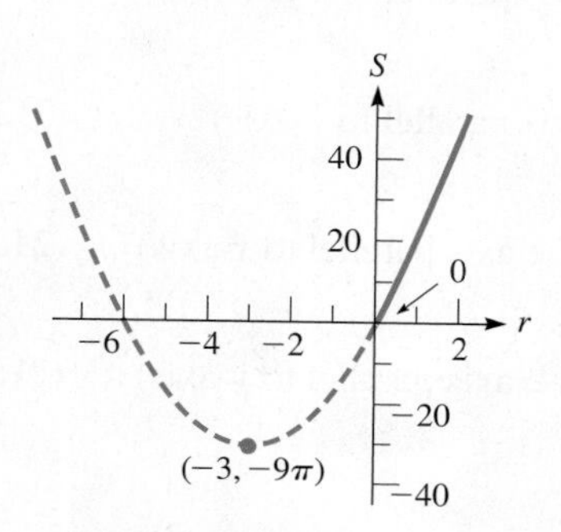

Fig. 21.88

This represents a parabola with vertex $(-3, -9\pi)$. See Fig. 21.88. Because $4p = 1/\pi$, $p = 1/(4\pi)$ and the focus is at $(-3, \frac{1}{4\pi} - 9\pi)$. The part of the graph for negative r is dashed because only positive values have meaning. ■

EXERCISES 21.7

In Exercises 1 and 2, make the given changes in the indicated examples of this section and then solve the resulting problem.

1. In Example 2, change $y + 2$ to $y - 2$ and change the sign before the second term from + to −. Then describe and sketch the curve.

2. In Example 3, change the fourth and fifth terms from $-4y - 4$ to $+ 6y - 9$ and then find the center.

In Exercises 3–10, describe the curve represented by each equation. Identify the type of curve and its center (or vertex if it is a parabola). Sketch each curve.

3. $(y - 2)^2 = 4(x + 1)$

4. $\dfrac{(x + 4)^2}{4} + (y - 1)^2 = 1$

5. $\dfrac{(x - 1)^2}{4} - \dfrac{(y - 2)^2}{25} = 1$

6. $(y + 5)^2 = -8(x - 2)$

7. $(x + 1)^2 + \dfrac{y^2}{36} = 1$

8. $\dfrac{(y - 4)^2}{49} - \dfrac{(x + 2)^2}{4} = 1$

9. $(x + 3)^2 = -12(y - 1)$

10. $\dfrac{x^2}{0.16} + \dfrac{(y + 1)^2}{0.25} = 1$

In Exercises 11–22, find the equation of each of the curves described by the given information.

11. Parabola: vertex $(-1, 3)$, focus $(-1, 6)$

12. Parabola: focus $(4, -4)$, directrix $y = -2$

13. Parabola: axis, directrix are coordinate axes, focus $(10, 0)$

14. Parabola: vertex $(4, 4)$, vertical directrix, passes through $(0, 1)$

15. Ellipse: center $(-2, 2)$, focus $(-5, 2)$, vertex $(-7, 2)$

16. Ellipse: center $(0, 3)$, focus $(12, 3)$, major axis 26 units

17. Ellipse: center $(-2, 1)$, vertex $(-2, 5)$, passes through $(0, 1)$

18. Ellipse: foci $(1, -2)$ and $(1, 10)$, minor axis 5 units

19. Hyperbola: vertex $(-1, 1)$, focus $(-1, 4)$, center $(-1, 2)$

20. Hyperbola: foci $(2, 1)$ and $(8, 1)$, conjugate axis 6 units

21. Hyperbola: vertices $(2, 1)$ and $(-4, 1)$, focus $(-6, 1)$

22. Hyperbola: center $(1, -4)$, focus $(1, 1)$, transverse axis 8 units

In Exercises 23–40, determine the center (or vertex if the curve is a parabola) of the given curve. Sketch each curve.

23. $x^2 + 2x - 4y - 3 = 0$

24. $y^2 - 2x - 2y - 9 = 0$

25. $x^2 + 4y = 24$

26. $x^2 + 4y^2 = 32y$

27. $4x^2 + 9y^2 + 24x = 0$

28. $2x^2 + y^2 + 8x = 8y$

29. $9x^2 - y^2 + 8y = 7$

30. $2x^2 - 4x = 9y - 2$

31. $5x^2 - 4y^2 + 20x + 8y = 4$

32. $0.04x^2 + 0.16y^2 = 0.01y$

33. $4x^2 - y^2 + 32x + 10y + 35 = 0$

34. $2x^2 + 2y^2 - 24x + 16y + 95 = 0$

35. $16x^2 + 25y^2 - 32x + 100y - 284 = 0$

36. $9x^2 - 16y^2 - 18x + 96y - 279 = 0$

37. $5x^2 - 3y^2 + 95 = 40x$

38. $5x^2 + 12y + 18 = 2y^2$

39. $9x^2 + 9y^2 + 14 = 6x + 24y$

40. $4y^2 + 29 = 15x + 12y$

In Exercises 41–54, solve the given problems.

41. Find the equation of the hyperbola with asymptotes $x - y = -1$ and $x + y = -3$ and vertex $(3, -1)$.

42. The circle $x^2 + y^2 + 4x - 5 = 0$ passes through the foci and the ends of the minor axis of an ellipse that has its major axis along the x-axis. Find the equation of the ellipse.

43. The vertex and focus of one parabola are, respectively, the focus and vertex of a second parabola. Find the equation of the first parabola, if $y^2 = 4x$ is the equation of the second.

44. Identify the curve represented by $4y^2 - x^2 - 6x - 2y = 14$ and view it on a graphing calculator.

45. What is the general form of the equation of a *family* of parabolas if each vertex and focus is on the x-axis?

46. What is the general form of the equation of a *family* of ellipses with foci on the y-axis if each passes through the origin?

47. If $(a,3)$ is a point on the parabola $y = x^2 + 2x$, what is a?

48. The vertical cross-section of a culvert under a road is elliptical. The culvert is 18 m wide and 12 m high. Find an equation to represent the perimeter of the culvert with the origin at road level and 2.0 m above the top of the culvert. See Fig. 21.89.

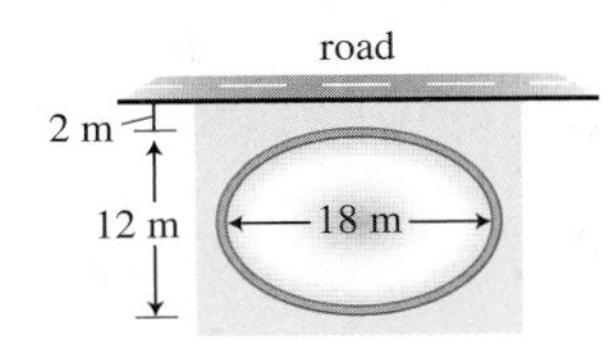

Fig. 21.89

49. An electric current (in A) is $i = 2 + \sin(2\pi t - \frac{\pi}{3})$. What is the equation for the current if the origin of the (t', i') system is taken as $(\frac{1}{6}, 2)$ of the (t, i) system?

50. The stopping distance d (in ft) of a car traveling at v mi/h is represented by $d = 0.05v^2 + v$. Where is the vertex of the parabola that represents d?

51. The stream from a fire hose follows a parabolic curve and reaches a maximum height of 60 ft at a horizontal distance of 95 ft from the nozzle. Find the equation that represents the stream, with the origin at the nozzle. Sketch the graph.

52. For a constant capacitive reactance and a constant resistance, sketch the graph of the impedance and inductive reactance (as abscissas) for an alternating-current circuit. (See Section 12.7.)

53. Two wheels in a friction drive assembly are equal ellipses, as shown in Fig. 21.90. They are always in contact, with the left wheel fixed in position and the right wheel able to move horizontally. Find the equation that can be used to represent the circumference of each wheel in the position shown.

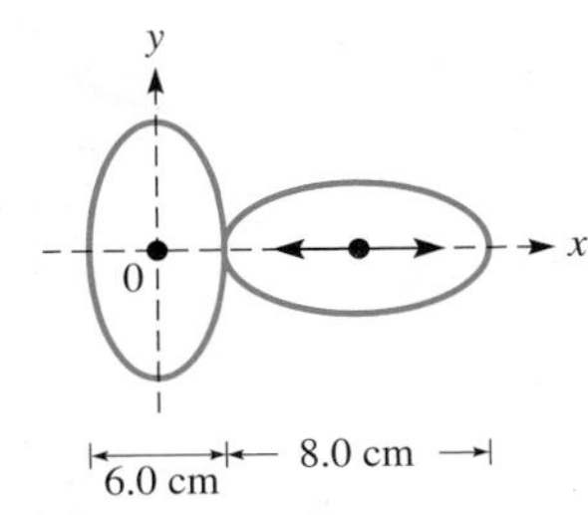

Fig. 21.90

54. An agricultural test station is to be divided into rectangular sections, each with a perimeter of 480 m. Express the area A of each section in terms of its width w and identify the type of curve represented. Sketch the graph of A as a function of w. For what value of w is A the greatest?

Answer to Practice Exercise

1. $C(2, -3), F(1, -3), F(3, -3)$

21.8 The Second-degree Equation

Second-degree Equation • Coefficients A, B, C Determine the Type of Curve • Conic Sections

The equations of the circle, parabola, ellipse, and hyperbola are all special cases of the same general equation. In this section, we discuss this equation and how to identify the particular form it takes when it represents a specific type of curve.

Each of these curves can be represented by a **second-degree equation** *of the form*

$$Ax^2 + Bxy + Cy^2 + Dx + Ey + F = 0 \tag{21.34}$$

The coefficients of the second-degree equation terms determine the type of curve that results. Recalling the discussions of the general forms of the equations of the circle, parabola, ellipse, and hyperbola from the previous sections of this chapter, Eq. (21.34) represents the indicated curve for given conditions of A, B, and C, as follows:

Identifying a Conic Section from the Second-degree Equation

1. If $A = C$, $B = 0$, a circle.
2. If $A \neq C$ (but they have the same sign), $B = 0$, an ellipse.
3. If A and C have different signs, $B = 0$, a hyperbola.
4. If $A = 0$, $C = 0$, $B \neq 0$, a hyperbola.
5. If either $A = 0$ or $C = 0$ (both not both), $B = 0$, a parabola.
(Special cases, such as a single point or no real locus, can also result.)

Another conclusion about Eq. (21.34) is that, if either $D \neq 0$ or $E \neq 0$ (or both), the center of the curve (or vertex of a parabola) is not at the origin. If $B \neq 0$, the axis of the curve has been rotated. We have considered only one such case (the hyperbola $xy = c$) so far in this chapter. In the following section, *rotation of axes* is taken up, and if $B \neq 0$, we will see that the type of curve depends on the value $B^2 - 4AC$. The following examples illustrate how the type of curve is identified from the equation according to the five criteria listed above.

EXAMPLE 1 Identify a circle from the equation

The equation $2x^2 = 3 - 2y^2$ represents a circle. This can be seen by putting the equation in the form of Eq. (21.34). This form is

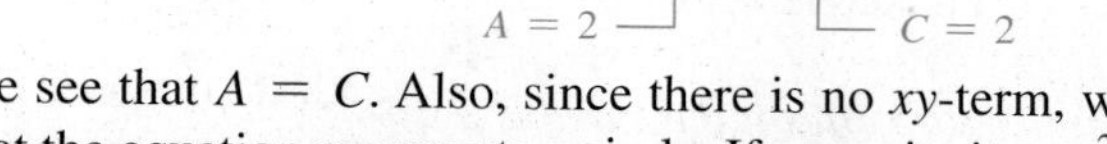

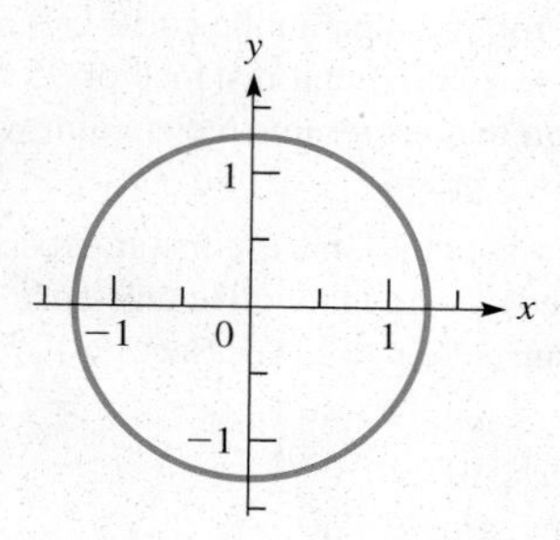

Fig. 21.91

We see that $A = C$. Also, since there is no xy-term, we know that $B = 0$. This means that the equation represents a circle. If we write it as $x^2 + y^2 = \frac{3}{2}$, we see that it fits the form of Eq. (21.12). The circle is shown in Fig. 21.91. ■

EXAMPLE 2 Identify an ellipse from the equation

The equation $3y^2 = 6y - x^2 + 3$ represents an ellipse. Before analyzing the equation, we should put it in the form of Eq. (21.34). For this equation, this form is

$$x^2 + 3y^2 - 6y - 3 = 0$$

$A = 1$ (coefficient of x^2), $C = 3$ (coefficient of y^2)

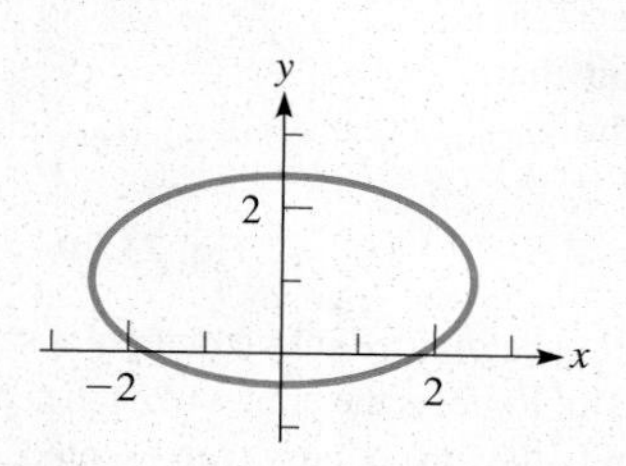

Fig. 21.92

Here, we see that $B = 0$, A and C have the same sign, and $A \neq C$. Therefore, it is an ellipse. The $-6y$ term indicates that the center of the ellipse is not at the origin. This ellipse is shown in Fig. 21.92. ■

EXAMPLE 3 Identify a hyperbola from the equation

Identify the curve represented by $2x^2 + 12x = y^2 - 14$. Determine the appropriate quantities for the curve, and sketch the graph.

Writing this equation in the form of Eq. (21.34), we have

$$2x^2 - y^2 + 12x + 14 = 0$$

$$A = 2 \qquad C = -1$$

We identify this equation as representing a hyperbola, because A and C have different signs and $B = 0$. We now write it in the standard form of a hyperbola:

$$2x^2 + 12x - y^2 = -14$$

$$2(x^2 + 6x \qquad) - y^2 = -14 \qquad \text{complete the square}$$

$$2(x^2 + 6x + 9) - y^2 = -14 + 18$$

$$2(x + 3)^2 - y^2 = 4$$

center is $(-3, 0)$

$$\frac{(x + 3)^2}{2} - \frac{y^2}{4} = 1 \quad \text{or} \quad \frac{x'^2}{2} - \frac{y'^2}{4} = 1$$

Fig. 21.93

Thus, we see that the center (h, k) of the hyperbola is the point $(-3, 0)$. Also, $a = \sqrt{2}$ and $b = 2$. This means that the vertices are $(-3 + \sqrt{2}, 0)$ and $(-3 - \sqrt{2}, 0)$, and the conjugate axis extends from $(-3, 2)$ to $(-3, -2)$. Also, $c^2 = 2 + 4 = 6$, which means that $c = \sqrt{6}$. The foci are $(-3 + \sqrt{6}, 0)$ and $(-3 - \sqrt{6}, 0)$. The graph is shown in Fig. 21.93. ■

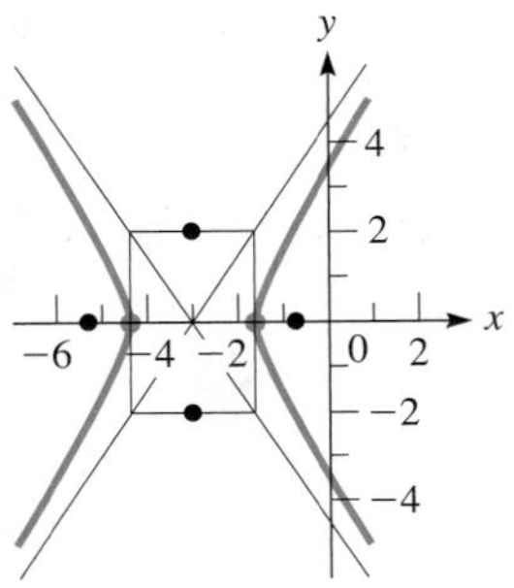

Practice Exercises

Identify the type of curve represented by each equation.

1. $2y^2 - 4y + 5 = 4x - x^2$

2. $x^2 + y^2 - 8y + 12x = x^2 - y^2$

EXAMPLE 4 Identify a parabola—calculator display

Identify the curve represented by $4y^2 - 23 = 4(4x + 3y)$ and find the appropriate important quantities. Then view it on a calculator.

Writing the equation in the form of Eq. (21.34), we have

$$4y^2 - 16x - 12y - 23 = 0$$

Therefore, we recognize the equation as representing a parabola, because $A = 0$ and $B = 0$. Now, writing the equation in the standard form of a parabola, we have

$$4y^2 - 12y = 16x + 23$$

$$4(y^2 - 3y \qquad) = 16x + 23 \qquad \text{complete the square}$$

$$4\left(y^2 - 3y + \frac{9}{4}\right) = 16x + 23 + 9$$

$$4\left(y - \frac{3}{2}\right)^2 = 16(x + 2)$$

vertex $\left(-2, \frac{3}{2}\right)$

$$\left(y - \frac{3}{2}\right)^2 = 4(x + 2) \quad \text{or} \quad y'^2 = 4x'$$

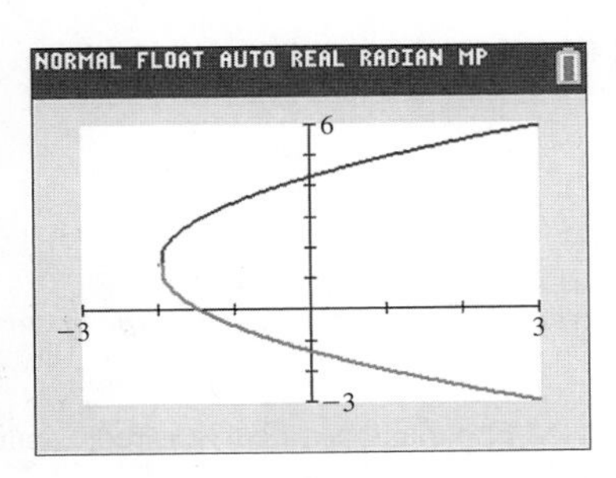

Fig. 21.94

We now note that the vertex is $(-2, 3/2)$ and that $p = 1$. This means that the focus is $(-1,3/2)$ and the directrix is $x = -3$.

To view the graph of this equation on a calculator, we first solve the equation for y and get $y = (3 \pm 4\sqrt{x + 2})/2$. Entering these two functions in the calculator, we get the view shown in Fig. 21.94. ■

In Chapter 14, when these curves were first introduced, they were referred to as **conic sections.** If a plane is passed through a cone, the intersection of the plane and the cone results in one of these curves; the curve formed depends on the angle of the plane with respect to the axis of the cone. This is shown in Fig. 21.95.

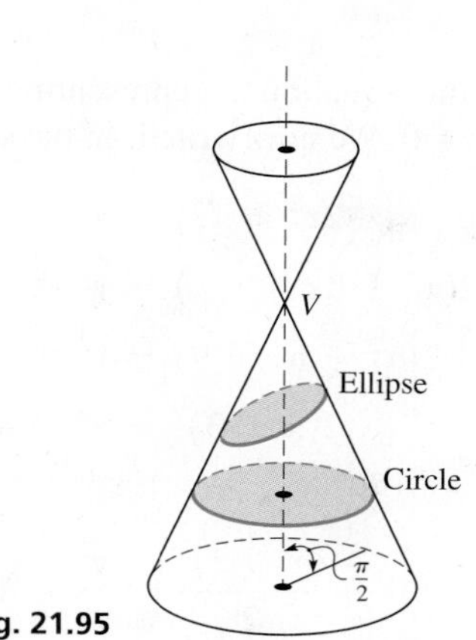

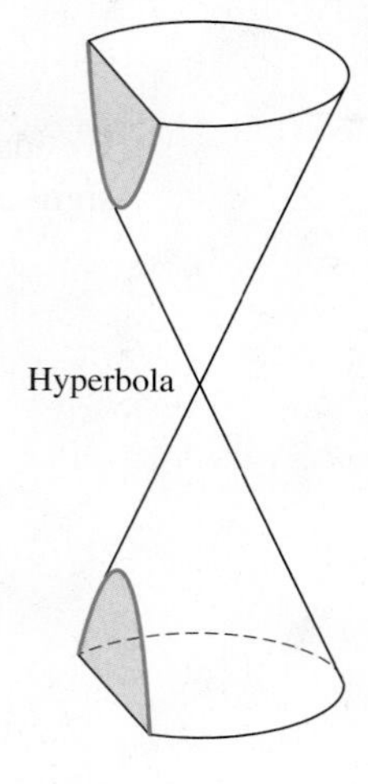

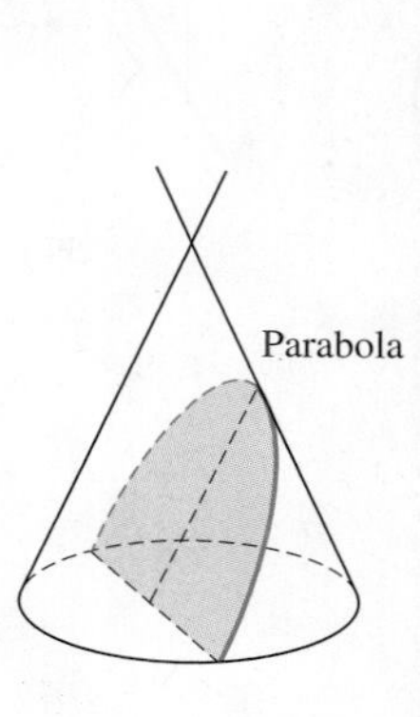

Fig. 21.95

EXERCISES 21.8

In Exercises 1 and 2, make the given changes in the indicated examples of this section and then solve the indicated problem.

1. In Example 1, change the $-$ before the $2y^2$ to $+$ and then determine what type of curve is represented.
2. In Example 3, change $y^2 - 14$ to $14 - y^2$ and then determine what type of curve is represented.

In Exercises 3–24, identify each of the equations as representing either a circle, a parabola, an ellipse, a hyperbola, or none of these.

3. $x^2 + 2y^2 - 2 = 0$
4. $x^2 - y - 5 = 0$
5. $2x^2 - y^2 - 1 = 0$
6. $y(y + x^2) = 4(x + y^2)$
7. $8x^2 + 2y^2 = 6y(1 - y)$
8. $x(x - 3) = y(1 - 2y^2)$
9. $2.2x^2 - (x + y) = 1.6$
10. $2x^2 + 4y^2 = y + 2x$
11. $x^2 = (y - 1)(y + 1)$
12. $3.2x^2 = 2.1y(1 - 2y)$
13. $36x^2 = 12y(1 - 3y) + 1$
14. $y = 3(1 - 2x)(1 + 2x)$
15. $y(3 - 2x) = x(5 - 2y)$
16. $x(13 - 5x) = 5y^2$
17. $2xy + x - 3y = 6$
18. $(y + 1)^2 = x^2 + y^2 - 1$
19. $2x(x - y) = y(3 + y - 2x)$
20. $15x^2 = x(x - 12) + 4y(y - 6)$
21. $(x + 1)^2 + (y + 1)^2 = 2(x + y + 1)$
22. $(2x + y)^2 = 4x(y - 2) - 16$
23. $x(y - x) = x^2 + x(y + 1) - y^2 + 1$
24. $4x(x - 1) = 2x^2 - 2y^2 + 3$

In Exercises 25–30, identify the curve represented by each of the given equations. Determine the appropriate important quantities for the curve and sketch the graph.

25. $x^2 = 8(y - x - 2)$
26. $x^2 = 6x - 4y^2 - 1$
27. $y^2 = 2(x^2 - 2x - 2y)$
28. $4x^2 + 4 = 9 - 8x - 4y^2$
29. $y^2 + 42 = 2x(10 - x)$
30. $x^2 - 4y = y^2 + 4(1 - x)$

In Exercises 31–36, identify the type of curve for each equation, and then view it on a calculator.

31. $x^2 + 2y^2 - 4x + 12y + 14 = 0$
32. $4y^2 - x^2 + 40y - 4x + 60 = 0$
33. $4(y^2 - 4x - 2) = 5(4y - 5)$
34. $2(2x^2 - y) = 8 - y^2$
35. $4(y^2 + 6y + 1) = x(x - 4) - 24$
36. $8x + 31 - xy = y(y - 2 - x)$

In Exercises 37–40, use the given values to determine the type of curve represented.

37. For the equation $x^2 + ky^2 = a^2$, what type of curve is represented if (a) $k = 1$, (b) $k < 0$, and (c) $k > 0$ $(k \neq 1)$?
38. In Eq. (21.34), if $A > C > 0$ and $B = D = E = F = 0$, describe the locus of the equation.
39. In Eq. (21.34), if $A = B = C = 0$, $D \neq 0$, $E \neq 0$, and $F \neq 0$, describe the locus of the equation.
40. In Eq. (21.34), if $A = -C \neq 0$, $B = D = E = 0$ and $F = C$, describe the locus of the equation if $C > 0$.

A flashlight emits a cone of light onto the floor. In Exercises 41–44, determine the type of curve for the perimeter of the lighted area on the floor depending on the position of the flashlight and cone as described. See Fig. 21.96.

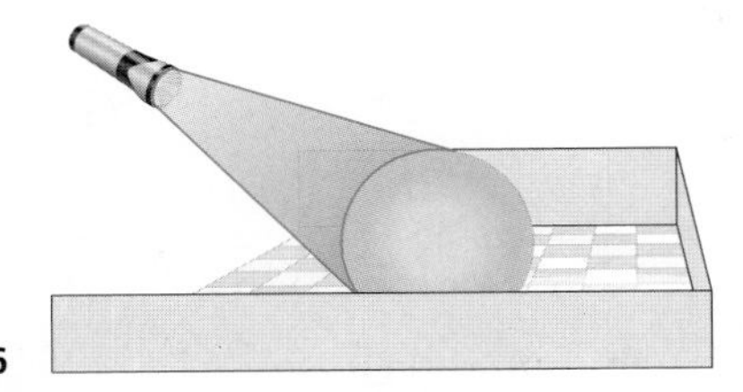

Fig. 21.96

41. The flashlight is directed at the floor and perpendicular to it.

42. The flashlight is directed toward the floor, but is not perpendicular to it.

43. The flashlight is parallel to the floor.

44. The flashlight is directed downward toward the floor such that the upper edge of the cone of light is parallel to the floor.

In Exercises 45–48, determine the type of curve from the given information.

45. The diagonal brace in a rectangular metal frame is 3.0 cm longer than the length of one of the sides. Determine the type of curve represented by the equation relating the lengths of the sides of the frame.

46. One circular solar cell has a radius that is 2.0 in. less than the radius r of a second circular solar cell. Determine the type of curve represented by the equation relating the total area A of both cells and r.

47. A supersonic jet creates a conical shock wave behind it. What type of curve is outlined on the surface of a lake by the shock wave if the jet is flying horizontally?

48. In Fig. 21.95, if the plane cutting the cones passes through the intersection of the upper and lower cones, what type of curve is the intersection of the plane and cones?

Answers to Practice Exercises

1. Ellipse **2.** Parabola

21.9 Rotation of Axes

Angle of Rotation of Axes • Eliminating the xy-term • Value of $B^2 - 4AC$ Determines Curve

In this chapter, we have discussed the circle, parabola, ellipse, and hyperbola, and in the last section, we showed how these curves are represented by the second-degree equation

$$Ax^2 + Bxy + Cy^2 + Dx + Ey + F = 0 \tag{21.34}$$

The discussion of these curves included their equations with the center (vertex of a parabola) at the origin. However, as noted in the last section, except for the special case of the hyperbola $xy = c$, we did not show what happens when the axes are *rotated* about the origin.

If a set of axes is rotated about the origin through an angle θ, as shown in Fig. 21.97, we say that there has been a **rotation of axes.** In this case, each point P in the plane has two sets of coordinates, (x, y) in the original system and (x', y') in the rotated system.

If we now let r equal the distance from the origin O to point P and let ϕ be the angle between the x'-axis and the line OP, we have

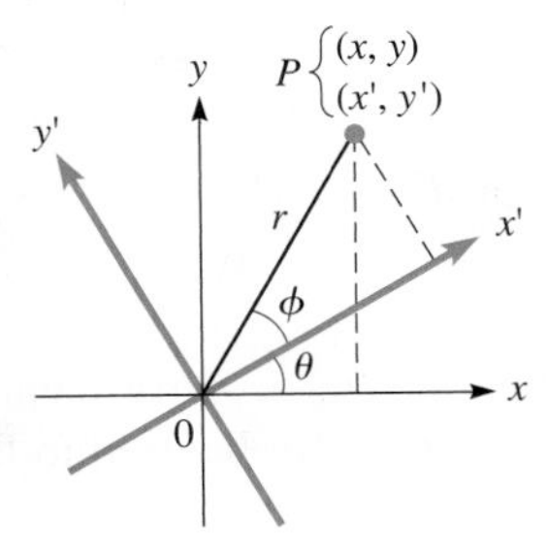

Fig. 21.97

$$x' = r\cos\phi \qquad y' = r\sin\phi \tag{21.35}$$

$$x = r\cos(\theta + \phi) \quad y = r\sin(\theta + \phi) \tag{21.36}$$

Using the cosine and sine of the sum of two angles, we can write Eqs. (21.36) as

$$\begin{aligned} x &= r\cos\phi\cos\theta - r\sin\phi\sin\theta \\ y &= r\cos\phi\sin\theta + r\sin\phi\cos\theta \end{aligned} \tag{21.37}$$

Now, using Eqs. (21.35), we have

$$\begin{aligned} x &= x'\cos\theta - y'\sin\theta \\ y &= x'\sin\theta + y'\cos\theta \end{aligned} \tag{21.38}$$

In our derivation, we have used the special case when θ is acute and P is in the first quadrant of both sets of axes. When simplifying equations of curves using Eqs. (21.38), we find that a rotation through a positive acute angle θ is sufficient. It can be shown, however, that Eqs. (21.38) hold for any θ and position of P.

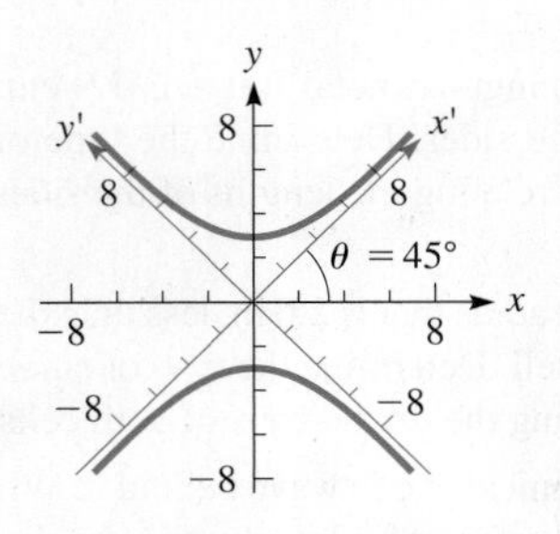

Fig. 21.98

EXAMPLE 1 Rotation through 45°

Transform $x^2 - y^2 + 8 = 0$ by rotating the axes through 45°.

When $\theta = 45°$, the rotation equations (21.38) become

$$x = x'\cos 45° - y'\sin 45° = \frac{x'}{\sqrt{2}} - \frac{y'}{\sqrt{2}}$$

$$y = x'\sin 45° + y'\cos 45° = \frac{x'}{\sqrt{2}} + \frac{y'}{\sqrt{2}}$$

Substituting into the equation $x^2 - y^2 + 8 = 0$ gives

$$\left(\frac{x'}{\sqrt{2}} - \frac{y'}{\sqrt{2}}\right)^2 - \left(\frac{x'}{\sqrt{2}} + \frac{y'}{\sqrt{2}}\right)^2 + 8 = 0$$

$$\frac{1}{2}x'^2 - x'y' + \frac{1}{2}y'^2 - \frac{1}{2}x'^2 - x'y' - \frac{1}{2}y'^2 + 8 = 0$$

$$x'y' = 4$$

The graph and both sets of axes are shown in Fig. 21.98. The original equation represents a hyperbola. In this example, we have shown that the $xy = c$ type of hyperbola is obtained by a 45° rotation of the standard form given by Eq. (21.34). ■

In Section 21.8, we saw that all conic sections can be represented by the second-degree equation $Ax^2 + Bxy + Cy^2 + Dx + Ey + F = 0$. When $B = 0$ (there is no xy-term), then the conic section will be aligned parallel to the x- or y-axis. However, when the xy-term is present in the equation, then the entire graph is rotated by some angle θ. In these cases, it is useful to rotate the axes by the angle θ to eliminate the xy-term. Then the conic section can be represented by one of the standard-form equations in the set of rotated axes.

By substituting Eqs. (21.38) into Eq. (21.34) and then simplifying, we have

$$(A\cos^2\theta + B\sin\theta\cos\theta + C\sin^2\theta)x'^2 + [B\cos 2\theta - (A - C)\sin 2\theta]x'y'$$
$$+ (A\sin^2\theta - B\sin\theta\cos\theta + C\cos^2\theta)y'^2 + (D\cos\theta + E\sin\theta)x'$$
$$+ (E\cos\theta - D\sin\theta)y' + F = 0$$

If there is to be no $x'y'$-term, its coefficient must be zero. This means that $B\cos 2\theta - (A - C)\sin 2\theta = 0$. Thus, the following equation can be used to find the angle of rotation.

Angle of Rotation

$$\tan 2\theta = \frac{B}{A - C} \quad (A \neq C) \tag{21.39}$$

Equation (21.39) gives the angle of rotation except when $A = C$. In this case, the coefficient of the $x'y'$-term is $B\cos 2\theta$, which is zero if $2\theta = 90°$. Thus,

$$\theta = 45° \quad (A = C) \tag{21.40}$$

Noting Eqs. (21.39) and (21.40), we see that a rotation of exactly 45° occurs if $A = C$. Therefore, when we check the equation, if $A \neq C$, we use Eq. (21.39). Consider the following example.

■ For reference:

Eq. (20.2) is $\cos\theta = \frac{1}{\sec\theta}$

Eq. (20.7) is $1 + \tan^2\theta = \sec^2\theta$

Eq. (20.26) is $\sin\frac{\alpha}{2} = \pm\sqrt{\frac{1 - \cos\alpha}{2}}$

Eq. (20.27) is $\cos\frac{\alpha}{2} = \pm\sqrt{\frac{1 + \cos\alpha}{2}}$

EXAMPLE 2 Rotating axes—eliminating the *xy*-term

By rotation of axes, transform $8x^2 + 4xy + 5y^2 = 9$ into a form without an xy-term. Identify and sketch the curve.

Here, $A = 8$, $B = 4$, and $C = 5$. Therefore, using Eq. (21.39), we have

$$\tan 2\theta = \frac{4}{8 - 5} = \frac{4}{3}$$

Because $\tan 2\theta$ is positive, we may take 2θ as an acute angle, which means θ is also acute. For the transformation, we need $\sin\theta$ and $\cos\theta$. We find these values by first finding the value of $\cos 2\theta$ and then using the half-angle formulas:

$$\cos 2\theta = \frac{1}{\sec 2\theta} = \frac{1}{\sqrt{1 + \tan^2 2\theta}} = \frac{1}{\sqrt{1 + (\frac{4}{3})^2}} = \frac{3}{5} \quad \text{using Eqs. (20.2) and (20.7)}$$

Now, using the half-angle formulas, Eqs. (20.26) and (20.27), we have

$$\sin\theta = \sqrt{\frac{1 - \cos 2\theta}{2}} = \sqrt{\frac{1 - \frac{3}{5}}{2}} = \frac{1}{\sqrt{5}} \qquad \cos\theta = \sqrt{\frac{1 + \cos 2\theta}{2}} = \sqrt{\frac{1 + \frac{3}{5}}{2}} = \frac{2}{\sqrt{5}}$$

Here, θ is about 26.6°. Now, substituting these values into Eqs. (21.38), we have

$$x = x'\left(\frac{2}{\sqrt{5}}\right) - y'\left(\frac{1}{\sqrt{5}}\right) = \frac{2x' - y'}{\sqrt{5}} \qquad y = x'\left(\frac{1}{\sqrt{5}}\right) + y'\left(\frac{2}{\sqrt{5}}\right) = \frac{x' + 2y'}{\sqrt{5}}$$

Now, substituting into the equation $8x^2 + 4xy + 5y^2 = 9$ gives

$$8\left(\frac{2x' - y'}{\sqrt{5}}\right)^2 + 4\left(\frac{2x' - y'}{\sqrt{5}}\right)\left(\frac{x' + 2y'}{\sqrt{5}}\right) + 5\left(\frac{x' + 2y'}{\sqrt{5}}\right)^2 = 9$$

$$8(4x'^2 - 4x'y' + y'^2) + 4(2x'^2 + 3x'y' - 2y'^2) + 5(x'^2 + 4x'y' + 4y'^2) = 45$$

$$45x'^2 + 20y'^2 = 45$$

$$\frac{x'^2}{1} + \frac{y'^2}{\frac{9}{4}} = 1$$

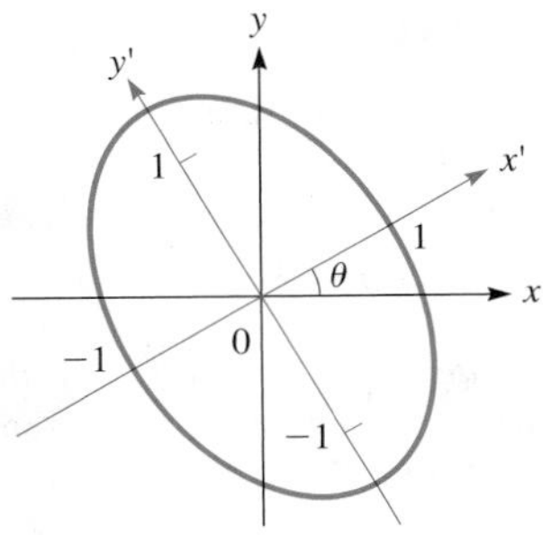

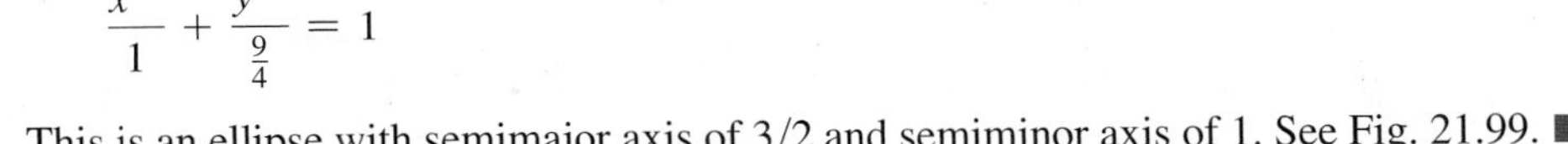

Fig. 21.99

This is an ellipse with semimajor axis of 3/2 and semiminor axis of 1. See Fig. 21.99. ■

In Example 2, $\tan 2\theta$ was positive, and we made 2θ and θ positive. If, when using Eq. (21.39), $\tan 2\theta$ is negative, we then make 2θ obtuse $(90° < 2\theta < 180°)$. In this case, $\cos 2\theta$ will be negative, but θ will be acute $(45° < \theta < 90°)$.

In Section 21.7, we showed the use of translation of axes in writing an equation in standard form if $B = 0$. In this section, we have seen how rotation of axes is used to eliminate the xy-term. It is possible that both a translation of axes and a rotation of axes are needed to write an equation in standard form.

In Section 21.8, we identified a conic section by inspecting the values of A and C when $B = 0$. If $B \neq 0$, these curves are identified as follows:

Identifying a Conic Section from the value of $B^2 - 4AC$

1. If $B^2 - 4AC = 0$, a parabola
2. If $B^2 - 4AC < 0$, an ellipse
3. If $B^2 - 4AC > 0$, a hyperbola

In some special cases, it is possible that a point, parallel lines, intersecting lines, or no curve will result.

Practice Exercises

Identify the type of curve represented by each equation.

1. $3x^2 - 2xy + 3y^2 = 8$
2. $x^2 - 4xy + y^2 = -5$
3. $4x^2 + 4xy + y^2 - 24x + 38y - 19 = 0$

EXAMPLE 3 Rotating axes—eliminating the xy-term

For the equation $16x^2 - 24xy + 9y^2 + 20x - 140y - 300 = 0$, identify the curve and simplify it to standard form. Sketch the graph and display it on a calculator.

With $A = 16$, $B = -24$, and $C = 9$, using Eq. (21.39), we have

$$\tan 2\theta = \frac{-24}{16 - 9} = -\frac{24}{7}$$

In this case, $\tan 2\theta$ is negative, and we take 2θ to be an obtuse angle. We then find that $\cos 2\theta = -7/25$. In turn, we find that $\sin\theta = 4/5$ and $\cos\theta = 3/5$. Here, θ is about 53.1°. Using these values in Eqs. (21.38), we find that

$$x = \frac{3x' - 4y'}{5} \qquad y = \frac{4x' + 3y'}{5}$$

Substituting these into the original equation and simplifying, we get

$$y'^2 - 4x' - 4y' - 12 = 0$$

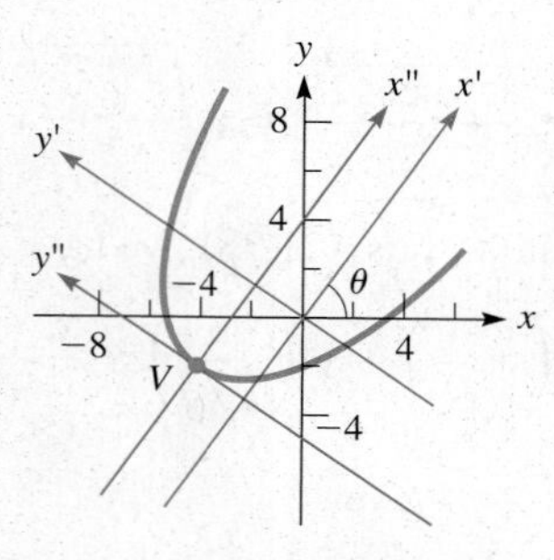

Fig. 21.100

This equation represents a parabola with its axis parallel to the x'-axis. The vertex is found by completing the square:

$$(y' - 2)^2 = 4(x' + 4)$$

The vertex is the point $(-4, 2)$ in the $x'y'$-rotated system. Therefore,

$$y''^2 = 4x''$$

is the equation in the $x''y''$-rotated and then translated system. The graph and the coordinate systems are shown in Fig. 21.100. ■

EXERCISES 21.9

In Exercises 1–4, transform the given equations by rotating the axes through the given angle. Identify and sketch each curve.

1. $x^2 - y^2 = 25, \theta = 45°$ **2.** $x^2 + y^2 = 16, \theta = 60°$

3. $8x^2 - 4xy + 5y^2 = 36, \theta = \tan^{-1}2$

4. $2x^2 + 24xy - 5y^2 = 8, \theta = \tan^{-1}\frac{3}{4}$

In Exercises 5–10, identify the type of curve that each equation represents by evaluating $B^2 - 4AC$.

5. $x^2 + 2xy + x - y - 3 = 0$

6. $8x^2 - 4xy + 2y^2 + 7 = 0$

7. $x^2 - 2xy + y^2 + 3y = 0$

8. $4xy + 3y^2 - 8x + 16y + 19 = 0$

9. $13x^2 + 10xy + 13y^2 + 6x - 42y - 27 = 0$

10. $x^2 - 4xy + 4y^2 + 36x + 28y + 24 = 0$

In Exercises 11–18, transform each equation to a form without an xy-term by a rotation of axes. Identify and sketch each curve. Then display each curve on a calculator.

11. $x^2 + 2xy + y^2 - 2x + 2y = 0$ **12.** $5x^2 - 6xy + 5y^2 = 32$

13. $xy = 8$ **14.** $5x^2 - 8xy + 5y^2 = 0$

15. $3x^2 + 4xy = 4$

16. $9x^2 - 24xy + 16y^2 - 320x - 240y = 0$

17. $11x^2 - 6xy + 19y^2 = 20$ **18.** $x^2 + 4xy - 2y^2 = 6$

In Exercises 19 and 20, transform each equation to a form without an xy-term by a rotation of axes. Then transform the equation to a standard form by a translation of axes. Identify and sketch each curve.

19. $16x^2 - 24xy + 9y^2 - 60x - 80y + 400 = 0$

20. $73x^2 - 72xy + 52y^2 + 100x - 200y + 100 = 0$

In Exercises 21–26, solve the given problems.

21. What curve does the value of $B^2 - 4AC$ indicate should result for the graph of $2x^2 + xy + y^2 = 0$? Is this the actual curve?

22. What curve does the value of $B^2 - 4AC$ indicate should result for the graph of $4x^2 - 4xy + y^2 = 0$? Is this the actual curve?

23. Find the $x'y'$ coordinates of the xy point $(-2, 6)$ rotated through 60°.

24. What is the $x'y'$ equation of the line $y = x$ when the axes are rotated through 45°.

25. What is the $x'y'$ equation of the function $y = 2^x$ when the axes are rotated through 90°?

26. An elliptical cam can be represented by the equation $x^2 - 3xy + 5y^2 - 13 = 0$. Through what angle is the cam rotated from its standard position?

Answers to Practice Exercises

1. Ellipse **2.** Hyperbola **3.** Parabola

21.10 Polar Coordinates

Pole • Polar Axis • Polar Coordinates • Converting Between Polar and Rectangular Coordinates

■ The Swiss mathematician Jakob Bernoulli (1654–1705) was among the first to make significant use of polar coordinates.

To this point, we have graphed all curves in the rectangular coordinate system. However, for certain types of curves, other coordinate systems are better adapted. We discuss one of these systems here.

Instead of designating a point by its x- and y-coordinates, we can specify its location by its radius vector and the angle the radius vector makes with the x-axis. Thus, the r and θ that are used in the definitions of the trigonometric functions can also be used as the coordinates of points in the plane. The important aspect of choosing coordinates is that, for each set of values, there must be only one point that corresponds to this set. We can see that this condition is satisfied by the use of r and θ as coordinates. *In* **polar coordinates,** *the origin is called the* **pole,** *and the half-line for which the angle is zero (equivalent to the positive x-axis) is called the* **polar axis.** The coordinates of a point are designated as (r, θ). We will use radians when measuring the value of θ. See Fig. 21.101.

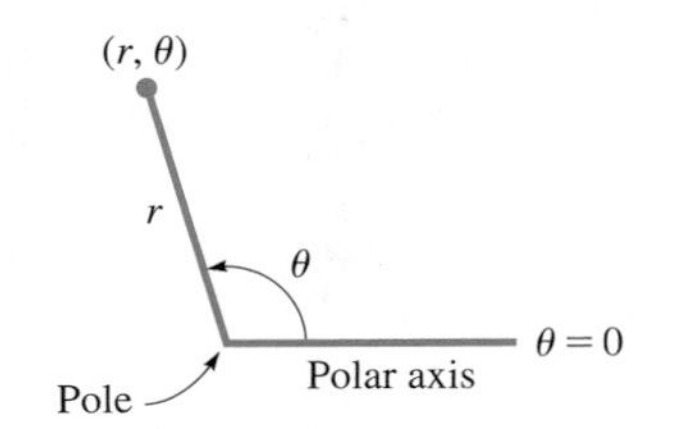

Fig. 21.101

When using polar coordinates, we generally label the lines for some of the values of θ; namely, those for $\theta = 0$ (the polar axis), $\theta = \pi/2$ (equivalent to the positive y-axis), $\theta = \pi$ (equivalent to the negative x-axis), $\theta = 3\pi/2$ (equivalent to the negative y-axis), and possibly others. In Fig. 21.102, these lines and those for multiples of $\pi/6$ are shown. Also, the circles for $r = 1$, $r = 2$, and $r = 3$ are shown in this figure.

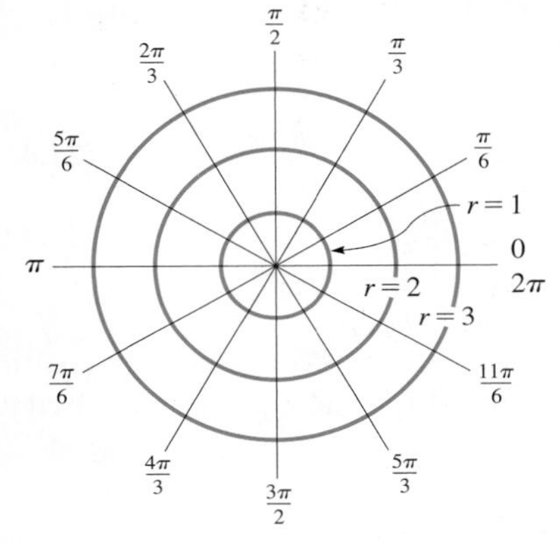

Fig. 21.102

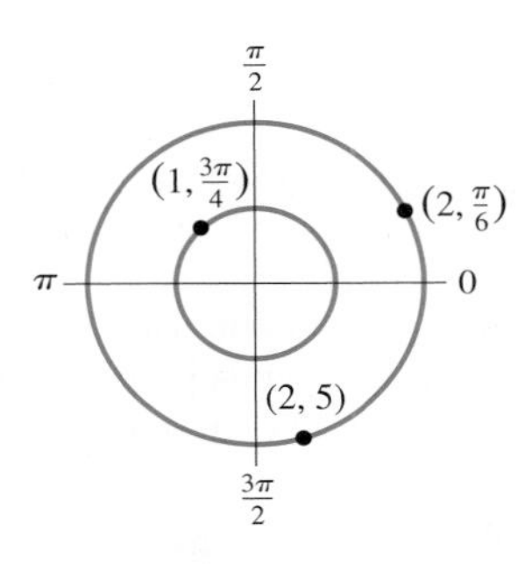

Fig. 21.103

EXAMPLE 1 Polar coordinates of a point

(a) If $r = 2$ and $\theta = \pi/6$, we have the point shown in Fig. 21.103. The point corresponds to $(\sqrt{3}, 1)$ in rectangular coordinates.

(b) The polar coordinate point $(1, 3\pi/4)$ is also shown. It is equivalent to $(-\sqrt{2}/2, \sqrt{2}/2)$ in rectangular coordinates.

(c) The polar coordinate point $(2, 5)$ is also shown. It is equivalent approximately to $(0.6, -1.9)$ in rectangular coordinates. Remember, the 5 is an angle in radians. ■

One difference between rectangular coordinates and polar coordinates is that, for each point in the plane, there are limitless possibilities for the polar coordinates of that point. For example, the point $(2, \frac{\pi}{6})$ can also be represented by $(2, \frac{13\pi}{6})$ because the angles $\frac{\pi}{6}$ and $\frac{13\pi}{6}$ are coterminal. We also remove one restriction on r that we imposed in the definition of the trigonometric functions. That is, r is allowed to take on positive and negative values. NOTE ▸ [If r is negative, θ is located as before, but the point is found r units from the pole ***on the opposite side*** from that on which it is positive.]

EXAMPLE 2 Polar coordinates—negative *r*

The polar coordinates $(3, 2\pi/3)$ and $(3, -4\pi/3)$ represent the same point. However, the point $(-3, 2\pi/3)$ is on the opposite side of the pole from $(3, 2\pi/3)$, 3 units from the pole. Another possible set of polar coordinates for the point $(-3, 2\pi/3)$ is $(3, 5\pi/3)$. These points are shown in Fig. 21.104. ■

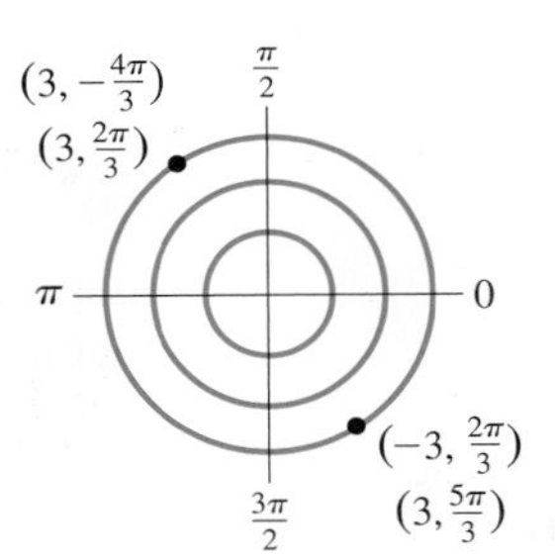

Fig. 21.104

NOTE ▸ [When locating and plotting a point in polar coordinates, it is generally best to first locate the terminal side of θ, and then measure r along this terminal side.] This is illustrated in the following example.

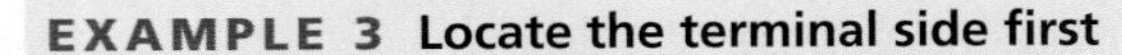

EXAMPLE 3 Locate the terminal side first

Plot the points $A(2, 5\pi/6)$ and $(-3.2, -2.4)$ in the polar coordinate system.

To locate A, we determine the terminal side of $\theta = 5\pi/6$ and then determine $r = 2$. See Fig. 21.105.

To locate B, we find the terminal side of $\theta = -2.4$, measuring clockwise from the polar axis (and recalling that $\pi = 3.14 = 180°$). Then we locate $r = -3.2$ on the opposite side of the pole. See Fig. 21.105.

We will find that points with negative values of r occur frequently when plotting curves in polar coordinates. ■

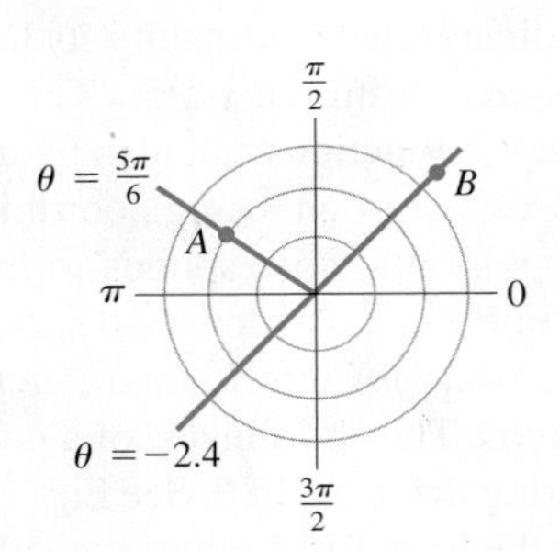

Fig. 21.105

POLAR AND RECTANGULAR COORDINATES

The relationships between the polar coordinates of a point and the rectangular coordinates of the same point come from the definitions of the trigonometric functions. Those most commonly used are (see Fig. 21.106)

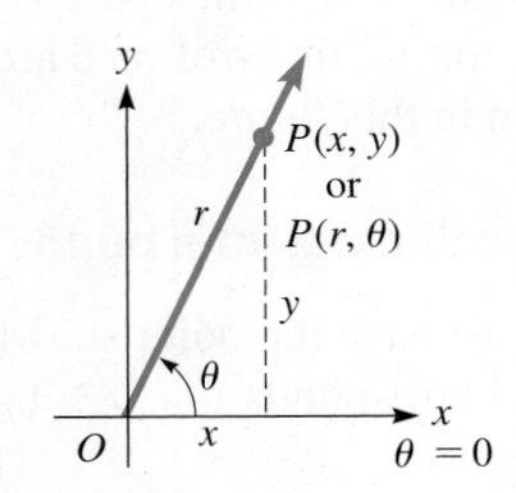

Fig. 21.106

$$x = r\cos\theta \quad y = r\sin\theta \tag{21.41}$$

$$\tan\theta = \frac{y}{x} \quad r = \sqrt{x^2 + y^2} \tag{21.42}$$

The following examples show the use of Eqs. (21.41) and (21.42) in changing coordinates in one system to coordinates in the other system. Also, these equations are used to transform equations from one system to the other.

EXAMPLE 4 Polar to rectangular coordinates

Using Eqs. (21.41), we can transform the polar coordinates $(4, \pi/4)$ into the rectangular coordinates $(2\sqrt{2}, 2\sqrt{2})$, because

$$x = 4\cos\frac{\pi}{4} = 4\left(\frac{\sqrt{2}}{2}\right) = 2\sqrt{2} \quad \text{and} \quad y = 4\sin\frac{\pi}{4} = 4\left(\frac{\sqrt{2}}{2}\right) = 2\sqrt{2}$$

See Fig. 21.107. ■

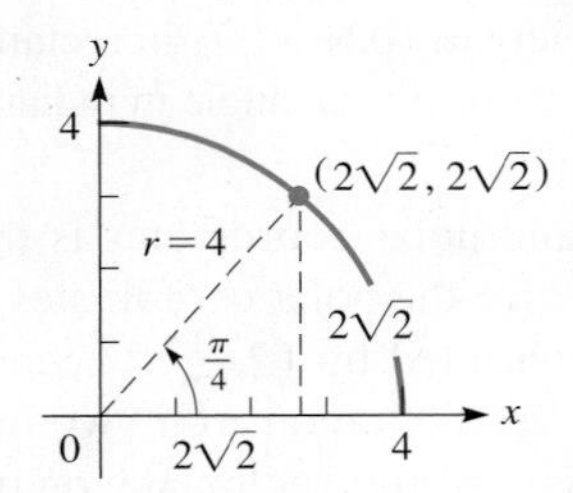

Fig. 21.107

EXAMPLE 5 Rectangular to polar coordinates

Using Eqs. (21.42), we can transform the rectangular coordinates $(3, -5)$ into polar coordinates, as follows.

$$\tan\theta = -\frac{5}{3}, \quad \theta = 5.25 \quad (\text{or } -1.03)$$

$$r = \sqrt{3^2 + (-5)^2} = 5.83$$

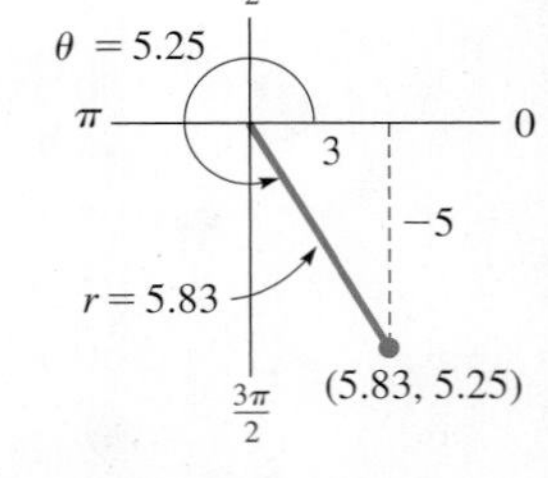

Fig. 21.108

We know that θ is a fourth-quadrant angle because x is positive and y is negative. Therefore, the point $(3, -5)$ in rectangular coordinates can be expressed as the point $(5.83, 5.25)$ in polar coordinates (see Fig. 21.108). Other polar coordinates for the point are also possible.

Calculators are programmed to make conversions between rectangular and polar coordinates. The display for the conversions of Examples 4 and 5 are shown in Fig. 21.109. ■

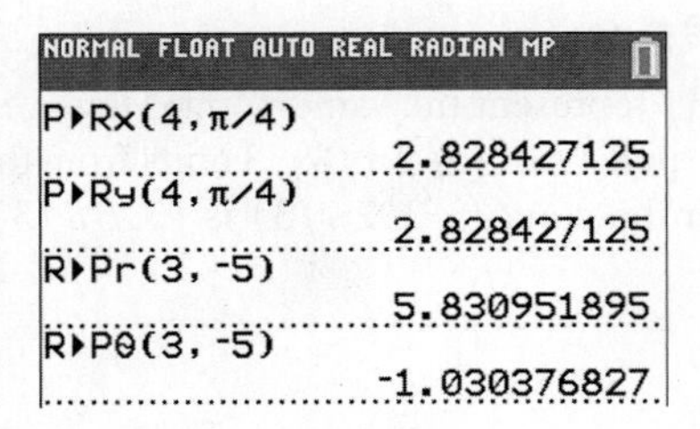

Fig. 21.109
Graphing calculator keystrokes: goo.gl/N6gGDY

Practice Exercise

1. Transform the polar coordinates $(4, 5\pi/6)$ into rectangular coordinates.

■ The cyclotron was invented in 1931 at the University of California. It was the first accelerator to deflect particles into circular paths.

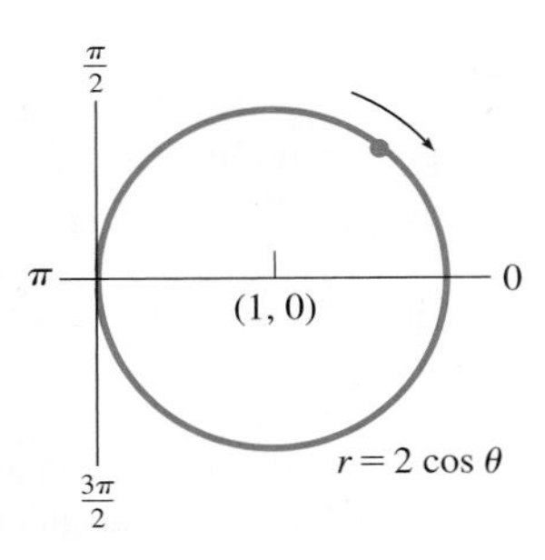

Fig. 21.110

EXAMPLE 6 Rectangular to polar equation—particle accelerator

If an electrically charged particle enters a magnetic field at right angles to the field, the particle follows a circular path. This fact is used in the design of nuclear particle accelerators.

A proton (positively charged) enters a magnetic field such that its path may be described by the rectangular equation $x^2 + y^2 = 2x$, where measurements are in meters. Find the polar equation of this circle.

We change this equation expressed in the rectangular coordinates x and y into an equation expressed in the polar coordinates r and θ by using the relations $r^2 = x^2 + y^2$ and $x = r\cos\theta$ as follows:

$$x^2 + y^2 = 2x \quad \text{rectangular equation}$$
$$r^2 = 2r\cos\theta \quad \text{substitute}$$
$$r = 2\cos\theta \quad \text{divided by } r$$

This is the polar equation of the circle, which is shown in Fig. 21.110. ■

EXAMPLE 7 Polar to rectangular equation of curve

Find the rectangular equation of the *rose* $r = 4\sin 2\theta$.

Using the trigonometric identity $\sin 2\theta = 2\sin\theta\cos\theta$ and Eqs. (21.41) and (21.42) leads to the solution:

$$r = 4\sin 2\theta \quad \text{polar equation}$$
$$= 4(2\sin\theta\cos\theta) = 8\sin\theta\cos\theta \quad \text{using identity}$$
$$\sqrt{x^2 + y^2} = 8\left(\frac{y}{r}\right)\left(\frac{x}{r}\right) = \frac{8xy}{r^2} = \frac{8xy}{x^2 + y^2} \quad \text{using Eqs. (21.41) and (21.42)}$$
$$x^2 + y^2 = \frac{64x^2y^2}{(x^2 + y^2)^2} \quad \text{squaring both sides}$$
$$(x^2 + y^2)^3 = 64x^2y^2 \quad \text{simplifying}$$

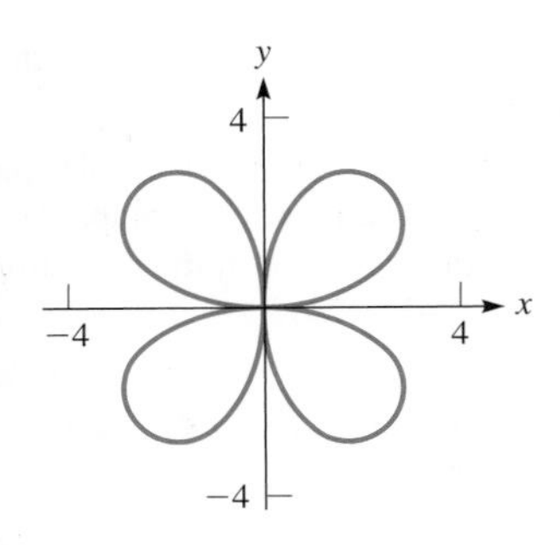

Fig. 21.111

Plotting the graph of this equation from the rectangular equation would be complicated. However, as we will see in the next section, plotting this graph in polar coordinates is quite simple. The curve is shown in Fig. 21.111. ■

Practice Exercise

2. Find the polar equation of the circle $x^2 + y^2 + 2x = 0$.

EXERCISES 21.10

In Exercises 1–4, make the given changes in the indicated examples of this section and then solve the indicated problems.

1. For both of the points plotted in Example 2, change $2\pi/3$ to $\pi/3$ and then find another set of coordinates for each point, similar to those shown for the points in the example.
2. In Example 4, change $\pi/4$ to $\pi/6$
3. In Example 5, change 3 to −3 and −5 to 5.
4. In Example 7, change sin to cos.

In Exercises 5–16, plot the given polar coordinate points on polar coordinate paper.

5. $\left(3, \frac{\pi}{6}\right)$ 6. $(2, \pi)$ 7. $\left(\frac{5}{2}, -\frac{2\pi}{5}\right)$

8. $\left(5, -\frac{\pi}{3}\right)$ 9. $\left(-8, \frac{7\pi}{6}\right)$ 10. $\left(-5, \frac{\pi}{4}\right)$

11. $\left(-3, -\frac{5\pi}{4}\right)$ 12. $\left(-4, -\frac{5\pi}{3}\right)$ 13. $(2, 2)$

14. $(-6, -6)$ 15. $(0.5, -8.4)$ 16. $(-2.2, 18.8)$

In Exercises 17–22, find a set of polar coordinates for each of the points for which the rectangular coordinates are given.

17. $(\sqrt{3}, 1)$ 18. $(-3, 3)$

19. $\left(-\frac{\sqrt{3}}{2}, -\frac{1}{2}\right)$ 20. $(-10, 8)$

21. $(0, 4)$ 22. $(-5, 0)$

In Exercises 23–28, find the rectangular coordinates for each of the points for which the polar coordinates are given.

23. $\left(8, \frac{4\pi}{3}\right)$ 24. $(4, -\pi)$

25. $(-3.0, -0.40)$ **26.** $(-3.0, 3.0)$

27. $(8.0, -8.0)$ **28.** $(\pi, 4.0)$

In Exercises 29–38, find the polar equation of each of the given rectangular equations.

29. $x = 3$ **30.** $y = -x$

31. $x + 2y + 3 = 0$ **32.** $x^2 + y^2 = 0.81$

33. $x^2 + (y - 2)^2 = 4$ **34.** $x^2 - y^2 = 0.01$

35. $x^2 + 4y^2 = 4$ **36.** $y^2 = 4x$

37. $x^2 + y^2 = 6y$ **38.** $xy = 9$

In Exercises 39–48, find the rectangular equation of each of the given polar equations. In Exercises 39–46, identify the curve that is represented by the equation.

39. $r = \sin\theta$ **40.** $r\sec\theta = 4$

41. $r\cos\theta = 4$ **42.** $r = -2\csc\theta$

43. $r = \dfrac{2}{\cos\theta - 3\sin\theta}$ **44.** $r = e^{r\cos\theta}\csc\theta$

45. $r = 4\cos\theta + 2\sin\theta$ **46.** $r\sin(\theta + \pi/6) = 3$

47. $r^2 = \sin 2\theta$ **48.** $r = \dfrac{3}{\sin\theta + 4\cos\theta}$

In Exercises 49–60, solve the given problems. All coordinates given are polar coordinates.

49. Is the point $(2, 3\pi/4)$ on the curve $r = 2\sin 2\theta$?

50. Is the point $(1/2, 3\pi/2)$ on the curve $r = \sin(\theta/3)$?

51. Show that the polar coordinate equation $r = a\sin\theta + b\cos\theta$ represents a circle by changing it to a rectangular equation.

52. Find the distance between the points $(3, \pi/6)$ and $(4, \pi/2)$ by using the law of cosines.

53. The center of a regular hexagon is at the pole with one vertex at $(2, \pi)$. What are the polar coordinates of the other vertices?

54. Find the distance between the points $(4, \pi/6)$ and $(5, 5\pi/3)$.

55. Under certain conditions, the x- and y-components of a magnetic field B are given by the equations

$$B_x = \frac{-ky}{x^2 + y^2} \quad \text{and} \quad B_y = \frac{kx}{x^2 + y^2}$$

Write these equations in terms of polar coordinates.

56. In designing a domed roof for a building, an architect uses the equation $x^2 + \dfrac{y^2}{k^2} = 1$, where k is a constant. Write this equation in polar form.

57. The shape of a swimming pool can be described by the polar equation $r = 5\cos\theta$ (dimensions in meters). Find the rectangular equation for the perimeter of the pool.

58. The polar equation of the path of a weather satellite of Earth is $r = \dfrac{4800}{1 + 0.14\cos\theta}$, where r is measured in miles. Find the rectangular equation of the path of this satellite. The path is an ellipse, with Earth at one of the foci.

59. The control tower of an airport is taken to be at the pole, and the polar axis is taken as due east in a polar coordinate graph. How far apart (in km) are planes, at the same altitude, if their positions on the graph are $(6.10, 1.25)$ and $(8.45, 3.74)$?

60. The perimeter of a certain type of machine part can be described by the equation $r = a\sin\theta + b\cos\theta\,(a > 0, b > 0)$. Explain why all such machine parts are circular.

Answers to Practice Exercises

1. $(-2\sqrt{3}, 2)$ **2.** $r = -2\cos\theta$

21.11 Curves in Polar Coordinates

Curve Sketched from Equations • Plotting Polar Coordinate Curves • Calculator Display of Polar Curves • Polar Curve Generated by Data

The basic method for finding a curve in polar coordinates is the same as in rectangular coordinates. We assume values of the independent variable—in this case, θ—and then find the corresponding values of the dependent variable r. These points are then plotted and joined, thereby forming the curve that represents the relation in polar coordinates.

Before using the basic method, it is useful to point out that certain basic curves can be sketched directly from the equation. This is done by noting the meaning of each of the polar coordinate variables, r and θ. This is illustrated in the following example.

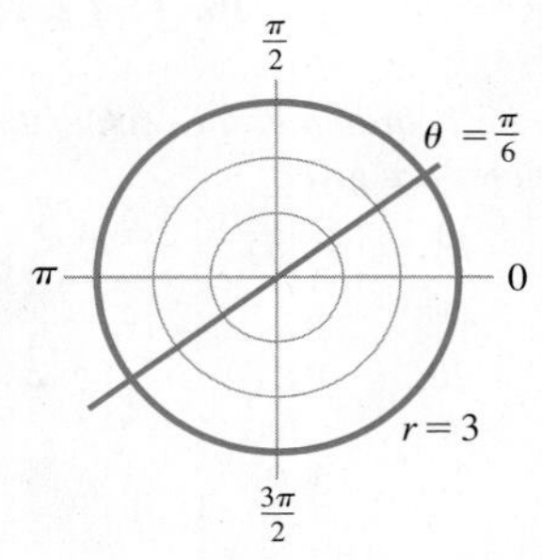

Fig. 21.112

EXAMPLE 1 Curves sketched from equations

(a) The graph of the polar equation $r = 3$ is a circle of radius 3, with center at the pole. This can be seen to be the case, since $r = 3$ for all possible values of θ. It is not necessary to find specific points for this circle, which is shown in Fig. 21.112.

(b) The graph of $\theta = \pi/6$ is a straight line through the pole. It represents all points for which $\theta = \pi/6$ for all possible values of r. This line is shown in Fig. 21.112. ■

Fig. 21.113

EXAMPLE 2 Plotting a cardioid

Plot the graph of $r = 1 + \cos\theta$.

First, for each of the chosen values of θ we calculate the value for r. Then, to plot these points we identify the ray that represents the value of θ, and then place the point the proper number of units for the value of r out on that ray. We then connect the points as shown in Fig. 21.113.

θ	0	$\frac{\pi}{4}$	$\frac{\pi}{2}$	$\frac{3\pi}{4}$	π	$\frac{5\pi}{4}$	$\frac{3\pi}{2}$	$\frac{7\pi}{4}$	2π
r	2	1.7	1	0.3	0	0.3	1	1.7	2
Point Number	1	2	3	4	5	6	7	8	9

We now see that the points are repeating, and it is unnecessary to find additional points. This curve is called a **cardioid**. ■

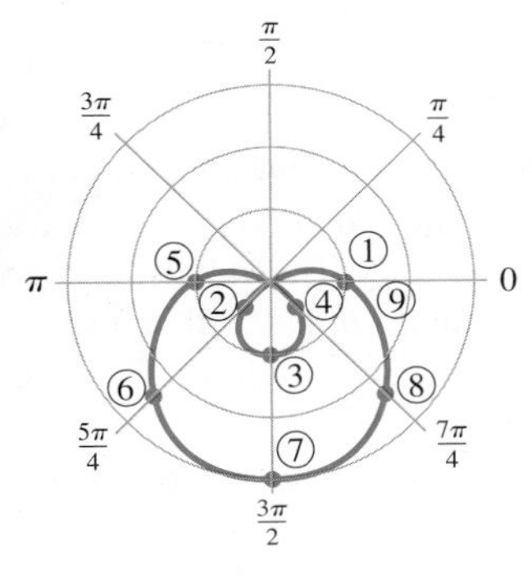

Fig. 21.114

EXAMPLE 3 Plotting a limaçon

Plot the graph of $r = 1 - 2\sin\theta$.

Choosing values of θ and then finding the corresponding values of r, we find the following table of values.

θ	0	$\frac{\pi}{4}$	$\frac{\pi}{2}$	$\frac{3\pi}{4}$	π	$\frac{5\pi}{4}$	$\frac{3\pi}{2}$	$\frac{7\pi}{4}$	2π
r	1	−0.4	−1	−0.4	1	2.4	3	2.4	1
Point Number	1	2	3	4	5	6	7	8	9

Practice Exercises

Determine the type of curve represented by each polar coordinate equation.

1. $\theta = 2\pi/5$ **2.** $r = -4$

NOTE ▸ [Particular care should be taken in plotting the points for which r is negative. We recall that when r is negative, the point is on the side *opposite* the pole from which it is positive.] This curve is known as a **limaçon** and is shown in Fig. 21.114. ■

EXAMPLE 4 Plotting a rose

Plot the graph of $r = 2\cos 2\theta$.

In finding values of r, we must be careful first to multiply the values of θ by 2 before finding the cosine of the angle. Also, for this reason, we take values of θ as multiples of $\pi/12$, so as to get enough useful points. The table of values follows:

θ	0	$\frac{\pi}{12}$	$\frac{\pi}{6}$	$\frac{\pi}{4}$	$\frac{\pi}{3}$	$\frac{5\pi}{12}$	$\frac{\pi}{2}$
r	2	1.7	1	0	−1	−1.7	−2

θ	$\frac{7\pi}{12}$	$\frac{2\pi}{3}$	$\frac{3\pi}{4}$	$\frac{5\pi}{6}$	$\frac{11\pi}{12}$	π
r	−1.7	−1	0	1	1.7	2

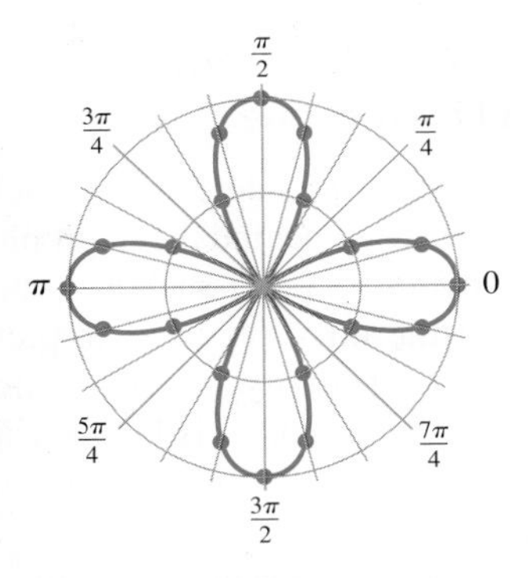

Fig. 21.115

For values of θ starting with π, the values of θ repeat. We have a four-leaf **rose,** as shown in Fig. 21.115. ■

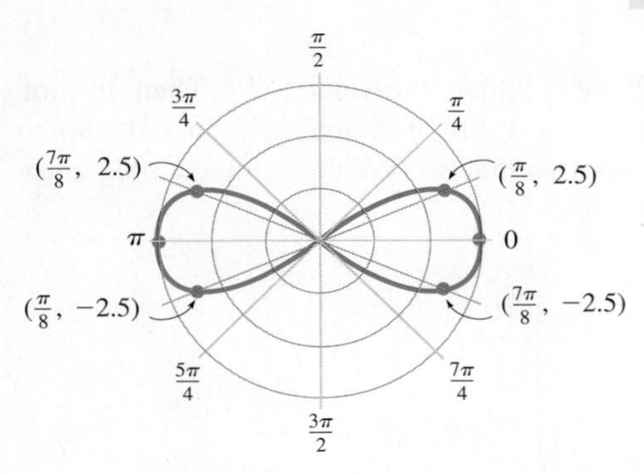

Fig. 21.116

EXAMPLE 5 Plotting a lemniscate

Plot the graph of $r^2 = 9\cos 2\theta$.

Choosing the indicated values of θ, we get the values of r as shown in the following table of values:

θ	0	$\frac{\pi}{8}$	$\frac{\pi}{4}$	...	$\frac{3\pi}{4}$	$\frac{7\pi}{8}$	π
r	± 3	± 2.5	0		0	± 2.5	± 3

There are no values of r corresponding to values of θ in the range $\pi/4 < \theta < 3\pi/4$, since twice these angles are in the second and third quadrants and the cosine is negative for such angles. The value of r^2 cannot be negative. Also, the values of r repeat for $\theta > \pi$. The figure is called a **lemniscate** and is shown in Fig. 21.116. ■

EXAMPLE 6 Calculator display of a polar curve

View the graph of $r = 1 - 2\cos\theta$ on a calculator.

Using the *mode* feature, a polar curve is displayed using the polar graph option or the parametric graph option, depending on the calculator. (Review the manual for the calculator.) The graph displayed will be the same with either method.

With the polar graph option, the function is entered directly. The values for the viewing window are determined by settings for x, y and the angles θ that will be used. These values are set in a manner similar to those used for parametric equations. (See page 317 for an example of graphing parametric equations.)

With the parametric graph option, to graph $r = f(\theta)$, we note that $x = r\cos\theta$ and $y = r\sin\theta$. This tells us that

$$x = f(\theta)\cos\theta$$
$$y = f(\theta)\sin\theta$$

Thus, for $r = 1 - 2\cos\theta$, by using

$$x = (1 - 2\cos\theta)\cos\theta$$
$$y = (1 - 2\cos\theta)\sin\theta$$

the graph can be displayed, as shown in Fig. 21.117. ■

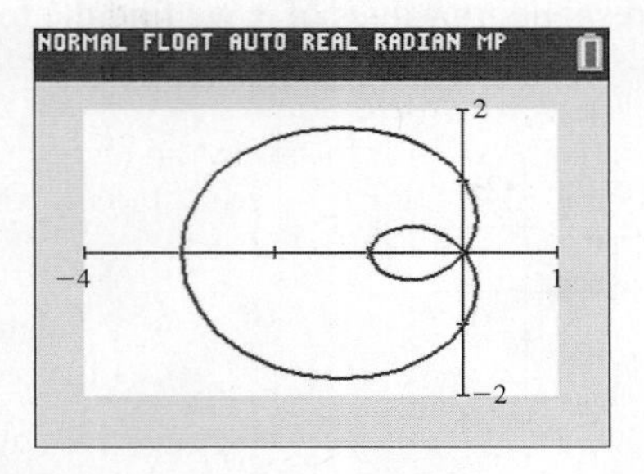

Fig. 21.117
Graphing calculator keystrokes: goo.gl/4B6EVr

■ To get a sense of how to interpret the decibel drops shown in Fig. 21.118, a whisper is about 40 dB lower than normal speech.

EXAMPLE 7 Polar graph from data—cardioid microphone

A cardioid microphone, commonly used for vocals or speech, is designed to pick up sound in front of the microphone but reject sound coming from behind the microphone. It gets its name from the fact that its polar sensitivity pattern is heart-shaped as shown in Fig. 21.118. The polar graph shows how sensitive the microphone is in picking up sound coming from different angles. The maximum sensitivity is at 0°, which is directly in front of the microphone. At other angles, the microphone is less sensitive, and therefore there is a decrease in decibels (dB). For example, at 30° to either side, there is about a 1-dB drop, and at 90°, the decrease is about 7 dB.

Polar patterns such as these are found by revolving a sound source in a circle around a microphone placed at the center and measuring the sound picked up by the microphone. Other types of microphones have circular or figure-eight shaped polar patterns.

Note that 0° is located straight upward instead of toward the right as is usually done in mathematics. Also notice that the angles increase as you rotate clockwise instead of the usual counterclockwise. ■

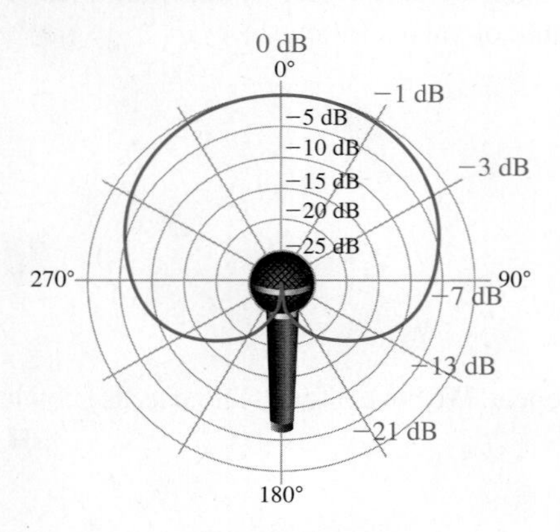

Fig. 21.118

EXERCISES 21.11

In Exercises 1–4, make the given changes in the indicated examples of this section and then make the indicated graphs.

1. In Example 1(b), change $\pi/6$ to $5\pi/6$.
2. In Example 3, change the $-$ before the $2\sin\theta$ to $+$.
3. In Example 4, change cos to sin.
4. In Example 7, what is the decibel loss when the sound approaches the microphone 120° from the point directly in front of the microphone?

In Exercises 5–32, plot the curves of the given polar equations in polar coordinates.

5. $r = 5$
6. $r = -2$
7. $\theta = 3\pi/4$
8. $\theta = -1.5$
9. $r = 4\sec\theta$
10. $r = 4\csc\theta$
11. $r = 2\sin\theta$
12. $r = -3\cos\theta$
13. $1 - r = \cos\theta$ (cardioid)
14. $r + 1 = \sin\theta$ (cardioid)
15. $r = 2 - \cos\theta$ (limaçon)
16. $r = 2 + 3\sin\theta$ (limaçon)
17. $r = 4\sin 2\theta$ (rose)
18. $r = 2\sin 3\theta$ (rose)
19. $r^2 = 4\sin 2\theta$ (lemniscate)
20. $r^2 = 2\cos\theta$
21. $r = 2^\theta$ (spiral)
22. $r = 1.5^{-\theta}$ (spiral)
23. $r = 4|\sin 3\theta|$
24. $r = 2\sin\theta\tan\theta$ (cissoid)
25. $r = \dfrac{3}{2 - \cos\theta}$ (ellipse)
26. $r = \dfrac{2}{1 - \cos\theta}$ (parabola)
27. $r - 2r\cos\theta = 6$ (hyperbola)
28. $3r - 2r\sin\theta = 6$ (ellipse)
29. $r = 4\cos\frac{1}{2}\theta$
30. $r = 2 + \cos 3\theta$
31. $r = 2[1 - \sin(\theta - \pi/4)]$
32. $r = 4\tan\theta$

In Exercises 33–42, view the curves of the given polar equations on a calculator.

33. $r = \theta \quad (-20 \le \theta \le 20)$
34. $r = 0.5^{\sin\theta}$
35. $r = 2\sec\theta + 1$
36. $r = 2\cos(\cos 2\theta)$
37. $r = 3\cos 4\theta$
38. $r\csc 5\theta = 3$
39. $r = 2\cos\theta + 3\sin\theta$
40. $r = 1 + 3\cos\theta - 2\sin\theta$
41. $r = \cos\theta + \sin 2\theta$
42. $2r\cos\theta + r\sin\theta = 2$

In Exercises 43–54, solve the given problems: sketch or display the indicated curves.

43. What is the graph of $\tan\theta = 1$? Verify by changing the equation to rectangular form.
44. Find the polar equation of the line through the polar points $(1, 0)$ and $(2, \pi/2)$.
45. Using a calculator, show that the curves $r = 2\sin\theta$ and $r = 2\cos\theta$ intersect at right angles. Proper *window* settings are necessary.
46. Using a calculator, determine what type of graph is displayed by $r = 3\sec^2(\theta/2)$.
47. Display the graph of $r = 5 - 4\cos\theta$ on a calculator, using (a) the polar curve mode, and (b) the parametric curve mode. (See Example 6).
48. Display the graph of $r = 4\sin 3\theta$ on a calculator, using (a) the polar curve mode, and (b) the parametric curve mode. (See Example 6).
49. An architect designs a patio shaped such that it can be described as the area within the polar curve $r = 4.0 - \sin\theta$, where measurements are in meters. Sketch the curve that represents the perimeter of the patio.
50. The radiation pattern of a certain television transmitting antenna can be represented by $r = 120(1 + \cos\theta)$, where distances (in km) are measured from the antenna. Sketch the radiation pattern.
51. The joint between two links of a robot arm moves in an elliptical path (in cm), given by $r = \dfrac{25}{10 + 4\cos\theta}$. Sketch the path.
52. A missile is fired at an airplane and is always directed toward the airplane. The missile is traveling at twice the speed of the airplane. An equation that describes the distance r between the missile and the airplane is $r = \dfrac{70\sin\theta}{(1 - \cos\theta)^2}$, where θ is the angle between their directions at all times. See Fig. 21.119. This is a *relative pursuit curve*. Sketch the graph of this equation for $\pi/4 \le \theta \le \pi$.

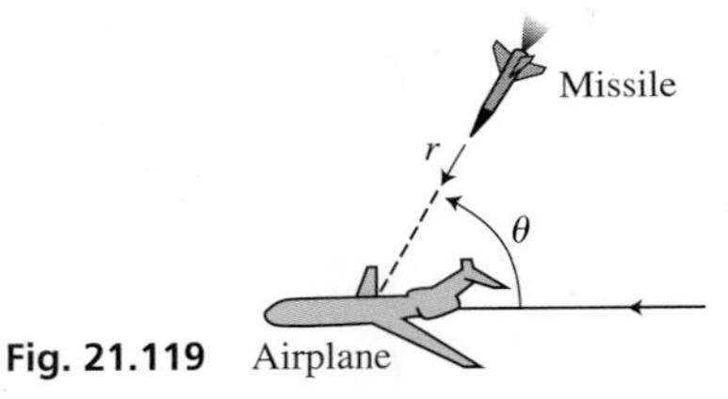

Fig. 21.119

53. In studying the photoelectric effect, an equation used for the rate R at which photoelectrons are ejected at various angles θ is $R = \dfrac{\sin^2\theta}{(1 - 0.5\cos\theta)^2}$. Sketch the graph.
54. Noting the graphs in Exercises 17, 18, 37, and 38, what conclusion do you draw about the value of n and the graph of $r = a\sin n\theta$ or $r = a\cos n\theta$?

Answers to Practice Exercises

1. Straight line 2. Circle

CHAPTER 21 KEY FORMULAS AND EQUATIONS

Distance formula	Fig. 21.2	$d = \sqrt{(x_2 - x_1)^2 + (y_2 - y_1)^2}$	**(21.1)**
Slope	Fig. 21.4	$m = \frac{y_2 - y_1}{x_2 - x_1}$	**(21.2)**
	Fig. 21.7	$m = \tan\alpha \quad (0° \le \alpha < 180°)$	**(21.3)**
	Fig. 21.9	$m_1 = m_2$ (for || lines)	**(21.4)**
	Fig. 21.10	$m_2 = -\frac{1}{m_1}$ or $m_1 m_2 = -1$ (for $\perp$ lines)	**(21.5)**
Straight line	Fig. 21.15	$y - y_1 = m(x - x_1)$	**(21.6)**
	Fig. 21.18	$x = a$	**(21.7)**
	Fig. 21.19	$y = b$	**(21.8)**
	Fig. 21.22	$y = mx + b$	**(21.9)**
		$Ax + By + C = 0$	**(21.10)**
Circle	Fig. 21.30	$(x - h)^2 + (y - k)^2 = r^2$	**(21.11)**
	Fig. 21.33	$x^2 + y^2 = r^2$	**(21.12)**
		$x^2 + y^2 + Dx + Ey + F = 0$	**(21.14)**
Parabola	Fig. 21.43	$y^2 = 4px$	**(21.15)**
	Fig. 21.46	$x^2 = 4py$	**(21.16)**
Ellipse	Fig. 21.58	$\frac{x^2}{a^2} + \frac{y^2}{b^2} = 1$	**(21.17)**
	Fig. 21.58	$a^2 = b^2 + c^2$	**(21.18)**
	Fig. 21.59	$\frac{y^2}{a^2} + \frac{x^2}{b^2} = 1$	**(21.19)**
Hyperbola	Fig. 21.71	$\frac{x^2}{a^2} - \frac{y^2}{b^2} = 1$	**(21.20)**
	Fig. 21.71	$c^2 = a^2 + b^2$	**(21.21)**
	Fig. 21.70	$y = \pm\frac{bx}{a}$ (asymptotes)	**(21.23)**
	Fig. 21.72	$\frac{y^2}{a^2} - \frac{x^2}{b^2} = 1$	**(21.24)**
	Fig. 21.78	$xy = c$	**(21.25)**
Translation of axes	Fig. 21.83	$x = x' + h$ and $y = y' + k$	**(21.26)**
		$x' = x - h$ and $y' = y - k$	**(21.27)**
Parabola, vertex (*h*, *k*)		$(y - k)^2 = 4p(x - h)$ (axis parallel to x-axis)	**(21.28)**
		$(x - h)^2 = 4p(y - k)$ (axis parallel to y-axis)	**(21.29)**

Ellipse, center (h, k)

$$\frac{(x-h)^2}{a^2} + \frac{(y-k)^2}{b^2} = 1 \quad \text{(major axis parallel to } x\text{-axis)} \qquad \textbf{(21.30)}$$

$$\frac{(y-k)^2}{a^2} + \frac{(x-h)^2}{b^2} = 1 \quad \text{(major axis parallel to } y\text{-axis)} \qquad \textbf{(21.31)}$$

Hyperbola, center (h, k)

$$\frac{(x-h)^2}{a^2} - \frac{(y-k)^2}{b^2} = 1 \quad \text{(transverse axis parallel to } x\text{-axis)} \qquad \textbf{(21.32)}$$

$$\frac{(y-k)^2}{a^2} - \frac{(x-h)^2}{b^2} = 1 \quad \text{(transverse axis parallel to } y\text{-axis)} \qquad \textbf{(21.33)}$$

Second-degree equation

$$Ax^2 + Bxy + Cy^2 + Dx + Ey + F = 0 \qquad \textbf{(21.34)}$$

Rotation of axes Fig. 21.97

$$x = x'\cos\theta - y'\sin\theta$$
$$y = x'\sin\theta + y'\cos\theta \qquad \textbf{(21.38)}$$

Angle of rotation

$$\tan 2\theta = \frac{B}{A - C} \quad (A \neq C) \qquad \textbf{(21.39)}$$

$$\theta = 45^\circ \quad (A = C) \qquad \textbf{(21.40)}$$

Polar coordinates Fig. 21.106

$$x = r\cos\theta \quad y = r\sin\theta \qquad \textbf{(21.41)}$$

$$\tan\theta = \frac{y}{x} \quad r = \sqrt{x^2 + y^2} \qquad \textbf{(21.42)}$$

CHAPTER 21 REVIEW EXERCISES

CONCEPT CHECK EXERCISES

*Determine each of the following as being either **true** or **false**. If it is false, explain why.*

1. The distance between $(4, -3)$ and $(3, -4)$ is $\sqrt{2}$.
2. $2y - 3x = 6$ is a straight line with intercepts $(2,0)$ and $(0,3)$.
3. The center of the circle $x^2 + y^2 + 2x + 4y + 5 = 0$ is $(1, 2)$.
4. The directrix of the parabola $x^2 = 8y$ is the line $y = 2$.
5. The vertices of the ellipse $9x^2 + 4y^2 = 36$ are $(2,0)$ and $(-2, 0)$.
6. The foci of the hyperbola $9x^2 - 16y^2 = 144$ are $(-5, 0)$ and $(5, 0)$.
7. The equation $x^2 = (y - 1)^2$ represents a hyperbola.
8. The equation $5x^2 - 8xy + 5y^2 = 9$ represents an ellipse.
9. The rectangular equation $x = 2$ represents the same curve as the polar equation $r = 2\sec\theta$.
10. The graph of the polar equation $\theta = \pi/4$ is a straight line.

PRACTICE AND APPLICATIONS

In Exercises 11–22, find the equation of the indicated curve, subject to the given conditions. Sketch each curve.

11. Straight line: passes through $(1, -7)$ with a slope of 4
12. Straight line: passes through $(-1, 5)$ and $(-2, -3)$
13. Straight line: perpendicular to $3x - 2y + 8 = 0$ and has a y-intercept of $(0, -1)$
14. Straight line: parallel to $2x - 5y + 1 = 0$ and has an x-intercept of $(2, 0)$
15. Circle: concentric with $x^2 + y^2 = 6x$, passes through $(4, -3)$
16. Circle: tangent to lines $x = 3$ and $x = 9$, center on line $y = 2x$
17. Parabola: focus $(-3, 0)$, vertex $(0, 0)$
18. Parabola: vertex $(0, 0)$, passes through $(1, 1)$ and $(-2, 4)$
19. Ellipse: vertex $(10, 0)$, focus $(8, 0)$, tangent to $x = -10$
20. Ellipse: center $(0, 0)$, passes through $(0, 3)$ and $(2, 1)$
21. Hyperbola: $V(0, 13)$, $C(0, 0)$, conj. axis of 24
22. Hyperbola: vertex $(0, 8)$, asymptotes $y = 2x$, $y = -2x$

In Exercises 23–36, find the indicated quantities for each of the given equations. Sketch each curve.

23. $x^2 + y^2 + 6x - 7 = 0$, center and radius
24. $2x^2 + 2y^2 + 4x - 8y - 15 = 0$, center and radius
25. $x^2 = -20y$, focus and directrix
26. $y^2 = 0.24x$, focus and directrix
27. $8x^2 + 2y^2 = 2$, vertices and foci
28. $2y^2 - 9x^2 = 18$, vertices and foci
29. $4x^2 - 25y^2 = 0.25$, vertices and foci
30. $4x^2 + 50y^2 = 1600$, vertices and foci
31. $x^2 - 8x - 4y - 16 = 0$, vertex and focus
32. $y^2 - 4x + 4y + 24 = 0$, vertex and directrix
33. $4x^2 + y^2 - 16x + 2y + 13 = 0$, center
34. $x^2 - 2y^2 + 4x + 4y + 6 = 0$, center
35. $x^2 - 2xy + y^2 + 4x + 4y = 0$, vertex
36. $9x^2 - 9xy + 21y^2 - 15 = 0$, center

In Exercises 37–44, plot the given curves in polar coordinates.

37. $r = 4(1 + \sin\theta)$
38. $r = 1 - 3\cos\theta$

39. $r = 4\cos 3\theta$
40. $r = 3\sin\theta - 4\cos\theta$

41. $r = \dfrac{3}{\sin\theta + 2\cos\theta}$
42. $r = \dfrac{1}{2(\sin\theta - 1)}$

43. $r = 2\sin\left(\dfrac{\theta}{2}\right)$
44. $r = 1 - \cos 2\theta$

In Exercises 45–48, find the polar equation of each of the given rectangular equations.

45. $y = 2x$
46. $2xy = 1$

47. $x^2 + xy + y^2 = 2$
48. $x^2 + (y + 3)^2 = 16$

In Exercises 49–52, find the rectangular equation of each of the given polar equations.

49. $r = 2\sin 2\theta$
50. $r^2 = 9\sin\theta$

51. $r = \dfrac{4}{2 - \cos\theta}$
52. $r\cos\theta = 4\tan\theta$

In Exercises 53–58, determine the number of real solutions of the given systems of equations by sketching the indicated curves. (See Section 14.1.)

53. $x^2 + y^2 = 9$
$4x^2 + y^2 = 16$

54. $y = e^x$
$x^2 - y^2 = 1$

55. $x^2 + y^2 - 4y - 5 = 0$
$y^2 - 4x^2 - 4 = 0$

56. $x^2 - 4y^2 + 2x - 3 = 0$
$y^2 - 4x - 4 = 0$

57. $y = 2\sin x$
$y = 2 - x^2$

58. $y = 4\ln x$
$xy = 6$

In Exercises 59–68, view the curves of the given equations on a calculator.

59. $x^2 + 3y + 2 - (1 + x)^2 = 0$
60. $y^2 = 4x + 6$

61. $2x^2 + 2y^2 + 4y - 3 = 0$
62. $2x^2 + (y - 3)^2 - 5 = 0$

63. $x^2 - 4y^2 + 4x + 24y - 48 = 0$

64. $x^2 + 2xy + y^2 - 3x + 8y = 0$

65. $r = 3\cos(3\theta/2)$
66. $r = 5 - 2\sin 4\theta$

67. $r = 2 - 3\csc\theta$
68. $r = 2\sin(\cos 3\theta)$

In Exercises 69–74, find the equation of the locus of a point $P(x, y)$ that moves as stated.

69. Always 4 units from $(3, -4)$

70. Passes through $(7, -5)$ with a constant slope of -2

71. The sum of its distances from $(1, -3)$ and $(7, -3)$ is 8.

72. The difference of its distances from $(3, -1)$ and $(3, -7)$ is 4.

73. Its distance from $y = 6$ always equals its distance to $(0, -6)$.

74. A standard form conic that passes through $(-3, 0)$ and $(0, 4)$.

In Exercises 75–120, solve the given problems.

75. Considering Eq. (21.30) of an ellipse, describe the graph if $a = b$.

76. Show that the ellipse $x^2 + 9y^2 = 9$ has the same foci as the hyperbola $x^2 - y^2 = 4$.

77. The points $(-2, -5)$, $(3, -3)$, and $(13, x)$ are collinear. Find x.

78. For the polar coordinate point $(-5, \pi/4)$, find another set of polar coordinates such that $r < 0$ and $-2\pi < \theta < 0$.

79. Find the distance between the polar coordinate points $(3, \pi/6)$ and $(6, -\pi/3)$.

80. Show that the parametric equations $y = \cot\theta$ and $x = \csc\theta$ define a hyperbola.

81. In two ways, show that the line segments joining $(-3, 11)$, $(2, -1)$, and $(14, 4)$ form a right triangle.

82. Find the equation of the circle that passes through $(3, -2)$, $(-1, -4)$, and $(2, -5)$.

83. What type of curve is represented by
$(x + jy)^2 + (x - jy)^2 = 2$? $(j = \sqrt{-1})$

84. For the ellipse in Fig. 21.120, show that the product of the slopes PA and PB is $-b^2/a^2$.

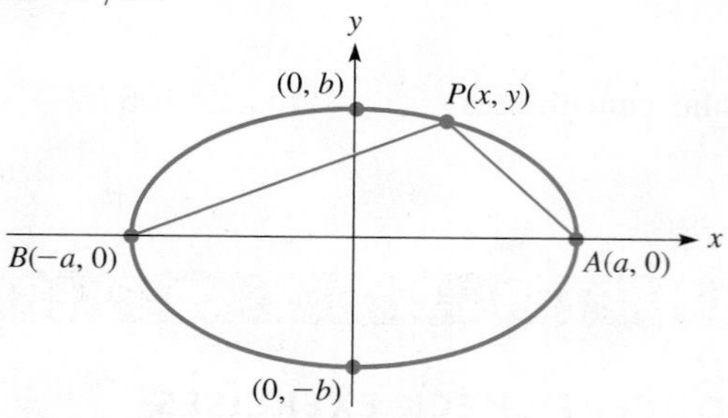

Fig. 21.120

85. Find the area of the square that can be inscribed in the ellipse $7x^2 + 2y^2 = 18$.

86. Using a graphing calculator, determine the number of points of intersection of the polar curves $r = 4|\cos 2\theta|$ and $r = 6\sin[\cos(\cos 3\theta)]$.

87. By means of the definition of a parabola, find the equation of the parabola with focus at $(3, 1)$ and directrix the line $y = -3$. Find the same equation by the method of translation of axes.

88. For what value(s) of k does $x^2 - ky^2 = 1$ represent an ellipse with vertices on the y-axis?

89. The total resistance R_T of two resistances in series in an electric circuit is the sum of the resistances. If a variable resistor R is in series with a 2.5-Ω resistor, express R_T as a function of R and sketch the graph.

90. The acceleration of an object is defined as the change in velocity v divided by the corresponding change in time t. Find the equation relating the velocity v and time t for an object for which the acceleration is 20 ft/s^2 and $v = 5.0$ ft/s when $t = 0$ s.

91. The velocity v of a crate sliding down a ramp is given by $v = v_0 + at$, where v_0 is the initial velocity, a is the acceleration, and t is the time. If $v_0 = 5.75$ ft/s and $v = 18.5$ ft/s when $t = 5.50$ s, find v as a function of t. Sketch the graph.

92. An airplane touches down when landing at 100 mi/h. Its velocity v while coming to a stop is given by $v = 100 - 20{,}000t$, where t is the time in hours. Sketch the graph of v vs. t.

93. It takes 2.010 kJ of heat to raise the temperature of 1.000 kg of steam by 1.000°C. In a steam generator, a total of y kJ is used to raise the temperature of 50.00 kg of steam from 100°C to T°C. Express y as a function of T and sketch the graph.

94. The temperature in a certain region is 27°C, and at an altitude of 2500 m above the region it is 12°C. If the equation relating the temperature T and the altitude h is linear, find the equation.

95. An elliptical tabletop is 4.0 m long and has a 3.0 m by 2.0 m rectangular design inscribed in it lengthwise. See Fig. 21.121. What is the area of the tabletop? (The area of an ellipse is $A = \pi ab$.)

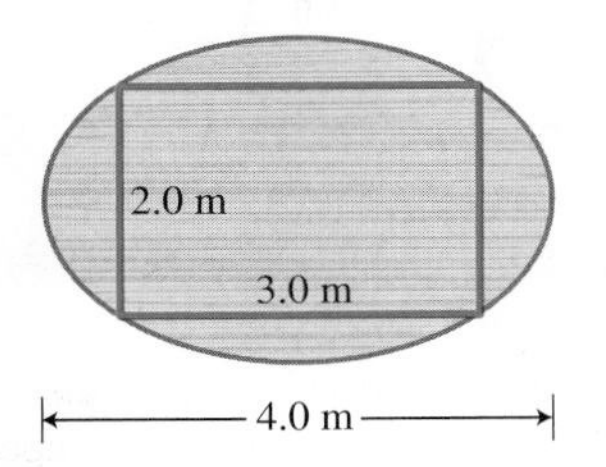

Fig. 21.121

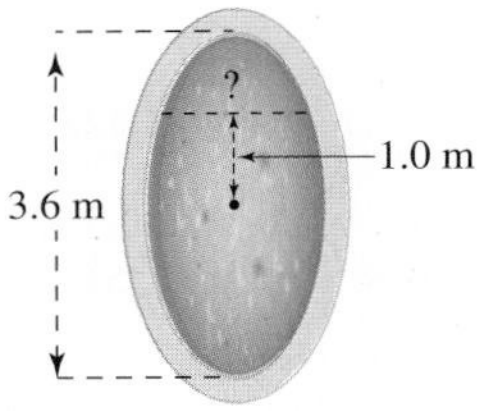

Fig. 21.122

96. An elliptical hot tub is twice as long as it is wide. If its length is 3.6 m, find the distance across the shorter span of the hot tub 1.0 m from the center. See Fig. 21.122.

97. The radar gun on a police helicopter 490 ft above a multilane highway is directed vertically down onto the highway. If the radar gun signal is cone-shaped with a vertex angle of 14°, what area of the highway is covered by the signal?

98. A circular wind turbine with a diameter of 90 m is attached to the top of a 110-m pole. Find the equation of the circle traced by the tips of the blades if the origin is at the bottom of the pole.

99. The arch of a small bridge across a stream is parabolic. If, at water level, the span of the arch is 80 ft and the maximum height above water level is 20 ft, what is the equation that represents the arch? Choose the most convenient point for the origin of the coordinate system.

100. A laser source is 2.00 cm from a spherical surface of radius 3.00 cm, and the laser beam is tangent to the surface. By placing the center of the sphere at the origin, and the source on the positive x-axis, find the equation of the line along which the beam shown in Fig. 21.123 is directed.

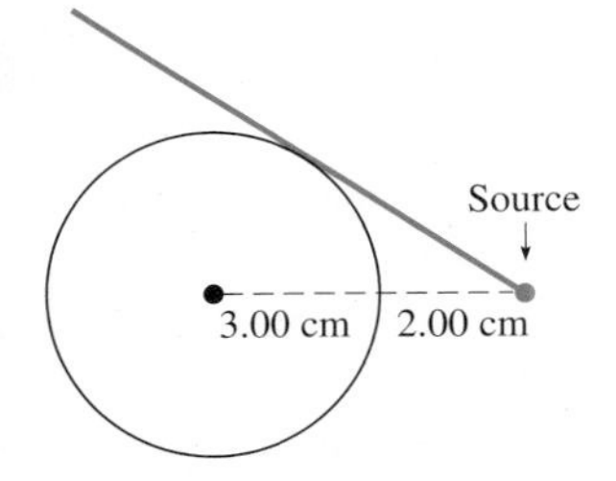

Fig. 21.123

101. A motorcycle cost $12,000 when new and then depreciated linearly $1250/year for four years. It then further depreciated linearly $1000/year until it had no resale value. Write the equation for the motorcycle's value V as a function of t and sketch the graph of $V = f(t)$.

102. The temperature of ocean water does not change with depth very much for about 300 m, and then as depth increases to about 1000 m, it decreases rapidly. Below 1000 m the temperature decreases very slowly with depth. A typical middle latitude approximation would be $T = 22$°C for the first 300 m of depth, then T decreases to 5°C at 1000 m, and then to 2°C at a depth of 5000 m. Graph T as a function of the depth d, assuming linear changes.

103. The top horizontal cross section of a dam is parabolic. The open area within this cross section is 80 ft across and 50 ft from front to back. Find the equation of the edge of the open area with the vertex at the origin of the coordinate system and the axis along the x-axis.

104. The *quality factor* Q of a series resonant electric circuit with resistance R, inductance L, and capacitance C is given by $Q = \frac{1}{R}\sqrt{\frac{L}{C}}$. Sketch the graph of Q and L for a circuit in which $R = 1000\ \Omega$ and $C = 4.00\ \mu$F.

105. At very low temperatures, certain metals have an electric resistance of zero. This phenomenon is called *superconductivity*. A magnetic field also affects the superconductivity. A certain level of magnetic field H_T, the threshold field, is related to the thermodynamic temperature T by $H_T/H_0 = 1 - (T/T_0)^2$, where H_0 and T_0 are specifically defined values of magnetic field and temperature. Sketch the graph of H_T/H_0 vs. T/T_0.

106. A rectangular parking lot is to have a perimeter of 600 m. Express the area A in terms of the width w and sketch the graph.

107. The electric power P (in W) supplied by a battery is given by $P = 12.0i - 0.500i^2$, where i is the current (in A). Sketch the graph of P vs. i.

108. The Colosseum in Rome is in the shape of an ellipse 188 m long and 156 m wide. Find the area of the Colosseum. ($A = \pi ab$ for an ellipse.)

109. A specialty electronics company makes an ultrasonic device to repel animals. It emits a 20–25 kHz sound (above those heard by people), which is unpleasant to animals. The sound covers an elliptical area starting at the device, with the longest dimension extending 120 ft from the device and the focus of the area 15 ft from the device. Find the area covered by the signal. ($A = \pi ab$)

110. A study indicated that the fraction f of cells destroyed by various dosages d of X-rays is given by the graph in Fig. 21.124. Assuming that the curve is a quarter-ellipse, find the equation relating f and d for $0 \leq f \leq 1$ and $0 < d \leq 10$ units.

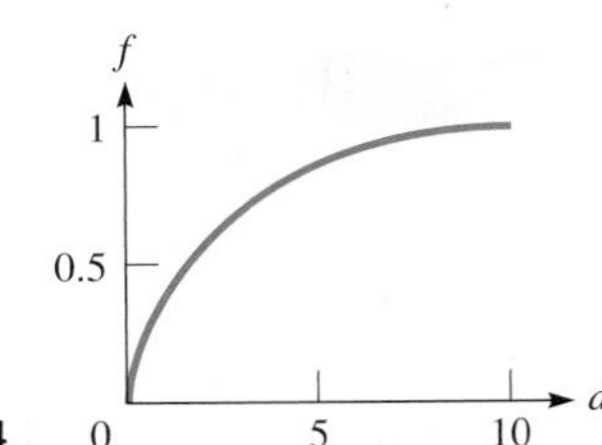

Fig. 21.124

111. A machine-part designer wishes to make a model for an elliptical cam by placing two pins in a design board, putting a loop of string over the pins, and marking off the outline by keeping the string taut. (Note that the definition of the ellipse is being used.) If the cam is to measure 10 cm by 6 cm, how long should the loop of string be and how far apart should the pins be?

112. Soon after reaching the vicinity of the moon, *Apollo 11* (the first spacecraft to land a man on the moon) went into an elliptical lunar orbit. The closest the craft was to the moon in this orbit was 70 mi, and the farthest it was from the moon was 190 mi. What was the equation of the path if the center of the moon was at one of the foci of the ellipse? Assume that the major axis is along the x-axis and that the center of the ellipse is at the origin. The radius of the moon is 1080 mi.

113. The vertical cross section of the cooling tower of a nuclear power plant is hyperbolic, as shown in Fig. 21.125. Find the radius r of the smallest circular horizontal cross section.

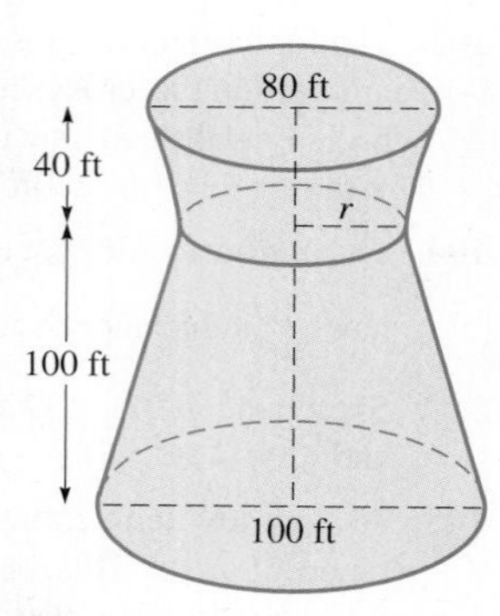

Fig. 21.125

114. Tremors from an earthquake are recorded at the California Institute of Technology (Pasadena, California) 36 s before they are recorded at Stanford University (Palo Alto, California). If the seismographs are 510 km apart and the shock waves from the tremors travel at 5.0 km/s, what is the curve on which lies the point where the earthquake occurred?

115. An electronic instrument located at point P records the sound of a rifle shot and the impact of the bullet striking the target at the same instant. Show that P lies on a branch of a hyperbola.

116. A 60-ft rope passes over a pulley 10 ft above the ground, and a crate on the ground is attached at one end. The other end of the rope is held at a level of 4 ft above the ground and is drawn away from the pulley. Express the height of the crate over the ground in terms of the distance the person is from directly below the crate. Sketch the graph of distance and height. See Fig. 21.126. (Neglect the thickness of the crate.)

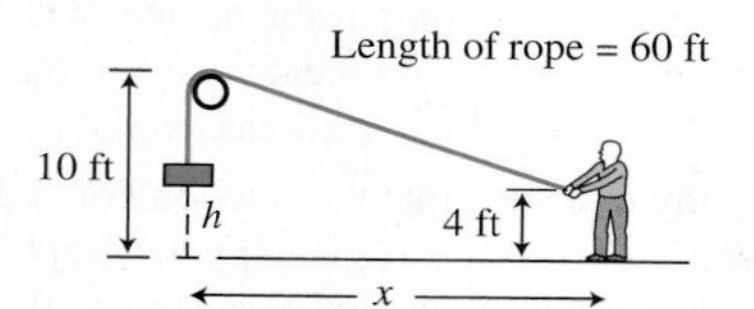

Fig. 21.126

117. A satellite at an altitude proper to make one revolution per day around the center of Earth will have for an excellent approximation of its projection on Earth of its path the curve $r^2 = R^2\cos 2(\theta + \frac{\pi}{2})$, where R is the radius of Earth. Sketch the path of the projection.

118. The vertical cross sections of two pipes as drawn on a drawing board are shown in Fig. 21.127. Find the polar equation of each.

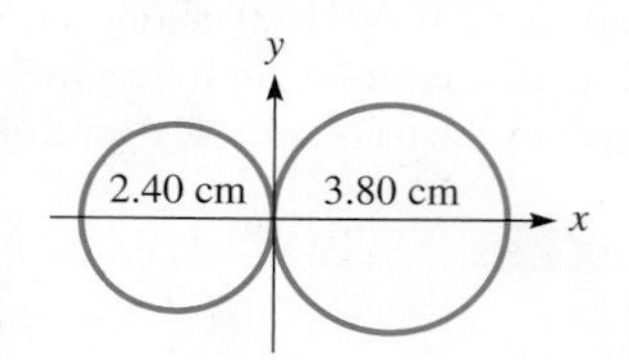

Fig. 21.127

119. The path of a certain plane is $r = 200(\sec\theta + \tan\theta)^{-5}/\cos\theta$, $0 < \theta < \pi/2$. Sketch the path and check it on a calculator.

120. The sound produced by a jet engine was measured at a distance of 100 m in all directions. The loudness of the sound d (in decibels) was found to be $d = 115 + 10\cos\theta$, where the 0° line for the angle θ is directed in front of the engine. Sketch the graph of d vs. θ in polar coordinates (use d as r).

121. Under a force that varies inversely as the square of the distance from an attracting object (such as the sun exerts on Earth), it can be shown that the equation of the path an object follows is given in general by

$$\frac{1}{r} = a + b\cos\theta$$

where a and b are constants for a particular path. First, transform this equation into rectangular coordinates. Then write one or two paragraphs explaining why this equation represents one of the conic sections, depending on the values of a and b. It is through this kind of analysis that we know the paths of the planets and comets are conic sections.

CHAPTER 21 PRACTICE TEST

As a study aid, we have included complete solutions for each Practice Test problem at the end of this book.

1. (a) Find the distance between $(4, -1)$ and $(6, 3)$. (b) Find the slope of the line perpendicular to the line segment joining the points in part (a).

2. Identify the type of curve represented by the equation $2(x^2 + x) = 1 - y^2$.

3. Sketch the graph of the straight line $4x - 2y + 5 = 0$ by finding its slope and y-intercept.

4. Find the polar equation of the curve whose rectangular equation is $x^2 = 2x - y^2$.

5. Find the vertex and the focus of the parabola $x^2 = -12y$. Sketch the graph.

6. Find the equation of the circle with center at $(-1, 2)$ and that passes through $(2, 3)$.

7. Find the equation of the straight line that passes through $(-4, 1)$ and $(2, -2)$.

8. Where is the focus of a parabolic reflector that is 12.0 cm across and 4.00 cm deep?

9. A hallway 16 ft wide has a ceiling whose cross section is a semi-ellipse. The ceiling is 10 ft high at the walls and 14 ft high at the center. Find the height of the ceiling 4 ft from each wall.

10. Plot the polar curve $r = 3 + \cos\theta$.

11. Find the center and vertices of the conic section $4y^2 - x^2 - 4x - 8y - 4 = 0$. Show completely the sketch of the curve.

12. (a) What type of curve is represented by $8x^2 - 4xy + 5y^2 = 36$?
(b) Through what angle must the curve in part (a) be rotated in order that there is no $x'y'$-term?

Introduction to Statistics

22

LEARNING OUTCOMES

After completion of this chapter, the student should be able to:

- Use bar graphs and pie charts to display qualitative data
- Summarize and display quantitative data using frequency distributions, histograms, stem-and-leaf plots, and time series plots
- Calculate measures of central tendency (mean, median, and mode)
- Calculate and interpret the standard deviation
- Apply the empirical rule
- Find relative frequencies involving normally distributed populations
- Plot and interpret $\bar{x}$, R, and p control charts
- Find the equation of the least squares regression line that best fits a given set of data
- Use a calculator to determine a nonlinear regression model to fit a set of data

After the invention of the steam engine in the late 1700s by the Scottish engineer James Watt, the production of machine-made goods became widespread during the 1800s. However, it was not until the 1920s that much attention was paid to the quality control of the goods being produced. In 1924, Walter Shewhart of Bell Telephone Laboratories used a statistical chart for controlling product variables; in the 1940s, quality control was used in much of wartime production.

Quality control is one of the modern uses of *statistics*, the branch of mathematics in which data are collected, analyzed, and interpreted. Today it is nearly impossible to read a newspaper or watch television news without seeing some type of study, in areas such as medicine or politics, that involves statistics. Other fields in which statistical methods are used include biology, physics, psychology, sociology, economics, business, education, and electronics.

The first significant use of statistics was made in the 1660s by John Graunt and in the 1690s by Edmund Halley (Halley's Comet), when each published some conclusions about the population in England based on mortality tables. There was little development of statistics until the 1800s, when statistical measures became more widely used. Examples are the scientist Francis Galton, who used statistics in the study of human heredity, and the nurse Florence Nightingale, who used statistical graphs to show that more soldiers died in the Crimean War (in the 1850s) from unsanitary conditions than from combat wounds.

This chapter is an introduction to some of the basic concepts and uses of statistics. We first consider certain basic statistical measures. Then a section is devoted to normal distributions, followed by a section on *control charts* that are used in statistical process control in industry. The final sections show how to start with a set of points on a graph and find an equation that best "fits" the data. This equation, which shows a basic relationship between the variables, can be useful in research.

◀ **Statistical analysis is used extensively in business and industry. In Section 22.5, we show how a *control chart* can be used to monitor the defects on DVDs.**

22.1 Graphical Displays of Data

Qualitative Data • Quantitative Data • Bar Graph • Pie Chart • Class • Frequency • Frequency Distribution • Relative Frequency • Histogram • Stem-and-Leaf Plot • Time Series Plot

There are many situations when we need to organize data so that important patterns can be seen and conclusions can be drawn. Statistics provides the tools for doing this. **Statistics is the science of collecting, describing, and interpreting data.** One important way of organizing data is by using statistical graphs. In this section, we will show some common graphs that are used to display data.

There are two main different kinds of data. **Qualitative data** are **categorical** in nature. Their values are nonnumeric, and they fit into one of a number of possible categories (like eye color, defective or not defective, and make of car). **Quantitative data,** on the other hand, have **numeric values** (like weight, income, and age). It is important to understand the difference between these two types of data because different graphs are used for each type.

GRAPHS FOR QUALITATIVE DATA

The two most common graphs used to display qualitative data are **bar graphs** and **pie charts.** NOTE ➧ [Bar graphs can always be used, but pie charts should only be used when the percentages add up to 100% and each member of the sample fits into only one category.]

The following two examples illustrate these graphs.

EXAMPLE 1 Bar chart—communication habits of high school students

In a survey of 1000 high school students, each student was asked which type of communication he or she uses daily to talk with friends and family. The **frequency** in the table below shows the **number of students** who used each form of communication. The **relative frequency** shows the **proportion of students as a percentage of the total** (this can also be given as a fraction or a decimal). Since there were a total of 1000 students, the relative frequency is found by dividing each frequency by 1000. Figure 22.1 shows a bar graph of these data.

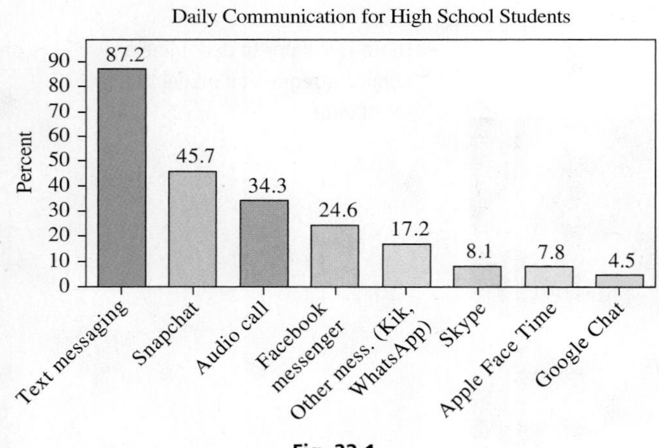

Fig. 22.1

Form of Communication	*Frequency*	*Relative Frequency (%)*
Text messaging	872	87.2
Snapchat	457	45.7
Audio call	343	34.3
Facebook messenger	246	24.6
Other messaging (Kik, WhatsApp)	172	17.2
Skype	81	8.1
Apple FaceTime	78	7.8
Google Chat	45	4.5

Bar graphs can show either frequency or relative frequency on the vertical axis (the one in Fig. 22.1 shows relative frequency). Either way, **the vertical scale must start at zero.** Otherwise, the graph is distorted and overexaggerates small differences between categories. NOTE ➧ [Note that a pie chart *cannot* be used in this example since the percentages add up to more than 100%. This is because a single student could use several of the forms of communication, not just one.] ■

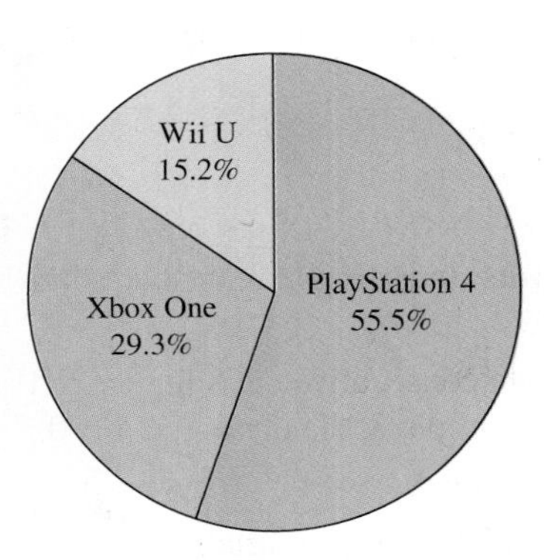

Fig. 22.2

EXAMPLE 2 Pie chart—market share among top video game consoles

In a certain month, the sales (in units sold) of the three leading video game consoles are shown in the table below. The number of units sold is the frequency. The relative frequency is found by dividing the number of units sold by 1,083,652. Figure 22.2 shows a pie chart of these data. The central angle of each slice is proportional to the frequency (or relative frequency). Note that use of the pie chart is appropriate here since the percentages add up to 100%.

Console	*Units Sold*	*Relative Frequency (%)*
PlayStation 4	601,529	55.5
Xbox One	317,421	29.3
Wii U	164,702	15.2
Total	1,083,652	100

■

GRAPHS FOR QUANTITATIVE DATA

There are several graphs that can be used to display quantitative data. We will discuss the **histogram**, **stem-and-leaf plot**, and **time series plot.** Histograms are important because they show how data are *distributed*, meaning the *shape* of the data. They are a graphical display of a **frequency distribution.** Stem-and-leaf plots are useful for displaying the original data values in ascending order in addition to showing the shape of the data. Time series plots show how a quantity changes over time. The following examples illustrate these graphs.

EXAMPLE 3 Frequency distribution and histogram—birth weight

The birth weights (in kg) for a sample of 32 infants are given below. Make a histogram of these data.

2.8	3.9	3.0	3.7	2.8	3.4	1.0	4.1
2.9	4.1	3.8	2.2	3.7	3.1	3.1	3.2
3.6	3.4	3.3	3.5	3.1	4.0	3.4	2.4
2.7	3.1	2.8	3.2	3.5	3.9	3.1	2.2

Birth Weight (kg)	*Frequency* (f)	*Relative Frequency* (%)
1.0–1.5	1	3.1
1.5–2.0	0	0.0
2.0–2.5	3	9.4
2.5–3.0	5	15.6
3.0–3.5	12	37.5
3.5–4.0	8	25.0
4.0–4.5	3	9.4
Total	32	100

We will begin by making a frequency distribution by grouping the data and then counting the number of data within each group. The groups are called **classes,** and the number of data within each class is called the **frequency.** A table that shows each of the classes along with the corresponding frequencies is called a **frequency distribution.** The **relative frequency** is found by dividing the frequency by the total number of data. We will adopt the somewhat common convention that ***the left endpoint, but not the right endpoint, is included in each class.*** The frequency distribution for birth weights is shown to the left, with the relative frequencies also included. A **histogram** is a graph that places the classes on the horizontal axis and uses rectangles to show either the frequency or relative frequency, which is plotted on the vertical axis. A frequency histogram for birth weights is shown in Fig. 22.3.

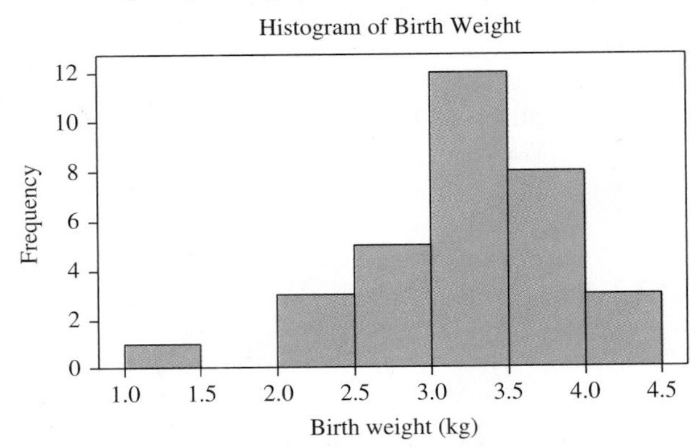

Fig. 22.3

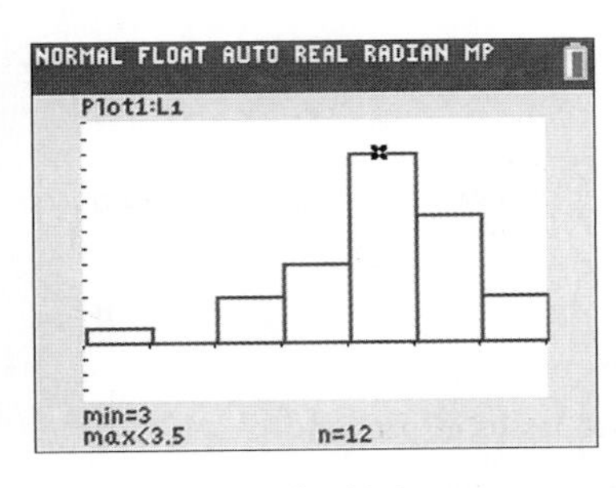

Fig. 22.4

The *stat plot* feature on a calculator can also be used to make a histogram. The classes may be adjusted with the *window* settings, and the *trace* feature can be used to toggle from bar to bar to observe the frequencies in each class. For example, the histogram shown in Fig. 22.4 shows there are 12 birth weights between 3 and 3.5 kg. ■

The following list contains some important points regarding histograms.

Notes Regarding Histograms

1. There should normally be between 5 and 15 classes.
2. The edges of the classes are called **class limits.** In Example 3, the class limits are 1.0, 1.5, 2.0, . . . , 4.5.
3. The **class width** is the difference between two consecutive class limits, and it should be the same for all the classes. In Example 3, the class width is 0.5.
4. The midpoint of each class is called the **class mark.**
5. The vertical scale of a histogram can measure frequency or relative frequency. Either way, it must start at zero.
6. The bars of a histogram should touch each other. Gaps should only occur when there are no data in a particular class.
7. Histograms show the *shape* of the data. Three common shapes are **symmetric** (bell shaped), **skewed to the left** (the longer tail extends to the left), and **skewed to the right** (the longer tail extends to the right). The histogram in Example 3 is skewed to the left.

Stem	*Leaf*
1	0
2	22478889
3	01111122344455677899
4	011

Fig. 22.5

Stem	*Leaf*
1	0
1	
2	224
2	78889
3	011111223444
3	55677899
4	011

Fig. 22.6

EXAMPLE 4 Stem-and-leaf plot—birth weights

In a stem-and-leaf plot, all but the last digit of each data value is called the stem and the last digit is the leaf. The stems are listed vertically with a line separating them from the leaves, which are listed in ascending order. A stem-and-leaf plot of the birth weights given in Example 3 is shown in Fig. 22.5. A "key" should be provided to show how to interpret the place values of the digits since decimal points are never included in this kind of chart. In this case, a typical key would be "leaf digit $= 0.1$" or "1|0 represents 1.0."

Sometimes it is useful to **split the stems.** This means that each stem will be listed twice, with the first stem including leaves 0–4 and the second including leaves 5–9. See Fig. 22.6.

NOTE ▸ [Note that the second stem of 1 is included even though there are no leaves. A stem-and-leaf plot is like a histogram on its side, and it should show that there is a bit of a gap between the data values 1.0 kg and 2.2 kg.] ■

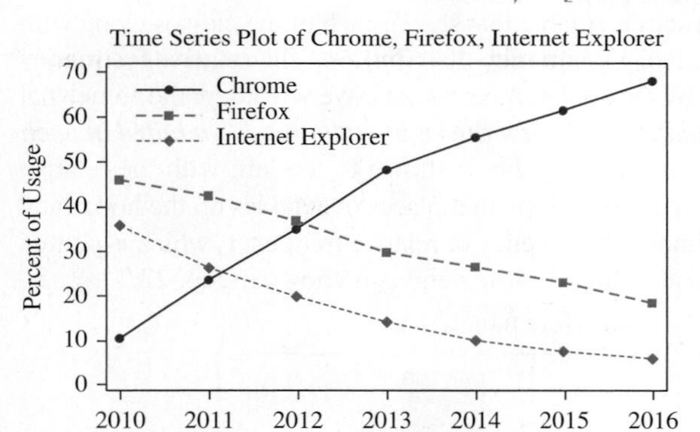

Fig. 22.7

EXAMPLE 5 Time series plot—Internet browser trends

A time series graph shows how one or more quantities change over time. The horizontal axis is scaled with time and the vertical axis with the value of one or more quantitative variables. Figure 22.7 shows the Web browser usage (as a percent of all Web browser usage) for the three leading browsers between the years 2010 and 2016. The graph clearly shows an upward trend for Chrome over this time period and a downward trend for both Firefox and Internet Explorer. In this type of graph, the vertical axis does not have to start at zero since it only shows how a quantity changes over time from a certain starting point. ■

EXERCISES 22.1

In Exercises 1 and 2, make the given changes to the indicated examples of this section and then solve the indicated problem.

1. In Example 3, change the class limits to 1.0, 2.0, 3.0, 4.0, and 5.0 and then make a table showing the frequencies and relative frequencies.
2. In Example 4, change the lowest data value from 1.0 to 0.9 and then make a stem-and-leaf plot with split stems.

In Exercises 3–6, indicate whether the variable is qualitative or quantitative.

3. The diameter of a bolt
4. A person's favorite genre of music
5. Whether or not a product passes inspection
6. The time it takes a worker to complete a task

In Exercises 7–10, use the following data. In a random sample, 500 college students were asked which social networks they use on a daily basis. The results are summarized below:

Social Network	*Frequency*
Facebook	305
Instagram	255
Twitter	175
Google +	115
Pinterest	80
Vine	80

7. Make a bar graph of these data showing frequency on the vertical axis.

8. Find the relative frequencies for each social network.

9. Is it appropriate to use a pie chart for these data? Explain why or why not.

10. Make a bar graph of these data showing relative frequency on the vertical axis.

In Exercises 11–14, use the following data. In a random sample, 800 smartphone owners were asked which type of smartphone they would choose with their next purchase (if they could only choose one). The results are summarized below:

Smartphone	*Frequency*
iPhone	320
Samsung	284
LG	82
Motorola	35
Other	79

11. Find the relative frequencies, rounded to the nearest tenth of a percent.

12. What is the sum of the relative frequencies found in Exercise 11? Why is this sum not exactly 100%?

13. Use the relative frequencies to make a pie chart of these data.

14. Make a bar graph of these data using the frequencies.

In Exercises 15–18, use the following data. In testing a new electric engine, an automobile company randomly selected 20 cars of a certain model and recorded the range (in mi) that the car could travel before the batteries needed recharging. The results are shown below.

143, 148, 146, 144, 149, 144, 150, 148, 148, 144
153, 146, 147, 146, 147, 149, 145, 151, 149, 148

15. Using the classes 141–144, 144–147, . . . , 153–156, form a frequency distribution table. Use the convention that the left endpoint is included in each class, but not the right endpoint.

16. Find the relative frequencies for the classes given in Exercise 15.

17. Draw a frequency histogram using the classes given in Exercise 15.

18. Draw a relative frequency histogram using the classes given in Exercise 15.

In Exercises 19–24, use the following data. In a random sample, 30 Android users were asked to record the number of apps that were installed on their phone. The resulting data are shown below:

112, 91, 101, 85, 76, 115, 93, 126, 78, 86, 105, 107, 58, 86, 109, 111, 103, 105, 97, 110, 92, 95, 107, 89, 101, 67, 103, 99, 93, 82

19. Make a stem-and-leaf plot of these data (without split stems).

20. Make a stem-and-leaf plot of these data using split stems.

21. Make a frequency distribution table using the class limits 50, 60, 70, . . . ,130.

22. Find the relative frequencies using the class limits given in Exercise 21.

23. Make a frequency histogram using the table in Exercise 21. Describe the shape.

24. Make a relative frequency histogram using the values found in Exercise 22.

In Exercises 25 and 26, use the following data. The dosage (in mR) given by a particular X-ray machine was measured 20 times, with the following readings:

4.25, 4.36, 3.96, 4.21, 4.44, 3.83, 4.37, 4.27, 4.33, 4.34, 4.15, 3.90, 4.41, 4.51, 4.18, 4.26, 4.29, 4.09, 4.36, 4.23

25. Using the class limits 3.80, 3.90, 4.00, . . . , 4.60, construct a frequency histogram of these data.

26. Make a stem-and-leaf plot of these data (without split stems).

In Exercises 27 and 28, use the following data. The blood alcohol level (in %) in the bloodstreams of those charged with DUI at a police check point were as follows:

0.15, 0.11, 0.13, 0.16, 0.12, 0.09. 0.10, 0.11, 0.13, 0.06, 0.12, 0.11, 0.09, 0.17, 0.14, 0.15, 0.11, 0.14, 0.12, 0.15, 0.10, 0.11, 0.13

27. Using the class limits 0.06, 0.08, 0.10, . . . , 0.18, make a frequency histogram of the data.

28. Find the relative frequencies (to the nearest tenth of a percent) for the classes in Exercise 27.

In Exercises 29 and 30, solve the given problems.

29. The data in the table show the global mean land-ocean temperature index (using a base period of 1951–1980) for various years. Make a time series graph of these data.

Year	1985	1990	1995	2000	2005	2010	2015
Temperature index (°C)	0.28	0.12	0.44	0.42	0.69	0.72	0.86

30. The data in the following table show the percentage of U.S. households that have a landline telephone for various years. Make a time series graph of these data.

Year	2006	2008	2010	2012	2014
Percentage (%)	84	78	69	60	53

22.2 Measures of Central Tendency

Median • Mean • Mode

Tables and graphical representations give a general description of data. However, it is often useful and convenient to find representative values for the location of the center of the distribution, and other numbers to give a measure of the deviation from this central value. In this way, we can obtain a numerical description of the data. We now discuss the values commonly used to measure the location of the center of the distribution. These are referred to as *measures of central tendency.*

The first of these measures of central tendency is the **median.** *The median is the middle number, that number for which there are as many above it as below it in the distribution.* If there is no middle number, the median is that number halfway between the two middle numbers of the distribution.

EXAMPLE 1 Median—odd or even number of values

Given the numbers 5, 2, 6, 4, 7, 4, 7, 2, 8, 9, 4, 11, 9, 1, 3, we first arrange them in numerical order. This arrangement is

middle number (arrow pointing to 5)

1, 2, 2, 3, 4, 4, 4, 5, 6, 7, 7, 8, 9, 9, 11

Because there are 15 numbers, the middle number is the eighth. Because the eighth number is 5, the median is 5.

If the number 11 is not included in this set of numbers and there are only 14 numbers in all, the median is that number halfway between the seventh and eighth numbers. Because the seventh is 4 and the eighth is 5, the median is 4.5. ■

■ The median may be expressed to one more significant digit than that given in the data.

EXAMPLE 2 Median—even number of values

In the distribution of birth weights in Example 3 of Section 22.1, the median is 3.2 kg. There are 32 values in all, and when listed in order, the 16th and 17th values are each 3.2 kg. The number halfway between the 16th and 17th values is the median. Because both are 3.2 kg, the median is 3.2 kg. ■

Another very widely applied measure of central tendency is the **mean.** *The mean is calculated by finding the sum of all the values and then dividing by the number of values.* (The mean is the number most people call the "average.").

EXAMPLE 3 Mean

The mean of the numbers given in Example 1 is determined by finding the sum of all the numbers and dividing by 15. Therefore, by letting $\bar{x}$ (read as "x bar") represent the mean, we have

$$\bar{x} = \frac{5 + 2 + 6 + 4 + 7 + 4 + 7 + 2 + 8 + 9 + 4 + 11 + 9 + 1 + 3}{15}$$

$$= \frac{82}{15} = 5.5$$

sum of values (arrow pointing to 82); number of values (arrow pointing to 15)

Thus, the mean is 5.5. (The mean is usually calculated to one more decimal place than was present in the original data.) ■

If we wish to find the mean of a large number of values, and if some of them appear more than once, the calculation can be simplified. The mean can be calculated by multiplying each value by its frequency, adding these results, and then dividing by the total number of values (the sum of the frequencies). Letting $\overline{x}$ represent the mean of the values $x_1, x_2, \ldots, x_n$, which occur with frequencies $f_1, f_2, \ldots, f_n$, respectively, we have

■ This is called a *weighted mean* because each value is given a weighting based on the number of times it occurs.

$$\overline{x} = \frac{x_1 f_1 + x_2 f_2 + \cdots + x_n f_n}{f_1 + f_2 + \cdots + f_n} \tag{22.1}$$

EXAMPLE 4 Mean using frequencies

Using Eq. (22.1) to find the mean of the numbers of Example 1, we first set up a table of values and their respective frequencies, as follows:

Values	1	2	3	4	5	6	7	8	9	11
Frequency	1	2	1	3	1	1	2	1	2	1

We now calculate the mean $\overline{x}$ by using Eq. (22.1):

multiply each value by its frequency and add results

$$\overline{x} = \frac{1(1) + 2(2) + 3(1) + 4(3) + 5(1) + 6(1) + 7(2) + 8(1) + 9(2) + 11(1)}{1 + 2 + 1 + 3 + 1 + 1 + 2 + 1 + 2 + 1}$$

sum of frequencies

$$= \frac{82}{15} = 5.5$$

We see that this agrees with the result of Example 3. ■

Summations such as those in Eq. (22.1) occur frequently in statistics and other branches of mathematics. In order to simplify writing these sums, the symbol Σ is used to indicate the process of summation. Σ is the Greek capital letter sigma and was first introduced in Chapter 19. Σx means the sum of the x's.

EXAMPLE 5 Summation symbol Σ

We can show the sum of the numbers $x_1, x_2, x_3, \ldots, x_n$ as

$$\sum x = x_1 + x_2 + x_3 + \cdots + x_n$$

If these numbers are 3, 7, 2, 6, 8, 4, and 9, we have

$$\sum x = 3 + 7 + 2 + 6 + 8 + 4 + 9 = 39$$ ■

Using the summation symbol $\sum$, we can write Eq. (22.1) for the mean as

$$\overline{x} = \frac{x_1 f_1 + x_2 f_2 + x_3 f_3 + \cdots + x_n f_n}{f_1 + f_2 + f_3 + \cdots + f_n} = \frac{\sum xf}{\sum f} \tag{22.1}$$

The summation notation $\sum x$ is an abbreviated form of the more general notation $\sum_{i=1}^{n} x_i$, which represents the sum $x_1 + x_2 + x_3 + \cdots + x_n$.

EXAMPLE 6 Mean using summation—smart devices

A sample of 50 people were asked how many smart devices they owned (including smartphones, tablets, and smartwatches). The grouped responses are shown below:

Number of devices, x	0	1	2	3	4
Frequency, f	6	21	16	5	2

We find the mean using Eq. (22.1):

$$\overline{x} = \frac{\sum xf}{\sum f} = \frac{0(6) + 1(21) + 2(16) + 3(5) + 4(2)}{50}$$

$$= \frac{76}{50} = 1.5 \text{ smart devices} \quad \text{(rounded off to tenths)}$$

■ The mean is one of a number of statistical measures that can be found on a calculator. The use of a calculator will be shown in the next section.

Another measure of central tendency is the **mode,** *which is the value that appears most frequently.* If two or more values appear with the same greatest frequency, each is a mode. If no value is repeated, there is no mode.

EXAMPLE 7 Mode

(a) The mode of the numbers in Example 1 is 4, since it appears three times and no other value appears more than twice.

(b) The modes of the numbers

1, 2, 2, 4, 5, 5, 6, 7

are 2 and 5, since each appears twice and no other number is repeated.

(c) There is no mode for the values

1, 2, 5, 6, 7, 9

since none of the values is repeated. ■

Practice Exercises

For the following numbers, find the indicated value.

16, 17, 16, 12, 14, 15, 14, 16, 15, 18

1. The median **2.** The mean

3. The mode

EXAMPLE 8 Measures of center—force of friction

To find the frictional force between two specially designed surfaces, the force to move a block with one surface along an inclined plane with the other surface is measured ten times. The results, with forces in newtons, are

2.2, 2.4, 2.1, 2.2, 2.5, 2.2, 2.4, 2.7, 2.1, 2.5

Find the mean, median, and mode of these forces.

To find the mean, we sum the values of the forces and divide this total by 10. This gives

$$\overline{F} = \frac{\sum F}{10} = \frac{2.2 + 2.4 + 2.1 + 2.2 + 2.5 + 2.2 + 2.4 + 2.7 + 2.1 + 2.5}{10}$$

$$= \frac{23.3}{10} = 2.33 \text{ N}$$

The median is found by arranging the values in order and finding the middle value. The values in order are

2.1, 2.1, 2.2, 2.2, 2.2, 2.4, 2.4, 2.5, 2.5, 2.7

Because there are ten values, we see that the fifth value is 2.2 and the sixth is 2.4. The value midway between these is 2.3, which is the median. Therefore, the median force is 2.3 N.

The mode is 2.2 N, because this value appears three times, which is more than any other value. ■

The mean is useful with many statistical methods and is used extensively. The median is also commonly used, particularly if there are many values and with some extreme values. The mode is occasionally used, especially for data that are approximately symmetric.

EXERCISES 22.2

In Exercises 1–4, delete the 5 from the data numbers given for Example 1 and then do the following with the resulting data.

1. Find the median.
2. Find the mean using the definition, as in Example 3.
3. Find the mean using Eq. (22.1), as in Example 4.
4. Find the mode, as in Example 7.

In Exercises 5–16, use the following sets of numbers.

A: 3, 6, 4, 2, 5, 4, 7, 6, 3, 4, 6, 4, 5, 7, 3
B: 25, 26, 23, 24, 25, 28, 26, 27, 23, 28, 25
C: 0.48, 0.53, 0.49, 0.45, 0.55, 0.49, 0.47, 0.55, 0.48, 0.57, 0.51, 0.46, 0.53, 0.50, 0.49, 0.53
D: 105, 108, 103, 108, 106, 104, 109, 104, 110, 108, 108, 104, 113, 106, 107, 106, 107, 109, 105, 111, 109, 108

In Exercises 5–8, determine the median of the numbers of the given set.

5. Set A
6. Set B
7. Set C
8. Set D

In Exercises 9–12, determine the mean of the numbers of the given set.

9. Set A
10. Set B
11. Set C
12. Set D

In Exercises 13–16, determine the mode of the numbers of the given set.

13. Set A
14. Set B
15. Set C
16. Set D

In Exercises 17–34, the required data are those in Exercises 22.1. Find the indicated measures of central tendency.

17. Median of the miles traveled in Exercise 15
18. Mean of the miles traveled in Exercise 15
19. Mode of the miles traveled in Exercise 15
20. Median of the number of apps in Exercise 19
21. Mean of the number of apps in Exercise 19
22. Mode of the number of apps in Exercise 19
23. Mean of X-ray dosages in Exercise 25
24. Median of X-ray dosages in Exercise 25
25. Mode of X-ray dosages in Exercise 25
26. Mean of the bloodstream alcohol percentages in Exercise 27
27. Median of the bloodstream alcohol percentages in Exercise 27
28. Mode of the bloodstream alcohol percentages in Exercise 27

In Exercises 29–42, find the indicated measure of central tendency.

29. The weekly salaries (in dollars) for the workers in a small factory are as follows:

    600, 750, 625, 575, 525, 700, 550,
    750, 625, 800, 700, 575, 600, 700

    Find the median and the mode of the salaries.
30. Find the mean salary for the salaries in Exercise 29.
31. In a particular month, the electrical usages, rounded to the nearest 100 kW · h (kilowatt-hours), of 1000 homes in a certain city were summarized as follows:

Usage	500	600	700	800	900	1000	1100	1200
No. Homes	22	80	106	185	380	122	90	15

    Find the mean of the electrical usage.
32. Find the median and mode of electrical usage in Exercise 31.
33. The diameters of a sample of fiber-optic cables were measured (to the nearest 0.0001 mm) with the following results.

Diameter (mm)	0.0057	0.0058	0.0059	0.0060	0.0061
No. Cables	18	36	50	65	31

    Find the mean of the diameters.
34. Find the median and mode of the diameters in Exercise 33.
35. A test of air pollution in a city gave the following readings of the concentration of sulfur dioxide (in parts per million) for 18 consecutive days:

    0.14, 0.18, 0.27, 0.19, 0.15, 0.22, 0.20, 0.18, 0.15,
    0.17, 0.24, 0.23, 0.22, 0.18, 0.32, 0.26, 0.17, 0.23

    Find the median and the mode of these readings.
36. Find the mean of the readings in Exercise 35.
37. The *midrange*, another measure of central tendency, is found by finding the sum of the lowest and the highest values and dividing this sum by 2. Find the midrange of the salaries in Exercise 29.
38. Find the midrange of the sulfur dioxide readings in Exercise 35. (See Exercise 37.)
39. Add $100 to each of the salaries in Exercise 29. Then find the median, mean, and mode of the resulting salaries. State any conclusion that might be drawn from the results.

40. Multiply each of the salaries in Exercise 29 by 2. Then find the median, mean, and mode of the resulting salaries. State any conclusion that might be drawn from the results.

41. Change the final salary in Exercise 29 to \$4000, with all other salaries being the same. Then find the mean of these salaries. State any conclusion that might be drawn from the result. (The \$4000 here is called an *outlier*, which is an extreme value.)

42. Find the median and mode of the salaries indicated in Exercise 41. State any conclusion that might be drawn from the results.

Answers to Practice Exercises

1. 15.5 **2.** 15.3 **3.** 16

22.3 Standard Deviation

Population • Sample • Standard Deviation of Sample • Calculator 1-Var Stats • Interpreting Standard Deviation

In using statistics, we generally collect a sample of data and draw certain conclusions about the complete collection of possible values. In statistics, *the complete collection of values (measurements, people, scores, etc.) is called the* **population,** *and a* **sample** *is a subset of the population.* In the applied examples and exercises of the previous two sections, we were using samples taken from larger populations.

We have discussed various ways of measuring the center of sample data. However, regardless of the measure that may be used, it does not tell us whether the values of the population tend to be grouped closely together or spread out over a large range of values. Therefore, we also need some measure of the deviation, or spread, of the values from the center. When taken together, measures of center and spread give a good summary of a set of data values.

In statistics, there are several measures of spread that may be defined. In this section, we discuss one that is very widely used: the *standard deviation.* The **standard deviation** s *of a set of* ***sample*** *values is defined by the equation*

$$s = \sqrt{\frac{\sum (x - \bar{x})^2}{n - 1}} \tag{22.2}$$

The definition of s shows that the following steps are used in computing its value.

Steps for Calculating Standard Deviation

1. Find the mean $\bar{x}$ of the numbers of the set.
2. Subtract the mean from each number of the set.
3. Square these differences.
4. Find the sum of these squares.
5. Divide this sum by $n - 1$.
6. Find the square root of this result.

The standard deviation s is a positive number. It is a *typical deviation from the mean,* regardless of whether the individual numbers are greater than or less than the mean. Numbers close together will have a small standard deviation, whereas numbers farther apart have a larger standard deviation. NOTE ▸ [Therefore, the standard deviation becomes larger as the spread of data increases.]

Following the steps shown above, we use Eq. (22.2) for the calculation of the standard deviation in the following example.

EXAMPLE 1 Standard deviation—using Eq. (22.2)

Find the standard deviation of the following numbers: 1, 5, 4, 2, 6, 2, 1, 1, 5, 3.

A table of the necessary values is shown below, and steps 1–6 are indicated:

x	$x - \bar{x}$ (step 2)	$(x - \bar{x})^2$ (step 3)
1	−2	4
5	2	4
4	1	1
2	−1	1
6	3	9
2	−1	1
1	−2	4
1	−2	4
5	2	4
3	0	0
30		32 (step 4)

$$\bar{x} = \frac{30}{10} = 3 \qquad \text{step 1}$$

$$\frac{\sum (x - \bar{x})^2}{n - 1} = \frac{32}{10 - 1} = \frac{32}{9} \qquad \text{step 5}$$

$$s = \sqrt{\frac{32}{9}} = 1.9 \qquad \text{step 6}$$

NOTE ➧ [In calculating the standard deviation, it is usually rounded off to one more decimal place than was present in the original data.] ■

It is possible to reduce the computational work required to find the standard deviation. Algebraically, it can be shown that the following equation is another form of Eq. (22.2) and therefore gives the same results.

$$s = \sqrt{\frac{n\left(\sum x^2\right) - \left(\sum x\right)^2}{n(n - 1)}} \tag{22.3}$$

Although the form of this equation appears more involved, it does reduce the amount of calculation that is necessary. Consider the following example.

EXAMPLE 2 Standard deviation—using Eq. (22.3)

Using Eq. (22.3), find s for the numbers in Example 1.

x	x^2
1	1
5	25
4	16
2	4
6	36
2	4
1	1
1	1
5	25
3	9
30	122

$$n = 10$$

$$\sum x^2 = 122$$

$$\left(\sum x\right)^2 = 30^2 = 900$$

$$s = \sqrt{\frac{10(122) - 900}{10(9)}} = 1.9$$

■

Practice Exercise

1. Find the standard deviation of the first eight numbers in Example 1.

Often in applied situations, a calculator or computer is used to find the standard deviation since the calculations can be quite extensive for large data sets. The following example illustrates the use of a calculator to find the basic descriptive statistics, including the standard deviation (shown as s_x), of a set of data.

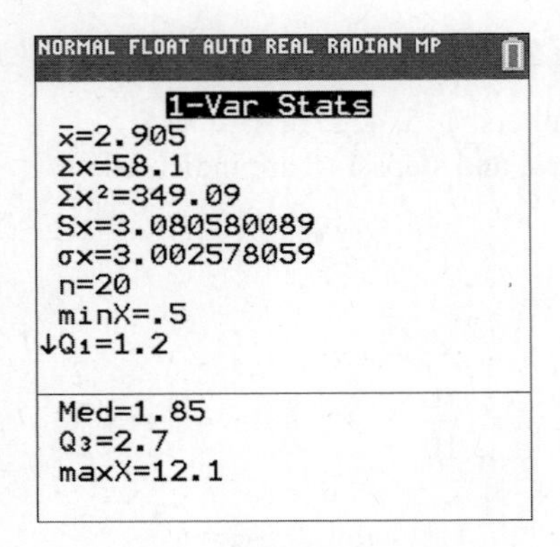

Fig. 22.8
Graphing calculator screenshots: goo.gl/Nc3XrZ

EXAMPLE 3 Calculator 1-Var Stats—cellular data usage

In a sample of 20 adults with smartphones, the amount of cellular data (in GB) they used in the previous month is shown below. Use a calculator to find the mean, median, and standard deviation of the data.

1.6	2.4	1.9	8.8	0.9	3.2	12.1	2.3	1.1	0.6
1.8	2.0	0.5	1.7	1.8	3.0	0.8	1.3	2.2	8.1

We first enter the data into the list L_1 on a calculator and then use the *1-Var Stats* feature (under *STAT, CALC*). The calculator display is shown in Fig. 22.8. The mean and standard deviation are 2.91 GB and 3.08 GB, respectively. By scrolling down, the median is shown to be 1.85 GB. Various other statistical values are also included. ■

The calculator can also be used to find the descriptive statistics for grouped data. The following example shows how this is done.

EXAMPLE 4 Calculator 1-Var Stats for grouped data—smart devices

The grouped data on smart device ownership from Example 6 in Section 22.2 is shown below. Use a calculator to find the mean, median, and standard deviation.

Number of devices, x	0	1	2	3	4
Frequency, f	6	21	16	5	2

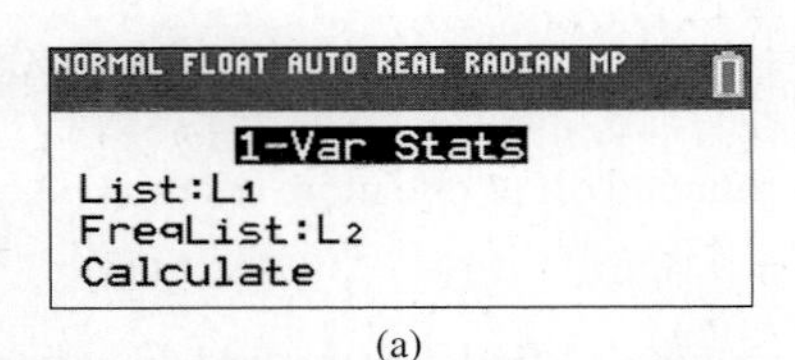

(a)

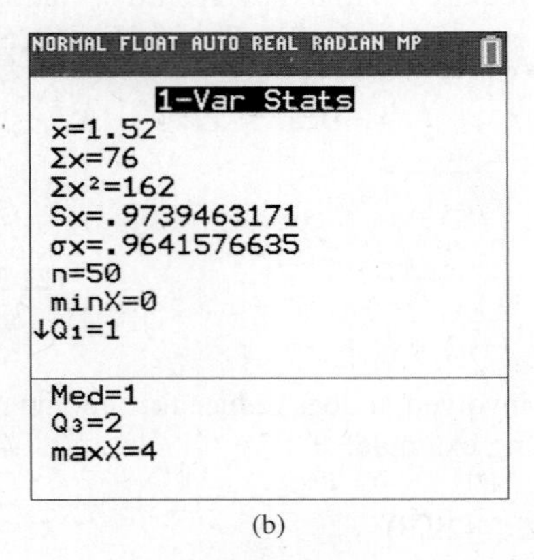

(b)

Fig. 22.9
Graphing calculator screenshots: goo.gl/QVV7SF

First, the values of x are entered into the list L_1 and the frequencies are entered into L_2. Then we use the *1-Var Stats* feature choosing L_1 as the list and L_2 as the frequency list as shown in Fig. 22.9(a). The calculator then computes the various statistical values of the grouped data. From Fig. 22.9(b), we see that the mean is 1.52 devices, the median is 1 device, and the standard deviation is 0.97 device. ■

We have shown how to calculate the standard deviation, but it is equally important to be able to interpret its value. ***The standard deviation is a typical distance between the individual data values and the mean.*** Some values in the data set will be farther away from the mean than the standard deviation, and some will be closer. Also, ***most of the data values in a data set will be within two standard deviations of the mean.*** It is somewhat rare for values to be more than two standard deviations away from the mean. The following example illustrates this.

EXAMPLE 5 Interpreting the standard deviation—assembly time

Suppose the mean amount of time it takes a worker to assemble a part is 8 min with a standard deviation of 1 min. Then two standard deviations below the mean is 6 min, and two standard deviations above the mean is 10 min. Therefore, the worker can almost always assemble the part in a time between 6 and 10 min. Very rarely will it take the worker less than 6 min or more than 10 min. ■

EXERCISES 22.3

In Exercises 1 and 2, in Example 1, change the first 1 to 6 and the first 2 to 7 and then find the standard deviation of the resulting data as directed.

1. Find s from the definition given by Eq. (22.2), as in Example 1.

2. Find s using Eq. (22.3), as in Example 2.

In Exercises 3–14, use the following sets of numbers. They are the same as those used in Exercises 22.2.

A: 3, 6, 4, 2, 5, 4, 7, 6, 3, 4, 6, 4, 5, 7, 3
B: 25, 26, 23, 24, 25, 28, 26, 27, 23, 28, 25
C: 0.48, 0.53, 0.49, 0.45, 0.55, 0.49, 0.47, 0.55, 0.48, 0.57, 0.51, 0.46, 0.53, 0.50, 0.49, 0.53
D: 105, 108, 103, 108, 106, 104, 109, 104, 110, 108, 108, 104, 113, 106, 107, 106, 107, 109, 105, 111, 109, 108

In Exercises 3–6, use Eq. (22.2) to find the standard deviation s for the indicated sets of numbers.

3. Set A **4.** Set B **5.** Set C **6.** Set D

In Exercises 7–10, use Eq. (22.3) to find the standard deviation s for the indicated sets of numbers.

7. Set A **8.** Set B **9.** Set C **10.** Set D

In Exercises 11–14, use the statistical feature of a calculator to find the mean and the standard deviation s for the indicated sets of numbers.

11. Set A **12.** Set B **13.** Set C **14.** Set D

In Exercises 15–22, find the standard deviation for the indicated sets of numbers. A calculator may be used.

15. The miles traveled in Exercise 15 of Section 22.1

16. The following data giving the mean number of days of rain for Vancouver, B.C., for the 12 months of the year

20, 17, 17, 14, 12, 11, 7, 8, 9, 16, 19, 22

17. The number of apps in Exercise 19 of Section 22.1

18. The X-ray dosages in Exercise 25 of Section 22.1

19. The alcohol percentages in Exercise 27 of Section 22.1

20. The salaries in Exercise 29 of Section 22.2

21. The electrical usages in Exercise 31 of Section 22.2

22. The diameters in Exercise 33 of Section 22.2

Answer to Practice Exercise

1. 2.0

22.4 Normal Distributions

Normal Distributions • Empirical Rule • *z*-Score • Standard Normal Distribution • Applications of Normal Distributions • Sampling Distribution of $\bar{x}$ • Standard Error

The distributions in the previous sections have been for a limited number of values. Let us now consider a very large population, such as the useable lifetime of all the AA batteries sold in the world in a year. It could have a large number of classes with many values within each class.

We would expect a histogram for this very large population to have its maximum frequency very near the mean and taper off to smaller frequencies on either side. It would probably be shaped something like the curve shown in Fig. 22.10.

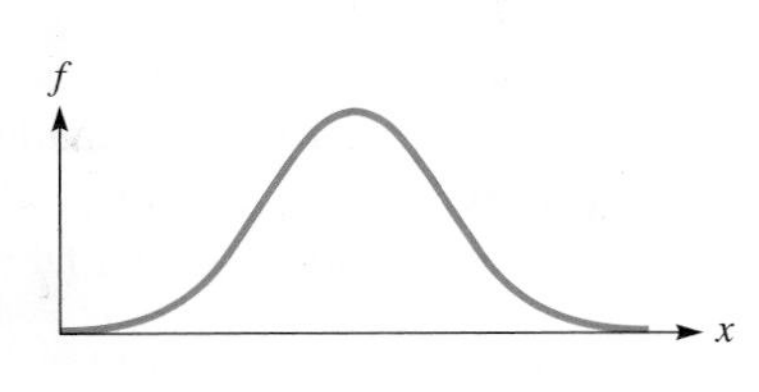

Fig. 22.10

The smooth bell-shaped curve in Fig. 22.10 shows the **normal distribution** of a population large enough that the distribution is considered to be *continuous*. Using advanced methods, its equation is found to be

$$y = \frac{e^{-(x-\mu)^2/2\sigma^2}}{\sigma\sqrt{2\pi}} \qquad (22.4)$$

■ The *population standard deviation* is given by

$$\sigma = \sqrt{\frac{\sum (x-\mu)^2}{N}}, \text{ where}$$

N is the number of elements in the population.

Here, **μ is the population mean** and **σ is the population standard deviation,** and π and e are the familiar numbers first used in Chapters 2 and 12, respectively.

From Eq. (22.4), we can see that any particular normal distribution depends on the values of μ and σ. The horizontal location of the curve depends on μ, and the spread of the curve depends on σ, but the bell shape remains. The next example illustrates the shapes of different normal distributions.

■ Note that normal distribution curves are symmetric with respect to the population mean μ.

EXAMPLE 1 Different normal distributions

In Fig. 22.11, for the left curve, $\mu = 10$ and $\sigma = 5$, whereas for the right curve, $\mu = 20$ and $\sigma = 10$.

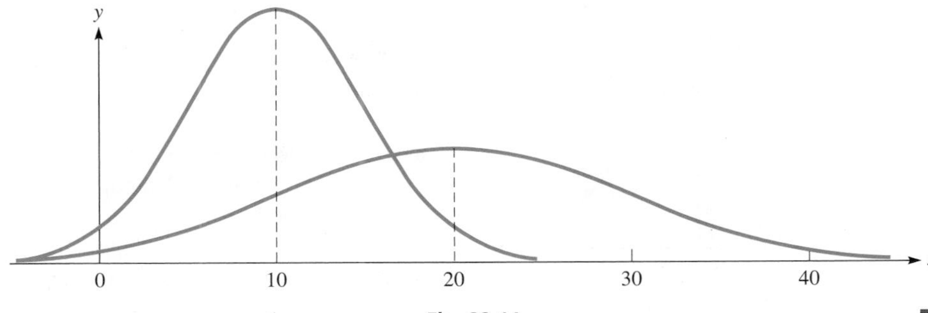

Fig. 22.11

■

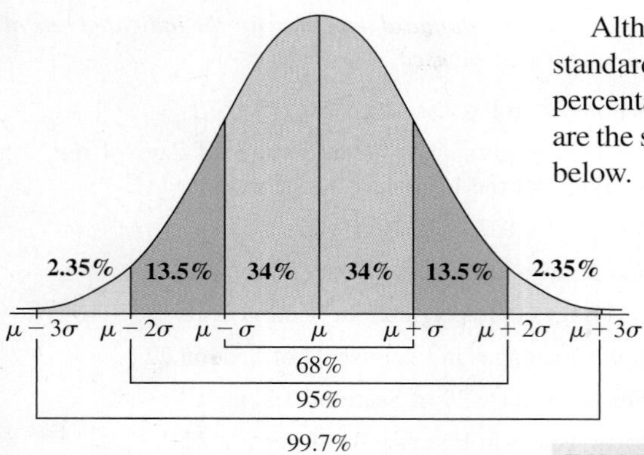

Fig. 22.12

■ The percentages given in the empirical rule can also be viewed as being areas under a normal curve. For example, the area under a normal curve within one standard deviation of the mean is 0.68 (or 68%).

Although there are many possible normal distributions (depending on the mean and standard deviation), **the total area under any normal curve is 1 (or 100%)**. Also, the percentages of the values that are within one, two, or three standard deviations of the mean are the same for all normal distributions. This fact is called the **empirical rule** and is stated below.

Empirical Rule

In any normal distribution, the percentages of the values that lie within one, two, and three standard deviations of the mean are about 68%, 95%, and 99.7%, respectively. See Fig. 22.12.

EXAMPLE 2 Empirical rule—achievement test scores

The scores on a certain achievement test are normally distributed with a mean of 500 and a standard deviation of 50. What percentage of the scores are (a) between 400 and 600, and (b) higher than 550?

(a) Since 400 is two standard deviations below the mean and 600 is two standard deviations above the mean, about 95% of all scores are between these values.

(b) The value 550 is one standard deviation above the mean. This means that about 34% (half of 68%) of the scores are between 500 and 550. Therefore, the percentage of scores that are higher than 550 is about 16% (50% − 34%). [Remember that the total area under any normal curve is 1 (or 100%). Because a normal curve is symmetric, the area on either side of the mean is 0.5 (or 50%).] ■

NOTE ➧

In order to find percentages of a normal distribution that lie between fractional increments of the standard deviation, we convert the given values to *z-scores* (or standard scores) and then use the *standard normal distribution*. [The **z-score** of a particular value *x* tells us the *number of standard deviations* the given value of *x* is above or below the mean.] It is calculated using the following formula:

NOTE ➧

$$\text{z-score} \quad z = \frac{x - \mu}{\sigma} \tag{22.5}$$

A positive z-score indicates the value is above the mean, whereas a negative z-score indicates it is below the mean.

EXAMPLE 3 z-scores—achievement test results

Using the mean and standard deviation given in Example 2, find and interpret the z-scores of achievement test results of (a) 610 and (b) 435.

(a) The test result of 610 has a z-score of $z = \frac{610 - 500}{50} = 2.2$. This test result is 2.2 standard deviations above the mean. This is a very high test score, which is rarely obtained.

(b) The test result of 435 has a z-score of $z = \frac{435 - 500}{50} = -1.3$. This test result is 1.3 standard deviations below the mean. This score is below the mean, but fairly typical. ■

NOTE ➧

By converting given values to z-scores, we can use the **standard normal distribution** to find areas (or percentages). [The standard normal distribution has a mean of 0 and a standard deviation of 1.] Like any normal curve, the total area under the curve is 1. However, the horizontal axis is labeled with z-scores, which can be interpreted as the number of standard deviations above or below the mean. Figure 22.13 shows the standard normal curve. Note that almost all of the area is between $z = -3$ and $z = 3$ (99.7% according to the empirical rule). Standard normal tables, such as the one given in Table 22.1 on the next page, can be used to find areas under the standard normal curve.

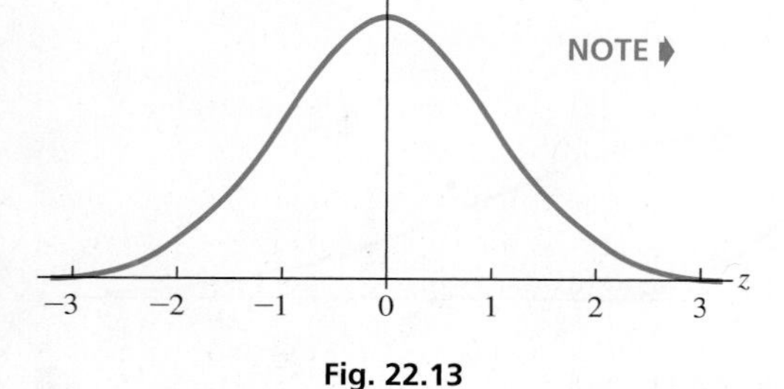

Fig. 22.13

NOTE ➧

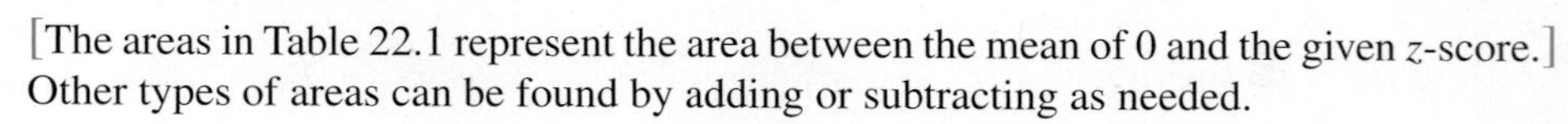
[The areas in Table 22.1 represent the area between the mean of 0 and the given z-score.] Other types of areas can be found by adding or subtracting as needed.

Table 22.1 Standard Normal (z) Distribution

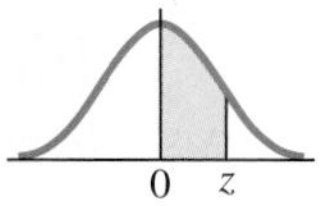

z	*Area*	z	*Area*	z	*Area*	z	*Area*	z	*Area*	z	*Area*
0.1	0.0398	0.6	0.2257	1.1	0.3643	1.6	0.4452	2.1	0.4821	2.6	0.4953
0.2	0.0793	0.7	0.2580	1.2	0.3849	1.7	0.4554	2.2	0.4861	2.7	0.4965
0.3	0.1179	0.8	0.2881	1.3	0.4032	1.8	0.4641	2.3	0.4893	2.8	0.4974
0.4	0.1554	0.9	0.3159	1.4	0.4192	1.9	0.4713	2.4	0.4918	2.9	0.4981
0.5	0.1915	1.0	0.3413	1.5	0.4332	2.0	0.4772	2.5	0.4938	3.0	0.4987

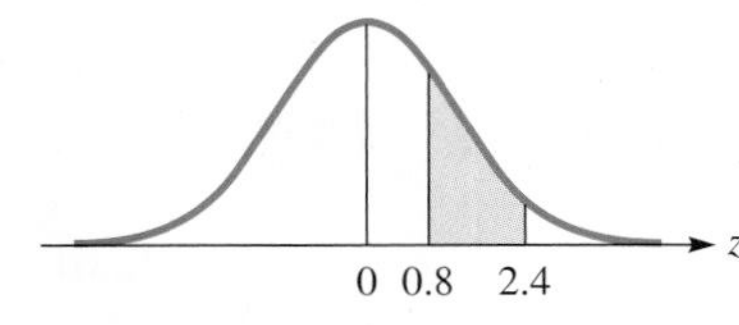

Fig. 22.14

Practice Exercise

1. Find the area under the standard normal curve between $z = -0.5$ and $z = 1$.

EXAMPLE 4 Finding area under the standard normal curve

Find the area under the standard normal curve between $z = 0.8$ and $z = 2.4$ as shown in Fig. 22.14.

Using Table 22.1, the area between $z = 0$ and $z = 2.4$ is 0.4918. The area between $z = 0$ and $z = 0.8$ is 0.2881. Therefore, the area between $z = 0.8$ and $z = 2.4$ is found by subtracting:

$$0.4918 - 0.2881 = 0.2037 \quad (20.37\% \text{ of the area})$$

In applications involving normal distributions, areas are found by first converting the values to z-scores and then using the standard normal distribution. This is illustrated in the next example.

EXAMPLE 5 Application of normal distribution—battery lifetimes

The lifetimes of a certain type of watch battery are normally distributed with a mean of 400 days and a standard deviation of 50 days. What percentage of batteries will last (a) between 360 days and 460 days, (b) more than 320 days, and (c) less than 280 days?

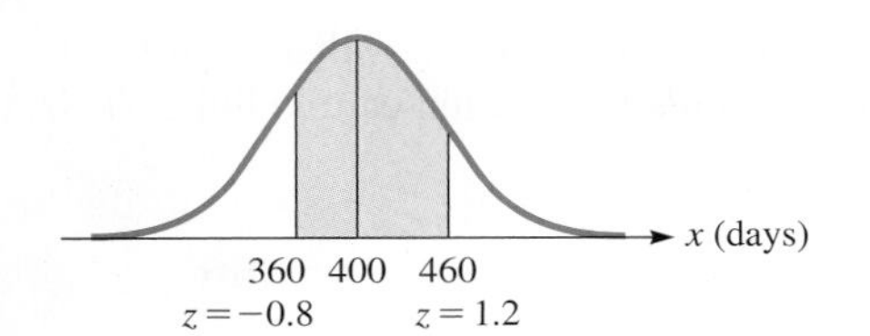

Fig. 22.15

(a) We first find the z-score of each value: $z = \dfrac{360 - 400}{50} = -0.8$ and $z = \dfrac{460 - 400}{50} = 1.2$ We wish to find the area between these two z-scores as shown in Fig. 22.15. Using Table 22.1, the area between $z = 0$ and $z = -0.8$ is 0.2881 and the area between $z = 0$ and $z = 1.2$ is 0.3849. Thus, the total area is found by adding: $0.2881 + 0.3849 = 0.6730$. This means that 67.3% of the batteries will last between 360 and 460 days.

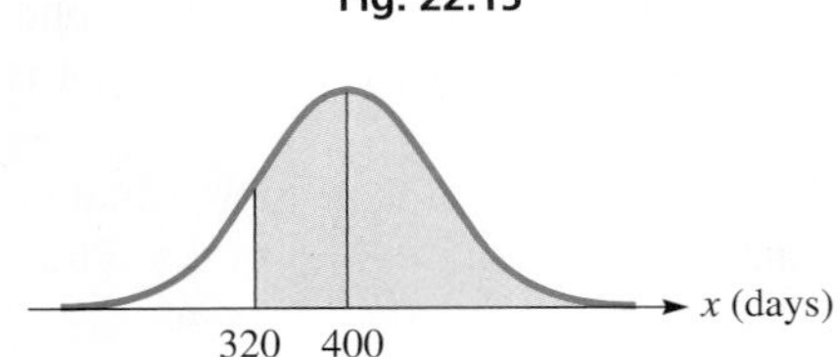

Fig. 22.16

(b) We begin by finding the z-score of 320 days: $z = \dfrac{320 - 400}{50} = -1.6$

We wish to find the area to the right of $z = -1.6$, shown in Fig. 22.16. The area between $z = 0$ and $z = -1.6$ is 0.4452. Since the area to the right of $z = 0$ is 0.5, we add to get the total area: $0.4452 + 0.5 = 0.9452$. This means that 94.52% of the batteries will last longer than 320 days.

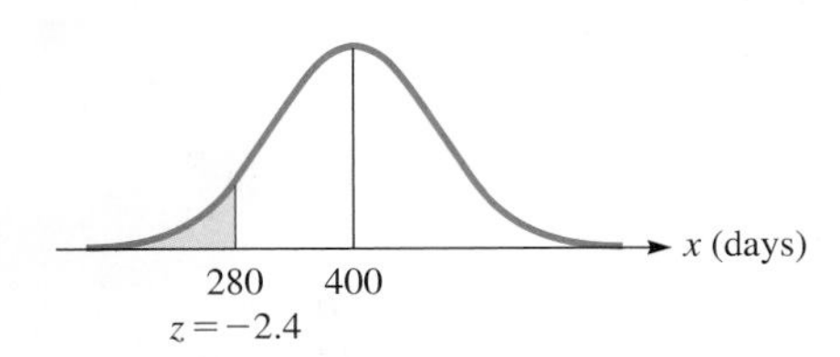

Fig. 22.17

(c) The z-score of 280 is $z = \dfrac{280 - 400}{50} = -2.4$. The area to the left of this (see Fig. 22.17) is $0.5 - 0.4918 = 0.0082$. Therefore, only 0.82% of the batteries will last fewer than 280 days.

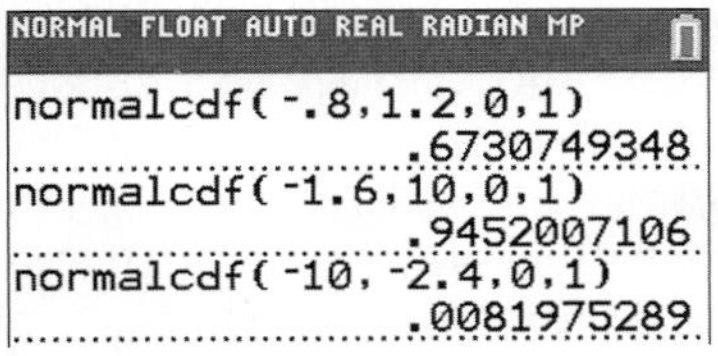

Instead of using Table 22.1, a calculator can also be used to find areas under the standard normal curve. The *normalcdf* feature calculates the area between a specified lower and upper bound. When using z-scores in the *normalcdf* feature, a 10 can be used as an upper bound for areas that continue indefinitely to the right. This is because there is virtually no area more than 10 standard deviations away from the mean. Similarly, a -10 can be used as a lower bound for areas that continue to the left. Figure 22.18 shows the calculator's evaluations of the areas in this example.

```
NORMAL FLOAT AUTO REAL RADIAN MP
normalcdf(-.8,1.2,0,1)
                  .6730749348
normalcdf(-1.6,10,0,1)
                  .9452007106
normalcdf(-10,-2.4,0,1)
                  .0081975289
```

Fig. 22.18

Graphing calculator screenshot: goo.gl/cnHTRz

THE SAMPLING DISTRIBUTION OF $\bar{x}$

Suppose we selected *all possible samples* of size n from a population with mean μ and standard deviation σ, and we calculated the sample mean $\bar{x}$ for *each* sample. We would have a *huge* number of sample means. The way in which these sample means are distributed is called the **sampling distribution of $\bar{x}$.** The following summarizes three very important facts about this sampling distribution.

Sampling Distribution of $\bar{x}$

1. The mean of the sampling distribution is μ. In other words, the mean of all possible sample means equals the population mean.
2. The standard deviation of the sampling distribution is

$$\sigma_{\bar{x}} = \frac{\sigma}{\sqrt{n}} \qquad (22.6)$$

   This is called the **standard error of the mean.** Note that the sample mean is less variable than individual members of the population, and as the sample size increases, the standard error decreases.
3. If the sample size is reasonably large (≥ 30), the sampling distribution is approximately normally distributed. This important fact is called the **central limit theorem.**

By knowing the mean, standard deviation, and shape of the sampling distribution, it is possible to find the likelihood that the *mean of a sample* falls within certain limits. This is illustrated in the following example.

EXAMPLE 6 Sampling distribution of $\bar{x}$—airline passenger weights

■ *The central limit theorem* can be used to determine the amount of weight a plane should be designed to carry.

The FAA standards assume that adult airline passengers and their carry-on bags have an average weight of 190 lb in the summer. If the standard deviation of these weights is 25 lb, find the likelihood that for a sample of 100 passengers, their *mean* weight is greater than 195 lb.

Since the sample size is $n = 100$, the sampling distribution of $\bar{x}$ is approximately normally distributed with a mean of 190 and a standard deviation of $\frac{25}{\sqrt{100}} = 2.5$. Thus, the z-score of 195 is

$$z = \frac{195 - 190}{2.5} = 2.0$$

We wish to find the area to the right of $z = 2.0$, which is $0.5 - 0.4772 = 0.0228$. In other words, 2.28% of samples of 100 passengers will have means over 195 lb. ■

Practice Exercise

2. In Example 6, find the likelihood that the mean weight of the 100 passengers is less than 197 lb.

EXERCISES 22.4

In Exercises 1–4, make the given changes in the indicated examples of this section and then solve the indicated problem.

1. In Example 1, change the second σ from 10 to 5 and then describe the curve that would result in terms of either or both curves shown in Fig. 22.11.
2. In Example 2, what percentage of the test scores are between 350 and 650?
3. In Example 3, find and interpret the z-score of an achievement test score of 630.
4. In Example 5(b), change 320 to 360 and then find the resulting percentage of batteries.

In Exercises 5–8, use the following information. If the weights of cement bags are normally distributed with a mean of 60 lb and a standard deviation of 1 lb, use the empirical rule to find the percent of the bags that weigh the following:

5. Between 58 lb and 62 lb
6. Between 59 lb and 61 lb
7. Less than 63 lb
8. More than 62 lb

In Exercises 9–12, use the following information. A standardized math test has a mean score of 200 and a standard deviation of 15. Find and interpret the z-scores of the following math test scores.

9. 218
10. 179
11. 164
12. 233

In Exercises 13–16, use the following data. It has been previously established that for a certain type of AA battery (when newly produced), the voltages are distributed normally with $\mu = 1.50$ V and $\sigma = 0.05$ V.

13. According to the empirical rule, what percent of the batteries have voltages between 1.45 V and 1.55 V?

14. What percent of the batteries have voltages between 1.52 V and 1.58 V?

15. What percent of the batteries have voltages below 1.54 V?

16. What percent of the batteries have voltages above 1.64 V?

In Exercises 17–24, use the following data. The lifetimes of a certain type of automobile tire have been found to be distributed normally with a mean lifetime of 100,000 km and a standard deviation of 10,000 km. Answer the following questions.

17. What percent of the tires will last between 85,000 km and 100,000 km?

18. What percent of the tires will last between 95,000 km and 115,000 km?

19. In a sample of 5000 of these tires, how many can be expected to last more than 118,000 km?

20. If the manufacturer guarantees to replace all tires that do not last 75,000 km, what percent of the tires may have to be replaced under this guarantee?

21. In a sample of 100 of these tires, find the likelihood that the mean lifetime for the sample is less than 98,200.

22. What happens to the standard error of the mean as n increases? Use the formula for the standard error to help explain your answer.

23. What percent of the samples of 100 of these tires should have a mean lifetime of more than 102,000 km?

24. If 144 of the tires are randomly selected, find the percent chance that the mean lifetime is more than 102,000.

In Exercises 25–30, solve the given problems,

25. With 75.8% of the area under the normal curve to the right of z, find the z-value.

26. With 21% of the area under the normal curve between z_1 and z_2, to the right of $z_1 = 0.8$, find z_2.

27. With 59% of the area under the normal curve between z_1 and z_2, to the left of $z_2 = 1.1$, find z_1.

28. With 5.8% of the area under the normal curve between z_1 and z_2, to the left of $z_2 = 2.0$, find z_1.

29. The residents of a city suburb live at a mean distance of 16.0 km from the center of the city, with a standard deviation of 4.0 km. What percent of the residents live between 12.0 km and 18.0 km of the center of the city?

30. For the data on the number of Android apps in Exercise 19 of Section 22.1, find the percent of the data that lie within one, two, and three standard deviations of the mean. Compare these percentages with the ones in the empirical rule.

Answer to Practice Exercises

1. 0.5328 **2.** 99.7%

22.5 Statistical Process Control

Control Charts • Central Line • Range • Upper and Lower Control Limits • $\bar{x}$ Chart • R Chart • p Chart

■ This is intended only as a brief introduction to this topic. A complete development requires at least a chapter in a statistics book.

One of the most important uses of statistics in industry is Statistical Process Control (SPC), which is used to maintain and improve product quality. Samples are tested during the production at specified intervals to determine whether the production process needs adjustment to meet quality requirements.

A particular industrial process is considered to be *in control* if it is stable and predictable, and sample measurements fall within upper and lower control limits. The process is *out of control* if it has an unpredictable amount of variation and there are sample measurements outside the control limits due to special causes.

EXAMPLE 1 Process—in control—out of control

The manufacturer of 1.5-V batteries states that the voltage of its batteries is no less than 1.45 V or greater than 1.55 V and has designed the manufacturing process to meet these specifications.

■ Minor variations may be expected, for example, from very small fluctuations in voltage, temperature, or material composition. Special causes resulting in an out-of-control process could include line stoppage, material defect, or an incorrect applied pressure.

If all samples of batteries that are tested have voltages that vary within expected control limits, the production process is *in control.*

However, if some samples have batteries with voltages out of the proper range, the process is *out of control*. This would indicate some special cause for the problem, such as an improperly operating machine or an impurity getting into the process. The process would probably be halted until the cause is determined. ■

CONTROL CHARTS

An important device used in SPC is the **control chart.** It is used to show a trend of a production characteristic over time. Samples are measured at specified intervals of time to see if the measurements are within the control limits. The measurements are plotted on a chart to check for trends and abnormalities in the production process.

In making a control chart, we must determine what the mean should be. For a stable process for which previous data are known, it can be based on a production specification or on previous data. For a new or recently modified process, it may be necessary to use present data, although the value may have to be revised for future charts. On a control chart, *this value is used as the population mean,* μ.

It is also necessary to establish the upper and lower control limits. The standard generally used is that 99.7% of the sample measurements should fall within these control limits. This assumes a normal distribution, and we note that this is within three standard deviations of the population mean. We will establish these limits by use of a table or a formula that has been made using statistical measures developed in a more complete coverage of quality control. This does follow the normal practice of using a formula or a more complete table in setting up the control limits.

In Fig. 22.19, we show a sample control chart, and in the examples that follow, we illustrate how control charts are made.

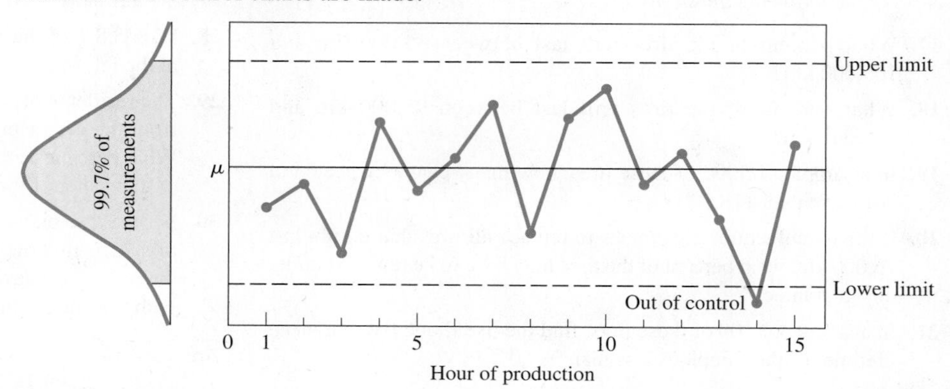

Fig. 22.19

EXAMPLE 2 Making $\bar{x}$ and R control charts

A pharmaceutical company makes a capsule of a prescription drug that contains 500 mg of the drug, according to the label. In a newly modified process of making the capsule, five capsules are tested every 15 min to check the amount of the drug in each capsule. Testing over a 5-h period gave the following results for the 20 subgroups of samples.

Subgroup	*Amount of Drug (in mg) of Five Capsules*					*Mean* $\bar{x}$	*Range R*
1	503	501	498	507	502	502.2	9
2	497	499	500	495	502	498.6	7
3	496	500	507	503	502	501.6	11
4	512	503	488	500	497	500.0	24
5	504	505	500	508	502	503.8	8
6	495	495	501	497	497	497.0	6
7	503	500	507	499	498	501.4	9
8	494	498	497	501	496	497.2	7
9	502	504	505	500	502	502.6	5
10	500	502	500	496	497	499.0	6
11	502	498	510	503	497	502.0	13
12	497	498	496	502	500	498.6	6
13	504	500	495	498	501	499.6	9
14	500	499	498	501	494	498.4	7
15	498	496	502	501	505	500.4	9
16	500	503	504	499	505	502.2	6
17	487	496	499	498	494	494.8	12
18	498	497	497	502	497	498.2	5
19	503	501	500	498	504	501.2	6
20	496	494	503	502	501	499.2	9
					Sums	9998.0	174
					Means	499.9	8.7

Practice Exercise

1. Is either the mean or range affected if subgroup 12 is 495, 498, 497, 503, 500?

As we noted from the table, *the **range R** of each sample is the difference between the highest value and the lowest value of the sample.*

From this table of values, we can make an $\bar{x}$ control chart and an R control chart. The $\bar{x}$ chart maintains a check on the average quality level, whereas the R chart maintains a check on the dispersion of the production process. These two control charts are often plotted together and referred to as the $\bar{x}$–R chart.

In order to define the **central line** of the $\bar{x}$ chart, which ideally is equivalent to the value of the population mean μ, we use the mean of the sample means $\bar{\bar{x}}$. For the central line of the R chart, we use $\bar{R}$. From the table, we see that

$$\bar{\bar{x}} = 499.9 \text{ mg} \quad \text{and} \quad \bar{R} = 8.7 \text{ mg}$$

Table 22.2 Control Chart Factors

n	d_2	A	A_2	D_1	D_2	D_3	D_4
5	2.326	1.342	0.577	0.000	4.918	0.000	2.115
6	2.534	1.225	0.483	0.000	5.078	0.000	2.004
7	2.704	1.134	0.419	0.205	5.203	0.076	1.924

The **upper control limit** (UCL) and the **lower control limit** (LCL) for each chart are defined in terms of the mean range $\bar{R}$ and an appropriate constant taken from a table of control chart factors. These factors, which are related to the sample size n, are determined by statistical considerations found in a more complete coverage of quality control. At the left is a brief table of control chart factors (Table 22.2).

The UCL and LCL for the $\bar{x}$ chart are found as follows:

$$\text{UCL}(\bar{x}) = \bar{\bar{x}} + A_2\bar{R} = 499.9 + 0.577(8.7) = 504.9 \text{ mg}$$

$$\text{LCL}(\bar{x}) = \bar{\bar{x}} - A_2\bar{R} = 499.9 - 0.577(8.7) = 494.9 \text{ mg}$$

The UCL and LCL for the R chart are found as follows:

$$\text{LCL}(R) = D_3\bar{R} = 0.000(8.7) = 0.0 \text{ mg}$$

$$\text{UCL}(R) = D_4\bar{R} = 2.115(8.7) = 18.4 \text{ mg}$$

Figure 22.20 shows an $\bar{x}$–R chart of these data constructed using the statistical software Minitab. The top graph is the $\bar{x}$ chart and the bottom graph is the R chart. Note that the central lines and control limits agree with the ones we calculated.

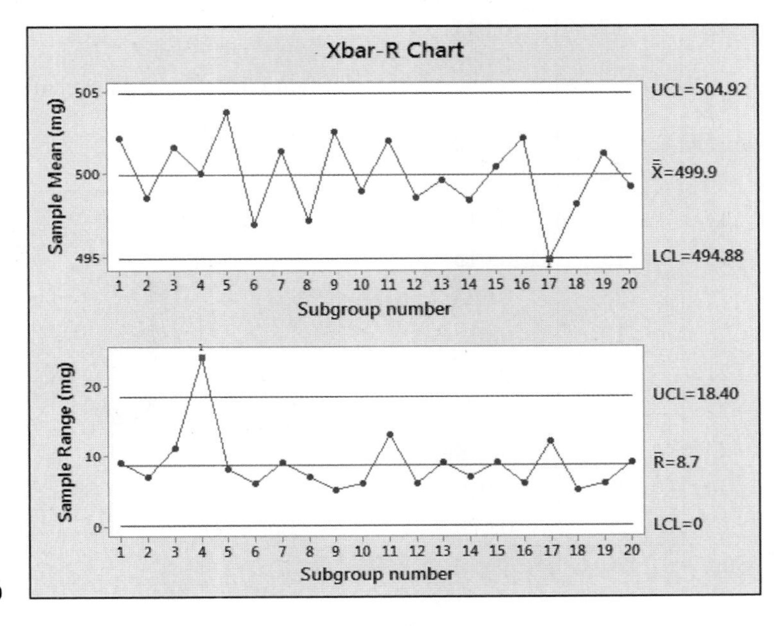

Fig. 22.20

This would be considered a *well-centered process* since $\bar{\bar{x}} = 499.9$ mg, which is very near the target value of 500.0 mg. We do note, however, that subgroup 17 was slightly outside the lower control limit and this might have been due to some special cause, such as the use of a substandard mixture of ingredients. We also note that the process was *out of control* due to some special cause of subgroup 4 since the range was above the upper control limit. ■

We should keep in mind that there are numerous considerations, including various human factors, that need to be taken into account when making and interpreting control charts. The coverage here is only a very brief introduction to this very important industrial use of statistics.

In Example 2, the weight (in mg) of a prescription drug was tested. Weight is a quantitative variable since its values are numeric.

Sometimes we wish to make a control chart for an **attribute,** which is a quality, or characteristic, that a sample member either has or doesn't have. Examples of attributes are color (acceptable or not), entries on a customer account (correct or incorrect), and defects (defective or not defective) in a product. To monitor an attribute in a production process, we get the *proportion* of defective parts by *dividing the number of defective parts in a sample by the total number of parts in the sample,* and then make a *p control chart.* This is illustrated in the following example.

■ See the chapter introduction.

EXAMPLE 3 Making a p control chart

Day	*Defective DVDs*	*Proportion Defective*
1	22	0.022
2	16	0.016
3	14	0.014
4	18	0.018
5	12	0.012
6	25	0.025
7	36	0.036
8	16	0.016
9	14	0.014
10	22	0.022
11	20	0.020
12	17	0.017
13	26	0.026
14	20	0.020
15	22	0.022
16	28	0.028
17	17	0.017
18	15	0.015
19	25	0.025
20	12	0.012
21	16	0.016
22	22	0.022
23	19	0.019
24	16	0.016
25	20	0.020
Sum	490	

The manufacturer of video discs has 1000 DVDs checked each day for defects (surface scratches, for example). The data for this procedure for 25 days are shown in the table at the left.

For the p control chart, the central line is the value of $\overline{p}$, which in this case is

$$\overline{p} = \frac{490}{25{,}000} = 0.0196$$

The control limits are each three standard deviations from $\overline{p}$. If n is the size of the subgroup, the standard deviation σ_p of a proportion is given by

$$\sigma_p = \sqrt{\frac{\overline{p}(1 - \overline{p})}{n}} = \sqrt{\frac{0.0196(1 - 0.0196)}{1000}} = 0.00438$$

Therefore, the control limits are

$$\text{UCL}(p) = 0.0196 + 3(0.00438) = 0.0327$$
$$\text{LCL}(p) = 0.0196 - 3(0.00438) = 0.0065$$

Figure 22.21 shows a p chart constructed using Minitab.

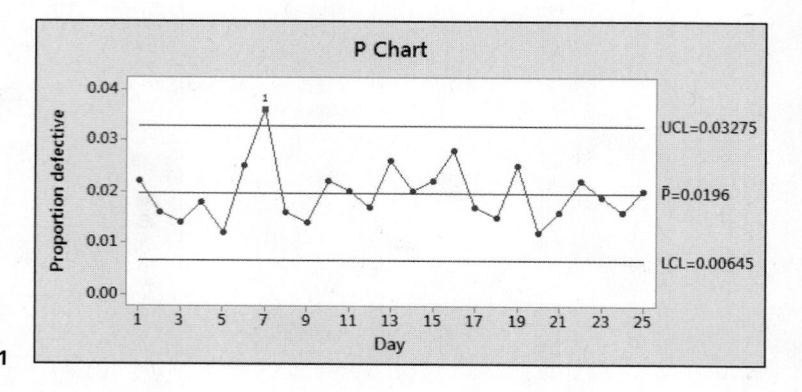

Fig. 22.21

According to the proportion mean of 0.0196, the process produces about 2% defective DVDs. We note that the process was out of control on day 7. An adjustment to the production process was probably made to remove the special cause of the additional defective DVDs. ■

Practice Exercise

2. In Example 3, change Day 9 datum from 14 to 24 defective parts. Then find UCL(p) and LCL(p).

EXERCISES 22.5

In Exercises 1–4, in Example 2, change the first subgroup to 497, 499, 502, 493, and 498 and then proceed as directed.

1. Find UCL $(\bar{x})$ and LCL $(\bar{x})$.

2. Find LCL (R) and UCL (R).

3. How would the $\bar{x}$ control chart differ from the top graph in Fig. 22.20?

4. How would the R control chart differ from the bottom graph in Fig. 22.20?

In Exercises 5–8, use the following data.

Five automobile engines are taken from the production line each hour and tested for their torque (in N · m) when rotating at a constant frequency. The measurements of the sample torques for 20 h of testing are as follows:

Hour	*Torques (in N · m) of Five Engines*				
1	366	352	354	360	362
2	370	374	362	366	356
3	358	357	365	372	361
4	360	368	367	359	363
5	352	356	354	348	350
6	366	361	372	370	363
7	365	366	361	370	362
8	354	363	360	361	364
9	361	358	356	364	364
10	368	366	368	358	360
11	355	360	359	362	353
12	365	364	357	367	370
13	360	364	372	358	365
14	348	360	352	360	354
15	358	364	362	372	361
16	360	361	371	366	346
17	354	359	358	366	366
18	362	366	367	361	357
19	363	373	364	360	358
20	372	362	360	365	367

5. Find the central line, UCL, and LCL for the mean.

6. Find the central line, UCL, and LCL for the range.

7. Plot the $\bar{x}$ chart.

8. Plot the R chart.

In Exercise 9–12, use the following data.

Five AC adaptors that are used to charge batteries of a cellular phone are taken from the production line each 15 min and tested for their direct-current output voltage. The output voltages for 24 sample subgroups are as follows:

Subgroup	*Output Voltages of Five Adaptors*				
1	9.03	9.08	8.85	8.92	8.90
2	9.05	8.98	9.20	9.04	9.12
3	8.93	8.96	9.14	9.06	9.00
4	9.16	9.08	9.04	9.07	8.97
5	9.03	9.08	8.93	8.88	8.95
6	8.92	9.07	8.86	8.96	9.04
7	9.00	9.05	8.90	8.94	8.93
8	8.87	8.99	8.96	9.02	9.03
9	8.89	8.92	9.05	9.10	8.93
10	9.01	9.00	9.09	8.96	8.98
11	8.90	8.97	8.92	8.98	9.03
12	9.04	9.06	8.94	8.93	8.92
13	8.94	8.99	8.93	9.05	9.10
14	9.07	9.01	9.05	8.96	9.02
15	9.01	8.82	8.95	8.99	9.04
16	8.93	8.91	9.04	9.05	8.90
17	9.08	9.03	8.91	8.92	8.96
18	8.94	8.90	9.05	8.93	9.01
19	8.88	8.82	8.89	8.94	8.88
20	9.04	9.00	8.98	8.93	9.05
21	9.00	9.03	8.94	8.92	9.05
22	8.95	8.95	8.91	8.90	9.03
23	9.12	9.04	9.01	8.94	9.02
24	8.94	8.99	8.93	9.05	9.07

9. Find the central line, UCL, and LCL for the mean.

10. Find the central line, UCL, and LCL for the range.

11. Plot the $\bar{x}$ chart.

12. Plot the R chart.

In Exercises 13–16, use the following information.

For a production process for which there is a great deal of data since its last modification, the population mean μ and population standard deviation σ are assumed known. For such a process, we have the following values (using additional statistical analysis):

$$\bar{x} \text{ chart: central line} = \mu, \text{UCL} = \mu + A\sigma, \text{LCL} = \mu - A\sigma$$

$$R \text{ chart: central line} = d_2\sigma, \text{UCL} = D_2\sigma, \text{LCL} = D_1\sigma$$

The values of A, d_2, D_2, and D_1 are found in the table of control chart factors in Example 2 (Table 22.2).

13. In the production of robot links and tests for their lenghs, it has been found that $\mu = 2.725$ in. and $\sigma = 0.032$ in. Find the central line, UCL, and LCL for the mean if the sample subgroup size is 5.

14. For the robot link samples of Exercise 13, find the central line, UCL, and LCL for the range.

15. After bottling, the volume of soft drink in six sample bottles is checked each 10 min. For this process $\mu = 750.0$ mL and $\sigma = 2.2$ mL. Find the central line, UCL, and LCL for the range.

16. For the bottling process of Exercise 15, find the central line, UCL, and LCL for the mean.

In Exercises 17 and 18, use the following data.

A telephone company rechecks the entries for 1000 of its new customers each week for name, address, and phone number. The data collected regarding the number of new accounts with errors, along with the proportion of these accounts with errors, is given in the following table for a 20-wk period:

Week	*Accounts with Errors*	*Proportion with Errors*
1	52	0.052
2	36	0.036
3	27	0.027
4	58	0.058
5	44	0.044
6	21	0.021
7	48	0.048
8	63	0.063
9	32	0.032
10	38	0.038
11	27	0.027
12	43	0.043
13	22	0.022
14	35	0.035
15	41	0.041
16	20	0.020
17	28	0.028
18	37	0.037
19	24	0.024
20	42	0.042
Total	738	

17. For the p chart, find the values for the central line, UCL, and LCL.

18. Plot the p chart.

In Exercises 19 and 20, use the following data.

The maker of electric fuses checks 500 fuses each day for defects. The number of defective fuses, along with the proportion of defective fuses for 24 days, is shown in the following table.

Day	*Number Defective*	*Proportion Defective*
1	26	0.052
2	32	0.064
3	37	0.074
4	16	0.032
5	28	0.056
6	31	0.062
7	42	0.084
8	22	0.044
9	31	0.062
10	28	0.056
11	24	0.048
12	35	0.070
13	30	0.060
14	34	0.068
15	39	0.078
16	26	0.052
17	23	0.046
18	33	0.066
19	25	0.050
20	25	0.050
21	32	0.064
22	23	0.046
23	34	0.068
24	20	0.040
Total	696	

19. For the p chart, find the values for the central line, UCL, and LCL.

20. Plot the p chart.

Answers to Practice Exercises

1. Mean: no change; range: 8.8

2. $\text{UCL}(p) = 0.0333$, $\text{LCL}(p) = 0.0067$

22.6 Linear Regression

Regression • Linear Regression • Method of Least Squares • Deviation • Least-squares Line • Using a Calculator • Interpolation • Extrapolation • Interpreting r and r^2

In the previous sections of this chapter, we have discussed various statistical methods for graphically and numerically summarizing the values of a single variable. In this section, we will show how to describe the relationship between *two different* quantitative variables, which allows us to predict the value of one from the other. There are many situations where data points from paired data form a general straight-line pattern but don't line up exactly along a straight line. In these cases, we wish to find the equation of a line that ***best fits*** the data points. The process of finding such a line is called **linear regression**.

In Chapters 5 and 21, we showed how the calculator can be used to find the equation of a regression line. In this section, we will show the mathematics behind the creation of this line and also discuss how to measure the strength of the linear relationship. We consider nonlinear regression in the next section.

EXAMPLE 1 Fitting a line to a set of points

All the students enrolled in a mathematics course took an entrance test. To study the reliability of this test as an indicator of future success, an instructor tabulated the test scores of ten students (selected at random), along with their course averages at the end of the course, and made a table of the data, which is shown below. The instructor then plotted the data and noticed that in general, the higher the test score, the higher the course grade. He wondered if there might be a straight line that would *fit* the data points reasonably well so that a student's success in the course could be predicted based on his or her entrance test score. Figure 22.22 shows two such possible lines.

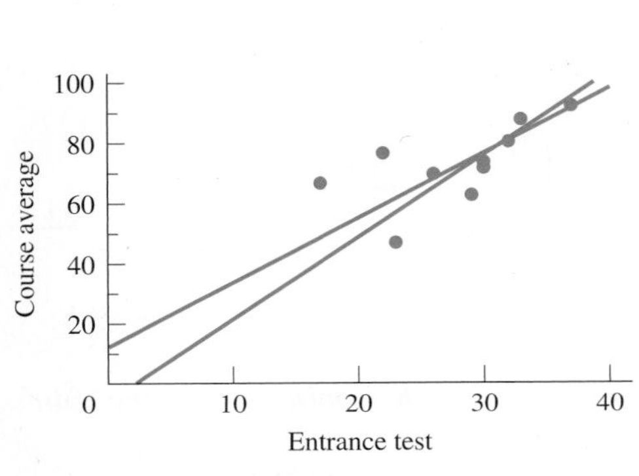

Fig. 22.22

Student	*Entrance Test Score, Based on 40*	*Course Average, Based on 100*
A	29	63
B	33	88
C	22	77
D	17	67
E	26	70
F	37	93
G	30	72
H	32	81
I	23	47
J	30	74

Since there are many ways of *visually* drawing a line through a set of points, we must establish *criteria* for determing which one fits the best. ■

There are a number of different methods of determining the straight line that best fits the given data points. We employ the method that is most widely used: the **method of least squares.** *The basic principle of this method is that the sum of the squares of the deviations of all data points from the best line (in accordance with this method) has the least value possible. By* **deviation,** *we mean the difference between the y-value of the line and the y-value for the point (of original data) for a particular value of x.*

EXAMPLE 2 Deviation and least squares line

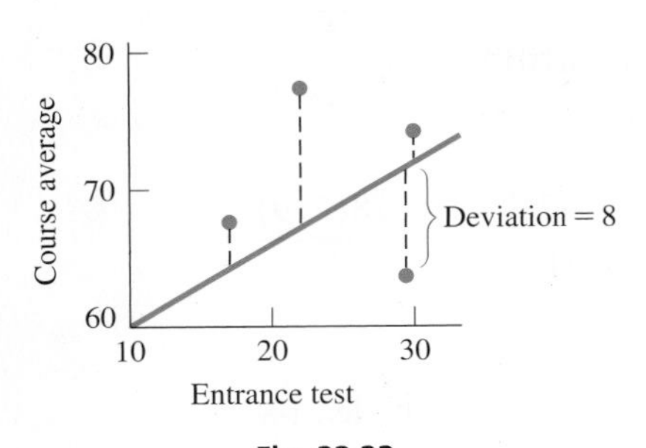

Fig. 22.23

In Fig. 22.23, the deviations of some of the points of Example 1 are shown. The point (29, 63) (student A of Example 1) has a deviation of 8 from the indicated line in the figure. Thus, we square the value of this deviation to obtain 64. In order to find the equation of the straight line that best fits the given points, the method of least squares requires that the sum of all such squares be a minimum.

Therefore, in applying this method of least squares, it is necessary to use the equation of a straight line and the coordinates of the points of the data. The deviations of all of these data points are determined, and these values are then squared. It is then necessary to determine the constants for the slope m and the y-intercept b in the equation of a straight line $y = mx + b$ for which the sum of the squared values is a minimum. To do this requires certain methods of advanced mathematics. ■

Using the methods that are required from advanced mathematics, it can be shown that *the equation of the* **least-squares line**

$$y = mx + b \tag{22.7}$$

is found by calculating the values of the slope m and the y-intercept b by using the formulas

$$m = \frac{n\sum xy - \left(\sum x\right)\left(\sum y\right)}{n\sum x^2 - \left(\sum x\right)^2} \tag{22.8}$$

and

$$b = \frac{\left(\sum x^2\right)\left(\sum y\right) - \left(\sum xy\right)\left(\sum x\right)}{n\sum x^2 - \left(\sum x\right)^2} \tag{22.9}$$

In the above equations, ***the x's and y's are the values of the coordinates of the points in the given data,*** and n is the number of points of data.

The following examples illustrate finding the least-squares line by use of Eqs. (22.8) and (22.9). Note that much of the work involved is finding the required sums for x, y, xy, and x^2. All of the calculations can be done on a calculator.

EXAMPLE 3 Finding equation of least-squares line—course averages

Find the least-squares line for the data of Example 1.

Here, the x-values will be the entrance-test scores and the y-values are the course averages.

x	y	xy	x^2
29	63	1,827	841
33	88	2,904	1089
22	77	1,694	484
17	67	1,139	289
26	70	1,820	676
37	93	3,441	1369
30	72	2,160	900
32	81	2,592	1024
23	47	1,081	529
30	74	2,220	900
279	732	20,878	8101

$$n = 10$$

$$m = \frac{10(20{,}878) - 279(732)}{10(8101) - 279^2} = 1.44$$

$$b = \frac{8101(732) - 20{,}878(279)}{10(8101) - 279^2} = 33.1$$

Thus, the equation of the least-squares line is $y = 1.44x + 33.1$.

The regression feature on a calculator can also be used to find the least-squares line. Figure 22.24(a) shows the regression equation, and Fig. 22.24(b) shows its graph through the scatterplot. Note that the calculator uses a to represent the slope of the regression line instead of m.

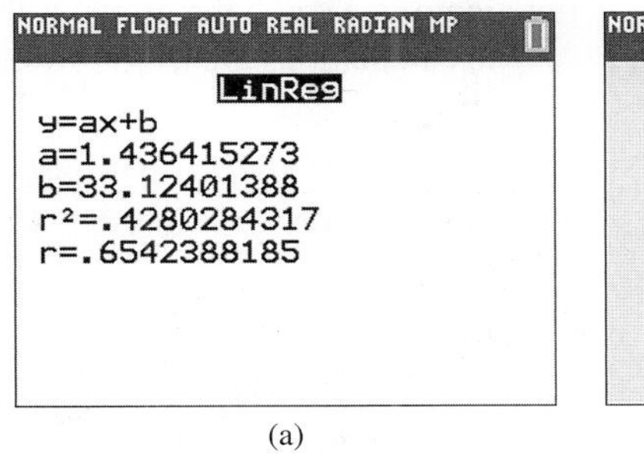

(a)

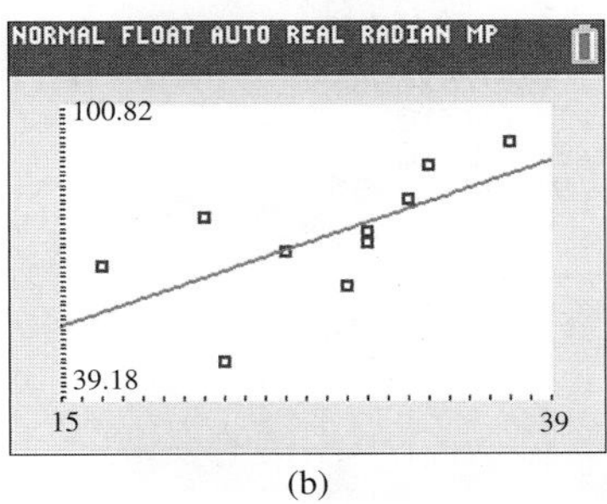

(b)

Fig. 22.24
Graphing calculator keystrokes:
goo.gl/wU04ct

NORMAL FLOAT AUTO REAL RADIAN MP
Y1(30)
76.21647207

Fig. 22.25
Graphing calculator keystrokes:
goo.gl/6Pr15m

The regression line can be used to predict the approximate course average based on the entrance test score. For example, to predict the course average for a student who scored 30 on the entrance test, we substitute 30 for x and evaluate:

$$y = 1.44(30) + 33.1$$

$$y = 76 \quad \text{rounded}$$

If we store the regression line in Y_1 when doing the regression, then we can predict the course average by entering $Y_1(30)$ as shown in Fig. 22.25. Thus, the predicted course average is 76. ■

NOTE ▸ [In regression, **interpolation** refers to predictions that are made using values *inside* the range of the given data and **extrapolation** refers to predictions made with values *outside* the range of the given data.] The prediction made in Example 3 was interpolation since the score of 30 is between the minimum and maximum entrance test scores.

MEASURING THE STRENGTH OF LINEAR CORRELATION: r and r^2

Note that the calculator screen in Fig. 22.24(a) shows values of r and r^2 in addition to the regression line. These two values measure how well the regression line fits the data. The **correlation coefficient r** is defined by $r = m\left(\frac{s_x}{x_y}\right)$, where s_x and s_y are the standard deviations of the x-values and y-values, respectively, and m is the slope of the regression line. NOTE ▸ [Because of its definition, the values of r always lie in the range $-1 \leq r \leq 1$. If r is close to 1 or -1, then there is strong correlation, which means the regression line fits the data points well.] Also, the sign of r is the same as the sign of the slope of the regression line. In Example 3, $r = 0.654$, which indicates a moderate level of positive correlation. The **coefficient of determination r^2** (the square of r) has an interesting interpretation. NOTE ▸ [When viewed as a percentage, it represents the percentage of variation in the y-variable that is explained by the linear model. The value of r^2 will always lie between 0% and 100%. The closer it is to 100%, the better the regression line fits the data points.] In Example 3, $r^2 = 0.428$, which means that 42.8% of the variation in the final course averages can be explained by their linear relationship with entrance-test scores. NOTE ▸ [To summarize, r-values close to 1 or -1 and r^2-values close to 100% indicate the regression model fits the data points very well.]

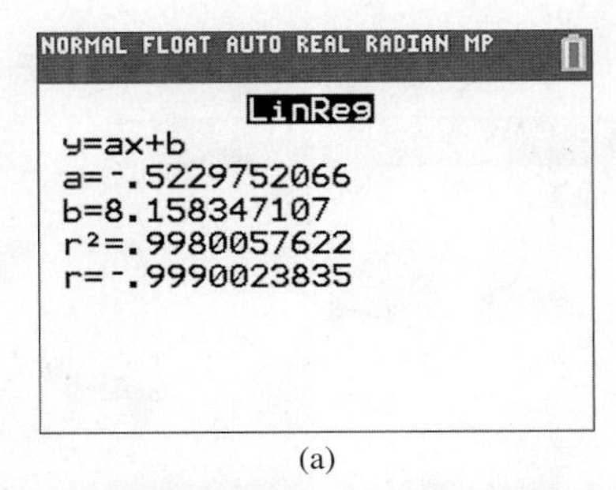

(a)

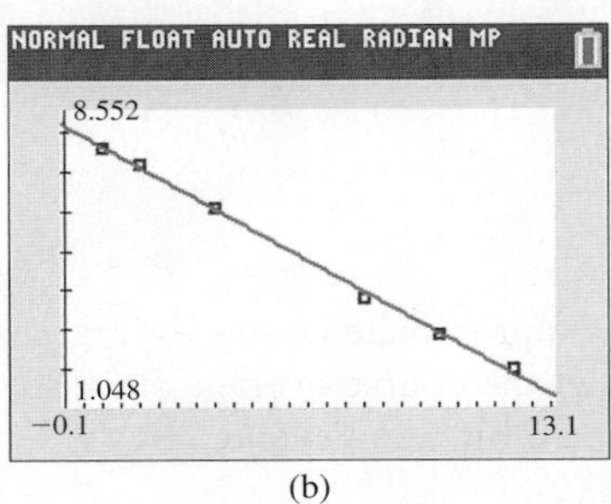

(b)

Fig. 22.26
Graphing calculator keystrokes:
goo.gl/LExw9S

EXAMPLE 4 Least-squares line—drug in bloodstream

In a research project to determine the amount of a drug that remains in the bloodstream after a given dosage, the amounts y (in mg of drug/dL of blood) were recorded after t h, as shown in the table below. Find the least-squares line for these data, expressing y as a function of t. Sketch the graph of the line and data points.

The calculations are as follows:

t (h)	1.0	2.0	4.0	8.0	10.0	12.0
y (mg/dL)	7.6	7.2	6.1	3.8	2.9	2.0

t	y	ty	t^2
1.0	7.6	7.6	1.0
2.0	7.2	14.4	4.0
4.0	6.1	24.4	16
8.0	3.8	30.4	64
10.0	2.9	29.0	100
12.0	2.0	24.0	144
37.0	29.6	129.8	329

$n = 6$

$$m = \frac{6(129.8) - 37.0(29.6)}{6(329) - 37.0^2} = -0.523$$

$$b = \frac{(329)(29.6) - (129.8)(37.0)}{6(329) - 37.0^2} = 8.16$$

The equation of the least-squares line is $y = -0.523t + 8.16$. This line is useful in determining the amount of the drug in the bloodstream at any given time. For example, using the regression line, the predicted amount of drug in the bloodstream after 13 hours is 1.4 mg/dL (extrapolation).

The calculator display of the regression line and its graph through the scatterplot are shown in Fig. 22.26(a) and (b), respectively. Note that r is very close to -1 and r^2 is close to 1 (or 100%). This indicates a very good fit. ■

CAUTION When making predictions, be careful to round only the final estimate, not the values in the regression equation. Small amounts of rounding in the regression equation can lead to fairly large errors in the estimates. When possible, use calculator stored regression equations to make predictions as shown in Example 3. ■

EXERCISES 22.6

In Exercises 1–14, find the equation of the least-squares line for the given data. Graph the line and data points on the same graph.

1.

x	1	2	3	4	5
y	3	7	9	9	12

2.

x	1	2	3	4	5	6	7
y	10	17	28	37	49	56	72

3.

x	20	26	30	38	48	60
y	160	145	135	120	100	90

4.

x	1	3	6	5	8	10	4	7	3	8
y	15	12	10	8	9	2	11	9	11	7

5. In Example 4, change the y (mg of drug/dL of blood) values to 8.7, 8.4, 7.7, 7.3, 5.7, 5.2. Then proceed to find y as a function of t, as in Example 4.

6. The speed v (in m/s) of sound was measured as a function of the temperature T (in °C) with the following results. Find v as a function of T.

T (°C)	0	10	20	30	40	50	60
v (m/s)	331	337	344	350	356	363	369

7. In an electrical experiment, the following data were found for the values of current and voltage for a particular element of the circuit. Find the voltage V as a function of the current i. Then predict the voltage if $i = 8.00$ mA. Is this interpolation or extrapolation?

Current (mA)	15.0	10.8	9.30	3.55	4.60
Voltage (V)	3.00	4.10	5.60	8.00	10.50

8. A particular muscle was tested for its speed of shortening as a function of the force applied to it. The results appear below. Find the speed as a function of the force. Then predict the speed if the force is 15.0 N. Is this interpolation or extrapolation?

Force (N)	60.0	44.2	37.3	24.2	19.5
Speed (m/s)	1.25	1.67	1.96	2.56	3.05

9. The altitude h (in m) of a rocket was measured at several positions at a horizontal distance x (in m) from the launch site, shown in the table. Find the least-squares line for h as a function of x.

x (m)	0	500	1000	1500	2000	2500
h (m)	0	1130	2250	3360	4500	5600

10. In testing an air-conditioning system, the temperature T in a building was measured during the afternoon hours with the results shown in the table. Find the least-squares line for T as a function of the time t from noon. Then predict the temperature when $t = 2.5$. Is this interpolation or extrapolation?

t (h)	0.0	1.0	2.0	3.0	4.0	5.0
T (°C)	20.5	20.6	20.9	21.3	21.7	22.0

11. The pressure p was measured along an oil pipeline at different distances from a reference point, with results as shown. Find the least-squares line for p as a function of x using a calculator. Then predict the pressure at a distance of $x = 500$ ft. Is this interpolation or extrapolation?

x (ft)	0	100	200	300	400
p (lb/in.^{2})	650	630	605	590	570

12. The heat loss L per hour through various thicknesses of a particular type of insulation was measured as shown in the table. Find the least-squares line for L as a function of t using a calculator.

t (in.)	3.0	4.0	5.0	6.0	7.0
L (Btu)	5900	4800	3900	3100	2450

13. In an experiment on the photoelectric effect, the frequency of light being used was measured as well as the stopping potential (the voltage just sufficient to stop the photoelectric effect) with the results given below. Use a calculator to find the least-squares line for V as a function of f. The frequency for $V = 0$ is known as the *threshold frequency*. From the graph determine the threshold frequency.

f (PHz)	0.550	0.605	0.660	0.735	0.805	0.880
V (V)	0.350	0.600	0.850	1.10	1.45	1.80

14. If gas is cooled under conditions of constant volume, it is noted that the pressure falls nearly proportionally as the temperature. If this were to happen until there was no pressure, the theoretical temperature for this case is referred to as *absolute zero*. In an elementary experiment, the following data were found for pressure and temperature under constant volume.

T (°C)	0.0	20	40	60	80	100
P(kPa)	133	143	153	162	172	183

Use a calculator to find the least-squares line for P as a function of T, and from the graph determine the value of absolute zero found in this experiment.

In Exercises 15–18, find and interpret the values of r and r^2 for the given data.

15. Exercise 7 **16.** Exercise 8 **17.** Exercise 10 **18.** Exercise 13

22.7 Nonlinear Regression

Nonlinear Regression • Types of Curves on a Calculator • Choosing an Appropriate Regression Model

The method of least-squares regression can be extended to cases where the data points tend to curve rather than follow a straight line. This is called **nonlinear regression.** The specific mathematical techniques and formulas used for nonlinear regression are beyond the scope of this chapter. However, a calculator can be used to find many different kinds of nonlinear regression models. NOTE [To choose an appropriate model, one should first plot the data to get a **scatterplot.** Then, depending on the shape of the points, an appropriate model can be chosen.] Figure 22.27 lists the types of regression available on most graphing calculators along with some examples of their shapes.

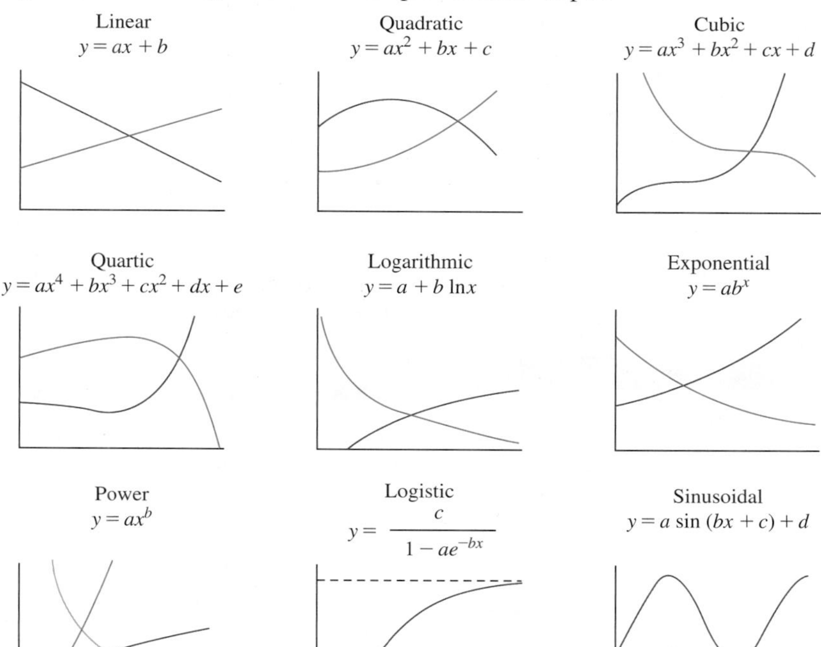

Fig. 22.27

When more than one type of model seems reasonable, we can try different types to see which one fits best. In nonlinear regression, the calculator still provides values of r, r^2, or R^2. Although these are not calculated using the same formulas as in linear regression, they still measure how well the regression model fits the data points. As before, values close to 1 (or -1 for r) indicate a good fit. NOTE ➧ [Practical considerations should also be taken into account when choosing a model. For example, if it is known that a quantity will continually decrease, one should try to avoid choosing a model that would eventually predict increases.] The following examples illustrate the process of choosing and determining a regression model.

EXAMPLE 1 Choosing a regression model—volume and pressure of a gas

In a physics experiment, the pressure P and volume V of a gas were measured at constant temperature. The resulting data are shown in the table below and plotted on a calculator in Fig. 22.28.

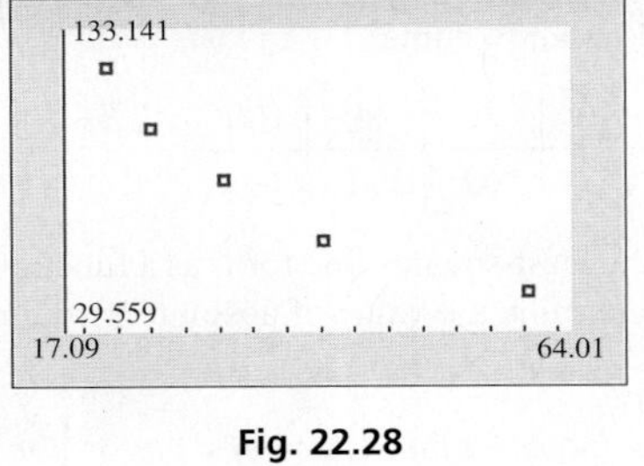

Fig. 22.28

Volume, V (cm^3)	21.0	25.0	31.8	41.1	60.1
Pressure, P (kPa)	120.0	99.2	81.3	60.6	42.7

Based on the curved downward trend of the points, the types of regression models that are reasonable to try here are logarithmic, exponential, and power. After trying all three, the power regression model clearly fits the data the best. The power regression equation is $y = 2382.57x^{-0.9831}$ as shown in Fig. 22.29(a). The graph of the model through the scatterplot is shown in Fig. 22.29(b). The fact that r is close to -1 and r^2 is close to 1 indicate the model fits the data very well.

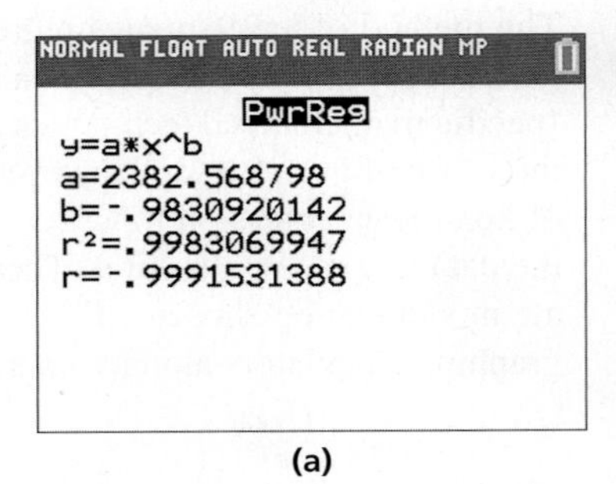

(a)

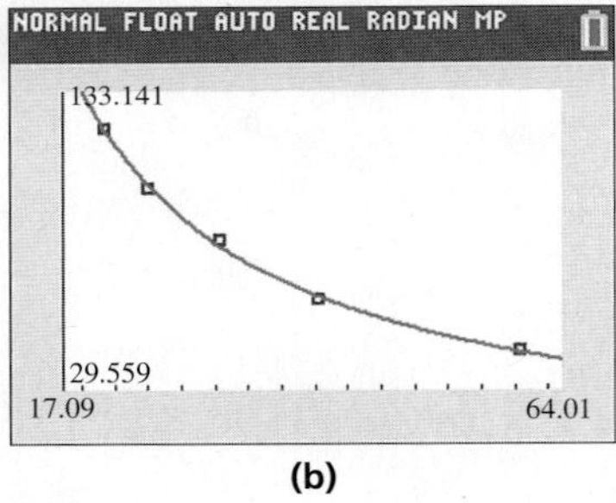

(b)

Fig. 22.29

Graphing calculator screenshots: goo.gl/GZYkRH

Using the variables in our problem, we can rewrite the regression equation as $P = 2382.57\,V^{-0.9831}$. This can be used to predict either variable when given the other. For example, to predict the pressure when the volume is 56.0 cm^3, we get $P = 2382.57(56.0)^{-0.9831} = 45.5$ kPa (interpolation). ■

EXAMPLE 2 Choosing a regression model—bacteria growth

In a medical research lab, an experiment was performed to measure the number N of bacteria present t min after the start of the experiment. The following table shows the resulting data, and the scatterplot is shown in Fig. 22.30.

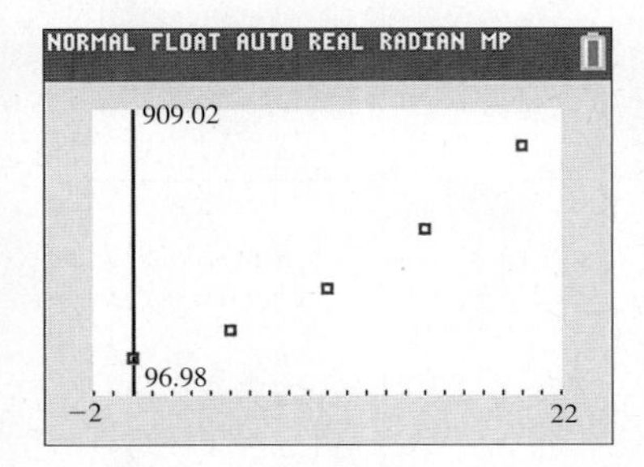

Fig. 22.30

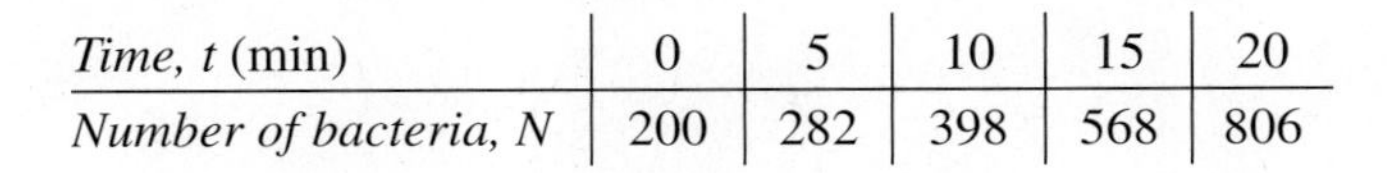

Time, t (min)	0	5	10	15	20
Number of bacteria, N	200	282	398	568	806

Because of the curved upward trend of the points, we will try both a quadratic and an exponential regression model. The results from a calculator are shown in Figs. 22.31 and 22.32. Both models provide a very good fit of the data points.

Quadratic regression model
$N = 1.046t^2 + 9.046t + 203.486$
$R^2 = 99.95\%$

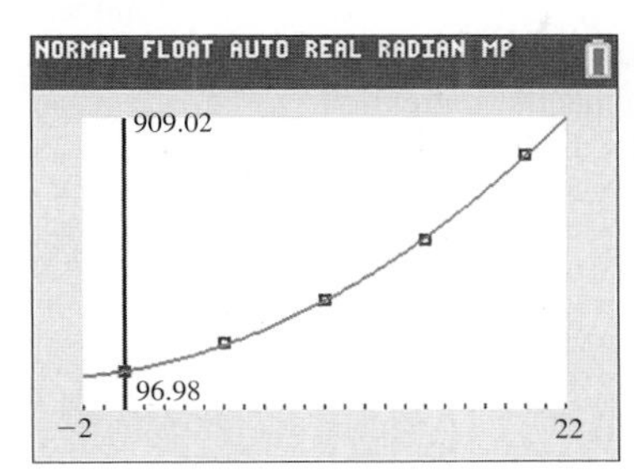

Fig. 22.31
Graphing calculator screenshots: goo.gl/FE0h4o

Exponential regression model
$N = 199.263(1.07225)^t$
$r^2 = 99.995\%$

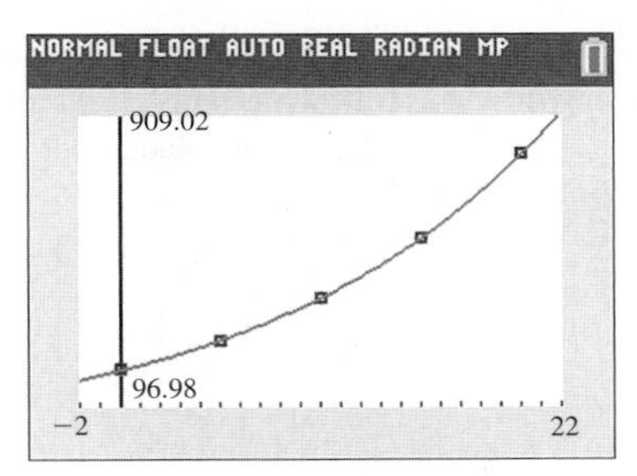

Fig. 22.32
Graphing calculator screenshots: goo.gl/nw97VL

Because both models fit the data well, either can be used. However, since it is known that bacteria grow exponentially, the exponential model will likely fit better in the long run. Suppose we wish to estimate the number of bacteria present after 25 min. Using the exponential model, we have $N = 199.263(1.07225)^{25} = 1140$ (extrapolation). ■

EXERCISES 22.7

In Exercises 1–12, use a calculator to find a regression model for the given data. Graph the scatterplot and regression model on the calculator. Use the regression model to make the indicated predictions.

1. Find an exponential regression model for the given data:

x	0	10	20	30	40
y	350	570	929	1513	2464

2. Find a logarithmic regression model for the given data:

x	1	4	8	12	16
y	2	9	11	14	15

3. In Example 1, change the V (volume of the gas) values to 19.9, 24.5, 29.4, 39.4, 56.0. Then find a power regression model as in Example 1.

4. The following data show the tensile strength (in 10^5 lb/in.^{2}) of brass (a copper-zinc alloy) that contains different percents of zinc.

Percent of Zinc	0	5	10	15	20	30
Tensile Strength	0.32	0.36	0.39	0.41	0.44	0.47

Find a logistic regression model for the data. Predict the tensile strength of brass that contains 25% zinc.

5. The following data were found for the distance y that an object rolled down an inclined plane in time t. Find a quadratic regression model for these data. Predict the distance at 2.5 s.

t (s)	1.0	2.0	3.0	4.0	5.0
y (cm)	6.0	23	55	98	148

6. The increase in length y of a certain metallic rod was measured in relation to particular increases x in temperature. Find a quadratic regression model for the given data.

x (°C)	50.0	100	150	200	250
y (cm)	1.00	4.40	9.40	16.4	24.0

7. The pressure p at which Freon, a refrigerant, vaporizes for temperature T is given in the following table. Find a quadratic regression model. Predict the vaporization pressure at 30°F.

T (° F)	0	20	40	60	80
p (lb/in.^{2})	23	35	49	68	88

8. A fraction f of annual hot-water loads at a certain facility are heated by solar energy. The fractions f for certain values of the collector area A are given in the following table. Find a power regression model for these data.

A (m^2)	0	12	27	56	90
f	0.0	0.2	0.4	0.6	0.8

9. The output torque (in J) of a certain engine was measured at various frequencies (in r/min) with the following results. Find a power regression model for these data. Predict the output torque for a frequency of 3500 r/min. Is this interpolation or extrapolation?

f (r/min)	500	1000	1500	2000	2500	3000
T (J)	220	102	77	50	43	30

10. The resonant frequency f of an electric circuit containing a 4-μF capacitor was measured as a function of the inductance L in the circuit. The following data were found. Find a power regression model for these data.

L (H)	1.0	2.0	4.0	6.0	9.0
f (Hz)	490	360	250	200	170

11. The displacement y of an object at the end of a spring at given times t is shown in the following table. Find an exponential regression model for this. Predict the displacement at 2.5 s. Is this interpolation or extrapolation?

t (s)	0.0	0.5	1.0	1.5	2.0	3.0
y (cm)	6.1	3.8	2.3	1.3	0.7	0.3

12. The average daily temperatures T (in °F) for each month in Minneapolis (National Weather Service records) are given in the following table. (1 = Jan, 2 = Feb, etc.).

Month, t	1	2	3	4	5	6	7	8	9	10	11	12
T (°F)	11	18	29	46	57	68	73	71	61	50	33	19

Find a sinusoidal regression model for these data.

CHAPTER 22 KEY FORMULAS AND EQUATIONS

Mean

$$\bar{x} = \frac{x_1 f_1 + x_2 f_2 + \cdots + x_n f_n}{f_1 + f_2 + \cdots + f_n} \tag{22.1}$$

Standard deviation

$$s = \sqrt{\frac{\sum (x - \bar{x})^2}{n - 1}} \tag{22.2}$$

$$s = \sqrt{\frac{n\left(\sum x^2\right) - \left(\sum x\right)^2}{n(n - 1)}} \tag{22.3}$$

Normal distribution

$$y = \frac{e^{-(x-\mu)^2/2\sigma^2}}{\sigma\sqrt{2\pi}} \tag{22.4}$$

Standard (z) score

$$z = \frac{x - \mu}{\sigma} \tag{22.5}$$

Standard error of $\bar{x}$

$$\sigma_{\bar{x}} = \frac{\sigma}{\sqrt{n}} \tag{22.6}$$

Least-squares lines

$$y = mx + b \tag{22.7}$$

$$m = \frac{n\sum xy - \left(\sum x\right)\left(\sum y\right)}{n\sum x^2 - \left(\sum x\right)^2} \tag{22.8}$$

$$b = \frac{\left(\sum x^2\right)\left(\sum y\right) - \left(\sum xy\right)\left(\sum x\right)}{n\sum x^2 - \left(\sum x\right)^2} \tag{22.9}$$

CHAPTER 22 REVIEW EXERCISES

CONCEPT CHECK EXERCISES

*Determine each of the following as being either **true** or **false**. If it is false, explain why.*

1. If the raw data includes 25 values, and a class includes 3 of these values, the relative frequency of this class is 12%.
2. The mean of the numbers 2, 3, 6, 4, 5 is 4.
3. The standard deviation of the numbers 2, 3, 6, 4, 5 is $\sqrt{5/2}$.
4. For a normal distribution for which $\mu = 10$ and $\sigma = 5$, the z-scores for $x = 20$ and $x = 30$ are $z = 1.5$ and $z = 2.5$.
5. If the least-squares line for a set of data is $y = 2x + 4$, the deviation of the data point (2, 6) is 4.
6. In linear regression, a value of r close to 0 indicates a good fit.

PRACTICE AND APPLICATIONS

In Exercises 7–14, use the following data. An airline's records showed that the percent of on-time flights each day for a 20-day period was as follows:

72, 75, 76, 70, 77, 73, 80, 75, 82, 85,
77, 78, 74, 86, 72, 77, 67, 78, 69, 80

7. Determine the median.
8. Determine the mode.
9. Determine the mean.
10. Determine the standard deviation.
11. Construct a frequency distribution table with class limits of 67, 71, 75, 79, 83, and 87.
12. Make a stem-and-leaf plot of these data using split stems.
13. Draw a histogram for the data using the table from Exercise 11.
14. Construct a relative frequency table for the data using the class limits given in Exercise 11.

In Exercises 15 and 16, use the following information. For the employees of a certain business, the highest education level is summarized below:

Education	*Frequency*
No high school diploma	4
High school diploma	8
2-year college degree	18
4-year college degree or higher	10

15. Use the frequencies to make a bar graph of these data.
16. Using relative frequencies, make a pie chart of these data.

In Exercises 17–20, use the following data. In testing the water supply of a town, the volume V (in mL/L) of settleable solids over a 12-day period were as follows (readings of $V < 0.15$ mL/L are considered acceptable): 0.11, 0.15, 0.16, 0.13, 0.14, 0.13, 0.12, 0.14, 0.15, 0.13, 0.13, 0.12.

17. Find the mean.
18. Find the median.
19. Find the standard deviation.
20. Draw a histogram.

In Exercises 21–24, use the following data. A sample of wind generators was tested for power output when the wind speed was 30 km/h. The following table gives the powers produced (to the nearest 10 W) and the number of generators for each power value.

Power (W)	650	660	670	680	690
No. Generators	3	2	7	12	27

Power (W)	700	710	720	730
No. Generators	34	15	16	5

21. Find the median.
22. Find the mean.
23. Find the standard deviation.
24. Find the mode.

In Exercises 25–28, use the following data. In an experiment to measure cosmic radiation, the number of cosmic rays was recorded for 200 different 5-s intervals. The grouped data are shown below:

No. of Cosmic Rays	0	1	2	3	4	5	6	7	8	9	10
Frequency	3	10	25	45	29	39	26	11	7	2	3

25. Find the mean.
26. Find the median.
27. Find the standard deviation.
28. Make a relative frequency table.

In Exercises 29–32, use the following information. IQ scores are normally distributed with a mean $\mu = 100$ and a standard deviation $\sigma = 15$.

29. According to the empirical rule, approximately what percentage of IQ scores are between 70 and 130?
30. According to the empirical rule, approximately what percentage of IQ scores are between 55 and 145?
31. Find and interpret the z-score of an IQ of 127.
32. Find the IQ score that has a z-score of -1.2.

In Exercises 33 and 34, use the following information. A company that makes electric light bulbs tests 500 bulbs each day for defects. The number of defective bulbs, along with the proportion of defective bulbs for 20 days, is shown in the following table.

Day	*Number Defective*	*Proportion Defective*
1	23	0.046
2	31	0.062
3	19	0.038
4	27	0.054
5	29	0.058
6	39	0.078
7	26	0.052
8	17	0.034
9	28	0.056
10	33	0.066
11	22	0.044
12	29	0.058
13	20	0.040
14	35	0.070
15	21	0.042
16	32	0.064
17	25	0.050
18	23	0.046
19	29	0.058
20	32	0.064
Total	540	

33. For a p chart, find the values of the central line, UCL, and LCL.

34. Plot a p chart.

In Exercises 35 and 36, use the following information. Five ball bearings are taken from the production line every 15 min and their diameters are measured. The diameters of the sample ball bearings for 16 successive subgroups are given in the following table.

Subgroup	*Diameters (mm) of Five Ball Bearings*				
1	4.98	4.92	5.02	4.91	4.93
2	5.03	5.01	4.94	5.06	5.07
3	5.05	5.03	5.00	5.02	4.96
4	5.01	4.92	4.91	4.99	5.03
5	4.92	4.97	5.02	4.95	4.94
6	5.02	4.95	5.01	5.07	5.15
7	4.93	5.03	5.02	4.96	4.99
8	4.85	4.91	4.88	4.92	4.90
9	5.02	4.95	5.06	5.04	5.06
10	4.98	4.98	4.93	5.01	5.00
11	4.90	4.97	4.93	5.05	5.02
12	5.03	5.05	4.92	5.03	4.98
13	4.90	4.96	5.00	5.02	4.97
14	5.09	5.04	5.05	5.02	4.97
15	4.88	5.00	5.02	4.97	4.94
16	5.02	5.09	5.03	4.99	5.03

35. Plot an $\bar{x}$ chart.

36. Plot an R chart.

In Exercises 37–40, solve the given problems.

37. Find the area under the standard normal curve between $z = -1.6$ and $z = 2.1$.

38. Find the area under the standard normal curve to the right of $z = 0.9$.

39. Find the two z-scores that bound the middle 83.84% of the area under the standard normal curve.

40. Find the z-score that has an area to the right of 0.0287 under the standard normal curve.

In Exercises 41–44, use the following data. After analyzing data for a long period of time, it was determined that the readings of an organic pollutant for an area are distributed normally with $\mu = 2.20\ \mu g/m^3$ and $\sigma = 0.50\ \mu g/m^3$.

41. What percentage of the readings are between $1.45\ \mu g/m^3$ and $2.50\ \mu g/m^3$?

42. What percentage of the readings are between $2.50\ \mu g/m^3$ and $3.50\ \mu g/m^3$?

43. What percentage of the readings are above $1.00\ \mu g/m^3$?

44. What percentage of the readings are below $2.00\ \mu g/m^3$?

In Exercises 45–54, find the indicated regression models. Sketch the curve and data points on the same graph.

45. In a certain experiment, the resistance R of a certain resistor was measured as a function of the temperature T. The data found are shown in the following table. Find the regression line that expresses R as a function of T.

T (°C)	0.0	20.0	40.0	60.0	80.0	100
R (Ω)	25.0	26.8	28.9	31.2	32.8	34.7

46. An air-pollution monitoring station took samples of air each hour during the later morning hours and tested each sample for the number n of parts per million (ppm) of carbon monoxide. The results are shown in the table, where t is the number of hours after 6 A.M. Find the regression line that expresses n as a function of t.

t (h)	0.0	1.0	2.0	3.0	4.0	5.0	6.0
n (ppm)	8.0	8.2	8.8	9.5	9.7	10.0	10.7

47. The *Mach number* of a moving object is the ratio of its speed to the speed of sound (740 mi/h). The following table shows the speed s of a jet aircraft, in terms of Mach numbers, and the time t after it starts to accelerate. Find the regression line that expresses s as a function of t.

t (min)	0.00	0.60	1.20	1.80	2.40	3.00
s (Mach number)	0.88	0.97	1.03	1.11	1.19	1.25

48. The distance s of a missile above the ground at time t after being released from a plane is given by the following table (see Fig. 22.33). Find the equation of the quadratic regression model for these data.

t (s)	0.0	3.0	6.0	9.0	12.0	15.0	18.0
s (m)	3000	2960	2820	2600	2290	1900	1410

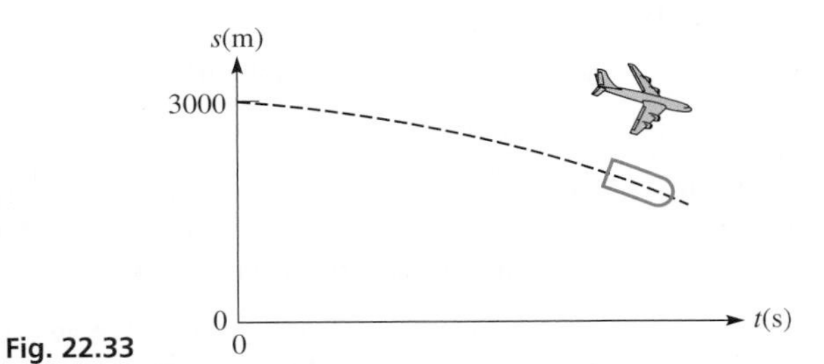

Fig. 22.33

49. In an experiment to determine the relation between the load x on a spring and the length y of the spring, the following data were found. Find the regression line that expresses y as a function of x.

Load (lb)	0.0	1.0	2.0	3.0	4.0	5.0
Length (in.)	10.0	11.2	12.3	13.4	14.6	15.9

50. In an elementary experiment that measured the wavelength L of sound as a function of the frequency f, the following results were obtained.

f (Hz)	240	320	400	480	560
L (cm)	140	107	81.0	70.0	60.0

Find the equation of the exponential regression model for these data that expresses L as a function of f.

51. For the data in Exercise 50, find equations for the quadratic, logarithmic, and power regression models. Which of these three models fits the data the *least* well?

52. The power P (in W) generated by a wind turbine was measured for various wind velocities v (in mi/h), as shown in the following table.

v (mi/h)	10	15	20	25	30	40
P (W)	75	250	600	1200	2100	4800

Find the equation of the cubic regression model for these data.

53. The vertical distance y of the cable of a suspension bridge above the surface of the bridge is measured at a horizontal distance x along the bridge from its center. See Fig. 22.34. The results are as follows:

x (m)	0	100	200	300	400	500
y (m)	15	17	23	33	47	65

Find the equation of the quadratic regression model for these data.

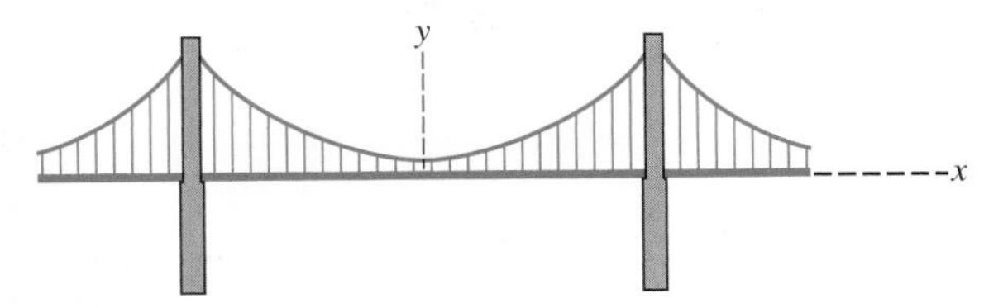

Fig. 22.34

54. The chlorine residual r (in ppm—parts per million) was measured in a swimming pool every 12 h after chemical treatment, with the following results:

t (in h)	0	12	24	36	48	60	72
r (ppm)	2.05	1.71	1.42	1.19	0.99	0.83	0.71

Find the equation of the exponential regression model for these data that expresses r as a function of t.

In Exercises 55–58, find the indicated regression model for the following data. Using aerial photography, the area A (in km²) of an oil spill as a function of the time t (in h) after the spill was found to be as follows:

t (h)	1.0	2.0	4.0	6.0	8.0	10.0
A (km^2)	1.4	2.5	4.7	6.8	8.8	10.2

55. Find the linear equation $y = ax + b$ to fit these data.

56. Find the quadratic equation $y = ax^2 + bx + c$ to fit these data.

57. Find the power equation $y = ax^b$ to fit these data.

58. Compare the values of the coefficient of correlation r, to determine whether the linear equation or power equation seems to fit the data better.

In Exercises 59 and 60, solve the given problems.

59. Show that Eqs. (22.8) and (22.9) satisfy the equation $\bar{y} = m\bar{x} + b$.

60. The sales (in millions of dollars) for a company is shown below. Make a time series plot of the data.

Year	2012	2013	2014	2015	2016
Sales	1.2	1.4	1.6	1.5	1.8

61. A research institute is planning a study of the effect of education on the income of workers. Explain what data should be collected and which of the measures discussed in this chapter would be useful in analyzing the data.

CHAPTER 22 PRACTICE TEST

As a study aid, we have included complete solutions for each Practice Test problem at the end of this book.

1. Of the 32 injuries that happened at a factory in a certain year, 14 were hand injuries, 8 were back injuries, 6 were eye injuries, and 4 were other injuries. Make a bar graph of these data.

In Problems 2–4, use the following set of numbers.

5, 6, 1, 4, 9, 5, 7, 3, 8, 10, 5, 8, 4, 9, 6

2. Find the median.
3. Find the mode.
4. Draw a histogram using class limits of 1, 3, 5, . . . , 11.
5. The amounts of energy produced by a solar panel on a sample of 10 days are given below. Make a stem-and-leaf plot of the data.

9.4 8.5 10.7 9.6 7.2 8.7 8.2 9.1 7.9 9.4

In Problems 6–8, use the following data. The thickness of 100 machine parts were measured (to the nearest 0.01 in.) and the grouped data is shown below:

Thickness (in.)	0.90	0.91	0.92	0.93	0.94	0.95	0.96
Number	3	9	31	38	12	5	2

6. Find the mean.
7. Find the standard deviation.
8. Make a relative frequency table.

In Problems 9–11, use the following information. At a certain customer call-in center, the customer hold time is normally distributed with a mean of 5.2 min and a standard deviation of 1.4 min.

9. According to the empirical rule, what percentage of hold times are between 3.8 min and 6.6 min?
10. Find and interpret the z-score of a hold time of 8.7 min.
11. What percent of hold times are less than 7.3 min?
12. Each hour, for 20 consecutive hours, a sample of 5 M&M packets are randomly selected and their weights are measured. Explain how these data can be used to make an R control chart.
13. Find the equation of the least-squares line for the points indicated in the following table. Graph the line and data points on the same graph.

x	1	3	5	7	9
y	5	11	17	20	27

14. The period of a pendulum as a function of its length was measured, giving the following results:

Length (ft)	1.00	3.00	5.00	7.00	9.00
Period (s)	1.10	1.90	2.50	2.90	3.30

Find the equation of the power regression model for these data.

The Derivative

23

Findings related to the motion of the planets and of projectiles in the early 1600s created interest in the mid-1600s as to the motion of objects. Noting, for example, that the velocity of a falling object changes from one instant to the next, just how fast an object is moving at a given instant, its *instantaneous velocity,* was of particular interest. It was also noted that the geometric problem of finding the slope of a curve *at a specific point* was really equivalent to finding the instantaneous velocity of an object, because each involved an *instantaneous rate of change.*

It was well known at the time how to find an *average velocity* (distance traveled divided by time taken) and a slope (difference in y-values divided by difference in x-values). However, these methods do not work in finding an instantaneous velocity or slope since it means dividing by zero. Therefore, a method for finding the slope of a tangent line to a curve at a given point was a major point of interest in mathematics in the 1600s.

This interest in motion and slope of a tangent line led to the development of *calculus*. In this chapter, we start developing *differential calculus,* which deals with finding the instantaneous rate of change of one quantity with respect to another. Other examples of an instantaneous rate of change are electric current (the rate of change of electric charge with respect to time), and the rate of change of light intensity with respect to the distance from the source. In Chapter 25, we will study *integral calculus,* which involves finding the function for which the rate of change is known.

Isaac Newton, the English mathematician and physicist, and Gottfried Leibnitz, a German mathematician and philosopher, are credited with the creation of the basic methods of calculus in the 1660s and 1670s. Others, including the French mathematician Pierre de Fermat, are known to have developed some of the topics related to calculus in the mid-1600s. In the 1700s and 1800s, many mathematicians further developed and refined the concepts of calculus.

The topic of this chapter, the *derivative,* is the basic concept of differential calculus that is used to measure an instantaneous rate of change. We will show some of the applications of the derivative in fields such as physical science and engineering in this chapter, and develop several important applications in the next chapter.

LEARNING OUTCOMES

After completion of this chapter, the student should be able to:

- Determine the continuity of a function at a single point or over an interval
- Define the concept of a limit
- Evaluate a limit of a function
- Find the slope of a function by taking the limit of the slopes of secant lines
- Find a derivative using the definition of the derivative
- Interpret instantaneous slope
- Use derivative formulas to calculate derivatives of polynomials, and powers, products, and quotients of functions
- Apply implicit differentiation
- Calculate higher derivatives of functions

◀ **The velocity of a diver continually changes at every instant in time. In Section 23.5, we show how the *derivative* of a function can be used to find the instantaneous velocity.**

23.1 Limits

Continuity • Limit of a Function • Limit as x Approaches Infinity

Before dealing with the rate of change of a function, we first take up the concept of a *limit*. We encountered a limit with infinite geometric series and with the asymptotes of a hyperbola. It is necessary to develop this concept further.

To help develop the concept of a limit, we first consider briefly the **continuity** of a function. *For a function to be* **continuous at a point,** *the function must exist at the point, and any small change in x produces only a small change in f(x).* In fact, the change in $f(x)$ can be made as small as we wish by restricting the change in x sufficiently, if the function is continuous. Also, *a function is said to be* **continuous over an interval** *if it is continuous at each point in the interval.*

NOTE ▸ [If the domain of a function includes a point, and other points on only one side of this point, it is *continuous at the point* if the definition of continuity holds for that part of the domain.]

EXAMPLE 1 Function continuous for all x

The function $f(x) = 3x^2$ is continuous for all values of x. That is, $f(x)$ is defined for all values of x, and a small change in x for any given value of x produces only a small change in $f(x)$. If we choose $x = 2$, and then let x change by 0.1, 0.01, and so on, we obtain the values in the following table:

x	2	2.1	2.01	2.001
$f(x)$	12	13.23	12.1203	12.012003
Change in x		0.1	0.01	0.001
Change in f(x)		1.23	0.1203	0.012003

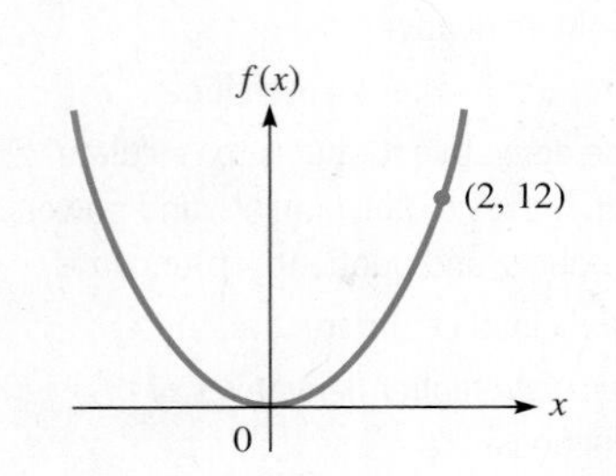

Fig. 23.1

We see that the change in $f(x)$ is smaller for smaller changes in x. This shows that $f(x)$ is continuous at $x = 2$. Because this type of result would be obtained for any other x we may chose, we see that $f(x)$ is continuous over the interval of all values of x. The graph of the function $f(x) = 3x^2$ is shown in Fig. 23.1. ■

EXAMPLE 2 Function discontinuous at $x = 2$

The function $f(x) = \frac{1}{x-2}$ is not continuous at $x = 2$. When we substitute 2 for x, we have division by zero. This means the function is not defined for the value $x = 2$. The condition of continuity—that the function must exist—is not satisfied. The graph of the function is shown in Fig. 23.2. ■

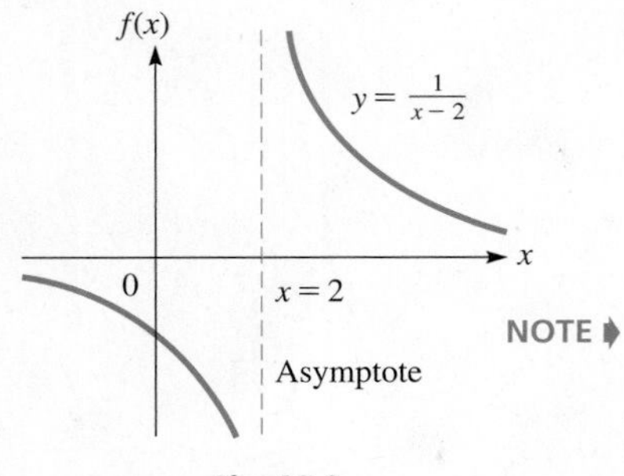

Fig. 23.2

NOTE ▸ [From a graphical point of view, a function that is continuous over an interval has no "breaks" in its graph over that interval. This means that the function is continuous over the interval if we can draw its graph without lifting the marker from the paper.] If the function is *discontinuous* for some value or values over an interval, a break occurs in the graph because the function is not defined or the definition of the function leads to an instantaneous "jump" in its values.

EXAMPLE 3 Continuity and graph of function

Looking at the graph of the function $f(x) = 3x^2$ in Fig. 23.1, we see that there are no breaks in the curve representing this function. In Example 1, we determined that this function is continuous for all values of x.

Now, looking at the graph of the function $f(x) = \frac{1}{x-2}$ in Fig. 23.2, we see that there is a break in the curve at $x = 2$. In Example 2, we determined that this function is not defined at $x = 2$, and is therefore not continuous at $x = 2$. We note that it is a hyperbola with an asymptote $x = 2$. ■

EXAMPLE 4 Continuous and discontinuous functions

(a) For the function represented in Fig. 23.3, the solid circle at $x = 1$ shows that the point is on the graph. Because it is continuous to the right of the point, it is also continuous at the point. Thus, the function is continuous for $x \geq 1$.

(b) The function represented by the graph in Fig. 23.4 is not continuous at $x = 1$. The function is defined (by the solid circle point) for $x = 1$. However, a small change from $x = 1$ may result in a change of at least 1.5 in $f(x)$, regardless of how small a change in x is made. The small change condition is not satisfied.

(c) The function represented by the graph in Fig. 23.5 is not continuous for $x = -2$. The open circle shows that the point is not part of the graph, and therefore $f(x)$ is not defined for $x = -2$.

■ Polynomials are continuous for all values x. The quotient of two continuous functions is continuous except at values of x that make the denominator zero. Roots of continuous functions are continuous throughout their domain.

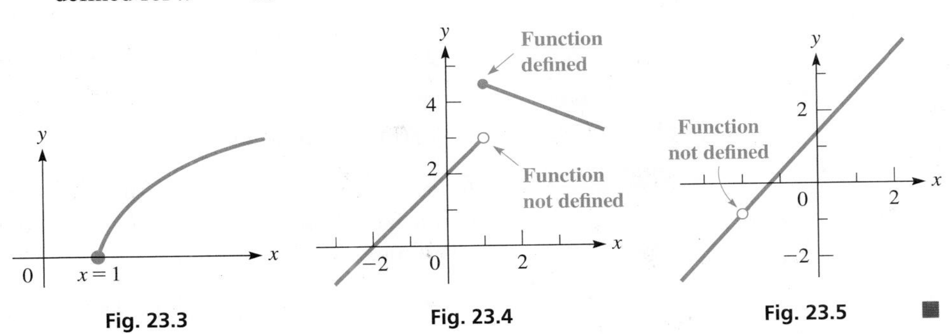

Fig. 23.3 Fig. 23.4 Fig. 23.5 ■

Practice Exercise

1. Determine the continuity of the function $f(x) = \frac{3}{x(x+3)}$.

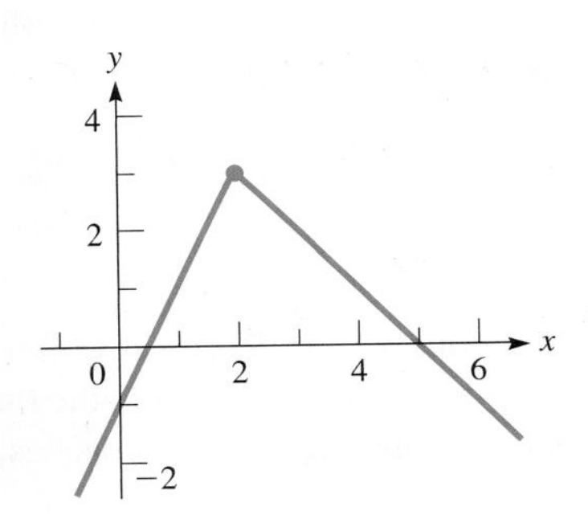

Fig. 23.6

EXAMPLE 5 Function defined differently over domain

(a) We can define the function in Fig. 23.4 as

$$f(x) = \begin{cases} x + 2 & \text{for } x < 1 \\ -\frac{1}{2}x + 5 & \text{for } x \geq 1 \end{cases}$$

where we note that the equation differs for different parts of the domain.

(b) The graph of the function

$$g(x) = \begin{cases} 2x - 1 & \text{for } x \leq 2 \\ -x + 5 & \text{for } x > 2 \end{cases}$$

is shown in Fig. 23.6. We see that it is a continuous function even though the equation for $x \leq 2$ is different from that for $x > 2$. ■

In our earlier discussions of infinite geometric series and the asymptotes of a hyperbola, we used the symbol $\rightarrow$, which means "approaches." **CAUTION** When we say that $x \rightarrow 2$, we mean that x may take on any value as close to 2 as desired, but it also distinctly means that x cannot be set equal to 2. ■

EXAMPLE 6 Behavior of function as x approaches 2

Consider the behavior of $f(x) = 2x + 1$ as $x \rightarrow 2$.

Because we are not to use $x = 2$, we use a calculator to set up tables in order to determine values of $f(x)$, as x gets close to 2:

x	1.000	1.500	1.900	1.990	1.999
$f(x)$	3.000	4.000	4.800	4.980	4.998

values approach 5

x	3.000	2.500	2.100	2.010	2.001
$f(x)$	7.000	6.000	5.200	5.020	5.002

We see that $f(x)$ approaches 5, as x approaches 2, from above 2 and from below 2. ■

LIMIT OF A FUNCTION

In Example 6, because $f(x) \to 5$ as $x \to 2$, the number 5 is called the limit of $f(x)$ as $x \to 2$. This leads to the meaning of the limit of a function. In general, *the* **limit of a function f(x)** *is that value which the function approaches as x approaches the given value a.* This is written as

$$\lim_{x \to a} f(x) = L \qquad (23.1)$$

where L is the value of the limit of the function. **CAUTION** Remember, in approaching a, x may come as arbitrarily close as desired to a, but x may not equal a. ■

An important conclusion can be drawn from the limit in Example 6. The function $f(x)$ is a continuous function, and $f(2)$ equals the value of the limit as $x \to 2$. In general, it is true that

if f(x) is continuous at x = a, then the limit as x → a equals f(a).

In fact, looking back at our definition of continuity, we see that this is what the definition means. That is, a function $f(x)$ is continuous at $x = a$ if *all three* of the following conditions are satisfied:

1. $f(a)$ exists **2.** $\lim_{x \to a} f(x)$ exists **3.** $\lim_{x \to a} f(x) = f(a)$

Although we can evaluate the limit for a continuous function as $x \to a$ by evaluating $f(a)$, it is possible that a function is not continuous at $x = a$ and that the limit exists and can be determined. Thus, we must be able to determine the value of a limit without finding $f(a)$. The following example illustrates the evaluation of such a limit.

EXAMPLE 7 Limit of function as x approaches 2

Find $\lim_{x \to 2} \frac{2x^2 - 3x - 2}{x - 2}$.

We note immediately that the function is not continuous at $x = 2$, for division by zero is indicated. Thus, we cannot evaluate the limit by substituting $x = 2$ into the function. Using a calculator to set up tables, we determine the value that $f(x)$ approaches, as x approaches 2 from either side (see Fig. 23.7):

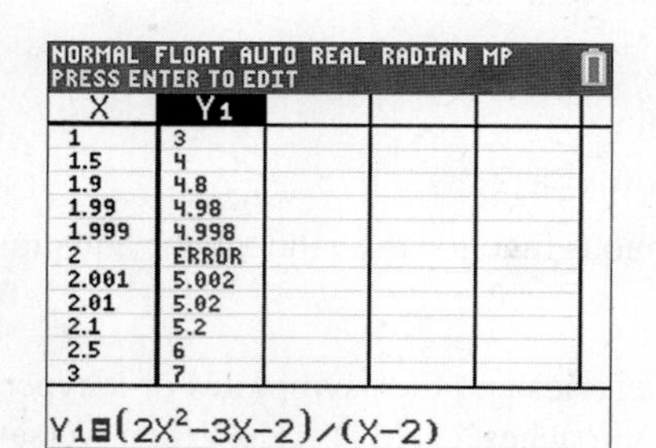

X	Y1
1	3
1.5	4
1.9	4.8
1.99	4.98
1.999	4.998
2	ERROR
2.001	5.002
2.01	5.02
2.1	5.2
2.5	6
3	7

Fig. 23.7
Graphing calculator keystrokes: goo.gl/EjwlNT

x	1.000	1.500	1.900	1.990	1.999
$f(x)$	3.000	4.000	4.800	4.980	4.998

x	3.000	2.500	2.100	2.010	2.001
$f(x)$	7.000	6.000	5.200	5.020	5.002

values approach 5

We see that the values obtained are identical to those in Example 6. Since $f(x) \to 5$ as $x \to 2$, we have

$$\lim_{x \to 2} \frac{2x^2 - 3x - 2}{x - 2} = 5$$

NOTE ▸ [Therefore, we see that *the limit exists* as $x \to 2$ although the function is not defined at $x = 2$.] ■

The reason that the functions in Examples 6 and 7 have the same limit as x approaches 2 is shown in the following example.

EXAMPLE 8 Limits equal—functions differ

The function $\frac{2x^2 - 3 - 2}{x - 2}$ in Example 7 is the same as the function $2x + 1$ in Example 6, except when $x = 2$. By factoring the numerator of the function of Example 7, and then canceling, we have

$$\frac{2x^2 - 3x - 2}{x - 2} = \frac{(2x + 1)(x - 2)}{x - 2} = 2x + 1$$

CAUTION The cancellation in this expression is valid as long as x does not equal 2 since we have division by zero at $x = 2$. ■ Also, in finding the limit as $x \to 2$, we do not use the value $x = 2$. Therefore,

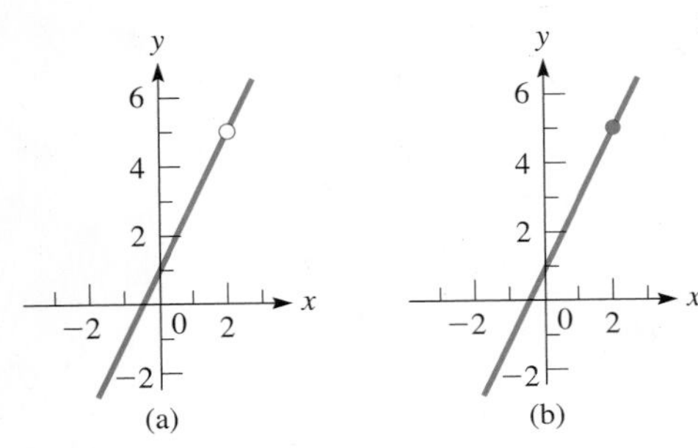

Fig. 23.8

$$\lim_{x \to 2} \frac{2x^2 - 3x - 2}{x - 2} = \lim_{x \to 2}(2x + 1) = 5$$

The limits of the two functions are equal because the limit depends on the behavior of the functions *near* $x = 2$, *not at* $x = 2$. The graphs of the two functions are shown in Fig. 23.8(a) and (b). We can see from the graphs that the limits are the same, although one of the functions is not continuous.

If $f(x) = 5$ for $x = 2$ is added to the definition of the function in Example 7, it is then the same as $2x + 1$, and its graph is that in Fig. 23.8(b). ■

The limit of the function in Example 7 was determined by calculating values near $x = 2$ and by means of an algebraic change in the function. This illustrates that limits may be found through the meaning and definition and through other procedures when the function is not continuous. The following example illustrates a function for which the limit does not exist as x approaches the indicated value.

EXAMPLE 9 Limit does not exist

In trying to find

$$\lim_{x \to 2} \frac{1}{x - 2}$$

we note that $f(x)$ is not defined for $x = 2$, since we would have division by zero. Therefore, we set up the following table to see how $f(x)$ behaves as $x \to 2$:

x	3	2.5	2.1	2.01	2.001
$f(x)$	1	2	10	100	1000

$f(x) \to +\infty$

x	1	1.5	1.9	1.99	1.999
$f(x)$	−1	−2	−10	−100	−1000

$f(x) \to -\infty$

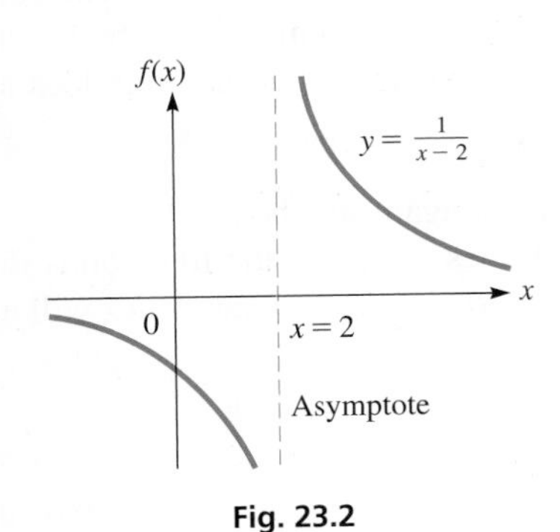

Fig. 23.2

We see that $f(x)$ gets larger as $x \to 2$ from above 2 and $f(x)$ gets smaller (large negative values) as $x \to 2$ from below 2. This may be written as $f(x) \to \infty$ as $x \to 2^+$ and $f(x) \to -\infty$ as $x \to 2^-$, but we must remember that ∞ is not a real number. Therefore, the limit as $x \to 2$ does not exist. The graph of this function is shown in Fig. 23.2, which is shown again for reference. ■

Practice Exercise

2. For $f(x) = \frac{2x^2 - 32}{x - 4}$, find $f(4)$ and $\lim_{x \to 4} f(x)$ if they exist.

The following examples further illustrate the evaluation of limits.

EXAMPLE 10 Limit for a continuous function

Find $\lim_{x \to 4} (x^2 - 7)$.

Because the function $x^2 - 7$ is continuous at $x = 4$, we may evaluate the limit by substituting $x = 4$ in the function. This method is called **direct substitution.** For $f(x) = x^2 - 7$, we have in this case $f(4) = 4^2 - 7 = 9$. This means that

$$\lim_{x \to 4} (x^2 - 7) = 9$$

■

EXAMPLE 11 Limit for a discontinuous function

Find $\lim_{t \to 2} \left(\frac{t^2 - 4}{t - 2} \right)$.

Because

$$\frac{t^2 - 4}{t - 2} = \frac{(t - 2)(t + 2)}{t - 2} = t + 2$$

is valid as long as $t \neq 2$, we find that

$$\lim_{t \to 2} \left(\frac{t^2 - 4}{t - 2} \right) = \lim_{t \to 2} (t + 2) = 4$$

Again, we do not have to be concerned with the fact that the cancellation is not valid for $t = 2$. The limit as $t \to 2$ is not affected by the behavior of the function at $t = 2$.

■

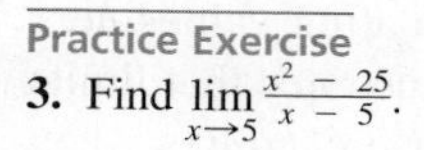

Practice Exercise

3. Find $\lim_{x \to 5} \frac{x^2 - 25}{x - 5}$.

EXAMPLE 12 Value of function exists—limit does not

Find $\lim_{x \to 1} f(x)$ if $f(x) = \begin{cases} x + 2 & \text{for } x < 1 \\ -\frac{1}{2}x + 5 & \text{for } x \geq 1 \end{cases}$

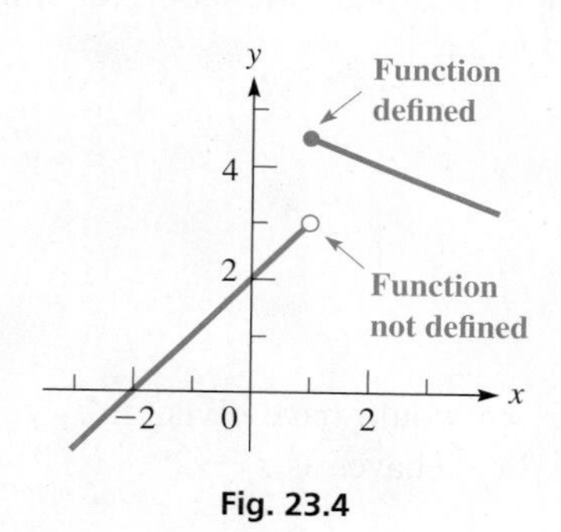

Fig. 23.4

The graph of this function is shown in Fig. 23.4. Although $f(1) = 4.5$, the limit as $x \to 1$ does not exist because $f(x)$ does not approach the same value as x approaches 1 from the left and right. (It approaches 3 from the left and 4.5 from the right.) The point of this example is that even if $f(a)$ exists, we cannot evaluate the limit simply by finding $f(a)$, unless $f(x)$ is continuous at $x = a$. In this case, $f(a)$ exists, but the limit does not exist.

■

Returning briefly to the discussion of continuity, we again see the need for all three conditions of continuity given on page 658. If $f(a)$ does not exist, the function is discontinuous at $x = a$, and if $f(a)$ does not equal $\lim_{x \to a} f(x)$, a small change in x will not result in a small change in $f(x)$.

LIMITS AS x APPROACHES INFINITY

Limits as x approaches infinity are also of importance, but we must be careful. **CAUTION** When dealing with these limits, we must remember that ∞ does not represent a real number and that algebraic operations may not be performed on it. ■

Therefore, when we write $x \to \infty$, we know that we are to consider values of x that are becoming larger and larger without bound. We encountered this concept in Chapter 19 when discussing infinite geometric series, and in Chapter 21 when discussing the asymptotes of a hyperbola. The following examples illustrate the evaluation of this type of limit for algebraic expressions.

EXAMPLE 13 Limit as value approaches infinity—engine efficiency

The efficiency of an engine is given by $E = 1 - Q_2/Q_1$, where Q_1 is the heat taken in and Q_2 is the heat ejected by the engine. ($Q_1 - Q_2$ is the work done by the engine.) If, in an engine cycle, $Q_2 = 500$ kJ, find E as Q_1 becomes large without bound.

We are to find

$$\lim_{Q_1 \to \infty}\left(1 - \frac{500}{Q_1}\right)$$

As Q_1 becomes larger and larger, $500/Q_1$ becomes smaller and smaller and approaches zero. This means $f(Q_1) \to 1$ as $Q_1 \to \infty$. Thus,

$$\lim_{Q_1 \to \infty}\left(1 - \frac{500}{Q_1}\right) = 1$$

We can verify our reasoning and the value of the limit by making a table of values for Q_1 and E as Q_1 becomes large:

Q_1	500	5000	50,000	500,000
E	0	0.9	0.99	0.999

values approach 1

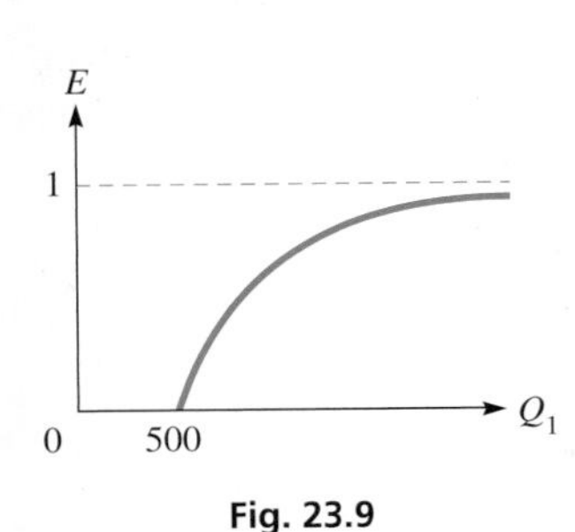

Fig. 23.9

Again, we see that $E \to 1$ as $Q_1 \to \infty$. See Fig. 23.9.

This is primarily a theoretical consideration, as there are obvious practical limitations as to how much heat can be supplied to an engine. An engine for which $E = 1$ would operate at 100% efficiency. ■

EXAMPLE 14 Limit as x approaches infinity

Find $\lim_{x \to \infty} \frac{x^2 + 1}{2x^2 + 3}$.

We note that as $x \to \infty$, both the numerator and the denominator become large without bound. Therefore, we use a calculator (see Fig. 23.10) to make a table to see how $f(x)$ behaves as x becomes very large:

x	1	10	100	1000
$f(x)$	0.4	0.4975369458	0.4999750037	0.49999975

values approach 0.5

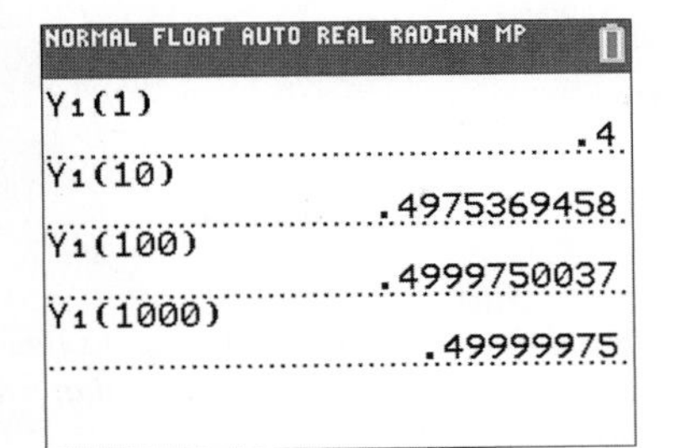

Fig. 23.10
Graphing calculator keystrokes: goo.gl/pEIYle

From this table, we see that $f(x) \to 0.5$ as $x \to \infty$.

NOTE ▸ This limit can also be found algebraically. [The key step is to divide both the numerator and the denominator of the function by x^2, which is the largest power of x that appears in either the numerator or the denominator.] By doing this, we have

$$\frac{x^2 + 1}{2x^2 + 3} = \frac{1 + \frac{1}{x^2}}{2 + \frac{3}{x^2}}$$

terms $\to 0$ as $x \to \infty$

Here, we see that $1/x^2$ and $3/x^2$ both approach zero as $x \to \infty$. This means that the numerator approaches 1 and the denominator approaches 2. Therefore,

$$\lim_{x \to \infty}\frac{x^2 + 1}{2x^2 + 3} = \lim_{x \to \infty}\frac{1 + \frac{1}{x^2}}{2 + \frac{3}{x^2}} = \frac{1}{2}$$

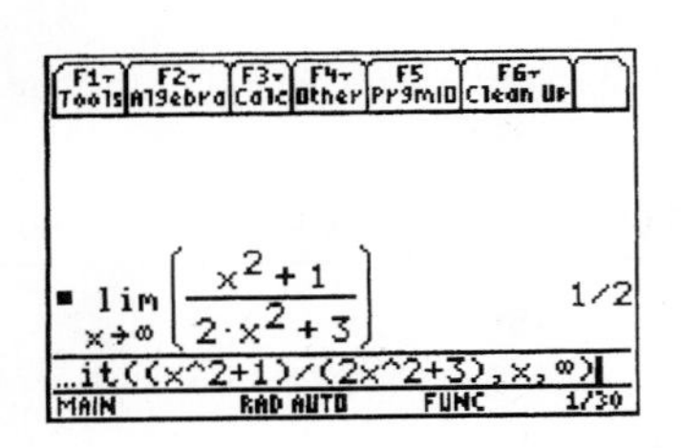

Fig. 23.11
TI-89 graphing calculator keystrokes: goo.gl/2XC42Y

Figure 23.11 shows a T1-89 graphing calculator screen on which the limit in this example is evaluated. ■

Practice Exercise

4. Find $\lim_{x \to \infty} \frac{x + 2}{2x - 7}$.

■ Calculus was not on a sound mathematical basis until limits were properly developed by the French mathematician Augustin-Louis Cauchy (1789–1857) and others in the mid-1800s.

The definitions and development of continuity and of a limit presented in this section are not mathematically rigorous. However, the development is consistent with a more rigorous development, and the *concept* of a limit is the principal concern.

EXERCISES 23.1

In Exercises 1–4, make the given changes in the indicated examples of this section. Then solve the resulting problems.

1. In Example 2, change the denominator to $x + 2$ and then determine the continuity.
2. In Example 8, change the numerator to $3x^2 - 5x - 2$ and find the resulting limit. Disregard references to Examples 6 and 7.
3. In Example 11, change the denominator to $t + 2$ and then find the limit as $t \to -2$.
4. In Example 14, change the numerator to $4x^2 + 1$ and find the resulting limit.

In Exercises 5–10, determine the values of x for which the function is continuous. If the function is not continuous, determine the reason.

5. $f(x) = 3x^2 - 98x$
6. $f(x) = \dfrac{3 - x}{9 + x^2}$
7. $f(x) = \dfrac{x + 4}{x^2 - x}$
8. $f(x) = \dfrac{2}{\sqrt{x + 3}}$
9. $f(x) = \dfrac{x}{\sqrt{x - 2}}$
10. $f(x) = \dfrac{3\sqrt{x + 5}}{x + 8}$

In Exercises 11–16, determine the values of x for which the function, as represented by the graphs in Fig. 23.12, is continuous. If the function is not continuous, determine the reason.

11.

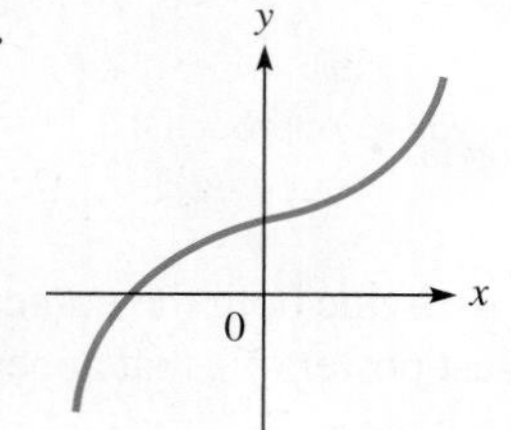

12.

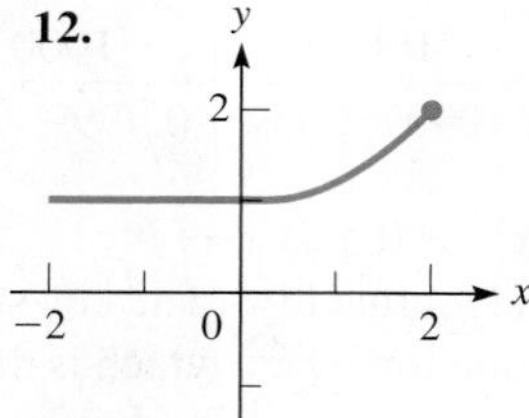

13.

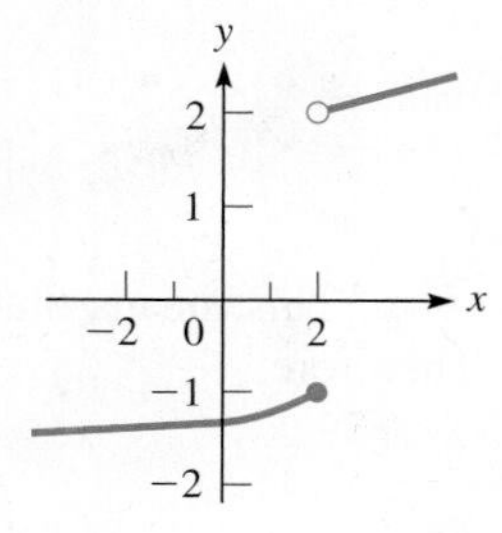

14.

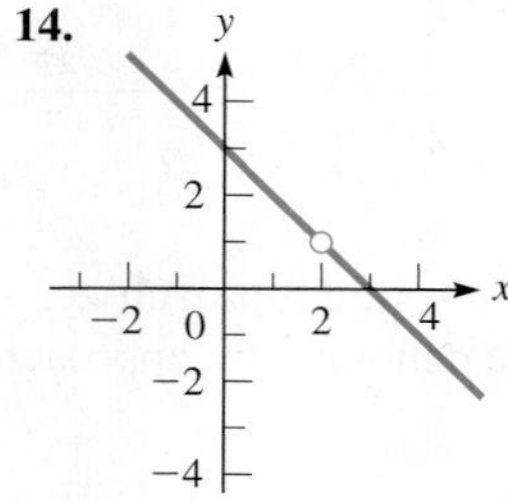

15.

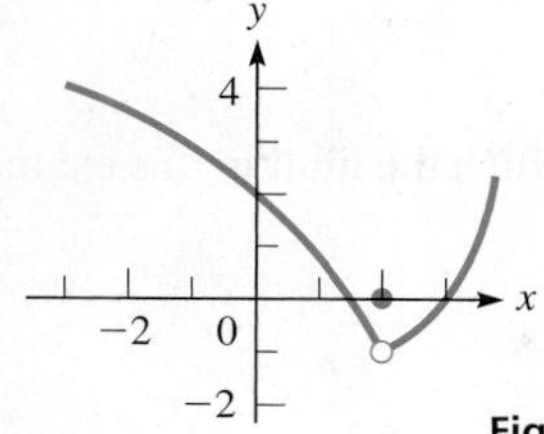

16.

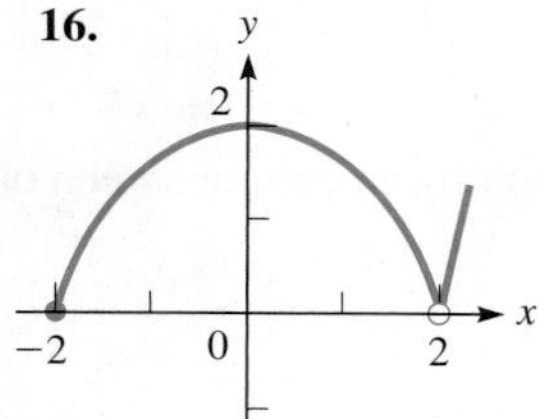

Fig. 23.12

In Exercises 17–20, for the function shown in the graph for the indicated exercise, find (a) $f(2)$, and (b) $\lim_{x\to 2} f(x)$, if they exist.

17. Exercise 13
18. Exercise 14
19. Exercise 15
20. Exercise 16

In Exercises 21–24, graph the function and determine the values of x for which the functions are continuous. Explain.

21. $f(x) = \begin{cases} x^2 & \text{for } x < 2 \\ 5 & \text{for } x \geq 2 \end{cases}$

22. $f(x) = \begin{cases} \dfrac{x^3 - x^2}{x - 1} & \text{for } x \neq 1 \\ 1 & \text{for } x = 1 \end{cases}$

23. $f(x) = \begin{cases} \dfrac{2x^2 - 18}{x - 3} & \text{for } x \neq 3 \\ 12 & \text{for } x = 3 \end{cases}$

24. $f(x) = \begin{cases} \dfrac{x + 2}{x^2 - 4} & \text{for } x < -2 \\ \dfrac{x}{8} & \text{for } x > -2 \end{cases}$

In Exercises 25–30, evaluate the indicated limits by evaluating the function for values shown in the table and observing the values that are obtained. Do not change the form of the function.

25. Find $\lim_{x\to 1} \dfrac{x^3 - x}{x - 1}$.

x	0.900	0.990	0.999	1.001	1.010	1.100
$f(x)$						

26. Find $\lim_{x\to -3} \dfrac{x^3 + 2x^2 - 2x + 3}{x + 3}$.

x	−3.100	−3.010	−3.001	−2.999	−2.990	−2.900
$f(x)$						

27. Find $\lim_{x\to 2} \dfrac{2 - \sqrt{x + 2}}{x - 2}$.

x	1.900	1.990	1.999	2.001	2.010	2.100
$f(x)$						

28. Find $\lim_{x\to 0} \dfrac{e^x - 1}{x}$.

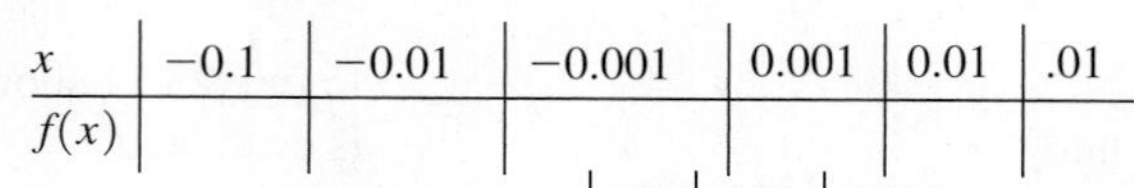

x	−0.1	−0.01	−0.001	0.001	0.01	.01
$f(x)$						

29. Find $\lim_{x\to\infty} \dfrac{2x + 1}{5x - 3}$.

x	10	100	1000
$f(x)$			

30. Find $\lim_{x\to\infty}\frac{1-x^2}{8x^2+5}$.

x	10	100	1000
$f(x)$			

In Exercises 31–50, evaluate the indicated limits algebraically as in Examples 10–14. Change the form of the function where necessary.

31. $\lim_{x\to3}(3x-2)$ **32.** $\lim_{x\to4}\sqrt{x^2-7}$ **33.** $\lim_{x\to0}\frac{6x^2+x}{x}$

34. $\lim_{v\to2}\frac{4v^2-8v}{v-2}$ **35.** $\lim_{x\to-1}\frac{x^2-1}{3x+3}$ **36.** $\lim_{x\to3}\frac{x^2-2x-3}{3-x}$

37. $\lim_{h\to3}\frac{h^3-27}{h-3}$ **38.** $\lim_{x\to1/3}\frac{9x-3}{3x^2+5x-2}$

39. $\lim_{x\to1}\frac{(2x-1)^2-1}{2x-2}$ **40.** $\lim_{x\to4}\frac{|x-4|}{x-4}$

41. $\lim_{p\to-1}\sqrt{p(p+1.3)}$ **42.** $\lim_{x\to1}(x-1)\sqrt{x^2-4}$

43. $\lim_{x\to1}\frac{\sqrt{x}-1}{x-1}$ **44.** $\lim_{x\to8}\frac{x-8}{\sqrt[3]{x}-2}$

45. $\lim_{h\to0}\frac{\sqrt{9+h}-3}{h}$ **46.** $\lim_{h\to0}\frac{(4+h)^2-16}{h}$

47. $\lim_{x\to\infty}\frac{3x^2+4.5}{x^2-1.5}$ **48.** $\lim_{x\to\infty}\frac{x-27}{7x+4}$

49. $\lim_{t\to\infty}\frac{\sqrt{t^2+16}}{t+1}$ **50.** $\lim_{x\to\infty}\frac{1-2x^2}{(4x+3)^2}$

In Exercises 51 and 52, evaluate the function at 0.1, 0.01, and 0.001 from both sides of the value it approaches. In Exercises 53 and 54, evaluate the function for values of x of 10, 100, and 1000. From these values, determine the limit. Then, by using an appropriate change of form, evaluate the limit algebraically and compare values.

51. $\lim_{x\to0}\frac{x^2-3x}{x}$ **52.** $\lim_{x\to3}\frac{2x^2-6x}{x-3}$

53. $\lim_{x\to\infty}\frac{2x^2+x}{x^2-3}$ **54.** $\lim_{x\to\infty}\frac{x^2+5}{\sqrt{64x^4+1}}$

In Exercises 55–60, solve the given problems involving limits.

55. Evaluate $\lim_{x\to2}\frac{x^3-8}{x-2}$ by first performing long division (or synthetic division) on $\frac{x^3-8}{x-2}$.

56. Draw the graph of a function that is discontinuous at $x = 2$, has a limit of 2 as $x\to2$, and has a value of 3 at $x = 2$.

57. A certain object, after being heated, cools at such a rate that its temperature T (in °C) decreases 10% each minute. If the object is originally heated to 100°C, find $\lim_{t\to10}T$ and $\lim_{t\to\infty}T$, where t is the time (in min).

58. The area A (in mm^2) of the pupil of a certain person's eye is given by $A=\frac{36+24b^3}{1+4b^3}$, where b is the brightness (in lumens) of the light source. Between what values does A vary if $b\geq0$?

59. Velocity can be found by dividing the displacement s of an object by the elapsed time t in moving through the displacement. In a certain experiment, the following values were measured for the displacements and elapsed times for the motion of an object. Determine the limiting value of the velocity as $t\to0$.

s (cm)	0.480000	0.280000	0.029800	0.0029980	0.00029998
t (s)	0.200000	0.100000	0.010000	0.0010000	0.00010000

60. A 5-Ω resistor and a variable resistor of resistance R are placed in parallel. The expression for the resulting resistance R_T is given by $R_T=\frac{5R}{5+R}$. Determine the limiting value of R_T as $R\to\infty$.

In Exercises 61–64, use a calculator to evaluate the indicated limits.

61. Approximate $\lim_{x\to2}\frac{2^x-4}{x-2}$ **62.** Approximate $\lim_{x\to1}\frac{4^x-4}{x-1}$

63. $\lim_{x\to0}(1+x)^{1/x}$ (Do you recognize the limiting value?)

64. $\lim_{x\to0}\frac{\sin x}{x}$ (Use radian mode.)

In Exercises 65–72, $\lim_{x\to a^-}f(x)$ means to find the limit as x approaches a from the left only, and $\lim_{x\to a^+}f(x)$ means to find the limit as x approaches a from the right only. These are called one-sided limits. *Solve the following problems.*

65. For the function displayed in Exercise 13, find:

(a) $\lim_{x\to2^-}f(x)$ (b) $\lim_{x\to2^+}f(x)$ (c) $\lim_{x\to2}f(x)$

66. For the function displayed in Exercise 16, find:

(a) $\lim_{x\to-2^-}f(x)$ (b) $\lim_{x\to-2^+}f(x)$ (c) $\lim_{x\to-2}f(x)$

67. Find $\lim_{x\to4^-}x\sqrt{16-x^2}$.

68. Explain why $\lim_{x\to0^+}2^{1/x}\neq\lim_{x\to0^-}2^{1/x}$.

69. For $f(x)=\frac{x}{|x|}$, find $\lim_{x\to0^-}f(x)$ and $\lim_{x\to0^+}f(x)$. Is $f(x)$ continuous at $x = 0$? Explain.

70. In Einstein's theory of relativity, the length L of an object moving at a velocity v is $L=L_0\sqrt{1-\frac{v^2}{c^2}}$, where c is the speed of light and L_0 is the length of the object at rest. Find $\lim_{v\to c^-}L$ and explain why a limit from the left is used.

71. Is there a difference between

$\lim_{x\to2^-}\frac{1}{x-2}$ and $\lim_{x\to2^+}\frac{1}{x-2}$?

72. Is there a difference between

$\lim_{x\to2^-}\frac{1}{\sqrt{x-2}}$ and $\lim_{x\to2^+}\frac{1}{\sqrt{x-2}}$?

Answers to Practice Exercises

1. Discontinuous at $x = -3$ and $x = 0$ **2.** Undefined, 16
3. 10 **4.** 0.5

23.2 The Slope of a Tangent to a Curve

Slope of Secant Line • $h = x_2 - x_1$ • Slope of Tangent Line

Having developed the basic operations with functions and the concept of a limit, we now turn our attention to a graphical interpretation of the rate of change of a function. This interpretation, basic to an understanding of calculus, deals with the slope of a line tangent to the curve of a function.

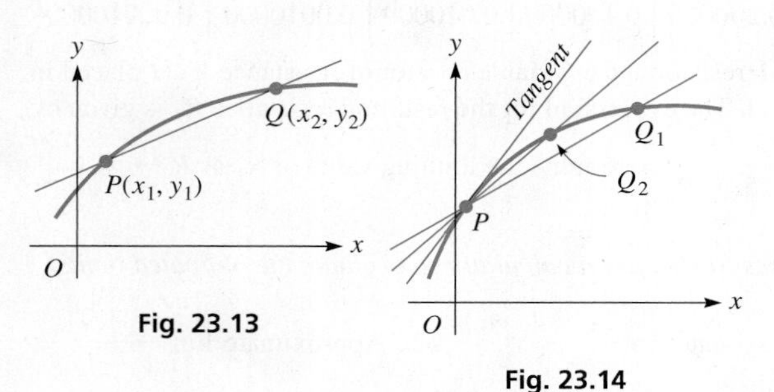

Fig. 23.13

Fig. 23.14

Consider the points $P(x_1, y_1)$ and $Q(x_2, y_2)$ in Fig. 23.13. From Chapter 21, we know that the slope of the line through these points is given by

$$m = \frac{y_2 - y_1}{x_2 - x_1}$$

This, however, represents the slope of the line through P and Q (a *secant line*) and no other line. If we now allow Q to be a point closer to P, the slope of PQ will more closely approximate the slope of a line drawn tangent to the curve at P (see Fig. 23.14). In fact, the closer Q is to P, the better this approximation becomes. It is not possible to allow Q to coincide with P, for then it would not be possible to define the slope of PQ in terms of two points.

NOTE ➧ [The slope of the tangent line, often referred to as the slope of the curve, is the *limiting value* of the slope of the secant line PQ as Q approaches P.]

EXAMPLE 1 Limit of slopes of secant lines

Find the slope of a line tangent to the curve $y = x^2 + 3x$ at the point $P(2, 10)$ by finding the limit of the slopes of the secant lines PQ as Q approaches P.

Let point Q have the x-values of 3.0, 2.5, 2.1, 2.01, and 2.001. Then, using a calculator, we tabulate the necessary values. Because P is the point $(2, 10)$, $x_1 = 2$ and $y_1 = 10$. Thus, using the values of x_2, we tabulate the values of y_2, $y_2 - 10$, $x_2 - 2$, and thereby the values of the slope m:

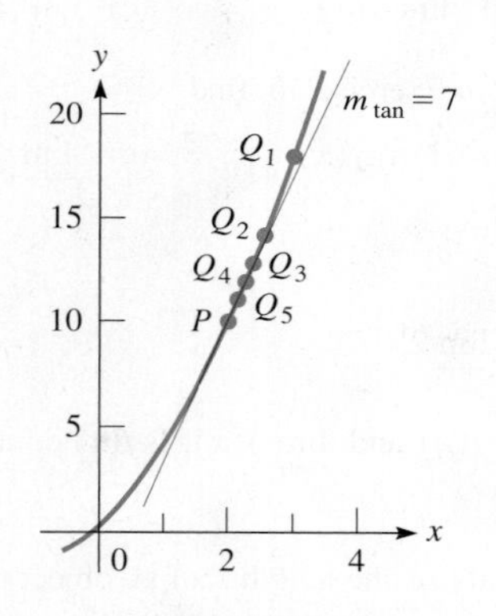

Fig. 23.15

Point	Q_1	Q_2	Q_3	Q_4	Q_5	P
x_2	3.0	2.5	2.1	2.01	2.001	2
y_2	18.0	13.75	10.71	10.0701	10.007001	10
$y_2 - 10$	8.0	3.75	0.71	0.0701	0.007001	
$x_2 - 2$	1.0	0.5	0.1	0.01	0.001	
$m = \frac{y_2 - 10}{x_2 - 2}$	8.0	7.5	7.1	7.01	7.001	

We see that the slope of PQ approaches the value of 7 as Q approaches P. Therefore, the slope of the tangent line at $(2, 10)$ is 7. See Fig. 23.15. ■

With the proper notation, it is possible to express the coordinates of Q in terms of the coordinates of P. By defining h as $x_2 - x_1$, we have

$$h = x_2 - x_1 \qquad \textbf{(23.2)}$$

$$x_2 = x_1 + h \qquad \textbf{(23.3)}$$

For the function $y = f(x)$, the point $P(x_1, y_1)$ can be written as $P(x_1, f(x_1))$ and the point $Q(x_2, y_2)$ can be written as $Q(x_1 + h, f(x_1 + h))$.

Using Eq. (23.3), along with the definition of slope, we can express the slope of PQ as

$$m_{PQ} = \frac{f(x_1 + h) - f(x_1)}{(x_1 + h) - x_1} = \frac{f(x_1 + h) - f(x_1)}{h} \tag{23.4}$$

By our previous discussion, as Q approaches P, the slope of the tangent line is more nearly *approximated* by Eq. (23.4).

EXAMPLE 2 Slope of tangent line at specific point

Find the slope of a line tangent to the curve of $y = x^2 + 3x$ at the point $(2, 10)$. (This is the same slope as calculated in Example 1.)

As in Example 1, point P has the coordinates $(2, 10)$. Thus, the coordinates of any other point Q on the curve can be expressed as $(2 + h, f(2 + h))$. See Fig. 23.16. The slope of PQ then becomes

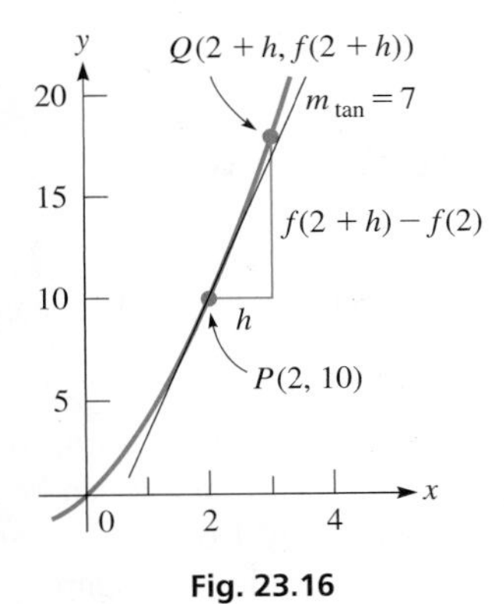

Fig. 23.16

$$\begin{aligned} m_{PQ} &= \frac{f(2 + h) - f(2)}{h} = \frac{[(2 + h)^2 + 3(2 + h)] - [2^2 + 3(2)]}{h} \\ &= \frac{(4 + 4h + h^2 + 6 + 3h) - (4 + 6)}{h} \\ &= \frac{7h + h^2}{h} = 7 + h \end{aligned}$$

From this expression, we can see that $m_{PQ} \rightarrow 7$ as $h \rightarrow 0$. Therefore, we can see that the slope of the tangent line is

$$m_{\tan} = \lim_{h \to 0} m_{PQ} = 7$$

We see that this result agrees with that found in Example 1. ■

Practice Exercise

1. Find the slope of a line tangent to the curve of $y = 4x - x^2$ at the point $(-1, -5)$.

EXAMPLE 3 Slope of tangent line at general point

Find the slope of a line tangent to the curve of $y = 4x - x^2$ at the point (x_1, y_1).

The points P and Q are $P(x_1, f(x_1))$ and $Q(x_1 + h, f(x_1 + h))$. Therefore,

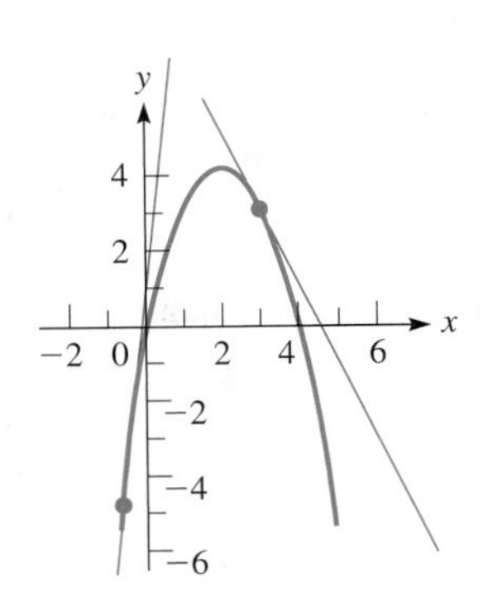

Fig. 23.17

$$\begin{aligned} m_{PQ} &= \frac{f(x_1 + h) - f(x_1)}{h} = \frac{[4(x_1 + h) - (x_1 + h)^2] - (4x_1 - x_1^2)}{h} \\ &= \frac{(4x_1 + 4h - x_1^2 - 2hx_1 - h^2) - (4x_1 - x_1^2)}{h} \\ &= \frac{4h - 2hx_1 - h^2}{h} = 4 - 2x_1 - h \end{aligned}$$

Here, we see that $m_{PQ} \rightarrow 4 - 2x_1$ as $h \rightarrow 0$. Therefore,

$$m_{\tan} = 4 - 2x_1$$

This method has an advantage over that used in Example 2. We now have a *general expression for the slope of a tangent line* for any value x_1. If $x_1 = -1$, $m_{\tan} = 6$ and if $x_1 = 3$, $m_{\tan} = -2$. These tangent lines are shown in Fig. 23.17. ■

EXAMPLE 4 Evaluating the slope of a tangent line

Find the expression for the slope of a line tangent to the curve of $y = x^3 + 2$ at the general point (x_1, y_1), and use this expression to find the slope when $x = 1/2$.

For $y = f(x)$, using points $P(x_1, f(x_1))$ and $Q(x_1 + h, f(x_1 + h))$, we have the following steps:

$$m_{PQ} = \frac{f(x_1 + h) - f(x_1)}{h} = \frac{[(x_1 + h)^3 + 2] - (x_1^3 + 2)}{h} \quad \text{using Eq. (23.4)}$$

$$= \frac{(x_1^3 + 3x_1^2h + 3x_1h^2 + h^3 + 2) - (x_1^3 + 2)}{h} = \frac{3x_1^2h + 3x_1h^2 + h^3}{h}$$

$$= 3x_1^2 + 3x_1h + h^2$$

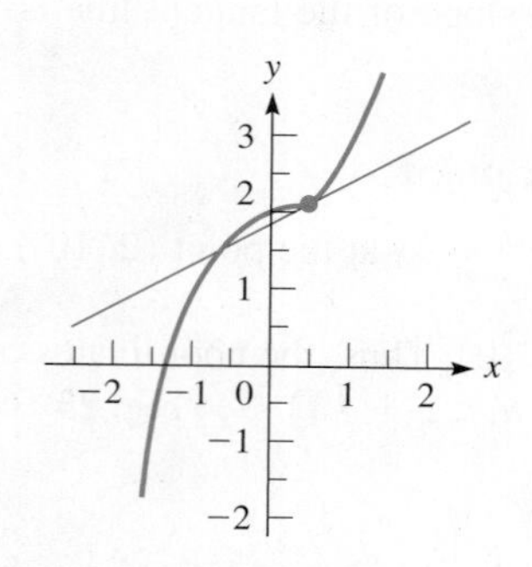

Fig. 23.18

As $h \to 0$, the expression above approaches the value $3x_1^2$. This means that

$$m_{\tan} = 3x_1^2$$

When $x_1 = 1/2$, we find that the slope of the tangent line is $3(1/4) = 3/4$. The curve and this tangent line are shown in Fig. 23.18. ■

Practice Exercise

2. Find the expression for the slope of a line tangent to the curve of $y = 4x^2$ at (x_1, y_1).

In interpreting this slope of a tangent line as the rate of change of a function, we see that if $h = 1$ and the corresponding value of $f(x + h) - f(x)$ is found, it can be said that $f(x)$ changes by the amount $f(x + h) - f(x)$ as x changes by 1 unit. If x changes by a lesser amount (h is less than 1), we can still calculate the ratio of the amount of change in $f(x)$ for the given value of h. Therefore, as long as x changes at all, there will be a corresponding change in $f(x)$. [This means that the ratio of the change in $f(x)$ to the change in x is the **average rate of change of $f(x)$ with respect to x**.]

NOTE ▶

Instantaneous Rate of Change

As $h \to 0$, the limit of the ratio of the change in $f(x)$ to the change in x is the instantaneous rate of change of $f(x)$ with respect to x.

EXAMPLE 5 Slope as instantaneous rate of change

In Example 1, consider points $P(2, 10)$ and $Q_2(2.5, 13.75)$. From P to Q_2, x changes by 0.5 unit and $f(x)$ changes by 3.75 units. This means *the average change in $f(x)$ for a 1-unit change in x is* $3.75/0.5 = 7.5$ units. However, this is not the rate at which $f(x)$ is changing with respect to x ***at*** most points within this interval.

At point P, the slope of 7 of the tangent line tells us that $f(x)$ is changing 7 units for a 1-unit change in x. However, ***this is an instantaneous rate of change at point P*** and tells the rate at which $f(x)$ is changing with respect to x at P. ■

EXERCISES 23.2

In Exercises 1 and 2, make the given changes in the indicated examples of this section and then find the indicated slopes.

1. In Example 2, change the point $(2, 10)$ to $(3, 18)$.
2. In Example 3, change the $4x$ to $3x$.

In Exercises 3–6, use the method of Example 1 to calculate the slope of the line tangent to the curve of each of the given functions. Let Q_1, Q_2, Q_3, and Q_4 have the indicated x-values. Sketch the curve and tangent lines.

3. $y = x^2$; P is $(2, 4)$; let Q have x-values of 1.5, 1.9, 1.99, 1.999.
4. $y = 1 - \frac{1}{2}x^2$; P is $(2, -1)$; let Q have x-values of 1.5, 1.9, 1.99, 1.999.
5. $y = 2x^2 + 5x$; P is $(-2, -2)$; let Q have x-values of -1.5, -1.9, -1.99, -1.999.
6. $y = x^3 + 1$; P is $(-1, 0)$; let Q have x-values of -0.5, -0.9, -0.99, -0.999.

In Exercises 7–10, use the method of Example 2 to calculate the slope of a line tangent to the curve of each of the functions $y = f(x)$ for the given point P. (These are the same functions and points as in Exercises 3–6.)

7. $y = x^2$; $P(2, 4)$
8. $y = 1 - \frac{1}{2}x^2$; $P(2, -1)$

9. $y = 2x^2 + 5x$; P is $(-2, -2)$.

10. $y = x^3 + 1$; P is $(-1, 0)$.

In Exercises 11–22, use the method of Example 3 to find a general expression for the slope of a tangent line to each of the indicated curves. Then find the slopes for the given values of x. Sketch the curves and tangent lines.

11. $y = x^2$; $x = -1$; $x = 2$

12. $y = 1 - \frac{1}{2}x^2$; $x = -3, x = 3$

13. $y = 2x^2 + 5x$; $x = -2, x = 0.5$

14. $y = 4.5 - 3x^2$; $x = -1, x = 0$

15. $y = x^2 + 3x + 2\pi$; $x = -3, x = 3$

16. $y = 2x^2 - 4x$; $x = 0.5, x = 1.5$

17. $y = 6x - x^2$; $x = -1, x = 1, x = 3$

18. $y = \frac{1}{3}x^3 - 5x$; $x = -3, x = 0, x = 3$

19. $y = 1.5x^4$; $x = -0.5, x = 0, x = 1$

20. $y = 4 - \frac{1}{8}x^4$; $x = -2, x = -1, x = 1, x = 2$

21. $y = x^5$; $x = -1, x = -0.5, x = 0.5, x = 1$

22. $y = \frac{1}{x}$; $x = -2, x = -1, x = 1, x = 2$

In Exercises 23–26, find the average rate of change of y with respect to x from P to Q. Then compare this with the instantaneous rate of change of y with respect to x at P by finding m_{tan} at P.

23. $y = x^2 + 2$; $P(2, 6)$, $Q(2.1, 6.41)$

24. $y = 1 - 2x^2$; $P(1, -1)$, $Q(1.1, -1.42)$

25. $y = 9 - x^3$; $P(2, 1)$, $Q(2.1, -0.261)$

26. $y = x^3 - 6x$; $P(3, 9)$, $Q(3.1, 11.191)$

In Exercises 27–30, find the point(s) where the slope of a tangent line to the given curve has the given value. In Exercises 31–34, solve the given problems.

27. $y = 2x^2$, $m_{tan} = -4$

28. $y = 8x - x^2$, $m_{tan} = 6$

29. $y = 12x - \frac{1}{3}x^3$; $m_{tan} = -4$

30. $y = x^3 + 3x^2$, $m_{tan} = 9$

31. Find the slope of a line perpendicular to the tangent of the curve of $y = 8 - 3x^2$ where $x = -1$.

32. Find the slopes of the tangent lines to the curve of $y = \frac{1}{3}x^3 + x$ at points where $x = 1$ and $x = 2$. Then find the acute angle between these lines at the point where they cross.

33. In a computer game, at one point an airplane is diving along the curve of $y = -2x^2 + 10$. What is the angle of the dive (with the vertical) when $x = 1.5$?

34. At an amusement park, a waterslide follows the curve of $y = 8/(x + 1)$, for $x = 0$ m to $x = 7$ m. What is the angle (with the horizontal) of the slide when $x = 4.0$ m?

Answers to Practice Exercises

1. 6 2. $m_{tan} = 8x_1$

23.3 The Derivative

Definition of Derivative • Differentiation • Procedure for Finding a Derivative • Calculator Evaluation of Derivative

In the preceding section, we found that we could find the slope of a line tangent to a curve at a point $(x_1, f(x_1))$ by calculating the limit (if it exists) of the difference $f(x_1 + h) - f(x_1)$ divided by h as $h \to 0$. We can write this as

$$m_{tan} = \lim_{h \to 0} \frac{f(x_1 + h) - f(x_1)}{h} \tag{23.5}$$

The limit on the right is defined as *the* **derivative** *of $f(x)$ at x_1*. This is one of the fundamental definitions of calculus.

Considering the limit at each point in the domain of $f(x)$, by letting $x_1 = x$ in Eq. (23.5), we have *the* **derivative** *of the function $f(x)$*:

Definition of the Derivative

$$f'(x) = \lim_{h \to 0} \frac{f(x + h) - f(x)}{h} \tag{23.6}$$

The process of finding a derivative is called **differentiation.**

A four-step procedure for finding the derivative of a function by use of the definition is outlined on the following page.

Procedure for Finding the Derivative of a Function

1. Find $f(x + h)$.
2. Subtract $f(x)$ from $f(x + h)$.
3. Divide the result of step 2 by h.
4. For the result of step 3, find the limit (if it exists) as $h \to 0$.

EXAMPLE 1 Using the definition to find a derivative

Find the derivative of $y = 2x^2 + 3x$ by using the definition.

With $y = f(x)$, using the above procedure to find the derivative $f'(x)$, we have the following:

$$f(x + h) = 2(x + h)^2 + 3(x + h) \qquad \text{step 1}$$

$$\begin{aligned} f(x + h) - f(x) &= 2(x + h)^2 + 3(x + h) - (2x^2 + 3x) \qquad \text{step 2} \\ &= 2x^2 + 4xh + 2h^2 + 3x + 3h - 2x^2 - 3x \\ &= 4xh + 3h + 2h^2 \end{aligned}$$

$$\frac{f(x + h) - f(x)}{h} = \frac{4hx + 3h + 2h^2}{h} = 4x + 3 + 2h \qquad \text{step 3}$$

$$\lim_{h \to 0} \frac{f(x + h) - f(x)}{h} = \lim_{h \to 0} (4x + 3 + 2h) = 4x + 3 \qquad \text{step 4}$$

$$f'(x) = 4x + 3$$

Fig. 23.19

Practice Exercise

1. Using the definition, find the derivative of $y = 5x - x^2$.

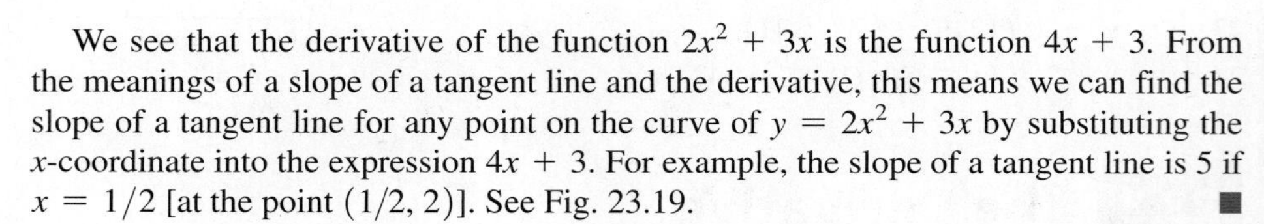

We see that the derivative of the function $2x^2 + 3x$ is the function $4x + 3$. From the meanings of a slope of a tangent line and the derivative, this means we can find the slope of a tangent line for any point on the curve of $y = 2x^2 + 3x$ by substituting the x-coordinate into the expression $4x + 3$. For example, the slope of a tangent line is 5 if $x = 1/2$ [at the point $(1/2, 2)$]. See Fig. 23.19. ■

Because the derivative of a function is itself a function, it is possible that it may not be defined for all values of x. *If the value x_0 is in the domain of the derivative, then the function is said to be differentiable at x_0.* The examples that follow illustrate functions that are not differentiable for all values of x.

EXAMPLE 2 Using the definition—derivative of a fraction

Find the derivative of $y = \dfrac{3}{x + 2}$ by using the definition.

$$f(x + h) = \frac{3}{x + h + 2} \qquad \text{step 1}$$

$$\begin{aligned} f(x + h) - f(x) &= \frac{3}{x + h + 2} - \frac{3}{x + 2} \qquad \text{step 2} \\ &= \frac{3(x + 2) - 3(x + h + 2)}{(x + h + 2)(x + 2)} = \frac{-3h}{(x + h + 2)(x + 2)} \end{aligned}$$

$$\frac{f(x + h) - f(x)}{h} = \frac{-3h}{h(x + h + 2)(x + 2)} = \frac{-3}{(x + h + 2)(x + 2)} \qquad \text{step 3}$$

$$\lim_{h \to 0} \frac{f(x + h) - f(x)}{h} = \lim_{h \to 0} \frac{-3}{(x + h + 2)(x + 2)} = \frac{-3}{(x + 2)^2} \qquad \text{step 4}$$

$$f'(x) = \frac{-3}{(x + 2)^2}$$

Note that neither the function nor the derivative is defined for $x = -2$. This means the function is not differentiable at $x = -2$. ■

In Example 2, it was necessary to combine fractions in the process of finding the derivative. Such algebraic operations must be done with care. **CAUTION** One of the more common sources of errors is the improper handling of fractions. ■

EXAMPLE 3 Derivative—proper handling of fractions

Find the derivative of $y = 4x^3 + \frac{5}{x}$ by using the definition.

$$f(x + h) = 4(x + h)^3 + \frac{5}{x + h}$$

$$f(x + h) - f(x) = 4(x + h)^3 + \frac{5}{x + h} - \left(4x^3 + \frac{5}{x}\right)$$

$$= 4(x^3 + 3x^2h + 3xh^2 + h^3) - 4x^3 + \frac{5}{x + h} - \frac{5}{x}$$

■ The algebra is handled most easily if the fractions are combined separately from the other terms.

$$= 4x^3 + 12x^2h + 12xh^2 + 4h^3 - 4x^3 + \frac{5x - 5(x + h)}{x(x + h)}$$

$$= 12x^2h + 12xh^2 + 4h^3 - \frac{5h}{x(x + h)}$$

$$\frac{f(x + h) - f(x)}{h} = \frac{12x^2h + 12xh^2 + 4h^3}{h} - \frac{5h}{hx(x + h)}$$

$$\lim_{h\to0}\frac{f(x + h) - f(x)}{h} = \lim_{h\to0}\left(12x^2 + 12xh + 4h^2 - \frac{5}{x(x + h)}\right) = 12x^2 - \frac{5}{x^2}$$

$$f'(x) = 12x^2 - \frac{5}{x^2}$$

Practice Exercise

2. Using the definition, find the derivative of $y = 3/x$.

Note that this function is not differentiable for $x = 0$. ■

There are notations other than $f'(x)$ that are used for the derivative. These other notations include y', D_xy, and $\frac{dy}{dx}$.

EXAMPLE 4 Evaluating a derivative

In Example 2, $y = \frac{3}{x + 2}$, and we found that the derivative is $\frac{-3}{(x + 2)^2}$. Therefore, we may write

$$y' = \frac{-3}{(x + 2)^2} \quad \text{or} \quad \frac{dy}{dx} = \frac{-3}{(x + 2)^2}$$

instead of $f'(x) = \frac{-3}{(x + 2)^2}$ as we did in Example 2.

If we wish to find the value of the derivative at some point, such as $(-1, 3)$, we write

$$\frac{dy}{dx} = \frac{-3}{(x + 2)^2}$$

$$\left.\frac{dy}{dx}\right|_{x=-1} = \frac{-3}{(-1 + 2)^2} = -3$$

Note that only the x-coordinate was needed to evaluate the derivative. Also note we also know that the slope of a tangent line to the curve at $(-1, 3)$ is -3. See Fig. 23.20. ■

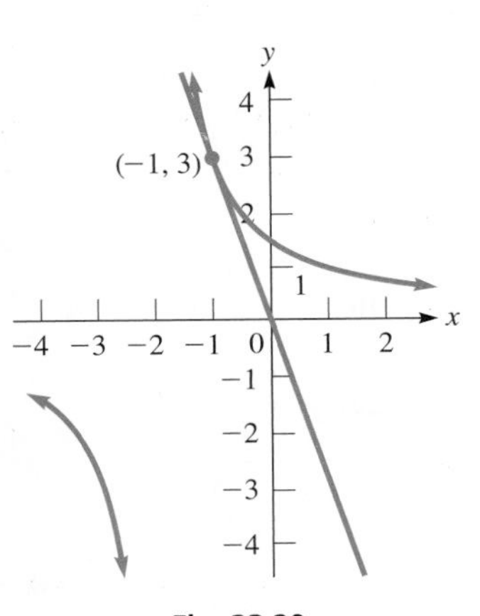

Fig. 23.20

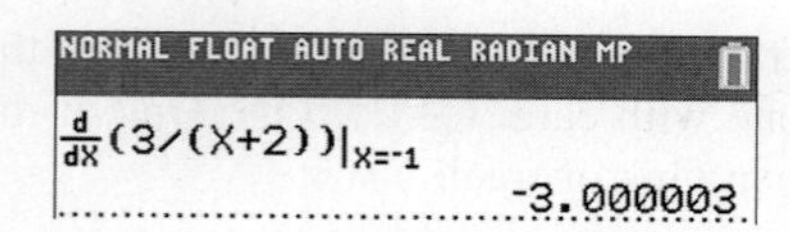

Fig. 23.21
Graphing calculator keystrokes: goo.gl/Gd90Ve

Most calculators have a feature for evaluating derivatives. Figure 23.21 shows the evaluation of Example 4 using the *numerical derivative* feature. The evaluation is done by numerical approximations, and the answer shown is to the default accuracy, but the accuracy can be adjusted with an additional entry of the derivative feature.

EXAMPLE 5 Using definition—derivative of square root

Find dy/dx for the function $y = \sqrt{x}$ by using the definition.

When finding this derivative, we will find that we have to rationalize the numerator of an expression involving radicals. See pages 340–341 regarding the rationalization of numerators or denominators that contain two terms.

$$f(x+h) = \sqrt{x+h}$$

$$f(x+h) - f(x) = \sqrt{x+h} - \sqrt{x}$$

rationalizing the numerator

$$\frac{f(x+h)-f(x)}{h} = \frac{\sqrt{x+h}-\sqrt{x}}{h} = \frac{(\sqrt{x+h}-\sqrt{x})(\sqrt{x+h}+\sqrt{x})}{h(\sqrt{x+h}+\sqrt{x})}$$

$$= \frac{(\sqrt{x+h})^2-(\sqrt{x})^2}{h(\sqrt{x+h}+\sqrt{x})} = \frac{x+h-x}{h(\sqrt{x+h}+\sqrt{x})} = \frac{1}{\sqrt{x+h}+\sqrt{x}}$$

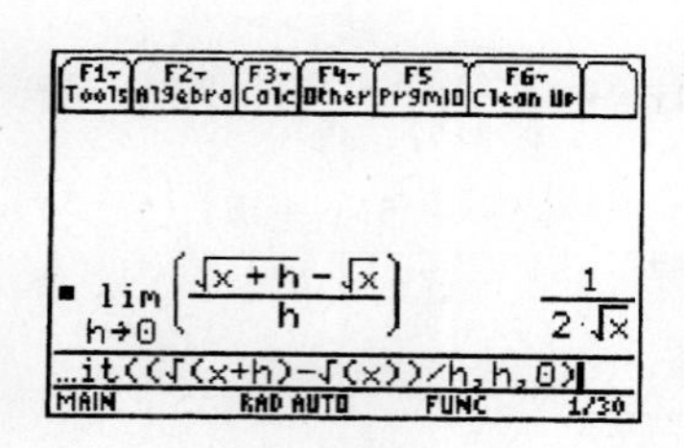

Fig. 23.22
TI-89 graphing calculator keystrokes: goo.gl/00x6b9

$$\lim_{h\to 0}\frac{f(x+h)-f(x)}{h} = \lim_{h\to 0}\frac{1}{\sqrt{x+h}+\sqrt{x}} = \frac{1}{2\sqrt{x}}$$

■ Figure 23.22 shows dy/dx on a TI-89 graphing calculator.

$$\frac{dy}{dx} = \frac{1}{2\sqrt{x}}$$

The domain of $f(x)$ is $x \geq 0$. However, since x appears in the denominator of the derivative, the domain of the derivative is $x > 0$. Thus, the function is differentiable for $x > 0$. ■

One might ask why, when finding a derivative, we take the limit as h approaches zero and not just let h equal zero. If we did this, we would find that the ratio $[f(x+h) - f(x)]/h$ is exactly $0/0$, which requires division by zero. As we know, this is an undefined operation, and therefore ***h cannot* equal *zero*.** However, it can equal any value as near zero as necessary. This idea is basic in the meaning of the word *limit*.

EXERCISES 23.3

In Exercises 1 and 2, make the given changes in the indicated examples of this section and then find the derivative by using the definition.

1. In Example 1, change $2x^2$ to $4x^2$.
2. In Example 2, change $x + 2$ in the denominator to $x - 2$.

In Exercises 3–24, find the derivative of each of the functions by using the definition.

3. $y = 3x - 1$
4. $y = 6x + 3$
5. $y = 1 - 7x$
6. $y = 2.3 - 5x$
7. $y = x^2 - 1$
8. $y = 16 - 3x^2$
9. $y = 3x^2$
10. $y = -\frac{1}{4}x^2$
11. $y = x^2 - 7x$
12. $y = x^2 + 4x$
13. $y = 8x - 2x^2$
14. $y = 3x - \frac{1}{2}x^2$
15. $y = x^3 + 2x$
16. $y = 2x^3$
17. $y = 5x^3 + 4$
18. $y = 2x - 4x^3$
19. $y = \dfrac{1}{x+2}$
20. $y = \dfrac{3}{5x+3}$
21. $y = x + \dfrac{4}{3x}$
22. $y = \dfrac{5x}{x-1}$
23. $y = \dfrac{2}{x^2}$
24. $y = \dfrac{8e^2}{x^2+4}$

In Exercises 25–28, find the derivative of each function by using the definition. Then evaluate the derivative at the given point. In Exercises 27 and 28, check your result using the derivative evaluation feature of a calculator.

25. $y = 3x^2 - 2x; (-1, 5)$

26. $y = 9x - x^3; (2, 10)$

27. $y = \frac{11}{3x + 2}; (3, 1)$

28. $y = x^2 - \frac{2}{x}; (-2, 5)$

In Exercises 29–32, find the derivative of each function by using the definition. Then determine the values for which the function is differentiable.

29. $y = 1 + \frac{2}{x}$

30. $y = \frac{5x}{x - 4}$

31. $y = \frac{3}{x^2 - 1}$

32. $y = \frac{4}{x^2 + 3}$

In Exercises 33–40, solve the given problems.

33. Find the point(s) on the curve of $y = x^2 - 4x$ for which the slope of a tangent line is 6.

34. Find the point(s) on the curve of $y = 1/(x + 1)$ for which the slope of the tangent line is -1.

35. At what point on the curve of $y = 2x^2 - 16x$ is there a tangent line that is horizontal?

36. At what point on the curve of $y = 9 - 2x^2$ is there a tangent line that is parallel to the line $12x - 2y + 7 = 0$?

37. Find dy/dx for $y = \sqrt{x + 1}$ by the method of Example 5. For what values of x is the function differentiable? Explain.

38. Find dy/dx for $y = \sqrt{x^2 + 3}$ by the method of Example 5.

39. A ski run follows the curve of $y = 0.01x^2 - 0.4x + 4$ from $x = 0$ m to $x = 20$ m. What is the angle between the ski run and the horizontal when $x = 10$ m?

40. The cross section of a hill can be approximated by the curve of $y = 0.3x - 0.00003x^3$ from $x = 0$ m to $x = 100$ m. The top of the hill is level. How high is the hill?

Answers to Practice Exercises

1. $y' = 5 - 2x$ **2.** $y' = -3/x^2$

23.4 The Derivative as an Instantaneous Rate of Change

Instantaneous Rate of Change • Instantaneous Velocity • Instantaneous Rate of Change of Dependent Variable with Respect to Independent Variable

In Section 23.2, we saw that the slope of a line tangent to a curve at point P was the limiting value of the slope of the line through points P and Q as Q approaches P. In Section 23.3, we defined the limit of the ratio $(f(x + h) - f(x))/h$ as $h \to 0$ as the derivative. Therefore, the first meaning we have given to *the derivative is the slope of a line tangent to a curve,* as we noted in Example 1 of Section 23.3. The following example further illustrates this meaning of the derivative.

EXAMPLE 1 Slope of a tangent line

Find the slope of the line tangent to the curve of $y = 4x - x^2$ at the point $(1,3)$.

As we have noted, from Sections 23.2 and 23.3, we know that we must first find the derivative and then evaluate it at the x-coordinate of the given point.

$$f(x + h) = 4(x + h) - (x + h)^2$$

$$f(x + h) - f(x) = 4(x + h) - (x + h)^2 - (4x - x^2)$$

$$= 4x + 4h - x^2 - 2xh - h^2 - 4x + x^2 = 4h - 2xh - h^2$$

$$\frac{f(x + h) - f(x)}{h} = \frac{4h - 2xh - h^2}{h} = 4 - 2x - h$$

$$\lim_{h \to 0} \frac{f(x + h) - f(x)}{h} = \lim_{h \to 0} (4 - 2x - h) = 4 - 2x$$

$$\frac{dy}{dx} = 4 - 2x$$

$$\left.\frac{dy}{dx}\right|_{x=1} = 4 - 2(1) = 2$$

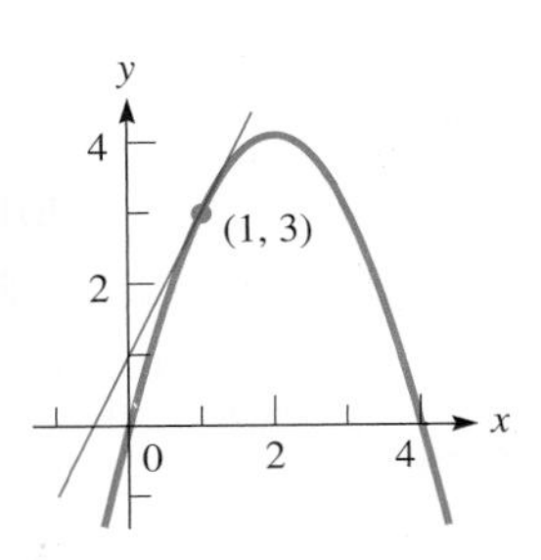

Fig. 23.23

The slope of the tangent line at $(1, 3)$ is 2. The curve and tangent line are shown in Fig. 23.23. ■

At the end of Section 23.2, we discussed the idea that the ratio $(f(x + h) - f(x))/h$ gives the average rate of change of $f(x)$ with respect to x. In defining the derivative as the limit of this ratio as $h \to 0$, it is a measure of the rate of change $f(x)$ with respect to x at point P. However, P may represent any point, which means the value of the derivative changes from one point on a curve to another point.

NOTE ➧ [Therefore, the derivative gives the instantaneous rate of change of $f(x)$ with respect to x.]

EXAMPLE 2 Rate of change of $f(x)$ for exact value of x

In Examples 1 and 2 of Section 23.2, $f(x)$ is changing at the rate of 7 units for a 1-unit change in x, ***when x equals exactly* 2.** In Example 3 of Section 23.2, $f(x)$ is increasing 6 units for a 1-unit change in x, ***when x equals exactly* −1**, and $f(x)$ is decreasing 2 units for a 1-unit increase in x, ***when x equals exactly* 3.** ■

This gives us a more general meaning of the derivative. If a functional relationship exists between any two variables, then one can be taken to be varying with respect to the other, and the derivative gives us the instantaneous rate of change. There are many applications of this principle, one of which is the velocity of an object. We consider now the case of motion along a straight line, called *rectilinear motion.*

The **average velocity** of an object is found by dividing the change in displacement by the time interval required for this change. As the time interval approaches zero, the limiting value of the average velocity gives the value of the **instantaneous velocity.** Using symbols for the derivative, *the instantaneous velocity of an object moving in rectilinear motion at a specified time t is given by*

$$v = \lim_{h \to 0} \frac{s(t + h) - s(t)}{h} \tag{23.7}$$

where $s(t)$ is the displacement as a function of the time t, and h is the time interval that approaches zero. In this case, the derivative has units of displacement divided by units of time, and we can denote it as ds/dt.

EXAMPLE 3 Instantaneous velocity—falling object

Find the instantaneous velocity when $t = 4$ s (exactly) of a falling object for which the distance s (in ft) fallen is the displacement, is given by $s = 16t^2$, by calculating average velocities between $t = 3.5$ s, 3.9 s, 3.99 s, 3.999 s and $t = 4$ s, and then noting the apparent limiting value as $h \to 0$.

The values of h are found by subtracting the given times from 4 s. Also, because $s = 256$ ft for $t = 4$ s, the differences in distance are found by subtracting the values of s for the given times from 256 ft. Therefore, we have

t(s)	3.5	3.9	3.99	3.999
s(ft)	196.0	243.36	254.7216	255.872016
$256 - s$ (ft)	60.0	12.64	1.2784	0.127984
$h = 4 - t$ (s)	0.5	0.1	0.01	0.001
$v = \dfrac{256 - s}{h}$(ft/s)	120.0	126.4	127.84	127.984

We can see that the value of v is approaching 128 ft/s, which is therefore the instantaneous velocity when $t = 4$ s. ■

EXAMPLE 4 Instantaneous velocity from the derivative

Find the expression for the instantaneous velocity of the object of Example 3, for which $s = 16t^2$, where s is the displacement (in ft) and t is the time (in s). Determine the instantaneous velocity for $t = 2$ s and $t = 4$ s.

The required expression is the derivative of s with respect to t.

$$f(t + h) = 16(t + h)^2$$

$$f(t + h) - f(t) = 16(t + h)^2 - 16t^2 = 32th + 16h^2$$

$$\frac{f(t + h) - f(t)}{h} = \frac{32th + 16h^2}{h} = 32t + 16h$$

$$v = \lim_{h \to 0} \frac{f(t + h) - f(t)}{h} = \lim_{h \to 0}(32t + 16h) = 32t$$

expression for instantaneous velocity

$$\left.\frac{ds}{dt}\right|_{t=2} = 32(2) = 64 \text{ ft/s} \quad \text{and} \quad \left.\frac{ds}{dt}\right|_{t=4} = 32(4) = 128 \text{ ft/s}$$

We see that the second result agrees with that found in Example 3. ■

Practice Exercise

1. Find the instantaneous velocity of an object moving such that $s = 6.0t^2$ for $t = 3.0$ s. Here, s is the displacement (in m) and t is the time (in s).

By finding $\lim_{h \to 0}(f(x + h) - f(x))/h$, we can find the instantaneous rate of change of $f(x)$ with respect to x. The expression $\lim_{h \to 0}(f(t + h) - f(t))/h$ gives the instantaneous velocity, or instantaneous rate of change of displacement with respect to time. [Generalizing, we can say that the derivative can be interpreted as the instantaneous rate of change of the dependent variable with respect to the independent variable.] This is true for any differentiable function, no matter what the variables represent.

NOTE ➧

EXAMPLE 5 Instantaneous rate of change—volume of a balloon

A spherical balloon is being inflated. Find the expression for the instantaneous rate of change of the volume with respect to the radius. Evaluate this instantaneous rate of change for a radius of 2.00 m.

$$V = f(r) = \tfrac{4}{3}\pi r^3 \qquad \text{volume of sphere}$$

$$f(r + h) = \tfrac{4}{3}\pi(r + h)^3 \qquad \text{find derivative}$$

$$f(r + h) - f(r) = \tfrac{4}{3}\pi(r + h)^3 - \tfrac{4}{3}\pi r^3$$

$$= \tfrac{4}{3}\pi(r^3 + 3r^2h + 3rh^2 + h^3 - r^3) = \tfrac{4}{3}\pi(3r^2h + 3rh^2 + h^3)$$

$$\frac{f(r + h) - f(r)}{h} = \frac{4\pi}{3}\left(\frac{3r^2h + 3rh^2 + h^3}{h}\right) = \frac{4\pi}{3}(3r^2 + 3rh + h^2)$$

$$\frac{dV}{dr} = \lim_{h \to 0} \frac{f(r + h) - f(r)}{h} = \lim_{h \to 0}\left(\frac{4\pi}{3}(3r^2 + 3rh + h^2)\right) = 4\pi r^2$$

$$\left.\frac{dV}{dr}\right|_{r=2.00 \text{ m}} = 4\pi(2.00)^2 = 16.0\pi = 50.3 \text{ m}^2$$

instantaneous rate of change when $r = 2.00$ m

The instantaneous rate of change of the volume with respect to the radius (dV/dr) for $r = 2.00$ m is 50.3 m^3/m (this way of showing the units is more meaningful).

As r increases, dV/dr also increases. This should be expected as the volume of a sphere varies directly as the cube of the radius. ■

EXAMPLE 6 Instantaneous rate of change of power

The power P produced by an electric current i in a resistor varies directly as the square of the current. Given that 1.2 W of power are produced by a current of 0.50 A in a certain resistor, find an expression for the instantaneous rate of change of power with respect to current. Evaluate this rate of change for $i = 2.5$ A.

We must first find the functional relationship between power and current, by solving the indicated problem in variation:

$$P = ki^2 \qquad 1.2 = k(0.50)^2 \qquad k = 4.8\text{ W/A}^2 \qquad P = 4.8i^2$$

Now, knowing the function, we may determine the expression for the instantaneous rate of change of P with respect to i by finding the derivative.

$$f(i + h) = 4.8(i + h)^2$$

$$f(i + h) - f(i) = 4.8(i + h)^2 - 4.8i^2 = 4.8(2ih + h^2)$$

$$\frac{f(i + h) - f(i)}{h} = \frac{4.8(2ih + h^2)}{h} = 4.8(2i + h)$$

expression for instantaneous rate of change

$$\frac{dP}{di} = \lim_{h \to 0} \frac{f(i + h) - f(i)}{h} = \lim_{h \to 0} [4.8(2i + h)] = 9.6i$$

$$\left.\frac{dP}{di}\right|_{i=2.5\text{ A}} = 9.6(2.5) = 24\text{ W/A}$$

instantaneous rate of change when $i = 2.5$ A

This tells us that when $i = 2.5$ A, the rate of change of power with respect to current is 24 W/A. Also, we see that the larger the current is, the greater is the rate of increase in power. This should be expected, since the power varies directly as the square of the current. ■

Practice Exercise

2. In Example 6, change 1.2 W of power to 1.6 W of power, and 0.50 A of current to 0.80 A of current, and then find the instantaneous rate of change of power with respect to current for $i = 3.0$ A.

EXERCISES 23.4

In Exercises 1 and 2, make the given changes in the indicated examples of this section and then solve the resulting problems.

1. In Example 4, change $16t^2$ to $48t - 16t^2$ and then evaluate the instantaneous velocity at the indicated times.

2. In Example 5, change "volume" in the second line to "surface area" and then evaluate the rate of change for $r = 2.00$ m.

In Exercises 3–6, find the slope of a line tangent to the curve of the given equation at the given point. Sketch the curve and the tangent line.

3. $y = x^2 - 1$; $(2, 3)$

4. $y = 2x - x^2$; $(-1, -3)$

5. $y = \dfrac{16}{3x + 1}$; $(-3, -2)$

6. $y = 4 - \dfrac{80}{x^2}$; $(4, -1)$

In Exercises 7–10, calculate the instantaneous velocity for the indicated value of the time (in s) of an object for which the displacement (in ft) is given by the indicated function. Use the method of Example 3 and calculate values of the average velocity for the given values of t and note the apparent limit as the time interval approaches zero.

7. $s = 4t + 10$; when $t = 3$; use values of t of 2.0, 2.5, 2.9, 2.99, 2.999

8. $s = 6 - 3t$; when $t = 4$; use values of t of 3.0, 3.5, 3.9, 3.99, 3.999

9. $s = 3t^2 - 4t$; when $t = 2$; use values of t of 1.0, 1.5, 1.9, 1.99, 1.999

10. $s = 120t - 16t^2$; when $t = 0.5$; use values of t of 0.4, 0.45, 0.49, 0.499, 0.4999

In Exercises 11–14, use the definition to find an expression for the instantaneous velocity of an object moving with rectilinear motion according to the given functions (the same as those for Exercises 7–10) relating s (in ft) and t (in s). Then calculate the instantaneous velocity for the given value of t.

11. $s = 4t + 10$; $t = 3$

12. $s = 6 - 3t$; $t = 4$

13. $s = 3t^2 - 4t$; $t = 2$

14. $s = 120t - 16t^2$; $t = 0.5$

In Exercises 15–22, use the definition to find an expression for the instantaneous velocity of an object moving with rectilinear motion according to the given functions relating s and t.

15. $s = 48t + 12$

16. $s = 3t^2 - 2t^3$

17. $s = 12t^2 - t^3$

18. $s = s_0 + v_0t + \frac{1}{2}at^2$ (s_0, v_0, and a are constants.)

19. $s = 3t - \dfrac{1}{3}\pi t^3$

20. $s = \dfrac{2t}{t + 4}$

21. $s = 7t^2 - \dfrac{2}{t + 1}$

22. $s = \dfrac{5}{t^2 + 1}$

In Exercises 23–26, use the definition to find an expression for the instantaneous acceleration of an object moving with rectilinear motion according to the given functions. The instantaneous acceleration of an object is defined as the instantaneous rate of change of the velocity with respect to time. Here, v is the velocity, s is the displacement, and t is the time.

23. $v = 6t^2 - 4t + 2$

24. $v = \sqrt{6t + 1}$

25. $s = t^3 + 15t$ (Find v, then find a.)

26. $s = s_0 + v_0t - \frac{1}{2}at^2$ (s_0, v_0, and a are constants.) (Find v, then find a.)

In Exercises 27–46, find the indicated instantaneous rates of change.

27. In Example 3, calculate v for $t = 4.5$ s, 4.1 s, 4.01 s, 4.001 s. This is finding the instantaneous velocity as $h \rightarrow 0$ through negative values of h.

28. In Example 4, calculate v for $t = 3.0$ s, 2.1 s, 2.01 s, 2.001 s. This is finding the instantaneous velocity as $h \rightarrow 0$ through negative values of h.

29. A metal circular ring is being cooled. Find the instantaneous rate at which the circumference changes with respect to the radius (measured in cm).

30. Liquid is poured into a tank with vertical sides and a square horizontal cross section of edge 6.25 in. Find the instantaneous rate of change of volume with respect to the depth h.

31. The distance s (in m) above the ground for a projectile fired vertically upward with a velocity of 44 m/s as a function of time t (in s) is given by $s = 44t - 4.9t^2$. Find t for $v = 0$.

32. For the projectile in Exercise 31, find v for $t = 4.0$ s and for $t = 5.0$ s. What conclusion can be drawn?

33. The electric current i at a point in an electric circuit is the instantaneous rate of change of the electric charge q that passes the point, with respect to the time t. Find i in a circuit for which $q = 30 - 2t$.

34. A load L (in N) is distributed along a beam 10 m long such that $L = 5x - 0.5x^2$, where x is the distance from one end of the beam. Find the expression for the instantaneous rate of change of L with respect to x.

35. A rectangular metal plate contracts while cooling. Find the expression for the instantaneous rate of change of the area A of the plate with respect to its width w, if the length of the plate is constantly three times as long as the width.

36. A circular oil spill is increasing in area. Find the difference in the rate of change of the area A of the spill with respect to the radius r for $r = 240$ m and $r = 480$ m.

37. The total power P (in W) transmitted by an AM radio station is given by $P = 500 + 250m^2$, where m is the modulation index. Find the instantaneous rate of change of P with respect to m for $m = 0.92$.

38. The bottom of a soft-drink can is being designed as an inverted spherical segment, the volume of which is $V = \frac{1}{6}\pi h^3 + 2.00\pi h$, where h is the depth (in cm) of the segment. Find the instantaneous rate of change of V with respect to h for $h = 0.60$ cm.

39. The total solar radiation H (in W/m^2) on a particular surface during an average clear day is given by $H = \dfrac{5000}{t^2 + 10}$, where t $(-6 \leq t \leq 6)$ is the number of hours from noon (6 A.M. is equivalent to $t = -6$ h). Find the instantaneous rate of change of H with respect to t at 3 P.M.

40. For the solar radiation in Exercise 39, find the average rate of change of H between 2 P.M. and 4 P.M. Compare with the instantaneous rate of change at 3 P.M.

41. The value (in thousands of dollars) of a certain car is given by the function $V = \dfrac{48}{t + 3}$, where t is measured in years. Find a general expression for the instantaneous rate of change of V with respect to t and evaluate this expression when $t = 3$ years.

42. For the car in Exercise 41, find the average rate of change of V between $t = 2$ years and $t = 4$ years. Compare with the instantaneous rate of change for $t = 3$ years.

43. Oil in a certain machine is stored in a conical reservoir, for which the radius and height are both 4 cm (see Fig. 23.24). Find the instantaneous rate of change of the volume V of oil in the reservoir with respect to the depth d of the oil.

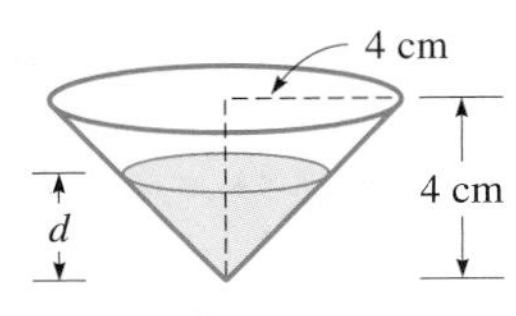

Fig. 23.24

44. The time t required to test a computer memory unit is directly proportional to the square of the number n of memory cells in the unit. For a particular type of unit, $n = 6400$ for $t = 25.0$ s. Find the instantaneous rate of change of t with respect to n for this type of unit for $n = 8000$.

45. A *holograph* (an image formed without using a lens) of concentric circles is formed. The radius r of each circle varies directly as the square root of the wavelength λ of the light used. If $r = 3.72$ cm for $\lambda = 592$ nm, find the expression for the instantaneous rate of change of r with respect to λ assuming both r and λ are measured in meters.

46. The force F between two electric charges varies inversely as the square of the distance r between them. For two charged particles, $F = 0.12$ N for $r = 0.060$ m. Find the instantaneous rate of change of F with respect to r for $r = 0.120$ m.

Answers to Practice Exercises

1. 36 m/s **2.** 15 W/A

23.5 Derivatives of Polynomials

Derivative of a Constant • Derivative of a Power of x • Derivative of Constant Times Function of x • Derivative of Sum of Functions of x

The task of finding a derivative can be considerably shortened by using the definition to derive certain basic formulas for derivatives. In this section, we derive the formulas that are used for finding the derivatives of polynomial functions of the form $f(x) = a_0x^n + a_1x^{n-1} + \cdots + a_n$.

First, we find the derivative of a constant. Letting $f(x) = c$ and then using the definition of a derivative, we find that $f(x + h) - f(x) = c - c = 0$. This means that

$$\lim_{h \to 0} \frac{f(x + h) - f(x)}{h} = \lim_{h \to 0} \left(\frac{0}{h}\right) = 0$$

NOTE ➧ [From this, we conclude that the derivative of a constant is zero.] Therefore, if $y = c$, $dy/dx = 0$. This can also be written as follows:

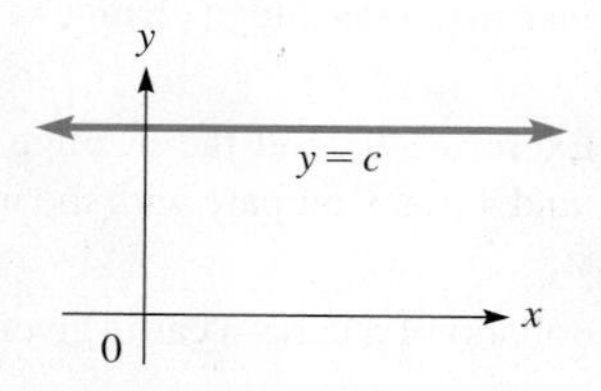

Fig. 23.25

Constant Rule

$$\frac{dc}{dx} = 0 \quad \textbf{(23.8)}$$

This rule is consistent with the fact that the constant function $y = c$ is a horizontal line, which has a slope of zero. See Fig. 23.25.

Next, we find the derivative of a positive integral power of x. If $f(x) = x^n$, where n is a positive integer, by using the binomial theorem we have

$$f(x + h) - f(x) = (x + h)^n - x^n = x^n + nx^{n-1}h + \frac{n(n-1)}{2}x^{n-2}h^2 + \cdots + h^n - x^n$$

$$= nx^{n-1}h + \frac{n(n-1)}{2}x^{n-2}h^2 + \cdots + h^n$$

$$\frac{f(x + h) - f(x)}{h} = nx^{n-1} + \frac{n(n-1)}{2}x^{n-2}h + \cdots + h^{n-1}$$

$$\lim_{h \to 0} \frac{f(x + h) - f(x)}{h} = nx^{n-1}$$

Thus, *the derivative of the nth power of x* can be found by using the following rule:

Power Rule

$$\frac{dx^n}{dx} = nx^{n-1} \quad \textbf{(23.9)}$$

EXAMPLE 1 Derivative of a constant

Find the derivative of the function $y = -5$.

Because -5 is a constant, applying Eq. (23.8), we have

$$\frac{dy}{dx} = \frac{d(-5)}{dx} = 0$$

■

EXAMPLE 2 Derivative of $y = x$

Find the derivative of the function $y = x$.

In using Eq. (23.9), we have $n = 1$ because $x = x^1$. This means

$$\frac{dy}{dx} = \frac{d(x)}{dx} = (1)x^{1-1} = (1)(x^0)$$

Because $x^0 = 1$, we have

$$\frac{dy}{dx} = 1$$

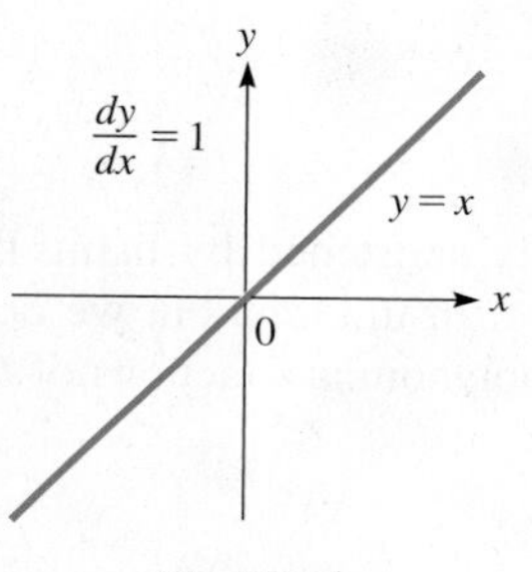

Fig. 23.26

Thus, the derivative of $y = x$ is 1, which means that the slope of the line $y = x$ is always 1. This is consistent with our previous discussion of the slope of a straight line. See Fig. 23.26.

■

EXAMPLE 3 Derivative of a power of r

Find the derivative of the function $v = r^{10}$.

Here, the dependent variable is v, and the independent variable is r. Therefore,

$$\frac{dv}{dr} = \frac{d(r^{10})}{dr} = 10r^{10-1}$$
$$= 10r^9$$

■

■ From here on, we will be using functions that are combinations of simpler functions. Therefore, we must denote some functions by symbols other than $f(x)$. We will often be using u and v.

Next, we find the derivative of a constant times a function of x. We denote this function as u, or to show directly that it is a function of x, as $u(x)$. In finding the derivative of cu with respect to x, we have

$$\frac{d}{dx}(cu) = \lim_{h\to 0}\frac{cu(x+h) - cu(x)}{h}$$
$$= c\lim_{h\to 0}\frac{u(x+h) - u(x)}{h} = c\frac{du}{dx}$$

Therefore, *the derivative of the product of a constant and a differentiable function of x is the product of the constant and the derivative of the function of x*. This is written as follows:

Constant Multiple Rule

$$\frac{d(cu)}{dx} = c\frac{du}{dx} \qquad (23.10)$$

EXAMPLE 4 Derivative of a constant times a power of x

Find the derivative of $y = 3x^2$.

In this case, $c = 3$ and $u = x^2$. Thus, $du/dx = 2x$. Therefore,

$$\frac{dx}{dy} = \frac{d(3x^2)}{dx} = 3\frac{d(x^2)}{dx} = 3(2x)$$
$$= 6x$$

■

Practice Exercise

1. Find the derivative of $y = 6x^3$.

If the types of functions for which we have found derivatives are added, the result is a polynomial function with more than one term. The derivative is found by letting $y = u + v$, where u and v are functions of x. Using the definition, we have

$$y = u(x) + v(x)$$
$$\frac{dy}{dx} = \frac{d}{dx}[u(x) + v(x)] = \lim_{h\to 0}\frac{[u(x+h) + v(x+h)] - [u(x) - v(x)]}{h}$$
$$= \lim_{h\to 0}\left[\frac{u(x+h) - u(x)}{h} + \frac{v(x+h) - v(x)}{h}\right]$$
$$= \lim_{h\to 0}\left[\frac{u(x+h) - u(x)}{h}\right] + \lim_{h\to 0}\left[\frac{v(x+h) - v(x)}{h}\right] = \frac{du}{dx} + \frac{dv}{dx}$$

This tells us that *the derivative of the sum of differentiable functions of x is the sum of the derivatives of the functions*. This is shown below.

Derivative of a Sum

$$\frac{d(u+v)}{dx} = \frac{du}{dx} + \frac{dv}{dx} \qquad (23.11)$$

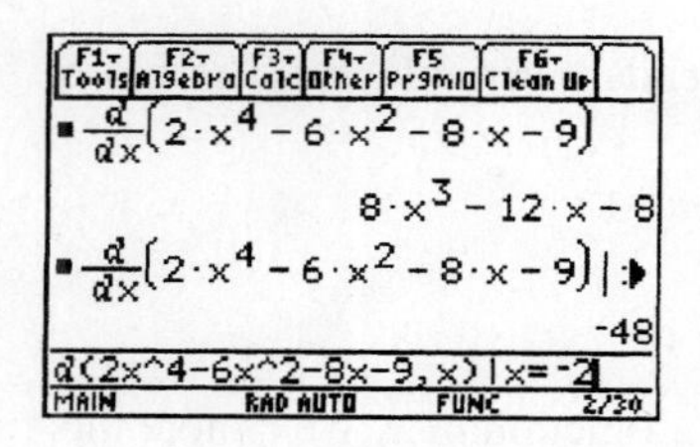

Fig. 23.27
TI-89 graphing calculator keystrokes: goo.gl/bGcKa6

EXAMPLE 5 Calculator evaluation of a derivative

Evaluate the derivative of $f(x) = 2x^4 - 6x^2 - 8x - 9$ at $(-2, 15)$.

First, finding the derivative, we have

$$f'(x) = \frac{d(2x^4)}{dx} - \frac{d(6x^2)}{dx} - \frac{d(8x)}{dx} - \frac{d(9)}{dx}$$

$$= 8x^3 - 12x - 8$$

We now evaluate this derivative for $x = -2$.

$$f'(x)|_{x=-2} = 8(-2)^3 - 12(-2) - 8 = -48$$

This evaluation on a TI-89 graphing calculator is shown in Fig. 23.27. ■

EXAMPLE 6 Slope of a tangent line

Find the slope of the line tangent to the curve of $y = 4x^7 - x^4$ at the point $(1, 3)$.

We must find the derivative and then evaluate it for the value $x = 1$:

$$\frac{dy}{dx} = 28x^6 - 4x^3 \quad \text{find derivative}$$

$$\left.\frac{dy}{dx}\right|_{x=1} = 28(1) - 4(1) = 24 \quad \text{evaluate derivative}$$

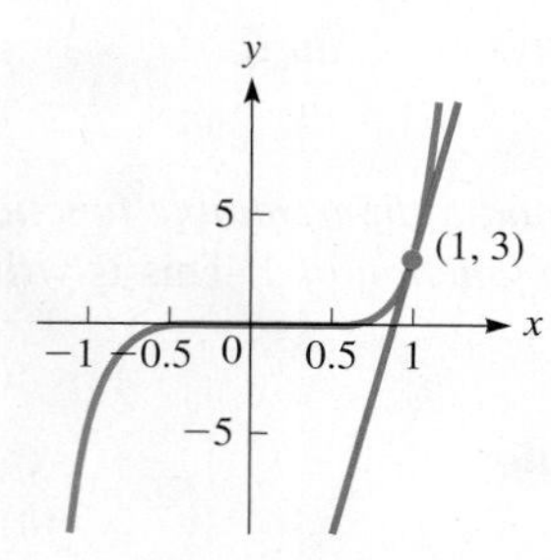

Fig. 23.28

Thus, the slope of the tangent line is 24. Again, we note that ***the substitution $x = 1$ must be made after the differentiation has been performed.*** The curve and the tangent line at $(1, 3)$ are shown in Fig. 23.28. ■

EXAMPLE 7 Evaluation of derivative—velocity of a diver

■ See the chapter introduction.

A diver jumped from a platform that is exactly 20 m above the water. Her height s (in m) above the water after t seconds is given by $s = -4.9t^2 + 4.6t + 20$. Find her instantaneous velocity after (a) 0.10 s (just after jumping), and (b) 2.5 s (just before impact with the water).

(a) $$\frac{ds}{dt} = -9.8t + 4.6$$

$$\left.\frac{ds}{dt}\right|_{t=0.10} = -9.8(0.10) + 4.6 = 3.6 \text{ m/s}$$

(b) $$\left.\frac{ds}{dt}\right|_{t=2.5} = -9.8(2.5) + 4.6 = -20 \text{ m/s}$$

Therefore, the diver is moving *upward* at a rate of 3.6 m/s shortly after jumping and she is moving *downward* (indicated by the *negative* velocity) at a rate if 20 m/s just before entering the water. ■

Practice Exercise

2. Find the expression for the instantaneous velocity if $s = 7t^4 - 4t + 5$, where s is the displacement and t is the time.

EXERCISES 23.5

In Exercises 1–4, make the given changes in the indicated examples of this section, and then solve the resulting problem.

1. In Example 3, change the exponent 10 to 9.
2. In Example 4, change the coefficient 3 to 4.
3. In Example 5, change $2x^4$ to $2x^3$ and $-8x$ to $+8x$.
4. In Example 6, change $4x^7 - x^4$ to $4x^4 - x^7$.

In Exercises 5–20, find the derivative of each of the given functions.

5. $y = x^5$
6. $y = x^{12}$
7. $f(x) = -4x^9$
8. $y = -7x^6$
9. $y = 5x^4 - 3\pi$
10. $s = 3t^5 - 14t$
11. $y = 4x^2 + 7x$
12. $y = 8x^3 - 1.5x^2$
13. $p = 5r^3 - 2r + 12$
14. $y = 6x^2 - 15x + 5e$
15. $y = 25x^8 - 34x^5 + 2^3 - x$
16. $u = 4v^4 - (12v^3 + 9v)$

17. $f(x) = -6x^7 + 5x^3 + \pi^2$

18. $y = 13x^4 - 6x^3 - 3(x - 8)$

19. $y = \frac{1}{3}x^3 + \frac{1}{2}x^2$

20. $f(z) = -\frac{1}{4}z^8 + \frac{1}{2}z^4 - 2^3$

In Exercises 21–24, evaluate the derivative of each of the given functions at the given point. In Exercises 23 and 24, check your result using the derivative evaluation feature of a calculator.

21. $y = 6x^2 - 8x + 1 \quad (2, 9)$

22. $s = 2t^3 - 5t^2 \quad (-1, -7)$

23. $y = 2x^3 + 9x - 7 \quad (-2, -41)$

24. $y = x^4 - 9x^2 - 5x \quad (3, -15)$

In Exercises 25–28, find the slope of a line tangent to the curve of each of the given functions for the given values of x.

25. $y = 2x^6 - 4x^2 \quad (x = -1)$

26. $y = 3x^3 - 9x \quad (x = 1)$

27. $y = 35x - 2x^4 \quad (x = 2)$

28. $y = x^4 - \frac{1}{2}x^2 + 2 \quad (x = -2)$

In Exercises 29–32, determine an expression for the instantaneous velocity of objects moving with rectilinear motion according to the functions given, if s represents displacement in terms of time t.

29. $s = 6t^5 - 5t + 2$

30. $s = 20 + 60t - 4.9t^2$

31. $s = 2 - (6t - 2t^3)$

32. $s = s_0 + v_0t + \frac{1}{2}at^2$

In Exercises 33–36, s represents the displacement, and t represents the time for objects moving with rectilinear motion, according to the given functions. Find the instantaneous velocity for the given times.

33. $s = 2t^3 - 4t^2; t = 4$

34. $s = 8t^2 - 2(10t + 6); t = 5$

35. $s = 120 + 80t - 16t^2; t = 2.5$

36. $s = 0.5t^4 - 1.5t^2 + 2.5; t = 3$

In Exercises 37–56, solve the given problems by finding the appropriate derivative.

37. For what value(s) of x is the tangent to the curve of $y = 3x^2 - 6x$ parallel to the x-axis? (That is, where is the slope zero?)

38. Find the value of a if the tangent to the curve of $y = ax^2 + 2x$ has a slope of -4 for $x = 2$.

39. For what point(s) on the curve of $y = 3x^2 - 4x$ is the slope of a tangent line equal to 8?

40. Explain why the curve $y = 5x^3 + 4x - 3$ does not have a tangent line with a slope less than 4.

41. Find the point at which a tangent line to the parabola $y = 2x^2 - 7x$ is perpendicular to the line $x - 3y = 16$.

42. Display the graphs of $y = x^2$ and its derivative on a calculator. State any conclusions you can draw from the relationship of the two graphs.

43. For what value(s) of x is the slope of a line tangent to the curve of $y = 4x^2 + 3x$ equal to the slope of a line tangent to the curve of $y = 5 - 2x^2$?

44. For what value(s) of t is the instantaneous velocity of an object moving according to $s = 5t - 2t^2$ equal to the instantaneous velocity of an object moving according to $s = 3t^2 + 4$?

45. The length of a certain metal tube is always 20 times the radius r. The tube is heated and the radius and volume V increase. Find dV/dr for $r = 3.0$ mm.

46. A rectangular solid block of ice is melting such that the height is always twice the edge of the square base. Find the expression for the instantaneous rate of change of surface area A with respect to the edge of the base e.

47. The electric power P (in W) as a function of the current i (in A) in a certain circuit is given by $P = 16i^2 + 60i$. Find the instantaneous rate of change of P with respect to i for $i = 0.75$ A.

48. The torque T on the arm of a robotic control mechanism varies directly as the cube of the diameter d of the arm. If $T = 850$ lb · in. for $d = 0.925$ in., find the expression for the instantaneous rate of change of T with respect to d.

49. The resistance R (in Ω) of a certain wire as a function of the temperature T (in °C) is given by $R = 16.0 + 0.450T + 0.0125T^2$. Find the instantaneous rate of change of R with respect to T when $T = 115$°C.

50. The deflection d of a diving board x m from the fixed end at the pool side is given by $d = kx^2(3L - x)$, where L is the length of the diving board and k is a positive constant. Find the expression for the instantaneous rate of change of d with respect to x.

51. The tensile strength S (in N) of a certain material as a function of the temperature T (in °C) is $S = 1600 - 0.000022T^2$. Find the instantaneous rate of change of S with respect to T for $T = 65$°C.

52. A tank containing 6000 L of water drains out in 30 min. The volume V of water in the tank after t min of draining is $V = 6000(1 - t/30)^2$. Find the instantaneous time rate of change of V after 15 min of draining.

53. The altitude h (in m) of a jet as a function of the horizontal distance x (in km) it has traveled is given by $h = 0.000104x^4 - 0.0417x^3 + 4.21x^2 - 8.33x$. Find the instantaneous rate of change of h with respect to x for $x = 120$ km.

54. The force F (in N) exerted by a cam on a lever is given by $F = x^4 - 12x^3 + 46x^2 - 60x + 25$, where x $(1 \le x \le 5)$ is the distance (in cm) from the center of rotation of the cam to the edge of the cam in contact with the lever (see Fig. 23.29). Find the instantaneous rate of change of F with respect to x when $x = 4.0$ cm.

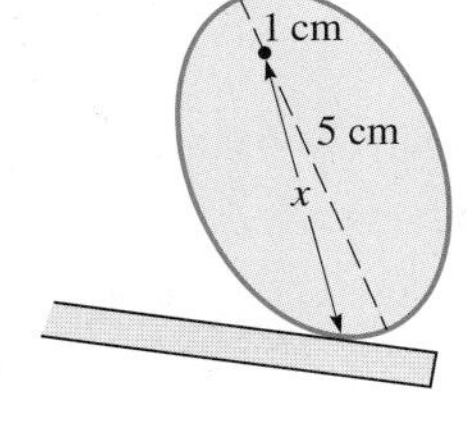

Fig. 23.29

55. Two ball bearings wear down such that the radius r of one is constantly 1.20 mm less than the radius of the other. Find the instantaneous rate of change of the total volume V_T of the two ball bearings with respect to r for $r = 3.30$ mm.

56. Equal squares of side x are to be cut from the corners of a 6.00 in. by 8.00 in. rectangular piece of cardboard. The sides are to be bent up and taped together to make an open top container. Find the instantaneous rate of change of the volume V of the container with respect to x for $x = 1.75$ in. See Fig. 23.30.

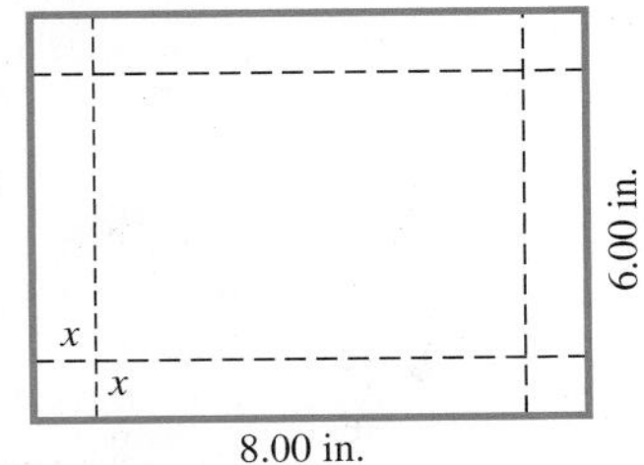

Fig. 23.30

Answers to Practice Exercises

1. $dy/dx = 18x^2$ **2.** $v = 28t^3 - 4$

23.6 Derivatives of Products and Quotients of Functions

Products and Quotients of Functions • Derivative of a Product • Derivative of a Quotient

In the previous section, we considered polynomial functions. For functions that are not polynomials, some are products of simpler functions, some are quotients of simpler functions, and others are powers of simpler functions. In this section, we develop the formula for the derivative of a product of functions and the formula for the quotient of functions.

EXAMPLE 1 Product and quotient of functions

The functions $f(x) = x^2 + 2$ and $g(x) = 3 - 2x$ can be combined to form functions of the type mentioned above, as we now illustrate:

$$p(x) = f(x)g(x) = (x^2 + 2)(3 - 2x) \quad \text{product of functions}$$

$$q(x) = \frac{g(x)}{f(x)} = \frac{3 - 2x}{x^2 + 2} \quad \text{quotient of functions}$$

$$F(x) = [g(x)]^2 = (3 - 2x)^2 \quad \text{power of a function}$$

■

If u and v are differentiable functions of x, the derivative of the function uv is found by letting $y = uv$, and applying the definition as follows:

$$y = u(x)v(x)$$

$$\frac{dy}{dx} = \frac{d}{dx}[u(x)v(x)] = \lim_{h\to 0}\frac{u(x + h)v(x + h) - u(x)v(x)}{h}$$

By adding and subtracting $u(x + h)v(x)$ in the numerator of the last fraction, we can put it in a form that includes the derivatives of $u(x)$ and $v(x)$. Therefore,

$$\frac{d}{dx}[u(x)v(x)] = \lim_{h\to 0}\frac{u(x + h)v(x + h) - u(x + h)v(x) + u(x + h)v(x) - u(x)v(x)}{h}$$

$$= \lim_{h\to 0}\left[u(x + h)\frac{v(x + h) - v(x)}{h} + v(x)\frac{u(x + h) - u(x)}{h}\right] = u(x)\frac{dv(x)}{dx} + v(x)\frac{du(x)}{dx}$$

We conclude that *the derivative of the product of two differentiable functions equals the first function times the derivative of the second function plus the second function times the derivative of the first function.* This is written below.

Product Rule

$$\frac{d(uv)}{dx} = u\frac{dv}{dx} + v\frac{du}{dx} \tag{23.12}$$

EXAMPLE 2 Derivative of a product of functions

Find the derivative of the product function in Example 1.

$$p(x) = (x^2 + 2)(3 - 2x) \qquad u = x^2 + 2 \qquad v = 3 - 2x$$

$$\frac{d(uv)}{dx} = u\frac{dv}{dx} + v\frac{du}{dx}$$

$$p'(x) = (x^2 + 2)(-2) + (3 - 2x)(2x) = -2x^2 - 4 + 6x - 4x^2$$

$$= -6x^2 + 6x - 4$$

We could have multiplied the functions first, and then taken the derivative as a polynomial. However, we will soon see functions for which the product rule must be used. ■

Practice Exercise

1. Find the derivative of $y = (3 - 2x^2)(x^4 - 1)$. Do not multiply factors together before finding the derivative.

We will now find the derivative of the quotient of two differentiable functions by applying the definition to the function $y = u/v$, as shown on the following page.

$$\frac{dy}{dx} = \frac{d}{dx}\left(\frac{u(x)}{v(x)}\right) = \lim_{h\to 0}\frac{\dfrac{u(x+h)}{v(x+h)} - \dfrac{u(x)}{v(x)}}{h} = \lim_{h\to 0}\frac{v(x)u(x+h) - u(x)v(x+h)}{hv(x+h)v(x)}$$

We can put the last fraction in a form that includes the derivatives of $u(x)$ and $v(x)$ by subtracting and adding $u(x)v(x)$ in the numerator. Therefore,

$$\frac{d}{dx}\left[\frac{u(x)}{v(x)}\right] = \lim_{h\to 0}\frac{v(x)u(x+h) - v(x)u(x) + v(x)u(x) - u(x)v(x+h)}{hv(x+h)v(x)}$$

$$= \lim_{h\to 0}\frac{v(x)\dfrac{u(x+h)-u(x)}{h} - u(x)\dfrac{v(x+h)-v(x)}{h}}{v(x+h)v(x)}$$

$$= \frac{v(x)\dfrac{du(x)}{dx} - u(x)\dfrac{dv(x)}{dx}}{[v(x)]^2}$$

Therefore, *the derivative of the quotient of two differentiable functions equals the denominator times the derivative of the numerator minus the numerator times the derivative of the denominator, all divided by the square of the denominator.*

Quotient Rule

$$\frac{d\left(\dfrac{u}{v}\right)}{dx} = \frac{v\dfrac{du}{dx} - u\dfrac{dv}{dx}}{v^2} \quad \textbf{(23.13)}$$

EXAMPLE 3 Derivative of a quotient of functions

Find the derivative of the quotient function in Example 1.

$$q(x) = \frac{3-2x}{x^2+2} \qquad u = 3 - 2x \qquad v = x^2 + 2$$

$$q'(x) = \frac{\overset{v}{(x^2+2)}\overset{du/dx}{(-2)} - \overset{u}{(3-2x)}\overset{dv/dx}{(2x)}}{\underset{v^2}{(x^2+2)^2}} = \frac{-2x^2 - 4 - 6x + 4x^2}{(x^2+2)^2}$$

$$= \frac{2(x^2 - 3x - 2)}{(x^2+2)^2}$$ ■

■ In the first expression for $q'(x)$, be careful not to cancel the factor of $(x^2 + 2)$, as it is not a factor of both terms of the numerator.

EXAMPLE 4 Derivative of a quotient—stress on a tube

The stress S on a hollow tube is given by

$$S = \frac{16DT}{\pi(D^4 - d^4)}$$

where T is the tension, D is the outer diameter, and d is the inner diameter of the tube. Find the expression for the instantaneous rate of change of S with respect to D, with the other values being constant.

We are to find the derivative of S with respect to D, and it is found as follows:

$$\frac{dS}{dD} = \frac{\pi(D^4 - d^4)(16T) - 16DT(\pi)(4D^3)}{\pi^2(D^4 - d^4)^2} = \frac{16\pi T(D^4 - d^4 - 4D^4)}{\pi^2(D^4 - d^4)^2}$$

$$= \frac{-16T(3D^4 + d^4)}{\pi(D^4 - d^4)^2}$$ ■

Practice Exercise

2. Find the derivative of $y = \dfrac{2x^2}{x^4 - 1}$.

EXAMPLE 5 Evaluation of a derivative

Evaluate the derivative of $y = \frac{3x^2 + x}{1 - 4x}$ at $(2, -2)$.

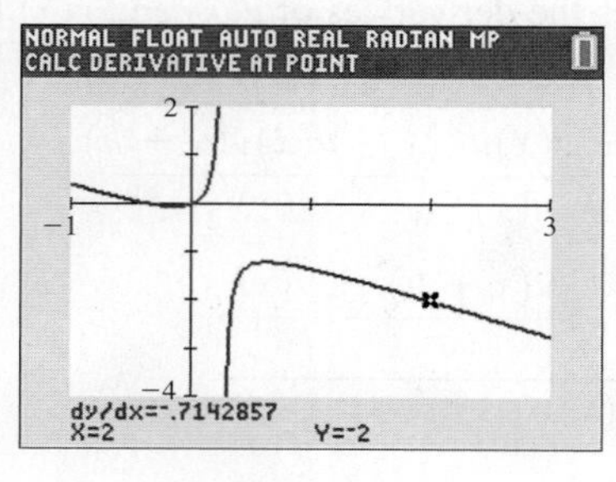

Fig. 23.31
Graphing calculator keystrokes: goo.gl/AKE5Mn

$$\frac{dy}{dx} = \frac{(1 - 4x)(6x + 1) - (3x^2 + x)(-4)}{(1 - 4x)^2}$$

$$= \frac{6x + 1 - 24x^2 - 4x + 12x^2 + 4x}{(1 - 4x)^2}$$

$$= \frac{-12x^2 + 6x + 1}{(1 - 4x)^2}$$

$$\left.\frac{dy}{dx}\right|_{x=2} = \frac{-12(2^2) + 6(2) + 1}{[1 - 4(2)]^2} = \frac{-35}{49}$$

$$= -\frac{5}{7}$$

The calculator evaluation using the dy/dx feature (graph-calc) is shown in Fig. 23.31. ■

EXERCISES 23.6

In Exercises 1 and 2, make the given changes in the indicated examples of this section and then solve the resulting problem.

1. In Example 2, change u to $5 - 3x^2$.
2. In Example 5, change the numerator to $3x^2 - x$ and then find and evaluate the derivative at $x = 2$.

In Exercises 3–8, find the derivative of each function by using the product rule. Do not find the product before finding the derivative.

3. $y = 6x(3x^2 - 5x)$
4. $y = 2x^3(3x^4 + x)$
5. $s = (3t + 2)(2t - 5)$
6. $f(x) = (3x - 2)(4x^2 + 3)$
7. $y = (x^4 - 3x^2 + 3)(1 - 2x^3)$
8. $y = (x^3 - 6x)(2 - 4x^3)$

In Exercises 9–12, find the derivative of each function by using the product rule. Then multiply out each function and find the derivative by treating it as a polynomial. Compare the results.

9. $y = (2x - 7)(5 - 2x)$
10. $f(s) = (5s^2 + 2)(2s^2 - 1)$
11. $V = (h^3 - 1)(2h^2 - h - 1)$
12. $y = (3x^2 - 4x + 1)(5 - 6x^2)$

In Exercises 13–24, find the derivative of each function by using the quotient rule.

13. $y = \frac{x}{8x + 3}$
14. $u = \frac{4}{v^2}$
15. $y = \frac{5}{2x^2 + 1}$
16. $R = \frac{5i + 9}{6i + 3}$
17. $y = \frac{6x^2}{3 - 2x}$
18. $y = \frac{e^2}{x(3x - 5)}$
19. $y = \frac{2x - 1}{3x^2 + 2}$
20. $P = \frac{8i^2}{4 - 3i}$
21. $f(x) = \frac{3x + 8}{x^2 + 4x + 2}$
22. $y = \frac{33x}{4x^5 - 3x - 4}$
23. $y = \frac{2x^2 - x - 1}{x^2(x + 2)}$
24. $y = \frac{x(3x^2 - 1)}{2x^2 - 5x + 4}$

In Exercises 25–32, evaluate the derivatives of the given functions for the given values of x. In Exercises 25–28, use the product rule. In Exercises 27, 28, 31, and 32, check your results using the derivative evaluation feature of a calculator.

25. $y = (3x - 1)(4 - 7x), x = 3$
26. $y = (3x^2 - 5)(2x^2 - 1), x = -1$
27. $y = (2x^2 - x + 1)(4 - 2x - x^2), x = -3$
28. $y = (4x^4 + 0.5x^2 + 1)(3x - 2x^2), x = 0.5$
29. $y = \frac{3x - 5}{2x + 3}, x = -2$
30. $y = \frac{2x^2 - 5x}{3x + 2}, x = 2$
31. $S = \frac{2n^3 - 3n + 8}{2n - 3n^4}, n = -1$
32. $y = \frac{2x^3 - x^2 - 2}{4x + 3}, x = 0.5$

In Exercises 33–58, solve the given problems by finding the appropriate derivatives.

33. Is it necessary to use the product rule to take the derivative of the function $y = \pi^2x^3$? Explain.
34. By use of the quotient rule, derive a formula for the derivative of the function $1/v(x)$.
35. Using the product rule, find the point(s) on the curve of $y = (2x^2 - 1)(1 - 4x)$ for which the tangent line is $y = 4x - 1$.
36. Do the curves of $y = x^2$ and $y = 1/x^2$ cross at right angles? Explain.

37. If $f(x)$ is a differentiable function, find an expression for the derivative of $y = x^2 f(x)$.

38. If $f(x)$ is a differentiable function, find an expression for the derivative of $y = f(x)/x^2$.

39. Find the derivative of $y = \dfrac{x^2 - 1}{x - 1}$ by (a) the quotient rule, and (b) by first simplifying the function.

40. Find the derivative of $y = [(3x - 1)(3x + 1)(x^2 - 4)]$ by using the product rule, and not first multiplying the factors. Check by first multiplying the factors.

41. Find the derivative of $y = \dfrac{x^2(1 - 2x)}{3x - 7}$ in each of the following two ways. (1) Do not multiply out the numerator before finding the derivative. (2) Multiply out the numerator before finding the derivative. Compare the results.

42. Find the derivative of $y = 4x^2 - \dfrac{1}{x - 1}$ in each of the following two ways. (1) Do not combine the terms over a common denominator before finding the derivative. (2) Combine the terms over a common denominator before finding the derivative. Compare the results.

43. Find the slope of a line tangent to the curve of the function $y = (4x + 1)(x^4 - 1)$ at the point $(-1, 0)$. Do not multiply the factors together before taking the derivative. Use the derivative evaluation feature of a calculator to check your result.

44. Find the slope of a line tangent to the curve of the function $y = (3x + 4)(1 - 4x)$ at the point $(2, -70)$. Do not multiply the factors together before taking the derivative. Use the derivative evaluation feature of a calculator to check your result.

45. For what value(s) of x is the slope of a tangent to the curve of $y = \dfrac{x}{x^2 + 1}$ equal to zero? View the graph on a calculator to verify the values found.

46. Determine the sign of the derivative of the function $y = \dfrac{2x - 1}{1 - x^2}$ for the following values of x: $-2, -1, 0, 1, 2$. Is the slope of a tangent line to this curve ever negative? View the graph on a calculator to verify your conclusion.

47. In the design of a rectangular container the area A (in cm^2) of the base is expressed as $A = 6x^2 - 11x - 10$, and the height h (in cm) is $h = 4x + 3$. Use the product rule to find the derivative of the volume V with respect to x for $x = 5.00$ cm.

48. The thermodynamic temperature T (in K) varies jointly as the pressure P (in Pa) and volume V (in m^3). Find the expression for dT/dt if $P = 1200(1 - 0.0025t)$ and $V = 2.50(1 + 0.0048t^2)$, where t is the time (in s).

49. If a constant current of 2 A passes through the *current divider* parallel resistors shown in Fig. 23.32, the current i is given by $i = 8R/(7R + 12)$, where R is a variable resistor. Find di/dR.

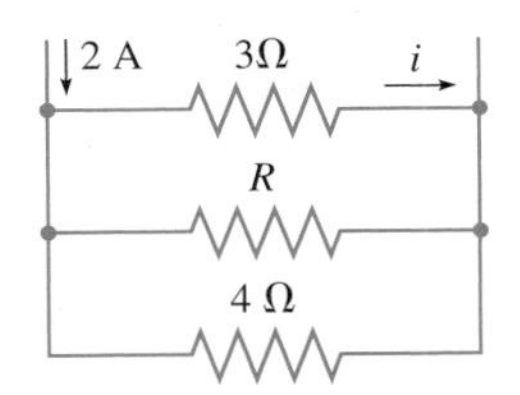

Fig. 23.32

50. The number n of dollars saved by increasing the fuel efficiency of e mi/gal to $e + 6$ mi/gal for a car driven 10,000 mi/year is $n = \dfrac{195{,}000}{e(e + 6)}$, if the cost of gas is \$3.25/gal. Find dn/de.

51. During each cycle, the vertical displacement s of the end of a robot arm is given by $s = (t^2 - 8t)(2t^2 + t + 1)$, where t is the time. Find the expression for the instantaneous velocity of the end of the robot arm. See Fig. 23.33.

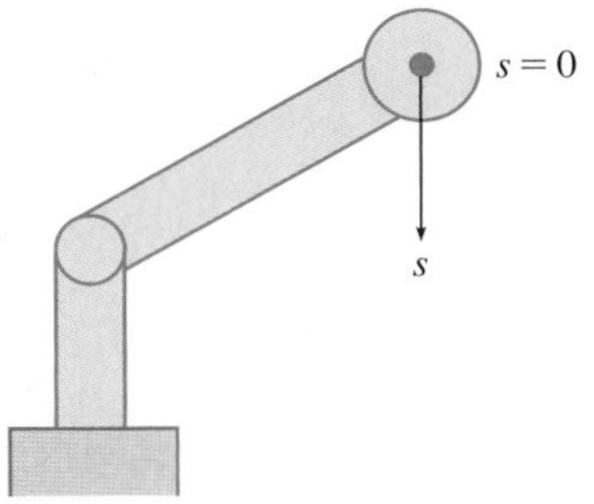

Fig. 23.33

52. The concentration c (in mg/L) of a certain drug in the bloodstream is found to be $c = 25t/(t^2 + 5)$, where t is the time (in h) after the drug is taken. Find dc/dt.

53. A computer, using data from a refrigeration plant, estimated that in the event of a power failure the temperature T (in °C) in the freezers would be given by $T = \dfrac{2t}{0.05t + 1} - 20$, where t is the number of hours after the power failure. Find the rate of change of temperature with respect to time after 6.0 h.

54. During a television commercial, a rectangular image appeared, increased in size, and then disappeared. If the length l (in cm) of the image, as a function of time t (in s) was $l = 6 - t$, and the width w (in cm) of the image was $w = t^2 + 4$, find the rate of change of the area of the rectangle with respect to time when $t = 5.00$ s.

55. The frictional radius r_f of a disc clutch is given by the equation $r_f = \dfrac{2(R^2 + Rr + r^2)}{3(R + r)}$, where R and r are the outer radius and the inner radius of the clutch, respectively. Find the derivative of r_f with respect to R with r constant.

56. In thermodynamics, an equation relating the thermodynamic temperature T, the pressure p, and the volume V of a gas is $T = \left(p + \dfrac{a}{V^2}\right)\left(\dfrac{V - b}{R}\right)$, where a, b, and R are constants. Find the derivative of T with respect to V, assuming p is constant.

57. The electric power P produced by a certain source is given by $P = \dfrac{E^2 r}{R^2 + 2Rr + r^2}$, where E is the voltage of the source, R is the resistance of the source, and r is the resistance in the circuit. Find the derivative of P with respect to r, assuming that the other quantities remain constant.

58. In the theory of lasers, the power P radiated is given by the equation $P = \dfrac{kf^2}{\omega^2 - 2\omega f + f^2 + a^2}$, where f is the field frequency and a, k, and ω are constants. Find the derivative of P with respect to f.

Answers to Practice Exercises

1. $dy/dx = -12x^5 + 12x^3 + 4x$ **2.** $dy/dx = \dfrac{-4x^5 - 4x}{(x^4 - 1)^2}$

23.7 The Derivative of a Power of a Function

Composite Function • Chain Rule • Derivative of a Power of a Function • Power Rule Extended to All Rational Exponents

In Example 1 of Section 23.6, we illustrated $y = (3 - 2x)^2$ as the power of a function of x, where $3 - 2x$ is the function. If we let $u = 3 - 2x$, we can write $y = u^2$, and in this way, y is a function of u, and u is a function of x. This means that *y is a function of a function of x, which is called a* **composite function.**

Because we will often need to find the derivative of a power of a function, which is a composite function, let us look at the derivative of $y = (3 - 2x)^2$.

EXAMPLE 1 Developing the chain rule

Find the derivative of $y = (3 - 2x)^2$.

We will use the product rule by treating $(3 - 2x)^2$ as $(3 - 2x)(3 - 2x)$. The reason for doing it this way will be shown.

$$y = (3 - 2x)^2 = (3 - 2x)(3 - 2x)$$

$$\frac{dy}{dx} = (3 - 2x)(-2) + (3 - 2x)(-2)$$

$$= 2(3 - 2x)(-2)$$

We want to leave the answer in this form in order to compare with a result we can get by letting $y = u^2$ and $u = 3 - 2x$:

$$y = u^2, \qquad \frac{dy}{du} = 2u$$

$$u = 3 - 2x, \qquad \frac{du}{dx} = -2$$

$$\left(\frac{dy}{du}\right)\left(\frac{du}{dx}\right) = 2u(-2) = 2(3 - 2x)(-2)$$

We see that this result is the same as the first result, and therefore for this function we see that $\frac{dy}{dx} = \left(\frac{dy}{du}\right)\left(\frac{du}{dx}\right)$. ■

The relationship shown in Example 1 is true for any differentiable composite function for which y is a function of u and u is a function of x.

Chain Rule

$$\frac{dy}{dx} = \frac{dy}{du}\frac{du}{dx} \tag{23.14}$$

Using Eq. (23.14) for $y = u^n$, where u is a differentiable function of x, we have

$$\frac{dy}{dx} = \frac{d(u^n)}{du}\frac{du}{dx}$$

This can be expressed as follows:

General Power Rule

$$\frac{du^n}{dx} = nu^{n-1}\left(\frac{du}{dx}\right) \tag{23.15}$$

We use Eq. (23.15) to find the derivative of a power of a differentiable function of x.

EXAMPLE 2 Derivative of a power of a function

Find the derivative of $y = (3 - 2x)^3$.

For this function, $n = 3$ and $u = 3 - 2x$. Therefore, $du/dx = -2$. This means

$$\frac{du^n}{dx} = n \quad u \quad n-1 \quad \left(\frac{du}{dx}\right)$$

$$\frac{dy}{dx} = 3(3 - 2x)^2(-2)$$

$$= -6(3 - 2x)^2$$

Practice Exercise

1. Find the derivative of $y = (5x + 2)^4$.

CAUTION A common type of error in finding this type of derivative is to omit the du/dx factor; in this case, it is the -2. The derivative is incomplete and therefore incorrect without this factor. ■ ■

EXAMPLE 3 Do not forget du/dx

Find the derivative of $p(x) = (1 - 3x^2)^4$.

In this example, $n = 4$ and $u = 1 - 3x^2$, and $du/dx = -6x$.

$$p'(x) = 4(1 - 3x^2)^3(-6x) = -24x(1 - 3x^2)^3$$

CAUTION We must not forget the $-6x$. ■ ■

EXAMPLE 4 Product rule combined with the power rule

Find the derivative of $y = 2x^3(3 - x^3)^4$.

Here, we must ***use the product rule in combination with the power rule.***

$$\frac{dy}{dx} = 2x^3[4(3 - x^3)^3(-3x^2)] + (3 - x^3)^4[2(3x^2)]$$

$$= -24x^5(3 - x^3)^3 + 6x^2(3 - x^3)^4 = 6x^2(3 - x^3)^3[-4x^3 + (3 - x^3)]$$

$$= 6x^2(3 - 5x^3)(3 - x^3)^3$$

■

Practice Exercise

2. Find the derivative of $y = 5x(2x + 7)^3$.

It is better to express the derivative in a factored, simplified form, because this is the form from which useful information may be found. In the next chapter, we will see that an analysis of the derivative has many uses. [Therefore, all derivatives should be in simplest algebraic form.]

NOTE ➧

To now, we have derived formulas for derivatives of differentiable functions of x raised to positive integer powers. We now show that these formulas are also valid for any rational number used as an exponent. If we raise each side of $y = u^{p/q}$ to the qth power, we have $y^q = u^p$. Applying the power rule, we have

$$qy^{q-1}\left(\frac{dy}{dx}\right) = pu^{p-1}\left(\frac{du}{dx}\right)$$

$$\frac{dy}{dx} = \frac{pu^{p-1}(du/dx)}{qy^{q-1}} = \frac{p}{q}\frac{u^{p-1}}{(u^{p/q})^{q-1}}\frac{du}{dx} = \frac{p}{q}\frac{u^{p-1}}{u^{p-p/q}}\frac{du}{dx}$$

$$= \frac{p}{q}u^{p-1-p+(p/q)}\frac{du}{dx}$$

Thus,

$$\frac{du^{p/q}}{dx} = \frac{p}{q}u^{(p/q)-1}\frac{du}{dx} \tag{23.16}$$

We see that in finding the derivative, we multiply the function by the rational exponent and subtract 1 from it to find the exponent of the function in the derivative. [This is the same rule as derived for positive integer exponents in the general power rule.]

NOTE ➧

In deriving Eqs. (23.15) and (23.16), we used the power rule, and we noted it was valid for positive integer exponents. We can show that the power rule is also valid for negative exponents by using the quotient rule on $1/x^n$, which is the same as x^{-n}. [Therefore, the general power rule can be extended to include all rational exponents, positive or negative.]

NOTE ▶

This, of course, includes all integer exponents, positive and negative. Also, we note that the power rule is a special case of the general power rule with $u = x$ (because $du/dx = 1$).

EXAMPLE 5 Derivative of a square root

We can now find the derivative of $y = \sqrt{x^2 + 1}$.

By using the general power rule and writing the square root as the fractional exponent 1/2, we can derive the result:

■ Note that we first rewrite the function in a different, more useful form. This is often an important step before taking the derivative.

$$y = (x^2 + 1)^{1/2}$$

$$\frac{dy}{dx} = \frac{1}{2}(x^2 + 1)^{-1/2}(2x)$$

$$= \frac{x}{(x^2 + 1)^{1/2}}$$

To avoid introducing apparently significant factors into the numerator, we do not usually rationalize such fractions. ■

Practice Exercise

3. Find the derivative of $y = \sqrt[3]{4 - 9x}$.

Having shown that we may use fractional exponents to find derivatives of roots of functions of x, we may also use them to find derivatives of roots of x itself. Consider the following example.

EXAMPLE 6 Derivative using a fractional exponent

Find the derivative of $y = 6\sqrt[3]{x^2}$.

We can write this function as $y = 6x^{2/3}$. In finding the derivative, we may use the power rule with $n = \frac{2}{3}$. This gives us

$$y = 6x^{2/3}$$

$$\frac{dy}{dx} = 6\left(\frac{2}{3}\right)x^{-1/3} = \frac{4}{x^{1/3}} \qquad \left(-\tfrac{1}{3} = \tfrac{2}{3} - 1\right)$$

We could also use the general power rule with $u = x$ and $n = \frac{2}{3}$. This give us

$$\frac{dy}{dx} = 6\left(\frac{2}{3}\right)x^{-1/3}(1) = \frac{4}{x^{1/3}} \qquad \left(\frac{du}{dx} = \frac{dx}{dx} = 1\right)$$

Note that the domain of the function is all real numbers, but the function is not differentiable for $x = 0$. ■

In the following examples, we illustrate the use of the power rule for the case in which n is a negative exponent. *Special care must be taken in the case of a negative exponent.*

EXAMPLE 7 Derivative using a negative exponent—acceleration of gravity

The acceleration due to gravity g on a satellite orbiting Earth varies inversely as the square of the distance r from the center of Earth. If $g = 8.7 \text{ m/s}^2$ for $r = 6.8 \times 10^6$ m, find the derivative of g with respect to r.

Because g varies inversely as the square of r, we have $g = k/r^2$. Using the given values, we have

$$8.7 \text{ m/s}^2 = \frac{k}{(6.8 \times 10^6)^2 \text{ m}^2}, \qquad k = 4.0 \times 10^{14} \text{ m}^3/\text{s}^2$$

which means that

$$g = \frac{4.0 \times 10^{14}}{r^2}$$

We could find the derivative by the quotient rule. However, when the numerator is constant, the derivative is easily found by using negative exponents:

$$g = \frac{4.0 \times 10^{14}}{r^2} = 4.0 \times 10^{14} r^{-2}$$

$$\frac{dg}{dr} = (4.0 \times 10^{14})(-2)(r^{-3}) \quad \longleftarrow -2 - 1 = -3$$

$$= -\frac{8.0 \times 10^{14}}{r^3} \text{ 1/s}^2$$

Here, we used the power rule directly. ■

EXAMPLE 8 Be careful using negative exponent

Find the derivative of $y = \frac{1}{(1 - 4x)^5}$.

The derivative is found as follows:

$$y = \frac{1}{(1 - 4x)^5} = (1 - 4x)^{-5} \quad \text{use negative exponent}$$

$$\frac{dy}{dx} = (-5)(1 - 4x)^{-6}(-4) \quad \text{use the general power rule}$$

$$= \frac{20}{(1 - 4x)^6} \quad \text{express result with positive exponent}$$

■

Practice Exercise

4. Find the derivative of $y = \frac{3}{(6x + 5)^4}$.

We now see the value of fractional exponents in calculus. They are useful in many algebraic operations, but they are almost essential in calculus. Without fractional exponents, it would be necessary to develop additional formulas to find the derivatives of radical expressions. In order to find the derivative of an algebraic function, we need only those formulas we have already developed. Often, it is necessary to combine these formulas, as we saw in Example 4. Actually, most derivatives are combinations. [The key step in finding the derivative is recognizing the *form of the function* with which you are dealing.] When you have recognized the form, completing the problem is only a matter of mechanics and algebra. You should now see the importance of being able to handle algebraic operations with ease.

NOTE ▸

In the next example, we again show the evaluation of a derivative and the calculator screen for this evaluation.

EXAMPLE 9 Evaluation of a derivative

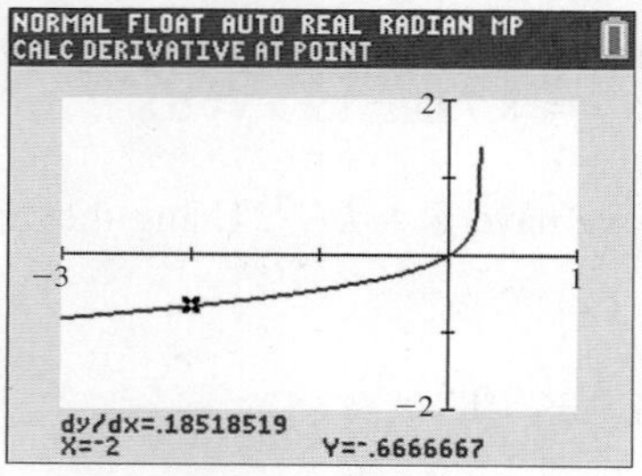

Evaluate the derivative of $y = \dfrac{x}{\sqrt{1-4x}}$ for $x = -2$.

Here, we have a quotient, and in order to find the derivative of this quotient, we must also use the power rule (and a derivative of a polynomial form). With sufficient practice in taking derivatives, we can recognize the rule to use almost automatically.

$$\frac{dy}{dx} = \frac{(1-4x)^{1/2}(1) - x(\frac{1}{2})(1-4x)^{-1/2}(-4)}{1-4x}$$

$$= \frac{(1-4x)^{1/2} + \dfrac{2x}{(1-4x)^{1/2}}}{1-4x} = \frac{\dfrac{(1-4x)^{1/2}(1-4x)^{1/2} + 2x}{(1-4x)^{1/2}}}{1-4x}$$

$$= \frac{(1-4x) + 2x}{(1-4x)^{1/2}(1-4x)} = \frac{1-2x}{(1-4x)^{3/2}}$$

Now, evaluating the derivative for $x = -2$, we have

$$\left.\frac{dy}{dx}\right|_{x=-2} = \frac{1-2(-2)}{[1-4(-2)]^{3/2}} = \frac{1+4}{(1+8)^{3/2}} = \frac{5}{9^{3/2}} = \frac{5}{27}$$

Using the dy/dx feature of a calculator, the graphical result is shown in Fig. 23.34 ($5/27 = 0.18518519$.) ■

Fig. 23.34
Graphing calculator keystrokes: goo.gl/GK0Qgs

EXERCISES 23.7

In Exercises 1–4, make the given changes in the indicated examples of this section and then find the derivatives.

1. In Example 3, change $1 - 3x^2$ to $2 + 3x^3$.
2. In Example 4, change $3 - x^3$ to $2 + x^5$.
3. In Example 5, change $x^2 + 1$ to $2 - 3x^2$.
4. In Example 8, change the exponent 5 to 3.

In Exercises 5–32, find the derivative of each of the given functions.

5. $y = 4\sqrt{x}$
6. $y = \sqrt[4]{x^3}$
7. $v = \dfrac{3}{5t^2}$
8. $y = \dfrac{2}{x^4}$
9. $y = \dfrac{3}{\sqrt[3]{x}} + 4x^2$
10. $y = \dfrac{55}{\sqrt[5]{x^2+3}}$
11. $y = x\sqrt{x} - \dfrac{6}{x}$
12. $f(x) = 2x^{-3} - 3x^{-2}$
13. $y = (4x^2 + 3)^5$
14. $y = (1 - 6x)^4$
15. $y = 2.25(7 - 4x^3)^8$
16. $s = 3(8t^2 - 7)^6$
17. $y = (2x^3 - 3)^{1/3}$
18. $y = 8(1 - 6x)^{1.5}$
19. $f(y) = \dfrac{3}{(4 - y^2)^4}$
20. $y = \dfrac{\pi^3}{\sqrt{1 - 3x}}$
21. $y = 4(2x^4 - 5)^{0.75}$
22. $r = 5(3\theta^6 - 4)^{2/3}$
23. $y = \sqrt[4]{1 - 8x^2}$
24. $y = 9\sqrt[3]{4x^6 + 2}$
25. $u = v\sqrt{8v + 5}$
26. $y = 8x^2(1 - 3x)^5$
27. $y = \dfrac{2\sqrt{1 - 6x}}{x^3}$
28. $R = \dfrac{5T^2}{\sqrt[3]{1 + 4T}}$
29. $y = \dfrac{6x\sqrt{x + 2}}{x + 4}$
30. $y = 8\sqrt{1 + \sqrt{x}}$
31. $f(R) = \sqrt{\dfrac{2R + 1}{4R + 1}}$
32. $y = \left(\dfrac{2x + 1}{3x - 2}\right)^2$

In Exercises 33–36, evaluate the derivatives of the given functions for the given values of x. In Exercises 35 and 36, check your results, using the derivative evaluation feature of a calculator.

33. $y = \sqrt{3x + 4}, x = 7$
34. $y = 6(4 - x^2)^{-1}, x = -1$
35. $y = \dfrac{\sqrt{x}}{1 - x}, x = 4$
36. $y = x^2\sqrt[3]{3x + 2}, x = 2$

In Exercises 37–58, solve the given problems by finding the appropriate derivatives.

37. Find the derivative of $y = 1/x^3$ as (a) a quotient and (b) a negative power of x and show that the results are the same.
38. Let $y = [u(x)]^2$ and find dy/dx, treating $[u(x)]^2$ as the product $u(x)u(x)$. (See Example 1.)
39. Find any values of x for which the derivative of $y = \dfrac{x^2}{\sqrt{x^2 + 1}}$ is zero. View the curve of the function on a calculator to verify the values found.

40. Find any values of x for which the derivative of $y = \dfrac{x}{\sqrt{4x-1}}$ is zero. View the curve of the function on a calculator to verify the values found.

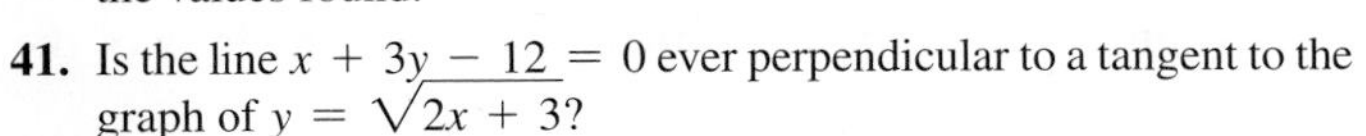

41. Is the line $x + 3y - 12 = 0$ ever perpendicular to a tangent to the graph of $y = \sqrt{2x+3}$?

42. Explain two different ways of taking the derivative of the function $y = \dfrac{1}{(x+4)^3}$. Which way seems easier?

43. Find the slope of a line tangent to the parabola $y^2 = 4x$ at the point $(1, 2)$. Use the derivative evaluation feature of a calculator to check your result.

44. A wheel that can be represented by $x^2 + y^2 = 25$ is rotating when a particle is ejected tangentially from the point $(4, 3)$. Find the slope of the line along which the particle traveled. Use the derivative feature of a calculator to check your result.

45. The lowest flying speed v (in ft/s) at which a certain airplane can fly varies directly as the square root of the wing load w (in lb/ft^2). If $v = 88$ ft/s for $w = 16$ lb/ft^2, find the derivative of v with respect to w.

46. During and after a period of rain, the depth h (in m) of water behind a certain dam was given by $h = 75(x+2)/(x+3)$, where x is the number of days after the start of the rain period. Find dh/dx for $x = 2.5$ days.

47. The displacement s (in cm) of a linkage joint of a robot is given by $s = (8t - t^2)^{2/3}$, where t is the time (in s). Find the velocity of the joint for $t = 6.25$ s.

48. Water is slowly rising in a horizontal drainage pipe. The width w of the water as a function of the depth h is $w = \sqrt{2rh - h^2}$, where r is the radius of the pipe. If $r = 6.00$ in., find dw/dh for $h = 2.25$ in. See Fig. 23.35.

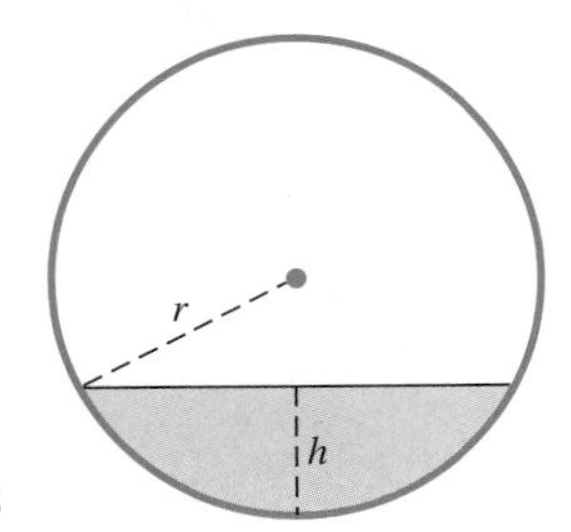

Fig. 23.35

49. When the volume of a gas changes very rapidly, an approximate relation is that the pressure P varies inversely as the 3/2 power of the volume. If P is 300 kPa when $V = 100$ cm^3, find the derivative of P with respect to V. Evaluate this derivative for $V = 100$ cm^3.

50. The power gain G of a certain antenna is inversely proportional to the square of the wavelength λ (in ft) of the carrier wave. If $G = 5.0 \times 10^4$ for $\lambda = 0.35$ ft, find the derivative of G with respect to λ for $\lambda = 0.35$ ft.

51. In deep water, the velocity of a wave is $v = k\sqrt{\dfrac{l}{a} + \dfrac{a}{l}}$, where a and k are constants and l is the length of the wave. For what value of l is $dv/dl = 0$?

52. Due to air friction, the drag F on a plane is $F = c_1v^2 + c_2v^{-2}$, where v is the plane's velocity and c_1 and c_2 are positive constants. For what values of v is $dF/dv = 0$?

53. The total solar radiation H (in W/m^2) on a certain surface during an average clear day is given by

$$H = \frac{4000}{\sqrt{t^6 + 100}} \quad (-6 < t < 6)$$

where t is the number of hours from noon. Find the rate at which H is changing with time at 4 P.M.

54. In determining the time for a laser beam to go from S to P (see Fig. 23.36), which are in different mediums, it is necessary to find the derivative of the time

$$t = \frac{\sqrt{a^2 + x^2}}{v_1} + \frac{\sqrt{b^2 + (c-x)^2}}{v_2}$$

with respect to x, where a, b, c, v_1, and v_2 are constants. Here, v_1 and v_2 are the velocities of the laser in each medium. Find this derivative.

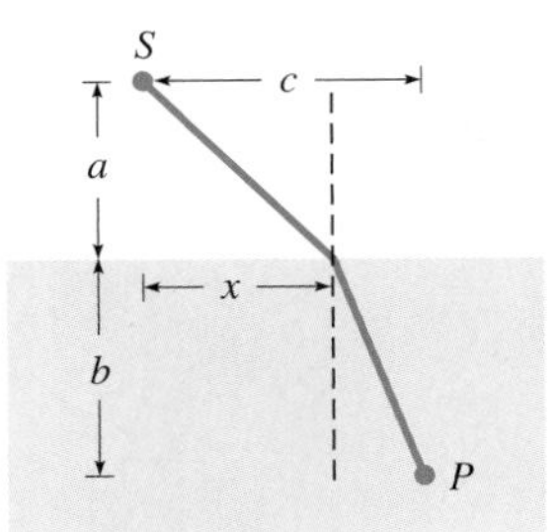

Fig. 23.36

55. The radio waveguide wavelength λ_r is related to its free-space wavelength λ by

$$\lambda_r = \frac{2a\lambda}{\sqrt{4a^2 - \lambda^2}}$$

where a is a constant. Find $d\lambda_r/d\lambda$.

56. The current I in a circuit containing a resistance R and an inductance L is found from the expression

$$I = \frac{V}{\sqrt{R^2 + (\omega L)^2}}$$

Find the expression for the instantaneous rate of change of current with respect to L, assuming that the other quantities remain constant.

57. The length l of a rectangular microprocessor chip is 2 mm longer than its width w. Find the derivative of the length of the diagonal D with respect to w.

58. The trapezoidal structure shown in Fig. 23.37 has an internal support of length l. Find the derivative of l with respect to x.

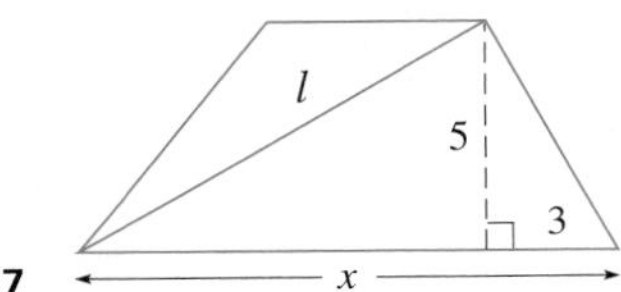

Fig. 23.37

Answers to Practice Exercises

1. $dy/dx = 20(5x+2)^3$ **2.** $dy/dx = 5(2x+7)^2(8x+7)$

3. $dy/dx = -3(4-9x)^{-2/3}$ **4.** $dy/dx = -72(6x+5)^{-5}$

23.8 Differentiation of Implicit Functions

Implicit Function • Differentiating Term by Term

To this point, the functions we have differentiated have been of the form $y = f(x)$. There are, however, occasions when we need to find the derivative of a function determined by an equation that does not express the dependent variable explicitly in terms of the independent variable.

An equation in which y is not expressed explicitly in terms of x may determine one or more functions. *Any such function, where y is defined implicitly as a function of x, is called an* **implicit function.** Some equations defining implicit functions may be solved to determine the explicit functions, and for others it is not possible to solve for the explicit functions. Also, not all such equations define y as a function of x for real values of x.

EXAMPLE 1 Illustrations of implicit functions

(a) The equation $3x + 4y = 5$ is an equation that defines a function, although it is not in explicit form. In solving for y as $y = -\frac{3}{4}x + \frac{5}{4}$, we have the explicit form of the function.

(b) The equation $y^2 + x = 3$ is an equation that defines two functions, although we do not have the explicit forms. When we solve for y, we obtain the explicit functions $y = \sqrt{3 - x}$ and $y = -\sqrt{3 - x}$.

(c) The equation $y^5 + xy^2 + 3x^2 = 5$ defines y as a function of x, although we cannot actually solve for the explicit algebraic form of the function.

(d) The equation $x^2 + y^2 + 4 = 0$ is not satisfied by any pair of real values of x and y. ■

Even when it is possible to determine the explicit form of a function given in implicit form, it is not always desirable to do so. In some cases, the implicit form is more convenient than the explicit form.

The derivative of an implicit function may be found directly without having to solve for the explicit function. [To find dy/dx when y is defined as an implicit function of x, we differentiate each term of the equation with respect to x, regarding y as a differentiable function of x. We then solve for dy/dx, which will usually be in terms of x and y.]

NOTE ▸

EXAMPLE 2 Implicit derivative

Find dy/dx if $y^2 + 2x^2 = 5$.

Here, we find the derivative of each term and then solve for dy/dx. Thus,

$$\frac{d(y^2)}{dx} + \frac{d(2x^2)}{dx} = \frac{d(5)}{dx}$$

$$2y^{2-1}\frac{dy}{dx} + 2\left(2x^{2-1}\frac{dx}{dx}\right) = 0$$

$$2y\frac{dy}{dx} + 4x = 0$$

$$\frac{dy}{dx} = -\frac{2x}{y}$$

■ For reference, the general power rule is $\frac{du^n}{dx} = nu^{n-1}\frac{du}{dx}$.

CAUTION The factor dy/dx arises from the derivative of the first term as a result of using the derivative of a power of a function of x (the general power rule). The factor dy/dx corresponds to the du/dx of the formula. ■ In the second term, no factor of dy/dx appears, because there are no y factors in the term. ■

EXAMPLE 3 Implicit derivative involving a product

Find dy/dx if $3y^4 + xy^2 + 2x^3 - 6 = 0$.

NOTE ▸ [In finding the derivative, we note that the second term is a *product*, and we must use the *product rule* for derivatives on it.] Thus, we have

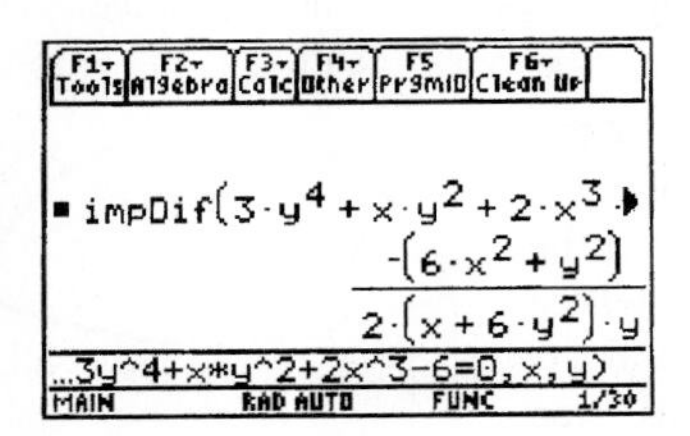

Fig. 23.38
TI-89 graphing calculator keystrokes: goo.gl/Y80Xn9

■ Figure 23.38 shows dy/dx on a TI-89 graphing calculator.

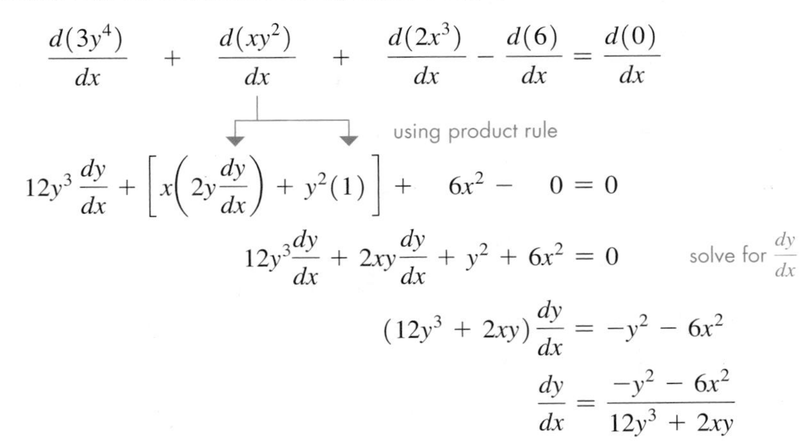

$$\frac{d(3y^4)}{dx} + \frac{d(xy^2)}{dx} + \frac{d(2x^3)}{dx} - \frac{d(6)}{dx} = \frac{d(0)}{dx}$$

using product rule

$$12y^3\frac{dy}{dx} + \left[x\left(2y\frac{dy}{dx}\right) + y^2(1)\right] + 6x^2 - 0 = 0$$

$$12y^3\frac{dy}{dx} + 2xy\frac{dy}{dx} + y^2 + 6x^2 = 0 \qquad \text{solve for } \frac{dy}{dx}$$

$$(12y^3 + 2xy)\frac{dy}{dx} = -y^2 - 6x^2$$

$$\frac{dy}{dx} = \frac{-y^2 - 6x^2}{12y^3 + 2xy}$$ ■

EXAMPLE 4 Implicit derivative—product and power

Find dy/dx if $2x^3y + (y^2 + x)^3 = x^4$.

In this case, we use the product rule on the first term and the power rule on the second term:

$$\frac{d(2x^3y)}{dx} + \frac{d(y^2+x)^3}{dx} = \frac{d(x^4)}{dx}$$

product rule power rule

$$2x^3\left(\frac{dy}{dx}\right) + y(6x^2) + 3(y^2+x)^2\left(2y\frac{dy}{dx} + 1\right) = 4x^3$$

$$2x^3\frac{dy}{dx} + 6x^2y + 3(y^2+x)^2\left(2y\frac{dy}{dx}\right) + 3(y^2+x)^2 = 4x^3$$

$$[2x^3 + 6y(y^2+x)^2]\frac{dy}{dx} = 4x^3 - 6x^2y - 3(y^2+x)^2$$

$$\frac{dy}{dx} = \frac{4x^3 - 6x^2y - 3(y^2+x)^2}{2x^3 + 6y(y^2+x)^2}$$ ■

Practice Exercise

1. Find dy/dx if $2y^3 + xy + 1 = 0$.

EXAMPLE 5 Slope of a tangent line

Find the slope of a tangent line to the curve $x^3 + y^3 - 9xy = 0$ at the point $(2, 4)$. This curve is known as a *folium*, which dates back to Descartes in the 1630s. See Fig. 23.39.

Here, we are to find dy/dx and evaluate it for $x = 2$ and $y = 4$.

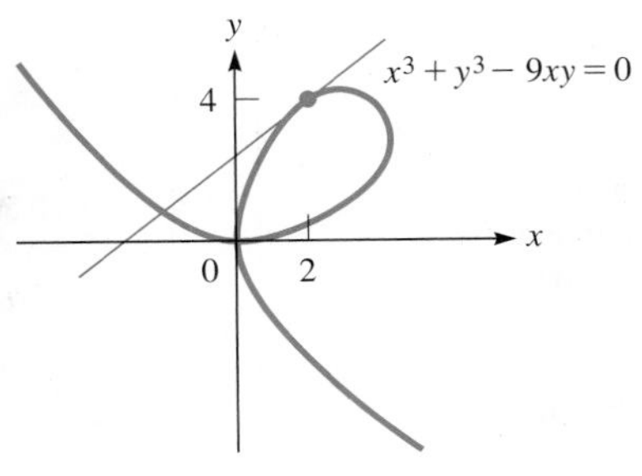

Fig. 23.39

$$\frac{d(x^3)}{dx} + \frac{d(y^3)}{dx} - \frac{d(9xy)}{dx} = \frac{d(0)}{dx}$$

$$3x^2 + 3y^2\frac{dy}{dx} - 9\left(x\frac{dy}{dx} + y\right) = 0$$

$$(3y^2 - 9x)\frac{dy}{dx} = 9y - 3x^2$$

$$\frac{dy}{dx} = \frac{3y - x^2}{y^2 - 3x} \qquad \left.\frac{dy}{dx}\right|_{(2,4)} = \frac{3(4) - 2^2}{4^2 - 3(2)} = \frac{8}{10} = \frac{4}{5}$$

Therefore, the slope of the tangent line at $(2, 4)$ is $4/5$. ■

EXERCISES 23.8

In Exercises 1 and 2, make the given changes in the indicated examples of this section and then find dy/dx.

1. In Example 2, change y^2 to y^3.

2. In Example 3, change xy^2 to x^2y.

In Exercises 3–22, find dy/dx by differentiating implicitly. When applicable, express the result in terms of x and y.

3. $3x + 2y = 5$

4. $6x - 3y = 4$

5. $4y - 3x^2 = x$

6. $x^5 - 5y = 6 - 4x^{3/2}$

7. $x^2 - 4y^2 - 9 = 0$

8. $\sqrt{x^2 + 2y^2} - 11 = 0$

9. $y^5 = x^2 - 1$

10. $x^{2/3} + y^{2/3} = 5$

11. $6y^{2/3} + y = x^2 - 4$

12. $2y^3 - y = 7 - x^4$

13. $y + 3xy - 4 = 0$

14. $xy^3 + 3y + x^2 = 2\pi^2$

15. $x^2 = \dfrac{x - y}{x + y}$

16. $y^2x - \dfrac{5y}{x + 1} + 3x = 4$

17. $\dfrac{3x^2}{y^2 + 1} + y = 3x + 1$

18. $\sqrt{xy} = \dfrac{x}{4} - \dfrac{1}{y^2}$

19. $(2y - x)^4 + x^2 = y + 3$

20. $(y^2 + 2)^3 = x^4y + e^2$

21. $2(x^2 + 1)^3 + \sqrt{y^2 + 1} = 17$

22. $(2x + 1)(1 - 3y) + y^2 = 13$

In Exercises 23–28, evaluate dy/dx at the given points.

23. $3x^3y^2 - 2y^3 = -4;\quad (1, 2)$

24. $2y + 5 - x^2 - y^3 = 0;\quad (2, -1)$

25. $5y^4 + 7 = x^4 - 3y;\quad (3, -2)$

26. $(xy - y^2)^{3/2} = 5y^2 + 3;\quad (5, 1)$

27. $xy^2 + 3x^2 - y^2 + 15 = 0;\quad (-1, 3)$

28. $2(x + y)^3 - y^2/x = 15;\quad (4, -2)$

In Exercises 29–44, solve the given problems by using implicit differentiation.

29. At what point(s) does the graph of $x^2 + y^2 = 4x$ have a horizontal tangent?

30. Show that if $P(x,y)$ is any point on the circle $x^2 + y^2 = a^2$, then a tangent line at P is perpendicular to a line through P and the origin.

31. Show that two tangents to the curve $x^2 + xy + y^2 = 7$ at the points where it crosses the x-axis are parallel.

32. At what point(s) is the tangent to the curve $y^2 = 2x^3$ perpendicular to the line $4x - 3y + 1 = 0$?

33. In an RLC circuit, the angular frequency ω at which the circuit resonates is given by $\omega^2 = 1/LC - R^2/L^2$. Find $d\omega/dL$.

34. Show that the graphs of $2x^2 + y^2 = 24$ and $y^2 = 8x$ are perpendicular at the point $(2, 4)$. Display the graphs on a calculator.

35. Find the slope of a line tangent to the curve of the implicit function $xy + y^2 + 2 = 0$ at the point $(-3, 1)$. Use the derivative evaluation feature of a calculator to check your result.

36. A lens is formed by cutting a cap (with a flat base) from a spherical piece of glass. The volume V of the lens is $V = \frac{1}{3}\pi h^2(3r - h)$, where r is the radius of the sphere and h is the thickness of the lens. See Fig. 23.40. If $V = 8\pi/3$, find dr/dh.

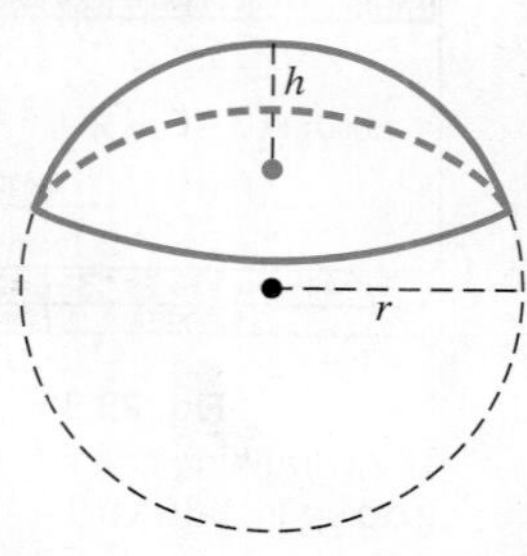

Fig. 23.40

37. The pressure P, volume V, and temperature T of a gas are related by $PV = n(RT + aP - bP/T)$, where a, b, n, and R are constants. For constant V, find dP/dT.

38. Oil moves through a pipeline such that the distance s it moves and the time t are related by $s^3 - t^2 = 7t$. Find the velocity of the oil for $s = 4.01$ m and $t = 5.25$ s.

39. The shelf support shown in Fig. 23.41 is 2.38 ft long. Find the expression for dy/dx in terms of x and y.

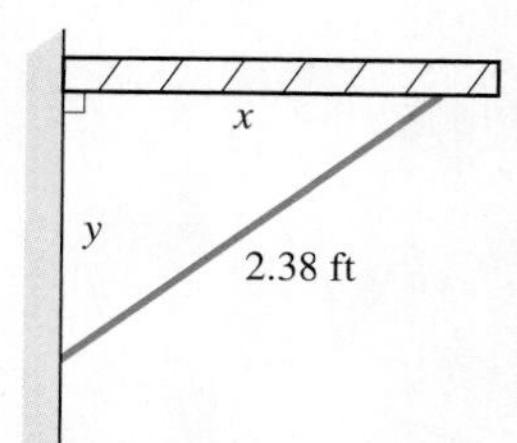

Fig. 23.41

40. An open (no top) right circular cylindrical container of radius r and height h has a total surface area of 940 cm^2. Find dr/dh in terms of r and h.

41. Two resistors, with resistances r and $r + 2$, are connected in parallel. Their combined resistance R is related to r by the equation $r^2 = 2rR + 2R - 2r$. Find dR/dr.

42. The polar moment of inertia I of a rectangular slab of concrete is given by $I = \frac{1}{12}(b^3h + bh^3)$, where b and h are the base and the height, respectively, of the slab. If I is constant, find the expression for db/dh.

43. A formula relating the length L and radius of gyration r of a steel column is $24C^3Sr^3 = 40C^3r^3 + 9LC^2r^2 - 3L^3$, where C and S are constants. Find dL/dr.

44. A computer is programmed to draw the graph of the implicit function $(x^2 + y^2)^3 = 64x^2y^2$ (see Fig. 23.42). Find the slope of a line tangent to this curve at $(2.00, 0.56)$ and at $(2.00, 3.07)$.

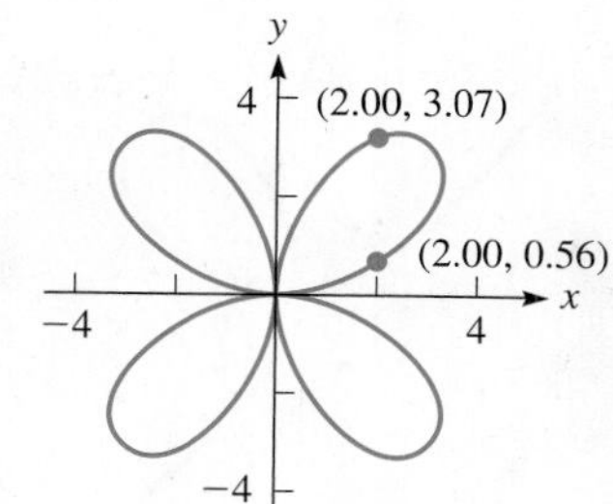

Fig. 23.42

Answer to Practice Exercise

1. $dy/dx = -y/(6y^2 + x)$

23.9 Higher Derivatives

First Derivative • Second Derivative • Higher Derivatives • Instantaneous Acceleration

Earlier we noted that the derivative of a function is itself a function. Therefore, we may take its derivative. In this section, we develop the concept and notation for the derivatives of a derivative, as well as show some of the applications.

The derivative of a function is called the **first derivative** *of the function. The derivative of the first derivative is called the* **second derivative.** Because the second derivative is a function, we may find its derivative, which is called the third derivative. We may continue to find the fourth derivative, fifth derivative, and so on (provided each derivative is defined). *The second derivative, third derivative, and so on, are known as* **higher derivatives.**

The notations used for higher derivatives follow closely those used for the first derivative. As shown in Section 23.3, the notations for the first derivative are y', D_xy, $f'(x)$, and dy/dx. The notations for the second derivative are y'', D_x^2y, $f''(x)$, and d^2y/dx^2. Similar notations are used for other higher derivatives.

EXAMPLE 1 Higher derivatives of a function

Find the higher derivatives of $y = 5x^3 - 2x$.

We find the first derivative as

$$\frac{dy}{dx} = 15x^2 - 2 \quad \text{or} \quad y' = 15x^2 - 2$$

Next, we obtain the second derivative by finding the derivative of the first derivative:

$$\frac{d^2y}{dx^2} = 30x \quad \text{or} \quad y'' = 30x$$

Continuing to find the successive derivatives, we have

$$\frac{d^3y}{dx^3} = 30 \quad \text{or} \quad y''' = 30$$

$$\frac{d^4y}{dx^4} = 0 \quad \text{or} \quad y^{(4)} = 0$$

Because the third derivative is a constant, the fourth derivative and all successive derivatives will be zero. This can be shown as $d^ny/dx^n = 0$ for $n \geq 4$. ■

Practice Exercise

1. Find the second derivative of $y = 4x^4 - 6x^2 + 8x$.

EXAMPLE 2 Higher derivatives of a function

Find the higher derivatives of $f(x) = x(x^2 - 1)^2$.

Using the product rule to find the first derivative, we have

$$\begin{aligned} f'(x) &= x(2)(x^2 - 1)(2x) + (x^2 - 1)^2(1) \\ &= (x^2 - 1)(4x^2 + x^2 - 1) = (x^2 - 1)(5x^2 - 1) \\ &= 5x^4 - 6x^2 + 1 \end{aligned}$$

■ For reference, the product rule is $\frac{d(uv)}{dx} = u\frac{dv}{dx} + v\frac{du}{dx}$.

Continuing to find the higher derivatives, we have

$$\begin{aligned} f''(x) &= 20x^3 - 12x \\ f'''(x) &= 60x^2 - 12 \\ f^{(4)}(x) &= 120x \\ f^{(5)}(x) &= 120 \\ f^{(n)}(x) &= 0 \qquad \text{for } n \geq 6 \end{aligned}$$

■ Note that when using the prime $(f'(x))$ notation the nth derivative may be shown as $f^{(n)}(x)$.

All derivatives after the fifth derivative are equal to zero. ■

EXAMPLE 3 Evaluation of second derivative

Evaluate the second derivative of $y = \frac{2}{1-x}$ for $x = -2$.

We write the function as $y = 2(1-x)^{-1}$ and then find the derivatives:

TI-89 graphing calculator keystrokes for Example 3: goo.gl/Fg6qWO

$$y = 2(1-x)^{-1}$$
$$\frac{dy}{dx} = 2(-1)(1-x)^{-2}(-1) = 2(1-x)^{-2}$$
$$\frac{d^2y}{dx^2} = 2(-2)(1-x)^{-3}(-1) = 4(1-x)^{-3} = \frac{4}{(1-x)^3}$$

Evaluating the second derivative for $x = -2$, we have

$$\left.\frac{d^2y}{dx^2}\right|_{x=-2} = \frac{4}{(1+2)^3} = \frac{4}{27}$$

The function is not differentiable for $x = 1$. Also, if we continue to find higher derivatives, the expressions will not become zero, as in Examples 1 and 2. ■

EXAMPLE 4 Second derivative of an implicit function

Find y'' for the implicit function defined by $2x^2 + 3y^2 = 6$.

Differentiating with respect to x, we have

$$2(2x) + 3(2yy') = 0$$
$$4x + 6yy' = 0 \quad \text{or} \quad 2x + 3yy' = 0$$

NOTE ➧ [Before differentiating again, we see that $3yy'$ is a *product*, and we note that the derivative of y' is y''.] Thus, differentiating again using the product rule, we have

differentiation of $3yy'$

$$2 + 3yy'' + 3y'(y') = 0$$
$$2 + 3yy'' + 3(y')^2 = 0$$

Now, solving Eq. (1) for y' and substituting this into Eq. (2), we have

$$y' = -\frac{2x}{3y}$$
$$2 + 3yy'' + 3\left(-\frac{2x}{3y}\right)^2 = 0$$
$$2 + 3yy'' + \frac{4x^2}{3y^2} = 0$$
$$6y^2 + 9y^3y'' + 4x^2 = 0$$
$$y'' = \frac{-4x^2 - 6y^2}{9y^3} = \frac{-2(2x^2 + 3y^2)}{9y^3}$$

Practice Exercise

2. Find the second derivative of $y = \frac{3}{x^2 + 4}$.

Because $2x^2 + 3y^2 = 6$, we have

$$y'' = \frac{-2(6)}{9y^3} = -\frac{4}{3y^3}$$ ■

As mentioned earlier, higher derivatives are useful in certain applications. This is particularly true of the second derivative. The first and second derivatives are used in the next chapter for several types of applications, and higher derivatives are used when we discuss infinite series in Chapter 30. An important technical application of the second derivative is shown in the example that follows.

In Section 23.4, we briefly discussed the instantaneous velocity of an object, and in the exercises we mentioned acceleration. From that discussion, recall that the instantaneous velocity is the rate of change of the displacement with respect to time, and that *the* **instantaneous acceleration** *is the rate of change of the instantaneous velocity with respect to time.* [Therefore, the acceleration is found from the *second derivative* of the displacement with respect to time.]

NOTE ▸

EXAMPLE 5 Instantaneous acceleration—rocket flight

For the first 12 s after launch, the height s (in m) of a certain rocket is given by $s = 10\sqrt{t^2 + 25} - 50$. Find the vertical acceleration of the rocket when $t = 10.0$ s.

Because the velocity is found from the first derivative and the acceleration is found from the second derivative, we must find the second derivative and evaluate it for $t = 10.0$ s.

$$s = 10\sqrt{t^4 + 25} - 50$$

$$v = \frac{ds}{dt} = 10\left(\frac{1}{2}\right)(t^4 + 25)^{-1/2}(4t^3) = \frac{20t^3}{(t^4 + 25)^{1/2}}$$

$$a = \frac{dv}{dt} = \frac{d^2s}{dt^2} = \frac{(t^4 + 25)^{1/2}(60t^2) - 20t^3\left(\frac{1}{2}\right)(t^4 + 25)^{-1/2}(4t^3)}{t^4 + 25}$$

$$= \frac{(t^4 + 25)(60t^2) - 40t^6}{(t^4 + 25)^{3/2}} = \frac{20t^6 + 1500t^2}{(t^4 + 25)^{3/2}}$$

$$= \frac{20t^2(t^4 + 75)}{(t^4 + 25)^{3/2}}$$

multiply numerator and denominator by $(t^4 + 25)^{1/2}$

Finding the value of the acceleration when $t = 10.0$ s, we have

$$a\big|_{t=10.0} = \frac{20(10.0)^2(10.0^4 + 75)}{(10.0^4 + 25)^{3/2}} = 20.1 \text{ m/s}^2$$

■

EXERCISES 23.9

In Exercises 1 and 2, make the given changes in the indicated examples of this section and then solve the resulting problem.

1. In Example 1, change $2x$ to $2x^2$.
2. In Example 3, in the denominator change $1 - x$ to $1 + 2x$.

In Exercises 3–10, find the first, second, and third derivatives of the given functions.

3. $y = x^3 + 7x^2$
4. $f(x) = 3x - x^4$
5. $f(x) = x^3 - 6x^4$
6. $s = 8t^5 + 5t^4$
7. $y = (1 - 2x)^4$
8. $f(x) = 4(3x + 2)^3$
9. $f(r) = r(4r + 9)^3$
10. $y = x(5x - 1)^3$

In Exercises 11–28, find the second derivative of each of the given functions.

11. $y = 2x^7 - x^6 - 3x$
12. $y = 6x - 2x^5$
13. $y = 5x + 8\sqrt{x}$
14. $r = 3\theta^2 - \dfrac{20}{\sqrt{\theta}}$
15. $f(x) = \sqrt[4]{8x - 3}$
16. $f(x) = 3\sqrt[3]{6x + 5}$
17. $f(p) = \dfrac{4.8\pi}{\sqrt{1 + 2p}}$
18. $f(x) = \dfrac{7.5}{\sqrt{3 - 4x}}$
19. $y = 2(2 - 5x)^4$
20. $y = \dfrac{1}{3}(4x + 1)^6$
21. $y = (3x^2 - 1)^5$
22. $y = 3(2x^3 + 3)^4$
23. $f(x) = \dfrac{2\pi^2}{6 - x}$
24. $f(R) = \dfrac{1 - 3R}{1 + 3R}$
25. $u = \dfrac{v^2}{4v + 15}$
26. $y = \dfrac{8x}{\sqrt{9 - x^2}}$
27. $x^2 - 4y^2 = 9$
28. $9x^2 + y^2 = 36$

In Exercises 29–34, evaluate the second derivative of the given function for the given value of x.

29. $f(x) = \sqrt{x^2 + 9}, x = 4$
30. $f(x) = x - \dfrac{2}{x^3}, x = -1$
31. $y = 3x^{2/3} - \dfrac{2}{x}, x = -8$
32. $y = 3(1 + 2x)^4, x = \dfrac{1}{2}$
33. $v = t(8 - t)^5, t = 2$
34. $y = \dfrac{9x}{2 - 3x}; x = -\dfrac{1}{3}$

In Exercises 35–38, find the acceleration of an object for which the displacement s (in m) is given as a function of the time t (in s) for the given value of t.

35. $s = 26t - 4.9t^2,\ t = 3.0$ s **36.** $s = 3(1 + 2t)^4,\ t = 0.500$ s

37. $s = \dfrac{16}{0.5t^2 + 1},\ t = 2$ s **38.** $s = 250\sqrt{6t + 7},\ t = 7.0$ s

In Exercises 39–52, solve the given problems by finding the appropriate derivatives.

39. Find the ordered pair (x, y) on the graph of $y = x^3 - 3x^2$ for which $f''(x) = 0$. (This is called an *inflection point.*)

40. Show that $\dfrac{d^6(x^6)}{dx^6} = 6!\quad (6! = 6 \times 5 \times 4 \times 3 \times 2 \times 1)$

41. What is the instantaneous rate of change of the first derivative of y with respect to x for $y = (1 - 2x)^4$ for $x = 1$?

42. What is the instantaneous rate of change of the first derivative of y with respect to x for $2xy + y = 1$ for $x = 0.5$?

43. If the population of a city is $P(t) = 8000(1 + 0.02t + 0.005t^2)$ (t is in years from 2010), what is the acceleration in the size of the population?

44. Find a second-degree polynomial such that $f(2) = 6, f'(2) = 3$, and $f''(2) = 2$.

45. Find a third-degree polynomial such that $f(-1) = 9, f'(-1) = 8$, $f''(-1) = -14$, and $f'''(-1) = 12$.

46. The potential V (in V) of a certain electric charge is given by $V = 6/(2t + 1)$, where t is the time (in s). Find d^2V/dt^2.

47. A bullet is fired vertically upward. Its distance s (in ft) above the ground is given by $s = 2250t - 16.1t^2$, where t is the time (in s). Find the acceleration of the bullet.

48. In testing the brakes on a new model car, it was found that the distance s (in ft) it traveled after the brakes were applied was given by $s = 57.6 - 1.20t^3$, where t is the time (in s). What were the velocity and acceleration for $t = 4.00$ s?

49. The voltage V induced in an inductor in an electric circuit is given by $V = L(d^2q/dt^2)$, where L is the inductance (in H). Find the expression for the voltage induced in a 1.60-H inductor if $q = \sqrt{2t + 1} - 1$.

50. How fast is the rate of change of solar radiation changing on the surface in Exercise 39 of Section 23.4 at 3 P.M.?

51. The deflection y (in m) of a 5.00-m beam as a function of the distance x (in m) from one end is $y = 0.0001(x^5 - 25x^2)$. Find the value of d^2y/dx^2 (the rate of change at which the slope of the beam changes) where $x = 3.00$ m.

52. The force F (in N) on an object is $F = 12\,dv/dt + 2.0v + 5.0$, where v is the velocity (in m/s) and t is the time (in s). If the displacement is $s = 25t^{0.60}$, find F for $t = 3.5$ s.

Answers to Practice Exercises

1. $y'' = 48x^2 - 12$ **2.** $y'' = \dfrac{18x^2 - 24}{(x^2 + 4)^3}$

CHAPTER 23 KEY FORMULAS AND EQUATIONS

Limit of function	$\lim_{x \to a} f(x) = L$	**(23.1)**
Difference in x-coordinates	$h = x_2 - x_1$	**(23.2)**
	$x_2 = x_1 + h$	**(23.3)**
Slope	$m_{PQ} = \dfrac{f(x_1 + h) - f(x_1)}{(x_1 + h) - x_1} = \dfrac{f(x_1 + h) - f(x_1)}{h}$	**(23.4)**
	$m_{\tan} = \lim_{h \to 0} \dfrac{f(x_1 + h) - f(x_1)}{h}$	**(23.5)**
Definition of derivative	$f'(x) = \lim_{h \to 0} \dfrac{f(x + h) - f(x)}{h}$	**(23.6)**
Instantaneous velocity	$v = \lim_{h \to 0} \dfrac{s(t + h) - s(t)}{h}$	**(23.7)**
Constant rule	$\dfrac{dc}{dx} = 0$	**(23.8)**
Power rule	$\dfrac{dx^n}{dx} = nx^{n-1}$	**(23.9)**
Constant multiple rule	$\dfrac{d(cu)}{dx} = c\dfrac{du}{dx}$	**(23.10)**
Derivative of a sum	$\dfrac{d(u + v)}{dx} = \dfrac{du}{dx} + \dfrac{dv}{dx}$	**(23.11)**

Product rule $$\frac{d(uv)}{dx} = u\frac{dv}{dx} + v\frac{du}{dx} \quad (23.12)$$

Quotient rule $$\frac{d\left(\frac{u}{v}\right)}{dx} = \frac{v\frac{du}{dx} - u\frac{dv}{dx}}{v^2} \quad (23.13)$$

Chain rule $$\frac{dy}{dx} = \frac{dy}{du}\frac{du}{dx} \quad (23.14)$$

General power rule $$\frac{du^n}{dx} = nu^{n-1}\left(\frac{du}{dx}\right) \quad (23.15)$$

$$\frac{du^{p/q}}{dx} = \frac{p}{q}u^{(p/q)-1}\frac{du}{dx} \quad (23.16)$$

CHAPTER 23 REVIEW EXERCISES

CONCEPT CHECK EXERCISES

*Determine each of the following as being either **true** or **false**. If it is false, explain why.*

1. $\lim_{x\to 3}\frac{x^2 - 3x}{x - 3} = 0$
2. In using the definition of the derivative, for the function $f(x) = 2x^2 - 4x$ it is necessary to find $\lim_{h\to 0}(4x - h - 4)$.
3. $\frac{d}{dx}(4x^3 - 3\pi^2 - x) = 12x^2 - 6\pi - 1$
4. $\frac{d}{dx}\left(\frac{x^2 + 1}{1 - x}\right) = -\frac{3x^2 - 2x - 1}{(1 - x)^2}$
5. $\frac{d}{dx}(3 - x)^3 = 3(3 - x)^2$
6. For the implicit function $3x^2y^2 = y + x^3$, $dy/dx = 6x^2y + 6xy^2 - 3x^2$
7. $\frac{d^2}{dx^2}(1 - 3x)^3 = 54(1 - 3x)$
8. The acceleration is the derivative of the displacement with respect to time.

PRACTICE AND APPLICATIONS

In Exercises 9–20, evaluate the given limits.

9. $\lim_{x\to 4}(8 - 3x)$
10. $\lim_{x\to 3}(2x^2 - 10)$
11. $\lim_{x\to -2}\frac{|x + 2|}{x + 2}$
12. $\lim_{x\to 3}\sqrt{2x^2 - 18}$
13. $\lim_{x\to 2}\frac{4x - 8}{x^2 - 4}$
14. $\lim_{x\to 5}\frac{x^2 - 25}{3x - 15}$
15. $\lim_{x\to 3}\frac{x^2 - 5x + 6}{x^2 - 2x - 3}$
16. $\lim_{x\to 0}\frac{(x - 3)^2 - 9}{x - 6}$
17. $\lim_{x\to\infty}\frac{2 + \frac{2}{x}}{3 - \frac{1}{x^2}}$
18. $\lim_{x\to\infty}\frac{3x^3 - 5x}{6x^2 + 3}$
19. $\lim_{x\to\infty}\frac{x - 2x^3}{(1 + x)^3}$
20. $\lim_{x\to\infty}\frac{\sqrt{4x^2 + 3}}{x + 5}$

In Exercises 21–28, use the definition to find the derivative of each of the given functions.

21. $y = 7 + 5x$
22. $y = 6x - 2$
23. $y = 6 - 2x^2$
24. $y = 12x^2 - x^3$
25. $y = \frac{2}{x^2}$
26. $y = \frac{3x}{1 - 4x}$
27. $y = \sqrt{x + 5}$
28. $y = \frac{6}{\sqrt{x}}$

In Exercises 29–44, find the derivative of each of the given functions.

29. $y = 2x^7 - 3x^2 + 5$
30. $y = 8x^7 - 2^5 - x$
31. $y = 4\sqrt{x} - \frac{3}{x} + \sqrt{3}$
32. $R = \frac{9}{T^2} - 8\sqrt[4]{T}$
33. $f(y) = \frac{12y}{1 - 5y}$
34. $y = \frac{2x - 1}{x^2 + 1}$
35. $y = (2 - 7x)^4$
36. $y = (2x^2 - 5x + 1)^6$
37. $y = \frac{3\pi}{(5 - 2x^2)^{3/4}}$
38. $f(Q) = \frac{70}{(3Q + 1)^3}$
39. $v = \frac{5t}{(t^2 + 2)^3}$
40. $y = (x - 1)^3(x^2 - 2)^2$
41. $y = \frac{\sqrt{4x + 3}}{2x}$
42. $R = \frac{\sqrt{t + 4}}{\sqrt{t - 4}}$
43. $(2x - 3y)^3 = x^2 - y$
44. $x^2y^2 = x^2 + y^2$

In Exercises 45–48, evaluate the derivatives of the given functions for the given values of x. Check your results, using the derivative evaluation feature of a calculator.

45. $y = \frac{4}{x} + 2\sqrt[3]{x}, x = 8$
46. $y = (3x - 5)^4, x = -2$
47. $y = 2x\sqrt{12x + 7}, x = 1.5$
48. $y = \frac{\sqrt{2x^2 + 1}}{3x}, x = 2$

In Exercises 49–52, find the second derivative of each of the given functions.

49. $y = 3x^4 - \frac{1}{x}$
50. $y = \sqrt{1 - 8x}$
51. $s = \frac{2 - 3t}{5 + 4t}$
52. $y = 2x(6x + 5)^4$

In Exercises 53–98, solve the given problems.

53. As x approaches 0 from the right, which of the functions $1/x$, $1/x^2$, and $1/\sqrt{x}$ increases most rapidly (all become infinite)?

54. The parabola $y = ax^2 + bx + c$ passes through (1, 2) and is tangent to the line $y = x$ at the origin. Find a, b, and c.

55. Find the acute angle between tangent lines to the parabolas $y = x^2$ and $y = (x - 2)^2$ at the point where they intersect.

56. Find the point(s) on the curve of $y = \dfrac{x}{x^2 + 1}$ where the tangent line is horizontal.

57. View the graph of $y = \dfrac{2(x^2 - 4)}{x - 2}$ on a calculator with *window* values such that y can be evaluated exactly for $x = 2$. [Xmin = −1 (or 0), Xmax = 4, Ymin = 0, Ymax = 10 will probably work.] Using the *trace* feature, determine the value of y for $x = 2$. Comment on the accuracy of the view and the value found.

58. A continuous function $f(x)$ is positive at $x = 0$ and negative for $x = 1$. How many solutions of $f(x) = 0$ are possible between $x = 0$ and $x = 1$? Explain.

59. The velocity v (in ft/s) of a weight falling in water is given by $v = \dfrac{6(t + 5)}{t + 1}$, where t is the time (in s). What are (a) the initial velocity and (b) the terminal velocity (as $t \to \infty$)?

60. Two lenses of focal lengths f_1 and f_2, separated by a distance d, are used in the study of lasers. The combined focal length f of this lens combination is $f = \dfrac{f_1 f_2}{f_1 + f_2 - d}$. If f_2 and d remain constant, find the limiting value of f as f_1 continues to increase in value.

61. Find the slope of a line tangent to the curve of $y = 7x^4 - x^3$ at $(-1, 8)$. Use the derivative evaluation feature of a calculator to check your result.

62. Find the slope of a line tangent to the curve of $y = \sqrt[3]{3 - 8x}$ at $(-3, 3)$. Use the derivative evaluation feature of a calculator to check your result.

63. Find the point(s) at which a tangent line to the graph of $y = 1/\sqrt{3x^2 + 3}$ is parallel to the x-axis.

64. Find the point(s) on the graph of $y = 2(1 - 3x)^2$ at which a tangent line is parallel to the line $y = -2x + 5$.

65. If \$5000 is invested at interest rate i, compounded quarterly, in two years it will grow to an amount A given by $A = 5000(1 + 0.25i)^8$. Find dA/dt.

66. The temperature T (in °C) of a rotating machine part that has been in operation for t hours is given by $T = \dfrac{100(t + 1)}{t + 5}$. Find dT/dt when $t = 4.0$ h.

67. Find the equations for (a) the velocity and (b) the acceleration if the displacement s (in m) of an object as a function of the time t (in s) is given by $s = \sqrt{1 + 8t}$.

68. Find the values of the velocity and acceleration for the object in Exercise 67 for $t = 3$ s.

69. The cable of a 200-m suspension bridge can be represented by $y = 0.0015x^2 + C$. At one point, the tension is directed along the line $y = 0.3x - 10$. Find the value of C.

70. The displacement s (in cm) of a piston during each 8-s cycle is given by $s = 8t - t^2$, where t is the time (in s). For what value(s) of t is the velocity of the piston 4 cm/s?

71. The reliability R of a computer system measures the probability that the system will be operating properly after t hours. For one system, $R = 1 - kt + \dfrac{k^2t^2}{2} - \dfrac{k^3t^3}{6}$, where k is a constant. Find the expression for the instantaneous rate of change of R with respect to t.

72. The distance s (in ft) traveled by a subway train after the brakes are applied is given by $s = 40t - 5t^2$, where t is the time (in s). How far does it travel in coming to a stop?

73. The gravitational force F between two objects m_1 and m_2 whose centers are at a distance r apart is $F = \dfrac{Gm_1m_2}{r^2}$, where G is a constant. Find dF/dr.

74. The velocity of an object moving with constant acceleration can be found from the equation $v = \sqrt{v_0^2 + 2as}$, where v_0 is the initial velocity, a is the acceleration, and s is the distance traveled. Find dv/ds.

75. The voltage induced in an inductor L is given by $E = L(dI/dt)$, where I is the current in the circuit and t is the time. Find the voltage induced in a 0.4-H inductor if the current I (in A) is related to the time (in s) by $I = t(0.01t + 1)^3$.

76. In studying the energy used by a mechanical robotic device, the equation $v = \dfrac{z}{\alpha(1 - z^2) - \beta}$ is used. If α and β are constants, find dv/dz.

77. The frictional radius r_f of a collar used in a braking system is given by $r_f = \dfrac{2(R^3 - r^3)}{3(R^2 - r^2)}$, where R is the outer radius and r is the inner radius. Find dr_f/dR if r is constant.

78. Water is being drained from a pond such that the volume V (in m^3) of water in the pond after t hours is given by $V = 5000(60 - t)^2$. Find the rate at which the pond is being drained after 4.00 h.

79. The energy output E of an electric heater is a function of the time t (in s) given by $E = t(1 + 2t)^2$ for $t < 10$ s. Find the power dE/dt (in W) generated by the heater for $t = 8.0$ s.

80. The amount n (in g) of a compound formed during a chemical change is $n = \dfrac{8t}{2t^2 + 3}$, where t is the time (in s). Find dn/dt for $t = 4.0$ s. What is the meaning of the result?

81. The deflection y of a 10-ft beam is $y = kx(x^4 + 450x^2 - 950)$, where k is a constant and x is the horizontal distance from one end. Find dy/dx.

82. The kinetic energy K (in J) of a rotating flywheel varies directly as the square of its angular velocity ω (in rad/s). If $K = 120$ J for $\omega = 75$ rad/s, find $dK/d\omega$ for $\omega = 140$ rad/s.

83. The frequency f of a certain electronic oscillator is given by $f = \dfrac{1}{2\pi\sqrt{C(L + 2)}}$, where C is a capacitance and L is an inductance. If C is constant, find df/dL.

84. The volume V of fluid produced in the retina of the eye in reaction to exposure to light of intensity I is given by $V = \dfrac{aI^2}{b - I}$, where a and b are constants. Find dV/dI.

85. The temperature T (in °C) in a freezer as a function of the time t (in h) is given by $T = \dfrac{10(1 - t)}{0.5t + 1}$. Find dT/dt.

86. Under certain conditions, the efficiency e (in %) of an internal combustion engine is given by

$$e = 100\left(1 - \frac{1}{(V_1/V_2)^{0.4}}\right)$$

where V_1 and V_2 are the maximum and minimum volumes of air in a cylinder, respectively. Assuming that V_2 is kept constant, find the expression for the instantaneous rate of change of efficiency with respect to V_1.

87. The deflection y of a cantilever beam (clamped at one end and free at the other end) is $y = \frac{w}{24EI}(6L^2x^2 - 4Lx^3 + x^4)$. Here, L is the length of the beam, and w, E, and I are constants. Find the first four derivatives of y with respect to x. (Each of these derivatives is useful in analyzing the properties of the beam.)

88. The number n of grams of a compound formed during a certain chemical reaction is given by $n = \frac{2t}{t + 1}$, where t is the time (in min). Evaluate d^2n/dt^2 (the rate of increase of the amount of the compound being formed) when $t = 4.00$ min.

89. The area of a rectangular patio is to be 75 m^2. Express the perimeter p of the patio as a function of its width w and find dp/dw.

90. A water tank is being designed in the shape of a right circular cylinder with a volume of 100 ft^3. Find the expression for the instantaneous rate of change of the total surface area A of the tank with respect to the radius r of the base.

91. Find the slope of a light ray perpendicular to the cross-section of a lens represented by $4x^2 - 3xy + y^2 = 14$ at $(-1, 2)$.

92. The power P (in μW) in a microprocessor circuit is the product of the voltage V (in mV) and current I (in mA). If V and I vary with time t $(0 \leq t \leq 0.0033 \text{ s})$ with $V = 10.0 - 60.0\sqrt{t}$ and $I = 0.025t$, find dP/dt.

93. Oil from an undersea well is leaking and forming a circular spill on the surface. Find the instantaneous rate of change of the area A of the spill with respect to the radius r, for $r = 1.8$ km.

94. A cylindrical metal container is being heated with the height h always twice the radius r. Find the expression for the instantaneous rate of change of the volume V with respect to r.

95. An arch over a walkway can be described by the first-quadrant part of the parabola $y = 4 - x^2$. In order to determine the size and shape of rectangular objects that can pass under the arch, express the area A of a rectangle inscribed under the parabola in terms of x. Find dA/dx.

96. A computer analysis showed that a specialized piece of machinery has a value (in dollars) given by $V = 1{,}500{,}000/(2t + 10)$, where t is the number of years after the purchase. Calculate the value of dV/dt and d^2V/dt^2 for $t = 5$ years. What is the meaning of these values?

97. An airplane flies over an observer with a velocity of 400 mi/h and at an altitude of 2640 ft. If the plane flies horizontally in a straight line, find the rate at which the distance x from the observer to the plane is changing 0.600 min after the plane passes over the observer. See Fig. 23.43.

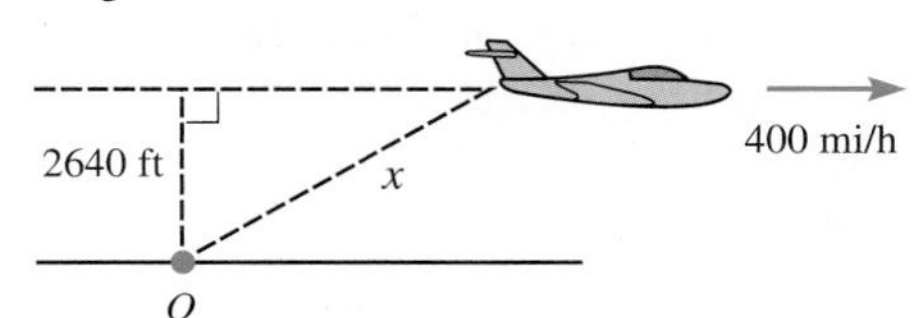

Fig. 23.43

98. The *radius of curvature* of $y = f(x)$ at the point (x, y) on the curve of $y = f(x)$ is given by $R = \frac{[1 + (y')^2]^{3/2}}{|y''|}$. A certain roadway follows the parabola $y = 1.2x - x^2$ for $0 < x < 1.2$, where x is measured in miles. Find R for $x = 0.2$ mi and $x = 0.6$ mi. See Fig. 23.44.

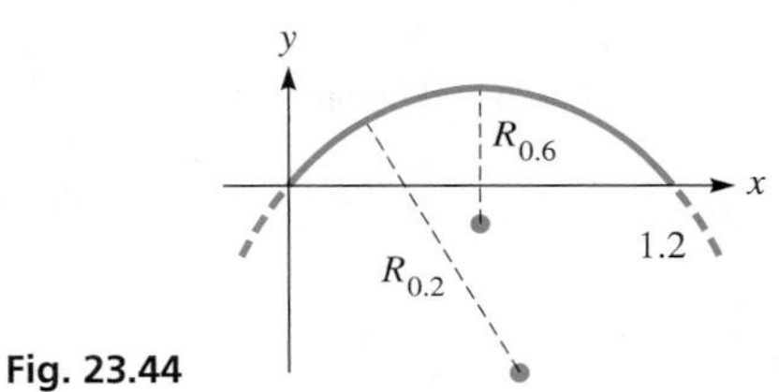

Fig. 23.44

99. An engineer designing military rockets uses computer simulation to find the path of a rocket as $y = f(x)$ and the path of an aircraft to be $y = g(x)$. Write two or three paragraphs explaining how the engineer can determine the angle at which the path of the rocket crosses the path of the aircraft.

CHAPTER 23 PRACTICE TEST

As a study aid, we have included complete solutions for each Practice Test problem at the end of this book.

1. Find $\lim\limits_{x \to 1} \frac{x^2 - x}{x^2 - 1}$.

2. Find $\lim\limits_{x \to \infty} \frac{1 - 4x^2}{x + 2x^2}$.

3. Find the slope of a line tangent to the curve of $y = 3x^2 - \frac{4}{x^2}$ at $(2, 11)$. Check your result using the derivative evaluation feature of a calculator. Write down the complete value shown on the calculator.

4. The displacement s (in cm) of a pumping machine piston in each cycle is given by $s = t\sqrt{10 - 2t}$, where t is the time (in s). Find the velocity of the piston for $t = 4.00$ s.

5. Find dy/dx: $y = 4x^6 - 2x^4 + \pi^3$

6. Find dy/dx: $y = 2x(5 - 3x)^4$

7. Find dy/dx: $(1 + y^2)^3 - x^2y = 7x$

8. Under certain conditions, due to the presence of a charge q, the electric potential V along a line is given by

$$V = \frac{kq}{\sqrt{x^2 + b^2}}$$

where k is a constant and b is the minimum distance from the charge to the line. Find the expression for the instantaneous rate of change of V with respect to x.

9. Find the second derivative of $y = \frac{2x}{3x + 2}$.

10. By using the definition, find the derivative of $y = 5x - 2x^2$ with respect to x.

24

Applications of the Derivative

LEARNING OUTCOMES

After completion of this chapter, the student should be able to:

- Find the equation of a line tangent or normal to a given curve
- Solve equations using Newton's method
- Find the velocity and acceleration of an object undergoing curvillinear motion
- Solve related rates problems
- Use derivatives to describe important features of the graph of a function such as maxima, minima, points of inflection, and concavity
- Sketch a curve using information about the function and its derivatives
- Solve applied maximum and minimum problems
- Use differentials to estimate errors in measurement
- Obtain the linear approximation of a function

▶ **In Section 24.7, we show how the derivative can be used to find the maximum revenue for a company that produces Bluetooth headphones.**

Following the work of Newton and Leibniz, the development of calculus proceeded rapidly but in a rather disorganized way.

Much of the progress in the late 1600s and early 1700s was due to a desire to solve applied problems, particularly in some areas of physics. These included problems such as finding velocities in more complex types of motion, accurately measuring time by use of a pendulum, and finding the equation of a uniform cable hanging under its own weight.

A number of mathematicians, most of whom also studied in various areas of physics, contributed to these advances in calculus. Among them was the Swiss mathematician Leonhard Euler, the most prolific mathematician of all time. Throughout the mid-to-later 1700s, he used the idea of a function to better organize the study of algebra, trigonometry, and calculus. In doing so, he fully developed the use of calculus on problems from physics in areas such as planetary motion, mechanics, and optics.

Euler had a nearly unbelievable memory and ability to calculate. At an early age, he memorized the entire *Aeneid* by the Roman poet Virgil and was able to recite it from memory at age 70. In his head he solved major problems related to the motion of the moon that Newton had not been able to solve. At one time, he was given two solutions to a problem that differed in the 50th decimal place, and he determined, in his head, which was correct. Although blind for the last 17 years of his life, it was one of his most productive periods. From memory, he dictated many of his articles (he wrote a total of over 70 volumes) until his sudden death in 1783.

We have noted some of the problems in technology in which the derivative plays a key role in the solution. Another important type of application is finding the maximum values or minimum values of functions. Such values are useful, for example, in finding the maximum possible income from production or the least amount of material needed in making a product. In this chapter, we consider several of these kinds of applications of the derivative.

24.1 Tangents and Normals

Tangent Line • Normal Line

The first application of the derivative we consider involves finding the equation of a line *tangent* to a given curve and the equation of a line *normal* (perpendicular) to a given curve.

TANGENT LINE

To find the equation of a line tangent to a curve at a given point, we first find the derivative, which is then evaluated at the point. This gives us the slope of the tangent line to the curve at the point. Then, by using the point-slope form of the equation of a straight line, we find the equation of the tangent line. The following examples illustrate the method.

EXAMPLE 1 Equation of a tangent line

Find the equation of the line tangent to the parabola $y = x^2 - 1$ at the point $(-2, 3)$.

Finding the derivative and evaluating it at $x = -2$, we have

$$\frac{dy}{dx} = 2x \quad \text{derivative}$$

$$\left.\frac{dy}{dx}\right|_{x=-2} = -4 \quad \text{evaluate derivative at } (-2, 3)$$

$$y - 3 = -4(x + 2) \quad \text{point-slope form of straight line}$$

$$y = -4x - 5 \quad \text{equation of tangent line}$$

The parabola and the tangent line are shown in Fig. 24.1. ■

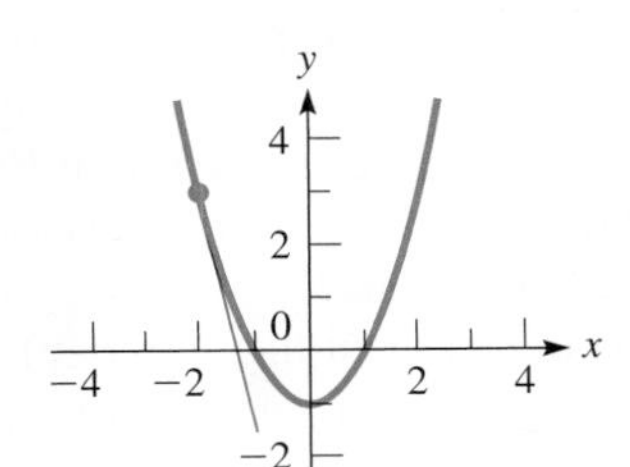

Fig. 24.1

EXAMPLE 2 Tangent line—implicit function

Find the equation of the line tangent to the ellipse $4x^2 + 9y^2 = 40$ at the point $(1, 2)$.

Using implicit differentiation, we have the following solution.

$$8x + 18yy' = 0 \quad \text{find derivative}$$

$$y' = -\frac{4x}{9y}$$

$$y'|_{(1,2)} = -\frac{4}{18} = -\frac{2}{9} \quad \text{evaluate derivative to find slope of tangent line}$$

$$y - 2 = -\frac{2}{9}(x - 1) \quad \text{point-slope form of tangent line}$$

$$9y - 18 = -2x + 2$$

$$2x + 9y - 20 = 0 \quad \text{general form of tangent line}$$

The ellipse and the tangent line $2x + 9y - 20 = 0$ are shown in Fig. 24.2. ■

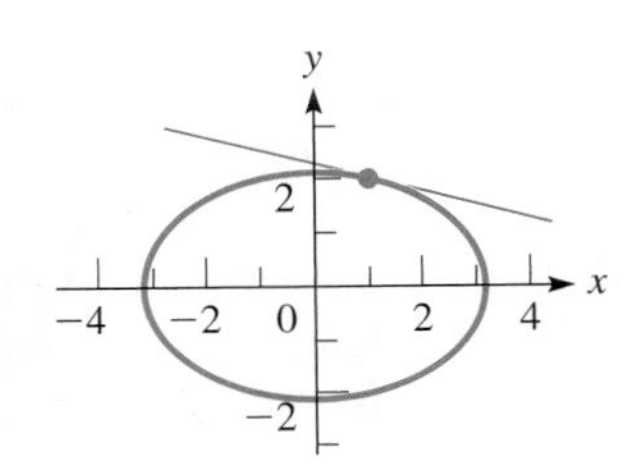

Fig. 24.2

NORMAL LINE

■ About 1700, the word *normal* was adapted from the Latin word *normals*, which was being used for *perpendicular*.

If we wish to obtain the equation of a line normal (perpendicular to a tangent) to a curve, recall that the slopes of perpendicular lines are negative reciprocals. Thus, the derivative is found and evaluated at the specified point. [Because this gives the slope of a tangent line, we take the negative reciprocal of this number to find the slope of the normal line.] Then by using the point-slope form of the equation of a straight line, we find the equation of the normal line. The following examples illustrate the method.

NOTE ▶

EXAMPLE 3 Equation of a normal line

Find the equation of a line normal to the hyperbola $y = 2/x$ at the point $(2, 1)$.

Taking the derivative and evaluating it for $x = 2$, we have

$$\frac{dy}{dx} = -\frac{2}{x^2}, \qquad \left.\frac{dy}{dx}\right|_{x=2} = -\frac{1}{2}$$

Therefore, the slope of the normal line at $(2, 1)$ is 2. The equation of the normal line is then

$$y - 1 = 2(x - 2)$$
$$y = 2x - 3$$

The hyperbola and normal line are shown in Fig. 24.3. ■

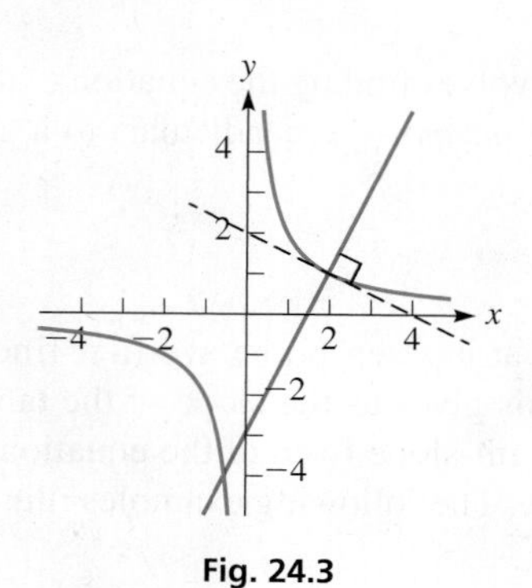

Fig. 24.3

There are many applications of tangents and normals in technology. One of these is shown in the following example. Others are shown in the exercises.

EXAMPLE 4 Normal line—parabolic solar reflector

In Fig. 24.4, the cross section of a parabolic solar reflector is shown, along with an incident ray of light and the reflected ray. The angle of incidence i is equal to the angle of reflection r where both angles are measured with respect to the normal to the surface. If the incident ray strikes at the point where the slope of the normal is -1 and the equation of the parabola is $4y = x^2$, what is the equation of the normal line?

If the slope of the normal line is -1, then the slope of a tangent line is $-(\frac{1}{-1}) = 1$. Therefore, we know that the value of the derivative at the point of reflection is 1. This allows us to find the coordinates of the point:

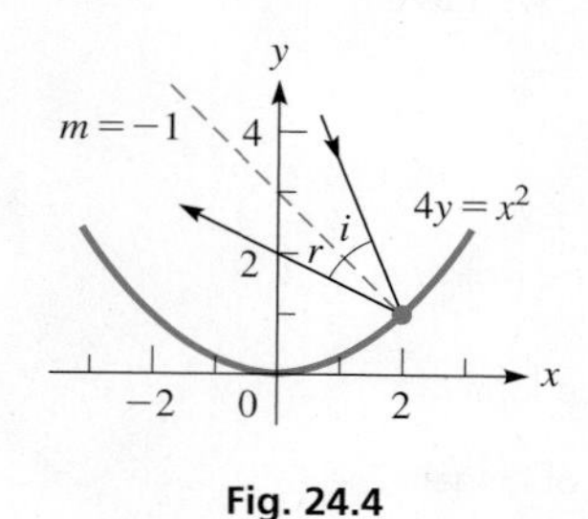

Fig. 24.4

$$4y = x^2$$

$$4\frac{dy}{dx} = 2x, \qquad \frac{dy}{dx} = \frac{1}{2}x \qquad \text{find derivative}$$

$$1 = \frac{1}{2}x \qquad \text{substitute } \frac{dy}{dx} = 1$$

$$x = 2$$

This means that the x-coordinate of the point of reflection is 2. We can find the y-coordinate by substituting $x = 2$ into the equation of the parabola. Thus, the point is $(2, 1)$. Because the slope is -1, the equation of the normal line is

$$y - 1 = (-1)(x - 2)$$
$$y = -x + 3$$

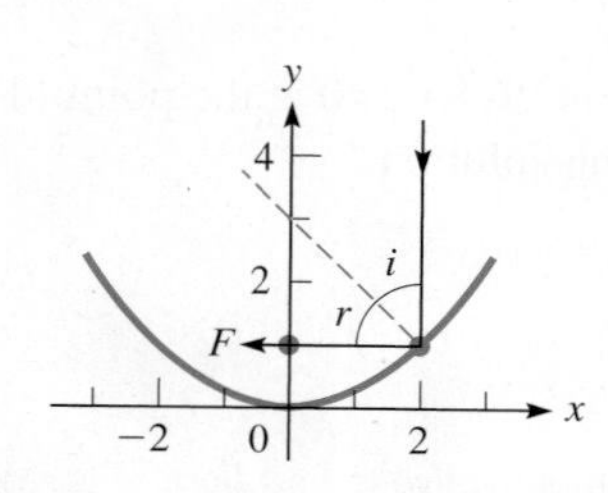

Fig. 24.5

If the incident ray is vertical, for which $i = 45°$ at the point $(2, 1)$, the reflected ray passes through $(0, 1)$, the focus of the parabola. See Fig. 24.5. This shows the important reflection property of a parabola that *any incident ray parallel to the axis passes through the focus.* We first noted this property in our discussion of the parabola in Section 21.4. ■

Practice Exercises

For the parabola $y = 4 - x^2$, at the point $(3, -5)$ find the equation of the line:

1. Tangent to the parabola.

2. Normal to the parabola.

EXERCISES 24.1

In Exercises 1 and 2, make the given changes in the indicated examples of this section and then solve the resulting problems.

1. In Example 2, change $4x^2 + 9y^2$ to $x^2 + 4y^2$, change 40 to 17, and then find the equation of the tangent line.

2. In Example 3, change $2/x$ to $3/(x + 1)$ and then find the equation of the normal line.

In Exercises 3–6, find the equations of the lines tangent to the indicated curves at the given points. In Exercises 3 and 6, sketch the curve and tangent line. In Exercises 4 and 5, use a calculator to view the curve and tangent line.

3. $y = x^2 + 2$ at $(2, 6)$

4. $y = \frac{1}{3}x^3 - 5x$ at $(3, -6)$

5. $y = \dfrac{1}{x^2 + 1}$ at $(-1, \frac{1}{2})$

6. $x^2 + y^2 = 25$ at $(-3, 4)$

In Exercises 7–10, find the equations of the lines normal to the indicated curves at the given points. In Exercises 7 and 10, sketch the curve and normal line. In Exercises 8 and 9, use a calculator to view the curve and normal line.

7. $y = 6x - 2x^2$ at $(2, 4)$ **8.** $y = 8 - x^3$ at $(-1, 9)$

9. $y^2(2 - x) = x^3$ at $(1, 1)$ **10.** $4x^2 - y^2 = 20$ at $(-3, -4)$

In Exercises 11–14, find the equations of the lines tangent or normal to the given curves and with the given slopes. View the curves and lines on a calculator.

11. $y = x^2 - 2x$, tangent line with slope 2

12. $y = \sqrt{2x - 9}$, tangent line with slope 1

13. $y = (2x - 1)^3$, normal line with slope $-\frac{1}{24}$, $x > 0$

14. $y = \frac{1}{2}x^4 + 1$, normal line with slope 4

In Exercises 15–30, solve the given problems involving tangent and normal lines.

15. Find the equations of the tangent and normal lines to the graph of $y = 2x^3 - 3x + 1$ at the point $(1, 0)$.

16. Find the equations of the tangent and normal lines to the graph of $y = 6x - x^3$ at the point $(2, 4)$.

17. Show that the line tangent to the graph of $y = x + 2x^2 - x^4$ at $(1, 2)$ is also tangent at $(-1, 0)$.

18. Show that the graphs of $y^2 = 4x + 4$ and $y^2 = 4 - 4x$ cross at right angles.

19. Suppose we wish to find the equation of the line normal to the graph of $y = 4x - x^2$ at the point $(2, 4)$. Explain why we can't find the slope by taking the negative reciprocal of the slope of the tangent line. What is the equation of the normal line?

20. Show that the curve $y = x^3 + 4x - 5$ has no normal line with a slope of $-1/3$.

21. Find the point of intersection between the tangent lines to the circle $x^2 + y^2 = 25$ at the points $(3, 4)$ and $(3, -4)$.

22. Where does the normal line to the parabola $y = x - x^2$ at $(1, 0)$ intersect the parabola other than at $(1, 0)$?

23. Heat flows normal to isotherms, curves along which the temperature is constant. Find the line along which heat flows through the point $(2, 1)$ and the isotherm is along the graph of $2x^2 + y^2 = 9$.

24. The sparks from an emery wheel to sharpen blades fly off tangent to the wheel. Find the equation along which sparks fly from a wheel described by $x^2 + y^2 = 25$, at $(3, 4)$.

25. A certain suspension cable with supports on the same level is closely approximated as being parabolic in shape. If the supports are 200 ft apart and the sag at the center is 30 ft, what is the equation of the line along which the tension acts (tangentially) at the right support? (Choose the origin of the coordinate system at the lowest point of the cable.)

26. In a video game, airplanes move from left to right along the path described by $y = 2 + 1/x$. They can shoot rockets tangent to the direction of flight at targets on the x-axis located at $x = 1, 2, 3,$ and 4. Will a rocket fired from $(1, 3)$ hit a target?

27. In an electric field, the lines of force are perpendicular to the curves of equal electric potential. In a certain electric field, a curve of equal potential is $y = \sqrt{2x^2 + 8}$. If the line along which the force acts on an electron has an inclination of 135°, find its equation.

28. A radio wave reflects from a reflecting surface in the same way as a light wave (see Example 4). A certain horizontal radio wave reflects off a parabolic reflector such that the reflected wave is 43.60° below the horizontal, as shown in Fig. 24.6. If the equation of the parabola is $y^2 = 8x$, what is the equation of the normal line through the point of reflection?

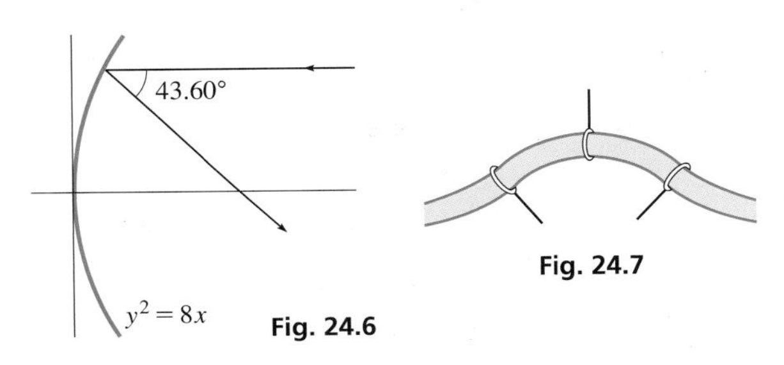

Fig. 24.6

Fig. 24.7

29. In designing a flexible tubing system, the supports for the tubing must be perpendicular to the tubing. If a section of the tubing follows the curve $y = \dfrac{4}{x^2 + 1}$ $(-2 \text{ dm} < x < 2 \text{ dm})$, along which lines must the supports be directed if they are located at $x = -1, x = 0,$ and $x = 1$? See Fig. 24.7.

30. On a particular drawing, a pulley wheel can be described by the equation $x^2 + y^2 = 100$ (units in cm). The pulley belt is directed along the lines $y = -10$ and $4y - 3x - 50 = 0$ when first and last making contact with the wheel. What are the first and last points on the wheel where the belt makes contact?

Answers to Practice Exercises

1. $6x + y = 13$ **2.** $x - 6y = 33$

24.2 Newton's Method for Solving Equations

Iterative Method • Newton's Method for Solving Equations

As we know, finding the roots of an equation $f(x) = 0$ is very important in mathematics and in many types of applications, and we have developed methods of solving many types of equations in the previous chapters. However, for a great many algebraic and nonalgebraic equations, there is no method for finding the roots exactly.

We have shown that the roots of an equation can be found with great accuracy on a calculator. In this section, we develop **Newton's method,** which uses the derivative to find approximately, but also with great accuracy, the real roots of many kinds of equations.

■ Another mathematical development by the English mathematician and physicist Issac Newton (1642–1727).

Newton's method is an **iterative method,** which starts with a reasonable guess for the root, and then yields a new and better approximation. This, in turn, is used to obtain a still better approximation, and so on until an approximate answer with the required degree of accuracy is obtained. Iterative methods are easily programmable for use on a computer.

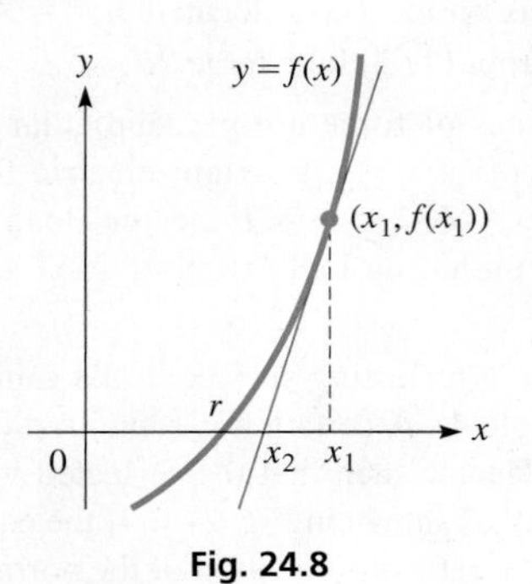

Fig. 24.8

Let us consider a section of the curve of $y = f(x)$ that (a) crosses the x-axis, (b) always has either a positive slope or a negative slope, and (c) has a slope that either becomes greater or becomes less as x increases. See Fig. 24.8. The curve in the figure crosses the x-axis at $x = r$, which means that $x = r$ is a root of the equation $f(x) = 0$. If x_1 is sufficiently close to r, a line tangent to the curve at $[x_1, f(x_1)]$ will cross the x-axis at a point $(x_2, 0)$, which is closer to r than is x_1.

We know that the slope of the tangent line is the value of the derivative at x_1, or $m_{\tan} = f'(x_1)$. Therefore, the equation of the tangent line is

$$y - f(x_1) = f'(x_1)(x - x_1)$$

For the point $(x_2, 0)$ on this line, we have

$$-f(x_1) = f'(x_1)(x_2 - x_1)$$

Solving for x_2, we have the formula

$$x_2 = x_1 - \frac{f(x_1)}{f'(x_1)} \quad \textbf{(24.1)}$$

NOTE ▸ [Here, x_2 is a second approximation to the root. We can then replace x_1 in Eq. (24.1) with x_2 and find a closer approximation, x_3. This process can be repeated as many times as needed to find the root to the required accuracy.] This method lends itself well to the use of a calculator or a computer for finding the root.

EXAMPLE 1 Using Newton's method

Find the root of $x^2 - 3x + 1 = 0$ between $x = 0$ and $x = 1$.

Here, $f(x) = x^2 - 3x + 1, f(0) = 1$, and $f(1) = -1$. This indicates that the root may be near the middle of the interval. Therefore, we choose $x_1 = 0.5$.

The derivative is

$$f'(x) = 2x - 3$$

Therefore, $f(0.5) = -0.25$ and $f'(0.5) = -2$, which gives us

$$x_2 = 0.5 - \frac{-0.25}{-2} = 0.375$$

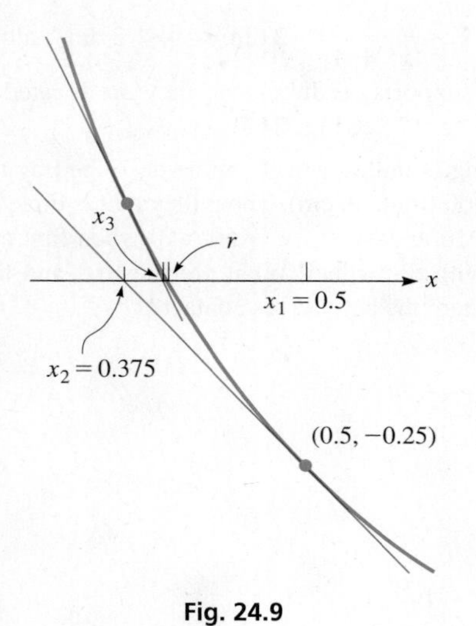

Fig. 24.9

This is a second approximation, which is closer to the actual value of the root. See Fig. 24.9. We can get an even better approximation, x_3, by using the method again with $x_2 = 0.375, f(0.375) = 0.015625$, and $f'(0.375) = -2.25$. This gives us

$$x_3 = 0.375 - \frac{0.015625}{-2.25} = 0.3819444$$

We can check this particular result by using the quadratic formula. This tells us the root is $x = 0.3819660$. Our result using Newton's method is good to three decimal places, and additional accuracy may be obtained by using the method again as many times as needed. ■

Practice Exercise

1. In Example 1, let $x_1 = 0.3$, and find x_2.

EXAMPLE 2 Newton's method—thickness of a water tank

A spherical water-storage tank holds 500.0 m^3. If the outside diameter is 10.0000 m, what is the thickness of the metal of which the tank is made?

Let x = the thickness of the metal. We know that the outside radius of the tank is 5.0000 m. Therefore, using the formula for the volume of a sphere, we have

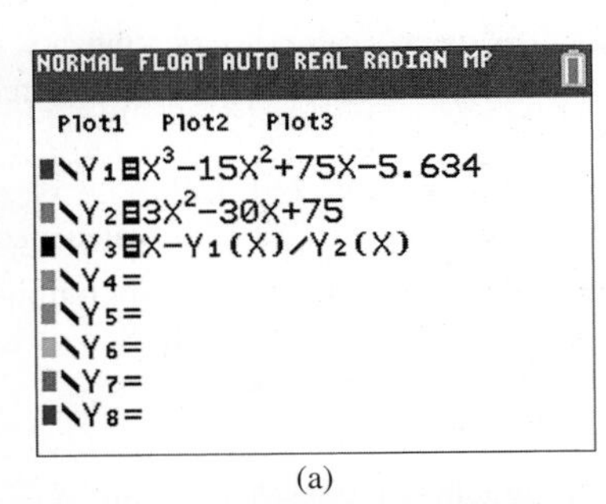

(a)

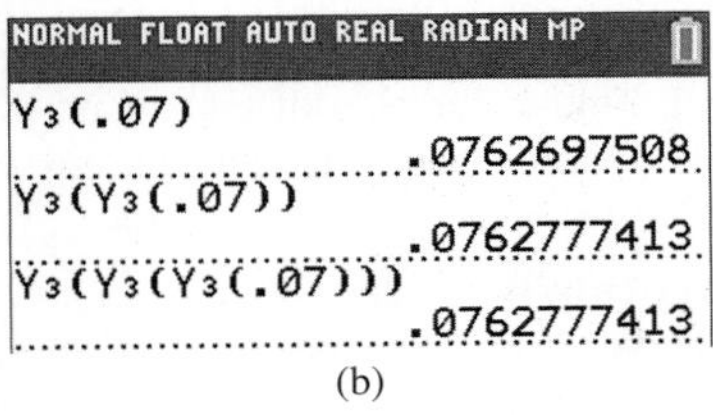

(b)

Fig. 24.10
Graphing calculator keystrokes: goo.gl/B3UFA1

$$\frac{4\pi}{3}(5.0000 - x)^3 = 500.0$$

$$125.0 - 75.00x + 15.00x^2 - x^3 = 119.366$$

$$x^3 - 15.00x^2 + 75.00x - 5.634 = 0$$

$$f(x) = x^3 - 15.00x^2 + 75.00x - 5.634$$

$$f'(x) = 3x^2 - 30.00x + 75.00$$

Because $f(0) = -5.634$ and $f(0.1) = 1.717$, the root may be closer to 0.1 than to 0.0. Therefore, we let $x_1 = 0.07$. Setting up a table, we have these values:

n	x_n	$f(x_n)$	$f'(x_n)$	$x_n - \frac{f(x_n)}{f'(x_n)}$
1	0.07	−0.457157	72.9147	0.0762697508
2	0.0762697508	−0.000581145	72.7293587	0.0762777413

Because $x_2 = x_3 = 0.0763$ to four decimal places, the thickness is 0.0763 m. This means the inside radius of the tank is 4.9237 m, and this value gives an inside volume of 500.0 m^3. As shown in Fig. 24.10(a) and (b), a calculator can be used to perform the iterations of Newton's method. ■

EXAMPLE 3 Graphically locating x_1

Solve the equation $x^2 - 1 = \sqrt{4x - 1}$.

We can see approximately where the root is by sketching the graphs of $y_1 = x^2 - 1$ and $y_2 = \sqrt{4x - 1}$ or by viewing the graphs on a calculator, as shown in Fig. 24.11. From this view, we see that they intersect between $x = 1$ and $x = 2$. Therefore, we choose $x_1 = 1.5$. With

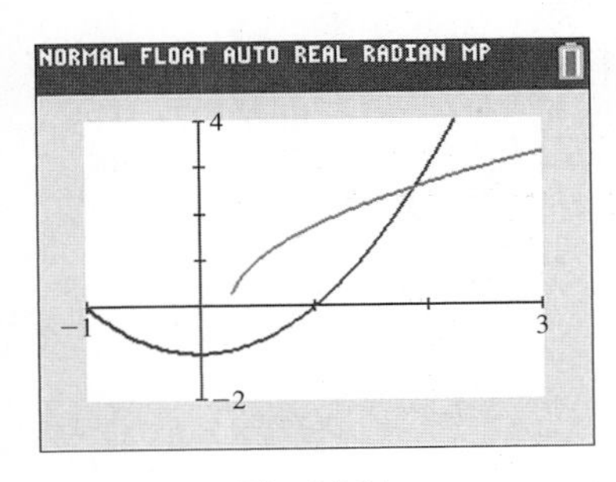

Fig. 24.11

$$f(x) = x^2 - 1 - \sqrt{4x - 1}$$

$$f'(x) = 2x - \frac{2}{\sqrt{4x - 1}}$$

we now find the values in the following table:

n	x_n	$f(x_n)$	$f'(x_n)$	$x_n - \frac{f(x_n)}{f'(x_n)}$
1	1.5	−0.98606798	2.1055728	1.9683134
2	1.9683134	0.25256859	3.1737598	1.8887332
3	1.8887332	0.00705269	2.9962957	1.8863794
4	1.8863794	0.00000620	2.9910265	1.8863773

Since $x_5 = x_4 = 1.88638$ to five decimal places, this is the required solution. (Here, rounded-off values of x_n are shown, although additional digits were carried and used.) This value can be verified on the calculator by using the *intersect* (or *zero*) feature. ■

EXERCISES 24.2

In Exercises 1–4, find the indicated roots of the given quadratic equations by finding x_3 from Newton's method. Compare this root with that obtained by using the quadratic formula.

1. In Example 1, change the middle term from $-3x$ to $-5x$ and use the same x_1.
2. $2x^2 - x - 2 = 0$ (between 1 and 2)
3. $3x^2 - 5x - 1 = 0$ (between -1 and 0)
4. $x^2 + 4x + 2 = 0$ (between -4 and -3)

In Exercises 5–16, find the indicated roots of the given equations to at least four decimal places by using Newton's method. Compare with the value of the root found using a calculator.

5. $x^3 - 6x^2 + 10x - 4 = 0$ (between 0 and 1)
6. $x^3 - 3x^2 - 2x + 3 = 0$ (between 0 and 1)
7. $x^3 - 6x^2 + 9x + 2 = 0$ (the real root)
8. $2x^3 + 2x^2 - 11x + 8 = 0$ (the real root)
9. $x^4 - x^3 - 3x^2 - x - 4 = 0$ (between 2 and 3)
10. $2x^4 - 2x^3 - 5x^2 - x - 3 = 0$ (between -2 and -1)
11. $x^4 - 2x^3 - 8x - 16 = 0$ (the negative root)
12. $3x^4 - 3x^3 - 11x^2 - x - 4 = 0$ (the negative root)
13. $2x^2 = \sqrt{2x + 1}$ (the positive real solution)
14. $x^3 = \sqrt{x + 1}$ (the real solution)
15. $x = \dfrac{1}{\sqrt{x + 2}}$ (the real solution)
16. $x^{3/2} = \dfrac{1}{2x + 1}$ (the real solution)

In Exercises 17–32, determine the required values by using Newton's method.

17. Find all the real roots of $x^3 - 2x^2 - 5x + 4 = 0$.
18. Find all the real roots of $x^4 - 2x^3 + 3x^2 + x - 7 = 0$.
19. Explain how to find $\sqrt[3]{4}$ by using Newton's method.
20. Explain why Newton's method does not work for finding the root of $x^3 - 3x = 5$ if x_1 is chosen as 1.
21. Use Newton's method to find an expression for x_{n+1}, in terms of x_n and a, for the equation $x^2 - a = 0$. Such an equation can be used to find $\sqrt{a}$.
22. In Appendix C at the back of this book, there is an explanation and example of Newton's method, which was copied directly from *Essays on Several Curious and Useful Subjects in Speculative and Mix'd Mathematicks* by Thomas Simpson (of Simpson's rule). It was published in London in 1740. Explain where the numerical error is in the example and what you think caused the error.
23. Use Newton's method on $f(x) = x^{1/3}$ with $x_1 = 1$. Calculate x_2, x_3, and x_4. What is happening as successive approximations are calculated?
24. To calculate reciprocals without dividing, a computer programmer applied Newton's method to the equation $1/x - a = 0$. Show that $x_2 = 2x_1 - ax_1^2$. From this, determine the expression for x_n.
25. Find the negative root of $3x^5 - x^4 - 12x^3 + 4x^2 + 12x - 4 = 0$ by choosing $x_1 = -1.4$. Then graph the polynomial on a calculator and note where this root appears.
26. See Example 7 in Section 15.3. Find x, the side of the square to be removed to form the box, using Newton's method to solve the equation $4x^3 - 136x^2 + 1147x - 2770 = 0$. Find x_2, using $x_1 = 4.4$.
27. The altitude h (in m) of a rocket is given by $h = -2t^3 + 84t^2 + 480t + 10$, where t is the time (in s) of flight. When does the rocket hit the ground?
28. A solid sphere of specific gravity s sinks in water to a depth h (in cm) given by $0.00926h^3 - 0.0833h^2 + s = 0$. Find h for $s = 0.786$.
29. A dome in the shape of a spherical segment is to be placed over the top of a sports stadium. If the radius r of the dome is to be 60.0 m and the volume V within the dome is 180,000 m^3, find the height h of the dome. See Fig. 24.12. $[V = \frac{1}{6}\pi h(h^2 + 3r^2).]$

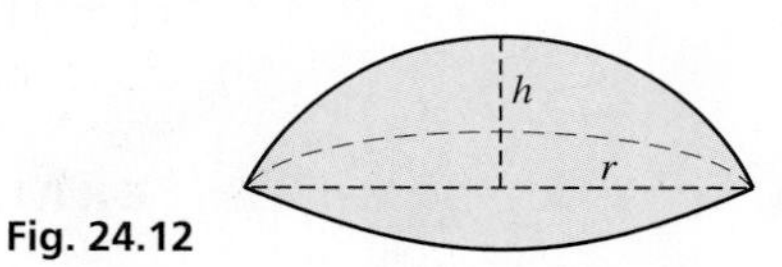

Fig. 24.12

30. An oil-storage tank has the shape of a right circular cylinder with a hemisphere at each end. See Fig. 24.13. If the volume of the tank is 1500 ft^3 and the length l is 12.0 ft, find the radius r.

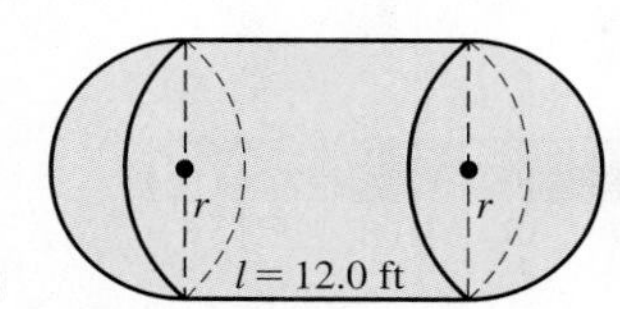

Fig. 24.13

Answer to Practice Exercise

1. $x_2 = 0.379167$

24.3 Curvilinear Motion

Vectors and Curvilinear Motion • Parameter • Parametric Form • Acceleration

When velocity was introduced in Section 23.4, the discussion was limited to rectilinear motion, or motion along a straight line. A more general discussion of velocity is necessary when we discuss the motion of an object in a plane. There are many important applications of motion in a plane, a principal one being the motion of a projectile.

An important concept in developing this topic is that of a vector. The necessary fundamentals related to vectors are taken up in Chapter 9. Although vectors can be used to represent many physical quantities, we will restrict our attention to their use in describing the velocity and acceleration of an object moving in a plane along a specified path. Such motion is called **curvilinear motion.**

In describing an object undergoing curvilinear motion, it is common to express the x- and y-coordinates of its position separately as functions of time. NOTE [Equations given in this form—that is, x and y both given in terms of a third variable (in this case, t)—are said to be in **parametric form,** which we encountered in Section 10.6. The third variable, t, is called the **parameter.**]

To find the velocity of an object whose coordinates are given in parametric form, we find its x-component of velocity v_x by determining dx/dt and its y-component of velocity v_y by determining dy/dt. These are then evaluated, and the resultant velocity is found from $v = \sqrt{v_x^2 + v_y^2}$. The direction in which the object is moving is found from $\tan\theta = v_y/v_x$.

EXAMPLE 1 Parametric form—resultant velocity

If the horizontal distance x that an object has moved is given by $x = 3t^2$ and the vertical distance y is given by $y = 1 - t^2$, find the resultant velocity when $t = 2$.

To find the resultant velocity, we must find v and θ, by first finding v_x and v_y. After the derivatives are found, they are evaluated for $t = 2$. Therefore,

$$v_x = \frac{dx}{dt} = 6t \qquad v_x\big|_{t=2} = 12 \qquad \text{find velocity components}$$

$$v_y = \frac{dy}{dt} = -2t \qquad v_y\big|_{t=2} = -4$$

$$v = \sqrt{12^2 + (-4)^2} = 12.6 \qquad \text{magnitude of velocity}$$

$$\tan\theta = \frac{-4}{12} \qquad \theta = -18.4^\circ \qquad \text{direction of motion}$$

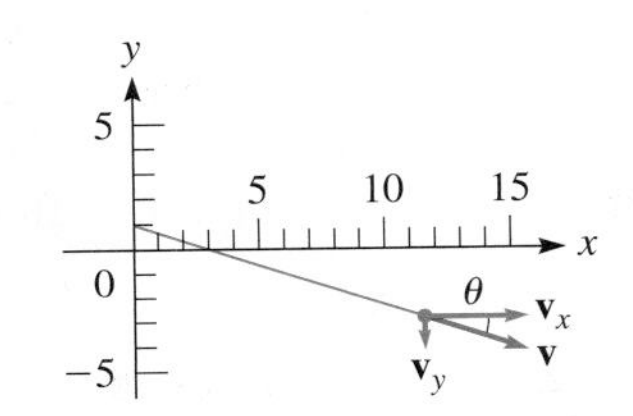

Fig. 24.14

The path and velocity vectors are shown in Fig. 24.14. ■

EXAMPLE 2 Parametric form—resultant velocity

Find the velocity and direction of motion when $t = 2$ of an object moving such that its x- and y-coordinates of position are given by $x = 1 + 2t$ and $y = t^2 - 3t$.

$$v_x = \frac{dx}{dt} = 2 \qquad v_x\big|_{t=2} = 2 \qquad \text{find velocity components}$$

$$v_y = \frac{dy}{dt} = 2t - 3 \qquad v_y\big|_{t=2} = 1$$

$$v\big|_{t=2} = \sqrt{2^2 + 1^2} = 2.24 \qquad \text{magnitude of velocity}$$

$$\tan\theta = \frac{1}{2} \qquad \theta = 26.6^\circ \qquad \text{direction of motion}$$

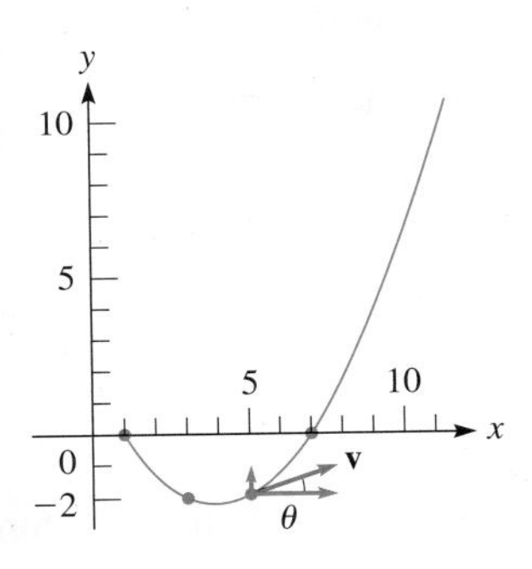

Fig. 24.15

These quantities are shown in Fig. 24.15. ■

CAUTION In these examples, note that we first find the necessary derivatives and then evaluate them. This procedure should always be followed. When a derivative is to be found, it is incorrect to take the derivative of the evaluated expression (which is a constant). ■

Acceleration *is the rate of change of velocity with respect to time.* Therefore, if the velocity, or its components, is known as a function of time, the acceleration of an object can be found by taking the derivative of the velocity with respect to time. If the displacement is known, the acceleration is found by finding the second derivative with respect to time. Finding the acceleration of an object is illustrated in the following example.

EXAMPLE 3 Parametric form—resultant acceleration

Find the magnitude and direction of the acceleration when $t = 2$ for an object that is moving such that its x- and y-coordinates of position are given by the parametric equations $x = t^3$ and $y = 1 - t^2$.

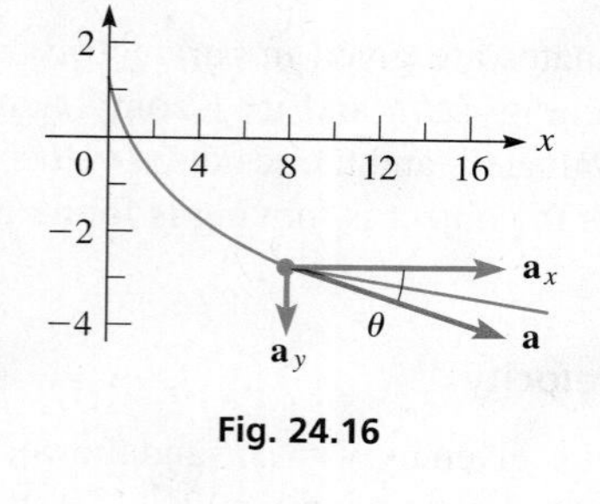

Fig. 24.16

$$v_x = \frac{dx}{dt} = 3t^2 \qquad a_x = \frac{dv_x}{dt} = \frac{d^2x}{dt^2} = 6t \qquad a_x|_{t=2} = 12$$

take second derivatives to find acceleration components

$$v_y = \frac{dy}{dt} = -2t \qquad a_y = \frac{dv_y}{dt} = \frac{d^2y}{dt^2} = -2 \qquad a_y|_{t=2} = -2$$

$$a|_{t=2} = \sqrt{12^2 + (-2)^2} = 12.2$$ magnitude of acceleration

$$\tan\theta = \frac{a_y}{a_x} = -\frac{2}{12} \qquad \theta = -9.5°$$ direction of acceleration

Practice Exercise

1. In Example 3, solve for the acceleration when $t = 2$, if $x = 0.8t^{5/2}$, instead of $x = t^3$.

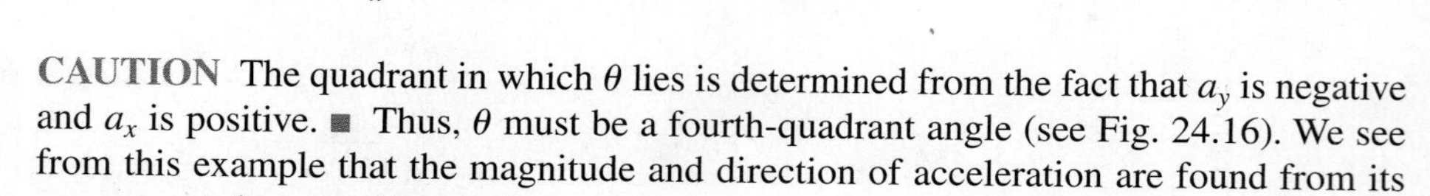

CAUTION The quadrant in which θ lies is determined from the fact that a_y is negative and a_x is positive. ■ Thus, θ must be a fourth-quadrant angle (see Fig. 24.16). We see from this example that the magnitude and direction of acceleration are found from its components just as with velocity. ■

We now summarize the equations used to find the velocity and acceleration of an object for which the displacement is a function of time. They indicate how to find the components, as well as the magnitude and direction, of each.

$$v_x = \frac{dx}{dt} \qquad v_y = \frac{dy}{dt} \qquad \text{velocity components} \tag{24.2}$$

$$a_x = \frac{dv_x}{dt} = \frac{d^2x}{dt^2} \qquad a_y = \frac{dv_y}{dt} = \frac{d^2y}{dt^2} \qquad \text{acceleration components} \tag{24.3}$$

$$v = \sqrt{v_x^2 + v_y^2} \qquad a = \sqrt{a_x^2 + a_y^2} \qquad \text{magnitude} \tag{24.4}$$

$$\tan\theta_v = \frac{v_y}{v_x} \qquad \tan\theta_a = \frac{a_y}{a_x} \qquad \text{direction} \tag{24.5}$$

■ For reference, Eq. (23.15) is $\frac{du^n}{dx} = nu^{n-1}\frac{du}{dx}$.

CAUTION If the curvilinear path an object follows is given with y as a function of x, the velocity (and acceleration) is found by taking derivatives of each term of the equation *with respect to time.* ■ It is assumed that both x and y are functions of time, although these functions are not stated. When finding derivatives, we must be careful in using the general power rule, Eq. (23.15), so that the factor du/dx is not neglected. In the following examples, we illustrate the use of Eqs. (24.2) to (24.5) in applied situations for which we know the equation of the path of the motion. Again, we must be careful to find the direction of the vector as well as its magnitude in order to have a complete solution.

EXAMPLE 4 Velocity at a point along a path

In a physics experiment, a small sphere is constrained to move along a parabolic path described by $y = \frac{1}{3}x^2$. If the horizontal velocity v_x is constant at 6.00 cm/s, find the velocity at the point (2.00, 1.33). See Fig. 24.17.

Because both y and x change with time, both can be considered to be functions of time. Therefore, we can take derivatives of $y = \frac{1}{3}x^2$ with respect to time.

$$\frac{dy}{dt} = \frac{1}{3}\left(2x\frac{dx}{dt}\right) \longleftarrow \frac{dx^2}{dt} = 2x\frac{dx}{dt} \qquad \longleftarrow \text{don't forget the } \frac{dx}{dt}$$

$$v_y = \frac{2}{3}xv_x \qquad \frac{dy}{dt} = v_y;\ \frac{dx}{dt} = v_x$$

$$v_y = \frac{2}{3}(2.00)(6.00) = 8.00 \text{ cm/s} \qquad \text{substituting}$$

$$v = \sqrt{6.00^2 + 8.00^2} = 10.0 \text{ cm/s} \qquad \text{magnitude: } v = \sqrt{v_x^2 + v_y^2}$$

$$\tan\theta = \frac{8.00}{6.00}, \quad \theta = 53.1^\circ \qquad \text{direction: } \tan\theta = \frac{v_y}{v_x}$$

■

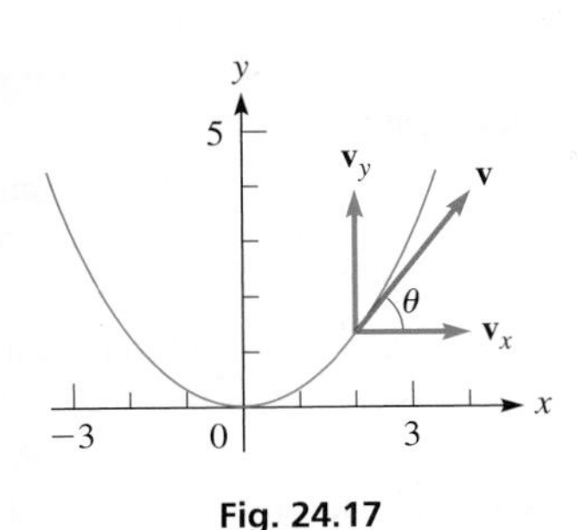

Fig. 24.17

EXAMPLE 5 Velocity and acceleration—rescue marker

A helicopter is flying at 18.0 m/s and at an altitude of 120 m when a rescue marker is released from it. The marker maintains a horizontal velocity and follows a path given by $y = 120 - 0.0151x^2$, as shown in Fig. 24.18. Find the magnitude and direction of the velocity and of the acceleration of the marker 3.00 s after release.

From the given information, we know that $v_x = dx/dt = 18.0$ m/s. Taking derivatives with respect to time leads to this solution:

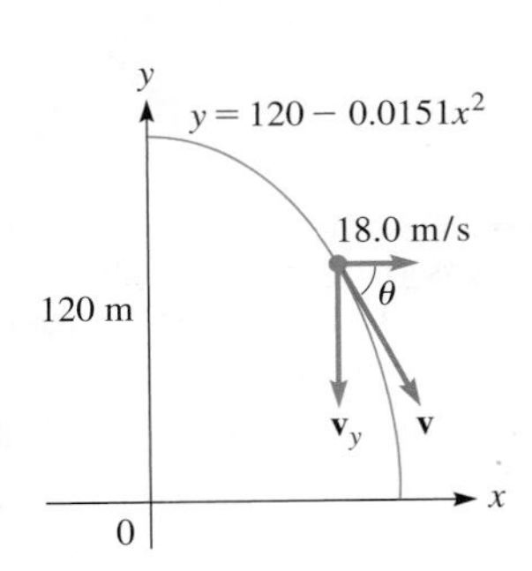

Fig. 24.18

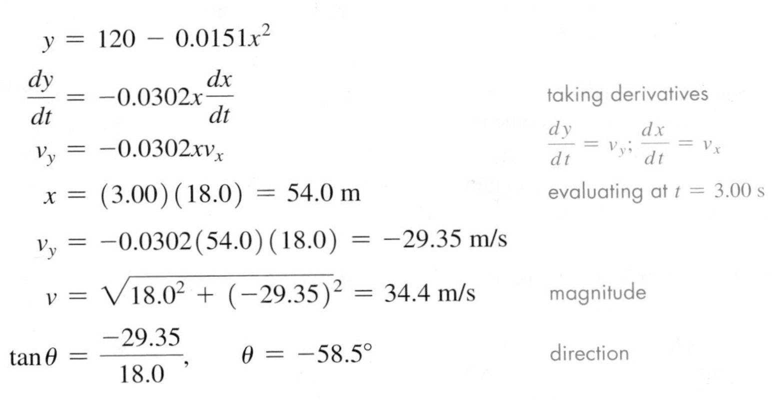

$$y = 120 - 0.0151x^2$$

$$\frac{dy}{dt} = -0.0302x\frac{dx}{dt} \qquad \text{taking derivatives}$$

$$v_y = -0.0302xv_x \qquad \frac{dy}{dt} = v_y;\ \frac{dx}{dt} = v_x$$

$$x = (3.00)(18.0) = 54.0 \text{ m} \qquad \text{evaluating at } t = 3.00 \text{ s}$$

$$v_y = -0.0302(54.0)(18.0) = -29.35 \text{ m/s}$$

$$v = \sqrt{18.0^2 + (-29.35)^2} = 34.4 \text{ m/s} \qquad \text{magnitude}$$

$$\tan\theta = \frac{-29.35}{18.0}, \quad \theta = -58.5^\circ \qquad \text{direction}$$

The velocity is 34.4 m/s and is directed at an angle of 58.5° below the horizontal.

To find the acceleration, we return to the equation $v_y = -0.0302xv_x$. Because v_x is constant, we can substitute 18.0 for v_x to get

$$v_y = -0.5436x$$

Again taking derivatives with respect to time, we have

$$\frac{dv_y}{dt} = -0.5436\frac{dx}{dt}$$

$$a_y = -0.5436v_x \qquad \frac{dv_y}{dt} = a_y;\ \frac{dx}{dt} = v_x$$

$$a_y = -0.5436(18.0) = -9.78 \text{ m/s}^2 \qquad \text{evaluating}$$

We know that v_x is constant, which means that $a_x = 0$. Therefore, the acceleration is 9.78 m/s² and is directed vertically downward. ■

EXERCISES 24.3

In Exercises 1 and 2, make the given changes in the indicated examples of this section and then solve the resulting problems.

1. In Example 1, change $x = 3t^2$ to $x = 4t^2$.
2. In Example 4, change $y = x^2/3$ to $y = x^2/4$ and $(2.00, 1.33)$ to $(2.00, 1.00)$.

In Exercises 3–6, given that the x- and y-coordinates of a moving particle are given by the indicated parametric equations, find the magnitude and direction of the velocity for the specific value of t. Sketch the curves and show the velocity and its components.

3. $x = 3t,\ y = 1 - t,\ t = 4$
4. $x = \dfrac{5t}{2t + 1},\ y = 0.1(t^2 + t),\ t = 2$
5. $x = t(2t + 1)^2,\ y = \dfrac{6}{\sqrt{4t + 3}},\ t = 0.5$
6. $x = \sqrt{1 + 2t},\ y = t - t^2,\ t = 4$

In Exercises 7–10, use the parametric equations and values of t in Exercises 3–6 to find the magnitude and direction of the acceleration in each case.

In Exercises 11–30, find the indicated velocities and accelerations.

11. A baseball is ejected horizontally toward home plate from a pitching machine on the mound with a velocity of 42.5 m/s. If y is the height of the ball above the ground, and t is the time (in s) after being ejected, $y = 1.5 - 4.9t^2$. What are the height and velocity of the ball when it crosses home plate in 0.43 s?
12. A section of a bike trail can be described by $y = 0.0016x^2$. On this section of the trail a bike maintains a constant $v_x = 650$ m/min. What is the bike's velocity when $x = 100$ m?
13. The water from a fire hose follows a path described by $y = 2.0 + 0.80x - 0.20x^2$ (units are in meters). If v_x is constant at 5.0 m/s, find the resultant velocity at the point $(4.0, 2.0)$.
14. A roller mechanism follows a path described by $y = \sqrt{4x + 1}$, where units are in feet. If $v_x = 2x$, find the resultant velocity (in ft/s) at the point $(2.0, 3.0)$.
15. A float is used to test the flow pattern of a stream. It follows a path described by $x = 0.2t^2$, $y = -0.1t^3$ (x and y in ft, t in min). Find the acceleration of the float after 2.0 min.
16. A radio-controlled model car is operated in a parking lot. The coordinates (in m) of the car are given by $x = 3.5 + 2.0t^2$ and $y = 8.5 + 0.25t^3$, where t is the time (in s). Find the acceleration of the car after 2.5 s.
17. A golfer drives a golf ball that moves according to the equations $x = 25t$ and $y = 36t - 4.9t^2$ (x and y in meters, t in seconds). Find the resultant velocity and acceleration of the golf ball for $t = 6.0$ s.
18. A package of relief supplies is dropped and moves according to the parametric equations $x = 45t$ and $y = -4.9t^2$ (x and y in m, t in s). Find the velocity and acceleration when $t = 3.0$ s.
19. A spacecraft moves along a path described by the parametric equations $x = 10(\sqrt{1 + t^4} - 1)$, $y = 40t^{3/2}$ for the first 100 s after launch. Here, x and y are measured in meters, and t is measured in seconds. Find the magnitude and direction of the velocity of the spacecraft 10.0 s and 100 s after launch.
20. An electron moves in an electric field according to the equations $x = 8.0/\sqrt{1 + t^2}$ and $y = 8.0t/\sqrt{1 + t^2}$ (x and y Mm and t in s). Find the velocity of the electron when $t = 0.5$ s.
21. In a computer game, an airplane starts at $(1.00, 4.00)$ (in cm) on the curve $y = 3.00 + x^{-1.50}$ and moves with a constant horizontal velocity of 1.20 cm/s. What is the plane's velocity after 0.500 s?
22. A person on a hoverboard is riding up a ramp and follows a path described by $y = 0.15x^{1.2}$. If v_x is constant at 0.50 m/s, find v_y when $x = 8$.
23. Find the resultant acceleration of the spacecraft in Exercise 19 for the specified times.
24. A ski jump is designed to follow the path given by the equations $x = 3.50t^2$ and $y = 20.0 + 0.120t^4 - 3.00\sqrt{t^4 + 1}$ $(0 \le t \le 4.00\text{ s})$ (x and y in m, t in s). Find the velocity and acceleration of a skier when $t = 4.00$ s. See Fig. 24.19.

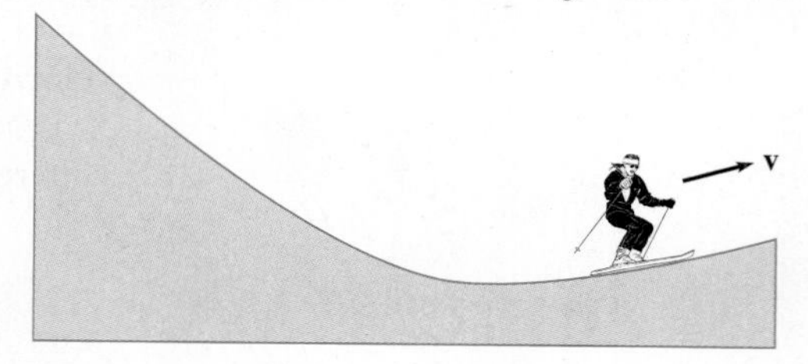

Fig. 24.19

25. A rocket follows a path given by $y = x - \frac{1}{90}x^3$ (distances in miles). If the horizontal velocity is given by $v_x = x$, find the magnitude and direction of the velocity when the rocket hits the ground (assume level terrain) if time is in minutes.
26. A ship is moving around an island on a route described by $y = 3x^2 - 0.2x^3$. If $v_x = 1.2$ km/h, find the velocity of the ship where $x = 3.5$ km.
27. A computer's hard disk is 3.50 in. in diameter and rotates at 3600 r/min. With the center of the disk at the origin, find the velocity components of a point on the rim for $x = 1.20$ in., if $y > 0$ and $v_x > 0$.
28. A robot arm joint moves in an elliptical path (horizontal major axis 8.0 cm, minor axis 4.0 cm, center at origin). For $y > 0$ and $-2\text{ cm} < x < 2\text{ cm}$, the joint moves such that $v_x = 2.5$ cm/s. Find its velocity for $x = -1.5$ cm.
29. An airplane ascends such that its gain h in altitude is proportional to the square root of the change x in horizontal distance traveled. If $h = 280$ m for $x = 400$ m and v_x is constant at 350 m/s, find the velocity at this point.
30. A meteor traveling toward Earth has a velocity inversely proportional to the square root of the distance from Earth's center. State how its acceleration is related to its distance from the center of Earth.

Answer to Practice Exercise

1. $a = 4.7,\ \theta = -25.2°$

24.4 Related Rates

Derivatives with Respect to Time • Rates of Change are Related

Often, variables vary with respect to time, and are therefore implicitly functions of time. If a relation is known to exist relating them, the rate of change with respect to time of one can be expressed in terms of the rate of change of the other(s). This is done by taking the derivative with respect to time of the expression relating the variables, even if t does not appear in the expression, as in Examples 4 and 5 of Section 24.3. Because the rates of change are related, this is referred to as a **related-rates** problem. Consider the following examples.

EXAMPLE 1 Related rates—voltage and temperature

The voltage E of a certain thermocouple as a function of the temperature T (in °C) is given by $E = 2.800T + 0.006T^2$. If the temperature is increasing at the rate of 1.00°C/min, how fast is the voltage increasing when $T = 100$°C?

Because we are asked to find the rate at which the voltage is changing, we first take derivatives with respect to time. This gives us

$$\frac{dE}{dt} = 2.800\frac{dT}{dt} + 0.012T\frac{dT}{dt}$$

$\frac{d}{dt}(0.006T^2) = 0.006\left(2T\frac{dT}{dt}\right)$

CAUTION We must be careful to include the factor dT/dt. ■ From the given information, we know that $dT/dt = 1.00$°C/min and that we wish to know dE/dt when $T = 100$°C. Thus,

$$\left.\frac{dE}{dt}\right|_{T=100} = 2.800(1.00) + 0.012(100)(1.00) = 4.00 \text{ V/min}$$

CAUTION The derivative must be taken before values are substituted. ■ In this problem, we are finding the rate at which the voltage is changing for a specified value of T. For other values of T, dE/dt would have different values. ■

EXAMPLE 2 Related rates—distances from a lens

The distance q that an image is from a certain lens in terms of p, the distance of the object from the lens, is given by

$$q = \frac{10p}{p - 10}$$

If the object distance is increasing at the rate of 0.200 cm/s, how fast is the image distance changing when $p = 15.0$ cm? See Fig. 24.20.

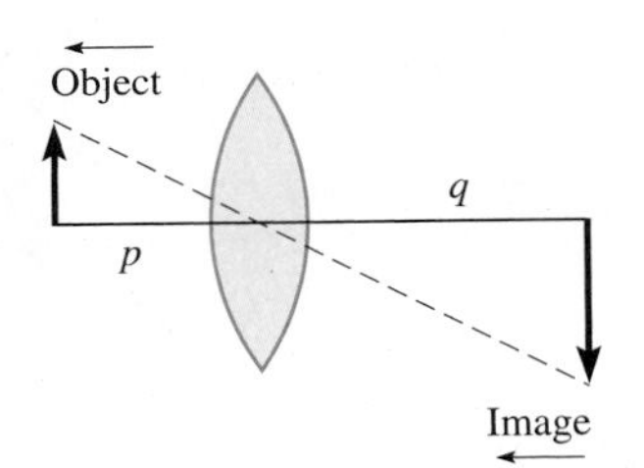

Fig. 24.20

Taking derivatives with respect to time, we have

don't forget the $\frac{dp}{dt}$

$$\frac{dq}{dt} = \frac{(p - 10)\left(10\frac{dp}{dt}\right) - 10p\left(\frac{dp}{dt}\right)}{(p - 10)^2} = \frac{-100\frac{dp}{dt}}{(p - 10)^2}$$

Now, substituting $p = 15.0$ and $dp/dt = 0.200$, we have

$$\left.\frac{dq}{dt}\right|_{p=15} = \frac{-100(0.200)}{(15.0 - 10)^2} = -0.800 \text{ cm/s}$$

Thus, the image distance is decreasing (the significance of the minus sign) at the rate of 0.800 cm/s when $p = 15.0$ cm. ■

Practice Exercise

1. In Example 2, change each 10 to 12, and then solve.

In many related-rate problems, the function is not given but must be set up according to the statement of the problem. The following examples illustrate this type of problem.

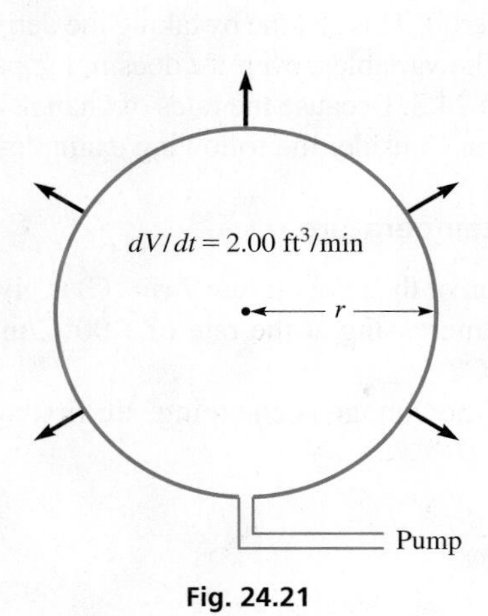

Fig. 24.21

EXAMPLE 3 Related rates—volume and radius of a sphere

A spherical balloon is being blown up such that its volume increases at the constant rate of 2.00 ft³/min. Find the rate at which the radius is increasing when it is 3.00 ft. See Fig. 24.21.

We are asked to find the relation between the rate of change of the volume of a sphere with respect to time and the corresponding rate of change of the radius with respect to time. Therefore, we are to ***take derivatives of the expression for the volume of a sphere with respect to time:***

$$V = \frac{4}{3}\pi r^3 \quad \text{volume of sphere}$$

$$\frac{dV}{dt} = 4\pi r^2\left(\frac{dr}{dt}\right) \quad \text{take derivatives with respect to time}$$

$$2.00 = 4\pi(3.00)^2\left(\frac{dr}{dt}\right) \quad \text{substitute } \frac{dV}{dt} = 2.00 \text{ ft}^3/\text{min and } r = 3.00 \text{ ft}$$

$$\left.\frac{dr}{dt}\right|_{r=3} = \frac{1}{18.0\pi} \quad \text{solve for } \frac{dr}{dt}$$

$$= 0.0177 \text{ ft/min}$$

■

EXAMPLE 4 Related rates—force and distance of a spacecraft

The force F of gravity of Earth on a spacecraft varies inversely as the square of the distance r of the spacecraft from the center of Earth. A particular spacecraft weighs 4500 N on the launchpad ($F = 4500$ N for $r = 6370$ km). Find the rate at which F changes later as the spacecraft moves away from Earth at the rate of 12 km/s, where $r = 8500$ km.

Noting Fig. 24.22 and setting up the equation, we have the following solution:

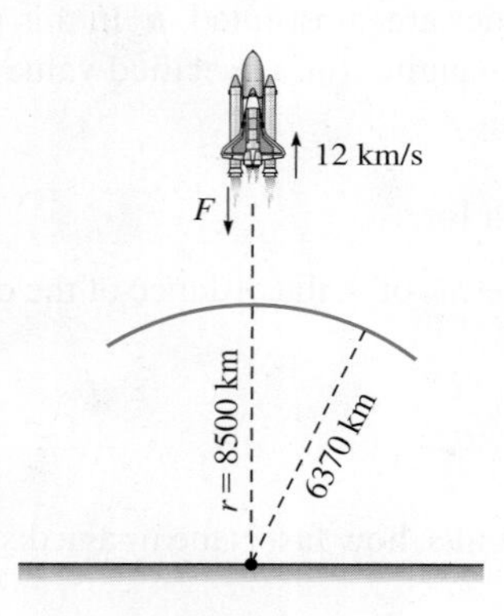

Fig. 24.22

$$F = \frac{k}{r^2} \quad \text{inverse variation}$$

$$4500 = \frac{k}{6370^2} \quad \text{substitute } F = 4500 \text{ N}, r = 6370 \text{ km}$$

$$k = 1.83 \times 10^{11} \text{ N}\cdot\text{km}^2 \quad \text{solve for } k$$

$$F = \frac{1.83 \times 10^{11}}{r^2} \quad \text{substitute for } k \text{ in equation}$$

$$\frac{dF}{dt} = (1.83 \times 10^{11})(-2)(r^{-3})\frac{dr}{dt} \quad \text{take derivatives with respect to time}$$

$$= \frac{-3.66 \times 10^{11}}{r^3}\frac{dr}{dt} \quad \text{evaluate derivate for } r = 8500 \text{ km}, dr/dt = 12 \text{ km/s}$$

$$\left.\frac{dF}{dt}\right|_{t=8500 \text{ km}} = \frac{-3.66 \times 10^{11}}{8500^3}(12)$$

$$= -7.2 \text{ N/s} \quad \text{gravitational force is decreasing}$$

■

These examples show the following method of solving a related-rates problem.

Steps for Solving Related-Rates Problems

1. Identify the variables and rates in the problem.
2. If possible, make a sketch showing the variables.
3. Determine the equation relating the variables.
4. Differentiate with respect to time.
5. Solve for the required rate.
6. Evaluate the required rate.

EXAMPLE 5 Related rates—distances

Two cruise ships leave Vancouver, British Columbia, at noon. Ship A travels west at 12.0 km/h (before leaving toward Alaska), and ship B travels south at 16.0 km/h (toward Seattle). How fast are they separating at 2 P.M.?

step 1

In Fig. 24.23, we let $x =$ the distance traveled by A and $y =$ the distance traveled by B. We can find the distance between them, z, from the Pythagorean theorem. Therefore, we are to find dz/dt for $t = 2.00$ h. Even though there are three variables, each is a function of time. This means we can find dz/dt by taking derivatives of each term with respect to time. This gives us

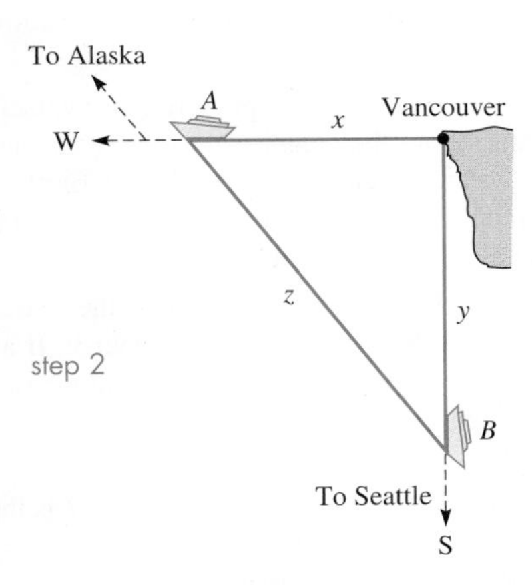

Fig. 24.23

$$z^2 = x^2 + y^2$$ using Pythagorean theorem (step 3)

$$2z\frac{dz}{dt} = 2x\frac{dx}{dt} + 2y\frac{dy}{dt}$$ taking derivatives with respect to time (step 4)

$$\frac{dz}{dt} = \frac{x(dx/dt) + y(dy/dt)}{z}$$ solve for $\frac{dz}{dt}$ (step 5)

At 2 P.M., we have the values (step 6)

$$x = 24.0 \text{ km}, \quad y = 32.0 \text{ km}, \quad z = 40.0 \text{ km}$$
$$dx/dt = 12.0 \text{ km/h}, \quad dy/dt = 16.0 \text{ km/h}$$

$d = rt$ and Pythagorean theorem from statement of problem

$$\left.\frac{dz}{dt}\right|_{z=40} = \frac{(24.0)(12.0) + (32.0)(16.0)}{40.0} = 20.0 \text{ km/h}$$ substitute values ■

EXERCISES 24.4

In Exercises 1 and 2, make the given changes in the indicated examples of this section and then solve the resulting problems.

1. In Example 1, change 0.006 to 0.012.
2. In Example 3, change "volume" to "surface area," change ft^3/min to ft^2/min.

In Exercises 3–6, assume that all variables are implicit functions of time t. Find the indicated rates.

3. $y = 5x^2 - 4x$; $dx/dt = 0.5$ when $x = 5$; find dy/dt.
4. $y = \sqrt{9 - 2x^2}$; $dy/dt = 3$ when $x = 2$ and $y = 1$; find dx/dt.
5. $x^2 + 3y^2 + 2y = 10$; $dx/dt = 2$ when $x = 3$ and $y = -1$; find dy/dt.
6. $z = 2x^2 - 3xy$; $dx/dt = -2$, $dy/dt = 3$ when $x = 1$ and $y = 4$; find dz/dt.

In Exercises 7–42, solve the problems in related rates.

7. The velocity v (in ft/s) of a pulse traveling in a certain string is a function of the tension T (in lb) in the string given by $v = 18\sqrt{T}$. Find dv/dt if $dT/dt = 0.20$ lb/s when $T = 25$ lb.
8. The force F (in lb) on the blade of a certain wind generator as a function of the wind velocity v (in ft/s) is given by $F = 0.0056v^2$. Find dF/dt if $dv/dt = 0.75$ ft/s^2 when $v = 28$ ft/s.
9. The electric resistance R (in Ω) of a certain resistor as a function of the temperature T (in°C) is $R = 4.000 + 0.003T^2$. If the temperature is increasing at the rate of 0.100°C/s, find how fast the resistance changes when $T = 150$°C.
10. The kinetic energy K (in J) of an object is given by $K = \frac{1}{2}mv^2$, where m is the mass (in kg) of the object and v is its velocity. If a 250-kg wrecking ball accelerates at 5.00 m/s^2, how fast is the kinetic energy changing when $v = 30.0$ m/s? (*Hint*: acceleration $= \frac{dv}{dt}$.)
11. The length L (in in.) of a pendulum is slowly decreasing at the rate of 0.100 in./s. How fast is the period T (in s) of the pendulum changing when $L = 16.0$ in., if the equation relating the period and length is $T = \pi\sqrt{L/96}$?
12. The voltage V that produces a current I (in A) in a wire of radius r (in in.) is $V = 0.030I/r^2$. If the current increases at 0.020 A/s in a wire of 0.040 in. radius, find the rate at which the voltage is increasing.
13. A plane flying at an altitude of 2.0 mi is at a direct distance $D = \sqrt{4.0 + x^2}$ from an airport control tower, where x is the horizontal distance to the tower. If the plane's speed is 350 mi/h, how fast is D changing when $x = 6.2$ mi? See Fig. 24.24.

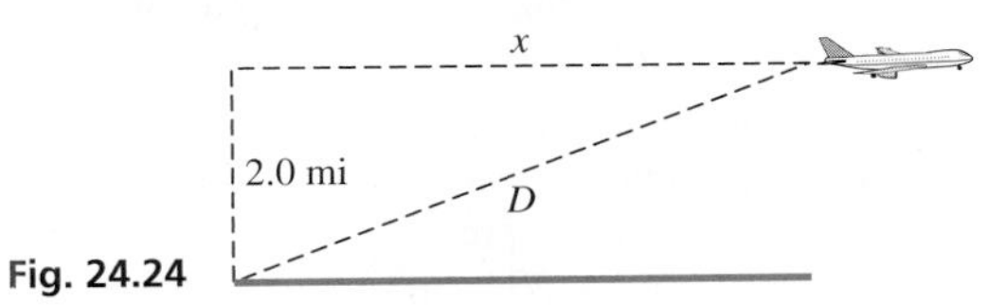

Fig. 24.24

14. A variable resistor R and an 8-Ω resistor in parallel have a combined resistance R_T given by $R_T = \frac{8R}{8 + R}$. If R is changing at 0.30 Ω/min, find the rate at which R_T is changing when $R = 6.0$ Ω.

15. The radius r (in m) of a ring of a certain holograph (an image produced without using a lens) is given by $r = \sqrt{0.4\lambda}$, where λ is the wavelength of the light being used. If λ is changing at the rate of 0.10×10^{-7} m/s when $\lambda = 6.0 \times 10^{-7}$ m, find the rate at which r is changing.

16. An Earth satellite moves in a path that can be described by $\frac{x^2}{2.80 \times 10^7} + \frac{y^2}{2.76 \times 10^7} = 1$, where x and y are in miles. If $dx/dt = 7750$ mi/h for $x = 2020$ mi and $y > 0$, find dy/dt.

17. The magnetic field B due to a magnet of length l at a distance r is given by $B = \frac{k}{[r^2 + (l/2)^2]^{3/2}}$, where k is a constant for a given magnet. Find the expression for $\frac{dB}{dt}$ in terms of $\frac{dr}{dt}$.

18. An approximate relationship between the pressure p and volume v of the vapor in a diesel engine cylinder is $pv^{1.4} = k$, where k is a constant. At a certain instant, $p = 4200$ kPa, $v = 75$ cm^3, and the volume is increasing at the rate of 850 cm^3/s. At what rate is the pressure changing at this instant?

19. A swimming pool with a rectangular surface 18.0 m long and 12.0 m wide is being filled at the rate of 0.80 m^3/min. At one end it is 1.0 m deep, and at the other end it is 2.5 m deep, with a constant slope between ends. How fast is the height of water rising when the depth of water at the deep end is 1.0 m?

20. An engine cylinder 15.0 cm deep is being bored such that the radius increases by 0.100 mm/min. How fast is the volume V of the cylinder changing when the diameter is 9.50 cm?

21. Fatty deposits have decreased the circular cross-sectional opening of a person's artery. A test drug reduces these deposits such that the radius of the opening increases at the rate of 0.020 mm/month. Find the rate at which the area of the opening increases when $r = 1.2$ mm.

22. A rectangular image 4.00 in. high on a computer screen is widening at the rate of 0.25 in./s. Find the rate at which the diagonal is increasing when the width is 6.50 in.

23. A metal cube dissolves in acid such that an edge of the cube decreases by 0.50 mm/min. How fast is the volume of the cube changing when the edge is 8.20 mm?

24. A metal sphere is placed in seawater to study the corrosive effect of seawater. If the surface area decreases at 35 cm^2/year due to corrosion, how fast is the radius changing when it is 12 cm?

25. A uniform layer of ice covers a spherical water-storage tank. As the ice melts, the volume V of ice decreases at a rate that varies directly as the surface area A. Show that the outside radius decreases at a constant rate.

26. A light in a garage is 9.50 ft above the floor and 12.0 ft behind the door. If the garage door descends vertically at 1.50 ft/s, how fast is the door's shadow moving toward the garage when the door is 2.00 ft above the floor?

27. One statement of Boyle's law is that the pressure of a gas varies inversely as the volume for constant temperature. If a certain gas occupies 650 cm^3 when the pressure is 230 kPa and the volume is increasing at the rate of 20.0 cm^3/min, how fast is the pressure changing when the volume is 810 cm^3?

28. The tuning frequency f of an electronic tuner is inversely proportional to the square root of the capacitance C in the circuit. If $f = 920$ kHz for $C = 3.5$ pF, find how fast f is changing at this frequency if $dC/dt = 0.3$ pF/s.

29. The shadow of a 24-m high building is increasing at the rate of 18 cm/min when the shadow is 18 m long. How fast is the distance from the top of the building to the end of the shadow increasing?

30. The acceleration due to the gravity g on a spacecraft is inversely proportional to its distance from the center of Earth. At the surface of Earth, $g = 32.2$ ft/s^2. Given that the radius of Earth is 3960 mi, how fast is g changing on a spacecraft approaching Earth at 4500 ft/s at a distance of 25,500 mi from the surface?

31. The intensity I of heat varies directly as the strength of the source and inversely as the square of the distance from the source. If an object approaches a heated object of strength 8.00 units at the rate of 50.0 cm/s, how fast is the intensity changing when it is 100 cm from the source?

32. The speed of sound v (in m/s) is $v = 331\sqrt{T/273}$, where T is the temperature (in K). If the temperature is 303 K (30°C) and is rising at 2.0°C/h, how fast is the speed of sound rising?

33. As a space shuttle moves into space, an astronaut's weight decreases. An astronaut weighing 650 N at sea level has a weight of $w = 650\left(\frac{6400}{6400 + h}\right)$ at h kilometers above sea level. If the shuttle is moving away from Earth at 6.0 km/s, at what rate is w changing when $h = 1200$ km?

34. The oil reservoir for the lubricating mechanism of a machine is in the shape of an inverted pyramid. It is being filled at the rate of 8.00 cm^3/s and the top surface is increasing at the rate of 6.00 cm^2/s. When the depth of oil is 6.50 cm and the top surface area is 22.5 cm^2, how fast is the level increasing? (*Hint*: The volume of a pyramid is given by $V = \frac{1}{3}Bh$, where B is the area of the base and h is the height).

35. Coffee is draining through a conical filter into a coffee pot at the rate of 18.0 cm.^{3}/min. If the filter is 15.0 cm in diameter and 15.0 cm deep, how fast is the level of coffee in the filter changing when the depth is 10.0 cm? (*Hint*: Use $V = \frac{1}{3}\pi r^2 h$.)

36. The top of a ladder 4.00 m long is slipping down a vertical wall at a constant rate of 0.75 m/s. How fast is the bottom of the ladder moving along the ground away from the wall when it is 2.50 m from the wall? See Fig. 24.25.

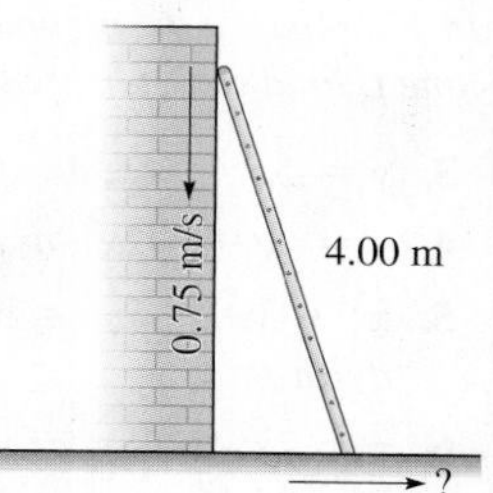

Fig. 24.25

37. A supersonic jet leaves an airfield traveling due east at 1600 mi/h. A second jet leaves the same airfield at the same time and travels 1800 mi/h along a line north of east such that it remains due north of the first jet. After a half-hour, how fast are the jets separating?

38. A car passes over a bridge at 15.0 m/s at the same time a boat passes under the bridge at a point 10.5 m directly below the car. If the boat is moving perpendicularly to the bridge at 4.0 m/s, how fast are the car and the boat separating 5.0 s later?

39. A rope attached to a boat is being pulled in at a rate of 10.0 ft/s. If the water is 20.0 ft below the level at which the rope is being drawn in, how fast is the boat approaching the wharf when 36.0 ft of rope are yet to be pulled in? See Fig. 24.26.

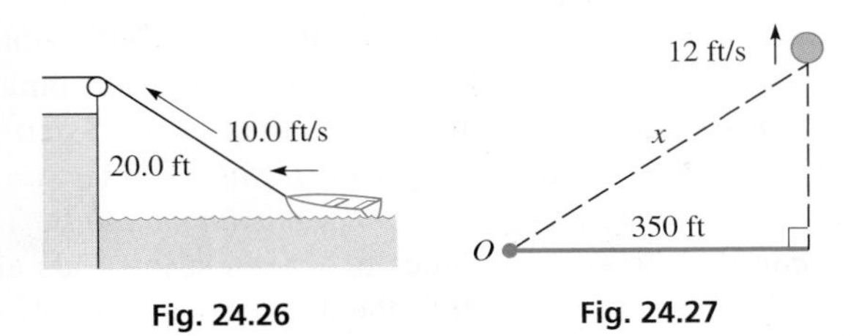

Fig. 24.26

Fig. 24.27

40. A weather balloon leaves the ground 350 ft from an observer and rises vertically at 12 ft/s. How fast is the line of sight from the observer to the balloon increasing when the balloon is 250 ft high? See Fig. 24.27.

41. A man 6.00 ft tall approaches a street light 15.0 ft above the ground at the rate of 5.00 ft/s. How fast is the end of the man's shadow moving when he is 10.0 ft from the base of the light? See Fig. 24.28.

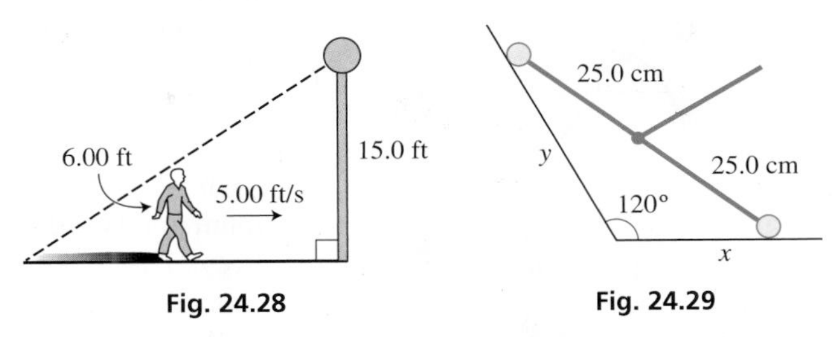

Fig. 24.28

Fig. 24.29

42. A roller mechanism, as shown in Fig. 24.29, moves such that the right roller is always in contact with the bottom surface and the left roller is always in contact with the left surface. If the right roller is moving to the right at 1.50 cm/s when $x = 10.0$ cm, how fast is the left roller moving? (*Hint*: Use the law of cosines.)

Answers to Practice Exercise

1. -3.20 cm/s

24.5 Using Derivatives in Curve Sketching

Function Increasing and Decreasing • Local Maximum and Minimum Points • First and Second Derivative Tests • Concave Up and Concave Down • Points of Inflection

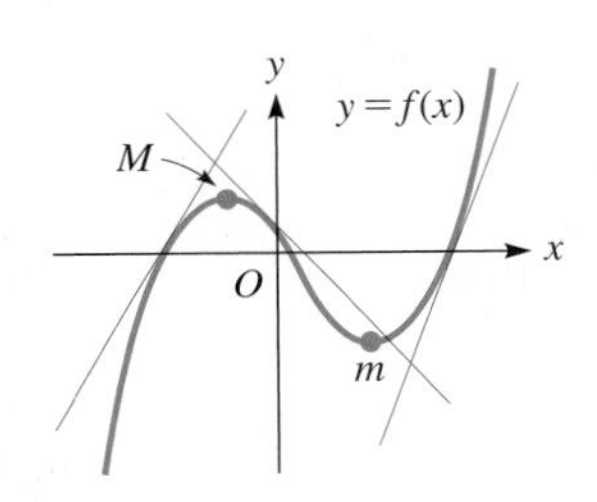

Fig. 24.30

For the function in Fig. 24.30, we see that as x increases (from left to right), y also increases until point M is reached. From M to m, y decreases. To the right of m, y again increases. Also, any tangent line left of M or right of m has a positive slope, and any tangent line between M and m has a negative slope. Because the derivative gives us the slope of a tangent line, we see that *as x increases, y increases if the derivative is positive and decreases if the derivative is negative.* This can be stated as

$$\boldsymbol{f(x) \text{ increases if } f'(x) > 0} \quad \text{and} \quad \boldsymbol{f(x) \text{ decreases if } f'(x) < 0}$$

It is always assumed that x is increasing. *Also, we assume in our present analysis that $f(x)$ and its derivatives are continuous over the indicated interval.*

EXAMPLE 1 Function increasing and decreasing

Find those values of x for which the function $f(x) = x^3 - 3x^2$ is increasing and those values for which it is decreasing.

We start by finding the values of x that make the **derivative equal to zero,** which are called the **critical values.** These are the only possible places where the function can change from increasing to decreasing or vice versa. Therefore, we use the critical values to subdivide the x-axis into **test intervals,** and then select a **test value** within each test interval to determine the sign of the derivative. This is shown below.

$$f'(x) = 3x^2 - 6x = 3x(x - 2)$$
$$\text{critical values: } x = 0, 2$$

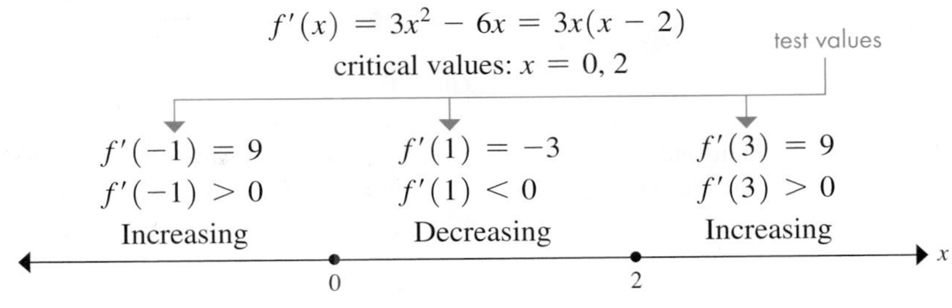

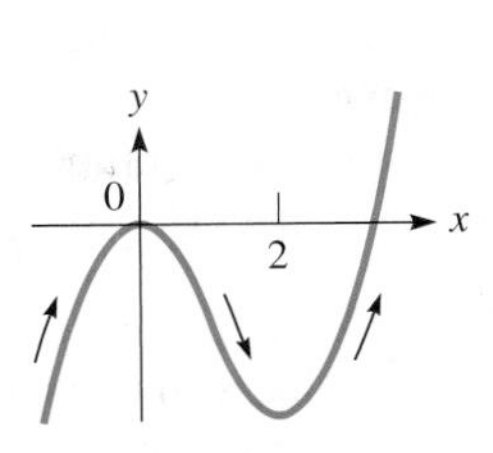

Fig. 24.31

Thus, $f(x)$ is increasing for $x < 0$ or $x > 2$, and decreasing for $0 < x < 2$. The graph of this function is shown in Fig. 24.31. ■

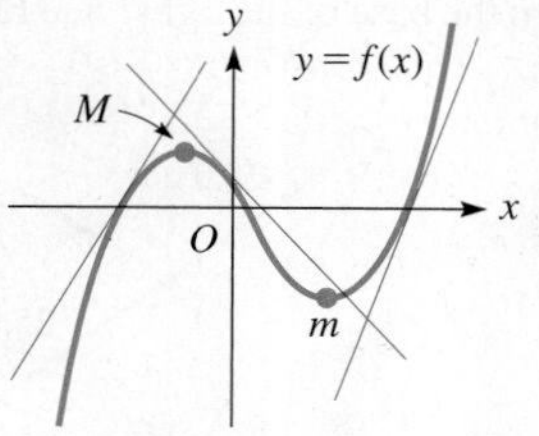

Fig. 24.30

LOCAL MAXIMUM AND MINIMUM POINTS

The points M and m in Fig. 24.30 are called a **local maximum point** and a **local minimum point,** respectively. This means that ***M has a greater y-value than any point near it,*** and that ***m has a smaller y-value than any point near it.*** This does not necessarily mean that M has the greatest y-value of any point on the curve, or that m has the least y-value of any point on the curve (that is why we use the word *local*).

An important characteristic of both M and m is that the ***derivative is zero at both points,*** meaning ***they occur at critical values.*** This leads us to the method of finding local minimum points and local maximum points (of continuous functions with continuous derivatives). *The derivative is found and set equal to zero to find the critical values. These critical values give us the x-coordinates of any possible local minimum or maximum points.*

As shown in Example 1, we can use the critical values to form test intervals and determine the sign of the derivative for each test interval. If the derivative changes from positive to negative, then there is a local maximum at that critical value. If the derivative changes from negative to positive, then there is a local minimum. If the sign of the derivative does not change, there is neither a maximum nor a minimum point. This is known as the **first-derivative test for maxima and minima.** In Fig. 24.32, a diagram for the first-derivative test is shown.

■ Local maximums and local minimums are also referred to as *relative maximums* and *relative minimums.*

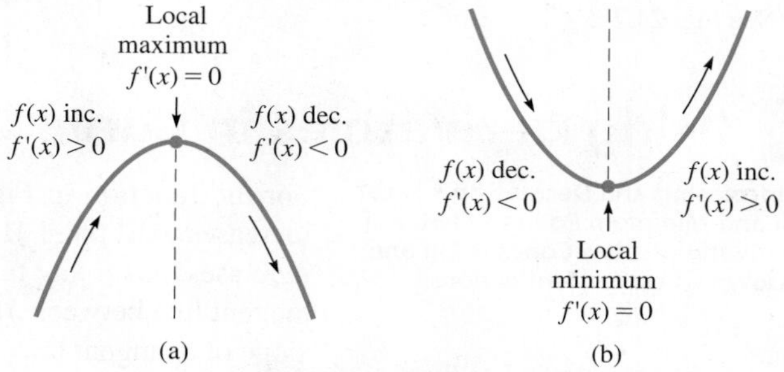

Fig. 24.32

First-Derivative Test for Maxima and Minima

EXAMPLE 2 Local maximum and minimum points

Find any local maximum points and any local minimum points of the graph of the function $y = 3x^5 - 5x^3$.

We will take the derivative, set it equal to zero, and use the critical values to find the values of x where the function is increasing or decreasing. Then we can determine any local maximum or minimum points. This is shown below.

$$y' = 15x^4 - 15x^2 = 15x^2(x^2 - 1) = 15x^2(x + 1)(x - 1)$$

critical values: $x = -1, 0, 1$

test values

$y'(-2) > 0$	$y'(-0.5) < 0$	$y'(0.5) < 0$	$y'(2) > 0$
Increasing	Decreasing	Decreasing	Increasing

-1 0 1 x

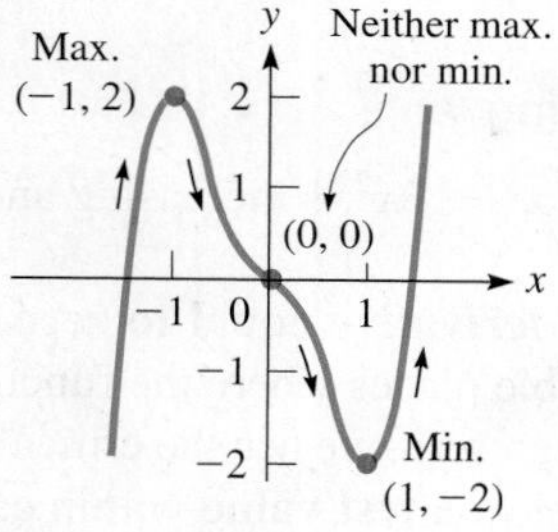

Fig. 24.33

There is a local maximum at $(-1, 2)$ (the function changes from increasing to decreasing) and a local minimum at $(1, -2)$ (the function changes from decreasing to increasing). Note that the y-values of these points are found by using the original function. There is neither a maximum nor a minimum at $(0, 0)$ since the sign of y' doesn't change at $x = 0$. The graph of $y = 3x^5 - 5x^3$ is shown in Fig. 24.33. ■

Having determined how to find local minimum points and local maximum points, we will now consider the rate at which the slope of a curve changes. This will lead us to finding other important points for the graph of the function.

Practice Exercise

1. Find the local maximum and minimum points on the graph of $y = x^3 - 3x^2 - 9x$.

Fig. 24.34

CONCAVITY AND INFLECTION POINTS

We now look again at the slope of a tangent drawn to a curve. In Fig. 24.34(a), consider the *change* in the values of the slope of a tangent at a point as the point moves from A to B. At A the slope is positive, and as the point moves toward M, the slope remains positive but becomes smaller until it becomes zero at M. To the right of M, the slope is negative and becomes more negative until it reaches I. Therefore, *from A to I, the slope continually decreases.* To the right of I, the slope remains negative but increases until it becomes zero again at m. To the right of m, the slope becomes positive and increases to point B. Therefore, *from I to B, the slope continually increases.* We say that *the curve is* **concave down** *from A to I and* **concave up** *from I to B.*

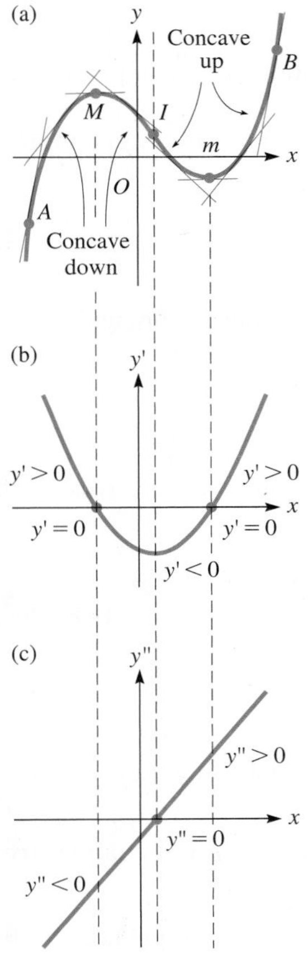

The curve in Fig. 24.34(b) is that of the derivative, and it therefore indicates the values of the slope of $f(x)$. If the slope changes, we are dealing with the rate of change of slope or the rate of change of the derivative. This function is the second derivative. The curve in Fig. 24.34(c) is that of the second derivative. We see that *where the second derivative of a function is* ***negative,*** *the slope is decreasing, or the curve is* ***concave down*** *(opens down). Where the second derivative is* ***positive,*** *the slope is increasing, or the curve is* ***concave up*** *(opens up).* This may be summarized as follows:

If $f''(x) > 0$, the curve is concave up.
If $f''(x) < 0$, the curve is concave down.

We can also now use this information in the determination of maximum and minimum points. By the nature of the definition of maximum and minimum points and of concavity, it is apparent that *a curve is concave down at a maximum point and concave up at a minimum point.* We can see these properties when we make a close analysis of the curve in Fig. 24.34. Therefore, at $x = a$,

$$\text{if } f'(a) = 0 \text{ and } f''(a) < 0,$$

then $f(x)$ has a local maximum at $x = a$, or

$$\text{if } f'(a) = 0 \text{ and } f''(a) > 0,$$

then $f(x)$ has a local minimum at $x = a$.

These statements comprise what is known as **the second-derivative test for maxima and minima.** This test is often easier to use than the first-derivative test. However, it can happen that $y'' = 0$ at a maximum or minimum point, and in such cases it is necessary that we use the first-derivative test.

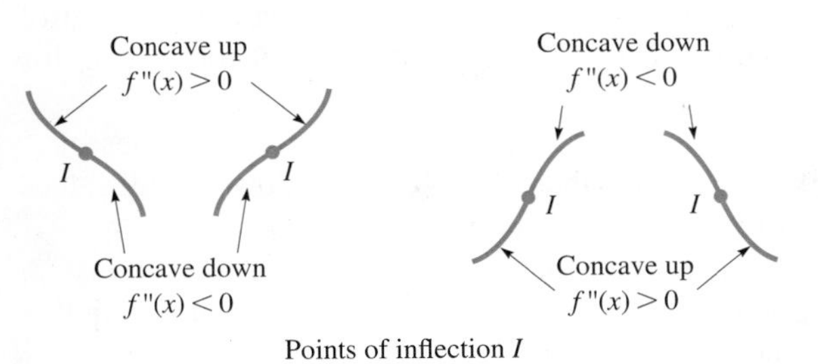

Fig. 24.35

CAUTION In using the second-derivative test, we should carefully note that $f''(x)$ is *negative* at a *local maximum* point, and $f''(x)$ is *positive* at a *local minimum* point. ■

The points at which the curve changes from concave up to concave down, or from concave down to concave up, are known as **points of inflection.** Thus, point I in Fig. 24.34(a) is a point of inflection. Inflection points are found by determining those values of x for which the second derivative changes sign. This is analogous to finding maximum and minimum points by the first-derivative test. In Fig. 24.35, various types of points of inflection are illustrated.

Throughout this analysis it has been assumed that $f(x)$ and its derivatives are continuous functions. To show that this is necessary, see Exercise 58 of this section.

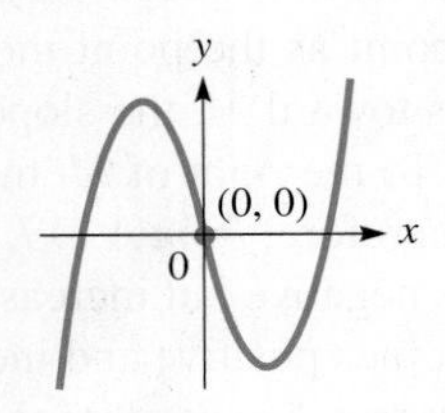

Fig. 24.36

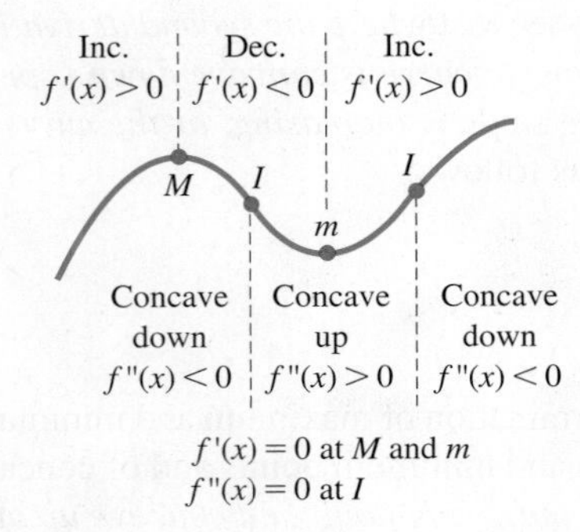

Fig. 24.37

EXAMPLE 3 Concavity—points of inflection

Determine the concavity and find any points of inflection on the graph of the function $y = x^3 - 3x$.

This requires an inspection and analysis of the second derivative. Therefore, we find the first two derivatives.

$$y' = 3x^2 - 3$$
$$y'' = 6x$$

The second derivative is positive where the function is concave up, and this occurs if $x > 0$. The curve is concave down for $x < 0$, because y'' is negative. Thus, $(0, 0)$ is a point of inflection, since the concavity changes there. The graph of $y = x^3 - 3x$ is shown in Fig. 24.36. ■

At this point, we summarize the information found from the derivatives of a function $f(x)$. See Fig. 24.37.

$f'(x) > 0$ where $f(x)$ increases; $f'(x) < 0$ where $f(x)$ decreases.

$f''(x) > 0$ where the graph of $f(x)$ is concave up; $f''(x) < 0$ where the graph of $f(x)$ is concave down.

If $f'(x) = 0$ at $x = a$, there is a local maximum point if $f'(x)$ changes from $+$ to $-$ or if $f''(a) < 0$.

If $f'(x) = 0$ at $x = a$, there is a local minimum point if $f'(x)$ changes from $-$ to $+$ or if $f''(a) > 0$.

If $f''(x) = 0$ at $x = a$, there is a point of inflection if $f''(x)$ changes from $+$ to $-$ or from $-$ to $+$.

The following examples illustrate how the above information is put together to obtain the graph of a function.

EXAMPLE 4 Sketching a curve using derivatives

Sketch the graph of $y = 6x - x^2$.

Finding the first two derivatives, we have

$$y' = 6 - 2x = 2(3 - x)$$
$$y'' = -2$$

We now note that $y' = 0$ for $x = 3$. For $x < 3$, we see that $y' > 0$, which means that y is increasing over this interval. Also, for $x > 3$, we note that $y' < 0$, which means that y is decreasing over this interval.

Because y' changes from positive on the left of $x = 3$ to negative on the right of $x = 3$, the curve has a maximum point where $x = 3$. Because $y = 9$ for $x = 3$, this maximum point is $(3, 9)$.

Because $y'' = -2$, this means that its value remains constant for all values of x. Therefore, there are no points of inflection, and the curve is concave down for all values of x. This also shows that the point $(3, 9)$ is a maximum point.

Summarizing, we know that y is increasing for $x < 3$, y is decreasing for $x > 3$, there is a maximum point at $(3, 9)$, and the curve is always concave down. Using this information, we sketch the curve shown in Fig. 24.38.

From the equation, we know this curve is a parabola. We could also find the maximum point from the material of Section 7.4 or Section 21.7. However, using derivatives we can find this kind of important information about the graphs of a great many types of functions. ■

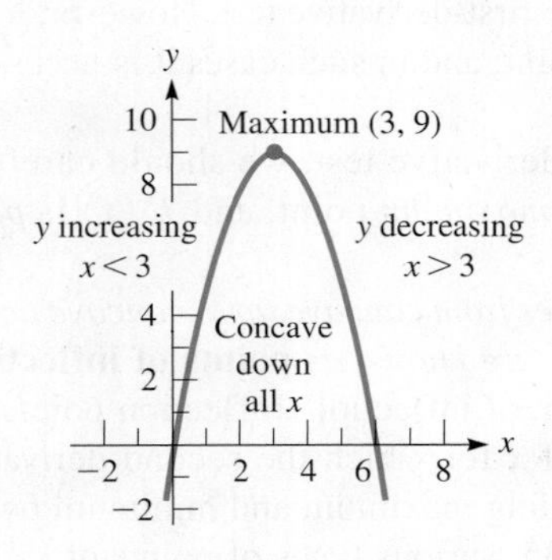

Fig. 24.38

Practice Exercise

2. Find the point(s) of inflection on the graph of $y = x^3 - 3x^2 - 9x$.

EXAMPLE 5 Sketching a curve using derivatives

Sketch the graph of $y = 2x^3 + 3x^2 - 12x$.

Using the first derivative, we have

$$y' = 6x^2 + 6x - 12 = 6(x + 2)(x - 1)$$
$$\text{critical values: } x = -2, 1$$

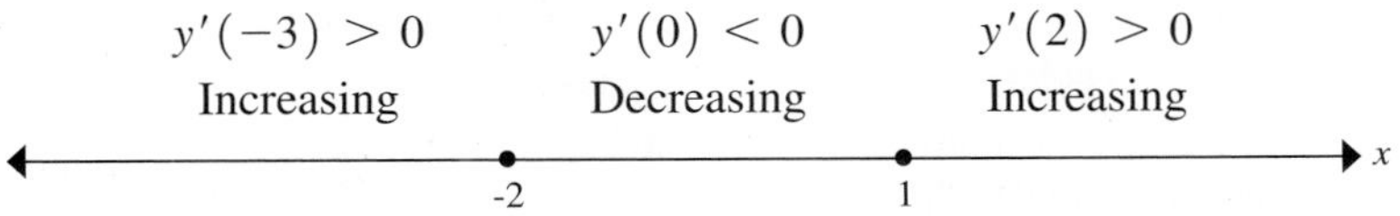

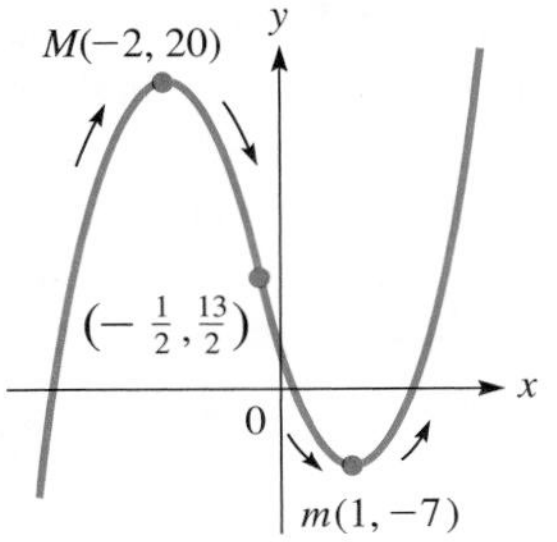

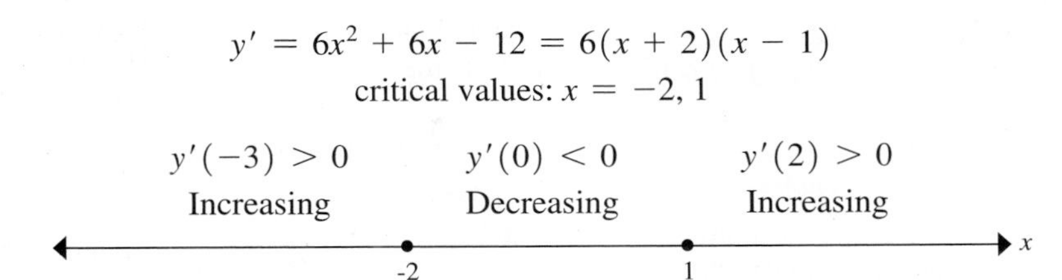

Fig. 24.39

Therefore, the function is increasing for $x < -2$ or $x > 1$, and decreasing for $-2 < x < 1$. There is a local maximum at $(-2, 20)$ and a local minimum at $(1, -7)$.

Now, using the second derivative, we get

$$y'' = 12x + 6$$
$$= 6(2x + 1)$$
$$\text{second order critical value: } x = -\tfrac{1}{2}$$

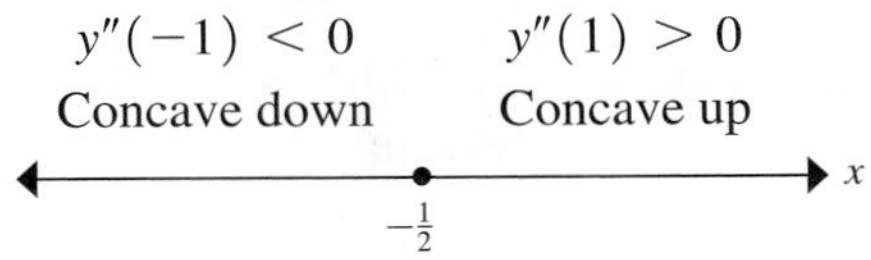

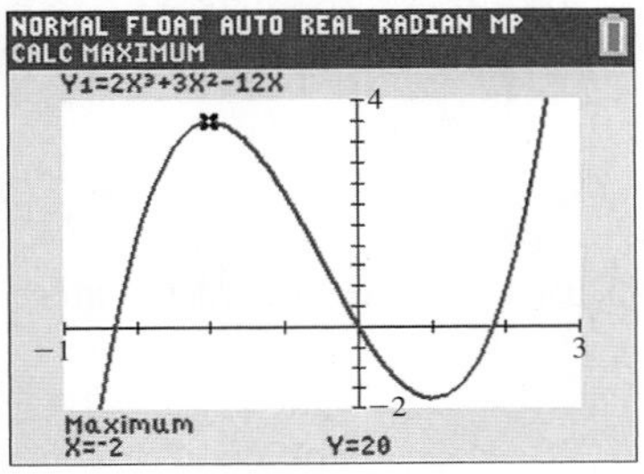

Fig. 24.40

Thus, the function is concave down for $x < -\frac{1}{2}$ and concave up for $x > -\frac{1}{2}$. Since the concavity changes at $x = -\frac{1}{2}$, there is an inflection point at $(-\frac{1}{2}, \frac{13}{2})$.

Finally, by locating the points $(-2, 20)$, $(-\frac{1}{2}, \frac{13}{2})$, and $(1, -7)$, we draw the curve *up* to $(-2, 20)$ and then *down* to $(-\frac{1}{2}, \frac{13}{2})$, with the curve *concave down*. ***Continuing* down, *but* concave up,** we draw the curve to $(1, -7)$, at which point we start *up* and continue up. We now know the key points and the shape of the curve. See Fig. 24.39. For more precision, additional points may be used. Figure 24.40 shows a graphical check of the local maximum at $(-2, 20)$. ■

EXAMPLE 6 Sketching a curve using derivatives

Sketch the graph of $y = x^5 - 5x^4$.

The first two derivatives are

$$y' = 5x^4 - 20x^3 = 5x^3(x - 4)$$
$$y'' = 20x^3 - 60x^2 = 20x^2(x - 3)$$

We now see that $y' = 0$ when $x = 0$ or $x = 4$. For $x = 0$, $y'' = 0$ also, which means ***we cannot use the second-derivative test*** for maximum and minimum points for $x = 0$ in this case. For $x = 4$, $y'' > 0$ $(+320)$, which means that $(4, -256)$ is a relative minimum point.

Next, we note that

$$y' > 0 \text{ for } x < 0 \text{ or } x > 4 \qquad \text{and} \qquad y' < 0 \text{ for } 0 < x < 4$$

Thus, by the first-derivative test, there is a local maximum point at $(0, 0)$. Also, y is increasing for $x < 0$ or $x > 4$ and decreasing for $0 < x < 4$.

The second derivative indicates that there is a point of inflection at $(3, -162)$. It also indicates that the curve is concave down for $x < 3$ $(x \neq 0)$ and concave up for $x > 3$. There is no point of inflection at $(0, 0)$ since the second derivative does not change sign at $x = 0$.

From this information, we sketch the curve in Fig. 24.41. ■

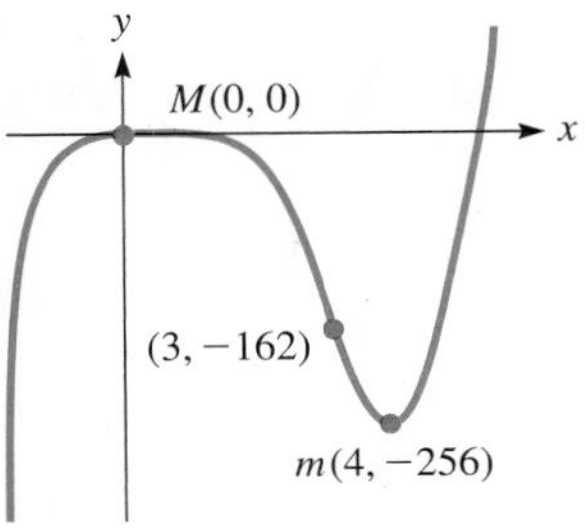

Fig. 24.41

EXERCISES 24.5

In Exercises 1–4, make the given changes in the indicated examples of this section and then solve the resulting problems.

1. In Example 1, change the − sign to a + sign and then find the values of x for which $f(x)$ is increasing and those for which it is decreasing.
2. In Example 2, change the $5x^3$ to $15x$ and then find the local maximum and minimum points.
3. In Example 3, change $3x$ to $6x^2$ and then determine the concavity and find any points of inflection.
4. In Example 4, change the $6x$ to $8x$ and then sketch the graph as in the example.

In Exercises 5–8, find those values of x for which the given functions are increasing and those values of x for which they are decreasing.

5. $y = x^2 + 2x$
6. $y = 2 + 27x - x^3$
7. $y = 2x^3 + 3x^2 - 36x - 10$
8. $y = x^4 - 6x^2$

In Exercises 9–12, find any local maximum or minimum points of the given functions. (These are the same functions as in Exercises 5–8.)

9. $y = x^2 + 2x$
10. $y = 2 + 27x - x^3$
11. $y = 2x^3 + 3x^2 - 36x - 10$
12. $y = x^4 - 6x^2$

In Exercises 13–16, find the values of x for which the given function is concave up, the values of x for which it is concave down, and any points of inflection. (These are the same functions as in Exercises 5–8.)

13. $y = x^2 + 2x$
14. $y = 2 + 27x - x^3$
15. $y = 2x^3 + 3x^2 - 36x - 10$
16. $y = x^4 - 6x^2$

In Exercises 17–20, sketch the graphs of the given functions by determining the appropriate information and points from the first and second derivatives (see Exercises 5–16). Use a calculator to check the graph.

17. $y = x^2 + 2x$
18. $y = 2 + 27x - x^3$
19. $y = 2x^3 + 3x^2 - 36x - 10$
20. $y = x^4 - 6x^2$

In Exercises 21–32, sketch the graphs of the given functions by determining the appropriate information and points from the first and second derivatives. Use a calculator to check the graph. In Exercises 27–32, use the calculator maximum-minimum feature to check the local maximum and minimum points.

21. $y = 12x - 2x^2$
22. $y = 20 + 16x - 4x^2$
23. $y = 2x^3 + 6x^2 - 5$
24. $y = x^3 - 9x^2 + 15x + 1$
25. $y = x^3 + 3x^2 + 3x + 2$
26. $y = x^3 - 12x + 12$
27. $y = 4x^3 - 24x^2 + 36x$
28. $y = x(x - 4)^3$
29. $y = 4x^3 - 3x^4 + 6$
30. $y = 20x^2 - x^5$
31. $y = x^5 - 5x$
32. $y = x^4 + 32x + 2$

In Exercises 33 and 34, view the graphs of y, y′, y″ together on a calculator. State how the graphs of y′ and y″ are related to the graph of y.

33. $y = x^3 - 12x$
34. $y = 24x - 9x^2 - 2x^3$

In Exercises 35–42, describe the indicated features of the given graphs.

35. Display the graph of $y = x^3 + cx$ for $c = -3, -1, 1, 3$ on a calculator. Describe how the graph changes as c varies.
36. Follow the instructions of Exercise 35 for the function $y = x^4 + cx^2$.
37. Display the graph of $y = 2x^5 - 7x^3 + 8x$ on a calculator. Describe the relative locations of the left local maximum point and the local minimum points.
38. Describe the following features of the graph in Fig. 24.42 between or at the points A, B, C, D, E, F, and G. (a) Increasing and decreasing, (b) local maximum and minimum points, (c) concavity, and (d) points of inflection.

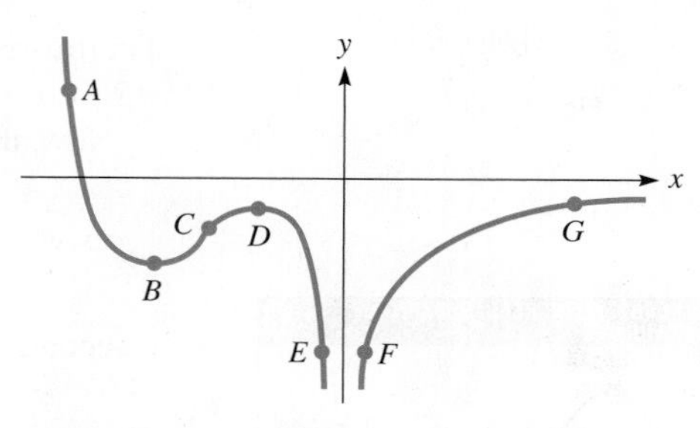

Fig. 24.42

39. For the curve in Fig. 24.43, list the points in the order in which their slopes increase.

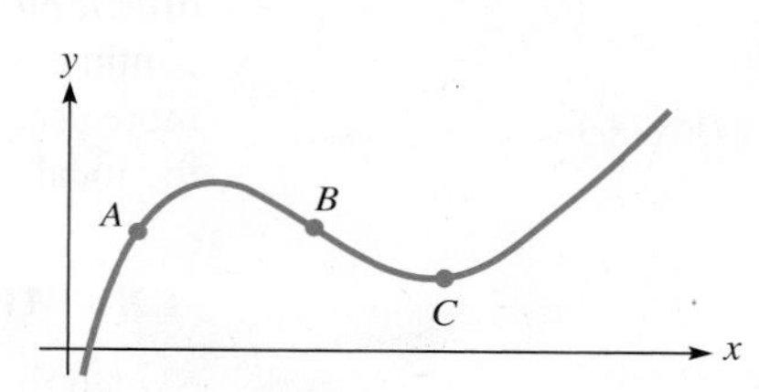

Fig. 24.43

40. For the curve in Fig. 24.43, list the points in the order in which the value of concavity increases.
41. The graph of $f'(x)$ is shown in Fig. 24.44. Describe how this graph can be used to determine the values of x for which $f(x)$ is increasing or decreasing. Also give the x-coordinates of any local maximum or minimum.

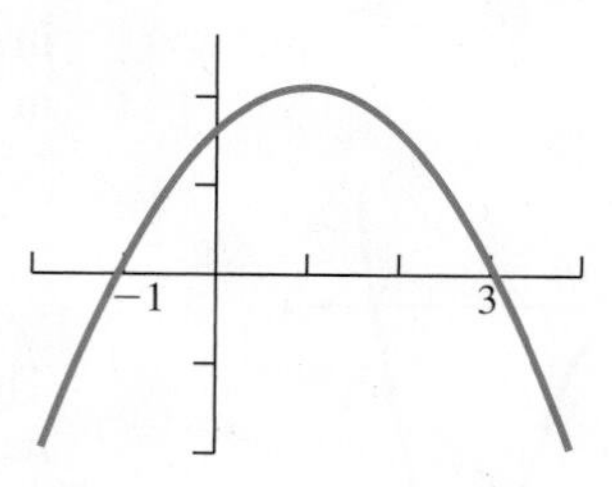

Fig. 24.44

42. Sketch a continuous curve $y = f(x)$, if $f(0) = -1, f'(x) < 0$ and $f''(x) < 0$ for $x < 0, f'(x) < 0$ and $f''(x) > 0$ for $x > 0$.

In Exercises 43–54, sketch the indicated curves by the methods of this section. You may check the graphs by using a calculator.

43. A batter hits a baseball that follows a path given by $y = x - 0.0025x^2$, where distances are in feet. Sketch the graph of the path of the baseball.

44. The angle θ (in degrees) of a robot arm with the horizontal as a function of the time t (in s) is given by $\theta = 10 + 12t^2 - 2t^3$. Sketch the graph for $0 \le t \le 6$ s.

45. The power P (in W) in a certain electric circuit is given by $P = 4i - 0.5i^2$. Sketch the graph of P vs. i.

46. A computer analysis shows that the thrust T (in kN) of an experimental rocket motor is $T = 20 + 9t - 4t^3$, where t is the time (in min) after the motor is activated. Sketch the graph for the first 2 min.

47. For the force F exerted on the cam on the lever in Exercise 54 on page 679, sketch the graph of F vs. x. (*Hint:* Use methods of Chapter 15 to analyze the first derivative.)

48. The solar-energy power P (in W) produced by a certain solar system does not rise and fall uniformly during a cloudless day because of the system's location. An analysis of records shows that $P = -0.45(2t^5 - 45t^4 + 350t^3 - 1000t^2)$, where t is the time (in h) during which power is produced. Show that, during the solar-power production, the production flattens (inflection) in the middle and then peaks before shutting down. (*Hint:* The solutions are integral.)

49. An electric circuit is designed such that the resistance R (in Ω) is a function of the current i (in mA) according to $R = 75 - 18i^2 + 8i^3 - i^4$. Sketch the graph if $R \ge 0$ and i can be positive or negative.

50. A horizontal 12-m beam is deflected by a load such that it can be represented by the equation $y = 0.0004(x^3 - 12x^2)$. Sketch the curve followed by the beam.

51. The altitude h (in ft) of a certain rocket is given by $h = -t^3 + 54t^2 + 480t + 20$, where t is the time (in s) of flight. Sketch the graph of $h = f(t)$.

52. An analysis of data showed that the mean density d (in mg/cm^3) of a calcium compound in the bones of women was given by $d = 0.00181x^3 - 0.289x^2 + 12.2x + 30.4$, where x represents the ages of women $(20 < x < 80 \text{ years})$. (A woman probably has osteoporosis if $d < 115$ mg/cm^3.) Sketch the graph.

53. A rectangular box is made from a piece of cardboard 8 in. by 12 in. by cutting equal squares from each corner and bending up the sides. See Fig. 24.45. Express the volume of the box as a function of the side of the square that is cut out and then sketch the curve of the resulting equation.

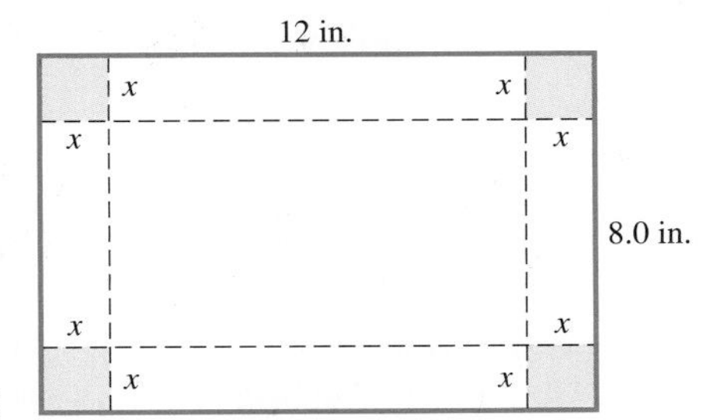

Fig. 24.45

54. A rectangular planter with a square end is to be made from 64 ft^2 of redwood. Express the volume of soil the planter can hold as a function of the side of the square of the end. Sketch the curve of the resulting function.

In Exercises 55–57, sketch a continuous curve that has the given characteristics.

55. $f(1) = 0; f'(x) > 0$ for all x; $f''(x) < 0$ for all x

56. $f(0) = 1; f'(x) < 0$ for all x; $f''(x) < 0$ for $x < 0; f''(x) > 0$ for $x > 0$

57. $f(-1) = 0; f(2) = 2; f'(x) < 0$ for $x < -1; f'(x) > 0$ for $x > -1; f''(x) < 0$ for $0 < x < 2; f''(x) > 0$ for $x < 0$ or $x > 2$

58. Display the graph of $f(x) = x^{2/3}$ for $-2 < x < 2$ on a calculator. Determine the continuity of $f(x)$, $f'(x)$, and $f''(x)$. Discuss the concavity of the curve in relation to the minimum point. (See the last paragraph on page 717.)

Answers to Practice Exercises

1. Max.$(-1, 5)$, Min.$(3, -27)$ **2.** Infl.$(1, -11)$

24.6 More on Curve Sketching

Intercepts • Symmetry • Behavior as x Becomes Large • Vertical Asymptotes • Domain and Range • Derivatives

When graphing functions in earlier chapters, we often used information that was obtainable from the function itself. We will now use this type of information, along with that found from the derivatives, to graph functions. We will find, in graphing any particular function, some types of information are of more value than others. The features we will consider in this section are as follows:

Features to Be Used in Graphing Functions

1. *Intercepts* Points for which the graph crosses (or is tangent to) each axis.
2. *Symmetry* For a review of symmetry, see Example 4 and the text before it on page 580.
3. *Behavior as x Becomes Large* We will find what happens to the function as $x \to +\infty$ and as $x \to -\infty$ in a way similar to that when we discussed the asymptotes of a hyperbola on page 594.
4. *Vertical Asymptotes* We can find vertical asymptotes by finding values that make factors in the denominator zero.
5. *Domain and Range* For a review of domain and range, see Section 3.2.
6. *Derivatives* We will find information from the first two derivatives as we did in Section 24.5.

EXAMPLE 1 Sketch a graph using function and derivatives

Sketch the graph of $y = \dfrac{8}{x^2 + 4}$.

Intercepts If $x = 0$, $y = 2$, which means $(0, 2)$ is an intercept. If $y = 0$, there is no corresponding value of x, because $8/(x^2 + 4)$ is a fraction greater than zero for all x. This also indicates that all points on the curve are above the x-axis.

Symmetry The curve is symmetric to the y-axis because

$$y = \frac{8}{(-x)^2 + 4} \text{ is the same as } y = \frac{8}{x^2 + 4}.$$

The curve is not symmetric to the x-axis because

$$-y = \frac{8}{x^2 + 4} \text{ is not the same as } y = \frac{8}{x^2 + 4}.$$

The curve is not symmetric to the origin because

$$-y = \frac{8}{(-x)^2 + 4} \text{ is not the same as } y = \frac{8}{x^2 + 4}.$$

The value in knowing the symmetry is that we should find those portions of the curve on either side of the y-axis reflections of the other. It is possible to use this fact directly or to use it as a check.

■ The curve for Example 1 is a special case (with $a = 1$) of the curve known as the *witch of Agnesi.* Its general form is

$$y = \frac{8a^3}{x^2 + 4a^2}$$

It is named for the Italian mathematician Maria Gaetana Agnesi (1718–1799). She wrote the first text that contained analytic geometry, differential and integral calculus, series (see Chapter 30), and differential equations (see Chapter 31). The word *witch* was used due to a mistranslation from Italian to English.

Behavior as x Becomes Large We note that as $x \to \infty$, $y \to 0$ since $8/(x^2 + 4)$ is always a fraction that is greater than zero but which becomes smaller as x becomes larger. Therefore, we see that $y = 0$ *is an asymptote.* From either the symmetry or the function, we also see that $y \to 0$ as $x \to -\infty$.

Vertical Asymptotes From the discussion of the hyperbola, recall that an asymptote is a line that a curve approaches. We have already noted that $y = 0$ is an asymptote for this curve. This asymptote, the x-axis, is a horizontal line. NOTE ▸ [Vertical asymptotes, if any exist, are found by determining those values of x for which the denominator of any term is zero.] Such a value of x makes y undefined. Because $x^2 + 4$ cannot be zero, this curve has no vertical asymptotes. The next example illustrates a curve that has a vertical asymptote.

Domain and Range Because the denominator $x^2 + 4$ cannot be zero, x can take on any value. This means the domain of the function is all values of x. Also, we have noted that $8/(x^2 + 4)$ is a fraction greater than zero. Since $x^2 + 4$ is 4 or greater, y is 2 or less. This tells us that the range of the function is $0 < y \leq 2$.

Derivatives We now determine what the derivatives can also tell us about the curve. We start with the first derivative.

$$y = \frac{8}{(x^2 + 4)} = 8(x^2 + 4)^{-1}$$

$$y' = -8(x^2 + 4)^{-2}(2x)$$

$$= \frac{-16x}{(x^2 + 4)^2}$$

Because $(x^2 + 4)^2$ is positive for all values of x, the sign of y' is determined by the numerator. Thus, we note that $y' = 0$ for $x = 0$ and that $y' > 0$ for $x < 0$ and $y' < 0$ for $x > 0$. The curve, therefore, is **increasing** for $x < 0$, is **decreasing** for $x > 0$, and has a maximum point at $(0, 2)$.

Now, finding the second derivative, we have

$$y'' = \frac{(x^2+4)^2(-16) + 16x(2)(x^2+4)(2x)}{(x^2+4)^4} = \frac{-16(x^2+4) + 64x^2}{(x^2+4)^3}$$

$$= \frac{48x^2 - 64}{(x^2+4)^3} = \frac{16(3x^2-4)}{(x^2+4)^3}$$

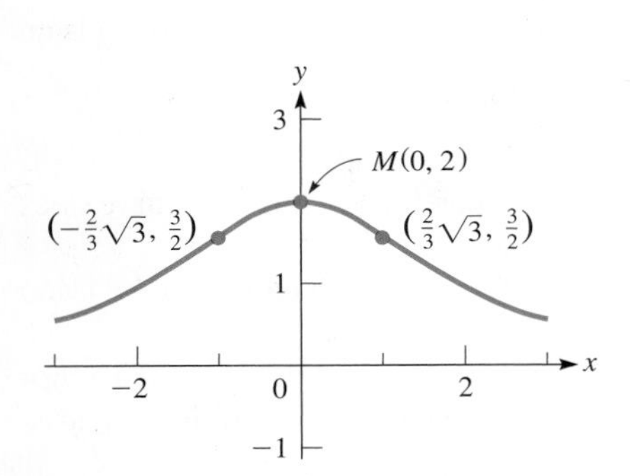

Fig. 24.46

We note that y'' is negative for $x = 0$, which confirms that $(0, 2)$ is a maximum point. Also, points of inflection are found for the values of x satisfying $3x^2 - 4 = 0$. Thus, $(-\frac{2}{3}\sqrt{3}, \frac{3}{2})$ and $(\frac{2}{3}\sqrt{3}, \frac{3}{2})$ are points of inflection. The curve is concave up if $x < -\frac{2}{3}\sqrt{3}$ or $x > \frac{2}{3}\sqrt{3}$, and the curve is concave down if $-\frac{2}{3}\sqrt{3} < x < \frac{2}{3}\sqrt{3}$.

Putting this information together, we sketch the curve shown in Fig. 24.46. Note that this curve could have been sketched primarily by use of the fact that $y \to 0$ as $x \to +\infty$ and as $x \to -\infty$ and the fact that a maximum point exists at $(0, 2)$. However, the other parts of the analysis, such as symmetry and concavity, serve as checks and make the curve more accurate. ■

EXAMPLE 2 Sketch a graph using function and derivatives

Sketch the graph of $y = x + \frac{4}{x}$.

Intercepts If we set $x = 0$, y is undefined. This means that the curve is not *continuous* at $x = 0$ and there are no y-intercepts. If we set $y = 0$, $x + 4/x = (x^2 + 4)/x$ cannot be zero because $x^2 + 4$ cannot be zero. Therefore, there are no intercepts. This may seem to be of little value, but we must realize *this curve does not cross either axis.* This will be of value when we sketch the curve in Fig. 24.47.

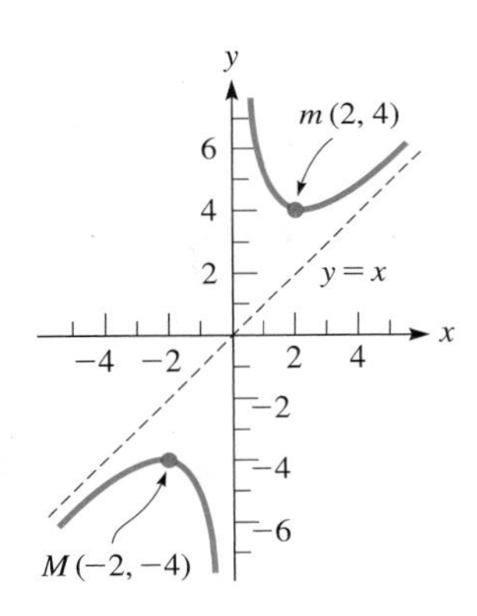

Fig. 24.47

Symmetry In testing for symmetry, we find that the curve is not symmetric to either axis. However, this curve does possess symmetry to the origin. This is determined by the fact that when $-x$ replaces x and at the same time $-y$ replaces y, the equation does not change.

Behavior as x Becomes Large As $x \to +\infty$ and as $x \to -\infty$, $y \to x$ because $4/x \to 0$. Thus, $y = x$ is an asymptote of the curve.

Vertical Asymptotes As we noted in Example 1, vertical asymptotes exist for values of x for which y is undefined. In this equation, $x = 0$ makes the second term on the right undefined, and therefore y is undefined. In fact, as $x \to 0$ from the positive side, $y \to +\infty$, and as $x \to 0$ from the negative side, $y \to -\infty$. This is derived from the sign of $4/x$ in each case.

Domain and Range Because x cannot be zero, the domain of the function is all x except zero. As for the range, the analysis from the derivatives will show it to be $y \leq -4$, $y \geq 4$.

Derivatives Finding the first derivative, we have

$$y' = 1 - \frac{4}{x^2} = \frac{x^2 - 4}{x^2}$$

The x^2 in the denominator indicates that the sign of the first derivative is the same as its numerator. The numerator is zero if $x = -2$ or $x = 2$. If $x < -2$ or $x > 2$, then $y' > 0$; and if $-2 < x < 2$, $x \neq 0$, $y' < 0$. Thus, y is increasing if $x < -2$ or $x > 2$, and also y is decreasing if $-2 < x < 2$, except at $x = 0$ (y is undefined). Also, $(-2, -4)$ is a local maximum point, and $(2, 4)$ is a local minimum point. The second derivative is $y'' = 8/x^3$. This cannot be zero, but it is negative if $x < 0$ and positive if $x > 0$. Thus, the curve is concave down if $x < 0$ and concave up if $x > 0$. Using this information, we have the curve shown in Fig. 24.47. ■

EXAMPLE 3 Sketch a graph using function and derivatives

Sketch the graph of $y = \dfrac{1}{\sqrt{1 - x^2}}$.

Intercepts If $x = 0$, $y = 1$. If $y = 0$, $1/\sqrt{1 - x^2}$ would have to be zero, but it cannot because it is a fraction with 1 as the numerator for all values of x. Thus, $(0, 1)$ is an intercept.

Symmetry The curve is symmetric to the y-axis.

Behavior as x Becomes Large The values of x cannot be considered beyond 1 or -1, for any value of $x < -1$ or $x > 1$ gives imaginary values for y. Thus, the curve does not exist for values of $x < -1$ or $x > 1$.

Vertical Asymptotes If $x = 1$ or $x = -1$, y is undefined. In each case, as $x \to 1$ and as $x \to -1$, $y \to +\infty$.

Domain and Range From the analysis of x becoming large and of the vertical asymptotes, we see that the domain is $-1 < x < 1$. Also, because $\sqrt{1 - x^2}$ is 1 or less, $1/\sqrt{1 - x^2}$ is 1 or more, which means the range is $y \geq 1$.

Derivatives

$$y' = -\frac{1}{2}(1 - x^2)^{-3/2}(-2x) = \frac{x}{(1 - x^2)^{3/2}}$$

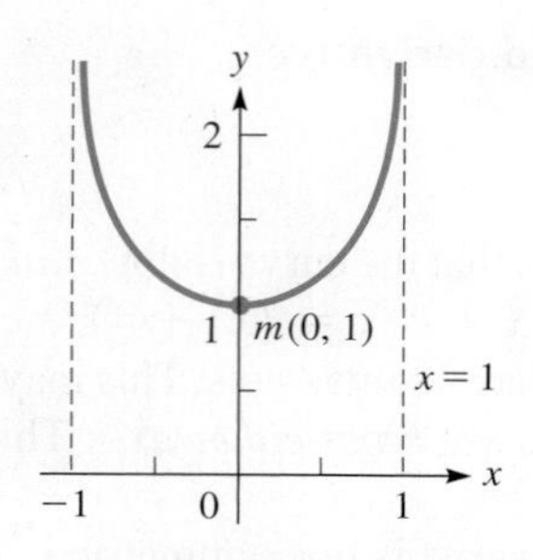

Fig. 24.48

We see that $y' = 0$ if $x = 0$. If $-1 < x < 0$, $y' < 0$, and also if $0 < x < 1$, $y' > 0$. Thus, the curve is decreasing if $-1 < x < 0$ and increasing if $0 < x < 1$. There is a minimum point at $(0, 1)$.

$$y'' = \frac{(1 - x^2)^{3/2} - x(\frac{3}{2})(1 - x^2)^{1/2}(-2x)}{(1 - x^2)^3} = \frac{(1 - x^2) + 3x^2}{(1 - x^2)^{5/2}}$$

$$= \frac{2x^2 + 1}{(1 - x^2)^{5/2}}$$

The second derivative cannot be zero because $2x^2 + 1$ is positive for all values of x. The second derivative is also positive for all permissible values of x, which means the curve is concave up for these values.

Using this information, we sketch the graph in Fig. 24.48. ■

Practice Exercise

1. Find the asymptotes of the curve

$$y = \frac{4}{x^2 - x}.$$

EXAMPLE 4 Sketch a graph using function and derivatives

Sketch the graph of $y = \dfrac{x}{x^2 - 4}$.

Intercepts If $x = 0$, $y = 0$, and if $y = 0$, $x = 0$. The only intercept is $(0, 0)$.

Symmetry The curve is not symmetric to either axis. However, because $-y = -x/[(-x)^2 - 4]$ is the same as $y = x/(x^2 - 4)$, it is symmetric to the origin.

Behavior as x Becomes Large As $x \to +\infty$ and as $x \to -\infty$, $y \to 0$. This means that $y = 0$ is an asymptote.

Vertical Asymptotes If $x = -2$ or $x = 2$, y is undefined. As $x \to -2$, $y \to -\infty$ if $x < -2$ because $x^2 - 4$ is positive, and $y \to +\infty$ if $x > -2$ since $x^2 - 4$ is negative. As $x \to 2$, $y \to -\infty$ if $x < 2$, and $y \to +\infty$ if $x > 2$.

Domain and Range The domain is all real values of x except -2 and 2. As for the range, if $x < -2$, $y < 0$ (the numerator is negative and the denominator is positive). If $x > 2$, $y > 0$ (both numerator and denominator are positive). Because $(0, 0)$ is an intercept, we see that the range is all values of y.

Derivatives

$$y' = \frac{(x^2 - 4)(1) - x(2x)}{(x^2 - 4)^2} = -\frac{x^2 + 4}{(x^2 - 4)^2}$$

Because $y' < 0$ for all values of x except -2 and 2, the curve is decreasing for all values in the domain.

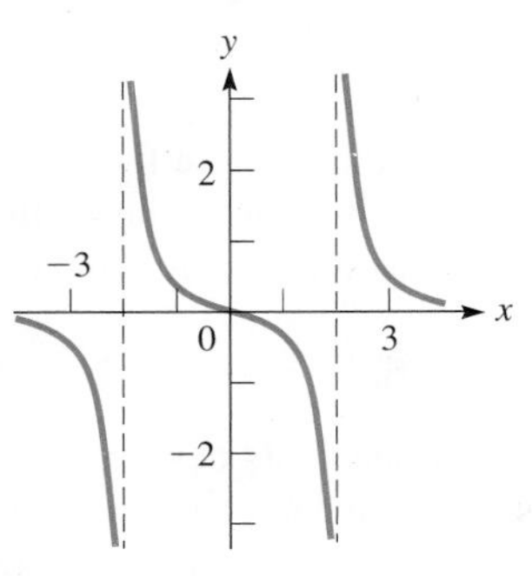

Fig. 24.49

Now, finding the second derivative, we have

$$y'' = -\frac{(x^2-4)^2(2x) - (x^2+4)(2)(x^2-4)(2x)}{(x^2-4)^4}$$

$$= -\frac{2x(x^2-4) - 4x(x^2+4)}{(x^2-4)^3} = \frac{2x^3+24x}{(x^2-4)^3} = \frac{2x(x^2+12)}{(x^2-4)^3}$$

The sign of y'' depends on x and $(x^2-4)^3$. If $x < -2$, $y'' < 0$. If $-2 < x < 0$, $y'' > 0$. If $0 < x < 2$, $y'' < 0$. If $x > 2$, $y'' > 0$. This means the curve is concave down for $x < -2$ or $0 < x < 2$ and is concave up for $-2 < x < 0$ or $x > 2$.

The curve is sketched in Fig. 24.49. ■

EXERCISES 24.6

In Exercises 1–18, use the method of the examples of this section to sketch the indicated curves. Use a calculator to check the graph.

1. In Example 2, change the + to − before $4/x$ and then proceed.

2. $y = \frac{4}{x^2}$ **3.** $y = \frac{2}{x+1}$ **4.** $y = \frac{x}{x-2}$

5. $y = x^2 + \frac{2}{x}$ **6.** $y = x + \frac{4}{x^2}$ **7.** $y = x - \frac{9}{x}$

8. $y = 3x + \frac{1}{x^3}$ **9.** $y = \frac{x^2}{x+1}$ **10.** $y = \frac{9x^2}{x^2+9}$

11. $y = \frac{3}{x^2-1}$ **12.** $y = \frac{x^2-1}{x^3}$ **13.** $y = \frac{4}{x} - \frac{4}{x^2}$

14. $y = 4x + \frac{1}{\sqrt{x}}$ **15.** $y = x\sqrt{1-x^2}$ **16.** $y = \frac{x-1}{x^2-2x}$

17. $y = \frac{9x}{9-x^2}$ **18.** $y = \frac{x^2-4}{x^2+4}$

In Exercises 19–32, solve the given problems.

19. Display the graph of $y = \frac{cx}{1+c^2x^2}$ on a calculator for $c = -3, -1, 1, 3$. Describe how the graph changes as c varies.

20. Display the graph of $y = x + \frac{4}{x^n}$ on a calculator for $n = 1, 2, 3, 4$. Describe how the graph changes as n varies.

21. Sketch a continuous curve such that $f(0) = 2$, $f'(x) > 0$, and $f''(x) < 0$ for all x, and $f(x) \to 4$ as $x \to +\infty$.

22. For a continuous function $y = f(x)$, if for all x, $f(x) > 0$, $f'(x) < 0$, and $f''(x) > 0$, what do you conclude about the graph of $f(x)$?

23. In Exercise 10, first divide the numerator by the denominator. Does this simplify any of the graphing features?

24. The concentration C (in mg/cm^3) of a certain drug in a patient's bloodstream is $C = \frac{0.15t}{t^2+1}$, where t is the time (in h) after the drug is administered. Sketch the graph.

25. The combined capacitance C_T (in μF) of a 6-μF capacitance and a variable capacitance C in series is given by $C_T = \frac{6C}{6+C}$. Display the graph on a calculator. What is the significance of the horizontal asymptote?

26. The number n of dollars saved by increasing the fuel efficiency of e mi/gal to $e + 6$ mi/gal for a car driven 10,000 mi/year is $n = \frac{195,000}{e(e+6)}$, if the cost of gas is \$3.25/gal. Sketch the graph.

27. The reliability R of a computer model is found to be $R = \frac{900}{\sqrt{t^2 + 810,000}}$, where t is the time of operation in hours. ($R = 1$ is perfect reliability, and $R = 0.5$ means there is a 50% chance of a malfunction.) Sketch the graph.

28. The electric power P (in W) produced by a source is given by $P = \frac{36R}{R^2 + 2R + 1}$, where R is the resistance in the circuit. Sketch the graph.

29. Assuming that a raindrop is always spherical and as it falls its radius increases from 1 mm to r mm, its velocity v (in mm/s) is $v = k\left(r - \frac{1}{r^3}\right)$. With $k = 1$, sketch the graph.

30. If a positive electric charge of $+q$ is placed between two negative charges of $-q$ that are two units apart, Coulomb's law states that the force F on the positive charge is

$$F = -\frac{kq^2}{x^2} + \frac{kq^2}{(x-2)^2},$$ where x is the distance from one of the negative charges. Let $kq^2 = 1$ and sketch the graph for $0 < x < 2$.

31. A cylindrical oil drum is to be made such that it will contain 20 kL. Sketch the area of sheet metal required for construction as a function of the radius of the drum.

32. A fence is to be constructed to enclose a rectangular area of 20,000 m^2. A previously constructed wall is to be used for one side. Sketch the length of fence to be built as a function of the side of the fence parallel to the wall. See Fig. 24.50.

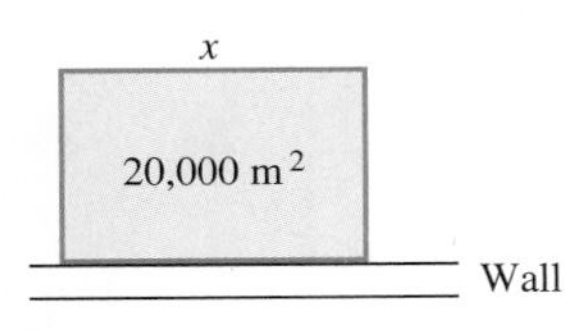

Fig. 24.50

Answer to Practice Exercise

1. $y = 0, x = 0, x = 1$

24.7 Applied Maximum and Minimum Problems

Steps in Solving Maximum and Minimum Problems • Setting Up and Solving Problems

Problems from various applied situations frequently occur that require finding a maximum or minimum value of some function. If the function is known, the methods we have already discussed can be used directly. This is discussed in the following example.

EXAMPLE 1 Finding maximum value—engine efficiency

An automobile manufacturer, in testing a new engine on one of its new models, found that the efficiency e (in %) of the engine as a function of the speed s (in km/h) of the car was given by $e = 0.768 - 0.00004s^3$. What is the maximum efficiency of the engine?

In order to find a maximum value, we find the derivative of e with respect to s:

■ The first gasoline-engine automobile was built by the German engineer Karl Benz (1844–1929) in the 1880s.

$$\frac{de}{ds} = 0.768 - 0.00012s^2$$

We then set the derivative equal to zero in order to find the value of s for which a maximum may occur:

$$\begin{aligned} 0.768 - 0.00012s^2 &= 0 \\ 0.00012s^2 &= 0.768 \\ s^2 &= 6400 \\ s &= 80.0 \text{ km/h} \end{aligned}$$

We know that s must be positive to have meaning in this problem. Therefore, the apparent solution of $s = -80$ is discarded. The second derivative is

$$\frac{d^2e}{ds^2} = -0.00024s$$

which is negative for any positive value of s. Therefore, we have a maximum for $s = 80.0$. Substituting $s = 80.0$ in the function for e, we obtain

$$e = 0.768(80.0) - 0.00004(80.0^3) = 61.44 - 20.48 = 40.96$$

The maximum efficiency is about 41.0% which occurs for $s = 80.0$ km/h. ■

In many problems for which a maximum or minimum value is to be found, the function is not given. To solve such a problem, we use these steps:

Steps in Solving Applied Maximum and Minimum Problems

1. Determine the quantity Q to be maximized or minimized.
2. If possible, draw a figure illustrating the problem.
3. Write an equation for Q in terms of another variable of the problem.
4. Take the derivative of the function in step 3.
5. Set the derivative equal to zero, and solve the resulting equation.
6. Check as to whether the value found in step 5 makes Q a maximum or a minimum. This might be clear from the statement of the problem, or it might require one of the derivative tests.
7. Be sure the stated answer is the one the problem required. Some problems require the maximum or minimum value, and others require values of other variables that give the maximum or minimum value.

CAUTION The principal difficulty that arises in these problems is finding the proper function. ■ We must carefully read the problem to find the information needed to set up the function. The following examples illustrate several types of stated problems involving maximum and minimum values.

EXAMPLE 2 Setting up equation—find maximum

Find the number that exceeds its square by the greatest amount.

The quantity to be maximized is the difference D between a number x and its square x^2. Therefore, the required function is

$$D = x - x^2$$

Because we want D to be a maximum, we find dD/dx, which is

$$\frac{dD}{dx} = 1 - 2x$$

Setting the derivative equal to zero and solving for x, we have

$$0 = 1 - 2x, \qquad x = \tfrac{1}{2}$$

The second derivative gives $d^2D/dx^2 = -2$, which tells us that the second derivative is always negative. This verifies that the solution $x = \frac{1}{2}$ corresponds to a maximum of the function. ■

EXAMPLE 3 Finding maximum area—rectangular corral

A rectangular corral is to be enclosed with 1600 ft of fencing. Find the maximum possible area of the corral.

There are limitless possibilities for rectangles of a perimeter of 1600 ft and differing areas. See Fig. 24.51. For example, if the sides are 700 ft and 100 ft, the area is 70,000 ft^2, or if the sides are 600 ft and 200 ft, the area is 120,000 ft^2. Therefore, we set up a function for the area of a rectangle in terms of its sides x and y:

$$A = xy$$

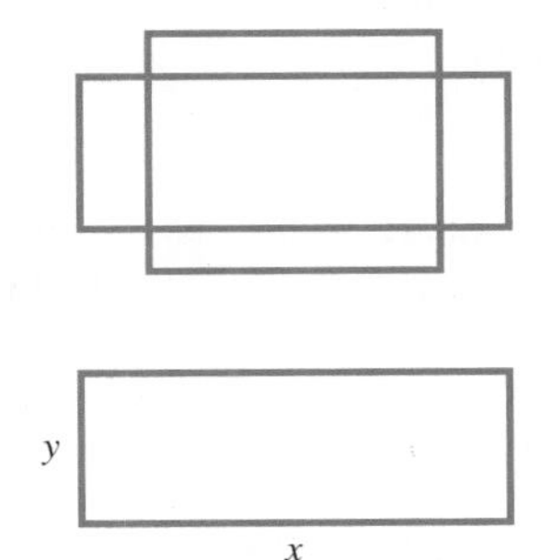

Fig. 24.51

Another important fact is that the perimeter of the corral is 1600 ft. Therefore, $2x + 2y = 1600$. Solving for y, we have $y = 800 - x$. By using this expression for y, we can express the area in terms of x only. This gives us

$$A = x(800 - x) = 800x - x^2$$

We complete the solution as follows:

$$\frac{dA}{dx} = 800 - 2x \qquad \text{take derivative}$$

$$800 - 2x = 0 \qquad \text{set derivative equal to zero}$$

$$x = 400 \text{ ft}$$

NOTE ▸ [By checking values of the derivative near 400 or by finding the second derivative, we can show that we have a maximum for $x = 400$.] This means that $x = 400$ ft and $y = 400$ ft give the maximum area of 160,000 ft^2 for the corral. ■

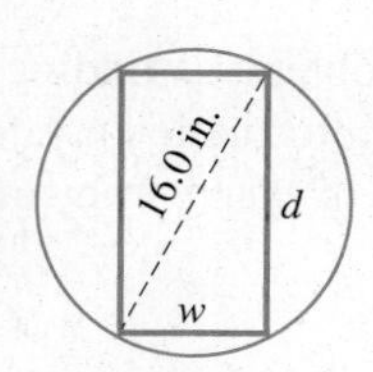

Fig. 24.52

EXAMPLE 4 Setting up equation using variation—strength of a beam

The strength S of a beam with a rectangular cross section is directly proportional to the product of the width w and the square of the depth d. Find the dimensions of the strongest beam that can be cut from a log with a circular cross section that is 16.0 in. in diameter. See Fig. 24.52.

The solution proceeds as follows:

$$S = kwd^2 \quad \text{direct variation}$$
$$d^2 = 256 - w^2 \quad \text{Pythagorean theorem}$$
$$S = kw(256 - w^2) \quad \text{substituting}$$
$$= k(256w - w^3) \quad S = f(w)$$
$$\frac{dS}{dw} = k(256 - 3w^2) \quad \text{take derivative}$$
$$0 = k(256 - 3w^2) \quad \text{set derivative equal to zero}$$
$$3w^2 = 256 \quad \text{solve for } w$$
$$w = \frac{16}{\sqrt{3}} = 9.24 \text{ in.}$$
$$d = \sqrt{256 - \frac{256}{3}} \quad \text{solve for } d$$
$$= 13.1 \text{ in.}$$

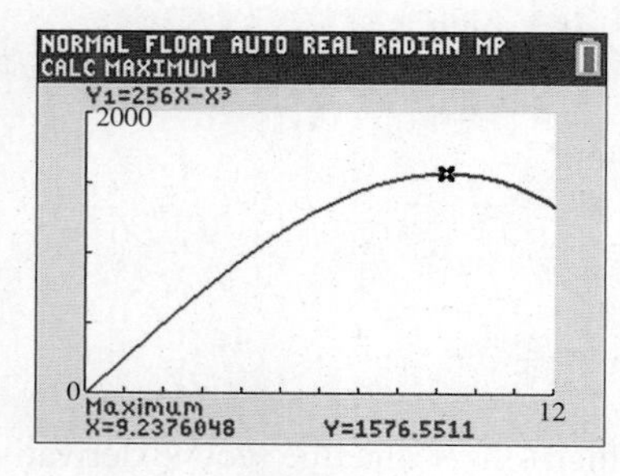

Fig. 24.53
Graphing calculator keystrokes: goo.gl/q19Qya

This means that the strongest beam is about 9.24 in. wide and 13.1 in. deep. Because $d^2S/dw^2 = -6kw$ and is negative for $w > 0$ (the only values with meaning in this problem), these dimensions give the maximum strength for the beam.

The solution can also be checked on a calculator. By graphing the equation $S = 256w - w^3$ using y for S, x for w, and $k = 1$ (*the value of k does not affect the solution*), we see in Fig. 24.53 that the strength is maximized for $w = 9.24$. ■

Practice Exercise

1. In Example 4, change 16.0 in. to 12.0 in. and solve the resulting problem.

EXAMPLE 5 Setting up equation using distance formula

Find the point on the parabola $y = x^2$ that is nearest to the point $(6, 3)$.

In this example, we must set up a function for the distance between a general point (x, y) on the parabola and the point $(6, 3)$. The relation is

$$D = \sqrt{(x - 6)^2 + (y - 3)^2}$$

However, to make it easier to take derivatives, we will square both sides of this expression. If a function is a minimum, then so is its square. We will also use the fact that the point (x, y) is on $y = x^2$ by replacing y by x^2. Thus, we have

$$D^2 = (x - 6)^2 + (x^2 - 3)^2 = x^2 - 12x + 36 + x^4 - 6x^2 + 9$$
$$= x^4 - 5x^2 - 12x + 45$$
$$\frac{dD^2}{dx} = 4x^3 - 10x - 12 \quad \text{take derivative}$$
$$0 = 2x^3 - 5x - 6 \quad \text{set derivative equal to zero}$$

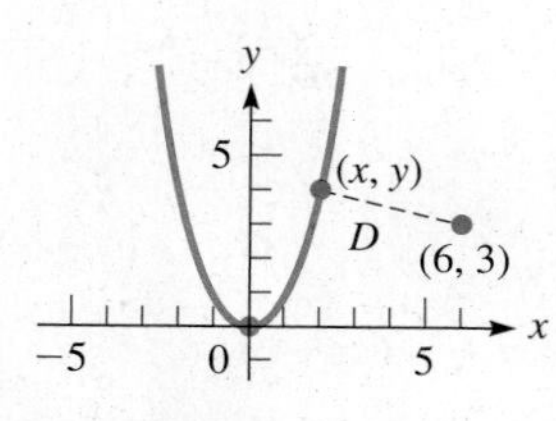

Fig. 24.54

Using synthetic division or some similar method, we find that the solution to this equation is $x = 2$. Thus, the required point on the parabola is $(2, 4)$. (See Fig. 24.54.) We can show that we have a minimum by analyzing the first derivative, by analyzing the second derivative, or by noting that points at much greater distances exist (therefore, it cannot be a maximum). ■

EXAMPLE 6 Maximizing revenue—Bluetooth headphones

■ See the chapter introduction.

A Bluetooth headphone company determines that it can sell 38,000 headphones per month if the price is \$150 for each unit. It also estimates that for each 1 dollar increase in the unit price, 200 fewer units can be sold. Under these conditions, what is the maximum possible revenue (income) and what price per unit gives this revenue?

If we let $x =$ the number of units sold and $p =$ the price, then

$$x = 38{,}000 - 200(p - 150)$$

Solving for p, we get

$$p = -0.005x + 340$$

The revenue R is the number of units sold times the price of each unit. Therefore,

$$R = x(-0.005x + 340) = -0.005x^2 + 340x$$

Taking the derivative and setting it equal to zero, we have

$$\frac{dR}{dx} = -0.01x + 340 \quad \text{take derivative}$$

$$0 = -0.01x + 340 \quad \text{set derivative equal to zero}$$

$$x = 34{,}000$$

We note that if $x < 34{,}000$, the derivative is positive, and if $x > 34{,}000$, the derivative is negative. Therefore, the revenue is maximized when $x = 34{,}000$ units. This in turn means that the maximum revenue is \$5,780,000 and the price per unit is \$170. These values are found by substituting $x = 34{,}000$ into the expression for R and for the price. ■

Practice Exercise

2. In Example 6, change 38,000 in the first line to 32,000 and solve the resulting problem.

EXAMPLE 7 Minimizing surface area—oil-storage tank

Find the dimensions of a 700-kL cylindrical oil-storage tank that can be made with the least cost of sheet metal, assuming there is no wasted sheet metal. See Fig. 24.55.

Analyzing the wording of the problem carefully, we see that we are to minimize the surface area of a right circular cylinder with a volume of 700 kL. Therefore, we set up expressions for the surface area and the volume $(700 \text{ kL} = 700 \text{ m}^3)$:

$$A = 2\pi r^2 + 2\pi rh \qquad V = 700 = \pi r^2 h$$

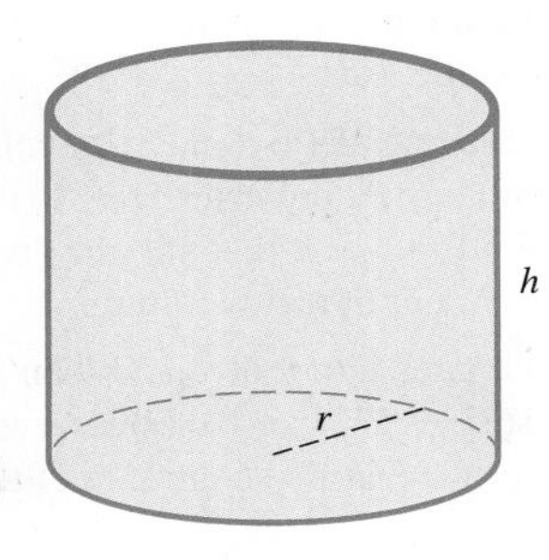

Fig. 24.55

We can express the equation for the area in terms of r only by solving the second equation for h and substituting in the first equation. This gives us

$$h = \frac{700}{\pi r^2}, \qquad A = 2\pi r^2 + 2\pi r\left(\frac{700}{\pi r^2}\right) = 2\pi r^2 + \frac{1400}{r}$$

In order to find the minimum value of A, we find dA/dr and set it equal to zero:

$$\frac{dA}{dr} = 4\pi r - \frac{1400}{r^2}, \qquad 4\pi r - \frac{1400}{r^2} = 0, \qquad \frac{4\pi r^3 - 1400}{r^2} = 0$$

$$4\pi r^3 - 1400 = 0, \qquad r^3 = \frac{1400}{4\pi} \quad \text{numerator must} = 0$$

$$r = \sqrt[3]{\frac{1400}{4\pi}} = 4.81 \text{ m} \qquad h = \frac{700}{\pi r^2} = 9.62 \text{ m}$$

Because dA/dr changes sign from negative to positive at $r = 4.81$ m, A is a minimum. ■

EXAMPLE 8 Minimizing a function—illuminance of light

The illuminance of a light source at any point equals the strength of the source divided by the square of the distance from the source. Two sources, of strengths 8 units and 1 unit, are 100 m apart. Determine at what point between them the illuminance is the least, assuming that the illuminance at any point is the sum of the illuminances of the two sources. See Fig. 24.56.

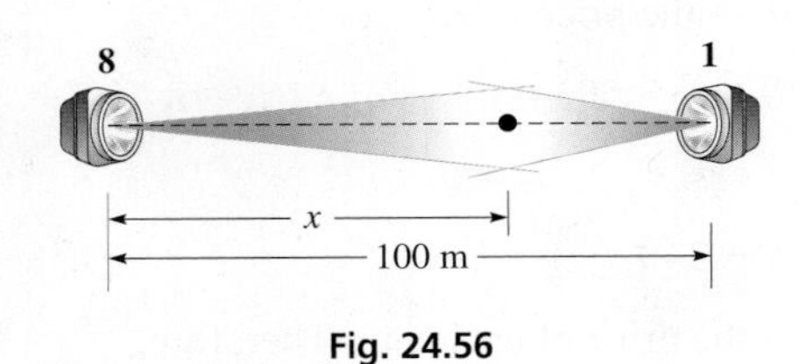

Fig. 24.56

Let $I =$ the sum of the illuminances and

$x =$ the distance from the source of strength 8

Then we find that

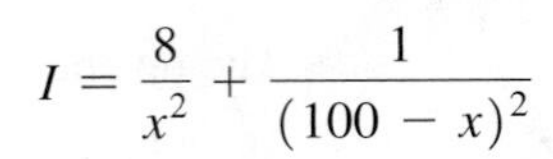

$$I = \frac{8}{x^2} + \frac{1}{(100 - x)^2}$$

is the function between the illuminance and the distance from the source of strength 8. We must now take a derivative of I with respect to x, set it equal to zero, and solve for x to find the point at which the illuminance is a minimum:

$$\frac{dI}{dx} = -\frac{16}{x^3} + \frac{2}{(100 - x)^3} = \frac{-16(100 - x)^3 + 2x^3}{x^3(100 - x)^3}$$

This function will be zero if the numerator is zero. Therefore, we have

$$2x^3 - 16(100 - x)^3 = 0 \quad \text{or} \quad x^3 = 8(100 - x)^3$$

Taking cube roots of each side, we have

$$x = 2(100 - x) \quad \text{or} \quad x = 66.7 \text{ m}$$

The illuminance is a minimum 66.7 m from the 8-unit source. ■

EXERCISES 24.7

In the following exercises, solve the given maximum and minimum problems.

1. In Example 3, change 1600 ft to 2400 ft and then proceed.
2. In Example 7, change 700 kL to 800 kL and then proceed.
3. The height (in ft) of a flare shot upward from the ground is given by $s = 112t - 16.0t^2$, where t is the time (in s). What is the greatest height to which the flare goes?
4. A small oil refinery estimates that its daily profit P (in dollars) from refining x barrels of oil is $P = 8x - 0.02x^2$. How many barrels should be refined for maximum daily profit, and what is the maximum profit?
5. The power output P (in W) of a certain 12-V battery is given by $P = 12I - 5.0I^2$, where I is the current (in A). Find the current for which the power is a maximum.
6. A rectangular enclosure will be constructed using 100 m of fencing. Find the dimensions that will yield a maximum area.
7. A company earns a weekly revenue R given by $R = xp$, where x is the number of units sold and p is the unit price. If p and x are related by the equation $p = 100 - 0.1x$, find the price that maximizes revenue.
8. If an airplane is moving at velocity v, the drag D on the plane is $D = av^2 + b/v^2$, where a and b are positive constants. Find the value(s) of v for which the drag is the least.
9. A company projects that its total savings S (in dollars) by converting to a solar-heating system with a solar-collector area A (in m^2) will be $S = 360A - 0.10A^3$. Find the area that should give the maximum savings and find the amount of the maximum savings.
10. The altitude h (in ft) of a jet that goes into a dive and then again turns upward is given by $h = 16t^3 - 240t^2 + 10{,}000$, where t is the time (in s) of the dive and turn. What is the altitude of the jet when it turns up out of the dive?
11. An alpha particle moves through a magnetic field along the parabolic path $y = x^2 - 4$. Determine the closest that the particle comes to the origin.
12. The electric potential V on the line $3x + 2y = 6$ is given by $V = 3x^2 + 2y^2$. At what point on this line is the potential a minimum?
13. In deep water, the velocity of a wave is $v = k\sqrt{\frac{l}{a} + \frac{a}{l}}$, where a and k are constants, and l is the length of the wave. What is the length of the wave that results in the minimum velocity?

14. The sum of the length l and width w of a rectangular table top is to be 240 cm. Determine l and w if the area of the table top is to be a maximum.

15. A rectangular hole is to be cut in a wall for a vent. If the perimeter of the hole is 48 in. and the length of the diagonal is a minimum, what are the dimensions of the hole?

16. When two electric resistors R_1 and R_2 are in series, their total resistance (the sum) is 32 Ω. If the same resistors are in parallel, their total resistance (the reciprocal of which equals the sum of the reciprocals of the individual resistances) is the maximum possible for two such resistors. What is the resistance of each?

17. A microprocessor chip is being designed with a given rectangular area A. Show that the chip with the minimum perimeter should be a square.

18. The rectangular animal display area in a zoo is enclosed by chain-link fencing and divided into two areas by internal fencing parallel to one of the sides. What dimensions will give the maximum area for the display if a total of 240 m of fencing are used?

19. What are the dimensions of the largest rectangular piece that can be cut from a semicircular metal sheet of diameter 14.0 cm?

20. A rectangular storage area is to be constructed alongside of a building. A security fence is required along the remaining three sides of the area. See Fig. 24.57. What is the maximum area that can be enclosed with 800 ft of fencing?

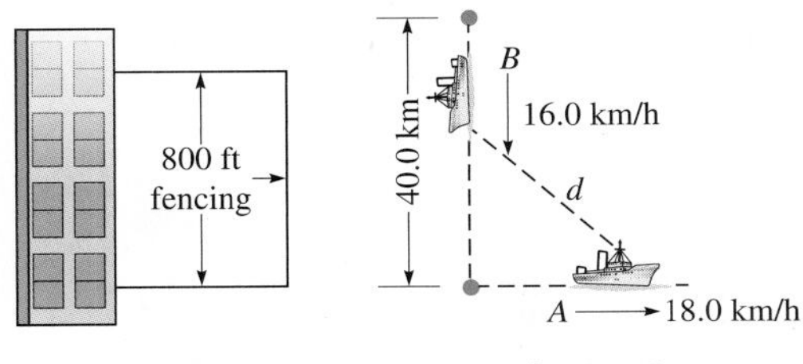

Fig. 24.57

Fig. 24.58

21. Ship A is traveling due east at 18.0 km/h as it passes a point 40.0 km due south of ship B, which is traveling due south at 16.0 km/h. See Fig. 24.58. How much later are the ships nearest each other?

22. An architect is designing a rectangular building in which the front wall costs twice as much per linear meter as the other three walls. The building is to cover 1350 m^2. What dimensions must it have such that the cost of the walls is a minimum?

23. A computer is programmed to display a slowly changing right triangle with its hypotenuse always equal to 12.0 cm. What are the legs of the triangle when it has its maximum area?

24. U.S. Postal Service regulations require that the length plus the girth (distance around) of a package not exceed 108 in. What are the dimensions of the largest (volume) rectangular box with square ends that can be mailed?

25. Referring to Exercise 24, what are the radius, length, and volume of the largest cylindrical package that may be sent through the mail?

26. An architect designs a window in the shape of a rectangle surmounted by an equilateral triangle. If the perimeter of the window is to be 6.00 m, what dimensions of the rectangle give the window the largest area?

27. The printed area of a rectangular poster is 384 cm^2, with margins of 4.00 cm on each side and margins of 6.00 cm at the top and bottom. Find the dimensions of the poster with the smallest area.

28. A conical funnel, with a very small opening, is being designed such that the slant height of the cone is 4.00 cm. What is the maximum volume of liquid that the funnel will be able to hold?

29. A culvert designed with a semicircular cross section of diameter 6.00 ft is redesigned to have an isosceles trapezoidal cross section by inscribing the trapezoid in the semicircle. See Fig. 24.59. What is the length of the bottom base b of the trapezoid if its area is to be maximum?

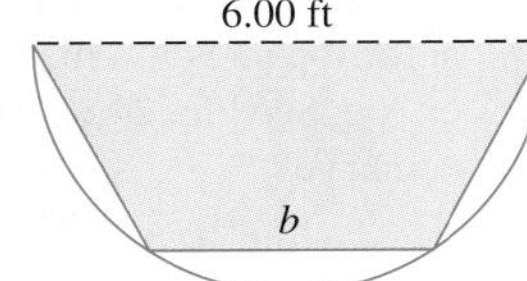

Fig. 24.59

30. A 36-cm-wide sheet of metal is bent into a rectangular trough as shown in Fig. 24.60. What dimensions give the maximum water flow?

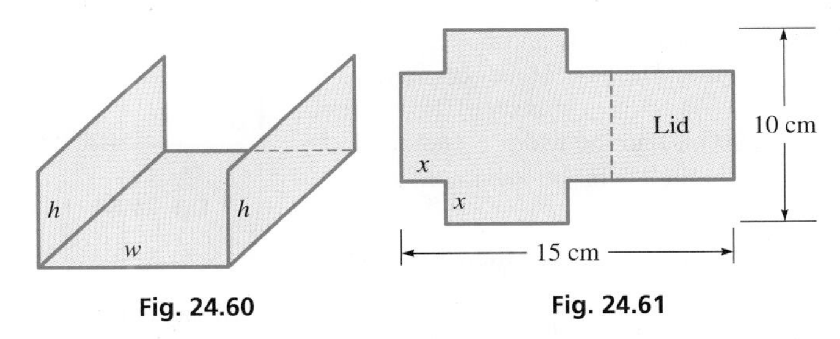

Fig. 24.60

Fig. 24.61

31. A box with a lid is to be made from a rectangular piece of cardboard 10 cm by 15 cm, as shown in Fig. 24.61. Two equal squares of side x are to be removed from one end, and two equal rectangles are to be removed from the other end so that the tabs can be folded to form the box with a lid. Find x such that the volume of the box is a maximum.

32. A lap pool (a pool for swimming laps) is designed to be seven times as long as it is wide. If the area of the sides and bottom is 980 ft^2, what are the dimensions of the pool if the volume of water it can hold is at maximum?

33. What is the maximum slope of the curve $y = 6x^2 - x^3$?

34. What is the minimum slope of the curve $y = x^5 - 10x^2$?

35. The deflection y of a beam of length L at a horizontal distance x from one end is given by $y = k(2x^4 - 5Lx^3 + 3L^2x^2)$, where k is a constant. For what value of x does the maximum deflection occur?

36. The electric power P (in W) produced by a certain battery is given by $P = \dfrac{144r}{(r + 0.6)^2}$, where r is the resistance in the circuit. For what value of r is the power a maximum?

37. A weight is suspended 50.0 cm below a ceiling by cables as shown in Fig. 24.62. Find the total minimum possible length of the cables.

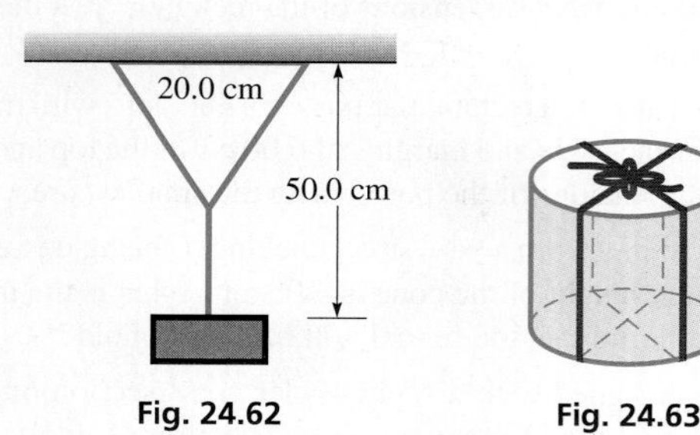

Fig. 24.62 Fig. 24.63

38. A cylindrical container is to be tied twice around vertically (crossed at the top and bottom) with 108 cm of string plus 8 cm for the knot on top. See Fig. 24.63. What is the maximum volume the container can have?

39. An airline requires that a carry-on bag has dimensions (length + width + height) that do not exceed 45 in. If a carry-on has a length 2.4 times the width, find the dimensions (to the nearest inch) of this type of carry-on that has the greatest volume.

40. A window is designed in the shape of a rectangle surmounted by a trapezoid, each of whose legs and upper base are one-half of the base of the rectangle. See Fig. 24.64. If the perimeter of the window is 5.70 m, find the width w and height h such that it lets in the maximum amount of light.

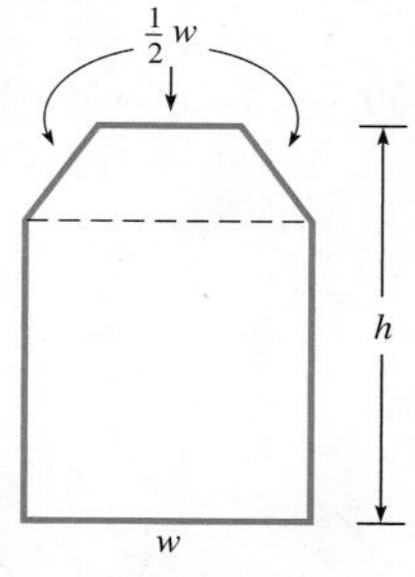

Fig. 24.64

41. For raising a load, the efficiency E (in %) of a screw with square threads is $E = \dfrac{100T(1 - fT)}{T + f}$, where f is the coefficient of friction and T is the tangent of the pitch angle of the screw. If $f = 0.25$, what acute angle makes E the greatest?

42. Computer simulation shows that the drag F (in N) on a certain airplane is $F = 0.00500v^2 + 3.00 \times 10^8/v^2$, where v is the velocity (in km/h) of the plane. For what velocity is the drag the least?

43. Factories A and B are 8.0 km apart, with factory B emitting eight times the pollutants into the air as factory A. If the number n of particles of pollutants is inversely proportional to the square of the distance from a factory, at what point between A and B is the pollution the least?

44. The potential energy E of an electric charge q due to another charge q_1 at a distance of r_1 is proportional to q_1 and inversely proportional to r_1. If charge q is placed directly between two charges of 2.00 nC and 1.00 nC that are separated by 10.0 mm, find the point at which the total potential energy (the sum due to the other two charges) of q is a minimum.

45. An open box is to be made from a square piece of cardboard whose sides are 8.00 in. long, by cutting equal squares from the corners and bending up the sides. Determine the side of the square that is to be cut out so that the volume of the box may be a maximum. See Fig. 24.65.

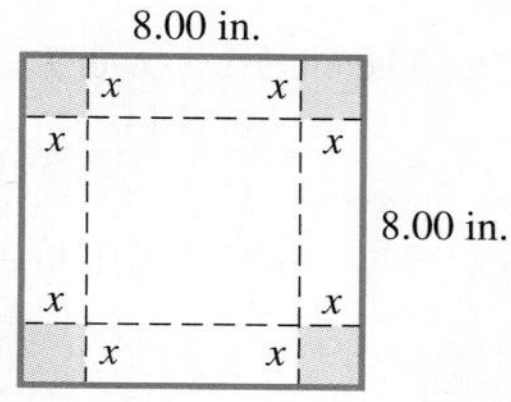

Fig. 24.65

46. A cone-shaped paper cup is to hold 100 cm^3 of water. Find the height and radius of the cup that can be made from the least amount of paper.

47. A race track 400 m long is to be built around an area that is a rectangle with a semicircle at each end. Find the length of the side of the rectangle adjacent to the track if the area of the rectangle is to be a maximum. See Fig. 24.66.

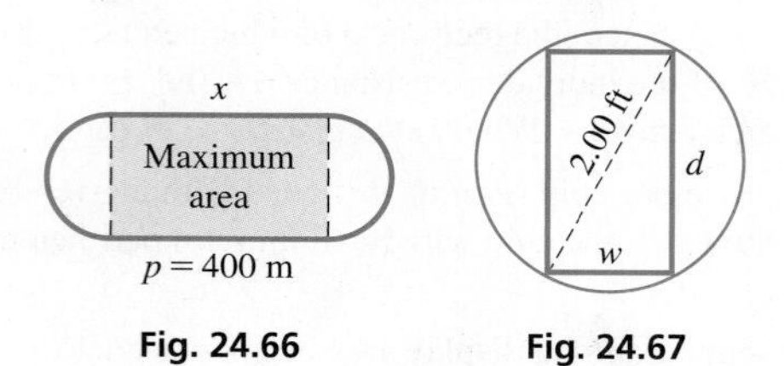

Fig. 24.66 Fig. 24.67

48. A beam of rectangular cross section is to be cut from a log 2.00 ft in diameter. The stiffness of the beam varies directly as the width and the cube of the depth. What dimensions will give the beam maximum stiffness? See Fig. 24.67.

49. An oil pipeline is to be built from a refinery to a tanker loading area. The loading area is 10.0 mi downstream from the refinery and on the opposite side of a river 2.5 mi wide. The pipeline is to run along the river and then cross to the loading area. If the pipeline costs \$50,000 per mile alongside the river and \$80,000 per mile across the river, find the point P (see Fig. 24.68) at which the pipeline should be turned to cross the river if construction costs are to be a minimum.

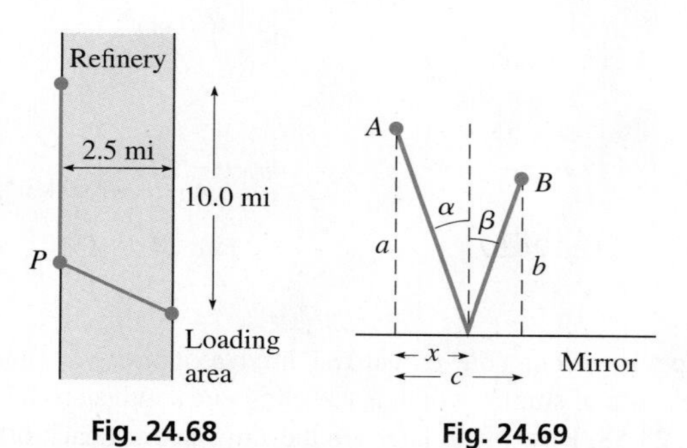

Fig. 24.68 Fig. 24.69

50. A light ray follows a path of least time. If a ray starts at point A (see Fig. 24.69) and is reflected off a plane mirror to point B, show that the angle of incidence α equals the angle of reflection β. (*Hint*: Set up the expression in terms of x, which will lead to $\sin\alpha = \sin\beta$.)

51. A rectangular building covering 7000 m^2 is to be built on a rectangular lot as shown in Fig. 24.70. If the building is to be 10.0 m from the lot boundary on each side and 20.0 m from the boundary in front and back, find the dimensions of the building if the area of the lot is a minimum.

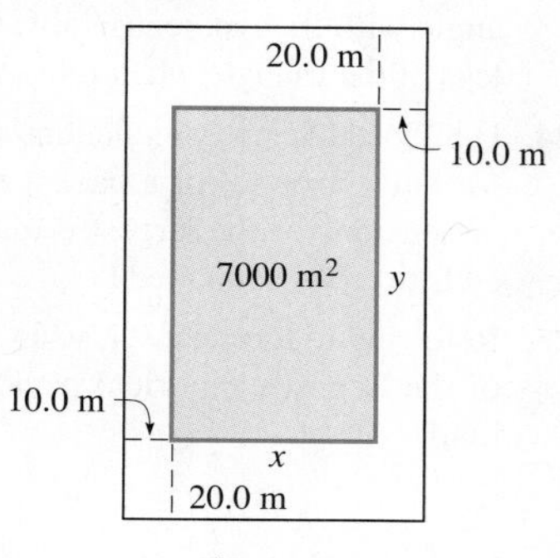

Fig. 24.70

52. On a computer simulation, a target is located at $(1.20, 7.00)$ (distances in km), and a rocket is fired along the path $y = 8.00 - 2.00x^2$. Find the minimum distance between the rocket's path and the target.

53. A company finds that there is a net profit of \$10 for each of the first 1000 units produced each week. For each unit over 1000 produced, there is 2 cents less profit per unit. How many units should be produced each week to net the greatest profit?

54. A cylindrical cup (no top) is designed to hold 375 cm^3 (375 mL). There is no waste in the material used for the sides. However, there is waste in that the bottom is made from a square $2r$ on a side. What are the most economical dimensions for a cup made under these conditions?

Answers to Practice Exercises

1. 6.93 in. by 9.80 in. 2. \$4,805,000 at \$155 per unit, 31,000 units

24.8 Differentials and Linear Approximations

Differential • Increment • Estimating Errors in Measurement • Linear Approximations

To this point, we have used the dy/dx notation for the derivative of y with respect to x, but we have not considered it to be a ratio. In this section, we define the quantities dy and dx, called *differentials*, such that their ratio is equal to the derivative. Then we show that differentials have applications in errors in measurement and in approximating values of functions. Also, in the next chapter, we use the differential notation in the development of integration, which is the inverse process of differentiation.

■ The symbol dy/dx for the derivatives was first used by Leibniz.

DIFFERENTIALS

We define the **differential** *of a function* $y = f(x)$ as

$$dy = f'(x)\,dx \tag{24.6}$$

In Eq. (24.6), dy is the *differential of* y, and dx is the *differential of* x. In this way, we can interpret the derivative as the ratio of the differential of y to the differential of x.

EXAMPLE 1 Differential of a polynomial

Find the differential of $y = 3x^5 - x$.

Because $f(x) = 3x^5 - x$, we find $f'(x) = 15x^4 - 1$. This means that

$$dy = (15x^4 - 1)dx$$

($15x^4 - 1$ is $f'(x)$; dx is the differential of x) ■

EXAMPLE 2 Differential of a rational expression

Find the differential of $s = \dfrac{4t}{t^2 + 4}$.

$$ds = \frac{(t^2 + 4)(4) - (4t)(2t)}{(t^2 + 4)^2}dt \quad \text{using derivative quotient rule}$$

$$= \frac{4t^2 + 16 - 8t^2}{(t^2 + 4)^2}dt = \frac{-4t^2 + 16}{(t^2 + 4)^2}dt$$

$$= \frac{-4(t^2 - 4)}{(t^2 + 4)^2}dt \quad \text{don't forget the } dt$$

■

Practice Exercise

1. Find the differential of $y = (2x - 1)^4$.

To understand more about differentials and their applications, we now introduce some useful notation. If a variable x changes value from x_1 to x_2, *the difference* $x_2 - x_1$ *is called the* **increment** *in* x. Traditionally, in calculus, this increment is denoted by Δx ("delta x"), which means $\Delta x = x_2 - x_1$. We must be careful to note that Δx is not a product of Δ and x, but a *single symbol* that represents the *change* in x.

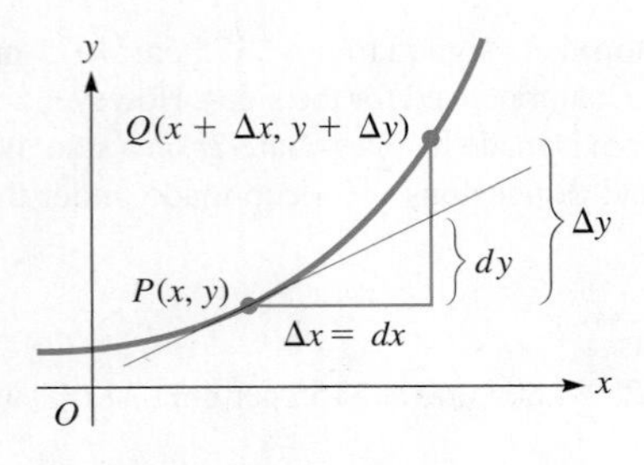

Fig. 24.71

By choosing $\Delta x = dx$, if dx is small, then the differential in y, dy, closely approximates the increment in y, Δy. This is the basis of the applications of the differential, and to understand this better, let us look at Fig. 24.71. We see that points $P(x, y)$ and $Q(x + \Delta x, y + \Delta y)$ lie on the graph of $f(x)$ and that the increments of x and y between the points are Δx and Δy. At point P, $f'(x) = dy/dx$, which means that the slope of a tangent line at P is indicated by dy/dx. With $dx = \Delta x$, we see that as dx becomes smaller, dy more nearly approximates Δy, which is the actual difference in the y-values. Therefore, *for small values of Δx, dy can be used to approximate Δy.*

EXAMPLE 3 Calculating the increment and differential

Calculate Δy and dy for $y = x^3 - 2x$ for $x = 3$ and $\Delta x = 0.1$.

First, we find Δy by calculating $f(3.1) - f(3)$. Therefore,

$$\Delta y = f(3.1) - f(3) = [3.1^3 - 2(3.1)] - [3^3 - 2(3)] = 2.591$$

The differential of y is

$$dy = (3x^2 - 2)dx$$

Because $dx = \Delta x$, we have

$$dy = [3(9) - 2](0.1) = 2.5$$

Thus, $\Delta y = 2.591$ and $dy = 2.5$. In this case, dy is very nearly equal to Δy. ■

ESTIMATING ERRORS IN MEASUREMENT

The fact that dy can be used to approximate Δy is useful in finding the error in a result from a measurement, if the data are in error, or the equivalent problem of finding the change in the result if a change is made in the data. Even though such changes can be found by using a calculator, the differential can be used to set up a general expression for the change of a particular function.

EXAMPLE 4 Calculating error in measurement

The edge of a cube of gold was measured to be 3.850 cm. From this value, the volume was found. Later, it was discovered that the measured value of the edge was 0.020 cm too small. By approximately how much was the volume in error?

The volume V of a cube, in terms of an edge e, is $V = e^3$. Because we wish to find the change in V for a given change in e, we want the value of dV for $e = 3.850$ cm and $de = 0.020$ cm.

First, finding the general expression for dV, we have

$$dV = 3e^2 de$$

Now, evaluating this expression for the given values, we have

$$dV = 3(3.850)^2(0.020) = 0.89 \text{ cm}^3$$

In this case, the volume was in error by about 0.89 cm^3. As long as de is small compared with e, we can closely approximate an error or change in the volume of a cube by calculating the value of $3e^2de$. ■

Practice Exercise

2. In Example 4, approximate the error in the total surface area A of the cube.

Often, when considering the error of a given value or result, *the actual numerical value of the error, the* **absolute error,** is not as important as its size in relation to the size of the quantity itself. *The ratio of the absolute error to the actual true size of the quantity is known as the* **relative error,** which is commonly expressed as a percent.

EXAMPLE 5 Absolute error and relative error

Referring to Example 4, we see that the absolute error in the edge was 0.020 cm. Since the actual length of the edge is 3.870 cm, the relative error in the edge was

$$\frac{de}{e} = \frac{0.020}{3.870} = 0.0052 = 0.52\%$$

The absolute error in the volume was 0.89 cm^3, and the actual value of the volume is $3.870^3 = 57.96$ cm^3. This means the relative error in the volume was

$$\frac{dV}{V} = \frac{0.89}{57.96} = 0.015 = 1.5\%$$ ■

LINEAR APPROXIMATIONS

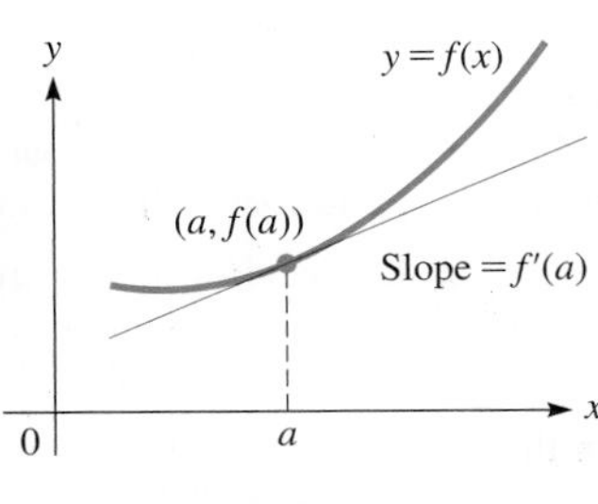

Fig. 24.72

Continuing our discussion related to Fig. 24.71, on the curve of the function $f(x)$ at the point $(a, f(a))$, the slope of a tangent line is $f'(a)$. See Fig. 24.72. For a point (x, y) on the tangent line, by using the point-slope form for the equation of a straight line, $y - y_1 = m(x - x_1)$, we have

$$f(x) = f(a) + f'(a)(x - a)$$

For the points on $y = f(x)$ near $x = a$, we can use the tangent line to approximate the function, as shown in Fig. 24.72. Therefore, the approximation $f(x) \approx L(x)$ *is the* ***linear approximation*** *of* $f(x)$ *near* $x = a$, *where the function*

$$L(x) = f(a) + f'(a)(x - a) \qquad \textbf{(24.7)}$$

is called the **linearization** *of* $f(x)$ *at* $x = a$.

EXAMPLE 6 Linearizing a function

Find the linearization of the function $f(x) = \sqrt{2x + 1}$ at $x = 4$. Use it to approximate $\sqrt{9.06}$.

The solution is as follows:

$$f(x) = \sqrt{2x + 1}$$

$$f'(x) = \tfrac{1}{2}(2x + 1)^{-1/2}(2) = \frac{1}{\sqrt{2x + 1}} \qquad \text{find the derivative}$$

$$f(4) = \sqrt{2(4) + 1} = 3 \qquad f'(4) = \frac{1}{\sqrt{2(4) + 1}} = \frac{1}{3} \qquad \text{evaluate } f(4) \text{ and } f'(4)$$

$$L(x) = 3 + \tfrac{1}{3}(x - 4) = \frac{x + 5}{3} \qquad \text{use Eq. (24.7)}$$

$$\sqrt{2x + 1} \approx \frac{x + 5}{3} \qquad 2x + 1 = 9.06.\ x = 4.03$$

$$\sqrt{9.06} \approx \frac{4.03 + 5}{3} = \frac{9.03}{3} = 3.01 \qquad \text{by calculator, } \sqrt{9.06} = 3.009983389$$

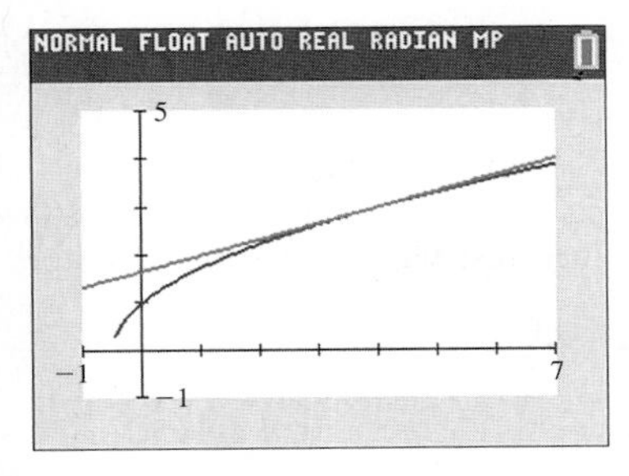

Fig. 24.73

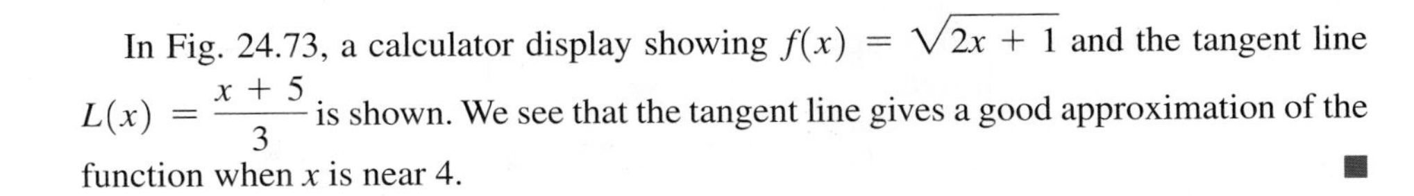

In Fig. 24.73, a calculator display showing $f(x) = \sqrt{2x + 1}$ and the tangent line $L(x) = \dfrac{x + 5}{3}$ is shown. We see that the tangent line gives a good approximation of the function when x is near 4. ■

EXERCISES 24.8

In Exercises 1–4, make the given changes in the indicated examples of this section and then solve the resulting problems.

1. In Example 2, change the t^2 to t^3.

2. In Example 3, change the $-$ before $2x$ to $+$.

3. In Example 5, change 0.020 to 0.025.

4. In Example 6, change $x = 4$ to $x = 12$; approximate $\sqrt{25.06}$.

In Exercises 5–16, find the differential of each of the given functions.

5. $y = x^4 + 3x$

6. $y = 3x^2 + 6$

7. $V = \frac{2}{r^5} + 3\pi^2$

8. $y = 2\sqrt{x} - \frac{1}{8x}$

9. $s = 3(t^2 - 5)^4$

10. $y = 5(4 + 3x)^{1/3}$

11. $y = \frac{12}{3x^2 + 1}$

12. $R = \frac{u}{\sqrt{1 + 2u}}$

13. $y = x(1 - x)^3$

14. $y = 6x\sqrt{1 - 4x}$

15. $y = \frac{x}{5x + 2}$

16. $y = \frac{3x + 1}{\sqrt{2x - 1}}$

In Exercises 17–20, find the values of Δy and dy for the given values of x and dx.

17. $y = 7x^2 + 4x, x = 4, \Delta x = 0.2$

18. $y = (x^2 + 2x)^3, x = 7, \Delta x = 0.02$

19. $y = x\sqrt{1 + 4x}, x = 12, \Delta x = 0.06$

20. $y = \frac{x}{\sqrt{6x - 1}}, x = 3.5, \Delta x = 0.025$

In Exercises 21–24, find the linearization $L(x)$ of the given functions for the given values of a. Display $f(x)$ and $L(x)$ on the same calculator screen.

21. $f(x) = x^2 + 2x, a = 0$

22. $f(x) = 6\sqrt[3]{x}, a = 8$

23. $f(x) = \frac{1}{2x + 1}, a = -1$

24. $g(x) = x\sqrt{2x + 8}, a = -2$

In Exercises 25–38, solve the given problems by finding the appropriate differential.

25. If a spacecraft circles Earth at an altitude of 250 km, how much farther does it travel in one orbit than an airplane that circles Earth at a low altitude? The radius of Earth is 6370 km.

26. Approximate the amount of paint needed to apply one coat of paint 0.50 mm thick on a hemispherical dome 55 m in diameter.

27. The radius of a circular manhole cover is measured to be 40.6 ± 0.05 cm (this means the possible error in the radius is 0.05 cm). Estimate the possible relative error in the area of the top of the cover.

28. The side of a square microprocessor chip is measured as 0.950 cm, and later it is measured as 0.952 cm. What is the difference in the calculations of the area due to the difference in the measurements of the side? See Fig. 24.74.

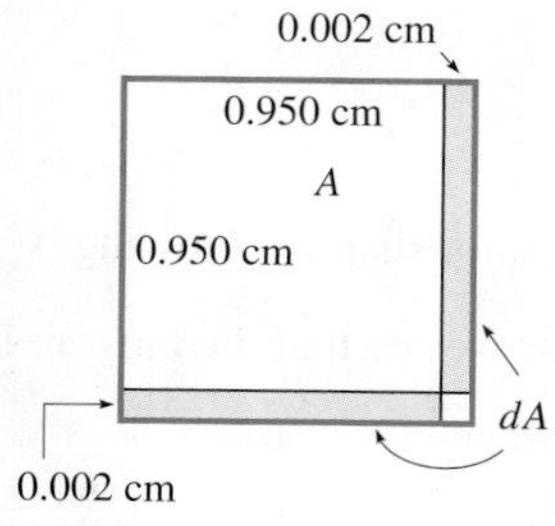

Fig. 24.74

29. The wavelength λ of light is inversely proportional to its frequency f. If $\lambda = 685$ nm for $f = 4.38 \times 10^{14}$ Hz, find the change in λ if f increases by 0.20×10^{14} Hz. (These values are for red light.)

30. The velocity of an object rolling down a certain inclined plane is given by $v = \sqrt{100 + 16x}$, where x is the distance (in ft) traveled along the plane by the object. What is the increase in velocity (in ft/s) of an object in moving from 20.0 ft to 20.5 ft along the plane? What is the relative change in the velocity?

31. If the diameter equals the height, what is the volume of the plastic that forms a closed cylindrical container for which the radius is 18.0 cm and the thickness is 2.00 mm?

32. The volume V of blood flowing through an artery is proportional to the fourth power of the radius r of the artery. Find how much a 5% increase in r affects V.

33. The radius r of a holograph is directly proportional to the square root of the wavelength λ of the light used. Show that $dr/r = \frac{1}{2}d\lambda/\lambda$.

34. The gravitational force F of Earth on an object is inversely proportional to the square of the distance r of the object from the center of Earth. Show that $dF/F = -2dr/r$.

35. Show that an error of 2% in the measurement of the radius of a DVD results in an error of approximately 4% in the calculation of the area.

36. Show that an error of 2% in the measurement of the radius of a ball bearing results in an error of approximately 6% in the calculation of the volume.

37. Calculate $\sqrt{4.05}$, using differentials.

38. Explain how to estimate 2.03^4 using differentials.

In Exercises 39–44, solve the given linearization problems.

39. Show that the linearization of $f(x) = (1 + x)^k$ at $x = 0$ is $L(x) = 1 + kx$.

40. Use the result shown in Exercise 39 to approximate the value of $f(x) = \frac{1}{\sqrt{1 + x}}$ near zero.

41. Linearize $f(x) = \sqrt{2 - x}$ for $a = 1$ and use it to approximate the value of $\sqrt{1.9}$.

42. Explain how to evaluate $\sqrt[3]{8.03}$, using linearization.

43. The capacitance C (in μF) in an element of an electronic tuner is given by $C = \frac{3.6}{\sqrt{1 + 2V}}$, where V is the voltage. Linearize C for $V = 4.0$ V.

44. A 16-Ω resistor is put in parallel with a variable resistor of resistance R. The combined resistance of the two resistors is $R_T = \frac{16R}{16 + R}$. Linearize R_T for $R = 4.0\ \Omega$.

Answers to Practice Exercises

1. $dy = 8(2x - 1)^3dx$ **2.** $dA = 0.92\text{ cm}^2$

CHAPTER 24 KEY FORMULAS AND EQUATIONS

Newton's method $$x_2 = x_1 - \frac{f(x_1)}{f'(x_1)} \quad (24.1)$$

Curvilinear motion $$v_x = \frac{dx}{dt} \qquad v_y = \frac{dy}{dt} \quad (24.2)$$

$$a_x = \frac{dv_x}{dt} = \frac{d^2x}{dt^2} \qquad a_y = \frac{dv_y}{dt} = \frac{d^2y}{dt^2} \quad (24.3)$$

$$v = \sqrt{v_x^2 + v_y^2} \qquad a = \sqrt{a_x^2 + a_y^2} \quad (24.4)$$

$$\tan\theta_v = \frac{v_y}{v_x} \qquad \tan\theta_a = \frac{a_y}{a_x} \quad (24.5)$$

Curve sketching and maximum and minimum values

$f'(x) > 0$ where $f(x)$ increases; $f'(x) < 0$ where $f(x)$ decreases.
$f''(x) > 0$ where the graph of $f(x)$ is concave up; $f''(x) < 0$ where the graph of $f(x)$ is concave down.
If $f'(x) = 0$ at $x = a$, there is a local maximum point if $f'(x)$ changes from + to − or if $f''(a) < 0$.
If $f'(x) = 0$ at $x = a$, there is a local minimum point if $f'(x)$ changes from − to + or if $f''(a) > 0$.
If $f''(x) = 0$ at $x = a$, there is a point of inflection if $f''(x)$ changes from + to − or from − to +.

Differential $$dy = f'(x)dx \quad (24.6)$$

Linearization $$L(x) = f(a) + f'(a)(x - a) \quad (24.7)$$

CHAPTER 24 REVIEW EXERCISES

CONCEPT CHECK EXERCISES

Determine each of the following as being either **true** *or* **false**. *If it is false, explain why.*

1. The slope of the line normal to the curve $y = 3x^2 - 5$ at $(1, 2)$ is $\frac{1}{6}$.
2. Newton's method can be used to find a root of $f(x)$ if a section of the curve is continuous.
3. If an object moves such that $x = \frac{1}{2}t^4$, then the acceleration component $a_x = 24$ when $x = 2$.
4. If $A = 4\pi r^2$, then $dA/dt = 8\pi r$.
5. The graph of $f(x) = 2x^4 - 4x^2$ has two points of inflection for $y = -2/3$.
6. An asymptote of the graph of $y = \frac{1}{x} - x$ is $y = x$.
7. The sum of a number and its reciprocal is never between -2 and 2.
8. For $y = 6x^3 - x$, $dy = 18x - 1$.

PRACTICE AND APPLICATIONS

In Exercises 9–14 find the equations of the tangent or normal lines. Use a calculator to view the curve and the line.

9. Find the equation of the line tangent to the parabola $y = 3x - x^2$ at the point $(-1, -4)$.
10. Find the equation of the line tangent to the curve $y = x^2 - \frac{6}{x}$ at the point $(3, 7)$
11. Find the equation of the line normal to $x^2 - 4y^2 = 9$ at the point $(5, 2)$.
12. Find the equation of the line normal to $y = \frac{4}{\sqrt{x - 2}}$ at the point $(6, 2)$.
13. Find the equation of the line tangent to the curve $y = \sqrt{x^2 + 3}$ and that has a slope of $\frac{1}{2}$.
14. Find the equation of the line normal to the curve $y = \frac{1}{2x + 1}$ and that has a slope of $\frac{1}{2}$ if $x \geq 0$.

In Exercises 15–20, find the indicated velocities and accelerations.

15. Given that the x- and y-coordinates of a moving particle are given as a function of time t by the parametric equations $x = \sqrt{t} + t, y = \frac{1}{12}t^3$, find the magnitude and direction of the velocity when $t = 4$.

16. If the x- and y-coordinates of a moving object as functions of time t are given by $x = 0.1t^2 + 1, y = \sqrt{4t + 1}$, find the magnitude and direction of the velocity when $t = 6$.

17. An object moves along the curve $y = 0.5x^2 + x$ such that $v_x = 0.5\sqrt{x}$. Find v_y at (2, 4).

18. A particle moves along the curve of $y = \dfrac{1}{x + 2}$ with a constant velocity in the x-direction of 4 cm/s. Find v_y at $(2, \frac{1}{4})$.

19. Find the magnitude and direction of the acceleration for the particle in Exercise 15.

20. Find the magnitude and direction of the acceleration for the particle in Exercise 18.

In Exercises 21–24, find the indicated roots of the given equations to at least four decimal places by use of Newton's method.

21. $x^3 - 3x^2 - x + 2 = 0$ (between 0 and 1)

22. $2x^3 - 4x^2 - 9 = 0$ (between 2 and 3)

23. $x^2 = \frac{3}{x}$ (the real solution)

24. $\sqrt{x} - 2 = \frac{1}{(x - 6)^2}$ (the real solution)

In Exercises 25–32, sketch the graphs of the given functions by information obtained from the function as well as information obtained from the derivatives. Use a calculator to check the graph.

25. $y = 4x^2 + 16x$

26. $y = x^3 + 2x^2 + x + 1$

27. $y = 9x - 3x^3$

28. $y = x(6 - x)^3$

29. $y = x^4 - 32x$

30. $y = x^5 - 20x^2 + 10$

31. $y = \dfrac{x^2}{\sqrt{x^2 - 9}}$

32. $y = x^3 + \dfrac{3}{x}$

In Exercises 33–36, find the differential of each of the given functions.

33. $y = 2x^3 + \dfrac{5}{x^2}$

34. $y = \dfrac{6}{(2x - 1)^2}$

35. $y = (x^2 - 3)^{1/3}$

36. $s = \sqrt{\dfrac{2 + t}{2 - t}}$

In Exercises 37 and 38, evaluate $\Delta y - dy$ for the given functions and values.

37. $y = 4x^3 - 12, x = 2, \Delta x = 0.1$

38. $y = 6x^2 - x, x = 3, \Delta x = 0.2$

In Exercises 39 and 40, find the linearization of the given functions for the given values of a.

39. $f(x) = \sqrt{x^2 + 7}, a = 3$

40. $f(x) = x^2(x + 1)^4, a = -2$

In Exercises 41–48, solve the given problems by finding the appropriate differentials.

41. A weather balloon 3.500 m in radius becomes covered with a uniform layer of ice 1.2 cm thick. What is the volume of the ice?

42. The total power P (in W) transmitted by an AM radio transmitter is $P = 460 + 230m^2$, where m is the modulation index. What is the change in power if m changes from 0.86 to 0.89?

43. A cylindrical silo with a flat top is 32.0 ft in diameter and is 32.0 ft high. By how much is the volume changed if the radius is increased by 1.0 ft and the height is unchanged?

44. A ski slope follows a path that can be represented by $y = 0.010x^2 - 0.86x + 24$. What is the change in the slope of the path when x changes from 26 m to 28 m?

45. The impedance Z of an electric circuit as a function of the resistance R and the reactance X is given by $Z = \sqrt{R^2 + X^2}$. Derive an expression of the relative error in impedance for an error in R and a given value of X.

46. Evaluate $\sqrt{8.94}$.

47. Evaluate 3.02^5.

48. Show that the relative error in the calculation of the volume of a spherical snowball is approximately three times the relative error in the measurement of the radius.

In Exercises 49–94, solve the given problems.

49. The parabolas $y = x^2 + 2$ and $y = 4x - x^2$ are tangent to each other. Find the equation of the line tangent to them at the point of tangency.

50. Find the equation of the line tangent to the curve of $y = x^4 - 8x$ and perpendicular to the line $4y - x + 5 = 0$.

51. For the cubic polynomial $f(x) = x^3 + bx^2 + cx + d$, find the relationship between b and c for which the graph of $f(x)$ has no local minimum or maximum points.

52. For the cubic polynomial of Exercise 51, find the x-coordinate of the one point of inflection.

53. Are the following statements always true? If not, give a counterexample. (a) If $f'(x) = 0$ at $x = a$, there is a local maximum or minimum point at $x = a$. (b) If $f''(b) = 0$ at $x = b$, there is a point of inflection at $x = b$.

54. The side s of a square on a TV screen is decreasing at the rate of 2.0 mm/s. How fast is the area decreasing when $s = 12.0$ cm?

55. A spherical metal object is ejected from an Earth satellite and reenters the atmosphere. It heats up (until it bursts) so that the radius r increases at the rate of 0.80 mm/s. How fast is the volume V changing when $r = 125$ mm?

56. A certain chemical reaction rate R (in mg/s) is dependent on the amount of chemical present. If $R = 12\sqrt{m}(27 - m)$, find m (in mg) for the maximum reaction rate R.

57. The deflection y (in m) of a beam at a horizontal distance x (in m) from one end is given by $y = k(x^4 - 30x^3 + 1000x)$, where k is a constant. Observing the equation and using Newton's method, find the values of x (to the nearest thousandth) where the deflection is zero.

58. The edges of a rectangular water tank are 3.00 ft, 5.00 ft, and 8.00 ft. By Newton's method, determine by how much each edge should be increased equally to double the volume of the tank.

59. A parachutist descends (after the parachute opens) in a path that can be described by $x = 8t$ and $y = -0.15t^2$, where distances are in meters and time is in seconds. Find the parachutist's velocity upon landing if the landing occurs when $t = 12$ s.

60. One of the curves on an automobile test track can be described by $y = 225/x$, where dimensions are in meters. Find the x- and y-coordinates of a car driving along this curve at the instant when $v_y = -2v_x$.

61. A person walks 250 ft north and turns west. Walking at the rate of 4.0 ft/s, at what rate is the distance between the person and the starting point increasing 1.0 min after turning west?

62. A glass prism for refracting light has equilateral triangular ends and vertical cross sections, and a volume of 45 cm^3. Find the edge of one of the ends such that the total surface area is a minimum.

63. In Fig. 24.75, the tension T supports the 40.0-N weight. The relation between the tension T and the deflection d is $d = \dfrac{1000}{\sqrt{T^2 - 400}}$. If the tension is increasing at 2.00 N/s when $T = 28.0$ N, how fast is the deflection (in cm) changing?

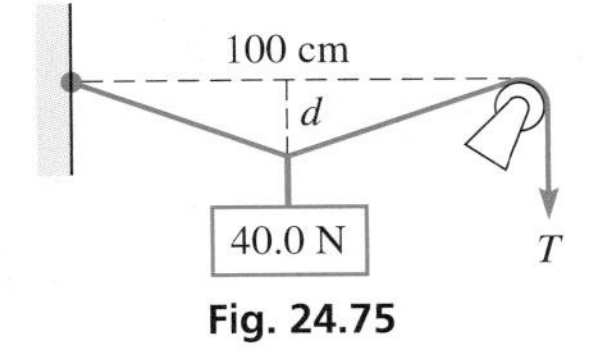

Fig. 24.75

64. The impedance Z (in Ω) in a particular electric circuit is given by $Z = \sqrt{48 + R^2}$, where R is the resistance. If R is increasing at a rate of 0.45 Ω/min for $R = 6.5\ \Omega$, find the rate at which Z is changing.

65. By using the methods of this chapter to graph $y = x^2 - \frac{2}{x}$, graph the solution of the inequality $x^2 > \frac{2}{x}$.

66. Display the graphs of $y_1 = x^2 - \frac{2}{x}$ and $y_2 = x^2$ on the same screen of a calculator, and explain why y_2 can be considered to be a nonlinear asymptote of y_1.

67. An analysis of the power output P (in kW/m^3) of a certain turbine showed that it depended on the flow rate r (in m^3/s) of water to the turbine according to the equation $P = 0.030r^3 - 2.6r^2 + 71r - 200$ $(6 \leq r \leq 30 \text{ m}^3/\text{s})$. Determine the rate for which P is a maximum.

68. The altitude h (in ft) of a certain rocket as a function of the time t (in s) after launching is given by $h = 1600t - 16t^2$. What is the maximum altitude the rocket attains?

69. Sketch the continuous curve having these characteristics:

$f(0) = 2$ $\quad f'(x) < 0$ for $x < 0$ $\quad f''(x) > 0$ for all x
$\quad f'(x) > 0$ for $x > 0$

70. Sketch a continuous curve having these characteristics:

$f(0) = 1$ $\quad f'(0) = 0$ $\quad f''(x) < 0$ for $x < 0$
$\quad f'(x) > 0$ for $|x| > 0$ $\quad f''(x) > 0$ for $x > 0$

71. A horizontal cylindrical oil tank (the length is parallel to the ground) of radius 6.00 ft is being emptied. Find how fast the width w of the oil surface is changing when the depth h is 1.50 ft and changing at the rate of 0.250 ft/min.

72. The current I (in A) in a circuit with a resistance R (in Ω) and a battery whose voltage is E and whose internal resistance is r (in Ω) is given by $I = E/(R + r)$. If R changes at the rate of 0.250 Ω/min, how fast is the current changing when $R = 6.25\ \Omega$, if $E = 3.10$ V and $r = 0.230\ \Omega$?

73. A field in the shape of the trapezoid (see Fig. 24.76) is to be enclosed with 1500 m of fencing. Find the dimensions of x and y such that the area of the field is a maximum.

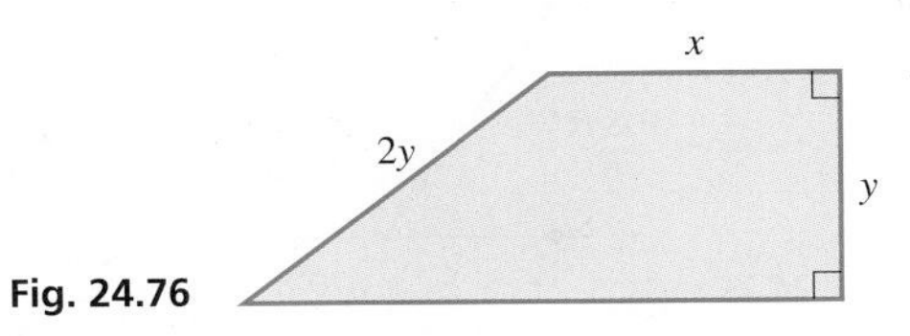

Fig. 24.76

74. A special insulation strip is to be sealed completely around three edges of a rectangular solar panel. If 200 cm of the strip are used, what is the maximum area of the panel?

75. A baseball diamond is a square 90.0 ft on a side. See Fig. 24.77. As a player runs from first base toward second base at 18.0 ft/s, at what rate is the player's distance from home plate increasing when the player is 40.0 ft from first base?

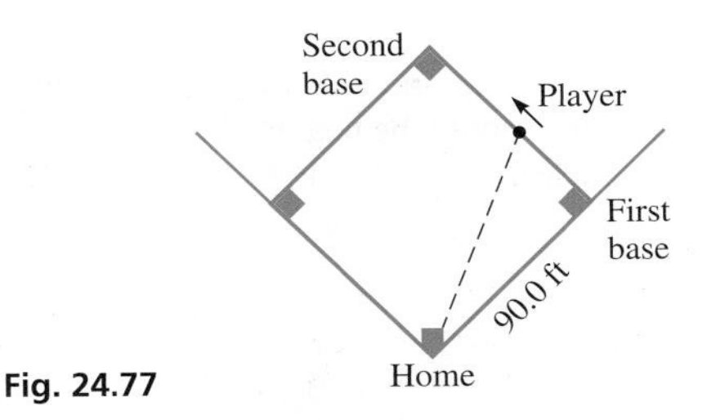

Fig. 24.77

76. A swimming pool with a rectangular surface of 1200 ft^2 is to have a cement border area that is 12.0 ft wide at each end and 8.00 ft wide at the sides. Find the surface dimensions of the pool if the total area covered is to be a minimum.

77. A study showed that the percent y of persons surviving burns to x percent of the body is given by $y = \dfrac{300}{0.0005x^2 + 2} - 50$. Linearize this function with $a = 50$ and sketch the graphs of y and $L(x)$.

78. A company estimates that the sales S (in dollars) of a new product will be $S = 5000t/(t + 4)^2$, where t is the time (in months) after it is put into production. Sketch the graph of S vs. t.

79. An airplane flying horizontally at 8000 ft is moving toward a radar installation at 680 mi/h. If the plane is directly over a point on the ground 5.00 mi from the radar installation, what is its actual speed? See Fig. 24.78.

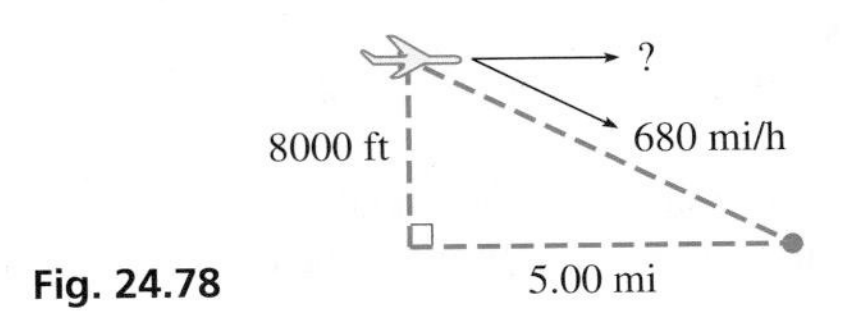

Fig. 24.78

80. The base of a conical machine part is being milled such that the height is decreasing at the rate of 0.050 cm/min. If the part originally had a radius of 1.0 cm and a height of 3.0 cm, how fast is the volume changing when the height is 2.8 cm?

81. The reciprocal of the total capacitance C_T of electrical capacitances in series equals the sum of the reciprocals of the individual capacitances. If the sum of two capacitances is 12 μF, find their values if their total capacitance in series is a maximum.

82. A cable is to be from point A to point B on a wall and then to point C. See Fig. 24.79. Where is B located if the total length of cable is a minimum?

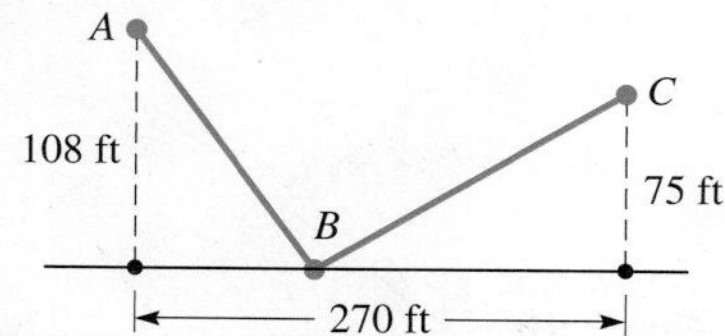

Fig. 24.79

83. A box with a square base and an open top is to be made of 27 ft^2 of cardboard. What is the maximum volume that can be contained within the box?

84. A car is traveling east at 90.0 km/h, and an airplane is traveling south at 450 km/h at an elevation of 2.00 km. At one instant the plane is directly above the car. At what rate are they separating 15.0 min later?

85. A person in a boat 4 km from the nearest point P on a straight shoreline wants to go to point A on the shoreline, 5 km from P. If the person can row at 3 km/h and walk at 5 km/h, at what point on the shoreline should the boat land in order that point A can be reached in the least time? See Fig. 24.80.

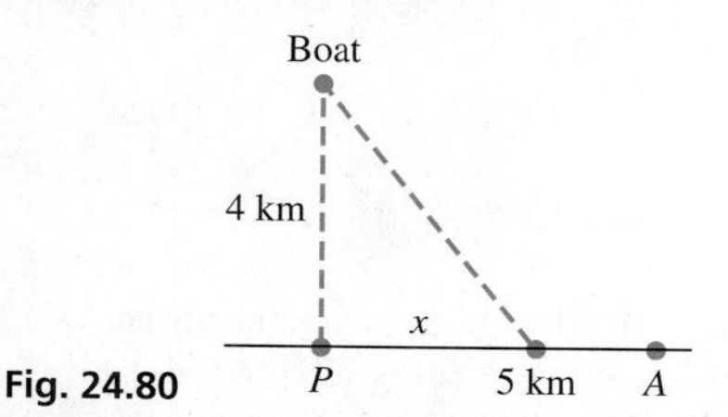

Fig. 24.80

86. A machine part is to be in the shape of a circular sector of radius r and central angle θ. Find r and θ if the area is one unit and the perimeter is a minimum. See Fig. 24.81.

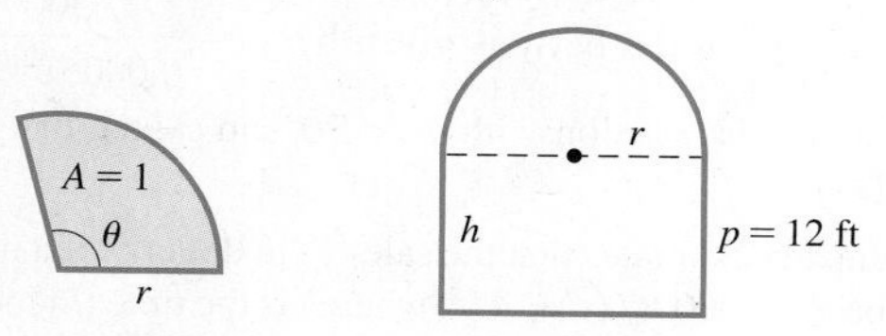

Fig. 24.81

Fig. 24.82

87. A Norman window has the form of a rectangle surmounted by a semicircle. Find the dimensions (radius of circular part and height of rectangular part) of the window that will admit the most light if the perimeter of the window is 12 ft. See Fig. 24.82.

88. An open drawer for small tools is to be made from a rectangular piece of heavy sheet metal 12.0 in. by 10.0 in., by cutting out equal squares from two corners and bending up the three sides, as shown in Fig. 24.83. Find the side of the square that should be cut out so that the volume of the drawer is a maximum.

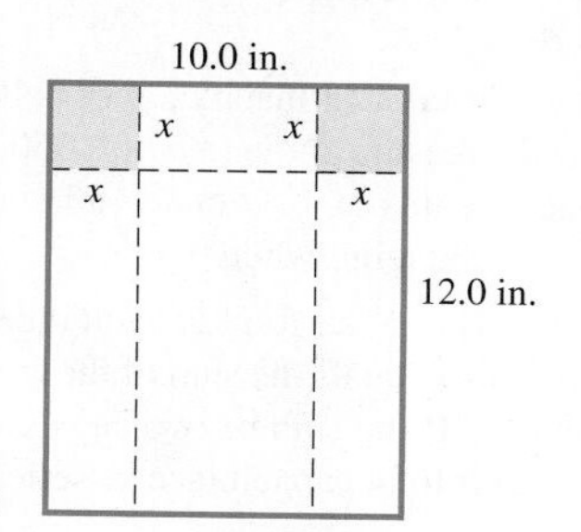

Fig. 24.83

89. A pile of sand in the shape of a cone has a radius that always equals the altitude. If 100 ft^3 of sand are poured onto the pile each minute, how fast is the radius increasing when the pile is 10.0 ft high?

90. A book is designed such that its (rectangular) pages have a 2.5-cm margin at the top and the bottom, 1.5-cm margins on each side, and a total page area of 320 cm^2. See Fig. 24.84. What are the dimensions of a page such that the area within the margins for print material and figures is a maximum?

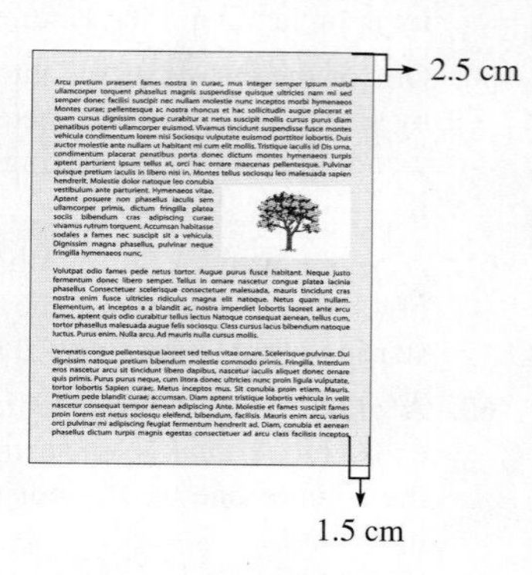

Fig. 24.84

91. A specially made cylindrical container is made of stainless steel sides and bottom and a silver top. If silver is ten times as expensive as stainless steel, what are the most economical dimensions of the container if it is to hold 314 cm^3?

92. A tax rate (on income or whatever) of 0% produces no revenue. Also, a tax rate of 100% would probably produce no revenue, since no one would choose to earn income that was completely taxed away. Economists have speculated as to what tax rate would produce the most revenue, and how revenue would vary with the rate. One proposal is that the revenue is proportional to the product of the tax rate x (in %) and $(100 - x)^{2/3}$. Under this proposal, what tax rate produces the greatest revenue r? [*Note:* The tax rate on the highest income bracket in the United States exceeded 90% (1944–45 and 1951–63). Also, this rate was below 30% (1988–90).]

93. A builder is designing a storage building with a total volume of 1350 ft^3, a rectangular base, and a flat roof. The width is to be 0.75 of the length. The cost per square foot is $6.00 for the floor, $9.00 for the sides, and $4.50 for the roof. What dimensions will minimize the cost of the building?

94. City B is 16.0 mi east and 12.0 mi north of city A. City A is 8.00 mi due south of a river that is 1.00 mi wide. A road is to be built between A and B that crosses straight across the river. See Fig. 24.85. Where should the bridge be located so that the road between A and B is as short as possible?

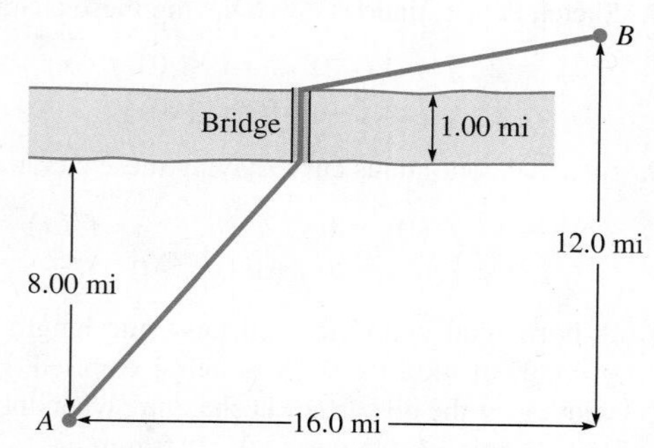

Fig. 24.85

95. A container manufacturer makes various sizes of closed cylindrical plastic containers for shipping liquid products. Write two or three paragraphs explaining how to determine the ratio of the height to radius of the container such that the least amount of plastic is used for each size. Include the reason why it is not necessary to specify the volume of the container in finding this ratio.

CHAPTER 24 PRACTICE TEST

As a study aid, we have included complete solutions for each Practice Test problem at the end of this book.

1. Find the equation of the line tangent to the curve $y = x^4 - 3x^2$ at the point $(1, -2)$.

2. For $y = 3x^2 - x$, evaluate (a) Δy, (b) dy, and (c) $\Delta y - dy$ for $x = 3$ and $\Delta x = 0.1$.

3. If the x- and y-coordinates of a moving object as functions of time are given by the parametric equations $x = 3t^2$, $y = 2t^3 - t^2$, find the magnitude and direction of the acceleration when $t = 2$.

4. The electric power (in W) produced by a certain source is given by $P = \frac{144r}{(r + 0.6)^2}$, where r is the resistance (in Ω) in the circuit. For what value of r is the power a maximum?

5. Find the root of the equation $x^2 - \sqrt{4x + 1} = 0$ between 1 and 2 to four decimal places by use of Newton's method. Use $x_1 = 1.5$ and find x_3.

6. Linearize the function $y = \sqrt{2x + 4}$ for $a = 6$.

7. Sketch the graph of $y = x^3 + 6x^2$ by finding the values of x for which the function is increasing, decreasing, concave up, and concave down and by finding any maximum points, minimum points, and points of inflection.

8. Sketch the graph of $y = \frac{4}{x^2} - x$ by finding the same information as required in Problem 7, as well as intercepts, symmetry, behavior as x becomes large, vertical asymptotes, and the domain and range.

9. Trash is being compacted into a cubical volume. The edge of the cube is decreasing at the rate of 0.50 ft/s. When an edge of the cube is 4.00 ft, how fast is the volume changing?

10. A rectangular field is to be fenced and then divided in half by a fence parallel to two opposite sides. If a total of 6000 m of fencing is used, what is the maximum area that can be fenced?

25

Integration

LEARNING OUTCOMES

After completion of this chapter, the student should be able to:

- Understand the concept of the antiderivative of a function
- Integrate basic functions such as constants, polynomials, and powers
- Evaluate the constant of integration
- Approximate the area under a curve using inscribed rectangles
- Evaluate a definite integral
- Find the area under a curve using a definite integral
- Use the trapezoidal rule and Simpson's rule to approximate a definite integral

Finding areas of geometric figures had been studied by the ancient Greeks, and they had found how to find the area of any polygon. Also, they had found that they could find areas of a curved figure by inscribing polygons in the figure and then letting the number of sides of the polygon increase. There was little more progress in finding such areas until the 1600s when analytic geometry was developed.

Several mathematicians of the 1600s studied both the area problem and the tangent problem that was discussed in Chapter 23. These included the French mathematician Pierre de Fermat and the English mathematician Isaac Barrow, both of whom developed a few formulas by which tangents and areas could be found. However, Newton and Leibniz found that these two problems were related and determined general methods of solving them. For these reasons, Newton and Leibniz are credited with the creation of calculus.

Finding tangents and finding areas appear to be very different, but they are closely related. As we will show, finding an area uses the inverse process of finding the slope of a tangent line, which we have shown can be interpreted as an instantaneous rate of change.

In physical and technical applications, we often find information that gives us the instantaneous rate of change of a variable. With such information, we have to reverse the process of differentiation in order to find the function when we know its derivative. This procedure is known as *integration*, which is the inverse process of differentiation.

This means that areas are found by integration. There are also many applications of integration in science and technology. A few of these applications will be illustrated in this chapter, and several specific applications will be developed in the next chapter.

▶ In Section 25.4, we will show how integration can be used to find the distance traveled by a light-rail train while coming to a stop.

25.1 Antiderivatives

Antidifferentiation • Antiderivative

We now show how to reverse the process of finding a derivative or a differential. *This reverse process is known as* **antidifferentiation.** In the next section, we formalize the process—it is only the basic idea that is the topic of this section.

EXAMPLE 1 Derivative known—find function

Find a function for which the derivative is $8x^3$. That is, find an antiderivative of $8x^3$.

When finding the derivative of a constant times a power of x, we multiply the constant coefficient by the power of x and reduce the power by 1. Therefore, in this case, the power of x must have been 4 before the differentiation was performed.

If we let the derivative function be $f(x) = 8x^3$ and then let its antiderivative function be $F(x) = ax^4$ [by increasing the power of x in $f(x)$ by 1], we can find the value of a by equating the derivative of $F(x)$ to $f(x)$. This gives us

$$F'(x) = 4ax^3 = 8x^3, \qquad 4a = 8, \qquad a = 2$$

This means that $F(x) = 2x^4$. ■

EXAMPLE 2 Antiderivative of a polynomial

Find the antiderivative of $v^2 + 2v$.

As for the v^2, we know that the power of v required in an antiderivative is 3. Also, to make the coefficient correct, we must multiply by $\frac{1}{3}$. The $2v$ should be recognized as the derivative of v^2. Therefore, we have as an antiderivative $\frac{1}{3}v^3 + v^2$. ■

NOTE ➧ [In Examples 1 and 2, we could add any constant to the antiderivative shown, and still have a correct antiderivative. This is true because the derivative of a constant is zero.] This is discussed in the next section, and we will not show any such constants in this section.

When we find an antiderivative of a function, we obtain another function. Thus, *we can define an* **antiderivative** *of the function $f(x)$ to be a function $F(x)$ such that $F'(x) = f(x)$.*

Practice Exercise

1. Find an antiderivative of $x^3 + 4x$.

EXAMPLE 3 Antiderivative—root and negative exponent

Find an antiderivative of the function $f(x) = \sqrt{x} - \dfrac{2}{x^3}$.

Because we wish to find an antiderivative of $f(x)$, we know that $f(x)$ is the derivative of the required function.

Considering the term $\sqrt{x}$, we first write it as $x^{1/2}$. To have x to the $\frac{1}{2}$ power in the derivative, we must have x to the $\frac{3}{2}$ power in the antiderivative. Knowing that the derivative of $x^{3/2}$ is $\frac{3}{2}x^{1/2}$, we write $x^{1/2}$ as $\frac{2}{3}(\frac{3}{2}x^{1/2})$. Thus, the first term of the antiderivative is $\frac{2}{3}x^{3/2}$.

As for the term $-2/x^3$, we write it as $-2x^{-3}$. This we recognize as the derivative of x^{-2}, or $1/x^2$.

This means that an antiderivative of the function

function $\qquad f(x) = \sqrt{x} - \dfrac{2}{x^3}$ ⟵ derivative

is the function

antiderivative ⟶ $F(x) = \dfrac{2}{3}x^{3/2} + \dfrac{1}{x^2}$ $\qquad$ function ■

A great many functions of which we must find an antiderivative are not polynomials or simple powers of x. It is these functions that may cause more difficulty in the general process of antidifferentiation. Pay special attention to the following examples, for they illustrate a type of problem that you will find to be very important.

EXAMPLE 4 Antiderivative of a power of a function

Find an antiderivative of the function $f(x) = 3(x^3 - 1)^2(3x^2)$.

Noting that we have a power of $x^3 - 1$ in the derivative, it is reasonable that the antiderivative may include a power of $x^3 - 1$. Because, in the derivative, $x^3 - 1$ is raised to the power 2, the antiderivative would then have $x^3 - 1$ raised to the power 3. Noting that the derivative of $(x^3 - 1)^3$ is $3(x^3 - 1)^2(3x^2)$, the desired antiderivative is

$$F(x) = (x^3 - 1)^3$$

Practice Exercise

2. Find an antiderivative of $5(4x - 3)^4(4)$.

CAUTION We note that the factor $3x^2$ does not appear in the antiderivative. ■ It is included when finding the derivative. Therefore, it must be present for $(x^3 - 1)^3$ to be a proper antiderivative, but must be excluded when finding an antiderivative. ■

EXAMPLE 5 Antiderivative of a power of a function

Find an antiderivative of the function $f(x) = (2x + 1)^{1/2}$.

Here, we note a power of $2x + 1$ in the derivative, which infers that the antiderivative has a power of $2x + 1$. Because, in finding a derivative, 1 is subtracted from the power of $2x + 1$, we should add 1 in finding the antiderivative. Thus, we should have $(2x + 1)^{3/2}$ as part of the antiderivative. Finding the derivative of $(2x + 1)^{3/2}$, we obtain $\frac{3}{2}(2x + 1)^{1/2}(2) = 3(2x + 1)^{1/2}$. This differs from the given derivative by the factor of 3. Thus, if we write $(2x + 1)^{1/2} = \frac{1}{3}[3(2x + 1)^{1/2}]$, we have the required antiderivative as

$$F(x) = \frac{1}{3}(2x + 1)^{3/2}$$

Checking, the derivative of $\frac{1}{3}(2x + 1)^{3/2}$ is $\frac{1}{3}\left(\frac{3}{2}\right)(2x + 1)^{1/2}(2) = (2x + 1)^{1/2}$. ■

EXERCISES 25.1

In Exercises 1–4, make the given changes in the indicated examples of this section and then solve the resulting problems.

1. In Example 1, change the coefficient 8 to 12.

2. In Example 2, change v^2 to v^3.

3. In Example 3, change $2/x^3$ to $3/x^4$.

4. In Example 5, change $2x$ to $4x$.

In Exercises 5–12, determine the value of a that makes F(x) an antiderivative of f(x).

5. $f(x) = 3x^2, F(x) = ax^3$

6. $f(x) = 5x^4, F(x) = ax^5$

7. $f(x) = 18x^5, F(x) = ax^6$

8. $f(x) = 40x^7, F(x) = ax^8$

9. $f(x) = 9\sqrt{x}, F(x) = ax^{3/2}$

10. $f(x) = 10x^{1/4}, F(x) = ax^{5/4}$

11. $f(x) = \frac{1}{x^2}, F(x) = \frac{a}{x}$

12. $f(x) = \frac{6}{x^4}, F(x) = \frac{a}{x^3}$

In Exercises 13–40, find antiderivatives of the given functions.

13. $f(x) = 5x^2$

14. $f(x) = 2x^3$

15. $f(t) = 6t^3 + 12$

16. $f(x) = 12x^5 + 6x$

17. $f(x) = 2x^{3/2} - 3x$

18. $f(x) = 6x^2 - 4x^{1/3}$

19. $f(x) = 4\sqrt{x} + 3$

20. $f(s) = 9\sqrt[3]{s} - 3s$

21. $f(x) = -\frac{7}{x^6} + \frac{1}{3^2}$

22. $f(x) = \frac{8}{x^5} - \pi$

23. $f(v) = 4v^4 + 3\pi^2$

24. $f(x) = \frac{1}{4\sqrt{x}} + \sqrt{3}$

25. $f(x) = 50x^{99} - 39x^{-79}$

26. $f(x) = 10x^{1/10} - 40x^{-19/20}$

27. $f(x) = x^2 - 4 + x^{-2}$

28. $f(x) = x\sqrt{x} - x^{-3}$

29. $f(x) = 6(2x + 1)^5(2)$

30. $f(R) = 3(R^2 + 1)^2(2R)$

31. $f(p) = 4(p^2 - 1)^3(2p)$

32. $f(x) = 5(2x^4 + 1)^4(8x^3)$

33. $f(x) = 4x^3(2x^4 + 1)^4$

34. $f(x) = 6x(1 - x^2)^7$

35. $f(x) = \frac{3}{2}(6x + 1)^{1/2}(6)$

36. $f(y) = \frac{5}{4}(1 - y)^{1/4}(-1)$

37. $f(x) = (3x + 1)^{1/3}$

38. $f(x) = (4x + 3)^{1/2}$

39. $f(x) = \frac{-12}{(2x + 1)^2}$

40. $f(s) = \frac{4s}{(1 - s^2)^3}$

In Exercises 41 and 42, answer the given questions.

41. Why is $(x + 5)^3$ a correct antiderivative of $3(x + 5)^2$, whereas $(2x + 5)^3$ is not a correct antiderivative of $3(2x + 5)^2$?

42. Is $\frac{1}{(x + 5)^3}$ a correct antiderivative of $\frac{1}{3(x + 5)^2}$?

Answers to Practice Exercises

1. $\frac{1}{4}x^4 + 2x^2$ **2.** $F(x) = (4x - 3)^5$

25.2 The Indefinite Integral

Indefinite Integral • Integration • Evaluating Constant of Integration

In the previous section, in developing the basic technique of finding an antiderivative, we noted that the results given are not unique. That is, we could have added any constant to the answers and the result would still have been correct. Again, this is the case because the derivative of a constant is zero.

EXAMPLE 1 Many antiderivatives for a given derivative

The derivatives of x^3, $x^3 + 4$, $x^3 - 7$, and $x^3 + 4\pi$ are all $3x^2$. This means that any of the functions listed, as well as innumerable others, would be a proper answer to the problem of finding an antiderivative of $3x^2$. ■

From Section 24.8, we know that the differential of a function $F(x)$ can be written as $d[F(x)] = F'(x)dx$. Therefore, because finding a differential of a function is closely related to finding the derivative, so is the antiderivative closely related to the process of finding the function for which the differential is known.

The notation used for finding the general form of the antiderivative, the **indefinite integral,** *is written in terms of the differential.* Thus, *the indefinite integral of a function* $f(x)$, *for which* $dF(x)/dx = f(x)$, *or* $dF(x) = f(x)dx$, *is defined as*

$$\int f(x)dx = F(x) + C \tag{25.1}$$

Here, $f(x)$ *is called the* **integrand,** $F(x) + C$ *is the indefinite integral, and C is an arbitrary constant, called the* **constant of integration.** It represents any of the constants that may be attached to an antiderivative to have a proper result. We must have additional information beyond a knowledge of the differential to assign a specific value to C. *The symbol* $\int$ *is the* **integral sign,** *and it indicates that the inverse of the differential is to be found. Determining the indefinite integral is called* **integration,** which we can see is essentially the same as finding an antiderivative.

EXAMPLE 2 Indefinite integral of a polynomial

In performing the integration

$$\int 5x^4\, dx = x^5 + C$$

constant of integration (points to C); integrand (points to $5x^4$); indefinite integral (points to $x^5 + C$)

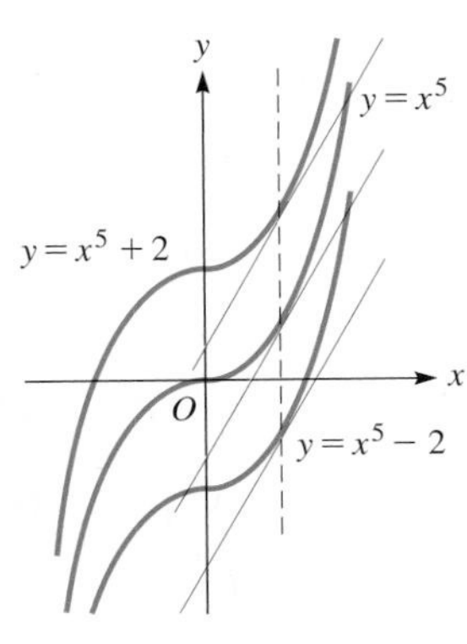

Fig. 25.1

we might think that the inclusion of this constant C would affect the derivative of the function x^5. However, the only effect of the C is to raise or lower the curve. The slope of $x^5 + 2$, $x^5 - 2$, or any function of the form $x^5 + C$, is the same for any given value of x. As Fig. 25.1 shows, tangents drawn to the curves are all parallel for the same value of x. ■

At this point, we shall derive some basic formulas for integration. Because we know $d(cu)/dx = c(du/dx)$, where u is a function of x and c is a constant, we can write

$$\int c\, du = c \int du = cu + C \tag{25.2}$$

Also, because the derivative of a sum of functions equals the sum of the derivatives, we write

$$\int (du + dv) = u + v + C \tag{25.3}$$

NOTE ▸ To find the differential of a power of a function, we multiply by the power, subtract 1 from it, and multiply by the differential of the function. [To find the integral, we reverse this procedure to get the **power rule for integration:**]

Power Rule for Integration

$$\int u^n\, du = \frac{u^{n+1}}{n+1} + C \qquad (n \neq -1) \tag{25.4}$$

We must be able to recognize the proper form and the component parts to use these formulas. Unless you do this and have a good knowledge of differentiation, you will have trouble using Eq. (25.4). Most of the difficulty, if it exists, arises from an improper identification of du.

EXAMPLE 3 Integration—power rule

Integrate: $\int 6x\, dx$.

We must identify u, n, du, and any multiplying constants. Noting that 6 is a multiplying constant, we identify x as u, which means dx must be du and $n = 1$.

$$\int 6x\, dx = 6\int x^1\, dx = 6\left(\frac{x^2}{2}\right) + C = 3x^2 + C$$

(Annotations: $\int u^n\, du$ points to $\int x^1\, dx$; u^{n+1} points to x^2; $n + 1$ points to 2; "do not forget the constant of integration" points to C.)

We see that our result checks since the differential of $3x^2 + C$ is $6x\, dx$. ■

EXAMPLE 4 Integration of a polynomial—power rule

Integrate: $\int (5x^3 - 6x^2 + 1)dx$.

Here, we must use a combination of Eqs. (25.2), (25.3), and (25.4). Therefore,

$$\begin{aligned}\int (5x^3 - 6x^2 + 1)dx &= \int 5x^3\, dx + \int (-6x^2)dx + \int dx \\ &= 5\int x^3\, dx - 6\int x^2\, dx + \int dx\end{aligned}$$

In the first integral, $u = x$, $n = 3$, and $du = dx$. In the second, $u = x$, $n = 2$, and $du = dx$. The third uses Eq. (25.2) directly, with $c = 1$ and $du = dx$. This means

$$\begin{aligned}5\int x^3\, dx - 6\int x^2\, dx + \int dx &= 5\left(\frac{x^4}{4}\right) - 6\left(\frac{x^3}{3}\right) + x + C \\ &= \frac{5}{4}x^4 - 2x^3 + x + C\end{aligned}$$

■

Practice Exercise

1. Integrate: $\int (6x^2 - 5)dx$.

EXAMPLE 5 Integration—root and negative exponent

Integrate: $\int\left(\sqrt{r} - \frac{1}{r^3}\right)dr$.

In order to use Eq. (25.4), we must first write $\sqrt{r} = r^{1/2}$ and $1/r^3 = r^{-3}$:

$$\int\left(\sqrt{r} - \frac{1}{r^3}\right)dr = \int r^{1/2}\,dr - \int r^{-3}\,dr = \frac{1}{\frac{3}{2}}r^{3/2} - \frac{1}{-2}r^{-2} + C$$

$$= \frac{2}{3}r^{3/2} + \frac{1}{2}r^{-2} + C = \frac{2}{3}r^{3/2} + \frac{1}{2r^2} + C$$ ■

EXAMPLE 6 Integration—power of a function

Integrate: $\int(x^2 + 1)^3(2x\,dx)$.

We first note that $n = 3$, for this is the power involved in the function being integrated. If $n = 3$, then $x^2 + 1$ must be u. If $u = x^2 + 1$, then $du = 2x\,dx$. Thus, the integral is in proper form for integration *as it stands*. Using the power rule for integration,

■ Again, we note that a good knowledge of differential forms is essential for the proper recognition of u and du.

$$\int(x^2 + 1)^3(2x\,dx) = \frac{(x^2 + 1)^4}{4} + C$$

CAUTION It must be emphasized that the entire quantity $(2x\,dx)$ must be equated to du. ■ Normally, u and n are recognized first, and then du is derived from u.

Showing the use of u directly, we can write the integration as

■ Other methods for integrating are discussed in Chapter 28. Also, many functions cannot be integrated algebraically.

$$\int(x^2 + 1)^3(2x\,dx) = \int u^3\,du = \frac{1}{4}u^4 + C = \frac{(x^2 + 1)^4}{4} + C$$ ■

EXAMPLE 7 Integration—power of a function

Integrate: $\int x^2\sqrt{x^3 + 2}\,dx$.

We first note that $n = \frac{1}{2}$ and u is then $x^3 + 2$. Because $u = x^3 + 2$, $du = 3x^2\,dx$. Now, we group $3x^2\,dx$ as du. Because there is no 3 in the integrand, we introduce one. In order not to change the numerical value, we also introduce $\frac{1}{3}$, normally before the integral sign. CAUTION In this way, we take advantage of the fact that a *constant* (and only a constant) factor may be moved across the integral sign. ■

$$\int x^2\sqrt{x^3 + 2}\,dx = \frac{1}{3}\int 3x^2\sqrt{x^3 + 2}\,dx = \frac{1}{3}\int\sqrt{x^3 + 2}\,(3x^2\,dx)$$

This is now in the proper form to use the power rule for integration. Rewriting the above integral in terms of u, we have

$$\frac{1}{3}\int\sqrt{x^3 + 2}\,(3x^2\,dx) = \frac{1}{3}\int\sqrt{u}\,du \qquad u = x^3 + 2,\ du = 3x^2dx$$

$$= \frac{1}{3}\int u^{1/2}\,du \qquad \sqrt{u} = u^{1/2}$$

$$= \frac{1}{3}\left(\frac{u^{3/2}}{3/2}\right) + C \qquad \text{integrate using } \int u^n\,du = \frac{u^{n+1}}{n+1} + C$$

$$= \frac{1}{3}\left(\frac{2}{3}\right)u^{3/2} + C \qquad \text{simplify}$$

$$= \frac{2}{9}(x^3 + 2)^{3/2} + C \qquad \text{back substitute: } u = x^3 + 2$$ ■

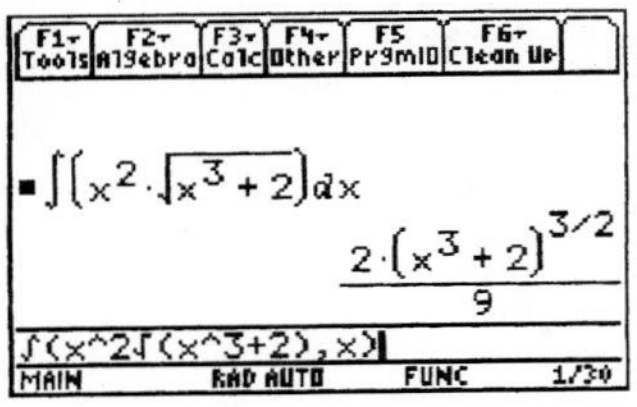

F1 Tools | F2 Algebra | F3 Calc | F4 Other | F5 PrgmIO | F6 Clean Up

$\blacksquare\ \int(x^2\cdot\sqrt{x^3+2})dx \qquad \frac{2\cdot(x^3+2)^{3/2}}{9}$

∫(x^2√(x^3+2),x)

MAIN RAD AUTO FUNC 1/30

Fig. 25.2
TI-89 graphing calculator keystrokes: goo.gl/9jntSd

■ A TI-89 calculator evaluation of this integral is shown in Fig. 25.2.

Practice Exercise

2. Integrate: $\int 8x\sqrt{1 - 2x^2}\,dx$.

EVALUATING THE CONSTANT OF INTEGRATION

To find the constant of integration, we need information such as a set of values that satisfy the function. A point through which the curve passes would provide the necessary information. This is illustrated in the following examples.

EXAMPLE 8 Evaluating the constant of integration

Find y in terms of x, given that $dy/dx = 3x - 1$ and the curve passes through $(1, 4)$.

The solution is as follows.

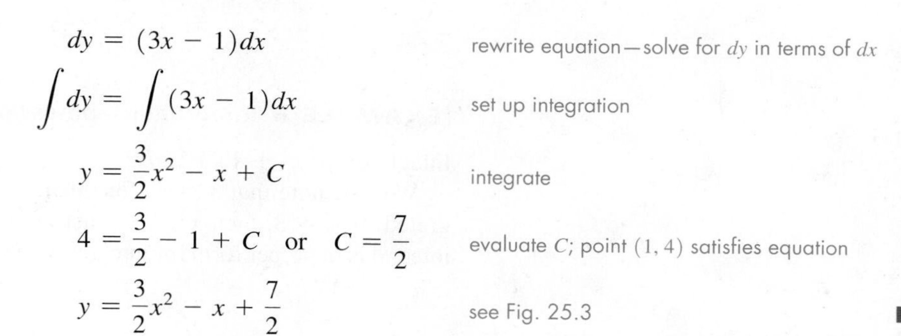

$$dy = (3x - 1)dx \qquad \text{rewrite equation—solve for } dy \text{ in terms of } dx$$

$$\int dy = \int (3x - 1)dx \qquad \text{set up integration}$$

$$y = \frac{3}{2}x^2 - x + C \qquad \text{integrate}$$

$$4 = \frac{3}{2} - 1 + C \quad \text{or} \quad C = \frac{7}{2} \qquad \text{evaluate } C\text{; point } (1, 4) \text{ satisfies equation}$$

$$y = \frac{3}{2}x^2 - x + \frac{7}{2} \qquad \text{see Fig. 25.3}$$

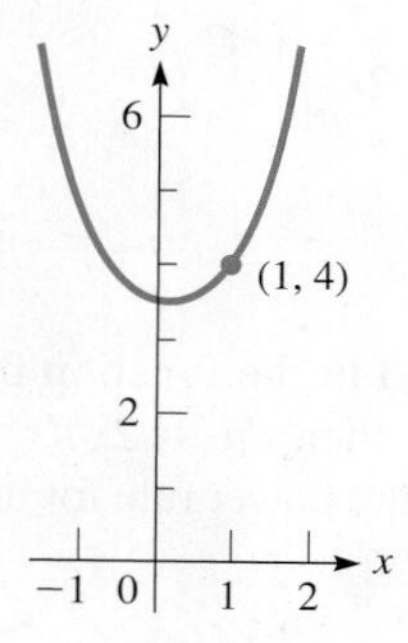

Fig. 25.3

EXAMPLE 9 Integral with negative exponent—robot arm displacement

The time rate of change of the displacement (velocity) of a robot arm is $ds/dt = 8t/(t^2 + 4)^2$. Find the expression for the displacement as a function of time if $s = -1$ m when $t = 0$ s.

First, we write $ds = \frac{8t\,dt}{(t^2+4)^2}$ and then integrate. To integrate the expression on the right, we note that $n = -2$, $u = t^2 + 4$, and $du = 2t\,dt$. This means we need a 2 with $t\,dt$ to form the proper du. In turn, this means we place a $\frac{1}{2}$ before the integral sign. Also, we place the 8 in front of the integral sign. Therefore,

$$\int ds = \int \frac{8t\,dt}{(t^2 + 4)^2} = \left(\frac{1}{2}\right)(8)\int (t^2 + 4)^{-2}(2t\,dt) \qquad \text{set up integration}$$

$$s = 4\left(\frac{1}{-1}\right)(t^2 + 4)^{-1} + C = \frac{-4}{t^2 + 4} + C \qquad \text{integrate; } -2 + 1 = -1$$

$$-1 = \frac{-4}{0 + 4} + C \quad \text{or} \quad C = 0 \qquad \text{evaluate } C \text{ given } s = -1 \text{ m when } t = 0 \text{ s}$$

$$s = \frac{-4}{t^2 + 4} \qquad \text{expression for displacement}$$

EXERCISES 25.2

In Exercises 1–4, make the given changes in the indicated examples of this section, and then solve the resulting problems.

1. In Example 3, change the coefficient 6 to 8.
2. In Example 5, change $\sqrt{r}$ to $\sqrt[3]{r}$.
3. In Example 6, change the power 3 to 4.
4. In Example 8, change $(1, 4)$ to $(2, 3)$.

In Exercises 5–36, integrate each of the given expressions.

5. $\int 2x\,dx$
6. $\int 5x^4\,dx$
7. $\int 6x^3\,dx$
8. $\int 0.6y^5\,dy$
9. $\int 8x^{3/2}\,dx$
10. $\int 6\sqrt[3]{x}\,dx$
11. $\int 9R^{-4}\,dR$
12. $\int \frac{4}{\sqrt{x}}dx$
13. $\int (x^7 - 3x^5)dx$

14. $\int (3 - 8x)dx$

15. $\int (9x^2 + x + 3^{-1})dx$

16. $\int x(x - 2)^2 dx$

17. $\int \left(\frac{t^2}{2} - \frac{2}{t^2}\right)dt$

18. $\int \frac{3x^2 - 4}{x^2}dx$

19. $\int \sqrt{x}(x^2 - 5x)dx$

20. $\int (3R\sqrt{R} - 5R^2)dR$

21. $\int (2x^{-2/3} + 9)dx$

22. $\int (x^{1/3} + x^{1/5} + x^{-1/7})dx$

23. $\int (1 + 2x^2)^2 dx$

24. $\int (x^2 + 4x + 4)^{1/3}\, dx$

25. $\int (x^2 - 1)^5(2x\, dx)$

26. $\int (t^3 - 2)^6(3t^2\, dt)$

27. $\int (x^4 + 3)^4(8x^3\, dx)$

28. $\int (1 - 2x)^{1/3}(4\, dx)$

29. $\int 40(2\theta^5 + 5)^7\theta^4\, d\theta$

30. $\int 6x^2(1 - x^3)^{4/3}\, dx$

31. $\int \sqrt{8x + 1}\, dx$

32. $\int \frac{8\, dV}{(0.3 + 2V)^3}$

33. $\int \frac{4x\, dx}{\sqrt{6x^2 + 1}}$

34. $\int \frac{2x^2\, dx}{\sqrt{2x^3 + 1}}$

35. $\int \frac{4z - 4}{\sqrt{z^2 - 2z}}dz$

36. $\int (x^2 - x)\left(x^3 - \frac{3}{2}x^2\right)^8 dx$

In Exercises 37–40, find y in terms of x.

37. $\frac{dy}{dx} = 6x^2$, curve passes through $(0, 2)$

38. $\frac{dy}{dx} = 8x + 1$, curve passes through $(-1, 4)$

39. $\frac{dy}{dx} = x^2(1 - x^3)^5$, curve passes through $(1, 5)$

40. $\frac{dy}{dx} = 2x^3(x^4 - 6)^4$, curve passes through $(2, 10)$

In Exercises 41–62, solve the given problems. In Exercises 41–46, explain your answers.

41. Is $\int 3x^2\, dx = x^3$?

42. Can $\int (x^2 - 1)^2\, dx$ be integrated with $u = x^2 - 1$?

43. Can $\int (4x^3 + 3)^5 x^4\, dx$ be integrated with $u = 4x^3 + 3$ and $du = x^4\, dx$?

44. Is $\int \sqrt{2x + 1}\, dx = \frac{2}{3}(2x + 1)^{3/2}$?

45. Is $\int 3(2x + 1)^2\, dx = (2x + 1)^3 + C$?

46. Is $\int x^{-2}\, dx = -\frac{1}{3}x^{-3} + C$?

47. Find an equation of the curve whose slope is $-x\sqrt{1 - 4x^2}$ and that passes through $(0, 7)$.

48. Find an equation of the curve whose slope is $\sqrt{6x - 3}$ and that passes through $(2, -1)$.

49. Find $f(x)$ if $f'(x) = 4x - 5$ and $f(-1) = 10$.

50. Find $f(x)$ if $f'(x) = 2/\sqrt{x}$ and $f(9) = 8$.

51. Find the general form of the function whose second derivative is $\sqrt{x}$.

52. Find the general form of the expression for the displacement s of an object if its acceleration is 9.8 m/s^2.

53. The velocity of a motorcycle driving on a straight highway is given by $v = \frac{ds}{dt} = 10 + t$ (in ft/s), where t is in seconds. Find an expression for the displacement s if $s = 0$ when $t = 0$.

54. The radius r (in ft) of a circular oil spill is increasing at the rate given by $\frac{dr}{dt} = \frac{10}{\sqrt{4t + 1}}$, where t is in minutes. Find the radius as a function of t, if t is measured form the time of the spill.

55. The rate of change of electric current i in a microprocessor circuit is given by $di/dt = 0.4t - 0.06t^2$. Find the expression for $i = f(t)$ if $i = 2\ \mu A$ when $t = 0$ s.

56. The rate of change of the frequency f of an electronic oscillator with respect to the inductance L is $df/dL = 80(4 + L)^{-3/2}$. Find f as a function of L if $f = 80$ Hz for $L = 0$ H.

57. The rate of change of the temperature T (in °C) from the center of a blast furnace to a distance r (in m) from the center is given by $dT/dr = -4500(r + 1)^{-3}$. Express T as a function of r if $T = 2500$°C for $r = 0$.

58. The rate of change of current i (in mA) in a circuit with a variable inductance is given by $di/dt = 300(5.0 - t)^{-2}$, where t (in ms) is the time since the circuit is closed. Find i as a function of t if $i = 300$ mA for $t - 2.0$ ms.

59. At a given site, the rate of change of the annual fraction f of energy supplied by solar energy with respect to the solar-collector area A (in m^2) is $\frac{df}{dA} = \frac{0.005}{\sqrt{0.01A + 1}}$. Find f as a function of A if $f = 0$ for $A = 0$ m^2.

60. An analysis of a company's records shows that in a day the rate of change of profit P (in dollars) in producing x generators is $\frac{dP}{dx} = \frac{600(30 - x)}{\sqrt{60x - x^2}}$. Find the profit in producing x generators if a loss of \$5000 is incurred if none are produced.

61. Find an equation of the curve for which the second derivative is 6. The curve passes through $(1, 2)$ with a slope of 8.

62. The second derivative of a function is $12x^2$. Explain how to find the function if its curve passes through the points $(1, 6)$ and $(2, 21)$. Find the function.

Answers to Practice Exercises

1. $2x^3 - 5x + C$ 2. $-\frac{4}{3}(1 - 2x^2)^{3/2} + C$

25.3 The Area Under a Curve

Sum Areas of Inscribed Rectangles • Limit of Areas of Inscribed Rectangles • Area Under Curve by Integration

In geometry, there are formulas and methods for finding the areas of regular figures. By means of integration, it is possible to find the area between curves for which we know the equations. The next example illustrates the basic idea behind the method.

EXAMPLE 1 Sum areas of inscribed rectangles

Approximate the area in the first quadrant to the left of the line $x = 4$ and under the parabola $y = x^2 + 1$. Here, "under" means between the curve and the x-axis. First, make this approximation by inscribing two rectangles of equal width under the parabola and finding the sum of the areas of these rectangles. Then, improve the approximation by repeating the process with eight inscribed rectangles.

The area to be approximated is shown in Fig. 25.4(a). The area with two rectangles inscribed under the curve is shown in Fig. 25.4(b). The first approximation of the area, admittedly small, can be found by adding the areas of the two rectangles. Both rectangles have a width of 2. The left rectangle is 1 unit high, and the right rectangle is 5 units high. Thus, the area of the two rectangles is

Practice Exercise

1. In Example 1, approximate the area by inscribing four rectangles.

$$A = 2(1 + 5) = 12$$

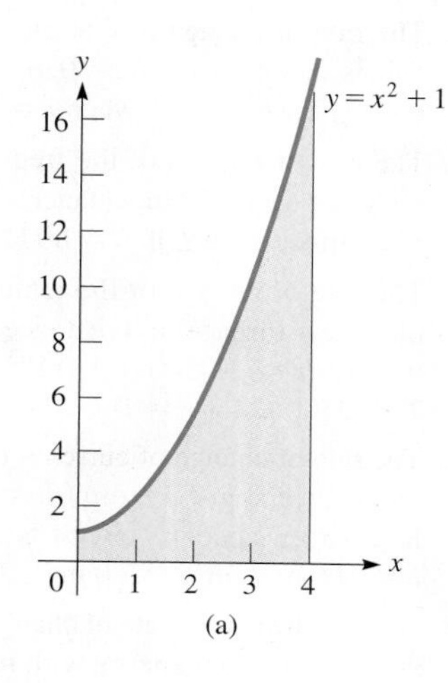

(a)

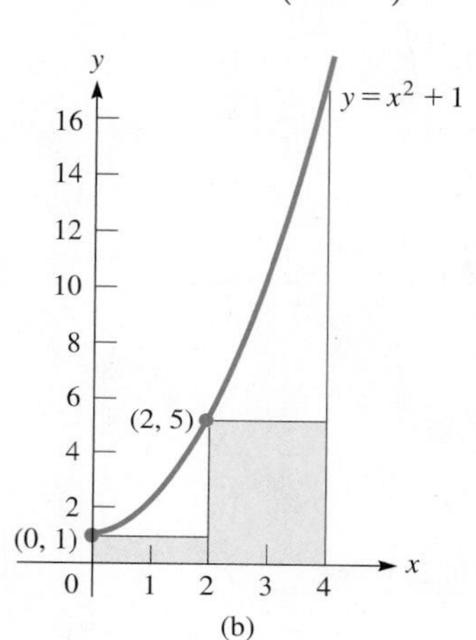

(b)

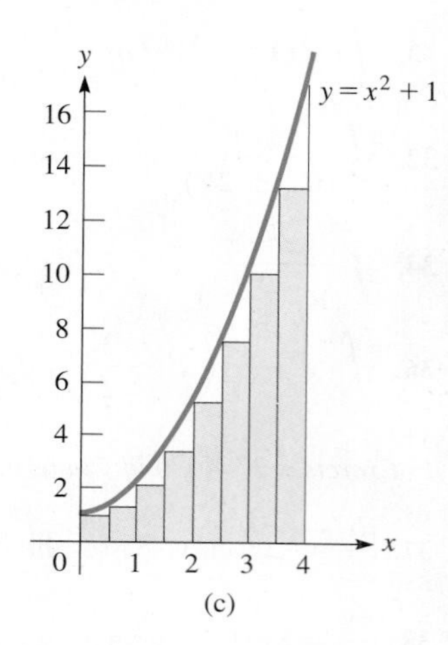

(c)

Fig. 25.4

Table 25.1

Number of Rectangles n	*Total Area of Rectangles*
8	21.5
100	25.0144
1,000	25.301344
10,000	25.330133

A much better approximation is found by inscribing eight rectangles as shown in Fig. 25.4(c). Each of these rectangles has a width of $\frac{1}{2}$. The leftmost rectangle has a height of 1. The next has a height of $\frac{5}{4}$, which is determined by finding y for $x = \frac{1}{2}$. The next rectangle has a height of 2, which is found by evaluating y for $x = 1$. Finding the heights of all rectangles and multiplying their sum by $\frac{1}{2}$ gives the area of the eight rectangles as

$$A = \frac{1}{2}\left(1 + \frac{5}{4} + 2 + \frac{13}{4} + 5 + \frac{29}{4} + 10 + \frac{53}{4}\right) = \frac{43}{2} = 21.5$$

An even better approximation could be obtained by inscribing more rectangles under the curve. The greater the number of rectangles, the more nearly the sum of their areas equals the area under the curve. See Table 25.1. By using integration later in this section, we determine the *exact* area to be $\frac{76}{3} = 25\frac{1}{3}$. ■

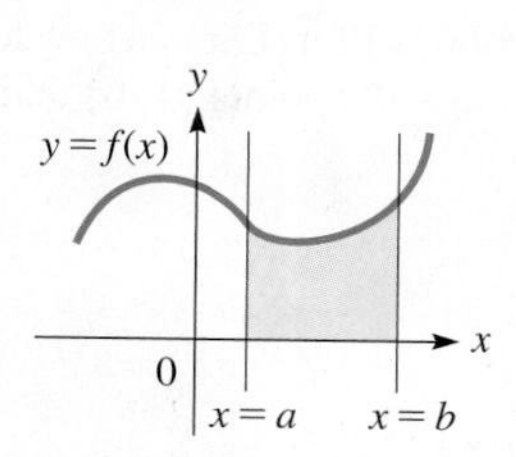

Fig. 25.5

We now develop the basic method used to find the area under a curve, which is the area bounded by the curve, the x-axis, and the lines $x = a$ and $x = b$. See Fig. 25.5. We assume here that $f(x)$ is never negative in the interval $a < x < b$. In Chapter 26, we will extend the method such that $f(x)$ may be negative.

In finding the area under a curve, we consider the sum of the areas of inscribed rectangles, as the number of rectangles is assumed to increase without bound. The reason for this last condition is that, as we saw in Example 1, as the number of rectangles increases, the approximation of the area is better.

EXAMPLE 2 Number of rectangles approaches infinity

Find the area under the straight line $y = 2x$, above the x-axis, and to the left of the line $x = 4$.

NOTE ➧ Because this figure is a right triangle, the area can easily be found. [However, the *method* we use here is the important concept.] We first subdivide the interval from $x = 0$ to $x = 4$ into n inscribed rectangles of Δx in width. The extremities of the intervals are labeled $a, x_1, x_2, \ldots, b(= x_n)$, as shown in Fig. 25.6, where

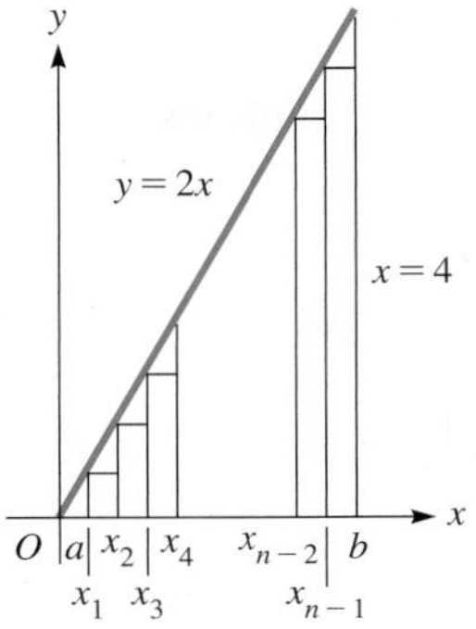

$$x_1 = \Delta x \qquad x_2 = 2\Delta x, \ldots \qquad x_{n-1} = (n-1)\Delta x \qquad b = n\Delta x$$

Fig. 25.6

The area of each of these n rectangles is as follows:

First: $f(a)\Delta x$, where $f(a) = f(0) = 2(0) = 0$ is the height.
Second: $f(x_1)\Delta x$, where $f(x_1) = 2(\Delta x) = 2\Delta x$ is the height.
Third: $f(x_2)\Delta x$, where $f(x_2) = 2(2\Delta x) = 4\Delta x$ is the height.
Fourth: $f(x_3)\Delta x$, where $f(x_3) = 2(3\Delta x) = 6\Delta x$ is the height.
⋮
Last: $f(x_{x-1})\Delta x$, where $f[(n-1)\Delta x] = 2(n-1)\Delta x$ is the height.

These areas are summed up as follows:

$$A_n = f(a)\Delta x + f(x_1)\Delta x + f(x_2)\Delta x + \cdots + f(x_{n-1})\Delta x \qquad (25.5)$$

$$= 0 + 2\Delta x(\Delta x) + 4\Delta x(\Delta x) + \cdots + 2[(n-1)\Delta x]\Delta x$$

$$= 2(\Delta x)^2[1 + 2 + 3 + \cdots + (n-1)]$$

Now, $b = n\Delta x$, or $4 = n\Delta x$, or $\Delta x = 4/n$. Thus,

$$A_n = 2\left(\frac{4}{n}\right)^2[1 + 2 + 3 + \cdots + (n-1)]$$

The sum of the arithmetic sequence $1 + 2 + 3 + \cdots + n - 1$ is

$$s = \frac{n-1}{2}(1 + n - 1) = \frac{n(n-1)}{2} = \frac{n^2 - n}{2}$$

Now, the expression for the sum of the areas can be written as

$$A_n = \frac{32}{n^2}\left(\frac{n^2 - n}{2}\right) = 16\left(1 - \frac{1}{n}\right)$$

This expression is an approximation of the actual area under consideration. The larger n becomes, the better the approximation. If we let $n \to \infty$ (which is equivalent to letting $\Delta x \to 0$), the limit of this sum will equal the area in question.

$$A = \lim_{n\to\infty} 16\left(1 - \frac{1}{n}\right) = 16 \qquad 1/n \to 0 \text{ as } n \to \infty$$

■ This checks with the geometric result.

NOTE ➧ [The area under the curve is the limit of the sum of the areas of the inscribed rectangles, as the number of rectangles approaches infinity.] ■

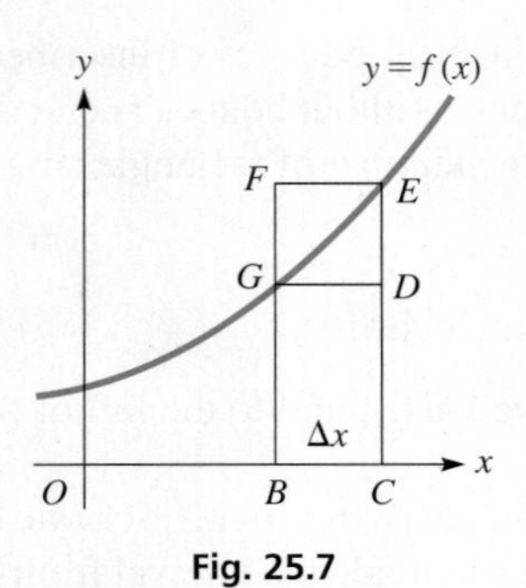

Fig. 25.7

The method indicated in Example 2 illustrates the interpretation of finding an area as a summation process, although it should not be considered as a proof. However, we will find that integration proves to be a much more useful method for finding an area. Let us now see how integration can be used directly.

Let ΔA represent the area $BCEG$ under the curve, as indicated in Fig. 25.7. We see that the following inequality is true for the indicated areas:

$$A_{BCDG} < \Delta A < A_{BCEF}$$

If the point G is now designated as (x, y) and E as $(x + \Delta x, y + \Delta y)$, we have $y\Delta x < \Delta A < (y + \Delta y)\Delta x$. Dividing through by Δx, we have

$$y < \frac{\Delta A}{\Delta x} < y + \Delta y$$

Now, we take the limit as $\Delta x \to 0$ (Δy then approaches zero). This results in

$$\frac{dA}{dx} = y \tag{25.6}$$

This is true because the left member of the inequality is y and the right member approaches y. Also, in the definition of the derivative, Eq. (23.6), $f(x + h) - f(x)$ is equivalent to ΔA, and h is equivalent to Δx, which means

$$\lim_{\Delta x \to 0} \frac{\Delta A}{\Delta x} = \frac{dA}{dx}$$

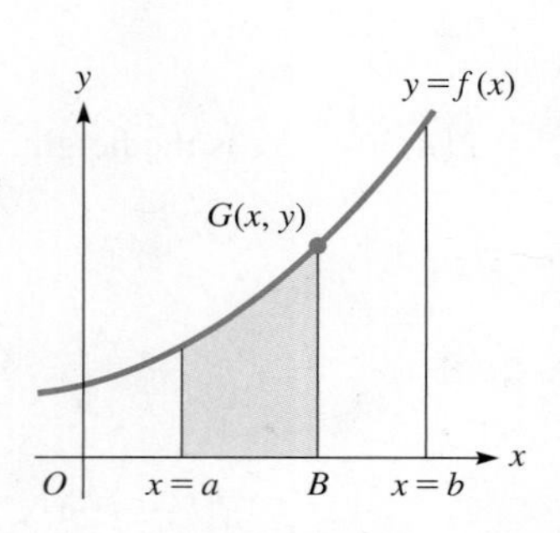

Fig. 25.8

We shall now use Eq. (25.6) to show the method of finding the complete area under a curve. We now let $x = a$ be the left boundary of the desired area and $x = b$ be the right boundary (Fig. 25.8). The area under the curve to the right of $x = a$ and bounded on the right by the line GB is now designated as $A_{a,x}$. From Eq. (25.6), we have

$$dA_{a,x} = [y\,dx]_a^x \qquad \text{or} \qquad A_{a,x} = \left[\int y\,dx\right]_a^x$$

where $[\quad]_a^x$ is the notation used to indicate the boundaries of the area. Thus,

$$A_{a,x} = \left[\int f(x)dx\right]_a^x = [F(x) + C]_a^x \tag{25.7}$$

But we know that if $x = a$, then $A_{aa} = 0$. Thus, $0 = F(a) + C$, or $C = -F(a)$. Therefore,

$$A_{a,x} = \left[\int f(x)dx\right]_a^x = F(x) - F(a) \tag{25.8}$$

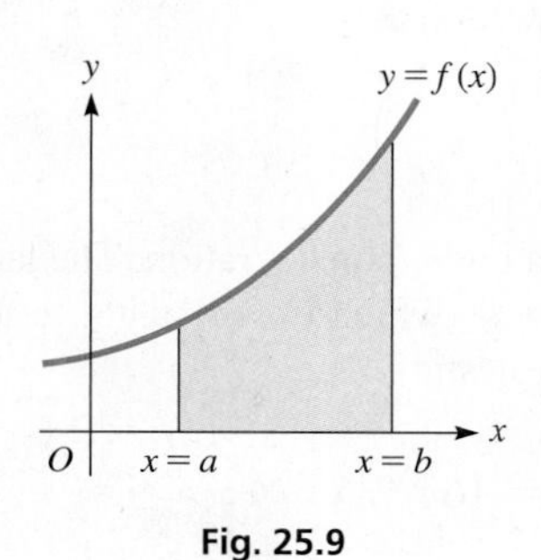

Fig. 25.9

Now, to find the area under the curve that reaches from a to b, we write

$$A_{a,b} = F(b) - F(a) \tag{25.9}$$

Thus, the area under the curve that reaches from a to b is given by

$$A_{a,b} = \left[\int f(x)dx\right]_a^b = F(b) - F(a) \tag{25.10}$$

NOTE ▶ [This shows that the area under the curve may be found by integrating the function $f(x)$ to find the function $F(x)$, which is then evaluated at each boundary value. The area is the difference between these values of $F(x)$.] See Fig. 25.9.

In Example 2, we found an area under a curve by finding the limit of the sum of the areas of the inscribed rectangles as the number of rectangles approaches infinity. Equation (25.10) expresses the area under a curve in terms of integration. We can now see that we have obtained an area by summation and also expressed it in terms of integration. Therefore, we conclude that

summations can be evaluated by integration.

Also, we have seen the connection between the problem of finding the slope of a tangent to a curve (differentiation) and the problem of finding an area under a curve (integration). We would not normally suspect that these two problems would have solutions that lead to reverse processes. We have also seen that the definition of integration has much more application than originally anticipated.

EXAMPLE 3 Area under curve by integration

Find the area under the curve of $y = x^2 + 1$ between the y-axis and the line $x = 4$. This is the same area that we illustrated in Example 1 and showed in Fig. 25.4(a). This figure is shown again here for reference.

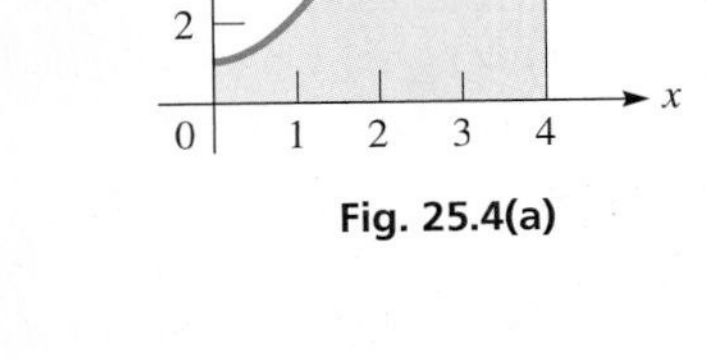

Fig. 25.4(a)

Using Eq. (25.10), we note that $f(x) = x^2 + 1$. This means that

$$\int (x^2 + 1)dx = \frac{1}{3}x^3 + x + C$$

Therefore, with $F(x) = \frac{1}{3}x^3 + x$, the area is given by

$$A_{0,4} = F(4) - F(0) \qquad \text{using Eq. (25.10)}$$

$$= \left[\frac{1}{3}(4^3) + 4\right] - \left[\frac{1}{3}(0^3) + 0\right] \qquad \text{evaluating } F(x) \text{ at } x = 4 \text{ and } x = 0$$

$$= \frac{1}{3}(64) + 4 = \frac{76}{3}$$

We note that 76/3 is a little more than 25 square units, and is therefore about 4 square units more than the value obtained using eight inscribed rectangles in Example 1. Therefore, from this result, we know that the *exact* area under the curve is $25\frac{1}{3}$, as stated at the end of Example 1. ■

Practice Exercise

2. In Example 3, change $x^2 + 1$ to $x^3 + 1$ and then find the area.

EXAMPLE 4 Area under curve by integration

Find the area under the curve $y = x^3$ that is between the lines $x = 1$ and $x = 2$.

In Eq. (25.10), $f(x) = x^3$. Therefore,

$$\int x^3\,dx = \underbrace{\frac{1}{4}x^4}_{F(x)} + C$$

$$A_{1,2} = F(2) - F(1) = \left[\frac{1}{4}(2^4)\right] - \left[\frac{1}{4}(1^4)\right] \qquad \text{using Eq. (25.10) and evaluating}$$

$$= 4 - \frac{1}{4} = \frac{15}{4}$$

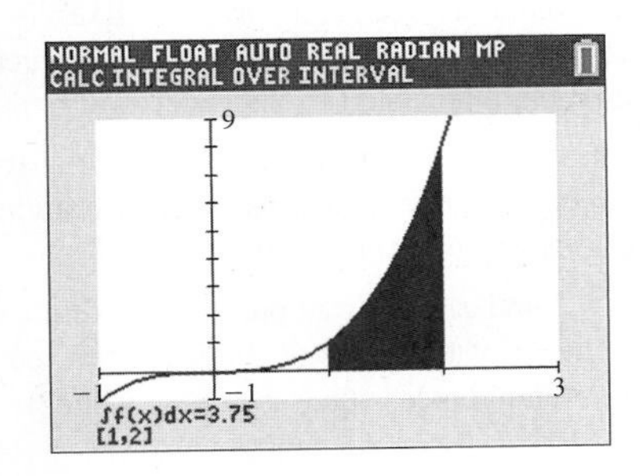

Fig. 25.10

Graphing calculator keystrokes: goo.gl/5Y788s

The calculated area of 15/4 is the exact area, not an approximation. Figure 25.10 shows a calculator check using the $\int f(x)dx$ feature. ■

Note that we do not have to include the constant of integration when finding areas. Any constant added to $F(x)$ cancels out when $F(a)$ is subtracted from $F(b)$.

EXERCISES 25.3

In Exercises 1–4, make the given changes in the indicated examples of this section and then solve the resulting problems.

1. In Example 1, change $x = 4$ to $x = 2$ and find the area of (a) two inscribed rectangles and (b) four inscribed rectangles.
2. In Example 3, change $x = 4$ to $x = 2$ and compare the results with those of Exercise 1.
3. In Example 4, change $x = 2$ to $x = 3$.
4. In Example 4, change $x = 1$ to $x = 2$ and $x = 2$ to $x = 3$. Note that the result added to the result of Example 4 is the same as the result for Exercise 3.

In Exercises 5–14, find the approximate area under the curves of the given equations by dividing the indicated intervals into n subintervals and then add up the areas of the inscribed rectangles. There are two values of n for each exercise and therefore two approximations for each area. The height of each rectangle may be found by evaluating the function for the proper value of x. See Example 1.

5. $y = 3x$, between $x = 0$ and $x = 3$, for (a) $n = 3$ $(\Delta x = 1)$, (b) $n = 10$ $(\Delta x = 0.3)$
6. $y = 2x + 1$, between $x = 0$ and $x = 2$, for (a) $n = 4$ $(\Delta x = 0.5)$, (b) $n = 10$ $(\Delta x = 0.2)$
7. $y = x^2$, between $x = 0$ and $x = 2$, for (a) $n = 5$ $(\Delta x = 0.4)$, (b) $n = 10$ $(\Delta x = 0.2)$
8. $y = 9 - x^2$, between $x = 2$ and $x = 3$, for (a) $n = 5$ $(\Delta x = 0.2)$, (b) $n = 10$ $(\Delta x = 0.1)$
9. $y = 4x - x^2$, between $x = 1$ and $x = 4$, for (a) $n = 6$, (b) $n = 10$
10. $y = 1 - x^2$, between $x = 0.5$ and $x = 1$, for (a) $n = 5$, (b) $n = 10$
11. $y = \dfrac{1}{x^2}$, between $x = 1$ and $x = 5$, for (a) $n = 4$, (b) $n = 8$
12. $y = \sqrt{x}$, between $x = 1$ and $x = 4$, for (a) $n = 3$, (b) $n = 12$
13. $y = \dfrac{1}{\sqrt{x + 1}}$, between $x = 3$ and $x = 8$, for (a) $n = 5$, (b) $n = 10$
14. $y = 2x\sqrt{x^2 + 1}$, between $x = 0$ and $x = 6$, for (a) $n = 6$, (b) $n = 12$

In Exercises 15–24, find the exact area under the given curves between the indicated values of x. The functions are the same as those for which approximate areas were found in Exercises 5–14.

15. $y = 3x$, between $x = 0$ and $x = 3$
16. $y = 2x + 1$, between $x = 0$ and $x = 2$
17. $y = x^2$, between $x = 0$ and $x = 2$
18. $y = 9 - x^2$, between $x = 2$ and $x = 3$
19. $y = 4x - x^2$, between $x = 1$ and $x = 4$
20. $y = 1 - x^2$, between $x = 0.5$ and $x = 1$
21. $y = \dfrac{1}{x^2}$, between $x = 1$ and $x = 5$
22. $y = \sqrt{x}$, between $x = 1$ and $x = 4$
23. $y = \dfrac{1}{\sqrt{x + 1}}$, between $x = 3$ and $x = 8$
24. $y = 2x\sqrt{x^2 + 1}$, between $x = 0$ and $x = 6$
Explain the reason for the difference between this result and the two values found in Exercise 14.

In developing the concept of the area under a curve, we first (in Examples 1 and 2) considered rectangles ***inscribed*** *under the curve. A more complete development also considers rectangles* ***circumscribed*** *above the curve and shows that the limiting area of the circumscribed rectangles equals the limiting area of the inscribed rectangles as the number of rectangles increases without bound. See Fig. 25.11 for an illustration of inscribed and circumscribed rectangles.*

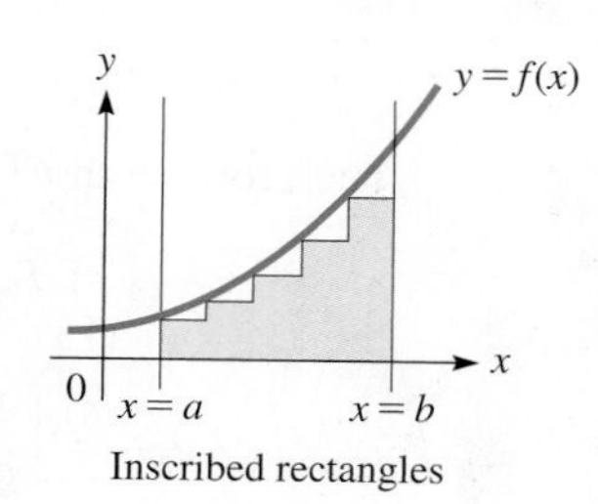

y y=f(x) 0 x=a x=b x

Circumscribed rectangles

Fig. 25.11

In Exercises 25–28, find the sum of the areas of 10 circumscribed rectangles for each curve and show that the exact area (as shown in Exercises 15–18) is between the sum of the areas of the circumscribed rectangles and the inscribed rectangles [as found in Exercises 5(b)–8(b)]. Also, note that the mean of the two sums is close to the exact value.

25. $y = 3x$ between $x = 0$ and $x = 3$ [compare with Exercises 5(b) and 15]. Why is the mean of the sums of the inscribed rectangles and circumscribed rectangles equal to the exact value?
26. $y = 2x + 1$ between $x = 0$ and $x = 2$ [compare with Exercises 6(b) and 16]. Why is the mean of the sums of the inscribed rectangles and circumscribed rectangles equal to the exact value?
27. $y = x^2$ between $x = 0$ and $x = 2$ [compare with Exercises 7(b) and 17]. Why is the mean of the sums of the inscribed rectangles and circumscribed rectangles greater than the exact value?
28. $y = 9 - x^2$ between $x = 2$ and $x = 3$ [compare with Exercises 8(b) and 18]. Why is the mean of the sums of the inscribed rectangles and circumscribed rectangles less than the exact value?

Answers to Practice Exercises

1. $A = 18$ 2. $A = 68$

25.4 The Definite Integral

Definite Integral - Result Is a Number • Lower and Upper Limits of Integration • Evaluation on a Calculator

Using reasoning similar to that in the preceding section, *we define the* **definite integral** *of a function f(x) as*

$$\int_a^b f(x)dx = F(b) - F(a) \tag{25.11}$$

where $F'(x) = f(x)$. *We call this a* **definite integral** *because the final result of integrating and evaluating is a* **number.** (The *indefinite* integral had an arbitrary constant in the result.) *The numbers a and b are called the* **lower and upper limit of integration,** *respectively. We can see that the value of a definite integral is found by evaluating the function (found by integration) at the upper limit and subtracting the value of this function at the lower limit.*

From Section 25.3, we know that this definite integral can be interpreted as the area under the curve of $y = f(x)$ from $x = a$ to $x = b$ (if $f(x) \geq 0$ for $a \leq x \leq b$), and in general as a summation. *This summation interpretation will be applied to many kinds of applied problems.*

■ That integration is equivalent to the limit of a sum is the reason that Leibniz (see page 655) used an elongated *S* for the integral sign. It stands for the Latin word for *sum.*

EXAMPLE 1 Definite integral—power of *x*

Evaluate the integral $\int_0^2 x^4\,dx$.

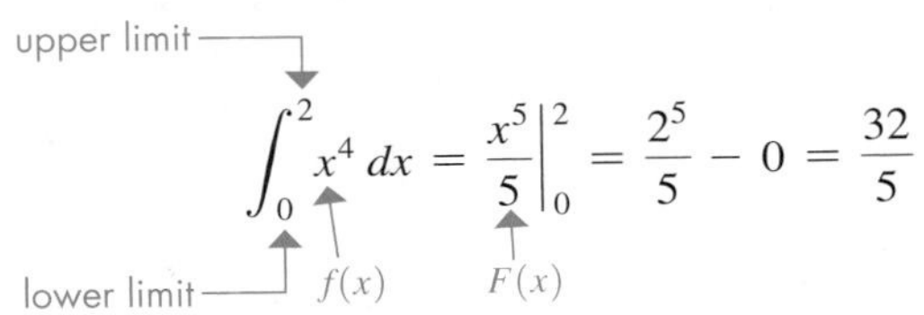

$$\int_0^2 x^4\,dx = \left.\frac{x^5}{5}\right|_0^2 = \frac{2^5}{5} - 0 = \frac{32}{5}$$

y
16
$y = x^4$
x
0
$x = 2$

Fig. 25.12

NOTE ▸ [Note that a vertical line—with the limits written at the top and the bottom—is the way the value is indicated after integration, but before evaluation.]

The area that this definite integral can represent is shown in Fig. 25.12. ■

EXAMPLE 2 Definite integral—negative exponent

Evaluate $\int_1^3 (x^{-2} - 1)dx$.

upper limit — subtract — lower limit

$$\int_1^3 (x^{-2} - 1)dx = -\frac{1}{x} - x\Big|_1^3 = \left(-\frac{1}{3} - 3\right) - (-1 - 1)$$

$$= -\frac{10}{3} + 2 = -\frac{4}{3}$$ ■

Practice Exercise

1. Evaluate: $\int_1^2 (4x - x^{-3})dx$.

EXAMPLE 3 Definite integral—power of a function

Evaluate $\int_0^1 5z(z^2 + 1)^5 dz$.

For purposes of integration, $n = 5$, $u = z^2 + 1$, and $du = 2z\,dz$. Hence,

$$\int_0^1 5z(z^2 + 1)^5\,dz = \frac{5}{2}\int_0^1 (z^2 + 1)^5(2z\,dz)$$

$$= \frac{5}{2}\left(\frac{1}{6}\right)(z^2 + 1)^6\Big|_0^1$$

$$= \frac{5}{12}(2^6 - 1^6) = \frac{5(63)}{12} = \frac{105}{4}$$ ■

EXAMPLE 4 Definite integral—radical in denominator

Evaluate $\displaystyle\int_{0.1}^{2.7} \frac{dx}{\sqrt{4x+1}}$.

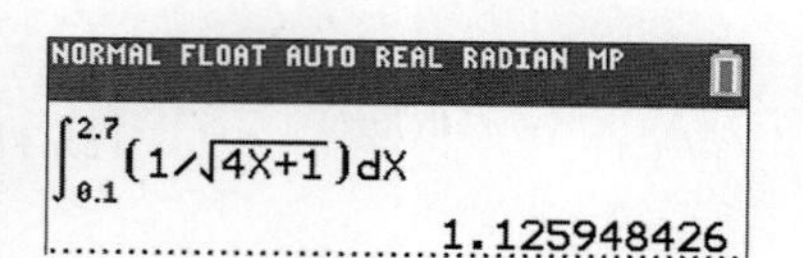

Fig. 25.13

Graphing calculator keystrokes: goo.gl/Yn1f1S

In order to integrate, we have $n = -\frac{1}{2}$, $u = 4x + 1$, and $du = 4\,dx$. Therefore,

$$\int_{0.1}^{2.7} \frac{dx}{\sqrt{4x+1}} = \int_{0.1}^{2.7} (4x+1)^{-1/2}dx = \frac{1}{4}\int_{0.1}^{2.7} (4x+1)^{-1/2}(4\,dx)$$

$$= \frac{1}{4}\left(\frac{1}{\frac{1}{2}}\right)(4x+1)^{1/2}\Big|_{0.1}^{2.7} = \frac{1}{2}(4x+1)^{1/2}\Big|_{0.1}^{2.7} \quad \text{integrate}$$

$$= \frac{1}{2}(\sqrt{11.8} - \sqrt{1.4}) = 1.126 \quad \text{evaluate}$$

Figure 25.13 shows a calculator evaluation of this integral using the *fnInt* feature. ■

Practice Exercise

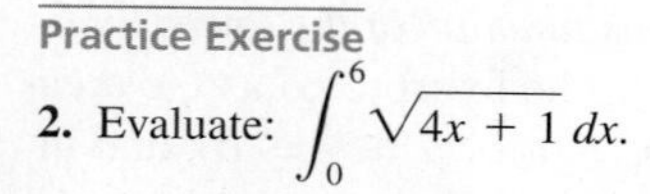

2. Evaluate: $\displaystyle\int_0^6 \sqrt{4x+1}\,dx$.

EXAMPLE 5 Integral; negative power of a function

Evaluate $\displaystyle\int_0^4 \frac{x+1}{(x^2+2x+2)^3}dx$.

For integrating, $n = -3$, $u = x^2 + 2x + 2$, and $du = (2x+2)dx$.

$$\int_0^4 (x^2+2x+2)^{-3}(x+1)dx = \frac{1}{2}\int_0^4 (x^2+2x+2)^{-3}[2(x+1)dx]$$

$$= \frac{1}{2}\left(\frac{1}{-2}\right)(x^2+2x+2)^{-2}\Big|_0^4 \quad \text{integrate}$$

$$= -\frac{1}{4}(16+8+2)^{-2} + \frac{1}{4}(0+0+2)^{-2} \quad \text{evaluate}$$

$$= \frac{1}{4}\left(-\frac{1}{26^2} + \frac{1}{2^2}\right) = \frac{1}{4}\left(\frac{1}{4} - \frac{1}{676}\right)$$

$$= \frac{1}{4}\left(\frac{168}{676}\right) = \frac{21}{338}$$

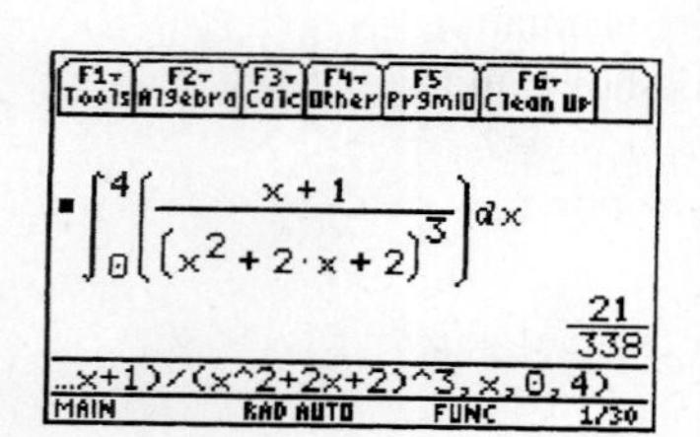

Fig. 25.14

TI-89 graphing calculator keystrokes: goo.gl/F0aE1x

The evaluation of this integral is shown on a TI-89 calculator in Fig. 25.14. ■

The following example illustrates an application of the definite integral. In Chapter 26, we will see that the definite integral has many applications in science and technology.

EXAMPLE 6 Definite integral—stopping distance of a train

■ See the chapter introduction.

The driver of a light-rail train traveling at 20.0 m/s (about 45 mi/h) applies the brakes, and the velocity then decreases by 1.25 m/s each second. Find the distance traveled by the train while coming to a stop.

Note that the velocity (in m/s) after t seconds is given by $v(t) = 20.0 - 1.25t$. By setting this equal to zero and solving for t, we find that it takes 16.0 s for the train to come to a stop. The distance d traveled during this time can be found by evaluating the definite integral of the velocity from 0 s to 16 s.

$$d = \int_0^{16} (20.0 - 1.25t)dt = \left(20.0\,t - \frac{1.25t^2}{2}\right)\Big|_0^{16} \quad \text{integrate}$$

$$= \left[20.0(16) - \frac{1.25(16)^2}{2}\right] - 0 = 160 \text{ m} \quad \text{evaluate}$$

Therefore, the train travels 160 m while coming to a stop. ■

The definition of the definite integral is valid regardless of the source of $f(x)$. That is, we may apply the definite integral whenever we want to sum a function in a manner similar to that which we use to find an area.

EXERCISES 25.4

In Exercises 1 and 2, make the given changes in the indicated examples of this section and then solve the resulting problems.

1. In Example 2, change the upper limit from 3 to 4.

2. In Example 4, change $4x$ to $2x$.

In Exercises 3–34, evaluate the given definite integrals.

3. $\int_0^1 4x\,dx$ **4.** $\int_0^2 3x^2\,dx$ **5.** $\int_1^4 x^{3/2}\,dx$

6. $\int_4^9 8(p^{3/2} - 3)dp$ **7.** $\int_3^6 \left(\frac{1}{\sqrt{x}} - 7\right)dx$ **8.** $\int_{1.2}^{1.6} \left(5 + \frac{6}{x^4}\right)dx$

9. $\int_{-1.6}^{0.7} (1 - x)^{1/3}\,dx$ **10.** $\int_1^5 9\sqrt{3v + 1}\,dv$

11. $\int_{-2}^{2} (T - 2)(T + 2)dT$ **12.** $\int_1^2 (3x^5 - 2x^3)dx$

13. $\int_1^8 (\sqrt[3]{x} - 2)dx$ **14.** $\int_{2.7}^{5.3} \left(\frac{1}{x\sqrt{x}} + 4\right)dx$

15. $\int_0^4 6(1 - \sqrt{x})^2\,dx$ **16.** $\int_1^4 \frac{y + 4}{\sqrt{y}}dy$

17. $\int_{-2}^{-1} 12x(4 - x^2)^3\,dx$ **18.** $\int_0^1 4x(3x^2 - 1)^3\,dx$

19. $\int_0^4 \frac{5x\,dx}{\sqrt{x^2 + 9}}$ **20.** $\int_{0.2}^{0.7} x^2(x^3 + 2)^{3/2}\,dx$

21. $\int_{2.75}^{3.25} \frac{dx}{\sqrt[3]{6x + 1}}$ **22.** $\int_2^6 \frac{8\,du}{\sqrt{4u + 1}}$

23. $\int_1^3 \frac{12x\,dx}{(2x^2 + 1)^3}$ **24.** $\int_{12.6}^{17.2} \frac{3\,dx}{(6x - 1)^2}$

25. $\int_3^7 \sqrt{16t^2 + 8t + 1}\,dt$ **26.** $\int_{-5}^{1} 6\sqrt{6 - 2x}\,dx$

27. $\int_0^2 2x(9 - 2x^2)^2\,dx$ **28.** $\int_{-1}^{2} V(V^3 + 1)dV$

29. $\int_{-1}^{2} \frac{8x - 2}{(2x^2 - x + 1)^3}dx$ **30.** $\int_2^3 \frac{8(x^2 + 1)}{(x^3 + 3x)^2}dx$

31. $\int_0^1 (x^2 + 3)(x^3 + 9x + 6)^2\,dx$

32. $\int_{-3}^{-2} (3x^2 - 2)\sqrt[3]{2x^3 - 4x + 1}\,dx$

33. $\int_{\sqrt{5}}^{3} 8z\sqrt[4]{z^4 + 8z^2 + 16}\,dz$

34. $\int_{-2}^{0} (\sqrt{2x + 4} - \sqrt[3]{3x + 8})dx$

In Exercises 35–54, solve the given problems.

35. Show that $\int_0^1 x^3\,dx + \int_1^2 x^3\,dx = \int_0^2 x^3\,dx$. In terms of area, explain the result.

36. Write $\int_1^8 f(x)\,dx - \int_5^8 f(x)\,dx$ as a single definite integral.

37. Evaluate $\int_{x=1}^{x=4} y\,dx$, when $y^2 = 4x\ (y > 0)$.

38. Show that $\int_1^3 4x\,dx = -\int_3^1 4x\,dx$.

39. Show that $\int_0^1 x\,dx > \int_0^1 x^2\,dx$ and $\int_1^2 x\,dx < \int_1^2 x^2\,dx$. In terms of area, explain the result.

40. Given that $\int_0^9 \sqrt{x}\,dx = 18$, evaluate $\int_0^9 2\sqrt{t}\,dt$.

41. Evaluate $\int_{-1}^{1} t^{2k}\,dt$, where k is a positive integer.

42. Evaluate the following integral, which arises in the study of electricity: $\int_0^L \frac{1}{EI}(-\frac{1}{2}wx^2)(-x)dx$.

43. If $\int_a^b f(x)dx = 4$, evaluate $\int_b^a f(x)dx$.

44. If $\int_1^7 f(x)dx = 16$ and $\int_{-4}^{7} f(x)dx = 8$, evaluate $\frac{1}{2}\int_{-4}^{1} f(x)dx$.

45. Evaluate $\int_{-3}^{3} |x - 1|\,dx$ by geometrically finding the area represented.

46. Evaluate $\int_{-2}^{2} \sqrt{4 - x^2}dx$ by geometrically finding the area represented.

47. Evaluate the following integral, which arises in the study of hydrodynamics: $\int_H^h \frac{Ay^{-1/2}\,dy}{a\sqrt{2g}}$.

48. It is estimated that a newly discovered oil field will produce oil at the rate of $\frac{dR}{dt} = \frac{(400t^2)^2}{t^3 + 20} + 10$ thousand barrels per year. How much oil can be expected from the field in the next ten years?

49. The work W (in ft · lb) in winding up an 80-ft cable is $W = \int_0^{80} (1000 - 5x)dx$. Evaluate W.

50. The total volume V of liquid flowing through a certain pipe of radius R is $V = k(R^2 \int_0^R r\,dr - \int_0^R r^3\,dr)$, where k is a constant. Evaluate V and explain why R, but not r, can be to the left of the integral sign.

51. The surface area A (in m^2) of a certain parabolic radio-wave reflector is $A = 4\pi\int_0^2 \sqrt{3x+9}\,dx$. Evaluate A.

52. The total force (in N) on the circular end of a water tank is $F = 19{,}600\int_0^5 y\sqrt{25-y^2}\,dy$. Evaluate F.

53. In finding the average electron energy in a metal at very low temperatures, the integral $\dfrac{3N}{2E_F^{3/2}}\displaystyle\int_0^{E_F} E^{3/2}\,dE$ is used. Evaluate this integral.

54. In finding the electric field E caused by a surface electric charge on a disk, the equation $E = k\displaystyle\int_0^R \frac{r\,dr}{(x^2+r^2)^{3/2}}$ is used. Evaluate the integral.

Answers to Practice Exercises

1. 45/8 **2.** 62/3

25.5 Numerical Integration: The Trapezoidal Rule

Numerical Integration • Trapezoidal Rule • Approximating Integrals

For data and functions that cannot be directly integrated by available methods, it is possible to develop numerical methods of integration. These numerical methods are of greater importance today because they are readily adaptable for use on a calculator or computer. There are a great many such numerical techniques for approximating the value of an integral. In this section, we develop one of these, the *trapezoidal rule.* In the following section, another numerical method is discussed.

We know from Sections 25.3 and 25.4 that we can interpret a definite integral as the area under a curve. We will therefore show how to approximate the value of the integral by approximating the appropriate area by a set of trapezoids. The basic idea here is very similar to that used when rectangles were inscribed under a curve. However, the use of trapezoids reduces the error and provides a better approximation.

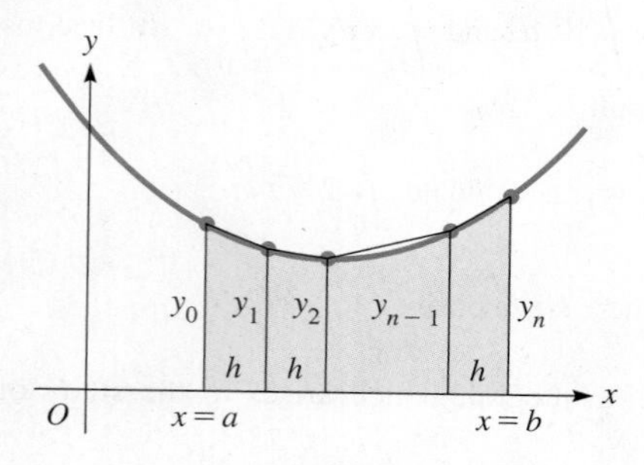

Fig. 25.15

The area to be found is subdivided into n intervals of equal width. Perpendicular lines are then dropped from the curve (or points, if only a given set of numbers is available). If the points on the curve are joined by straight-line segments, the area of successive parts under the curve is approximated by finding the areas of the trapezoids formed. However, if these points are not too far apart, the approximation will be very good (see Fig. 25.15). From geometry, recall that the area of a trapezoid equals one-half the product of the sum of the bases times the altitude. For these trapezoids, the bases are the y-coordinates, and the altitudes are h. Therefore, when we indicate the sum of these trapezoidal areas, we have

$$A_T = \frac{1}{2}(y_0+y_1)h + \frac{1}{2}(y_1+y_2)h + \frac{1}{2}(y_2+y_3)h + \cdots$$
$$+ \frac{1}{2}(y_{n-2}+y_{n-1})h + \frac{1}{2}(y_{n-1}+y_n)h$$

We note, when this addition is performed, that the result is

$$A_T = h\left(\frac{1}{2}y_0 + y_1 + y_2 + \cdots + y_{n-1} + \frac{1}{2}y_n\right) \tag{25.12}$$

The y-values to be used either are derived from the function $y = f(x)$ or are the y-coordinates of a set of data.

Because A_T approximates the area under the curve, it also approximates the value of the definite integral. By factoring out $\frac{1}{2}$, we get what is known as the **trapezoidal rule.**

Trapezoidal Rule

$$\int_a^b f(x)\,dx \approx \frac{h}{2}(y_0 + 2y_1 + 2y_2 + \cdots + 2y_{n-1} + y_n) \tag{25.13}$$

The trapezoidal rule is really the same rule that we used in Chapter 2 to measure irregular geometric areas (see page 73). Because we know that the value we get when evaluating a definite integral can be interpreted as the area under the curve of a function, we can now use the trapezoidal rule to find the approximate value of a definite integral.

Whenever the curve of the function being integrated is concave up, the approximation segments are above the curve and each trapezoid has slightly more area than the corresponding area under the curve (see Fig. 25.16(a) on the next page). If the curve is

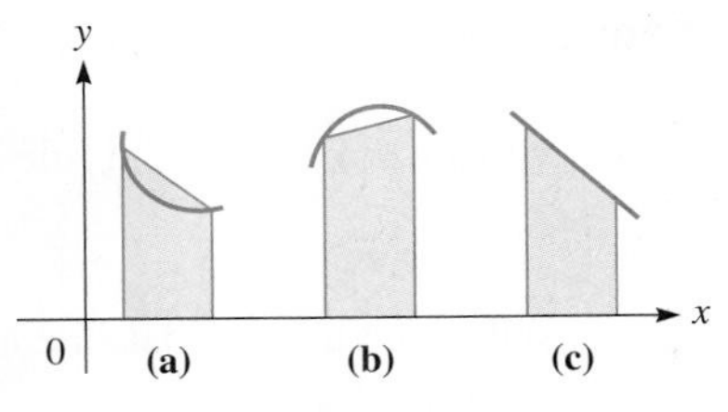

Fig. 25.16

concave down, the approximating segments are below the curve, and each trapezoid has slightly less area than the corresponding area under the curve (see Fig. 25.16(b)). For a straight line segment, the trapezoidal rule gives the exact value (see Fig. 25.16(c)).

EXAMPLE 1 Trapezoidal rule—approximating an integral

Approximate the value of $\int_1^3 \frac{1}{x}dx$ by the trapezoidal rule. Let $n = 4$.

We are to approximate the area under $y = 1/x$ from $x = 1$ to $x = 3$ by dividing the area into four trapezoids. Figure 25.17 shows the graph. Using the trapezoidal rule, we have $f(x) = 1/x$, and

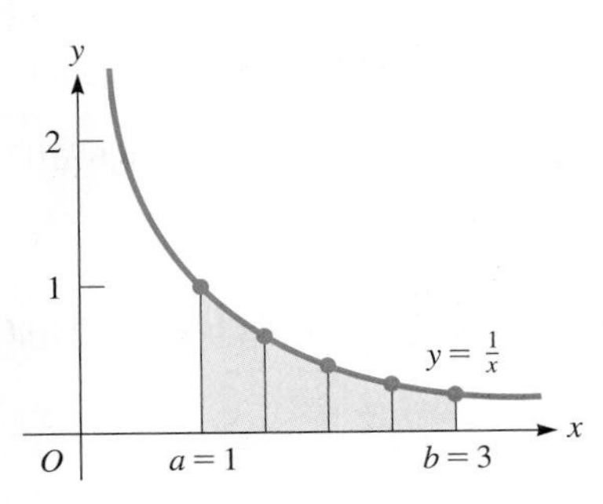

Fig. 25.17

$$h = \frac{3-1}{4} = \frac{1}{2} \qquad y_0 = f(a) = f(1) = 1$$

$$y_1 = f\left(\frac{3}{2}\right) = \frac{2}{3} \qquad y_2 = f(2) = \frac{1}{2}$$

$$y_3 = f\left(\frac{5}{2}\right) = \frac{2}{5} \qquad y_n = y_4 = f(b) = f(3) = \frac{1}{3}$$

$$A_T = \frac{1/2}{2}\left[1 + 2\left(\frac{2}{3}\right) + 2\left(\frac{1}{2}\right) + 2\left(\frac{2}{5}\right) + \frac{1}{3}\right]$$

$$= \frac{1}{4}\left(\frac{15 + 20 + 15 + 12 + 5}{15}\right) = \frac{1}{4}\left(\frac{67}{15}\right) = \frac{67}{60}$$

Practice Exercise

1. In Example 1, use the trapezoidal rule with $n = 3$.

Therefore,

$$\int_1^3 \frac{1}{x}dx \approx \frac{67}{60} = 1.12$$

Table 25.2

Number of Trapezoids n	*Total Area of Trapezoids*
4	1.1166667
100	1.0986419
1,000	1.0986126
10,000	1.0986123

We cannot perform this integration directly by methods developed up to this point. As we increase the number of trapezoids, the value becomes more accurate. See Table 25.2. The actual value to seven decimal places is 1.0986123. ■

EXAMPLE 2 Trapezoidal rule—calculator evaluation

Approximate the value of $\int_0^1 \sqrt{x^2+1}\,dx$ by the trapezoidal rule. Let $n = 5$.

Figure 25.18 shows the graph. In this example,

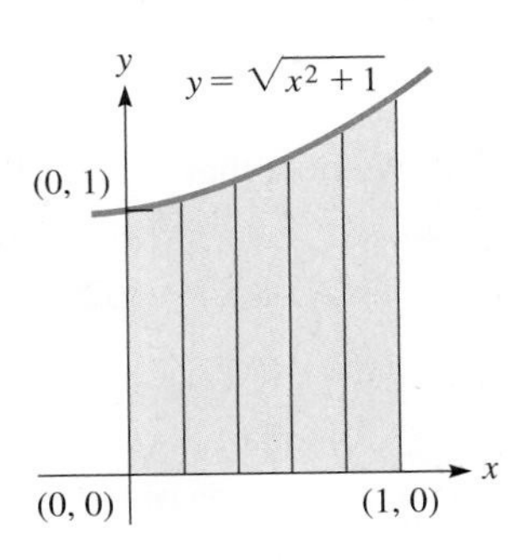

Fig. 25.18

$$h = \frac{1-0}{5} = 0.2$$

$$y_0 = f(0) = 1 \qquad y_1 = f(0.2) = \sqrt{1.04} = 1.0198039$$

$$y_2 = f(0.4) = \sqrt{1.16} = 1.0770330 \qquad y_3 = f(0.6) = \sqrt{1.36} = 1.1661904$$

$$y_4 = f(0.8) = \sqrt{1.64} = 1.2806248 \qquad y_5 = f(1) = \sqrt{2.00} = 1.4142136$$

Hence, we have

$$A_T = \frac{0.2}{2}[1 + 2(1.0198039) + 2(1.0770330) + 2(1.1661904) + 2(1.2806248) + 1.4142136]$$

$$= 1.150 \quad \text{(rounded off)}$$

This means that

$$\int_0^1 \sqrt{x^2+1}\,dx \approx 1.150 \quad \text{the actual value is 1.148 to three decimal places}$$

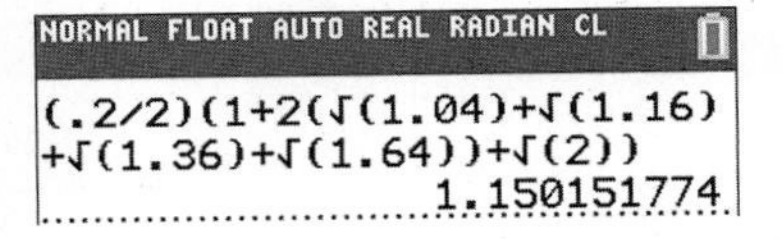

Fig. 25.19

The entire calculation can be done on a calculator without tabulating values, as shown in the display in Fig. 25.19. ■

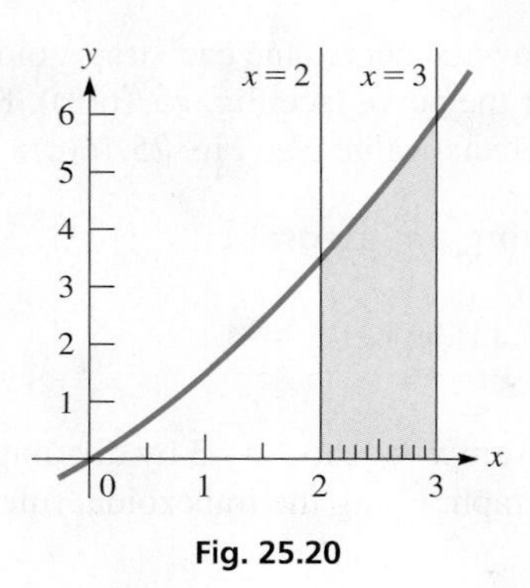

Fig. 25.20

EXAMPLE 3 Trapezoidal rule—calculator verification

Approximate the value of the integral $\int_2^3 x\sqrt{x+1}\,dx$ by using the trapezoidal rule. Use $n = 10$.

In Fig. 25.20, the graph of the function and the area used in the trapezoidal rule are shown. From the given values, we have $h = \frac{3-2}{10} = 0.1$. Therefore,

$y_0 = f(2) = 2\sqrt{3} = 3.4641016$ $\qquad y_1 = f(2.1) = 2.1\sqrt{3.1} = 3.6974315$

$y_2 = f(2.2) = 2.2\sqrt{3.2} = 3.9354796$ $\qquad y_3 = f(2.3) = 2.3\sqrt{3.3} = 4.1781575$

$y_4 = f(2.4) = 2.4\sqrt{3.4} = 4.4253813$ $\qquad y_5 = f(2.5) = 2.5\sqrt{3.5} = 4.6770717$

$y_6 = f(2.6) = 2.6\sqrt{3.6} = 4.9331531$ $\qquad y_7 = f(2.7) = 2.7\sqrt{3.7} = 5.1935537$

$y_8 = f(2.8) = 2.8\sqrt{3.8} = 5.4582048$ $\qquad y_9 = f(2.9) = 2.9\sqrt{3.9} = 5.7270411$

$y_{10} = f(3) = 3\sqrt{4} = 6.0000000$

$$A_T = \frac{0.1}{2}[3.4641016 + 2(3.6974315) + \cdots + 2(5.7270411) + 6.0000000]$$
$$= 4.6958$$

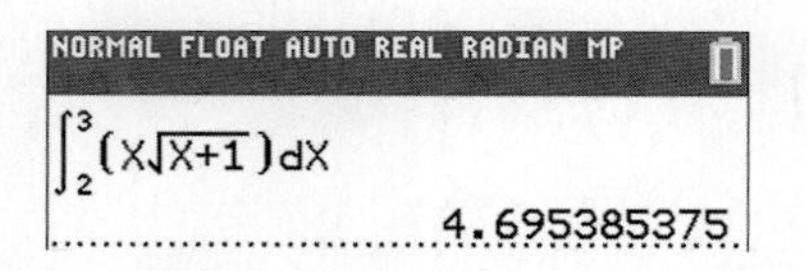

Fig. 25.21
Graphing calculator keystrokes: goo.gl/n8LgjT

Therefore, $\int_2^3 x\sqrt{x+1}\,dx \approx 4.6958$. We see from the calculator display shown in Fig. 25.21 that the value is 4.6954 to four decimal places. ■

EXAMPLE 4 Trapezoidal rule with empirical data—area of a park

In estimating the area of a proposed city park bounded by three straight streets, two parallel and perpendicular to the third, and a river (see Fig. 25.22), measurements were taken at 10.0-m intervals of the distance to the river, and the values found are in the following table. Find the area of the proposed park, using the trapezoidal rule.

x (m)	0	10	20	30	40	50
y (m)	56.8	67.5	73.2	73.5	68.8	62.4

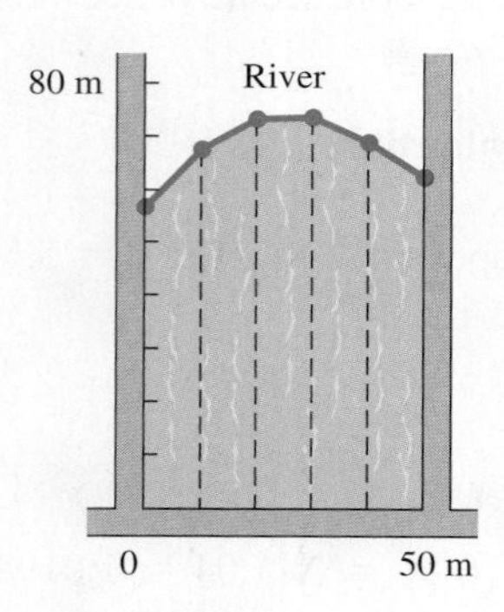

Fig. 25.22

To find the area, we use the values directly from the table. We note that $h = 10.0$ m.

$$A = \frac{10.0}{2}[56.8 + 2(67.5) + 2(73.2) + 2(73.5) + 2(68.8) + 62.4]$$
$$= 3426 \text{ m}^2$$

Rounding off to three significant digits, the accuracy of the data, the area of the proposed park is about 3430 m^2.

Although we do not know the mathematical form of the function, we can state that

$$\int_0^{50.0} f(x)dx \approx 3430$$ ■

EXERCISES 25.5

In Exercises 1 and 2, make the given changes in the indicated examples of this section and then solve the resulting problems.

1. In Example 1, change n from 4 to 2.
2. In Example 3, change n from 10 to 5.

In Exercises 3–6, (a) approximate the value of each of the given integrals by use of the trapezoidal rule, using the given value of n, and (b) check by direct integration.

3. $\int_0^2 2x^2\,dx,\ n = 4$
4. $\int_0^1 (1 - x^2)dx,\ n = 3$
5. $\int_1^4 (1 + \sqrt{x})dx,\ n = 6$
6. $\int_3^8 \sqrt{1 + x}\,dx,\ n = 5$

In Exercises 7–14, approximate the value of each of the given integrals by use of the trapezoidal rule, using the given value of n.

7. $\int_2^3 \frac{1}{2x}dx,\ n = 2$
8. $\int_2^6 \frac{dx}{x + 3},\ n = 4$
9. $\int_0^5 \sqrt{25 - x^2}\,dx,\ n = 5$
10. $\int_0^2 \sqrt{x^3 + 1}\,dx,\ n = 4$
11. $\int_1^5 \frac{1}{x^2 + x}dx,\ n = 10$
12. $\int_2^4 \frac{1}{x^2 + 1}dx,\ n = 10$
13. $\int_0^4 2^x\,dx,\ n = 12$
14. $\int_0^{1.5} 10^x\,dx,\ n = 15$

In Exercises 15 and 16, approximate the values of the integrals defined by the given sets of points.

15. $\int_2^{14} y\,dx$

x	2	4	6	8	10	12	14
y	0.67	2.34	4.56	3.67	3.56	4.78	6.87

16. $\int_{1.4}^{3.2} y\,dx$

x	1.4	1.7	2.0	2.3	2.6	2.9	3.2
y	0.18	7.87	18.23	23.53	24.62	20.93	20.76

In Exercises 17–22, solve each given problem by using the trapezoidal rule.

17. Explain why the approximate value of the integral in Exercise 5 is less than the exact value.
18. $\int_0^2 \sqrt{4 - x^2}\,dx = \pi$. Approximate the value of the integral with $n = 8$. Compare with π.
19. $\int_0^3 \frac{dx}{2x + 2} = \ln 2$. Approximate the value of the integral with $n = 6$. Compare with ln 2.
20. A force F that a distributed electric charge has on a point charge is $F = k\int_0^2 \frac{dx}{(4 + x^2)^{3/2}}$, where x is the distance along the distributed charge and k is a constant. With $n = 8$, evaluate F in terms of k.
21. The length L (in ft) of telephone wire needed (considering the sag) between two poles exactly 100 ft apart is $L = 2\int_0^{50}\sqrt{6.4 \times 10^{-7}x^2 + 1}\,dx$. With $n = 10$, evaluate L (to six significant digits).
22. The rate $\frac{dA}{dt}$ (in standard pollution index per hour) of a pollutant put into the air by a smokestack is given by $\frac{dA}{dt} = \frac{150}{1 + 0.25(t - 4.0)^2} + 25$, where t is the time (in h) after 6 A.M. With $n = 6$, estimate the total amount of the pollutant put into the air between 6 A.M. and noon.

Answer to Practice Exercise

1. 1.13

25.6 Simpson's Rule

Parabolic Arcs • Simpson's Rule • Number of Intervals Must Be Even

The numerical method of integration developed in this section is also readily programmable for use on a computer or easily usable with the necessary calculations done on a calculator. It is obtained by interpreting the definite integral as the area under a curve, as we did in developing the trapezoidal rule, and by approximating the curve by a set of parabolic arcs. The use of parabolic arcs, rather than chords as with the trapezoidal rule, usually gives a better approximation.

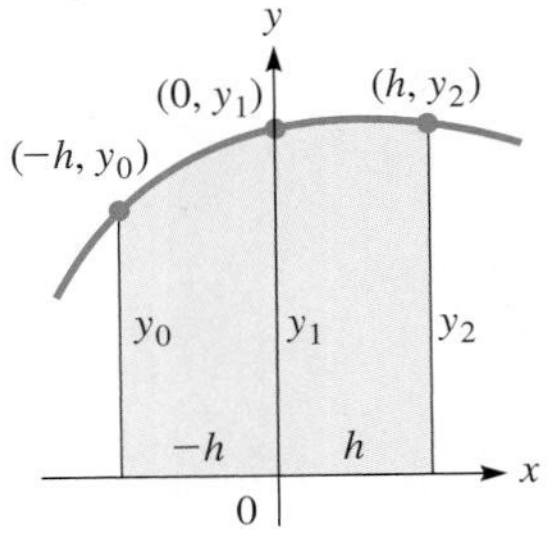

First finding the area under a parabolic arc, the curve in Fig. 25.23 represents the parabola $y = ax^2 + bx + c$ with points $(-h, y_0)$, $(0, y_1)$, and (h, y_2). The area is

$$A = \int_{-h}^{h} y\,dx = \int_{-h}^{h}(ax^2 + bx + c)dx = \frac{ax^3}{3} + \frac{bx^2}{2} + cx\Bigg|_{-h}^{h}$$

$$= \frac{2}{3}ah^3 + 2ch$$

$$A = \frac{h}{3}(2ah^2 + 6c) \tag{25.14}$$

Fig. 25.23

The coordinates of the three points also satisfy the equation $y = ax^2 + bx + c$. This means that

$$y_0 = ah^2 - bh + c$$
$$y_1 = c$$
$$y_2 = ah^2 + bh + c$$

By finding the sum of $y_0 + 4y_1 + y_2$, we have

$$y_0 + 4y_1 + y_2 = 2ah^2 + 6c \qquad \textbf{(25.15)}$$

Substituting Eq. (25.15) into Eq. (25.14), we have

■ Note that the area depends only on the distance h and the three y-coordinates.

$$A = \frac{h}{3}(y_0 + 4y_1 + y_2) \qquad \textbf{(25.16)}$$

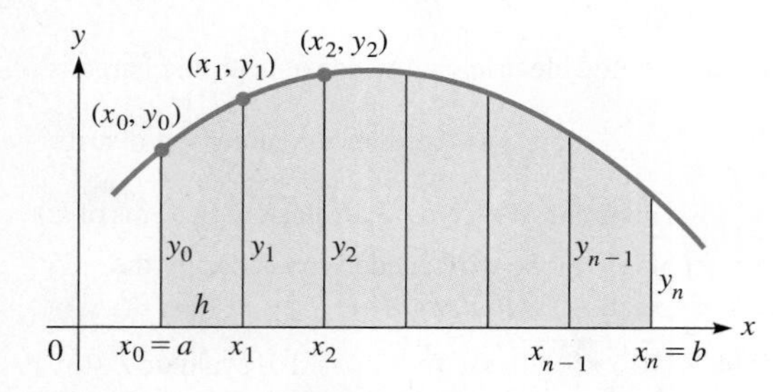

Fig. 25.24

Now, let us consider the area under the curve in Fig. 25.24. If a parabolic arc is passed through the points (x_0, y_0), (x_1, y_1), and (x_2, y_2), we may use Eq. (25.16) to approximate the area under the curve between x_0 and x_2. We again note that the distance h is the difference in the x-coordinates. Therefore, the area under the curve between x_0 and x_2 is

$$A_1 = \frac{h}{3}(y_0 + 4y_1 + y_2)$$

We note that the base of A_1 extends from x_0 to x_2. If we now find the area A_2, the base of which extends from x_2 to x_4, by finding the area under the parabola and through the three points (x_2, y_2), (x_3, y_3), and (x_4, y_4), we find it to be

$$A_2 = \frac{h}{3}(y_2 + 4y_3 + y_4)$$

The sum of these areas is

$$A_1 + A_2 = \frac{h}{3}(y_0 + 4y_1 + 2y_2 + 4y_3 + y_4) \qquad \textbf{(25.17)}$$

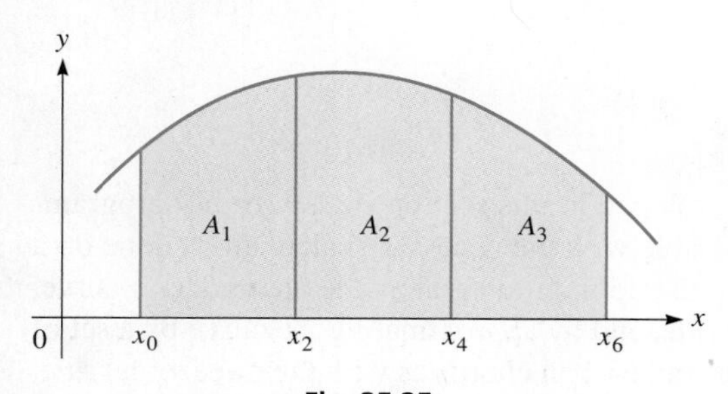

Fig. 25.25

We can continue this procedure by finding A_3, the base of which extends from x_4 to x_6, and then finding the area $A_1 + A_2 + A_3$, as shown in Fig. 25.25. This procedure can continue to include additional areas under the parabola. However, we must note that each time we add another area, we include *two more subintervals,* which implies the following. **CAUTION** When using this method, the number of subintervals n of width h *must be even.* ■

Generalizing on Eq. (25.17) and recalling that the value of the definite integral is the area under the curve, we have what is known as **Simpson's rule.**

Simpson's Rule

$$\int_a^b f(x)\,dx \approx \frac{h}{3}(y_0 + 4y_1 + 2y_2 + 4y_3 + 2y_4 + \cdots + 4y_{n-1} + y_n) \qquad \textbf{(25.18)}$$

■ Although named for the English mathematician Thomas Simpson (1710–1761), he did not discover the rule. It was well known when he included it in some of his many books on mathematics.

As with the trapezoidal rule, we used Simpson's rule in Chapter 2 to measure irregular areas (see page 74).

EXAMPLE 1 Simpson's rule—approximating an integral

Approximate the value of the integral $\int_0^1 \frac{dx}{x+1}$ by Simpson's rule. Let $n = 2$.

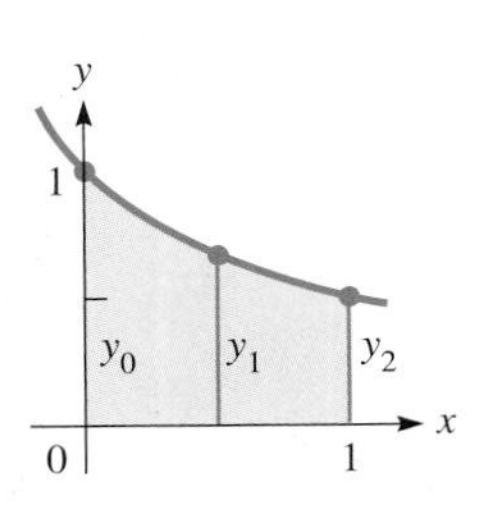

Fig. 25.26

In Fig. 25.26, the graph of the function and the area used are shown. We are to approximate the integral by using Simpson's rule. We therefore note that $f(x) = 1/(x+1)$. Also, $x_0 = a = 0$, $x_1 = 0.5$, and $x_2 = b = 1$. This is due to the fact that $n = 2$ and $h = 0.5$ because the total interval is 1 unit (from $x = 0$ to $x = 1$). Therefore,

$$y_0 = \frac{1}{0+1} = 1.0000 \qquad y_1 = \frac{1}{0.5+1} = 0.6667 \qquad y_2 = \frac{1}{1+1} = 0.5000$$

Substituting, we have

$$\int_0^1 \frac{dx}{x+1} \approx \frac{0.5}{3}[1.0000 + 4(0.6667) + 0.5000]$$
$$= 0.694$$

To three decimal places, the actual value of the integral is 0.693. We will consider the method of integrating this function in a later chapter. ■

Practice Exercise

1. In Example 1, use Simpson's rule with $n = 4$.

When we use the trapezoidal rule or Simpson's rule to approximate the value of a definite integral, we get a better approximation by using more subdivisions. In the previous example we used only two subdivisions. By using more subdivisions, a more accurate approximation could be obtained. [This is further use of the method we used in developing the value of the definite integral as an area by letting the number of inscribed rectangles become unlimited to find the exact result.] In the following example we use ten subdivisions, and find the result is accurate to eight significant digits, which is usually much more accuracy than is needed.

NOTE ➧

EXAMPLE 2 Simpson's rule—approximating an integral

Approximate the value of $\int_2^3 x\sqrt{x+1}\,dx$ by Simpson's rule. Use $n = 10$.

Because the necessary values for this function are shown in Example 3 of Section 25.5, we shall simply tabulate them here. ($h = 0.1$.) See Fig. 25.27.

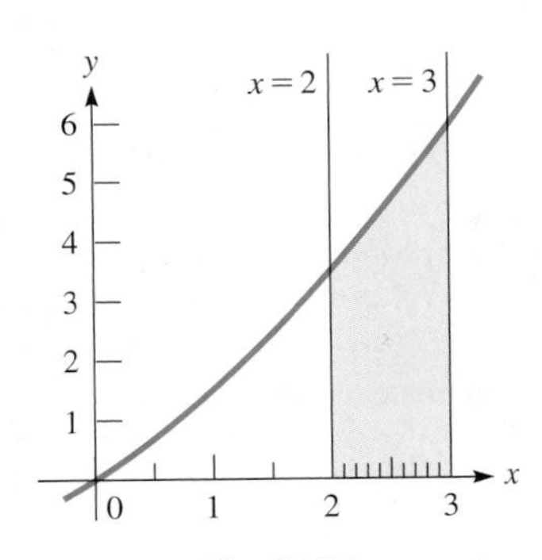

Fig. 25.27

$y_0 = 3.4641016$ $y_1 = 3.6974315$ $y_2 = 3.9354796$ $y_3 = 4.1781575$
$y_4 = 4.4253813$ $y_5 = 4.6770717$ $y_6 = 4.9331531$ $y_7 = 5.1935537$
$y_8 = 5.4582048$ $y_9 = 5.7270411$ $y_{10} = 6.0000000$

Therefore, we evaluate the integral as follows:

$$\int_2^3 x\sqrt{x+1}\,dx \approx \frac{0.1}{3}[3.4641016 + 4(3.6974315) + 2(3.9354796)$$
$$+ 4(4.1781575) + 2(4.4253813) + 4(4.6770717)$$
$$+ 2(4.9331531) + 4(5.1935537) + 2(5.4582048)$$
$$+ 4(5.7270411) + 6.0000000]$$
$$= \frac{0.1}{3}(140.86156) = 4.6953854$$

This result agrees with the actual value to the eight significant digits shown. The value we obtained with the trapezoidal rule was 4.6958. ■

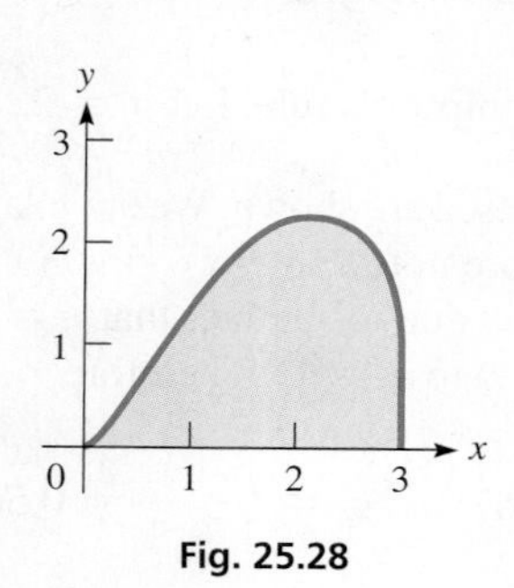

Fig. 25.28

EXAMPLE 3 Simpson's rule—area of an aircraft stabilizer

One side of an aircraft's rear stabilizer is shown in Fig. 25.28. If its area A (in m^2) is represented as $A = \int_0^3 (3x^2 - x^3)^{0.6}\,dx$, find the area, using Simpson's rule, with $n = 6$.

Noting that $f(x) = (3x^2 - x^3)^{0.6}$, $a = 0$, $b = 3$, and $h = \dfrac{3-0}{6} = 0.5$, we have

$$y_0 = f(0) = [3(0)^2 - 0^3]^{0.6} = 0$$
$$y_1 = f(0.5) = [3(0.5)^2 - 0.5^3]^{0.6} = 0.7542720$$
$$y_2 = f(1) = [3(1)^2 - 1^3]^{0.6} = 1.5157166$$
$$y_3 = f(1.5) = [3(1.5)^2 - 1.5^3]^{0.6} = 2.0747428$$
$$y_4 = f(2) = [3(2)^2 - 2^3]^{0.6} = 2.2973967$$
$$y_5 = f(2.5) = [3(2.5)^2 - 2.5^3]^{0.6} = 1.9811165$$
$$y_6 = f(3) = [3(3)^2 - 3^3]^{0.6} = 0$$

$$A = \int_0^3 (3x^2 - x^3)^{0.6}\,dx \approx \frac{0.5}{3}[0 + 4(0.7542720) + 2(1.5157166)$$
$$+ 4(2.0747428) + 2(2.2973967) + 4(1.9811165) + 0]$$
$$= 4.4777920 \text{ m}^2$$

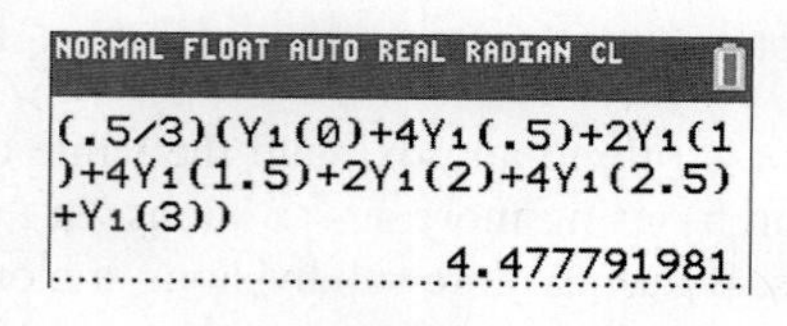

Fig. 25.29
Graphing calculator keystrokes: goo.gl/X9mNh0

Thus, the area of one side of the stabilizer is 4.478 m^2 (rounded off).

As with the trapezoidal rule, since the calculation can be done completely on a calculator, it is not necessary to record the above values. Using $y = (3x^2 - x^3)^{0.6}$, the display in Fig. 25.29 shows the calculator evaluation of the approximate area. ■

EXERCISES 25.6

In Exercises 1 and 2, make the given changes in the indicated examples of this section, and then solve the indicated problems.

1. In Example 1, change the denominator $x + 1$ to $x + 2$ and then find the approximate value of the integral.

2. In Example 2, change n such that $h = 0.2$ and explain why Simpson's rule cannot be used.

In Exercises 3–6, (a) approximate the value of each of the given integrals by use of Simpson's rule, using the given value of n, and (b) check by direct integration.

3. $\int_0^2 (1 + x^3)dx,\ n = 2$

4. $\int_0^8 x^{1/3}\,dx,\ n = 4$

5. $\int_1^4 (2x + \sqrt{x})dx,\ n = 6$

6. $\int_0^2 x\sqrt{x^2 + 1}\,dx,\ n = 4$

In Exercises 7–12, approximate the value of each of the given integrals by use of Simpson's rule, using the given values of n. Exercises 8–10 are the same as Exercises 10–12 of Section 25.5.

7. $\int_0^5 \sqrt{25 - x^2}\,dx,\ n = 4$

8. $\int_0^2 \sqrt{x^3 + 1}\,dx,\ n = 4$

9. $\int_1^5 \dfrac{1}{x^2 + x}dx,\ n = 10$

10. $\int_2^4 \dfrac{1}{x^2 + 1}dx,\ n = 10$

11. $\int_{-4}^5 (2x^4 + 1)^{0.1}\,dx,\ n = 6$

12. $\int_0^{2.4} \dfrac{dx}{(4 + \sqrt{x})^{3/2}},\ n = 8$

In Exercises 13 and 14, approximate the values of the integrals defined by the given sets of points by using Simpson's rule. These are the same as Exercises 15 and 16 of Section 25.5.

13. $\int_2^{14} y\,dx$

x	2	4	6	8	10	12	14
y	0.67	2.34	4.56	3.67	3.56	4.78	6.87

14. $\int_{1.4}^{3.2} y\,dx$

x	1.4	1.7	2.0	2.3	2.6	2.9	3.2
y	0.18	7.87	18.23	23.53	24.62	20.93	20.76

In Exercises 15–18, solve the given problems, using Simpson's rule. Exercises 15 and 16 are the same as 18 and 19 of Section 25.5.

15. $\int_0^2 \sqrt{4 - x^2}\,dx = \pi$. Approximate the value of the integral with $n = 8$. Compare with π.

16. $\int_0^3 \dfrac{dx}{2x + 2} = \ln 2$. Approximate the value of the integral with $n = 6$. Compare with ln 2.

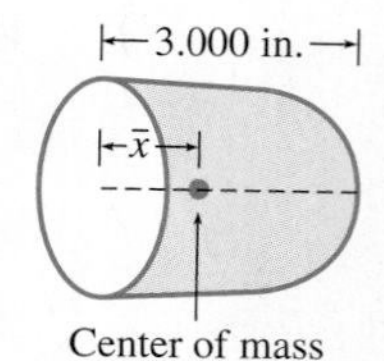

Fig. 25.30

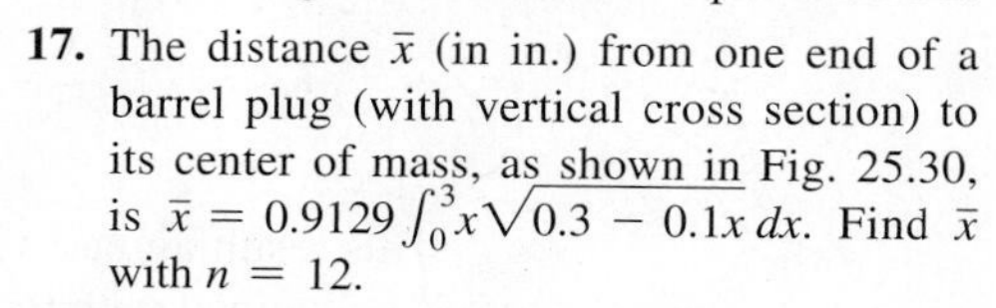

17. The distance $\bar{x}$ (in in.) from one end of a barrel plug (with vertical cross section) to its center of mass, as shown in Fig. 25.30, is $\bar{x} = 0.9129\int_0^3 x\sqrt{0.3 - 0.1x}\,dx$. Find $\bar{x}$ with $n = 12$.

18. The average value of the electric current i_{av} (in A) in a circuit for the first 4 s is $i_{av} = \frac{1}{4}\int_0^4 (4t - t^2)^{0.2}\,dt$. Find i_{av} with $n = 10$.

Answer to Practice Exercise

1. 0.693

CHAPTER 25 KEY FORMULAS AND EQUATIONS

Indefinite integral $$\int f(x)\,dx = F(x) + C \quad (25.1)$$

Integrals $$\int c\,du = c\int du = cu + C \quad (25.2)$$

$$\int (du + dv) = u + v + C \quad (25.3)$$

Power rule for integration $$\int u^n\,du = \frac{u^{n+1}}{n+1} + C \qquad (n \neq -1) \quad (25.4)$$

Area under a curve $$A_{ab} = \left[\int f(x)\,dx\right]_a^b = F(b) - F(a) \quad (25.10)$$

Definite integral $$\int_a^b f(x)\,dx = F(b) - F(a) \quad (25.11)$$

Trapezoidal rule $$\int_a^b f(x)\,dx \approx \frac{h}{2}(y_0 + 2y_1 + 2y_2 + \cdots + 2y_{n-1} + y_n) \quad (25.13)$$

Simpson's rule $$\int_a^b f(x)\,dx \approx \frac{h}{3}(y_0 + 4y_1 + 2y_2 + 4y_3 + 2y_4 + \cdots + 4y_{n-1} + y_n) \quad (25.18)$$

CHAPTER 25 REVIEW EXERCISES

CONCEPT CHECK EXERCISES

*Determine each of the following as being either **true** or **false**. If it is false, explain why.*

1. $\int (3x^2+1)^5 dx = \frac{1}{6}(3x^2+1)^6 + C$
2. $\int_0^2 \sqrt{4x+1} = \frac{13}{3}$
3. The area under the curve $y = 3x^2$ between $x = 1$ and $x = 2$ is 8.
4. Either the trapezoidal rule or Simpson's rule can be used to approximate $\int_0^{10} \sqrt{x^4+1}\,dx$ with $h = 2$.

PRACTICE AND APPLICATIONS

In Exercises 5–30, evaluate the given integrals.

5. $\int (4x^3 - x)\,dx$
6. $\int (5 + 3t^2)\,dt$
7. $\int \sqrt{u}(u^2+8)\,du$
8. $\int x(x - 3x^4)\,dx$
9. $\int_1^4 \left(\frac{\sqrt{x}}{2} + \frac{2}{\sqrt{x}}\right)dx$
10. $\int_1^2 \left(x + \frac{4}{x^2}\right)dx$
11. $\int \frac{x^2+4x}{\sqrt{x}}\,dx$
12. $\int \frac{6x^3 - 4\sqrt{x}}{x}\,dx$
13. $\int_0^2 5x(4-x)\,dx$
14. $\int_0^1 2t(2t+1)^2\,dt$
15. $\int \left(5^2 + \frac{6}{x^3}\right)dx$
16. $\int \left(3\sqrt{x} + \frac{1}{2\sqrt{x}} - \frac{1}{4}\right)dx$
17. $\int_{-2}^5 \frac{dx}{\sqrt[3]{x^2+6x+9}}$
18. $\int_{0.35}^{0.85} x(\sqrt{1-x^2}+1)\,dx$
19. $\int \frac{10\,dn}{(9-5n)^3}$
20. $\int \frac{2}{x^2}\sqrt{1+\frac{1}{x}}\,dx$
21. $\int 3(7-2x)^{3/4}\,dx$
22. $\int \frac{x}{(x^2+1)^2}\,dx$
23. $\int_0^2 \frac{3x\,dx}{\sqrt[3]{1+2x^2}}$
24. $\int_1^6 \frac{12\,dx}{(3x-2)^{3/4}}$
25. $\int 4x^2(1-2x^3)^4\,dx$
26. $\int 3R^3(1-5R^4)\,dR$
27. $\int \frac{(2-3x^2)\,dx}{(2x-x^3)^2}$
28. $\int \frac{2x^2-6}{\sqrt{6+9x-x^3}}\,dx$
29. $\int_1^3 (x^2+x+2)(2x^3+3x^2+12x)\,dx$
30. $\int_0^2 (4x+18x^2)(x^2+3x^3)^2\,dx$

In Exercises 31–46, solve the given problems.

31. Find an equation of the curve that passes through $(-1, 3)$ for which the slope is given by $3 - x^2$.

32. Find an equation of the curve that passes through $(1, -2)$ for which the slope is $x(x^2 + 1)^2$.

33. Perform the integration $\int(1 - 2x)dx$ (a) term by term, labeling the constant of integration as C_1, and then (b) by letting $u = 1 - 2x$, using the general power rule and labeling the constant of integration as C_2. Is $C_1 = C_2$? Explain.

34. Following the methods (a) and (b) in Exercise 33, perform the integration $\int(3x + 2)dx$. In (b) let $u = 3x + 2$. Is $C_1 = C_2$? Explain.

35. Write $\int_3^8 F(v)dv - \int_4^8 F(v)dv$ as a single definite integral.

36. Show that $\int_{-3}^0 x^3\,dx = -\int_0^3 x^3\,dx$.

37. Show that $\int_2^9 \sqrt{3x-2}\,dx = \int_2^6 \sqrt{3x-2}\,dx + \int_6^9 \sqrt{3x-2}\,dx$.

38. Evaluate $\int_0^4 (3x - 4)dx$ by (a) geometrically finding the area represented, and (b) by integration.

39. Find the general form of the function whose second derivative is $1/\sqrt{6x + 5}$.

40. Find an equation of the curve for which the second derivative is $-6x$ if the curve passes through $(-1, 3)$ with a slope of -3.

41. Show that $\int_0^1 x^3\,dx = \int_1^2 (x - 1)^3\,dx$. In terms of area, explain this result.

42. If $\int_0^1 [f(x) - g(x)]dx = 3$ and $\int_0^1 g(x)dx = -1$, find the value of $\int_0^1 2f(x)dx$.

43. Use Eq. (25.10) to find the area under $y = 6x - 1$ between $x = 1$ and $x = 3$.

44. Use Eq. (25.10) to find the first-quadrant area under $y = 8x - x^4$.

45. Given that $f(x)$ is continuous, $f(x) > 0$, and $f''(x) < 0$ for $a \le x \le b$, explain why the exact value of $\int_a^b f(x)dx$ is greater than the approximate value found by use of the trapezoidal rule.

46. It is shown in more advanced works that when evaluating $\int_a^b f(x)dx$, the maximum error in using Simpson's rule is $\frac{M(b - a)^5}{180n^4}$, where M is the greatest absolute value of the fourth derivative of $f(x)$ for $a \le x \le b$. Evaluate the maximum error for the integral $\int_2^3 \frac{dx}{2x}$ with $n = 4$.

In Exercises 47 and 48, solve the given problems by using the trapezoidal rule. In Exercises 49 and 50, solve the given problems using Simpson's rule.

47. Approximate $\int_1^3 \frac{dx}{2x - 1}$ with $n = 4$.

48. The streamflow F (in m³/s) passing through a cross section of a stream is found by evaluating $F = \int_0^L h(x)dx$, where L is the width of the stream cross section and $h(x)$ is the product of the depth and velocity of the stream x m from the bank. From the following table, estimate F with $L = 24$ m.

x(m)	0.0	3.0	6.0	9.0	12.0	15.0	18.0	21.0	24.0
$h(x)$	0.00	0.59	0.13	0.34	0.76	0.65	0.29	0.07	0.00

49. Approximate $\int_1^3 \frac{dx}{2x - 1}$ with $n = 4$ (see Exercise 47).

50. The velocity v (in km/h) of a car was recorded at 1-min intervals as shown. Estimate the distance traveled by the car.

t (min)	0	1	2	3	4	5	6	7	8	9	10
v (km/h)	60	62	65	69	72	74	76	77	77	75	76

In Exercises 51–58, use the function $y = x\sqrt[3]{2x^2 + 1}$ and approximate the area under the curve between $x = 1$ and $x = 4$ by the indicated method. See Fig. 25.31.

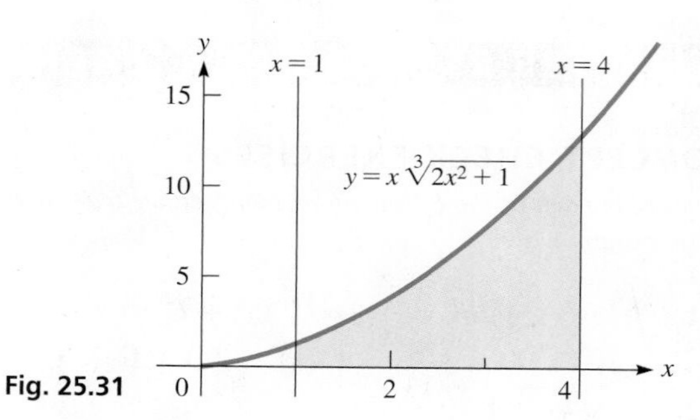

Fig. 25.31

51. Find the sum of the areas of three inscribed rectangles.

52. Find the sum of the areas of six inscribed rectangles.

53. Use the trapezoidal rule with $n = 3$.

54. Use the trapezoidal rule with $n = 6$.

55. Use Simpson's rule with $n = 2$.

56. Use Simpson's rule with $n = 3$ or explain why it can't be done.

57. Use Simpson's rule with $n = 6$.

58. Use integration (for the exact area).

In Exercises 59–62, find the area of the archway, as shown in Fig. 25.32, by the indicated method. The archway can be described as the area bounded by the elliptical arc $y = 4 + \sqrt{1 + 8x - 2x^2}$, $x = 0$, $x = 4$, and $y = 0$, where dimensions are in meters.

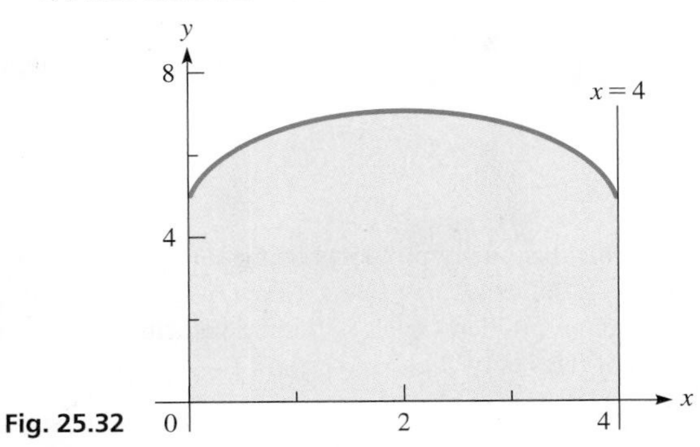

Fig. 25.32

59. Find the sum of the areas of eight inscribed rectangles.

60. Use the trapezoidal rule with $n = 8$.

61. Use Simpson's rule with $n = 8$.

62. Use the integration evaluation feature on a calculator.

In Exercises 63–68, solve the given problems by integration.

63. The charge Q (in C) transmitted in a certain electric circuit in 3 s is given by $Q = \int_0^3 6t^2\, dt$. Evaluate Q.

64. The velocity ds/dt (in m/s) of a projectile is $ds/dt = -9.8t + 16$. Find the displacement s of the object after 4.0 s if the initial displacement is 48 m.

65. The deflection y of a certain beam at a distance x from one end is given by $dy/dx = k(2L^3 - 12Lx + 2x^4)$, where k is a constant and L is the length of the beam. Find y as a function of x if $y = 0$ for $x = 0$.

66. The total electric charge Q on a charged sphere is given by $Q = k\int\left(r^2 - \frac{r^3}{R}\right)dr$, where k is a constant, r is the distance from the center of the sphere, and R is the radius of the sphere. Find Q as a function of r if $Q = Q_0$ for $r = R$.

67. Part of the deck of a boat is the parabolic area shown in Fig. 25.33. The area A (in m²) is $A = 2\int_0^5 \sqrt{5 - y}\, dy$. Evaluate A.

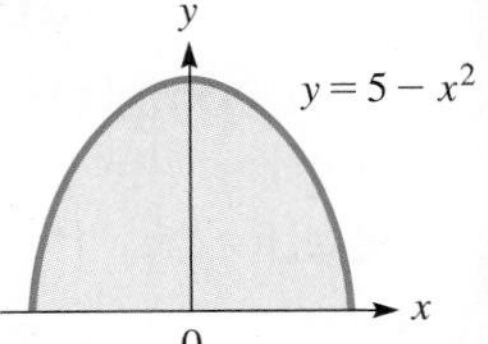

Fig. 25.33

68. The distance s (in in.) through which a cam follower moves in 4 s is $s = \int_0^4 t\sqrt{4 + 9t^2}\, dt$. Evaluate s.

69. A computer science student is writing a program to find a good approximation for the value of π by using the formula $A = \pi r^2$ for a circle. The value of π is to be found by approximating the area of a circle with a given radius. Write two or three paragraphs explaining how the value of π can be approximated in this way. Include any equations and values that may be used, but do not actually make the calculations.

CHAPTER 25 PRACTICE TEST

As a study aid, we have included complete solutions for each Practice Test problem at the end of this book.

1. Find an antiderivative of $f(x) = 2x - (1 - x)^4$.

2. Integrate: $\int 6x\sqrt{1 - 2x^2}\, dx$.

3. Find y in terms of x if $dy/dx = (6 - x)^4$ and the curve passes through $(5, 2)$.

4. Approximate the area under $y = \frac{4}{x + 2}$ between $x = 1$ and $x = 4$ (above the x-axis) by inscribing six rectangles and finding the sum of their areas.

5. Evaluate $\int_1^4 \frac{4dx}{x + 2}$ by using the trapezoidal rule with $n = 6$.

6. Evaluate the definite integral of Problem 5 by using Simpson's rule with $n = 6$.

7. The total electric current i (in A) to pass a point in a circuit between $t = 1$ s and $t = 3$ s is $i = \int_1^3 \left(t^2 + \frac{1}{t^2}\right)dt$. Evaluate i.

26

Applications of Integration

LEARNING OUTCOMES

After completion of this chapter, the student should be able to:

- Solve application problems involving indefinite integrals, including velocity and displacement and voltage across a capacitor
- Find the area between curves
- Obtain the volume of a solid of revolution
- Find the center of mass, the moment of inertia, and the radius of gyration of a flat plate and of a solid of revolution
- Solve application problems involving definite intergrals, including work, liquid pressure, and average value of a function

With the development of calculus, many problems being studied in the 1600s and later were much more easily solved. As we saw in Chapters 23 and 24, differential calculus led to the solution of problems such as finding velocities, maximum and minimum values, and various types of rates of change.

Integral calculus also led to the solution of many types of problems that were being studied. As Newton and Leibniz developed the methods of integral calculus, they were interested in finding areas and the problems that could be solved by finding these areas. They were also very interested in applications in which the rate of change was known and therefore led to the relation between the variables being studied.

One of the problems being studied in the 1600s by the French mathematician and physicist Blaise Pascal was that of the pressure within a liquid and the force on the walls of the container due to this pressure. Because pressure is a measure of the force on an area, finding the force became essentially solving an area problem. Also at that time, many mathematicians and physicists were studying various kinds of motion, such as motion along a curved path and the motion of a rotating object. When information about the velocity is known, the solution is found by integrating. Later in the 1800s, because electric current is the time rate change of electric charge, the current could be found by integrating known expressions for the charge.

As it turns out, integration is useful in many areas of science, engineering, and technology. It has important applications in areas such as electricity, mechanics, architecture, machine design, and business, as well as other areas of physics and geometry.

In the first section of this chapter, we present some important applications of the indefinite integral, with emphasis on the motion of an object and the voltage across a capacitor. In the remaining sections, we show uses of the definite integral related to geometry, mechanics, work by a variable force, and force due to liquid pressure.

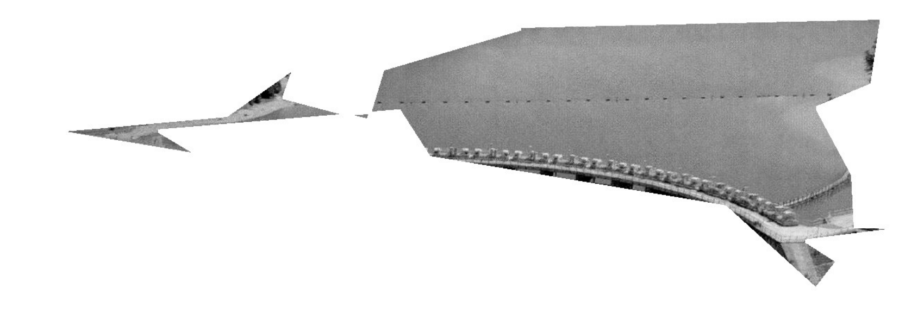

▶ **In Section 26.6, integration is used to find the force that water exerts on the floodgate of a dam.**

26.1 Applications of the Indefinite Integral

Velocity and Displacement • Voltage Across a Capacitor

VELOCITY AND DISPLACEMENT

We first apply integration to the problem of finding the displacement and velocity as functions of time, when we know the relationship between acceleration and time, and certain values of displacement and velocity. As shown in Section 25.2, these values are needed for finding the constants of integration that are introduced.

■ Velocity as a first derivative and acceleration as a second derivative were introduced in Chapter 23.

Recalling that the acceleration a of an object is given by $a = dv/dt$, we can find the expression for the velocity v in terms of a, the time t, and the constant of integration. Therefore, we write $dv = a\,dt$, or

$$v = \int a\,dt \tag{26.1}$$

If the acceleration is constant, we have

$$v = at + C_1 \tag{26.2}$$

In general, Eq. (26.1) is used to find the velocity as a function of time when we know the acceleration as a function of time. Because the case of constant acceleration is often encountered, Eq. (26.2) can often be used.

EXAMPLE 1 Given the acceleration—find the velocity

Find the expression for the velocity if $a = 12t$, given that $v = 8$ when $t = 1$.

Using Eq. (26.1), we have

$$v = \int (12t)\,dt = 6t^2 + C_1$$

Substituting the known values, we have $8 = 6 + C_1$, or $C_1 = 2$. This means that

$$v = 6t^2 + 2$$

■

EXAMPLE 2 Constant acceleration—find the velocity of a falling object

For an object falling under the influence of gravity, the acceleration due to gravity is essentially constant. Its value is -32 ft/s^2. [The negative sign is chosen so that all quantities directed up are positive and all quantities directed down are negative.] Find the expression for the velocity of an object under the influence of gravity if $v = v_0$ when $t = 0$.

NOTE ▸

We write

$$\begin{aligned} v &= \int (-32)\,dt && \text{substitute } a = -32 \text{ in Eq. (26.1)} \\ &= -32t + C_1 && \text{integrate} \\ v_0 &= -32(0) + C_1 && \text{substitute given values} \\ C_1 &= v_0 && \text{solve for } C_1 \\ v &= v_0 - 32t && \text{substitute} \end{aligned}$$

The velocity v_0 is called the *initial velocity*. If the object is given an initial upward velocity of 100 ft/s, $v_0 = 100$ ft/s. If the object is dropped, $v_0 = 0$. If the object is given an initial downward velocity of 100 ft/s, $v_0 = -100$ ft/s. ■

Once we have the expression for velocity, we can then integrate to find the expression for displacement s in terms of the time. Because $v = ds/dt$, we can write $ds = v\,dt$, or

$$s = \int v\,dt \tag{26.3}$$

EXAMPLE 3 Find the initial velocity of a ball

A ball is thrown vertically from the top of a building 200 ft high and hits the ground 5.0 s later. What initial velocity was the ball given?

Measuring vertical distances from the ground, we know that $s = 200$ ft when $t = 0$ and that $v = v_0 - 32t$. Thus,

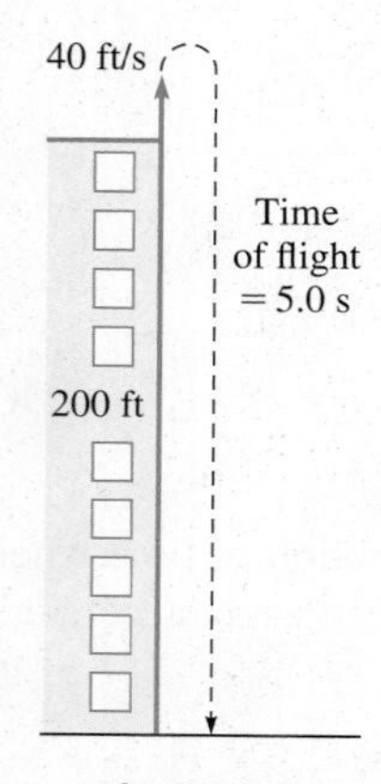

Fig. 26.1

$$s = \int (v_0 - 32t)\,dt = v_0t - 16t^2 + C \quad \text{integrate}$$

$$200 = v_0(0) - 16(0) + C, \qquad C = 200 \quad \text{evaluate } C$$

$$s = v_0t - 16t^2 + 200$$

We also know that $s = 0$ when $t = 5.0$ s. Thus,

$$0 = v_0(5.0) - 16(5.0)^2 + 200 \quad \text{substitute given values}$$

$$5.0v_0 = 200$$

$$v_0 = 40 \text{ ft/s}$$

This means that the initial velocity was 40 ft/s upward. See Fig. 26.1. ■

EXAMPLE 4 Given acceleration—find the displacement of a spacecraft

During the initial stage of launching a spacecraft vertically, the acceleration a (in m/s^2) of the spacecraft is $a = 6t^2$. Find the height s of the spacecraft after 6.0 s if $s = 12$ m for $t = 0.0$ s and $v = 16$ m/s for $t = 2.0$ s.

First, we use Eq. (26.1) to get an expression for the velocity:

$$v = \int 6t^2\,dt = 2t^3 + C_1 \quad \text{integrate}$$

$$16 = 2(2.0)^3 + C_1, \qquad C_1 = 0 \quad \text{evaluate } C_1$$

$$v = 2t^3$$

We now use Eq. (26.3) to get an expression for the displacement:

$$s = \int 2t^3\,dt = \tfrac{1}{2}t^4 + C_2 \quad \text{integrate}$$

$$12 = \tfrac{1}{2}(0.0)^4 + C_2, \qquad C_2 = 12 \quad \text{evaluate } C_2$$

$$s = \tfrac{1}{2}t^4 + 12$$

Now, finding s for $t = 6.0$ s, we have $s = \frac{1}{2}(6.0)^4 + 12 = 660$ m ■

Practice Exercise

1. In Example 4, change the acceleration to $a = 4$ m/s^2 and then find the height under the same given conditions.

VOLTAGE ACROSS A CAPACITOR

The second basic application of the indefinite integral we will discuss comes from the field of electricity. By definition, *the current i in an electric circuit equals the time rate of change of the charge q (in coulombs) that passes a given point in the circuit,* or

$$i = \frac{dq}{dt} \tag{26.4}$$

Rewriting this expression $dq = i\,dt$ and integrating both sides, we have

$$q = \int i\,dt \tag{26.5}$$

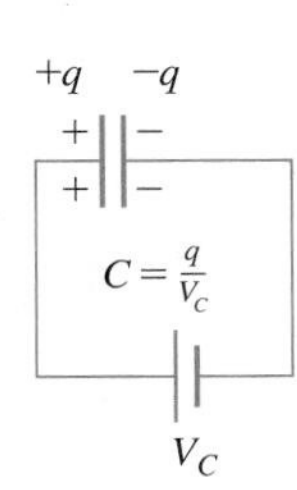

Fig. 26.2

Now, the voltage V_C across a capacitor with capacitance C (see Fig. 26.2) is given by $V_C = q/C$. By combining equations, the voltage V_C is given by

$$V_C = \frac{1}{C}\int i\,dt \tag{26.6}$$

Here, V_C is measured in volts, C in farads, i in amperes, and t in seconds.

EXAMPLE 5 Given the current—find the electric charge

■ Here, C represents coulombs and is not the C for capacitance or the constant of integration.

The current i (in A) in an electric circuit as a function of time t (in s) is given by $i = 6t^2 + 4$. Find an expression for the electric charge q (in C) that passes a point in the circuit as a function of t. If $q = 0$ C when $t = 0$ s, determine the total charge that passes the point in 2.0 s.

Because $q = \int i\,dt$, we have

$$\begin{aligned} q &= \int (6t^2 + 4)\,dt && \text{substitute into Eq. (26.5)} \\ &= 2t^3 + 4t + C && \text{integrate} \end{aligned}$$

This is the required expression of charge as a function of time. When $t = 0$, $q = C$, which means the constant of integration represents the initial charge, or the charge that passed a given point before the timing started. Using q_0 to represent this charge, we have

$$q = 2t^3 + 4t + q_0$$

Returning to the second part of the problem, we note that $q_0 = 0$. Evaluating q for $t = 2.0$ s, we have $q = 2(2.0)^3 + 4(2.0) = 24$ C, which is the charge that passes any point in 2.0 s. ■

Practice Exercise

2. In Example 5, change the current to $i = 12t + 6$, and then find the charge under the same given conditions.

EXAMPLE 6 Find the voltage across a capacitor

The voltage across a 5.0-μF capacitor is zero. What is the voltage after 20 ms if a current of 75 mA charges the capacitor?

Because the current is 75 mA, we know that $i = 0.075\text{ A} = 7.5 \times 10^{-2}$ A. **CAUTION** We see that we must use the proper power of 10 that corresponds to each prefix. ■ Because $5.0\ \mu\text{F} = 5.0 \times 10^{-6}$ F, we have

$$\begin{aligned} V_C &= \frac{1}{5.0 \times 10^{-6}}\int 7.5 \times 10^{-2} dt && \text{substituting into Eq. (26.6)} \\ &= (1.5 \times 10^4)\int dt = (1.5 \times 10^4)t + C_1 && \text{integrate} \end{aligned}$$

From the given information, we know that $V_C = 0$ when $t = 0$. Thus,

$$0 = (1.5 \times 10^4)(0) + C_1 \quad \text{or} \quad C_1 = 0 \qquad \text{evaluate } C_1$$

This means that

$$V_C = (1.5 \times 10^4)t$$

Evaluating this expression for $t = 20 \times 10^{-3}$ s, we have

$$\begin{aligned} V_C &= (1.5 \times 10^4)(20 \times 10^{-3}) \\ &= 30 \times 10 = 300\text{ V} \end{aligned}$$

■

EXAMPLE 7 Find the capacitance of a capacitor

A certain capacitor has 100 V across it. At this instant, a current $i = 0.06t^{1/2}$ is sent through the circuit. After 0.25 s, the voltage across the capacitor is 140 V. What is the capacitance?

Substituting $i = 0.06t^{1/2}$, we find that

$$V_C = \frac{1}{C}\int (0.06t^{1/2}dt) = \frac{0.06}{C}\int t^{1/2}dt \quad \text{using Eq. (26.6)}$$

$$= \frac{0.04}{C}t^{3/2} + C_1 \quad \text{integrate}$$

From the given information, we know that $V_C = 100$ V when $t = 0$. Thus,

$$100 = \frac{0.04}{C}(0) + C_1 \quad \text{or} \quad C_1 = 100\ V \quad \text{evaluate } C_1$$

$$V_C = \frac{0.04}{C}t^{3/2} + 100 \quad \text{substituting 100 for } C_1$$

We also know that $V_C = 140$ V when $t = 0.25$ s. Therefore,

$$140 = \frac{0.04}{C}(0.25)^{3/2} + 100$$

$$40 = \frac{0.04}{C}(0.125)$$

$$C = 1.25 \times 10^{-4}\ \text{F} = 125\ \mu\text{F}$$

■

EXERCISES 26.1

1. In Example 3, change 5.0 s to 2.5 s and then solve the resulting problem.
2. In Example 7, change $0.06t^{1/2}$ to $0.06t$ and then solve the resulting problem.
3. What is the velocity (in ft/s) of a sandbag 1.5 s after it is released from a hot-air balloon that is rising at 12 ft/s? (*Hint:* The acceleration of gravity is -32 ft/s^2.)
4. A beach ball is rolled up a shallow slope with an initial velocity of 18 ft/s. If the acceleration of the ball is 3.0 ft/s^2 down the slope, find the velocity of the ball after 8.0 s.
5. A conveyor belt 8.00 m long moves at 0.25 m/s. If a package is placed at one end, find its displacement from the other end as a function of time.
6. During each cycle, the velocity v (in mm/s) of a piston is $v = 6t - 6t^2$, where t is the time (in s). Find the displacement s of the piston after 0.75 s if the initial displacement is zero.
7. The velocity (in km/h) of a plane flying into an increasing headwind is $v = 50(12 - t)$, where t is the time (in h). How far does the plane travel in a 2.0-h trip?
8. A cyclist goes downhill for 15 min with a velocity $v = 40 + 64t$ (in km/h), and then maintains the speed at the bottom for another 30 min. How far does the cyclist go in the 45 min?
9. A car crosses an intersection as a fire engine approaches the intersection at 64 ft/s. How far does the fire engine travel while stopping to avoid a collision, if its acceleration is $-8.0t$?
10. In designing a highway, a civil engineer must determine the length of a highway on-ramp for cars going onto the ramp at 25 km/h and entering the highway at 95 km/h in 12.0 s. What minimum length should the on-ramp be?
11. While in the barrel of a tennis ball machine, the acceleration a (in ft/s^2) of a ball is $a = 90\sqrt{1 - 4t}$, where t is the time (in s). If $v = 0$ for $t = 0$, find the velocity of the ball as it leaves the barrel at $t = 0.25$ s.
12. A person skis down a slope with an acceleration (in m/s^2) given by $a = \frac{600t}{(60 + 0.5t^2)^2}$, where t is the time (in s). Find the skier's velocity as a function of time if $v = 0$ when $t = 0$.
13. A certain Chevrolet Corvette goes from 0 mi/h to 60.0 mi/h (88.0 ft/s) in 3.60 s. Assuming constant acceleration, how far (in ft) does it travel in this time?
14. The engine of a lunar lander is cut off when the lander is 5.0 m above the surface of the moon and descending at 2.0 m/s. If the acceleration due to gravity on the moon is 1.6 m/s^2, what is the speed of the lander just before it touches the surface?

15. A catapult can launch a plane from the deck of an aircraft carrier from 0 to 260 km/h in 2.0 s. How many g's is the average acceleration for such a launch? ($1 \text{ g} = 9.8 \text{ m/s}^2$.)

16. A stone is thrown straight up from the edge of a 45.0-m-high cliff. A loose stone at the edge of the cliff falls off 1.50 s later. What is the vertical velocity of the first stone, if the two stones reach the ground below at the same time?

17. What must be the nozzle velocity of the water from a fire hose if it is to reach a point 90 ft directly above the nozzle?

18. An arrow is shot from 5.0 ft above the top of a hill with a vertical upward velocity of 108 ft/s. If it strikes the plain below after 7.5 s, how high is the hill?

19. A truck driver traveling at 84 ft/s suddenly sees a bicyclist going in the same direction 120 ft ahead. Because of oncoming traffic the driver slams on the brakes and decelerates at 12 ft/s^2. If the cyclist continues on at 32 ft/s, will the truck hit the bicycle? Solve using a calculator to graph distances traveled.

20. A hoist mechanism raises a crate with an acceleration (in m/s^2) $a = \sqrt{1 + 0.2t}$, where t is the time in seconds. Find the displacement of the crate as a function of time if $v = 0$ m/s and $s = 2$ m for $t = 0$ s.

21. The electric current in a microprocessor circuit is 0.230 μA. How many coulombs pass a given point in the circuit in 1.50 ms?

22. The voltage across a 0.10-μF capacitor in a microwave oven is zero. What is the voltage after being charged by a 0.25 μA current for 3.0 μs?

23. In an amplifier circuit, the current i (in A) changes with time t (in s) according to $i = 0.06t\sqrt{1 + t^2}$. If 0.015 C of charge has passed a point in the circuit at $t = 0$, find the total charge to have passed the point at $t = 0.25$ s.

24. The current i (in μA) in a DVD player circuit is given by $i = 6.0 - 0.50t$, where t is the time (in μs) and $0 \le t \le 30$ μs. If $q_0 = 0$ C, for what value of t (other than $t = 0$) is $q = 0$ C? What interpretation can be given to this result?

25. The voltage across a 2.5-μF capacitor in a copying machine is zero. What is the voltage after 12 ms if a current of 25 mA charges the capacitor?

26. The voltage across an 8.50-nF capacitor in an FM receiver circuit is zero. Find the voltage after 2.00 μs if a current (in mA) $i = 0.042t$ charges the capacitor.

27. The voltage across a 3.75-μF capacitor in a television circuit is 4.50 mV. Find the voltage after 0.565 ms if a current (in μA) $i = \sqrt[3]{1 + 6t}$ further charges the capacitor.

28. A current $i = t/\sqrt{t^2 + 1}$ (in A) is sent through an electric dryer circuit containing a previously uncharged (zero voltage) 2.0-μF capacitor. How long does it take for the capacitor voltage to reach 120 V?

29. The angular velocity ω is the time rate of change of the angular displacement θ of a rotating object. See Fig. 26.3. In testing the shaft of an engine, its angular velocity is $\omega = 16t + 0.50t^2$, where t is the time (in s) of rotation. Find the angular displacement through which the shaft goes in 10.0 s.

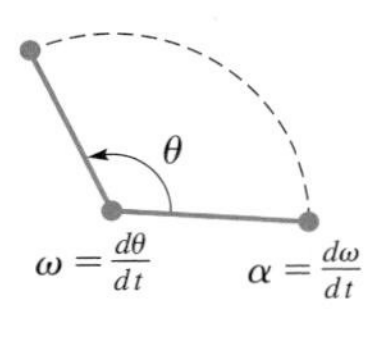

Fig. 26.3

30. The angular acceleration α is the time rate of change of angular velocity ω of a rotating object. See Fig. 26.3. When starting up, the angular acceleration of a helicopter blade is $\alpha = \sqrt{8t + 1}$. Find the expression for θ if $\omega = 0$ and $\theta = 0$ for $t = 0$.

31. An inductor in an electric circuit is essentially a coil of wire in which the voltage is affected by a changing current. By definition, the voltage caused by the changing current is given by $V_L = L(di/dt)$, where L is the inductance (in H). If $V_L = 12.0 - 0.2t$ for a 3.0-H inductor, find the current in the circuit after 20 s if the initial current was zero.

32. If the inner and outer walls of a container are at different temperatures, the rate of change of temperature with respect to the distance from one wall is a function of the distance from the wall. Symbolically, this is stated as $dT/dx = f(x)$, where T is the temperature. If x is measured from the outer wall, at 20°C, and $f(x) = 72x^2$, find the temperature at the inner wall if the container walls are 0.5 cm thick.

33. Show that a 1-mA constant current increases the voltage on a 1-μF capacitor by 1 V in 1 ms.

34. The rate of change of the vertical deflection y with respect to the horizontal distance x from one end of a beam is a function of x. For a particular beam, the function is $k(x^5 + 1350x^3 - 7000x^2)$, where k is a constant. Find y as a function of x if $y = 0$ when $x = 0$.

35. Freshwater is flowing into a brine solution, with an equal volume of mixed solution flowing out. The amount of salt in the solution decreases, but more slowly as time increases. Under certain conditions, the time rate of change of mass of salt (in g/min) is given by $-1/\sqrt{t + 1}$. Find the mass m of salt as a function of time if 1000 g were originally present. Under these conditions, how long would it take for all the salt to be removed?

36. A holograph of a circle is formed. The rate of change of the radius r of the circle with respect to the wavelength λ of the light used is inversely proportional to the square root of λ. If $dr/d\lambda = 3.55 \times 10^4$ and $r = 4.08$ cm for $\lambda = 574$ nm, find r as a function of λ.

Answers to Practice Exercises

1. 132 m 2. 36 C

26.2 Areas by Integration

Summing Elements of Area • Vertical Elements • Horizontal Elements

In Section 25.3, we introduced the method of finding the area under a curve by integration. We also showed that the area can be found by a summation process on the rectangles inscribed under the curve, which means that integration can be interpreted as a summation process. *The applications of the definite integral use this summation interpretation of the integral.* We now develop a general procedure for finding the area for which the bounding curves are known by summing the areas of inscribed rectangles and using integration for the summation.

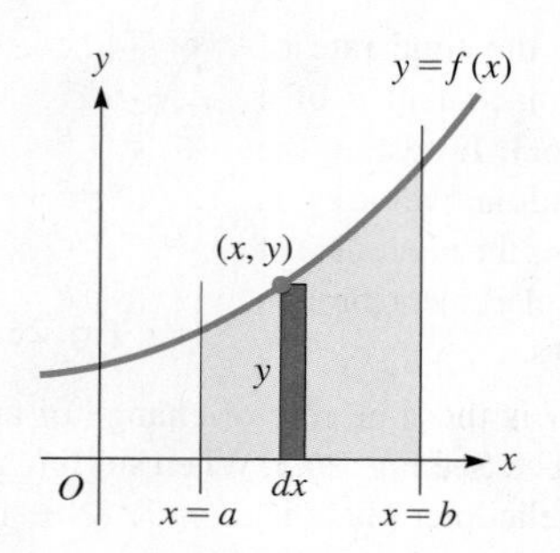

Fig. 26.4

The first step is to make a sketch of the area. Next, a representative **element of area** dA (a typical rectangle) is drawn. In Fig. 26.4, the width of the element is dx. The length of the element is determined by the y-coordinate (of the vertex of the element) of the point on the curve. Thus, the length is y. The area of this element is $y\,dx$, which in turn means that $dA = y\,dx$, or

$$A = \int_a^b y\,dx = \int_a^b f(x)\,dx \tag{26.7}$$

This equation states that the elements are to be summed (this is the meaning of the integral sign) from a (the left boundary) to b (the right boundary).

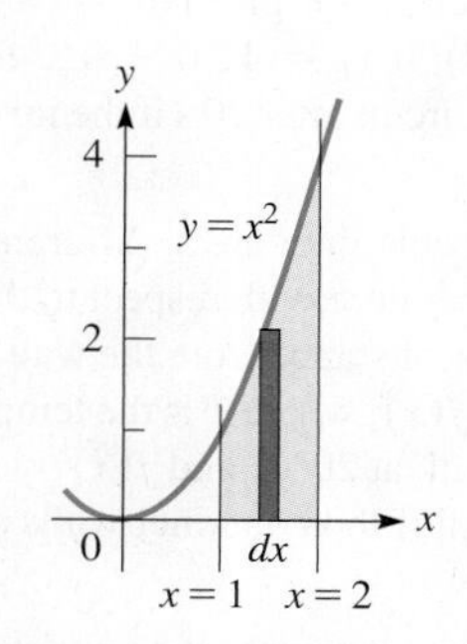

Fig. 26.5

EXAMPLE 1 Find area using vertical elements

Find the area bounded by $y = x^2$, $x = 1$, and $x = 2$.

This area is shown in Fig. 26.5. The rectangle shown is the representative element, and its area is $y\,dx$. The elements are to be summed from $x = 1$ to $x = 2$.

sum — right boundary

$$A = \int_1^2 y\,dx$$

left boundary — area of element

$$A = \int_1^2 y\,dx = \int_1^2 x^2\,dx \qquad \text{substitute } x^2 \text{ for } y$$

$$= \frac{1}{3}x^3\Big|_1^2 = \frac{1}{3}(8) - \frac{1}{3}(1) \qquad \text{integrate and evaluate}$$

$$= \frac{7}{3}$$

■

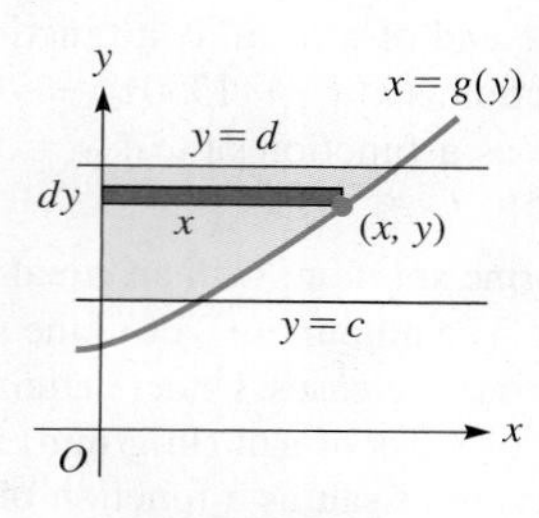

Fig. 26.6

In Figs. 26.4 and 26.5 the elements of area are vertical. Some problems are simplified using horizontal elements. With horizontal elements, the length (longest dimension) is measured in terms of the x-coordinate of the point on the curve, and the width is dy. In Fig. 26.6, the area of the element is $x\,dy$, which means $dA = x\,dy$, or

$$A = \int_c^d x\,dy = \int_c^d g(y)\,dy \tag{26.8}$$

In using Eq. (26.8), the elements are summed from c (the lower boundary) to d (the upper boundary).

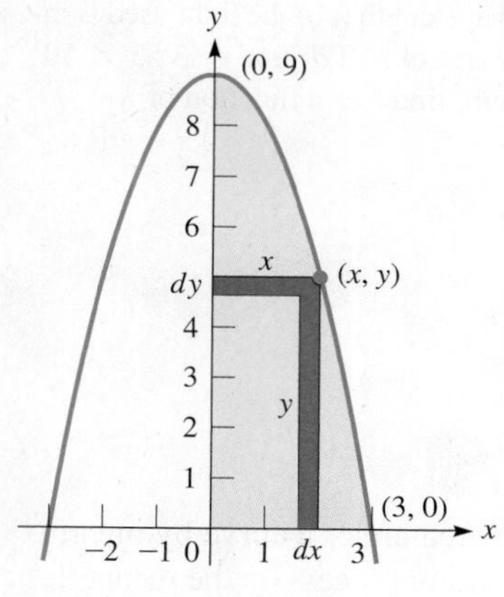

Fig. 26.7

EXAMPLE 2 Vertical and horizontal elements of area

Find the area in the first quadrant bounded by $y = 9 - x^2$. See Fig. 26.7.

vertical element of length y and width dx — horizontal element of length x and width dy

$$A = \int_0^3 y\,dx \qquad \text{sum of areas of elements} \qquad A = \int_0^9 x\,dy$$

$$= \int_0^3 (9 - x^2)dx \qquad \text{substitute for } y\text{; substitute for } x \qquad = \int_0^9 \sqrt{9 - y}\,dy = -\int_0^9 (9 - y)^{1/2}(-dy)$$

$$= \left(9x - \frac{x^3}{3}\right)\Big|_0^3 \qquad \text{integrate} \qquad = -\frac{2}{3}(9 - y)^{3/2}\Big|_0^9$$

$$= (27 - 9) - 0 = 18 \qquad \text{evaluate} \qquad = -\frac{2}{3}(9 - 9)^{3/2} + \frac{2}{3}(9 - 0)^{3/2} = 18$$

Note that the limits of integration are 0 to 3 for the vertical elements, and 0 to 9 for the horizontal elements. They are chosen so that the summation is done in a positive direction. [As we have noted, vertical elements are summed from left to right, and horizontal elements are summed from bottom to top.]

NOTE ▸

■

Practice Exercise

1. Find the area in the first quadrant bounded by $y = 4 - x^2$.

The choice of vertical or horizontal elements is determined by (1) which one leads to the simplest solution or (2) the form of the resulting integral. In some problems, it makes little difference which is chosen. However, in other cases, the choice of using either a horizontal or a vertical element greatly affects the difficulty of the solution.

AREA BETWEEN TWO CURVES

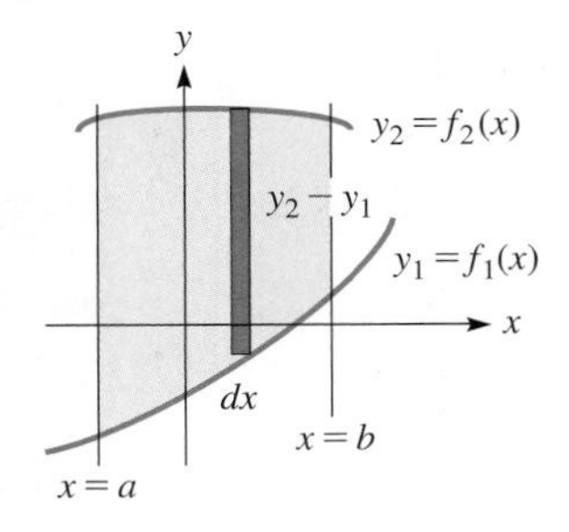

Fig. 26.8

It is also possible to find the area between two curves when one of the curves is not an axis. In such a case, the length of the element of area becomes the difference in the y- or x-coordinates, depending on whether a vertical element or a horizontal element is used.

In Fig. 26.8, by using vertical elements, the element of area is bounded on the bottom by $y_1 = f_1(x)$ and on the top by $y_2 = f_2(x)$. The length of the element is $y_2 - y_1$, and its width is dx. Thus, the area is

$$A = \int_a^b (y_2 - y_1)dx \tag{26.9}$$

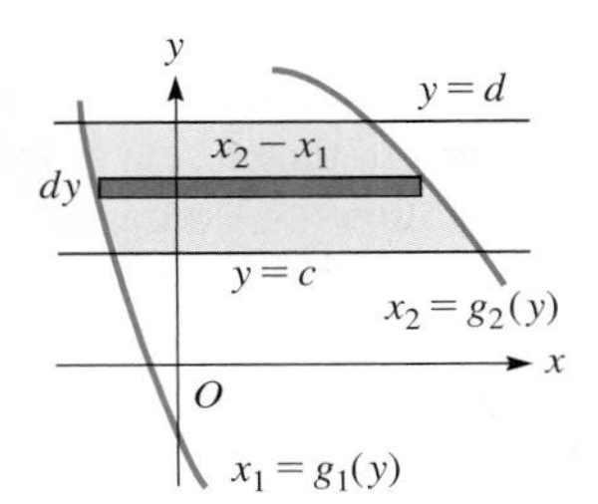

Fig. 26.9

In Fig. 26.9, by using horizontal elements, the element of area is bounded on the left by $x_1 = g_1(y)$ and on the right by $x_2 = g_2(y)$. The length of the element is $x_2 - x_1$, and its width is dy. Thus, the area is

$$A = \int_c^d (x_2 - x_1)dy \tag{26.10}$$

The following examples show the use of Eqs. (26.9) and (26.10) to find the indicated areas.

EXAMPLE 3 Area between curves—vertical elements

Find the area bounded by $y = x^2$ and $y = x + 2$.

First, by sketching each curve, we see that the area to be found is that shown in Fig. 26.10. The points of intersection of these curves are found by solving the equations simultaneously. The solution for the x-values is shown at the left. We then find the y-coordinates by substituting into either equation. The substitution shows the points of intersection to be $(-1, 1)$ and $(2, 4)$.

■ $x^2 = x + 2$
$x^2 - x - 2 = 0$
$(x + 1)(x - 2) = 0$
$x = -1, 2$

Here, we choose vertical elements, because they are all bounded at the top by the line $y = x + 2$ and at the bottom by the parabola $y = x^2$. If we were to choose horizontal elements, the bounding curves are different above $(-1, 1)$ from below this point. Choosing horizontal elements would then require two separate integrals for solution. Therefore, using vertical elements, we have

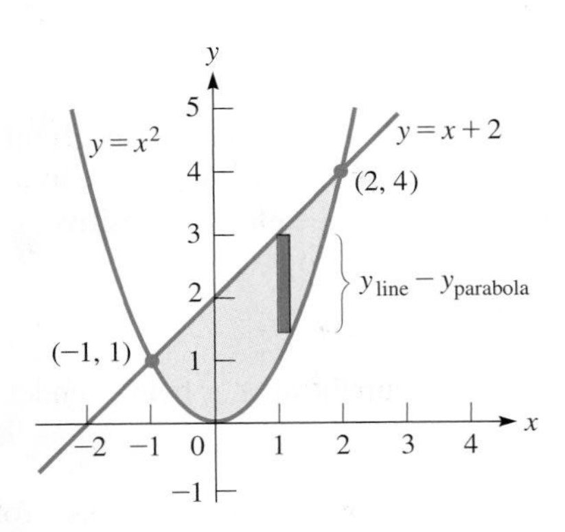

Fig. 26.10

$$A = \int_{-1}^{2} (y_{\text{line}} - y_{\text{parabola}})dx \quad \text{using Eq. (26.9)}$$

$$= \int_{-1}^{2} (x + 2 - x^2)dx = \left(\frac{x^2}{2} + 2x - \frac{x^3}{3}\right)\Bigg|_{-1}^{2}$$

$$= \left(2 + 4 - \frac{8}{3}\right) - \left(\frac{1}{2} - 2 + \frac{1}{3}\right)$$

$$= \frac{10}{3} + \frac{7}{6} = \frac{27}{6}$$

$$= \frac{9}{2}$$

■

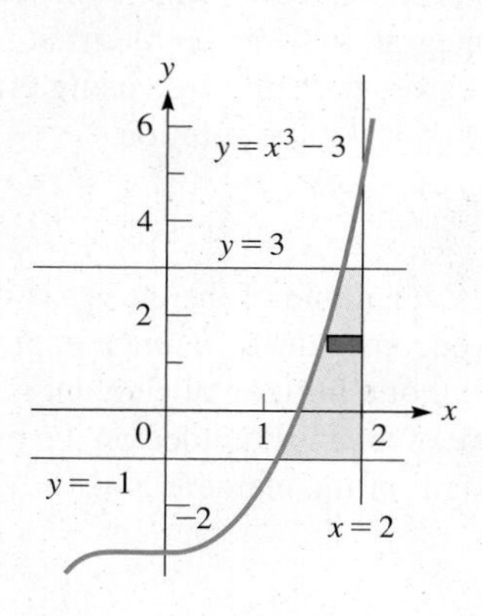

Fig. 26.11

EXAMPLE 4 Area between curves—horizontal elements

Find the area bounded by the curve $y = x^3 - 3$ and the lines $x = 2$, $y = -1$, and $y = 3$.

Sketching the curve and lines, we show the area in Fig. 26.11. Horizontal elements are better, because they avoid having to evaluate the area in two parts. Therefore, we have

$$A = \int_{-1}^{3} (x_{\text{line}} - x_{\text{cubic}})dy = \int_{-1}^{3} (2 - \sqrt[3]{y + 3})dy \quad \text{using Eq. (26.10)}$$

$$= 2y - \frac{3}{4}(y + 3)^{4/3}\Big|_{-1}^{3} = \left[6 - \frac{3}{4}(6^{4/3})\right] - \left[-2 - \frac{3}{4}(2^{4/3})\right]$$

$$= 8 - \frac{9}{2}\sqrt[3]{6} + \frac{3}{2}\sqrt[3]{2} = 1.713$$

As we see, the choice of horizontal elements leads to the limits of integration −1 and 3. If we had chosen vertical elements, the limits would have been $\sqrt[3]{2}$ and $\sqrt[3]{6}$ for the area to the left of $(\sqrt[3]{6}, 3)$, and $\sqrt[3]{6}$ and 2 to the right of this point. ■

CAUTION It is important to set up the element of area so that its length is positive. If the difference is taken incorrectly, the result will show a negative area. Getting positive lengths can be ensured for vertical elements if we subtract y of the lower curve from y of the upper curve. For horizontal elements, we should subtract x of the left curve from x of the right curve. ■ This important point is illustrated in the following example.

EXAMPLE 5 Area below the x-axis

Find the area bounded by $x^3 - 3x - 2$ and the x-axis.

Sketching the graph, we find that $y = x^3 - 3x - 2$ has a maximum point at $(-1, 0)$, a minimum point at $(1, -4)$, and an intercept at $(2, 0)$. The curve is shown in Fig. 26.12, and we see that the area is *below* the x-axis. Using vertical elements, we see that the top is the x-axis $(y = 0)$ and the bottom is the curve $y = x^3 - 3x - 2$. Therefore, we have

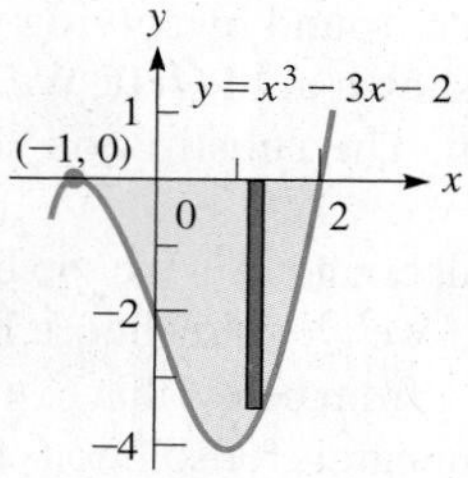

Fig. 26.12

$$A = \int_{-1}^{2} [0 - (x^3 - 3x - 2)]dx = \int_{-1}^{2} (-x^3 + 3x + 2)dx$$

$$= -\frac{1}{4}x^4 + \frac{3}{2}x^2 + 2x\Big|_{-1}^{2}$$

$$= \left[-\frac{1}{4}(2^4) + \frac{3}{2}(2^2) + 2(2)\right] - \left[-\frac{1}{4}(-1)^4 + \frac{3}{2}(-1)^2 + 2(-1)\right]$$

$$= \frac{27}{4} = 6.75$$

NOTE ▸ [If we had simply set up the area as $A = \int_{-1}^{2}(x^3 - 3x - 2)\,dx$, we would have found $A = -6.75$. The negative sign shows that the area is below the x-axis.] Again, we avoid any complications with negative areas by making the length of the element positive.

Also, note that because

$$0 - (x^3 - 3x - 2) = -(x^3 - 3x - 2)$$

an area bounded on top by the x-axis can be found by setting up the area as being "under" the curve and using the negative of the function. ■

Practice Exercise

2. Find the area bounded by $y = x^2 - 4$ and the x-axis.

NOTE ▸ [We must be very careful if the bounding curves of an area cross.] In such a case, for part of the area one curve is above the area, and for a different part of the area this same curve is below the area. When this happens, *two integrals must be used* to find the area. The following example illustrates the necessity of using this procedure.

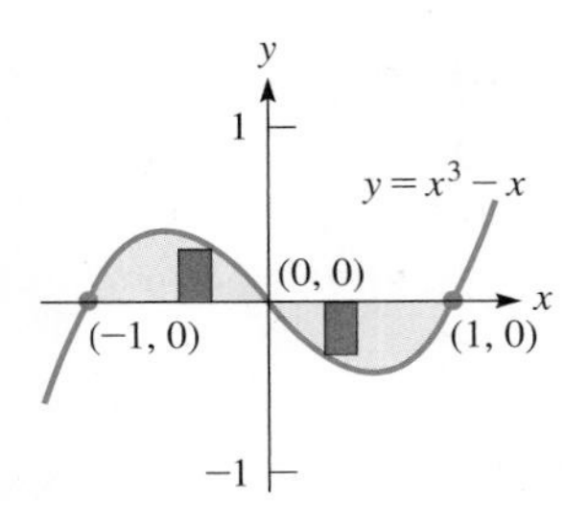

Fig. 26.13

EXAMPLE 6 Area above and below the x-axis

Find the area between $y = x^3 - x$ and the x-axis.

We note from Fig. 26.13 that the area to the left of the origin is above the axis and the area to the right is below. If we find the area from

$$A = \int_{-1}^{1} (x^3 - x)\,dx = \frac{x^4}{4} - \frac{x^2}{2}\Big|_{-1}^{1}$$
$$= \left(\frac{1}{4} - \frac{1}{2}\right) - \left(\frac{1}{4} - \frac{1}{2}\right) = 0$$

we see that the apparent area is zero. From the figure, we know this is not correct. Noting that the y-values (of the area) are negative to the right of the origin, we set up the integrals

$$A = \int_{-1}^{0} (x^3 - x)dx + \int_{0}^{1} [0 - (x^3 - x)]dx$$
$$= \left(\frac{x^4}{4} - \frac{x^2}{2}\right)\Big|_{-1}^{0} + \left(-\frac{x^4}{4} + \frac{x^2}{2}\right)\Big|_{0}^{1}$$
$$= \left[0 - \left(\frac{1}{4} - \frac{1}{2}\right)\right] + \left[\left(-\frac{1}{4} + \frac{1}{2}\right) - 0\right] = \frac{1}{2}$$

■

EXAMPLE 7 Area under a curve—finding total solar energy

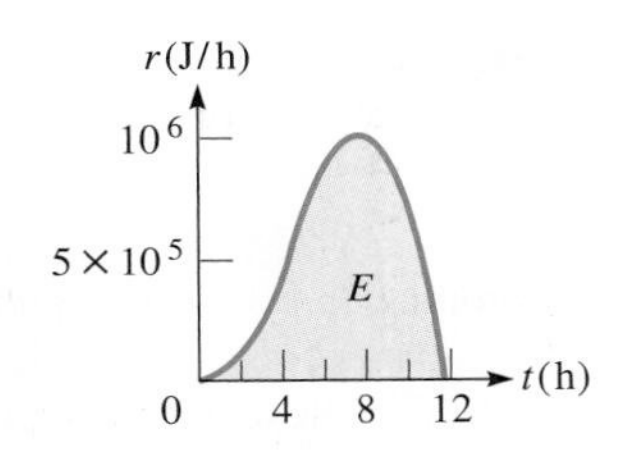

Fig. 26.14

Measurements of solar radiation on a particular surface indicate that the rate r (in J/h) at which solar energy is received between 8 A.M. and 8 P.M. on a certain day is given by $r = 3600(12t^2 - t^3)$, where t is the time (in h) after 8 A.M. Because r is a rate, we write $r = dE/dt$, where E is the energy (in J) received at the surface. Thus, $dE = 3600(12t^2 - t^3)dt$, and we find the total energy over the 12-h period by evaluating the integral.

$$E = 3600 \int_{0}^{12} (12t^2 - t^3)dt$$

This integral can be interpreted as being the area under $f(t) = 3600(12t^2 - t^3)$ from $t = 0$ to $t = 12$, as shown in Fig. 26.14. Evaluating this integral, we have

$$E = 3600 \int_{0}^{12} (12t^2 - t^3)dt = 3600\left(4t^3 - \frac{1}{4}t^4\right)\Big|_{0}^{12}$$
$$= 3600\left[4(12^3) - \frac{1}{4}(12^4) - 0\right]$$
$$= 6.22 \times 10^6 \text{ J}$$

Therefore, 6.22 MJ of energy were received in 12 h.

A calculator evaluation of this definite integral is shown in Fig. 26.15. ■

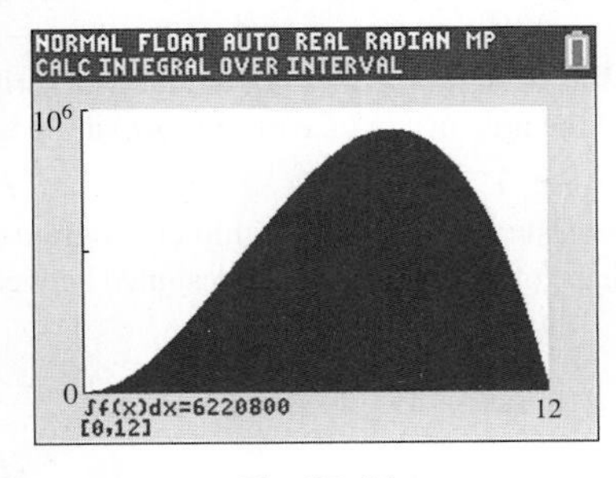

Fig. 26.15
Graphing calculator keystrokes: goo.gl/kYr5Dv

EXERCISES 26.2

In Exercises 1 and 2, make the given changes in the indicated examples of this section and then find the resulting areas.

1. In Example 1, change $x = 2$ to $x = 3$.
2. In Example 3, change $y = x + 2$ to $y = 2x$.

In Exercises 3–28, find the areas bounded by the indicated curves.

3. $y = 4x, y = 0, x = 1$
4. $y = 3x + 3, y = 0, x = 2$
5. $y = 8 - 2x^2, y = 0$
6. $y = \frac{1}{2}x^2 + 2; x = 0, y = 4\ (x > 0)$
7. $y = x^2 - 4, y = 0, x = -4$
8. $y = 4x^2 - 6x, y = 0$
9. $y = 3/x^2, y = 0, x = 2, x = 3$
10. $y = 16 - x^2, y = 0, x = -2, x = 3$

11. $y = 4\sqrt{x}, x = 0, y = 1, y = 3$

12. $y = 3\sqrt{x + 1}, x = 0, y = 6$

13. $y = 2/\sqrt{x}, x = 0, y = 1, y = 4$

14. $x = y^2 - 4y, x = 0$

15. $y = 6 - 3x, x = 0, y = 0, y = 3$

16. $y = x, y = 3 - x, x = 0$

17. $y = x - 4\sqrt{x}, y = 0$

18. $y = 2x^3 - x^4, y = 0$

19. $y = x^2, y = 2 - x, x = 0 \quad (x \geq 0)$

20. $y = x^2, y = 2 - x, y = 1$

21. $y = x^4 - 8x^2 + 16, y = 16 - x^4$

22. $y = \sqrt{x - 1}, y = 3 - x, y = 0$

23. $y = x^2 + 5x, y = 3 - x^2$

24. $y = x^3, y = x^2 + 4, x = -1$

25. $y = \frac{1}{2}x^5, x = -1, x = 2, y = 0$

26. $y = x^2 + 2x - 8, y = x + 4$

27. $y = 4 - x^2, y = 4x - x^2, x = 0, x = 2$

28. $y = x^2\sqrt{1 - x^3}, y = 0$

In Exercises 29–38, solve the given problems.

29. Describe a region for which the area is found by evaluating the integral $\int_1^2 (2x^2 - x^3)dx$.

30. Although the integral $\int_{-2}^{2} \sqrt{4 - x^2}\,dx$ cannot be integrated by methods we have developed to this point, by recognizing the region represented, it can be evaluated. Evaluate this integral.

31. Use integration to find the area of the triangle with vertices (0, 0), (4, 4), and (10, 0).

32. Show that the area bounded by the parabola $y = x^2$ and the line $y = b\ (b > 0)$ is two-thirds of the area of the rectangle that circumscribes it.

33. Show that the curve $y = x^n (n > 0)$ divides the unit square bounded by $x = 0, y = 0, x = 1$, and $y = 1$ into regions with areas in the ratio of $n/1$.

34. Why can the integral $\int_a^2 (2 + x - x^2)dx$ be used to find the area bounded by $x = a, y = 0$, and $y = 2 + x - x^2$ if $a = -1$, but not if $a = -2$?

35. Find the area of the parallelogram with vertices at (0, 0), (2, 0), (2, 1), and (4, 1) by integration. Show any integrals you set up.

36. Set up the integrals (do not evaluate) for the upper of the two areas bounded by $y = 4 - x^2, y = 3x$, and $y = 4 - 2x$, using vertical elements of area.

37. Find the value of c such that the region bounded by $y = x^2$ and $y = 4$ is divided by $y = c$ into two regions of equal area.

38. Find the positive value of c such that the region bounded by $y = x^2 - c^2$ and $y = c^2 - x^2$ has an area of 576.

In Exercises 39–42, find the areas bounded by the indicated curves, using (a) vertical elements and (b) horizontal elements.

39. $y = 8x, x = 0, y = 4$

40. $y = x^3, x = 0, y = 3$

41. $y = x^4, y = 8x$

42. $y = 4x, y = x^3 + 1$

In Exercises 43–50, some applications of areas are shown.

43. Certain physical quantities are often represented as an area under a curve. By definition, power is the time rate of change of performing work. Thus, $p = dw/dt$, or $dw = p\,dt$. If $p = 12t - 4t^2$, find the work (in J) performed in 3 s by finding the area under the curve of p vs. t. See Fig. 26.16. Round the answer to three significant digits.

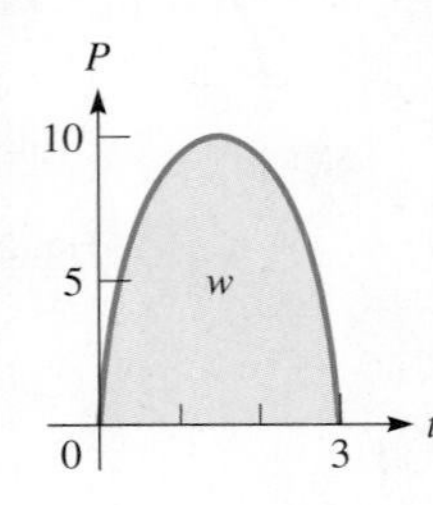

Fig. 26.16

44. The total electric charge Q (in C) to pass a point in the circuit from time t_1 to t_2 is $Q = \int_{t_1}^{t_2} i\,dt$, where i is the current (in A). Find Q if $t_1 = 1$ s, $t_2 = 4$ s, and $i = 0.0032t\sqrt{t^2 + 1}$.

45. Because the displacement s, velocity v, and time t of a moving object are related by $s = \int v\,dt$, it is possible to represent the change in displacement as an area. A rocket is launched such that its vertical velocity v (in km/s) as a function of time t (in s) is $v = 1 - 0.01\sqrt{2t + 1}$. Find the change in vertical displacement from $t = 10$ s to $t = 100$ s.

46. The total cost C (in dollars) of production can be interpreted as an area. If the cost per unit C' (in dollars per unit) of producing x units is given by $100/(0.01x + 1)^2$, find the total cost of producing 100 units by finding the area under the curve of C' vs x.

47. A cam is designed such that one face of it is described as being the area between the curves $y = x^3 - 2x^2 - x + 2$ and $y = x^2 - 1$ (units in cm). Show that this description does not uniquely describe the face of the cam. Find the area of the face of the cam, if a complete description requires that $x \leq 1$.

48. Using CAD (computer-assisted design), an architect programs a computer to sketch the shape of a swimming pool designed between the curves

$$y = \frac{800x}{(x^2 + 10)^2} \qquad y = 0.5x^2 - 4x \qquad x = 8$$

(dimensions in m). Find the area of the surface of the pool.

49. A coffee-table top is designed to be the region between $y = 0.25x^4$ and $y = 12 - 0.25x^4$ (dimensions in dm). What is the area of the table top?

50. A window is designed to be the area between a parabolic section and a straight base, as shown in Fig. 26.17. What is the area of the window?

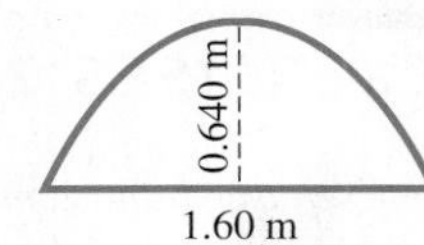

Fig. 26.17

Answers to Practice Exercises

1. 16/3 **2.** 32/3

26.3 Volumes by Integration

Summing Elements of Volume • Disk • Cylindrical Shell

Consider a region in the xy-plane and its representative element of area, as shown in Fig. 26.18(a). When the region is revolved about the x-axis, it is said to generate a **solid of revolution,** which is also shown in the figure. We now show methods of finding volumes of solids that are generated in this way.

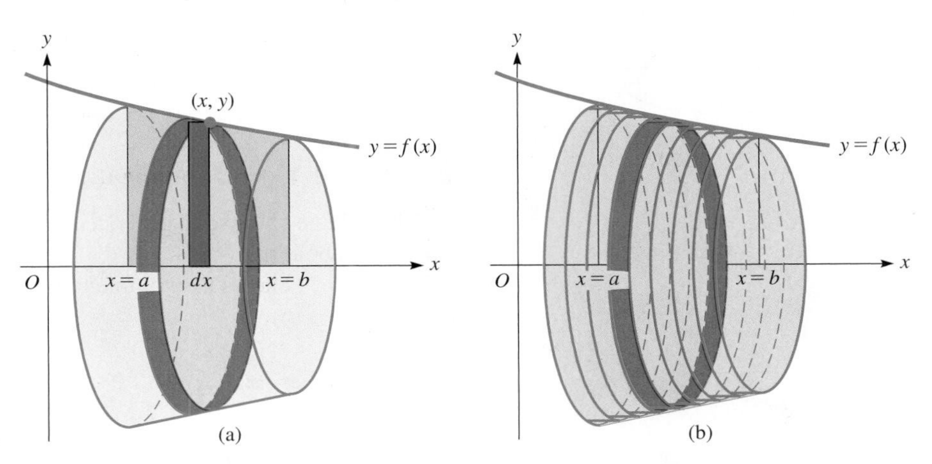

Fig. 26.18

As the region revolves about the x-axis, so does its representative element, which generates a solid for which the volume is known—an infinitesimally thin cylindrical **disk.** The volume of a right circular cylinder is π times its radius squared times its height (in this case, the thickness) of the cylinder. Because the element is revolved about the x-axis, the y-coordinate of the point on the curve that touches the element is the radius. Also, the thickness is dx. This disk, the representative **element of volume,** has a volume of $dV = \pi y^2 dx$. Summing these elements of volume from left to right, as shown in Fig. 26.18(b), we have for the total volume

■ This is the sum of the volumes of the disks whose thickness approach zero as the number of disks approaches infinity.

$$V = \pi \int_a^b y^2 dx = \pi \int_a^b [f(x)]^2\, dx \qquad \textbf{(26.11)}$$

The element of volume is a **disk,** and *by use of Eq. (26.11), we can find the volume of the solid generated by a region bounded by the x-axis, which is revolved about the x-axis.*

EXAMPLE 1 Find volume using vertical disks

Find the volume of the solid generated by revolving the region bounded by $y = x^2$, $x = 2$, and $y = 0$ about the x-axis. See Fig. 26.19.

From the figure, we see that the radius of the disk is y and its thickness is dx. The elements are summed from left ($x = 0$) to right ($x = 2$):

$$V = \pi \int_0^2 y^2 dx \qquad \text{using Eq. (26.11)}$$

$$= \pi \int_0^2 (x^2)^2 dx = \pi \int_0^2 x^4 dx \qquad \text{substitute } x^2 \text{ for } y$$

$$= \frac{\pi}{5} x^5 \Big|_0^2 = \frac{32\pi}{5} \qquad \text{integrate and evaluate}$$

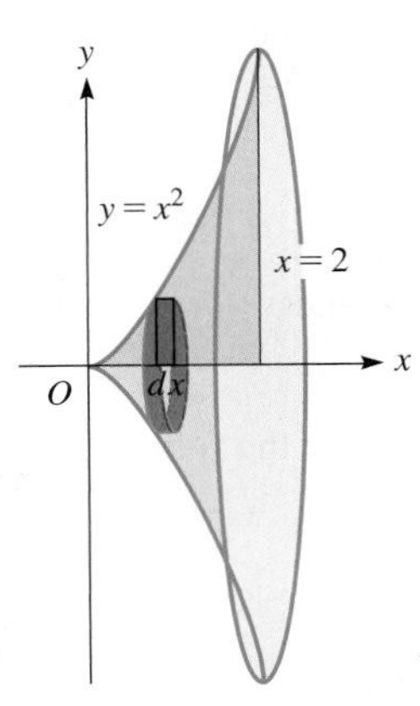

Fig. 26.19

NOTE ▸ [Because π is used in Eq. (26.11), it is common to leave results in terms of π. In applied problems, a decimal result would normally be given.] ■

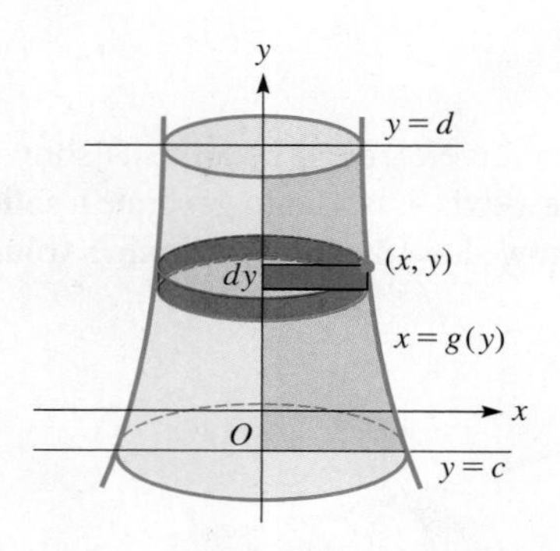

Fig. 26.20

If a region bounded by the y-axis is revolved about the y-axis, the volume of the solid generated is given by

$$V = \pi \int_c^d x^2\, dy \tag{26.12}$$

In this case, the radius of the element of volume, a ***disk,*** is the x-coordinate of the point on the curve, and the thickness of the disk is dy, as shown in Fig. 26.20. One should always be careful to identify the radius and the thickness properly.

EXAMPLE 2 Find volume using horizontal disks

Find the volume of the solid generated by revolving the region bounded by $y = 2x$, $y = 6$, and $x = 0$ about the y-axis.

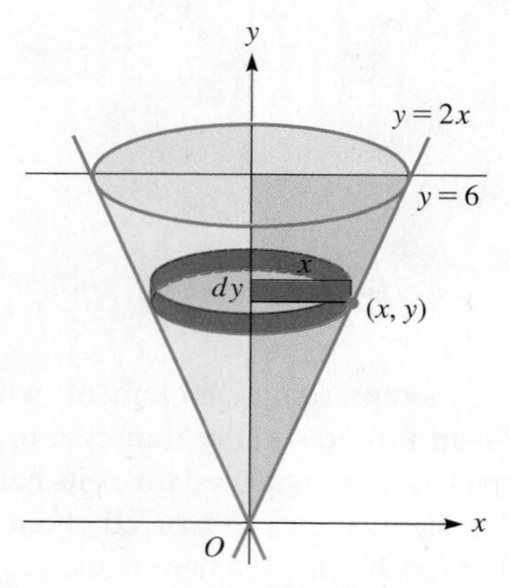

Fig. 26.21

Figure 26.21 shows the volume to be found. Note that the radius of the disk is x and its thickness is dy.

$$V = \pi \int_0^6 x^2 dy \qquad \text{using Eq. (26.12)}$$

$$= \pi \int_0^6 \left(\frac{y}{2}\right)^2 dy = \frac{\pi}{4}\int_0^6 y^2 dy \qquad \text{substituting } \frac{y}{2} \text{ for } x$$

$$= \frac{\pi}{12} y^3 \Big|_0^6 = 18\pi \qquad \text{integrate and evaluate}$$

Since this volume is a right circular cone, it is possible to check the result:

$$V = \frac{1}{3}\pi r^2 h = \frac{1}{3}\pi(3^2)(6) = 18\pi$$

■

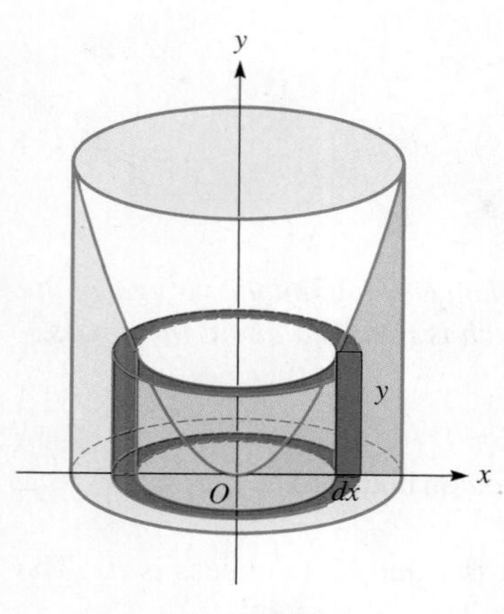

Fig. 26.22

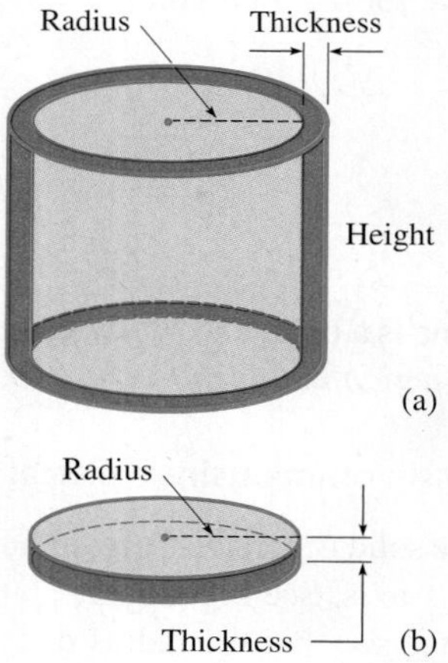

Radius
Thickness
Height
(a)
Radius
Thickness
(b)

Fig. 26.23

If the region in Fig. 26.22 is revolved about the y-axis, the element of the area $y\ dx$ generates a different element of volume from that when it is revolved about the x-axis. In Fig. 26.22, this element of volume is a **cylindrical shell.** *The total volume is made up of an infinite number of concentric shells.* When the volumes of these shells are summed, we have the total volume generated. Thus, we must now find the approximate volume dV of the representative shell. By finding the circumference of the base and multiplying this by the height, we obtain an expression for the surface area of the shell. Then, by multiplying this by the thickness of the shell, we find its volume. The volume of the representative ***shell*** shown in Fig. 26.23(a) is given below.

Shell

$$dV = 2\pi(\text{radius}) \times (\text{height}) \times (\text{thickness}) \tag{26.13}$$

Similarly, the volume of a ***disk*** ([see Fig. 26.23(b)]) is given by the following:

Disk

$$dV = \pi(\text{radius})^2 \times (\text{thickness}) \tag{26.14}$$

NOTE ▶ [When using disks, the representative rectangle is drawn *perpendicular* to the axis of revolution, and when using shells, it is drawn *parallel* to this axis.]

It is generally better to remember the formulas for the elements of volume in the general forms given in Eqs. (26.13) and (26.14), and not in the specific forms such as Eqs. (26.11) and (26.12) (both of these use *disks*). If we remember the formulas in this way, we can readily apply these methods to finding any such volume of a solid of revolution.

EXAMPLE 3 Find volume using cylindrical shells

Use the method of cylindrical shells to find the volume of the solid generated by revolving the first-quadrant region bounded by $y = 4 - x^2, x = 0$, and $y = 0$ about the y-axis.

From Fig. 26.24, we identify the radius, the height, and the thickness of the shell:

$$\text{radius} = x \qquad \text{height} = y \qquad \text{thickness} = dx$$

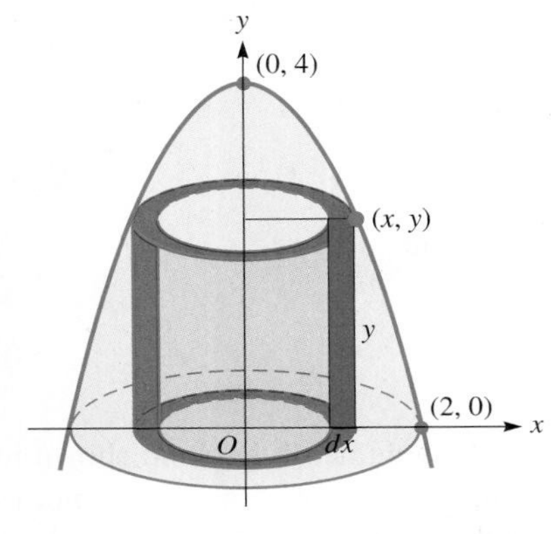

Fig. 26.24

The fact that the elements of area that generate the shells go from $x = 0$ ***to*** $x = 2$ ***determines the limits of integration as 0 and 2.*** Therefore,

$$V = 2\pi \int_0^2 xy\, dx \quad \text{using Eq. (26.13)}$$

(radius: x; height: y; thickness: dx)

$$= 2\pi \int_0^2 x(4 - x^2)dx = 2\pi \int_0^2 (4x - x^3)dx \quad \text{substitute } 4 - x^2 \text{ for } y$$

$$= 2\pi\left(2x^2 - \frac{1}{4}x^4\right)\Bigg|_0^2 \quad \text{integrate}$$

$$= 8\pi \quad \text{evaluate}$$

■

We can find the volume shown in Example 3 by using disks, as we show in the following example.

EXAMPLE 4 Same volume using horizontal disks

Use the method of disks to find the volume indicated in Example 3.

From Fig. 26.25, we identify the radius and the thickness of the disk:

$$\text{radius} = x \qquad \text{thickness} = dy$$

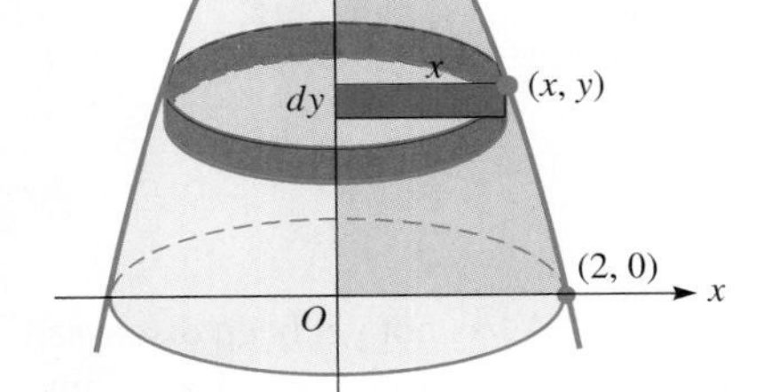

Fig. 26.25

Because the elements of area that generate the disks go from $y = 0$ ***to*** $y = 4$***, the limits of integration are 0 and 4.*** Thus,

$$V = \pi \int_0^4 x^2\, dy \quad \text{using Eq. (26.14)}$$

(radius: x; thickness: dy)

$$= \pi \int_0^4 (4 - y)dy \quad \text{substitute } \sqrt{4 - y} \text{ for } x$$

$$= \pi\left(4y - \frac{1}{2}y^2\right)\Bigg|_0^4 \quad \text{integrate}$$

$$= 8\pi \quad \text{evaluate}$$

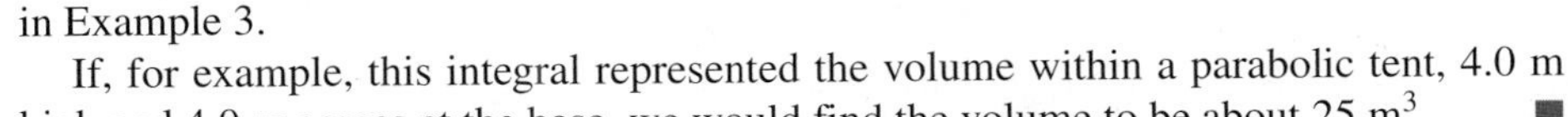

We see that the volume of 8π using disks agrees with the result we obtained using shells in Example 3.

If, for example, this integral represented the volume within a parabolic tent, 4.0 m high and 4.0 m across at the base, we would find the volume to be about 25 m^3. ■

Practice Exercise

1. Find the volume of the solid generated by revolving the first-quadrant region bounded by $y = 4 - x$ about the x-axis. Use disks.

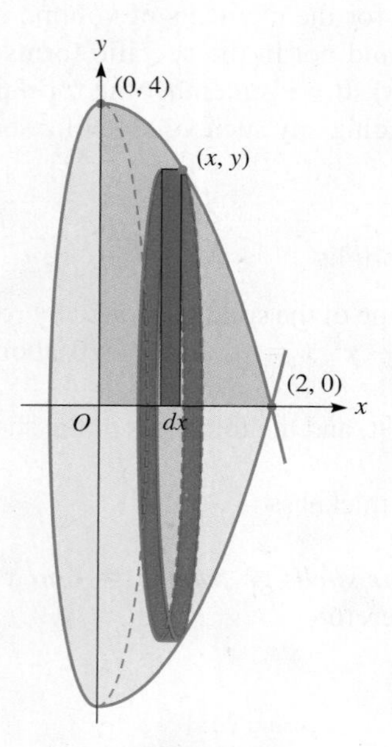

Fig. 26.26

EXAMPLE 5 Find volume using vertical disks

Using disks, find the volume of the solid generated by revolving the first-quadrant region bounded by $y = 4 - x^2$, $x = 0$, and $y = 0$ about the x-axis. (This is the same region as used in Examples 3 and 4.)

For the disk in Fig. 26.26, we have

$$\text{radius} = y \qquad \text{thickness} = dx$$

and the limits of integration are $x = 0$ and $x = 2$. This gives us

$$\begin{aligned} V &= \pi \int_0^2 y^2 dx && \text{using Eq. (26.14)} \\ &= \pi \int_0^2 (4 - x^2)^2 dx && \text{substitute } 4 - x^2 \text{ for } y \\ &= \pi \int_0^2 (16 - 8x^2 + x^4) dx \\ &= \pi\left(16x - \frac{8}{3}x^3 + \frac{1}{5}x^5\right)\Bigg|_0^2 && \text{integrate} \\ &= \frac{256\pi}{15} && \text{evaluate} \end{aligned}$$

■

We now show how to set up the integral to find the volume of the solid shown in Example 5 by using cylindrical shells. As it turns out, we are not able at this point to integrate the expression that arises, but we are still able to set up the proper integral.

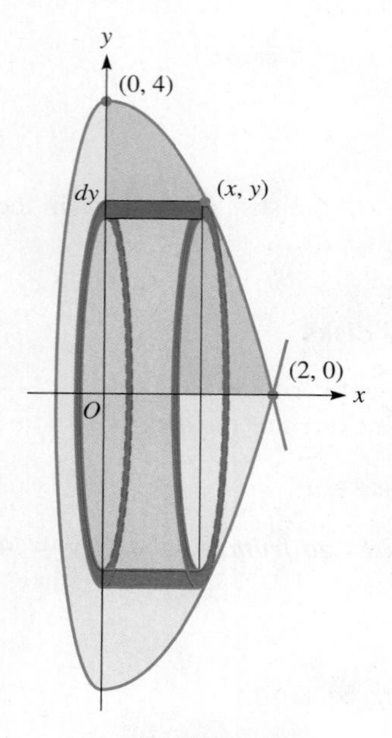

Fig. 26.27

EXAMPLE 6 Same volume using cylindrical shells

Use the method of cylindrical shells to find the volume indicated in Example 5.

From Fig. 26.27, we see for the shell we have

$$\text{radius} = y \qquad \text{height} = x \qquad \text{thickness} = dy$$

Because the elements go from $y = 0$ to $y = 4$, the limits of integration are 0 and 4. Hence,

$$\begin{aligned} V &= 2\pi \int_0^4 xy\, dy && \text{using Eq. (26.13)} \\ &= 2\pi \int_0^4 \sqrt{4 - y}\,(y\, dy) && \text{substitute } \sqrt{4 - y} \text{ for } x \\ &= \frac{256\pi}{15} \end{aligned}$$

The method of performing the integration $\int \sqrt{4 - y}\,(y\, dy)$ has not yet been discussed. We present the answer here for the reader's information to show that the volume found in this example is the same as that found in Example 5. ■

Practice Exercise

2. Find the volume of the solid generated by revolving the first-quadrant region bounded by $y = 4 - x$ about the x-axis. Use shells.

In the next example, we show how to find the volume of the solid generated if a region is revolved about a line other than one of the axes. We will see that a proper choice of the radius, height, and thickness for Eq. (26.13) leads to the result.

EXAMPLE 7 Rotate region about a line other than x- or y-axis

Find the volume of the solid generated if the region in Examples 3–6 is revolved about the line $x = 2$.

Shells are convenient, because the volume of a shell can be expressed as a single integral. We can find the radius, height, and thickness of the shell from Fig. 26.28. CAUTION Carefully note that **the radius is not x but is $2 - x$**, because the region is revolved about the line $x = 2$. ■ This means

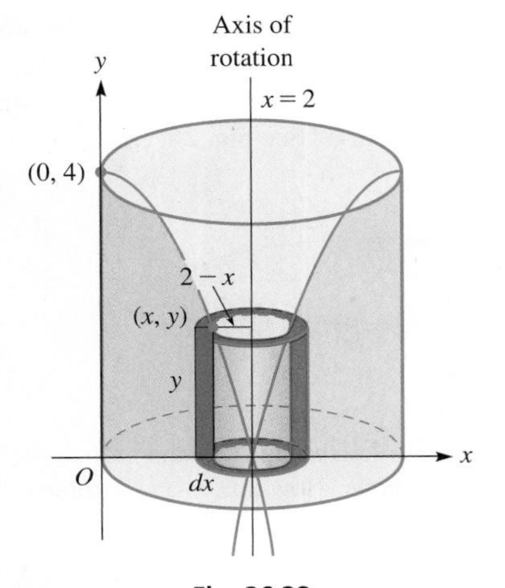

Fig. 26.28

$$\text{radius} = 2 - x \qquad \text{height} = y \qquad \text{thickness} = dx$$

Because the elements that generate the shells go from $x = 0$ to $x = 2$, the limits of integration are 0 and 2. This means we have

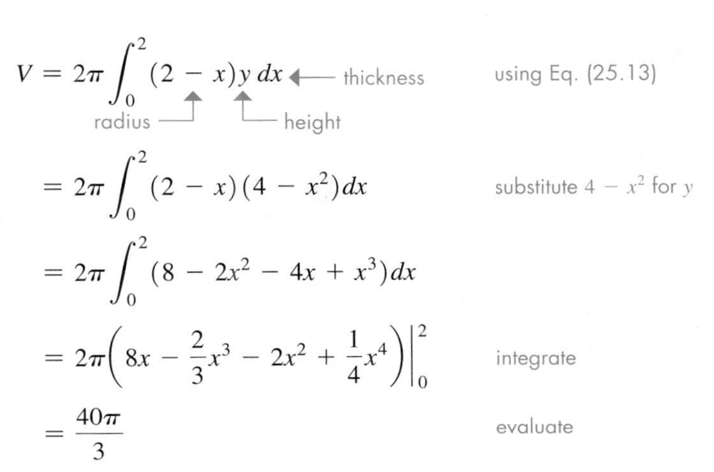

$$V = 2\pi \int_0^2 (2 - x)y\,dx \quad \text{using Eq. (25.13)}$$

(radius: $2 - x$; height: y; thickness: dx)

$$= 2\pi \int_0^2 (2 - x)(4 - x^2)dx \quad \text{substitute } 4 - x^2 \text{ for } y$$

$$= 2\pi \int_0^2 (8 - 2x^2 - 4x + x^3)dx$$

$$= 2\pi\left(8x - \frac{2}{3}x^3 - 2x^2 + \frac{1}{4}x^4\right)\Bigg|_0^2 \quad \text{integrate}$$

$$= \frac{40\pi}{3} \quad \text{evaluate}$$

■

EXERCISES 26.3

In Exercises 1 and 2, make the given changes in the indicated examples of this section and then find the indicated volumes.

1. In Example 1, change $y = x^2$ to $y = x^3$.

2. In Example 3, change $y = 4 - x^2$ to $y = 4 - x$.

In Exercises 3–6, find the volume generated by revolving the region bounded by $y = 4 - 2x$, $x = 0$, and $y = 0$ about the indicated axis, using the indicated element of volume.

3. x-axis (disks) **4.** y-axis (disks)

5. y-axis (shells) **6.** x-axis (shells)

In Exercises 7–16, find the volume generated by revolving the regions bounded by the given curves about the x-axis. Use the indicated method in each case.

7. $y = 2x, y = 0, x = 3$ (disks)

8. $y = \sqrt{x}, x = 0, y = 2$ (shells)

9. $y = 3\sqrt{x}, y = 0, x = 4$ (disks)

10. $y = 4x - x^2, y = 0$ (disks)

11. $y = x^3, y = 8, x = 0$ (shells)

12. $y = x^2 + 1, y = x + 1$ (shells)

13. $y = x^2 + 1, x = 0, x = 3, y = 0$ (disks)

14. $y = 6 - x - x^2, x = 0, y = 0$ (quadrant I), (disks)

15. $y = \sqrt{x}, y = 1, y = 2, x = 0$ (shells)

16. $y = x^4, x = 0, y = 1, y = 8$ (shells)

In Exercises 17–26, find the volume generated by revolving the regions bounded by the given curves about the y-axis. Use the indicated method in each case.

17. $y = 2x^{1/3}, x = 0, y = 2$ (disks)

18. $y = \sqrt{x^2 - 1}, y = 0, x = 3$ (shells)

19. $y = 3\sqrt{x}, x = 0, y = 3$ (disks)

20. $y^2 = x, y = 5, x = 0$ (disks)

21. $x^2 - 4y^2 = 4, x = 3$ (shells)

22. $y = 3x^2 - x^3, y = 0$ (shells)

23. $y = \sqrt{x}, y = 1, y = 2, x = 0$ (disks)

24. $x^2 + 4y^2 = 4$ (quadrant I), (disks)

25. $y = \sqrt{4 - x^2}$ (quadrant I), (shells)

26. $y = 8 - x^3, x = 0, y = 0$ (shells)

In Exercises 27–40, find the indicated volumes by integration.

27. Describe a region that is revolved about the x-axis to generate a

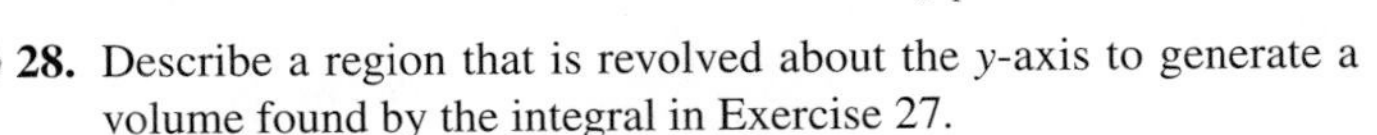

volume found by evaluating the integral $\pi \int_1^2 x^3\,dx$.

28. Describe a region that is revolved about the y-axis to generate a volume found by the integral in Exercise 27.

29. Find the volume generated if the region bounded by $y = 4 - x^2$, $y = 4$, and $x = 2$ is revolved about the line $x = 2$.

30. Find the volume generated if the region bounded by $y = \sqrt{x}$ and $y = x/2$ is revolved about the line $y = 4$.

31. Derive the formula for the volume of a right circular cone of radius r and height h by revolving the area bounded by $y = (r/h)x$, $y = 0$, and $x = h$ about the x-axis.

32. Explain how to derive the formula for the volume of a sphere by using the disk method.

33. The base of a floor floodlight can be represented as the area bounded by $x = y^2$ and $x = 25$ (in cm). The casing has all vertical *square* cross sections, with an edge in the xy-plane. See Fig. 26.29. The volume within the casing can be found by evaluating $\int_0^{25} A(x)dx$, where $A(x)$ represents the area (in terms of x) of a representative square cross section. Find this volume.

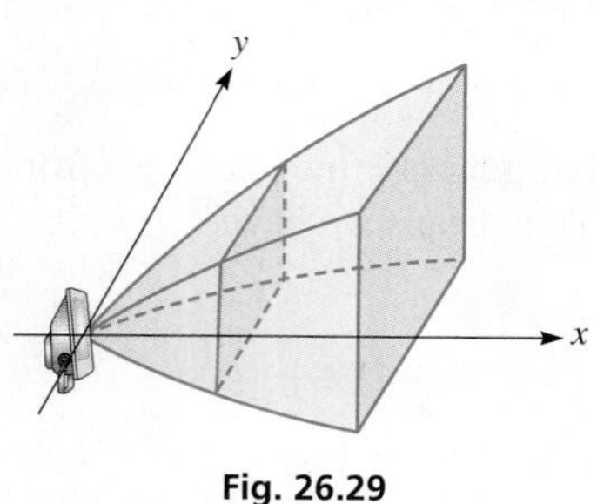

Fig. 26.29

34. The oil in a spherical tank 40.0 ft in diameter is 15.0 ft deep. How much oil is in the tank?

35. A *drumlin* is an oval hill composed of relatively soft soil that was deposited beneath glacial ice (the campus of Dutchess Community College in Poughkeepsie, New York, is on a drumlin). Computer analysis showed that the surface of a certain drumlin can be approximated by $y = 10(1 - 0.0001x^2)$ revolved 180° about the x-axis from $x = -100$ to $x = 100$ (see Fig. 26.30). Find the volume (in m^3) of this drumlin.

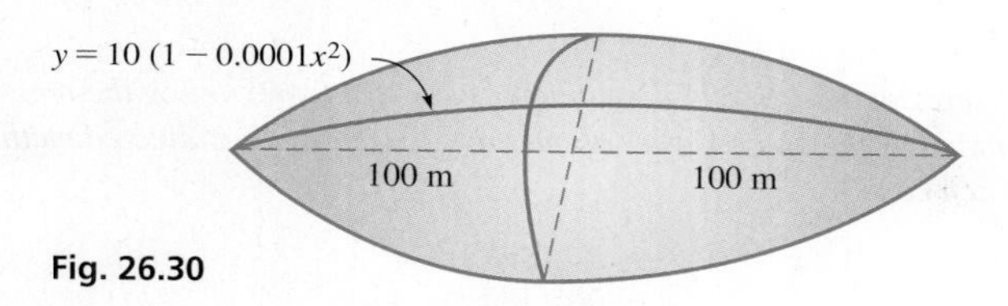

Fig. 26.30

36. If the area bounded by $y = 0$, $y = 2x$, and $x = 5$ is rotated about each axis, which volume is greater?

37. The ball used in Australian football is elliptical. Find its volume if it is 275 mm long and 170 mm wide.

38. A commercial airship used for outdoor advertising has a helium-filled balloon in the shape of an ellipse revolved about its major axis. If the balloon is 124 ft long and 36.0 ft in diameter, what volume of helium is required to fill it? See Fig. 26.31.

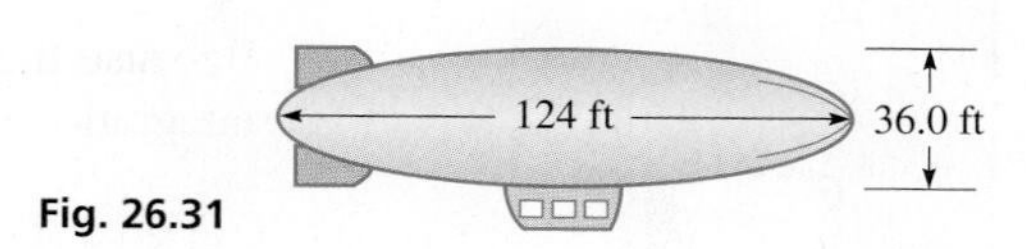

Fig. 26.31

39. A hole 2.00 cm in diameter is drilled through the center of a spherical lead weight 6.00 cm in diameter. How much lead is removed? See Fig. 26.32.

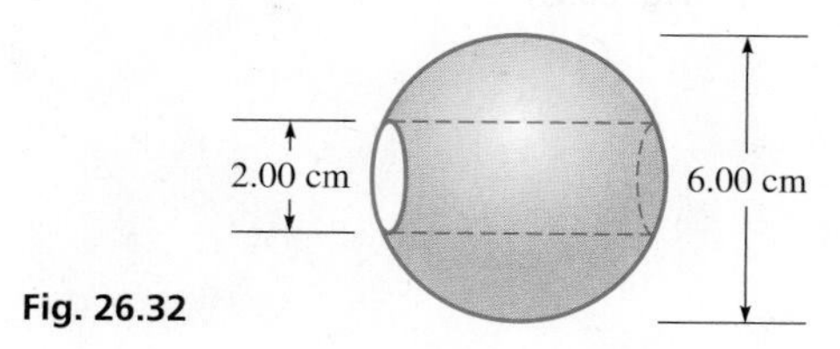

Fig. 26.32

40. All horizontal cross sections of a keg 4.00 ft tall are circular, and the sides of the keg are parabolic. The diameter at the top and bottom is 2.00 ft, and the diameter in the middle is 3.00 ft. Find the volume that the keg holds.

Answers to Practice Exercises

1. $64\pi/3$ **2.** $64\pi/3$

26.4 Centroids

Center of Mass • Centroid • Centroid of Thin Flat Plate • Centroid of Solid of Revolution

In the study of mechanics, a very important property of an object is its center of mass. In this section, we explain the meaning of center of mass and then show how integration is used to determine the center of mass for regions and solids of revolution.

If a mass m is at a distance d from a specified point O, the **moment** *of the mass about O is defined as md.* If several masses $m_1, m_2, \ldots, m_n$ are at distances $d_1, d_2, \ldots, d_n$, respectively, from point O, the total moment (as a group) about O is defined as $m_1d_1 + m_2d_2 + \cdots + m_nd_n$. *The* **center of mass** *is that point $\overline{d}$ units from O at which all the masses could be concentrated to get the same total moment.* Therefore $\overline{d}$ is defined by the equation

$$m_1d_1 + m_2d_2 + \cdots + m_nd_n = (m_1 + m_2 + \cdots + m_n)\overline{d} \quad \textbf{(26.15)}$$

The moment of a mass is a measure of its tendency to rotate about a point. A weight far from the point of balance of a long rod is more likely to make the rod turn than if the same weight were placed near the point of balance. It is easier to open a door if you push near the doorknob than if you push near the hinges. This is the type of physical property that the moment of mass measures.

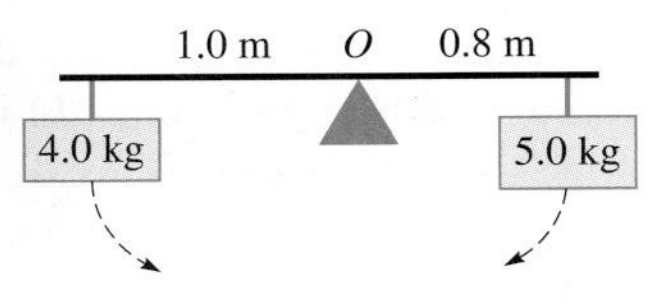

Fig. 26.33

EXAMPLE 1 Balancing masses

One of the simplest and most basic illustrations of moments and center of mass is seen in balancing a long rod (of negligible mass) with masses of two different sizes, one on either side of the balance point.

In Fig. 26.33, a mass of 5.0 kg is hung from the rod 0.8 m to the right of point O. We see that this 5.0-kg mass tends to turn the rod clockwise. A mass placed on the opposite side of O will tend to turn the rod counterclockwise. Neglecting the mass of the rod, in order to balance the rod at O, the moments must be equal in magnitude but opposite in sign. Therefore, a 4.0-kg mass would have to be placed 1.0 m to the left.

Thus, with $d_1 = 0.8$ m, $d_2 = -1.0$ m, we can find $\overline{d}$ by solving the equation below:

$$(5.0 + 4.0)\overline{d} = 5.0(0.8) + 4.0(-1.0) = 4.0 - 4.0$$
$$\overline{d} = 0.0 \text{ m}$$

The center of mass of the combination of the 5.0-kg mass and the 4.0-kg mass is at O. CAUTION Note that we must use **directed distances** in finding moments. ■ ■

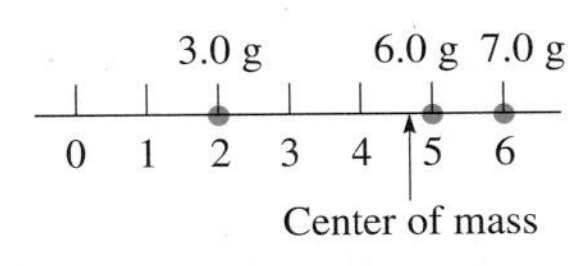

Fig. 26.34

EXAMPLE 2 Center of mass of three masses

A mass of 3.0 g is placed at (2.0, 0) on the x-axis (distances in cm). Another mass of 6.0 g is placed at (5.0, 0), and a third mass of 7.0 g is placed at (6.0, 0). See Fig. 26.34. Find the center of mass of these three masses.

Taking the reference point as the origin, we find $d_1 = 2.0$ cm, $d_2 = 5.0$ cm, and $d_3 = 6.0$ cm. Thus, $m_1d_1 + m_2d_2 + m_3d_3 = (m_1 + m_2 + m_3)\overline{d}$ becomes

$$3.0(2.0) + 6.0(5.0) + 7.0(6.0) = (3.0 + 6.0 + 7.0)\overline{d} \quad \text{or} \quad \overline{d} = 4.9 \text{ cm}$$

This means that the center of mass of the three masses is at (4.9, 0). Therefore, a mass of 16.0 g placed at this point has the same moment as the three masses as a unit. ■

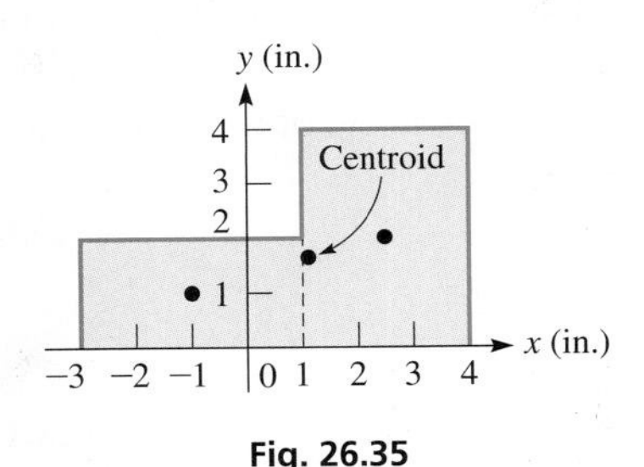

Fig. 26.35

EXAMPLE 3 Center of mass of a metal plate

Find the center of mass of the flat metal plate that is shown in Fig. 26.35.

We first note that the center of mass is not *on* either axis. This can be seen from the fact that the major portion of the area is in the first quadrant. *We will therefore measure the moments with respect to each axis to find the point that is the center of mass. This point is also called the* **centroid** *of the plate.*

The easiest method of finding this centroid is to divide the plate into rectangles, as indicated by the dashed line in Fig. 26.35, and assume that we may consider the mass of each rectangle to be concentrated at its center. In this way, the center of the left rectangle is at $(-1.0, 1.0)$ (distances in in.), and the center of the right rectangle is at (2.5, 2.0). The mass of each rectangle area, assumed uniform, is proportional to its area. The area of the left rectangle is 8.0 in.^{2}, and that of the right rectangle is 12.0 in.^{2}. Thus, taking moments with respect to the y-axis, we have

$$8.0(-1.0) + 12.0(2.5) = (8.0 + 12.0)\overline{x}$$

where $\overline{x}$ is the x-coordinate of the centroid. Solving for $\overline{x}$, we have $\overline{x} = 1.1$ in.

Now, taking moments with respect to the x-axis, we have

$$8.0(1.0) + 12.0(2.0) = (8.0 + 12.0)\overline{y}$$

where $\overline{y}$ is the y-coordinate of the centroid. Thus, $\overline{y} = 1.6$ in. This means that the coordinates of the centroid, the center of mass, are (1.1, 1.6). This may be interpreted as meaning that a plate of this shape would balance on a single support under this point. As an approximate check, we note from the figure that this point appears to be a reasonable balance point for the plate. ■

■ Since the center of mass does not depend on the density of the metal, we have assumed the constant of proportionality to be 1.

CENTROID OF A THIN, FLAT PLATE BY INTEGRATION

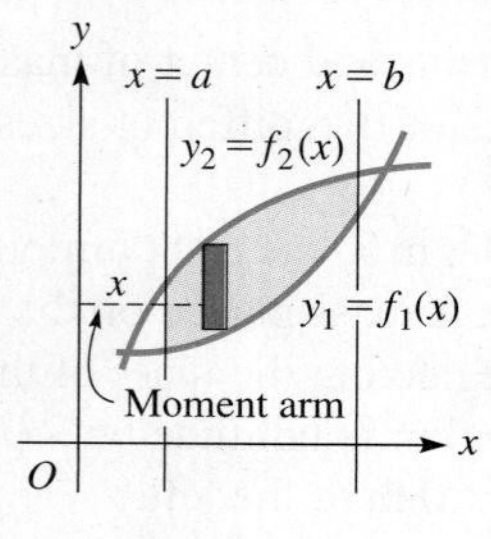

Fig. 26.36

If a thin, flat plate covers the region bounded by $y_1 = f_1(x)$, $y_2 = f_2(x)$, $x = a$, and $x = b$, as shown in Fig. 26.36, the moment of the mass of the element of area about the y-axis is given by $(kdA)x$, where k is the mass per unit area. In this expression, kdA is the mass of the element, and x is its distance (moment arm) from the y-axis. The element dA may be written as $(y_2 - y_1)dx$, which means that the moment may be written as $kx(y_2 - y_1)dx$. If we then sum up the moments of all the elements and express this as an integral (which, of course, means sum), we have $k\int_a^b x(y_2 - y_1)dx$. If we consider all the mass of the plate to be concentrated at one point $\bar{x}$ units from the y-axis, the moment would be $(kA)\bar{x}$, where kA is the mass of the entire plate and $\bar{x}$ is the distance the center of mass is from the y-axis. By the previous discussion, these two expressions should be equal. This means $k\int_a^b x(y_2 - y_1)dx = kA\bar{x}$. Because k appears on each side of the equation, we divide it out (we are assuming that the mass per unit area is constant). The area A is found by the integral $\int_a^b (y_2 - y_1)dx$. Therefore, the x-coordinate of the centroid of the plate is given by

$$\bar{x} = \frac{\int_a^b x(y_2 - y_1)dx}{\int_a^b (y_2 - y_1)dx} \tag{26.16}$$

Equation (26.16) gives us the x-coordinate of the centroid of the plate if vertical elements are used.

CAUTION Note that the two integrals in Eq. (26.16) must be evaluated separately. We cannot cancel out the apparent common factor $y_2 - y_1$, and we cannot combine quantities and perform only one integration. The two integrals must be evaluated separately first. Then any possible cancellations of factors common to the numerator and the denominator may be made. ■

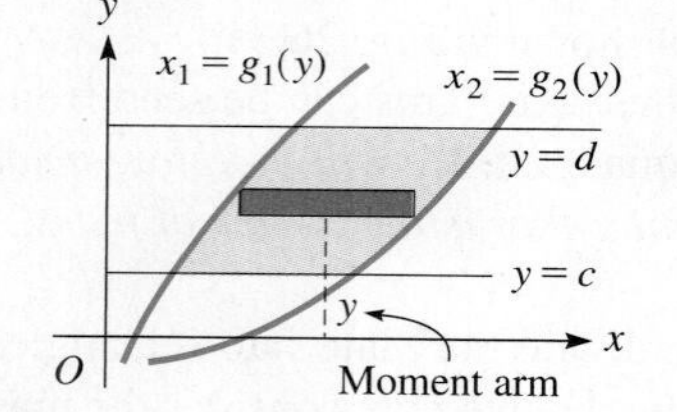

Fig. 26.37

Following the same reasoning that we used in developing Eq. (26.16), for a thin plate covering the region bounded by the functions $x_1 = g_1(y)$, $x_2 = g_2(y)$, $y = c$, and $y = d$, as shown in Fig. 26.37, the y-coordinate of the centroid of the plate is given by the equation

$$\bar{y} = \frac{\int_c^d y(x_2 - x_1)dy}{\int_c^d (x_2 - x_1)dy} \tag{26.17}$$

In this equation, horizontal elements are used.

NOTE ➧ [In applying Eqs. (26.16) and (26.17), we should keep in mind that each denominator gives the area of the plate. Once we have found this area, we may use it for both $\bar{x}$ and $\bar{y}$. In this way, we can avoid having to set up and perform one of the indicated integrations.] Also, in finding the coordinates of the centroid, we should look for and utilize any symmetry the region may have.

EXAMPLE 4 Centroid of a thin plate by integration

Find the coordinates of the centroid of a thin plate covering the region bounded by the parabola $y = x^2$ and the line $y = 4$.

Fig. 26.38

We sketch a graph indicating the region and an element of area (see Fig. 26.38). The curve is a parabola whose axis is the y-axis. [Because the region is symmetric to the y-axis, the centroid must be on this axis.] This means that the x-coordinate of the centroid is zero, or $\bar{x} = 0$. To find the y-coordinate of the centroid, we have

NOTE ▸

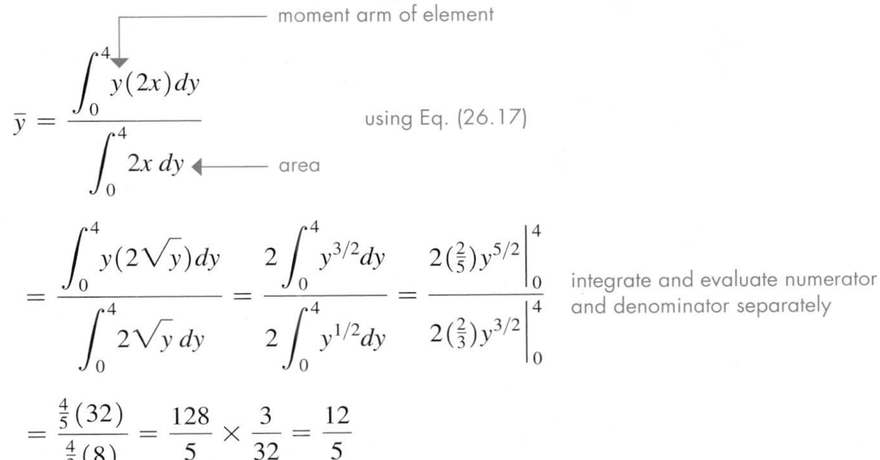

$$\bar{y} = \frac{\int_0^4 y(2x)\,dy}{\int_0^4 2x\,dy}$$

$$= \frac{\int_0^4 y(2\sqrt{y})\,dy}{\int_0^4 2\sqrt{y}\,dy} = \frac{2\int_0^4 y^{3/2}dy}{2\int_0^4 y^{1/2}dy} = \frac{2\left(\frac{2}{5}\right)y^{5/2}\Big|_0^4}{2\left(\frac{2}{3}\right)y^{3/2}\Big|_0^4}$$

$$= \frac{\frac{4}{5}(32)}{\frac{4}{3}(8)} = \frac{128}{5} \times \frac{3}{32} = \frac{12}{5}$$

The coordinates of the centroid are $\left(0, \frac{12}{5}\right)$. This plate would balance if a single pointed support were to be put under this point. ■

Practice Exercise

1. In Example 4, change $y = 4$ to $y = 1$ and solve the resulting problem.

EXAMPLE 5 Centroid of a triangular plate by integration

Find the coordinates of the centroid of an isosceles right triangular plate with side a. See Fig. 26.39.

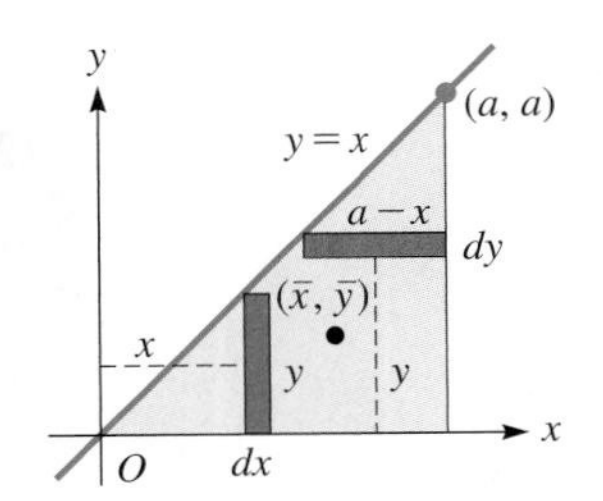

Fig. 26.39

We must first set up the region in the xy-plane. The choice shown in Fig. 26.39 is to place the triangle with one vertex at the origin and the right angle on the x-axis. Because each side is a, the hypotenuse passes through the point (a, a). The equation of the hypotenuse is $y = x$. The x-coordinate of the centroid is found by using Eq. (26.16):

$$\bar{x} = \frac{\int_0^a xy\,dx}{\int_0^a y\,dx} = \frac{\int_0^a x(x)\,dx}{\int_0^a x\,dx} = \frac{\int_0^a x^2dx}{\frac{1}{2}x^2\Big|_0^a} = \frac{\frac{1}{3}x^3\Big|_0^a}{\frac{a^2}{2}} = \frac{\frac{a^3}{3}}{\frac{a^2}{2}} = \frac{2a}{3}$$

The y-coordinate of the centroid is found by using Eq. (26.17):

$$\bar{y} = \frac{\int_0^a y(a-x)\,dy}{\frac{a^2}{2}} = \frac{\int_0^a y(a-y)\,dy}{\frac{a^2}{2}} = \frac{\int_0^a (ay - y^2)\,dy}{\frac{a^2}{2}}$$

$$= \frac{\frac{ay^2}{2} - \frac{y^3}{3}\Big|_0^a}{\frac{a^2}{2}} = \frac{\frac{a^3}{6}}{\frac{a^2}{2}} = \frac{a}{3}$$

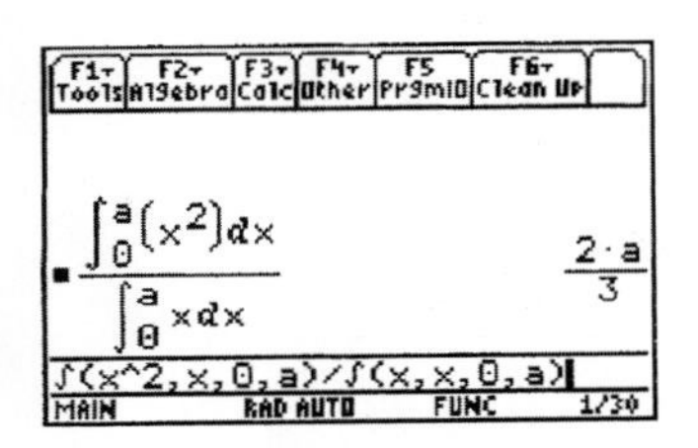

Fig. 26.40
TI-89 graphing calculator keystrokes:
goo.gl/GW6Sw1

Thus, the coordinates of the centroid are $\left(\frac{2}{3}a, \frac{1}{3}a\right)$. The results indicate that the center of mass is $\frac{1}{3}a$ units from each of the equal sides. A TI-89 calculator evaluation of the x-coordinate of this centroid is shown is Fig. 26.40. ■

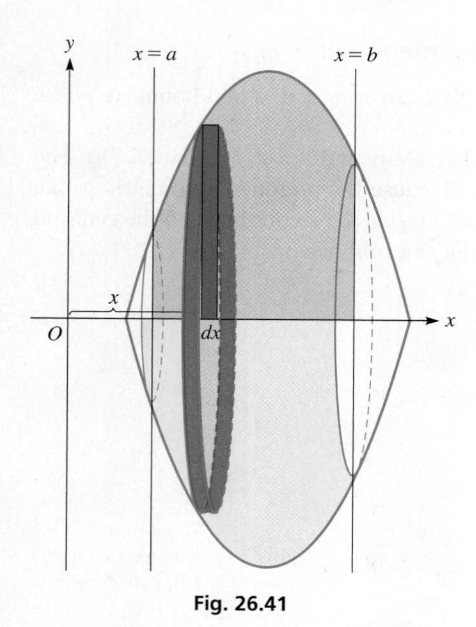

Fig. 26.41

CENTROID OF A SOLID OF REVOLUTION

Another figure for which we wish to find the centroid is a solid of revolution. If the density of the solid is constant, the centroid is on the axis of revolution. The problem that remains is to find just where on the axis the centroid is located.

If a region bounded by the x-axis is revolved about the x-axis, as shown in Fig. 26.41, a vertical element of area generates a disk element of volume. The center of mass of the disk is at its center, and we may consider its mass concentrated there. The moment about the y-axis of a typical element is $x(k)(\pi y^2 dx)$, where x is the moment arm, k is the density, and $\pi y^2\, dx$ is the volume. The sum of the moments of the elements can be expressed as an integral; it equals the volume times the density times the x-coordinate of the centroid of the volume. Because π and the density k would appear in both the numerator and denominator, they cancel and need not be written. Therefore,

$$\overline{x} = \frac{\int_a^b xy^2 dx}{\int_a^b y^2 dx} \tag{26.18}$$

is the equation for the x-coordinate of the centroid of a solid of revolution about the x-axis.

In the same manner, we may find that *the y-coordinate of the centroid of a solid of revolution about the y-axis is*

$$\overline{y} = \frac{\int_c^d yx^2 dy}{\int_c^d x^2 dy} \tag{26.19}$$

EXAMPLE 6 Centroid of a solid by integration

Find the coordinates of the centroid of the volume generated by revolving the first-quadrant region under the curve $y = 4 - x^2$ about the y-axis as shown in Fig. 26.42.

Because the curve is rotated about the y-axis, $\overline{x} = 0$. The y-coordinate is

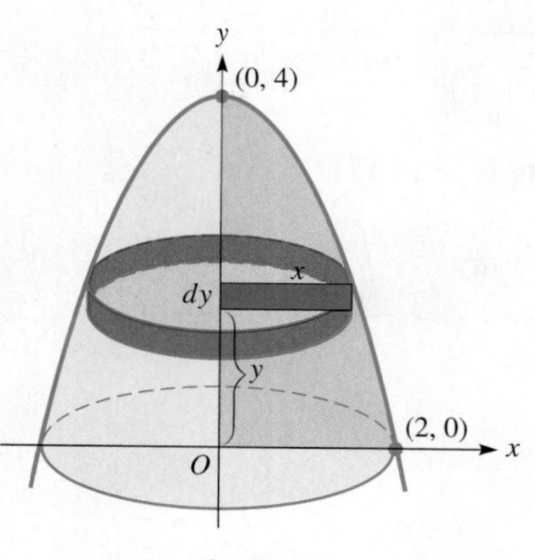

Fig. 26.42

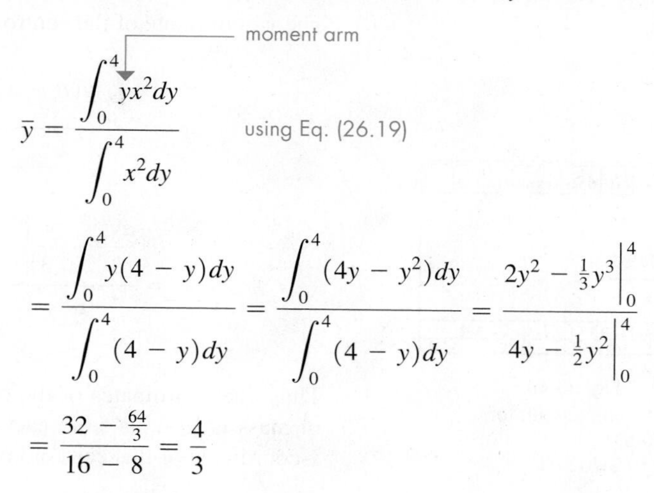

$$\overline{y} = \frac{\int_0^4 yx^2 dy}{\int_0^4 x^2 dy}$$

$$= \frac{\int_0^4 y(4-y)dy}{\int_0^4 (4-y)dy} = \frac{\int_0^4 (4y-y^2)dy}{\int_0^4 (4-y)dy} = \frac{2y^2 - \frac{1}{3}y^3\Big|_0^4}{4y - \frac{1}{2}y^2\Big|_0^4}$$

$$= \frac{32 - \frac{64}{3}}{16 - 8} = \frac{4}{3}$$

The coordinates of the centroid are $(0, \frac{4}{3})$. ■

Fig. 26.43

EXAMPLE 7 Centroid of a right circular cone—machine parts

A company makes solid conical machine parts of various sizes. Show that the centroid of every part of any size is in the same relative position within the part. We can do this by finding the centroid of any right circular cone of radius r and height h.

To generate a right circular cone, we revolve a right triangle about one of its legs (see Fig. 26.43). Placing a leg of length h along the x-axis, we revolve the right triangle whose hypotenuse is given by $y = (a/h)x$ about the x-axis. Therefore,

$$\overline{x} = \frac{\int_0^h xy^2 dx}{\int_0^h y^2 dx} \quad \text{using Eq. (26.18)} \quad (x \text{ is the moment arm})$$

$$= \frac{\int_0^h x\left[\left(\frac{a}{h}\right)x\right]^2 dx}{\int_0^h \left[\left(\frac{a}{h}\right)x\right]^2 dx} = \frac{\left(\frac{a^2}{h^2}\right)\left(\frac{1}{4}x^4\right)\Big|_0^h}{\left(\frac{a^2}{h^2}\right)\left(\frac{1}{3}x^3\right)\Big|_0^h} = \frac{3}{4}h$$

The centroid of every part is on the central axis $\frac{3}{4}$ of the way from the vertex to the base. ■

EXERCISES 26.4

In Exercises 1 and 2, make the given changes in the indicated examples of this section and then find the coordinates of the centroid.

1. In Example 4, change $y = x^2$ to $y = |x|$ ($y = x$ for $x \geq 0$, and $y = -x$ for $x < 0$).
2. In Example 6, change $y = 4 - x^2$ to $y = 4 - x$.

In Exercises 3–6, find the center of mass (in cm) of the particles with the given masses located at the given points on the x-axis.

3. 5.0 g at (1.0, 0), 8.5 g at (4.2, 0), 3.6 g at (7.3, 0)
4. 2.3 g at (1.3, 0), 6.5 g at (5.8, 0), 1.2 g at (9.5, 0)
5. 31 g at (−3.5, 0), 24 g at (0, 0), 15 g at (2.6, 0), 84 g at (3.7, 0)
6. 550 g at (−42, 0), 230 g at (−27, 0), 470 g at (16, 0), 120 g at (22, 0)

In Exercises 7–10, find the coordinates (to 0.01 in.) of the centroids of the uniform flat-plate machine parts shown.

7.

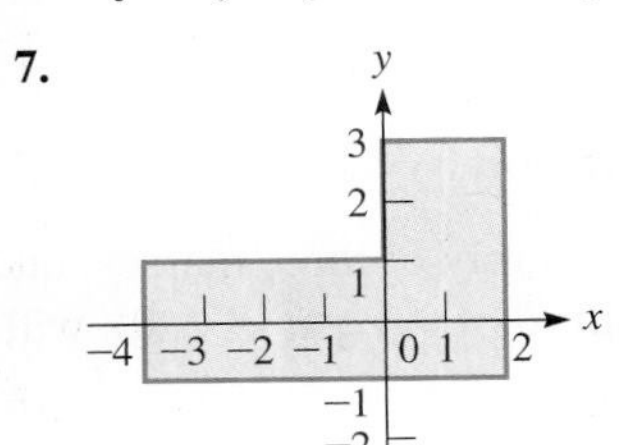

8.

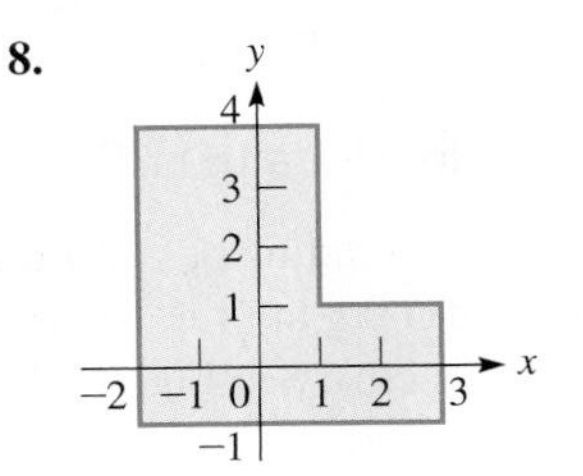

9.

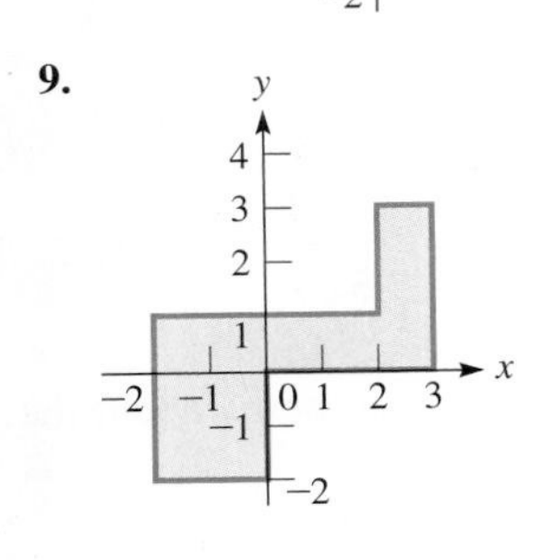

10.

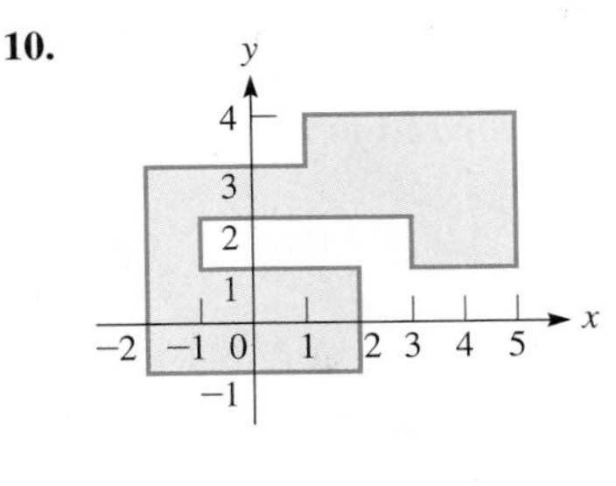

In Exercises 11–34, find the coordinates of the centroids of the given figures. In Exercises 11–22, each region is covered by a thin, flat plate.

11. The region bounded by $y = x^2$ and $y = 2$
12. The semicircular region in Fig. 26.44.

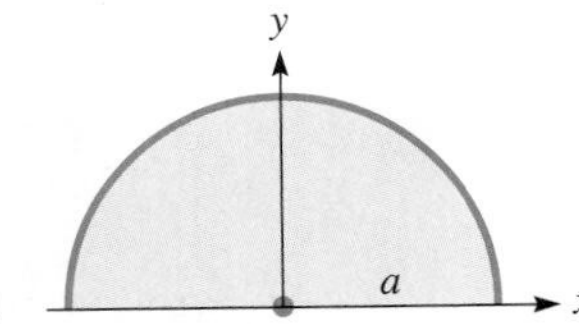

Fig. 26.44

13. The region bounded by $y = 4 - x$ and the axes
14. The region bounded by $y = x^3$, $x = 2$, and the x-axis
15. The region bounded by $y = 4x^2$ and $y = 2x^3$
16. The region bounded by $y^2 = x$, $y = 2$, and $x = 0$
17. The region bounded by $y = 2(x + 1)$, $y = 3x + 2$, and $y = 8$
18. The region bounded by $y = x^{2/3}$, $x = 8$, and $y = 0$
19. The region bounded by $y = 4 - 2x$, $x = 2$, and $y = 4$
20. The region bounded by $y = \sqrt{x}$, $y = 0$, and $x = 9$
21. The region bounded by $x^2 = 4py$ and $y = a$ if $p > 0$ and $a > 0$
22. The region above the x-axis, bounded by the ellipse with vertices $(a, 0)$ and $(-a, 0)$, and minor axis $2b$. (The area of an ellipse is πab.)
23. The solid generated by revolving the region bounded by $y = x^3$, $y = 0$, and $x = 1$ about the x-axis
24. The solid generated by revolving the region bounded by $y = 2 - 2x$, $x = 0$, and $y = 0$ about the y-axis
25. The solid generated by revolving the region in the first quadrant bounded by $y^2 = 4x$, $x = 0$, and $y = 2$ about the y-axis

26. The solid generated by revolving the region bounded by $y = x^2$, $x = 2$, and the x-axis about the x-axis

27. The solid generated by revolving the region in the first quadrant bounded by $y^2 = 4x$ and $x = 1$ about the x-axis

28. The solid generated by revolving the region in the first quadrant bounded by $x^2 - y^2 = 9$, $y = 4$, and the x-axis about the y-axis

29. A sailboat has a right-triangular sail with a horizontal base 3.0 m long and a vertical side of 4.5 m high. Where is the centroid of the sail?

30. A lens with semielliptical vertical cross sections and circular horizontal cross sections is shown in Fig. 26.45. For proper installation in an optical device, its centroid must be known. Locate its centroid.

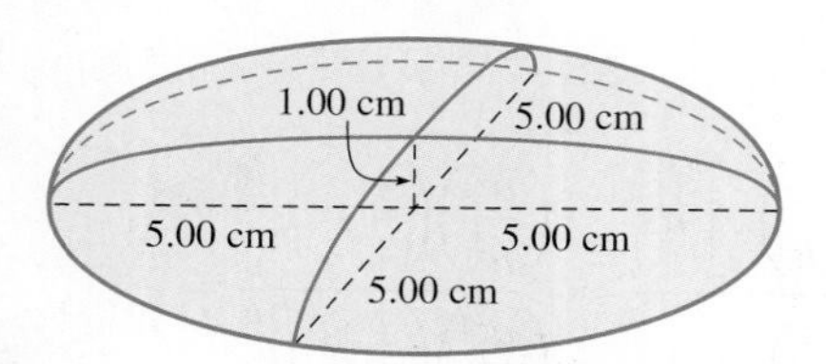

Fig. 26.45

31. Find the location of the centroid of a hemisphere of radius a. Using this result, locate the centroid of the northern hemisphere of Earth for which the radius is 6370 km.

32. A sanding machine disc can be described as the solid generated by rotating the region bounded by $y^2 = 4/x$, $y = 1$, $y = 2$, and the y-axis about the y-axis (measurements in in.). Locate the centroid of the disc.

33. A highway marking pylon has the shape of a frustum of a cone. Find its centroid if the radii of its bases are 5.00 cm and 20.0 cm and the height between bases is 60.0 cm.

34. A floodgate is in the shape of an isosceles trapezoid. Find the location of the centroid of the floodgate if the upper base is 20 m, the lower base is 12 m, and the height between bases is 6.0 m. See Fig. 26.46.

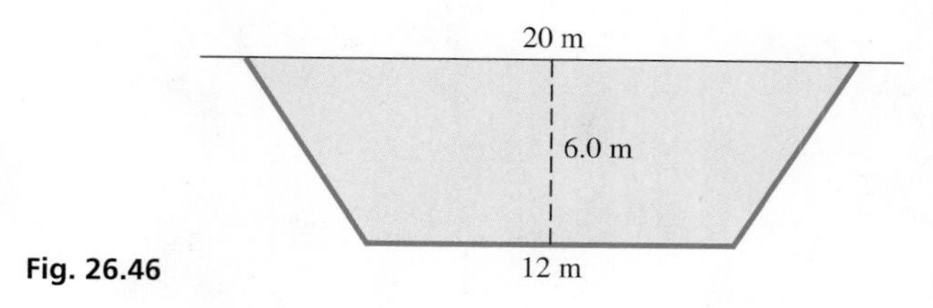

Fig. 26.46

Answer to Practice Exercise

1. $(0, \frac{3}{5})$

26.5 Moments of Inertia

Radius of Gyration • Moment of Inertia of Thin, Flat Plate • Moment of Inertia of Solid

Important in the rotational motion of an object is its **moment of inertia,** which is analogous to the mass of a moving object. In each case, *the moment of inertia or mass is the measure of the tendency of the object to resist a change in motion.*

Suppose that a particle of mass m is rotating about some point: We define its moment of inertia as md^2, where d is the distance from the particle to the point. If a group of particles of masses $m_1, m_2, \ldots, m_n$ are rotating about an axis, as shown in Fig. 26.47, the moment of inertia I with respect to the axis of the group is

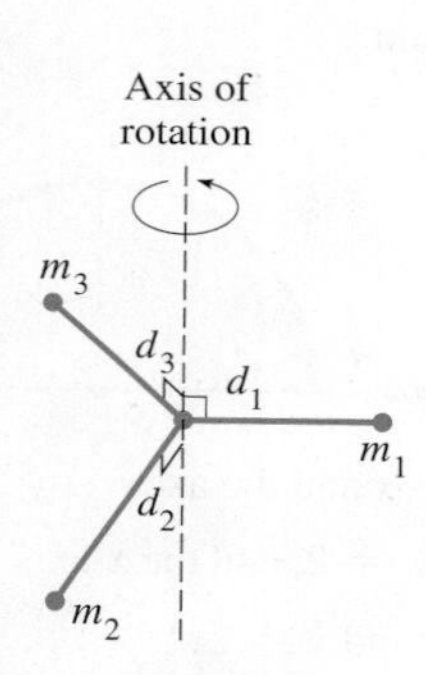

Fig. 26.47

$$I = m_1d_1^2 + m_2d_2^2 + \cdots + m_nd_n^2$$

where the d's are the respective distances of the particles from the axis. If all the masses were at the same distance R from the axis of rotation, so that the total moment of inertia were the same, we would have

$$m_1d_1^2 + m_2d_2^2 + \cdots + m_nd_n^2 = (m_1 + m_2 + \cdots + m_n)R^2 \tag{26.20}$$

where ***R* is called the radius of gyration.**

EXAMPLE 1 Moment of inertia and radius of gyration

Find the moment of inertia and the radius of gyration of the array of three masses, one of 3.0 g at $(-2.0, 0)$, another of 5.0 g at (1.0, 0), and the third of 4.0 g at (4.0, 0), with respect to the origin (distances in cm). See Fig. 26.48.

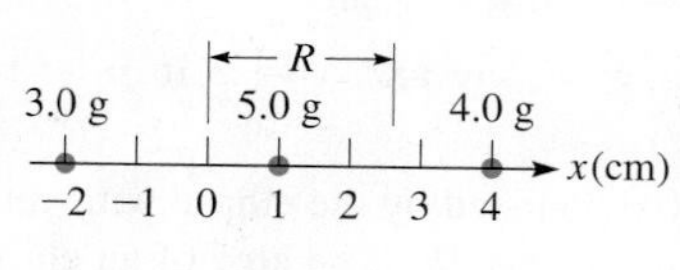

Fig. 26.48

The moment of inertia of the array is

$$I = 3.0(-2.0)^2 + 5.0(1.0)^2 + 4.0(4.0)^2 = 81 \text{ g}\cdot\text{cm}^2$$

The radius of gyration is found from $I = (m_1 + m_2 + m_3)R^2$. Thus,

$$81 = (3.0 + 5.0 + 4.0)R^2, \qquad R^2 = \frac{81}{12}, \qquad R = 2.6 \text{ cm}$$

Therefore, a mass of 12.0 g placed at (2.6, 0) has the same rotational inertia about the origin as the array of masses as a unit. ■

Practice Exercise

1. In Example 1, interchange the positions of the 3.0-g and 4.0-g masses, and calculate R.

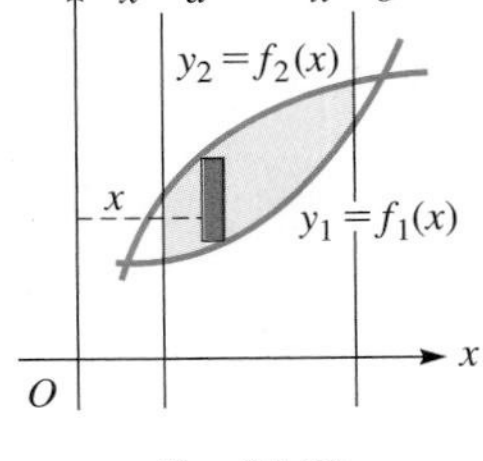

Fig. 26.49

MOMENT OF INERTIA OF A THIN, FLAT PLATE

If a thin, flat plate covering the region is bounded by the curves of the functions $y_1 = f_1(x)$, $y_2 = f_2(x)$ and the lines $x = a$ and $x = b$, as shown in Fig. 26.49, the moment of inertia of this plate with respect to the y-axis, I_y, is given by the sum of the moments of inertia of the individual elements. The mass of each element is $k(y_2 - y_1)dx$, where k is the mass per unit area and $(y_2 - y_1)dx$ is the area of the element. The distance of the element from the y-axis is x. Representing this sum as an integral, we have

$$I_y = k\int_a^b x^2(y_2 - y_1)dx \qquad \textbf{(26.21)}$$

NOTE ➧ [To find the radius of gyration of the plate with respect to the y-axis, R_y, first find the moment of inertia, divide this by the mass of the plate, and take the square root of this result.]

In the same manner, the moment of inertia of a thin plate, with respect to the x-axis, bounded by $x_1 = g_1(y)$ and $x_2 = g_2(y)$ as shown in Fig. 26.50 is given by

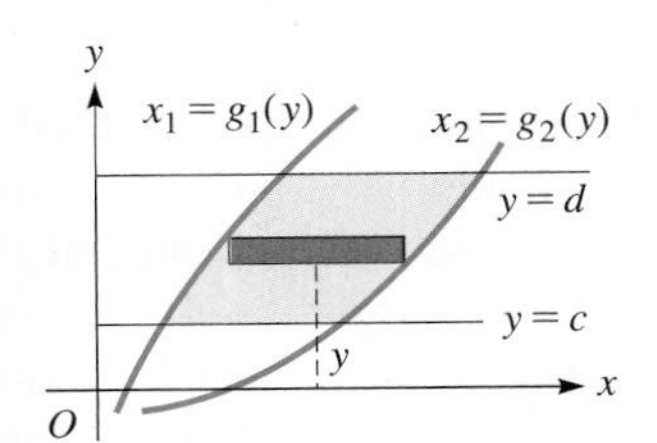

Fig. 26.50

$$I_x = k\int_c^d y^2(x_2 - x_1)dy \qquad \textbf{(26.22)}$$

We find the radius of gyration of the plate with respect to the x-axis, R_x, in the same manner as we find it with respect to the y-axis.

EXAMPLE 2 Moment of inertia of a plate

Find the moment of inertia and radius of gyration of the plate covering the region bounded by $y = 4x^2$, $x = 1$, and the x-axis with respect to the y-axis.

We find the moment of inertia of this plate (see Fig. 26.51) as follows:

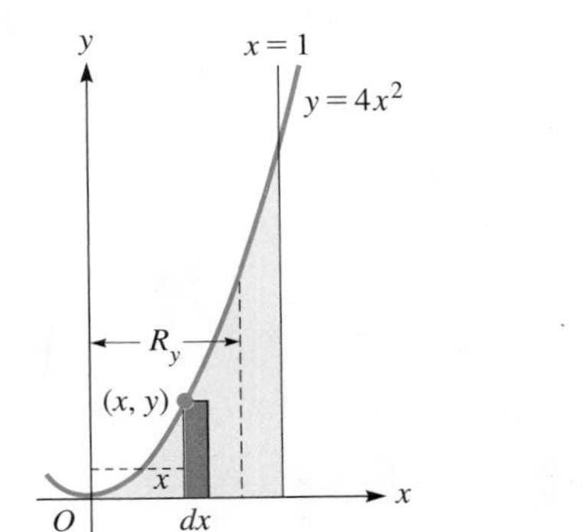

Fig. 26.51

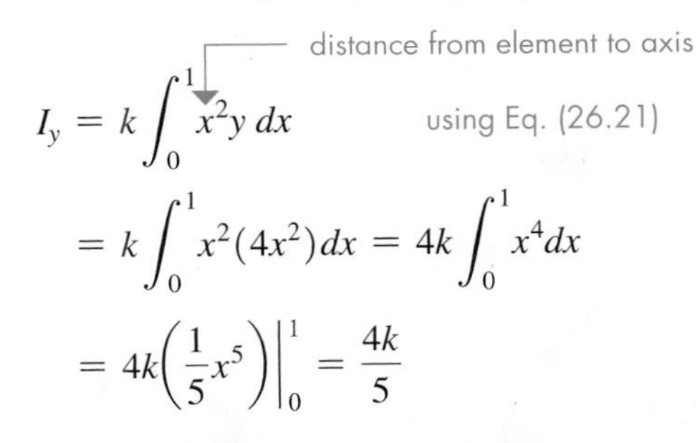

$$I_y = k\int_0^1 x^2 y\,dx$$

$$= k\int_0^1 x^2(4x^2)dx = 4k\int_0^1 x^4 dx$$

$$= 4k\left(\frac{1}{5}x^5\right)\Big|_0^1 = \frac{4k}{5}$$

To find the radius of gyration, we first determine the mass of the plate:

$$m = k\int_0^1 y\,dx = k\int_0^1 (4x^2)dx \qquad m = kA$$

$$= 4k\left(\frac{1}{3}x^3\right)\Big|_0^1 = \frac{4k}{3}$$

$$R_y^2 = \frac{I_y}{m} = \frac{4k}{5} \times \frac{3}{4k} = \frac{3}{5} \qquad R_y^2 = I_y/m$$

$$R_y = \sqrt{\frac{3}{5}} = \frac{\sqrt{15}}{5} = 0.775$$

NOTE ➧ [Therefore, if all of the mass of the plate were at a distance of 0.775 from the y-axis, the moment of inertia about the y-axis is the same as the moment of inertia of the plate itself.] ■

EXAMPLE 3 Moment of inertia of a triangular plate

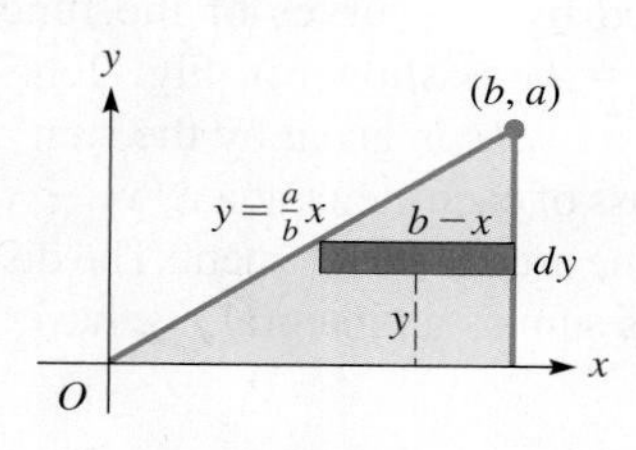

Fig. 26.52

Find the moment of inertia of a right triangular plate with sides a and b with respect to side b. Assume that $k = 1$.

Placing the triangle as shown in Fig. 26.52, we see that the equation of the hypotenuse is $y = (a/b)x$. The moment of inertia is

distance from element to axis

$$I_x = \int_0^a y^2(b - x)dy \quad \text{using Eq. (26.22)}$$

$$= \int_0^a y^2\left(b - \frac{b}{a}y\right)dy = b\int_0^a \left(y^2 - \frac{1}{a}y^3\right)dy$$

$$= b\left(\frac{1}{3}y^3 - \frac{1}{4a}y^4\right)\Bigg|_0^a = b\left(\frac{a^3}{3} - \frac{a^3}{4}\right) = \frac{ba^3}{12}$$

■

MOMENT OF INERTIA OF A SOLID

In applications, among the most important moments of inertia are those of solids of revolution. NOTE ➧ [Because all parts of an element of mass should be at the same distance from the axis, the most convenient element of volume to use is the cylindrical shell.] In Fig. 26.53, if the region bounded by the curves $y_1 = f_1(x)$, $y_2 = f_2(x)$, $x = a$, and $x = b$ is revolved about the y-axis, the moment of inertia of the element of volume is $k[2\pi x(y_2 - y_1)dx](x^2)$, where k is the density, $2\pi x(y_2 - y_1)dx$ is the volume of the element, and x^2 is the square of the distance from the y-axis. Expressing the sum of the elements as an integral, *the moment of inertia of the solid with respect to the y-axis, I_y, is*

■ Note carefully that Eq. (26.23) gives the moment of inertia with respect to the y-axis and that $(y_2 - y_1)$ is the height of the shell (see Fig. 26.53).

$$I_y = 2\pi k \int_a^b (y_2 - y_1)x^3 dx \tag{26.23}$$

The radius of gyration R_y is found by determining (1) the moment of inertia, (2) the mass of the solid, and (3) the square root of the quotient of the moment of inertia divided by the mass.

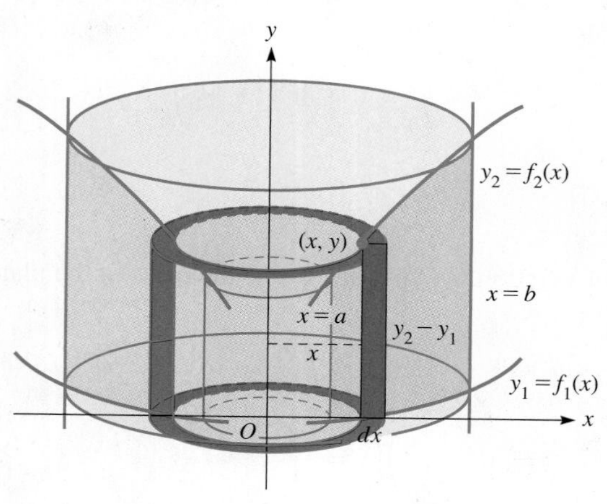

Fig. 26.53

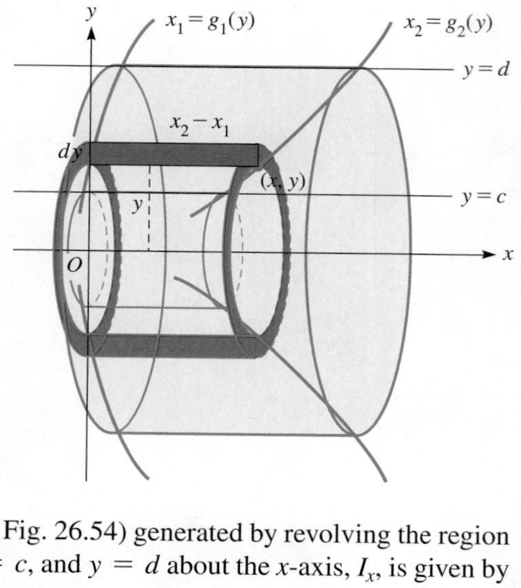

Fig. 26.54

The moment of inertia of the solid (see Fig. 26.54) generated by revolving the region bounded by $x_1 = g_1(y)$, $x_2 = g_2(y)$, $y = c$, and $y = d$ about the x-axis, I_x, is given by

■ Note carefully that Eq. (26.24) gives the moment of inertia with respect to the x-axis and that $(x_2 - x_1)$ is the height of the shell (see Fig. 26.54).

$$I_x = 2\pi k \int_c^d (x_2 - x_1)y^3 dy \tag{26.24}$$

The radius of gyration with respect to the x-axis, R_x, is found in the same manner as R_y.

EXAMPLE 4 Moment of inertia of a solid

Find the moment of inertia and radius of gyration with respect to the x-axis of the solid generated by revolving the region bounded by the curves of $y^3 = x$, $y = 2$, and the y-axis about the x-axis. See Fig. 26.55.

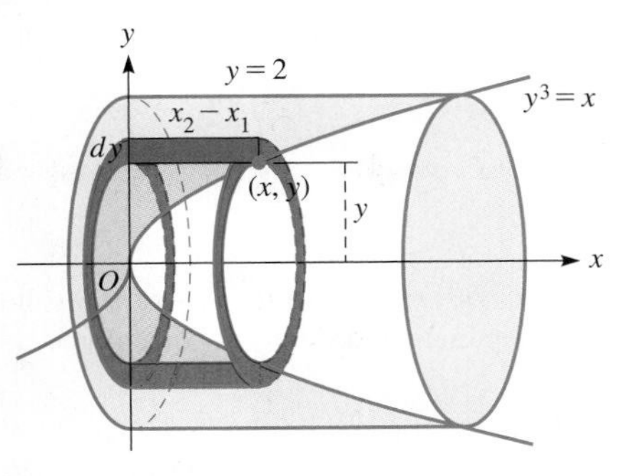

Fig. 26.55

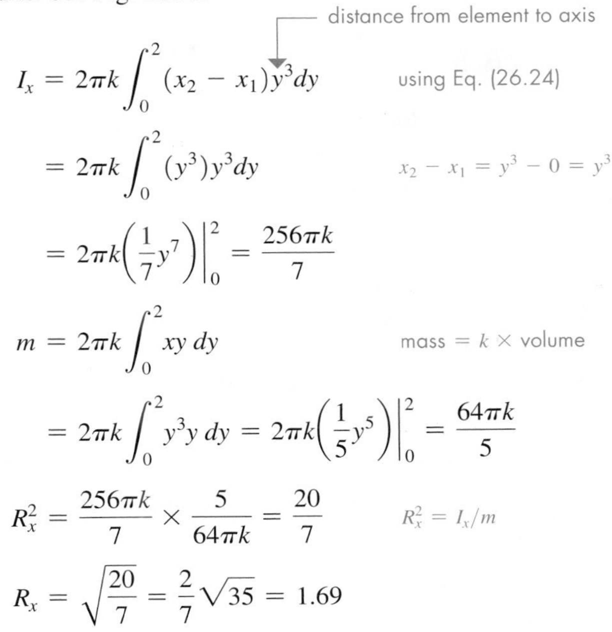

$$I_x = 2\pi k \int_0^2 (x_2 - x_1) y^3 dy \qquad \text{using Eq. (26.24)}$$

(distance from element to axis: y)

$$= 2\pi k \int_0^2 (y^3) y^3 dy \qquad x_2 - x_1 = y^3 - 0 = y^3$$

$$= 2\pi k \left(\frac{1}{7} y^7\right)\Bigg|_0^2 = \frac{256\pi k}{7}$$

$$m = 2\pi k \int_0^2 xy\, dy \qquad \text{mass} = k \times \text{volume}$$

$$= 2\pi k \int_0^2 y^3 y\, dy = 2\pi k\left(\frac{1}{5}y^5\right)\Bigg|_0^2 = \frac{64\pi k}{5}$$

$$R_x^2 = \frac{256\pi k}{7} \times \frac{5}{64\pi k} = \frac{20}{7} \qquad R_x^2 = I_x/m$$

$$R_x = \sqrt{\frac{20}{7}} = \frac{2}{7}\sqrt{35} = 1.69$$

■

EXAMPLE 5 Moment of inertia of a solid disk

As noted at the beginning of this section, the moment of inertia is important when studying the rotational motion of an object. For this reason, the moments of inertia of various objects are calculated, and the formulas tabulated. Such formulas are usually expressed in terms of the mass of the object.

Among the objects for which the moment of inertia is important is a solid disk. Find the moment of inertia of a disk with respect to its axis and in terms of its mass.

To generate a disk (see Fig. 26.56), we rotate the region bounded by the axes, $x = r$, and $y = b$, about the y-axis. We then have

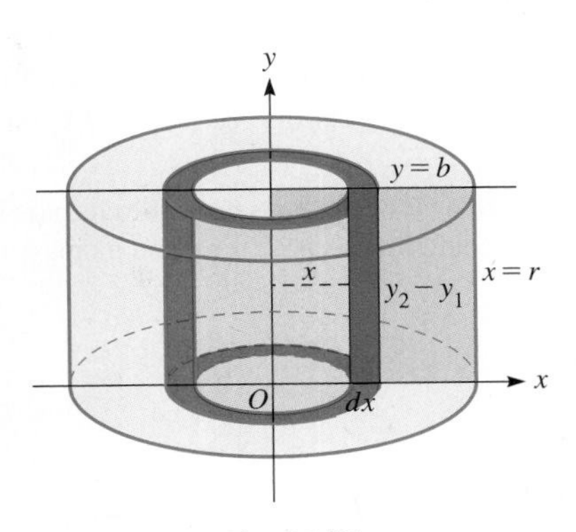

Fig. 26.56

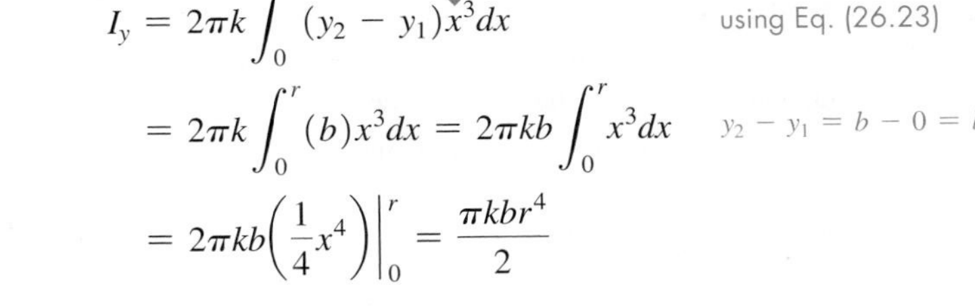

$$I_y = 2\pi k \int_0^r (y_2 - y_1) x^3 dx \qquad \text{using Eq. (26.23)}$$

(distance from element to axis: x)

$$= 2\pi k \int_0^r (b) x^3 dx = 2\pi k b \int_0^r x^3 dx \qquad y_2 - y_1 = b - 0 = b$$

$$= 2\pi k b\left(\frac{1}{4}x^4\right)\Bigg|_0^r = \frac{\pi k b r^4}{2}$$

The mass of the disk is $k(\pi r^2)b$. Rewriting the expression for I_y, we have

$$I_y = \frac{(\pi k b r^2) r^2}{2} = \frac{mr^2}{2}$$

The rotation of a solid disk is important in the design and use of objects such as flywheels, pulley wheels, train wheels, engine rotors, various rollers such as are in printing presses, various types of machine tools, and numerous others. ■

Due to the limited methods of integration available at this point, we cannot integrate the expressions for the moments of inertia of circular areas or of a sphere. These will be introduced in Section 28.8 in the exercises, by which point the proper method of integration will have been developed.

EXERCISES 26.5

In Exercises 1 and 2, make the given changes in the indicated examples of this section, and then solve the resulting problems.

1. In Example 2, change $y = 4x^2$ to $y = 4x$.
2. In Example 4, change $y^3 = x$ to $y^2 = x$.

In Exercises 3–6, find the moment of inertia (in g · cm²) and the radius of gyration (in cm) with respect to the origin of each of the given arrays of masses located at the given points on the x-axis.

3. 4.2 g at (1.7, 0), 3.2 g at (3.5, 0)
4. 3.4 g at (−1.5, 0), 6.0 g at (2.1, 0), 2.6 g at (3.8, 0)
5. 82.0 g at (−3.80, 0), 90.0 g at (0.00, 0), 62.0 g at (5.50, 0)
6. 564 g at (−45.0, 0), 326 g at (−22.5, 0), 720 g at (15.4, 0), 205 g at (64.0, 0)

In Exercises 7–28, find the indicated moment of inertia or radius of gyration.

7. Find the moment of inertia of a plate covering the region bounded by $x = -1$, $x = 1$, $y = 0$, and $y = 1$ with respect to the x-axis.
8. Find the radius of gyration of a plate covering the region bounded by $x = 2$, $x = 4$, $y = 0$, and $y = 4$, with respect to the y-axis.
9. Find the moment of inertia of a plate covering the first-quadrant region bounded by $y^2 = x$, $x = 9$, and the x-axis with respect to the x-axis.
10. Find the moment of inertia of a plate covering the region bounded by $y = 2x$, $x = 1$, $x = 2$, and the x-axis with respect to the y-axis.
11. Find the radius of gyration of a plate covering the region bounded by $y = x^3$, $x = 3$, and the x-axis with respect to the y-axis.
12. Find the radius of gyration of a plate covering the first-quadrant region bounded by $y^2 = 1 - x$ with respect to the x-axis.
13. Find the radius of gyration of a plate covering the region bounded by $y = x^2$, $x = 3$, and the x-axis with respect to the x-axis.
14. Find the radius of gyration of a plate covering the region bounded by $y^2 = x^3$, $y = 8$, and the y-axis with respect to the y-axis.
15. Find the radius of gyration of the plate of Exercise 14 with respect to the x-axis.
16. Find the radius of gyration of a plate covering the first-quadrant region bounded by $x = 1$, $y = 2 - x$, and the y-axis with respect to the y-axis.
17. Find the moment of inertia with respect to its axis of the solid generated by revolving the region bounded by $y^2 = 4x$, $y = 2$, and the y-axis about the x-axis.
18. Find the radius of gyration with respect to its axis of the solid generated by revolving the first-quadrant region under the curve $y = 4 - x^2$ about the y-axis.
19. Find the radius of gyration with respect to its axis of the solid generated by revolving the region bounded by $y = 4x - x^2$ and the x-axis about the y-axis.
20. Find the radius of gyration with respect to its axis of the solid generated by revolving the region bounded by $y = 2x$ and $y = x^2$ about the y-axis.
21. Find the moment of inertia of the triangular sail in Exercise 29 of Section 26.4, with respect to the vertical side in terms of its mass.
22. Find the moment of inertia of a rectangular sheet of metal of sides a and b with respect to side a. Express the result in terms of the mass of the metal sheet.
23. A top in the shape of an inverted right circular cone has a base radius r (at the top), height h, and mass m. Find the moment of inertia of the cone with respect to its axis (the height) in terms of its mass and radius. See Fig. 26.57.

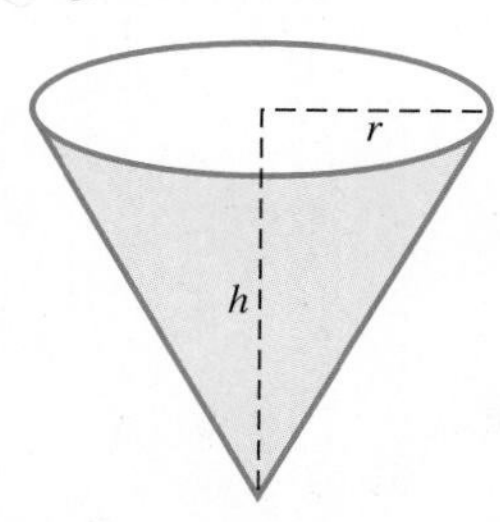

Fig. 26.57

24. Find the moment of inertia in terms of its mass of a circular hoop of radius r and of negligible thickness with respect to its center.
25. A rotating drill head is in the shape of a right circular cone. Find the moment of inertia of the drill head with respect to its axis if its radius is 0.600 cm, its height is 0.800 cm, and its mass is 3.00 g. (See Exercise 23.)
26. Find the moment of inertia (in kg · m²) of a rectangular door 2 m high and 1 m wide with respect to its hinges if $k = 3$ kg/m². (See Exercise 22.)
27. Find the moment of inertia of a flywheel with respect to its axis if its inner radius is 4.0 cm, its outer radius is 6.0 cm, and its mass is 1.2 kg. See Fig. 26.58.

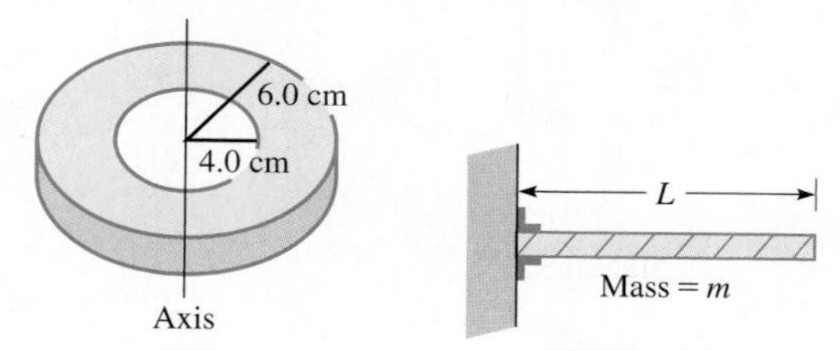

Fig. 26.58 Fig. 26.59

28. A cantilever beam is supported only at its left end, as shown in Fig. 26.59. Explain how to find the formula for the moment of inertia of this beam with respect to a vertical axis through its left end if its length is L and its mass is m. (Consider the mass to be distributed evenly along the beam. This is not an area or volume type of problem.) Find the formula for the moment of inertia.

Answer to Practice Exercise

1. $R = 2.4$ cm

26.6 Other Applications

Work Done by a Variable Force • Force Due to Liquid Pressure • Average Value of a Function

We have seen that the definite integral is used to find the exact measure of the sum of products in which one factor is an increment that approaches a limit of zero. This makes the definite integral a powerful mathematical tool in that a great many applications can be expressed in this form. The following examples show three more applications of the definite integral, and others are shown in the exercises.

WORK BY A VARIABLE FORCE

In physics, **work** *is defined as the product of a constant force times the distance through which it acts.* When we consider the work done in stretching a spring, the first thing we recognize is that the more the spring is stretched, the greater is the force necessary to stretch it. Thus, the force varies. However, if we are stretching the spring a distance Δx, where we are considering the limit as $\Delta x \rightarrow 0$, the force can be considered as approaching a constant over Δx. Adding the product of force$_1$ times Δx_1, force$_2$ times Δx_2, and so forth, we see that the total is the sum of these products. Thus, the work can be expressed as a definite integral in the form

$$W = \int_a^b f(x)\,dx \qquad (26.25)$$

where $f(x)$ is the force as a function of the distance the spring is stretched. CAUTION The limits a and b refer to the initial and final distances the spring is stretched from its normal length. ■

One problem remains: We must find the function $f(x)$ From physics, we learn that the force required to stretch a spring is proportional to the amount it is stretched (Hooke's law). If a spring is stretched x units from its normal length, then $f(x) = kx$. From conditions stated for a particular spring, the value of k may be determined. Thus, $W = \int_a^b kx\,dx$ is the formula for finding the total work done in stretching a spring. Here, a and b are the initial and final amounts the spring is stretched from its natural length.

■ Named for the English physicist Robert Hooke (1635–1703)

EXAMPLE 1 Work done stretching a spring

A spring of natural length 12 in. requires a force of 6.0 lb to stretch it 2.0 in. See Fig. 26.60. Find the work done in stretching it 6.0 in.

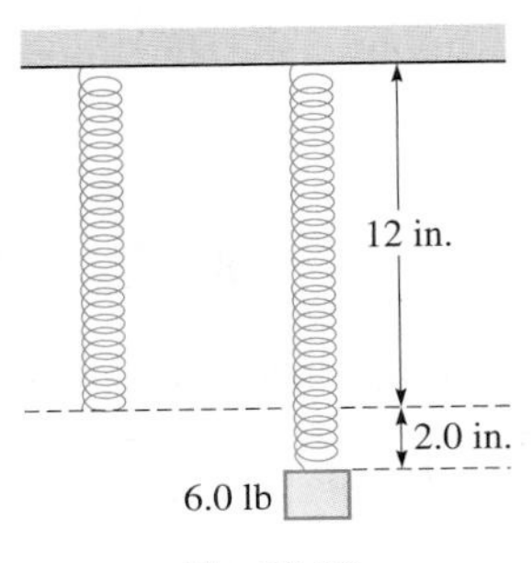

Fig. 26.60

From Hooke's law, we find the constant k for the spring, and therefore $f(x)$ as

$$f(x) = kx, \qquad 6.0 = k(2.0), \qquad k = 3.0 \text{ lb/in.}$$
$$= 3.0x$$

Since the spring is to be stretched 6.0 in., $a = 0$ (it starts unstretched) and $b = 6.0$ (it is 6.0 in. longer than its normal length). Therefore, the work done in stretching it is

$$W = \int_0^{6.0} 3.0x\,dx = 1.5x^2 \Big|_0^{6.0} \qquad \text{using Eq. (26.25)}$$
$$= 54 \text{ lb} \cdot \text{in.}$$

■

Problems involving work by a variable force arise in many fields of technology. On the following page is an example from electricity that deals with the motion of an electric charge through an electric field created by another electric charge.

Electric charges are of two types, designated as positive and negative. A basic law is that charges of the same sign repel each other and charges of opposite signs attract each other. *The force between charges is proportional to the product of their charges, and inversely proportional to the square of the distance between them.*

The force $f(x)$ between electric charges is therefore given by

$$f(x) = \frac{kq_1q_2}{x^2} \tag{26.26}$$

when q_1 and q_2 are the charges (in coulombs), x is the distance (in meters), the force is in newtons, and $k = 9.0 \times 10^9\ \text{N}\cdot\text{m}^2/\text{C}^2$. For other systems of units, the numerical value of k is different. We can find the work done when electric charges move toward each other or when they separate by use of Eq. (26.26) in Eq. (26.25).

EXAMPLE 2 Work done in moving α-particles

Find the work done when two α-particles, $q = 0.32$ aC each, move until they are 10 nm apart, if they were originally separated by 1.0 m.

From the given information, we have for each α-particle

$$q = 0.32\ \text{aC} = 0.32 \times 10^{-18}\ \text{C} = 3.2 \times 10^{-19}\ \text{C}$$

Because the particles start 1.0 m apart and are moved to 10 nm apart, $a = 1.0$ m and $b = 10 \times 10^{-9}\ \text{m} = 10^{-8}$ m. The work done is

$f(x)$ from Eq. (26.26)

$$W = \int_{1.0}^{10^{-8}} \frac{9.0 \times 10^9(3.2 \times 10^{-19})^2}{x^2}dx \quad \text{using Eq. (26.25)}$$

$$= 9.2 \times 10^{-28}\int_{1.0}^{10^{-8}} \frac{dx}{x^2} = 9.2 \times 10^{-28}\left(-\frac{1}{x}\right)\Bigg|_{1.0}^{10^{-8}}$$

$$= -9.2 \times 10^{-28}(10^8 - 1) = -9.2 \times 10^{-20}\ \text{J}$$

Because $10^8 \gg 1$, where $\gg$ means "much greater than," the 1 may be neglected in the calculation. NOTE ▸ [The minus sign in the result means that work must be done on the system to move the particles *toward each other*. If free to move, they tend to separate.] ■

The following is another type of problem involving work by a variable force.

EXAMPLE 3 Work done in winding up a cable

Find the work done in winding up 60.0 ft of a 100-ft cable that weighs 4.00 lb/ft. See Fig. 26.61.

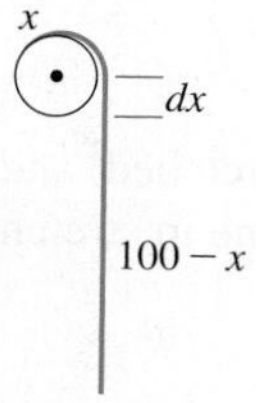

First, we let x denote the length of cable that has been wound up at any time. Then the force required to raise the remaining cable equals the weight of the cable that has not yet been wound up. This weight is the product of the unwound cable length, $100 - x$, and its weight per unit length, 4.00 lb/ft, or

$$f(x) = 4.00(100 - x)$$

x
dx
100 − x

Fig. 26.61

Since 60.0 ft of cable are to be wound up, $a = 0$ (none is initially wound up) and $b = 60.0$ ft. The work done is

$$W = \int_0^{60.0} 4.00(100 - x)dx \quad \text{using Eq. (26.25)}$$

$$= \int_0^{60.0} (400 - 4.00x)dx = 400x - 2.00x^2\Big|_0^{60.0} = 16{,}800\ \text{ft}\cdot\text{lb}$$ ■

Practice Exercise

1. Find the work done in winding up 30.0 ft of a 50.0-ft cable that weighs 2.00 lb/ft.

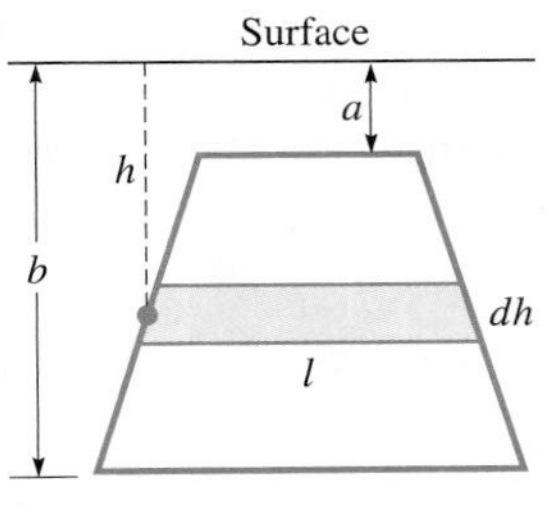

Fig. 26.62

FORCE DUE TO LIQUID PRESSURE

The second application of integration in this section deals with the force due to liquid pressure. The force F on an area A at the depth h in a liquid of density w is $F = whA$. Let us assume that the plate shown in Fig. 26.62 is submerged vertically in water. Using integration to sum the forces on the elements of area, *the total force on the plate is given by*

$$F = w\int_a^b lh\,dh \qquad (26.27)$$

Here, l is the length of the element of area, h is the depth of the element of area, w is the weight per unit volume of the liquid, a is the depth of the top, and b is the depth of the bottom of the area on which the force is exerted.

EXAMPLE 4 Force of water on the floodgate of a dam

■ See the chapter introduction.

A vertical rectangular floodgate of a dam is 5.00 ft wide and 4.00 ft high. Find the force on the floodgate if its upper edge is 3.00 ft below the surface of the water. See Fig. 26.63.

Each element of area of the floodgate has a length of 5.00 ft, which means that $l = 5.00$ ft. Because the top of the gate is 3.0 ft below the surface, $a = 3.00$ ft, and since the gate is 4.00 ft high, $b = 7.00$ ft. Using $w = 62.4$ lb/ft^3, we have the force on the gate as

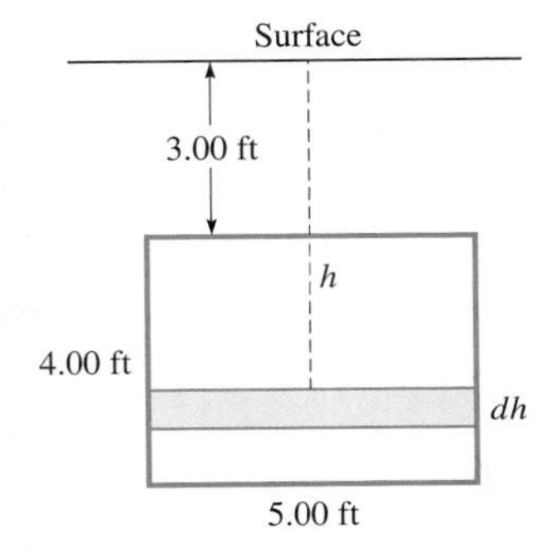

Fig. 26.63

$$\begin{aligned} F &= 62.4\int_{3.00}^{7.00} 5.00h\,dh \qquad \text{using Eq. (26.27)} \\ &= 312\int_{3.00}^{7.00} h\,dh \\ &= 156h^2\Big|_{3.00}^{7.00} = 156(49.0 - 9.00) \\ &= 6240 \text{ lb} \end{aligned}$$

■

Practice Exercise

2. In Example 4, find the force on the floodgate if the gate is 2.00 ft high.

EXAMPLE 5 Force on the end of a tank of water

The vertical end of a tank full of water is in the shape of a right triangle as shown in Fig. 26.64. What is the force on the end of the tank?

NOTE ▶ In setting up the figure, it is convenient to use coordinate axes. [It is also convenient to have the y-axis directed downward, because we integrate from the top to the bottom of the area.] The equation of the line OA is $y = \frac{1}{2}x$. Thus, we see that the length of an element of area of the end of the tank is $4.0 - x$, the depth of the element of area is y, the top of the tank is $y = 0$, and the bottom is $y = 2.0$ m. Therefore, the force on the end of the tank is $(w = 9800 \text{ N/m}^3)$

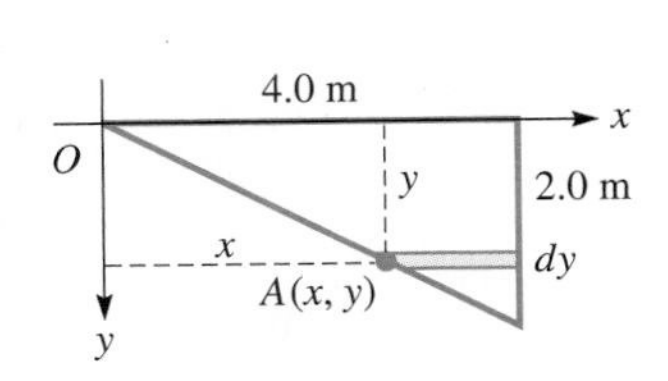

Fig. 26.64

$$\begin{aligned} F &= 9800\int_0^{2.0} \overbrace{(4.0 - x)}^{\text{length}}\overbrace{(y)}^{\text{depth}}(dy) \qquad \text{using Eq. (26.27)} \\ &= 9800\int_0^{2.0} (4.0 - 2y)(y\,dy) \\ &= 19{,}600\int_0^{2.0} (2.0y - y^2)dy = 19{,}600\left(1.0y^2 - \frac{1}{3}y^3\right)\Big|_0^{2.0} \\ &= 26{,}000 \text{ N} \qquad \text{rounded to two significant digits} \end{aligned}$$

■

AVERAGE VALUE OF A FUNCTION

The third application of integration shown in this section is that of the *average value* of a function. In general, an average is found by summing up the quantities to be averaged and then dividing by the total number of them. Generalizing on this and using integration for the summation, *the* **average value of a function** *y with respect to x from x = a to x = b is given by*

$$y_{av} = \frac{\int_a^b y\,dx}{b-a} \tag{26.28}$$

The following examples illustrate applications of the average value of a function.

EXAMPLE 6 Average value of velocity

The velocity v (in ft/s) of an object falling under the influence of gravity as a function of time t (in s) is given by $v = 32t$. What is the average velocity of the object with respect to time for the first 3.0 s?

In this case, we want the average value of the function $v = 32t$ from $t = 0$ s to $t = 3.0$ s. This gives us

$$v_{av} = \frac{\int_0^{3.0} v\,dt}{3.0-0} \quad \text{using Eq. (26.28)}$$

$$= \frac{\int_0^{3.0} 32t\,dt}{3.0} = \frac{16t^2}{3.0}\Bigg|_0^{3.0}$$

$$= 48 \text{ ft/s}$$

NOTE ➧ [This result can be interpreted as meaning that an average velocity of 48 ft/s for 3.0 s would result in the same distance, 144 ft, being traveled by the object as that with the variable velocity.] Because $s = \int v\,dt$, the numerator represents the distance traveled. ■

EXAMPLE 7 Average value of electric power

The power P (in W) developed in a certain resistor as a function of the current i (in A) is $P = 6.0i^2$. What is the average power with respect to the current as the current changes from 2.0 A to 5.0 A?

In this case, we are to find the average value of the function P from $i = 2.0$ A to $i = 5.0$ A. This average value of P is

$$P_{av} = \frac{\int_{2.0}^{5.0} P di}{5.0-2.0} \quad \text{using Eq. (26.28)}$$

$$= \frac{6.0\int_{2.0}^{5.0} i^2 dt}{3.0} = \frac{2.0i^3}{3.0}\Bigg|_{2.0}^{5.0} = \frac{2.0(125-8.0)}{3.0} = 78 \text{ W}$$ ■

NOTE ➧ [In general, it might be noted that the average value of y with respect to x is that value of y which, when multiplied by the length of the interval for x, gives the same area as that under the curve of y as a function of x.]

EXERCISES 26.6

1. In Example 4, find the force on the floodgate if the upper edge is 2.00 ft below the surface.

2. In Example 6, find the average velocity of the object with respect to time between 3.00 s and 6.00 s.

3. The spring of a spring balance is 8.0 in. long when there is no weight on the balance, and it is 9.5 in. long with 6.0 lb hung from the balance. How much work is done in stretching it from 8.0 in. to a length of 10.0 in?

4. How much work is done in stretching the spring of Exercise 3 from a length of 10.0 in. to 12.0 in.?

5. A 160-lb person compresses a bathroom scale 0.080 in. If the scale obeys Hooke's law, how much work is done compressing the scale if a 180-lb person stands on it?

6. A force F of 25 N on the spring in the lever-spring mechanism shown in Fig. 26.65 stretches the spring by 16 mm. How much work is done by the 25-N force in stretching the spring?

7. An electron has a 1.6×10^{-19} C negative charge. How much work is done in separating two electrons from 1.0 pm to 4.0 pm?

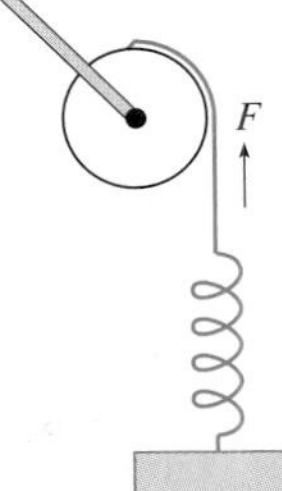

Fig. 26.65

8. How much work is done in separating an electron (see Exercise 7) and an oxygen nucleus, which has a positive charge of 1.3×10^{-18} C, from a distance of 2.0 μm to a distance of 1.0 m?

9. The gravitational force (in lb) of attraction between two objects is given by $F = k/x^2$, where x is the distance between the objects. If the objects are 10 ft apart, find the work required to separate them until they are 100 ft apart. Express the result in terms of k.

10. Find the work done in winding up (a) 30 m of a 40-m rope on which the force of gravity is 6.0 N/m, and (b) all of the rope.

11. A 1500-lb elevator is suspended on cables that together weigh 12 lb/ft. How much work is done in raising the elevator from the basement to the top floor, a distance of 24 ft?

12. A chain is being unwound from a winch. The force of gravity on it is 12.0 N/m. When 20 m have been unwound, how much work is done by gravity in unwinding another 30 m?

13. At liftoff, a rocket weighs 32.5 tons, including the weight of its fuel. During the first (vertical) stage of ascent, fuel is consumed at the rate of 1.25 tons per 1000 ft of ascent. How much work is done in lifting the rocket to an altitude of 12,000 ft?

14. While descending, a 550-N weather balloon enters a zone of freezing rain in which ice forms on the balloon at the rate of 7.50 N per 100 m of descent. Find the work done on the balloon during the first 1000 m of descent through the freezing rain.

15. A meteorite is 75,000 km from the center of Earth and falls to the surface of Earth. From Newton's law of gravity, the force of gravity varies inversely as the square of the distance between the meteorite and the center of Earth. Find the work done by gravity if the meteorite weighs 160 N at the surface, and the radius of Earth is 6400 km.

16. A rectangular swimming pool full of water is 16.0 ft wide, 40.0 ft long, and 5.00 ft deep. Find the work done in pumping the water from the pool to a level 2.00 ft above the top of the pool.

17. Compare the work done in emptying the top half of the water in the swimming pool in Exercise 16 with the work in emptying the bottom half.

18. Find the work done in pumping the water out of the top of a cylindrical tank 3.00 ft in radius and 10.0 ft high, given that the tank is initially full and water weighs 62.4 lb/ft^3. [*Hint:* If horizontal slices dx ft thick are used, each element weighs $62.4(\pi)(3.00^2)dx$ lb, and each element must be raised $10 - x$ ft, if x is the distance from the base to the element (see Fig. 26.66). In this way, the force, which is the weight of the slice, and the distance through which the force acts are determined. Thus, the products of force and distance are summed by integration.]

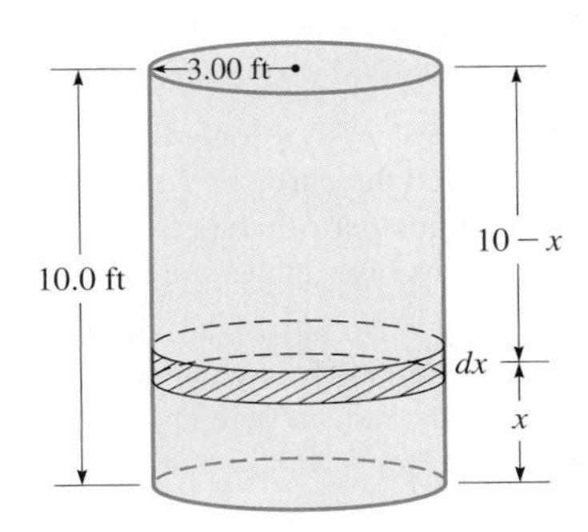

Fig. 26.66

19. How much work is done in drinking a full glass of lemonade (density equal to that of water) if the glass is cylindrical with a radius of 0.050 m and height of 0.100 m, and the top of the straw is 0.050 m above the top of the glass?

20. A hemispherical tank (with the flat end up) of radius 10.0 ft is full of water. Find the work done in pumping the water out of the top of the tank. [See Exercise 18. This problem is similar, except that the weight of each element is 62.4π (radius)^2 (thickness), where the radius of each element is different. If we let x be the radius of an element and y be the distance the element must be raised. We have $62.4\pi x^2\, dy$ with $x^2 + y^2 = 100$.]

21. One end of a spa is a vertical rectangular wall 12.0 ft wide. What is the force exerted on this wall by the water if it is 2.50 ft deep?

22. Find the force on one side of a cubical container 6.0 cm on an edge if the container is filled with mercury. The density of mercury is 133 kN/m^3.

23. A rectangular sea aquarium observation window is 10.0 ft wide and 5.00 ft high. What is the force on this window if the upper edge is 4.00 ft below the surface of the water? The density of seawater is 64.0 lb/ft^3.

24. A horizontal tank has vertical circular ends, each with a radius of 4.00 ft. It is filled to a depth of 4.00 ft with oil of density 60.0 lb/ft^3. Find the force on one end of the tank.

25. A swimming pool is 6.00 m wide and 15.0 m long. The bottom has a constant slope such that the water is 1.00 m deep at one end and 2.00 m deep at the other end. Find the force of the water on one of the sides of the pool.

26. Find the force on the lower half of the wall at the deep end of the swimming pool in Exercise 25.

27. A small dam is in the shape of the area bounded by $y = x^2$ and $y = 20$ (distances in ft). Find the force on the area below $y = 4$ if the surface of the water is at the top of the dam.

28. The tank on a tanker truck has vertical elliptical ends with the major axis horizontal. The major axis is 8.00 ft and the minor axis 6.00 ft. Find the force on one end of the tank when it is half-filled with fuel oil of density 50.0 lb/ft^3.

29. A watertight cubical box with an edge of 2.00 m is suspended in water such that the top surface is 1.00 m below water level. Find the total force on the top of the box and the total force on the bottom of the box. What meaning can you give to the difference of these two forces?

30. Find the force on the region bounded by $x = 2y - y^2$ and the y-axis if the upper point of the area is at the surface of the water. All distances are in feet.

31. The electric current i (in μA) as a function of time t (in μs) for a certain circuit is given by $i = 0.4t - 0.1t^2$. Find the average value of the current with respect to time for the first 4.0 μs.

32. The temperature T (in °C) recorded in a city during a given day approximately followed the curve of $T = 0.00100t^4 - 0.280t^2 + 25.0$, where t is the number of hours from noon $(-12 \text{ h} \leq t \leq 12 \text{ h})$. What was the average temperature during the day?

33. The efficiency e (in %) of an automobile engine is given by $e = 0.768s - 0.00004s^3$, where s is the speed (in km/h) of the car. Find the average efficiency with respect to the speed for $s = 30.0$ km/h to $s = 90.0$ km/h.

34. Find the average value of the volume of a sphere with respect to the radius. Explain the meaning of the result.

35. The length of arc s of a curve from $x = a$ to $x = b$ is

$$s = \int_a^b \sqrt{1 + \left(\frac{dy}{dx}\right)^2}\, dx$$

The cable of a bridge can be described by the equation $y = 0.04x^{3/2}$ from $x = 0$ to $x = 100$ ft. Find the length of the cable. See Fig. 26.67.

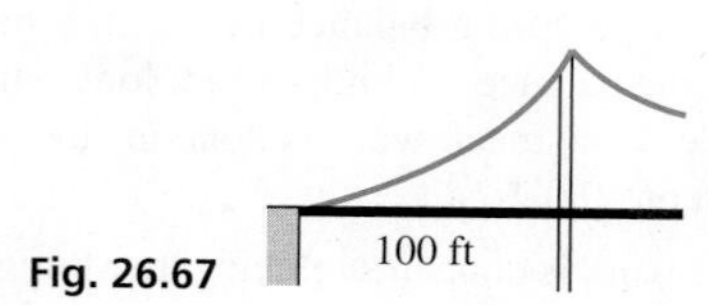

Fig. 26.67

36. A rocket takes off in a path described by the equation $y = \frac{2}{3}(x^2 - 1)^{3/2}$. Find the distance traveled by the rocket for $x = 1.0$ km to $x = 3.0$ km. (See Exercise 35.)

37. The area of a surface of revolution from $x = a$ to $x = b$ is

$$S = 2\pi \int_a^b y \sqrt{1 + \left(\frac{dy}{dx}\right)^2}\, dx$$

Find the formula for the lateral surface area of a right circular cone of radius r and height h.

38. The grinding surface of a grinding machine can be described as the surface generated by rotating the curve $y = 0.2x^3$ from $x = 0$ to $x = 2.0$ cm about the x-axis. Find the grinding surface area. (See Exercise 37.)

Answers to Practice Exercises

1. 2100 ft · lb **2.** 2500 lb

CHAPTER 26 KEY FORMULAS AND EQUATIONS

Velocity	$v = \int a\, dt$	**(26.1)**
Velocity for constant acceleration	$v = at + C_1$	**(26.2)**
Displacement	$s = \int v\, dt$	**(26.3)**
Electric current	$i = \frac{dq}{dt}$	**(26.4)**
Electric charge	$q = \int i\, dt$	**(26.5)**
Voltage across capacitor	$V_C = \frac{1}{C}\int i\, dt$	**(26.6)**

Area

$$A = \int_a^b y\,dx = \int_a^b f(x)\,dx \tag{26.7}$$

$$A = \int_c^d x\,dy = \int_c^d g(y)\,dy \tag{26.8}$$

$$A = \int_a^b (y_2 - y_1)\,dx \tag{26.9}$$

$$A = \int_c^d (x_2 - x_1)\,dy \tag{26.10}$$

Volume

$$V = \pi \int_a^b y^2 dx = \pi \int_a^b [f(x)]^2 dx \tag{26.11}$$

$$V = \pi \int_c^d x^2 dy \tag{26.12}$$

Shell

$$dV = 2\pi(\text{radius}) \times (\text{height}) \times (\text{thickness}) \tag{26.13}$$

Disk

$$dV = \pi(\text{radius})^2 \times (\text{thickness}) \tag{26.14}$$

Center of mass

$$m_1 d_1 + m_2 d_2 + \cdots + m_n d_n = (m_1 + m_2 + \cdots + m_n)\overline{d} \tag{26.15}$$

Centroid of flat plate

$$\overline{x} = \frac{\int_a^b x(y_2 - y_1)\,dx}{\int_a^b (y_2 - y_1)\,dx} \tag{26.16}$$

$$\overline{y} = \frac{\int_c^d y(x_2 - x_1)\,dy}{\int_c^d (x_2 - x_1)\,dy} \tag{26.17}$$

Centroid of solid of revolution

$$\overline{x} = \frac{\int_a^b xy^2 dx}{\int_a^b y^2 dx} \tag{26.18}$$

$$\overline{y} = \frac{\int_c^d yx^2 dy}{\int_c^d x^2 dy} \tag{26.19}$$

Radius of gyration

$$m_1 d_1^2 + m_2 d_2^2 + \cdots + m_n d_n^2 = (m_1 + m_2 + \cdots + m_n)R^2 \tag{26.20}$$

Moment of inertia of flat plate

$$I_y = k \int_a^b x^2 (y_2 - y_1)\,dx \tag{26.21}$$

$$I_x = k \int_c^d y^2 (x_2 - x_1)\,dy \tag{26.22}$$

Moment of inertia of solid of revolution

$$I_y = 2\pi k \int_a^b (y_2 - y_1)x^3 dx \tag{26.23}$$

$$I_x = 2\pi k \int_c^d (x_2 - x_1)y^3 dy \tag{26.24}$$

Work

$$W = \int_a^b f(x)\,dx \tag{26.25}$$

Force between electric charges

$$f(x) = \frac{kq_1 q_2}{x^2} \tag{26.26}$$

Force due to liquid pressure

$$F = w \int_a^b lh\,dh \tag{26.27}$$

Average value of a function

$$y_{\text{av}} = \frac{\int_a^b y\,dx}{b - a} \tag{26.28}$$

CHAPTER 26 REVIEW EXERCISES

CONCEPT CHECK EXERCISES

Determine each of the following as being either **true** *or* **false.** *If it is false, explain why.*

1. If the acceleration of an object moving along a horizontal straight line is 8 m/s^2, then it moves 8 m more each second than in the previous second.

2. To find the area bounded by the curves $y = 2x^2$ and $y = x + 1$, vertical elements of area lead to a simpler solution than horizontal elements.

3. For the volume generated by revolving the region bounded by $y = 4 - 2x$ and the axes about the y-axis,

$$V = 2\pi \int_0^2 (4x - 2x^2)dx.$$

4. For the centroid of a thin plate covering the region bounded by $y = 4x^2$ and $y = 4$, $\bar{y} = \int_0^4 y^{3/2}\,dy$.

5. For a vertical rectangular floodgate, 4 m wide and 2 m high, with its top edge at the surface of the water, the force on the floodgate is $F = w\int_0^2 4h\,dh$.

6. The average value of the function $P = 4.0v^3$ from $v = 10$ to $v = 20$ is given by $\int_{10}^{20} 4.0v^3\,dv$.

PRACTICE AND APPLICATIONS

7. A pitcher releases a baseball horizontally at 45.0 m/s. How far does it drop while traveling 17.1 m to home plate?

8. If the velocity v (m/s) of a subway train after the brakes are applied can be expressed as $v = \sqrt{400 - 20t}$, where t is the time in seconds, how far does it travel in coming to a stop?

9. A weather balloon is rising at the rate of 20 ft/s when a small metal part drops off. If the balloon is 200 ft high at this instant, when will the part hit the ground?

10. A float is dropped into a river at a point where it is flowing at 5.0 ft/s. How far does the float travel in 30 s if it accelerates downstream at 0.020 ft/s^2?

11. A golf ball is putted straight for the hole with an initial velocity of 2.50 m/s and acceleration of -0.750 m/s^2. Will the ball make it to the hole, which is 4.20 m away?

12. What must be the acceleration of a plane taking off from the flight deck of an aircraft carrier if the takeoff velocity is 68 m/s and the available deck length is 180 m?

13. A quarterback throws a *Hail Mary* pass with a vertical velocity of 64 ft/s and a horizontal velocity of 45 ft/s. Will the pass reach the back of the end zone 180 ft (60 yd) away within 3.00 ft of the height from which it was thrown?

14. What is the initial vertical velocity of a baseball that just reaches the ceiling of an indoor stadium that is 65 m high?

15. The electric current i (in A) in a circuit as a function of the time t (in s) is $i = 0.25(2\sqrt{t} - t)$. Find the total charge to pass a point in the circuit in 2.0 s.

16. The current i (in A) in a certain electric circuit is given by $i = \sqrt{1 + 4t}$, where t is the time (in s). Find the charge that passes a given point from $t = 1.0$ s to $t = 3.0$ s if $q_0 = 0$.

17. The voltage across a 5.5-nF capacitor in an FM radio receiver is zero. What is the voltage after 25 μs if a current of 12 mA charges the capacitor?

18. The initial voltage across a capacitor is zero, and $V_C = 2.50$ V after 8.00 ms. If a current $i = t/\sqrt{t^2 + 1}$, where i is the current (in A) and t is the time (in s), charges the capacitor, find the capacitance C of the capacitor.

19. The distribution of weight on a cable is not uniform. If the slope of the cable at any point is given by $dy/dx = 20 + 0.025x^2$ and if the origin of the coordinate system is at the lowest point, find the equation that gives the curve described by the cable.

20. The time rate of change of the reliability R (in %) of a computer system is $dR/dt = -2.5(0.05t + 1)^{-1.5}$, where t is the time (in h). If $R = 100$ for $t = 0$, find R for $t = 100$ h.

21. Find the area between $y = \sqrt{5 - x}$, $y = x - 5$, and $x = 0$.

22. Find the area between $y = 4x^2 - 2x^3$ and the x-axis.

23. Find the area bounded by $y^2 = 2x$ and $y = x - 4$.

24. Find the area bounded by $y = 6/(x + 3)^2$, $y = 0$, $x = -9$, and $x = -5$.

25. Find the area between $y = x^2 + 1$ and $y = x^3 - 2x^2 + 1$.

26. Find the area between $y = x^2 + 8$ and $y = 3x^2$.

27. Show that the curve $y = x^{2n}$ $(n > 0)$ divides the square bounded by $x = 0$, $y = 0$, $x = 1$, and $y = 1$ into two regions, the areas of which are in the ratio $2n/1$.

28. Find the value of a such that the line $x = a$ bisects the area under the curve $y = 1/x^2$ between $x = 1$ and $x = 4$.

29. Find the volume generated by revolving the region bounded by $y = 3 + x^2$ and the line $y = 4$ about the x-axis.

30. Find the volume generated by revolving the region bounded by $y = 8x - x^4$ and the x-axis about the x-axis.

31. Find the volume generated by revolving the region bounded by $y = x^3 - 4x^2$ and the x-axis about the y-axis.

32. Find the volume generated by revolving the region bounded by $y = x$ and $y = 3x - x^2$ about the y-axis.

33. Find the volume generated by revolving an ellipse about its major axis.

34. A hole of radius 1.00 cm is bored along the diameter of a sphere of radius 4.00 cm. Find the volume of the material that is removed from the sphere.

35. Find the center of mass of the following array of four masses in the xy-plane (distances in cm): 60 g at (4, 4), 160 g at $(-3, 6)$, 70 g at $(-5, -4)$, 130 g at $(3, -5)$.

36. Find the centroid of the flat-plate machine part shown in Fig. 26.68. Each section is uniform, and the mass of the section to the right of the y-axis is twice that of the section to the left.

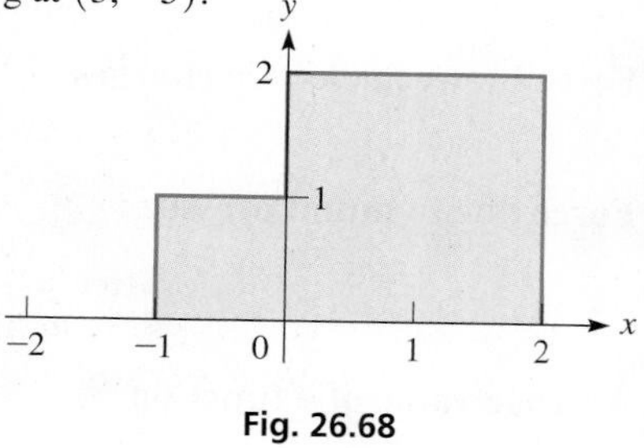

Fig. 26.68

37. Find the centroid of a flat plate covering the region bounded by $y^2 = x^3$ and $y = 3x$.

38. Find the centroid of a flat plate covering the region bounded by $y = 2x - 4$, $x = 1$, and $y = 0$.

39. Find the centroid of the volume generated by revolving the region bounded by $y = \sqrt{x}$, $x = 1$, $x = 4$, and $y = 0$ about the x-axis.

40. Find the centroid of the volume generated by revolving the region in the first quadrant bounded by $yx^4 = 1$, $y = 1$, and $y = 4$ about the y-axis.

41. Find the moment of inertia of a flat plate covering the region bounded by $y = 3x - x^2$ and $y = x$ with respect to the y-axis.

42. Find the radius of gyration of a flat plate covering the region bounded by $y = 8 - x^3$, and the axes, with respect to the y-axis.

43. Find the moment of inertia with respect to its axis of a lead bullet that is defined by revolving the region bounded by $y = 3.00x^{0.10}$, $x = 0$, $x = 20.0$, and $y = 0$ about the x-axis (all measurements in mm). The density of lead is $0.0114\ \text{g/mm}^3$.

44. Find the radius of gyration with respect to its axis of a rotating machine part that can be defined by revolving the region bounded by $y = 1/x$, $x = 1.00$, $x = 4.00$, and $y = 0.25$ (all measurements in in.) about the x-axis.

45. A pail and its contents weigh 80 lb. The pail is attached to the end of a 100-ft rope that weighs 10 lb and is hanging vertically. How much work is done in winding up the rope with the pail attached?

46. The gravitational force (in lb) of Earth on a satellite (the weight of the satellite) is given by $F = 10^{11}/x^2$, where x is the vertical distance (in mi) from the center of Earth to the satellite. How much work is done in moving the satellite from Earth's surface to an altitude of 2000 mi? The radius of Earth is 3960 mi.

47. A decorative glass table-top is designed to be the region between the curves of $y = 0.0001x^4$ and $y = 110 - 0.0001x^4$. Find the area (in cm^2) of the table top.

48. A level putting green at a golf course can be approximated as the area bounded by $y = 0.003x^3 - 2x$ and $y = 1.5x - 0.001x^3$ for $x \geq 0$. Find the area (in m^2) of the green.

49. In a video game, a large rock is propelled up a slope at 28 ft/s, but is accelerating down the slope at 18 ft/s^2. What is the velocity of the rock after 5.0 s?

50. The vertical end of a trough is made of a right triangular section of concrete and a similar section of wood, as shown in Fig. 26.69. Find the force on each section if the trough is full of water.

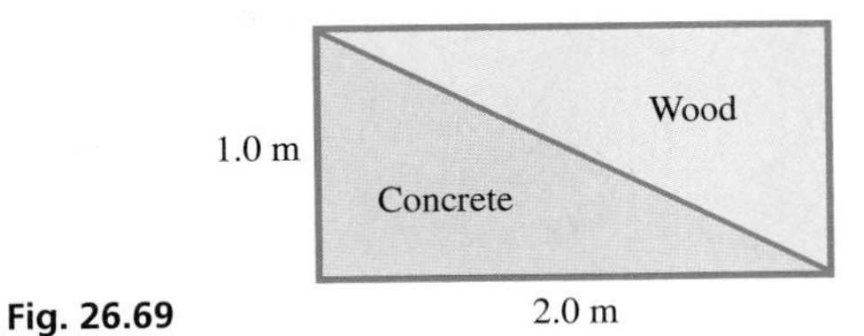

Fig. 26.69

51. An acoustic horn can be represented by revolving the region bounded by $y = x^{3/2}$, $y = 0$, and $x = 4$ about the x-axis. Find the volume (in cm^3) of the horn.

52. A sea-life observation boat has a vertical rectangular window 3.50 m wide and 1.60 m high. What is the total force on the window if the top is 2.50 m below the water surface? ($w = 10.1\ \text{kN/m}^3$)

53. The rear stabilizer of a certain aircraft can be described as the region under the curve $y = 3x^2 - x^3$, as shown in Fig. 26.70. Find the x-coordinate (in m) of the centroid of the stabilizer.

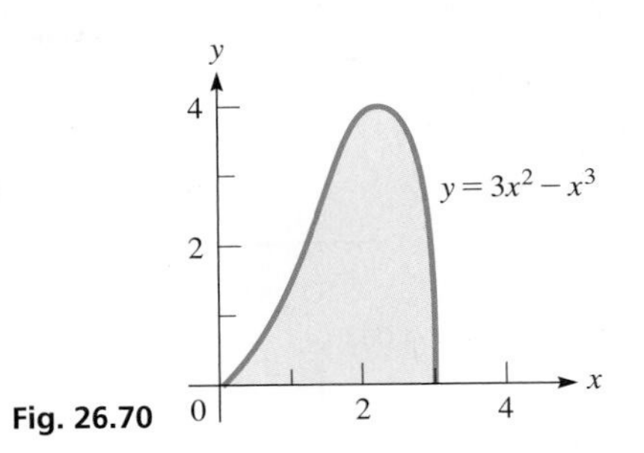

Fig. 26.70

54. The diameter of a circular swimming pool is 32 ft, and the sides are 5.0 ft high. If the depth of the water is 4.0 ft, how much work is done in pumping all of the water out over the side?

55. The nose cone of a rocket has the shape of a semiellipse revolved about its major axis, as shown in Fig. 26.71. What is the volume of the nose cone?

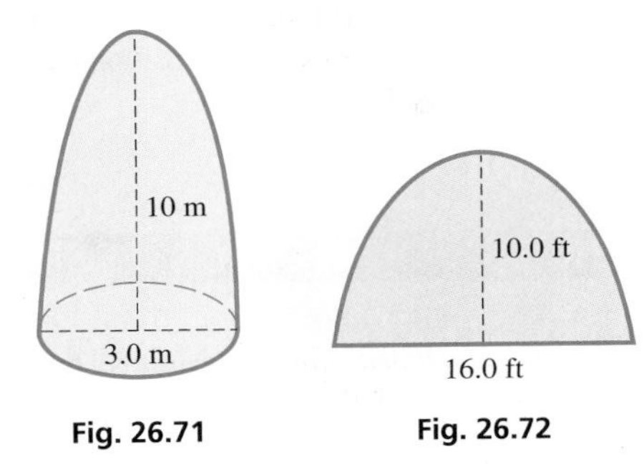

Fig. 26.71 Fig. 26.72

56. The deck area of a boat is a parabolic section as shown in Fig. 26.72. What is the area of the deck?

57. The vertical ends of a fuel storage tank have a parabolic bottom section and a triangular top section, as shown in Fig. 26.73. What volume does the tank hold?

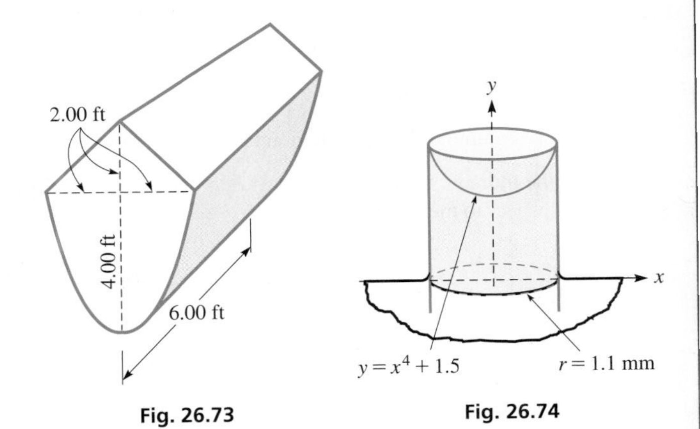

Fig. 26.73 Fig. 26.74

58. The capillary tube shown in Fig. 26.74 has circular horizontal cross sections of inner radius 1.1 mm. What is the volume of the liquid in the tube above the level of liquid outside the tube if the top of the liquid in the center vertical cross section is described by the equation $y = x^4 + 1.5$, as shown?

59. A cylindrical chemical waste-holding tank 4.50 ft in radius has a depth of 3.25 ft. Find the total force on the circular side of the tank when it is filled with liquid with a density of 68.0 lb/ft^3.

60. A section of a dam is in the shape of a right triangle. The base of the triangle is 6.00 ft and is on the surface of the water. If the triangular section goes to a depth of 4.00 ft, find the force on it. See Fig. 26.75.

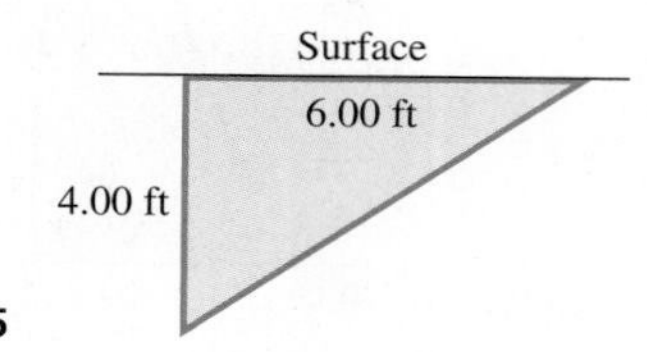

Fig. 26.75

61. The temperature T (in °C) recorded during a day followed the curve $T = 5.00 \times 10^{-4}t^4 - 0.0240t^3 + 0.300t^2 + 10.0$, where t is the number of hours after midnight $(0 \leq t \leq 24)$. What was the average temperature during that day?

62. The mass of Earth is 5.98×10^{24} kg, and the mass of the moon is 7.36×10^{22} kg. Assuming all of the mass of each is at its center, find the center of mass of the Earth-moon system, if their centers are 3.82×10^8 m apart. Compare this position with the radius of Earth, which is 6.37×10^6 m.

63. The velocity v of blood that flows in a blood vessel with radius R is given by $v = k(R^2 - r^2)$, where r is the distance from the central axis of the vessel and k is a constant. Find the average velocity of blood for $0 \leq r \leq R$.

64. A horizontal straight section of pipe is supported at its center by a vertical wire as shown in Fig. 26.76. Find the formula for the moment of inertia of the pipe with respect to an axis along the wire if the pipe is of length L and mass m.

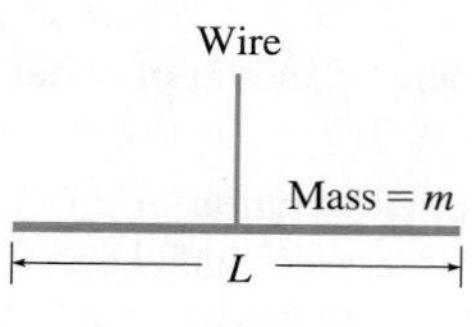

Fig. 26.76

65. The float for a certain valve control has a circular top of radius a. All cross sections of the float perpendicular to a fixed diameter of the top are squares. Write one or two paragraphs explaining how to derive the formula that gives the volume of the float. What is the formula?

CHAPTER 26 PRACTICE TEST

As a study aid, we have included complete solutions for each Practice Test problem at the end of this book.

In Problems 1–3, use the region bounded by $y = \frac{1}{4}x^2$, $y = 0$, *and* $x = 2$.

1. Find the area.

2. Find the coordinates of the centroid of a flat plate that covers the region.

3. Find the volume if the given region is revolved about the x-axis.

In Problems 4 and 5, use the first-quadrant region bounded by $y = x^2$, $x = 0$, *and* $y = 9$.

4. Find the volume if the given region is revolved about the x-axis.

5. Find the moment of inertia of a flat plate that covers the region, with respect to the y-axis.

6. The current i (in A) in an electric circuit is given by $i = 5.0 - 0.20t$, where t is time (in s). If 0 C of charge have passed a given point in the circuit when $t = 0$, how many coulombs pass the point after 2.0 s?

7. The velocity v of an object as a function of the time t is $v = 60 - 4t$. Find the expression for the displacement s if $s = 10$ for $t = 0$.

8. The natural length of a spring is 8.0 cm. A force of 12 N stretches it to a length of 10.0 cm. How much work is done in stretching it from a length of 10.0 cm to a length of 14.0 cm?

9. A vertical rectangular floodgate is 6.00 ft wide and 2.00 ft high. Find the force on the gate if its upper edge is 1.00 ft below the surface of the water ($w = 62.4$ lb/ft^3).

10. What is the average value of $f(x) = x^2 + 2x - 3$ for $0 \leq x \leq 6$?

Differentiation of Transcendental Functions

LEARNING OUTCOMES

After completion of this chapter, the student should be able to:

- Find the derivative of expressions involving trigonometric and inverse trigonometric functions
- Find the derivative of expressions involving logarithmic and exponential functions
- Find limits of indeterminate forms using L'Hospital's rule
- Solve application problems involving derivatives of transcendental functions

While studying vibrations in a rod in the mid-1700s, the Swiss mathematician Euler noted that the trigonometric functions arose naturally as solutions to equations in which derivatives appeared. This was the first treatment of the trigonometric functions as functions of numbers essentially as we do today. Later, in 1755, Euler wrote a textbook on differential calculus in which he included differentiation of the trigonometric, inverse trigonometric, logarithmic, and exponential functions. Euler called these functions "transcendental," as "they transcend the power of algebraic methods." These functions are not algebraic in that they cannot be expressed using basic algebraic operations (addition, subtraction, multiplication, division, and taking roots) on polymomials.

Logarithms were first developed as a tool for calculation. In establishing calculus, Newton and Leibniz did some formulation of the logarithmic and exponential functions. However, the calculus of the transcendental functions was formulated mostly in the 1700s by Euler and a number of other mathematicians. This led to the rapid progress made later in many technical and scientific areas. For example, transcendental functions, along with their derivatives and integrals, have been of great importance in the development of the fields of electricity and electronics in the 1800s, 1900s, and 2000s, particularly with respect to alternating current.

In this chapter, we develop formulas for the derivatives of these transcendental functions; and in the next chapter, we will take up integration that involves these functions. Other areas in which we will show important applications include harmonic motion, rocket motion, monetary interest calculations, population growth, acoustics, and optics.

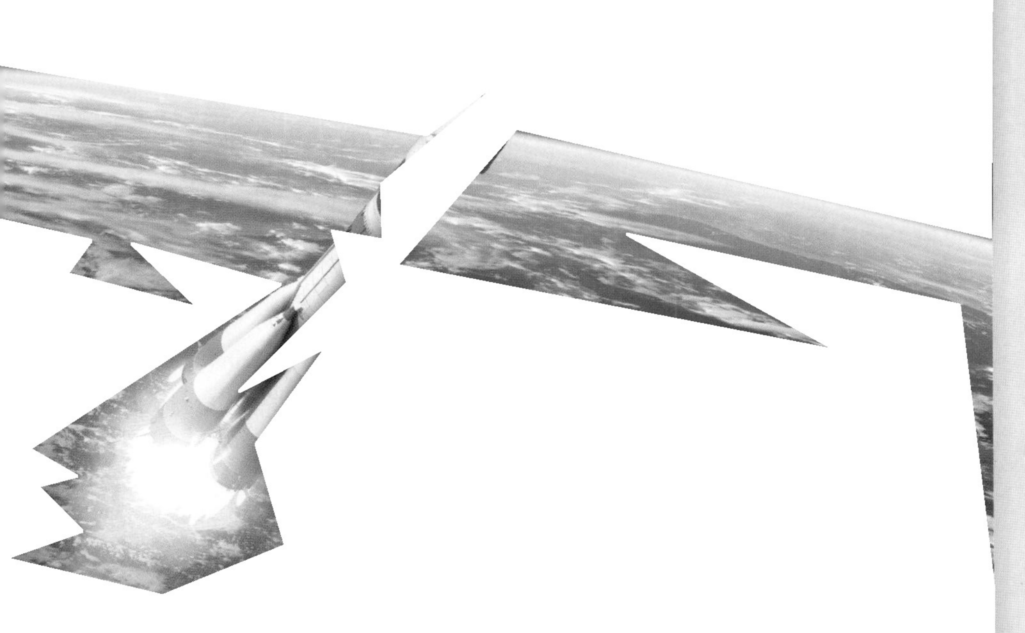

◀ In Sections 27.4 and 27.8, we use derivatives of transcendental functions in analyzing the motion of a rocket.

27.1 Derivatives of the Sine and Cosine Functions

Limit of $\sin\theta/\theta$ as $\theta \to 0$ • Derivative of $\sin u$ • Derivative of $\cos u$

We now find the derivative of the sine function. We will then be able to use it in finding the derivatives of the other trigonometric and inverse trigonometric functions.

Let $y = \sin x$, where x is expressed in radians. If x changes by an amount h, from the definition of the derivative, we have

$$\frac{dy}{dx} = \lim_{h\to 0}\frac{\sin(x+h) - \sin x}{h}$$

■ For reference, Eq. (20.18) is $\sin x - \sin y = 2\sin\frac{1}{2}(x-y)\cos\frac{1}{2}(x+y)$.

Referring now to Eq. (20.18), we have

$$\frac{dy}{dx} = \lim_{h\to 0}\frac{2\sin\frac{1}{2}(x+h-x)\cos\frac{1}{2}(x+h+x)}{h}$$
$$= \lim_{h\to 0}\frac{\sin(h/2)\cos(x+h/2)}{h/2}$$

Looking ahead to the next step of letting $h \to 0$, we see that the numerator and denominator both approach zero. This situation is precisely the same as that in which we were finding the derivatives of the algebraic functions. To find the limit, we must find

$$\lim_{h\to 0}\frac{\sin(h/2)}{h/2}$$

because these are the factors that cause the numerator and denominator to approach zero.

In finding this limit, we let $\theta = h/2$ for convenience of notation. This means that we are to determine $\lim_{\theta\to 0}\frac{\sin\theta}{\theta}$. Of course, it would be convenient to know before proceeding if this limit does actually exist. Therefore, by using a calculator, we can develop a table of values of $\frac{\sin\theta}{\theta}$ as θ becomes very small:

θ (radians)	0.5	0.1	0.05	0.01	0.001
$\frac{\sin\theta}{\theta}$	0.9588511	0.9983342	0.9995834	0.9999833	0.9999998

We see from this table that the limit of $\frac{\sin\theta}{\theta}$, as $\theta \to 0$, appears to be 1.

In order to prove that $\lim_{\theta\to 0}\frac{\sin\theta}{\theta} = 1$, we use a geometric approach. Considering Fig. 27.1, we see that the following inequality is true, assuming that θ is in radians:

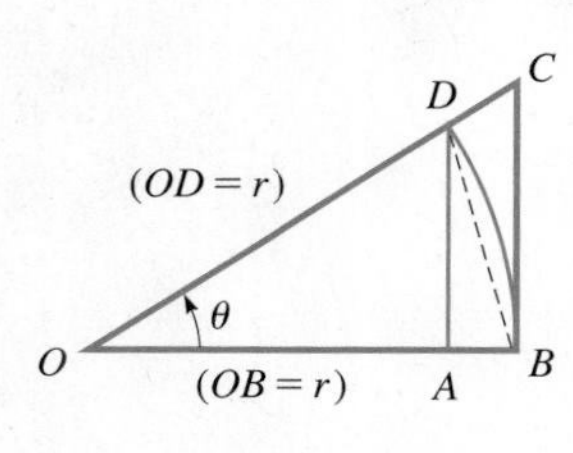

Fig. 27.1

$$\text{Area triangle } OBD < \text{area sector } OBD < \text{area triangle } OBC$$

$$\frac{1}{2}r(r\sin\theta) < \frac{1}{2}r^2\theta < \frac{1}{2}r(r\tan\theta) \quad \text{or} \quad \sin\theta < \theta < \tan\theta$$

Remembering that we want to find the limit of $(\sin\theta)/\theta$, we next divide through by $\sin\theta$ and then take reciprocals:

$$1 < \frac{\theta}{\sin\theta} < \frac{1}{\cos\theta} \quad \text{or} \quad 1 > \frac{\sin\theta}{\theta} > \cos\theta$$

When we consider the limit as $\theta \to 0$, we see that the left member remains 1 and the right member approaches 1. Thus, $(\sin\theta)/\theta$ must approach 1. This means

$$\lim_{\theta\to 0}\frac{\sin\theta}{\theta} = \lim_{h\to 0}\frac{\sin(h/2)}{h/2} = 1 \tag{27.1}$$

Using this result in the expression for dy/dx, we have

$$\lim_{h \to 0}\left[\cos(x + h/2)\frac{\sin(h/2)}{h/2}\right] = \cos x$$

$$\frac{dy}{dx} = \cos x \tag{27.2}$$

To find the derivative of $y = \sin u$, where u is a function of x, we use the chain rule, which we repeat here for reference:

$$\frac{dy}{dx} = \frac{dy}{du}\frac{du}{dx} \tag{27.3}$$

Therefore, for $y = \sin u$, $dy/du = \cos u$, and we have

$$\frac{d(\sin u)}{dx} = \cos u \frac{du}{dx} \tag{27.4}$$

EXAMPLE 1 Derivative of sin *u*

(a) Find the derivative of $y = \sin 2x$.

In this example, $u = 2x$. Therefore, using Eq. (27.4),

$$\frac{dy}{dx} = \frac{d(\sin 2x)}{dx} = \cos 2x\frac{d(2x)}{dx} = (\cos 2x)(2) \quad \leftarrow \frac{du}{dx}$$

$$= 2\cos 2x$$

(b) Find the derivative of $y = 2\sin(x^2)$.

In this example, $u = x^2$, which means that $du/dx = 2x$. This means

$$\frac{dy}{dx} = 2[\cos(x^2)](2x) \quad \text{using Eq. (27.4)}$$

$$= 4x\cos(x^2)$$

CAUTION It is important here, just as it is in finding the derivatives of powers of all functions, to remember to include the factor du/dx. ■ ■

Practice Exercise

1. Find the derivative of $y = 3\sin(4x + 1)$.

EXAMPLE 2 Derivative of a power of sin *u*

Find the derivative of $r = \sin^2\theta$.

This example is a combination of the use of the general power rule and the derivative of the sine function. Because $\sin^2\theta$ means $(\sin\theta)^2$, in using the general power rule with $u = \sin\theta$, we have

$$\frac{dr}{d\theta} = 2(\sin\theta)\frac{d\sin\theta}{d\theta} \quad \text{using } \frac{du^n}{dx} = nu^{n-1}\left(\frac{du}{dx}\right)$$

$$= 2\sin\theta\cos\theta \quad \text{using } \frac{d(\sin u)}{dx} = \cos u\frac{du}{dx}$$

$$= \sin 2\theta \quad \text{using identity } \sin 2\alpha = 2\sin\alpha\cos\alpha$$

■

In order to find the derivative of the cosine function, we write it in the form $\cos u = \sin(\frac{\pi}{2} - u)$. Thus, if $y = \sin(\frac{\pi}{2} - u)$, we have

$$\frac{dy}{dx} = \cos\left(\frac{\pi}{2} - u\right)\frac{d(\frac{\pi}{2} - u)}{dx} = \cos\left(\frac{\pi}{2} - u\right)\left(-\frac{du}{dx}\right)$$
$$= -\cos\left(\frac{\pi}{2} - u\right)\frac{du}{dx}$$

Because $\cos(\frac{\pi}{2} - u) = \sin u$, we have

$$\frac{d(\cos u)}{dx} = -\sin u\frac{du}{dx} \tag{27.5}$$

EXAMPLE 3 Derivative of cos u—power in an amplifier

The electric power p developed in a resistor of an amplifier circuit is $p = 25\cos^2 120\pi t$, where t is the time. Find the expression for the time rate of change of power.

From Chapter 23, we know that we are to find the derivative dp/dt. Therefore,

$$p = 25\cos^2 120\pi t$$
$$\frac{dp}{dt} = 25(2\cos 120\pi t)\frac{d\cos 120\pi t}{dt} \qquad \text{using } \frac{du^n}{dx} = nu^{n-1}\left(\frac{du}{dx}\right)$$
$$= 50\cos 120\pi t(-\sin 120\pi t)\frac{d(120\pi t)}{dt} \qquad \text{using } \frac{d(\cos u)}{dx} = -\sin u\frac{du}{dx}$$
$$= (-50\cos 120\pi t\sin 120\pi t)(120\pi)$$
$$= -6000\pi\cos 120\pi t\sin 120\pi t$$
$$= -3000\pi\sin 240\pi t \qquad \text{using } \sin 2\alpha = 2\sin\alpha\cos\alpha$$ ■

EXAMPLE 4 Derivative of a root of cos u

Find the derivative of $y = \sqrt{1 + \cos 2x}$.

$$y = (1 + \cos 2x)^{1/2}$$
$$\frac{dy}{dx} = \frac{1}{2}(1 + \cos 2x)^{-1/2}\frac{d(1 + \cos 2x)}{dx} \qquad \text{using } \frac{du^n}{dx} = nu^{n-1}\left(\frac{du}{dx}\right)$$
$$= \frac{1}{2}(1 + \cos 2x)^{-1/2}(-\sin 2x)(2) \qquad \text{using } \frac{d(\cos u)}{dx} = -\sin u\frac{du}{dx}$$
$$= -\frac{\sin 2x}{\sqrt{1 + \cos 2x}}$$ ■

Practice Exercise

2. Find the derivative of $y = 5(1 + \cos x^2)^2$.

EXAMPLE 5 Differential of product sin u cos u

Find the differential of $y = \sin 2x\cos x^2$.

From Section 24.8, recall that the differential of a function $y = f(x)$ is $dy = f'(x)dx$. Thus, using the derivative product rule and the derivatives of the sine and cosine functions, we arrive at the following result:

$$y = \sin 2x\cos x^2 \qquad y = (\sin 2x)(\cos x^2)$$
$$dy = [\sin 2x(-\sin x^2)(2x) + \cos x^2(\cos 2x)(2)]dx$$
$$= (-2x\sin 2x\sin x^2 + 2\cos 2x\cos x^2)dx$$ ■

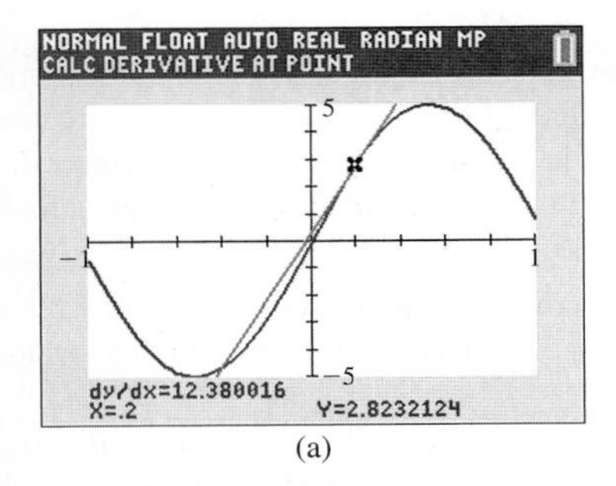

(a)

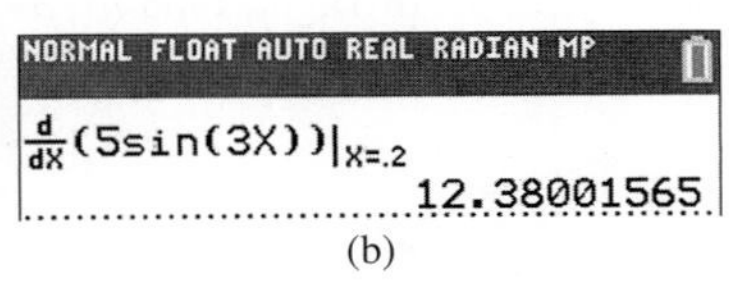

(b)

Fig. 27.2
Graphing calculator keystrokes:
goo.gl/bjZA8T

EXAMPLE 6 Slope of a tangent line

Find the slope of a line tangent to the curve of $y = 5\sin 3x$, where $x = 0.2$.

Here, we are to find the derivative of $y = 5\sin 3x$ and then evaluate the derivative for $x = 0.2$. Therefore, we have the following:

$$y = 5\sin 3x$$

$$\frac{dy}{dx} = 5(\cos 3x)(3) = 15\cos 3x \qquad \text{find derivative}$$

$$\left.\frac{dy}{dx}\right|_{x=0.2} = 15\cos 3(0.2) = 15\cos 0.6 \qquad \text{evaluate}$$

$$= 12.38$$

In evaluating the slope, we must remember that $x = 0.2$ means the ***values are in radians.*** Therefore, the slope is 12.38.

The calculator screen in Fig. 27.2 (a) shows the function, the tangent line, and the slope at $x = 0.2$. The screen in Fig. 27.2 (b) shows the slope at $x = 0.2$ using the *numerical derivative* feature. ■

EXERCISES 27.1

In Exercises 1 and 2, make the given changes in the indicated examples of this section and then find the derivatives.

1. In Example 2, in the given function, change θ to $2\theta^2$.

2. In Example 4, in the given function, change $2x$ to x^2.

In Exercises 3–34, find the derivatives of the given functions.

3. $y = \sin(3x + 2)$

4. $y = 3\sin 4x$

5. $y = 2\sin(2x^3 - 1)$

6. $s = 5\sin(7 - 3t)$

7. $y = 9\cos\frac{4}{3}x$

8. $y = \cos(1 - 2x)$

9. $y = 3x + 2\cos(3x - \pi)$

10. $y = x^2 - \cos(1 - 3x)$

11. $r = \sin^2 5\pi\theta$

12. $y = 3\sin^3(2x^4 + 1)$

13. $y = 3\cos^3(5x + 2)$

14. $y = 4\cos^2\sqrt{x}$

15. $y = 4x\sin 3x$

16. $v = 6t^2\sin 3\pi t$

17. $y = 3x^3\cos 5x$

18. $y = 0.5\theta\cos(2\theta + \pi/4)$

19. $u = 3\sin v^2\cos 5v$

20. $y = \sin 3x\cos 4x$

21. $y = \sqrt{1 + \sin 4x}$

22. $y = (3x - \cos^2 x)^4$

23. $r = \dfrac{\sin(3t - \pi/3)}{2t}$

24. $T = \dfrac{4z + 3}{\sin \pi z}$

25. $y = \dfrac{2\cos x^2}{3x - 1}$

26. $y = \dfrac{\cos^2 3x}{1 + 2\sin^2 2x}$

27. $y = 2\sin^2 3x\cos 2x$

28. $y = \cos^3 4x\sin^2 2x$

29. $s = \sin(3\sin 2t)$

30. $z = 0.2\cos(4\sin 3\phi)$

31. $y = \sin^3 x - \cos 2x$

32. $y = x\sin x + \cos x$

33. $p = \dfrac{1}{\sin s} + \dfrac{1}{\cos s}$

34. $y = 2x\sin x + 2\cos x - x^2\cos x$

35. $\cos y = x^2 + y$

36. $y\sin x = 2y$

In Exercises 37–60, solve the given problems.

37. Using a calculator: (a) display the graph of $y = (\sin x)/x$ to verify that $(\sin\theta)/\theta \to 1$ as $\theta \to 0$, and (b) verify the values for $(\sin\theta)/\theta$ in the table on page 806.

38. Evaluate $\lim_{\theta\to 0}(\tan\theta)/\theta$. [Use the fact that $\lim_{\theta\to 0}(\sin\theta)/\theta = 1$.]

39. On a calculator, find the values of (a) $\cos 1.0000$ and (b) $(\sin 1.0001 - \sin 1.0000)/0.0001$. Compare the values and give the meaning of each in relation to the derivative of the sine function where $x = 1$.

40. On a calculator, find the values of (a) $-\sin 1.0000$ and (b) $(\cos 1.0001 - \cos 1.0000)/0.0001$. Compare the values and give the meaning of each in relation to the derivative of the cosine function where $x = 1$.

41. On the graph of $y = \sin x$ in Fig. 27.3, draw tangent lines at the indicated points and estimate the slopes of these tangent lines (the slope at the x-intercepts is ± 1). Then plot the values of these slopes for the same values of x and join the points with a smooth curve. Compare the resulting curve with $y = \cos x$. (Note the meaning of the derivative as the slope of a tangent line.)

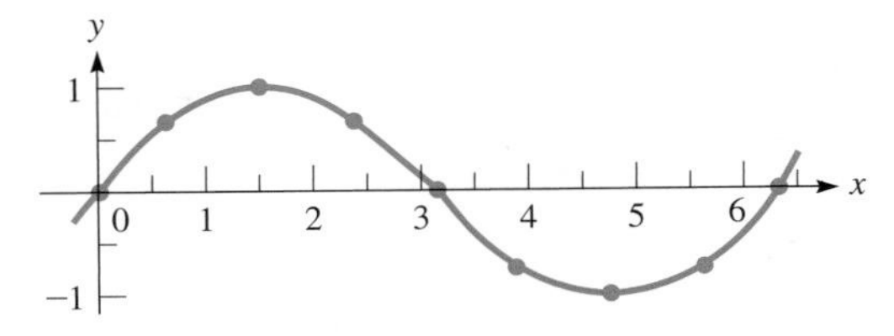

Fig. 27.3

42. Display the graphs of $y_1 = \cos x$ and $y_n = \dfrac{\sin(x + h) - \sin x}{h}$ on the same screen of a calculator for $0 \le x \le 2\pi$. For $n = 2$, let $h = 0.5$; for $n = 3$, let $h = 0.1$. (You might try some even smaller values of h.) What do these curves show?

43. Find the derivative $\frac{dy}{dx}$ of the implicit function $\sin(xy) + \cos 2y = x^2$.

44. Find the derivative $\frac{dy}{dx}$ of the implicit function $x\cos 2y + \sin x \cos y = 1$.

45. Show that $\frac{d^4 \sin x}{dx^4} = \sin x$.

46. Show that $y = A\sin kx + B\cos kx$ satisfies the equation $y'' + k^2 y = 0$.

47. Find the derivative of each member of the identity $\cos 2x = 2\cos^2 x - 1$ and thereby obtain another trigonometric identity.

48. Find values of x for which the following curves have horizontal tangents: (a) $y = x + \sin x$ (b) $y = 4x + \cos \pi x$

49. Use differentials to estimate the value of sin 31°.

50. Find the linearization $L(x)$ of the function $f(x) = \sin(\cos x)$ for $a = \pi/2$.

51. Find the slope of a line tangent to the curve of $y = \frac{2\sin 3x}{x}$, where $x = 0.15$. Verify the result by using the *derivative-evaluating* feature of a calculator.

52. An object is oscillating vertically on the end of a spring such that its displacement d (in cm) is $d = 2.5\cos 16t$, where t is the time (in s). What is the acceleration of the object after 1.5 s?

53. The displacement d (in mm) of a piano wire as a function of the time t (in s) is $d = 3.0 \sin 188t \cos 188t$. How fast is the displacement changing when $t = 2.0$ ms?

54. A water slide at an amusement park follows the curve (y in m) $y = 2.0 + 2.0\cos(0.53x + 0.40)$ for $0 \le x \le 5.0$ m. Find the angle with the horizontal of the slide for $x = 2.5$ m.

55. The blade of a saber saw moves vertically up and down, and its displacement y (in cm) is given by $y = 1.85 \sin 36\pi t$, where t is the time (in s). Find the velocity of the blade for $t = 0.0250$ s.

56. The current i (in A) in an amplifier circuit as a function of the time t (in s) is given by $i = 0.10\cos(120\pi t + \pi/6)$. The voltage caused by the changing current is given by $V_L = L(di/dt)$, where L is the inductance (in H). Find the expression for the voltage across a 2.0-mH inductor in the circuit.

57. In testing a heat-seeking rocket, it is always moving directly toward a remote-controlled aircraft. At a certain instant, the distance r (in km) from the rocket to the aircraft is $r = \frac{100}{1 - \cos\theta}$, where θ is the angle between their directions of flight. Find $dr/d\theta$ for $\theta = 120°$. See Fig. 27.4.

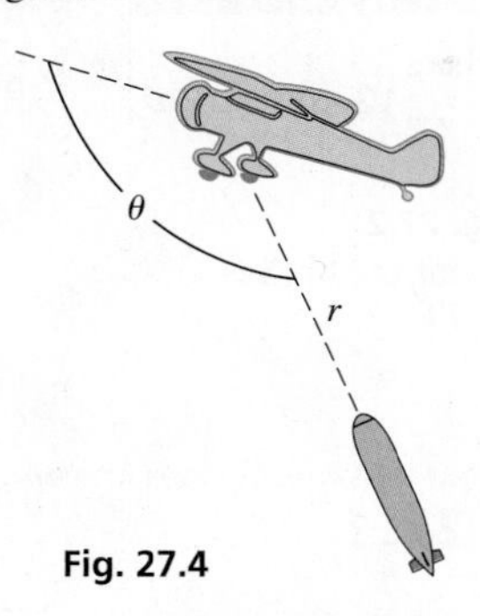

Fig. 27.4

58. The number N of reflections of a light ray passing through an optic fiber of length L and diameter d is $N = \frac{L\sin\theta}{d\sqrt{n^2 - \sin^2\theta}}$. Here, n is the index of refraction of the fiber, and θ is the angle between the light ray and the fiber's axis. Find $dN/d\theta$.

59. In the study of the transmission of light, the equation $T = \frac{A}{1 + B\sin^2(\theta/2)}$ arises. Find $dT/d\theta$.

60. A conical paper cup has a slant height of 10.0 cm. Express the volume V of the cup in terms of $\theta/2$, where θ is the vertex angle at the bottom of the cup. Then find $dV/d\theta$.

Answers to Practice Exercises

1. $y' = 12\cos(4x + 1)$ **2.** $y' = -20x\sin x^2(1 + \cos x^2)$

27.2 Derivatives of the Other Trigonometric Functions

Derivatives of tan *u*, cot *u*, sec *u*, csc *u*

We obtain the derivative of tan u by expressing tan u as sin u/cos u. Therefore, letting $y = \sin u/\cos u$, by employing the quotient rule, we have

$$\frac{dy}{dx} = \frac{\cos u[\cos u(du/dx)] - \sin u[-\sin u(du/dx)]}{\cos^2 u}$$

$$= \frac{\cos^2 u + \sin^2 u}{\cos^2 u}\frac{du}{dx} = \frac{1}{\cos^2 u}\frac{du}{dx} = \sec^2 u\frac{du}{dx}$$

$$\frac{d(\tan u)}{dx} = \sec^2 u\frac{du}{dx} \tag{27.6}$$

We find the derivative of cot u by letting $y = \cos u/\sin u$, again using the quotient rule.

$$\frac{dy}{dx} = \frac{\sin u[-\sin u(du/dx)] - \cos u[\cos u(du/dx)]}{\sin^2 u}$$
$$= \frac{-\sin^2 u - \cos^2 u}{\sin^2 u}\frac{du}{dx}$$

$$\frac{d(\cot u)}{dx} = -\csc^2 u\frac{du}{dx} \quad (27.7)$$

To obtain the derivative of sec u, we let $y = 1/\cos u$. Then,

$$\frac{dy}{dx} = -(\cos u)^{-2}\left[(-\sin u)\left(\frac{du}{dx}\right)\right] = \frac{1}{\cos u}\frac{\sin u}{\cos u}\frac{du}{dx}$$

$$\frac{d(\sec u)}{dx} = \sec u \tan u\frac{du}{dx} \quad (27.8)$$

We obtain the derivative of csc u by letting $y = 1/\sin u$. And so,

$$\frac{dy}{dx} = -(\sin u)^{-2}\left(\cos u\frac{du}{dx}\right) = -\frac{1}{\sin u}\frac{\cos u}{\sin u}\frac{du}{dx}$$

$$\frac{d(\csc u)}{dx} = -\csc u \cot u\frac{du}{dx} \quad (27.9)$$

EXAMPLE 1 Derivative of a power of sec u

Find the derivative of $y = 3\sec^2 4x$.

Using the general power rule (or just power rule, for short) and Eq. (27.8), we have

$$\frac{dy}{dx} = 3(2)(\sec 4x)\frac{d(\sec 4x)}{dx} \qquad \text{using } \frac{du^n}{dx} = nu^{n-1}\frac{du}{dx}$$
$$= 6(\sec 4x)(\sec 4x \tan 4x)(4) \qquad \text{using } \frac{d\sec u}{dx} = \sec u\tan u\frac{du}{dx}$$
$$= 24\sec^2 4x\tan 4x$$

■

Practice Exercise

1. Find the derivative of $y = 3\tan 8x$.

EXAMPLE 2 Derivative of a product with csc u

Find the derivative of $y = t\csc^3 2t$.

Using the power rule, the product rule, and Eq. (27.9), we have

$$\frac{dy}{dt} = t(3\csc^2 2t)(-\csc 2t\cot 2t)(2) + (\csc^3 2t)(1)$$
$$= \csc^3 2t(-6t\cot 2t + 1)$$

■

EXAMPLE 3 Derivative of a power with tan u and sec u

Find the derivative of $y = (\tan 2x + \sec 2x)^3$.

Using the power rule and Eqs. (27.6) and (27.8), we have

$$\frac{dy}{dx} = 3(\tan 2x + \sec 2x)^2[\sec^2 2x(2) + \sec 2x\tan 2x(2)]$$
$$= 3(\tan 2x + \sec 2x)^2(2\sec 2x)(\sec 2x + \tan 2x)$$
$$= 6\sec 2x(\tan 2x + \sec 2x)^3$$

■

Practice Exercise

2. Find the derivative of $y = 5(\cot 3x + \csc 3x)^2$.

EXAMPLE 4 Differential with $\sin u$ and $\tan u$

Find the differential of $r = \sin 2\theta \tan \theta^2$.

NOTE ➧ [Here, we are to find the derivative of the given function and multiply by $d\theta$.] Therefore, using the product rule along with the derivatives of the sine and tangent functions, we have

$$dr = [(\sin 2\theta)(\sec^2\theta^2)(2\theta) + (\tan\theta^2)(\cos 2\theta)(2)]d\theta$$
$$= (2\theta \sin 2\theta \sec^2\theta^2 + 2\cos 2\theta \tan\theta^2)d\theta \quad \text{don't forget the } d\theta$$ ■

EXAMPLE 5 Derivative of an implicit function

Find dy/dx if $\cot 2x - 3\csc xy = y^2$.

In finding the derivative of this implicit function, we must be careful not to forget the factor dy/dx when it occurs. The derivative is found as follows:

$$\cot 2x - 3\csc xy = y^2$$
$$(-\csc^2 2x)(2) - 3(-\csc xy \cot xy)\left(x\frac{dy}{dx} + y\right) = 2y\frac{dy}{dx}$$
$$3x\csc xy \cot xy\frac{dy}{dx} - 2y\frac{dy}{dx} = 2\csc^2 2x - 3y\csc xy \cot xy$$
$$\frac{dy}{dx} = \frac{2\csc^2 2x - 3y\csc xy \cot xy}{3x\csc xy \cot xy - 2y}$$ ■

EXAMPLE 6 Evaluation of a derivative at a point

Evaluate the derivative of $y = \dfrac{2x}{1 - \cot 3x}$, where $x = 0.25$.

Finding the derivative, we have

$$\frac{dy}{dx} = \frac{(1 - \cot 3x)(2) - 2x(\csc^2 3x)(3)}{(1 - \cot 3x)^2}$$
$$= \frac{2 - 2\cot 3x - 6x\csc^2 3x}{(1 - \cot 3x)^2}$$

Now, substituting $x = 0.25$, we have

$$\left.\frac{dy}{dx}\right|_{x=0.25} = \frac{2 - 2\cot 0.75 - 6(0.25)\csc^2 0.75}{(1 - \cot 0.75)^2}$$
$$= -626.0$$

NOTE ➧ In using the calculator, we note again that we must have it in ***radian mode.*** [To evaluate the above expression for dy/dx, we must use the reciprocals of tan 0.75 and sin 0.75 for cot 0.75 and csc 0.75, respectively.] ■

EXERCISES 27.2

In Exercises 1 and 2, make the given changes in the indicated examples of this section and then find the derivatives.

1. In Example 1, in the given function, change $4x$ to x^2.
2. In Example 4, in the given function, change θ^2 to 3θ.

In Exercises 3–34, find the derivatives of the given functions.

3. $y = 2\tan 4x$
4. $y = 3\tan(3x + 2)$
5. $y = \cot(2\pi - 3\theta)$
6. $y = 3\cot 6x$
7. $u = 3\sec 5v$
8. $y = \sec\sqrt{1 - 4x}$
9. $y = 4x^3 - 3\csc\sqrt{2x + 3}$
10. $h = 0.5\csc(3 - 2\pi t)$
11. $R = 3\tan^5(2t)$
12. $y = 3x^4 + 2\tan^2(x^2)$
13. $y = 2\cot^4(\frac{1}{2}x + \pi)$
14. $p = 3\cot^2(4 - 3r^2)$
15. $y = \sqrt{\sec 4x}$
16. $y = 0.8\sec^3 5u$
17. $y = 3\csc^4(7x - \pi/2)$
18. $y = 7\csc^2(9x^3)$
19. $r = t^2\tan 0.5t$
20. $y = 3x\sec(2\pi x - 1)$
21. $y = 4\cos x \csc x^2$
22. $y = \frac{1}{2}\sin 2x \sec x$
23. $y = \dfrac{2\csc 3x}{x^2}$
24. $u = \dfrac{5\cot 0.25z}{2z}$

25. $y = \dfrac{2\cos 4x}{1 + \cot 3x}$

26. $y = \dfrac{\tan^2 3x}{2 + \sin(x^2 + 1)}$

27. $y = \frac{1}{3}\tan^3 x - \tan x$

28. $y = 4\csc 4x - 2\cot 4x$

29. $r = \tan(\sin 2\pi\theta)$

30. $y = x\tan x + \sec^2 2x$

31. $y = \sqrt{2x + \tan 4x}$

32. $V = (4 - \csc^2 3r)^3$

33. $x\sec y - 2y = \sin 2x$

34. $3\cot(x + y) = \cos y^2$

In Exercises 35–38, find the differentials of the given functions.

35. $y = 4\tan^2 3x$

36. $y = 2.5\sec^3 2t$

37. $y = \tan 4x\sec 4x$

38. $y = 2x\cot 3x$

In Exercises 39–54, solve the given problems.

39. On a calculator, find the values of (a) $\sec^2 1.0000$ and (b) $(\tan 1.0001 - \tan 1.0000)/0.0001$. Compare the values and give the meaning of each in relation to the derivative of $\tan x$ where $x = 1$.

40. Display the graphs of $y_1 = \sec^2 x$ and $y_n = [\tan(x + h) - \tan x]/h$ on the same screen of a calculator for $-\pi/2 < x < \pi/2$. For $n = 2$, let $h = 0.5$, for $n = 3$, let $h = 0.1$. (You might try some even smaller values of h.) What do these curves show?

41. (a) Display the graph of $y = \tan x$ on a calculator, and using the *derivative* feature, evaluate dy/dx for $x = 1$. (b) Display the graph of $y = \sec^2 x$ and evaluate y for $x = 1$. [Compare the values in parts (a) and (b).]

42. Follow the instructions in Exercise 41, using the graphs of $y = \sec x$ and $y = \sec x\tan x$.

43. Find the derivative of each member of the identity $1 + \tan^2 x = \sec^2 x$ and show that the results are equal.

44. Find the points where a tangent to the curve of $y = \tan x$ is parallel to the line $y = 2x$ if $0 < x < 2\pi$.

45. Find the slope of a line tangent to the curve of $y = 2\cot 3x$ where $x = \pi/12$. Verify the result by using the *numerical derivative* feature of a calculator.

46. Find the slope of a line normal to the curve of $y = \csc\sqrt{2x + 1}$ where $x = 0.45$. Verify the result by using the *numerical derivative* feature of a calculator.

47. Show that $y = 2\tan x - \sec x$ satisfies $\dfrac{dy}{dx} = \dfrac{2 - \sin x}{\cos^2 x}$.

48. For the spring mechanism in Exercise 31 of Section 10.4, find db/dA. (Note that a and angle B are constants.)

49. For the cantilever column in Exercise 32 of Section 10.4, find x' $(= dx/dL)$. (We use x' because of the constant d in the equation.)

50. A helicopter takes off such that its height h (in ft) above the ground is $h = 25\sec 0.16t$ for the first 8.0 s of flight. What is its vertical velocity after 6.0 s?

51. The vertical displacement y (in cm) of the end of an industrial robot arm for each cycle is $y = 2t^{1.5} - \tan 0.1t$, where t is the time (in s). Find its vertical velocity for $t = 15$ s.

52. The electric charge q (in C) passing a given point in a circuit is given by $q = t\sec\sqrt{0.2t^2 + 1}$, where t is the time (in s). Find the current i (in A) for $t = 0.80$ s. $(i = dq/dt.)$

53. An observer to a rocket launch was 1000 ft from the takeoff position. The observer found the angle of elevation of the rocket as a function of time to be $\theta = 3t/(2t + 10)$. Therefore, the height h (in ft) of the rocket was $h = 1000\tan\dfrac{3t}{2t + 10}$. Find the time rate of change of height after 5.0 s. See Fig. 27.5.

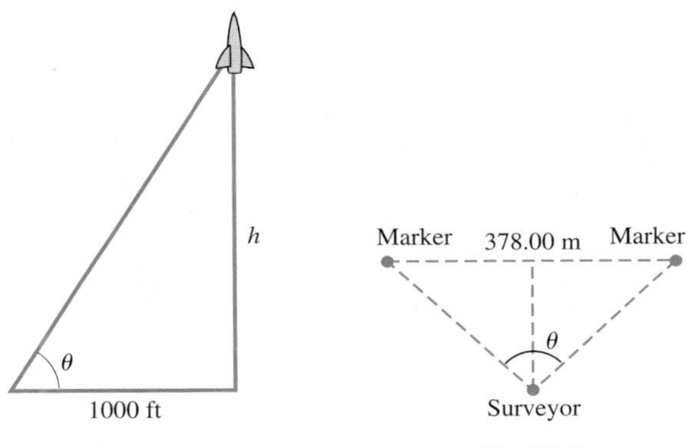

Fig. 27.5

Fig. 27.6

54. A surveyor measures the distance between two markers to be 378.00 m. Then, moving along a line equidistant from the markers, the distance d from the surveyor to each marker is $d = 189.00\csc\frac{1}{2}\theta$, where θ is the angle between the lines of sight to the markers. See Fig. 27.6. By using differentials, find the change in d if θ changes from 98.20° to 98.45°.

Answers to Practice Exercises

1. $y' = 24\sec^2 8x$ **2.** $y' = -30\csc 3x(\cot 3x + \csc 3x)^2$

27.3 Derivatives of the Inverse Trigonometric Functions

Derivatives of $\sin^{-1} u$, $\cos^{-1} u$, $\tan^{-1} u$

To obtain the derivative of $y = \sin^{-1} u$, we first solve for u in the form $u = \sin y$, and then take the derivative with respect to x. This results in $\frac{du}{dx} = \cos y\frac{dy}{dx}$. Solving this for dy/dx, we have

$$\frac{dy}{dx} = \frac{1}{\cos y}\frac{du}{dx} = \frac{1}{\sqrt{1 - \sin^2 y}}\frac{du}{dx} = \frac{1}{\sqrt{1 - u^2}}\frac{du}{dx}$$

We choose the positive square root because $\cos y > 0$ for $-\frac{\pi}{2} < y < \frac{\pi}{2}$, which is the range of the defined values of $\sin^{-1} u$. Therefore, we obtain the following result:

$$\frac{d(\sin^{-1} u)}{dx} = \frac{1}{\sqrt{1 - u^2}}\frac{du}{dx} \tag{27.10}$$

■ Note that the derivative of the inverse sine function is an algebraic function.

EXAMPLE 1 Derivative of $\sin^{-1} u$

Find the derivative of $y = \sin^{-1} 4x$.

$$\frac{dy}{dx} = \frac{1}{\sqrt{1 - (4x)^2}}(4) = \frac{4}{\sqrt{1 - 16x^2}} \quad \text{using Eq. (27.10)}$$

(with $u = 4x$ and $\frac{du}{dx} = 4$) ■

We find the derivative of the inverse cosine function by letting $y = \cos^{-1} u$ and by following the same procedure as that used in finding the derivative of $\sin^{-1} u$:

$$u = \cos y, \qquad \frac{du}{dx} = -\sin y \frac{dy}{dx}$$

$$\frac{dy}{dx} = -\frac{1}{\sin y}\frac{du}{dx} = -\frac{1}{\sqrt{1 - \cos^2 y}}\frac{du}{dx}$$

$$\frac{d(\cos^{-1} u)}{dx} = -\frac{1}{\sqrt{1 - u^2}}\frac{du}{dx} \tag{27.11}$$

Practice Exercise

1. Find the derivative of $y = 5\sin^{-1} x^2$.

The positive square root is chosen here because $\sin y > 0$ for $0 < y < \pi$, which is the range of the defined values of $\cos^{-1} u$. We note that the derivative of the inverse cosine is the negative of the derivative of the inverse sine.

By letting $y = \tan^{-1} u$, solving for u, and taking derivatives, we find the derivative of the inverse tangent function:

$$u = \tan y, \qquad \frac{du}{dx} = \sec^2 y \frac{dy}{dx}, \qquad \frac{dy}{dx} = \frac{1}{\sec^2 y}\frac{du}{dx} = \frac{1}{1 + \tan^2 y}\frac{du}{dx}$$

$$\frac{d(\tan^{-1} u)}{dx} = \frac{1}{1 + u^2}\frac{du}{dx} \tag{27.12}$$

We can see that the derivative of the inverse tangent is an algebraic function also.

The inverse sine, inverse cosine, and inverse tangent prove to be of the greatest importance in applications and in further development of mathematics. Therefore, the formulas for the derivatives of the other inverse functions are not presented here, although they are included in the exercises.

EXAMPLE 2 Derivative of $\cos^{-1} u$—forces on a sign

A 20-N force acts on a sign as shown in Fig. 27.7. Express the angle θ as a function of the x-component, F_x, of the force, and then find the expression for the instantaneous rate of change of θ with respect to F_x.

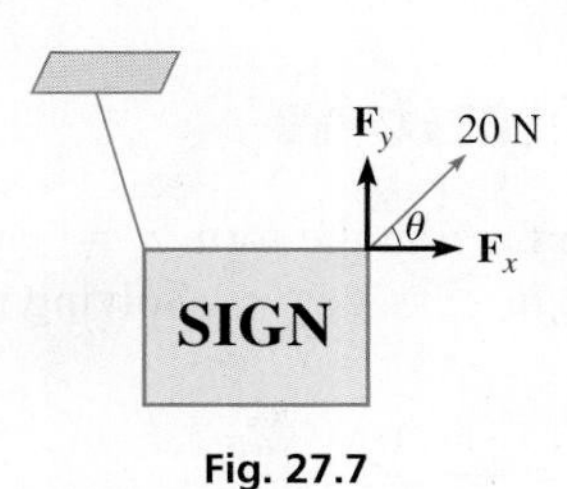

Fig. 27.7

From the figure, we see that $F_x = 20\cos\theta$. Solving for θ, we have $\theta = \cos^{-1}(F_x/20)$. To find the instantaneous rate of change of θ with respect to F_x, we are to take the derivative $d\theta/dF_x$:

$$\theta = \cos^{-1}\frac{F_x}{20} = \cos^{-1} 0.05F_x$$

$$\frac{d\theta}{dF_x} = -\frac{1}{\sqrt{1 - (0.05F_x)^2}}(0.05) \quad \text{using Eq. (27.11)}$$

$$= \frac{-0.05}{\sqrt{1 - 0.0025F_x^2}}$$

■

EXAMPLE 3 Derivative of $\tan^{-1}u$

Find the derivative of $y = (x^2 + 1)\tan^{-1}x - x$.

Using the product rule along with the derivative of the inverse tangent function on the first term, we have

$$\frac{dy}{dx} = (x^2 + 1)\left(\frac{1}{1 + x^2}\right)(1) + (\tan^{-1}x)(2x) - 1$$

using Eq. (27.12)

$$= 2x\tan^{-1}x$$

Practice Exercise

2. Find the derivative of $y = (\tan^{-1}3x)^2$

EXAMPLE 4 Derivative of $\sin^{-1}u$

Find the derivative of $y = x\sin^{-1}2x + \frac{1}{2}\sqrt{1 - 4x^2}$.

■ For reference, the *general power rule* is $\frac{du^n}{dx} = nu^{n-1}\left(\frac{du}{dx}\right)$.

using Eq. (27.10) — using the general power rule

$$\frac{dy}{dx} = x\left(\frac{2}{\sqrt{1 - 4x^2}}\right) + \sin^{-1}2x + \frac{1}{2}\left(\frac{1}{2}\right)(1 - 4x^2)^{-1/2}(-8x)$$

$$= \frac{2x}{\sqrt{1 - 4x^2}} + \sin^{-1}2x - \frac{2x}{\sqrt{1 - 4x^2}}$$

$$= \sin^{-1}2x$$

EXAMPLE 5 Slope of a tangent line

Find the slope of a tangent to the curve of $y = \frac{\tan^{-1}x}{x^2 + 1}$, where $x = 3.60$.

Here, we are to find the derivative and then evaluate it for $x = 3.60$.

$$\frac{dy}{dx} = \frac{(x^2 + 1)\left(\frac{1}{1 + x^2}\right)(1) - (\tan^{-1}x)(2x)}{(x^2 + 1)^2} \quad \text{take derivative}$$

$$= \frac{1 - 2x\tan^{-1}x}{(x^2 + 1)^2}$$

$$\left.\frac{dy}{dx}\right|_{x=3.60} = \frac{1 - 2(3.60)(\tan^{-1}3.60)}{(3.60^2 + 1)^2} = -0.0429 \quad \text{evaluate}$$

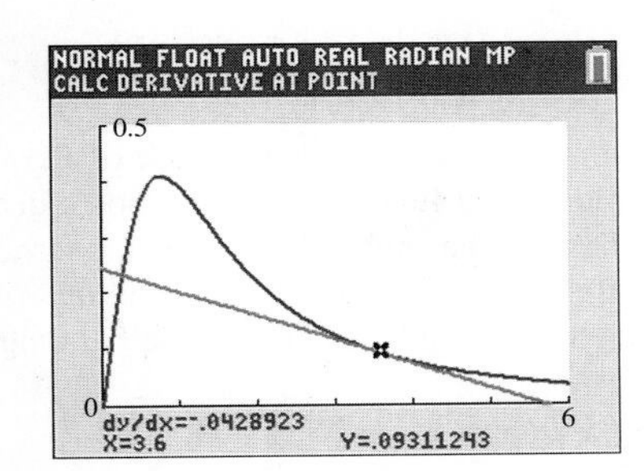

Fig. 27.8

Graphing calculator keystrokes: goo.gl/ErCk6r

Therefore, the slope of the tangent line is -0.0429. Note that the calculator must be in the radian mode when evaluating. In Fig. 27.8, the function, the tangent line, and the slope at $x = 3.60$ are shown.

EXERCISES 27.3

In Exercises 1 and 2, make the given changes in the indicated examples of this section and then find the derivatives.

1. In Example 1, in the given function, change $4x$ to x^2.
2. In Example 3, in the given function, change $(x^2 + 1)\tan^{-1}x$ to $(4x^2 + 1)\tan^{-1}2x$.

In Exercises 3–34, find the derivatives of the given functions.

3. $y = \sin^{-1}7x$
4. $R = 3\sin^{-1}(4 - t^2)$
5. $y = 3\sin^{-1}x^2$
6. $y = \sin^{-1}\sqrt{3 - 2x}$
7. $y = 2.4\cos^{-1}2t$
8. $\theta = 0.2\cos^{-1}5t$
9. $y = 6\cos^{-1}\sqrt{2 - x}$
10. $y = 3\cos^{-1}(x^2 + 0.5)$
11. $V = 8\tan^{-1}\sqrt{s}$
12. $y = \tan^{-1}(1 - 2x)$
13. $y = 6x\tan^{-1}(1/x)$
14. $w = 4u^2\tan^{-1}\pi u^4$
15. $y = 5x\sin^{-1}2x + \sqrt{1 - 4x^2}$
16. $y = x^2\cos^{-1}x + \sqrt{1 - x^2}$
17. $v = 0.4u\tan^{-1}2u$
18. $y = (x^2 + 1)\sin^{-1}4x$
19. $T = \frac{R}{\sin^{-1}3R}$
20. $\theta = \frac{\tan^{-1}2r}{\pi r}$
21. $y = \frac{\cos^{-1}x}{x^3}$
22. $y = \frac{4x^2 + 1}{\tan^{-1}2x}$
23. $y = 2(\cos^{-1}4x)^3$
24. $r = 0.5(\sin^{-1}3t)^4$
25. $u = [\sin^{-1}(4t + 3)]^2$
26. $y = \sqrt{\sin^{-1}(x - 1)}$
27. $y = \tan^{-1}\left(\frac{1 - t}{1 + t}\right)$
28. $p = \frac{3}{\cos^{-1}2w}$

29. $y = \frac{1}{1 + 4x^2} - \tan^{-1}2x$

30. $y = \sin^{-1}x - \sqrt{1 - x^2}$

31. $y = 3(4 - \cos^{-1}2x)^3$

32. $\sin^{-1}(x + y) + y = x^2$

33. $2\tan^{-1}y + x^2 = 3x$

34. $y = \sqrt{2\pi - \sin^{-1}4x}$

In Exercises 35–54, solve the given problems.

35. On a calculator, find the values of (a) $1/\sqrt{1 - 0.5^2}$ and (b) $(\sin^{-1}0.5001 - \sin^{-1}0.5000)/0.0001$. Compare the values and give the meaning of each in relation to the derivative of $\sin^{-1}x$ where $x = 0.5$.

36. Display the graphs of $y_1 = \frac{1}{\sqrt{1 - x^2}}$ and $y_n = \frac{\sin^{-1}(x + h) - \sin^{-1}x}{h}$ on the same screen of a calculator for $-1.2 < x < 1.2$. For $n = 2$, let $h = 0.5$; for $n = 3$, let $h = 0.1$. (You might try even smaller values for h.) What do these curves show?

37. Find the differential of the function $y = (\sin^{-1}x)^3$.

38. Find the linearization $L(x)$ of the function $f(x) = 2x\cos^{-1}x$ for $a = 0$.

39. Find the slope of a line tangent to the curve of $y = x/\tan^{-1}x$ at $x = 0.80$. Verify the result by using the *numerical derivative* feature of a calculator.

40. Explain what is wrong with a problem that requires finding the derivative of $y = \sin^{-1}(x^2 + 1)$.

41. Find the second derivative of $y = x\tan^{-1}x$.

42. Find the point(s) at which the line normal to $y = 2\sin^{-1}0.5x$ is parallel to the line $y = 1 - x$.

43. Use a calculator to display the graphs of $y = \sin^{-1}x$ and $y = 1/\sqrt{1 - x^2}$. By roughly estimating slopes of tangent lines of $y = \sin^{-1}x$, note that $y = 1/\sqrt{1 - x^2}$ gives reasonable values for the derivative of $y = \sin^{-1}x$.

44. Use a calculator to display the graphs of $y = \tan^{-1}x$ and $y = 1/(1 + x^2)$. By roughly estimating slopes of tangent lines of $y = \tan^{-1}x$, note that $y = 1/(1 + x^2)$ gives reasonable values for the derivative of $y = \tan^{-1}x$.

45. Find the second derivative of the function $y = \tan^{-1}2x$.

46. Show that $\frac{d(\cot^{-1}u)}{dx} = -\frac{1}{1 + u^2}\frac{du}{dx}$.

47. Show that $\frac{d(\sec^{-1}u)}{dx} = \frac{1}{\sqrt{u^2(u^2 - 1)}}\frac{du}{dx}$.

48. Show that $\frac{d(\csc^{-1}u)}{dx} = -\frac{1}{\sqrt{u^2(u^2 - 1)}}\frac{du}{dx}$.

49. In the analysis of the waveform of an AM radio wave, the equation $t = \frac{1}{\omega}\sin^{-1}\frac{A - E}{mE}$ arises. Find dt/dm, assuming that the other quantities are constant.

50. An equation that arises in the theory of solar collectors is $\alpha = \cos^{-1}\frac{2f - r}{r}$. Find the expression for $d\alpha/dr$ if f is constant.

51. When an alternating current passes through a series *RLC* circuit, the voltage and current are out of phase by angle ϕ (see Section 12.7). Here $\phi = \tan^{-1}[(X_L - X_C)/R]$, where X_L and X_C are the reactances of the inductor and capacitor, respectively, and R is the resistance. Find $d\phi/dX_C$ for constant X_L and R.

52. When passing through glass, a light ray is refracted (bent) such that the angle of refraction r is given by $r = \sin^{-1}[(\sin i)/\mu]$. Here, i is the angle of incidence, and μ is the index of refraction of the glass (see Fig. 27.9). For different types of glass, μ differs. Find the expression for dr for a constant value of i.

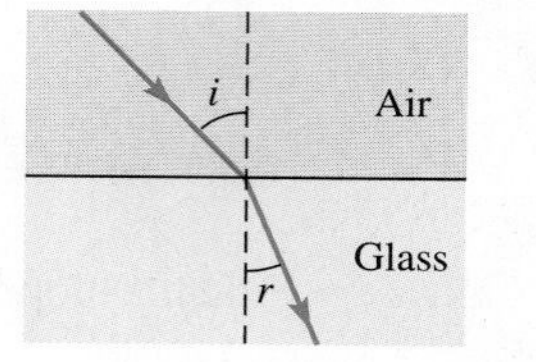

Fig. 27.9

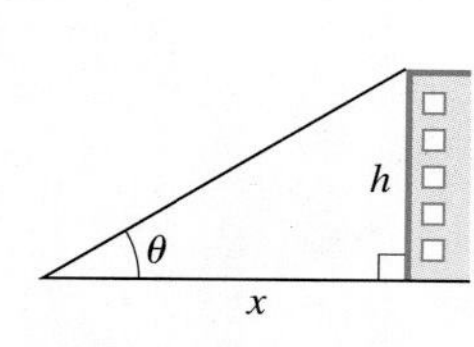

Fig. 27.10

53. As a person approaches a building of height h, the angle of elevation of the top of the building is a function of the person's distance from the building. Express the angle of elevation θ in terms of h and the distance x from the building and then find $d\theta/dx$. Assume the person's height is negligible to that of the building. See Fig. 27.10.

54. A triangular metal frame is designed as shown in Fig. 27.11. Express angle A as a function of x and evaluate dA/dx for $x = 6$ cm.

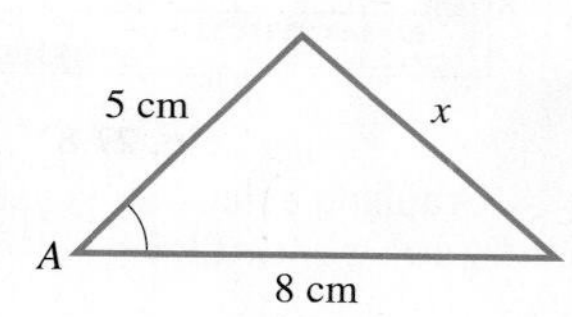

Fig. 27.11

Answers to Practice Exercises

1. $y' = 10x/\sqrt{1 - x^4}$ **2.** $y' = (6\tan^{-1}3x)/(1 + 9x^2)$

27.4 Applications

Tangents and Normals • Newton's Method • Time Rates of Change • Curve Sketching • Maximum and Minimum Problems • Differentials

With our development of the formulas for the derivatives of the trigonometric and inverse trigonometric functions, it is now possible to use these derivatives in the same manner as we applied the derivatives of algebraic functions. We can now use these functions in the types of applications listed at the left as shown in the examples and exercises of this section.

EXAMPLE 1 Sketching a curve

Sketch the curve $y = \sin^2 x - \frac{x}{2}(0 \leq x \leq 2\pi)$.

First, by setting $x = 0$, we see that the only easily obtainable intercept is $(0, 0)$. Replacing x by $-x$ and y by $-y$, we find that the curve is not symmetric to either axis or to the origin. Also, since x does not appear in a denominator, there are no vertical asymptotes. We are considering only the restricted domain $0 \leq x \leq 2\pi$.

In order to find the information from the derivatives, we write

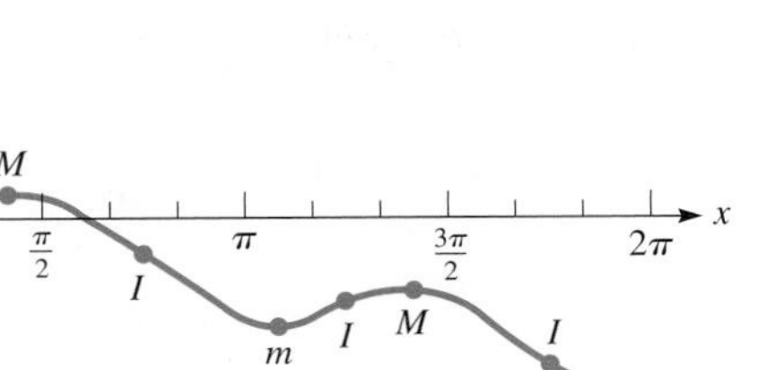

Fig. 27.12

$$\frac{dy}{dx} = 2\sin x\cos x - \frac{1}{2} = \sin 2x - \frac{1}{2}$$

$$\frac{d^2y}{dx^2} = 2\cos 2x$$

Local maximum and minimum points will occur for $\sin 2x = 1/2$. Thus, we have possible local maximum and minimum points for

$$2x = \frac{\pi}{6}, \frac{5\pi}{6}, \frac{13\pi}{6}, \frac{17\pi}{6}, \quad \text{or} \quad x = \frac{\pi}{12}, \frac{5\pi}{12}, \frac{13\pi}{12}, \frac{17\pi}{12}$$

Now, using the second derivative, we find that d^2y/dx^2 is positive for $x = \frac{\pi}{12}$ and $x = \frac{13\pi}{12}$ and is negative for $x = \frac{5\pi}{12}$ and $x = \frac{17\pi}{12}$. Thus, the maximum points are $(\frac{5\pi}{12}, 0.279)$ and $(\frac{17\pi}{12}, -1.29)$. Minimum points are $(\frac{\pi}{12}, -0.064)$ and $(\frac{13\pi}{12}, -1.63)$. Inflection points occur for $\cos 2x = 0$, or

$$2x = \frac{\pi}{2}, \frac{3\pi}{2}, \frac{5\pi}{2}, \frac{7\pi}{2}, \quad \text{or} \quad x = \frac{\pi}{4}, \frac{3\pi}{4}, \frac{5\pi}{4}, \frac{7\pi}{4}$$

Therefore, the points of inflection are $(\frac{\pi}{4}, 0.11)$, $(\frac{3\pi}{4}, -0.68)$, $(\frac{5\pi}{4}, -1.46)$, and $(\frac{7\pi}{4}, -2.25)$. Using this information, we sketch the curve in Fig. 27.12. ■

EXAMPLE 2 Solving an equation using Newton's method

By using Newton's method, solve the equation $2x - 1 = 3\cos x$.

First, we locate the required root approximately by sketching $y_1 = 2x - 1$ and $y_2 = 3\cos x$. Using the calculator view shown in Fig. 27.13, we see that they intersect between $x = 1$ and $x = 2$, near $x = 1.2$. Therefore, using $x_1 = 1.2$, with

$$f(x) = 2x - 1 - 3\cos x$$

$$f'(x) = 2 + 3\sin x$$

we use Eq. (24.1), which is

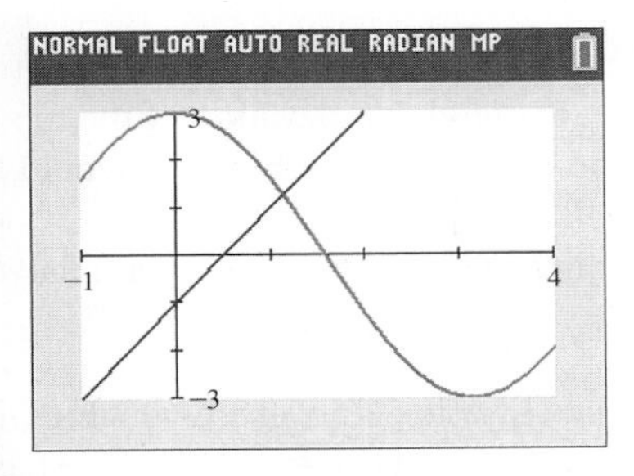

Fig. 27.13

$$x_2 = x_1 - \frac{f(x_1)}{f'(x_1)}$$

To find x_2, we have

$$f(x_1) = 2(1.2) - 1 - 3\cos 1.2 = 0.3129267$$

$$f'(x_1) = 2 + 3\sin 1.2 = 4.7961173$$

$$x_2 = 1.2 - \frac{0.3129267}{4.7961173} = 1.1347542$$

Finding the next approximation, we find $x_3 = 1.1342366$, which is accurate to the value shown. Again, when using the calculator, it is not necessary to list the values of $f(x_1)$ and $f'(x_1)$, as the complete calculation can be done directly on the calculator (see Example 2 of Section 24.2). ■

EXAMPLE 3 Finding a maximum value—lumber area

Logs with a circular cross section 4.0 ft in diameter are cut in half lengthwise. Find the largest rectangular cross-sectional area that can then be cut from one of the halves.

From Fig. 27.14, we see that $x = 2.0\cos\theta$ and $y = 2.0\sin\theta$, which means the area of the rectangle inscribed within the semicircular area is

$$A = (2x)y = 2(2.0\cos\theta)(2.0\sin\theta) = 8.0\cos\theta\sin\theta$$
$$= 4.0\sin 2\theta \quad \text{using identity } \sin 2\alpha = 2\sin\alpha\cos\alpha$$

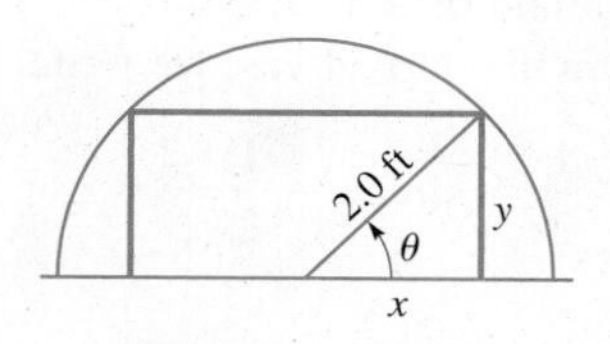

Fig. 27.14

Now, taking the derivative and setting it equal to zero, we have

$$\frac{dA}{d\theta} = (4.0\cos 2\theta)(2) = 8.0\cos 2\theta$$

$$8.0\cos 2\theta = 0, \quad 2\theta = \frac{\pi}{2}, \quad \theta = \frac{\pi}{4} \quad \cos\frac{\pi}{2} = 0$$

(Using $2\theta = \frac{3\pi}{2}$, $\theta = \frac{3\pi}{4}$ leads to the same solution.) Because the minimum area is zero, we have the maximum area when $\theta = \frac{\pi}{4}$, and this maximum area is

$$A = 4.0\sin 2\left(\frac{\pi}{4}\right) = 4.0\sin\frac{\pi}{2} = 4.0 \text{ ft}^2 \quad \sin\frac{\pi}{2} = 1$$

Therefore, the largest rectangular cross-sectional area is 4.0 ft^2. ■

EXAMPLE 4 Related rates—rocket velocity

■ See the chapter introduction.

A rocket is taking off vertically at a distance of 6500 ft from an observer. If, when the angle of elevation is 38.4°, it is changing at 5.0°/s, how fast is the rocket ascending?

From Fig. 27.15, letting $x =$ the height of the rocket, we see that $\tan\theta = x/6500$, or $\theta = \tan^{-1}(x/6500)$. Taking derivatives *with respect to time*, we have

$$\frac{d\theta}{dt} = \frac{1}{1 + (x/6500)^2}\frac{dx/dt}{6500} = \frac{6500\, dx/dt}{6500^2 + x^2}$$

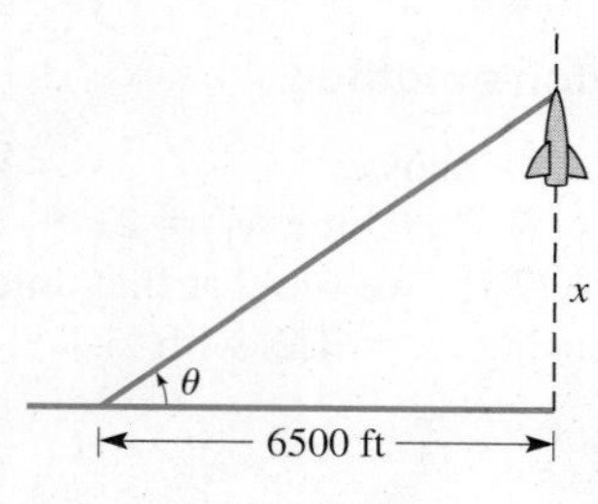

Fig. 27.15

We must remember to express angles in radians. This means $d\theta/dt = 5.0°/s = 0.0873$ rad/s. Substituting this value and $x = 6500\tan 38.4° = 5150$ ft, we have

$$0.0873 = \frac{6500\, dx/dt}{6500^2 + 5150^2}$$
$$\frac{dx}{dt} = 924 \text{ ft/s}$$

■

EXAMPLE 5 Differential—error in calculated building height

On level ground, 180 ft from the base of a building, the angle of elevation of the top of the building is 30.00°. What error in calculating the height h of the building would be caused by an error of 0.25° in the angle?

From Fig. 27.16, $h = 180\tan\theta$. To find the error in h, we must find the differential dh.

$$dh = 180\sec^2\theta\, d\theta$$

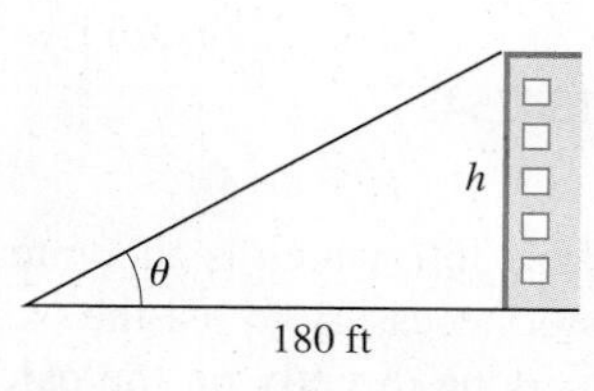

Fig. 27.16

The possible error in θ is 0.25°, which in *radian measure* is $0.25\pi/180$, which is the value we should use in the calculation of dh. We can use the value $\theta = 30.00°$ if we have the calculator set in degree mode. Calculating dh, we have

$$dh = \frac{180(0.25\pi/180)}{\cos^2 30.00°} = 1.05 \text{ ft}$$

An error of 0.25° in the angle results in an error of over 1 ft in the calculated value of the height. In using the calculator, we divide by $\cos^2 30.00°$ because $\sec\theta = 1/\cos\theta$. ■

EXAMPLE 6 Curvilinear motion—parametric equations

A particle is rotating so that its x- and y-coordinates are given by $x = \cos 2t$ and $y = \sin 2t$. Find the magnitude and direction of its velocity when $t = \pi/8$.

Taking derivatives with respect to time, we have

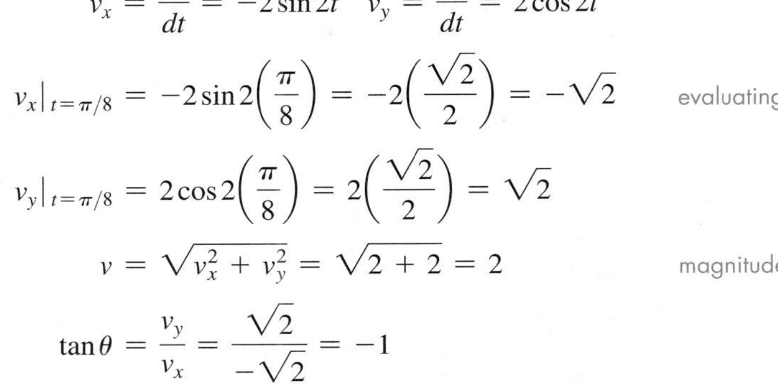

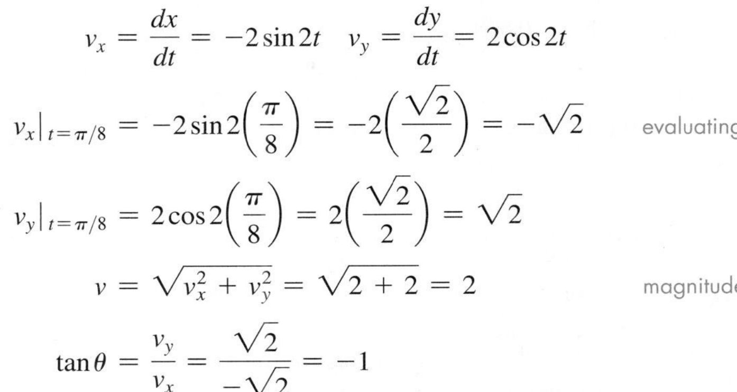

$$v_x = \frac{dx}{dt} = -2\sin 2t \quad v_y = \frac{dy}{dt} = 2\cos 2t$$

$$v_x\big|_{t=\pi/8} = -2\sin 2\left(\frac{\pi}{8}\right) = -2\left(\frac{\sqrt{2}}{2}\right) = -\sqrt{2} \qquad \text{evaluating}$$

$$v_y\big|_{t=\pi/8} = 2\cos 2\left(\frac{\pi}{8}\right) = 2\left(\frac{\sqrt{2}}{2}\right) = \sqrt{2}$$

$$v = \sqrt{v_x^2 + v_y^2} = \sqrt{2 + 2} = 2 \qquad \text{magnitude}$$

$$\tan\theta = \frac{v_y}{v_x} = \frac{\sqrt{2}}{-\sqrt{2}} = -1$$

Fig. 27.17

Because v_x is negative and v_y is positive, $\theta = 135°$. Note that in this example θ is the angle, in standard position, between the horizontal and the resultant velocity.

Plotting the curve shows that it is a circle (see Fig. 27.17). This means the particle is moving counterclockwise around a circle. Also, we can see that it is a circle by squaring x and y and adding. This gives $x^2 + y^2 = \cos^2 2t + \sin^2 2t = 1$, which is a circle of radius 1. ■

Practice Exercise

1. In Example 6, find the velocity when $t = 5\pi/8$.

EXERCISES 27.4

In Exercises 1 and 2, make the given changes in the indicated examples of this section and then solve the resulting problems.

1. In Example 1, change $\sin^2 x$ to $\sin x$.
2. In Example 6, change $\cos 2t$ to $3 \cos 2t$, and $\sin 2t$ to $2 \sin 2t$.

In Exercises 3–40, solve the given problems.

3. Show that the slopes of the sine and cosine curves are negatives of each other at the points of intersection.
4. Show that the tangent curve is always increasing (when the tangent is defined).
5. Show that the curve of $y = \tan^{-1}x$ is always increasing.
6. Sketch the graph of $y = \sin x + \cos x$ $(0 \le x \le 2\pi)$. Check the graph on a calculator.
7. Sketch the graph of $y = x - \tan x$ $(-\frac{\pi}{2} < x < \frac{\pi}{2})$. Check the graph on a calculator.
8. Sketch the graph of $y = 2\sin x + \sin 2x$ $(0 \le x \le 2\pi)$. Check the graph on a calculator.
9. Find the equation of the line tangent to the curve of $y = x\sin^{-1}x$ at $x = 0.50$.
10. Find the equation of the line normal to the curve of $y = 3\tan x^2$ at $x = 0.25$.
11. By Newton's method, find the positive root of the equation $x^2 - 4\sin x = 0$. Verify the result by using the *zero* feature of a calculator.
12. By Newton's method, find the smallest positive root of the equation $\tan x = 2x$. Verify the result by using the *zero* feature of a calculator.
13. Find the minimum value of the function $y = 6\cos x - 8\sin x$.
14. Find the maximum value of the function $y = \tan^{-1}(1 + x) + \tan^{-1}(1 - x)$.
15. Power P is the time rate of change of work W. Find the equation for the power in a circuit for which $W = 8\sin^2 2t$.
16. The phase shift ϕ in a certain electric circuit with a resistance R and variable capacitance C is $\phi = \tan^{-1}\omega RC$. Find the equation for the instantaneous rate of change of ϕ with respect to C.
17. In studying water waves, the vertical displacement y (in ft) of a wave was determined to be $y = 0.50\sin 2t + 0.30\cos t$, where t is the time (in s). Find the velocity and the acceleration for $t = 0.40$ s.
18. At 30°N latitude, the number of hours h of daylight each day during the year is given approximately by the equation $h = 12.1 + 2.0\sin\left[\frac{\pi}{6}(x - 2.7)\right]$, where x is measured in months ($x = 0.5$ is Jan. 15, etc.). Find the date of the longest day and the date of the shortest day. (Cities near 30°N are Houston, Texas, and Cairo, Egypt.)
19. Find the time rate of change of the horizontal component T_x of the constant 46.6-lb tension shown in Fig. 27.18 if $d\theta/dt = 0.36°$/s for $\theta = 14.2°$.

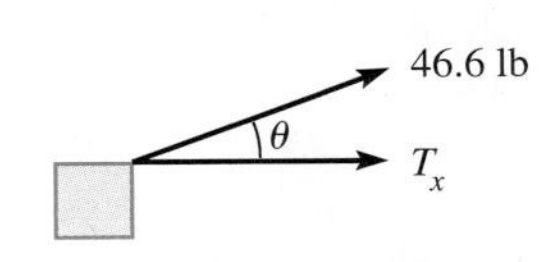

Fig. 27.18

20. The *apparent power* P_a (in W) in an electric circuit whose power is P and whose impedance phase angle is ϕ is given by $P_a = P\sec\phi$. Given that P is constant at 12 W, find the time rate of change of P_a if ϕ is changing at the rate of 0.050 rad/min, when $\phi = 40.0°$.

21. A point on the outer edge of a 38.0-cm wheel can be described by the equations $x = 19.0\cos 6\pi t$ and $y = 19.0\sin 6\pi t$. Find the velocity of the point for $t = 0.600$ s.

22. A machine is programmed to move an etching tool such that the position (in cm) of the tool is given by $x = 2\cos 3t$ and $y = \cos 2t$, where t is the time (in s). Find the velocity of the tool for $t = 4.1$ s.

23. Find the acceleration of the tool of Exercise 22 for $t = 4.1$ s.

24. The volume V (in m^3) of water used each day by a community during the summer is found to be $V = 2500 + 480\sin(\pi t/90)$, where t is the number of the summer day, and $t = 0$ is the first day of summer. On what summer day is the water usage the greatest?

25. A person observes an object dropped from the top of a building 100 ft away. If the top of the building is 200 ft above the person's eye level, how fast is the angle of elevation of the object changing after 1.0 s? (The distance the object drops is given by $s = 16t^2$.) See Fig. 27.19.

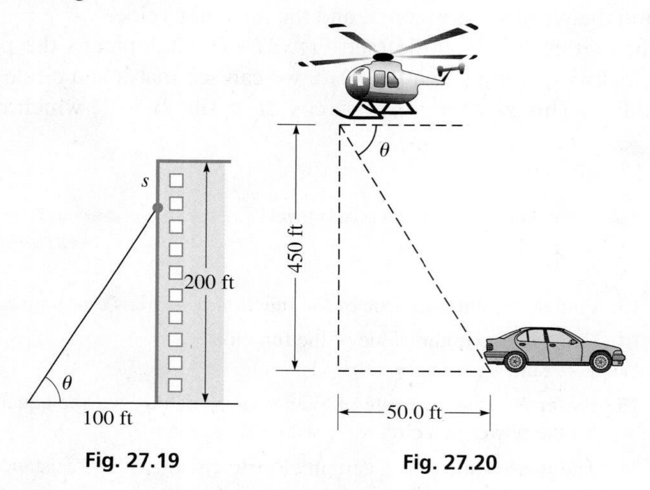

Fig. 27.19 Fig. 27.20

26. A car passes directly under a police helicopter 450 ft above a straight and level highway. After the car travels another 50.0 ft, the angle of depression of the car from the helicopter is decreasing at 0.215 rad/s. What is the speed of the car? See Fig. 27.20.

27. A searchlight is 225 ft from a straight wall. As the beam moves along the wall, the angle between the beam and the perpendicular to the wall is increasing at the rate of 1.5°/s. How fast is the length of the beam increasing when it is 315 ft long? See Fig. 27.21.

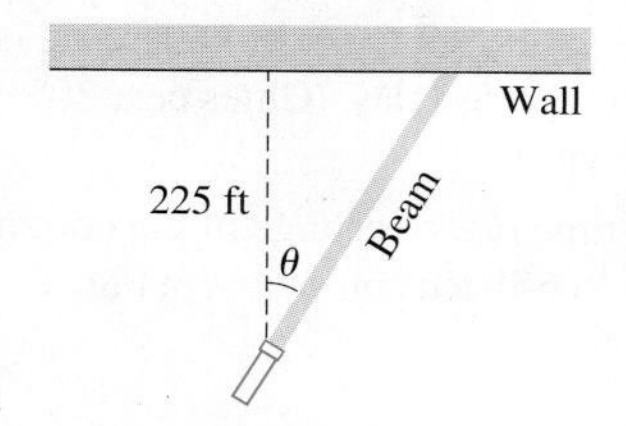

Fig. 27.21

28. In a modern hotel, where the elevators are directly observable from the lobby area (and a person can see from the elevators), a person in the lobby observes one of the elevators rising at the rate of 12.0 ft/s. If the person was 50.0 ft from the elevator when it left the lobby, how fast is the angle of elevation of the line of sight to the elevator increasing 10.0 s later?

29. A crate of weight w is being pulled along a level floor by a force F that is at an angle θ with the floor. The force is given by $F = \dfrac{0.25w}{0.25\sin\theta + \cos\theta}$. Find θ for the minimum value of F.

30. The electric power p (in W) developed in a resistor in an FM receiver circuit is $p = 0.0307\cos^2 120\pi t$, where t is the time (in s). Linearize p for $t = 0.0010$s.

31. When an astronaut views the horizon of Earth from a spacecraft at an altitude of 610 km, the angle θ in Fig. 27.22 is found to be $65.8° \pm 0.5°$. Use differentials to approximate the possible error in the astronaut's calculation of Earth's radius.

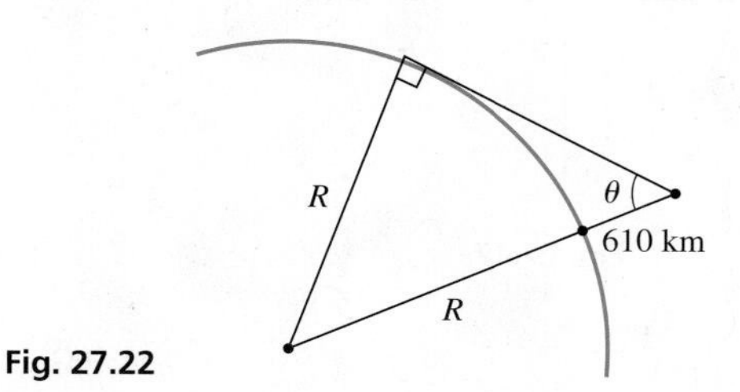

Fig. 27.22

32. A surveyor measures two sides and the included angle of a triangular parcel of land to be 82.04 m, 75.37 m, and 38.38°. What error is caused in the calculation of the third side by an error of 0.15° in the angle?

33. The volume V (in L) of air in a person's lungs during one normal cycle of inhaling and exhaling at any time t is $V = 0.48(1.2 - \cos 1.26t)$. What is the maximum flow rate (in L/s) of air?

34. To connect the four vertices of a square with the minimum amount of electric wire requires using the wiring pattern shown in Fig. 27.23. Find θ for the total length of wire $(L = 4x + y)$ to be a minimum.

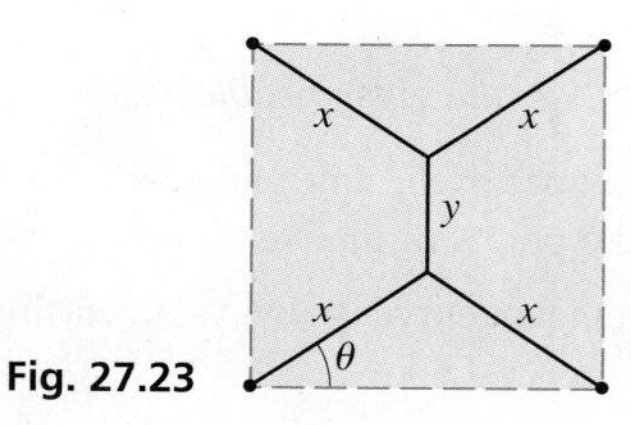

Fig. 27.23

35. The strength S of a rectangular beam is directly proportional to the product of its width w and the square of its depth d. Use trigonometric functions to find the dimensions of the strongest beam that can be cut from a circular log 16.0 in. in diameter. (See Example 4 on page 728.)

36. An architect is designing a window in the shape of an isosceles triangle with a perimeter of 60 in. What is the vertex angle of the window of greatest area?

37. A camera is on the starting line of a drag race 15.0 m from a racing car. After 1.5 s the car has traveled 30.0 m and the camera is rotating at 0.75 rad/s while filming the car. What is the speed of the car at this time?

38. What is the vertex angle at the bottom of an ice cream cone such that the cone holds a given amount of ice cream (within the cone itself) and the cone requires the least possible surface? (*Hint:* Set up equations using half the angle.)

39. A wall is 6.0 ft high and 4.0 ft from a building. What is the length of the shortest pole that can touch the building and the ground beyond the wall? (*Hint:* From Fig. 27.24, it can be shown that $y = 6.0\csc\theta + 4.0\sec\theta$.)

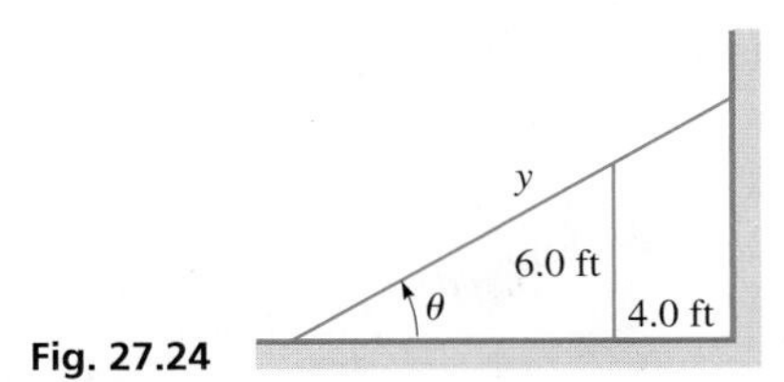

Fig. 27.24

40. The television screen at a sports arena is vertical and 8.0 ft high. The lower edge is 25.0 ft above an observer's eye level. If the best view of the screen is obtained when the angle subtended by the screen at eye level is a maximum, how far from directly below the screen must the observer's eye be? See Fig. 27.25.

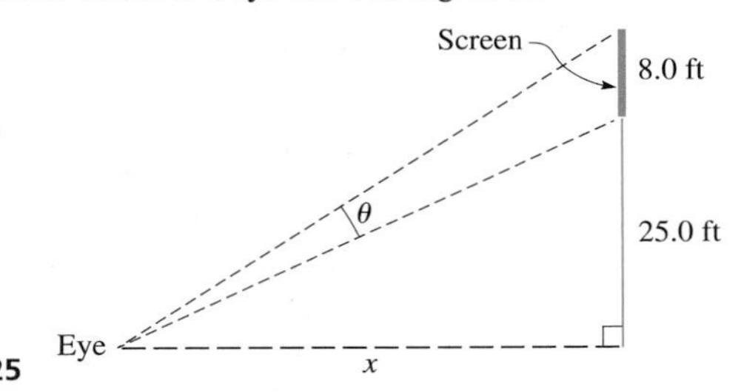

Fig. 27.25

Answer to Practice Exercise

1. $v = 2, \theta = 315°$

27.5 Derivative of the Logarithmic Function

Limit of $(1 + h/x)^{x/h}$ as $h \to 0$ • Derivative of $\log_b u$ • Derivative of $\ln u$ • Using Properties of Logarithms

Using the definition of a derivative, we next find the derivative of the logarithmic function. Therefore, if letting $y = \log_b x$, we have

$$\frac{dy}{dx} = \lim_{h\to 0}\frac{\log_b(x + h) - \log_b x}{h} = \lim_{h\to 0}\frac{\log_b \frac{x+h}{x}}{h}$$
$$= \lim_{h\to 0}\frac{1}{x}\frac{x}{h}\log_b\left(1 + \frac{h}{x}\right)$$
$$= \frac{1}{x}\lim_{h\to 0}\log_b\left(1 + \frac{h}{x}\right)^{x/h}$$

(We multiplied and divided by x for purposes of evaluating the limit, as we now show.)

$$\lim_{h\to 0}\left(1 + \frac{h}{x}\right)^{x/h}$$

We can see that the exponent becomes unbounded, but the number being raised to this exponent approaches 1. Therefore, we will investigate this limiting value.

To approximate the value, we graph the function $y = (1 + t)^{1/t}$ (for purposes of graphing, we let $h/x = t$). Constructing a table of values, we then graph this function in Fig. 27.26.

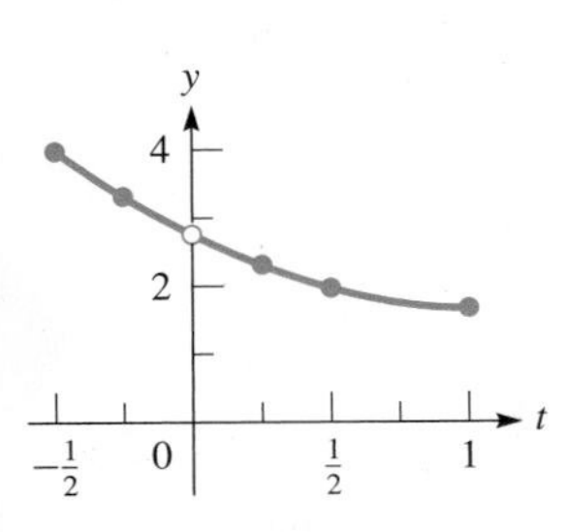

Fig. 27.26

t	−0.5	−0.25	+0.25	+0.50	+1.00
y	4.00	3.16	2.44	2.25	2.00

Only these values are shown, because we are interested in the y-value corresponding to $t = 0$. We see from the graph that this value is approximately 2.7. Choosing very small values of t, we may obtain these values:

t	0.1	0.01	0.001	0.0001
y	2.5937	2.7048	2.7169	2.71815

By methods developed in Chapter 30, it can be shown that this value is about 2.7182818.

NOTE ▸ [The limiting value is the irrational number e.] This is the same number used in the exponential form of a complex number in Chapter 12 and as the base of natural logarithms in Chapter 13.

Returning to the derivative of the logarithmic function, we have

$$\frac{dy}{dx} = \lim_{h\to 0}\left[\frac{1}{x}\log_b\left(1 + \frac{h}{x}\right)^{x/h}\right] = \frac{1}{x}\log_b e$$

Now, for $y = \log_b u$, where u is a function of x, using the chain rule, we have

■ Since $\log_b e = \frac{1}{\ln b}$, Eq. (27.13) can also be written as $\frac{d(\log_b u)}{dx} = \frac{1}{(\ln b)u}\frac{du}{dx}$.

$$\frac{d(\log_b u)}{dx} = \frac{1}{u}\log_b e\frac{du}{dx} \quad (27.13)$$

At this point, we see that if we choose e as the base of a system of logarithms, the above formula becomes

$$\frac{d(\ln u)}{dx} = \frac{1}{u}\frac{du}{dx} \quad (27.14)$$

The choice of e as the base b makes $\log_e e = 1$; thus, this factor does not appear in Eq. (27.14). We now see why the number e is chosen as the base for the natural logarithm. The notation $\ln u$ is the same as that used in Chapter 13 for natural logarithms.

EXAMPLE 1 Derivative of log u

Find the derivative of $y = \log 4x$.

Using Eq. (27.13), we find

$$\frac{dy}{dx} = \frac{1}{4x}(\log e)(4)$$

($4x$ is u; 4 is $\frac{du}{dx}$)

$$= \frac{1}{x}\log e$$

$\log e = 0.4343$ ■

EXAMPLE 2 Derivative of ln u

Find the derivative of $s = \ln 3t^4$.

Using Eq. (27.14), we have (with $u = 3t^4$)

$$\frac{ds}{dt} = \frac{1}{3t^4}(12t^3)$$

($12t^3$ is $\frac{du}{dt}$)

$$= \frac{4}{t}$$

■

Practice Exercise

1. Find the derivative of $y = 2\ln 5x^2$.

EXAMPLE 3 Derivative of ln tan u

Find the derivative of $y = \ln \tan 4x$.

Using Eq. (27.14), along with the derivative of the tangent, we have

$$\frac{dy}{dx} = \frac{1}{\tan 4x}(\sec^2 4x)(4)$$

($(\sec^2 4x)(4)$ is $\frac{d\tan 4x}{dx}$)

$$= \frac{\cos 4x}{\sin 4x}\frac{4}{\cos^2 4x}$$

using trigonometric relations

$$= \frac{1}{\sin 4x}\frac{4}{\cos 4x} = 4\csc 4x\sec 4x$$

A TI-89 calculator evaluation of this derivative is shown in Fig. 27.27. ■

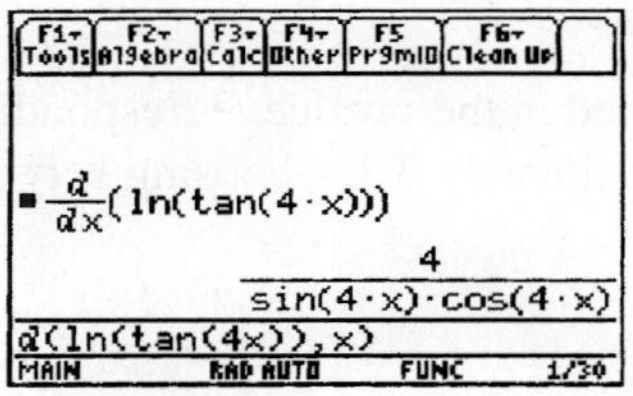

Fig. 27.27
TI-89 graphing calculator keystrokes: goo.gl/nsEn29

NOTE ➧ [Often, finding the derivative of a logarithmic function is simplified by using the properties of logarithms to simplify the logarithmic expression before taking the derivative.]

EXAMPLE 4 Derivative using properties of logarithms

Find the derivative of $y = \ln\frac{x-1}{x+1}$.

In this example, it is easier to find the derivative if we write y in the form

$$y = \ln(x-1) - \ln(x+1)$$

by using the property $\log_b\left(\frac{x}{y}\right) = \log_b x - \log_b y$. Hence,

$$\frac{dy}{dx} = \frac{1}{x-1} - \frac{1}{x+1} = \frac{x+1-x+1}{(x-1)(x+1)}$$

$$= \frac{2}{x^2-1}$$ ■

EXAMPLE 5 Derivative of $\ln u^3$ and $\ln^3 u$

(a) Find the derivative of $y = \ln(1-2x)^3$.

First, using the property $\log_b x^n = n\log_b x$, we rewrite the equation as $y = 3\ln(1-2x)$. Then we have

$$\frac{dy}{dx} = 3\left(\frac{1}{1-2x}\right)(-2) = \frac{-6}{1-2x}$$

(b) Find the derivative of $y = \ln^3(1-2x)$.

First, we note that

$$y = \ln^3(1-2x) = [\ln(1-2x)]^3$$

where $\ln^3(1-2x)$ is usually the preferred notation.

NOTE ➧ [Next, we must be careful to distinguish this function from that in part (a). For $y = \ln^3(1-2x)$, it is the logarithm of $1-2x$ that is being cubed, whereas for $y = \ln(1-2x)^3$, it is $1-2x$ that is being cubed.]

Now, finding the derivative of $y = \ln^3(1-2x)$, we have

$$\frac{dy}{dx} = 3[\ln^2(1-2x)]\left(\frac{1}{1-2x}\right)(-2) \quad \text{using } \frac{du^n}{dx} = nu^{n-1}\left(\frac{du}{dx}\right)$$

$$= -\frac{6\ln^2(1-2x)}{1-2x}$$

(arrow: $\frac{1}{1-2x}(-2)$ is $\frac{d\ln(1-2x)}{dx}$) ■

Practice Exercise

2. Find the derivative of $y = \ln\frac{4x}{x+4}$.

EXAMPLE 6 Derivative of $\ln u$ using properties

Evaluate the derivative of $y = \ln[(\sin 2x)(\sqrt{x^2+1})]$ for $x = 0.375$.

First, using the properties of logarithms, we rewrite the function as

$$y = \ln\sin 2x + \frac{1}{2}\ln(x^2+1)$$

Now, we have

$$\frac{dy}{dx} = \frac{1}{\sin 2x}(\cos 2x)(2) + \frac{1}{2}\left(\frac{1}{x^2+1}\right)(2x) \quad \text{take the derivative}$$

$$= 2\cot 2x + \frac{x}{x^2+1}$$

$$\left.\frac{dy}{dx}\right|_{x=0.375} = 2\cot 0.750 + \frac{0.375}{0.375^2+1} = 2.48 \quad \text{evaluate}$$ ■

EXERCISES 27.5

In Exercises 1 and 2, make the given changes in the indicated examples of this section and then find the derivatives.

1. In Example 3, in the given function, change tan to cos.
2. In Example 4, in the given function, change $x - 1$ to x^2.

In Exercises 3–34, find the derivatives of the given functions.

3. $y = \log x^2$
4. $y = \log_2 6x$
5. $y = 4\log_5(3 - x)$
6. $y = \log_7(x^2 + 4)$
7. $u = 2\ln(3 - x)^4$
8. $y = 2\ln(3x^2 - 1)$
9. $y = 2\ln\tan 2x$
10. $s = 3\ln\sin^2 t$
11. $R = \ln\sqrt{4T + 1}$
12. $y = \ln(4x - 3)^3$
13. $y = \ln(x - x^2)^3$
14. $s = 3\ln^2(7t^3 - 1)$
15. $v = 3(t + \ln t^2)^2$
16. $y = 6x^2\ln 5x$
17. $y = 3x\ln(6 - x)^2$
18. $y = \dfrac{8\ln(2x + 1)}{x}$
19. $y = \ln(\ln x)$
20. $y = \ln\dfrac{2x}{1 + x}$
21. $r = 0.5\ln\cos(\pi\theta^2)$
22. $y = \ln(x\sqrt{x + 1})$
23. $y = \sin\ln x$
24. $y = \tan^{-1}(\ln 2x + \ln x)$
25. $u = 3v\ln^2 2v$
26. $h = 0.1s\ln^4 s$
27. $y = \ln(x\tan x)$
28. $y = \ln(x + \sqrt{x^2 - 1})$
29. $r = \ln\dfrac{v^2}{v + 2}$
30. $y = \sqrt{x + \ln 3 + \ln x}$
31. $y = x - \ln^2(x + y)$
32. $y = \ln(x + \ln x)$
33. $\ln\left(\dfrac{x}{y}\right) = x$
34. $x\ln y = y^2$

In Exercises 35–56, solve the given problems.

35. On a calculator, find the value of $(\ln 2.0001 - \ln 2.0000)/0.0001$ and compare it with 0.5. Give the meanings of the value found and 0.5 in relation to the derivative of ln x, where $x = 2$.

36. Display the graphs of $y_1 = \dfrac{1}{x}$ and $y_n = \dfrac{\ln(x + h) - \ln x}{h}$ on the same calculator screen for $0 < x < 10$. For $n = 2$, let $h = 0.5$; for $n = 3$, let $h = 0.1$. (You might try smaller values of h.) What do these curves show?

37. Using a calculator, (a) display the graph of $y = (1 + x)^{1/x}$ to verify that $(1 + x)^{1/x} \rightarrow e\ (\approx 2.718)$ as $x \rightarrow 0$ and (b) verify the values for $(1 + x)^{1/x}$ in the tables on page 821.

38. (a) Display the graph of $y = \ln x$ on a calculator, and using the derivative feature, evaluate dy/dx for $x = 2$. (b) Display the graph of $y = 1/x$, and evaluate y for $x = 2$. (c) Compare the values in parts (a) and (b).

39. Given that $\ln\sin 45° = -0.3466$, use differentials to approximate $\ln\sin 44°$.

40. Find the second derivative of the function $y = x^2\ln x$.

41. Evaluate the derivative of $y = \ln\sqrt{1 - 4x^2}$ at $x = \frac{1}{4}$.

42. Evaluate the derivative of $y = \ln\sqrt{\dfrac{2x + 1}{3x + 1}}$, where $x = 2.75$.

43. Find the linearization $L(x)$ for the function $f(x) = 2\ln\tan x$ for $a = \pi/4$.

44. Find the differential of the function $y = 6\log_x 2$.

45. Find the slope of a line tangent to the curve of $y = \tan^{-1}2x + \ln(4x^2 + 1)$, where $x = 0.625$. Verify the result by using the *numerical derivative* feature of a calculator.

46. Find the slope of a line tangent to the curve of $y = x\ln 3x$ at $x = 4$. Verify the result by using the *numerical derivative* feature of a calculator.

47. Find the derivative of $y = x^x$ by first taking logarithms of each side of the equation. Explain why the power rule cannot be used to find the derivative of this function.

48. Find the derivative of $y = (\sin x)^x$ by first taking logarithms of each side of the equation. Explain why the power rule cannot be used to find the derivative of this function.

49. Find the derivatives of $y_1 = \ln(x^2)$ and $y_2 = 2\ln x$, and evaluate these derivatives for $x = -1$. Explain your results.

50. The inductance L (in μH) of a coaxial cable is given by $L = 0.032 + 0.15\log(a/x)$, where a and x are the radii of the outer and inner conductors, respectively. For constant a, find dL/dx.

51. If the loudness b (in decibels) of a sound of intensity I is given by $b = 10\log(I/I_0)$, where I_0 is a constant, find the expression for db/dt in terms of dI/dt.

52. The time t for a particular computer system to process N bits of data is directly proportional to N ln N. Find the expression for dt/dN.

53. When a tractor-trailer turns a right-angle corner, the rear wheels follow a curve known as a *tractrix*, the equation for which is

$$y = \ln\left(\frac{1 + \sqrt{1 + x^2}}{x}\right) - \sqrt{1 - x^2}.$$ Find dy/dx.

54. When designing a computer to sort files on a hard disk, the equation $y = xA\log_x A$ arises. If A is constant, find dy/dx.

55. When air friction is considered, the time t (in s) it takes a certain falling object to attain a velocity v (in ft/s) is given by $t = 5\ln\dfrac{16}{16 - 0.1v}$. Find dt/dv for $v = 100$ ft/s.

56. The electric potential V at a point P at a distance x from an electric charge distributed along a wire of length $2a$ (see Fig. 27.28) is

$$V = k\ln\frac{\sqrt{a^2 + x^2} + a}{\sqrt{a^2 + x^2} - a},$$

where k is a constant. Find the expression for the electric field E, where $E = -dV/dx$.

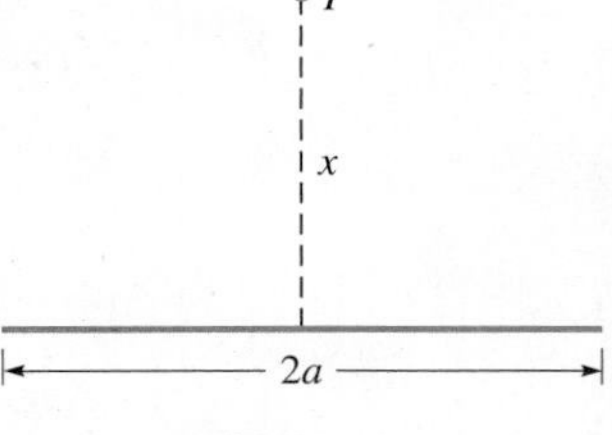

Fig. 27.28

Answers to Practice Exercises

1. $y' = 4/x$ 2. $y' = 4/(x^2 + 4x)$

27.6 Derivative of the Exponential Function

Derivative of b^x • Derivative of e^x • Using Laws of Exponents

To obtain the derivative of the exponential function, we let $y = b^u$ and then take natural logarithms of both sides:

$$\ln y = \ln b^u = u \ln b$$
$$\frac{1}{y}\frac{dy}{dx} = \ln b \frac{du}{dx}$$
$$\frac{dy}{dx} = y \ln b \frac{du}{dx}$$

$$\frac{d(b^u)}{dx} = b^u \ln b\left(\frac{du}{dx}\right) \quad \textbf{(27.15)}$$

If we let $b = e$, Eq. (27.15) becomes

$$\frac{d(e^u)}{dx} = e^u\left(\frac{du}{dx}\right) \quad \textbf{(27.16)}$$

The simplicity of Eq. (27.16) compared with Eq. (27.15) again shows the advantage of choosing e as the base of natural logarithms. It is for this reason that e appears so often in applications of calculus.

EXAMPLE 1 Derivative of e^x

Find the derivative of $y = e^x$.

Using Eq. (27.16), we have

$$\frac{dy}{dx} = e^x(1) = e^x$$

(the factor (1) is $\frac{du}{dx}$)

We see that the derivative of the function e^x equals itself. This exponential function is widely used in applications of calculus. ■

Note carefully that the power rule is used with a variable raised to a constant exponent, whereas Eqs. (27.15) and (27.16) are used with a constant raised to a variable exponent.

(variable u, constant n) $$\frac{du^n}{dx} = nu^{n-1}\left(\frac{du}{dx}\right)$$ (constant b, variable u) $$\frac{db^u}{dx} = b^u \ln b\left(\frac{du}{dx}\right)$$

In the following example, we must note carefully this difference when choosing the rule to find the derivative.

EXAMPLE 2 Derivative of u^2 and 2^u

Find the derivatives of $y = (4x)^2$ and $y = 2^{4x}$.

Using the power rule, we have

$$y = (4x)^2$$
$$\frac{dy}{dx} = 2(4x)^1(4)$$
$$= 32x$$

(the factor (4) is $\frac{du}{dx}$)

Using Eq. (27.15), we have

$$y = 2^{4x}$$
$$\frac{dy}{dx} = 2^{4x}(\ln 2)(4)$$
$$= (4 \ln 2)(2^{4x})$$

(the factor (4) is $\frac{du}{dx}$) ■

EXAMPLE 3 Derivative of ln cos e^u

Find the derivative of $y = \ln\cos e^{2x}$.

Here, we use Eq. (27.16) and the derivatives of the logarithmic and cosine functions.

$$\frac{dy}{dx} = \frac{1}{\cos e^{2x}}\frac{d\cos e^{2x}}{dx} \qquad \text{using } \frac{d\ln u}{dx} = \frac{1}{u}\frac{du}{dx}$$

$$= \frac{1}{\cos e^{2x}}(-\sin e^{2x})\frac{de^{2x}}{dx} \qquad \text{using } \frac{d\cos u}{dx} = -\sin u\frac{du}{dx}$$

$$= -\frac{\sin e^{2x}}{\cos e^{2x}}(e^{2x})(2) \qquad \text{using } \frac{de^u}{dx} = e^u\frac{du}{dx}$$

$$= -2e^{2x}\tan e^{2x} \qquad \text{using } \frac{\sin\theta}{\cos\theta} = \tan\theta$$

■

Practice Exercise

1. Find the derivative of $y = \ln(e^{3x} + 1)$.

EXAMPLE 4 Derivative of a product

Find the derivative of $r = \theta e^{\tan\theta}$.

Here, we use Eq. (27.16) with the derivatives of a product and the tangent:

$$\frac{dr}{d\theta} = \theta e^{\tan\theta}(\sec^2\theta) + e^{\tan\theta}(1)$$

$$= e^{\tan\theta}(\theta\sec^2\theta + 1)$$

■

Practice Exercise

2. Find the derivative of $y = 8e^{4x}\cos x$.

EXAMPLE 5 Using laws of exponents

Find the derivative of $y = (e^{1/x})^2$.

In this example, we use the power rule and Eq. (27.16):

$$\frac{dy}{dx} = 2(e^{1/x})(e^{1/x})\left(-\frac{1}{x^2}\right) \qquad \text{using } \frac{du^n}{dx} = nu^{n-1}\left(\frac{du}{dx}\right)$$

(arrow pointing to $(e^{1/x})\left(-\frac{1}{x^2}\right)$: $\frac{du}{dx}$)

$$= \frac{-2(e^{1/x})^2}{x^2} = \frac{-2e^{2/x}}{x^2}$$

NOTE ▸ [We can also solve this problem by using the laws of exponents and writing the function as $y = e^{2/x}$.] This change in form simplifies the steps for finding the derivative:

(arrow pointing to $\left(-\frac{2}{x^2}\right)$: $\frac{du}{dx}$)

$$\frac{dy}{dx} = e^{2/x}\left(-\frac{2}{x^2}\right) = \frac{-2e^{2/x}}{x^2} \qquad \text{using } \frac{d(e^u)}{dx} = e^u\left(\frac{du}{dx}\right)$$

■

EXAMPLE 6 Slope of a tangent line

Find the slope of a line tangent to the curve of $y = \frac{3e^{2x}}{x^2 + 1}$, where $x = 1.275$.

Here, we are to evaluate the derivative for $x = 1.275$. The solution is as follows:

$$\frac{dy}{dx} = \frac{(x^2 + 1)(3e^{2x})(2) - 3e^{2x}(2x)}{(x^2 + 1)^2} \qquad \text{take the derivative}$$

$$= \frac{6e^{2x}(x^2 - x + 1)}{(x^2 + 1)^2}$$

$$\left.\frac{dy}{dx}\right|_{x=1.275} = \frac{6e^{2(1.275)}(1.275^2 - 1.275 + 1)}{(1.275^2 + 1)^2} = 15.05 \qquad \text{evaluate}$$

■

EXAMPLE 7 Derivative of a power

Find the derivative of $y = (3e^{4x} + 4x^2 \ln x)^3$.

Using the general power rule, the derivative of the exponential function, the product rule, and the derivative of a logarithm, we have

$$\frac{dy}{dx} = 3(3e^{4x} + 4x^2 \ln x)^2 \left[12e^{4x} + 4x^2\left(\frac{1}{x}\right) + (\ln x)(8x)\right]$$
$$= 3(3e^{4x} + 4x^2 \ln x)^2 (12e^{4x} + 4x + 8x \ln x)$$ ■

EXERCISES 27.6

In Exercises 1 and 2, make the given changes in the indicated examples of this section and then solve the resulting problems.

1. In Example 3, in the given function, change cos to sin.

2. In Example 6, in the given function, change $x^2 + 1$ to $x + 1$.

In Exercises 3–32, find the derivatives of the given functions.

3. $y = 4^{6x}$

4. $y = 10^{x^2}$

5. $y = 6e^{\sqrt{x}}$

6. $r = 0.3e^{\theta^2}$

7. $y = 4e^t(e^{2t} - e^t)$

8. $y = 0.6 \ln(e^{5x} + 3^4)$

9. $R = Te^{-3T}$

10. $y = 5x^2 e^{2x}$

11. $y = xe^{\tan 2x}$

12. $y = 4e^x \sin \frac{1}{2}x$

13. $r = \dfrac{2(e^{2s} - e^{-2s})}{e^{2s}}$

14. $u = \dfrac{e^{0.5v}}{2v}$

15. $y = e^{-3x} \sec 4x$

16. $y = (\cos^{-1} 2x)(e^{x^2 - 1})$

17. $y = \dfrac{2e^{3x}}{4x + 3}$

18. $y = \dfrac{7 \ln 3x}{e^{2x} + 8}$

19. $y = 0.5 \ln(e^{t^2} + 4)$

20. $p = (3e^{2n} + e^2)^3$

21. $y = (2e^{2x})^3 \sin x^2$

22. $y = (e^{3/x} \cos x)^2$

23. $u = 4\sqrt{\ln 2t + e^{2t}}$

24. $y = (2e^{x^2} + x^2)^3$

25. $y = e^{xy}$

26. $y = 4e^{-2/x} \ln y + 1$

27. $y = e^{2x} \ln x^3$

28. $r = 0.4e^{2\theta} \ln \cos \theta$

29. $I = \ln \sin 2e^{6t}$

30. $y = 6 \tan e^{x+1}$

31. $y = 2 \sin^{-1} e^{2x}$

32. $w = 5 \tan^{-1} e^{3x}$

In Exercises 33–54, solve the given problems.

33. On a calculator, find the values of (a) e and (b) $(e^{1.0001} - e^{1.0000})/0.0001$. Compare the values and give the meaning of each in relation to the derivative of e^x, where $x = 1$.

34. Display the graphs of $y_1 = e^x$ and $y_n = \dfrac{e^{x+h} - e^x}{h}$ on the same calculator screen for $0 < x < 3$. For $n = 2$, let $h = 0.5$; for $n = 3$, let $h = 0.1$. (You might try smaller values of h.) What do these curves show?

35. Display the graph of $y = e^x$ on a calculator. Using the *derivative* feature, evaluate dy/dx for $x = 2$ and compare with the value of y for $x = 2$.

36. Find a formula for the nth derivative of $y = ae^{bx}$.

37. Find the slope of a line tangent to the curve of $y = e^{-x/2} \cos 4x$ for $x = 0.625$. Verify the result by using the *numerical derivative* feature of a calculator.

38. Find the slope of a line tangent to the curve of $y = \dfrac{e^{-x}}{1 + \ln 4x}$ for $x = 1.842$. Verify the result by using the *numerical derivative* feature of a calculator.

39. Find the differential of the function $y = \dfrac{12e^{4x}}{x + 6}$.

40. Find the linearization of the function $f(x) = \dfrac{6e^{4x}}{2x + 3}$ for $a = 0$.

41. Use a calculator to display the graph of $y = e^x$. By roughly estimating slopes of tangent lines, note that it is reasonable that these values are equal to the y-coordinates of the points at which these estimates are made. (*Remember:* For $y = e^x$, $dy/dx = e^x$ also.)

42. Use a calculator to display the graphs of $y = e^{-x}$ and $y = -e^{-x}$. By roughly estimating slopes of tangent lines of $y = e^{-x}$, note that $y = -e^{-x}$ gives reasonable values for the derivative of $y = e^{-x}$.

43. Show that $y = xe^{-x}$ satisfies the equation $(dy/dx) + y = e^{-x}$.

44. Show that $y = e^{-x} \sin x$ satisfies the equation

$$\frac{d^2y}{dx^2} + 2\frac{dy}{dx} + 2y = 0.$$

45. For what values of m does the function $y = ae^{mx}$ satisfy the equation $y'' + y' - 6y = 0$?

46. For what values of m does the function $y = ae^{mx}$ satisfy the equation $y'' + 4y = 0$?

47. If $y = Ae^{kx} + Be^{-kx}$, show that $y'' = k^2 y$.

48. The average energy consumption C (in MJ/year) of a certain model of refrigerator-freezer is approximately $C = 5350e^{-0.0748t} + 1800$, where t is measured in years, with $t = 0$ corresponding to 1990, and a newer model is produced each year. Assuming the function is continuous, use differentials to estimate the reduction of the 2012 model from that of the 2011 model.

49. The reliability R $(0 \leq R \leq 1)$ of a certain computer system is given by $R = e^{-0.0002t}$, where t is the time of operation (in h). Find dR/dt for $t = 1000$ h.

50. A thermometer is taken from a freezer at $-16°C$ and placed in a room at $24°C$. The temperature T of the thermometer as a function of the time t (in min) after removal is given by $T = 8.0(3.0 - 5.0e^{-0.50t})$. How fast is the temperature changing when $t = 6.0$ min?

51. For the electric circuit shown in Fig. 27.29, the current i (in A) is given by $i = 4.42e^{-66.7t}\sin 226t$, where t is the time (in s). Find the expression for di/dt.

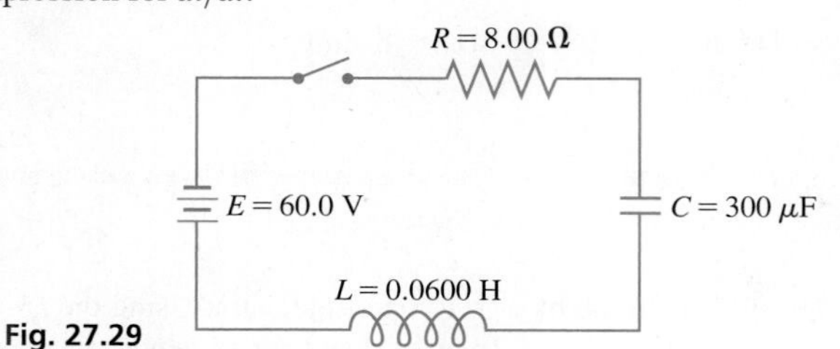

Fig. 27.29

52. Under certain assumptions of limitations to population growth, the population P (in billions) of the world is given by the *logistic equation* $P = \dfrac{10}{1 + 0.65e^{-0.036t}}$, where t is the number of years after the year 2010. Find the expression for dP/dt.

In Exercises 55–58, use the following information.
The **hyperbolic sine** *of u is defined as*

$$\sinh u = \frac{1}{2}(e^{u} - e^{-u}).$$

Figure 27.30 shows the graph of $y = \sinh x$.

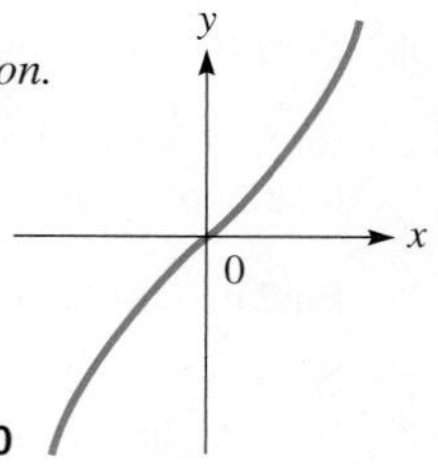

Fig. 27.30

The **hyperbolic cosine** *of u is defined as*

$$\cosh u = \frac{1}{2}(e^{u} + e^{-u}).$$

Figure 27.31 shows the graph of $y = \cosh x$.

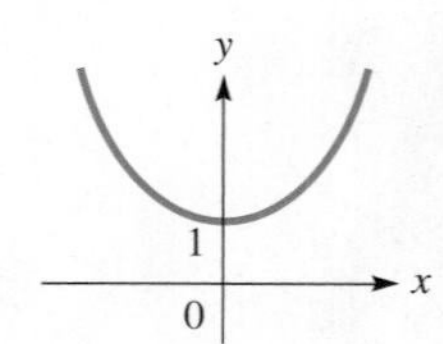

Fig. 27.31

These functions are called hyperbolic *functions since, if $x = \cosh u$ and $y = \sinh u$, x and y satisfy the equation of the hyperbola $x^2 - y^2 = 1$.*

53. Verify the fact that the exponential expressions for the hyperbolic sine and hyperbolic cosine given above satisfy the equation of the hyperbola.

54. Show that $\sinh u$ and $\cosh u$ satisfy the identity. $\cosh^2 u - \sinh^2 u = 1$.

55. Show that $\dfrac{d}{dx}\sinh u = \cosh u\dfrac{du}{dx}$ and $\dfrac{d}{dx}\cosh u = \sinh u\dfrac{du}{dx}$ where u is a function of x.

56. Show that $\dfrac{d^2\sinh x}{dx^2} = \sinh x$ and $\dfrac{d^2\cosh x}{dx^2} = \cosh x$

Answers to Practice Exercises

1. $y' = 3e^{3x}/(e^{3x} + 1)$ **2.** $y' = 8e^{4x}(4\cos x - \sin x)$

27.7 L'Hospital's Rule

Indeterminate Forms 0/0 and ∞/∞ • L'Hospital's Rule • Indeterminate Form 0 • ∞

In Section 27.1, using a geometric method, we found that $\lim_{x\to 0}\dfrac{\sin\theta}{\theta} = 1$. Now, having developed the derivatives of the transcendental functions, we can use a more general method to evaluate such limits, where both numerator and denominator approach zero.

For the limit of $u(x)/v(x)$ as $x \to a$, *if both $u(x)$ and $v(x)$ approach zero, the limit is called an* **indeterminate form of the type *0/0*.** *If $u(x)$ and $v(x)$ approach infinity, the limit is called an* **indeterminate form of the type ∞/∞.**

To find these limits, we use ***L'Hospital's rule,*** which is a method of finding the limit (if it exists) of a quotient by using derivatives. The proof of L'Hospital's rule can be found in texts covering advanced methods in calculus.

■ Named for the French mathematician, the Marquis de l'Hospital (1661–1704).

L'Hospital's Rule

If $u(x)$ and $v(x)$ are differentiable such that $v'(x)$ is not zero, and if $\lim_{x\to a} u(x) = 0$ and $\lim_{x\to a} v(x) = 0$ or $\lim_{x\to a} u(x) = \pm\infty$ and $\lim_{x\to a} v(x) = \pm\infty$, then

$$\lim_{x\to a}\frac{u(x)}{v(x)} = \lim_{x\to a}\frac{u'(x)}{v'(x)}$$

EXAMPLE 1 L'Hospital's rule—indeterminate form 0/0

Using L'Hospital's rule, find $\lim_{\theta \to 0} \frac{\sin\theta}{\theta}$.

This, of course, is the same limit we found in Section 27.1 when deriving the derivative of the sine function. Noting again that

$$\lim_{\theta \to 0} \sin\theta = 0 \quad \text{and} \quad \lim_{\theta \to 0} \theta = 0$$

we see that this is of the indeterminate form 0/0. Applying L'Hospital's rule, we have

$$\lim_{\theta \to 0} \frac{\sin\theta}{\theta} = \lim_{\theta \to 0} \frac{\frac{d}{d\theta}\sin\theta}{\frac{d}{d\theta}\theta} = \lim_{\theta \to 0} \frac{\cos\theta}{1} = \frac{1}{1} = 1$$

This agrees with the result in Section 27.1. ■

Practice Exercise

1. Find $\lim_{x \to 0} \frac{1 - e^{4x}}{2x}$.

EXAMPLE 2 L'Hospital's rule—indeterminate form ∞/∞

Evaluate $\lim_{x \to \infty} \frac{3e^{2x}}{5x}$.

Here, we see that $3e^{2x} \to \infty$ and $5x \to \infty$ as $x \to \infty$, which means the limit is of the indeterminate form ∞/∞. Therefore, using L'Hospital's rule, we have

$$\lim_{x \to \infty} \frac{3e^{2x}}{5x} = \lim_{x \to \infty} \frac{\frac{d}{dx}3e^{2x}}{\frac{d}{dx}5x} = \lim_{x \to \infty} \frac{6e^{2x}}{5} = \infty$$

This means that the limit does not exist. ■

CAUTION In using L'Hospital's Rule find the limit of the quotient of derivatives, not the limit of the derivative of the quotient. ■

Practice Exercise

2. Find $\lim_{x \to \infty} \frac{\ln x}{x}$.

EXAMPLE 3 L'Hospital's rule—limit as $t \to \pi/2$

Evaluate $\lim_{t \to \pi/2} \frac{1 - \sin t}{\cos t}$.

Here, $\lim_{t \to \pi/2} (1 - \sin t) = 0$ and $\lim_{t \to \pi/2} \cos t = 0$, which means the limit is of the indeterminate form 0/0. Using L'Hospital's rule, we have

$$\lim_{t \to \pi/2} \frac{1 - \sin t}{\cos t} = \lim_{t \to \pi/2} \frac{\frac{d}{dt}(1 - \sin t)}{\frac{d}{dt}\cos t} = \lim_{t \to \pi/2} \frac{-\cos t}{-\sin t} = \frac{0}{-1} = 0$$

■

EXAMPLE 4 Using L'Hospital's rule twice

Using L'Hospital's rule, find $\lim_{x \to \infty} \frac{x^2 + 1}{2x^2 + 3}$.

This is the same limit we found algebraically in Example 14 of Section 23.1.

Noting that $(x^2 + 1) \to \infty$ and $(2x^2 + 3) \to \infty$ as $x \to \infty$, this limit is of the indeterminate form ∞/∞. Therefore, applying L'Hospital's rule, we have

$$\lim_{x \to \infty} \frac{x^2 + 1}{2x^2 + 3} = \lim_{x \to \infty} \frac{\frac{d}{dx}(x^2 + 1)}{\frac{d}{dx}(2x^2 + 3)} = \lim_{x \to \infty} \frac{2x}{4x} = \lim_{x \to \infty} \frac{1}{2} = \frac{1}{2}$$

After applying L'Hospital's rule, we were able to algebraically simplify the expression. However, because $2x \to \infty$ and $4x \to \infty$ as $x \to \infty$, this last limit also is of the indeterminate form ∞/∞ and we can use L'Hospital's rule again. This gives us

$$\lim_{x \to \infty} \frac{2x}{4x} = \lim_{x \to \infty} \frac{\frac{d}{dx}2x}{\frac{d}{dx}4x} = \lim_{x \to \infty} \frac{2}{4} = \frac{1}{2}$$

We note that this result agrees with the result of Example 14 of Section 23.1. ■

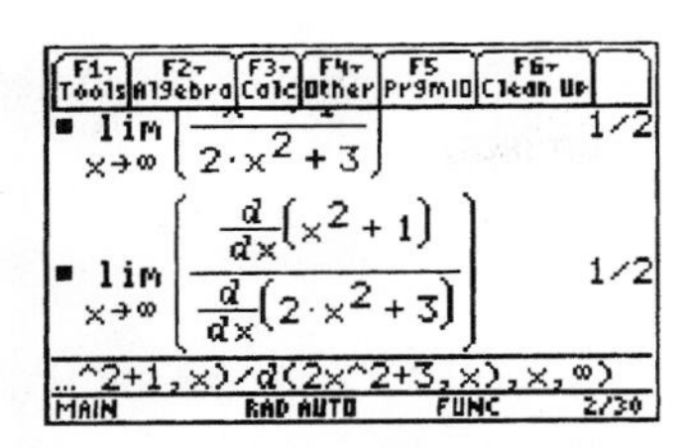

Fig. 27.32
TI-89 graphing calculator keystrokes: goo.gl/J94CSe

■ On a TI-89 calculator, the limit of this function found directly, and using L'Hospital's rule, is shown in Fig. 27.32.

EXAMPLE 5 Be careful to check for indeterminate form

Find $\lim\limits_{x\to\pi^-} \dfrac{2\sin x}{1 - \cos x}$.

Applying L'Hospital's rule, we have

$$\lim_{x\to\pi^-} \frac{2\sin x}{1-\cos x} = \lim_{x\to\pi^-} \frac{\frac{d}{dx}2\sin x}{\frac{d}{dx}(1-\cos x)} = \lim_{x\to\pi^-} \frac{2\cos x}{\sin x} = -\infty$$

CAUTION However, **this result is INCORRECT.** Checking back to the original expression, we note that **it does not fit the indeterminate form 0/0.** As $x \to \pi^-$ (x approaches π from values below π), $2\sin x$ approaches 0, but $1 - \cos x$ approaches 2. Therefore, we cannot apply L'Hospital's rule to this limit. **Always check to see that the limit fits the indeterminate form of 0/0 or ∞/∞ before using L'Hospital's rule.** ■

Because this function is continuous as $x \to \pi^-$, we have

$$\lim_{x\to\pi^-} \frac{2\sin x}{1-\cos x} = \frac{2\sin\pi}{1-\cos\pi} = \frac{0}{1-(-1)} = 0$$ ■

We can use a calculator to check the value of a limit found by use of L'Hospital's rule. The following example illustrates this.

EXAMPLE 6 Calculator verification of limit

Find $\lim\limits_{x\to\pi} \dfrac{1+\cos x}{(\pi - x)^2}$ and verify the value of the limit using the *Table* feature on a calculator. Because $\cos\pi = -1$, we see that both the numerator and the denominator approach zero as x approaches π. Therefore, we have

$$\lim_{x\to\pi} \frac{1+\cos x}{(\pi-x)^2} = \lim_{x\to\pi} \frac{\frac{d}{dx}(1+\cos x)}{\frac{d}{dx}(\pi-x)^2} = \lim_{x\to\pi} \frac{-\sin x}{-2(\pi-x)}$$

indeterminate form 0/0; use L'Hospital's rule again

$$= \lim_{x\to\pi} \frac{\frac{d}{dx}(-\sin x)}{\frac{d}{dx}[-2(\pi-x)]} = \lim_{x\to\pi} \frac{-\cos x}{2} = \frac{-(-1)}{2} = \frac{1}{2}$$

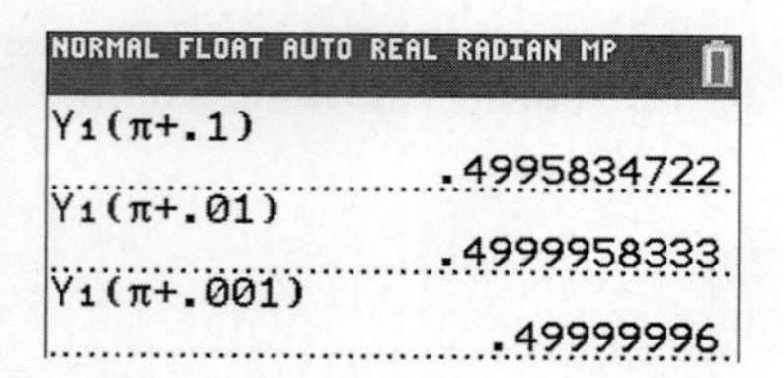

Fig. 27.33
Graphing calculator keystrokes: goo.gl/d61uMB

To check this limit on a calculator, we first let $Y_1 = (1 + \cos X)/(\pi - X)^2$. We then evaluate the function for x-values that are closer and closer to π as shown in Fig. 27.33. We can see that the function values are approaching 1/2, which verifies the limit found using L'Hospital's rule. ■

There are indeterminate forms other than 0/0 and ∞/∞. One of these is illustrated in the following example, and others are covered in the exercises.

EXAMPLE 7 Indeterminate form 0 · ∞

■ L'Hospital's rule can be used for indeterminant forms of any combination of $\dfrac{\pm\infty}{\pm\infty}$ as well as 0/0.

Find $\lim\limits_{x\to 0^+} x^2 \ln x$.

As $x \to 0^+$, we note that $x^2 \to 0$ and $\ln x \to -\infty$. This is the indeterminate form $0 \cdot -\infty$. By writing x^2 as $1/(1/x^2)$ we can change this indeterminate form to $-\infty/\infty$. Making this change, and then applying L'Hospital's rule, we have

$$\lim_{x\to 0^+} x^2 \ln x = \lim_{x\to 0^+} \frac{\ln x}{\frac{1}{x^2}}$$

$$= \lim_{x\to 0^+} \frac{\frac{d}{dx}\ln x}{\frac{d}{dx}x^{-2}} = \lim_{x\to 0^+} \frac{\frac{1}{x}}{-\frac{2}{x^3}} = \lim_{x\to 0^+}\left(-\frac{x^2}{2}\right) = 0$$ ■

EXERCISES 27.7

In Exercises 1 and 2, make the given changes in the indicated examples of this section and then solve the resulting problems.

1. In Example 2, change the denominator $5x$ to $5 \ln x$.
2. In Example 5, change the denominator $1 - \cos x$ to $1 + \cos x$.

In Exercises 3–36, evaluate each limit (if it exists). Use L'Hospital's rule (if appropriate).

3. $\lim_{x \to 3} \frac{x - 3}{x^2 - 9}$
4. $\lim_{x \to 1} \frac{x^5 - 1}{x^3 - 1}$
5. $\lim_{\theta \to 0} \frac{\tan\theta}{\theta}$
6. $\lim_{x \to \infty} \frac{e^x}{x^2}$
7. $\lim_{x \to \infty} \frac{x \ln x}{x + \ln x}$
8. $\lim_{x \to 1} \frac{\ln x}{x - 1}$
9. $\lim_{t \to \pi/4} \frac{1 - \sin 2t}{\frac{\pi}{4} - t}$
10. $\lim_{x \to 0} \frac{\ln \cos x}{x}$
11. $\lim_{x \to 0^+} \frac{\ln x}{x^{-1}}$
12. $\lim_{\theta \to \pi/2} \frac{1 + \sec\theta}{\tan\theta}$
13. $\lim_{x \to 0} \frac{\sin x - x}{x^3}$
14. $\lim_{x \to \infty} \frac{x^2 + x}{e^x + 1}$
15. $\lim_{x \to 1} \frac{\sin \pi x}{x - 1}$
16. $\lim_{x \to 0} \frac{\tan^{-1} x}{x}$
17. $\lim_{x \to 0} x \cot x$
18. $\lim_{z \to \infty} z e^{-z}$
19. $\lim_{x \to 0} x \ln \sin x$
20. $\lim_{x \to \pi/2} (1 - \sin x) \tan x$
21. $\lim_{x \to 0} \frac{2 \sin x}{5e^x}$
22. $\lim_{x \to \infty} \frac{e^{2x} - 1}{4x + 1}$
23. $\lim_{x \to \infty} \frac{1 + e^{2x}}{2 + \ln x}$
24. $\lim_{x \to \pi/2} \frac{\frac{\pi}{2} - x}{1 + \sin x}$
25. $\lim_{x \to \infty} \frac{\ln x^3}{x^2}$
26. $\lim_{x \to \pi/2} \frac{2 \cos x}{x - \frac{\pi}{2}}$
27. $\lim_{x \to 0} \frac{2x + \sin 3x}{x - \sin 3x}$
28. $\lim_{x \to 0} \frac{\sin^{-1} x}{\tan^{-1} x}$
29. $\lim_{x \to 1} \frac{\ln x}{\sin 2\pi x}$
30. $\lim_{t \to \infty} \frac{\ln \ln t}{\ln t}$
31. $\lim_{x \to 0^+} (\sin x)(\ln x)$
32. $\lim_{x \to 0^+} (\sin^{-1} x)(\ln x)$
33. $\lim_{x \to 2} \frac{x^3 - 8}{x^5 - 32}$
34. $\lim_{x \to 1} \frac{x^5 + 2x^3 - 3}{x^7 - 3x^2 + 2}$
35. $\lim_{x \to 0} \frac{\sqrt{1 - x} - \sqrt{1 + x}}{x}$
36. $\lim_{x \to 0} \frac{e^x + e^{-x} - 2}{1 - \cos 2x}$

In Exercises 37–48, solve the given problems.

37. Another indeterminate form is $\infty - \infty$. Often, it is possible to make an algebraic or trigonometric change in the function so that it will take on the form of a $0/0$ or ∞/∞ indeterminate form. Find $\lim_{\theta \to \frac{\pi}{2}^-} (\sec\theta - \tan\theta)$.

38. Find $\lim_{x \to 0^+} \left(\frac{1}{\sin x} - \frac{1}{x}\right)$. (See Exercise 37.)

39. Three other indeterminate forms are 0^0, ∞^0, and 1^∞. For the function $y = [f(x)]^{g(x)}$ and the following limits, we have the indicated indeterminate form:

 $\lim_{x \to a} f(x) = 0$ and $\lim_{x \to a} g(x) = 0$ (indet. form 0^0)

 $\lim_{x \to a} f(x) = \infty$ and $\lim_{x \to a} g(x) = 0$ (indet. form ∞^0)

 $\lim_{x \to a} f(x) = 1$ and $\lim_{x \to a} g(x) = \infty$ (indet. form 1^∞)

 By taking the logarithm of $y = [f(x)]^{g(x)}$, we have $\ln y = g(x) \ln f(x)$ and in each case the right-hand member is a type that can be solved by L'Hospital's rule. By knowing the limit of $\ln y$, we can find the limit of y.

 Find $\lim_{x \to 0} x^x$.

40. Find $\lim_{x \to 0} \left(\frac{1}{x}\right)^{\sin x}$ (See Exercise 39.)

41. Find $\lim_{x \to \infty} (1 + x^2)^{1/x}$ (See Exercise 39.)

42. Find $\lim_{\theta \to \pi/2} (\sin\theta)^{\tan\theta}$ (See Exercise 39.)

43. Is $\lim_{x \to 3} \frac{x^2 - 9}{x - 1} = \lim_{x \to 3} \frac{2x}{1} = 6$? Explain.

44. Is $\lim_{x \to 3} \frac{x^3 - 3x^2 + x - 3}{x^2 - 9} = \lim_{x \to 3} \frac{3x^2 - 6x + 1}{2x} = \frac{5}{3}$? Explain.

45. Explain why L'Hospital's rule cannot be applied to $\lim_{x \to \infty} x \sin x$.

46. By inspection, find $\lim_{x \to \frac{\pi}{2}^-} (\cos x)^{\tan x}$. What is the form of this limit? (Note that it is not one of the indeterminate forms noted in this section.)

47. If the force resisting the fall of an object of mass m through the atmosphere is directly proportional to the velocity v, then the velocity at time t is $v = \frac{mg}{k}(1 - e^{-kt/m})$, where g is the acceleration due to gravity and k is a positive constant. Find $\lim_{k \to 0^+} v$.

48. In Exercise 47, find $\lim_{m \to \infty} v$.

Answers to Practice Exercises

1. -2 **2.** 0

27.8 Applications

Curve Sketching • Newton's Method • Time-Rate-of-Change Problems

The following examples show applications of the logarithmic and exponential functions to the types of applications shown at the left. These and other applications are included in the exercises.

EXAMPLE 1 Sketching a graph

Sketch the graph of the function $y = x\ln x$.

First, we note that x cannot be zero since $\ln x$ is not defined at $x = 0$. Because $\ln 1 = 0$, we have an intercept at $(1, 0)$. There is no symmetry to the axes or origin, and there are no vertical asymptotes. Also, because $\ln x$ is defined only for $x > 0$, the domain is $x > 0$.

Finding the first two derivatives, we have

$$\frac{dy}{dx} = x\left(\frac{1}{x}\right) + \ln x = 1 + \ln x \qquad \frac{d^2y}{dx^2} = \frac{1}{x}$$

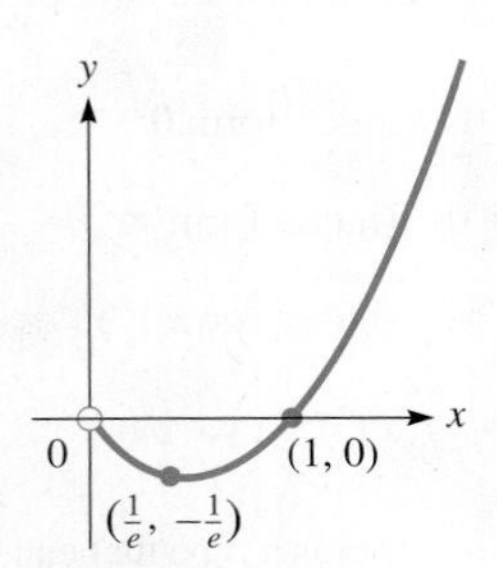

Fig. 27.34

The first derivative is zero if $\ln x = -1$, or $x = e^{-1}$. The second derivative is positive for this value of x. Thus, there is a minimum point at $(1/e, -1/e)$. Because the domain is $x > 0$, the second derivative indicates that the curve is always concave up. In turn, we now see that the range of the function is $y \geq -1/e$. The graph is shown in Fig. 27.34.

NOTE ▸ [Using L'Hospital's rule, we find $\lim_{x\to 0} x\ln x = 0$. Therefore, the curve approaches the origin as x approaches zero. However, the origin is not included on the graph of the function since ln 0 is undefined.] ■

EXAMPLE 2 Sketching a graph

Sketch the graph of the function $y = e^{-x}\cos x$ $(0 \leq x \leq 2\pi)$.

This curve has intercepts for all values for which $\cos x$ is zero. Those values in the domain $0 \leq x \leq 2\pi$ for which $\cos x = 0$ are $x = \frac{\pi}{2}$ and $x = \frac{3\pi}{2}$. The factor e^{-x} is always positive, and $e^{-x} = 1$ for $x = 0$, which means $(0, 1)$ is also an intercept. There is no symmetry to the axes or the origin, and there are no vertical asymptotes.

Next, finding the first derivative, we have

$$\frac{dy}{dx} = -e^{-x}\sin x - e^{-x}\cos x = -e^{-x}(\sin x + \cos x)$$

Setting the derivative equal to zero, since e^{-x} is always positive, we have

$$\sin x + \cos x = 0, \qquad \tan x = -1, \qquad x = \frac{3\pi}{4}, \frac{7\pi}{4}$$

Now, finding the second derivative, we have

$$\frac{d^2y}{dx^2} = -e^{-x}(\cos x - \sin x) - e^{-x}(-1)(\sin x + \cos x) = 2e^{-x}\sin x$$

The sign of the second derivative depends only on $\sin x$. Therefore,

$$\frac{d^2y}{dx^2} > 0 \quad \text{for} \quad x = \frac{3\pi}{4} \qquad \text{and} \qquad \frac{d^2y}{dx^2} < 0 \quad \text{for} \quad x = \frac{7\pi}{4}$$

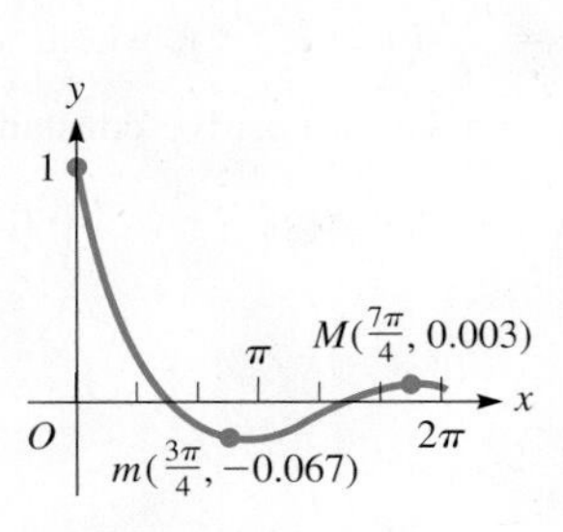

Fig. 27.35

This means that $(\frac{3\pi}{4}, -0.067)$ is a minimum and $(\frac{7\pi}{4}, 0.003)$ is a maximum.

Also, from the second derivative, points of inflection occur for $x = 0, \pi$, and 2π because $\sin x = 0$ for these values. The graph is shown in Fig. 27.35. ■

EXAMPLE 3 Solving an equation using Newton's method

Find the root of the equation $e^{2x} - 4\cos x = 0$ that lies between 0 and 1, by using Newton's method.

Here,

$$f(x) = e^{2x} - 4\cos x$$
$$f'(x) = 2e^{2x} + 4\sin x$$

This means that $f(0) = -3$ and $f(1) = 5.2$. Therefore, we choose $x_1 = 0.5$. Using Eq. (24.1), which is

$$x_2 = x_1 - \frac{f(x_1)}{f'(x_1)}$$

we have these values:

$$f(x_1) = e^{2(0.5)} - 4\cos 0.5 = -0.7920484$$
$$f'(x_1) = 2e^{2(0.5)} + 4\sin 0.5 = 7.3542658$$
$$x_2 = 0.5 - \frac{-0.7920484}{7.3542658} = 0.6076992$$

Using the method again, we find $x_3 = 0.5979751$, which is correct to three decimal places. ■

Practice Exercise

1. In Example 3, choose $x_1 = 0.7$ and then find x_2.

EXAMPLE 4 Time rate of change—population growth

One model for population growth is that the population P at time t is given by $P = P_0e^{kt}$, where P_0 is the *initial* population (at $t = 0$, when timing starts for the population being considered) and k is a constant. Show that the instantaneous time rate of change of population is directly proportional to the population present at time t.

To find the time rate of change, we find the derivative dP/dt:

$$\frac{dP}{dt} = (P_0e^{kt})(k) = kP_0e^{kt}$$
$$= kP \qquad \text{since } P = P_0e^{kt}$$

Thus, we see that population growth increases as population increases. ■

EXAMPLE 5 Acceleration of a rocket

A rocket is moving such that the only force acting on it is due to gravity and its mass is decreasing (because of the use of fuel) at a constant rate r. If it moves vertically, its velocity v as a function of time t is given by

■ See the chapter introduction.

$$v = v_0 - gt - k\ln\left(1 - \frac{rt}{m_0}\right)$$

where v_0 is the initial velocity, g is the acceleration due to gravity, t is the time, m_0 is the initial mass, and k is a constant. Determine the expression for the acceleration.

Because acceleration is the time rate of change of the velocity, we must find dv/dt. Therefore,

$$\frac{dv}{dt} = -g - k\frac{1}{1 - \frac{rt}{m_0}}\left(\frac{-r}{m_0}\right) = -g + \frac{km_0}{m_0 - rt}\left(\frac{r}{m_0}\right)$$
$$= \frac{kr}{m_0 - rt} - g$$ ■

EXERCISES 27.8

In Exercises 1 and 2, make the given changes in the indicated examples of this section and then solve the resulting problems.

1. In Example 2, in the given function, change cos to sin.

2. In Example 3, in the given function, change e^{2x} to $2e^x$.

In Exercises 3–14, sketch the graphs of the given functions. Check each by displaying the graph on a calculator.

3. $y = \ln \cos x$

4. $y = \dfrac{2\ln x}{x}$

5. $y = 3xe^{-x}$

6. $y = \dfrac{e^x}{x}$

7. $y = \ln\dfrac{e}{x^2 + 1}$

8. $y = \dfrac{2x}{\ln x}$

9. $y = 4e^{-x^2}$

10. $y = 4x - e^x$

11. $y = 8\ln x - 2x$

12. $y = e^{-x}\sin x$

13. $y = \frac{1}{2}(e^x - e^{-x})$ (See Exercise 53 of Section 27.6.)

14. $y = \frac{1}{2}(e^x + e^{-x})$ (See Exercise 53 of Section 27.6.)

In Exercises 15–46, solve the given problems by finding the appropriate derivative.

15. Find the values of x for which the graphs of $y_1 = e^{2/x^2}$ and $y_2 = e^{-2/x^2}$ are increasing and decreasing. Explain why they differ as they do.

16. Find the values of x for which the graphs of $y_1 = \ln(x^2 + 1)$ and $y_2 = \ln(x^2 - 1)$ are concave up and concave down. Explain why they differ as they do.

17. Find the equation of the line tangent to the curve of $y = x^2\ln x$ at the point $(1, 0)$.

18. Find the equation of the line tangent to the curve of $y = \tan^{-1}2x$, where $x = 1$.

19. Find the equation of the line normal to the curve of $y = 2\sin\frac{1}{2}x$, where $x = 3\pi/2$.

20. Find the equation of the line normal to the curve of $y = e^{2x}/x$, at $x = 1$.

21. By Newton's method, solve the equation $x^2 - 3 + \ln 4x = 0$. Check the result by using the *zero* feature of a calculator.

22. By Newton's method, find the value of x for which $y = e^{\cos x}$ is minimum for $0 < x < 2\pi$. Check the result by displaying the graph on a calculator and then finding the minimum point.

23. The electric current i (in A) through an inductor of 0.50 H as a function of time t (in s) is $i = e^{-5.0t}\sin 120\pi t$. The voltage across the inductor is given by $V_L = L(di/dt)$, where L is the inductance (in H). Find the voltage across the inductor for $t = 1.0$ ms.

24. A computer analysis showed that the population density D (in persons/km^2) at a distance r (in km) from the center of a city is approximately $D = 200(1 + 5e^{-0.01r^2})$ if $r < 20$ km. At what distance from the city center does the decrease in population density (dD/dr) itself start to decrease?

25. The power supply P (in W) in a satellite is $P = 100e^{-0.005t}$, where t is measured in days. Find the time rate of change of power after 100 days.

26. The number N of atoms of radium at any time t is given in terms of the number at $t = 0$, N_0, by $N = N_0e^{-kt}$. Show that the time rate of change of N is proportional to N.

27. A metal bar is heated, and then allowed to cool. Its temperature T (in °C) is found to be $T = 15 + 75e^{-0.25t}$, where t (in min) is the time of cooling. Find the time rate of change of temperature after 5.0 min.

28. The insulation resistance R (in Ω/m) of a shielded cable is given by $R = k\ln(r_2/r_1)$. Here r_1 and r_2 are the inner and outer radii of the insulation. Find the expression for dR/dr_2 if k and r_1 are constant.

29. The vapor pressure p and thermodynamic temperature T of a gas are related by the equation $\ln p = \dfrac{a}{T} + b\ln T + c$, where a, b, and c are constants. Find the expression for dp/dT.

30. The charge q on a capacitor in a circuit containing a capacitor of capacitance C, a resistance R, and a source of voltage E is given by $q = CE(1 - e^{-t/RC})$. Show that this equation satisfies the equation $R\dfrac{dq}{dt} + \dfrac{q}{C} = E$.

31. Assuming that force is proportional to acceleration, show that a particle moving along the x-axis, so that its displacement $x = ae^{kt} + be^{-kt}$, has a force acting on it which is proportional to its displacement.

32. The radius of curvature at a point on a curve is given by

$$R = \frac{[1 + (dy/dx)^2]^{3/2}}{d^2y/dx^2}$$

A roller mechanism moves along the path defined by $y = \ln \sec x$ $(-1.5\,\text{dm} \le x \le 1.5\,\text{dm})$. Find the radius of curvature of this path for $x = 0.85$ dm.

33. Sketch the graph of $y = \ln \sec x$, marking that part which is the path of the roller mechanism of Exercise 32.

34. In an electronic device, the maximum current density i_m as a function of the temperature T is given by $i_m = AT^2e^{k/T}$, where A and k are constants. Find the expression for a small change in i_m for a small change in T.

35. The energy E (in J) dissipated by a certain resistor after t seconds is given by $E = \ln(t + 1) - 0.25t$. At what time is the energy dissipated the greatest?

36. In a study of traffic control, the number n of vehicles on a certain section of a highway from 2 p.m. to 8 p.m. was found to be $n = 200(1 + t^3e^{-t})$, where t is the number of hours after 2 p.m. At what time is the number of vehicles the greatest?

37. The curve given by $y = \dfrac{1}{\sqrt{2\pi}}e^{-x^2/2}$, called the *standard normal curve*, is very important in statistics. Show that this curve has inflection points at $x = \pm 1$.

38. The reliability R $(0 \le R \le 1)$ of a certain computer system after t hours of operation is found from $R = 3e^{-0.004t} - 2e^{-0.006t}$. Use Newton's method to find how long the system operates to have a reliability of 0.8 (80% probability that there will be no system failure).

39. An object on the end of a spring is moving so that its displacement (in cm) from the equilibrium position is given by $y = e^{-0.5t}(0.4\cos 6t - 0.2\sin 6t)$. Find the expression for the velocity of the object. What is the velocity when $t = 0.26$ s? The motion described by this equation is called *damped harmonic motion.*

40. A package of weather instruments is propelled into the air to an altitude of about 7 km. A parachute then opens, and the package returns to the surface. The altitude y of the package as a function of the time t (in min) is given by $y = \dfrac{10t}{e^{0.4t} + 1}$. Find the vertical velocity of the package for $t = 8.0$ min.

41. The speed s of signaling by use of a certain communications cable is directly proportional to $x^2 \ln x^{-1}$, where x is the ratio of the radius of the core of the cable to the thickness of the surrounding insulation. For what value of x is s a maximum?

42. A computer is programmed to inscribe a series of rectangles in the first quadrant under the curve of $y = e^{-x}$. What is the area of the largest rectangle that can be inscribed?

43. The relative number N of gas molecules in a container that are moving at a velocity v can be shown to be $N = av^2e^{-bv^2}$, where a and b are constants. Find v for the maximum N.

44. A missile is launched and travels along a path that can be represented by $y = \sqrt{x}$. A radar tracking station is located 2.00 km directly behind the launch pad. Placing the launch pad at the origin and the radar station at $(-2.00, 0)$, find the largest angle of elevation required of the radar to track the missile.

45. A connecting rod 4 cm long connects a piston to a crank 2 cm in radius (see Fig. 27.36). The acceleration of the piston is given by $a = 2w^2(\cos\theta + 0.5\cos 2\theta)$. If the angular velocity w is constant, find θ for maximum or minimum a.

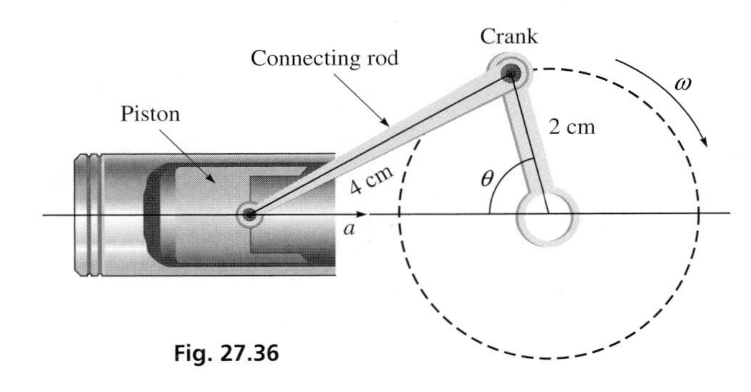

Fig. 27.36

46. The St. Louis Gateway Arch (see Fig. 27.46 on page 839) has a shape that is given approximately by (measurements in m) $y = -19.46(e^{x/38.92} + e^{-x/38.92}) + 230.9$. What is the maximum height of the arch?

Answer to Practice Exercise

1. $x_2 = 0.6068$

CHAPTER 27 KEY FORMULAS AND EQUATIONS

Limit of $\dfrac{\sin\theta}{\theta}$ as $\theta \to 0$

$$\lim_{\theta\to 0}\frac{\sin\theta}{\theta} = \lim_{h\to 0}\frac{\sin(h/2)}{h/2} = 1 \qquad (27.1)$$

Chain rule

$$\frac{dy}{dx} = \frac{dy}{du}\frac{du}{dx} \qquad (27.3)$$

Derivatives

$$\frac{d(\sin u)}{dx} = \cos u\frac{du}{dx} \qquad (27.4)$$

$$\frac{d(\cos u)}{dx} = -\sin u\frac{du}{dx} \qquad (27.5)$$

$$\frac{d(\tan u)}{dx} = \sec^2 u\frac{du}{dx} \qquad (27.6)$$

$$\frac{d(\cot u)}{dx} = -\csc^2 u\frac{du}{dx} \qquad (27.7)$$

$$\frac{d(\sec u)}{dx} = \sec u\tan u\frac{du}{dx} \qquad (27.8)$$

$$\frac{d(\csc u)}{dx} = -\csc u\cot u\frac{du}{dx} \qquad (27.9)$$

$$\frac{d(\sin^{-1} u)}{dx} = \frac{1}{\sqrt{1-u^2}}\frac{du}{dx} \qquad (27.10)$$

$$\frac{d(\cos^{-1}u)}{dx} = -\frac{1}{\sqrt{1-u^2}}\frac{du}{dx} \quad (27.11)$$

$$\frac{d(\tan^{-1}u)}{dx} = \frac{1}{1+u^2}\frac{du}{dx} \quad (27.12)$$

$$\frac{d(\log_b u)}{dx} = \frac{1}{u}\log_b e\frac{du}{dx} \quad (27.13)$$

$$\frac{d(\ln u)}{dx} = \frac{1}{u}\frac{du}{dx} \quad (27.14)$$

$$\frac{d(b^u)}{dx} = b^u \ln b\frac{du}{dx} \quad (27.15)$$

$$\frac{d(e^u)}{dx} = e^u\frac{du}{dx} \quad (27.16)$$

CHAPTER 27 REVIEW EXERCISES

CONCEPT CHECK EXERCISES

*Determine each of the following as being either **true** or **false**. If it is false, explain why.*

1. For $y = \sin^3 2x$, $\frac{dy}{dx} = 3\sin^2 2x\cos 2x$
2. For $y = 2\tan^4 2x$, $\frac{dy}{dx} = 16\tan^3 2x\sec^2 2x$
3. For $y = 6\tan^{-1}2x$, $\frac{dy}{dx} = \frac{6}{1+4x^2}$
4. The curve of $y = \cos^{-1}x$ decreases for all values of x.
5. For $y = \ln 2x$, $\frac{dy}{dx} = \frac{1}{x}$.
6. For $y = e^{2x}$, $\frac{dy}{dx} = e^{2x}$.
7. For $y = 2^{5x}$, $\frac{dy}{dx} = 5x(2^{5x-1})$.
8. $\lim_{x\to 0}\frac{e^{3x}-1}{x} = 3$

PRACTICE AND APPLICATIONS

In Exercises 9–48, find the derivative of the given functions.

9. $y = 3\cos(4x-1)$
10. $y = 4\sec(1-x^3)$
11. $u = 0.2\tan\sqrt{3-2v}$
12. $y = 5\sin(1-6x)$
13. $y = \csc^2(3x+2)$
14. $r = \cot^2 5\pi\theta$
15. $y = 3\cos^4 x^2$
16. $y = 2\sin^3\sqrt{x}$
17. $y = (e^{x-3})^2$
18. $y = 0.5e^{\sin 2x}\cos 2x$
19. $y = 3x\ln(x^2+1)$
20. $R = \ln(3+\sin T^2)$
21. $y = 10\tan^{-1}(x/5)$
22. $y = 0.4t^2\cos^{-1}(2\pi t+1)$
23. $\theta = \ln\sin^{-1}0.1t$
24. $y = \sin(\tan^{-1}x)$
25. $y = \sqrt{\csc 4x + \cot 4x}$
26. $y = 3\cos^2(\tan 3x)$
27. $y = 7\ln(x-e^{-x})^2$
28. $h = 4\ln\sqrt[3]{\sin 6\theta}$
29. $y = \frac{\cos^2 x}{e^{3x}+\pi^2}$
30. $y = \sqrt{\frac{1+\cos 2x}{2}}$
31. $v = \frac{u^2}{\tan^{-1}2u}$
32. $y = \frac{\sin^{-1}x}{4x}$
33. $y = \frac{\ln\csc x^2}{x}$
34. $u = 0.25\ln\tan e^{8x}$
35. $y = 3\ln^2(4+\sin 2x)$
36. $y = \frac{\ln(3+\sin\pi t)}{2t}$
37. $L = 0.1e^{-2t}\sec\pi t$
38. $y = 5e^{3x}\ln x$
39. $y = \sqrt{\sin 2x + e^{4x}}$
40. $x + y\ln 2x = y^2$
41. $\tan^{-1}\frac{y}{x} = x^2e^y$
42. $3y + \ln x + \ln y = 2 + x^2$
43. $r = 0.5t(e^{2t}+1)(e^{-2t}-1)$
44. $y = (\ln 4x - \tan 4x)^3$
45. $e^x\ln xy + y = e^x$
46. $y = x(\sin^{-1}x)^2$
47. $y = x\cos^{-1}x - \sqrt{1-x^2}$
48. $W = \ln(4s^2+1) + \tan^{-1}2s$

In Exercises 49–52, sketch the graphs of the given functions. Check each by displaying the graph on a calculator.

49. $y = x - \cos 0.5x$
50. $y = 4\sin x + \cos 2x$
51. $y = x(\ln x)^2$
52. $y = 2\ln(3+x)$

In Exercises 53–56, find the equations of the indicated tangent or normal lines.

53. Find the equation of the line tangent to the curve of $y = 4\cos^2(x^2)$ at $x = 1$.
54. Find the equation of the line tangent to the curve of $y = \ln\cos x$ at $x = \frac{\pi}{6}$.

55. Find the equation of the line normal to the curve of $y = e^{x^2}$ at $x = \frac{1}{2}$.

56. Find the equation of the line normal to the curve of $y = \tan^{-1}4x$ at $x = 1/2$.

In Exercises 57–62, find the indicated limits by use of L'Hospital's rule.

57. $\lim_{x\to 0}\frac{\sin 2x}{\sin 3x}$

58. $\lim_{x\to 0}\frac{xe^x}{1 - e^x}$

59. $\lim_{x\to \pi^+}\frac{\sin x}{x - \pi}$

60. $\lim_{t\to +\infty}\frac{x^4 + 5x^2 + 1}{3x^4 + 4}$

61. $\lim_{x\to \infty}\frac{\ln x}{\sqrt[3]{x}}$

62. $\lim_{x\to \pi}(x - \pi)\cot x$

In Exercises 63–112, solve the given problems.

63. Find the derivative of each member of the identity $\sin^2 x + \cos^2 x = 1$ and show that the results are equal.

64. Find the derivative of each member of the identity $\sin(x + 1) = \sin x\cos 1 + \cos x\sin 1$ and show that the results are equal.

65. If $y = \sin 3x$, show that $\frac{d^2y}{dx^2} = -9y$.

66. If $y = e^{5x}(a + bx)$, show that $\frac{d^2y}{dx^2} - 10\frac{dy}{dx} + 25y = 0$.

67. By Newton's method, solve the equation $e^x - x^2 = 0$. Check the solution by displaying the graph on a calculator and using the *intersect* (or *zero*) feature.

68. By Newton's method, solve the equation $x^2 = \tan^{-1}x$. Check the solution by displaying the graph on a calculator and using the *intersect* (or *zero*) feature.

69. Find the values of x for which the graph of $y = e^x - 2e^{-x}$ is concave up.

70. Find the values of x for which the graph of $y = 2e^{\sin(x/2)}$ has maximum or minimum points.

71. What is the slope of a line tangent to the curve $y = x + 2\sin x$, where $x = \pi/3$?

72. The distance s traveled by a motorboat in t seconds after the engine is cut off is given by $s = k^{-1}(\ln kv_0t + 1)$, where v_0 is the velocity of the boat at the time the engine was cut off. Find ds/dt.

73. In finding the path of a certain plane, the equation $\ln r = \ln a - \ln\cos\theta - (v/w)\ln(\sec\theta + \tan\theta)$ is used. Find $dr/d\theta$.

74. The outline of an archway can be described as the area (in m^2) bounded by $y = 3e^{-x^2}$, $x = -1$, $x = 1$, and $y = 0$. What is the cross-sectional area of the largest rectangular object that can pass through the archway? Noting the exact solution, identify the points where the upper corners touch the archway.

75. A surveyor measures two sides and the included angle of a triangular tract of land to be 315.2 m, 464.3 m, and 65.5°. What is the error caused in calculating the third side if the angle is in error by 0.2°?

76. For the paper cup in Exercise 60 of Section 27.1, find the vertex angle θ for which the volume is a maximum.

77. If a block is placed on a plane inclined with the horizontal at an angle θ such that the block stays at the same position, the coefficient of friction μ is given by $\mu = \tan\theta$. Use differentials to find the change in μ if θ changes from 18° to 20°.

78. The tensile strength S (in N) of a plastic is tested and found to change with the temperature T (in °C) according to the equation $S = 1800\ln(T + 25) - 40T + 8600$. For what temperature is the tensile strength the greatest?

79. If a 200-N crate is dragged along a horizontal floor by a force F (in N) acting along a rope at an angle θ with the floor, the magnitude of F is given by $F = \frac{200\mu}{\mu\sin\theta + \cos\theta}$, where μ is the coefficient of friction. Evaluate the instantaneous rate of change of F with respect to θ when $\theta = 15°$ if $\mu = 0.20$.

80. Find the date of the maximum number of hours of daylight for cities at 40°N. The hours h of daylight is approximately $h = 12.2 + 2.8\sin\left[\frac{\pi}{6}(x - 2.7)\right]$, where x is measured in months ($x = 0.5$ is Jan. 15, etc.).

81. Periodically a robot moves a part vertically y cm in an automobile assembly line. If y as a function of the time t (in s) is $y = 0.75(\sec\sqrt{0.15t} - 1)$, find the velocity at which the part is moved after 5.0 s of each period.

82. The length L (in m) of the shadow of a tree 15 m tall is $L = 15\cot\theta$, where θ is the angle of elevation of the sun. Approximate the change in L if θ changes from 50° to 52°.

83. An analysis of temperature records for Sydney, Australia, indicates that the average daily temperature (in °C) during the year is given approximately by $T = 17.2 + 5.2\cos\left[\frac{\pi}{6}(x - 0.50)\right]$, where x is measured in months ($x = 0.5$ is Jan. 15, etc.). What is the *daily* time rate of change of temperature on March 1? (*Hint*: 12 months/365 days = 0.033 month/day = dx/dt.)

84. An Earth-orbiting satellite is launched such that its altitude (in mi) is given by $y = 150(1 - e^{-0.05t})$, where t is the time (in min). Find the vertical velocity of the satellite for $t = 10.0$ min.

85. Power P is the time rate of change of doing work W. If work is being done in an electric circuit according to $W = 25\sin^2 2t$, find P as a function of t.

86. A bank account that has continuous compounding at an annual interest rate r has a value $A = A_0e^{rt}$, where A_0 is the original amount invested, and t is the time in years. Show that the account grows at a rate proportional to A. (Continuous compounding means interest is being added continually, rather than after a specified length of time.)

87. In determining how to divide files on the hard disk of a computer, we can use the equation $n = xN\log_x N$. Sketch the graph of n as a function of x for $1 < x \le 10$ if $N = 8$.

88. Under certain conditions, the potential V (in V) due to a magnet is given by $V = -k\ln\left(1 + \frac{L}{x}\right)$, where L is the length of the magnet and x is the distance from the point where the potential is measured. Find the expression for dV/dx.

89. In the theory of making images by holography, an expression used for the light-intensity distribution is $I = kE_0^2\cos^2\frac{1}{2}\theta$, where k and E_0 are constant and θ is the phase angle between two light waves. Find the expression for $dI/d\theta$.

90. Neglecting air resistance, the range R of a bullet fired at an angle θ with the horizontal is $R = \frac{v_0^2}{g}\sin 2\theta$, where v_0 is the initial velocity and g is the acceleration due to gravity. Find θ for the maximum range. See Fig. 27.37.

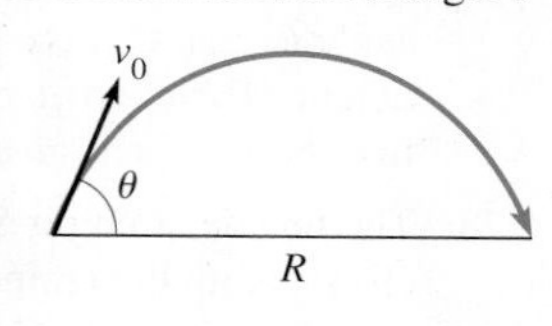

Fig. 27.37

91. In the design of a cone-type clutch, an equation that relates the cone angle θ and the applied force F is $\theta = \sin^{-1}(Ff/R)$, where R is the frictional resistance and f is the coefficient of friction. For constant R and f, find $d\theta/dF$.

92. If inflation makes the dollar worth 5% less each year, then the value of \$100 in t years will be $V = 100(0.95)^t$. What is the approximate change in the value during the fourth year?

93. An object attached to a cord of length l, as shown in Fig. 27.38, moves in a circular path. The angular velocity ω is given by $\omega = \sqrt{g/(l\cos\theta)}$. By use of differentials, find the approximate change in ω if θ changes from 32.50° to 32.75°, given that $g = 9.800$ m/s^2 and $l = 0.6375$ m.

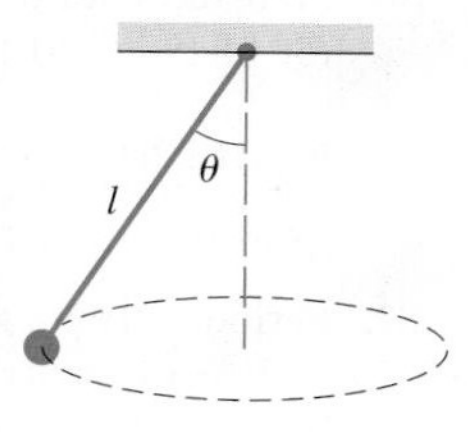

Fig. 27.38

94. An analysis of samples of air for a city showed that the number of parts per million p of sulfur dioxide on a certain day was $p = 0.05\ln(2 + 24t - t^2)$, where t is the hour of the day. Using differentials, find the approximate change in the amount of sulfur dioxide between 10 A.M. and noon.

95. According to Newton's law of cooling (Isaac Newton, again), the rate at which a body cools is proportional to the difference in temperature between it and the surrounding medium. By use of this law, the temperature T (in °F) of an engine coolant as a function of the time t (in min) is $T = 80 + 120(0.5)^{0.200t}$. The coolant was initially at 200°F, and the air temperature was 80°F. Linearize this function for $t = 5.00$ min and display the graphs of $T = f(t)$ and $L(t)$ on a calculator.

96. The charge q on a certain capacitor in an amplifier circuit as a function of time t is given by $q = e^{-0.1t}(0.2\sin 120\pi t + 0.8\cos 120\pi t)$. The current i in the circuit is the instantaneous time rate of change of the charge. Find the expression for i as a function of t.

97. A new 2017 car was purchased for \$32,000. Due to depreciation, its projected value V is given by $V = 32{,}000e^{-0.14t}$, where t is the number of years after 2017. Find the instantaneous rate at which the value is changing after (a) one year and (b) five years. Round to two significant digits.

98. A football is thrown horizontally (very little arc) at 56 ft/s parallel to the sideline. A TV camera is 92 ft from the path of the football. Find $d\theta/dt$, the rate at which the camera must turn to follow the ball when $\theta = 15°$. See Fig. 27.39.

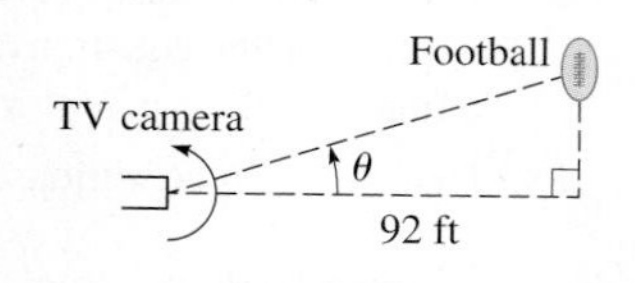

Fig. 27.39

99. An architect designs an arch of height y (in m) over a walkway by the curve of the equation $y = 3.00e^{-0.500x^2}$. What are the dimensions of the largest rectangular passage area under the arch?

100. A force P (in lb) at an angle θ above the horizontal drags a 50-lb box across a level floor. The coefficient of friction between the floor and the box is constant and equals 0.20. The magnitude of the force P is given by $P = \frac{(0.20)(50)}{0.20\sin\theta + \cos\theta}$. Find θ such that P is a minimum.

101. A jet is flying at 880 ft/s directly away from the control tower of an airport. If the jet is at a constant altitude of 6800 ft, how fast is the angle of elevation of the jet from the control tower changing when it is 13.0°?

102. The current i in an electric circuit with a resistance R and an inductance L is $i = i_0e^{-Rt/L}$, where i_0 is the initial current. Show that the time rate of change of the current is directly proportional to the current.

103. A silo constructed as shown in Fig. 27.40 is to hold 2880 m^3 of silage when completely full. It can be shown (can you?) that the surface area S (in m^2) (not including the base) is $S = 640 + 81\pi(\csc\theta - \frac{2}{3}\cot\theta)$. Find θ such that S is a minimum.

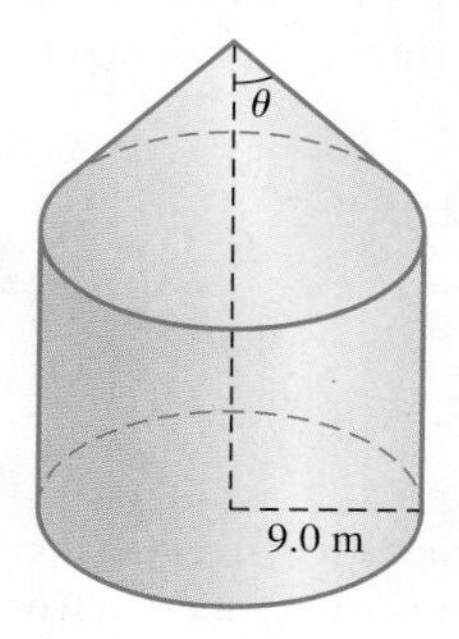

Fig. 27.40

104. Light passing through a narrow slit forms patterns of light and dark (see Fig. 27.41). The intensity I of the light at an angle θ is given by $I = I_0\left[\frac{\sin(k\sin\theta)}{k\sin\theta}\right]^2$, where k and I_0 are constants. Show that the maximum and minimum values of I occur for $k\sin\theta = \tan(k\sin\theta)$.

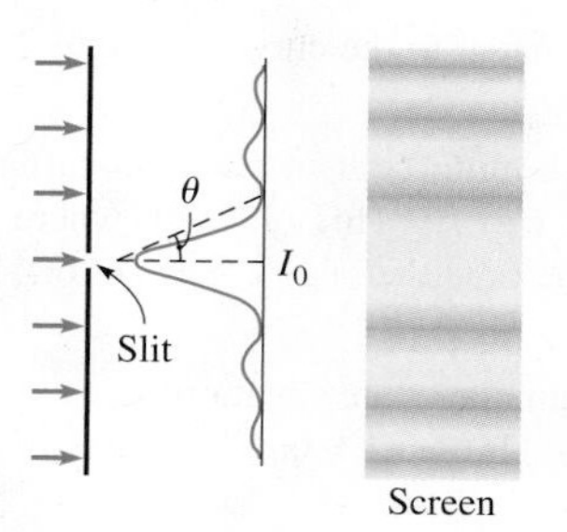

Fig. 27.41

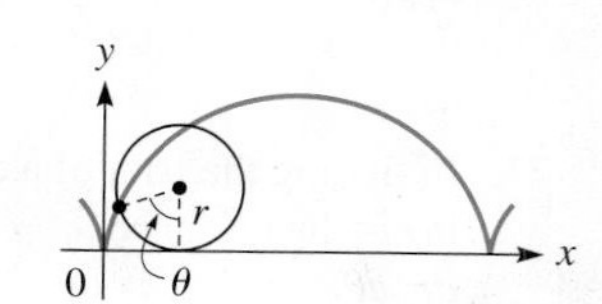

Fig. 27.42

105. When a wheel rolls along a straight line, a point P on the circumference traces a curve called a *cycloid.* See Fig. 27.42. The parametric equations of a cycloid are $x = r(\theta - \sin\theta)$ and $y = r(1 - \cos\theta)$. Find the velocity of the point on the rim of a wheel for which $r = 5.500$ cm and $d\theta/dt = 0.12$ rad/s for $\theta = 35.0°$.

106. In the study of atomic spectra, it is necessary to solve the equation $x = 5(1 - e^{-x})$ for x. Use Newton's method to find the solution.

107. The illuminance from a point source of light varies directly as the cosine of the angle of incidence (measured from the perpendicular) and inversely as the square of the distance r from the source. How high above the center of a circle of radius 10.0 in. should a light be placed so that illuminance at the circumference will be a maximum? See Fig. 27.43.

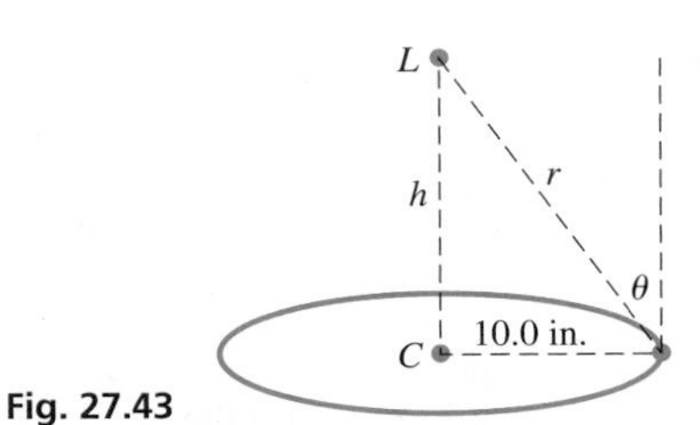

Fig. 27.43

108. A Y-shaped metal bracket is to be made such that its height is 10.0 cm and its width across the top is 6.00 cm. What shape will require the least amount of material? See Fig. 27.44.

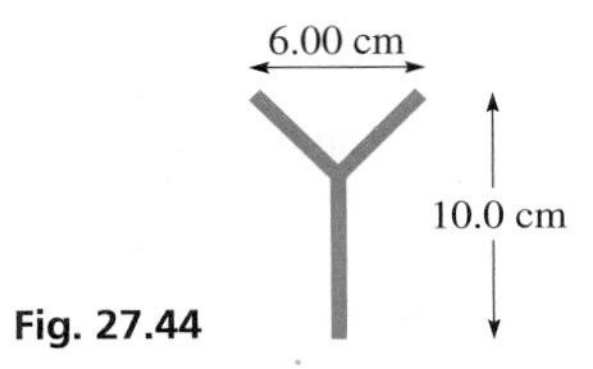

Fig. 27.44

109. A gutter is to be made from a sheet of metal 12 in. wide by turning up strips of width 4 in. along each side to make equal angles θ with the vertical. Sketch a graph of the cross-sectional area A as a function of θ. See Fig. 27.45.

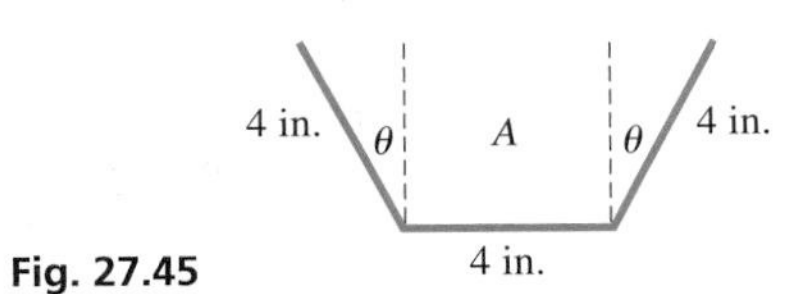

Fig. 27.45

110. The displacement y (in cm) of a weight on a spring in water is given by $y = 3.0te^{-0.20t}$, where t is the time (in s). What is the maximum displacement? (For this type of displacement, the motion is called *critically damped,* as the weight returns to its equilibrium position as quickly as possible without oscillating.)

111. Show that the equation of the hyperbolic cosine function $y = \frac{H}{w}\cosh\frac{wx}{H}$ (w and H are constants) satisfies the equation

$$\frac{d^2y}{dx^2} = \frac{w}{H}\sqrt{1 + \left(\frac{dy}{dx}\right)^2}$$

(see Exercise 53 of Section 27.6). A *catenary* is the curve of a uniform cable hanging under its own weight and is in the shape of a hyperbolic cosine curve. Also, this shape (inverted) was chosen for the St. Louis Gateway Arch (shown in Fig. 27.46) and makes the arch self-supporting.

Fig. 27.46

112. A conical filter is made from a circular piece of wire mesh of radius 24.0 cm by cutting out a sector with central angle θ and then taping the cut edges of the remaining piece together (see Fig. 27.47). What is the maximum possible volume the resulting filter can hold?

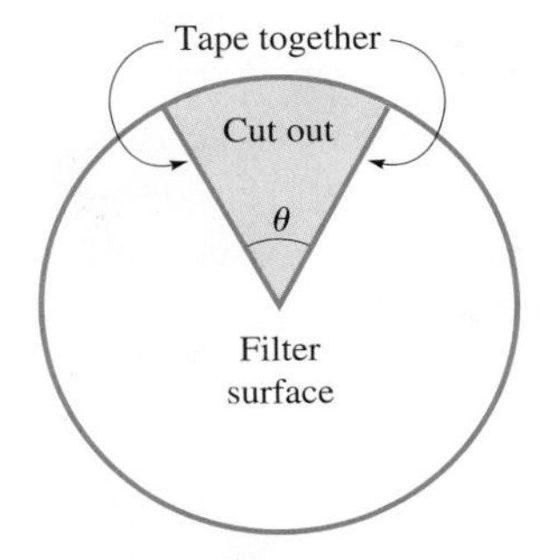

Fig. 27.47

113. To find the area of the largest rectangular microprocessor chip with a perimeter of 40 mm, it is possible to use either an algebraic function or a trigonometric function. Write two or three paragraphs to explain how each type of function can be used to find the required area.

CHAPTER 27 PRACTICE TEST

As a study aid, we have included complete solutions for each Practice Test problem at the end of this book.

In Problems 1–3, find the derivative of each of the functions.

1. $y = \tan^3 2x + \tan^{-1} 2x$

2. $y = 2(3 + \cot 4x)^3$

3. $y\sec 2x = \sin^{-1} 3y$

4. Find the differential of the function $y = \dfrac{\cos^2(3x + 1)}{x}$.

5. Find the slope of a tangent to the curve of $y = \ln\dfrac{2x - 1}{1 + x^2}$ for $x = 2$.

6. Find the expression for the time rate of change of electric current that is given by the equation $i = 8e^{-t}\sin 10t$, where t is the time.

7. Sketch the graph of the function $y = xe^x$.

8. A balloon leaves the ground 250 ft from an observer and rises at the rate of 5.0 ft/s. How fast is the angle of elevation of the balloon increasing after 8.0 s?

9. Find $\lim\limits_{x\to 0}\dfrac{\tan^{-1}x}{2\sin x}$ using L'Hospital's rule.

28

Methods of Integration

LEARNING OUTCOMES

After completion of this chapter, the student should be able to:

- Evaluate integrals by any of the following methods:
  - the power rule for integration
  - integration using basic logarithmic and exponential forms
  - integration using basic trigonometric and inverse trigonometric forms
  - integration by parts
  - trigonometric substitution
  - integration by partial fractions
  - using tables
- Solve application problems involving integration

In developing calculus, mathematicians saw that integration and differentiation were inverse processes, and that many integrals could be formed by finding the antiderivative. However, many of the integrals that arose mathematically and from the study of mechanical systems did not fit a form from which the antiderivative could be found directly. This led to the creation of numerous methods to integrate various types of functions.

Some of these methods of integration had been developed by the early 1700s, and were used by many mathematicians, including Newton and Leibniz. By the mid-1700s, many of these methods were included in textbooks. One of these texts was written by the Italian mathematician Maria Agnesi. Her text included analytic geometry, differential calculus, integral calculus, and some advanced topics and was noted for clear and organized explanations with many examples. Another important set of textbooks was written by Euler, who wrote separate texts on precalculus topics, differential calculus, and integral calculus. These texts were also noted for clear and organized presentations and were widely used until the early 1800s. Euler's integral calculus text included most of the methods of integration presented in this chapter.

Being able to integrate functions by using special methods, as well as by directly using antiderivatives, made integral calculus much more useful in developing many areas of geometry, science, and technology in the 1800s and 1900s. In earlier chapters, we have noted a number of applications of integration, and additional examples are found in the examples and exercises of this chapter.

In this chapter, we expand the use of the power rule for integration for use with integrands that include transcendental functions and several special methods of integration for integrands that do not directly fit standard forms. In using all forms of integration, *recognition of the integral form* is of great importance.

▶ **In Section 28.5, we show an application of integration that is important in the design of electronic equipment and electric appliances.**

28.1 The Power Rule for Integration

Integration Using Power Rule • Recognizing u, n, and du • Special Care with du

The first formula for integration that we will discuss is the power rule, and we will expand its use to include transcendental integrands. It was first introduced with the integration of basic algebraic forms in Chapter 25 and is repeated here for reference.

$$\int u^n\,du = \frac{u^{n+1}}{n+1} + C \qquad (n \neq -1) \tag{28.1}$$

In applying Eq. (28.1) to transcendental integrands, as well as with algebraic integrands, ***we must properly recognize the quantities u, n, and du.*** This requires familiarity with the differential forms of Chapters 23 and 27.

EXAMPLE 1 Trigonometric integrand

Integrate: $\int \sin^3 x \cos x\,dx$.

Because $d(\sin x) = \cos x\,dx$, we note that this integral fits the form of Eq. (28.1) for $u = \sin x$. Thus, with $u = \sin x$, we have $du = \cos x\,dx$, which means that this integral is of the form $\int u^3\,du$. Therefore, the integration can now be completed:

$$\int \sin^3 x \cos x\,dx = \int \sin^3 x(\cos x\,dx) \qquad u = \sin x \quad (\cos x\,dx \text{ is } du)$$

$$= \frac{1}{4}\sin^4 x + C \quad \leftarrow \text{do not forget the constant of integration}$$

CAUTION We note here that the factor $\cos x$ is a necessary part of the du in order to have the proper form of integration and therefore does not appear in the final result. ■

We check our result by finding the derivative of $\frac{1}{4}\sin^4 x + C$, which is

$$\frac{d}{dx}\left(\frac{1}{4}\sin^4 x + C\right) = \frac{1}{4}(4)\sin^3 x \cos x$$

$$= \sin^3 x \cos x$$

■

Practice Exercise

1. Integrate: $\int \cos^2 x \sin x\,dx$.

EXAMPLE 2 Trigonometric integrand

Integrate: $\int 2\sqrt{1 + \tan\theta}\sec^2\theta\,d\theta$.

Here, we note that $d(\tan\theta) = \sec^2\theta\,d\theta$, which means that the integral fits the form of Eq. (28.1) with

$$u = 1 + \tan\theta \qquad du = \sec^2\theta\,d\theta \qquad n = \tfrac{1}{2}$$

The integral is of the form $\int u^{1/2}\,du$. Thus,

$$\int 2\sqrt{1 + \tan\theta}(\sec^2\theta\,d\theta) = 2\int (1 + \tan\theta)^{1/2}(\sec^2\theta\,d\theta) \quad (u = 1 + \tan\theta,\ du = \sec^2\theta\,d\theta)$$

$$= 2\left(\frac{2}{3}\right)(1 + \tan\theta)^{3/2} + C$$

$$= \frac{4}{3}(1 + \tan\theta)^{3/2} + C$$

■

EXAMPLE 3 Logarithmic integral

Integrate: $\int \ln x\left(\frac{dx}{x}\right)$.

By noting that $d(\ln x) = \frac{dx}{x}$, we have: $u = \ln x \quad du = \frac{dx}{x} \quad n = 1$

This means that the integral is of the form $\int u\,du$. Thus,

$$\int \ln x\left(\frac{dx}{x}\right) = \frac{1}{2}(\ln x)^2 + C = \frac{1}{2}\ln^2 x + C$$ ■

Practice Exercise

2. Integrate: $\int \frac{2\ln 2x\,dx}{x}$.

EXAMPLE 4 Inverse trigonometric integrand

Find the value of $\int_0^{0.5} \frac{\sin^{-1}x}{\sqrt{1-x^2}}dx$.

For purposes of integrating: $u = \sin^{-1}x \quad du = \frac{dx}{\sqrt{1-x^2}} \quad n = 1$

$$\int_0^{0.5} \frac{\sin^{-1}x}{\sqrt{1-x^2}}dx = \int_0^{0.5} \sin^{-1}x\left(\frac{dx}{\sqrt{1-x^2}}\right) \qquad \int u\,du$$

$$= \frac{(\sin^{-1}x)^2}{2}\Bigg|_0^{0.5} \qquad \text{integrate}$$

$$= \frac{\left(\frac{\pi}{6}\right)^2}{2} - 0 = \frac{\pi^2}{72} \qquad \text{evaluate}$$ ■

EXAMPLE 5 Exponential integrand—area under a curve

Find the first-quadrant area bounded by $y = \frac{e^{2x}}{\sqrt{e^{2x}+1}}$ and $x = 1.5$.

The area is shown in the calculator display in Fig. 28.1. Using a representative element of area $y\,dx$, the area is found by evaluating the integral

$$\int_0^{1.5} \frac{e^{2x}\,dx}{\sqrt{e^{2x}+1}}$$

For the purpose of integration, $n = -\frac{1}{2}$, $u = e^{2x} + 1$, and $du = 2e^{2x}\,dx$. Therefore,

$$\int_0^{1.5} (e^{2x}+1)^{-1/2}e^{2x}\,dx = \frac{1}{2}\int_0^{1.5} (e^{2x}+1)^{-1/2}(2e^{2x}\,dx)$$

$$= \frac{1}{2}(2)(e^{2x}+1)^{1/2}\Bigg|_0^{1.5} = (e^{2x}+1)^{1/2}\Bigg|_0^{1.5}$$

$$= \sqrt{e^3+1} - \sqrt{2} = 3.178$$

This result agrees with the use of the $\int f(x)\,dx$ calculator feature shown in Fig. 28.1. ■

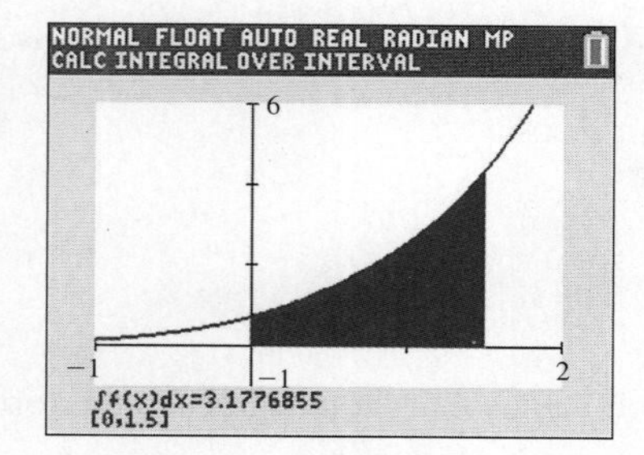

Fig. 28.1
Graphing calculator keystrokes: goo.gl/aB2UIP

EXERCISES 28.1

In Exercises 1 and 2, make the given changes in the indicated examples of this section and then solve the given problems.

1. In Example 1, change $\sin^3 x\cos x$ to $\cos^3 x\sin x$ and then integrate.
2. In Example 3, change $\ln x$ to $\ln x^2$ and then integrate.

In Exercises 3–28, integrate each of the functions.

3. $\int (x^2+1)^3(2x\,dx)$
4. $\int (x^3-2)^6(3x^2\,dx)$
5. $\int \sin^4 x\cos x\,dx$
6. $\int \cos^5 x(-\sin x\,dx)$

7. $\int 0.4\sqrt{\cos\theta}\sin\theta\, d\theta$

8. $\int 8\sin^{1/3}x\cos x\, dx$

9. $\int 4\tan^2 x\sec^2 x\, dx$

10. $\int 3\sec^3 x(\sec x\tan x)dx$

11. $\int_0^{\pi/8} \frac{\cos 2x}{\csc 2x}dx$

12. $\int_{\pi/6}^{\pi/4} 3\sqrt{\cot x}\csc^2 x\, dx$

13. $\int (\sin^{-1}x)^3\left(\frac{dx}{\sqrt{1-x^2}}\right)$

14. $\int \frac{20(\cos^{-1}2t)^4\, dt}{\sqrt{1-4t^2}}$

15. $\int \frac{5\tan^{-1}5x}{25x^2+1}dx$

16. $\int \frac{\sin^{-1}4x\, dx}{\sqrt{1-16x^2}}$

17. $\int [\ln(x+1)]^2\frac{dx}{x+1}$

18. $\int 0.8(3+2\ln u)^3\frac{du}{u}$

19. $\int_0^{1/2} \frac{\ln(2x+3)}{4x+6}dx$

20. $\int_1^e \frac{(1-2\ln x)dx}{4x}$

21. $\int 3(4+e^x)^3 e^x\, dx$

22. $\int 2\sqrt{1-e^{-x}}(-e^{-x}\, dx)$

23. $\int \frac{4e^{2t}\, dt}{(1-e^{2t})^3}$

24. $\int \frac{(1+3e^{-2x})^4\, dx}{6e^{2x}}$

25. $\int (1+\sec^2 x)^4(\sec^2 x\tan x\, dx)$

26. $\int (e^x+e^{-x})^{1/4}(e^x-e^{-x})\, dx$

27. $\int_{\pi/6}^{\pi/4} (1+\cot x)^2\csc^2 x\, dx$

28. $\int_{\pi/3}^{\pi/2} \frac{2\sin\theta\, d\theta}{\sqrt{1+\cos\theta}}$

In Exercises 29–32, rewrite the given integrals so that they fit the form $\int u^n\, du$, and identify u, n, and du.

29. $\int \sec^5 x\sin x\, dx$

30. $\int \frac{\tan^3 x\, dx}{\cos^2 x}$

31. $\int \frac{dx}{x\ln^2 x}$

32. $\int \frac{e^{-1/x}}{x^2}dx$

In Exercises 33–44, solve the given problems by integration.

33. Find the first-quadrant area under the curve of $y = \ln^2 x/x$ from $x = 1$ to $x = 4$.

34. Find the area under $y = 6\sin^2 x\cos x$ from $x = 0$ to $x = \pi/2$.

35. Find the volume generated when the first-quadrant area bounded by $y = e^x$ and $x = 2$ is rotated about the x-axis. [*Hint:* After setting up the integral, rewrite $(e^x)^2$ as $(e^x)(e^x)$.]

36. Find the volume generated if the first-quadrant region bounded by $y = e^{2x}$ and $x = 1$ is revolved about the x-axis. See the hint in Exercise 35.

37. Find the area under the curve $y = \frac{1+\tan^{-1}2x}{1+4x^2}$ from $x = 0$ to $x = 2$.

38. Find the first-quadrant area bounded by $y = \frac{\ln(4x+1)}{4x+1}$ and $x = 5$.

39. The general expression for the slope of a given curve is $(\ln x)^2/x$. If the curve passes through $(1, 2)$, find its equation.

40. Find an equation of the curve for which $dy/dx = (1+\tan 2x)^2\sec^2 2x$ if the curve passes through $(2, 1)$.

41. In the development of the expression for the total pressure P on a wall due to molecules with mass m and velocity v striking the wall, the equation $P = mnv^2\int_0^{\pi/2}\sin\theta\cos^2\theta\, d\theta$ is found. The symbol n represents the number of molecules per unit volume, and θ represents the angle between a perpendicular to the wall and the direction of the molecule. Find the expression for P.

42. The solar energy E passing through a hemispherical surface per unit time, per unit area, is $E = 2\pi I\int_0^{\pi/2}\cos\theta\sin\theta\, d\theta$, where I is the solar intensity and θ is the angle at which it is directed (from the perpendicular). Evaluate this integral.

43. After an electric power interruption, the current i in a circuit is given by $i = 3(1-e^{-t})^2(e^{-t})$, where t is the time. Find the expression for the total electric charge q to pass a point in the circuit if $q = 0$ for $t = 0$.

44. A space vehicle is launched vertically from the ground such that its velocity v (in km/s) is given by $v = [\ln^2(t^3+1)]\frac{t^2}{t^3+1}$, where t is the time (in s). Find the altitude of the vehicle after 10.0 s.

Answers to Practice Exercises

1. $-\frac{1}{3}\cos^3 x + C$ **2.** $\ln^2 2x + C$

28.2 The Basic Logarithmic Form

Integration of du/u • Result Is Absolute Value of u

The power rule for integration, Eq. (28.1), is valid for all values of n except $n = -1$. If n were set equal to -1, this would cause the result to be undefined. When we obtained the derivative of the logarithmic function, we found

$$\frac{d(\ln u)}{dx} = \frac{1}{u}\frac{du}{dx}$$

This means the differential of the logarithmic form is $d(\ln u) = du/u$. Reversing the process, we then determine that $\int du/u = \ln u + C$. In other words, when the exponent of the expression being integrated is -1, the expression is a logarithmic form.

Logarithms are defined only for positive numbers. Thus, $\int du/u = \ln u + C$ is valid if $u > 0$. If $u < 0$, then $-u > 0$. In this case, $d(-u) = -du$, or $\int(-du)/(-u) = \ln(-u) + C$. However, $\int du/u = \int(-du)/(-u)$. [These results can be combined into a single form using the absolute value of u.] Therefore,

NOTE ▶

$$\int \frac{du}{u} = \ln|u| + C \tag{28.2}$$

We will refer to Eq. (28.2) as the **log rule for integration.**

EXAMPLE 1 Algebraic integrand

Integrate: $\int \frac{dx}{x+1}$.

Because $d(x + 1) = dx$, this integral fits the form of Eq. (28.2) with $u = x + 1$ and $du = dx$. Therefore, we have

$$\int \frac{dx}{x+1} = \ln|x+1| + C$$

(annotations: dx ← du; $x + 1$ ← u) ■

EXAMPLE 2 Algebraic integrand—cooling object

Newton's law of cooling states that the rate at which an object cools is directly proportional to the difference in its temperature T and the temperature of the surrounding medium. By use of this law, the time t (in min) a hot plate from a dishwasher takes to cool from 80°C to 50°C in air at 20°C is found to be

$$t = -9.8\int_{80}^{50} \frac{dT}{T - 20}$$

Find the value of t.

We see that the integral fits Eq. (28.2) with $u = T - 20$ and $du = dT$. Thus,

$$\begin{aligned} t &= -9.8\int_{80}^{50} \frac{dT}{T - 20} && \text{(} dT \leftarrow du\text{; } T-20 \leftarrow u\text{)} \\ &= -9.8\ln|T - 20|\Big|_{80}^{50} && \text{integrate} \\ &= -9.8(\ln 30 - \ln 60) && \text{evaluate} \\ &= -9.8\ln\frac{30}{60} = -9.8\ln(0.50) && \ln x - \ln y = \ln\frac{x}{y} \\ &= 6.8 \text{ min} \end{aligned}$$

■

Practice Exercise

1. Integrate: $\int \frac{3\,dx}{5 - 2x}$.

EXAMPLE 3 Trigonometric integrand

Integrate: $\int \frac{\cos x}{\sin x}dx$.

We note that $d(\sin x) = \cos x\,dx$. This means that this integral fits the form of Eq. (28.2) with $u = \sin x$ and $du = \cos x\,dx$. Thus,

$$\begin{aligned} \int \frac{\cos x}{\sin x}dx &= \int \frac{\cos x\,dx}{\sin x} && \text{(} \cos x\,dx \leftarrow du\text{; } \sin x \leftarrow u\text{)} \\ &= \ln|\sin x| + C \end{aligned}$$

■

EXAMPLE 4 Algebraic integrand—note exponent

Integrate: $\int \frac{x\,dx}{4 - x^2}$.

This integral fits the form of Eq. (28.2) with $u = 4 - x^2$ and $du = -2x\,dx$. This means that we must introduce a factor of -2 into the numerator and a factor of $-\frac{1}{2}$ before the integral. Therefore,

$$\int \frac{x\,dx}{4 - x^2} = -\frac{1}{2}\int \frac{-2x\,dx}{4 - x^2} \begin{matrix}\leftarrow du \\ \leftarrow u\end{matrix}$$

$$= -\frac{1}{2}\ln|4 - x^2| + C$$

CAUTION We should note that if the quantity $4 - x^2$ were raised to any power other than 1, we would have to employ the power rule for integration. ■ For example,

$$\int \frac{x\,dx}{(4 - x^2)^2} = -\frac{1}{2}\int \frac{-2x\,dx}{(4 - x^2)^2} \begin{matrix}\leftarrow du \\ \leftarrow u^2\end{matrix}$$

$$= -\frac{1}{2}\frac{(4 - x^2)^{-1}}{-1} + C = \frac{1}{2(4 - x^2)} + C$$

■

EXAMPLE 5 Exponential integrand

Integrate: $\int \frac{e^{4x}\,dx}{1 + 3e^{4x}}$.

Because $d(1 + 3e^{4x})/dx = 12e^{4x}$, we see that we can use the log rule of integration with $u = 1 + 3e^{4x}$ and $du = 12e^{4x}\,dx$. Therefore, we write

$$\int \frac{e^{4x}\,dx}{1 + 3e^{4x}} = \frac{1}{12}\int \frac{12e^{4x}\,dx}{1 + 3e^{4x}} \quad \text{introduce factors of 12}$$

$$= \frac{1}{12}\ln|1 + 3e^{4x}| + C \quad \text{integrate}$$

$$= \frac{1}{12}\ln(1 + 3e^{4x}) + C \quad 1 + 3e^{4x} > 0 \text{ for all } x$$

■

Practice Exercise

2. Integrate: $\int \frac{\sin x\,dx}{1 - \cos x}$.

EXAMPLE 6 Trigonometric integrand—definite integral

Evaluate: $\int_0^{\pi/8} \frac{\sec^2 2\theta}{1 + \tan 2\theta}\,d\theta$.

Because $d(1 + \tan 2\theta) = 2\sec^2 2\theta\,d\theta$, we can use the log rule of integration with $u = 1 + \tan 2\theta$ and $du = 2\sec^2 2\theta\,d\theta$. Therefore, we have

$$\int_0^{\pi/8} \frac{\sec^2 2\theta}{1 + \tan 2\theta}\,d\theta = \frac{1}{2}\int_0^{\pi/8} \frac{2\sec^2 2\theta\,d\theta}{1 + \tan 2\theta} \quad \text{introduce factors of 2}$$

$$= \frac{1}{2}\ln|1 + \tan 2\theta|\Big|_0^{\pi/8} \quad \text{integrate}$$

$$= \frac{1}{2}(\ln|1 + 1| - \ln|1 + 0|) \quad \text{evaluate}$$

$$= \frac{1}{2}(\ln 2 - \ln 1) = \frac{1}{2}(\ln 2 - 0)$$

$$= \frac{1}{2}\ln 2$$

■

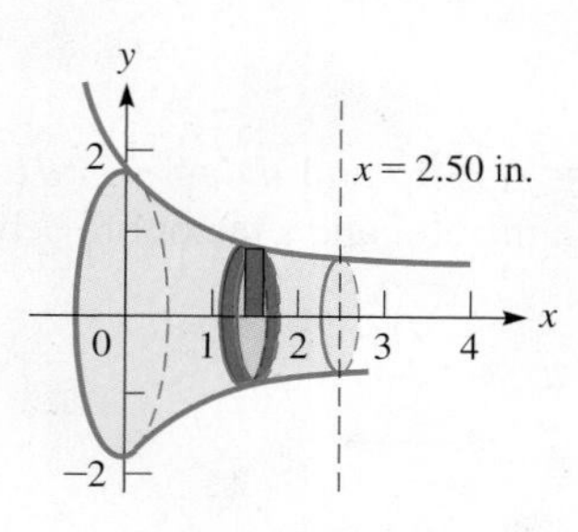

Fig. 28.2

EXAMPLE 7 Find volume using logarithmic form

Find the volume within the piece of tapered tubing shown in Fig. 28.2, which can be described as the volume generated by revolving the region bounded by the curve of $y = \frac{3}{\sqrt{4x+3}}$, $x = 2.50$ in., and the axes about the x-axis.

The volume can be found by setting up only one integral by using a disk element of volume, as shown. The volume is found as follows:

$$V = \pi\int_0^{2.50} y^2\,dx = \pi\int_0^{2.50}\left(\frac{3}{\sqrt{4x+3}}\right)^2 dx$$

$$= \pi\int_0^{2.50}\frac{9\,dx}{4x+3} = \frac{9\pi}{4}\int_0^{2.50}\frac{4\,dx}{4x+3}$$

$$= \frac{9\pi}{4}\ln(4x+3)\Big|_0^{2.50} = \frac{9\pi}{4}(\ln 13.0 - \ln 3)$$

$$= \frac{9\pi}{4}\ln\frac{13.0}{3} = 10.4 \text{ in.}^3$$

■

EXERCISES 28.2

In Exercises 1 and 2, make the given changes in the indicated examples of this section and then solve the given problems.

1. In Example 1, change $x + 1$ to $2x + 1$ and then integrate.
2. In Example 3, interchange $\cos x$ and $\sin x$ and then integrate.

In Exercises 3–30, integrate each of the given functions.

3. $\int \frac{dx}{1+4x}$
4. $\int \frac{dx}{1-4x}$
5. $\int \frac{2x\,dx}{4-3x^2}$
6. $\int \frac{y^2}{y^3-1}dy$
7. $\int_0^2 \frac{6\,dx}{8-3x}$
8. $\int_{-1}^3 \frac{8x^3\,dx}{x^4+1}$
9. $\int \frac{0.4\csc^2 2\theta\,d\theta}{\cot 2\theta}$
10. $\int \frac{9\sin 3x}{\cos 3x}dx$
11. $\int_0^{\pi/2} \frac{\cos x\,dx}{1+\sin x}$
12. $\int_0^{\pi/4} \frac{2\sec^2 x\,dx}{4+\tan x}$
13. $\int \frac{e^{-x}}{1-e^{-x}}dx$
14. $\int \frac{15e^{3x}}{1-e^{3x}}dx$
15. $\int \frac{1+e^x}{x+e^x}dx$
16. $\int \frac{4e^{2t}}{e^{2t}+4}dt$
17. $\int \frac{8\sec x\tan x\,dx}{1+4\sec x}$
18. $\int \frac{3\csc x\cot x}{\csc x+2}dx$
19. $\int_1^3 \frac{1+x}{4x+2x^2}dx$
20. $\int_1^2 \frac{4x+6x^2}{x^2+x^3}dx$
21. $\int \frac{dr}{r\ln r}$
22. $\int \frac{6\,dx}{x(1+2\ln x)}$
23. $\int \frac{2+\sec^2 x}{2x+\tan x}dx$
24. $\int \frac{x+\cos 2x}{x^2+\sin 2x}dx$
25. $\int \frac{16\,dx}{\sqrt{1-2x}}$
26. $\int \frac{4x\,dx}{(1+x^2)^2}$
27. $\int \frac{x+2}{x^2}dx$
28. $\int \frac{3v^2-2v}{v^2}dv$
29. $\int_0^{\pi/12} \frac{\sec^2 3x}{4+\tan 3x}dx$
30. $\int_1^2 \frac{x^2+1}{x^3+3x}dx$

In Exercises 31–50, solve the given problems by integration.

31. Find the area bounded by $y = \frac{1}{x+1}$, $x = 0$, $y = 0$, and $x = 2$. See Fig. 28.3.

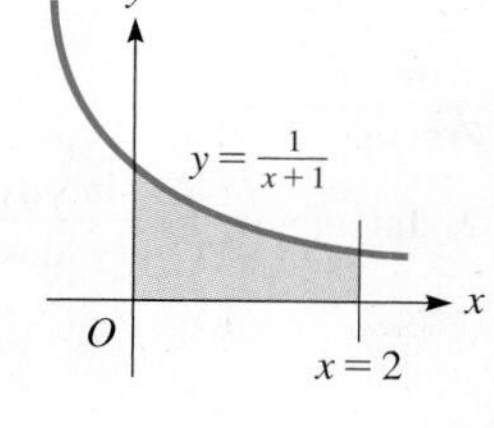

Fig. 28.3

32. Evaluate $\int_1^2 x^{-1}\,dx$ and $\int_2^4 x^{-1}\,dx$. Give a geometric interpretation of these two results.
33. Integrate $\int \frac{x-4}{x+4}dx$ by first using algebraic division to change the form of the integrand.
34. Integrate $\int \sec x\,dx$ by first multiplying the integrand by $\frac{\sec x + \tan x}{\sec x + \tan x}$.
35. Find the volume generated by revolving the region bounded by $y = 1/(x^2+1)$, $x = 0$, $x = 1$, and $y = 0$ about the y-axis. Use shells.
36. Show that $\int_0^3 \frac{dx}{2x+2} = \ln 2$.

37. The general expression for the slope of a curve is $\frac{\sin x}{3 + \cos x}$. If the curve passes through the point $(\pi/3, 2)$, find its equation.

38. Find the volume of the solid generated by revolving the region bounded by $y = \frac{2}{\sqrt{3x + 1}}$, $x = 0$, $x = 3.5$, and $y = 0$ about the x-axis.

39. If $a > 0$ and $b > 0$, show that $\int_1^a \frac{du}{u} + \int_1^b \frac{du}{u} = \int_1^{ab} \frac{du}{u}$.

40. If $x > 0$, find $f(x)$ if $f''(x) = x^{-2}$, $f(1) = 0$, and $f(2) = 0$.

41. A marathon runner's speed (in km/h) is $v = \frac{12.0}{0.200t + 1}$. How far does the runner go in 3.00 h?

42. The population P of elk on a refuge is changing at a rate of $\frac{dP}{dt} = \frac{25.0}{1.00 + 0.100t}$, where t is the time in years. If the original population (when $t = 0$) was 125 elk, find the population 5 years later.

43. The acceleration (in m/s^2) of a rolling ball is $a = 8(t + 1)^{-1}$. Find its velocity for $t = 4.0$ s if its initial velocity is zero.

44. The pressure P (in kPa) and volume V (in cm^3) of a gas are related by $PV = 8600$. Find the average value of P from $V = 75\text{ cm}^3$ to $V = 95\text{ cm}^3$.

45. A hot metal rod with an initial temperature of 425°C is placed in a room with temperature 20°C. The time t (in min) required for the temperature T of the rod to cool to 175°C is given by $t = 8.72 \int_{175}^{425} \frac{1}{T - 20.0} dT$. Find this time.

46. In determining the temperature that is absolute zero (0 K, or about −273°C), the equation $\ln T = -\int \frac{dr}{r - 1}$ is used. Here, T is the thermodynamic temperature and r is the ratio between certain specific vapor pressures. If $T = 273.16$ K for $r = 1.3361$, find T as a function of r (if $r > 1$ for all T).

47. The time t and electric current i for a certain circuit with a voltage E, a resistance R, and an inductance L is given by $t = L \int \frac{di}{E - iR}$. If $t = 0$ for $i = 0$, integrate and express i as a function of t.

48. Conditions are often such that a force proportional to the velocity tends to retard the motion of an object moving through a resisting medium. Under such conditions, the acceleration of a certain object moving down an inclined plane is given by $20 - v$. This leads to the equation $t = \int \frac{dv}{20 - v}$. If the object starts from rest, find the expression for the velocity as a function of time.

49. An architect designs a wall panel that can be described as the first-quadrant area bounded by $y = \frac{50}{x^2 + 20}$ and $x = 3.00$. If the area of the panel is 6.61 m^2, find the x-coordinate (in m) of the centroid of the panel.

50. The power used by a robotic drilling machine is given by $p = 3 \int \frac{\sin \pi t}{2 + \cos \pi t} dt$, where t is the time. Find $p = f(t)$.

Answers to Practice Exercises

1. $-\frac{3}{2}\ln|5 - 2x| + C$ 2. $\ln|1 - \cos x| + C$

28.3 The Exponential Form

Integration of $e^u\,du$

In deriving the derivative for the exponential function, we obtained the result $de^u/dx = e^u(du/dx)$. This means that the differential of the exponential form is $d(e^u) = e^u\,du$. Reversing this form to find the proper form of the integral for the exponential function, we have

$$\int e^u\,du = e^u + C \tag{28.3}$$

EXAMPLE 1 Algebraic u

Integrate: $\int xe^{x^2}\,dx$.

Because $d(x^2) = 2x\,dx$, we can write this integral in the form of Eq. (28.3) with $u = x^2$ and $du = 2x\,dx$. Thus,

$$\int xe^{x^2}\,dx = \frac{1}{2}\int e^{x^2}(2x\,dx)$$

(with $u = x^2$ and $du = 2x\,dx$ indicated)

$$= \frac{1}{2}e^{x^2} + C$$

■

Practice Exercise

1. Integrate: $\int 6e^{-3x}\,dx$.

EXAMPLE 2 Algebraic u—current in an RL circuit

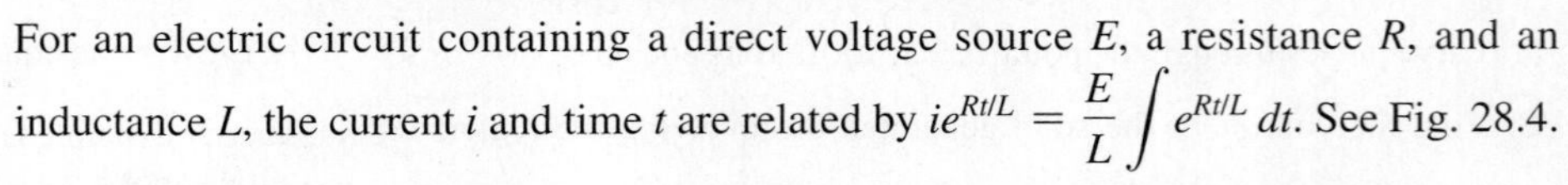

For an electric circuit containing a direct voltage source E, a resistance R, and an inductance L, the current i and time t are related by $ie^{Rt/L} = \frac{E}{L}\int e^{Rt/L}\,dt$. See Fig. 28.4. If $i = 0$ for $t = 0$, perform the integration and then solve for i as a function of t.

For this integral, we see that $u = \frac{Rt}{L}$, which means that $du = \frac{R\,dt}{L}$. The solution is then as follows:

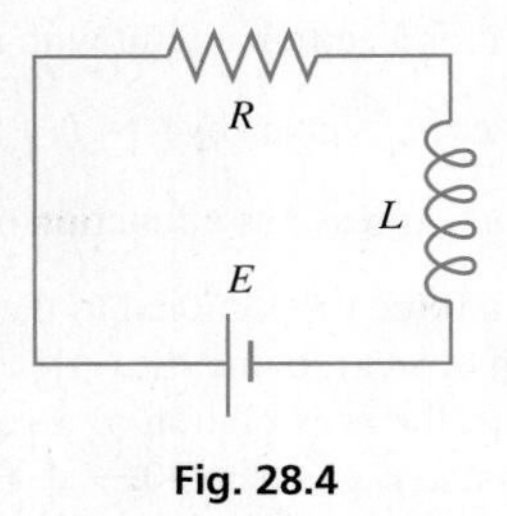

Fig. 28.4

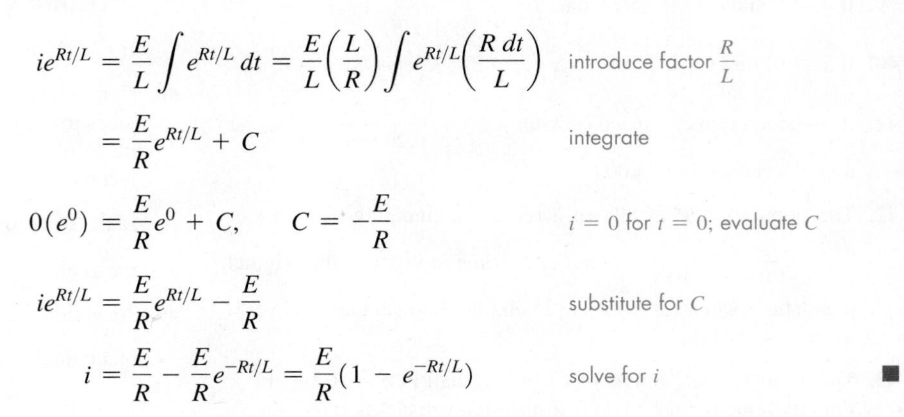

$$ie^{Rt/L} = \frac{E}{L}\int e^{Rt/L}\,dt = \frac{E}{L}\left(\frac{L}{R}\right)\int e^{Rt/L}\left(\frac{R\,dt}{L}\right) \qquad \text{introduce factor } \frac{R}{L}$$

$$= \frac{E}{R}e^{Rt/L} + C \qquad \text{integrate}$$

$$0(e^0) = \frac{E}{R}e^0 + C, \quad C = -\frac{E}{R} \qquad i = 0 \text{ for } t = 0\text{; evaluate } C$$

$$ie^{Rt/L} = \frac{E}{R}e^{Rt/L} - \frac{E}{R} \qquad \text{substitute for } C$$

$$i = \frac{E}{R} - \frac{E}{R}e^{-Rt/L} = \frac{E}{R}(1 - e^{-Rt/L}) \qquad \text{solve for } i$$

■

EXAMPLE 3 e^u in denominator

Integrate: $\int \frac{dx}{e^{3x}}$.

NOTE ▸ [This integral can be put in proper form by writing it as $\int e^{-3x}\,dx$.] In this form, $u = -3x$, $du = -3\,dx$. Thus,

$$\int \frac{dx}{e^{3x}} = \int e^{-3x}\,dx = -\frac{1}{3}\int e^{-3x}(-3\,dx)$$

$$= -\frac{1}{3}e^{-3x} + C$$

■

EXAMPLE 4 Proper form using laws of exponents

Integrate: $\int \frac{4e^{3x} - 3e^x}{e^{x+1}}dx$.

By using the laws of exponents, this integral can be put in proper form for integration, and then integrated as follows:

$$\int \frac{4e^{3x} - 3e^x}{e^{x+1}}dx = \int \frac{4e^{3x}}{e^{x+1}}dx - \int \frac{3e^x}{e^{x+1}}dx$$

$$= 4\int e^{3x-(x+1)}\,dx - 3\int e^{x-(x+1)}\,dx \qquad \text{using } \frac{a^m}{a^n} = a^{m-n}$$

$$= 4\int e^{2x-1}\,dx - 3\int e^{-1}\,dx$$

$$= \frac{4}{2}\int e^{2x-1}(2\,dx) - \frac{3}{e}\int dx$$

$$= 2e^{2x-1} - \frac{3}{e}x + C$$

■

EXAMPLE 5 Trigonometric u—definite integral

Evaluate: $\int_0^{\pi/2}(\sin 2\theta)(e^{\cos 2\theta})\,d\theta$.

With $u = \cos 2\theta$, $du = -2\sin 2\theta\,d\theta$, we have

$$\int_0^{\pi/2}(\sin 2\theta)(e^{\cos 2\theta})\,d\theta = -\frac{1}{2}\int_0^{\pi/2}(e^{\cos 2\theta})(-2\sin 2\theta\,d\theta)$$

$$= -\frac{1}{2}e^{\cos 2\theta}\Big|_0^{\pi/2} \qquad \text{integrate}$$

$$= -\frac{1}{2}\left(\frac{1}{e} - e\right) = 1.175 \qquad \text{evaluate}$$

NORMAL FLOAT AUTO REAL RADIAN MP

$\int_0^{\pi/2}(\sin(2X)e^{\cos(2X)})dX$

1.175201194

Fig. 28.5

Graphing calculator keystrokes: goo.gl/bzDLPC

Figure 28.5 shows a calculator check for this result. ■

EXAMPLE 6 Algebraic u—find equation of a curve

Find the equation of the curve for which $\frac{dy}{dx} = \frac{e^{\sqrt{x+1}}}{\sqrt{x+1}}$ if the curve passes through (0, 1).

The solution of this problem requires that we integrate the given function and then evaluate the constant of integration. Hence,

$$dy = \frac{e^{\sqrt{x+1}}}{\sqrt{x+1}}dx \qquad \int dy = \int \frac{e^{\sqrt{x+1}}}{\sqrt{x+1}}dx$$

For purposes of integrating the right-hand side,

$$u = \sqrt{x+1} \quad \text{and} \quad du = \frac{1}{2\sqrt{x+1}}dx$$

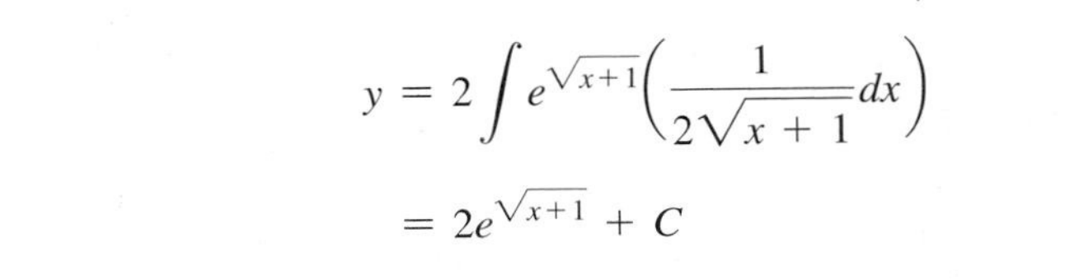

$$y = 2\int e^{\sqrt{x+1}}\left(\frac{1}{2\sqrt{x+1}}dx\right)$$

$$= 2e^{\sqrt{x+1}} + C$$

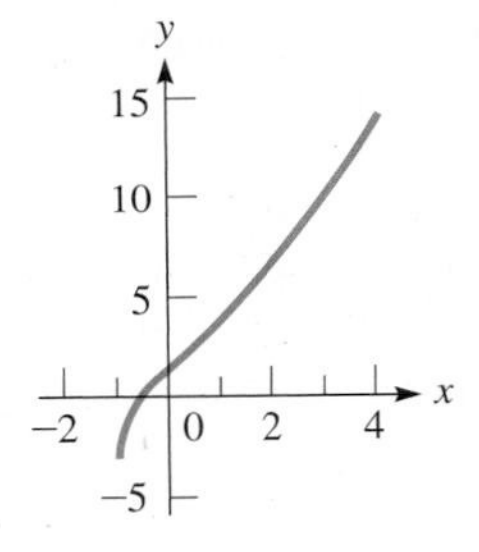

Fig. 28.6

Letting $x = 0$ and $y = 1$, we have $1 = 2e + C$, or $C = 1 - 2e$. This means that the equation is

$$y = 2e^{\sqrt{x+1}} + 1 - 2e \qquad \text{see Fig. 28.6}$$ ■

Practice Exercise

2. Integrate: $\int \frac{e^{\sin x}}{\sec x}dx$.

EXERCISES 28.3

In Exercises 1 and 2, make the given changes in the indicated examples of this section and then solve the given problems.

1. In Example 1, change x to x^2 and x^2 to x^3 and then integrate.

2. In Example 4, change e^{x+1} to e^{x-1} and then integrate.

In Exercises 3–28, integrate each of the given functions.

3. $\int e^{7x}(7\,dx)$

4. $\int e^{x^4}(4x^3\,dx)$

5. $\int 4e^{2x+5}\,dx$

6. $\int 2e^{-4x}\,dx$

7. $\int_{-2}^{2} 6e^{s/2}\,ds$

8. $\int_1^2 3e^{4x}\,dx$

9. $\int 6x^2e^{x^3}\,dx$

10. $\int \frac{8x}{e^{x^2}}dx$

11. $\int_1^4 \frac{e^{\sqrt{x}}}{2\sqrt{x}}dx$

12. $\int_0^1 4(\ln e^u)e^{-2u^2}\,du$

13. $\int 14(\sec\theta\tan\theta)e^{2\sec\theta}\,d\theta$

14. $\int (\sec^2 x)e^{\tan x}\,dx$

15. $\int 3e^y\sqrt{1 + e^y}\,dy$

16. $\int \frac{4\,dx}{x^2e^{1/x}}$

17. $\int_1^3 3e^{2x}(e^{-2x} - 1)dx$

18. $\int_0^{0.5} \frac{3e^{3x}}{e^{x-1}}dx$

19. $\int \frac{e^{3x} + 1}{e^x}dx$

20. $\int \frac{4\,dx}{e^{\sin x}\sec x}$

21. $\int \frac{e^{\tan^{-1}2x}}{8x^2 + 2}dx$

22. $\int \frac{e^{\sin^{-1}x}}{\sqrt{1 - x^2}}dx$

23. $\int \frac{e^{\cos 3x}\,dx}{\csc 3x}$

24. $\int (e^x - e^{-x})^2\,dx$

25. $\int_0^{\pi} (\sin 2x)e^{\cos^2 x}\, dx$

26. $\int_0^2 \frac{e^{2t}\, dt}{\sqrt{e^{2t}+4}}$

27. $\int \frac{6e^x\, dx}{e^x + 1}$

28. $\int \frac{4e^x\, dx}{(1-e^x)^2}$

In Exercises 29–44, solve the given problems by integration.

29. Find the area bounded by $y = 3e^x$, $x = 0$, $y = 0$, and $x = 2$.

30. Find the area bounded by $x = a$, $x = b$, $y = 0$, and $y = e^x$, assuming that $a < b$.

31. Integrate $\int 2e^{x^2 + \ln x}\, dx$. (*Hint:* Use the property $x^{n+m} = x^n x^m$. Then use the property $e^{\ln x} = x$.)

32. Integrate: $\int e^{x+\ln x} \frac{dx}{x}$. See Exercise 31.

33. Find the volume generated by revolving the region bounded by $y = e^{x^2}$, $x = 1$, $y = 0$, and $x = 2$ about the y-axis. See Fig. 28.7.

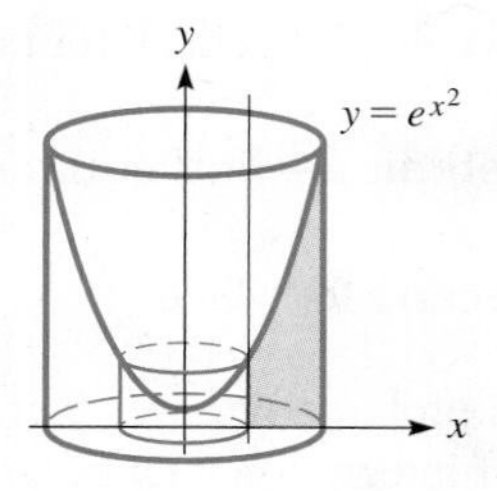

Fig. 28.7

34. Find an equation of the curve for which $dy/dx = 8e^{4x}$ if the curve passes through (0, 6).

35. Find the average value of $y = 4e^{x/2}$ from $x = 0$ to $x = 4$.

36. Find the moment of inertia with respect to the y-axis of a flat plate that covers the first-quadrant region bounded by $y = e^{x^3}$, $x = 1$, and the axes.

37. Using Eq. (27.15), show that $\int b^u\, du = \frac{b^u}{\ln b} + C$ $(b > 0, b \neq 1)$.

38. Find the first-quadrant area bounded by $y = 2^x$ and $x = 3$. See Exercise 37.

39. A marathon runner's speed (in km/h) is $v = 12.0e^{-t/6.00}$. This runner is in the same race as the runner in Exercise 41 of Section 28.2. Who is ahead (a) after 3.00 h, (b) after 4.00 h?

40. The energy consumption rate (in MW/year) in a certain city is projected to be given by $\frac{dP}{dt} = 775e^{-0.0500t}$, where P is the power consumption and t is the number of years after 2017. Find the total projected energy consumption between 2017 and 2022. (*Hint:* Integrate from $t = 0$ to $t = 5$.)

41. For an electric circuit containing a voltage source E, a resistance R, and a capacitance C, an equation relating the charge q on the capacitor and the time t is $qe^{t/RC} = \frac{E}{R}\int e^{t/RC}\, dt$. See Fig. 28.8. If $q = 0$ for $t = 0$, perform the integration and then solve for q as a function of t.

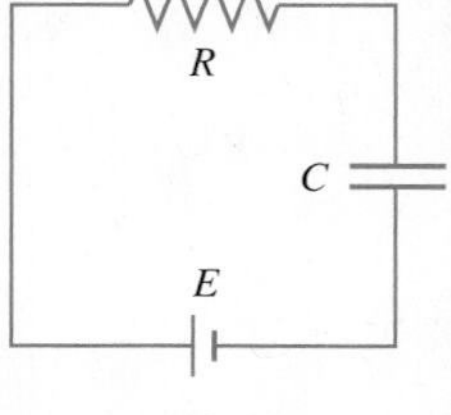

Fig. 28.8

42. In the theory dealing with energy propagation of lasers, the equation $E = a\int_0^{I_0} e^{-Tx}\, dx$ is used. Here, a, I_0, and T are constants. Evaluate this integral.

43. The St. Louis Gateway Arch has a shape that is given approximately by (measurements in m) $y = -19.46(e^{x/38.92} + e^{-x/38.92}) + 230.9$. Find the area under the Arch by determining the area bounded by this curve and the x-axis.

44. The force F (in lb) exerted by a robot programmed to staple carton sections together is given by $F = 6\int e^{\sin \pi t} \cos \pi t\, dt$, where t is the time (in s). Find F as a function of t if $F = 0$ for $t = 1.5$ s.

Answers to Practice Exercises

1. $-2e^{-3x} + C$ **2.** $e^{\sin x} + C$

28.4 Basic Trigonometric Forms

Integration of the Six Trigonometric Functions and Four Trigonometric Integrals That Directly Give Trigonometric Functions

By noting the formulas for differentiating the six trigonometric functions, and the antiderivatives that are found using them, we have the following six integration formulas:

$$\int \sin u\, du = -\cos u + C \tag{28.4}$$

$$\int \cos u\, du = \sin u + C \tag{28.5}$$

$$\int \sec^2 u\, du = \tan u + C \tag{28.6}$$

$$\int \csc^2 u\, du = -\cot u + C \tag{28.7}$$

$$\int \sec u \tan u\, du = \sec u + C \tag{28.8}$$

$$\int \csc u \cot u\, du = -\csc u + C \tag{28.9}$$

EXAMPLE 1 Integration of $\sec^2 u\, du$

Integrate: $\int x\sec^2 x^2\, dx$.

With $u = x^2$, $du = 2x\, dx$, we have

$$\int x\sec^2 x^2\, dx = \frac{1}{2}\int (\sec^2 \underbrace{x^2}_{u})\overbrace{(2x\, dx)}^{du}$$

$$= \frac{1}{2}\tan x^2 + C \qquad \text{using Eq. (28.6)}$$

■

Practice Exercise

1. Integrate: $\int \sin 5x\, dx$.

EXAMPLE 2 Integration of $\tan u \sec u\, du$

Integrate: $\int \frac{\tan 2x}{\cos 2x} dx$.

NOTE ➧ [By using the basic identity $\sec\theta = 1/\cos\theta$, we can transform this integral into the form $\int \sec 2x \tan 2x\, dx$.] In this form, $u = 2x$, $du = 2\, dx$. Therefore,

$$\int \frac{\tan 2x}{\cos 2x} dx = \int \sec 2x \tan 2x\, dx = \frac{1}{2}\int \sec \underbrace{2x}_{u} \tan \underbrace{2x}_{u} \overbrace{(2\, dx)}^{du}$$

$$= \frac{1}{2}\sec 2x + C \qquad \text{using Eq. (28.8)}$$

■

Practice Exercise

2. Integrate: $\int 6\csc^2 3x\, dx$.

EXAMPLE 3 Integration of $\cos u\, du$—velocity and displacement

The vertical velocity v (in cm/s) of the end of a vibrating rod is given by $v = 80\cos 20\pi t$, where t is the time in seconds. Find the vertical displacement y (in cm) as a function of t if $y = 0$ for $t = 0$.

Because $v = dy/dt$, we have the following solution:

$$\frac{dy}{dt} = 80\cos 20\pi t$$

$$\int dy = \int 80\cos 20\pi t\, dt = \frac{80}{20\pi}\int (\cos 20\pi t)(20\pi\, dt) \qquad \text{set up integration}$$

$$y = \frac{4}{\pi}\sin 20\pi t + C \qquad \text{using Eq. (28.5)}$$

$$0 = 1.27\sin 0 + C, \quad C = 0 \qquad \text{evaluate } C$$

$$y = 1.27\sin 20\pi t \qquad \text{solution}$$

■

To find the integrals for the other trigonometric functions, we must change them to a form for which the integral can be determined by methods previously discussed. We can accomplish this by using the basic trigonometric relations.

The formula for $\int \tan u\, du$ is found by expressing the integral in the form $\int (\sin u/\cos u)du$. We recognize this as being a logarithmic form, where the u of the logarithmic form is cos u in this integral. The differential of cos u is $-\sin u\, du$. Therefore, we have

$$\int \tan u\, du = \int \frac{\sin u}{\cos u} du = -\int \frac{-\sin u\, du}{\cos u} = -\ln|\cos u| + C$$

The formula for $\int \cot u\, du$ is found by writing it in the form $\int (\cos u/\sin u)du$. In this manner, we obtain the result

$$\int \cot u\, du = \int \frac{\cos u}{\sin u} du = \int \frac{\cos u\, du}{\sin u} = \ln|\sin u| + C$$

The formula for $\int \sec u\, du$ is found by writing it in the form

$$\int \frac{\sec u(\sec u + \tan u)}{\sec u + \tan u} du$$

We see that this form is also a logarithmic form, because

$$d(\sec u + \tan u) = (\sec u \tan u + \sec^2 u)\, du$$

The right side of this equation is the expression appearing in the numerator of the integral. Thus,

$$\int \sec u\, du = \int \frac{\sec u(\sec u + \tan u) du}{\sec u + \tan u} = \int \frac{\sec u \tan u + \sec^2 u}{\sec u + \tan u} du$$
$$= \ln|\sec u + \tan u| + C$$

To obtain the formula for $\int \csc u\, du$, we write it in the form

$$\int \frac{\csc u(\csc u - \cot u) du}{\csc u - \cot u}$$

Thus, we have

$$\int \csc u\, du = \int \frac{\csc u(\csc u - \cot u)}{\csc u - \cot u} du = \int \frac{(-\csc u \cot u + \csc^2 u) du}{\csc u - \cot u}$$
$$= \ln|\csc u - \cot u| + C$$

Summarizing these results, we have the following integrals:

$$\int \tan u\, du = -\ln|\cos u| + C \quad \textbf{(28.10)}$$

$$\int \cot u\, du = \ln|\sin u| + C \quad \textbf{(28.11)}$$

$$\int \sec u\, du = \ln|\sec u + \tan u| + C \quad \textbf{(28.12)}$$

$$\int \csc u\, du = \ln|\csc u - \cot u| + C \quad \textbf{(28.13)}$$

■ An equivalent form of Eq. (28.13) is $\int \csc u\, du = -\ln|\csc u + \cot u| + C$.

EXAMPLE 4 Integration of tan *u du*

Integrate: $\int \tan 4\theta\, d\theta$.

Noting that $u = 4\theta$, $du = 4\, d\theta$, we have

$$\int \tan 4\theta\, d\theta = \frac{1}{4}\int \tan 4\theta(4\, d\theta) \quad \text{introducing factors of 4}$$
$$= -\frac{1}{4}\ln|\cos 4\theta| + C \quad \text{using Eq. (28.10)}$$
■

Practice Exercise

3. Integrate: $\int 4x \cot x^2\, dx$.

EXAMPLE 5 Integration of sec *u du*

Integrate: $\int \frac{\sec e^{-x}\, dx}{e^x}$.

In this integral, $u = e^{-x}$, $du = -e^{-x}\, dx$. Therefore,

$$\int \frac{\sec e^{-x}\, dx}{e^x} = -\int (\sec e^{-x})(-e^{-x}\, dx) \quad \text{introducing } - \text{ sign}$$
$$= -\ln|\sec e^{-x} + \tan e^{-x}| + C \quad \text{using Eq. (28.12)}$$
■

EXAMPLE 6 Integration of csc u and cot u du

Evaluate: $\int_{\pi/6}^{\pi/4} \frac{1 + \cos x}{\sin x} dx$.

$$\int_{\pi/6}^{\pi/4} \frac{1 + \cos x}{\sin x} dx = \int_{\pi/6}^{\pi/4} \csc x \, dx + \int_{\pi/6}^{\pi/4} \cot x \, dx \quad \frac{1}{\sin x} = \csc x, \ \frac{\cos x}{\sin x} = \cot x$$

$$= \ln|\csc x - \cot x| \Big|_{\pi/6}^{\pi/4} + \ln|\sin x| \Big|_{\pi/6}^{\pi/4} \quad \text{integrating}$$

$$= \ln|\sqrt{2} - 1| - \ln|2 - \sqrt{3}| + \ln\left|\frac{1}{2}\sqrt{2}\right| - \ln\left|\frac{1}{2}\right| \quad \text{evaluating}$$

$$= \ln\frac{(\frac{1}{2}\sqrt{2})(\sqrt{2} - 1)}{(\frac{1}{2})(2 - \sqrt{3})} = \ln\frac{2 - \sqrt{2}}{2 - \sqrt{3}} = 0.782$$

■

EXERCISES 28.4

In Exercises 1 and 2, make the given changes in the indicated examples of this section and then solve the given problems.

1. In Example 1, change $\sec^2 x^2$ to $\sec^2 x^3$. What other change must be made in the integrand to have a result of $\tan x^3 + C$?
2. In Example 4, change tan to cot and then integrate.

In Exercises 3–26, integrate each of the given functions.

3. $\int \cos 2x \, dx$
4. $\int 4\sin(2 - x) dx$
5. $\int \sec^2 3\theta \, d\theta$
6. $\int 4\csc 8x \cot 8x \, dx$
7. $\int 2\sec 4x \tan 4x \, dx$
8. $\int \frac{\csc^2(e^{-x}) dx}{e^x}$
9. $\int_{0.5}^{1} x^2 \cot x^3 \, dx$
10. $\int_0^1 6\sin\frac{1}{2}t \sec\frac{1}{2}t \, dt$
11. $\int 2\theta^2 \sec(\theta^3) d\theta$
12. $\int (2\cos^2 4x - 1) dx$
13. $\int \frac{\sin(1/x)}{2x^2} dx$
14. $\int \frac{3 \, dx}{\sin 4x}$
15. $\int_0^{\pi/6} \frac{dx}{\cos^2 2x}$
16. $\int_0^1 \frac{2e^s \, ds}{\sec e^s}$
17. $\int \frac{e^x}{\sin(e^x)} dx$
18. $\int \frac{\sin 2x}{2\cos^2 x} dx$
19. $\int \sqrt{\tan^2 2x + 1} \, dx$
20. $\int 5(\tan u)(\ln \cos u) \, du$
21. $\int \cos 3x \sec^2 3x \, dx$
22. $\int \frac{1 - \cot^2 x}{\cos^2 x} dx$
23. $\int \frac{\sin^2 5x + \cos^2 5x}{\tan 5x} dx$
24. $\int \frac{1 + \sec^2 x}{x + \tan x} dx$
25. $\int_0^{\pi/9} \sin 3x(\csc 3x + \sec 3x) dx$
26. $\int_{\pi/4}^{\pi/3} (1 + \sec x)^2 \, dx$

In Exercises 27–38, solve the given problems by integration.

27. Find the area bounded by $y = 2\tan x$, $x = \frac{\pi}{4}$, and $y = 0$.
28. Find the area under the curve $y = \sin x$ from $x = 0$ to $x = \pi$.
29. Although $\int \frac{dx}{1 + \sin x}$ does not appear to fit a form for integration, show that it can be integrated by multiplying the numerator and the denominator by $1 - \sin x$.
30. Integrate $\int \frac{1 - \sin x}{1 + \cos x} dx$ by multiplying the numerator and the denominator by $1 - \cos x$.
31. Integrate $\int \sec^2 x \tan x \, dx$ (a) with $u = \tan x$, and (b) with $u = \sec x$. Explain why the answers appear to be different.
32. Evaluate $\int_a^{a+2\pi} \sin x \, dx$ for any real value of a. Show an interpretation of the result in terms of the area under a curve.
33. Find the volume generated by revolving the region bounded by $y = \sec x$, $x = 0$, $x = \frac{\pi}{3}$, and $y = 0$ about the x-axis.
34. The volume under a tent can be described as being generated by revolving the region bounded by $y = 4.00\cos(0.150x^2)$, $x = 0$, $y = 0$, for $x < 3$ about the y-axis. Find the volume (in m^3).
35. The angular velocity ω (in rad/s) of a pendulum is $\omega = -0.25\sin 2.5t$. Find the angular displacement θ as a function of t if $\theta = 0.10$ for $t = 0$.
36. If the current i (in A) in a certain electric circuit is given by $i = 110\cos 377t$, find the expression for the voltage across a 500-μF capacitor as a function of time. The initial voltage is zero. Show that the voltage across the capacitor is 90° out of phase with the current.
37. A fin on a wind-direction indicator has a shape that can be described as the region bounded by $y = \tan x^2$, $y = 0$, and $x = 1$. Find the x-coordinate (in m) of the centroid of the fin if its area is 0.3984 m^2.
38. A force is given as a function of the distance from the origin as $F = \frac{2 + \tan x}{\cos x}$. Express the work done by this force as a function of x if $W = 0$ for $x = 0$.

Answers to Practice Exercises

1. $-\frac{1}{5}\cos 5x + C$ 2. $-2\cot 3x + C$ 3. $2\ln|\sin x^2| + C$

28.5 Other Trigonometric Forms

Odd Power of $\sin u$ or $\cos u$ • Even Power of $\sin u$ or $\cos u$ • Power of $\tan u$, $\cot u$, $\sec u$, $\csc u$ • Root-Mean-Square Value of a Function

By use of trigonometric relations developed in Chapter 20, it is possible to transform many integrals involving powers of the trigonometric functions into integrable form. We now show the relationships that are useful for these integrals.

$$\cos^2 x + \sin^2 x = 1 \quad \textbf{(28.14)}$$

$$1 + \tan^2 x = \sec^2 x \quad \textbf{(28.15)}$$

$$1 + \cot^2 x = \csc^2 x \quad \textbf{(28.16)}$$

$$2\cos^2 x = 1 + \cos 2x \quad \textbf{(28.17)}$$

$$2\sin^2 x = 1 - \cos 2x \quad \textbf{(28.18)}$$

To integrate a product of powers of the sine and cosine, we use Eq. (28.14) ***if at least one of the powers is odd.*** The method is based on transforming the integral so that it is made up of powers of either the sine or cosine and the first power of the other. In this way, this first power becomes a factor of du.

EXAMPLE 1 Integration with an odd power of $\sin u$

Integrate: $\int \sin^3 x \cos^2 x\, dx$.

Because $\sin^3 x = \sin^2 x \sin x = (1 - \cos^2 x)\sin x$, we can write this integral with powers of $\cos x$ along with $-\sin x$, which is the necessary du for this integral. Therefore,

$$\int \sin^3 x \cos^2 x\, dx = \int (1 - \cos^2 x)(\sin x)(\cos^2 x)dx \qquad \text{using Eq. (28.14)}$$

$$= \int (\cos^2 x - \cos^4 x)(\sin x\, dx)$$

$$= -\int \cos^2 x(\underbrace{-\sin x\, dx}_{du}) + \int \cos^4 x(\underbrace{-\sin x\, dx}_{du})$$

$$= -\frac{1}{3}\cos^3 x + \frac{1}{5}\cos^5 x + C \quad \blacksquare$$

Practice Exercise

1. Integrate: $\int \sin^3 x\, dx$.

EXAMPLE 2 Integration with an odd power of $\cos u$

Integrate: $\int \cos^5 2x\, dx$.

Because $\cos^5 2x = \cos^4 2x \cos 2x = (1 - \sin^2 2x)^2 \cos 2x$, it is possible to write this integral with powers of $\sin 2x$ along with $\cos 2x\, dx$. Thus, with the introduction of a factor of 2, $(\cos 2x)(2\, dx)$ is the necessary du for this integral. Thus,

$$\int \cos^5 2x\, dx = \int (1 - \sin^2 2x)^2 \cos 2x\, dx \qquad \text{using Eq. (28.14)}$$

$$= \int (1 - 2\sin^2 2x + \sin^4 2x)\cos 2x\, dx$$

$$= \int \cos 2x\, dx - \int 2\sin^2 2x \cos 2x\, dx + \int \sin^4 2x \cos 2x\, dx$$

$$= \frac{1}{2}\int \cos 2x(\underbrace{2\, dx}_{du}) - \int \sin^2 2x(\underbrace{2\cos 2x\, dx}_{du}) + \frac{1}{2}\int \sin^4 2x(\underbrace{2\cos 2x\, dx}_{du})$$

$$= \frac{1}{2}\sin 2x - \frac{1}{3}\sin^3 2x + \frac{1}{10}\sin^5 2x + C \quad \blacksquare$$

CAUTION In products of powers of the sine and cosine, if the powers to be integrated are even, we use Eqs. (28.17) and (28.18) to transform the integral. ■ Those most commonly met are $\int \cos^2 u\, du$ and $\int \sin^2 u\, du$. Consider the following examples.

EXAMPLE 3 Integration with an even power of sin u

Integrate: $\int \sin^2 2x\, dx$.

Using Eq. (28.18) in the form $\sin^2 2x = \frac{1}{2}(1 - \cos 4x)$, this integral can be transformed into a form that can be integrated. (Here, we note ***the x of Eq. (28.18) is treated as 2x*** for this integral.) Therefore, we write

$$\int \sin^2 2x\, dx = \int \left[\frac{1}{2}(1 - \cos 4x)\right] dx \qquad \sin^2 x = \frac{1}{2}(1 - \cos 2x)$$

$$= \frac{1}{2}\int dx - \frac{1}{8}\int \cos 4x(4\, dx)$$

$$= \frac{x}{2} - \frac{1}{8}\sin 4x + C$$

■

NOTE ▶ *To integrate even powers of the secant, powers of the tangent, or products of the secant and tangent, we use Eq. (28.15) to transform the integral.* [In transforming, we look for **powers of the tangent with $\sec^2 x$,** which becomes part of du, or **powers of the secant along with $\sec x \tan x$,** which becomes part of du in this case.] Similar transformations are made when we integrate powers of the cotangent and cosecant, with the use of Eq. (28.16).

EXAMPLE 4 Integration with a power of sec u

Integrate: $\int \sec^3 t \tan t\, dt$.

By writing $\sec^3 t \tan t$ as $\sec^2 t(\sec t \tan t)$, we can use the $\sec t \tan t\, dt$ as the du of the integral. Thus,

$$\int \sec^3 t \tan t\, dt = \int (\sec^2 t)(\underbrace{\sec t \tan t\, dt}_{du}) \qquad u = \sec t$$

$$= \frac{1}{3}\sec^3 t + C$$

■

Practice Exercise

2. Integrate: $\int \sec^4 x\, dx$.

EXAMPLE 5 Integration with a power of tan u

Integrate: $\int \tan^5 x\, dx$.

Because $\tan^5 x = \tan^3 x \tan^2 x = \tan^3 x(\sec^2 x - 1)$, we can write this integral with powers of $\tan x$ along with $\sec^2 x\, dx$. Thus, $\sec^2 x\, dx$ becomes the necessary du of the integral. It is necessary to replace $\tan^2 x$ with $\sec^2 x - 1$ twice during the integration. Therefore,

$$\int \tan^5 x\, dx = \int \tan^3 x(\sec^2 x - 1)dx \qquad \text{using } \tan^2 x = \sec^2 x - 1$$

$$= \int \tan^3 x(\sec^2 x\, dx) - \int \tan^3 x\, dx$$

$$= \frac{1}{4}\tan^4 x - \int \tan x(\sec^2 x - 1)dx \qquad \text{using } \tan^2 x = \sec^2 x - 1$$

$$= \frac{1}{4}\tan^4 x - \int \tan x(\sec^2 x\, dx) + \int \tan x\, dx$$

$$= \frac{1}{4}\tan^4 x - \frac{1}{2}\tan^2 x - \ln|\cos x| + C$$

■

EXAMPLE 6 Integration of another trigonometric form

Integrate: $\displaystyle\int_0^{\pi/4} \frac{\tan^3 x}{\sec^3 x} dx.$

Of several possible ways in which the integrand can be transformed into an integrable form, among the easiest is the following:

$$\int_0^{\pi/4} \frac{\tan^3 x}{\sec^3 x} dx = \int_0^{\pi/4} \frac{\sin^3 x}{\cos^3 x} \frac{1}{\sec^3 x} dx \qquad \tan x = \sin x/\cos x$$

$$= \int_0^{\pi/4} \sin^3 x \, dx = \int_0^{\pi/4} (1 - \cos^2 x) \sin x \, dx \qquad \cos x \sec x = 1$$

$$= \int_0^{\pi/4} \sin x \, dx - \int_0^{\pi/4} \cos^2 x \sin x \, dx$$

$$= -\cos x + \frac{1}{3}\cos^3 x \Big|_0^{\pi/4} \qquad \text{integrate}$$

$$= -\frac{\sqrt{2}}{2} + \frac{1}{3}\left(\frac{\sqrt{2}}{2}\right)^3 - \left(-1 + \frac{1}{3}\right) \qquad \text{evaluate}$$

$$= \frac{8 - 5\sqrt{2}}{12} = 0.0774$$

■

EXAMPLE 7 Root-mean-square value—effective current of a plasma TV

The *root-mean-square value* of a function with respect to x is defined by the following equation:

■ See the chapter introduction.

$$y_{\text{rms}} = \sqrt{\frac{1}{T}\int_0^T y^2 \, dx} \qquad \textbf{(28.19)}$$

Usually, the value of T that is of importance is the period of the function. Find the root-mean-square value of the electric current i (in A) used by a plasma TV, for which $i = 3.75 \cos 120\pi t$, for one period.

■ In most countries, the rms voltage is 240 V. In the United States and Canada, it is 120 V.

The period is $\frac{2\pi}{120\pi} = \frac{1}{60.0}$ s. Thus, we must find the square root of the integral

$$\frac{1}{1/60.0}\int_0^{1/60.0} (3.75 \cos 120\pi t)^2 \, dt = 844 \int_0^{1/60.0} \cos^2 120\pi t \, dt$$

Evaluating this integral, we have

$$844 \int_0^{1/60.0} \cos^2 120\pi t \, dt = 422 \int_0^{1/60.0} (1 + \cos 240\pi t) dt$$

$$= 422t \Big|_0^{1/60.0} + \frac{422}{240\pi}\int_0^{1/60.0} \cos 240\pi t (240\pi \, dt)$$

$$= 7.033 + \frac{422}{240\pi} \sin 240\pi t \Big|_0^{1/60.0} = 7.033$$

This means the root-mean-square current is

$$i_{\text{rms}} = \sqrt{7.033} = 2.65 \text{ A}$$

This value of the current, often referred to as the *effective current*, is the value of direct current that would produce the same quantity of heat energy in the same time. It is important in the design of electronic equipment and electric appliances. ■

EXERCISES 28.5

In Exercises 1 and 2, answer the given questions related to the indicated examples of this section.

1. In Example 3, change $2x$ to $3x$ and then integrate.
2. In Example 5, change the exponent of 5 to a 3 and then integrate.

In Exercises 3–34, integrate each of the given functions.

3. $\int \sin^2 x \cos x \, dx$
4. $\int \sin x \cos^5 x \, dx$
5. $\int \sin^3 2x \, dx$
6. $\int 3\cos^3 T \, dT$
7. $\int \tan^2 x \sec^2 x \, dx$
8. $\int \tan x \sec^3 x \, dx$
9. $\int_0^{\pi/2} \sin^4 x \cos^3 x \, dx$
10. $\int_{\pi/3}^{\pi/2} 10 \sin t(1 - \cos 2t)^2 \, dt$
11. $\int \sin^2 x \, dx$
12. $\int 2\cos^2 2x \, dx$
13. $\int \sin^3 2\theta \cos^2 2\theta \, d\theta$
14. $\int_0^1 (1 - \cos^2 4x) dx$
15. $\int \tan^2 x \, dx$
16. $\int \frac{6\cot^2 y}{\tan y} dy$
17. $\int_0^{\pi/4} \frac{\tan x}{\cos^4 x} dx$
18. $\int \frac{\csc^4 4x}{\tan 4x} dx$
19. $\int \tan^4 2x \, dx$
20. $\int 4\cot^4 x \, dx$
21. $\int 0.5 \sin s \sin 2s \, ds$
22. $\int \sqrt{\tan x} \sec^4 x \, dx$
23. $\int (\sin x + \cos x)^2 \, dx$
24. $\int (\tan 2x + \cot 2x)^2 \, dx$
25. $\int \frac{1 - \cot x}{\sin^4 x} dx$
26. $\int \frac{(\sin u + \sin^2 u)^2}{\sec u} du$
27. $\int_{\pi/6}^{\pi/4} \cot^5 p \, dp$
28. $\int_{\pi/6}^{\pi/3} \frac{2 \, dx}{1 + \sin x}$
29. $\int \sec^6 x \, dx$
30. $\int \tan^7 x \, dx$
31. $\int \frac{\sin 2x}{\cos^3 x} dx$
32. $\int \frac{\sec^2 t \tan t}{4 + \sec^2 t} dt$
33. $\int \frac{\sec e^{-x}}{e^x} dx$
34. $\int_0^{\pi/4} \sqrt{1 + \cos 4x} \, dx$

In Exercises 35–52, solve the given problems by integration.

35. Using the identity $\sin\alpha\cos\beta = \frac{1}{2}[\sin(\alpha + \beta) + \sin(\alpha - \beta)]$, integrate $\int \sin 4x \cos 5x \, dx$.
36. Using the identity $\cos\alpha\cos\beta = \frac{1}{2}[\cos(\alpha + \beta) + \cos(\alpha - \beta)]$, integrate $\int \cos 3x \cos 4x \, dx$.
37. Find the volume generated by revolving the region bounded by $y = \sin x$ and $y = 0$, from $x = 0$ to $x = \pi$, about the x-axis.
38. Find the volume generated by revolving the region bounded by $y = \tan^3(x^2)$, $y = 0$, and $x = \frac{\pi}{4}$ about the y-axis.
39. Find the area bounded by $y = \sin^4 x$, $y = \cos^4 x$, and $x = 0$ in the first quadrant. (*Hint:* Use factoring to simplify the integrand as much as possible.)
40. The length of an arc along a function is given by $s = \int_a^b \sqrt{1 + \left(\frac{dy}{dx}\right)^2} dx$. Find the arc length of the curve $y = \ln \cos x$ from $x = 0$ to $x = \frac{\pi}{3}$.
41. Show that $\int \sin x \cos x \, dx$ can be integrated in two ways. Explain the difference in the answers.
42. For $n > 0$, show that $\int \tan x \sec^n x \, dx = \frac{1}{n}\sec^n x + C$.
43. Show that $\int_0^{\pi} \sin^2 nx \, dx = \frac{1}{2}\pi$, where n is any positive integer.
44. The velocity v (in cm/s) of an object is $v = \cos^2 \pi t$. How far does the object move in 4.0 s?
45. The acceleration a (in ft/s^2) of an object is $a = \sin^2 t \cos t$. If the object starts at the origin with a velocity of 6 ft/s, what is its position at time t?
46. Find the root-mean-square current in a circuit from $t = 0$ s to $t = 0.50$ s if $i = i_0 \sin t \sqrt{\cos t}$.
47. In the study of the rate of radiation by an accelerated charge, the following integral must be evaluated: $\int_0^{\pi} \sin^3 \theta \, d\theta$. Find the value of the integral.
48. In finding the volume of a special O-ring for a space vehicle, the integral $\int \frac{\sin^2\theta}{\cos^2\theta} d\theta$ must be evaluated. Perform this integration.
49. For a voltage $V = 340 \sin 120\pi t$, show that the root-mean-square voltage for one period is 240 V.
50. For a current $i = i_0 \sin \omega t$, show that the root-mean-square current for one period is $i_0/\sqrt{2}$.
51. In the analysis of the intensity of light from a certain source, the equation $I = A\int_{-a/2}^{a/2} \cos^2 [b\pi(c - x)] \, dx$ is used. Here, A, a, b, and c are constants. Evaluate this integral. (The simplification is quite lengthy.)
52. In the study of the lifting force L due to a stream of fluid passing around a cylinder, the equation $L = k\int_0^{2\pi} (a \sin\theta + b \sin^2\theta - b \sin^3\theta) \, d\theta$ is used. Here, k, a, and b are constants and θ is the angle from the direction of flow. Evaluate the integral.

Answers to Practice Exercises

1. $-\cos x + \frac{1}{3}\cos^3 x + C$ 2. $\tan x + \frac{1}{3}\tan^3 x + C$

28.6 Inverse Trigonometric Forms

Inverse Sine Form • Inverse Tangent Form • Comparison of Inverse Sine, Inverse Tangent, Power, and Logarithmic Forms

Using Eq. (27.10), we can find the differential of $\sin^{-1}(u/a)$, where a is constant:

$$d\left(\sin^{-1}\frac{u}{a}\right) = \frac{1}{\sqrt{1-(u/a)^2}}\frac{du}{a} = \frac{a}{\sqrt{a^2-u^2}}\frac{du}{a} = \frac{du}{\sqrt{a^2-u^2}}$$

■ For reference, Eq. (27.10) is $\frac{d(\sin^{-1}u)}{dx} = \frac{1}{\sqrt{1-u^2}}\frac{du}{dx}$.

Noting this differentiation formula, and the antiderivative of the result, we have the important integration formula:

$$\int \frac{du}{\sqrt{a^2-u^2}} = \sin^{-1}\frac{u}{a} + C \tag{28.20}$$

By finding the differential of $\tan^{-1}(u/a)$, we have

$$d\left(\tan^{-1}\frac{u}{a}\right) = \frac{1}{1+(u/a)^2}\frac{du}{a} = \frac{a^2}{a^2+u^2}\frac{du}{a} = \frac{a\,du}{a^2+u^2}$$

Again, noting the antiderivative, we have another important integration formula.

$$\int \frac{du}{a^2+u^2} = \frac{1}{a}\tan^{-1}\frac{u}{a} + C \tag{28.21}$$

This shows one of the principal uses of the inverse trigonometric functions: They provide a solution to the integration of important algebraic functions.

EXAMPLE 1 Inverse sine form

Integrate: $\int \frac{dx}{\sqrt{9-x^2}}$.

This integral fits the form of Eq. (28.20) with $u = x$, $du = dx$, and $a = 3$. Thus,

$$\int \frac{dx}{\sqrt{9-x^2}} = \int \frac{dx}{\sqrt{3^2-x^2}}$$
$$= \sin^{-1}\frac{x}{3} + C$$ ■

Practice Exercise

1. Integrate: $\int \frac{dx}{\sqrt{9-4x^2}}$.

EXAMPLE 2 Inverse tangent form—liquid flow rate

The volume flow rate Q (in m^3/s) of a constantly flowing liquid is given by $Q = 24\int_0^2 \frac{dx}{6+x^2}$, where x is the distance from the center of flow. Find the value of Q.

For the integral, we see that it fits Eq. (28.21) with $u = x$, $du = dx$, and $a = \sqrt{6}$.

$$Q = 24\int_0^2 \frac{dx}{6+x^2} = 24\int_0^2 \frac{dx}{(\sqrt{6})^2+x^2}$$
$$= \frac{24}{\sqrt{6}}\tan^{-1}\frac{x}{\sqrt{6}}\Big|_0^2$$
$$= \frac{24}{\sqrt{6}}\left(\tan^{-1}\frac{2}{\sqrt{6}} - \tan^{-1}0\right)$$
$$= 6.71\ m^3/s$$ ■

EXAMPLE 3 Inverse sine form

Integrate: $\int \frac{dx}{\sqrt{25 - 4x^2}}$.

This integral fits the form of Eq. (28.20) with $u = 2x$, $du = 2\,dx$, and $a = 5$. Thus, in order to have the proper du, we must include a factor of 2 in the numerator, and therefore we also place a $\frac{1}{2}$ before the integral. This leads to

$$\int \frac{dx}{\sqrt{25 - 4x^2}} = \frac{1}{2}\int \frac{2\,dx}{\sqrt{5^2 - (2x)^2}}$$

($2\,dx$ ← du; $2x$ ← u)

$$= \frac{1}{2}\sin^{-1}\frac{2x}{5} + C$$ ■

EXAMPLE 4 Complete the square to fit $\tan^{-1}u$ form

Integrate: $\int_{-1}^{3} \frac{dx}{x^2 + 6x + 13}$.

At first glance, it does not appear that this integral fits any of the forms presented up to this point. However, by writing the denominator in the form $(x^2 + 6x + 9) + 4 = (x + 3)^2 + 2^2$, we recognize that $u = x + 3$, $du = dx$, and $a = 2$. Thus,

$$\int_{-1}^{3} \frac{dx}{x^2 + 6x + 13} = \int_{-1}^{3} \frac{dx}{(x + 3)^2 + 2^2}$$

(dx ← du; $x + 3$ ← u)

$$= \frac{1}{2}\tan^{-1}\frac{x + 3}{2}\Big|_{-1}^{3} \quad \text{integrate}$$

$$= \frac{1}{2}(\tan^{-1}3 - \tan^{-1}1) \quad \text{evaluate}$$

$$= 0.2318$$

Now, we can see the use of completing the square when we are transforming integrals into proper form. ■

Practice Exercise

2. Integrate: $\int \frac{dx}{4 + 9x^2}$.

EXAMPLE 5 Inverse tangent and logarithmic forms

Integrate: $\int \frac{2r + 5}{r^2 + 9}dr$.

By writing this integral as the sum of two integrals, we may integrate each of these separately:

$$\int \frac{2r + 5}{r^2 + 9}dr = \int \frac{2r\,dr}{r^2 + 9} + \int \frac{5\,dr}{r^2 + 9}$$

The first integral is a logarithmic form, and the second is an inverse tangent form. For the first, $u = r^2 + 9$, $du = 2r\,dr$. For the second, $u = r$, $du = dr$, $a = 3$.

$$\int \frac{2r\,dr}{r^2 + 9} + 5\int \frac{dr}{r^2 + 9} = \ln|r^2 + 9| + \frac{5}{3}\tan^{-1}\frac{r}{3} + C$$ ■

CAUTION The inverse trigonometric integral forms show very well the importance of *proper recognition of the form of the integral*. It is important that these forms are not confused with those of the power rule or the logarithmic form. ■

EXAMPLE 6 Inverse sine, power, and logarithmic forms

The integral $\int \frac{dx}{\sqrt{1 - x^2}}$ is of the inverse sine form with $u = x$, $du = dx$, and $a = 1$. Thus,

$$\int \frac{dx}{\sqrt{1 - x^2}} = \sin^{-1} x + C$$

The integral $\int \frac{x\,dx}{\sqrt{1 - x^2}}$ is not of the inverse sine form due to the factor of x in the numerator. It is integrated by use of the general power rule, with $u = 1 - x^2$, $du = -2x\,dx$, and $n = -\frac{1}{2}$. Thus,

$$\int \frac{x\,dx}{\sqrt{1 - x^2}} = -\sqrt{1 - x^2} + C$$

The integral $\int \frac{x\,dx}{1 - x^2}$ is of the basic logarithmic form with $u = 1 - x^2$ and $du = -2x\,dx$. If $1 - x^2$ is raised to any power other than 1 in the denominator, we would use the general power rule. To be of the inverse sine form, we would have the square root of $1 - x^2$ and no factor of x, as in the first illustration. Thus,

$$\int \frac{x\,dx}{1 - x^2} = -\frac{1}{2}\ln|1 - x^2| + C$$ ■

EXAMPLE 7 $\sin^{-1} u$, $\tan^{-1} u$, u^n, and $\ln u$ forms

CAUTION Note carefully the form of u and du In each of the forms of Example 7. ■

The following integrals are of the form indicated.

$\int \frac{dx}{1 + x^2}$	Inverse tangent form	$u = x$, $du = dx$
$\int \frac{x\,dx}{1 + x^2}$	Logarithmic form	$u = 1 + x^2$, $du = 2x\,dx$
$\int \frac{x\,dx}{\sqrt{1 + x^2}}$	General power form	$u = 1 + x^2$, $du = 2x\,dx$
$\int \frac{dx}{1 + x}$	Logarithmic form	$u = 1 + x$, $du = dx$
$\int \frac{x\,dx}{\sqrt{1 - x^4}}$	Inverse sine form	$u = x^2$, $du = 2x\,dx$
$\int \frac{x\,dx}{1 + x^4}$	Inverse tangent form	$u = x^2$, $du = 2x\,dx$ ■

There are a number of integrals whose forms appear to be similar to those in Examples 6 and 7, but which do not fit the forms we have discussed. They include

$$\int \frac{dx}{\sqrt{x^2 - 1}} \qquad \int \frac{dx}{\sqrt{1 + x^2}} \qquad \int \frac{dx}{1 - x^2} \qquad \int \frac{dx}{x\sqrt{1 + x^2}}$$

We will develop methods to integrate some of these forms, and all of them can be integrated by the tables discussed in Section 28.11.

EXERCISES 28.6

In Exercises 1 and 2, answer the given questions related to the indicated examples of this section.

1. In Example 1, what change must be made in the integrand in order to have a result of $\sqrt{9 - x^2} + C$?

2. In Example 2, what change must be made in the integrand in order that the integration would lead to a result of $\ln(6 + x^2) + C$?

In Exercises 3–30, integrate each of the given functions.

3. $\int \frac{dx}{\sqrt{4 - x^2}}$

4. $\int \frac{dx}{\sqrt{49 - x^2}}$

5. $\int \frac{12\,dx}{64 + x^2}$

6. $\int \frac{6p^2\,dp}{4 + p^6}$

7. $\int \frac{8x\,dx}{\sqrt{1 - 16x^4}}$

8. $\int_0^1 \frac{2dx}{\sqrt{9 - 4x^2}}$

9. $\int_0^2 \frac{3e^{-t}\,dt}{1 + 9e^{-2t}}$

10. $\int_1^3 \frac{4\,dx}{49 + 4x^2}$

11. $\int_0^{0.4} \frac{2\,dx}{\sqrt{4 - 5x^2}}$

12. $\int \frac{dx}{2\sqrt{x}\sqrt{1 - x}}$

13. $\int \frac{8x\,dx}{9x^2 + 16}$

14. $\int \frac{y\,dy}{4\sqrt{25 - 16y^2}}$

15. $\int_1^e \frac{3\,du}{u[1 + (\ln u)^2]}$

16. $\int_0^1 \frac{4x\,dx}{1 + x^4}$

17. $\int \frac{2e^x\,dx}{\sqrt{1 - e^{2x}}}$

18. $\int \frac{\sec^2 x\,dx}{\sqrt{9 - \tan^2 x}}$

19. $\int \frac{dT}{T^2 + 2T + 2}$

20. $\int \frac{2\,dx}{x^2 + 8x + 20}$

21. $\int \frac{4\,dx}{\sqrt{-4x - x^2}}$

22. $\int \frac{0.3ds}{\sqrt{2s - s^2}}$

23. $\int_{\pi/6}^{\pi/2} \frac{2\cos 2\theta\,d\theta}{1 + \sin^2 2\theta}$

24. $\int_{-4}^{0} \frac{dx}{x^2 + 4x + 5}$

25. $\int \frac{2 - x}{\sqrt{4 - x^2}}dx$

26. $\int \frac{3x^3 - 2x}{1 + 16x^4}dx$

27. $\int \frac{\sin^{-1} x}{\sqrt{1 - x^2}}dx$

28. $\int \frac{dx}{e^x + e^{-x}}$

29. $\int \frac{x^2 + 3x^5}{1 + x^6}dx$

30. $\int \frac{x\tan^{-1} x^2}{1 + x^4}dx$

In Exercises 31–34, identify the form of each integral as being inverse sine, inverse tangent, logarithmic, or general power, as in Examples 6 and 7. Do not integrate. In each part (a), explain how the choice was made.

31. (a) $\int \frac{2\,dx}{4 + 9x^2}$ (b) $\int \frac{2\,dx}{4 + 9x}$ (c) $\int \frac{2x\,dx}{\sqrt{4 + 9x^2}}$

32. (a) $\int \frac{2x\,dx}{4 - 9x^2}$ (b) $\int \frac{2\,dx}{\sqrt{4 - 9x}}$ (c) $\int \frac{2x\,dx}{4 + 9x^2}$

33. (a) $\int \frac{2x\,dx}{\sqrt{4 - 9x^2}}$ (b) $\int \frac{2\,dx}{\sqrt{4 - 9x^2}}$ (c) $\int \frac{2\,dx}{4 - 9x}$

34. (a) $\int \frac{2\,dx}{9x^2 + 4}$ (b) $\int \frac{2x\,dx}{\sqrt{9x^2 - 4}}$ (c) $\int \frac{2x\,dx}{9x^2 - 4}$

In Exercises 35–46, solve the given problems by integration.

35. Explain how to integrate $\int \frac{dx}{\sqrt{x}(1 + x)}$. What is the result?

36. Integrate $\int \sqrt{\frac{1 + x}{1 - x}}dx$ by first multiplying the numerator and denominator of the fraction under the radical by $1 + x$.

37. Find the area bounded by $y(1 + x^2) = 1, x = 0, y = 0$, and $x = 2$. See Fig. 28.9.

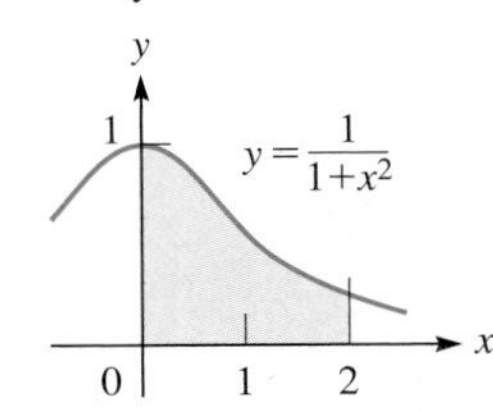

Fig. 28.9

38. Find the circumference of the circle $x^2 + y^2 = r^2$ by using the formula for arc length:

$$s = \int_a^b \sqrt{1 + \left(\frac{dy}{dx}\right)^2}\,dx$$

39. Find the area bounded by $y = \frac{2}{\sqrt{4 - x^2}}$ and $y = \frac{4}{4 + x^2} + 1$. Use a calculator to find the points of intersection.

40. A ball is rolling such that its velocity v (in cm/s) as a function of time t (in s) is $v = \frac{0.45}{0.25t^2 + 1}$. How far does it move in 10.0 s?

41. To find the electric field E from an electric charge distributed uniformly over the entire xy-plane at a distance d from the plane, it is necessary to evaluate the integral $kd\int \frac{dx}{d^2 + x^2}$. Here, x is the distance from the origin to the element of charge. Perform the indicated integration.

42. An oil-storage tank can be described as the volume generated by revolving the region bounded by $y = 24/\sqrt{16 + x^2}, x = 0, y = 0$, and $x = 3$ about the x-axis. Find the volume (in m^3) of the tank.

43. In dealing with the theory for simple harmonic motion, it is necessary to solve the equation $\frac{dx}{\sqrt{A^2 - x^2}} = \sqrt{\frac{k}{m}}dt$ (k, m, and A are constants). Determine the solution if $x = x_0$ when $t = 0$.

44. During each cycle, the velocity v (in ft/s) of a robotic welding device is given by $v = 2t - \frac{12}{2 + t^2}$, where t is the time (in s). Find the expression for the displacement s (in ft) as a function of t if $s = 0$ for $t = 0$.

45. Find the moment of inertia with respect to the y-axis for a flat plate covering the region bounded by $y = 1/(1 + x^6)$, the x-axis, $x = 1$, and $x = 2$.

46. Find the average value of the function $y = \frac{1}{\sqrt{1 - x^2}}$ for $0 \le x \le \frac{1}{2}$.

Answers to Practice Exercises

1. $\frac{1}{2}\sin^{-1}\frac{2x}{3} + C$ 2. $\frac{1}{6}\tan^{-1}\frac{3x}{2} + C$

28.7 Integration by Parts

Integration by Parts • Identifying u and dv • Repeated Use of Method

There are many methods of transforming integrals into forms that can be integrated by one of the basic formulas. In the preceding sections, we saw that completing the square and trigonometric identities can be used for this purpose. In this section and the following one, we develop two general methods. The method of *integration by parts* is discussed in this section.

Because the derivative of a product of functions is found by use of the formula

$$\frac{d(uv)}{dx} = u\frac{dv}{dx} + v\frac{du}{dx}$$

the differential of a product of functions is given by $d(uv) = u\,dv + v\,du$. Integrating both sides of this equation, we have $uv = \int u\,dv + \int v\,dv$. Solving for $\int u\,dv$, we obtain

$$\int u\,dv = uv - \int v\,du \tag{28.22}$$

Integration by use of Eq. (28.22) is called **integration by parts.**

EXAMPLE 1 Algebraic—trigonometric integrand

Integrate: $\int x \sin x\,dx$.

This integral does not fit any of the previous forms we have discussed, because neither x nor $\sin x$ can be made a factor of a proper du. However, by choosing $u = x$ and $dv = \sin x\,dx$, integration by parts may be used. Thus,

$$u = x \qquad dv = \sin x\,dx$$

By finding the differential of u and integrating dv, we find du and v. This gives us

$$du = dx \qquad v = -\cos x + C_1$$

Now, substituting in Eq. (28.22), we have

$$\begin{aligned}
\int \underset{u}{(x)}\,\underset{dv}{(\sin x\,dx)} &= \underset{u}{(x)}\,\underset{v}{(-\cos x + C_1)} - \int \underset{v}{(-\cos x + C_1)}\,\underset{du}{(dx)} \\
&= -x\cos x + C_1 x + \int \cos x\,dx - \int C_1\,dx \\
&= -x\cos x + C_1 x + \sin x - C_1 x + C \\
&= -x\cos x + \sin x + C
\end{aligned}$$

Other choices of u and dv may be made, but they are not useful. For example, if we choose $u = \sin x$ and $dv = x\,dx$, then $du = \cos x\,dx$ and $v = \frac{1}{2}x^2 + C_2$. This makes $\int v\,du = \int (\frac{1}{2}x^2 + C_2)(\cos x\,dx)$, which is more complex than the integrand of the original problem.

NOTE ▶ [We also note that the constant C_1 that was introduced when we integrated dv does not appear in the final result. This constant will always cancel out, and therefore we will not show any constant of integration when finding v.] ■

Practice Exercise

1. Integrate: $\int x \sec^2 x\,dx$.

As in Example 1, there is often more than one choice as to the part of the integrand that is selected to be u and the part that is selected to be dv. There are no set rules that may be stated for the best choice of u and dv, but two guidelines may be stated.

Guidelines for Choosing u and dv

1. The quantity u is normally chosen such that du/dx is of simpler form than u.
2. The differential dv is normally chosen such that $\int dv$ is easily integrated.

Working examples, and thereby gaining experience in methods of integration, is the best way to determine when this method should be used and how to use it.

EXAMPLE 2 Algebraic integrand

Integrate: $\int x\sqrt{1-x}\,dx$.

We see that this form does not fit the power rule, for $x\,dx$ is not a factor of the differential of $1-x$. By choosing $u = x$ and $dv = \sqrt{1-x}\,dx$, we have $du/dx = 1$, and v can readily be determined. Thus,

$$u = x \qquad dv = \sqrt{1-x}\,dx = (1-x)^{1/2}\,dx$$

$$du = dx \qquad v = -\frac{2}{3}(1-x)^{3/2}$$

Substituting in Eq. (28.22), we have

$$\underbrace{\int}\ \overset{u}{x}\,\overset{dv}{[(1-x)^{1/2}\,dx]} = \overset{u}{x}\overset{v}{\left[-\frac{2}{3}(1-x)^{3/2}\right]} - \int \overset{v}{\left[-\frac{2}{3}(1-x)^{3/2}\right]}\overset{du}{dx}$$

At this point, we see that we can complete the integration. Thus,

$$\begin{aligned}\int x(1-x)^{1/2}\,dx &= -\frac{2x}{3}(1-x)^{3/2} + \frac{2}{3}\int(1-x)^{3/2}\,dx\\ &= -\frac{2x}{3}(1-x)^{3/2} + \frac{2}{3}\left(-\frac{2}{5}\right)(1-x)^{5/2} + C\\ &= -\frac{2}{3}(1-x)^{3/2}\left[x + \frac{2}{5}(1-x)\right] + C\\ &= -\frac{2}{15}(1-x)^{3/2}(2+3x) + C\end{aligned}$$

■

EXAMPLE 3 Algebraic—logarithmic integrand

Integrate: $\int\sqrt{x}\ln x\,dx$.

For this integral, we have

$$u = \ln x \qquad dv = x^{1/2}\,dx$$

$$du = \frac{1}{x}dx \qquad v = \frac{2}{3}x^{3/2}$$

$$\begin{aligned}\int\sqrt{x}\ln x\,dx &= \frac{2}{3}x^{3/2}\ln x - \frac{2}{3}\int x^{1/2}\,dx\\ &= \frac{2}{3}x^{3/2}\ln x - \frac{4}{9}x^{3/2} + C\end{aligned}$$

■

EXAMPLE 4 Inverse sine integrand

Integrate: $\int \sin^{-1} x\, dx$.

We write

$$u = \sin^{-1} x \qquad dv = dx$$
$$du = \frac{dx}{\sqrt{1-x^2}} \qquad v = x$$
$$\int \sin^{-1} x\, dx = x\sin^{-1} x - \int \frac{x\, dx}{\sqrt{1-x^2}} = x\sin^{-1} x + \frac{1}{2}\int \frac{-2x\, dx}{\sqrt{1-x^2}}$$
$$= x\sin^{-1} x + \sqrt{1-x^2} + C$$

■

EXAMPLE 5 Algebraic—exponential integrand

In a certain electric circuit, the current i (in A) is given by the equation $i = te^{-t}$, where t is the time (in s). Find the charge q to pass a point in the circuit between $t = 0$ s and $t = 1.0$ s.

Because $i = \frac{dq}{dt}$, we have $\frac{dq}{dt} = te^{-t}$. Thus, with $q = \int_0^{1.0} te^{-t}\, dt$, we are to solve for q. For this integral,

$$u = t \qquad dv = e^{-t}\, dt$$
$$du = dt \qquad v = -e^{-t}$$
$$q = \int_0^{1.0} te^{-t}\, dt = -te^{-t}\Big|_0^{1.0} + \int_0^{1.0} e^{-t}\, dt = -te^{-t} - e^{-t}\Big|_0^{1.0}$$
$$= -e^{-1.0} - e^{-1.0} + 1.0 = 0.26 \text{ C}$$

Therefore, 0.26 C passes a given point in the circuit. ■

EXAMPLE 6 Integration using method twice

Integrate: $\int e^x \sin x\, dx$.

Let $u = \sin x$, $dv = e^x\, dx$, $du = \cos x\, dx$, $v = e^x$.

$$\int e^x \sin x\, dx = e^x \sin x - \int e^x \cos x\, dx$$

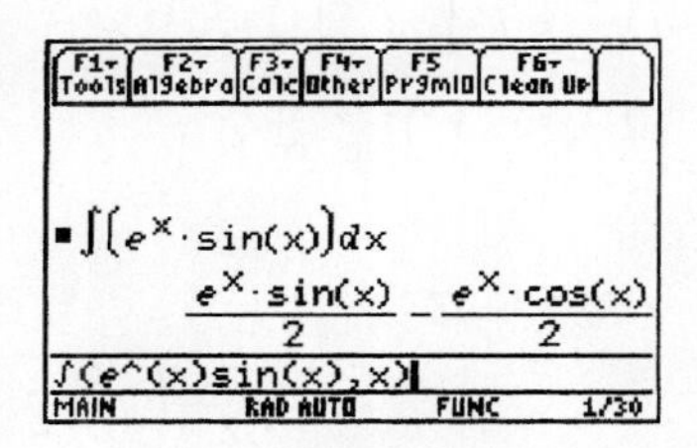

Fig. 28.10
TI-89 graphing calculator keystrokes: goo.gl/C8KdDa

■ This integration is shown on a TI-89 calculator in Fig. 28.10.

At first glance, it appears that we have made no progress in applying the method of integration by parts. We note, however, that when we integrated $\int e^x \sin x\, dx$, part of the result was a term of $\int e^x \cos x\, dx$. This implies that ***if*** $\int e^x \cos x\, dx$ ***were integrated, a term of*** $\int e^x \sin x\, dx$ ***might result.*** Thus, the method of integration by parts is now applied to the integral $\int e^x \cos x\, dx$:

$$u = \cos x \quad dv = e^x\, dx \quad du = -\sin x\, dx \quad v = e^x$$

And so $\int e^x \cos x\, dx = e^x \cos x + \int e^x \sin x\, dx$. Substituting this expression into the expression for $\int e^x \sin x\, dx$, we obtain

$$\int e^x \sin x\, dx = e^x \sin x - \left(e^x \cos x + \int e^x \sin x\, dx\right)$$
$$= e^x \sin x - e^x \cos x - \int e^x \sin x\, dx$$
$$2\int e^x \sin x\, dx = e^x(\sin x - \cos x) + 2C \qquad \text{add } \int e^x \sin x\, dx \text{ to both sides}$$
$$\int e^x \sin x\, dx = \frac{e^x}{2}(\sin x - \cos x) + C \qquad \text{divide both sides by 2}$$

Thus, by combining integrals of like form, we obtain the desired result. ■

Practice Exercise

2. Identify u, dv, du, and v for the integral $\int e^{-x} \cos x\, dx$.

EXERCISES 28.7

In Exercises 1 and 2, answer the given questions related to the indicated examples of this section.

1. In Example 2, do the choices $u = \sqrt{1-x}$ and $dv = x\,dx$ work for this integral? Explain.
2. In Example 6, do the choices $u = e^x$ and $dv = \sin x\,dx$ work for this integral? Explain.

In Exercises 3–26, integrate each of the given functions.

3. $\int \theta\cos\theta\,d\theta$
4. $\int x\sin 2x\,dx$
5. $\int 4xe^{2x}\,dx$
6. $\int 3xe^x\,dx$
7. $\int 3x\sec^2 x\,dx$
8. $\int_0^{\pi/4} x\sec x\tan x\,dx$
9. $\int 2\tan^{-1}x\,dx$
10. $\int \ln s\,ds$
11. $\int_{-3}^{0}\frac{4t\,dt}{\sqrt{1-t}}$
12. $\int x\sqrt{x+1}\,dx$
13. $\int x\ln x\,dx$
14. $\int 8x^2\ln 4x\,dx$
15. $\int \frac{\ln x}{x^3}dx$
16. $\int_0^1 r^2e^{2r}\,dr$
17. $\int_0^{\pi/2} e^x\cos x\,dx$
18. $\int 2e^{-x}\sin 2x\,dx$
19. $\int \sin x\ln\cos x\,dx$
20. $\int x\sin^{-1}x^2\,dx$
21. $\int_0^{\ln 2} xe^x\,dx$
22. $\int x\tan^{-1}x\,dx.$
23. $\int (x+4)^8(x+5)dx$
24. $\int \cos x\ln(\sin x)dx$
25. $\int \cos(\ln x)dx$
26. $\int \sec^3 x\,dx$

In Exercises 27–42, solve the given problems by integration.

27. To integrate $\int e^{-\sqrt{x}}\,dx$, the only choices possible of $u = e^{-\sqrt{x}}$ and $dv = dx$ do not work. However, if we first let $t = \sqrt{x}$, $dt = dx/2\sqrt{x}$, the integration can be done by parts. Perform this integration.
28. To integrate $\int x\ln(x+1)\,dx$, the substitution $t = x+1$, $dt = dx$ leads to an integral that can be done readily by parts. Perform this integration in this way.
29. Find the area bounded by $y = xe^{-x}$, $y = 0$, and $x = 2$. See Fig. 28.11.
30. Find the area bounded by $y = 2(\ln x)/x^2$, $y = 0$, and $x = 3$.

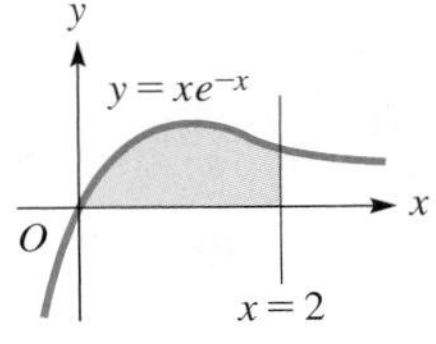

Fig. 28.11

31. Find the volume generated by revolving the region bounded by $y = \tan^2 x$, $y = 0$, and $x = 0.5$ about the y-axis.
32. Find the volume generated by revolving the region bounded by $y = \sin x$ and $y = 0$ from $x = 0$ to $x = \pi$ about the y-axis.
33. Find the x-coordinate of the centroid of a flat plate covering the region bounded by $y = \cos x$ and $y = 0$ for $0 \le x \le \pi/2$.
34. Find the moment of inertia with respect to its axis of the solid generated by revolving the region bounded by $y = e^x$, $x = 1$, and the coordinate axes about the y-axis.
35. Find the root-mean-square value of the function $y = \sqrt{\sin^{-1}x}$ between $x = 0$ and $x = 1$. (See Example 7 of Section 28.5.)
36. The general expression for the slope of a curve is $dy/dx = x^3\sqrt{1+x^2}$. Find an equation of the curve if it passes through the origin.
37. Find the area bounded by $y = x\sin x$, the x-axis, (a) between 0 and π, (b) between π and 2π, (c) between 2π and 3π. Note the pattern.
38. The displacement y (in cm) of a weight on a spring is given by $y = 4e^{-t}\cos t\,(t \ge 0)$. Find the average value of the displacement for the interval $0 \le t \le 2\pi$ s.
39. Computer simulation shows that the velocity v (in ft/s) of a test car is $v = t^3/\sqrt{t^2+1}$ from $t = 0$ to $t = 8.0$ s. Find the expression for the distance traveled by the car in t seconds.
40. The nose cone of a rocket has the shape of the solid that is generated by revolving the region bounded by $y = \ln x$, $y = 0$, and $x = 9.5$ about the x-axis. Find the volume (in m^3) of the nose cone. See Fig. 28.12.

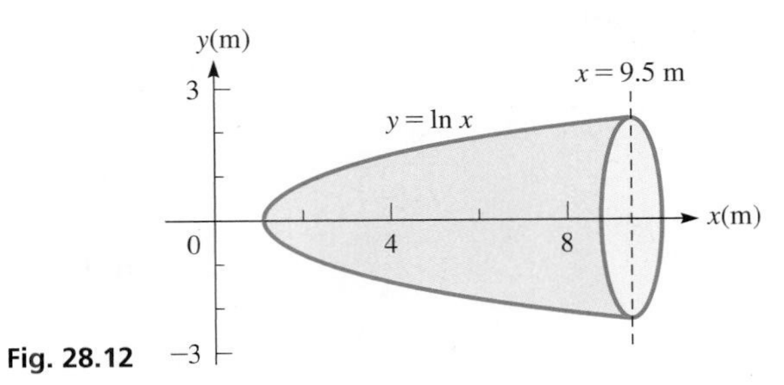

Fig. 28.12

41. The current in a given circuit is given by $i = e^{-2t}\cos t$. Find an expression for the amount of charge that passes a given point in the circuit as a function of the time, if $q_0 = 0$.
42. In finding the average length $\bar{x}$ (in nm) of a certain type of large molecule, we use the equation $\bar{x} = \lim_{b\to\infty}[0.1\int_0^b x^3e^{-x^2/8}\,dx]$. Evaluate the integral and then use a calculator to show that $\bar{x} \to 3.2$ nm as $b \to \infty$.

Answer to Practice Exercises

1. $x\tan x + \ln|\cos x| + C$
2. $u = \cos x$, $dv = e^{-x}\,dx$, $du = -\sin x\,dx$, $v = -e^{-x}$

28.8 Integration by Trigonometric Substitution

Transform Algebraic Integrals into Trigonometric Integrals

In this section, we show how trigonometric relations are useful in integrating certain types of algebraic integrals. Substitutions based on Eqs. (28.14), (28.15), and (28.16), which are shown in the margin, are particularly useful for integrating expressions involving radicals. Consider the following examples.

■ For reference, Eqs. (28.14), (28.15), and (28.16) are

$\cos^2 x + \sin^2 x = 1$
$1 + \tan^2 x = \sec^2 x$
$1 + \cot^2 x = \csc^2 x$

EXAMPLE 1 For $\sqrt{a^2 - x^2}$, let $x = a\sin\theta$

Integrate: $\displaystyle\int \frac{dx}{x^2\sqrt{1 - x^2}}$.

If we let $x = \sin\theta$, then $\sqrt{1 - \sin^2\theta} = \cos\theta$, and the integral can be transformed into a trigonometric integral. Carefully ***replacing all factors of the integral with expressions in terms of θ***, we have $x = \sin\theta$, $\sqrt{1 - x^2} = \cos\theta$, and $dx = \cos\theta\, d\theta$. Therefore,

$$\int \frac{dx}{x^2\sqrt{1 - x^2}} = \int \frac{\cos\theta\, d\theta}{\sin^2\theta\sqrt{1 - \sin^2\theta}} \quad \text{substituting}$$

$$= \int \frac{\cos\theta\, d\theta}{\sin^2\theta\cos\theta} = \int \csc^2\theta\, d\theta \quad \text{using trigonometric relations}$$

$$= -\cot\theta + C \quad \text{using Eq. (28.7)}$$

CAUTION We have now performed the integration, but the answer we now have is in terms of θ, and we must express the result in terms of x. ■ Making a triangle with an angle θ such that $\sin\theta = x/1$ (see Fig. 28.13), we may express any of the trigonometric functions in terms of x. (This is the method used with inverse trigonometric functions.) Thus,

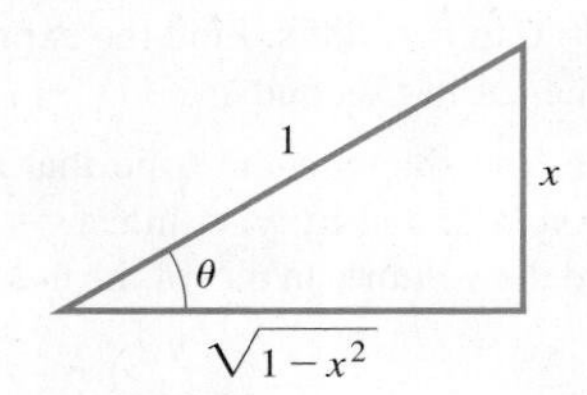

Fig. 28.13

$$\cot\theta = \frac{\sqrt{1 - x^2}}{x}$$

Therefore, the result of the integration becomes

$$\int \frac{dx}{x^2\sqrt{1 - x^2}} = -\cot\theta + C = -\frac{\sqrt{1 - x^2}}{x} + C$$ ■

EXAMPLE 2 For $\sqrt{a^2 + x^2}$, let $x = a\tan\theta$

Integrate: $\displaystyle\int \frac{dx}{\sqrt{x^2 + 4}}$.

If we let $x = 2\tan\theta$, the radical in this integral becomes

$$\sqrt{x^2 + 4} = \sqrt{4\tan^2\theta + 4} = 2\sqrt{\tan^2\theta + 1} = 2\sqrt{\sec^2\theta} = 2\sec\theta$$

Therefore, with $x = 2\tan\theta$ and $dx = 2\sec^2\theta\, d\theta$, we have

$$\int \frac{dx}{\sqrt{x^2 + 4}} = \int \frac{2\sec^2\theta\, d\theta}{\sqrt{4\tan^2\theta + 4}} = \int \frac{2\sec^2\theta\, d\theta}{2\sec\theta} \quad \text{substituting}$$

$$= \int \sec\theta\, d\theta = \ln|\sec\theta + \tan\theta| + C \quad \text{using Eq. (28.12)}$$

$$= \ln\left|\frac{\sqrt{x^2 + 4}}{2} + \frac{x}{2}\right| + C = \ln\left|\frac{\sqrt{x^2 + 4} + x}{2}\right| + C \quad \text{see Fig. 28.14}$$

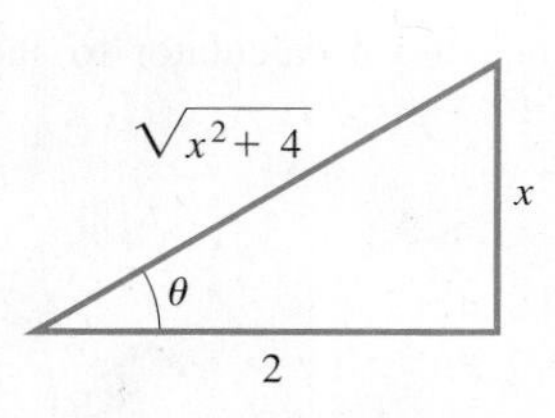

Fig. 28.14

This answer is acceptable, but by using the properties of logarithms, we have

$$\ln\left|\frac{\sqrt{x^2 + 4} + x}{2}\right| + C = \ln\left|\sqrt{x^2 + 4} + x\right| + (C - \ln 2)$$

$$= \ln\left|\sqrt{x^2 + 4} + x\right| + C'$$ ■

■ Combining constants, as in $C' = C - \ln 2$, is a common practice in integration problems.

EXAMPLE 3 For $\sqrt{x^2 - a^2}$, let $x = a\sec\theta$

Integrate: $\displaystyle\int \frac{2\,dx}{x\sqrt{x^2 - 9}}$.

If we let $x = 3\sec\theta$, the radical in this integral becomes

$$\sqrt{x^2 - 9} = \sqrt{9\sec^2\theta - 9} = 3\sqrt{\sec^2\theta - 1} = 3\sqrt{\tan^2\theta} = 3\tan\theta$$

Therefore, with $x = 3\sec\theta$ and $dx = 3\sec\theta\tan\theta\,d\theta$, we have

$$\int \frac{2\,dx}{x\sqrt{x^2 - 9}} = 2\int \frac{3\sec\theta\tan\theta\,d\theta}{3\sec\theta\sqrt{9\sec^2\theta - 9}} = 2\int \frac{\tan\theta\,d\theta}{3\tan\theta}$$

$$= \frac{2}{3}\int d\theta = \frac{2}{3}\theta + C = \frac{2}{3}\sec^{-1}\frac{x}{3} + C$$

It is not necessary to refer to a triangle to express the result in terms of x. The solution is found by solving $x = 3\sec\theta$ for θ, as indicated. ■

These examples show that by making the proper substitution, we can integrate algebraic functions by using the equivalent trigonometric forms. In summary, for the indicated radical form, the following **trigonometric substitutions** are used:

$$\begin{array}{lll} \text{For } \sqrt{a^2 - x^2}, & \text{use} & x = a\sin\theta \\ \text{For } \sqrt{a^2 + x^2}, & \text{use} & x = a\tan\theta \\ \text{For } \sqrt{x^2 - a^2}, & \text{use} & x = a\sec\theta \end{array} \qquad \textbf{(28.23)}$$

Practice Exercise

1. What substitution should be used to integrate $\displaystyle\int \frac{2\,dx}{3x^2\sqrt{4 + x^2}}$?

EXAMPLE 4 Trigonometric substitution—arc length of a robot arm

The joint between two links of a robot arm moves back and forth along the curve defined by $y = 3.0\ln x$ from $x = 1.0$ in. to $x = 4.0$ in. Find the distance the joint moves in one cycle.

■ The arc length s along a curve is given by

$$s = \int_a^b \sqrt{1 + \left(\frac{dy}{dx}\right)^2}\,dx$$

The required distance is found by using the equation for arc length, which is shown in the margin. Therefore, we find the derivative as $dy/dx = 3.0/x$, which means the total distance s moved in one cycle is

$$s = 2\int_{1.0}^{4.0} \sqrt{\left[1 + \left(\frac{3.0}{x}\right)^2\right]}\,dx = 2\int_{1.0}^{4.0} \frac{\sqrt{x^2 + 9.0}}{x}dx$$

To integrate, we make the substitution $x = 3.0\tan\theta$ and $dx = 3.0\sec^2\theta\,d\theta$. Thus,

$$\int \frac{\sqrt{x^2 + 9.0}}{x}dx = \int \frac{\sqrt{9.0\tan^2\theta + 9.0}}{3.0\tan\theta}(3.0\sec^2\theta\,d\theta) = 3.0\int \frac{\sec\theta}{\tan\theta}(1 + \tan^2\theta)\,d\theta$$

$$= 3.0\left(\int \frac{\sec\theta}{\tan\theta}d\theta + \int \tan\theta\sec\theta\,d\theta\right) = 3.0\left(\int \csc\theta\,d\theta + \int \tan\theta\sec\theta\,d\theta\right)$$ see Fig. 28.15

$$= 3.0(\ln|\csc\theta - \cot\theta| + \sec\theta) + C = 3.0\left[\ln\left|\frac{\sqrt{x^2 + 9.0}}{x} - \frac{3.0}{x}\right| + \frac{\sqrt{x^2 + 9.0}}{3.0}\right] + C$$

Limits have not been included, due to the change in variables. Evaluating, we have

$$s = 2\int_{1.0}^{4.0} \frac{\sqrt{x^2 + 9.0}}{x}dx = 6.0\left[\ln\left|\frac{\sqrt{x^2 + 9.0} - 3.0}{x}\right| + \frac{\sqrt{x^2 + 9.0}}{3.0}\right]\Bigg|_{1.0}^{4.0}$$

$$= 6.0\left[\left(\ln 0.50 - \ln\frac{\sqrt{10.0} - 3.0}{1.0}\right) + \frac{5.0}{3.0} - \frac{\sqrt{10.0}}{3.0}\right] = 10.4\text{ in.}$$ ■

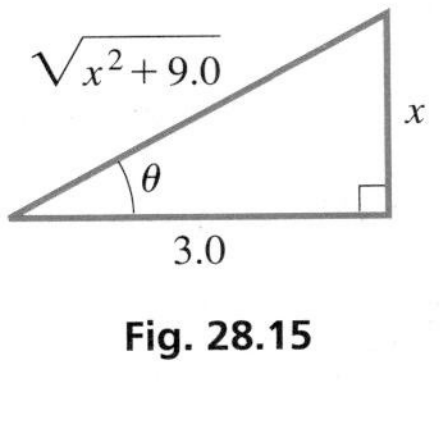

$\sqrt{x^2+9.0}$, x, θ, 3.0

Fig. 28.15

EXERCISES 28.8

In Exercises 1 and 2, answer the given equations related to the indicated examples of this section.

1. In Example 1, how must the integrand be changed in order to have a result of $\sin^{-1}x + C$?

2. In Example 2, how must the integrand be changed in order to have a result of $\tan^{-1}(x/2) + C$?

In Exercises 3–8, give the proper trigonometric substitution and find the transformed integral, but do not integrate.

3. $\int \frac{\sqrt{9-x^2}}{x^2}dx$ **4.** $\int \frac{dx}{\sqrt{x^2-16}}$ **5.** $\int \frac{dx}{x^2\sqrt{x^2+1}}$

6. $\int \frac{dx}{(1-x^2)^{3/2}}$ **7.** $\int \frac{dx}{x\sqrt{x^2-1}}$ **8.** $\int \sqrt{25+x^2}\,dx$

In Exercises 9–26, integrate each of the given functions.

9. $\int \frac{\sqrt{1-x^2}}{x^2}dx$ **10.** $\int_0^4 \frac{9\,dt}{(t^2+9)^{3/2}}$

11. $\int \frac{2\,dx}{\sqrt{x^2-36}}$ **12.** $\int \frac{\sqrt{x^2-25}}{x}dx$

13. $\int \frac{6\,dz}{z^2\sqrt{z^2+9}}$ **14.** $\int \frac{6\,dx}{x\sqrt{4-x^2}}$

15. $\int \frac{4\,dx}{(4-x^2)^{3/2}}$ **16.** $\int \frac{6p^3\,dp}{\sqrt{9+p^2}}$

17. $\int_0^{0.5} \frac{x^3\,dx}{\sqrt{1-x^2}}$ **18.** $\int_4^5 \frac{\sqrt{x^2-16}}{x^2}dx$

19. $\int \frac{5\,dx}{\sqrt{x^2+2x+2}}$ **20.** $\int \frac{dx}{\sqrt{x^2+4x}}$

21. $\int_{2.5}^3 \frac{dy}{y\sqrt{4y^2-9}}$ **22.** $\int \sqrt{16-x^2}\,dx$

23. $\int \frac{dx}{\sqrt{16-x^2}}$ **24.** $\int \frac{dx}{x^2\sqrt{4-x^2}}$

25. $\int \frac{2\,dx}{\sqrt{e^{2x}-1}}$ **26.** $\int \frac{12\sec^2 u\,du}{(4-\tan^2 u)^{3/2}}$

In Exercises 27–38, solve the given problems by integration.

27. Perform the integration $\int x\sqrt{1-x^2}\,dx$ (a) by using the power formula, and (b) by trigonometric substitution. Compare results.

28. Perform the integration $\int \frac{x\,dx}{x^2+1}$ (a) by using the logarithmic formula, and (b) by trigonometric substitution. Compare results.

29. Find the area of a quarter circle of radius 2 by integrating $\int_0^2 \sqrt{4-x^2}\,dx.$

30. Find the area bounded by $y = \frac{1}{x^2\sqrt{x^2-1}}, x = \sqrt{2}, x = \sqrt{5},$ and $y = 0$. See Fig. 28.16.

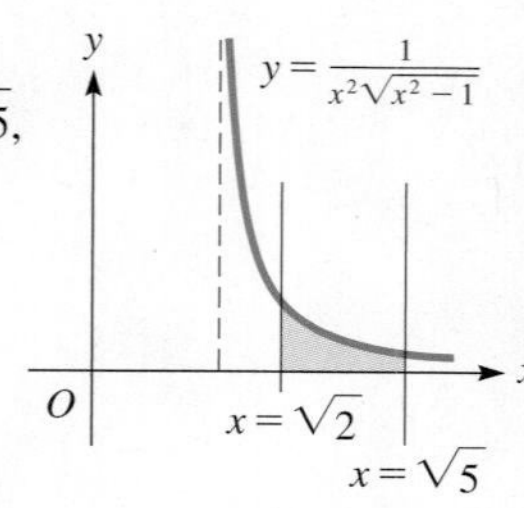

Fig. 28.16

31. Find the moment of inertia with respect to the y-axis of a flat plate covering the first-quadrant region under the circle $x^2 + y^2 = a^2$ in terms of its mass.

32. Find the moment of inertia of a sphere of radius a with respect to its axis in terms of its mass.

33. A marathon runner's speed (in km/h) is $v = \frac{48}{0.25t^2+4}$. This runner is in the same race as the runners in Exercise 41 of Section 28.2 and Exercise 39 of Section 28.3. (a) Who is ahead after 3.00 h? (b) Who is first, second, and third after 4.00 h?

34. The perimeter of the rudder of a boat can be described as the region in the fourth quadrant bounded by $y = -0.5x^2\sqrt{4-x^2}$ and the x-axis. Find the area of one side of the rudder.

35. Find the x-coordinate of the centroid of the boat rudder in Exercise 34.

36. The vertical cross section of a highway culvert is defined by the region within the ellipse $1.00x^2 + 9.00y^2 = 9.00$, where dimensions are in meters. Find the area of the cross section of the culvert.

37. If an electric charge Q is distributed along a straight wire of length $2a$, the electric potential V at a point P, which is at a distance b from the center of the wire, is $V = kQ\int_{-a}^{a}\frac{dx}{\sqrt{b^2+x^2}}$. Here, k is a constant and x is the distance along the wire. Evaluate the integral.

38. An electric insulating ring for a machine part can be described as the volume generated by revolving the region bounded by $y = x^2\sqrt{x^2-4}$, $y = 0$, and $x = 2.5$ cm about the y-axis. Find the volume (in cm^3) of material in the ring.

Certain algebraic integrals can be transformed into integrable form with the appropriate ***algebraic substitution.*** *For an expression of the form $(ax+b)^{p/q}$, a substitution of the form $u = (ax+b)^{1/q}$ may put it into an integrable form. In Exercises 39–42, use this type of substitution for the given integrals.*

39. $\int x\sqrt{x+1}\,dx$ **40.** $\int x\sqrt[3]{8-x}\,dx$

41. $\int x(x-4)^{2/3}\,dx$ **42.** $\int \frac{x^2\,dx}{(4x+1)^{5/2}}$

Answer to Practice Exercise

1. $x = 2\tan\theta$

28.9 Integration by Partial Fractions: Nonrepeated Linear Factors

Method of Partial Fractions • Nonrepeated Linear Factors • Solution by Substitution • Solution by Equating Coefficients

We have seen how the *derivative of a product* and *trigonometric identities* are used to write integrals into a form that can be integrated. In this section and the next, we show an algebraic method by which integrands that are fractions can also be changed into an integrable form.

In algebra, we combine fractions into a single fraction by means of addition. However, if we wish to integrate an expression that contains a rational function, in which both numerator and denominator are polynomials, it is often advantageous to reverse the process of addition and express the rational function as the sum of simpler fractions.

EXAMPLE 1 Illustrating partial fractions

In attempting to integrate $\int \frac{7-x}{x^2+x-2}dx$, we find that it does not fit any of the standard forms in this chapter. However, we can show that

$$\frac{7-x}{x^2+x-2} = \frac{2}{x-1} - \frac{3}{x+2}$$

This means that

$$\int \frac{7-x}{x^2+x-2}dx = \int \frac{2\,dx}{x-1} - \int \frac{3\,dx}{x+2}$$
$$= 2\ln|x-1| - 3\ln|x+2| + C$$

We see that by writing the fraction in the original integrand as the sum of the simpler fractions, each resulting integrand can be integrated. ■

In Example 1, we saw that the integral is readily determined once the fraction $(7-x)/(x^2+x-2)$ is replaced by the simpler fractions. In this section and the next, we describe how certain rational functions can be expressed in terms of simpler fractions and thereby be integrated. This technique is called the **method of partial fractions.**

NOTE ▸ [In order to express the rational function $f(x)/g(x)$ in terms of simpler partial fractions, the degree of the numerator $f(x)$ must be less than that of the denominator $g(x)$.] If this is not the case, we divide the numerator by the denominator until the remainder is of the proper form. Then the denominator $g(x)$ is factored into a product of linear and quadratic factors. The method of determining the partial fractions depends on the factors that are obtained. In advanced algebra, the form of the partial fractions is shown (but we shall not show the proof here).

There are four cases for the types of factors of the denominator. They are (1) *nonrepeated linear factors,* (2) *repeated linear factors,* (3) *nonrepeated quadratic factors,* and (4) *repeated quadratic factors.* In this section, we consider the case of nonrepeated linear factors, and the other cases are discussed in the next section.

NONREPEATED LINEAR FACTORS

For the case of nonrepeated linear factors, we use the fact that *corresponding to each linear factor $ax+b$, occurring once in the denominator, there will be a partial fraction of the form*

$$\frac{A}{ax+b}$$

where A is a constant to be determined. The following examples illustrate the method.

EXAMPLE 2 Find partial fractions—integrate

Integrate: $\int \frac{7-x}{x^2+x-2}dx$. (This is the same integral as in Example 1. Here, we see how the partial fractions are found.)

First, we note that the degree of the numerator is 1 (the highest-power term is x) and that of the denominator is 2 (the highest-power term is x^2). Because the degree of the denominator is higher, we may proceed to factoring it. Thus,

$$\frac{7-x}{x^2+x-2} = \frac{7-x}{(x-1)(x+2)}$$

There are two linear factors, $(x-1)$ and $(x+2)$, in the denominator, and they are different. This means that there are two partial fractions. Therefore, we write

$$\frac{7-x}{(x-1)(x+2)} = \frac{A}{x-1} + \frac{B}{x+2} \quad \textbf{(1)}$$

We are to determine constants A and B so that Eq. (1) is an identity. In finding A and B, we clear Eq. (1) of fractions by multiplying both sides by $(x-1)(x+2)$.

$$7-x = A(x+2) + B(x-1) \quad \textbf{(2)}$$

Equation (2) is also an identity, which means that there are two ways of determining the values of A and B.

NOTE ▸ *Solution by substitution:* [Because Eq. (2) is an identity, **it is true for any value of x.**] Thus, in turn we pick $x=-2$ and $x=1$, for each of these values makes a factor on the right equal to zero, and the values of B and A are easily found. Therefore,

$$\text{For } x=-2: \quad 7-(-2) = A(-2+2) + B(-2-1)$$
$$9 = -3B \qquad B=-3$$
$$\text{For } x=1: \quad 7-1 = A(1+2) + B(1-1)$$
$$6 = 3A, \qquad A = 2$$

NOTE ▸ *Solution by equating coefficients:* [Because Eq. (2) is an identity, another way of finding the constants A and B is to **equate coefficients of like powers of x from each side.**] Thus, writing Eq. (2) as

$$7-x = (2A-B) + (A+B)x$$

we have

$$2A - B = 7 \qquad \text{equating constants}$$
$$A + B = -1 \qquad \text{equating coefficients of } x$$

Now, using the values $A=2$ and $B=-3$ (as found at the left), we have

$$\frac{7-x}{(x-1)(x+2)} = \frac{2}{x-1} - \frac{3}{x+2}$$

Therefore, the integral is found as in Example 1.

$$\int \frac{7-x}{x^2+x-2}dx = \int \frac{7-x}{(x-1)(x+2)}dx = \int \frac{2\,dx}{x-1} - \int \frac{3\,dx}{x+2}$$
$$= 2\ln|x-1| - 3\ln|x+2| + C$$

Using the properties of logarithms, we may write this as

$$\int \frac{7-x}{x^2+x-2}dx = \ln\left|\frac{(x-1)^2}{(x+2)^3}\right| + C$$

■

■ The method of partial fractions essentially reverses the process of combining fractions over a common denominator. By determining that

$$\frac{7-x}{x^2+x-2} = \frac{2}{x-1} - \frac{3}{x+2}$$

reverses the process by which

$$\frac{2}{x-1} - \frac{3}{x+2} = \frac{2(x+2)-3(x-1)}{(x-1)(x+2)} = \frac{7-x}{x^2+x-2}$$

■
$$2A - B = 7$$
$$A + B = -1$$
$$3A = 6$$
$$A = 2$$
$$2(2) - B = 7$$
$$-B = 3$$
$$B = -3$$

Practice Exercise

1. Find the partial fractions for $\frac{3x+11}{x^2-2x-3}$.

EXAMPLE 3 Find partial fractions—integrate

Integrate: $\displaystyle\int \frac{6x^2 - 14x - 11}{(x+1)(x-2)(2x+1)}dx.$

The denominator is factored and is of degree 3 (when multiplied out, the highest-power term of x is x^3). This means we have three nonrepeated linear factors.

$$\frac{6x^2 - 14x - 11}{(x+1)(x-2)(2x+1)} = \frac{A}{x+1} + \frac{B}{x-2} + \frac{C}{2x+1}$$

Multiplying through by $(x+1)(x-2)(2x+1)$, we have

$$6x^2 - 14x - 11 = A(x-2)(2x+1) + B(x+1)(2x+1) + C(x+1)(x-2)$$

We now substitute the values of 2, $-\frac{1}{2}$, and -1 for x. Again, these are chosen because they make factors of the coefficients of A, B, or C equal to zero.

For $x = 2$: $\quad 6(4) - 14(2) - 11 = A(0)(5) + B(3)(5) + C(3)(0), \quad B = -1$

For $x = -\frac{1}{2}$: $\quad 6\left(\frac{1}{4}\right) - 14\left(-\frac{1}{2}\right) - 11 = A\left(-\frac{5}{2}\right)(0) + B\left(\frac{1}{2}\right)(0) + C\left(\frac{1}{2}\right)\left(-\frac{5}{2}\right), \quad C = 2$

For $x = -1$: $\quad 6(1) - 14(-1) - 11 = A(-3)(-1) + B(0)(-1) + C(0)(-3), \quad A = 3$

Therefore,

$$\int \frac{6x^2 - 14x - 11}{(x+1)(x-2)(2x+1)}dx = \int \frac{3\,dx}{x+1} - \int \frac{dx}{x-2} + \int \frac{2\,dx}{2x+1}$$

$$= 3\ln|x+1| - \ln|x-2| + \ln|2x+1| + C_1 = \ln\left|\frac{(2x+1)(x+1)^3}{x-2}\right| + C_1$$

Here, we have let the constant of integration be C_1 since we used C as the numerator of the third partial fraction. ■

EXAMPLE 4 Find partial fractions—integrate

Integrate: $\displaystyle\int \frac{2x^4 - x^3 - 9x^2 + x - 12}{x^3 - x^2 - 6x}dx.$

CAUTION Because the numerator is of a higher degree than the denominator, we must first divide the numerator by the denominator. ■ This gives

$$\frac{2x^4 - x^3 - 9x^2 + x - 12}{x^3 - x^2 - 6x} = 2x + 1 + \frac{4x^2 + 7x - 12}{x^3 - x^2 - 6x}$$

We must now express this rational function in terms of its partial fractions.

$$\frac{4x^2 + 7x - 12}{x^3 - x^2 - 6x} = \frac{4x^2 + 7x - 12}{x(x+2)(x-3)} = \frac{A}{x} + \frac{B}{x+2} + \frac{C}{x-3}$$

Clearing fractions, we have

$$4x^2 + 7x - 12 = A(x+2)(x-3) + Bx(x-3) + Cx(x+2)$$

Now, substituting the values -2, 3, and 0 in for x, we obtain the values of $B = -1$, $C = 3$, and $A = 2$, respectively. Therefore,

$$\int \frac{2x^4 - x^3 - 9x^2 + x - 12}{x^3 - x^2 - 6x}dx = \int \left(2x + 1 + \frac{2}{x} - \frac{1}{x+2} + \frac{3}{x-3}\right)dx$$

$$= x^2 + x + 2\ln|x| - \ln|x+2| + 3\ln|x-3| + C_1$$

$$= x^2 + x + \ln\left|\frac{x^2(x-3)^3}{x+2}\right| + C_1$$

■

In Example 2, we showed two ways of finding the values of A and B. The method of substitution is generally easier to use with linear factors, and we used it in Examples 3 and 4. However, as we will see in the next section, the method of equating coefficients can be very useful for other cases with partial fractions.

EXERCISES 28.9

In Exercises 1 and 2, make the given changes in the integrands of the indicated examples of this section, and then find the resulting fractions to be used in the integration. Do not integrate.

1. In Example 2, change the numerator to $10 - x$.

2. In Example 3, change the numerator to $x^2 - 12x - 10$.

In Exercises 3–6, write out the form of the partial fractions, similar to that shown in Eq. (1) of Example 2, that would be used to perform the indicated integrations. Do not evaluate the constants.

3. $\int \frac{3x+2}{x^2+x}dx$

4. $\int \frac{9-x}{x^2+2x-3}dx$

5. $\int \frac{x^2-6x-8}{x^3-4x}dx$

6. $\int \frac{2x^2-5x-7}{x^3+2x^2-x-2}dx$

In Exercises 7–24, integrate each of the given functions.

7. $\int \frac{x+3}{(x+1)(x+2)}dx$

8. $\int \frac{x+2}{x(x+1)}dx$

9. $\int \frac{8\,dx}{x^2-4}$

10. $\int \frac{p-9}{2p^2-3p+1}dp$

11. $\int \frac{x^2+3}{x^2+3x}dx$

12. $\int \frac{x^3}{x^2+3x+2}dx$

13. $\int_0^1 \frac{2t+4}{3t^2+5t+2}dt$

14. $\int_1^3 \frac{x-1}{4x^2+x}dx$

15. $\int \frac{4x^2-10}{x(x+1)(x-5)}dx$

16. $\int \frac{4x^2+21x+6}{(x+2)(x-3)(x+4)}dx$

17. $\int \frac{12x^2-4x-2}{4x^3-x}dx$

18. $\int_1^2 \frac{x^3+7x^2+9x+2}{x(x^2+3x+2)}dx$

19. $\int_2^3 \frac{dR}{R^3-R}$

20. $\int \frac{2x^3+x-1}{x^3+x^2-4x-4}dx$

21. $\int \frac{dV}{(V^2-4)(V^2-9)}$

22. $\int \frac{x^3+2x}{x^2+x-2}dx$

23. $\int \frac{2x\,dx}{x^4-3x^3+2x^2}$

24. $\int \frac{e^x\,dx}{e^{2x}+3e^x+2}$

In Exercises 25–36, solve the given problems by integration.

25. Derive the general formula $\int \frac{du}{u(a+bu)} = -\frac{1}{a}\ln\frac{a+bu}{u} + C$.

26. Derive the general formula $\int \frac{du}{u^2-a^2} = \frac{1}{2a}\ln\frac{u-a}{u+a} + C$.

27. Integrate $\int \frac{dx}{x-\sqrt[3]{x}}$ by first letting $x = u^3$.

28. To integrate $\int \frac{x^2\,dx}{(x-2)(x+3)}$, explain why A and B cannot be found if we let $\frac{x^2}{(x-2)(x+3)} = \frac{A}{x-2} + \frac{B}{x+3}$.

29. Integrate: $\int \frac{\cos\theta}{\sin^2\theta + 2\sin\theta - 3}d\theta$. (*Hint:* First, make the substitution $u = \sin\theta$.)

30. Find the first-quadrant area bounded by $y = 1/(x^3+3x^2+2x)$, $x = 1$, and $x = 3$.

31. Find the volume generated if the region of Exercise 30 is revolved about the y-axis.

32. Find the x-coordinate of the centroid of a flat plate that covers the region bounded by $y(x^2-1) = 1$, $y = 0$, $x = 2$, and $x = 4$.

33. Find an equation of the curve that passes through (1, 0), and the general expression for the slope is $(3x+5)/(x^2+5x)$.

34. The current i (in A) as a function of the time t (in s) in a certain electric circuit is given by $i = (4t+3)/(2t^2+3t+1)$. Find the total charge that passes through a point during the first second.

35. The force F (in N) applied by a stamping machine in making a certain computer part is $F = 4x/(x^2+3x+2)$, where x is the distance (in cm) through which the force acts. Find the work done by the force from $x = 0$ to $x = 0.500$ cm.

36. Under specified conditions, the time t (in min) required to form x grams of a substance during a chemical reaction is given by $t = \int dx/[(4-x)(2-x)]$. Find the equation relating t and x if $x = 0$ g when $t = 0$ min.

Answer to Practice Exercise

1. $\frac{3x+11}{x^2-2x-3} = \frac{5}{x-3} - \frac{2}{x+1}$

28.10 Integration by Partial Fractions: Other Cases

Repeated Linear Factors • The Only Factor Is Repeated • Nonrepeated Quadratic Factors • Repeated Quadratic Factors

In the previous section, we introduced the method of partial fractions and considered the case of nonrepeated linear factors. In this section, we develop the use of partial fractions for the cases of repeated linear factors and nonrepeated quadratic factors. We also briefly discuss the case of repeated quadratic factors.

REPEATED LINEAR FACTORS

For repeated linear factors, we use the fact that *corresponding to each linear factor $ax + b$ that occurs n times in the denominator there will be n partial fractions*

■ We usually use capital letters A, B, C, D, E, etc. in the numerators of the partial fractions.

$$\frac{A_1}{ax + b} + \frac{A_2}{(ax + b)^2} + \cdots + \frac{A_n}{(ax + b)^n}$$

where $A_1, A_2, \ldots, A_n$ are constants to be determined.

EXAMPLE 1 Two factors—one repeated

Integrate: $\int \frac{dx}{x(x + 3)^2}$.

Here, we see that the denominator has a factor of x and two factors of $x + 3$. For the factor of x, we use a partial fraction as in the previous section, for it is a nonrepeated factor. For the factor $x + 3$, we need two partial fractions, one with a denominator of $x + 3$ and the other with a denominator of $(x + 3)^2$. Thus, we write

$$\frac{1}{x(x + 3)^2} = \frac{A}{x} + \frac{B}{x + 3} + \frac{C}{(x + 3)^2}$$

Multiplying each side by $x(x + 3)^2$, we have

$$1 = A(x + 3)^2 + Bx(x + 3) + Cx \qquad \textbf{(1)}$$

Substituting the values -3 and 0 in for x, we have

$$\text{For } x = -3: \quad 1 = A(0^2) + B(-3)(0) + (-3)C, \qquad C = -\frac{1}{3}$$

$$\text{For } x = 0: \quad 1 = A(3^2) + B(0)(3) + C(0), \qquad A = \frac{1}{9}$$

Because no other numbers make a factor in Eq. (1) equal to zero, we either choose some other value of x or equate coefficients of some power of x in Eq. (1). Because Eq. (1) is an identity, we may choose any value of x. With $x = 1$, we have

$$1 = A(4^2) + B(1)(4) + C(1)$$
$$1 = 16A + 4B + C$$

Using the known values of A and C, we have

$$1 = 16\left(\frac{1}{9}\right) + 4B - \frac{1}{3}, \qquad B = -\frac{1}{9}$$

This means that

$$\frac{1}{x(x + 3)^2} = \frac{\frac{1}{9}}{x} + \frac{-\frac{1}{9}}{x + 3} + \frac{-\frac{1}{3}}{(x + 3)^2}$$

or

$$\int \frac{dx}{x(x + 3)^2} = \frac{1}{9}\int \frac{dx}{x} - \frac{1}{9}\int \frac{dx}{x + 3} - \frac{1}{3}\int \frac{dx}{(x + 3)^2}$$
$$= \frac{1}{9}\ln|x| - \frac{1}{9}\ln|x + 3| - \frac{1}{3}\left(\frac{1}{-1}\right)(x + 3)^{-1} + C_1$$
$$= \frac{1}{9}\ln\left|\frac{x}{x + 3}\right| + \frac{1}{3(x + 3)} + C_1$$

The properties of logarithms are used to write the final form of the answer. ■

Practice Exercise

1. Find the partial fractions for $\frac{2}{x(x - 1)^2}$.

EXAMPLE 2 Two factors—one repeated

Integrate: $\int \frac{3x^3 + 15x^2 + 21x + 15}{(x-1)(x+2)^3} dx$.

First, we set up the partial fractions as

$$\frac{3x^3 + 15x^2 + 21x + 15}{(x-1)(x+2)^3} = \frac{A}{x-1} + \frac{B}{x+2} + \frac{C}{(x+2)^2} + \frac{D}{(x+2)^3}$$

Next, we clear fractions:

$$3x^3 + 15x^2 + 21x + 15 = A(x+2)^3 + B(x-1)(x+2)^2 + C(x-1)(x+2) + D(x-1) \quad \textbf{(1)}$$

For $x = 1$: $\quad 3 + 15 + 21 + 15 = 27A, \quad 54 = 27A, \quad A = 2$

For $x = -2$: $\quad 3(-8) + 15(4) + 21(-2) + 15 = -3D, \quad 9 = -3D, \quad D = -3$

To find B and C, we equate coefficients of powers of x. Therefore, we write Eq. (1) as

$$3x^3 + 15x^2 + 21x + 15 = (A+B)x^3 + (6A + 3B + C)x^2 + (12A + C + D)x + (8A - 4B - 2C - D)$$

Coefficients of x^3: $\quad 3 = A + B, \quad 3 = 2 + B, \quad B = 1$

Coefficients of x^2: $\quad 15 = 6A + 3B + C, \quad 15 = 12 + 3 + C, \quad C = 0$

$$\frac{3x^3 + 15x^2 + 21x + 15}{(x-1)(x+2)^3} = \frac{2}{x-1} + \frac{1}{x+2} + \frac{0}{(x+2)^2} + \frac{-3}{(x+2)^3}$$

$$\int \frac{3x^3 + 15x^2 + 21x + 15}{(x-1)(x+2)^3} dx = 2\int \frac{dx}{x-1} + \int \frac{dx}{x+2} - 3\int \frac{dx}{(x+2)^3}$$

$$= 2\ln|x-1| + \ln|x+2| - 3\left(\frac{1}{-2}\right)(x+2)^{-2} + C_1$$

$$= \ln|(x-1)^2(x+2)| + \frac{3}{2(x+2)^2} + C_1$$

■

EXAMPLE 3 The only factor is repeated

Integrate: $\int \frac{x\,dx}{(x-2)^3}$.

This could be integrated by first setting up the appropriate partial fractions. However, the solution is more easily found by using the substitution $u = x - 2$. Using this, we have

$$u = x - 2 \qquad x = u + 2 \qquad dx = du$$

$$\int \frac{x\,dx}{(x-2)^3} = \int \frac{(u+2)(du)}{u^3} = \int \frac{du}{u^2} + 2\int \frac{du}{u^3}$$

$$= \int u^{-2}\,du + 2\int u^{-3}\,du = \frac{1}{-u} + \frac{2}{-2}u^{-2} + C$$

$$= -\frac{1}{u} - \frac{1}{u^2} + C = -\frac{u+1}{u^2} + C$$

$$= -\frac{x-2+1}{(x-2)^2} + C = \frac{1-x}{(x-2)^2} + C$$

■

NONREPEATED QUADRATIC FACTORS

For the case of nonrepeated quadratic factors, we use the fact that *corresponding to each irreducible quadratic factor* $ax^2 + bx + c$ *that occurs once in the denominator there is a partial fraction of the form*

$$\frac{Ax + B}{ax^2 + bx + c}$$

where A and B are constants to be determined. (Here, an *irreducible quadratic factor* is one that cannot be further factored into linear factors involving only real numbers.) This case is illustrated in the following examples.

EXAMPLE 4 Two factors—one quadratic

Integrate: $\int \frac{4x + 4}{x^3 + 4x} dx$.

In setting up the partial fractions, we note that the denominator factors as $x^3 + 4x = x(x^2 + 4)$. Here, the factor $x^2 + 4$ cannot be further factored. This means we have

$$\frac{4x + 4}{x^3 + 4x} = \frac{4x + 4}{x(x^2 + 4)} = \frac{A}{x} + \frac{Bx + C}{x^2 + 4}$$

Clearing fractions, we have

$$\begin{aligned} 4x + 4 &= A(x^2 + 4) + Bx^2 + Cx \\ &= (A + B)x^2 + Cx + 4A \end{aligned}$$

Equating coefficients of powers of x gives us

$$\begin{aligned} &\text{For } x^2\text{:} \quad 0 = A + B \\ &\text{For } x\text{:} \quad 4 = C \\ &\text{For constants:} \quad 4 = 4A, \qquad A = 1 \end{aligned}$$

Therefore, we easily find that $B = -1$ from the first equation. This means that

$$\frac{4x + 4}{x^3 + 4x} = \frac{1}{x} + \frac{-x + 4}{x^2 + 4}$$

and

$$\begin{aligned} \int \frac{4x + 4}{x^3 + 4x} dx &= \int \frac{1}{x} dx + \int \frac{-x + 4}{x^2 + 4} dx \\ &= \int \frac{1}{x} dx - \int \frac{x\,dx}{x^2 + 4} + \int \frac{4\,dx}{x^2 + 4} \\ &= \ln|x| - \frac{1}{2}\ln|x^2 + 4| + 2\tan^{-1}\frac{x}{2} + C_1 \end{aligned}$$

We could use the properties of logarithms to combine the first two terms of the answer. Doing so, we have the following result.

$$\int \frac{4x + 4}{x^3 + 4x} dx = \ln\frac{|x|}{\sqrt{x^2 + 4}} + 2\tan^{-1}\frac{x}{2} + C_1$$ ■

Practice Exercise

2. Find the partial fractions for $\frac{x^2 - 2}{x^3 + 2x}$.

EXAMPLE 5 Repeated linear factor—quadratic factor

Integrate: $\displaystyle\int \frac{x^3 + 3x^2 + 2x + 4}{x^2(x^2 + 2x + 2)}dx.$

In the denominator, we have a repeated linear factor, x^2, and a quadratic factor. Therefore,

$$\frac{x^3 + 3x^2 + 2x + 4}{x^2(x^2 + 2x + 2)} = \frac{A}{x} + \frac{B}{x^2} + \frac{Cx + D}{x^2 + 2x + 2}$$

$$\begin{aligned} x^3 + 3x^2 + 2x + 4 &= Ax(x^2 + 2x + 2) + B(x^2 + 2x + 2) + Cx^3 + Dx^2 \\ &= (A + C)x^3 + (2A + B + D)x^2 + (2A + 2B)x + 2B \end{aligned}$$

Equating coefficients, we find that

For constants:	$2B = 4,$	$B = 2$	
For x:	$2A + 2B = 2,$	$A + B = 1,$	$A = -1$
For x^2:	$2A + B + D = 3,$	$-2 + 2 + D = 3,$	$D = 3$
For x^3:	$A + C = 1,$	$-1 + C = 1,$	$C = 2$

$$\frac{x^3 + 3x^2 + 2x + 4}{x^2(x^2 + 2x + 2)} = -\frac{1}{x} + \frac{2}{x^2} + \frac{2x + 3}{x^2 + 2x + 2}$$

$$\begin{aligned} \int \frac{x^3 + 3x^2 + 2x + 4}{x^2(x^2 + 2x + 2)}dx &= -\int \frac{dx}{x} + 2\int \frac{dx}{x^2} + \int \frac{2x + 3}{x^2 + 2x + 2}dx \\ &= -\ln|x| - 2\left(\frac{1}{x}\right) + \int \frac{2x + 2 + 1}{x^2 + 2x + 2}dx \\ &= -\ln|x| - \frac{2}{x} + \int \frac{2x + 2}{x^2 + 2x + 2}dx + \int \frac{dx}{(x^2 + 2x + 1) + 1} \\ &= -\ln|x| - \frac{2}{x} + \ln|x^2 + 2x + 2| + \tan^{-1}(x + 1) + C_1 \end{aligned}$$

■ Partial fractions can be found on a TI-89 graphing calculator by using the *expand* feature.

CAUTION Note the manner in which the integral with the quadratic denominator was handled for the purpose of integration. ■ First, the numerator, $2x + 3$, was written in the form $(2x + 2) + 1$ so that we could fit the logarithmic form with the $2x + 2$. Then we completed the square in the denominator of the final integral so that it then fit an inverse tangent form. ■

REPEATED QUADRATIC FACTORS

Finally, considering the case of repeated quadratic factors, we use the fact that *corresponding to each irreducible quadratic factor $ax^2 + bx + c$ that occurs n times in the denominator there will be n partial fractions*

$$\frac{A_1x + B_1}{ax^2 + bx + c} + \frac{A_2x + B_2}{(ax^2 + bx + c)^2} + \cdots + \frac{A_nx + B_n}{(ax^2 + bx + c)^n}$$

where $A_1, A_2, \ldots, A_n, B_1, B_2, \ldots, B_n$ are constants to be determined. The procedures that lead to the solution are the same as those for the other cases. Exercises 23 and 24 in the following set are solved by using these partial fractions for repeated quadratic factors.

EXERCISES 28.10

In Exercises 1–4, make the given changes in the integrands of the indicated examples of this section. Then write out the equation that shows the partial fractions that would be used for the integration.

1. In Example 1, change the numerator from 1 to 2.
2. In Example 2, change the denominator to $(x - 1)^2(x + 2)^2$.
3. In Example 4, change the denominator to $x^3 - 9x^2$.
4. In Example 5, change the denominator to $x^2(x^2 + 3x + 2)$.

In Exercises 5–24, integrate each of the given functions.

5. $\int \frac{1}{x^2(x+1)}dx$
6. $\int \frac{x}{(x-2)^2}dx$
7. $\int \frac{x-8}{x^3-4x^2+4x}dx$
8. $\int \frac{dT}{T^3-T^2}$
9. $\int \frac{2\,dx}{x^2(x^2-1)}$
10. $\int_1^3 \frac{3x^3+8x^2+10x+2}{x(x+1)^3}dx$
11. $\int_1^2 \frac{2s\,ds}{(s-3)^3}$
12. $\int \frac{x\,dx}{(x+2)^4}$
13. $\int \frac{x^3-2x^2-7x+28}{(x+1)^2(x-3)^2}dx$
14. $\int \frac{4\,dx}{(x+1)^2(x-1)^2}$
15. $\int_0^2 \frac{x^2+x+5}{(x+1)(x^2+4)}dx$
16. $\int \frac{v^2+v-2}{(v^2+1)(3v-1)}dv$
17. $\int \frac{5x^2-3x+2}{x^3-2x^2}dx$
18. $\int \frac{3x^2+2x+4}{x^3+x^2+4x+4}dx$
19. $\int \frac{5x^2+8x+16}{x^2(x^2+4x+8)}dx$
20. $\int \frac{2x^2+x+3}{(x^2+2)(x-1)}dx$
21. $\int \frac{10x^3+40x^2+22x+7}{(4x^2+1)(x^2+6x+10)}dx$
22. $\int_3^4 \frac{5x^3-4x}{x^4-16}dx$
23. $\int \frac{-x^3+x^2+x+3}{(x+1)(x^2+1)^2}dx$
24. $\int \frac{2r^3}{r^4+2r^2+1}dr$

In Exercises 25–34, solve the given problems by integration.

25. For the integral of Example 3, set up the integration by partial fractions, and then integrate. Compare results with Example 3.
26. Integrate: $\int \frac{\cos x\,dx}{\sin x + \sin^3 x}$. (*Hint:* Begin by making the substitution $u = \sin x$.)
27. Find the area bounded by $y = \frac{x-3}{x^3+x^2}$, $y = 0$, and $x = 1$.
28. Find the volume generated by revolving the first-quadrant region bounded by $y = 4/(x^4 + 6x^2 + 5)$ and $x = 2$ about the y-axis.
29. Find the volume generated by revolving the first-quadrant region bounded by $y = x/(x + 3)^2$ and $x = 3$ about the x-axis.
30. The work done (in J) in moving a crate through a distance of 10.0 m is $W = \int_0^{10} \frac{2400\,ds}{(s+1)(s^2+4)^2}$. Evaluate W.
31. Under certain conditions, the velocity v (in m/s) of an object moving along a straight line as a function of the time t (in s) is given by $v = \frac{t^2+14t+27}{(2t+1)(t+5)^2}$. Find the distance traveled by the object during the first 2.00 s.
32. By a computer analysis, the electric current i (in A) in a certain circuit is given by $i = \frac{0.0010(7t^2+16t+48)}{(t+4)(t^2+16)}$, where t is the time (in s). Find the total charge that passes a point in the circuit in the first 0.250 s.
33. Find the x-coordinate of the centroid of a flat plate covering the region bounded by $y = 4/(x^3 + x)$, $x = 1$, $x = 2$, and $y = 0$.
34. The slope of a curve is given by $\frac{dy}{dx} = \frac{29x^2+36}{4x^4+9x^2}$. Find the equation of the curve if it passes through (1, 5).

Answers to Practice Exercises

1. $\frac{2}{x(x-1)^2} = \frac{2}{x} - \frac{2}{x-1} + \frac{2}{(x-1)^2}$ 2. $\frac{x^2-2}{x^3+2x} = -\frac{1}{x} + \frac{2x}{x^2+2}$

28.11 Integration by Use of Tables

Proper Recognition of the Form, Variables, and Constants • Proper Identification of *u* and *du*

In this chapter, we have introduced certain basic integrals and have also brought in some methods of reducing other integrals to these basic forms. Often, this transformation and integration requires a number of steps to be performed, and therefore integrals are tabulated for reference. The integrals found in tables have been derived by using the methods introduced thus far, as well as many other methods that can be used. Therefore, an understanding of the basic forms and some of the basic methods is very useful in finding integrals from tables. Such an understanding forms a basis for proper recognition of the forms that are used in the tables, as well as the types of results that may be expected. Therefore, ***the use of the tables depends on proper recognition of the form and the variables and constants of the integral.*** The following examples illustrate the use of the table of integrals found in Appendix D. More extensive tables are available in other sources.

■ For reference, Formula 6 is

$$\int \frac{u\,du}{\sqrt{a+bu}} = -\frac{2(2a-bu)\sqrt{a+bu}}{3b^2}$$

EXAMPLE 1 Integral fits the form of Formula 6

Integrate: $\int \frac{x\,dx}{\sqrt{2+3x}}$.

We first note that this integral fits the form of Formula 6 of Appendix D, with $u = x$, $a = 2$, and $b = 3$. Therefore,

$$\int \frac{x\,dx}{\sqrt{2+3x}} = -\frac{2(4-3x)\sqrt{2+3x}}{27} + C$$

■

■ For reference, Formula 18 is

$$\int \frac{\sqrt{a^2-u^2}}{u}\,du = \sqrt{a^2-u^2} - a\ln\left(\frac{a+\sqrt{a^2-u^2}}{u}\right)$$

EXAMPLE 2 Integral fits the form of Formula 18

Integrate: $\int \frac{\sqrt{4-9x^2}}{x}dx$.

This fits the form of Formula 18, with proper identification of constants; $u = 3x$, $du = 3\,dx$, $a = 2$. Hence,

$$\int \frac{\sqrt{4-9x^2}}{x}dx = \int \frac{\sqrt{4-9x^2}}{3x}3\,dx$$
$$= \sqrt{4-9x^2} - 2\ln\left(\frac{2+\sqrt{4-9x^2}}{3x}\right) + C$$

■

■ For reference, Formula 37 is

$$\int \sec^n u\,du = \frac{\sec^{n-2}u\tan u}{n-1} + \frac{n-2}{n-1}\int \sec^{n-2}u\,du$$

EXAMPLE 3 Integral fits the form of Formula 37

Integrate: $\int 5\sec^3 2x\,dx$.

This fits the form of Formula 37; $n = 3$, $u = 2x$, $du = 2\,dx$. And so,

$$\int 5\sec^3 2x\,dx = 5\left(\frac{1}{2}\right)\int \sec^3 2x(2\,dx)$$
$$= \frac{5}{2}\frac{\sec 2x\tan 2x}{2} + \frac{5}{2}\left(\frac{1}{2}\right)\int \sec 2x(2\,dx)$$

To complete this integral, we use the basic form $\int \sec u\,du = \ln|\sec u + \tan u| + C$. Thus, we get

$$\int 5\sec^3 2x\,dx = \frac{5\sec 2x\tan 2x}{4} + \frac{5}{4}\ln|\sec 2x + \tan 2x| + C$$

■

EXAMPLE 4 Area—integral fits the form of Formula 46

Find the area bounded by $y = x^2\ln 2x$, $y = 0$, and $x = e$.

From Fig. 28.17, we see that the area is

$$A = \int_{0.5}^{e} x^2\ln 2x\,dx$$

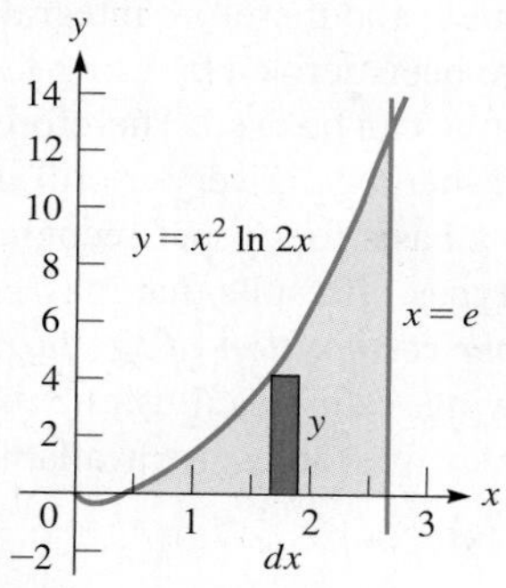

This integral fits the form of Formula 46 in Appendix D if $u = 2x$. Thus, we have

$$A = \frac{1}{8}\int_{0.5}^{e}(2x)^2\ln 2x(2\,dx) = \frac{1}{8}(2x)^3\left[\frac{\ln 2x}{3} - \frac{1}{9}\right]\Bigg|_{0.5}^{e}$$
$$= e^3\left(\frac{\ln 2e}{3} - \frac{1}{9}\right) - \frac{1}{8}\left(\frac{\ln 1}{3} - \frac{1}{9}\right) = e^3\left(\frac{3\ln 2e - 1}{9}\right) + \frac{1}{72}$$
$$= 9.118$$

■

Fig. 28.17

NOTE ▸ [The proper identification of u and du is the key step in the use of tables.] Therefore, for the integrals in the following example, the proper u and du, along with the appropriate formula from the table, are identified, but the integrations are not performed.

EXAMPLE 5 Identify the formula, u, and du

(a) $\int x\sqrt{1 - x^4}\,dx$ $u = x^2$, $du = 2x\,dx$ Formula 15

(b) $\int \frac{(4x^6 - 9)^{3/2}}{x}dx$ $u = 2x^3$, $du = 6x^2\,dx$ Formula 22

introduce a factor of x^2 into numerator and denominator

(c) $\int x^3 \sin x^2\,dx$ $u = x^2$, $du = 2x\,dx$ Formula 47 ■

Practice Exercise

1. Which formula should be used to integrate $\int \frac{2\,dx}{x\sqrt{4 + x^4}}$?

EXERCISES 28.11

In Exercises 1 and 2, make the given changes in the indicated examples of this section, and then state which formula from Appendix D would be used to complete the integration.

1. In Example 1, change the denominator to $(2 + 3x)^2$.

2. In Example 2, in the numerator, change $-$ to $+$.

In Exercises 3–8, identify u, du, and the formula from Appendix D that would be used to complete the integration. Do not integrate.

3. $\int \frac{4\,dy}{3y\sqrt{1 + 2y}}$

4. $\int \frac{x\,dx}{\sqrt{x^4 - 16}}$

5. $\int \frac{x\,dx}{(4 - x^4)^{3/2}}$

6. $\int x^5 \ln x^3\,dx$

7. $\int x\cos^2(x^2)dx$

8. $\int \frac{ds}{s(s^4 - 1)^{3/2}}$

In Exercises 9–52, integrate each function by using the table in Appendix D.

9. $\int \frac{3x\,dx}{2 + 5x}$

10. $\int \frac{4x\,dx}{1 + 2x + x^2}$

11. $\int_2^7 4x\sqrt{2 + x}\,dx$

12. $\int \frac{dx}{x^2 - 4}$

13. $\int \frac{8\,dy}{(y^2 + 4)^{3/2}}$

14. $\int_0^{\pi/3} 3\sin^3 x\,dx$

15. $\int \sin 2x \sin 3x\,dx$

16. $\int 6\sin^{-1} 3x\,dx$

17. $\int \frac{\sqrt{4x^2 - 9}}{x}dx$

18. $\int \frac{(9x^2 + 16)^{3/2}}{x}dx$

19. $\int \cos^5 4x\,dx$

20. $\int \tan^{-1} 2x\,dx$

21. $\int 6r\tan^{-1} r^2\,dr$

22. $\int 5xe^{4x}\,dx$

23. $\int_1^2 (4 - x^2)^{3/2}\,dx$

24. $\int \frac{3\,dx}{9 - 16x^2}$

25. $\int \frac{dx}{2x\sqrt{x^2 + \frac{1}{4}}}$

26. $\int \frac{\sqrt{9 + x^2}}{x}dx$

27. $\int \frac{8\,dx}{x\sqrt{1 - 4x^2}}$

28. $\int \sqrt{5 - 16x^2}\,dx$

29. $\int_0^{\pi/12} \sin\theta\cos 5\theta\,d\theta$

30. $\int_0^2 x^2 e^{3x}\,dx$

31. $\int 6x^5 \cos x^3\,dx$

32. $\int 15\sin^3 t\cos^2 t\,dt$

33. $\int \frac{2x\,dx}{(1 - x^4)^{3/2}}$

34. $\int \frac{dx}{x - 4x^2}$

35. $\int_1^3 \frac{\sqrt{3 + 5x^2}\,dx}{x}$

36. $\int_0^1 \frac{\sqrt{9 - 4x^2}}{x}dx$

37. $\int x^3 \ln x^2\,dx$

38. $\int \frac{1.2u\,du}{u^2\sqrt{u^4 - 9}}$

39. $\int \frac{9x^2\,dx}{(x^6 - 1)^{3/2}}$

40. $\int x^7\sqrt{x^4 + 4}\,dx$

41. $\int t^2(t^6 + 1)^{3/2}\,dt$

42. $\int \frac{\sqrt{3 + 4x^2}\,dx}{x}$

43. $\int \sin^3 4x\cos^3 4x\,dx$

44. $\int 6\cot^4 2\,dx$

45. A good representation of the cables between towers of the 2280-m section of the Golden Gate Bridge is $y = 0.000370x^2$ for $-1140 \le x \le 1140$, where x and y are in meters. Find the length of the cables (see Exercise 35 of Section 26.6).

46. The design of a rotor can be represented as the volume generated by rotating the area bounded by $y = 3.00 \ln x$, $x = 3.00$, and the x-axis. Find its radius of gyration about the y-axis.

47. Find the area of an ellipse with a major axis $2a$ and a minor axis $2b$.

48. The voltage across a 5.0-μF capacitor in an electric circuit is zero. What is the voltage after 5.00 μs if a current i (in mA) as a function of the time t (in s) given by $i = \tan^{-1} 2t$ charges the capacitor?

49. Find the force (in lb) on the region bounded by $x = 1/\sqrt{1+y}$, $y = 0$, $y = 3$, and the y-axis, if the surface of the water is at the upper edge of the area.

50. If 6.00 g of a chemical are placed in water, the time t (in min) it takes to dissolve half of the chemical is given by $t = 560 \int_3^6 \frac{dx}{x(x+4)}$, where x is the amount of undissolved chemical at any time. Evaluate t.

51. The dome of a sports arena is the surface generated by revolving $y = 20.0 \cos 0.0196x$ $(0 \le x \le 80.0 \text{ m})$ about the y-axis. Find the volume within the dome.

52. If an electric charge Q is distributed along a wire of length $2a$, the force F exerted on an electric charge q placed at point P is $F = kqQ \int \frac{b\,dx}{(b^2 + x^2)^{3/2}}$. Integrate to find F as a function of x.

Answer to Practice Exercise

1. Formula 11

CHAPTER 28 KEY FORMULAS AND EQUATIONS

Integrals

$$\int u^n\,du = \frac{u^{n+1}}{n+1} + C \quad (n \ne -1) \tag{28.1}$$

$$\int \frac{du}{u} = \ln|u| + C \tag{28.2}$$

$$\int e^u\,du = e^u + C \tag{28.3}$$

$$\int \sin u\,du = -\cos u + C \tag{28.4}$$

$$\int \cos u\,du = \sin u + C \tag{28.5}$$

$$\int \sec^2 u\,du = \tan u + C \tag{28.6}$$

$$\int \csc^2 u\,du = -\cot u + C \tag{28.7}$$

$$\int \sec u \tan u\,du = \sec u + C \tag{28.8}$$

$$\int \csc u \cot u\,du = -\csc u + C \tag{28.9}$$

$$\int \tan u\,du = -\ln|\cos u| + C \tag{28.10}$$

$$\int \cot u\,du = \ln|\sin u| + C \tag{28.11}$$

$$\int \sec u\,du = \ln|\sec u + \tan u| + C \tag{28.12}$$

$$\int \csc u\,du = \ln|\csc u - \cot u| + C \tag{28.13}$$

Trigonometric identities	$\cos^2 x + \sin^2 x = 1$	**(28.14)**
	$1 + \tan^2 x = \sec^2 x$	**(28.15)**
	$1 + \cot^2 x = \csc^2 x$	**(28.16)**
	$2\cos^2 x = 1 + \cos 2x$	**(28.17)**
	$2\sin^2 x = 1 - \cos 2x$	**(28.18)**
Root-mean-square value	$y_{rms} = \sqrt{\frac{1}{T}\int_0^T y^2\,dx}$	**(28.19)**
Integrals	$\int \frac{du}{\sqrt{a^2 - u^2}} = \sin^{-1}\frac{u}{a} + C$	**(28.20)**
	$\int \frac{du}{a^2 + u^2} = \frac{1}{a}\tan^{-1}\frac{u}{a} + C$	**(28.21)**
Integration by parts	$\int u\,dv = uv - \int v\,du$	**(28.22)**
Trigonometric substitutions	For $\sqrt{a^2 - x^2}$ use $x = a\sin\theta$ For $\sqrt{a^2 + x^2}$ use $x = a\tan\theta$ For $\sqrt{x^2 - a^2}$ use $x = a\sec\theta$	**(28.23)**

CHAPTER 28 REVIEW EXERCISES

CONCEPT CHECK EXERCISES

*Determine each of the following as being either **true** or **false**. If it is false, explain why.*

1. $\int \frac{dx}{1 + 2x} = \ln|1 + 2x| + C$

2. $4\int e^{2x}\,dx = 2e^{2x}$

3. $6\int \tan 3x\,dx = 2\ln|\cos 3x| + C$

4. To integrate $\int \sin^4 x\cos^3 x\,dx$, first let $\sin^4 x = \sin^3 x\sin x$.

5. To integrate $\int \frac{dx}{\sqrt{1 - 16x^2}}$, first let $u = 4x$.

6. To integrate $\int x\sin 2x\,dx$, first let $u = x$ and $dv = \sin 2x\,dx$.

7. To integrate $\int \frac{2\,dx}{\sqrt{x^2 + 1}}$, first let $x = \sec\theta$.

8. To integrate $\int \frac{3\,dx}{x^2 + x - 2}$, first let $\frac{3}{x^2 + x - 2} = \frac{A}{x + 2} + \frac{B}{x - 1}$.

PRACTICE AND APPLICATIONS

In Exercises 9–50, integrate the given functions without using a table of integrals.

9. $\int e^{-8x}\,dx$

10. $\int e^{\cos 2x}\sin x\cos x\,dx$

11. $\int \frac{dx}{x(\ln 2x)^2}$

12. $\int_1^8 y^{1/3}\sqrt{y^{4/3} + 9}\,dy$

13. $\int_0^{\pi/2} \frac{4\cos\theta\,d\theta}{1 + \sin\theta}$

14. $\int \frac{\sec^2 x\,dx}{2 + \tan x}$

15. $\int \frac{20\,dx}{25 + 49x^2}$

16. $\int \frac{4\,dx}{\sqrt{1 - 4x^2}}$

17. $\int_0^{\pi/2} \cos^3 2\theta\,d\theta$

18. $\int_0^{\pi/18} \sec^3 6x\tan 6x\,dx$

19. $\int_0^2 \frac{x\,dx}{4 + x^2}$

20. $\int_1^e \frac{\ln v^2\,dv}{\ln e^v}$

21. $\int (\sin t + \cos t)^2\sin t\,dt$

22. $\int \frac{\sin^3 x\,dx}{\sqrt{\cos x}}$

23. $\int \frac{e^x\,dx}{1 + e^{2x}}$

24. $\int \frac{p + 25}{p^2 - 25}dp$

25. $\int 6\sec^4 3x\,dx$

26. $\int \frac{(1 - \cos^2\theta)d\theta}{1 + \cos 2\theta}$

27. $\int \frac{2x^2 + 6x + 1}{2x^3 - x^2 - x}dx$

28. $\int \frac{4 - e^{\sqrt{x}}}{\sqrt{x}\,e^{\sqrt{x}}}dx$

29. $\int \frac{3x\,dx}{4 + x^4}$

30. $\int_1^3 \frac{12dR}{\sqrt{R}(1 + R)}$

31. $\int \frac{4\,dx}{\sqrt{4x^2 - 9}}$

32. $\int \frac{6x^2\,dx}{\sqrt{9 - x^2}}$

33. $\int \frac{e^{2x}\,dx}{\sqrt{e^{2x}+1}}$

34. $\int \frac{x^2-2x+3}{(x-1)^3}dx$

35. $\int \frac{2x^2+3x+18}{x^3+9x}dx$

36. $\int_{1/2}^{e/2} \frac{(4+\ln 2u)^3\,du}{u}$

37. $\int_0^{\pi/6} 3\sin^2 3\phi\,d\phi$

38. $\int 16\sin^4 x\,dx$

39. $\int 2x\csc^2 2x\,dx$

40. $\int x\tan^{-1}x\,dx$

41. $\int \frac{3u^2-6u-2}{3u^3+u^2}du$

42. $\int \frac{R^2+3}{R^4+3R^2+2}dR$

43. $\int 8e^{2x}\cos e^{2x}\,dx$

44. $\int \frac{3\,dx}{x^2+6x+10}$

45. $\int_1^e \frac{3\cos(\ln x)dx}{x}$

46. $\int_1^3 \frac{2\,dx}{x^2-2x+5}$

47. $\int \frac{u^2-8}{u+3}du$

48. $\int \frac{1}{x(\ln x)^2}dx$

49. $\int \frac{\sin x\cos^2 x}{5+\cos^2 x}dx$

50. $\int \frac{e^{2x}\,dx}{16+e^{4x}}$

In Exercises 51–94, solve the given problem by integration.

51. For the integral $\int \frac{dx}{x\sqrt{2-x^2}}$, using a trigonometric substitution, find the transformed integral but do not integrate.

52. For the integral $\int \frac{dx}{\sqrt{x^2+4x+3}}$, using a trigonometric substitution, find the transformed integral but do not integrate.

53. Perform the integration $\int e^{\ln 4x}\,dx$ and $\int \ln e^{4x}\,dx$. Compare results.

54. Perform the integration $\int \sin 6x \sin 5x\,dx$ using Eq. (20.16) on page 546.

55. For the integral $\int \frac{\sqrt{16+x^8}}{x}dx$, identify the formula found in Appendix D, u, and du that would be used for the integration. Do not integrate.

56. For the integral $\int x^9\sqrt{x^5+8}\,dx$, identify the formula in Appendix D, u, and du, that would be used for the integration. Do not integrate.

57. Show that $\int e^x(e^x+1)^2\,dx$ can be integrated in two ways. Explain the difference in the answers.

58. Show that $\int \frac{1}{x}(1+\ln x)dx$ can be integrated in two ways. Explain the difference in the answers.

59. Integrate $\int \sin\sqrt{x}\,dx$ by first making the substitution $u=\sqrt{x}$ and then using integration by parts.

60. Integrate $\int \frac{dx}{1+e^x}$ by first rewriting the integrand. (*Hint:* It is possible to multiply the numerator and the denominator by an appropriate expression.)

61. The integral $\int \frac{x}{\sqrt{x^2+4}}dx$ can be integrated in more than one way. Explain what methods can be used and which is simpler.

62. Find an equation of the curve for which $dy/dx = e^x(2-e^x)^2$, if the curve passes through (0, 4).

63. Find an equation of the curve for which $dy/dx = 3\sec^4 dx$, if the curve passes through the origin.

64. Find an equation of the curve for which $\frac{dy}{dx} = \frac{\sqrt{4+x^2}}{x^4}$, if the curve passes through (2, 1).

65. Show that $\int \sin 2x\,dx$ can be integrated in three ways. Explain the difference in the answers.

66. Find the area bounded by $y=\sin 2x$, $y=\cos 2x$, and $x=0$ in the first quadrant.

67. Find the area bounded by $y=4e^{2x}$, $x=1.5$, and the axes.

68. Find the area bounded by $y=x/(1+x)^2$, the x-axis, and the line $x=4$.

69. Find the area inside the circle $x^2+y^2=25$ and to the right of the line $x=3$.

70. Find the area bounded by $y=x\sqrt{x+4}$, $y=0$, and $x=5$.

71. Find the area bounded by $y=\tan^{-1}2x$, $x=2$, and the x-axis.

72. In polar coordinates, the area A bounded by the curve $r=f(\theta)$, $\theta=\alpha$, and $\theta=\beta$ is found by evaluating the integral $A=\frac{1}{2}\int_\alpha^\beta r^2\,d\theta$. Find the area bounded by $\alpha=0$, $\beta=\pi/2$, and $r=e^\theta$.

73. Find the volume generated by revolving the region bounded by $y=xe^x$, $y=0$, and $x=2$ about the y-axis.

74. Find the volume generated by revolving about the y-axis the region bounded by $y=x+\sqrt{x+1}$, $x=3$, and the axes.

75. Find the volume of the solid generated by revolving the region bounded by $y=e^x\sin x$ and the x-axis between $x=0$ and $x=\pi$ about the x-axis.

76. The rudder of a boat has the shape of the area bounded by the x-axis, $y=-2\ln x$, and $x=2$. Find the centroid (in m).

77. The wire brace in a sunshade has the shape of the curve $y=\ln\sin x$ from $x=0.5$ to $x=2.5$. Find the length (in m) of the brace. (See Exercise 35 of Section 26.6.)

78. By integration, find the surface area of a tennis ball. The diameter is 6.20 cm. Check using the geometric formula. (See Exercise 37 of Section 26.6.)

79. The force F (in lb) on a nail by a hammer is $F=5/(1+2t)$, where t is the time (in s). The impulse I of the force from $t=0$ s to $t=0.25$ s is $I=\int_0^{0.25}F\,dt$. Find the impulse.

80. The acceleration of a parachutist is given by $dv/dt = g-kv$, where k is a constant depending on the resisting force due to air friction. Find v as a function of the time t.

81. The change in the thermodynamic entity of entropy ΔS may be expressed as $\Delta S = \int (c_v/T)dt$, where c_v is the heat capacity at constant volume and T is the temperature. For increased accuracy, c_v is often given by the equation $c_v = a+bT+cT^2$, where a, b, and c are constants. Express ΔS as a function of temperature.

82. A certain type of chemical reaction leads to the equation $dt = \frac{dx}{k(a-x)(b-x)}$, where a, b, and k are constants. Solve for t as a function of x.

83. An electric transmission line between two towers has a shape given by $y = 16.0(e^{x/32} + e^{-x/32})$. Find the length of transmission line if the towers are 50.0 m apart (from $x = -25.0$ m to $x = 25.0$ m). See Exercise 35 of Section 26.6 for the length-of-arc formula.

84. An object at the end of a spring is immersed in liquid. Its velocity (in cm/s) is then described by the equation $v = 2e^{-2t} + 3e^{-5t}$, where t is the time (in s). Such motion is called *overdamped.* Find the displacement s as a function of t if $s = -1.6$ cm for $t = 0$.

85. When we consider the resisting force of the air, the velocity v (in ft/s) of a falling brick in terms of the time t (in s) is given by $dv/(32 - 0.5v) = dt$. If $v = 0$ when $t = 0$, find v as a function of t.

86. The power delivered to an electric circuit is given by $P = ei$, where e and i are the instantaneous voltage and the instantaneous current in the circuit, respectively. The mean power, averaged over a period $2\pi/\omega$, is given by $P_{av} = \dfrac{\omega}{2\pi}\displaystyle\int_0^{2\pi/\omega} ei\,dt$. If $e = 20\cos 2t$ and $i = 3\sin 2t$, find the average power over a period of $\pi/4$.

87. Find the root-mean-square value for one period of the electric current i if $i = 2\sin t$.

88. In atomic theory, when finding the number n of atoms per unit volume of a substance, we use the equation $n = A\int_0^{\pi} e^{\alpha\cos\theta}\sin\theta\,d\theta$. Perform the indicated integration.

89. Find the volume within the piece of tubing in an oil distribution line shown in Fig. 28.18. All cross sections are circular.

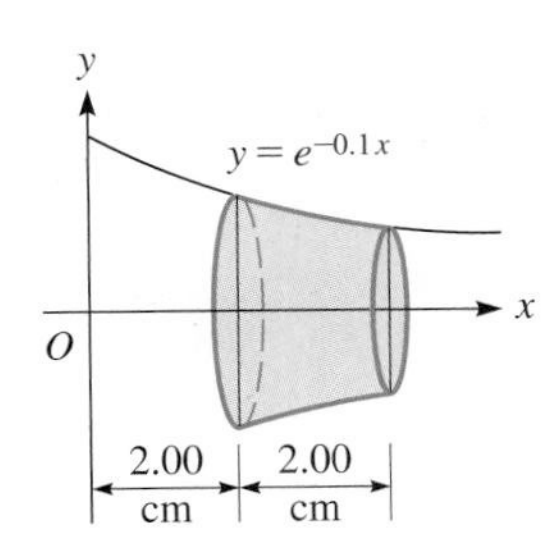

Fig. 28.18

90. In the study of the effects of an electric field on molecular orientation, the integral $\int_0^{\pi}(1 + k\cos\theta)\cos\theta\sin\theta\,d\theta$ is used. Evaluate this integral.

91. In finding the lift of the air flowing around an airplane wing, we use the integral $\int_{-\pi/2}^{\pi/2}\theta^2\cos\theta\,d\theta$. Evaluate this integral.

92. A metal plate has a shape shown in Fig. 28.19. Find the x-coordinate of the centroid of the plate.

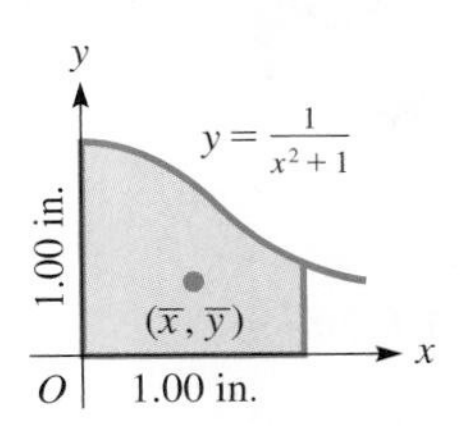

Fig. 28.19

93. The nose cone of a space vehicle is to be covered with a heat shield. The cone is designed such that a cross section x feet from the tip and perpendicular to its axis is a circle of radius $1.5x^{2/3}$ feet. Find the surface area of the heat shield if the nose cone is 16.0 ft long. See Fig. 28.20. (See Exercise 37 of Section 26.6.)

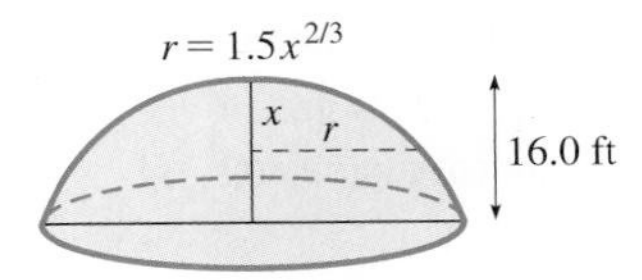

Fig. 28.20

94. A window has a shape of a semiellipse, as shown in Fig. 28.21. What is the area of the window?

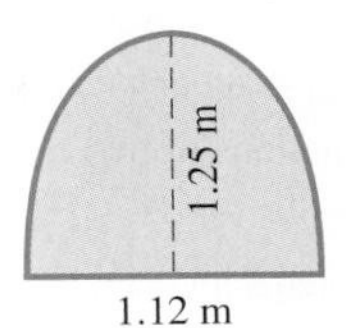

Fig. 28.21

95. The side of a blade designed to cut leather at a shoe factory can be described as the region bounded by $y = 4\cos^2 x$ and $y = 4$ from $x = 0$ to $x = 3.14$. Write two or three paragraphs explaining how this area of the blade may be found by integrating the appropriate integral by either of two methods of this chapter.

CHAPTER 28 PRACTICE TEST

As a study aid, we have included complete solutions for each Practice Test problem at the end of this book.

In Problems 1–7, evaluate the given integrals.

1. $\displaystyle\int(\sec x - \sec^3 x\tan x)\,dx$

2. $\displaystyle\int\sin^3 x\,dx$

3. $\displaystyle\int\tan^3 2x\,dx$

4. $\displaystyle\int\cos^2 4\theta\,d\theta$

5. $\displaystyle\int\frac{dx}{x^2\sqrt{4 - x^2}}$

6. $\displaystyle\int xe^{-2x}\,dx$

7. $\displaystyle\int\frac{x^3 + 5x^2 + x + 2}{x^4 + x^2}\,dx$

8. The electric current in a certain circuit is given by $i = \displaystyle\int\frac{6t + 1}{4t^2 + 9}\,dt$, where t is the time. Integrate and find the resulting function if $i = 0$ for $t = 0$.

9. Find the first-quadrant area bounded by $y = \dfrac{1}{\sqrt{16 - x^2}}$ and $x = 3$.

29

Partial Derivatives and Double Integrals

LEARNING OUTCOMES

After completion of this chapter, the student should be able to:

- Set up and evaluate a function of two independent variables
- Use traces and sections to sketch the graph of a surface in the rectangular coordinate system in three dimensions
- Convert points and equations from rectangular to cylindrical coordinates and vice versa
- Find the first- and second-order partial derivatives of a function of several variables
- Evaluate double integrals
- Understand the geometric interpretation of partial derivatives and double integrals of functions of two independent variables
- Solve application problems involving functions of two independent variables, their partial derivatives, and their integrals

To this point, we have been dealing with functions that have a single independent variable. There are, however, numerous applications in which functions with more than one independent variable are used. Although a number of these functions involve three or more independent variables, many involve only two, and we shall be concerned primarily with these.

In the first two sections of this chapter, we establish the meaning of a function of two independent variables, and discuss how the graph of this type of function is shown. In the last two sections, we develop some of the basic concepts of the calculus of these functions, primarily *partial derivatives* and *double integrals.* In finding the partial derivative, we take the derivative of the function with respect to one independent variable, holding the other constant. Similarly, when integrating an expression with two differentials, we integrate with respect to one, holding the other constant, and then integrate with respect to the other.

While studying problems in motion and optics in the early 1700s, mathematicians including Leibniz developed the meaning and use of partial derivatives. One of the first uses of double integrals was by Leibniz in solving a problem presented to him in 1692. In the 1700s, the development of double integrals continued, and in 1769, Euler gave the first detailed explanation of them.

Partial derivatives and double integrals have applications in many areas of science and technology, including acoustics, electricity, electronics, mechanics, product design, and wave motion. Some of these applications are shown in the examples and exercises of this chapter.

▶ **In Section 29.3, we see how partial derivatives can be used to find the slope of the roof (in different directions) of the Saddledome arena in Calgary, Alberta.**

29.1 Functions of Two Variables

Functions of Two Independent Variables • $z = f(x, y)$ Notation • Restrictions on x and y

Many familiar formulas express one variable in terms of two or more other variables. The following example illustrates one from geometry.

EXAMPLE 1 Function of two variables—surface area of a cylinder

The total surface area A of a right circular cylinder is a function of the radius r and the height h of the cylinder. That is, the area will change if either or both of these change. The formula for the total surface area is

$$A = 2\pi r^2 + 2\pi rh$$

We say that A is a function of r and h. See Fig. 29.1. ■

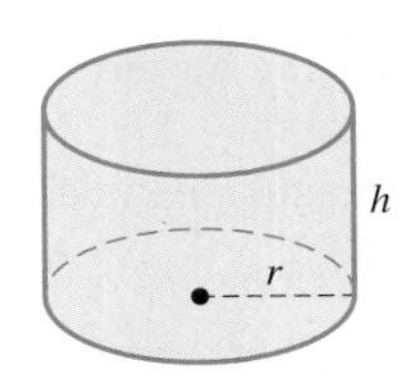

Fig. 29.1

We define a function of two variables as follows: *If z is uniquely determined for given values of x and y, then z is a function of x and y.* The notation used is similar to that used for one independent variable. It is $z = f(x, y)$, where both x and y are independent variables. NOTE ➧ [Therefore, it follows that $f(a, b)$ means "the value of the function when $x = a$ and $y = b$.]

EXAMPLE 2 Illustrating $z = f(x, y)$ notation

If $f(x, y) = 3x^2 + 2xy - y^3$, find $f(-1, 2)$.

Substituting -1 for x and 2 for y, we have

$$f(-1, 2) = 3(-1)^2 + 2(-1)(2) - (2)^3$$
$$= 3 - 4 - 8 = -9$$ ■

EXAMPLE 3 Function of two variables—electric current

For a certain electric circuit, the current i (in A), in terms of the voltage E and resistance R (in Ω) is given by

$$i = \frac{E}{R + 0.25}$$

Find the current for $E = 1.50$ V and $R = 1.20$ Ω and for $E = 1.60$ V and $R = 1.05$ Ω.

Substituting the first values, we have

$$i = \frac{1.50}{1.20 + 0.25} = 1.03 \text{ A}$$

For the second pair of values, we have

$$i = \frac{1.60}{1.05 + 0.25} = 1.23 \text{ A}$$

For this circuit, the current generally changes if either or both of E and R change. ■

EXAMPLE 4 Using $f(x, y)$ notation

If $f(x, y) = 2xy^2 - y$, find $f(x, 2x) - f(x, x^2)$.

We note that in each evaluation, the x factor remains as x, but that we are to substitute $2x$ for y and subtract the function for which x^2 is substituted for y.

$$\begin{aligned} f(x, 2x) - f(x, x^2) &= [2x(2x)^2 - (2x)] - [2x(x^2)^2 - x^2] \\ &= [8x^3 - 2x] - [2x^5 - x^2] \\ &= 8x^3 - 2x - 2x^5 + x^2 \\ &= -2x^5 + 8x^3 + x^2 - 2x \end{aligned}$$

This type of difference of functions is important in Section 29.4. ■

Practice Exercise

1. If $f(x, y) = 4xy^2 - 3x^2y$, find $f(1, 4) - f(x, 2x)$.

NOTE ▸ [Restricting values of the function to real numbers means that values of either x or y or both that lead to division by zero or to imaginary values for z are not permissible.]

EXAMPLE 5 Restrictions on independent variables

If $f(x, y) = \dfrac{3xy}{(x - y)(x + 3)}$, all values of x and y are permissible except those for which $x = y$ and $x = -3$. Each of these would indicate division by zero.

If $f(x, y) = \sqrt{4 - x^2 - y^2}$, neither x nor y may be greater than 2 in absolute value, and the sum of their squares may not exceed 4. Otherwise, imaginary values of the function would result. ■

Following is an example of setting up a function of two variables from stated conditions. As with word problems with one unknown, although no general rules can be given for this procedure, a careful analysis of the statement should lead to the required function.

EXAMPLE 6 Function of two variables—fish tank surface area

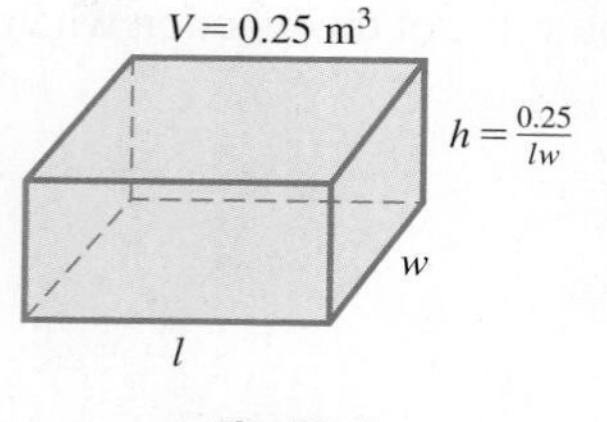

Fig. 29.2

An open rectangular fish tank is to hold 0.25 m^3 of water when completely full. Express the total surface area of glass required to make the tank as a function of the length and width of the tank.

An "open" tank is one that has no top. Therefore, the surface area S of the tank is

$$S = lw + 2lh + 2hw$$

where l is the length, w the width, and h the height of the tank (see Fig. 29.2). However, this equation contains three independent variables. Using the condition that the volume of water is 0.25 m^3, we have $lwh = 0.25$. Because we wish to have only l and w, we solve this equation for h and find that $h = 0.25/lw$. Substituting for h in the equation for the surface area, we have

$$S = lw + 2l\left(\frac{0.25}{lw}\right) + 2\left(\frac{0.25}{lw}\right)w$$

$$= lw + \frac{0.50}{w} + \frac{0.50}{l} \quad \text{the required function}$$ ■

EXERCISES 29.1

In Exercises 1 and 2, make the given changes in the indicated examples of this section and then solve the resulting problems.

1. In Example 2, change $f(-1, 2)$ to $f(-2, 1)$.
2. In Example 4, change $f(x, 2x) - f(x, x^2)$ to $f(2x, x) - f(x^2, x)$.

In Exercises 3–8 determine the indicated function.

3. Express the volume V of a right circular cylinder as a function of the radius r and height h.
4. Express the length of a diagonal of a rectangle as a function of the length and the width.
5. A cylindrical can is to be made to contain a volume V. Express the total surface area (including the top) of the can as a function of V and the radius of the can.
6. The angle between two forces $\mathbf{F}_1$ and $\mathbf{F}_2$ is 30°. Express the magnitude of the resultant $\mathbf{R}$ in terms of F_1 and F_2. See Fig. 29.3.

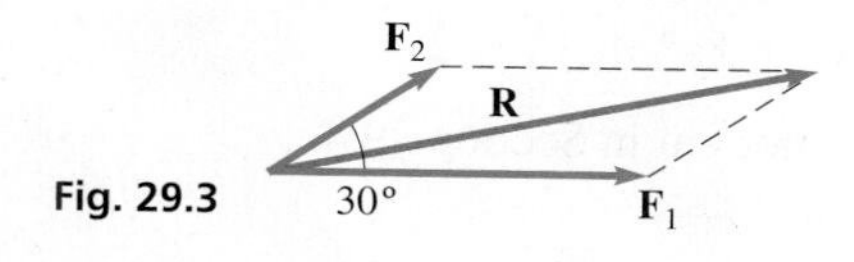

Fig. 29.3

7. A right circular cylinder is to be inscribed in a sphere of radius r. Express the volume of the cylinder as a function of the height h of the cylinder and r. See Fig. 29.4.

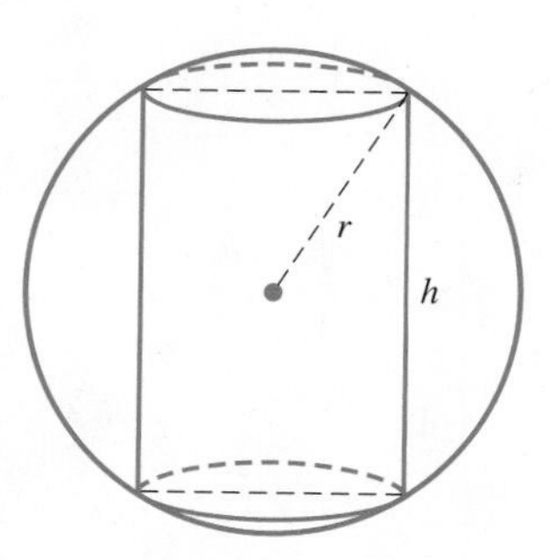

Fig. 29.4

8. An office-furniture-leasing firm charges a monthly fee F plus \$100 for each week the furniture is leased. Express the total monthly charge T as a function of F and the number of weeks w the furniture is leased.

In Exercises 9–24, evaluate the given functions.

9. $f(x, y) = 2x - 6y$; find $f(0, -4)$, and $f(-3, 2)$

10. $F(x, y) = x^2 - 5y + y^2$; find $F(2, -2)$, and $F(-3, -3)$

11. $g(r, s) = r - 2rs - r^2s$; find $g(-2, 1)$, and $g(0, -5)$

12. $f(r, \theta) = 2r(r\tan\theta - \sin 2\theta)$; find $f(3, \pi/4)$, and $f(3, 9\pi/4)$

13. $Y(y, t) = \dfrac{2 - 3y}{t - 1} + 2y^2t$; find $Y(2, -1)$, and $Y(y, 2)$

14. $f(r, t) = r^3 - 3r^2t + 3rt^2 - t^3$; find $f(-2, -3)$, and $f(3, t)$

15. $X(x, t) = -6xt + xt^2 - t^3$; find $X(x, -t)$, and $X(t, 2x)$

16. $g(y, z) = 2yz^2 - 6y^2z - y^2z^2$; find $g(y, 2y)$, and $g(2y, -z)$

17. $H(p, q) = p - \dfrac{p - 2q^2 - 5q}{p + q}$; find $H(p, q + k)$

18. $g(x, z) = z\tan^{-1}(x^2 + xz)$; find $g(-x, z)$

19. $f(x, y) = x^2 - 2xy - 4x$; find $f(x + h, y + k) - f(x, y)$

20. $g(y, z) = 4yz - z^3 + 4y$; find $g(y + 1, z + 2) - g(y, z)$

21. $f(x, y) = xy + x^2 - y^2$; find $f(x, x) - f(x, 0)$

22. $f(x, y) = 4x^2 - xy - 2y$; find $f(x, x^2) - f(x, 1)$

23. $g(y, z) = 3y^3 - y^2z + 5z^2$; find $g(3z^2, z) - g(z, z)$

24. $X(x, t) = 2x - \dfrac{t^2 - 2x^2}{x}$; find $X(2t, t) - X(2t^2, t)$

In Exercises 25–28, determine which values of x and y, if any, are not permissible. In Exercise 27, explain your answer.

25. $f(x, y) = \dfrac{\sqrt{y}}{2x}$

26. $f(x, y) = \dfrac{x^2 - 4y^2}{x^2 + 9}$

27. $f(x, y) = \sqrt{x^2 - x^2y + y^2 - y^3}$

28. $f(x, y) = \dfrac{1}{xy - y}$

In Exercises 29–44, solve the given problems.

29. The momentum M of an object is the product of its mass m and its velocity v. What is the momentum of a 0.160-kg hockey puck moving at 45.0 m/s (about 100 mi/h)?

30. A baseball pitcher's earned run average (ERA) is found by taking 9 times the number of earned runs r and dividing by the number of innings pitched i. Express the ERA as a function of r and i. Then find the ERA (to 2 decimal places) for Clayton Kershaw of the Dodgers, who in 2015 had 55 earned runs against him in 232 innings pitched.

31. Find the volume of a conical pile of sand if $r = 3.5$ m and $h = 1.5$ m.

32. The centripetal acceleration a of an object moving in a circular path is $a = v^2/R$, where v is the velocity of the object and R is the radius of the circle. What is the centripetal acceleration of an object moving at 6 ft/s in a circular path of radius 4 ft?

33. The pressure p (in Pa) of a gas as a function of its volume V and temperature T is $p = nRT/V$. If $n = 3.00$ mol and $R = 8.31$ J/mol · K, find p for $T = 300$ K and $V = 50.0$ m^3.

34. The power P (in W) used by a kitchen appliance is $P = i^2R$, where i is the current and R is the resistance. Find the power used if $i = 4.0$ A and $R = 2.4\ \Omega$.

35. The atmospheric temperature T near ground level in a certain region is $T = ax^2 + by^2$, where a and b are constants. What type of curve is each isotherm (along which the temperature is constant) in this region?

36. The pressure p exerted by a force F on an area A is $p = F/A$. If a given force is doubled on an area that is 2/3 of a given area, what is the ratio of the initial pressure to the final pressure?

37. The current i (in A) in a certain electric circuit is a function of the time t (in s) and a variable resistor R (in Ω), given by $i = \dfrac{6.0\sin 0.01t}{R + 0.12}$. Find i for $t = 0.75$ s and $R = 1.50\ \Omega$.

38. The reciprocal of the image distance q from a lens as a function of the object distance p and the focal length f of the lens is $\dfrac{1}{q} = \dfrac{1}{f} - \dfrac{1}{p}$. Find the image distance of an object 20 cm from a lens whose focal length is 5 cm.

39. A spherical cap (the smaller part of a sphere cut through by a plane) has a volume $V = \dfrac{1}{3}\pi h^2(3r - h)$, where r is the radius of the sphere and h is the height of the cap. What is the volume of Earth above 30° north latitude ($r = 6380$ km)? (*Hint:* 30° north latitude is 30° above the plane passing through the equator measured from the center of Earth.)

40. A planner for an electronics supplier plots a grid on a map and finds that its three biggest customers are located at A (0, 0), B (60, 0), and C (20, 50), where units are in km. If the supplier builds a warehouse at the point $W(x, y)$, express the sum S of the distances from W to A, B, and C as a function of x and y. See Fig. 29.5. (*Hint:* Use the distance formula given in Section 21.1.)

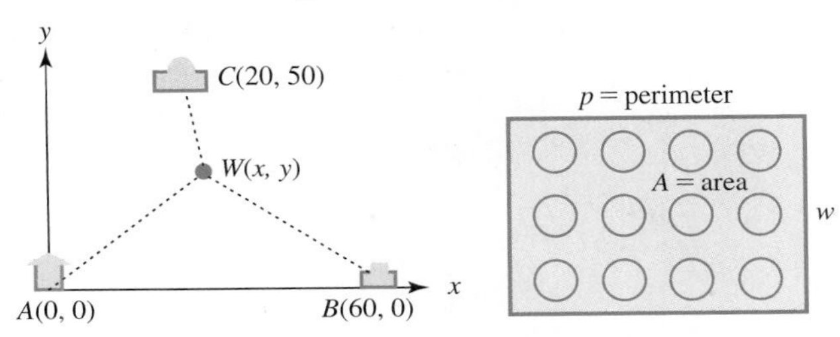

Fig. 29.5

Fig. 29.6

41. A rectangular solar cell panel has a perimeter p and a width w. Express the area A of the panel in terms of p and w and evaluate the area for $p = 250$ cm and $w = 55$ cm. See Fig. 29.6.

42. A gasoline storage tank is in the shape of a right circular cylinder with a hemisphere at each end, as shown in Fig. 29.7. Express the volume V of the tank in terms of r and h and then evaluate the volume for $r = 3.75$ ft and $h = 12.5$ ft.

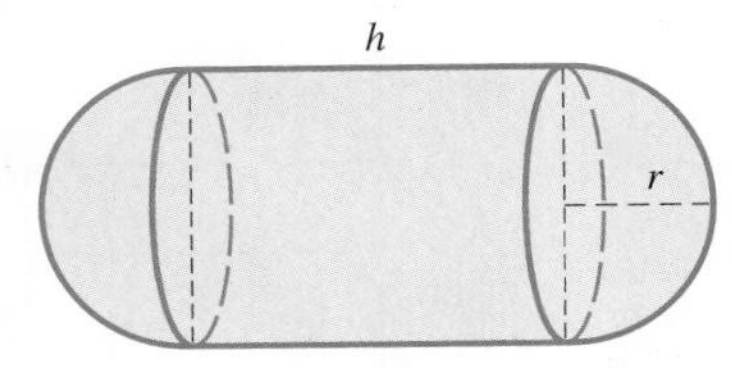

Fig. 29.7

43. The crushing load L of a pillar varies as the fourth power of its radius r and inversely as the square of its length l. Express L as a function of r and l for a pillar 20 ft tall and 1 ft in diameter that is crushed by a load of 20 tons.

44. The resonant frequency f (in Hz) of an electric circuit containing an inductance L and capacitance C is inversely proportional to the square root of the product of the inductance and the capacitance. If the resonant frequency of a circuit containing a 4-H inductor and a 64-μF capacitor is 10 Hz, express f as a function of L and C.

Answer to Practice Exercise

1. $f(1, 4) - f(x, 2x) = 52 - 10x^3$

29.2 Curves and Surfaces in Three Dimensions

Rectangular Coordinates • Octant • Plane • Surfaces and Curves • Trace • Section • Cylindrical Coordinates

We will now undertake a brief description of the graphical representation of a function of two variables. We shall show first a method of representation in the rectangular coordinate system in two dimensions. The following example illustrates the method.

EXAMPLE 1 Representing $f(x, y)$ in two dimensions

In order to represent $z = 2x^2 + y^2$, we will assume various values of z and sketch the curve of the resulting equation in the xy-plane. For example, if $z = 2$ we have

$$2x^2 + y^2 = 2$$

We recognize this as an ellipse with its major axis along the y-axis and vertices at $(0, \sqrt{2})$ and $(0, -\sqrt{2})$. The ends of the minor axis are at $(1, 0)$ and $(-1, 0)$. However, the ellipse $2x^2 + y^2 = 2$ represents the function $z = 2x^2 + y^2$ only for the value of $z = 2$. If $z = 4$, we have $2x^2 + y^2 = 4$, which is another ellipse. In fact, for all positive values of z, an ellipse is the resulting curve. Negative values of z are not possible, because neither x^2 nor y^2 may be negative. Figure 29.8 shows the ellipses that are obtained by using the indicated values of z. ■

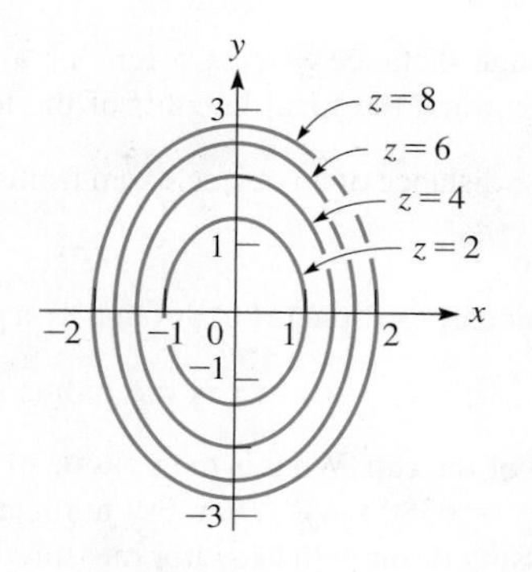

Fig. 29.8

The method of representation illustrated in Example 1 is useful if only a few specific values of z are to be used, or at least if the various curves do not intersect in such a way that they cannot be distinguished. If a general representation of z as a function of x and y is desired, it is necessary to use three coordinate axes, one each for x, y, and z. The most widely applicable system of this kind is to place a third coordinate axis at right angles to each of the x- and y-axes. In this way, we employ three dimensions for the representation.

The three mutually perpendicular coordinate axes—the x-axis, the y-axis, and the z-axis—are the basis of the **rectangular coordinate system in three dimensions.** Together they form three mutually perpendicular planes in space: the xy-plane, the yz-plane, and the xz-plane. To every point in space of the coordinate system is associated the set of numbers (x, y, z), called an **ordered triple.** The point at which the axes meet is the *origin.* The positive directions of the axes are indicated in Fig. 29.9. *That part of space in which all values of the coordinates are positive is called the* **first octant.** The other seven octants are not numbered in any particular way.

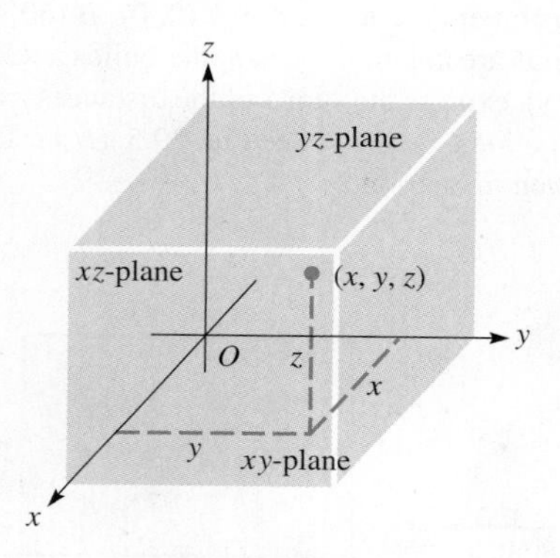

Fig. 29.9

EXAMPLE 2 Representing (x, y, z) in rectangular coordinates

Represent the point $(2, 4, 3)$ in rectangular coordinates.

We draw a line 4 units long from the point $(2, 0, 0)$ on the x-axis in the xy-plane, parallel to the y-axis. This locates the point $(2, 4, 0)$. From this point, a line 3 units long is drawn vertically upward, and this locates the point $(2, 4, 3)$. See the points and dashed lines in Fig. 29.10.

This point $(2, 4, 3)$ may also be located by starting from $(0, 4, 0)$ on the y-axis, then proceeding 2 units *parallel* to the x-axis to $(2, 4, 0)$, and then proceeding vertically 3 units upward to $(2, 4, 3)$.

In may also be located by starting from the point $(0, 0, 3)$ on the z-axis, although it is generally preferred to start on the x-axis, or possibly the y-axis. ■

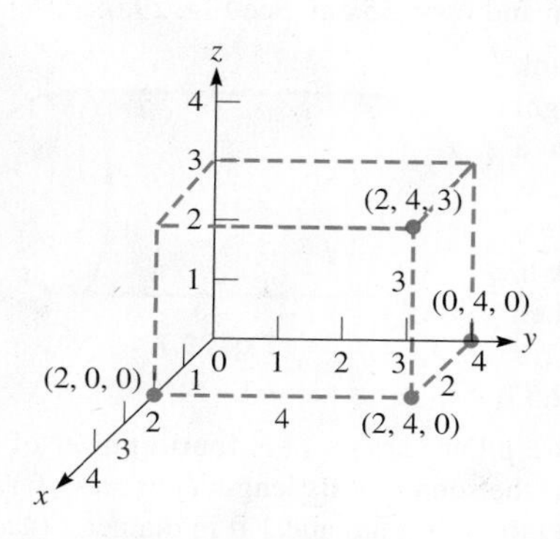

Fig. 29.10

We now show certain basic techniques by which three-dimensional figures may be drawn. We start by showing the general equation of a plane.

In Chapters 5 and 21, we showed that the graph of the equation $Ax + By + C = 0$ in two dimensions is a straight line. In the following example, we will verify that *the graph of the equation shown below is a* **plane** *in three dimensions.*

Equation of a Plane

$$Ax + By + Cz + D = 0 \tag{29.1}$$

EXAMPLE 3 Illustrating the graph of a plane

Show that the graph of the linear equation $2x + 3y + z - 6 = 0$ in the rectangular coordinate system of three dimensions is a plane.

NOTE ➧ [If we let any of the three variables take on a specific value, we obtain a linear equation in the other two variables.] For example, the point $(\frac{1}{2}, 1, 2)$ satisfies the equation and therefore lies on the graph of the equation. For $x = \frac{1}{2}$, we have

$$3y + z - 5 = 0$$

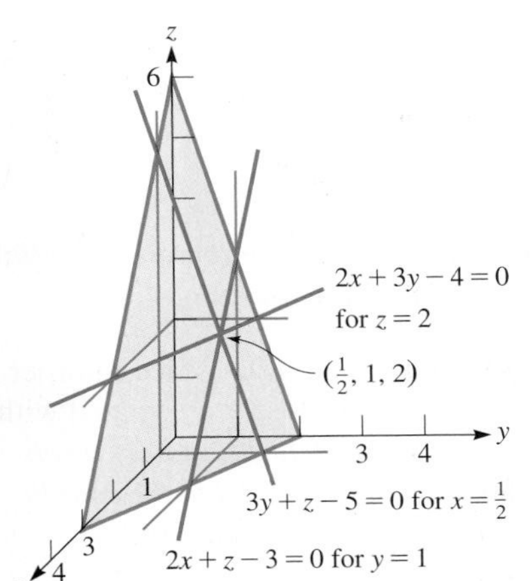

Fig. 29.11

which is the equation of a straight line. This means that all pairs of values of y and z that satisfy this equation, along with $x = \frac{1}{2}$, satisfy the given equation. Thus, for $x = \frac{1}{2}$, the straight line $3y + z - 5 = 0$ lies on the graph of the equation.

For $z = 2$, we have

$$2x + 3y - 4 = 0$$

which is also a straight line. By similar reasoning, this line lies on the graph of the equation. Because two lines through a point define a plane, these lines through $(\frac{1}{2}, 1, 2)$ define a plane. This plane is the graph of the equation (see Fig. 29.11).

For any point on the graph of this equation, there is a straight line on the graph that is parallel to one of the coordinate planes. Therefore, there are intersecting straight lines through the point. Thus, the graph is a plane. A similar analysis can be made for any equation of the same linear form. ■

NOTE ➧ [Because we know that the graph of an equation of the form of Eq. (29.1) is a plane, its graph can be found by determining its three intercepts, and the plane can then be represented by drawing in the lines between these intercepts.] If the plane passes through the origin, by letting two of the variables in turn be zero, two straight lines that define the plane are found.

EXAMPLE 4 Sketching the graph of a plane

Sketch the graph of $3x - y + 2z - 4 = 0$.

As in the case of two-dimensional graphs, the intercepts of a graph in three dimensions are those points where it crosses the respective axes. Therefore, by letting two of the variables at a time equal zero, we obtain the intercepts. For the graph of the given equation, the intercepts are $(\frac{4}{3}, 0, 0)$, $(0, -4, 0)$, and $(0, 0, 2)$. These points are located (see Fig. 29.12), and lines drawn between them to represent the plane. ■

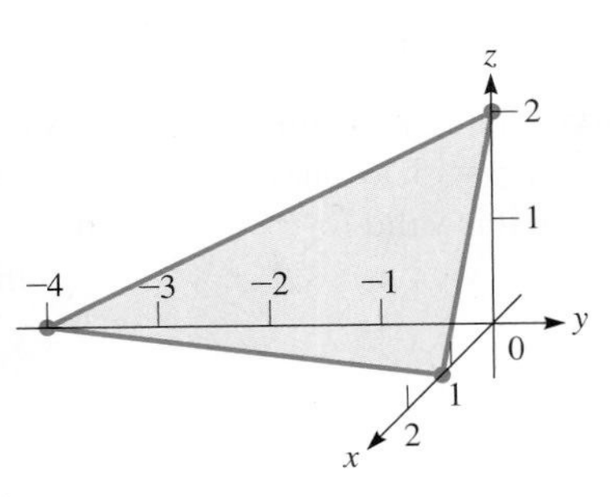

Fig. 29.12

The graph of an equation in three variables, which is essentially equivalent to a function with two independent variables, is a **surface** *in space.* This is seen for the plane and will be verified for other equations in the examples that follow.

The intersection of two surfaces is a **curve** *in space.* This has been seen in Examples 3 and 4, because the intersections of the given planes and the coordinate planes are lines (which in the general sense are curves). *We define the* **traces** *of a surface to be the curves of intersection of the surface and the coordinate planes.* The traces of a plane are those lines drawn between the intercepts to represent the plane. Many surfaces may be sketched by finding their traces and intercepts.

EXAMPLE 5 Using intercepts and traces to sketch a graph

Find the intercepts and traces for the graph of the equation $z = 4 - x^2 - y^2$, and then sketch the graph.

The intercepts of the graph of the equation are (2, 0, 0), $(-2, 0, 0)$, (0, 2, 0), $(0, -2, 0)$, and (0, 0, 4).

Because the traces of a surface lie within the coordinate planes, for each trace one of the variables is zero. Thus, ***by letting each variable in turn be zero, we find the trace of the surface in the plane of the other two variables.*** Therefore, the traces of this surface are

in the yz-plane:	$z = 4 - y^2$	(a parabola)
in the xz-plane:	$z = 4 - x^2$	(a parabola)
in the xy-plane:	$x^2 + y^2 = 4$	(a circle)

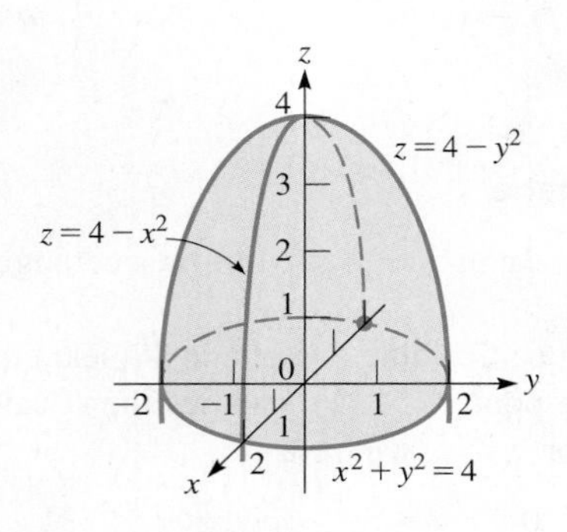

Fig. 29.13

Using the intercepts and sketching the traces, we obtain the surface represented by the equation as shown in Fig. 29.13. This figure is called a **circular paraboloid.** ■

There are numerous techniques for analyzing the equation of a surface in order to obtain its graph. Another that we shall discuss here, which is closely associated with a trace, is that of a **section.** *By assuming a specific value of one of the variables, we obtain an equation in two variables, the graph of which lies in a plane parallel to the coordinate plane of the two variables.* The following example illustrates sketching a surface by use of intercepts, traces, and sections.

EXAMPLE 6 Using traces and sections to sketch a graph

Sketch the graph of $4x^2 + y^2 - z^2 = 4$.

The intercepts are (1, 0, 0), $(-1, 0, 0)$, (0, 2, 0), and $(0, -2, 0)$. We note that there are no intercepts on the z-axis, for this would necessitate $z^2 = -4$.

The traces are

in the yz-plane:	$y^2 - z^2 = 4$	(a hyperbola)
in the xz-plane:	$4x^2 - z^2 = 4$	(a hyperbola)
in the xy-plane:	$4x^2 + y^2 = 4$	(an ellipse)

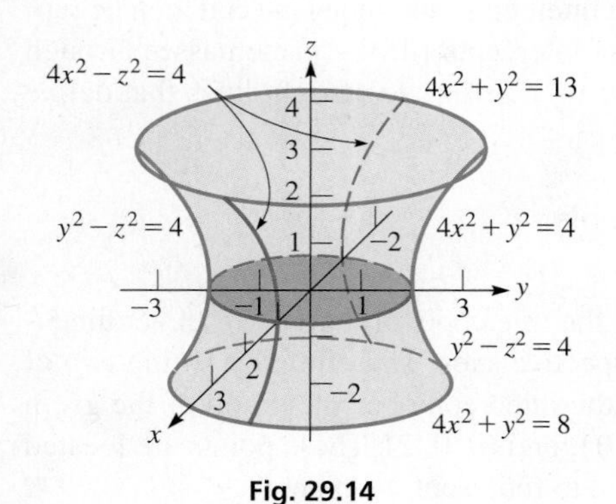

Fig. 29.14

The surface is reasonably defined by these curves, but by assuming suitable values of z, we may indicate its shape better. For example, if $z = 3$, we have $4x^2 + y^2 = 13$, which is an ellipse. In using it, we must remember that it is valid for $z = 3$ and therefore should be drawn 3 units above the xy-plane. If $z = -2$, we have $4x^2 + y^2 = 8$, which is also an ellipse. Thus, we have the following sections:

for $z = 3$:	$4x^2 + y^2 = 13$	(an ellipse)
for $z = -2$:	$4x^2 + y^2 = 8$	(an ellipse)

Other sections could be found, but these are sufficient to obtain a good sketch of the graph (see Fig. 29.14). The figure is called an **elliptic hyperboloid.**

The shape of a nuclear cooling tower is a circular hyperboloid, which is a special case of an elliptical hyperboloid. ■

Practice Exercise

1. What is the trace of $x^2 + 4y^2 - 4z^2 = 4$ in the (a) xy-plane? (b) xz-plane?

Having developed the rectangular coordinate system in three dimensions, we can compare the graph of a function using two dimensions and three dimensions. The next example shows the surface for the function of Example 1.

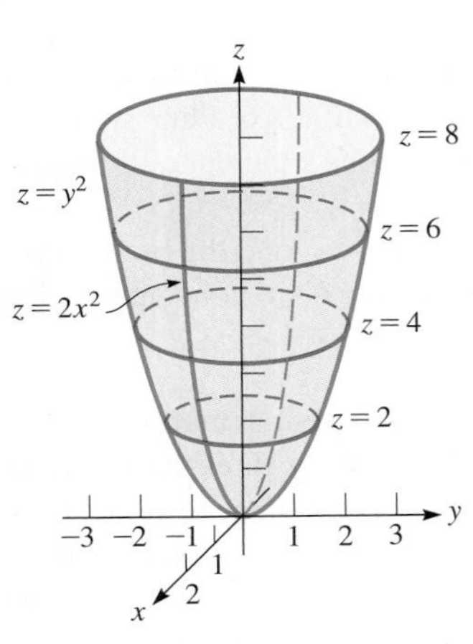

Fig. 29.15

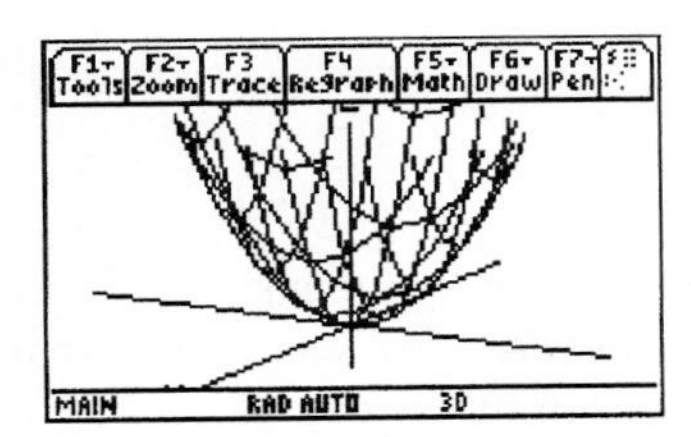

Fig. 29.16
TI-89 graphing calculator keystrokes: goo.gl/6UwV2I

EXAMPLE 7 Comparing 2D graph and 3D graph

Sketch the graph of $z = 2x^2 + y^2$.

The only intercept is (0, 0, 0). The traces are

in the yz-plane:	$z = y^2$	(a parabola)
in the xz-plane:	$z = 2x^2$	(a parabola)
in the xy-plane:	the origin	

The trace in the xy-plane is only the point of origin, because $2x^2 + y^2 = 0$ may be written as $y^2 = -2x^2$, which is true only for $x = 0$ and $y = 0$.

To get a better graph, we should use some positive values for z. As we noted in Example 1, negative values of z cannot be used. Because we used $z = 2$, $z = 4$, $z = 6$, and $z = 8$ in Example 1, we shall use these values here. Therefore,

for $z = 2$:	$2x^2 + y^2 = 2$	for $z = 4$:	$2x^2 + y^2 = 4$
for $z = 6$:	$2x^2 + y^2 = 6$	for $z = 8$:	$2x^2 + y^2 = 8$

Each of these sections is an ellipse. The surface, called an **elliptic paraboloid,** is shown in Fig. 29.15. Compare with Fig. 29.8.

Figure 29.16 shows the graph of this function on a TI-89 calculator. ■

Example 7 illustrates how *topographic maps* may be drawn. These maps represent three-dimensional terrain in two dimensions. For example, if Fig. 29.15 represents an excavation in the surface of Earth, then Fig. 29.8 represents the curves of constant elevation, or *contours,* with equally spaced elevations measured from the bottom of the excavation.

An equation with only two variables may represent a surface in space. Because only two variables are included in the equation, the surface is independent of the other variable. Another interpretation is that all sections, for all values of the variable not included, are the same.

[That is, all sections parallel to the coordinate plane of the included variables are the same as the trace in that plane.]

■ Whether an equation in two variables, like $x + y = 2$, represents a two- or three-dimensional graph must be determined from the context of the problem.

EXAMPLE 8 Comparing a linear equation in 2D and 3D

Sketch the graph of $x + y = 2$ in the rectangular coordinate system in three dimensions and in two dimensions.

Because z does not appear in the equation, we can consider the equation to be $x + y + 0z = 2$. Therefore, we see that *for any value of z,* the section is the straight line $x + y = 2$. Thus, the graph is a plane as shown in Fig. 29.17(a). The graph as a straight line in two dimensions is shown in Fig. 29.17(b).

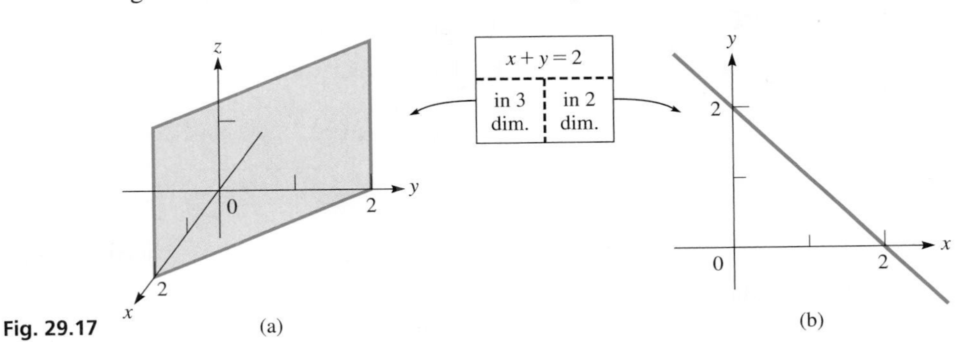

Fig. 29.17 (a) (b) ■

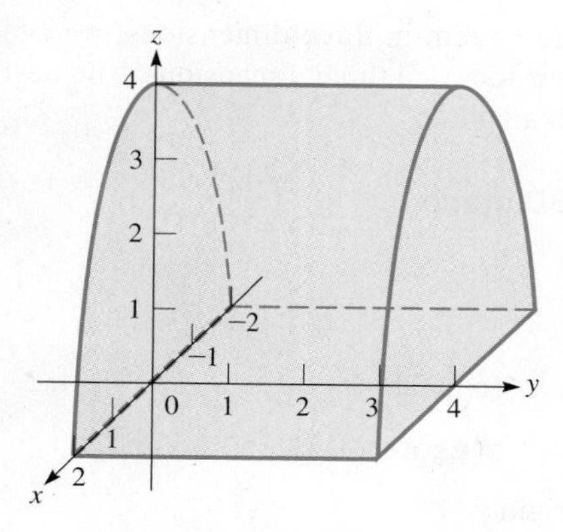

Fig. 29.18

EXAMPLE 9 Sketching a cylindrical surface

The graph of the equation $z = 4 - x^2$ in three dimensions is a surface whose sections, for all values of y, are given by the parabola $z = 4 - x^2$. The surface is shown in Fig. 29.18.

This type of surface is known as a **cylindrical surface.** In general, a cylindrical surface is one that can be generated by a line moving parallel to a fixed line while passing through a plane curve. This is not the same as a right circular cylinder, although a right circular cylinder is an example of a cylindrical surface.

It must be realized that most of the figures shown extend beyond the ranges indicated by the traces and sections. However, these traces and sections are convenient for representing and visualizing these surfaces. ■

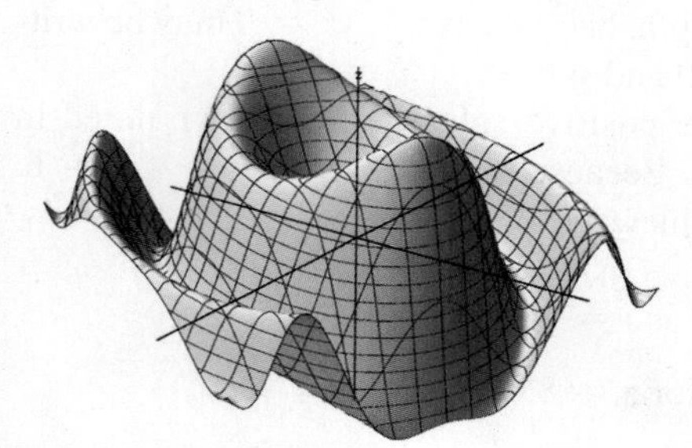
Fig. 29.19

There are various computer programs, and some graphing calculators (as we have shown for the TI-89 in Example 7) that can be used to display three-dimensional surfaces. They generally use sections in planes that are perpendicular to both the x- and y-axes. Such computer or calculator-drawn surfaces are of great value for visualizing all types of surfaces. They are especially useful for those of a complex nature that can be very difficult to draw by hand. Figure 29.19 shows a computer-generated graph of

$$z = \frac{\sin(2x^2 + y^2)}{x^2 + 1}$$

CYLINDRICAL COORDINATES

NOTE ▸ Another set of coordinates that can be used to display a three-dimensional figure are ***cylindrical coordinates,*** *which are polar coordinates and the z-axis combined.* [In using cylindrical coordinates, every point in space is designated by the coordinates (r, θ, z), as shown in Fig. 29.20.] The equations relating the rectangular coordinates (x, y, z) and the cylindrical coordinates (r, θ, z) of a point are

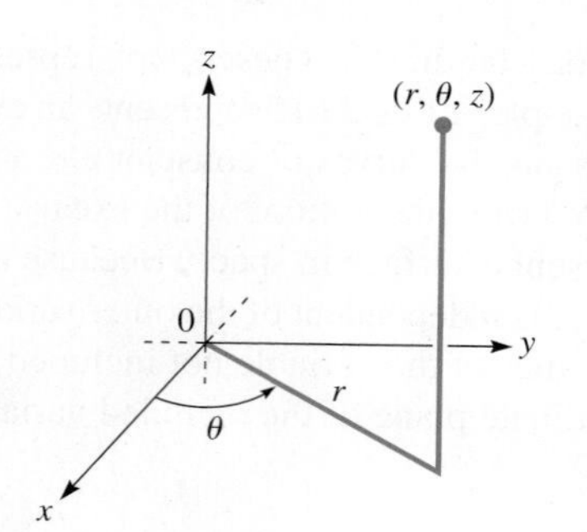

Fig. 29.20

$$x = r\cos\theta \qquad y = r\sin\theta \qquad z = z$$
$$r^2 = x^2 + y^2 \qquad \tan\theta = \frac{y}{x} \tag{29.2}$$

The following examples illustrate the use of cylindrical coordinates.

EXAMPLE 10 Plotting points in cylindrical coordinates

(a) Plot the point with cylindrical coordinates $(4, 2\pi/3, 2)$ and find the corresponding rectangular coordinates.

This point is shown in Fig. 29.21. From Eqs. (29.2), we have

$$x = 4\cos\frac{2\pi}{3} = 4\left(-\frac{1}{2}\right) = -2 \qquad y = 4\sin\frac{2\pi}{3} = 4\left(\frac{\sqrt{3}}{2}\right) = 2\sqrt{3} \qquad z = 2$$

Therefore, the rectangular coordinates of the point are $(-2, 2\sqrt{3}, 2)$.

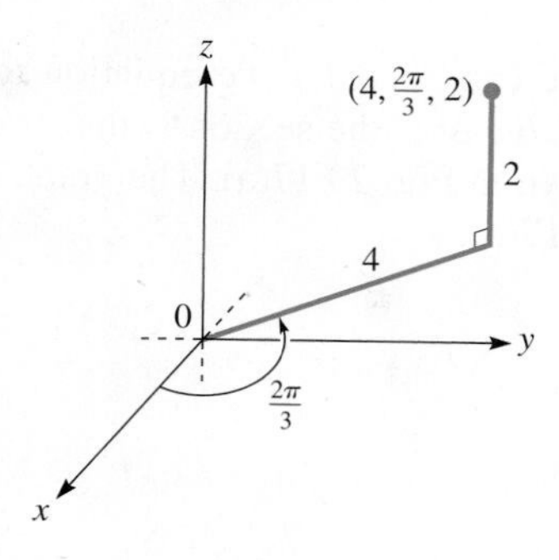

Fig. 29.21

(b) Find the cylindrical coordinates of the point with rectangular coordinates $(2, -2, 6)$.

Using Eqs. (29.2), we have

$$r = \sqrt{2^2 + (-2)^2} = 2\sqrt{2} \qquad z = 6$$

$$\tan\theta = \frac{-2}{2} = -1 \quad \text{(quadrant IV)} \qquad \theta = 2\pi - \frac{\pi}{4} = \frac{7\pi}{4}$$

The cylindrical coordinates are $(2\sqrt{2}, 7\pi/4, 6)$. As with polar coordinates, the value of θ can be $7\pi/4 + 2n\pi$, where n is an integer. ■

Practice Exercise

2. Find the cylindrical coordinates of the point with rectangular coordinates $(1, \sqrt{3}, -2)$.

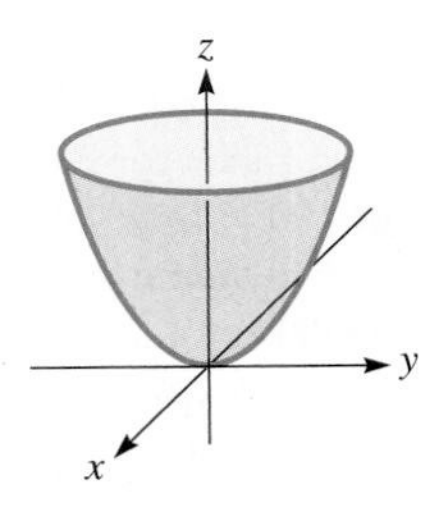

Fig. 29.22

EXAMPLE 11 Sketching a surface in cylindrical coordinates

Find the equation for the surface $z = 4x^2 + 4y^2$ in cylindrical coordinates and sketch the surface.

Factoring the 4 from the terms on the right, we have $z = 4(x^2 + y^2)$. We can now use the fact that $x^2 + y^2 = r^2$ to write this equation in cylindrical coordinates (r, θ, z) as $z = 4r^2$. From this equation, we see that $z \geq 0$, and that as r increases, we have circular sections (parallel to the xy-plane) of increasing radius. Therefore, we see that it is a **circular paraboloid.** See Fig. 29.22. ■

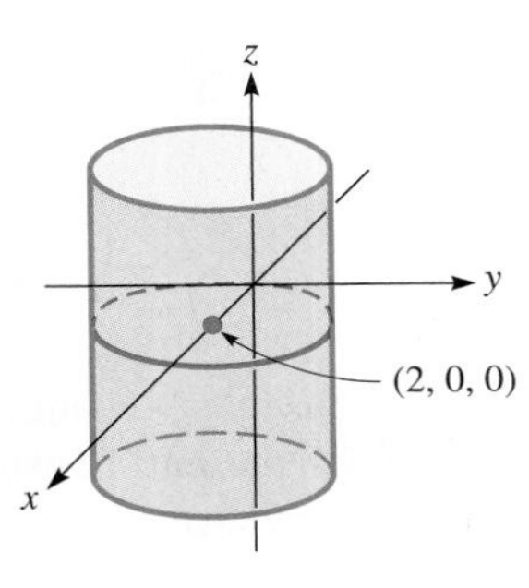

Fig. 29.23

EXAMPLE 12 Sketching a surface in cylindrical coordinates

Find the equation for the surface $r = 4\cos\theta$ in rectangular coordinates and sketch the surface.

If we multiply each side of the equation by r, we obtain $r^2 = 4r\cos\theta$. Now, from Eqs. (29.2), we have $r^2 = x^2 + y^2$ and $r\cos\theta = x$, which means that $x^2 + y^2 = 4x$, or

$$x^2 - 4x + 4 + y^2 = 4$$
$$(x - 2)^2 + y^2 = 4$$

We recognize this as a cylinder that has as its trace in the xy-plane a circle with center at (2, 0, 0), as shown in Fig. 29.23. Since there is no z in the equation, sections for all values of z are circles. ■

If the center of the trace of a circular cylinder in the xy-plane is at the origin, the equation of the cylinder in cylindrical coordinates is simply $r = a$. For this reason, these coordinates are called "cylindrical" coordinates.

EXERCISES 29.2

In Exercises 1 and 2, make the given changes in the indicated examples of this section and then solve the resulting problems.

1. In Example 4, change the −4 to +6.

2. In Example 6, change the − sign before z^2 to +.

In Exercises 3 and 4, use the method of Example 1 and show the graphs of the given equations for the given values of z.

3. $z = x^2 + y^2, z = 1, z = 4, z = 9$

4. $z = y - x^2, z = 0, z = 2, z = 4$

In Exercises 5–26, sketch the graphs of the given equations in the rectangular coordinate system in three dimensions.

5. $x + y + 2z - 4 = 0$

6. $2x - y - z + 6 = 0$

7. $4x - 2y + z - 8 = 0$

8. $3x + 3y - 2z - 6 = 0$

9. $z = y - 2x - 2$

10. $z = x - 4y$

11. $x + 2y = 4$

12. $2x - 3z = 6$

13. $x^2 + y^2 + z^2 = 4$

14. $2x^2 + 2y^2 + z^2 = 8$

15. $z = 4 - 4x^2 - y^2$

16. $z = x^2 + y^2$

17. $z = 2x^2 + y^2 + 2$

18. $x^2 + y^2 - 4z^2 = 4$

19. $x^2 - y^2 - z^2 = 9$

20. $z = \sqrt{9x^2 + 4y^2}$

21. $x^2 + y^2 = 16$

22. $4z = x^2$

23. $y^2 + 9z^2 = 9$

24. $xy = 2$

25. $z = \dfrac{1}{x^2 + y^2}$

26. $x^2 + y^2 + z^2 - 4z = 0$

In Exercises 27–30, use a calculator or computer to display the graphs of the given equations.

27. $z = \ln(x^2 + y^2)$

28. $z = 2\sin\sqrt{2x^2 + y^2}$

29. $z = y^4 - 4y^2 - 2x^2$

30. $z = 4e^{-x^2} + 4e^{-4y^2}$

In Exercises 31–38, perform the indicated operations involving cylindrical coordinates.

31. Find the rectangular coordinates of the points whose cylindrical coordinates are (a) $(3, \pi/4, 5)$, (b) $(2, \pi/2, 3)$.

32. Find the rectangular coordinates of the points whose cylindrical coordinates are (a) $(4, \pi/3, 2)$, (b) $(5, \pi, -4)$.

33. Find the cylindrical coordinates of the points whose rectangular coordinates are (a) $(\sqrt{3}, 1, 7)$, (b) $(0, 4, 1)$.

34. Find the cylindrical coordinates of the points whose rectangular coordinates are (a) $(1, \sqrt{3}, 6)$ (b) $(1, -1, -5)$

35. Describe the surface for which the cylindrical coordinate equation is (a) $r = 2$, (b) $\theta = 2$, (c) $z = 2$.

36. Write the equation $x^2 + y^2 + 4z^2 = 4$ in cylindrical coordinates and sketch the surface.

37. Write the equation $r^2 = 4z$ in rectangular coordinates and sketch the surface.

38. Write the equation $r = 2\sin\theta$ in rectangular coordinates and sketch the surface.

In Exercises 39–46, sketch the indicated curves and surfaces.

39. Curves that represent a constant temperature are called *isotherms*. The temperature at a point (x, y) of a flat plate is t (°C), where $t = 4x - y^2$. In two dimensions, draw the isotherms for $t = -4, 0, 8$.

40. At a point (x, y) in the xy-plane, the electric potential V (in volts) is given by $V = y^2 - x^2$. Draw the lines of equal potential for $V = -9, 0, 9$.

41. An electric charge is so distributed that the electric potential at all points on an imaginary surface is the same. Such a surface is called an *equipotential surface*. Sketch the graph of the equipotential surface whose equation is $2x^2 + 2y^2 + 3z^2 = 6$.

42. The surface of a small hill can be roughly approximated by the equation $z(2x^2 + y^2 + 100) = 1500$, where the units are meters. Draw the surface of the hill and the contours for $z = 3$ m, $z = 6$ m, $z = 9$ m, $z = 12$ m, and $z = 15$ m.

43. The pressure p (in kPa), volume V (in m^3), and temperature T (in K) for a certain gas are related by the equation $p = T/2V$. Sketch the p-V-T surface by using the z-axis for p, the x-axis for V, and the y-axis for T. Use units of 100 K for T and 10 m for V. Sections must be used for this surface, *a thermodynamic surface,* because none of the variables may equal zero.

44. Sketch the line in space defined by the intersection of the planes $x + 2y + 3z - 6 = 0$ and $2x + y + z - 4 = 0$.

45. Sketch the graph of $x^2 + y^2 - 2y = 0$ in three dimensions and in two dimensions.

46. Sketch the curve in space defined by the intersection of the surfaces $x^2 + (z - 1)^2 = 1$ and $z = 4 - x^2 - y^2$.

Answers to Practice Exercises

1. (a) $x^2 + 4y^2 = 4$ (ellipse) (b) $x^2 - 4z^2 = 4$ (hyperbola)
2. $(2, \pi/3, -2)$

29.3 Partial Derivatives

Meaning of Partial Derivative • Notation • Geometric Interpretation • Partial Derivatives of Higher Order

In Chapter 23, when showing the derivative to be the instantaneous rate of change of one variable with respect to another, only one independent variable was present. To extend the derivative to functions of two (or more) variables, we find the derivative of the function with respect to one variable, while holding the other variable(s) constant.

If $z = f(x, y)$ and y is held constant, z becomes a function of only x. The derivative of $f(x, y)$ with respect to x is termed the **partial derivative** *of z with respect to x. Similarly, if x is held constant, the derivative of $f(x, y)$ with respect to y is the* **partial derivative** *of z with respect to y.* The notations for the partial derivative of $z = f(x, y)$ with respect to x include

■ The symbol ∂ was introduced by the German mathematician Carl Jacobi (1804–1851).

$$\frac{\partial z}{\partial x} \qquad \frac{\partial f}{\partial x} \qquad f_x \qquad \frac{\partial}{\partial x}f(x, y) \qquad f_x(x, y)$$

Similarly, $\partial z/\partial y$ denotes the partial derivative of z with respect to y. In speaking, this is often shortened to "the partial of z with respect to y."

EXAMPLE 1 Finding partial derivatives

If $z = 4x^2 + xy - y^2$, find $\partial z/\partial x$ and $\partial z/\partial y$.

Finding the partial derivatives of z, we have

$$z = 4x^2 + xy - y^2$$

$$\frac{\partial z}{\partial x} = 8x + y \quad \text{treat } y \text{ as a constant}$$

$$\frac{\partial z}{\partial y} = x - 2y \quad \text{treat } x \text{ as a constant}$$

■

EXAMPLE 2 Finding partial derivatives

If $z = \dfrac{x \ln y}{x^2 + 1}$, find $\partial z/\partial x$ and $\partial z/\partial y$.

$$\frac{\partial z}{\partial x} = \frac{(x^2 + 1)(\ln y) - (x \ln y)(2x)}{(x^2 + 1)^2} = \frac{(1 - x^2)\ln y}{(1 + x^2)^2}$$

$$\frac{\partial z}{\partial y} = \left(\frac{x}{x^2 + 1}\right)\left(\frac{1}{y}\right) = \frac{x}{y(x^2 + 1)}$$

We note that in finding $\partial z/\partial x$, it is necessary to use the quotient rule, because x appears in both numerator and denominator. However, when finding $\partial z/\partial y$, the only derivative needed is that of $\ln y$. ■

Practice Exercise

1. If $z = 4x^2 + x \sin y$, find $\partial z/\partial x$ and $\partial z/\partial y$.

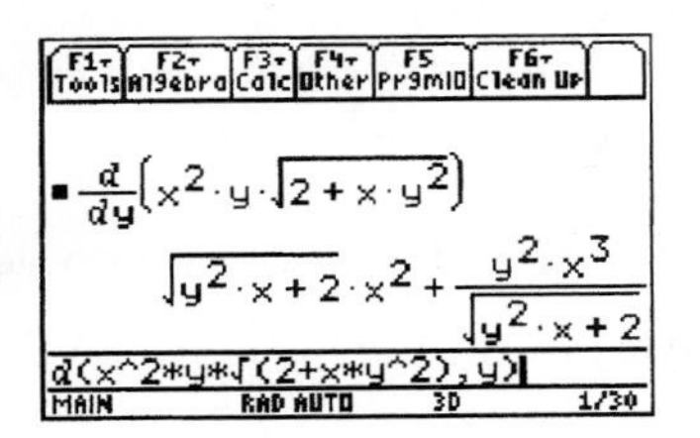

Fig. 29.24
TI-89 graphing calculator keystrokes: goo.gl/zEoWqP

EXAMPLE 3 Evaluating a partial derivative

For the function $f(x, y) = x^2y\sqrt{2 + xy^2}$, find $f_y(2, 1)$.

The notation $f_y(2, 1)$ means the partial derivative of f with respect to y, evaluated for $x = 2$ and $y = 1$. Thus, first finding $f_y(x, y)$, we have

$$f(x, y) = x^2y(2 + xy^2)^{1/2}$$

$$f_y(x, y) = x^2y\left(\frac{1}{2}\right)(2 + xy^2)^{-1/2}(2xy) + (2 + xy^2)^{1/2}(x^2)$$

$$= \frac{x^3y^2}{(2 + xy^2)^{1/2}} + x^2(2 + xy^2)^{1/2} \quad \text{see Fig. 29.24}$$

$$= \frac{x^3y^2 + x^2(2 + xy^2)}{(2 + xy^2)^{1/2}} = \frac{2x^2 + 2x^3y^2}{(2 + xy^2)^{1/2}}$$

$$f_y(2, 1) = \frac{2(4) + 2(8)(1)}{(2 + 2)^{1/2}} = 12$$

■

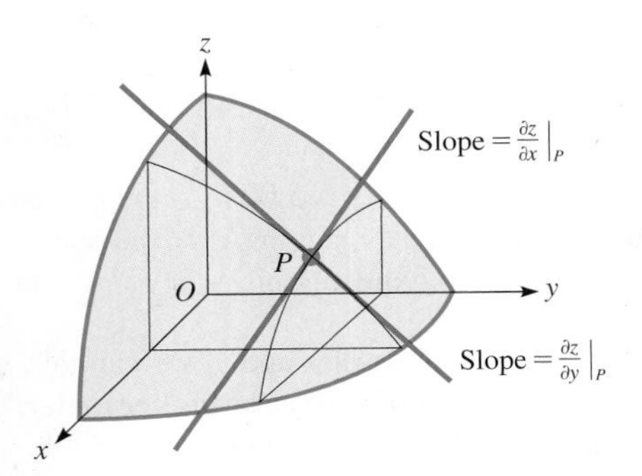

Fig. 29.25

To determine the geometric interpretation of a partial derivative, assume that $z = f(x, y)$ is the surface shown in Fig. 29.25. Choosing a point P on the surface, we then draw a plane through P parallel to the xz-plane. On this plane through P, the value of y is constant. The intersection of this plane and the surface is the curve as indicated. *The partial derivative of z with respect to x represents the slope of a line tangent to this curve.* When the values of the coordinates of point P are substituted into the expression for this partial derivative, it gives the slope of the tangent line at that point. In the same way, the partial derivative of z with respect to y, evaluated at P, gives the slope of the line tangent to the curve that is found from the intersection of the surface and the plane parallel to the yz-plane through P.

EXAMPLE 4 Slopes of lines tangent to a surface—Saddledome roof

■ See the chapter introduction.

The saddle-shaped roof surface of the Saddledome arena in Calgary, Alberta, is given approximately by the equation $z = \frac{1}{327}y^2 - \frac{1}{763}x^2$, with the origin at the center of the roof and all measurements in meters. Find the slopes of the tangent lines at the point (50.0, 40.0, 1.62) that are parallel to the xz- and yz-planes (see Fig. 29.26).

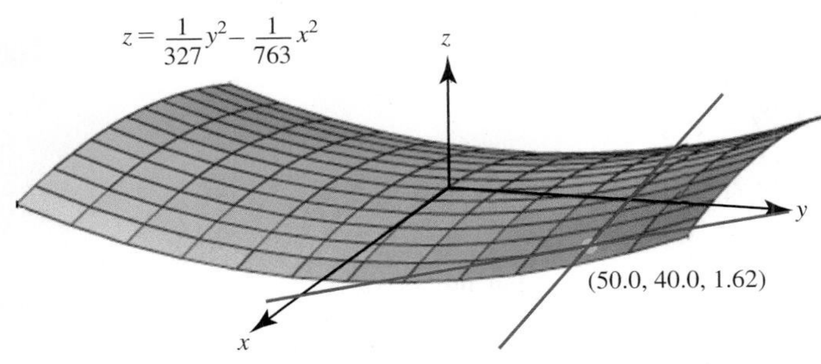

Fig. 29.26

Finding the partial derivatives with respect to x and y, we have

$$\frac{\partial z}{\partial x} = -\frac{2}{763}x \quad \text{and} \quad \frac{\partial z}{\partial y} = \frac{2}{327}y$$

These partial derivatives, when evaluated at the point (50.0, 40.0, 1.62), give us the slopes of the tangent lines parallel to the xz- and yz-planes, respectively:

$$\left.\frac{\partial z}{\partial x}\right|_{(50.0,40.0,1.62)} = -\frac{2}{763}(50.0) = -0.131 \quad \text{slope of tangent parallel to } xz\text{-plane}$$

$$\left.\frac{\partial z}{\partial y}\right|_{(50.0,40.0,1.62)} = \frac{2}{327}(40.0) = 0.245 \quad \text{slope of tangent parallel to } yz\text{-plane}$$

Therefore, in the x-direction, the roof is descending by 0.131 m per horizontal meter and in the y-direction, it is rising by 0.245 m per horizontal meter.

Note that the traces of this surface in the xz- and yz-planes are both parabolas (one opening upward and one opening downward) and sections for nonzero values of z are hyperbolas. Therefore, this surface is called a **hyperbolic paraboloid.** ■

■ Other structures with saddle-shaped roofs include the Stadium of Peace and Friendship in Athens, Greece, and the Velodrome at the Olympic Park near London, England.

The general interpretation of the partial derivative follows that of a derivative of a function with one independent variable. *The partial derivative* $f_x(x_0, y_0)$ *is the instantaneous rate of change of the function* $f(x, y)$ *with respect to* x, *with* y *held constant at the value of* y_0. This is illustrated in the following example.

EXAMPLE 5 Partial derivatives—rates of change of wind turbine power

The power P (in W) in the wind hitting a certain style of wind turbine is given by

$$P = 0.750r^2v^3,$$

where r is the length of the blades (in m) and v is the wind velocity (in m/s). Find and interpret $\frac{\partial P}{\partial r}$ and $\frac{\partial P}{\partial v}$ when $r = 30.0$ m and $v = 5.00$ m/s.

In finding and evaluating the partial derivatives, we have

$$\frac{\partial P}{\partial r} = 1.50rv^3 \qquad \left.\frac{\partial P}{\partial r}\right|_{r=30.0,\ v=5.00} = 1.50(30.0)(5.00)^3 = 5625 \text{ W/m}$$

$$\frac{\partial P}{\partial v} = 2.25r^2v^2 \qquad \left.\frac{\partial P}{\partial v}\right|_{r=30.0,\ v=5.00} = 2.25(30.0)^2(5.00)^2 = 50{,}625 \text{ W/(m/s)}$$

■ The values of these partial derivatives show us that an increase of 1 m/s in the wind velocity has a much larger impact on the power than an increase of 1 m in the length of the blades.

Therefore, when the blades are 30.0 m long and the wind velocity is 5.00 m/s, the power is increasing at a rate of 5625 W per meter of increase in the blade length. Also, the power is increasing at a rate of 50,625 W for each m/s increase in wind velocity. ■

Because the partial derivatives $\partial f/\partial x$ and $\partial f/\partial y$ are functions of x and y, we can take partial derivatives of each of them. This gives rise to **partial derivatives of higher order,** in a manner similar to the higher derivatives of a function of one independent variable. *The possible* **second-order partial derivatives** *of a function* $f(x, y)$ *are*

$$\frac{\partial^2 f}{\partial x^2} = \frac{\partial}{\partial x}\left(\frac{\partial f}{\partial x}\right) \qquad \frac{\partial^2 f}{\partial y^2} = \frac{\partial}{\partial y}\left(\frac{\partial f}{\partial y}\right)$$

$$\frac{\partial^2 f}{\partial x \partial y} = \frac{\partial}{\partial x}\left(\frac{\partial f}{\partial y}\right) \qquad \frac{\partial^2 f}{\partial y \partial x} = \frac{\partial}{\partial y}\left(\frac{\partial f}{\partial x}\right)$$

EXAMPLE 6 Second-order partial derivatives

Find the second-order partial derivatives of $z = x^3y^2 - 3xy^3$.

First, we find $\partial z/\partial x$ and $\partial z/\partial y$:

$$\frac{\partial z}{\partial x} = 3x^2y^2 - 3y^3 \qquad \frac{\partial z}{\partial y} = 2x^3y - 9xy^2$$

Therefore, we have the following second-order partial derivatives:

$$\frac{\partial^2 z}{\partial x^2} = \frac{\partial}{\partial x}\left(\frac{\partial z}{\partial x}\right) = 6xy^2 \qquad \frac{\partial^2 z}{\partial y^2} = \frac{\partial}{\partial y}\left(\frac{\partial z}{\partial y}\right) = 2x^3 - 18xy$$

$$\frac{\partial^2 z}{\partial x \partial y} = \frac{\partial}{\partial x}\left(\frac{\partial z}{\partial y}\right) = 6x^2y - 9y^2 \qquad \frac{\partial^2 z}{\partial y \partial x} = \frac{\partial}{\partial y}\left(\frac{\partial z}{\partial x}\right) = 6x^2y - 9y^2$$ ■

Practice Exercise

2. If $z = 2x^2y - 5x^2y^4$, find $\frac{\partial^2 z}{\partial x \partial y}$.

In Example 6, we note that

$$\frac{\partial^2 z}{\partial x \partial y} = \frac{\partial^2 z}{\partial y \partial x} \qquad \textbf{(29.3)}$$

In general, this is true if the function and partial derivatives are continuous.

EXAMPLE 7 Equality of second-order partial derivatives

For $f(x, y) = \tan^{-1}\dfrac{y}{x^2}$, show that $\dfrac{\partial^2 f}{\partial x \partial y} = \dfrac{\partial^2 f}{\partial y \partial x}$.

Finding $\partial f/\partial x$ and $\partial f/\partial y$, we have

$$\frac{\partial f}{\partial x} = \frac{1}{1 + \left(\dfrac{y}{x^2}\right)^2}\left(\frac{-2y}{x^3}\right) = \frac{x^4}{x^4 + y^2}\left(-\frac{2y}{x^3}\right) = \frac{-2xy}{x^4 + y^2}$$

$$\frac{\partial f}{\partial y} = \frac{1}{1 + \left(\dfrac{y}{x^2}\right)^2}\left(\frac{1}{x^2}\right) = \frac{x^4}{x^4 + y^2}\left(\frac{1}{x^2}\right) = \frac{x^2}{x^4 + y^2}$$

Now, we show that $\partial^2 f/\partial x\, \partial y$ and $\partial^2/\partial y\, \partial x$ are equal.

$$\frac{\partial^2 f}{\partial x \partial y} = \frac{(x^4 + y^2)(2x) - x^2(4x^3)}{(x^4 + y^2)^2} = \frac{-2x^5 + 2xy^2}{(x^4 + y^2)^2}$$

$$\frac{\partial^2 f}{\partial y \partial x} = \frac{(x^4 + y^2)(-2x) - (-2xy)(2y)}{(x^4 + y^2)^2} = \frac{-2x^5 + 2xy^2}{(x^4 + y^2)^2}$$

■

EXERCISES 29.3

In Exercises 1 and 2, make the given changes in the indicated examples of this section and then solve the resulting problems.

1. In Example 2, change x^2 to y^2.
2. In Example 4, change the point to (40.0, 50.0, 5.55) and then find the slopes of the two tangent lines.

In Exercises 3–26, find the partial derivative of the dependent variable or function with respect to each of the independent variables.

3. $z = 9x^6 - 3y^5$
4. $z = 6y^2 + 8x^4$
5. $z = e^{3x} - \sin y$
6. $z = \cos 2x + y^3$
7. $z = 5x + 4x^2y$
8. $z = 3x^2y^3 - 3x + 4y$
9. $f(x, y) = xe^{-2y}$
10. $z = 3y\cos^4 2x$
11. $f(x, y) = \dfrac{2 + \cos x}{1 - \sec 3y}$
12. $f(x, y) = \dfrac{\tan^{-1} 4y}{2 + x^2}$
13. $\phi = r\sqrt{1 + 2rs}$
14. $w = uv^2\sqrt[3]{1 - u^3}$
15. $z = (3x^2 + xy^3)^4$
16. $f(x, y) = (2xy - x^2)^5$
17. $z = \sin x^2 y$
18. $y = \tan^{-1}\left(\dfrac{8x}{t^2}\right)$
19. $y = \ln(r^2 + 6s)$
20. $u = e^{3x+2y}$
21. $f(x, y) = \dfrac{2\sin^3 2x}{1 - 3y}$
22. $f(x, y) = \dfrac{3x + \ln y}{x^2 + y^2}$
23. $z = \sin x + \cos xy - \cos y$
24. $t = 2re^{rs^2} - \tan(2r + s)$
25. $f(x, y) = e^x \cos xy + e^{-2x}\tan y$
26. $u = \ln\dfrac{y^2}{x - y} + e^{-x}(\sin y - \cos 2y)$

In Exercises 27–30, evaluate the indicated partial derivatives at the given points.

27. $z = 3xy - x^2,\ \left.\dfrac{\partial z}{\partial x}\right|_{(1, -2, -7)}$
28. $z = x^2\cos 4y,\ \left.\dfrac{\partial z}{\partial y}\right|_{(2, \frac{\pi}{2}, 4)}$
29. $z = x\sqrt{x^2 - y^2},\ \left.\dfrac{\partial z}{\partial x}\right|_{(5, 3, 20)}$
30. $z = e^y \ln xy,\ \left.\dfrac{\partial z}{\partial y}\right|_{(e, 1, e)}$

In Exercises 31–34, find all second-order partial derivatives.

31. $z = 2xy^3 - 3x^2y$
32. $F(x, y) = y\ln(x + 2y)$
33. $z = \dfrac{x}{y} + e^x \sin y$
34. $f(x, y) = \dfrac{2 + \cos y}{1 + x^2}$

In Exercises 35–50, solve the given problems.

35. The surface area of a cone as a function of its radius r and height h is $A = \pi r^2 + \pi r\sqrt{r^2 + h^2}$. Find $\partial A/\partial r$ and $\partial A/\partial h$.
36. Given the law of cosines $c^2 = a^2 + b^2 - 2ab\cos C$, if $C = 60°$, find $\partial c/\partial a$ and $\partial c/\partial b$.
37. The displacement y of a wave in a string as a function of its position x and time t is $y = \sin(\pi x)\sin(\pi t/2)$. Find $\partial y/\partial x$ and $\partial y/\partial t$.
38. The kinetic energy E of a moving object is given by $E = \frac{1}{2}mv^2$, where m is the mass (in kg) and v is the velocity (in m/s). Find $\dfrac{\partial E}{\partial m}$ and $\dfrac{\partial E}{\partial v}$.
39. Find the slope of a line tangent to the surface $z = 9 - x^2 - y^2$ and parallel to the yz-plane that passes through (1, 2, 4). Repeat the instructions for the line through (2, 2, 1). Draw an appropriate figure.
40. A metal plate in the shape of a circular segment of radius r expands by being heated. Express the width w (straight dimension) as a function of r and the height h. Then find both $\partial w/\partial r$ and $\partial w/\partial h$.

41. Two resistors R_1 and R_2, placed in parallel, have a combined resistance R_T given by $\frac{1}{R_T} = \frac{1}{R_1} + \frac{1}{R_2}$. Find $\frac{\partial R_T}{\partial R_1}$.

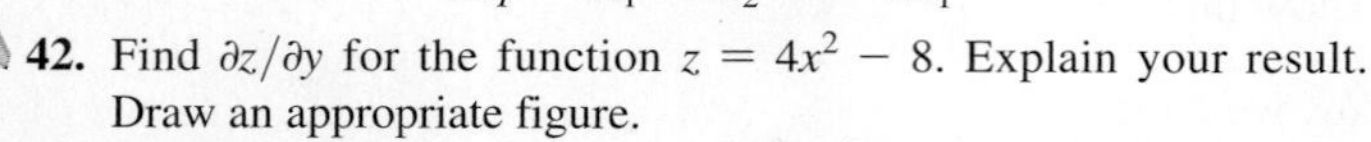

42. Find $\partial z/\partial y$ for the function $z = 4x^2 - 8$. Explain your result. Draw an appropriate figure.

43. A metallic machine part contracts while cooling. It is in the shape of a hemisphere attached to a cylinder, as shown in Fig. 29.27. Find the rate of change of volume with respect to r when $r = 2.65$ cm and $h = 4.20$ cm.

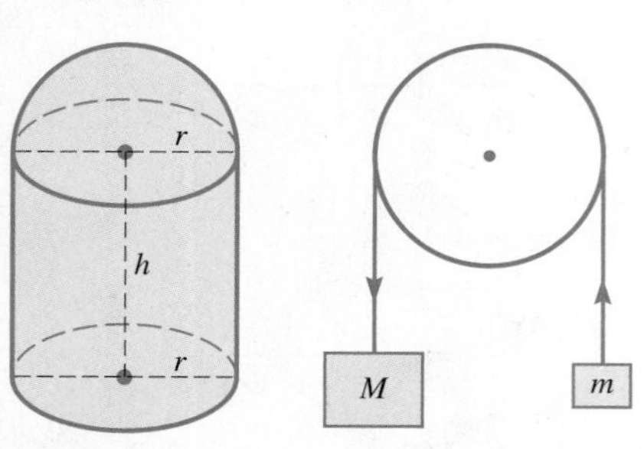

Fig. 29.27 Fig. 29.28

44. Two masses M and m are attached as shown in Fig. 29.28. If $M > m$, the downward acceleration a of mass M is given by $a = \frac{M - m}{M + m}g$, where g is the acceleration due to gravity. Show that $M\frac{\partial a}{\partial M} + m\frac{\partial a}{\partial m} = 0$.

45. In quality testing, a rectangular sheet of vinyl is stretched. Express the length of the diagonal d of the sheet as a function of the sides x and y. Find the rate of change of d with respect to x for $x = 6.50$ ft if y remains constant at 4.75 ft.

46. If an observer and a source of sound are moving toward or away from each other, the observed frequency of sound is different from that emitted. This is known as the *Doppler effect*. The equation relating the frequency f_0 the observer hears and the frequency f_s emitted by the source (a constant) is $f_0 = f_s\left(\frac{v + v_0}{v - v_s}\right)$, where v is the velocity of sound in air (a constant), v_0 is the velocity of the observer, and v_s is the velocity of the source. Show that $f_s\frac{\partial f_0}{\partial v_s} = f_0\frac{\partial f_0}{\partial v_0}$. Explain the meaning of $\partial f_0/\partial v_s$.

47. The *mutual conductance* (in $1/\Omega$) of a certain electronic device is defined as $g_m = \partial i_b/\partial e_c$. Under certain circumstances, the current i_b (in μA) is given by $i_b = 50(e_b + 5e_c)^{1.5}$. Find g_m when $e_b = 200$ V and $e_c = -20$ V.

48. The *amplification factor* of the electronic device of Exercise 47 is defined as $\mu = -\partial e_b/\partial e_c$. For the device of Exercise 47, under the given conditions, find the amplification factor.

49. The temperature u in a metal bar depends on the distance x from one end and the time t. Show that $u(x, t) = 5e^{-t}\sin 4x$ satisfies the *one-dimensional heat-conduction equation*. $\frac{\partial u}{\partial t} = k\frac{\partial^2 u}{\partial x^2}$, where k is called the *diffusivity*. Here $k = 1/16$.

50. The displacement y at any point in a taut, flexible string depends on the distance x from one end of the string and the time t. Show that $y(x, t) = 2\sin 2x\cos 4t$ satisfies the *wave equation* $\frac{\partial^2 y}{\partial t^2} = a^2\frac{\partial^2 y}{\partial x^2}$ with $a = 2$.

Answers to Practice Exercises

1. $\partial z/\partial x = 8x + \sin y$, $\partial z/\partial y = x\cos y$ **2.** $\frac{\partial^2 z}{\partial x \partial y} = 4x - 40xy^3$

29.4 Double Integrals

Meaning of Double Integral • Notation • Volume Under a Surface

We now turn our attention to integration in the case of a function of two variables. The analysis has similarities to that of partial differentiation, in that an operation is performed while holding one of the independent variables constant.

If $z = f(x, y)$ and we wish to integrate with respect to x and y, we first consider either x or y constant and integrate with respect to the other. After this integral is evaluated, we then integrate with respect to the variable first held constant. We shall now define this type of integral and then give an appropriate geometric interpretation.

If $z = f(x, y)$ the **double integral** *of the function over x and y is defined as*

$$\int_a^b \left[\int_{g(x)}^{G(x)} f(x, y)\,dy\right] dx$$

NOTE ▸ [Note that the limits on the inner integral are functions of x, and those on the outer integral are explicit values of x. **While the inner integral is evaluated first, x is held constant,** and this results in an integral with x only.] Then this integral is evaluated. Since it is customary not to include the brackets in writing a double integral, we write

$$\int_a^b \left[\int_{g(x)}^{G(x)} f(x, y)\,dy\right] dx = \int_a^b \int_{g(x)}^{G(x)} f(x, y)\,dy\,dx \qquad (29.4)$$

EXAMPLE 1 Evaluating a double integral

Evaluate $\int_0^1 \int_{x^2}^{x} xy\, dy\, dx.$

First, we integrate the inner integral with y as the variable and x as a constant.

treat as constant

$$\int_{x^2}^{x} xy\, dy = \left(x\frac{y^2}{2}\right)\Big|_{x^2}^{x} = x\left(\frac{x^2}{2} - \frac{x^4}{2}\right) = \frac{1}{2}(x^3 - x^5)$$

This means

$$\int_0^1 \int_{x^2}^{x} xy\, dy\, dx = \int_0^1 \frac{1}{2}(x^3 - x^5)dx = \frac{1}{2}\left(\frac{x^4}{4} - \frac{x^6}{6}\right)\Big|_0^1$$

$$= \frac{1}{2}\left(\frac{1}{4} - \frac{1}{6}\right) - \frac{1}{2}(0) = \frac{1}{24}$$ ■

Practice Exercise

1. Evaluate: $\int_0^1 \int_0^x 2xy\, dy\, dx.$

EXAMPLE 2 Evaluating a double integral

Evaluate $\int_0^{\pi/2} \int_0^{\sin y} e^{2x} \cos y\, dx\, dy.$

Because the inner differential is dx *(the inner limits must then be functions of y, which may be constant)*, we first integrate with x as the variable and y as a constant. The second integration is with y as the variable.

$$\int_0^{\pi/2} \int_0^{\sin y} e^{2x} \cos y\, dx\, dy = \int_0^{\pi/2} \left[\frac{1}{2}e^{2x}\cos y\right]_0^{\sin y} dy$$

$$= \frac{1}{2}\int_0^{\pi/2} (e^{2\sin y}\cos y - \cos y)dy$$

$$= \frac{1}{2}\left[\frac{1}{2}e^{2\sin y} - \sin y\right]_0^{\pi/2} = \frac{1}{2}\left(\frac{1}{2}e^2 - 1\right) - \frac{1}{2}\left(\frac{1}{2} - 0\right)$$

$$= \frac{1}{4}e^2 - \frac{1}{2} - \frac{1}{4} = \frac{1}{4}(e^2 - 3) = 1.097$$ ■

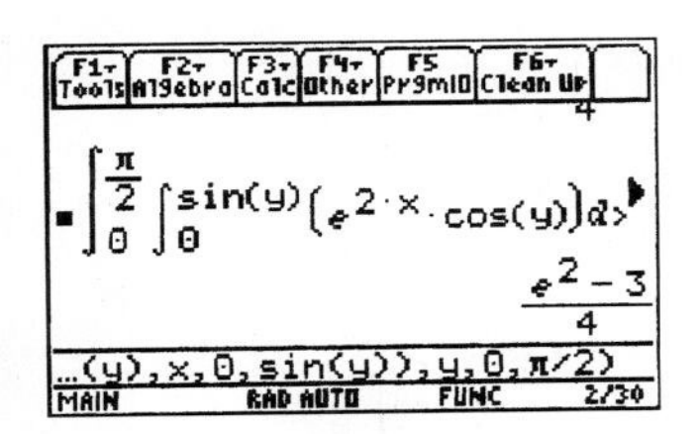

Fig. 29.29

TI-89 graphing calculator keystrokes: goo.gl/EB4IT4

■ This double integral is shown on a TI-89 calculator in Fig. 29.29.

For the geometric interpretation of a double integral, consider the surface shown in Fig. 29.30(a). An **element of volume** (dimensions of dx, dy, and z) extends from the xy-plane to the surface. With x a constant, sum (integrate) these elements of volume from the left boundary, $y = g(x)$, to the right boundary, $y = G(x)$. Now, the volume of the vertical slice is a function of x, as shown in Fig. 29.30(b). By summing (integrating) the volumes of these slices from $x = a$ ($x = 0$ in the figure) to $x = b$, we have the complete volume as shown in Fig. 29.30(c).

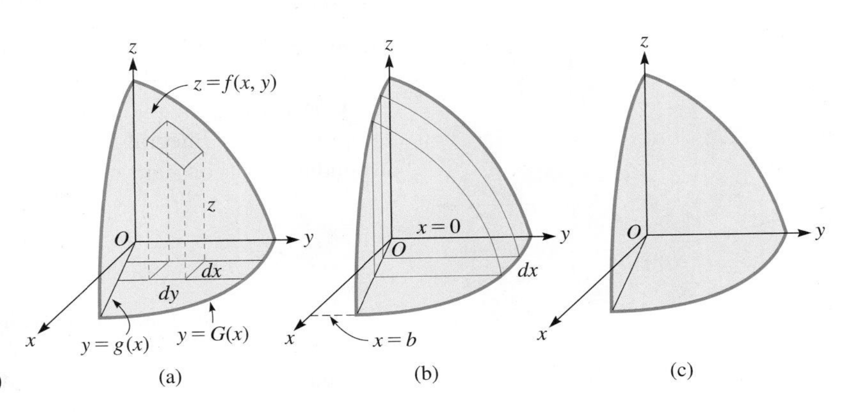

Fig. 29.30

Thus, *we may interpret a double integral as the* **volume under a surface,** in the same way as the integral was interpreted as the area of a plane figure. This allows us to find the volume of a more general figure, as this volume is not necessarily a volume of revolution, as discussed in Section 26.3.

EXAMPLE 3 Volume under a plane

Find the volume that is in the first octant and under the plane $x + 2y + 4z - 8 = 0$. See Fig. 29.31.

This figure is a tetrahedron, for which $V = \frac{1}{3}Bh$. Assuming the base is in the xy-plane, $B = \frac{1}{2}(4)(8) = 16$, and $h = 2$. Therefore, $V = \frac{1}{3}(16)(2) = \frac{32}{3}$ cubic units. We shall use this value to check the one we find by double integration.

To find $z = f(x, y)$, we solve the given equation for z. Thus,

$$z = \frac{8 - x - 2y}{4}$$

Next, we must find the limits on y and x. Choosing to integrate over y first, we see that y goes from $y = 0$ to $y = (8 - x)/2$. ***This last limit is the trace of the surface in the xy-plane.*** Next, we note that x goes from $x = 0$ to $x = 8$. Therefore, we set up and evaluate the integral:

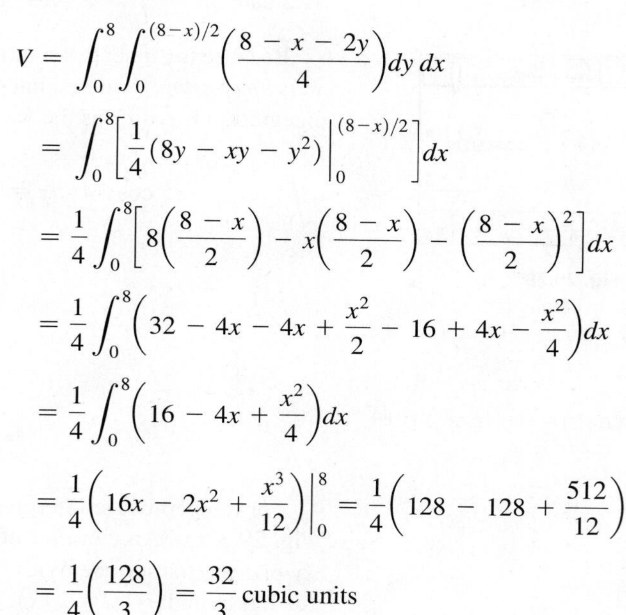

$$\begin{aligned}
V &= \int_0^8 \int_0^{(8-x)/2} \left(\frac{8 - x - 2y}{4}\right) dy\, dx \\
&= \int_0^8 \left[\frac{1}{4}(8y - xy - y^2)\Big|_0^{(8-x)/2}\right] dx \\
&= \frac{1}{4}\int_0^8 \left[8\left(\frac{8 - x}{2}\right) - x\left(\frac{8 - x}{2}\right) - \left(\frac{8 - x}{2}\right)^2\right] dx \\
&= \frac{1}{4}\int_0^8 \left(32 - 4x - 4x + \frac{x^2}{2} - 16 + 4x - \frac{x^2}{4}\right) dx \\
&= \frac{1}{4}\int_0^8 \left(16 - 4x + \frac{x^2}{4}\right) dx \\
&= \frac{1}{4}\left(16x - 2x^2 + \frac{x^3}{12}\right)\Big|_0^8 = \frac{1}{4}\left(128 - 128 + \frac{512}{12}\right) \\
&= \frac{1}{4}\left(\frac{128}{3}\right) = \frac{32}{3} \text{ cubic units}
\end{aligned}$$

Fig. 29.31

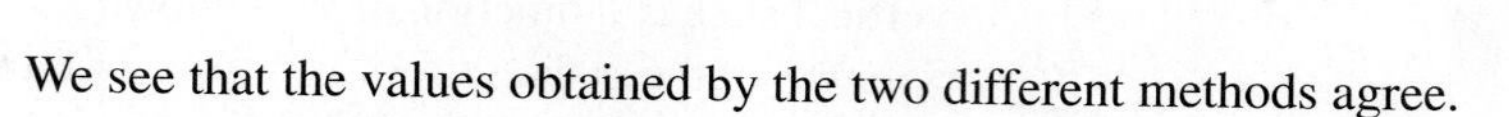

We see that the values obtained by the two different methods agree. ■

EXAMPLE 4 Volume under a surface

Find the volume above the xy-plane, below the surface $z = xy$, and enclosed by the cylinder $y = x^2$ and the plane $y = x$.

Constructing the figure, shown in Fig. 29.32, we note that $z = xy$ is the desired function of x and y. Integrating over y first, the limits on y are $y = x^2$ to $y = x$. The corresponding limits on x are $x = 0$ to $x = 1$. Therefore, the double integral to be evaluated is

$$V = \int_0^1 \int_{x^2}^{x} xy\, dy\, dx$$

This integral has already been evaluated in Example 1 of this section, and we can now see the geometric interpretation of that integral. Using the result from Example 1, we see that the required volume is $1/24$ cubic unit. ■

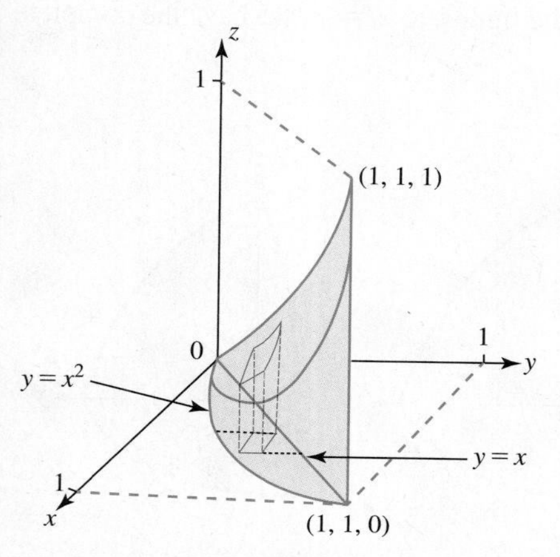

Fig. 29.32

EXAMPLE 5 Volume under a surface

Find the volume in the first octant that is under the surface $z = 4 - x^2 - y^2$, and is between the cylinder $x^2 = 3y$ and the plane $y = 1$. See Fig. 29.33.

Setting up the integration such that we integrate over x first, we have

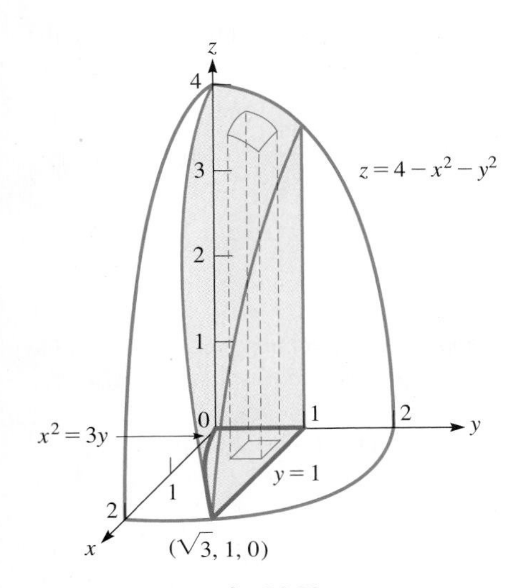

Fig. 29.33

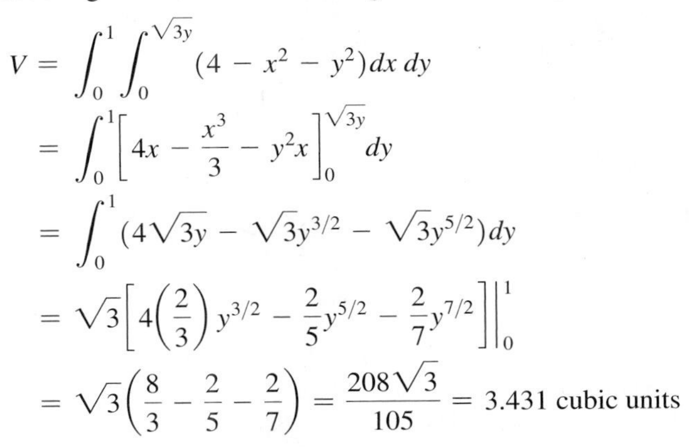

$$V = \int_0^1 \int_0^{\sqrt{3y}} (4 - x^2 - y^2)dx\,dy$$

$$= \int_0^1 \left[4x - \frac{x^3}{3} - y^2x\right]_0^{\sqrt{3y}} dy$$

$$= \int_0^1 (4\sqrt{3y} - \sqrt{3}y^{3/2} - \sqrt{3}y^{5/2})dy$$

$$= \sqrt{3}\left[4\left(\frac{2}{3}\right)y^{3/2} - \frac{2}{5}y^{5/2} - \frac{2}{7}y^{7/2}\right]\Bigg|_0^1$$

$$= \sqrt{3}\left(\frac{8}{3} - \frac{2}{5} - \frac{2}{7}\right) = \frac{208\sqrt{3}}{105} = 3.431 \text{ cubic units}$$

If we integrate over y first, we arrive at the same result evaluating the double integral

$$V = \int_0^{\sqrt{3}} \int_{x^2/3}^{1} (4 - x^2 - y^2)dy\,dx$$ ■

We can now see how to find many different kinds of volumes. In Chapter 2, we used basic geometric formulas to find volumes of regular-shaped objects. In Chapter 26, we used calculus to find various volumes of revolution. Here, we see that double integrals allow us to find volumes of many irregular shapes. We can apply these calculus methods to find volumes such as under tents, sections of a pipe cut at an angle (see Exercise 30), irregular swimming pools, specialized containers, and atmospheric regions.

EXERCISES 29.4

In Exercises 1 and 2, make the given changes in the indicated examples of this section and then solve the resulting problems.

1. In Example 1, change the integrand xy to $(x + y)$.
2. In Example 2, delete the e^{2x}.

In Exercises 5–18, evaluate the given double integrals.

3. $\int_0^2 \int_0^3 x^2\,dx\,dy$
4. $\int_0^2 \int_0^2 4y^3\,dy\,dx$
5. $\int_2^4 \int_0^1 xy^2\,dx\,dy$
6. $\int_0^2 \int_0^1 \frac{y}{(xy + 1)^2}dx\,dy$
7. $\int_1^2 \int_0^{y^2} xy^2\,dx\,dy$
8. $\int_0^4 \int_1^{\sqrt{y}} (x - y)dx\,dy$
9. $\int_0^1 \int_0^{\sqrt{1-x^2}} y\,dy\,dx$
10. $\int_4^9 \int_0^x \sqrt{x - y}\,dy\,dx$
11. $\int_0^{\pi/6} \int_{\pi/3}^{y} \sin x\,dx\,dy$
12. $\int_0^{\sqrt{3}} \int_{x^2/3}^{1} (4 - x^2)dy\,dx$
13. $\int_1^e \int_1^y \frac{1}{x}dx\,dy$
14. $\int_{-1}^1 \int_1^{e^x} \frac{1}{xy}dy\,dx$
15. $\int_1^2 \int_0^x yx^3e^{xy^2}\,dy\,dx$
16. $\int_0^{\pi/6} \int_0^1 y\sin x\,dy\,dx$
17. $\int_0^{\ln 3} \int_0^x e^{2x+3y}\,dy\,dx$
18. $\int_0^{1/2} \int_y^{y^2} \frac{dx\,dy}{\sqrt{y^2 - x^2}}$

In Exercises 19–28, find the indicated volumes by double integration.

19. The first-octant volume under the plane $x + y + z - 4 = 0$
20. The first-octant volume under the surface $z = y^2$ and bounded by the planes $x = 2$ and $y = 3$
21. The volume above the xy-plane and under the surface $z = 4 - x^2 - y^2$
22. The volume above the xy-plane, below the surface $z = x^2 + y^2$, and inside the cylinder $x^2 + y^2 = 4$
23. The first-octant volume bounded by the xy-plane, the planes $x = y$, $y = 2$, and $z = 2 + x^2 + y^2$
24. The volume bounded by the planes $x + 3y + 2z - 6 = 0$, $2x = y$, $x = 0$, and $z = 0$
25. The first-octant volume under the plane $z = x + y$ and inside the cylinder $x^2 + y^2 = 9$
26. The volume above the xy-plane and bounded by the cylinders $x = y^2$, $y = 8x^2$, and $z = x^2 + 1$. Integrate over y first and then check by integrating over x first.
27. Evaluate $A = \int_{-\pi/6}^{\pi/6} \int_{\sqrt{2}}^{2\sqrt{\cos 2\theta}} r\,dr\,d\theta$, the area outside the circle $r = \sqrt{2}$ and inside the lemniscate $r^2 = 4\cos 2\theta$, using polar coordinates.
28. Evaluate the *triple integral* $\int_{-2}^2 \int_1^2 \int_1^e \frac{xy^2}{z}dz\,dx\,dy$ by integrating over z, x, and y in that order (similar to a double integral).

29. A wedge is to be made in the shape shown in Fig. 29.34 (all vertical cross sections are equal right triangles). By double integration, find the volume of the wedge.

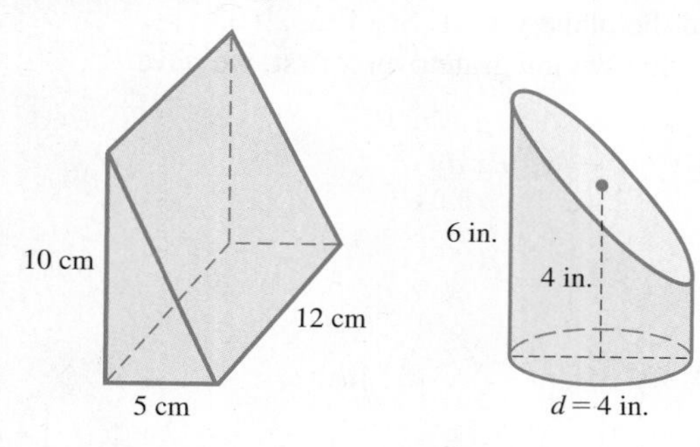

Fig. 29.34

Fig. 29.35

30. A circular piece of pipe is cut as shown in Fig. 29.35. Find the volume within the pipe. Describe how to set up the coordinate system in order to determine the required volume.

In Exercises 31 and 32, draw the appropriate figure.

31. Draw the appropriate figure that has a volume given by the integral

$$\int_0^{1/2}\int_{x^2}^{1}(4 - x - 2y)\,dy\,dx$$

32. Repeat Exercise 31 for the integral

$$\int_1^2\int_0^{2-y}\sqrt{1 + x^2 + y^2}\,dx\,dy$$

In Exercises 33 and 34, by noting the given limits, rewrite the double integral with the order of integration reversed. Evaluate Exercise 34.

33. $\displaystyle\int_0^1\int_{2x}^2 f(x, y)\,dy\,dx$ **34.** $\displaystyle\int_0^1\int_{2x}^2 e^{y^2}\,dy\,dx$

Answer to Practice Exercise

1. 1/4

CHAPTER 29 KEY FORMULAS AND EQUATIONS

Equation of a plane $Ax + By + Cz + D = 0$ **(29.1)**

Cylindrical coordinates $x = r\cos\theta \quad y = r\sin\theta \quad z = z$

$r^2 = x^2 + y^2 \quad \tan\theta = \dfrac{y}{x}$ **(29.2)**

Partial derivatives $\dfrac{\partial^2 z}{\partial x\,\partial y} = \dfrac{\partial^2 z}{\partial y\,\partial x}$ **(29.3)**

Double integral $\displaystyle\int_a^b\left[\int_{g(x)}^{G(x)} f(x, y)\,dy\right]dx = \int_a^b\int_{g(x)}^{G(x)} f(x, y)\,dy\,dx$ **(29.4)**

CHAPTER 29 REVIEW EXERCISES

CONCEPT CHECK EXERCISES

Determine each of the following as being either ***true*** *or* ***false****. If it is false, explain why.*

1. If $f(x, y) = \dfrac{2x^2y - y^2}{2xy}$, then $f(y^2, x) = \dfrac{2xy^4 - x^2}{2x^2y}$.

2. In graphing $2x^2 - y^2 + z^2 = 2$, the trace in the xz-plane is $2x^2 + z^2 = 2$.

3. If $z = x^2\sin xy + y\sin y$, then $\dfrac{\partial z}{\partial y} = x^3\sin xy + \sin x + y\cos y$

4. $\displaystyle\int_2^4\int_0^1 2xy\,dx\,dy = 6$

PRACTICE AND APPLICATIONS

In Exercises 5–8, evaluate the given functions.

5. $f(x, y) = 4xy^3 - y^2$, find $f(-4, 1)$ and $f(1, -2)$

6. $f(r, \theta) = r^2\cos 2\theta - r\sin\theta$, find $f(3, \pi)$ and $f\left(-2, \dfrac{\pi}{2}\right)$

7. $f(s, t) = \dfrac{4t - st}{s}$, find $f(2, s^2)$

8. $f(x, y) = \dfrac{4x}{xy - 2}$, find $f(x^2, 2x)$

In Exercises 9–12, sketch the graphs of the given equations in the rectangular coordinate system in three dimensions.

9. $x - y + 2z - 4 = 0$ **10.** $2y + 3z = 6$

11. $z = x^2 + 4y^2$ **12.** $x^2 + y^2 - 4z^2 - 4 = 0$

In Exercises 13–22, find the partial derivatives of the given functions with respect to each of the independent variables.

13. $z = 5x^3y^2 - 2xy^4$ **14.** $z = 2x\sqrt{y} - x^2y$

15. $z = \sqrt{x^2 - 3y^2}$ **16.** $u = \dfrac{r\tan 2s}{(r - 3s)^2}$

17. $z = \dfrac{e^{2y}}{x + y}$ **18.** $z = x(y^2 + xy + 2)^4$

19. $u = \sin x\ln(x + y)$ **20.** $q = p\ln(r + 1) - \dfrac{rp}{r + 1}$

21. $z = \cos^{-1}\sqrt{x + y}$ **22.** $z = ye^{x^2y}\sin(2x - y)$

In Exercises 23–26, find all of the second-order partial derivatives of the given functions.

23. $z = 3x^2y - y^3 + 2xy$

24. $z = 4x\sqrt{2y + 1} + y^2$

25. $r = 4e^s \cos 2t - 2te^{-s}$

26. $z = u^2 \ln v - \dfrac{2e^u}{v}$

In Exercises 27–34, evaluate each of the given double integrals.

27. $\displaystyle\int_0^2 \int_1^2 (3y + 2xy)\,dx\,dy$

28. $\displaystyle\int_2^7 \int_0^1 x\sqrt{2 + x^2y}\,dx\,dy$

29. $\displaystyle\int_0^3 \int_1^x (x + 2y)\,dy\,dx$

30. $\displaystyle\int_1^2 \int_0^{\pi/4} r\sec^2\theta\,d\theta\,dr$

31. $\displaystyle\int_0^1 \int_0^{2x} x^2e^{xy}\,dy\,dx$

32. $\displaystyle\int_{\pi/4}^{\pi/2} \int_1^{\sqrt{\cos\theta}} r\sin\theta\,dr\,d\theta$

33. $\displaystyle\int_1^e \int_1^x \frac{\ln y}{xy}\,dy\,dx$

34. $\displaystyle\int_1^3 \int_0^x \frac{2}{x^2 + y^2}\,dy\,dx$

In Exercises 35–54, solve the given problems.

35. Sketch the surface representing $z = \sqrt{x^2 + 4y^2}$.

36. For the function of Exercise 35, find the equation of a line tangent to the surface at $(2, 1, 2\sqrt{2})$ that is parallel to the yz-plane.

37. Sketch the surface representing $z = e^{x+y}$.

38. For the function of Exercise 37, find the volume in the first octant under the surface and inside the planes $x = 1$ and $y = x$.

39. Describe the surface for which the cylindrical coordinate equation is (a) $\theta = 3$, (b) $z = r^2$.

40. Write the cylindrical coordinate equation $r = 2(\sin\theta + \cos\theta)$ in rectangular coordinates and sketch the surface.

41. In a simple series electric circuit, with two resistors r and R connected across a voltage source E, the voltage v across r is $v = rE/(r + R)$. Assuming E to be constant, find $\partial v/\partial r$ and $\partial v/\partial R$.

42. For a gas, the volume expansivity is defined as $\beta = \dfrac{1}{V}\dfrac{\partial V}{\partial T}$, where V is the volume and T is the temperature of the gas. If V is a function of T and the pressure p is given by $V = a + bT/p - c/T^2$, where a, b, and c are constants, find β.

43. In the theory dealing with transistors, the current gain α of a transistor is defined as $\alpha = \dfrac{\partial i_c}{\partial i_e}$, where i_c is the collector current and i_e is the emitter current. If i_c is a function of i_e and the collector voltage v_c given by $i_c = i_e(1 - e^{-2v_c})$, find α if v_c is 2 V.

44. A tank in the shape of a hemisphere is being filled with water. The volume V of water in the tank is given by $V = \frac{1}{3}\pi h^2(3r - h)$, where h is the height of the water and r is the radius of the hemisphere. Find $\dfrac{\partial V}{\partial h}$ if $r = 1.50$ m and $h = 0.500$ m.

45. The period T of the pendulum as a function of its length l and the acceleration due to gravity g is given by $T = 2\pi\sqrt{l/g}$. Show that $\partial T/\partial l = T/(2l)$.

46. The volume of a right circular cone of radius r and height h is given by $V = \frac{1}{3}\pi r^2h$. Show that $\partial V/\partial r = 2V/r$.

47. Is a square the region of integration for $\displaystyle\int_3^6 \int_{-1}^2 dy\,dx$?

48. The area bounded by $y = 0$, $y = 2x$, and $x = 2$ is shaded on a calculator screen. Does $\displaystyle\int_0^4 \int_{y/2}^2 dx\,dy$ represent the area?

49. In analyzing the frictional resistance of a bearing, the integral $M = \dfrac{k}{\pi(b^2 - a^2)}\displaystyle\int_0^{2\pi} \int_a^b r^2\,dr\,d\theta$ arises. Evaluate this integral.

50. An *isothermal process* is one during which the temperature does not change. If the volume V, pressure p, and temperature T of an ideal gas are related by the equation $pV = nRT$, where n and R are constants, find the expression for $\partial p/\partial V$, which is the rate of change of pressure with respect to volume for an isothermal process.

51. For the ideal gas of Exercise 50, show that

$$\left(\frac{\partial V}{\partial T}\right)\left(\frac{\partial T}{\partial p}\right)\left(\frac{\partial p}{\partial V}\right) = -1.$$

52. Find the volume in the first octant below the plane $x + y + z - 6 = 0$ and inside the cylinder $y = 4 - x^2$.

53. Find the first-octant volume bounded by $x^2 + y^2 = 16$ and $x + z = 8$. Describe each of the bounding surfaces.

54. Find the first-octant volume below the surface $z = 4 - y^2$ and inside the plane $x + y = 2$.

55. An architectural design student determined that the area of a patio could be described by the double integral $\displaystyle\int_0^4 \int_0^{\sqrt{y}} f(x, y)\,dx\,dy$. Write the integration with the order of integration interchanged. Write one or two paragraphs to explain your method on interchanging the order of integration.

CHAPTER 29 PRACTICE TEST

As a study aid, we have included complete solutions for each Practice Test problem at the end of this book.

1. Given $f(x, y) = \dfrac{2y}{x^2 - y^2}$, find $f(-1, 3)$.

2. Sketch the surface representing the function $z = 4 - x^2 - 4y^2$.

3. Given $z = xe^{2xy}$, find $\dfrac{\partial z}{\partial x}$ and $\dfrac{\partial z}{\partial y}$.

4. Given $z = 3x^3y + 2x^2y^4$, find $\dfrac{\partial^2 z}{\partial x \partial y}$.

5. Evaluate: $\displaystyle\int_0^2 \int_{x^2}^{2x} (x^3 + 4y)\,dy\,dx$

6. Evaluate: $\displaystyle\int_1^{\ln 8} \int_0^{\ln y} e^{x+y}\,dx\,dy$

7. Find the volume in the first octant bounded by the coordinate planes and the cylinders $x^2 + y^2 = 9$ and $y^2 + z^2 = 9$.

8. The fundamental frequency of vibration f of a string varies directly as the square root of the tension T and inversely as the length L. If a string 60 cm long is under a tension of 65 N and has a fundamental frequency of 30 Hz, find the partial derivative of f with respect to T and evaluate it for the given values.

30

Expansion of Functions in Series

LEARNING OUTCOMES

After completion of this chapter, the student should be able to:

- Find the terms of sequences and series
- Decide whether a geometric series is convergent or divergent and, if convergent, find its sum
- Find the Maclaurin series expansion of a function
- Use algebraic or calculus procedures on known series to obtain other series expansions
- Find the Taylor series expansion of a function
- Approximate the value of functions by using series
- Find the Fourier series expansion of a periodic function
- Find the half-range Fourier series of a function
- Solve application problems using series

In the mid-1600s, mathematicians found that transcendental functions can be represented by polynomials and that by using these polynomials it was possible to more easily calculate the values of these functions. In this chapter, we show how a given function may be expressed in terms of a polynomial and how this polynomial is used to evaluate the function.

The polynomials that we will develop are known as *power series,* and they can be expressed with an unlimited number of terms. Although first noted for their usefulness in calculating values of transcendental functions, many mathematicians, including Newton, used power series extensively to further develop various areas of mathematics. In fact, the French mathematician Joseph-Louis Lagrange (1736–1813) attempted to make power series the basis for the development of all methods in calculus.

Another type of series was developed by the French physicist and mathematician Jean Baptiste Joseph Fourier (1768–1830) in the study of heat conduction. In 1822, he showed that a function can be expressed in a series of sine and cosine terms. Today, these series are very important in the study of electricity and electronics. They are also useful in the study of mechanical vibrations, signal processing, and other applications that are periodic in nature. We study these series in the last two sections of this chapter.

We see again that a concept, first used to ease calculation, became very important in the later development of mathematics. Also, a concept developed for the study of heat, long before the advent of electronics, has become important in electronics.

▶ Fourier series are used in electronic synthesizers to generate waveforms that mimic various musical instruments. In Section 30.7, we see how a square wave, which produces an organ-like sound, can be created by adding different sine waves together.

30.1 Infinite Series

Sequence • Term • Partial Sum • Convergent • Divergent • Geometric Series

In addition to arithmetic sequences and geometric sequences, there are many other ways of generating sequences of numbers. For example, the squares of the integers 1, 4, 9, 16, 25 . . . form a sequence. In general, *a* **sequence** *(or* **infinite sequence***) is an infinite succession of numbers. Each of the numbers is a* **term** *of the sequence.* Each term of the sequence is associated with a positive integer, although at times it is convenient to associate the first term with zero (or some specified positive integer). We shall use a_n to designate the term of the sequence corresponding to the integer n.

EXAMPLE 1 Given a_n—find terms of a series

Find the first three terms of the sequence for which the general term $a_n = 2n + 1$, $n = 1, 2, 3, \ldots$.

Substituting the values of n, we obtain the values

$$a_1 = 2(1) + 1 = 3 \qquad a_2 = 2(2) + 1 = 5 \qquad a_3 = 2(3) + 1 = 7$$

Therefore, we have the sequence 3, 5, 7, ...

Given $a_n = 2n + 1$ for $n = 0, 1, 2, \ldots$, the sequence is 1, 3, 5, ... ■

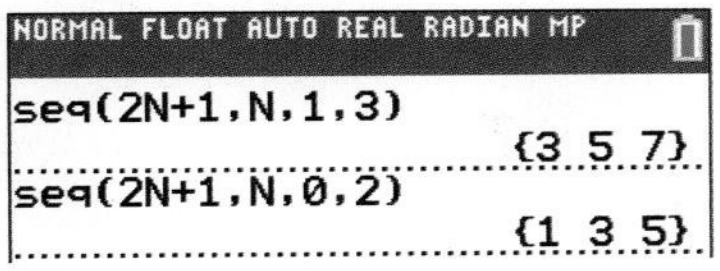

```
NORMAL FLOAT AUTO REAL RADIAN MP
seq(2N+1,N,1,3)
                    {3 5 7}
seq(2N+1,N,0,2)
                    {1 3 5}
```

Fig. 30.1
Graphing calculator keystrokes: goo.gl/zDXSku

■ Figure 30.1 shows a calculator display of the sequence in Example 1.

As we stated in Chapter 19, *the indicated sum of the terms of a sequence is called an* **infinite series.** Thus, for the sequence

$$a_1, a_2, a_3, \ldots, a_n, \ldots$$

the associated infinite series is

$$a_1 + a_2 + a_3 + \cdots + a_n + \cdots$$

Using the summation sign Σ to indicate the sum, we have the following:

Infinite Series

$$\sum_{n=1}^{\infty} a_n = a_1 + a_2 + a_3 + \cdots + a_n + \cdots \qquad (30.1)$$

Practice Exercise

1. Find the first three terms of the sequence for which $a_n = \dfrac{n+3}{n^2+1}$, $n = 1, 2, 3, \ldots$

We define the sum of an infinite series in terms of a limit. For the infinite series of Eq. (30.1), we let S_n represent the sum of the first n terms. Therefore,

$$S_1 = a_1$$
$$S_2 = a_1 + a_2$$
$$S_3 = a_1 + a_2 + a_3$$
$$S_n = a_1 + a_2 + a_3 + \cdots + a_n$$

The numbers $S_1, S_2, S_3, \ldots, S_n, \ldots$ form a sequence. *Each term of this sequence is called a* **partial sum.** *We say that the infinite series, Eq. (30.1), is* **convergent** *and has the* **sum** *S given by*

$$S = \lim_{n\to\infty} S_n = \lim_{n\to\infty} \sum_{i=1}^{n} a_i \qquad (30.2)$$

if this limit exists. If the limit does not exist, the series is **divergent.**

EXAMPLE 2 Partial sums—convergent series

For the infinite series

$$\sum_{n=0}^{\infty} \frac{1}{5^n} = \frac{1}{5^0} + \frac{1}{5^1} + \frac{1}{5^2} + \cdots + \frac{1}{5^n} + \cdots$$

the first six partial sums are

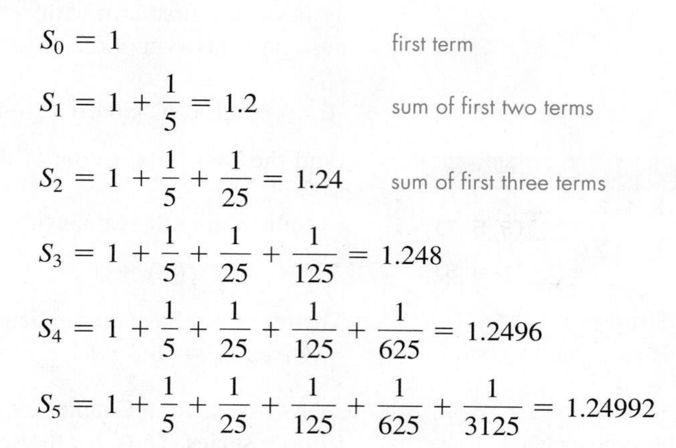

$$S_0 = 1 \qquad \text{first term}$$

$$S_1 = 1 + \frac{1}{5} = 1.2 \qquad \text{sum of first two terms}$$

$$S_2 = 1 + \frac{1}{5} + \frac{1}{25} = 1.24 \qquad \text{sum of first three terms}$$

$$S_3 = 1 + \frac{1}{5} + \frac{1}{25} + \frac{1}{125} = 1.248$$

$$S_4 = 1 + \frac{1}{5} + \frac{1}{25} + \frac{1}{125} + \frac{1}{625} = 1.2496$$

$$S_5 = 1 + \frac{1}{5} + \frac{1}{25} + \frac{1}{125} + \frac{1}{625} + \frac{1}{3125} = 1.24992$$

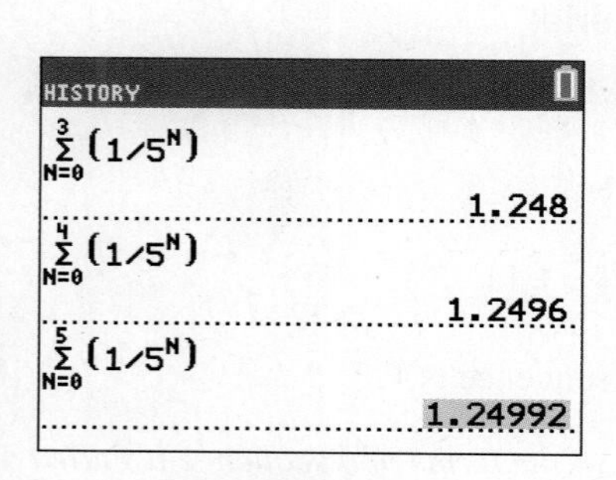

Fig. 30.2
Graphing calculator keystrokes: goo.gl/sNnkQv

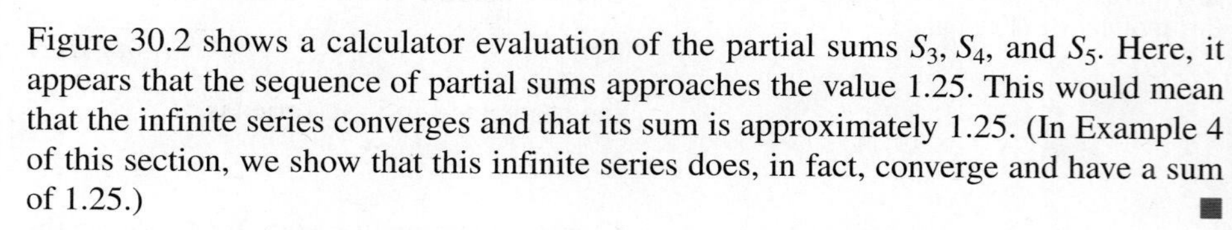

Figure 30.2 shows a calculator evaluation of the partial sums S_3, S_4, and S_5. Here, it appears that the sequence of partial sums approaches the value 1.25. This would mean that the infinite series converges and that its sum is approximately 1.25. (In Example 4 of this section, we show that this infinite series does, in fact, converge and have a sum of 1.25.) ■

EXAMPLE 3 Divergent series

(a) The infinite series

$$\sum_{n=1}^{\infty} 5^n = 5 + 5^2 + 5^3 + \cdots + 5^n + \cdots$$

is a divergent series. The first four partial sums are

$$S_1 = 5 \qquad S_2 = 30 \qquad S_3 = 155 \qquad S_4 = 780$$

Obviously, they are increasing without bound.

(b) The infinite series

$$\sum_{n=0}^{\infty} (-1)^n = 1 + (-1) + 1 + (-1) + \cdots + (-1)^n + \cdots$$

has as its first five partial sums

$$S_0 = 1 \qquad S_1 = 0 \qquad S_2 = 1 \qquad S_3 = 0 \qquad S_4 = 1$$

The values of these partial sums do not approach a limiting value and, therefore, the series diverges. ■

Because convergent series are those that have a value associated with them, they are the ones that are of primary use to us. However, generally, it is not easy to determine whether a given series is convergent, and many types of tests have been developed for this purpose. These tests for convergence may be found in most textbooks that include the more advanced topics in calculus.

One important series for which we are able to determine the convergence, and its sum if convergent, is the geometric series. For this series, the nth partial sum is

$$S_n = a_1 + a_1r + a_1r^2 + \cdots + a_1r^{n-1}$$

where r is the fixed number by which we multiply a given term to get the next term. In Chapter 19, we determined that if $|r| < 1$, the sum S of the infinite geometric series is

$$S = \lim_{n\to\infty} S_n = \frac{a_1}{1 - r} \tag{30.3}$$

If $r = 1$, we see that the series is $a_1 + a_1 + a_1 + \cdots + a_1 + \cdots$ and is, therefore, divergent. If $r = -1$, the series is $a_1 - a_1 + a_1 - a_1 + \cdots$ and is also divergent. If $|r| > 1$, $\lim_{n\to\infty} r^n$ is unbounded. Therefore, ***the geometric series is convergent only if*** $|r| < 1$ *and has the value given by Eq. (30.3).*

EXAMPLE 4 Geometric series

Show that the infinite series

$$\sum_{n=0}^{\infty} \frac{1}{5^n} = \frac{1}{5^0} + \frac{1}{5^1} + \frac{1}{5^2} + \cdots + \frac{1}{5^n} + \cdots$$

is convergent and find its sum. This is the same series as in Example 2.

This is a geometric series with $r = \frac{1}{5}$. Because $|r| < 1$, the series converges. The sum is

$$S = \frac{1}{1 - \frac{1}{5}} = \frac{1}{\frac{4}{5}} = \frac{5}{4} = 1.25 \quad \text{using Eq. (30.3)}$$

■

Practice Exercise

2. Show that the infinite series $\sum_{n=0}^{\infty} \frac{2}{3^n}$ is convergent and find its sum.

EXERCISES 30.1

In Exercises 1 and 2, make the given changes in the indicated examples of this section and then solve the given problems.

1. In Example 3(a), change 5^n to 0.5^n. What other changes occur?

2. In Example 4, change $n = 0$ to $n = 1$. What is the value of S?

In Exercises 3–6, give the first four terms of the sequences for which a_n is given.

3. $a_n = n^3, \quad n = 1, 2, 3, \ldots$

4. $a_n = \dfrac{2^{n+1}}{n!}, n = 1, 2, 3, \ldots$

5. $a_n = \dfrac{n}{n + 1}, \quad n = 0, 1, 2, \ldots$

6. $a_n = \dfrac{n^2 + 1}{2n + 1}, \quad n = 0, 1, 2, \ldots$

In Exercises 7–10, give (a) the first four terms of the sequence for which a_n is given and (b) the first four terms of the infinite series associated with the sequence.

7. $a_n = \left(\dfrac{2}{3}\right)^n, \quad n = 1, 2, 3, \ldots$

8. $a_n = \dfrac{1}{n} + \dfrac{1}{n + 1}, \quad n = 1, 2, 3, \ldots$

9. $a_n = \cos\dfrac{n\pi}{2}, \quad n = 0, 1, 2, \ldots$

10. $a_n = \dfrac{n^n}{n!}, \quad n = 2, 3, 4, \ldots$

In Exercises 11–14, find the nth term of the given infinite series for which $n = 1, 2, 3, \ldots$.

11. $\dfrac{1}{2} + \dfrac{1}{3} + \dfrac{1}{4} + \dfrac{1}{5} + \cdots$

12. $\dfrac{1}{2} + \dfrac{1}{4} + \dfrac{1}{8} + \dfrac{1}{16} + \cdots$

13. $\dfrac{1}{2 \times 3} - \dfrac{1}{3 \times 4} + \dfrac{1}{4 \times 5} - \dfrac{1}{5 \times 6} + \cdots$

14. $\dfrac{1}{\sqrt{2}} - \dfrac{1}{2} + \dfrac{\sqrt{2}}{4} - \dfrac{1}{4} + \cdots$

In Exercises 15–24, find the first five partial sums of the given series and determine whether the series appears to be convergent or divergent. If it is convergent, find its approximate sum.

15. $1 + \dfrac{1}{8} + \dfrac{1}{27} + \dfrac{1}{64} + \dfrac{1}{125} + \cdots$

16. $1 + 2 + 5 + 10 + 17 + \cdots$

17. $1 + \dfrac{1}{2} + \dfrac{2}{3} + \dfrac{3}{4} + \dfrac{4}{5} + \cdots$

18. $\dfrac{1}{3} - \dfrac{1}{9} + \dfrac{1}{27} - \dfrac{1}{81} + \dfrac{1}{243} - \cdots$

19. $\sum_{n=0}^{\infty} \sqrt{n}$ **20.** $\sum_{n=1}^{\infty} \frac{2}{n(n+1)}$ **21.** $\sum_{n=1}^{\infty} \frac{2n+1}{n^2(n+1)^2}$

22. $\sum_{n=1}^{\infty} \frac{n^2}{2n^2+1}$ **23.** $\sum_{n=1}^{\infty} \frac{\sin n}{4^n}$ **24.** $\sum_{n=3}^{\infty} \frac{\ln n}{e^n}$

In Exercises 25–32, test each of the given geometric series for convergence or divergence. Find the sum of each series that is convergent.

25. $1 + 2 + 4 + \cdots + 2^n + \cdots$

26. $1 + \frac{1}{2} + \frac{1}{4} + \cdots + \frac{1}{2^n} + \cdots$

27. $1 - \frac{1}{3} + \frac{1}{9} - \cdots + \left(-\frac{1}{3}\right)^n + \cdots$

28. $1 - \frac{3}{2} + \frac{9}{4} - \cdots + \left(-\frac{3}{2}\right)^n + \cdots$

29. $10 + 9 + 8.1 + 7.29 + 6.561 + \cdots$

30. $4 + 1 + \frac{1}{4} + \frac{1}{16} + \frac{1}{64} + \cdots$

31. $512 - 64 + 8 - 1 + \frac{1}{8} - \cdots$

32. $16 + 12 + 9 + \frac{27}{4} + \frac{81}{16} + \cdots$

In Exercises 33 and 34, find the values of x for which the given series converge.

33. $\sum_{n=0}^{\infty} (x-4)^n$ **34.** $\sum_{n=2}^{\infty} \frac{x^n}{5^n}$

In Exercises 35–38, determine the convergence or divergence of the given sequence. If a_n is the term of a sequence and $f(x)$ exists for $x \geq 1$ such that $f(n) = a_n$, then $\lim_{x\to\infty} f(x) = L$ means $a_n \to L$ as $n \to \infty$. This lets us analyze convergence or divergence by using the equivalent continuous function. Therefore, if applicable, L'Hospital's rule may be used.

35. $a_n = 2 + \frac{2}{n}$ **36.** $a_n = \frac{5n^2 - 2n + 3}{2n^2 + 3n - 1}$

37. $a_n = \frac{e^n}{n^3}$ **38.** $a_n = \frac{\ln n}{n^2}$

In Exercises 39–48, solve the given problems as indicated.

39. The repeating decimal 0.151515 . . . can be expressed as $\frac{15}{100} + \frac{15}{10{,}000} + \frac{15}{1{,}000{,}000} + \cdots$. Find the sum of this series.

40. Using a calculator, (a) take successive square roots of 0.01 and then (b) take successive square roots of 100. From these sequences of square roots, state any general conclusions that might be drawn.

41. Referring to Chapter 19, we see that the sum of the first n terms of a geometric sequence is

$$S_n = \frac{a_1(1-r^n)}{1-r} \quad (r \neq 1) \qquad \text{Eq. (19.6)}$$

where a_1 is the first term and r is the common ratio. We can visualize the corresponding infinite series by graphing the function $f(x) = a_1(1-r^x)/(1-r)\,(r \neq 1)$ (or using a calculator that can graph a sequence). The graph represents the sequence of partial sums for values where $x = n$, since $f(n) = S_n$.

Use a calculator to visualize the first five partial sums of the series

$$\frac{1}{2} + \frac{1}{4} + \frac{1}{8} + \cdots$$

What value does the infinite series approach? (*Remember:* Only points for which x is an integer have real meaning.)

42. Write out the first four terms of the series $\sum_{n=1}^{\infty}\left(\frac{1}{n} - \frac{1}{n+1}\right)$. Then find the first four partial sums. Does this series appear to converge? Why or why not?

43. When two dice are rolled repeatedly, the probability of rolling a sum of 4 before rolling a sum of 7 is given by $\frac{3}{36}\left[1 + \frac{27}{36} + \left(\frac{27}{36}\right)^2 + \left(\frac{27}{36}\right)^3 + \cdots\right]$. Find this probability.

44. If an electric discharge is passed through hydrogen gas, a spectrum of isolated parallel lines, called the Balmer series, is formed. See Fig. 30.3. The wavelengths λ (in nm) of the light for these lines is given by the formula

$$\frac{1}{\lambda} = 1.097 \times 10^{-2}\left(\frac{1}{2^2} - \frac{1}{n^2}\right) \qquad (n = 3, 4, 5, \ldots)$$

Find the wavelengths of the first three lines and the shortest wavelength of all the lines of the series.

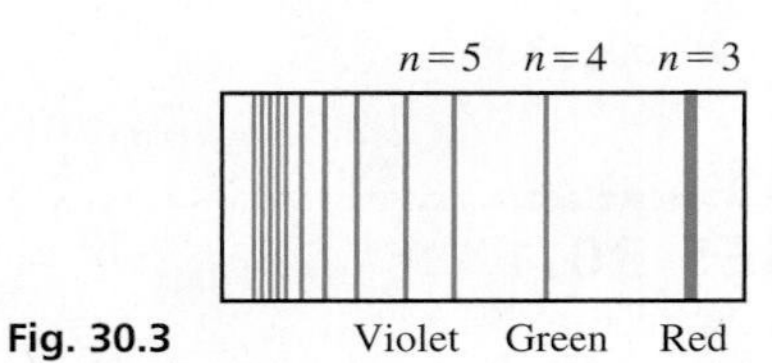

Fig. 30.3

45. Use geometric series to show that $\sum_{n=0}^{\infty} x^n = \frac{1}{1-x}$ for $|x| < 1$.

46. Use geometric series to show that $\sum_{n=0}^{\infty} (-1)^n x^n = \frac{1}{1+x}$ for $|x| < 1$.

47. If term a_1 is given along with a rule to find term a_{n+1} from term a_1, the sequence is said to be defined *recursively*. If $a_1 = 2$ and $a_{n+1} = (n+1)a_n$, find the first five terms of the sequence.

48. A sequence is defined recursively (see Exercise 47) by $x_1 = \frac{N}{2}, x_{n+1} = \frac{1}{2}\left(x_n + \frac{N}{x_n}\right)$. With $N = 10$, find x_6 and compare the value with $\sqrt{10}$. It can be seen that $\sqrt{N}$ can be approximated using this recursion sequence.

Answers to Practice Exercises

1. 2, 1, 3/5, . . . **2.** Converges since $|r| = \frac{1}{3} < 1$; $S = 3$

30.2 Maclaurin Series

Power Series Expansion • Interval of Convergence • Maclaurin Series Expansion

In this section, we develop a very important basic polynomial form of a function. Before developing the method using calculus, we will review how this can be done for some functions algebraically.

EXAMPLE 1 Algebraic function represented by a series

$$\begin{array}{r} 1 + \frac{x}{2} \\ 2 - x\overline{)2} \\ \underline{2 - x} \\ x \\ \underline{x - \frac{x^2}{2}} \\ \frac{x^2}{2} \end{array}$$

By using long division (as started at the left), we have

$$\frac{2}{2-x} = 1 + \frac{1}{2}x + \frac{1}{4}x^2 + \cdots + \left(\frac{1}{2}x\right)^{n-1} + \cdots \quad (1)$$

where n is the number of the term of the expression on the right. Because x represents a number, the right-hand side of Eq. (1) becomes a geometric series.

From Eq. (30.3), we know that the sum of a geometric series with first term a_1 and common ratio r converges to the sum

$$S = \frac{a_1}{1-r}$$

if $|r| < 1$.

If $x = 1$, the right-hand side of Eq. (1) is

$$1 + \frac{1}{2} + \frac{1}{4} + \cdots + \left(\frac{1}{2}\right)^{n-1} + \cdots$$

For this series, $r = \frac{1}{2}$ and $a_1 = 1$, which means that the series converges and $S = 2$. If $x = 3$, the right-hand side of Eq. (1) is

$$1 + \frac{3}{2} + \frac{9}{4} + \cdots + \left(\frac{3}{2}\right)^{n-1} + \cdots$$

which diverges since $r > 1$. Referring to the left side of Eq. (1), we see that it also equals 2 when $x = 1$. Thus, we see that the two sides agree for $x = 1$, but that the series diverges for $x = 3$. In fact, as long as $|x| < 2$, the series will converge to the value of the function on the left. From this we conclude that the series on the right properly represents the function on the left, as long as $|x| < 2$. ■

From Example 1, we see that an algebraic function may be properly represented by a function of the following form, called a **power series expansion** *of the function* $f(x)$.

Power Series

$$f(x) = a_0 + a_1x + a_2x^2 + \cdots + a_nx^n + \cdots \quad (30.4)$$

The problem now arises as to whether or not functions in general may be represented in this form. If such a representation were possible, it would provide a means of evaluating the transcendental functions for making tables of values, or developing programs for use in computers and calculators. Also, because a power series expansion is in the form of a polynomial, it makes algebraic operations much simpler due to the properties of polynomials. A further study of calculus shows many other uses of power series.

In Example 1, we saw that the function could be represented by a power series as long as $|x| < 2$. That is, if we substitute any value of x in this interval into the series and also into the function, the series will converge to the value of the function. [This interval of values for which the series converges is called the **interval of convergence.**]

NOTE ▸

EXAMPLE 2 Interval of convergence

In Example 1, the interval of convergence for the series

$$1 + \frac{1}{2}x + \frac{1}{4}x^2 + \cdots + \left(\frac{1}{2}x\right)^{n-1} + \cdots$$

is $|x| < 2$. We saw that the series converges for $x = 1$, with $S = 2$, and that the value of the function is 2 for $x = 1$. This verifies that $x = 1$ is in the interval of convergence.

Also, we saw that the series diverges for $x = 3$, which verifies that $x = 3$ is not in the interval of convergence. ■

At this point, we will assume that unless otherwise noted, the functions with which we will be dealing may be properly represented by a power-series expansion (it takes more advanced methods to prove that this is generally possible), for appropriate intervals of convergence. We will find that the methods of calculus are very useful in developing the method of general representation. Thus, writing a general power series, along with the first few derivatives, we have

$$f(x) = a_0 + a_1x + a_2x^2 + a_3x^3 + a_4x^4 + a_5x^5 + \cdots + a_nx^n + \cdots$$

$$f'(x) = a_1 + 2a_2x + 3a_3x^2 + 4a_4x^3 + 5a_5x^4 + \cdots + na_nx^{n-1} + \cdots$$

$$f''(x) = 2a_2 + 2(3)a_3x + 3(4)a_4x^2 + 4(5)a_5x^3 + \cdots + (n-1)na_nx^{n-2} + \cdots$$

$$f'''(x) = 2(3)a_3 + 2(3)(4)a_4x + 3(4)(5)a_5x^2 + \cdots + (n-2)(n-1)na_nx^{n-3} + \cdots$$

$$f^{\text{iv}}(x) = 2(3)(4)a_4 + 2(3)(4)(5)a_5x + \cdots + (n-3)(n-2)(n-1)na_nx^{n-4} + \cdots$$

NOTE ▸ [Regardless of the values of the constants a_n for any power series, ***if*** $\boldsymbol{x = 0}$, ***the left and right sides must be equal,*** and all the terms on the right are zero except the first.] Thus, setting $x = 0$ in each of the above equations, we have

$$f(0) = a_0 \qquad f'(0) = a_1 \qquad f''(0) = 2a_2$$

$$f'''(0) = 2(3)a_3 \qquad f^{\text{iv}}(0) = 2(3)(4)a_4$$

Solving each of these for the constants a_n, we have

$$a_0 = f(0) \qquad a_1 = f'(0) \qquad a_2 = \frac{f''(0)}{2!} \qquad a_3 = \frac{f'''(0)}{3!} \qquad a_4 = \frac{f^{\text{iv}}(0)}{4!}$$

Substituting these into the expression for $f(x)$, we get the following equation, which is known as the **Maclaurin series expansion** *of a function.*

Maclaurin Series

$$f(x) = f(0) + f'(0)x + \frac{f''(0)x^2}{2!} + \frac{f'''(0)x^3}{3!} + \cdots + \frac{f^n(0)x^n}{n!} + \cdots \qquad \textbf{(30.5)}$$

■ Named for the Scottish mathematician Colin Maclaurin (1698–1746).

For a function to be represented by a Maclaurin expansion, the function and all of its derivatives must exist at $x = 0$. Also, we note that the factorial notation introduced in Section 19.4 is used in writing the Maclaurin series expansion.

As we mentioned earlier, one of the uses we will make of series expansions is that of determining the values of functions for particular values of x. If x is sufficiently small, successive terms become smaller and smaller and the series will converge rapidly. This is considered in the sections that follow.

The following examples illustrate Maclaurin expansions for algebraic, exponential, and trigonometric functions.

EXAMPLE 3 Maclaurin series for an algebraic function

Find the first four terms of the Maclaurin series expansion of $f(x) = \dfrac{2}{2 - x}$.

■ Compare with Example 1.

$$f(x) = \frac{2}{2-x} \qquad f(0) = 1 \qquad f''(x) = \frac{4}{(2-x)^3} \qquad f''(0) = \frac{1}{2}$$

$$f'(x) = \frac{2}{(2-x)^2} \qquad f'(0) = \frac{1}{2} \qquad f'''(x) = \frac{12}{(2-x)^4} \qquad f'''(0) = \frac{3}{4}$$

find derivatives and evaluate each at $x = 0$

$$f(x) = 1 + \frac{1}{2}x + \frac{1}{2}\left(\frac{x^2}{2!}\right) + \frac{3}{4}\left(\frac{x^3}{3!}\right) + \cdots$$

using Eq. (30.5)

$$\frac{2}{2-x} = 1 + \frac{1}{2}x + \frac{1}{4}x^2 + \frac{1}{8}x^3 + \cdots$$

■

Practice Exercise

1. Find the first four terms of the Maclaurin series expansion for $f(x) = \dfrac{1}{1 + x}$.

EXAMPLE 4 Maclaurin series for an exponential function

Find the first four terms of the Maclaurin series expansion of $f(x) = e^{-x}$.

$$f(x) = e^{-x} \qquad f(0) = 1 \qquad f''(x) = e^{-x} \qquad f''(0) = 1$$

$$f'(x) = -e^{-x} \qquad f'(0) = -1 \qquad f'''(x) = -e^{-x} \qquad f'''(0) = -1$$

find derivatives and evaluate each at $x = 0$

$$f(x) = 1 + (-1)x + 1\left(\frac{x^2}{2!}\right) + (-1)\left(\frac{x^3}{3!}\right) + \cdots$$

using Eq. (30.5)

$$e^{-x} = 1 - x + \frac{x^2}{2!} - \frac{x^3}{3!} + \cdots$$

■

EXAMPLE 5 Maclaurin series for a trigonometric function

Find the first three nonzero terms of the Maclaurin series expansion of $f(x) = \sin 2x$.

$$f(x) = \sin 2x \qquad f(0) = 0 \qquad f'''(x) = -8\cos 2x \qquad f'''(0) = -8$$

$$f'(x) = 2\cos 2x \qquad f'(0) = 2 \qquad f^{\text{iv}}(x) = 16\sin 2x \qquad f^{\text{iv}}(0) = 0$$

$$f''(x) = -4\sin 2x \qquad f''(0) = 0 \qquad f^{\text{v}}(x) = 32\cos 2x \qquad f^{\text{v}}(0) = 32$$

$$f(x) = 0 + 2x + 0 + (-8)\frac{x^3}{3!} + 0 + 32\frac{x^5}{5!} + \cdots$$

$$\sin 2x = 2x - \frac{4}{3}x^3 + \frac{4}{15}x^5 - \cdots$$

This series is called an alternating series, since every other term is negative.

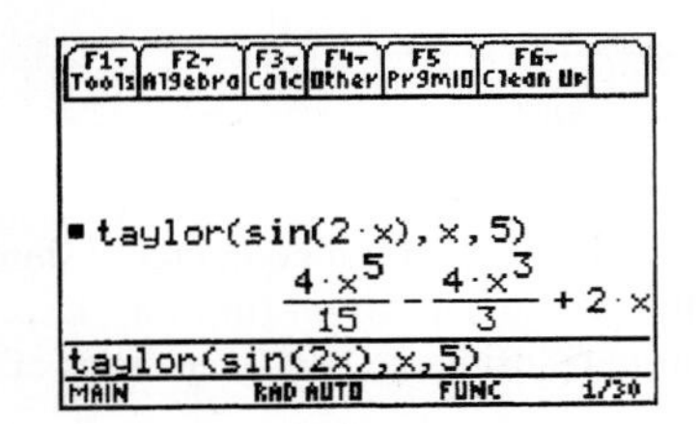

Fig. 30.4
TI-89 graphing calculator keystrokes: goo.gl/6dxpG7

In Fig. 30.4, the TI-89 calculator uses a *Taylor series* (discussed in Section 30.5) because a Maclaurin series is a special case (expansion about $x = 0$) of a Taylor series. ■

EXAMPLE 6 Maclaurin series—critically damped motion

Frictional forces in the spring shown in Fig. 30.5 are just sufficient so that the lever does not oscillate after being depressed. Such motion is called *critically damped.* The displacement y as a function of the time t for one case is $y = (1 + t)e^{-t}$. To study the motion for small values of t, a Maclaurin expansion of $y = f(t)$ is to be used. Find the first four terms of the expansion.

$$f(t) = (1 + t)e^{-t} \qquad f(0) = 1$$

$$f'(t) = (1 + t)e^{-t}(-1) + e^{-t} = -te^{-t} \qquad f'(0) = 0$$

$$f''(t) = te^{-t} - e^{-t} \qquad f''(0) = -1$$

$$f'''(t) = -te^{-t} + e^{-t} + e^{-t} = 2e^{-t} - te^{-t} \qquad f'''(0) = 2$$

$$f^{\text{iv}}(t) = -2e^{-t} + te^{-t} - e^{-t} = te^{-t} - 3e^{-t} \qquad f^{\text{iv}}(0) = -3$$

$$f(t) = 1 + 0 + (-1)\frac{t^2}{2!} + 2\frac{t^3}{3!} + (-3)\frac{t^4}{4!} + \cdots$$

$$(1 + t)e^{-t} = 1 - \frac{t^2}{2} + \frac{t^3}{3} - \frac{t^4}{8} + \cdots$$

■

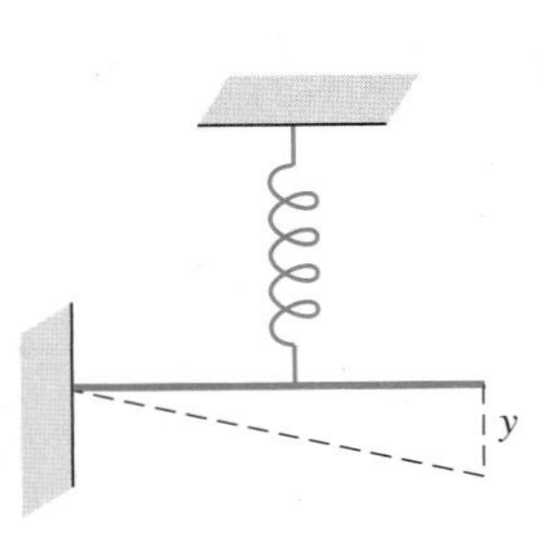

Fig. 30.5

EXERCISES 30.2

In Exercises 1 and 2, make the given changes in the indicated examples of this section and then find the resulting series.

1. In Example 3, in $f(x)$, change the denominator to $2 + x$.
2. In Example 5, in $f(x)$, change $2x$ to $(-2x)$.

In Exercises 3–20, find the first three nonzero terms of the Maclaurin expansion of the given functions.

3. $f(x) = e^x$
4. $f(x) = \sin x$
5. $f(x) = \cos x$
6. $f(x) = \ln(1 + x)$
7. $f(x) = \sqrt{1 + x}$
8. $f(x) = \sqrt[3]{1 + x}$
9. $f(x) = e^{-2x}$
10. $f(x) = \dfrac{1}{\sqrt{1 + x}}$
11. $f(x) = \sin 3x$
12. $f(x) = \cos \pi x$
13. $f(x) = \dfrac{1}{1 - x}$
14. $f(x) = \dfrac{1}{(1 + x)^2}$
15. $f(x) = \ln(1 - 2x)$
16. $f(x) = (4 + x)^{3/2}$
17. $f(x) = \cos^2 x$
18. $f(x) = \ln(1 + 4x)^2$
19. $f(x) = \sin(x + \frac{\pi}{4})$
20. $f(x) = (2x - 1)^2$

In Exercises 21–28, find the first two nonzero terms of the Maclaurin expansion of the given functions.

21. $f(x) = \tan^{-1} x$
22. $f(x) = \cos x^2$
23. $f(x) = \tan x$
24. $f(x) = \sec x$
25. $f(x) = \ln \cos x$
26. $f(x) = xe^{\sin x}$
27. $f(x) = \sqrt{1 + \sin x}$
28. $f(x) = xe^{-x^2}$

In Exercises 29–44, solve the given problems.

29. Use long division to find a series expansion for $f(x) = \dfrac{1}{1 - x}$. Compare the results with Exercise 13.
30. Use long division to find a series expansion for $f(x) = \dfrac{1}{(1 + x)^2}$. Compare the results with Exercise 14.
31. Is it possible to find a Maclaurin expansion for (a) $f(x) = \csc x$ or (b) $f(x) = \ln x$? Explain.
32. Is it possible to find a Maclaurin expansion for (a) $f(x) = \sqrt{x}$ or (b) $f(x) = \sqrt{1 + x}$? Explain.
33. Find the first three nonzero terms of the Maclaurin expansion for (a) $f(x) = e^x$ and (b) $f(x) = e^{x^2}$. Compare these expansions.
34. By finding the Maclaurin expansion of $f(x) = (1 + x)^n$, derive the first four terms of the binomial series, which is Eq. (19.10). Its interval of convergence is $|x| < 1$ for all values of n.
35. If $f(x) = e^{3x}$, compare the Maclaurin expansion with the linearization for $a = 0$.
36. The hyperbolic sine function is defined as $\sinh x = \frac{1}{2}(e^x - e^{-x})$. Find the Maclaurin series for $y = \sinh x$.
37. The hyperbolic cosine function is defined as $\cosh x = \frac{1}{2}(e^x + e^{-x})$. Find the Maclaurin series for $y = \cosh x$.
38. Find the Maclaurin series for $f(x) = \cos^2 x$, by using the identity $\cos^2 x = \frac{1}{2}(1 + \cos 2x)$. Compare the result with that of Exercise 17.
39. If $f(x) = x^2$, show that this function is obtained when a Maclaurin expansion is found.
40. If $f(x) = x^4 + 2x^2$, show that this function is obtained when a Maclaurin expansion is found.
41. The displacement y (in cm) of an object hung vertically from a spring and allowed to oscillate is given by the equation $y = 4e^{-0.2t}\cos t$, where t is the time (in s). Find the first three terms of the Maclaurin expansion of this function.
42. For the circuit shown in Fig. 30.6, after the switch is closed, the transient current i (in A) is given by $i = 2.5(1 + e^{-0.1t})$. Find the first three terms of the Maclaurin expansion of this function.

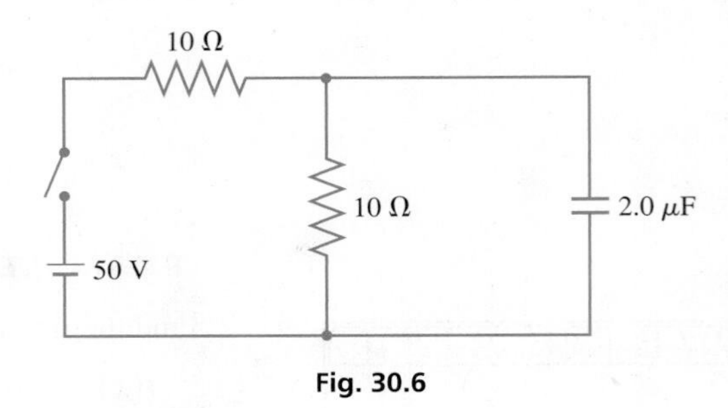

Fig. 30.6

43. The reliability R $(0 \leq R \leq 1)$ of a certain computer system is $R = e^{-0.001t}$, where t is the time of operation (in min). Express $R = f(t)$ in polynomial form by using the first three terms of the Maclaurin expansion.
44. In the analysis of the optical paths of light from a narrow slit S to a point P as shown in Fig. 30.7, the law of cosines is used to obtain the equation

$$c^2 = a^2 + (a + b)^2 - 2a(a + b)\cos\frac{s}{a}$$

where s is part of the circular arc $\widehat{AB}$. By using two nonzero terms of the Maclaurin expansion of $\cos\frac{s}{a}$, simplify the right side of the equation. (In finding the expansion, let $x = \frac{s}{a}$ and then substitute back into the expansion.)

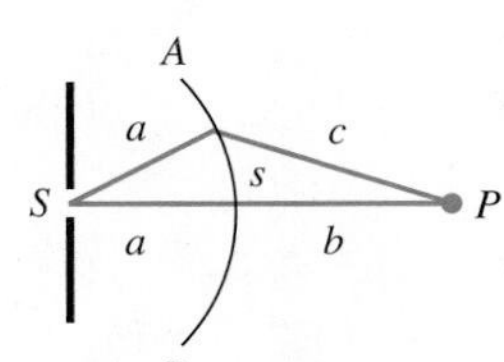

Fig. 30.7

Answer to Practice Exercise

1. $\dfrac{1}{1 + x} = 1 - x + x^2 - x^3 + \cdots$

30.3 Operations with Series

Form New Series from Known Series • Using Functional Notation • Using Algebraic Operations • Differentiating and Integrating • Accuracy of Series

The series found in Exercises 3 to 6 and 32 (the binomial series) of Section 30.2 are of particular importance. They are used to evaluate exponential functions, trigonometric functions, logarithms, powers, and roots, as well as develop other series. For reference, we give them here with their intervals of convergence.

$$e^x = 1 + x + \frac{x^2}{2!} + \frac{x^3}{3!} + \cdots \quad (\text{all } x) \tag{30.6}$$

$$\sin x = x - \frac{x^3}{3!} + \frac{x^5}{5!} - \cdots \quad (\text{all } x) \tag{30.7}$$

$$\cos x = 1 - \frac{x^2}{2!} + \frac{x^4}{4!} - \cdots \quad (\text{all } x) \tag{30.8}$$

$$\ln(1 + x) = x - \frac{x^2}{2} + \frac{x^3}{3} - \frac{x^4}{4} + \cdots \quad (|x| < 1) \tag{30.9}$$

$$(1 + x)^n = 1 + nx + \frac{n(n-1)}{2!}x^2 + \cdots \quad (|x| < 1) \tag{30.10}$$

In the next section, we will see how to use these series in finding values of functions. In this section, we see how new series are developed by using the above basic series, and we also show other uses of series.

When we discussed functions in Chapter 3, we mentioned functions such as $f(2x)$ and $f(-x)$. [By using functional notation and the preceding series, we can find the series expansions of many other series without using direct expansion.] This can often save time in finding a desired series.

NOTE ▸

EXAMPLE 1 Series formed using functional notation

Find the Maclaurin expansion of e^{2x}.

From Eq. (30.6), we know the expansion of e^x. Hence,

$$f(x) = 1 + x + \frac{x^2}{2!} + \frac{x^3}{3!} + \cdots$$

Because $e^{2x} = f(2x)$, we have

$$f(2x) = 1 + (2x) + \frac{(2x)^2}{2!} + \frac{(2x)^3}{3!} + \cdots \qquad \text{in } f(x)\text{, replace } x \text{ by } 2x$$

$$e^{2x} = 1 + 2x + 2x^2 + \frac{4x^3}{3} + \cdots$$

■

EXAMPLE 2 Series formed using functional notation

Find the Maclaurin expansion of $\sin x^2$.

From Eq. (30.7), we know the expansion of $\sin x$. Therefore,

$$f(x) = x - \frac{x^3}{3!} + \frac{x^5}{5!} - \cdots$$

$$f(x^2) = (x^2) - \frac{(x^2)^3}{3!} + \frac{(x^2)^5}{5!} - \cdots \qquad \text{in } f(x)\text{, replace } x \text{ by } x^2$$

$$\sin x^2 = x^2 - \frac{x^6}{3!} + \frac{x^{10}}{5!} - \cdots$$

Direct expansion of this series is quite lengthy. ■

Practice Exercise

1. Using the Maclaurin series for $\ln(1 + x)$, find the first four terms of the Maclaurin expansion of $\ln(1 - 2x)$.

The basic algebraic operations may be applied to series in the same manner they are applied to polynomials. That is, we may add, subtract, multiply, or divide series in order to obtain other series. The interval of convergence for the resulting series is that which is common to those of the series being used. The multiplication of series is illustrated in the following example.

EXAMPLE 3 Series formed by multiplication

Multiply the series expansion for e^x by the series expansion for $\cos x$ to obtain the series expansion for $e^x \cos x$.

Using the series expansion for e^x and $\cos x$ as shown in Eqs. (30.6) and (30.8), we have the following indicated multiplication:

$$e^x \cos x = \left(1 + x + \frac{x^2}{2!} + \frac{x^3}{3!} + \frac{x^4}{4!} + \cdots\right)\left(1 - \frac{x^2}{2!} + \frac{x^4}{4!} - \cdots\right)$$

Using the distributive property, we have the following result, considering through the x^4-terms in the product.

$$1\left(1 - \frac{x^2}{2!} + \frac{x^4}{4!}\right) \quad x\left(1 - \frac{x^2}{2!}\right) \quad \frac{x^2}{2!}\left(1 - \frac{x^2}{2!}\right) \quad \left(\frac{x^3}{3!} + \frac{x^4}{4!}\right)(1)$$

$$e^x \cos x = 1 - \frac{x^2}{2} + \frac{x^4}{24} + x - \frac{x^3}{2} + \frac{x^2}{2} - \frac{x^4}{4} + \frac{x^3}{6} + \frac{x^4}{24} + \cdots$$

$$= 1 + x - \frac{1}{3}x^3 - \frac{1}{6}x^4 + \cdots$$

■

It is also possible to use the operations of differentiation and integration to obtain series expansions, although the proof of this is found in more advanced texts. Consider the following example.

EXAMPLE 4 Series formed by differentiating

Show that by differentiating the series for $\ln(1 + x)$ term by term, the result is the same as the series for $\frac{1}{1 + x}$.

The series for $\ln(1 + x)$ is shown in Eq. (30.9) as

$$\ln(1 + x) = x - \frac{x^2}{2} + \frac{x^3}{3} - \frac{x^4}{4} + \cdots$$

Differentiating, we have

$$\frac{1}{1 + x} = 1 - \frac{2x}{2} + \frac{3x^2}{3} - \frac{4x^3}{4} + \cdots$$

$$= 1 - x + x^2 - x^3 + \cdots$$

Using the binomial expansion for $\frac{1}{1 + x} = (1 + x)^{-1}$, we have

$$(1 + x)^{-1} = 1 + (-1)x + \frac{(-1)(-2)}{2!}x^2 + \frac{(-1)(-2)(-3)}{3!}x^3 + \cdots$$

using Eq. (30.10) with $n = -1$

$$= 1 - x + x^2 - x^3 + \cdots$$

We see that the results are the same.

■

Practice Exercise

2. Show that by differentiating the series for e^{-x} term by term, the result is the same as the series for $-e^{-x}$.

We can use algebraic operations on series to verify that the definition of the exponential form of a complex number, $re^{j\theta} = r(\cos\theta + j\sin\theta)$, is consistent with other definitions. The only assumption required here is that the Maclaurin expansions for e^x, $\sin x$, and $\cos x$ are also valid for complex numbers. This is shown in advanced calculus. Thus,

$$e^{j\theta} = 1 + j\theta + \frac{(j\theta)^2}{2!} + \frac{(j\theta)^3}{3!} + \cdots = 1 + j\theta - \frac{\theta^2}{2!} - j\frac{\theta^3}{3!} + \cdots \quad (30.11)$$

$$j\sin\theta = j\theta - j\frac{\theta^3}{3!} + \cdots \quad (30.12)$$

$$\cos\theta = 1 - \frac{\theta^2}{2!} + \cdots \quad (30.13)$$

■ Eq. (30.14) is known as Euler's Formula. If $\theta = \pi$, we have $e^{j\pi} = -1$, which can be written as

$$e^{j\pi} + 1 = 0$$

This equation connects the five fundamental numbers $e, j, \pi, 1$, and 0, and it has been called a "beautiful" equation by mathematicians.

When we add the terms of Eq. (30.12) to those of Eq. (30.13), the result is the series given in Eq. (30.11). Thus,

$$e^{j\theta} = \cos\theta + j\sin\theta \quad (30.14)$$

By multiplying both sides of Eq. (30.14) by r, we get the exponential form of a complex number: $re^{j\theta} = r(\cos\theta + j\sin\theta)$.

An additional use of power series is now shown. Many integrals that occur in practice cannot be integrated by methods given in the preceding chapters. However, power series can be very useful in giving excellent approximations to some definite integrals.

EXAMPLE 5 Using series for integration—area of a cutting blade

The shape of a special cutting blade can be described by the region bounded by the axes, the line $x = 0.500$ cm, and the curve $y = \sqrt{1 + x^3}$. Find the area of the blade (in cm^2).

From Fig. 30.8, we see that the area is

$$A = \int_0^{0.5} \sqrt{1 + x^3}\, dx$$

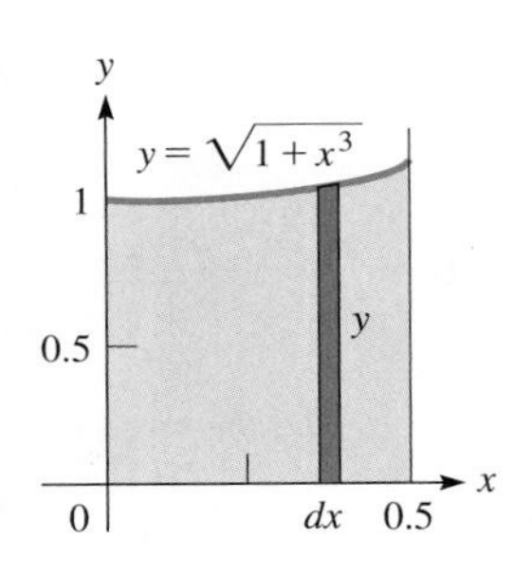

Fig. 30.8

This integral does not fit any form we have used. However, its value can be closely approximated by using the binomial expansion for $\sqrt{1 + x^3}$ and then integrating.

Using the binomial expansion to find the first three terms of the expansion for $\sqrt{1 + x^3}$, we have

$$\sqrt{1 + x^3} = (1 + x^3)^{0.5} = 1 + 0.5x^3 + \frac{0.5(-0.5)}{2}(x^3)^2 + \cdots$$
$$= 1 + 0.5x^3 - 0.125x^6 + \cdots$$

Substituting in the integral, we have

$$A = \int_0^{0.5} (1 + 0.5x^3 - 0.125x^6 + \cdots)dx$$
$$= x + \frac{0.5}{4}x^4 - \frac{0.125}{7}x^7 + \cdots \Big|_0^{0.5}$$
$$= 0.5 + 0.0078125 - 0.0001395 + \cdots = 0.507673 + \cdots$$

We can see that each of the terms omitted was very small. Because the data given ($x = 0.500$ cm) is accurate to three significant digits, we conclude that the area of the blade is $A = 0.508\ \text{cm}^2$. ■

EXAMPLE 6 Using series for integration

Evaluate: $\int_0^{0.1} e^{-x^2}\,dx$.

$$e^{-x^2} = 1 + (-x^2) + \frac{(-x^2)^2}{2!} + \cdots \quad \text{using Eq. (30.6)}$$

$$\int_0^{0.1} e^{-x^2}\,dx = \int_0^{0.1}\left(1 - x^2 + \frac{x^4}{2} - \cdots\right)dx \quad \text{substitute}$$

$$= \left(x - \frac{x^3}{3} + \frac{x^5}{10} - \cdots\right)\Bigg|_0^{0.1} \quad \text{integrate}$$

$$= 0.1 - \frac{0.001}{3} + \frac{0.00001}{10} = 0.0996677 \quad \text{evaluate}$$

This answer is correct to the indicated accuracy. ■

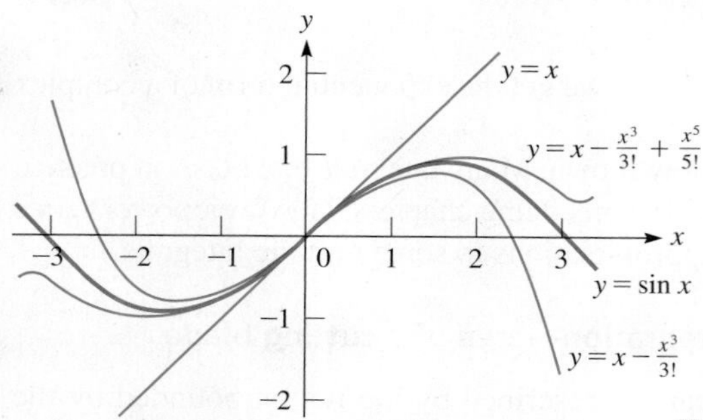

Fig. 30.9

The question of accuracy now arises. The integrals just evaluated indicate that the more terms used, the greater the accuracy of the result. To graphically show the accuracy involved, Fig. 30.9 depicts the graphs of $y = \sin x$ and the graphs of

$$y = x \qquad y = x - \frac{x^3}{3!} \qquad y = x - \frac{x^3}{3!} + \frac{x^5}{5!}$$

which are the first three approximations of $y = \sin x$. We can see that each term added gives a better fit to the curve of $y = \sin x$. Also, this gives a graphical representation of the meaning of a series expansion.

We have just shown that the more terms included, the more accurate the result. For small values of x, a Maclaurin series gives good accuracy with very few terms. In this case, the series *converges* rapidly, as we mentioned earlier. [For this reason, a Maclaurin series is of particular use for small values of x.] For larger values of x, usually a function is expanded in a Taylor series (see Section 30.5). Of course, if we omit any term in a series, there is some error in the calculation.

NOTE ▸

EXERCISES 30.3

In Exercises 1 and 2, make the given changes in the indicated examples of this section, and then find the resulting series.

1. In Example 1, change e^{2x} to e^{2x^2}.
2. In Example 3, change e^x to e^{-x}.

In Exercises 3–10, find the first four nonzero terms of the Maclaurin expansions of the given functions by using Eqs. (30.6) to (30.10).

3. $f(x) = e^{3x}$
4. $f(x) = e^{-2x}$
5. $f(x) = \sin\frac{1}{2}x$
6. $f(x) = \sin(-x^4)$
7. $f(x) = x\cos 4x$
8. $f(x) = \sqrt{1 - x^4}$
9. $f(x) = \ln(1 + x^2)$
10. $f(x) = x^2\ln(1 - x)$

In Exercises 11–16, evaluate the given integrals by using three terms of the appropriate series.

11. $\int_0^1 \sin x^2\,dx$
12. $\int_0^{0.4} \sqrt[4]{1 - 2x^2}\,dx$
13. $\int_0^{0.5} e^{-\sqrt{x}}\,dx$
14. $\int_{0.1}^{0.2} \frac{\cos x - 1}{x}\,dx$
15. $\int_0^{0.2} \frac{\ln(1 + x)}{x}\,dx$
16. $\int_0^{0.5} \frac{dx}{\sqrt{1 + x^2}}$

In Exercises 17–30, find the indicated series by the given operation.

17. Find the first four terms of the Maclaurin expansion of the function $f(x) = \frac{2}{1 - x^2}$ by adding the terms of the series for the functions $\frac{1}{1 - x}$ and $\frac{1}{1 + x}$.
18. Find the first four nonzero terms of the expansion of the function $f(x) = \frac{1}{2}(e^x - e^{-x})$ by subtracting the terms of the appropriate series. The result is the series for $\sinh x$. (See Exercise 53 of Section 27.6.)
19. Find the first three terms of the expansion for $e^x \sin x$ by multiplying the proper expansions together, term by term.
20. Find the first three nonzero terms of the expansion for $f(x) = \tan x$ by dividing the series for $\sin x$ by that for $\cos x$.
21. By using the properties of logarithms and the series for $\ln(1 + x)$, find the series for $x^2\ln(1 - x)^2$.

22. By using the properties of logarithms and the series for $\ln(1 + x)$, find the series for $\ln\frac{1+x}{1-x}$.

23. Find the first three terms of the expansion for $\ln(1 + \sin x)$ by using the expansions for $\ln(1 + x)$ and $\sin x$.

24. Show that by differentiating term by term the expansion for $\sin x$, the result is the expansion for $\cos x$.

25. Show that by differentiating term by term the expansion for e^x, the result is also the expansion for e^x.

26. Find the expansion for $\sin x + x\cos x$ by differentiating term by term the expansion for $x \sin x$.

27. Show that by integrating term by term the expansion for $\cos x$, the result is the expansion for $\sin x$.

28. Show that by integrating term by term the expansion for $-1/(1 - x)$ (see Exercise 13 of Section 30.2), the result is the expansion for $\ln(1 - x)$.

29. By multiplication of series, find the first three terms of the expansion for the displacement of the oscillating object in Exercise 41 of Section 30.2.

30. By using the series for e^x, find the first three terms of the expansion of the electric current given in Exercise 42 of Section 30.2.

In Exercises 31–42, solve the given problems.

31. Evaluate $\int_0^1 e^x\,dx$ directly and compare the result obtained by using four terms of the series for e^x and then integrating.

32. Evaluate $\lim_{x\to 0}\frac{\sin x}{x}$ by using the series expansion for $\sin x$. Compare the result with Eq. (27.1).

33. Evaluate $\lim_{x\to 0}\frac{\sin x - x}{x^3}$ by using the expansion for $\sin x$.

34. Find the approximate area bounded by $y = \sin x$, $y = 0$, and $x = \pi/6$ by using two terms of the expansion for $\sin x$. Compare the result with that found by direct integration.

35. Find the approximate value of the area bounded by $y = x^2e^x$, $x = 0.2$, and the x-axis by using three terms of the appropriate Maclaurin series.

36. Use series to evaluate $\lim_{x\to 0}\frac{e^x - (1 + x)}{x^2}$.

37. Use series to evaluate $\lim_{x\to 0}\frac{\ln(1 + x^2)}{x^2}$. Compare your answer to the one you get by using L'Hospital's rule.

38. Find the approximate area under the graph of $y = \frac{1}{\sqrt{2\pi}}e^{-x^2/2}$ (the standard normal curve) from $x = -1$ to $x = 1$ by using three terms of the appropriate series. See Fig. 30.10. (According to the *empirical rule*, this area is approximately 68%.)

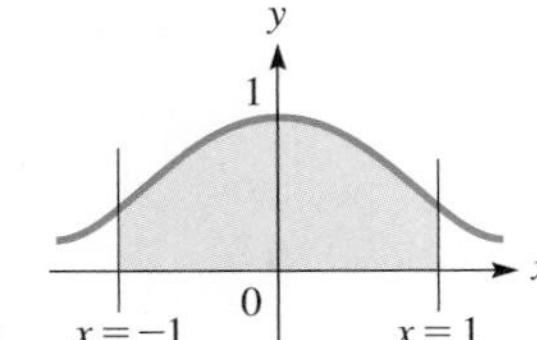

Fig. 30.10

39. The *Fresnel integral* $\int_0^x \cos t^2\,dt$ is used in the analysis of beam displacements (and in optics). Evaluate this integral for $x = 0.2$ by using two terms of the appropriate series.

40. The dome of a sports arena is designed as the surface generated by revolving the curve of $y = 20.0\cos 0.0196x$ $(0 \le x \le 80.0\text{ m})$ about the y-axis. Find the volume within the dome by using three terms of the appropriate series.

41. In the theory of relativity, the equation for energy E is $E = mc^2\left(1 - \frac{v^2}{c^2}\right)^{-1/2}$, where m is the mass of the object, v is its velocity, and c is the speed of light. Treating v as the variable, use Eq. (30.10) to find the first three terms of the power series for E. (If you include only the first term, you should get the famous formula $E = mc^2$.)

42. The charge q on a capacitor in a certain electric circuit is given by $q = ce^{-at}\sin 6at$, where t is the time. By multiplication of series, find the first four nonzero terms of the expansion for q.

In Exercises 43–46, use a calculator to display (a) the given function and (b) the first three series approximations of the function in the same display. Each display will be similar to that in Fig. 30.9 for the function $y = \sin x$ and its first three approximations. Be careful in choosing the appropriate window values.

43. $y = e^x$

44. $y = \cos x$

45. $y = \ln(1 + x)$ $\quad(|x| < 1)$

46. $y = \sqrt{1 + x}$ $\quad(|x| < 1)$

Answers to Practice Exercises

1. $\ln(1 - 2x) = -2x - 2x^2 - \frac{8}{3}x^3 - 4x^4 - \cdots$

2. $-e^{-x} = -1 + x - \frac{x^2}{2!} + \frac{x^3}{3!} - \cdots$

30.4 Computations by Use of Series Expansions

Approximating Values of Algebraic, Trigonometric, Exponential, and Logarithmic Functions • Approximating Error in a Calculation

As we mentioned at the beginning of the previous section, power-series expansions can be used to compute numerical values of exponential functions, trigonometric functions, logarithms, powers, and roots. By including a sufficient number of terms in the expansion, we can calculate these values to any degree of accuracy that may be required.

It is through such calculations that tables of values can be made, and decimal approximations of numbers such as e and π can be found. Also, many of the values found on a calculator or a computer are calculated by using series expansions that have been programmed into the chip which is in the calculator or computer.

EXAMPLE 1 Exponential value

Calculate the value of $e^{0.1}$.

In order to evaluate $e^{0.1}$, we substitute 0.1 for x in the expansion for e^x. The more terms that are used, the more accurate a value we can obtain. The limit of the partial sums would be the actual value. However, since $e^{0.1}$ is irrational, we cannot express the exact value in decimal form.

Therefore, the value is found as follows:

$$e^x = 1 + x + \frac{x^2}{2!} + \cdots \qquad \text{Eq. (30.6)}$$

$$e^{0.1} = 1 + 0.1 + \frac{(0.1)^2}{2} + \cdots \qquad \text{substitute 0.1 for } x$$

$$= 1.105 \qquad \text{using 3 terms}$$

Using a calculator, we find that $e^{0.1} = 1.105170918$, which shows that our answer is valid to the accuracy shown. ■

Practice Exercise

1. Using three terms of the appropriate series, calculate the value of $e^{0.06}$.

EXAMPLE 2 Trigonometric value

CAUTION When using series to find trigonometric values, the angle must be expressed in radians. ■

Calculate the value of sin 2°.

In finding trigonometric values, we must be careful to ***express the angle in radians.*** Thus, the value of sin 2° is found as follows:

$$\sin x = x - \frac{x^3}{3!} + \cdots \qquad \text{Eq. (30.7)}$$

$$\sin 2° = \left(\frac{\pi}{90}\right) - \frac{(\pi/90)^3}{6} + \cdots \qquad 2° = \frac{\pi}{90} \text{ rad}$$

$$= 0.0348994963 \qquad \text{using 2 terms}$$

A calculator gives the value 0.0348994967. Here, we note that the second term is much smaller than the first. In fact, a good approximation of 0.0349 can be found by using just one term. We now see that $\sin\theta \approx \theta$ for small values of θ, as we noted in Section 8.4. ■

EXAMPLE 3 Trigonometric value

Calculate the value of cos 0.5429.

Because the angle is expressed in radians, we have

$$\cos 0.5429 = 1 - \frac{0.5429^2}{2} + \frac{0.5429^4}{4!} - \cdots \qquad \text{using Eq. (30.8)}$$

$$= 0.8562495 \qquad \text{using 3 terms}$$

A calculator shows that $\cos 0.5429 = 0.8562140824$. Because the angle is not small, additional terms are needed to obtain this accuracy. With one more term, the value 0.8562139 is obtained. ■

Practice Exercise

2. Using two terms of the appropriate series, calculate the value of cos 2°.

EXAMPLE 4 Logarithmic value

Calculate the value of ln 1.2.

$$\ln(1 + x) = x - \frac{x^2}{2} + \frac{x^3}{3} - \cdots \qquad \text{Eq. (30.9)}$$

$$\ln 1.2 = \ln(1 + 0.2)$$

$$= 0.2 - \frac{(0.2)^2}{2} + \frac{(0.2)^3}{3} - \cdots = 0.1827$$

To four significant digits, $\ln(1.2) = 0.1823$. One more term is required to obtain this accuracy. ■

We now illustrate the use of series in finding the amount of error in a calculation that could result from a certain measurement error. We also discussed this as an application of differentials. However, a series allows us to estimate the error with much more accuracy than with differentials since more terms can be included.

EXAMPLE 5 Approximation of error—velocity of a falling object

The velocity v of an object that has fallen h feet is $v = 8.00\sqrt{h}$. Find the approximate error in calculating the velocity of an object that has fallen 100.0 ft with a possible error of 2.0 ft.

If we ***let*** $\boldsymbol{v = 8.00\sqrt{100.0 + x}}$***, where*** $\boldsymbol{x}$ ***is the error in*** $\boldsymbol{h}$, we may express v as a Maclaurin expansion in x:

$$f(x) = 8.00(100.0 + x)^{1/2} \qquad f(0) = 80.0$$

$$f'(x) = 4.00(100.0 + x)^{-1/2} \qquad f'(0) = 0.400$$

$$f''(x) = -2.00(100.0 + x)^{-3/2} \qquad f''(0) = -0.00200$$

Therefore,

$$v = 8.00\sqrt{100.0 + x} = 80.0 + 0.400x - 0.00100x^2 + \cdots$$

Because the calculated value of v for $x = 0$ is 80.0, the error E in the value of v is

$$E = 0.400x - 0.00100x^2 + \cdots$$

Calculating, the error for $x = 2.0$ is

$$E = 0.400(2.0) - 0.00100(4.0) = 0.800 - 0.0040 = 0.796 \text{ ft/s}$$

The value 0.800 is that which is found using differentials. The additional terms are corrections to this term. The additional term in this case shows that the first term is a good approximation to the error. Although this problem can be done numerically, a series solution allows us to find the error for any value of x. ■

EXERCISES 30.4

In Exercises 1 and 2, make the given changes in the indicated examples of this section, and then solve the resulting problems.

1. In Example 1, change $e^{0.1}$ to $e^{-0.1}$.

2. In Example 4, change ln 1.2 to ln 0.8.

In Exercises 3–20, calculate the value of each of the given functions. Use the indicated number of terms of the appropriate series. Compare with the value found directly on a calculator.

3. $e^{0.2}$ (3) **4.** 1.01^{-1} (4) **5.** $\sin 0.1$ (2)

6. $\cos 0.05$ (2) **7.** e (7) **8.** $1/\sqrt{e}$ (5)

9. $\cos \pi°$ (2) **10.** $\sin(-4°)$ (3) **11.** $\ln 1.4$ (4)

12. $\ln 0.95$ (4) **13.** $\sin 0.3625$ (3) **14.** $\cos 1$ (4)

15. $\ln 0.8461$ (5) **16.** $\ln 0.9493^{-1}$ (3) **17.** 1.032^6 (3)

18. 0.9982^8 (3) **19.** $1.1^{-0.2}$ (3) **20.** 0.96^{-1} (3)

In Exercises 21–24, use a series to approximate the value of each of the given functions. In Exercises 21 and 22, use the expansion for $\sqrt{1 + x}$, *and in Exercises 23 and 24, use the expansion for* $\sqrt[3]{1 + x}$. *Use three terms of the appropriate series.*

21. $\sqrt{1.1076}$ **22.** $\sqrt{0.7915}$

23. $\sqrt[3]{0.9628}$ **24.** $\sqrt[3]{1.1392}$

In Exercises 25–28, calculate the maximum error of the values calculated in the indicated exercises. If a series is alternating (every other term is negative), the maximum possible error in the calculated value is the value of the first term omitted.

25. Exercise 5 **26.** Exercise 4

27. Exercise 9 **28.** Exercise 11

In Exercises 29–40, solve the given problems by using series expansions.

29. Evaluate $\sqrt{3.92}$ by noting that $\sqrt{3.92} = \sqrt{4 - 0.08} = 2\sqrt{1 - 0.02}$.

30. Evaluate sin 32° by first finding the expansion for $\sin(x + \pi/6)$.

31. We can evaluate π by use of $\frac{1}{4}\pi = \tan^{-1}\frac{1}{2} + \tan^{-1}\frac{1}{3}$ along with the series for $\tan^{-1}x$. The first three terms are $\tan^{-1}x = x - \frac{1}{3}x^3 + \frac{1}{5}x^5$. Using these terms, expand $\tan^{-1}\frac{1}{2}$ and $\tan^{-1}\frac{1}{3}$ and approximate the value of π.

32. Use the fact that $\frac{1}{4}\pi = \tan^{-1}\frac{1}{7} + 2\tan^{-1}\frac{1}{3}$ to approximate the value of π. (See Exercise 31.)

33. Explain why $e^x > 1 + x + \frac{1}{2}x^2$ for $x > 0$.

34. Using a calculator, determine how many terms of the expansion for $\ln(1 + x)$ are needed to give the value of ln 1.3 accurate to five decimal places.

35. The time t (in years) for an investment to increase by 10% when the interest rate is 6% is given by $t = \frac{\ln 1.1}{0.06}$. Evaluate this expression by using the first four terms of the appropriate series.

36. The period T of a pendulum of length L is given by

$$T = 2\pi\sqrt{\frac{L}{g}}\left(1 + \frac{1}{4}\sin^2\frac{\theta}{2} + \frac{9}{64}\sin^4\frac{\theta}{2} + \cdots\right)$$

where g is the acceleration due to gravity and θ is the maximum angular displacement. If $L = 1.000$ m and $g = 9.800$ m/s^2, calculate T for $\theta = 10.0°$ (a) if only one term (the 1) of the series is used and (b) if two terms of the indicated series are used. In the second term, estimate $\sin^2(\theta/2)$ by using the first term of its series expansion.

37. The current in a circuit containing a resistance R, an inductance L, and a battery whose voltage is E is given by the equation $i = \frac{E}{R}(1 - e^{-Rt/L})$, where t is the time. Approximate this expression by using the first three terms of the appropriate exponential series. Under what conditions will this approximation be valid?

38. The image distance q from a certain lens as a function of the object distance p is given by $q = 20p/(p - 20)$. Find the first three nonzero terms of the expansion of the right side. From this expression, calculate q for $p = 2.00$ cm and compare it with the value found by substituting 2.00 in the original expression.

39. An empty underground cubical tank was later filled with water. The amount of water needed to fill the tank was 30.0 m^3 with a possible error of 1.00 m^3. Use a series to estimate the error in calculating the length of one side, s, of the tank $(s = V^{1/3})$. [*Hint*: Find a series for $(30 + x)^{1/3}$.]

40. The efficiency E (in %) of an internal combustion engine in terms of its compression ratio r is given by $E = 100(1 - r^{-0.40})$. Determine the possible approximate error in the efficiency for a compression ratio measured to be 6.00 with a possible error of 0.50. [*Hint:* Set up a series for $(6 + x)^{-0.40}$.]

Answers to Practice Exercises

1. 1.0618 **2.** 0.9993908

30.5 Taylor Series

Taylor Series Expansion • More General than Maclaurin Series • Choice of the Value of *a*

To obtain accurate values of a function for values of x that are not close to zero, it is usually necessary to use many terms of a Maclaurin expansion. However, we can use another type of series, called a **Taylor series,** *which is a more general expansion than a Maclaurin expansion.* Also, functions for which a Maclaurin series may not be found may have a Taylor series.

The basic assumption in formulating a Taylor expansion is that a function may be expanded in a polynomial of the form

$$f(x) = c_0 + c_1(x - a) + c_2(x - a)^2 + \cdots \tag{30.15}$$

Following the same line of reasoning as in deriving the Maclaurin expansion, we may find the constants $c_0, c_1, c_2, \ldots$. That is, derivatives of Eq. (30.15) are taken, and the function and its derivatives are evaluated at $x = a$. This leads to the following equation, which is called the **Taylor series expansion** *of a function.*

■ Named for the English mathematician Brook Taylor (1685–1731).

Taylor Series

$$f(x) = f(a) + f'(a)(x - a) + \frac{f''(a)(x - a)^2}{2!} + \cdots \tag{30.16}$$

The number a is called the **center of expansion** of the Taylor series (in a Maclaurin series, $a = 0$). A Taylor series converges rapidly for values of x that are close to a, and this is illustrated in Examples 3 and 4.

EXAMPLE 1 Taylor series for e^x

Expand $f(x) = e^x$ in a Taylor series with $a = 1$.

$$\begin{aligned} f(x) &= e^x & f(1) &= e & &\text{find derivatives and evaluate each at } x = 1 \\ f'(x) &= e^x & f'(1) &= e \\ f'(x) &= e^x & f'(1) &= e \\ f''(x) &= e^x & f''(1) &= e \end{aligned}$$

$$f(x) = e + e(x-1) + e\frac{(x-1)^2}{2!} + e\frac{(x-1)^3}{3!} + \cdots \quad \text{using Eq. (30.16)}$$

$$e^x = e\left[1 + (x-1) + \frac{(x-1)^2}{2} + \frac{(x-1)^3}{6} + \cdots\right]$$

This series can be used in evaluating e^x for values of x near 1. ■

Practice Exercise

1. Expand $f(x) = e^x$ in a Taylor series with $a = 3$.

EXAMPLE 2 Taylor series for $\sqrt{x}$

Expand $f(x) = \sqrt{x}$ in powers of $(x - 4)$.

Another way of stating this is to find the Taylor series for $f(x) = \sqrt{x}$, with $a = 4$. Thus,

$$\begin{aligned} f(x) &= x^{1/2} & f(4) &= 2 & &\text{find derivatives and evaluate each at } x = 4 \\ f'(x) &= \frac{1}{2x^{1/2}} & f'(4) &= \frac{1}{4} \\ f''(x) &= -\frac{1}{4x^{3/2}} & f''(4) &= -\frac{1}{32} \\ f'''(x) &= \frac{3}{8x^{5/2}} & f'''(4) &= \frac{3}{256} \end{aligned}$$

$$f(x) = 2 + \frac{1}{4}(x-4) - \frac{1}{32}\frac{(x-4)^2}{2!} + \frac{3}{256}\frac{(x-4)^3}{3!} - \cdots \quad \text{using Eq. (30.16)}$$

$$\sqrt{x} = 2 + \frac{(x-4)}{4} - \frac{(x-4)^2}{64} + \frac{(x-4)^3}{512} - \cdots$$

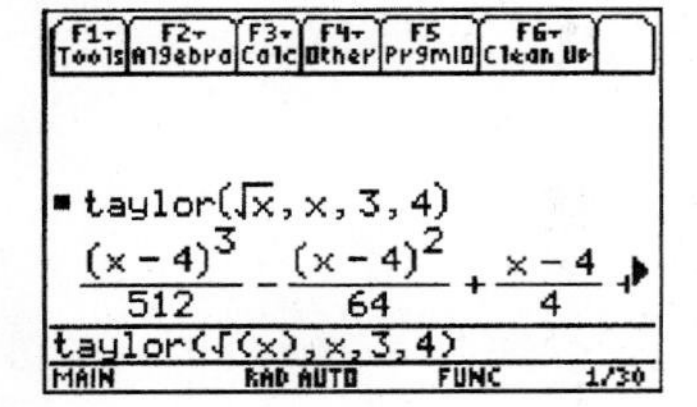

Fig. 30.11
TI-89 graphing calculator keystrokes: goo.gl/OpyGqz

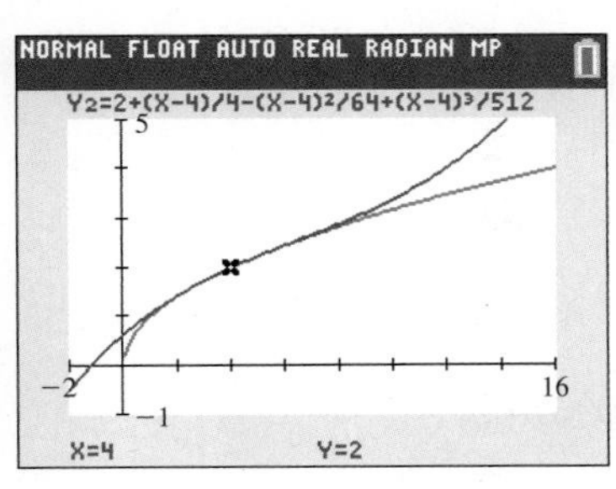

Fig. 30.12

This series would be used to evaluate square roots of numbers near 4.

In Fig. 30.11, the TI-89 calculator display of this expansion is shown. In Fig. 30.12, the original function $f(x) = \sqrt{x}$ is graphed (in red) along with the first four terms of the Taylor series (in blue). We see that each curve passes through (4, 2), and they have nearly equal values of y for values of x near 4. ■

In the last section, we evaluated functions by using Maclaurin series. In the following examples, we use Taylor series to evaluate functions.

EXAMPLE 3 Evaluating a square root using a Taylor series

By using Taylor series, evaluate $\sqrt{4.5}$.

Using the four terms of the series found in Example 2, we have

$$\begin{aligned} \sqrt{4.5} &= 2 + \frac{(4.5-4)}{4} - \frac{(4.5-4)^2}{64} + \frac{(4.5-4)^3}{512} \quad \text{substitute 4.5 for } x \\ &= 2 + \frac{(0.5)}{4} - \frac{(0.5)^2}{64} + \frac{(0.5)^3}{512} \\ &= 2.121337891 \end{aligned}$$

The value found directly on a calculator is 2.121320344. Therefore, the value found by these terms of the series expansion is correct to four decimal places. ■

In Example 3, note that successive terms become small rapidly. If a value of x is chosen such that $x - a$ is not close to zero, the successive terms may not become small rapidly, and many terms may be required. Therefore, ***we should choose the value of a as conveniently close as possible to the x-values that will be used.*** Also, we should note that a Maclaurin expansion for $\sqrt{x}$ cannot be used since the derivatives of $\sqrt{x}$ are not defined for $x = 0$.

EXAMPLE 4 Evaluating sine value using Taylor series

Calculate the approximate value of sin 29° by using three terms of the appropriate Taylor expansion.

Because the value of sin 30° is known to be $\frac{1}{2}$, we let $a = \frac{\pi}{6}$ (remember, we must use values expressed in radians) when we evaluate the expansion for $x = 29°$ [when expressed in radians, the quantity $(x - a)$ is $-\frac{\pi}{180}$ (equivalent to $-1°$)]. This means that its numerical values are small and become smaller when it is raised to higher powers. Therefore,

$$f(x) = \sin x \qquad f\left(\frac{\pi}{6}\right) = \frac{1}{2} \qquad \text{find derivatives and evaluate each at } x = \frac{\pi}{6}$$

$$f'(x) = \cos x \qquad f'\left(\frac{\pi}{6}\right) = \frac{\sqrt{3}}{2}$$

$$f''(x) = -\sin x \qquad f''\left(\frac{\pi}{6}\right) = -\frac{1}{2}$$

$$f(x) = \frac{1}{2} + \frac{\sqrt{3}}{2}\left(x - \frac{\pi}{6}\right) - \frac{1}{4}\left(x - \frac{\pi}{6}\right)^2 - \cdots \qquad \text{using Eq. (30.16)}$$

$$\sin x = \frac{1}{2} + \frac{\sqrt{3}}{2}\left(x - \frac{\pi}{6}\right) - \frac{1}{4}\left(x - \frac{\pi}{6}\right)^2 - \cdots \qquad f(x) = \sin x$$

$$\sin 29° = \sin\left(\frac{\pi}{6} - \frac{\pi}{180}\right) \qquad 29° = 30° - 1° = \frac{\pi}{6} - \frac{\pi}{180}$$

$$= \frac{1}{2} + \frac{\sqrt{3}}{2}\left(\frac{\pi}{6} - \frac{\pi}{180} - \frac{\pi}{6}\right) - \frac{1}{4}\left(\frac{\pi}{6} - \frac{\pi}{180} - \frac{\pi}{6}\right)^2 - \cdots \qquad \text{substitute } \frac{\pi}{6} - \frac{\pi}{180} \text{ for } x$$

$$= \frac{1}{2} + \frac{\sqrt{3}}{2}\left(-\frac{\pi}{180}\right) - \frac{1}{4}\left(-\frac{\pi}{180}\right)^2 - \cdots$$

$$= 0.4848088509$$

The value found directly on a calculator is 0.4848096202. ■

EXERCISES 30.5

In Exercises 1 and 2, make the given changes in the indicated examples of this section, and then solve the resulting problems.

1. In Example 2, change $(x - 4)$ to $(x - 1)$.
2. In Example 4, change sin 29° to sin 31°.

In Exercises 3–10, evaluate the given functions by using the series developed in the examples of this section.

3. $e^{1.2}$
4. $e^{0.7}$
5. $\sqrt{4.3}$
6. $\sqrt{3.5}$
7. sin 33°
8. sin 28°
9. $\sin\frac{9\pi}{60}$
10. $\sqrt[6]{27}$

In Exercises 11–22, find the first three nonzero terms of the Taylor expansion for the given function and given value of a.

11. e^{-x} $(a = 2)$
12. $\cos x$ $(a = \frac{\pi}{4})$
13. $\sin x$ $(a = \frac{\pi}{3})$
14. $\ln x$ $(a = 3)$
15. $\sqrt[3]{x}$ $(a = 8)$
16. $\frac{1}{x}$ $(a = 2)$
17. $\tan x$ $(a = \frac{\pi}{4})$
18. $\ln \sin x$ $(a = \frac{\pi}{2})$
19. $e^x \sin x$ $(a = \frac{\pi}{2})$
20. xe^{-x} $(a = -1)$
21. $\frac{1}{x + 2}$ $(a = 3)$
22. $\frac{1}{(1 + x)^2}$ $(a = -2)$

In Exercises 23–30, evaluate the given functions by using three terms of the appropriate Taylor series.

23. e^{π}
24. ln 3.1
25. $\sqrt{9.3}$
26. 2.056^{-1}
27. $\sqrt[3]{8.3}$
28. tan 46°
29. sin 61°
30. $\cos\frac{7\pi}{30}$

In Exercises 31–38, solve the given problems.

31. By completing the steps indicated before Eq. (30.16) in the text, complete the derivation of Eq. (30.16).

32. Find the first three terms of the Taylor expansion of $f(x) = \ln x$ with $a = 1$. Compare this Taylor expansion with the linearization $L(x)$ of $f(x)$ with $a = 1$. Compare the graphs of $f(x)$, $L(x)$, and the Taylor expansion on a calculator.

33. Show that the polynomial $2x^3 + x^2 - 3x + 5$ can be written as $2(x-1)^3 + 7(x-1)^2 + 5(x-1) + 5$.

34. Calculate $\sqrt{3}$ using the series in Example 2 and compare with the value using the series in Exercise 1. Which is the better approximation?

35. Calculate $\sin 31°$ by using three terms of the Maclaurin expansion for $\sin x$. Also, calculate $\sin 31°$ by using three terms of the Taylor expansion in Example 4 (see Exercise 2). Compare the accuracy of the values obtained with that found directly on a calculator.

36. Referring to Eq. (30.16), show that a Taylor series can be expressed in the form

$$f(a+h) = f(a) + f'(a)h + \frac{f''(a)}{2!}h^2 + \cdots$$

37. The current i in a certain electric circuit is $i = 6\sin \pi t$. Write the first three terms of the Taylor series of this function about $t = \pi/2$.

38. In the analysis of the electric potential of an electric charge distributed along a straight wire of length L, the expression $\ln\frac{x+L}{x}$ is used. Find three terms of the Taylor expansion of this expression in powers of $(x - L)$.

In Exercises 39–42, use a calculator to display (a) the function in the indicated exercise of this set and (b) the first two terms of the Taylor series found for that exercise in the same display. Describe how closely the graph in part (b) fits the graph in part (a). Use the given values of x for Xmin and Xmax.

39. Exercise 13 ($\sin x$), $x = 0$ to $x = 2$

40. Exercise 15 ($\sqrt[3]{x}$), $x = 0$ to $x = 16$

41. Exercise 16 ($1/x$), $x = 0$ to $x = 4$

42. Exercise 17 ($\tan x$), $x = 0$ to $x = 1.5$

Answer to Practice Exercise

1. $f(x) = e^3\left[1 + (x-3) + \frac{(x-3)^2}{2} + \frac{(x-3)^3}{6} + \cdots\right]$

30.6 Introduction to Fourier Series

Periodic Functions • Fourier Series Is a Series of Sines and Cosines • Formulas for Coefficients

We have already shown how transcendental functions can be represented by Taylor series, which contain an infinite number of terms of the form $c_n(x-a)^n$, where a is the center of the expansion. In this section, we discuss another way of representing a function, again using an infinite number of terms, provided that the function is periodic. A **periodic function** is one for which $f(x+P) = f(x)$, where P is the **period.** We noted in Chapter 10 that all of the trigonometric functions are periodic. There are numerous applied problems that involve periodic functions, including alternating-current voltages, mechanical oscillations, and sound waves. The main focus of this section is to show how periodic functions like these, which often have complicated waveforms, can be expressed as an ***infinite sum of sine and cosine waves***, called a **Fourier series.**

■ Named for the French mathematician and physicist Jean Baptiste Joseph Fourier (1768–1830).

We will begin by discussing how to find a Fourier series for a function with a period of 2π. (In Section 30.7, we will show how to find a Fourier series for a function with a period different than 2π.) Since both $\sin(nx)$ and $\cos(nx)$ repeat every 2π units for integer values of n, a function with a period of 2π can be represented by a series of the following form, where a_n and b_n are constant coefficients.

Fourier Series with Period 2π

$$f(x) = a_0 + a_1\cos x + a_2\cos 2x + \cdots + a_n\cos nx + \cdots$$
$$+ b_1\sin x + b_2\sin 2x + \cdots + b_n\sin nx + \cdots \quad \textbf{(30.17)}$$

Although we will not show the derivation (it is provided online at the URL in the margin), the coefficients can be shown to be given by the following integrals.

■ Derivations for the coefficients given in Eqs. (30.18)–(30.20) can be found at goo.gl/YnPs27.

$$a_0 = \frac{1}{2\pi}\int_{-\pi}^{\pi} f(x)\,dx \quad \textbf{(30.18)}$$

$$a_n = \frac{1}{\pi}\int_{-\pi}^{\pi} f(x)\cos nx\,dx \quad \textbf{(30.19)}$$

$$b_n = \frac{1}{\pi}\int_{-\pi}^{\pi} f(x)\sin nx\,dx \quad \textbf{(30.20)}$$

By finding the values of these coefficients and inserting them into Eq. (30.17), we obtain the Fourier series for a given function.

If we use some, but not all, of the terms of the series, we will get an approximation of the function. The more terms we use, the better the approximation will be. In essence, we are *building* a complicated waveform by using the basic ingredients of sine and cosine waves. Fourier series provide us a good approximation over a greater interval than Maclaurin and Taylor series, which are only accurate at x-values near the center of the expansion (0 for Maclaurin and a for Taylor series). With a reasonable number of terms included, a Fourier series can accurately represent a function throughout its entire domain.

The following examples illustrate the method of finding Fourier series for various functions.

EXAMPLE 1 Fourier series for a square wave function

Find the Fourier series for the square wave function

$$f(x) = \begin{cases} -1 & -\pi \le x < 0 \\ 1 & 0 \le x < \pi \end{cases}$$

[Many of the functions we shall expand in Fourier series are discontinuous (not continuous) like this one. See Section 23.1 for a discussion of continuity.]

Because $f(x)$ is defined differently for the intervals of x indicated, ***it requires two integrals for each coefficient:***

using Eq. (30.18)
$$a_0 = \frac{1}{2\pi}\int_{-\pi}^{0}(-1)dx + \frac{1}{2\pi}\int_{0}^{\pi}(1)dx = -\frac{x}{2\pi}\bigg|_{-\pi}^{0} + \frac{x}{2\pi}\bigg|_{0}^{\pi} = -\frac{1}{2} + \frac{1}{2} = 0$$

using Eq. (30.19)
$$a_n = \frac{1}{\pi}\int_{-\pi}^{0}(-1)\cos nx\,dx + \frac{1}{\pi}\int_{0}^{\pi}(1)\cos nx\,dx = -\frac{1}{n\pi}\sin nx\bigg|_{-\pi}^{0} + \frac{1}{n\pi}\sin nx\bigg|_{0}^{\pi} = 0 + 0 = 0$$

for all values of n, since $\sin n\pi = 0$;

using Eq. (30.20) with $n = 1$
$$\begin{aligned} b_1 &= \frac{1}{\pi}\int_{-\pi}^{0}(-1)\sin x\,dx + \frac{1}{\pi}\int_{0}^{\pi}(1)\sin x\,dx = \frac{1}{\pi}\cos x\bigg|_{-\pi}^{0} - \frac{1}{\pi}\cos x\bigg|_{0}^{\pi} \\ &= \frac{1}{\pi}(1+1) - \frac{1}{\pi}(-1-1) = \frac{4}{\pi} \end{aligned}$$

using Eq. (30.20) with $n = 2$
$$\begin{aligned} b_2 &= \frac{1}{\pi}\int_{-\pi}^{0}(-1)\sin 2x\,dx + \frac{1}{\pi}\int_{0}^{\pi}(1)\sin 2x\,dx = \frac{1}{2\pi}\cos 2x\bigg|_{-\pi}^{0} - \frac{1}{2\pi}\cos 2x\bigg|_{0}^{\pi} \\ &= \frac{1}{2\pi}(1-1) - \frac{1}{2\pi}(1-1) = 0 \end{aligned}$$

using Eq. (30.20) with $n = 3$
$$\begin{aligned} b_3 &= \frac{1}{\pi}\int_{-\pi}^{0}(-1)\sin 3x\,dx + \frac{1}{\pi}\int_{0}^{\pi}(1)\sin 3x\,dx = \frac{1}{3\pi}\cos 3x\bigg|_{-\pi}^{0} - \frac{1}{3\pi}\cos 3x\bigg|_{0}^{\pi} \\ &= \frac{1}{3\pi}(1+1) - \frac{1}{3\pi}(-1-1) = \frac{4}{3\pi} \end{aligned}$$

In general, if n is even, $b_n = 0$, and if n is odd, then $b_n = 4/n\pi$. Therefore,

$$f(x) = \frac{4}{\pi}\sin x + \frac{4}{3\pi}\sin 3x + \frac{4}{5\pi}\sin 5x + \cdots = \frac{4}{\pi}\left(\sin x + \frac{1}{3}\sin 3x + \frac{1}{5}\sin 5x + \cdots\right)$$

A graph of the function as defined, and the curve found by using the first three terms of the Fourier series, are shown in Fig. 30.13.

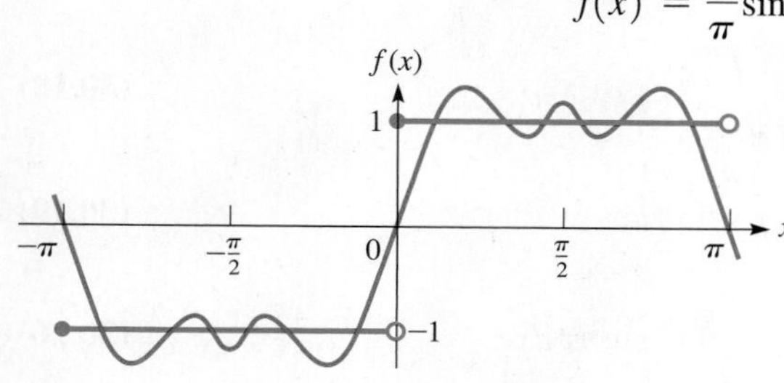

Fig. 30.13

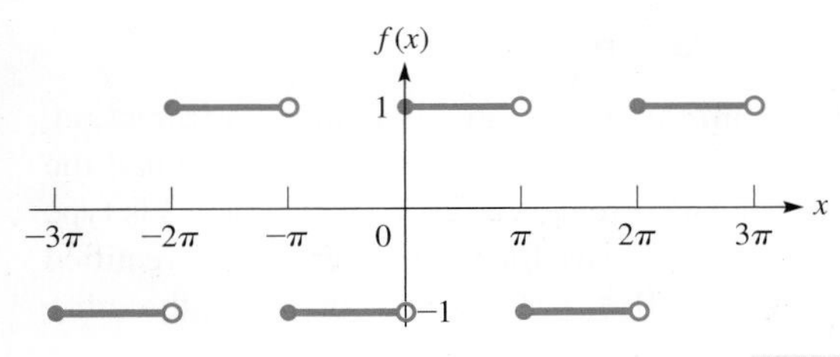

Fig. 30.14

Since functions found by Fourier series have a period of 2π, they can represent functions with this period. If the function $f(x)$ were defined to be periodic with period 2π, with the same definitions as originally indicated, we would graph the function as shown in Fig. 30.14. The Fourier series representation would follow it as in Fig. 30.13. If more terms were used, the fit would be closer. ■

EXAMPLE 2 Finding Fourier series

Find the Fourier series for the function

$$f(x) = \begin{cases} 1 & -\pi \le x < 0 \\ x & 0 \le x < \pi \end{cases}$$

For the periodic function, let $f(x + 2\pi) = f(x)$ for all x.

A graph of three periods of this function is shown in Fig. 30.15.

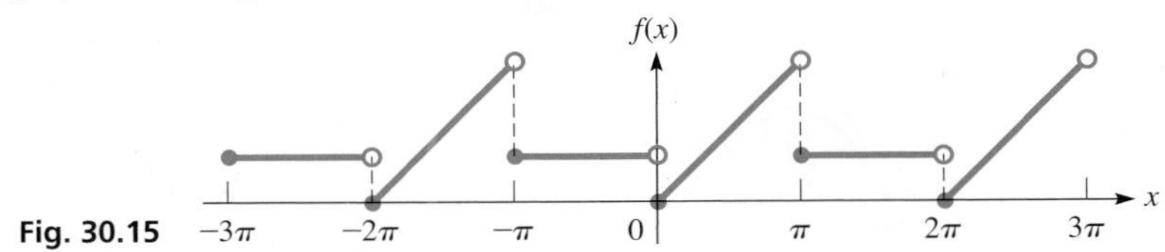

Fig. 30.15

Now, finding the coefficients, we have

$$a_0 = \frac{1}{2\pi}\int_{-\pi}^{0} dx + \frac{1}{2\pi}\int_0^{\pi} x\,dx = \frac{x}{2\pi}\Big|_{-\pi}^{0} + \frac{x^2}{4\pi}\Big|_0^{\pi}$$ using Eq. (30.18)

$$= \frac{1}{2} + \frac{\pi}{4} = \frac{2+\pi}{4}$$

$$a_1 = \frac{1}{\pi}\int_{-\pi}^{0}\cos x\,dx + \frac{1}{\pi}\int_0^{\pi} x\cos x\,dx$$ using Eq. (30.19) with $n = 1$

$$= \frac{1}{\pi}\sin x\Big|_{-\pi}^{0} + \frac{1}{\pi}(\cos x + x\sin x)\Big|_0^{\pi} = -\frac{2}{\pi}$$

$$a_2 = \frac{1}{\pi}\int_{-\pi}^{0}\cos 2x\,dx + \frac{1}{\pi}\int_0^{\pi} x\cos 2x\,dx$$ using Eq. (30.19) with $n = 2$

$$= \frac{1}{2\pi}\sin 2x\Big|_{-\pi}^{0} + \frac{1}{4\pi}(\cos 2x + 2x\sin 2x)\Big|_0^{\pi} = 0$$

$$a_3 = \frac{1}{\pi}\int_{-\pi}^{0}\cos 3x\,dx + \frac{1}{\pi}\int_0^{\pi} x\cos 3x\,dx$$ using Eq. (30.19) with $n = 3$

$$= \frac{1}{3\pi}\sin 3x\Big|_{-\pi}^{0} + \frac{1}{9\pi}(\cos 3x + 3x\sin 3x)\Big|_0^{\pi} = -\frac{2}{9\pi}$$

$$b_1 = \frac{1}{\pi}\int_{-\pi}^{0}\sin x\,dx + \frac{1}{\pi}\int_0^{\pi} x\sin x\,dx$$ using Eq. (30.20) with $n = 1$

$$= -\frac{1}{\pi}\cos x\Big|_{-\pi}^{0} + \frac{1}{\pi}(\sin x - x\cos x)\Big|_0^{\pi} = \frac{\pi - 2}{\pi}$$

$$b_2 = \frac{1}{\pi}\int_{-\pi}^{0}\sin 2x\,dx + \frac{1}{\pi}\int_0^{\pi} x\sin 2x\,dx$$ using Eq. (30.20) with $n = 2$

$$= -\frac{\cos 2x}{2\pi}\Big|_{-\pi}^{0} + \frac{\sin 2x - 2x\cos 2x}{4\pi}\Big|_0^{\pi} = -\frac{1}{2}$$

Therefore, the first few terms of the Fourier series are

$$f(x) = \frac{2+\pi}{4} - \frac{2}{\pi}\cos x - \frac{2}{9\pi}\cos 3x - \cdots + \left(\frac{\pi - 2}{\pi}\right)\sin x - \frac{1}{2}\sin 2x + \cdots$$ ■

Practice Exercise

1. In Example 2, in the definition of $f(x)$, replace 1 with 0 and find a_0.

EXAMPLE 3 Fourier series for a half-wave rectifier

Certain electronic devices allow an electric current to pass through in only one direction. When an alternating current is applied to the circuit, the current exists for only half the cycle. Figure 30.16 is a representation of such a current as a function of time. This type of electronic device is called a *half-wave rectifier*. Derive the Fourier series for a rectified wave for which half is defined by $f(t) = \sin t\ (0 \le t \le \pi)$ and for which the other half is defined by $f(t) = 0$.

In finding the Fourier coefficients, we first find a_0 as

$$a_0 = \frac{1}{2\pi}\int_0^{\pi} \sin t\, dt = \frac{1}{2\pi}(-\cos t)\Big|_0^{\pi} = \frac{1}{2\pi}(1 + 1) = \frac{1}{\pi}$$

In the previous example, we evaluated each of the coefficients individually. Here, we show how to set up a general expression for a_n and another for b_n. Once we have determined these, we can substitute values of n in the formula to obtain the individual coefficients:

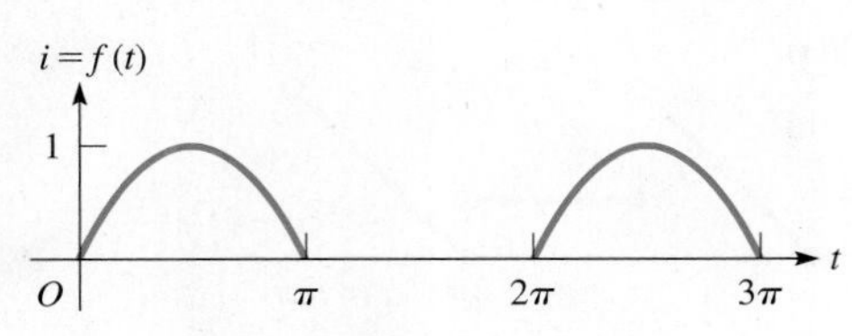

Fig. 30.16

$$a_n = \frac{1}{\pi}\int_0^{\pi} \sin t \cos nt\, dt = -\frac{1}{2\pi}\left[\frac{\cos(1-n)t}{1-n} + \frac{\cos(1+n)t}{1+n}\right]_0^{\pi}$$

$$= -\frac{1}{2\pi}\left[\frac{\cos(1-n)\pi}{1-n} + \frac{\cos(1+n)\pi}{1+n} - \frac{1}{1-n} - \frac{1}{1+n}\right]$$

See Formula 40 in the table of integrals in Appendix D. It is valid for all values of n except $n = 1$. Now, we write

$$a_1 = \frac{1}{\pi}\int_0^{\pi} \sin t \cos t\, dt = \frac{1}{2\pi}\sin^2 t\Big|_0^{\pi} = 0$$

$$a_2 = -\frac{1}{2\pi}\left(\frac{-1}{-1} + \frac{-1}{3} - \frac{1}{-1} - \frac{1}{3}\right) = -\frac{2}{3\pi}$$

TI-89 graphing calculator keystrokes for finding the Fourier series coefficients in Example 3: goo.gl/S1ah4v

$$a_3 = -\frac{1}{2\pi}\left(\frac{1}{-2} + \frac{1}{4} - \frac{1}{-2} - \frac{1}{4}\right) = 0$$

$$a_4 = -\frac{1}{2\pi}\left(\frac{-1}{-3} + \frac{-1}{5} - \frac{1}{-3} - \frac{1}{5}\right) = -\frac{2}{15\pi}$$

$$b_n = \frac{1}{\pi}\int_0^{\pi} \sin t \sin nt\, dt = \frac{1}{2\pi}\left[\frac{\sin(1-n)t}{1-n} - \frac{\sin(1+n)t}{1+n}\right]_0^{\pi}$$

$$= -\frac{1}{2\pi}\left[\frac{\sin(1-n)\pi}{1-n} - \frac{\sin(1+n)\pi}{1+n}\right]$$

See Formula 39 in Appendix D. It is valid for all values of n except $n = 1$.

Therefore, we have

$$b_1 = \frac{1}{\pi}\int_0^{\pi} \sin t \sin t\, dt = \frac{1}{\pi}\int_0^{\pi} \sin^2 t\, dt = \frac{1}{2\pi}(t - \sin t \cos t)\Big|_0^{\pi} = \frac{1}{2}$$

We see that $b_n = 0$ if $n > 1$, since each is evaluated in terms of the sine of a multiple of π.

Therefore, the Fourier series for the rectified wave is

$$f(t) = \frac{1}{\pi} + \frac{1}{2}\sin t - \frac{2}{\pi}\left(\frac{1}{3}\cos 2t + \frac{1}{15}\cos 4t + \cdots\right)$$

The graph of these terms of the Fourier series is shown in the calculator display in Fig. 30.17. ■

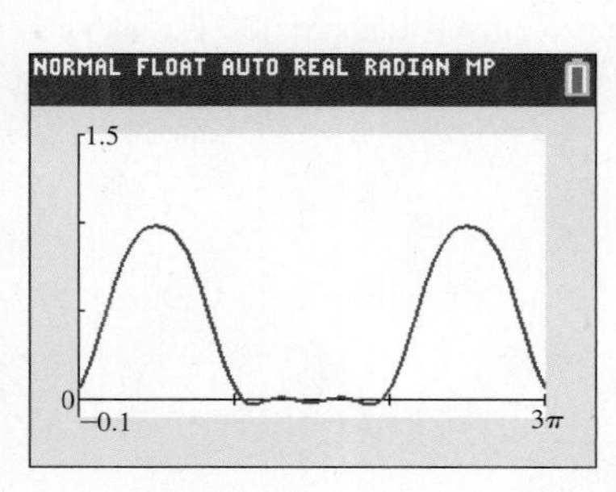

Fig. 30.17

All the types of periodic functions included in this section (as well as many others) may actually be seen on an oscilloscope when the proper signal is set into it. In this way, the oscilloscope may be used to analyze the periodic nature of such phenomena as sound waves and electric currents.

EXERCISES 30.6

In Exercises 1 and 2, make the given changes in Example 1 of this section and then find the resulting Fourier series.

1. Change the -1 to -2, and the 1 to 2.
2. Change the -1 to 0.

In Exercises 3–14, find at least three nonzero terms (including a_0 and at least two cosine terms and two sine terms if they are not all zero) of the Fourier series for the given functions, and sketch at least three periods of the function.

3. $f(x) = \begin{cases} 1 & -\pi \le x < 0 \\ 0 & 0 \le x < \pi \end{cases}$

4. $f(x) = \begin{cases} 0 & -\pi \le x < -\frac{\pi}{2}, \frac{\pi}{2} \le x < \pi \\ 2 & -\frac{\pi}{2} \le x < \frac{\pi}{2} \end{cases}$

5. $f(x) = \begin{cases} 1 & -\pi \le x < 0 \\ 2 & 0 \le x < \pi \end{cases}$

6. $f(x) = \begin{cases} 0 & -\pi \le x < 0, \dfrac{\pi}{2} < x < \pi \\ 1 & 0 \le x \le \dfrac{\pi}{2} \end{cases}$

7. $f(x) = \begin{cases} 0 & -\pi \le x < 0 \\ x & 0 \le x < \pi \end{cases}$

8. $f(x) = x \quad -\pi \le x < \pi$

9. $f(x) = \begin{cases} -1 & -\pi \le x < 0 \\ 0 & 0 \le x < \dfrac{\pi}{2} \\ 1 & \dfrac{\pi}{2} \le x < \pi \end{cases}$

10. $f(x) = x^2 \quad -\pi \le x < \pi$

11. $f(x) = \begin{cases} -x & -\pi \le x < 0 \\ x & 0 \le x < \pi \end{cases}$

12. $f(x) = \begin{cases} 0 & -\pi \le x < 0 \\ x^2 & 0 \le x < \pi \end{cases}$

13. $f(x) = e^x \quad -\pi \le x < \pi$

14. $f(x) = \begin{cases} \pi + x & -\pi \le x < 0 \\ \pi - x & 0 < x < \pi \end{cases}$

In Exercises 15–20, use a graphing calculator to display the terms of the Fourier series given in the indicated example or answer for the indicated exercise. Compare with the sketch of the function. For each calculator display, use Xmin = −8 and Xmax = 8.

15. Example 1
16. Example 2
17. Exercise 5
18. Exercise 7
19. Exercise 11
20. Exercise 10

In Exercises 21–24, solve the given problems.

21. The periodic force F (in N) applied in testing a spring system can be represented by $F = 0$ for $-\pi \le t < 0$ and $F = t^2 + t$ for $0 < t < \pi$, where the time t is in seconds. Find the Fourier series that represents this force.

22. Another representation for a half-wave rectifier (see Example 3) is $f(t) = \cos t\ (-\pi/2 \le t < \pi/2), f(t) = 0$ $(-\pi < t < -\pi/2, \pi/2 < t < \pi)$. Find the Fourier series for this half-wave rectifier.

23. Find the Fourier expansion of the electronic device known as a *full-wave rectifier.* This is found by using as the function for the current $f(t) = -\sin t$ for $-\pi \le t \le 0$ and $f(t) = \sin t$ for $0 < t \le \pi$. The graph of this function is shown in Fig. 30.18. The portion of the curve to the left of the origin is dashed because from a physical point of view we can give no significance to this part of the wave, although mathematically we can derive the proper form of the Fourier expansion by using it.

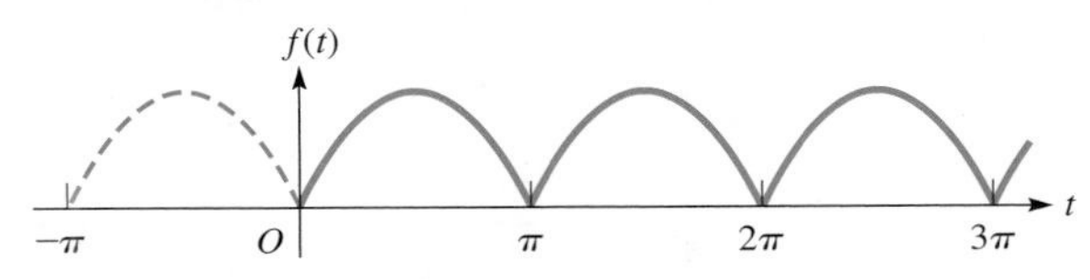

Fig. 30.18

24. The loudness L (in decibels) of a certain siren as a function of time t (in s) can be described by the function

$$\begin{aligned} L &= 0 & -\pi &\le t < 0 \\ L &= 100t & 0 &\le t < \pi/2 \\ L &= 100(\pi - t) & \pi/2 &\le t < \pi \end{aligned}$$

with a period of 2π seconds (where only positive values of t have physical significance). Find a_0, the first nonzero cosine term, and the first two nonzero sine terms of the Fourier expansion for the loudness of the siren. See Fig. 30.19.

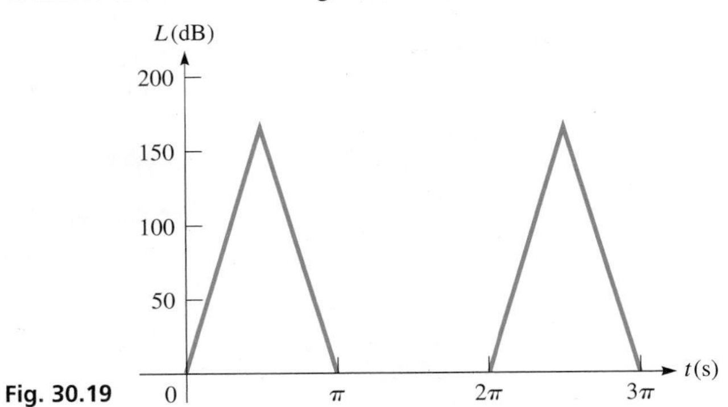

Fig. 30.19

Answer to Practice Exercise

1. $a_0 = \dfrac{\pi}{4}$

30.7 More About Fourier Series

Even and Odd Functions • Constant Added to a Fourier Series • Fourier Series with Period 2L • Half-range Expansions

When finding the Fourier expansion of some functions, it may turn out that all the sine terms evaluate to be zero or that all the cosine terms evaluate to be zero. In fact, in Example 1 of Section 30.6, we see that all of the cosine terms were zero and that the expansion for the square wave contained only sine terms. We now show how to quickly determine if an expansion will contain only sine terms, or only cosine terms.

EVEN FUNCTIONS AND ODD FUNCTIONS

In Chapter 21 (Section 21.3), we showed that when $-x$ replaces x in a function $f(x)$, and the function does not change, the curve of the function is symmetric to the y-axis. *Such a function is called an* **even function.**

EXAMPLE 1 $\cos x$ is an even function

We can show that the function $y = \cos x$ is an even function by using the Maclaurin expansions for $\cos x$ and $\cos(-x)$. These are

$$\cos x = 1 - \frac{x^2}{2} + \frac{x^4}{24} - \cdots$$

$$\cos(-x) = 1 - \frac{(-x)^2}{2} + \frac{(-x)^4}{24} \cdots = 1 - \frac{x^2}{2} + \frac{x^4}{24} - \cdots$$

Because the expansions are the same, $\cos x$ is an even function. ■

NOTE ▸ [Because $\cos x$ is an even function and all of its terms are even functions, it follows that an ***even function*** will have a Fourier series that contains only ***cosine*** terms (and possibly a constant term).]

EXAMPLE 2 Fourier series shows $f(x)$ is an even function

The Fourier series for the function

$$f(x) = \begin{cases} 0 & -\pi \le x < -\pi/2,\ \pi/2 \le x < \pi \\ 1 & -\pi/2 \le x < \pi/2 \end{cases}$$

is $f(x) = \frac{1}{2} + \frac{2}{\pi}\left(\cos x - \frac{1}{3}\cos 3x + \frac{1}{5}\cos 5x - \cdots\right)$. We see that $f(x) = f(-x)$, which means it is an even function. We also see that its Fourier series expansion contains only cosine terms (and a constant). Thus, *when finding the Fourier series, we do not have to find any sine terms.* The graph of $f(x)$ in Fig. 30.20 shows its symmetry to the y-axis. ■

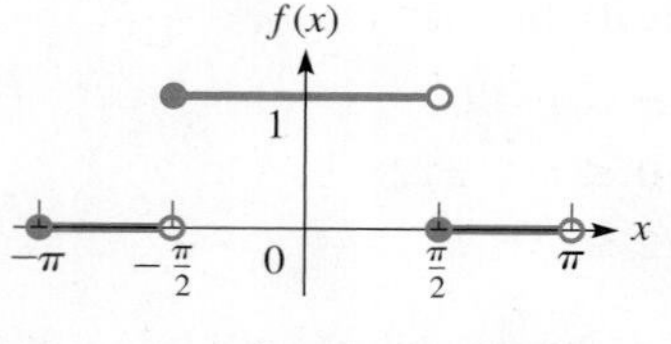

Fig. 30.20

Again referring to Chapter 21, we recall that if $-x$ replaces x and $-y$ replaces y at the same time, and the function does not change, then the function is symmetric to the origin. *Such a function is called an* **odd function.**

EXAMPLE 3 $\sin x$ is an odd function

We can show that the function $y = \sin x$ is an odd function by using the Maclaurin expansions for $\sin x$ and $-\sin(-x)$ [the $-$ sign before $\sin(-x)$ is equivalent to making y negative]. These are

$$\sin x = x - \frac{x^3}{6} + \frac{x^5}{120} - \cdots$$

$$-\sin(-x) = -\left[(-x) - \frac{(-x)^3}{6} + \frac{(-x)^5}{120} - \cdots\right] = x - \frac{x^3}{6} + \frac{x^5}{120} - \cdots$$

Because $\sin x = -\sin(-x)$, $\sin x$ is an odd function. ■

Practice Exercise

Determine whether the following functions are even or odd or neither:

1. $f(x) = x^{1/3}$

2. $f(x) = \sin^2 x$

NOTE ➧ [Because $\sin x$ is an odd function and all its terms are odd functions, it follows that an ***odd function*** will have a Fourier series that contains only ***sine*** terms (and no constant term).]

EXAMPLE 4 Fourier series shows $f(x)$ is an odd function

As we showed in Example 1 of Section 30.6, the Fourier series for the function

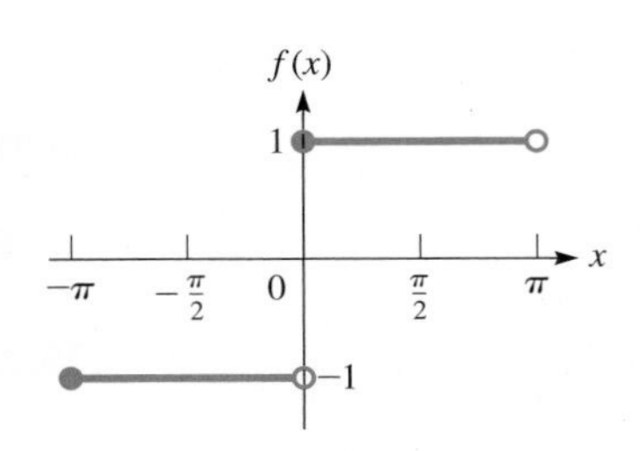

Fig. 30.21

$$f(x) = \begin{cases} -1 & -\pi \le x < 0 \\ 1 & 0 \le x < \pi \end{cases}$$

is $f(x) = \dfrac{4}{\pi}\left(\sin x + \dfrac{1}{3}\sin 3x + \dfrac{1}{5}\sin 5x + \cdots\right)$. We can see that $f(-x) = -f(-x)$, which means $f(x)$ is an odd function. We also see that its Fourier series expansion contains only sine terms. Therefore, *when finding the Fourier series, we do not have to find any cosine terms.* The graph of $f(x)$ in Fig. 30.21 shows its symmetry to the origin. ■

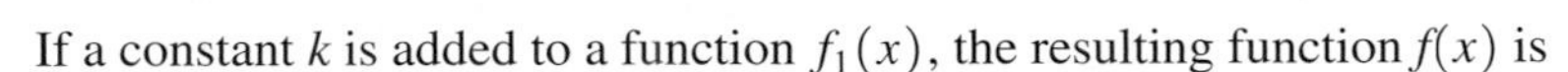

If a constant k is added to a function $f_1(x)$, the resulting function $f(x)$ is

$$f(x) = k + f_1(x)$$

NOTE ➧ [Therefore, if we know the Fourier series expansion for $f_1(x)$, the Fourier series expansion of $f(x)$ is found by adding k to the Fourier series expansion of $f_1(x)$.]

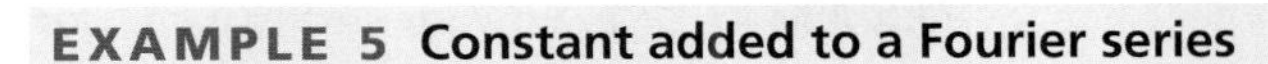

EXAMPLE 5 Constant added to a Fourier series

The values of the function

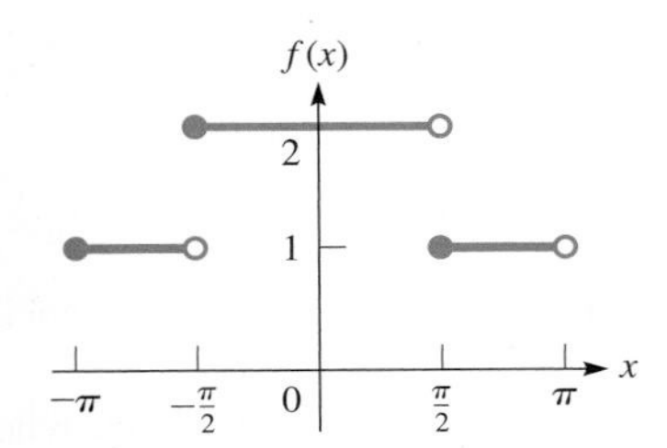

Fig. 30.22

$$f(x) = \begin{cases} 1 & -\pi \le x < -\pi/2,\ \pi/2 \le x < \pi \\ 2 & -\pi/2 \le x < \pi/2 \end{cases}$$

are all 1 greater than those of the function of Example 2. Therefore, denoting the function of Example 2 as $f_1(x)$, we have $f(x) = 1 + f_1(x)$. This means that the Fourier series for $f(x)$ is

$$\begin{aligned} f(x) &= 1 + \left[\frac{1}{2} + \frac{2}{\pi}\left(\cos x - \frac{1}{3}\cos 3x + \frac{1}{5}\cos 5x - \cdots\right)\right] \\ &= \frac{3}{2} + \frac{2}{\pi}\left(\cos x - \frac{1}{3}\cos 3x + \frac{1}{5}\cos 5x - \cdots\right) \end{aligned}$$

In Fig. 30.22, we see that the graph of $f(x)$ is shifted up vertically by 1 unit from the graph of $f_1(x)$ in Fig. 30.20. This is equivalent to a vertical translation of axes. We also note that $f(x)$ is an even function. ■

EXAMPLE 6 Constant subtracted from a Fourier series

The values of the function

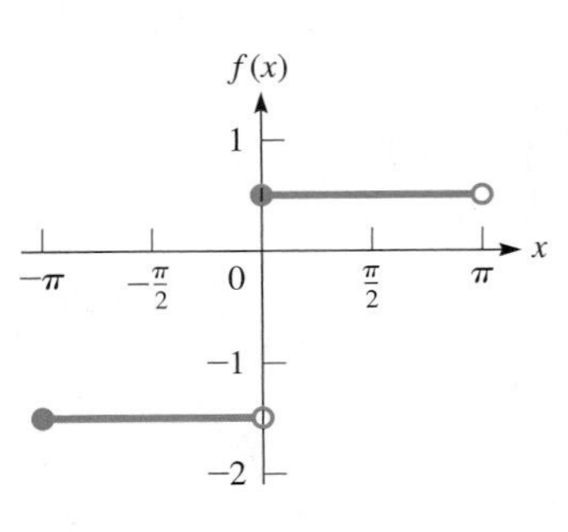

Fig. 30.23

$$f(x) = \begin{cases} -\frac{3}{2} & -\pi \le x < 0 \\ \frac{1}{2} & 0 \le x < \pi \end{cases}$$

are all $\frac{1}{2}$ less than those of the function of Example 4. Therefore, denoting the function of Example 4 as $f_1(x)$, we have $f(x) = -\frac{1}{2} + f_1(x)$. This means that the Fourier series for $f(x)$ is

$$f(x) = -\frac{1}{2} + \frac{4}{\pi}\left(\sin x - \frac{1}{3}\sin 3x + \frac{1}{5}\sin 5x - \cdots\right)$$

In Fig. 30.23, we see that the graph of $f(x)$ is shifted vertically down by $\frac{1}{2}$ unit from the graph of $f_1(x)$ in Fig. 30.21. Although $f(x)$ is not an odd function, it would be an odd function if the origin were translated to $(0, -\frac{1}{2})$. ■

FOURIER SERIES WITH PERIOD $2L$

To this point, we have discussed how to find a Fourier series for a function with a period of 2π, defined over the interval $x = -\pi$ to $x = \pi$. We now show how to find such a series for a function with a period of $2L$, defined from $x = -L$ to $x = L$. Since functions of the form $\sin(n\pi x/L)$ and $\cos(n\pi x/L)$ repeat every $2L$ units, the Fourier series will consist of these types of terms as shown below:

Fourier Series with Period $2L$

$$f(x) = a_0 + a_1\cos(\pi x/L) + a_2\cos(2\pi x/L) + \cdots + a_n\cos(n\pi x/L) + \cdots$$
$$+ b_1\sin(\pi x/L) + b_2\sin(2\pi x/L) + \cdots + b_n\sin(n\pi x/L) + \cdots \quad \textbf{(30.21)}$$

$$a_0 = \frac{1}{2L}\int_{-L}^{L} f(x)\,dx \quad \textbf{(30.22)}$$

$$a_n = \frac{1}{L}\int_{-L}^{L} f(x)\cos\frac{n\pi x}{L}\,dx \quad \textbf{(30.23)}$$

$$b_n = \frac{1}{L}\int_{-L}^{L} f(x)\sin\frac{n\pi x}{L}\,dx \quad \textbf{(30.24)}$$

EXAMPLE 7 Fourier series with period of 8—synthesizer square wave

A musical note (B, octave 2) played on a synthesizer has a square wave that is given approximately by the function

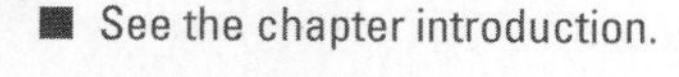

■ See the chapter introduction.

$$f(t) = \begin{cases} 0 & -4 \le t < 0 \\ 2 & 0 \le t < 4 \end{cases}$$

where t is in ms and the period is 8 ms. See Fig. 30.24. Find the Fourier series expansion of this function.

Because the period is 8 ms, $L = 4$ ms. Next, we note that $f(t) = 1 + f_1(t)$, where $f_1(t)$ is an odd function [from the definition of $f(t)$, and from Fig. 30.24 we can see the symmetry to the point (0,1)]. Therefore, *the constant is* 1 *and there are no cosine terms* in the Fourier series for $f(t)$. Now, finding the sine terms, we have

$$b_n = \frac{1}{4}\int_{-4}^{0} 0\sin\frac{n\pi t}{4}\,dt + \frac{1}{4}\int_{0}^{4} 2\sin\frac{n\pi t}{4}\,dt \qquad \text{using Eq. (30.24)}$$

$$= \frac{1}{2}\left(\frac{4}{n\pi}\right)\int_{0}^{4}\sin\frac{n\pi t}{4}\left(\frac{n\pi\,dt}{4}\right) = -\frac{2}{n\pi}\cos\frac{n\pi t}{4}\bigg|_{0}^{4}$$

$$= -\frac{2}{n\pi}(\cos n\pi - \cos 0) = \frac{2}{n\pi}(1 - \cos n\pi)$$

$$b_1 = \frac{2}{\pi}[1 - (-1)] = \frac{4}{\pi} \qquad b_2 = \frac{2}{2\pi}(1 - 1) = 0$$

$$b_3 = \frac{2}{3\pi}[1 - (-1)] = \frac{4}{3\pi} \qquad b_4 = \frac{2}{4\pi}(1 - 1) = 0$$

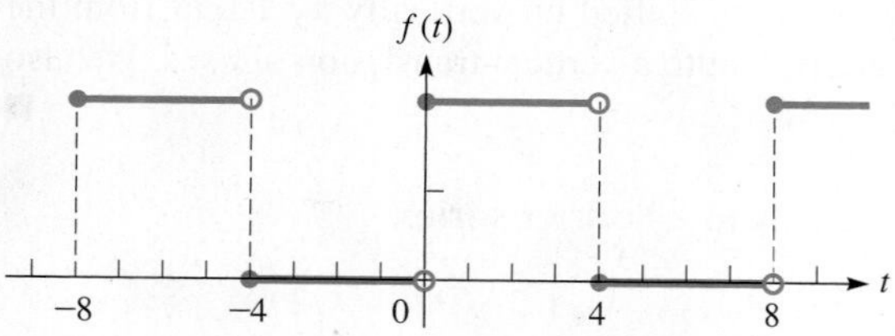

Fig. 30.24

Therefore, the Fourier series is

$$f(t) = 1 + \frac{4}{\pi}\sin\frac{\pi t}{4} + \frac{4}{3\pi}\sin\frac{3\pi t}{4} + \cdots$$

■

EXAMPLE 8 Fourier series with period of 2

Find the Fourier series for the function

$$f(x) = x^2 \quad -1 \le x < 1$$

for which the period is 2. See Fig. 30.25.

Fig. 30.25

Because the period is 2, $L = 1$. Next, we note that $f(x) = f(-x)$, which means it is an even function. Therefore, there are no sine terms in the Fourier series. Finding the constant and the cosine terms, we have

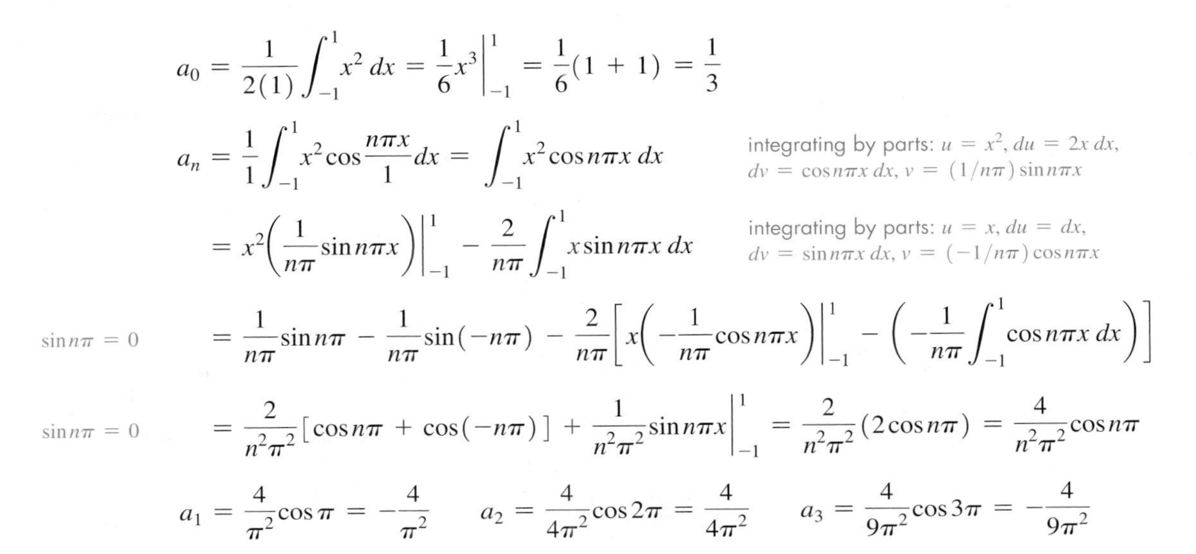

$$a_0 = \frac{1}{2(1)}\int_{-1}^{1} x^2\,dx = \frac{1}{6}x^3\Big|_{-1}^{1} = \frac{1}{6}(1+1) = \frac{1}{3}$$

$$a_n = \frac{1}{1}\int_{-1}^{1} x^2 \cos\frac{n\pi x}{1}\,dx = \int_{-1}^{1} x^2 \cos n\pi x\,dx$$

integrating by parts: $u = x^2$, $du = 2x\,dx$, $dv = \cos n\pi x\,dx$, $v = (1/n\pi)\sin n\pi x$

$$= x^2\left(\frac{1}{n\pi}\sin n\pi x\right)\Big|_{-1}^{1} - \frac{2}{n\pi}\int_{-1}^{1} x \sin n\pi x\,dx$$

integrating by parts: $u = x$, $du = dx$, $dv = \sin n\pi x\,dx$, $v = (-1/n\pi)\cos n\pi x$

$\sin n\pi = 0$

$$= \frac{1}{n\pi}\sin n\pi - \frac{1}{n\pi}\sin(-n\pi) - \frac{2}{n\pi}\left[x\left(-\frac{1}{n\pi}\cos n\pi x\right)\Big|_{-1}^{1} - \left(-\frac{1}{n\pi}\int_{-1}^{1}\cos n\pi x\,dx\right)\right]$$

$\sin n\pi = 0$

$$= \frac{2}{n^2\pi^2}[\cos n\pi + \cos(-n\pi)] + \frac{1}{n^2\pi^2}\sin n\pi x\Big|_{-1}^{1} = \frac{2}{n^2\pi^2}(2\cos n\pi) = \frac{4}{n^2\pi^2}\cos n\pi$$

$$a_1 = \frac{4}{\pi^2}\cos\pi = -\frac{4}{\pi^2} \qquad a_2 = \frac{4}{4\pi^2}\cos 2\pi = \frac{4}{4\pi^2} \qquad a_3 = \frac{4}{9\pi^2}\cos 3\pi = -\frac{4}{9\pi^2}$$

Therefore, the Fourier series is

$$f(x) = \frac{1}{3} - \frac{4}{\pi^2}\left(\cos\pi x - \frac{1}{4}\cos 2\pi x + \frac{1}{9}\cos 3\pi x - \cdots\right)$$ ■

Practice Exercise

3. Find the Fourier series for the function $f(x) = x^2 + 2 \quad -1 \le x < 1$.

HALF-RANGE EXPANSIONS

We have seen that the Fourier series expansion for an even function contains only cosine terms (and possibly a constant), and the expansion of an odd function contains only sine terms. It is also possible to specify a function to be even or odd, such that the expansion will contain only cosine terms or only sine terms.

Considering the symmetry of an even function, the area under the curve from $-L$ to 0 is the same as the area under the curve from 0 to L (see Fig. 30.26). This means the value of the integral from $-L$ to 0 equals the value of the integral from 0 to L. Therefore, the value of the integral from $-L$ to L equals twice the value of the integral from 0 to L, or

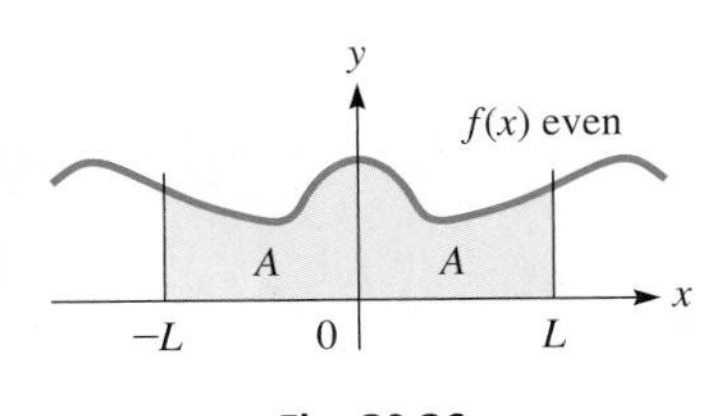

Fig. 30.26

$$\int_{-L}^{L} f(x)dx = 2\int_{0}^{L} f(x)dx \quad f(x)\text{ even}$$

Therefore, to obtain the Fourier coefficients for an expression from $-L$ to L for an even function, we can multiply the coefficients obtained using Eqs. (30.22) and (30.23) from 0 to L by 2. Similar reasoning shows that the Fourier coefficients for an expansion from $-L$ to L for an odd function may be found by multiplying the coefficients obtained using Eq. (30.24) from 0 to L by 2.

NOTE ▸ [A **half-range Fourier cosine series** is a series that contains only cosine terms, and a **half-range Fourier sine series** is a series that contains only sine terms.] To find the half-range expansion for a function $f(x)$, it is defined for interval 0 to L (*half* of the interval from $-L$ to L) and then specified as odd or even, thereby clearly defining the function in the interval from $-L$ to 0. This means that *the Fourier coefficients for a half-range cosine series are given by*

$$a_0 = \frac{1}{L}\int_0^L f(x)dx \quad \text{and} \quad a_n = \frac{2}{L}\int_0^L f(x)\cos\frac{n\pi x}{L}dx \qquad (n = 1, 2, \dots) \qquad \textbf{(30.25)}$$

Similarly, *the Fourier coefficients for a half-range sine series are given by*

$$b_n = \frac{2}{L}\int_0^L f(x)\sin\frac{n\pi x}{L}dx \qquad (n = 1, 2, \dots) \qquad \textbf{(30.26)}$$

EXAMPLE 9 Half-range cosine series

Find $f(x) = x$ in a half-range cosine series for $0 \leq x < 2$.

Because we are to have a cosine series, we extend the function to be an even function with its graph as shown in Fig. 30.27. The red portion between $x = 0$ and $x = L$ shows the given function as defined, and blue portions show the extension that makes it an even function. Now, by use of Eqs. (30.25), we find the Fourier expansion coefficients, with $L = 2$.

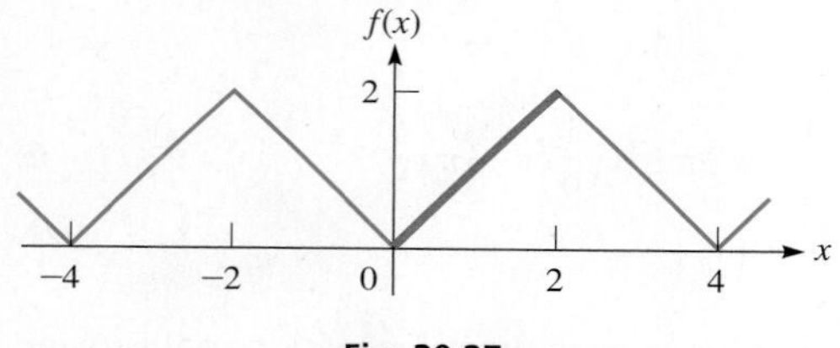

Fig. 30.27

$$a_0 = \frac{1}{2}\int_0^2 x\,dx = \frac{1}{4}x^2\Big|_0^2 = 1$$

$$a_n = \frac{2}{2}\int_0^2 x\cos\frac{n\pi x}{2}dx = x\left(\frac{2}{n\pi}\sin\frac{n\pi x}{2}\right) - \left(\frac{-4}{n^2\pi^2}\cos\frac{n\pi x}{2}\right)\Big|_0^2$$

$$= \frac{4}{n^2\pi^2}(\cos n\pi - 1) \qquad (n \neq 0)$$

If n is even, $\cos n\pi - 1 = 0$. Therefore, we evaluate a_n for the odd values of n, and find the expansion is

$$f(x) = 1 - \frac{8}{\pi^2}\left(\cos\frac{\pi x}{2} + \frac{1}{9}\cos\frac{3\pi x}{2} + \frac{1}{25}\cos\frac{5\pi x}{2} + \cdots\right)$$ ■

EXAMPLE 10 Half-range sine series

Expand $f(x) = x$ in a half-range sine series for $0 \leq x < 2$.

Because we are to have a sine series, we extend the function to be an odd function with its graph as shown in Fig. 30.28. Again, the red portion shows the given function as defined, and blue portions show the extension that makes it an odd function. By using Eq. (30.26), we find the Fourier expansion coefficients, with $L = 2$.

f(x)
2
−2
−4
−2
0
2
4
x

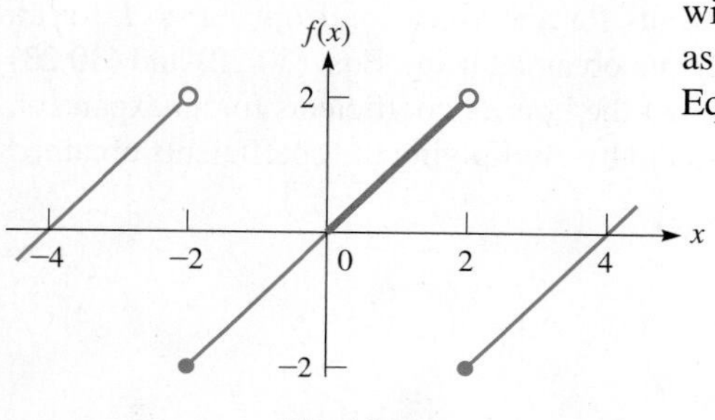

Fig. 30.28

$$b_n = \frac{2}{2}\int_0^2 x\sin\frac{n\pi x}{2}dx$$

$$= x\left(\frac{-2}{n\pi}\cos\frac{n\pi x}{2}\right) - \left(\frac{-4}{n^2\pi^2}\sin\frac{n\pi x}{2}\right)\Big|_0^2 = -\frac{4}{n\pi}\cos n\pi$$

$$f(x) = \frac{4}{\pi}\left(\sin\frac{\pi x}{2} - \frac{1}{2}\sin\pi x + \frac{1}{3}\sin\frac{3\pi x}{2} - \cdots\right)$$ ■

EXERCISES 30.7

In Exercises 1–4, write the Fourier series for each function by comparing it to an appropriate function given in an example of this section. Do not use any of the formulas for a_0, a_n, or b_n.

1. $f(x) = \begin{cases} 2 & -\pi \le x < -\pi/2, \pi/2 \le x < \pi \\ 3 & -\pi/2 \le x < \pi/2 \end{cases}$

2. $f(x) = \begin{cases} -\frac{1}{2} & -\pi \le x < 0 \\ \frac{3}{2} & 0 \le x < \pi \end{cases}$

3. $f(x) = \begin{cases} -2 & -4 \le x < 0 \\ 0 & 0 \le x < 4 \end{cases}$

4. $f(x) = \begin{cases} -\frac{1}{3} & -\pi \le x < -\pi/2, \pi/2 \le x < \pi \\ \frac{2}{3} & -\pi/2 \le x < \pi/2 \end{cases}$

In Exercises 5–12, determine whether the given function is even, odd, or neither. One period is defined for each function.

5. $f(x) = \begin{cases} 5 & -3 \le x < 0 \\ 0 & 0 \le x < 3 \end{cases}$ **6.** $f(x) = \begin{cases} -1 & -2 \le x < 0 \\ 1 & 0 \le x < 2 \end{cases}$

7. $f(x) = \begin{cases} 2 & -1 \le x < 1 \\ 0 & -2 \le x < -1, 1 \le x < 2 \end{cases}$

8. $f(x) = \begin{cases} 0 & -2 \le x < 0, 1 \le x < 2 \\ 1 & 0 \le x < 1 \end{cases}$

9. $f(x) = |x| \quad -4 \le x < 4$ **10.** $f(x) = \begin{cases} 0 & -1 \le x < 0 \\ e^x & 0 \le x < 1 \end{cases}$

11. $f(x) = x^2 \sin x \quad -3 \le x < 3$

12. $f(x) = x\cos 2x \quad -4 \le x < 4$

In Exercises 13–16, determine whether the Fourier series of the given functions will include only sine terms, only cosine terms, or both sine terms and cosine terms.

13. $f(x) = 2 - x \quad -4 \le x < 4$

14. $f(x) = \cos(\sin x) \quad -\pi \le x < \pi$

15. $f(x) = \begin{cases} 0 & -\pi \le x < 0 \\ \cos x & 0 \le x < \pi \end{cases}$

16. $f(x) = \begin{cases} -3 & -3 \le x < 0 \\ 0 & 0 \le x < 3 \end{cases}$

In Exercises 17–22, find at least three nonzero terms (including a_0 and at least two cosine terms and two sine terms if they are not all zero) of the Fourier series for the function from the indicated exercise of this section. Sketch at least three periods of the function.

17. Exercise 5 **18.** Exercise 6 **19.** Exercise 7

20. Exercise 8 **21.** Exercise 9 **22.** Exercise 10

In Exercises 23–28, solve the given problems.

23. Expand $f(x) = 1$ in a half-range sine series for $0 \le x < 4$.

24. Expand $f(x) = 1\ (0 \le x < 2)$, $f(x) = 0\ (2 \le x < 4)$ in a half-range cosine series for $0 \le x < 4$.

25. Expand $f(x) = x^2$ in a half-range cosine series for $0 \le x < 2$.

26. Expand $f(x) = x^2$ in a half-range sine series for $0 \le x < 2$.

27. Each pulse of a pulsating force F of a pressing machine is 8 N. The force lasts for 1 s, followed by a 3-s pause. Thus, it can be represented by $F = 0$ for $-2 \le t < 0$ and $1 \le t < 2$, and $F = 8$ for $0 \le t < 1$, with a period of 4 s (only positive values of t have physical significance). Find the Fourier series for the force.

28. A pulsating electric current i (in mA) with a period of 2 s can be described by $i = e^{-t}$ for $-1 \le t < 1$ s for one period (only positive values of t have physical significance). Find the Fourier series that represents this current.

Answers to Practice Exercises

1. odd **2.** even **3.** $f(x) = \frac{7}{3} - \frac{4}{\pi^2}(\cos \pi x - \frac{1}{4}\cos 2\pi x + \cdots)$

CHAPTER 30 KEY FORMULAS AND EQUATIONS

Infinite series $$\sum_{n=1}^{\infty} a_n = a_1 + a_2 + a_3 + \cdots + a_n + \cdots \quad (30.1)$$

Sum of series $$S = \lim_{n\to\infty} S_n = \lim_{n\to\infty} \sum_{i=1}^{n} a_i \quad (30.2)$$

Sum of geometric series $$S = \lim_{n\to\infty} S_n = \frac{a_1}{1 - r} \quad (30.3)$$

Power series $$f(x) = a_0 + a_1 x + a_2 x^2 + \cdots + a_n x^n + \cdots \quad (30.4)$$

Maclaurin series $$f(x) = f(0) + f'(0)x + \frac{f''(0)x^2}{2!} + \frac{f'''(0)x^3}{3!} + \cdots + \frac{f^n(0)x^n}{n!} + \cdots \quad (30.5)$$

Special series

$$e^x = 1 + x + \frac{x^2}{2!} + \frac{x^3}{3!} + \cdots \quad (\text{all } x) \tag{30.6}$$

$$\sin x = x - \frac{x^3}{3!} + \frac{x^5}{5!} - \cdots \quad (\text{all } x) \tag{30.7}$$

$$\cos x = 1 - \frac{x^2}{2!} + \frac{x^4}{4!} - \cdots \quad (\text{all } x) \tag{30.8}$$

$$\ln(1 + x) = x - \frac{x^2}{2} + \frac{x^3}{3} - \frac{x^4}{4} + \cdots \quad (|x| < 1) \tag{30.9}$$

$$(1 + x)^n = 1 + nx + \frac{n(n-1)}{2!}x^2 + \cdots \quad (|x| < 1) \tag{30.10}$$

Taylor series

$$f(x) = f(a) + f'(a)(x - a) + \frac{f''(a)(x - a)^2}{2!} + \cdots \tag{30.16}$$

Fourier series with period 2π

$$f(x) = a_0 + a_1\cos x + a_2\cos 2x + \cdots + a_n\cos nx + \cdots + b_1\sin x + b_2\sin 2x + \cdots + b_n\sin nx + \cdots \tag{30.17}$$

$$a_0 = \frac{1}{2\pi}\int_{-\pi}^{\pi} f(x)dx \tag{30.18}$$

$$a_n = \frac{1}{\pi}\int_{-\pi}^{\pi} f(x)\cos nx\, dx \tag{30.19}$$

$$b_n = \frac{1}{\pi}\int_{-\pi}^{\pi} f(x)\sin nx\, dx \tag{30.20}$$

Fourier series with period $2L$

$$f(x) = a_0 + a_1\cos(\pi x/L) + a_2\cos(2\pi x/L) + \cdots + a_n\cos(n\pi x/L) + \cdots + b_1\sin(\pi x/L) + b_2\sin(2\pi x/L) + \cdots + b_n\sin(n\pi x/L) + \cdots \tag{30.21}$$

$$a_0 = \frac{1}{2L}\int_{-L}^{L} f(x)\, dx \tag{30.22}$$

$$a_n = \frac{1}{L}\int_{-L}^{L} f(x)\cos\frac{n\pi x}{L}dx \tag{30.23}$$

$$b_n = \frac{1}{L}\int_{-L}^{L} f(x)\sin\frac{n\pi x}{L}dx \tag{30.24}$$

Half-range expansions

$$a_0 = \frac{1}{L}\int_0^L f(x)dx \quad \text{and} \quad a_n = \frac{2}{L}\int_0^L f(x)\cos\frac{n\pi x}{L}dx \quad (n = 1, 2, \ldots) \tag{30.25}$$

$$b_n = \frac{2}{L}\int_0^L f(x)\sin\frac{n\pi x}{L}dx \quad (n = 1, 2, \ldots) \tag{30.26}$$

CHAPTER 30 REVIEW EXERCISES

CONCEPT CHECK EXERCISES

*Determine each of the following as being either **true** or **false**. If it is false, explain why.*

1. The series $6 + 4 + \frac{8}{3} + \cdots$ is convergent.
2. The sum of the series $1 + \frac{1}{4} + \frac{1}{16} + \cdots$ is $\frac{4}{3}$.
3. $f(x) = e^{2x} = 1 + 2x + 2x^2 + \frac{4}{3}x^3 + \cdots$
4. Using three terms of the Maclaurin series, to four decimal places, $\ln 1.1 = 0.1053$.
5. The Talyor expansion of $f(x)$ with $a = 2$ is $f(x) = f(2) + f'(2)(x - 2) + f''(2)(x - 2) + \cdots$
6. For the Fourier expansion of $f(x)$, $a_0 = \frac{1}{\pi}\int_{-\pi}^{\pi} f(x)dx$

PRACTICE AND APPLICATIONS

In Exercises 7–16, find the first three nonzero terms of the Maclaurin expansion of the given functions.

7. $f(x) = \frac{1}{1 + e^x}$
8. $f(x) = e^{\cos x}$
9. $f(x) = \sin 2x^2$
10. $f(x) = \frac{1}{(1 - x)^2}$
11. $f(x) = (x + 1)^{1/3}$
12. $f(x) = \frac{x^2}{1 + x^2}$
13. $f(x) = \sin^{-1}x$
14. $f(x) = \frac{1}{1 - \sin x}$
15. $f(x) = \cos(a + x)$
16. $f(x) = \ln(a + x)$

In Exercises 17–28, approximate the value of each of the given functions. Use three terms of the appropriate series.

17. $e^{-0.2}$
18. $\ln(1.10)$
19. $\sqrt[3]{1.3}$
20. $\sin 3.5°$
21. 1.086^{-1}
22. 0.9839^{10}
23. $\ln 1.2237^{-1}$
24. $\cos(-0.1376)$
25. $\tan 43.62°$
26. $\sqrt[4]{260}$
27. $\sqrt{148}$
28. $\cos\frac{47\pi}{180}$

In Exercises 29 and 30, evaluate the given integrals by using three terms of the appropriate series.

29. $\int_{0.1}^{0.2} \frac{\cos x}{\sqrt{x}}dx$
30. $\int_{0}^{0.1} \sqrt[3]{1 + x^2}dx$

In Exercises 31–34, find the first three terms of the Taylor expansion for the given function and value of a.

31. $\cos x \quad (a = \pi/3)$
32. $\ln \cos x \quad (a = \pi/4)$
33. $\sin^{-1}x \quad \left(a = \frac{1}{2}\right)$
34. $xe^x \quad (a = -1)$

In Exercises 35–38, write the Fourier series for each function by comparing it to an appropriate function in an example of either Section 30.6 or 30.7. One period is given for each function. Do not use any formulas for a_0, a_n, or b_n.

35. $f(x) = \begin{cases} 0 & -\pi \le x < 0 \\ x - 1 & 0 \le x < \pi \end{cases}$
36. $f(x) = x^2 - 1 \quad -1 \le x < 1$
37. $f(x) = \begin{cases} \pi - 1 & -4 \le x < 0 \\ \pi + 1 & 0 \le x < 4 \end{cases}$
38. $f(x) = \begin{cases} 1 & -\pi \le x < 0 \\ 1 + \sin x & 0 \le x < \pi \end{cases}$

In Exercises 39–42, find at least three nonzero terms (including a_0 and at least two cosine terms and two sine terms if they are not all zero) of the Fourier series for the given function. One period is given for each function. Sketch three periods of the function.

39. $f(x) = \begin{cases} 0 & -\pi \le x < -\pi/2, \pi/2 \le x < \pi \\ 1 & -\pi/2 \le x < \pi/2 \end{cases}$
40. $f(x) = \begin{cases} -x & -\pi \le x < 0 \\ 0 & 0 \le x < \pi \end{cases}$
41. $f(x) = x \quad -2 \le x < 2$
42. $f(x) = \begin{cases} -2 & -3 \le x < 0 \\ 2 & 0 \le x < 3 \end{cases}$

In Exercises 43–80, solve the given problems.

43. Test the series $1000 + 800 + 640 + 512 + \cdots$ for convergence or divergence. If convergent, find its sum.
44. Test the series $1 + 1.1 + 1.21 + 1.331 + \cdots$ for convergence or divergence. If convergent, find its sum.
45. Find the sum of the series $64 + 48 + 36 + 27 + \cdots$.
46. Find the first five partial sums of the series $\sum_{n=1}^{\infty}\frac{n}{3n + 1}$ and determine whether it appears to be convergent or divergent.
47. Express the integration of the indefinite integral $\int \sin(x^2)dx$ as an infinite series.
48. Express the integration of the indefinite integral $\int (1 + x^4)^{-1}dx$ as an infinite series.
49. Find the first three terms of the Taylor series for $f(x) = \tan x$ with $a = \pi/4$.
50. Use the result of Exercise 49 to approximate tan 46°. Compare this with the answer obtained from a calculator.
51. Differentiate the series found in Exercise 49 to find the Taylor expansion for $\sec^2 x$ with $a = \pi/4$.
52. Use the series for $(1 - x^2)^{-1/2}$ to find the Maclaurin series for $\sin^{-1}x$.
53. Use the series expansion for $\cos x$ to evaluate $\lim_{x\to 0}\frac{1 - \cos x}{x^2}$.
54. Find the first three nonzero terms of the Maclaurin expansion of the function $\sin x + x\cos x$ by differentiating the expansion term by term for $x \sin x$.

55. Using the properties of logarithms and Eq. (30.9), find four terms of the Maclaurin expansion of $\ln(1 + x)^4$.

56. By multiplication of series, show that the first two terms of the Maclaurin series for 2 sin x cos x are the same as those of the series for sin $2x$.

57. Find the first four nonzero terms of the expansion for $\sin^2 x$ by using the identity $\sin^2 x = \frac{1}{2}(1 - \cos 2x)$ and the series for cos x.

58. Evaluate the integral $\int_0^1 x \sin x\, dx$ (a) by methods of Chapter 28 and (b) using three terms of the series for sin x. Compare results.

59. Use the first three terms of the Maclaurin series for $e^{\sqrt{x}}$ to approximate $e^{\sqrt{0.1}}$.

60. Show that $\frac{d}{dx}[\ln(1 + x)] = \frac{1}{1 + x}$ by comparing the series expansion of both sides of the equation.

61. Find the first four terms of the power series for ln x by replacing x with $(x - 1)$ in the power series for $\ln(1 + x)$. Compare this with the Taylor series for ln x found by direct expansion with $a = 1$.

62. Show that the Maclaurin expansions for cos x and $\cos(-x)$ are the same.

63. Expand $f(x) = x^2$ in a half-range cosine series for $0 < x \leq 1$.

64. Expand $f(x) = 2 - x$ in a half-range sine series for $0 < x \leq 2$.

65. Calculate $e^{0.9}$ by using four terms of the Maclaurin expansion for e^x. Also, calculate $e^{0.9}$ by using the first three terms of the Taylor expansion in Example 1 of Section 30.5. Compare the accuracy of the values obtained with that found directly on a calculator.

66. Find the volume generated by revolving the region bounded by $y = e^{-x}$, $y = 0$, $x = 0$, and $x = 0.1$ about the y-axis by using three terms of the appropriate series.

67. Find the approximate area between the curve of $y = \frac{x - \sin x}{x^2}$ and the x-axis between $x = 0.1$ and $x = 0.2$.

68. Find the approximate value of the moment of inertia with respect to its axis of the solid generated by revolving the smaller region bounded by $y = \sin x$, $x = 0.3$, and the x-axis about the y-axis. Use two terms of the appropriate series.

69. Find three terms of the Maclaurin series for $\tan^{-1} x$ by integrating the series for $1/(1 + x^2)$, term by term.

70. The current i in an electric circuit containing a resistance R and an inductance L is given by $i = Ie^{-Rt/L}$, where I is the current at $t = 0$. Express i as an infinite series.

71. A piano wire vibrates with a displacement y (in mm) given by $y = 3.2\cos 880\pi t$, where t is the time (in s). Express y as an infinite series.

72. The displacement y (in m) of a water wave as a function of the time t (in s) is $y = 0.5\sin 0.5t - 0.2\sin 0.4t$. Find the first three terms of the Maclaurin series for the displacement.

73. The number N of radioactive nuclei in a radioactive sample is $N = N_0 e^{-\lambda t}$. Here, t is the time, N_0 is the number at $t = 0$, and λ is the *decay constant.* By using four terms of the appropriate series, express the right side of this equation as a polynomial.

74. The vertical displacement y of a mass at the end of a spring is given by $y = \sin 3t - \cos 2t$, where t is the time. By subtraction of series, find the first four nonzero terms of the series for y.

75. The electric potential V at a distance x along a certain surface is given by $V = \ln\frac{1 + x}{1 - x}$. Find the first four terms of the Maclaurin series for V.

76. If a mass M is hung from a spring of mass m, the ratio of the masses is $m/M = k\omega \tan k\omega$, where k is a constant and ω is a measure of the frequency of vibration. By using two terms of the appropriate series, express m/M as a polynomial in terms of ω.

77. In the study of electromagnetic radiation, the expression $\frac{N_0}{1 - e^{-k/T}}$ is used. Here, T is the thermodynamic temperature, and N_0 and k are constants. Show that this expression can be written as $N_0(1 + e^{-k/T} + e^{-2k/T} + \cdots)$. (*Hint:* Let $x = e^{-k/T}$.)

78. In the analysis of reflection from a spherical mirror, it is necessary to express the x-coordinate on the surface shown in Fig. 30.29 in terms of the y-coordinate and the radius R. Using the equation of the semicircle shown, solve for x (note that $x \leq R$). Then express the result as a series. (Note that the first approximation gives a parabolic surface.)

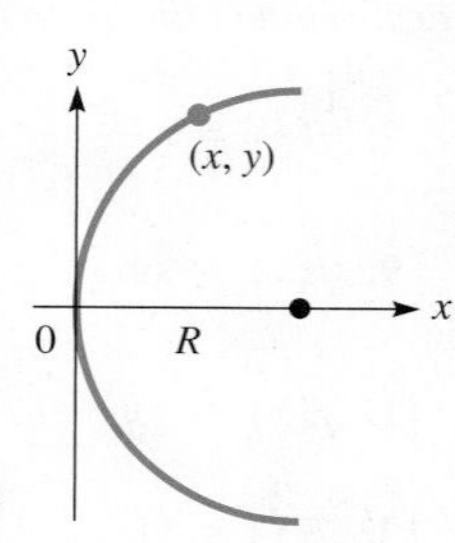

Fig. 30.29

79. A certain electric current is pulsating so that the current as a function of time is given by $f(t) = 0$ if $-\pi \leq t < 0$ and $\pi/2 < t < \pi$. If $0 < t < \pi/2$, $f(t) = \sin t$. Find the Fourier expansion for this pulsating current and sketch three periods.

80. The force F applied to a spring system as a function of the time t is given by $F = t/\pi$ if $0 \leq t \leq \pi$ and $F = 0$ if $\pi < t < 2\pi$. If the period of the force is 2π, find the first few terms of the Fourier series that represents the force.

81. A computer science class is assigned to write a program to find the Maclaurin series for $\sin^2 x$, using only the series for sin x and/or cos x, and any algebraic, trigonometric, or calculus procedures. Write a paragraph or two explaining two different ways this can be done.

CHAPTER 30 PRACTICE TEST

As a study aid, we have included complete solutions for each Practice Test problem at the end of this book.

1. By direct expansion, find the first four nonzero terms of the Maclaurin expansion for $f(x) = (1 + e^x)^2$.

2. Find the first three nonzero terms of the Taylor expansion for $f(x) = \cos x$, with $a = \pi/3$.

3. Evaluate ln 0.96 by using four terms of the expansion for $\ln(1 + x)$.

4. Find the first three nonzero terms of the expansion for $f(x) = \frac{1}{\sqrt{1 - 2x}}$ by using the binomial series.

5. Evaluate $\int_0^1 x \cos x\, dx$ by using three terms of the appropriate series.

6. An electric current is pulsating such that it is a function of the time with a period of 2π. If $f(t) = 2$ for $0 \leq t < \pi$ and $f(t) = 0$ for the other half-cycle, find the first three nonzero terms of the Fourier series for this current.

7. $f(x) = x^2 + 2$ for $-2 \leq x < 2$ (period $= 4$). Is $f(x)$ an even function, an odd function, or neither? If $F(x)$ represents the Fourier series for $f(x)$, express the Fourier series for the function $g(x) = x^2 - 1$ for $-2 \leq x < 2$ (period $= 4$) in terms of $F(x)$. Do *not* use integration to derive specific terms for either series.

Differential Equations

31

Many of the physical problems being studied in the 1700s, such as velocity and light, led to equations that involved derivatives or differentials. These equations are called *differential equations*. Therefore, solving these differential equations became a very important topic of mathematical development during the eighteenth century.

In this chapter, we study some of the basic methods of solving differential equations. Many of these methods were first developed by the famous Swiss mathematician Leonhard Euler (1707–1783). He is undoubtedly the most prolific mathematician of all time, in that his work in mathematics and other fields filled over 75 large volumes.

The final topic covered in this chapter is the solution of differential equations by *Laplace transforms*. They are named for the French mathematician Pierre Laplace (1749–1827). Actually, Laplace had devised a mathematical method in the late 1700s that the English electric engineer Oliver Heaviside (1850–1925) refined and developed into its present useful form in the late 1800s. The Laplace transform is particularly useful for solving problems involving electric circuits and mechanical systems.

Here, we see again that a field of mathematics was developed in response to the need for solving real-life physical problems. Once developed, this type of mathematics was usefully applied 100 years later in electricity, an area of study that did not exist when the method was first devised.

We actually solved a few simple differential equations in earlier chapters when we started a solution with the expression for the slope of a tangent line or the velocity of an object. Also, we have noted the applications of differential equations in electric circuits, mechanical systems, and the study of light. Other areas of application include chemical reactions, interest calculations, changes in pressure and temperature, population growth, forces on beams and structures, and nuclear energy.

LEARNING OUTCOMES

After completion of this chapter, the student should be able to:

- Show that a function is a solution of a differential equation
- Solve first-order differential equations by separation of variables or by recognizing integrating combinations
- Solve first-order linear differential equations using an integrating factor
- Solve first-order differential equations numerically by Euler's method or by the Runge-Kutta method
- Solve homogeneous linear differential equations of higher order
- Solve higher-order nonhomogeneous linear differential equations by the method of undetermined coefficients
- Find the Laplace transform and the inverse Laplace transform of a function
- Solve differential equations using Laplace transforms
- Solve application problems involving differential equations

◀ In Section 31.6, we show how differential equations are used in the study of historical events by using the method of carbon dating.

31.1 Solutions of Differential Equations

Differential Equation • Order • Degree • General Solution • Particular Solution • Show Equation Is a Solution

A **differential equation** *is an equation that contains derivatives or differentials.* Most differential equations we shall consider contain first and/or second derivatives, although some will have higher derivatives. *An equation that contains only first derivatives is called a* **first-order** *differential equation. An equation that contains second derivatives, and possibly first derivatives, is called a* **second-order** *differential equation.* In general, *the* **order** *of the differential equation is that of the highest derivative in the equation, and the* **degree** *is the highest power of that derivative.*

EXAMPLE 1 Order and degree of differential equations

(a) The equation $dy/dx + x = y$ is a first-order differential equation because it contains only a first derivative. It is of the first degree since this derivative is raised to the first power.

(b) The equations $\dfrac{d^2y}{dx^2} + y = 3x^2$ and $\dfrac{d^2y}{dx^2} + 2\dfrac{dy}{dx} = x$ are second-order differential equations because each contains a second derivative and no higher derivatives. They are both of the first degree since the second derivatives are raised to the first power.

(c) The equation $\dfrac{d^2y}{dx^2} + \left(\dfrac{dy}{dx}\right)^4 - y = 6$ is a differential equation of the second order and first degree. That is, the highest derivative that appears is the second, and it is raised to the first power. ■

■ Note carefully that $\dfrac{d^2y}{dx^2}$ is the second derivative. $\left(\dfrac{dy}{dx}\right)^2$ is the square of the first derivative.

In our discussion of differential equations, we will restrict our attention to equations of the first degree.

A **solution** *of a differential equation is a relation between the variables that satisfies the differential equation.* That is, when this relation is substituted into the differential equation, an algebraic identity results. *A solution containing a number of independent arbitrary constants equal to the order of the differential equation is called the* **general solution** *of the equation. When specific values are given to at least one of these constants, the solution is called a* **particular solution.**

EXAMPLE 2 Independent arbitrary constants

Any coefficients that are not specified numerically after like terms have been combined are independent arbitrary constants. In the expression $c_1x + c_2 + c_3x$, there are only two arbitrary constants, since the x-terms may be combined; $c_2 + c_4x$ is an equivalent expression with $c_4 = c_1 + c_3$. ■

EXAMPLE 3 Illustration of a general solution

$y = c_1e^{-x} + c_2e^{2x}$ is the general solution of the differential equation

$$\frac{d^2y}{dx^2} - \frac{dy}{dx} = 2y$$

The order of this differential equation is 2, and there are two independent arbitrary constants in the solution. The equation $y = 4e^{-x}$ is a particular solution. It can be derived from the general solution by letting $c_1 = 4$ and $c_2 = 0$. Each of these solutions can be shown to satisfy the differential equation by taking two derivatives and substituting. ■

To solve a differential equation, we have to find some method of transforming the equation so that each term may be integrated. Some of these methods will be considered after this section. NOTE ▶ [The purpose here is to show that a given equation is a solution of the differential equation by taking the required derivatives and showing that an *identity* results after substitution.]

EXAMPLE 4 Showing an equation is a general solution

Show that $y = c_1 \sin x + c_2 \cos x$ is the general solution of the differential equation $y'' + y' = 0$.

The function and its first two derivatives are

$$y = c_1 \sin x + c_2 \cos x$$
$$y' = c_1 \cos x - c_2 \sin x$$
$$y'' = -c_1 \sin x - c_2 \cos x$$

Substituting these into the differential equation, we have

$$y'' \quad + \quad y \quad = 0$$
$$(-c_1 \sin x - c_2 \cos x) + (c_1 \sin x + c_2 \cos x) = 0 \quad \text{or} \quad 0 = 0$$

We know that this must be the general solution, because there are two independent arbitrary constants and the order of the differential equation is 2. ■

Practice Exercise

1. Show that $y = 2 + ce^{2x}$ is a solution of $y'' - 2y' = 0$. Is it the general solution?

EXAMPLE 5 Showing an equation is a particular solution

Show that $y = 3x + x^2$ is a solution of the differential equation $xy' - y = x^2$.

Taking one derivative and substituting in the differential equation, we have

$$y = 3x + x^2$$

particular solution – no arbitrary constants

$$xy' - y = x^2$$

$$y' = 3 + 2x$$

$$x(3 + 2x) - (3x + x^2) = x^2 \quad \text{or} \quad x^2 = x^2$$ ■

EXERCISES 31.1

In Exercises 1 and 2, show that the indicated solutions are, in fact, solutions of the differential equations in the indicated examples.

1. In Example 3, two solutions are shown for the given differential equation. Show that each is a solution.
2. In Example 4, show that $y_1 = c \sin x + 5 \cos x$ and $y_2 = 2 \sin x - 3 \cos x$ are solutions of the given differential equation.

In Exercises 3–6, determine whether the given equation is the general solution or a particular solution of the given differential equation.

3. $\frac{dy}{dx} + 2xy = 0, \quad y = e^{-x^2}$
4. $y' \ln x - \frac{y}{x} = 0, \quad y = c \ln x$
5. $y'' + 3y' - 4y = 3e^x, \quad y = c_1 e^x + c_2 e^{-4x} + \frac{3}{5}xe^x$
6. $\frac{d^2y}{dx^2} + 4y = 8, \quad y = c \sin 2x + 3 \cos 2x + 2$

In Exercises 7–10, show that each function $y = f(x)$ is a solution of the given differential equation.

7. $\frac{dy}{dx} - y = 1; \quad y = e^x - 1, \quad y = 5e^x - 1$
8. $\frac{dy}{dx} = 2xy^2; \quad y = -\frac{1}{x^2}, \quad y = -\frac{1}{x^2 + c}$
9. $y'' + 4y = 0; \quad y = 3 \cos 2x, \quad y = c_1 \sin 2x + c_2 \cos 2x$
10. $y'' = 2y'; \quad y = 3e^{2x}, \quad y = 2e^{2x} - 5$

In Exercises 11–30, show that the given equation is a solution of the given differential equation.

11. $\frac{dy}{dx} = 2x, \quad y = x^2 + 1$
12. $xy' = 2y, \quad y = cx^2$
13. $\frac{dy}{dx} = 1 - 3x^2, \quad y = 2 + x - x^3$
14. $\frac{dy}{dx} = 3y + 2x, \quad y = ce^{3x} - \frac{2}{3}x - \frac{2}{9}$
15. $y' + 2y = 2x, \quad y = ce^{-2x} + x - \frac{1}{2}$
16. $y'' = 6x + 2, \quad y = x^3 + x^2 + c$
17. $y'' + 9y = 4 \cos x, \quad 2y = \cos x$
18. $y' = y \sec x, \quad y = \sec x + \tan x$
19. $x^2y' + y^2 = 0, \quad xy = cx + cy$
20. $xy' - 3y = x^2, \quad y = cx^3 - x^2$
21. $x\frac{d^2y}{dx^2} + \frac{dy}{dx} = 0, \quad y = c_1 \ln x + c_2$
22. $y'' + 4y = 10e^x, \quad y = c_1 \sin 2x + c_2 \cos 2x + 2e^x$
23. $y' + y = 2 \cos x, \quad y = \sin x + \cos x - e^{-x}$
24. $(x + y) - xy' = 0, \quad y = x \ln x - cx$
25. $y'' - 3y' + 2y = 3, \quad y = c_1 e^x + c_2 e^{2x} + 3/2$
26. $xy'' + y' = 16x^3, \quad y = x^4 + c_1 + c_2 \ln x$

27. $\frac{d^3y}{dx^3} = \frac{d^2y}{dx^2}$, $y = c_1 + c_2x + c_3e^x$

28. $2xyy' + x^2 = y^2$, $x^2 + y^2 = cx$

29. $(y')^2 + xy' = y$, $y = cx + c^2$

30. $x^4(y')^2 - xy' = y$, $y = c^2 + \frac{c}{x}$

In Exercises 31–34, determine whether or not each of the given functions is a solution of the differential equation $y'' - 2y' - 3y = -4e^x$.

31. $y = e^x + e^{-x}$

32. $y = e^{3x} + e^{-x}$

33. $y = e^x + e^{2x}$

34. $y = 2e^{-x} + e^x$

In Exercises 35–38, solve the given problems.

35. The general solution of the differential equation $y'' - y'/x = 3x$ is $y = x^3 + c_1x^2 + c_2$. Find the particular solution if the graph of the solution passes through the point $(0, -4)$.

36. Find the particular solution of the differential equation in Exercise 35, if the graph of the solution passes through $(0, -4)$ and $(2, 8)$.

37. A differential equation that arises in the study of radioactivity is $dN/dt = kN$. Show that $N = N_0e^{kt}$ is the general solution.

38. Show that the electric charge $q = 0.01(1 - \cos 316t)$ in a circuit, where t represents time satisfies the equation $\frac{d^2q}{dt^2} + 10^5q = 10^3$.

Answer to Practice Exercise

1. $(4ce^{2x}) - 2(2ce^{2x}) = 0$, No

31.2 Separation of Variables

Separation of Variables • Each Term with Only One Variable • Solutions with Logarithms • Finding Particular Solutions

We will now solve differential equations of the first order and first degree. Of the many methods for solving such equations, a few are presented in this and the next two sections. The first of these is *the method of* **separation of variables.**

A differential equation of the first order and first degree contains the first derivative to the first power. That is, it may be written as $dy/dx = f(x, y)$. This type of equation is more commonly expressed in its differential form,

$$M(x, y)dx + N(x, y)dy = 0 \tag{31.1}$$

where $M(x, y)$ and $N(x, y)$ may represent constants, functions of either x or y, or functions of x and y.

To solve an equation of the form of Eq. (31.1), we must integrate. However, if $M(x, y)$ contains y, the first term cannot be integrated. Also, if $N(x, y)$ contains x, the second term cannot be integrated. If it is possible to rewrite Eq. (31.1) as

$$A(x)dx + B(y)dy = 0 \tag{31.2}$$

where $A(x)$ does not contain y and $B(y)$ does not contain x, then *we may find the solution by integrating each term and adding the constant of integration.* (In rewriting Eq. (31.1), if division is used, the solution is not valid for values that make the divisor zero.) Many differential equations can be solved in this way.

EXAMPLE 1 Separation of variables—divide by x

Solve the differential equation $dx - 4xy^3dy = 0$.

We can write this equation as

$$(1)dx + (-4xy^3)dy = 0$$

which means that $M(x, y) = 1$ and $N(x, y) = -4xy^3$.

We must remove the x from the coefficient of dy ***without introducing y into the coefficient of dx.*** We do this by dividing each term by x, which gives us

$$\frac{dx}{x} - 4y^3\,dy = 0$$

It is now possible to integrate each term. Performing this integration, we have

$$\ln|x| - y^4 = c$$

The constant of integration c becomes the arbitrary constant of the solution. ■

In Example 1, we showed the integration of dx/x as $\ln|x|$, which follows our discussion in Section 28.2. We know $\ln|x| = \ln x$ if $x > 0$ and $\ln|x| = \ln(-x)$ if $x < 0$. NOTE ➧ [Because we know the values being used when we find a particular solution, we generally will not use the absolute value notation when integrating logarithmic forms.] We would show the integration of dx/x as $\ln x$, with the understanding that we know $x > 0$ (as is the case in many applications). When using negative values of x, we would express it as $\ln(-x)$.

EXAMPLE 2 Separation of variables—divide by $y(x^2 + 1)$

Solve the differential equation $xy\,dx + (x^2 + 1)dy = 0$.

In order to integrate each term, it is necessary to divide each term by $y(x^2 + 1)$. When this is done, we have

$$\frac{x\,dx}{x^2 + 1} + \frac{dy}{y} = 0$$

Integrating, we obtain the solution

$$\frac{1}{2}\ln(x^2 + 1) + \ln y = c$$

■ For reference, Eq. (13.9) is $\log_b x^n = n\log_b x$ and Eq. (13.7) is $\log_b xy = \log_b x + \log_b y$.

It is possible to make use of the properties of logarithms to make the form of this solution neater. If we write the constant of integration as $\ln c_1$, rather than c, we have $\frac{1}{2}\ln(x^2 + 1) + \ln y = \ln c_1$. Multiplying through by 2 and using the property of logarithms given by Eq. (13.9), we have $\ln(x^2 + 1) + \ln y^2 = \ln c_1^2$. Next, using the property of logarithms given by Eq. (13.7), we then have $\ln(x^2 + 1)y^2 = \ln c_1^2$, which means

$$(x^2 + 1)y^2 = c_1^2$$

This form of the solution is more compact and generally would be preferred. NOTE ➧ [However, any expression that represents a constant may be chosen as the constant of integration and will lead to a correct solution.] In checking answers, we must remember that a different choice of constant will lead to a different form of the solution. Thus, two different-appearing answers may both be correct. *Often there is more than one reasonable choice of a constant, and different forms of the solution may be expected.* ■

Practice Exercise

1. Find the general solution of the differential equation $dx + 2y\sec x\,dy = 0$.

EXAMPLE 3 Separation of variables—divide by θ and multiply by dt

Solve the differential equation $\dfrac{d\theta}{dt} = \dfrac{\theta}{t^2 + 4}$.

The solution proceeds as follows:

$$\frac{d\theta}{\theta} = \frac{dt}{t^2 + 4} \quad \text{separate variables by multiplying by } dt \text{ and dividing by } \theta$$

$$\ln\theta = \frac{1}{2}\tan^{-1}\frac{t}{2} + \frac{c}{2} \quad \text{integrate}$$

$$2\ln\theta = \tan^{-1}\frac{t}{2} + c$$

$$\ln\theta^2 = \tan^{-1}\frac{t}{2} + c$$

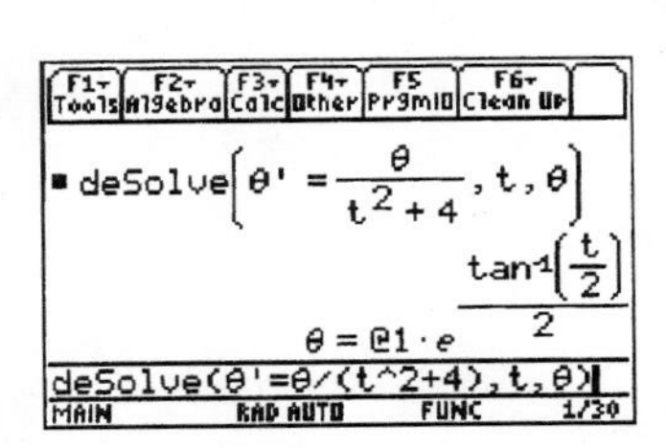

Fig. 31.1
TI-89 graphing calculator keystrokes: goo.gl/yRw6bv

■ Figure 31.1 shows the solution of this differential equation on a calculator. Note that the solution is solved for θ. Also note the symbol used for the constant.

Note the form of the result using $c/2$ as the constant of integration. The form would differ somewhat had we chosen c as the constant.

The choice of $\ln c$ as the constant of integration (on the left) is also reasonable. It would lead to the result $2\ln c\theta = \tan^{-1}(t/2)$. ■

EXAMPLE 4 Separation of variables—divide by $e^x \sin y$

Solve the differential equation $2e^{3x}\sin y\,dx + e^x \csc y\,dy = 0$.

To separate variables, we divide by $e^x \sin y$ to remove $\sin y$ from the first term and e^x from the second term. Using properties of trigonometric and exponential functions, we have

$$\frac{2e^{3x}\sin y\,dx}{e^x \sin y} + \frac{e^x \csc y\,dy}{e^x \sin y} = 0 \quad \text{divide by } e^x \sin y$$

$$2e^{2x}dx + \csc^2 y\,dy = 0 \quad \text{variables separated}$$

$$e^{2x}(2dx) + \csc^2 y\,dy = 0 \quad \text{form for integrating}$$

$$e^{2x} - \cot y = c \quad \text{integrate}$$

■

Practice Exercise

2. In Example 4, find the particular solution if $x = 0$ when $y = \pi/4$.

FINDING PARTICULAR SOLUTIONS

To find a particular solution, we need information that allows us to evaluate the constant of integration. We now show how to find particular solutions, and graphically show the difference between the general solution and a particular solution.

EXAMPLE 5 Finding a particular solution

Solve the equation $(x^2 + 1)^2\,dy + 4x\,dx = 0$, subject to the condition that $x = 1$ when $y = 3$.

$$dy + \frac{4x\,dx}{(x^2+1)^2} = 0 \quad \text{dividing by } (x^2+1)^2$$

$$y - \frac{2}{x^2+1} = c \quad \text{integrating}$$

$$y = \frac{2}{x^2+1} + c \quad \text{general solution}$$

$$3 = \frac{2}{1+1} + c, \quad \text{use } x = 1,\ y = 3 \text{ to evaluate } c;\ c = 2$$

$$y = \frac{2}{x^2+1} + 2 \quad \text{particular solution}$$

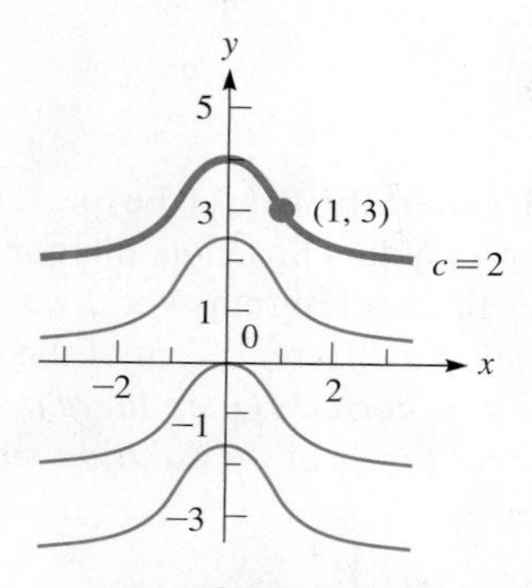

Fig. 31.2

The general solution defines a *family* of curves, one member of the family for each value of c. A few of these curves are shown in Fig. 31.2. When c is specified as in the particular solution, we have the specific (red) curve shown. ■

EXAMPLE 6 Particular solution—form of the constant

Find the particular solution in Example 2, given that $x = 0$ when $y = e$.

Using the solution $\frac{1}{2}\ln(x^2 + 1) + \ln y = c$, we have

$$\tfrac{1}{2}\ln(0+1) + \ln e = c, \qquad \tfrac{1}{2}\ln 1 + 1 = c, \qquad c = 1$$

$$\tfrac{1}{2}\ln(x^2+1) + \ln y = 1 \quad \text{substitute } c = 1$$

$$\ln(x^2+1) + 2\ln y = 2$$

$$\ln y^2(x^2+1) = 2 \quad \text{using properties of logarithms}$$

$$y^2(x^2+1) = e^2 \quad \text{exponential form of particular solution}$$

Using the general solution $(x^2 + 1)y^2 = c_1^2$, we have

$$(0+1)e^2 = c_1^2, \qquad c_1^2 = e^2$$

$$y^2(x^2+1) = e^2$$

NOTE ▸ This is precisely the same solution as above. [This shows that the choice of the form of the constant does not affect the final result, and the constant is truly arbitrary.] ■

EXERCISES 31.2

In Exercises 1 and 2, make the given changes in the indicated examples of this section and then solve the resulting differential equations.

1. In Example 2, change the first term to $2xy\,dx$.

2. In Example 4, change the second term to $e^{2x}\csc y\,dy$.

In Exercises 3–34, solve the given differential equations.

3. $\frac{dy}{dx} = \frac{x}{y^2}$

4. $e^y\frac{dy}{dx} = 4x$

5. $dy = k(20 + y)dt$

6. $\frac{dy}{dt} = \sqrt{3 - t}$

7. $\frac{ds}{dt} = \frac{\cos t}{s - 1}$

8. $\frac{dy}{d\theta} = \frac{2\theta + \sec^2\theta}{2y}$

9. $\frac{dp}{dx} = \sqrt{\frac{p}{x}}$

10. $\frac{1}{x}dy = \frac{e^{x^2}}{y^2}dx$

11. $\frac{dy}{t^2} = ydt$

12. $e^x(y' - 1) = 1$

13. $2xdx + dy = 0$

14. $y^2dy + x^3dx = 0$

15. $y^2dx + dy = 0$

16. $ydt + tdy = 3tydt$

17. $\frac{dV}{dP} = -\frac{V}{P^2}$

18. $\frac{2\,dy}{dx} = \frac{y(x + 1)}{x}$

19. $x^2 + (x^3 + 5)y' = 0$

20. $xyy' + \sqrt{1 + y^2} = 0$

21. $dy + \ln x y dx = (4x + \ln y)dx$

22. $r\sqrt{1 - \theta^2}\frac{dr}{d\theta} = \theta + 4$

23. $e^{x^2}\,dy = x\sqrt{1 - y}dx$

24. $\sqrt{1 + 4x^2}\,dy = y^3xdx$

25. $e^{x+y}\,dx + dy = 0$

26. $e^{2x}\,dy + e^x\,dx = 4\,dx$

27. $y' - y = 4$

28. $9\,ds - s^2\,dt = 9\,dt$

29. $x\frac{dy}{dx} = y^2 + y^2\ln x$

30. $(yx^2 + y)\frac{dy}{dx} = \tan^{-1}x$

31. $y\tan x\,dx + \cos^2 x\,dy = 0$

32. $\sin x \sec y\,dx = dy$

33. $yx^2dx = ydx - x^2dy$

34. $y\sqrt{1 - x^2}dy + 2dx = 0$

In Exercises 35–40, find the particular solution of the given differential equation for the indicated values.

35. $\frac{dy}{dx} + yx^2 = 0$; $x = 0$ when $y = 1$

36. $\frac{ds}{dt} = \sec s$; $t = 0$ when $s = 0$

37. $y' = (1 - y)\cos x$; $x = \pi/6$ when $y = 0$

38. $x\,dy = y\ln y\,dx$; $x = 2$ when $y = e$

39. $y^2e^x\,dx + e^{-x}\,dy = y^2\,dx$; $x = 0$ when $y = 2$

40. $2y\cos y\,dy - \sin y\,dy = y\sin y\,dx$; $x = 0$ when $y = \pi/2$

In Exercises 41–44, solve the given problems.

41. On a wildlife refuge, the deer population grows at a rate of 10% per year due to reproduction. However, approximately 20 deer are hit and killed by cars each year. Therefore, the rate of growth is given by $\frac{dP}{dt} = 0.1P - 20$, where P is the population of deer and t is the time in years. Express P as a function of t.

42. Explain why more than one function can be a solution of a given differential equation. Find three different functions that are solutions of $y' = y$.

43. The temperature reading T (in °F) at time t (in s) of a thermometer initially reading 100°F and then placed in water at 10°F is found by solving the equation $dT + 0.15(T - 10)dt = 0$. Solve for T as a function of t.

44. The depth h (in m) of water that is being drained from a tank changes with time t (in min) according to $dh = -0.0214\sqrt{h}\,dt$. Find h as a function of time if $h = 8.0$ m for $t = 16$ min.

Answers to Practice Exercises

1. $\sin x + y^2 = c$ **2.** $\cot y = e^{2x}$

31.3 Integrating Combinations

Combinations of Differentials Integrated as a Unit • Basic Combinations

For different equations that cannot be solved by separation of variables, other methods have been developed. One is based on the fact that *certain combinations of differentials can be integrated as a unit.* The following differentials show some of the possible combinations:

$$d(xy) = xdy + ydx \tag{31.3}$$

$$d(x^2 + y^2) = 2(xdx + ydy) \tag{31.4}$$

$$d\left(\frac{y}{x}\right) = \frac{xdy - ydx}{x^2} \tag{31.5}$$

$$d\left(\frac{x}{y}\right) = \frac{ydx - xdy}{y^2} \tag{31.6}$$

Equation (31.3) suggests that if the combination $x\,dy + y\,dx$ occurs in a differential equation, we should look for a function of xy as a solution. Equation (31.4) suggests that if the combination $x\,dx + y\,dy$ occurs, we should look for a function of $x^2 + y^2$. Equations (31.5) and (31.6) suggest that if either of the combinations $x\,dy - y\,dx$ or $y\,dx - x\,dy$ occurs, we should look for a function of y/x or x/y.

EXAMPLE 1 Recognizing the differential of xy

Solve the differential equation $x\,dy + y\,dx + xy\,dy = 0$.

By dividing through by xy, we have

$$\frac{x\,dy + y\,dx}{xy} + dy = 0$$

The left term is the differential of xy divided by xy. Thus, it integrates to $\ln xy$.

$$\frac{d(xy)}{xy} + dy = 0$$

for which the solution is

$$\ln xy + y = c$$

■

EXAMPLE 2 Recognizing the differential of y/x

Solve the differential equation $y\,dx - x\,dy + x\,dx = 0$.

The combination of $y\,dx - x\,dy$ suggests that this equation might make use of either Eq. (31.5) or (31.6). This would require dividing through by x^2 or y^2. If we divide by y^2, the last term cannot be integrated, but ***division by x^2 still allows integration of the last term.*** Performing this division, we obtain

$$\frac{y\,dx - x\,dy}{x^2} + \frac{dx}{x} = 0$$

This left combination is the negative of Eq. (31.5). Thus, we have

$$-d\left(\frac{y}{x}\right) + \frac{dx}{x} = 0$$

for which the solution is $-\dfrac{y}{x} + \ln x = c$. ■

Practice Exercise

1. Find the general solution of the differential equation $(4x - y)dx = x\,dy$.

EXAMPLE 3 Two combinations in the same equation

Find the general solution of the differential equation

$$(x^3 + xy^2 + 2y)dx + (y^3 + x^2y + 2x)dy = 0$$

which satisfies the condition that $x = 1$ when $y = 0$.

Regrouping the terms of the equation, we have

$$x(x^2 + y^2)dx + y(x^2 + y^2)dy + 2(y\,dx + x\,dy) = 0$$

Factoring $x^2 + y^2$ from each of the first two terms gives

$$(x^2 + y^2)(x\,dx + y\,dy) + 2(y\,dx + x\,dy) = 0$$

$$\frac{1}{2}(x^2 + y^2)\overbrace{(2x\,dx + 2y\,dy)}^{d(x^2+y^2)} + 2\overbrace{(y\,dx + x\,dy)}^{d(xy)} = 0$$

$$\frac{1}{2}\left(\frac{1}{2}\right)(x^2 + y^2)^2 + 2xy + \frac{c}{4} = 0 \quad \text{integrating}$$

$$(x^2 + y^2)^2 + 8xy + c = 0$$

■

NOTE ▶ [The use of integrating combinations depends on proper recognition of the forms.] It may take two or three arrangements to find the combination that leads to the solution. Of course, many equations cannot be arranged so as to give integrable combinations in all terms.

EXAMPLE 4 Particular solution—differential of $x^2 + y^2$

Find the particular solution of the differential equation $(x^2 + y^2 + x)dx + y\,dy = 0$ that satisfies the condition $x = 1$ when $y = 0$.

Regrouping the terms of this equation, we have

$$(x^2 + y^2)dx + (x\,dx + y\,dy) = 0 \qquad \text{divide each term by } x^2 + y^2$$

$$dx + \frac{x\,dx + y\,dy}{x^2 + y^2} = 0$$

The right term now can be put in the form of du/u (with $u = x^2 + y^2$) by multiplying each of the terms of the numerator by 2. This leads to

$$dx + \left(\frac{1}{2}\right)\frac{2x\,dx + 2y\,dy}{x^2 + y^2} = 0 \qquad d(x^2 + y^2) = 2x\,dx + 2y\,dy$$

$$x + \frac{1}{2}\ln(x^2 + y^2) = \frac{c}{2} \quad \text{or} \quad 2x + \ln(x^2 + y^2) = c$$

Using the condition that $x = 1$ when $y = 0$, $2(1) + \ln(1^2 + 0^2) = c$, or $c = 2$. The particular solution is then $2x + \ln(x^2 + y^2) = 2$. ■

EXERCISES 31.3

In Exercises 1 and 2, make the given changes in the indicated examples of this section and then solve the resulting differential equations.

1. In Example 1, change the third term to $2xy^2\,dy$.
2. In Example 2, change the third term to $2\,dx$.

In Exercises 3–18, solve the given differential equations.

3. $x\,dy + y\,dx + x\,dx = 0$
4. $(2y + t)dy + y\,dt = 0$
5. $y\,dx - x\,dy + x^3dx = 2dx$
6. $x\,dy - y\,dx + y^2dx = 0$
7. $t^3\,dr + rt^2\,dt = t\,dr - r\,dt$
8. $(y + \sec xy)dx + x\,dy = 0$
9. $\sin x\,dy = (1 - y\cos x)dx$
10. $(y^2 + xe^{x/y})dy = ye^{x/y}\,dx$
11. $\sqrt{x^2 + y^2}dx - 2y\,dy = 2x\,dx$
12. $R\,dR + (R^2 + T^2 + T)dT = 0$
13. $\tan(x^2 + y^2)dy + x\,dx + y\,dy = 0$
14. $(x^2 + y^3)^2\,dy + 2x\,dx + 3y^2\,dy = 0$
15. $y\,dy + (y^2 - x^2)dx = x\,dx$ (Explain your solution.)
16. $e^{x+y}(dx + dy) + 4x\,dx = 0$
17. $10x\,dy + 5y\,dx + 3y\,dy = 0$
18. $2(u\,dv + v\,du)\ln uv + 3u^3v\,du = 0$

In Exercises 19–24, find the particular solutions to the given differential equations that satisfy the given conditions.

19. $2(x\,dy + y\,dx) + 3x^2dx = 0$; $x = 1$ when $y = 2$
20. $t\,dt + s\,ds = 2(t^2 + s^2)dt$; $t = 1$ when $s = 0$
21. $y\,dx - x\,dy = y^3\,dx + y^2x\,dy$; $x = 2$ when $y = 4$
22. $e^{x/y}(x\,dy - y\,dx) = y^4\,dy$; $x = 0$ when $y = 2$
23. $2\csc(xy)dx + x\,dy + y\,dx = 0$; $x = 0$ when $y = \pi/2$
24. $\sqrt[3]{x^2 + y^2}dy = 3(x\,dx + y\,dy)$; $x = 0$ when $y = 8$

In Exercises 25 and 26, rewrite each equation such that each resulting term or combination is in an integrable form.

25. $e^{-x}\,dy - 2y\,dy = ye^{-x}\,dx$
26. $\cos(x + 2y)dx + 2\cos(x + 2y)dy - x\,dy = y\,dx$

In Exercises 27 and 28, solve the given equations by letting $u = y/x$.

27. $dy = \left(\frac{y}{x} + \frac{y^2}{x^2}\right)dx$
28. $dy = \left(\frac{x - y}{x}\right)dx$

Answer to Practice Exercise

1. $2x^2 - xy = c$

31.4 The Linear Differential Equation of the First Order

Linear Differential Equation • P and Q Functions of x only • Identifying P, Q, and $e^{\int P\,dx}$

■ Another common form of Eq. (31.7) is $\frac{dy}{dx} + Py = Q$.

There is one type of differential equation of the first order and first degree for which an integrable combination can always be found. *It is the* **linear differential equation** *of the first order and is of the form*

$$dy + Py\,dx = Q\,dx \tag{31.7}$$

where ***P and Q are functions of x only.*** This type of equation occurs widely in applications.

If each side of Eq. (31.7) is multiplied by $e^{\int P\,dx}$, it becomes integrable, because the left side becomes of the form du with $u = ye^{\int P\,dx}$ and the right side is a function of x only. This is shown by finding the differential of $ye^{\int P\,dx}$. Thus,

$$d(ye^{\int P\,dx}) = e^{\int P\,dx}(dy + Py\,dx)$$

In finding the differential of $\int P\,dx$, we use the fact that, by definition, these are reverse processes. Thus, $d(\int P\,dx) = P\,dx$. Therefore, if each side is multiplied by $e^{\int P\,dx}$, the left side may be immediately integrated to get $ye^{\int P\,dx}$, and the right-side integration may be indicated. The solution becomes

$$ye^{\int P\,dx} = \int Qe^{\int P\,dx}\,dx + c \tag{31.8}$$

EXAMPLE 1 Note the logarithmic form in the solution

Solve the differential equation $dy + \left(\frac{2}{x}\right)y\,dx = 4x\,dx$.

This equation fits the form of Eq. (31.7) with $P = 2/x$ and $Q = 4x$. The first expression to find is $e^{\int P\,dx}$. In this case, this is

$$e^{\int (2/x)dx} = e^{2\ln x} = e^{\ln x^2} = x^2 \quad \text{see text comments following example}$$

The left side integrates to yx^2, while the right side becomes $\int 4x(x^2)dx$. Thus,

$$ye^{\int P\,dx} = \int Qe^{\int P\,dx}\,dx + c$$

$$y(x^2) = \int (4x)(x^2)dx + c \quad \text{using Eq. (31.8)}$$

$$yx^2 = \int 4x^3\,dx + c = x^4 + c \quad \text{integrating}$$

$$y = x^2 + cx^{-2}$$

■

As in Example 1, in finding the factor $e^{\int P\,dx}$, we often obtain an expression of the form $e^{\ln u}$. [From Section 13.5, we know this can be simplified by using the property $\boldsymbol{e^{\ln u} = u}$.]

NOTE ▸

[Also, in finding $e^{\int P\,dx}$, the constant of integration in the exponent $\int P\,dx$ can always be taken as zero, as we did in Example 1.] To show why this is so, let $P = 2/x$ as in Example 1:

NOTE ▸

$$e^{\int (2/x)dx} = e^{\ln x^2 + c} = (e^{\ln x^2})(e^c) = x^2e^c$$

The solution to the differential equation, as given in Eq. (30.8), is then

$$y(x^2)(e^c) = \int 4x(x^2)(e^c)dx + c_1e^c$$

Regardless of the value of c, the factor e^c can be divided out. Therefore, it is convenient to let $c = 0$ and have $e^c = 1$.

EXAMPLE 2 Note the logarithmic form in the solution

Solve the differential equation $x\,dy - 3y\,dx = x^3\,dx$.

Putting this equation in the form of Eq. (31.7) by dividing through by x gives $dy - (3/x)y\,dx = x^2\,dx$. Here, $P = -3/x$, $Q = x^2$, and the factor $e^{\int P\,dx}$ becomes

$$e^{\int(-3/x)dx} = e^{-3\ln x} = e^{\ln x^{-3}} = x^{-3}$$

Therefore,

$$ye^{\int P\,dx} = \int Qe^{\int P\,dx}\,dx + c$$

$$yx^{-3} = \int x^2(x^{-3})\,dx + c \quad \text{using Eq. (31.8)}$$

$$= \int x^{-1}\,dx + c = \ln x + c$$

$$y = x^3(\ln x + c)$$

■

Practice Exercise

1. Find the general solution of the differential equation $x\,dy - 2y\,dx = 2x^4\,dx$.

EXAMPLE 3 Solution uses integration by parts or tables

Solve the differential equation $dy + y\,dx = x\,dx$.

Here, $P = 1$, $Q = x$, and $e^{\int P\,dx} = e^{\int(1)dx} = e^x$. Therefore,

$$ye^x = \int xe^x\,dx + c = e^x(x - 1) + c \quad \text{using Eq. (31.8) and integrating by ports or tables}$$

$$y = x - 1 + ce^{-x}$$

■

EXAMPLE 4 Logarithmic and trigonometric forms

Solve the differential equation $\cos x\dfrac{dy}{dx} = 1 - y\sin x$.

Writing this in the form of Eq. (31.7), we have

$$dy + y\tan x\,dx = \sec x\,dx \quad \text{dividing by } \cos x \text{ and multiplying by } dx$$

Thus, with $P = \tan x$, we have

$$e^{\int P\,dx} = e^{\int \tan x\,dx} = e^{-\ln\cos x} = \sec x \quad \text{see the first NOTE after Example 1}$$

$$y\sec x = \int \sec^2 x\,dx = \tan x + c \quad \text{using Eq. (31.8)}$$

$$y = \sin x + c\cos x$$

■

EXAMPLE 5 Finding a particular solution

For the differential equation $dy = (1 - 2y)x\,dx$, find the particular solution such that $x = 0$ when $y = 2$.

The solution proceeds as follows:

$$dy + 2xy\,dx = x\,dx \quad \text{form of Eq. (31.7)}$$

$$e^{\int P\,dx} = e^{\int 2x\,dx} = e^{x^2} \quad \text{find } e^{\int P\,dx}$$

$$ye^{x^2} = \int xe^{x^2}\,dx \quad \text{using Eq. (31.8)}$$

$$ye^{x^2} = \frac{1}{2}e^{x^2} + c \quad \text{general solution}$$

$$(2)(e^0) = \frac{1}{2}(e^0) + c, \quad 2 = \frac{1}{2} + c, \quad c = \frac{3}{2} \quad x = 0,\ y = 2;\ \text{evaluate } c$$

$$ye^{x^2} = \frac{1}{2}e^{x^2} + \frac{3}{2} \quad \text{substitute } c = \frac{3}{2}$$

$$y = \frac{1}{2}(1 + 3e^{-x^2}) \quad \text{particular solution}$$

■

EXERCISES 31.4

In Exercises 1 and 2, make the given changes in the indicated examples of this section and then solve the resulting differential equations.

1. In Example 1, change the right side to 3 dx.
2. In Example 3, change the right side to 2 dx.

In Exercises 3–28, solve the given differential equations.

3. $dy + y\,dx = e^{-x}dx$
4. $dy + 3y\,dx = e^{-3x}dx$
5. $dy + 2y\,dx = 2e^{-4x}dx$
6. $di + i\,dt = e^{-t}\cos t\,dt$
7. $\frac{dy}{dx} - 2y = 4$
8. $2\frac{dy}{dx} = 5 - 6y$
9. $dy = 3x^2(2 - y)dx$
10. $x\,dy + 3y\,dx = dx$
11. $2x\,dy + y\,dx = 8x^3dx$
12. $3x\,dy - y\,dx = 9x\,dx$
13. $dr + r\cot\theta\,d\theta = d\theta$
14. $y' = x^2y + 3x^2$
15. $\sin x\frac{dy}{dx} = 1 - y\cos x$
16. $\frac{dv}{dt} - \frac{v}{t} = 4\ln t$
17. $y' + y = x + e^x$
18. $y' + 2y = \sin x$
19. $ds = (te^{4t} + 4s)dt$
20. $y' - 2y = 2e^{2x}$
21. $y' = x^3(1 - 4y)$
22. $y' + y\tan x = -\sin x$
23. $x\frac{dy}{dx} = y + (x^2 - 1)^2$
24. $dy = dt - \frac{y\,dt}{(1 + t^2)\tan^{-1}t}$
25. $\sqrt{1 + x^2}\,dy + x(1 + y)dx = 0$
26. $(1 + x^2)dy + xy\,dx = x\,dx$
27. $\tan\theta\frac{dr}{d\theta} - r = \tan^2\theta$
28. $y' + y = y^2$ (Solve by letting $y = 1/u$ and solving the resulting linear equation for u.)

In Exercises 29 and 30, solve the given differential equations. Explain how each can be solved using either of two different methods.

29. $y' = 2(1 - y)$
30. $x\,dy = (2x - y)dx$

In Exercises 31–36, find the indicated particular solutions of the given differential equations.

31. $\frac{dy}{dx} + 2y = e^{-x}$; $x = 0$ when $y = 1$
32. $dq - 4q\,du = 2\,du$; $q = 2$ when $u = 0$
33. $y' + 2y\cot x = 4\cos x$; $x = \pi/2$ when $y = 1/3$
34. $y'\sqrt{x} + \frac{1}{2}y = e^{\sqrt{x}}$; $x = 1$ when $y = 3$
35. $(\sin x)y' + y = \tan x$; $x = \pi/4$ when $y = 0$
36. $f(x)dy + 2yf'(x)dx = f(x)f'(x)dx$; $f(x) = -1$ when $y = 3$

In Exercises 37–42, solve the given problems.

37. The differential equation $y' + P(x)y = Q(x)y^2$ is not linear. Show that the substitution $u = y^{-1}$ will transform it into a linear equation.
38. An equation used in the analysis of rocket motion is $m\,dv + kv\,dt = 0$, where m and k are positive constants. Solve this equation for v as a function of t in two ways.
39. In a certain national forest, dead leaves fall and accumulate on the ground at the rate of 20 kg per square meter per year. Once on the ground, the leaves decompose at the rate of 80% per year. This leads to the differential equation $\frac{dL}{dt} = 20 - 0.8L$, where L is the amount of leaves on the ground (in kg/m^2) and t is the time in years. Express L as a function of t.
40. If A dollars are placed in an account that pays 5% interest, compounded *continuously,* and A dollars are added to the account each year, the number of dollars n in the account after t years is given by $dn/dt = A + 0.05n$. Find n for $A = \$2000$ and $t = 5$ years.
41. A certain country has a population of 30 million at time $t = 0$. Because of reproduction, the population grows by 2.0% annually. However, due to anticipated increasing emigration, $0.3e^{0.05t}$ million people are expected to leave the country in year t. Therefore, the rate of change in the population is given by $\frac{dP}{dt} = 0.02P - 0.3e^{0.05t}$, where P is the population in millions and t is in years. Express the population of the country as a function of time. Use a calculator to view the graph of this function for $0 \le t \le 30$, and describe one key feature of the graph.
42. A drug is given to a patient intravenously at a rate of 0.5 mg per hour. The person's body continuously removes 2.0% of the drug from the blood per hour through absorption. If y is the amount of the drug (in mg) in the blood at time t (in h), then $\frac{dy}{dt} = 0.5 - 0.02y$. (a) Express y as a function of t if $y = 0$ when $t = 0$. (b) What is the limiting value of y as $t \to \infty$?

Answer to Practice Exercise

1. $y = x^4 + cx^2$

31.5 Numerical Solutions of First-order Equations

Euler's Method • Runge–Kutta Method

Many differential equations do not have exact solutions. Therefore, in this section, we show one basic method and one more advanced method of solving such equations numerically.

EULER'S METHOD

■ Named for the Swiss mathematician Leonhard Euler (1707–1783).

To find an approximate solution to a differential equation of the form $dy/dx = f(x, y)$, that passes through a known point (x_0, y_0), we write the equation as $dy = f(x, y)dx$ and then approximate dy as $y_1 - y_0$, and replace dx with Δx. From Section 24.8, we recall that Δy closely approximates dy for a small dx and that $dx = \Delta x$. This gives us

$$y_1 = y_0 + f(x_0, y_0)\Delta x \quad \text{and} \quad x_1 = x_0 + \Delta x$$

Therefore, we now know another point (x_1, y_1) that is on (or very nearly on) the curve of the solution. We can now repeat this process using (x_1, y_1) as a known point to obtain a next point (x_2, y_2). Continuing this process, we can get a series of points that are approximately on the solution curve. The method is called ***Euler's method.***

EXAMPLE 1 Euler's method

For the differential equation $dy/dx = x + y$, use Euler's method to find the y-values of the solution for $x = 0$ to $x = 0.5$ with $\Delta x = 0.1$, if the curve of the solution passes through (0, 1).

■ A graphing calculator program that gives numerical values of the solution of a differential equation using Euler's method is at goo.gl/xt4mFA

Using the method outlined above, we have $x_0 = 0$, $y_0 = 1$, and

$$y_1 = 1 + (0 + 1)(0.1) = 1.1 \qquad \text{and} \qquad x_1 = 0 + 0.1 = 0.1$$

This tells us that the curve passes (or nearly passes) through the point (0.1, 1.1). Assuming this point is correct, we use it to find the next point on the curve.

$$y_2 = 1.1 + (0.1 + 1.1)(0.1) = 1.22 \qquad \text{and} \qquad x_2 = 0.1 + 0.1 = 0.2$$

Therefore, the next approximate point is (0.2, 1.22). Continuing this process, we find a set of points that would approximately satisfy the function that is the solution of the differential equation. Tabulating results, we have the following table.

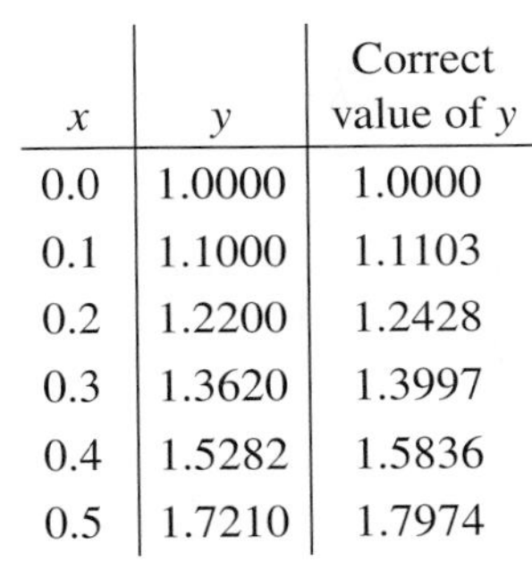

x	y	Correct value of y
0.0	1.0000	1.0000
0.1	1.1000	1.1103
0.2	1.2200	1.2428
0.3	1.3620	1.3997
0.4	1.5282	1.5836
0.5	1.7210	1.7974

the values shown have been rounded off, although more digits were carried in the calculations

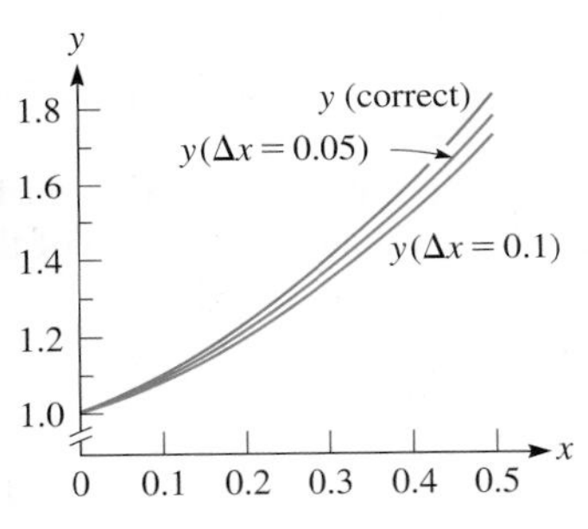

Fig. 31.3

In this case, we are able to find the correct values because the equation can be written as $dy/dx - y = x$, and the solution is $y = 2e^x - x - 1$. Although numerical methods are generally used with equations that cannot be solved exactly, we chose this equation so that we could compare values obtained with known values.

We can see that as x increases, the error in y increases. More accurate values can be found by using smaller values of Δx. In Fig. 31.3 the solution curves using $\Delta x = 0.1$ and $\Delta x = 0.05$ are shown along with the correct values of y.

Euler's method is easy to use and understand, but it is less accurate than other methods. We will show one of the more accurate methods in the next example. ■

RUNGE–KUTTA METHOD

■ Named for the German mathematicians Carl Runge (1856–1927) and Martin Kutta (1867–1944).

For more accurate numerical solutions of a differential equation, the ***Runge–Kutta method*** is often used. Starting at a first point (x_0, y_0), the coordinates of the second point (x_1, y_1) are found by using a weighted average of the slopes calculated at the points where $x = x_0$, $x = x_0 + \frac{1}{2}\Delta x$, and $x = x_0 + \Delta x$. The formulas for y_1 and x_1 are

$$y_1 = y_0 + \frac{1}{6}H(J + 2K + 2L + M) \qquad \text{and} \qquad x_1 = x_0 + H \qquad (\text{for convenience, } H = \Delta x)$$

where

$$J = f(x_0, y_0)$$
$$K = f(x_0 + 0.5H, y_0 + 0.5HJ)$$
$$L = f(x_0 + 0.5H, y_0 + 0.5HK)$$
$$M = f(x_0 + H, y_0 + HL)$$

We have used uppercase letters to correspond to calculator use. Traditional sources normally use h for H and a lowercase letter (such as k) with subscripts for J, K, L, and M, and express 0.5 as 1/2.

■ A graphing calculator program that gives numerical values of the solution of a differential equation using the Range–Kutta method is at goo.gl/6WXHeL

As with Euler's method, once (x_1, y_1) is determined, we use the formulas again to find (x_2, y_2) by replacing (x_0, y_0) with (x_1, y_1). The following example illustrates the use of the Runge–Kutta method.

x	y
0.0	0.0
0.1	0.0050125208
0.2	0.0202013395
0.3	0.0460278455
0.4	0.0832868181
0.5	0.1331460062

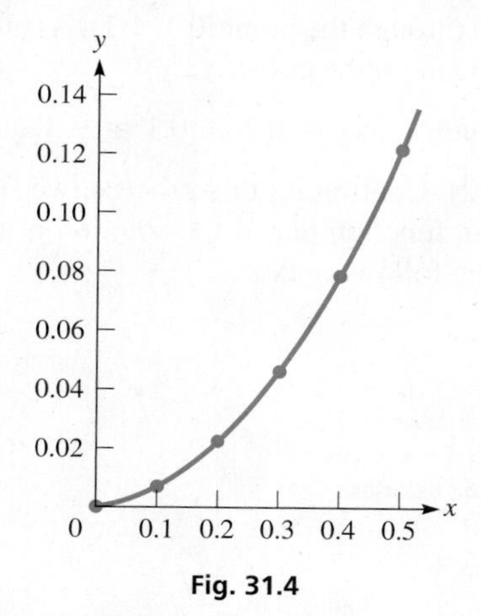

Fig. 31.4

EXAMPLE 2 Runge–Kutta method

For the differential equation $dy/dx = x + \sin xy$, use the Runge–Kutta method to find y-values of the solution for $x = 0$ to $x = 0.5$ with $\Delta x = 0.1$, if the curve of the solution passes through (0, 0).

Using the formulas and method outlined above, we have the following solution, with calculator notes to the right of the equations. Also, calculator symbols are used on the right sides of the equations to indicate the way in which they should be entered.

$x_0 = 0$ store as X

$y_0 = 0$ store as Y

$H = 0.1$ store as H

$J = X + \sin XY = 0$ store as J

$K = X + 0.5H + \sin[(X + 0.5H)(Y + 0.5HJ)] = 0.05$ store as K

$L = X + 0.5H + \sin[(X + 0.5H)(Y + 0.5HK)] = 0.050125$ store as L

$M = X + H + \sin[(X + H)(Y + HL)] = 0.10050125$ store as M

$y_1 = Y + (H/6)(J + 2K + 2L + M) = 0.0050125208$ store as Y

$x_1 = X + H = 0.1$ store as X

We now use (x_1, y_1) as we just used (x_0, y_0) to get the next point (x_2, y_2), which is then used to find (x_3, y_3), and so on. A table showing calculator values and a graph of these values is shown in Fig. 31.4. ■

EXERCISES 31.5

In Exercises 1–8, use Euler's method to find y-values of the solution for the given values of x and Δx, if the curve of the solution passes through the given point. Check the results against known values by solving the differential equations exactly. Plot the graphs of the solutions in Exercises 1–4.

1. $\dfrac{dy}{dx} = x + 1$; $x = 0$ to $x = 1$; $\Delta x = 0.2$; $(0, 1)$
2. $\dfrac{dy}{dx} = \sqrt{2x + 1}$; $x = 0$ to $x = 1.2$; $\Delta x = 0.3$; $(0, 2)$
3. $\dfrac{dy}{dx} = y(0.4x + 1)$; $x = -0.2$ to $x = 0.3$; $\Delta x = 0.1$; $(-0.2, 2)$
4. $\dfrac{dy}{dx} = y + e^x$; $x = 0$ to $x = 0.5$; $\Delta x = 0.1$; $(0, 0)$
5. The differential equation of Exercise 1 with $\Delta x = 0.1$
6. The differential equation of Exercise 2 with $\Delta x = 0.1$
7. The differential equation of Exercise 3 with $\Delta x = 0.05$
8. The differential equation of Exercise 4 with $\Delta x = 0.05$

In Exercises 9–14, use the Runge–Kutta method to find y-values of the solution for the given values of x and Δx, if the curve of the solution passes through the given point.

9. $\dfrac{dy}{dx} = xy + 1$; $x = 0$ to $x = 0.4$; $\Delta x = 0.1$; $(0, 0)$
10. $\dfrac{dy}{dx} = x^2 + y^2$; $x = 0$ to $x = 0.4$; $\Delta x = 0.1$; $(0, 1)$
11. $\dfrac{dy}{dx} = e^{xy}$; $x = 0$ to $x = 1$; $\Delta x = 0.2$; $(0, 0)$
12. $\dfrac{dy}{dx} = \sqrt{1 + xy}$; $x = 0$ to $x = 0.2$; $\Delta x = 0.05$; $(0, 1)$
13. $\dfrac{dy}{dx} = \cos(x + y)$; $x = 0$ to $x = 0.6$; $\Delta x = 0.1$; $(0, \pi/2)$
14. $\dfrac{dy}{dx} = y + \sin x$; $x = 0.5$ to $x = 1.0$; $\Delta x = 0.1$; $(0.5, 0)$

In Exercises 15–18, solve the given problems.

15. In Example 1, use Euler's method to find the y-values from $x = 0$ to $x = 3$ with $\Delta x = 1$. Compare with the values found using the exact solution. Comment on the use of Euler's method in finding the value of y for $x = 3$.
16. For the differential equation $dy/dx = x + 1$, if the curve of the solution passes through (0, 0), calculate the y-value for $x = 0.04$ with $\Delta x = 0.01$. Find the exact solution, and compare the result using three terms of the Maclaurin series that represents the solution.
17. An electric circuit contains a 1-H inductor, a 2-Ω resistor, and a voltage source of $\sin t$. The resulting differential equation relating the current i and the time t is $di/dt + 2i = \sin t$. Find i after 0.5 s by Euler's method with $\Delta t = 0.1$ s if the initial current is zero. Solve the equation exactly and compare the values.
18. An object is being heated such that the rate of change of the temperature T (in °C) with respect to time t (in min) is $dT/dt = \sqrt[3]{1 + t^3}$. Find T for $t = 5$ min by using the Runge–Kutta method with $\Delta t = 1$ min, if the initial temperature is 0°C.

31.6 Elementary Applications

Geometric Applications • Electrical, Mechanical, and Other Technical Applications

The differential equations of the first order and first degree we have discussed thus far have numerous applications in geometry and the various fields of technology. In this section, we illustrate some of these applications.

EXAMPLE 1 Given the slope—find the equation

The slope of a given curve is given by the expression $6xy$. Find the equation of the curve if it passes through the point (2, 1).

Because the slope is $6xy$, the differential equation for the curve is

$$\frac{dy}{dx} = 6xy$$

We now want to find the particular solution of this equation for which $x = 2$ when $y = 1$. The solution follows:

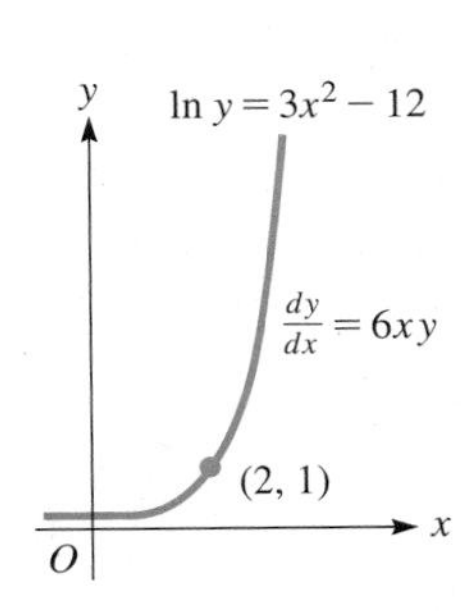

Fig. 31.5

$$\frac{dy}{y} = 6x\,dx \quad \text{separate variables}$$

$$\ln y = 3x^2 + c \quad \text{general solution}$$

$$\ln 1 = 3(2^2) + c \quad \text{evaluate } c$$

$$0 = 12 + c, \qquad c = -12$$

$$\ln y = 3x^2 - 12 \quad \text{particular solution}$$

The graph of this solution is shown in Fig. 31.5. ■

EXAMPLE 2 Orthogonal trajectories

A curve that intersects all members of a family of curves at right angles is called an **orthogonal trajectory** *of the family.* Find the equations of the orthogonal trajectories of the parabolas $x^2 = cy$. As before, each value of c gives us a particular member of the family.

The derivative of the given equation is $dy/dx = 2x/c$. This equation contains the constant c, which depends on the point (x, y) on the parabola. ***Eliminating this constant between the equations of the parabolas and the derivative,*** we have

$$c = \frac{x^2}{y} \qquad \frac{dy}{dx} = \frac{2x}{c} = \frac{2x}{x^2/y} \quad \text{or} \quad \frac{dy}{dx} = \frac{2y}{x}$$

This equation gives a general expression for the slope of any of the members of the family. For a curve to be perpendicular, its slope must equal the negative reciprocal of this expression, or the slope of the orthogonal trajectories must be

$$\left.\frac{dy}{dx}\right|_{OT} = -\frac{x}{2y} \quad \longleftarrow \text{this equation must not contain the constant } c$$

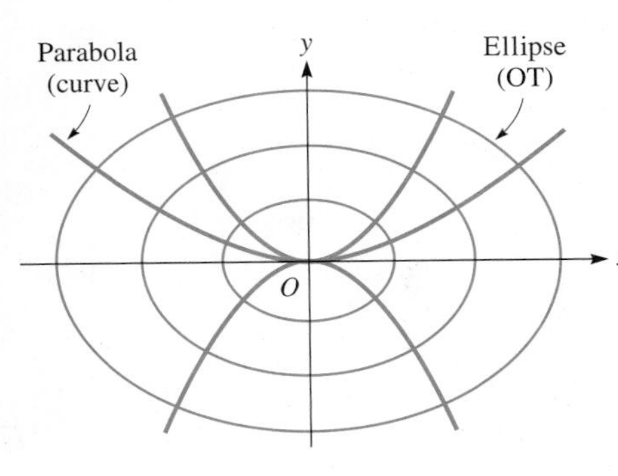

Fig. 31.6

Solving this differential equation gives the family of orthogonal trajectories.

$$2y\,dy = -x\,dx$$

$$y^2 = -\frac{x^2}{2} + \frac{k}{2}$$

$$2y^2 + x^2 = k \quad \text{orthogonal trajectories}$$

Thus, the orthogonal trajectories are ellipses. Note in Fig. 31.6 that each parabola intersects each ellipse at right angles. ■

Practice Exercise

1. Find the equation of the orthogonal trajectories of the curves $y^2 = cx^3$.

EXAMPLE 3 Radioactivity—carbon dating

Radioactive elements decay at rates that are proportional to the amount of the element present. Carbon-14 decays such that one-half of an original amount decays into other forms in about 5730 years. By measuring the proportion of carbon-14 in remains at an ancient site, the approximate age of the remains can be determined. This method, called *carbon dating,* is used to determine the dates of prehistoric events.

■ See the chapter introduction.

The analysis of some wood at the site of the ancient city of Troy showed that the concentration of carbon-14 was 67.8% of the concentration that new wood would have. Determine the equation relating the amount of carbon-14 present with the time and then determine the age of the wood at the site of Troy.

■ Radioactivity was discovered in 1898 by the French physicist Henri Becquerel (1852–1908). Carbon dating was developed in 1947 by the U.S. chemist Willard Libby (1908–1980).

Let N_0 be the original amount and N be the amount present at any time t (in years). The rate of decay can be expressed as a derivative. Therefore, since the rate of change is proportional to N, we have the equation

$$\frac{dN}{dt} = kN$$

Solving this differential equation, we have

$$\begin{aligned}
\frac{dN}{N} &= k\,dt && \text{separate variables} \\
\ln N &= kt + \ln c && \text{general solution} \\
\ln N_0 &= k(0) + \ln c && N = N_0 \text{ for } t = 0 \\
c &= N_0 && \text{solve for } c \\
\ln N &= kt + \ln N_0 && \text{substitute } N_0 \text{ for } c \\
\ln N - \ln N_0 &= kt && \text{use properties of logarithms} \\
\ln\frac{N}{N_0} &= kt \\
N &= N_0 e^{kt} && \text{exponential form}
\end{aligned}$$

Now, using the condition that one-half of carbon-14 decays in 5730 years, we have $N = N_0/2$ when $t = 5730$ years. This gives

■ The time it takes for one-half of an isotope (a specific form of an element) to decay into other forms is called the "half-life" of the isotope.

$$\frac{N_0}{2} = N_0 e^{5730k} \quad \text{or} \quad \frac{1}{2} = (e^k)^{5730}$$

$$e^k = 0.5^{1/5730}$$

Therefore, the equation relating N and t is

$$N = N_0(0.5)^{t/5730}$$

Because the present concentration is $N = 0.678N_0$, we can solve for t:

$$0.678N_0 = N_0(0.5)^{t/5730} \quad \text{or} \quad 0.678 = 0.5^{t/5730}$$

$$\ln 0.678 = \ln 0.5^{t/5730}, \qquad \ln 0.678 = \frac{t}{5730}\ln 0.5 \qquad \text{solving an exponential equation}$$

$$t = 5730\frac{\ln 0.678}{\ln 0.5} = 3210 \text{ years}$$

Therefore, the wood at the Troy site is about 3210 years old. ■

EXAMPLE 4 Electric circuit

The general equation relating the current i, voltage E, inductance L, capacitance C, and resistance R in a simple electric circuit (see Fig. 31.7) is

$$L\frac{di}{dt} + Ri + \frac{q}{C} = E \tag{31.9}$$

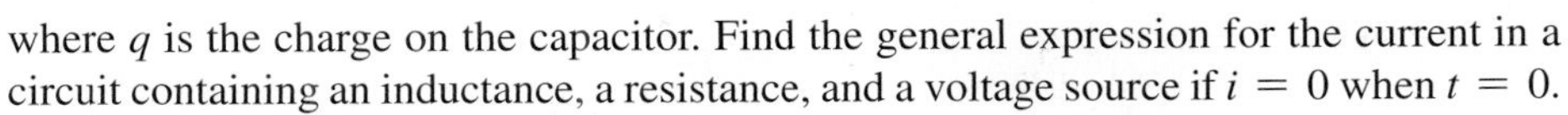

where q is the charge on the capacitor. Find the general expression for the current in a circuit containing an inductance, a resistance, and a voltage source if $i = 0$ when $t = 0$.

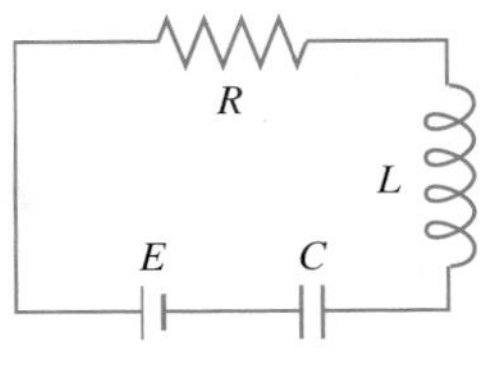

Fig. 31.7

The differential equation for this circuit is

$$L\frac{di}{dt} + Ri = E$$

Using the method of the linear differential equation of the first order, we have the equation

$$di + \frac{R}{L}i\,dt = \frac{E}{L}dt$$

The factor $e^{\int P\,dt}$ is $e^{\int (R/L)dt} = e^{(R/L)t}$. This gives

$$ie^{(R/L)t} = \frac{E}{L}\int e^{(R/L)t}\,dt = \frac{E}{R}e^{(R/L)t} + c$$

Letting the current be zero for $t = 0$, we have $c = -E/R$. The result is

$$ie^{(R/L)t} = \frac{E}{R}e^{(R/L)t} - \frac{E}{R}$$

$$i = \frac{E}{R}(1 - e^{-(R/L)t})$$

(We can see that $i \to E/R$ as $t \to \infty$. In practice, the exponential term becomes negligible very quickly.) ■

EXAMPLE 5 Mixing problem—salt solution

Fifty gallons of brine originally containing 20.0 lb of salt are in a tank into which 2.00 gal of water run each minute with the same amount of mixture running out each minute. How much salt is in the tank after 10.0 min?

NOTE ▸ Let $x =$ the number of pounds of salt in the tank after t minutes. [Each gallon of brine contains $x/50$ lb of salt, and in time dt, $2dt$ gal of mixture leave the tank with $(x/50)(2\,dt)$ lb of salt.] The amount of salt that is leaving may also be written as $-dx$ (the minus sign is included to show that x is decreasing). Thus,

$$-dx = \frac{2x\,dt}{50} \quad \text{or} \quad \frac{dx}{x} = -\frac{dt}{25}$$

This leads to $\ln x = -(t/25) + \ln c$. Using the fact that $x = 20$ lb when $t = 0$, we find that $\ln 20 = \ln c$, or $c = 20$. Therefore,

$$x = 20e^{-t/25}$$

is the general expression for the amount of salt in the tank at time t. Therefore, when $t = 10.0$ min, we have

$$x = 20e^{-10/25} = 20e^{-0.4} = 20(0.670) = 13.4 \text{ lb}$$

There are 13.4 lb of salt in the tank after 10.0 min. (Although the data were given with three significant digits, we did not use all significant digits in writing the equations that were used.) ■

EXAMPLE 6 Motion in a resisting medium

An object moving through (or across) a resisting medium often experiences a retarding force approximately proportional to the velocity as well as the force that causes the motion. An example is a ball falling due to the force of gravity, and air resistance produces a retarding force. Applying Newton's laws of motion (from physics) to the ball leads to the equation

■ There is at least some resistance to the motion of any object moving through a medium (not a vacuum). The exact nature of the resistance is not usually known.

$$m\frac{dv}{dt} = F - kv \tag{31.10}$$

where m is the mass of the object, v is the velocity of the object, t is the time, F is the force causing the motion, and k ($k > 0$) is a constant. The quantity kv is the retarding force.

■ Slug is the unit of mass when force is expressed in pounds.

We assume that these conditions hold for a falling object whose mass is 1.0 slug and experiences a force (its own weight) of 32.0 lb. The object starts from rest, and the air causes a retarding force numerically equal to 0.200 times the velocity.

Substituting in Eq. (31.10) and then solving the differential equations, we have

$$\frac{dv}{dt} = 32 - 0.2\,v$$

$$\frac{dv}{32 - 0.2v} = dt \qquad \text{separate variables}$$

$$-5\ln(32 - 0.2v) = t - 5\ln c \qquad \text{integrate}$$

$$\ln(32 - 0.2v) = -\frac{t}{5} + \ln c \qquad \text{solve for } v$$

$$\ln\frac{32 - 0.2v}{c} = -\frac{t}{5}$$

$$32 - 0.2v = ce^{-t/5}$$

$$0.2v = 32 - ce^{-t/5}$$

$$v = 5(32 - ce^{-t/5}) \qquad \text{general solution}$$

■ The data were given to three significant digits, but not all significant digits were used in writing the equations.

Because the object started from rest, $v = 0$ when $t = 0$. Thus,

$$0 = 5(32 - c) \quad \text{or} \quad c = 32 \qquad \text{evaluate } c$$

$$v = 160(1 - e^{-t/5}) \qquad \text{particular solution}$$

$$= 160(1 - e^{-1.00}) = 160(1 - 0.368) \qquad \text{evaluating } v \text{ for } t = 5.00\text{ s}$$

$$= 101 \text{ ft/s}$$

After 5.00 s, the velocity is 101 ft/s. Without the air resistance, the velocity would be about 160 ft/s. ■

EXERCISES 31.6

In Exercises 1–4, make the given changes in the indicated examples of this section, and then solve the resulting problems.

1. In Example 2, change $x^2 = cy$ to $y^2 = cx$.
2. In Example 4, if the term $L\frac{di}{dt}$ is deleted (no inductance) in Eq. (31.9), find the expression for the charge in the circuit. (There is a capacitance C, and $i = dq/dt$.) The initial charge is zero.
3. In Example 5, change 2.00 gal to 1.00 gal.
4. In Example 6, change 0.200 to 0.100 (for the retarding force).

In Exercises 5–8, find the equation of the curve for the given slope and point through which it passes. Use a calculator to display the curve.

5. Slope given by $2x/y$; passes through (2, 3)
6. Slope given by $-y/(x + y)$; passes through $(-1, 3)$
7. Slope given by $y + x$; passes through (0, 1)
8. Slope given by $-2y + e^{-x}$; passes through (0, 2)

In Exercises 9–12, find the equation of the orthogonal trajectories of the curves for the given equations. Use a graphing calculator to display at least two members of the family of curves and at least two of the orthogonal trajectories.

9. The exponential curves $y = ce^x$

10. The hyperbolas $x^2 - y^2 = a^2$

11. The curves $y = c(\sec x + \tan x)$

12. The family of circles, all with centers at the origin

In Exercises 13–52, solve the given problems by solving the appropriate differential equation.

13. The isotope cobalt-60, with half-life of 5.27 years, is used in treating cancerous tumors. What percent of an initial amount remains after 2.00 years?

14. In 1986, the world's worst nuclear accident occurred in Chernobyl, Ukraine. Since then, over 20,000 people have died from the radioactivity of Cesium 137, which has a half-life of 30.1 years. What percent of Cesium 137 released in 1986 remained in 2016?

15. A possible health hazard in the home is radon gas. It is radioactive and about 90.0% of an original amount disintegrates in 12.7 days. Find the half-life of radon gas. (The problem with radon is that it is a gas and is being continually produced by the radioactive decay of minute amounts of radioactive radium found in the soil and rocks of an area.)

16. Noting Example 3, another element used to date more recent events is helium-3, which has a half-life of 12.3 years. If a building has 5.0% of helium-3 that a new building would have, about how old is the building?

17. A radioactive element leaks from a nuclear power plant at a constant rate r, and it decays at a rate that is proportional to the amount present. Find the relation between the amount N present in the environment in which it leaks and the time t, if $N = 0$ when $t = 0$.

18. Most use of the pesticide DDT was banned in the United States in 1972. It has been found that 19% of an initial amount of DDT is degraded into harmless products in 10 years. If no DDT has been used in an area since 1972, in what year will the concentration become only 20%, of the 1972 amount?

19. A person takes an 81-mg aspirin tablet, and 8.0 h later there are 51 mg of aspirin in the person's bloodstream. What is the half-life of this aspirin?

20. One model of animal growth is that the rate of growth is proportional to remaining growth expected to develop. If a certain animal has a length x (in m) at age t (in years), find the equation relating the animal's length as a function of t, if the expected length is L.

21. The growth of the population P of a nation with a constant immigration rate I may be expressed as $\frac{dP}{dt} = kP + I$, where t is in years. If the population of Canada in 2015 was 35.9 million and about 0.240 million immigrants enter Canada each year, what will the population of Canada be in 2025, given that the growth rate k is about 0.800% (0.00800) annually?

22. Assuming that the natural environment of Earth is limited and that the maximum population it can sustain is M, the rate of growth of the population P is given by the *logistic* differential equation $\frac{dP}{dt} = kP(M - P)$. Using this equation for Earth, if $P = 7.4$ billion in 2016, $k = 0.00040$, and $M = 25$ billion, what will be the population of Earth in 2026?

23. The rate of change of the radial stress S on the walls of a pipe with respect to the distance r from the axis of the pipe is given by $r\frac{dS}{dr} = 2(a - S)$, where a is a constant. Solve for S as a function of r.

24. The velocity v of a meteor approaching Earth is given by $v\frac{dv}{dr} = -\frac{GM}{r^2}$, where r is the distance from the center of Earth, M is the mass of Earth, and G is a universal gravitational constant. If $v = 0$ for $r = r_0$, solve for v as a function of r.

25. Assume that the rate at which highway construction increases is directly proportional to the total mileage M of all highways already completed at time t (in years). Solve for M as a function of t if $M = 5250$ mi for a certain county when $t = 0$ and $M = 5460$ mi for $t = 2.00$ years.

26. The marginal profit function gives the change in the total profit P of a business due to a change in the business, such as adding new machinery or reducing the size of the sales staff. A company determines that the marginal profit dP/dx is $e^{-x^2} - 2Px$, where x is the amount invested in new machinery. Determine the total profit (in thousands of dollars) as a function of x, if $P = 0$ for $x = 0$.

27. According to Newton's law of cooling, the rate at which a body cools is proportional to the difference in temperature between it and the surrounding medium. Assuming Newton's law holds, how long will it take a cup of hot water, initially at 200°F, to cool to 100°F if the room temperature is 80.0°F, if it cools to 140°F in 5.0 min?

28. An object whose temperature is 100°C is placed in a medium whose temperature is 20°C. The temperature of the object falls to 50°C in 10 min. Express the temperature T of the object as a function of time t (in min). (Assume it cools according to Newton's law of cooling as stated in Exercise 27.)

29. If interest in a bank account is compounded continuously, the amount grows at a rate that is proportional to the amount present in the account. Determine the amount in an account after one year if $1000 is placed in the account and it pays 4% interest per year, compounded continuously.

30. The rate of change in the intensity I of light below the surface of the ocean with respect to the depth y is proportional to I. If the intensity at 15 ft is 50% of the intensity I_0 at the surface, at what depth is the intensity 15% of I_0?

31. For a DNA sample in a liquid containing a solute of constant concentration c_0, the rate at which the concentration $c(t)$ of solute in the sample changes is proportional to $c_0 - c(t)$. Find $c(t)$ if $c(0) = 0$.

32. In a town of N persons, during a flu epidemic, it was determined that the rate dS/dt at which persons were being infected was proportional to the product of the number S of infected persons and the number $N - S$ of healthy persons. Find S as a function of t.

33. If the current in an RL circuit with a voltage source E is zero when $t = 0$ (see Example 4), show that $\lim_{t\to\infty} i = E/R$. See Fig. 31.8.

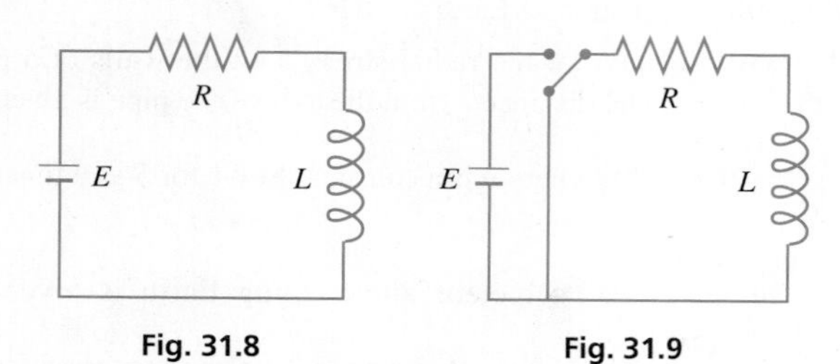

Fig. 31.8 Fig. 31.9

34. If a circuit contains only an inductance and a resistance, with $L = 2.0$ H and $R = 30\ \Omega$, find the current i as a function of time t if $i = 0.020$ A when $t = 0$. See Fig. 31.9.

35. An amplifier circuit contains a resistance R, an inductance L, and a voltage source $E \sin \omega t$. Express the current in the circuit as a function of the time t if the initial current is zero.

36. A radio transmitter circuit contains a resistance of 2.0 Ω, a variable inductor of $100 - t$ henrys, and a voltage source of 4.0 V. Find the current i in the circuit as a function of the time t for $0 \leq t \leq 100$ s if the initial current is zero.

37. If a circuit contains only a resistance and a capacitance C, find the equation relating the charge q on the capacitor in terms of the time t if $i = dq/dt$ and $q = q_0$ when $t = 0$. See Fig. 31.10.

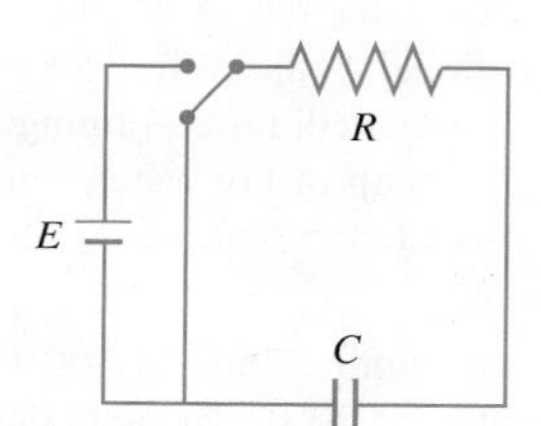

Fig. 31.10

38. One hundred gallons of brine originally containing 30 lb of salt are in a tank into which 5.0 gal of water run each minute. The same amount of mixture from the tank leaves each minute. How much salt is in the tank after 20 min?

39. An object falling under the influence of gravity has a variable acceleration given by $32 - v$, where v represents the velocity. If the object starts from rest, find an expression for the velocity in terms of the time. Also, find the limiting value of the velocity (find $\lim_{t\to\infty} v$).

40. In a ballistics test, a bullet is fired into a sandbag. The acceleration of the bullet within the sandbag is $-40\sqrt{v}$, where v is the velocity (in ft/s). When will the bullet stop if it enters the sandbag at 950 ft/s?

41. A boat with a mass of 10 slugs is being towed at 8.0 mi/h. The tow rope is then cut, and a motor that exerts a force of 20 lb on the boat is started. If the water exerts a retarding force that numerically equals twice the velocity, what is the velocity of the boat 3.0 min later?

42. A parachutist is falling at a rate of 196 ft/s when her parachute opens. With the parachute open, the air resists the fall with a force equal to $0.5v^2$. Find the velocity as a function of time, where t is the number of seconds after the parachute opens. The person and equipment have a combined mass of 5.00 slugs (weight is 160 lb).

43. Assuming a person expends 18 calories per pound of their weight each day, one model for weight loss is given by $\frac{dw}{dt} = \frac{1}{3500}(I_c - 18w)$, where w is the person's weight (in lb) and I_c is the constant daily intake of calories. If a person that originally weighs 185 lb goes on a diet and limits their daily calorie intake to 2880, express the person's weight as a function of time.

44. In studying the flow of water in a stream, it is found that an object follows the hyperbolic path $y(x + 1) = 10$ such that $(t + 1)dx = (x - 2)dt$. Find x and y (in ft) in terms of time t (in s) if $x = 4$ ft and $y = 2$ ft for $t = 0$.

45. The rate of change of air pressure p (in lb/ft^2) with respect to height h (in ft) is approximately proportional to the pressure. If the pressure is 15.0 lb/in.^{2} when $h = 0$ and $p = 10.0$ lb/in.^{2} when $h = 9800$ ft, find the expression relating pressure and height.

46. Water flows from a vertical cylindrical storage tank through a hole of area A at the bottom of the tank. The rate of flow is $4.8A\sqrt{h}$, where h is the distance (in ft) from the surface of the water to the hole. If h changes from 9.0 ft to 8.0 ft in 16 min, how long will it take the tank to empty? See Fig. 31.11.

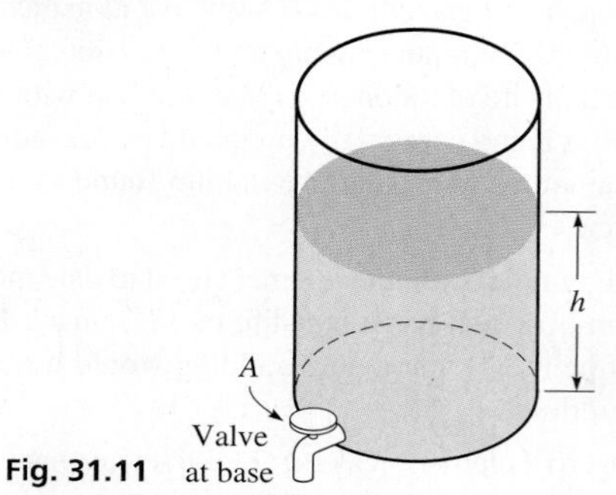

Fig. 31.11

47. Assume that the rate of depreciation of an object is proportional to its value at any time t. If a car costs \$33,000 new and its value 3 years later is \$19,700, what is its value 11 years after it was purchased?

48. Assume that sugar dissolves at a rate proportional to the undissolved amount. If there are initially 525 g of sugar and 225 g remain after 4.00 min, how long does it take to dissolve 375 g?

49. Fresh air is being circulated into a room whose volume is 4000 ft^3. Under specified conditions the number of cubic feet x of carbon dioxide present at any time t (in min) is found by solving the differential equation $dx/dt = 1 - 0.25x$. Find x as a function of t if $x = 12$ ft^3 when $t = 0$.

50. Moisture evaporates from a surface at a rate proportional to the amount of moisture present at any time. If 75% of the moisture evaporates from a certain surface in 1.00 h, how long did it take for 50% to evaporate?

51. On a certain weather map, the *isobars* (curves of equal barometric pressure) are given by $y = e^{x/2} + c$. Find the equation of the orthogonal trajectories (curves that show the wind direction), and display a few of each on a calculator.

52. The lines of equal potential in a field of force are all at right angles to the lines of force. In an electric field of force caused by charged particles, the lines of force are given by $x^2 + y^2 = cx$. Find the equation of the lines of equal potential. Use a graphing calculator to view a few members of the lines of force and those of equal potential.

Answer to Practice Exercise

1. $2x^2 + 3y^2 = k$

31.7 Higher-order Homogeneous Equations

*n*th-Order Linear Differential Equation • Operator D • Homogeneous and Nonhomogeneous Equations • Second-order Homogeneous Equations with Constant Coefficients • Auxiliary Equation

Another important type of differential equation is the linear differential equation of higher order with constant coefficients. First, we shall briefly describe the general higher-order equation and the notation we shall use with this type of equation.

The general **linear differential equation of the** ***n*****th order** *is of the form*

$$a_0\frac{d^n y}{dx^n} + a_1\frac{d^{n-1}y}{dx^{n-1}} + \cdots + a_{n-1}\frac{dy}{dx} + a_n y = b \qquad (31.11)$$

where the a's and b are either functions of x or constants.

For convenience of notation, the *n*th derivative with respect to the independent variable will be denoted by D^n. Here, D is called the **operator,** because it denotes the *operation* of differentiation. Using this notation with x as the independent variable, Eq. (31.11) becomes

$$a_0D^n y + a_1D^{n-1}y + \cdots + a_{n-1}Dy + a_n y = b \qquad (31.12)$$

NOTE ▸ [If $b = 0$, the general linear equation is called **homogeneous,** and if $b \neq 0$, it is called **nonhomogeneous.**] Both types of equations have important applications.

EXAMPLE 1 Illustrating the operator *D*

Using the operator form of Eq. (31.12), the differential equation

$$\frac{d^3y}{dx^3} - 3\frac{d^2y}{dx^2} + 4\frac{dy}{dx} - 2y = e^x \sec x$$

is written as

$$D^3y - 3D^2y + 4Dy - 2y = e^x \sec x$$

This equation is nonhomogeneous since $b = e^x \sec x$. ■

Although the a's may be functions of x, we shall restrict our attention to the cases in which they are constants. We shall, however, consider both homogeneous equations and nonhomogeneous equations. Also, because second-order linear equations are the most commonly found in elementary applications, we shall devote most of our attention to them. The methods used to solve second-order equations may be applied to equations of higher order, and we shall consider certain of these higher-order equations.

SECOND-ORDER HOMOGENEOUS EQUATIONS WITH CONSTANT COEFFICIENTS

Using the operator notation, *a second-order, linear, homogeneous differential equation with constant coefficients is one of the form*

$$a_0D^2y + a_1Dy + a_2y = 0 \qquad (31.13)$$

where the a's are constants. The following example indicates the kind of solution we should expect for this type of equation.

EXAMPLE 2 Solution of a second-order equation

Solve the differential equation $D^2y - Dy - 2y = 0$.

First, we put this equation in the form $(D^2 - D - 2)y = 0$. This is another way of saying that we are to take the second derivative of y, subtract the first derivative, and finally subtract twice the function. This expression may now be factored as $(D - 2)(D + 1)y = 0$. (We will not develop the algebra of the operator D. However, most such algebraic operations can be shown to be valid.) This formula tells us to find the first derivative of the function and add this to the function. Then twice this result is to be subtracted from the derivative of this result. If we let $z = (D + 1)y$, which is valid because $(D + 1)y$ is a function of x, we have $(D - 2)z = 0$. This equation is easily solved by separation of variables. Thus,

$$\frac{dz}{dx} - 2z = 0 \qquad \frac{dz}{z} - 2\,dx = 0 \qquad \ln z - 2x = \ln c_1$$

$$\ln\frac{z}{c_1} = 2x \quad \text{or} \quad z = c_1e^{2x}$$

Replacing z by $(D + 1)y$, we have

$$(D + 1)y = c_1e^{2x}$$

This is a linear equation of the first order. Then,

$$dy + y\,dx = c_1e^{2x}\,dx$$

The factor $e^{\int P\,dx}$ is $e^{\int dx} = e^x$. And so,

$$ye^x = \int c_1e^{3x}\,dx = \frac{c_1}{3}e^{3x} + c_2 \qquad \text{using Eq. (31.8)}$$

$$y = c_1'e^{2x} + c_2e^{-x}$$

NOTE ▸ where $c_1' = \frac{1}{3}c_1$. [This example indicates that solutions of the form e^{mx} result for this equation.] ■

Based on the result of Example 2, assume that an equation of the form of Eq. (31.13) has a particular solution ce^{mx}. Substituting this into Eq. (31.13) gives

$$a_0cm^2e^{mx} + a_1cme^{mx} + a_2ce^{mx} = 0$$

Because the exponential function $e^{mx} > 0$ for all real x, this equation will be satisfied if m is a root of the following equation, called the **auxiliary equation** of Eq. (31.13).

Auxiliary Equation

$$a_0m^2 + a_1m + a_2 = 0 \qquad \textbf{(31.14)}$$

NOTE ▸ [Note that the auxiliary equation may be formed directly from Eq. (31.13) by replacing the second derivative with m^2, the first derivative with m, and the original function with 1.] If the auxiliary equation has two *distinct* roots m_1 and m_2, then the general solution of Eq. (31.13) is given by the following:

General Solution for Distinct Roots m_1 and m_2

$$y = c_1e^{m_1x} + c_2e^{m_2x} \qquad \textbf{(31.15)}$$

We see that this is in agreement with the results of Example 2.

EXAMPLE 3 Using the auxiliary equation

Solve the differential equation $D^2y - 5y + 6y = 0$.

From this operator form of the differential equation, we write the auxiliary equation

$$m^2 - 5m + 6 = 0$$
$$(m - 3)(m - 2) = 0 \quad \text{solving auxiliary equation}$$
$$m_1 = 3 \qquad m_2 = 2$$
$$y = c_1e^{3x} + c_2e^{2x} \quad \text{using Eq. (31.15)}$$

It makes no difference which constant is written with each exponential term. ■

EXAMPLE 4 Auxiliary equation with a root of zero

Solve the differential equation $y'' = 6y'$.

We first rewrite this equation using the D notation for derivatives. Also, we want to write it in the proper form of a homogeneous equation. This gives us

$$D^2y - 6\,Dy = 0$$

Proceeding with the solution, we have

$$m^2 - 6m = 0 \quad \text{auxiliary equation}$$
$$m(m - 6) = 0 \quad \text{solve for } m$$
$$m_1 = 0 \qquad m_2 = 6$$
$$y = c_1e^{0x} + c_2e^{6x} \quad \text{using Eq. (31.15)}$$

Because $e^{0x} = 1$, we have

$$y = c_1 + c_2e^{6x} \quad \text{general solution}$$

■

Practice Exercise

1. Solve the differential equation $2D^2y - Dy - 3y = 0$.

EXAMPLE 5 Third-order equation

Solve the differential equation $2\frac{d^3y}{dx^3} + \frac{d^2y}{dx^2} - 7\frac{dy}{dx} = 0$.

Although this is a third-order equation, the *method* of solution is the same as in the previous examples. Using the D-notation for derivatives, we have

$$2\,D^3y + D^2y - 7\,Dy = 0$$
$$2m^3 + m^2 - 7m = 0, \quad \text{auxiliary equation}$$
$$m(2m^2 + m - 7) = 0$$

We can see that one root is $m = 0$. The quadratic factor is not factorable, but we can find the roots from it by using the quadratic formula. This gives

$$m = \frac{-1 \pm \sqrt{1 + 56}}{4} = \frac{-1 \pm \sqrt{57}}{4}$$

NOTE ▸ [Because there are three roots, there are three arbitrary constants.] Following the solution indicated by Eq. (31.15), we have

$$y = c_1e^{0x} + c_2e^{(-1+\sqrt{57})x/4} + c_3e^{(-1-\sqrt{57})x/4}$$

Again, $e^{0x} = 1$. Also, factoring $e^{-x/4}$ from the second and third terms, we have

$$y = c_1 + e^{-x/4}(c_2e^{x\sqrt{57}/4} + c_3e^{-x\sqrt{57}/4})$$

■

In all examples and exercises of this section, all the roots of the auxiliary equation are different, and they do not include complex numbers. If the auxiliary equation has repeated or complex roots, the solutions have a different form. These cases are discussed in the following section.

31.8 Auxiliary Equation with Repeated or Complex Roots

Repeated Roots • Complex Roots • Roots Repeated More Than Once • Repeated Complex Roots

In solving higher-order homogeneous differential equations in the previous section, we purposely avoided repeated or complex roots of the auxiliary equation. In this section, we develop the solutions for such equations. The following example indicates the type of solution that results from the case of repeated roots.

EXAMPLE 1 Solution for a double root

Solve the differential equation $D^2y - 4Dy + 4y = 0$.

Using the method of Example 2 of the previous section, we have the following steps:

$$(D^2 - 4D + 4)y = 0, \quad (D - 2)(D - 2)y = 0, \quad (D - 2)z = 0$$

where $z = (D - 2)y$. The solution to $(D - 2)z = 0$ is found by separation of variables. And so,

$$\frac{dz}{dx} - 2z = 0 \qquad \frac{dz}{z} - 2\,dx = 0$$

$$\ln z - 2x = \ln c_1 \quad \text{or} \quad z = c_1e^{2x}$$

Substituting back, we have $(D - 2)y = c_1e^{2x}$, which is a linear equation of the first order. Then

$$dy - 2y\,dx = c_1e^{2x}\,dx \qquad e^{\int -2\,dx} = e^{-2x}$$

This leads to

$$ye^{-2x} = c_1\int dx = c_1x + c_2 \qquad \text{or} \qquad y = c_1xe^{2x} + c_2e^{2x}$$

NOTE ▸ [This example indicates the type of solution that results when the auxiliary equation has repeated roots.] ■

Based on the above example, the solution to the equation $a_0D^2y + a_1Dy + a_2y = 0$ when the auxiliary equation has a *double root m* is given by the following:

General Solution for a Double Root m

$$y = e^{mx}(c_1 + c_2x) \tag{31.16}$$

In Example 1, the auxiliary equation is $m^2 - 4m + 4 = 0$, or in factored form, $(m - 2)^2 = 0$. Since 2 is a double root, the general solultion given by Eq. (31.16) is $y = e^{2x}(c_1 + c_2x)$. This is equivalent to the solution found in Example 1.

EXAMPLE 2 Solving an equation with a double root

Solve the differential equation $(D + 2)^2y = 0$.

The solution is as follows:

$$(m + 2)^2 = 0 \qquad \text{auxiliary equation}$$
$$m = -2, -2$$
$$y = e^{-2x}(c_1 + c_2x) \qquad \text{using Eq. (31.16)}$$

■

EXAMPLE 3 Solving an equation with a double root

Solve the differential equation $\frac{d^2y}{dx^2} + 25y = 10\frac{dy}{dx}$.

$$D^2y + 25y = 10Dy \quad \text{using operator } D \text{ notation}$$
$$D^2y - 10Dy + 25y = 0 \quad \text{put in proper form with 0 on right}$$
$$m^2 - 10m + 25 = 0 \quad \text{auxiliary equation}$$
$$(m - 5)^2 = 0 \quad \text{solve for } m$$
$$m = 5, 5 \quad \text{double root}$$
$$y = e^{5x}(c_1 + c_2x) \quad \text{using Eq. (31.16)}$$

■

Practice Exercise

1. Solve the differential equation $D^2y + 8Dy + 16y = 0$.

When the auxiliary equation has complex roots, it can be solved by the method of the previous section and the solution can be put in a more useful form. For complex roots of the auxiliary equation $m = \alpha \pm j\beta$, the solution is of the form

$$y = c_1e^{(\alpha+j\beta)x} + c_2e^{(\alpha-j\beta)x} = e^{\alpha x}(c_1e^{j\beta x} + c_2e^{-j\beta x})$$

Using the exponential form of a complex number, Eq. (12.11), we have

$$\begin{aligned} y &= e^{\alpha x}[c_1\cos\beta x + jc_1\sin\beta x + c_2\cos(-\beta x) + jc_2\sin(-\beta x)] \\ &= e^{\alpha x}(c_1\cos\beta x + c_2\cos\beta x + jc_1\sin\beta x - jc_2\sin\beta x) \\ &= e^{\alpha x}(c_3\cos\beta x + c_4\sin\beta x) \quad c_3 = c_1 + c_2,\ c_4 = jc_1 - jc_2 \end{aligned}$$

Therefore, if the auxiliary equation has *complex roots* of the form $\alpha \pm j\beta$, the solution to Eq. (31.13) is as follows:

General Solution for Complex Roots $m = \alpha \pm j\beta$

$$y = e^{\alpha x}(c_1\sin\beta x + c_2\cos\beta x) \quad \textbf{(31.17)}$$

EXAMPLE 4 Solving an equation with complex roots

Solve the differential equation $D^2y - Dy + y = 0$.

We have the following solution:

$$m^2 - m + 1 = 0 \quad \text{auxiliary equation}$$
$$m = \frac{1 \pm j\sqrt{3}}{2} \quad \text{complex roots}$$
$$\alpha = \frac{1}{2} \qquad \beta = \frac{\sqrt{3}}{2} \quad \text{identify } \alpha \text{ and } \beta$$
$$y = e^{x/2}\left(c_1\sin\frac{\sqrt{3}}{2}x + c_2\cos\frac{\sqrt{3}}{2}x\right) \quad \text{using Eq. (31.17)}$$

■

EXAMPLE 5 Third-order equation—real and complex roots

Solve the differential equation $D^3y + 4Dy = 0$.

This is a third-order equation, which means there are three arbitrary constants in the general solution.

$$m^3 + 4m = 0, \qquad m(m^2 + 4) = 0 \quad \text{auxiliary equation}$$
$$m_1 = 0 \qquad m_2 = 2j \qquad m_3 = -2j \quad \text{three roots, two of them complex}$$
$$\alpha = 0 \qquad \beta = 2 \quad \text{identify } \alpha \text{ and } \beta \text{ for complex roots}$$
$$y = c_1e^{0x} + e^{0x}(c_2\sin 2x + c_3\cos 2x) \quad \text{using Eqs. (31.15) and (31.17)}$$
$$= c_1 + c_2\sin 2x + c_3\cos 2x \quad e^0 = 1$$

■

Practice Exercise

2. Solve the differential equation $D^2y + 2Dy + 5y = 0$.

EXAMPLE 6 Complex roots—particular solution

Solve the differential equation $y'' - 2y' + 12y = 0$, if $y' = 2$ and $y = 1$ when $x = 0$.

$$D^2y - 2Dy + 12y = 0 \qquad \text{using operator } D \text{ notation}$$

$$m^2 - 2m + 12 = 0 \qquad \text{auxiliary equation}$$

$$m = \frac{2 \pm \sqrt{4 - 48}}{2} = 1 \pm j\sqrt{11} \qquad \text{complex roots: } \alpha = 1,\ \beta = \sqrt{11}$$

$$y = e^x(c_1 \cos\sqrt{11}x + c_2 \sin\sqrt{11}x) \qquad \text{general solution}$$

Using the condition that $y = 1$ when $x = 0$, we have

$$1 = e^0(c_1 \cos 0 + c_2 \sin 0) \quad \text{or} \quad c_1 = 1$$

Since $y' = 2$ when $x = 0$, we find the derivative and then evaluate c_2.

$$y' = e^x(c_1 \cos\sqrt{11}x + c_2 \sin\sqrt{11}x - \sqrt{11}c_1 \sin\sqrt{11}x + \sqrt{11}c_2 \cos\sqrt{11}x)$$

$$2 = e^0(\cos 0 + c_2 \sin 0 - \sqrt{11}\sin 0 + \sqrt{11}c_2 \cos 0) \qquad y' = 2 \text{ when } x = 0$$

$$2 = 1 + \sqrt{11}c_2, \quad c_2 = \frac{1}{11}\sqrt{11} \qquad \text{solve for } c_2$$

$$y = e^x\left(\cos\sqrt{11}x + \frac{1}{11}\sqrt{11}\sin\sqrt{11}x\right) \qquad \text{particular solution}$$

■

NOTE ➧ [If the root of the auxiliary equation is repeated more than once—for example, a triple root—an additional term with another arbitrary constant and the next higher power of x is added to the solution for each additional root. Also, if a pair of complex roots is repeated, an additional term with a factor of x and another arbitrary constant is added for each root of the pair.] These are illustrated in the following two examples.

EXAMPLE 7 Root repeated more than once

For the differential equation $D^3y + 3D^2y + 3Dy + y = 0$, the auxiliary equation is

$$m^3 + 3m^2 + 3m + 1 = 0, \quad (m + 1)^3 = 0$$

Each of the three roots is $m = -1$. The equation is a third-order equation, which means there are three arbitrary constants. Therefore, the general solution is

$$y = e^{-x}(c_1 + c_2x + c_3x^2)$$

■

EXAMPLE 8 Repeated complex roots

For the differential equation $D^4y + 8D^2y + 16y = 0$, the auxiliary equation is

$$m^4 + 8m^2 + 16 = 0, \quad (m^2 + 4)^2 = 0$$

With two factors of $m^2 + 4$, the roots are $2j$, $2j$, $-2j$, and $-2j$. The fourth-order equation, and four roots, indicate four arbitrary constants. Because $e^{0x} = 1$, the general solution is

$$y = (c_1 + c_2x)\sin 2x + (c_3 + c_4x)\cos 2x$$

■

Knowing the various types of possible solutions, it is possible to determine the differential equation if the solution is known. Consider the following example.

EXAMPLE 9 Determine an equation from the solution

(a) A solution of $y = c_1e^x + c_2e^{2x}$ indicates an auxiliary equation with roots of $m_1 = 1$ and $m_2 = 2$. Thus, the auxiliary equation is $(m - 1)(m - 2) = 0$, and the simplest form of the differential equation is $D^2y - 3Dy + 2y = 0$.

(b) A solution of $y = e^{2x}(c_1 + c_2x)$ indicates repeated roots $m_1 = m_2 = 2$ of the auxiliary equation $(m - 2)^2 = 0$, and a differential equation $D^2y - 4Dy + 4y = 0$. ■

EXERCISES 31.8

In Exercises 1–4, make the given changes in the indicated examples of this section, and then solve the resulting differential equations.

1. In Example 3, change the sign of the term $10\frac{dy}{dx}$.
2. In Example 3, delete the term $10\frac{dy}{dx}$.
3. In Example 3, delete the term $25y$.
4. In Example 4, change the $-$ sign to $+$.

In Exercises 5–32, solve the given differential equations.

5. $\frac{d^2y}{dx^2} - 2\frac{dy}{dx} + y = 0$
6. $\frac{d^2y}{dx^2} - 6\frac{dy}{dx} + 9y = 0$
7. $D^2y + 12\,Dy + 36y = 0$
8. $16\,D^2y + 8\,Dy + y = 0$
9. $\frac{d^2y}{dx^2} + 9y = 0$
10. $\frac{d^2y}{dx^2} + y = 0$
11. $D^2y + Dy + 2y = 0$
12. $D^2y + 4y = 2\,Dy$
13. $D^4y - y = 0$
14. $4\,D^2y = 12\,Dy - 9y$
15. $4\,D^2y + y = 0$
16. $9\,D^2y + 4y = 0$
17. $16y'' - 24y' + 9y = 0$
18. $9y'' - 24y' + 16y = 0$
19. $25y'' + 4y = 0$
20. $y'' + 5y = 4y'$
21. $2\,D^2y + 5y = 4\,Dy$
22. $D^2y + 4\,Dy + 6y = 0$
23. $25y'' + 16y = 40y'$
24. $9y''' + 0.6y'' + 0.01y' = 0$
25. $2\,D^2y - 3\,Dy - y = 0$
26. $D^2y - 5\,Dy = 14y$
27. $3\,D^2y + 12\,Dy = 2y$
28. $36\,D^2y = 25y$
29. $D^3y - 6\,D^2y + 12\,Dy - 8y = 0$
30. $D^4y - 2\,D^3y + 2\,D^2y - 2Dy + y = 0$
31. $D^4y + 2\,D^2y + y = 0$
32. $16\,D^4y - y = 0$

In Exercises 33–36, find the particular solutions of the given differential equations that satisfy the given conditions.

33. $y'' + 2y' + 10y = 0$; $y = 0$ when $x = 0$ and $y = e^{-\pi/6}$ when $x = \pi/6$
34. $9\,D^2y + 16y = 0$; $Dy = 0$ and $y = 2$ when $x = \pi/2$
35. $D^2y + 16y = 8\,Dy$; $Dy = 2$ and $y = 4$ when $x = 0$
36. $D^4y + 3\,D^3y + 2\,D^2y = 0$; $y = 0$ and $Dy = 4$ and $D^2y = -8$ and $D^3y = 16$ when $x = 0$

In Exercises 37–40, find the simplest form of the second-order homogeneous linear differential equation that has the given solution. In Exercises 38 and 39, explain how the equation is found.

37. $y = c_1e^{3x} + c_2e^{-3x}$
38. $y = c_1e^{3x} + c_2xe^{3x}$
39. $y = c_1\cos 3x + c_2\sin 3x$
40. $y = c_1e^{2x}\cos x + c_2e^{2x}\sin x$

In Exercises 41 and 42, solve the given problems.

41. The displacement y (in cm) of an object at the end of a spring is described by the equation $d^2y/dt^2 + 4\,dy/dt + 4y = 0$, where t is the time (in s). Find $y = f(t)$ if $f(0) = 0$ and $f(1) = 0.50$ cm.
42. Following the method of Example 1, solve the differential equation $D^2y + 4y = 0$. Do not use Eq. (31.17).

Answers to Practice Exercises

1. $y = e^{-4x}(c_1 + c_2x)$ 2. $y = e^{-x}(c_1\sin 2x + c_2\cos 2x)$

31.9 Solutions of Nonhomogeneous Equations

Complementary Solution • Particular Solution • Method of Undetermined Coefficients • Special Case

We now consider the solution of a nonhomogeneous linear equation of the form

$$a_0D^2y + a_1Dy + a_2y = b \tag{31.18}$$

where the a's are constants and b is a function of x or is a constant. When the solution is substituted into the left side, we must obtain b. Solutions found from the methods of Sections 31.7 and 31.8 give zero when substituted into the left side, but they do contain the arbitrary constants necessary in the solution. If we could find a particular solution that when substituted into the left side produced b, it could be added to the solution containing the arbitrary constants. Therefore, *the solution is of the form*

$$y = y_c + y_p \tag{31.19}$$

where y_c, called the **complementary solution,** *is obtained by solving the corresponding homogeneous equation and where y_p is the* **particular solution** *necessary to produce the expression b of Eq. (31.18).* It should be noted that y_p satisfies the differential equation, but it has no arbitrary constants and therefore cannot be the general solution. The arbitrary constants are part of y_c.

EXAMPLE 1 Complementary and particular solutions

The differential equation $D^2y - Dy - 6y = e^x$ has the solution

$$y = c_1e^{3x} + c_2e^{-2x} - \frac{1}{6}e^x$$

where the complementary solution y_c and particular solution y_p are

$$y_c = c_1e^{3x} + c_2e^{-2x} \qquad y_p = -\frac{1}{6}e^x$$

The complementary solution y_c is obtained by solving the corresponding homogeneous equation $D^2y - Dy - 6y = 0$, and we shall discuss below the method of finding y_p. Again, we note that y_c contains the arbitrary constants and y_p contains the expression needed to produce the e^x on the right. Therefore both are needed to have the complete general solution. ■

By inspecting the form of b on the right side of the equation, we can find the form that the particular solution must have. Because a combination of the particular solution and its derivatives must form the function b,

y_p is an expression that contains all possible forms of b and its derivatives.

The method that is used to find the exact form of y_p is called the **method of undetermined coefficients.**

EXAMPLE 2 Forms of particular solutions

(a) If the function b is $4x$, we choose the particular solution y_p to be of the form $y_p = A + Bx$. The Bx-term is included to account for the $4x$, and the A is included because the derivative of $4x$ is a constant.

(b) If the function b is e^{2x}, we choose the form of the particular solution to be $y_p = Ce^{2x}$. *Because all derivatives of e^{2x} are a constant times e^{2x}, no other terms are needed in y_p.*

(c) If the function b is $4x + e^{2x}$, we choose the form of the particular solution to be $y_p = A + Bx + Ce^{2x}$. ■

EXAMPLE 3 Forms of particular solutions

(a) If b is of the form $x^2 + e^{-x}$, we choose the particular solution to be of the form $y_p = A + Bx + Cx^2 + Ee^{-x}$.

(b) If b is of the form $xe^{-2x} - 5$, we choose the form of the particular solution to be $y_p = Ae^{-2x} + Bxe^{-2x} + C$.

(c) If b is of the form $x\sin x$, we choose the form of the particular solution to be $y_p = A\sin x + B\cos x + Cx\sin x + Ex\cos x$. All these types of terms occur in the derivatives of $x\sin x$. ■

Practice Exercise

1. Find the form of y_p if b is of the form $x + \cos x$.

EXAMPLE 4 Forms of particular solutions

(a) If b is of the form $e^x + xe^x$, we should then choose y_p to be of the form $y_p = Ae^x + Bxe^x$. These terms occur for xe^x and its derivatives. [Because the form of the e^x-term of b is already included in Ae^x, we do not include another e^x-term in y_p.]

NOTE ▸

(b) In the same way, if b is of the form $2x + 4x^2$, we choose the form of y_p to be $y_p = A + Bx + Cx^2$. *These are the only forms that occur in either $2x$ or $4x^2$ and their derivatives.* ■

NOTE ▸

Once we have determined the form of y_p, we have to find the numerical values of the coefficients $A, B, \ldots$. [The *method of undetermined coefficients* is to substitute the chosen form of y_p into the differential equation and equate the coefficients of like terms.]

EXAMPLE 5 Solving a nonhomogeneous equation

Solve the differential equation $D^2y - Dy - 6y = e^x$.

In this case, the solution of the auxiliary equation $m^2 - m - 6 = 0$ gives us the roots $m_1 = 3$ and $m_2 = -2$. Thus,

$$y_c = c_1e^{3x} + c_2e^{-2x}$$

The proper form of y_c is $y_p = Ae^x$. This means that $Dy_p = Ae^x$ and $D^2y_p = Ae^x$. Substituting y_p and its derivatives into the differential equation, we have

$$Ae^x - Ae^x - 6Ae^x = e^x$$

To produce equality, the coefficients of e^x must be the same on each side of the equation. Thus,

$$-6A = 1 \quad \text{or} \quad A = -1/6$$

Therefore, $y_p = -\frac{1}{6}e^x$. This gives the complete solution $y = y_c + y_p$,

$$y = c_1e^{3x} + c_2e^{-2x} - \frac{1}{6}e^x \qquad \text{see Example 1}$$

This solution checks when substituted into the original differential equation. ■

EXAMPLE 6 Solving a nonhomogeneous equation

Solve the differential equation $D^2y + 4y = x - 4e^{-x}$.

In this case, we have $m^2 + 4 = 0$, which give us $m_1 = 2j$ and $m_2 = -2j$. Therefore, $y_c = c_1 \sin 2x + c_2 \cos 2x$.

The proper form of the particular solution is $y_p = A + Bx + Ce^{-x}$. Finding two derivatives and then substituting into the differential equation gives

$$y_p = A + Bx + Ce^{-x} \qquad Dy_p = B - Ce^{-x} \qquad D^2y_p = Ce^{-x}$$

$$D^2y + 4y = x - 4e^{-x} \qquad \text{differential equation}$$

$$(Ce^{-x}) + 4(A + Bx + Ce^{-x}) = x - 4e^{-x} \qquad \text{substituting}$$

$$Ce^{-x} + 4A + 4Bx + 4Ce^{-x} = x - 4e^{-x}$$

$$(4A) + (4B)x + (5C)e^{-x} = 0 + (1)x + (-4)e^{-x} \qquad \text{note coefficients}$$

Equating the constants, and the coefficients of x and e^{-x} on either side gives

$$4A = 0 \quad 4B = 1 \quad 5C = -4$$
$$A = 0 \quad B = 1/4 \quad C = -4/5$$

This means that the particular solution is

$$y_p = \frac{1}{4}x - \frac{4}{5}e^{-x}$$

In turn, this tells us that the complete solution is

$$y = c_1 \sin 2x + c_2 \cos 2x + \frac{1}{4}x - \frac{4}{5}e^{-x}$$

Substitution into the original differential equation verifies this solution. ■

Practice Exercise

2. Find y_p for the differential equation $D^2y + 4y = 8xe^{2x}$.

EXAMPLE 6 Particular solution

Solve the differential equation $D^2y - 2\,Dy - 15y = 0$ and find the particular solution that satisfies the conditions $Dy = 2$ and $y = -1$, when $x = 0$. (It is necessary to give two conditions because there are two constants to evaluate.)

We have

$$m^2 - 2m - 15 = 0, \qquad (m-5)(m+3) = 0$$
$$m_1 = 5 \qquad m_2 = -3$$
$$y = c_1e^{5x} + c_2e^{-3x}$$

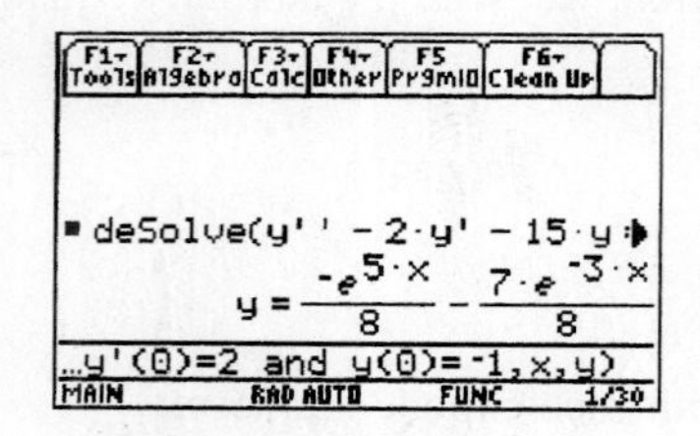

Fig. 31.12
TI-89 graphing calculator keystrokes: goo.gl/evjzRF

■ Figure 31.12 shows the particular solution of this differential equation on a calculator.

This equation is the general solution. In order to evaluate the constants c_1 and c_2, ***we use the given conditions to find two simultaneous equations in c_1 and c_2.*** These are then solved to determine the particular solution. Thus,

$$y' = 5c_1e^{5x} - 3c_2e^{-3x}$$

Using the given conditions in the general solution and its derivative, we have

$$c_1 + c_2 = -1 \qquad y = -1 \text{ when } x = 0$$
$$5c_1 - 3c_2 = 2 \qquad Dy = 2 \text{ when } x = 0$$

The solution to this system of equations is $c_1 = -\frac{1}{8}$ and $c_2 = -\frac{7}{8}$. The particular solution becomes

$$y = -\frac{1}{8}e^{5x} - \frac{7}{8}e^{-3x} \quad \text{or} \quad 8y + e^{5x} + 7e^{-3x} = 0$$ ■

EXERCISES 31.7

In Exercises 1 and 2, make the given changes in the indicated examples of this section and then solve the resulting differential equations.

1. In Example 3, delete the $+6y$ term.

2. In Example 4, add $16y$ to the right side.

In Exercises 3–26, solve the given differential equations.

3. $\frac{d^2y}{dx^2} - \frac{dy}{dx} - 6y = 0$

4. $\frac{d^2y}{dx^2} + \frac{dy}{dx} = 0$

5. $3\frac{d^2y}{dx^2} + 4\frac{dy}{dx} + y = 0$

6. $\frac{d^2y}{dx^2} - 2\frac{dy}{dx} - 8y = 0$

7. $D^2y - 3Dy = 0$

8. $D^2y = 25y$

9. $2\,D^2y - 3y = Dy$

10. $D^2y + 7\,Dy + 6y = 0$

11. $3\,D^2y + 12y = 20\,Dy$

12. $4\,D^2y + 12\,Dy = 7y$

13. $3y'' + 8y' - 3y = 0$

14. $8y'' + 6y' = 9y$

15. $3y'' + 2y' - y = 0$

16. $2y'' - 7y' + 6y = 0$

17. $2\frac{d^2y}{dx^2} - 4\frac{dy}{dx} + y = 0$

18. $\frac{d^2y}{dx^2} + \frac{dy}{dx} = 5y$

19. $4\,D^2y - 3\,Dy = 2y$

20. $2\,D^2y - 3\,Dy - y = 0$

21. $y'' = 3y' + y$

22. $5y'' - y' = 3y$

23. $y'' + y' = 8y$

24. $8y'' = y' + y$

25. $2\,D^2y + 5aDy - 12a^2 = 0 \quad (a > 0)$

26. $3k^4D^2y + 14k^2Dy - 5y = 0$

In Exercises 27–30, find the particular solutions of the given differential equations that satisfy the given conditions.

27. $D^2y - 4\,Dy - 21y = 0$; $Dy = 0$ and $y = 2$ when $x = 0$

28. $4\,D^2y - Dy = 0$; $Dy = 2$ and $y = 4$ when $x = 0$

29. $D^2y - Dy = 12y$; $y = 0$ when $x = 0$, and $y = 1$ when $x = 1$

30. $2\,D^2y + 5\,Dy = 0$; $y = 0$ when $x = 0$, and $y = 2$ when $x = 1$

In Exercises 31–34, solve the given third- and fourth-order differential equations.

31. $y''' - 2y'' - 3y' = 0$

32. $D^3y - 6\,D^2y + 11\,Dy - 6y = 0$

33. $D^4y - 5\,D^2y + 4y = 0$

34. $D^4y - D^3y - 9\,D^2y + 9\,Dy = 0$

In Exercises 35–38, solve the given problems.

35. The voltage v at a distance s along a transmission line is given by $d^2v/ds^2 = a^2v$, where a is called the *attenuation constant*. Solve for v as a function of s.

36. The displacement y (in cm) of an object at the end of a spring is described by the equation $d^2y/dt^2 + 2\,dy/dt + 4y = 0$, where t is the time (in s). Find $y = f(t)$ if $f(0) = 0$ and $f(1) = 2.00$ cm.

37. Following the method of Example 2, solve the differential equation $D^2y + 4Dy = 0$. Do not use Eqs. (31.14) and (31.15).

38. Following the method of Example 2, solve the differential equation $D^2y - 3Dy + 2y = 0$. Do not use Eqs. (31.14) and (31.15).

Answer to Practice Exercise

1. $y = c_1e^{-x} + c_2e^{3x/2}$

EXAMPLE 7 Solving a nonhomogeneous equation

Solve the differential equation $D^3y - 3\,D^2y + 2\,Dy = 10\sin x$.

$$m^3 - 3m^2 + 2m = 0, \quad m(m-1)(m-2) = 0, \quad m_1 = 0 \quad m_2 = 1 \quad m_3 = 2 \qquad \text{auxiliary equation}$$

$$y_c = c_1 + c_2e^x + c_3e^{2x} \qquad \text{complementary solution}$$

We now find the particular solution:

$$y_p = A\sin x + B\cos x \qquad \text{particular solution form}$$

$$Dy_p = A\cos x - B\sin x \quad D^2y_p = -A\sin x - B\cos x \quad D^3y_p = -A\cos x + B\sin x \qquad \text{find three derivatives}$$

$$(-A\cos x + B\sin x) - 3(-A\sin x - B\cos x) + 2(A\cos x - B\sin x) = 10\sin x \qquad \text{substitute into differential equation}$$

$$(3A - B)\sin x + (A + 3B)\cos x = 10\sin x$$

$$3A - B = 10 \quad A + 3B = 0 \qquad \text{equate coefficients}$$

The solution of this system is $A = 3, B = -1$.

$$y_p = 3\sin x - \cos x \qquad \text{particular solution}$$

$$y = c_1 + c_2e^x + c_3e^{2x} + 3\sin x - \cos x \qquad \text{complete general solution}$$

This solution checks when substituted into the differential equation. ■

EXAMPLE 8 Particular solution

Find the particular solution of $y'' + 16y = 2e^{-x}$ if $Dy = -2$ and $y = 1$ when $x = 0$.

In this case, we must not only find y_c and y_p, but we must also evaluate the constants of y_c from the given conditions. The solution is as follows:

$$D^2y + 16y = 2e^{-x} \qquad \text{operator } D \text{ form}$$

$$m^2 + 16 = 0, \quad m = \pm 4j \qquad \text{auxiliary equation}$$

$$y_c = c_1\sin 4x + c_2\cos 4x \qquad \text{complete general solution}$$

$$y_p = Ae^{-x} \qquad \text{particular solution form}$$

$$Dy_p = -Ae^{-x}, \qquad D^2y_p = Ae^{-x}$$

$$Ae^{-x} + 16Ae^{-x} = 2e^{-x} \qquad \text{substituting}$$

$$17Ae^{-x} = 2e^{-x}, \qquad A = \frac{2}{17} \qquad \text{equate coefficients}$$

$$y_p = \frac{2}{17}e^{-x}$$

$$y = c_1\sin 4x + c_2\cos 4x + \frac{2}{17}e^{-x} \qquad \text{complete general solution}$$

We now evaluate c_1 and c_2 from the given conditions:

$$Dy = 4c_1\cos 4x - 4c_2\sin 4x - \frac{2}{17}e^{-x}$$

$$1 = c_1(0) + c_2(1) + \frac{2}{17}(1), \qquad c_2 = \frac{15}{17} \qquad y = 1 \text{ when } x = 0$$

$$-2 = 4c_1(1) - 4c_2(0) - \frac{2}{17}(1), \quad c_1 = -\frac{8}{17} \qquad Dy = -2 \text{ when } x = 0$$

$$y = -\frac{8}{17}\sin 4x + \frac{15}{17}\cos 4x + \frac{2}{17}e^{-x} \qquad \text{required particular solution}$$

This solution checks when substituted into the differential equation. ■

A SPECIAL CASE

■ From Eq. (13.18), we note that the function b is the function on the right side of the differential equation.

It may happen that a term of the proposed y_p and a term of y_c are *similar terms*. Because any term of y_c gives zero when substituted in the differential equation, so will that term of the proposed y_p. This means the proposed y_p must be modified.

CAUTION If a term of the proposed y_p is similar to a term of y_c, any term of the proposed y_p, included to account for the similar term of the function b, must be multiplied by the smallest possible integer power of x such that any resulting term y_p is not similar to the term of y_c. ■

The following example shows that this is not as involved as it sounds.

EXAMPLE 9 Term of proposed y_p similar to a term of y_c

Solve the differential equation $D^2y - 2Dy + y = x + e^x$.

We find that the auxiliary equation and complementary solution are

$$m^2 - 2m + 1 = 0, \qquad (m-1)^2 = 0 \qquad m_1 = 1 \qquad m_2 = 1$$
$$y_c = e^x(c_1 + c_2x)$$

Based on the function b on the right side, the *proposed* form of y_p is

$$y_p = A + Bx + Ce^x \qquad \text{proposed form}$$

We now note that the term Ce^x is similar to the term c_1e^x of y_c. Therefore, we must multiply the term Ce^x by the smallest power of x such that it is not similar to any term of y_c. If we multiply by x, the term becomes similar to c_2xe^x. Therefore, we must multiply Ce^x by x^2 such that y_p is

■ Note that the $A + Bx$ are not multiplied by x^2 since they are not included in y_p to account for the e^x in the function b.

$$y_p = A + Bx + Cx^2e^x \qquad \text{correct modified form}$$

Using this form of y_p, we now complete the solution.

$$Dy_p = B + Cx^2e^x + 2Cxe^x \quad D^2y_p = Cx^2e^x + 2Cxe^x + 2Ce^x + 2Cxe^x = 2Ce^x + 4Cxe^x + Cx^2e^x$$
$$(2Ce^x + 4Cxe^x + Cx^2e^x) - 2(B + Cx^2e^x + 2Cxe^x) + (A + Bx + Cx^2e^x) = x + e^x$$
$$(A - 2B) + Bx + 2Ce^x = x + e^x$$
$$A - 2B = 0 \qquad B = 1 \qquad 2C = 1 \qquad A = 2 \qquad B = 1 \qquad C = 1/2$$
$$y_p = 2 + x + \tfrac{1}{2}x^2e^x$$
$$y = e^x(c_1 + c_2x) + 2 + x + \tfrac{1}{2}x^2e^x$$
■

EXERCISES 31.9

In Exercises 1–4, make the given changes in the indicated examples of this section and then solve the given problems.

1. In Example 3(a), add $2x$ to the form b and then determine the form of y_p.
2. In Example 5, change e^x to e^{2x} and then find the solution.
3. In Example 6, on the right side, change x to $\sin x$, and then find the proper form of y_c and y_p.
4. In Example 8, change the right side to $x^2 + xe^x$ and then find the proper form for y_p.

In Exercises 5–16, solve the given differential equations. The form of y_p is given.

5. $D^2y - Dy - 2y = 4$ (Let $y_p = A$.)
6. $D^2y - Dy - 6y = 4x$ (Let $y_p = A + Bx$.)
7. $D^2y - y = 2 + x^2$ (Let $y_p = A + Bx + Cx^2$.)
8. $D^2y + 4Dy + 3y = 2 + e^x$ (Let $y_p = A + Be^x$.)
9. $y'' - 3y' = 2e^x + xe^x$ (Let $y_p = Ae^x + Bxe^x$.)
10. $y'' + y' - 2y = 8 + 4x + 2xe^{2x}$ (Let $y_p = A + Bx + Ce^{2x} + Exe^{2x}$.)
11. $9D^2y - y = \sin x$ (Let $y_p = A\sin x + B\cos x$.)
12. $D^2y + 4y = \sin x + 4$ (Let $y_p = A + B\sin x + C\cos x$.)
13. $\dfrac{d^2y}{dx^2} - 2\dfrac{dy}{dx} + y = 2x + x^2 + \sin 3x$ (Let $y_p = A + Bx + Cx^2 + E\sin 3x + F\cos 3x$.)
14. $D^2y - y = e^{-x}$ (Let $y_p = Axe^{-x}$.)
15. $D^2y + 4y = -12\sin 2x$ (Let $y_p = Ax\sin 2x + Bx\cos 2x$.)
16. $y'' - 2y' + y = 3 + e^x$ (Let $y_p = A + Bx^2e^x$.)

In Exercises 17–32, solve the given differential equations.

17. $\frac{d^2y}{dx^2} - \frac{dy}{dx} - 30y = 10$

18. $2\frac{d^2y}{dx^2} + 11\frac{dy}{dx} - 6y = 8x$

19. $3\frac{d^2y}{dx^2} - \frac{dy}{dx} - 4y = 5e^{3x}$

20. $\frac{d^2y}{dx^2} + 4y = 2\sin 3x$

21. $D^2y - 4y = \sin x + 2\cos x$

22. $6D^2y + Dy - y = e^x - e^{-x}$

23. $D^2y + y = 4 + \sin 2x$

24. $D^2y - Dy + y = x + \sin x$

25. $D^2y + 5Dy + 4y = xe^x + 4$

26. $3D^2y + Dy - 2y = 4 + 2x + e^x$

27. $y''' - y' = \sin 2x$

28. $D^4y - y = x$

29. $D^2y + y = \cos x$

30. $4y'' - 4y' + y = 4e^{x/2}$

31. $D^2y + 2Dy = 8x + e^{-2x}$

32. $D^3y - Dy = 4e^{-x} + 3e^{2x}$

In Exercises 33–36, find the particular solution of each differential equation for the given conditions.

33. $D^2y - Dy - 6y = 5 - e^x$; $Dy = 4$ and $y = 2$ when $x = 0$

34. $3y'' - 10y' + 3y = xe^{-2x}$; $y' = -\frac{9}{35}$ and $y = -\frac{13}{35}$ when $x = 0$

35. $y'' + y = x + \sin 2x$; $y' = 1$ and $y = 0$ when $x = \pi$

36. $D^2y - 2\,Dy + y = xe^{2x} - e^{2x}$; $Dy = 4$ and $y = -2$ when $x = 0$

In Exercises 37–40, solve the given problems.

37. Solve the first-order equation $Dy - y = x^2$ by the method of undetermined coefficients. For this equation, why is this method easier than the method of Section 31.4?

38. Show that the equation $(D^2 + 1)y = x^3$, subject to the conditions that $y = 0$ for $x = 0$ and $x = \pi$, has no solution.

39. Show that $y = (c + x)\sin x$ satisfies the differential equation $D^2y + y = 2\cos x$ for any value of c.

40. The displacement y (in cm) of an object at the end of a spring is described by the equation $d^2y/dt^2 + 4\,dy/dt + 4y = 0$, where t is the time (in s). If $f(0) = 0$ and $f(1) = 0.50$ cm, solve for $y = f(t)$.

Answers to Practice Exercises

1. $y_p = A + Bx + C\sin x + D\cos x$ **2.** $y_p = -\frac{1}{2}e^{2x} + xe^{2x}$

31.10 Applications of Higher-order Equations

Simple Harmonic Motion • Damped Simple Harmonic Motion • Electric Circuits • Deflection of Beams

We now show important applications of second-order differential equations to simple harmonic motion and simple electric circuits. Also, we will show an application of a fourth-order differential equation to the deflection of a beam.

EXAMPLE 1 Simple harmonic motion

Simple harmonic motion may be defined as motion in a straight line for which the acceleration is proportional to the displacement and in the opposite direction. Examples of this type of motion are a weight on a spring, a simple pendulum, and an object bobbing in water. If x represents the displacement, d^2x/dt^2 is the acceleration.

Using the definition of simple harmonic motion, we have

$$\frac{d^2x}{dt^2} = -k^2x$$

(We chose k^2 for convenience of notation in the solution.) We write this equation in the form

$$D^2x + k^2x = 0 \qquad \text{here, } D = d/dt$$

The roots of the auxiliary equation are kj and $-kj$, and the solution is

$$x = c_1 \sin kt + c_2 \cos kt$$

This solution indicates an oscillating motion, which is known to be the case. If, for example, $k = 4$ and we know that $x = 2$ and $Dx = 0$ (which means the velocity is zero) for $t = 0$, we have

$$Dx = 4c_1 \cos 4t - 4c_2 \sin 4t$$

$$2 = c_1(0) + c_2(1) \qquad x = 2 \text{ for } t = 0$$

$$0 = 4c_1(1) - 4c_2(0) \qquad Dx = 0 \text{ for } t = 0$$

which gives $c_1 = 0$ and $c_2 = 2$. Therefore,

$$x = 2\cos 4t$$

is the equation relating the displacement and time; Dx is the velocity and D^2x is the acceleration. See Fig. 31.13. ■

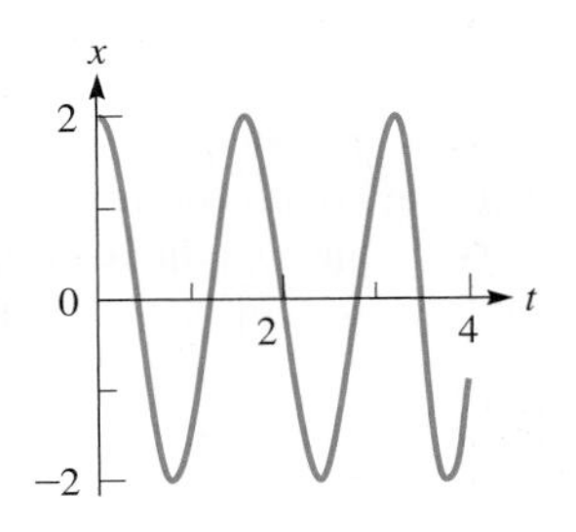

Fig. 31.13

Practice Exercise

1. In Example 1, find the solution if $x = 0$ and $Dx = 2$ when $t = 0$.

EXAMPLE 2 Damped simple harmonic motion

■ Newton's second law states that the net force acting on an object is equal to its mass times its acceleration. (This is one of Newton's best-known contributions to physics.)

In practice, an object moving with simple harmonic motion will in time cease to move due to unavoidable frictional forces. A "freely" oscillating object has a retarding force that is approximately proportional to the velocity. The differential equation for this case is $D^2x = -k^2x - bDx$. This results from applying (from physics) Newton's second law of motion (see the margin note at the left). Again, using the operator $D^2x = d^2x/dt^2$, the term D^2x represents the acceleration of the object, the term $-k^2x$ is a measure of the restoring force (of the spring, for example), and the term $-bDx$ represents the retarding (damping) force. This equation can be written as

$$D^2x + bDx + k^2x = 0$$

The auxiliary equation is $m^2 + bm + k^2 = 0$, for which the roots are

$$m = \frac{-b \pm \sqrt{b^2 - 4k^2}}{2}$$

If $k = 3$ and $b = 4$, $m = -2 \pm j\sqrt{5}$, which means the solution is

$$\mathbf{x = e^{-2t}(c_1 \sin \sqrt{5}t + c_2 \cos \sqrt{5}t)} \qquad \textbf{(1)}$$

Here, $4k^2 > b^2$, and this case is called **underdamped.** In this case, the object oscillates as the amplitude becomes smaller.

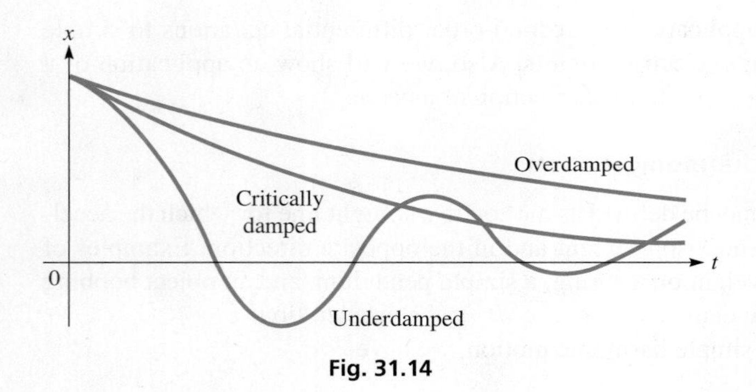

Fig. 31.14

If $k = 2$ and $b = 5$, $m = -1, -4$, which means the solution is

$$\mathbf{x = c_1e^{-t} + c_2e^{-4t}} \qquad \textbf{(2)}$$

Here, $4k^2 < b^2$, and the motion is called **overdamped.** Note that the motion is not oscillatory, since no sine or cosine terms appear. In this case, the object returns slowly to equilibrium without oscillating.

If $k = 2$ and $b = 4$, $m = -2, -2$, which means the solution is

$$\mathbf{x = e^{-2t}(c_1 + c_2t)} \qquad \textbf{(3)}$$

Here, $4k^2 = b^2$, and the motion is called **critically damped.** Again the motion is not oscillatory. In this case, there is just enough damping to prevent any oscillations. The object returns to equilibrium in the minimum time.

See Fig. 31.14, in which Eqs. (1), (2), and (3) are represented in general. The actual values depend on c_1 and c_2, which in turn depend on the conditions imposed on the motion. ■

EXAMPLE 3 Underdamped harmonic motion

In testing the characteristics of a particular type of spring, it is found that a weight of 16.0 lb stretches the spring 1.60 ft when the weight and spring are placed in a fluid that resists the motion with a force equal to twice the velocity. If the weight is brought to rest and then given a velocity of 12.0 ft/s, find the equation of motion. See Fig. 31.15.

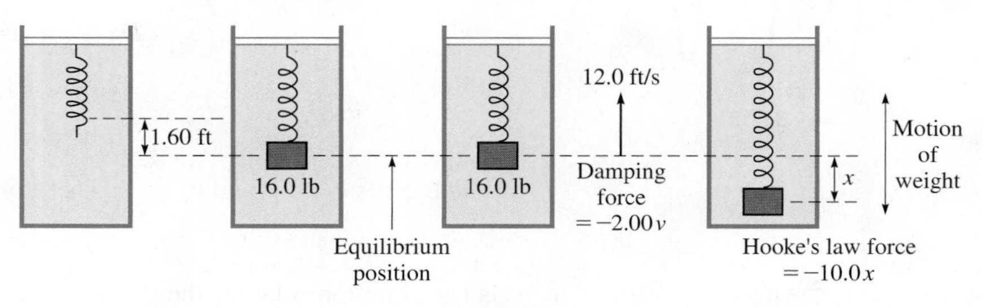

Fig. 31.15

In order to find the equation of motion, we use Newton's second law of motion (see Example 2). The weight (one force) at the end of the spring is offset by the equilibrium position force exerted by the spring, in accordance with Hooke's law (see Section 26.6). Therefore, the net force acting on the weight is the sum of the Hooke's law force due to the displacement from the equilibrium position and the resisting force. Using Newton's second law, we have

mass × acceleration = resisting force + Hooke's law force

$$mD^2x = -2.00\,Dx - kx$$

The mass of an object is its weight divided by the acceleration due to gravity. The weight is 16.0 lb, and the acceleration due to gravity is 32.0 ft/s². Thus, the mass m is

$$m = \frac{16.0 \text{ lb}}{32.0 \text{ ft/s}^2} = 0.500 \text{ slug}$$

where the slug is the unit of mass if the weight is in pounds.

The constant k for the Hooke's law force is found from the fact that the spring stretches 1.60 ft for a force of 16.0 lb. Thus, using Hooke's law,

$$16.0 = k(1.60), \quad k = 10.0 \text{ lb/ft}$$

This means that the differential equation to be solved is

$$0.500D^2x + 2.00\,Dx + 10.0x = 0$$

or

$$1.00D^2x + 4.00\,Dx + 20.0x = 0$$

Solving this equation, we have

$$1.00m^2 + 4.00m + 20.0 = 0 \qquad \text{auxiliary equation}$$

$$m = \frac{-4.00 \pm \sqrt{16.0 - 4(20.0)(1.00)}}{2.00}$$

$$= -2.00 \pm 4.00j \qquad \text{complex roots}$$

$$x = e^{-2.00t}(c_1 \cos 4.00t + c_2 \sin 4.00t) \qquad \text{general solution}$$

Because the weight started from the equilibrium position with a velocity of 12.0 ft/s, we know that $x = 0$ and $Dx = 12.0$ for $t = 0$. Thus,

$$0 = e^0(c_1 + 0c_2) \quad \text{or} \quad c_1 = 0 \qquad x = 0 \text{ for } t = 0$$

Thus, because $c_1 = 0$, we have

$$x = c_2 e^{-2.00t} \sin 4.00t$$

$$Dx = c_2 e^{-2.00t}(\cos 4.00t)(4.00) + c_2 \sin 4.00t(e^{-2.00t})(-2.00)$$

$$12.0 = c_2 e^0(1)(4.00) + c_2(0)(e^0)(-2.00) \qquad Dx = 12.0 \text{ for } t = 0$$

$$c_2 = 3.00$$

This means that the equation of motion is

$$x = 3.00e^{-2.00t} \sin 4.00t$$

The motion is underdamped; the graph is shown in Fig. 31.16. ■

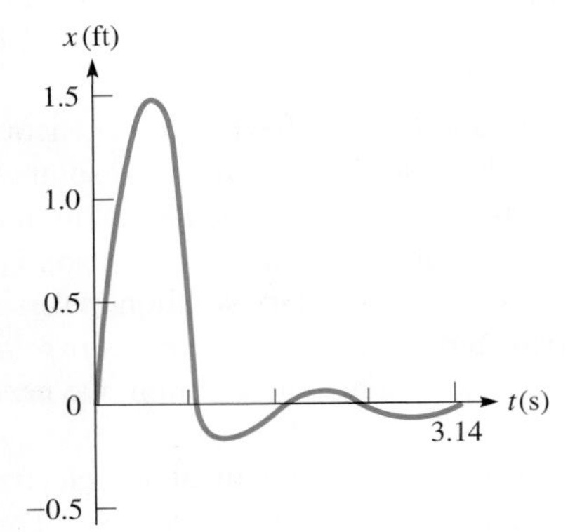

Fig. 31.16

In Example 3, the force was given in pounds, and the mass was therefore expressed in slugs. If metric units are used, it is common to give the mass of the object in kilograms, and its weight is therefore in newtons. The weight, which is needed to determine the constant k in the Hooke's law force, is found by multiplying the mass, in kilograms, by the acceleration of gravity, 9.80 m/s^2.

It is possible to have an additional force acting on a weight such as the one in Example 3. For example, a vibratory force may be applied to the support of the spring. In such a case, called ***forced vibrations,*** this additional external force is added to the other net force. This means that the added force $F(t)$ becomes a nonzero function on the right side of the differential equation, and we must then solve a nonhomogeneous equation.

EXAMPLE 4 Electric circuits

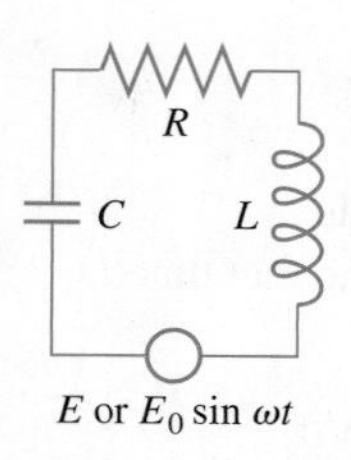

Fig. 31.17

The impressed voltage in the electric circuit shown in Fig. 31.17 equals the sum of the voltages across the components of the circuit. For this circuit with a resistance R, an inductance L, a capacitance C, and a voltage source E, we have

$$L\frac{d^2q}{dt^2} + R\frac{dq}{dt} + \frac{q}{C} = E \tag{31.20}$$

By definition, q represents the electric charge, $dq/dt = i$ is the current, and d^2q/dt^2 is the time rate of change of current. This equation may be written as

$$LD^2q + RDq + q/C = E$$

The auxiliary equation is $Lm^2 + Rm + 1/C = 0$. The roots are

$$m = \frac{-R \pm \sqrt{R^2 - 4L/C}}{2L} = -\frac{R}{2L} \pm \sqrt{\frac{R^2}{4L^2} - \frac{1}{LC}}$$

If we let $a = R/2L$ and $\omega = \sqrt{1/LC - R^2/4L^2}$, we have (assuming complex roots, which corresponds to realistic values of R, L, and C)

$$q_c = e^{-at}(c_1 \sin \omega t + c_2 \cos \omega t)$$

This indicates an oscillating charge, or an alternating current. However, the exponential term usually is such that the current dies out rapidly unless there is a source of voltage in the circuit. ■

If there is no source of voltage in the circuit of Example 4, we have a homogeneous differential equation to solve. If we have a constant voltage source, the particular solution is of the form $q_p = A$. If there is an alternating voltage source, the particular solution is of the form $q_p = A \sin \omega_1 t + B \cos \omega_1 t$, where ω_1 is the angular velocity of the source.

NOTE ▸ [After a very short time, the exponential factor in the complementary solution makes it negligible. For this reason, it is referred to as the **transient** term, and *the particular solution is the* **steady-state** *solution.*] Therefore, to find the steady-state solution, we need find only the particular solution.

It should be noted that the complementary solutions of the mechanical and electric cases are of identical form. There is also an equivalent mechanical case to that of an impressed sinusoidal voltage source in the electric case. This arises in the case of forced vibrations, when an external force affecting the vibrations is applied to the system. Thus, we may have transient and steady-state solutions to mechanical and other nonelectric situations.

EXAMPLE 5 Electric circuit

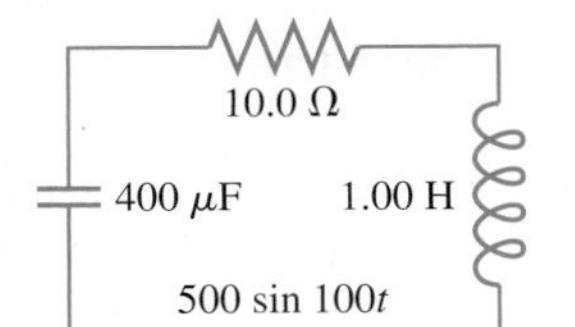

Fig. 31.18

Find the steady-state solution for the current in a circuit containing the following elements: $C = 400\ \mu\text{F}$, $L = 1.00$ H, $R = 10.0\ \Omega$, and a voltage source of $500 \sin 100t$. See Fig. 31.18.

This means the differential equation to be solved is

$$\frac{d^2q}{dt^2} + 10\frac{dq}{dt} + \frac{10^4}{4}q = 500 \sin 100t$$

Because we wish to find the steady-state solution, we must find q_p, from which we may find i_p by finding a derivative. The solution now follows:

$$q_p = A \sin 100t + B \cos 100t \qquad \text{particular solution form}$$

$$\frac{dq_p}{dt} = 100A \cos 100t - 100B \sin 100t$$

$$\frac{d^2q_p}{dt} = -10^4 A \sin 100t - 10^4 B \cos 100t$$

$$-10^4 A \sin 100t - 10^4 B \cos 100t + 10^3 A \cos 100t - 10^3 B \sin 100t \qquad \text{substitute into differential equation}$$
$$+ \frac{10^4}{4}A \sin 100t + \frac{10^4}{4}B \cos 100t = 500 \sin 100t$$

$$(-0.75 \times 10^4 A - 10^3 B) \sin 100t + (-0.75 \times 10^4 B + 10^3 A) \cos 100t = 500 \sin 100t$$
$$-7.5 \times 10^3 A - 10^3 B = 500 \qquad \text{equate coefficients of } \sin 100t$$
$$10^3 A - 7.5 \times 10^3 B = 0 \qquad \text{equate coefficients of } \cos 100t$$

Solving these equations, we obtain

$$B = -8.73 \times 10^{-3} \quad \text{and} \quad A = -65.5 \times 10^{-3}$$

Therefore,

$$q_p = -65.5 \times 10^{-3} \sin 100t - 8.73 \times 10^{-3} \cos 100t$$

$$i_p = \frac{dq_p}{dt} = -6.55 \cos 100t + 0.87 \sin 100t$$

which is the required solution. (We assumed three significant digits for the data but did not use all of them in most equations of the solution.) ■

The solutions to the second-order differential equations for the applications of simple harmonic motion and electric circuits generally include sines and cosines, because of the oscillatory nature of these applications. We now consider problems involving the **deflections of beams,** which involve fourth-order differential equations and algebraic functions.

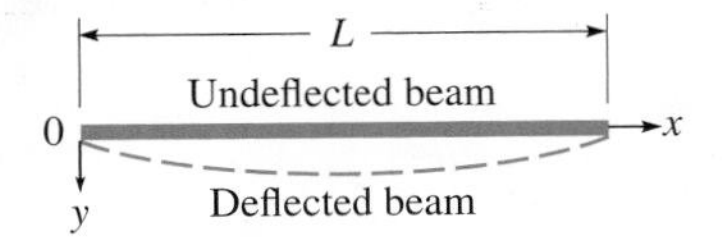

Fig. 31.19

In the study of the strength of materials and elasticity, it is shown that the deflection y of a beam of length L satisfies the differential equation $\boldsymbol{EI\, d^4y/dx^4 = w(x)}$, where EI is a measure of the stiffness of the beam and $w(x)$ is the weight distribution along the beam. See Fig. 31.19. Since this is a fourth-order equation, it is necessary to specify four conditions to obtain a solution. These conditions are determined by the way in which the ends, where $x = 0$ and where $x = L$, are held. For an end held in the specified manner, these conditions are: *clamped:* $y = 0$ and $y' = 0$; *hinged:* $y = 0$ and $y'' = 0$; and *free:* $y'' = 0$ and $y''' = 0$ ($y' = 0$ indicates no change in alignment; $y'' = 0$ indicates no curvature; $y''' = 0$ indicates no shearing force). Because the conditions are given for specific positions, this kind of problem is called a *boundary value problem.* Consider the following example.

EXAMPLE 6 Deflection of a beam

A uniform beam of length L is hinged at both ends and has a constant load distribution of w due to its own weight. Find the deflection y of the beam in terms of the distance x from one end of the beam.

Using the differential equation given on the previous page, we have $EI\, d^4y/dx^4 = w$. For convenience in the solution, let $k = w/EI$. Thus, the solution is as follows:

$$D^4y = k$$

$$m^4 = 0 \qquad m_1 = m_2 = m_3 = m_4 = 0$$

Because the four roots of the auxiliary equation are equal,

$$y_c = c_1 + c_2x + c_3x^2 + c_4x^3$$

The form of y_p indicates that we must multiply the *proposed* $y_p = A$ by x^4 so that it is not similar to any of the terms of y_c. This gives us $y_p = Ax^4$. Therefore, we now find the general solution as follows:

$$y_p = Ax^4 \qquad Dy_p = 4Ax^3 \qquad D^2y_p = 12Ax^2 \qquad D^3y_p = 24Ax \qquad D^4y_p = 24A$$

$$24A = k, \qquad A = k/24$$

$$y = c_1 + c_2x + c_3x^2 + c_4x^3 + \frac{k}{24}x^4 \quad \text{general solution}$$

NOTE ▸ From the discussion about beams before this example, we now find four boundary conditions in order to evaluate the four constants in y_c. [For a beam hinged at both ends, we know that $y = 0$ and $D^2y = 0$ for both $x = 0$ and $x = L$.] Therefore, we now find four derivatives, use these conditions, and thereby find y_c.

find derivatives

$$Dy = c_2 + 2c_3x + 3c_4x^2 + \frac{k}{6}x^3$$

$$D^2y = 2c_3 + 6c_4x + \frac{k}{2}x^2$$

$$D^3y = 6c_4 + kx$$

$$D^4y = k$$

use conditions to evaluate constants

$$\text{At } x = 0, y = 0: \quad c_1 = 0; \qquad \text{At } x = 0, D^2y = 0: \quad c_3 = 0$$

$$\text{At } x = L, D^2y = 0: \quad 0 = 6c_4L + \frac{kL^2}{2}; \quad c_4 = -\frac{kL}{12}$$

$$\text{At } x = L, y = 0: \quad 0 = c_2L + c_4L^3 + \frac{kL^4}{24}$$

$$0 = c_2L + \left(-\frac{kL}{12}\right)L^3 + \frac{kL^4}{24}; \quad c_2 = \frac{kL^3}{24}$$

Therefore, substituting the values of the four constants in the general solution above, the particular solution that satisfies these conditions is

$$y = \frac{kL^3}{24}x - \frac{kL}{12}x^3 + \frac{kx^4}{24} = \frac{k}{24}(L^3x - 2Lx^3 + x^4)$$

$$= \frac{w}{24EI}(L^3x - 2Lx^3 + x^4) \quad k = w/EI$$

■

EXERCISES 31.10

In Exercises 1 and 2, make the given changes in the indicated examples of this section and then solve the resulting problems.

1. In Example 1, change the conditions that $x = 2$ and $Dx = 0$ for $t = 0$ to $x = 2$ and $Dx = 4$ for $t = 0$.

2. In Example 5, change the voltage source to $500 \cos 100t$.

In Exercises 3–28, solve the given problems.

3. An object moves with simple harmonic motion according to $D^2x + 0.2Dx + 100x = 0$, $D = d/dt$. Find the displacement as a function of time, subject to the conditions $x = 4$ and $Dx = 0$ when $t = 0$.

4. What must be the value of b so that the motion of an object given by the equation $D^2x + bDx + 100x = 0$ is critically damped?

5. For the motion of the object in Exercise 4, (a) what values of b give underdamped motion?; (b) what values of b give overdamped motion?

6. When the angular displacement θ of a pendulum is small (less than about 6°), the pendulum moves with simple harmonic motion closely approximated by $D^2\theta + \frac{g}{l}\theta = 0$. Here, $D = d/dt$, g is the acceleration due to gravity, and l is the length of the pendulum. Find θ as a function of time (in s) if $g = 9.8 \text{ m/s}^2$, $l = 1.0 \text{ m}$, $\theta = 0.1$, and $D\theta = 0$ when $t = 0$. Sketch the curve.

7. For the pendulum in Exercise 6, what values, if any, of the length l give a critically or overdamped solution? Explain.

8. A block of wood floating in oil is depressed from its equilibrium position such that its equation of motion is $D^2y + 8Dy + 3y = 0$, where y is the displacement (in in.) and $D = d/dt$. Find its displacement after 12 s if $y = 6.0$ in. and $Dy = 0$ when $t = 0$.

9. A car suspension is depressed from its equilibrium position such that its equation of motion is $D^2y + b\,Dy + 25y = 0$, where y is the displacement and $D = d/dt$. What must be the value of b if the motion is critically damped?

10. In an electric circuit, if a capacitor discharges through a negligible resistance, the current i is related to the time t by the equation $d^2i/dt^2 = -a^2i$, where a is a constant. Find the frequency of the current if $a = 1000$.

11. For an elastic band that is stretched vertically, with one end fixed and a mass m at the other end, the displacement s of the mass is given by $m\frac{d^2s}{dt^2} = -\frac{mg}{e}(s - L)$, where L is the natural length of the band and e is the elongation due to the weight mg. Find s if $s = s_0$ and $ds/dt = 0$ when $t = 0$.

12. A mass of 0.820 kg stretches a given spring by 0.250 m. The mass is pulled down 0.150 m below the equilibrium position and released. Find the equation of motion of the mass if there is no damping.

13. A 4.00-lb weight stretches a certain spring 0.125 ft. With this weight attached, the spring is pulled 3.00 in. longer than its equilibrium length and released. Find the equation of the resulting motion, assuming no damping.

14. Find the solution for the spring of Exercise 13 if a damping force numerically equal to the velocity is present.

15. Find the solution for the spring of Exercise 13 if no damping is present but an external force of $4 \sin 2t$ is acting on the spring.

16. Find the solution for the spring of Exercise 13 if the damping force of Exercise 14 and the impressed force of Exercise 15 are both acting.

17. Find the equation relating the charge and the time in an electric circuit with the following elements: $L = 0.200$ H, $R = 8.00\ \Omega$, $C = 1.00\ \mu\text{F}$, and $E = 0$. In this circuit, $q = 0$ and $i = 0.500$ A when $t = 0$.

18. For a given electric circuit, $L = 2$ mH, $R = 0$, $C = 50$ nF, and $E = 0$. Find the equation relating the charge and the time if $q = 10^5\ C$ and $i = 0$ when $t = 0$.

19. For a given circuit, $L = 0.100$ H, $R = 0$, $C = 100\ \mu\text{F}$, and $E = 100$ V. Find the equation relating the charge and the time if $q = 0$ and $i = 0$ when $t = 0$.

20. Find the relation between the current and the time for the circuit of Exercise 19.

21. For a radio tuning circuit, $L = 0.500$ H, $R = 10.0\ \Omega$, $C = 200\ \mu\text{F}$, and $E = 120 \sin 120\pi t$. Find the equation relating the charge and time.

22. Find the steady-state current for the circuit of Exercise 21.

23. In a given electric circuit $L = 8.00$ mH, $R = 0$, $C = 0.500\ \mu\text{F}$, and $E = 20.0e^{-200t}$ mV. Find the relation between the current and the time if $q = 0$ and $i = 0$ for $t = 0$.

24. Find the current as a function of time for a circuit in which $L = 0.400$ H, $R = 60.0\ \Omega$, $C = 0.200\ \mu\text{F}$, and $E = 0.800e^{-100t}$ V, if $q = 0$ and $i = 5.00$ mA for $t = 0$.

25. Find the steady-state current for a circuit with $L = 1.00$ H, $R = 5.00\ \Omega$, $C = 150\ \mu\text{F}$, and $E = 120 \sin 100t$ V.

26. Find the steady-state solution for the current in an electric circuit containing the following elements: $C = 20.0\ \mu\text{F}$, $L = 2.00$ H, $R = 20.0\ \Omega$, and $E = 200 \sin 10t$.

27. A *cantilever* beam is clamped at the end $x = 0$ and is free at the end $x = L$. Find the equation for the deflection y of the beam in terms of the distance x from one end if it has a constant load distribution of w due to its own weight. See Fig. 31.20.

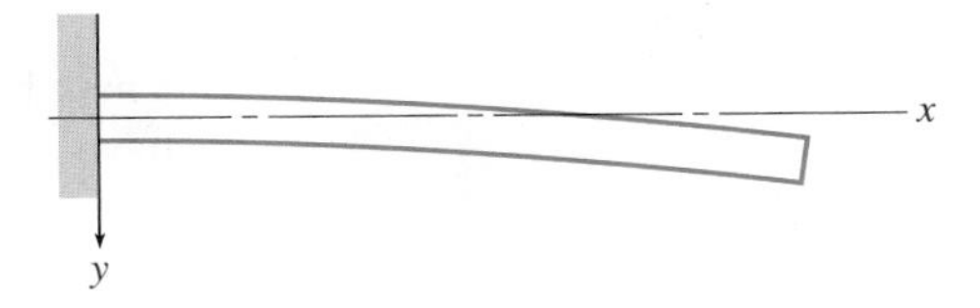

Fig. 31.20

28. A beam 10 m in length is hinged at both ends and has a variable load distribution of $w = kEIx$, where $k = 7.2 \times 10^{-4}/\text{m}$ and x is the distance from one end. Find the equation of the deflection y in terms of x.

Answer to Practice Exercise

1. $x = \frac{1}{2}\sin 4t$

31.11 Laplace Transforms

Definition of Laplace Transform • Improper Integral • Table of Laplace Transforms • Linearity Property • Inverse Transforms

Laplace transforms *provide an algebraic method of obtaining a* ***particular*** *solution of a differential equation from stated initial conditions.* Because this is frequently what is wanted, Laplace transforms are often used in engineering and electronics. The treatment in this text is intended only an as introduction to Laplace transforms.

The Laplace transform of a function $f(t)$ is defined as the function $F(s)$ as

■ Named for the French mathematician and astronomer Pierre Laplace (1749–1827).

$$F(s) = \int_0^{\infty} e^{-st}f(t)\,dt \tag{31.21}$$

By writing the transform as $F(s)$, we show that the result of integrating and evaluating is a function of s. To denote that we are dealing with "the Laplace transform of the function $f(t)$," the notation $\mathscr{L}(f)$ is used. Thus,

$$F(s) = \mathscr{L}(f) = \int_0^{\infty} e^{-st}f(t)\,dt \tag{31.22}$$

We shall see that both notations are quite useful.

In Eqs. (31.21) and (31.22), we note that the upper limit is ∞, which means it is unbounded. This integral is one type of what is known as an **improper integral.** In evaluating at the upper limit, it is necessary to find the limit of the resulting function as the upper limit approaches infinity. This may be shown as

$$\lim_{c\to\infty}\int_0^{c} e^{-st}f(t)\,dt$$

where we substitute c for t in the resulting function and determine the limit as $c \to \infty$. This also means that the Laplace transform, $F(s)$, is defined only for those values of s for which the limit is defined.

EXAMPLE 1 Finding a transform from the definition

Find the Laplace transform of the function $f(t) = t, t > 0$.

By the definition of the Laplace transform

$$\mathscr{L}(f) = \mathscr{L}(t) = \int_0^{\infty} e^{-st}t\,dt$$

This may be integrated by parts or by Formula (44) in Appendix D. Using the formula, we have

$$\mathscr{L}(t) = \int_0^{\infty} te^{-st}\,dt = \lim_{c\to\infty}\int_0^{c} te^{-st}\,dt = \lim_{c\to\infty}\frac{e^{-st}(-st-1)}{s^2}\bigg|_0^c$$

$$= \lim_{c\to\infty}\left[\frac{e^{-sc}(-sc-1)}{s^2}\right] + \frac{1}{s^2}$$

■ L'Hospital's rule is discussed in Section 27.7.

Now, for $s > 0$, as $c \to \infty$, $e^{-sc} \to 0$ and $sc \to \infty$. However, $e^{-sc} \to 0$ much faster than $sc \to \infty$. Using L'Hospital's rule, the above limit can be shown to be zero, and therefore,

$$\mathscr{L}(t) = \frac{1}{s^2} \quad \text{defined for } s > 0$$

■

EXAMPLE 2 Finding a transform from the definition

Find the Laplace transform of the function $f(t) = \cos at$.

By definition,

$$\mathcal{L}(f) = \mathcal{L}(\cos at) = \int_0^{\infty} e^{-st}\cos at\,dt$$

Using Formula (50) in Appendix D, we have

$$\begin{aligned}\mathcal{L}(\cos at) &= \int_0^{\infty} e^{-st}\cos at\,dt = \lim_{c\to\infty}\int_0^{c} e^{-st}\cos at\,dt\\ &= \lim_{c\to\infty}\frac{e^{-st}(-s\cos at + a\sin at)}{s^2+a^2}\Bigg|_0^c\\ &= \lim_{c\to\infty}\frac{e^{-sc}(-s\cos ac + a\sin ac)}{s^2+a^2} - \left(-\frac{s}{s^2+a^2}\right)\\ &= 0 + \frac{s}{s^2+a^2} = \frac{s}{s^2+a^2} \qquad (s > 0)\end{aligned}$$

Therefore, the Laplace transform of the function $\cos at$ is

$$\mathcal{L}(\cos at) = \frac{s}{s^2+a^2}$$

NOTE ➧ [In both examples, the resulting transform was an algebraic function of s]. ■

We now present Table 31.1, a short table of Laplace transforms. They are sufficient for our work in this chapter. More complete tables are available in many references.

Table 31.1 Laplace Transforms

	$f(t) = \mathcal{L}^{-1}(F)$	$\mathcal{L}(f) = F(s)$		$f(t) = \mathcal{L}^{-1}(F)$	$\mathcal{L}(f) = F(s)$
1.	1	$\frac{1}{s}$	11.	te^{-at}	$\frac{1}{(s+a)^2}$
2.	$\frac{t^{n-1}}{(n-1)!}$	$\frac{1}{s^n}(n = 1, 2, 3, \ldots)$	12.	$t^{n-1}e^{-at}$	$\frac{(n-1)!}{(s+a)^n}$
3.	e^{-at}	$\frac{1}{s+a}$	13.	$e^{-at}(1-at)$	$\frac{s}{(s+a)^2}$
4.	$1 - e^{-at}$	$\frac{a}{s(s+a)}$	14.	$[(b-a)t+1]e^{-at}$	$\frac{s+b}{(s+a)^2}$
5.	$\cos at$	$\frac{s}{s^2+a^2}$	15.	$\sin at - at\cos at$	$\frac{2a^3}{(s^2+a^2)^2}$
6.	$\sin at$	$\frac{a}{s^2+a^2}$	16.	$t\sin at$	$\frac{2as}{(s^2+a^2)^2}$
7.	$1 - \cos at$	$\frac{a^2}{s(s^2+a^2)}$	17.	$\sin at + at\cos at$	$\frac{2as^2}{(s^2+a^2)^2}$
8.	$at - \sin at$	$\frac{a^3}{s^2(s^2+a^2)}$	18.	$t\cos at$	$\frac{s^2-a^2}{(s^2+a^2)^2}$
9.	$e^{-at} - e^{-bt}$	$\frac{b-a}{(s+a)(s+b)}$	19.	$e^{-at}\sin bt$	$\frac{b}{(s+a)^2+b^2}$
10.	$ae^{-at} - be^{-bt}$	$\frac{s(a-b)}{(s+a)(s+b)}$	20.	$e^{-at}\cos bt$	$\frac{s+a}{(s+a)^2+b^2}$

An important property of transforms is the **linearity property,**

$$\mathcal{L}[af(t) + bg(t)] = a\mathcal{L}(f) + b\mathcal{L}(g) \quad \textbf{(31.23)}$$

We state this property here because it determines that the transform of a sum of functions is the sum of the transforms. This is of definite importance when dealing with a sum of functions. This property is a direct result of the definition of the Laplace transform.

Another Laplace transform important to the solution of a differential equation is the transform of the derivative of a function. Let us first find the Laplace transform of the first derivative of a function.

By definition,

$$\mathcal{L}(f') = \int_0^\infty e^{-st}f'(t)\,dt$$

To integrate by parts, let $u = e^{-st}$ and $dv = f'(t)\,dt$, so $du = -se^{-st}dt$ and $v = f(t)$ (the integral of the derivative of a function is the function). Therefore,

$$\mathcal{L}(f') = e^{-st}f(t)\big|_0^\infty + s\int_0^\infty e^{-st}f(t)dt$$

$$= 0 - f(0) + s\mathcal{L}(f)$$

It is noted that the integral in the second term on the right is the Laplace transform of $f(t)$ by definition. Therefore, *the Laplace transform of the first derivative of a function is*

$$\mathcal{L}(f') = s\mathcal{L}(f) - f(0) \quad \textbf{(31.24)}$$

Applying the same analysis, we may find *the Laplace transform of the second derivative of a function. It is*

$$\mathcal{L}(f'') = s^2\mathcal{L}(f) - sf(0) - f'(0) \quad \textbf{(31.25)}$$

Here, it is necessary to integrate by parts twice to derive the result. The transforms of higher derivatives are found in a similar manner.

Equations (31.24) and (31.25) allow us to express the transform of each derivative in terms of s and the transform itself. This is illustrated in the following example.

EXAMPLE 3 Linearity property—transforms of derivatives

Given that $f(0) = 0$ and $f'(0) = 1$, express the transform of $f''(t) - 2f'(t)$ in terms of s and the transform of $f(t)$.

By using the linearity property and the transforms of the derivatives, we have

$$\mathcal{L}[f''(t) - 2f'(t)] = \mathcal{L}(f'') - 2\mathcal{L}(f') \quad \text{using Eq. (31.23)}$$

$$= [s^2\mathcal{L}(f) - sf(0) - f'(0)] - 2[s\mathcal{L}(f) - f(0)] \quad \text{using Eqs. (31.25) and (31.24)}$$

$$= [s^2\mathcal{L}(f) - s(0) - 1] - 2[s\mathcal{L}(f) - 0] \quad \text{substitute given values}$$

$$= (s^2 - 2s)\mathcal{L}(f) - 1$$

■

Practice Exercise

1. Given that $f(0) = 1$ and $f'(0) = 0$, express the transform of $f''(t) - f'(t)$ in terms of s and the transform of $f(t)$.

NOTE ▸ [In the next section we will show how the properties of Laplace transforms, and their derivatives, lead directly to particular solutions of differential equations when the given conditions involve the values of the function and derivatives for $t = 0$.]

INVERSE TRANSFORMS

If the Laplace transform of a function is known, it is then possible to find the function by finding the **inverse transform,**

$$\mathscr{L}^{-1}(F) = f(t) \qquad \textbf{(31.26)}$$

where $\mathscr{L}^{-1}$ denotes the inverse transform.

EXAMPLE 4 Inverse transform from Table 31.1

If $F(s) = \frac{s}{s^2 + a^2}$, from Transform (5) of Table 13.1, we see that

$$\mathscr{L}^{-1}(F) = \mathscr{L}^{-1}\left(\frac{s}{s^2 + a^2}\right) = \cos at$$

$$f(t) = \cos at$$

■

EXAMPLE 5 Inverse transform from Table 31.1

If $(s^2 - 2s)\mathscr{L}(f) - 1 = 0$, then

$$\mathscr{L}(f) = \frac{1}{s^2 - 2s} \quad \text{or} \quad F(s) = \frac{1}{s(s-2)}$$

Therefore, we have

$$f(t) = \mathscr{L}^{-1}(F) = \mathscr{L}^{-1}\left[\frac{1}{s(s-2)}\right] \qquad \text{inverse transform}$$

$$= -\frac{1}{2}\mathscr{L}^{-1}\left[\frac{-2}{s(s-2)}\right] \qquad \text{fit form of Transform (4)}$$

$$= -\frac{1}{2}(1 - e^{2t}) \qquad \text{use Transform (4)}$$

■

Practice Exercise

2. Find $f(t)$ if $F(s) = \frac{6}{s^2 + 9}$.

The introduction of the factor -2 in Example 3 illustrates that it often takes some algebra steps to get $F(s)$ to match the proper form in Table 13.1. Another algebraic step that may be useful is *completing the square.* For a review of this algebraic method, see Section 7.2. The following example illustrates its use in finding an inverse transform.

EXAMPLE 6 Inverse transform—completing the square

If $F(s) = \frac{s + 5}{s^2 + 6s + 10}$, then

$$\mathscr{L}^{-1}(F) = \mathscr{L}^{-1}\left[\frac{s + 5}{s^2 + 6s + 10}\right]$$

It appears that this function does not fit any of the forms given. However,

$$s^2 + 6s + 10 = (s^2 + 6s + 9) + 1 = (s + 3)^2 + 1$$

By writing $F(s)$ as

$$F(s) = \frac{(s + 3) + 2}{(s + 3)^2 + 1} = \frac{s + 3}{(s + 3)^2 + 1} + \frac{2}{(s + 3)^2 + 1}$$

we can find the inverse of each term. Therefore,

$$\mathscr{L}^{-1}(F) = e^{-3t}\cos t + 2e^{-3t}\sin t \qquad \text{using Transforms (20) and (19)}$$

$$f(t) = e^{-3t}(\cos t + 2\sin t)$$

■

The following example shows how *partial fractions* can be used to find the inverse transform of $F(s)$. For a review of the method of expressing a given algebraic fraction in terms of partial fractions, refer to Sections 28.9 and 28.10.

EXAMPLE 7 Inverse transform—partial fractions

If $F(s) = \dfrac{5s^2 - 17s + 32}{s^3 - 8s^2 + 16s}$, then

$$\mathscr{L}^{-1}(F) = \mathscr{L}^{-1}\left[\frac{5s^2 - 17s + 32}{s^3 - 8s^2 + 16s}\right]$$

To fit forms in Table 13.1, we will now use partial fractions.

$$\frac{5s^2 - 17s + 32}{s^3 - 8s^2 + 16s} = \frac{5s^2 - 17s + 32}{s(s-4)^2} = \frac{A}{s} + \frac{B}{s-4} + \frac{C}{(s-4)^2}$$ factor of s, repeated factor $s - 4$

$$5s^2 - 17s + 32 = A(s-4)^2 + Bs(s-4) + Cs$$ multiply each side by $s(s-4)^2$

$$s = 0: \quad 32 = 16A, \quad A = 2$$

$$s = 4: \quad 5(4^2) - 17(4) + 32 = 4C, \quad C = 11$$

$$s^2 \text{ terms:} \quad 5 = A + B, \quad 5 = 2 + B, \quad B = 3$$

$$\mathscr{L}^{-1}(F) = \mathscr{L}^{-1}\left[\frac{2}{s} + \frac{3}{s-4} + \frac{11}{(s-4)^2}\right]$$ substitute in $F(s)$

$$\mathscr{L}^{-1}(F) = f(t) = 2 + 3e^{4t} + 11te^{4t}$$ using Transforms (1), (3), (11) ■

EXERCISES 31.11

In Exercises 1–4, make the given changes in the indicated examples of this section, and then solve the resulting problems.

1. In Example 1, change the function $f(t)$. Let $f(t) = 1$.
2. In Example 2, change the function $f(t)$. Let $f(t) = \sin at$.
3. In Example 3, interchange the values of $f(0)$ and $f'(0)$.
4. In Example 4, in the function $F(s)$, change the numerator to a.

In Exercises 5–12, find the transforms of the given functions by use of the table.

5. $f(t) = e^{3t}$
6. $f(t) = 1 - \cos 2t$
7. $f(t) = 5t^3e^{-2t}$
8. $f(t) = 8e^{-3t}\sin 4t$
9. $f(t) = \cos 2t - \sin 2t$
10. $f(t) = 2t\sin 3t + e^{-3t}\cos t$
11. $f(t) = 3 + 2t\cos 3t$
12. $f(t) = t^3 - 3te^{-t}$

In Exercises 13–16, express the transforms of the given expressions in terms of s and $\mathscr{L}(f)$.

13. $y'' + y', f(0) = 0, f'(0) = 0$
14. $y'' - 3y', f(0) = 2, f'(0) = -1$
15. $2y'' - y' + y, f(0) = 1, f'(0) = 0$
16. $y'' - 3y' + 2y, f(0) = -1, f'(0) = 2$

In Exercises 17–28, find the inverse transforms of the given functions of s.

17. $F(s) = \dfrac{2}{s^3}$
18. $F(s) = \dfrac{6}{s^2 + 4}$
19. $F(s) = \dfrac{15}{2s + 6}$
20. $F(s) = \dfrac{3}{s^4 + 4s^2}$
21. $F(s) = \dfrac{1}{s^3 + 3s^2 + 3s + 1}$
22. $F(s) = \dfrac{s^2 - 1}{s^4 + 2s^2 + 1}$
23. $F(s) = \dfrac{s + 2}{(s^2 + 9)^2}$
24. $F(s) = \dfrac{s + 3}{s^2 + 4s + 13}$
25. $F(s) = \dfrac{4s^2 - 8}{(s+1)(s-2)(s-3)}$
26. $F(s) = \dfrac{3s + 1}{(s-1)(s^2 + 1)}$
27. $F(s) = \dfrac{2s + 3}{s^2 - 2s + 5}$
28. $F(s) = \dfrac{3s^4 + 3s^3 + 6s^2 + s + 1}{s^5 + s^3}$ (Explain your method of solution.)

In Exercises 29 and 30, find the indicated Laplace transforms:

29. For the Laplace transform $F(s)$ of the function $f(t)$, it can be shown that $\mathscr{L}\{tf(t)\} = -\frac{d}{ds}F(s)$. Verify this relationship by deriving Transform 11 from Transform 3.
30. Using the equation given in Exercise 29, derive the transform for $t^2\cos at$ from Transform 18.

Answers to Practice Exercises

1. $(s^2 - s)\mathscr{L}(f) - s + 1$ 2. $2\sin 3t$

31.12 Solving Differential Equations by Laplace Transforms

Solutions are Particular Solutions • Change Differential Equation into Algebraic Form

We will now show how certain differential equations can be solved by using Laplace transforms. *It must be remembered that these solutions are the* **particular** *solutions of the equations subject to the given conditions.* The necessary operations were developed in the preceding section. The following examples illustrate the method.

EXAMPLE 1 Solution of a first-order equation

Using Laplace transforms, solve the differential equation $2y' - y = 0$, if $y(0) = 1$. (Note that we are using y to denote the function.)

Taking transforms of each term in the equation, we have

$$\mathcal{L}(2y') - \mathcal{L}(y) = \mathcal{L}(0)$$

$$2\mathcal{L}(y') - \mathcal{L}(y) = 0$$

$\mathcal{L}(0) = 0$ by direct use of the definition of the transform. Now, using Eq. (31.24), $\mathcal{L}(y') = s\mathcal{L}(y) - y(0)$, we have

$$2[s\mathcal{L}(y) - 1] - \mathcal{L}(y) = 0 \qquad y(0) = 1$$

Solving for $\mathcal{L}(y)$, we obtain

$$2s\mathcal{L}(y) - \mathcal{L}(y) = 2$$

$$\mathcal{L}(y) = \frac{2}{2s - 1} = \frac{1}{s - \frac{1}{2}}$$

Finding the inverse transform, we have

$$y = e^{t/2} \qquad \text{using Transform (3)}$$

You should check this solution with that obtained by methods developed earlier. Also, it should be noted that the solution was essentially an algebraic one. This points out the power and usefulness of Laplace transforms. [We are able to change a differential equation into an *algebraic form,* which can in turn be translated into the solution of the differential equation.] Thus, we can solve a differential equation by using algebra and specific algebraic forms. ■

NOTE ▸

EXAMPLE 2 Solution of a second-order equation

Using Laplace transforms, solve the differential equation $y'' + 2y' + 2y = 0$, if $y(0) = 0$ and $y'(0) = 1$.

Using the same steps as outlined in Example 1, we have

$$\mathcal{L}(y'') + 2\mathcal{L}(y') + 2\mathcal{L}(y) = 0 \qquad \text{take transforms}$$

$$[s^2\mathcal{L}(y) - sy(0) - y'(0)] + 2[s\mathcal{L}(y) - y(0)] + 2\mathcal{L}(y) = 0 \qquad \text{using Eqs. (31.25) and (31.24)}$$

$$[s^2\mathcal{L}(y) - s(0) - 1] + 2[s\mathcal{L}(y) - 0] + 2\mathcal{L}(y) = 0 \qquad \text{substitute given values}$$

$$s^2\mathcal{L}(y) - 1 + 2s\mathcal{L}(y) + 2\mathcal{L}(y) = 0$$

$$(s^2 + 2s + 2)\mathcal{L}(y) = 1 \qquad \text{solve for } \mathcal{L}(y)$$

$$\mathcal{L}(y) = \frac{1}{s^2 + 2s + 2} = \frac{1}{(s + 1)^2 + 1} \qquad \text{fit transform form}$$

$$y = e^{-t}\sin t \qquad \text{take inverse transform using Transform (19)}$$ ■

Practice Exercise

1. In Example 2, find the solution if $y(0) = 1$ and $y'(0) = 0$.

EXAMPLE 3 Solution of a second-order equation

Solve the differential equation $y'' + y = \cos t$, if $y(0) = 1$ and $y'(0) = 2$.

$$\mathscr{L}(y'') + \mathscr{L}(y) = \mathscr{L}(\cos t) \qquad \text{take transforms}$$

$$[s^2\mathscr{L}(y) - s(1) - 2] + \mathscr{L}(y) = \frac{s}{s^2 + 1} \qquad \text{using Eq. (31.25) and Transform (5)}$$

$$(s^2 + 1)\mathscr{L}(y) = \frac{s}{s^2 + 1} + s + 2$$

$$\mathscr{L}(y) = \frac{s}{(s^2 + 1)^2} + \frac{s}{s^2 + 1} + \frac{2}{s^2 + 1}$$

$$y = \frac{t}{2}\sin t + \cos t + 2\sin t \qquad \text{using Transforms (16), (5), (6)}$$

■

EXAMPLE 4 Application—simple harmonic motion

A spring is stretched 1 ft by a weight of 16 lb (mass of 1/2 slug). The medium resists the motion with a force of $4v$, where v is the velocity of the motion. The differential equation describing the displacement y of the weight is

$$\frac{1}{2}\frac{d^2y}{dt^2} + 4\frac{dy}{dt} + 16y = 0 \qquad \text{see Example 3 of Section 31.10}$$

Find y as a function of time t, if $y(0) = 1$ and $dy/dx = 0$ for $t = 0$.

Clearing fractions and denoting derivatives by y'' and y', we have the following differential equation and solution.

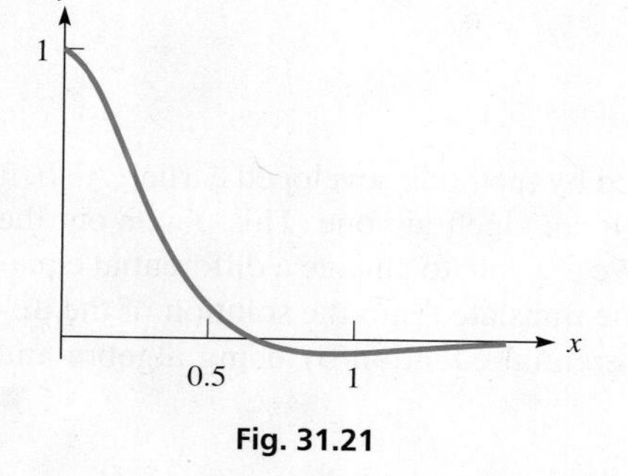

Fig. 31.21

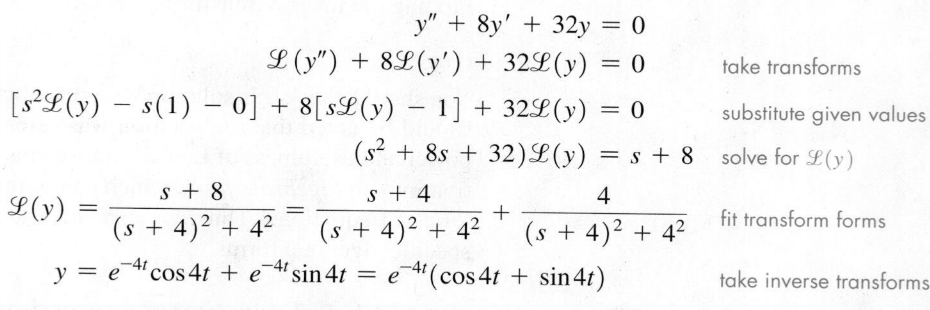

$$y'' + 8y' + 32y = 0$$

$$\mathscr{L}(y'') + 8\mathscr{L}(y') + 32\mathscr{L}(y) = 0 \qquad \text{take transforms}$$

$$[s^2\mathscr{L}(y) - s(1) - 0] + 8[s\mathscr{L}(y) - 1] + 32\mathscr{L}(y) = 0 \qquad \text{substitute given values}$$

$$(s^2 + 8s + 32)\mathscr{L}(y) = s + 8 \qquad \text{solve for } \mathscr{L}(y)$$

$$\mathscr{L}(y) = \frac{s + 8}{(s + 4)^2 + 4^2} = \frac{s + 4}{(s + 4)^2 + 4^2} + \frac{4}{(s + 4)^2 + 4^2} \qquad \text{fit transform forms}$$

$$y = e^{-4t}\cos 4t + e^{-4t}\sin 4t = e^{-4t}(\cos 4t + \sin 4t) \qquad \text{take inverse transforms}$$

The graph of this solution is shown in Fig. 31.21.

■

EXAMPLE 5 Application—electric current

The initial current in the circuit shown in Fig. 31.22 is zero. Find the current as a function of the time t.

Setting up the differential equation, and then solving it, we have

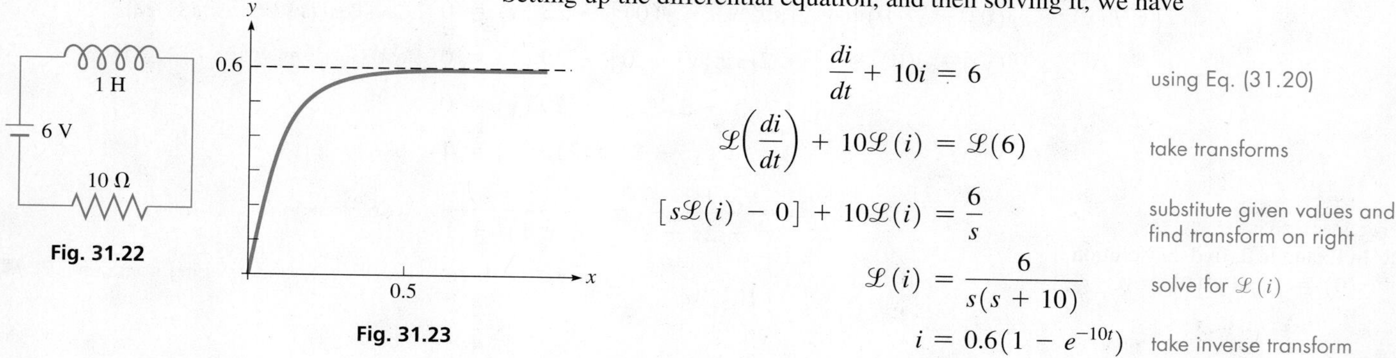

Fig. 31.22

Fig. 31.23

$$\frac{di}{dt} + 10i = 6 \qquad \text{using Eq. (31.20)}$$

$$\mathscr{L}\left(\frac{di}{dt}\right) + 10\mathscr{L}(i) = \mathscr{L}(6) \qquad \text{take transforms}$$

$$[s\mathscr{L}(i) - 0] + 10\mathscr{L}(i) = \frac{6}{s} \qquad \text{substitute given values and find transform on right}$$

$$\mathscr{L}(i) = \frac{6}{s(s + 10)} \qquad \text{solve for } \mathscr{L}(i)$$

$$i = 0.6(1 - e^{-10t}) \qquad \text{take inverse transform}$$

The graph of this solution is shown in Fig. 31.23.

■

EXAMPLE 6 Application—electric current

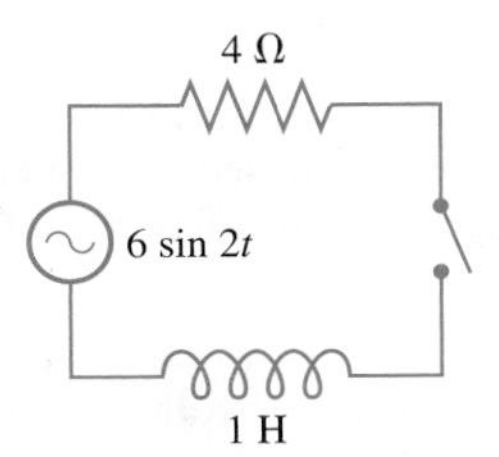

Fig. 31.24

An electric circuit in an FM radio transmitter contains a 1-H inductor and a 4-Ω resistor. It is being tested using a voltage source of 6 sin $2t$. If the initial current is zero, find the current i as a function of time t. See Fig. 31.24.

The solution is as follows:

$$(1)Di + 4i = 6\sin 2t \quad \text{differential equation, } D = d/dt$$

$$\mathcal{L}(Di) + 4\mathcal{L}(i) = 6\mathcal{L}(\sin 2t) \quad \text{take transforms}$$

$$[s\mathcal{L}(i) - 0] + 4\mathcal{L}(i) = \frac{6(2)}{s^2 + 4} \quad i(0) = 0$$

$$\mathcal{L}(i) = \frac{12}{(s + 4)(s^2 + 4)} = \frac{A}{s + 4} + \frac{Bs + C}{s^2 + 4} \quad \text{use partial fractions}$$

$$12 = A(s^2 + 4) + B(s^2 + 4s) + C(s + 4)$$

■ $s = 0$: $12 = 4A + 4C, 3 = A + C$
s terms: $0 = 4B + C$
s^2 terms: $0 = A + B$

The equations that give us the following values are shown in the margin at the left.

$$A = 0.6 \quad B = -0.6 \quad C = 2.4$$

$$\mathcal{L}(i) = 0.6\left(\frac{1}{s + 4}\right) - 0.6\left(\frac{s}{s^2 + 4}\right) + 1.2\left(\frac{2}{s^2 + 4}\right) \quad \text{fit transform forms}$$

$$i = 0.6e^{-4t} - 0.6\cos 2t + 1.2\sin 2t \quad \text{take inverse transforms}$$

This is checked by showing that $i(0) = 0$ and that it satisfies the original equation. ■

EXAMPLE 7 Application—electric current

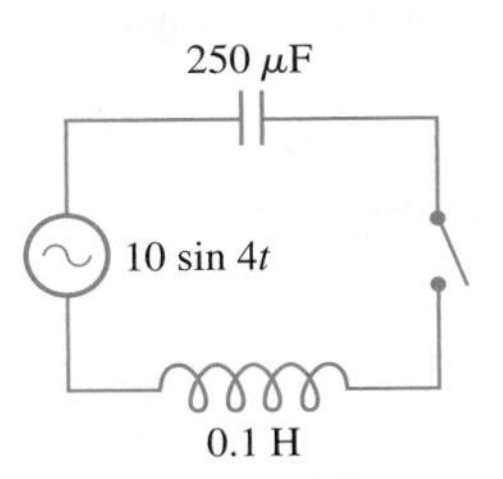

Fig. 31.25

An electric circuit contains a 0.1-H inductor, a 250-μF capacitor, a voltage source of 10 sin $4t$, and negligible resistance (assume $R = 0$). See Fig. 31.25. If the initial charge on the capacitor is zero, and the initial current is also zero, find the current in the circuit as a function of the time t.

The solution is as follows:

$$0.1D^2q + \frac{1}{250 \times 10^{-6}}q = 10\sin 4t \quad \text{differential equation, } D = d/dt$$

$$D^2q + 40{,}000q = 100\sin 4t$$

$$\mathcal{L}(D^2q) + 40{,}000\mathcal{L}(q) = 100\mathcal{L}(\sin 4t) \quad \text{take transforms}$$

$$[s^2\mathcal{L}(q) - sq(0) - Dq(0)] + 40{,}000\mathcal{L}(q) = \frac{400}{s^2 + 16} \quad q(0) = 0, D(q) = 0$$

$$\mathcal{L}(q) = \frac{400}{(s^2 + 200^2)(s^2 + 16)} = \frac{As + B}{s^2 + 200^2} + \frac{Cs + E}{s^2 + 16} \quad \text{use partial fractions}$$

$$400 = (As + B)(s^2 + 16) + (Cs + E)(s^2 + 200^2)$$

■ $s = 0$: $400 = 16B + 200^2E$
s terms: $0 = 16A + 200^2C$
s^2 terms: $0 = B + E$
s^3 terms: $0 = A + C$

The equations that give us the following values are shown in the margin at the left.

$$A = 0 \quad B = -0.010 \quad C = 0 \quad E = 0.010$$

$$\mathcal{L}(q) = \frac{0.010}{s^2 + 16} - \frac{0.010}{s^2 + 200^2} = \frac{0.010}{4}\left(\frac{4}{s^2 + 16}\right) - \frac{0.010}{200}\left(\frac{200}{s^2 + 200^2}\right)$$

$$q = 0.0025\sin 4t - 5.0 \times 10^{-5}\sin 200t \quad \text{take inverse transforms}$$

$$i = 0.010\cos 4t - 0.010\cos 200t \quad \text{take derivative}$$

■

EXERCISES 31.12

In Exercises 1–4, make the given changes in the indicated examples of this section, and then solve the resulting problems.

1. In Example 1, change the function $y(0)$ from 1 to 2.
2. In Example 2, change the function $y(0)$ from 0 to 1.
3. In Example 3, interchange the values of $y(0)$ and $y'(0)$.
4. In Example 5, change the initial current to 1 A.

In Exercises 5–38, solve the given differential equations by Laplace transforms. The function is subject to the given conditions.

5. $y' + y = 0, y(0) = 1$
6. $y' - 2y = 0, y(0) = 2$
7. $2y' - 3y = 0, y(0) = -1$
8. $y' + 2y = 1, y(0) = 0$
9. $y' + 3y = e^{-3t}, y(0) = 1$
10. $y' + 2y = te^{-2t}, y(0) = 0$
11. $y'' + 4y = 0, y(0) = 0, y'(0) = 1$
12. $9y'' - 4y = 0, y(0) = 1, y'(0) = 0$
13. $4y'' + 4y' + 5y = 0, \quad y(0) = 1, y'(0) = -1/2$
14. $y'' + 2y' + y = 0, y(0) = 0, y'(0) = -2$
15. $y'' - 4y' + 5y = 0, y(0) = 1, y'(0) = 2$
16. $4y'' + 4y' + y = 0, y(0) = 1, y'(0) = 0$
17. $y'' + y = 1, y(0) = 1, y'(0) = 1$
18. $9y'' + 4y = 2t, y(0) = 0, \quad y'(0) = 0$
19. $y'' + 2y' + y = 3te^{-t}, y(0) = 4, y'(0) = 2$
20. $2y'' + 8y = 3\sin 2t, y(0) = 0, y'(0) = 0$
21. $y'' - 4y = 10e^{3t}, y(0) = 5, y'(0) = 0$
22. $y'' - 2y' + y = e^{2t}, y(0) = 1, y'(0) = 3$
23. $y'' - 4y = 3\cos t, y(0) = 0, y'(0) = 0$
24. $2y'' + y' - y = \sin 3t, \quad y(0) = 0, y'(0) = 0$
25. A constant force of 6 lb moves a 2-slug mass through a medium that resists the motion with a force equal to the velocity v. The equation relating the velocity and the time is $2\frac{dv}{dt} = 6 - v$. Find v as a function of t if the object starts from rest.
26. A pendulum moves with simple harmonic motion according to the differential equation $D^2\theta + 20\theta = 0$, where θ is the angular displacement and $D = d/dt$. Find θ as a function of t if $\theta = 0$ and $D\theta = 0.40$ rad/s when $t = 0$.
27. The end of a certain vibrating metal rod oscillates according to $D^2y + 6400y = 0$ (assuming no damping), where $D^2 = d^2/dt^2$. If $y = 4$ mm and $Dy = 0$ when $t = 0$, find the equation of motion.
28. If there is a retarding force of $0.2Dy$ to the motion of the rod in Exercise 27, find the equation of motion.
29. A 50-Ω resistor, a 4.0-μF capacitor, and a 40-V battery are connected in series. Find the charge on the capacitor as a function of time t if the initial charge is zero.
30. A 2-H inductor, an 80-Ω resistor, and an 8-V battery are connected in series. Find the current in the circuit as a function of time if the initial current is zero.
31. A 10-H inductor, a 40-μF capacitor, and a voltage supply whose voltage is given by 100 sin $50t$ are connected in series in an electric circuit. Find the current as a function of the time if the initial charge on the capacitor is zero and the initial current is zero.
32. A 20-mH inductor, a 40-Ω resistor, a 50-μF capacitor, and a voltage source of $100e^{-1000t}$ are connected in series in an electric circuit. Find the charge on the capacitor as a function of time t, if $q = 0$ and $i = 0$ when $t = 0$.
33. The weight on a spring undergoes forced vibrations according to the equation $D^2y + 9y = 18\sin 3t$. Find its displacement y as a function of the time t, if $y = 0$ and $Dy = 0$ when $t = 0$.
34. A spring is stretched 1 m by a 20-N weight. The spring is stretched 0.5 m below the equilibrium position with the weight attached and then released. If it is in a medium that resists the motion with a force equal to $12v$, where v is the velocity, find the displacement y of the weight as a function of the time.
35. For the electric circuit shown in Fig. 31.26, find the current as a function of the time t if the initial current is zero.

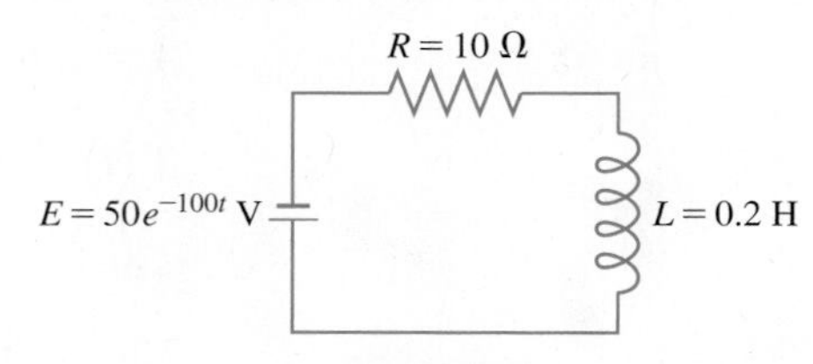

Fig. 31.26

36. For the electric circuit shown in Fig. 31.27, find the current as a function of time t if the initial charge on the capacitor is zero and the initial current is zero.

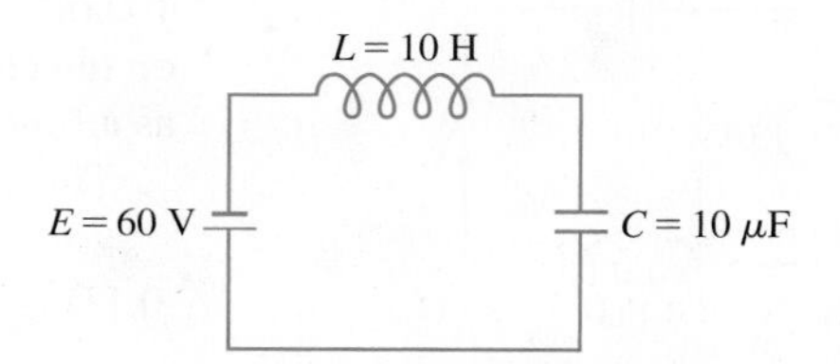

Fig. 31.27

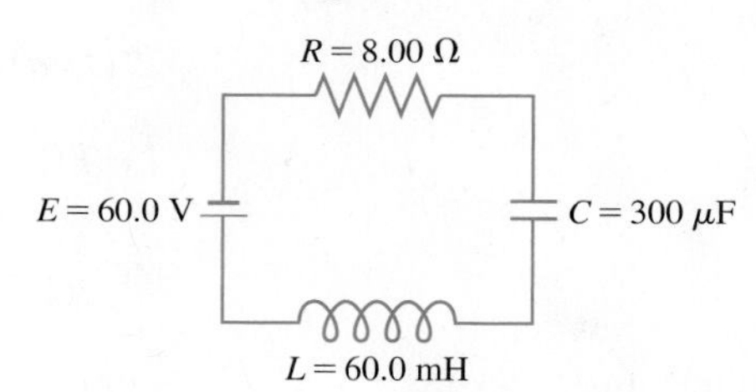

Fig. 31.28

37. For the electric circuit shown in Fig. 31.28, find the current as a function of the time t, if the initial charge on the capacitor is zero and the initial current is zero.
38. For the beam in Example 6 of Section 31.10, find the deflection y as a function of x using Laplace transforms. The Laplace transform of the fourth derivative y^{iv} is given by

$$\mathcal{L}(f^{iv}) = s^4\mathcal{L}(f) - s^3f(0) - s^2f'(0) - sf''(0) - f'''(0)$$

Also, since $y'(0)$ and $y'''(0)$ are not given, but are constants, assume $y'(0) = a$, and $y'''(0) = b$. It is then possible to evaluate a and b to obtain the solution.

Answer to Practice Exercise

1. $y = e^{-t}(\sin t + \cos t)$

CHAPTER 31 KEY FORMULAS AND EQUATIONS

Separation of variables

$$M(x, y)dx + N(x, y)dy = 0 \tag{31.1}$$

$$A(x)dx + B(y)dy = 0 \tag{31.2}$$

Integrating combinations

$$d(xy) = x\,dy + y\,dx \tag{31.3}$$

$$d(x^2 + y^2) = 2(x\,dx + y\,dy) \tag{31.4}$$

$$d\left(\frac{y}{x}\right) = \frac{x\,dy - y\,dx}{x^2} \tag{31.5}$$

$$d\left(\frac{x}{y}\right) = \frac{y\,dx - x\,dy}{y^2} \tag{31.6}$$

Linear differential equation of first order

$$dy + Py\,dx = Q\,dx \tag{31.7}$$

$$ye^{\int P\,dx} = \int Qe^{\int P\,dx}\,dx + c \tag{31.8}$$

Electric circuit

$$L\frac{di}{dt} + Ri + \frac{q}{C} = E \tag{31.9}$$

Motion in resisting medium

$$m\frac{dv}{dt} = F - kv \tag{31.10}$$

General linear differential equation

$$a_0\frac{d^ny}{dx^n} + a_1\frac{d^{n-1}y}{dx^{n-1}} + \cdots + a_{n-1}\frac{dy}{dx} + a_ny = b \tag{31.11}$$

$$a_0D^ny + a_1\,D^{n-1}y + \cdots + a_{n-1}\,Dy + a_ny = b \tag{31.12}$$

Homogeneous linear differential equation

$$a_0D^2y + a_1\,Dy + a_2y = 0 \tag{31.13}$$

Auxiliary equation

$$a_0m^2 + a_1m + a_2 = 0 \tag{31.14}$$

Distinct roots

$$y = c_1e^{m_1x} + c_2e^{m_2x} \tag{31.15}$$

Repeated roots

$$y = e^{mx}(c_1 + c_2x) \tag{31.16}$$

Complex roots

$$y = e^{\alpha x}(c_1\sin\beta x + c_2\cos\beta x) \tag{31.17}$$

Nonhomgeneous linear differential equation

$$a_0D^2y + a_1Dy + a_2y = b \tag{31.18}$$

$$y = y_c + y_p \tag{31.19}$$

Electric circuit

$$L\frac{d^2q}{dt^2} + R\frac{dq}{dt} + \frac{q}{C} = E \tag{31.20}$$

Laplace transforms

$$F(s) = \int_0^\infty e^{-st}f(t)\,dt \tag{31.21}$$

$$F(s) = \mathcal{L}(f) = \int_0^\infty e^{-st}f(t)\,dt \tag{31.22}$$

$$\mathcal{L}[af(t) + bg(t)] = a\mathcal{L}(f) + b\mathcal{L}(g) \tag{31.23}$$

$$\mathcal{L}(f') = s\mathcal{L}(f) - f(0) \tag{31.24}$$

$$\mathcal{L}(f'') = s^2\mathcal{L}(f) - sf(0) - f'(0) \tag{31.25}$$

Inverse transform

$$\mathcal{L}^{-1}(F) = f(t) \tag{31.26}$$

CHAPTER 31 REVIEW EXERCISES

CONCEPT CHECK EXERCISES

Determine each of the following as being either **true** *or* **false.** *If it is false, explain why.*

1. $y = ce^{-x} + 2e^{2x}$ is a solution of the differential equation $D^2y - Dy = 2y$.
2. Multiplying each term of the differential equation $e^{\sin x}\cot x\,dx + \csc x\,dy = 0$ by $\sin x$ will lead to the solution.
3. Combining terms properly and dividing through by x^2 will lead to the solution of the differential equation $x\,dy - y\,dx - 2x\,dx = 0$.
4. Dividing the terms of the differential equation $x\,dy + 3y\,dx = x^2\,dx$ by x will lead to the solution.
5. The solution of the differential equation $\frac{d^2y}{dx^2} - 6\frac{dy}{dx} + 9y = 0$ is $y = c_1e^{3x} + c_2e^{3x}$.
6. The form of the particular solution y_p of the differential equation $D^2y - 4Dy + 4y = e^{2x}$ is $y_p = c_1e^{2x} + c_2x^2e^{2x}$.
7. Given that $f(0) = 0$ and $f'(0) = 1$, the Laplace transform of $f''(t) - f'(t)$, in terms of the transform $f(t)$ is $(s^2 - s)\mathcal{L}(f) - 1$.
8. The inverse transform of $F(s) = \frac{s}{s^2 + 9}$ is $\sin 3t$.

PRACTICE AND APPLICATIONS

In Exercises 9–40, find the general solution to the given differential equations.

9. $4xy^3dx + (x^2 + 1)dy = 0$
10. $\frac{dy}{dx} = e^{x-y}$
11. $\sin 2x\,dx + y\sin x\,dy = \sin x\,dx$
12. $x\,dy + y\,dx = y\,dy$
13. $2D^2y + Dy = 0$
14. $2D^2y - 5Dy + 2y = 0$
15. $16y'' - 8y' + y = 0$
16. $y'' + 2y' + 2y = 0$
17. $(x + y)dx + (x + y^3)dy = 0$
18. $R\ln L\,dL = L\,dR$
19. $V\frac{dP}{dV} - 5P = V^2$
20. $dy - 2y\,dx = (x - 2)e^xdx$
21. $dy = 2y\,dx + y^2dx$
22. $x^2y\,dy = (1 + x)\csc y\,dx$
23. $D^2y + 2Dy + 6y = 0$
24. $4D^2y - 4Dy + y = 0$
25. $y' + 4y = 2e^{-2x}$
26. $2uv\,du = (2v - \ln v)dv$
27. $\sin x\frac{dy}{dx} + y\cos x + x = 0$
28. $y\,dy = (x^2 + y^2 - x)dx$
29. $2\frac{d^2s}{dt^2} + \frac{ds}{dt} - 3s = 6$
30. $\frac{d^2y}{dx^2} + 6\frac{dy}{dx} + 9y = 3x$
31. $y'' + y' - y = 2e^x$
32. $4D^3y + 9Dy = xe^x$
33. $9D^2y - 18Dy + 8y = 16 + 4x$
34. $9y'' + 4y = 4\cos 2x$
35. $D^3y - D^2y + 9Dy - 9y = \sin x$
36. $y'' + y' = e^x + \cos 2x$
37. $y'' - 7y' - 8y = 2e^{-x}$
38. $3y'' - 6y' = 4 + xe^x$
39. $D^2y + 25y = 50\cos 5x$
40. $D^2y + 4y = 8x\sin 2x$

In Exercises 41–48, find the indicated particular solution of the given differential equations.

41. $3y' = 2y\cot x$; $x = \frac{\pi}{2}$ when $y = 2$
42. $T\,dV - V\,dT = V^3dV$; $T = 1$ when $V = 3$
43. $y' = 4x - 2y$; $x = 0$ when $y = -2$
44. $xy^2dx + e^xdy = 0$; $x = 0$ when $y = 2$
45. $\frac{d^2v}{dt^2} + \frac{dv}{dt} + 4v = 0$, $\frac{dv}{dt} = \sqrt{15}$, $v = 0$ when $t = 0$
46. $5y'' + 7y' - 6y = 0$; $y' = 10$, $y = 2$ when $x = 0$
47. $D^2y + 4Dy + 4y = 4\cos x$; $Dy = 1, y = 0$ when $x = 0$
48. $y'' - 2y' + y = e^x + x$; $y = 0$, $y' = 0$ when $x = 0$

In Exercises 49–58, solve the given differential equations by using Laplace transforms, where the function is subject to the given conditions.

49. $4y' - y = 0, y(0) = 1$
50. $2y' - y = 4, y(0) = 1$
51. $y' - 3y = e^t, y(0) = 0$
52. $y' + 2y = e^{-2t}, y(0) = 2$
53. $y'' - 6y' + 9y = t, y(0) = 0, y'(0) = 1$
54. $y'' + 2y' + 5y = 10\cos t, y(0) = 2, y'(0) = 1$
55. $y'' + y = 0, y(0) = 0, y'(0) = -4$
56. $y'' + 4y' + 5y = 0, y(0) = 1, y'(0) = 1$
57. $16y'' + 9y = 3e^x, y(0) = 0, y'(0) = 0$
58. $y'' - 2y' + y = e^x + x, y(0) = 0, y'(0) = 1$

In Exercises 59–102, solve the given problems.

59. Use Euler's method to find the y-values of the solution of the equation $dy/dx = 1 + y^2$ from $x = 0$ to $x = 0.4$, with $\Delta x = 0.1$, if the curve passes through (0, 0).
60. Solve the equation of Exercise 59, subject to the same conditions, using the Runge–Kutta method.
61. Find the particular solution of the equation $dy/dx - 2y = e^{3x}$, if $y = 1$ when $x = 0$, (a) as a first-order linear equation, and (b) using Laplace transforms.
62. Find the particular solution of the equation $D^2y - 4y = 2 - 8x$, if $y = 0$ and $y' = 0$ when $x = 0$, (a) by the method of undetermined coefficients, and (b) by using Laplace transforms.
63. Solve the initial value problem $\frac{d^2y}{dt^2} + y = 1, y(0) = y'(0) = 0$ using (a) Laplace transforms, and (b) the method of undetermined coefficients.
64. The current i (in A) in an electric circuit changes with time t (in s) according to the equation $di + 10i\,dt = 6\,dt$. Find i as a function of t if the initial current is zero.

65. An object moves along a hyperbolic path described by $xy = 1$, such that $dx/dt = 2t$. Express x and y in terms of t if $x = 1$, $y = 1$ when $t = 0$.

66. An object moves along a parabolic path described by $y = x^2 + x$, such that $dx/dt = 4t + 1$. Express x and y in terms of t, if both x and y are zero when $t = 0$.

67. The time rate of change of volume of an evaporating substance is proportional to the surface area. Express the radius of an evaporating sphere of ice as a function of time. Let $r = r_0$ when $t = 0$. (*Hint:* Express both V and A in terms of the radius r.)

68. An insulated tank is filled with a solution containing radioactive cobalt. Due to the radioactivity, energy is released and the temperature T (in °F) of the solution rises with the time t (in h). The following equation expresses the relation between temperature and time for a specific case:

$$56{,}600 = 262(T - 70) + 20{,}200\frac{dT}{dt}$$

If the initial temperature is 70°F, what is the temperature 24 h later?

69. In a certain chemical reaction, the velocity of the reaction is proportional to the mass m of the chemical that remains unchanged. If m_0 is the initial mass and dm/dt is the velocity of the reaction, find m as a function of the time t.

70. Under proper conditions, bacteria grow at a rate proportional to the number present. In a certain culture, there were 10^4 bacteria present at a given time, and there were 3.0×10^5 bacteria present after 10 h. How many were present after 5.0 h?

71. An object with a mass of 1.00 kg slides down a long inclined plane. The effective force of gravity is 4.00 N, and the motion is retarded by a force numerically equal to the velocity. If the object starts from rest, what is the velocity (in m/s) 4.00 s later?

72. A 192-lb object falls from rest under the influence of gravity. Find the equation for the velocity at any time t (in s) if the air resists the motion with a force numerically equal to twice the velocity.

73. A particle is moving along a path $y = f(x)$ such that the slope of the path is $y/(y - x)$, and the path passes through the point $(-1, 2)$. Find the equation of the path $(y > 0)$.

74. On a certain weather map, the lines indicating equal temperature (*isotherms*) are given by $y = x^3 + c$. Find the equation of the orthogonal trajectories of the isotherms, the curves that show the direction of heat flow.

75. After 10.0 s, it is noted that 15.9% of the radioactive isotope neon-23 has decayed. Find the half-life of neon-23.

76. The isotope iodine-131, with a half-life of 8.04 days, is used in nuclear medicine to study the thyroid gland. What percent of an original amount of iodine-131 remains after 21 days?

77. Radioactive potassium-40 with a half-life of 1.28×10^9 years is used for dating rock samples. If a given rock sample has 75% of its original amount of potassium-40, how old is the rock?

78. When a gas undergoes an adiabatic change (no gain or loss of heat), the rate of change of pressure with respect to volume is directly proportional to the pressure and inversely proportional to the volume. Express the pressure in terms of the volume.

79. Under ideal conditions, the natural law of population change is that the population increases at a rate proportional to the population at any time. Under these conditions, project the population of the world in 2025, if it reached 6.9 billion in 2010, and 7.3 billion in 2015.

80. A spherical balloon is being blown up such that its volume V increases at a rate proportional to its surface area. Show that this leads to the differential equation $dV/dt = kV^{2/3}$ and solve for V as a function of t.

81. Find the orthogonal trajectories of the family of curves $y = cx^5$.

82. Find the equation of the curves for which their normals at all points are in the direction with the lines connecting the points and the origin.

83. Find the temperature after 1.0 h of an object originally 100°C, if it cools to 90°in 5.0 min in air that is at 20°C. (See Exercise 27 of Section 31.6.)

84. If a circuit contains a resistance R, a capacitance C, and a source of voltage E, express the charge q on the capacitor as a function of time.

85. A 2-H inductor, a 40-Ω resistor, and a 20-V battery are connected in series. Find the current in the circuit as a function of time if the initial current is zero.

86. A hollow cylinder moves vertically up and down in water according to the equation $D^2y + 6.5y = 0$, where $D = d/dt$. Find the displacement y as a function of the time t, if $y = 8.0$ cm when $t = 0$ s.

87. A certain spring stretches 0.50 m by a 40-N weight. With this weight suspended on it, the spring is stretched 0.50 m beyond the equilibrium position and released. Find the equation of the resulting motion if the medium in which the weight is suspended retards the motion with a force equal to 16 times the velocity. Classify the motion as underdamped, critically damped, or overdamped. Explain your choice.

88. The end of a vibrating rod moves according to the equation $D^2y + 0.2Dy + 4000y = 0$, where y is the displacement and $D = d/dt$. Find y as a function of t if $y = 3.00$ cm and $Dy = -0.300$ cm/s when $t = 0$.

89. A 0.5-H inductor, a 6-Ω resistor, and a 20-mF capacitor are connected in series with a generator for which $E = 24\sin 10t$. Find the charge on the capacitor as a function of time if the initial charge and initial current are zero.

90. A 5.00-mH inductor and a 10.0-μF capacitor are connected in series with a voltage source of $0.200e^{-200t}$ V. Find the charge on the capacitor as a function of time if $q = 0$ and $i = 4.00$ mA when $t = 0$.

91. Find the equation for the current as a function of time if a resistor of 20 Ω, an inductor of 4 H, a capacitor of 100 μF, and a battery of 100 V are in series. The initial charge on the capacitor is 10 mC, and the initial current is zero.

92. If an electric circuit contains an inductance L, a capacitor with a capacitance C, and a sinusoidal source of voltage $E_0 \sin \omega t$, express the charge q on the capacitor as a function of the time. Assume $q = 0$, $i = 0$ when $t = 0$.

93. The differential equation relating the current and time for a certain electric circuit is $2\,di/dt + i = 12$. Solve this equation by use of Laplace transforms, given that the initial current is zero. Evaluate the current for $t = 0.300$ s.

94. A 6-H inductor and a 30-Ω resistor are connected in series with a voltage source of $10 \sin 20t$. Find the current as a function of time if the initial current is zero. Use Laplace transforms.

95. A 0.25-H inductor, a 4.0-Ω resistor, and a 100-μF capacitor are connected in series. If the initial charge on the capacitor is 400 μC and the initial current is zero, find the charge on the capacitor as a function of time. Use Laplace transforms.

96. An inductor of 0.5 H, a resistor of 6 Ω, and a capacitor of 200 μF are connected in series. If the initial charge on the capacitor is 10 mC and the initial current is zero, find the charge on the capacitor as a function of time after the switch is closed. Use Laplace transforms.

97. Air containing 20% oxygen passes into a 5.00-L container initially filled with 100% oxygen. A uniform mixture of the air and oxygen then passes from the container at the same rate. What volume of oxygen is in the container after 5.00 L of air have passed into it?

98. When a circular disk of mass m and radius r is suspended by a wire at the center of one of its flat faces and the disk is twisted through an angle θ, torsion in the wire tends to turn the disk back in the opposite direction. The differential equation for this case is $\frac{1}{2}mr^2\frac{d^2\theta}{dt^2} = -k\theta$, where k is a constant. Determine the equation of motion if $\theta = \theta_0$ and $d\theta/dt = \omega_0$ when $t = 0$. See Fig. 31.29.

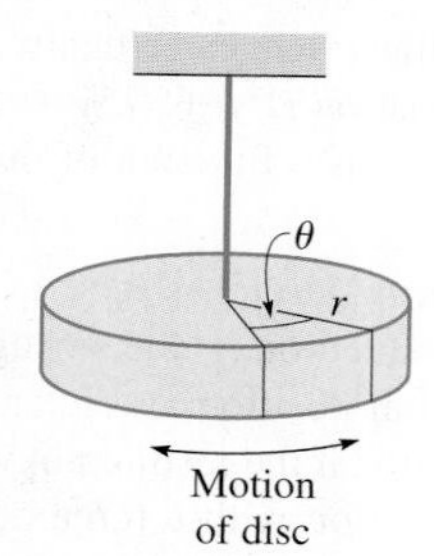

Fig. 31.29

99. An 8.0-lb weight (1/4-slug mass) stretches a spring 6.0 in. An external force of $\cos 8t$ is applied to the spring. Express the displacement y of the object as a function of time if the initial displacement and velocity are zero. Use Laplace transforms.

100. A spring is stretched 1.00 m by a mass of 5.00 kg (assume the weight to be 50.0 N). Find the displacement y of the object as a function of time if $y(0) = 1$ m and $dy/dt = 0$ when $t = 0$. Use Laplace transforms.

101. The approximate differential equation relating the displacement y of a beam at a horizontal distance x from one end is $EI\frac{d^2y}{dx^2} = M$, where E is the modulus of elasticity, I is the moment of inertia of the cross section of the beam perpendicular to its axis, and M is the bending moment at the cross section. If $M = 2000x - 40x^2$ for a particular beam of length L for which $y = 0$ when $x = 0$ and when $x = L$, express y in terms of x. Consider E and I as constants.

102. The gravitational acceleration of an object is inversely proportional to the square of its distance r from the center of Earth. Use the chain rule $\left(\frac{dy}{dx} = \frac{dy}{du}\frac{du}{dx}\right)$ to show that the acceleration is $\frac{dv}{dt} = v\frac{dv}{dr}$, where $v = \frac{dr}{dt}$ is the velocity of the object. Then solve for v as a function of r if $dv/dt = -g$ and $v = v_0$ for $r = R$, where R is the radius of Earth. Finally, show that a spacecraft must have a velocity of at least $v_0 = \sqrt{2gR}$ ($= 7$ mi/s) in order to escape from Earth's gravitation. (Note the expression for v^2 as $r \rightarrow \infty$.)

103. An electric circuit contains an inductor L, a resistor R, and a battery of voltage E. The initial current in the circuit is zero. Write three or four paragraphs explaining how the differential equation for the current in the circuit is solved using (a) separation of variables, (b) the linear differential equation of the first order, and (c) Laplace transforms.

CHAPTER 31 PRACTICE TEST

As a study aid, we have included complete solutions for each Practice Test problem at the end of this book.

In Problems 1–6, find the general solution of each of the given differential equations.

1. $x\frac{dy}{dx} + 2y = 4$

2. $y'' + 2y' + 5y = 0$

3. $x\,dx + y\,dy = x^2dx + y^2dx$

4. $2D^2y - Dy = 2\cos x$

5. $\frac{d^2y}{dx^2} - 4\frac{dy}{dx} + 4y = 3x$

6. $D^2y - 2Dy - 8y = 4e^{-2x}$

7. Find the particular solution of the differential equation $(xy + y)\frac{dy}{dx} = 2$, if $y = 2$ when $x = 0$.

8. If interest in a bank account is compounded continuously, the amount grows at a rate that is proportional to the amount present. Derive the equation for the amount A in an account with continuous compounding in which the initial amount is A_0 and the interest rate is r as a function of the time t after A_0 is deposited.

9. Using Laplace transforms, solve the differential equation $y'' + 9y = 9$, if $y(0) = 0$ and $y'(0) = 1$.

10. Using Laplace transforms, solve the differential equation $D^2y - Dy - 2y = 12$, if $y(0) = 0$ and $y'(0) = 0$.

11. Find the equation for the current as a function of the time (in s) in a circuit containing a 2-H inductance, an 8-Ω resistor, and a 6-V battery in series, if $i = 0$ when $t = 0$.

12. A 16-lb weight stretches a certain spring 0.5 ft. With this weight attached, the spring is pulled 0.3 ft longer than its equilibrium length and released. Find the equation of the resulting motion, assuming no damping. (The acceleration due to gravity is 32 ft/s^2.)

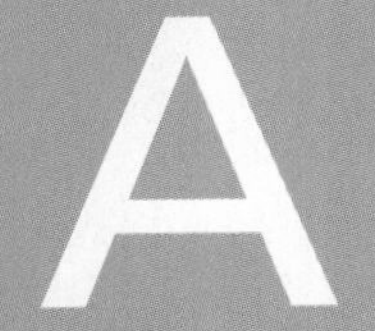

APPENDIX A Solving Word Problems

Drill-type problems require a working knowledge of the methods presented, and some algebraic steps to change the algebraic form may be required to complete the solution. *Word problems*, however, require a proper interpretation of the statement of the problem before they can be put in a form for solution.

We have to put word problems in symbolic form in order to solve them, and it is this procedure that most students find difficult. Because such problems require more than going through a certain routine, they demand more analysis and appear to be more difficult. Among the reasons for the student's difficulty at solving word problems are (1) unsuccessful previous attempts at solving word problems, leading the student to believe that all word problems are "impossible," (2) a poorly organized approach to the solution, and (3) failure to read the problem carefully, thereby having an improper and incomplete interpretation of the statement given. These can be overcome with proper attitude and care.

NOTE ➧ [A specific procedure for solving word problems is shown on page 46, when word problems are first covered in our study of algebra.] There are over 120 completely worked examples of word problems (as well as numerous other examples that show a similar analysis) throughout this text, illustrating proper interpretations and approaches to these problems.

RISERS

The procedure shown on page 46 is similar to that used by most instructors and texts. One of the variations that a number of instructors use is called *RISERS*. This is a word formed from the first letters (an *acronym*) of the words that outline the procedure. These are *Read, Imagine, Sketch, Equate, Relate*, and *Solve*. We now briefly outline this procedure here.

Read the statement of the problem carefully.

Imagine. Take time to get a mental image of the situation described.

Sketch a figure.

Equate, *on the sketch*, the known and unknown quantities.

Relate the known and unknown quantities with an equation.

Solve the equation.

There are problems where a *sketch* may simply be words and numbers placed so that we may properly *equate* the known and unknown quantities. For example, in Example 2 on page 47, the *sketch, equate*, and *relate* steps might look like this:

sketch	34 lights		1000 W
	25-W lights	40-W lights	
equate	x	$34 - x$	
power (watts)	$25x$	$40(34 - x)$	1000
relate	$25x$	$+ 40(34 - x)$	$= 1000$

If you follow the method on page 46, or this **RISERS** variation, or any appropriate step-by-step method, and *write out the solution neatly*, you will find that word problems lend themselves to solution more readily than you have previously found.

APPENDIX

B Units of Measurement

Scientific and technical calculations often involve numbers that include **units of measurement.** It is very important to be able to work with these numbers algebraically. Performing operations on numbers with units and converting between units are discussed in Section 1.4. Included in that section are tables of conversion factors and metric prefixes as well as several examples and exercises involving units and unit conversions. This appendix is intended to provide an overview of the types of units used in this book, including the quantities they measure, their symbols, and how some units are derived from others.

■ The abbreviation SI, used throughout the world, comes from the French *Le Système International d'Unitès,* the official name of the metric system.

There are two basic systems of units, the **SI metric system (International System of Units)** and the **U.S. customary system.** The SI metric system (which includes the meter, kilogram, and newton, for example) is used worldwide in all scientific work and in business related to international commerce. Most countries in the world use the SI metric system for all measurements in science, industry, and everyday activities.

The U.S. customary system (which includes the foot, pound, and mile, for example) is the system historically used by the United States. In the 1970s and 1980s, there was a trend in the United States to change to the metric system in order to align itself with most other countries. However, that trend lost its steam in the 1990s, and today, both the metric system and the U.S. customary system are used. For this reason, this book uses both SI and U.S. customary units in the examples and exercises. Technicians and engineers must be able to use both systems and convert units from one system to the other.

In each system, universally accepted **base units** are used to measure certain fundamental quantities. For example, the base units for length are the foot (U.S.) and the meter (SI). The base unit for time is the second in both systems. There is one case where the base units used in the two systems measure different, but related, fundamental quantities. The U.S. customary system uses the pound as the base unit for *force,* whereas the SI system uses the kilogram as the base unit for *mass*. Since force and mass are two different quantities, it is not really proper to convert from one to the other. However, this conversion is frequently done by using the fact that 1 kilogram exerts a force of 2.205 pounds near Earth's surface (this conversion would not be valid anywhere except near Earth's surface).

In the SI system, there are seven different base units, which are shown in boldface type in Table B.1 on the next page. Other units are expressed in terms of the base units. For example, the SI unit for force is the newton (N), which is defined by $1\text{ N} = 1\frac{\text{m}\cdot\text{kg}}{\text{s}^2}$. Here, the newton is expressed in terms of the base units meter (m), kilogram (kg), and second (s). Units such as this are called **derived units.** Not all combinations of base units are given special names. For example, the U.S. customary unit for acceleration is $\frac{\text{ft}}{\text{s}^2}$ (without any special name). Table B.1 gives a listing of all the commonly used units, their symbols, and how each of the derived units are expressed in terms of other units.

Table B.1 Quantities and Their Associated Units

		Unit				
		U.S. Customary		*Metric (SI)*		
Quantity	*Quantity Symbol*	*Name*	*Symbol*	*Name*	*Symbol*	*In Terms of Other SI Units*
Length	*s*	foot	ft	**meter**	m	
Mass	*m*	slug		**kilogram**	kg	
Force	*F*	pound	lb	newton	N	m · kg/s^2
Time	*t*	second	s	**second**	s	
Area	*A*		ft^2		m^2	
Volume	*V*		ft^3		m^3	
Capacity	*V*	gallon	gal	liter	L	(1 L = 1 dm^3)
Velocity	*v*		ft/s		m/s	
Acceleration	*a*		ft/s^2		m/s^2	
Density	*d*, *ρ*		lb/ft^3		kg/m^3	
Pressure	*p*		lb/ft^2	pascal	Pa	N/m^2
Energy, work	*E*, *W*		ft · lb	joule	J	N · m
Power	*P*	horsepower	hp	watt	W	J/s
Period	*T*		s		s	
Frequency	*f*		1/s	hertz	Hz	1/s
Angle	*θ*	radian	rad	radian	rad	
Electric current	*I*, *i*	ampere	A	**ampere**	A	
Electric charge	*q*	coulomb	C	coulomb	C	A · s
Electric potential	*V E*	volt	V	volt	V	J/(A · s)
Capacitance	*C*	farad	F	farad	F	s/Ω
Inductance	*L*	henry	H	henry	H	Ω · s
Resistance	*R*	ohm	Ω	ohm	Ω	V/A
Thermodynamic temperature	*T*			**kelvin**	K	(temp. interval
Temperature	*T*	degrees Fahrenheit	°F	degrees Celsius	°C	1°C = 1 K)
Quantity of heat	*Q*	British thermal unit	Btu	joule	J	
Amount of substance	*n*			**mole**	mol	
Luminous intensity	*I*	candlepower	cp	**candela**	cd	

Special Notes:

1. The SI base units are shown in boldface type.
2. Other units of time, along with their symbols, that are used in this book are the minute (min), hour (h), and day (d).
3. This table includes most of the units used in this book. Occasionally, other units will be noted and used.
4. In addition to the radian, the units degree (°) and revolution (r) are commonly used to measure angles or rotations.
5. Other common U.S. units that are used in this text are the inch (in.), yard (yd), mile (mi), ounce (oz), quart (qt), ton, and acre.
6. Unit symbols are case sensitive. They should be written exactly as shown in the table.
7. The dot symbol is used to show units are multiplied.

APPENDIX

C Newton's Method

■ At the right is an explanation and example of Newton's method for solving equations. They were copied directly from *Essays on Several Curious and Useful Subjects in Speculative and Mix'd Mathematicks* by Thomas Simpson (of Simpson's rule). It was published in London in 1740.

■ See Exercise 22 in Section 24.2.

A new Method for the Solution of Equations in Numbers.

CASE I.

When only one Equation is given, and one Quantity (x) to be determined.

TAKE the Fluxion of the given Equation (be it what it will) ſuppoſing, x, the unknown, to be the variable Quantity; and having divided the whole by $\dot{x}$, let the Quotient be repreſented by A. Eſtimate the Value of x pretty near the Truth, ſubſtituting the ſame in the Equation, as alſo in the Value of A, and let the Error, or reſulting Number in the former, be divided by this numerical Value of A, and the Quotient be ſubtracted from the ſaid former Value of x; and from thence will ariſe a new Value of that Quantity much nearer to he Truth than the former, wherewith proceeding as before, another new Value may be had, and ſo another, &c. 'till we arrive to any Degree of Accuracy deſired.

EXAMPLE I.

LET $300x - x^3 - 1000$ be given $= 0$; to find a Value of x. From $300\dot{x} - 3x^2\dot{x}$, the Fluxion of the given Equation, having expunged $\dot{x}$, (*Caſe* I.) there will be $300 - 3xx = A$: And, becauſe it appears by Inſpection, that the Quantity $300x - x^3$, when x is $= 3$, will be leſs, and when $x = 4$, greater than 1000, I eſtimate x at 3.5, and ſubſtitute inſtead thereof, both in the Equation and in the Value of A, finding the Error in the former $= 7.125$, and the Value of the latter $= 263.25$: Wherefore, by taking $\frac{7.125}{263.25} = .027$ from 3.5 there will remain 3.473 for a new Value of x; with which proceeding as before, the next Error, and the next Value of A, will come out .00962518, and 263.815 reſpectively; and from thence the third Value of $x = 3.47296351$; which is true, at leaſt, to 7 or 8 Places.

APPENDIX

D A Table of Integrals

The basic forms of Chapter 28 are not included. The constant of integration is omitted.

Forms containing $a + bu$ and $\sqrt{a + bu}$

1. $$\int \frac{u\,du}{a + bu} = \frac{1}{b^2}[(a + bu) - a\ln(a + bu)]$$

2. $$\int \frac{du}{u(a + bu)} = -\frac{1}{a}\ln\frac{a + bu}{u}$$

3. $$\int \frac{u\,du}{(a + bu)^2} = \frac{1}{b^2}\left[\frac{a}{a + bu} + \ln(a + bu)\right]$$

4. $$\int \frac{du}{u(a + bu)^2} = \frac{1}{a(a + bu)} - \frac{1}{a^2}\ln\frac{a + bu}{u}$$

5. $$\int u\sqrt{a + bu}\,du = -\frac{2(2a - 3bu)(a + bu)^{3/2}}{15b^2}$$

6. $$\int \frac{u\,du}{\sqrt{a + bu}} = -\frac{2(2a - bu)\sqrt{a + bu}}{3b^2}$$

7. $$\int \frac{du}{u\sqrt{a + bu}} = \frac{1}{\sqrt{a}}\ln\left(\frac{\sqrt{a + bu} - \sqrt{a}}{\sqrt{a + bu} + \sqrt{a}}\right), \quad a > 0$$

8. $$\int \frac{\sqrt{a + bu}}{u}du = 2\sqrt{a + bu} + a\int \frac{du}{u\sqrt{a + bu}}$$

Forms containing $\sqrt{u^2 \pm a^2}$ and $\sqrt{a^2 - u^2}$

9. $$\int \frac{du}{u^2 - a^2} = \frac{1}{2a}\ln\frac{u - a}{u + a}$$

10. $$\int \frac{du}{\sqrt{u^2 \pm a^2}} = \ln(u + \sqrt{u^2 \pm a^2})$$

11. $$\int \frac{du}{u\sqrt{u^2 + a^2}} = -\frac{1}{a}\ln\left(\frac{a + \sqrt{u^2 + a^2}}{u}\right)$$

12. $$\int \frac{du}{u\sqrt{u^2 - a^2}} = \frac{1}{a}\sec^{-1}\frac{u}{a}$$

13. $$\int \frac{du}{u\sqrt{a^2 - u^2}} = -\frac{1}{a}\ln\left(\frac{a + \sqrt{a^2 - u^2}}{u}\right)$$

14. $$\int \sqrt{u^2 \pm a^2}\,du = \frac{u}{2}\sqrt{u^2 \pm a^2} \pm \frac{a^2}{2}\ln(u + \sqrt{u^2 \pm a^2})$$

15. $$\int \sqrt{a^2 - u^2}\,du = \frac{u}{2}\sqrt{a^2 - u^2} + \frac{a^2}{2}\sin^{-1}\frac{u}{a}$$

16. $$\int \frac{\sqrt{u^2 + a^2}}{u}du = \sqrt{u^2 + a^2} - a\ln\left(\frac{a + \sqrt{u^2 + a^2}}{u}\right)$$

17. $\displaystyle\int \frac{\sqrt{u^2 - a^2}}{u}\,du = \sqrt{u^2 - a^2} - a\sec^{-1}\frac{u}{a}$

18. $\displaystyle\int \frac{\sqrt{a^2 - u^2}}{u}\,du = \sqrt{a^2 - u^2} - a\ln\left(\frac{a + \sqrt{a^2 - u^2}}{u}\right)$

19. $\displaystyle\int (u^2 \pm a^2)^{3/2}\,du = \frac{u}{4}(u^2 \pm a^2)^{3/2} \pm \frac{3a^2u}{8}\sqrt{u^2 \pm a^2} + \frac{3a^4}{8}\ln(u + \sqrt{u^2 \pm a^2})$

20. $\displaystyle\int (a^2 - u^2)^{3/2}\,du = \frac{u}{4}(a^2 - u^2)^{3/2} + \frac{3a^2u}{8}\sqrt{a^2 - u^2} + \frac{3a^4}{8}\sin^{-1}\frac{u}{a}$

21. $\displaystyle\int \frac{(u^2 + a^2)^{3/2}}{u}\,du = \frac{1}{3}(u^2 + a^2)^{3/2} + a^2\sqrt{u^2 + a^2} - a^3\ln\left(\frac{a + \sqrt{u^2 + a^2}}{u}\right)$

22. $\displaystyle\int \frac{(u^2 - a^2)^{3/2}}{u}\,du = \frac{1}{3}(u^2 - a^2)^{3/2} - a^2\sqrt{u^2 - a^2} + a^3\sec^{-1}\frac{u}{a}$

23. $\displaystyle\int \frac{(a^2 - u^2)^{3/2}}{u}\,du = \frac{1}{3}(a^2 - u^2)^{3/2} - a^2\sqrt{a^2 - u^2} + a^3\ln\left(\frac{a + \sqrt{a^2 - u^2}}{u}\right)$

24. $\displaystyle\int \frac{du}{(u^2 \pm a^2)^{3/2}} = \pm\frac{u}{a^2\sqrt{u^2 \pm a^2}}$

25. $\displaystyle\int \frac{du}{(a^2 - u^2)^{3/2}} = \frac{u}{a^2\sqrt{a^2 - u^2}}$

26. $\displaystyle\int \frac{du}{u(u^2 + a^2)^{3/2}} = \frac{1}{a^2\sqrt{u^2 + a^2}} - \frac{1}{a^3}\ln\left(\frac{a + \sqrt{u^2 + a^2}}{u}\right)$

27. $\displaystyle\int \frac{du}{u(u^2 - a^2)^{3/2}} = -\frac{1}{a^2\sqrt{u^2 - a^2}} - \frac{1}{a^3}\sec^{-1}\frac{u}{a}$

28. $\displaystyle\int \frac{du}{u(a^2 - u^2)^{3/2}} = \frac{1}{a^2\sqrt{a^2 - u^2}} - \frac{1}{a^3}\ln\left(\frac{a + \sqrt{a^2 - u^2}}{u}\right)$

Trigonometric forms

29. $\displaystyle\int \sin^2 u\,du = \frac{u}{2} - \frac{1}{2}\sin u\cos u$

30. $\displaystyle\int \sin^3 u\,du = -\cos u + \frac{1}{3}\cos^3 u$

31. $\displaystyle\int \sin^n u\,du = -\frac{1}{n}\sin^{n-1}u\cos u + \frac{n-1}{n}\int \sin^{n-2}u\,du$

32. $\displaystyle\int \cos^2 u\,du = \frac{u}{2} + \frac{1}{2}\sin u\cos u$

33. $\displaystyle\int \cos^3 u\,du = \sin u - \frac{1}{3}\sin^3 u$

34. $\displaystyle\int \cos^n u\,du = \frac{1}{n}\cos^{n-1}u\sin u + \frac{n-1}{n}\int \cos^{n-2}u\,du$

35. $\displaystyle\int \tan^n u\,du = \frac{\tan^{n-1}u}{n-1} - \int \tan^{n-2}u\,du$

36. $\displaystyle\int \cot^n u\,du = -\frac{\cot^{n-1}u}{n-1} - \int \cot^{n-2}u\,du$

37. $\displaystyle\int \sec^n u\,du = \frac{\sec^{n-2}u\tan u}{n-1} + \frac{n-2}{n-1}\int \sec^{n-2}u\,du$

38. $\displaystyle\int \csc^n u\,du = \frac{\csc^{n-2}u\cot u}{n-1} + \frac{n-2}{n-1}\int \csc^{n-2}u\,du$

39. $\displaystyle\int \sin au\sin bu\,du = \frac{\sin(a-b)u}{2(a-b)} - \frac{\sin(a+b)u}{2(a+b)}$

40. $\displaystyle\int \sin au\cos bu\,du = -\frac{\cos(a-b)u}{2(a-b)} - \frac{\cos(a+b)u}{2(a+b)}$

41. $\displaystyle\int \cos au\cos bu\,du = \frac{\sin(a-b)u}{2(a-b)} + \frac{\sin(a+b)u}{2(a+b)}$

42. $\displaystyle\int \sin^m u\cos^n u\,du = \frac{\sin^{m+1}u\cos^{n-1}u}{m+n} + \frac{n-1}{m+n}\int \sin^m u\cos^{n-2}u\,du$

43. $\displaystyle\int \sin^m u\cos^n u\,du = -\frac{\sin^{m-1}u\cos^{n+1}u}{m+n} + \frac{m-1}{m+n}\int \sin^{m-2}u\cos^n u\,du$

Other forms

44. $\displaystyle\int ue^{au}\,du = \frac{e^{au}(au-1)}{a^2}$

45. $\displaystyle\int u^2e^{au}\,du = \frac{e^{au}}{a^3}(a^2u^2 - 2au + 2)$

46. $\displaystyle\int u^n\ln u\,du = u^{n+1}\left(\frac{\ln u}{n+1} - \frac{1}{(n+1)^2}\right)$

47. $\displaystyle\int u\sin u\,du = \sin u - u\cos u$

48. $\displaystyle\int u\cos u\,du = \cos u + u\sin u$

49. $\displaystyle\int e^{au}\sin bu\,du = \frac{e^{au}(a\sin bu - b\cos bu)}{a^2+b^2}$

50. $\displaystyle\int e^{au}\cos bu\,du = \frac{e^{au}(a\cos bu + b\sin bu)}{a^2+b^2}$

51. $\displaystyle\int \sin^{-1}u\,du = u\sin^{-1}u + \sqrt{1-u^2}$

52. $\displaystyle\int \tan^{-1}u\,du = u\tan^{-1}u - \frac{1}{2}\ln(1+u^2)$

Photo Credits

Cover Daniel Schoenen/Image Broker/Alamy Stock Photo

Chapter 1 Page 1, Jason Larkin

Chapter 2 Page 54, Marshall Ikonography/Alamy Stock Photo

Chapter 3 Page 85, Majeczka/Shutterstock

Chapter 4 Page 113, Stocktrek Images, Inc./Alamy Stock Photo

Chapter 5 Page 140, Stoupa/Shutterstock

Chapter 6 Page 180, Jürgen Fälchle/Fotolia

Chapter 7 Page 219, Luciano mortula/Fotolia

Chapter 8 Page 240, 3dsculptor/Fotolia

Chapter 9 Page 263, Jakub Cejpek/Fotolia

Chapter 10 Page 299, Rannev/Shutterstock

Chapter 11 Page 323, Penka Todorova Vitkova/Shutterstock

Chapter 12 Page 345, Africa Studio/Fotolia

Chapter 13 Page 373, Rawpixel.com/Shutterstock

Chapter 14 Page 413, Allyn J. Washington

Chapter 14 Page 403, Stanca Sanda/Alamy Stock Photo

Chapter 15 Page 420, Michele Bosi

Chapter 16 Page 439, Monty Rakusen/Cultura Creative/Alamy Stock Photo

Chapter 17 Page 470, Maxoidos/Fotolia

Chapter 18 Page 499, Syda Productions/Fotolia

Chapter 19 Page 514, Keith Brofsky/Getty Images

Chapter 20 Page 535, Allyn J. Washington

Chapter 21 Page 568, Greg801/iStock/Getty Images

Chapter 22 Page 621, Real Deal Photo/Shutterstock

Chapter 23 Page 655, Microgen/Fotolia

Chapter 23 Page 672, Allyn J. Washington

Chapter 24 Page 700, StockLite/Shutterstock

Chapter 25 Page 742, Katatonia/Fotolia

Chapter 26 Page 768, Emberose/Fotolia

Chapter 27 Page 805, 3dsculptor/Fotolia

Chapter 28 Page 840, Dmitrimaruta/Fotolia

Chapter 29 Page 884, Daniel Meissner/ImageBROKER/Alamy Stock Photo

Chapter 30 Page 904, Chris Tefme/Fotolia

Chapter 31 Page 937, iStock/Getty Images

Appendix C A.4 Essays on Several Curious and Useful Subjects in Speculative and Mix'd Mathematicks by Thomas Simpson. Publshed by H. Woodfall, jun.

Screenshots from Texas Instruments. Courtesy of Texas Instruments Inc.

Screenshots from Minitab. Courtesy of Minitab Corporation.

Answers to Odd-Numbered Exercises and Chapter Review Exercises

Because statements will vary for writing exercises, answers here are in abbreviated form. Answers are not included for end-of-chapter writing exercises.

Exercises 1.1, page 5

1. Change $\frac{5}{1}$ to $\frac{-7}{1}$ and $\frac{-19}{1}$ to $\frac{12}{1}$. **3.** Change $2 > -4$, 2 is to the right of -4, to $-6 < -4$, -6 is to the left of -4. Also the figure must be changed to show -6 to left of -4.

5. integer, rational, real; imag.

7. irrational, real; rational, real

9. 3, 3, $\frac{\pi}{4}$, not real

11. $6 < 8$ **13.** $\pi < 3.1416$ **15.** $-4 < -|-3|$

17. $-\frac{2}{3} > -\frac{3}{4}$ **19.** $\frac{1}{3}, -\frac{\sqrt{3}}{4}, \frac{b}{y}$

21.

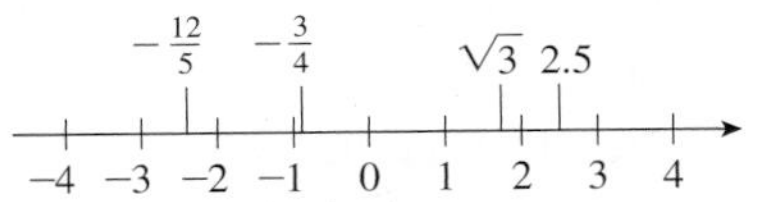

23. No, $|0| = 0$ **25.** The number itself **27.** True

29. $-3.1, -|-3|, -1, \sqrt{5}, \pi, |-8|, 9$

31. (a) Positive integer (b) Negative integer (c) Positive rational number less than 1

33. (a) Yes (b) Yes **35.** (a) To right of origin (b) To left of -4

37. Between 0 and 1 **39.** a = any real number; $b = 0$

41. 0.0008 F **43.** $N = 1000an$

45. Yes; -20 is to right of -30

Exercises 1.2, page 10

1. 20 **3.** -4 **5.** -3 **7.** 6 **9.** -3 **11.** -28

13. 35 **15.** 20 **17.** 40 **19.** -1 **21.** -32

23. -17 **25.** Undefined **27.** 20 **29.** 16 **31.** -6

33. -24 **35.** -6 **37.** -1

39. Commutative law of multiplication **41.** Distributive law

43. Associative law of addition

45. Associative law of multiplication **47.** d **49.** b **51.** =

53. (a) Positive (b) Negative

55. Correct. (For $x > 0$, x is positive; for $x < 0$, $-x$ is positive.)

57. (a) Negative reciprocals of each other (b) They may not be equal.

59. -0.29 **61.** -2.4 kW·h **63.** -2°C **65.** 10 V

67. 100 m + 200 m = 200 m + 100 m; commutative law of addition

69. 7(25 min + 15 min); distributive law

Exercises 1.3, page 16

1. 0.390 has 3 sig. digits; the zero is not needed for proper location of the decimal point.

3. 76 **5.** 8 is exact; 55 is approx. **7.** 24 and 1440 are exact

9. 1 and 9 are approx. **11.** 3, 4, 3 **13.** 3, 3, 2 **15.** 1, 5, 6

17. 1, 5, 4 **19.** (a) 0.010 (b) 30.8 **21.** (a) Same (b) 78.0

23. (a) 0.004 (b) Same **25.** (a) 4.94 (b) 4.9

27. (a) -50.9 (b) -51 **29.** (a) 5970 (b) $6\bar{0}00$

31. (a) 0.945 (b) 0.94 **33.** 12.20 **35.** -6.06 **37.** 13.4

39. 5.57 **41.** 2.91 **43.** 15.8788 **45.** -204.2

47. 2.745 MHz, 2.755 MHz

49. Too many sig. digits; time has only 2 sig. digits

51. (a) 19.3 (b) 27 **53.** (a) 2 (b) 2 (c) -2 (d) 0 (e) Undefined

55. (a) $3.1416 > \pi$; (b) $\frac{22}{7} > \pi$

57. (a) $\frac{1}{3} = 0.33333333\ldots$ (b) $\frac{5}{11} = 0.4545454545\ldots$ (c) $\frac{2}{5} = 0.400000000$ (0 repeats)

59. 95.3 MJ **61.** 6330 g **63.** 59.14% **65.** (a) 30 (b) 27.3

Exercises 1.4 , page 22

1. x^6 **3.** x^7 **5.** $2b^6$ **7.** m^2 **9.** $-\frac{1}{7n^4}$ **11.** P^8

13. $a^{30}T^{60}$ **15.** $\frac{8}{b^3}$ **17.** $\frac{x^8}{16}$ **19.** 1 **21.** -3

23. $\frac{1}{6}$ **25.** R^2 **27.** $-t^{14}$ **29.** $-L^2$ **31.** $\frac{1}{8}$

33. 1 **35.** $\frac{a}{x^2}$ **37.** $\frac{64s^6}{g^2}$ **39.** $\frac{x^3}{64a^3}$ **41.** $\frac{5n^3}{T}$

43. -53 **45.** 253 **47.** -0.421 **49.** 114 **51.** Yes

53. 625 **55.** 1 **57.** $\frac{G^2k^5T^5}{h}$ **59.** $\frac{r}{6}$ **61.** $3212.27

63. 2, 2.5937, 2.7048, 2.7169 **65.** 277 ft **67.** 85,600 ft/min^2

69. 14.9 L **71.** 38.0 in.^{2} **73.** 12,800 ft/min **75.** 39.6 cm

77. 90,700 L/h **79.** 1240 km/h **81.** 101,000 Pa

Exercises 1.5, page 26

1. 8060 **3.** 45,000 **5.** 0.00201 **7.** 3.23 **9.** 18.6

11. 4×10^3 **13.** 8.7×10^{-3} **15.** 6.09×10^8 **17.** 5.28×10^{-2}

19. 5.6×10^{13} **21.** 2.2×10^8 **23.** 35.6×10^6 **25.** 97.3×10^{-3}

27. 475×10^{-9} **29.** 3.2×10^{-34} **31.** 1.728×10^{87}

33. 7.31×10^{10} **35.** 1.59×10^7 **37.** 9.965×10^{-3}

39. 3.38×10^{16} **41.** 5.0×10^8 tweets **43.** 3×10^{-6} W

45. 1.2×10^9 Hz **47.** 1.2×10^{10} m^2 **49.** 0.0000000000016 W

51. (a) 2.3×10^3 W, 2.3 kW (b) 230×10^{-3} W, 230 mW (c) 2.3×10^6 W, 2.3 MW (d) 230×10^{-6} W, 230 μW

53. (a) 1×10^{100} (b) $10^{10^{100}}$ **55.** 1.3×10^7 m **57.** 4.2×10^{-8} s

59. 2.3×10^7 m **61.** 3.32×10^{-18} kg **63.** 3.433 Ω

Exercises 1.6, page 28

1. v **3.** 12 **5.** 7 **7.** -11 **9.** -8 **11.** 0.3

13. 5 **15.** 3 **17.** 5 **19.** 47 **21.** 53 **23.** $3\sqrt{2}$

25. $20\sqrt{3}$ **27.** $4\sqrt{21}$ **29.** $2\sqrt{5}$ **31.** -4 **33.** 7

35. 10 **37.** $3\sqrt{10}$ **39.** 9.24 **41.** 0.9029

43. (a) 60.00 (b) 84 **45.** (a) 0.0388 (b) 0.0246 **47.** 60 mi/h

49. 1450 m/s **51.** 59.9 in. **53.** 670 km/h

55. no, not true if $a < 0$ **57.** (a) 12.9 (b) -0.598 **59.** 50.0 Hz

Exercises 1.7, page 33

1. $3x - 3y$ **3.** $4ax + 5s$ **5.** $8x$ **7.** $y + 4x$ **9.** $-5s$
11. $5F - 3T - 2$ **13.** $-a^2b - a^2b^2$ **15.** $p - 6$
17. $9x - v - 7$ **19.** $5a - 5$ **21.** $-5a + 2$ **23.** $-2t + 5u$
25. $7r + 8s$ **27.** $-50 + 19j$ **29.** $7n - 7$ **31.** $18 - 2t^2$
33. $6a$ **35.** $2aZ + 1$ **37.** $3z - 5$ **39.** $8p - 5q$
41. $-4x^2 + 22$ **43.** $7V^2 - 3$ **45.** $-6t + 13$ **47.** $4Z - 24R$
49. $2D + d$ **51.** $3B - 2\alpha$ **53.** $12x + 200$
55. (a) $x^2 + 2y + 2a - b$ (b) $3x^2 - 4y + 2a + b$
57. No; did not change signs correctly
59. Yes; $|a - b| = |-(b - a)| = |b - a|$

Exercises 1.8, page 35

1. $-8s^8t^{13}$ **3.** $x^2 - 5x + 6$ **5.** a^3x **7.** $-a^4c^3x^3$
9. $-8a^3x^5$ **11.** $i^3R + 2i^3$ **13.** $-3s^3 + 15st$ **15.** $5m^3n + 15m^2n$
17. $-3M^2 - 3MN + 6M$ **19.** $x^4ty + x^3ty^4$ **21.** $x^2 + 2x - 15$
23. $2x^2 + 9x - 5$ **25.** $y^2 - 64$ **27.** $6a^2 - 7ab + 2b^2$
29. $6s^2 + 11st - 35t^2$ **31.** $2x^3 + 5x^2 - 2x - 5$
33. $x^2 - 4xy + 4y^2 - 16$ **35.** $2a^2 - 16a - 18$
37. $18T^2 - 15T - 18$ **39.** $-2L^3 + 6L^2 + 8L$
41. $9x^2 - 42x + 49$ **43.** $x_1^2 + 6x_1x_2 + 9x_2^2$
45. $x^2y^2z^2 - 4xyz + 4$ **47.** $2x^2 + 32x + 128$
49. $-x^3 + 2x^2 + 5x - 6$ **51.** $6T^3 + 9T^2 - 6T$
53. (a) $49 \neq 9 + 16$ (b) $1 \neq 9 - 16$
55. $(x + y)^3 = x^3 + 3x^2y + 3xy^2 + y^3$
57. $P + 0.02Pr + 0.0001Pr^2$ **59.** $2w^2 + 30w + 100$
61. $3R^2 - 4RX$ **63.** $n^2 + 200n + 10{,}000$ **65.** $R_1^2 - R_2^2$

Exercises 1.9, page 38

1. $\frac{3}{y^3}$ **3.** $3x - 2$ **5.** $-4x^2y$ **7.** $\frac{4t^4}{r^2}$ **9.** $3x^2z$
11. a **13.** $a^2 + 2y$ **15.** $t - 2rt^2$ **17.** $q + 2p - 4q^3$
19. $\frac{2L}{R} - R$ **21.** $-\frac{1}{2b} + \frac{1}{a} - \frac{3a}{2b^2}$ **23.** $3y^n - 2ay$
25. $x + 5$ **27.** $2x + 1$ **29.** $x - 1$ **31.** $4x^2 - x - 1, R = -3$
33. $Z - 2 - \frac{1}{4Z + 3}$ **35.** $x^2 + x - 6$ **37.** $2a^2 + 8$
39. $y^2 - 3y + 9$ **41.** $x - y$ **43.** $t - 2$ **45.** -5
47. $\frac{x^4 + 1}{x + 1} = x^3 - x^2 + x - 1 + \frac{2}{x + 1}$ **49.** $\frac{V_1T_2}{T_1}$
51. $A + \frac{\mu^2E^2}{2A} - \frac{\mu^4E^4}{8A^3}$ **53.** $\frac{GMm}{R}$
55. $s^2 + 2s + 6 + \frac{16s + 16}{s^2 - 2s - 2}$

Exercises 1.10, page 42

1. $-9, -15, -36, -4$ **3.** $\frac{1}{8}$ **5.** 9 **7.** -1 **9.** -10
11. 20 **13.** -5 **15.** -2 **17.** 4 **19.** $-\frac{7}{2}$ **21.** -8
23. $-\frac{2}{3}$ **25.** 4 **27.** -2.5 **29.** 0 **31.** 8
33. 11 or -11 **35.** -1 or 4 **37.** 9.5 **39.** -2.1 **41.** 5.7
43. 0.85 **45.** (a) Identity (b) Conditional equation
47. $x = 3.75$ **49.** \$800 **51.** 120°C **53.** 750 gal
55. 98 kW · h

Exercises 1.11, page 44

1. $\frac{v - v_0}{t}$ **3.** $\frac{V_0 + bTV_0 - V}{bV_0}$ **5.** $\frac{E}{I}$ **7.** $g_2 - rL$
9. $\frac{12B}{TWL}$ **11.** $\frac{p - p_a}{dh}$ **13.** $\frac{mv^2}{F_c}$ **15.** $5TS_T - 0.25dT$
17. $\frac{0.3t - ct^2}{c}$ **19.** $Tv - c$ **21.** $\frac{K_1m_1 - K_2m_1}{K_2}$
23. $\frac{2mg - 2am}{a}$ **25.** $\frac{C_0^2 - C_1^2}{2C_1^2}$ **27.** $\frac{rA - N}{r}$
29. $100T_1 - 100T_2$ **31.** $\frac{Q_1 + PQ_1}{P}$ **33.** $\frac{N + N_2 - N_2T}{T}$
35. $\frac{L - \pi r_2 - 2x_1 - x_2}{\pi}$ **37.** $\frac{gJP + V_1^2}{V_1}$ **39.** $\frac{Cd(k_1 + k_2)}{2Ak_1k_2}$
41. $\frac{CN - NV}{C}$ **43.** 12.3 L **45.** 32.3°C **47.** 3.22 Ω
49. $\frac{d - 4v_2 - 2v_1}{v_1}$

Exercises 1.12, page 48

1. 16 25-W lights, 15 40-W lights **3.** 2.500 h
5. \$29,500, \$34,500
7. 3.2 million the first year, 3.7 million the second year
9. 50 acres at \$200/acre, 90 acres at \$300/acre **11.** 60 ppm/h
13. \$85, \$95 **15.** $-2.3\ \mu$A, $-4.6\ \mu$A, $6.9\ \mu$A
17. 6.9 km, 9.5 km **19.** 240 ec, 195 ec+, 120 de
21. 6.5 s **23.** 900 m **25.** 84.2 km/h, 92.2 km/h
27. 390 s, first car **29.** 64 mi from A **31.** 4 L
33. 79 km/h

Review Exercises for Chapter 1, page 50

Concept Check Exercises

1. F **2.** T **3.** F **4.** F **5.** T **6.** F **7.** F
8. T **9.** F **10.** F **11.** F **12.** F

Practice and Application

13. -10 **15.** -20 **17.** -22 **19.** -25 **21.** -4
23. 5 **25.** $4r^2t^4$ **27.** $-\frac{24t}{m^2n}$ **29.** $\frac{8T^3}{N}$ **31.** $3\sqrt{5}$
33. (a) 1 (b) 8000 **35.** (a) 4 (b) 9.1 **37.** 18.0
39. 1.3×10^{-4} **41.** 923,000 J **43.** 59 ft/s
45. 10,100,000 J/min **47.** $-a - 2ab$ **49.** $7LC - 3$
51. $2x^2 + 9x - 5$ **53.** $x^2 + 16x + 64$ **55.** $hk - 3h^2k^4$
57. $7R - 6r$ **59.** $13xy - 10z$ **61.** $2x^3 - x^2 - 7x - 3$
63. $-3x^2y + 24xy^2 - 48y^3$ **65.** $-9p^2 + 3pq + 18p^2q$
67. $\frac{6q}{p} - 2 + \frac{3q^4}{p^3}$ **69.** $2x - 5$ **71.** $x^2 - 2x + 3$
73. $4x^3 - 2x^2 + 6x, R = -1$ **75.** $15r - 3s - 3t$
77. $y^2 + 5y - 1, R = 4$ **79.** $-\frac{9}{2}$ **81.** $\frac{21}{10}$ **83.** $-\frac{7}{3}$
85. 3 **87.** $-\frac{19}{5}$ **89.** 1.0 **91.** 6×10^{13} bytes
93. 1.54×10^{10} km **95.** 25,300,000,000,000 mi
97. 0.000000000001 W/m^2 **99.** 0.15 Bq/L **101.** $\frac{V}{\pi r^2}$
103. $\frac{PL^2}{\pi^2I}$ **105.** $\frac{Rr - Pp}{Q}$ **107.** $\frac{d + A}{A}$ **109.** $\frac{N_1 + TN_3 - N_3}{T}$
111. $\frac{RH + AT_1}{A}$ **113.** $\frac{d - 3kbx^2 + kx^3}{3kx^2}$ **115.** 8.2×10^8
117. 111 m **119.** 0.0188 Ω **121.** $4x + 4a$
123. $4t + 4h - 2t^2 - 4th - 2h^2$ **125.** Yes (18 to 0)
127. Equation is an identity. **129.** (a) -4, (b) 12 **131.** 4

133. $x^3 - 3x^2y + 3xy^2 - y^3 = -y^3 + 3y^2x - 3yx^2 + x^3$

135. 4×10^{-7} **137.** 190 kw **139.** 81 ft · lb

141. \$59, \$131 **143.** 160 cm^3, 80 cm^3, 320 cm^3

145. 1900 Ω, 3100 Ω **147.** 6.7 h

149. 15.3 h after second ship enters Lake Superior

151. 400 L, 600 L **153.** 290 ft^2

Exercises 2.1, page 57

1. 90° **3.** 4 **5.** $\angle EBD, \angle DBC$ **7.** $\angle ABC$ **9.** 25°

11. *BD, BC* **13.** 140° **15.** 40° **17.** 30° **19.** 62°

21. 28° **23.** 118° **25.** 134° **27.** 44° **29.** 46°

31. 4.53 m **33.** 3.40 m **35.** $\angle BHC, \angle CGD$

37. $\angle BCH, \angle CHG\ \angle HGC, \angle DCG$ **39.** $\angle AHG, \angle HGE$

41. 130° **43.** 133° **45.** 5.38 cm **47.** 180°

49. Sum of angles is 180°.

Exercises 2.2, page 63

1. 65° **3.** 7.02 m **5.** 56° **7.** 48° **9.** 6.9 ft^2

11. 38,400 cm^2 **13.** 4.41 ft^2 **15.** 0.390 m^2 **17.** 976 cm

19. 64.5 cm **21.** 5 in. **23.** 26.6 ft **25.** 522 cm

27. 67° **29.** 227.2 cm **31.** 45°, 135°

33. Equilateral triangle **35.** All three triangles are similar.

37. $\angle K = \angle N = 90°$; $\angle LMK = \angle OMN$;
$\angle KLM = \angle NOM$; $\Delta MKL \sim \Delta MNO$

39. 8 **41.** Yes **43.** 65° **45.** 7.2 ft **47.** 60 ft^2

49. 12.1 ft **51.** 23 ft **53.** 7.5 m, 9.0 m **55.** 9.6 ft

57. 20.0 mi

Exercises 2.3, page 67

1. Trapezoid **3.** 8900 ft^2 **5.** 340 m **7.** 33.24 in.

9. 12.8 m **11.** 2.144 ft **13.** 41 mm^2 **15.** 23.5 $in.^2$

17. 9.3 m^2 **19.** 2.000 ft^2 **21.** $p = 4a + 2b$

23. $A = bh + a^2$ **25.** It is a rectangle. **27.** 288 cm^2

29. 360° **31.** $A = a(b + c) = ab + ac$; distributive law

33. The diagonal always divides the rhombus into two congruent triangles. All outer sides are always equal.

35. (a) 344 m (b) 996 m^2

37. 1470 ft^2 **39.** 0.84 gal **41.** $w = 36.7$ cm, $h = 22.9$ cm

43. 3.04 km^2

45. 360°. A diagonal divides a quadrilateral into two triangles, and the interior angles of each triangle are 180°.

Exercises 2.4, page 70

1. 18° **3.** $p = 11.6$ in., $A = 8.30$ $in.^2$

5. (a) *AD* (b) *AF*

7. (a) $AF \perp OE$ (b) ΔOEC

9. 1730 ft **11.** 72.6 mm **13.** 0.0285 yd^2 **15.** 4.26 m^2

17. 128 cm^2 **19.** 25° **21.** 25° **23.** 120° **25.** 40°

27. 0.393 rad **29.** 2.185 rad **31.** $p = \frac{1}{2}\pi r + 2r$

33. $p = \sqrt{2r^2} + \frac{1}{2}\pi r$ **35.** All are on the same diameter.

37. 10.3 $in.^2$ **39.** π is the ratio of circumference to diameter.

41. $A = \frac{1}{4}\pi r^2 - \frac{1}{2}r^2$ **43.** 383 km **45.** No

47. 24,900 mi **49.** $\frac{4}{9}$ **51.** 35.7 in. **53.** 7/9 **55.** 73.8 in.

57. Horizontally and opposite to original direction

Exercises 2.5, page 75

1. Simpson's rule. Arcs give better estimate.

3. Simpson's rule should be more accurate in that it accounts better for the arcs between points on the upper curve.

5. Exact **7.** 84 m^2 **9.** 4.9 ft^2 **11.** 9.8 mi^2

13. 12,300 km^2 **15.** 120,000 ft^2 **17.** 8100 ft^2

19. 2.73 $in.^2$ The trapezoids are inside the boundary and do not include some of the area.

21. 2.98 $in.^2$ The ends of the areas are curved so that they can get closer to the boundary.

Exercises 2.6, page 78

1. The volume is six times as much.

3. 771 cm^3 **5.** 336 ft^3 **7.** 3.99×10^6 mm^2 **9.** 4.671 yd^3

11. 20,500 cm^2 **13.** 0.25 $in.^3$ **15.** 153,000 mm^3 **17.** 3.358 m^2

19. 0.072 yd^3 **21.** 72.3 cm^2 **23.** $V = \frac{1}{6}\pi d^3$ **25.** $\frac{6}{1}$

27. $\frac{4}{1}$ **29.** 604 $in.^2$ **31.** 0.00034 mi^3 = 5.0×10^7 ft^3

33. 426 m^3 **35.** 3.3×10^6 yd^3 **37.** 2,350,000 ft^3

39. 1560 mm^2 **41.** 447 $in.^3$ **43.** 1.10 $in.^3$ **45.** 2.7%

Review Exercises for Chapter 2, page 81

Concept Check Exercises

1. F **2.** T **3.** F **4.** T **5.** T **6.** T

Practice and Applications

7. 32° **9.** 32° **11.** 37 **13.** 700 **15.** 0.630

17. 36.5 **19.** 25.5 mm **21.** 0.0118 ft^2

23. 235 mm **25.** 3320 $in.^2$ **27.** 6190 cm^3 **29.** 160,000 ft^3

31. 55,700 cm^3 **33.** 1.62 m^2 **35.** 101 $in.^2$ **37.** 25°

39. 65° **41.** 53° **43.** 2.4

45. $p = \pi a + b + \sqrt{4a^2 + b^2}$ **47.** $A = ab + \frac{1}{2}\pi a^2$

49. yes **51.** n^2; $A = \pi(nr)^2 = n^2(\pi r^2)$ **53.** $\frac{a}{d} = \frac{b}{c}$

55. 74° **57.** 7.8 m **59.** $1\overline{0}$ cm **61.** $1\overline{0}$ m **63.** 6.0 m

65. 3.73 mi **67.** 26,200 mi **69.** 30 ft^2 **71.** 1.0×10^6 m^2

73. 190 m^3 **75.** 10 ft **77.** 1000 m **79.** 159 gal

81. $w = 132$ cm, $h = 74.5$ cm **83.** $r = \frac{3}{2}h$

Exercises 3.1, page 88

1. -13 **3.** $f(T-10) = 9.1 + 0.08T + 0.001T^2$

5. (a) $A = \pi r^2$ **(b)** $A = \frac{1}{4}\pi d^2$ **7.** $d = \sqrt[3]{\frac{6V}{\pi}}$

9. $A = \frac{1}{2}d^2, d = \sqrt{2A}$ **11.** $A = 4r^2 - \pi r^2$ **13.** $7, -9$

15. $6, 6$ **17.** $\frac{3-2\pi^2}{2\pi}, -\frac{1}{2}$ **19.** $\frac{1}{9}a - \frac{1}{3}a^2, 0$

21. $3s^2 - 11s + 16, 3s^2 + s + 8$ **23.** -6

25. $62.9, 260$ **27.** -299.7 **29.** $f(x) = 3x^2 - 4$

31. $f(x) = x^3 - 7x$ **33.** Square the input and add 2.

35. Cube the input and subtract this from 6 times the input.

37. Multiply by 3 the sum of twice the input added to 5.

39. $A = 5e^2, f(e) = 5e^2$

41. $A = 5200 - 120\,t, f(t) = 5200 - 120t$

43. 10.4 m **45.** 75 ft, $2v + 0.2v^2$, 240 ft, 240 ft

47. 31 lb/in.^{2} **49.** $d = 120t + 65$

51. (a) $f[f(x)]$ means "function of the function of x". **(b)** $8x^4$

Exercises 3.2, page 92

1. $-x^2 + 2$ is never greater than 2. The range is all real numbers $f(x) \le 2$.

3. 4 mA, 0 mA

5. Domain: all real numbers; range: all real numbers

7. Domain: all real numbers except 0; range: all real numbers except 0

9. Domain: all real numbers $s \ge 2$, range: all real numbers $f(s) \ge 0$

11. Domain all real numbers $h \ge 0$; range: all real numbers $H(h) \ge 1$; $\sqrt{h}$ cannot be negative

13. Domain: all real numbers; range: all real numbers $y \ge 0$

15. All real numbers $y > 2$ **17.** All real numbers $D \ge 0, D \ne 2$

19. $y = \frac{x+3}{x}$; domain all x except 0 **21.** 2, 2, not defined

23. 2, 0.75 **25.** $d = 110 + 40t$ **27.** $w = 5500 - 2t$

29. $m = 0.5h - 390$ **31.** $C = 5l + 250$

33. (a) $y = 3000 - 0.25x$ (b) 2900 L

35. $y = -\frac{1}{2}x + 95$ **37.** $A = \frac{1}{16}p^2 + \frac{(60-p)^2}{4\pi}$

39. $d = \sqrt{14{,}400 + h^2}$; domain: $h \ge 0$; range: $d \ge 120\ m$

41. $s = \frac{300}{t-3}$; domain: $t > 3$; range: $s > 0$ (upper limits depend on truck)

43. Domain is all values of $C > 0$, with some upper limit depending on the circuit.

45. $m = \begin{cases} 0.5h - 390 \text{ for } h > 1000 \\ 110 \text{ for } 0 \le h \le 1000 \end{cases}$

47. (a) $V = 4w^2 - 24w + 32$ (b) $w \ge 4$ in. **49.** 2

51. All real numbers $y \ge 2$.

Exercises 3.3, page 95

1. (0, 1)

3. (2. 1), $(-1, 2)$, $(-2, -3)$

5.

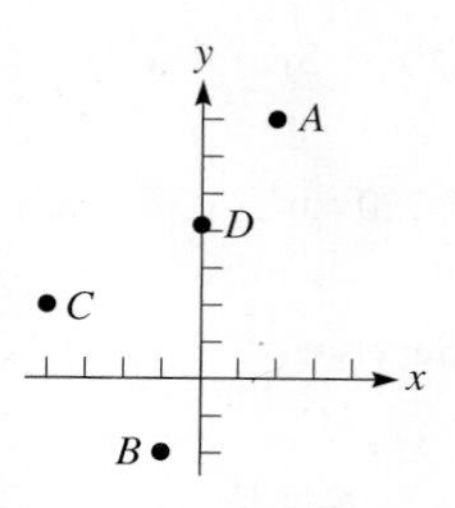

7. Scalene triangle **9.** Rectangle **11.** $(6, -2)$

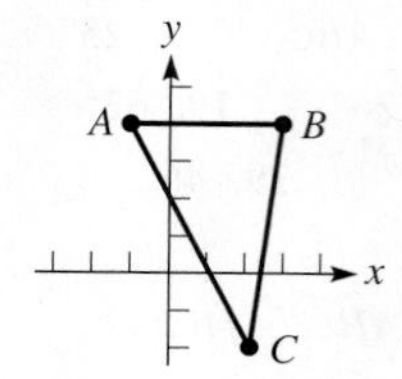

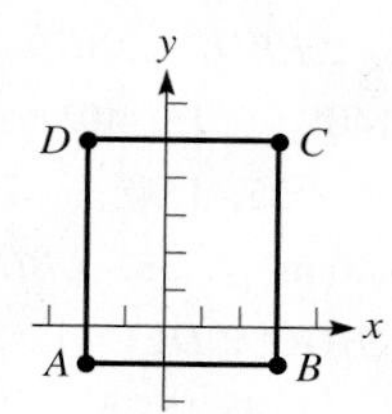

13. $(3, -6)$

15. Quadrant II **17.** Quadrant III

19. On a line parallel to the y-axis, 1 unit to the right

21. On a line parallel to the x-axis, 3 units above

23. On a line through the origin that bisects the first and third quadrants

25. 0 **27.** To the right of the y-axis

29. To the left of a line that is parallel to the y-axis, 1 unit to its left

31. In first or third quadrant **33.** On either axis

35. Third quadrant **37.** $\sqrt{32}$

Exercises 3.4, page 100

1.

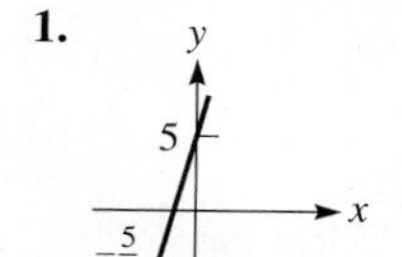

3.

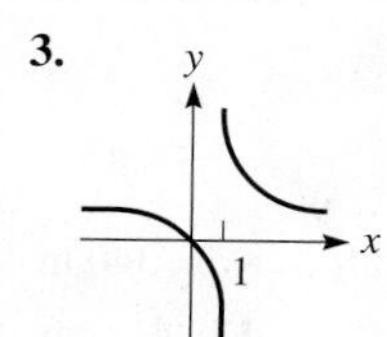

5.

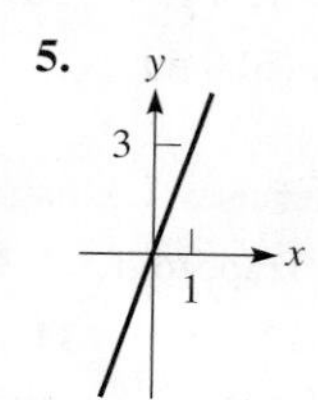

7.

9.

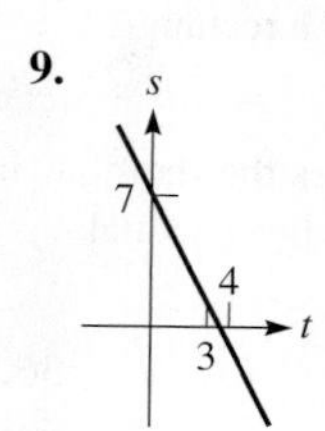

11.

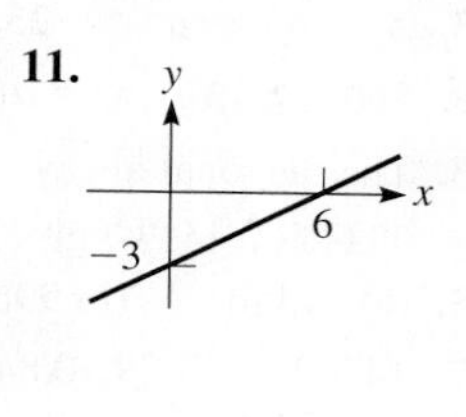

13.

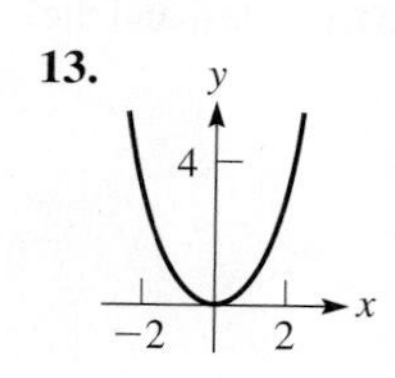

15.

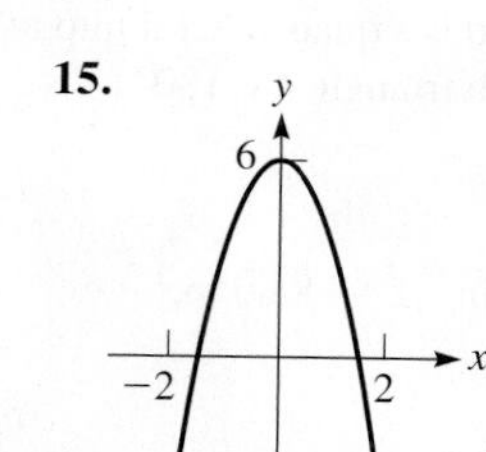

17.

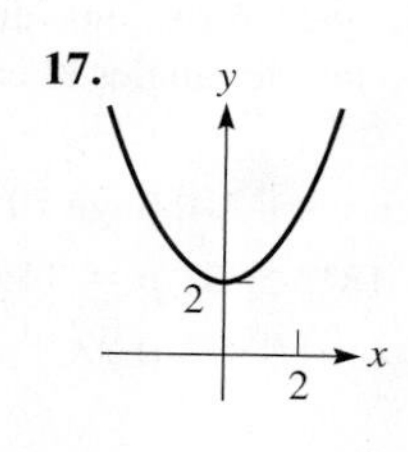

19.
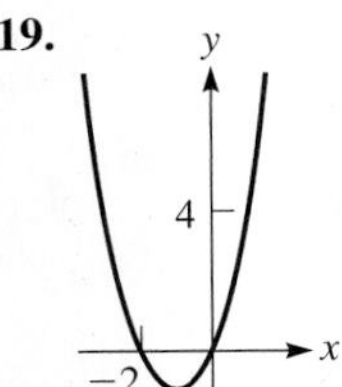

21.
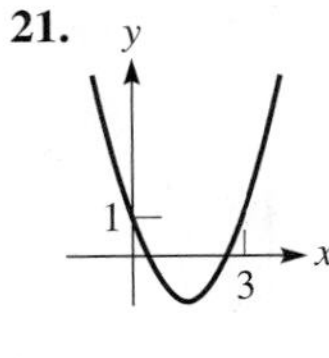

23.
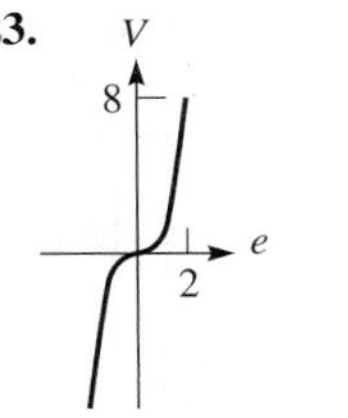

25.
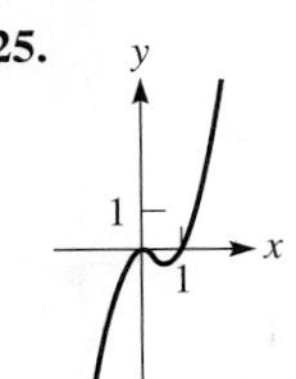

27.
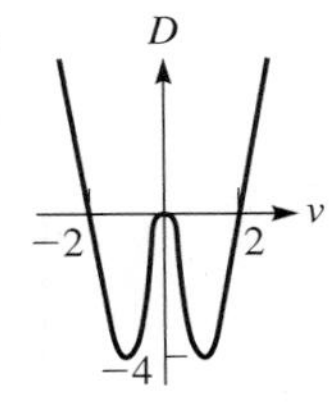

29.
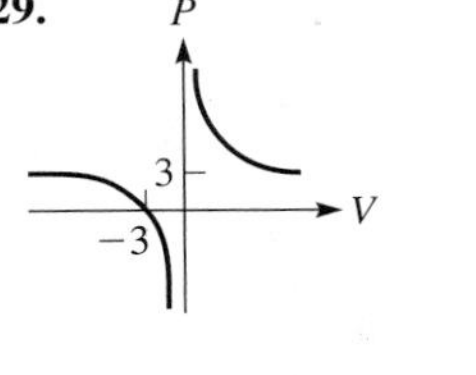

31. **33.**
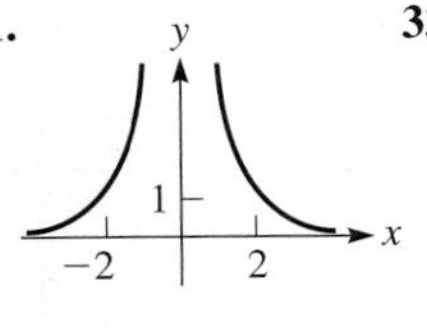
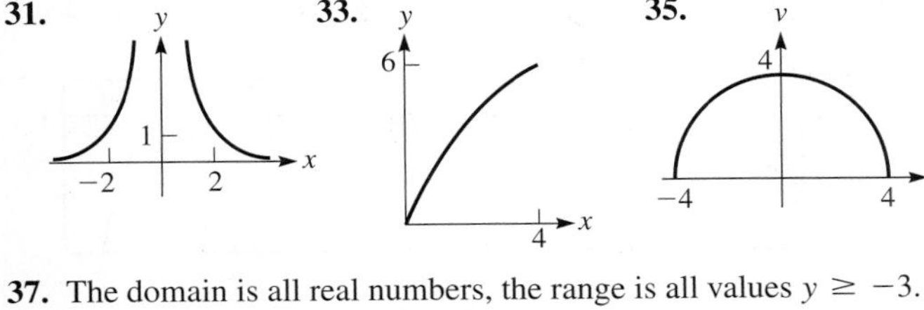

35.
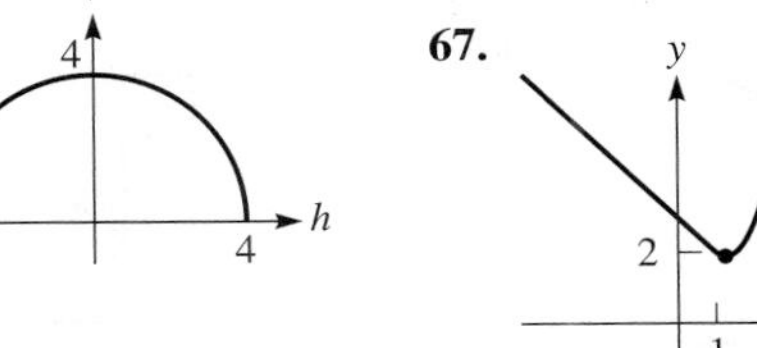

37. The domain is all real numbers, the range is all values $y \geq -3$.

39. The domain is all values $x \geq 2$, the range is all values $y \geq 0$.

41.
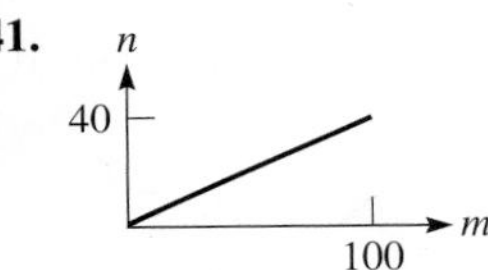

43.
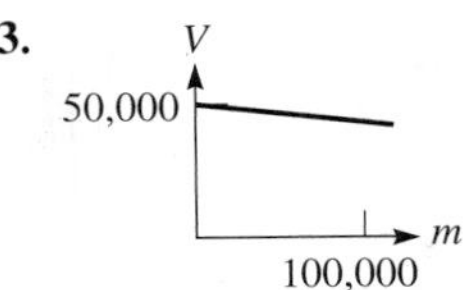

45.
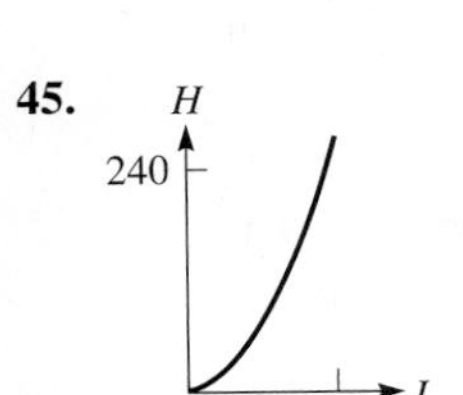

47.
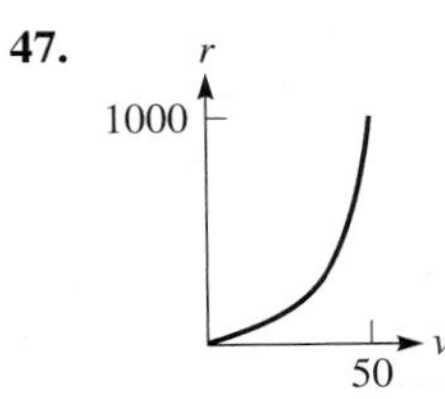
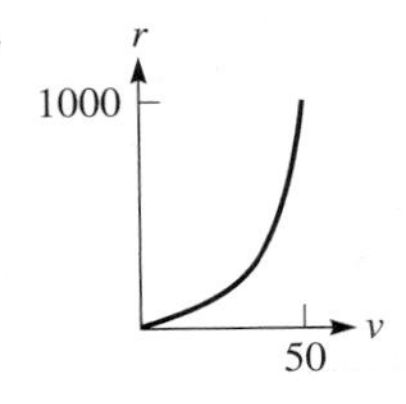
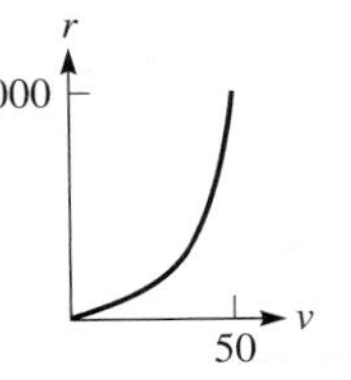

49.
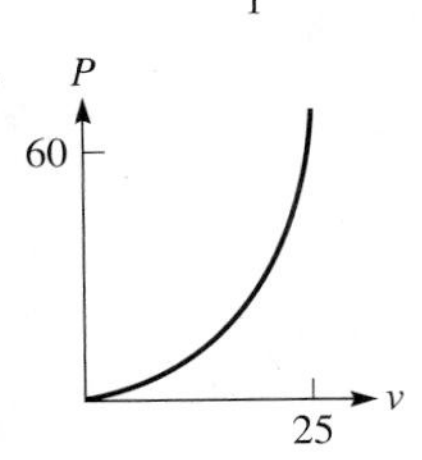

51.
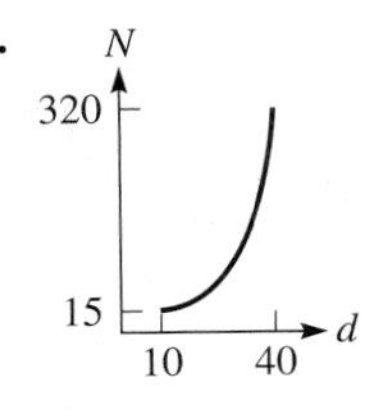

53.
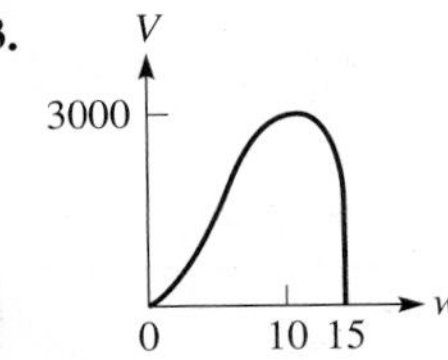

55. $A = 100w - w^2$
$30 \text{ m} \leq w \leq 70 \text{ m}$
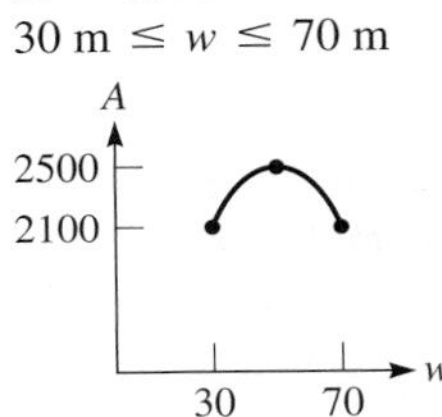

57.
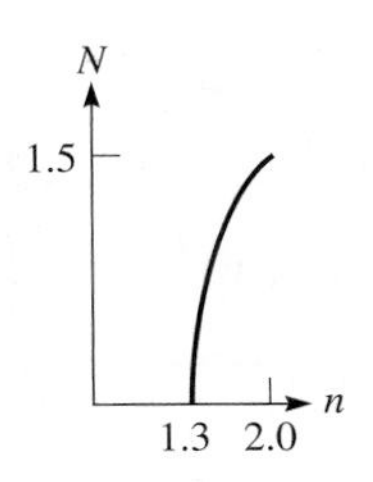

59. Only integer values of n have meaning.
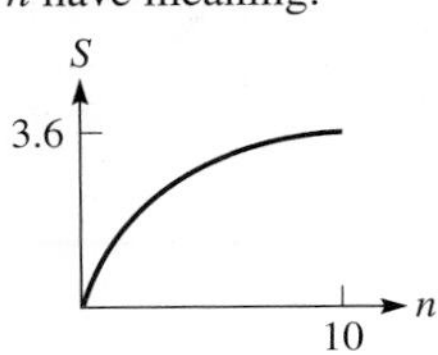

61. No. $f(1) = 2$, but $f(2)$ is not determined.

63.
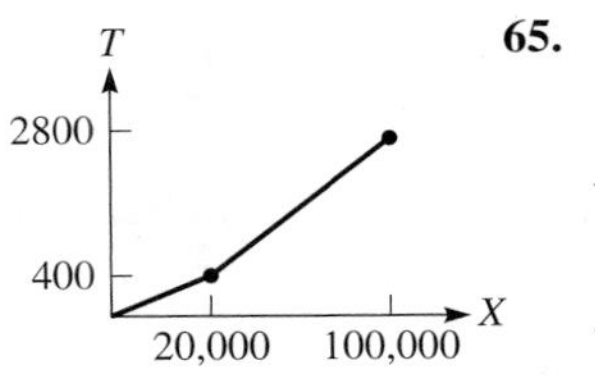

65.
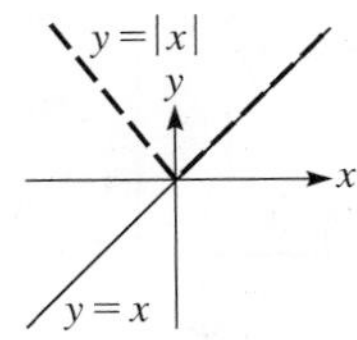

$y = x$ is the same as $y = |x|$ for $x \geq 0$
$y = |x|$ is the same as $y = -x$ for $x < 0$.

67.
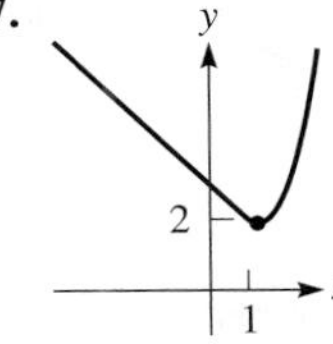

69. They are the same except $x = 2$ is not in the domain of (b).
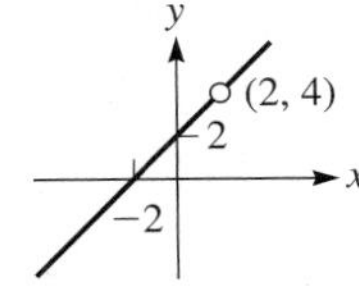

71. Yes **73.** No

Exercises 3.5, page 106

1. $-2.414, 0.414$

3.
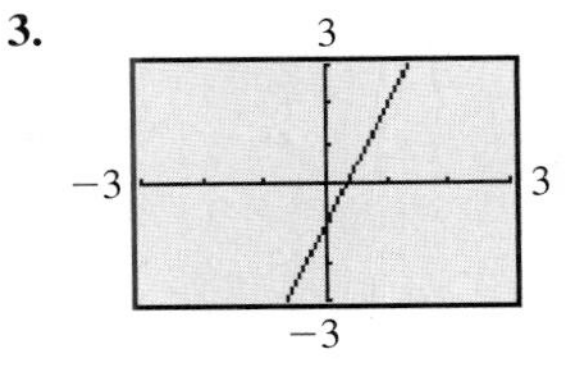

5.
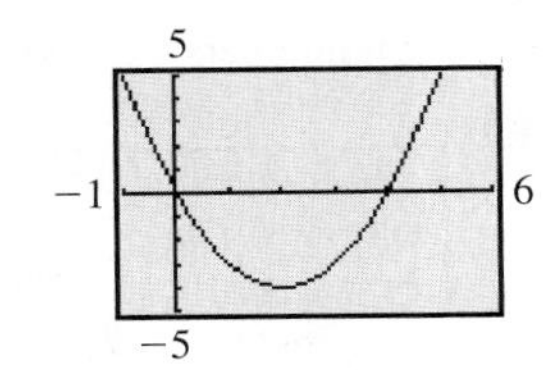

7.
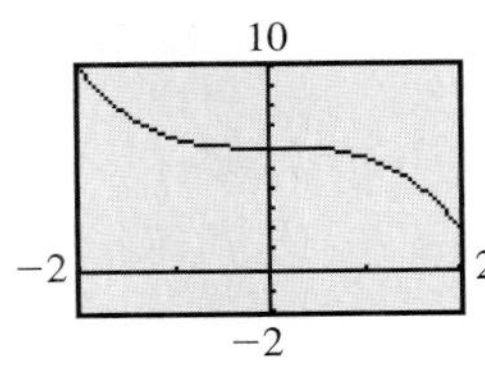

9.
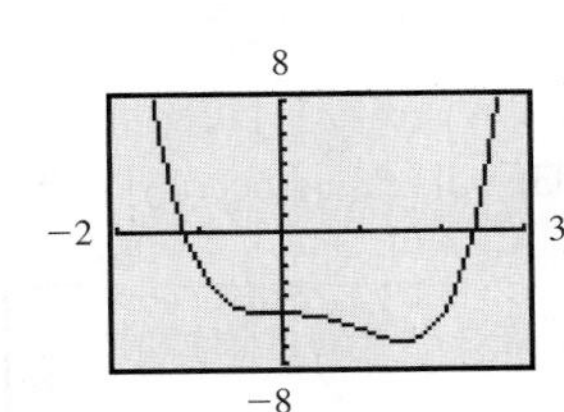

11.
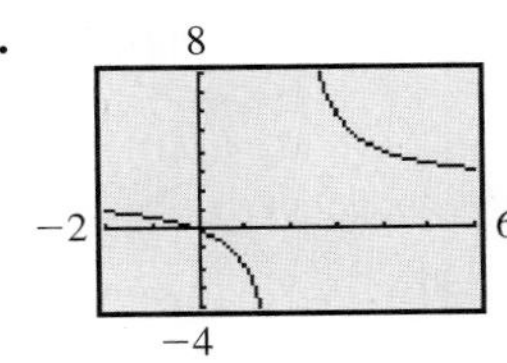

13.
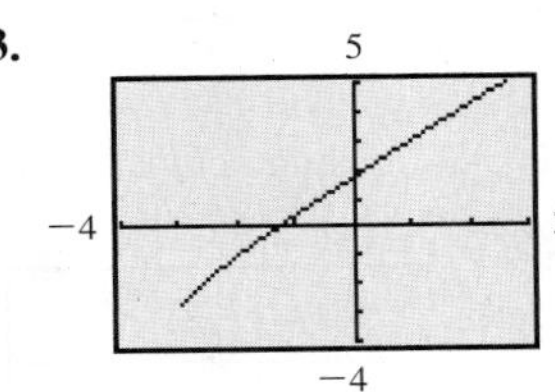

15.

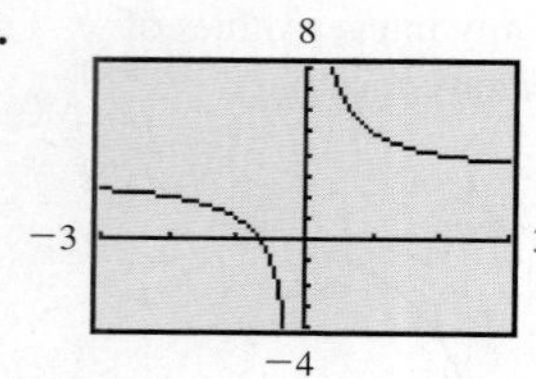

17.

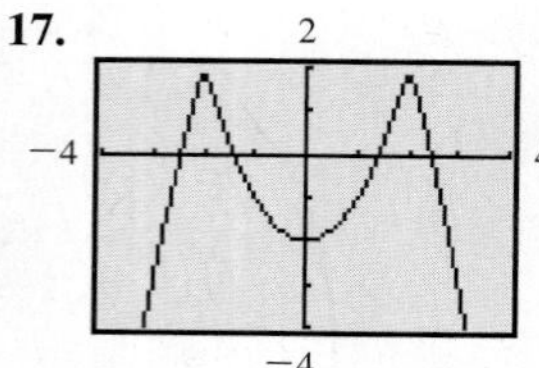

19. −2.791, 1.791

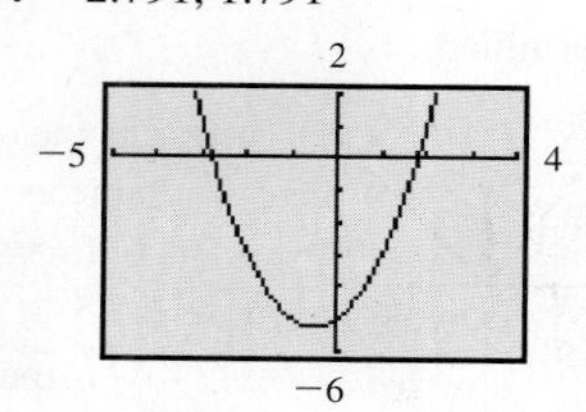

21. 2.104

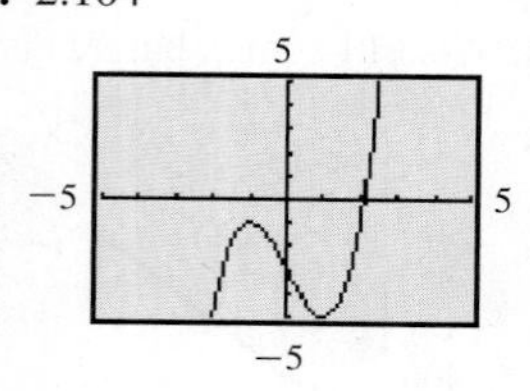

23. 4.321

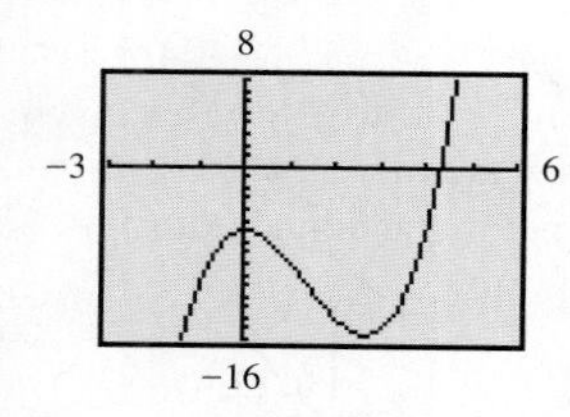

25. 1.4

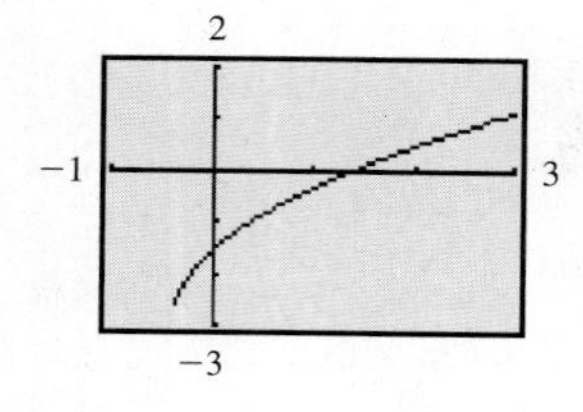

27. No values

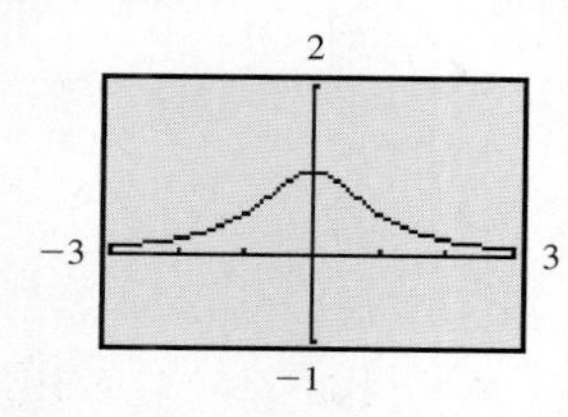

29. All real numbers $y \leq 7$

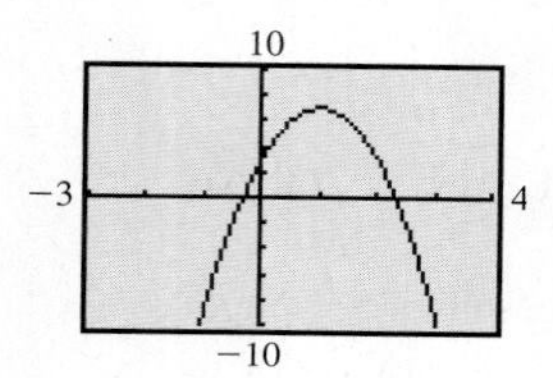

31. All real numbers $-5 \leq y \leq 5$

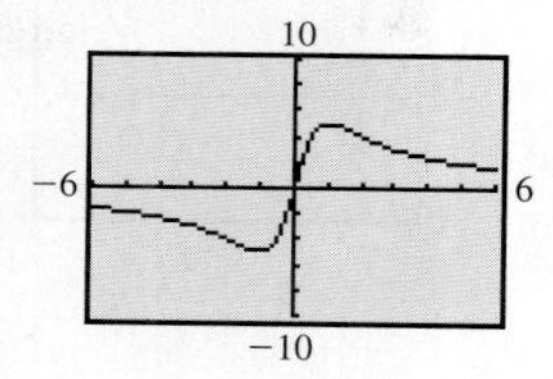

33. All real numbers $y > 0$ or $y \leq -1$

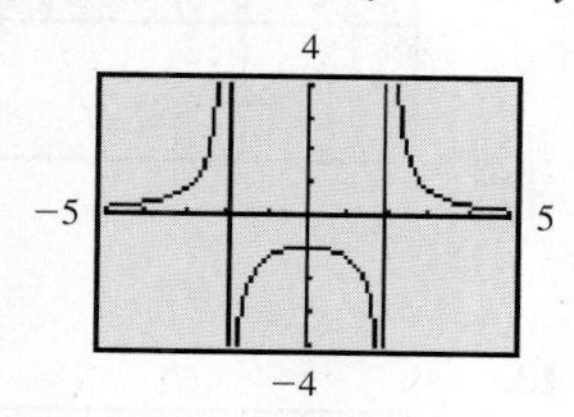

35. All real numbers $y \geq 0$ or $y \leq -4$

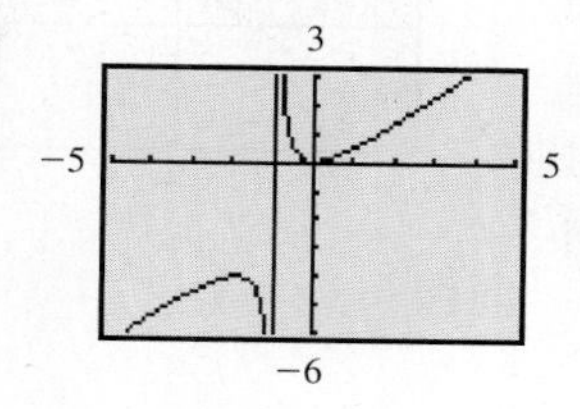

37. All real numbers $Y(y) > 3.46$ (approx.)

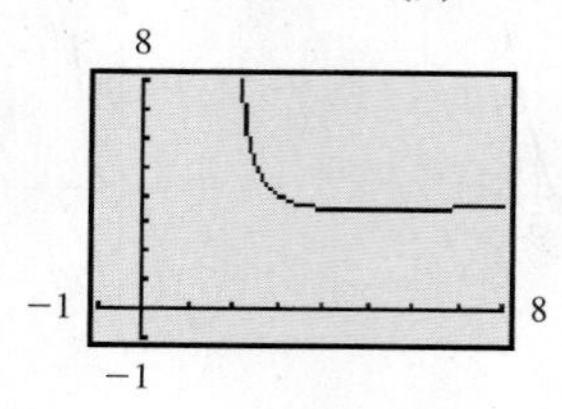

39. All real numbers

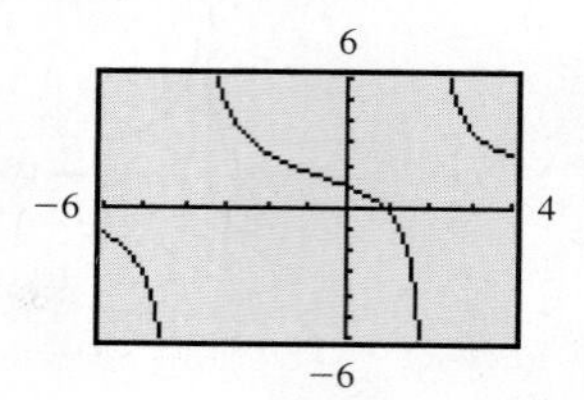

41. $y = 3x + 1$

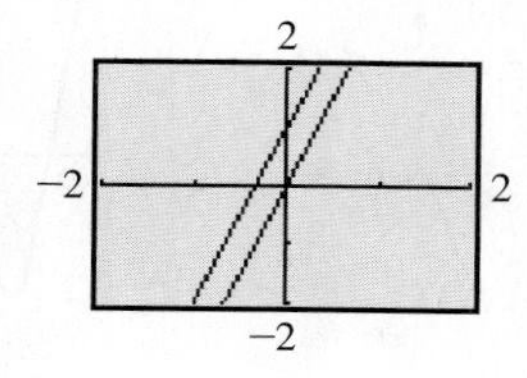

43. $y = \sqrt{x - 3}$

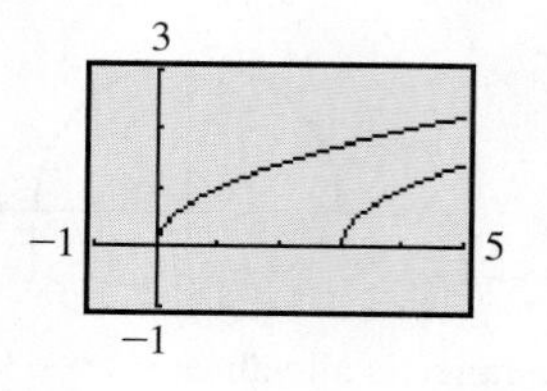

45. $y = -2(x + 2)^2 - 3$

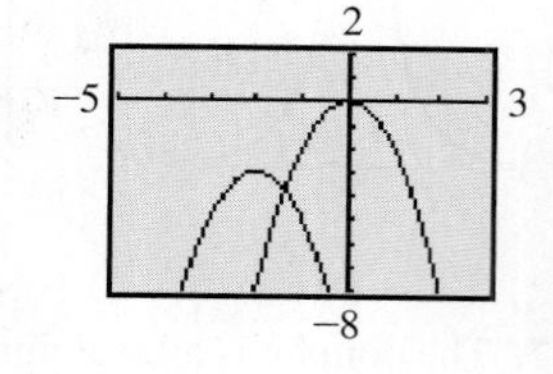

47. $y = \sqrt{2x + 3} + 1$

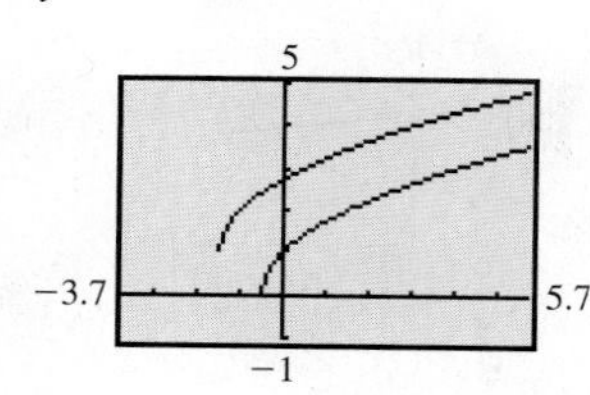

49.

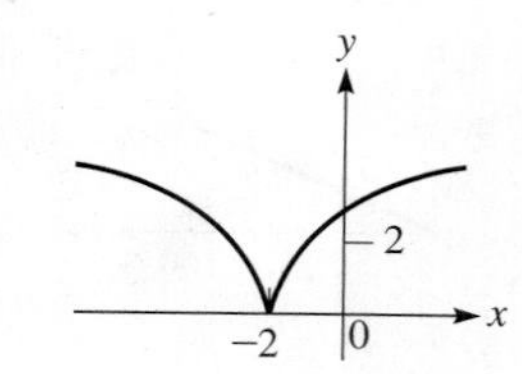

51.

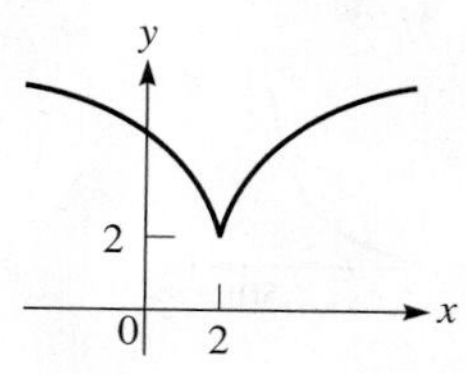

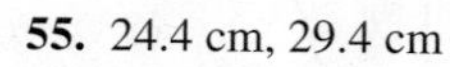

53. 6.0 V

55. 24.4 cm, 29.4 cm

57. 18 cm, 30 cm

59. 67 ft

61. 6.51 h

63. 0.25 ft/min

65. Graph for (b)

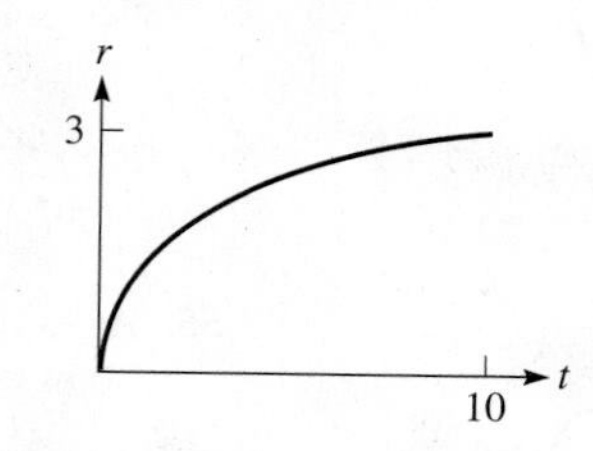

67. A curve with negative c is inverted from the curve with positive c.

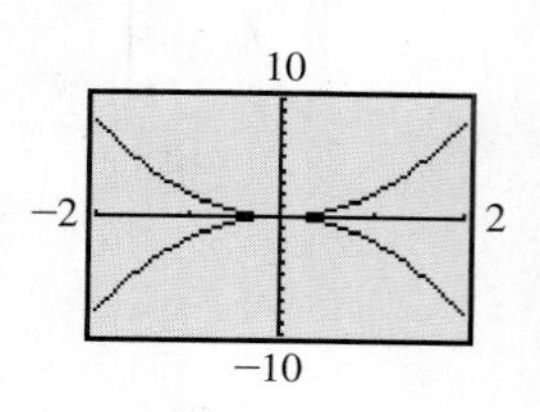

Exercises 3.6, page 109

1.

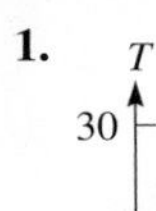

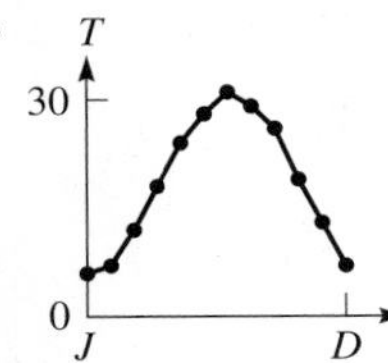

3.

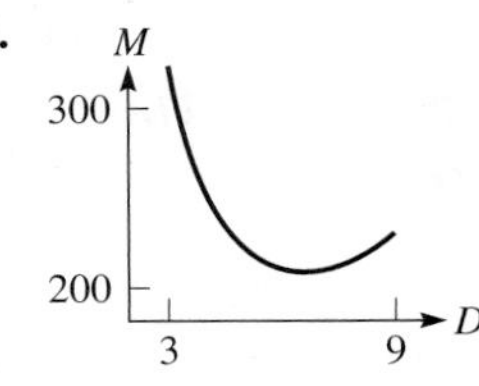

5.

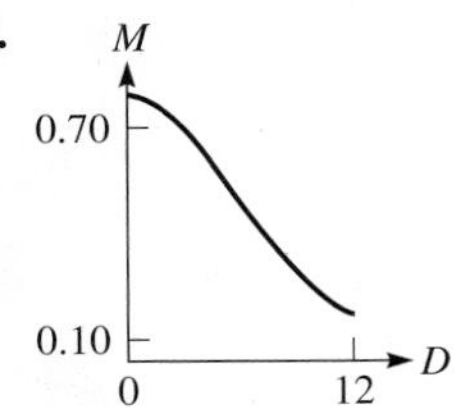

7.

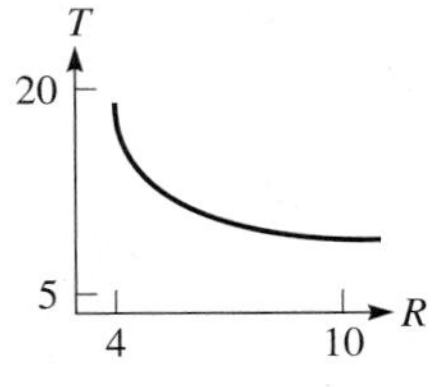

9. 132.1°C **11.** 0.7 min **13.** (a) 1.5 mm (b) 3.2 mm

15. 0.30 H **17.** 7.2% **19.** (a) 3.4 ft (b) 17 ft^3/s

21. 14 ft^3/s **23.** 0.34

25. 76 m^2 **27.** 130.3°C

29. 36 ft^3/s

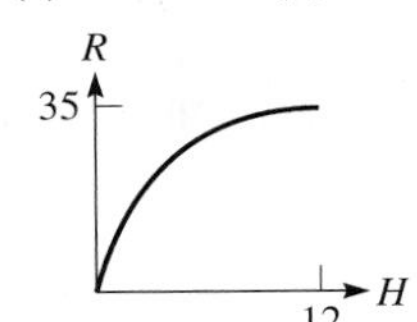

Review Exercises for Chapter 3, page 110

Concept Check Exercises

1. F **2.** F **3.** F **4.** F **5.** T **6.** T

Practice and Applications

7. $A = 4\pi t^2$ **9.** $y = 50.000 - 1.25x$ **11.** 16, −47

13. 3, $\sqrt{1 - 4h} + 2$ **15.** $h^3 + 11h^2 + 36h$ **17.** −4

19. −3.67, 16.7 **21.** 0.16503, −0.21476

23. Domain: all real numbers; range: all real numbers $f(x) \geq 1$

25. Domain: all real numbers $t > -4$; range: all real numbers $g(t) > 0$

27. Domain: all real numbers except 5; range: all real numbers $f(n) > 1$

29.

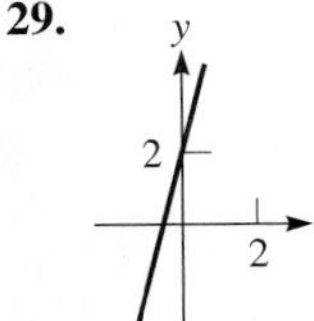

31.

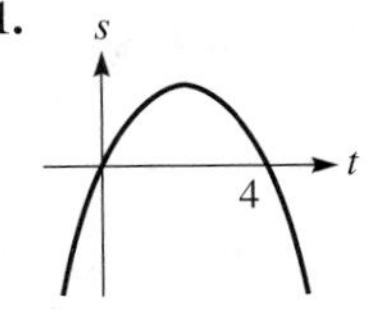

33.

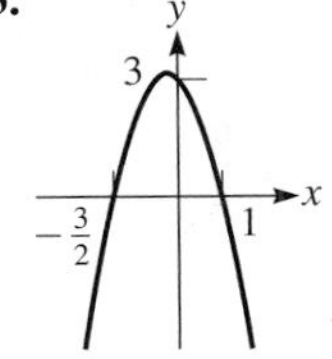

35.

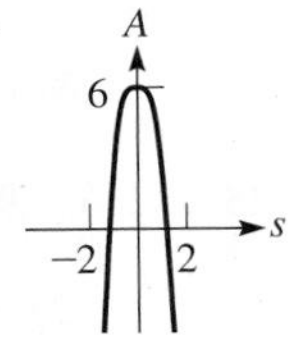

37.

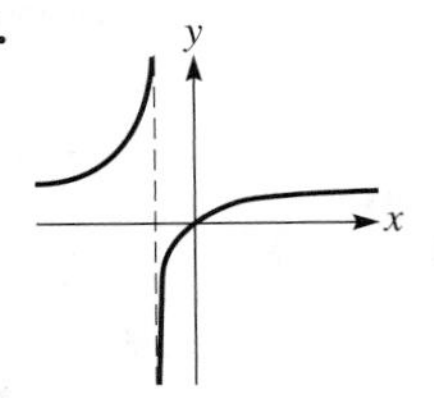

39. 0.4

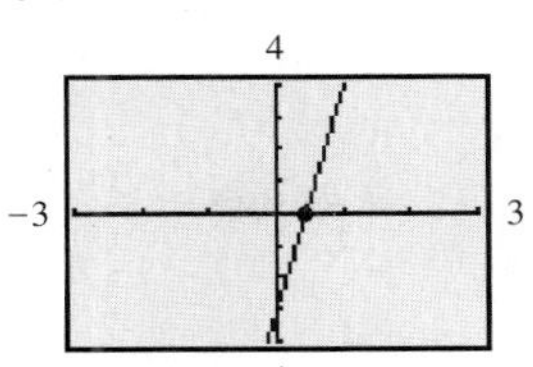

41. 0.2, 5.8

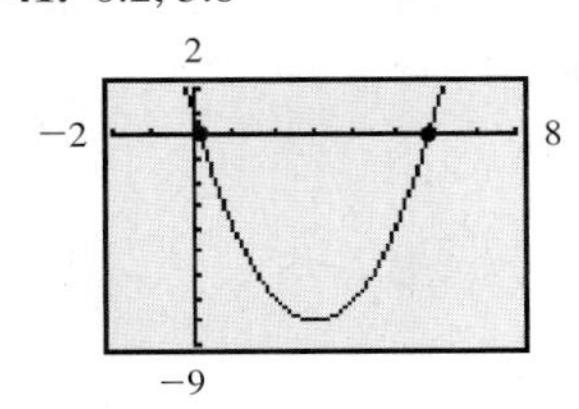

43. 1.4

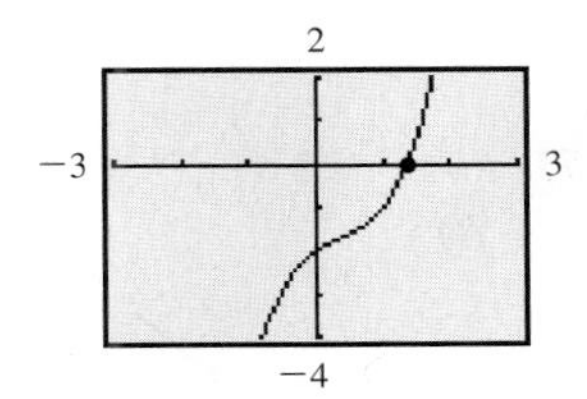

45. −4.1, 1.0

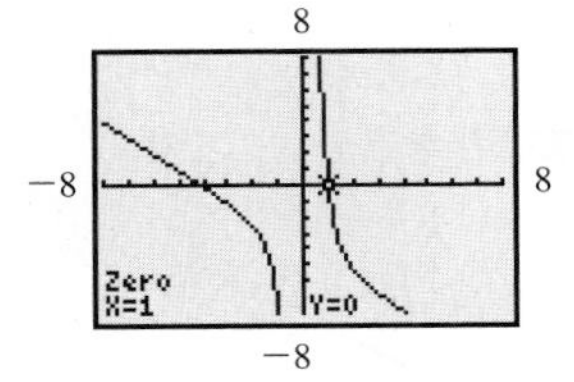

47. All real numbers $y \geq -6.25$

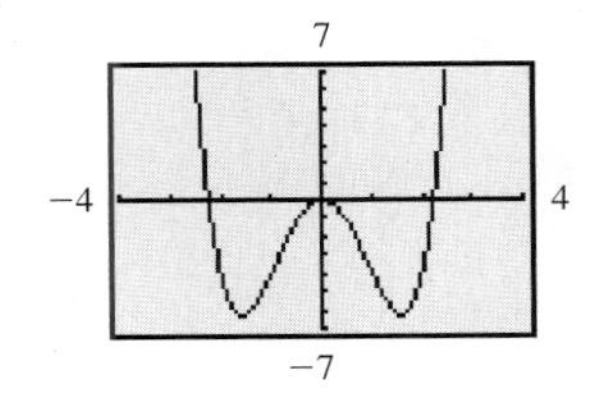

49. All real numbers $y \leq -2.83$ or $y \geq 2.83$

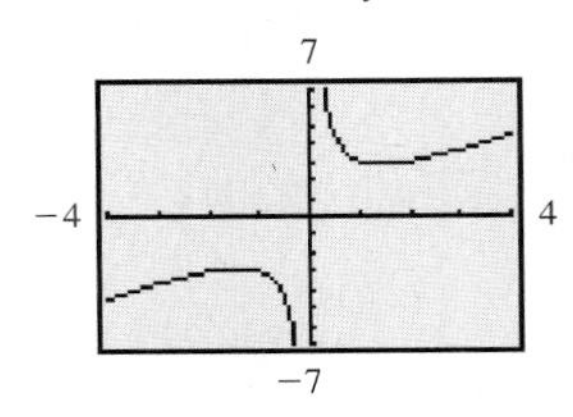

51. Either a or b is positive, the other is negative.

53. $(1, \sqrt{3})$ or $(1, -\sqrt{3})$ **55.** In any quadrant (not on an axis)

57. Many possibilities (two shown) **59.** $y = \sqrt{x + 1} + 1$

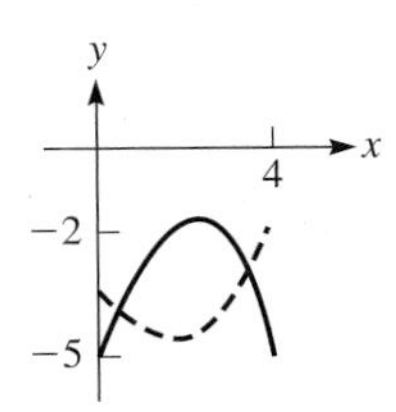

61. They are reflections of each other across the y-axis.

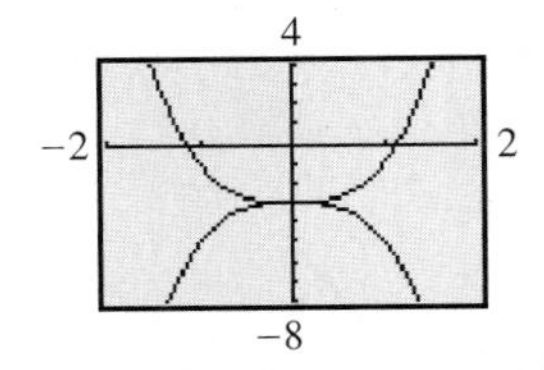

63. Yes; no two x-values are the same. **65.** $A = \frac{1}{2}\pi s^2$

67. 13.4 **69.** 72.0°

71.

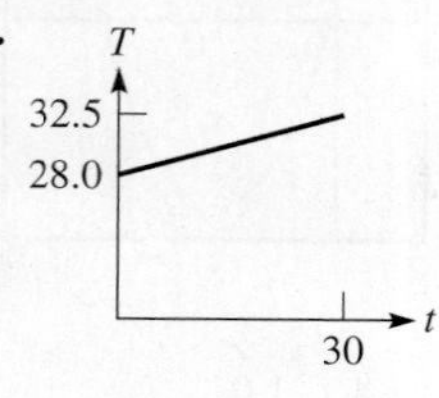

73.

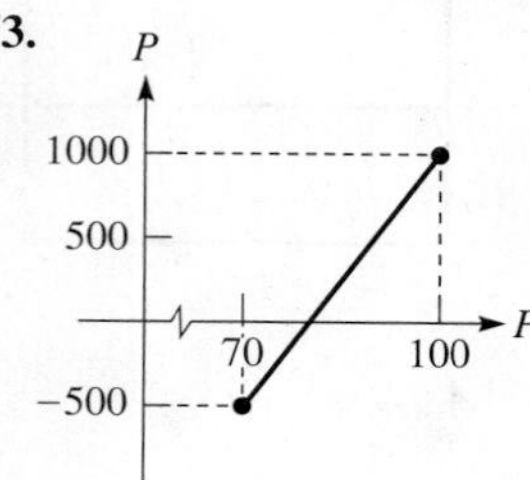

75. $L = 2\pi r + 12$

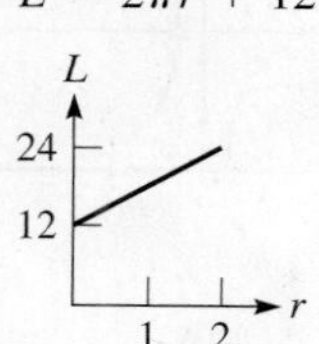

77. $f(T)$

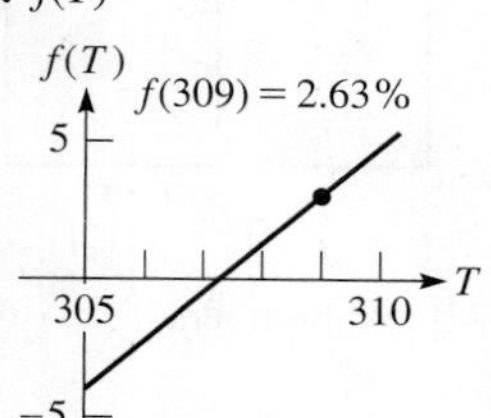

79.

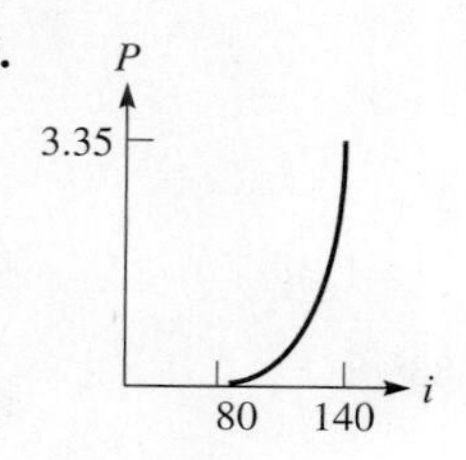

81.

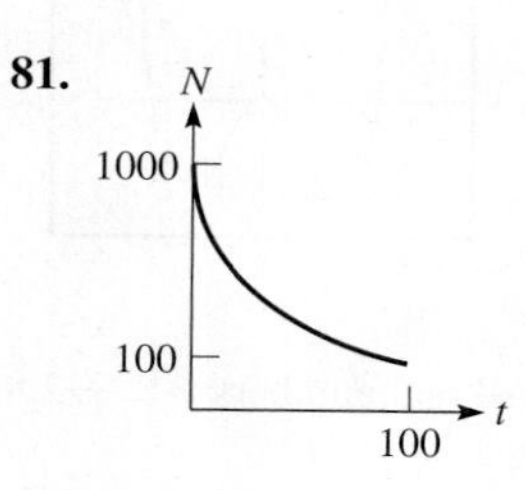

83.

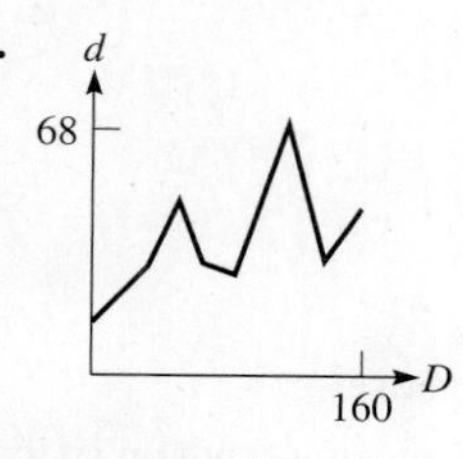

85. 11.3 ft

87. 3.5 h

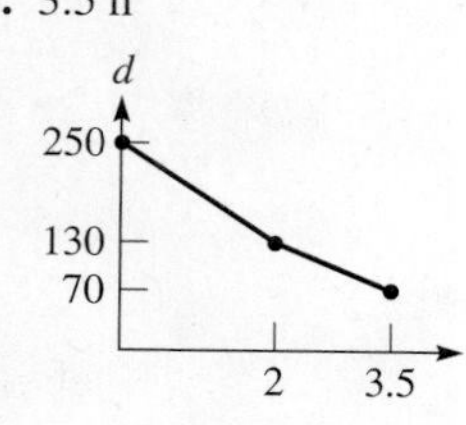

89. 33°C

91. 1.03 ft

93. 15.5 ft

95. 6.5 h

Exercises 4.1, page 116

1. 865.6° **3.** −135° (answers may vary)

5.

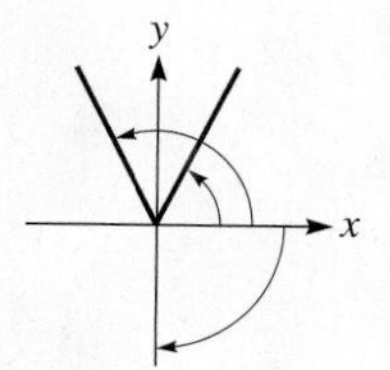

7.

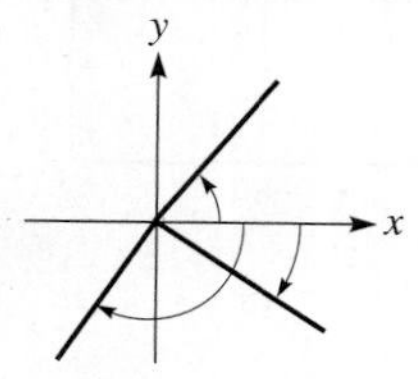

9. 485°, −235° **11.** 210°, −510° **13.** 638.1°, −81.9°

15. 38.67° **17.** 254.79° **19.** 72.02° **21.** −235.49°

23. 1.48 **25.** 6.72 **27.** 47° 30′ **29.** −5° 37′

31. 15.20° **33.** 301.27°

35.

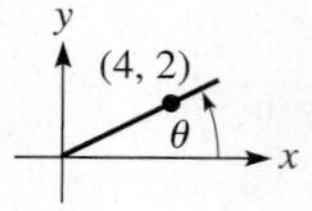

37.

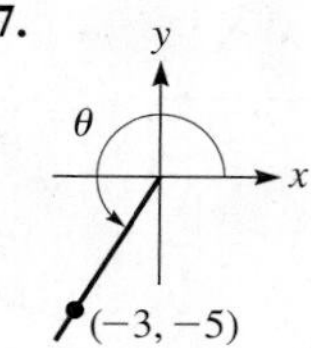

39.

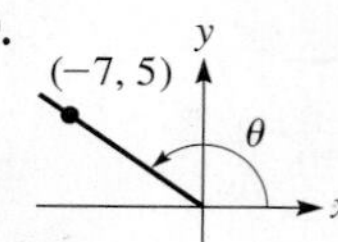

41.

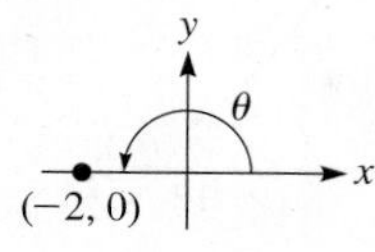

43. I, IV **45.** I, quadrantal **47.** I, II **49.** III, I

51. 21.710° **53.** 86°16′26″ **55.** −1260°

Exercises 4.2, page 120

1. $\sin\theta = \frac{3}{5}, \cos\theta = \frac{4}{5}, \tan\theta = \frac{3}{4}, \cot\theta = \frac{4}{3},$ $\sec\theta = \frac{5}{4}, \csc\theta = \frac{5}{3}$

3. $\sin\theta = \frac{4}{5}, \cos\theta = \frac{3}{5}, \tan\theta = \frac{4}{3}, \cot\theta = \frac{3}{4},$ $\sec\theta = \frac{5}{3}, \csc\theta = \frac{5}{4}$

5. $\sin\theta = \frac{8}{17}, \cos\theta = \frac{15}{17}, \tan\theta = \frac{8}{15}, \cot\theta = \frac{15}{8},$ $\sec\theta = \frac{17}{15}, \csc\theta = \frac{17}{8}$

7. $\sin\theta = \frac{40}{41}, \cos\theta = \frac{9}{41}, \tan\theta = \frac{40}{9}, \cot\theta = \frac{9}{40},$ $\sec\theta = \frac{41}{9}, \csc\theta = \frac{41}{40}$

9. $\sin\theta = \frac{35}{37}, \cos\theta = \frac{12}{37}, \tan\theta = \frac{35}{12},$ $\cot\theta = \frac{12}{35}, \sec\theta = \frac{37}{12}, \csc\theta = \frac{37}{35}$

11. $\sin\theta = \frac{\sqrt{15}}{4}, \cos\theta = \frac{1}{4}, \tan\theta = \sqrt{15}, \cot\theta = \frac{1}{\sqrt{15}},$ $\sec\theta = 4, \csc\theta = \frac{4}{\sqrt{15}}$

13. $\sin\theta = \frac{1}{\sqrt{2}}, \cos\theta = \frac{1}{\sqrt{2}}, \tan\theta = 1, \cot\theta = 1,$ $\sec\theta = \sqrt{2}, \csc\theta = \sqrt{2}$

15. $\sin\theta = \frac{2}{\sqrt{29}}, \cos\theta = \frac{5}{\sqrt{29}}, \tan\theta = \frac{2}{5}, \cot\theta = \frac{5}{2},$ $\sec\theta = \frac{\sqrt{29}}{5}, \csc\theta = \frac{\sqrt{29}}{2}$

17. $\sin\theta = 0.808, \cos\theta = 0.589, \tan\theta = 1.37, \cot\theta = 0.729,$ $\sec\theta = 1.70, \csc\theta = 1.24$

19. $\frac{5}{13}, \frac{12}{5}$ **21.** $\frac{2}{\sqrt{5}}, \sqrt{5}$ **23.** 0.882, 1.33 **25.** 0.246, 3.94

27. $\sin\theta = \frac{4}{5}, \tan\theta = \frac{4}{3}$ **29.** $\tan\theta = \frac{1}{3}, \sec\theta = \frac{\sqrt{10}}{3}$ **31.** 1

33. $\sin\theta = 0.96, \csc\theta = \frac{25}{24}$ **35.** $\tan A$ **37.** 1

39. $\cos\theta = \sqrt{1 - y^2}$ **41.** $\sec\theta$

Exercises 4.3, page 123

1. 20.65° **3.** 71.0° **5.** 0.572 **7.** 1.58 **9.** 0.9626
11. 1.0 **13.** 0.6283 **15.** 6.00 **17.** 116.9 **19.** 0.07063
21. 70.97° **23.** 65.70° **25.** 17.6° **27.** 49.453°
29. 22.86° **31.** 86.5° **33.** not possible **35.** 7.0°
37. $\sin 40° = 0.64$, $\cos 40° = 0.77$, $\tan 40° = 0.84$, $\cot 40° = 1.19$, $\sec 40° = 1.31$, $\csc 40° = 1.56$
39. $\sin 15° = 0.26$, $\cos 15° = 0.97$, $\tan 15° = 0.27$, $\cot 15° = 3.73$, $\sec 15° = 1.04$, $\csc 15° = 3.86$
41. $0.9556 = 0.9556$ **43.** $2.747 = 2.747$
45. y is always less than r.
47. For a given r, x decreases as θ increases from 0° to 90°.
49. $\frac{x}{r} + \frac{y}{r} = \frac{x+y}{r}$; $x + y > r$
51. 0.8885 **53.** 0.93614 **55.** 32.1°
57. 83.4 cm **59.** 48.6°

Exercises 4.4, page 128

1. $\sin A = 0.868$, $\cos A = 0.496$, $\tan A = 1.75$, $\sin B = 0.496$, $\cos B = 0.868$, $\tan B = 0.571$

3.

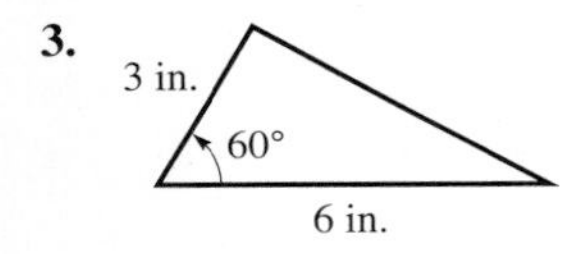

5.

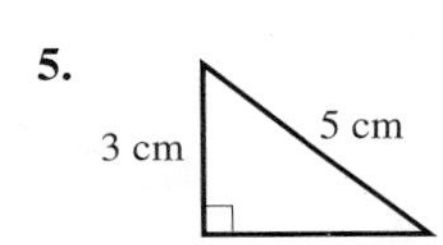

7. $B = 12.2°$, $b = 1450$, $c = 6850$
9. $A = 25.8°$, $B = 64.2°$, $b = 311$
11. $A = 57.9°$, $a = 202$, $b = 126$
13. $A = 84.7°$, $a = 877$, $B = 5.3°$
15. $a = 30.21$, $B = 57.90°$, $b = 48.16$
17. $A = 52.15°$, $B = 37.85°$, $c = 71.85$
19. $A = 52.5°$, $b = 0.661$, $c = 1.09$
21. $A = 15.82°$, $a = 0.5239$, $c = 1.922$
23. $A = 65.886°$, $B = 24.114°$, $c = 648.46$
25. $a = 0.75835$, $c = 14.612$, $B = 87.025°$
27. $a = 0.0959$, $b = 0.0162$, $A = 80.44°$
29. Cannot solve: need one more given part.
31. 4.45 **33.** 40.24° **35.** 39.2° **37.** 919
39. 108π **41.** 1776 ft

Exercises 4.5, page 131

1. 2090 ft **3.** 50.7 ft **5.** 8.0° **7.** 0.4°
9. $Z = 19.2\ \Omega$, $\phi = 51.3°$
11. 25.3 ft **13.** 0.21 mi **15.** 850.1 cm
17. 26.6°, 63.4°, 90.0° **19.** 23.5° **21.** 8.1° **23.** 3.4°
25. 3.07 cm **27.** 651 ft **29.** 30.2° **31.** 865,000 mi
33. 4.1° **35.** 47.3 m **37.** 2100 ft
39. $A = a(b + a\cos\theta)\sin\theta$
41. $d = 2x\tan(0.5\,\theta)$ **43.** 35.3°

Review Exercises for Chapter 4, page 135

Concept Check Exercises

1. F **2.** T **3.** F **4.** T **5.** T **6.** T

Practice and Applications

7. 407.0°, −313.0° **9.** 142.5°, −577.5° **11.** 31.9°
13. −83.35° **15.** 17° 30′ **17.** 749° 45′
19. $\sin\theta = \frac{7}{25}$, $\cos\theta = \frac{24}{25}$, $\tan\theta = \frac{7}{24}$, $\cot\theta = \frac{24}{7}$, $\sec\theta = \frac{25}{24}$, $\csc\theta = \frac{25}{7}$
21. $\sin\theta = \frac{1}{\sqrt{2}}$, $\cos\theta = \frac{1}{\sqrt{2}}$, $\tan\theta = 1$, $\cot\theta = 1$, $\sec\theta = \sqrt{2}$, $\csc\theta = \sqrt{2}$
23. 0.923, 2.40 **25.** 0.447, 1.12 **27.** 0.945, 1.06
29. 0.952 **31.** 13.24 **33.** 1.24 **35.** 0 **37.** 31.8°
39. 57.57° **41.** 12.25° **43.** 88.3° **45.** 8.00°
47. not possible **49.** $\left(\frac{\sqrt{21}}{5}, \frac{2}{5}\right)$
51. $a = 1.83$, $B = 73.0°$, $c = 6.27$
53. $A = 51.5°$, $B = 38.5°$, $c = 104$
55. $B = 52.5°$, $b = 15.6$, $c = 19.7$
57. $A = 31.61°$, $a = 4.006$, $B = 58.39°$
59. $a = 0.6292$, $B = 40.33°$, $b = 0.5341$
61. 4.6 **63.** 61.2 **65.** 10.5
67. $x/h = \cot A$, $y/h = \cot B$, $c = x + y = h\cot A + h\cot B$

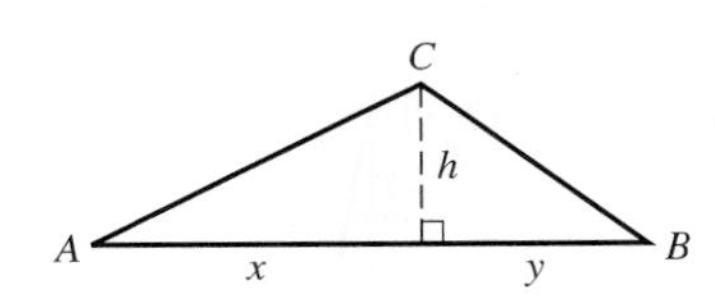

69. $c = 2r\sin(\theta/2)$ **71.** 22.6°
73. 71 ft, 35 ft **75.** 80.3° **77.** 12.0°
79. (a) $A = \frac{1}{2}bh = \frac{1}{2}b(a\sin C) = \frac{1}{2}ab\sin C$ (b) 679.2 m^2
81. 4.92 km **83.** 9.77 ft^2 **85.** 135 ft^2 **87.** 4.43 m
89. 34 m **91.** 56% **93.** 10.2 in. **95.** 4.71 min
97. 1.83 km **99.** 30.8° **101.** 464 m **103.** 73.3 cm

Exercises 5.1, page 146

1. $x = -3$, $y = -10$
3. The slope is $\frac{2}{3}$ and the y-intercept is $(0, -\frac{4}{3})$.
5. Linear **7.** Not linear **9.** Yes; no **11.** Yes; yes
13. $3, -\frac{21}{2}$ **15.** $\frac{1}{4}, -0.6$ **17.** 4 **19.** −5
21. $\frac{2}{7}$ **23.** $\frac{1}{2}$

25. **27.**

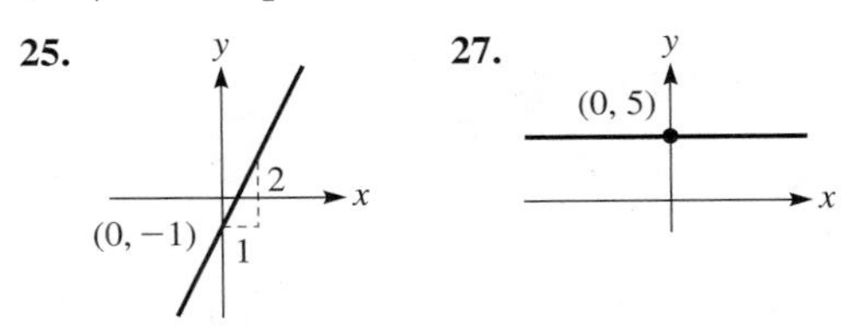

29.

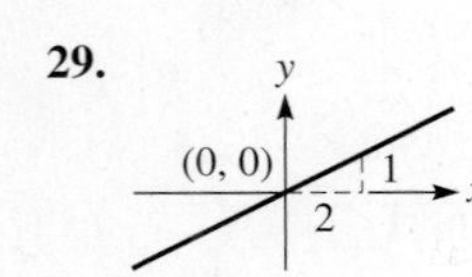

31.

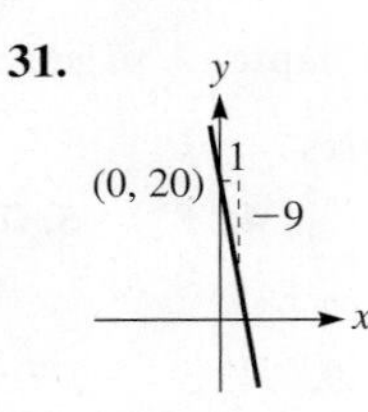

33. $m = -2, b = 1$

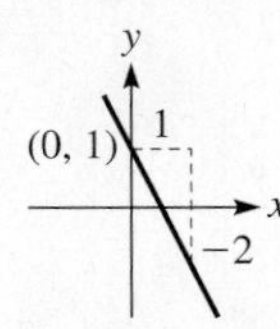

35. $m = 1, b = -4$

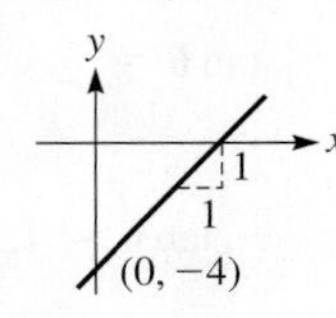

37. $m = \frac{5}{2}, b = -20$

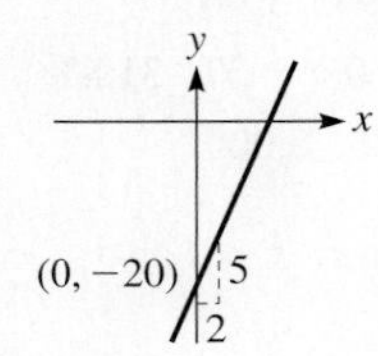

39. $m = -\frac{3}{5}, b = \frac{3}{8}$

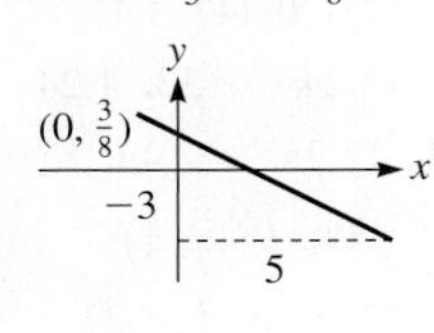

41.

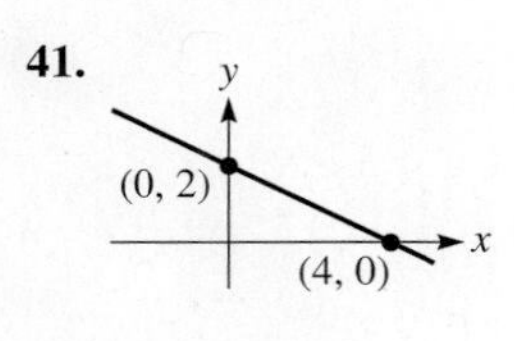

43.

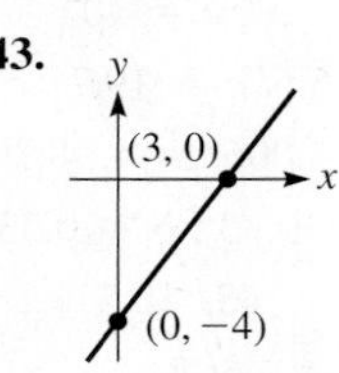

45.

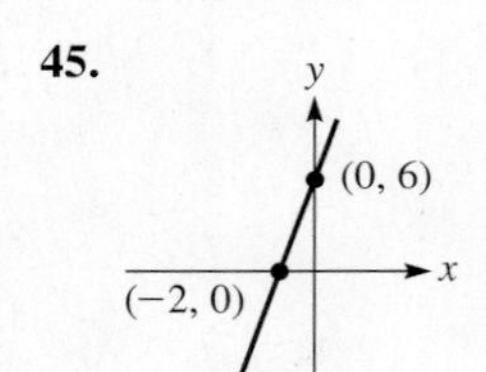

47.

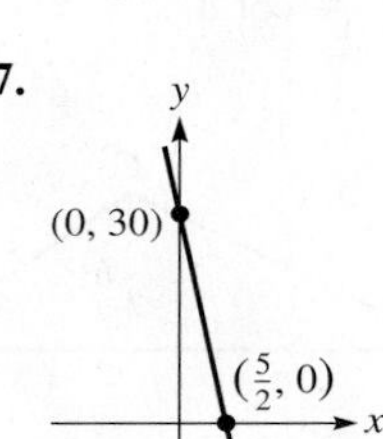

49. $s = 32.5d - 84.714$, 62 kN/mm

51. $v = 5.689d - 45.582$, 39.8 ft^3 **53.** $(a, 0), (0, b)$ **55.** No

57.

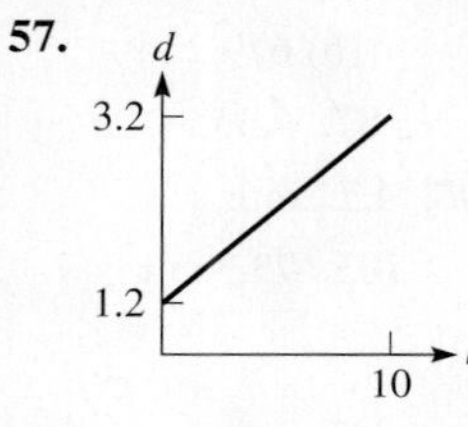

59.

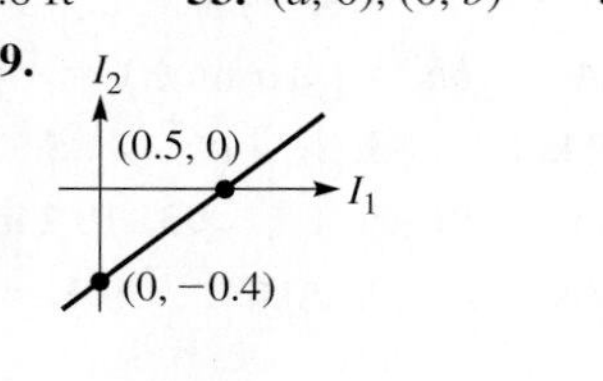

61. $h = 2500 - 150t$

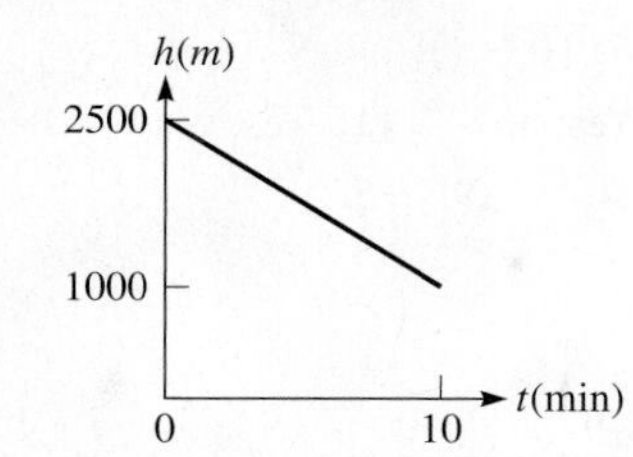

Exercises 5.2, page 150

1. Yes **3.** The system is dependent. **5.** Yes **7.** No
9. No **11.** Yes **13.** $x = 3.0, y = 1.0$ **15.** $x = 3.0, y = 0.0$
17. $x = 2.2, y = -0.3$ **19.** $x = -0.9, y = -2.3$
21. $s = 1.1, t = -1.7$ **23.** $x = 0.0, y = 3.0$
25. $x = -14.0, y = -5.0$ **27.** $r_1 = 4.0, r_2 = 7.5$
29. $x = -3.600, y = -1.400$ **31.** $x = 1.111, y = 0.841$
33. Dependent **35.** $t = -1.887, v = -3.179$
37. $x = 1.500, y = 4.500$ **39.** Inconsistent **41.** Yes
43. No **45.** 50N, 47N **47.** $p = 220$ km/h, $w = 40$ km/h
49. \$18, \$4 **51.** $m = 3, b \neq -7$

Exercises 5.3, page 157

1. $x = -3, y = -3$ **3.** $x = \frac{16}{13}, y = -\frac{2}{13}$
5. $x = 1, y = -2$ **7.** $V = 7, p = 3$
9. $x = -1, y = -4$ **11.** $x = \frac{1}{2}, y = 2$
13. $x = 1, y = \frac{1}{2}$ **15.** $x = 3, y = 1$
17. $x = -1, y = -2$ **19.** $t = -\frac{1}{3}, y = 2$
21. Inconsistent **23.** $x = -\frac{14}{5}, y = -\frac{16}{5}$
25. $x = \frac{1}{2}, y = -4$ **27.** $t = -\frac{2}{3}, y = 0$
29. $x = \frac{3}{5}, y = \frac{1}{5}$ **31.** $C = -1, V = -2$
33. $A = -1, B = 3$ **35.** $a = \frac{1}{3}, b = -7$
37. $x = 2.38, y = 0.45$ **39.** $s = 3.9, t = 3.2$
41. $x = \frac{17}{3}, y = \frac{1}{6}$ **43.** $f(x) = 2x - 3$
45. $x = 1, y = 2$ **47.** $V_1 = 9.0$ V, $V_2 = 6.0$ V
49. $x = 6250$ L, $y = 3750$ L **51.** 62
53. $W_r = 8120$ N, $W_f = 9580$ N **55.** $t_1 = 32$ s, $t_2 = 20$ s
57. 34 at \$900/month, 20 at \$1250/month
59. $A = 150{,}000, B = 148{,}000$ **61.** 7.00 kW, 9.39 kW
63. Incorrect conclusion or error in sales figures; system of equations is inconsistent.
65. $a \neq b$ **67.** $(-3, -4), (9, 0)$

Exercises 5.4, Page 163

1. 86 **3.** $x = -13, y = -27$ **5.** −4 **7.** 29
9. 108 **11.** 9300 **13.** 1.083 **15.** −256
17. $-a^2 + a + 2$ **19.** $x = 3, y = 1$ **21.** $x = -1, y = -2$
23. $t = -\frac{1}{3}, y = 2$ **25.** Inconsistent **27.** $x = -\frac{14}{5}, y = -\frac{16}{5}$
29. $x = 2, y = 0.4$ **31.** $s = 3.9, t = 3.2$
33. $x = -11.2, y = -9.26$ **35.** $x = -1.0, y = -2.0$
37. 0 **39.** 0 **41.** $F_1 = 15$ lb, $F_2 = 6.0$ lb
43. $x = 18.4$ gal, $y = 17.6$ gal
45. 160 3-bedroom homes, 80 4-bedroom homes
47. 210 phones, 110 detectors **49.** $L = 1.30$ m, $W = 0.802$ m
51. \$2000,6% **53.** 2.5 h, 2.1 h
55. $d = 835$ ft, $h = 327$ ft

Exercises 5.5, page 167

1. $x = 1, y = -4, z = \frac{1}{3}$ **3.** $x = 2, y = -1, z = 1$
5. $x = 4, y = -3, z = 3$ **7.** $l = \frac{1}{2}, w = \frac{2}{3}, h = \frac{1}{6}$
9. $x = \frac{2}{3}, y = -\frac{1}{3}, z = 1$ **11.** $x = \frac{4}{15}, y = -\frac{3}{5}, z = \frac{1}{3}$

13. $x = \frac{3}{4}, y = 1, z = -\frac{1}{2}$ **15.** $x = -3, y = 2, z = 1$
17. $I_1 = 3.62, I_2 = 5.11, I_3 = 1.49$ **19.** $f(x) = x^2 - 3x + 5$
21. $P = 800$ h, $M = 125$ h, $I = 225$ h
23. $F_1 = 9.43$ N, $F_2 = 8.33$ N, $F_3 = 1.67$ N
25. $A = 22.5°, B = 45.0°, C = 112.5°$
27. $A - 19{,}400, B - 20{,}000, C - 19{,}900$
29. MA − 260, MS − 120, PhD − 40 **31.** 70 lb, 100 lb, 30 lb
33. Unlimited: $x = -10, y = -6, z = 0$
35. No solution

Exercises 5.6, page 173

1. −38 **3.** 7 **5.** 651 **7.** −340 **9.** 202
11. 18,270 **13.** 0.128 **15.** $x = -1, y = 2, z = 0$
17. $x = 2, y = -1, z = 1$ **19.** $x = 4, y = -3, z = 3$
21. $l = \frac{1}{2}, w = \frac{2}{3}, h = \frac{1}{6}$ **23.** $x = \frac{2}{3}, y = -\frac{1}{3}, z = 1$
25. $x = \frac{4}{15}, y = -\frac{3}{5}, z = \frac{1}{3}$ **27.** $p = -2, q = \frac{2}{3}, r = \frac{1}{3}$
29. Collinear **31.** −35 (sign changes) **33.** 35 (no change)
35. $A = 125$ N, $B = 60$ N, $F = 75$ N
37. $s_0 = 2$ ft, $v_0 = 5$ ft/s, $a = 4$ ft/s^2
39. $\angle A = 120°; \angle B = 40°; \angle L = 80°$
41. $V = 5.0 + 0.5T + 0.1T^2$ **43.** 79% Ni, 16% Fe, 5% Mo
45. 30.9 mi/h, 45.9 mi/h, 551 mi/h

Review Exercises for Chapter 5, page 175

Concept Check Exercises

1. F **2.** T **3.** F **4.** F **5.** F **6.** T
7. F **8.** F

Practice and Applications

9. −17 **11.** −1485 **13.** −4 **15.** $\frac{2}{7}$
17. $m = -2, b = 4$ **19.** $m = 4, b = -\frac{5}{2}$

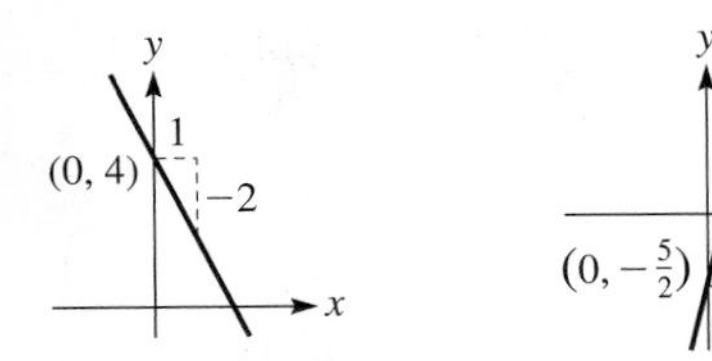

y
x
$(0, -\frac{5}{2})$
4
1

21. $x = 2.0, y = 0.0$ **23.** $A = 2.2, B = 2.7$
25. $x = 1.5, y = -1.9$ **27.** $M = 1.5, N = 0.4$
29. $x = 1, y = 2$ **31.** $x = \frac{1}{2}, y = -2$
33. $i = 2, v = -\frac{1}{3}$ **35.** $x = -\frac{6}{19}, y = \frac{36}{19}$
37. $x = \frac{43}{39}, y = \frac{7}{13}$ **39.** $x = 1, y = 2$
41. $x = \frac{1}{2}, y = -2$ **43.** $i = 2, v = -\frac{1}{3}$
45. $x = -\frac{6}{19}, y = \frac{36}{19}$ **47.** $x = 1.10, y = 0.54$
49. Best choices: 41, 42 **51.** Best choices: 47, 48
53. −115 **55.** 230.08 **57.** $x = 2, y = -1, z = 1$
59. $r = 3, s = -1, t = \frac{3}{2}$ **61.** $x = -0.17, y = 0.16, z = 2.4$
63. $x = -3, y = 4, z = -1.5$ **65.** $x = 2, y = -1, z = 1$
67. $r = 3, s = -1, t = \frac{3}{2}$ **69.** $x = -0.17, y = 0.16, z = 2.4$
71. 4 **73.** $-\frac{15}{7}$ **75.** $y = -0.007x + 54.570$, 30 mpg
77. −6 **79.** $-\frac{4}{3}$ **81.** $F_1 = 230$ lb, $F_2 = 24$ lb, $F_3 = 200$ lb
83. $p_1 = 42\%, p_2 = 58\%$ **85.** \$40,000, \$4000
87. 28 tons of 6% copper, 14 tons of 2.4% copper
89. $a = 440$ m·°C, $b = 9.6$°C
91. 17.5 tons of 18.0 gal/ton, 24.5 tons of 30.0 gal/ton
93. 34 at \$900/month, 20 at \$1250/month
95. 22,800 km/h, 1400 km/h **97.** $R_1 = 0.50\ \Omega, R_2 = 1.5\ \Omega$
99. $L = 10$ lb, $w = 40$ lb **101.** $A = 75°, B = 65°, C = 40°$
103. 68 MB, 24 MB, 48 MB **105.** 4.3 N, 1.7 N

Exercises 6.1, page 185

1. $2ax(2x - 1)$ **3.** $5(x + 3)(x - 3)$ **5.** $T^2 - 36$
7. $16x^2 - 25y^2$ **9.** $7(x + y)$ **11.** $5(a - 1)$ **13.** $3x(x - 3)$
15. $7b(bh - 4)$ **17.** $24n(3n^2 + 1)$ **19.** $2(x + 2y - 4z)$
21. $3ab(b - 2 + 4b^2)$ **23.** $4pq(3q - 2 - 7q^2)$
25. $2(a^2 - b^2 + 2c^2 - 3d^2)$ **27.** $(x + 3)(x - 3)$
29. $4(5 + a)(5 - a)$ **31.** $36a^4 + 1$(prime)
33. $2(9s + 5t)(9s - 5t)$ **35.** $(12n - 13p^2)(12n + 13p^2)$
37. $(x + y + 3)(x + y - 3)$ **39.** $2(2 + x)(2 - x)$
41. $300(x + 3z)(x - 3z)$ **43.** $2(I - 1)(I - 5)$
45. $(x^2 + 4)(x + 2)(x - 2)$
47. $x^2(x^4 + 1)(x^2 + 1)(x + 1)(x - 1)$
49. $\frac{3 + b}{2 - b}$ **51.** $\frac{3}{2(t - 1)}$ **53.** $\frac{5}{2}$
55. $(3 + b)(x - y)$ **57.** $(a - b)(a + x)$
59. $(x + 2)(x - 2)(x + 3)$ **61.** $(x - y)(x + y + 1)$
63. $8^8 = 16{,}777{,}216$
65. $n^2 + n = n(n + 1)$; product of two consecutive positive integers is a positive integer.
67. $2(Q^2 + 1)$ **69.** $s(9 - s)(9 + s)$
71. $r(R - r)(R + r)$ **73.** $r^2(\pi - 2)$
75. $4\pi(r_2 + r_1)(r_2 - r_1)$ **77.** $\frac{i_2R_2}{R_1 + R_2}$
79. $\frac{5Y}{3(3S - Y)}$ **81.** $\frac{ER}{A(T_0 - T_1)}$

Exercises 6.2, page 192

1. $(x + 3)(x + 1)$ **3.** $(2x - 1)(x - 5)$
5. $x^2 - 14x + 49$ **7.** $a^2 + 6ab + 9b^2$
9. $(x + 1)(x + 3)$ **11.** $(s - 7)(s + 6)$
13. $(t + 8)(t - 3)$ **15.** $(x + 4)^2$
17. $(a - 3b)^2$ **19.** $(3x + 1)(x - 2)$
21. $4(3y + 1)(y - 3)$ **23.** $(2s + 11)(s + 1)$
25. $(3z - 1)(z - 6)$ **27.** $(2t - 3)(t + 5)$
29. $(3t - 4u)(t - u)$ **31.** $(6x - 5)(x + 1)$
33. $(9x - 2y)(x + y)$ **35.** $3(2m + 5)^2$
37. $2(2x - 3)^2$ **39.** $(3t - 4)(3t - 1)$

41. $(8b^3 - 1)(b^3 + 4)$ **43.** $(4p - q)(p - 6q)$
45. $(12x - y)(x + 4y)$ **47.** $2(x - 1)(x - 6)$
49. $2x(2x^2 - 1)(x^2 + 4)$ **51.** $ax(x + 6a)(x - 2a)$
53. $(4x^n - 3)(x^n + 4)$ **55.** $16(t - 1)(t - 4)$
57. $(d^2 - 2)(d^2 - 8)$ **59.** $(3Q + 10)(Q - 3)$
61. $(V - nB)^2$ **63.** $wx^2(x - 2L)(x - 3L)$
65. $Ad(3u - v)(u - v)$ **67.** 12, −12
69. −19, −11, −9, 9, 11, 19 **71.** $-16(t - 4)(t + 2)$

Exercises 6.3, page 194

1. $(x - 2)(x^2 + 2x + 4)$ **3.** $(x + 1)(x^2 - x + 1)$
5. $(y - 5)(y^2 + 5y + 25)$ **7.** $(3 - t)(9 + 3t + t^2)$
9. $(2a - 3b)(4a^2 + 6ab + 9b^2)$ **11.** $4(x + 2)(x^2 - 2x + 4)$
13. $7n^2(n - 1)(n^2 + n + 1)$ **15.** $6x^3y(9 - y^3)$
17. $x^3y^3(x + y)(x^2 - xy + y^2)$ **19.** $3a^2(a^2 + 1)(a + 1)(a - 1)$
21. $\frac{4}{3}\pi(R - r)(R^2 + Rr + r^2)$ **23.** $27L^3(L + 2)(L^2 - 2L + 4)$
25. $(a + b + 4)(a^2 + 2ab + b^2 - 4a - 4b + 16)$
27. $(2 - x)(2 + x)(16 + 4x^2 + x^4)$
29. $[(x - 1)(x^2 + x + 1)]^2$ **31.** $4x(2 - x)(4 + 2x + x^2)$
33. $D(D - d)(D^2 + Dd + d^2)$
35. $QH(H + Q)(H^2 - HQ + Q^2)$
37. Expanding $(a + b)(a^2 - ab + b^2)$ verifies that it equals $a^3 + b^3$.
39. $(x + y)(x - y)(x^4 + x^2y^2 + y^4)$
41. $(n^3 + 1) = (n + 1)(n^2 - n + 1)$; If $n > 1$, both factors are integers and greater than 1.

Exercises 6.4, page 198

1. $\frac{x - 6}{x - 2}$ **3.** $\frac{14}{21}$ **5.** $\frac{3a^2x}{3ay}$ **7.** $\frac{2x - 4}{x^2 + x - 6}$
9. $\frac{ax + ay}{x^2 - y^2}$ **11.** $\frac{7}{11}$ **13.** $\frac{2xy}{4y^2}$ **15.** $\frac{4}{R + 2}$
17. $\frac{s - 5}{2s - 1}$ **19.** $9xy$ **21.** $a^2 - 16$ **23.** $2x$
25. 1 **27.** $\frac{5}{9}$ **29.** $\frac{3x}{4}$ **31.** $\frac{1}{5a}$ **33.** $\frac{2(a - b)}{2a - b}$
35. $\frac{4x^2 + 1}{(2x + 1)(2x - 1)}$ (cannot be reduced) **37.** $3x$
39. $\frac{1}{2y^2}$ **41.** $\frac{x - 5}{x + 5}$ **43.** $\frac{2w^2 - 1}{w^2 + 8}$ **45.** $\frac{5x + 4}{x(x + 3)}$
47. $\frac{(N + 2)(N^2 + 4)}{8}$ **49.** $\frac{1}{2t + 1}$ **51.** $\frac{x + 3}{x - 3}$ **53.** $-\frac{x + y}{2}$
55. $\frac{n + 1}{n - 1}$ **57.** $\frac{(x + 5)(x - 3)}{(5 - x)(x + 3)}$ **59.** $\frac{x^2 - xy + y^2}{2}$
61. $\frac{2x}{9x^2 - 3x + 1}$
63. (a) $\frac{x^2(x + 2)}{x^2 + 4}$ (b) $\frac{x^2}{(x - 2)(x + 2)}$ Numerator and denominator have no common factor. In each, x^2 is not a factor of the denominator.
65. (a) $\frac{(x - 2)(x + 1)}{x(x - 1)}$ (b) $\frac{x - 2}{x}$ Numerator and denominator have no common factor. In each, x is not a factor of the numerator.
67. No **69.** $\frac{v + v_o}{t}$ **71.** $u + v$ **73.** $\frac{E^2(R - r)}{(R + r)^3}$

Exercises 6.5, page 203

1. $\frac{(x - 2)(x + 3)}{3x - 1}$ **3.** $\frac{1}{15}$ **5.** $6xy$ **7.** $\frac{7}{18}$ **9.** $\frac{y^2}{b^2}$ **11.** $\frac{2y}{5}$
13. $3(u + v)(u - v)$ **15.** $\frac{50}{3(a + 4)}$ **17.** $\frac{x^2 - 3}{x^2}$ **19.** $\frac{3x}{5a}$
21. $\frac{(x + 1)(x - 1)(x - 4)}{4(x + 2)}$ **23.** $\frac{x^2}{a + x}$ **25.** $\frac{15}{4}$
27. $\frac{3(x - 5)}{(x - 2)(x + 1)}$ **29.** $\frac{2(3T + N)(5V - 6)}{(V - 7)(4T + N)}$ **31.** $\frac{7x^4}{3a^4}$
33. $\frac{4t(2t - 1)(t + 5)}{(2t + 1)^2}$ **35.** $\frac{x + y}{2}$ **37.** $(x + y)(3p + 7q)$
39. $\frac{x - 2}{6x}$ **41.** $\frac{2(2x - 1)}{5(2 - x)}$ **43.** $\frac{\pi}{2}$
45. $\frac{c(\lambda + \lambda_0)(\lambda - \lambda_0)}{\lambda^2 + \lambda_0^2}$
47. $v = \sqrt{\frac{GM}{r}}$

Exercises 6.6, page 208

1. $12a^2b^3$ **3.** $\frac{12}{s(s + 4)}$ **5.** 2 **7.** $\frac{8}{x}$ **9.** $\frac{5}{4}$
11. $\frac{x^2 + 7}{3x}$ **13.** $\frac{ax - b}{x^2}$ **15.** $\frac{30 + ax^2}{25x^3}$ **17.** $\frac{14 - a^2}{10a}$
19. $\frac{xy + 3x + 2y}{4xy}$ **21.** $y - 5$ **23.** $\frac{3x + 8}{6(x - 1)}$
25. $\frac{5 - 3x}{2x(x + 1)}$ **27.** $\frac{-3}{4(s - 3)}$ **29.** $\frac{7R + 6}{3(R + 3)(R - 3)}$
31. $\frac{2x - 5}{(x - 4)^2}$ **33.** $\frac{-4}{(v + 1)(v - 3)}$ **35.** $\frac{9x^2 + x - 2}{(3x - 1)(x - 4)}$
37. $\frac{-2t^3 + 3t^2 + 15t}{(t - 3)(t + 2)(t + 3)^2}$ **39.** $\frac{-2w^3 + w^2 - w}{(w + 1)(w^2 - w + 1)}$
41. $\frac{3}{1 - x}$ **43.** $\frac{x - y}{y}$ **45.** $-\frac{(3x + 4)(x - 1)}{2x}$
47. $\frac{h}{(x + 1)(x + h + 1)}$ **49.** $\frac{-2hx - h^2}{x^2(x + h)^2}$ **51.** $\frac{y^2 - rx + r^2}{r^2}$
53. $\frac{2a - 1}{a^2}$ **55.** $\frac{a(a + 2)}{a + 1}$ **57.** $\frac{a + b}{\frac{1}{a} + \frac{1}{b}} = ab$
59. $\frac{2mn}{m^2 + n^2}$ **61.** $\frac{3(H - H_0)}{4\pi H}$ **63.** $\frac{n(2n + 1)}{2(n + 2)(n - 1)}$
65. $\frac{P^2(9x^2 + L^2)}{4L^4}$ **67.** $\frac{sL + R}{s^2LC + sRC + 1}$
69. $\frac{\omega^2L^2 + \omega^4R^2C^2L^2 - 2\omega^2R^2CL + R^2}{\omega^2R^2L^2}$
71. $\frac{10v}{(v + w)(v - w)}$

Exercises 6.7, page 213

1. $\frac{2}{b - 1}$ **3.** $\frac{arv^2}{2aM - rM}$ **5.** 4 **7.** −3 **9.** $\frac{11}{3}$
11. $\frac{16}{21}$ **13.** −9 **15.** −5 **17.** $-\frac{2}{13}$ **19.** $\frac{15}{2}$
21. −2 **23.** $-\frac{1}{2}$ **25.** $\frac{51}{8}$ **27.** 7 **29.** $\frac{37}{6}$
31. No solution **33.** $\frac{2}{3}$ **35.** $\frac{3b}{1 - 2b}$
37. $\frac{(2b - 1)(b + 6)}{2(b - 1)}$ **39.** $\frac{2s - 2s_0 - v_0t}{t}$ **41.** $\frac{40V - 240}{78 - 5.0V}$
43. $\frac{jX}{1 - g_mZ}$ **45.** $\frac{PV^3 - bPV^2 + aV - ab}{RV^2}$ **47.** $\frac{CC_2 + CC_3 - C_2C_3}{C_2 - C}$
49. $\frac{fnR_2 - fR_2}{R_2 + f - fn}$ **51.** 3.1 h **53.** 3.2 min **55.** 220 m
57. 6 m^3 **59.** Mach 1.6 **61.** 80 km/h **63.** 2.2 V
65. $A = 3, B = -2$

Review Exercises for Chapter 6, page 215

Concept Check Exercises

1. F **2.** T **3.** F **4.** F **5.** F **6.** T **7.** F **8.** F

Practice and Applications

9. $12ax + 15a^2$ **11.** $9a^2 - 16b^2$ **13.** $b^2 + 6b - 16$

15. $5(s + 4t)$ **17.** $a^2(x^2 + 1)$

19. $b^x(Wb + 12)(Wb - 12)$ **21.** $(4x + 8 + t^2)(4x + 8 - t^2)$

23. $4(3t - 1)^2$ **25.** $(5t + 1)^2$ **27.** $(x + 7)(x - 6)$

29. $(t + 3)(t - 3)(t^2 + 4)$ **31.** $(2k - 9)(k + 4)$

33. $2(2x + 5)(2x - 7)$ **35.** $(5b - 1)(2b + 5)$

37. $4(x + 2y)(x - 2y)$ **39.** $2(5 - 2y^2)(25 + 10y^2 + 4y^4)$

41. $(2x + 3)(4x^2 - 6x + 9)$ **43.** $(a - 3)(b^2 + 1)$

45. $(x + 5)(n - x + 5)$ **47.** $\frac{16x^2}{3a^2}$ **49.** $\frac{3x + 1}{x - 1}$

51. $\frac{16}{5x(x - y)}$ **53.** $-\frac{6}{L - 5}$ **55.** $\frac{x + 2}{2x(7x - 1)}$

57. $\frac{1}{x - 1}$ **59.** $\frac{16x - 15}{36x^2}$ **61.** $\frac{5y + 6}{2xy}$ **63.** $\frac{2(3a + 1)}{a(a + 2)}$

65. $\frac{4x^2 - x + 1}{2x(x + 3)(x - 1)}$ **67.** $\frac{12x^2 - 7x - 4}{2(x - 1)(x + 1)(4x - 1)}$

69. $\frac{x^3 + 6x^2 - 2x + 2}{x(x - 1)(x + 3)}$

71.

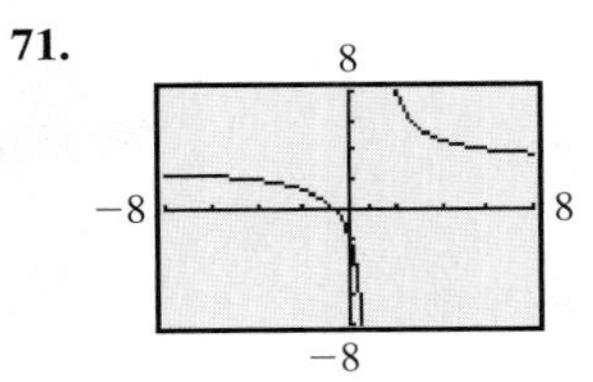

73.

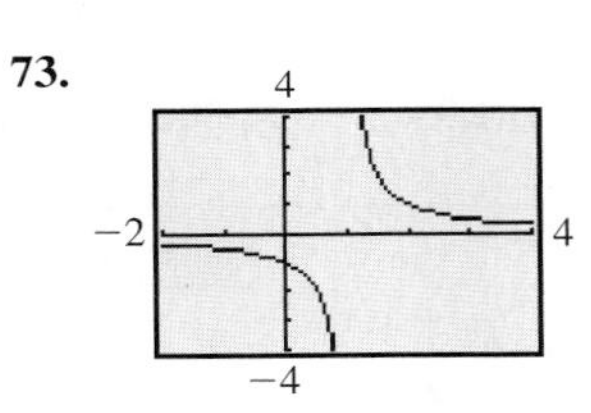

75. 2 **77.** $-\frac{(a - 1)^2}{2a}$ **79.** $\frac{10}{3}$ **81.** -6

83. (a) Sign changes (b) No change in sign

85. $\frac{1}{4}[(x + y)^2 - (x - y)^2]$

$= \frac{1}{4}(x^2 + 2xy + y^2 - x^2 + 2xy - y^2)$

$= \frac{1}{4}(4xy) = xy$

87. $PbL^2 - Pb^3$ **89.** $4b^2 + 4bn\lambda - 4b\lambda + n^2\lambda^2 - 2n\lambda^2 + \lambda^2$

91. $(c + R)(T_2 - T_1)$ **93.** $R(3R - 4r)$

95. $8n^6 + 36n^5 + 66n^4 + 63n^3 + 33n^2 + 9n + 1$

97. $10aT - 10at + aT^2 - 2aTt + at^2$

99. $4(3x^2 + 12x + 16)$ **101.** $\frac{12\pi^2 wv^2 D}{gn^2 t}$

103. $\frac{R^2 + r^2}{2}$ **105.** $\frac{120 - 60d^2 + 5d^4 - d^6}{120}$

107. $\frac{4k^2 + k - 2}{4k(k - 1)}$ **109.** $\frac{4r^3 - 3ar^2 - a^3}{4r^3}$

111. $\frac{c^2u^2 - 2gc^2x}{1 - c^2u^2 + 2gc^2x}$ **113.** $\frac{W}{g(h_2 - h_1)}$ **115.** $\frac{RHw}{w - RH}$

117. $\frac{p^2}{2E - 2V_0 - (m + M)V^2}$ **119.** $\frac{-mb^2s^2 - kL^2}{b^2 s}$ **121.** $\frac{m}{sF - F_0}$

123. 3.4 h **125.** 600 Hz **127.** 11.3 **129.** 18 Ω

Exercises 7.1, page 224

1. $-\frac{1}{3}, -2$ **3.** $a = 1, b = -2, c = -4$

5. Not quadratic, no x^2-term **7.** $a = 1, b = -1, c = 0$

9. Not quadratic, has y^3-term **11.** $5, -5$ **13.** $\frac{3}{2}, -\frac{3}{2}$

15. $-2, 7$ **17.** $3, 4$ **19.** $0, \frac{5}{2}$ **21.** $\frac{1}{2}, -\frac{1}{2}$ **23.** $\frac{1}{3}, 4$

25. $-1, \frac{3}{7}$ **27.** $\frac{2}{3}, \frac{3}{2}$ **29.** $\frac{1}{2}, -\frac{3}{2}$ **31.** $-\frac{2b}{3}, \frac{b}{2}$ **33.** $\frac{5}{2}, -\frac{9}{2}$

35. $0, -2$ **37.** $-a - b, -a + b$ **39.** $-a - b, -a + b$

41. $x = 3$ or $x = \frac{1}{2}, 3 + \frac{1}{2} = \frac{-(-7)}{2} = \frac{7}{2}$ **43.** 4

45. 2 A, -6 A (if current is negative) **47.** 2 A.M., 10 A.M.

49. $2x^2 - 5x + 2 = 0$ **51.** $-1, 0, 1$ **53.** $\frac{3}{2}, 4$ **55.** $-2, \frac{1}{2}$

57. 3 N/cm, 6 N/cm **59.** 30 km/h, 40 km/h

Exercises 7.2, page 227

1. $-3 \pm \sqrt{17}$ **3.** $-5, 5$ **5.** $-\sqrt{7}, \sqrt{7}$

7. $-\sqrt{3}, \sqrt{3}$ **9.** $-3, 7$ **11.** $-3 \pm \sqrt{7}$

13. $3, -5$ **15.** $-2, -1$ **17.** $3 \pm \sqrt{5}$

19. $-2 \pm \sqrt{10}$ **21.** $-3, \frac{1}{2}$ **23.** $\frac{1}{6}(3 \pm \sqrt{33})$

25. $\frac{1}{4}(1 \pm \sqrt{17})$ **27.** $\frac{1}{5}(5 \pm \sqrt{5})$ **29.** $-\frac{1}{3}, -\frac{1}{3}$

31. $-b \pm \sqrt{b^2 - c}$ **33.** 5°C, 15°C **35.** 18 in.

Exercises 7.3, page 230

1. $-3, -2$ **3.** $3, -5$ **5.** $-2, -1$ **7.** $\frac{1}{2}(5 \pm \sqrt{13})$

9. $-5, 3$ **11.** $-3, \frac{1}{2}$ **13.** $\frac{1}{6}(3 \pm \sqrt{33})$ **15.** $\frac{1}{4}(1 \pm \sqrt{17})$

17. $-\frac{8}{5}, \frac{5}{6}$ **19.** No real roots **21.** $\frac{1}{6}(1 \pm \sqrt{109})$

23. $\pm\frac{9}{5}$ **25.** $\frac{3}{4}, -\frac{5}{8}$ **27.** $-0.54, 0.74$ **29.** $-0.26, 2.43$

31. $-c \pm \sqrt{c^2 + 1}$ **33.** $\frac{b + 1 \pm \sqrt{-3b^2 + 2b + 4b^2a + 1}}{2b^2}$

35. Imaginary, unequal **37.** Real, irrational, unequal **39.** 4

41. $-2, -1, 1, 2$

43. $b^2 - 4ac = 729$ is a perfect square, so the equation can be solved by factoring.

45. 4.376 cm **47.** 0.58 L **49.** 3.1 km, 5.1 km

51. $\frac{1}{2}(1 + \sqrt{5}) = 1.618$ **53.** $\frac{-R \pm \sqrt{R^2 - 4L/C}}{2L}$

55. 1.6 cm **57.** 2.6 ft **59.** 2.25 cm **61.** 11.0 m by 23.8 m

Exercises 7.4, page 235

1.

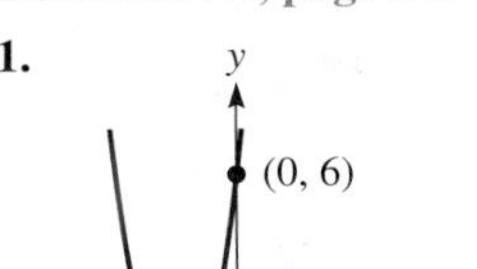

3.

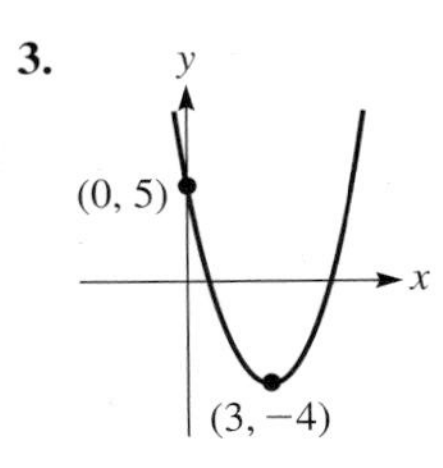

5.

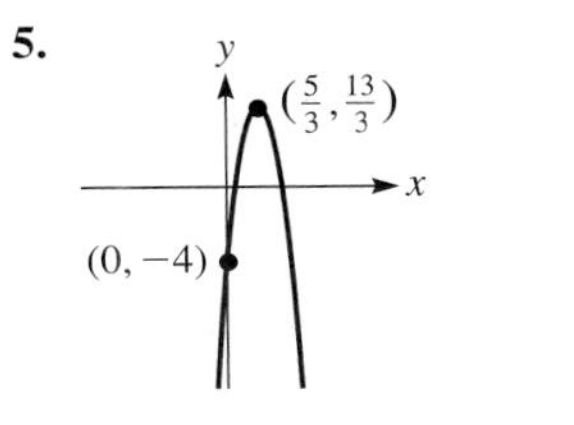

7.

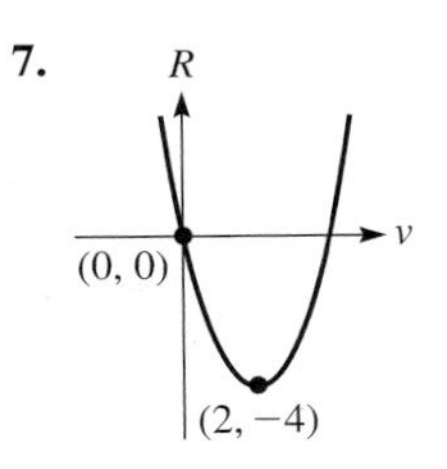

9.

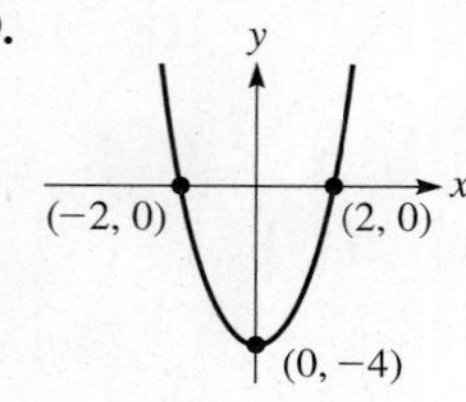

11.

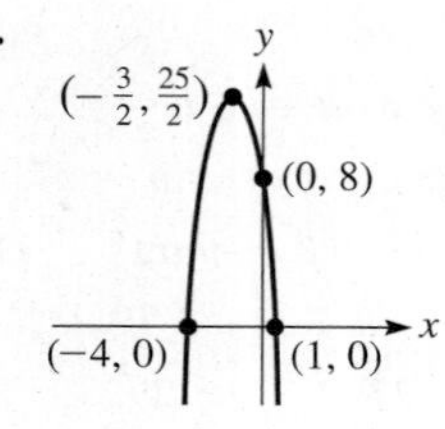

13.

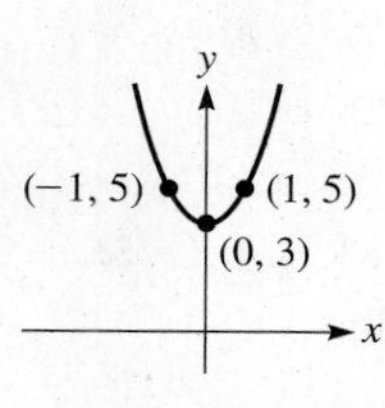

15.

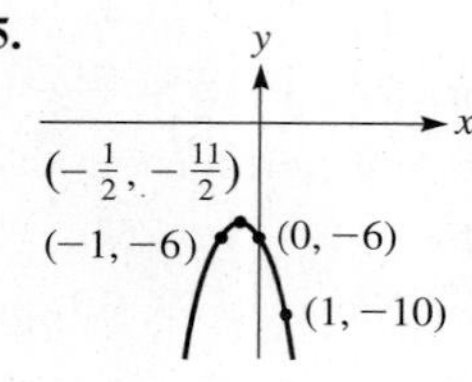

17. −1.87, 1.87 **19.** 0.74, 2.26 **21.** No real roots

23. −2.82, 1.07

25.

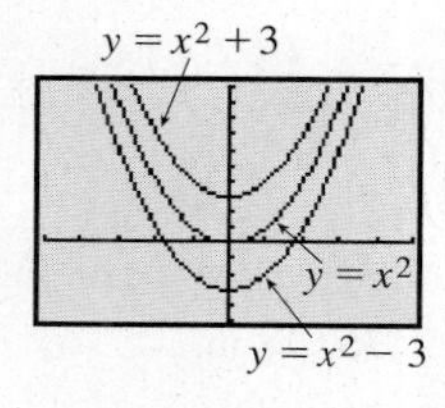

(a) The parabola $y = x^2 + 3$ is 3 units above, and the parabola $y = x^3 - 3$ is 3 units below.

(b) All parabolas open up.

27.

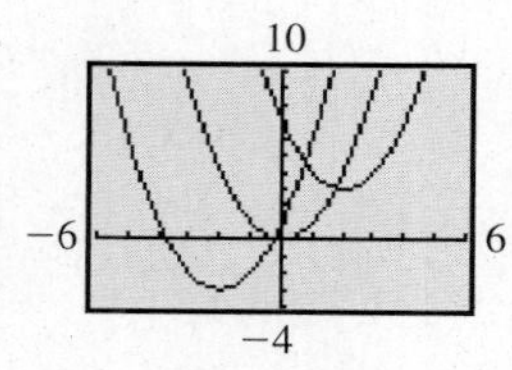

(a) Parabola (b) is 2 units to the right and 3 units up. Parabola (c) is 2 units to the left and 3 units down.

(b) All parabolas open up.

29.

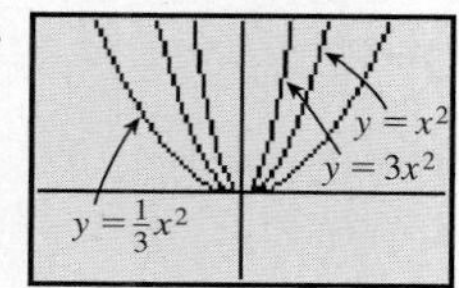

(a) There are no shifts.

(b) The parabola $y = 3x^2$ rises more quickly, and the parabola $y = \frac{1}{3}x^2$ rises more slowly.

31. (1.28, 19.94) **33.** $y = 0.2x^2$

35. −2 **37.** $y = x^2 + 2x - 3$

39.

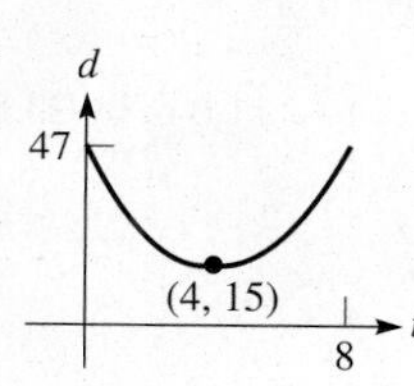

41. $h = w = 192$ m

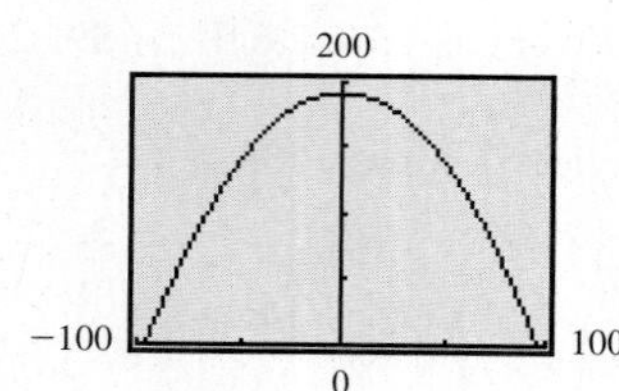

43.

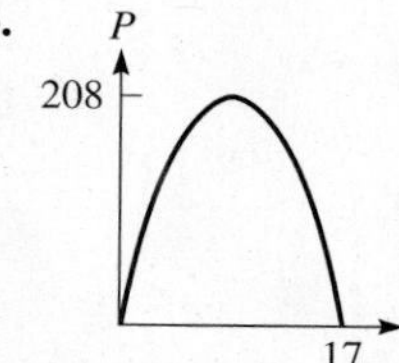

45. 6.9 s **47.** 4.53 cm

49. 173 ft by 116 ft, or 77 ft by 259 ft

Review Exercises for Chapter 7, page 237

Concept Check Exercises

1. F **2.** F **3.** T **4.** T

Practice and Applications

5. −4, 1 **7.** 3, 7 **9.** $\frac{1}{3}, -4$ **11.** $\frac{1}{2}, \frac{5}{3}$ **13.** $0, \frac{9}{2}$

15. $-\frac{3}{2}, \frac{7}{2}$ **17.** −8, 12 **19.** $-1 \pm \sqrt{7}$ **21.** $-4, \frac{9}{2}$

23. $\frac{1}{8}(3 \pm \sqrt{41})$ **25.** $\frac{1}{42}(-23 \pm \sqrt{-4091})$

27. $\frac{1}{6}(-2 \pm \sqrt{58})$ **29.** $\pm\sqrt{5}$ **31.** $-2 \pm 2\sqrt{2}$

33. $\frac{1}{3}(-4 \pm \sqrt{10})$ **35.** $-1, \frac{3}{4}$ **37.** No real roots

39. No real roots **41.** $\frac{-3 \pm \sqrt{9 + 4a^2}}{2a}$ **43.** −5, 6

45. $\frac{1}{4}(1 \pm \sqrt{33})$ **47.** $3 \pm \sqrt{7}$ **49.** 0 (3 is not a solution)

51.

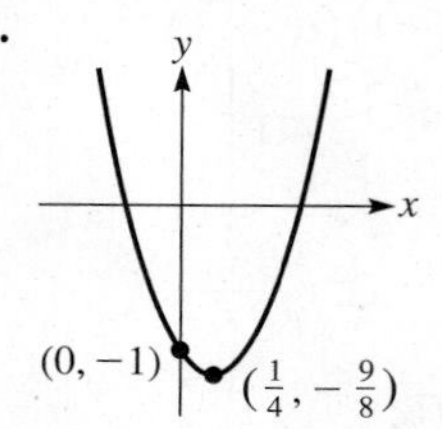

53.

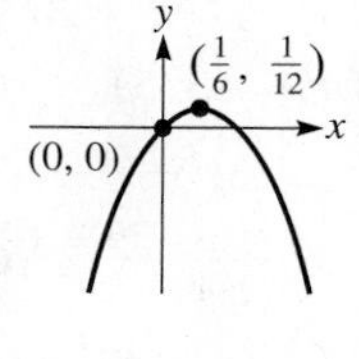

55. −3.26, 2.76 **57.** No real roots **59.** $x = -4$

61. $0, L$ **63.** 1.57 **65.** 17 **67.** 202°F

69. 0.6 s, 1.9 s **71.** 8000 **73.** $\frac{-\pi h \pm \sqrt{\pi^2 h^2 + 2\pi A}}{2\pi}$

75. $\frac{1 + r \pm \sqrt{(1 + r)^2 - 4rp_2}}{2r}$

77.

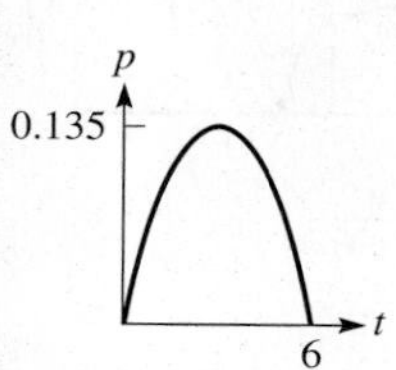

79. 9.9 cm **81.** 61 m, 81 m

83. 15.3 cm, 11.3 cm, 4.00 cm **85.** 29.4 in., 52.3 in. **87.** 25

89.

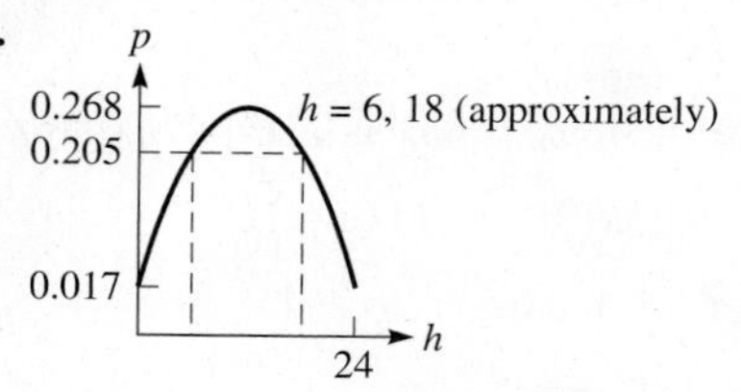

Exercises 8.1, page 243

1. (a) −, +, −, −, +, − (b) +, −, +, +, +, −

3. −, + **5.** −, − **7.** +, + **9.** −, +

11. +, − **13.** −, + **15.** +, +

17. $\sin\theta = \frac{1}{\sqrt{5}}$, $\cos\theta = \frac{2}{\sqrt{5}}$, $\tan\theta = \frac{1}{2}$, $\cot\theta = 2$,

$\sec\theta = \frac{1}{2}\sqrt{5}$, $\csc\theta = \sqrt{5}$

19. $\sin\theta = -\frac{3}{\sqrt{13}}$, $\cos\theta = -\frac{2}{\sqrt{13}}$, $\tan\theta = \frac{3}{2}$, $\cot\theta = \frac{2}{3}$,

$\sec\theta = -\frac{1}{2}\sqrt{13}$, $\csc\theta = -\frac{1}{3}\sqrt{13}$

21. $\sin\theta = \frac{12}{13}$, $\cos\theta = -\frac{5}{13}$, $\tan\theta = -\frac{12}{5}$, $\cot\theta = -\frac{5}{12}$, $\sec\theta = -\frac{13}{5}$, $\csc\theta = \frac{13}{12}$

23. $\sin\theta = -\frac{2}{\sqrt{29}}$, $\cos\theta = \frac{5}{\sqrt{29}}$, $\tan\theta = -\frac{2}{5}$, $\cot\theta = -\frac{5}{2}$, $\sec\theta = \frac{1}{5}\sqrt{29}$, $\csc\theta = -\frac{1}{2}\sqrt{29}$

25. I, II **27.** II, IV **29.** II, III **31.** II **33.** II **35.** IV **37.** I **39.** II **41.** − **43.** −

Exercises 8.2, page 248

1. $\sin 200° = -\sin 20° = -0.3420$, $\tan 150° = -\tan 30° = -0.5774$, $\cos 265° = -\cos 85° = -0.0872$, $\cot 300° = -\cot 60° = -0.5774$, $\sec 344° = \sec 16° = 1.040$, $\sin 397° = \sin 37° = 0.6018$

3. $\sin 25°$; $-\cos 40°$ **5.** $-\tan 75°$; $-\csc 32°$ **7.** $\sec 65°$; $-\sin 20°$ **9.** $-\sin 15° = -0.26$ **11.** $-\cos 73.7° = -0.281$ **13.** $\sec 31.67° = 1.175$ **15.** $\tan 70.9° = 2.89$ **17.** 0.459 **19.** −0.7620 **21.** −3.910 **23.** −0.6188 **25.** 237.99°, 302.01° **27.** 66.40°, 293.60° **29.** 91°, 271° **31.** 119.5° **33.** 263° **35.** 312.94° **37.** 299.24° **39.** 0.5 **41.** −1 **43.** −0.7003 **45.** −0.777 **47.** $<$ **49.** = **51.** −0.9659 **53.** $\cot\theta$ **55.** 0 **57.** 0.0183 A **59.** 12.6 in.

Exercises 8.3, page 252

1. 160° **3.** $\frac{\pi}{12}, \frac{2\pi}{3}$ **5.** $\frac{5\pi}{12}, \frac{11\pi}{6}$ **7.** $\frac{7\pi}{6}, \frac{11\pi}{20}$ **9.** $4\pi, -\frac{\pi}{20}$ **11.** 108°, 270° **13.** 100°, 315° **15.** 70°, 150° **17.** −12°, 27° **19.** 1.47 **21.** 4.40 **23.** −5.821 **25.** 8.351 **27.** 43.0° **29.** 195.2° **31.** 710° **33.** −940.8° **35.** 0.7071 **37.** 3.732 **39.** −0.8660 **41.** −8.327 **43.** 0.906 **45.** −0.890 **47.** −2.09 **49.** 0.149 **51.** 0.3141, 2.827 **53.** 2.932, 6.074 **55.** 0.8309, 5.452 **57.** 2.442, 3.841 **59.** 1.25 **61.** −0.3827 **63.** $-\cot\theta$ **65.** 612 mil **67.** 10.1 rad **69.** 0.030 ft · lb **71.** 2900 m

Exercises 8.4, page 257

1. 2.36 in. **3.** 14.4 m/s **5.** 4.48 in. **7.** 385 mm **9.** 0.1849 mi^2 **11.** 48 cm^2 **13.** 0.0647 ft **15.** 8.21 m **17.** 0.0431 mi **19.** 16.7 in. **21.** 32.73 min past noon **23.** 4240 ft^2 **25.** 0.52 rad/s **27.** 34.73 m^2 **29.** 2.30 ft **31.** 22.6 m^2 **33.** 369 m^3 **35.** 0.4 km **37.** 1070 ft/min **39.** 5370 in./min **41.** 3.87 km/s **43.** 72.0 r/min **45.** 25.7 ft **47.** 1400 in./min **49.** 649 r/min **51.** 8800 ft/min **53.** 250 rad **55.** 940 m/s **57.** 14.9 m^3 **59.** Ratios become very close to 1 **61.** 7.13×10^7 mi

Review Exercises for Chapter 8, page 260

Concept Check Exercises

1. T **2.** T **3.** F **4.** T **5.** F **6.** T

Practice and Applications

7. $\sin\theta = \frac{4}{5}$, $\cos\theta = \frac{3}{5}$, $\tan\theta = \frac{4}{3}$, $\cot\theta = \frac{3}{4}$, $\sec\theta = \frac{5}{3}$, $\csc\theta = \frac{5}{4}$

9. $\sin\theta = -\frac{2}{\sqrt{53}}$, $\cos\theta = \frac{7}{\sqrt{53}}$, $\tan\theta = -\frac{2}{7}$, $\cot\theta = -\frac{7}{2}$, $\sec\theta = \frac{\sqrt{53}}{7}$, $\csc\theta = -\frac{\sqrt{53}}{2}$

11. $-\cos 48°$, $\tan 14°$ **13.** $-\sin 71°$, $\sec 15°$ **15.** $\frac{2\pi}{9}, \frac{17\pi}{20}$ **17.** $\frac{34\pi}{15}, -\frac{9\pi}{8}$ **19.** 252°, 130° **21.** 12°, 330° **23.** 32.1° **25.** −2067° **27.** 1.78 **29.** 0.3534 **31.** −11.10 **33.** $3\pi/2$ **35.** $-5\pi/3$ **37.** −0.539 **39.** −0.47 **41.** −1.080 **43.** −0.0488 **45.** −1.64 **47.** 4.140 **49.** −0.5878 **51.** −0.8660 **53.** 0.5569 **55.** 1.197 **57.** 10.30°, 190.30° **59.** 118.23°, 241.77° **61.** 0.5759, 5.707 **63.** 4.187, 5.238 **65.** 227.8° **67.** 246.78° **69.** 10.8 in. **71.** 3.23 = 185.3° **73.** 14.2 m **75.** 2120 in.^{2} **77.** 0

79. $A = \frac{1}{2}r^2\theta - \frac{1}{2}r(r\sin\theta) = \frac{1}{2}r^2(\theta - \sin\theta)$; 3.66 cm^2

81. (a) 1040 m^2 (b) 137.5 m **83.** 0.0562 W **85.** 0.800 = 45.8° **87.** 20.4 r/min **89.** 2200 mi/h **91.** (a) 4150 mi (b) 6220 mi; great circle route shortest of all routes **93.** 4710 cm/s **95.** 24.4 ft^2 **97.** 1.81×10^6 cm/s **99.** 138 m^2 **101.** 3.58×10^5 km

Exercises 9.1, page 267

1. **3.**

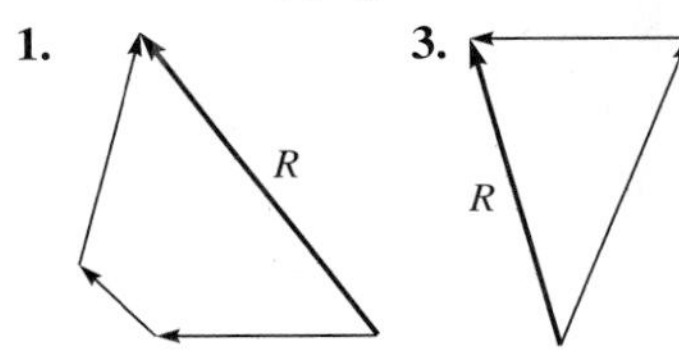

5. (a) Scalar because no direction is given. (b) Vector because direction is given.

7. (a) Vector: magnitude and direction are specified. (b) Scalar: only magnitude is specified.

9. **11.** **13.**

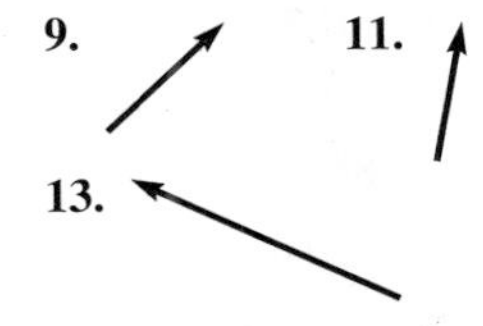

15. 5.6, 50° **17.** 4.3, 156°

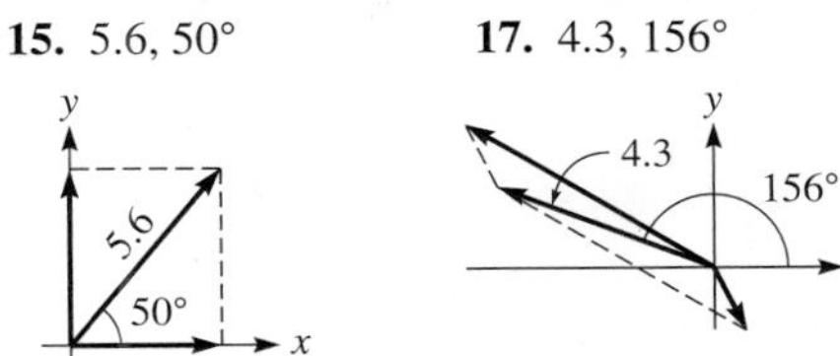

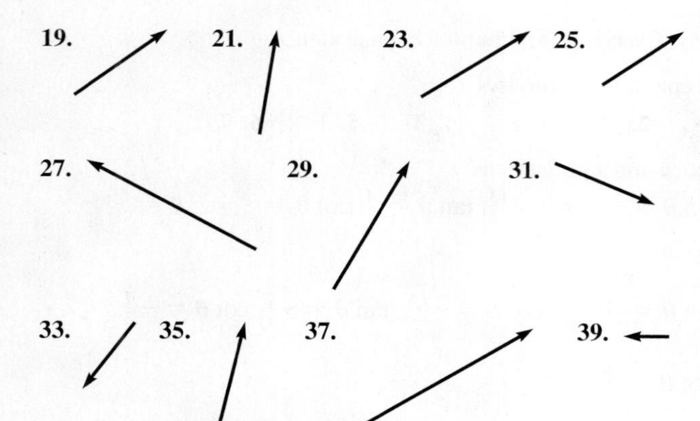

41.

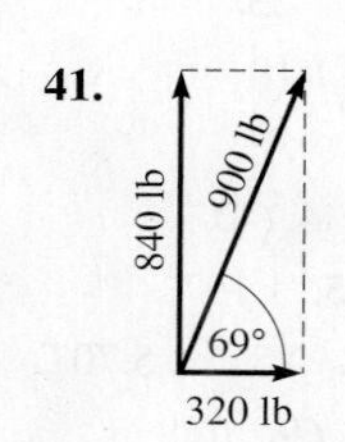

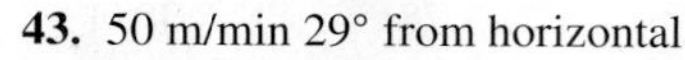

43. 50 m/min 29° from horizontal

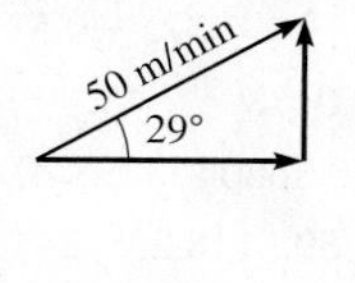

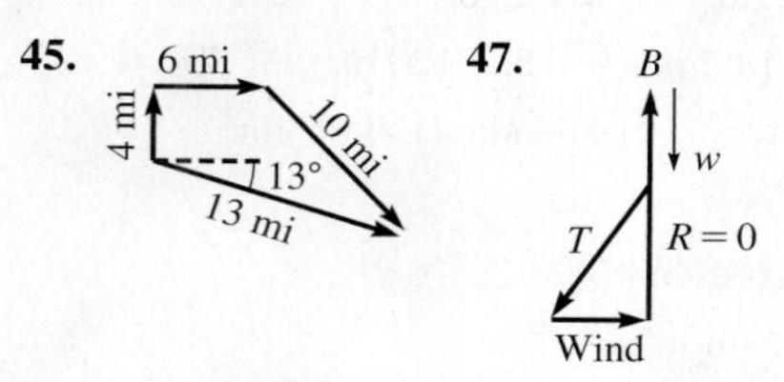

Exercises 9.2, page 270

1. $V_x = -11.6$ m, $V_y = -8.46$ m **3.** $A_x = 0, A_y = -375.4$
5. 662, 352 **7.** −349, −664 **9.** −750, 0
11. 3.82 lb, 15.3 lb **13.** −62.9 m/s, 44.1 m/s
15. 0, -8.17 ft/s^2 **17.** −2.53 mN, −0.788 mN
19. −31.83 ft, 21.61 ft **21.** 23.9 km/h, 7.43 km/h
23. 2780 lb, 170 lb **25.** 116 km/h **27.** 17°
29. 89 N, −190 N **31.** 75 lb
33. 0.57(km/h)/m, 0.48(km/h)/m

Exercises 9.3, page 276

1. $R = 1650, \theta = 73.8°$ **3.** $R = 24.2, \theta = 52.6°$ with A
5. $R = 7.781, \theta = 66.63°$ with A **7.** $R = 10.0, \theta = 58.8°$
9. $R = 2.74, \theta = 111.0°$ **11.** $R = 2130, \theta = 107.7°$
13. $R = 7052, \theta = 349.82°$ **15.** $R = 29.2, \theta = 10.8°$
17. $R = 521, \theta = 257.4°$ **19.** $R = 5920, \theta = 88.4°$
21. $R = 27.27, \theta = 33.14°$ **23.** $R = 50.2, \theta = 50.3°$
25. $R = 0.242, \theta = 285.9°$ **27.** $R = 532$ m, $\theta = 95.7°$
29. $R = 235$ lb, $\theta = 121.7°$
31. 15,000 lb, 15° from 5500-lb force
33. 4000 N **35.** 24.3 kg · m/s

Exercises 9.4, page 281

1. 47.10 mi, 10.27° N of E
3. 8.20 lb, 33.3° from the 6.85-lb force
5. 11,800 N, 47.3° above horizontal
7. 3070 ft, 17.8° S of W **9.** 229.4 ft, 72.82° N of E
11. 940 lb, 34° from 780 lb force
13. 21.9 km/h, 34.8° S of E **15.** 9.6 m/s^2, 7.0° from vertical
17. 175 lb, 9.3° above horizontal
19. 540 km/h, 6° from direction of plane
21. $T_1 = 420$ N, $T_2 = 359$ N **23.** 27.0 km, bearing 192.4°
25. 0.94 km, 65° S of E **27.** 9.6 km/h **29.** 910 N
31. $T = 389$ lb, $F = 275$ lb **33.** 138 mi, 65.0° N of E
35. 79.0 m/s, 11.3° from direction of plane, 75.6° from vertical
37. $T_1 = 517$ lb, $T_2 = 659$ lb

Exercises 9.5, page 288

1. $b = 76.01, c = 50.01, C = 40.77°$
3. $b = 38.1, C = 66.0°, c = 46.1$
5. $a = 2800, b = 2620, C = 108.0°$
7. $B = 12.20°, C = 149.57°, c = 7.448$
9. $a = 11{,}050, A = 149.70°, C = 9.57°$
11. $A = 99.4°, b = 55.1, c = 24.4$
13. $A = 68.01°, a = 5520, c = 5376$
15. $A_1 = 61.36°, C_1 = 70.51°, c_1 = 5.628; A_2 = 118.64°, C_2 = 13.23°, c_2 = 1.366$
17. $A_1 = 107.3°, a_1 = 5280, C_1 = 41.3°; A_2 = 9.9°, a_2 = 952, C_2 = 138.7°$
19. No solution **21.** 373 m **23.** 9.9° **25.** 1490 ft
27. 880 N **29.** 20.7 m **31.** 13.94 cm **33.** 27,300 km
35. 77.3° with bank downstream **37.** 309 ft

Exercises 9.6, page 293

1. $A = 14.0°, B = 21.0°, c = 107$
3. $A = 50.3°, B = 75.7°, c = 6.31$
5. $A = 70.9°, B = 11.1°, c = 4750$
7. $A = 34.72°, B = 40.67°, C = 104.61°$
9. $A = 18.21°, B = 22.28°, C = 139.51°$
11. $A = 6.0°, B = 16.0°, c = 1150$
13. $A = 82.3°, b = 2160, C = 11.4°$
15. $A = 36.24°, B = 39.09°, a = 97.22$
17. $A = 137.9°, B = 33.7°, C = 8.4°$
19. $b = 3700, C = 25°, c = 2400$
21. 11.2 ft **23.** $c^2 = a^2 + b^2$ **25.** 291 mi
27. 4.96 km **29.** 47.8° **31.** 57.3°, 141.7°
33. 16.5 mm **35.** 5.09 km/h **37.** 17.8 mi **39.** 13.6 m

Review Exercises for Chapter 9, page 295

Concept Check Exercises

1. T **2.** F **3.** F **4.** F **5.** F **6.** T

Practice and Applications

7. $A_x = 65.8, A_y = 29.3$ **9.** $A_x = -4.648, A_y = -3.327$
11. $R = 602, \theta = 57.1°$ with A

13. $R = 6239, \theta = 31.91°$ with A **15.** $R = 965, \theta = 8.6°$
17. $R = 26.12, \theta = 146.03°$ **19.** $R = 71.93, \theta = 336.50°$
21. $R = 0.9942, \theta = 359.57°$ **23.** $b = 181, C = 64.0°, c = 175$
25. $A = 21.2°, b = 34.8, c = 51.5$
27. $A = 39.88°, a = 5194, C = 30.03°$
29. $A_1 = 54.8°, a_1 = 12.7, B_1 = 68.6°; A_2 = 12.0°, a_2 = 3.24, B_2 = 111.4°$
31. $A = 32.3°, b = 267, C = 17.7°$
33. $A = 176.4°, B = 1.1°, c = 5.41$
35. $a = 1782, b = 1920, C = 16.00°$
37. $A = 37°, B = 25°, C = 118°$
39. $A = 20.6°, B = 35.6°, C = 123.8°$
41. 1300 m **43.** $A_t = \frac{1}{2}ab; b = \frac{a \sin B}{\sin A}$; substitute
45. −155.7 lb, 81.14 lb **47.** 2100 ft/s
49. $T_1 = 441$ lb, $T_2 = 298$ lb **51.** $R = 2700$ lb, $\theta = 107°$
53. 6.1 m/s, 35° with horizontal **55.** 0.11 N
57. 1.52 pm **59.** 2.30 m, 2.49 m **61.** 0.039 km
63. 32,900 mi **65.** 95 m, 14° E of S **67.** 2.65 km
69. 190 mi **71.** 2510 ft or 3370 ft (ambiguous)
73. 810 N, 36° N of E

Exercises 10.1, page 302

1. $y = 3 \cos x$

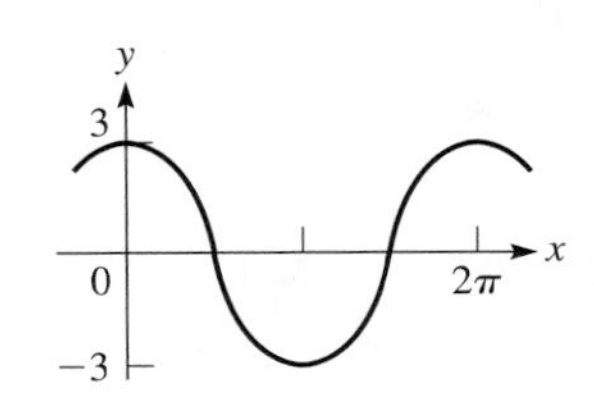

3. 0, −0.7, −1, −0.7, 0, 0.7, 1, 0.7, 0, −0.7, −1, −0.7, 0, 0.7, 1, 0.7, 0

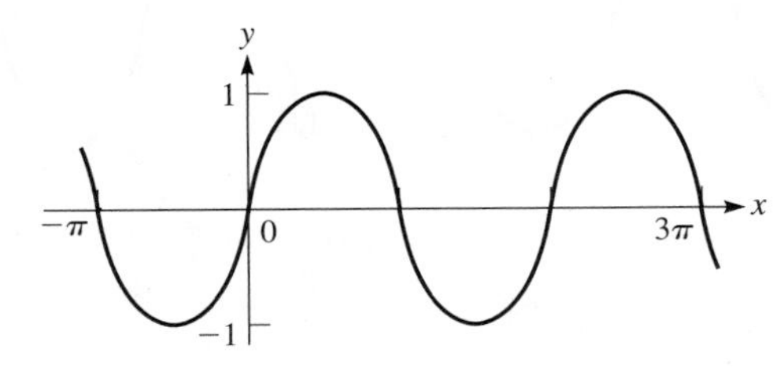

5. 3, 2.1, 0, −2.1, −3, −2.1, 0, 2.1, 3, 2.1, 0, −2.1, −3, −2.1, 0, 2.1, 3

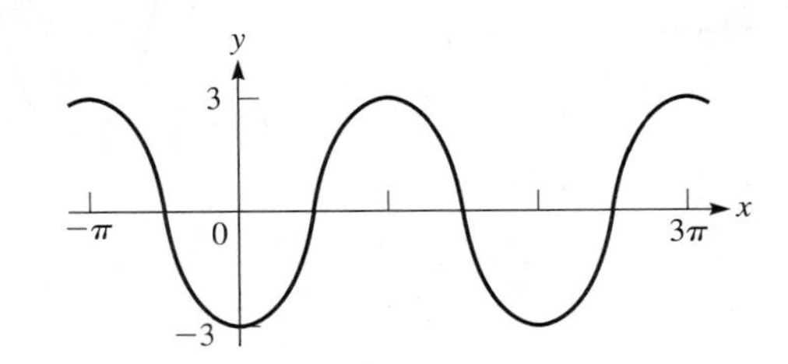

7. Amplitude = 3

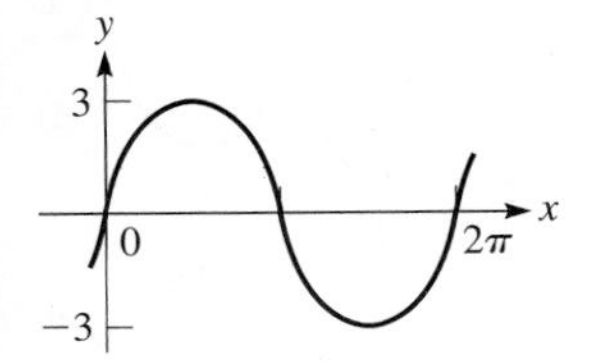

9. Amplitude = $\frac{5}{2}$

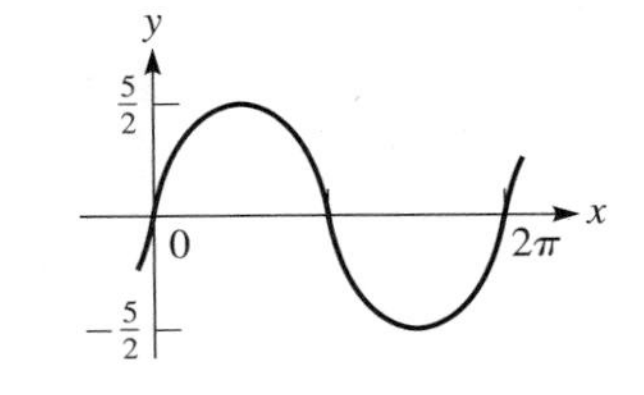

11. Amplitude = 200

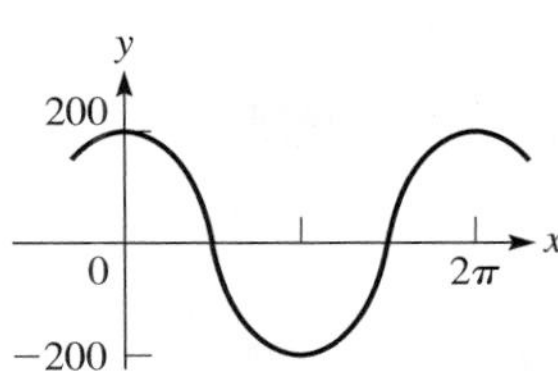

13. Amplitude = 0.8

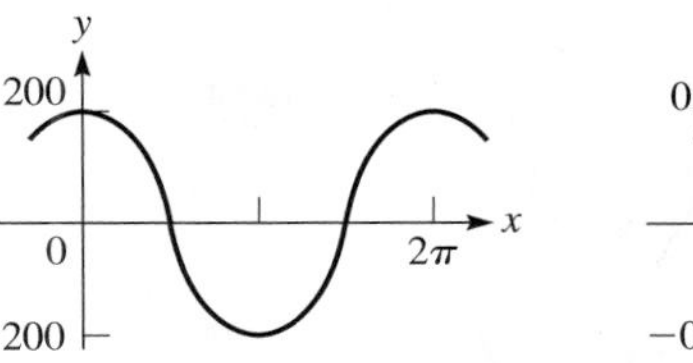

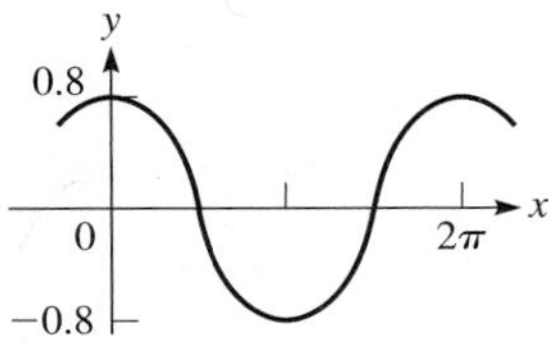

15. Amplitude = 1

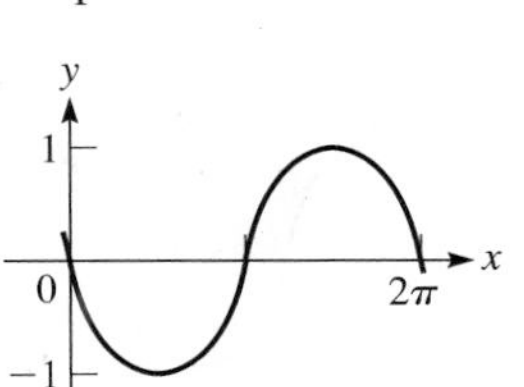

17. Amplitude = 1500

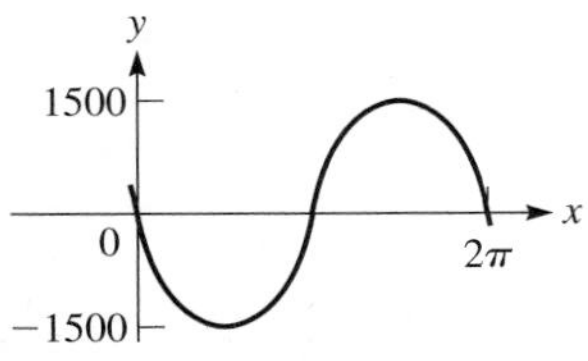

19. Amplitude = 4

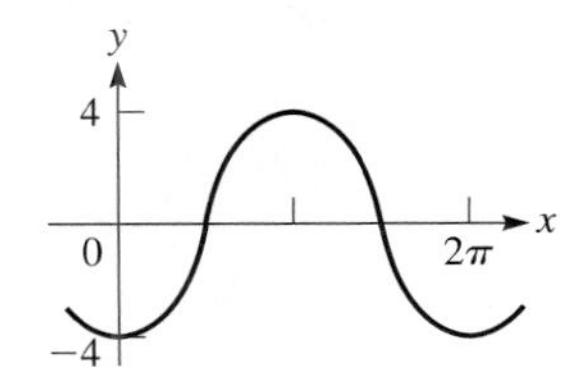

21. Amplitude = 50

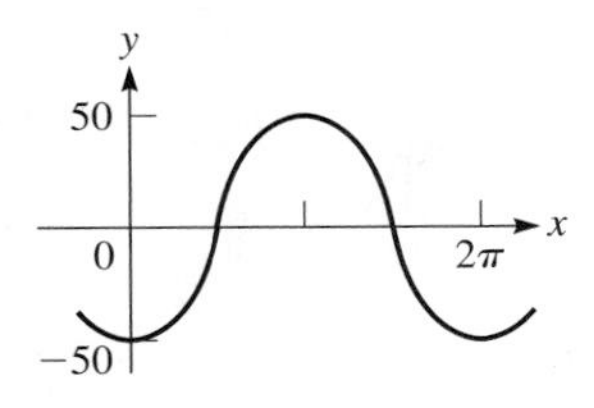

23. 0, 0.84, 0.91, 0.14, −0.76, −0.96, −0.28, 0.66

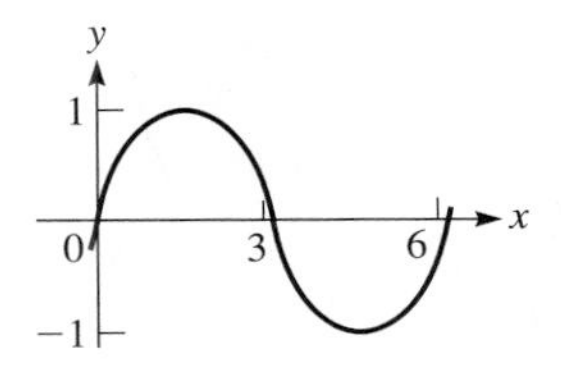

25. −12, −6.48, 4.99, 11.88, 7.84, −3.40, −11.52, −9.05

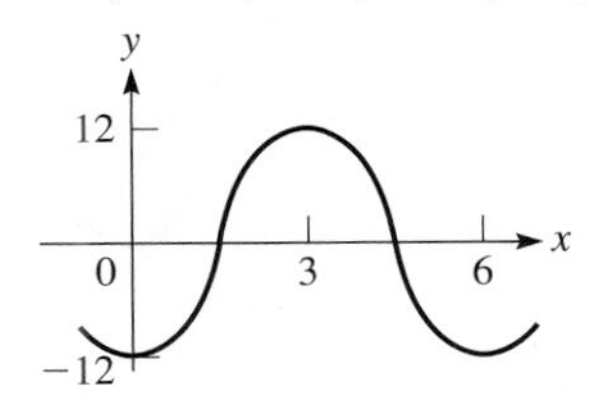

27. $y = -2 \sin x$

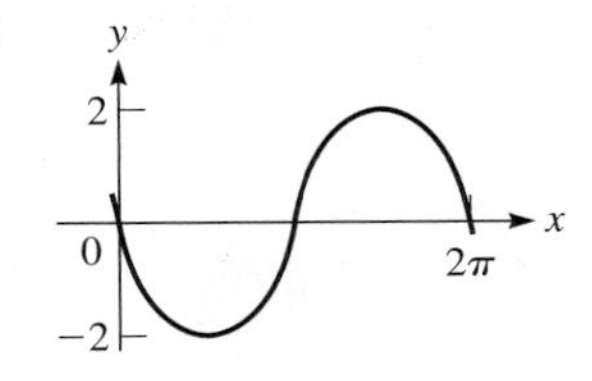

29.

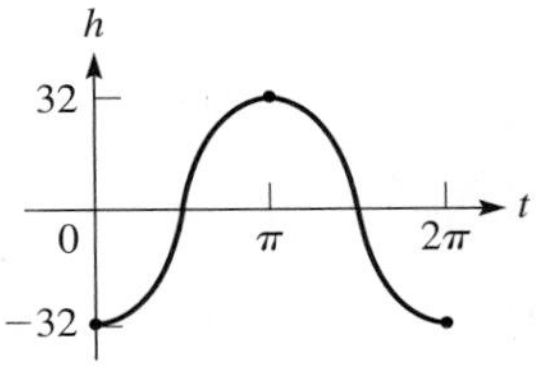

31.

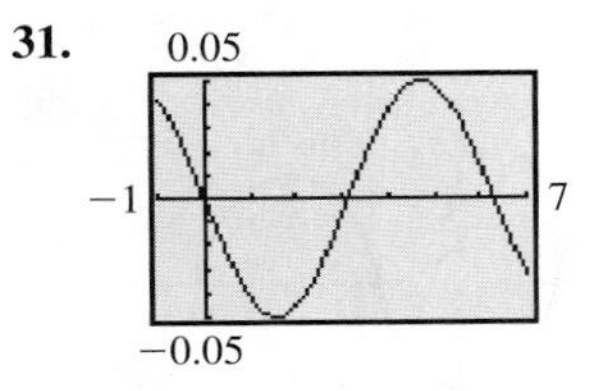

33. $y = 4 \sin x$ **35.** $y = -1.5 \cos x$
37. $y = -2.50 \sin x$ **39.** $y = -2.50 \cos x$

Exercises 10.2, page 305

1. $y = 3 \sin 6x$

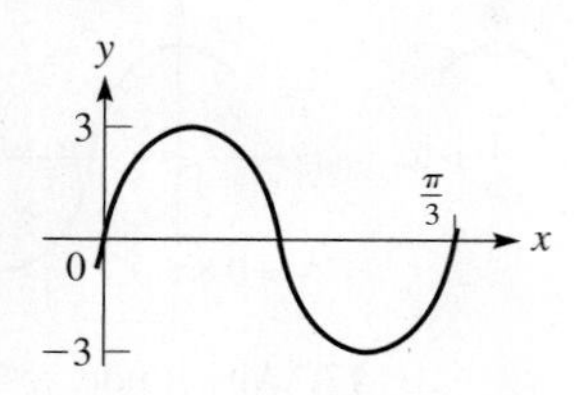

3. $2, \frac{\pi}{3}$

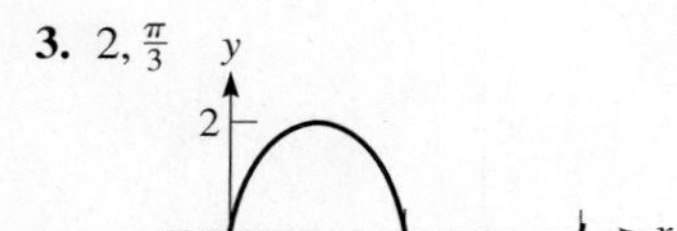

5. $3, \frac{\pi}{4}$

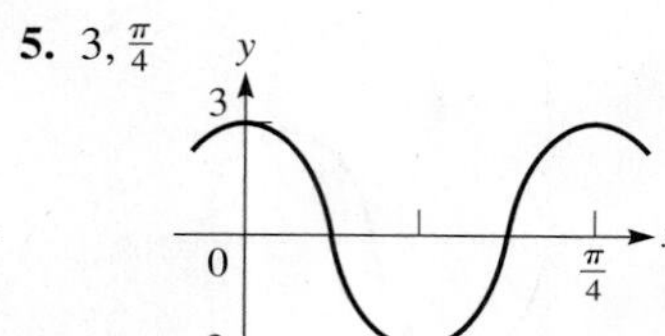

7. $2, \frac{\pi}{6}$

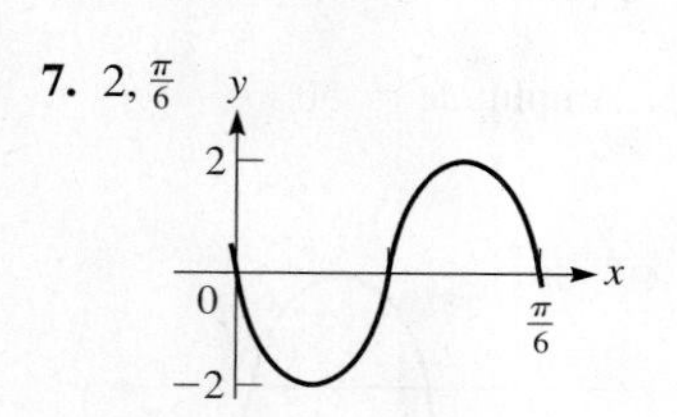

9. $1, \frac{\pi}{8}$

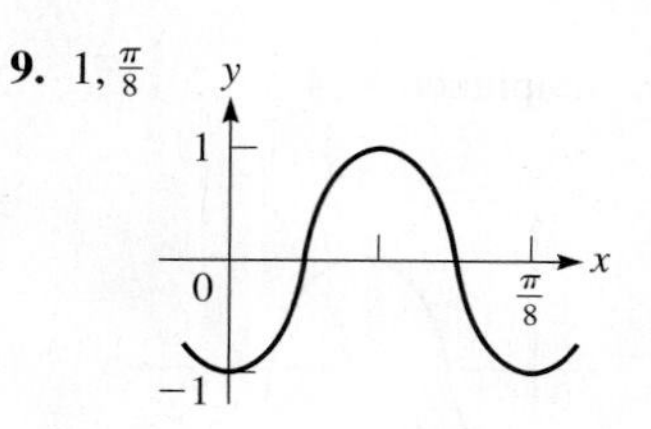

11. 520, 1

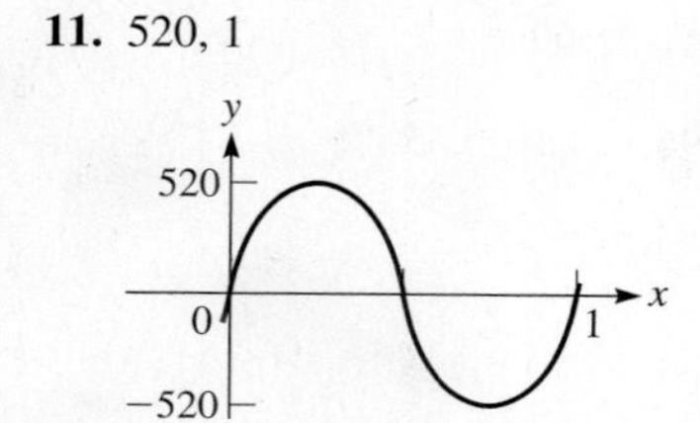

13. $3, \frac{1}{2}$

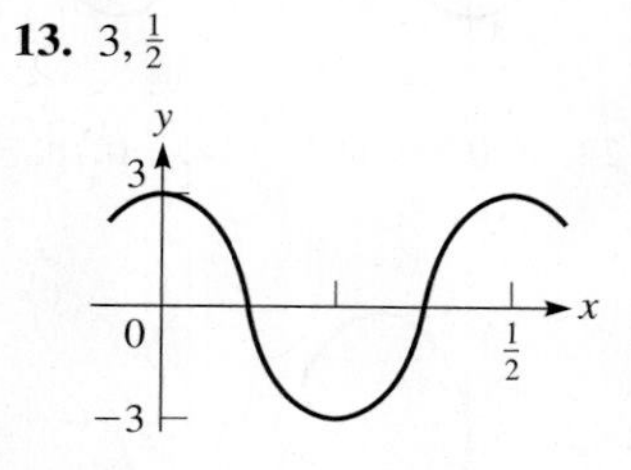

15. $15, 6\pi$

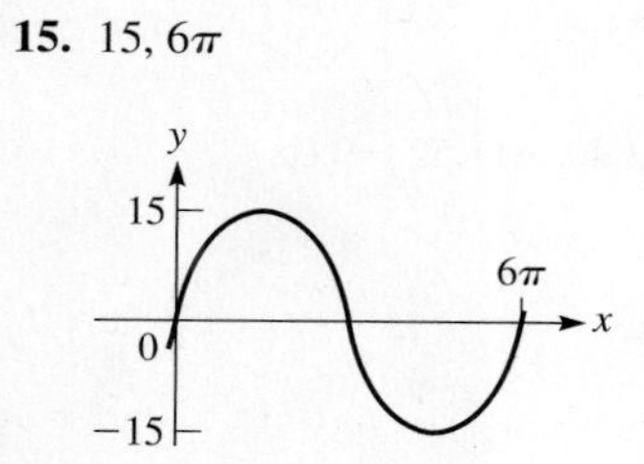

17. $\frac{1}{2}, 3\pi$

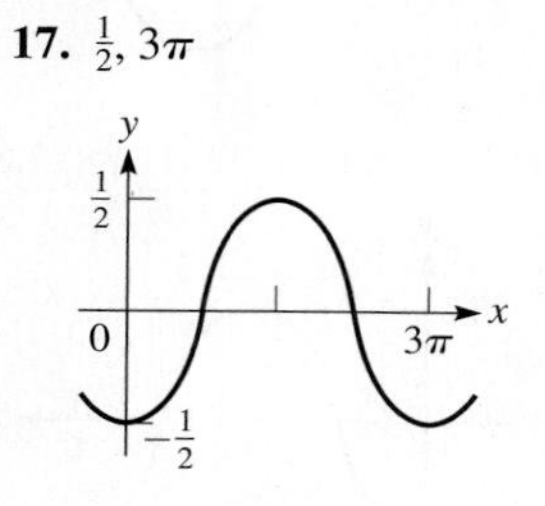

19. 0.4, 9

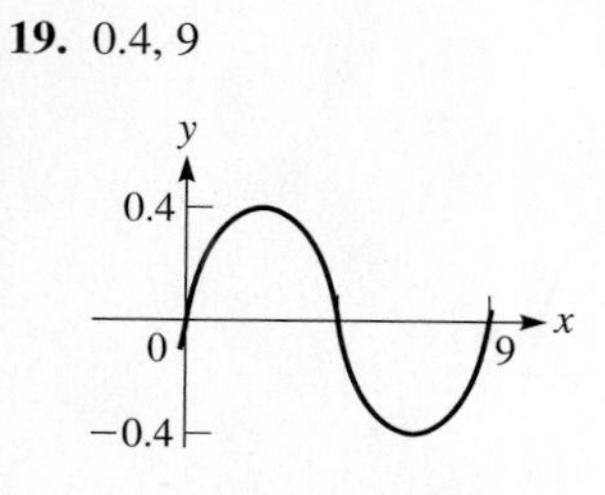

21. $3.3, \frac{2}{\pi}$

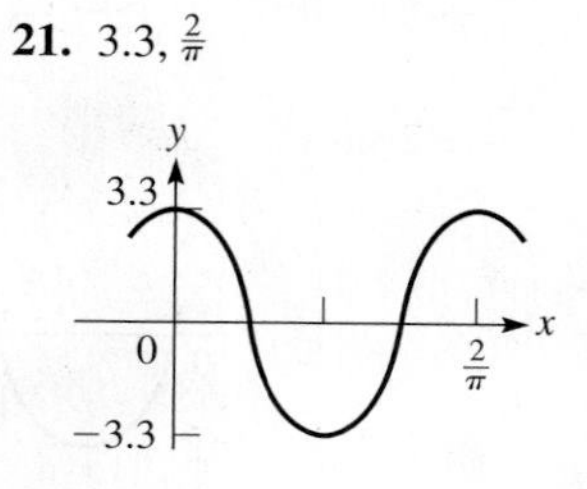

23. $y = \sin 6x$ **25.** $y = \sin 6\pi x$

27. $y = -3 \sin 2x$

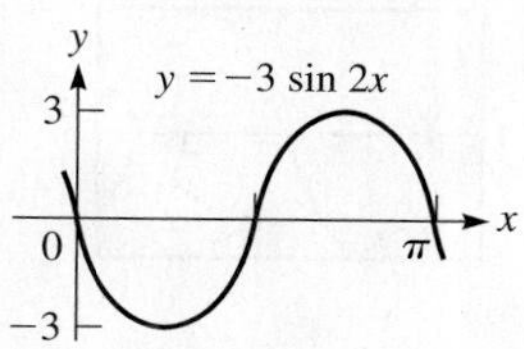

29.

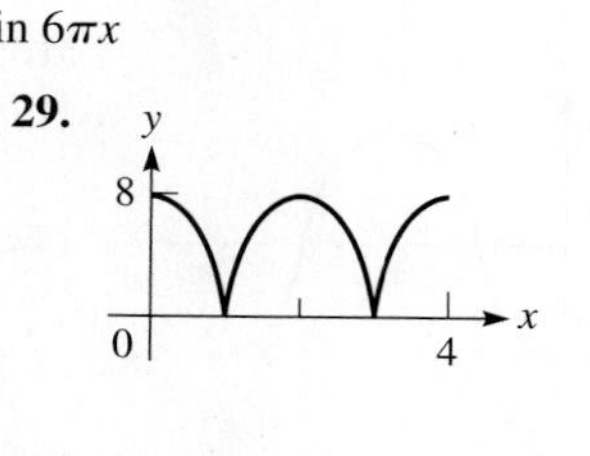

31. $2 \sin 3x = -2 \sin(-3x)$

33. 2π

35. $y = -2 \sin 2x$

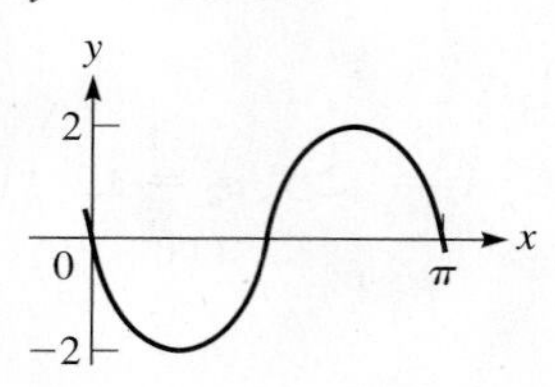

37. 105 MHz

39.

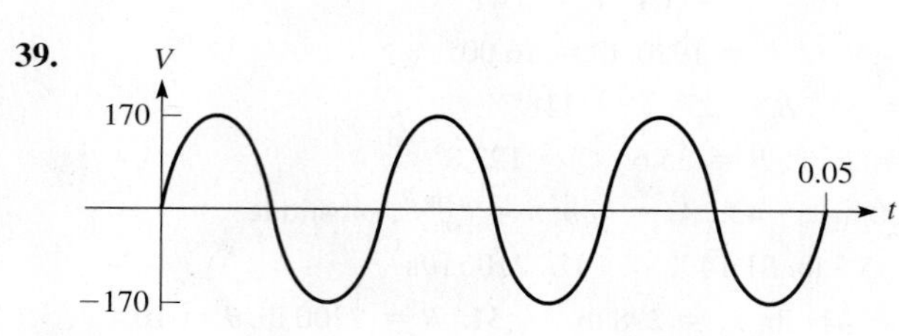

41.

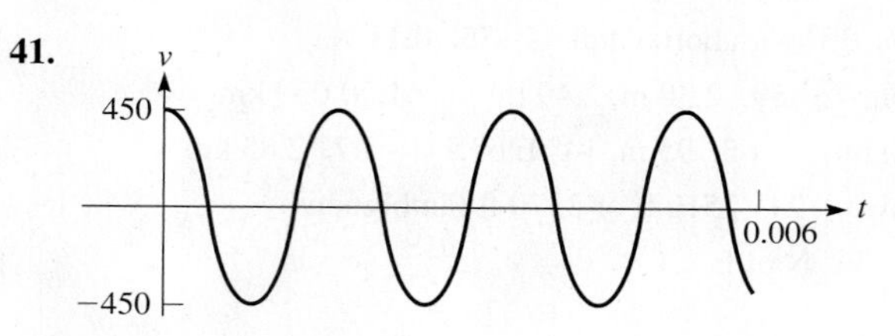

43. $y = \frac{1}{2} \cos 2x$ **45.** $y = -4 \sin \pi x$

Exercises 10.3, page 309

1. $y = -\cos\left(2x - \frac{\pi}{6}\right)$

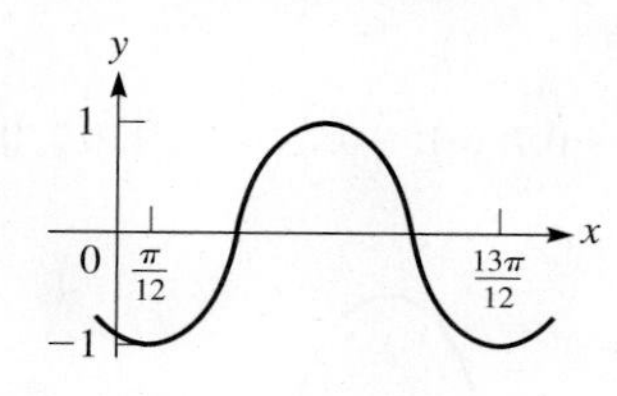

3. $1, 2\pi, \frac{\pi}{6}$

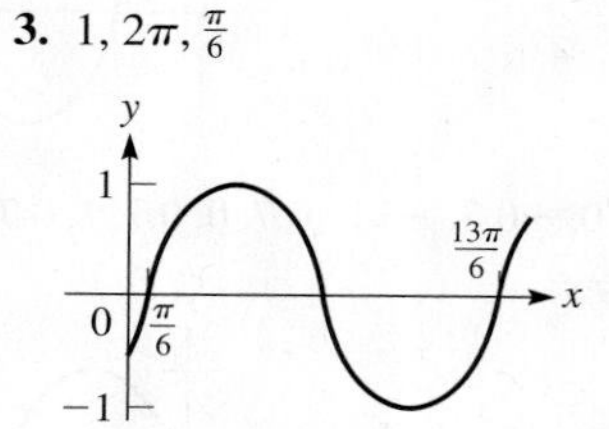

5. $1, 2\pi, -\frac{\pi}{6}$

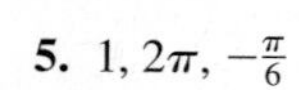

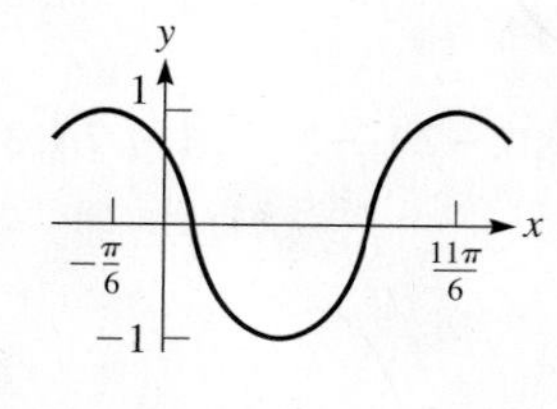

7. $0.2, \pi, -\frac{\pi}{4}$

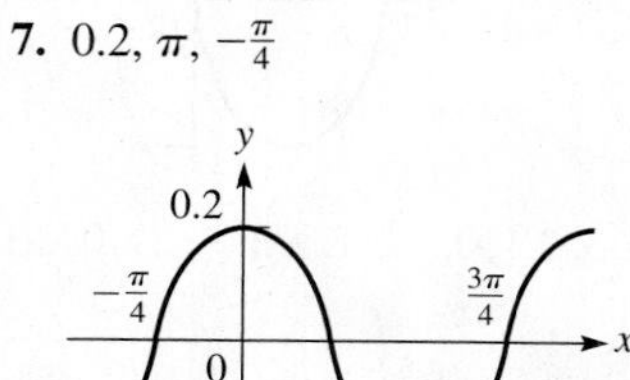

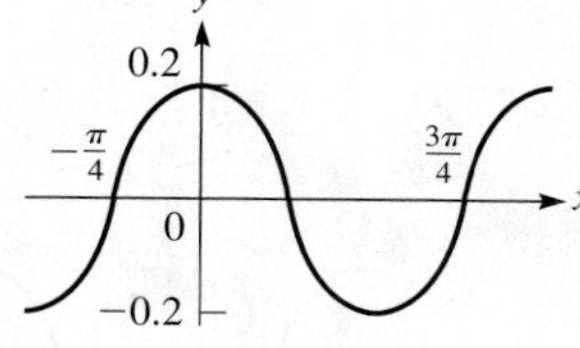

9. $1, \pi, \frac{\pi}{2}$

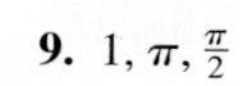

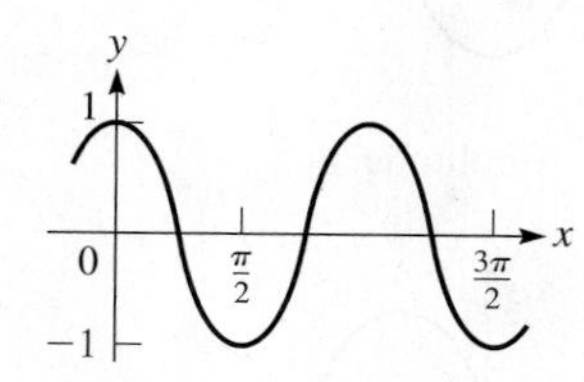

11. $\frac{1}{2}, 4\pi, \frac{\pi}{2}$

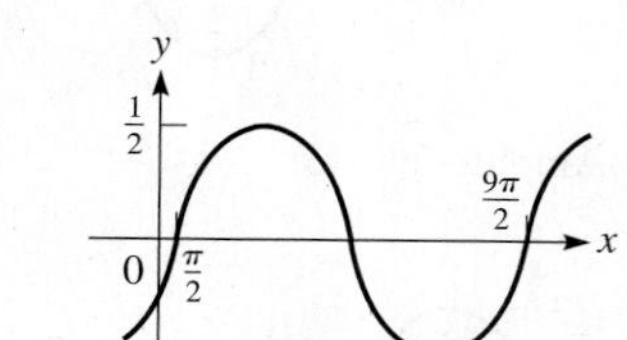

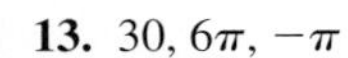

13. 30, 6π, $-\pi$

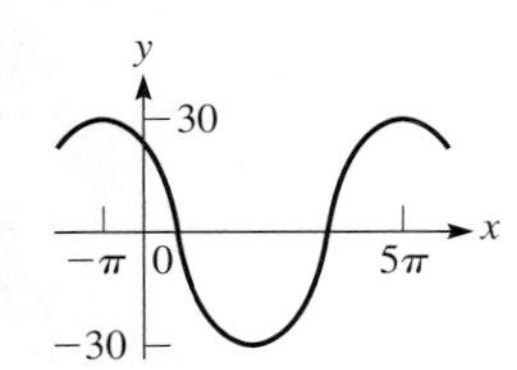

15. 1, 2, $-\frac{1}{8}$

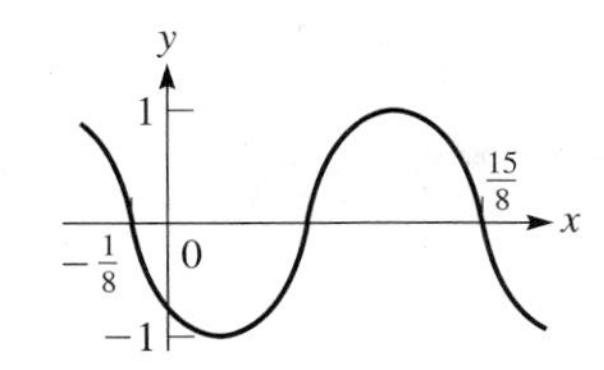

17. 0.08, $\frac{1}{2}$, $\frac{1}{20}$

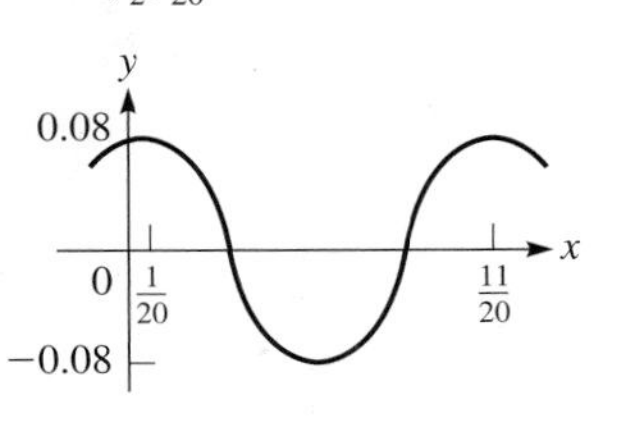

19. 0.6, 1, $\frac{1}{2\pi}$

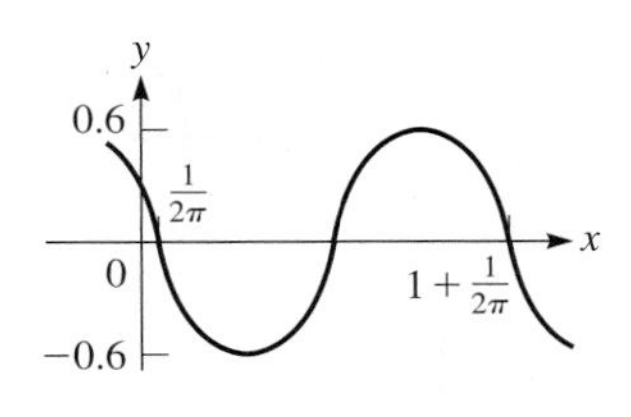

21. 40, $\frac{2}{3}$, $-\frac{1}{3\pi}$

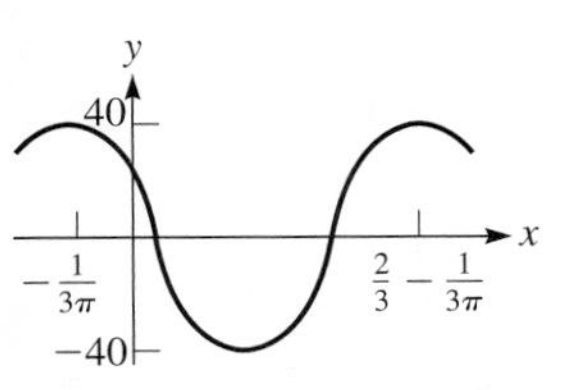

23. 1, $\frac{2}{\pi}$, $\frac{1}{\pi}$

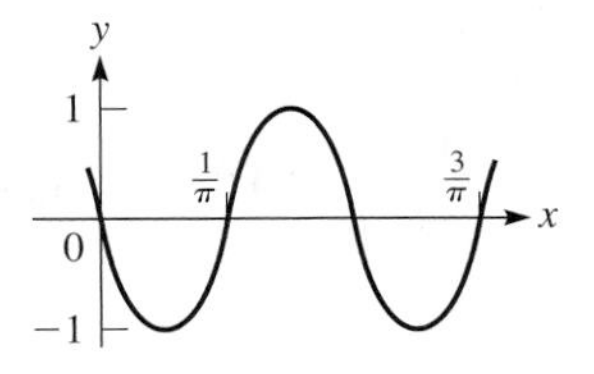

25. $\frac{3}{2}$, 2, $-\frac{\pi}{6}$

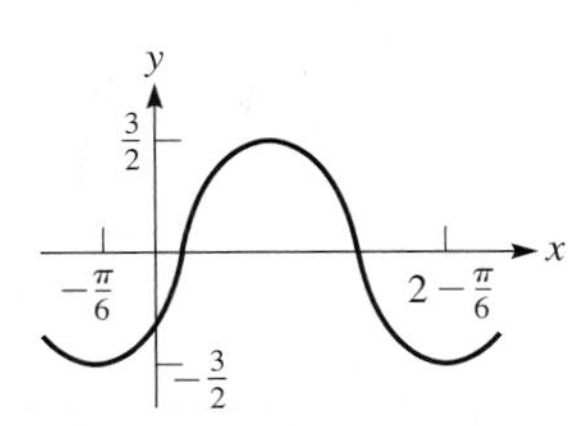

27. $y = 4\sin(\frac{2}{3}x + \frac{\pi}{6})$ **29.** $y = 12\cos(4\pi x - \frac{\pi}{2})$

31.

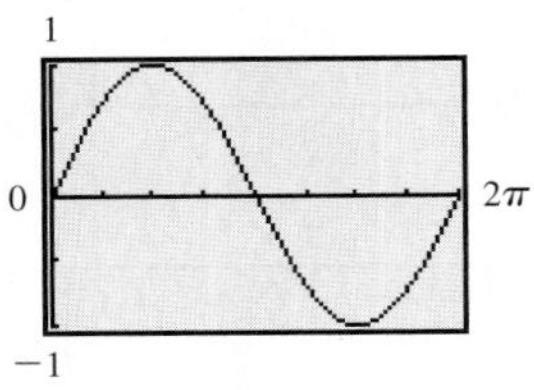

33. $y_1 = y_2$

35. $v = 18\sin(120\pi t - 60°)$, $\frac{1}{360}$ s

37.

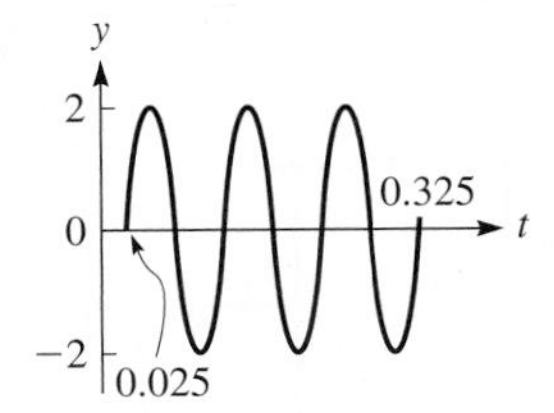

39.

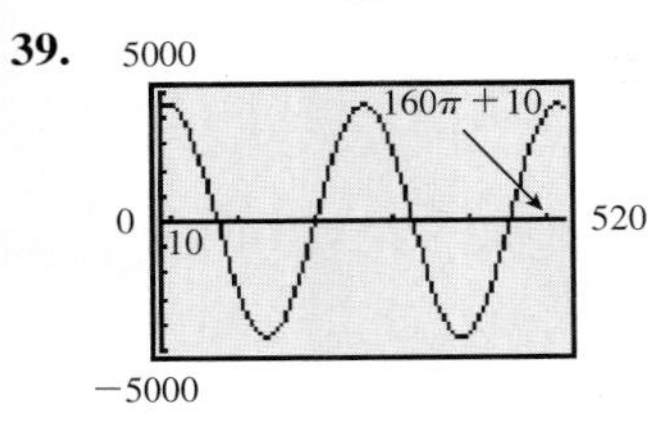

41. $y = 5\sin(\frac{\pi}{8}x + \frac{\pi}{8})$. As shown: amplitude $= 5$; period $= \frac{2\pi}{b} = 16$, $b = \frac{\pi}{8}$; displacement $= -\frac{c}{b} = -1$, $c = \frac{\pi}{8}$

43. $y = -0.8\cos 2x$. As shown: amplitude $= |-0.8| = 0.8$; period $= \frac{2\pi}{b} = \pi$, $b = 2$; displacement $= -\frac{c}{b} = 0$, $c = 0$

Exercises 10.4, page 312

1. $y = 5\cot 2x$

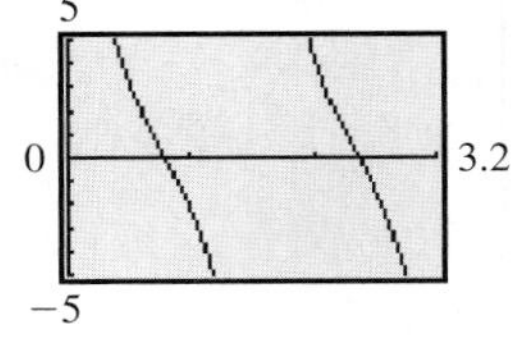

3. Undef., -1.7, -1, -0.58, 0, 0.58, 1, 1.7, undef., -1.7, -1, -0.58, 0

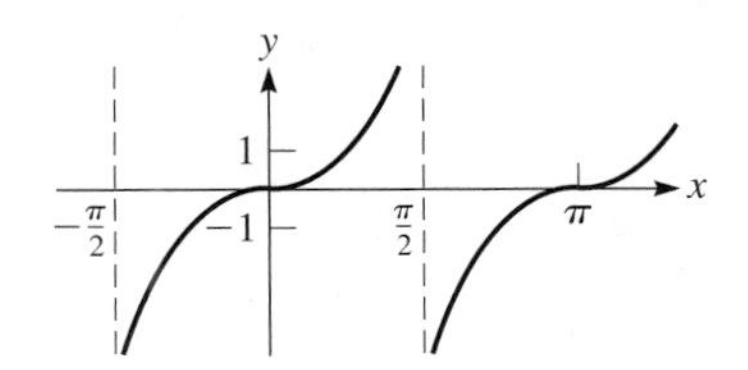

5. Undef., 2, 1.4, 1.2, 1, 1.2, 1.4, 2, undef., -2, -1.4, -1.2, -1

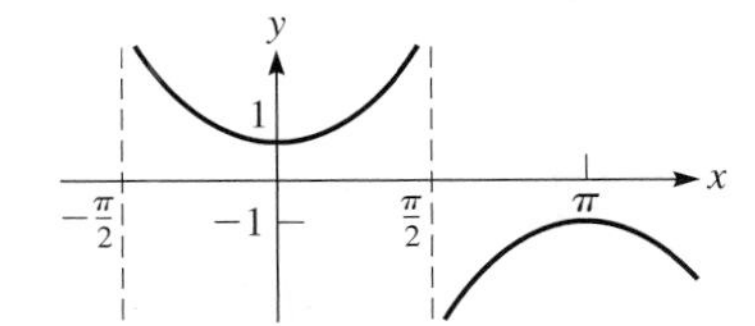

7.

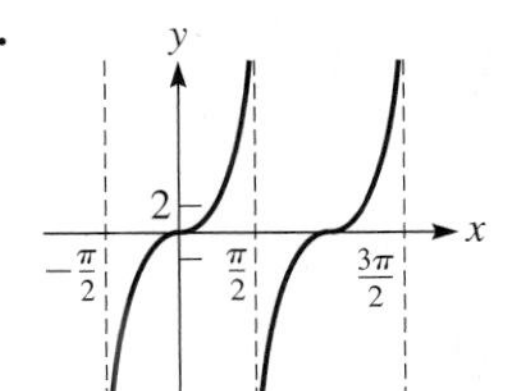

9.

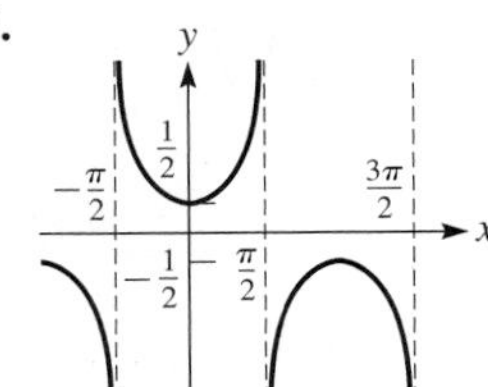

11.

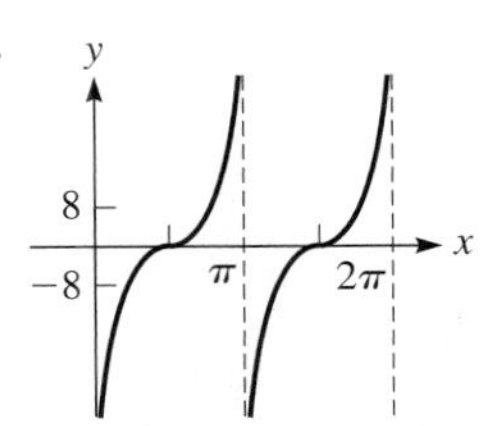

13.

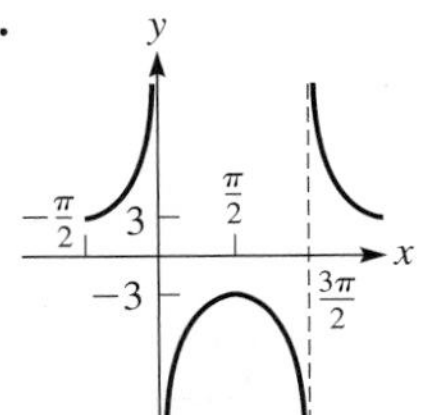

15.

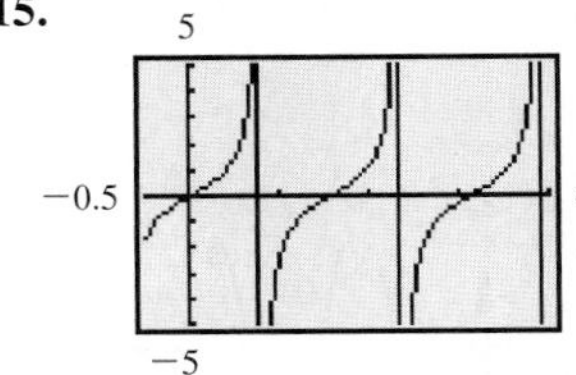

17.

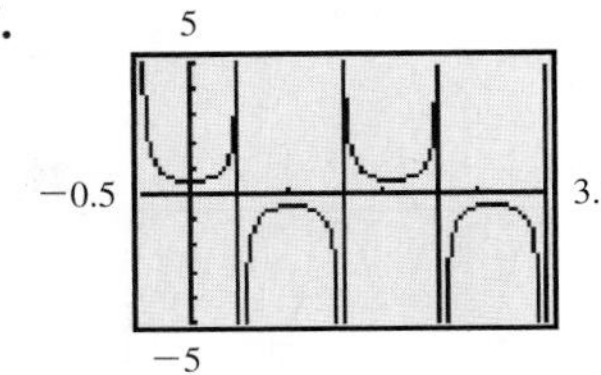

19.

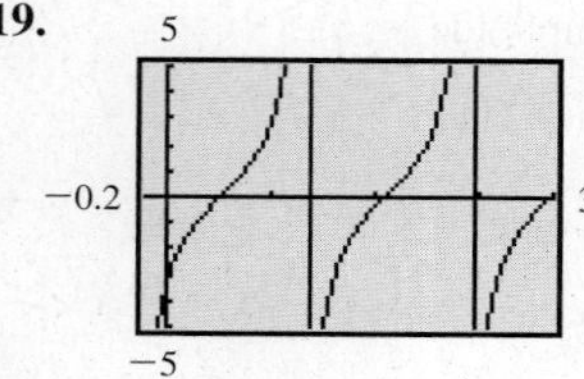

21.

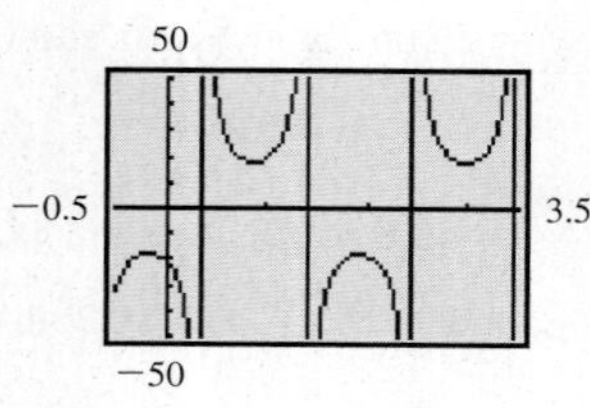

23.

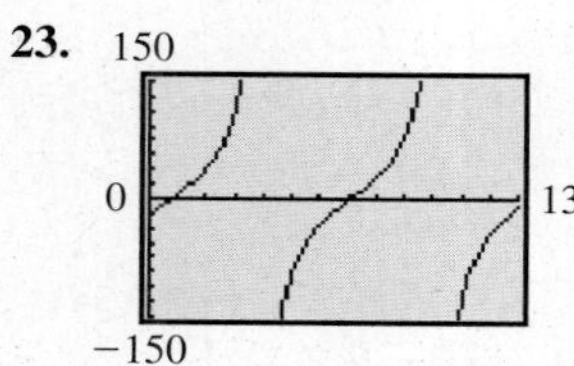

25. (a) $\tan x$ approaches ∞
(b) $\tan x$ approaches $-\infty$

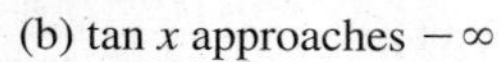

27. $y = -3 \sec(x/2)$

29.

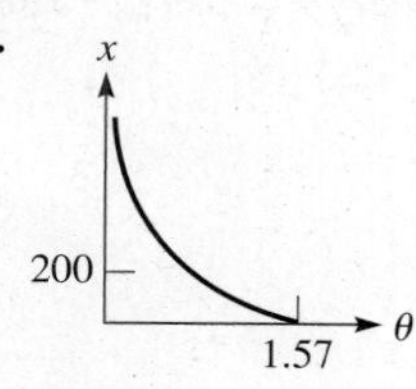

31.

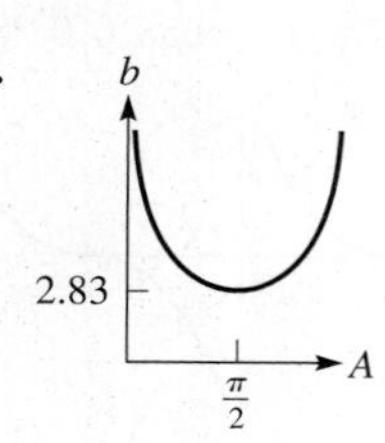

Exercises 10.5, page 314

1. $d = R \cos \omega t$ **3.** $\frac{1}{40}$ min **5.** $\frac{1}{20}$ s

7.

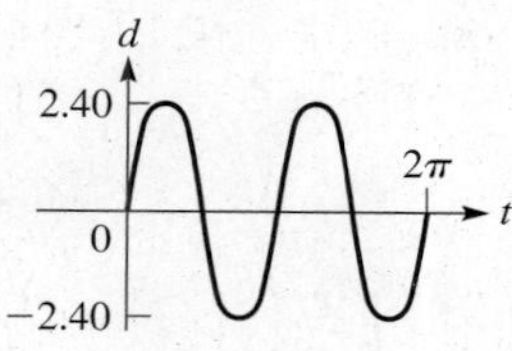

9.

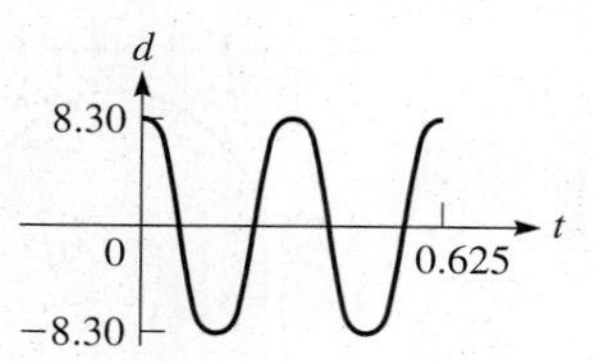

11.

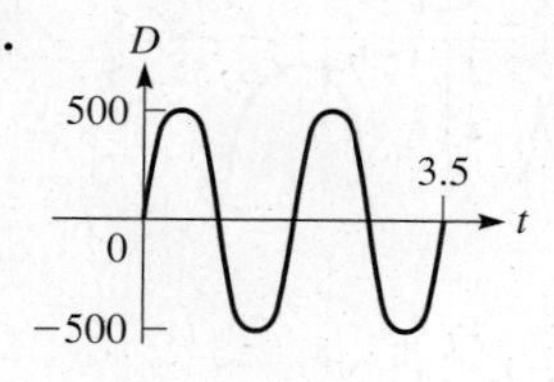

13.

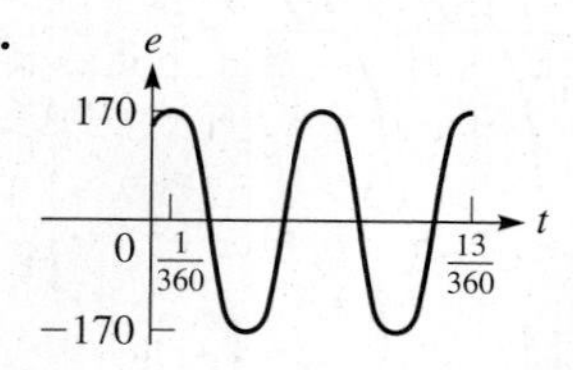

15.

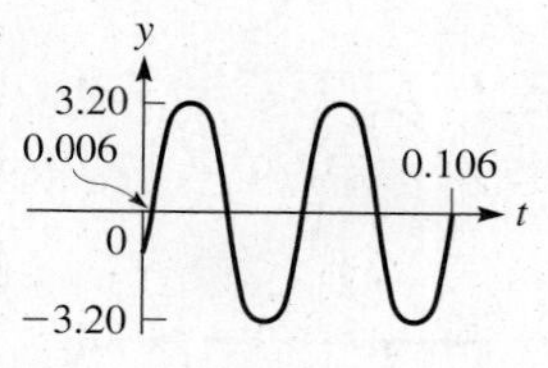

17.

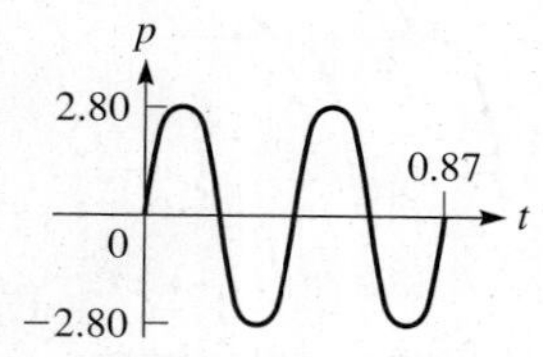

19.

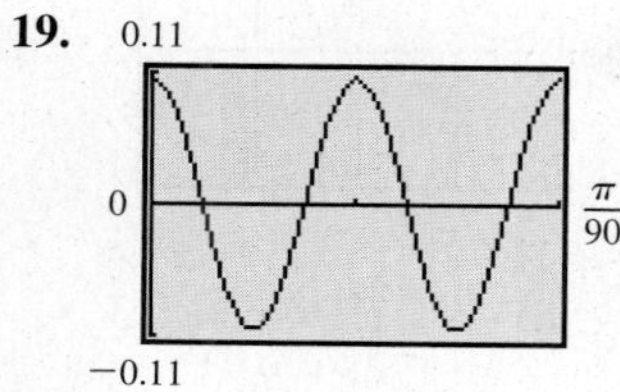

21.

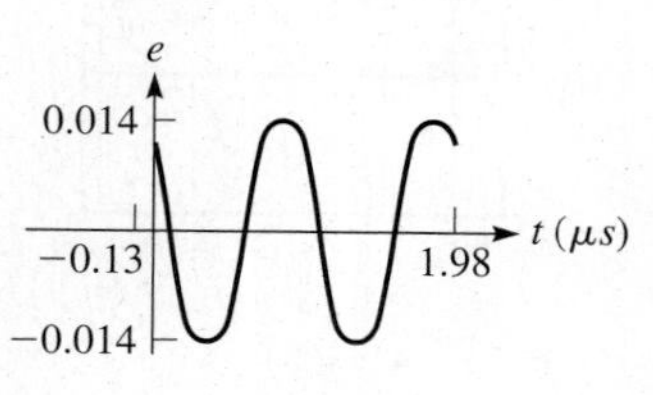

23. (a)

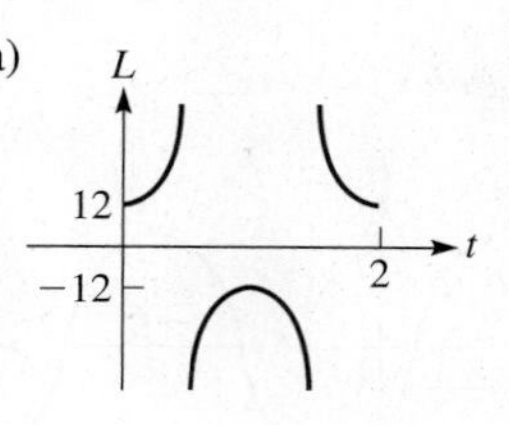

(b) $L > 0$, above x-axis

25. $y = 12 \sin 36\pi t$

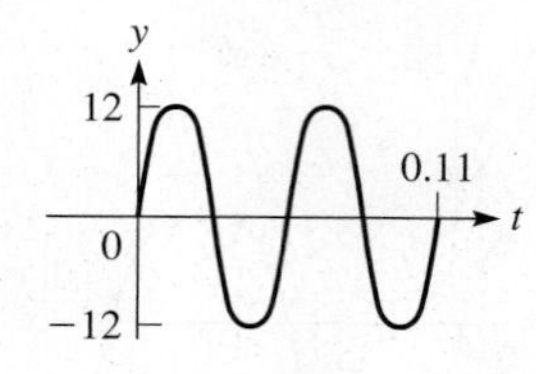

Exercises 10.6, page 318

1.

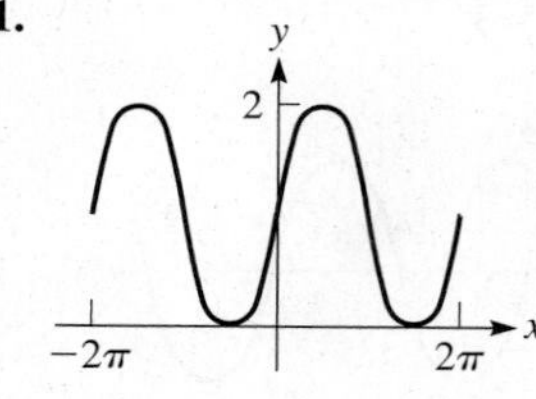

3.

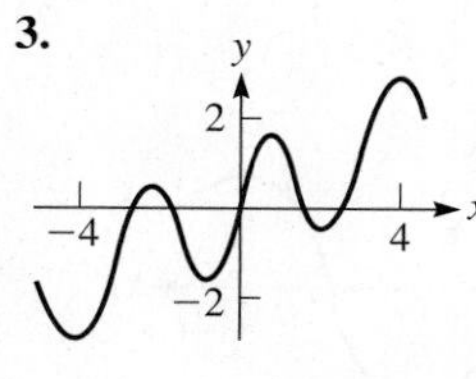

5.

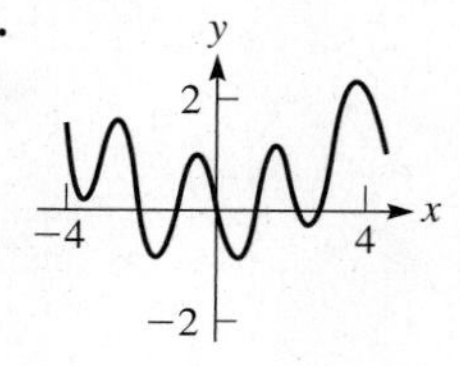

7.

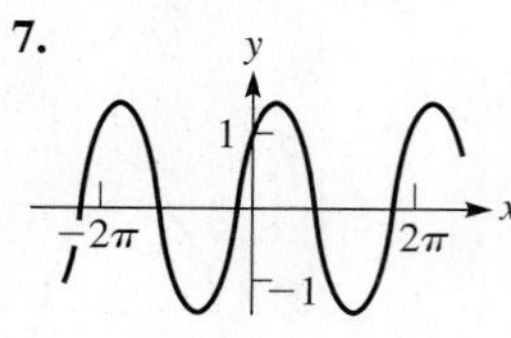

9.

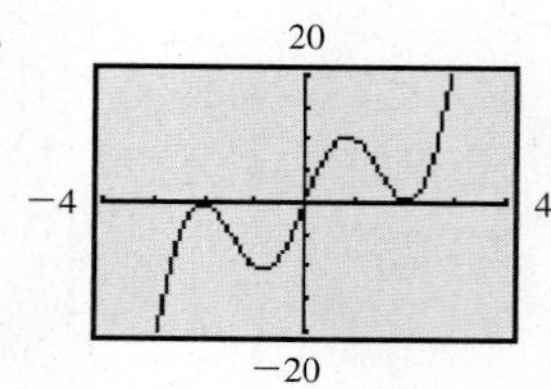

11.

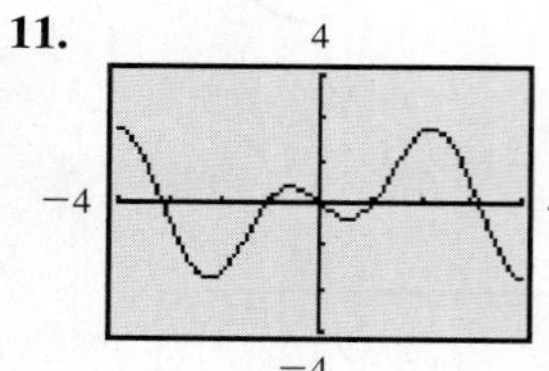

13.

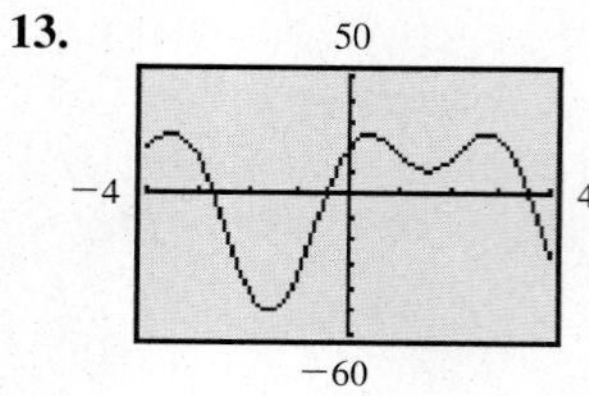

15.

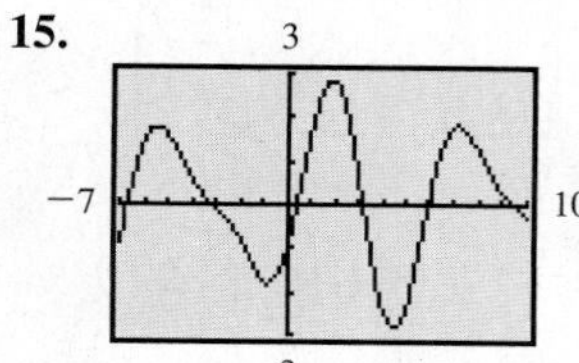

17.

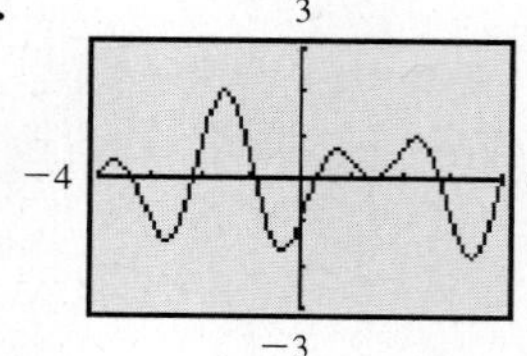

19.

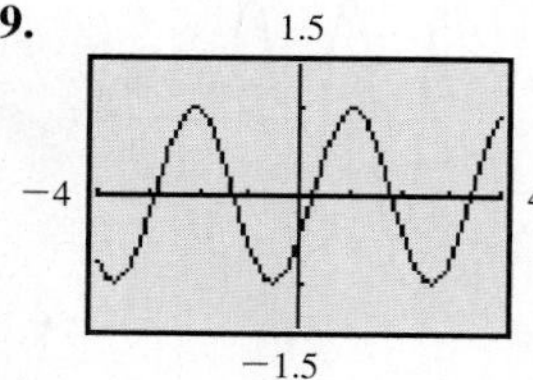

21.

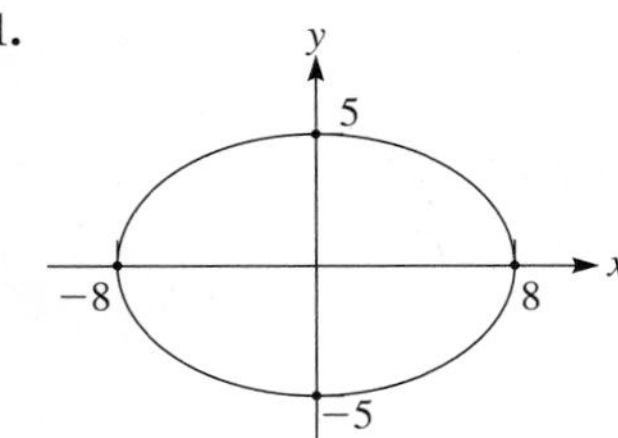

23.

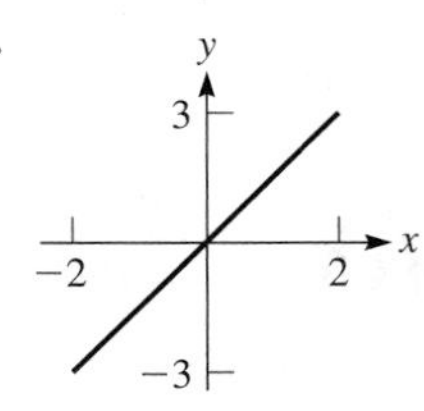

25.

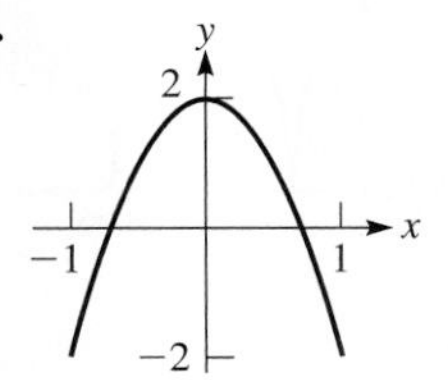

27.

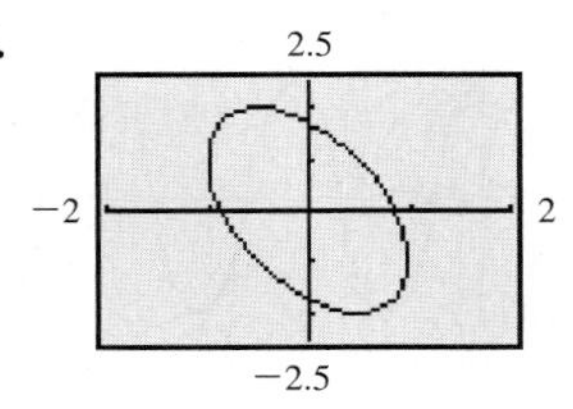

29.

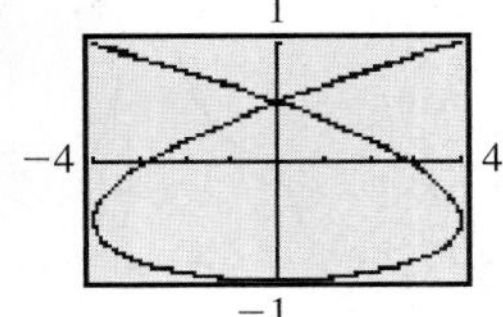

31.

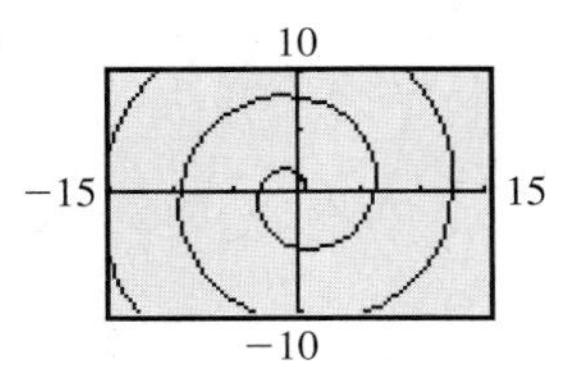

33.

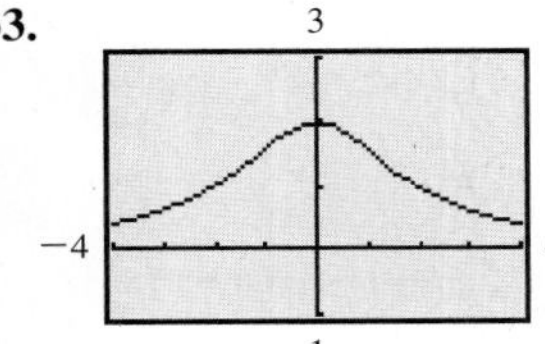

35.

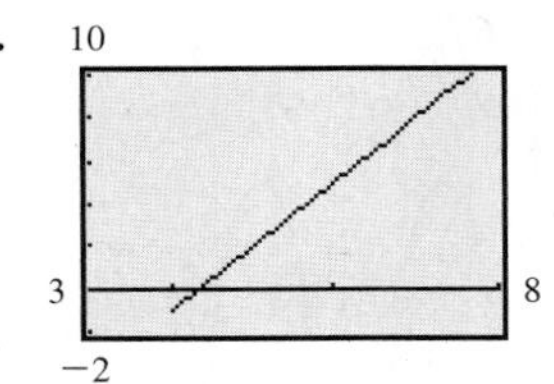

37.

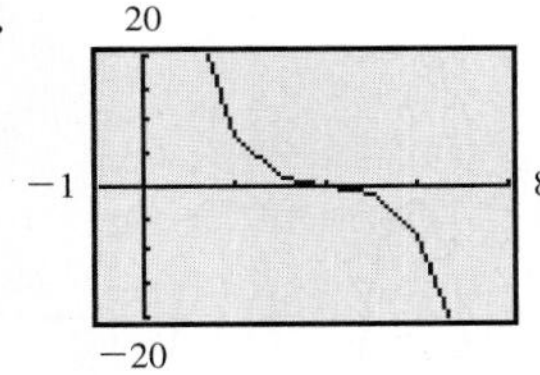

39.

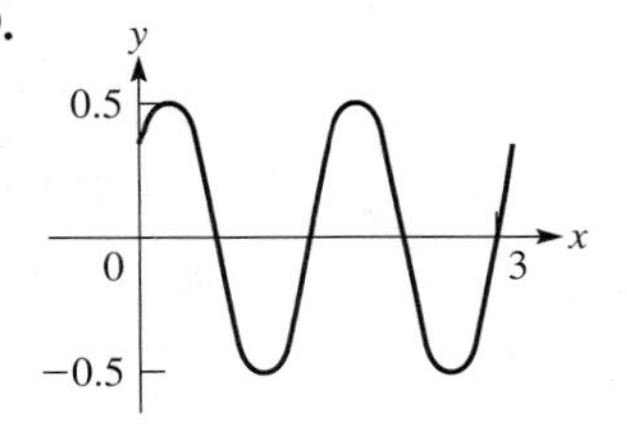

41.

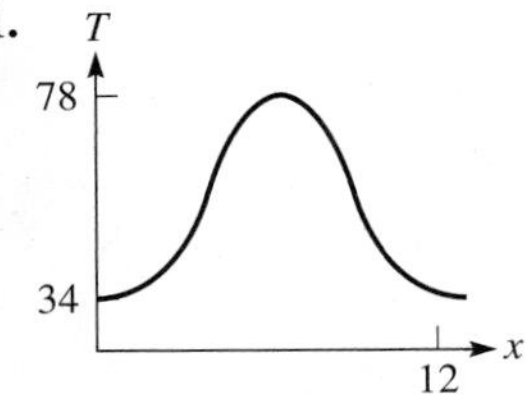

43. $p = 100 + 20 \cos 2\pi t$

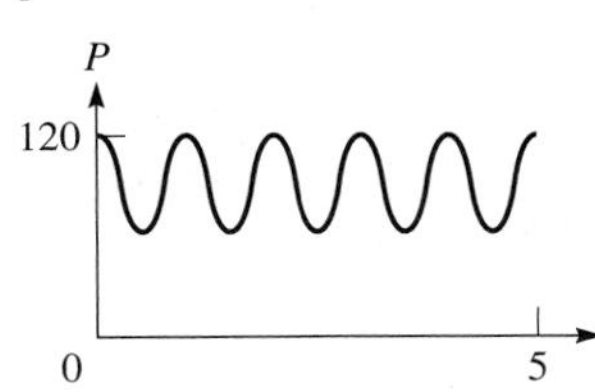

45.

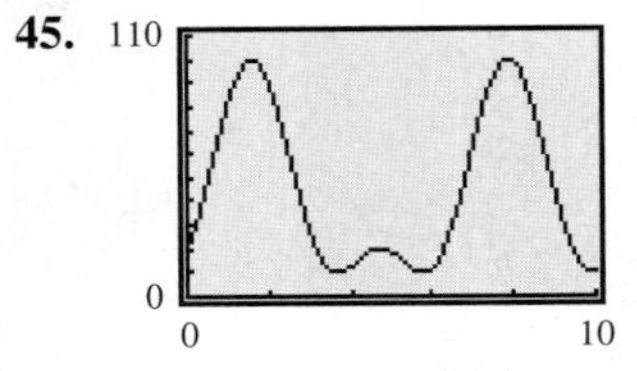

47.

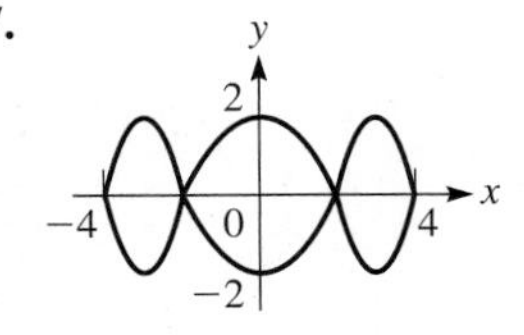

49.

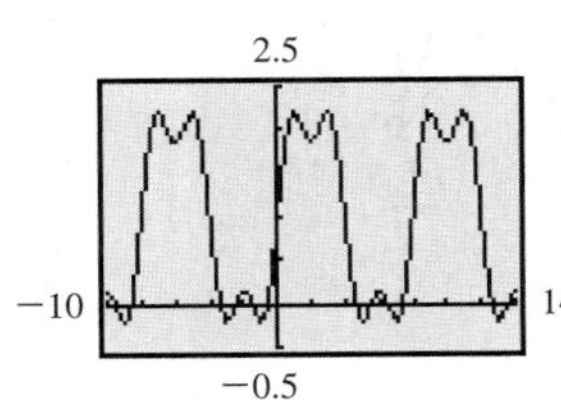

Review Exercises for Chapter 10, page 320

Concept Check Exercises

1. T **2.** T **3.** F **4.** F **5.** F **6.** T

Practice and Applications

7.

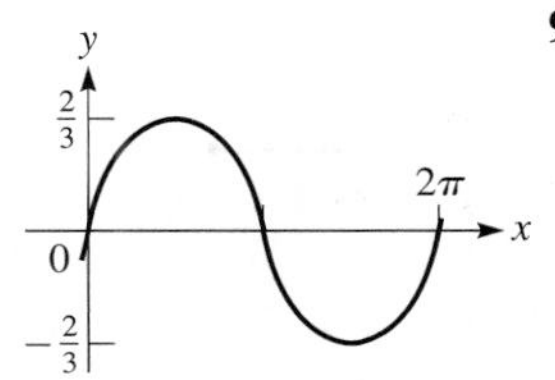

9.

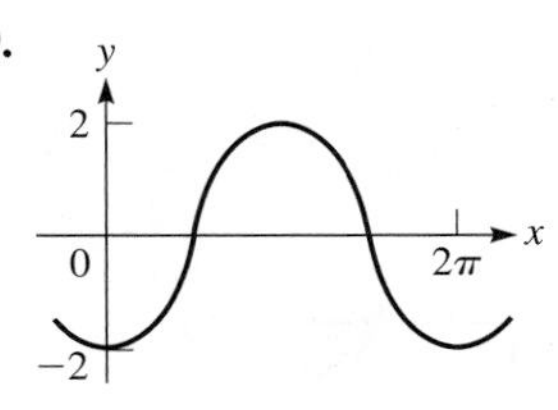

11.

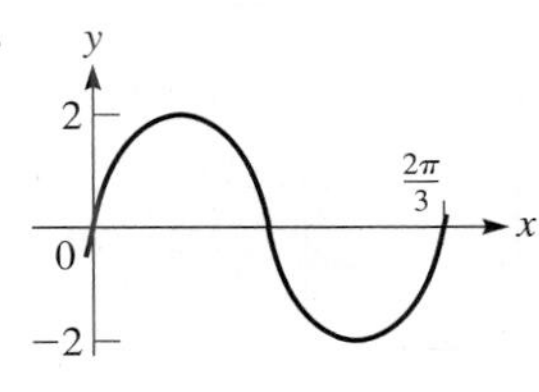

13.

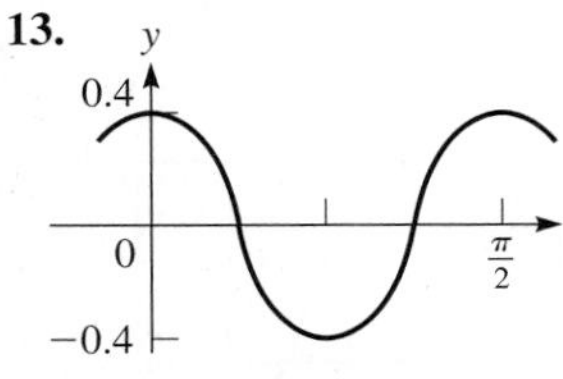

15.

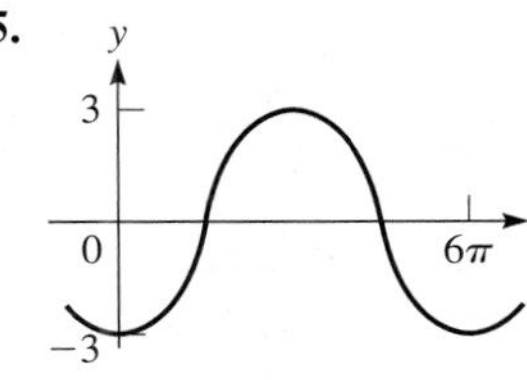

17.

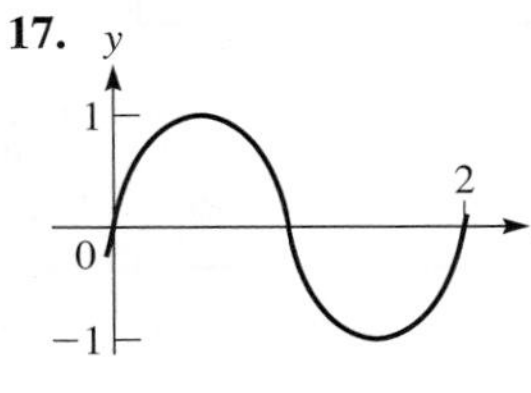

19.

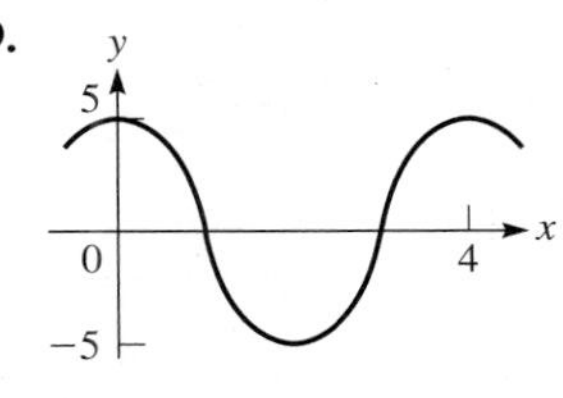

21.

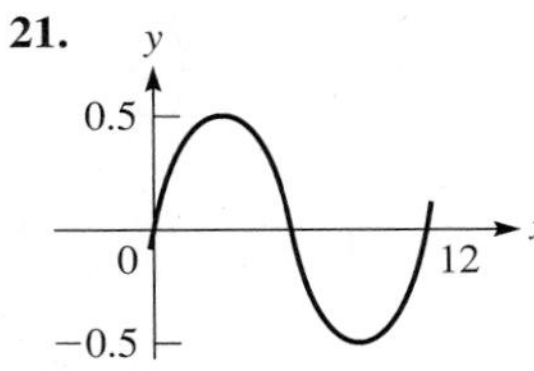

23.

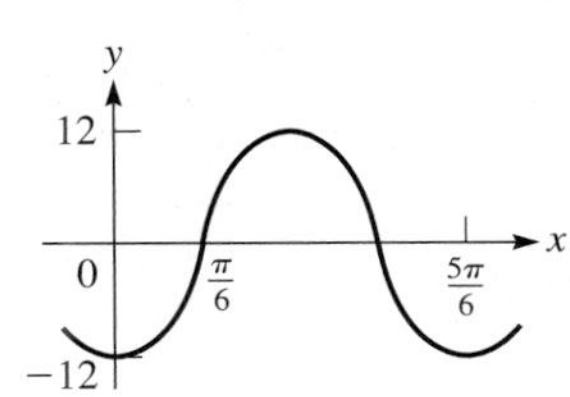

25.

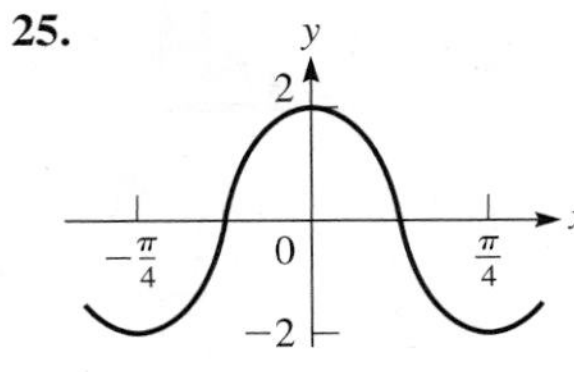

27.

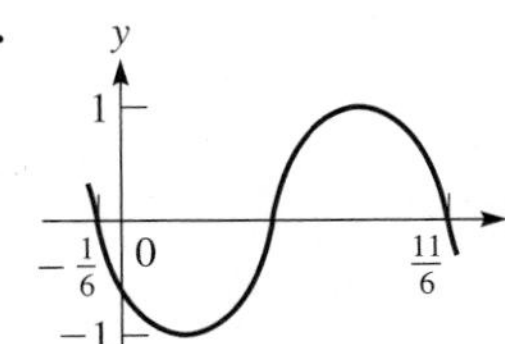

29.

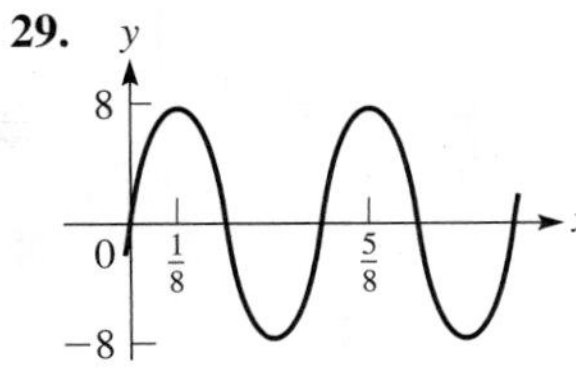

31.

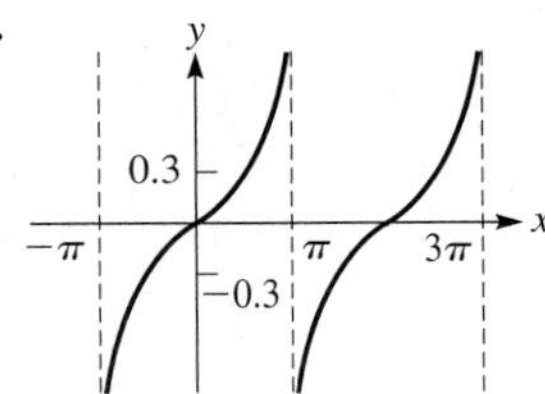

33.

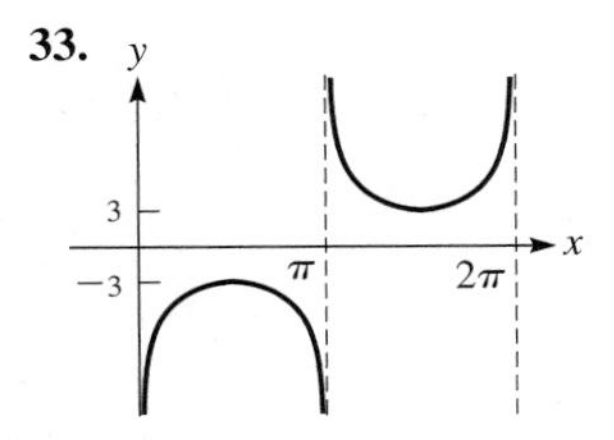

35.

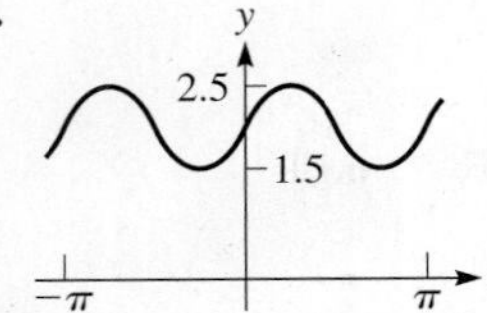

37.

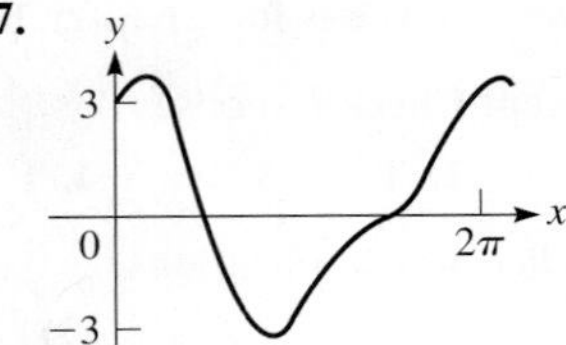

39.

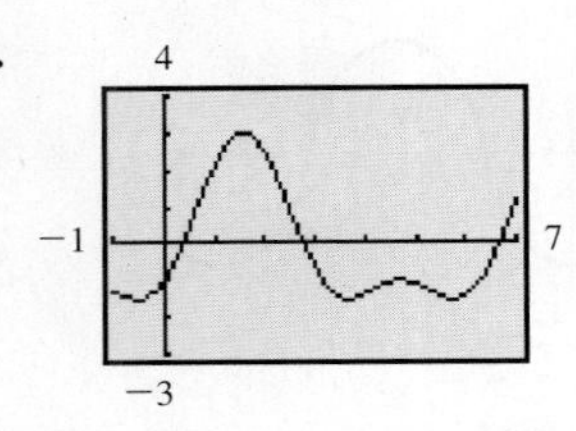

41.

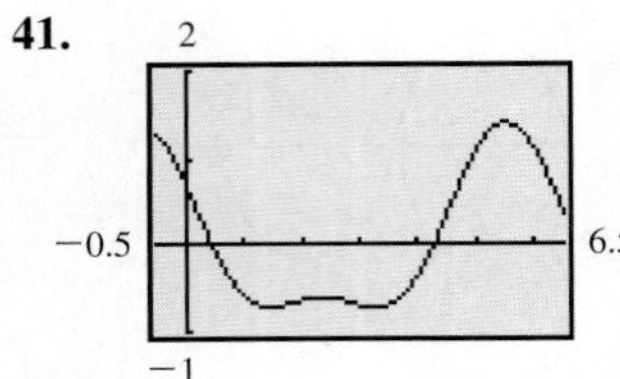

43.

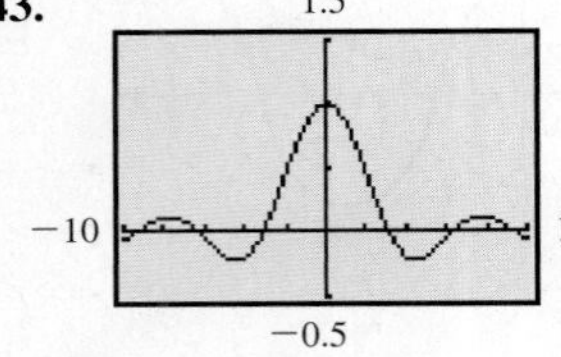

45.

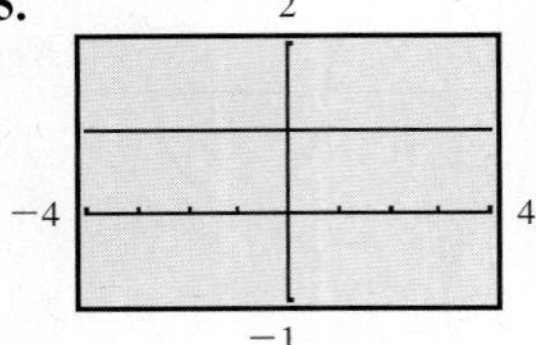

47. $y = 2\sin(2x + \frac{\pi}{2})$

49. $y = \cos(\frac{\pi}{4}x - \frac{3\pi}{4})$

51.

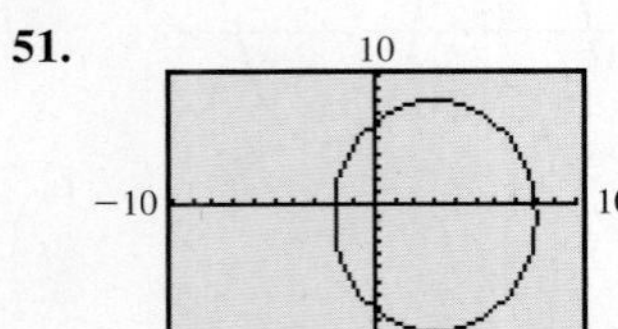

53.

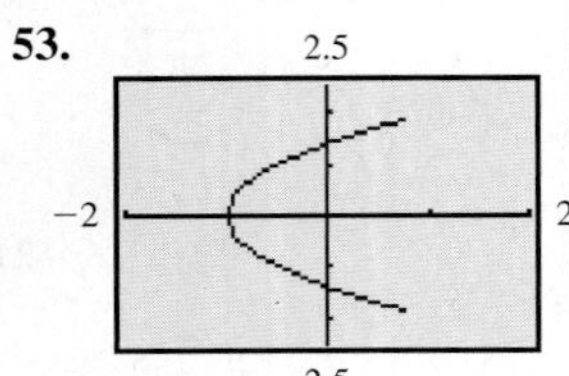

55.

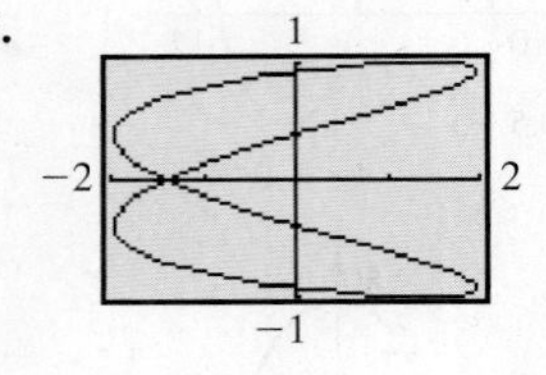

57.

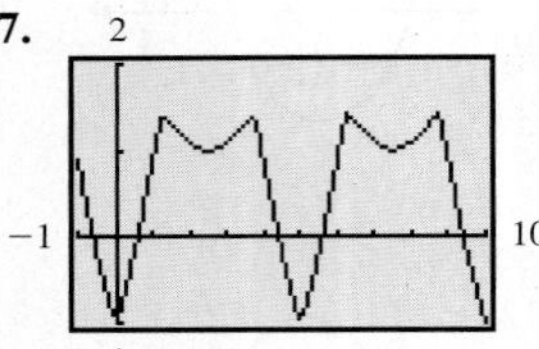

59.

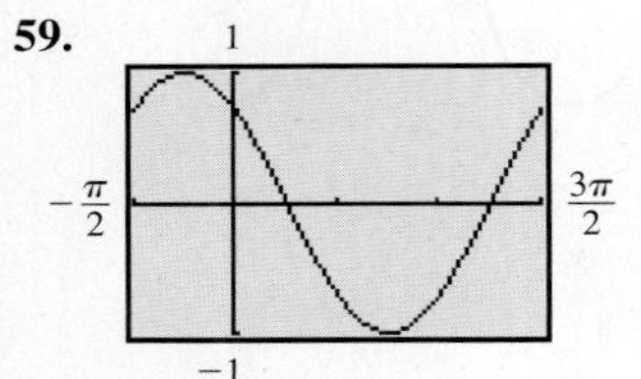

61. 4π

63. $y = 3\sin x$

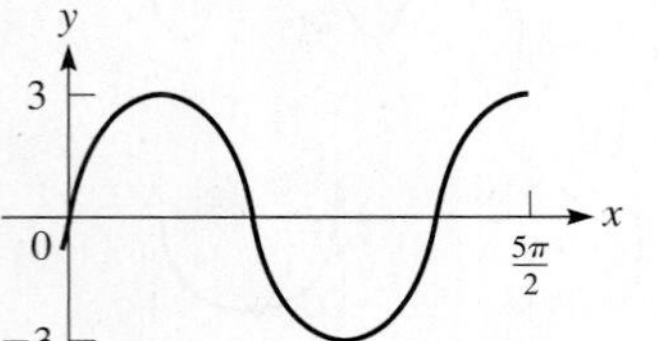

65. $y = 3\cos 3x$

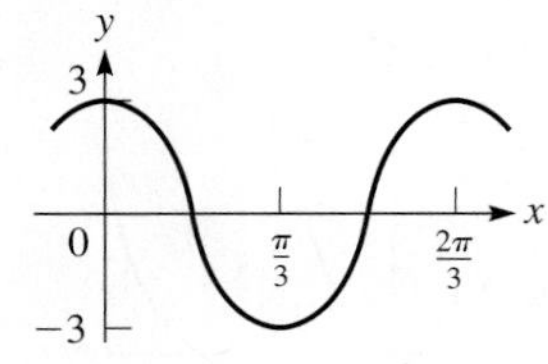

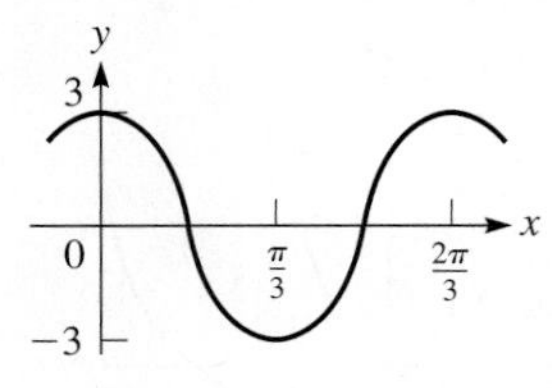

67. $y = 3\sin(\pi x + 0.25\pi)$

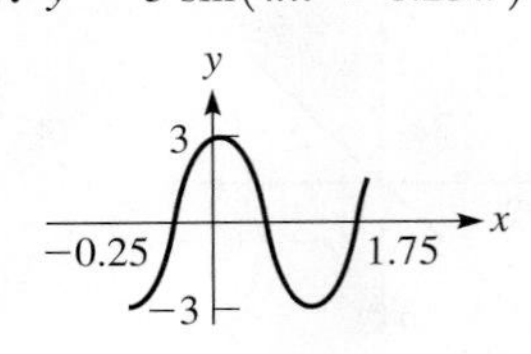

69.

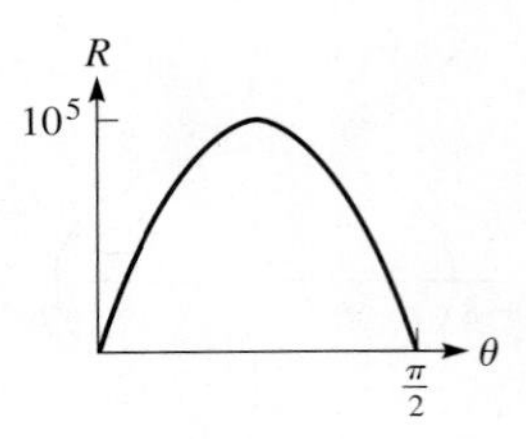

71.

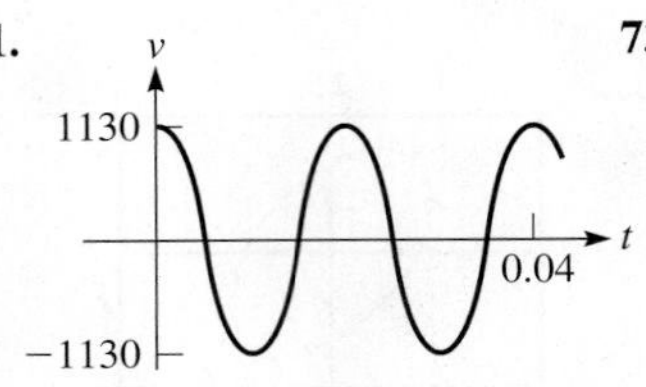

73.

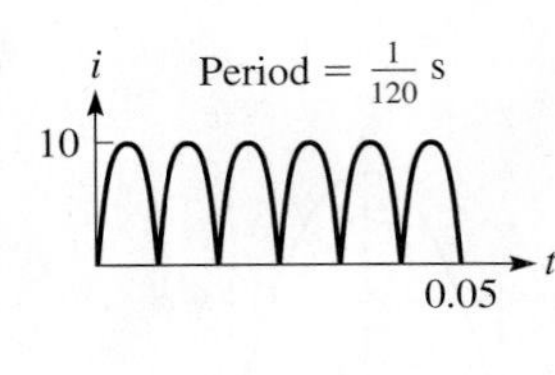

75.

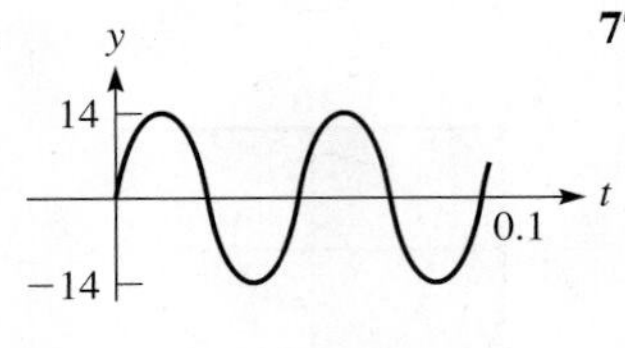

77.

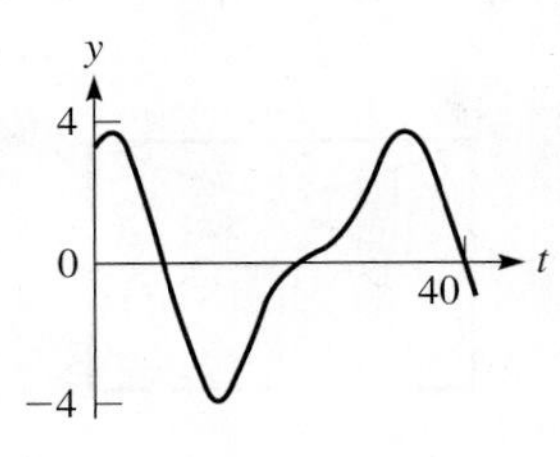

79. $y = 67.5(1 - \cos\frac{\pi t}{15})$

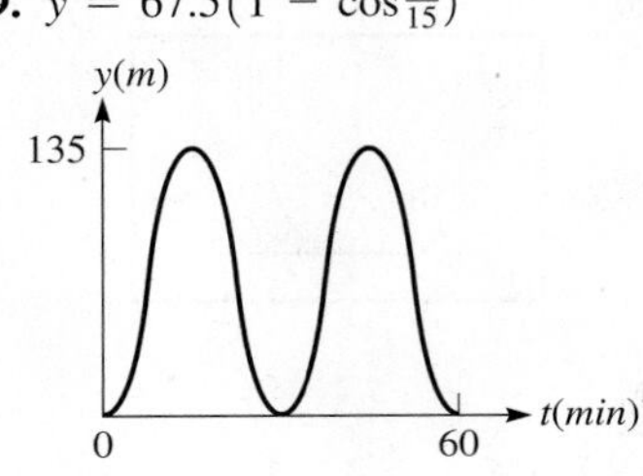

81.

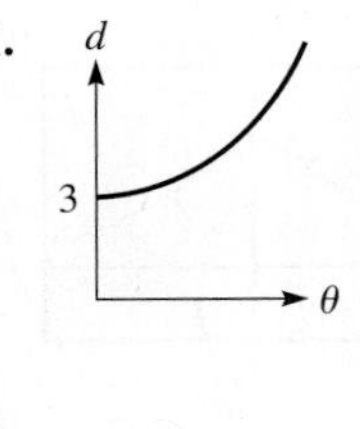

83.

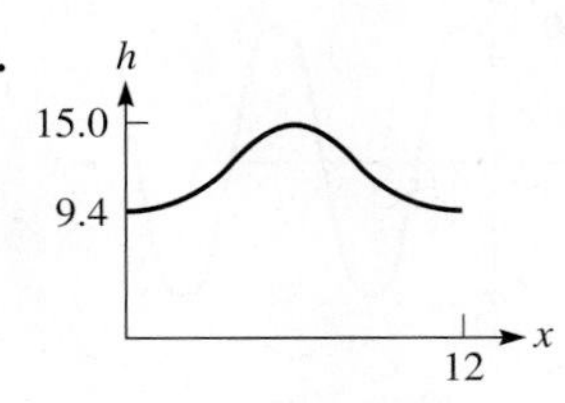

85.

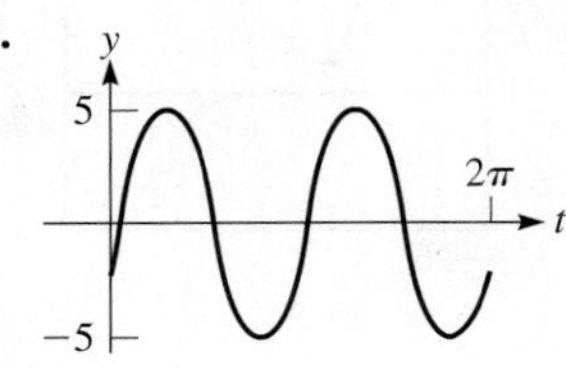

87.

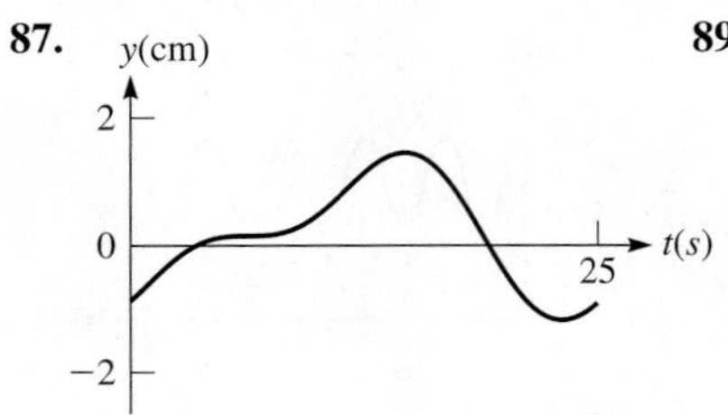

89.

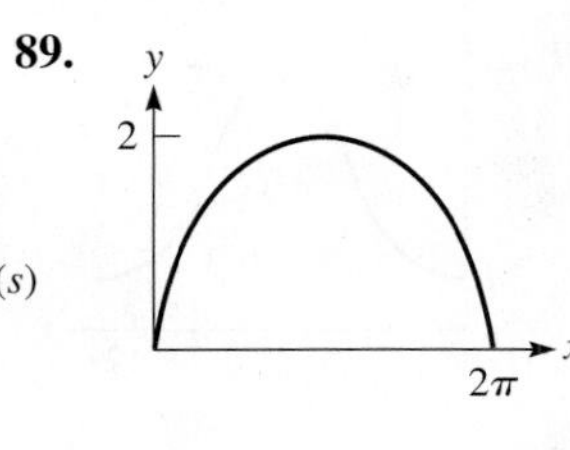

91.

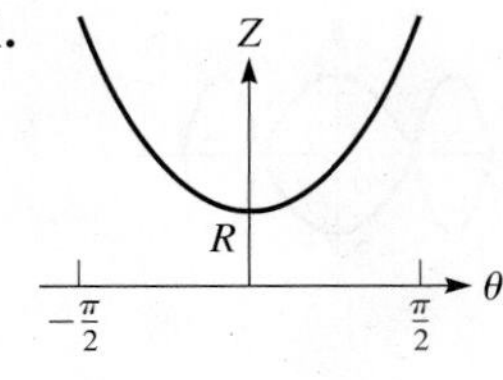

93.

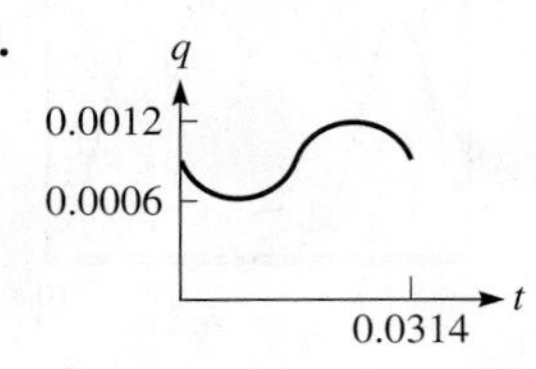

Exercises 11.1, page 327

1. $\frac{y^2}{4x^2}$ **3.** $\frac{3}{416}$ **5.** x^5 **7.** $\frac{2}{a^4}$ **9.** c^9 **11.** x^5y^2

13. $\frac{1}{125}$ **15.** $\frac{4\pi^2}{x^2}$ **17.** $\frac{2n^2}{5a}$ **19.** 1 **21.** -9 **23.** $\frac{3}{x^2}$

25. $\frac{a^3}{343x^3}$ **27.** $\frac{n^{12}}{16}$ **29.** $\frac{3}{a^3b^6}$ **31.** $\frac{1}{a+b}$ **33.** $\frac{4x^3+2y^2}{x^3y^2}$

35. $\frac{8}{3a^n}$ **37.** $\frac{b^3}{432a}$ **39.** $\frac{4}{t^4V^4}$ **41.** $\frac{3a^6+81}{a^8}$ **43.** $\frac{10}{9}$

45. $\frac{R_1R_2}{R_1+R_2}$ **47.** $\frac{4n^2-4n+1}{n^4}$ **49.** $\frac{8}{99}$ **51.** $\frac{x+y}{xy}$

53. $\frac{t^2+t+2}{t^2}$ **55.** $\frac{2D}{D^2-1}$ **57.** No

59. (a) 4^5 (b) 2^{10}

61. (a) $\left(\frac{a}{b}\right)^{-n} = \frac{1}{\left(\frac{a}{b}\right)^n} = \frac{1}{\frac{a^n}{b^n}} = \frac{b^n}{a^n} = \left(\frac{b}{a}\right)^n$

(b) 303.55182 = 303.55182

63. $n = 3$ **65.** Yes **67.** $16(4^a)$ **69.** 7 **71.** N·m

73. J/s^3 **75.** $m{\cdot}s^{-1} = (m{\cdot}s^{-2})^1(s)^1$; So $p = r = 1$ **77.** \$368.33

Exercises 11.2, page 331

1. 16 **3.**

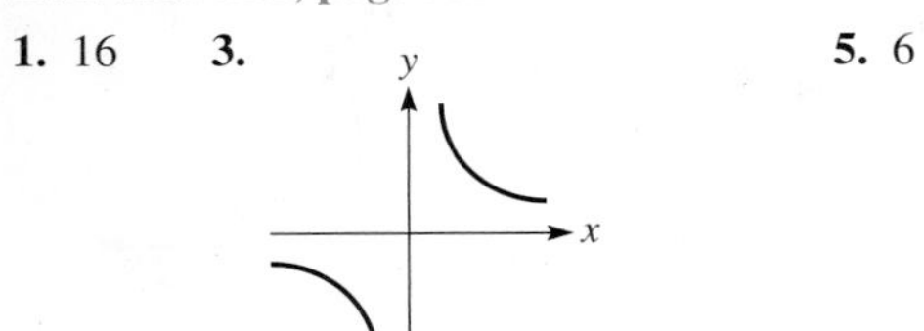

5. 6

7. 3 **9.** 1000 **11.** $\frac{1}{4}$ **13.** $\frac{1}{16}$ **15.** 25 **17.** 81

19. -200 **21.** $\frac{3}{5}$ **23.** -2 **25.** $\frac{39}{1000}$ **27.** 2.059

29. 0.53891 **31.** $m^{5/6}$ **33.** $\frac{-4}{y^{9/10}}$ **35.** $\frac{1}{x^{3/2}}$ **37.** $2ab^2$

39. $\frac{1}{8a^3b^{9/4}}$ **41.** $a^{1/12}$ **43.** $\frac{4x}{(4x^2+1)^{1/2}}$ **45.** $-y$

47. $\frac{T}{(T+2)^{1/2}}$ **49.** $\frac{a^2+1}{a^4}$ **51.** $\frac{a+1}{a^{1/2}}$ **53.** $\frac{5x^2-2x}{(2x-1)^{1/2}}$

55. **57.**

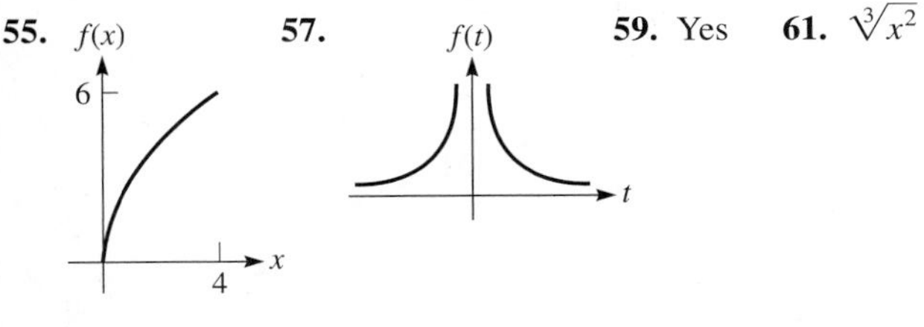

59. Yes **61.** $\sqrt[3]{x^2}$

63. If $(A/S)^{-1/4} = 0.5 = 1/2$, then $(A/S)^{1/4} = 2$. Raise each to the fourth power and get $A/S = 16$.

65. $R = \frac{T^{2/3}}{k^{1/3}} - d$ **67.** (a) $r = \left(\frac{3V}{\pi}\right)^{1/3}$ (b) 4.92 cm

69. 37.1 mg **71.** 229 km

Exercises 11.3, page 335

1. $ab^2\sqrt{a}$ **3.** $\frac{\sqrt[4]{54}}{3}$ **5.** $5\sqrt{3}$ **7.** $6\sqrt{3}$ **9.** $xy^2\sqrt{y}$

11. $x^2yz^3\sqrt{y}$ **13.** $4R^2V^2\sqrt{5RT}$ **15.** $2\sqrt[3]{2}$ **17.** $2\sqrt[5]{3}$

19. $2\sqrt[3]{a^2}$ **21.** $2st\sqrt[4]{4r^3t}$ **23.** -2 **25.** $P\sqrt[3]{V}$

27. $\frac{2\sqrt{3}}{3}$ **29.** $\frac{1}{2}\sqrt{6}$ **31.** $\frac{1}{2}\sqrt[3]{6}$ **33.** $\frac{1}{3}\sqrt[5]{27}$ **35.** $2\sqrt{5}$

37. 2 **39.** 200 **41.** 2000 **43.** $\sqrt{3y}$ **45.** $\frac{1}{2}\sqrt{2}$

47. $\sqrt[3]{2}$ **49.** $\sqrt[8]{n}$ **51.** $\frac{2u\sqrt{7uv}}{v^3}$ **53.** $4\sqrt{13}$ **55.** $\frac{\sqrt{6x}}{3c^2}$

57. $\frac{3}{4}\sqrt{2}$ **59.** $\frac{\sqrt{xy(x^2+y^2)}}{xy}$ **61.** $\frac{\sqrt{C^2-4}}{C+2}$ **63.** $\sqrt{a^2+b^2}$

65. $3x - 1$ **67.** $\sqrt[6]{a^3}$; $\sqrt[6]{b^2}$; $\sqrt[6]{c}$

69. **71.** $f = \frac{\sqrt{\mu T}}{2\mu L}$

73. $\sqrt{2ag}$ **75.** $\frac{8Af\sqrt{f^2+f_0^2}}{\pi^2(f^2+f_0^2)}$

Exercises 11.4, page 338

1. $16\sqrt{5}$ **3.** $8\sqrt{7}$ **5.** $\sqrt{5} - \sqrt{7}$ **7.** $11\sqrt{2} - 3\sqrt{3}$

9. $3\sqrt{5}$ **11.** $-4t\sqrt{3}$ **13.** $-3\sqrt{2y}$ **15.** $19\sqrt{7}$

17. $9\sqrt{2}$ **19.** $23\sqrt{3R} - 6\sqrt{2R}$ **21.** $\frac{5\sqrt{10}}{2}$ **23.** $-9\sqrt{2}$

25. $13\sqrt[3]{3}$ **27.** $\sqrt[4]{2}$ **29.** $(5a - 2b^2)\sqrt{ab}$

31. $(3 - 2a)\sqrt{10}$ **33.** $(2b - a)\sqrt[3]{3a^2b}$ **35.** $\frac{(a^2 - c^3)\sqrt{ac}}{a^2c^3}$

37. $\frac{(a - 2b)\sqrt[3]{ab^2}}{ab}$ **39.** $\frac{-2V\sqrt{T^2 - V^2}}{T^2 - V^2}$ **41.** $8\sqrt{x+2}$

43. $15\sqrt{3} - 11\sqrt{5} = 1.3840144$ **45.** $\frac{1}{6}\sqrt{6} = 0.4082483$

47. $\frac{7}{2}\sqrt[3]{2} = 4.4097237$ **49.** $3\sqrt{3}$

51. Positive, $\sqrt{1000} = 10\sqrt{10}$, $\sqrt{11} > \sqrt{10}$

53. $6\sqrt{2} + 2\sqrt{6}$ **55.** $18 + 3\sqrt{2} = 22.2$ ft

Exercises 11.5, page 341

1. $3\sqrt{10} - 16$ **3.** $\sqrt{3} - \sqrt{2}$ **5.** $\sqrt{35}$ **7.** $2\sqrt{3}$

9. 2 **11.** 48 **13.** $6\sqrt{5}$ **15.** $3\sqrt{2} - \sqrt{15}$ **17.** -1

19. $-12 + 15\sqrt{10}$ **21.** $66 + 13\sqrt{11x} - 5x$

23. $a\sqrt{b} + c\sqrt{ac}$ **25.** $\frac{2 - 3\sqrt{6}}{2}$ **27.** $x - 8\sqrt{xy} + 16y$

29. $2a - 3b + 2\sqrt{2ab}$ **31.** $\sqrt[6]{72}$ **33.** $\frac{1}{4}(\sqrt{7} - \sqrt{3})$

35. $\frac{1}{11}(\sqrt{7} + 3\sqrt{2} - 6 - \sqrt{14})$ **37.** $\frac{1}{17}(-56 + 9\sqrt{15})$

39. $\frac{6x + 6\sqrt{5x}}{x - 5}$ **41.** 1 **43.** $\frac{2 - R - R^2}{R}$

45. $-\frac{\sqrt{x^2 - y^2} + \sqrt{x^2 + xy}}{y}$ **47.** $a - 1 + \sqrt{a(a-2)}$

49. $-1 - \sqrt{66} = -9.1240384$

51. $-\frac{16 + 5\sqrt{30}}{26} = -1.6686972$ **53.** $\frac{2x+1}{\sqrt{x}}$ **55.** $\frac{5x^2+2x}{\sqrt{2x+1}}$

57. $\frac{-1}{5\sqrt{2} - 4\sqrt{5}}$ **59.** $\frac{1}{\sqrt{x+h} + \sqrt{x}}$

61. Yes — if either x or y is zero and the other is ≥ 0.

63. $(1 - \sqrt{2})^2 - 2(1 - \sqrt{2}) - 1$
$= 1 - 2\sqrt{2} + 2 - 2 + 2\sqrt{2} - 1 = 0$

65. $a = c$ **67.** $\frac{\sqrt[3]{x} - 1}{x - 1}$

69. $[\frac{1}{2}(\sqrt{b^2 - 4k^2} - b)]^2 + b[\frac{1}{2}(\sqrt{b^2 - 4k^2} - b)] + k^2$
$= \frac{1}{4}(b^2 - 4k^2) - \frac{b}{2}\sqrt{b^2 - 4k^2} + \frac{1}{4}b^2$
$+ \frac{b}{2}\sqrt{b^2 - 4k^2} - \frac{1}{2}b^2 + k^2 = 0$

71. $\frac{2500 - 50\sqrt{V}}{2500 - V}$ **73.** $2Q\sqrt{\sqrt{2} + 1}$ **75.** $\frac{\sqrt{LC - R^2C^2}}{LC}$

Review Exercises for Chapter 11, page 343

Concept Check Exercises

1. F **2.** F **3.** F **4.** T **5.** T **6.** T

Practice and Applications

7. $\frac{8}{x^3}$ **9.** $\frac{9d^3}{c}$ **11.** 375 **13.** $\frac{1}{1000}$ **15.** $\frac{t^4}{9}$ **17.** -28

19. $64a^4b$ **21.** $-8m^9n^6$ **23.** $\frac{6C - 4L^2}{CL^2}$ **25.** $\frac{3y}{x + 3y}$

27. $\frac{b}{ab - 3}$ **29.** $\frac{(x^3y^3 - 1)^{1/3}}{y}$ **31.** $\frac{1}{W - H}$ **33.** $\frac{-2(x + 1)}{(x - 1)^3}$

35. $2\sqrt{15}$ **37.** $b^2c\sqrt{ab}$ **39.** $3ab^2\sqrt{a}$ **41.** $\frac{2t\sqrt{21st}}{u}$

43. $\frac{5\sqrt{2s}}{2s}$ **45.** $\frac{2\sqrt{6}}{9}$ **47.** $mn^2\sqrt[4]{8m^2n}$ **49.** $\sqrt{2}$ **51.** 0

53. $-7\sqrt{7}$ **55.** $3ax\sqrt{2x}$ **57.** $(2a + b)\sqrt[3]{a}$

59. $150 - 25\sqrt{7}$ **61.** $4\sqrt{3} - 4\sqrt{5}$

63. $6 - 51B - 7\sqrt{17B}$ **65.** $42 - 7\sqrt{7a} - 3a$

67. $36x - 48\sqrt{xy} + 16y$ **69.** $\frac{6x + \sqrt{3xy}}{12x - y}$ **71.** $-\frac{8 + \sqrt{6}}{29}$

73. $\frac{13 - 2\sqrt{35}}{29}$ **75.** $\frac{6x - 13a\sqrt{x} + 5a^2}{9x - 25a^2}$ **77.** $\sqrt{4b^2 + 1}$

79. $9x - 4x^2$ **81.** $1 + 6^{1/2}$ **83.** $\frac{15 - 2\sqrt{15}}{4}$

85. $51 - 7\sqrt{105} = -20.728655$

87. $\sqrt{\sqrt{2} - 1}(\sqrt{2} + 1) = \sqrt{(\sqrt{2} - 1)(\sqrt{2} + 1)^2}$
$= \sqrt{(\sqrt{2} - 1)(3 + 2\sqrt{2})} = \sqrt{\sqrt{2} + 1} = 1.5537740$

89. $12\sqrt{2}$ **91.** $\frac{33}{4} - 2\sqrt{3}$ **93.** 1.676%

95. (a) $v = k(P/W)^{1/3}$ (b) $v = \frac{k\sqrt[3]{PW^2}}{W}$ **97.** $\frac{100(1 + 2\sqrt{x})}{x}$

99. $\frac{c}{(c^2 - v^2)^{1/2}}$ **101.** $\frac{1}{\sqrt{A + h} + \sqrt{A}}$ **103.** $6\sqrt{2}$ cm

105. $\frac{\sqrt{LC_1C_2(C_1 + C_2)}}{2\pi LC_1C_2}$

Exercises 12.1, page 348

1. 6 **3.** -1 **5.** $9j$ **7.** $-2j$ **9.** $0.6j$ **11.** $8j\sqrt{2}$

13. $\frac{1}{2}j\sqrt{7}$ **15.** $-2\pi j\sqrt{\pi}$ **17.** (a) -7 (b) 7

19. (a) 4 (b) -4 **21.** $-\frac{3}{5}$ **23.** $10j$ **25.** (a) 1 (b) -1

27. 0 **29.** $-2j$ **31.** $-j$ **33.** $2 + 3j$ **35.** $-7j$

37. $2 + 2j$ **39.** $-2 + 3j$ **41.** $3\sqrt{2} - 2j\sqrt{2}$ **43.** -1

45. -15 **47.** $-\frac{1}{2} + \frac{\sqrt{2}}{2}$ **49.** $j - 2$

51. (a) $6 + 7j$ (b) $8 - j$ **53.** (a) $-2j$ (b) -4

55. $x = 2, y = -2$ **57.** $x = 10, y = -6$ **59.** $x = -2, y = 3$

61. $4j\sqrt{2}, -4j\sqrt{2}$ **63.** $-1 + j\sqrt{6}, -1 - j\sqrt{6}$

65. (a) No change (b) Sign changes **67.** Yes **69.** 0

71. Yes; imag. part is zero. **73.** Each is x.

Exercises 12.2, page 351

1. $1 - 5j$ **3.** $\frac{1}{25}(29 + 22j)$ **5.** $5 - 8j$ **7.** $-25 + 10j$

9. $-0.23 + 0.86j$ **11.** $-36 + 21j$ **13.** $7 + 49j$

15. $22 + 3j$ **17.** $-18j\sqrt{2}$ **19.** $22 - 7j$ **21.** $3\sqrt{7} + 3j$

23. $-40 - 42j$ **25.** 73 **27.** $\frac{1}{29}(-30 + 12j)$ **29.** $-\frac{1}{3}(1 + j)$

31. $\frac{1}{11}(-13 + 8j\sqrt{2})$ **33.** $\frac{-1 + 3j}{5}$ **35.** $\frac{-45}{13} + \frac{48j}{13}$

37. $-8j$ **39.** $-80j$ **41.** $-\frac{7}{8} + 4j$

43. $(-1 - j)^2 + 2(-1 - j) + 2 = 1 + 2j - 1 - 2 - 2j + 2 = 0$

45. 4 **47.** 10 **49.** $\frac{3}{10} + \frac{1}{10}j$ **51.** $-1 + j$ **53.** -4

55. $\frac{1}{10}(11 + 27j)$ **57.** $281 + 35.2j$ volts

59. $0.016 + 0.037j$ amperes **61.** (a) Real (b) Pure imaginary

63. Product is sum of squares of two real numbers
$(a + bj)(a - bj) = a^2 + b^2$

Exercises 12.3, page 353

1. $3 - j$ **3.** **5.**

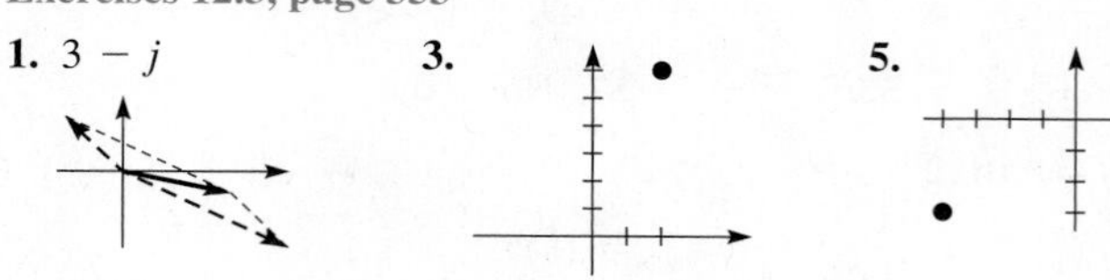

7. $-3j$ **9.** $5 + 4j$ **11.** $8 + j$

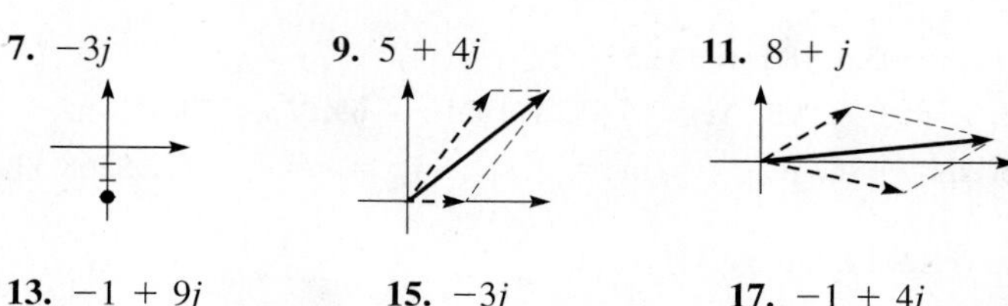

13. $-1 + 9j$ **15.** $-3j$ **17.** $-1 + 4j$

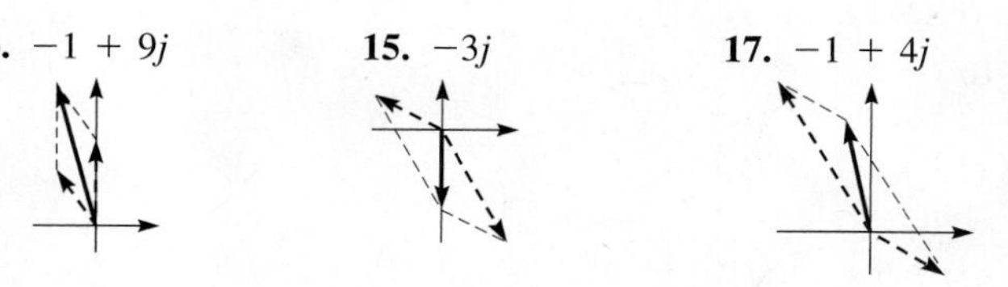

19. $-180 + 150j$ **21.** $9 + 4j$ **23.** $4 + 2j$

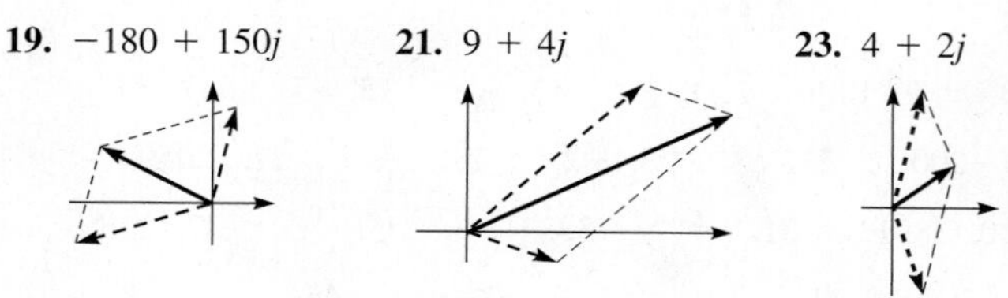

25. $-2j$ **27.**

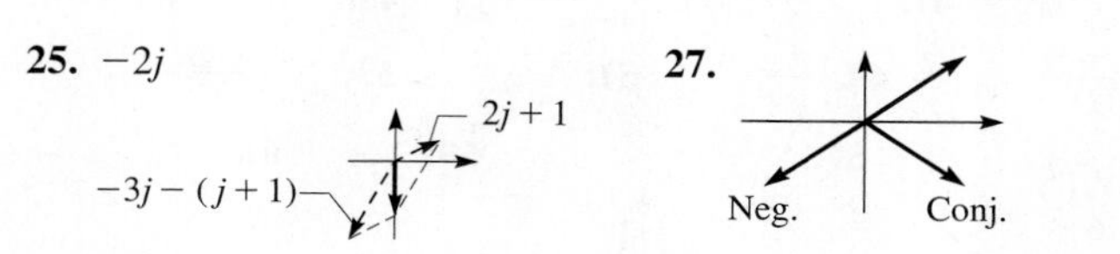

29. **31.** $-9 + 3j$

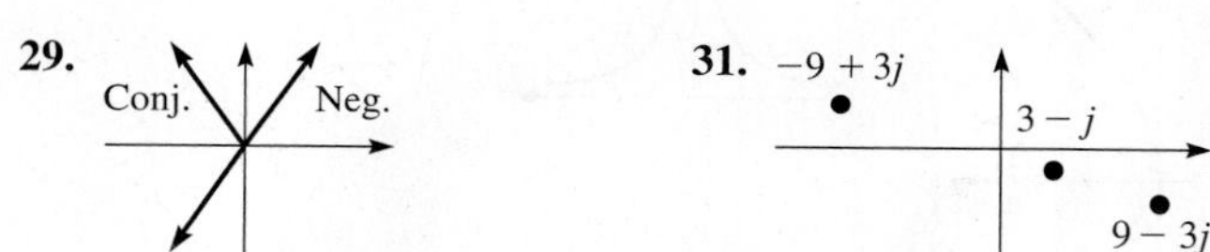

33. A complex number is on the opposite side of the real axis from its conjugate.

Conj.

35. Subtracting the conjugate from the complex number results in an imaginary number.

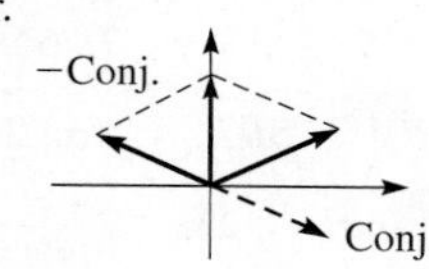

37. $90 - 15j$ lb

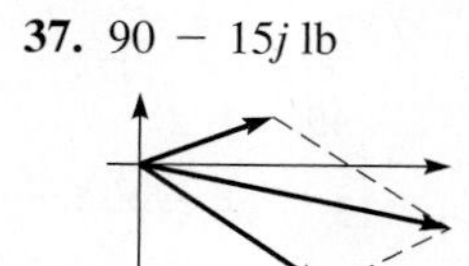

Exercises 12.4, page 356

1. $5(\cos 126.9^\circ + j \sin 126.9^\circ)$

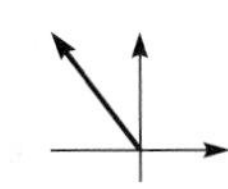

3. $10(\cos 36.9^\circ + j \sin 36.9^\circ)$

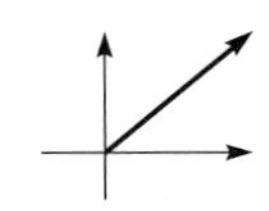

5. $50(\cos 306.9^\circ + j \sin 306.9^\circ)$

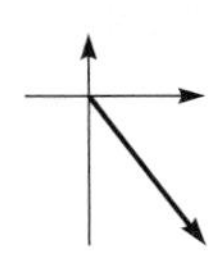

7. $3.61(\cos 123.7^\circ + j \sin 123.7^\circ)$

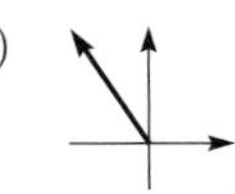

9. $0.60(\cos 204^\circ + j \sin 204^\circ)$

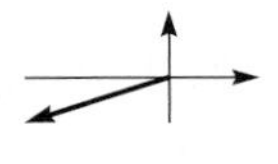

11. $2(\cos 60^\circ + j \sin 60^\circ)$

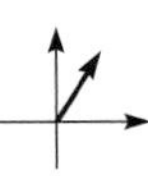

13. $8.062(\cos 295.84^\circ + j \sin 295.84^\circ)$

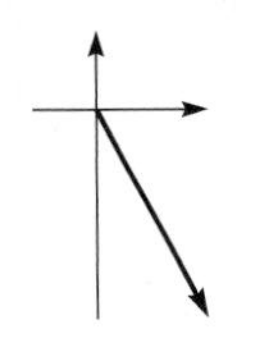

15. $3(\cos 180^\circ + j \sin 180^\circ)$ **17.** $9(\cos 90^\circ + j \sin 90^\circ)$

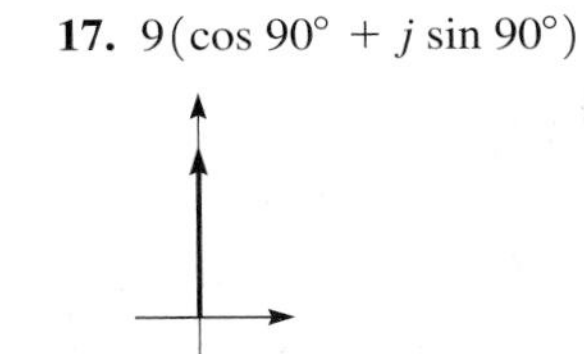

19. $2.94 + 4.05j$ **21.** $-139 + 80.0j$ **23.** $-1.85 - 2.36j$

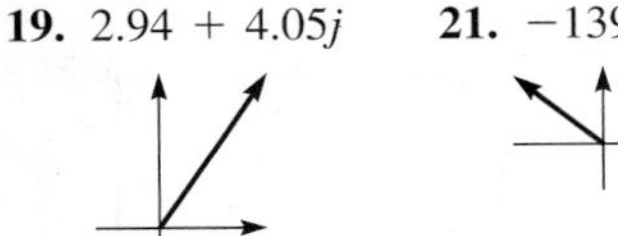

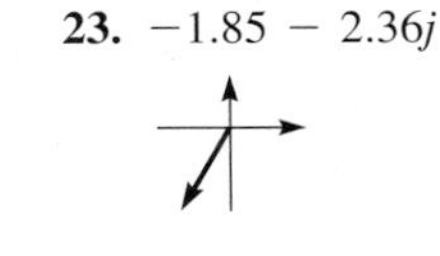

25. 0.08 **27.** $-120j$

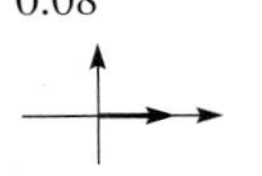

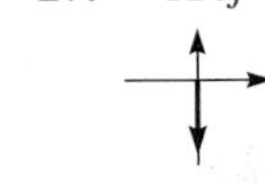

29. $-4.71 + 0.595j$ **31.** $-0.6052 - 0.7096j$ **33.** $7.32j$

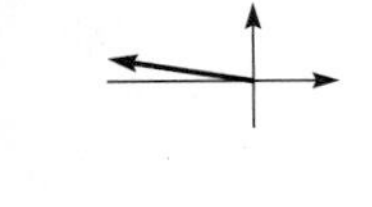

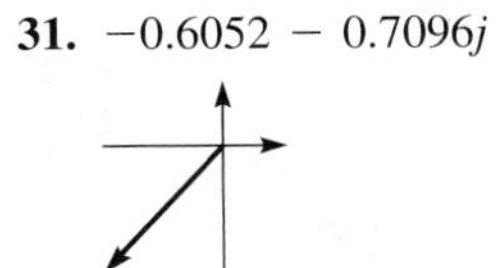

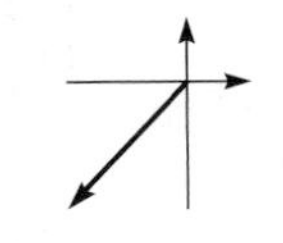

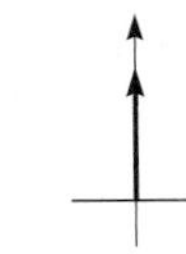

35. $-6.961 + 86.14j$ **37.** 180°

39. conj. of $r(\cos\theta + j\sin\theta) = r(\cos\theta - j\sin\theta)$
$= r(\cos(-\theta) + j\sin(-\theta))$
$= r\angle -\theta$

41. $3.03\angle 339.5^\circ$ kV **43.** $2.51 + 12.1j$ V/m

Exercises 12.5, page 358

1. $8.50e^{3.95j}$ **3.** $3.00e^{1.05j}$ **5.** $0.450e^{4.93j}$ **7.** $375.5e^{-1.666j}$
9. $0.515e^{3.46j}$ **11.** $4.06e^{-1.07j} = 4.06e^{5.21j}$ **13.** $9245e^{5.172j}$
15. $5.00e^{5.36j}$ **17.** $36.1e^{2.55j}$ **19.** $6.37e^{0.386j}$ **21.** $825.7e^{3.836j}$
23. $3.00\angle 28.6^\circ$; $2.63 + 1.44j$ **25.** $464\angle 106.0^\circ$; $-128 + 446j$
27. $3.20\angle 50.0^\circ$; $2.06 + 2.45j$
29. $1724\angle 137.0^\circ$; $-1261 + 1176j$
31. $-18.2 + 9.95j$, $20.7\angle 151.3^\circ$
33. $-89.1 - 2750j$; $2750\angle 268.1^\circ$ **35.** $6.43e^{0.952j}$
37. $3.91e^{0.285j}$ ohms; $3.91\ \Omega$ **39.** $3.17 \times 10^{-4}\, e^{0.478j}\ 1/\Omega$

Exercises 12.6, page 363

1. $5.09(\cos 258.7^\circ + j \sin 258.7^\circ)$
3. $613(\cos 281.5^\circ + j \sin 281.5^\circ)$ **5.** $8(\cos 80^\circ + j \sin 80^\circ)$
7. $3(\cos 250^\circ + j \sin 250^\circ)$ **9.** $2(\cos 35^\circ + j \sin 35^\circ)$
11. $2.4\angle 170^\circ$ **13.** $0.008(\cos 105^\circ + j \sin 105^\circ)$
15. $256(\cos 0^\circ + j \sin 0^\circ)$ **17.** $0.8\angle 273^\circ$ **19.** $5\angle 87^\circ$
21. $1.73\angle 79.8^\circ$ **23.** $11{,}750\angle 115.91^\circ$
25. $65.0(\cos 345.7^\circ + j \sin 345.7^\circ) = 63 - 16j$
27. $61.4(\cos 343.9^\circ + j \sin 343.9^\circ)$; $59 - 17j$
29. $2.21(\cos 71.6^\circ + j \sin 71.6^\circ)$; $\frac{7}{10} + \frac{21}{10}j$
31. $3.85(\cos 120.5^\circ + j \sin 120.5^\circ) = \frac{10}{169}(-33 + 56j)$
33. $625(\cos 212.5^\circ + j \sin 212.5^\circ) = -527 - 336j$
35. $2(\cos 30^\circ + j \sin 30^\circ)$; $2(\cos 210^\circ + j \sin 210^\circ)$
37. $-0.364 + 1.67j$, $-1.26 - 1.15j$, $1.63 - 0.520j$
39. $1.10 + 0.455j$, $-1.10 - 0.455j$ **41.** $1, -1, j, -j$
43. $3j$, $-\frac{3}{2}(\sqrt{3} + j)$, $\frac{3}{2}(\sqrt{3} - j)$
45. $1.62 + 1.18j$, $-0.618 + 1.90j$, -2, $-0.618 - 0.190j$, $1.62 - 1.18j$
47. $-5, \frac{5}{2} + j\frac{5\sqrt{3}}{2}, \frac{5}{2} - j\frac{5\sqrt{3}}{2}$
49. $[\frac{1}{2}(1 - j\sqrt{3})]^3 = \frac{1}{8}[1 - 3(j\sqrt{3}) + 3(j\sqrt{3})^2$
$-(j\sqrt{3})^3] = \frac{1}{8}[1 - 3j\sqrt{3} - 9 + 3j\sqrt{3}]$
$= \frac{1}{8}(-8) = -1$
51. $-1, \frac{1}{2} + j\frac{\sqrt{3}}{2}, \frac{1}{2} - j\frac{\sqrt{3}}{2}$ **53.** $p = 0.479\angle 40.5^\circ$ watts
55. $43.3\angle 165^\circ$, 43.3 V

Exercises 12.7, page 369

1. $V_R = 24.0$ V, $V_L = 32.0$ V, $V_{RL} = 40.0$ V, $\theta = 53.1^\circ$ (voltage leads current) **3.** 12.9 V **5.** (a) 2850 Ω (b) 37.9° (c) 16.4 V **7.** (a) 14.6 Ω (b) −90.0°
9. (a) 47.8 Ω (b) 19.8°

11. 38.0 V **13.** 54.5 Ω, −62.3° **15.** 0.682 H
17. 2.08×10^5 Hz **19.** 1.30×10^{-11} F = 13.0 pF
21. 1.02 mW **23.** $21.4 - 33.9j$

Review Exercises for Chapter 12, page 371

Concept Check Exercises

1. F **2.** T **3.** T **4.** T **5.** T **6.** F

Practice and Applications

7. $10 - j$ **9.** $6 + 2j$ **11.** $25 + 30j$ **13.** $33 + 6j$
15. $\frac{1}{85}(21 + 18j)$ **17.** $-2 - 3j$ **19.** $\frac{1}{10}(-12 + 9j)$
21. $x = -3, y = -2$ **23.** $x = \frac{10}{13}, y = \frac{11}{13}$
25. $3 + 11j$ **27.** $4 + 8j$

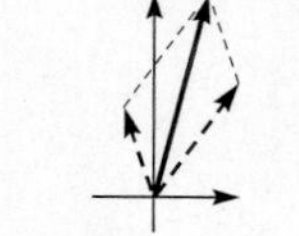

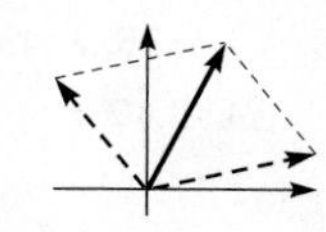

29. $1.41(\cos 315° + j \sin 315°) = 1.41e^{5.50j}$
31. $80.1(\cos 254.1° + j \sin 254.1°) = 80.1e^{4.43j}$
33. $4.67(\cos 76.8° + j \sin 76.8°); 4.67e^{1.34j}$
35. $5000\angle 0°$, $5000e^{0j}$ **37.** $-1.41 - 1.41j$
39. $-2.789 + 4.162j$ **41.** $0.19 - 0.59j$
43. $26.31 - 6.427j$ **45.** $1.94 + 0.495j$
47. $-728.1 + 1017j$ **49.** $15(\cos 84° + j \sin 84°)$
51. $20\angle 263°$ **53.** $\frac{8}{9}(\cos 0° + j \sin 0°) = \frac{8}{9}$
55. $14.29\angle 133.61°$ **57.** $1.26\angle 59.7°$
59. $9682\angle 249.52°$ **61.** $1024(\cos 160° + j \sin 160°)$
63. $343\angle 331.5°$ **65.** $32(\cos 270° + j \sin 270°) = -32j$
67. $\frac{625}{2}(\cos 270° + j \sin 270°) = -\frac{625}{2}j$
69. $1 + j\sqrt{3}, -2, 1 - j\sqrt{3}$
71. $0.383 + 0.924j, -0.924 + 0.383j, -0.383 - 0.924j,$ $0.924 - 0.383j$ **73.** $40 + 9j, 41(\cos 12.7° + j \sin 12.7°)$
75. $-15.0 - 10.9j, 18.5(\cos 216.0° + j \sin 216.0°)$
77. $15 - 16j$ **79.** $4 + 5j, 4 - 5j$
81. $1 - j$, Yes; $-1 - j$, No **83.** $\frac{1}{2}$ **85.** $2 + \frac{9}{2}j$
87. $5.83e^{5.74j}$ **89.** 60 V **91.** −21.6° **93.** 22.9 Hz
95. $5500(\cos 53.1° + j \sin 53.1°)$ N **97.** $\frac{u - j\omega n}{u^2 + \omega^2 n^2}$
99. $e^{j\pi} = \cos \pi + j \sin \pi = -1$

Exercises 13.1, page 375

1. $-\frac{1}{4}$ **3.** 11.7 **5.** 0.00677 **7.** (a) Yes (b) Yes
9. (a) No (base cannot be negative) (b) Yes **11.** 2
13. $\frac{1}{16}$ **15.** $\frac{1}{8}$

17.

19.

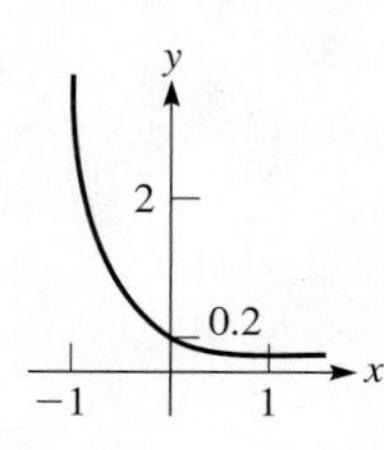

21.

23.

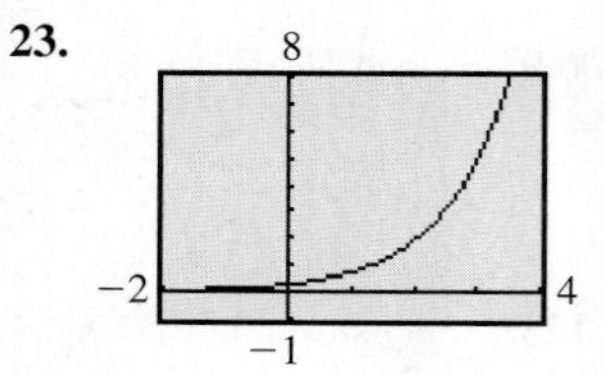

25.

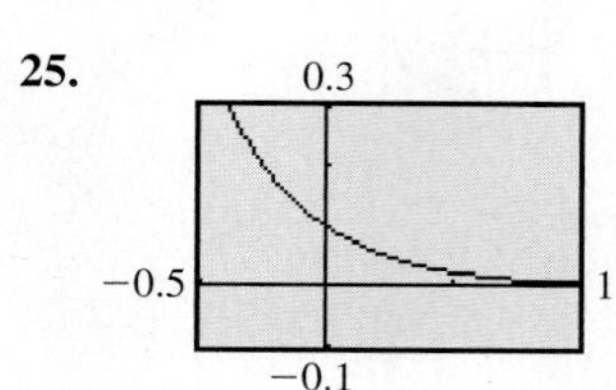

27.

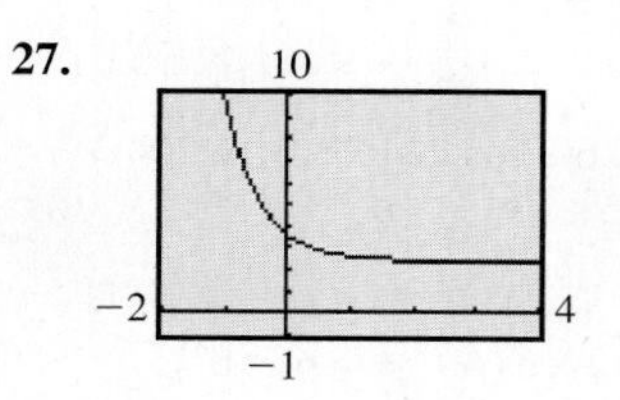

29.

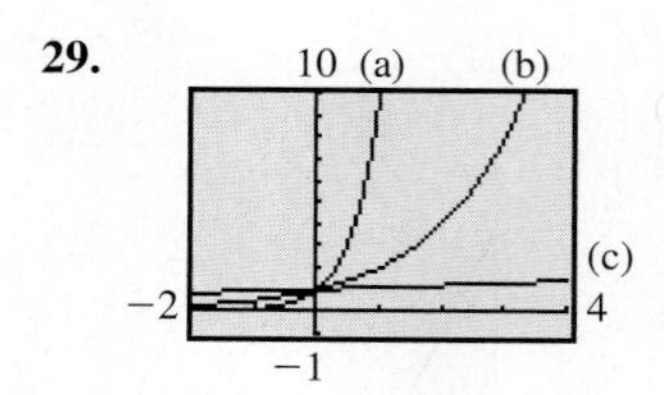

31. 4 **33.** $f(c + d) = b^{c+d} = b^c b^d = f(c) \cdot f(d)$
35. **37.** −0.7667, 2, 4 **39.** $e^k - 1$

41. $303.88 **43.** 0.0012 mA **45.**

Exercises 13.2, page 379

1. $32^{4/5} = 16$ in logarithmic form is $\frac{4}{5} = \log_{32} 16$.
3. If $x = 16$, $y = \log_4 16$ means $y = 2$, since $4^2 = 16$. If $x = \frac{1}{16}$, $y = \log_4(\frac{1}{16})$ means $y = -2$, since $4^{-2} = \frac{1}{16}$.
5. $\log_3 81 = 4$ **7.** $\log_7 343 = 3$ **9.** $\log_4(\frac{1}{16}) = -2$
11. $\log_2(\frac{1}{64}) = -6$ **13.** $\log_8 2 = \frac{1}{3}$ **15.** $\log_{1/4}(\frac{1}{16}) = 2$
17. $32 = 2^5$ **19.** $9 = 9^1$ **21.** $5 = 25^{1/2}$ **23.** $3 = 243^{1/5}$
25. $0.01 = 10^{-2}$ **27.** $16 = (0.5)^{-4}$ **29.** 2 **31.** −2
33. 343 **35.** $\frac{9}{4}$ **37.** $\sqrt{3}$ **39.** $\frac{1}{64}$ **41.** 0.2 **43.** −4

45.

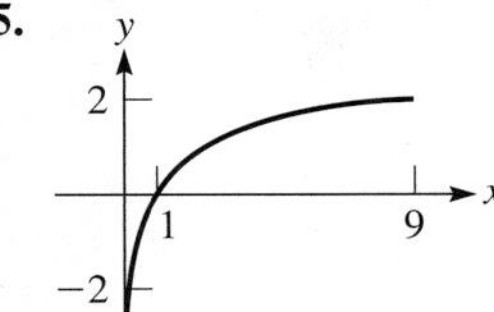

47.

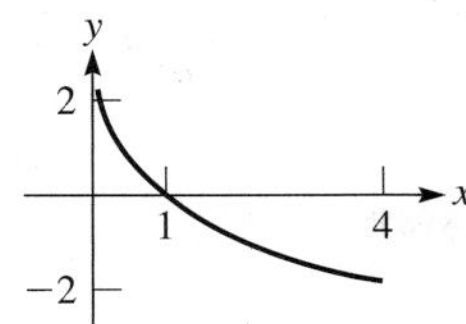

49.

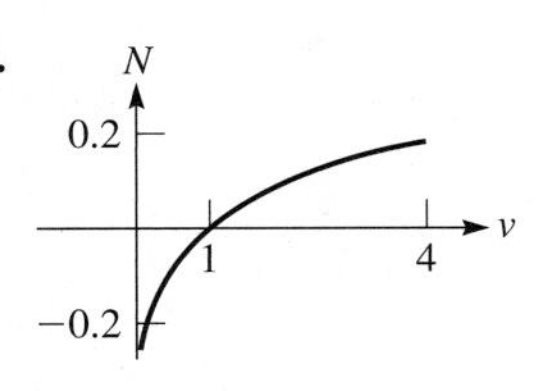

51.

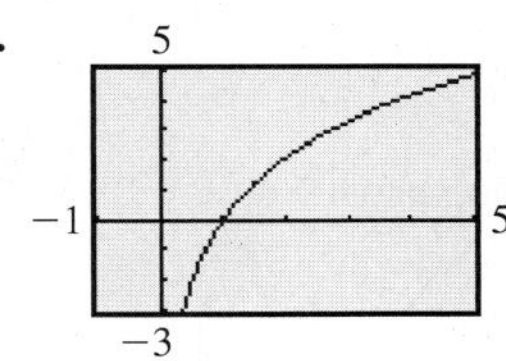

53.

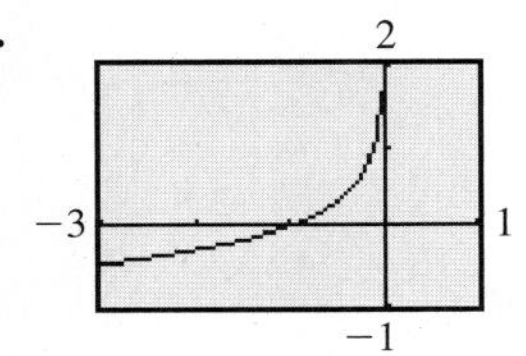

55. (a) -3
(b) No value (x cannot be negative)

57. (a) $\frac{1}{2}$
(b) Not defined

59. $\sqrt{2}$

61.

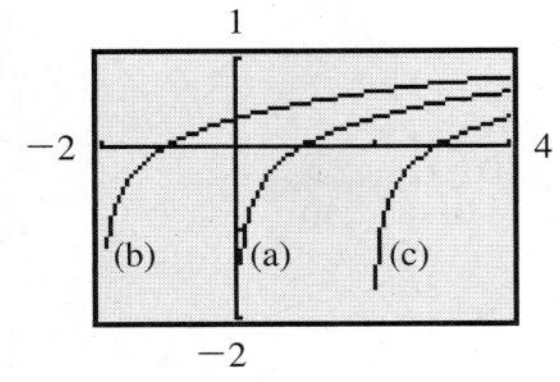

63. 0.0102, 2.376

65. $x < 2$

67. $b_2 = b_1 10^{0.4(m_1 - m_2)}$

69. $N = N_0 e^{-kt}$

71. $V_1 = V_2 e^{-W/k}$

73.

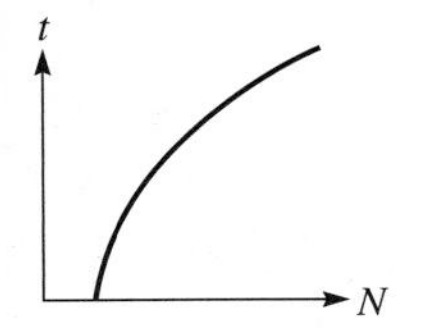

75. -0.132

77.

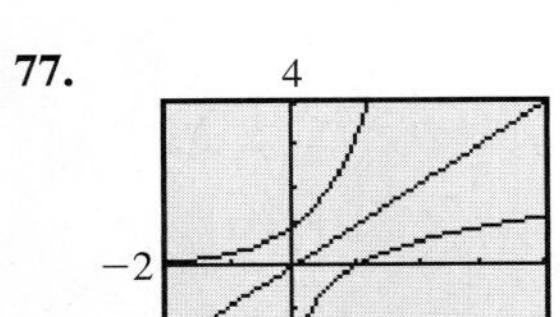

79.

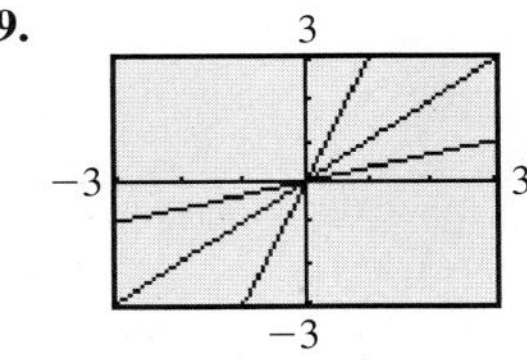

Exercises 13.3, page 384

1. $\log_4 3 + \log_4 7$ **3.** $\log_4\left(\frac{5}{x}\right)$ **5.** 3 **7.** $\frac{1}{2}$

9. $\log_5 3 + \log_5 11$ **11.** $\log_7 9 - \log_7 2$ **13.** $3 \log_2 a$

15. $\log_6 a + \log_6 b + 2 \log_6 c$ **17.** $5 \log_5 t$

19. $\frac{1}{2} \log_2 x - 2 \log_2 a$ **21.** $\log_b ac$ **23.** $\log_5 3$

25. $\log_b x^{5/2}$ **27.** $\log_e \frac{4\pi^3}{3}$ **29.** -5 **31.** 2.5 **33.** 3

35. 8 **37.** $2 + \log_3 2$ **39.** $2 - \log_2 7$ **41.** $\frac{1}{2}(1 + \log_3 2)$

43. $3 + \log_{10} 3$ **45.** $y = 2x$ **47.** $y = \frac{3x}{5}$ **49.** $y = \frac{49}{x^3}$

51. $y = 2(2ax)^{1/5}$ **53.** $y = \frac{2}{x}$ **55.** $y = \frac{x^2}{25}$

57. $\log_{10} x + \log_{10} 3 = \log_{10} 3x$

59. $y = \log_e(e^2 x) = 2 \log_e e + \log_e x = 2 + \log_e x$

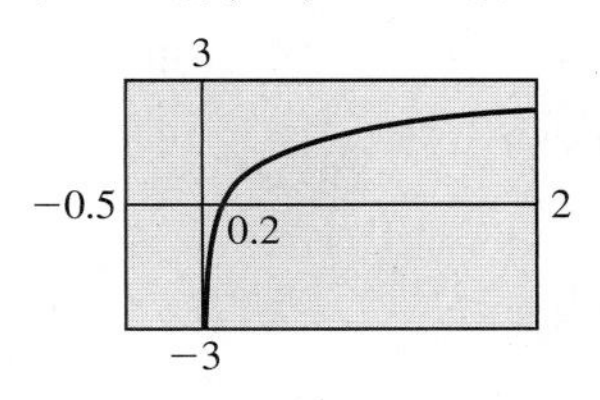

61. 8 **63.** $2x$

65. $y_1 = y_2$

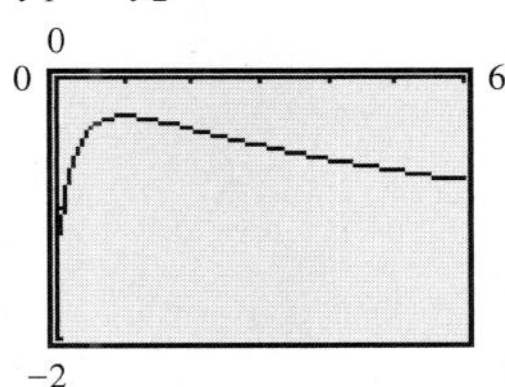

67. $D = ae^{cr^2 - br}$ **69.** $T = 90.0e^{-0.23t}$

Exercises 13.4, page 387

1. -0.4372 **3.** 2.444 **5.** 6.966 **7.** -0.2787

9. -0.149 **11.** 1.219 **13.** 18.1 **15.** 0.06090

17. 2000.4 **19.** 0.0057882 **21.** 5.21×10^{226}

23. 2.32×10^{-120} **25.** $1.1461 - 0.3010 = 0.8451$

27. $1.9085 = 1.9085$ **29.** 8.9542 **31.** -13.89

33. 10^8 K **35.** 0.45 **37.** $\log_b x$ **39.** 0.23

41. 9.2 **43.** $2^{400} = 2.58 \times 10^{120}$

Exercises 13.5, page 390

1. 5.298 **3.** 3.258 **5.** 0.4460 **7.** -4.91623

9. 2.03 **11.** 4.806 **13.** 1.795 **15.** 4.332 **17.** 0.3322

19. -0.1136 **21.** -17.39066 **23.** 8.94 **25.** 1.0085

27. 0.4757 **29.** 6.20×10^{-11}

31.

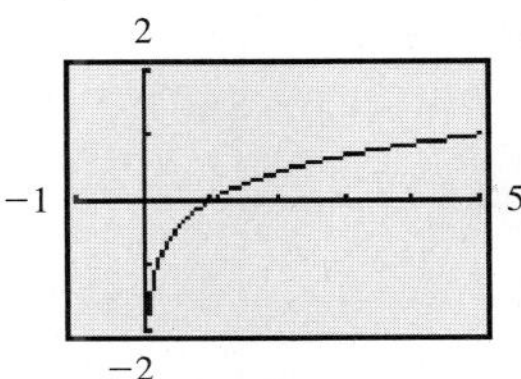

33.

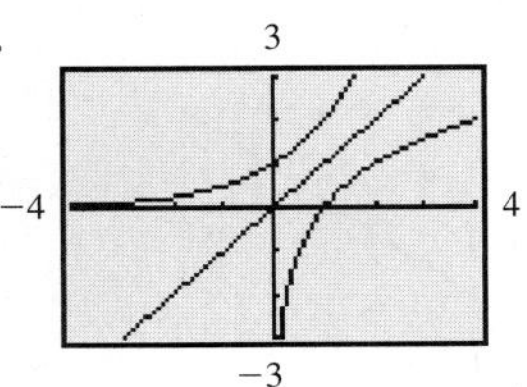

35. $1.6094 + 2.0794 = 3.6889$ **37.** $4.3944 = 4.3944$ **39.** 3

41. 10 **43.** $\ln 80 = 2x + y$ **45.** $2 + \frac{1}{2}\ln(1 - x)$

47. 2.45×10^9 Hz **49.** 8.155% **51.** 0.384 s **53.** 21.7 s

Exercises 13.6, page 394

1. -0.535 **3.** 4 **5.** 3.53 **7.** 0.587 **9.** 0.479

11. 1.85 **13.** $0, \frac{\ln 0.6}{\ln 2} = -0.737$ **15.** 2, 4 **17.** $\frac{1}{4}$

19. 17 **21.** 3 **23.** -0.162 **25.** 250 **27.** 4

29. 2 **31.** 1.42 **33.** -0.1042 **35.** 0.2031

37. 0.9739 **39.** 10.87 **41.** 1.585 **43.** $\frac{1}{125}$

45. ± 0.9624 **47.** 0.0025 **49.** -1.21 **51.** 2030

53. Noon of previous day (36 h earlier) **55.** 1.72×10^{-5}

57. $10^{8.3} = 2.0 \times 10^{8}$

59. $c = 15e^{-0.20t}$

61. $P = P_0(0.999)^t$

63. $x = 3.353$ **65.** ± 3.7 m

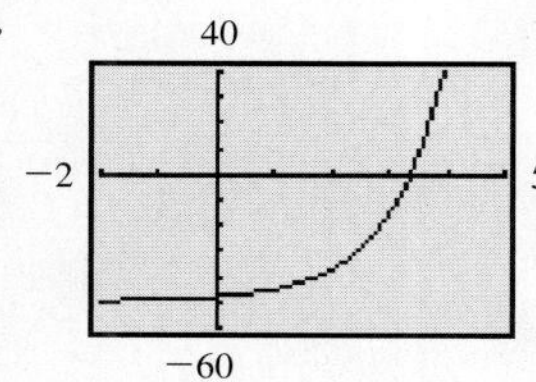

Exercises 13.7, page 398

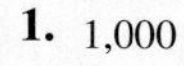

1.

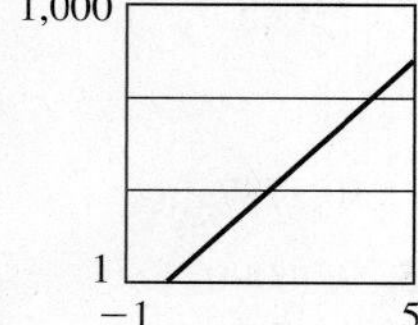

3.

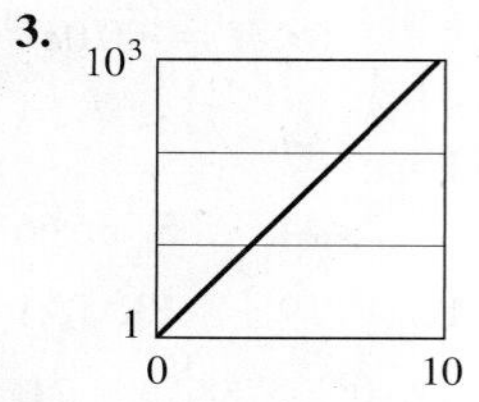

5.

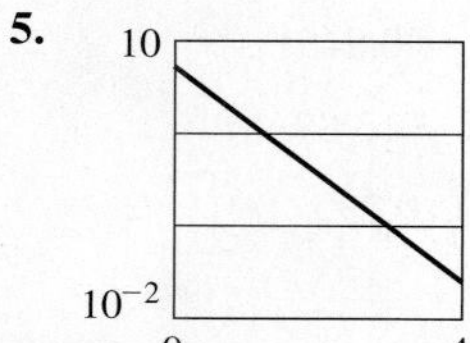

7.

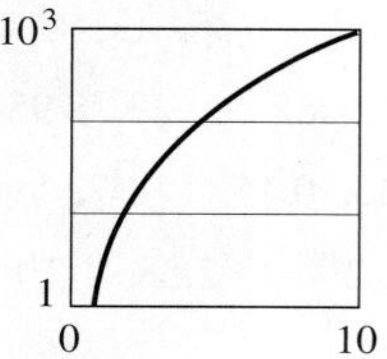

9.

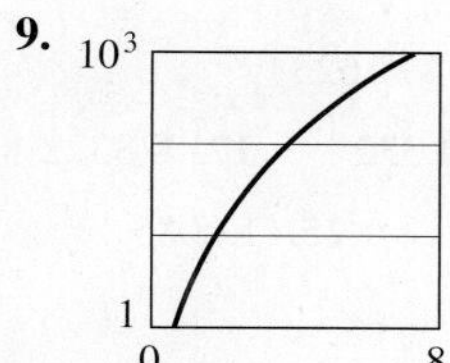

11.

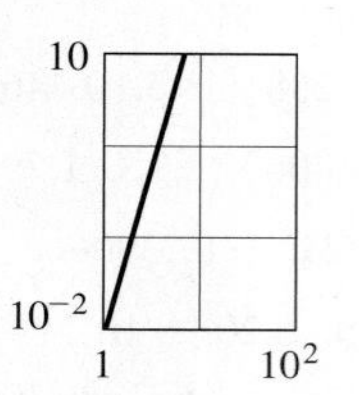

13.

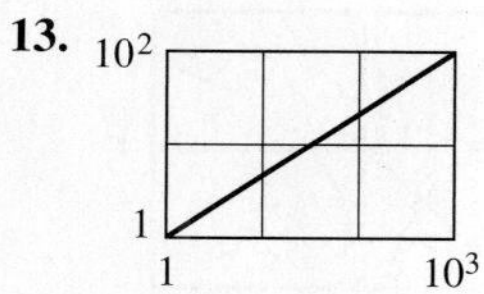

15.

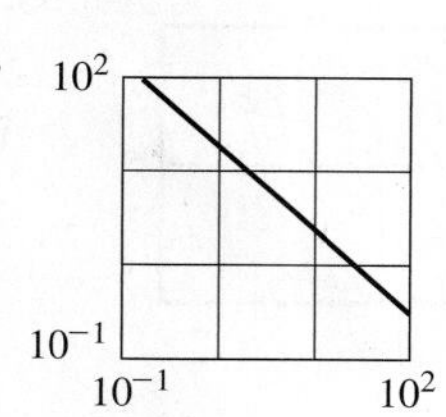

17.

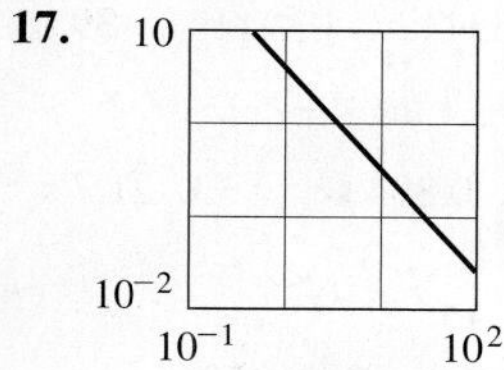

19. Semilog paper

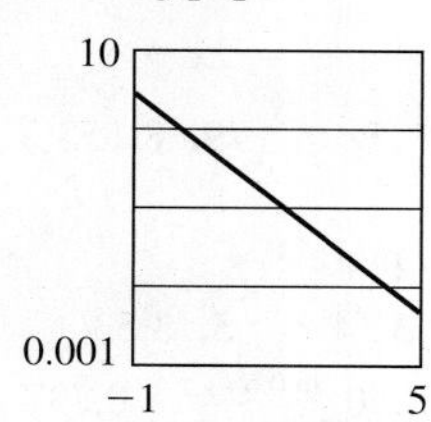

21. Log-log paper

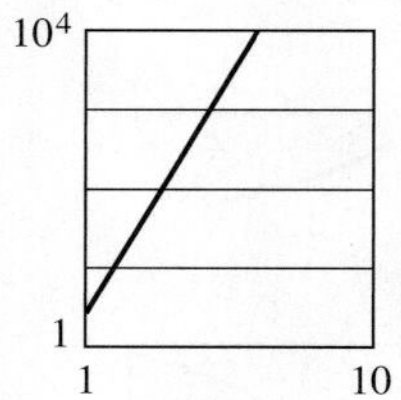

23. Semilog paper

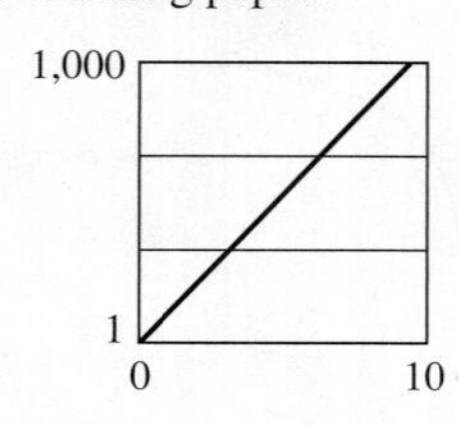

25. Log-log paper

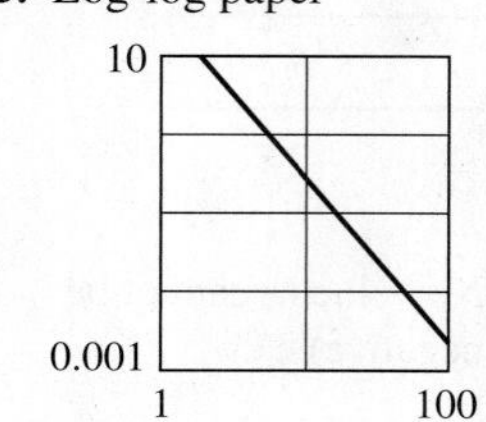

27. (a)

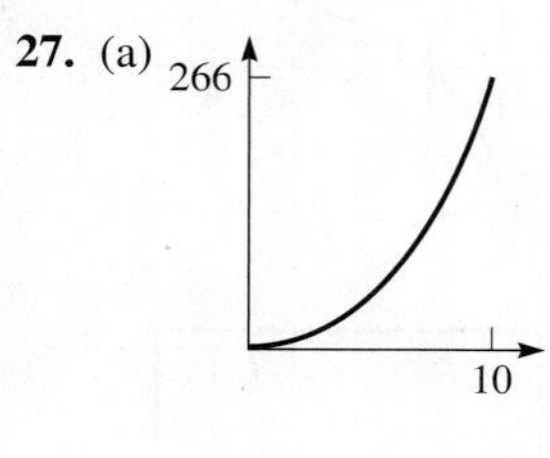

27. (b)

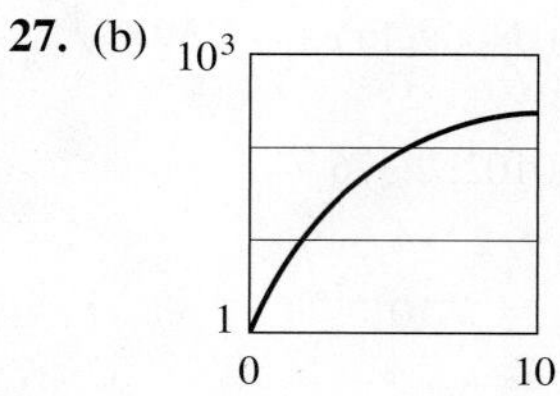

29.

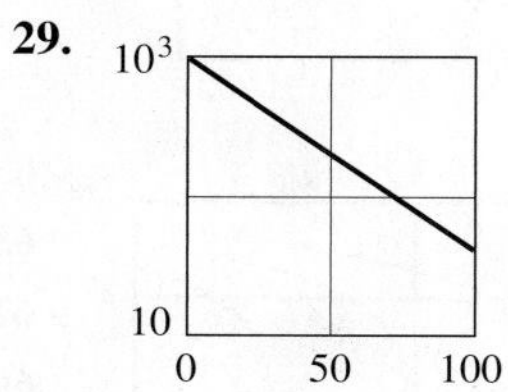

31.

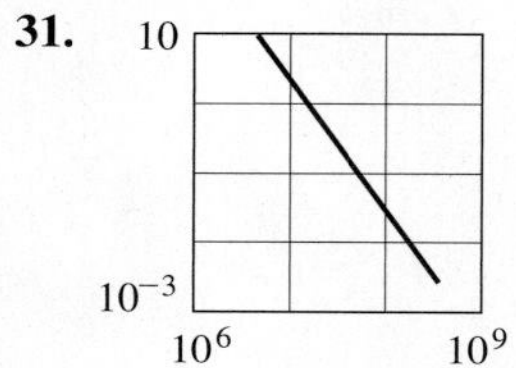

33.

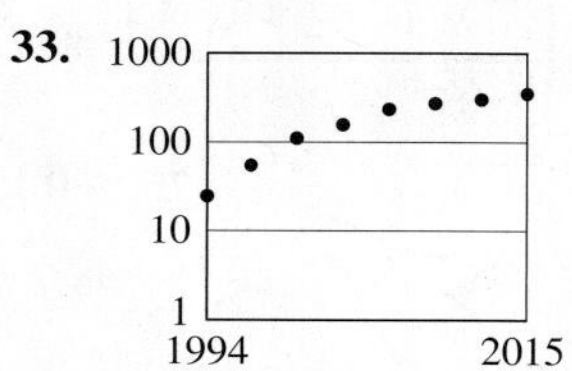

35.

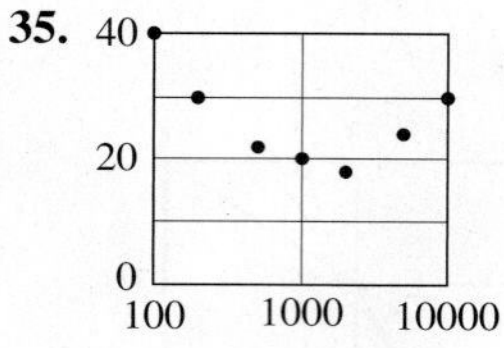

37.

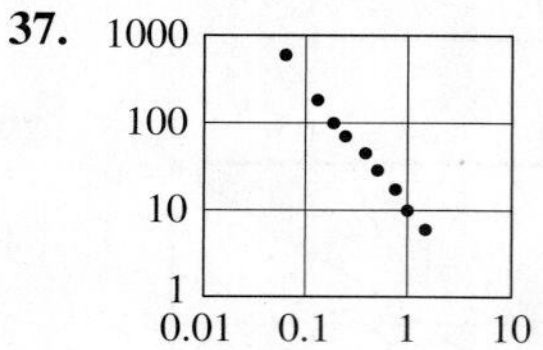

39.

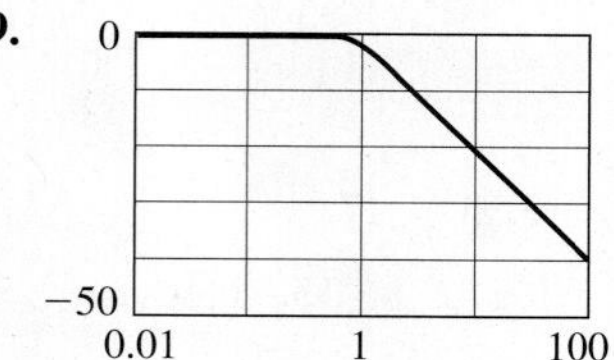

Review Exercises for Chapter 13, page 400

Concept Check Exercises

1. T **2.** F **3.** F **4.** T **5.** F **6.** T

Practice and Applications

7. 10,000 **9.** $\frac{1}{5}$ **11.** -6 **13.** $\frac{2}{3}$ **15.** 6 **17.** 1000

19. $1 + \log_3 2 + \log_3 x$ **21.** $2 \log_3 t$ **23.** $3 + \log_2 7$

25. $1 + \frac{1}{2}\log_4 3$ **27.** $2 - \log_3 x$ **29.** $3 + 4\log_{10} x$

31. $y = \frac{4}{x}$ **33.** $y = e^{2/3}x$ **35.** $y = \sqrt{7x}$ **37.** $y = \frac{15}{x}$

39. $y = 8x^3$ **41.** $y = x^3$

43.

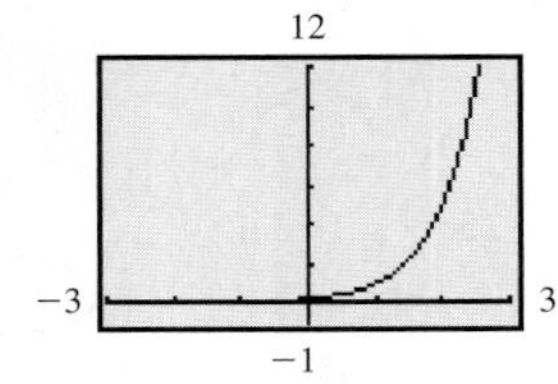

45.

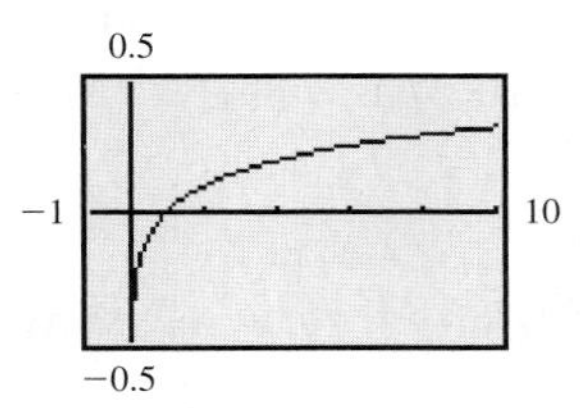

47.

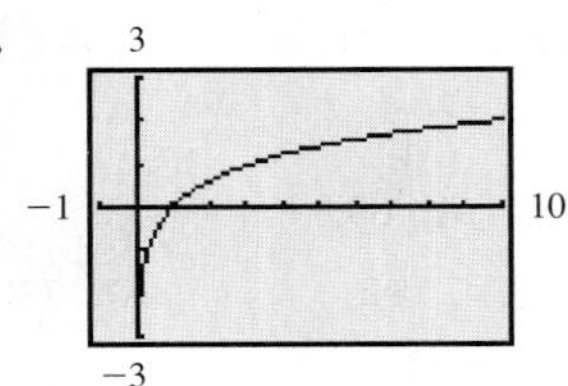

49.

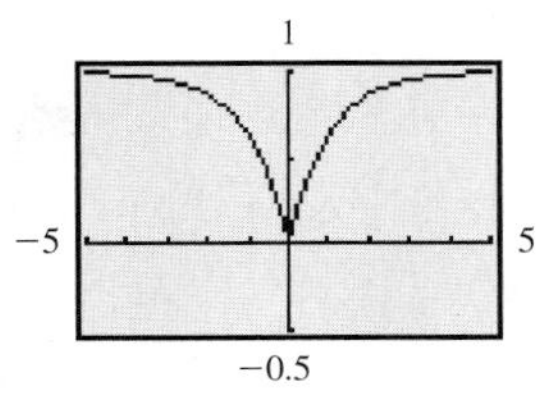

51. −1.084 **53.** 4.476 **55.** 5.814 **57.** −3.46

59. 0.805 **61.** 4.30 **63.** 2 **65.** −0.2

67.

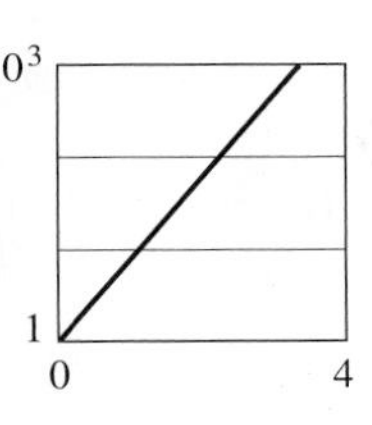

69.

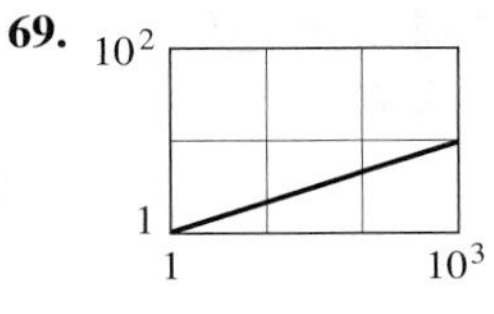

71. 4 **73.** 12 **75.** 77.1 **77.** 0 **79.** 2

81. 0.706 **83.** 1.28×10^{-5} or 3.925 **85.** 4.593×10^6

87. (a) $n = \frac{\log(A/A_0)}{\log(1 + r)}$ (b) 17.7 years **89.** $I = I_0e^{-\beta h}$

91. 5.4h **93.** 2040 **95.**

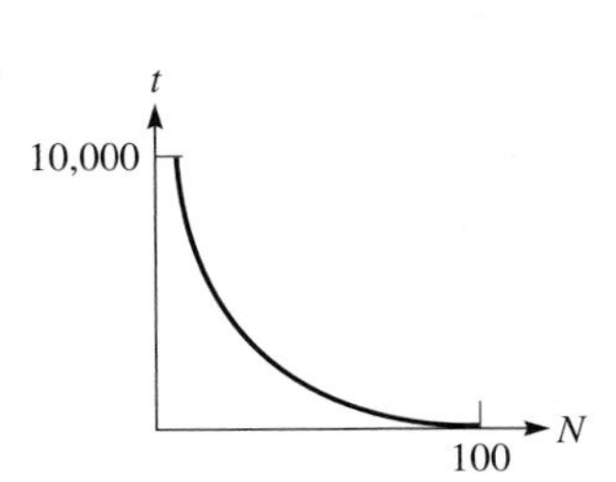

97. $\sin\theta = \frac{l\omega^2}{3g}$ **99.** $R = 2^{C/B} - 1$ **101.** 910 times brighter

103. 1.17 min **105.** 0.813 cm **107.** $-\frac{1}{k}\ln\left(\frac{T - T_1}{T_0 - T_1}\right)$

109. $n = 20e^{-0.04t}$ **111.**

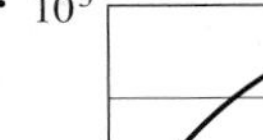

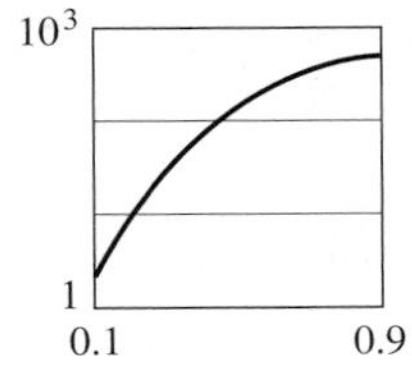

Exercises 14.1, page 406

1. $y = 3x^2 + 6x$ **3.** $x = -1.2, y = 0.3; x = 0.8, y = 2.3$

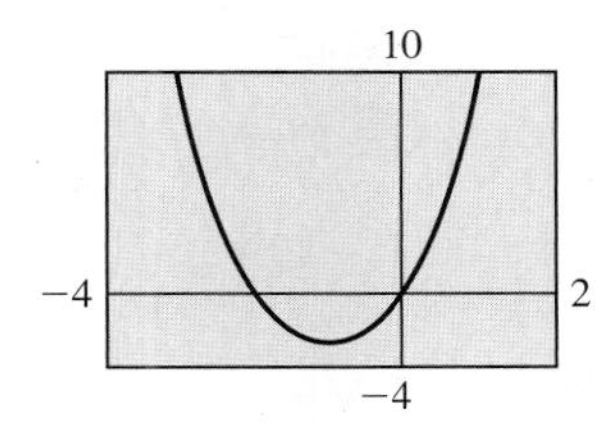

5. $x = 3.6, y = 1.8; x = -3.6, y = -1.8$

7. $x = 0.0, y = -2.0; x = 2.7, y = -0.7$

9. $x = 1.5, y = 0.2$

11. $x = 2.2, y = 9.1$

13. $x = -2.7, y = 4.2; x = 2.7, y = 4.2$

15. No real solutions

17. $x = -2.8, y = -1.0; x = 2.8, y = 1.0;$
$x = 2.8, y = -1.0; x = -2.8, y = 1.0$

19. $x = -0.8, y = 2.5; x = 2.6, y = 0.7$

21. $x = 0.0, y = 0.0; x = 1.9, y = 0.9$

23. $x = 0.5, y = 0.6$

25. $x = 4.0, y = 3.0$

27. $x = 3.6, y = 1.0$

29. $x = -2.1, y = 4.2; x = 1.1, y = 1.1$

31.

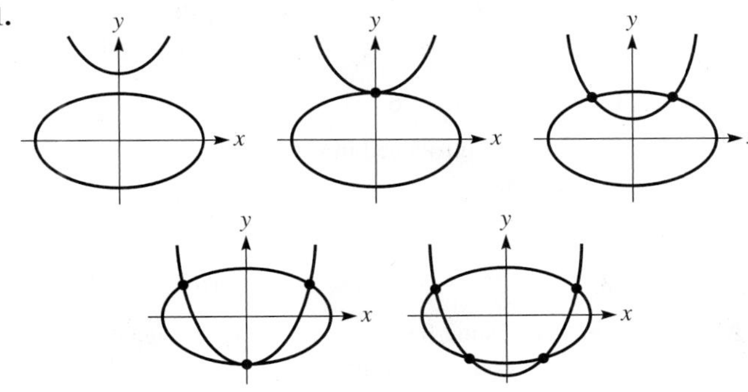

33. 2.0 h, 3.0 h; 2.4 h, 2.6 h **35.** 2.2 A, 0.9 A **37.** No

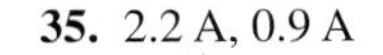

Exercises 14.2, page 409

1. $x = 2, y = 0; x = \frac{10}{3}, y = -\frac{8}{3}$

3. $x = -\sqrt{6}, y = -\sqrt{3}; x = -\sqrt{6}, y = \sqrt{3};$
$x = \sqrt{6}, y = -\sqrt{3}; x = \sqrt{6}, y = \sqrt{3}$

5. $x = -2, y = 5; x = 4, y = 17$

7. $x = -\frac{19}{5}, y = \frac{17}{5}; x = 5, y = -1$ **9.** $x = 1, y = 0$

11. $x = \frac{2}{7}(3 + \sqrt{2}), y = \frac{2}{7}(-1 + 2\sqrt{2});$
$x = \frac{2}{7}(3 - \sqrt{2}), y = \frac{2}{7}(-1 - 2\sqrt{2})$

13. $w = 3, h = 3$ **15.** $x = -\frac{3}{2}, y = -\frac{8}{3}; x = 2, y = 2$

17. $x = -5, y = 25; x = 5, y = 25$

19. $x = 1, y = 3; x = -1, y = 3$

21. $D = 1, R = 0; D = -1, R = 0; D = \frac{1}{2}\sqrt{6}, R = \frac{1}{2};$
$D = -\frac{1}{2}\sqrt{6}, R = \frac{1}{2}$

23. $x = \sqrt{19}, y = \sqrt{6}; x = \sqrt{19}, y = -\sqrt{6};$
$x = -\sqrt{19}, y = \sqrt{6}; x = -\sqrt{19}, y = -\sqrt{6}$

25. $x = -5, y = -2; x = -5, y = 2; x = 5, y = -2; x = 5, y = 2$

27. $x = -3, y = 2; x = -1, y = -2$ **29.** $x = a, y = b$

31. $x = 50$ mi, $h = 25$ mi **33.** 2.00 cm, 4.00 cm

35. 16.6 m, 13.0 cm **37.** 2.19 m, 0.21 m **39.** 12, 10

41. 12 in., 18 in. **43.** 3720 cm^2 **45.** 80 mi/h

Exercises 14.3, page 413

1. $\pm\frac{1}{2}j\sqrt{2}, \pm 2$ **3.** $-3, -1, 1, 3$ **5.** $-\frac{3}{2}, \frac{1}{3}$

7. $-\frac{1}{2}, \frac{1}{2}, \frac{1}{6}j\sqrt{6}, -\frac{1}{6}j\sqrt{6}$ **9.** $1, \frac{9}{4}$ **11.** $\frac{64}{729}, 1$

13. $-27, 125$ **15.** 81 **17.** 26 **19.** $-2, -1, 3, 4$

21. 18 **23.** $\frac{3}{4}, \frac{4}{3}$ **25.** $1, -1, j\sqrt{2}, -j\sqrt{2}$ **27.** 0, ln 2

29. $\pm 2, \pm 4$ **31.** 1, 4 **33.** 10, 100 **35.** $-2, -1, 1, 2$

37. $R_1 = 2.62\ \Omega, R_2 = 1.62\ \Omega$ **39.** 0.610

41. 52.3 in., 29.5 in.

Exercises 14.4, page 417

1. $\frac{13}{12}$ **3.** 10 **5.** 12 **7.** 3 **9.** $\frac{2}{3}$ **11.** -1

13. 15 **15.** 4, 9 **17.** ± 6

19. 12 (Extraneous root introduced in squaring both sides of $\sqrt{x+4} = x - 8$.)

21. 16 **23.** $0, \frac{1}{9}$ **25.** $-1, 7$ **27.** 0 **29.** 5

31. 25 **33.** 8 **35.** 4 **37.** 4 **39.** 16

41. $x = 2$; For $x = 5$, squaring $3 - x$ is squaring a negative number. A different extraneous root is introduced.

43. $-r \pm \sqrt{r^2 + R^2}$ **45.** $r_1^2 = (kC + A - \sqrt{R_1^2 - R_2^2})^2 + r_2^2$

47. \$228,000 **49.** 9.2 km **51.** 1.6 mi

Review Exercises for Chapter 14, page 418

Concept Check Exercises

1. F **2.** F **3.** F **4.** F

Practice and Applications

5. $x = -1.3, y = 5.2; x = 1.1, y = 3.9$

7. $x = 2.0, y = 0.0; x = 1.6, y = 0.6$

9. $x = -0.7, y = 1.4; x = 0.7, y = 1.4$

11. $x = -2, y = 7; x = 2, y = 7$

13. $x = 0.0, y = 0.0; x = 3.3, y = 4.3$

15. $x = 0, y = 0; x = 2, y = 16$

17. $L = \sqrt{3}, R = 1; L = -\sqrt{3}, R = 1$

19. $u = \frac{1}{12}(1 + \sqrt{97}), v = \frac{1}{18}(5 - \sqrt{97});$
$u = \frac{1}{12}(1 - \sqrt{97}), v = \frac{1}{18}(5 + \sqrt{97})$

21. $x = 7, y = 5; x = 7, y = -5; x = -7, y = 5;$
$x = -7, y = -5$

23. $x = -2, y = 2; x = \frac{2}{3}, y = \frac{10}{9}$ **25.** $-4, -2, 2, 4$

27. 1, 16 **29.** $\frac{1}{3}, -\frac{1}{7}$ **31.** 1, 1.65 **33.** $-\frac{3}{2}, 2$ **35.** $\frac{1}{4}, 49$

37. $\sqrt{3}, -\sqrt{3}, \frac{1}{2}j\sqrt{3}, -\frac{1}{2}j\sqrt{3}$ **39.** 6 **41.** 8 **43.** $\frac{9}{16}$

45. $\frac{1}{3}(11 - 4\sqrt{15})$ **47.** 2 **49.** 4 **51.** 5 **53.** $-1, 2$

55. $x = a + 1, y = a$ **57.** 25

59. $x = -2, y = -3; x = 3, y = 2$ **61.** 69.3 cm

63. $m = \frac{1}{2}(-y \pm \sqrt{2s^2 - y^2})$ **65.** 0.75 s, 1.5 s **67.** 230 ft

69. 20 in., 26 in., 26 in., or 27.6 in., 22.2 in., 22.2 in.

71. 3.0 mm, 1.0 mm **73.** 34 mm, 52 mm

75. 0.353 cm, 6.29 cm **77.** 26 mi/h, 32 mi/h

Exercises 15.1, page 425

1. -25 **3.** Coefficients 1 0 0 -4 11, $R = -26$ **5.** 4

7. 33 **9.** -43 **11.** 30 **13.** 16 **15.** 38 **17.** Yes

19. No **21.** Yes **23.** $x^2 + 3x + 2, R = 0$

25. $x^2 - 2x + 5, R = -16$

27. $p^5 + 2p^4 + 4p^3 + 2p^2 + 2p + 4, R = 2$

29. $x^6 + 2x^5 + 4x^4 + 8x^3 + 16x^2 + 32x + 64, R = 0$

31. $x^3 + x^2 + 2x + 1, R = 0$ **33.** No **35.** No **37.** Yes

39. No **41.** Yes **43.** Yes **45.** $2x^2 - 5x + 1 - \frac{8}{x+4}$

47. $(4x^3 + 8x^2 - x - 2) \div (2x - 1) = 2x^2 + 5x + 2$; no, because the coefficient of x in $2x - 1$ is 2, not 1.

49. -7 **51.** $x^2 - (3 + j)x + 3j, R = 0$

53. Yes, because both functions cross the x-axis at the same values.

55. $3x^2 - 8ax + 5a^2$ **57.** (a) No (b) Yes **59.** Yes

Exercises 15.2, page 430

(Note: Unknown roots listed)

1. $-1, -1, 1, 1, 1$ **3.** $-3, -2, 2$ **5.** $5, 3j, -3j$ **7.** $-1, 4$

9. $-2, -2$ **11.** $-j, \frac{2}{5}$ **13.** $3, 2 - j$ **15.** $-1, \frac{1}{2}$

17. $-2, 1$ **19.** $1 - j, -2, \frac{1}{2}$ **21.** $j, -j$ **23.** $-j, -1, 1$

25. $-2j, -3, 3$

27. If j is a root, $-j$ must also be a root, and $-j$ is not one of the roots.

29. -7 **31.** $-1, \frac{1}{2}(1 \pm \sqrt{13})$

Exercises 15.3, page 435

1. No more than two positive roots and three negative roots

3. $-4, -2, 1$ **5.** $-3, -1, 2$

7. $0.1, \frac{1}{6}(-3 + j\sqrt{15}), \frac{1}{6}(-3 - j\sqrt{15})$ **9.** $-1, \frac{1}{2}, 2$

11. $-2, -2, 2 \pm \sqrt{3}$ **13.** $-2, \frac{2}{5}, 1 + j\sqrt{3}, 1 - j\sqrt{3}$

15. $-3, -\frac{2}{3}, -\frac{1}{2}, \frac{1}{2}$ **17.** $-3, -1, -1, 2, 2$

19. $2, 2, 2, \frac{1}{2}j, -\frac{1}{2}j$

21. $-2.17, 0.39, 1.78$ **23.** $-3.01, -1.49, -0.33, 0.33$

25. $x = -2, y = -28; x = 2 + \sqrt{3}, y = 20 + 12\sqrt{3}$: $x = 2 - \sqrt{3}, y = 20 - 12\sqrt{3}$

27. $-\frac{3}{4}, \sqrt{5}, -\sqrt{5}$

29. $x = \frac{1}{2}, x = 8$ **31.** 1.8 s, 4.5 s **33.** $0, L$ **35.** 19.5%

37. 1.23 cm or 2.14 cm **39.** 3.0 mm, 4.0 mm; 5.0 mm, 6.0 mm

41. $(4, -7)$

43. Two positive, one negative; zero positive, one negative, two nonreal complex

Review Exercises for Chapter 15, page 436

Concept Check Exercises

1. F **2.** F **3.** F **4.** F

Practice and Applications

5. 4 **7.** -80 **9.** Yes **11.** No

13. $x^2 + 5x + 10, R = 11$ **15.** $2x^2 - 7x + 10, R = -17$

17. $x^3 - 3x^2 - 2x + 10, R = -4$

19. $2m^4 + 10m^3 + 2m^2 + 11m + 55, R = 266$ **21.** Yes

23. No, Yes **25.** (unlisted roots) $-2, 1$

27. (unlisted roots) 3, 4

29. (unlisted roots) $-1 + j\sqrt{2}, -1 - j\sqrt{2}$

31. (unlisted roots) $-j, -\frac{3}{2}, \frac{1}{2}$

33. (unlisted roots) $-2, 2$

35. (unlisted roots) $-2 - j, \frac{1}{2}(-1 \pm j\sqrt{3})$

37. $-4, 1, 2$ **39.** $-\frac{1}{2}, 1, 1$ **41.** $-1, -\frac{1}{2}, \frac{5}{3}$

43. $\frac{1}{2}, \frac{3}{2}, -1 + \sqrt{2}, -1 - \sqrt{2}$

45. $f(x) = (x + 2)(x + 1)(x - 4)(3x - 4)$ **47.** 1, 3, 5

49. 2 or 4 **51.** Use k in synthetic division, $k = 4$.

53. $x^3 - 5x^2 + x - 5$ **55.** $x = -1, y = -2; x = 2, y = 1$

57. 11.7 ft **59.** April **61.** 0.75 cm **63.** 2 in.

65. 5.1 m, 8.3 m **67.** $h = 6.75$ ft, $w = 3.60$ ft

69. $A = 6.16 \text{ m}^2$

Exercises 16.1, page 443

1. $\begin{bmatrix} 5 & 7 & -1 & 9 \\ 6 & 4 & 1 & 12 \end{bmatrix}$ **3.** $a = 1, b = -3, c = 4, d = 7$

5. $x = -2, y = 5, z = -9, r = 48, s = 4, t = -1$

7. $C = -3, D = 6, E = 2$

9. Elements cannot be equated; different number of rows.

11. $\begin{bmatrix} 5 & 10 \\ 0 & -6 \end{bmatrix}$ **13.** $\begin{bmatrix} -5 & 0 \\ -4 & 71 \\ 11 & -8 \end{bmatrix}$ **15.** $\begin{bmatrix} 9 & 9 \\ -5 & -11 \end{bmatrix}$

17. Cannot be added **19.** $\begin{bmatrix} -8 & -5 & -1 \\ 6 & -6 & 3 \end{bmatrix}$

21. $\begin{bmatrix} 9 & 1 & 8 \\ -8 & 12 & -14 \end{bmatrix}$ **23.** $\begin{bmatrix} 11 & -7 & 22 \\ -12 & 24 & -36 \end{bmatrix}$

25. $\begin{bmatrix} -15 & 21 \\ -21 & 9 \end{bmatrix}$ **27.** $\begin{bmatrix} 18 & -9 \\ 12 & -15 \end{bmatrix}$

29. $\begin{bmatrix} 20 & 31 & -25 \\ -10 & -6 & 35 \end{bmatrix}$ **31.** $\begin{bmatrix} 18 & -63 \\ 56 & -1 \end{bmatrix}$ **33.** $\begin{bmatrix} -42 & -60 \\ 38 & 56 \end{bmatrix}$

35. $A + B = B + A = \begin{bmatrix} 3 & 1 & 0 & 7 \\ 5 & -3 & -2 & 5 \\ 10 & 10 & 8 & 0 \end{bmatrix}$

37. $-(A - B) = B - A = \begin{bmatrix} 5 & -3 & -6 & -7 \\ 5 & 3 & 0 & -3 \\ -8 & 12 & 8 & 4 \end{bmatrix}$

39. $v_w = 31.0$ km/h, $v_p = 249$ km/h

41. $\begin{bmatrix} 288 & 225 & 0 & 0 \\ 186 & 132 & 72 & 0 \\ 0 & 105 & 204 & 234 \end{bmatrix}$

43. $B = \begin{bmatrix} 15 & 25 & 10 & 25 \\ 10 & 10 & 10 & 45 \end{bmatrix}, J = \begin{bmatrix} 15 & 30 & 3 & 3 \\ 0 & 100 & 2 & 2 \end{bmatrix}$

Exercises 16.2, page 447

1. $\begin{bmatrix} 5 & 6 & 7 & -10 \\ -9 & 0 & -3 & 12 \\ 1 & 12 & 11 & -8 \end{bmatrix}$ **3.** $\begin{bmatrix} -8 & -12 \end{bmatrix}$ **5.** $\begin{bmatrix} 435 \\ 320 \end{bmatrix}$

7. $\begin{bmatrix} -10 & \frac{63}{2} \\ -\frac{70}{3} & -31 \\ 36 & 34 \end{bmatrix}$ **9.** $\begin{bmatrix} 33 & -22 \\ 31 & -12 \\ 15 & 13 \\ 50 & -41 \end{bmatrix}$ **11.** $\begin{bmatrix} -15 & 15 & -20 \\ 17 & 5 & -14 \end{bmatrix}$

13. $\begin{bmatrix} -35.78 & 44.28 \\ -45.28 & 49.49 \end{bmatrix}$

15. $AB = [40], BA = \begin{bmatrix} -1 & 3 & -8 \\ 5 & -15 & 40 \\ 7 & -21 & 56 \end{bmatrix}$

17. $AB = \begin{bmatrix} 45 \\ 327 \end{bmatrix}$, BA not defined **19.** $AI = IA = A$

21. $AI = IA = A$ **23.** $B = A^{-1}$ **25.** $B = A^{-1}$ **27.** Yes

29. No **31.** $A^2 = A$ **33.** $B^3 = B$

35. $\begin{bmatrix} ac + bd & ad + bc \\ bc + ad & bd + ac \end{bmatrix}$

37. $\begin{bmatrix} -1 & 0 \\ 0 & -1 \end{bmatrix}\begin{bmatrix} -1 & 0 \\ 0 & -1 \end{bmatrix} = \begin{bmatrix} 1 & 0 \\ 0 & 1 \end{bmatrix}$

39. $A^2 - I = (A + I)(A - I) = \begin{bmatrix} 15 & 28 \\ 21 & 36 \end{bmatrix}$

41. 45.2% **43.** $\begin{bmatrix} 0 & -j \\ j & 0 \end{bmatrix}\begin{bmatrix} 0 & -j \\ j & 0 \end{bmatrix} = \begin{bmatrix} 1 & 0 \\ 0 & 1 \end{bmatrix}$

45. $v_2 = v_1, i_2 = -v_1/R + i_1$ **47.** Ellipse

Exercises 16.3, page 452

1. $\begin{bmatrix} -\frac{5}{2} & \frac{3}{2} \\ -2 & 1 \end{bmatrix}$ **3.** $\begin{bmatrix} 2 & \frac{3}{2} \\ 1 & 1 \end{bmatrix}$ **5.** $\begin{bmatrix} -\frac{1}{3} & \frac{1}{6} \\ \frac{2}{15} & \frac{1}{30} \end{bmatrix}$ **7.** $\begin{bmatrix} \frac{3}{4} & \frac{1}{2} \\ -\frac{1}{4} & 0 \end{bmatrix}$

9. $\begin{bmatrix} -\frac{5}{33} & -\frac{3}{22} \\ \frac{13}{165} & \frac{1}{11} \end{bmatrix}$ **11.** $\begin{bmatrix} 5 & -2 \\ -2 & 1 \end{bmatrix}$ **13.** $\begin{bmatrix} -\frac{1}{2} & -2 \\ \frac{1}{2} & 1 \end{bmatrix}$

15. $\begin{bmatrix} 2 & -5 \\ 1 & -2 \end{bmatrix}$ **17.** $\begin{bmatrix} -18 & -7 & 5 \\ -3 & -1 & 1 \\ -5 & -2 & 1 \end{bmatrix}$ **19.** $\begin{bmatrix} 2 & 4 & \frac{7}{2} \\ -1 & -2 & -\frac{3}{2} \\ 1 & 1 & \frac{1}{2} \end{bmatrix}$

21. $\begin{bmatrix} -1.5 & 2 \\ -0.25 & 0.5 \end{bmatrix}$ **23.** $\begin{bmatrix} 2 & 4 & 3.5 \\ -1 & -2 & -1.5 \\ 1 & 1 & 0.5 \end{bmatrix}$

25. $\begin{bmatrix} 2.5 & -2 & -2 \\ -1 & 1 & 1 \\ 1.75 & -1.5 & -1 \end{bmatrix}$ **27.** $\begin{bmatrix} 1 & 2 & 3 & 1 \\ 1 & 3 & 3 & 2 \\ 2 & 4 & 3 & 3 \\ 1 & 1 & 1 & 1 \end{bmatrix}$

29. $\begin{bmatrix} 11 & 14 \\ 6.5 & 10 \end{bmatrix}$ **31.** $\begin{bmatrix} -4 & 14 \\ -11 & 29 \end{bmatrix}$ **33.** $\begin{bmatrix} 17 & -17 & 6 \\ 6 & -5 & 2 \\ -25 & 29 & -11 \end{bmatrix}$

35. $\begin{vmatrix} 1 & 1 \\ 1 & 1 \end{vmatrix} = 0$ means the inverse does not exist.

37. $\frac{1}{ad - bc}\begin{bmatrix} ad - bc & -ba + ab \\ cd - dc & -bc + ad \end{bmatrix} = \begin{bmatrix} 1 & 0 \\ 0 & 1 \end{bmatrix}$

39. $(1, -1), (-1, 1), (1, 1)$

41. $v_1 = (a_{22}i_1 - a_{12}i_2)/(a_{11}a_{22} - a_{12}a_{21})$

$v_2 = (-a_{21}i_1 + a_{11}i_2)/(a_{11}a_{22} - a_{12}a_{21})$

43. MATH CAN BE FUN

Exercises 16.4, page 456

1. $x = 2, y = -3$ **3.** $x = \frac{9}{2}, y = 5$ **5.** $x = 13, y = -3$

7. $x = -4, y = 2, z = -1$ **9.** $x = -\frac{3}{2}, y = -2$

11. $x = 1.6, y = -2.5$ **13.** $x = 2, y = -4, z = 1$

15. $x = 2, y = -\frac{1}{2}, z = 3$ **17.** $x = 2, y = 2$

19. $x = \frac{1}{3}, y = -2$ **21.** $x = 1, y = -2, z = -3$

23. $u = 2, v = -5, w = 4$ **25.** $x = 2, y = 3, z = -2, t = 1$

27. $v = 2, w = -1, x = \frac{1}{2}, y = \frac{3}{2}, z = -3$

29. $x = \pm 2, y = -2$

31. $x = 1, y = -2$; The three lines meet at the point $(1, -2)$.

33. $A = 1180$ N, $B = 1860$ N **35.** 10 V, 8 V

37. 40 mL of 20% solution, 8 mL of 50% solution

39. 6.4 L, 1.6 L, 2.0 L

Exercises 16.5, page 460

1. $x = \frac{11}{7}, y = \frac{6}{7}$ **3.** $x = 2, y = 1$

5. $x = 1.7, y = 1.6$ **7.** Inconsistent

9. $x = -2, y = -\frac{2}{3}, z = \frac{1}{3}$ **11.** $w = 1, x = 0, y = -2, z = 3$

13. Unlimited; $x = 0, y = -1, z = 1; x = 11, y = 0, z = -6$

15. Inconsistent **17.** $x = 4, y = \frac{2}{3}, z = -2$

19. $x = \frac{1}{2}, y = -\frac{3}{2}$ **21.** $x = \frac{1}{6}, y = -\frac{1}{2}$

23. Inconsistent **25.** $x = 1, y = 3, z = 2$

27. $r = -\frac{8}{7}, s = -\frac{8}{9}, t = \frac{83}{63}, u = -\frac{22}{21}$

29. $x = \frac{c_1b_2 - c_2b_1}{a_1b_2 - a_2b_1}, y = \frac{a_1c_2 - a_2c_1}{a_1b_2 - a_2b_1}$

31. 1080 km, 1290 km **33.** 250 parts/h, 220 parts/h, 180 parts/h

Exercises 16.6, page 464

1. 13 **3.** −90 **5.** 0 **7.** −40 **9.** 120 **11.** 22

13. 57 **15.** −13 **17.** −36 **19.** 0

21. $x = -1, y = 0, z = 2, t = 1$

23. $x = 1, y = 2, z = -1, t = 3$

25. $x = 2, y = -1, z = -1, t = 3$

27. $D = 1, E = 2, F = -1, G = -2$ **29.** 0 **31.** 8

33. $\frac{33}{16}$ A, $\frac{11}{8}$ A, $-\frac{5}{8}$ A, $-\frac{15}{8}$ A, $-\frac{15}{16}$ A **35.** 20.8 km^2

37. ppm SO_2: 0.5, NO: 0.3, NO_2: 0.2, CO: 5.0

Review Exercises for Chapter 16, page 466

Concept Check Exercises

1. F **2.** T **3.** F **4.** T **5.** F **6.** F

Practice and Applications

7. $a = 2, b = -3$

9. $x = 2, y = -3, z = 4, a = 5, b = -1, c = 6$

11. $x = -1, y = \frac{1}{2}, a = -\frac{1}{2}, b = -\frac{3}{2}$

13. $\begin{bmatrix} 1 & -3 \\ 8 & -5 \\ -8 & -2 \\ 3 & -10 \end{bmatrix}$ **15.** $\begin{bmatrix} -3 & 3 \\ 0 & -7 \\ 2 & -2 \\ -1 & -4 \end{bmatrix}$ **17.** $\begin{bmatrix} 7 & -6 \\ -4 & 20 \\ -1 & 6 \\ 1 & 15 \end{bmatrix}$

19. $\begin{bmatrix} 0 & 0 \\ 0 & 0 \end{bmatrix}$ **21.** $\begin{bmatrix} 0.30 & 0.06 & 0 \\ 0.02 & -0.08 & 0.10 \\ -0.01 & -0.17 & 0.20 \end{bmatrix}$ **23.** $\begin{bmatrix} -2 & \frac{5}{2} \\ -1 & 1 \end{bmatrix}$

25. $\begin{bmatrix} -\frac{7}{6} & \frac{1}{6} \\ -\frac{2}{3} & -\frac{4}{3} \end{bmatrix}$ **27.** $\begin{bmatrix} 11 & 10 & 3 \\ -4 & -4 & -1 \\ 3 & 3 & 1 \end{bmatrix}$ **29.** $\begin{bmatrix} \frac{1}{2} & -\frac{1}{2} & -1 \\ -3 & 2 & 1 \\ -4 & 3 & 2 \end{bmatrix}$

31. $x = -3, y = 1$ **33.** $x = 10, y = -15$

35. $u = -1, v = -3, w = 0$ **37.** $x = 1, y = \frac{1}{2}, z = -\frac{1}{3}$

39. $x = -3, y = 1$ **41.** $u = -1, v = -3, w = 0$

43. $x = 1, y = \frac{1}{2}, z = -\frac{1}{3}$ **45.** Unlimited number of solutions

47. $u = -1, v = -3, w = 0$ **49.** $x = 1, y = \frac{1}{2}, z = -\frac{1}{3}$

51. $x = 3, y = 1, z = -1$ **53.** $x = 1, y = 2, z = -3, t = 1$

55. $x = -\frac{1}{3}, y = 3, z = \frac{2}{3}, t = -4$

57. $r = \frac{1}{2}, s = -7, t = -\frac{3}{4}, u = 5, v = \frac{3}{2}$

59. $\begin{bmatrix} 1 & 0 \\ 15 & 16 \end{bmatrix}, \begin{bmatrix} 1 & 0 \\ 63 & 64 \end{bmatrix}, \begin{bmatrix} 1 & 0 \\ 255 & 256 \end{bmatrix}$

61. $B^3 = \begin{bmatrix} 1 & 0 & 0 \\ 0 & 1 & 0 \\ 0 & 0 & 1 \end{bmatrix}$ **63.** 33 **65.** 323

67. 33 **69.** 323

71. $N^{-1} = -N = \begin{bmatrix} 0 & 1 \\ -1 & 0 \end{bmatrix}$

73. $\begin{bmatrix} n & 1+n \\ 1-n & -n \end{bmatrix}\begin{bmatrix} n & 1+n \\ 1-n & -n \end{bmatrix}$

$= \begin{bmatrix} n^2 + (1-n^2) & (n+n^2) - (n+n^2) \\ (n-n^2) - (n-n^2) & (1-n^2) + n^2 \end{bmatrix}$

$= \begin{bmatrix} 1 & 0 \\ 0 & 1 \end{bmatrix}$

75. $\begin{bmatrix} 8 & 10 \\ 6 & 8 \end{bmatrix}$

77. $(A+B)(A-B) = \begin{bmatrix} -6 & 2 \\ 4 & 2 \end{bmatrix}, A^2 - B^2 = \begin{bmatrix} -10 & -4 \\ 8 & 6 \end{bmatrix}$

79. $\begin{bmatrix} \frac{1}{2} & \frac{1}{3} \\ 0 & \frac{1}{6} \end{bmatrix} = \frac{1}{2}\begin{bmatrix} 1 & \frac{2}{3} \\ 0 & \frac{1}{3} \end{bmatrix}$

81. $R_1 = 4\ \Omega, R_2 = 6\ \Omega$

83. $F = 303$ lb, $T = 175$ lb

85. $R_1 = 4\ \Omega, R_2 = 6\ \Omega$

87. $F = 303$ lb, $T = 175$ lb

89. 0.22 h after police pass intersection

91. 30 g, 50 g, 20 g

93. $\begin{bmatrix} 12{,}000 & 24{,}000 & 4{,}000 \\ 15{,}000 & 8{,}000 & 30{,}000 \end{bmatrix} + \begin{bmatrix} 15{,}000 & 12{,}000 & 2{,}000 \\ 20{,}000 & 3{,}000 & 22{,}000 \end{bmatrix}$

$= \begin{bmatrix} 27{,}000 & 36{,}000 & 6{,}000 \\ 35{,}000 & 11{,}000 & 52{,}000 \end{bmatrix}$

95. $(R_1 + R_2)i_1 - R_2 i_2 = 6$

$-R_2 i_1 + (R_1 + R_2)i_2 = 0$

Exercises 17.1, page 474

1. $x + 1 < 0$ is true for all values of x less than -1; $x < -1$.

3. $-2 > -4$, multiplied by -3 gives $6 < 12$; divided by -2 gives $1 < 2$.

5. $9 < 14$ **7.** $16 < 36$ **9.** $-4 > -9$ **11.** $16 < 81$

13. $x > -2$ **15.** $x \le 38$ **17.** $1 < x < 7$

19. $x < -5$ or $x \ge 3$ **21.** $x < 1$ or $3 < x \le 5$

23. $-2 < x < 2$ or $3 \le x < 4$

25. x is greater than 0 and less than or equal to 9.

27. x is less than -10, or greater than or equal to 10 and less than 20.

29. 3 **31.** -1 0.5

33. 0 5 **35.** -3 5

37. -1 1 4 **39.** -3 -1 0.5 3

41. -5 **43.** 3 5 8 10

45. Absolute inequality **47.** No, $a - b < 0$

49. $|x| + |y| > |x + y|$

51. From step (4) to step (5), both sides divided by log 0.5, which is negative. Sign in step (5) should be $>$. **53.** $130 \le d \le 144$ yd

55. $2000 \le M \le 1{,}000{,}000$ **57.** $18{,}000 < v < 25{,}000$ mi/h

2000 1,000,000 0 18,000 25,000

59. $630 \le \omega \le 1530$ r/min **61.** $5 \le t \le 6$ h

Exercises 17.2, page 479

1. $x \le 5$ **3.** $1 < x < \frac{9}{2}$ **5.** $x > -1$

1 $\frac{9}{2}$ -1

7. $x < 64$ **9.** $x \le -2$

64 -2

11. $y < -2$ **13.** $x \le \frac{5}{2}$

-2 $\frac{5}{2}$

15. $x < -\frac{177}{20}$ **17.** $x > -1.80$

$-\frac{177}{20}$ -1.80

19. $L > -\frac{7}{9}$ **21.** $-1 \le x \le 1$

$-\frac{7}{9}$

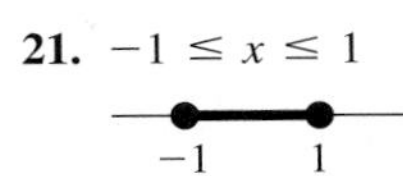

23. $2 < x \le 5$ **25.** $-3 \le x < -1$

2 5

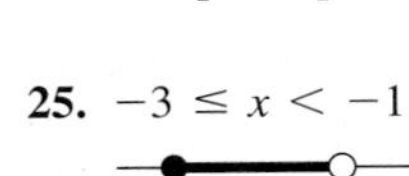

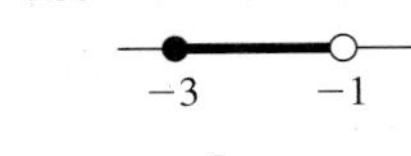

27. No values **29.** $x < \frac{5}{2}$

0

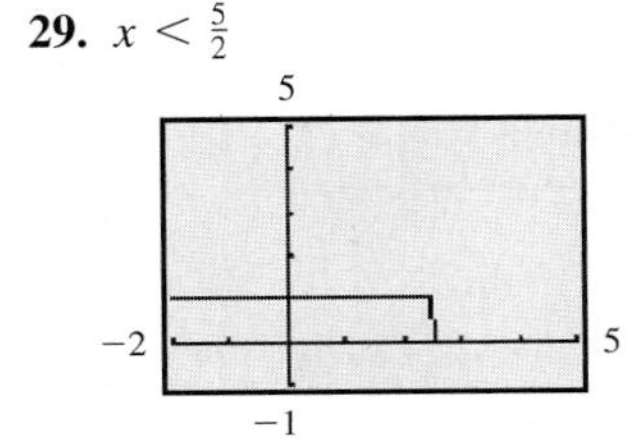

31. $x \ge -1$ **33.** $-2 < t < 2$

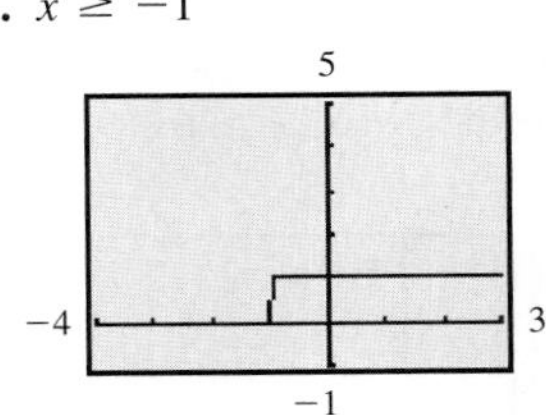

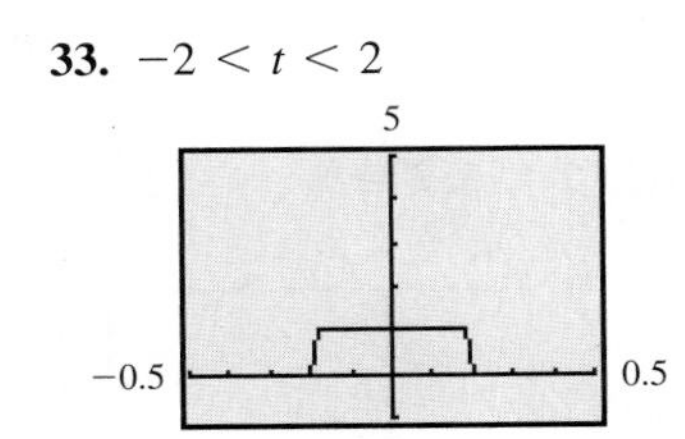

35. $x > -\frac{1}{2}$ **37.** $x \le 0$

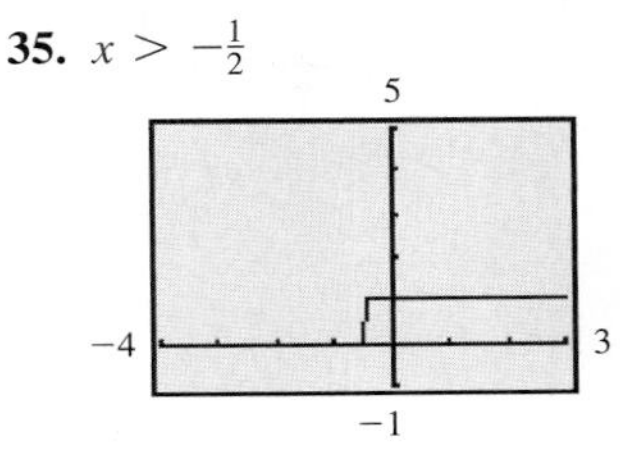

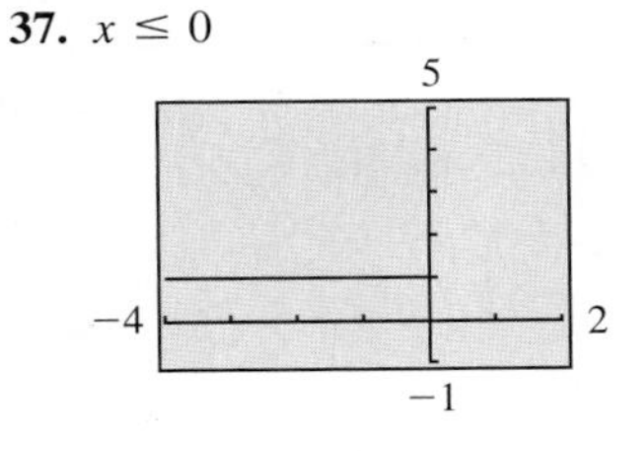

39. $x \ge 5$

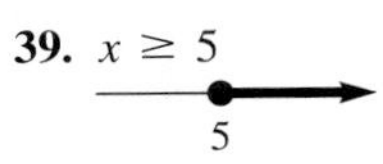

41. $-6 < k < 6$

43. $a = 3, b = 11$

45. $=$

47. $3.2\% < I < 4.8\%$

49. $n > 35$ h

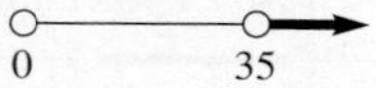

51. $50° < F < 68°$

53. $0.54 \text{ m} < w < 0.81 \text{ m}$

55. $0.4 < t < 2.6$ h

57. $0 \le x \le 500$
$200 \le y \le 700$

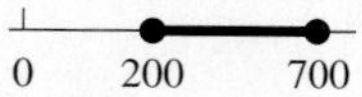

59. 2 min $\le$ stop times $\le$ 4 min

Exercises 17.3, page 484

1. $x < 1$ or $x > 3$

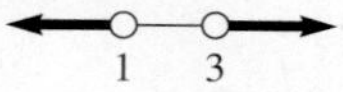

3. $-4 < x < 4$

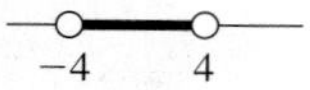

5. $x \le 0, x \ge 2$

7. $-4 \le x \le \frac{3}{2}$

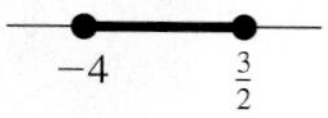

9. $x = -2$

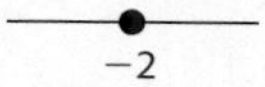

11. All R

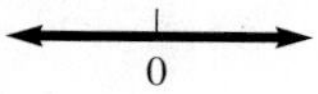

13. $x < -2, 0 < x < 1$

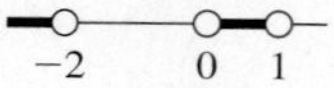

15. $-2 \le s \le -1, s \ge 1$

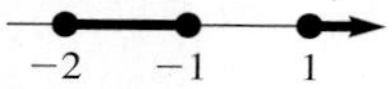

17. $-6 < x \le \frac{3}{2}$

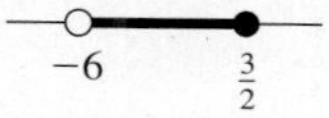

19. $-5 < x < -1, x > 7$

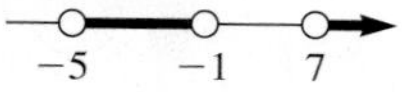

21. $x < -3$

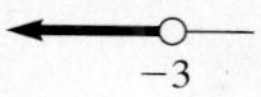

23. $x \le -2, x \ge \frac{1}{3}$

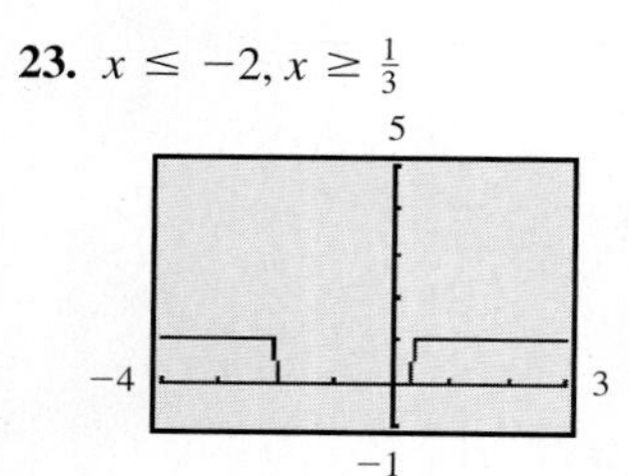

25. $T < 3, T \ge 8$

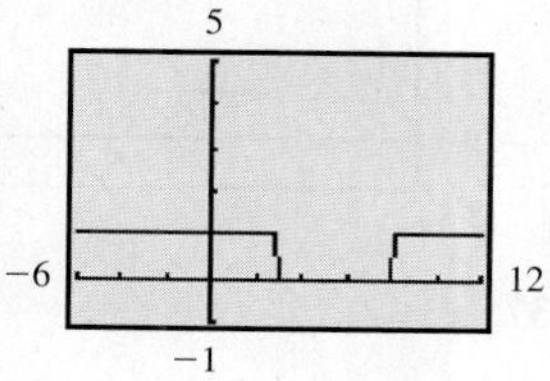

27. $-1 < x < \frac{3}{4}, x \ge 6$

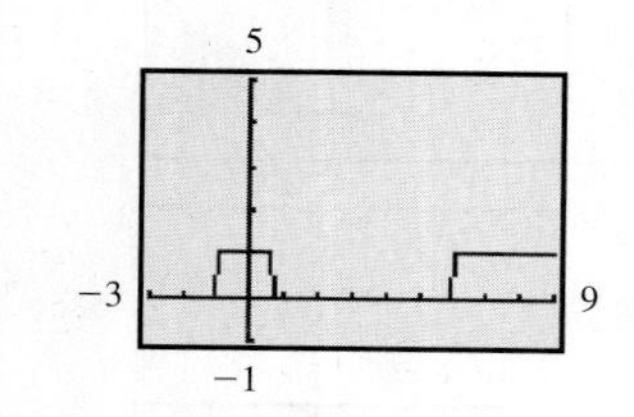

29. $x < 0, 0 < x < 2, 4 < x < 5, x > 9$

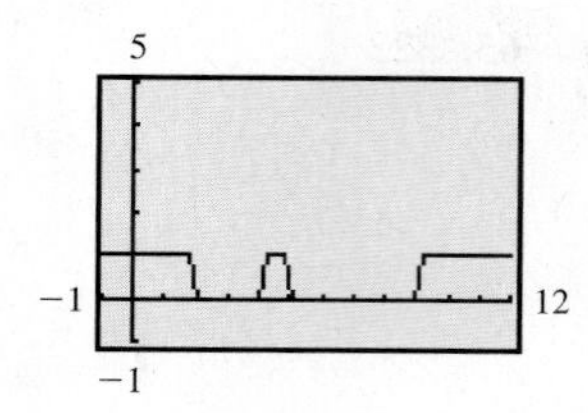

31. $x \le -1, x \ge 4$

33. $-2 \le x \le 0$

35. $x > 1.52$

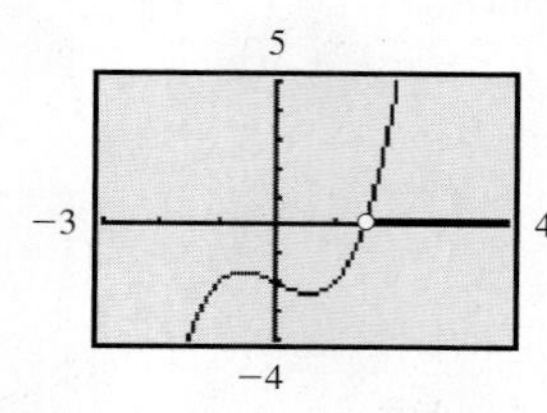

37. $-1.39 < x < -0.43$

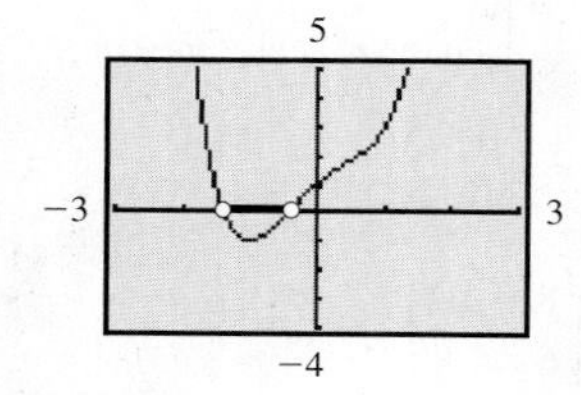

39. $x < -1.69, x > 2.00$

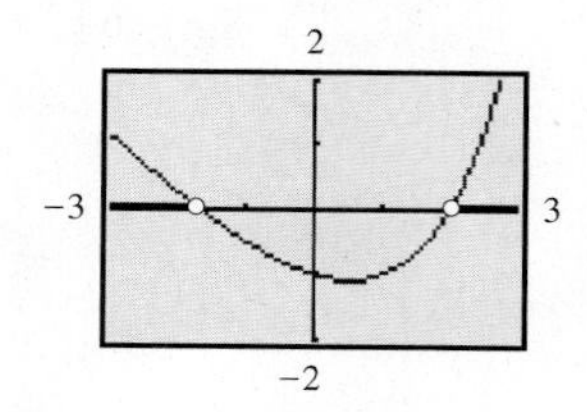

41. $x < -4.43, -3.11 < x < -1.08, x > 3.15$

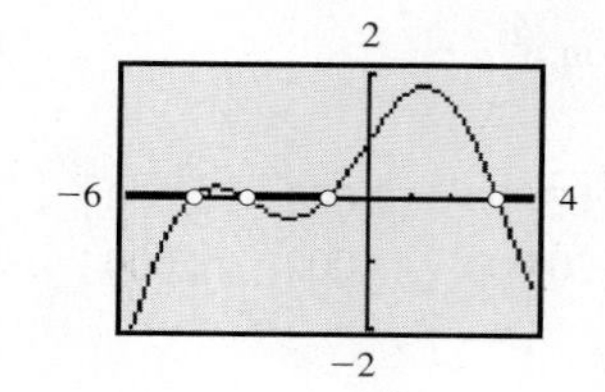

43. No; not true if $0 \le x \le 1$

45. $x^2 - 3x - 4 > 0$

47. $x < 3.113, x < \frac{3 \log 3 + 2 \log 2}{2 \log 3 - \log 2}$

49. If $b > a, a < x < b$; If $a > b, b < x < a$

51. $2 < t < 3$ min

53. $5 \le t \le 20$ weeks

55. $C_1 > \frac{4}{3}\,\mu\text{F}$

57. $h > 2640$ km

59. $h \ge 2$ cm

61. $0 \le t < 0.92$ h

Exercises 17.4, page 488

1. $-2 < x < 3$

3. $3 < x < 5$

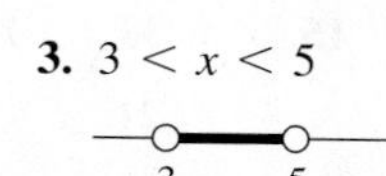

5. $x < -2, x > \frac{2}{5}$

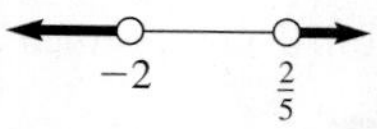

7. $\frac{1}{6} \le x \le \frac{3}{2}$

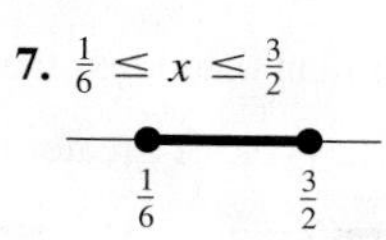

9. $x < 0, x > \frac{3}{2}$

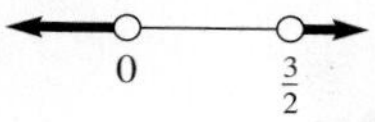

11. $-26 < x < 24$

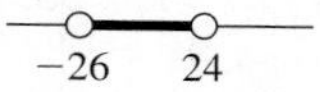

13. $-6.55 \le x \le -1.95$

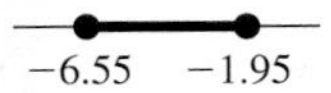

$-6.55 \quad -1.95$

15. $x < -18, x > 66$

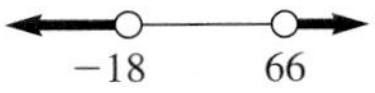

$-18 \quad 66$

17. $1 < x < 2$

$1 \quad 2$

19. $x \le \frac{1}{10}, x \ge \frac{7}{10}$

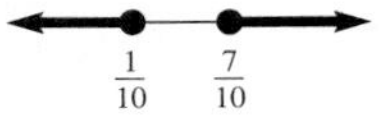

$\frac{1}{10} \quad \frac{7}{10}$

21. $-15 < R < \frac{35}{3}$

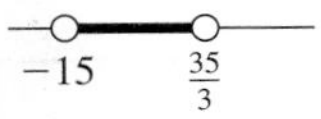

$-15 \quad \frac{35}{3}$

23. $x \le 8.4, x \ge 17.6$

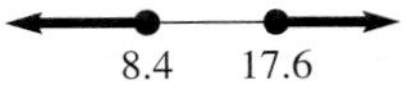

$8.4 \quad 17.6$

25. $1 < x < 4$

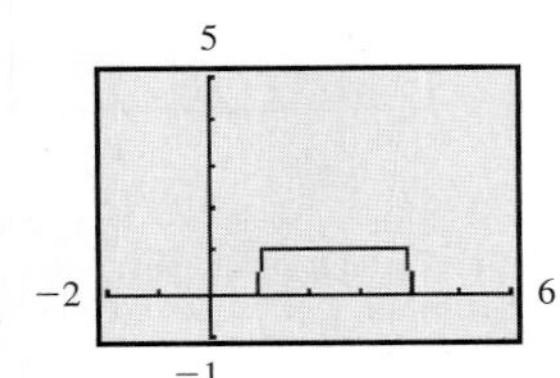

27. $x \le 6, x \ge 10$

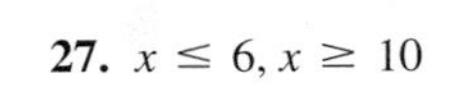

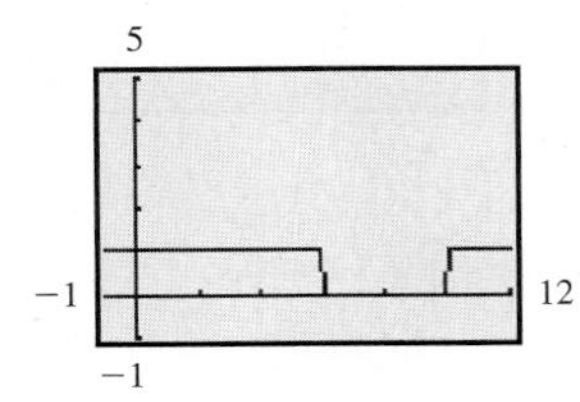

29. $x < -3, -2 < x < 1, x > 2$

31. $-3 < x < -2, 1 < x < 2$

33. No solution

35. $5 < x < 8$ **37.** $a = 1, b = 9$ **39.** 4 km, 50 km

41. $|p - 2{,}000{,}000| \le 200{,}000$; production is at least 1,800,000 barrels, but not greater than 2,200,000 barrels.

43. $|T - 70| \le 20$

45. $|d - 3.765| \le 0.002$ **47.** $0.020 < i < 0.040$ A

Exercises 17.5, page 491

1. $y < 3 - x$

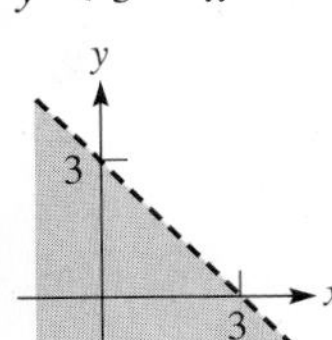

3.

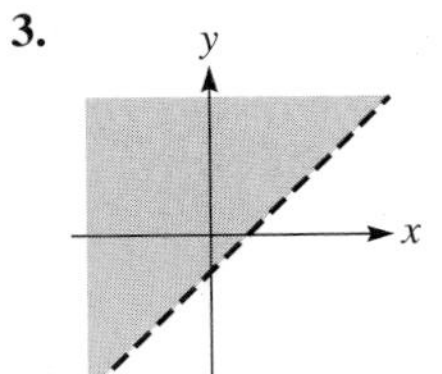

5.

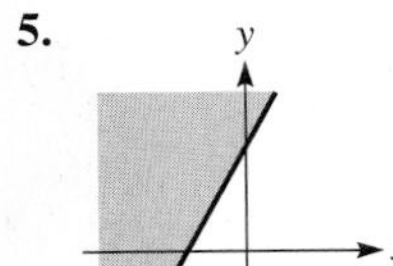

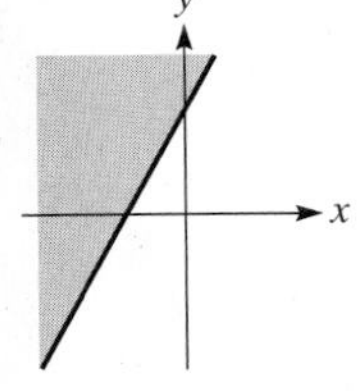

7.

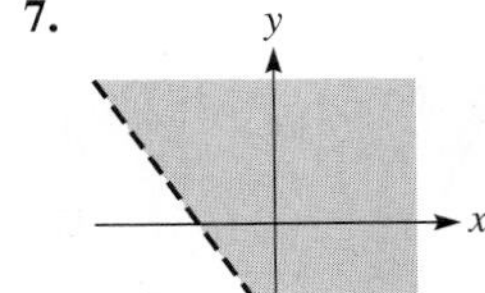

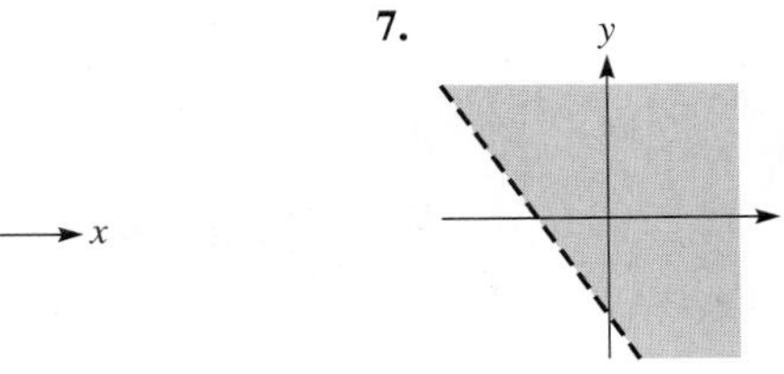

9.

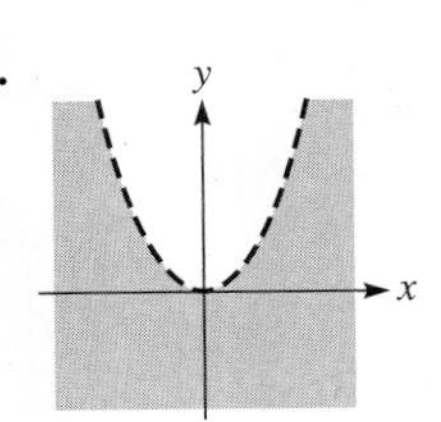

11.

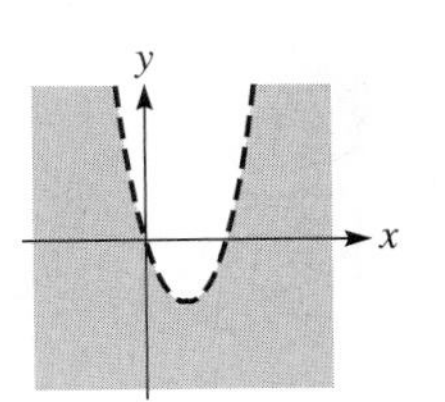

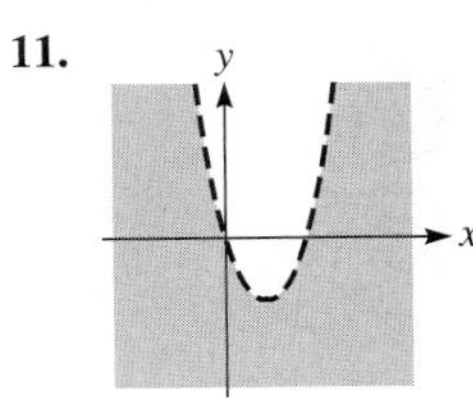

13.

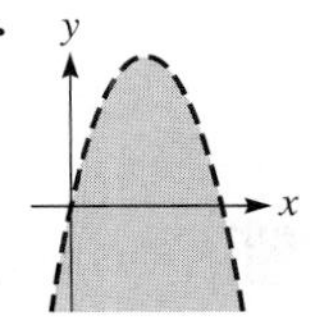

15.

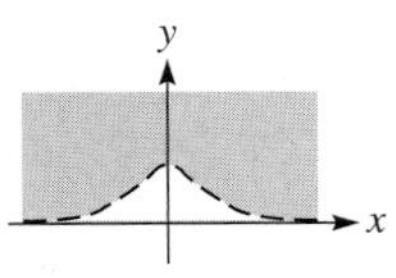

17.

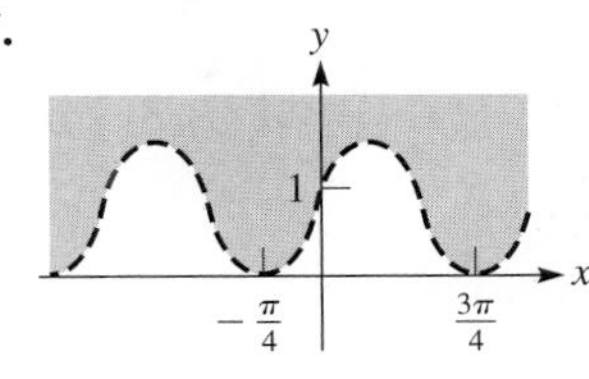

19.

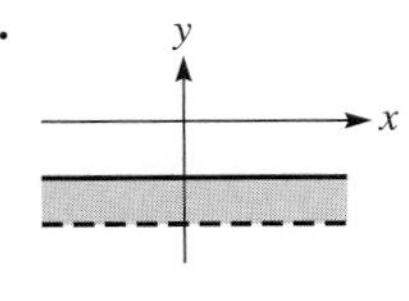

21.

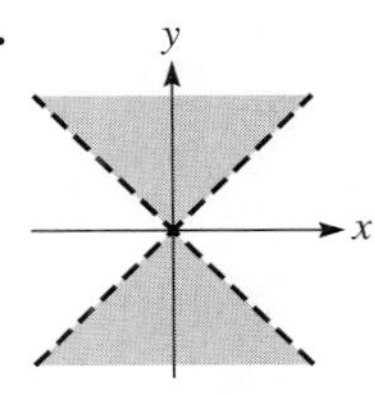

23.

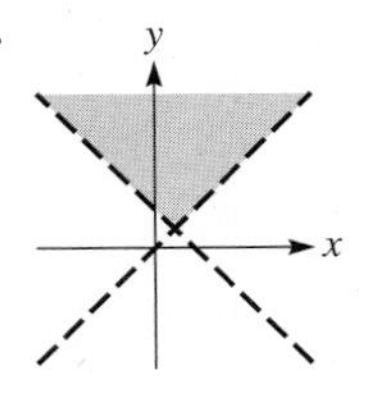

25.

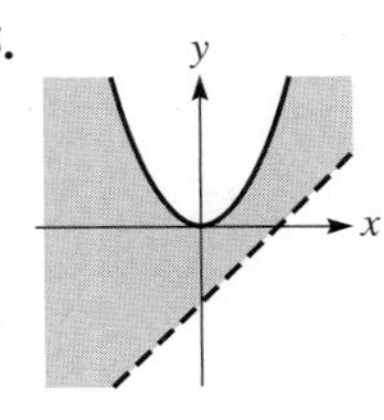

27.

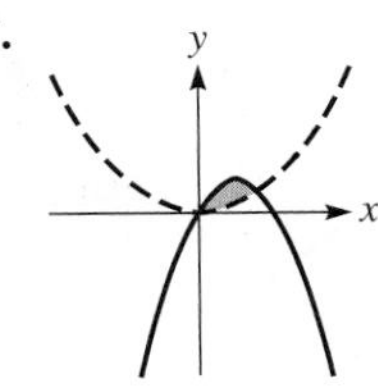

29.

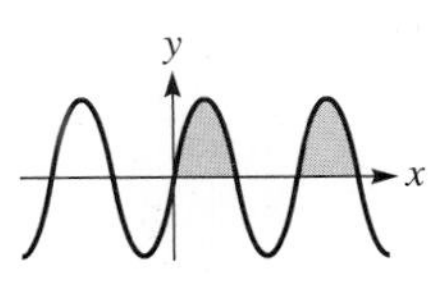

31.

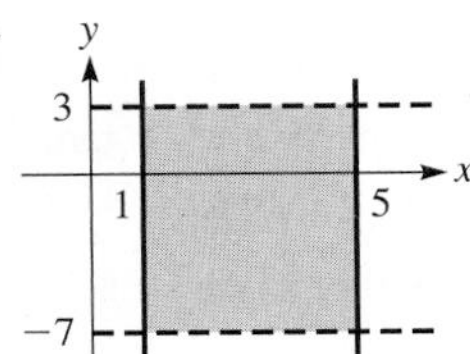

33.

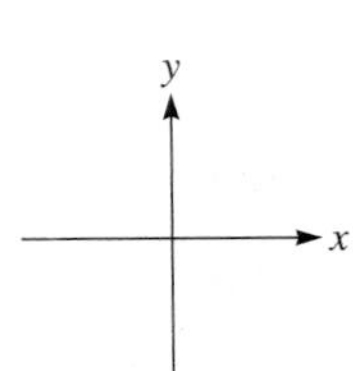

35.

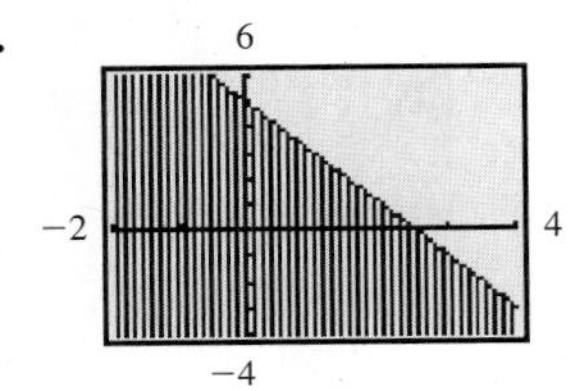

37.

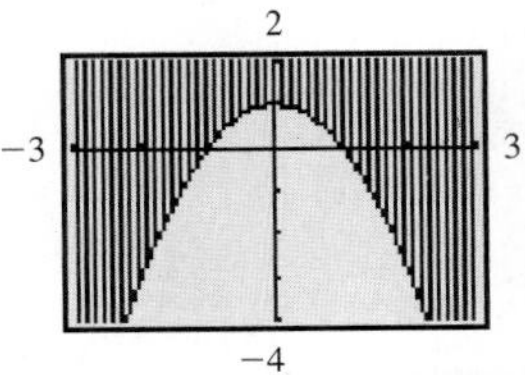

39.

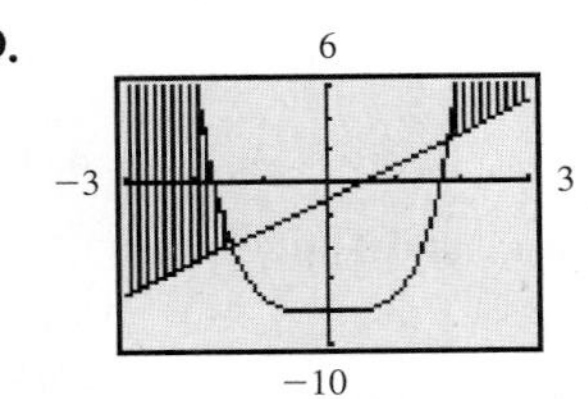

41.

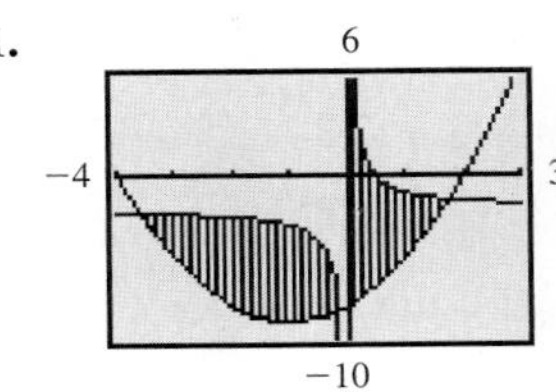

43.

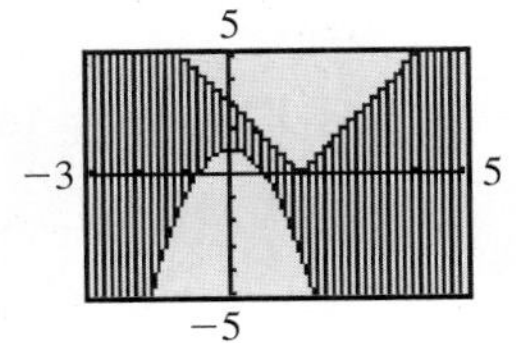

45. $y < 3x + 4$ **47.** Above

49.

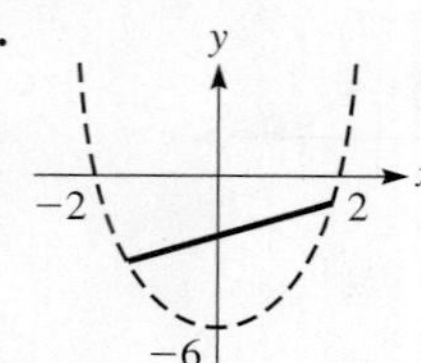

51.

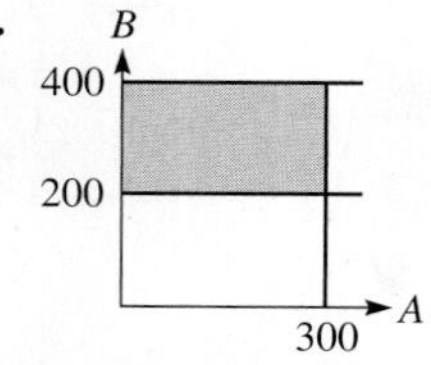

53.

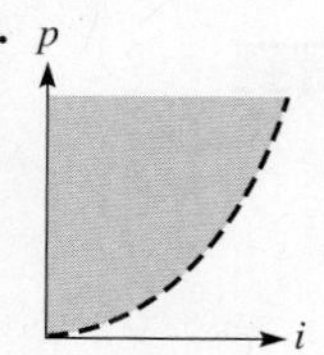

55.

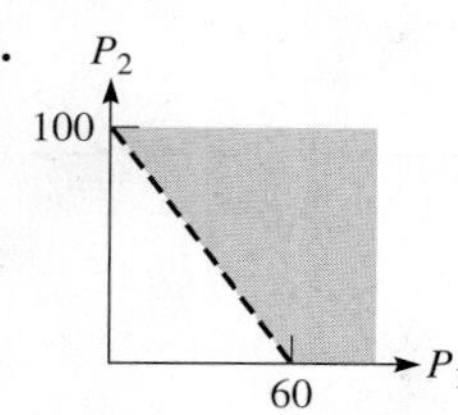

Exercises 17.6, page 495

1. Max $F = 18$ at (0,6)

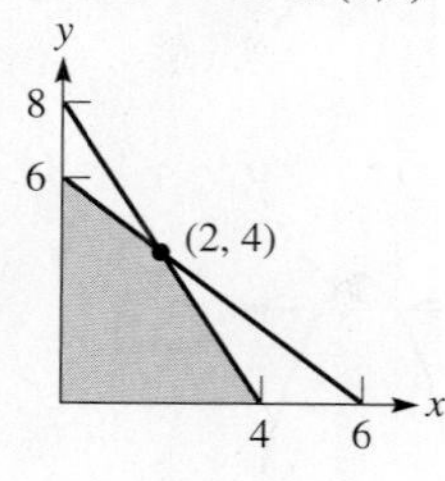

3. Max $F = 55$ at (10, 7)
Min $F = 0$ at (0, 0)

5. Max $P = 15$ at (3, 0)

7. Max $P = 30$ at (6, 0)

9. Min $C = 27$ at (3, 2)

11. Min $F = 7$ at (1, 2)
Max $F = \frac{47}{3}$ at $\left(\frac{5}{3}, \frac{14}{3}\right)$

13. Max $P = 26$ at $\left(\frac{5}{2}, 1\right)$ **15.** Min $C = 40$ at (4,4)

17. 60 newspaper ads, 40 radio ads

19. 40 business models
60 graphing models

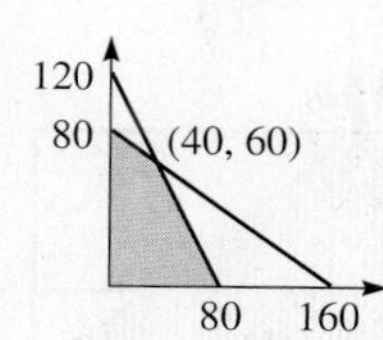

21. $4\frac{2}{7}$ oz of A
$2\frac{6}{7}$ oz of B

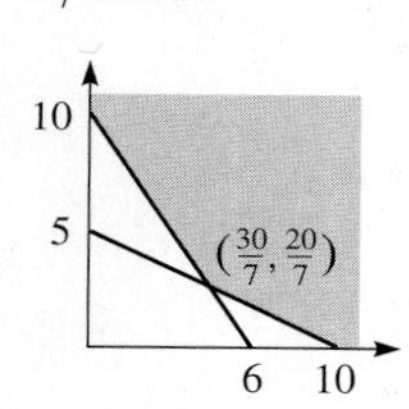

Review Exercises for Chapter 17, page 496

Concept Check Exercises

1. F **2.** F **3.** T **4.** F **5.** F **6.** T

Practice and Applications

7. $x > 6$

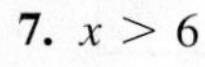

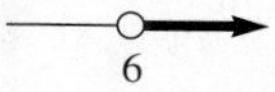

9. $\frac{5}{2} < x < 6$

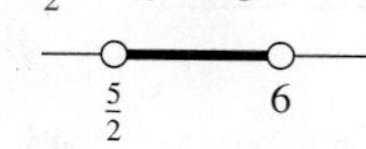

11. $-2 < x < \frac{1}{5}$

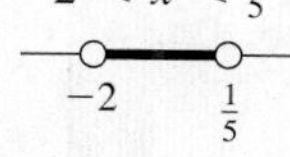

13. $n < -\frac{7}{3}$ or $n > \frac{5}{2}$

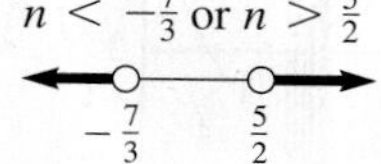

15. $x < -4, \frac{1}{2} < x < 3$

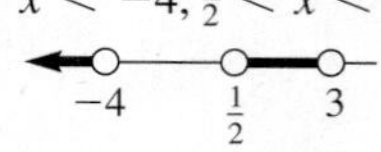

17. $x < 0, x > 4$

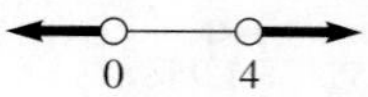

19. $-2 \le x \le \frac{2}{3}$

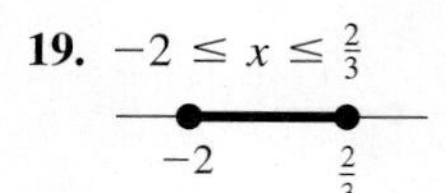

21. $x < -\frac{4}{5}, x > 2$

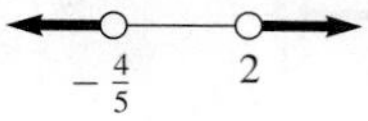

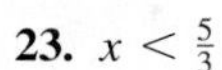

23. $x < \frac{5}{3}$

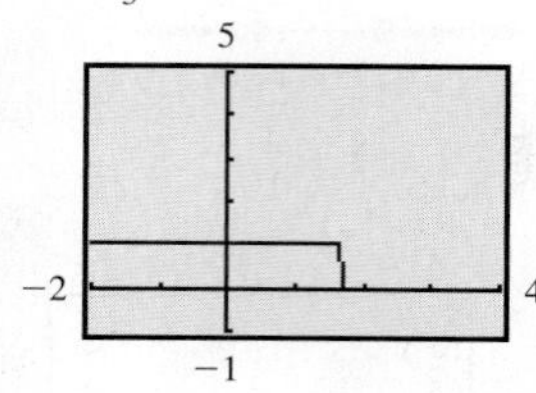

25. $2 \le n < \frac{11}{4}$

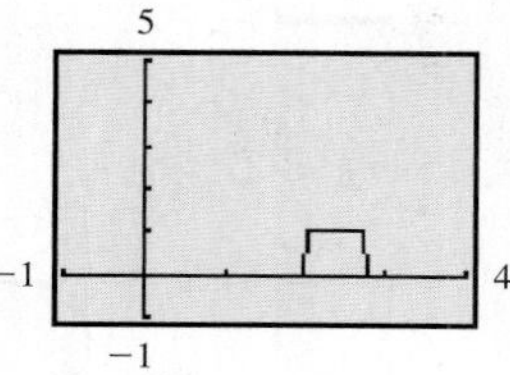

27. $R < -\frac{1}{2}, R \ge 8$

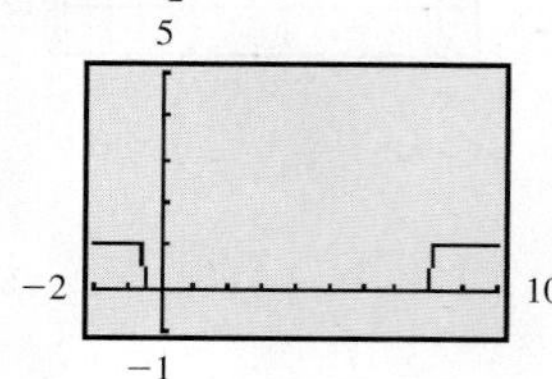

29. $x < -18, x > 78$

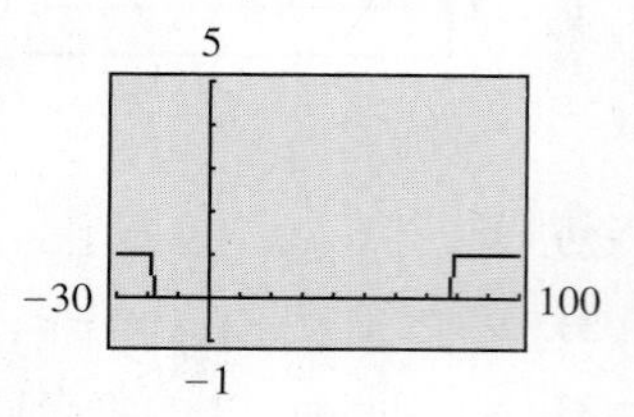

31. $x < -0.68$

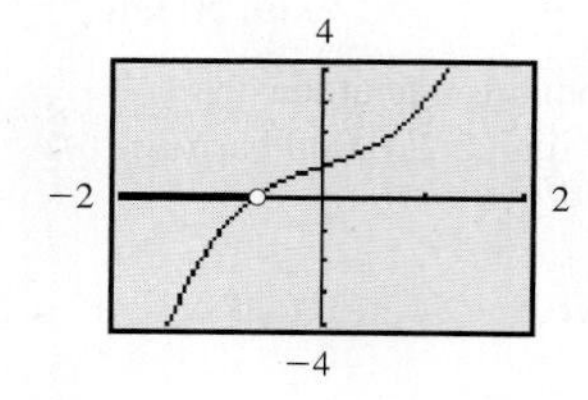

33. $t < 0.69$

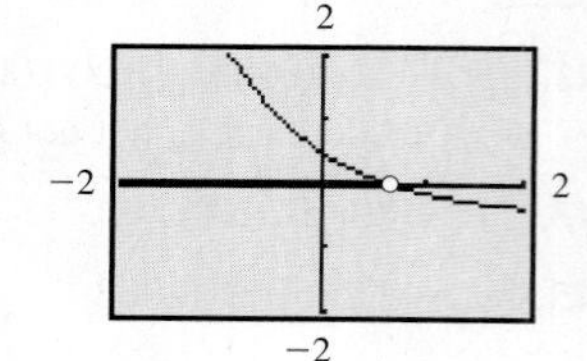

35.

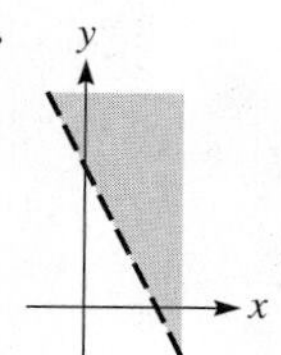

37.

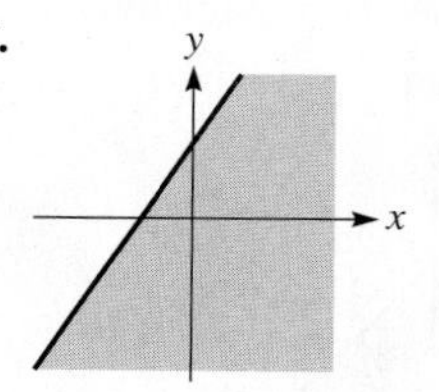

39.

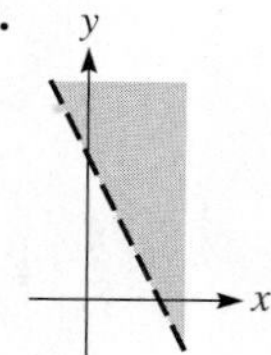

41.

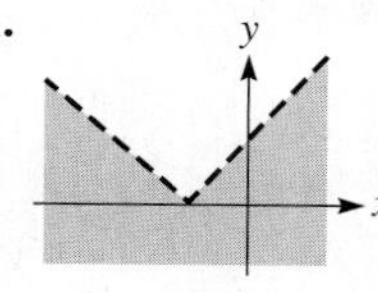

43.

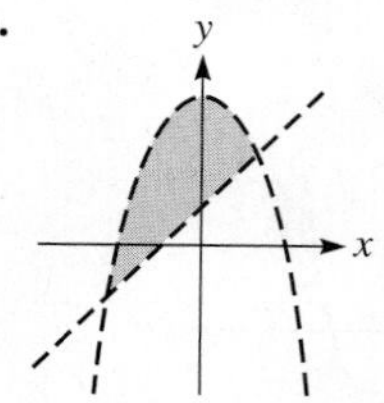

45.

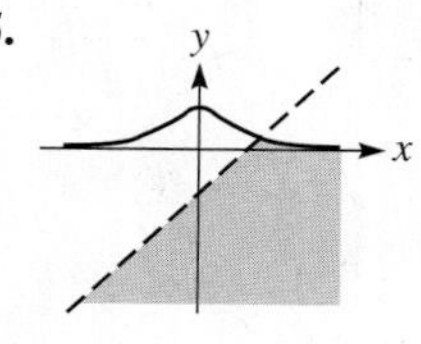

47.

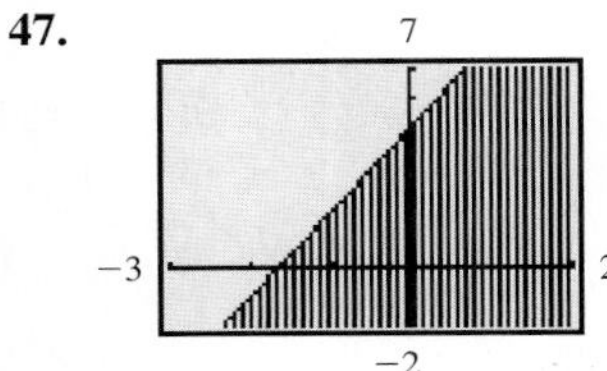

49.

25
−2
10
−5

51.

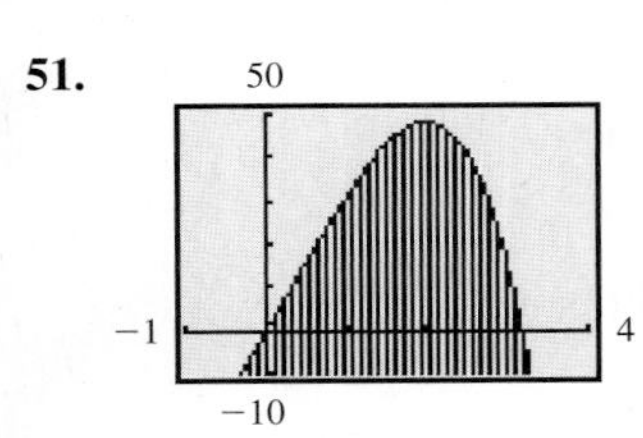

53.

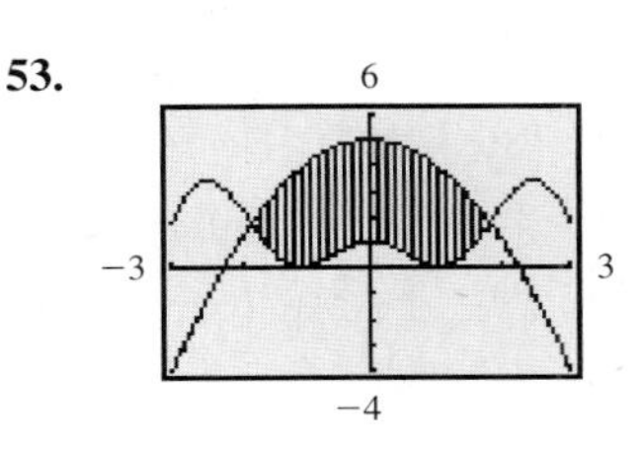

55. $x \le 4$ **57.** $x \le -3, x \ge 0$ **59.** Max $P = 29$ at (1,3)

61. Min $C = \frac{65}{8}$ at $(\frac{3}{8}, \frac{7}{4})$ **63.** a and b must have different signs.

65. $x < -3$ or $x > 2$ **67.** $x^2 - 2x - 15 < 0$

69.

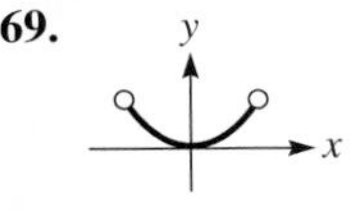

71. $f(x) > 0$ for $x < 2$ or $x > 3$
$f(x) < 0$ for $2 < x < 3$
$f(x) = 0$ for $x = 2, x = 3$
$f(0) = 6, f(5) = 6$

73. No values satisfy both inequalities

75. $0 < x \le 8.0$ cm **77.** Between \$5 and \$12.50

79. $d > 39.5$ m **81.** $168{,}000 \le B \le 400{,}000$ Btu

83. $0.456 < i < 0.816$ A **85.** $r > 5.7$

87.

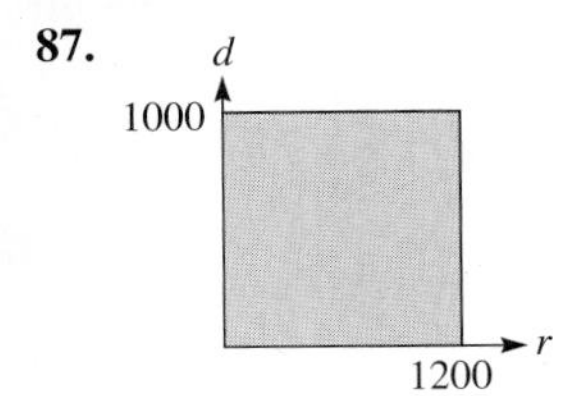

89. 300 regular
150 deluxe
(Max P = \$4650)

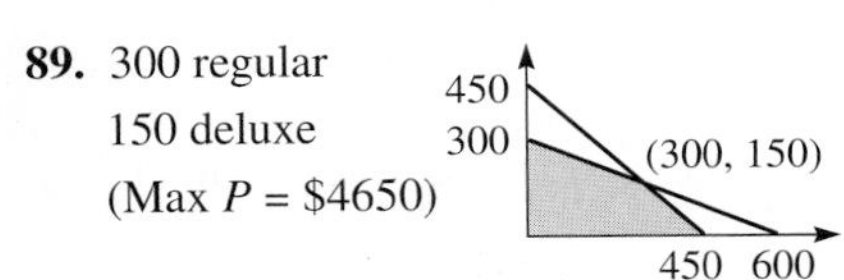

Exercises 18.1, page 502

1. $\frac{3}{2}$ **3.** 4 **5.** $\frac{4}{3}$ **7.** 2 **9.** 50 **11.** 23.14

13. 0.27 = 27% **15.** 0.41 **17.** 0.025 μF **19.** 5.25

21. 7.80 **23.** 3.5% **25.** 863 kg **27.** 0.103 m^3

29. 1.44 Ω **31.** 5.63 $in.^2$ **33.** 0.335 hp **35.** 1210 km

37. 12.5 m/s **39.** 23,400 cm^3 **41.** 1.2% **43.** 8.8 m

45. 290 $in.^2$, 870 $in.^2$ **47.** 19.67 kg **49.** 17,500 chips

51. Traditional screen area is 12% greater than HDTV screen

Exercises 18.2, page 507

1. $C = kd, k = \pi$ **3.** 667 Hz **5.** $v = kr$

7. $R = \frac{k}{d^2}$ **9.** $P = \frac{k}{\sqrt{A}}$ **11.** $S = kwd^3$

13. A varies directly as the square of r.

15. f varies directly as L and inversely as the square root of m.

17. $V = 3H^2$ **19.** $p = \frac{16q}{r^3}$ **21.** 125 **23.** 19.3

25. 1800 **27.** 2.56×10^5

29. $A = k_1x, B = k_2x; A + B = (k_1 + k_2)x$

31. 34,200 kW **33.** 209 kJ **35.** halved **37.** 2.40 in.

39. 560 kJ **41.** 1.4 h **43.** (a) Inverse (b) $a = 60/m$

45. 10,200 hp **47.** $F = 0.00523Av^2$ **49.** 0.536 N

51. 480 m/s **53.** $R = \frac{(4.44 \times 10^{-5})l}{A}$

55. (a) $p = kT$ (b) 265 kPa **57.** 80.0 W **59.** $G = \frac{(5.9 \times 10^4)d^2}{x^2}$

61. -6.57 ft/s^2 **63.** 0.288 W/m^2

Review Exercises for Chapter 18, page 510

Concept Check Exercises

1. T **2.** F **3.** F **4.** T

Practice and Applications

5. 0.28 **7.** 1.5 **9.** 3.1417 **11.** 4.8 **13.** 8.36 $lb/in.^2$

15. 2260 J/g **17.** 4.5%

19. $\frac{a+b}{b} = \frac{a}{b} + \frac{b}{b} = \frac{a}{b} + 1 = \frac{c}{d} + 1 = \frac{c}{d} + \frac{d}{d} = \frac{c+d}{d}$

21. 92 mi **23.** 2.82×10^6 J = 2.82 MJ **25.** 240

27. 310 mL **29.** 71.9 ft **31.** 4500, 7500 **33.** 30.2 kg

35. 115 bolts, 207 bolts **37.** $y = 3x^2$ **39.** $v = \frac{128x}{y^3}$

41. 11.7 in. **43.** 10 cm **45.** $R = 0.94\,A$

47. 1500 bacteria/h **49.** $F = \frac{5500}{L}$ **51.** 2.22s **53.** 96 ft/s

55. 4.3 μC **57.** 22.5 hp **59.** 144 ft **61.** 1.4

63. 48.7 Hz **65.** 2.99×10^8 m/s **67.** 5940 lb

69. 3.26 cm **71.** 78 m **73.** 150 ft **75.** 150%

77. \$125.00, \$600.13 **79.** 4800 Btu **81.** 0.023 W/m^2

Exercises 19.1, page 517

1. -3 **3.** 4, 6, 8, 10, 12 **5.** 2.5, 1.5, 0.5. -0.5, -1.5

7. 22 **9.** $-\frac{16}{3}$ **11.** 30.9 **13.** $49b$ **15.** 544

17. $-\frac{85}{2}$ **19.** $n = 6, S_6 = 150$ **21.** $d = -\frac{2}{19}, a_{20} = -\frac{1}{3}$

23. $a_1 = 19, a_{30} = 106$ **25.** $n = 62, S_n = -486.7$

27. $a_1 = 360, d = 40, S_{10} = 5400$

29. Yes, $d = \ln 2$; $a_5 = \ln 48$ **31.** $a_n = 2n + 1$

33. $2b - c, b, c, 2c - b, 3c - 2b$ **35.** 5050 **37.** -8

39. 1800° **41.** 2700 m^2 **43.** 195 logs **45.** 10 rows

47. 13 years, \$11,700 **49.** 304 ft, 1600 ft

51. Marketing company, \$3000 **53.** $S_n = \frac{1}{2}n[2a_1 + (n-1)d]$

55. $a_1 = 1, a_n = n; S_n = \frac{n}{2}(a_1 + a_n) = \frac{n}{2}(1 + n) = \frac{1}{2}n(n + 1)$

Exercises 19.2, page 521

1. 243 **3.** 6400, 1600, 400, 100, 25 **5.** $\frac{1}{6}, \frac{1}{2}, \frac{3}{2}, \frac{9}{2}, \frac{27}{2}$

7. 128 **9.** $\frac{1}{125}$ **11.** $\frac{100}{729}$ **13.** 1 **15.** $\frac{341}{8}$

17. 762 **19.** $\frac{3(2^{10} - k^{10})}{16(2 + k)}$ **21.** $a_6 = 64, S_6 = \frac{1365}{16}$

23. $a_1 = 16, a_5 = 81$ **25.** $a_1 = 1, r = 3$

27. $n = 7, S_n = \frac{58{,}593}{625}$ **29.** Yes; $r = 3^x$; $a_{20} = 3^{19x+1}$ **31.** 5

33. $\frac{x^2[1 - (-x)^n]}{x + 1}$ **35.** 1.4% **37.** 1.09 mA **39.** \$366.76

41. 5800°C **43.** 43% **45.** \$13,947 **47.** \$671,088.64

49. The 5% bonus, \$1,609,564.06 more **51.** 0.25 ft

53. Yes, the ratio is squared.

55. 8, 1, −6; 8, 16, 24; 1, −6, 36; 16, 24, 36

Exercises 19.3, page 525

1. $\frac{32}{7}$ **3.** 8 **5.** 0.2 **7.** $\frac{400}{21}$ **9.** 32 **11.** $\frac{10{,}000}{9999}$

13. $\frac{1}{2}(5 + 3\sqrt{3})$ **15.** $\frac{1}{3}$ **17.** $\frac{1}{2}$ **19.** $\frac{40}{99}$

21. $\frac{91}{3330}$ **23.** $\frac{13}{110}$ **25.** $\frac{5}{11}$ **27.** 350 gal **29.** 346 g

31. 100 m **33.** 81.8% **35.** $-\frac{1}{4}$

Exercises 19.4, page 530

1. $32x^5 + 240x^4 + 720x^3 + 1080x^2 + 810x + 243$

3. $t^3 + 12t^2 + 48t + 64$

5. $16x^4 - 96x^3 + 216x^2 - 216x + 81$ **7.** 8445.96301

9. $n^5 + 10\pi^3n^4 + 40\pi^6n^3 + 80\pi^9n^2 + 80\pi^{12}n + 32\pi^{15}$

11. $64a^6 - 192a^5b^2 + 240a^4b^4 - 160a^3b^6 + 60a^2b^8 - 12ab^{10} + b^{12}$

13. $625x^4 - 1500x^3 + 1350x^2 - 540x + 81$

15. $64a^6 + 192a^5 + 240a^4 + 160a^3 + 60a^2 + 12a + 1$

17. $x^{10} + 20x^9 + 180x^8 + 960x^7 + \cdots$

19. $128a^7 - 448a^6 + 672a^5 - 560a^4 + \cdots$

21. $x^6 - 48x^{11/2}y + 1056x^5y^2 - 14{,}080x^{9/2}y^3 + \cdots$

23. $b^{40} + 10b^{37} + \frac{95}{2}b^{34} + \frac{285}{2}b^{31} + \cdots$ **25.** 1.338

27. 1.015 **29.** $1 + 8x + 28x^2 + 56x^3 + \cdots$

31. $1 + 6x + 27x^2 + 108x^3 + \cdots$

33. $1 + \frac{1}{2}x - \frac{1}{8}x^2 + \frac{1}{16}x^3 - \cdots$

35. $\frac{1}{3}[1 + \frac{1}{2}x + \frac{3}{8}x^2 + \frac{5}{16}x^3 + \cdots]$

37. (a) 3.557×10^{14} (b) 5.109×10^{19} (c) 8.536×10^{15} (d) 2.480×10^{96}

39. $n! = n(n - 1)(n - 2)(\cdots)(2)(1) = n \times (n - 1)!$; for $n = 1$, $1! = 1 \times 0!$ Since $1! = 1$, $0!$ must $= 1$.

41. $56a^3b^5$ **43.** $10{,}264{,}320x^8b^4$ **45.** 8

47. $n!$ contains factors 2 and 5. **49.** 0.171 **51.** 2.45

53. $V = A(1 - 5r + 10r^2 - 10r^3 + 5r^4 - r^5)$

55. $1 - \frac{x}{a} + \frac{x^3}{2a^3} - \cdots$

57. $(1-t)^4P_0 + 4(1-t)^3tP_1 + 6(1-t)^2t^2P_2 + 4(1-t)t^3P_3 + t^4P_4$, $0 \le t \le 1$

Review Exercises for Chapter 19, page 531

Concept Check Exercises

1. F **2.** T **3.** F **4.** T

Practice and Applications

5. 81 **7.** $\frac{4}{3125}$ **9.** 66.5 **11.** $\frac{16}{243}$ **13.** $\frac{195}{2}$

15. $\frac{5461}{512}$ **17.** 81 **19.** 32 **21.** −15 **23.** −0.25

25. 186 **27.** 2.7 **29.** 51 **31.** $\frac{1}{33}$ **33.** $\frac{4}{55}$

35. $x^4 - 20x^3 + 150x^2 - 500x + 625$

37. $x^{10} + 20x^8 + 160x^6 + 640x^4 + 1280x^2 + 1024$

39. $a^{10} + 30a^9e + 405a^8e^2 + 3240a^7e^3 + \cdots$

41. $p^{18} - \frac{3}{2}p^{16}q + p^{14}q^2 - \frac{7}{18}p^{12}q^3 + \cdots$

43. $1 + 12x + 66x^2 + 220x^3 + \cdots$

45. $1 + \frac{1}{2}x^2 - \frac{1}{8}x^4 + \frac{1}{16}x^6 - \cdots$

47. $1 - \frac{1}{2}a^2 - \frac{1}{8}a^4 - \frac{1}{16}a^6 - \cdots$

49. $\frac{1}{8} + \frac{3}{4}x + 3x^2 + 10x^3 + \cdots$

51. 1,001,000 **53.** $4b - 3a$ **55.** No **57.** $a_n = -3n - 2$

59. 0.7156 **61.** 5.475 **63.** 81 **65.** $3, -\frac{1}{2}$ **67.** 11th

69. 12.6 mm **71.** 302.9 in. **73.** \$4700

75. 4.4×10^9 in. = 69,000 mi **77.** 624 ft **79.** 100 cm

81. \$47,340.80 **83.** \$6.40 **85.** 4.0°C

87. $1 + \frac{1}{2}am^2 + \frac{1}{8}am^4$ **89.** 21 years **91.** Five applications

93. 22 years **95.** $S = \frac{n}{2}(a_1 + a_n)$; $a_{\text{mid}} = \frac{a_1 + a_n}{2} = \frac{2S}{2n} = \frac{S}{n}$

97. Yes; term to term ratios are equal.

Exercises 20.1, page 541

(*Note*: "Answers" to trigonometric identities are intermediate steps of suggested reductions of the left member.)

1. $\sin x = \frac{\tan x}{\sec x} = \frac{\frac{\sin x}{\cos x}}{\frac{1}{\cos x}} = \frac{\sin x}{\cos x} \cdot \frac{\cos x}{1} = \sin x$

3. $0.7813 = \frac{0.6157}{0.7880}$ **5.** $(-\frac{1}{2}\sqrt{3})^2 + (-\frac{1}{2})^2 = \frac{3}{4} + \frac{1}{4} = 1$

7. $\sin x - 1$ **9.** $\cot\theta - 2\cos\theta$ **11.** $\sec^2 u$

13. $\tan x \sec x$ **15.** $\sin t \cos t$ **17.** $\cot^2 y(\csc^2 y + 1)$

19. $\frac{\sin x}{\frac{\sin x}{\cos x}} = \frac{\sin x}{1}\left(\frac{\cos x}{\sin x}\right)$ **21.** $\sin x\left(\frac{1}{\cos x}\right)$ **23.** $\csc^2 x(\sin^2 x)$

25. $\sin x(\csc^2 x) = (\sin x)(\csc x)(\csc x) = \sin x\left(\frac{1}{\sin x}\right)\csc x$

27. $\cos\theta\left(\frac{\cos\theta}{\sin\theta}\right) + \sin\theta = \frac{\cos^2\theta + \sin^2\theta}{\sin\theta} = \frac{1}{\sin\theta}$

29. $\cot\theta(\sec^2\theta - 1) = \cot\theta\tan^2\theta = (\cot\theta\tan\theta)\tan\theta$

31. $\frac{\sin x}{\cos x} + \frac{\cos x}{\sin x} = \frac{\sin^2 x + \cos^2 x}{\cos x \sin x} = \frac{1}{\cos x \sin x}$ **33.** $(1 - \sin^2 x) - \sin^2 x$

35. $\frac{\sin\theta}{\frac{1}{\sin\theta}} + \frac{\cos\theta}{\frac{1}{\cos\theta}} = \sin^2\theta + \cos^2\theta$

37. $(2\sin^2 x - 1)(\sin^2 x - 1) = (2\sin^2 x - 1)(-\cos^2 x)$

39. $\cot x$ **41.** $\sin x$ **43.** $\sec x$ **45.** $\cos x$

47. $\sin x = \frac{\sqrt{\sec^2 x - 1}}{\sec x}$ **49.** $\tan x = \frac{1}{\sqrt{\csc^2 x - 1}}$

51.

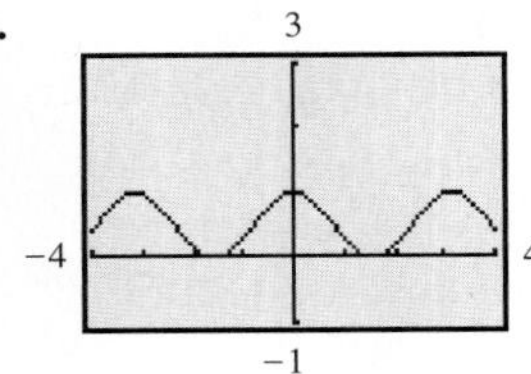

53.

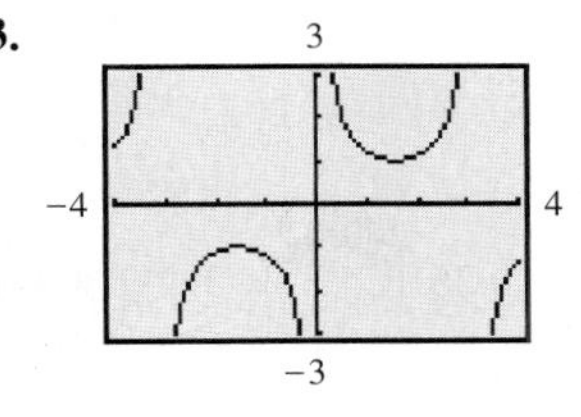

55. Yes

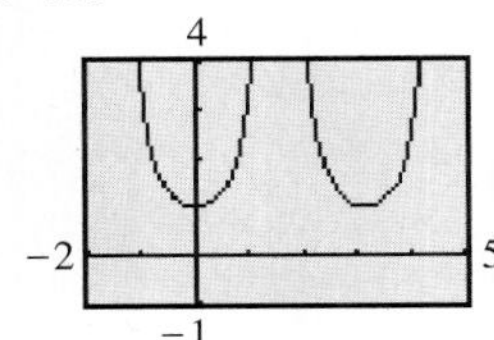

57. No

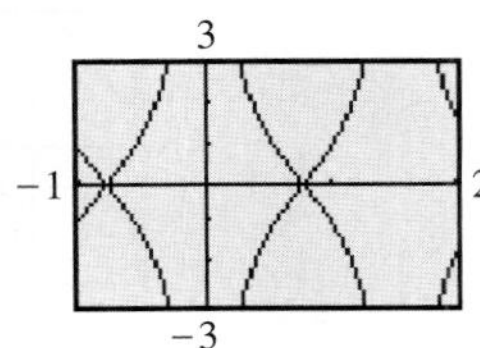

59. $0 = \cos A \cos B \cos C + \sin A \sin B.$

$\cos C = -\frac{\sin A \sin B}{\cos A \cos B}$

61. $l = a \csc\theta + a \sec\theta = a\left(\frac{1}{\sin\theta} + \frac{\tan\theta}{\sin\theta}\right)$

63. Replacing $\tan\theta$ with $\sin\theta/\cos\theta$ gives $\sin^2\theta/\cos\theta + \cos\theta$. Using LCD of $\cos\theta$ gives $(\sin^2\theta + \cos^2\theta)/\cos\theta$ which equals $\sec\theta$.

65. $\sin^2 x - \sin^2 x \sec^2 x + \cos^2 x + \cos^2 x \sec^4 x$

$= \sin^2 x - \tan^2 x + \cos^2 x + \sec^2 x = 2$

67. 2 **69.** $\left(\frac{r}{x}\right)^2 + \left(\frac{r}{y}\right)^2 = \frac{r^2(x^2 + y^2)}{x^2y^2}$

71. $\sqrt{1 - \cos^2\theta} = \sqrt{\sin^2\theta}$

73. $\sqrt{4 + 4\tan^2\theta} = 2\sqrt{1 + \tan^2\theta}$

Exercises 20.2, page 546

1. $\frac{16}{65}$

3. $\sin 105° = \sin 60° \cos 45° + \cos 60° \sin 45°$

$= \frac{\sqrt{3}}{2}\frac{\sqrt{2}}{2} + \frac{1}{2}\frac{\sqrt{2}}{2} = 0.9659$

5. $\cos 15° = \cos(60° - 45°) = \cos 60° \cos 45° + \sin 60° \sin 45°$

$= \left(\frac{1}{2}\right)\left(\frac{1}{2}\sqrt{2}\right) + \left(\frac{1}{2}\sqrt{3}\right)\left(\frac{1}{2}\sqrt{2}\right)$

$= \frac{1}{4}\sqrt{2} + \frac{1}{4}\sqrt{6} = \frac{1}{4}(\sqrt{2} + \sqrt{6}) = 0.9659$

7. $-\frac{33}{65}$ **9.** $-\frac{56}{65}$ **11.** $\sin 3x$ **13.** $\cos 4x$ **15.** $\cos x$

17. $\tan x$ **19.** 0 **21.** 1 **23.** 0

25. $(\sin x \cos y + \cos x \sin y)(\sin x \cos y - \cos x \sin y)$

$= \sin^2 x \cos^2 y - \cos^2 x \sin^2 y$

$= \sin^2 x(1 - \sin^2 y) - (1 - \sin^2 x)\sin^2 y = \sin^2 x - \sin^2 y$

27. $(\cos\alpha\cos\beta - \sin\alpha\sin\beta) + (\cos\alpha\cos\beta + \sin\alpha\sin\beta)$

$= 2\cos\alpha\cos\beta$

29.

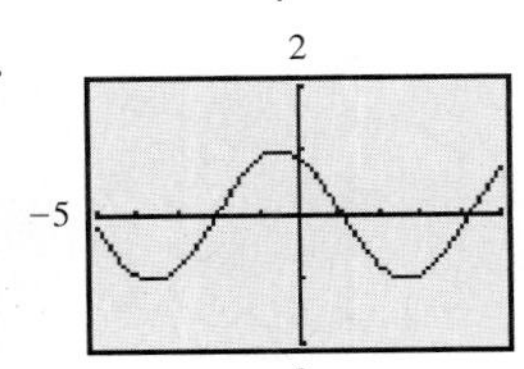

31.

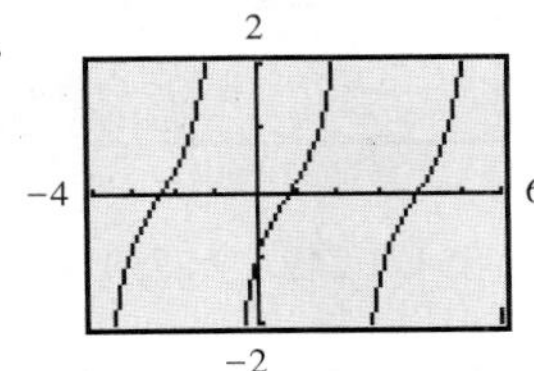

33, 35, 37, 39. Use the indicated method. **41.** $\frac{1}{2}$

43. $\frac{\sin(x + x)}{\sin x} = \frac{\sin x \cos x + \cos x \sin x}{\sin x}$

$= \frac{2 \sin x \cos x}{\sin x}$

45. Using Eq. (20.9), $\sin 75° = \sin(30° + 45°)$. Using Eq. (20.11), $\sin 75° = \sin(135° - 60°)$. (Other angles are also possible.)

47. $I(\cos\theta \cos 30° - \sin\theta \sin 30°)$

$+I(\cos\theta \cos 150° - \sin\theta \sin 150°)$

$+I(\cos\theta \cos 270° - \sin\theta \sin 270°)$

$= I[\frac{1}{2}\sqrt{3}\cos\theta - \frac{1}{2}\sin\theta - \frac{1}{2}\sqrt{3}\cos\theta - \frac{1}{2}\sin\theta$

$+ 0 - (-\sin\theta)] = 0$

49.

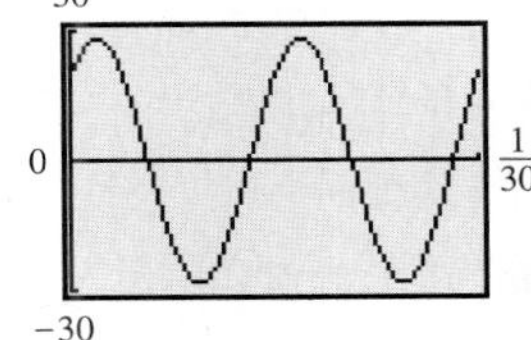

51. $d_0 \sin(\omega t + \alpha) = d_0(\sin\omega t \cos\alpha + \cos\omega t \sin\alpha)$

53. $\tan\alpha(R + \cos\beta) = \sin\beta, R = \frac{\sin\beta - \tan\alpha\cos\beta}{\tan\alpha}$

$= \frac{\sin\beta\cos\alpha - \cos\beta\sin\alpha}{\cos\alpha\tan\alpha}$

Exercises 20.3, page 549

1. $-\sqrt{3}$ **3.** $-\frac{24}{25}$

5. $\sin 60° = \sin 2(30°) = 2 \sin 30° \cos 30°$

$= 2\left(\frac{1}{2}\right)\left(\frac{1}{2}\sqrt{3}\right) = \frac{1}{2}\sqrt{3}$

7. $\tan 120° = \frac{2\tan 60°}{1 - \tan^2 60°} = \frac{2\sqrt{3}}{1 - (\sqrt{3})^2} = -\sqrt{3}$

9. $\sin 100° = 2 \sin 50° \cos 50° = 0.9848$

11. $\cos 96° = \cos^2 48° - \sin^2 48° = -0.1045285$

13. $\tan\frac{2\pi}{7} = \frac{2\tan\frac{\pi}{7}}{1 - \tan^2\frac{\pi}{7}}$

$= 1.254$

15. $\frac{24}{25}$ **17.** $-\sqrt{3}$ **19.** $3 \sin 10x$ **21.** $\cos 8x$ **23.** $\cos x$

25. $-4\cos 4x$ **27.** $2\sin 3\theta$ **29.** 2

31. $\cos^2\alpha - (1 - \cos^2\alpha)$

33. $\frac{\cos x - (\sin x/\cos x)\sin x}{1/\cos x} = \cos^2 x - \sin^2 x$

35. $\frac{2\sin\theta\cos\theta}{1 + 2\cos^2\theta - 1} = \frac{\sin\theta}{\cos\theta}$

37. $1 - (1 - 2\sin^2 2\theta) = \frac{2}{\csc^2\theta}$

39. $\ln\frac{1 - \cos 2x}{1 + \cos 2x} = \ln\frac{2\sin^2 x}{2\cos^2 x} = \ln\tan^2 x$

41.

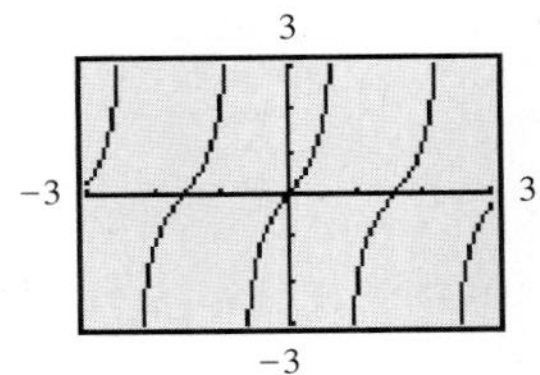

43.

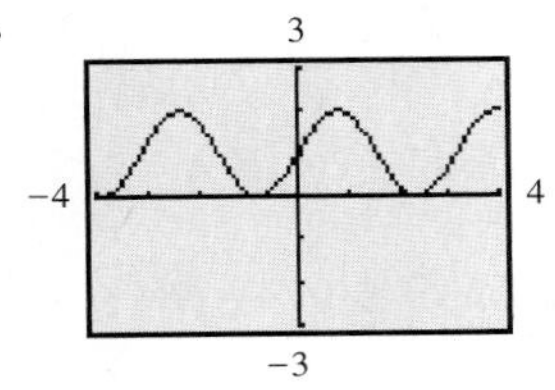

45. $3\sin x - 4\sin^3 x$ **47.** $8\cos^4 x - 8\cos^2 x + 1$

49. 1 **51.** $\log\cos 2x$

53. amp. = 2, per. = π; write equation as $y = 2(2\sin x\cos x) = 2\sin 2x$.

55. $\sqrt{\sin^2 x + 2\sin x\cos x + \cos^2 x} = \sqrt{1 + \sin 2x}$

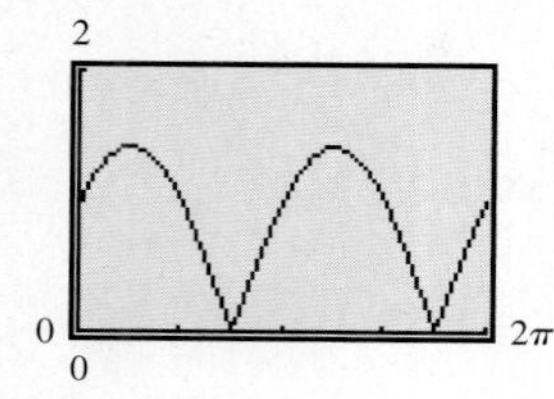

57. 474 m **59.** $R = v\left(\frac{2v\sin\alpha}{g}\right)\cos\alpha = \frac{v^2(2\sin\alpha\cos\alpha)}{g}$

61. $vi\sin\omega t\sin\left(\omega t - \frac{\pi}{2}\right)$

$= vi\sin\omega t\left(\sin\omega t\cos\frac{\pi}{2} - \cos\omega t\sin\frac{\pi}{2}\right)$

$= vi\sin\omega t[-(\cos\omega t)(1)] = -\frac{1}{2}vi(2\sin\omega t\cos\omega t)$

Exercises 20.4, page 553

1. $\cos 57° = 0.544639035$

3. $\cos 15° = \cos\frac{1}{2}(30°) = \sqrt{\frac{1 + \cos 30°}{2}} = \sqrt{\frac{1.8660}{2}}$ $= 0.9659$

5. $\sin 105° = \sin\frac{1}{2}(210°) = \sqrt{\frac{1 - \cos 210°}{2}}$ $= \sqrt{\frac{1.8660}{2}} = 0.9659$

7. $\cos\frac{3\pi}{8} = \cos\frac{1}{2}\left(\frac{3\pi}{4}\right) = \sqrt{\frac{1 + \cos\frac{3\pi}{4}}{2}}$ $= \sqrt{\frac{0.29289}{2}} = 0.3827$

9. $\sin 118° = 0.8829476$ **11.** $\cos 48° = 0.6691$

13. $\pm\sin 4x$ **15.** $\pm\cos 3x$ **17.** $\pm 2\sqrt{2}\sin 5\theta$

19. 1 **21.** $\frac{1}{26}\sqrt{26}$ **23.** 0.1414 **25.** $\pm\sqrt{\frac{2\sec\alpha}{\sec\alpha - 1}}$

27. $\tan\frac{1}{2}\alpha = \frac{1 - \cos\alpha}{\sin\alpha} = \frac{\sin\alpha}{1 + \cos\alpha}$

29. $\frac{1 - \cos\alpha}{2\sin\frac{1}{2}\alpha} = \frac{1 - \cos\alpha}{2\sqrt{\frac{1}{2}(1 - \cos\alpha)}} = \sqrt{\frac{1 - \cos\alpha}{2}}$

31. $2\sqrt{2\frac{(1 + \cos x)(1 + \cos x)}{2(2)(1 + \cos x)}} = \frac{2(1 + \cos x)}{2}\cdot\sqrt{\frac{2}{1 + \cos x}}$

33.

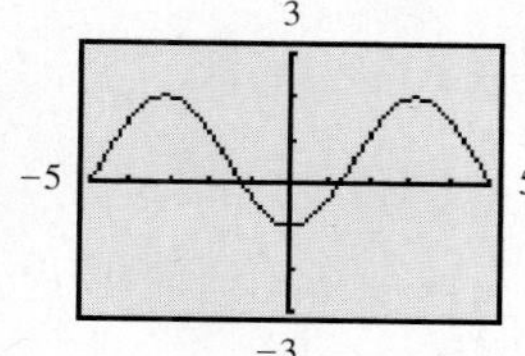

35.

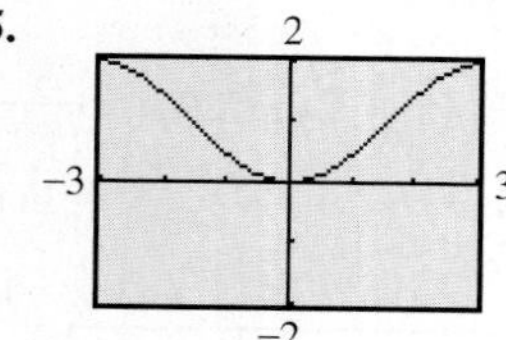

37. $\pm\frac{24}{7}$ **39.** $\sqrt{3 - 2\sqrt{2}}$ **41.** $\frac{1}{5}\sqrt{5}$ **43.** $4\sin\frac{\theta}{2}$

45. $\sin\omega t = \pm\sqrt{\frac{1 - \cos 2\omega t}{2}}$, $\sin^2\omega t = \frac{1}{2}(1 - \cos 2\omega t)$

47. $\frac{\sqrt{\frac{1 - \cos(A + \phi)}{2}}}{\sqrt{\frac{1 - \cos A}{2}}} = \sqrt{\frac{1 - \cos(A + \phi)}{1 - \cos A}}$

Exercises 20.5, page 557

1. $\frac{\pi}{4}, \frac{5\pi}{4}$ **3.** $0, \frac{\pi}{3}, \frac{5\pi}{3}$ **5.** $\frac{\pi}{2}$ **7.** $\frac{\pi}{3}, \frac{5\pi}{3}$ **9.** $\frac{\pi}{3}, \frac{2\pi}{3}, \frac{4\pi}{3}, \frac{5\pi}{3}$

11. $0, \frac{\pi}{6}, \frac{5\pi}{6}, \pi$ **13.** $\frac{\pi}{2}, \frac{3\pi}{2}$ **15.** $\frac{\pi}{4}, \frac{3\pi}{4}, \frac{5\pi}{4}, \frac{7\pi}{4}$

17. 0.2618, 1.309, 3.403, 4.451

19. $0, \pi$ **21.** $\frac{3\pi}{4} = 2.36, \frac{7\pi}{4} = 5.50$

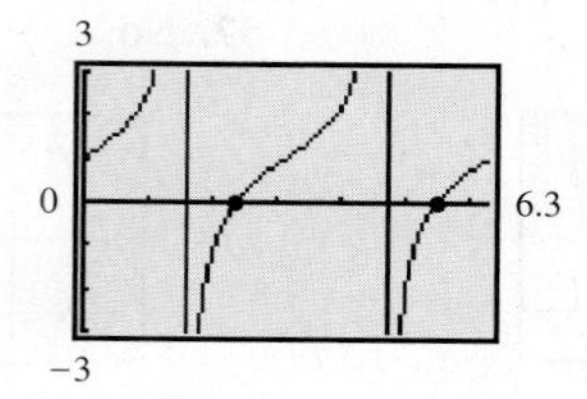

23. 1.9823, 4.3009

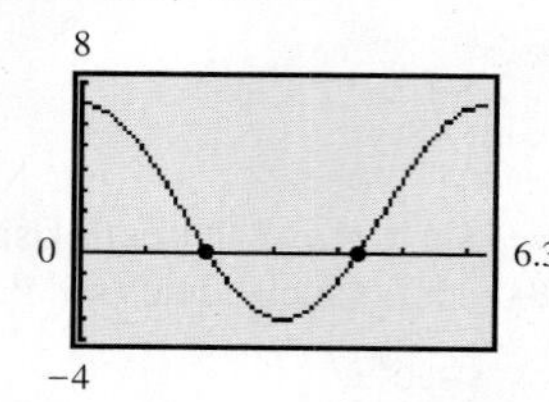

25. $\frac{\pi}{3} = 1.05, \frac{2\pi}{3} = 2.09,$

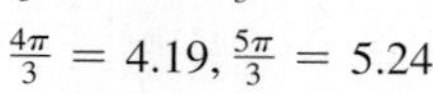

$\frac{4\pi}{3} = 4.19, \frac{5\pi}{3} = 5.24$

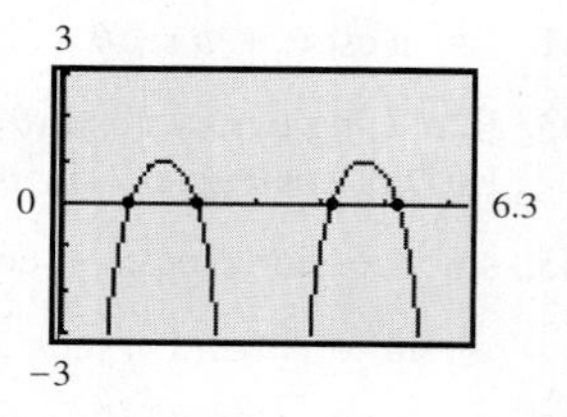

27. $\frac{\pi}{12} = 0.26, \frac{\pi}{4} = 0.79, \frac{5\pi}{12} = 1.31, \frac{3\pi}{4} = 2.36,$ $\frac{13\pi}{12} = 3.40, \frac{5\pi}{4} = 3.93, \frac{17\pi}{12} = 4.45, \frac{7\pi}{4} = 5.50$

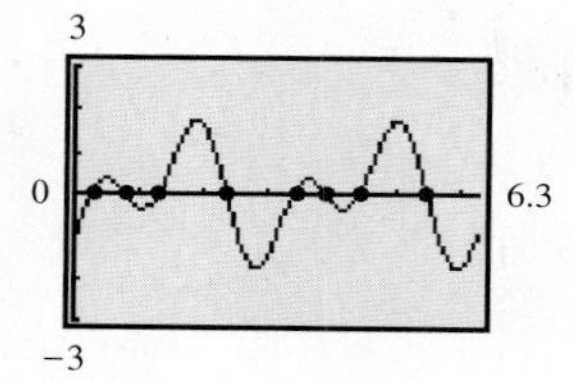

29. $0 = 0.00, \frac{\pi}{3} = 1.05, \pi = 3.14, \frac{5\pi}{3} = 5.24$

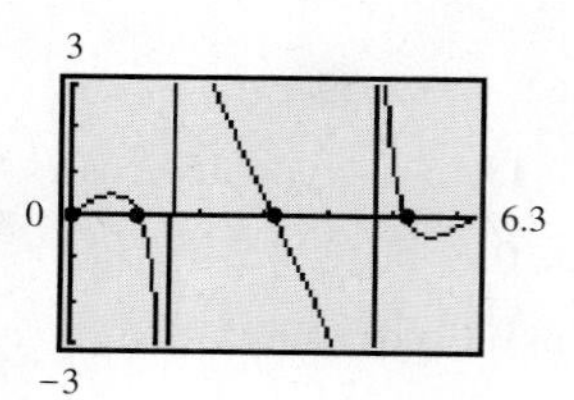

31. 3.569, 5.856

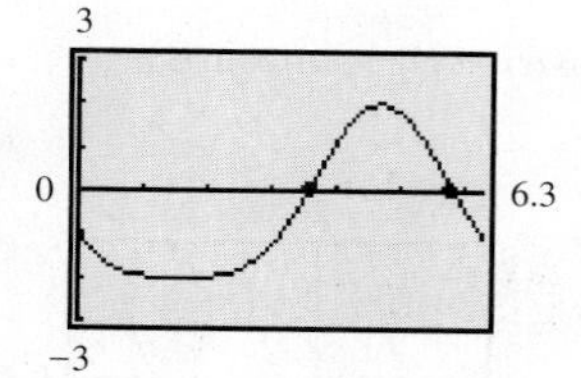

33. 0.7854, 1.249, 3.927, 4.391

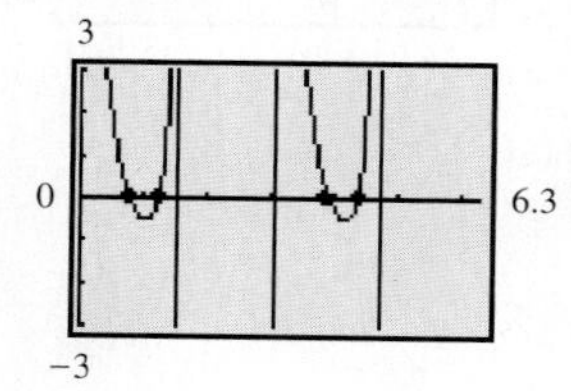

35. $\frac{3\pi}{8} = 1.18, \frac{7\pi}{8} = 2.75, \frac{11\pi}{8} = 4.32, \frac{15\pi}{8} = 5.89$

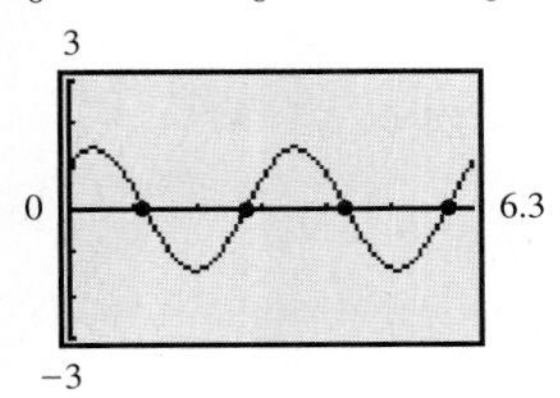

37. $0, \frac{\pi}{2}, \pi, \frac{3\pi}{2}$

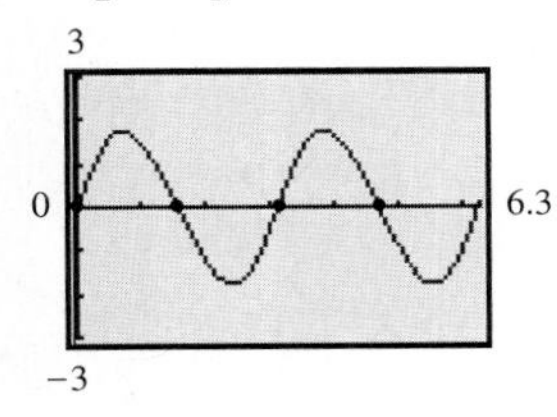

39. $0, \frac{\pi}{2}, \pi, \frac{3\pi}{2}$

41. No: $\cot\theta > 1, \sec\theta > 1, \csc\theta > 1$ for $0 < \theta < \frac{\pi}{2}$

43. $\theta = 0, r = 0; \theta = \frac{\pi}{3}, r = \frac{\sqrt{3}}{2}; \theta = \pi, r = 0; \theta = \frac{5\pi}{3}, r = -\frac{\sqrt{3}}{2}$

45. 30°, 60°, 90° **47.** 37.8° **49.** 10.2 s, 15.7 s, 21.2 s, 47.1 s

51. 6.56×10^{-4} **53.** 300.0 N, 400.0 N

55. −2.28, 0.00, 2.28 **57.** 0.29, 0.95

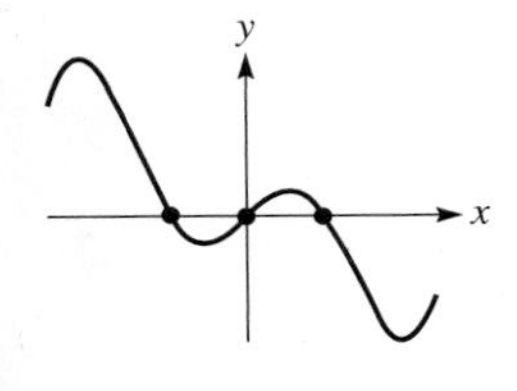

59. 2.10 **61.** 1.08

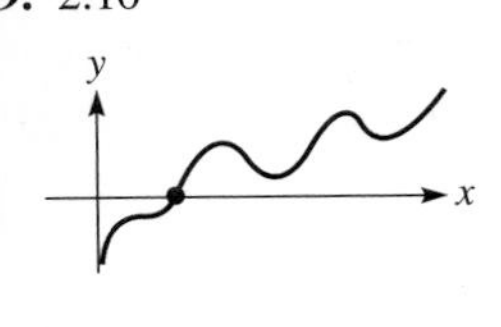

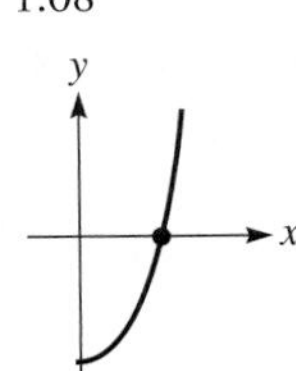

Exercises 20.6, page 563

1. y is the angle whose tangent is $3A$. **3.** 0

5. y is the angle whose tangent is $5x$.

7. y is twice the angle whose sine is x.

9. y is five times the angle whose cosine is $2x - 1$.

11. $\frac{\pi}{3}$ **13.** $\frac{\pi}{4}$ **15.** $-\frac{\pi}{3}$ **17.** No value **19.** $-\frac{\pi}{4}$

21. $\frac{1}{2}\sqrt{3}$ **23.** $\frac{\pi}{4}$ **25.** $\frac{\sqrt{26}}{26}$ **27.** −1 **29.** −1.2303

31. 0.0219 **33.** −1.2389 **35.** −0.2239 **37.** $x = \frac{1}{3}\sin^{-1} y$

39. $x = 4\tan y$ **41.** $x = 1 - \cos(1 - y)$ **43.** $\frac{x}{\sqrt{1 - x^2}}$

45. $\frac{\sqrt{x^2 - 16}}{x}$ **47.** $\frac{3x}{\sqrt{9x^2 - 1}}$ **49.** $2x\sqrt{1 - x^2}$

51. No: $\sin^{-1}(\sin x) = x$ for $-\frac{\pi}{2} \le x \le \frac{\pi}{2}$.

53. A = area of sector − area of triangle

55. $t = \frac{1}{\omega}\left(\sin^{-1}\frac{i}{I_m} - \alpha - \phi\right)$

57. $\sin\left(\sin^{-1}\frac{3}{5} + \sin^{-1}\frac{5}{13}\right) = \frac{3}{5}\cdot\frac{12}{13} + \frac{4}{5}\cdot\frac{5}{13} = \frac{56}{65}$ **59.** $-\frac{\sqrt{2}}{10}$

61. $xy + \sqrt{1 - x^2 - y^2 + x^2y^2}$ **63.** $\frac{\pi}{2}$ **65.** 0

67. $\sin^{-1}\left(\frac{a}{c}\right)$ **69.** $\tan^{-1}\left(\frac{b\tan B}{a}\right)$

71. Let y = height to top of pedestal;
$\tan\alpha = \frac{151 + y}{d}, \tan\beta = \frac{y}{d}; \tan\alpha = \frac{151 + d\tan\beta}{d}$

73. $\theta = \tan^{-1}\left(\frac{y + 50}{x}\right) - \tan^{-1}\left(\frac{y}{x}\right)$

75. $\theta = \pi - \tan^{-1}\left(\frac{50}{x}\right) - \tan^{-1}\left(\frac{25}{35 - x}\right)$

Review Exercises for Chapter 20, page 565

Concept Check Exercises

1. F **2.** T **3.** F **4.** F **5.** F **6.** T

Practice and Applications

7. $\sin(90° + 30°) = \sin 90°\cos 30° + \cos 90°\sin 30°$
$= (1)(\frac{1}{2}\sqrt{3}) + (0)(\frac{1}{2}) = \frac{1}{2}\sqrt{3}$

9. $\sin(360° - 45°) = \sin 360°\cos 45° - \cos 360°\sin 45°$
$= 0(\frac{1}{2}\sqrt{2}) - 1(\frac{1}{2}\sqrt{2}) = -\frac{1}{2}\sqrt{2}$

11. $\cos(2\frac{\pi}{2}) = \cos^2\frac{\pi}{2} - \sin^2\frac{\pi}{2} = 0 - 1^2 = -1$

13. $\tan 2(30°) = \frac{2\tan 30°}{1 - \tan^2 30°} = \frac{2(\sqrt{3}/3)}{1 - (\sqrt{3}/3)^2} = \sqrt{3}$

15. $\sin 50° = 0.76604$ **17.** $\sin(\frac{\pi}{7}) = 0.43388$

19. $\cos 215° = -0.8192$ **21.** $\tan 36° = 0.72654$

23. $\sin 5x$ **25.** $2\sin 14x$ **27.** $2\cos 12x$ **29.** $2\cos x$

31. $-\frac{\pi}{2}$ **33.** 0.5298 **35.** $-\frac{1}{3}\sqrt{3}$ **37.** $-\frac{\pi}{3}$

39. $\sec^2 y - \tan^2 y = 1$

41. $\sin x\csc x - \sin^2 x = 1 - \sin^2 x = \cos^2 x$

43. $\frac{(\sec^2 x - 1)(\sec^2 x + 1)}{\tan^2 x} = \sec^2 x + 1$

45. $2\left(\frac{1}{\sin 2x}\right)\left(\frac{\cos x}{\sin x}\right) = 2\left(\frac{1}{2\sin x\cos x}\right)\left(\frac{\cos x}{\sin x}\right)$
$= \frac{1}{\sin^2 x} = \csc^2 x$

47. $\frac{\cos^2\theta}{\sin^2\theta}$ **49.** $\frac{1}{2}(2\sin\frac{\theta}{2}\cos\frac{\theta}{2})$ **51.** $\cot x$ **53.** $\sec x$

55. $\sin x$ **57.** $\sin x$

59.

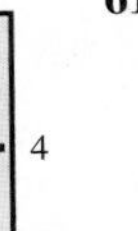

61.

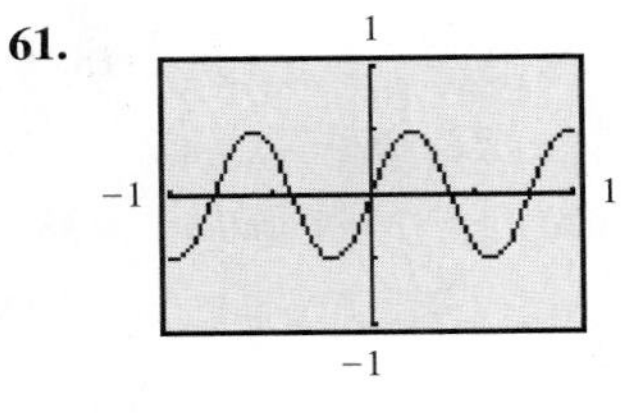

63.

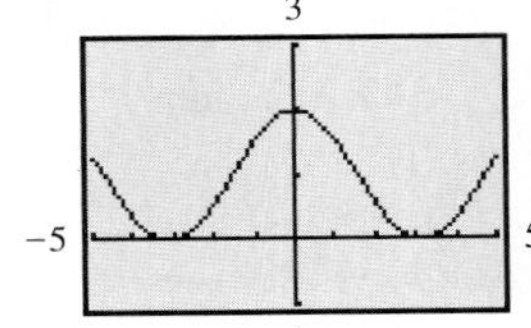

65.

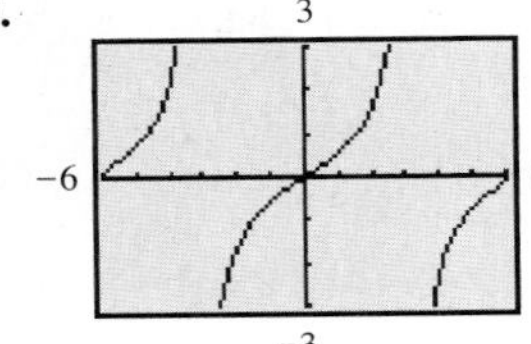

67. $x = \frac{1}{2}\cos^{-1}\frac{1}{2}y$ **69.** $x = \frac{1}{5}\sin\frac{1}{3}\left(\frac{1}{4}\pi - y\right)$

71. 1.2925, 4.4341 **73.** $\frac{\pi}{6}, \frac{5\pi}{6}, \frac{7\pi}{6}, \frac{11\pi}{6}$ **75.** $0, \frac{\pi}{3}, \frac{5\pi}{3}$

77. $0, \frac{2\pi}{3}$ **79.** 0.3142, 1.571, 2.827, 4.084, 4.712, 5.341

81. 0

83. Identify:

$$\tan x + \frac{1}{\tan x} = \frac{\tan^2 x + 1}{\tan x} = \sec^2 x(\cot x) = \frac{\sec^2 x \cos x}{\sin x}$$

85. $\frac{\pi}{2}, \pi$ **87.** 1.56, 2.16, 3.46 **89.** -2.31, 1.14

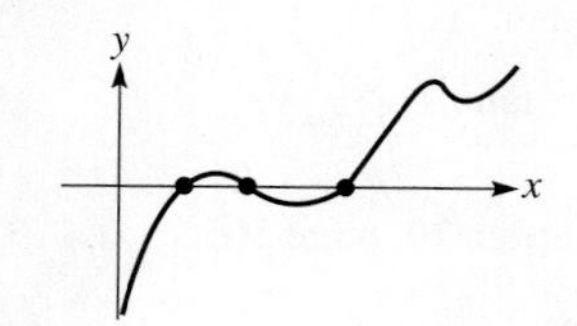

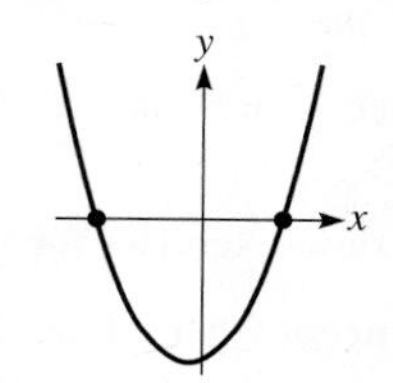

91. $\frac{1}{x}$ **93.** $\frac{\sqrt{25 - x^2}}{5}$ **95.** $\frac{\sqrt{1 - x^2} - xy}{\sqrt{1 + y^2}}$ **97.** $2\sqrt{1 - \cos^2\theta}$

99. $\frac{\tan\theta}{\sqrt{1 + \tan^2\theta}} = \frac{\tan\theta}{\sec\theta}$ **101.** $\frac{24}{25}$ **103.** $\frac{3\sqrt{10}}{10}$

105. $(\cos\theta + j\sin\theta)^2 = (\cos^2\theta - \sin^2\theta) + j(2\sin\theta\cos\theta)$

107. -1 **109.** $\frac{\pi}{2} < x < \frac{3\pi}{2}$

111. $C\left(\frac{A}{C}\sin 2t + \frac{B}{C}\cos 2t\right)$

$= C(\cos\alpha\sin 2t + \sin\alpha\cos 2t)$

113. $\frac{2ab}{c^2}$

115. $R = \sqrt{(A\cos\theta - B\sin\theta)^2 + (A\sin\theta + B\cos\theta)^2}$

$= \sqrt{A^2(\cos^2\theta + \sin^2\theta) + B^2(\sin^2\theta + \cos^2\theta)}$

117. $\frac{k}{2}\cdot\frac{1}{\sin^2\frac{\theta}{2}} = \frac{k}{2}\cdot\frac{1}{\frac{1 - \cos\theta}{2}}$

119. $\theta = \alpha + R\sin\omega t$

121. $\frac{\cos 2\alpha}{2\cos^2\alpha}$

123. $p = VI\cos\omega t[\cos(\phi + \omega t)]$

125. (a) $\theta = \tan^{-1}\left(\frac{x + 53.3}{y}\right) - \tan^{-1}\left(\frac{x}{y}\right)$ (b) 14.8°

127. 13.6 ft **129.** 54.7°

Exercises 21.1, page 572

1. $\sqrt{61}$ **3.** 150° **5.** $2\sqrt{29}$ **7.** 3 **9.** 55

11. $2\sqrt{41}$ **13.** 2.86 **15.** $\frac{5}{2}$ **17.** Undefined **19.** $-\frac{3}{4}$

21. $\frac{1}{9}$ **23.** 0.747 **25.** $\frac{1}{3}\sqrt{3}$ **27.** -0.0664 **29.** 20.0°

31. 98.5° **33.** Parallel **35.** Perpendicular **37.** -2

39. -3 **41.** Two sides equal $2\sqrt{10}$. **43.** $m_1 = \frac{5}{12}, m_2 = \frac{4}{3}$

45. 10 **47.** $4\sqrt{10} + 4\sqrt{2} = 18.3$ **49.** (1.5)

51. $(-2.8, 4.2)$ **53.** $x^2 + y^2 = 9$ **55.** $m_1 = 1, m_2 = -1$

57. $\left(-\frac{11}{3}, 0\right)$ **59.** $\left(0, -\frac{7}{9}\right)$ **61.** No **63.** 66.7 mm

65. 1700 km

Exercises 21.2, page 577

1. $x + 2y - 2 = 0$ **3.** $2, \left(0, \frac{5}{2}\right)$

5. $4x - y + 20 = 0$ **7.** $7x - 2y - 24 = 0$

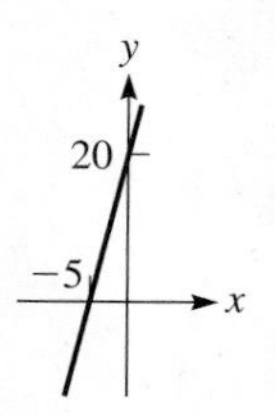

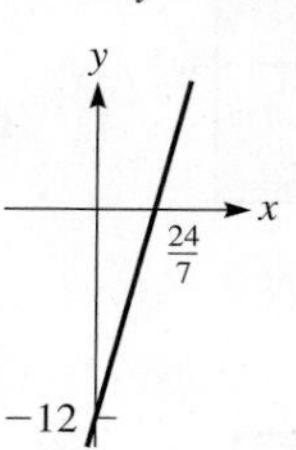

9. $x - y + 19 = 0$ **11.** $y = -2.7$

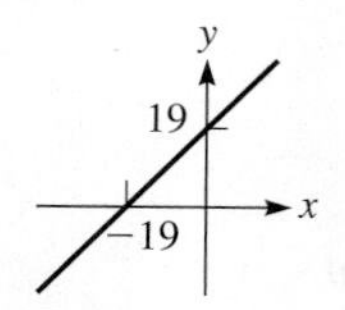

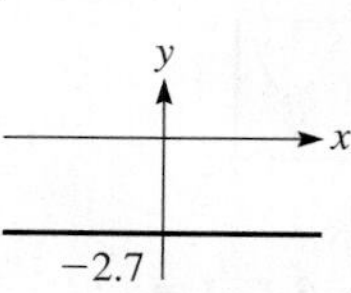

13. $x = -3$ **15.** $x - 3y - 7 = 0$

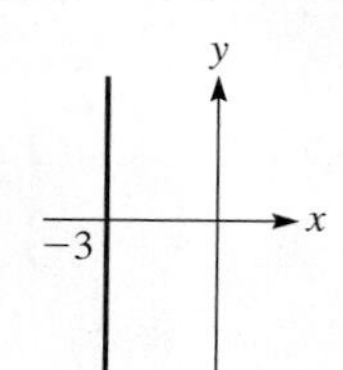

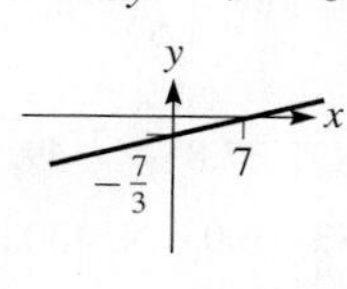

17. $x + y - 7 = 0$ **19.** $3x + y - 18 = 0$

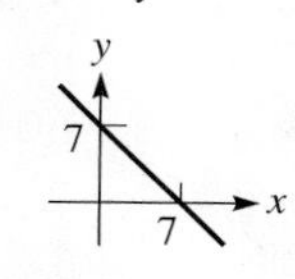

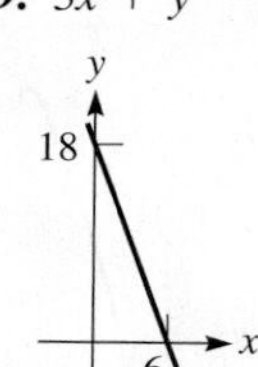

21. $y = 4x - 8; m = 4, (0, -8)$

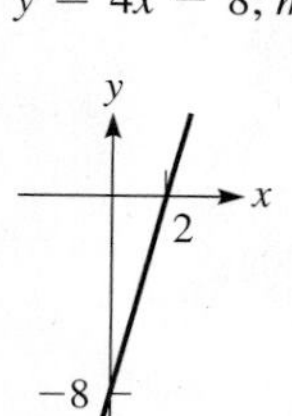

23. $y = -\frac{3}{5}x + 2; m = -\frac{3}{5}, (0, 2)$

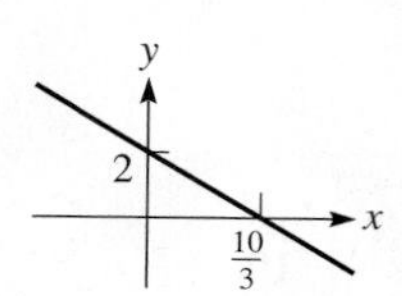

25. $y = \frac{3}{2}x - \frac{1}{2}; m = \frac{3}{2}, \left(0, -\frac{1}{2}\right)$

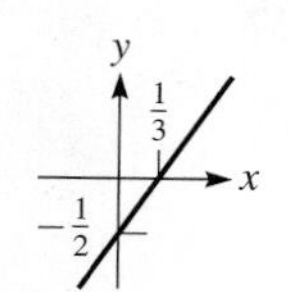

27. $y = 3.5x + 0.5;\ m = 3.5,\ (0, 0.5)$

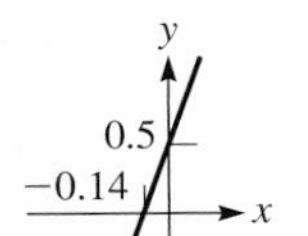

29. Parallel

31. Perpendicular **33.** Neither **35.** Perpendicular

37. -2

39. The slope of the first line is 3. A line perpendicular to it has a slope of $-\frac{1}{3}$. The slope of the second line is $-\frac{k}{3}$, so $k = 1$.

41. 5

43. $m_1 = -\frac{4}{5}, m_2 = \frac{2}{3}, m_3 = \frac{2}{3}; m_4 = -\frac{4}{5}; m_1 = m_4, m_2 = m_3$

45. $m_1 = -\frac{a}{b}, m_2 = -\frac{a}{b}, m_1 = m_2; m_3 = -\frac{a}{b}, m_4 = \frac{b}{a}, m_3 = 1/m_4$

47. All parallel, $m = 2$ **49.** $y - mx - b = 0$

51. $F = \frac{9}{4}R + 32$ **53.** $v = 0.607T + 331$

55. $T = \frac{4}{3}x + 3$; If the distance increases by 1 cm, the temperature increases by $\frac{4}{3}$°C.

57. $w = 30 - 2t$

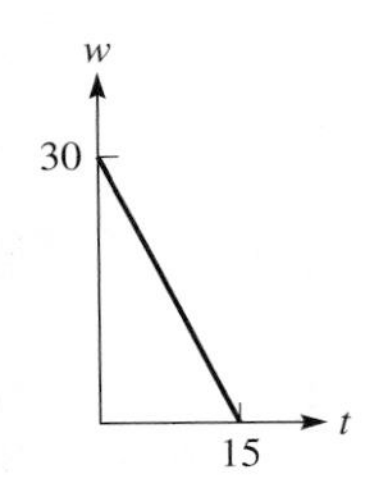

59. $y = 10^{-5}(2.4 - 5.6x)$

61. $n = \frac{7}{6}t + 10$; at 6:30, $n = 10$; at 8.30, $n = 150$; For each minute that passes, $\frac{7}{6}$ more cars pass that point.

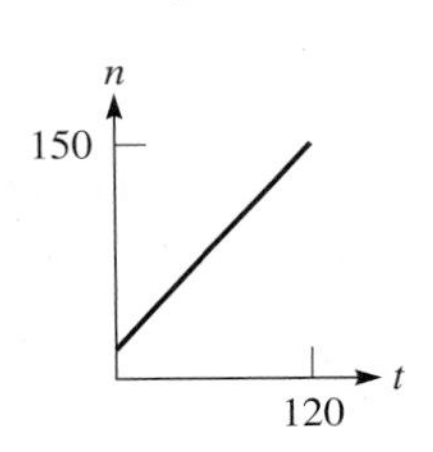

63. $p = 9.789d + 101.286$; 444 kPa; interpolation

65.

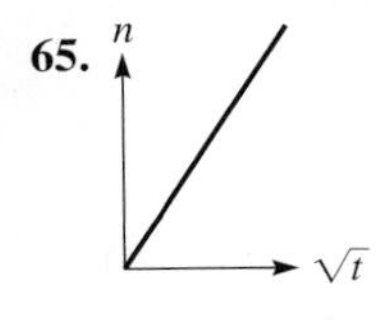

67.

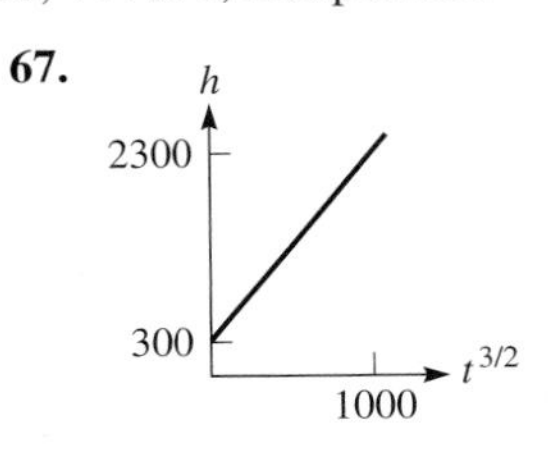

Exercises 21.3, page 582

1. $C(1, -1), r = 4$ **3.** $C(3, 4), r = 7$

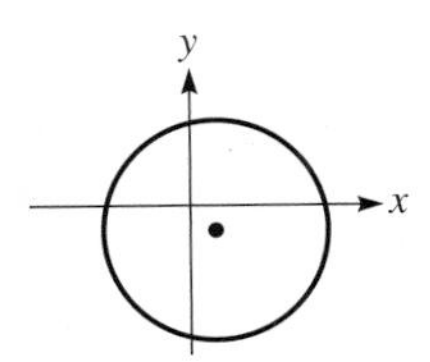

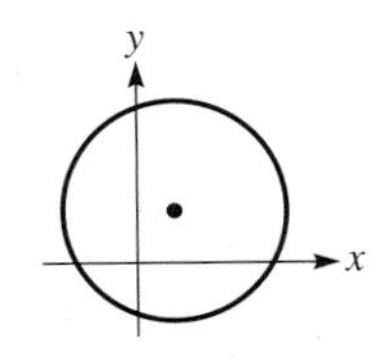

5. $(2, 1), r = 5$ **7.** $C(-1, 0), r = \frac{11}{2}$ **9.** $x^2 + y^2 = 9$

11. $(x - 2)^2 + (y - 3)^2 = 16$, or $x^2 + y^2 - 4x - 6y - 3 = 0$

13. $(x - 12)^2 + (y + 15)^2 = 324$, or $x^2 + y^2 - 24x + 30y + 45 = 0$

15. $(x + 3)^2 + (y - 4)^2 = 25$ **17.** $(x - 2)^2 + (y - 1)^2 = 8$

19. $(x + 3)^2 + (y - 5)^2 = 25$, or $x^2 + y^2 + 6x - 10y + 9 = 0$

21. $(x + 2)^2 + (y - 2)^2 = 4$, or $x^2 + y^2 + 4x - 4y + 4 = 0$

23. $x^2 + y^2 = 2$

25. $(0, 3), r = 2$

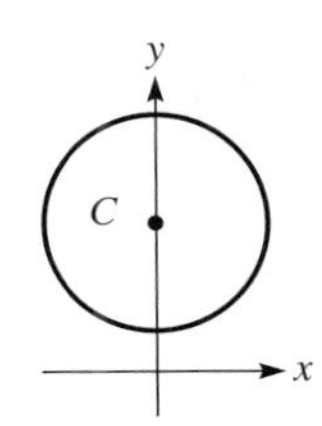

27. $(-1, 5), r = \frac{9}{2}$

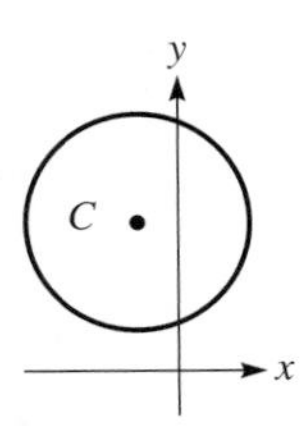

29. $(1, 0), r = 3$

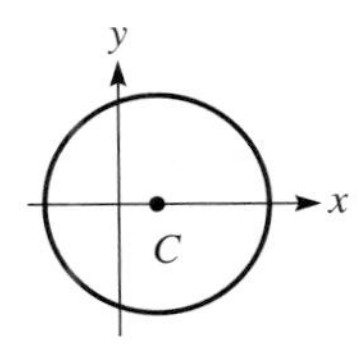

31. $(-2.1, 1.3), r = 3.1$

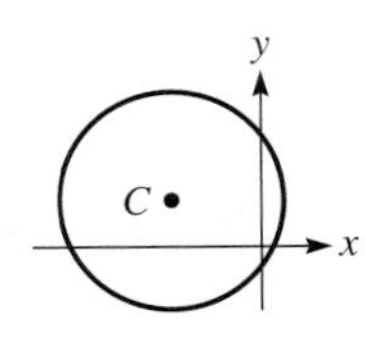

33. $(0, 2), r = \frac{5}{2}$

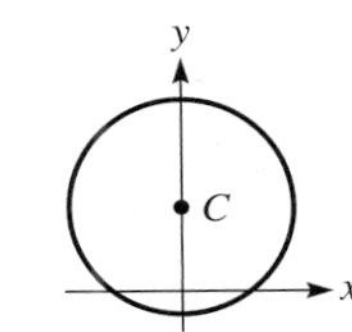

35. $(1, 2), r = \frac{1}{2}\sqrt{22}$

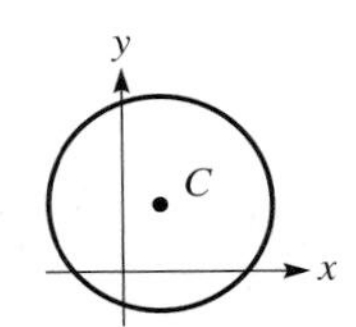

37. Symmetric to both axes and origin

39. Symmetric to y-axis **41.** 64 sq. units **43.** (1,0), (5,0)

45. 16π unit^2 **47.** $(7, 0), (-1, 0)$

49. The x- and y-axes may not be scaled equally.

51.

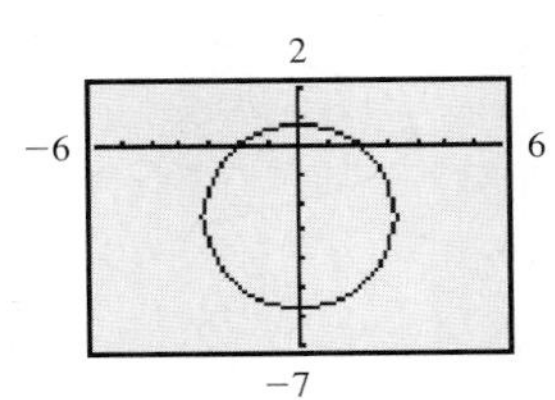

53. (a) Semicircle
(b) Semicircle
(c) Yes, there is only one value of y for each x in the domain.

55. Outside

57. $p > 0$, circle; $p = 0$, point; $p < 0$, does not exist

59. 0.0912 in. **61.** 2.82 in.

63. $x^2 + y^2 = 0.0100$ **65.** $x^2 + (y + 7)^2 = 1$

67. $(x - 500 \times 10^{-6})^2 + y^2 = 0.16 \times 10^{-6}$

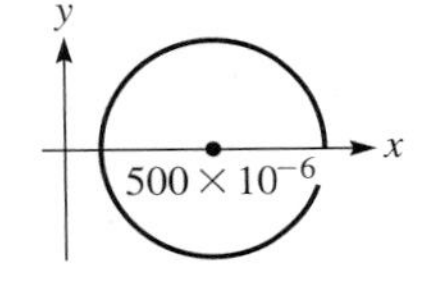

Exercises 21.4, page 586

1. $F(5, 0), x = -5$

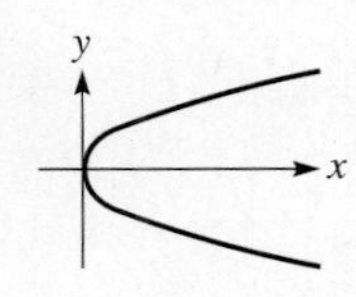

3. $F(0, -\frac{3}{2}), y = \frac{3}{2}$

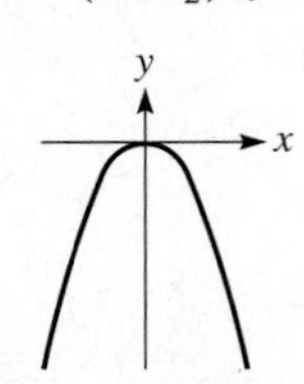

5. $F(1, 0), x = -1$

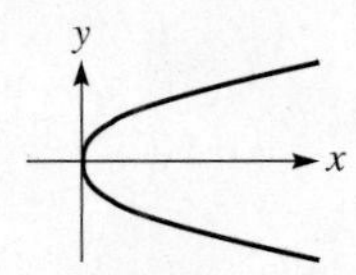

7. $F(-16, 0), x = 16$

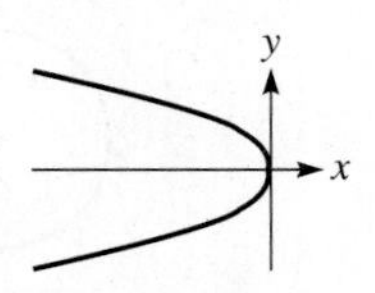

9. $F(0, 18), y = -18$

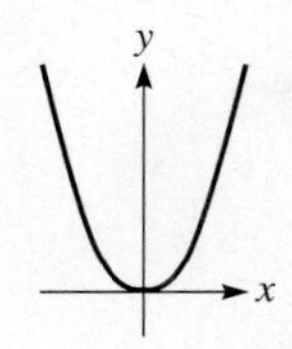

11. $F(0, -1), y = 1$

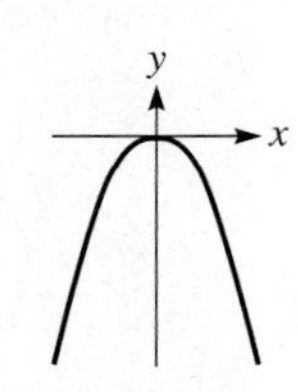

13. $F(\frac{5}{8}, 0), x = -\frac{5}{8}$

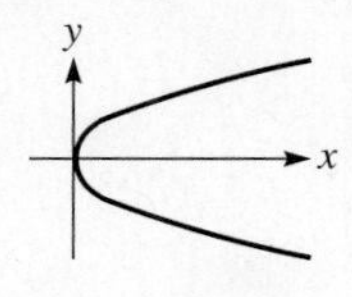

15. $F(0, \frac{25}{48}), y = -\frac{25}{48}$

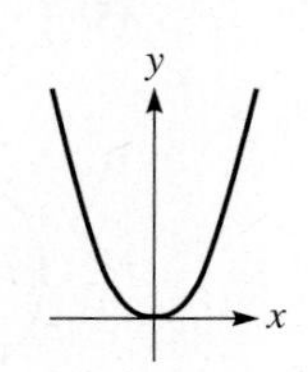

17. $y^2 = 12x$ **19.** $x^2 = -2y$ **21.** $x^2 = 0.64y$

23. $x^2 = 336y$ **25.** $x^2 = \frac{1}{8}y$ **27.** $y^2 = \frac{25}{3}x$

29. $y^2 = 3x$

31. $(0, 0), (8, -4)$

33. $y^2 - 2y - 12x + 37 = 0$

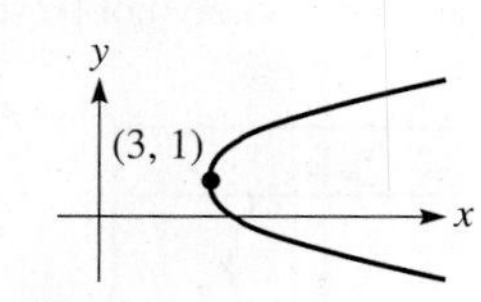

35.

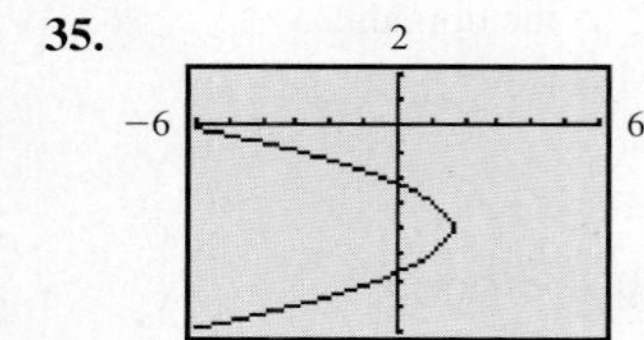

37.

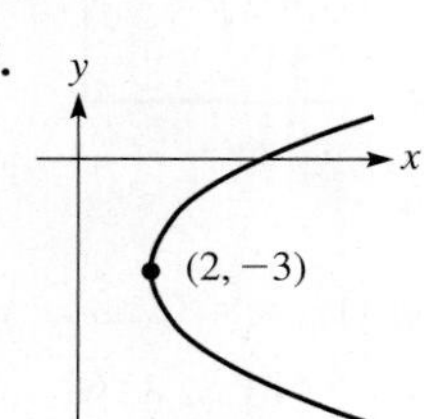

39. 4p **41.** $y^2 = 2x$ **43.** 32.0 cm **45.** 59.0 ft

47. 10^7 km **49.** 57.6 m

51. No (a course correction is necessary)

53. 2.16 cm from vertex

55.

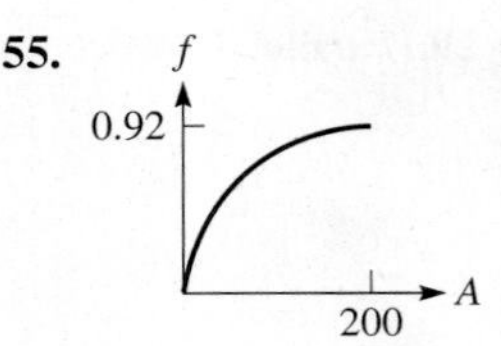

57. $y^2 = 8x$ or $x^2 = 8y$ with vertex midway between island and shore

Exercises 21.5, page 592

1. $V(0, 6), V(0, -6)$, ends minor axis $(5, 0), (-5, 0)$ foci $(0, \sqrt{11}), (0, -\sqrt{11})$

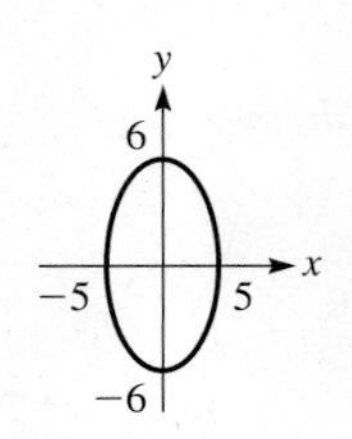

3. $V(2, 0), V(-2, 0), F(\sqrt{3}, 0), F(-\sqrt{3}, 0)$

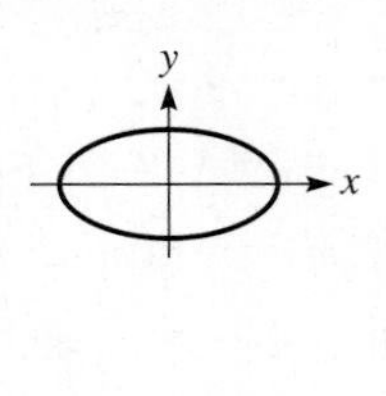

5. $V(0, 12), V(0, -12), F(0, \sqrt{119}), F(0, -\sqrt{119})$

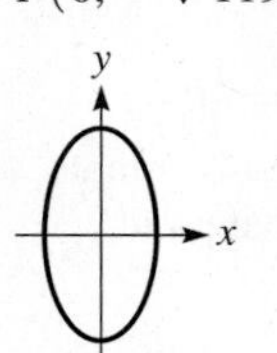

7. $V(\frac{5}{2}, 0), V(-\frac{5}{2}, 0), F(\frac{\sqrt{209}}{6}, 0), F(-\frac{\sqrt{209}}{6}, 0)$

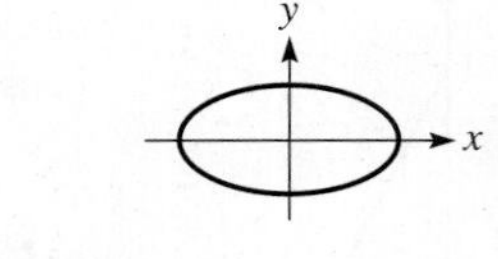

9. $V(9, 0), V(-9, 0), F(3\sqrt{5}, 0), F(-3\sqrt{5}, 0)$

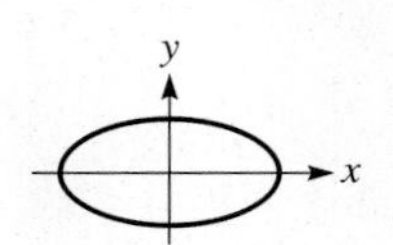

11. $V(0, \frac{7}{2}\sqrt{2}), V(0, -\frac{7}{2}\sqrt{2}), F(0, \frac{1}{2}\sqrt{82}), F(0, -\frac{1}{2}\sqrt{82})$

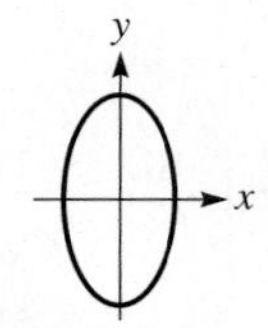

13. $V(0, 4), V(0, -4), F(0, \sqrt{14}), F(0, -\sqrt{14})$

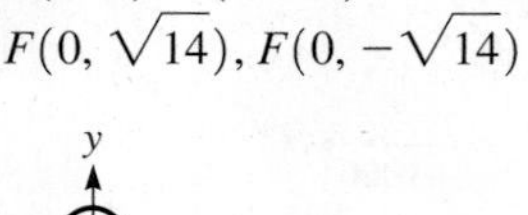

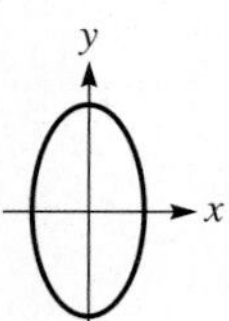

15. $V(0.25, 0), V(-0.25, 0), F(0.23, 0), F(-0.23, 0)$

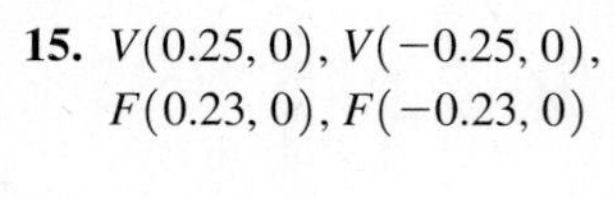

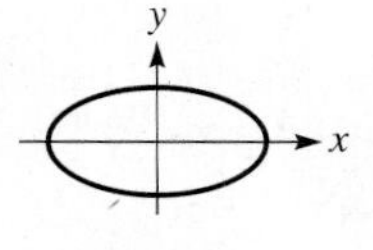

17. $\frac{x^2}{225} + \frac{y^2}{144} = 1$, or $144x^2 + 225y^2 = 32{,}400$

19. $\frac{x^2}{225} + \frac{y^2}{289} = 1$, or $289x^2 + 225y^2 = 65{,}025$

21. $\frac{x^2}{208} + \frac{y^2}{144} = 1$, or $144x^2 + 208y^2 = 29{,}952$

23. $\frac{x^2}{64} + \frac{15y^2}{144} = 1$, or $3x^2 + 20y^2 = 192$

25. $\frac{x^2}{5} + \frac{y^2}{20} = 1$, or $4x^2 + y^2 = 20$

27. $\frac{x^2}{100} + \frac{y^2}{64} = 1$, or $16x^2 + 25y^2 = 1600$

29. $(1, -2), (1, 2)$

31. $16x^2 + 25y^2 - 32x - 50y - 359 = 0$

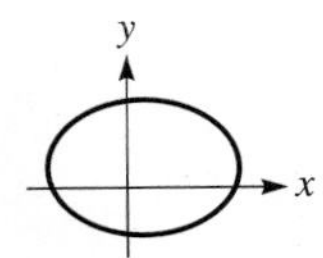

33.

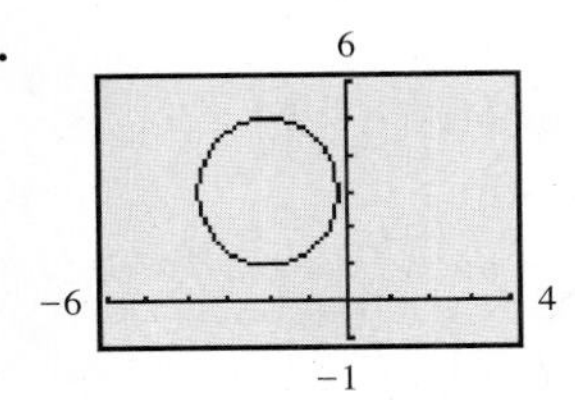

35.

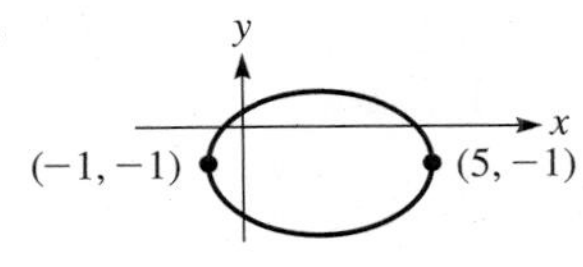

37. Write equation as $\frac{x^2}{1} + \frac{y^2}{1/k} = 1$. Thus, $\sqrt{\frac{1}{k}} > 1$, or $0 < k < 1$.

39. $2x^2 + 3y^2 - 8x - 4 = 2x^2 + 3(-y)^2 - 8x - 4$

41. $\sin^2 t + \cos^2 t = \frac{x^2}{4} + \frac{y^2}{9} = 1$

43. $\frac{x^2}{256} + \frac{y^2}{25} = 1$

45. $i_1^2 + 4i_2^2 = 32$ **47.** 0.44 **49.** $7x^2 + 16y^2 = 112$

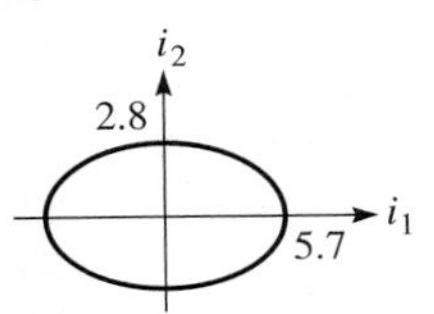

51. 27.5m **53.** 13 ft **55.** 843 ft^3

Exercises 21.6, page 597

1. $V(0, -4), V(0, 4)$,
conj. axis $(-2, 0), (2,0)$
$F(0, -2\sqrt{5}), F(0, 2\sqrt{5})$

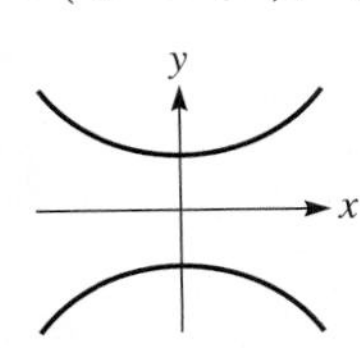

3. $V(5, 0), V(-5, 0)$,
$F(13, 0), F(-13, 0)$

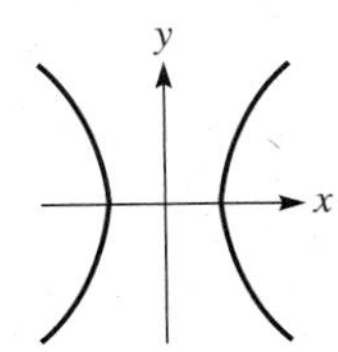

5. $V(0, 3), V(0, -3), F(0, \sqrt{10})$,
$F(0, -\sqrt{10})$

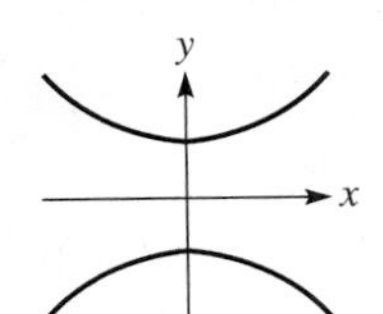

7. $V(-\frac{5}{2}, 0), V(\frac{5}{2}, 0)$,
$F(-\frac{1}{2}\sqrt{41}, 0)$,
$F(\frac{1}{2}\sqrt{41}, 0)$

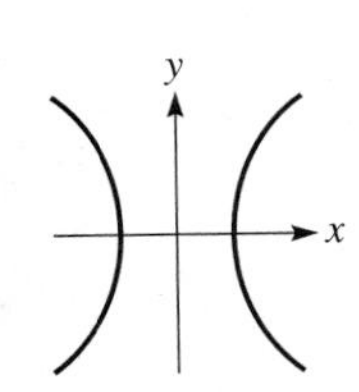

9. $V(\frac{2}{3}, 0), V(-\frac{2}{3}, 0), F(\frac{2}{3}\sqrt{10}, 0)$
$F(-\frac{2}{3}\sqrt{10}, 0)$

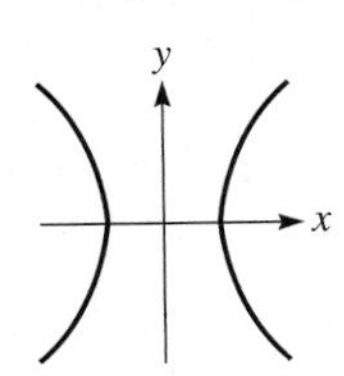

11. $V(0, \sqrt{10}), V(0, -\sqrt{10})$,
$F(0, \sqrt{14}), F(0, -\sqrt{14})$

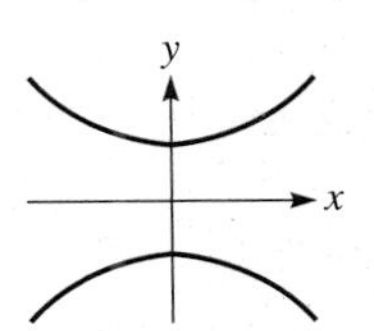

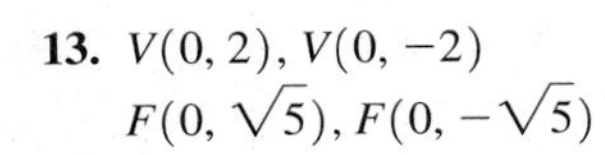

13. $V(0, 2), V(0, -2)$
$F(0, \sqrt{5}), F(0, -\sqrt{5})$

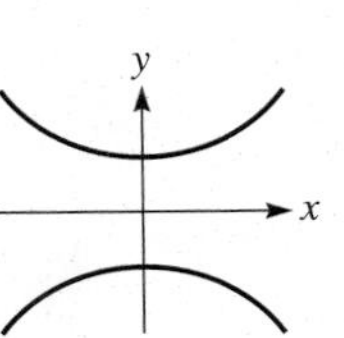

15. $V(0.4, 0), V(-0.4, 0)$
$F(0.9, 0), F(-0.9, 0)$

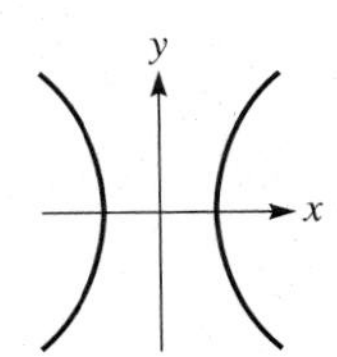

17. $\frac{x^2}{9} - \frac{y^2}{16} = 1$, or $16x^2 - 9y^2 = 144$

19. $\frac{y^2}{100} - \frac{x^2}{576} = 1$, or $144y^2 - 25x^2 = 14{,}400$

21. $\frac{x^2}{1} - \frac{y^2}{3} = 1$, or $3x^2 - y^2 = 3$

23. $\frac{x^2}{5} - \frac{y^2}{4} = 1$, or $4x^2 - 5y^2 = 20$

25. $x^2 - \frac{y^2}{4} = 1$, or $4x^2 - y^2 = 4$

27. $\frac{x^2}{36} - \frac{y^2}{64} = 1$, or $16x^2 - 9y^2 = 576$

29.

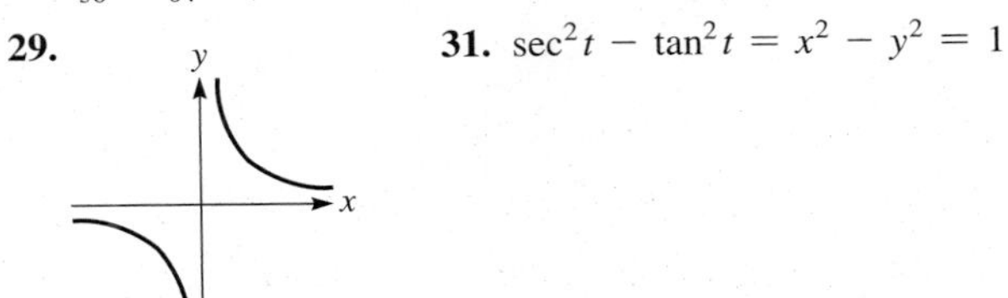

31. $\sec^2 t - \tan^2 t = x^2 - y^2 = 1$

33. $(-2, -3), (-2, 3), (2, -3), (2, 3)$

35. $9x^2 - 16y^2 - 108x + 64y + 116 = 0$

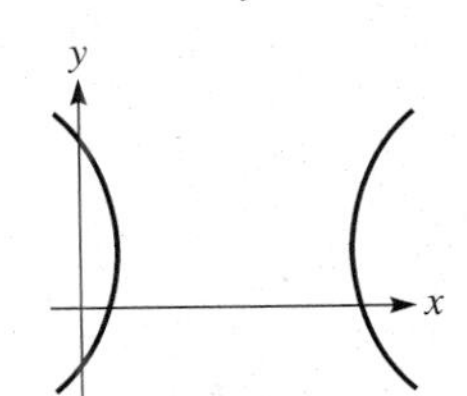

37.

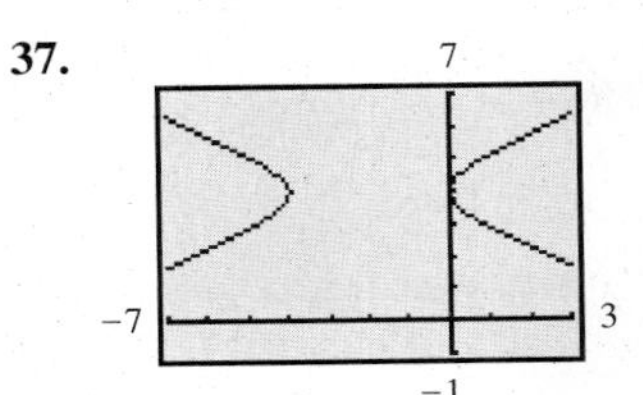

39.

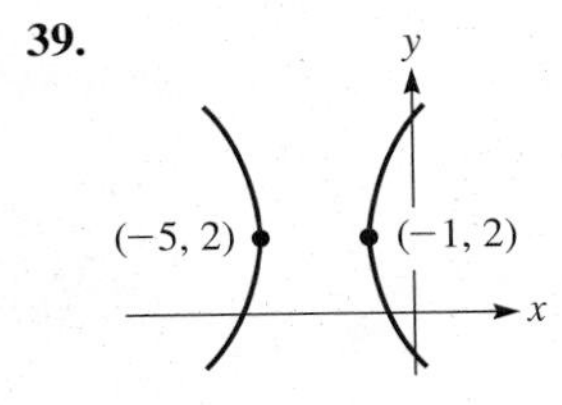

41. $\frac{y^2}{16} - \frac{x^2}{9} = 1$ **43.** $9x^2 - 16y^2 = 144$

45. (a) $8x^2 - y^2 = 32$,
(b) $5x^2 - 4y^2 = 80$

47. $3y^2 - x^2 = 27$

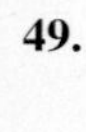

49.

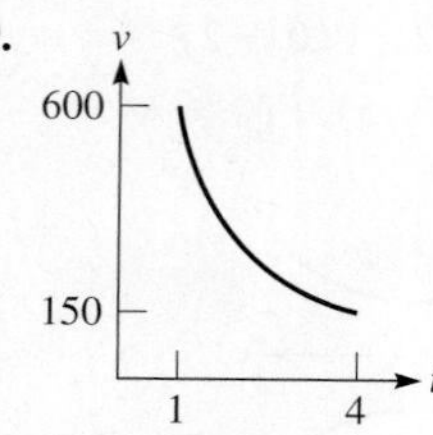

51. $i = 6.00/R$

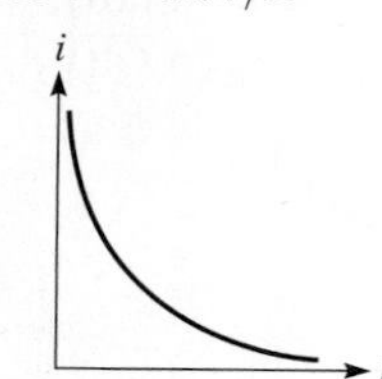

53.

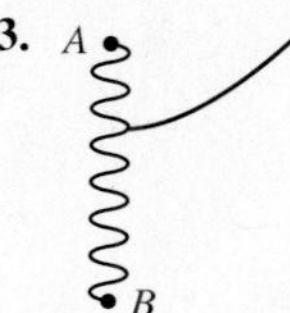

Exercises 21.7, page 601

1. Hyperbola, $C(3, 2)$ transverse axis parallel to x-axis; $a = 5, b = 3$

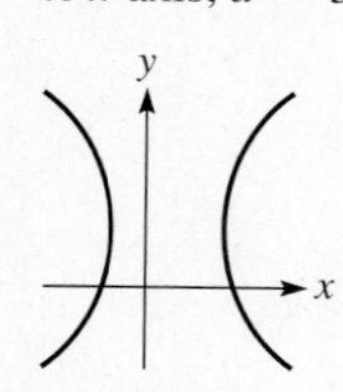

3. Parabola, $(-1, 2)$

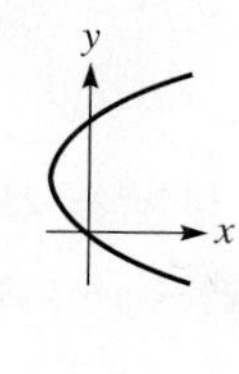

5. Hyperbola, $(1, 2)$

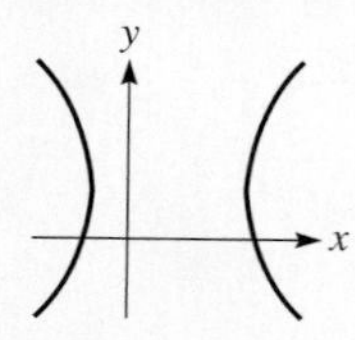

7. Ellipse, $(-1, 0)$

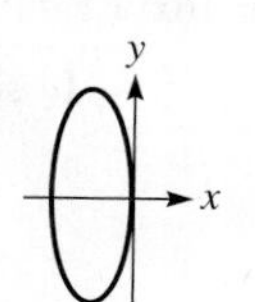

9. Parabola, $(-3, 1)$

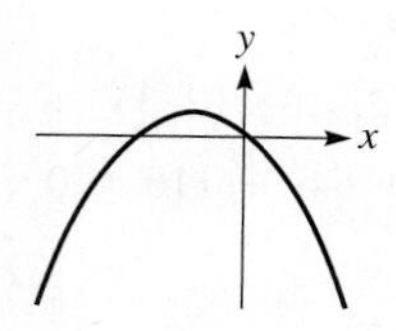

11. $(x + 1)^2 = 12(y - 3)$ or $x^2 + 2x - 12y + 37 = 0$

13. $y^2 = 20(x - 5)$

15. $\frac{(x + 2)^2}{25} + \frac{(y - 2)^2}{16} = 1$, or $16x^2 + 25y^2 + 64x - 100y - 236 = 0$

17. $\frac{(y - 1)^2}{16} + \frac{(x + 2)^2}{4} = 1$, or $4x^2 + y^2 + 16x - 2y + 1 = 0$

19. $\frac{(y - 2)^2}{1} - \frac{(x + 1)^2}{3} = 1$, or $x^2 - 3y^2 + 2x + 12y - 8 = 0$

21. $\frac{(x + 1)^2}{9} - \frac{(y - 1)^2}{16} = 1$, or $16x^2 - 9y^2 + 32x + 18y - 137 = 0$

23. Parabola, $(-1, -1)$

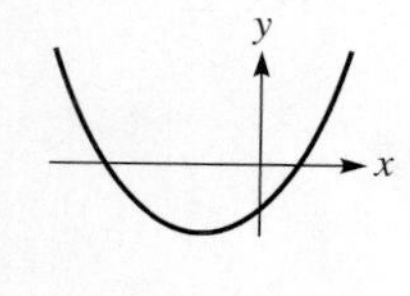

25. Parabola, $(0, 6)$

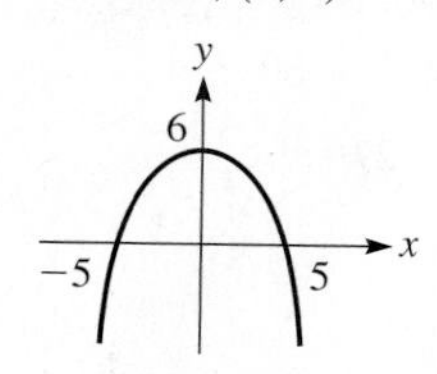

27. Ellipse, $(-3, 0)$

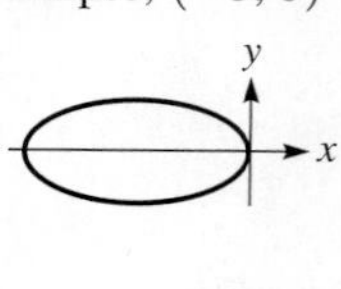

29. Hyperbola, $(0, 4)$

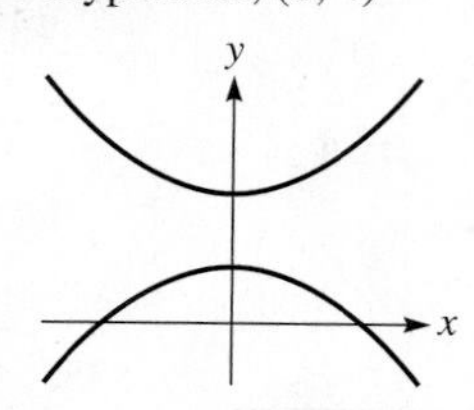

31. Hyperbola, $(-2, 1)$

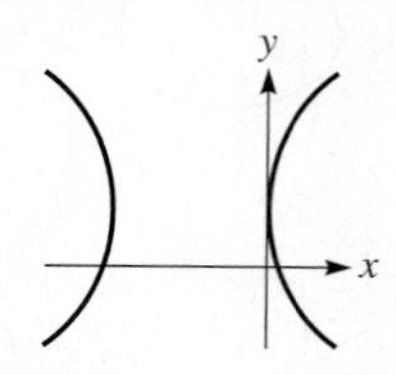

33. Hyperbola, $(-4, 5)$

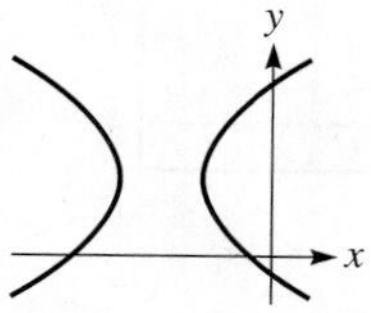

35. Ellipse, $(1, -2)$

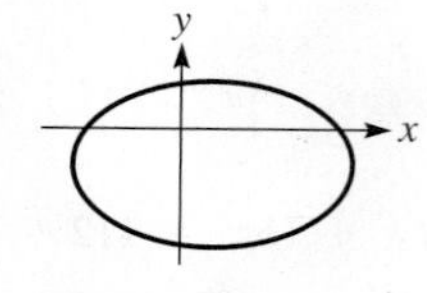

37. Hyperbola, $(4, 0)$

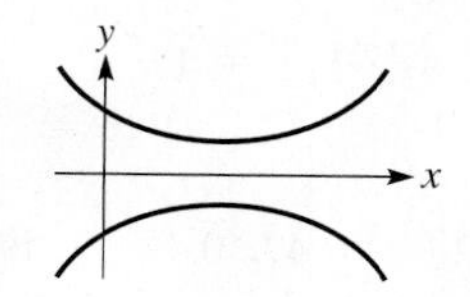

39. Circle, $(\frac{1}{3}, \frac{4}{3})$

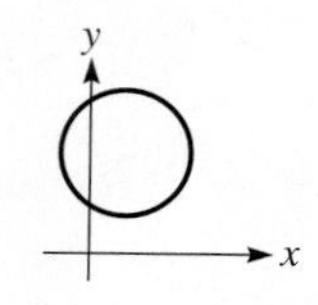

41. $x^2 - y^2 + 4x - 2y - 22 = 0$ **43.** $y^2 + 4x - 4 = 0$

45. $y^2 = 4p(x - h)$ **47.** 1 or -3 **49.** $i' = \sin 2\pi t'$

51. $(x - 95)^2 = -\frac{95^2}{60}(y - 60)$

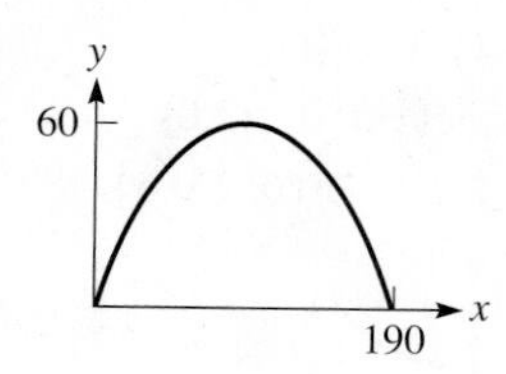

53. $\frac{x^2}{9.0} + \frac{y^2}{16} = 1, \frac{(x - 7.0)^2}{16} + \frac{y^2}{9.0} = 1$

Exercises 21.8, page 604

1. Hyperbola **3.** Ellipse **5.** Hyperbola **7.** Circle

9. Parabola **11.** Hyperbola **13.** Circle

15. None (straight line) **17.** Hyperbola **19.** Hyperbola

21. None (point at origin) **23.** Hyperbola

25. Parabola: $V(-4, 0)$: $F(-4, 2)$

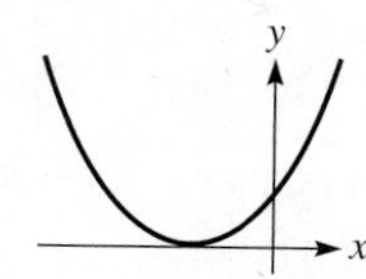

27. Hyperbola: $C(1, -2)$; $V(1, -2 \pm \sqrt{2})$

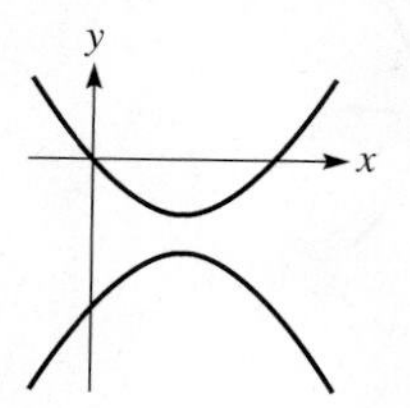

29. Ellipses; $C(5, 0)$; $V(5, \pm 2\sqrt{2})$

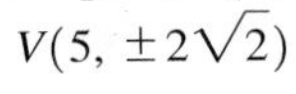

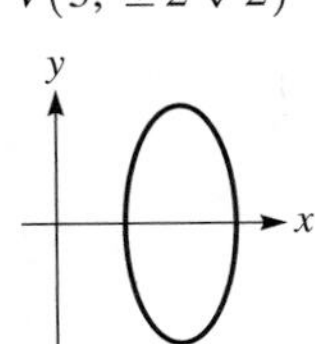

31. Ellipse

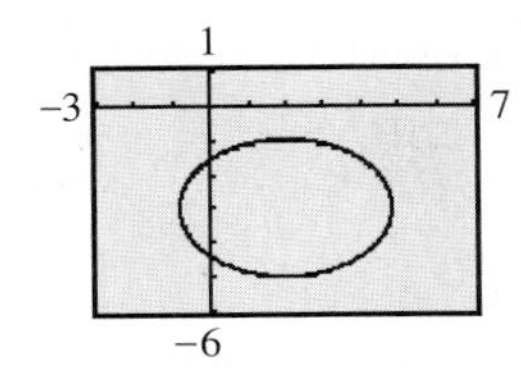

33. Parabola

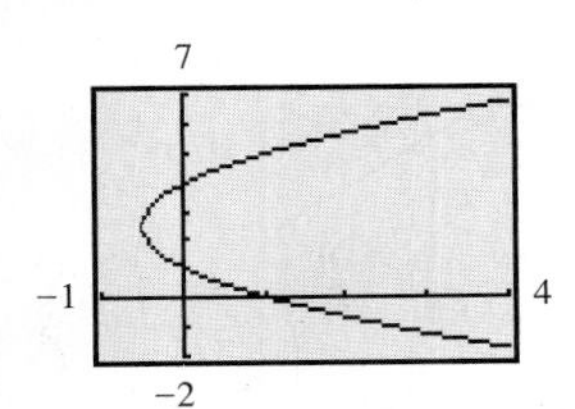

35. Hyperbola

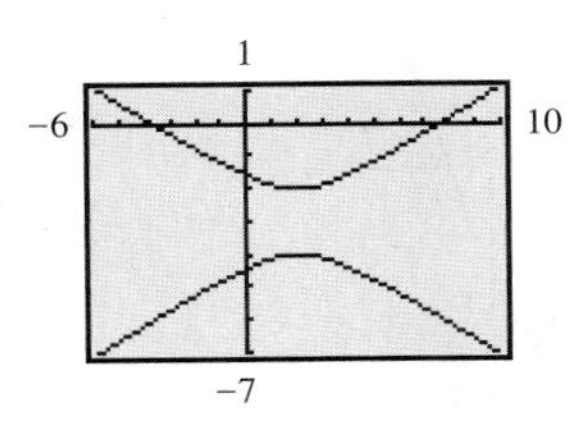

Enter both functions (from quadratic formula).

37. (a) Circle (b) Hyperbola (c) Ellipse

39. Straight line **41.** Circle **43.** Hyperbola

45. Parabola **47.** One branch of a hyperbola

Exercises 21.9, page 608

1. Hyperbola; $2x'y' + 25 = 0$

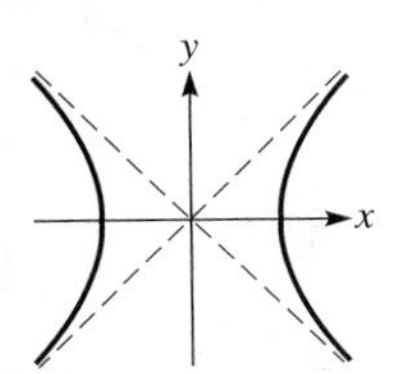

3. Ellipse; $4x'^2 + 9y'^2 = 36$

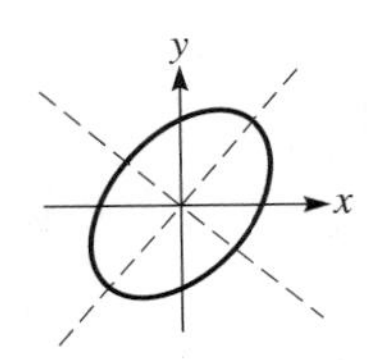

5. Hyperbola **7.** Parabola **9.** Ellipse

11. Parabola; $x'^2 + \sqrt{2}y' = 0$

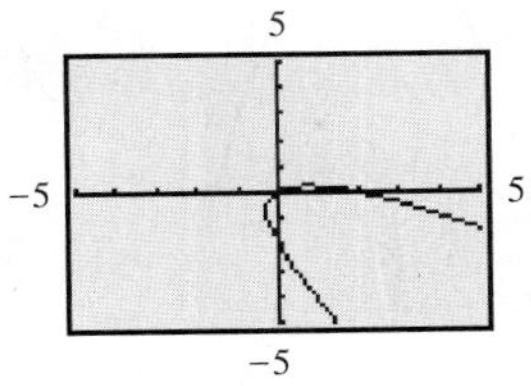

13. $y'^2 - x'^2 = 16$

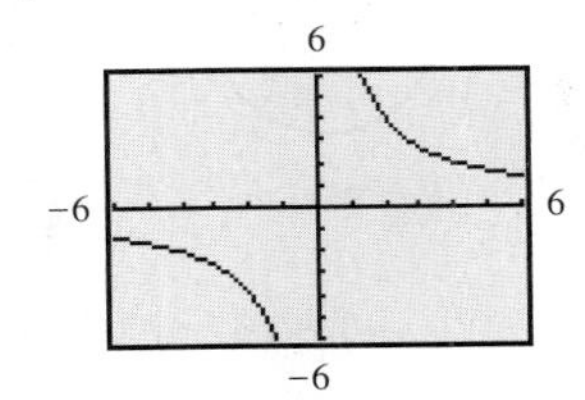

15. Hyperbola; $4x'^2 - y'^2 = 4$

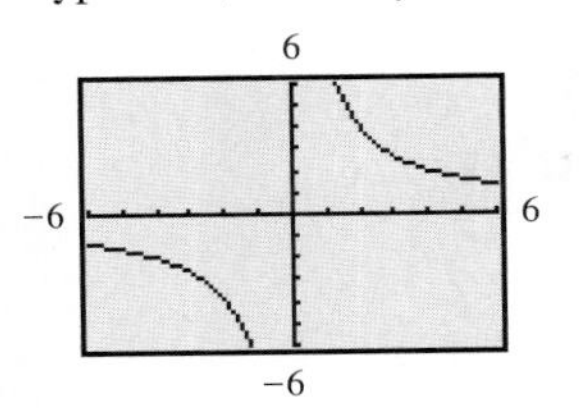

17. Ellipse; $x'^2 + 2y'^2 = 2$

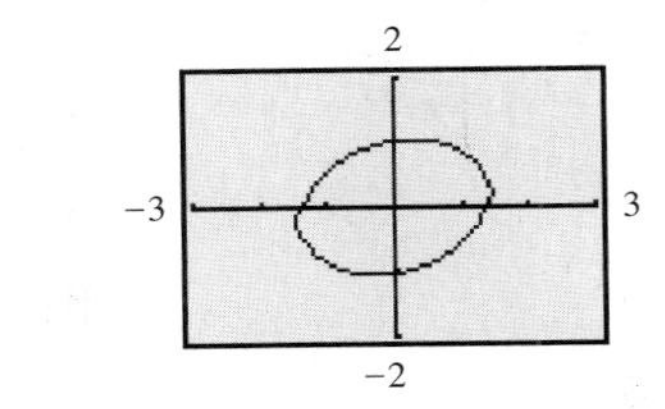

19. Parabola
$y'^2 - 4x' + 16 = 0$
$y''^2 = 4x''$

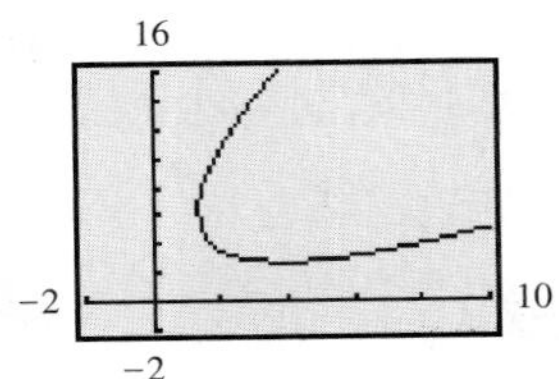

21. Ellipse. The single point (0, 0). **23.** $(-1 + 3\sqrt{3}, 3 + \sqrt{3})$

25. $y' = -\log_2 x'$

Exercises 21.10, page 611

1. $(3, \frac{-5\pi}{3}), (3, \frac{4\pi}{3})$

3. (5.83, 2.11)

5.

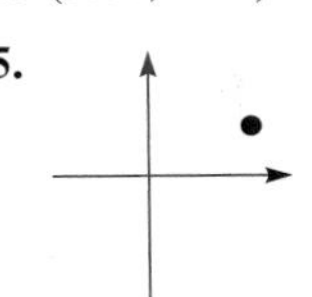

7.

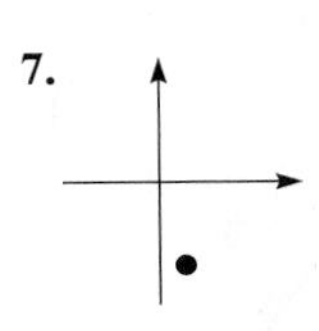

9.

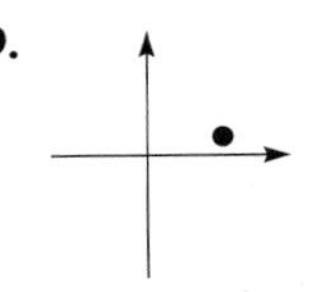

11. 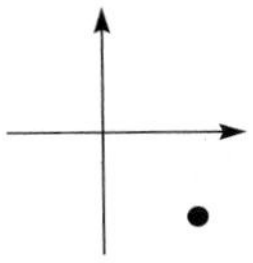 **13.** 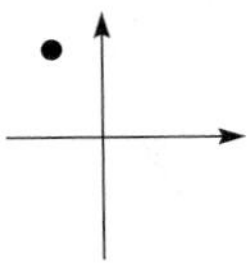 **15.**

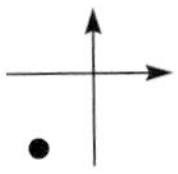

17. $(2, \frac{\pi}{6})$ **19.** $(1, \frac{7\pi}{6})$ **21.** $(4, \frac{\pi}{2})$ **23.** $(-4, -4\sqrt{3})$

25. (−2.76, 1.17) **27.** (−1.16, −7.91) **29.** $r = 3\sec\theta$

31. $r = \frac{-3}{\cos\theta + 2\sin\theta}$ **33.** $r = 4\sin\theta$

35. $r^2 = \frac{4}{1 + 3\sin^2\theta}$ **37.** $r = 6\sin\theta$

39. $x^2 + y^2 - y = 0$, circle **41.** $x = 4$, straight line

43. $x - 3y - 2 = 0$, straight line

45. $x^2 + y^2 - 4x - 2y = 0$, circle

47. $(x^2 + y^2)^2 = 2xy$ **49.** Yes [as $(-2, \frac{7\pi}{4})$]

51. $x^2 + y^2 - bx - ay = 0$

53. $(2, 0), (2, \frac{\pi}{3}), (2, \frac{2\pi}{3}), (2, \frac{4\pi}{3}), (2, \frac{5\pi}{3})$

55. $B_x = -\frac{k\sin\theta}{r}, B_y = \frac{k\cos\theta}{r}$ **57.** $x^2 + y^2 - 5x = 0$

59. 13.8 km

Exercises 21.11, page 615

1. $\theta = \frac{5\pi}{6}$ **3.** $r = 2\sin 2\theta$

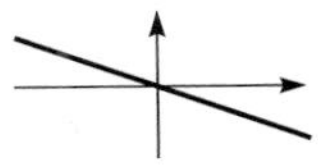

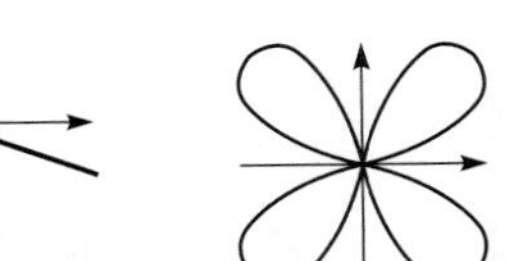

5.

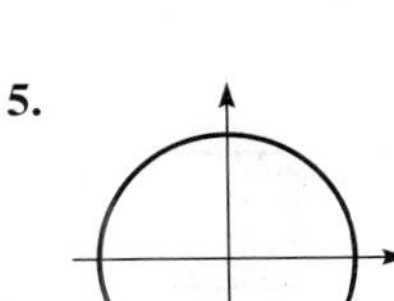

7.

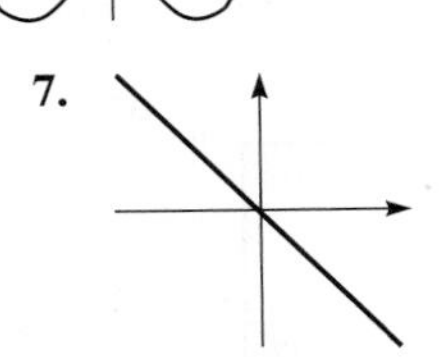

9. 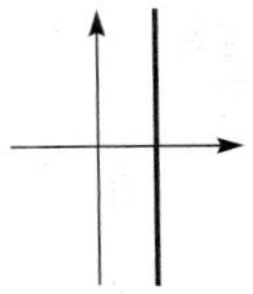 **11.** 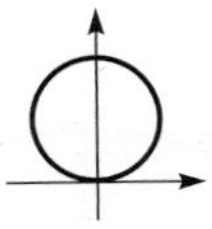 **13.**

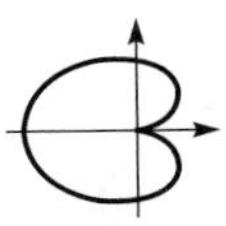

15.
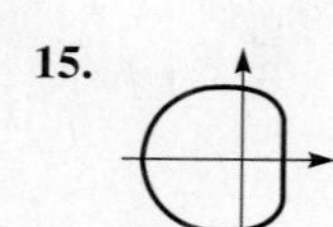

17.
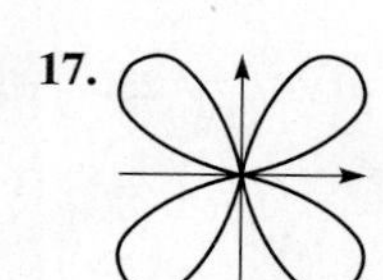

19.
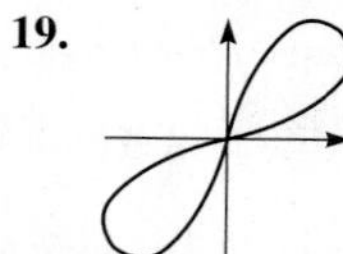

21.
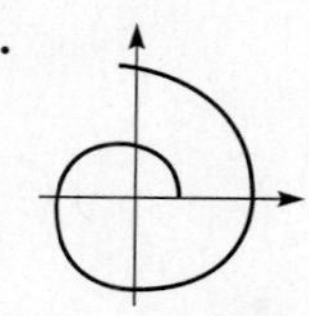

23.
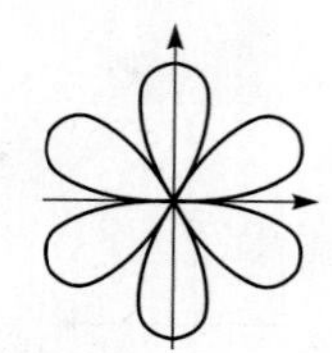

25.
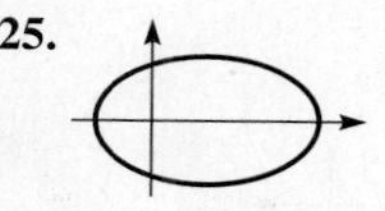

27.
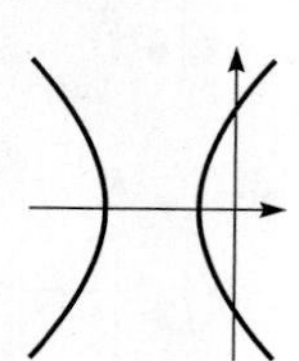

29.
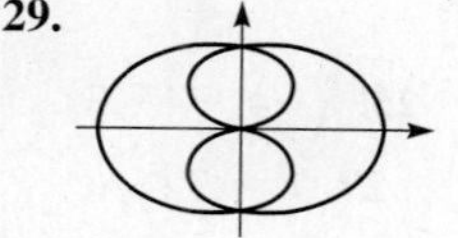

31.
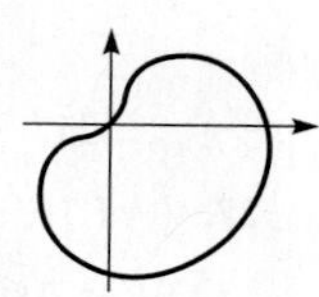

33.
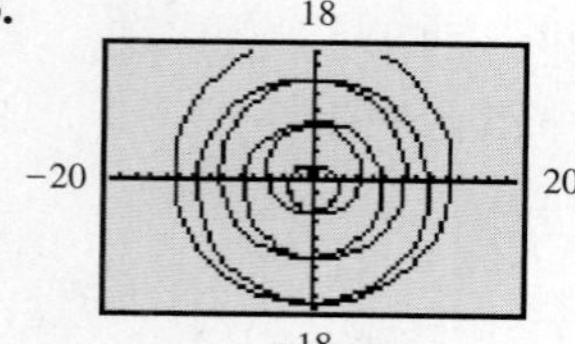

35.
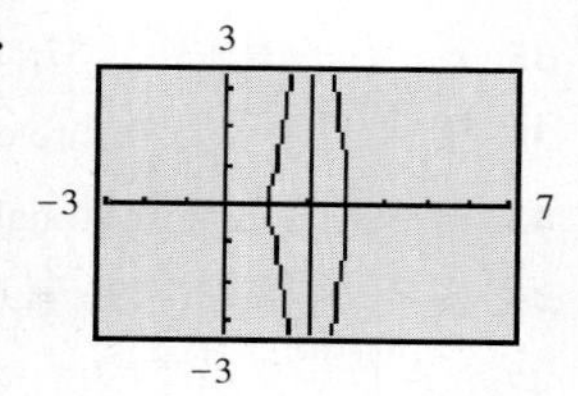

37.
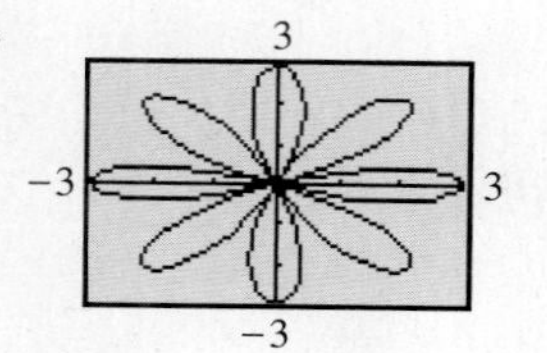

39.
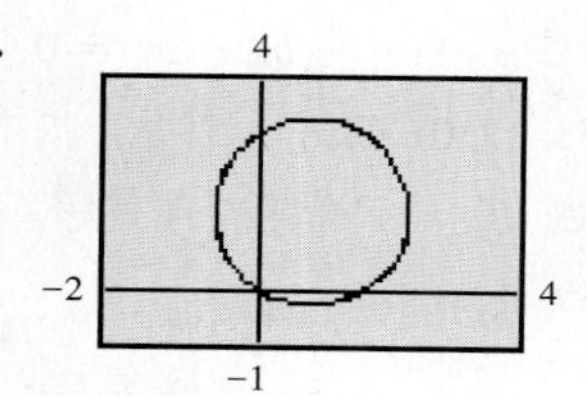

41.
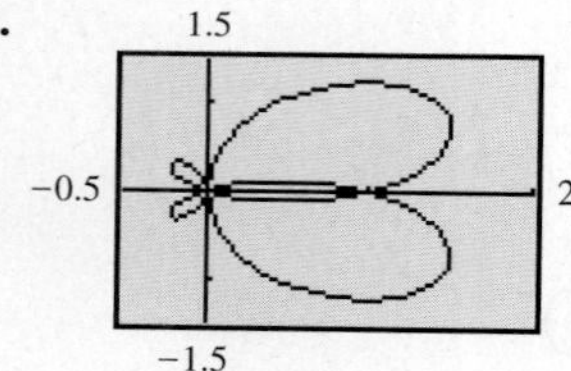

43. Straight line $y = x$
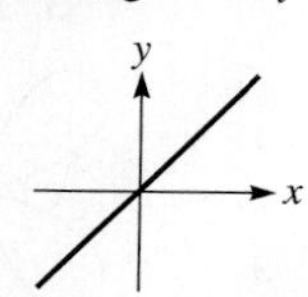

45.
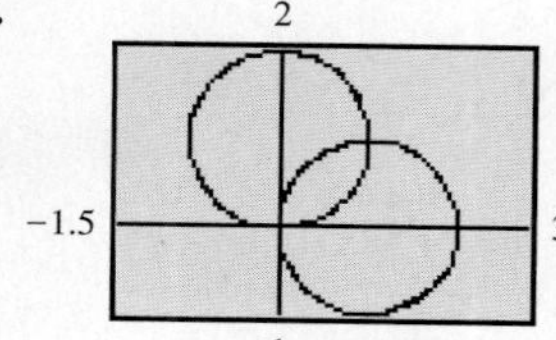

47.
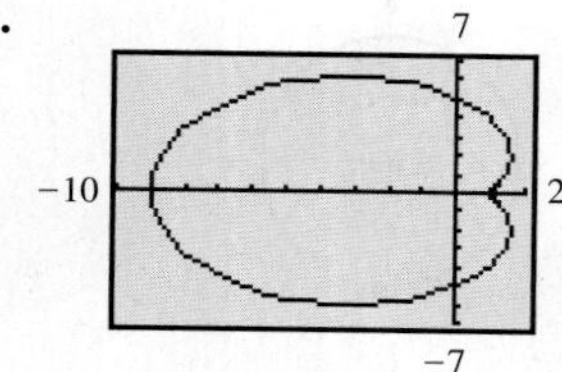

49.
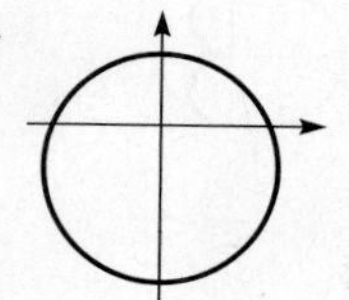

51.
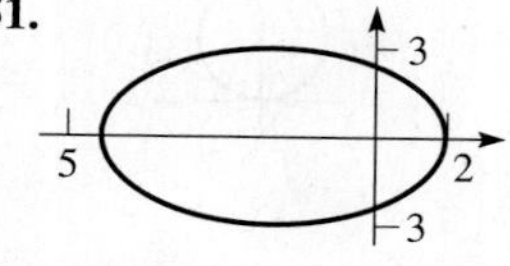

53.
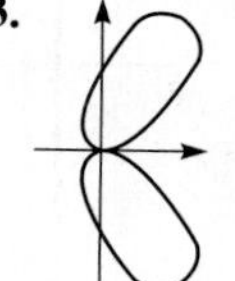

Review Exercises for Chapter 21, page 617

Concept Check Exercises

1. T **2.** F **3.** F **4.** F **5.** F **6.** T **7.** F
8. T **9.** T **10.** T

Practice and Applications

11. $4x - y - 11 = 0$
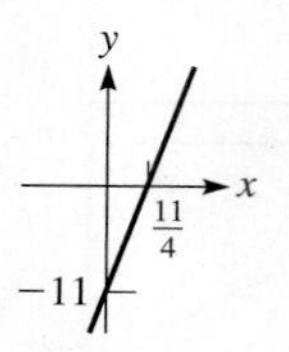

13. $2x + 3y + 3 = 0$

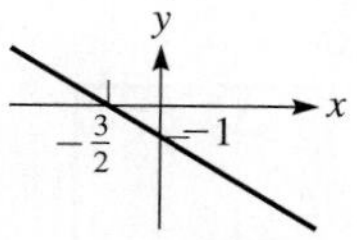

y
x
$-\frac{3}{2}$
−1

15. $x^2 - 6x + y^2 - 1 = 0$
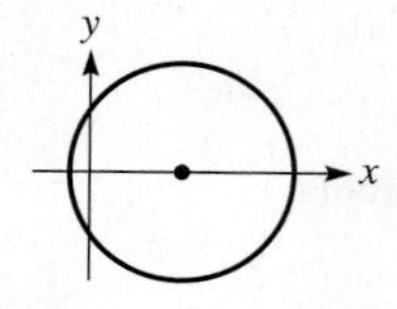

17. $y^2 = -12x$
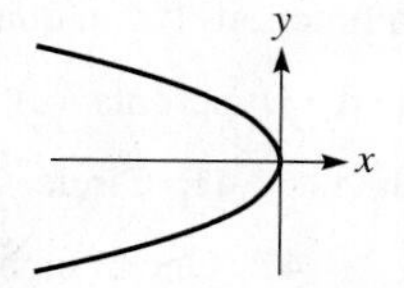

19. $9x^2 + 25y^2 = 900$
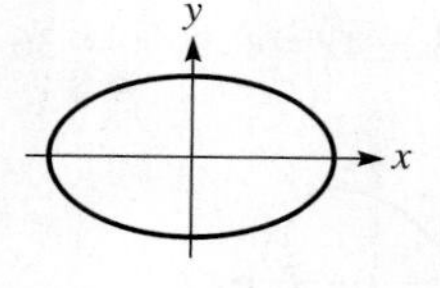

21. $144y^2 - 169x^2 = 24{,}336$
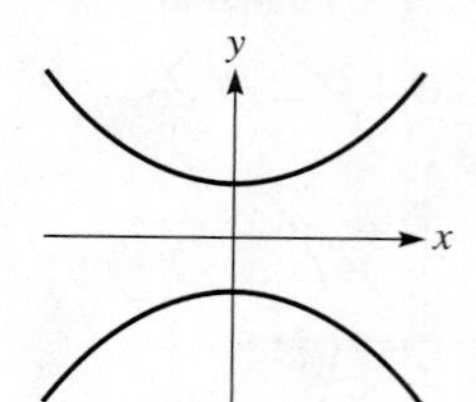

23. $(-3, 0), r = 4$
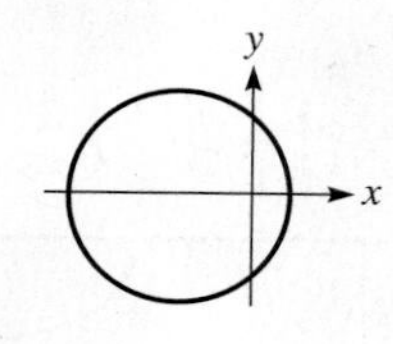

25. $(0, -5), y = 5$
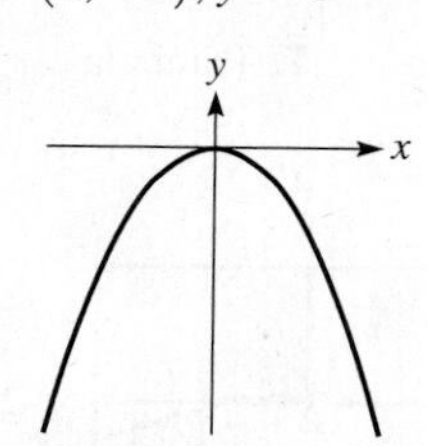

27. $V(0, 1), V(0, -1), F(0, \frac{1}{2}\sqrt{3}), F(0, -\frac{1}{2}\sqrt{3})$
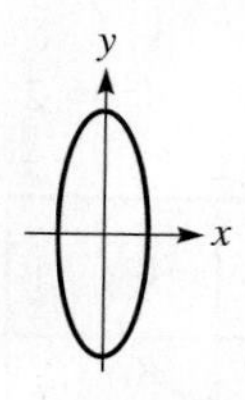

29. $V(\frac{1}{4}, 0), V(-\frac{1}{4}, 0), F(\frac{\sqrt{29}}{20}, 0), F(-\frac{\sqrt{29}}{20}, 0)$
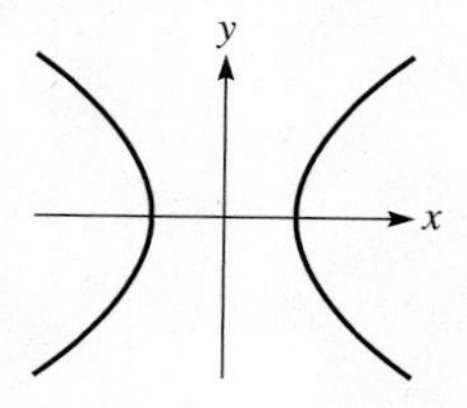

31. $V(4, -8), F(4, -7)$

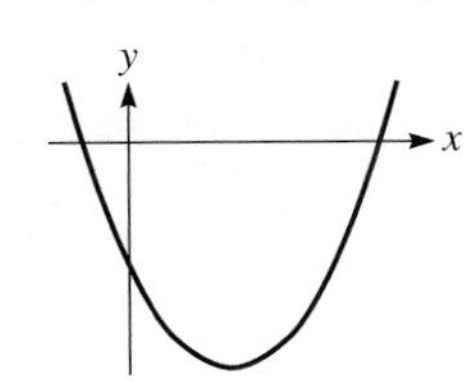

33. $(2, -1)$

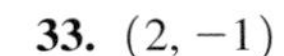

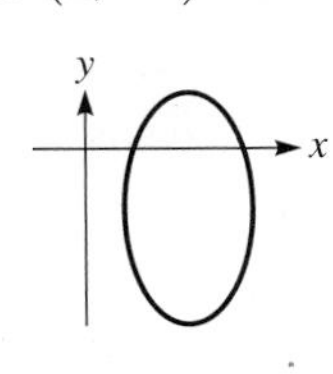

35. $(0, 0)$

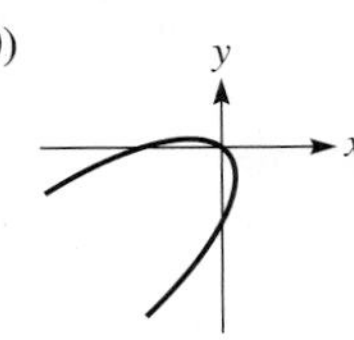

37.

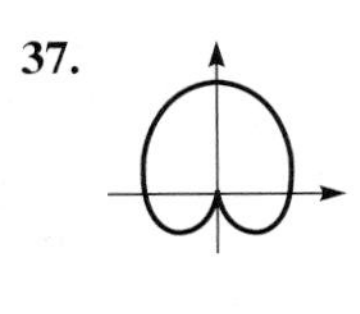

39.

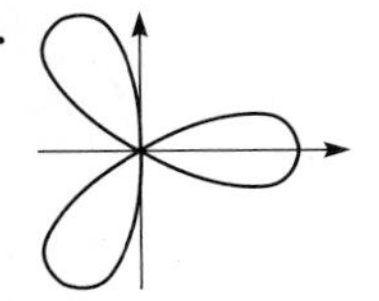

41.

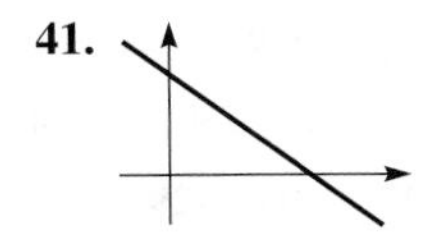

43.

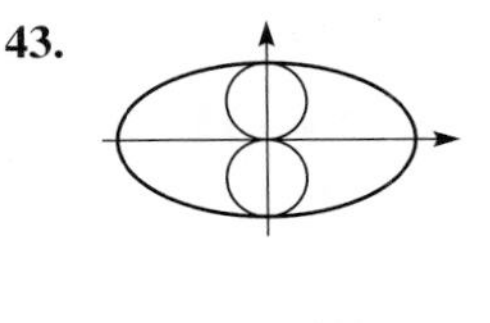

45. $\theta = \tan^{-1}2 = 1.11$

47. $r^2 = \frac{2}{1 + \sin\theta\cos\theta}$ **49.** $(x^2 + y^2)^3 = 16x^2y^2$

51. $3x^2 + 4y^2 - 8x - 16 = 0$ **53.** 4 **55.** 2 **57.** 2

59.

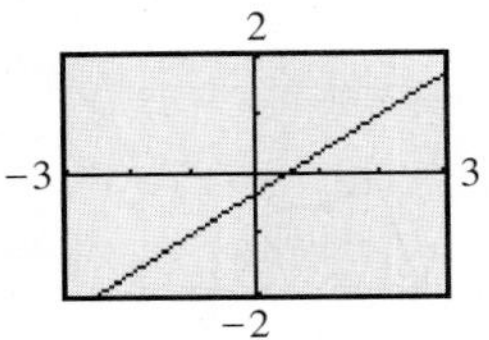

61.

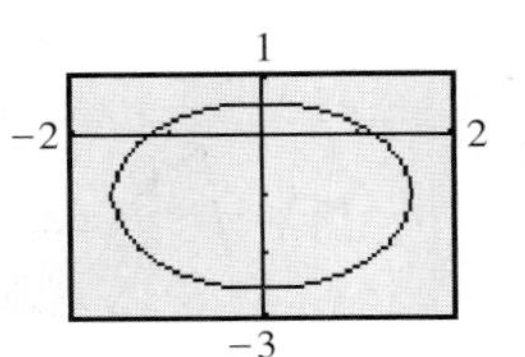

63.

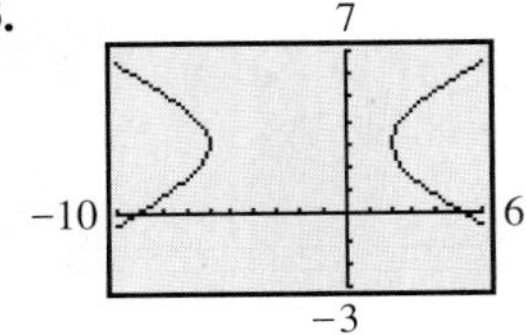

65.

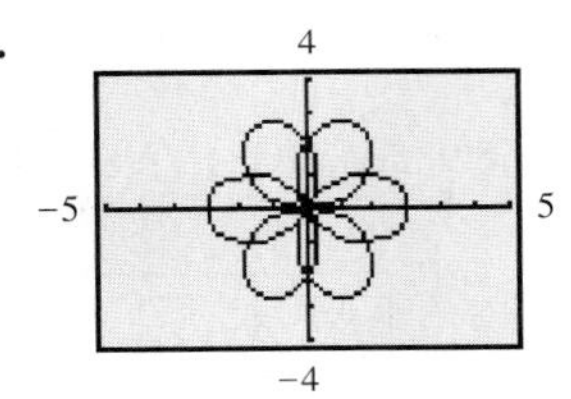

67.

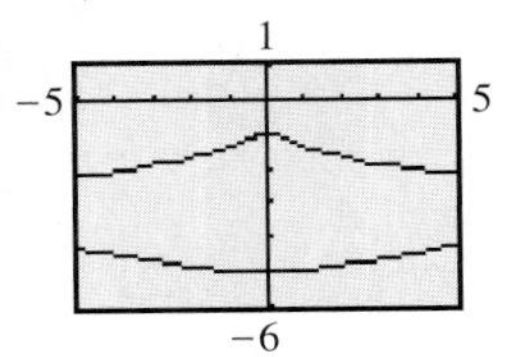

69. $x^2 + y^2 - 6x + 8y + 9 = 0$

71. $\frac{(x-4)^2}{16} + \frac{(y+3)^2}{7} = 1$ or

$7x^2 + 16y^2 - 56x + 96y + 144 = 0$

73. $x^2 = -24y$ **75.** Circle **77.** 1 **79.** 6.71

81. $m_1 = -\frac{12}{5}, m_2 = \frac{5}{12}, d_1^2 = 169, d_2^2 = 169, d_3^2 = 338$

83. Hyperbola **85.** 8

87. $x^2 - 6x - 8y + 1 = 0$ **89.** $R_T = R + 2.5$

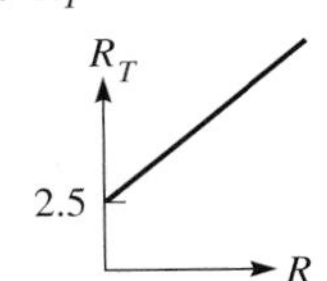

91. $v = 5.75 + 2.32t$

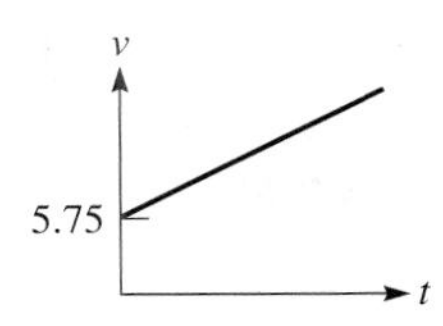

93. $y = 100.5T - 10050$

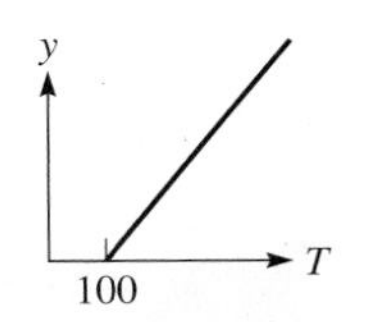

95. 9.5m^2 **97.** $11{,}000 \text{ ft}^2$ **99.** $y = -\frac{1}{80}x^2$

101. $v = 12{,}000 - 1250t (0 \le t \le 4 \text{ years})$

$y = 7000 - 1000(t - 4)(4 < t \le 11 \text{ years})$

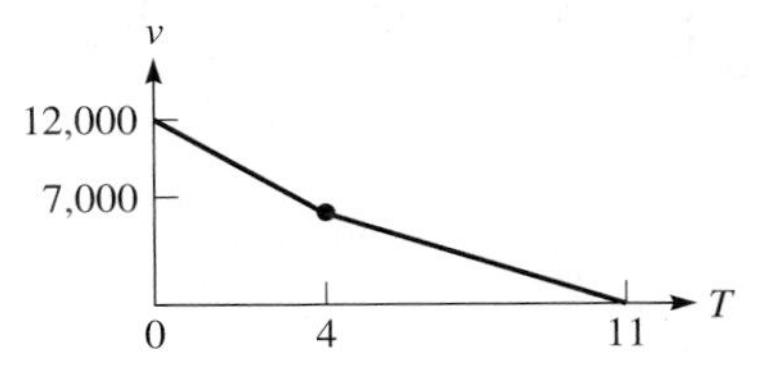

103. $y^2 = 32x$

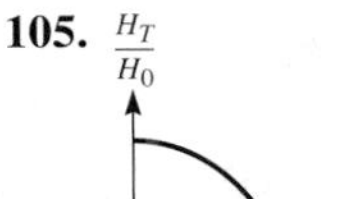

105. **107.**

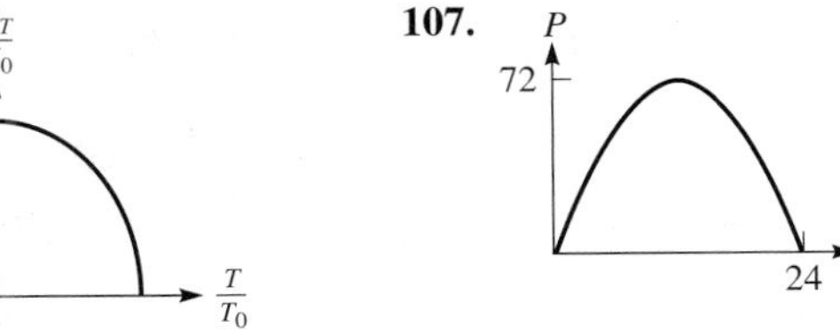

109. 7500 ft^2 **111.** 18 cm, 8 cm **113.** 37.8 ft

115. Dist. from rifle to P − dist. from target to P = constant (related to dist. from rifle to target).

117.

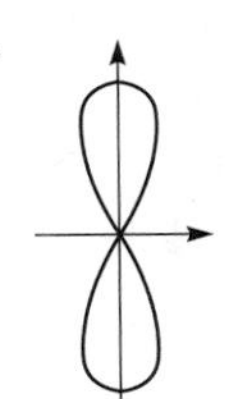

119.

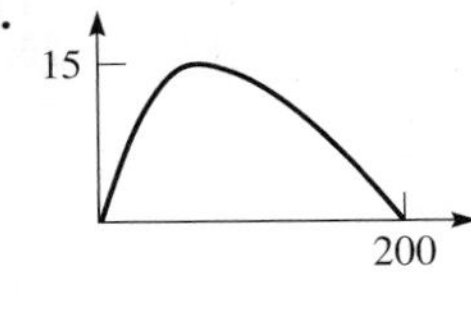

Exercises 22.1, page 624

1.

Weight	Freq.	Rel. freq. (%)
1.0–2.0	1	3.1
2.0–3.0	8	25
3.0–4.0	20	62.5
4.0–5.0	3	9.4

3. Quantitative **5.** Qualitative

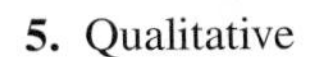

7.

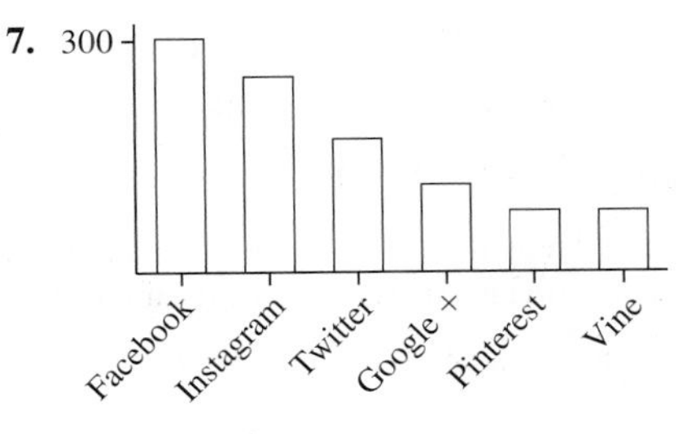

9. No. The percentages do not add up to 100%.

11. 40%, 35.5%, 10.3%, 4.4%, 9.9%

13.

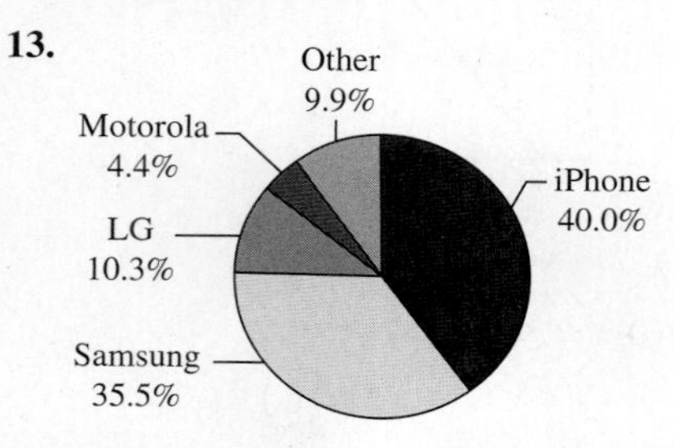

15.

Range (mi)	141–144	144–147	147–150	150–153	153–156
f	1	7	9	2	1

17.

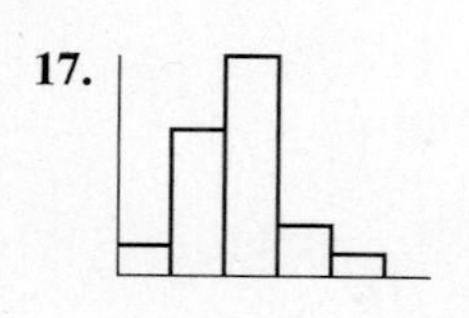

19.

Stem	Leaf
5	8
6	7
7	68
8	25669
9	1233579
10	113355779
11	0125
12	6

21.

Apps	50–60	60–70	70–80	80–90	90–100	100–110	110–120	120–130
f	1	1	2	5	7	9	4	1

23.

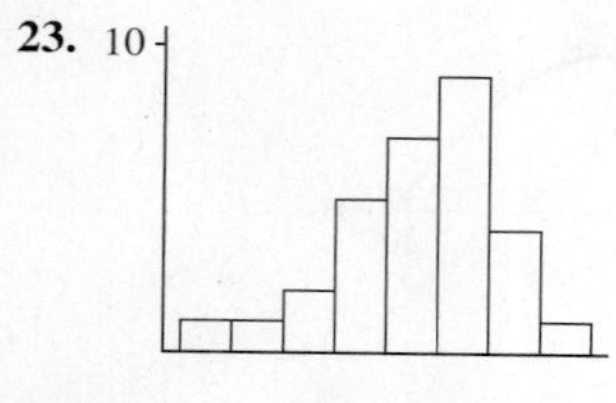

25.

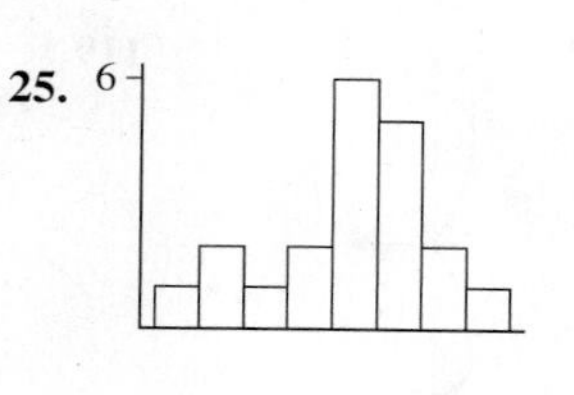

27.

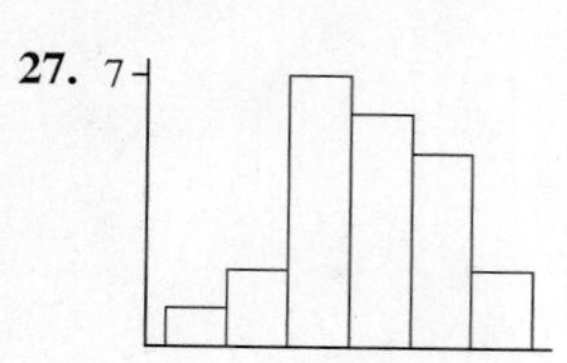

29.

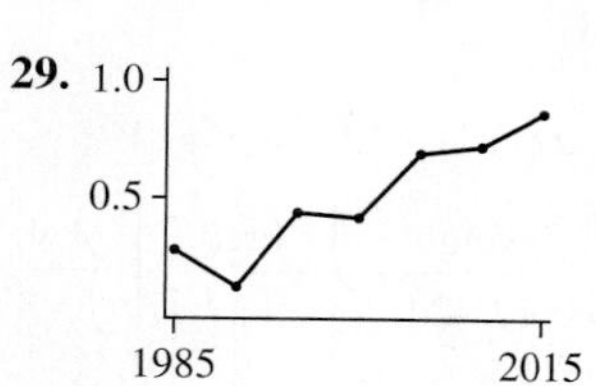

Exercises 22.2, page 629

1. 5 **3.** 5.5 **5.** 4 **7.** 0.495 **9.** 4.6 **11.** 0.505

13. 4 **15.** 0.49, 0.53 **17.** 147.5 mi **19.** 148 mi

21. 96.1 **23.** 4.237 mR **25.** 4.36 mR **27.** 0.12%

29. \$625, \$700 **31.** 862 kW · h **33.** 0.00593 mm

35. 0.195, 0.18 **37.** \$663

39. \$725, \$748, \$800; if each value is increased by the same amount, the median, mean, and mode are also increased by this amount.

41. \$884; an outlier can make the mean a poor measure of the center of the distribution.

Exercises 22.3, page 632

1. 2.2 **3.** 1.55 **5.** 0.035 **7.** 1.55 **9.** 0.035

11. 4.6, 1.55 **13.** 0.505, 0.035 **15.** 2.6 mi **17.** 14.7 apps

19. 0.026% **21.** 147 kW · h

Exercises 22.4, page 636

1. The peak would be as high as for the left curve, and it would be centered as for the right curve.

3. 2.6; It is 2.6 standard deviations above the mean.

5. 95% **7.** 99.85%

9. 1.2; It is 1.2 standard deviations above the mean.

11. −2.4; It is 2.4 standard deviations below the mean.

13. 68% **15.** 78.81% **17.** 43.32% **19.** 180 **21.** 3.59%

23. 2.28% **25.** −0.7 **27.** −0.6 **29.** 53.28%

Exercises 22.5, page 641

1. $\mathrm{UCL}(\bar{x}) = 504.7$ mg, $\mathrm{LCL}(\bar{x}) = 494.7$ mg

3. The first point would be below the $\bar{x}$ line at 499.7. The UCL and LCL lines would be 0.2 unit lower.

5. $\bar{\bar{x}} = 361.8$ N · m, UCL = 368.9 N · m, LCL = 354.6 N · m

7.

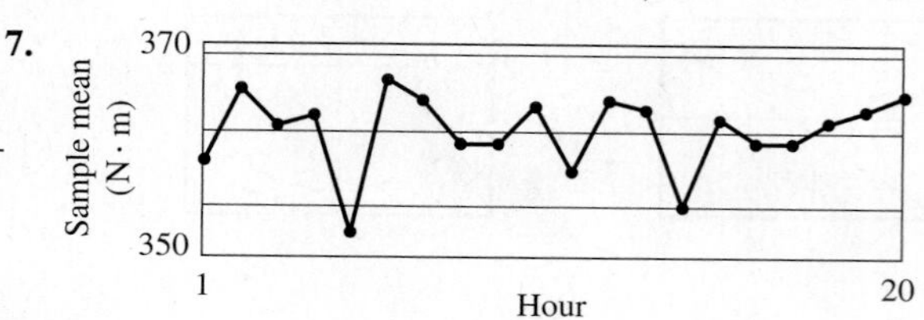

9. $\bar{\bar{x}} = 8.986$ V, UCL = 9.081 V, LCL = 8.890 V

11.

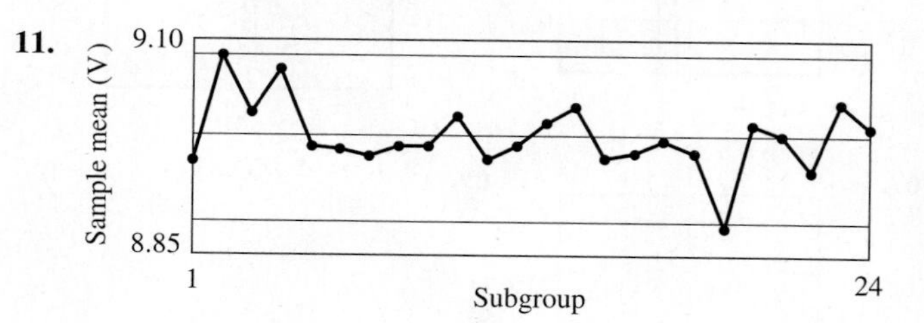

13. $\mu = 2.725$ in., UCL = 2.768 in., LCL = 2.682 in.

15. $\mu_R = 5.57$ mL, UCL = 11.17 mL, LCL = 0.00 mL

17. $\bar{p} = 0.0369$, UCL = 0.0548, LCL = 0.0190

19. $\bar{p} = 0.0580$, UCL = 0.0894, LCL = 0.0266

Exercises 22.6 , page 646

1. $y = 2x + 2$

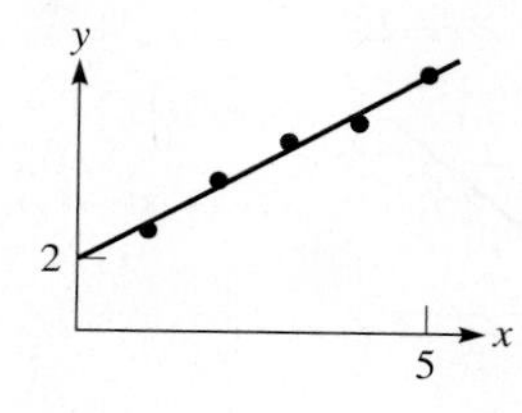

3. $y = -1.77x + 191$

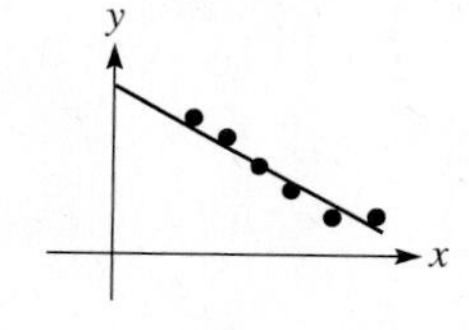

5. $y = -0.308t + 9.07$

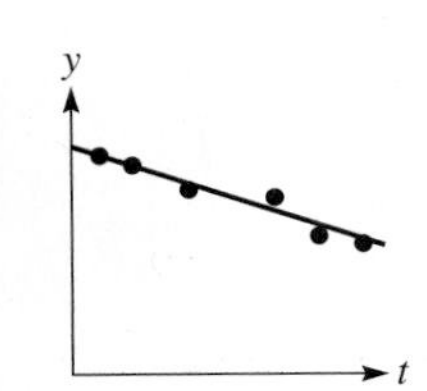

7. $V = -0.590i + 11.3$; 6.62 V; interpolation

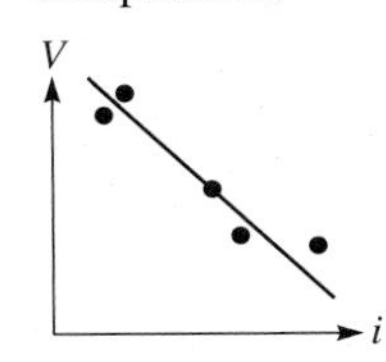

9. $h = 2.24x + 5.2$

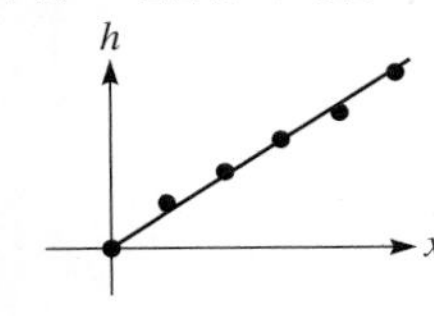

11. $p = -0.200x + 649$; 549 lb/in.^{2}; extrapolation

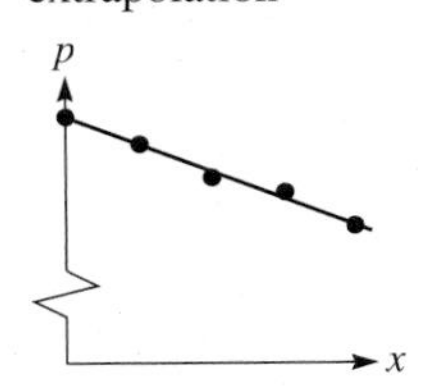

13. $V = 4.32f - 2.03$
$f_0 = 0.470$ PHz

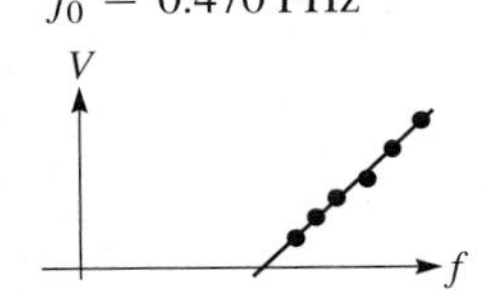

15. $r = -0.913$ (strong negative correlation)
$r^2 = 0.833$ (83.3% of the variation in voltage is explained by the model)

17. $r = 0.989$ (strong positive correlation)
$r^2 = 0.977$ (97.7% of the variation in temperature is explained by the model)

Exercises 22.7, page 649

1. $y = 349.998\,(1.050)^x$ **3.** $P = 2445.949x^{-1.006}$

5. $y = 5.5t^2 + 2.9t - 3.2$; 38 cm

7. $p = 0.00375T^2 + 0.515T + 23$; 42 lb/in.^{2}

9. $T = 170{,}188.7x^{-1.068}$; 28 J; extrapolation

11. $y = 6.161\,(0.357)^t$; 0.5 cm; interpolation

Review Exercises for Chapter 22, page 651

Concept Check Exercises

1. T **2.** T **3.** T **4.** F **5.** F **6.** F

Practice and Applications

7. 76.5 **9.** 76.2

11.

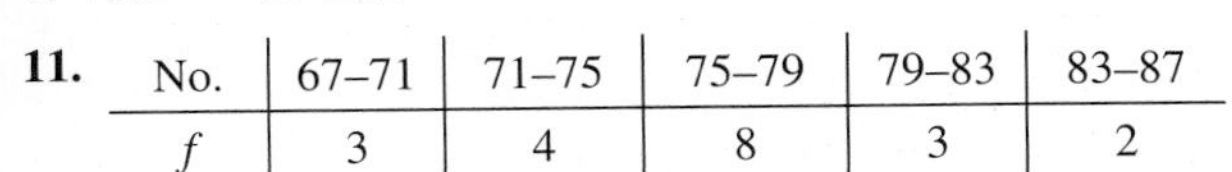

No.	67–71	71–75	75–79	79–83	83–87
f	3	4	8	3	2

13.

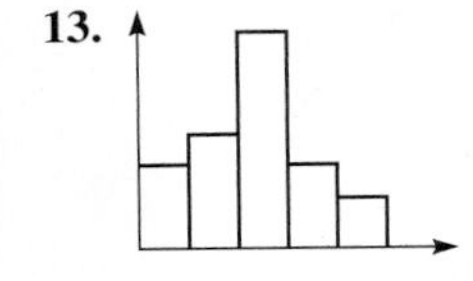

15.

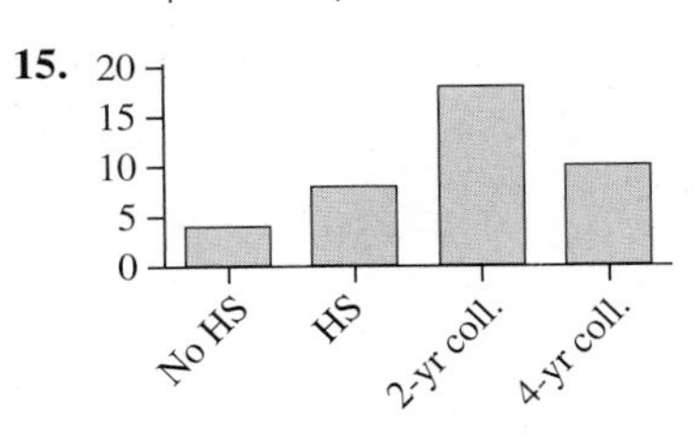

17. 0.134 mL/L **19.** 0.014 mL/L **21.** 700 W

23. 17.3 W **25.** 4.2 **27.** 2.0 **29.** 95%

31. 1.8; An IQ of 127 is 1.8 standard deviations above the mean.

33. $\bar{p} = 0.0540$, UCL $= 0.0843$, LCL $= 0.0237$

35.

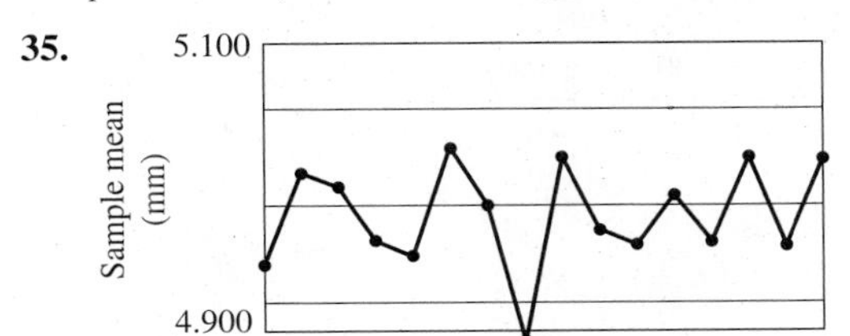

37. 0.9273 **39.** -1.4, 1.4 **41.** 65.89%

43. 99.18% **45.** $R = 0.0983T + 25.0$

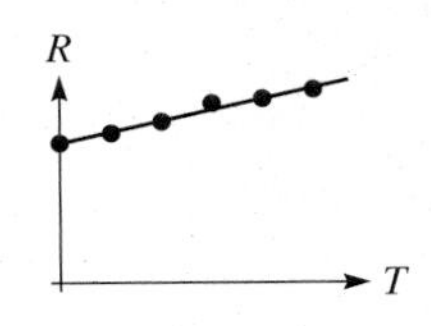

47. $s = 0.123t + 0.887$

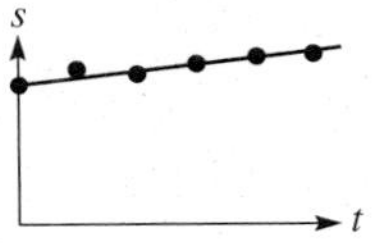

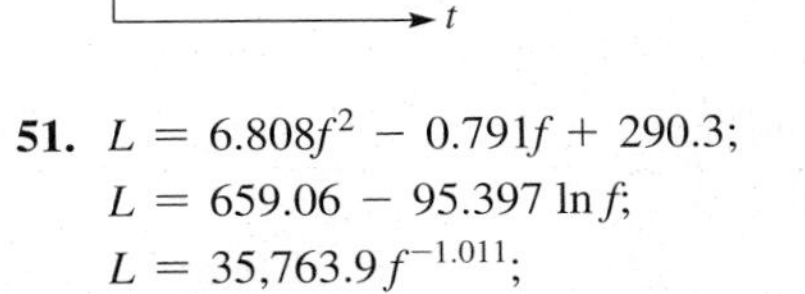

49. $y = 1.17x + 9.99$

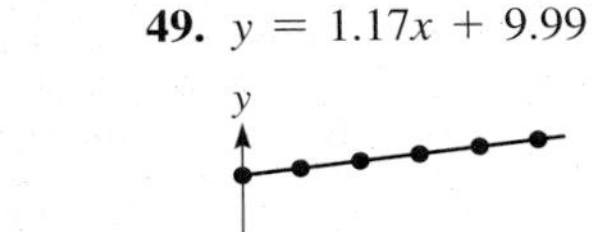

51. $L = 6.808f^2 - 0.791f + 290.3$;
$L = 659.06 - 95.397 \ln f$;
$L = 35{,}763.9 f^{-1.011}$;
logarithmic

53. $y = 0.0002x^2 + 15$

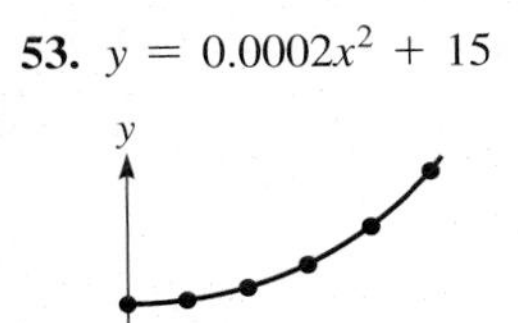

55. $y = 0.997x + 0.581$ **57.** $y = 1.39x^{0.878}$

59. Substitute expressions for m, b, and $\bar{x}$. Simplify to $\bar{y} = \bar{y}$.

Exercises 23.1, page 662

1. Not cont. at $x = -2$ **3.** -4 **5.** Cont. all x

7. Not cont. $x = 0$ and $x = 1$, div. by zero

9. Cont. $x > 2$; function not defined **11.** Cont. all x

13. Not cont. $x = 2$, small change

15. Not cont. $x = 2$, $f(2) \neq \lim_{x \to 2} f(x)$

17. **(a)** -1, **(b)** $\lim_{x \to 2} f(x)$ does not exist

19. **(a)** 0, **(b)** $\lim_{x \to 2} f(x) = -1$

21. Not cont. $x = 2$, small change **23.** Cont. all x

25.

x	0.900	0.990	0.999	1.001
$f(x)$	1.7100	1.9701	1.9970	2.0030

x	1.010	1.100
$f(x)$	2.0301	2.3100

$\lim_{x \to 1} f(x) = 2$

27.

x	1.900	1.990	1.999	2.001
$f(x)$	−0.2516	−0.2502	−0.25002	−0.24998

x	2.010	2.100
$f(x)$	−0.2498	−0.2485

$\lim_{x\to 2} f(x) = -0.25$

29.

x	10	100	1000
$f(x)$	0.4468	0.4044	0.4004

$\lim_{x\to\infty} f(x) = 0.4$

31. 7 **33.** 1 **35.** $-\frac{2}{3}$ **37.** 27 **39.** 2

41. Does not exist **43.** $\frac{1}{2}$ **45.** $\frac{1}{6}$ **47.** 3 **49.** 1

51.

x	−0.1	−0.01	−0.001	0.001
$f(x)$	−3.1	−3.01	−3.001	−2.999

x	0.01	0.1
$f(x)$	−2.99	−2.9

$\lim_{x\to 0} f(x) = -3$

53.

x	10	100	1000
$f(x)$	2.1649	2.0106	2.0010

$\lim_{x\to\infty} f(x) = 2$

55. $\lim_{x\to 2}(x^2 + 2x + 4) = 12$

57. 34.9°C, 0°C **59.** 3 cm/s **61.** 2.77 **63.** e

65. **(a)** −1, **(b)** 2, **(c)** Does not exist **67.** 0

69. −1; +1; no, $\lim_{x\to 0^+} f(x) \neq \lim_{x\to 0^-} f(x)$

71. Yes limits are $-\infty$ and $+\infty$

Exercises 23.2, page 666

1. 9 **3.** (Slopes) 3.5, 3.9, 3.99, 3.999; $m = 4$ **5.** (Slopes) −2, −2.8, −2.98, −2.998; $m = -3$

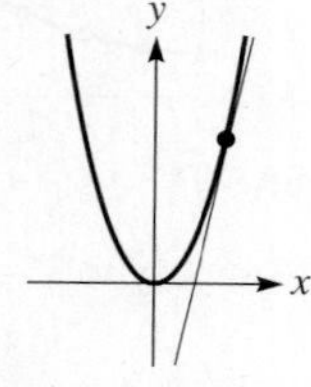

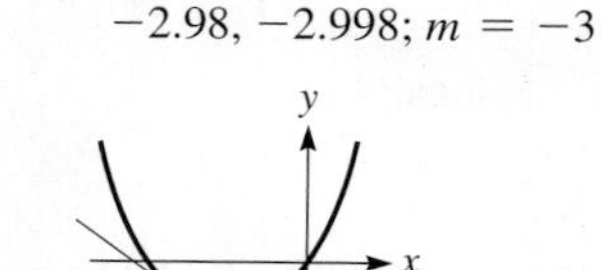

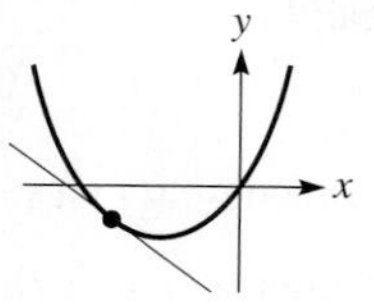

7. 4 **9.** −3 **11.** $m_{tan} = 2x_1$; −2, 4

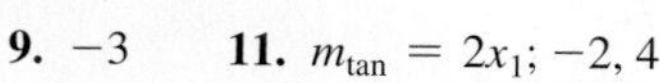

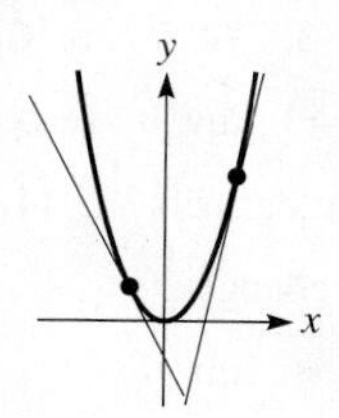

13. $m_{tan} = 4x_1 + 5$; −3, 7 **15.** $m_{tan} = 2x_1 + 3$; −3, 9

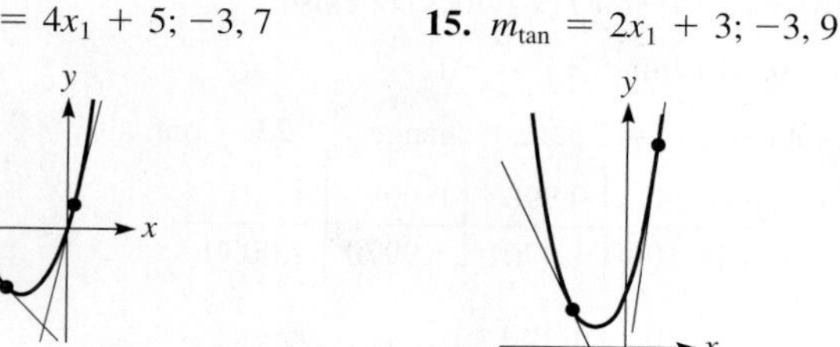

17. $m_{tan} = 6 - 2x_1$; 8, 4, 0

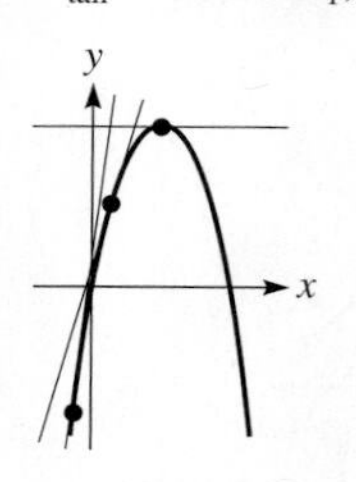

19. $m_{tan} = 6x_1^3$; −0.75, 0, 6

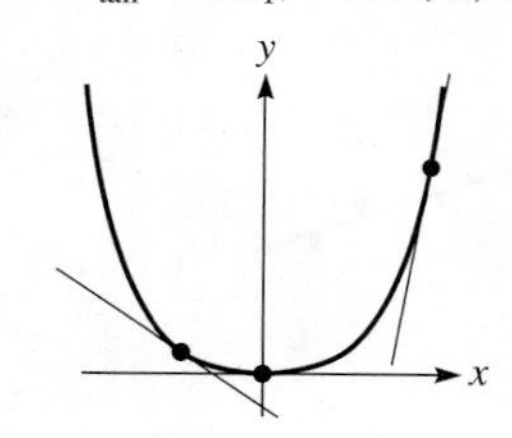

21. $m_{tan} = 5x_1^4$; 5, −0.31, 0.31, 5

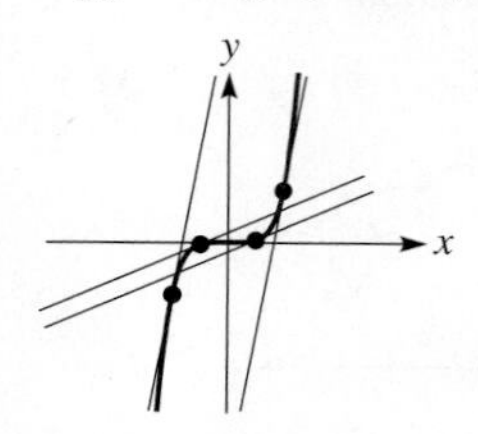

23. av. ch. = 4.1, $m_{tan} = 4$

25. av. ch. = −12.61, $m_{tan} = -12$ **27.** (−1, 2)

29. $(4, \frac{80}{3}), (-4, -\frac{80}{3})$ **31.** $-\frac{1}{6}$ **33.** 9.5°

Exercises 23.3, page 670

1. $8x + 3$ **3.** 3 **5.** −7 **7.** $2x$ **9.** $6x$

11. $2x - 7$ **13.** $8 - 4x$ **15.** $3x^2 + 2$ **17.** $15x^2$

19. $-\frac{1}{(x + 2)^2}$ **21.** $1 - \frac{4}{3x^2}$ **23.** $-\frac{4}{x^3}$ **25.** $6x - 2$; −8

27. $-\frac{33}{(3x + 2)^2}$; $-\frac{3}{11}$ **29.** $-\frac{2}{x^2}$; all real numbers except 0

31. $\frac{-6x}{(x^2 - 1)^2}$, all real numbers except −1 and 1

33. (5, 5) **35.** (4, −32)

37. $\frac{1}{2\sqrt{x + 1}}$; differentiable for $x > -1$ because derivative and function are both defined for these values

39. 11.3°

Exercises 23.4, page 674

1. $v = 48 - 32t$; −16 ft/s, −80 ft/s

3. $m = 4$ **5.** $m = -\frac{3}{4}$

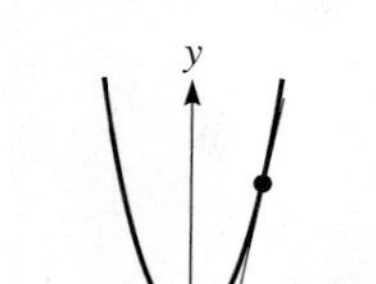

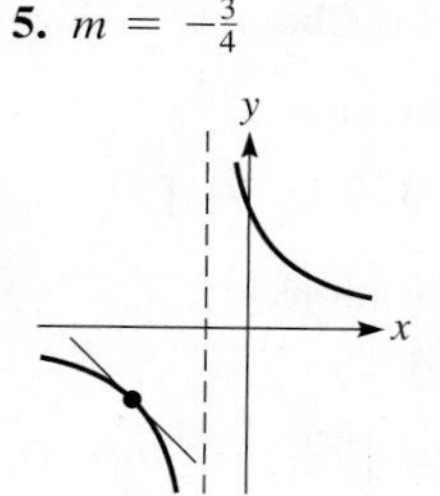

7. 4.00, 4.00, 4.00, 4.00, 4.00; $\lim_{t\to 3} v = 4$ ft/s

9. 5, 6.5, 7.7, 7.97, 7.997,; $\lim_{t\to 2} v = 8$ ft/s **11.** 4; 4 ft/s

13. $6t - 4$; 8 ft/s **15.** 48 **17.** $24t - 3t^2$ **19.** $3 - \pi t^2$

21. $14t + \frac{2}{(t + 1)^2}$ **23.** $12t - 4$ **25.** $6t$

27. v (ft/s): 136, 129.6, 128.16, 128.016; The limit is 128 ft/s.

29. 2π cm/cm

31. 4.5 s **33.** -2 **35.** $6w$ **37.** 460 W

39. $-83.1\ \text{W}/(\text{m}^2\cdot\text{h})$ **41.** $-\frac{48}{(t+3)^2}$; $-\$1300$/year

43. πd^2 **45.** $24.2/\sqrt{\lambda}$

Exercises 23.5, page 678

1. $9r^8$ **3.** $6x^2 - 12x + 8$; 56 **5.** $5x^4$ **7.** $-36x^8$

9. $20x^3$ **11.** $8x + 7$ **13.** $15r^2 - 2$

15. $200x^7 - 170x^4 - 1$ **17.** $-42x^6 + 15x^2$ **19.** $x^2 + x$

21. 16 **23.** 33 **25.** -4 **27.** -29 **29.** $30t^4 - 5$

31. $-6 + 6t^2$ **33.** 64 **35.** 0 **37.** 1 **39.** $(2, 4)$

41. $(1, -5)$ **43.** $-\frac{1}{4}$ **45.** 1700 mm² **47.** 84 W/A

49. 3.33 Ω/°C **51.** -0.00286 N/°C

53. -80.5 m/km **55.** 391 mm²

Exercises 23.6, page 682

1. $18x^2 - 18x - 10$

3. $6x(6x - 5) + (3x^2 - 5x)(6) = 54x^2 - 60x$

5. $(3t + 2)(2) + (2t - 5)(3) = 12t - 11$

7. $(x^4 - 3x^2 + 3)(-6x^2) + (1 - 2x^3)(4x^3 - 6x)$
$= -14x^6 + 30x^4 + 4x^3 - 18x^2 - 6x$

9. $(2x - 7)(-2) + (5 - 2x)(2) = -8x + 24$

11. $(h^3 - 1)(4h - 1) + (2h^2 - h - 1)(3h^2)$
$= 10h^4 - 4h^3 - 3h^2 - 4h + 1$

13. $\frac{3}{(8x+3)^2}$ **15.** $-\frac{20x}{(2x^2+1)^2}$ **17.** $\frac{36x - 12x^2}{(3-2x)^2}$

19. $\frac{-6x^2 + 6x + 4}{(3x^2+2)^2}$ **21.** $\frac{-3x^2 - 16x - 26}{(x^2 + 4x + 2)^2}$

23. $\frac{-2x^3 + 2x^2 + 5x + 4}{x^3(x+2)^2}$ **25.** -107 **27.** 75 **29.** 19

31. -5.64 **33.** No. π^2 is a constant. **35.** $(0, -1)$

37. $x^2f'(x) + 2xf(x)$ **39.** 1 (except or $x = 1$)

41. (1) $\frac{-12x^3 + 45x^2 - 14x}{(3x-7)^2}$ (2) $\frac{-12x^3 + 45x^2 - 14x}{(3x-7)^2}$

43. 12 **45.** $1, -1$ **47.** 1470 cm²

49. $\frac{96}{(7R+12)^2}$ **51.** $8t^3 - 45t^2 - 14t - 8$ **53.** 1.2°C/h

55. $\frac{2R(R+2r)}{3(R+r)^2}$ **57.** $\frac{E^2(R-r)}{(R+r)^3}$

Exercises 23.7, page 688

1. $36x^2(2 + 3x^3)^3$ **3.** $-\frac{3x}{(2-3x^2)^{1/2}}$ **5.** $\frac{2}{x^{1/2}}$ **7.** $-\frac{6}{5t^3}$

9. $-\frac{1}{x^{4/3}} + 8x$ **11.** $\frac{3}{2}x^{1/2} + \frac{6}{x^2}$ **13.** $40x(x^2+3)^4$

15. $-216x^2(7 - 4x^3)^7$ **17.** $\frac{2x^2}{(2x^3-3)^{2/3}}$ **19.** $\frac{24y}{(4-y^2)^5}$

21. $\frac{24x^3}{(2x^4-5)^{0.25}}$ **23.** $\frac{-4x}{(1-8x^2)^{3/4}}$ **25.** $\frac{12v+5}{(8v+5)^{1/2}}$

27. $\frac{6(5x-1)}{x^4(1-6x)^{1/2}}$ **29.** $\frac{3(x^2+12x+16)}{(x+4)^2(x+2)^{1/2}}$

31. $\frac{-1}{\sqrt{2R+1}(4R+1)^{3/2}}$ **33.** $\frac{3}{10}$ **35.** $\frac{5}{36}$

37. $\frac{x^3(0) - 1(3x^2)}{x^6} = -3x^{-4}$ **39.** $x = 0$

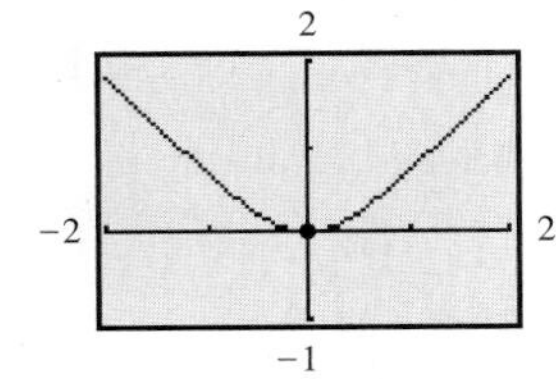

41. Yes, at $\left(-\frac{13}{9}, \frac{1}{3}\right)$ **43.** 1 **45.** $\frac{11}{\sqrt{w}}$ **47.** -1.35 cm/s

49. $\frac{-450{,}000}{V^{5/2}}$, -4.50 kPa/cm³ **51.** $l = a$

53. $-45.2\ \text{W}/(\text{m}^2\cdot\text{h})$ **55.** $\frac{8a^3}{(4a^2 - x^2)^{3/2}}$

57. $\frac{2(w+1)}{(2w^2+4w+4)^{1/2}}$

Exercises 23.8, page 692

1. $-\frac{4x}{3y^2}$ **3.** $-\frac{3}{2}$ **5.** $\frac{6x+1}{4}$ **7.** $\frac{x}{4y}$ **9.** $\frac{2x}{5y^4}$

11. $\frac{2x}{4y^{-1/3}+1}$ **13.** $\frac{-3y}{3x+1}$ **15.** $\frac{y - x^3 - 2x^2y - xy^2}{x}$

17. $\frac{3(y^2+1)(y^2-2x+1)}{(y^2+1)^2 - 6x^2y}$ **19.** $\frac{4(2y-x)^3 - 2x}{8(2y-x)^3 - 1}$

21. $-\frac{12x(x^2+1)^2(y^2+1)^{1/2}}{y}$ **23.** 3 **25.** $-\frac{108}{157}$

27. $\frac{1}{4}$ **29.** $(2, 2)$, $(2, -2)$

31. At $(\sqrt{7}, 0)$ and $(-\sqrt{7}, 0)$, $m_{\tan} = -2$

33. $\frac{2R^2C - L}{2\omega CL^3}$ **35.** 1 **37.** $\frac{nRT^2 + bnP}{VT^2 - anT^2 + bnT}$ **39.** $-\frac{x}{y}$

41. $\frac{r - R + 1}{r + 1}$ **43.** $\frac{2C^2r(12CSr - 20Cr - 3L)}{3(C^2r^2 - L^2)}$

Exercises 23.9, page 695

1. $y' = 15x^2 - 4x$, $y'' = 30x - 4$, $y''' = 30$, $y^{(n)} = 0\ (n \geq 4)$

3. $y' = 3x^2 + 14x$, $y'' = 6x + 14$, $y''' = 6$

5. $f'(x) = 3x^2 - 24x^3$, $f''(x) = 6x - 72x^2$,
$f'''(x) = 6 - 144x$

7. $y' = -8(1 - 2x)^3$, $y'' = 48(1 - 2x)^2$,
$y''' = -192(1 - 2x)$

9. $f'(r) = (16r + 9)(4r + 9)^2$, $f''(r) = 24(8r + 9)(4r + 9)$,
$f'''(r) = 96(16r + 27)$

11. $84x^5 - 30x^4$ **13.** $-\frac{2}{x^{3/2}}$ **15.** $-\frac{12}{(8x-3)^{7/4}}$

17. $\frac{14.4\pi}{(1+2p)^{5/2}}$ **19.** $600(2 - 5x)^2$

21. $30(27x^2 - 1)(3x^2 - 1)^3$ **23.** $\frac{4\pi^2}{(6-x)^3}$

25. $\frac{450}{(4v+15)^3}$ **27.** $-\frac{9}{16y^3}$ **29.** $\frac{9}{125}$ **31.** $-\frac{13}{384}$

33. -4320 **35.** -9.8 m/s² **37.** 3.0 m/s²

39. $(1, -2)$ **41.** 48 **43.** $\frac{d^2P}{dt^2} = 80$

45. $y = 2x^3 - x^2 + 12$ **47.** -32.2 ft/s^2

49. $-\frac{1.60}{(2t + 1)^{3/2}}$ **51.** 0.049

Review Exercises for Chapter 23, page 697

Concept Check Exercises

1. F **2.** F **3.** F **4.** F **5.** F **6.** F **7.** T **8.** F

Practice and Applications

9. -4 **11.** Does not exist. **13.** 1 **15.** $\frac{1}{4}$ **17.** $\frac{2}{3}$

19. -2 **21.** 5 **23.** $-4x$ **25.** $-\frac{4}{x^3}$ **27.** $\frac{1}{2\sqrt{(x + 5)}}$

29. $14x^6 - 6x$ **31.** $\frac{2}{x^{1/2}} + \frac{3}{x^2}$ **33.** $\frac{12}{(1 - 5y)^2}$

35. $-28(2 - 7x)^3$ **37.** $\frac{9\pi x}{(5 - 2x^2)^{7/4}}$ **39.** $\frac{5(2 - 5t^2)}{(t^2 + 2)^4}$

41. $\frac{-2x - 3}{2x^2(4x + 3)^{1/2}}$ **43.** $\frac{2x - 6(2x - 3y)^2}{1 - 9(2x - 3y)^2}$ **45.** $\frac{5}{48}$

47. 13.6 **49.** $36x^2 - \frac{2}{x^3}$ **51.** $\frac{184}{(5 + 4t)^3}$ **53.** $\frac{1}{x^2}$

55. 53.1°

57. It appears to be 8, but using *trace*, there is no value shown for $x = 2$.

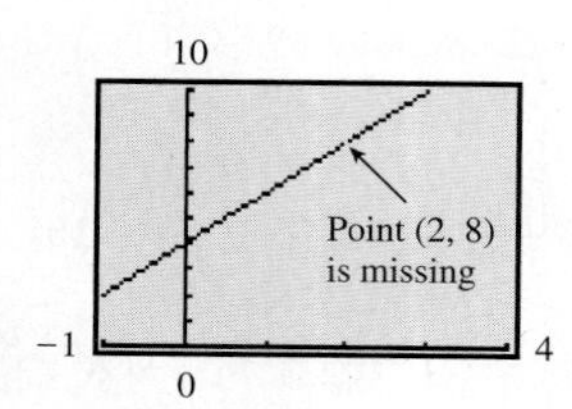

59. (a) 30 ft/s (b) 6 ft/s **61.** -31 **63.** $(0, \frac{1}{3}\sqrt{3})$

65. $10{,}000(1 + 0.25t)^7$ **67.** (a) $v = \frac{4}{(1 + 8t)^{1/2}}$

(b) $a = \frac{-16}{(1 + 8t)^{3/2}}$

69. 5 **71.** $-k + k^2t - \frac{1}{2}k^3t^2$

73. $-\frac{2G\,m_1m_2}{r^3}$ **75.** $0.4(0.01t + 1)^2(0.04t + 1)$

77. $\frac{2R(R + 2r)}{3(R + r)^2}$ **79.** 830 W **81.** $5k(x^4 + 270x^2 - 190)$

83. $-\frac{1}{4\pi\sqrt{C}(L + 2)^{3/2}}$ **85.** $\frac{-15}{(0.5t + 1)^2}$

87. $y' = \frac{w}{6EI}(3L^2x - 3Lx^2 + x^3)$

$y'' = \frac{w}{2EI}(L - x)^2$

$y''' = \frac{w}{EI}(x - L)$

$y^{\text{iv}} = \frac{w}{EI}$

89. $p = 2w + \frac{150}{w}, \frac{dp}{dw} = 2 - \frac{150}{w^2}$ **91.** $-\frac{1}{2}$

93. 11.3 km^2/km **95.** $A = 4x - x^3, \frac{dA}{dx} = 4 - 3x^2$

97. 397 mi/h

Exercises 24.1, page 702

1. $x + 8y - 17 = 0$

3. $4x - y - 2 = 0$

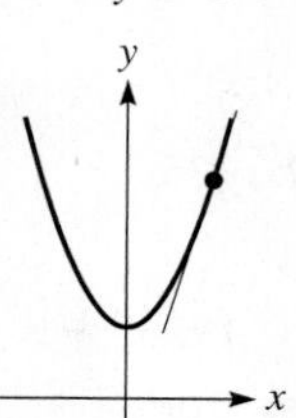

5. $2y - x - 2 = 0$

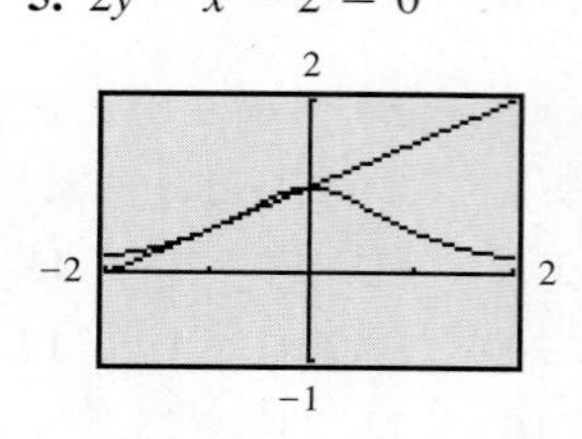

7. $x - 2y + 6 = 0$

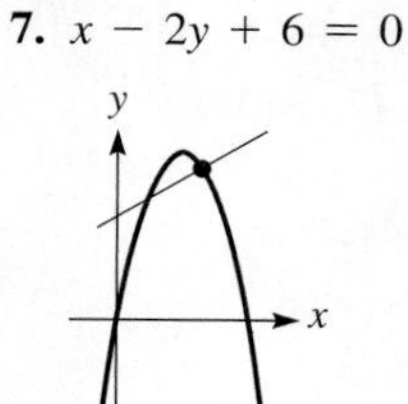

9. $x + 2y - 3 = 0$

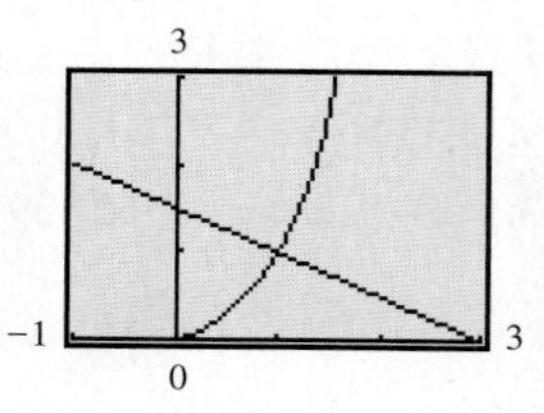

11. $y = 2x - 4$

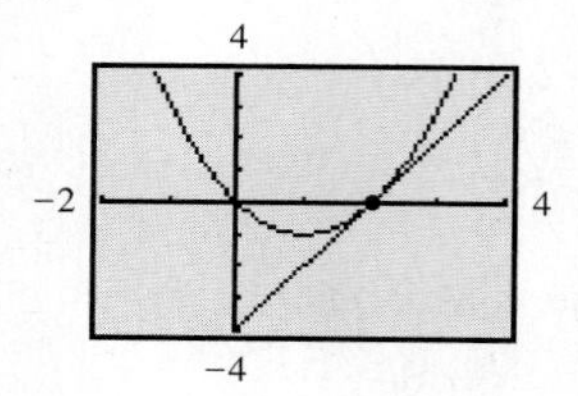

13. $y - 8 = -\frac{1}{24}(x - \frac{3}{2})$, *or* $2x + 48y - 387 = 0$

15. $y = 3x - 3; y = -\frac{1}{3}x + \frac{1}{3}$

17. The line is $y - x - 1 = 0$.

19. It causes division by zero; $x = 2$ **21.** $(\frac{25}{3}, 0)$

23. $x - 4y + 2 = 0$ **25.** $3x - 5y - 150 = 0$

27. $x + y - 6 = 0$

29. $x + 2y - 3 = 0, x = 0, x - 2y + 3 = 0$

Exercises 24.2, page 706

1. $x_3 = 0.2086$; quadratic formula: 0.2087 (to four decimal places)

3. $x_3 = -0.1805$; quadratic formula: -0.1805 (to four decimal places)

5. 0.5857864 **7.** -0.1958

9. 2.5615528 **11.** -1.2360680 **13.** 0.9175433

15. 0.6180340 **17.** -1.8557725, 0.6783628, 3.1774097

19. Find the real root of $x^3 - 4 = 0$; 1.5874011

21. $x_{n+1} = \frac{1}{2}x_n + \frac{a}{2x_n}$

23. $x_2 = -2, x_3 = 4, x_4 = -8$; Successive approximations form geometric sequence, but do not converge to result of 0.

25. -1.4142 (tangent) **27.** 47 s **29.** 29.5 m

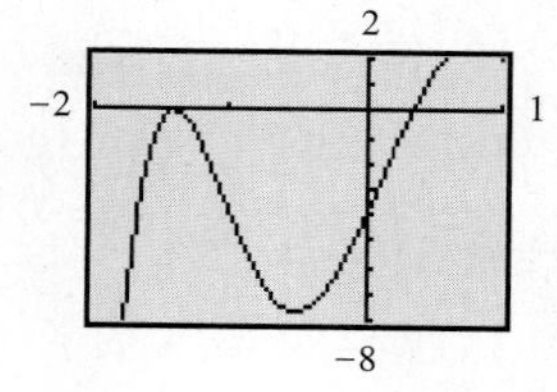

Exercises 24.3, page 710

1. 16.5, $-14.0°$

3. 3.16, 341.6°   **5.** 8.07, 352.4°

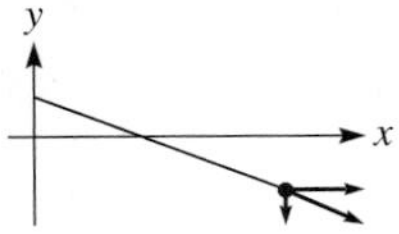

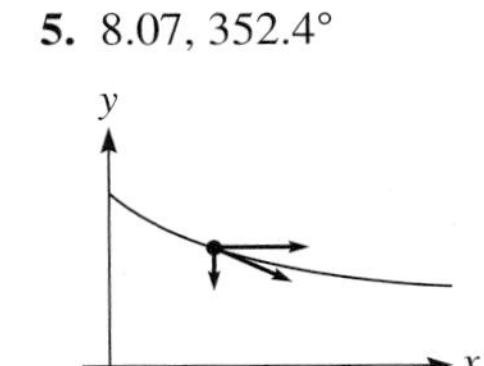

7. $a = 0$ **9.** 20.0, 3.7°

11. $y = 0.59$ m, $v = 42.7$ m/s, 5.7° below horizontal

13. 6.4 m/s, 321°   **15.** 1.3 ft/min^2, 288°

17. 34 m/s, 318°; 9.8 m/s^2, 270°

19. 276 m/s, 43.5°; 2090 m/s, 16.7°   **21.** 1.32 cm/s, $-24.9°$

23. 22.1 m/s^2, 25.4°; 20.2 m/s^2, 8.5°   **25.** 21.2 mi/min, 296.6°

27. $x^2 + y^2 = 1.75^2$; $v_x = 28{,}800$ in./min, $v_y = -27{,}100$ in./min

29. 370 m/s, 19°

Exercises 24.4, page 713

1. 5.20 V/min   **3.** 23   **5.** 3   **7.** 0.36 ft/s   **9.** 0.0900 Ω/s

11. -0.00401 s/s   **13.** 330 mi/h   **15.** 4.1×10^{-6} m/s

17. $\frac{dB}{dt} = \frac{-3kr(dr/dt)}{[r^2 + (l/2)^2]^{5/2}}$ **19.** 0.0056 m/min

21. 0.15 mm^2/month   **23.** -101 mm^3/min

25. $\frac{dV}{dt} = kA$; $\frac{dr}{dt} = k$ **27.** -4.6 kpa/min

29. 11 cm/min   **31.** $I = \frac{8k}{x^2}$; $\frac{dI}{dt} = 0.000800k$ unit/s

33. -0.432 N/s   **35.** -0.23 cm/min   **37.** 820 mi/h

39. 12.0 ft/s   **41.** 8.33 ft/s

Exercises 24.5, page 720

1. Inc. $x < -2$, $x > 0$, dec. $-2 < x < 0$

3. Conc. down $x < 2$, conc. up $x > 2$, infl. $(2, -16)$

5. Inc. $x > -1$, dec. $x < -1$

7. Inc. $x < -3$, $x > 2$; dec. $-3 < x < 2$

9. Min. $(-1, -1)$ **11.** Max. $(-3, 71)$; min. $(2, -54)$

13. Conc. up all x

15. Conc. up, $x > -\frac{1}{2}$; conc. down, $x < -\frac{1}{2}$; infl. $(-\frac{1}{2}, \frac{17}{2})$

17.

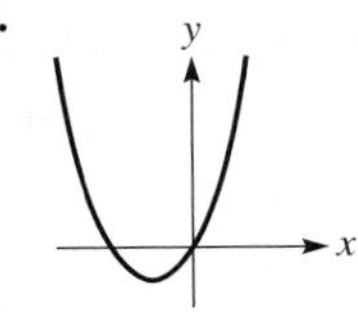

19.

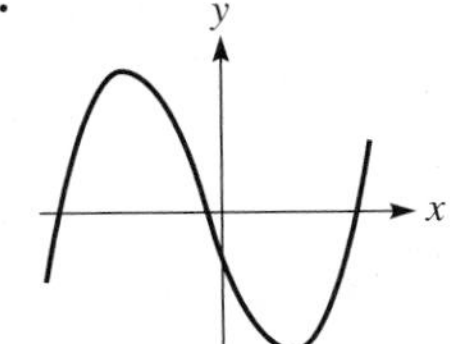

21. Max. $(3, 18)$, conc. down all x

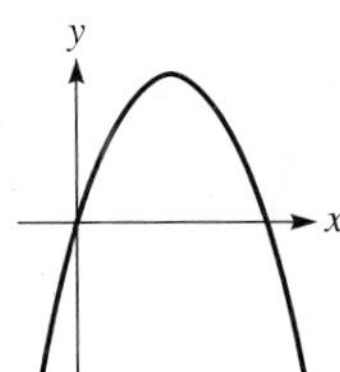

23. Max. $(-2, 3)$, min. $(0, -5)$, infl. $(-1, -1)$

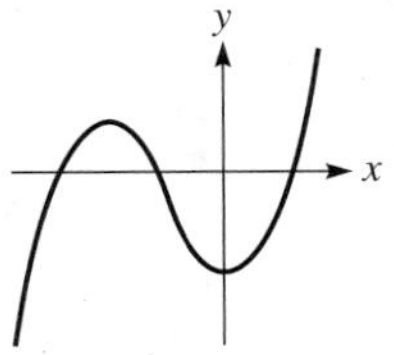

25. No max. or min., infl. $(-1, 1)$

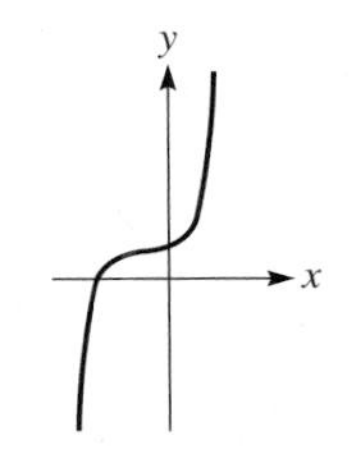

27. Max. $(1, 16)$, min. $(3, 0)$, infl. $(2, 8)$

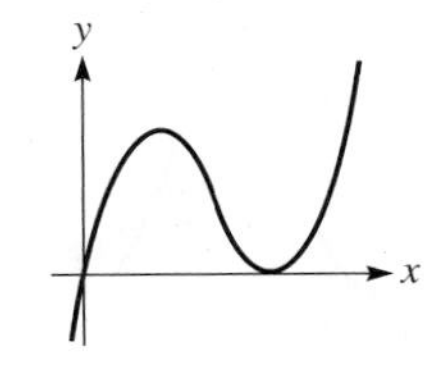

29. Max. $(1, 7)$, infl. $(0, 6)$, $(\frac{2}{3}, \frac{178}{27})$

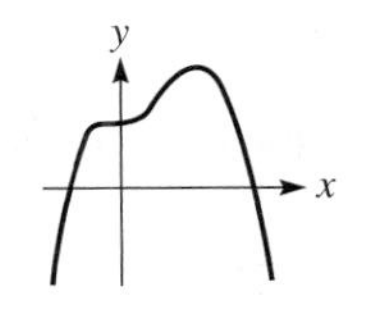

31. Max. $(-1, 4)$, min. $(1, -4)$, infl. $(0, 0)$

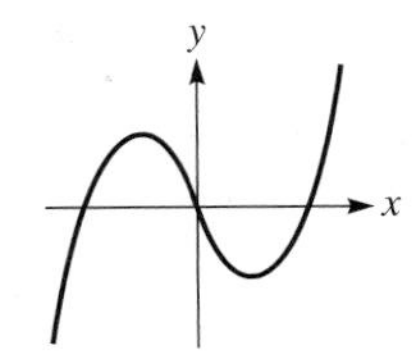

33. Where $y' > 0$, y inc.
$y' = 0$, y has a max. or min.
$y' < 0$, y dec.
$y'' > 0$, y conc. up
$y'' = 0$, y has infl.
$y'' < 0$, y conc. down

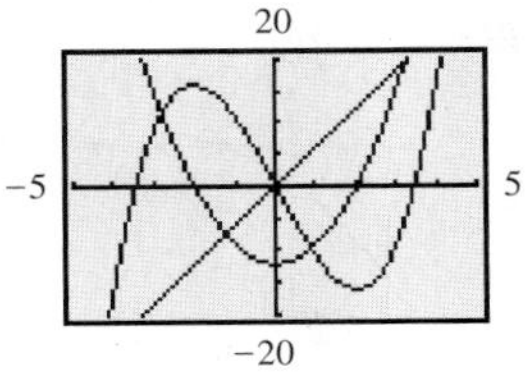

20, −5, 5, −20

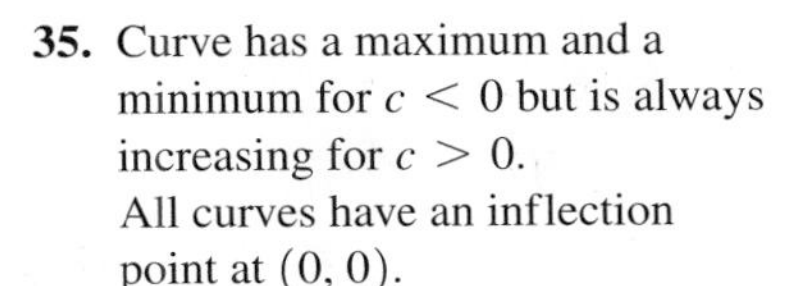

35. Curve has a maximum and a minimum for $c < 0$ but is always increasing for $c > 0$.
All curves have an inflection point at $(0, 0)$.

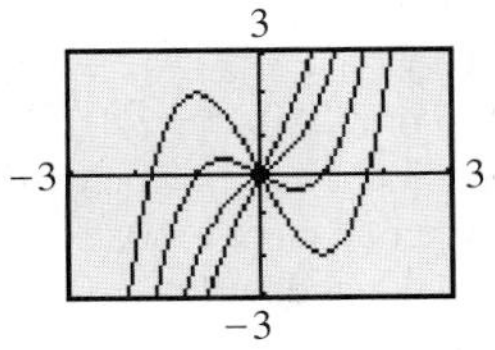

3, −3, 3, −3

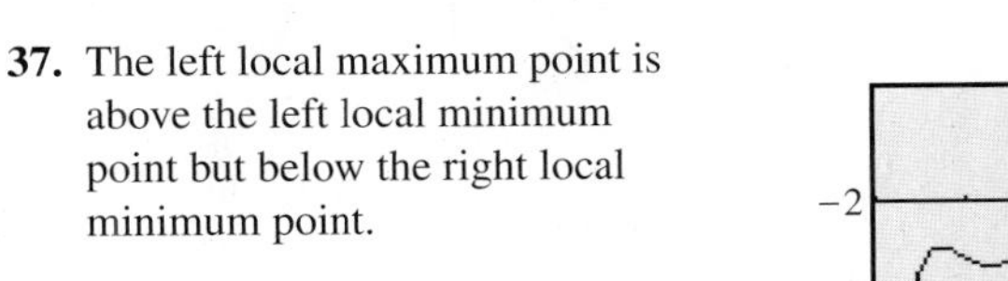

37. The left local maximum point is above the left local minimum point but below the right local minimum point.

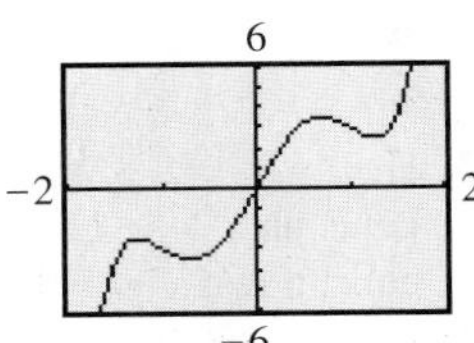

39. B, C, A

41. Dec. $x < -1$, $x > 3$ $(f'(x) < 0)$;
incr. $-1 < x < 3$ $(f'(x) > 0)$;
local min. at $x = -1$; local max. at $x = 3$

43. Max. (200, 100)

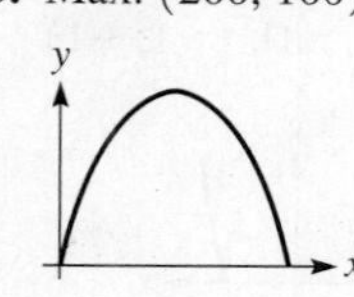

45. Max. (4, 8)

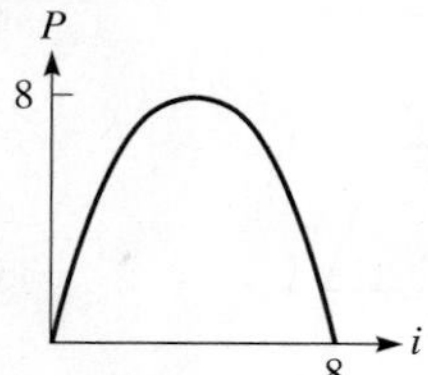

47. Max. (3, 16)

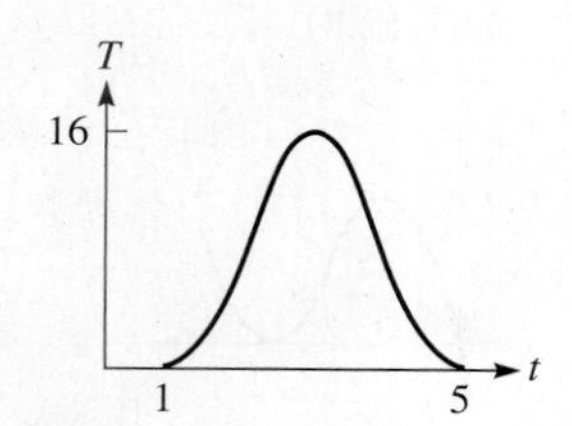

49. Max. (0, 75), infl. (1, 64), (3, 48)

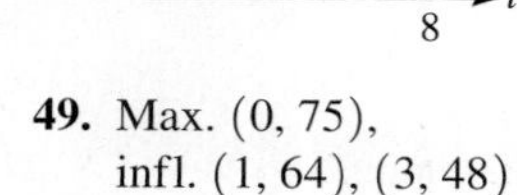

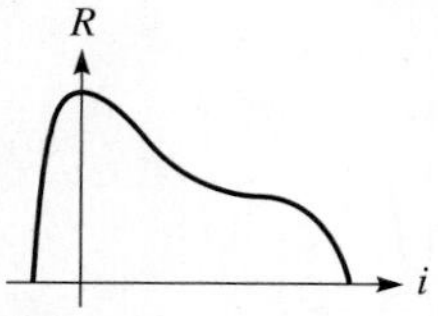

51. Max. (40, 41,620), infl. (18, 20,324)

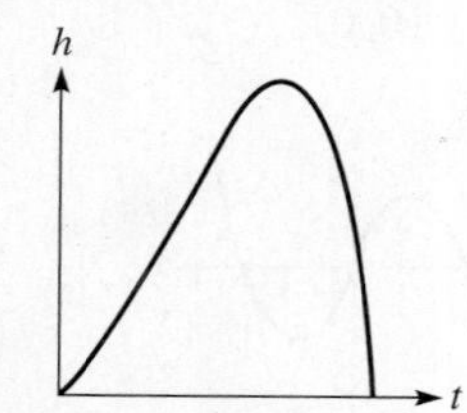

53. $V = 4x^3 - 40x^2 + 96x$, max. (1.57, 67.6)

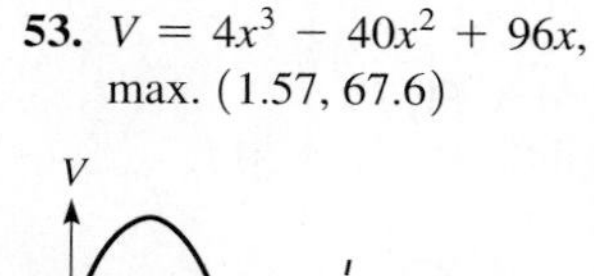

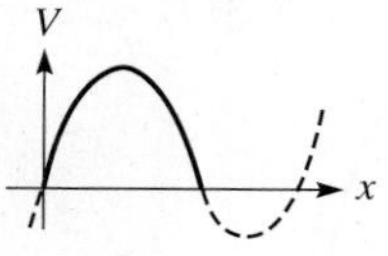

55.

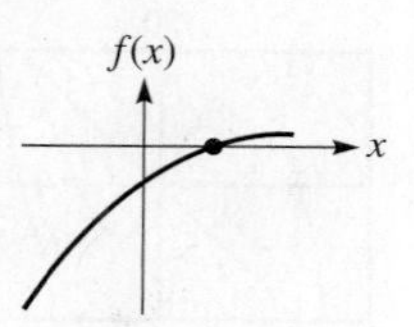

57.

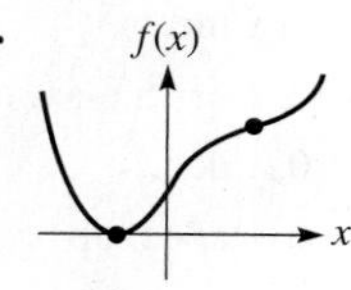

Exercises 24.6, page 725

1. $y = x - \frac{4}{x}$
int. (2, 0), (−2, 0);
inc. all x, except $x = 0$;
conc. up $x < 0$,
conc. down $x > 0$

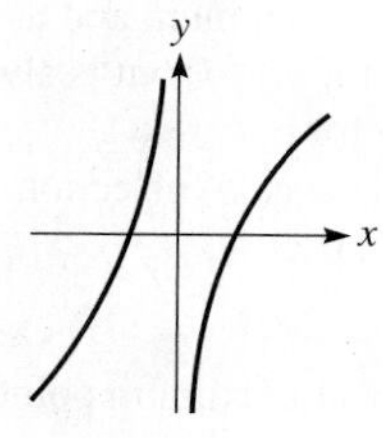

3. Dec. $x < -1, x > -1$,
conc. up $x > -1$,
conc. down $x < -1$,
int. (0, 2), asym. $x = -1, y = 0$

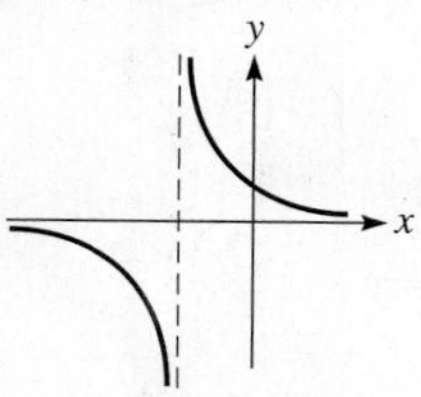

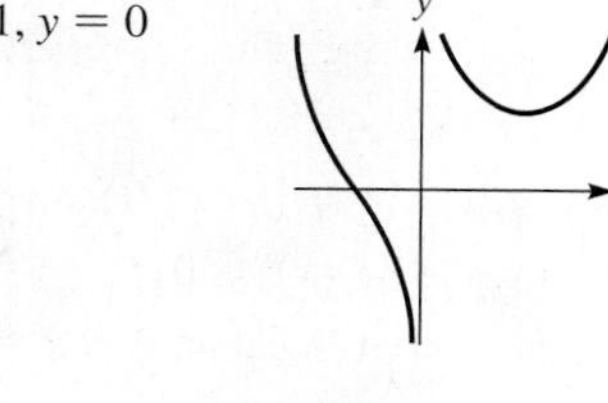

5. Int. $(-\sqrt[3]{2}, 0)$, min. (1, 3),
infl. $(-\sqrt[3]{2}, 0)$, asym. $x = 0$

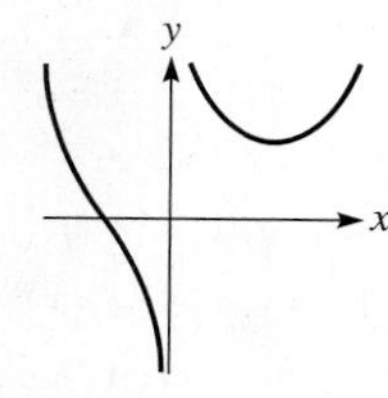

7. Int. (3, 0), (−3, 0),
asym. $x = 0, y = x$,
conc. up $x < 0$,
conc. down $x > 0$

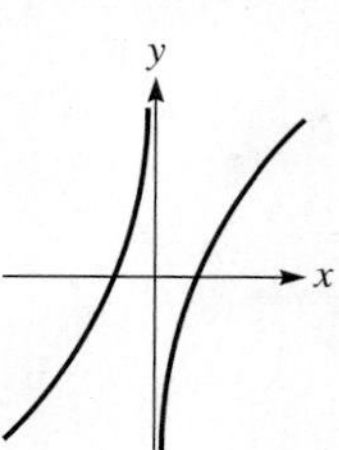

9. Int. (0, 0), max. (−2, −4),
min. (0, 0), asym. $x = -1$

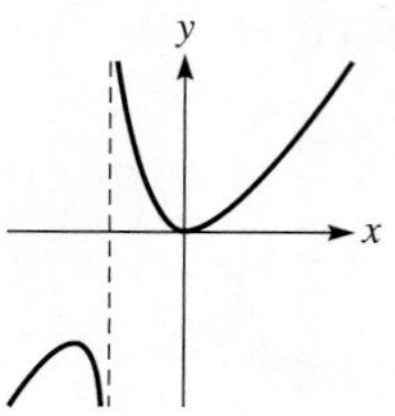

11. Int. (0, −1),
max. (0, −1),
asym. $x = 1$,
$x = -1, y = 0$

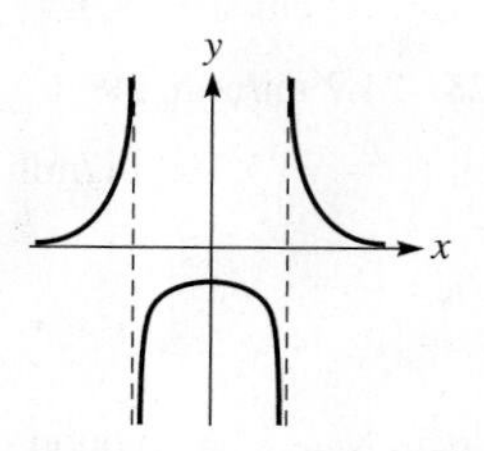

13. Int. (1, 0), max. (2, 1),
infl. $(3, \frac{8}{9})$, asym.
$x = 0, y = 0$

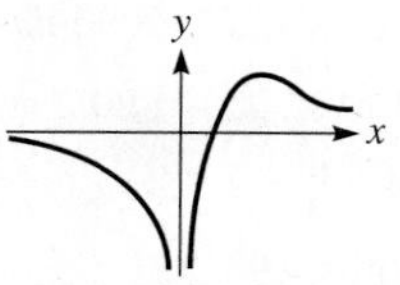

15. Int. (0, 0), (1, 0), (−1, 0),
max. $(\frac{1}{2}\sqrt{2}, \frac{1}{2})$,
min. $(-\frac{1}{2}\sqrt{2}, -\frac{1}{2})$

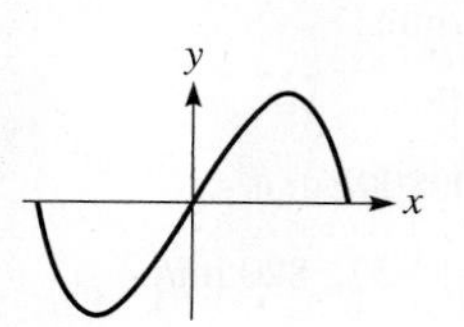

17. Int. (0, 0),
infl. (0, 0), asym.
$x = -3$,
$x = 3, y = 0$

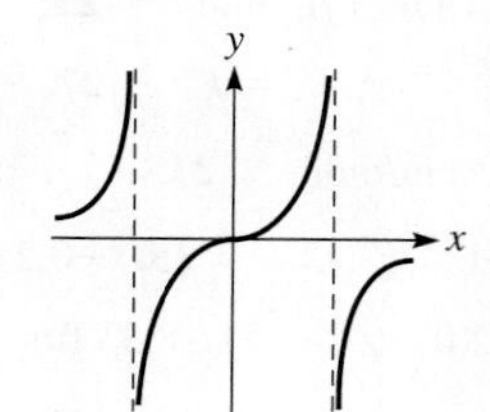

19. As c goes from −3 to +3, the graph goes from second and fourth quadrants to the third and first quadrants.

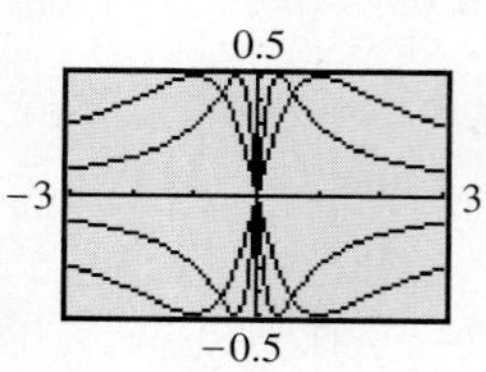

21.

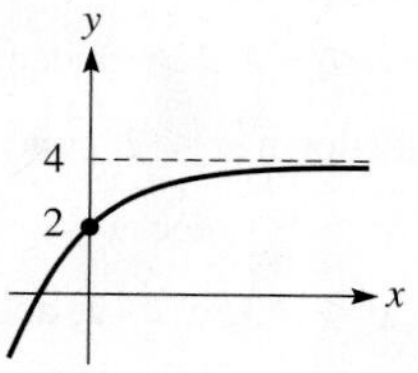

23. Dividing helps determine behavior as x becomes large; the range.

25. Int. (0, 0), asym. $C_T = 6$,
inc. $C \geq 0$,
conc. down $C \geq 0$
asym. $C_T = 6$ represents the limiting value of C_T.

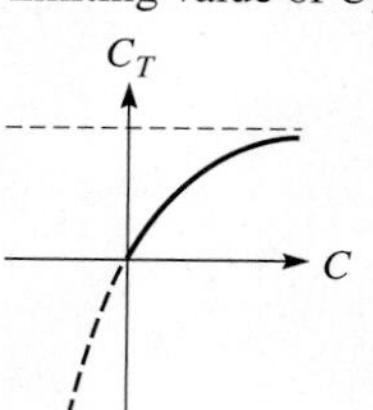

27. Int. (0, 1), max. (0, 1)
infl. (636, 0.82),
asym. $R = 0$

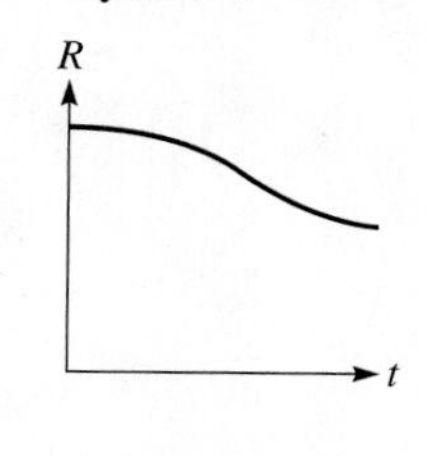

29. Int. $(1, 0)$, $(-1, 0)$; inc, all r, except $r = 0$; conc. up $r < 0$, conc. down $r > 0$

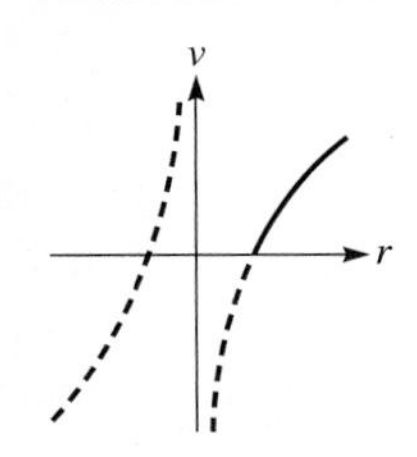

31. $A = 2\pi r^2 + \frac{40}{r}$, min. $(1.47, 40.8)$, asym. $r = 0$

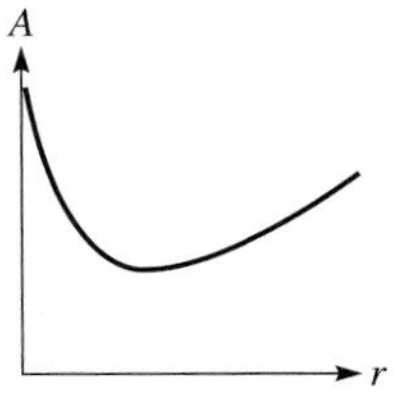

Exercises 24.7, page 730

1. 360,000 ft^2 **3.** 196 ft **5.** 1.2 A **7.** \$50

9. 35 m^2, \$8300 **11.** 1.94 units **13.** a

15. 12 in. by 12 in.

17. $xy = A, p = 2x + 2y; p = 2x + \frac{2A}{x}, \frac{dp}{dx} = 2 - \frac{2A}{x^2}, \frac{dp}{dx} = 0$
$x = \sqrt{A}, y = \sqrt{A}; x = y$ (square)

19. 4.95 cm, 9.90 cm **21.** 1.1 h **23.** 8.49 cm, 8.49 cm

25. $r = 11.5$ in., $l = 36.0$ in., $V = 14{,}850$ in^3.

27. 24 cm, 36 cm **29.** 3.00 ft **31.** 2 cm **33.** 12

35. $0.58L$ **37.** 67.3 cm **39.** 15 in., 21 in., 9 in.

41. 38.0° **43.** 2.7 km from A **45.** 1.33 in.

47. 100 m **49.** 8.0 mi from refinery **51.** 59.2 m, 118 m

53. 1250

Exercises 24.8, page 736

1. $\frac{-8(t^3 - 2)}{(t^3 + 4)^2}dt$

3. $\frac{de}{e} = 0.0065 = 0.65\%; \frac{dV}{V} = 0.019 = 1.9\%$

5. $(4x^3 + 3)dx$ **7.** $\frac{-10dr}{r^6}$ **9.** $24t(t^2 - 5)^3 dt$

11. $\frac{-72x dx}{(3x^2 + 1)^2}$ **13.** $(1 - x)^2(1 - 4x)dx$ **15.** $\frac{2dx}{(5x + 2)^2}$

17. 12.28, 12 **19.** 0.6264903, 0.6257

21. $L(x) = 2x$

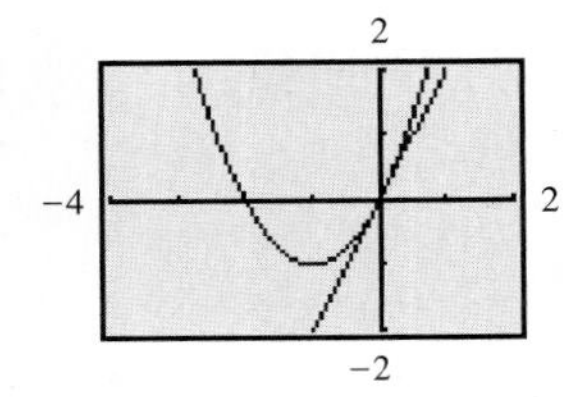

23. $L(x) = -2x - 3$

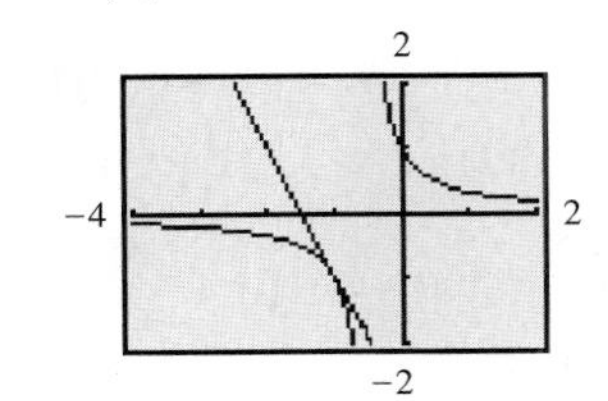

25. 1570 km **27.** 0.25% **29.** −31 nm **31.** 1220 cm^3

33. $\frac{dr}{r} = \frac{kd\lambda/(2\lambda^{1/2})}{k\lambda^{1/2}}$ **35.** $\frac{dA}{A} = \frac{2dr}{r}$ **37.** 2.0125

39. $L(x) = 1^k + k(1^{k-1})(x - 0) = 1 + kx$

41. $L(x) = -\frac{1}{2}x + \frac{3}{2}$; 1.45 **43.** $L(V) = 1.73 - 0.13V$

Review Exercises for Chapter 24, page 737

Concept Check Exercises

1. F **2.** F **3.** F **4.** F **5.** F **6.** F **7.** T **8.** F

Practice and Applications

9. $5x - y + 1 = 0$

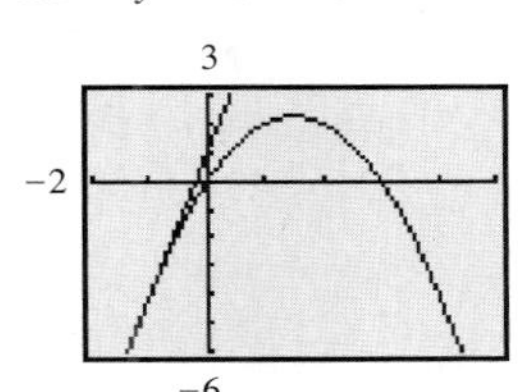

11. $8x + 5y - 50 = 0$

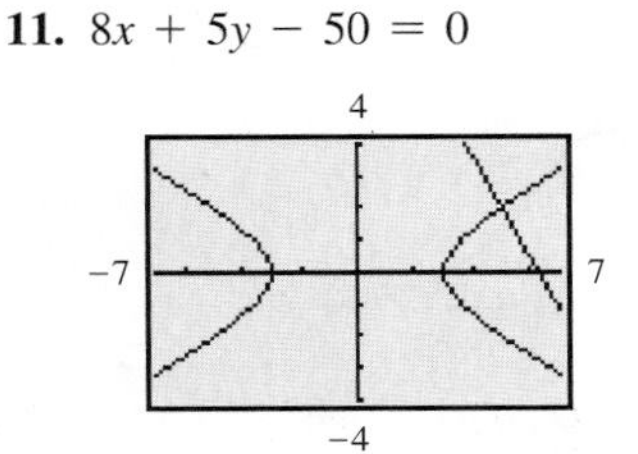

13. $x - 2y + 3 = 0$

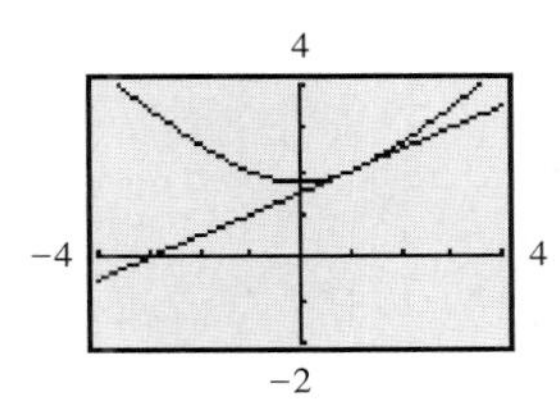

15. 4.19, 72.6° **17.** 2.12

19. 2.00, 90.9° **21.** 0.7458983 **23.** 1.4422

25. Min. $(-2, -16)$, conc. up all x

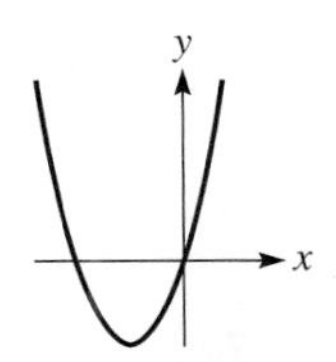

27. Int. $(0, 0)$, $(\pm\sqrt{3}, 0)$; max. $(1, 6)$, min. $(-1, -6)$; infl. $(0, 0)$

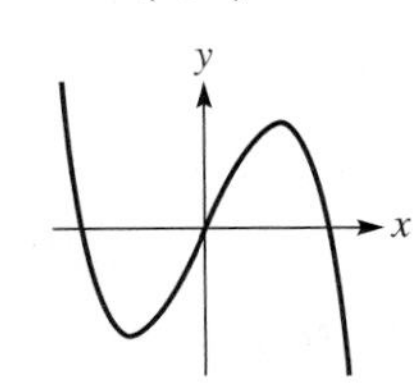

29. Min. $(2, -48)$; conc. up $x < 0, x > 0$

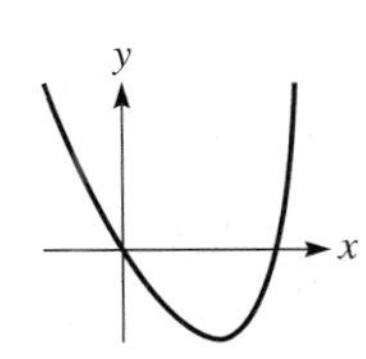

31. Min. $(\pm 3\sqrt{2}, 6)$; asym. $x = \pm 3$; conc. up $x < -3, x > 3$

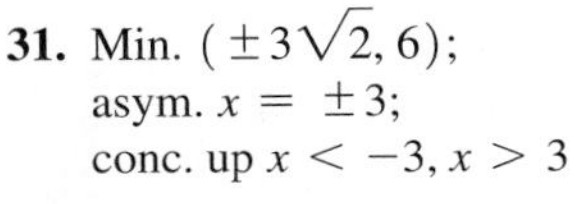

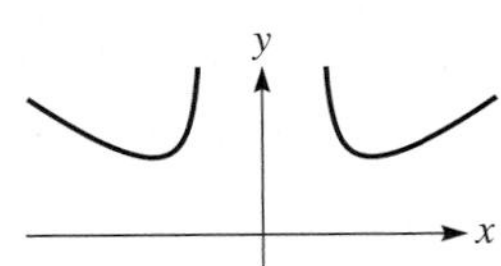

33. $\left(6x^2 - \frac{10}{x^3}\right)dx$ **35.** $\frac{2x\,dx}{3(x^2 - 3)^{2/3}}$ **37.** 0.244

39. $L(x) = \frac{3}{4}x + \frac{7}{4}$ **41.** 1.85 m^3 **43.** 3200 ft^3

45. $\frac{RdR}{R^2 + X^2}$ **47.** 251.1 **49.** $2x - y + 1 = 0$

51. $b^2 < 3c$ **53.** (a) False, $f(x) = x^3$; (b) False, $f(x) = x^4$

55. 1.6×10^5 mm^3/s **57.** 0.0 m, 6.527 m **59.** 8.8 m/s, 336°

61. 2.8 ft/s **63.** −7.44 cm/s **65.** 0 1.26

67. 22 m^3/s **69.**

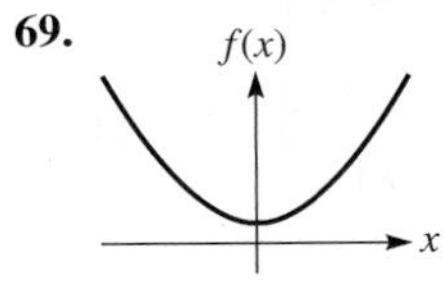

71. −0.567 ft/min

73. $x = 160, y = 250$ **75.** 7.31 ft/s

77. Max. (0, 100); infl. (37, 63); int. (0, 100), (89, 0) $L(x) = 113 - 1.42x$ **79.** 710 mi/h

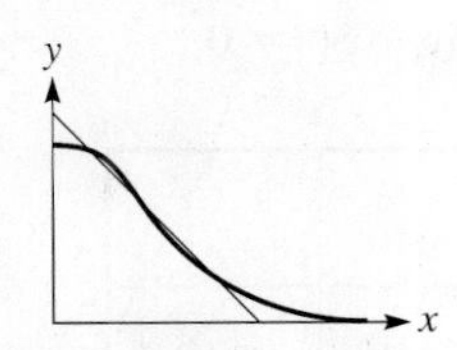

81. 6 μF, 6 μF **83.** 13.5 ft^3 **85.** 3 km

87. $r = h = 1.68$ ft **89.** 0.318 ft/min

91. $r = 2.09$ cm; $h = 23.0$ cm

93. $l = 15.3$ ft, $w = 11.5$ ft, $h = 7.66$ ft

Exercises 25.1, page 744

1. $F(x) = 3x^4$ **3.** $F(x) = \frac{2}{3}x^{3/2} + \frac{1}{x^3}$ **5.** 1 **7.** 3

9. 6 **11.** −1 **13.** $\frac{5}{3}x^3$ **15.** $\frac{3}{2}t^4 + 12t$

17. $\frac{4}{5}x^{5/2} - \frac{3}{2}x^2$ **19.** $\frac{8}{3}x^{3/2} + 3x$ **21.** $\frac{7}{5x^5} + \frac{x}{9}$

23. $\frac{4}{5}v^5 + 3\pi^2 v$ **25.** $\frac{1}{2}x^{100} + \frac{1}{2}x^{-78}$ **27.** $\frac{1}{3}x^3 - 4x - \frac{1}{x}$

29. $(2x + 1)^6$ **31.** $(p^2 - 1)^4$ **33.** $\frac{1}{10}(2x^4 + 1)^5$

35. $(6x + 1)^{3/2}$ **37.** $\frac{1}{4}(3x + 1)^{4/3}$ **39.** $\frac{6}{2x + 1}$

41. $(2x + 5)^3$ is antiderivative of $6(2x + 5)^2$. Factor of 2 from $2x + 5$ is needed.

Exercises 25.2, page 748

1. $4x^2 + C$ **3.** $\frac{(x^2 + 1)^5}{5} + C$ **5.** $x^2 + C$ **7.** $\frac{3}{2}x^4 + C$

9. $\frac{16}{5}x^{5/2} + C$ **11.** $-\frac{3}{R^3} + C$ **13.** $\frac{1}{8}x^8 - \frac{1}{2}x^6 + C$

15. $3x^3 + \frac{1}{2}x^2 + \frac{1}{3}x + C$ **17.** $\frac{1}{6}t^3 + \frac{2}{t} + C$

19. $\frac{2}{7}x^{7/2} - 2x^{5/2} + C$ **21.** $6x^{1/3} + 9x + C$

23. $x + \frac{4}{3}x^3 + \frac{4}{5}x^5 + C$ **25.** $\frac{1}{6}(x^2 - 1)^6 + C$

27. $\frac{2}{5}(x^4 + 3)^5 + C$ **29.** $\frac{1}{2}(2\theta^5 + 5)^8 + C$

31. $\frac{1}{12}(8x + 1)^{3/2} + C$ **33.** $\frac{2}{3}\sqrt{6x^2 + 1} + C$

35. $4\sqrt{z^2 - 2z} + C$ **37.** $y = 2x^3 + 2$

39. $y = 5 - \frac{1}{18}(1 - x^3)^6$

41. No. The result should include the constant of integration.

43. No. With $u = 4x^3 + 3$, $du = 12x^2dx$, and the x^4 would have to be x^2.

45. No. There should be a factor of 1/2 in the result.

47. $12y = 83 + (1 - 4x^2)^{3/2}$ **49.** $f(x) = 2x^2 - 5x + 3$

51. $\frac{4}{15}x^{5/2} + C_1x + C_2$ **53.** $s = 10t + \frac{1}{2}t^2$

55. $i = 0.2t^2 - 0.02t^3 + 2$ **57.** $T = 2250(r + 1)^{-2} + 250$

59. $f = \sqrt{0.01A + 1} - 1$ **61.** $y = 3x^2 + 2x - 3$

Exercises 25.3, page 754

1. (a) 3 (b) $\frac{15}{4}$ **3.** 20 **5.** 9, 12.15 **7.** 1.92, 2.28

9. 7.625, 8.208 **11.** 0.464, 0.5995 **13.** 1.92, 1.96

15. 13.5 **17.** $\frac{8}{3}$ **19.** 9 **21.** 0.8 **23.** 2

25. 14.85; the extra area above $y = 3x$ using circumscribed rectangles is the same as the omitted area under $y = 3x$ using inscribed rectangles.

27. 3.08; the extra area above $y = x^2$ using circumscribed rectangles is greater than the omitted area under $y = x^2$ using inscribed rectangles.

Exercises 25.4, page 757

1. $-\frac{9}{4}$ **3.** 2 **5.** $\frac{62}{5}$ **7.** $2\sqrt{6} - 2\sqrt{3} - 21 = -19.57$

9. 2.53 **11.** $-\frac{32}{3}$ **13.** $-\frac{11}{4}$ **15.** 8 **17.** $-\frac{243}{2}$

19. 10 **21.** $\frac{1}{4}(20.5^{2/3} - 17.5^{2/3}) = 0.19$

23. $\frac{176}{1083} = 0.1625$ **25.** 84 **27.** $\frac{364}{3}$ **29.** $\frac{33}{784} = 0.0421$

31. $\frac{3880}{9} = 431.1$ **33.** $\frac{8}{3}(13\sqrt{13} - 27) = 53.0$

35. $\frac{1}{4} + \frac{15}{4} = 4$; under $y = x^3$, the area from $x = 0$ to $x = 1$ plus the area from $x = 1$ to $x = 2$ equals the area from $x = 0$ to $x = 2$.

37. $\frac{28}{3}$

39. $\frac{1}{2} > \frac{1}{3}$ and $\frac{3}{2} < \frac{7}{3}$; From $x = 0$ to $x = 1$ the area under $y = x$ is greater than the area under $y = x^2$. Between $x = 1$ and $x = 2$, the area under $y = x$ is less than the area under $y = x^2$.

41. $\frac{2}{2k + 1}$ **43.** −4 **45.** 10 **47.** $\frac{A\sqrt{2}}{a\sqrt{g}}(\sqrt{h} - \sqrt{H})$

49. 64,000 ft · lb **51.** 86.8 m^2 **53.** $\frac{3NE_F}{5}$

Exercises 25.5, page 761

1. $\frac{7}{6}$ **3.** $\frac{11}{2} = 5.50$, $\frac{16}{3} = 5.33$ **5.** 7.661, $\frac{23}{3} = 7.667$

7. 0.2042 **9.** 18.98 **11.** 0.5205 **13.** 21.74 **15.** 45.36

17. The tops of all trapezoids are below $y = 1 + \sqrt{x}$.

19. 0.703, ln 2 = 0.693 **21.** 100.027 ft

Exercises 25.6, page 764

1. 0.406 **3.** (a) 6 (b) 6 **5.** (a) 19.67 (b) 19.67

7. 19.27 **9.** 0.5114 **11.** 13.147 **13.** 44.63

15. 3.121, $\pi = 3.142$ **17.** 1.191 in.

Review Exercises for Chapter 25, page 765

Concept Check Exercises

1. F **2.** T **3.** F **4.** F

Practice and Applications

5. $x^4 - \frac{1}{2}x^2 + C$ **7.** $\frac{2}{7}u^{7/2} + \frac{16}{3}u^{3/2} + C$ **9.** $\frac{19}{3}$

11. $\frac{2}{5}x^{5/2} + \frac{8}{3}x^{3/2} + C$ **13.** $\frac{80}{3}$ **15.** $25x - \frac{3}{x^2} + C$ **17.** 3

19. $\frac{1}{(9 - 5n)^2} + C$ **21.** $-\frac{6}{7}(7 - 2x)^{7/4} + C$ **23.** $\frac{9}{8}(3\sqrt[3]{3} - 1)$

25. $-\frac{2}{15}(1 - 2x^3)^5 + C$ **27.** $-\frac{1}{2x - x^3} + C$ **29.** $\frac{3350}{3}$

31. $y = 3x - \frac{1}{3}x^3 + \frac{17}{3}$

33. (a) $x - x^2 + C_1$

(b) $-\frac{1}{4}(1 - 2x)^2 + C_2 = x - x^2 + C_2 - \frac{1}{4}$; $C_1 = C_2 - \frac{1}{4}$

35. $\int_3^4 F(v)dv$ **37.** $\frac{112}{9} + \frac{122}{9} = 26$

39. $\frac{1}{27}(6x + 5)^{3/2} + C_1x + C_2$

41. $0.25 = 0.25$; $y = x^3$ shifted 1 unit to the right is $y = (x - 1)^3$. Therefore, areas are the same.

43. 22

45. The graph of $f(x)$ is concave down. This means the tops of the trapezoids are below $y = f(x)$.

47. 0.842 **49.** 0.811 **51.** 13.6 **53.** 19.3016

55. 19.0417 **57.** 19.0354 **59.** 24.68 m^2 **61.** 25.81 m^2

63. 54 C **65.** $y = k(2L^3x - 6Lx^2 + \frac{2}{5}x^5)$ **67.** 14.9 m^2

Exercises 26.1, page 772

1. −40 ft/s **3.** −36 ft/s **5.** $s = 8.00 - 0.25t$

7. 1100 km **9.** 170 ft **11.** 15 ft/s **13.** 158 ft

15. 3.7 g's **17.** 76 ft/s **19.** No **21.** 0.345 nC

23. 0.017 C **25.** 120 V **27.** 4.65 mV **29.** 970 rad

31. 66.7 A **33.** $V_C(t) = \frac{1}{0.001}t + V_0$; $V_C(0.001) = 1 + V_0$

35. $m = 1002 - 2\sqrt{t + 1}$, 2.51×10^5 min

Exercises 26.2, page 777

1. $\frac{26}{3}$ **3.** 2 **5.** $\frac{64}{3}$ **7.** $\frac{32}{3}$ **9.** $\frac{1}{2}$ **11.** $\frac{13}{24}$

13. 3 **15.** $\frac{9}{2}$ **17.** $\frac{128}{3}$ **19.** $\frac{7}{6}$ **21.** $\frac{256}{15}$ **23.** $\frac{343}{24}$

25. $\frac{65}{12}$ **27.** 4

29. The area bounded by $x = 1$, $y = 2x^2$, and $y = x^3$ or the area bounded by $x = 1$, $y = 0$, and $y = 2x^2 - x^3$.

31. 20 **33.** $\frac{A_1}{A_2} = \frac{1 - \frac{1}{n+1}}{\frac{1}{n+1}} = \frac{n}{n+1}\cdot\frac{n+1}{1} = \frac{n}{1}$

35. 2 **37.** $4^{2/3}$ **39.** 1 **41.** $\frac{48}{5}$ **43.** 18.0 J

45. 80.8 km **47.** 4 cm^2 **49.** 42.5 dm^2

Exercises 26.3, page 783

1. $\frac{128\pi}{7}$ **3.** $\frac{32\pi}{3}$ **5.** $\frac{16\pi}{3}$ **7.** 36π **9.** 72π **11.** $\frac{768\pi}{7}$

13. $\frac{348\pi}{5}$ **15.** $\frac{15\pi}{2}$ **17.** $\frac{2\pi}{7}$ **19.** $\frac{3\pi}{5}$ **21.** $\frac{10\pi}{3}\sqrt{5}$

23. $\frac{31\pi}{5}$ **25.** $\frac{16\pi}{3}$

27. The region bounded by $y = x^{3/2}$, $y = 0$, $x = 1$, and $x = 2$

29. $\frac{8}{3}\pi$ **31.** $\frac{1}{3}\pi r^2h$ **33.** 1250 cm^3 **35.** 16,800 m^3

37. 4.16×10^6 mm^3 **39.** 18.3 cm^3

Exercises 26.4, page 789

1. $(0, \frac{8}{3})$ **3.** 3.9 cm **5.** 1.6 cm **7.** (−0.5 in., 0.5 in.)

9. (0.32 in., 0.23 in.) **11.** $(0, \frac{6}{5})$ **13.** $(\frac{4}{3}, \frac{4}{3})$

15. (1.20, 5.49) **17.** $(\frac{5}{3}, 6)$ **19.** $(\frac{4}{3}, \frac{8}{3})$ **21.** $(0, \frac{3}{5}a)$

23. $(\frac{7}{8}, 0)$ **25.** $(0, \frac{5}{3})$ **27.** $(\frac{2}{3}, 0)$

29. From right angle: horizontally 1.0 m, vertically 1.5 m

31. 2390 km above center

33. 19.3 cm from larger base

Exercises 26.5, page 794

1. $I_y = k$, $R_y = \frac{1}{2}\sqrt{2}$ **3.** 51 g · cm^2, 2.6 cm

5. 3060 g · cm^2, 3.62 cm **7.** $\frac{2}{3}k$ **9.** $\frac{162}{5}k$ **11.** $\sqrt{6}$

13. $\frac{9}{7}\sqrt{7}$ **15.** $\frac{8}{11}\sqrt{55}$ **17.** $\frac{16}{3}\pi k$ **19.** $\frac{4}{5}\sqrt{10}$

21. $1.5m$ **23.** $\frac{3}{10}mr^2$ **25.** 0.324 g · cm^2

27. 31.2 kg · cm^2

Exercises 26.6, page 799

1. 4990 lb **3.** 8.0 lb · in. **5.** 8.1 lb · in.

7. 1.7×10^{-16} J **9.** $0.09k$ ft · lb **11.** 39,000 ft · lb

13. 3.00×10^5 ft · ton **15.** 9.4×10^5 N · km

17. top: 3.24×10^5 ft · lb

bottom: 5.74×10^5 ft · lb

19. 0.770 Nm **21.** 2340 lb **23.** 20,800 lb

25. 1.72×10^5 N **27.** 11,700 lb

29. 3.92×10^4 N, 1.18×10^5 N buoyant force **31.** 0.27 μA

33. 35.3% **35.** 109 ft

37. $S = \pi r\sqrt{r^2 + h^2}$

Review Exercises for Chapter 26, page 802

Concept Check Exercises

1. F **2.** T **3.** T **4.** F **5.** T **6.** F

Practice and Applications

7. 0.71 m **9.** 4.2 s **11.** No **13.** Yes **15.** 0.44 C

17. 55 V **19.** $y = 20x + \frac{1}{120}x^3$ **21.** 20.0 **23.** 18

25. $\frac{27}{4}$ **27.** $A_{top}/A_{bot} = (\frac{2n}{2n+1})/(\frac{1}{2n+1}) = \frac{2n}{1}$

29. $\frac{48}{5}\pi$ **31.** $\frac{512}{5}\pi$ **33.** $\frac{4}{3}\pi ab^2$ **35.** (−0.5 cm, 0.6 cm)

37. $(\frac{30}{7}, \frac{45}{4})$ **39.** $(\frac{14}{5}, 0)$ **41.** $\frac{8}{5}k$ **43.** 68.7 g · mm^2

45. 8500 ft · lb **47.** 4790 cm^2 **49.** −62 ft/s

51. 200 cm^3 **53.** 1.8 m **55.** 47 m^3 **57.** 88 ft^3

59. 10,200 lb **61.** 17.8°C **63.** $\frac{2kR^2}{3}$

Exercises 27.1, page 809

1. $8\theta\sin 2\theta^2\cos 2\theta^2 = 4\theta\sin 4\theta^2$ **3.** $3\cos(3x + 2)$

5. $12x^2\cos(2x^3 - 1)$ **7.** $-12\sin\frac{4}{3}x$ **9.** $3 - 6\sin(3x - \pi)$

11. $10\pi\sin 5\pi\theta\cos 5\pi\theta = 5\pi\sin 10\pi\theta$

13. $-45\cos^2(5x + 2)\sin(5x + 2)$ **15.** $4\sin 3x + 12x\cos 3x$

17. $9x^2\cos 5x - 15x^3\sin 5x$

19. $6v\cos(5v)\cos v^2 - 15\sin(5v)\sin v^2$ **21.** $\frac{2\cos 4x}{\sqrt{1 + \sin 4x}}$

23. $\frac{3t\cos(3t - \pi/3) - \sin(3t - \pi/3)}{2t^2}$ **25.** $\frac{4x(1 - 3x)\sin x^2 - 6\cos x^2}{(3x - 1)^2}$

27. $4\sin 3x(3\cos 3x\cos 2x - \sin 3x\sin 2x)$

29. $6\cos 2t\cos(3\sin 2t)$ **31.** $3\sin^2 x\cos x + 2\sin 2x$

33. $-\frac{\cos s}{\sin^2 s} + \frac{\sin s}{\cos^2 s}$ **35.** $\frac{-2x}{1 + \sin y}$

37. (a)

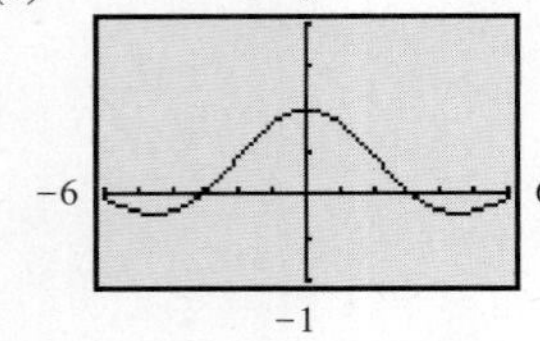

(b) See the table.

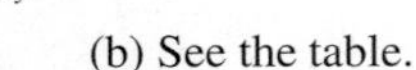

39. (a) 0.5403023, value of derivative

(b) 0.5402602, slope of secant line

41. Resulting curve is $y = \cos x$. **43.** $\frac{2x - y\cos xy}{x\cos xy - 2\sin 2y}$

45. $\frac{d\sin x}{dx} = \cos x$, $\frac{d^2\sin x}{dx^2} = -\sin x$,

$\frac{d^3\sin x}{dx^3} = -\cos x$, $\frac{d^4\sin x}{dx^4} = \sin x$

47. $\sin 2x = 2\sin x\cos x$ **49.** 0.515 **51.** −2.646

53. 410 mm/s **55.** −199 cm/s **57.** −38.5 km/°

59. $-\frac{AB\sin\theta}{2[1 + B\sin^2(\frac{\theta}{2})]^2}$

Exercises 27.2, page 812

1. $12x\sec^2 x^2\tan x^2$ **3.** $8\sec^2 4x$ **5.** $3\csc^2(2\pi - 3\theta)$

7. $15\sec 5v\tan 5v$ **9.** $12x^2 + \frac{3}{\sqrt{2x + 3}}\csc\sqrt{2x + 3}\cot\sqrt{2x + 3}$

11. $30\tan^4 2t\sec^2 2t$ **13.** $-4\cot^3(\frac{1}{2}x + \pi)\csc^2(\frac{1}{2}x + \pi)$

15. $2\tan 4x\sqrt{\sec 4x}$ **17.** $-84\csc^4(7x - \frac{\pi}{2})\cot(7x - \frac{\pi}{2})$

19. $0.5t^2\sec^2 0.5t + 2t\tan 0.5t$

21. $-4\csc x^2(2x\cos x\cot x^2 + \sin x)$ **23.** $\frac{-6x\csc 3x\cot 3x - 4\csc 3x}{x^3}$

25. $\frac{2(-4\sin 4x - 4\sin 4x\cot 3x + 3\cos 4x\csc^2 3x)}{(1 + \cot 3x)^2}$ **27.** $\sec^2 x(\tan^2 x - 1)$

29. $2\pi\cos 2\pi\theta\sec^2(\sin 2\pi\theta)$ **31.** $\frac{1 + 2\sec^2 4x}{\sqrt{2x + \tan 4x}}$

33. $\frac{2\cos 2x - \sec y}{x\sec y\tan y - 2}$ **35.** $24\tan 3x\sec^2 3x\,dx$

37. $4\sec 4x(\tan^2 4x + \sec^2 4x)dx$

39. (a) 3.4255188, value of derivative

(b) 3.4260524, slope of secant line

41. (a)

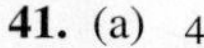

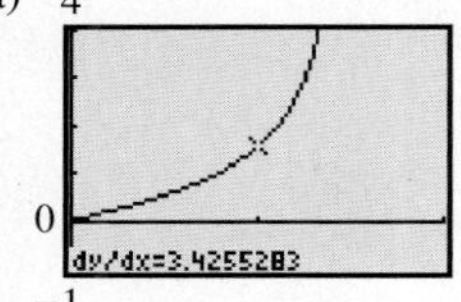

(b)

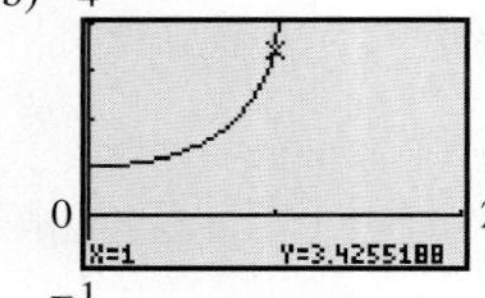

43. $2\tan x\sec^2 x = 2\sec x(\sec x\tan x)$ **45.** −12

47. $2\sec^2 x - \sec x\tan x = \frac{2}{\cos^2 x} - \frac{\sin x}{\cos^2 x}$

49. $dk\sec(kL)\tan(kL)$ **51.** −8.4 cm/s **53.** 140 ft/s

Exercises 27.3, page 815

1. $\frac{2x}{\sqrt{1 - x^4}}$ **3.** $\frac{7}{\sqrt{1 - 49x^2}}$ **5.** $\frac{6x}{\sqrt{1 - x^4}}$

7. $\frac{-4.8}{\sqrt{1 - 4t^2}}$ **9.** $\frac{3}{\sqrt{(x - 1)(2 - x)}}$ **11.** $\frac{4}{\sqrt{s}(1 + s)}$

13. $\frac{-6x}{x^2 + 1} + 6\tan^{-1}(\frac{1}{x})$ **15.** $\frac{6x}{\sqrt{1 - 4x^2}} + 5\sin^{-1}2x$

17. $\frac{0.8u}{1 + 4u^2} + 0.4\tan^{-1}2u$ **19.** $\frac{\sqrt{1 - 9R^2}\sin^{-1}3R - 3R}{\sqrt{1 - 9R^2}(\sin^{-1}3R)^2}$

21. $\frac{-x - 3\sqrt{1 - x^2}\cos^{-1}x}{\sqrt{1 - x^2}x^4}$ **23.** $\frac{-24(\cos^{-1}4x)^2}{\sqrt{1 - 16x^2}}$

25. $\frac{4\sin^{-1}(4t + 3)}{\sqrt{-(4t^2 + 6t + 2)}}$ **27.** $\frac{-1}{t^2 + 1}$ **29.** $\frac{-2(2x + 1)^2}{(1 + 4x^2)^2}$

31. $\frac{18(4 - \cos^{-1}2x)^2}{\sqrt{1 - 4x^2}}$ **33.** $\frac{(1 + y^2)(3 - 2x)}{2}$

35. (a) 1.1547005, value of derivative

(b) 1.1547390, slope of secant line

37. $\frac{3(\sin^{-1}x)^2dx}{\sqrt{1 - x^2}}$

39. 0.41

41. $\frac{2}{(x^2 + 1)^2}$

43.

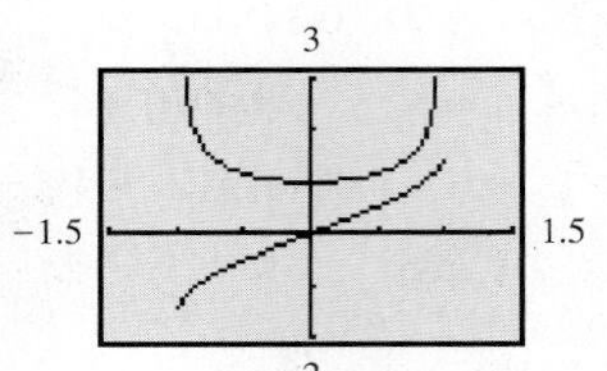

45. $\frac{-16x}{(1 + 4x^2)^2}$

47. Let $y = \sec^{-1}u$; solve for u; take derivatives; substitute.

49. $\frac{E - A}{\omega m\sqrt{m^2E^2 - (A - E)^2}}$ **51.** $-\frac{R}{R^2 + (X_L - X_C)^2}$

53. $\theta = \tan^{-1}\frac{h}{x}$; $\frac{d\theta}{dx} = \frac{-h}{x^2 + h^2}$

Exercises 27.4, page 819

1. Max. $(\frac{\pi}{3}, 0.34)$;

min. $(\frac{5\pi}{3}, -3.48)$;

infl. $(0, 0)$, $(\pi, -\frac{\pi}{2})$

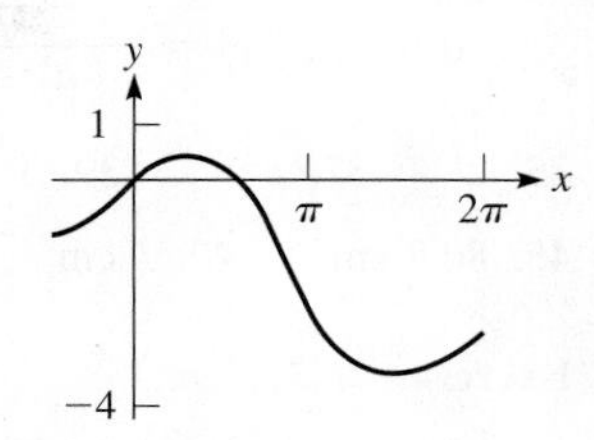

3. $d\sin x/dx = \cos x$ and $d\cos x/dx = -\sin x$, and $\sin x = \cos x$ at points of intersection.

5. $\frac{1}{x^2 + 1}$ is always positive.

7.

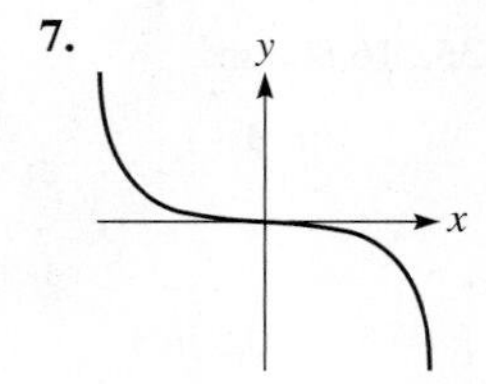

Dec. $x > 0, x < 0$;

infl. $(0, 0)$;

asym. $x = \frac{\pi}{2}, x = -\frac{\pi}{2}$

9. $y = 1.10x - 0.29$ **11.** 1.9337538 **13.** -10

15. $32 \sin 2t \cos 2t$ **17.** 0.58 ft/s, -1.7 ft/s^2 **19.** -0.072 lb/s

21. 358 cm/s, 18.0° **23.** 17.4 cm/s^2, 176° **25.** -0.073 rad/s

27. 8.08 ft/s **29.** 14.0° **31.** 280 km **33.** 0.60 L/s

35. $w = 9.24$ in., $d = 13.1$ in. **37.** 56 m/s **39.** 14 ft

Exercises 27.5, page 824

1. $-4\tan 4x$ **3.** $\frac{2\log e}{x}$ **5.** $\frac{4\log_5 e}{x-3}$ **7.** $\frac{8}{x-3}$

9. $\frac{4\sec^2 2x}{\tan 2x} = 4\sec 2x \csc 2x$ **11.** $\frac{2}{4T+1}$ **13.** $\frac{3(1-2x)}{x-x^2}$

15. $\frac{6(t+2)(t+\ln t^2)}{t}$ **17.** $6\ln(6-x) - \frac{6x}{6-x}$ **19.** $\frac{1}{x\ln x}$

21. $-\pi\theta\tan(\pi\theta^2)$ **23.** $\frac{\cos \ln x}{x}$ **25.** $6\ln 2v + 3\ln^2 2v$

27. $\frac{x\sec^2 x + \tan x}{x\tan x}$ **29.** $\frac{v+4}{v(v+2)}$ **31.** $\frac{x+y-2\ln(x+y)}{x+y+2\ln(x+y)}$

33. $y(\frac{1}{x} - 1)$

35. 0.5 is value of derivative; 0.4999875 is slope of secant line.

37. (a) (b) See the table.

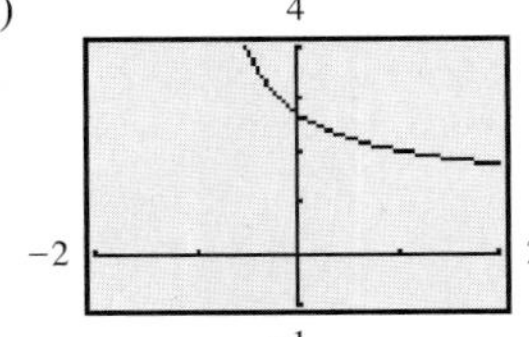

39. -0.3641 **41.** $-\frac{4}{3}$ **43.** $L(x) = 4x - \pi$ **45.** 2.73

47. $x^x(\ln x + 1)$. In Eq. (23.15) the exponent is constant. For x^x, both the base and the exponent are variables.

49. $\frac{dy_1}{dx}\Big|_{x=-1} = \frac{2}{x}\Big|_{x=-1} = -2[\ln(x^2)$ defined for all x, $x \neq 0]$
$\frac{dy_2}{dx}\Big|_{x=-1} = \frac{2}{x}\Big|_{x=-1}$ does not exist ($\ln x$ not defined for $x \leq 0$)

51. $\frac{10\log e}{I}\frac{dI}{dt}$ **53.** $\frac{x}{x^2+1+\sqrt{1+x^2}} - \frac{1}{x} + \frac{x}{\sqrt{1-x^2}}$

55. 0.083 s^2/ft.

Exercises 27.6, page 827

1. $2e^{2x}\cot e^{2x}$ **3.** $(6\ln 4)4^{6x}$ **5.** $\frac{3e^{\sqrt{x}}}{\sqrt{x}}$ **7.** $4e^{2t}(3e^t - 2)$

9. $e^{-3T}(1 - 3T)$ **11.** $e^{\tan 2x}(1 + 2x\sec^2 2x)$ **13.** $8e^{-4s}$

15. $e^{-3x}\sec 4x(4\tan 4x - 3)$ **17.** $\frac{2e^{3x}(12x+5)}{(4x+3)^2}$ **19.** $\frac{te^{t^2}}{e^{t^2}+4}$

21. $16e^{6x}(x\cos x^2 + 3\sin x^2)$ **23.** $\frac{2(1+2te^{2t})}{t\sqrt{\ln 2t + e^{2t}}}$ **25.** $\frac{ye^{xy}}{1-xe^{xy}}$

27. $\frac{3e^{2x}}{x} + 6e^{2x}\ln x$ **29.** $12e^{6t}\cot 2e^{6t}$ **31.** $\frac{4e^{2x}}{\sqrt{1-e^{4x}}}$

33. (a) 2.7182818, value of derivative
(b) 2.7184177, slope of secant line

35.

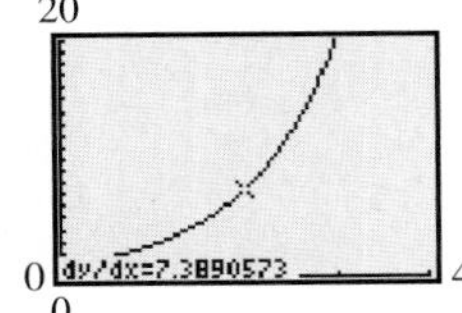

37. -1.458

39. $\frac{12e^{4x}(4x+23)dx}{(x+6)^2}$ **41.**

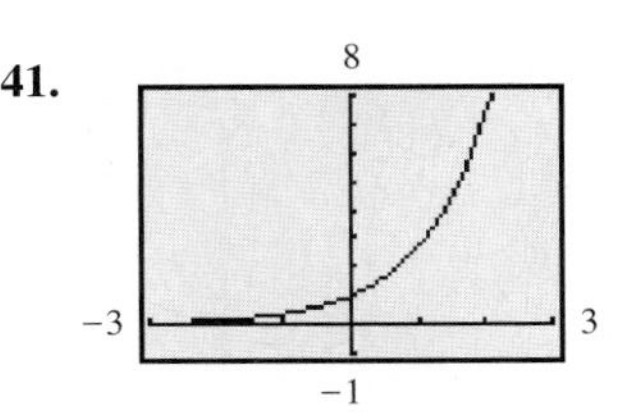

43. $(-xe^{-x} + e^{-x}) + (xe^{-x}) = e^{-x}$ **45.** $-3, 2$

47. $Ak^2e^{kt} + Bk^2e^{-kx} = k^2(Ae^{kx} + Be^{-kx})$ **49.** -0.000164/h

51. $e^{-66.7t}(999\cos 226t - 295\sin 226t)$

53. Substitute and simplify.

55. $\frac{d}{dx}\left[\frac{1}{2}(e^u - e^{-u})\right] = \frac{1}{2}(e^u + e^{-u})\frac{du}{dx}$;
$\frac{d}{dx}\left[\frac{1}{2}(e^u + e^{-u})\right] = \frac{1}{2}(e^u - e^{-u})\frac{du}{dx}$

Exercises 27.7, page 831

1. ∞ **3.** $\frac{1}{6}$ **5.** 1 **7.** ∞ **9.** 0 **11.** 0 **13.** $-\frac{1}{6}$

15. $-\pi$ **17.** 1 **19.** 0 **21.** 0 **23.** ∞ **25.** 0

27. $-\frac{5}{2}$ **29.** $\frac{1}{2\pi}$ **31.** 0 **33.** $\frac{3}{20}$ **35.** -1 **37.** 0

39. 1 **41.** 1 **43.** No-not indeterminant form

45. $\sin x$ varies between 1 and -1 as $x \to \infty$. **47.** gt

Exercises 27.8, page 834

1. Int. $(0, 0)$, $(\pi, 0)$, $(2\pi, 0)$;
max. $(\frac{\pi}{4}, 0.322)$;
min. $(\frac{5\pi}{4}, -0.014)$;
infl. $(\frac{\pi}{2}, 0.208)$

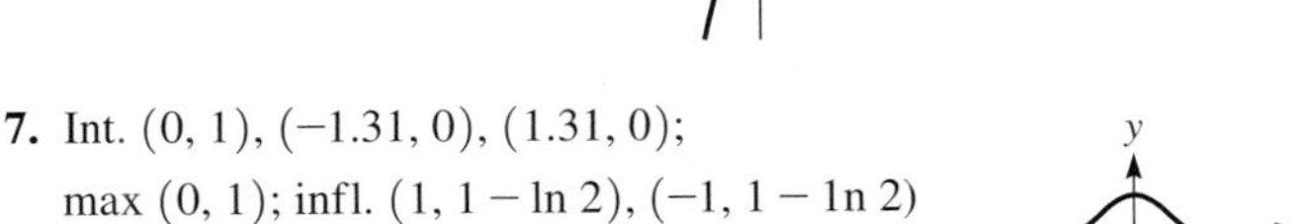

3. Int. $(0, 0)$, max. $(0, 0)$,
not defined for $\cos x < 0$,
asym. $x = -\frac{1}{2}\pi, \frac{1}{2}\pi, \ldots$

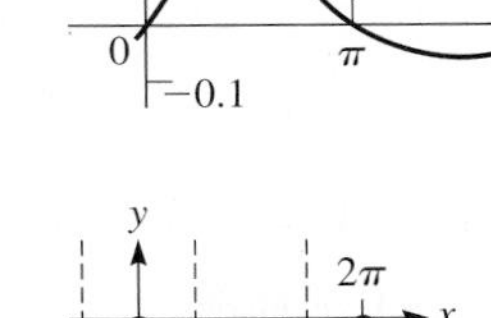

5. Int. $(0, 0)$, max. $(1, \frac{3}{e})$,
infl. $(2, \frac{6}{e^2})$ asym. $y = 0$

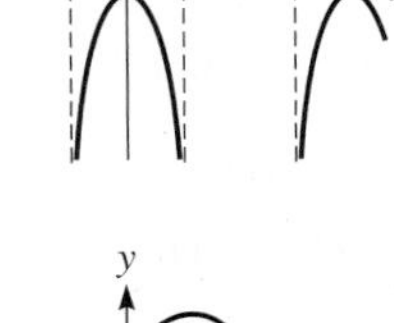

7. Int. $(0, 1)$, $(-1.31, 0)$, $(1.31, 0)$;
max $(0, 1)$; infl. $(1, 1 - \ln 2)$, $(-1, 1 - \ln 2)$

9. Int. $(0, 4)$; max. $(0, 4)$;
infl. $(\frac{1}{2}\sqrt{2}, \frac{4}{e}\sqrt{e})$, $(-\frac{1}{2}\sqrt{2}, \frac{4}{e}\sqrt{e})$;
asym. $y = 0$

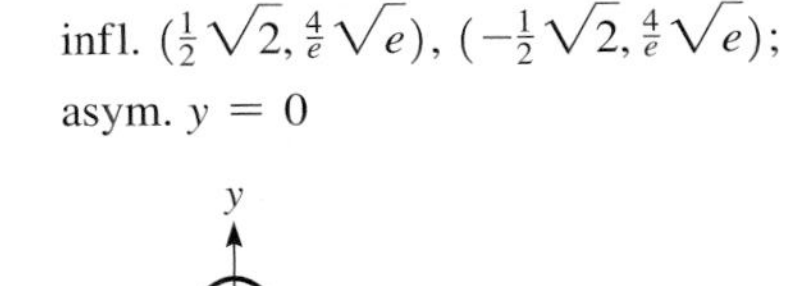

11. Max. (4, 3.09), asym. $x = 0$

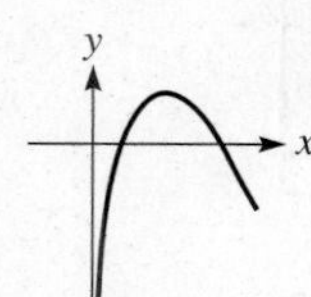

13. Int. (0, 0), infl. (0, 0), inc. all x

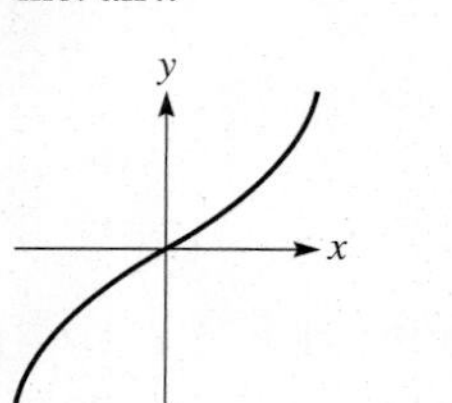

15. y_1 inc. $x < 0$, y_1 dec. $x > 0$; y_2 inc. $x > 0$, y_1 dec. $x < 0$; y_1 and y_2 are reciprocals.

17. $y = x - 1$ **19.** $2\sqrt{2}x - 2y + 2\sqrt{2} - 3\pi\sqrt{2} = 0$

21. 1.1973338 **23.** 170 V **25.** −0.303 W/day

27. −5.4°C/min **29.** $\frac{p(-a + bT)}{T^2}$ **31.** $a = k^2x$

33. Int. (0, 0); min. (0, 0). $(2\pi, 0), \ldots$; asym. $x = -\frac{\pi}{2}, \frac{\pi}{2}, \frac{3\pi}{2}, \ldots$

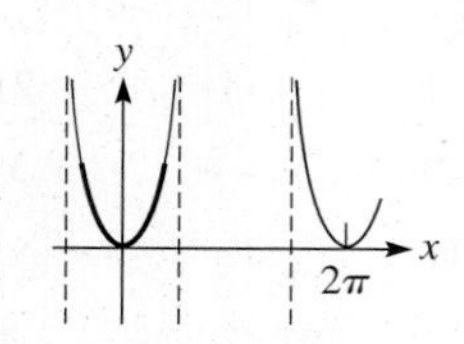

35. 3.0 s

37. $y'' = \frac{1}{\sqrt{2\pi}}(e^{-x^2/2})(x^2 - 1)$, $y'' = 0$ if $x = \pm 1$

39. $v = -e^{-0.5t}(1.4\cos 6t + 2.3\sin 6t)$, −2.03 cm/s

41. $1/\sqrt{e} = 0.607$ **43.** $\frac{1}{\sqrt{b}}$

45. Max. 0°, 360°; min. 120°, 240°

49. Infl. (π, π), $(3\pi, 3\pi)$

51. Max. $(e^{-2}, 4e^{-2})$, min. (1, 0), infl. (e^{-1}, e^{-1})

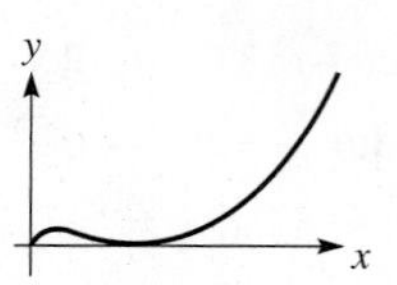

53. $7.27x + y - 8.44 = 0$ **55.** $2x + 2.57y - 4.30 = 0$

57. $\frac{2}{3}$ **59.** −1 **61.** 0

63. $2\sin x\cos x - 2\cos x\sin x = 0$ **65.** $-9\sin 3x = -9(\sin 3x)$

67. −0.7034674 **69.** $x > \frac{1}{2}\ln 2$ (= 0.3466) **71.** 2

73. $r(\tan\theta - \frac{v}{w}\sec\theta)$ **75.** 1.06 m **77.** 0.039 **79.** 2.5 N

81. 0.12 cm/s **83.** −0.064°C/day **85.** $50\sin 4t$

87. $n = \frac{(8\ln 8)x}{\ln x}$, min. $(e, 8e\ln 8)$, asym. $x = 1$

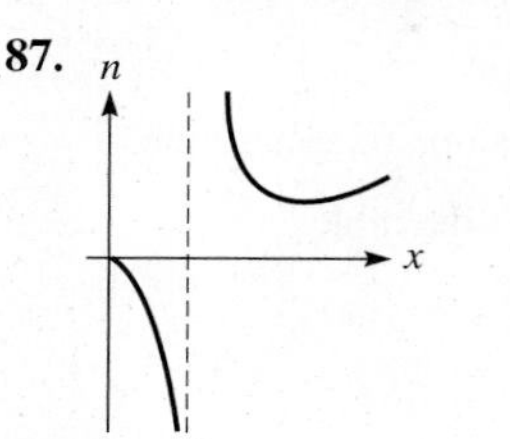

89. $-kE_0^2\cos\frac{1}{2}\theta\sin\frac{1}{2}\theta$ **91.** $\frac{f}{\sqrt{R^2 - F^2f^2}}$

93. 0.005934 rad/s **95.** $L(t) = 181.6 - 8.32t$

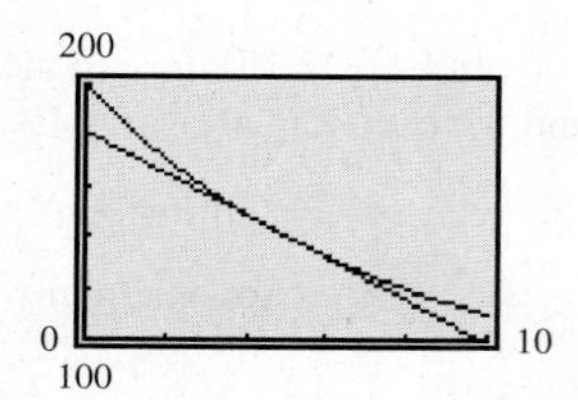

97. −3900 dollars/year; −2200 dollars/year

99. 2.00 m, 1.82 m **101.** −0.0065 rad/s **103.** 48.2°

105. 0.40 cm/s, 72.5° **107.** 7.07 in.

109. $A = 16\cos\theta(1 + \sin\theta)$ Max. $(\frac{\pi}{6}, 20.8)$

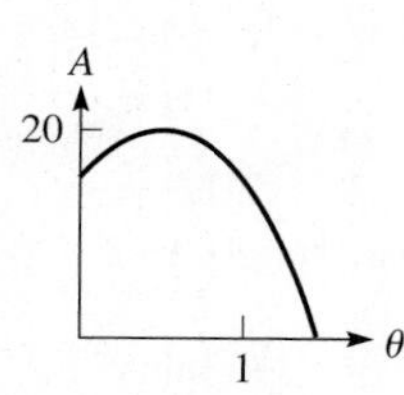

111. $\frac{w}{H}\cosh\frac{wx}{H} = \frac{w}{H}\sqrt{1 + \sinh^2\frac{wx}{H}}$

Review Exercises for Chapter 27, page 836

Concept Check Exercises

1. F **2.** T **3.** F **4.** F **5.** T **6.** F **7.** F **8.** T

Practice and Applications

9. $-12\sin(4x - 1)$ **11.** $\frac{-0.2\sec^2\sqrt{3 - 2v}}{\sqrt{3 - 2v}}$

13. $-6\csc^2(3x + 2)\cot(3x + 2)$ **15.** $-24x\cos^3 x^2\sin x^2$

17. $2e^{2(x-3)}$ **19.** $3\left[\frac{2x^2}{x^2 + 1} + \ln(x^2 + 1)\right]$

21. $\frac{50}{25 + x^2}$ **23.** $\frac{0.1}{\sin^{-1}0.1t\sqrt{1 - 0.01t^2}}$

25. $(-2\csc 4x)\sqrt{\csc 4x + \cot 4x}$ **27.** $\frac{14(1 + e^{-x})}{x - e^{-x}}$

29. $-\frac{2\sin x\cos x}{e^{3x} + \pi^2} - \frac{3e^{3x}\cos^2 x}{(e^{3x} + \pi^2)^2}$ **31.** $\frac{2u(1 + 4u^2)(\tan^{-1}2u) - 2u^2}{(1 + 4u^2)(\tan^{-1}2u)^2}$

33. $-\frac{2x^2\cot x^2 + \ln\csc x^2}{x^2}$ **35.** $\frac{12\cos 2x\ln(4 + \sin 2x)}{4 + \sin 2x}$

37. $0.1e^{-2t}\sec\pi t(-2 + \pi\tan\pi t)$ **39.** $\frac{\cos 2x + 2e^{4x}}{\sqrt{\sin 2x + e^{4x}}}$

41. $\frac{2x^3e^y + 2xy^2e^y + y}{x - x^4e^y - x^2y^2e^y}$ **43.** $0.5[e^{-2t}(1 - 2t) - e^{2t}(1 + 2t)]$

45. $\frac{ye^x(x - x\ln xy - 1)}{x(y + e^x)}$ **47.** $\cos^{-1}x$

Exercises 28.1, page 842

1. $-\frac{1}{4}\cos^4 x + C$ **3.** $\frac{1}{4}(x^2+1)^4 + C$ **5.** $\frac{1}{5}\sin^5 x + C$

7. $-\frac{4}{15}(\cos\theta)^{3/2} + C$ **9.** $\frac{4}{3}\tan^3 x + C$ **11.** $\frac{1}{8}$

13. $\frac{1}{4}(\sin^{-1}x)^4 + C$ **15.** $\frac{1}{2}(\tan^{-1}5x)^2 + C$

17. $\frac{1}{3}[\ln(x+1)]^3 + C$ **19.** 0.0894 **21.** $\frac{3}{4}(4+e^x)^4 + C$

23. $\frac{1}{(1-e^{2t})^2} + C$ **25.** $\frac{1}{10}(1+\sec^2 x)^5 + C$ **27.** 4.1308

29. $-\int \frac{-\sin x\, dx}{\cos^5 x}$; $u = \cos x$, $du = -\sin x\, dx$, $n = -5$

31. $\int \frac{1}{\ln^2 x}\frac{dx}{x}$, $u = \ln x$, $du = \frac{dx}{x}$, $n = -2$

33. 0.888 **35.** $\frac{\pi}{2}(e^4 - 1)$ **37.** 1.102

39. $y = \frac{1}{3}(\ln x)^3 + 2$ **41.** $\frac{1}{3}mnv^2$ **43.** $q = (1 - e^{-t})^3$

Exercises 28.2, page 846

1. $\frac{1}{2}\ln|2x+1| + C$ **3.** $\frac{1}{4}\ln|1+4x| + C$

5. $-\frac{1}{3}\ln|4-3x^2| + C$ **7.** $2\ln 4 = 2.773$

9. $-0.2\ln|\cot 2\theta| + C$ **11.** $\ln 2 = 0.693$

13. $\ln|1-e^{-x}| + C$ **15.** $\ln|x+e^x| + C$

17. $2\ln|1+4\sec x| + C$ **19.** $\frac{1}{4}\ln 5 = 0.402$

21. $\ln|\ln r| + C$ **23.** $\ln|2x + \tan x| + C$

25. $-16\sqrt{1-2x} + C$ **27.** $\ln|x| - \frac{2}{x} + C$

29. $\frac{1}{3}\ln(\frac{5}{4}) = 0.0744$ **31.** $\ln 3 = 1.10$

33. $\int \frac{x-4}{x+4}dx = \int (1 - \frac{8}{x+4})dx = x - 8\ln|x+4| + C$

35. $\pi\ln 2 = 2.18$ **37.** $y = \ln(\frac{3.5}{3+\cos x}) + 2$

39. $\ln a + \ln b = \ln ab$ **41.** 28.2 km **43.** 12.9 m/s

45. 8.38 min **47.** $i = \frac{E}{R}(1 - e^{-Rt/L})$ **49.** 1.41 m

Exercises 28.3, page 849

1. $\frac{1}{3}e^{x^3} + C$ **3.** $e^{7x} + C$ **5.** $2e^{2x+5} + C$

7. 28.2 **9.** $2e^{x^3} + C$ **11.** $e^2 - e = 4.67$

13. $7e^{2\sec\theta} + C$ **15.** $2(1+e^y)^{3/2} + C$

17. $6 - \frac{3(e^6 - e^2)}{2} = -588.06$ **19.** $\frac{1}{2}e^{2x} - e^{-x} + C$

21. $\frac{1}{4}e^{\tan^{-1}2x} + C$ **23.** $-\frac{1}{3}e^{\cos 3x} + C$ **25.** 0

27. $6\ln(e^x+1) + C$ **29.** $3e^2 - 3 = 19.2$

31. $e^{x^2} + C$ **33.** $\pi(e^4 - e) = 163$

35. $2(e^2 - 1) = 12.8$ **37.** $\ln b\int b^u\, du = b^u + C_1$

39. (a) Second runner by 0.1 km
(b) First runner by 0.3 km

41. $q = EC(1 - e^{-t/RC})$ **43.** 26,610 m^2

Exercises 28.4, page 853

1. Change xdx to $3x^2\,dx$. **3.** $\frac{1}{2}\sin 2x + C$

5. $\frac{1}{3}\tan 3\theta + C$ **7.** $\frac{1}{2}\sec 4x + C$ **9.** 0.6365

11. $\frac{2}{3}\ln|\sec\theta^3 + \tan\theta^3| + C$ **13.** $\frac{1}{2}\cos(\frac{1}{x}) + C$

15. $\frac{1}{2}\sqrt{3}$ **17.** $\ln|\csc e^x - \cot e^x| + C$ **19.** $\frac{1}{2}\ln|\sec 2x + \tan 2x| + C$

21. $\frac{1}{3}\ln|\sec 3x + \tan 3x| + C$ **23.** $\frac{1}{5}\ln|\sin 5x| + C$

25. $\frac{1}{9}\pi + \frac{1}{3}\ln 2 = 0.580$ **27.** $\ln 2 = 0.693$

29. Integral changes to $\int(\sec^2 x - \sec x\tan x)dx$

31. (a) $\frac{1}{2}\tan^2 x + C_1$, (b) $\frac{1}{2}\sec^2 x + C_2$; $C_1 = C_2 + \frac{1}{2}$

33. $\pi\sqrt{3} = 5.44$ **35.** $\theta = 0.10\cos 2.5t$ **37.** 0.7726 m

Exercises 28.5, page 857

1. $\frac{x}{2} - \frac{1}{12}\sin 6x + C$ **3.** $\frac{1}{3}\sin^3 x + C$

5. $-\frac{1}{2}\cos 2x + \frac{1}{6}\cos^3 2x + C$ **7.** $\frac{1}{3}\tan^3 x + C$

9. $\frac{2}{35}$ **11.** $\frac{1}{2}x - \frac{1}{4}\sin 2x + C$

13. $-\frac{1}{6}\cos^3 2\theta + \frac{1}{10}\cos^5 2\theta + C$

15. $\tan x - x + C$ **17.** $\frac{3}{4}$

19. $\frac{1}{6}\tan^3 2x - \frac{1}{2}\tan 2x + x + C$ **21.** $\frac{1}{3}\sin^3 s + C$

23. $x - \frac{1}{2}\cos 2x + C$

25. $\frac{1}{4}\cot^4 x - \frac{1}{3}\cot^3 x + \frac{1}{2}\cot^2 x - \cot x + C$

27. $1 + \frac{1}{2}\ln 2 = 1.347$ **29.** $\frac{1}{5}\tan^5 x + \frac{2}{3}\tan^3 x + \tan x + C$

31. $2\sec x + C$ **33.** $-\ln|\sec e^{-x} + \tan e^{-x}| + C$

35. $\frac{1}{2}\cos x - \frac{1}{18}\cos 9x + C$ **37.** $\frac{1}{2}\pi^2 = 4.935$

39. 0.5

41. $\int \sin x\cos x\, dx = \frac{1}{2}\sin^2 x + C_1 = -\frac{1}{2}\cos^2 x + C_2$; $C_2 = C_1 + \frac{1}{2}$

43. $\int_0^{\pi}\sin^2 nx\, dx = \frac{1}{2}\int_0^{\pi}(1 - \cos 2nx)dx = \frac{1}{2}x - \frac{1}{4n}\sin 2nx\Big|_0^{\pi} = \frac{\pi}{2}$

45. $s = 6t - \frac{1}{9}\sin^2 t\cos t - \frac{2}{9}\cos t + \frac{2}{9}$

47. $\frac{4}{3}$ **49.** $V = \sqrt{\frac{1}{1/60.0}\int_0^{1/60.0}(340\sin 120\pi t)^2\, dt} = 240$ V

51. $\frac{aA}{2} + \frac{A}{2b\pi}\sin ab\pi\cos 2bc\pi$

Exercises 28.6, page 861

1. Change dx to $-x\,dx$. **3.** $\sin^{-1}\frac{1}{2}x + C$ **5.** $\frac{3}{2}\tan^{-1}\frac{1}{8}x + C$

7. $\sin^{-1}4x^2 + C$ **9.** 0.8634 **11.** 0.415

13. $\frac{4}{9}\ln(9x^2 + 16) + C$ **15.** 2.356 **17.** $2\sin^{-1}e^x + C$

19. $\tan^{-1}(T+1) + C$ **21.** $4\sin^{-1}\frac{1}{2}(x+2) + C$

23. -0.714 **25.** $2\sin^{-1}(\frac{1}{2}x) + \sqrt{4-x^2} + C$

27. $\frac{1}{2}(\sin^{-1}x)^2 + C$ **29.** $\frac{1}{3}\tan^{-1}x^3 + \frac{1}{2}\ln(x^6+1) + C$

31. (a) Inverse tangent, $\int \frac{du}{a^2 + u^2}$ where $u = 3x$, $du = 3\,dx$, $a = 2$; numerator cannot fit du of denominator. Positive $9x^2$ leads to inverse tangent form.

(b) logarithmic, $\int \frac{du}{u}$ where $u = 4 + 9x$, $du = 9\,dx$

(c) general power, $\int u^{-1/2}\,du$ where $u = 4 + 9x^2$, $du = 18x\,dx$

33. (a) General power, $\int u^{-1/2}\,du$ where $u = 4 - 9x^2$, $du = -18x\,dx$; numerator can fit du of denominator. Square root becomes $-1/2$ power. Does not fit inverse sine form.
(b) Inverse sine, $\int \frac{du}{\sqrt{a^2 - u^2}}$ where $u = 3x$, $du = 3\,dx$, $a = 2$
(c) Logarithmic, $\int \frac{du}{u}$ where $u = 4 - 9x$, $du = -9\,dx$

35. Form fits inverse tangent integral with $u = \sqrt{x}$, $du = \frac{dx}{2\sqrt{x}}$. Result is $2\tan^{-1}\sqrt{x} + C$.

37. $\tan^{-1}2 = 1.11$ **39.** 2.19 **41.** $k\tan^{-1}\frac{x}{d} + C$

43. $\sin^{-1}\frac{x}{A} = \sqrt{\frac{k}{m}}\,t + \sin^{-1}\frac{x_0}{A}$ **45.** $0.22k$

Exercises 28.7, page 865

1. No. Integral $\int v\,du$ is more complex than the given integral.

3. $\cos\theta + \theta\sin\theta + C$ **5.** $2xe^{2x} - e^{2x} + C$

7. $3x\tan x + 3\ln|\cos x| + C$ **9.** $2x\tan^{-1}x - 2\ln\sqrt{1 + x^2} + C$

11. $-\frac{32}{3}$ **13.** $\frac{1}{2}x^2\ln x - \frac{1}{4}x^2 + C$

15. $-\frac{\ln x}{2x^2} - \frac{1}{4x^2} + C$ **17.** $\frac{1}{2}(e^{\pi/2} - 1) = 1.91$

19. $\cos x(1 - \ln\cos x) + C$ **21.** $2\ln 2 - 1$

23. $\frac{1}{10}(x + 4)^{10} + \frac{1}{9}(x + 4)^9 + C$

25. $\frac{1}{2}x[\cos(\ln x) + \sin(\ln x)] + C$

27. $-2e^{-\sqrt{x}}(\sqrt{x} + 1) + C$ **29.** $1 - \frac{3}{e^2} = 0.594$

31. 0.1104 **33.** $\frac{1}{2}\pi - 1 = 0.571$ **35.** 0.756

37. (a) π (b) 3π (c) 5π; Answers are odd multiples of π.

39. $s = \frac{1}{3}[(t^2 - 2)\sqrt{t^2 + 1} + 2]$

41. $q = \frac{1}{5}[e^{-2t}(\sin t - 2\cos t) + 2]$

Exercises 28.8, page 868

1. Delete the x^2 before the radical in the denominator.

3. $x = 3\sin\theta$, $\int \cot^2\theta\,d\theta$ **5.** $x = \tan\theta$, $\int \csc\theta\cot\theta\,d\theta$

7. $x = \sec\theta$, $\int d\theta$ **9.** $-\frac{\sqrt{1 - x^2}}{x} - \sin^{-1}x + C$

11. $2\ln|x + \sqrt{x^2 - 36}| + C$ **13.** $-\frac{2\sqrt{z^2 + 9}}{3z} + C$

15. $\frac{x}{\sqrt{4 - x^2}} + C$ **17.** $\frac{16 - 9\sqrt{3}}{24} = 0.017$

19. $5\ln|\sqrt{x^2 + 2x + 2} + x + 1| + C$ **21.** 0.03997

23. $\sin^{-1}(\frac{x}{4}) + C$ **25.** $2\sec^{-1}e^x + C$

27. (a) $-\frac{1}{3}(1 - x^2)^{3/2} + C$, (b) $-\frac{1}{3}(1 - x^2)^{3/2} + C$

29. π **31.** $\frac{1}{4}ma^2$

33. (a) third runner
(b) third runner, first runner, second runner

35. 1.36 **37.** $kQ\ln\frac{\sqrt{a^2 + b^2} + a}{|\sqrt{a^2 + b^2} - a|}$

39. $\frac{2}{15}(3x - 2)(x + 1)^{3/2} + C$ **41.** $\frac{3}{40}(5x + 12)(x - 4)^{5/3} + C$

Exercises 28.9, page 872

1. $\frac{3}{x - 1} - \frac{4}{x + 2}$ **3.** $\frac{A}{x} + \frac{B}{x + 1}$

5. $\frac{A}{x} + \frac{B}{x + 2} + \frac{C}{x - 2}$ **7.** $\ln\left|\frac{(x + 1)^2}{x + 2}\right| + C$

9. $2\ln\left|\frac{x - 2}{x + 2}\right| + C$ **11.** $x + \ln\left|\frac{x}{(x + 3)^4}\right| + C$

13. 1.057 **15.** $\ln\left|\frac{x^2(x - 5)^3}{x + 1}\right| + C$

17. $\frac{1}{2}\ln\left|\frac{x^4(2x + 1)^3}{2x - 1}\right| + C$ **19.** $\frac{1}{2}\ln(\frac{32}{27}) = 0.08495$

21. $\frac{1}{60}\ln\left|\frac{(V + 2)^3(V - 3)^2}{(V - 2)^3(V + 3)^2}\right| + C$ **23.** $-\frac{1}{2}\ln\left|\frac{(x - 1)^4}{x^2(x - 2)^2}\right| + C$

25. $\frac{1}{u(a + bu)} = \frac{A}{u} + \frac{B}{a + bu}$; $1 = A(a + bu) + Bu$;
$A = \frac{1}{a}$, $B = -\frac{b}{a}$; Then integrate $\frac{1}{a}\int \frac{1}{u}\,du - \frac{b}{a}\int \frac{1}{a + bu}\,du$

27. $\frac{3}{2}\ln|x^{2/3} - 1| + C$ **29.** $-\frac{1}{4}\ln\left|\frac{\sin\theta + 3}{\sin\theta - 1}\right| + C$

31. $2\pi\ln\frac{6}{5} = 1.146$ **33.** $y = \ln\left|\frac{x(x + 5)^2}{36}\right|$ **35.** $0.163\ \text{N}\cdot\text{cm}$

Exercises 28.10, page 877

1. $\frac{2}{x(x + 3)^2} = \frac{A}{x} + \frac{B}{x + 3} + \frac{C}{(x + 3)^2}$

3. $\frac{4x + 4}{x^3 - 9x^2} = \frac{A}{x} + \frac{B}{x^2} + \frac{C}{x - 9}$ **5.** $\ln|\frac{x+1}{x}| - \frac{1}{x} + C$

7. $\frac{3}{x - 2} + 2\ln|\frac{x - 2}{x}| + C$ **9.** $\frac{2}{x} + \ln|\frac{x - 1}{x + 1}| + C$

11. $-\frac{5}{4}$ **13.** $-\frac{2}{x + 1} - \frac{1}{x - 3} + \ln|x + 1| + C$

15. $\frac{1}{8}\pi + \ln 3 = 1.491$ **17.** $\frac{1}{x} + \ln(|x|(x - 2)^4) + C$

19. $-\frac{2}{x} + \frac{3}{2}\tan^{-1}\frac{x + 2}{2} + C$

21. $\frac{1}{4}\ln(4x^2 + 1) + \ln|x^2 + 6x + 10| + \tan^{-1}(x + 3) + C$

23. $\tan^{-1}x + \frac{x}{x^2 + 1} + \ln|x + 1| - \frac{1}{2}\ln|x^2 + 1| + C$

25. $\frac{1 - x}{(x - 2)^2} + C$ **27.** $2 + 4\ln\frac{2}{3} = 0.3781$

29. $\frac{\pi}{72} = 0.0436$ **31.** 0.919 m **33.** 1.369

Exercises 28.11, page 879

1. Formula 3 **3.** Formula 7; $u = y$, $du = dy$

5. Formula 25; $u = x^2$, $du = 2x\,dx$

7. Formula 32; $u = x^2$, $du = 2x\,dx$

9. $\frac{3}{25}[2 + 5x - 2\ln|2 + 5x|] + C$ **11.** $\frac{3544}{15} = 236.3$

13. $\frac{2y}{\sqrt{y^2 + 4}} + C$ **15.** $\frac{1}{2}\sin x - \frac{1}{10}\sin 5x + C$

17. $\sqrt{4x^2 - 9} - 3\sec^{-1}(\frac{2x}{3}) + C$

19. $\frac{1}{20}\cos^4 4x\sin 4x + \frac{1}{5}\sin 4x - \frac{1}{15}\sin^3 4x + C$

21. $3r^2\tan^{-1}r^2 - \frac{3}{2}\ln(1 + r^4) + C$

23. $\frac{1}{4}(8\pi - 9\sqrt{3}) = 2.386$ **25.** $-\ln\left(\frac{1 + \sqrt{4x^2 + 1}}{2x}\right) + C$

27. $-8\ln\left(\frac{1 + \sqrt{1 - 4x^2}}{2x}\right) + C$ **29.** 0.0208

31. $2(\cos x^3 + x^3\sin x^3) + C$ **33.** $\frac{x^2}{\sqrt{1 - x^4}} + C$

35. 4.892 **37.** $\frac{1}{4}x^4(\ln x^2 - \frac{1}{2}) + C$ **39.** $-\frac{3x^3}{\sqrt{x^6 - 1}} + C$

41. $\frac{t^3}{12}(t^6 + 1)^{3/2} + \frac{t^3}{8}\sqrt{t^6 + 1} + \frac{1}{8}\ln(t^3 + \sqrt{t^6 + 1}) + C$

43. Using Formula 42: $\frac{1}{48}\sin^4 4x\,(2\cos^2 4x + 1) + C$;
Using Formula 43: $-\frac{1}{48}(1 + 2\sin^2 4x)\cos^4 4x + C$

45. 2530 m **47.** πab **49.** 208 lb **51.** 187,000 m^3

Review Exercises for Chapter 28, page 881

Concept Check Exercises

1. F **2.** F **3.** F **4.** F **5.** T **6.** T
7. F **8.** T

Practice and Applications

9. $-\frac{1}{8}e^{-8x} + C$ **11.** $-\frac{1}{\ln 2x} + C$ **13.** $4\ln 2 = 2.773$

15. $\frac{4}{7}\tan^{-1}\frac{7}{5}x + C$ **17.** 0 **19.** $\frac{1}{2}\ln 2 = 0.3466$

21. $\frac{2}{3}\sin^3 t - \cos t + C$ **23.** $\tan^{-1}e^x + C$

25. $\frac{2}{3}\tan^3 3x + 2\tan 3x + C$ **27.** $\ln\left|\frac{(x - 1)^3}{x(2x + 1)}\right| + C$

29. $\frac{3}{4}\tan^{-1}\frac{x^2}{2} + C$ **31.** $2\ln|2x + \sqrt{4x^2 - 9}| + C$

33. $\sqrt{e^{2x} + 1} + C$ **35.** $2\ln|x| + \tan^{-1}\frac{x}{3} + C$

37. $\frac{\pi}{4} = 0.7854$ **39.** $-x\cot 2x + \frac{1}{2}\ln|\sin 2x| + C$

41. $\frac{2}{u} + \ln|3u + 1| + C$ **43.** $4\sin e^{2x} + C$

45. $3\sin 1 = 2.524$ **47.** $\frac{1}{2}u^2 - 3u + \ln|u + 3| + C$

49. $\sqrt{5}\tan^{-1}\left(\frac{\sqrt{5}\cos x}{5}\right) - \cos x + C$

51. $x = \sqrt{2}\sin\theta; \frac{1}{\sqrt{2}}\int \csc\theta\, d\theta$ **53.** $2x^2 + C, 2x^2 + C$

55. Formula 16, $u = x^4$, $du = 4x^3\,dx$

57. $\frac{1}{3}(e^x + 1)^3 + C_1 = \frac{1}{3}e^{3x} + e^{2x} + e^x + C_2; C_2 = C_1 + \frac{1}{3}$

59. $-2\sqrt{x}\cos\sqrt{x} + 2\sin\sqrt{x} + C$

61. Power rule with $u = x^2 + 4$, $du = 2x\,dx$, and $n = -1/2$ is easier to use than a trigonometric substitution with $x = 2\tan\theta$.

63. $y = \tan^3 x + 3\tan x$

65. $\int \sin 2x\, dx = \frac{1}{2}\int \sin 2x(2\, dx) = 2\int \sin x(\cos x\, dx) =$
$-2\int \cos x(-\sin x\, dx)$; constants are different.

67. $2(e^3 - 1) = 38.17$ **69.** 11.18

71. $\frac{1}{4}(8\tan^{-1}4 - \ln 17) = 1.943$ **73.** $4\pi(e^2 - 1) = 80.29$

75. $\frac{1}{8}\pi(e^{2\pi} - 1) = 209.9$ **77.** 2.47 m **79.** 1.01 lb · s

81. $\Delta S = a\ln T + bT + \frac{1}{2}cT^2 + C$ **83.** 55.2 m

85. $v = 64(1 - e^{-0.5t})$ **87.** $\sqrt{2}$ **89.** 3.47 cm^3

91. $\frac{1}{2}\pi^2 - 4$ **93.** 644 ft^2

Exercises 29.1, page 886

1. 7 **3.** $V = \pi r^2 h$ **5.** $A = \frac{2V}{r} + 2\pi r^2$

7. $V = \frac{1}{4}\pi h(4r^2 - h^2)$ **9.** 24, −18 **11.** −2, 0

13. $-6, 2 - 3y + 4y^2$ **15.** $6xt + xt^2 + t^3, -12xt - 4x^2t - 8x^3$

17. $\frac{p^2 + pq + kp - p + 2q^2 + 4kq + 2k^2 + 5q + 5k}{p + q + k}$

19. $2hx - 2kx - 2hy + h^2 - 2hk - 4h$ **21.** 0

23. $81z^6 - 9z^5 - 2z^3$ **25.** $x \neq 0, y \geq 0$ **27.** $y \leq 1$

29. 7.20 kg · m/s **31.** 19 m^3 **33.** 150 Pa

35. for a, b, and T with same sign: circle if $a = b$, ellipse if $a \neq b$: for a and b of different signs, hyperbola

37. 0.028 A **39.** 1.70×10^{11} km^3

41. $A = \frac{pw - 2w^2}{2}$, 3850 cm^2

43. $L = \frac{(1.28 \times 10^5)r^4}{l^2}$

Exercises 29.2, page 893

1.
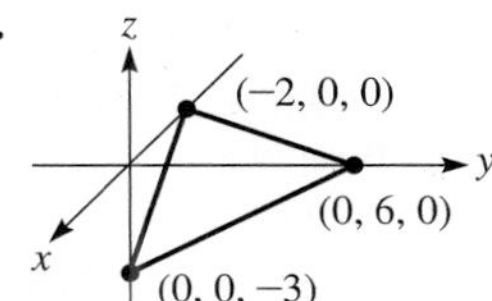

3.
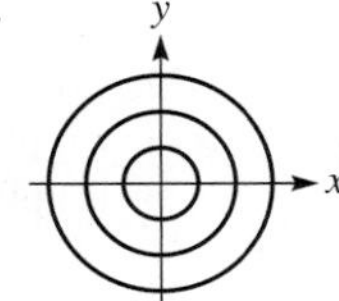

5.
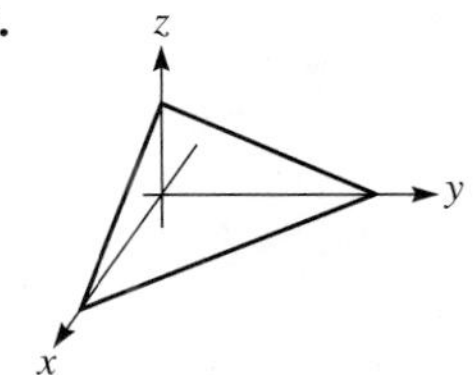

7.
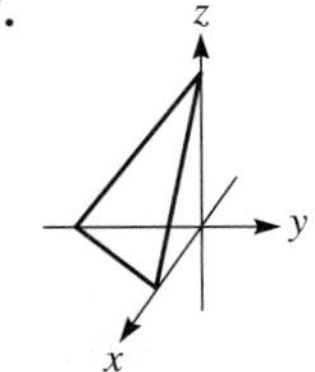

9.
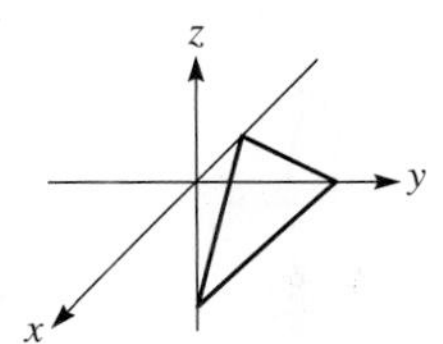

11.
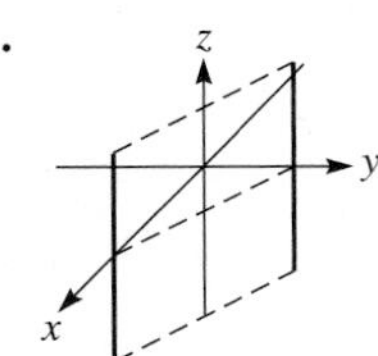

13.
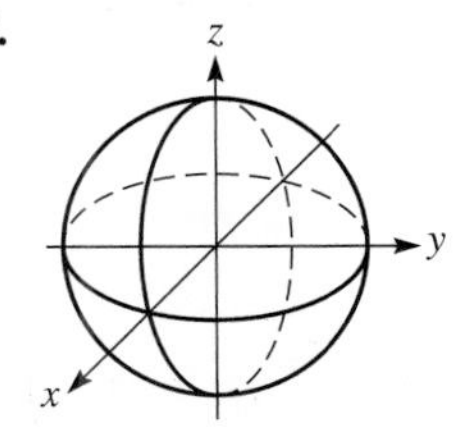

15.
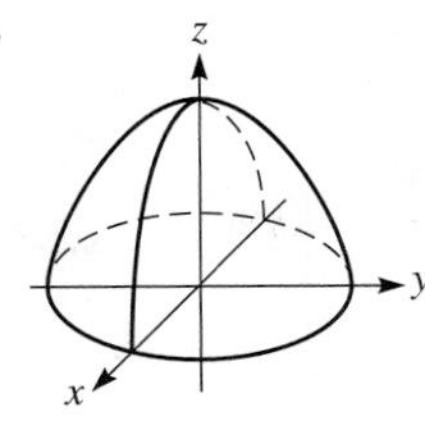

17.
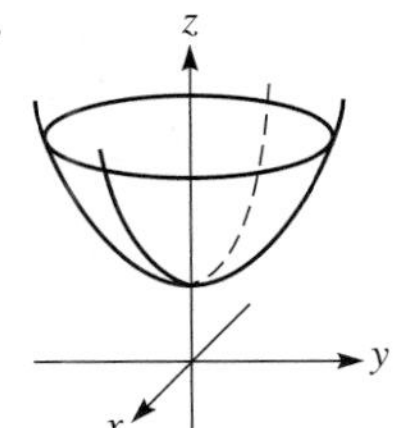

19.
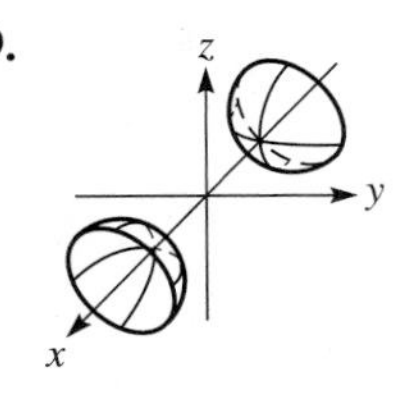

21.
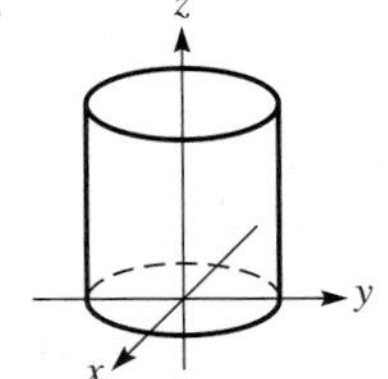

23.
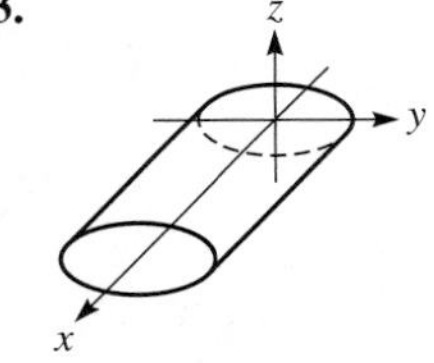

25.

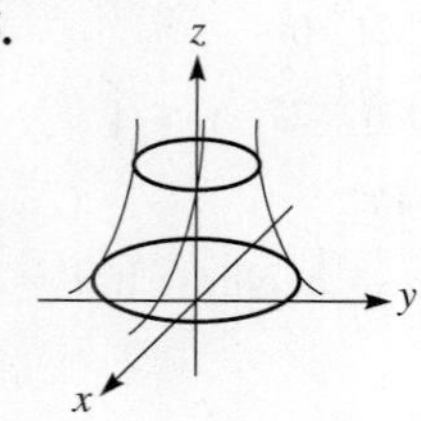

27.

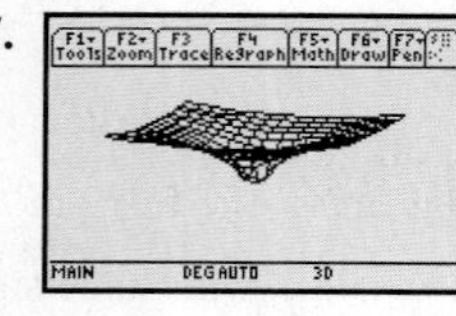

29.

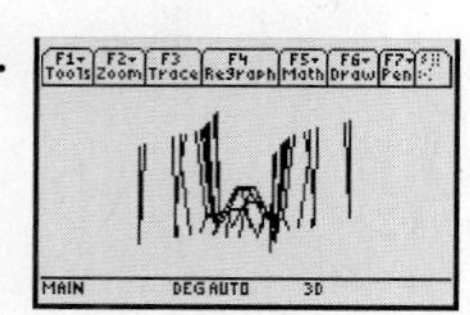

31. (a) $(\frac{3}{2}\sqrt{2}, \frac{3}{2}\sqrt{2}, 5)$
(b) $(0, 2, 3)$

33. (a) $(2, \frac{\pi}{6}, 7)$
(b) $(4, \frac{\pi}{2}, 1)$

35. (a) cylinder, axis is z-axis, $r = 2$
(b) plane $\theta = 2$ for all r and z
(c) plane $z = 2$ for all r and θ

37. $x^2 + y^2 = 4z$

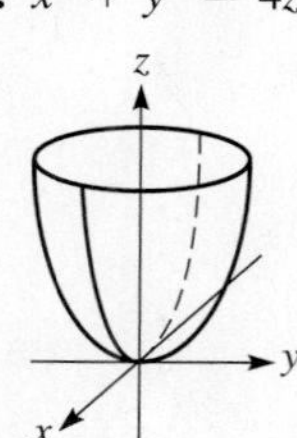

39.

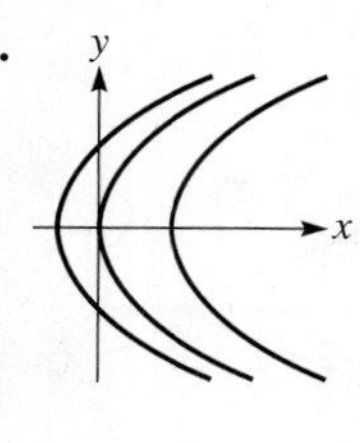

41.

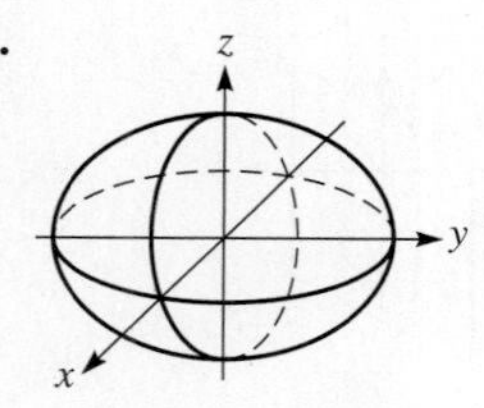

43.

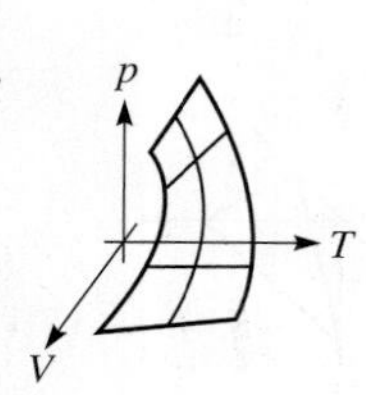

45.

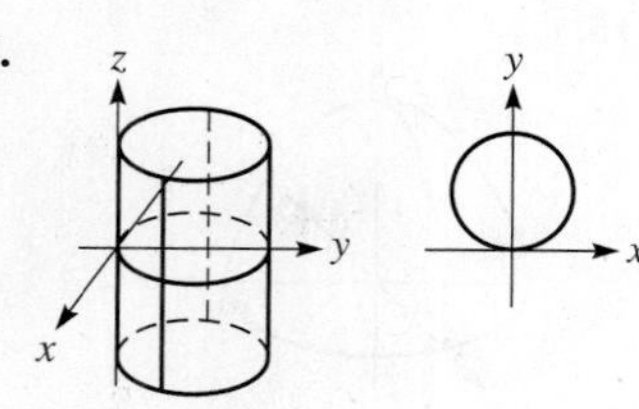

Exercises 29.3, page 897

1. $\frac{\partial z}{\partial x} = \frac{\ln y}{y^2 + 1}, \frac{\partial z}{\partial y} = \frac{x(y^2 + 1) - 2xy^2 \ln y}{y(y^2 + 1)^2}$

3. $\frac{\partial z}{\partial x} = 54x^5, \frac{\partial z}{\partial y} = -15y^4$

5. $\frac{\partial z}{\partial x} = 3e^{3x}, \frac{\partial z}{\partial y} = -\cos y$

7. $\frac{\partial z}{\partial x} = 5 + 8xy, \frac{\partial z}{\partial y} = 4x^2$

9. $\frac{\partial f}{\partial x} = e^{-2y}, \frac{\partial f}{\partial y} = -2xe^{-2y}$

11. $\frac{\partial f}{\partial x} = -\frac{\sin x}{1 - \sec 3y}, \frac{\partial f}{\partial y} = \frac{3(2 + \cos x)\sec 3y \tan 3y}{(1 - \sec 3y)^2}$

13. $\frac{\partial \phi}{\partial r} = \frac{1 + 3rs}{\sqrt{1 + 2rs}}, \frac{\partial \phi}{\partial s} = \frac{r^2}{\sqrt{1 + 2rs}}$

15. $\frac{\partial z}{\partial x} = 4(6x + y^3)(3x^2 + xy^3)^3, \frac{\partial z}{\partial y} = 12xy^2\,(3x^2 + xy^3)^3$

17. $\frac{\partial z}{\partial x} = 2xy\cos x^2y, \frac{\partial z}{\partial y} = x^2\cos x^2y$

19. $\frac{\partial y}{\partial r} = \frac{2r}{r^2 + 6s}, \frac{\partial y}{\partial s} = \frac{6}{r^2 + 6s}$

21. $\frac{\partial f}{\partial x} = \frac{12\sin^2 2x \cos 2x}{1 - 3y}, \frac{\partial f}{\partial y} = \frac{6\sin^3 2x}{(1 - 3y)^2}$

23. $\frac{\partial z}{\partial x} = \cos x - y\sin xy, \frac{\partial z}{\partial y} = -x\sin xy + \sin y$

25. $\frac{\partial f}{\partial x} = e^x(\cos xy - y\sin xy) - 2e^{-2x}\tan y,$
$\frac{\partial f}{\partial y} = -xe^x\sin xy + e^{-2x}\sec^2 y$

27. -8 **29.** $\frac{41}{4}$

31. $\frac{\partial^2 z}{\partial x^2} = -6y, \frac{\partial^2 z}{\partial y^2} = 12xy, \frac{\partial^2 z}{\partial x\,\partial y} = \frac{\partial^2 z}{\partial y\,\partial x} = 6y^2 - 6x$

33. $\frac{\partial^2 z}{\partial x^2} = e^x \sin y, \frac{\partial^2 z}{\partial y^2} = \frac{2x}{y^3} - e^x \sin y,$
$\frac{\partial^2 z}{\partial x\,\partial y} = \frac{\partial^2 z}{\partial y\,\partial x} = -\frac{1}{y^2} + e^x \cos y$

35. $\frac{\partial A}{\partial r} = 2\pi r + \pi\sqrt{r^2 + h^2} + \frac{\pi r^2}{\sqrt{r^2 + h^2}}, \frac{\partial A}{\partial h} = \frac{\pi rh}{\sqrt{r^2 + h^2}}$

37. $\frac{\partial y}{\partial x} = \pi\cos\pi x \sin\frac{\pi t}{2}; \frac{\partial y}{\partial t} = \frac{\pi}{2}\sin\pi x \cos\frac{\pi t}{2}$

39. $-4, -4$ **41.** $(\frac{R_2}{R_1 + R_2})^2$ **43.** 114 cm^2

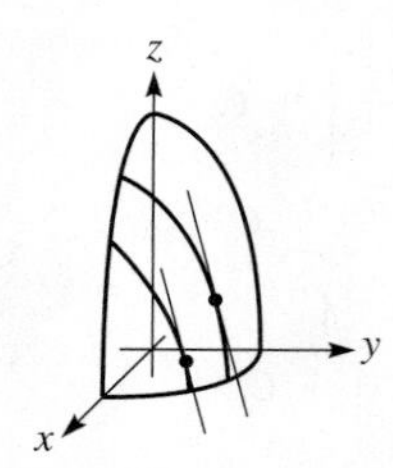

45. 0.807 **47.** $3.75 \times 10^{-3}\,1/\Omega$

49. $-5e^{-t}\sin 4x = \frac{1}{16}(-80e^{-t}\sin 4x)$

Exercises 29.4, page 901

1. $\frac{3}{20}$ **3.** 18 **5.** $\frac{28}{3}$ **7.** $\frac{127}{14}$ **9.** $\frac{1}{3}$ **11.** $\frac{\pi - 6}{12}$

13. 1 **15.** 495.2 **17.** $\frac{74}{5}$ **19.** $\frac{32}{3}$ **21.** 8π **23.** $\frac{28}{3}$

25. 18 **27.** $\sqrt{3} - \frac{\pi}{3}$ **29.** 300 cm^3 **31.**

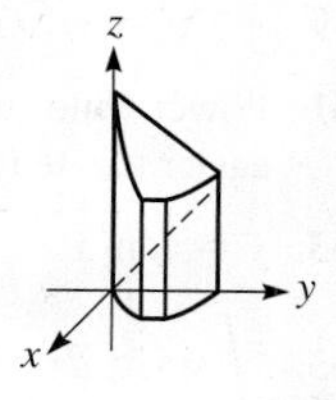

33. $\int_0^2 \int_0^{y/2} f(x, y)\,dx\,dy$

Review Exercises for Chapter 29, page 902

Concept Check Exercises

1. F **2.** T **3.** F **4.** T

Practice and Applications

5. $-17, -36$ **7.** s^2

9.

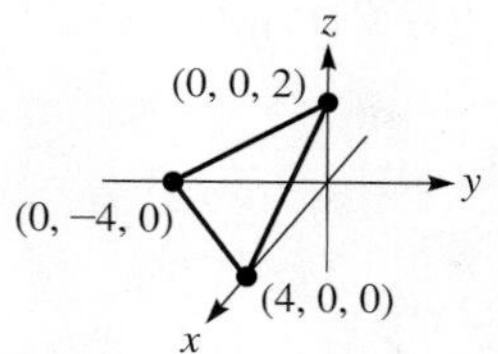

11.

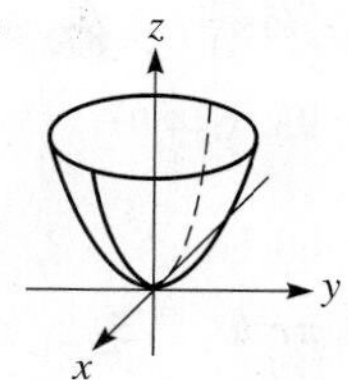

13. $\frac{\partial z}{\partial x} = 15x^2y^2 - 2y^4, \frac{\partial z}{\partial y} = 10x^3y - 8xy^3$

15. $\frac{\partial z}{\partial x} = \frac{x}{\sqrt{x^2 - 3y^2}}, \frac{\partial z}{\partial y} = \frac{-3y}{\sqrt{x^2 - 3y^2}}$

17. $\frac{\partial z}{\partial x} = -\frac{e^{2y}}{(x + y)^2}$

$\frac{\partial z}{\partial y} = \frac{e^{2y}(2x + 2y - 1)}{(x + y)^2}$

19. $\frac{\partial u}{\partial x} = \cos x \ln(x + y) + \frac{\sin x}{x + y},$

$\frac{\partial u}{\partial y} = \frac{\sin x}{x + y}$

21. $\frac{\partial z}{\partial x} = \frac{\partial z}{\partial y} = \frac{-1}{2\sqrt{(x + y)(1 - x - y)}}$

23. $\frac{\partial^2 z}{\partial x^2} = 6y, \frac{\partial^2 z}{\partial y^2} = -6y, \frac{\partial^2 z}{\partial x \partial y} = 6x + 2$

25. $\frac{\partial^2 r}{\partial s^2} = 4e^s \cos 2t - 2te^{-s}, \frac{\partial^2 r}{\partial t^2} = -16e^s \cos 2t,$

$\frac{\partial^2 r}{\partial s \partial t} = -8e^s \sin 2t + 2e^{-s}$

27. 12 **29.** $\frac{21}{2}$ **31.** $\frac{e^2 - 3}{4} = 1.097$ **33.** $\frac{1}{6}$

35.

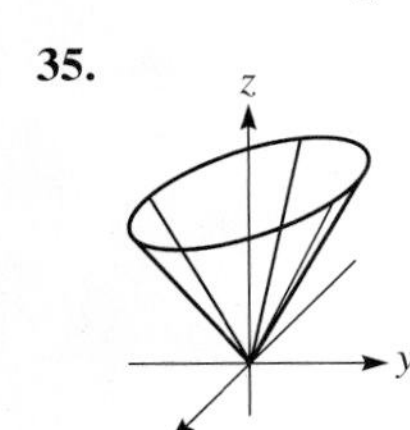

37.

39. (a) $\theta = 3$ represents a line through the origin with an inclination of 172°.

(b) $z = r^2$ represents a circular paraboloid.

41. $\frac{\partial v}{\partial r} = \frac{ER}{(r + R)^2}, \frac{\partial v}{\partial R} = \frac{-rE}{(r + R)^2}$ **43.** 0.982

45. $\frac{\pi}{\sqrt{gl}} = \frac{2\pi\sqrt{l/g}}{2l}$ **47.** Yes **49.** $\frac{2k}{3}\left(\frac{b^2 + ab + a^2}{b + a}\right)$

51. $\left(\frac{\partial V}{\partial T}\right)\left(\frac{\partial T}{\partial p}\right)\left(\frac{\partial p}{\partial V}\right) = -\left(\frac{nR}{p}\right)\left(\frac{V}{nR}\right)\left(\frac{pV}{V^2}\right) = -1$

53. $32(\pi - \frac{2}{3}) = 79.20$. The bounding surfaces are a cylinder with axis along the z-axis, and a plane parallel to the y-axis.

Exercises 30.1, page 907

1. Converges; $S_1 = 0.5, S_2 = 0.75, S_3 = 0.875, S_4 = 0.9375$

3. 1, 8, 27, 64 **5.** $0, \frac{1}{2}, \frac{2}{3}, \frac{3}{4}$

7. (a) $\frac{2}{3}, \frac{4}{9}, \frac{8}{27}, \frac{16}{81}$ (b) $\frac{2}{3} + \frac{4}{9} + \frac{8}{27} + \frac{16}{81}$

9. (a) 1, 0, −1, 0 (b) 1 + 0 − 1 + 0

11. $a_n = \frac{1}{n + 1}$ **13.** $a_n = \frac{(-1)^{n+1}}{(n + 1)(n + 2)}$

15. 1, 1.125, 1.1620370, 1.1776620, 1.1856620; convergent; 1.2

17. 1, 1.5, 2.1666667, 2.9166667, 3.7166667; divergent

19. 0, 1, 2.4142, 4.1463, 6.1463; divergent

21. 0.75, 0.8888889, 0.9375, 0.96, 0.9722222; convergent; 1

23. 0.2103677, 0.2671988, 0.2694038, 0.2664476, 0.2655111; convergent, 0.26

25. Divergent **27.** Convergent, $S = \frac{3}{4}$

29. Convergent, $S = 100$ **31.** Convergent, $S = \frac{4096}{9}$

33. $3 < x < 5$ **35.** Convergent **37.** Divergent **39.** $\frac{5}{33}$

41. 1

43. $\frac{1}{3}$

45. $r = x; S = \frac{x^0}{1 - x} = \frac{1}{1 - x}$ **47.** 2, 4, 12, 48, 240

Exercises 30.2, page 912

1. $\frac{2}{2 + x} = 1 - \frac{1}{2}x + \frac{1}{4}x^2 - \frac{1}{8}x^3 + \cdots$

3. $1 + x + \frac{1}{2}x^2 + \cdots$ **5.** $1 - \frac{1}{2}x^2 + \frac{1}{24}x^4 - \cdots$

7. $1 + \frac{1}{2}x - \frac{1}{8}x^2 + \cdots$ **9.** $1 - 2x + 2x^2 - \cdots$

11. $3x - \frac{9}{2}x^3 + \frac{81}{40}x^5 - \cdots$ **13.** $1 + x + x^2 + \cdots$

15. $-2x - 2x^2 - \frac{8}{3}x^3 - \cdots$ **17.** $1 - x^2 + \frac{1}{3}x^4 - \cdots$

19. $\frac{\sqrt{2}}{2}(1 + x - \frac{1}{2}x^2 - \cdots)$ **21.** $x - \frac{1}{3}x^3 + \cdots$

23. $x + \frac{1}{3}x^3 + \cdots$ **25.** $-\frac{1}{2}x^2 - \frac{1}{12}x^4 - \cdots$

27. $1 + \frac{1}{2}x - \cdots$ **29.** $1 + x + x^2 + \cdots$

31. No. Functions are not defined at $x = 0$.

33. $e^x = 1 + x + \frac{x^2}{2} + \cdots, e^{x^2} = 1 + x^2 + \frac{x^4}{2} + \cdots$

35. $f(x) = 1 + 3x + \frac{9}{2}x^2 + \cdots$; $L(x) = 1 + 3x$; The linearization is the first two terms of the Maclaurin expansion.

37. $1 + \frac{x^2}{2!} + \frac{x^4}{4!} + \cdots$ **39.** $f(x) = x^2$

41. $4 - 0.8t - 1.92t^2 + \cdots$

43. $R = e^{-0.001t} = 1 - 0.001t + (5 \times 10^{-7})t^2 - \cdots$

Exercises 30.3, page 916

1. $e^{2x^2} = 1 + 2x^2 + 2x^4 + \frac{4}{3}x^6 + \cdots$

3. $1 + 3x + \frac{9}{2}x^2 + \frac{9}{2}x^3 + \cdots$

5. $\frac{x}{2} - \frac{x^3}{2^3 3!} + \frac{x^5}{2^5 5!} - \frac{x^7}{2^7 7!} + \cdots$

7. $x - 8x^3 + \frac{32}{3}x^5 - \frac{256}{45}x^7 + \cdots$

9. $x^2 - \frac{1}{2}x^4 + \frac{1}{3}x^6 - \frac{1}{4}x^8 + \cdots$ **11.** 0.3103

13. 0.327 **15.** 0.191 **17.** $2(1 + x^2 + x^4 + x^6 + \cdots)$

19. $x + x^2 + \frac{1}{3}x^3 + \cdots$ **21.** $-2x^3 - x^4 - \frac{2x^5}{3} - \frac{x^6}{2} - \cdots$

23. $x - \frac{1}{2}x^2 + \frac{1}{6}x^3 - \cdots$

25. $\frac{d}{dx}(1 + x + \frac{1}{2}x^2 + \frac{1}{6}x^3 + \cdots) = 1 + x + \frac{1}{2}x^2 + \cdots$

27. $\int \cos x\, dx = x - \frac{x^3}{3!} + \cdots$ **29.** $4 - 0.8t - 1.92t^2 - \cdots$

31. $\int_0^1 e^x dx = 1.7182818,$

$\int_0^1 (1 + x + \frac{1}{2}x^2 + \frac{1}{6}x^3)dx = 1.7083333$

33. $-\frac{1}{6}$ **35.** 0.003099

37. 1; L'Hospital's rule gives the same answer **39.** 0.199968

41. $E = mc^2 + \frac{1}{2}mv^2 + \frac{3}{8}m\frac{v^4}{c^2} + \cdots$

43.

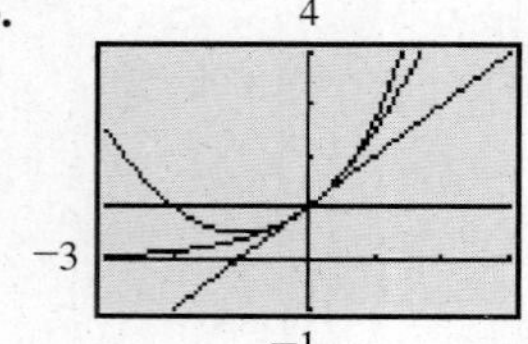

45.

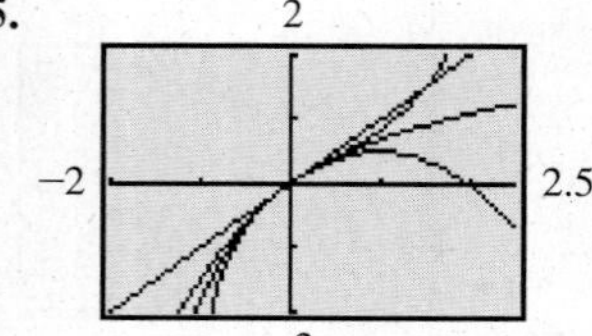

Exercises 30.4, page 919

1. 0.905 **3.** 1.22, 1.2214028 **5.** 0.0998333, 0.0998334

7. 2.7180556, 2.7182818 **9.** 0.99849677, 0.99849715

11. 0.3349333, 0.3364722 **13.** 0.3546130, 0.3546129

15. −0.16711517, −0.16711772 **17.** 1.20736, 1.20803

19. 0.9812, 0.98111850 **21.** 1.0523528 **23.** 0.9874462

25. 8.3×10^{-8} **27.** 3.77×10^{-7} **29.** 1.9799 **31.** 3.146

33. The terms of the expansion for e^x after those on the right side of the inequality have a positive value.

35. 1.59 years **37.** $i = \frac{E}{L}\left(t - \frac{Rt^2}{2L}\right)$; small values of t

39. 0.034 m

Exercises 30.5, page 922

1. $\sqrt{x} = 1 + \frac{1}{2}(x-1) - \frac{1}{8}(x-1)^2 + \frac{1}{16}(x-1)^3 - \cdots$

3. 3.32 **5.** 2.074 **7.** 0.5447 **9.** 0.45397

11. $e^{-2}\left[1 - (x-2) + \frac{(x-2)^2}{2!} - \cdots\right]$

13. $\frac{1}{2}\left[\sqrt{3} + (x - \frac{1}{3}\pi) - \frac{\sqrt{3}}{2!}(x - \frac{1}{3}\pi)^2 - \cdots\right]$

15. $2 + \frac{1}{12}(x-8) - \frac{1}{288}(x-8)^2 + \cdots$

17. $1 + 2(x - \frac{1}{4}\pi) + 2(x - \frac{1}{4}\pi)^2 + \cdots$

19. $e^{\pi/2}\left[1 + (x - \frac{\pi}{2}) - \frac{1}{3}(x - \frac{\pi}{2})^3 + \cdots\right]$

21. $\frac{1}{5} - \frac{x-3}{25} + \frac{(x-3)^2}{125} - \cdots$ **23.** 23.1308

25. 3.0496 **27.** 2.0247 **29.** 0.87462

31. Use the indicated method.

33. $2x^3 + x^2 - 3x + 5 = 5 + 5(x-1) + 14\frac{(x-1)^2}{2} + 12\frac{(x-1)^3}{6}$

35. 0.5150408, 0.5150388, 0.5150381

37. $6\sin\frac{\pi^2}{2} + 6\pi\cos\frac{\pi^2}{2}(t - \frac{\pi}{2}) - 3\pi^2\sin\frac{\pi^2}{2}(t - \frac{\pi}{2})^2 + \cdots$

39. Graph of part (b) fits well near $x = \pi/3$.

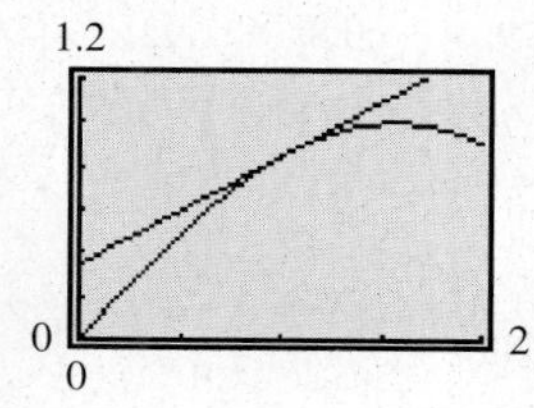

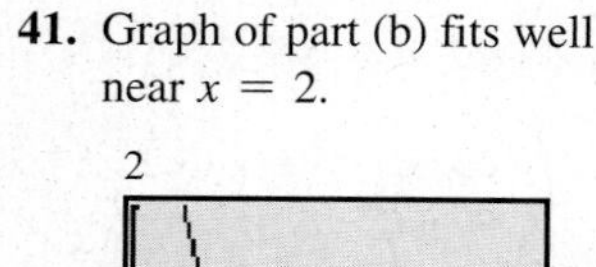

41. Graph of part (b) fits well near $x = 2$.

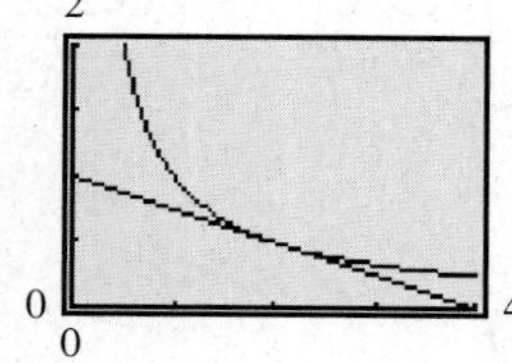

Exercises 30.6, page 927

1. $f(x) = \frac{8}{\pi}\left(\sin x + \frac{1}{3}\sin 3x + \frac{1}{5}\sin 5x + \cdots\right)$

3. $f(x) = \frac{1}{2} - \frac{2}{\pi}\sin x - \frac{2}{3\pi}\sin 3x - \cdots$

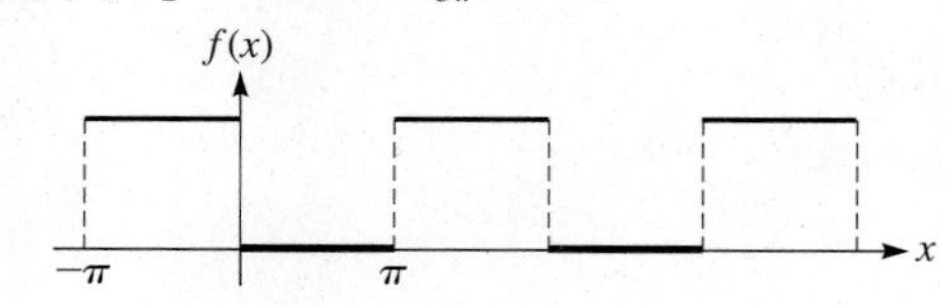

5. $f(x) = \frac{3}{2} + \frac{2}{\pi}\sin x + \frac{2}{3\pi}\sin 3x + \cdots$

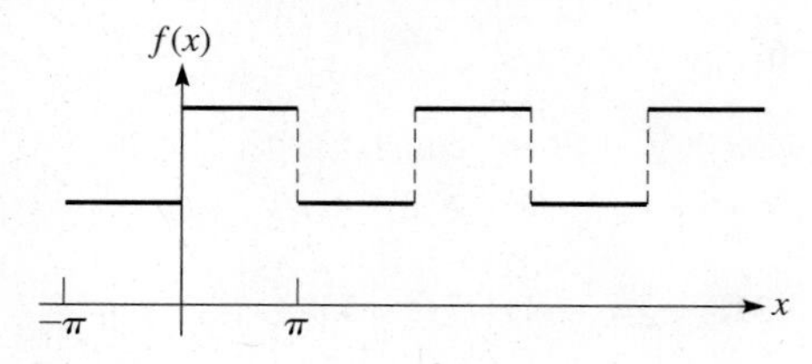

7. $f(x) = \frac{\pi}{4} - \frac{2}{\pi}\left(\cos x + \frac{1}{9}\cos 3x + \cdots\right) + \left(\sin x - \frac{1}{2}\sin 2x + \cdots\right)$

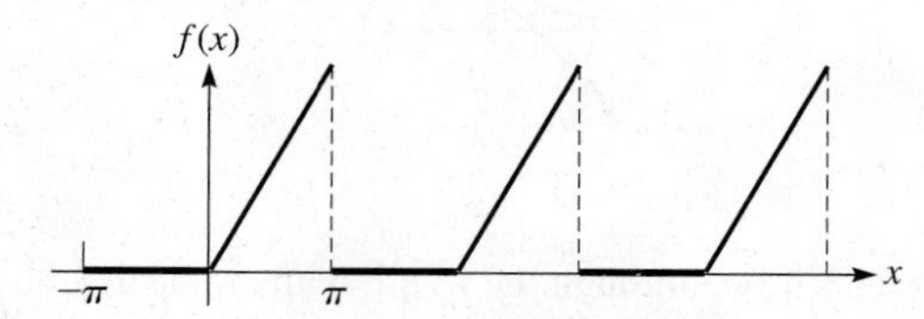

9. $f(x) = -\frac{1}{4} - \frac{1}{\pi}\cos x + \frac{1}{3\pi}\cos 3x - \cdots + \frac{3}{\pi}\sin x - \frac{1}{\pi}\sin 2x + \cdots$

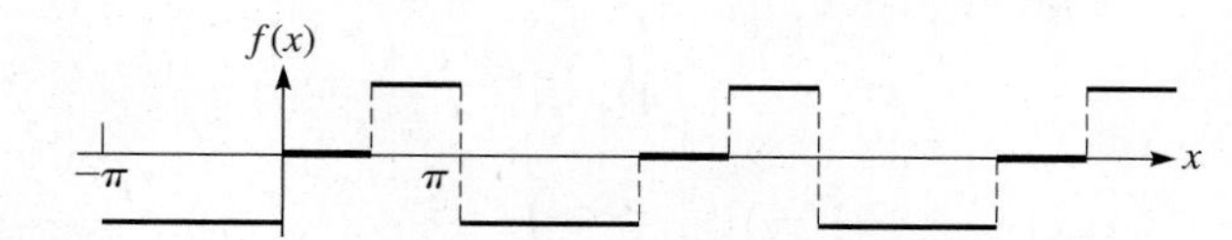

11. $f(x) = \frac{\pi}{2} - \frac{4}{\pi}\cos x - \frac{4}{9\pi}\cos 3x - \cdots$

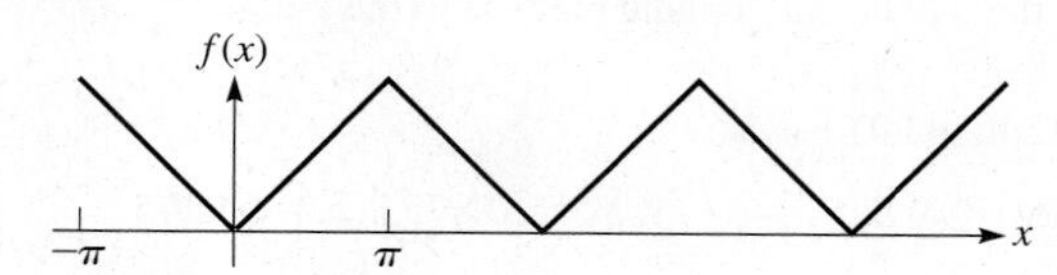

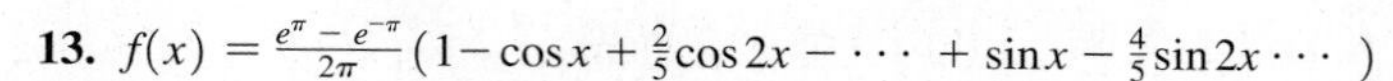

13. $f(x) = \frac{e^{\pi} - e^{-\pi}}{2\pi}\left(1 - \cos x + \frac{2}{5}\cos 2x - \cdots + \sin x - \frac{4}{5}\sin 2x \cdots\right)$

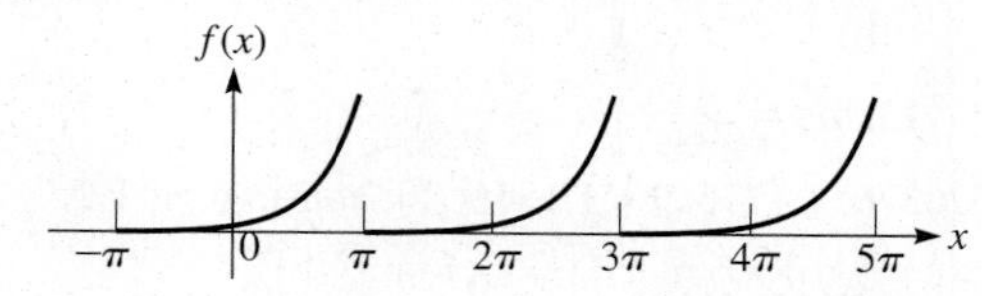

15.

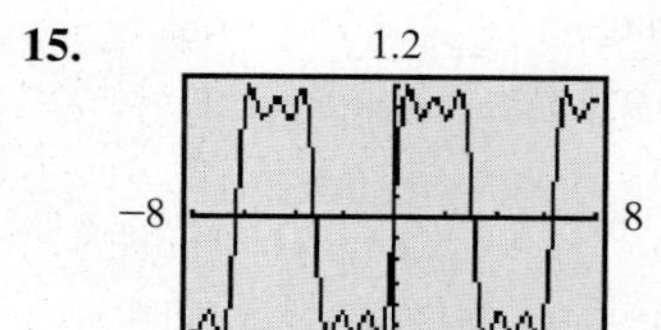

17.

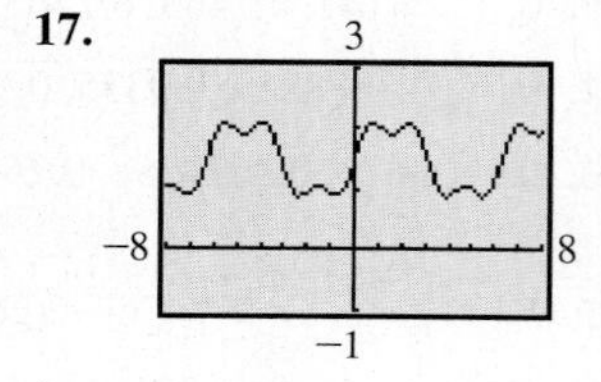

19.

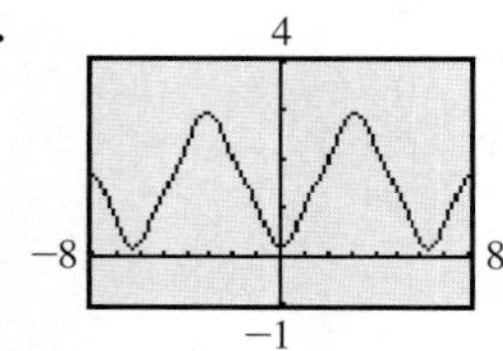

21. $f(x) = \frac{\pi(2\pi + 3)}{12} - \frac{2\pi + 2}{\pi}\cos t + \frac{1}{2}\cos 2t \ldots$
$+ \frac{\pi^2 + \pi - 4}{\pi}\sin t - \frac{\pi + 1}{2}\sin 2t \ldots$

23. $f(t) = \frac{2}{\pi} - \frac{4}{3\pi}\cos 2t - \frac{4}{15\pi}\cos 4t - \cdots$

Exercises 30.7, page 933

1. $f(x) = \frac{5}{2} + \frac{2}{\pi}(\cos x - \frac{1}{3}\cos 3x + \frac{1}{5}\cos 5x - \cdots)$

3. $f(x) = -1 + \frac{4}{\pi}\sin\frac{\pi x}{4} + \frac{4}{3\pi}\sin\frac{3\pi x}{4} + \cdots$

5. Neither **7.** Even **9.** Even **11.** Odd

13. Sine terms **15.** Sine terms and cosine terms

17. $f(x) = \frac{5}{2} - \frac{10}{\pi}(\sin\frac{\pi x}{3} + \frac{1}{3}\sin \pi x - \cdots) + \cdots$

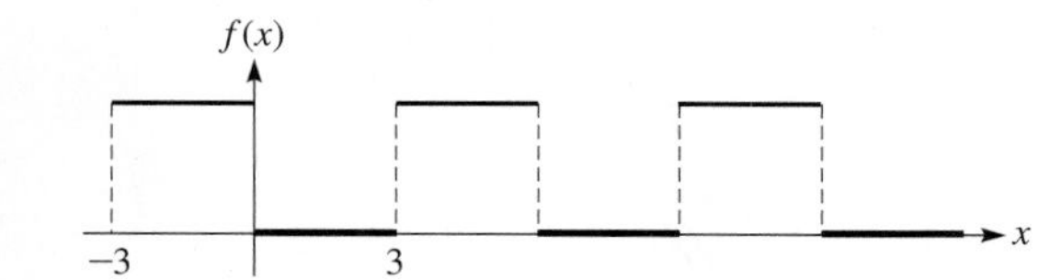

19. $f(x) = 1 + \frac{4}{\pi}\cos\frac{\pi x}{2} - \frac{4}{3\pi}\cos\frac{3\pi x}{2} + \cdots$

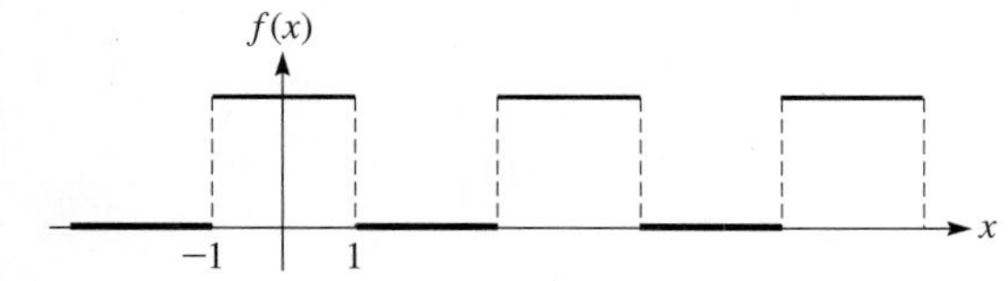

21. $f(x) = 2 - \frac{16}{\pi^2}(\cos\frac{\pi x}{4} + \frac{1}{9}\cos\frac{3\pi x}{4} + \cdots)$

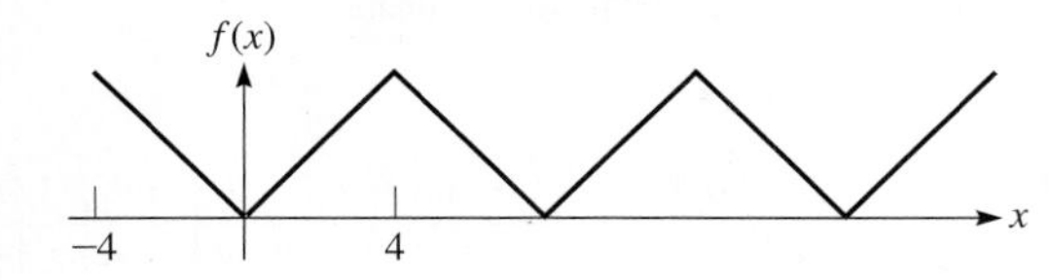

23. $f(x) = \frac{4}{\pi}(\sin\frac{\pi x}{4} + \frac{1}{3}\sin\frac{3\pi x}{4} + \frac{1}{5}\sin\frac{5\pi x}{4} + \cdots)$

25. $f(x) = \frac{4}{3} - \frac{16}{\pi^2}(\cos\frac{\pi x}{2} - \frac{1}{4}\cos \pi x + \frac{1}{9}\cos\frac{3\pi x}{2} - \cdots)$

27. $f(t) = 2 + \frac{8}{\pi}(\cos\frac{\pi}{2}t - \frac{1}{3}\cos\frac{3\pi}{2}t + \cdots + \sin\frac{\pi}{2}t$
$+ \sin \pi t + \frac{1}{3}\sin\frac{3\pi}{2}t + \cdots)$

Review Exercises for Chapter 30, page 935

Concept Check Exercises

1. T **2.** T **3.** T **4.** F **5.** F **6.** F

Practice and Applications

7. $\frac{1}{2} - \frac{1}{4}x + \frac{1}{48}x^3 - \cdots$ **9.** $2x^2 - \frac{4}{3}x^6 + \frac{4}{15}x^{10} - \cdots$

11. $1 + \frac{1}{3}x - \frac{1}{9}x^2 + \cdots$ **13.** $x + \frac{1}{6}x^3 + \frac{3}{40}x^5 + \cdots$

15. $\cos a - (\sin a)x - (\cos a)\frac{x^2}{2} + \cdots$ **17.** 0.82

19. 1.09 **21.** 0.9214 **23.** −0.2024 **25.** 0.95299

27. 12.1655 **29.** 0.259

31. $\frac{1}{2} - \frac{1}{2}\sqrt{3}(x - \frac{1}{3}\pi) - \frac{1}{4}(x - \frac{1}{3}\pi)^2 + \cdots$

33. $\frac{\pi}{6} + \frac{2}{\sqrt{3}}(x - \frac{1}{2}) + \frac{2}{3\sqrt{3}}(x - \frac{1}{2})^2 + \cdots$

35. $f(x) = \frac{\pi - 2}{4} - \frac{2}{\pi}(\cos x + \frac{1}{9}\cos 3x + \cdots)$
$+ (\frac{\pi - 2}{\pi})\sin x - \frac{1}{2}\sin 2x + \cdots$

37. $f(x) = \pi + \frac{4}{\pi}(\sin\frac{\pi x}{4} + \frac{1}{3}\sin\frac{3\pi x}{4} + \cdots)$

39. $f(x) = \frac{1}{2} + \frac{2}{\pi}(\cos x - \frac{1}{3}\cos 3x + \cdots)$

41. $f(x) = \frac{4}{\pi}(\sin\frac{\pi x}{2} - \frac{1}{2}\sin \pi x + \frac{1}{3}\sin\frac{3\pi x}{2} - \cdots)$

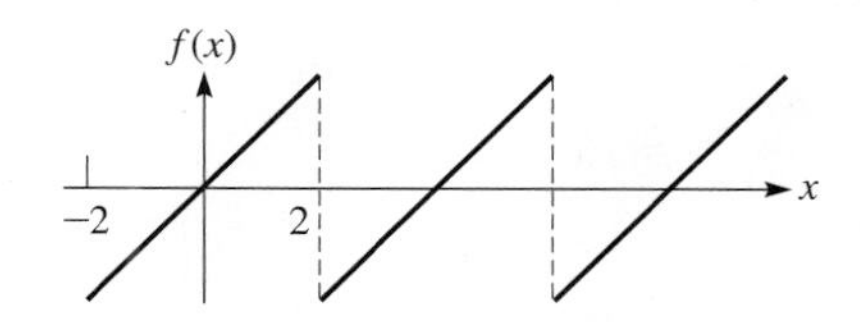

43. Convergent, $S = 5000$ **45.** $S = 256$

47. $\frac{1}{3}x^3 - \frac{1}{42}x^7 + \cdots$

49. $1 + 2(x - \frac{\pi}{4}) + 2(x - \frac{\pi}{2})^2 + \cdots$

51. $2 + 4(x - \frac{\pi}{4}) + \cdots$ **53.** $\frac{1}{2}$

55. $4x - 2x^2 + \frac{4}{3}x^3 - x^4 + \cdots$

57. $x^2 - \frac{1}{3}x^4 + \frac{2}{45}x^6 - \frac{1}{315}x^8 + \cdots$ **59.** 1.366

61. $(x - 1) - \frac{(x-1)^2}{2} + \frac{(x-1)^3}{3} - \frac{(x-1)^4}{4}$; Taylor series found by direct expansion is the same.

63. $\frac{1}{3} + \frac{4}{\pi^2}(-\cos \pi x + \frac{1}{4}\cos 2\pi x - \frac{1}{9}\cos 3\pi x + \cdots)$

65. 2.4265, 2.4600, 2.4596031 **67.** 0.00249688

69. $x - \frac{x^3}{3} + \frac{x^5}{5} - \cdots$

71. $3.2[1 - \frac{(880\pi t)^2}{2} + \frac{(880\pi t)^4}{24} - \cdots]$

73. $N_0(1 - \lambda t + \frac{\lambda^2 t^2}{2} - \frac{\lambda^3 t^3}{6})$

75. $2x + \frac{2}{3}x^3 + \frac{2}{5}x^5 + \frac{2}{7}x^7 + \cdots$

77. $N_0[1 + e^{-k/T} + (e^{-k/T})^2 + \cdots]$
$= N_0(1 + e^{-k/T} + e^{-2k/T} + \cdots)$

79. $f(t) = \frac{1}{2\pi} + \frac{1}{\pi}(\frac{1}{2}\cos t - \frac{1}{3}\cos 2t + \cdots)$
$+ \frac{1}{4}\sin t + \frac{2}{3\pi}\sin 2t + \cdots$

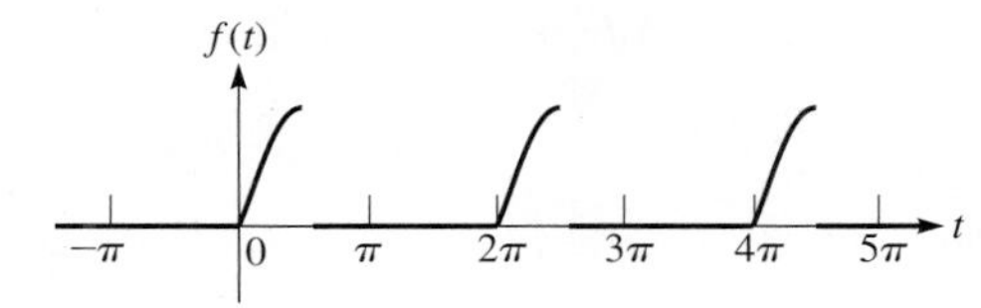

Exercises 31.1, page 939

1. $(c_1e^{-x} + 4c_2e^{2x}) - (-c_1e^{-x} + 2c_2e^{2x}) = 2(c_1e^{-x} + c_2e^{2x})$; $(4e^{-x}) - (-4e^{-x}) = 2(4e^{-x})$

3. Particular solution **5.** General solution

(The following "answers" are the unsimplified expressions obtained by substituting functions and derivatives.)

7. $e^x - (e^x - 1) = 1$; $5e^x - (5e^x - 1) = 1$

9. $-12\cos 2x + 4(3\cos 2x) = 0$; $(-4c_1\sin 2x - 4c_2\cos 2x) + 4(c_1\sin 2x + c_2\cos 2x) = 0$

11. $2x = 2x$ **13.** $1 - 3x^2 = 1 - 3x^2$

15. $(-2ce^{-2x} + 1) + 2(ce^{-2x} + x - \frac{1}{2}) = 2x$

17. $-\frac{1}{2}\cos x + \frac{9}{2}\cos x = 4\cos x$

19. $x^2[-\frac{c^2}{(x-c)^2}] + [\frac{cx}{(x-c)}]^2 = 0$

21. $x(-\frac{c_1}{x^2}) + \frac{c_1}{x} = 0$

23. $(\cos x - \sin x + e^{-x}) + (\sin x + \cos x - e^{-x}) = 2\cos x$

25. $(c_1e^x + 4c_2e^{2x}) - 3(c_1e^x + 2c_2e^{2x}) + 2(c_1e^x + c_2e^{2x} + \frac{3}{2}) = 3$

27. $c_3e^x = c_3e^x$ **29.** $c^2 + cx = cx + c^2$ **31.** Solution

33. Not a solution **35.** $y = x^3 + c_1x - 4$

37. $kN_0e^{kt} = kN_0e^{kt}$

Exercises 31.2, page 943

1. $y(x^2 + 1) = c$ **3.** $y^3 = \frac{3}{2}x^2 + c$

5. $\ln(20 + y) = kt + c$ **7.** $\frac{1}{2}s^2 - s = \sin t + c$

9. $\sqrt{p} = \sqrt{x} + c$ **11.** $3\ln y = t^3 + c$ **13.** $y = c - x^2$

15. $x - \frac{1}{y} = c$ **17.** $\ln V = \frac{1}{p} + c$ **19.** $\ln(x^3 + 5) + 3y = c$

21. $y = 2x^2 + x - x\ln x + c$ **23.** $4\sqrt{1 - y} = e^{-x^2} + c$

25. $e^x - e^{-y} = c$ **27.** $\ln(y + 4) = x + c$

29. $y(1 + \ln x)^2 + cy + 2 = 0$ **31.** $\tan^2 x + 2\ln y = c$

33. $x^2 + 1 + x\ln y + cx = 0$ **35.** $3\ln y + x^3 = 0$

37. $2\ln(1 - y) = 1 - 2\sin x$ **39.** $e^{2x} - \frac{2}{y} = 2(e^x - 1)$

41. $P = 200 + ce^{0.1t}$ **43.** $T = 10 + 90e^{-0.15t}$

Exercises 31.3, page 945

1. $\ln xy + y^2 = c$ **3.** $2xy + x^2 = c$ **5.** $x^3 - 2y = cx - 4$

7. $t^2r - r = ct$ **9.** $y\sin x = x + c$ **11.** $2\sqrt{x^2 + y^2} = x + c$

13. $y = c - \frac{1}{2}\ln\sin(x^2 + y^2)$

15. $\ln(y^2 - x^2) + 2x = c$; subtract $x dx$ from each side and divide through by $y^2 - x^2$.

17. $5xy^2 + y^3 = c$ **19.** $2xy + x^3 = 5$ **21.** $2x = 2xy^2 - 15y$

23. $2x + 1 = \cos xy$ **25.** $(e^{-x}\,dy - ye^{-x}\,dx) - 2y\,dy = 0$

27. $y = -\frac{x}{\ln|x| + c}$

Exercises 31.4, page 948

1. $y = x + cx^{-2}$ **3.** $y = e^{-x}(x + c)$

5. $y = -e^{-4x} + ce^{-2x}$ **7.** $y = -2 + ce^{2x}$

9. $y = ce^{-x^3} + 2$ **11.** $y = \frac{8}{7}x^3 + \frac{c}{\sqrt{x}}$

13. $r = -\cot\theta + c\csc\theta$ **15.** $y = (x + c)\csc x$

17. $y = x + \frac{1}{2}e^x - 1 + ce^{-x}$ **19.** $2s = e^{4t}(t^2 + c)$

21. $y = \frac{1}{4} + ce^{-x^4}$ **23.** $3y = x^4 - 6x^2 - 3 + cx$

25. $y = ce^{-\sqrt{x^2+1}} - 1$ **27.** $r = \sin\theta[\ln(\tan\theta + \sec\theta)] + c\sin\theta$

29. Can solve by separation of variables: $\frac{dy}{1 - y} = 2\,dx$. Can also solve as linear differential equation of first order: $dy + 2y\,dx = 2\,dx$; $y = 1 + ce^{-2x}$

31. $y = e^{-x}$ **33.** $y = \frac{4}{3}\sin x - \csc^2 x$

35. $y(\csc x - \cot x) = \ln\frac{(\sqrt{2} - 1)(\csc 2x - \cot 2x)}{\csc x - \cot x}$

37. $u' - P(x)u = -Q(x)$ **39.** $L = 25 + ce^{-0.8t}$

41. $P = -10e^{0.05t} + 40e^{0.02t}$; The population increases and then decreases.

Exercises 31.5, page 950

1.

x	0.0	0.2	0.4	0.6	0.8	1.0
y	1.00	1.20	1.44	1.72	2.04	2.40

$y = \frac{1}{2}x^2 + x + 1$

y
Exact
(1, 2.5)
(1, 2.4)
Approximate
x

3.

x	−0.2	−0.1	0.0	0.1	0.2	0.3
y	2.0000	2.1840	2.3937	2.6330	2.9069	3.2208

$y = 2.4233e^{0.2x^2+x}$

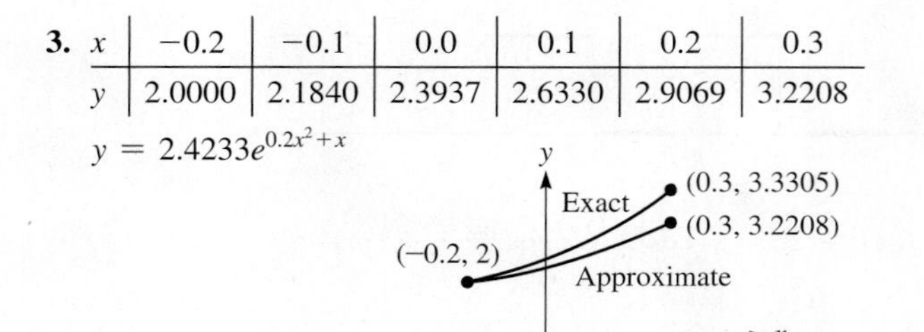

5.

x	0.0	0.1	0.2	0.3	0.4	0.5	0.6	0.7	0.8	0.9	1.0
y	1.00	1.10	1.21	1.33	1.46	1.60	1.75	1.91	2.08	2.26	2.45

7. (Not all values shown)

x	−0.2	−0.1	0.0	0.1	0.2	0.3
y	2.000	2.1903	2.4079	2.6573	2.9436	3.2732

9.

x	0.0	0.1	0.2	0.3	0.4
y	0.0000	0.1003	0.2027	0.3092	0.4220

11.

x	0.0	0.2	0.4	0.6	0.8	1.0
y	0.0000	0.2027	0.4232	0.6884	1.0588	1.7722

13.

x	0.0	0.1	0.2	0.3	0.4	0.5	0.6
y	1.5708	1.5660	1.5521	1.5302	1.5011	1.4656	1.4244

15. $y_3 = 12$; $y_{actual} = 36.2$. Euler method is too inaccurate for larger values of x or Δx.

17. $i_{approx} = 0.0804\text{A}$, $i_{exact} = 0.0898\text{A}$

Exercises 31.6, page 954

1. $y^2 + 2x^2 = k$ **3.** 16.4 lb

5. $y^2 = 2x^2 + 1$

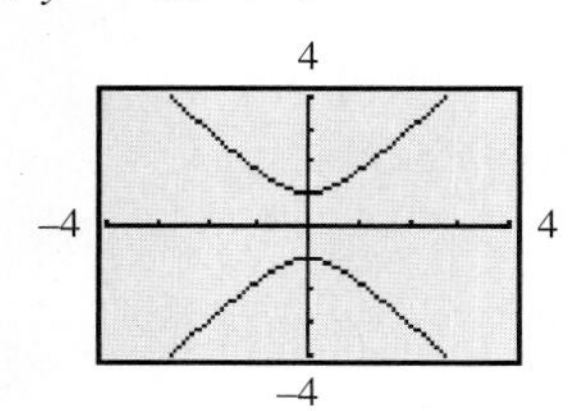

7. $y = 2e^x - x - 1$

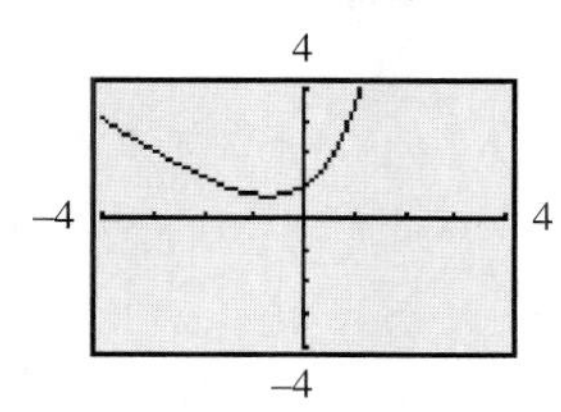

9. $y^2 = c - 2x$

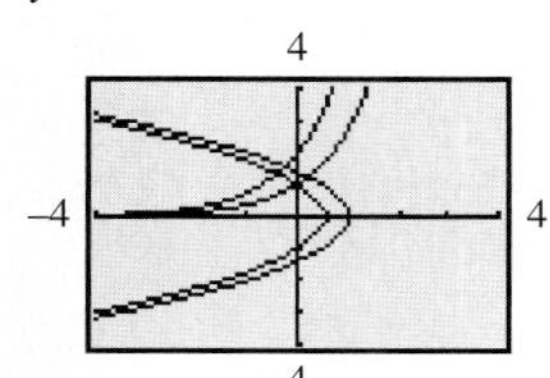

11. $y^2 = c - 2\sin x$

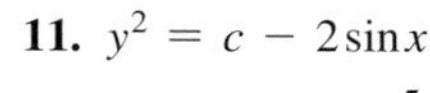

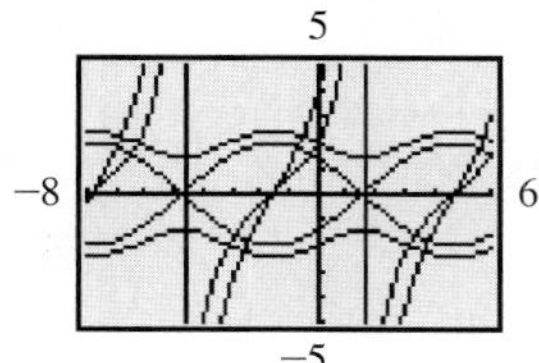

13. 76.9% **15.** 3.82 days **17.** $N = \frac{r}{k}(1 - e^{-kt})$

19. 12 h **21.** 41.4 million **23.** $S = a + \frac{c}{r^2}$

25. $5250e^{0.0196t}$ **27.** 13 min **29.** \$1040.81

31. $c = c_0(1 - e^{-kt})$ **33.** $\lim\limits_{t\to\infty} \frac{E}{R}(1 - e^{-Rt/L}) = \frac{E}{R}$

35. $i = \frac{E}{R^2 + \omega^2 L^2}(R\sin\omega t - \omega L\cos\omega t + \omega L e^{-Rt/L})$

37. $q = q_0 e^{-t/RC}$ **39.** $v = 32(1 - e^{-t}), 32$ **41.** 11 ft/s

43. $w = 160 + 25e^{-18t/3500}$ **45.** $p = 15(0.667)^{10^{-4}h}$

47. \$4980

49. $x = 4(1 + 2e^{-0.25t})$

51. $y = 4e^{-x/2} + c_1$

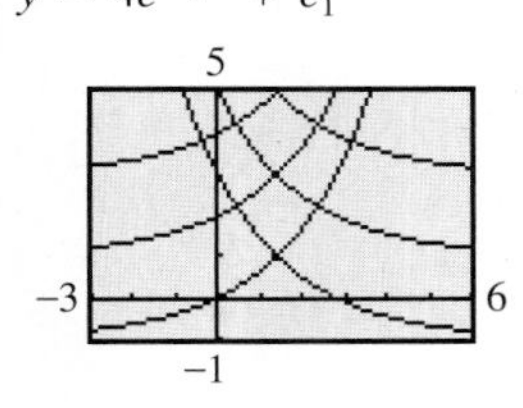

Exercises 31.7, page 960

1. $y = c_1 + c_2 e^{5x}$ **3.** $y = c_1 e^{3x} + c_2 e^{-2x}$

5. $y = c_1 e^{-x} + c_2 e^{-x/3}$ **7.** $y = c_1 + c_2 e^{3x}$

9. $y = c_1 e^{3x/2} + c_2 e^{-x}$ **11.** $y = c_1 e^{6x} + c_2 e^{2x/3}$

13. $y = c_1 e^{x/3} + c_2 e^{-3x}$ **15.** $y = c_1 e^{x/3} + c_2 e^{-x}$

17. $y = e^x(c_1 e^{x\sqrt{2}/2} + c_2 e^{-x\sqrt{2}/2})$

19. $y = e^{3x/8}(c_1 e^{x\sqrt{41}/8} + c_2 e^{-x\sqrt{41}/8})$

21. $y = e^{3x/2}(c_1 e^{x\sqrt{13}/2} + c_2 e^{-x\sqrt{13}/2})$

23. $y = e^{-x/2}(c_1 e^{x\sqrt{33}/2} + c_2 e^{-x\sqrt{33}/2})$

25. $y = c_1 e^{3ax/2} + c_2 e^{-4ax}$ **27.** $y = \frac{1}{5}(3e^{7x} + 7e^{-3x})$

29. $y = \frac{e^3}{e^7 - 1}(e^{4x} - e^{-3x})$ **31.** $y = c_1 + c_2 e^{-x} + c_3 e^{3x}$

33. $y = c_1 e^x + c_2 e^{-x} + c_3 e^{2x} + c_4 e^{-2x}$

35. $v = c_1 e^{as} + c_2 e^{-as}$ **37.** $y = c_1 + c_2 e^{-4x}$

Exercises 31.8, page 964

1. $y = e^{-5x}(c_1 + c_2 x)$ **3.** $y = c_1 + c_2 e^{10x}$

5. $y = e^x(c_1 + c_2 x)$ **7.** $y = e^{-6x}(c_1 + c_2 x)$

9. $y = c_1 \sin 3x + c_2 \cos 3x$

11. $y = e^{-x/2}(c_1 \sin\frac{1}{2}\sqrt{7}x + c_2\cos\frac{1}{2}\sqrt{7}x)$

13. $y = c_1 e^x + c_2 e^{-x} + c_3 \sin x + c_4 \cos x$

15. $y = c_1 \sin\frac{1}{2}x + c_2\cos\frac{1}{2}x$ **17.** $y = e^{3x/4}(c_1 + c_2 x)$

19. $y = c_1 \sin\frac{2}{5}x + c_2 \cos\frac{2}{5}x$

21. $y = e^x(c_1\cos\frac{1}{2}\sqrt{6}x + c_2\sin\frac{1}{2}\sqrt{6}x)$

23. $y = e^{4x/5}(c_1 + c_2 x)$

25. $y = e^{3x/4}(c_1 e^{x\sqrt{17}/4} + c_2 e^{-x\sqrt{17}/4})$

27. $y = e^{-2x}(c_1 e^{x\sqrt{42}/3} + c_2 e^{-x\sqrt{42}/3})$

29. $y = e^{2x}(c_1 + c_2 x + c_3 x^2)$

31. $y = (c_1 + c_2 x)\sin x + (c_3 + c_4 x)\cos x$ **33.** $y = e^{-x}\sin 3x$

35. $y = e^{4x}(4 - 14x)$ **37.** $D^2y - 9y = 0$

39. $D^2y + 9y = 0$. The sum of cos 3x and sin 3x with no exponential factor indicates imaginary roots with $\alpha = 0$ and $\beta = 3$.

41. $y = 3.7t\, e^{-2.0t}$

Exercises 31.9, page 968

1. $y_p = A + Bx + Cx^2 + Ee^{-x}$

3. $y_c = c_1 \sin 2x + c_2 \cos 2x$; $y_p = A\sin x + B\cos x + Ce^{-x}$

5. $y = c_1 e^{2x} + c_2 e^{-x} - 2$ **7.** $y = c_1 e^{-x} + c_2 e^x - 4 - x^2$

9. $y = c_1 + c_2 e^{3x} - \frac{3}{4}e^x - \frac{1}{2}xe^x$

11. $y = c_1 e^{x/3} + c_2 e^{-x/3} - \frac{1}{10}\sin x$

13. $y = (c_1 + c_2 x)e^x + 10 + 6x + x^2 - \frac{2}{25}\sin 3x + \frac{3}{50}\cos 3x$

15. $y = c_1 \sin 2x + c_2 \cos 2x + 3x\cos 2x$

17. $y = c_1 e^{-5x} + c_2 e^{6x} - \frac{1}{3}$ **19.** $y = c_1 e^{4x/3} + c_2 e^{-x} + \frac{1}{4}e^{3x}$

21. $y = c_1 e^{2x} + c_2 e^{-2x} - \frac{1}{5}\sin x - \frac{2}{5}\cos x$

23. $y = c_1 \sin x + c_2 \cos x - \frac{1}{3}\sin 2x + 4$

25. $y = c_1 e^{-x} + c_2 e^{-4x} - \frac{7}{100}e^x + \frac{1}{10}xe^x + 1$

27. $y = c_1 + c_2 e^x + c_3 e^{-x} + \frac{1}{10}\cos 2x$

29. $y = c_1 \sin x + c_2 \cos x + \frac{1}{2}x\sin x$

31. $y = c_1 + c_2 e^{-2x} + 2x^2 - 2x - \frac{1}{2}xe^{-2x}$

33. $y = \frac{1}{6}(11e^{3x} + 5e^{-2x} + e^x - 5)$

35. $y = -\frac{2}{3}\sin x + \pi\cos x + x - \frac{1}{3}\sin 2x$

37. $y = ce^x - x^2 - 2x - 2$; If solved as a first-order linear equation, integration is more complex.

39. $(2\cos x - y) + y = 2\cos x$

Exercises 31.10, page 975

1. $x = \sin 4t + 2\cos 4t$ **3.** $x = e^{-0.1t}(0.04\sin 10t + 4\cos 10t)$

5. (a) $b < 20$ (b) $b > 20$ **7.** No values **9.** 10

11. $s = (s_0 - L)\cos\sqrt{\frac{g}{\ell}}t + L$ **13.** $y = 0.250\cos 16.0t$

15. $y = 0.250\cos 16.0t + 0.127\sin 2.00t - 0.016\sin 16.0t$

17. $q = 2.24 \times 10^{-4}e^{-20t}\sin 2240t$ **19.** $q = 0.01(1 - \cos 316t)$

21. $q = e^{-10t}(c_1\sin 99.5t + c_2\cos 99.5t)$
$-1.81 \times 10^{-3}\sin 120\pi t$
$-1.03 \times 10^{-4}\cos 120\pi t$

23. $i = 10^{-6}[2.00\cos(1.58 \times 10^4 t) + 158\sin(1.58 \times 10^4 t) - 2.00e^{-200t}]$

25. $i_p = 0.528\sin 100t - 3.52\cos 100t$

27. $y = \frac{w}{24EI}(6L^2x^2 - 4Lx^3 + x^4)$

35. $y = c_1e^x + c_2\sin 3x + c_3\cos 3x - \frac{1}{16}(\sin x + \cos x)$

37. $y = c_1e^{-x} + c_2e^{8x} - \frac{2}{9}xe^{-x}$

39. $y = c_1\sin 5x + c_2\cos 5x + 5x\sin 5x$ **41.** $y^3 = 8\sin^2 x$

43. $y = 2x - 1 - e^{-2x}$ **45.** $v = 2e^{-t/2}\sin(\frac{1}{2}\sqrt{15}t)$

47. $y = \frac{1}{25}[16\sin x + 12\cos x - 3e^{-2x}(4 + 5x)]$

49. $y = e^{t/4}$ **51.** $y = \frac{1}{2}(e^{3t} - e^t)$

53. $y = \frac{2}{27} + \frac{1}{9}t - \frac{2}{27}e^{3t} + \frac{10}{9}te^{3t}$ **55.** $y = -4\sin t$

57. $y = \frac{1}{25}(3e^x - 3\cos\frac{3x}{4} - 4\sin\frac{3x}{4})$

59.

x	0	0.1	0.2	0.3	0.4
y	0	0.100	0.201	0.305	0.414

61. (a) and (b) $y = e^{3x}$ **63.** $y = 1 - \cos t$; $y = 1 - \cos t$

65. $x = t^2 + 1$; $y = \frac{1}{t^2 + 1}$ **67.** $r = r_0 + kt$

69. $m = m_0 e^{kt}$ $(k < 0)$ **71.** 3.93 m/s

73. $y^2 - 2xy - 8 = 0$ **75.** 40.0 s

77. 5.31×10^8 years **79.** 8.2 billion

81. $5y^2 + x^2 = c$ **83.** 36.1°C

85. $i = 0.5(1 - e^{-20t})$

87. $y = 0.25e^{-2t}(2\cos 4t + \sin 4t)$, underdamped

89. $q = e^{-6t}(0.4\cos 8t + 0.3\sin 8t) - 0.4\cos 10t$

91. $i = 0$ **93.** $i = 12(1 - e^{-t/2})$; $i(0.3) = 1.67$ A

95. $q = 10^{-4}e^{8t}(4.0\cos 200t + 0.16\sin 200t)$ **97.** 2.47 L

99. $y = 0.25t\sin 8t$

101. $y = \frac{10}{3EI}[100x^3 - x^4 + xL^2(L - 100)]$

Exercises 31.11, page 980

1. $F(s) = \int_0^\infty e^{-st}dt = -\frac{1}{s}e^{-st}\Big|_0^\infty = \frac{1}{s}$

3. $(s^2 - 2s)\mathcal{L}(f) - s + 2$ **5.** $\frac{1}{s - 3}$ **7.** $\frac{30}{(s + 2)^4}$

9. $\frac{s - 2}{s^2 + 4}$ **11.** $\frac{3}{s} + \frac{2(s^2 - 9)}{(s^2 + 9)^2}$ **13.** $(s^2 + s)\mathcal{L}(f)$

15. $(2s^2 - s + 1)\mathcal{L}(f) - 2s + 1$ **17.** t^2

19. $\frac{15}{2}e^{-3t}$ **21.** $\frac{1}{2}t^2e^{-t}$ **23.** $\frac{1}{54}(9t\sin 3t + 2\sin 3t - 6t\cos 3t)$

25. $-\frac{1}{3}e^{-t} - \frac{8}{3}e^{2t} + 7e^{3t}$ **27.** $\frac{1}{2}e^t(4\cos 2t + 5\sin 2t)$

29. $-\frac{d}{ds}\left(\frac{1}{s + a}\right) = \frac{1}{(s + a)^2}$

Exercises 31.12, page 984

1. $y = 2e^{t/2}$ **3.** $y = \frac{t}{2}\sin t + \sin t + 2\cos t$

5. $y = e^{-t}$ **7.** $y = -e^{3t/2}$ **9.** $y = (1 + t)e^{-3t}$

11. $y = \frac{1}{2}\sin 2t$ **13.** $y = e^{-t/2}\cos t$ **15.** $y = e^{2t}\cos t$

17. $y = 1 + \sin t$ **19.** $y = e^{-t}(\frac{1}{2}t^3 + 6t + 4)$

21. $y = 2e^{3t} + 3e^{-2t}$ **23.** $\frac{3}{10}e^{-2t} + \frac{3}{10}e^{2t} - \frac{3}{5}\cos t$

25. $v = 6(1 - e^{-t/2})$ **27.** $y = 4\cos 80t$

29. $q = 1.6 \times 10^{-4}(1 - e^{-5000t})$ **31.** $i = 5t\sin 50t$

33. $y = \sin 3t - 3t\cos 3t$ **35.** $i = 5.0e^{-50t} - 5.0e^{-100t}$

37. $i = 4.42e^{-66.7t}\sin 226t$

Review Exercises for Chapter 31, page 986

Concept Check Exercises

1. T **2.** T **3.** T **4.** T **5.** F **6.** F **7.** T **8.** F

Practice and Applications

9. $2\ln(x^2 + 1) - \frac{1}{2y^2} = c$ **11.** $y^2 = 2x - 4\sin x + c$

13. $y = c_1 + c_2e^{-x/2}$ **15.** $y = (c_1 + c_2x)e^{x/4}$

17. $2x^2 + 4xy + y^4 = c$ **19.** $P = cV^5 - \frac{1}{3}V^2$

21. $y = c(y + 2)e^{2x}$ **23.** $y = e^{-x}(c_1\sin\sqrt{5}x + c_2\cos\sqrt{5}x)$

25. $y = e^{-2x} + ce^{-4x}$ **27.** $y = \frac{1}{2}(c - x^2)\csc x$

29. $s = c_1e^t + c_2e^{-3t/2} - 2$

31. $y = e^{-x/2}(c_1e^{x\sqrt{5}/2} + c_2e^{-x\sqrt{5}/2}) + 2e^x$

33. $y = c_1e^{2x/3} + c_2e^{4x/3} + \frac{1}{2}x + \frac{25}{8}$

Solutions to Practice Test Problems

Chapter 1

1. $\sqrt{9+16} = \sqrt{25} = 5$

2. $\frac{(7)(-3)(-2)}{(-6)(0)}$ is undefined (division by zero).

3. $\frac{3.372 \times 10^{-3}}{7.526 \times 10^{12}} = 4.480 \times 10^{-16}$

```
3.372E-3/7.526E1
2
     4.48046771E-16
```

4. $\frac{(+6)(-2) - 3(-1)}{5-2} = \frac{-12-(-3)}{3} = \frac{-12+3}{3}$

$= \frac{-9}{3} = -3$

5. $\frac{346.4 - 23.5}{287.7} - \frac{0.944^3}{(3.46)(0.109)} = -1.108$

```
(346.4-23.5)/287
.7-.944^3/(3.46*
.109)
        -1.10820764
```

6. $(2a^0b^{-2}c^3)^{-3} = 2^{-3}a^{0(-3)}b^{(-2)(-3)}c^{3(-3)} = 2^{-3}a^0b^6c^{-9}$

$= \frac{b^6}{8c^9}$

7. $(2x+3)^2 = (2x+3)(2x+3)$

$= 2x(2x) + 2x(3) + 3(2x) + 3(3)$

$= 4x^2 + 6x + 6x + 9$

$= 4x^2 + 12x + 9$

8. $3m^2(am - 2m^3) = 3m^2(am) + 3m^2(-2m^3)$

$= 3am^3 - 6m^5$

9. $\frac{8a^3x^2 - 4a^2x^4}{-2ax^2} = \frac{8a^3x^2}{-2ax^2} - \frac{4a^2x^4}{-2ax^2} = -4a^2 - (-2ax^2)$

$= -4a^2 + 2ax^2$

10.

$$\begin{array}{r l}
 3x - 5 & \text{(quotient)} \\
2x - 1\overline{)6x^2 - 13x + 7} & \\
6x^2 - 3x & \\
\hline
-10x + 7 & \\
-10x + 5 & \\
\hline
2 & \text{(remainder)}
\end{array}$$

11. $(2x-3)(x+7)$

$= 2x(x) + 2x(7) + (-3)(x) + (-3)(7)$

$= 2x^2 + 14x - 3x - 21 = 2x^2 + 11x - 21$

12. $3x - [4x - (3 - 2x)] = 3x - [4x - 3 + 2x]$

$= 3x - [6x - 3]$

$= 3x - 6x + 3 = -3x + 3$

13. $5y - 2(y - 4) = 7$

$5y - 2y + 8 = 7$

$3y + 8 - 8 = 7 - 8$

$3y = -1$

$y = -1/3$

14. $3(x-3) = x - (2 - 3d)$

$3x - 9 = x - 2 + 3d$

$3x - 9 - x + 9 = x - 2 + 3d - x + 9$

$2x = 3d + 7$

$x = \frac{3d+7}{2}$

15. $245 \text{ lb/ft}^3 = \frac{245 \text{ lb}}{1 \text{ ft}^3} \times \frac{1 \text{ kg}}{2.205 \text{ lb}} \times \frac{1 \text{ ft}^3}{28.32 \text{ L}}$

$= \frac{245(1)(1) \text{ kg}}{1(2.205)(28.32) \text{ L}} = 3.92 \text{ kg/L}$

16. $0.0000036 = 3.6 \times 10^{-6}$ (six places to right)

17.

$-\pi$	-3	0.3	$\sqrt{2}$	$\lvert -4 \rvert$	(order)
-3.14	-3	0.3	1.41	4	(value)

18. $3(5+8) = 3(5) + 3(8)$ illustrates the distributive law.

19. (a) 5, (b) 3.0 (zero is significant)

20. Evaluation: $1000(1 + 0.05/2)^{2(3)} = 1000(1.025)^6 = \$1159.69$

21. $8(100 - x)^2 + x^2 = 8(100 - x)(100 - x) + x^2$

$= 8(10{,}000 - 200x + x^2) + x^2$

$= 80{,}000 - 1600x + 8x^2 + x^2$

$= 80{,}000 - 1600x + 9x^2$

22. $L = L_0[1 + \alpha(t_2 - t_1)]$

$L = L_0[1 + \alpha t_2 - \alpha t_1]$

$L = L_0 + \alpha L_0 t_2 - \alpha L_0 t_1$

$L - L_0 + \alpha L_0 t_1 = \alpha L_0 t_2$

$t_2 = \frac{L - L_0 + \alpha L_0 t_1}{\alpha L_0}$

23. Let n = number of pounds of second alloy

$0.3(20) + 0.8n = 0.6(n + 20)$

$6 + 0.8n = 0.6n + 12$

$0.2n = 6$

$n = 30$ lb

Chapter 2

1. $\angle 1 + \angle 3 + 90° = 180°$ (sum of angles of a triangle)
$\angle 3 = 52°$ (vertical angles)
$\angle 1 = 180° - 90° - 52° = 38°$

2. $\angle 2 + \angle 4 = 180°$ (straight angle)
$\angle 4 = 52°$ (corresponding angles)
$\angle 2 + 52° = 180°$
$\angle 2 = 128°$

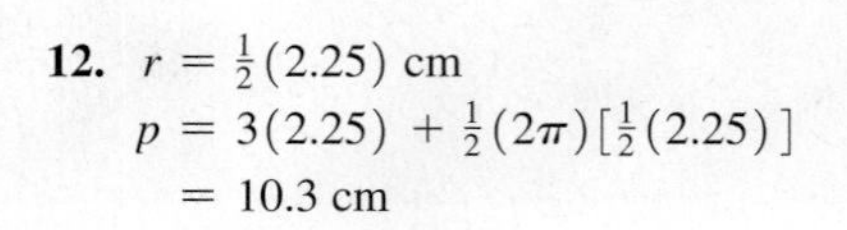

3.

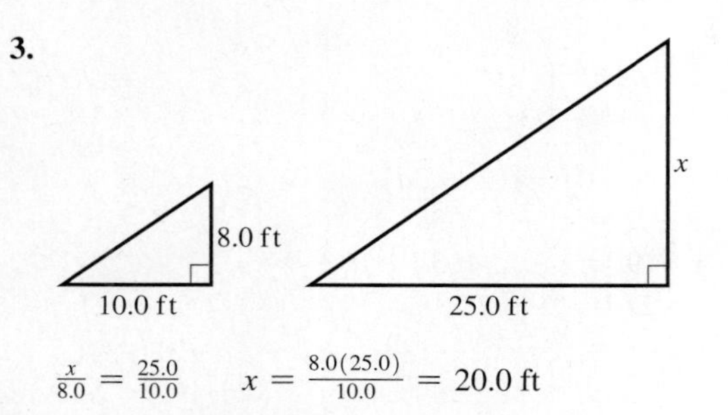

$\frac{x}{8.0} = \frac{25.0}{10.0} \qquad x = \frac{8.0(25.0)}{10.0} = 20.0 \text{ ft}$

4. Use Hero's formula:
$s = \frac{1}{2}(24.6 + 36.5 + 40.7) = 50.9 \text{ cm}$
$A = \sqrt{50.9(50.9 - 24.6)(50.9 - 36.5)(50.9 - 40.7)}$
$= 443 \text{ cm}^2$

5. $d^2 = 125^2 + 170$
$d = \sqrt{125^2 + 170^2}$
$= 211 \text{ ft}$

d, 125 ft, 170 ft

6. $A = \frac{1}{2}h(b_1 + b_2)$
$= \frac{1}{2}(2.76)(9.96 + 4.70) = 20.2 \text{ m}^2$

7. Let m = mass of block
(a) $m = 0.92 \times 10^3\,(0.40^3) = 59 \text{ kg}$
(b) $A = 6(0.40^2) = 0.96 \text{ m}^2$

8. $c = 2\pi r \qquad A = 4\pi r^2$
$21.0 = 2\pi r \qquad = 4\pi\left(\frac{10.5}{\pi}\right)^2 = \frac{4(10.5^2)}{\pi}$
$r = \frac{10.5}{\pi} \qquad = 140 \text{ cm}^2$

9. $V = \frac{1}{3}\pi r^2 h = \frac{1}{3}\pi(2.08^2)(1.78)$
$= 8.06 \text{ m}^3$

10. $\angle ACO + 64° = 90°$ (tangent perpendicular to radius)
$\angle ACO = 26°$
$\angle A = \angle ACO = 26°$ (isosceles triangle)
$\frac{1}{2}\overline{CD} = 26°$ (intercepted arc)
$\overline{CD} = 52°$
$\angle 1 = 52°$ (central angle)

64°, C, 2, 1, A, O, D, B

11. $\angle CBO + \angle 1 + 90° = 180°$ (sum of angles of triangle)
$\angle CBO + 52° + 90° = 180°$
$\angle CBO = 180° - 52° - 90° = 38°$
$\angle 2 + \angle CBO = 180°$ (straight angle)
$\angle 2 + 38° = 180°$
$\angle 2 = 142°$

12. $r = \frac{1}{2}(2.25) \text{ cm}$
$p = 3(2.25) + \frac{1}{2}(2\pi)\left[\frac{1}{2}(2.25)\right]$
$= 10.3 \text{ cm}$

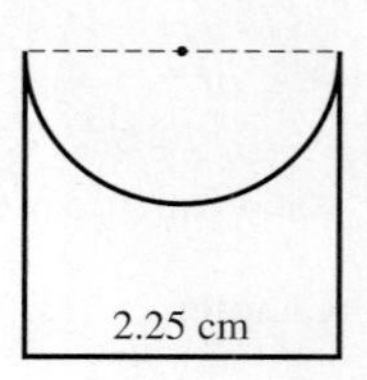

13. $A = 2.25^2 - \frac{1}{2}\pi\left[\frac{1}{2}(2.25)\right]^2 = 3.07 \text{ cm}^2$

14. $A = \frac{1}{2}(50)[0 + 2(90) + 2(145) + 2(260)$
$+ 2(205) + 2(110) + 20]$
$= 41{,}000 \text{ ft}^2$

Chapter 3

1. $f(x) = \frac{8}{x} - 2x^2$

(a) $f(-4) = \frac{8}{-4} - 2(-4)^2 = -2 - 2(16) = -2 - 32 = -34$
(b) $f(x - 4) = \frac{8}{x - 4} - 2(x - 4)^2 = \frac{8}{x - 4} - 2(x^2 - 8x + 16)$
$= \frac{8}{x - 4} - 2x^2 + 16x - 32$

2. $m = 2000 - 10t$

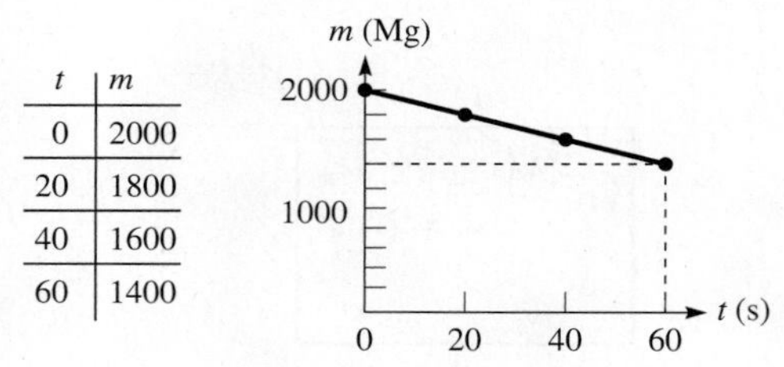

t	m
0	2000
20	1800
40	1600
60	1400

3. $f(x) = 4 - 2x$
$y = 4 - 2x$
$y = 4 - 2(-1) = 6$
$y = 4 - 2(0) = 4$
$y = 4 - 2(1) = 2$
$y = 4 - 2(2) = 0$
$y = 4 - 2(3) = -2$
$y = 4 - 2(4) = -4$

x	y
−1	6
0	4
1	2
2	0
3	−2
4	−4

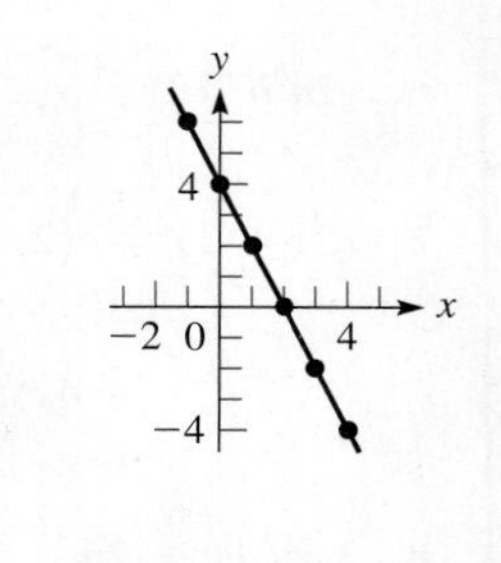

4. $y = 2x^2 - 3x - 3$

2, −2, 3, −5

$x = -0.7$ and $x = 2.2$

5. $y = \sqrt{4 + 2x}$
$y = \sqrt{4 + 2(-2)} = 0$
$y = \sqrt{4 + 2(-1)} = 1.4$
$y = \sqrt{4 + 2(0)} = 2$
$y = \sqrt{4 + 2(1)} = 2.4$
$y = \sqrt{4 + 2(2)} = 2.8$
$y = \sqrt{4 + 2(4)} = 3.5$

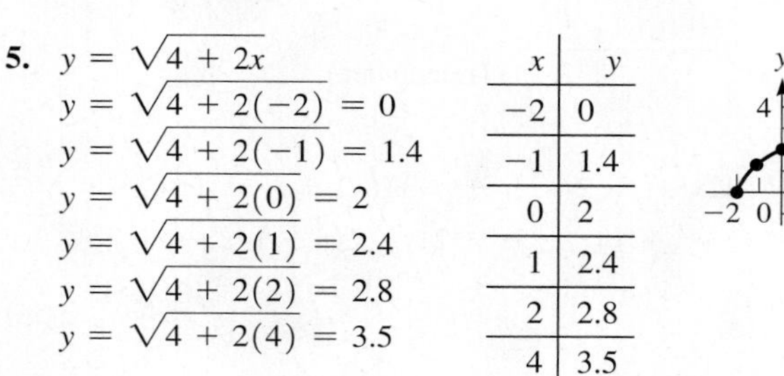

x	y
−2	0
−1	1.4
0	2
1	2.4
2	2.8
4	3.5

6. On negative x-axis
7. $f(x) = \sqrt{6 - x}$
   Domain: $x \leq 6$; x cannot be greater than 6 to have real values of $f(x)$.
   Range: $f(x) \geq 0$; $\sqrt{6 - x}$ is the principal square root of $6 - x$ and cannot be negative.
8. Shifting $y = 2x^2 - 3$ to the right 1 and up 3 gives
   $y = 2(x - 1)^2 - 3 + 3 = 2(x - 1)^2 = 2x^2 - 4x + 2.$
9. Range: $y < -8.9$ or $y > 0.9$

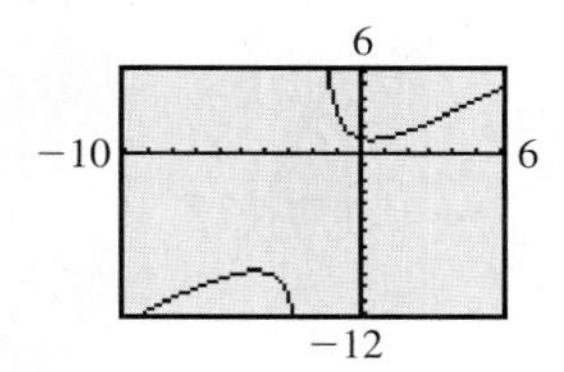

10. Let r = radius of circular part
    square **semicircle**
    $A = (2r)(2r) + \frac{1}{2}(\pi r^2)$
    $= 4r^2 + \frac{1}{2}\pi r^2$

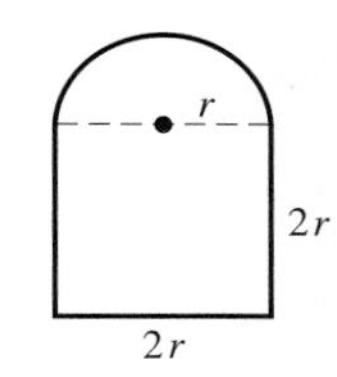

11.

Voltage V	10.0	20.0	30.0	40.0	50.0	60.0
Current i	145	188	220	255	285	315

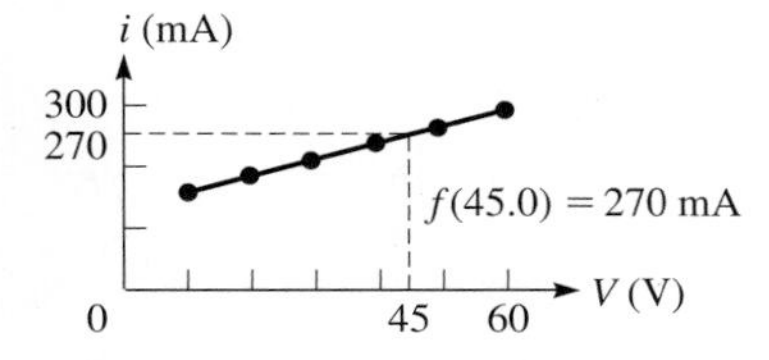

12. $V = 20.0$ V for $i = 188$ mA.
    $V = 30.0$ V for $i = 220$ mA.
    $\frac{x}{10} = \frac{12}{32}$, $x = 3.8$ (rounded off)
    $V = 20.0 + 3.8 = 23.8$ V
    for $i = 200$ mA

$$10\left[x\left[\begin{array}{cc} V & i \\ 20.0 & 188 \\ ? & 200 \end{array}\right]12 \quad \begin{array}{cc} 30.0 & 220 \end{array}\right]32$$

Chapter 4

1. $-165° + 360° = 195°$
   $-165° - 360° = -525°$
2. $39' = \left(\frac{39}{60}\right)° = 0.65°$
   $37°39' = 37.65°$
3. $\tan 73.8° = 3.44$

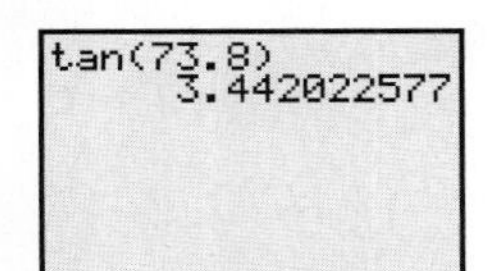

4. $\cos\theta = 0.3726$; $\theta = 68.12°$

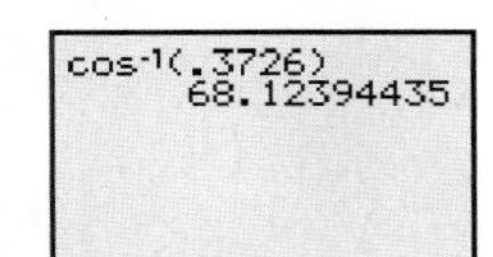

5. Let x = distance from course to east
   $\frac{x}{22.62} = \sin 4.05°$
   $x = 22.62 \sin 4.05°$
   $= 1.598$ km

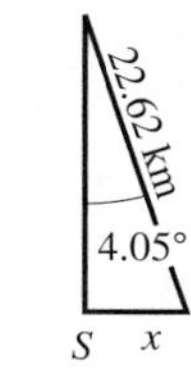

6. $\sin\theta = \frac{2}{3}$
   $x = \sqrt{3^2 - 2^2} = \sqrt{5}$
   $\tan\theta = \frac{2}{\sqrt{5}}$

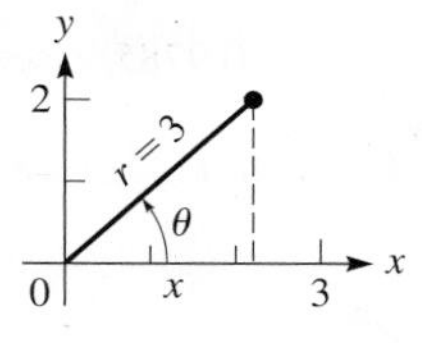

7. $\tan\theta = 1.294$; $\csc\theta = 1.264$

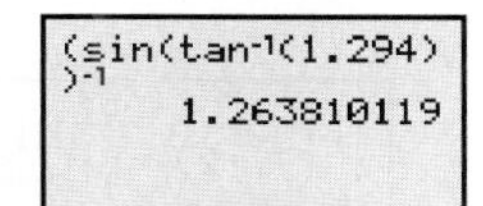

8. $B = 90° - 37.4° = 52.6°$
   $\frac{a}{52.8} = \tan 37.4°$
   $a = 52.8 \tan 37.4°$
   $= 40.4$
   $\frac{52.8}{c} = \cos 37.4°$
   $c = \frac{52.8}{\cos 37.4°}$
   $= 66.5$

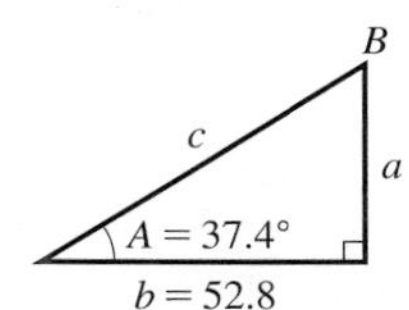

9. $2.49^2 + b^2 = 3.88^2$
   $b = \sqrt{3.88^2 - 2.49^2} = 2.98$
   $\sin A = \frac{2.49}{3.88}$
   $A = \sin^{-1}\left(\frac{2.49}{3.88}\right)$
   $A = 39.9°$ $B = 50.1°$

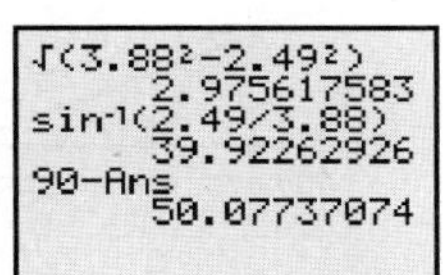

B, $c = 3.88$, $a = 2.49$, A, b

10. 12.0, 12.0, 42.0°, 42.0°, s

    $\frac{s/2}{12.0} = \cos 42.0°$
    $s = 24.0 \cos 42.0° = 17.8$

11. h, 9, θ, 40

    $h = \sqrt{9^2 + 40^2} = 41$
    $\sin\theta = \frac{9}{41}$, $\cos\theta = \frac{40}{41}$
    $\frac{\sin\theta}{\cos\theta} = \frac{9/41}{40/41} = \frac{9}{40}$

12. $\lambda = d \sin\theta$

$= 30.05 \sin 1.167°$

$= 0.6120\ \mu\text{m}$

13. $r = \sqrt{5^2 + 2^2} = \sqrt{29}$

$\sin\theta = \frac{2}{\sqrt{29}} = 0.3714 \quad \csc\theta = \frac{\sqrt{29}}{2} = 2.693$

$\cos\theta = \frac{5}{\sqrt{29}} = 0.9285 \quad \sec\theta = \frac{\sqrt{29}}{5} = 1.077$

$\tan\theta = \frac{2}{5} = 0.4000 \quad \cot\theta = \frac{5}{2} = 2.500$

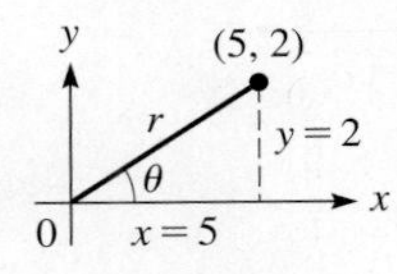

14. Let x = new length of ramp

$\sin 4.50° = \frac{2.50}{x}$

$x = \frac{2.50}{\sin 4.50°} = 31.9$ ft

added length $= 31.9 - 9.5 = 22.4$ ft

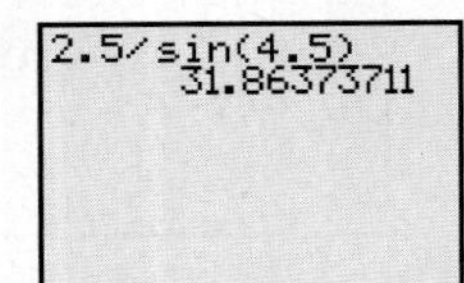

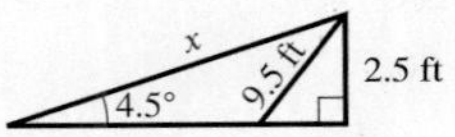

15. Distance between points is $x - y$.

$\frac{18.525}{x} = \tan 13.500° \qquad \frac{18.525}{y} = \tan 21.375°$

$x - y = \frac{18.525}{\tan 13.500°} - \frac{18.525}{\tan 21.375°} = 29.831$ m

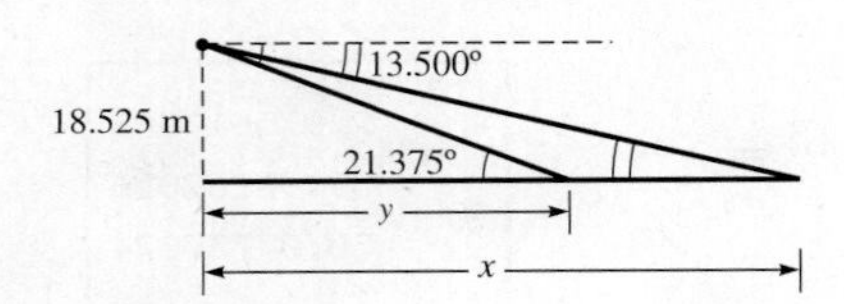

Chapter 5

1. Points $(2, -5)$ and $(-1, 4)$

$m = \frac{4 - (-5)}{-1 - 2} = \frac{9}{-3} = -3$

2. $2x + 5y = 11 \quad 2(-2) + 5(3) = 11$ OK

$y - 5x = 12 \quad 3 - 5(-2) = 13$ No

$x = -2, y = 3$ Not a solution. Values do not satisfy second equation.

3. $\begin{vmatrix} -1 & 3 & -2 \\ 4 & -3 & 0 \\ 5 & -4 & 2 \end{vmatrix} \begin{matrix} -1 & 3 \\ 4 & -3 \\ 5 & -4 \end{matrix} = 6 + 0 + 32 - 30 - 0 - 24 = -16$

4. $x + 2y = 5 \quad 4y = 3 - 2(5 - 2y)$

$4y = 3 - 2x \quad 4y = 3 - 10 + 4y$

$x = 5 - 2y \quad 0 = -7$

Inconsistent

5. $3x - 2y = 4$

$2x + 5y = -1$

$x = \frac{\begin{vmatrix} 4 & -2 \\ -1 & 5 \end{vmatrix}}{\begin{vmatrix} 3 & -2 \\ 2 & 5 \end{vmatrix}} = \frac{20 - 2}{15 - (-4)} = \frac{18}{19}$

$y = \frac{\begin{vmatrix} 3 & 4 \\ 2 & -1 \end{vmatrix}}{19} = \frac{-3 - 8}{19} = -\frac{11}{19}$

6. $2x + y = 4$

$y = -2x + 4$

$m = -2 \quad b = 4$

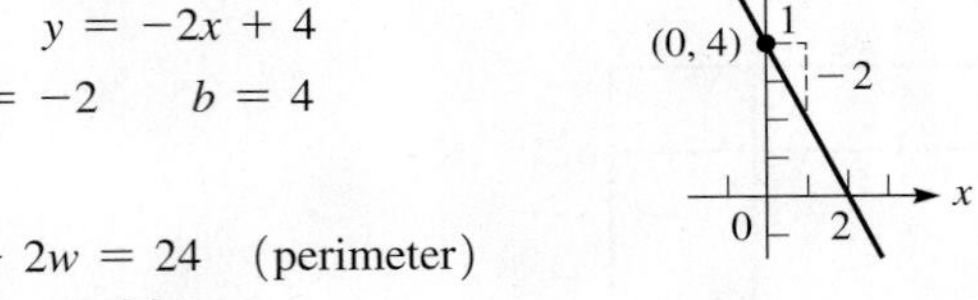

7. $2l + 2w = 24$ (perimeter)

$l = w + 6.0$

$l + w = 12$

$l - w = 6.0$

$2l = 18$

$l = 9$ km

$9 = w + 6$

$w = 3$ km

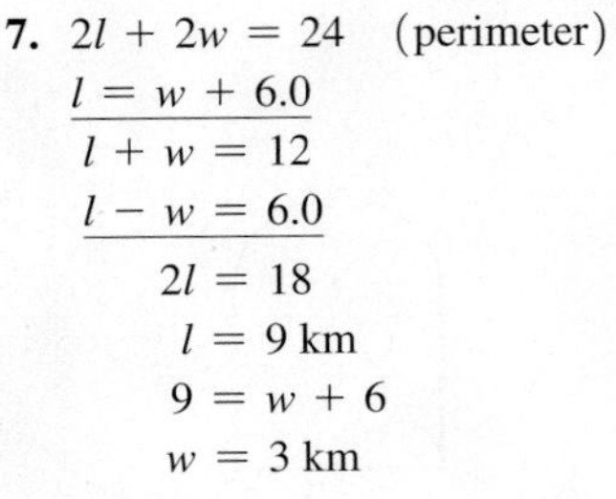

8. $6N - 2P = 13$

$4N + 3P = -13$

$18N - 6P = 39$

$8N + 6P = -26$

$26N = 13$

$N = \frac{1}{2}$

$6(\frac{1}{2}) - 2P = 13$

$-2P = 10$

$P = -5$

9. $C = 310 - 2d$

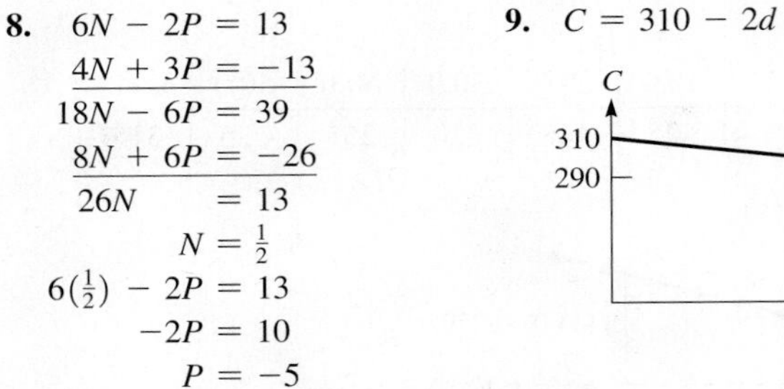

10. $2x - 3y = 6 \qquad 4x + y = 4$

$x = 0: y = -2 \qquad x = 0: y = 4$

$y = 0: x = 3 \qquad y = 0: x = 1$

Int: $(0, -2), (3,0) \qquad$ Int: $(0,4), (1,0)$

$x = 1.3 \quad y = -1.1$

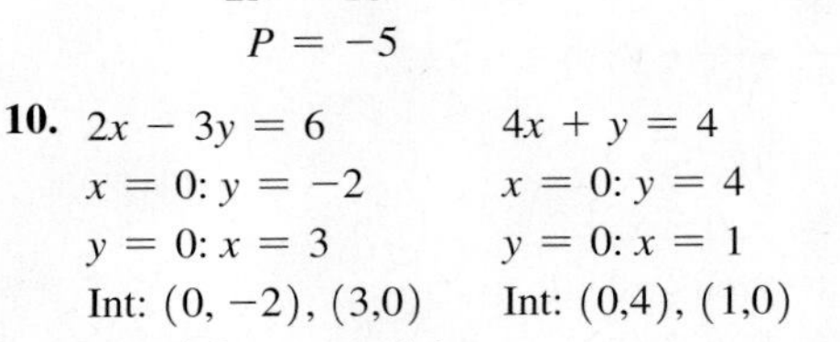

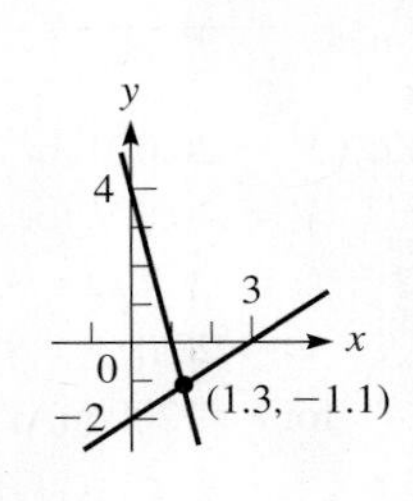

11. $x + y + z = 4$

$x + y - z = 6$

$2x + y + 2z = 5$

$2x + 2y = 10$

$4x + 3y = 17$

$-y = -3$

$y = 3$

$x = 2$

$z = -1$

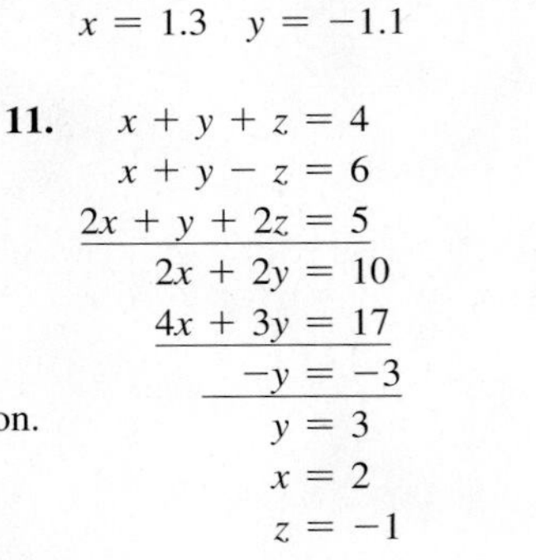

12. Let $x =$ vol. of first alloy
$y =$ vol. of second alloy
$z =$ vol. of third alloy

$$\begin{aligned} x + y + z &= 100 \quad \text{total vol.}\\ 0.6x + 0.5y + 0.3z &= 40 \quad \text{copper}\\ 0.3x + 0.3y &= 15 \quad \text{zinc}\\ \hline x + y + z &= 100\\ 6x + 5y + 3z &= 400\\ 3x + 3y &= 150\\ \hline 3x + 2y &= 100\\ 3x + 3y &= 150\\ \hline y &= 50\text{ cm}^3\\ x &= 0\text{ cm}^3\\ z &= 50\text{ cm}^3 \end{aligned}$$

13.
$$\begin{aligned} 3x + 2y - z &= 4\\ 2x - y + 3z &= -2\\ x \quad\quad + 4z &= 5 \end{aligned}$$

$$y = \frac{\begin{vmatrix} 3 & 4 & -1\\ 2 & -2 & 3\\ 1 & 5 & 4 \end{vmatrix}}{\begin{vmatrix} 3 & 2 & -1\\ 2 & -1 & 3\\ 1 & 0 & 4 \end{vmatrix}}$$

$$= \frac{-24 + 12 - 10 - 2 - 45 - 32}{-12 + 6 + 0 - 1 - 0 - 16}$$

$$= \frac{-101}{-23} = \frac{101}{23}$$

14.
$$\begin{aligned} I_1 + I_2 - I_3 &= 0\\ 3I_1 - 4I_2 &= 10\\ 4I_2 + 5I_3 &= 5 \end{aligned}$$

$$\left[\begin{array}{ccc|c} 1 & 1 & -1 & 0\\ 3 & -4 & 0 & 10\\ 0 & 4 & 5 & 5 \end{array}\right]$$

$$\left[\begin{array}{ccc|c} 1 & 0 & 0 & 2.34\\ 0 & 1 & 0 & -0.745\\ 0 & 0 & 1 & 1.60 \end{array}\right]$$

$I_1 = 2.34$ A
$I_2 = -0.745$ A
$I_3 = 1.60$ A

Chapter 6

1.
$$\begin{aligned} 2x(2x - 3)^2 &= 2x[(2x)^2 - 2(2x)(3) + 3^2]\\ &= 2x(4x^2 - 12x + 9)\\ &= 8x^3 - 24x^2 + 18x \end{aligned}$$

2. $\frac{1}{R} = \frac{1}{R_1 + r} + \frac{1}{R_2}$

$$\frac{RR_2(R_1 + r)}{R} = \frac{RR_2(R_1 + r)}{R_1 + r} + \frac{RR_2(R_1 + r)}{R_2}$$

$$\begin{aligned} R_2(R_1 + r) &= RR_2 + R(R_1 + r)\\ R_1R_2 + rR_2 &= RR_2 + RR_1 + rR\\ R_1R_2 - RR_1 &= RR_2 + rR - rR_2\\ R_1(R_2 - R) &= RR_2 + rR - rR_2\\ R_1 &= \frac{RR_2 + rR - rR_2}{R_2 - R} \end{aligned}$$

3. $\frac{2x^2 + 5x - 3}{2x^2 + 12x + 18} = \frac{(2x - 1)(x + 3)}{2(x + 3)^2} = \frac{2x - 1}{2(x + 3)}$

4. $4x^2 - 16y^2 = 4(x^2 - 4y^2) = 4(x + 2y)(x - 2y)$

5.
$$\begin{aligned} pb^3 + 8a^3p &= p(b^3 + 8a^3)\\ &= p(b + 2a)(b^2 - 2ab + 4a^2) \end{aligned}$$

6.
$$\begin{aligned} 2a - 4T - ba + 2bT &= 2(a - 2T) - b(a - 2T)\\ &= (2 - b)(a - 2T) \end{aligned}$$

7.
$$\begin{aligned} 36x^2 + 14x - 16 &= 2(18x^2 + 7x - 8)\\ &= 2(9x + 8)(2x - 1) \end{aligned}$$

8.
$$\begin{aligned} \frac{3}{4x^2} - \frac{2}{x^2 - x} - \frac{x}{2x - 2} &= \frac{3}{4x^2} - \frac{2}{x(x - 1)} - \frac{x}{2(x - 1)}\\ &= \frac{3(x - 1) - 2(4x) - x(2x^2)}{4x^2(x - 1)}\\ &= \frac{3x - 3 - 8x - 2x^3}{4x^2(x - 1)}\\ &= \frac{-2x^3 - 5x - 3}{4x^2(x - 1)} \end{aligned}$$

9.
$$\begin{aligned} \frac{x^2 + x}{2 - x} \div \frac{x^2}{x^2 - 4x + 4} &= \frac{x^2 + x}{2 - x} \times \frac{x^2 - 4x + 4}{x^2}\\ &= \frac{x(x + 1)}{2 - x} \times \frac{(x - 2)^2}{x^2}\\ &= -\frac{x(x + 1)(x - 2)^2}{(x - 2)(x^2)}\\ &= -\frac{(x + 1)(x - 2)}{x} \end{aligned}$$

10.
$$\begin{aligned} \frac{1 - \frac{3}{2x + 2}}{\frac{x}{5} - \frac{1}{2}} &= \frac{\frac{2(x + 1) - 3}{2(x + 1)}}{\frac{2x - 5}{10}}\\ &= \frac{2x + 2 - 3}{2(x + 1)} \times \frac{10}{2x - 5}\\ &= \frac{5(2x - 1)}{(x + 1)(2x - 5)} \end{aligned}$$

11. Let $t =$ time working together

$\frac{t}{12} + \frac{t}{16} = 1$

LCD of 12 and 16 is 48.

$$\begin{aligned} \frac{48t}{12} + \frac{48t}{16} &= 48\\ 4t + 3t &= 48\\ t = \frac{48}{7} &= 6.9\text{ days} \end{aligned}$$

12. $\frac{3}{2x^2 - 3x} + \frac{1}{x} = \frac{3}{2x - 3}$ LCD $= x(2x - 3)$

$$\frac{3x(2x - 3)}{x(2x - 3)} + \frac{x(2x - 3)}{x} = \frac{3x(2x - 3)}{2x - 3}$$

$$3 + (2x - 3) = 3x$$

$$x = 0$$

No solution due to division by zero in first two terms

Chapter 7

1.
$$\begin{aligned} 2x^2 + 5x &= 12\\ 2x^2 + 5x - 12 &= 0\\ (2x - 3)(x + 4) &= 0 \end{aligned}$$

$2x - 3 = 0 \quad x + 4 = 0$

$x = \frac{3}{2} \quad$ or $\quad x = -4$

2.
$$\begin{aligned} x^2 &= 3x + 5\\ x^2 - 3x - 5 &= 0 \end{aligned}$$

$a = 1, \quad b = -3, \quad c = -5$

$$\begin{aligned} x &= \frac{-(-3) \pm \sqrt{(-3)^2 - 4(1)(-5)}}{2(1)}\\ &= \frac{3 \pm \sqrt{29}}{2} \end{aligned}$$

3. $4x^2 - 5x - 3 = 0$

On calculator

$Y_1 = 4X^2 - 5X - 3$

From zero feature

$x = -0.44$ or $x = 1.69$

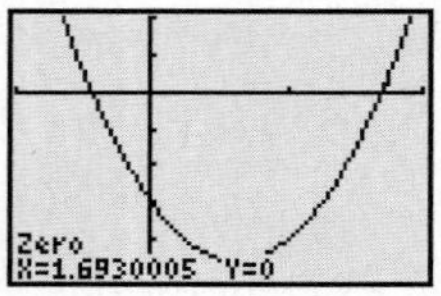

4. $2x^2 - x = 6 - 2x(3 - x)$

$2x^2 - x = 6 - 6x + 2x^2$

$5x = 6$ (not quadratic)

$x = \frac{6}{5}$

5. $\frac{3}{x} - \frac{2}{x + 2} = 1 \quad \text{LCD} = x(x + 2)$

$\frac{3x(x + 2)}{x} - \frac{2x(x + 2)}{x + 2} = x(x + 2)$

$3(x + 2) - 2x = x^2 + 2x$

$3x + 6 - 2x = x^2 + 2x$

$0 = x^2 + x - 6$

$(x + 3)(x - 2) = 0$

$x = -3, 2$

6. $y = 2x^2 + 8x + 5$

$\frac{-b}{2a} = \frac{-8}{2(2)} = -2$

$y = 2(-2)^2 + 8(-2) + 5 = -3$

Min. pt. $(a > 0)$ is $(-2, -3)$,

$c = 5$, y-intercept is $(0, 5)$.

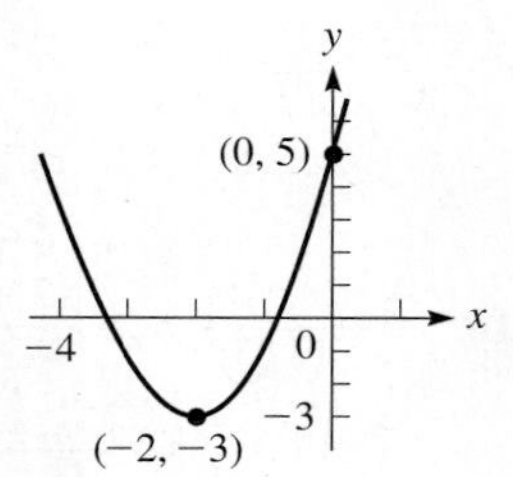

7. $P = EI - RI^2$

$RI^2 - EI + P = 0$

$I = \frac{-(-E) \pm \sqrt{(-E)^2 - 4RP}}{2R}$

$= \frac{E \pm \sqrt{E^2 - 4RP}}{2R}$

8. $x^2 = 6x + 9$

$x^2 - 6x - 9 = 0$

$x^2 - 6x = 9$

$x^2 - 6x + 9 = 9 + 9$

$(x - 3)^2 = 18$

$x - 3 = \pm\sqrt{18}$

$x = 3 \pm 3\sqrt{2}$

9. Let w = width of window

h = height of window

$2w + 2h = 8.4$ (perimeter)

$w + h = 4.2 \quad h = 4.2 - w$

$w(4.2 - w) = 3.8$ (area)

$-w^2 + 4.2w = 3.8$

$w^2 - 4.2w + 3.8 = 0$

$w = \frac{-(-4.2) \pm \sqrt{(-4.2)^2 - 4(1)(3.8)}}{2}$

$= 2.88, 1.32$

$w = 1.3$ m, $h = 2.9$ m or

$w = 2.9$ m, $h = 1.3$ m

10. $y = x^2 - 8x + 8$

$a = 1, b = -8$

$\frac{-b}{2a} = \frac{-(-8)}{2(1)} = 4$

$y = 4^2 - 8(4) + 8 = -8$

$V(4, -8)$, $c = 8$, y-int $(0, 8)$

For $x = 8$, $y = 8$

$y = 0$ for $x = 1.2, 6.8$

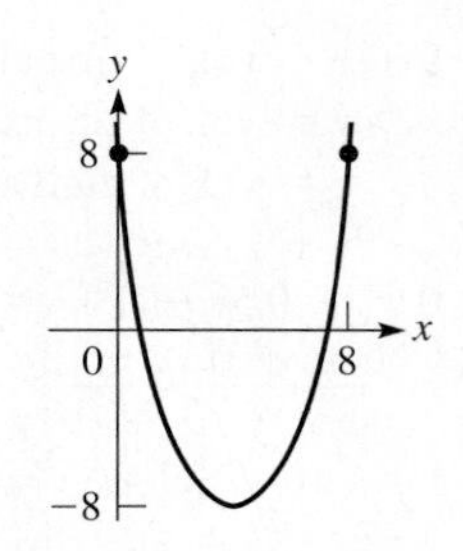

Chapter 8

1. $150° = \left(\frac{\pi}{180}\right)(150) = \frac{5\pi}{6}$ **2.** (a) + (b) − (c) +

3. $\sin 205° = -\sin(205° - 180°) = -\sin 25°$

4. $x = -9, \quad y = 12$

$r = \sqrt{(-9)^2 + 12^2} = 15$

$\sin\theta = \frac{12}{15} = \frac{4}{5}$

$\sec\theta = \frac{15}{-9} = -\frac{5}{3}$

5. $r = 2.80$ ft

$\omega = 2200 \text{ r/min} = (2200 \text{ r/min})(2\pi \text{ rad/r})$

$= 4400\pi$ rad/min

$v = \omega r = (4400\pi)(2.80) = 39{,}000$ ft/min

6. $3.572 = 3.572\left(\frac{180°}{\pi}\right) = 204.7°$

7. $\tan\theta = 0.2396$

$\theta_{ref} = 13.47°$

$\theta = 13.47°$ or

$\theta = 180° + 13.47° = 193.47°$

8. $\cos\theta = -0.8244$, $\csc\theta < 0$;

$\cos\theta$ negative, $\csc\theta$ negative,

θ in third quadrant

$\theta_{ref} = 0.6017, \quad \theta = 3.7432$

9. θ is in quadrant IV since $\sin\theta < 0$ and $\cos\theta > 0$. Since $\sin\theta = -\frac{6}{7}$, we will use $y = -6$ and $r = 7$. Thus, $x = \sqrt{7^2 - (-6)^2} = \sqrt{13}$. So $\tan\theta = \frac{y}{x} = -\frac{6}{\sqrt{13}}$.

10. $s = r\theta$, $32.0 = 8.50\theta$

$\theta = \frac{32.0}{8.50} = 3.76$ rad

$A = \frac{1}{2}\theta r^2 = \frac{1}{2}(3.76)(8.50)^2$

$= 136 \text{ ft}^2$

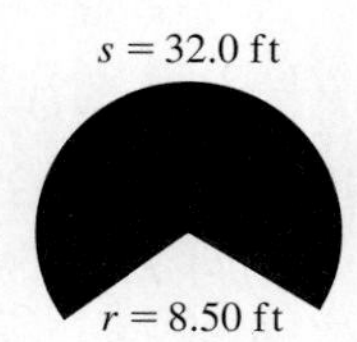

11. $A = \frac{1}{2}\theta r^2$

$A = 38.5 \text{ cm}^2$, $d = 12.2$ cm, $r = 6.10$ cm

$38.5 = \frac{1}{2}\theta(6.10)^2$, $\theta = \frac{2(38.5)}{6.10^2} = 2.07$

$s = r\theta$, $s = 6.10(2.07) = 12.6$ cm

Chapter 9

1.

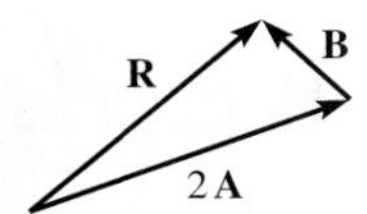

2. $b^2 = a^2 + c^2 - 2ac\cos B$

$b = \sqrt{22.5^2 + 30.9^2 - 2(22.5)(30.9)\cos 78.6^\circ}$

$= 34.4$

3. $x^2 = 36.50^2 + 21.38^2 - 2(36.50)(21.38)\cos 45.00^\circ$

$x = 26.19 \text{ m} \qquad \frac{21.38}{\sin\alpha} = \frac{26.19}{\sin 45.00^\circ}$

$\sin\alpha = \frac{21.38\sin 45.00^\circ}{26.19}, \qquad \alpha = 35.26^\circ$

$\theta = 45.00^\circ - 35.26^\circ = 9.74^\circ$

Displacement is 26.19 m, 9.74° N of E.

4. $C = 180^\circ - (18.9^\circ + 104.2^\circ) = 56.9^\circ$

$\frac{c}{\sin C} = \frac{a}{\sin A}$

$c = \frac{426\sin 56.9^\circ}{\sin 18.9^\circ} = 1100$

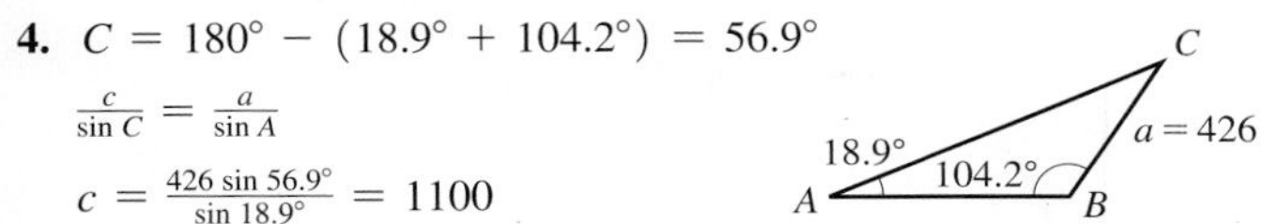

5. Since a is longest side, find A first.

$a^2 = b^2 + c^2 - 2bc\cos A$

$\cos A = \frac{b^2 + c^2 - a^2}{2bc} = \frac{3.29^2 + 8.44^2 - 9.84^2}{2(3.29)(8.44)}$

$A = 105.4^\circ$

$\frac{b}{\sin B} = \frac{a}{\sin A}$

$\sin B = \frac{b\sin A}{a} = \frac{3.29\sin 105.4^\circ}{9.84}$

$B = 18.8^\circ$

$C = 180^\circ - (105.4^\circ + 18.8^\circ) = 55.8^\circ$

6. $R_x = -235, R_y = 152$

$R = \sqrt{(-235)^2 + 152^2} = 280$

$\tan\theta_{ref} = \left|\frac{152}{-235}\right| = \frac{152}{235}, \qquad \theta_{ref} = 32.9^\circ$

Quad. II: $\theta = 180^\circ - 32.9^\circ = 147.1^\circ$

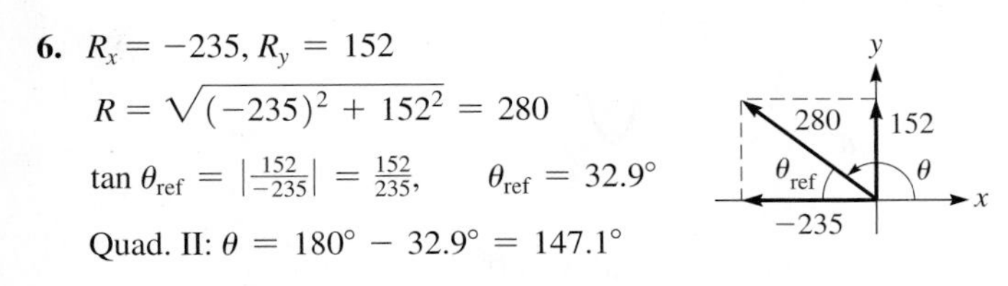

7. Let the force be **F**.

$F_x = 871\cos 284.3^\circ = 215 \text{ kN}$

$F_y = 871\sin 284.3^\circ = -844 \text{ kN}$

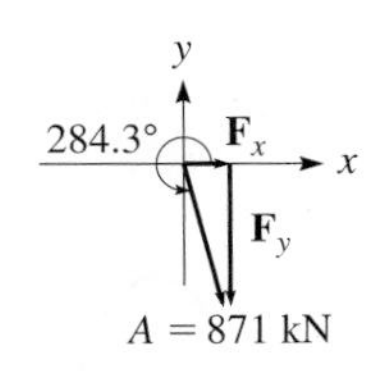

8. $\frac{63.0}{\sin 148.5^\circ} = \frac{42.0}{\sin A} \qquad \frac{x}{\sin 11.1^\circ} = \frac{63.0}{\sin 148.5^\circ}$

$\sin A = \frac{42.0\sin 148.5^\circ}{63.0} \qquad x = \frac{63.0\sin 11.1^\circ}{\sin 148.5^\circ}$

$A = 20.4^\circ \qquad = 23.2 \text{ mi}$

$C = 180^\circ - (148.5^\circ + 20.4^\circ) = 11.1^\circ$

9. $A_x = 449\cos 74.2^\circ \qquad B_x = 285\cos 208.9^\circ$

$A_y = 449\sin 74.2^\circ \qquad B_y = 285\sin 208.9^\circ$

$R_x = A_x + B_x = 449\cos 74.2^\circ + 285\cos 208.9^\circ$

$= -127.3$

$R_y = A_y + B_y = 449\sin 74.2^\circ + 285\sin 208.9^\circ$

$= 294.3$

$R = \sqrt{(-127.3)^2 + 294.3^2} = 321$

$\tan\theta_{ref} = \frac{294.3}{127.3}, \qquad \theta_{ref} = 66.6^\circ, \qquad \theta = 113.4^\circ$

θ is in second quadrant, since R_x is negative and R_y is positive.

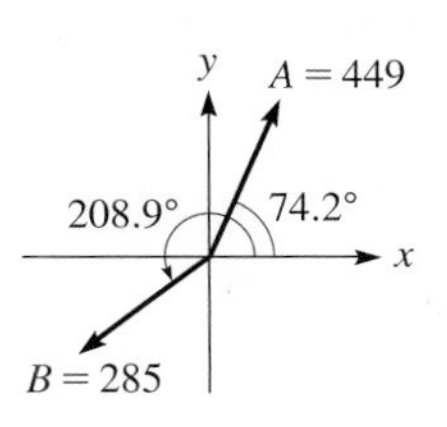

10. $29.6\sin 36.5^\circ = 17.6$

$17.6 < 22.3 < 29.6$ means two solutions.

$\frac{29.6}{\sin B} = \frac{22.3}{\sin 36.5^\circ}, \qquad \sin B = \frac{29.6\sin 36.5^\circ}{22.3}$

$B_1 = 52.1^\circ \qquad C_1 = 180^\circ - 36.5^\circ - 52.1^\circ = 91.4^\circ$

$B_2 = 180^\circ - 52.1^\circ = 127.9^\circ,$

$C_2 = 180^\circ - 36.5^\circ - 127.9^\circ = 15.6^\circ$

$\frac{c_1}{\sin 91.4^\circ} = \frac{22.3}{\sin 36.5^\circ}, \qquad c_1 = \frac{22.3\sin 91.4^\circ}{\sin 36.5^\circ} = 37.5$

$\frac{c_2}{\sin 15.6^\circ} = \frac{22.3}{\sin 36.5^\circ}, \qquad c_2 = \frac{22.3\sin 15.6^\circ}{\sin 36.5^\circ} = 10.1$

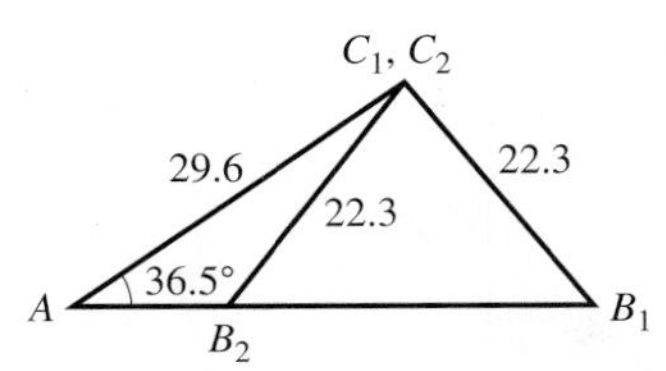

11. Sum of x-components = 0:

$185\cos 305.6^\circ + T_2\cos 90^\circ + T_1\cos 221.7^\circ = 0$

$T_1 = \frac{-185\cos 305.6^\circ}{\cos 221.7^\circ} = 144 \text{ N}$

Sum of y-components = 0:

$185\sin 305.6^\circ + T_2\sin 90^\circ + T_1\sin 221.7^\circ = 0$

$185\sin 305.6^\circ + T_2 + (144.24)\sin 221.7^\circ = 0$

$-T_2 = -246.38$

$T_2 = 246 \text{ N}$

Chapter 10

1. $y = -3\sin(4\pi x - \pi/3)$

$a = -3, b = 4\pi, c = -\pi/3$ amplitude $= |-3| = 3$

period $= \frac{2\pi}{4\pi} = \frac{1}{2}$ displacement $= -\dfrac{-\frac{\pi}{3}}{4\pi} = \frac{1}{12}$

2. $y = 0.5\cos\frac{\pi}{2}x$

Amp. $= 0.5$, disp. $= 0$,

per. $= \frac{2\pi}{\pi/2} = 4$

x	0	1	2	3	4
y	0.5	0	−0.5	0	0.5

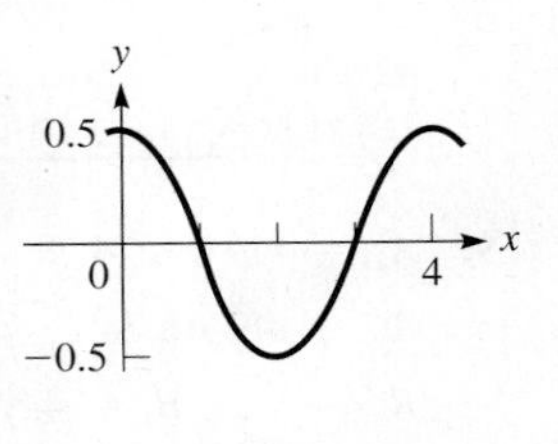

3. $y = 2 + 3\sin x$

For $y_1 = 3\sin x$,

amp. $= 3$, per. $= 2\pi$,

disp. $= 0$

x	0	$\frac{\pi}{2}$	π	$\frac{3\pi}{2}$	2π
$y_1 = 3\sin x$	0	3	0	−3	0
$y = 2 + 3\sin x$	2	5	2	−1	2

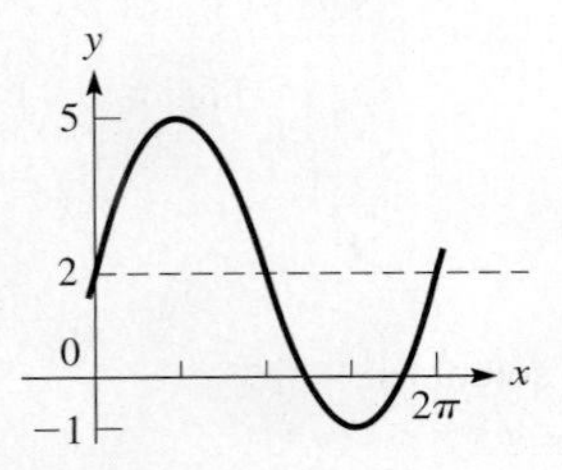

4. $y = 3\sec x$

$\sec x = \frac{1}{\cos x}$

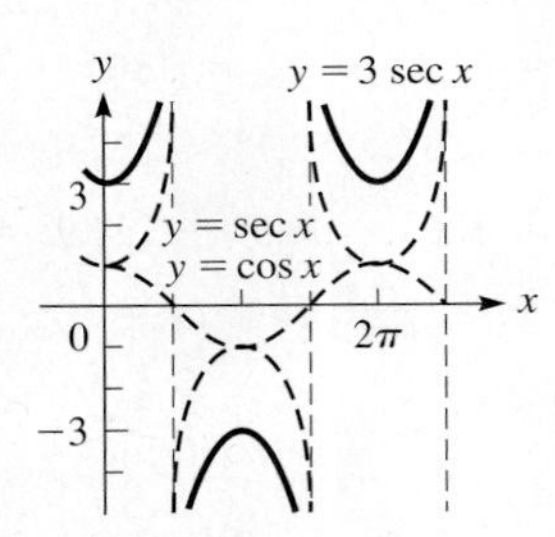

5. $y = 2\sin(2x - \frac{\pi}{3})$

Amp. $= 2$, per. $= \frac{2\pi}{2} = \pi$

disp. $= -\frac{-\pi/3}{2} = \frac{\pi}{6}$

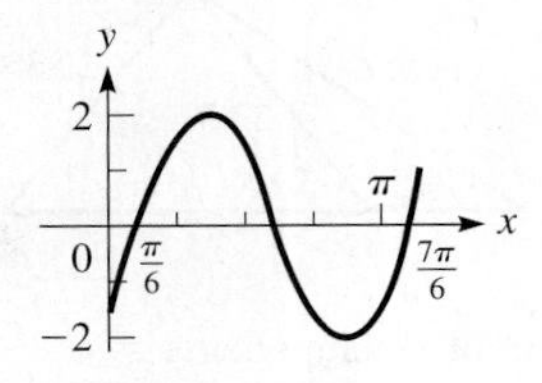

x	$\frac{\pi}{6}$	$\frac{5\pi}{12}$	$\frac{2\pi}{3}$	$\frac{11\pi}{12}$	$\frac{7\pi}{6}$
y	0	2	0	−2	0

$\frac{\pi}{6} + \frac{\pi}{4} = \frac{5\pi}{12}, \frac{\pi}{6} + \frac{\pi}{2} = \frac{2\pi}{3}, \frac{\pi}{6} + \frac{3\pi}{4} = \frac{11\pi}{12}$

6. $y = A\cos\frac{2\pi}{T}t$

$A = 0.200$ in., $T = 0.100$ s,

$y = 0.200\cos 20\pi t$,

amp. $= 0.200$ in., per. $= 0.100$ s

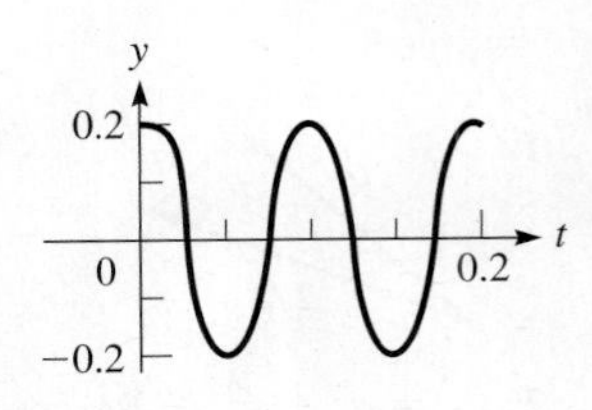

7. $y = 2\sin x + \cos 2x$

For $y_1 = 2\sin x$,

amp. $= 2$, per. $= 2\pi$, disp. $= 0$

For $y_2 = \cos 2x$,

amp. $= 1$, per. $= \frac{2\pi}{2} = \pi$, disp. $= 0$

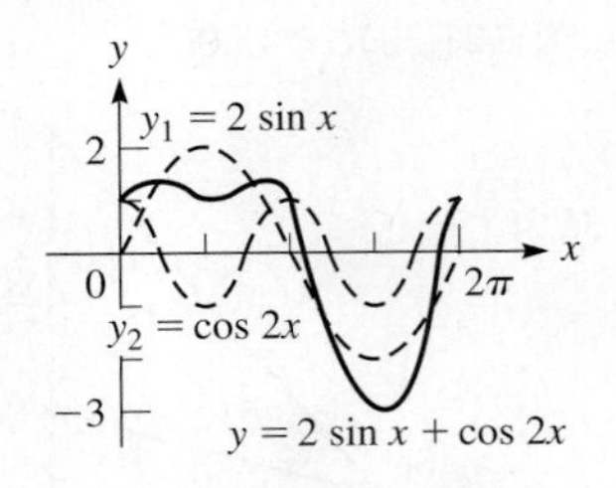

8. $x = \sin\pi t, y = 2\cos 2\pi t$

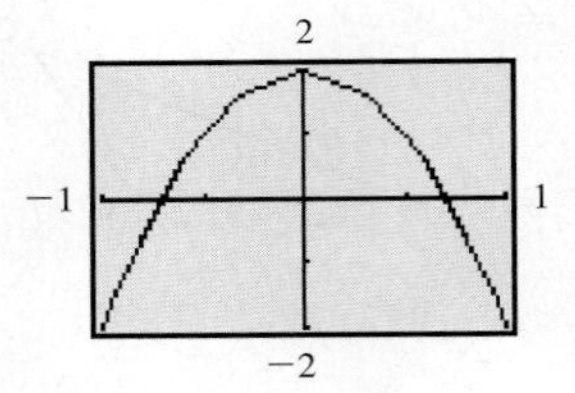

9. $d = R\sin(\omega t + \frac{\pi}{6})$

$\omega = 2.00$ rad/s

$d = R\sin(2.00t + \frac{\pi}{6})$

Amp. $= R$, per. $= \dfrac{2\pi}{2.00} = \pi$ s $= 3.14$ s,

disp. $= \dfrac{-\pi/6}{2.00} = -\dfrac{\pi}{12}$ s $= -0.26$ s

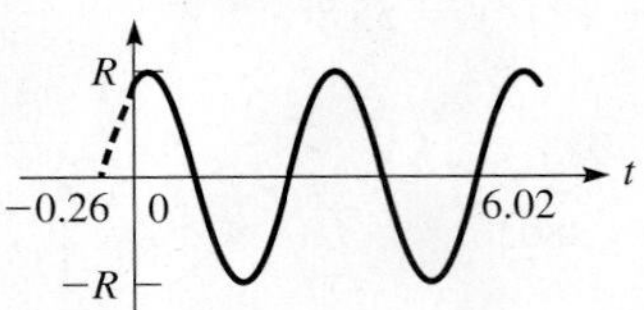

10. $y = 2\sin bx$

$(\frac{\pi}{3}, 2)$, is first max. with $x > 0$

period $= \frac{2\pi}{b}, \frac{1}{4}(\frac{2\pi}{b}) = \frac{\pi}{3}, \frac{\pi}{2b} = \frac{\pi}{3}, b = \frac{3}{2}$

$y = 2\sin\frac{3x}{2}$

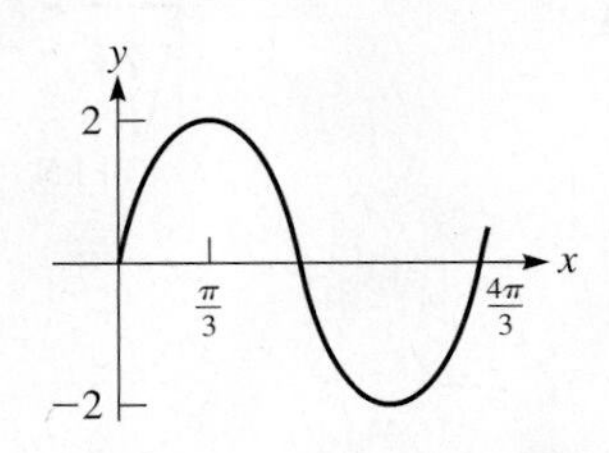

Chapter 11

1. $-5y^0 = -5(1) = -5$

2. $(3\pi x^{-4})^{-2} = (3\pi)^{-2}(x^{-4})^{-2} = \frac{x^8}{(3\pi)^2} = \frac{x^8}{9\pi^2}$

3. $\frac{100^{3/2}}{8^{-2/3}} = (100^{3/2})(8^{2/3}) = [(100^{1/2})^3][(8^{1/3})^2]$
$= (10^3)(2^2) = 4000$

4. $(as^{-1/3}t^{3/4})^{12} = a^{12}s^{(-1/3)(12)}t^{(3/4)(12)} = a^{12}s^{-4}t^9 = \frac{a^{12}t^9}{s^4}$

5. $2\sqrt{20} - \sqrt{125} = 2\sqrt{4 \times 5} - \sqrt{25 \times 5}$
$= 2(2\sqrt{5}) - 5\sqrt{5}$
$= 4\sqrt{5} - 5\sqrt{5} = -\sqrt{5}$

6. $(2x^{-1} + y^{-2})^{-1} = \frac{1}{2x^{-1} + y^{-2}} = \frac{1}{\frac{2}{x} + \frac{1}{y^2}} = \frac{1}{\frac{2y^2 + x}{xy^2}}$
$= \frac{xy^2}{2y^2 + x}$

7. $(\sqrt{2x} - 3\sqrt{y})^2 = (\sqrt{2x})^2 - 2\sqrt{2x}(3\sqrt{y}) + (3\sqrt{y})^2$
$= 2x - 6\sqrt{2xy} + 9y$

8. $\sqrt[3]{\sqrt[4]{4}} = \sqrt[12]{4} = \sqrt[12]{2^2} = 2^{2/12} = 2^{1/6} = \sqrt[6]{2}$

9. $\frac{3 - 2\sqrt{2}}{2\sqrt{x}} = \frac{3 - 2\sqrt{2}}{2\sqrt{x}} \times \frac{\sqrt{x}}{\sqrt{x}} = \frac{3\sqrt{x} - 2\sqrt{2x}}{2x}$

10. $\sqrt{27a^4b^3} = \sqrt{9 \times 3 \times (a^2)^2(b^2)(b)} = 3a^2b\sqrt{3b}$

11. $2\sqrt{2}(3\sqrt{10} - \sqrt{6}) = 2\sqrt{2}(3\sqrt{10}) - 2\sqrt{2}(\sqrt{6})$
$= 6\sqrt{20} - 2\sqrt{12}$
$= 6(2\sqrt{5}) - 2(2\sqrt{3})$
$= 12\sqrt{5} - 4\sqrt{3}$

12. $(2x + 3)^{1/2} + (x + 1)(2x + 3)^{-1/2}$
$= (2x + 3)^{1/2} + \frac{x + 1}{(2x + 3)^{1/2}}$
$= \frac{(2x + 3)^{1/2}(2x + 3)^{1/2} + x + 1}{(2x + 3)^{1/2}} = \frac{2x + 3 + x + 1}{(2x + 3)^{1/2}}$
$= \frac{3x + 4}{(2x + 3)^{1/2}}$

13. $\left(\frac{4a^{-1/2}b^{3/4}}{b^{-2}}\right)\left(\frac{b^{-1}}{2a}\right) = \frac{4a^{-1/2}b^{3/4 - 1}}{2ab^{-2}} = \frac{2b^{2 - 1/4}}{a^{1 + 1/2}} = \frac{2b^{7/4}}{a^{3/2}}$

14. $\frac{2\sqrt{15} + \sqrt{3}}{\sqrt{15} - 2\sqrt{3}} = \frac{2\sqrt{15} + \sqrt{3}}{\sqrt{15} - 2\sqrt{3}} \times \frac{\sqrt{15} + 2\sqrt{3}}{\sqrt{15} + 2\sqrt{3}}$
$= \frac{2(15) + 4\sqrt{45} + \sqrt{45} + 2(3)}{15 - 4(3)}$
$= \frac{36 + 15\sqrt{5}}{3}$
$= 12 + 5\sqrt{5}$

15. $\frac{3^{-1/2}}{2} = \frac{1}{2 \times 3^{1/2}} = \frac{1}{2\sqrt{3}}$
$= \frac{\sqrt{3}}{2\sqrt{3}\sqrt{3}} = \frac{\sqrt{3}}{6}$

16. $0.220N^{-1/6} \quad N = 64 \times 10^6$
$0.220(64 \times 10^6)^{-1/6} = \frac{0.220}{(64 \times 10^6)^{1/6}} = \frac{0.220}{2 \times 10} = 0.011$

Chapter 12

1. $(3 - \sqrt{-4}) + (5\sqrt{-9} - 1) = (3 - 2j) + [5(3j) - 1]$
$= 3 - 2j + 15j - 1$
$= 2 + 13j$

2. $(2\angle 130^\circ)(3\angle 45^\circ) = (2)(3)\angle 130^\circ + 45^\circ = 6\angle 175^\circ$

3. $2 - 7j$: $r = \sqrt{2^2 + (-7)^2} = \sqrt{53} = 7.28$
$\tan\theta = \frac{-7}{2} = -3.500, \quad \theta_{ref} = 74.1^\circ, \quad \theta = 285.9^\circ$
$2 - 7j = 7.28(\cos 285.9^\circ + j\sin 285.9^\circ)$
$= 7.28\angle 285.9^\circ$

4. (a) $-\sqrt{-64} = -(8j) = -8j$
(b) $-j^{15} = -j^{12}j^3 = (-1)(-j) = j$

5. $(4 - 3j) + (-1 + 4j) = 3 + j$

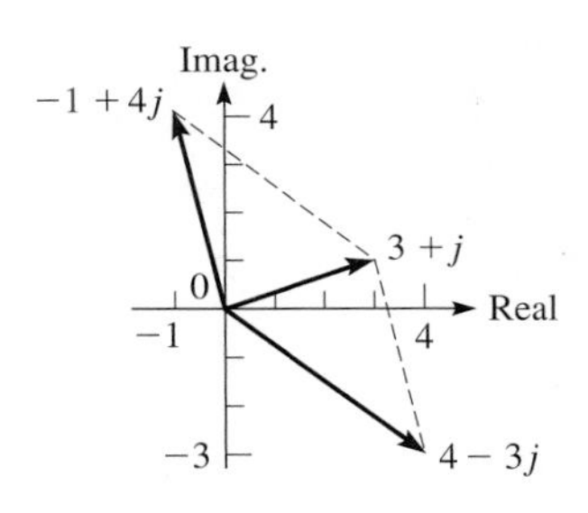

6. $\frac{2 - 4j}{5 + 3j} = \frac{(2 - 4j)(5 - 3j)}{(5 + 3j)(5 - 3j)} = \frac{10 - 26j + 12j^2}{25 - 9j^2}$
$= \frac{10 - 26j - 12}{25 - 9(-1)} = \frac{-2 - 26j}{34} = -\frac{1 + 13j}{17}$

7. $2.56(\cos 125.2^\circ + j\sin 125.2^\circ) = 2.56e^{2.185j}$
$125.2^\circ = \frac{125.2\pi}{180} = 2.185 \text{ rad}$

8. $R = 3.50\ \Omega, X_L = 6.20\ \Omega, X_C = 7.35\ \Omega$
$|Z| = \sqrt{R^2 + (X_L - X_C)^2}$
$= \sqrt{3.50^2 + (6.20 - 7.35)^2} = 3.68\ \Omega$
$\tan\phi = \frac{X_L - X_C}{R} = \frac{6.20 - 7.35}{3.50} = \frac{-1.15}{3.50}$
$\phi = -18.2^\circ$

9. $3.47 - 2.81j = 4.47e^{5.60j}$
$R = \sqrt{3.47^2 + (-2.81)^2} = 4.47$
$\tan\theta = \frac{-2.81}{3.47}, \quad \theta_{ref} = 39.0^\circ$
$\theta = 321.0^\circ = 5.60 \text{ rad}$

10. $x + 2j - y = yj - 3xj$
$x - y + 3xj - yj = -2j$
$(x - y) + (3x - y)j = 0 - 2j$
$x - y = 0$
$3x - y = -2$
$2x = -2$
$x = -1, \quad y = -1$

11. $L = 8.75 \text{ mH} = 8.75 \times 10^{-3} \text{ H}$
$f = 600 \text{ kHz} = 6.00 \times 10^5 \text{ Hz}$
$2\pi fL = \frac{1}{2\pi fC}$
$C = \frac{1}{(2\pi f)^2L} = \frac{1}{(2\pi)^2(6.00 \times 10^5)^2(8.75 \times 10^{-3})}$
$= 8.04 \times 10^{-12} = 8.04 \text{ pF}$

12. $j = 1(\cos 90° + j\sin 90°)$

$j^{1/3} = 1^{1/3}(\cos\frac{90°}{3} + j\sin\frac{90°}{3})$

$= \cos 30° + j\sin 30° = 0.8660 + 0.5000j$

$= 1^{1/3}(\cos\frac{90° + 360°}{3} + j\sin\frac{90° + 360°}{3})$

$= \cos 150° + j\sin 150° = -0.8660 + 0.5000j$

$= 1^{1/3}(\cos\frac{90° + 360°}{3} + j\sin\frac{90° + 360°}{3})$

$= \cos 270° + j\sin 270° = -j$

Cube roots of j: $0.8660 + 0.5000j$

$-0.8660 + 0.5000j$

$-j$

Chapter 13

1. $\log_9 x = -\frac{1}{2}$

$x = 9^{-1/2}$

$= \frac{1}{9^{1/2}} = \frac{1}{3}$

2. $\log_3 x - \log_3 2 = 2$

$\log_3 \frac{x}{2} = 2$

$\frac{x}{2} = 3^2$

$x = 2(3^2) = 18$

3. $\log_x 64 = 3$

$64 = x^3$

$4^3 = x^3$

$x = 4$

4. $3^{3x+1} = 8$

$(3x + 1)\log 3 = \log 8$

$3x + 1 = \frac{\log 8}{\log 3}$

$x = \frac{1}{3}(\frac{\log 8}{\log 3} - 1) = 0.298$

5. $y = 2\log_4 x$

x	$\frac{1}{4}$	1	4	16
y	-2	0	2	4

$\log_4 \frac{1}{4} = -1, \log_4 16 = 2$

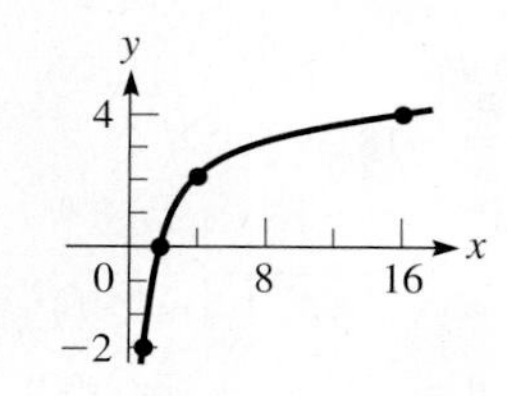

6. $y = 2(3^x)$

x	-1	0	1	2
y	0.7	2	6	18

x	3	4	5
y	54	162	486

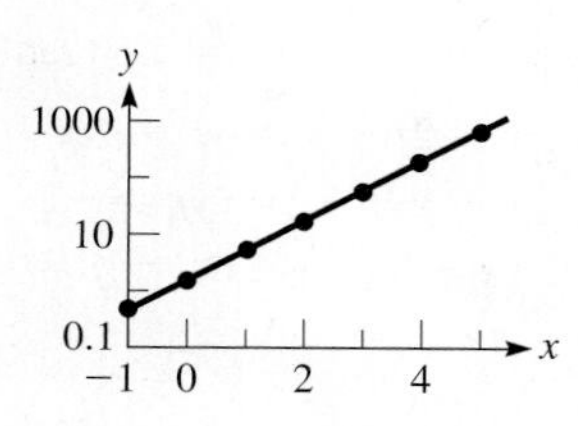

7. $\log_5(\frac{4a^3}{7}) = \log_5 4a^3 - \log_5 7$

$= \log_5 4 + \log_5 a^3 - \log_5 7$

$= \log_5 4 + 3\log_5 a - \log_5 7$

$= 2\log_5 2 + 3\log_5 a - \log_5 7$

8. $3\log_7 x - \log_7 y = 2$

$\log_7 x^3 - \log_7 y = 2$

$\log_7 \frac{x^3}{y} = 2 \qquad \frac{x^3}{y} = 7^2$

$x^3 = 49y \quad y = \frac{1}{49}x^3$

9. $\ln i - \ln I = -t/RC$

$\ln\frac{i}{I} = -t/RC$

$\frac{i}{I} = e^{-t/RC} \qquad i = Ie^{-t/RC}$

10. $\frac{2\ln 0.9523}{\log 6066} = -0.02584$

11. $\log_b x = \frac{\log_a x}{\log_a b}$

$\log_5 732 = \frac{\log 732}{\log 5} = 4.098$

12. $A = A_0e^{0.08t} \qquad A = 2A_0$

$2A_0 = A_0e^{0.08t} \qquad 2 = e^{0.08t}$

$\ln 2 = \ln e^{0.08t} = 0.08t \qquad t = \frac{\ln 2}{0.08} = 8.66$ years

Chapter 14

1. $x^{1/2} - 2x^{1/4} = 3$

Let $y = x^{1/4}$

$y^2 - 2y - 3 = 0$

$(y - 3)(y + 1) = 0$

$y = 3, -1$

$x^{1/4} \neq -1$

$x^{1/4} = 3, \qquad x = 81$

Check: $81^{1/2} - 2(81^{1/4}) = 3$

$9 - 6 = 3$

Solution: $x = 81$

2. $3\sqrt{x - 2} - \sqrt{x + 1} = 1$

$3\sqrt{x - 2} = 1 + \sqrt{x + 1}$

$9(x - 2) = 1 + 2\sqrt{x + 1} + (x + 1)$

$8x - 20 = 2\sqrt{x + 1}$

$4x - 10 = \sqrt{x + 1}$

$16x^2 - 80x + 100 = x + 1$

$16x^2 - 81x + 99 = 0$

$x = \frac{81 \pm \sqrt{81^2 - 4(16)(99)}}{32}$

$= \frac{81 \pm 15}{32} = 3, \frac{33}{16}$

Check:

$x = 3$: $3\sqrt{3 - 2} - \sqrt{3 + 1} \stackrel{?}{=} 1$

$3 - 2 = 1$

$x = \frac{33}{16}$: $3\sqrt{\frac{33}{16} - 2} - \sqrt{\frac{33}{16} + 1} \stackrel{?}{=} 1$

$\frac{3}{4} - \frac{7}{4} \neq 1$

Solution: $x = 3$

3. $x^4 - 17x^2 + 16 = 0$

Let $y = x^2$

$$y^2 - 17y + 16 = 0$$
$$(y - 1)(y - 16) = 0$$
$$y = 1, 16$$
$$x^2 = 1, 16$$
$$x = -1, 1, -4, 4$$

All values check.

4. $x^2 - 2y = 5$

$\underline{2x + 6y = 1}$

$$y = \tfrac{1}{2}(x^2 - 5)$$
$$2x + 6(\tfrac{1}{2})(x^2 - 5) = 1$$
$$3x^2 + 2x - 16 = 0$$
$$(3x + 8)(x - 2) = 0$$
$$x = -\tfrac{8}{3}, 2$$
$$x = -\tfrac{8}{3}: y = \tfrac{1}{2}(\tfrac{64}{9} - 5) = \tfrac{19}{18}$$
$$x = 2: y = \tfrac{1}{2}(4 - 5) = -\tfrac{1}{2}$$
$$x = -\tfrac{8}{3}, y = \tfrac{19}{18}$$

or $x = 2, y = -\frac{1}{2}$

5. $\sqrt[3]{2x + 5} = 5$

$$2x + 5 = 125$$
$$2x = 120$$
$$x = 60$$

6. $v = \sqrt{v_0^2 + 2gh}$

$$v^2 = v_0^2 + 2gh$$
$$2gh = v^2 - v_0^2$$
$$h = \frac{v^2 - v_0^2}{2g}$$

7. $x^2 - y^2 = 4 \qquad xy = 2$

$y = \pm\sqrt{x^2 - 4} \qquad y = \frac{2}{x}$

x	y
± 2	0
± 3	± 2.2
± 4	± 3.5

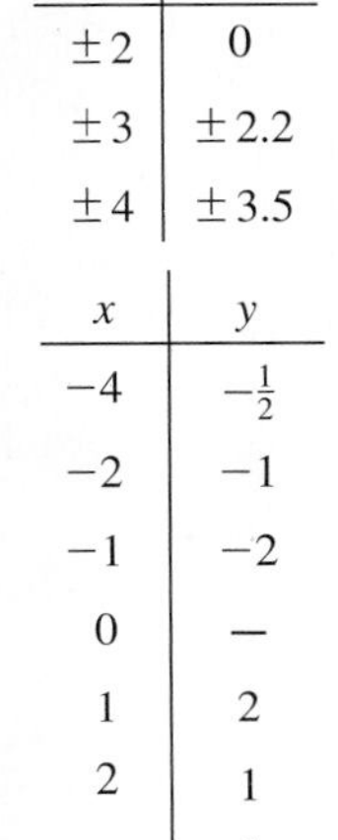

x	y
-4	$-\frac{1}{2}$
-2	-1
-1	-2
0	—
1	2
2	1
4	$\frac{1}{2}$

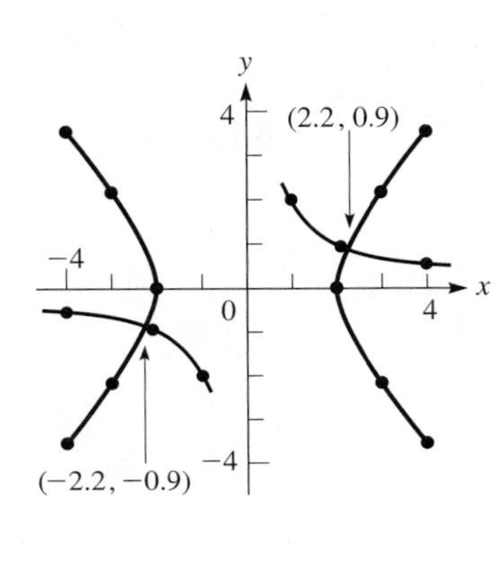

$x = 2.2, \quad y = 0.9; \quad x = -2.2, \quad y = -0.9$

8. Let l = length, w = width

$$2l + 2w = 14$$
$$\underline{lw = 10}$$
$$l + w = 7$$
$$l = 7 - w$$
$$(7 - w)w = 10$$
$$7w - w^2 = 10$$
$$w^2 - 7w + 10 = 0$$
$$(w - 5)(w - 2) = 0$$
$$w = 5, 2$$

$l = 5.00$ ft, $w = 2.00$ ft

(or $l = 2.00$ ft, $w = 5.00$ ft)

Chapter 15

1. $f(x) = 2x^3 + 3x^2 + 7x - 6$

$$f(-3) = 2(-3)^3 + 3(-3)^2 + 7(-3) - 6$$
$$= -54 + 27 - 21 - 6 = -54$$

$f(-3) \neq 0, \quad -3$ is not a zero

2. $x^4 - 2x^3 - 7x^2 + 20x - 12 = 0$

1	−2	−7	20	−12	⌊2
	2	0	−14	12	
1	0	−7	6		⌊2
	2	4	−6		
1	2	−3			

$x^2 + 2x - 3 = (x + 3)(x - 1)$

Other roots: $x = -3, 1$

3. $(x^3 - 5x^2 + 4x - 9) \div (x - 3)$

1	−5	4	−9	⌊3
	3	−6	−6	
1	−2	−2	−15	

Quotient: $x^2 - 2x - 2$

Remainder $= -15$

4. $f(x) = 2x^4 + 15x^3 + 23x^2 - 16$

$2x + 1 = 2(x + \frac{1}{2})$

2	15	23	0	−16	⌊$-\frac{1}{2}$
	−1	−7	−8	4	
2	14	16	−8	−12	

Remainder is not zero;

$2x + 1$ is not a factor.

5. $(x^3 + 4x^2 + 7x - 9) \div (x + 4)$

$$f(x) = x^3 + 4x^2 + 7x - 9$$
$$f(-4) = (-4)^3 + 4(-4)^2 + 7(-4) - 9$$
$$= -64 + 64 - 28 - 9 = -37$$

Remainder $= -37$

6. $2x^4 - x^3 + 5x^2 - 4x - 12 = 0$

$f(x) = 2x^4 - x^3 + 5x^2 - 4x - 12$

$f(-x) = 2x^4 + x^3 + 5x^2 + 4x - 12$

$n = 4$; 4 roots

No more than 3 positive roots

One negative root

Rational roots: factors of 12 divided by factors of 2

Possible rational roots:

$\pm 1, \pm 2, \pm 3, \pm 4, \pm 6, \pm 12, \pm\frac{1}{2}, \pm\frac{3}{2}$

$$\begin{array}{rrrrr|l} 2 & -1 & 5 & -4 & -12 & \underline{2} \\ & 4 & 6 & 22 & 36 & \\ \hline 2 & 3 & 11 & 18 & 24 & \end{array}$$

2 is too large.

$$\begin{array}{rrrrr|l} 2 & -1 & 5 & -4 & -12 & \underline{\tfrac{3}{2}} \\ & 3 & 3 & 12 & 12 & \\ \hline 2 & 2 & 8 & 8 & 0 & \underline{-1} \\ & -2 & 0 & -8 & & \\ \hline 2 & 0 & 8 & 0 & & \end{array}$$

$2x^2 + 8 = 0, \quad x^2 + 4 = 0, \quad x = \pm 2j$

roots: $\frac{3}{2}, -1, 2j, -2j$

7. $y = kx^2(x^3 + 436x - 4000)$

$y = 0, kx^2(x^3 + 436x - 4000) = 0$

$x = 0, x^3 + 436x - 4000 = 0$

$$\begin{array}{rrrr|l} 1 & 0 & 436 & -4000 & \underline{8} \\ & 8 & 64 & 4000 & \\ \hline 1 & 8 & 500 & 0 & \end{array}$$

$y = 0$ for $x = 0$ ft, $x = 8$ ft

8. Let $x =$ length of edge.

$V = x^3, \quad 2V = (x + 1)^3$

$2x^3 = x^3 + 3x^2 + 3x + 1$

$x^3 - 3x^2 - 3x - 1 = 0$

$y = x^3 - 3x^2 - 3x - 1$

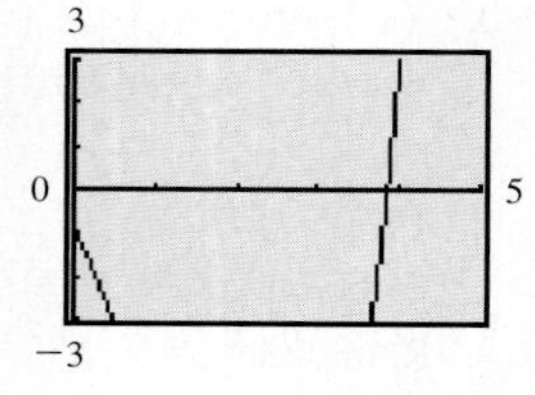

$x = 3.8$ mm

Chapter 16

1. $A = \begin{bmatrix} 3 & -1 & 4 \\ 2 & 0 & -2 \end{bmatrix} \quad B = \begin{bmatrix} 1 & 4 & 5 \\ -1 & -2 & 3 \end{bmatrix}$

$2B = \begin{bmatrix} 2 & 8 & 10 \\ -2 & -4 & 6 \end{bmatrix}$

$A - 2B = \begin{bmatrix} 3-2 & -1-8 & 4-10 \\ 2+2 & 0+4 & -2-6 \end{bmatrix}$

$= \begin{bmatrix} 1 & -9 & -6 \\ 4 & 4 & -8 \end{bmatrix}$

2. $\begin{bmatrix} 2x & x-y & z \\ x+z & 2y & y+z \end{bmatrix} = \begin{bmatrix} 6 & -2 & 4 \\ a & b & c \end{bmatrix}$

$2x = 6 \quad x - y = -2 \quad z = 4$

$x = 3 \quad 3 - y = -2$

$y = 5$

$x + z = a \quad 2y = b \quad y + z = c$

$3 + 4 = a \quad 2(5) = b \quad 5 + 4 = c$

$a = 7 \quad b = 10 \quad c = 9$

3. $CD = \begin{bmatrix} 1 & 0 & 4 \\ 2 & -2 & 1 \\ -1 & 3 & 2 \end{bmatrix}\begin{bmatrix} 2 & -2 \\ 4 & -5 \\ 6 & 1 \end{bmatrix}$

$= \begin{bmatrix} 2+0+24 & -2+0+4 \\ 4-8+6 & -4+10+1 \\ -2+12+12 & 2-15+2 \end{bmatrix}$

$= \begin{bmatrix} 26 & 2 \\ 2 & 7 \\ 22 & -11 \end{bmatrix}$

$DC = \begin{bmatrix} 2 & -2 \\ 4 & -5 \\ 6 & 1 \end{bmatrix}\begin{bmatrix} 1 & 0 & 4 \\ 2 & -2 & 1 \\ -1 & 3 & 2 \end{bmatrix}$ not defined, since D has 2 columns and C has 3 rows

4. $A = \begin{bmatrix} 2 & -5 \\ 1 & -2 \end{bmatrix} \quad B = \begin{bmatrix} -2 & 5 \\ -1 & 2 \end{bmatrix}$

$AB = \begin{bmatrix} 2 & -5 \\ 1 & -2 \end{bmatrix}\begin{bmatrix} -2 & 5 \\ -1 & 2 \end{bmatrix} = \begin{bmatrix} -4+5 & 10-10 \\ -2+2 & 5-4 \end{bmatrix}$

$= \begin{bmatrix} 1 & 0 \\ 0 & 1 \end{bmatrix} = I$

$BA = \begin{bmatrix} -2 & 5 \\ -1 & 2 \end{bmatrix}\begin{bmatrix} 2 & -5 \\ 1 & -2 \end{bmatrix} = \begin{bmatrix} -4+5 & 10-10 \\ -2+2 & 5-4 \end{bmatrix}$

$= \begin{bmatrix} 1 & 0 \\ 0 & 1 \end{bmatrix} = I$

$AB = BA = I, B = A^{-1}$

5. $\left[\begin{array}{rrr|rrr} 1 & 0 & 4 & 1 & 0 & 0 \\ 2 & -2 & 1 & 0 & 1 & 0 \\ -1 & 3 & 2 & 0 & 0 & 1 \end{array}\right] \to \left[\begin{array}{rrr|rrr} 1 & 0 & 4 & 1 & 0 & 0 \\ 0 & -2 & -7 & -2 & 1 & 0 \\ 0 & 3 & 6 & 1 & 0 & 1 \end{array}\right] \to$

$\left[\begin{array}{rrr|rrr} 1 & 0 & 4 & 1 & 0 & 0 \\ 0 & 1 & \frac{7}{2} & 1 & -\frac{1}{2} & 0 \\ 0 & 3 & 6 & 1 & 0 & 1 \end{array}\right] \to \left[\begin{array}{rrr|rrr} 1 & 0 & 4 & 1 & 0 & 0 \\ 0 & 1 & \frac{7}{2} & 1 & -\frac{1}{2} & 0 \\ 0 & 0 & -\frac{9}{2} & -2 & \frac{3}{2} & 1 \end{array}\right] \to$

$\left[\begin{array}{rrr|rrr} 1 & 0 & 4 & 1 & 0 & 0 \\ 0 & 1 & \frac{7}{2} & 1 & -\frac{1}{2} & 0 \\ 0 & 0 & 1 & \frac{4}{9} & -\frac{1}{3} & -\frac{2}{9} \end{array}\right] \to \left[\begin{array}{rrr|rrr} 1 & 0 & 0 & -\frac{7}{9} & \frac{4}{3} & \frac{8}{9} \\ 0 & 1 & 0 & -\frac{5}{9} & \frac{2}{3} & \frac{7}{9} \\ 0 & 0 & 1 & \frac{4}{9} & -\frac{1}{3} & -\frac{2}{9} \end{array}\right]$

$C^{-1} = \begin{bmatrix} -\frac{7}{9} & \frac{4}{3} & \frac{8}{9} \\ -\frac{5}{9} & \frac{2}{3} & \frac{7}{9} \\ \frac{4}{9} & -\frac{1}{3} & -\frac{2}{9} \end{bmatrix}$

6. $2x - 3y = 11$

$x + 2y = 2$

$$A = \begin{bmatrix} 2 & -3 \\ 1 & 2 \end{bmatrix} \quad A^{-1} = \begin{bmatrix} \frac{2}{7} & \frac{3}{7} \\ -\frac{1}{7} & \frac{2}{7} \end{bmatrix} \quad C = \begin{bmatrix} 11 \\ 2 \end{bmatrix}$$

$$A^{-1}C = \begin{bmatrix} \frac{2}{7} & \frac{3}{7} \\ -\frac{1}{7} & \frac{2}{7} \end{bmatrix}\begin{bmatrix} 11 \\ 2 \end{bmatrix} = \begin{bmatrix} \frac{22}{7} + \frac{6}{7} \\ -\frac{11}{7} + \frac{4}{7} \end{bmatrix} = \begin{bmatrix} 4 \\ -1 \end{bmatrix}$$

$x = 4 \qquad y = -1$

7.

$2x - 3y = 11$	
$x + 2y = 2$	
$x + 2y = 2$	interchange equations
$2x - 3y = 11$	
$x + 2y = 2$	subtract 2 times first equation
$-7y = 7$	from second equation
$x + 2y = 2$	divide second equation by -7
$y = -1$	
$x + 2(-1) = 2$	substitute $y = -1$ in first equation
$x = 4$	

8.

$$\begin{vmatrix} 2 & -4 & -3 \\ -3 & 6 & 2 \\ 5 & -1 & 5 \end{vmatrix} = 2\begin{vmatrix} 6 & 2 \\ -1 & 5 \end{vmatrix} - (-4)\begin{vmatrix} -3 & 2 \\ 5 & 5 \end{vmatrix} + (-3)\begin{vmatrix} -3 & 6 \\ 5 & -1 \end{vmatrix}$$

$$= 2[30 - (-2)] + 4[-15 - 10] - 3[3 - 30]$$

$$= 2(32) + 4(-25) - 3(-27) = 64 - 100 + 81 = 45$$

9.

$$A = \begin{bmatrix} 7 & -2 & 1 \\ 2 & 3 & -4 \\ 4 & -5 & 2 \end{bmatrix} \qquad C = \begin{bmatrix} 6 \\ 6 \\ 10 \end{bmatrix}$$

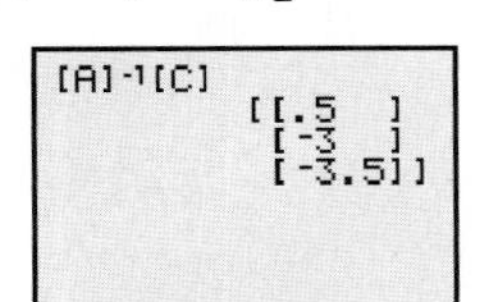

$x = 0.5 \qquad y = -3 \qquad z = -3.5$

10. Let A = price of stock A

B = price of stock B

$50A + 30B = 2600$

$30A + 40B = 2000$

$5A + 3B = 260$ (divide both equations by 10)

$3A + 4B = 200$

Let C = coefficient matrix

$$C = \begin{bmatrix} 5 & 3 \\ 3 & 4 \end{bmatrix}, \quad \begin{vmatrix} 5 & 3 \\ 3 & 4 \end{vmatrix} = 20 - 9 = 11$$

$$C^{-1} = \frac{1}{11}\begin{bmatrix} 4 & -3 \\ -3 & 5 \end{bmatrix} = \begin{bmatrix} \frac{4}{11} & -\frac{3}{11} \\ -\frac{3}{11} & \frac{5}{11} \end{bmatrix}$$

$$\begin{bmatrix} A \\ B \end{bmatrix} = \begin{bmatrix} \frac{4}{11} & -\frac{3}{11} \\ -\frac{3}{11} & \frac{5}{11} \end{bmatrix}\begin{bmatrix} 260 \\ 200 \end{bmatrix} = \begin{bmatrix} \frac{4}{11}(260) - \frac{3}{11}(200) \\ -\frac{3}{11}(260) + \frac{5}{11}(200) \end{bmatrix}$$

$$= \begin{bmatrix} 40 \\ 20 \end{bmatrix}$$

$A = \$40 \qquad B = \$20$

Chapter 17

1. $x < 0, y > 0$

2. $\frac{-x}{2} \geq 3$

$-x \geq 6$

$x \leq -6$

3. $3x + 1 < -5$

$3x < -6$

$x < -2$

4. $-1 < 1 - 2x < 5$

$-2 < -2x < 4$

$1 > x > -2$

$-2 < x < 1$

5. $\frac{x^2 + x}{x - 2} \leq 0, \frac{x(x + 1)}{x - 2} \leq 0$

Interval	$\frac{x(x+1)}{x-2}$	Sign
$x < -1$	$\frac{- -}{-}$	$-$
$-1 < x < 0$	$\frac{- +}{-}$	$+$
$0 < x < 2$	$\frac{+ +}{-}$	$-$
$x > 2$	$\frac{+ +}{+}$	$+$

Solution: $x \leq -1$ or $0 \leq x < 2$

(x cannot equal 2)

6. $|2x + 1| \geq 3$

$2x + 1 \geq 3$ or $2x + 1 \leq -3$

$2x \geq 2$ or $2x \leq -4$

$x \geq 1$ or $x \leq -2$

7. $|2 - 3x| < 8$

$-8 < 2 - 3x < 8$

$-10 < -3x < 6$

$\frac{10}{3} > x > -2$

$-2 < x < \frac{10}{3}$

8.

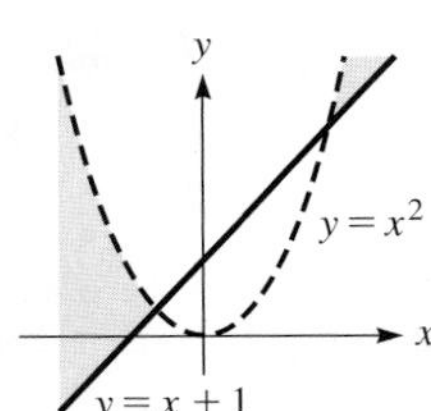

9. If $\sqrt{x^2 - x - 6}$ is real, then $x^2 - x - 6 \geq 0$.

$(x - 3)(x + 2) \geq 0$

$x \leq -2$ or $x \geq 3$

Interval	$(x - 3)(x + 2)$	Sign
$x < -2$	$(-)(-)$	$+$
$-2 < x < 3$	$(-)(+)$	$-$
$x > 3$	$(+)(+)$	$+$

10. Let w = width, l = length

$l = w + 20$

$wl \geq 4800$

$w(w + 20) \geq 4800$

$w^2 + 20w - 4800 \geq 0$

$(w + 80)(w - 60) \geq 0$

$w \geq 60$ m

11. Let A = length of type A wire

B = length of type B wire

$0.10A + 0.20B < 5.00$

$A + 2B < 50$

12. $|\lambda - 550 \text{ nm}| < 150 \text{ nm}$ (within 150 nm of $\lambda = 550$ nm)

13. $x^2 > 12 - x$

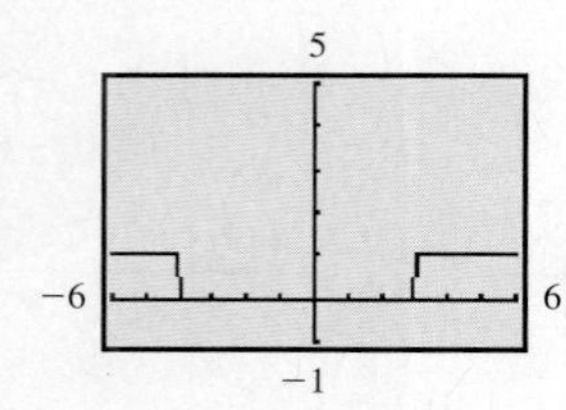

$x < -4, x > 3$

14. $P = 5x + 3y$

$x \geq 0, y \geq 0$

$2x + 3y \leq 12$

$4x + y \leq 8$

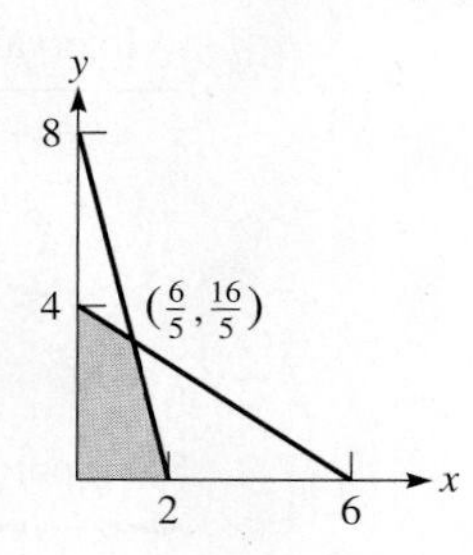

Vertices:	(0,0)	(2,0)	$(\frac{6}{5}, \frac{16}{5})$	(0,4)
$P = 5x + 3y$	0	10	15.6	12

Max. value of $P = 15.6$ at $(\frac{6}{5}, \frac{16}{5})$

Chapter 18

1. $\frac{180\text{ s}}{4\text{ min}} = \frac{180\text{ s}}{240\text{ s}} = \frac{3}{4}$

2. $\frac{1.8\text{ m}}{20.0\text{ mm}} = \frac{x}{14.5\text{ mm}}$

$20.0x = 1.8(14.5)$

$x = 1.3\text{ m}$

3. $L = kT$

$L = 2.7\text{ cm}$ for $T = 150°C$

$2.7 = k(150)$

$k = 0.018\text{ cm/°C}$

$L = 0.018T$

4. Let x = length of the image (in in.)

$\frac{1.00\text{ in.}}{2.54\text{ cm}} = \frac{x}{7.24\text{ cm}}$

$x = \frac{7.24}{2.54} = 2.85\text{ in.}$

5. $2l + 2w = 210.0$

$l = 105.0 - w$

$\frac{l}{w} = \frac{7}{3}$

$\frac{105.0 - w}{w} = \frac{7}{3}$

$315.0 - 3w = 7w$

$10w = 315.0$

$w = 31.5\text{ in.}$

$l = 105.0 - 31.5$

$= 73.5\text{ in.}$

6. $p = kdh$

Using values of water.

$1.96 = k(1000)(0.200)$

$k = 0.00980\text{ kPa}\cdot\text{m}^2/\text{kg}$

For alcohol,

$p = 0.00980(800)(0.300)$

$= 2.35\text{ kPa}$

7. Let L_1 = crushing load of first pillar

L_2 = crushing load of second pillar

$L_2 = \frac{kr_2^4}{l_2^2} \qquad L_1 = \frac{k(2r_2)^4}{(3l_2)^2}$

$$\frac{L_1}{L_2} = \frac{\frac{k(2r_2)^4}{(3l_2)^2}}{\frac{kr_2^4}{l_2^2}} = \frac{16kr_2^4}{9l_2^2} \times \frac{l_2^2}{kr_2^4} = \frac{16}{9}$$

Chapter 19

1. (a) $a_1 = 8, d = -1/2$; arith. seq. 8, 15/2, 7, 13/2, 6, . . .

(b) $a_1 = 8, r = -1/2$; geom. seq. 8, −4, 2, −1, 1/2, . . .

2. $6, -2, \frac{2}{3}, \ldots;$

geometric sequence

$a_1 = 6 \qquad r = \frac{-2}{6} = -\frac{1}{3}$

$$S_7 = \frac{6[1 - (-\frac{1}{3})^7]}{1 - (-\frac{1}{3})}$$

$$= \frac{6(1 + \frac{1}{3^7})}{\frac{4}{3}} = \frac{1094}{243}$$

3. $d = (x + 6) - 4$ and $d = (3x + 2) - (x + 6)$

$d = x + 2$ and $d = 2x - 4$

So $x + 2 = 2x - 4$

$x = 6$

Thus, $d = 8$.

$a_{10} = 4 + 9(8) = 76$

$S_{10} = \frac{10}{2}(4 + 76) = 400$

4. $0.454545\ldots = 0.45 + 0.0045 + 0.000045 + \cdots$

$a = 0.45 \quad r = 0.01$

$S = \frac{0.45}{1 - 0.01} = \frac{0.45}{0.99} = \frac{5}{11}$

5. $\sqrt{1 - 4x} = (1 - 4x)^{1/2}$

$$= 1 + (\tfrac{1}{2})(-4x) + \frac{\frac{1}{2}(\frac{1}{2} - 1)}{2}(-4x)^2 + \cdots$$

$$= 1 - 2x - 2x^2 + \cdots$$

6. $(2x - y)^5 = (2x)^5 + 5(2x)^4(-y) + \frac{5(4)}{2}(2x)^3(-y)^2$

$+ \frac{5(4)(3)}{2(3)}(2x)^2(-y)^3 + \frac{5(4)(3)(2)}{2(3)(4)}(2x)(-y)^4 + (-y)^5$

$= 32x^5 - 80x^4y + 80x^3y^2 - 40x^2y^3 + 10xy^4 - y^5$

7. $5\% = 0.05$

Value after 1 year is

$2500 + 2500(0.05) = 2500(1.05)$

$V_{20} = 2500(1.05)^{20}$

$= \$6633.24$

8. $2 + 4 + \cdots + 200$

$a_1 = 2, \quad a_{100} = 200, \quad n = 100$

$S_{100} = \frac{100}{2}(2 + 200) = 10{,}100$

9. Ball falls 8.00 ft, rises 4.00 ft, falls 4.00 ft, etc.

Distance $= 8.00 + (4.00 + 4.00) + (2.00 + 2.00) + \cdots$

$= 8.00 + 8.00 + 4.00 + 2.00 + \cdots$

$= 8.00 + \frac{8.00}{1 - 0.5} = 8.00 + 16.0 = 24.0\text{ ft}$

Chapter 20

1. $\sec\theta - \frac{\tan\theta}{\csc\theta} = \frac{1}{\cos\theta} - \frac{\frac{\sin\theta}{\cos\theta}}{\frac{1}{\sin\theta}} = \frac{1}{\cos\theta} - \frac{\sin^2\theta}{\cos\theta}$

$= \frac{1 - \sin^2\theta}{\cos\theta} = \frac{\cos^2\theta}{\cos\theta} = \cos\theta$

$\cos\theta = \cos\theta$

2. $\sin 2x + \sin x = 0$

$2\sin x\cos x + \sin x = 0$

$\sin x(2\cos x + 1) = 0$

$\sin x = 0 \qquad \cos x = -\frac{1}{2}$

$x = 0, \pi, \frac{2\pi}{3}, \frac{4\pi}{3}$

3. $\theta = \sin^{-1}x$

$\cos\theta = \cos(\sin^{-1}x)$

$= \sqrt{1 - x^2}$

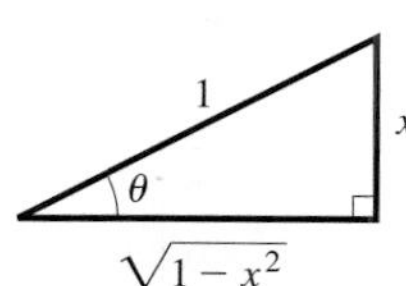

4. $i = 8.00e^{-20t}(1.73\cos 10.0t - \sin 10.0t)$

Because $8.00e^{-20t}$ will not equal zero, $i = 0$ if

$1.73\cos 10.0t - \sin 10.0t = 0$

$1.73 = \tan 10.0t$

$10.0t = \tan^{-1}1.73$

$t = 0.105$ s

5. $\frac{\tan\alpha + \tan\beta}{\tan\alpha - \tan\beta} = \frac{\frac{\sin\alpha}{\cos\alpha} + \frac{\sin\beta}{\cos\beta}}{\frac{\sin\alpha}{\cos\alpha} - \frac{\sin\beta}{\cos\beta}} = \frac{\sin\alpha\cos\beta + \sin\beta\cos\alpha}{\sin\alpha\cos\beta - \sin\beta\cos\alpha}$

$= \frac{\sin(\alpha + \beta)}{\sin(\alpha - \beta)}$

6. $\cot^2 x - \cos^2 x = \frac{\cos^2 x}{\sin^2 x} - \cos^2 x =$

$\cos^2 x\,(\csc^2 x - 1) = \cos^2 x\cot^2 x;$

$\cos^2 x\cot^2 x = \cos^2 x\cot^2 x$

7. $\sin x = -\frac{3}{5}$

$\cos x = \frac{4}{5}$

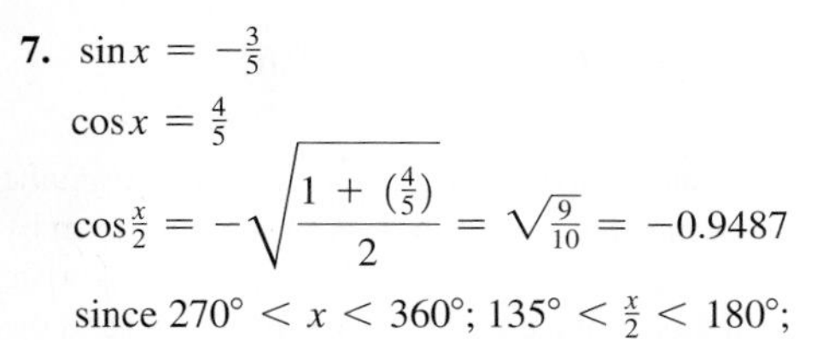

$\cos\frac{x}{2} = -\sqrt{\frac{1 + (\frac{4}{5})}{2}} = \sqrt{\frac{9}{10}} = -0.9487$

since $270° < x < 360°$; $135° < \frac{x}{2} < 180°$;

$\frac{x}{2}$ is in second quadrant, where $\cos\frac{x}{2}$ is negative.

8. $I = I_0\sin 2\theta\cos 2\theta$

$\frac{2I}{I_0} = 2\sin 2\theta\cos 2\theta = \sin 4\theta$

$4\theta = \sin^{-1}\frac{2I}{I_0}$

$\theta = \frac{1}{4}\sin^{-1}\frac{2I}{I_0}$

9. $y = x - 2\cos x - 5$

$x = 3.02, 4.42, 6.77$

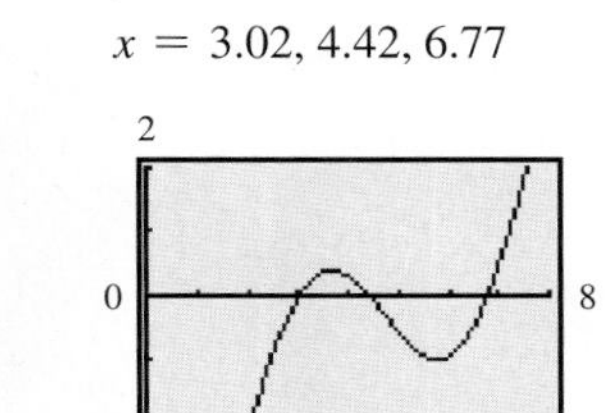

10. $\cos^{-1}x = -\tan^{-1}(-1)$

$\tan^{-1}(-1) = -\frac{\pi}{4}$

$\cos^{-1}x = -(-\frac{\pi}{4}) = \frac{\pi}{4}$

$x = \cos\frac{\pi}{4} = \frac{1}{2}\sqrt{2}$

Chapter 21

1. Points: $(4, -1), (6, 3)$

(a) $d = \sqrt{(6 - 4)^2 + [3 - (-1)]^2} = \sqrt{4 + 16} = 2\sqrt{5}$

(b) Between points $m = \frac{3 - (-1)}{6 - 4} = 2$; perp. line, $m = -\frac{1}{2}$

2. $2(x^2 + x) = 1 - y^2$

$2x^2 + y^2 + 2x - 1 = 0$

$A \neq C$ (same sign)

$B = 0$: ellipse

3. $4x - 2y + 5 = 0$

$y = 2x + \frac{5}{2}$

$m = 2 \quad b = \frac{5}{2}$

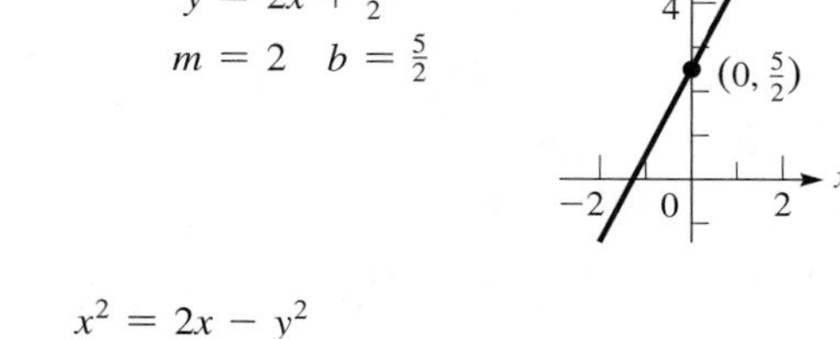

4. $x^2 = 2x - y^2$

$x^2 + y^2 = 2x$

$r^2 = 2r\cos\theta$

$r = 2\cos\theta$

5. $x^2 = -12y$

$4p = -12, \quad p = -3$

$V(0, 0) \quad F(0, -3)$

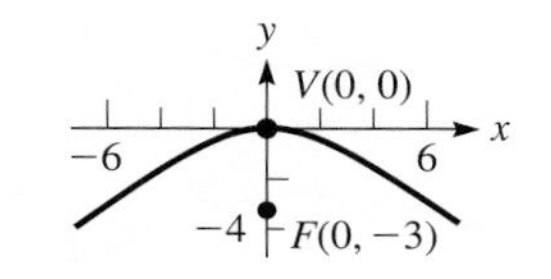

6. Center $(-1, 2)$; $h = -1, k = 2$

$r = \sqrt{(2 + 1)^2 + (3 - 2)^2}$

$= \sqrt{10}$

$(x + 1)^2 + (y - 2)^2 = 10$

or

$x^2 + y^2 + 2x - 4y - 5 = 0$

7. $m = \frac{-2 - 1}{2 + 4} = -\frac{1}{2}$

$y - 1 = -\frac{1}{2}(x + 4)$

$2y - 2 = -x - 4$

$x + 2y + 2 = 0$

8. $x^2 = 4py$

$(6.00)^2 = 4p(4.00)$

$p = 2.25$

Focus is 2.25 cm from vertex.

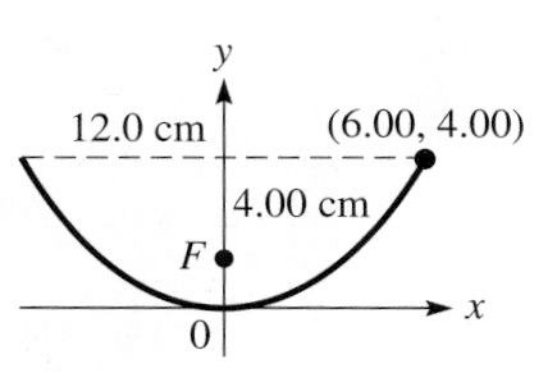

9. $a = 8 \qquad b = 4$

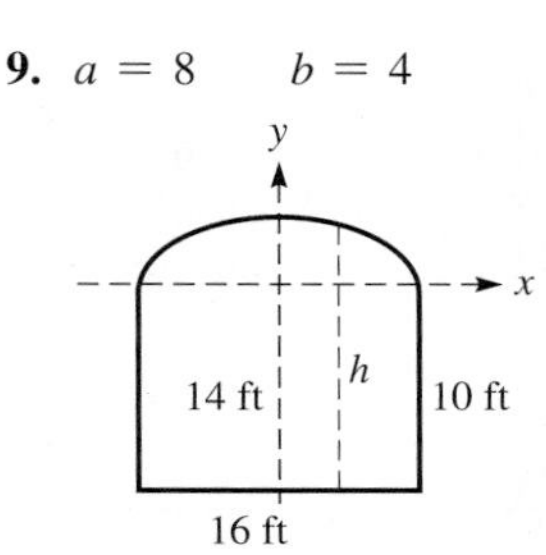

$\frac{x^2}{64} + \frac{y^2}{16} = 1$

Find y for $x = 4$ ft.

$\frac{16}{64} + \frac{y^2}{16} = 1$

$y^2 = 12 \qquad y = 3.5$ ft

$h = 10.0 + 3.5 = 13.5$ ft

10. $r = 3 + \cos\theta$

θ	0	$\frac{\pi}{4}$	$\frac{\pi}{2}$	$\frac{3\pi}{4}$	π	$\frac{5\pi}{4}$	$\frac{3\pi}{2}$	$\frac{7\pi}{4}$	2π
r	4.0	3.7	3.0	2.3	2.0	2.3	3.0	3.7	4.0

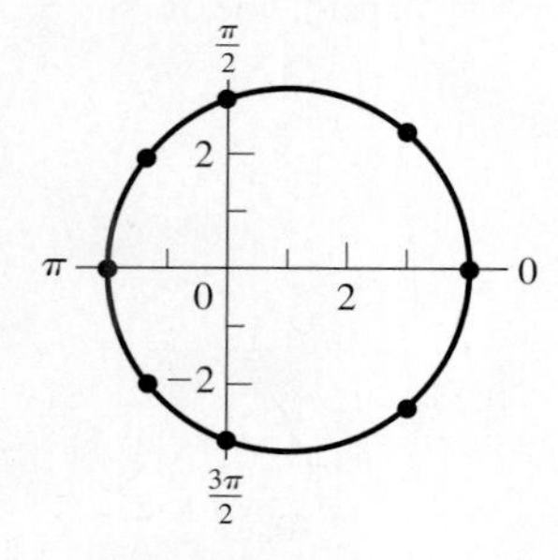

11. $4y^2 - x^2 - 4x - 8y - 4 = 0$

$4(y^2 - 2y \qquad) - (x^2 + 4x \qquad) = 4$

$4(y^2 - 2y + 1) - (x^2 + 4x + 4) = 4 + 4 - 4$

$\frac{(y-1)^2}{1^2} - \frac{(x+2)^2}{2^2} = 1$

$C(-2, 1) \qquad V(-2, 0) \qquad V(-2, 2)$

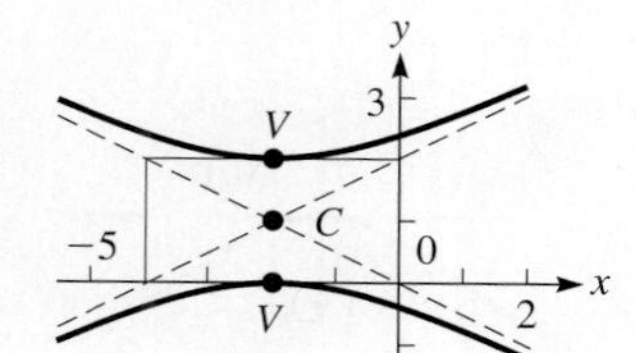

12. Equation of curve: $8x^2 - 4xy + 5y^2 = 36$; $A = 8$, $B = -4$, $C = 5$

(a) $B^2 - 4AC = (-4)^2 - 4(8)(5) = 16 - 160 < 0$; ellipse

(b) $\tan 2\theta = \frac{B}{A - C} = \frac{-4}{8 - 5} = -\frac{4}{3}$;

$\tan^{-1}\left(\frac{4}{3}\right) = 53.13°$;

2θ is obtuse with a reference angle of $53.13°$;

$2\theta = 126.87°$;

$\theta = 63.4°$

Chapter 22

1.

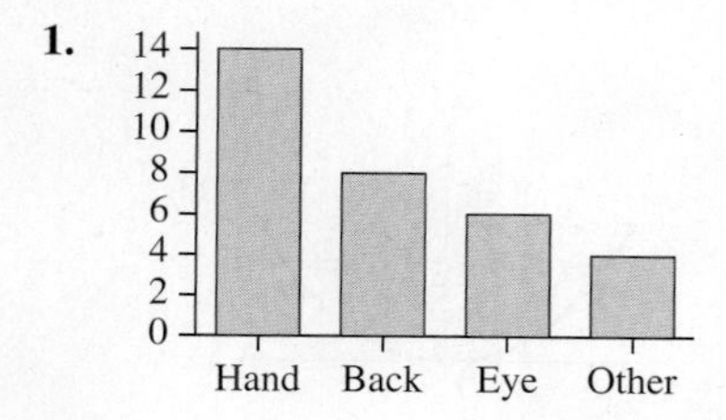

2.

Number	1	2	3	4	5	6	7	8	9	10
Frequency	1	0	1	2	3	2	1	2	2	1

$\Sigma f = 15$

Median is eighth number; median = 6.

3. 5 appears three times, and no other number appears more than twice; mode = 5.

4.

Number	1–3	3–5	5–7	7–9	9–11
Frequency	1	3	5	3	3

(left endpoint included)

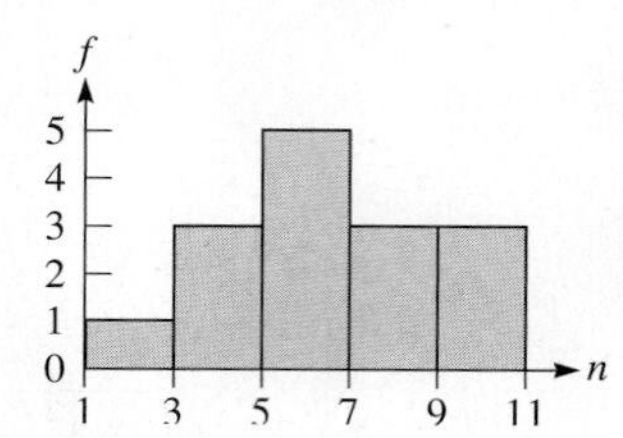

5.

Stem	Leaf
7	2 9
8	2 5 7
9	1 4 4 6
10	7

6. Let t = thickness

$\bar{t} = \frac{3(0.90) + 9(0.91) + 31(0.92) + 38(0.93) + 12(0.94) + 5(0.95) + 2(0.96)}{3 + 9 + 31 + 38 + 12 + 5 + 2}$

$= \frac{92.7}{100} = 0.927$ in.

7. Using 1-Var stats with thicknesses in L1 and frequencies in L2, $s = 0.0117$ in.

8.

t (in.)	0.90	0.91	0.92	0.93	0.94	0.95	0.96
%	3	9	31	38	12	5	2

9. 68% (within 1 standard deviation of the mean)

10. $z = \frac{8.7 - 5.2}{1.4} = 2.5$

A hold time of 8.7 min is 2.5 standard deviations above the mean.

11. $z = \frac{7.3 - 5.2}{1.4} = 1.5$

$0.5000 + 0.4332 = 0.9332$ (total area to the left)

$= 93.32\%$

12. Find the range of each subgroup (subtract the lowest value from the largest value). Find the mean R of these ranges (divide their sum by 20). This is the value of the central line. Multiply R by the appropriate control chart factors to obtain the upper control limit (UCL) and the lower control limit (LCL). Draw the central line, the UCL line, and the LCL line on a graph. Plot the points for each R corresponding to its subgroup and join successive points by straight-line segments.

13.

x	y	xy	x^2
1	5	5	1
3	11	33	9
5	17	85	25
7	20	140	49
9	27	243	81
25	80	506	165

$n = 5$

$m = \frac{5(506) - (25)(80)}{5(165) - 25^2}$

$= 2.65$

$b = \frac{165(80) - (506)(25)}{5(165) - 25^2}$

$= 2.75$

$y = 2.65x + 2.75$

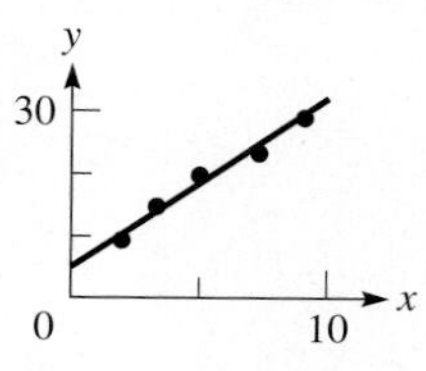

14. Using the regression features on a calculator, $y = 1.101x^{0.501}$.

Chapter 23

1. $\lim_{x\to 1}\frac{x^2-x}{x^2-1} = \lim_{x\to 1}\frac{x(x-1)}{(x+1)(x-1)} = \lim_{x\to 1}\frac{x}{x+1} = \frac{1}{2}$

2. $\lim_{x\to\infty}\frac{1-4x^2}{x+2x^2} = \lim_{x\to\infty}\frac{\frac{1}{x^2}-4}{\frac{1}{x}+2} = -2$

3. $y = 3x^2 - \frac{4}{x^2}$

$\frac{dy}{dx} = 6x + \frac{8}{x^3}$

$\frac{dy}{dx}\Big|_{x=2} = 6(2) + \frac{8}{2^3}$

$= 13$

$m_{\tan} = 13$

```
nDeriv(3X²-4/X²,
X,2)
          13.0000005
```

4. $s = t\sqrt{10-2t}$

$v = \frac{ds}{dt} = t(\frac{1}{2})(10-2t)^{-1/2}(-2) + (10-2t)^{1/2}(1)$

$= \frac{-t}{(10-2t)^{1/2}} + (10-2t)^{1/2} = \frac{10-3t}{(10-2t)^{1/2}}$

$v\big|_{t=4.00} = \frac{10-3(4.00)}{[10-2(4.00)]^{1/2}} = \frac{10-12.00}{2.00^{1/2}} = -1.41\text{ cm/s}$

5. $y = 4x^6 - 2x^4 + \pi^3$

$\frac{dy}{dx} = 4(6x^5) - 2(4x^3)$ π^3 is constant

$= 24x^5 - 8x^3$

6. $y = 2x(5-3x)^4$

$\frac{dy}{dx} = 2x[4(5-3x)^3(-3)] + (5-3x)^4(2)$

$= -24x(5-3x)^3 + 2(5-3x)^4$

$= 2(5-3x)^3(-12x+5-3x)$

$= 2(5-15x)(5-3x)^3$

$= 10(1-3x)(5-3x)^3$

7. $(1+y^2)^3 - x^2y = 7x$

$3(1+y^2)^2(2yy') - x^2y' - y(2x) = 7$

$6y(1+y^2)^2y' - x^2y' = 7 + 2xy$

$y' = \frac{7+2xy}{6y(1+y^2)^2 - x^2}$

8. $V = \frac{kq}{\sqrt{x^2+b^2}} = kq(x^2+b^2)^{-1/2}$

$\frac{dV}{dx} = kq(-\frac{1}{2})(x^2+b^2)^{-3/2}(2x) = \frac{-kqx}{(x^2+b^2)^{3/2}}$

9. $y = \frac{2x}{3x+2}$

$\frac{dy}{dx} = \frac{(3x+2)(2) - 2x(3)}{(3x+2)^2} = \frac{4}{(3x+2)^2} = 4(3x+2)^{-2}$

$\frac{d^2y}{dx^2} = -2(4)(3x+2)^{-3}(3) = \frac{-24}{(3x+2)^3}$

10. $y = 5x - 2x^2$

$f(x+h) = 5(x+h) - 2(x+h)^2$

$f(x+h) - f(x) = 5(x+h) - 2(x+h)^2 - (5x - 2x^2)$

$= 5h - 4xh - 2h^2$

$\frac{f(x+h)-f(x)}{h} = \frac{5h - 4xh - 2h^2}{h} = 5 - 4x - 2h$

$\lim_{h\to 0}\frac{f(x+h)-f(x)}{h} = 5 - 4x$

Chapter 24

1. $y = x^4 - 3x^2$

$\frac{dy}{dx} = 4x^3 - 6x$

$\frac{dy}{dx}\Big|_{x=1} = 4(1^3) - 6(1)$

$= -2$

$y - (-2) = -2(x-1)$

$y = -2x$

2. $y = 3x^2 - x$

$\Delta y = f(3.1) - f(3)$

$[3(3.1)^2 - 3.1] - [3(3)^2 - 3] = 1.73$

$dy = (6x-1)dx$

$= [6(3) - 1](0.1) = 1.7$

$\Delta y - dy = 0.03$

3. $x = 3t^2$ $y = 2t^3 - t^2$

$v_x = \frac{dx}{dt} = 6t$ $v_y = \frac{dy}{dt} = 6t^2 - 2t$

$a_x = \frac{dv_x}{dt} = \frac{d^2x}{dt^2} = 6$ $a_y = \frac{dv_y}{dt} = \frac{d^2y}{dt^2} = 12t - 2$

$a_x\big|_{t=2} = 6$ $a_y\big|_{t=2} = 12(2) - 2 = 22$

$a\big|_{t=2} = \sqrt{6^2 + 22^2} = 22.8$ $\tan\theta = \frac{22}{6}$, $\theta = 74.7°$

4. $P = \frac{144r}{(r+0.6)^2}$

$\frac{dP}{dr} = \frac{144[(r+0.6)^2(1) - r(2)(r+0.6)(1)]}{(r+0.6)^4}$

$= \frac{144[(r+0.6) - 2r]}{(r+0.6)^3} = \frac{144(0.6-r)}{(r+0.6)^3}$

$\frac{dP}{dr} = 0$; $0.6 - r = 0$, $r = 0.6\ \Omega$

$(r < 0.6, \frac{dP}{dr} > 0; r > 0.6, \frac{dP}{dr} < 0)$

5. $x^2 - \sqrt{4x+1} = 0$; $f(x) = x^2 - \sqrt{4x+1}$

$f'(x) = 2x - \frac{1}{2}(4x+1)^{-1/2}(4) = 2x - \frac{2}{(4x+1)^{1/2}}$

n	x_n	$f(x_n)$	$f'(x_n)$	$x_n - \frac{f(x_n)}{f'(x_n)}$
1	1.5	−0.3957513	2.2440711	1.6763542
2	1.6763542	0.0343001	2.6322118	1.6633233

$x_3 = 1.6633$

6. $y = \sqrt{2x+4}$, $a = 6$

$\frac{dy}{dx} = \frac{1}{2}(2x+4)^{-1/2}(2) = \frac{1}{\sqrt{2x+4}}$

$\frac{dy}{dx}\Big|_{x=6} = \frac{1}{\sqrt{2(6)+4}} = \frac{1}{4}$

$f(6) = \sqrt{2(6)+4} = 4$

$L(x) = 4 + \frac{1}{4}(x-6) = \frac{1}{4}x + \frac{5}{2}$

7. $y = x^3 + 6x^2$

$y' = 3x^2 + 12x = 3x(x + 4)$

$y'' = 6x + 12 = 6(x + 2)$

$x < -4$	y inc.	Max. $(-4, 32)$
$-4 < x < 0$	y dec.	Min. $(0, 0)$
$x > 0$	y inc.	Infl. $(-2, 16)$
$x < -2$	y conc. down	
$x > -2$	y conc. up	

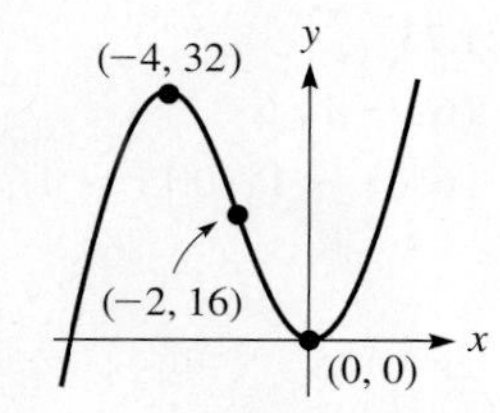

8. $y = \frac{4}{x^2} - x$

$y' = -\frac{8}{x^3} - 1 = -\frac{8 + x^3}{x^3}$

$y'' = \frac{24}{x^4}$

$x < -2$	y dec.
$-2 < x < 0$	y inc.
$x > 0$	y dec., conc. up
$x < 0$	y conc. up

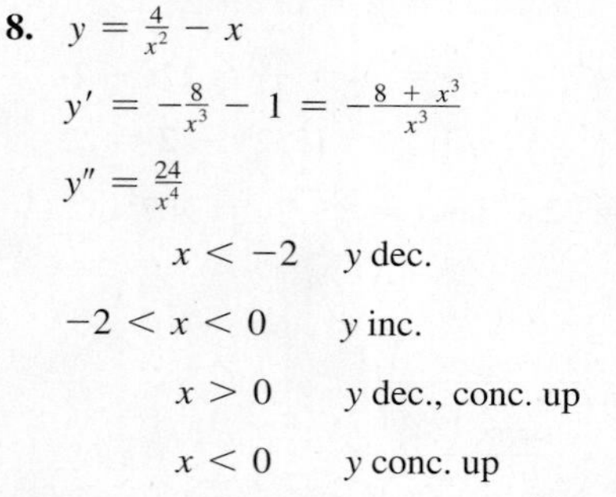

Min. $(-2, 3)$, no infl., int. $(\sqrt[3]{4}, 0)$, sym. none; as $x \to \pm\infty$, $y \to -x$, asym. $y = -x$, $x = 0$. Domain: all real x except 0; range: all real y.

9. Let V = volume of cube

e = edge of cube

$V = e^3 \qquad \frac{dV}{dt} = 3e^2\frac{de}{dt}$

$\frac{dV}{dt}\Big|_{e=4.00} = 3(4.00)^2(-0.50)$

$= -24 \text{ ft}^3/\text{s}$

10.

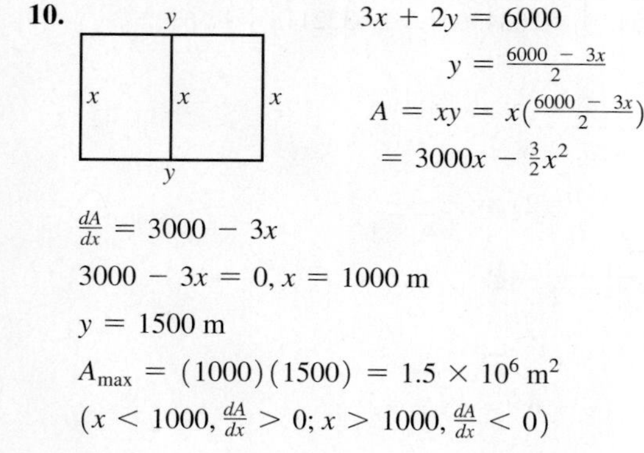

$3x + 2y = 6000$

$y = \frac{6000 - 3x}{2}$

$A = xy = x\left(\frac{6000 - 3x}{2}\right)$

$= 3000x - \frac{3}{2}x^2$

$\frac{dA}{dx} = 3000 - 3x$

$3000 - 3x = 0, x = 1000 \text{ m}$

$y = 1500 \text{ m}$

$A_{max} = (1000)(1500) = 1.5 \times 10^6 \text{ m}^2$

$(x < 1000, \frac{dA}{dx} > 0; x > 1000, \frac{dA}{dx} < 0)$

Chapter 25

1. Power of x required for $2x$ is 2. Therefore, multiply by 1/2. Antiderivative of $2x = \frac{1}{2}(2x^2) = x^2$. Power of $(1 - x)^4$ required is 5. Derivative of $(1 - x)^5$ is $5(1 - x)^4(-1)$. Writing $-(1 - x)^4$ as $\frac{1}{5}[5(1 - x)^4(-1)]$, the antiderivative of $-(1 - x)^4$ is $\frac{1}{5}(1 - x)^5$. Therefore, an antiderivative of $2x - (1 - x)^4$ is $x^2 + \frac{1}{5}(1 - x)^5$.

2. $\int 6x\sqrt{1 - 2x^2}\,dx = \int 6x(1 - 2x^2)^{1/2}\,dx$

$u = 1 - 2x^2 \qquad du = -4x\,dx \qquad n = \frac{1}{2} \qquad n + 1 = \frac{3}{2}$

$\int 6x(1 - 2x^2)^{1/2}\,dx = -\frac{6}{4}\int (1 - 2x^2)^{1/2}(-4x\,dx)$

$= -\left(\frac{3}{2}\right)\left(\frac{2}{3}\right)(1 - 2x^2)^{3/2} + C$

$= -(1 - 2x^2)^{3/2} + C$

3. $\frac{dy}{dx} = (6 - x)^4, dy = (6 - x)^4\,dx$

$\int dy = \int (6 - x)^4\,dx$

$y = -\int (6 - x)^4(-dx) = -\frac{1}{5}(6 - x)^5 + C$

$2 = -\frac{1}{5}(6 - 5) + C, C = \frac{11}{5}$

$y = -\frac{1}{5}(6 - x)^5 + \frac{11}{5}$

4. $y = \frac{4}{x + 2} \quad n = 6 \quad \Delta x = \frac{4 - 1}{6} = \frac{1}{2}$

$A = \frac{1}{2}\left(\frac{8}{7} + 1 + \frac{8}{9} + \frac{4}{5} + \frac{8}{11} + \frac{2}{3}\right) = 2.613$

x	1	$\frac{3}{2}$	2	$\frac{5}{2}$	3	$\frac{7}{2}$	4
y	$\frac{4}{3}$	$\frac{8}{7}$	1	$\frac{8}{9}$	$\frac{4}{5}$	$\frac{8}{11}$	$\frac{2}{3}$

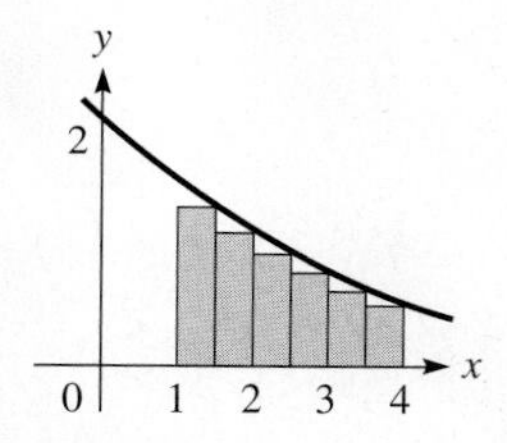

5. (See values for Problem 4.)

$\int_1^4 \frac{4\,dx}{x + 2} = \frac{1}{4}\left[\frac{4}{3} + 2\left(\frac{8}{7}\right) + 2(1) + 2\left(\frac{8}{9}\right)\right.$

$\left. + 2\left(\frac{4}{5}\right) + 2\left(\frac{8}{11}\right) + \frac{2}{3}\right]$

$= 2.780$

6. (See values for Problem 4.)

$\int_1^4 \frac{4\,dx}{x + 2} = \frac{1}{6}\left[\frac{4}{3} + 4\left(\frac{8}{7}\right) + 2(1) + 4\left(\frac{8}{9}\right)\right.$

$\left. + 2\left(\frac{4}{5}\right) + 4\left(\frac{8}{11}\right) + \frac{2}{3}\right] = 2.773$

7. $i = \int_1^3 \left(t^2 + \frac{1}{t^2}\right)dt = \frac{1}{3}t^3 - \frac{1}{t}\Big|_1^3$

$= \frac{1}{3}(27) - \frac{1}{3} - \left(\frac{1}{3} - 1\right) = 9.3 \text{ A}$

Chapter 26

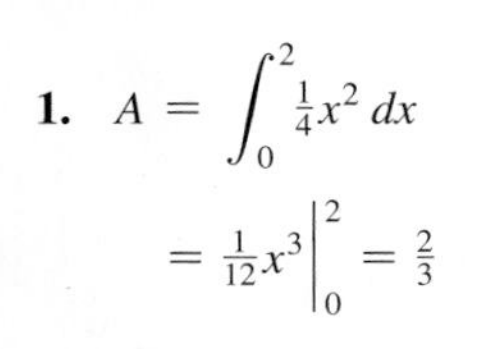

1. $A = \int_0^2 \frac{1}{4}x^2\,dx$

$= \frac{1}{12}x^3\Big|_0^2 = \frac{2}{3}$

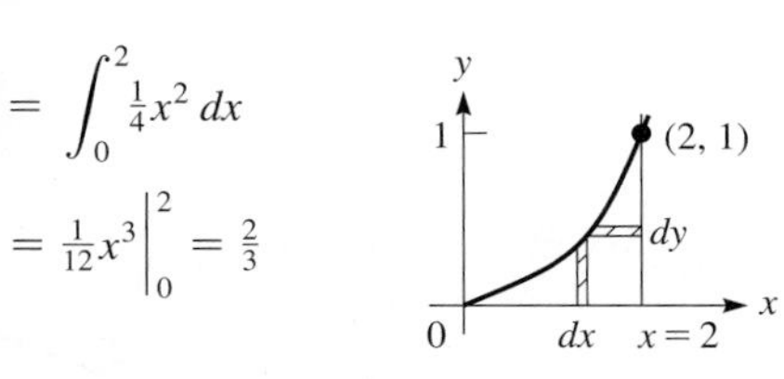

2. $\bar{x} = \dfrac{\int_0^2 x\left(\frac{1}{4}x^2\right)dx}{\frac{2}{3}} = \dfrac{\frac{1}{4}\int_0^2 x^3\,dx}{\frac{2}{3}} = \dfrac{\frac{1}{16}x^4\Big|_0^2}{\frac{2}{3}} = \dfrac{1}{\frac{2}{3}} = \frac{3}{2}$

$\bar{y} = \dfrac{\int_0^1 y(2 - 2\sqrt{y})\,dy}{\frac{2}{3}} = \dfrac{2\int_0^1 (y - y^{3/2})\,dy}{\frac{2}{3}}$

$= \dfrac{2\left(\frac{1}{2}y^2 - \frac{2}{5}y^{5/2}\right)\Big|_0^1}{\frac{2}{3}}$

$= \dfrac{1 - \frac{4}{5}}{\frac{2}{3}} = \frac{1}{5} \times \frac{3}{2} = \frac{3}{10}$

3. $V = \pi\int_0^2 \left(\frac{1}{4}x^2\right)^2 dx = \frac{\pi}{16}\int_0^2 x^4\,dx = \frac{\pi}{80}x^5\Big|_0^2$

$= \frac{32\pi}{80} = \frac{2\pi}{5}$

4. $V = \pi\int_0^3 9^2 dx - \pi\int_0^3 (x^2)^2 dx = 81\pi x\Big|_0^3 - \frac{\pi}{5}x^5\Big|_0^3$

$= 243\pi - \frac{243\pi}{5} = \frac{972\pi}{5}$

or $V = 2\pi\int_0^9 xy\,dy = 2\pi\int_0^9 y^{1/2}y\,dy = \frac{4\pi}{5}y^{5/2}\Big|_0^9$

$= \frac{4\pi}{5}(3^5 - 0) = \frac{972\pi}{5}$

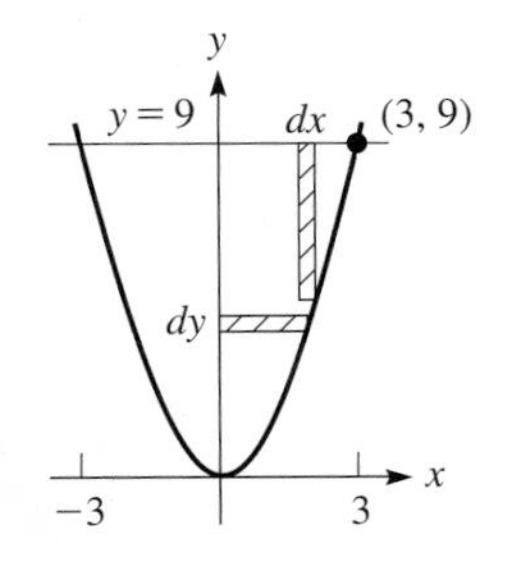

5. $I_y = k\int_0^3 x^2(9 - x^2)\,dx = k\int_0^3 (9x^2 - x^4)\,dx$

$= k\left(3x^3 - \frac{1}{5}x^5\right)\Big|_0^3 = k\left(81 - \frac{243}{5}\right) = \frac{162k}{5}$

6. $q = \int (5.0 - 0.20t)\,dt$

$q = 5.0t - 0.10t^2 + C$

$0 = C$

$q = 5.0t - 0.10t^2$

At $t = 2.0$, $q = 5.0(2.0) - 0.10(2.0)^2 = 9.6$ C

7. $s = \int (60 - 4t)\,dt = 60t - 2t^2 + C$

$s = 10$ for $t = 0$, $10 = 60(0) - 2(0^2) + C$, $C = 10$

$s = 60t - 2t^2 + 10$

8. $F = kx$, $12 = k(2.0)$, $k = 6.0$ N/cm

$W = \int_{2.0}^{6.0} 6.0x\,dx = 3.0x^2\Big|_{2.0}^{6.0} = 3.0(36.0 - 4.0)$

$= 96$ N · cm

9. $F = 62.4\int_{1.00}^{3.00} 6.00h\,dh = (62.4)(6.00)\frac{1}{2}h^2\Big|_{1.00}^{3.00}$

$= (62.4)(3.00)(9.00 - 1.00) = 1500$ lb

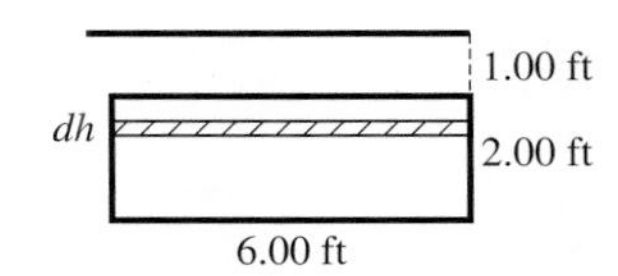

10. $\frac{1}{6}\int_0^6 (x^2 + 2x - 3)\,dx = \frac{1}{6}\left(\frac{1}{3}x^3 + x^2 - 3x\right)\Big|_0^6 = 15$

Chapter 27

1. $y = \tan^3 2x + \tan^{-1} 2x$

$\frac{dy}{dx} = 3(\tan^2 2x)(\sec^2 2x)(2) + \frac{2}{1 + (2x)^2}$

$= 6\tan^2 2x\sec^2 2x + \frac{2}{1 + 4x^2}$

2. $y = 2(3 + \cot 4x)^3$

$\frac{dy}{dx} = 2(3)(3 + \cot 4x)^2(-\csc^2 4x)(4)$

$= -24(3 + \cot 4x)^2\csc^2 4x$

3. $y\sec 2x = \sin^{-1} 3y$

$y(\sec 2x\tan 2x)(2) + (\sec 2x)(y') = \dfrac{3y'}{\sqrt{1 - (3y)^2}}$

$\sqrt{1 - 9y^2}\sec 2x(2y\tan 2x + y') = 3y'$

$y' = \dfrac{2y\sqrt{1 - 9y^2}\sec 2x\tan 2x}{3 - \sqrt{1 - 9y^2}\sec 2x}$

4. $y = \dfrac{\cos^2(3x + 1)}{x}$

$dy = \dfrac{x\{2\cos(3x + 1)[-\sin(3x + 1)(3)]\} - \cos^2(3x + 1)(1)}{x^2}dx$

$= \dfrac{-6x\cos(3x + 1)\sin(3x + 1) - \cos^2(3x + 1)}{x^2}dx$

5. $y = \ln\dfrac{2x - 1}{1 + x^2}$

$= \ln(2x - 1) - \ln(1 + x^2)$

$\frac{dy}{dx} = \frac{2}{2x - 1} - \frac{2x}{1 + x^2}$

$m_{\tan} = \frac{dy}{dx}\Big|_{x=2} = \frac{2}{4 - 1} - \frac{4}{1 + 4} = \frac{2}{3} - \frac{4}{5} = -\frac{2}{15}$

6. $i = 8e^{-t}\sin 10t$

$\frac{di}{dt} = 8[(e^{-t}\cos 10t)(10) + (\sin 10t)(-e^{-t})]$

$= 8e^{-t}(10\cos 10t - \sin 10t)$

7. $y = xe^x$

$y' = xe^x + e^x = e^x(x + 1)$

$y'' = xe^x + e^x + e^x = e^x(x + 2)$

$x < -1$ y dec.

$x > -1$ y inc.

$x < -2$ y conc. down

$x > -2$ y conc. up

Int. $(0, 0)$, min. $(-1, -e^{-1})$

infl. $(-2, -2e^{-2})$, asym. $y = 0$

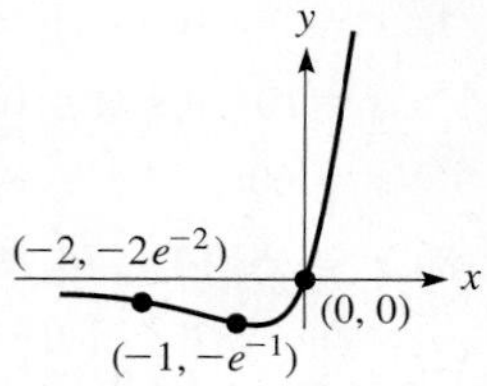

y, x, $(-2, -2e^{-2})$, $(0, 0)$, $(-1, -e^{-1})$

8. For $t = 8.0$ s, $x = 40$ ft

$\theta = \tan^{-1}\frac{x}{250}$

$$\frac{d\theta}{dt} = \frac{1}{1 + \frac{x^2}{250^2}}\frac{dx/dt}{250} = \frac{250^2}{250^2 + x^2}\frac{dx/dt}{250}$$

$$\frac{d\theta}{dt}\Big|_{t=8.0} = \frac{250}{250^2 + 40^2}(5.0) = 0.020 \text{ rad/s}$$

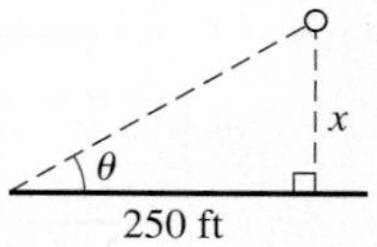

θ, x, 250 ft

9. Function is 0/0 indeterminate form.

$$\lim_{x\to 0}\frac{\tan^{-1}x}{2\sin x} = \lim_{x\to 0}\frac{\frac{d}{dx}\tan^{-1}x}{\frac{d}{dx}(2\sin x)} = \lim_{x\to 0}\frac{\frac{1}{1 + x^2}}{2\cos x} = \frac{1}{2}$$

Chapter 28

1. $\int (\sec x - \sec^3 x\tan x)\, dx$

$$= \int \sec x\, dx - \int \sec^2 x(\sec x\tan x)\, dx$$

$$= \ln|\sec x + \tan x| - \tfrac{1}{3}\sec^3 x + C$$

2. $\int \sin^3 x\, dx = \int \sin^2 x\sin x\, dx$

$$= \int (1 - \cos^2 x)\sin x\, dx$$

$$= \int \sin x\, dx - \int \cos^2 x\sin x\, dx$$

$$= -\cos x + \tfrac{1}{3}\cos^3 x + C$$

3. $\int \tan^3 2x\, dx = \int \tan 2x(\tan^2 2x)\, dx$

$$= \int \tan 2x(\sec^2 2x - 1)\, dx$$

$$= \tfrac{1}{2}\int \tan 2x\sec^2 2x(2\, dx) - \tfrac{1}{2}\int \tan 2x(2\, dx)$$

$$= \tfrac{1}{4}\tan^2 2x + \tfrac{1}{2}\ln|\cos 2x| + C$$

4. $\int \cos^2 4\theta\, d\theta = \tfrac{1}{2}\int (1 + \cos 8\theta)\, d\theta$

$$= \tfrac{1}{2}\int d\theta + \tfrac{1}{16}\int \cos 8\theta(8\, d\theta)$$

$$= \tfrac{1}{2}\theta + \tfrac{1}{16}\sin 8\theta + C$$

5. Let $x = 2\sin\theta$,

$dx = 2\cos\theta\, d\theta$.

$$\int \frac{dx}{x^2\sqrt{4 - x^2}} = \int \frac{2\cos\theta\, d\theta}{4\sin^2\theta\sqrt{4 - 4\sin^2\theta}}$$

$$= \tfrac{1}{4}\int \frac{\cos\theta\, d\theta}{\sin^2\theta\sqrt{\cos^2\theta}} = \tfrac{1}{4}\int \csc^2\theta\, d\theta$$

$$= -\tfrac{1}{4}\cot\theta + C$$

$$= -\frac{\sqrt{4 - x^2}}{4x} + C$$

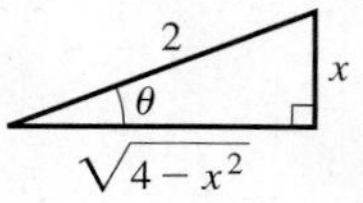

2, x, θ, $\sqrt{4 - x^2}$

6. $\int xe^{-2x}\, dx$; $u = x$, $du = dx$, $dv = e^{-2x}\, dx$, $v = -\frac{1}{2}e^{-2x}$

$$\int xe^{-2x}\, dx = x(-\tfrac{1}{2}e^{-2x}) - \int (-\tfrac{1}{2}e^{-2x})dx$$

$$= -\tfrac{1}{2}xe^{-2x} - \tfrac{1}{4}e^{-2x} + C$$

7. $\int \frac{x^3 + 5x^2 + x + 2}{x^4 + x^2}dx$

$$\frac{x^3 + 5x^2 + x + 2}{x^4 + x^2} = \frac{x^3 + 5x^2 + x + 2}{x^2(x^2 + 1)}$$

$$= \frac{A}{x} + \frac{B}{x^2} + \frac{Cx + D}{x^2 + 1}$$

$x^3 + 5x^2 + x + 2 = Ax(x^2 + 1) + B(x^2 + 1) + Cx^3 + Dx^2$

x^3-terms: $1 = A + C$ $\quad$ x^2-terms: $5 = B + D$

x-terms: $1 = A$ $\quad$ $x = 0$; $2 = B$

$A = 1, B = 2, C = 0, D = 3$

$$\int \frac{x^3 + 5x^2 + x + 2}{x^4 + x^2}dx = \int \frac{dx}{x} + \int \frac{2\, dx}{x^2} + \int \frac{3\, dx}{x^2 + 1}$$

$$= \ln|x| - \tfrac{2}{x} + 3\tan^{-1}x + C$$

8. $i = \int \frac{6t + 1}{4t^2 + 9}dt = \int \frac{6t\, dt}{4t^2 + 9} + \int \frac{dt}{4t^2 + 9}$

$$= \tfrac{6}{8}\int \frac{8t\, dt}{4t^2 + 9} + \tfrac{1}{2}\int \frac{2\, dt}{9 + (2t)^2}$$

$$= \tfrac{3}{4}\ln(4t^2 + 9) + \tfrac{1}{2}(\tfrac{1}{3})\tan^{-1}\tfrac{2t}{3} + C$$

$i = 0$ for $t = 0$; $0 = \frac{3}{4}\ln 9 + \frac{1}{6}\tan^{-1}0 + C$, $C = -\frac{3}{4}\ln 9$

$i = \frac{3}{4}\ln(4t^2 + 9) + \frac{1}{6}\tan^{-1}\frac{2t}{3} - \frac{3}{4}\ln 9$

$$= \tfrac{3}{4}\ln\frac{4t^2 + 9}{9} + \tfrac{1}{6}\tan^{-1}\tfrac{2t}{3}$$

9. $y = \frac{1}{\sqrt{16 - x^2}}$; $A = \int_0^3 y\, dx$

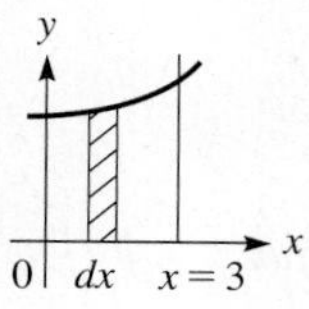

$$= \int_0^3 \frac{dx}{\sqrt{16 - x^2}}$$

$$= \sin^{-1}\tfrac{x}{4}\Big|_0^3$$

$$= \sin^{-1}\tfrac{3}{4} = 0.8481$$

y, x, 0, dx, $x = 3$

Chapter 29

1. $f(x, y) = \frac{2y}{x^2 - y^2}; f(-1, 3) = \frac{2(3)}{(-1)^2 - 3^2} = \frac{6}{-8} = -\frac{3}{4}$

2. $z = 4 - x^2 - 4y^2$

Intercepts: $(2, 0, 0), (-2, 0, 0), (0, 1, 0). (0, -1, 0), (0, 0, 4)$

Traces:

in yz-plane: $z = 4 - 4y^2$ (parabola)

in xz-plane: $z = 4 - x^2$ (parabola)

in xy-plane: $x^2 + 4y^2 = 4$ (ellipse)

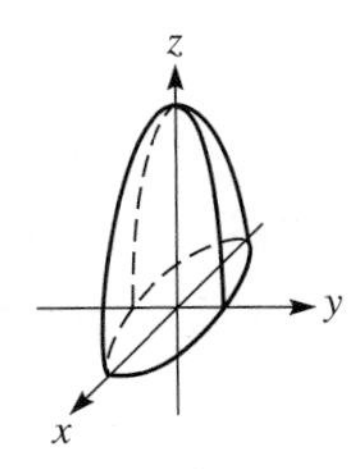

3. $z = xe^{2xy}; \frac{\partial z}{\partial x} = x(e^{2xy})(2y) + e^{2xy}(1) = e^{2xy}(2xy + 1)$

$\frac{\partial z}{\partial y} = xe^{2xy}(2x) = 2x^2e^{2xy}$

4. $z = 3x^3y + 2x^2y^4; \frac{\partial z}{\partial y} = 3x^3(1) + 2x^2(4y^3) = 3x^3 + 8x^2y^3$

$\frac{\partial^2 z}{\partial x \partial y} = 9x^2 + (16x)y^3 = 9x^2 + 16xy^3$

5. $$\int_0^2 \int_{x^2}^{2x} (x^3 + 4y)dy\, dx = \int_0^2 (x^3y + 2y^2)\Big|_{x^2}^{2x} dx$$

$$= \int_0^2 (2x^4 + 8x^2 - x^5 - 2x^4)dx = \int_0^2 (8x^2 - x^5)dx$$

$$= \left(\tfrac{8}{3}x^3 - \tfrac{1}{6}x^6\right)\Big|_0^2 = \tfrac{8}{3}(8) - \tfrac{1}{6}(64) = \tfrac{64}{3} - \tfrac{32}{3} = \tfrac{32}{3}$$

6. $$\int_1^{\ln 8} \int_0^{\ln y} e^{x+y} dx\, dy = \int_1^{\ln 8} e^{x+y}\Big|_0^{\ln y} dy$$

$$= \int_1^{\ln 8} (e^{y+\ln y} - e^y)dy$$

$$= \int_1^{\ln 8} (e^y e^{\ln y} - e^y)dy = \int_1^{\ln 8} (ye^y - e^y)dy$$

$$= e^y(y - 1) - e^y\Big|_1^{\ln 8} = e^{\ln 8}(\ln 8 - 1) - e^{\ln 8} + e$$

$$= 8(\ln 8 - 1) - 8 + e = 8\ln 2^3 - 8 - 8 + e$$

$$= e - 16 + 24\ln 2 = 3.354$$

7. $$V = \int_0^3 \int_0^{\sqrt{9-y^2}} \sqrt{9 - y^2}\, dx\, dy$$

$$= \int_0^3 x\sqrt{9 - y^2}\Big|_0^{\sqrt{9-y^2}} dy$$

$$= \int_0^3 (9 - y^2)dy$$

$$= 9y - \tfrac{1}{3}y^3|_0^3 = 18$$

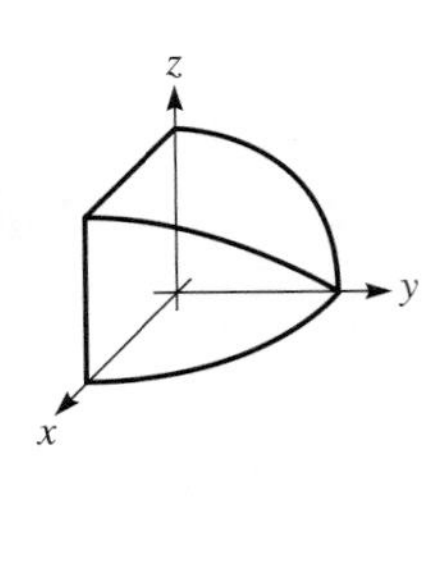

8. $f = \frac{k\sqrt{T}}{L}$; 30 Hz $= \frac{k\sqrt{65\text{ N}}}{60\text{ cm}}; k = \frac{1800}{\sqrt{65}}\frac{\text{Hz}\cdot\text{cm}}{\text{N}^{1/2}}$

$$\frac{\partial f}{\partial T} = \frac{k}{L}\frac{1}{2\sqrt{T}} = \frac{1800}{L\sqrt{65}}\frac{1}{2\sqrt{T}} = \frac{900}{L\sqrt{65T}}$$

$$\frac{\partial f}{\partial T}\Big|_{L=60,\,T=65} = \frac{900}{60\sqrt{65 \times 65}} = \frac{15}{65} = 0.23 \text{ Hz/N}$$

Chapter 30

1. $f(x) = (1 + e^x)^2 \qquad f(0) = (1 + 1)^2 = 4$

$f'(x) = 2(1 + e^x)(e^x) \qquad f'(0) = 2(1 + 1)(1) = 4$

$= 2e^x + 2e^{2x}$

$f''(x) = 2e^x + 4e^{2x} \qquad f''(0) = 2(1) + 4(1) = 6$

$f'''(x) = 2e^x + 8e^{2x} \qquad f'''(0) = 2(1) + 8(1) = 10$

$(1 + e^x)^2 = 4 + 4x + \frac{6}{2}x^2 + \frac{10}{6}x^3 + \cdots$

$= 4 + 4x + 3x^2 + \frac{5}{3}x^3 + \cdots$

2. $f(x) = \cos x \qquad f(\frac{\pi}{3}) = \frac{1}{2}$

$f'(x) = -\sin x \qquad f'(\frac{\pi}{3}) = -\frac{\sqrt{3}}{2}$

$f''(x) = -\cos x \qquad f''(\frac{\pi}{3}) = -\frac{1}{2}$

$$\cos x = \tfrac{1}{2} - \tfrac{\sqrt{3}}{2}\left(x - \tfrac{\pi}{3}\right) - \frac{\frac{1}{2}\left(x - \frac{\pi}{3}\right)^2}{2} + \cdots$$

$$= \tfrac{1}{2}\left[1 - \sqrt{3}\left(x - \tfrac{\pi}{3}\right) - \tfrac{1}{2}\left(x - \tfrac{\pi}{3}\right)^2 + \cdots\right]$$

3. $\ln(1 + x) = x - \frac{x^2}{2} + \frac{x^3}{3} - \frac{x^4}{4} + \cdots$

$\ln 0.96 = \ln(1 - 0.04)$

$$= -0.04 - \frac{(-0.04)^2}{2} + \frac{(-0.04)^3}{3} - \frac{(-0.04)^4}{4}$$

$= -0.0408220$

4. $f(x) = \frac{1}{\sqrt{1 - 2x}} = (1 - 2x)^{-1/2}$

$(1 + x)^n = 1 + nx + \frac{n(n - 1)}{2}x^2 + \cdots$

$$(1 - 2x)^{-1/2} = 1 + \left(-\tfrac{1}{2}\right)(-2x) + \frac{-\frac{1}{2}\left(-\frac{3}{2}\right)}{2}(-2x)^2 + \cdots$$

$= 1 + x + \frac{3}{2}x^2 + \cdots$

5. $$\int_0^1 x\cos x\, dx = \int_0^1 x\left(1 - \tfrac{x^2}{2} + \tfrac{x^4}{24}\right)dx$$

$$= \int_0^1 \left(x - \tfrac{x^3}{2} + \tfrac{x^5}{24}\right)dx$$

$$= \tfrac{1}{2}x^2 - \tfrac{x^4}{8} + \tfrac{x^6}{144}\Big|_0^1$$

$$= \tfrac{1}{2} - \tfrac{1}{8} + \tfrac{1}{144} - 0 = 0.3819$$

6. $f(t) = 0 \qquad -\pi \le t < 0$

$f(t) = 2 \qquad 0 \le t < \pi$

$$a_0 = \frac{1}{2\pi}\int_0^{\pi} 2\, dt = \tfrac{1}{\pi}t\Big|_0^{\pi} = 1$$

$$a_n = \tfrac{1}{\pi}\int_0^{\pi} 2\cos nt\, dt = \tfrac{2}{n\pi}\sin nt\Big|_0^{\pi}$$

$= 0$ for all n

$$b_n = \tfrac{1}{\pi}\int_0^{\pi} 2\sin nt\, dt = \tfrac{2}{n\pi}(-\cos nt)\Big|_0^{\pi}$$

$= \frac{2}{n\pi}(1 - \cos n\pi)$

$b_1 = \frac{2}{\pi}(1 + 1) = \frac{4}{\pi} \quad b_2 = \frac{2}{2\pi}(1 - 1) = 0$

$b_3 = \frac{2}{3\pi}(1 + 1) = \frac{4}{3\pi}$

$f(t) = 1 + \frac{4}{\pi}\sin t + \frac{4}{3\pi}\sin 3t + \cdots$

7. $f(x) = x^2 + 2, f(-x) = (-x)^2 + 2 = x^2 + 2, -f(-x) \ne f(x)$

Since $f(x) = f(-x), f(x)$ is an even function.

Since $f(x) \ne -f(-x), f(x)$ is not an odd function.

$g(x) = x^2 - 1, g(x) = f(x) - 3$

Fourier series for $g(x)$ is $F(x) - 3$.

Chapter 31

1. $x\frac{dy}{dx} + 2y = 4$

$dy + \frac{2}{x}y\,dx = \frac{4}{x}dx$

$e^{\int \frac{2}{x}dx} = e^{2\ln x} = x^2$

$yx^2 = \int \frac{4}{x}x^2\,dx = \int 4x\,dx = 2x^2 + c$

$y = 2 + \frac{c}{x^2}$

2. $y'' + 2y' + 5y = 0$

$m^2 + 2m + 5 = 0$

$m = \frac{-2 \pm \sqrt{4 - 20}}{2} = -1 \pm 2j$

$y = e^{-x}(c_1 \sin 2x + c_2 \cos 2x)$

3. $x\,dx + y\,dy = x^2\,dx + y^2\,dx = (x^2 + y^2)\,dx$

$\frac{x\,dx + y\,dy}{x^2 + y^2} = dx$

$\frac{1}{2}\ln(x^2 + y^2) = x + \frac{1}{2}c \qquad \ln(x^2 + y^2) = 2x + c$

4. $2D^2y - Dy = 2\cos x$

$2m^2 - m = 0 \qquad m = 0, \frac{1}{2}$

$y_c = c_1 + c_2e^{x/2}$

$y_p = A\sin x + B\cos x$

$Dy_p = A\cos x - B\sin x$

$D^2y_p = -A\sin x - B\cos x$

$2(-A\sin x - B\cos x) - (A\cos x - B\sin x) = 2\cos x$

$-2A + B = 0 \qquad -2B - A = 2 \qquad A = -\frac{2}{5} \qquad B = -\frac{4}{5}$

$y = c_1 + c_2e^{x/2} - \frac{2}{5}\sin x - \frac{4}{5}\cos x$

5. $\frac{d^2y}{dx^2} - 4\frac{dy}{dx} + 4y = 3x$

$m^2 - 4m + 4 = 0 \qquad m = 2, 2$

$y_c = (c_1 + c_2x)e^{2x}$

$y_p = A + Bx \quad Dy_p = B \quad D^2y_p = 0$

$0 - 4B + 4(A + Bx) = 3x$

$4A - 4B = 0 \quad 4B = 3$

$B = \frac{3}{4} \quad A = \frac{3}{4}$

$y = (c_1 + c_2x)e^{2x} + \frac{3}{4} + \frac{3}{4}x$

6. $D^2y - 2Dy - 8y = 4e^{-2x}$

$m^2 - 2m - 8 = 0 \quad (m + 2)(m - 4) = 0 \quad m = -2, 4$

$y_c = c_1e^{-2x} + c_2e^{4x}$

$y_p = Axe^{-2x}$ (factor of x necessary due to first term of y_c)

$Dy_p = Ae^{-2x} - 2Axe^{-2x}$

$D^2y_p = -2Ae^{-2x} - 2Ae^{-2x} + 4Axe^{-2x} = -4Ae^{-2x} + 4Axe^{-2x}$

$(-4Ae^{-2x} + 4Axe^{-2x}) - 2(Ae^{-2x} - 2Axe^{-2x})$

$-8Axe^{-2x} = 4e^{-2x}$

$-6Ae^{-2x} = 4e^{-2x} \quad A = -\frac{2}{3}$

$y = c_1e^{-2x} + c_2e^{4x} - \frac{2}{3}xe^{-2x}$

7. $(xy + y)\frac{dy}{dx} = 2$

$y(x + 1)dy = 2\,dx$

$y\,dy = \frac{2\,dx}{x + 1}$

$\frac{1}{2}y^2 = 2\ln(x + 1) + c$

$y = 2 \quad \text{when} \quad x = 0$

$\frac{1}{2}(4) = 2\ln + c, \quad c = 2$

$\frac{1}{2}y^2 = 2\ln(x + 1) + 2$

$y^2 = 4\ln(x + 1) + 4$

8. $\frac{dA}{dt} = rA$

$\frac{dA}{A} = r\,dt$

$\ln A = rt + \ln c$

$\ln\frac{A}{c} = rt$

$A = ce^{rt}$

$A_0 = ce^0$

$c = A_0$

$A = A_0e^{rt}$

9. $y'' + y = 9$

$y(0) = 0 \quad y'(0) = 1$

$\mathcal{L}(y'') + 9\mathcal{L}(y) = \mathcal{L}(9)$

$s^2\mathcal{L}(y) - s(0) - 1 + 9\mathcal{L}(y) = \frac{9}{s}$

$(s^2 + 9)\mathcal{L}(y) = \frac{9}{s} + 1$

$\mathcal{L}(y) = \frac{9}{s(s^2 + 9)} + \frac{1}{s^2 + 9}$

$y = 1 - \cos 3t + \frac{1}{3}\sin 3t$

10. $D^2y - Dy - 2y = 12 \quad y(0) = 0, y'(0) = 0$

$\mathcal{L}(y'') - \mathcal{L}(y') - 2\mathcal{L}(y) = \mathcal{L}(12)$

$s^2\mathcal{L}(y) - s(0) - 0 - [s\mathcal{L}(y) - 0] - 2\mathcal{L}(y) = \frac{12}{s}$

$\mathcal{L}(y)(s^2 - s - 2) = \frac{12}{s}$

$\mathcal{L}(y) = \frac{12}{s(s^2 - s - 2)} = \frac{12}{s(s + 1)(s - 2)}$

$= \frac{A}{s} + \frac{B}{s + 1} + \frac{C}{s - 2}$

$12 = A(s + 1)(s - 2) + Bs(s - 2) + Cs(s + 1)$

$s = 0: \quad 12 = -2A, A = -6$

$s = -1: \quad 12 = 3B, B = 4$

$s = 2: \quad 12 = 6C, C = 2$

$\mathcal{L}(y) = -\frac{6}{s} + \frac{4}{s + 1} + \frac{2}{s - 2}$

$y = -6 + 4e^{-t} + 2e^{2t}$

11. $L\frac{d^2q}{dt^2} + R\frac{dq}{dt} = E \qquad L\frac{di}{dt} + Ri = E$

$L = 2\text{ H} \quad R = 8\ \Omega \quad E = 6\text{ V}$

$2\frac{di}{dt} + 8i = 6 \quad di + 4i\,dt = 3\,dt$

$e^{\int 4\,dt} = e^{4t} \quad ie^{4t} = \int 3e^{4t}\,dt = \frac{3}{4}e^{4t} + c$

$i = 0 \text{ for } t = 0 \quad 0 = \frac{3}{4} + c \quad c = -\frac{3}{4}$

$ie^{4t} = \frac{3}{4}e^{4t} - \frac{3}{4}$

$i = \frac{3}{4}(1 - e^{-4t}) = 0.75(1 - e^{-4t})$

12. $F = kx;\ 16 = k(0.5),\ k = 32\text{ lb/ft}$

$m = \frac{16\text{ lb}}{32\text{ ft/s}^2} = 0.5\text{ slug}$

$mD^2x = -kx, \quad 0.5D^2x + 32x = 0$

$D^2x + 64x = 0$

$m^2 + 64 = 0; \quad m = \pm 8j$

$x = c_1 \sin 8t + c_2 \cos 8t$

$Dx = 8c_1 \cos 8t - 8c_2 \sin 8t$

$x = 0.3\text{ ft} \quad D_x = 0 \quad \text{for} \quad t = 0$

$0.3 = c_1 \sin 0 + c_2 \cos 0, \quad c_2 = 0.3$

$0 = 8c_1 \cos 0 - 8c_2 \sin 0, \quad c_1 = 0$

$x = 0.3 \cos 8t$

INDEX OF APPLICATIONS

Biology and Medical Science

Business and Finance

Chemistry

Civil Engineering

Computer

Construction

Design

Electricity and Electronics

Energy Technology

Environmental Science

Fire Science

Geodesy, Geology, Seismology

Hydrodynamics and Hydrostatics

Machine Technology

Materials

Measurement

Meteorology

Motion

General

Rectilinear and Curvilinear

Navigation

Nuclear and Atomic Physics

Optics

Petroleum

Photography

Physics

Police Science

Refrigeration and Air Conditioning

Space Technology

Statics

Strength of Materials

Surveying

Thermal

Thermodynamics

Wastewater Technology

Index

Notes

Notes

Notes

Notes

Notes

BASIC CURVES

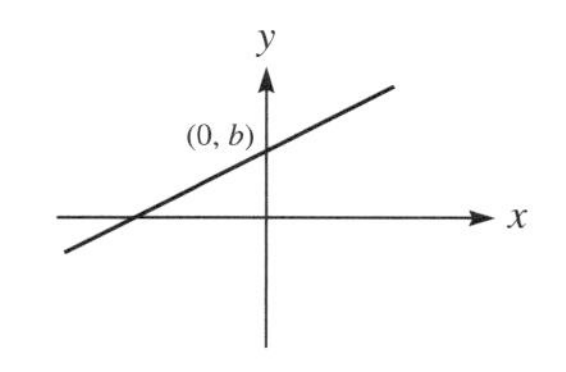

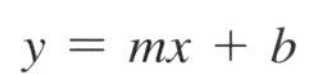
$y = mx + b$

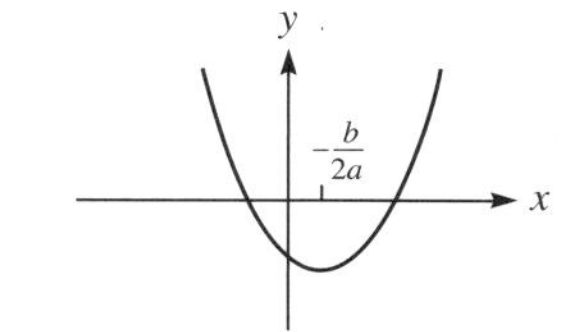

$y = ax^2 + bx + c \quad (a > 0)$

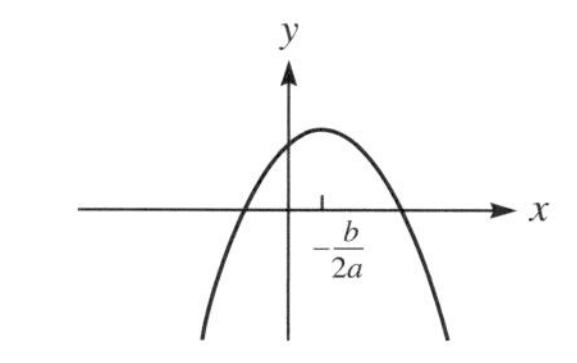

$y = ax^2 + bx + c \quad (a < 0)$

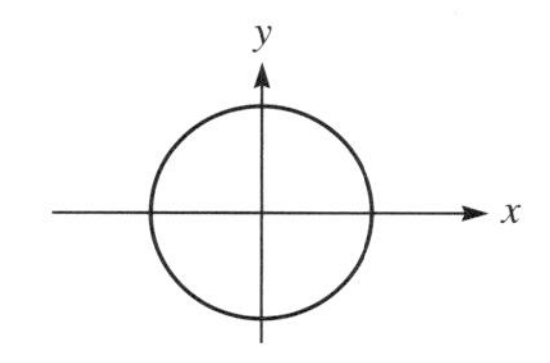

$x^2 + y^2 = a^2$

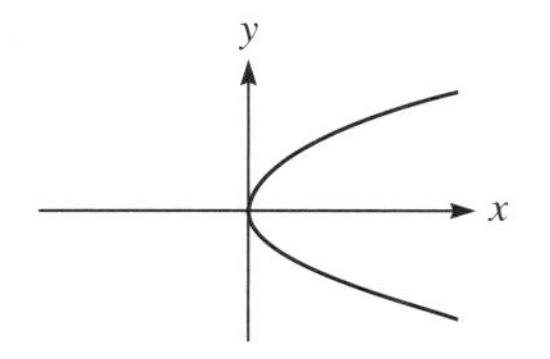

$y^2 = 4px \quad (p > 0)$

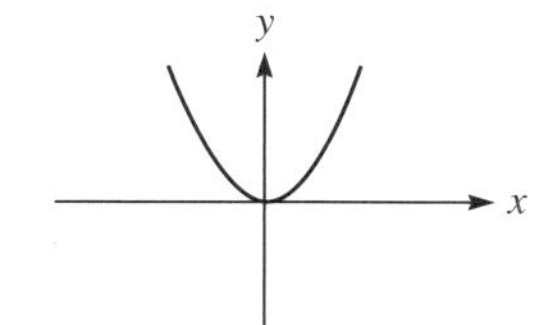

$x^2 = 4py \quad (p > 0)$

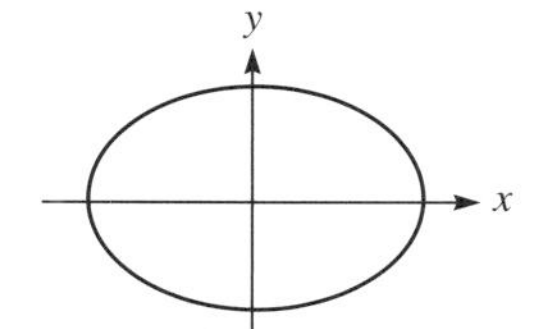

$\frac{x^2}{a^2} + \frac{y^2}{b^2} = 1$

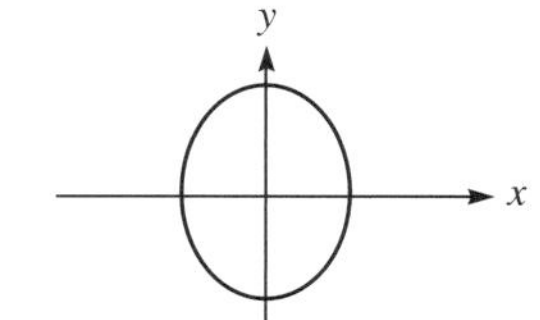

$\frac{y^2}{a^2} + \frac{x^2}{b^2} = 1$

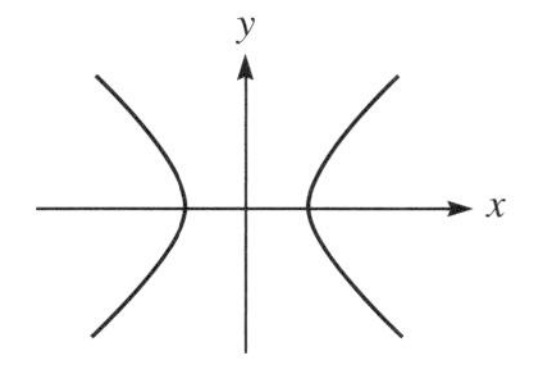

$\frac{x^2}{a^2} - \frac{y^2}{b^2} = 1$

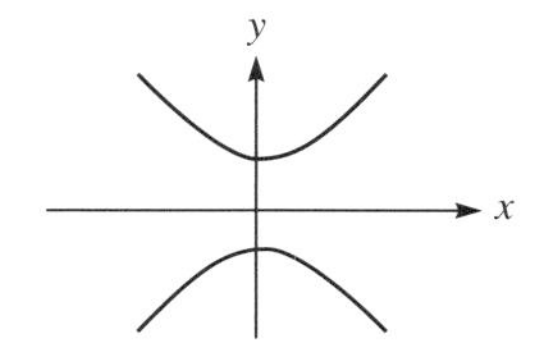

$\frac{y^2}{a^2} - \frac{x^2}{b^2} = 1$

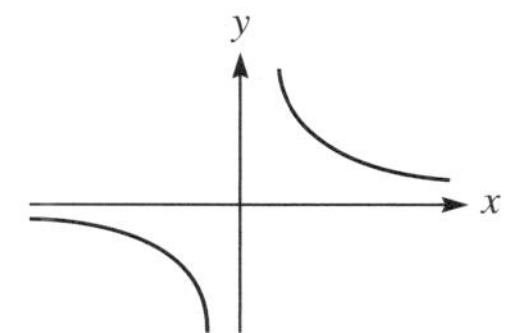

$xy = a \quad (a > 0)$

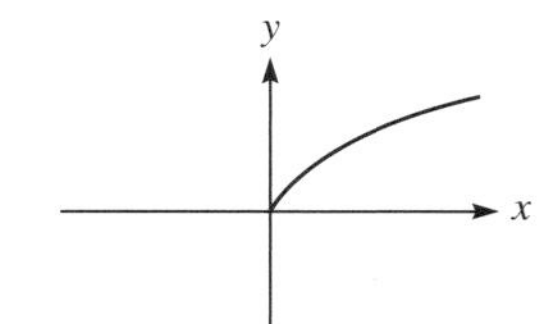

$y = a\sqrt{x} \quad (a > 0)$

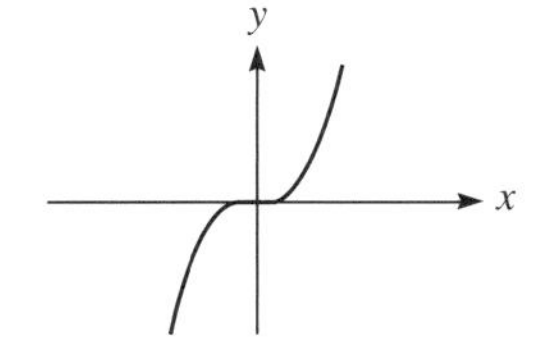

$y = ax^3 \quad (a > 0)$

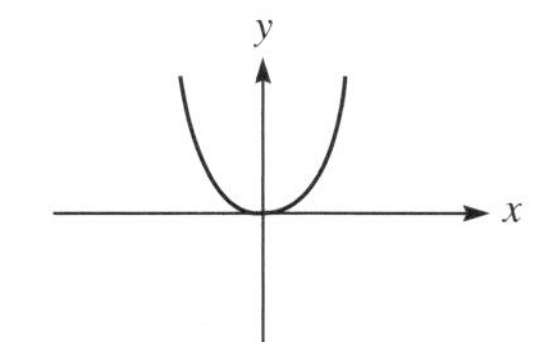

$y = ax^4 \quad (a > 0)$

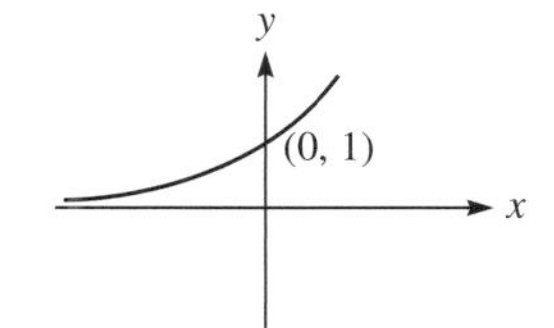

$y = b^x \quad (b > 1)$

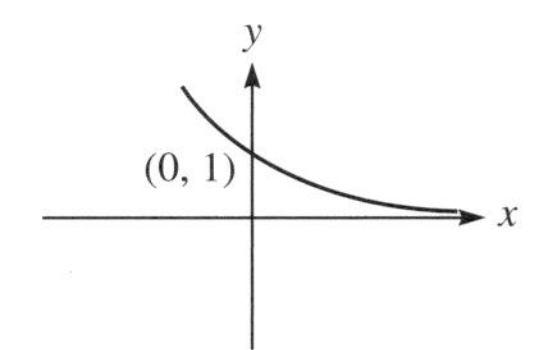

$y = b^{-x} \quad (b > 1)$

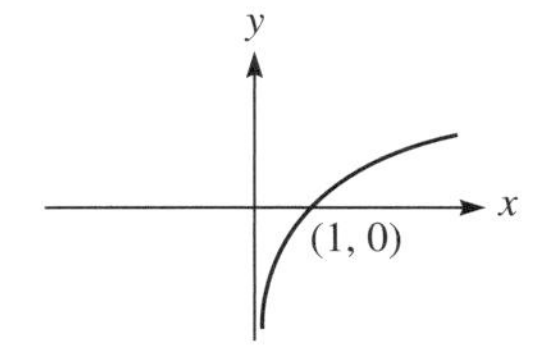

$y = \log_b x$

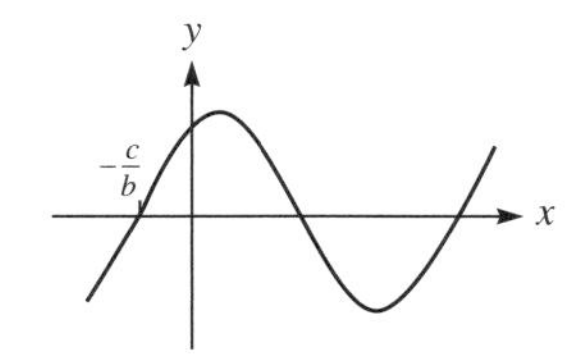

$y = a\sin(bx + c)$
$(a > 0, c > 0)$

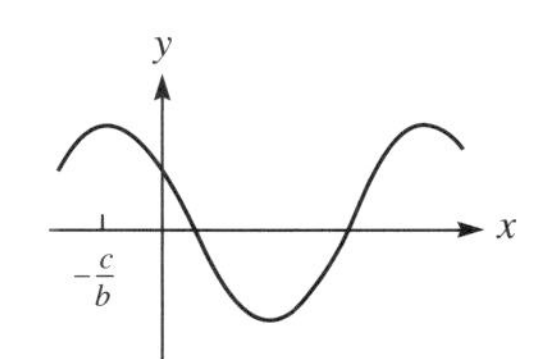

$y = a\cos(bx + c)$
$(a > 0, c > 0)$

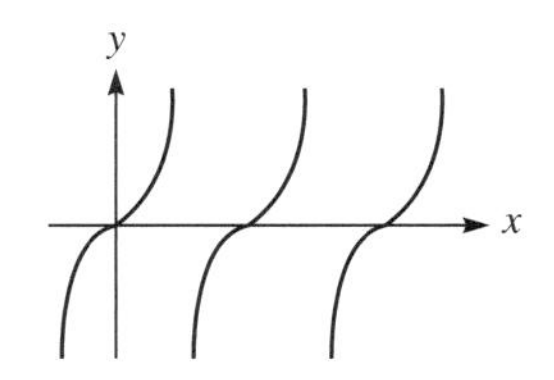

$y = a\tan x$
$(a > 0)$

ALGEBRA

Exponents and Radicals

$$a^m \cdot a^n = a^{m+n}$$

$$\frac{a^m}{a^n} = a^{m-n} \quad a \neq 0$$

$$(a^m)^n = a^{mn}$$

$$(ab)^n = a^n b^n$$

$$\left(\frac{a}{b}\right)^n = \frac{a^n}{b^n} \quad b \neq 0$$

$$a^0 = 1 \quad a \neq 0$$

$$a^{-n} = \frac{1}{a^n} \quad a \neq 0$$

$$a^{m/n} = \sqrt[n]{a^m} = (\sqrt[n]{a})^m$$

$$\sqrt{ab} = \sqrt{a}\sqrt{b}$$

Special Products

$$a(x + y) = ax + ay$$

$$(x + y)(x - y) = x^2 - y^2$$

$$(x + y)^2 = x^2 + 2xy + y^2$$

$$(x - y)^2 = x^2 - 2xy + y^2$$

Quadratic Equation and Formula

$$ax^2 + bx + c = 0$$

$$x = \frac{-b \pm \sqrt{b^2 - 4ac}}{2a}$$

Properties of Logarithms

$$\log_b x + \log_b y = \log_b xy$$

$$\log_b x - \log_b y = \log_b\left(\frac{x}{y}\right)$$

$$n \log_b x = \log_b (x^n)$$

Complex Numbers

$$\sqrt{-a} = j\sqrt{a} \quad (a > 0)$$

$$x + yj = r(\cos\theta + j\sin\theta)$$

$$= re^{j\theta} = r\underline{/\theta}$$

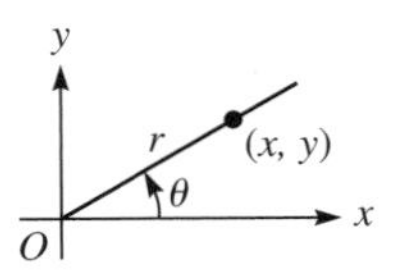

Variation

Direct variation: $y = kx$
Inverse variation: $y = k/x$

GEOMETRIC FORMULAS

Triangle

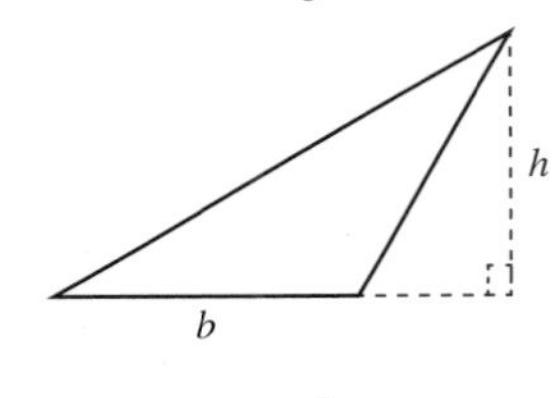

$$A = \tfrac{1}{2}bh$$

Circle

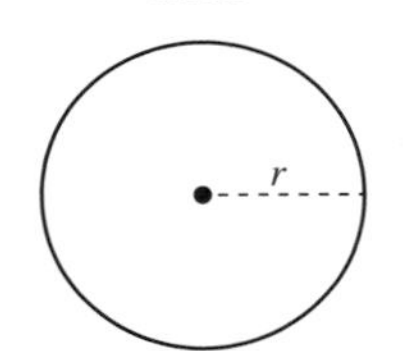

$$A = \pi r^2$$
$$c = 2\pi r$$

Parallelogram

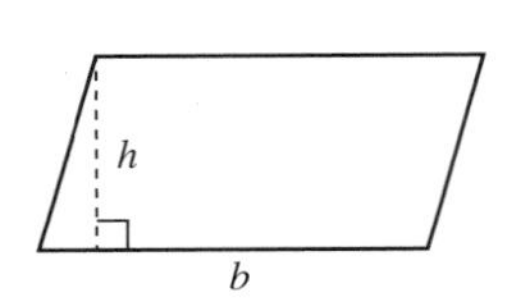

$$A = bh$$

Trapezoid

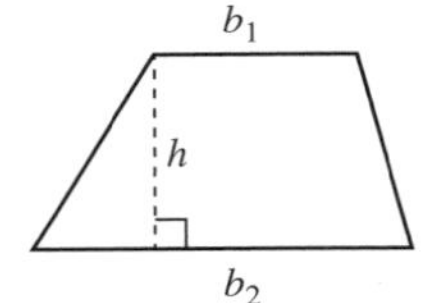

$$A = \tfrac{1}{2}h(b_1 + b_2)$$

Pythagorean Theorem

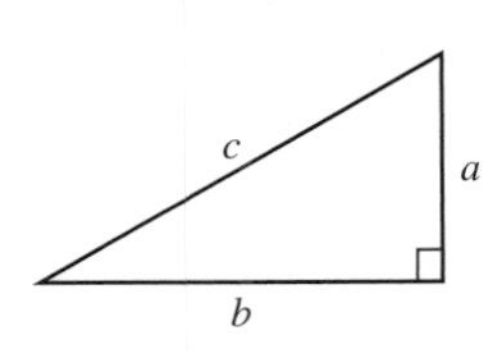

$$c^2 = a^2 + b^2$$

Cylinder

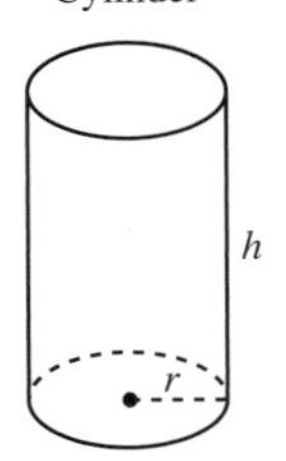

$$V = \pi r^2 h$$
$$A = 2\pi rh + 2\pi r^2$$

Cone

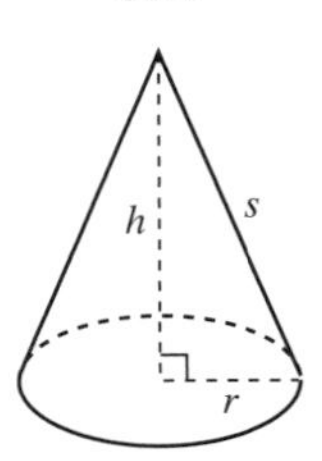

$$V = \tfrac{1}{3}\pi r^2 h$$
$$A = \pi rs + \pi r^2$$

Sphere

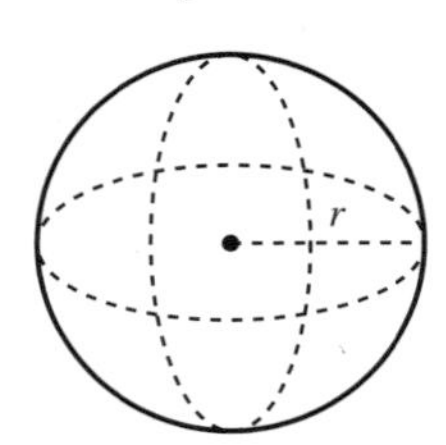

$$V = \tfrac{4}{3}\pi r^3$$
$$A = 4\pi r^2$$